2014 Graduate Programs

in Physics, Astronomy, and Related Fields

 American Institute of Physics

College Park, Maryland

Copyright 2013 by American Institute of Physics
One Physics Ellipse
College Park, MD 20740-3843
Tel. (516) 576-2388
E-mail: jzhu@aip.org
http://www.aip.org

To order additional copies of the book, contact info@gradschoolshopper.com.

International Standards Book Number: 978-0-7354-1177-7

International Standards Serial Number: 1533-5445

Printed in the United States of America

CONTENTS

Part I
United States: Geographic Listing of Graduate Programs

Part II
International: Geographic Listing of Graduate Programs

Foreword

2014 Graduate Programs in Physics, Astronomy, and Related Fields provides information on graduate programs in North America (U.S and Canada). This 49th edition also features several physics graduate programs in China, Hong Kong, Taiwan, and Mexico

The great majority of U.S. physics and astronomy doctoral programs and many master's programs are featured. A number of physics-related fields are also listed, including nuclear engineering, electrical engineering, chemical physics, materials science, meteorology, geophysics, medical physics, oceanography, and acoustics.

I thank the department chairs and administrative assistants for supplying information on their graduate and research programs. Their cooperation and assistance has enabled AIP to produce this resource and other products that benefit the physics and astronomy communities.

2014 Graduate Programs in Physics, Astronomy, and Related Fields is a companion resource to GradschoolShopper.com, a one-stop website for researching graduate programs in physics, astronomy, and related fields. GradschoolShopper.com adds valuable dimension to the print publication: searchability, accessibility, and convenience. I invite you to visit www.GradschoolShopper.com.

I hope that students, their advisers, and others interested in graduate science education will find this publication useful. I welcome your suggestions for improvements. We are committed to making every edition a better resource.

H. Frederick Dylla
Executive Director and CEO
American Institute of Physics

Introduction

2014 Graduate Programs in Physics, Astronomy and Related Fields is designed to provide easily accessible, comprehensive information on graduate programs and research in physics and fields based upon the principles of physics. Students planning graduate study, faculty advisers and others interested in comparative information on graduate programs and physics research will find the information contained in this book valuable. This is the 49th annual edition of the book.

This year departments from all over the United States as well as several top departments in China, Hong Kong, Taiwan, Mexico, and Canada were approached to submit their updated information through an enhanced online process, extending the amount of information supplied and displayed in the printed book and in the accompanying website, GradSchool-Shopper.com.

The content and format of this edition have remained the same as the previous edition. The information in this book is displayed, as much as possible, in tabular format to make it easier to compare information and to make the department profiles more robust. The presentation format has also improved to accommodate additional data.

The enhanced Web-based data-collection process allows the departments to have greater control over their information. As a result, the book and accompanying website are the most comprehensive and up-to-date source of information about graduate programs in physics, astronomy and related fields available in North America as well as several top programs in East Asia.

Sources of Funding

It should be noted that the same approach was used to fund the preparation and distribution of this edition as for the previous edition. Listed graduate departments paid a listing charge to cover the cost of preparing the book, maintaining the website, and distributing information about the resource widely, including a copy of this book to all departments in North America offering at least a bachelor's degree in physics, astronomy, electrical engineering, or nuclear engineering. Almost all physics and astronomy doctoral programs in the United States and many Master's programs are included. AIP anticipates that the number of featured departments will grow

in the future due to the evolving nature of the resource, which will be continually enhanced.

Organization of the Book

Each entry in the book describes the graduate program offered by an academic department at an institution of higher learning in North America as well as several top graduate programs in East Asia. Entries are organized alphabetically by state or province and within each state or province alphabetically by the name of the institution.

Finding Departments

There are three mechanisms by which a user can locate the listing of a department at a particular institution. First, if the country, state or province in which the institution is located is known, the entry can be found relatively quickly from Appendix I. Appendix II provides an alphabetical listing of institutions and departments. Appendix III lists institutions and departments by field and highest degree offered.

Additionally, the www.GradSchoolShopper.com website can be used to locate particular departments within various parameters. A search based on criteria such as the availability of a Master's degree, specific research specialties and research budgets will produce a list of relevant departments and point to their data. The departmental profiles then can be viewed online or in the book.

Comprehensiveness

All known departments in North America as well as several top departments in East Asia with programs leading to a Ph.D. or Master's degree in physics, astronomy or a physics-related field were invited to submit entries. Additional departments, including medical physics, geophysics, chemical physics, materials science, electrical engineering, nuclear engineering, meteorology and oceanography departments that have physics-oriented research programs were also invited to submit entries. The response was excellent. The great majority of U.S. physics doctoral programs and astronomy programs as well as six top East Asian physics graduate programs responded, with 208 departments included in the book.

PART I

UNITED STATES

Geographic Listing of Graduate Programs

ALABAMA AGRICULTURAL AND MECHANICAL UNIVERSITY

DEPARTMENT OF PHYSICS, CHEMISTRY & MATHEMATICS

Normal, Alabama 35762
http://www.physics.aamu.edu

General University Information
President: Dr. Andrew Hugine, Jr.
Dean of Graduate School: Dr. Vann Newkirk
University website: http://www.aamu.edu
Control: Public
Setting: Urban
Total Faculty: 250
Total Graduate Faculty: 79
Total number of Students: 4,853
Total number of Graduate Students: 760

Department Information
Department Chairman: M. D. Aggarwal, Chair
Department Contact: Mary Strong, Administrative Secretary
  Total full-time faculty: 11
  Total number of full-time equivalent positions: 10
  Full-Time Graduate Students: 22
  First-Year Graduate Students: 6
  Female First-Year Students: 3

Department Address
P.O. Box 1268
4900 North Meridian Street
Normal, AL 35762
Phone: (256) 372-5305
Fax: (256) 372-5622
E-mail: mary.strong@aamu.edu
Website: http://www.physics.aamu.edu

ADMISSIONS

Admission Contact Information
Address admission inquiries to: Dean, School of Graduate Studies, Alabama A&M University, P.O. Box 998, Normal, AL 35762
Phone: (256) 372-5266
E-mail: gradschool@aamu.edu
Admissions website: http://www.aamu.edu/academics/gradstudies/Pages/default.aspx

Application deadlines
Fall admission:
U.S. students: July 15   *Int'l. students*: July 15
Spring admission:
U.S. students: November 1   *Int'l. students*: November 1

Application fee
U.S. students: $45   *Int'l. students*: $45

Admissions information
For Fall of 2012:
  Number of applicants: 8
  Number admitted: 5
  Number enrolled: 4

Admission requirements
Bachelor's degree requirements: Bachelor's degree in Physical Science/Electrical Engineering Materials Science/optics is required.
Minimum undergraduate GPA: 3.0

GRE requirements
The GRE is required.
  Quantitative score: 140
  Verbal score: 146
No analytical writing or Mean GRE score required.

Advanced GRE requirements
The Advanced GRE is not required.
No Physics GRE Exam.

TOEFL requirements
The TOEFL exam is required for students from non-English-speaking countries.
  iBT score: 61

Other admissions information
Undergraduate preparation assumed: Physics: Halliday, Resnick, and Krane, Physics Part I & II; Modern Physics: Beiser, Concepts of Modern Physics; Mechanics: Arya, Introduction to Classical Mechanics; Methods in Mathematical Physics: Boas, Mathematical Methods in the Physical Sciences; Electricity and Magnetism: Lorrain, Corson, and Lorrain, Electric Fields and Waves.

TUITION

Tuition year 2013–14:
Tuition for in-state residents
  Full-time students: $337 per credit
Tuition for out-of-state residents
  Full-time students: $674 per credit
$337 is per semester hour for In-state residents. $674 is per semester hour for Out-of-state residents.
Credit hours per semester to be considered full-time: 9
Deferred tuition plan: No
Health insurance: Yes, $112.00.
Other academic fees: $650.
Academic term: Semester
Number of first-year students who received full tuition waivers: 3

Teaching Assistants, Research Assistants, and Fellowships
Number of first-year
  Teaching Assistants: 2
  Research Assistants: 4
Average stipend per academic year
  Teaching Assistant: $12,000
  Research Assistant: $15,000
  Fellowship student: $18,000

FINANCIAL AID

Application deadlines
Fall admission:
U.S. students: March 1   *Int'l. students*: March 1
Spring admission:
U.S. students: September 1   *Int'l. students*: September 1

Loans
Loans are available for U.S. students.
Loans are not available for international students.
GAPSFAS application required: Yes
FAFSA application required: No

For further information

Address financial aid inquiries to: Dean, School of Graduate Studies, Alabama A&M University, Normal, AL 35762.
Phone: (256) 372-5400
E-mail: financialaid@aamu.edu
Financial aid website: http://www.aamu.edu/admissions/fincialaid/pages/default.aspx

HOUSING

Availability of on-campus housing
Single students: Yes
Married students: No

For further information

Address housing inquiries to: Housing Office, P.O. Box 630, Alabama A&M University, Normal, AL 35762.
Phone: (256) 372-5797
E-mail: housing@aamu.edu
Housing aid website: http://www.aamu.edu/campuslife/living-on-campus/residentiallife/pages/default.aspx

Table A—Faculty, Enrollments, and Degrees Granted

Research Specialty	2010–11 Faculty	Enrollment Fall 2011		Number of Degrees Granted 2011–12 (2012-13)		
		Master's	Doctorate	Master's	Terminal Master's	Doctorate
Materials Science, Metallurgy	–	6	8	3(-)	–	3(-)
Optics	–	4	4	–	–	–
Other	–	1	1	1(-)	–	1(-)
Total	–	11	13	2(-)	1(-)	1(-)
Full-time Grad. Stud.	–	10	11	–	–	–
First-year Grad. Stud.	–	3	3	–	–	–

GRADUATE DEGREE REQUIREMENTS

Master's: For admission to the Master of Science program in physics, applicants must have received a Bachelor's degree from a recognized university with a major in any of the physical sciences or electrical engineering or materials science or optics and must have an overall GPA of 3.00 (based on a 4.00 system). Also, students with Bachelor's degrees in optical science, optical engineering programs will be eligible for admission into the optics program and with materials science, or materials engineering programs into materials program. Students with a degree in an area other than physics may be required to take prerequisite undergraduate physics courses. Thesis Option The students must complete at least 24 semester hours of course work with a minimum of 12 hours in area of concentration, and write a thesis (6 semester hours credit) on an approved topic under the supervision of a thesis advisor, and satisfactorily defend the finding of the thesis before a committee of faculty appointed by the department and appointed by the Dean of Graduate Studies. Non-Thesis Option The students must complete at least 30 semester hours of course work with at least 15 of these being in the area of concentration and pass a comprehensive examination given by the Department.

Doctorate: The program is open for admission to students who satisfy the general criteria for admission to graduate-school and who also meet the departmental requirements for admission to the graduate program in the specialization of choice. The applicants with a B.S. in Physics must have an overall GPA of 3.30 (based on a 4.00 system) in the area of concentration and also must have a GRE score of 50 percent in the applicant's major area.

These applicants, as well as applicants with Master's degrees, must pass the various examinations described later. Graduates with a major in any of the physical sciences and a minor in physics, as well as graduates in electrical engineering, are eligible for conditional admission. Such students may be required to take additional courses in physics to attain regular status. Students from non-English speaking countries are required to demonstrate proficiency in English via the University's English Competency test for graduate students. A minimum score of 550 on the Test for English as a Foreign Language (TOEFL) WILL BE REQUIRED FOR ADMISSION. Applicants who hold an M.S. Degree in the particular specialization, namely optics or materials science, will be granted provisional admission based on their performance at the Master's level as evidenced by the corresponding transcripts and also based on letters of recommendation from the departmental faculty where they graduated. Such applicants also must have a minimum GPA of 3.30 (based on a 4.00 system) in the major area. Persons holding the M.S. in traditional physics or electrical engineering or chemistry may be eligible for admission subject to the condition given above in this paragraph. Those students may be required to complete some Master's level courses. However, credit will be given only for courses which are in the list of required or optional courses for the specialization to which the applicant will be admitted. In order to earn the Ph.D. degree, a graduate student must earn a total of at least 60 semester hours of credit, with 45 hours in the area of specialization (optics or materials science) and 15 in the general area of physics. In addition to this, a student must pass a departmental qualifying examination before completing 24 semester hours of graduate credits and must also pass a departmental comprehensive examination before being admitted to the Ph.D. candidacy. Also, the student must do research on an approved topic, earn 12 semester hours of credit for the dissertation, and defend the findings of research before a committee of faculty members. A student cannot register for more than 6 credit hours of dissertation during a given semester. A student may skip the M.S. and proceed to the Ph.D. program. There is no foreign language requirement for the degree, but all students will be required to show proficiency in the use of computers. A student must pass three examinations in the following sequence before the degree is awarded: 1. All students seeking a Ph.D. must pass a qualifying examination before completing 24 semester hours of graduate credits. A person who has been admitted on the basis of a Master's degree may take the qualifying examination after the first semester in the program. 2. All students must take a written departmental candidacy examination in the area of specialization before filing for candidacy. This examination must be passed at least nine months before the expected graduation date. A student is considered as a Ph.D. candidate only after passing the departmental examination.

Thesis: Theses may be written in absentia.

SPECIAL EQUIPMENT, FACILITIES, OR PROGRAMS

The Department houses 20 labs, a Center for the Irradiation of Materials, Center for Nonlinear Optics & Materials, a Departmental Library, a glass shop, a Learning Resource Center, machine shop, graduate student office to 15 graduate students. Argon laser (6 W, 15 W C.W. and Ring dye laser)

High pressure CO_2 Laser (Continuously tunable)

Nd:YAG Nanosecond Laser (second and third harmonics)

244 nm and 365 nm UV Lasers

Nd:YAG Laser pumped Optical Parametric Oscillator

Keithley 2400 Photovoltaic Solar Cell's Testing Station

Quad Tech 7600 LCR Meter, Czochralski Crystal Pullers with automatic diameter control, Bridgman Growth Furnaces for Organic and Inorganic Crystals, Joel-6200 Scanning Electron Microscope, NEC Pelletron Ion Beam Accelerator (2X2.0 MeV) Leitz Microscope with heating stage up to 1400C,Optical and Metallurgical Microscopes, Automatic C-V Data Acquisition Station, DTA-1700 Differential Thermal Analyzer, GenRad 1620 Capacitance Measuring Assembly, Ferroelectric Materials Characterization set-up, Automatic C-V Data Acquisition Station, Furnaces and Temperature Controllers, X-Band Electron Spin Resonance, Solution Crystal Growth System, X-Ray Diffraction Unit (Norelco), Excimer Laser and a tunable dye laser, Perkin Elmer 5100 Atomic Absorption Spectrometer, Varian-Cary 3E Spectrophotometer, Nd:YAG Laser pumped Optical Parametric Oscillator, 0.5 m Monochromator fitted with a CCD Detector, PDA1 Dye Amplifier (Spectra Physics), Bomem DA8 FT-IR Spectrophotometer, Hamamatsu Streak Camera,PDA1 Dye Amplifier (Spectra Physics), Heterodyne Optical Profilometer

Table B—Separately Budgeted Research Expenditures by Source of Support

Source of Support	Departmental Research	Physics-related Research Outside Department
Federal government	$2,950,000	
State/local government		
Non-profit organizations	$10,000	
Business and industry		
Other		$150,000
Total	$2,960,000	$150,000

Table C—Separately Budgeted Research Expenditures by Research Specialty

Research Specialty	No. of Grants	Expenditures ($)
Materials Science, Metallurgy	3	$800,000
Optics	3	$600,000
Other	2	$150,000
Total	8	$1,550,000

FACULTY

Professor

Aggarwal, M. D., Ph.D., University of Calcutta, 1974. Chairman of Department *Applied Physics, Materials Science, Metallurgy*. Crystal growth and characterization.
Dokhanian, Mostafa, Ph.D., Alabama A&M University, 1999. *Applied Physics, Optics*. Optics; applied physics.
Edwards, Matthew E., Ph.D., Howard University, 1977. *Applied Physics, Materials Science, Metallurgy*. Materials science/condensed matter, laser optics.
Reddy, B. R., Ph.D., Indian Institute of Technology, Kampur, 1981. *Applied Physics*. Laser spectroscopy.
Sharma, A., Ph.D., Columbia University, 1982. *Applied Physics, Materials Science, Metallurgy*. Optics.
Zhang, T. X., Ph.D., Nagoya University, 1995. Space science, physics.

Associate Professor

Batra, A. K., Ph.D., Indian Institute of Technology, Delhi, 1981. Materials science, physics.

Assistant Professor

Edwards, Vernessa, Ph.D., Alabama A&M University, 2004. Applied physics.
Guggilla, Padmaja, Ph.D., Alabama A&M University, 2007. Applied physics.
Schamschula, M., Ph.D., University of Alabama, Huntsville, 1994. Optics.

Emeritus

Lal, R. B., Ph.D., Agra University, 1963. Solid state physics; materials science; crystal growth.
Tan, A., Ph.D., University of Alabama, Huntsville, 1979. Space science.

Research Professor

Bhatnagar, V. P., Ph.D., Delphi University, Delhi, 1968. Space physics.
Kukhtarev, N., Ph.D., Institute of Physics, Kiev, Ukraine, 1973. *Optics*.

Research Associate Professor

Curley, Michael, Ph.D., Alabama A&M University, 1997. Applied physics.
Kukhtareva, Tanya, M.S., Kiev University, Ukraine, 1972.
Wladislaw, Lyatsky, Ph.D., University of St. Petersburg, 1968.

Adjunct Professor

Hathaway, David H., Ph.D., University of Colorado, 1979. Astrophysics.
Koshak, William J., Ph.D., University of Arizona. Atmospheric physics.
Nash-Stevenson, Dr. Shelia, Ph.D., Alabama A&M University, 1993. *Optics*. Applied physics.
Phanord, Diendonne D., Ph.D., University of Illinois, 1988. Mathematical physics.
Ruffin, Dr. Paul, Ph.D., University of Alabama, 1986. Optics, physics.
Wu, Shi T., Ph.D., University of Colorado, 1967. Aerospace engineering science.

Visiting Professor

Johnson, R. Barry, Ph.D., Southeastern Institute of Technology. Optical systems.

DEPARTMENTAL RESEARCH SPECIALTIES AND STAFF

Theoretical

Atmosphere, Space Physics, Cosmic Rays. Aggarwal, Matthew Edwards, Sharma.
Materials Science, Metallurgy. Aggarwal, Matthew Edwards, Sharma.
Optics.
Other.

Experimental

Atmosphere, Space Physics, Cosmic Rays.
Materials Science, Metallurgy.
Nano Science and Technology.

View additional information about this department at
www.gradschoolshopper.com

AUBURN UNIVERSITY

DEPARTMENT OF PHYSICS

Auburn, Alabama 36849

http://www.auburn.edu/academic/cosam/departments/physics/

General University Information
President: Jay Gogue
Dean of Graduate School: George Flowers
University website: http://www.auburn.edu/
Control: Public
Setting: Rural
Total Faculty: 1,176
Total Graduate Faculty: 1,000
Total number of Students: 25,134
Total number of Graduate Students: 3,881

Department Information
Department Chairman: James D. Hanson, Chair
Department Contact: James D. Hanson, Chair
  Total full-time faculty: 19
  Total number of full-time equivalent positions: 23
  Full-Time Graduate Students: 54
  First-Year Graduate Students: 14
  Female First-Year Students: 2
  Total Post Doctorates: 6

Department Address
206 Allison Lab
Auburn, AL 36849
Phone: (334) 844-4264
Fax: (334) 844-4613
E-mail: jdhanson@auburn.edu
Website: http://www.auburn.edu/academic/cosam/departments/physics/

ADMISSIONS

Admission Contact Information
Address admission inquiries to: Chairman, Physics Graduate Admissions Committee, 206 Allison Lab, Auburn University, AL 36849
Phone: (334) 844-4264
E-mail: jdhanson@auburn.edu
Admissions website: http://www.auburn.edu/academic/cosam/departments/physics/grad/applying.htm

Application deadlines
Fall admission:
U.S. students: March 15   *Int'l. students*: January 15

Application fee
There is no application fee required.

Admissions information
For Fall of 2013:
  Number of applicants: 100
  Number admitted: 20
  Number enrolled: 14

Admission requirements
Bachelor's degree requirements: Bachelor's degree in physics or related major is required.
Minimum undergraduate GPA: 3.0

GRE requirements
The GRE is required.

Advanced GRE requirements
The Advanced GRE is not required.

TOEFL requirements
The TOEFL exam is required for students from non-English-speaking countries.
  PBT score: 550
Minimum accepted computer-based exam (CBT) score: 213.

Other admissions information
Additional requirements: A combined total greater than 1,200 is desired.
Undergraduate preparation assumed: Upper division mechanics, electricity and magnetism, quantum mechanics, and thermal physics. Students sometimes take these courses during their first year as graduate students.

TUITION

Tuition year 2013-14:
Tuition for in-state residents
  Full-time students: $459 per credit
Tuition for out-of-state residents
  Full-time students: $1,357 per credit
Students with assistantships pay no tuition.
Credit hours per semester to be considered full-time: 9
Deferred tuition plan: No
Other academic fees: Yes.

Teaching Assistants, Research Assistants, and Fellowships
  Number of first-year
    Teaching Assistants: 14

FINANCIAL AID

Application deadlines
Fall admission:
U.S. students: March 15   *Int'l. students*: January 15

Loans
Loans are not available for U.S. students.
Loans are not available for international students.
GAPSFAS application required: No
FAFSA application required: No

For further information
Address financial aid inquiries to: Chairman, Physics Graduate Admissions Committee.
Phone: (334) 844-4264
E-mail: jdhanson@auburn.edu
Financial aid website: http://www.auburn.edu/academic/cosam/departments/physics/grad/graduate-application.htm

HOUSING

Availability of on-campus housing
Single students: Yes
Married students: No

For further information
Address housing inquiries to: Director of Housing, Burton Hall. Kim Trupp.
Phone: (334) 844-4580
E-mail: truppki@auburn.edu

Table A—Faculty, Enrollments, and Degrees Granted

Research Specialty	2013-14 Faculty	Enrollment Fall 2013		Number of Degrees Granted 2012-2013 (2008-2013)		
		Master's	Doctorate	Master's	Terminal Master's	Doctorate
Astrophysics	1	–	–	–	–	–
Atomic, Molecular, & Optical Physics	5	–	–	–	–	1(5)
Condensed Matter Physics	5	–	–	–	–	–(13)
Plasma and Fusion	8	–	–	–	–	–(8)
Total	19	–	54	10(36)	3(19)	8(-)
Full-time Grad. Stud.	–	–	54	–	–	–
First-year Grad. Stud.	–	–	14	–	–	–

GRADUATE DEGREE REQUIREMENTS

Master's: Each student must complete 30 hours of course work with a minimum grade average of 3.0/4.0. No specific residency period is stipulated. A thesis option and nonthesis option are available.

Doctorate: Each student must complete 60 hours of course work including dissertation and research hours. Each candidate must maintain a grade point average of 3.0/4.0 or better. Passage of a general doctoral examination is required of all students. A dissertation based upon original research must be completed and defended in a final oral examination.

SPECIAL EQUIPMENT, FACILITIES, OR PROGRAMS

Materials characterization facilities including low energy ion accelerator; surface science lab; magnetic fusion laboratory with Compact Toroidal Hybrid device; laboratory plasma devices; dusty plasma laboratory; atomic physics laboratory; 90 node computer cluster.

Table B—Separately Budgeted Research Expenditures by Source of Support

Source of Support	Departmental Research	Physics-related Research Outside Department
Federal government	$3,500,000	
State/local government	$1,500,000	
Non-profit organizations		
Business and industry		
Other		
Total	$5,000,000	

FACULTY

Professor

Bozack, Michael J., Ph.D., University of Oregon, 1985. Surface physics.

Giordano, Nicholas J, Ph.D., Yale University, 1977. Dean. The physics of nanostructures and mesoscopic systems; musical acoustics and the physics of the piano; computational neuroscience; computational physics.

Hanson, James D., Ph.D., University of Maryland, 1982. Plasma physics.

Hinata, S., Ph.D., University of Illinois, 1973. Space physics.

Landers, Allen, Ph.D., Kansas State University, 1999. Atomic and molecular physics.

Lin, Yu, Ph.D., University of Alaska, 1993. Space physics.

Oks, Eugene, Ph.D., Moscow Physical Technological Institute, Moscow, 1975. Plasma physics; atomic and molecular physics; nonlinear dynamics.

Park, Minseo, Ph.D., North Carolina State University, 1998. Solid state physics.

Perez, Joseph D., Ph.D., University of Maryland, 1968. Space and plasma physics.

Pindzola, Michael S., Ph.D., University of Virginia, 1975. Atomic and molecular physics.

Thomas, Edward, Ph.D., Auburn University, 1996. Experimental plasma physics.

Associate Professor

Dong, Jianjun, Ph.D., Ohio University, 1998. Condensed matter theory; computational physics.

Konopka, Uwe, Ph.D., Ruhr-University of Bochum, 2000. Plasma physics.

Loch, Stuart, Ph.D., University of Strathclyde, 2001. Atomic and molecular physics.

Maurer, David A., Ph.D., Columbia University, 2000. Plasma physics.

Assistant Professor

Dhar, Sarit, Ph.D., Vanderbilt University, 2005. *Condensed Matter Physics*.

Fogle, Michael R., Ph.D., University of Stockholm, 2004. Experimental atomic physics.

Liu, Kaijun, Ph.D., Cornell University, 2007. Plasma physics.

DEPARTMENTAL RESEARCH SPECIALTIES AND STAFF

Theoretical

Astrophysics. Magnetic fields in stars and galaxies; structure of solar atmosphere.

Atomic, Molecular, & Optical Physics. Photon interactions with atoms, electron scattering by atoms and molecules. Stark Broadening, line shape analysis.

Condensed Matter Physics. Electronic structure; charge transport; dielectric breakdown. Surface dynamics.

Other. Magnetospheric plasma physics.

Plasma and Fusion. MHD equilibrium and equilibrium reconstruction; magnetic field configurations; particle dynamics.

Experimental

Atomic, Molecular, & Optical Physics. Synchotron studies using imaging techniques; electron spectroscopy; momentum imaging.

Condensed Matter Physics. Surface kinetics; gas-surface interactions; thermal desorption; electron-stimulated desorption; low-work function surfaces; wetting and adhesion; corrosion phenomena.

Condensed Matter Physics. Wide bandgap semiconductors; epitaxy; solid state switches; surface properties of solids.

Plasma and Fusion. Magnetic confinement of fusion plasma, plasma diagnostics; plasma spectroscopy; plasma-heating; laboratory simulation of space plasmas; dusty plasmas.

THE UNIVERSITY OF ALABAMA AT BIRMINGHAM

DEPARTMENT OF PHYSICS

Birmingham, Alabama 35294-1170
http://www.uab.edu/physics

General University Information
President: Ray L. Watts
Dean of Graduate School: Brian D. Noe
University website: http://www.uab.edu
Control: Public
Setting: Urban
Total Faculty: 2,289
Total Graduate Faculty: 1,574
Total number of Students: 17,999
Total number of Graduate Students: 5,663

Department Information
Department Chairman: David L. Shealy, Chair
Department Contact: Amanda Holt, Administrative Associate
Total full-time faculty: 16
Total number of full-time equivalent positions: 19
Full-Time Graduate Students: 35
First-Year Graduate Students: 5
Total Post Doctorates: 2

Department Address
1720 2nd Avenue South
Campbell Hall 310
Birmingham, AL 35294-1170
Phone: (205) 934-8190
Fax: (205) 934-8042
E-mail: physics@uab.edu
Website: http://www.uab.edu/physics

ADMISSIONS

Admission Contact Information
Address admission inquiries to: Graduate School Office, HUC
511, 1720 2nd Avenue South, Birmingham, AL 35294-1150
Phone: (205) 934-8227
E-mail: gradschool@uab.edu
Admissions website: www.uab.edu/graduate/

Application deadlines
Fall admission:
U.S. students: July 7   *Int'l. students*: April 1

Application fee
U.S. students: $35   *Int'l. students*: $60

Admissions information
For Fall of 2012:
Number of applicants: 25
Number admitted: 19
Number enrolled: 5

Admission requirements
Bachelor's degree requirements: Bachelor's degree in physics
is required.

GRE requirements
The GRE is required.

Advanced GRE requirements
The Advanced GRE is recommended.

TOEFL requirements
The TOEFL exam is required for students from non-English-
speaking countries.
PBT score: 550

Other admissions information
Additional requirements: The average GRE score for 2012–2013
admissions was 306 (total).
Undergraduate preparation assumed: Halliday and Resnick &
Walker, Fundamentals of Physics; Thornton & Rex, Modern
Physics; Morin, Introduction to Classical Mechanics; Grif-
fiths, Introduction to Electrodynamics; Reif, Fundamentals of
Statistical and Thermal Physics, Berkeley Course Vol. 5; Li-
boff, Introductory Quantum Mechanics.

TUITION

Tuition year 2012–13:
Tuition for in-state residents
Full-time students: $335 per credit
Tuition for out-of-state residents
Full-time students: $788 per credit
Total fees are included in cost of first hour of tuition, which is
$530 for in-state students and $983 for out-of-state students.
Credit hours per semester to be considered full-time: 9
Deferred tuition plan: No
Health insurance: Yes, $1,704.00.
Academic term: Semester
Number of first-year students who received full tuition waivers: 4

Teaching Assistants, Research Assistants, and Fellowships
Number of first-year
Teaching Assistants: 3
Research Assistants: 1
Average stipend per academic year
Teaching Assistant: $20,500

FINANCIAL AID

Application deadlines
Fall admission:
U.S. students: March 1

Loans
Loans are available for U.S. students.
Loans are available for international students.
GAPSFAS application required: No
FAFSA application required: Yes

For further information
Address financial aid inquiries to: HUC 317, 1720 2nd Avenue
South, Birmingham, AL 35294-1150.
Phone: (205) 934-8223
E-mail: finaid@uab.edu
Financial aid website: www.uab.edu/students/paying-for-college

HOUSING

Availability of on-campus housing
Single students: Yes
Married students: No

For further information
Address housing inquiries to: Student Housing and Residential Life, DNMH 101, 1720 2nd Avenue South, Birmingham, AL 35294-1230.
Phone: (205) 934-2092
E-mail: StudentHousing@uab.edu
Housing aid website: www.uab.edu/students/housing

Table A—Faculty, Enrollments, and Degrees Granted

Research Specialty	2012-13 Faculty	Enrollment Fall 2012 Master's	Enrollment Fall 2012 Doctorate	Number of Degrees Granted 2012–13 (2008–12) Master's	Number of Degrees Granted 2012–13 (2008–12) Terminal Master's	Number of Degrees Granted 2012–13 (2008–12) Doctorate
Astrophysics	1	–	1	1(-)	-(1)	-(2)
Biophysics	2	2	2	1(1)	1(-)	-(1)
Condensed Matter Physics	8	1	20	3(16)	–	2(6)
Optics	4	1	8	1(7)	–	1(3)
Physics and other Science Education	1	–	–	–	–	–
Total	16	4	31	6(24)	1(1)	3(12)
Full-time Grad. Stud.	–	4	31	–	–	–
First-year Grad. Stud.	–	–	5	–	–	–

GRADUATE DEGREE REQUIREMENTS

Master's: 30 semester hours of credit with thesis; minimum B (3.0 average); no residency requirements. Thesis is optional with approval of faculty. An additional "interdisciplinary track" for an M.S. degree with thesis option is also offered to non-physics majors and requires a minimum of 12 hours of graduate-level courses offered by other departments.

Doctorate: Minimum residence of three full-time academic years or equivalent periods of part-time enrollment with minimum GPA of B (3.0). Pass: oral placement examination on basic physics concepts; comprehensive examination covering the areas of classical mechanics, quantum mechanics, electromagnetic theory, and two selected topics from thermodynamics/statistical mechanics, optics, or solid-state physics in no more than two attempts; oral examination on area of research specialization; oral defense of written dissertation proposal; and oral final defense of dissertation. In addition, there is an "applied physics track" for the Ph.D. degree that requires students to complete successfully a sequence of core graduate physics classes in classical mechanics, electromagnetism, quantum mechanics, statistical mechanics, and scientific communication seminars totaling 14 credit hours; 12 credit hours of elective courses in applied physics; 3 credit hours of an applied physics training course; and dissertation research hours.

Thesis: Thesis may be written in absentia.

SPECIAL EQUIPMENT, FACILITIES, OR PROGRAMS

The department has active research programs in applied and theoretical astrophysics, biophysics, condensed-matter physics, materials science, nanophysics, optics, lasers, and laser spectroscopy. Opportunities exist for interaction with major government laboratories, including NASA AMES Research Center, Jet Propulsion Laboratory, NASA Goddard Space Flight Center, NASA Marshall Space Flight Center, Advanced Photon Source (APS) at Argonne National Laboratory, National Synchrotron Light Source (NSLS) at Brookhaven National Laboratory, Lawrence Livermore National Laboratory, Oak Ridge National Laboratory, Sandia National Laboratory, Naval Research Laboratory, Wright Patterson Air Force Base Air Force Research Laboratory, National Cancer Institute at the National Institutes of Health, National High Magnetic Field Laboratory (Tallahassee, FL), and Center for Integrated Nanotechnology-Los Alamos National Laboratory.

Table B—Separately Budgeted Research Expenditures by Source of Support

Source of Support	Departmental Research	Physics-related Research Outside Department
Federal government	$2,365,443	
State/local government	$329,985	
Non-profit organizations		
Business and industry		
Other		
Total	$2,695,428	

Table C—Separately Budgeted Research Expenditures by Research Specialty

Research Specialty	No. of Grants	Expenditures ($)
Astrophysics	1	$43,582
Biophysics	3	$196,498
Condensed Matter Physics	18	$1,147,304
Optics	8	$985,731
Other	6	$322,313
Total	36	$2,695,428

FACULTY

Professor

Lawson, Chris M., Ph.D., Oklahoma State University, 1981. Executive Director, Alabama Experimental Program to Stimulate Competitive Research (ALEPSCoR); Director, Center for Optical Sensors and Spectroscopies (COSS). *Optics.* Nonlinear optics; fiber optics; optical sensors; optical coherence imaging and tomography.

Mirov, Sergey B., Ph.D., Lebedev Physical Institute, Moscow, 1983. University Professor of Physics; Co-Director, Center for Optical Sensors and Spectroscopies. *Optics.* Experimental quantum electronics; solid-state lasers; physics of color centers; laser spectroscopy.

Shealy, David L., Ph.D., University of Georgia, 1973. Chair of Department of Physics; Director of Research Computing, UAB IT. *Computational Physics, Optics.* Geometrical optics; laser beam shaping optics; radiative transfer; caustic and optical aberration theory.

Vohra, Yogesh K., Ph.D., Bombay University, 1980. University Scholar; Director of Physics Graduate Program; Director, Center for Nanoscale Materials & Biointegrations: www.uab.edu/cnmb. *Applied Physics, Condensed Matter Physics.* High-pressure physics; synthesis and characterization of diamond crystals and thin films; nanostructured ceramic and polymeric biomaterials.

Wenger, Lowell E., Ph.D., Purdue University, 1975. *Condensed Matter Physics.* Synthesis and characterization of magnetic materials and nanostructures; superconductivity.

Zvanut, Mary Ellen, Ph.D., Lehigh University, 1988. *Condensed Matter Physics.* Electrical studies and EPR studies of insulators and semiconductors; microelectronics and optoelectronics.

Associate Professor

Camata, Renato P., Ph.D., California Institute of Technology, 1998. Director of Undergraduate Physics Program. *Applied Physics, Condensed Matter Physics.* Synthesis and properties of metal and semiconductor nanoparticles; nanostructured materials; aerosol strategies in nanomaterials fabrication; pulsed laser deposition of thin films and nanostructured materials.

Harrison, Joseph G., Ph.D., University of Wisconsin-Madison, 1981. *Computational Physics, Condensed Matter Physics.* Solid-state theory; atomic and molecular physics; MRI modeling; chemical kinetics; simulation of nanoparticle-facilitated hyperthermia.

Kawai, Ryoichi, Ph.D., Waseda University, 1985. *Condensed Matter Physics.* Condensed-matter theory; biophysics theory; materials physics theory; computational physics; complex systems.

Nordlund, Thomas M., Ph.D., University of Illinois, 1977. *Biophysics.* Structural dynamics of DNA and proteins; protein-DNA recognition; picosecond fluorescence; laser tweezers; biomolecule-nanoparticle interactions.

Stanishevsky, Andrei V., Ph.D., Belarus Academy of Sciences, 1996. Focused ion beam micro- and nanofabrication; PVD thin films deposition, characterization, and application; nanoparticle research.

Wang, Xujing, Ph.D., Texas A&M University, 1995. *Biophysics.* Theoretical physics; network theory; biophysics; theoretical and mathematical biology; genetics.

Assistant Professor

Catledge, Aaron S., Ph.D., University of Alabama at Birmingham, 1999. *Condensed Matter Physics.* Synthesis and properties of nanostructured super-hard materials; chemical vapor deposition (CVD) of diamond films and novel nanostructured coatings for biomedical implants; composite scaffolds for tissue engineering; mechanical properties.

Hilton, David J., Ph.D., Cornell University, 2002. *Applied Physics, Optics.* Ultrafast spectroscopy and ultrashort pulse generations; ultrafast terahertz spectroscopy; correlated electron materials; superconductivity; high-magnetic field spectroscopy; magnetic semiconductors; complex functional nanomaterials; materials in extreme environments.

Emeritus

Agresti, David G., Ph.D., California Institute of Technology, 1967. Professor Emeritus. *Astrophysics, Condensed Matter Physics.*

Bauman, Robert P., Ph.D., University of Pittsburgh, 1954. Professor Emeritus.

Martin, James C., Ph.D., Georgia Institute of Technology, 1978. Conformations of biological macromolecules; laser light scattering; optical pattern recognition; Raman spectroscopy.

Wills, Edward L., Ph.D., University of Virginia, 1968. Research Associate Professor Emeritus.

Young, John H., Ph.D., Clark University, 1969. Professor Emeritus.

Research Assistant Professor

Fedorov, Vladimir V., Ph.D., Lisano International, 1999. Physical and mathematical science; coherent and laser spectroscopic characterization of doped laser materials; solid-state lasers; laser spectroscopy for molecular-sensing applications.

Martyshkin, Dmitri V., Ph.D., University of Alabama at Birmingham, 2004. Development and spectroscopic characterization of doped laser materials; solid-state lasers; laser spectroscopy for molecular-sensing applications.

Tsoi, Georgiy, Ph.D., Ukraine Academy of Sciences, 1984. Physics and mathematics; physical and quantum electronics.

Adjunct Faculty

Gerakines, Perry A., Ph.D., Rensselaer Polytechnic Institute, 1998. *Astrophysics.* Astrophysics; interstellar molecules; interstellar dust; laboratory astrophysics; infrared astrophysics; comets; planetary science; origin of life; observational astronomy.

Instructor

Devore, Todd, Ph.D., University of Alabama at Birmingham, 1999. Coordinator of Undergraduate Laboratories. *Computational Physics, Physics and other Science Education.* Computational physics.

Mohr, Rob, Ph.D., University of Alabama, 2001. *Astronomy.* Computational applications to theoretical astrophysical problems.

DEPARTMENTAL RESEARCH SPECIALTIES AND STAFF

Theoretical

Astrophysics. Computer modeling of astrochemical processes and planetary data from Mars and the outer solar system; origin of the solar system; impact. Agresti, Gerakines, Hilton.

Biophysics. Macromolecular structure; assembly and dynamics by computer modeling. Harrison, Kawai, Nordlund, Wang.

Condensed Matter Physics. Low-dimensional systems; defects in insulators and semiconductors; positron states in condensed matter; simulation of chemical vapor deposition processes; computational electromagnetics; surface adsorption; ab initio molecular dynamics simulations; computational algorithms applicable to massively parallel computers; quantum Monte Carlo simulations; non-equilibrium statistical mechanics; stochastic processes. Agresti, Camata, Harrison, Kawai, Shealy, Wang.

Optics. Laser physics; laser spectroscopy; fiber, laser, soft x-ray/UV optics; geometrical optics; nonlinear optics; laser beam shaping; optical design; caustic and optical aberration theory. Hilton, Mirov, Shealy.

Experimental

Astrophysics. Astrochemistry of cosmic ices and complex interstellar molecules; molecular evolution and precursors of life; hydrothermal systems; instruments for in situ planetary science and life search; participation in the Mars Rover exploration missions; mass extinctions and Pre-Cambrian paleontology; bringing to bear tools such as Mössbauer, uv/vis/ir, Raman, and mass spectroscopies, XRD, and chemical analysis. Agresti, Gerakines, Mohr.

Biophysics. DNA and protein structure and function via continuous and time-resolved fluorescence spectroscopy and molecular calculations; fiber-optic biosensors; TIRF; FRET; transient kinetics of molecular interactions; energy transfer and photophysics of sunscreens; spectroscopy and imaging of assembly and interactions between biomolecules and nanoparticles. Camata, Catledge, Fedorov, Mirov, Nordlund, Vohra.

Condensed Matter Physics. EPR studies of bulk crystals and thin films; optical Mössbauer effect; design and construction of portable Mössbauer spectrometer for use in extraterrestrial studies; high-pressure physics; electrical studies of semiconducting and insulating materials; electrical and optical properties of bulk synthetic diamond and diamond thin films; radiational defects in crystals; optical properties of laser crystals; time-resolved laser spectroscopy; synthesis and characterization of metallic, semiconducting, and magnetic materials/nanostructures; superconductivity; aerosol strategies. Agresti, Camata, Catledge, Fedorov, Gerakines, Hilton, Lawson, Martyshkin, Mirov, Stanishevsky, Tsoi, Vohra, Wenger, Zvanut.

Materials Science, Metallurgy. Nanostructured materials; carbon nanotube synthesis and properties; nanoscale direct writing and patterning; nanocomposite biomaterials. Camata, Catledge, Stanishevsky, Vohra, Wenger.

Optics. Laser optics; laser resonators; solid-state laser materials; tunable lasers; laser spectroscopy; UV holographic projection processing of materials; physiological optics; nonlinear optics and nonlinear optical materials; diamond windows for optical spectroscopy; fiber optics; optical sensors; optical imaging; optical coherence; tomography. Camata, Catledge, Fedorov, Hilton, Lawson, Martyshkin, Mirov, Shealy, Stanishevsky, Vohra.

View additional information about this department at
www.gradschoolshopper.com

THE UNIVERSITY OF ALABAMA

DEPARTMENT OF PHYSICS AND ASTRONOMY

Tuscaloosa, Alabama 35487-0324
http://physics.ua.edu

General University Information
President: Judy Bonner
Dean of Graduate School: David Francko
University website: http://www.ua.edu
Control: Public
Setting: Suburban
Total Faculty: 1,211
Total Graduate Faculty: 801
Total number of Students: 30,232
Total number of Graduate Students: 4,726

Department Information
Department Chairman: Raymond E. White III, Chair
Department Contact: Nancy Pekera, Administrative Secretary
Total full-time faculty: 26
Total number of full-time equivalent positions: 26
Full-Time Graduate Students: 43
First-Year Graduate Students: 7
Total Post Doctorates: 14

Department Address
514 University Boulevard
Tuscaloosa, AL 35487-0324
Phone: (205) 348-5050
Fax: (205) 348-5051
E-mail: npekera@ua.edu
Website: http://physics.ua.edu

ADMISSIONS

Admission Contact Information
Address admission inquiries to: Graduate School Office, Box 870118, Tuscaloosa, AL 35487-0118
Phone: (877) 824-7237
E-mail: graduate.school@ua.edu
Admissions website: http://graduate.ua.edu/applicants.html

Application deadlines
Fall admission:
U.S. students: February 15   *Int'l. students*: February 15
Spring admission:
U.S. students: November 1   *Int'l. students*: June 1

Application fee
U.S. students: $50   *Int'l. students*: $60

Admissions information
For Fall of 2013:
Number of applicants: 62
Number admitted: 19
Number enrolled: 10

Admission requirements
Bachelor's degree requirements: Bachelor's degree in physics is required.
Minimum undergraduate GPA: 3.0

GRE requirements
The GRE is required.
A score of at least 300 on revised GRE is required, or a score of at least 1000 on previous GRE general test.

Advanced GRE requirements
The Advanced GRE is not required.

TOEFL requirements
The TOEFL exam is required for students from non-English-speaking countries.
PBT score: 550
iBT score: 79

Other admissions information
Undergraduate preparation assumed: Halliday and Resnick, Fundamentals of Physics; Serway, Moses, and Moyer, Modern Physics; Symon, Mechanics; Reitz, Milford, Foundation of Electromagnetic Theory; Eisberg, Resnick, Quantum Physics of Atoms; etc.

TUITION
Tuition year 2012–13:
Tuition for in-state residents
   Full-time students: $4,600 per semester
Tuition for out-of-state residents
   Full-time students: $11,475 per semester
Credit hours per semester to be considered full-time: 9
Deferred tuition plan: Yes
Health insurance: Available

Academic term: Semester
Number of first-year students who received full tuition waivers: 7

Teaching Assistants, Research Assistants, and Fellowships

Number of first-year
   Teaching Assistants: 4
   Research Assistants: 2
   Fellowship students: 1
Average stipend per academic year
   Teaching Assistant: $17,280
   Research Assistant: $17,280
   Fellowship student: $15,000

FINANCIAL AID

Application deadlines
Fall admission:
U.S. students: February 15   *Int'l. students*: February 15

Loans
Loans are available for U.S. students.
Loans are available for international students.
GAPSFAS application required: No
FAFSA application required: No

For further information
Address financial aid inquiries to: Office of Financial Aid, Box 870162, 106 Student Services Center, The University of Alabama, Tuscaloosa, AL 35487.
Phone: (855) 469-2262
Financial aid website: http://financialaid.ua.edu/

HOUSING

Availability of on-campus housing
Single students: Yes
Married students: Yes

For further information
Address housing inquiries to: Director of Housing, Residential Life, Box 870399, Tuscaloosa, AL 35487.
Housing aid website: http://housing.ua.edu/

Table A—Faculty, Enrollments, and Degrees Granted

Research Specialty	2012–13 Faculty	Enrollment Fall 2012 Master's	Enrollment Fall 2012 Doctorate	Number of Degrees Granted 2012–13 (2008–13) Master's	Number of Degrees Granted 2012–13 (2008–13) Terminal Master's	Number of Degrees Granted 2012–13 (2008–13) Doctorate
Astronomy	4	2	4	–(1)	–(3)	–(3)
Astrophysics	2	–	1	–	–	–
Condensed Matter Physics	9	2	15	1(6)	2(7)	1(11)
High Energy Physics	4	–	10	–(1)	–	–(1)
Particles and Fields	7	2	7	1(7)	–	1(5)
Non-specialized	–	–	–	–(1)	–	–(1)
Total	26	6	37	2(16)	2(10)	2(21)
Full-time Grad. Stud.	–	6	37	–	–	–
First-year Grad. Stud.	–	–	5	–	–	–

GRADUATE DEGREE REQUIREMENTS

Master's: Plan I: 24 graduate semester hours in an approved program with satisfactory performance required; "B" average; one semester in residence; master's examination required; thesis required; no language requirement. Plan II: 30 graduate semester hours in an approved program with satisfactory performance required; master's examination required; thesis not required; no language requirement.

Doctorate: A minimum of 48 graduate semester hours required in an approved program with satisfactory performance; one academic year in residence required; oral preliminary examination required; dissertation and dissertation examination required.

Thesis: Thesis may be written in absentia.

SPECIAL EQUIPMENT, FACILITIES, OR PROGRAMS

Facilities include well-equipped laboratories for research in condensed-matter physics, high-energy physics, and image processing. Supporting facilities include a machine shop, electronics shop, computer workstations, and direct access to the campus mainframe computer and the Alabama supercomputer. Faculty and students participate in the Center for Materials for Information Technology and the Tri-Campus Material Science Ph.D. Program. We are members of the SARA Telescope consortium, which operates a 0.9 meter telescope at Kitt Peak, Arizona and a 0.6 meter telescope at Cerro Tololo in Chile.

Table B—Separately Budgeted Research Expenditures by Source of Support

Source of Support	Departmental Research	Physics-related Research Outside Department
Federal government	$4,500,000	
State/local government		
Non-profit organizations		
Business and industry		
Other		
Total	$4,500,000	

Table C—Separately Budgeted Research Expenditures by Research Specialty

Research Specialty	No. of Grants	Expenditures ($)
Astrophysics	12	$600,000
Atomic, Molecular, & Optical Physics	1	$55,000
Condensed Matter Physics	23	$3,115,000
Particles and Fields	7	$730,000
Total	43	$4,500,000

FACULTY

Professor

Busenitz, Jerome K., Ph.D., University of Illinois, 1985. *Particles and Fields*. Experimental elementary particle physics.

Buta, Ronald J., Ph.D., University of Texas, Austin, 1984. *Astronomy*. Galaxy morphology and catalogs.

Butler, William H., Ph.D., University of California, San Diego, 1969. *Condensed Matter Physics*. Theoretical condensed-matter physics.

Harms, Benjamin C., Ph.D., Florida State University, 1969. *Particles and Fields*. Theoretical particle physics.

Harrell, J. W., Ph.D., University of North Carolina, Chapel Hill, 1969. *Condensed Matter Physics*. Experimental condensed-matter physics.

Keel, William C., Ph.D., University of California, Santa Cruz, 1982. *Astronomy*. Galactic nuclei, jets, and galaxy interactions.

Mankey, Gary J., Ph.D., Pennsylvania State University, 1992. *Condensed Matter Physics*. Experimental condensed-matter physics.

Piepke, Andreas, Ph.D., Heidelberg University, 1990. *Particles and Fields*. Experimental elementary particle physics.

Sarker, S. K., Ph.D., Cornell University, 1980. *Condensed Matter Physics*. Theoretical condensed-matter physics.

Schad, Rainer, Ph.D., University of Hannover, 1991. *Condensed Matter Physics*. Experimental condensed-matter physics.

Stancu, Ion, Rice University, 1990. *Particles and Fields*. Experimental elementary particle physics.

Stern, Allen, Ph.D., Syracuse University, 1980. *Particles and Fields*. Theoretical particle physics.

White, Raymond E., Ph.D., University of Virginia, 1986. *Astronomy, Astrophysics*. Dynamics and hydrodynamics in galaxies and galaxy clusters.

Associate Professor

LeClair, Patrick R., Ph.D., Eindhoven University of Technology, 2002. *Condensed Matter Physics*. Experimental condensed-matter physics.

Mewes, Tim, Ph.D., University of Kaisersleutern, 2002. *Condensed Matter Physics*. Experimental condensed-matter physics.

Mryasov, Oleg, Ph.D., Russian Academy of Sciences, 1993. *Condensed Matter Physics*. Theoretical condensed-matter physics.

Assistant Professor

Bailin, Jeremy, Ph.D., University of Arizona, 2004. *Astrophysics*. Galaxy formation and evolution.

Henderson, Conor, Ph.D., Massachusetts Institute of Technology, 2005. *Particles and Fields*. Experimental particle physics.

Irwin, Jimmy, Ph.D., University of Virginia, 1997. *Astronomy*. Accreting black holes and neutron stars.

Mewes, Claudia K.A., Ph.D., University of Kaiserslautern, 2004. *Condensed Matter Physics*.

Okada, Nobuchika, Ph.D., Tokyo Metropolitan University, 1998. *Cosmology & String Theory, Particles and Fields*. Physics beyond the Standard Model.

Rumerio, Paolo G., Ph.D., Northwestern University, 2003. *Particles and Fields*.

Toale, Patrick A., Ph.D., University of Colorado Boulder, 2004. *Particles and Fields*.

Townsley, Dean M., Ph.D., University of California, Santa Barbara, 2004. *Astrophysics*. White dwarf supernovae.

Williams, Dawn R., Ph.D., University of California, Los Angeles, 2004. *Particles and Fields*. Experimental particle astrophysics.

Professor Emeritus

Alexander, Chester, Ph.D., Duke University, 1968. *Condensed Matter Physics*. Experimental condensed-matter and chemical physics.

Byrd, Gene G., Ph.D., University of Texas, Austin, 1974. *Astrophysics*. Theoretical astrophysics.

Clavelli, Louis J., Ph.D., University of Chicago, 1967. *Particles and Fields*. Theoretical particle physics.

Coulter, Philip W., Ph.D., Stanford University, 1965. *Particles and Fields*. Theoretical particle physics.

Fujiwara, Hideo, Ph.D., University of Tokyo, 1969. *Condensed Matter Physics*. Experimental condensed-matter physics.

Hardee, Philip E., Ph.D., University of Maryland, 1976. *Astrophysics*. Theoretical and observational astrophysics.

Izatt, Jerald R., Ph.D., Johns Hopkins University, 1960. *Atomic, Molecular, & Optical Physics*. Experimental nonlinear optics and lasers.

Jones, Stanley T., Ph.D., University of Illinois, 1970. *Physics and other Science Education*. Physics education.

Sulentic, Jack W., Ph.D., SUNY, Albany, 1975. *Astronomy*. Observational astrophysics.

Tipping, Richard H., Ph.D., Pennsylvania State University, 1969. *Atomic, Molecular, & Optical Physics*. Theoretical physics; molecular spectroscopy.

Visscher, Pieter B., Ph.D., University of California, Berkeley, 1971. *Condensed Matter Physics*. Theoretical condensed-matter physics; computer simulation.

Adjunct Professor

Biermann, Peter L., Ph.D., University of Gottingen, 1971. Theoretical astrophysics.

Crocker, Deborah A., Ph.D., University of Virginia, 1987. Observational astrophysics.

Gupta, Arunava, Ph.D., Stanford University, 1980. Experimental condensed-matter physics.

Pandey, Raghvendra K., Ph.D., University of Cologne, 1967. Experimental condensed-matter physics.

DEPARTMENTAL RESEARCH SPECIALTIES AND STAFF

Theoretical

Astrophysics. Galactic dynamics; galaxy formation; galactic structure; extragalactic astronomy; high-energy astrophysics; stellar evolution; supernovae. Bailin, Biermann, Townsley.

Condensed Matter Physics. Electronic structure of solids; magnetic properties; hierarchical and renormalization-group methods; magnetic lattice models. Butler, Claudia Mewes, Mryasov, Sarker, Visscher.

Particles and Fields. Supersymmetry phenomenology; field theory; quantum black holes; particle astrophysics. Biermann, Harms, Okada, Stern.

Experimental

Astronomy. Black holes; galaxy evolution; galaxy morphology; spectroscopy of AGN; galaxy clusters; globular clusters; x-ray astronomy; x-ray binaries. Buta, Irwin, Keel, White.

Condensed Matter Physics. Magnetic materials and thin films; nanoparticles spintronics. Gupta, Harrell, LeClair, Mankey, Tim Mewes, Pandey, Schad.

Particles and Fields. Detector research and development; neutrino physics; particle astrophysics. Busenitz, Henderson, Piepke, Rumerio, Stancu, Toale, Williams.

View additional information about this department at
www.gradschoolshopper.com

UNIVERSITY OF ALASKA, FAIRBANKS

PHYSICS DEPARTMENT

Fairbanks, Alaska 99775-5920
http://www.uaf.edu/physics/graduate-programs/

General University Information
President: Patrick Gamble
Dean of Graduate School: John C. Eichelberger
University website: http://www.uaf.edu
Control: Public
Setting: Urban
Total Faculty: 1,024
Total number of Students: 11,149
Total number of Graduate Students: 1,117

Department Information
Department Chairman: Curt Szuberla, Chair
Department Contact: Dr. Renate Wackerbauer, Graduate
   Program Coordinator
   Total full-time faculty: 13
   Total number of full-time equivalent positions: 8
   Full-Time Graduate Students: 28
   First-Year Graduate Students: 4
   Total Post Doctorates: 2

Department Address
900 Yukon Drive, REIC 102
UAF Physics
Fairbanks, AK 99775-5920
Phone: (907) 474-6108
Fax: (907) 474-6130
E-mail: physics@uaf.edu
Website: http://www.uaf.edu/physics/graduate-programs/

ADMISSIONS

Admission Contact Information
Address admission inquiries to: Admissions, PO Box 757480,
   Fairbanks, AK 99775-7480
Phone: (800) 478-1823
E-mail: admissions@uaf.edu
Admissions website: http://www.uaf.edu/admissions

Application deadlines
Fall admission:
U.S. students: June 1   *Int'l. students*: March 3
Spring admission:
U.S. students: October 15   *Int'l. students*: September 1

Application fee
U.S. students: $60   *Int'l. students*: $60

Admissions information
For Fall of 2012:
   Number of applicants: 16
   Number admitted: 8
   Number enrolled: 4

Admission requirements
Bachelor's degree requirements: Physics or related field is re-
   quired.
Minimum undergraduate GPA: 3.0

GRE requirements
The GRE is required.

Advanced GRE requirements
The Advanced GRE is recommended.

TOEFL requirements
The TOEFL exam is required for students from non-English-
   speaking countries.
   PBT score: 550
   iBT score: 80
Must meet minimum TOEFL score to apply and be considered
   by department.

Other admissions information
Additional requirements: 3 letters of recommendation, transcripts, resume, statement of goals, GRE PHYS recom-
   mended.
Undergraduate preparation assumed: Classical Mechanics,
   Computational Physics, Electricity and Magnetism, Quantum
   Mechanics, Optics, Differential equations, linear algebra, nu-
   merical analysis and complex analysis; Mechanics: Taylor,
   Classical Mechanics; Electricity and Magnetism: Griffith, In-
   tro to Electrodynamics; Quantum mechanics: Griffith, Quantum mechanics; Optics: Hecht, Optics; Statistical Physics:
   Reif, Statistical and Thermal Physics.

TUITION
Tuition year 2012–13:
Tuition for in-state residents
   Full-time students: $3,447 per semester
   Part-time students: $383 per credit
Tuition for out-of-state residents
   Full-time students: $7,047 per semester
   Part-time students: $783 per credit
Credit hours per semester to be considered full-time: 9
Deferred tuition plan: Yes
Health insurance: Yes, included with TA/RA positions.
Other academic fees: Approximately $525/semester. These are
   mandatory fees incurred by UAF. www.uaf.edu/register/
   expenses/fee-chart/
Academic term: Semester
Number of first-year students who received full tuition waivers: 4

Teaching Assistants, Research Assistants, and Fellowships
Number of first-year
   Teaching Assistants: 10
   Fellowship students: 1
Average stipend per academic year
   Teaching Assistant: $17,987
   Research Assistant: $17,987
   Fellowship student: $20,845

FINANCIAL AID

Application deadlines
Fall admission:
U.S. students: June 1   *Int'l. students*: March 1
Spring admission:
U.S. students: September 1   *Int'l. students*: October 15

Loans

Loans are available for U.S. students.
Loans are not available for international students.
GAPSFAS application required: No
FAFSA application required: No

For further information

Address financial aid inquiries to: UAF Financial Aid, 101 Eielson Building, P.O. Box 756360, Fairbanks, AK 99775-6360.
Phone: (907) 474-7256 Toll free: (888) 474-7256
E-mail: financialaid@uaf.edu
Financial aid website: http://www.uaf.edu/finaid/

HOUSING

Availability of on-campus housing

Single students: Yes
Married students: Yes

For further information

Address housing inquiries to: UAF Department of Residence Life, 732 Yukon Drive, P.O. Box 756860, Fairbanks, AK 99775.
Phone: (907) 474-7247
E-mail: housing@uaf.edu
Housing aid website: http://www.uaf.edu/reslife/

Table A—Faculty, Enrollments, and Degrees Granted

Research Specialty	2012-13 Faculty	Enrollment Fall 2012		Number of Degrees Granted 2012-2013 (2008-2013)		
		Master's	Doctorate	Master's	Terminal Master's	Doctorate
Atmospheric Infrasound	2	2	2	1(3)	–	-(1)
Condensed Matter Physics	1	–	–	-(1)	–	–
Geophysics	1	1	2	-(1)	–	–
Nonlinear Dynamics and Complex Systems	2	4	1	1(2)	–	-(2)
Space Physics	6	4	11	-(3)	–	2(5)
Theoretical Physics	1	–	–	–	–	–
Other	–	1	–	-(1)	–	–
Total	**13**	**12**	**16**	**2(11)**	**–**	**2(8)**
Full-time Grad. Stud.	–	12	16	–	–	–
First-year Grad. Stud.	–	3	1	–	–	–

GRADUATE DEGREE REQUIREMENTS

Master's: The Master's degree in Physics is offered with a concentration in Physics, or Computational Physics, or Space Physics. (a) Thesis Option: The minimum number of credits that must be earned is 30 semester hours. A maximum of 12 credits may be devoted to thesis or to thesis and research. At least 21 credits in any master's program, including thesis and research, must be at the 600 level. A maximum of 9 semester hours of credit from another institution may be transferred to UAF and applied toward a master's degree upon approval of the student's advisory committee and the dean of the college or school in which the student is enrolled. A thesis is required and an oral defense of the thesis must be taken in conjunction with a comprehensive/final examination. The examining committees shall consist of the candidate's advisory committee. (b) Project Option: A non-thesis Master's degree requires a minimum of 33 credits. Three credits

must be devoted to a short research project resulting in a written report. Coursework requirements, totaling a maximum of 30 credits, are the same as those listed for part (a) Thesis Option. A student may apply for admission to candidacy for a specific master's degree if he/she is in good standing and has satisfied the following requirements: the student must have (1) satisfactorily completed at least 9 credits of graduate study at UAF, (2) received approval for the provisional thesis title if a thesis is required, and (3) received approval of the finalized Graduate Study Plan. All work toward the fulfillment of a master's degree must be completed in 7 years.

Doctorate: We offer a PhD degree in Physics and a PhD degree in Space Physics. In both cases, the degree of Doctor of Philosophy is granted for proven ability and scholarly attainment. There are no fixed credit requirements for this degree. However, coursework will be set by individual student's background, coursework requirements for passing Ph.D. comprehensive examination, and research requirements. The student chooses a major line of study and, with the advice of his/her advisory committee, such lines of study in related fields as are necessary for achievement of a thorough and scholarly knowledge of his/her subject. The committee and the student will prepare the student's graduate study plan for the degree, which, including applicable and acceptable work transferred from other institutions, shall represent approximately 3 full years of study beyond the bachelor's degree. Admission to graduate study does not imply admission to candidacy for a degree. The student should seek admission to candidacy approximately 1 year before completing the requirements for the doctorate. A student may be accepted as a candidate by his/her advisory committee after (1) completing the full-time equivalent of 2 academic years of graduate study, (2) completing at least 1 semester in residence at UAF, (3) finalizing the graduate study plan, (4) obtaining approval by the advisory committee of the title and synopsis of the dissertation, and (5) passing a written comprehensive examination administered by the Department. The dissertation, which is expected to represent the equivalent of at least 1 full academic year's work at the University of Alaska Fairbanks, must be a substantial contribution to knowledge. After submitting the dissertation, the candidate must pass an oral examination supporting the dissertation. The examining committee will consist of the student's advisory committee supplemented by additional examiners, including one from outside the candidate's college or school representing the Office of the Graduate School. All work toward the fulfillment of a Ph.D. degree must be completed within 10 years.

Other Degrees: Interdisciplinary Degrees (M.S. or Ph.D.). Students can create their own Master's or Doctorate degree program by combining course work in more than one discipline. They have the responsibility to design the program, organize a committee of faculty members to serve as advisors and make sure that it conforms to rigorous academic standards. The proposed program must differ significantly from and may not substitute for an existing UAF graduate degree program. The student may select no more than one half of his/her program credits from one existing graduate degree program.

SPECIAL EQUIPMENT, FACILITIES, OR PROGRAMS

The University houses a number of research centers and institutes that provide support facilities and services for graduate student research. A majority of department faculty hold joint appointments with the Geophysical Institute. Many are also affiliated with the International Arctic Research Center and/or the Arctic Region Supercomputing Center. Examples of facilities are: 1152 CPU processor Cray XK6m and 2816 processor Penguin Computing Cluster; the Poker Flat Research Range with an active rocket launching facility; research and a major state-of-the-art satellite remote-sensing facility to receive, process, and analyze synthetic aperture radar (SAR) data from European Space Agency, Japanese, Canadian, and U.S. spacecraft, machine, and electronic shops; a network of field sites including stations in Antarctica, Spitzbergen, and sites throughout Alaska; specialized optical, radiation and infrasound instrumentation.

FACULTY

Professor

Newman, David E., Ph.D., University of Wisconsin-Madison, 1993. *Computational Physics, Nonlinear Dynamics and Complex Systems, Plasma and Fusion.* Complex systems, turbulence, nonlinear dynamics, fusion plasma physics.

Olson, John, Ph.D., University of California, Los Angeles, 1970. *Acoustics, Atmosphere, Space Physics, Cosmic Rays.* Plasma wave propagation, atmospheric infrasound, digital signal processing, magnetospheric physics.

Otto, Antonius, Ph.D., Ruhr-Universität Bochum, 1987. *Atmosphere, Space Physics, Cosmic Rays, Computational Physics.* Space plasma theory and simulation.

Truffer, Martin, Ph.D., University of Alaska, Fairbanks, 1999. *Computational Physics, Geophysics.* Glacier dynamics, application of geophysical and borehole techniques to glaciology, numerical modeling of ice flow.

Associate Professor

Chowdhury, Ataur, Ph.D., Clark University, 1985. *Condensed Matter Physics.* Condensed matter physics.

Conde, Mark, Ph.D., University of Adelaide, 1991. *Astronomy, Atmosphere, Space Physics, Cosmic Rays.* Auroral processes and space weather.

Delamere, Peter A., Ph.D., University of Alaska, Fairbanks, 1998. *Atmosphere, Space Physics, Cosmic Rays, Computational Physics.* Comparative magnetospheric physics, numerical simulation of space plasmas using hybrid and multi-fluid techniques.

Makarevich, Roman, Ph.D., University of Saskatchewan, Saskatoon, 2003. *Atmosphere, Space Physics, Cosmic Rays.* Radio remote sensing of the ionosphere and magnetosphere.

Ng, Chung-Sang, Ph.D., Auburn University, Alabama, 1994. *Atmosphere, Space Physics, Cosmic Rays, Computational Physics, Plasma and Fusion.* Theoretical and computational plasma physics, with applications in space and fusion plasmas.

Price, Channon P., Ph.D., University of California, Santa Barbara, 1981. *Astrophysics, Quantum Foundations, Theoretical Physics.* Theoretical physics, astrophysics, quantum information.

Szuberla, Curt, Ph.D., University of Alaska, Fairbanks, 1997. *Acoustics, Atmosphere, Space Physics, Cosmic Rays.* Atmospheric infrasound, digital signal processing.

Wackerbauer, Renate A., Ph.D., Ludwig Maximilian University, Munich, 1995. *Computational Physics, Nonlinear Dynamics and Complex Systems.* Complex systems, nonlinear dynamics and chaos, modeling of biological systems and Arctic sea ice.

Assistant Professor

Zhang, Hui, Ph.D., Boston University, 2008. *Atmosphere, Space Physics, Cosmic Rays.* Space Plasma Physics, Magnetospheric Physics, Solar Wind-Magnetosphere Interaction.

Affiliate Professor

Bailey, Scott, Ph.D., University of Colorado, 1995. aeronomy of the atmosphere.

Bristow, William, Ph.D., University of Alaska, Fairbanks, 1992. *Atmosphere, Space Physics, Cosmic Rays, Electrical Engineering.* Space physics and upper atmospheric physics.

Carreras, Benjamin A., Ph.D., Valencia University, Spain, 1968. *Nonlinear Dynamics and Complex Systems, Plasma and Fusion.* Fusion plasma physics, complex systems, turbulence and transport.

Collins, Richard, Ph.D., University of Illinois, 1994. *Climate/Atmospheric Science.* Laser studies of the atmospheres.

Hampton, Donald, Ph.D., University of Alaska, Fairbanks, 1996. Optical Science Mgr, Poker Flat Research Range. *Atmosphere, Space Physics, Cosmic Rays, Climate/Atmospheric Science, Optics, Systems Science/Engineering.*

Heavner, Matthew, Ph.D., University of Alaska, Fairbanks, 2000. *Atmosphere, Space Physics, Cosmic Rays, Climate/Atmospheric Science.* Sensor network applications for geophysics, bioacoustics, and research experience for undergraduates.

Lummerzheim, Dirk, Ph.D., University of Alaska, Fairbanks, 1987. *Atmosphere, Space Physics, Cosmic Rays.* Penetration of auroral electrons and protons into the atmosphere and the subsequent optical emissions.

Pettit, Erin, Ph.D., University of Washington, 2003. Assistant Professor of Geophysics. *Energy Sources & Environment, Geophysics.* glacier dynamics with applications ranging from paleoclimatology to ice/ocean interactions.

Sanchez, Raul, Ph.D., Universidad Complutense de Madrid (Spain), 1997. *Nonlinear Dynamics and Complex Systems, Plasma and Fusion.* Computational physics, complex systems, turbulent transport, fusion plasma physics.

Simpson, William, Ph.D., Stanford University, 1995. Assoc. Professor of Physical Chemistry. *Chemical Physics, Climate/Atmospheric Science.* spectroscopy to study environmental/atmospheric chemistry.

Weingartner, Tom, Ph.D., North Carolina State University, 1990. Professor, Physical Oceanography. *Marine Science/Oceanography.* Physical oceanography of Alaskan continental shelves and slopes; Interdisciplinary marine research; Wind- and buoyancy-forced shelf circulation systems.

DEPARTMENTAL RESEARCH SPECIALTIES AND STAFF

Theoretical

Computational Physics. Computational plasma physics with applications in space and fusion plasmas, hybrid and multi-fluid techniques for space plasmas, numerical modeling of ice flow, nonlinear dynamics and chaos, turbulence, modeling of complex systems. Delamere, Newman, Ng, Otto, Truffer, Wackerbauer.

Geophysics. Glacier dynamics, numerical modeling of ice flow. Truffer.

Nonlinear Dynamics and Complex Systems. Complex systems, nonlinear dynamics and chaos, turbulence, network dynamics, modeling and time series analysis. Newman, Wackerbauer.

Space Physics. Theoretical and computational space plasma physics, magnetospheric physics, solar wind - magnetosphere interaction, auroral processes, comparative magnetospheric physics. Delamere, Ng, Otto.

Theoretical Physics. Diverse theoretical studies in astrophysics, quantum information. Price.

Experimental

Acoustics. Atmospheric Infrasound, Digital signal analysis, installation and operation of infrasonic arrays across the world. Olson, Szuberla.

Condensed Matter Physics. Semiconductors. Chowdhury.

Geophysics. Glacier dynamics, application of geophysical and borehole techniques to glaciology. Truffer.
Other.
Space physics. Space plasma physics, magnetospheric physics, solar wind - magnetosphere interaction, auroral processes and space weather; active launching of sounding rockets, radio remote sensing of the ionosphere and magnetosphere, Satellite Data analysis. Conde, Makarevich, Zhang.

View additional information about this department at
www.gradschoolshopper.com

ARIZONA STATE UNIVERSITY

DEPARTMENT OF PHYSICS

Tempe, Arizona 85287-1504
http://physics.asu.edu

General University Information
President: Michael M. Crow
Dean of Graduate School: Andrew Webber
University website: http://www.asu.edu
Control: Public
Setting: Urban
Total Faculty: 2,655
Total Graduate Faculty: 995
Total number of Students: 73,378
Total number of Graduate Students: 13,996

Department Information
Department Chairman: Peter Bennett, Chair
Department Contact: Araceli Vizcarra, Program Coordinator
Total full-time faculty: 48
Total number of full-time equivalent positions: 48
Full-Time Graduate Students: 121
First-Year Graduate Students: 43
Female First-Year Students: 8
Total Post Doctorates: 5

Department Address
P.O. Box 871504
Bateman Physical Sciences F-Wing
Tempe, AZ 85287-1504
Phone: (480) 965-3561
Fax: (480) 965-7954
E-mail: physics.grad@asu.edu
Website: http://physics.asu.edu

ADMISSIONS

Admission Contact Information
Address admission inquiries to: Director of Graduate Admissions, Arizona State University, 1120 South Cady Mall, Interdisciplinary Building, B Wing, Suite 285, Graduate Enrollment Services, Room 170, Mail Code 1003, Tempe, AZ 85287-1003
Phone: (480) 965-6113
E-mail: grad-ges@asu.edu
Admissions website: http://graduate.asu.edu/

Application deadlines
Fall admission:
U.S. students: January 15   *Int'l. students*: January 15
Spring admission:
U.S. students: September 1   *Int'l. students*: September 1

Application fee
U.S. students: $70   *Int'l. students*: $90

Admissions information
For Fall of 2013:
  Number of applicants: 148
  Number admitted: 61
  Number enrolled: 32

Admission requirements
Bachelor's degree requirements: Bachelor's degree in physics or a closely related program is required with a minimum undergraduate GPA of 3.0 specified and a minimum Junior-Senior GPA of 3.0.
Minimum undergraduate GPA: 3.0

GRE requirements
The GRE is required.
  Quantitative score: 150
  Verbal score: 150
  Analytical score: 3
  Mean GRE score range (25th–75th percentile): 1240-1380
Mean GRE scores have been calculated using the old grading scale.

Advanced GRE requirements
The Advanced GRE is required.
  Minimum accepted Advanced GRE score: 600
  Mean Advanced GRE score range (25th–75th percentile): 650-800
Undergraduate/Graduate GPA plus GRE scores considered to determine admission. Minimum and Mean Physics GRE scores have been calculated using the old grading scale.

TOEFL requirements
The TOEFL exam is required for students from non-English-speaking countries.
  PBT score: 550
  iBT score: 84

Other admissions information

Additional requirements: Minimum expected GRE scores: 500 verbal, 650 quantitative, 500 analytic. The GRE Physics subject exam is highly recommended, although not mandatory.

Undergraduate preparation assumed: Sturge, Statistical and Thermal Physics; Griffiths, Introduction to Electrodynamics; Arfken, Mathematical Methods for Physicists; Fowles and Cassiday, Analytic Mechanics; Griffiths, Introduction to Quantum Mechanics; Zettili, Quantum Mechanics, Concepts and Applications.

TUITION

Tuition year 2012–13:

Tuition for in-state residents
Full-time students: $10,515 annual
Part-time students: $868 per credit

Tuition for out-of-state residents
Full-time students: $18,928 annual
Part-time students: $1,177 per credit

See Graduate Catalog for more detailed information. Nonresident tuition is waived for students with Graduate Assistantships.

Credit hours per semester to be considered full-time: 9

Deferred tuition plan: No

Health insurance: Available at the cost of $1,956 per year.

Other academic fees: Health and Wellness Fee; Technology Fee; Recreation Fee; FA Trust Fee; Tempe Arizona Student Associate Fee; Student Programs Fee.

Academic term: Semester

Number of first-year students who received full tuition waivers: 19

Teaching Assistants, Research Assistants, and Fellowships

Number of first-year
Teaching Assistants: 16
Research Assistants: 1
Fellowship students: 2

Average stipend per academic year
Teaching Assistant: $15,631
Research Assistant: $15,631
Fellowship student: $22,500

FINANCIAL AID

Application deadlines

Fall admission:
U.S. students: June 30   *Int'l. students*: June 30

Loans

Loans are not available for U.S. students.
Loans are not available for international students.

GAPSFAS application required: No

FAFSA application required: No

For further information

Address financial aid inquiries to: Graduate Admissions, Interdisciplinary B-Wing, Room-285.

Phone: (480) 965-3521

E-mail: grad-financial@asu.edu

Financial aid website: http://graduate.asu.edu/financing

HOUSING

Availability of on-campus housing

Single students: Yes
Married students: No

For further information

Address housing inquiries to: Residence Life, Student Services Bldg.

Phone: (480) 965-3515

E-mail: housing@asu.edu

Housing aid website: http://www.asu.edu/housing/

Table A—Faculty, Enrollments, and Degrees Granted

Research Specialty	2012-13 Faculty	Enrollment Fall 2012		Number of Degrees Granted 2012-13		
		Master's	Doctorate	Master's	Terminal Master's	Doctorate
Biophysics	11	–	19	–	–	2(-)
Nano Science and Technology	21	14	45	–	5(-)	11(-)
Particles and Fields	12	–	23	3(-)	–	5(-)
Physics and other Science Education	4	6	–	–	3(-)	–
Total	48	20	87	3(-)	8(-)	18(-)
Full-time Grad. Stud.	–	20	19	–	–	–
First-year Grad. Stud.	–	14	19	–	–	–

GRADUATE DEGREE REQUIREMENTS

Master's: Thirty semester hours required with 3.0 average; at least 18 semester hours resident credit; no language requirement. Final written examination required.

Doctorate: Eighty-four semester hours with 3.0 average; at least two semesters continuous full time residence after first year (30 semester hours); at least 30 semester hours earned in residence, with no language requirement. Written, oral comprehensive examination; thesis prospectus; dissertation required; final oral examination required.

Other Degrees: Interdepartmental program in science education, natural sciences, and mathematics for teachers. Does not replace usual teacher certification requirements. Thirty semester hours required with 3.0 average; at least 18 semester hours resident credit; no language requirement; not more than 21 semester hours credit (including thesis or project credit) from any one area of natural sciences or mathematics may be applied toward this degree. Thesis or project required; final examination, written, oral, or both is required. Professional Science Masters in Nanoscience Interdisciplinary program spanning physics, chemistry, and biochemistry, materials, and electrical engineering. 30 credit hours either on accelerated 12-month track or part-time 24-month track. 15 hours of core courses (including applied project) and 15 hours elective courses. No thesis requirement.

Thesis: Thesis may be written in absentia.

SPECIAL EQUIPMENT, FACILITIES, OR PROGRAMS

Optional M.S. programs in interdisciplinary physics, technical physics, and physics teaching. Optional Ph.D. program in applied physics. Experimental and theoretical diffraction physics and electron microscopy; experimental and theoretical sub-atomic physics; experimental and theoretical condensed matter and materials physics; observational and theoretical astrophysics; science education; interdisciplinary Center for Solid State Science; Center for High-Resolution Electron Microscopy; Surface Science and Ion Beam Analysis facilities; access to University of Arizona and other national and international facilities.

Table B—Separately Budgeted Research Expenditures by Source of Support

Source of Support	Departmental Research	Physics-related Research Outside Department
Federal government	$7,495,089	$987,081
State/local government	$8,026	$41,174
Non-profit organizations	$32,338	
Business and industry	$193,943	$64,648
Other	$99,461	$4,998
Total	**$7,828,857**	**$1,097,901**

Table C—Separately Budgeted Research Expenditures by Research Specialty

Research Specialty	No. of Grants	Expenditures ($)
Biophysics	170	$4,417,160
Nano Science and Technology	116	$3,187,948
Particles and Fields	23	$1,289,238
Physics and other Science Education	3	$32,412
Total	**312**	**$8,926,758**

FACULTY

Chair Professor

Bennett, Peter A., Ph.D., University of Wisconsin-Madison, 1980. Department Chair. *Condensed Matter Physics*. Experimental surface physics; epitaxial growth.

Professor

Alarcon, Ricardo O., Ph.D., Ohio University, 1985. Department Associate Chair. *Astrophysics, Nuclear Physics, Particles and Fields*. Experimental intermediate-energy nuclear physics.

Belitsky, Andrei V., Ph.D., Bogoliubov Laboratory of Theoretical Physics, Joint Institute for Nuclear Research, Dubna, Russia, 1996. *Particles and Fields*. Elementary particle physics, field theory, string theory.

Chamberlin, Ralph V., Ph.D., University of California, Los Angeles, 1984. Graduate Program Director. *Condensed Matter Physics*. Experimental condensed matter physics; dynamics of complex systems; nanothermodynamics.

Comfort, Joseph R., Ph.D., Yale University, 1968. *Nuclear Physics, Particles and Fields*. Experimental nuclear physics; low- and medium-energy nuclear reactions and spectroscopy; reaction mechanisms; nuclear structure models.

Davies, Paul C. W., Ph.D., University College London, 1970. Director of Beyond Center for Fundamental Concepts in Science & Co-Director of Cosmology Initiative. *Astrophysics*. Astrophysics and cosmology.

Doak, R. Bruce, Ph.D., Massachusetts Institute of Technology, 1981. *Condensed Matter Physics*. Helium scattering studies of surface phonon physics; time-resolved measurements of dynamical processes at surfaces; tunable ultracold He beams; novel He beam scattering experiments.

Drucker, Jeff, Ph.D., University of California, Santa Barbara, 1986. *Condensed Matter Physics*. Synthesis and characterization of nanostructured electronic materials for novel opto electronic and photonic applications.

Krauss, Lawrence M., Ph.D., Massachusetts Institute of Technology, 1982. Foundation Professor, and Inaugural Director of the Origins Initiative & Co-Director of the Cosmology Initiative. *Cosmology & String Theory, Particles and Fields*. Elementary particle physics and cosmology.

Lebed, Richard F., Ph.D., University of California, Berkeley, 1994. *Particles and Fields*. Elementary particle theory. Hadronic physics, fundamental symmetries.

Lindsay, Stuart M., Ph.D., University of Manchester, 1976. Director of the Center for Single Molecule Biophysics. *Biophysics*. Biophysics and nanoscale physics; scanning probe microscopy and nanofabrication; molecular electronics.

Liu, Jingyue, Ph.D., Arizona State University, 1990. Director of the Professional Science Masters-Nanoscience Program. *Nano Science and Technology*. Nanomaterials for catalysis and application of high resolution electron microscopy to nanomaterials.

Matyushov, Dmitry, Ph.D., Kiev State University, Ukraine, 1987. Director of the Center of Biological Physics. *Biophysics*. Biophysics, electron transfer, theoretical chemistry.

Mauskopf, Philip, Ph.D., University of California, Berkeley, 1997. *Astrophysics, Particles and Fields*. Cosmology and astronomical measurements at millimeter/sub-millimeter wavelengths.

McCartney, Martha, Ph.D., Arizona State University, 1989. *Condensed Matter Physics*. Electron microscopy techniques, leading expert on electron holography.

Menendez, José, Ph.D., Stüttgart University, 1985. *Condensed Matter Physics*. Experimental condensed matter physics.

Nelson, Alan C., Ph.D., University of California, Berkeley, 1980. *Biophysics*. Biophysics in cancer early detection, imaging devices, image reconstructions and diagnostic knowledge generation.

Nemanich, Robert J., Ph.D., University of Chicago, 1976. Department Chair. *Condensed Matter Physics*. Experimental surface science and nanoscience.

Ponce, Fernando A., Ph.D., Stanford University, 1997. *Materials Science, Metallurgy, Nano Science and Technology*. Microscopic properties of electronic materials for applications in microelectronics, photonics and optoelectronics.

Rez, Peter, Ph.D., University of Oxford, 1976. *Biophysics*. Electron diffraction and microscopy; medical physics; biophysics; solid state theory.

Ritchie, Barry, Ph.D., University of South Carolina, 1979. Vice Provost. *Nuclear Physics*. Experimental medium-energy nuclear physics.

Schmidt, Kevin E., Ph.D., University of Illinois at Urbana-Champaign, 1979. *Condensed Matter Physics*. Theoretical solid state physics; computational many-body theory.

Smith, David J., Ph.D., University of Melbourne, 1978. *Condensed Matter Physics*. Electron diffraction and high-resolution electron microscopy; electron holography; magnetic materials, and semiconductors.

Spence, John C. H., Ph.D., University of Melbourne, 1974. *Condensed Matter Physics*. Ultrahigh resolution electron microscopy; electron channeling; electron microscope contrast theory; excitations in solids by inelastic electron scattering; surface physics; STM; biophysics; x-ray imaging; nanolithography by stem.

Thorpe, Michael F., Other, University of Oxford, 1968. *Biophysics, Condensed Matter Physics*. Theoretical molecular biophysics and soft condensed matter physics.

Treacy, Mike, Ph.D., University of Cambridge, 1980. Director of Undergraduate Programs. *Condensed Matter Physics*. Diffraction physics, complex materials. Fluctuation microscopy of disordered materials; enumeration of hypothetical framework materials; diffraction phenomena.

Tsen, Kong-Thon, Ph.D., Purdue University, 1983. *Condensed Matter Physics*. Experimental condensed matter physics.

Tsong, Ignatius S. T., Ph.D., University of London, 1970. *Condensed Matter Physics*. Surface physics; epitaxial thin-film growth processes; wide band gap semiconductors.

Vachaspati, Tanmay, Ph.D., Tufts University, 1985. Director of the Cosmology Initiative. *Particles and Fields*. Theoretical cosmology, particle physics, and gravitational physics.

Venables, John A., Ph.D., University of Cambridge, 1961. *Condensed Matter Physics*. Electron microscopy and surface physics; atomic processes in adsorption and crystal growth; modeling and graduate education, using the Internet.

Associate Professor

Culbertson, Robert J., Ph.D., Pennsylvania State University, 1979. Director of the Masters of Natural Science Program. *Condensed Matter Physics, Physics and other Science Education*. Studies of surface modification and characterization of materials, crystal surfaces, and interfaces using ion beams.

Lunardini, Cecilia, Ph.D., SISSA, Trieste, 2001. *Astrophysics, Particles and Fields*. Supernovae neutrinos, neutrino-matter interactions, particle astrophysics.

Marzke, Robert, Ph.D., Columbia University, 1966. *Condensed Matter Physics*. Experimental solid state and chemical physics; NMR, studies of molten ceramics, electrolytes, catalysts and collagen. Biomechanics of hand function.

Ozkan, S. Banu, Ph.D., Bogazici University, Istanbuly, 2001. *Biophysics*. Biophysics, modeling, protein folding dynamics.

Parikh, Maulik, Ph.D., Princeton University, 1998. *Cosmology & String Theory*. Theoretical physics; black holes; cosmology; classical and quantum gravity; string theory.

Ros, Robert, Ph.D., University of Basel, 2000. *Biophysics*. Nano-biophysics, structural biology, molecular recognition.

Shumway, John B., Ph.D., University of Illinois at Urbana-Champaign, 1999. *Condensed Matter Physics*. Theoretical condensed matter physics; quantum dots; semiconductors; path-integrals.

Assistant Professor

Beckstein, Oliver, Ph.D., University of Oxford, 2005. *Biophysics*. Biophysics.

Chen, Tingyong, Ph.D., Johns Hopkins University, 2006. *Condensed Matter Physics*. Experimental condensed matter physics; magnetism, superconductivity, nanostructures and nano-materials.

Easson, Damien A., Ph.D., Brown University, 2002. *Particles and Fields*. Particle cosmology; cosmology of the early universe; quantum aspects of gravity.

Qing, Quan, Ph.D., Peking University, 2006. *Biophysics*.

Vaiana, Sara M., Ph.D., University of Palermo, 2004. *Biophysics*. Biophysics, protein self-assembly.

Emeritus

Bauer, Ernst, Ph.D., Universitat Munchen, 1955. Distinguished Research Professor. *Condensed Matter Physics*. Surface and thin film physics, in particular surface electron growth.

Dow, John D., Ph.D., University of Rochester, 1967. *Condensed Matter Physics*. Theoretical solid state physics; theory of materials; experimental scanning tunneling microscopy and surface physics.

Herbots, Nicole, Ph.D., U. Catholique de Louvain, Belgium, 1984. *Condensed Matter Physics*. Synthesis of new thin-film heterostructures by combined ion and molecular deposition and characterization by a wide variety of modern analysis techniques.

Hestenes, David O., Ph.D., University of California, Los Angeles, 1963. *Physics and other Science Education*. Theoretical foundations of physics; relativistic electron theory; physics education research and development.

Jacob, Richard J., Ph.D., University of Utah, 1963. *Nuclear Physics, Particles and Fields*. Elementary particle theory; intermediate-energy hadronic interactions.

Kaufmann, William B., Ph.D., University of California, Berkeley, 1968. *Nuclear Physics, Particles and Fields*. Theoretical intermediate-energy; nuclear and elementary particle physics.

Mayer, James, Ph.D., Purdue University, 1960. *Condensed Matter Physics*. Ion beam analysis; ion implantation in silicon thin film reactions.

Page, John B., Ph.D., University of Utah, 1966. *Condensed Matter Physics*. Condensed matter theory; dynamical localization in nonlinear lattices, first-principles studies of fullerenes and fullerene polymers; resonance Raman scattering; phonons and electron-phonon interactions in solids.

Sankey, Otto F., Ph.D., Washington University, 1979. *Biophysics, Condensed Matter Physics*. Theoretical solid state physics; molecular electronics; biophysics.

Research Professor

Jiang, Nan, Ph.D., The University of Birmingham, 1998. *Biophysics*.

Weierstall, Uwe, Ph.D., Eberhard Karls University Tübingen, 1994. *Condensed Matter Physics*. Multidisciplinary research based on electron and X-ray diffraction.

Lecturer

Adams, Gary, Ph.D., Arizona State University, 1992. First principles simulations of semiconductor surfaces and fullerenes, fullerene derivatives and carbon nanotubes.

Covatto, Carl, Ph.D., Arizona State University, 2002. Focused on calculating the evolution of grain size distributions in the outflows of cool starts.

Hakhoyan, Armen, Ph.D., Ukrainian National Academy of Sciences, 1986. *Physics and other Science Education*. Instructional resource development and teaching.

Makarova, Darya V., Ph.D., Herzen State Pedagogical University of Russia, 2006. *Physics and other Science Education*. Theory and methods of physics teaching.

DEPARTMENTAL RESEARCH SPECIALTIES AND STAFF

Theoretical

Biological Physics and Biophysics. The biological physics group studies biological systems from the molecular to the cell level. With improved experimental data, biology is becoming much more quantitative. At ASU, we are researching the underlying principles involved in the machinery of living things and searching for unifying themes both within and between organisms in an interdisciplinary environment. Biological physics at ASU is a leader in this area and welcomes inquiries from prospective physics graduate students who would like to join one of our exciting research areas. Beckstein, Lindsay, Matyushov, Nelson, Qing, Rez, Ros, Sankey, Spence, Thorpe, Vaiana, Weierstall.

Cosmology, Particle Physics and Astrophysics. Cosmology, Particle Physics and Astrophysics research at ASU specializes in several areas. Major focus is upon particles whose constituents interact so strongly that their interactions cannot be handled using perturbative techniques, and is upon key experiments designed to test nature's fundamental symmetries. For example, the structure of hadrons (particles composed of quarks and gluons, which interact by means of the quantum field theory called quantum chromodynamics [QCD]), is one of our areas of theoretical and experimental specialty. Alarcon, Belitsky, Comfort, Davies, Easson, Krauss, Lebed, Lunardini, Mauskopf, Parikh, Ritchie, Vachaspati.

Education and Societal Impact. Physics interacts with society in many important ways. Within the university, the physics department teaches many undergraduate classes needed for future engineers and for many other professions. The general studies program involves most of our faculty and our graduate students who serve as teaching assistants. We offer professional degrees through the Master of Natural Science (M.N.S.) and Professional Science Master (P.S.M.) programs. Adams, Covatto, Culbertson, Hakhoyan, Makarova.

Nanoscience and Materials Physics. At the nanometer length scale, materials and structures behave differently, offering exciting new opportunities for scientific discoveries as well as technological advances. Our faculty are working to define the cutting edge in many aspects of nanoscale physics. ASU is well known for its John M. Cowley Center for High Resolution Electron Microscopy, where researchers use and develop new techniques for probing structural, magnetic, electronic and optical properties at the nanoscale. Bauer, Chamberlin, Chen, Culbertson, Doak, Dow, Drucker, Herbots, Liu, Marzke, McCartney, Menendez, Nemanich, Page, Ponce, Rez, Schmidt, Shumway, Smith, Treacy, Tsen, Tsong, Venables.

Experimental

Biological Physics and Biophysics. The biological physics group studies biological systems from the molecular to the cell level. With improved experimental data, biology is becoming much more quantitative. At ASU, we are researching the underlying principles involved in the machinery of living things and searching for unifying themes both within and between organisms in an interdisciplinary environment. Biological physics at ASU is a leader in this area and welcomes inquiries from prospective physics graduate students who would like to join one of our exciting research areas. Beckstein, Lindsay, Matyushov, Nelson, Qing, Rez, Ros, Sankey, Spence, Thorpe, Vaiana, Weierstall.

Cosmology, Particle Physics and Astrophysics. Cosmology, Particle Physics, and Astrophysics research at ASU specializes in several areas. Major focus is upon particles whose constituents interact so strongly that their interactions cannot be handled using perturbative techniques, and is upon key experiments designed to test nature's fundamental symmetries. For example, the structure of hadrons (particles composed of quarks and gluons, which interact by means of the quantum field theory called quantum chromodynamics [QCD]), is one of our areas of theoretical and experimental specialty. Alarcon, Belitsky, Comfort, Davies, Easson, Krauss, Lebed, Lunardini, Mauskopf, Parikh, Ritchie, Vachaspati.

Education and Societal Impact. Physics interacts with society in many important ways. Within the university, the physics department teaches many undergraduate classes needed for future engineers and for many other professions. The general studies program involves most of our faculty and our graduate students who serve as teaching assistants. We offer professional degrees through the Master of Natural Science (M.N.S.) and Professional Science Master (P.S.M.) programs. Adams, Covatto, Hakhoyan, Makarova.

Nanoscience and Materials Physics. At the nanometer length scale, materials and structures behave differently, offering exciting new opportunities for scientific discoveries as well as technological advances. Our faculty are working to define the cutting edge in many aspects of nanoscale physics. ASU is well known for its John M. Cowley Center for High Resolution Electron Microscopy, where researchers use and develop new techniques for probing structural, magnetic, electronic and optical properties at the nanoscale. Bauer, Chamberlin, Chen, Culbertson, Doak, Dow, Drucker, Herbots, Liu, Marzke, McCartney, Menendez, Nemanich, Page, Ponce, Rez, Schmidt, Shumway, Smith, Treacy, Tsen, Tsong, Venables, Weierstall.

View additional information about this department at
www.gradschoolshopper.com

NORTHERN ARIZONA UNIVERSITY

PHYSICS & ASTRONOMY

Flagstaff, Arizona 86011
http://www.physics.nau.edu

General University Information
President: John Haeger
Dean of Graduate School: Ramona Mellott
University website: http://home.nau.edu
Control: Public
Setting: Urban
Total Faculty: 1,496
Total number of Students: 25,364
Total number of Graduate Students: 4,614

Department Information
Department Chairman: Stephen C. Tegler, Chair
Department Contact: Jamie Housholder, Administrative
  Associate
  Total full-time faculty: 14

Total number of full-time equivalent positions: 14
Full-Time Graduate Students: 15
First-Year Graduate Students: 9
Female First-Year Students: 3
Total Post Doctorates: 2

Department Address
PO Box 6010
Flagstaff, AZ 86011
Phone: (928) 523-2661
Fax: (928) 523-1371
E-mail: astro.physics@nau.edu
Website: http://www.physics.nau.edu

ADMISSIONS

Admission Contact Information
Address admission inquiries to: NAU Graduate College, P.O.
 Box 4125, Flagstaff, AZ 86011-4125
Phone: (928) 523-4348
E-mail: graduate@nau.edu
Admissions website: http://home.nau.edu/gradcol

Application deadlines
Fall admission:
U.S. students: January 15   *Int'l. students*: January 15

Application fee
U.S. students: $65

Admissions information
For Fall of 2013:
 Number of applicants: 22
 Number admitted: 15
 Number enrolled: 9

Admission requirements
Bachelor's degree requirements: Bachelors degree in Physics,
 Astronomy, Chemistry, or a related field is required.
Minimum undergraduate GPA: 3.0

GRE requirements
The GRE is not required.

Advanced GRE requirements
The Advanced GRE is recommended.

TOEFL requirements
The TOEFL exam is required for students from non-English-
 speaking countries.
 iBT score: 80

Other admissions information
Additional requirements: The minimum acceptable score suggested for admission is quantitative 625.
 No minimum score is specified. The average GRE scores for
 admissions were not specified. The average GRE Advanced
 for admissions was not specified.
Undergraduate preparation assumed: Although preparation will
 vary, we expect at least one semester each of upper-division
 mechanics, electricity and magnetism, quantum mechanics,
 and advanced laboratory.

TUITION

Tuition year 2012–13:
Tuition for in-state residents
 Full-time students: $7,560 annual
Tuition for out-of-state residents
 Full-time students: $18,654 annual
Academic term: Semester
Number of first-year students who received full tuition waivers: 9

Teaching Assistants, Research Assistants, and Fellowships
 Number of first-year
 Teaching Assistants: 9
 Average stipend per academic year
 Teaching Assistant: $12,390
 Research Assistant: $13,300

FINANCIAL AID

Application deadlines
Fall admission:
U.S. students: March 1

Loans
Loans are available for U.S. students.
Loans are not available for international students.
GAPSFAS application required: No
FAFSA application required: Yes

For further information
Address financial aid inquiries to: Office of Student Financial
 Aid, Northern Arizona University, Box 4108, Flagstaff, AZ
 86011.
Phone: (928) 523-4951
E-mail: Financial.Aid@nau.edu
Financial aid website: http://www.nau.edu/FinAid/Welcome

HOUSING

Availability of on-campus housing
 Single students: Yes
 Married students: Yes

For further information
Address housing inquiries to: Office of Residence Life, Northern
 Arizona University, Box 6100, Flagstaff, AZ 86011.
Phone: (928) 523-3978
E-mail: Residence.Life@nau.edu
Housing aid website: http://www.nau.edu/Residence-Life/

Table A—Faculty, Enrollments, and Degrees Granted

Research Specialty	2012-13 Faculty	Enrollment Fall 2012		Number of Degrees Granted 2012-13 (2008-13)		
		Master's	Doctorate	Master's	Terminal Master's	Doctorate
Astrophysics	5	7	–	6(16)	–	–
Condensed Matter Physics	2	2	–	1(5)	–	–
Optics	1	5	–	4(4)	–	–
Physics and other Science Education	2	–	–	–	–	–
Other	3	4	–	3(10)	–	–
Total	13	18	–	14(35)	–	–
Full-time Grad. Stud.	–	18	–	–	–	–
First-year Grad. Stud.	–	4	–	–	–	–

GRADUATE DEGREE REQUIREMENTS

Master's: Thirty-six hours of graduate courses, including both
 thesis and non-thesis options. There is no foreign language
 requirement. Qualifying and comprehensive exams are not required. This program can be interdisciplinary, integrating a
 broad range of subject areas to enhance student opportunities
 in the industrial or research world. Individual programs may
 be customized to meet specific student needs.
Thesis: Thesis may be written in absentia.

SPECIAL EQUIPMENT, FACILITIES, OR PROGRAMS

Our Department has access to a campus observatory, ultra-high
vacuum laboratories, PHI x-ray photoelectron spectroscopy instrumentation, scanning force microscopy, infrared spectroscopy
(Nicolet), mass spectroscopy, Unix workstation laboratory, and
computational physics laboratory. Local astrophysical facilities
include Lowell Observatory, U.S. Naval Observatory, and NAU
has access to the University of Arizona telescopes.

Table B—Separately Budgeted Research Expenditures by Source of Support

Source of Support	Departmental Research	Physics-related Research Outside Department
Federal government	$1,089,688	
State/local government		
Non-profit organizations		
Business and industry	$9,000	
Other		
Total	$1,098,688	

Table C—Separately Budgeted Research Expenditures by Research Specialty

Research Specialty	No. of Grants	Expenditures ($)
Astrophysics	19	$970,232
Condensed Matter Physics	1	$35,192
Optics	2	$42,780
Physics and other Science Education	–	$41,484
Total	**22**	**$1,089,688**

FACULTY

Professor

Barlow, Nadine, Ph.D., University of Arizona, 1987. Planetary astronomy.

Delinger, William G., Ph.D., University of Iowa, 1972. Solid state, electronic instrumentation, solar energy, computers.

Dillingham, T. Randall, Ph.D., Kansas State University, 1983. Surface physics, x-ray photoelectron spectroscopy, surface chemistry of ices.

Eastwood, Kathleen DeGioia, Ph.D., University of Wyoming, 1982. Optical and infrared astronomy; star formation, interstellar medium, stellar populations.

Tegler, Stephen C., Ph.D., Arizona State University, 1989. Optical and infrared astronomy, Kuiper belt objects, icy dwarf planets, laboratory astrophysics.

Associate Professor

Bowman, Gary E., Ph.D., University of Notre Dame, 2000. Foundations of quantum mechanics.

James, Mark C., Ph.D., Kansas State University, 2003. Science education.

Koerner, David, Ph.D., California Institute of Technology, 1994. Origin of planetary systems.

Trilling, David, Ph.D., University of Arizona, 1999. Planetary science.

Assistant Professor

Mann, Christopher J., Ph.D., University of South Florida, 2006. Optics, interferometry, digital holography.

Lecturer

Cole, David M., Ph.D., Texas A&M University, 1997. Magnetic resonance imaging, science education.

Dolle, Ethan, Ph.D., University of Arizona, 2009. Physics beyond the Standard Model, particle phenomenology, particle dark matter.

DEPARTMENTAL RESEARCH SPECIALTIES AND STAFF

Theoretical

High Energy Physics. Physics beyond the Standard Model, particle phenomenology, particle dark matter. Dolle.

Quantum Foundations. Foundations of quantum mechanics, Bohmian mechanics, quantum chaos, decoherence and the classical limit. Bowman.

Experimental

Astrophysics. Planetary science, origins of solar systems, star formation and evolution, optical and infrared astronomy, unmanned space missions. Barlow, Eastwood, Koerner, Tegler, Trilling.

Condensed Matter Physics. Inorganic/Organic materials, semiconductor physics, microsensor development, polymer physics. Delinger, Dillingham.

Optics. Interferometry, digital holography. Mann.

Physics and other Science Education. Science education, action-based research. Cole, James.

View additional information about this department at
www.gradschoolshopper.com

UNIVERSITY OF ARIZONA

DEPARTMENT OF ASTRONOMY

Tucson, Arizona 85721-0065
http://www.as.arizona.edu

General University Information

President: Ann Weaver Hart
Dean of Graduate School: Andrew Carnie
University website: http://www.arizona.edu
Control: Public
Setting: Urban
Total Faculty: 3,022
Total Graduate Faculty: No Formal Graduate Faculty

Total number of Students: 40,223
Total number of Graduate Students: 8,658

Department Information

Department Chairman: Buell T. Jannuzi, Head
Department Contact: Michelle Cournoyer, Program Coordinator, Senior

Total full-time faculty: 32
Full-Time Graduate Students: 43
First-Year Graduate Students: 7
Female First-Year Students: 3
Total Post Doctorates: 21

Department Address
933 North Cherry Avenue
Tucson, AZ 85721-0065
Phone: (520) 621-2288
Fax: (520) 621-1532
E-mail: michelle@email.arizona.edu
Website: http://www.as.arizona.edu

ADMISSIONS

Admission Contact Information
Address admission inquiries to: Chairman, Admissions Committee, Steward Observatory, The University of Arizona.
Phone: (520) 621-2288
E-mail: gradadmissions@as.arizona.edu
Admissions website: http://www.as.arizona.edu

Application deadlines
Fall admission:
U.S. students: January 10   *Int'l. students*: December 1

Application fee
U.S. students: $75   *Int'l. students*: $75

Admissions information
For Fall of 2013:
Number of applicants: 109
Number admitted: 20
Number enrolled: 3

Admission requirements
Bachelor's degree requirements: Bachelor's degree in astronomy or physics is preferred.
Minimum undergraduate GPA: 3.5

GRE requirements
The GRE is required.

Advanced GRE requirements
The Advanced GRE is required.
The Physics GRE is required.

TOEFL requirements
The TOEFL exam is required for students from non-English-speaking countries.
PBT score: 550
iBT score: 79

Other admissions information
Additional requirements: No minimum acceptable score is specified. The average GRE scores for 2013 admissions were: verbal–157; quantitative–159; total–316. The average GRE Physics score for 2013 admissions was 670.
Undergraduate preparation assumed: Normal preparation in mathematics through advanced calculus; physics, including classical mechanics, electromagnetism, quantum mechanics, and thermodynamics.

TUITION
Tuition year 2013-14:
Tuition for in-state residents
Full-time students: $11,525 annual
Tuition for out-of-state residents
Full-time students: $27,398 annual
Part-time students: $1,559 per credit

See graduate catalog for more detailed information. Nonresident tuition is waived for students with graduate assistantships.
Credit hours per semester to be considered full-time: 6
Deferred tuition plan: No
Health insurance: Available at the cost of 2,016 per year.
Other academic fees: Registration: Average total cost with fees (In-State) $5,254 (Out-of-State) $13,035.
Academic term: Semester

Teaching Assistants, Research Assistants, and Fellowships
Number of first-year
Research Assistants: 3
Fellowship students: 4
Average stipend per academic year
Teaching Assistant: $18,088
Research Assistant: $18,088
Fellowship student: $18,088
Summer salary: $12,087.

FINANCIAL AID

Application deadlines
Fall admission:
U.S. students: January 10   *Int'l. students*: December 1

Loans
Loans are available for U.S. students.
Loans are not available for international students.
GAPSFAS application required: No
FAFSA application required: Yes

For further information
Address financial aid inquiries to: Admissions Chairman.
Phone: (520) 621-2288
E-mail: gradadmissions@as.arizona.edu
Financial aid website: http://grad.arizona.edu/financial-resources

HOUSING

Availability of on-campus housing
Single students: Yes
Married students: Yes

For further information
Address housing inquiries to: Department of Student Housing.
Phone: 520-621-6501
E-mail: housing@life.arizona.edu
Housing aid website: http://www.life.arizona.edu/home/

Table A—Faculty, Enrollments, and Degrees Granted

Research Specialty	2012-13 Faculty	Enrollment Fall 2012 Master's	Enrollment Fall 2012 Doctorate	Number of Degrees Granted 2012-13 (2008-13) Master's	Terminal Master's	Doctorate
Astronomy	32	–	43	–	2(5)	8(26)
Total	32	–	43	–	2(5)	8(26)
Full-time Grad. Stud.	–	–	43	–	–	–
First-year Grad. Stud.	–	–	7	–	–	–

GRADUATE DEGREE REQUIREMENTS
Master's: Thirty semester hours credit. Thesis with final oral exam required. A GPA of 3.0 (A = 4) must be maintained. There is a 30-week residence requirement. No language requirement or written comprehensive or qualifying examination. Normal admission is only to the Ph.D. program.

Doctorate: Successful completion of course requirements; second year Independent Research; Ph.D. qualifying exam; and dissertation defense.

Thesis: Thesis may be written in absentia.

SPECIAL EQUIPMENT, FACILITIES, OR PROGRAMS

The primary research telescopes of the Steward Observatory include the 6.5-m MMT located on the summit of Mt. Hopkins in the Santa Rita Mountains, the 90-in. (2.3-m) Ritchey-Chrétien reflector on Kitt Peak, the 61-in. (1.55-m) Cassegrain reflector on Mt. Bigelow in the Santa Catalina Mountains, and the Vatican Advanced Technology Telescope (VATT) on Mt. Graham. The VATT has a 1.8-m f/1.0 primary mirror fabricated at the Steward Observatory Mirror Lab; Steward Observatory staff receive 25% of the scheduled observing time on this telescope. The 6.5-m telescope of the MMT Observatory is jointly operated by The University of Arizona and the Smithsonian Astrophysical Observatory.

Steward Observatory is a partner in the twin 6.5-m Magellan telescopes at Las Campanas Observatory in Chile (built with Mirror Lab mirrors), giving students and researchers access to the southern skies. The observatory is currently involved in developing several large (8.4-m) optical telescopes using spin cast mirror technology. The first of these, the Large Binocular Telescope (LBT) on Mt. Graham (two 8.4-m mirrors on a common mount) started normal operation in 2009. The first two 8.4-m mirrors have already been cast for the Giant Magellan Telescope (GMT) and are currently being polished. The GMT will consist of seven 8.4-m mirrors with an equivalent circular aperture of 21.5-m to be located at Las Campanas Observatory. A fourth 8.4-m mirror was cast in 2008 for the Large Synopic Survey Telescope (LSST). A fifth 8.4-m mirror was cast in 2012 for GMT and a sixth will be cast for the GMT in 2013. The GMT would be, by far, the world's largest.

These major telescopes are equipped with a wide variety of instrumentation and detectors and are supported by several smaller telescopes used for teaching or special research projects. A 6.5-m mirror was cast in 2009 for a joint telescope project to be located on Cerro San Pedro Martir in Baja, California, in collaboration with Mexico, Smithsonian, UCB, and UCSC. The main areas of research at the observatory include extragalactic and galactic astronomy, with major specializations in the areas of quasars, active galactic nuclei, degenerate stars, infrared sources, formation of stars and planetary systems, interstellar medium, novae, and radio galaxies.

Observational work is concentrated in the optical and infrared, but also includes work at radio, ultraviolet, and x-ray wavelengths using other facilities. The observatory operates the Heinrich Hertz 10-m Submillimeter Telescope (SMT) on Mt. Graham for work at millimeter and submillimeter wavelengths, and a 12-m radio telescope on Kitt Peak.

The research programs also include a wide range of observational and theoretical studies in astrophysics, and an involvement in astronomy in space, such as the Spitzer Space Telescope, the Hubble Space Telescope (HST), and the James Webb Space Telescope (JWST). Laboratory work includes astrobiology, astrochemistry, development of special instruments, CCD and CMOS imaging systems, high frequency receivers for work at submillimeter wavelengths, including observations at the South Pole (Antarctica): both from ground and high altitude balloons. The Center for Astronomical Adaptive Optics (CAAO) is developing diffraction limited imaging systems for use with the LBT, Magellan, and GMT. An AO system is in regular operation on the MMT and a second AO system has been installed on the LBT. A third AO system is being installed on one of the 6.5-m telescopes at Las Campanas.

Table B—Separately Budgeted Research Expenditures by Source of Support

Source of Support	Departmental Research	Physics-related Research Outside Department
Federal government	$49,409,459	
State/local government		
Non-profit organizations		
Business and industry		
Other	$9,680,619	
Total	**$59,090,078**	

Table C—Separately Budgeted Research Expenditures by Research Specialty

Research Specialty	No. of Grants	Expenditures ($)
Astronomy	293	$59,090,078
Total	293	$59,090,078

FACULTY

Professor

Angel, J. Roger P., Ph.D., University of Oxford, 1967. *Astronomy*. High-energy astrophysics; instrumentation; polarization, solar energy.

Arnett, David, Ph.D., Yale University, 1965. *Astronomy*. Nuclear relativistic and computational astrophysics; stellar and galactic evolution.

Bechtold, Jill, Ph.D., University of Arizona, 1985. *Astronomy*. Quasars; the intergalactic medium.

Bieging, John H., Ph.D., California Institute of Technology, 1974. *Astronomy*. Interstellar medium; radio astronomy; stellar evolution.

Brown, Robert H., Ph.D., University of Hawaii, 1982. *Astronomy*. Kuiper belt objects; brown dwarfs; extra solar planets; IR instrumentation.

Burge, James, Ph.D., University of Arizona, 1993. *Optics*. Astronomical instrumentation, space optics, adaptive optics.

Close, Laird, Ph.D., University of Arizona, 1995. *Astronomy, Optics*. Adaptive optics; star/planet formation; IR instrumentation.

Fan, Xiaohui, Ph.D., Princeton University, 2000. *Astronomy*. Cosmology; quasars; galaxy formation; brown dwarfs.

Impey, Christopher, Ph.D., University of Edinburgh, 1981. *Astronomy*. Quasars; observational cosmology.

Jannuzi, Buell T., Ph.D., University of Arizona, 1990. Chairman of the Department. *Astronomy*. Observational cosmology, quasar absorption line systems, active galaxies, instrumentation for surveys.

Jokipii, J. R., Ph.D., California Institute of Technology, 1965. *Astronomy, Planetary Science*. Theoretical astrophysics; plasma physics.

Melia, Fulvio, Ph.D., Massachusetts Institute of Technology, 1985. *Astronomy, Astrophysics*. Galactic center; relativistic jets and active galactic nuclei; accretion-disk coronae; cataclysmic variables.

Olszewski, Edward, Ph.D., University of Washington, 1982. *Astronomy*. Dwarf galaxies; Magellanic clouds.

Psaltis, Dimitrios, Ph.D., University of Illinois, 1998. *Astronomy, Astrophysics, Physics and other Science Education*. Theoretical astrophysics and gravitation.

Rieke, George H., Ph.D., Harvard University, 1969. *Astronomy*. Stellar and nonstellar infrared observational studies.

Rieke, Marcia, Ph.D., Massachusetts Institute of Technology, 1976. *Astronomy*. Infrared astronomy; instrumentation.

Sarcevic, Ina, Ph.D., University of Minnesota, 1986. *Astronomy, Physics and other Science Education*. Theoretical particle and nuclear astrophysics.

Sasian, José, Ph.D., University of Arizona, 1988. *Astronomy, Optics*. Optical design, optical fabrication and testing, optical design for astronomical instrumentation.

Strittmatter, Peter A., Ph.D., University of Cambridge, 1966. *Astronomy*. Extragalactic radio resources; speckle interferometry; quasars, accretion disks.

Thompson, Rodger I., Ph.D., Massachusetts Institute of Technology, 1970. *Astronomy*. Infrared spectroscopy; molecular astrophysics; stellar evolution, nucleosynthesis, star formation, galaxy evolution, and cosmology.

Walker, Christopher, Ph.D., University of Arizona, 1988. *Astronomy*. Protostellar objects; submillimeter wave instrumentation.

Zabludoff, Ann, Ph.D., Harvard University, 1993. *Astronomy*. Dependence of galaxy evolution on environment, type, merger history, and mass; structure and evolution of galaxy clusters and groups.

Zaritsky, Dennis, Ph.D., University of Arizona, 1991. *Astronomy*. Magellanic clouds; the distribution and nature of dark matter; large scale structure; interstellar and intergalactic dust; galaxy evolution.

Ziurys, Lucy, Ph.D., University of California, Berkeley, 1984. *Astronomy*. Laboratory microwave spectroscopy; astrochemistry, astrobiology.

Associate Professor

Davé, Romeel, Ph.D., University of California, Santa Cruz, 1998. *Astronomy*. Galaxy formation; cosmology; intergalactic medium; numerical hydrodynamics.

Eisner, Joshua, Ph.D., California Institute of Technology, 2005. *Astronomy*. Galactic astronomy and star formation; extrasolar planets.

Hinz, Philip, Ph.D., University of Arizona, 2001. *Astronomy*. Instrumentation; adaptive optics; nulling interferometry; exosolar planets.

Ozel, Feryal, Ph.D., Harvard University, 2002. *Astronomy, Astrophysics*. Theoretical astrophysics.

Pinto, Philip A., Ph.D., University of California, Santa Cruz, 1988. *Astronomy*. Supernovae, radiative transfer, cosmic distance scale.

Poss, Richard L., Ph.D., University of Georgia, 1986. *Astronomy*. History of astronomy; astronomy and the arts.

Prather, Edward E., Ph.D., University of Maine, 2000. *Astronomy, Physics and other Science Education*. Research on college teaching and learning in astronomy, physics, and space science; professional development of college faculty.

Assistant Professor

Apai, Daniel, Ph.D., Heidelberg University, 2004. *Astronomy*. Exoplanets; planet formation; astrobiology; circumstellar disks.

Guyon, Oliver, Ph.D., University of Paris VI, 2002. *Astronomy*. Instrumentation; extrasolar planets.

Marrone, Daniel, Ph.D., Harvard University, 2006. *Astronomy*. Extragalactic astronomy and cosmology; instrumentation; active galactic nuclei; star formation.

Robertson, Brant E., Ph.D., Harvard University, 2006. *Astronomy, Astrophysics, Theoretical Physics*. Theoretical topics related to galaxy and structure formation, the nature of dark matter, hydrodynamics and numerical stimulation methodologies.

Shirley, Yancy, Ph.D., University of Texas, Austin, 2002. *Astronomy*. Galactic astronomy and star formation; extragalactic astronomy; stellar astronomy; astrobiology.

Smith, Nathan, Ph.D., University of Minnesota, 2002. *Astronomy*. Massive stars; supernovae; eruptive stars; circumstellar material; star formation; observational astronomy; spectroscopy and high-resolution imaging.

Emeritus

Cocke, William J., Ph.D., Cornell University, 1964. *Astronomy*. Relativity; turbulence; pulsating radio sources; speckle interferometry.

Hoffmann, William F., Ph.D., Princeton University, 1962. *Astronomy*. Infrared astronomy from balloons; instrumentation, nulling interferometry.

Pacholczyk, Andrzej G., Ph.D., Warsaw University, 1961. *Astronomy, Astrophysics*. Radio-astrophysics; Seyfert galaxies.

Tifft, William G., Ph.D., California Institute of Technology, 1958. *Astronomy*. Space astronomy; galaxies.

Woolf, Neville J., Ph.D., University of Manchester, 1959. *Astronomy*. Circumstellar matter; infrared astronomy, astrobiology.

Senior Research Scientist

Codona, Johanan, Ph.D., University of California, San Diego, 1985. *Astronomy*. Adaptive optics; mathematical modeling in wave propagation.

Hubeny, Ivan, Ph.D., Charles University, Prague, 1977. *Astronomy*. Theory of stellar and planetary atmospheres; radiation transfer; accretion disks.

Lesser, Michael, Ph.D., University of Arizona, 1988. *Astronomy*. Astronomical instrumentation; CCD development.

West, Steven, Ph.D., University of Arizona, 1987. *Astronomy*.

Astronomer

Axelrod, Timothy, Ph.D., University of California, Santa Cruz, 1980. *Astronomy*. Supernovae, microlensing.

Egami, Eiichi, Ph.D., University of Hawaii, 1995. *Astronomy*. Infrared astronomy, extragalactic astronomy; cosmology.

Fleming, Thomas, Ph.D., University of Arizona, 1988. *Astronomy*. Low mass stars; white dwarfs; x-ray astronomy; nearby stars.

Green, Richard, Ph.D., California Institute of Technology, 1977. *Astronomy*. Quasars; black holes.

Hart, Michael, Ph.D., University of Arizona, 1991. *Astronomy, Optics*. Astronomical instrumentation; adaptive optics; infrared detectors.

Hill, John M., Ph.D., University of Arizona, 1984. *Astronomy*. Large telescope design; clusters of galaxies, instrumentation.

McCarthy, Donald W., Ph.D., University of Arizona, 1976. *Astronomy*. Low mass stars; brown dwarfs; infrared speckle interferometry.

Schneider, Glenn, Ph.D., University of Florida, 1985. *Astronomy*. Infrared astronomy; high contrast imaging; brown dwarfs and exosolar planets; star formation.

Smith, Paul S., Ph.D., University of New Mexico, 1986. *Astronomy*. Active galactic nuclei; observing techniques; polarimetry.

Sykes, Mark, Ph.D., University of Arizona, 1986. *Astronomy*. Interplanetary dust; asteroids; thermal modeling.

Williams, G. Grant, Ph.D., Clemson University, 2000. *Astronomy*. Massive stars; supernovae, gamma-ray bursts; instrumentation and polarimetry.

Associate Astronomer

Green, Elizabeth M., Ph.D., University of Texas, 1981. *Astronomy*. Stellar evolution; binaries; subdwarf B stars.

Hege, E. Keith, Ph.D., Rensselaer Polytechnic Institute, 1965. *Astronomy*. Speckle interferometry; optical instrumentation.

Kim, Serena, Ph.D., Stony Brook University, 2002. *Astronomy*. Star and planet formation.

Kulesa, Craig A., Ph.D., University of Arizona, 2002. *Astronomy*. Submillimeter wave instrumentation; life cycle of the interstellar medium; star formation.

Milne, Peter, Ph.D., Clemson University, 1998. *Astronomy*. Multi-wavelength emission from supernovae.

Stansberry, John A., Ph.D., University of Arizona, 1995. *Astronomy, Planetary Science*. Brown dwarfs; debris disks; planetary science.

Su, Kate, Ph.D., University of Calgary, 2000. *Astronomy*. Space-based optical and infrared observations on dusty objects.

Weiner, Benjamin, Ph.D., Rutgers University, 1998. *Astronomy*. Galaxy structure and evolution, kinematics, star formation.

Willmer, Christopher N. A., Ph.D., Observatorio Nacional, Brazil, 1990. *Astronomy, Astrophysics*. Extragalactic astrophysics; galaxy evolution; dwarf galaxies; infrared astronomy.

Assistant Astronomer

Cunha, Katia, Ph.D., Observatorio Nacional/MCT Rio de Janeiro, 1993. *Astronomy*. High-resolution spectroscopy; stellar abundances; metallicity gradients and chemical evolution.

Frye, Brenda, Ph.D., University of California, Berkeley, 1999. *Astronomy*. Extragalactic astronomy and cosmology, galaxy evolution and gravitational lensing.

Halfen, DeWayne T., Ph.D., University of Arizona, 2006. *Astronomy*. Astrochemistry, laboratory astrophysics and radio astronomy.

DEPARTMENTAL RESEARCH SPECIALTIES AND STAFF

Theoretical

Theoretical Astronomy. Arnett, Cocke, Davé, Hubeny, Jokipii, Melia, Ozel, Pacholczyk, Psaltis, Robertson, Sarcevic, Weiner.

Experimental

Active Galaxies and QSO's. Angel, Bechtold, Frye, Impey, Jannuzi, Melia, Pacholczyk, George Rieke, Marcia Rieke, Strittmatter, Woolf.

Astrobiology, Astrochemistry and Origin of Life. Apai, Eisner, Halfen, Woolf, Ziurys.

Astronomy Education and Outreach. Fleming, Frye, Impey, McCarthy, Poss, Prather.

Atomic and Molecular Astrophysics. Bieging, Thompson, Walker, Ziurys.

Cosmology. Cocke, Davé, Egami, Fan, Impey, Jannuzi, Robertson, Thompson, Tifft, Zabludoff, Zaritsky.

Galaxies and Galactic Evolution. Arnett, Bechtold, Davé, Fan, Frye, Hill, Impey, Jannuzi, Jokipii, Marrone, Olszewski, George Rieke, Marcia Rieke, Robertson, Thompson, Tifft, Willmer, Zabludoff, Zaritsky.

Instrumentation. Angel, Bechtold, Brown, Burge, Close, Eisner, Guyon, Hart, Hege, Hill, Hinz, Hoffmann, Jannuzi, Lesser, McCarthy, George Rieke, Marcia Rieke, Strittmatter, Thompson, Walker, Zaritsky.

Large Telescopes. Angel, Eisner, Hart, Hill, Jannuzi, Woolf.

Solar System and Planetary Astronomy. Apai, Brown, Close, Jokipii, Sykes.

Star Formation. Close, Eisner, Hoffmann, Schneider, Shirley, Thompson, Walker.

Stellar Astronomy. Arnett, Bieging, Brown, Close, Fleming, McCarthy, Melia, Olszewski, George Rieke, Marcia Rieke, Shirley, Nathan Smith, Strittmatter.

Stellar and Galactic Dynamics. Olszewski, Nathan Smith, Zaritsky.

Submillimeter (aka Terahertz). Kulesa, Marrone, Shirley, Walker, Ziurys.

Supernovae and Nucleosynthesis. Arnett, Axelrod, Pinto.

View additional information about this department at
www.gradschoolshopper.com

UNIVERSITY OF ARIZONA

DEPARTMENT OF PHYSICS

Tucson, Arizona 85721
http://www.physics.arizona.edu

General University Information

President: Ann Weaver Hart
Dean of Graduate School: Andrew Comrie
University website: http://www.arizona.edu
Control: Public
Setting: Urban
Total Faculty: 2,619
Total Graduate Faculty: 1,594
Total number of Students: 37,036
Total number of Graduate Students: 8,215

Department Information

Department Chairman: Sumit Mazumdar, Chair
Department Contact: Lisa Shapouri, Program Coordinator
Total full-time faculty: 29
Full-Time Graduate Students: 86
First-Year Graduate Students: 18
Female First-Year Students: 4
Total Post Doctorates: 12

Department Address
1118 E. 4th Street
Tucson, AZ 85721
Phone: (520) 621-2290
Fax: (520) 621-4721
E-mail: lshapour@email.arizona.edu
Website: http://www.physics.arizona.edu

ADMISSIONS

Admission Contact Information
Address admission inquiries to: Univ. of Arizona, Dept. of Physics, Graduate Coordinator, 1118 E. 4th St., Tucson, AZ 85721
Phone: (520) 621-2290
E-mail: lshapour@email.arizona.edu
Admissions website: http://www.physics.arizona.edu/physics/graduate.php?page=application_procedure

Application deadlines
Fall admission:
U.S. students: January 1   *Int'l. students*: December 1

Application fee
U.S. students: $75   *Int'l. students*: $75
http://www.physics.arizona.edu/physics/graduate.php?page=application_procedure. No Spring Admissions, except under special circumstances.

Admissions information
For Fall of 2013:
  Number of applicants: 192
  Number admitted: 66
  Number enrolled: 18

Admission requirements
Bachelor's degree requirements: Bachelor's degree in physics or a related field is required.
Minimum undergraduate GPA: 3.0

GRE requirements
The GRE is required.
No minimum score has been set.

Advanced GRE requirements
The Advanced GRE is required.
No minimum score has been set.

TOEFL requirements
The TOEFL exam is required for students from non-English-speaking countries.
  PBT score: 550
  iBT score: 80

Other admissions information
Undergraduate preparation assumed: Tipler, Modern Physics; (Phys. 242); Symon, Mechanics (Phys. 321); Wangsness, Griffiths, Electromagnetic Fields (Phys. 331/332); Hecht and Zajac, Optics (Phys. 320); Fermi, Thermodynamics; Callen, Thermodynamics (Phys. 325); Liboff, Griffiths, Quantum Physics (Phys. 371); Gasiorowicz, Quantum Physics (Phys. 472); Herzberg, Kodak, Beveridge, Atomic Spectra and Atomic Structure (Phys. 473/474); Kittel, Introduction to Solid State Physics (Phys. 460); Arfken, Mathematical Methods for Physics (Phys. 475); Melissinos, Experiments in Modern Physics; Young, Statistical Treatment of Experimental Data; Taylor, An Introduction to Error Analysis (all 3 optional); (Phys. 381, 481).

TUITION

Tuition year 2013-14:
Tuition for in-state residents
  Full-time students: $5,419 per semester
  Part-time students: $4,613 per semester
Tuition for out-of-state residents
  Full-time students: $12,900 per semester
  Part-time students: $8,647 per semester
Graduate Assistants receive a Non-Resident Tuition Waiver and a Tuition Remission. Average cost for a Graduate Assistant is $371.00-$467.00/semester.
Credit hours per semester to be considered full-time: 9
Deferred tuition plan: Yes
Health insurance: Yes, 1286.00 per year, free with assistantship.
Other academic fees: Miscellaneous fees ranging from $371.00 to $467.00 depending on the number of units taken per semester.
Academic term: Semester

Teaching Assistants, Research Assistants, and Fellowships
Number of first-year
  Teaching Assistants: 16
  Research Assistants: 1
Average stipend per academic year
  Teaching Assistant: $15,042
  Research Assistant: $15,042

FINANCIAL AID

Application deadlines
Fall admission:
U.S. students: January 1   *Int'l. students*: December 1

Loans
Loans are available for U.S. students.
Loans are available for international students.
GAPSFAS application required: No
FAFSA application required: No

For further information
Address financial aid inquiries to: Office of Student Financial Aid, 203 Admin. Building, University of Arizona.
Phone: (520) 621-1858
Financial aid website: http://www.financialaid.arizona.edu

HOUSING

Availability of on-campus housing
  Single students: Yes
  Married students: Yes

For further information
Address housing inquiries to: Department of Residence Life, Administration Building.
Phone: (520) 621-6501
E-mail: housing@life.arizona.edu
Housing aid website: http://www.life.arizona.edu

Table A—Faculty, Enrollments, and Degrees Granted

Research Specialty	2012-13 Faculty	Enrollment Fall 2013		Number of Degrees Granted 2012-13		
		Master's	Doctorate	Master's	Terminal Master's	Doctorate
Astrophysics	1	–	13	–	–	2(-)
Atmosphere, Space Physics, Cosmic Rays	1	–	3	–	–	–
Atomic, Molecular, & Optical Physics	3	–	27	–	–	2(-)
Biophysics	2	–	4	–	1(-)	2(-)
Condensed Matter Physics	7	–	15	–	–	2(-)
High Energy Physics	9	–	16	–	1(-)	3(-)
Medical, Health Physics	1	8	–	–	1(-)	–
Nuclear Physics	4	–	4	–	–	4(-)
Physics and other Science Education	1	–	–	–	–	–
Non-specialized	–	–	4	–	–	–
Total	29	8	86	–	3(-)	5(-)
Full-time Grad. Stud.	–	8	86	–	–	–
First-year Grad. Stud.	–	4	14	–	–	–

GRADUATE DEGREE REQUIREMENTS

Master's: In order to qualify for the M.S. degree, the student must complete the qualifying examination. In addition, the student must complete at least 30 units of graduate work, at least 15 of which must be in physics. The student must pass the written comprehensive exam at the M.S. level and maintain a 3.0 average. No foreign language required. The student must also satisfy one of the following options: 1. Write a thesis (for which up to six units may be allowed) and pass an oral examination. 2. Take six additional graduate credits in physics and pass a final oral examination. Under this option, the student completes a total of at least 21 units in physics out of the total of 30 units required. 3. Pass the written and oral parts of the comprehensive examination for the Ph.D. Students are not normally admitted into the MS program; students with bachelors degrees who seek the Ph.D. as the terminal degree should apply directly for admission into the Ph.D. program. Such students may then earn the M.S. degree en route to the Ph.D. Professional Science Masters in Medical Physics: This degree requires completion of at least 36 units of graduate work, including 6 units of internship, 12 units of physics, up to 6 units of business, and 9 units of other specialty courses.

Doctorate: The Ph.D. degree requires completion of 36 units of graduate work in physics, with 12 additional units in the minor, with a 3.0/4.0 grade average (exclusive of dissertation credits); passing the qualifying and comprehensive examinations; submitting a dissertation based on independent research; and defending this dissertation in a final examination. Note that independent study research can be used to earn a substantial portion of the required graduate units.

Other Degrees: Interdisciplinary work with other departments is possible, especially in the areas of materials and surface physics, chemical physics, astrophysics, space physics, optical science, medical physics, biophysics, applied math, and mathematical physics.

Thesis: Thesis may be written in absentia.

SPECIAL EQUIPMENT, FACILITIES, OR PROGRAMS

Our graduate program is dedicated to research at the cutting edge of discovery. We also have strong interdisciplinary initiatives, with campus-wide programs in theoretical astrophysics, lunar and planetary sciences, geosciences, optical sciences, biophysics, and applied mathematics. Additional connections to biology, chemistry, and medicine provide additional research opportunities. Moreover, in addition to a traditional Ph.D. program, we are also one of the few institutions nationwide to have a Professional Science Masters (PSM) program in medical physics. Graduate students have excellent resources at their disposal.

General facilities include a shop for students and staff. An extensive collection of physics journals and more than 5,000 physics books are located in the science library building. The University is connected to Internet2 for high-speed connectivity to the NSF supercomputer centers, other universities, and other supercomputer centers. The University also operates a large SGI supercomputer, numerous UNIX servers, Windows PC and Macintosh public access sites, and a computerized library information system (sabio.arizona.edu). The Physics Department has a large number of Linux PCs, Windows PCs, and laser printers (color and b/w) available for general department use.

Facilities and equipment for particular research areas in physics include a 3-MV tandem accelerator, spectrographs, spectrometers, superconducting magnets for solid state physics, electron microscopes, electron probes (including Auger, RHEED, and LEED), a materials-processing laboratory, thin-film molecular beam epitaxy (MBE) and sputter epitaxy equipment, thin-film x-ray diffraction facilities, an atomic force microscope, optical tweezers and single molecule detection for biophysics, cryogenic systems covering the entire temperature range of 300 K down to 0.02 K, a computerized Mössbauer spectroscopy system, ultrahigh-resolution infrared and visible lasers, atomic-beam machines, and an observatory for work in experimental relativity and solar seismology. The high-energy experimental group does work at Fermilab and CERN. There is an active program in theoretical physics covering atomic, nuclear, condensed matter, astrophysics, and high-energy theory. The Departments of Physics and Geosciences jointly operate a facility, sponsored by the National Science Foundation, that uses accelerators for radioisotope dating.

Table B—Separately Budgeted Research Expenditures by Source of Support

Source of Support	Departmental Research	Physics-related Research Outside Department
Federal government	$4,463,120	
State/local government	$121,597	
Non-profit organizations	$14,000	
Business and industry		
Other		
Total	$4,598,717	

Table C—Separately Budgeted Research Expenditures by Research Specialty

Research Specialty	No. of Grants	Expenditures ($)
Astrophysics	1	$30,000
Atomic, Molecular, & Optical Physics	5	$267,626
Condensed Matter Physics	10	$1,062,055
Nuclear Physics	3	$406,966
Optics	5	$441,128
Particles and Fields	9	$2,469,932
Physics and other Science Education	0	
Total	33	$4,677,707

FACULTY

Professor

Barrett, Bruce R., Ph.D., Stanford University, 1967. *Nuclear Physics*. Nuclear many-body theory; microscopic shell model calculations; three-nucleon interactions; nuclear collective motion.

Cheu, Elliott, Ph.D., Cornell University, 1991. Associate Dean, College of Science. *High Energy Physics*. Experimental high energy.

Dienes, Keith R., Ph.D., Cornell University, 1991. *High Energy Physics*. Theoretical high-energy physics.

Jacquod, Philippe, Ph.D., University of Neuchatel, 1997. *Condensed Matter Physics*. Condensed matter theory.

Johns, Ken, Ph.D., Rice University, 1986. *High Energy Physics*. Experimental high energy.

Lebed, Andrei, Ph.D., Landau Institute for Theoretical Physics, 1986. *Condensed Matter Physics*. Theoretical condensed matter.

Mazumdar, Sumit, Ph.D., Princeton University, 1980. Department Head. *Condensed Matter Physics*. Condensed matter theory.

Melia, Fulvio, Ph.D., Massachusetts Institute of Technology, 1985. *Astrophysics*. Theoretical astrophysics.

Meystre, Pierre, Ph.D., Swiss Federal Institute of Technology, 1974. *Optics*. Theoretical quantum optics.

Rafelski, Johann, Ph.D., University of Frankfurt, 1973. *Nuclear Physics, Particles and Fields*. Relativistic nuclear theory; laser particle physics; nuclear cosmology.

Rutherfoord, John P., Ph.D., Cornell University, 1968. *High Energy Physics*. Experimental high-energy physics.

Sarcevic, Ina, Ph.D., University of Minnesota, 1986. *High Energy Physics, Nuclear Physics*. High-energy theory, nuclear theory.

Shupe, Michael A., Ph.D., Tufts University, 1976. *High Energy Physics*. Experimental high-energy physics.

Stafford, Charles A., Ph.D., Princeton University, 1992. Condensed matter theory.

Toussaint, Douglas, Ph.D., Princeton University, 1978. *High Energy Physics*. High-energy theory.

van Kolck, Ubirajara, Ph.D., University of Texas, Austin, 1993. *Nuclear Physics*. Effective field theories in particle, nuclear, atomic, and astro physics.

Zhang, Shufeng, Ph.D., New York University, 1991. *Condensed Matter Physics*. Condensed matter theory.

Associate Professor

Cronin, Alexander D., Ph.D., University of Washington, 1999. *Atomic, Molecular, & Optical Physics*. Experimental atomic physics.

Fleming, Sean, Ph.D., Northwestern University, 1995. *Nuclear Physics*. Nuclear theory, high-energy theory.

LeRoy, Brian, Ph.D., Harvard University, 2003. *Condensed Matter Physics*. Experimental nanoscience.

Manne, Srinivas, Ph.D., University of California, Santa Barbara, 1994. *Condensed Matter Physics*. Experimental condensed matter.

Sandhu, Arvinder, Ph.D., Tata Institute of Fundamental Research, 2004. *Atomic, Molecular, & Optical Physics*. Experimental atomic, molecular, and optical physics.

Su, Shufang, Ph.D., Massachusetts Institute of Technology, 2000. *High Energy Physics*. Theoretical high-energy physics.

Varnes, Erich W., Ph.D., University of California, Berkeley, 1997. *High Energy Physics*. Experimental high-energy physics.

Visscher, Koen, Ph.D., University of Amsterdam, 1993. *Biophysics*. Biophysics.

Wolgemuth, Charles W., Ph.D., University of Arizona, 2000. *Biophysics*. Biological physics.

Assistant Professor

Wang, Weigang, Ph.D., University of Delaware, 2008. *Condensed Matter Physics*. Experimental condensed matter.

Emeritus

Bickel, William S., Ph.D., Pennsylvania State University, 1965. *Biophysics, Optics*. Optical biophysics; atomic physics; physics of music.

Bowen, Theodore, Ph.D., University of Chicago, 1954. Experimental elementary particle and cosmic-ray physics; medical physics.

Chambers, Robert H., Ph.D., Carnegie Mellon University, 1957. Science education; dislocation dynamics; internal friction.

Donahue, Douglas J., Ph.D., University of Wisconsin-Madison, 1952. Atomic and nuclear physics; accelerator mass spectrometry.

Emrick, Roy M., Ph.D., University of Illinois, 1960. Experimental solid-state physics.

Garcia, Jose D., Ph.D., University of Wisconsin-Madison, 1966. Collision theory; atomic physics; physics education.

Hill, Henry A., Ph.D., University of Minnesota, 1957. Astrophysics and experimental general relativity.

Hsieh, Ke Chiang, Ph.D., University of Chicago, 1969. Experimental cosmic-ray and space physics.

Huffman, Donald R., Ph.D., University of California, Riverside, 1966. Optical properties of solids; astrophysics.

Jenkins, Edgar W., Ph.D., Columbia University, 1962. Experimental elementary particle physics.

Just, Kurt W., Ph.D., Berlin University, 1954. Gauge theories of gravity and particles.

Kessler, John O., Ph.D., Columbia University, 1953. Applied physics; biophysics and fluids.

Kilkson, Rein, Ph.D., Yale University, 1956. Molecular biophysics.

Kohler, Sigurd, Uppsala, University of Sweden, 1959. Nuclear many-body theory; heavy-ion collisions.

Mahmoud, Hormoz M., Ph.D., Indiana University, 1953. Field theory.

McIntyre, Laurence C., University of Wisconsin-Madison, 1965. Ion-beam analysis.

Scadron, Michael D., Ph.D., University of California, Berkeley, 1964. Theoretical elementary particle physics.

Stark, Royal W., Ph.D., Case Western Reserve University, 1962. Experimental solid-state physics.

Stoner, John O., Ph.D., Princeton University, 1964. Experimental atomic spectroscopy; thin-film optics.

Thews, Robert L., Ph.D., Massachusetts Institute of Technology, 1966. High-energy and nuclear theory.

Tomizuka, Carl T., University of Illinois, 1954. Experimental solid-state physics.

Vuillemin, Joseph J., Ph.D., University of Chicago, 1965. Experimental solid-state physics.

Wangsness, Roald K., Ph.D., Stanford University, 1950. Science education; statistical mechanics; electromagnetic theory.

Weaver, Albert B., Ph.D., University of Chicago, 1952. Cosmic rays.

Wing, William H., Ph.D., University of Michigan, 1968. Experimental atomic physics and quantum optics.

Research Professor

Beck, Warren, Ph.D., University of Minnesota, 1988. AMS Facility.

Burr, George S., Ph.D., University of Chicago, 1990. Geophysical sciences.

Curtis, Charles, Ph.D., University of Arizona, 1978.

Hodgins, Gregory, Ph.D., University of Oxford, 1999.

Research Associate Professor

Loch, Peter, Ph.D., University of Hamburg, 1992.

Research Assistant Professor

Biddulph, Dana, Ph.D., University of Arizona, 2004.
Cheng, Li, Ph.D., University of Arizona, 2000.

Adjunct Professor

Barker, Delmar, Ph.D., University of Arizona, 1974.
Coon, Sidney, Nuclear Theory.
Denison, Arthur B., Research Corporation Technologies.
Gallardo, Juan C., Ret., Experimental High Energy.
Pattison, John.
Radziemski, Leon, Research Corporation Technologies.
Timmes, Francis, Arizona State University. Theoretical Astrophysics.
Weekes, Trevor, Fred Lawrence Whipple Observatory.
Wiener, Richard J., Research Corporation Technologies.

Affiliate Professor

Özel, Feryal, Ph.D., Harvard University, 2002. *Astrophysics.* Theoretical astrophysics.
Adamowicz, Ludwik, Ph.D., Institute of Physical Chemistry of the Polish Academy of Sciences, 1977. *Chemical Physics.* Theoretical chemistry, chemical physics.
Anderson, Brian, Ph.D., Stanford University, 1999. *Atomic, Molecular, & Optical Physics.* Experimental Bose-Einstein condensation.
Binder, Rudolf, Ph.D., University of Dortmund, 1988. *Atomic, Molecular, & Optical Physics.* AMO physics.
Brown, Michael, Ph.D., University of California, Santa Cruz, 1975. *Biophysics, Chemical Physics.* Physical chemistry and biochemistry, nuclear magnetic resonance spectroscopy.
Falco, Charles M., Ph.D., University of California, Irvine, 1974. *Condensed Matter Physics, Optics.* Condensed matter physics, optics.
Jessen, Poul, Ph.D., University of Aarhus, 1993. *Atomic, Molecular, & Optical Physics.* Atomic and optical physics.
Jull, Timothy, Ph.D., University of Bristol, 1976. *Geophysics.* Geosciences.
Kennedy, Thomas G., Ph.D., University of Virginia, 1984. Mathematical physics.
Lunine, Jonathan L., Ph.D., California Institute of Technology, 1985. Theoretical astrophysics.
Maier, Robert S., Ph.D., Rutgers State University of New Jersey, 1983. *Applied Mathematics.* Applied Mathematics, mathematical physics.
Miyashita, Osamu, Ph.D., Kyoto University, 2000. *Biophysics.* Biochemistry and molecular biophysics.
Novodvorsky, Ingrid, Ph.D., University of Arizona, 1993. *Physics and other Science Education.*
Pinto, Philip A., Ph.D., University of California, Santa Cruz, 1988. *Astronomy.* Astronomy.
Psaltis, Dimitrios, Ph.D., University of Illinois at Urbana-Champaign, 1998. *Astrophysics.* Theoretical astrophysics.
Restrepo, Juan, Ph.D., Pennsylvania State University, 1992. *Applied Mathematics.* Mathematics, mathematical physics.
Tabor, Michael, University of Bristol, 1990. *Applied Mathematics.*
Wright, Ewan, Ph.D., Heriot-Watt University, 1983. *Optics.* Theory and stimulation of light string propagation in air.
Xin, Hao, Ph.D., Massachusetts Institute of Technology, 2000. *Condensed Matter Physics.* Microwave and millimeter wave technology.

Affiliate Associate Professor

Trouard, Theodore P., Ph.D., University of Virginia, 1992. *Biophysics.* Biomedical engineering.

Lecturer

Jackson, Shawn S., Washington University, St. Louis, 1988. *Astrophysics.* Astrophysics.
Milsom, John A., Ph.D., Northwestern University, 1996. *Astrophysics.* Astrophysics.

DEPARTMENTAL RESEARCH SPECIALTIES AND STAFF

Theoretical

Astrophysics. Black holes, compact objects, AGN's particle astrophysics, cosmology. Özel, Melia, Milsom, Psaltis, Rafelski, Sarcevic.
Atomic, Molecular, & Optical Physics. Ultracold atoms, Bose condensation and atom lasers, cavity QED, ion-atom collision theory, atomic structure calculations. Binder, Meystre, Wright.
Condensed Matter Physics. Strongly correlated electron systems, mesoscopic and nanoscopic materials, disordered systems, nonlinear dynamics, quantum chaos, quantum spin systems, superconductivity. Adamowicz, Jacquod, Lebed, Mazumdar, Stafford.
High Energy Physics. Theory and phenomenology of strong and electroweak interactions, lattice QCD, neutrino physics, Higgs physics, physics beyond the standard model, supersymmetry, grand unification, extra dimensions, and string theory. Dienes, Fleming, Sarcevic, Su, Toussaint.
Nuclear Physics. Nuclear many-body theory, heavy ion collisions, quark-gluon substructure of nuclear matter, effective field theories of nuclear forces, chiral perturbation theory, Soft-collinear effective theory, heavy-quark systems, laser-induced particle phenomena. Barrett, Fleming, Kohler, Rafelski, van Kolck.
Physics and other Science Education. Secondary school teacher preparation, physics education theory and modeling. Novodvorsky.

Experimental

Atomic, Molecular, & Optical Physics. Cold atom trapping, Bose-Einstein condensates, atom interferometry, quantum computation, attosecond spectroscopy. Anderson, Cronin, Jessen, Sandhu.
Biophysics. Theory and experiment of cell motility, chemotaxis, pattern formation by swimming cells, membrane proteins and membrane biophysics, solid-state NMR, nucleic acid structure, single-molecule biophysics, optical tweezers, molecular dynamics simulations. Brown, Kessler, Miyashita, Visscher, Wolgemuth.
Condensed Matter Physics. Self-assembly at solid/liquid interfaces, fullerenes, graphene, scanning tunneling microscopy, magnetic nanostructures, ultrafast optical spectroscopy. Huffman, LeRoy, Manne, Sandhu, Wang.
High Energy Physics. Accelerator-based experiments at Fermilab and CERN, including: direct CP violation in kaon decay, search for Higgs, top quark physics, jets, search for extra dimensions, electroweak physics. Cheu, Johns, Rutherfoord, Shupe, Varnes.
Other. Radiocarbon dating, cosmic ray effects in terrestrial materials, climate change, geophysical studies. Beck, Jull.

THE UNIVERSITY OF ARIZONA

COLLEGE OF OPTICAL SCIENCES

Tucson, Arizona 85721
http://www.optics.arizona.edu

General University Information
President: Ann Weaver Hart
Dean of Graduate School: Andrew Comrie
University website: http://www.arizona.edu/
Control: Public
Setting: Urban
Total Faculty: 2,554
Total number of Students: 38,876
Total number of Graduate Students: 7,162

Department Information
Department Chairman: Thomas L. Koch, Dean
Department Contact: Carl F. Maes, Associate Dean, Academic
  Programs
  Total full-time faculty: 58
  Total number of full-time equivalent positions: 30
  Full-Time Graduate Students: 217
  First-Year Graduate Students: 56
  Female First-Year Students: 18
  Total Post Doctorates: 18

Department Address
1630 E. University Boulevard
Tucson, AZ 85721
Phone: (520) 621-4112
Fax: (520) 626-1480
E-mail: admissions@optics.arizona.edu
Website: http://www.optics.arizona.edu

ADMISSIONS

Admission Contact Information
Address admission inquiries to: Dr. Carl F. Maes, Associate Dean
  of Academic Programs, College of Optical Sciences
Phone: (520) 621-4112
E-mail: admissions@optics.arizona.edu
Admissions website: http://www.optics.arizona.edu/academics/
  for-prospective-students/admissions/graduate

Application deadlines
Fall admission:
U.S. students: January 1   *Int'l. students*: January 1
Spring admission:
U.S. students: November 1   *Int'l. students*: November 1

Application fee
U.S. students: $75   *Int'l. students*: $75
Ph.D. applications are not accepted for spring semester admis-
sions.

Admissions information
For Fall of 2012:
  Number of applicants: 243
  Number admitted: 106
  Number enrolled: 56

Admission requirements
Bachelor's degree requirements: A bachelor's degree in engi-
  neering, physics, mathematics or optics is required.
Minimum undergraduate GPA: 3.0

GRE requirements
The GRE is required.
GRE subject is not required.

Advanced GRE requirements
The Advanced GRE is not required.

TOEFL requirements
The TOEFL exam is required for students from non-English-
  speaking countries.
  PBT score: 600
  iBT score: 79
A TOEFL IELTS score of 7 is required.

Other admissions information
Additional requirements: The minimum acceptable GRE score
  for admission is analytical: 65 percent; quantitative: 75 percent.
  The average GRE scores for 2012-2013 admissions were verbal:
  64 percent; quantitative: 82 percent; analytical: 45 percent.
Undergraduate preparation assumed: Applicants should hold a
  bachelor's degree in optics, engineering, physics, mathematics
  or a related field. Before beginning graduate-level coursework,
  students should have taken four semesters of advanced mathematics, including calculus, vector calculus and differential equations. A course in linear algebra is also recommended.

TUITION

Tuition year 2012–13:
Tuition for in-state residents
  Full-time students: $5,100 per semester
Tuition for out-of-state residents
  Full-time students: $12,806 per semester
A limited number of graduate tuition scholarships and fellow-
  ships are available. Applicants from WRGP-participating
  states may qualify for in-state tuition. For more information,
  see http://www.optics.arizona.edu/academics/funding.
Credit hours per semester to be considered full-time: 9
Deferred tuition plan: Yes
Health insurance: Available at the cost of $1,664 per year.
Other academic fees: Additional mandatory fees are $935 per
  academic year.
Academic term: Semester
Number of first-year students who received full tuition waivers: 16

Teaching Assistants, Research Assistants, and Fellowships
Number of first-year
  Research Assistants: 28
  Fellowship students: 3
Average stipend per academic year
  Teaching Assistant: $15,912
  Research Assistant: $15,912
  Fellowship student: $10,000
Students with graduate research and teaching assistantships
qualify for in-state tuition. They are required to enroll in a
minimum of six units for full-time status (as opposed to the
nine units required of other students). Assistantships include
student health insurance coverage.

FINANCIAL AID

Application deadlines
Fall admission:
U.S. students: January 1   *Int'l. students*: January 1
Spring admission:
U.S. students: November 1   *Int'l. students*: November 1

Loans
Loans are available for U.S. students.
Loans are not available for international students.
GAPSFAS application required: No
FAFSA application required: Yes

For further information
Address financial aid inquiries to: Dr. Carl F. Maes, Associate Dean of Academic Programs, College of Optical Sciences, P.O. Box 210094, Tucson, AZ 85721-0094.
Phone: (520) 626-8837
E-mail: carl.maes@optics.arizona.edu
Financial aid website: http://www.optics.arizona.edu/academics/funding

HOUSING

Availability of on-campus housing
Single students: Yes
Married students: Yes

For further information
Address housing inquiries to: Department of Residence Life.
Phone: (520) 626-0336
E-mail: laaldea@life.arizona.edu
Housing aid website: http://www.life.arizona.edu/home/

Table A—Faculty, Enrollments, and Degrees Granted

Research Specialty	2012-13 Faculty	Enrollment Fall 2012		Number of Degrees Granted 2012-2013 (2006-13)		
		Master's	Doctorate	Master's	Terminal Master's	Doctorate
Optics	33	99	180	29(312)	29(163)	26(153)
Total	33	99	180	29(312)	29(163)	26(153)
Full-time Grad. Stud.	–	49	169	–	–	–
First-year Grad. Stud.	–	30	23	–	–	–

GRADUATE DEGREE REQUIREMENTS

Master's: Thesis option: a minimum of 32 units of graduate credit in optics or optics-related courses, including eight units of OPTI 910 (thesis) and at least two optics laboratory courses, and a final oral examination based primarily on the thesis. Nonthesis option: a minimum of 35 units of graduate credit in optics or optics-related courses, including at least two units of optics laboratory courses; three units' credit for demonstrated competence in written communication, either for writing an acceptable master's report or successfully completing an appropriate course in technical writing; and a final oral examination, based primarily on the subject matter of the courses taken. A cumulative GPA of 3.0 is required for a M.S. degree to be awarded. The M.S. in Optical Sciences can be completed by distance with one visit to campus. The Master of Science in Photonics Communication has a core curriculum: http://www.optics.arizona.edu/academics/degree-programs/master-science-photonic-communications-engineering. The M.S. in Optical Sciences and MBA Dual Degree offers the opportunity to earn two degrees concur-rently: http://www.optics.arizona.edu/academics/degree-programs/master-science-optical-sciences-and-mba-dual-degree.

Doctorate: The equivalent of six semesters of full-time graduate coursework is required, including at least two optics laboratory courses, for a total of 54 units. The equivalent of two semesters of full-time study must be spent in residence, and 30 units of graduate credit must be earned at the University of Arizona. Students are also required to earn 18 dissertation units. To receive a Ph.D., students must maintain a cumulative GPA of 3.0. A foreign language is not required. Required examinations are the written and oral comprehensive examinations (usually completed during the fifth or sixth semester), the dissertation proposal examination and the final oral examination.

Other Degrees: The Graduate Professional Certificate in Optical Sciences requires completion of 15 units of graduate-level course work in optics.

Thesis: Theses and dissertations for master's and doctoral degrees may be written in absentia.

SPECIAL EQUIPMENT, FACILITIES, OR PROGRAMS

The College of Optical Sciences is recognized internationally for its innovative and unusually comprehensive research programs. Research encompasses a broad set of technologies and techniques for exploiting the properties and applications of light, touching virtually every field of science and industry. The extensive research facilities at the college provide the resources for both theoretical and applied research programs in all areas related to optics and the optical sciences. We continually refine and upgrade our facilities to expand our research capabilities and programs.

Table B—Separately Budgeted Research Expenditures by Source of Support

Source of Support	Departmental Research	Physics-related Research Outside Department
Federal government	$17,468,787	
State/local government	$2,218,156	
Non-profit organizations	$228,635	
Business and industry	$2,344,710	
Other	$279,460	
Total	$22,539,748	

Table C—Separately Budgeted Research Expenditures by Research Specialty

Research Specialty	No. of Grants	Expenditures ($)
Optics	122	$31,000,000
Total	122	$31,000,000

FACULTY

Professor
Angel, J. Roger P., Ph.D., University of Oxford, 1967. *Astronomy, Atmosphere, Space Physics, Cosmic Rays, Atomic, Molecular, & Optical Physics, Optics, Other*. Adaptive optics; instrumentation; extrasolar planets; telescope design and optical fabrication; interferometry.

Armstrong, Neal, Ph.D., University of New Mexico, 1974. *Atomic, Molecular, & Optical Physics, Chemical Physics, Optics, Other*. New molecular materials through self-assembly and patterning; interface characterization through surface analysis and scanning probe microscopies. Electrochemistry. Chemical sensors. Analytical chemistry.

Banerjee, Bhaskar, Other, University of London, 1983. *Medical, Health Physics*. Optical detection of gastrointestinal cancer using native fluorophores; receptor-targeted imaging of gastrointestinal cancer.

Barrett, Harrison H., Ph.D., Harvard University, 1969. *Medical, Health Physics, Nuclear Engineering, Optics, Other*. Inverse problems in medicine; applications of statistical decision theory; medical imaging; three-dimensional reconstruction; nuclear medicine.

Barton, Jennifer, Ph.D., University of Texas, Austin, 1998. *Medical, Health Physics, Other*. Optical imaging (optical coherence tomography); laser-tissue interaction; bioinstrumentation.

Binder, Rolf, Ph.D., Universität Dortmund, 1988. *Applied Physics, Optics, Physics of Beams, Other*. Theoretical investigations of the optical properties of semiconductor structures and modeling of semiconductor lasers.

Bloembergen, Nicolaas, Ph.D., University of Leiden, 1948. Nobel laureate in physics, 1981. Nuclear and electronic magnetic resonance; solid-state masers and lasers; nonlinear optics; spectroscopy.

Burge, James H., Ph.D., University of Arizona, 1993. *Astronomy, Atmosphere, Space Physics, Cosmic Rays, Optics, Other*. Optical system engineering; optical designl opto-mechanics; pointing and tracking; detectors; cryogenic systems; optical testing and precision metrology; aspheric surfaces; ultra-lightweight mirrors for space; diffractive optics; stellar interferometry; astronomical instrumentation.

Chipman, Russell, Ph.D., University of Arizona, 1987. *Optics, Other*. Optical polarization; ophthalmic optics.

Clarkson, Eric, Ph.D., Arizona State University, 1985. *Medical, Health Physics, Optics, Other*. Image science; mathematical optics; medical imaging; inverse problems; image quality assessment.

Cvijetic, Milorad, Ph.D., University of Belgrade, 1984. *Electrical Engineering, Optics, Other*. High-speed optical communication systems (high speed DWDM systems and all optical switching schemes solution transmission coherent detection); multilayer optical networks; advanced optical access networks: high-speed PON with OFDMA; optical-wireless integration of high-speed systems and networks' high-capacity photonic networks modeling for major North American backbone/regional/access optical networks.

Dallas, William John, Ph.D., University of California, San Diego, 1973. *Electrical Engineering, Medical, Health Physics, Optics, Other*. Picture archiving and communications systems; electrical current imaging from biomagnetic field measurements; medical image processing; image display; cardiac imaging.

Denninghoff, Kurt R., Ph.D., Vanderbilt University School of Medicine, 1987. Also an M.D. *Medical, Health Physics, Other*. Emergency medicine; ophthalmic technologies.

Dereniak, Eustace L., Ph.D., University of Arizona, 1976. *Electrical Engineering, Optics, Physics of Beams, Other*. Infrared-radiation detection; imaging spectrometers; cryogenically cooled detector/electronics technology; CCD and CMOS devices; infrared detectors using charge-transfer concepts; image processing of infrared sensors.

Falco, Charles, Ph.D., University of California, Irvine, 1974. Chair of Condensed Matter Physics. *Energy Sources & Environment, Optics, Other*. Metallic superlattices; x-ray optics; magnetism; magneto-optics; far-IR detector materials; superconductivity; nucleation and epitaxy of thin films; multilayered materials and superlattices.

Fallahi, Mahmoud, Ph.D., Toulouse University/CNRS, 1988. *Atomic, Molecular, & Optical Physics, Electrical Engineer-*

ing, Nano Science and Technology, Optics, Other. High-power semiconductor lasers; DFB/DBR lasers; grating-assisted integrated optics; photonic integrated circuits; optical communication. Wavelength multiplexers and demultiplexers. Wavelength filters. Solgel-semiconductor integration. WDM components. Sensors, design and microfabrication. Nanofabrication and nanostructures. Solgel PIC.

Furenlid, Lars, Ph.D., Georgia Institute of Technology, 1988. *Medical, Health Physics, Optics, Other*. Development and application of novel detectors; optical configurations; readout electronics; data-processing methods for biomedical imaging systems with special emphasis on biological questions related to cancer, cardiovascular, and neurodegenerative diseases.

Gmitro, Arthur, Ph.D., University of Arizona, 1982. *Medical, Health Physics, Optics, Other*. Magnetic resonance imaging; optics in medicine; optical computing.

Greivenkamp, John E., Ph.D., University of Arizona, 1980. *Optics, Other*. Ophthalmic and visual optics; ophthalmic instrumentation and measurements; interferometry and optical testing of aspheric surfaces; optical fabrication; optical system design; optical metrology systems; distance measurement systems; sampled imaging theory; optics of electronic imaging systems.

Jessen, Poul S., Ph.D., University of Aarhus, 1993. Chair of Quantum Information & Control. *Atomic, Molecular, & Optical Physics, Physics of Beams, Other*. Quantum state preparation; coherent control; quantum tunneling and transport phenomena in optical lattices; quantum computation. Macroscopic superposition states and coherent evolution in dissipative quantum systems. Few-body quantum state preparation in optical lattices. Matter-wave equivalent of the laser/micromaser. Atom-optics in nano-fabrication. Laser cooling, trapping and manipulation of atoms and ions.

Khitrova, Galina, Ph.D., New York University, 1986. *Nonlinear Dynamics and Complex Systems, Optics, Other*. Nonlinear and quantum optics of semiconductor microcavities; fundamental studies of quantum confined semiconductors.

Koch, Thomas, Ph.D., California Institute of Technology, 1982. Dean of the College of Optical Sciences Professor of Electrical and Computer Engineering. *Electrical Engineering*. Semiconductor optoelectronics; optical fiber communications; photonic integrated circuits; silicon photonics.

Kostuk, Raymond K., Ph.D., Stanford University, 1986. *Electrical Engineering, Optics, Other*. Holographic techniques, systems, and materials; ion-exchange waveguide devices-interfacing to polymer and PBG layers; fiber-optic systems including OCDMA, error-correction codes, and all-optical network issues; medical imaging sensors including OCT and holographic filtering of coherent image data.

Maes, Carl F., Ph.D., University of Arizona, 2003. Associate Dean for Academic Programs, College of Optical Sciences. *Atomic, Molecular, & Optical Physics, Electromagnetism, Optics, Other*. Theory and simulations of laser resonators; mechanical effects of light on atoms and molecules in high-Q cavities; electromagnetic wave propagation, aberrations, and adaptive optics; experimental work with alkali and solid-state lasers, laser line-width measurement, and adaptive optics operations.

Mansuripur, Masud, Ph.D., Stanford University, 1981. Chair of Optical Data Storage. *Electrical Engineering, Electromagnetism, Optics, Other*. Optical data storage; magneto-optics; optics of polarized light in systems of high numerical aperture; magnetic and magneto-optical properties of thin solid films; magnetization dynamics; integrated optics for optical heads (data storage systems); information theory; optical signal processing; biological data storage; erbium-doped fiber amplifiers and lasers.

Marcellin, Michael, Ph.D., Texas A&M, 1987. *Computational Physics*. Digital communication and data storage systems; image and video compression; image processing; digital signal processing.

Marmorstein, Alan, Ph.D., State of University of New York, Brooklyn, 1994. Professor of Ophthalmology. *Medical, Health Physics*. Etiology and mechanisms of age-related macular degeneration.

Mazumdar, Sumit, Ph.D., Princeton University, 1980. *Engineering Physics/Science, Nonlinear Dynamics and Complex Systems, Optics, Polymer Physics/Science, Other*. Linear and nonlinear optical properties of organic conjugated molecules and polymers; effects of strong Coulomb correlations, excitons, and multiexcitons in organic systems; organic optoelectronic devices; strong Coulomb interactions and broken symmetries; charge and spin density waves and superconductivity in organic and inorganic materials with emphasis on organic charge-transfer solids and transition metal oxides.

Meystre, Pierre, Ph.D., Ecole Polytechnique Federale, 1974. Director of Biosphere 2. *Atomic, Molecular, & Optical Physics, Electrical Engineering, Energy Sources & Environment, Optics, Other*. Theoretical quantum optics; statistical properties of radiation; laser theory; nonlinear optics; atomic physics; ultracold atoms; Bose condensation and atom lasers; cavity quantum electrodynamics; atom optics.

Miller, Joseph M., Ph.D., Northeastern Ohio Universities College of Medicine, 1985. *Medical, Health Physics, Optics, Other*. The effect of astigmatism on visual development; non-invasive assessment of buried optical elements.

Milster, Thomas, Ph.D., University of Arizona, 1987. *Optics, Other*. Improving data density and signal readout quality in optical storage systems; design and implementation of novel storage modalities; signal detection and components.

Moloney, Jerome, Ph.D., University of Western Ontario, 1977. *Electrical Engineering, Nonlinear Dynamics and Complex Systems, Optics, Physics of Beams, Other*. Mathematical modeling and simulation of photonics systems; fundamental theory of semiconductor lasers; modeling high-power femtosecond atmospheric light strings; nonlinear theory of partial differential equations and chaos synchronization in extended complex spatiotemporal interacting systems; algorithm development for large-scale computational photonics systems simulations.

Neifeld, Mark A., Ph.D., California Institute of Technology, 1991. *Electrical Engineering, Optics, Other*. Nontraditional imaging; pattern recognition and neural networks; parallel coding and signal processing; volume optical storage; multiple-quantum-well photonics.

Nofziger, Michael, Ph.D., University of Arizona, 1995. Outreach Coordinator, College of Optical Sciences. *Optics, Physics and other Science Education, Other*.

Norwood, Robert, Ph.D., University of Pennsylvania, 1988. *Optics, Polymer Physics/Science, Other*. Electro-optic polymers and devices; photorefractive polymers; sol-gels; materials for linear and nonlinear photonic crystals; organic light-emitting diodes; solar cells; sensors.

Peyghambarian, Nasser, Ph.D., Indiana University, 1982. Chair, Photonics and Lasers; Director, ERC NSF CIAN. *Electrical Engineering, Nonlinear Dynamics and Complex Systems, Optics, Physics and other Science Education, Polymer Physics/Science, Other*. Optical telecommunication; fiber optics; fiber amplifiers and fiber lasers; integrated optics; femtosecond laser spectroscopy and dynamics of optical phenomena in semiconductors and organic materials; nonlinear photonics and high speed optical switching; characterization of optical materials in terms of speed and nonlinearities; polymer optoelectronics; photorefractive polymers; organic light-emitting diodes and lasers.

Potter, Kelly Simmons, Ph.D., University of Arizona, 1994. *Electrical Engineering, Optics, Other*. Elements for integrated optical systems using both optically active and passive novel photowritable materials; examination of single and multiphoton processes leading to both linear and nonlinear response in optical materials as a result of exposure to either ionizing or non-ionizing radiation; research into the impact of defect physics on material optical behavior, waveguide device design, and optical device performance.

Potter Jr., Barrett G., University of Florida, 1991. Professor of Materials Science and Engineering. *Materials Science, Metallurgy, Nano Science and Technology*. Synthesis and study of glass, ceramic and molecular hybrid materials for photonic and electronic applications: optically driven molecular assembly strategies; nanostructured photovoltaic energy conversion materials; photoactivated phenomena in glass and hybrid thin films; solution and physical vapor phase deposition of thin films and nanocomposites (oxides, inorganic-organic hybrids); thermal stability of complex oxide optical materials; environmental sensing; optical behavior of rare-earth-doped matrices; semiconductor quantum-dot ensembles; optical spectroscopy.

Sasian, José M., Ph.D., University of Arizona, 1988. *Astronomy, Atomic, Molecular, & Optical Physics, Medical, Health Physics, Optics, Other*. Lens and mirror design; optical fabrication; optomechanics; illumination optics; optical instrumentation for astronomy and biomedical sciences; conformal optics; microlithography; novel optical systems.

Schwiegerling, James, Ph.D., University of Arizona, 1995. *Medical, Health Physics, Optics, Other*. Visual system modeling with raytracing software; corneal topographic analysis for disease detection and visual performance assessment following surgery; optimization of refractive surgery techniques and development of ophthalmic instrumentation.

Seraphin, Supapan, Ph.D., Arizona State University, 1990. Green chemistry approach to synthesizing nanoparticles for waste water treatment and CO_2.

Shadman, Farhang, Ph.D., University of California, Berkeley, 1972. Regents Professor of Chemical & Environmental Engineering; Director of the Semiconductor Research Corporation Engineering Research Center for Environmentally Benign Semiconductor Manufacturing. *Chemical Physics, Energy Sources & Environment, Other*. Applications of chemical reaction engineering in semiconductor and opto-electronics manufacturing; advanced materials processing and environmental contamination control.

Simmons, Joseph H., Ph.D., Catholic University of America, 1969. Professor of Materials & Science Engineering; Head of the Department of Materials Science & Engineering. *Energy Sources & Environment, Materials Science, Metallurgy, Nonlinear Dynamics and Complex Systems, Other*. Quantum-size effects in the optical properties of semiconductor clusters; optical properties and carrier dynamics in wide-gap semiconductors; photosensitivity in glass films; nonlinear optical behavior of materials and glasses; optical spectroscopy of materials at the nanoscale level; molecular dynamics simulations. Non-linear viscous flow and rheological behavior of molten glasses.

Tyo, J. Scott, Ph.D., University of Pennsylvania, 1997. *Applied Mathematics, Electromagnetism, Optics, Other*. Processing of high-dimensional spectropolarimetric data; investigation of statistical properties of hyperspectral imagery; optimization of polarimetric sensors for remote sensing applications; integration of polarimetric and spectral sensors and information; spectropolarimetry to improve imaging in scattering media; fusion of multidimensional data into intelligible images; design of UWB antennas and antenna arrays; development of sensors, measurement techniques, and processing strategies for UWB EM measurements; generation and radiation of electromagnetic transients.

Uhlmann, Donald R., Ph.D., Harvard University, 1963. *Materials Science, Metallurgy, Other*. Sol-gel synthesis of ceramics nanocomposites; optical materials; hybrid silicone materials and devices; ferroelectric and pyroelectric materials and devices; kinetic processes in materials.

Walker, Christopher K., Ph.D., University of Arizona, 1988. *Astronomy, Optics, Other*. Star formation; millimeter instrumentation.

Wright, Ewan M., Ph.D., Heriot-Watt University, 1983. *Applied Physics, Atomic, Molecular, & Optical Physics, Engineering Physics/Science, Low Temperature Physics, Nano Science and Technology, Nonlinear Dynamics and Complex Systems, Optics, Physics and other Science Education, Physics of Beams, Other*. Theory and simulation of light-string propagation in air; electromagnetic pulse emission from light-string-induced plasmas; supercontinuum generation and self-guiding in condensed media; Bose-Einstein condensation in atomic vapors; mean-field theory and beyond for small BECs; quantum theory of 1-D gases in the Tonks-Girardeau regime; theory of 1-D atom waveguides and interferometers; theory of light-induced waveguides for cold atoms; optically bound matter; theory of anyon matter in planar chiral nanostructures.

Ziolkowski, Richard, Ph.D., University of Illinois at Urbana-Champaign, 1980. *Applied Mathematics, Electromagnetism, Other*. Application of new mathematical and numerical methods to linear and nonlinear problems dealing with the interaction of acoustic and electromagnetic waves with complex media, metamaterials, and realistic structures.

Associate Professor

Anderson, Brian P., Ph.D., Stanford University, 1999. *Applied Physics, Atomic, Molecular, & Optical Physics, Physics of Beams, Other*. Atomic gas Bose-Einstein condensates held in combined optical and magnetic potentials; vortices, solitons, and superfluid dynamics; atom interferometry, phase transitions, and BEC manipulation using tailored optical potentials.

Djordjevic, Ivan B., Ph.D., University of Nis, 1999. *Electrical Engineering, Other*. Optical networks; error control coding, constrained coding, and coded modulation; turbo equalization; OFDM applications; quantum error correction.

Hua, Hong, Ph.D., Beijing Institute of Technology, 1999. *Medical, Health Physics, Optics, Other*. Development of 2-D and 3-D display systems, imaging systems, tracking systems, and interaction methods; stereoscopic displays; human-computer interface techniques; virtual and augmented environments.

Jacquod, Philippe, Ph.D., University of Neuchâtel, 1997. *Other*.

Jones, Ronald Jason, Ph.D., University of New Mexico, 2001. *Atomic, Molecular, & Optical Physics, Electrical Engineering, Optics, Other*. Ultrafast laser science; femtosecond frequency combs; extreme nonlinear light/matter interactions and generation; optical frequency metrology; high-resolution spectroscopy.

Kolesik, Miroslav, Ph.D., Slovak Academy of Sciences, 1992. *Nonlinear Dynamics and Complex Systems, Physics of Beams, Other*. Semiconductor laser simulation; femtosecond light-matter interactions; computational nonlinear optics.

Kupinski, Matthew, Ph.D., University of Chicago, 2000. *Applied Mathematics, Medical, Health Physics, Statistical & Thermal Physics, Other*. Task-based assessment of image quality for both tumor detection and parameter estimation tasks; statistical characteristics of images and the objects being imaged; imaging hardware optimization; human-observer models for image analysis.

Liang, Rongguang, Ph.D., University of Arizona, 2001. *Medical, Health Physics, Optics, Other*. Biomedical imaging; optical design.

Pau, Stanley, Ph.D., Stanford University, 1996. *Computer Science, Electrical Engineering, Nano Science and Technology, Optics, Other*. Micro-optics, MEMS/NEMS for imaging and sensing applications; optical lithography and novel techniques for nanofabrication; microfabricated neutral atom trap and ion trap for mass spectrometry and quantum computing; microfluidic and microfabricated chemical reactor.

Takashima, Yuzuru, Ph.D., Stanford University, 2007. *Electrical Engineering, Optics, Other*. Optical systems for high-density bit-based holographic data storage systems.

Utzinger, Urs, Ph.D., Federal Institute of Technology, 1995. *Medical, Health Physics, Optics, Other*. Optical tissue spectroscopy; optical biosignatures and bioinstrumentation.

Visscher, Koen, Ph.D., University of Amsterdam, 1993. *Applied Physics, Optics, Other*. Regulation of gene expression by mechanical force; RNA structure; optical tweezers.

Assistant Professor

Ashok, Amit, Ph.D., University of Arizona, 2008. *Electrical Engineering, Optics, Other*. Optical imaging and sensing; physical optics; statistical inference and information theory.

Cronin, Alexander, Ph.D., University of Washington, 1999. *Atomic, Molecular, & Optical Physics, Optics, Other*. Atom interferometry.

Gehm, Michael, Ph.D., Duke University, 2003. *Atomic, Molecular, & Optical Physics, Electrical Engineering, Optics, Other*. Computational sensing; compressive sensing; optical sensors; spectroscopy; imaging; spectral imaging; terahertz technology; rapid prototyping; volumetric optical components; optical physics.

Guyon, Olivier, Ph.D., Pierre-and-Marie-Curie University, 2002. Assistant Professor of Astronomy. *Astronomy*. Innovative techniques for extrasolar planets; new wavefront sensing techniques for adaptive optics; Phase-Induced Amplitude Apodization coronograph; high-contrast imaging techniques.

Kieu, Khanh, Ph.D., The University of Arizona, 2007. Quantum electronics, ultra-fast lasers, and nonlinear optics; optical fiber technologies such as fiber lasers, fiber optical sensors, nonlinear effects and devices in waveguiding structures; developing new advanced components and instruments.

Peng, Leilei, Ph.D., Purdue University, 2005. *Optics, Other*. High-speed Fourier transform fluorescence spectrometer for hyperspectral imaging.

Sandhu, Arvinder S., Ph.D., Tata Institute of Fundamental Research, 2005. *Atomic, Molecular, & Optical Physics, Engineering Physics/Science, High Energy Physics, Other*. High-harmonic generation in hollow waveguides/XUV sources; generation and applications of attosecond pulse trains; carrier envelope phase stabilization techniques for grating-based Ti: Sapphire ultrashort pulse amplifier systems; cold target recoil ion momentum spectroscopy (a.k.a. reaction microscope) for coincidence imaging of molecular or atomic fragments; time-resolved experiments to probe ultrafast dynamics in atoms, molecules and plasmas using X-ray and electrons in both high-average-power and high-peak-power regime.

Witte, Russell, Ph.D., Arizona State University, 2002. *Medical, Health Physics, Optics, Other*. Electrical brain mapping during neurosurgery (epilepsy, brain cancer); novel paradigms to diagnose and treat neurological disorders (Alzheimer's); neural prostheses; functional electrical stimulation; neural engineering: muscular disease, sports injury, rehabilitation, fatigue, and aging.

Professor Emeritus

Strickland, Robin, Ph.D., Sheffield University, 1979. Professor of Electrical and Computer Engineering. *Computer Science, Electrical Engineering, Other*. Digital image processing; computer vision; signal processing.

Wyant, James C., Ph.D., University of Rochester, 1968. Founding Dean of College of Optical Sciences. *Optics, Other*. Implementation of microcomputers and software to interferometric techniques for optical measurements, in particular for optical testing. Testing of supersmooth optical surfaces. Measurement of optically rough surfaces. Testing of complex aspheric surfaces. Development of commercial optical test equipment based on phase-shifting interferometry.

Research Professor

Barber, Bradford H., Ph.D., University of Arizona, 1976. *Medical, Health Physics, Nuclear Engineering, Nuclear Physics, Optics, Other*. Radiation physics; radiation detectors; dosimetry; ultra-high-resolution imaging gamma-ray detectors.

Biggar, Stuart F., Ph.D., University of Arizona, 1990. *Optics, Other*. Spacecraft and aircraft optical sensors; optical system design, evaluation, and absolute radiometric calibration; reflectance measurement.

Kost, Alan, Ph.D., University of Texas, Austin, 1983. Administrative Director, CIAN Engineering Research Center. *Electrical Engineering, Optics, Other*. Nano-photonics (photonic integrated circuits, fiber optics, and nonlinear optical materials); optical data storage.

Kwong, Nai-Hang, Ph.D., California Institute of Technology, 1983. *Quantum Foundations*. Microscopic theory of nonlinear optics of semiconductor quantum well structures; quantum mechanics of interacting electrons and/or excitons in the semiconductor; electromagnetically induced transparency in quantum wells, Raman coherences, four-wave mixing in microcavities and optically induced polarization shift in quantum well Bragg structures; conceptual issues such as the bosonic aspects of quantum well excitons.

LaComb, Lloyd, Ph.D., Stanford University, 1989. *Optics*. holography, interferometry, spectroscopy, ellipsometry, reflectometry, image formation, image processing, image analysis and correction, fiber optics, lasers, film thickness measurement, surface measurements.

Wissinger, John, Ph.D., Massachusetts Institute of Technology, 1994. *Computer Science*. optical communications networks, network monitoring and control, distributed sensor networks, distributed inference and learning.

Research Associate Professor

Dubin, Matthew, Ph.D., University of Arizona, 2002. *Optics, Systems Science/Engineering, Other*. Optical systems and engineering; innovation; design, prototyping, and use of systems involving interferometry, alignment, imaging, optomechanical design, display design, and illumination combining conceptual and analytical skills with system design and practical implementation.

Hader, Jorg, Ph.D., Philipp University of Marburg, 1997. semiconductor many-body physics; carrier recombination processes in semiconductors, nonequilibrium dynamics in semiconductor lasers, modeling of vertical external cavity surface emitting lasers, microscopic optoelectronics of wide bandgap nitride-based and mid- to far-infrared semiconductor lasers and laser diodes.

Kaneda, Yushi, Ph.D., University of Tokyo, 1998. *Nonlinear Dynamics and Complex Systems*. Solid-state lasers, optically pumped semiconductor lasers, fiber lasers, nonlinear frequency conversion.

Kupinski, Meredith K., Ph.D., University of Arizona, 2008. *Optics, Physics and other Science Education, Other*.

Lesser, Michael, Ph.D., University of Arizona, 1988. Director, the University of Arizona Imaging Technology Laboratory. Optimization of scientific CCD detectors, particularly for astronomical and industrial use; illuminating, packaging, backside charging, and antireflection coating of CCDs and CMOS imagers. Astronomical instrumentation. Visible and ultraviolet imaging and spectroscopy and associated software technologies.

Martin, Hubert M., Ph.D., University of Cambridge, 1983. Project Scientist at the Steward Observatory. *Optics, Other*. Fabrication and testing of large astronomical mirrors; active and adaptive optics.

McClain, Stephen, Ph.D., Cornell University, 1992. Polarization modeling, design, and analysis; optical system design and lens design with CODE V, Zemax, and ASAP; polarization measurement, including design, construction, and calibration of polarimeters; optical system development for monitors, TVs, and projectors; polarization of critical fiber optic devices; colorimetry in display applications.

Polynkin, Pavel, Ph.D., Texas A&M University, 2000. High-intensity ultrafast laser science; plasma generation in intense laser fields; material processing with ultrafast lasers; free-space communications through turbid media; nonlinear wavelength conversion; optical fiber lasers and amplifiers; optical sensors.

Yushi, Kaneda, Ph.D., University of Tokyo, 1998. Solid-state lasers; optically pumped semiconductor lasers; fiber lasers; nonlinear frequency conversion.

Zhao, Chunyu, Ph.D., University of Arizona, 2002. Optical system engineering; optical testing; optical design; aberration theory.

Research Assistant Professor

Balakrishnan, Kaushik, Ph.D., Southern Illinois University, 2008. *Materials Science, Metallurgy, Nano Science and Technology, Optics*. Self-assembled structures and functional nanostructured materials for photonic, optoelectronic and electronic, and energy applications; photorefractive sensitizers for holographic applications; bio-nano composite materials.

Blanche, Pierre-Alexandre, Ph.D., University of Liege, 1999. *Nano Science and Technology, Optics*. Photorefractive materials and their applications; holography and recording materials; technical holographic applications; Dispersive Volume Phase Holographic Gratings and holographic optical elements; photovoltaic materials; nano-structuring and nanoparticles for optics and photonics; nonlinear optics; space instrumentation.

Czapla-Myers, Jeffrey, Ph.D., University of Arizona, 2006. *Optics, Other*. Preflight and post-launch radiometric calibration of airborne and satellite sensors; design and testing of field and laboratory radiometers.

He, Jun, Ph.D., University of Virginia, 2008. *Computer Science, Optics, Other*. Architectural, algorithmic, and performance aspects of communication networks; optical access network control and management system that aggregates the newly developed optical devices and components and satisfy the networking requirement in integrated access networks; multifunctional autonomous sensor networks consisting of spatially distributed self-sustainable sensor nodes that can monitor locally various environmental parameters; cross-layer optimization of joint routing and resource allocation for optical networks impaired by physical degradations.

Kim, Dae Wook, Ph.D., University of Arizona, 2009. *Electrical Engineering, Optics*. Large precision optics fabrication using Computer Controlled Optical Surfacing (CCOS) process; optical testing for large optical components using computer generated holograms, laser tracker, interferometer, etc.; optical system design and manufacturing; open-source data analysis and visualization S/W platform development for optical engineering.

Parks, Robert E., M.A., Williams College, 1966. Optical testing and test instrumentation; optical fabrication methods.

Smith, Gregory A., Ph.D., The University of Arizona, 2006. *Engineering Physics/Science, Optics*. polarization software design and polarization engineering; hands-on design for quantum physics; optical simulation and analysis.

Stone, Robert, Ph.D., University of Arizona, 1997. Design and analysis of complex mechanical systems subject to static, dynamic, or thermal loads; structural design and analysis of spaceborne telescopes; welded structures, pneumatic and hydraulic systems, optical components, and support structures; material characterization of cellular foam materials for lightweight optic and space applications.

Su, Peng, Ph.D., University of Arizona, 2008. Optical metrology for testing aspheric surfaces: swing arm profilometer, CGH test, interferometry, large flat test, and scanning pentaprism test. Optical design. Diffractive optics. Adaptive optics. Global warming. Solar reflector. Optical system engineering.

Zhao, Ming, Ph.D., Purdue University, 2009. *Biophysics, Engineering Physics/Science, Medical, Health Physics, Optics*. Fluorescence microscopy and spectroscopy; biosensors and microarray immunoassays.

Zhu, Xiushan, Ph.D., The University of Arizona, 2008. *Nano Science and Technology, Optics*. optical fibers, fiber devices, fiber lasers and solid-state lasers; novel materials for fiber laser and amplifiers, high power ZBLAN fiber lasers, single-frequency fiber lasers and amplifiers, advanced fiber devices and nonlinear optics.

Adjunct Professor

Koch, Stephan W., Ph.D., University of Frankfurt, 1979. *Atomic, Molecular, & Optical Physics, Nano Science and Technology, Theoretical Physics*. condensed matter theory; optical and electronic properties of semiconductors; many-body interactions; semiconductor quantum optics; quantum confinement in solids; coherent and ultrafast phenomena; semiconductor laser theory; microcavity and photonic crystal effects.

Postdoctoral Research Associate

Scheller, Maik, Ph.D., Philipp University of Marburg, 2011. *Atomic, Molecular, & Optical Physics*. Nonlinear frequency conversion; optically pumped semiconductor lasers; terahertz generation and detection.

DEPARTMENTAL RESEARCH SPECIALTIES AND STAFF

Theoretical

Image Science. http://www.optics.arizona.edu/research/faculty-specialties/image-science Ashok, Barrett, Chipman, Matthew Kupinski, Peng.

Optical Engineering. http://www.optics.arizona.edu/research/faculty-specialties/optical-engineering Burge, Dereniak, Greivenkamp, Hua, Liang, Milster, Sasian, Schwiegerling, Takashima, Tyo.

Optical Physics. http://www.optics.arizona.edu/research/faculty-specialties/optical-physics Anderson, Binder, Bloembergen, Jessen, Jones, Khitrova, Kolesik, Maes, Wright.

Photonics. http://www.optics.arizona.edu/research/faculty-specialties/photonics Cvijetic, Falco, Fallahi, Thomas Koch, Mansuripur, Norwood, Pau, Peyghambarian.

View additional information about this department at www.gradschoolshopper.com

---

UNIVERSITY OF ARKANSAS

DEPARTMENT OF PHYSICS

Fayetteville, Arkansas 72701

http://www.uark.edu/depts/physics/apps/

General University Information

President: Donald Bobbitt
Dean of Graduate School: Todd Shields
University website: http://uark.edu
Control: Public
Setting: Urban
Total Faculty: 1,203
Total Graduate Faculty: 784
Total number of Students: 24,537
Total number of Graduate Students: 3,777

Department Information

Department Chairman: Julio Gea-Banacloche, Chair
Department Contact: Dianne Melahn, Office Manager
 Total full-time faculty: 20
 Total number of full-time equivalent positions: 27

Full-Time Graduate Students: 49
First-Year Graduate Students: 12
Female First-Year Students: 1
Total Post Doctorates: 11

Department Address

825 W. Dickson Street
PHYS-226
Fayetteville, AR 72701
Phone: (479) 575-2506
Fax: (479) 575-4580
E-mail: physics@uark.edu
Website: http://www.uark.edu/depts/physics/apps/

ADMISSIONS

Admission Contact Information
Address admission inquiries to: Department of Physics
Phone: (479) 575-2506
E-mail: physics@uark.edu
Admissions website: http://admissions.uark.edu/

Application deadlines
Fall admission:
U.S. students: January 15   *Int'l. students*: January 15
Spring admission:
U.S. students: September 30   *Int'l. students*: September 30

Application fee
U.S. students: $40   *Int'l. students*: $50
Fee waived if approved by department. Late applications will
be considered only if positions are available.

Admissions information
For Fall of 2013:
  Number of applicants: 90
  Number admitted: 21
  Number enrolled: 9

Admission requirements
Bachelor's degree requirements: Bachelor's degree in physics
is preferred with a minimum undergraduate GPA of 3.0
(A = 4) or 3.2 on last 60 hours credit.
Minimum undergraduate GPA: 3.0

GRE requirements
The GRE is recommended.

Advanced GRE requirements
The Advanced GRE is recommended.

TOEFL requirements
The TOEFL exam is required for students from non-English-
speaking countries.
  iBT score: 80

Other admissions information
Additional requirements: A minimum score of 26 on the speaking
section is required for a graduate teaching assistantship.
Undergraduate preparation assumed: Candidates should have an
undergraduate degree with the equivalent of a 30-hour major
in physics including intermediate level courses in mechanics,
electricity and magnetism, thermal physics, quantum mechanics, and mathematics through differential equations.

TUITION

Tuition year 2012–13:
Tuition for in-state residents
  Full-time students: $349.47 per credit
  Part-time students: $349.47 per credit
Tuition for out-of-state residents
  Full-time students: $826.77 per credit
  Part-time students: $826.77 per credit
Students generally take 6-9 hours.
Credit hours per semester to be considered full-time: 6
Deferred tuition plan: No
Health insurance: Available at the cost of ~1000 per year.
Other academic fees: $40 Application fee, $50 for international
  $575.49 for international students $35.08 University Fees
  $10.19 College Fees $607 Books, supplies + Lab fees
Academic term: Semester
Number of first-year students who received full tuition waivers: 9

Teaching Assistants, Research Assistants, and Fellowships
Number of first-year
  Teaching Assistants: 12
Average stipend per academic year
  Teaching Assistant: $15,500
  Research Assistant: $15,500
  Fellowship student: $25,300

FINANCIAL AID

Application deadlines
Fall admission:
U.S. students: July 1
Spring admission:
U.S. students: December 1   *Int'l. students*: December 1

Loans
Loans are not available for U.S. students.
Loans are not available for international students.
GAPSFAS application required: No
FAFSA application required: No

For further information
Address financial aid inquiries to: Graduate Admissions Com-
mittee, Department of Physics.
Phone: (479) 575-2506
E-mail: physics@uark.edu
Financial aid website: http://finaid.uark.edu/

HOUSING

Availability of on-campus housing
  Single students: Yes
  Married students: No

For further information
Address housing inquiries to: University Housing, 900 Hotz Hall.
Phone: (479) 575-3951
E-mail: housing@uark.edu
Housing aid website: http://housing.uark.edu/

Table A—Faculty, Enrollments, and Degrees Granted

Research Specialty	2012-13 Faculty	Enrollment Fall 2012		Number of Degrees Granted 2013 (2007-12)		
		Mas-ter's	Doc-torate	Mas-ter's	Terminal Master's	Doc-torate
Astronomy	3	1	3	2(3)	–(4)	–
Atomic, Molecular, & Optical Physics	5	4	15	–(1)	–(1)	–(7)
Biophysics	3	–	2	–(1)	1(2)	2(1)
Condensed Matter Physics	–	–	15	2(6)	–(8)	2(5)
Physics and other Science Education	2	–	13	–(5)	–(5)	–
Non-specialized	1	–	1	–	–	1(-)
Total	20	10	40	4(16)	1(20)	5(13)
Full-time Grad. Stud.	–	5	44	–	–	–
First-year Grad. Stud.	–	–	12	–	–	–

GRADUATE DEGREE REQUIREMENTS

Master's: Choose either a 30 credit thesis path or a 36 credit non-thesis path. A GPA of 3.0 (A = 4) must be maintained. There is a 30-week residency requirement. Core courses plus physics electives. No language requirement or written comprehensive or qualifying examination.

Doctorate: 40 semester hours of coursework at the graduate level and 18 hours of doctoral dissertations. A GPA of 3.0 (A = 4) must be maintained. Residency of two consecutive semesters required after admission to candidacy. No language requirement. Written and oral candidacy examination required by third semester of graduate work. Dissertation with final oral examination.

Other Degrees: M.A. degree. Education track. 30 semester hours credit. A GPA of 3.0 (A = 4) must be maintained. There is a 30-week residency requirement. No language requirement or written comprehensive or qualifying examination. No thesis. Written report with oral final examination.

Thesis: Thesis may be written in absentia.

SPECIAL EQUIPMENT, FACILITIES, OR PROGRAMS

Research facilities include well-equipped research laboratories in quantum optics, laser spectroscopy, nonlinear optics, high pressure physics, surface physics, nanoscience, biophysics and computer graphics. The laboratories possess a complete range of equipment including ultra-violet, visible and infrared gas lasers, ultra-high stability CW dye and solid-state laser systems, femtosecond and ultra-high power pulsed solid state laser systems, and a SQUID magnetometer. A $1.5M molecular beam epitaxy (MBE) facility is available in the department for fabricating and characterizing semiconductor heterostructures. Excellent sample characterization and research facilities equipped with X-ray diffractometer (XRD), scanning electron microscope (SEM), atomic force microscope (AFM), Raman spectrometer, etc. are available at the High Density Electronics Center HiDEC at the University.

Table B—Separately Budgeted Research Expenditures by Source of Support

Source of Support	Departmental Research	Physics-related Research Outside Department
Federal government	$3,523,262	
State/local government	$100,921	
Non-profit organizations		
Business and industry		
Other		
Total	**$3,624,183**	

Table C—Separately Budgeted Research Expenditures by Research Specialty

Research Specialty	No. of Grants	Expenditures ($)
Astronomy	3	$238,190
Biophysics	4	$63,397
Condensed Matter Physics	24	$1,959,649
Physics and other Science Education	4	$1,213,786
Quantum Foundations	3	$149,161
Total	**38**	**$3,624,183**

FACULTY

Chair Professor

Gea-Banacloche, Julio, Ph.D., New Mexico State University, 1985. Department Chair. *Optics*. Theoretical quantum optics; quantum information.

Professor

Bellaiche, Laurent, Ph.D., University of Paris XI, 1994. *Computational Physics, Condensed Matter Physics*. Developing and/or using direct first-principles methods to calculate properties of ferroelectrics, semiconductors, magnetic compounds and low dimensional systems.

Harter, William G., Ph.D., University of California, Irvine, 1967. *Atomic, Molecular, & Optical Physics*. Theoretical physics; molecular dynamics; computer graphics.

Lacy, Claud H., Ph.D., University of Texas, Austin, 1978. Department vice chair. *Astronomy, Astrophysics*. Astronomy; eclipsing binaries; near-earth asteroids; binary supermassive black holes.

Salamo, Gregory J., Ph.D., City University of New York, 1974. Distinguished Professor; Basore Professorship; Director of the Nano Material Science and Engineering Building. *Atomic, Molecular, & Optical Physics, Biophysics, Condensed Matter Physics, Nano Science and Technology, Optics*. Lasers; quantum optics; quantum structures; growing semiconductors; optical computing; nanofabrication.

Singh, Surendra P., Ph.D., University of Rochester, 1982. *Atomic, Molecular, & Optical Physics, Nonlinear Dynamics and Complex Systems, Optics*. Quantum optics; lasers.

Stewart, Gay, Ph.D., University of Illinois, 1994. *Physics and other Science Education*. Physics education; educational engineering.

Thibado, Paul M., Ph.D., University of Pennsylvania, 1994. *Condensed Matter Physics, Nano Science and Technology*. Condensed matter physics; surface physics.

Vyas, Reeta, Ph.D., State University of New York at Buffalo, 1984. *Nonlinear Dynamics and Complex Systems, Nuclear Physics, Optics*. Nuclear theory; quantum optics; interaction of simple atomic systems with nonclassical light.

Xiao, Min, Ph.D., University of Texas, Austin, 1988. Distinguished Professor; Twenty-First Century Chair in Nanotechnology. *Atomic, Molecular, & Optical Physics, Optics*. Quantum/nonlinear optics with multi-level systems and optical properties of semiconductor nanostructures.

Associate Professor

Fu, Huaxiang, Ph.D., Fudan University, 1994. *Condensed Matter Physics*. Theoretical solid-state physics.

Li, Jiali, Ph.D., City University of New York, 1999. *Biophysics, Condensed Matter Physics, Nano Science and Technology*. Condensed matter physics; bio/nano physics; nanofabrication, nanoscale materials science, and single DNA and protein detection.

Oliver, William F., Ph.D., University of Colorado, 1987. *Condensed Matter Physics*. Condensed matter under extreme conditions.

Stewart, John, Ph.D., University of Illinois, 1994. Clinical Associate Professor. *Condensed Matter Physics, Physics and other Science Education*. Condensed matter physics, physics education; educational engineering.

Tchakhalian, Jak, Ph.D., University of British Columbia, 2002. *Condensed Matter Physics, Nano Science and Technology*. Condensed matter physics; multiscale phenomena in artificial complex oxide nanostructures.

Assistant Professor

Barraza-Lopez, Salvador, Ph.D., University of Illinois at Urbana-Champaign, 2006. *Condensed Matter Physics*. Quantum transport; graphene.

Kennefick, Daniel, Ph.D., California Institute of Technology, 1997. *Astrophysics*. Astrophysics; supermassive black holes; Einstein.

Kennefick, Julia, Ph.D., California Institute of Technology, 1995. *Astronomy*. Astronomy; observational cosmology; quasars at high redshifts.

Shew, Woodrow L., Ph.D., University of Maryland, 2004. *Neuroscience/Neuro Physics*.

Research Professor

Vickers, Ken, M.S., University of Arkansas, 1978. Director, Microelectronics and Photonics Program. *Engineering Physics/Science, Nano Science and Technology*. Advanced materials and devices.

DEPARTMENTAL RESEARCH SPECIALTIES AND STAFF

Theoretical

Atomic, Molecular, & Optical Physics. Molecular spectroscopy; quantum optics; quantum information with atomic and optical systems. Gea-Banacloche, Harter, Singh, Vyas.

Condensed Matter Physics. First-principles calculations of material properties; quantum transport. Barraza-Lopez, Bellaiche, Fu.

Neuroscience/Neuro Physics. Testing predictions from statistical physics in living neural networks, focusing on predictions that have implications for brain function. Such predictions are many, arising from decades of theoretical work, but experiments are needed to move these predictions from the realm of speculation to the realm of practical importance and concrete measurements. Actual experiments in Dr. Shew's lab focus on sensory processing and how information arriving from the senses is integrated into the ongoing neural activity of the cortex. The results of these experiments have the potential to provide insight on brain disorders such as autism and Down's syndrome. Shew.

Optics. Quantum optics and electronics; nonlinear optics; laser theory. Gea-Banacloche, Singh, Vyas.

Physics and other Science Education. Physics education research; science teacher training. Gay Stewart, John Stewart.

Experimental

Astronomy. Astronomy and astrophysics: eclipsing binaries; supermassive black holes. Daniel Kennefick, Julia Kennefick, Lacy.

Biophysics. Nanopore physics; surfaces, biopolymers; optical tweezers; bioptical and electrical phenomena, neural network, membranes. Li, Shew.

Condensed Matter Physics. Optical properties of solids; thin films; electronic transport; quantum well structures; surface physics; molecular beam epitaxy (MBE); scanning tunneling spectroscopy; Raman scattering, Brillouin scattering, high-pressure physics. Oliver, Salamo, Tchakhalian, Thibado.

Optics. Self-induced transparency; coherence and fluctuation in lasers; nonlinear optics; photorefraction; photon statistics; spectroscopy, Raman scattering, Brillouin scattering. Salamo, Singh, Xiao.

View additional information about this department at
www.gradschoolshopper.com

CALIFORNIA STATE UNIVERSITY, LONG BEACH

DEPARTMENT OF PHYSICS & ASTRONOMY

Long Beach, California 90840-9505
http://www.csulb.edu/depts/physics/

General University Information

President: Donald J. Para
Dean of Graduate School: Cécile Lindsay
University website: http://www.csulb.edu/
Control: Public
Setting: Urban
Total Faculty: 2,319
Total number of Students: 34,875
Total number of Graduate Students: 5,584

Department Information

Department Chairman: Chuhee Kwon, Chair
Department Contact: Irene Howard, Administrative Coordinator
Total full-time faculty: 13
Total number of full-time equivalent positions: 24
Full-Time Graduate Students: 41
First-Year Graduate Students: 17
Female First-Year Students: 4

Department Address

1250 Bellflower Boulevard
Long Beach, CA 90840-9505
Phone: (562) 985-7925
Fax: (562) 985-7924
E-mail: irene.howard@csulb.edu
Website: http://www.csulb.edu/depts/physics/

ADMISSIONS

Admission Contact Information

Address admission inquiries to: Professor Andreas Bill
Phone: (562) 985-8616
E-mail: Andreas.Bill@csulb.edu
Admissions website: http://www.csulb.edu/divisions/aa/projects/grad/

Application deadlines

Fall admission:
U.S. students: June 1   *Int'l. students*: April 1
Spring admission:
U.S. students: November 1   *Int'l. students*: October 1

Application fee

U.S. students: $55   *Int'l. students*: $55
Application Website http://www.csumentor.edu

Admissions information

For Fall of 2013:
  Number enrolled: 17

Admission requirements

Bachelor's degree requirements: Bachelor degree in Physics is preferred. Other bachelor degrees are accepted under certain conditions. Contact the graduate advisor.
Minimum undergraduate GPA: 2.5

GRE requirements

The GRE is not required.
Applicants are encouraged to indicate their GRE score if they have one.

Advanced GRE requirements

The Advanced GRE is not required.
Applicants are encouraged to indicate their GRE score if they have one.

TOEFL requirements

The TOEFL exam is required for students from non-English-speaking countries.
  PBT score: 550
  iBT score: 80

TUITION

Tuition year 2012–13:
Tuition for in-state residents
  Full-time students: $3,764 per semester
  Part-time students: $2,348 per semester
Tuition for out-of-state residents
  Full-time students: $7,112 per semester
  Part-time students: $4,580 per semester
The information above is for 9 units (full-time) or 6 units (part-time). However, for US citizens and permanent residents, a graduate student with 6 units is considered a full-time graduate student. For international graduate students, 9 units is considered a full-time graduate student.
Credit hours per semester to be considered full-time: 6
Deferred tuition plan: Yes
Health insurance: Available at the cost of $759 per year.
Academic term: Semester

Teaching Assistants, Research Assistants, and Fellowships

Number of first-year
  Teaching Assistants: 11
Average stipend per academic year
  Teaching Assistant: $9,120
TA for 1 lab section is $2,280 for one semester. The maximum teaching load is 4 lab sections per semester.

FINANCIAL AID

Application deadlines

Fall admission:
U.S. students: March 1

Loans

Loans are available for U.S. students.
Loans are not available for international students.
GAPSFAS application required: No
FAFSA application required: Yes

For further information

Address financial aid inquiries to: CSULB Office of Financial Aid, In person: 101 Brotman Hall.
Phone: (562) 985-8403
Financial aid website: http://www.csulb.edu/depts/enrollment/financial_aid/

HOUSING

Availability of on-campus housing

Single students: Yes
Married students: No

For further information

Address housing inquiries to: Housing and Residential Life, California State University, Long Beach, 1250 Bellflower Boulevard, MS 8701, Long Beach, CA 90840-8701.
Phone: (562) 985-4187
E-mail: housing@csulb.edu
Housing aid website: http://www.csulb.edu/divisions/students/housing/

SPECIAL EQUIPMENT, FACILITIES, OR PROGRAMS

Many experimental facilities are housed in the department, including a Physical Properties Measurement System (PPMS), Apertureless Near-field Scanning Optical Microscopy (ANSOM), multi-target Sputtering systems, Magneto-Optical Kerr Effect (MOKE), X-ray diffraction system, Photo-current measurement systems, Vibrating Sample Magnetometer, Atomic Force Microscope, etc.

The Department also has a modern Computational Physics Laboratory.

Most equipment has been funded by various external grants, including NSF, DoD, and the Army Research Laboratory.

FACULTY

Professor

Anwar, Zahur, Ph.D., University of British Columbia, 1964.
Bill, Andreas, Ph.D., University of Stuttgart, 1995. *Applied Physics, Computational Physics, Condensed Matter Physics, Low Temperature Physics, Nano Science and Technology, Solid State Physics, Theoretical Physics*. Superconductivity, magnetism, condensed matter theory, quantum mechanics, crystallization of solids. http://www.csulb.edu/~pjaikuma/Compuphys/comp.html.
Hintzen, Paul, Ph.D., University of Arizona, 1975. *Astronomy*.
Hlousek, Zvonimir T., Ph.D., Brown University, 1987. *High Energy Physics, Particles and Fields, Physics and other Science Education, Theoretical Physics*.
Kenealy, Patrick, Ph.D., University of Notre Dame, 1967. *Physics and other Science Education*.
Kwon, Chuhee, Ph.D., University of Maryland, College Park, 1995. *Applied Physics, Condensed Matter Physics, Low Temperature Physics, Materials Science, Metallurgy, Nano Science and Technology, Physics and other Science Education, Solid State Physics*.
Leung, Alfred, Ph.D., University of California, Los Angeles. *Optics*.
Papp, Zoltan, Ph.D., University of Debrecen, 1986. *Atomic, Molecular, & Optical Physics, Computational Physics, High Energy Physics, Nuclear Physics, Theoretical Physics*. few body systems, computational physics, quantum mechanics. http://www.csulb.edu/~pjaikuma/Compuphys/comp.html.
Pickett, Galen T., Ph.D., University of Chicago, 1995. *Computational Physics, Condensed Matter Physics, Mechanics, Phys-*

ics and other Science Education, Polymer Physics/Science, Surface Physics, Theoretical Physics. Condensed Matter Theory, Membranes. http://www.csulb.edu/~pjaikuma/Compuphys/comp.html.

Rajpoot, Subhash, Ph.D., Imperial College, London, 1979. *Cosmology & String Theory, High Energy Physics, Particles and Fields, Theoretical Physics.* Unification of all interactions. Study of supersymmetry, supergravity, superstring theory, and super membrane theory.

Associate Professor

Gredig, Thomas, Ph.D., University of Minnesota, Twin Cities, 2002. *Applied Physics, Condensed Matter Physics, Materials Science, Metallurgy, Nano Science and Technology, Solar Physics, Solid State Physics.* Organic Semiconductors, Thin Film Crystallography, Nanomagnetism.

Gu, Jiyeong, Ph.D., Seoul National University, 1998. *Applied Physics, Condensed Matter Physics, Low Temperature Physics, Materials Science, Metallurgy, Nano Science and Technology, Solid State Physics.* Thin Films, Nanomagnetism, Superconductivity.

Assistant Professor

Abate, Yohannes, Ph.D., University of Iowa, Iowa City, 2006. *Applied Physics, Nano Science and Technology, Optics.* Near-field Scanning Optical Microscopy (s-SNOM), plasmonics, semiconductor quantum dots, strongly correlated materials, metamaterials.

Jaikumar, Prashanth, Ph.D., Stony Brook University, 2002. *Astrophysics, Computational Physics, Nuclear Physics, Relativity & Gravitation, Theoretical Physics.* Nuclear Astrophysics, Quantum Chromodynamics (QCD), Neutron Stars, Gravitational waves. http://www.csulb.edu/~pjaikuma/Compuphys/comp.html Computational Physics: http://www.csulb.edu/~pjaikuma/Compuphys/comp.html.

Peterson, Michael, Ph.D., Pennsylvania State University, 2005. *Computational Physics, Condensed Matter Physics, Nano Science and Technology, Solid State Physics, Theoretical Physics.* Strongly Correlated Systems, Quantum Topological Phases, Condensed Matter Theory, Quantum Mechanics. http://www.csulb.edu/~pjaikuma/Compuphys/comp.html.

Instructor

Stankovic, Jasmina, M.S., University of Nis, 1991.

Lecturer

Abachi, Shahriar, Ph.D., University of California, Los Angeles, 1985. *High Energy Physics, Particles and Fields.* LHC (Large hadron Collider at CERN), supersymmetry.

Chuang, Kuan-Wen, Ph.D., University of California, Riverside, 1990.

Geier, Montserrat, Ph.D..

Nishino, Hitoshi, Ph.D., University of Tokyo, 1981. *Cosmology & String Theory, High Energy Physics, Particles and Fields, Theoretical Physics.* Unification of all interactions. Study of supersymmetry, supergravity, superstring theory, and super membrane theory.

Sharma, Deepak.

DEPARTMENTAL RESEARCH SPECIALTIES AND STAFF

Theoretical

Astrophysics. Gravitational waves, nuclear astrophysics. Hintzen, Jaikumar.

Computational Physics. This is an overarching specialty for most of our theoretical work. We provide training in computational physics for any path chosen by students but have research projects in condensed matter, nuclear, and particle physics. Bill, Jaikumar, Papp, Peterson, Pickett.

Condensed Matter Physics. We study phenomena in superconductivity, magnetism, quantum Hall systems, topological phases, strongly correlated system, and low-dimensional systems. Bill, Peterson, Pickett.

High Energy Physics. Abachi, Hlousek, Jaikumar, Nishino, Papp, Rajpoot.

Low Temperature Physics. superconductivity. Bill.

Nuclear Physics. We study the quantum mechanics of few-body systems, and nuclear astrophysics. Jaikumar, Papp.

Solid State Physics. Bill, Peterson.

Experimental

Applied Physics. Most of our research in experimental physics studies fundamental properties of matter but have applications in mind. Abate, Gredig, Gu, Kwon, Leung.

Condensed Matter Physics. We have a strong emphasis in condensed matter experiments and materials science. Abate, Gredig, Gu, Kwon.

Nano Science and Technology. We have state-of-the-art experiments, including near field microscopy and sputtering machines to study nano particles, thin films and combinations thereof. Abate, Gredig, Gu, Kwon.

Optics. We have a state-of-the-art near field microscopy group. Abate, Leung.

Solar Physics. We study solar cells built of organic semiconductors. Gredig.

View additional information about this department at
www.gradschoolshopper.com

SAN DIEGO STATE UNIVERSITY

DEPARTMENT OF PHYSICS

San Diego, California 92182-1233
http://www.physics.sdsu.edu

General University Information
President: Elliot Hirshman
Dean of Graduate School: Radmilla Prislin (Interim)
University website: http://www.sdsu.edu
Control: Public
Setting: Urban
Total Faculty: 1,795
Total Graduate Faculty: 894
Total number of Students: 30,541
Total number of Graduate Students: 4,800

Department Information
Department Chairman: Usha Sinha, Chair
Department Contact: Tim Bonner, Department Buyer
  Total full-time faculty: 9
  Total number of full-time equivalent positions: 9
  Full-Time Graduate Students: 34
  First-Year Graduate Students: 17
  Female First-Year Students: 3

Department Address
5500 Campanile Drive
San Diego, CA 92182-1233
Phone: (619) 594-6165
Fax: (619) 594-1263
E-mail: tbonner@mail.sdsu.edu
Website: http://www.physics.sdsu.edu

ADMISSIONS

Admission Contact Information
Address admission inquiries to: Admission Office
Admissions website: http://www.physics.sdsu.edu

Application deadlines
Fall admission:
U.S. students: February 3   *Int'l. students*: February 3

Application fee
U.S. students: $55   *Int'l. students*: $55

Admissions information
For Fall of 2011:
  Number of applicants: 56
  Number admitted: 39
  Number enrolled: 21

Admission requirements
Bachelor's degree requirements: Bachelor's degree in physics, engineering, or mathematics is required.
Minimum undergraduate GPA: 2.85

GRE requirements
The GRE is required.
  Quantitative score: 150
  Verbal score: 150
  Analytical score: 3
We will also review applicants with lower GRE scores than those stated above.

Advanced GRE requirements
The Advanced GRE is not required.

TOEFL requirements
The TOEFL exam is required for students from non-English-speaking countries.
PBT score: 550

TUITION

Tuition year 2013-14:
Tuition for in-state residents
  Full-time students: $4,016 per semester
  Part-time students: $2,600 per semester
Tuition for out-of-state residents
  Full-time students: $372 per credit
  Part-time students: $372 per credit
Credit hours per semester to be considered full-time: 9
Deferred tuition plan: Yes
Health insurance: Not available.
Other academic fees: 0–6 units, $2,686; 6.1 or more units, $4,231.
Academic term: Semester
Number of first-year students who received full tuition waivers: 2

Teaching Assistants, Research Assistants, and Fellowships
Number of first-year
  Teaching Assistants: 10
  Research Assistants: 5
Average stipend per academic year
  Teaching Assistant: $15,900
  Research Assistant: $16,000

FINANCIAL AID

Loans
Loans are available for U.S. students.
Loans are not available for international students.
GAPSFAS application required: Yes
FAFSA application required: No

For further information
Address financial aid inquiries to: Office of Financial Aid and Scholarships, 5500 Campanile Drive, San Diego, CA 92182-7436.
Phone: (619) 594-6323
Financial aid website: http://www.sa.sdsu.edu/fao

HOUSING

Availability of on-campus housing
Single students: Yes
Married students: No

For further information

Address housing inquiries to: Office of Housing Administration, 5500 Campanile Drive, San Diego, CA 92182-1802.
Phone: (619) 594-5742
E-mail: oha@mail.sdsu.edu
Housing aid website: http://www.sdsu.edu/housing

Table A—Faculty, Enrollments, and Degrees Granted

Research Specialty	2011–12 Faculty	Enrollment Fall 2011		Number of Degrees Granted 2011–12 (2010–11)		
		Master's	Doctorate	Master's	Terminal Master's	Doctorate
Computational Physics	1	–	–	4(-)	–	–
Condensed Matter Physics	1	4	–	–	1(6)	–
Energy Sources & Environment	1	–	–	–	–(1)	–
Medical, Health Physics	2	32	–	–	4(4)	–
Nuclear Physics	2	4	–	–	4(2)	–
Optics	2	6	–	–	3(7)	–
Non-specialized	–	2	–	–	2(1)	–
Other	1	2	–	–	2(2)	–
Total	10	50	–	–	16(23)	–
Full-time Grad. Stud.	–	25	–	–	–	–
First-year Grad. Stud.	–	11	–	–	–	–

GRADUATE DEGREE REQUIREMENTS

Master's: Master of Science (Physics): The student must complete a graduate program in which 18 of the 30 units must include graduate course work in quantum mechanics, statistical mechanics, classical mechanics, electricity and magnetism, and thesis. The remaining units must be approved by the graduate advisor. There is a minimum of 24 units in residence. The student must pass a final oral examination on his/her thesis and maintain a 3.0 GPA. There is no foreign language requirement. Master of Science (Medical Physics): The student must complete a graduate program in which 18 of the 30 units must include graduate course work in radiological physics, nuclear medicine physics, nuclear instrumentation and radiation biology, and thesis. Other requirements are as for the graduate MS degree in Physics. Master of Art (Physics): The student must complete a graduate program in which 21 of the 30 units include graduate course work in electricity and magnetism, quantum mechanics, statistical mechanics, and classical mechanics. The remaining units must be approved by the students' graduate committee. The student must pass a comprehensive examination and maintain a 3.0 GPA. There is a minimum of 24 units in residence. Master of Art (Medical Physics): The student must complete a graduate program in which 12 of the 30 units must include graduate course work in radiological physics, nuclear medicine physics, nuclear instrumentation, and radiation biology. Other requirements are as for the graduate MA degree in Physics.

Thesis: Thesis may be written in absentia; however, residency is required.

SPECIAL EQUIPMENT, FACILITIES, OR PROGRAMS

Condensed matter physics laboratory, nuclear instrumentation laboratory, radiation therapy treatment planning systems, laser optics, holography facility; electro-optics measurements laboratory; image processing facility; materials laboratory, ultrafast laser; computational physics lab; Beowulf cluster.

Table B—Separately Budgeted Research Expenditures by Source of Support

Source of Support	Departmental Research	Physics-related Research Outside Department
Federal government	$5,815,740	
State/local government		
Non-profit organizations		
Business and industry	$250,000	
Other		
Total	$6,065,740	

Table C—Separately Budgeted Research Expenditures by Research Specialty

Research Specialty	No. of Grants	Expenditures ($)
Condensed Matter Physics	5	$276,452
Energy Sources & Environment	1	$900,000
Medical, Health Physics	2	$700,000
Optics	2	$190,500
Physics and other Science Education	3	$3,082,788
Computational Physics	2	$666,000
Total	15	$5,815,740

FACULTY

Professor

Davis, Jeffrey A., Ph.D., Cornell University, 1970. *Optics*. Optics; optical pattern recognition; computer generated holograms.

Goldberg, Fred M., Ph.D., University of Michigan, 1971. *Physics and other Science Education*. Physics education.

Johnson, Calvin W., Ph.D., University of Washington, 1989. *Astrophysics, Computational Physics, Nuclear Physics, Theoretical Physics*. Theoretical nuclear physics; astrophysics.

Morris, Richard H., Ph.D., University of California, Berkeley, 1957. *Electromagnetism, Optics*. Electromagnetic theory; laser physics.

Papin, Patrick J., Ph.D., University of California, Los Angeles, 1985. *Medical, Health Physics*. Radiation dosimetry; medical imaging.

Roeder, Stephen B.W., Ph.D., University of Wisconsin-Madison, 1968. *Applied Physics*. Magnetic resonance; scientific instrumentation.

Sinha, Usha, Ph.D., Indian Institute of Science, Bangalore, 1985. *Medical, Health Physics*. Medical physics; magnetic resonance imaging.

Sweedler, Alan R., Ph.D., University of California, San Diego, 1970. *Energy Sources & Environment*. Physical and environmental science; computer modeling of energy systems.

Torikachvili, Milton S., Ph.D., University of Campinas, 1978. *Condensed Matter Physics*. Experimental condensed matter physics.

Weber, Fridolin, Ph.D., University of Munich, Germany, 1992. *Astrophysics, Nuclear Physics, Theoretical Physics*. Theoretical nuclear physics; astrophysics.

Associate Professor

Anderson, Matt, Ph.D., University of Oregon, 1998. *Optics*. Ultrafast laser physics.

Baljon, Arlette R.C., Ph.D., University of Chicago, 1993. *Biophysics, Computational Physics*. Computational soft condensed matter physics.

Assistant Professor

Tambasco, Mauro, Ph.D., University of Western Ontario, 2002. *Medical, Health Physics*.

DEPARTMENTAL RESEARCH SPECIALTIES AND STAFF

Theoretical

Computational Physics. We have specialties in astrophysics, medical imaging, nuclear and polymers. SDSU also offers a Ph.D. in Computational Science. Baljon, Johnson, Weber.

Experimental

Condensed Matter Physics. Specializing in materials and superconductivity. Roeder, Torikachvili.

Medical, Health Physics. Medical image processing and imaging informatics, magnetic resonance imaging, radiation therapy physics, dose calculations. CAMPEP accredited. Option for clinical internships. Papin, Sinha, Tambasco.

Optics. Strong connection to industry. Anderson, Davis, Morris.

View additional information about this department at www.gradschoolshopper.com

---

SAN FRANCISCO STATE UNIVERSITY

DEPARTMENT OF PHYSICS AND ASTRONOMY

San Francisco, California 94132

http://physics.sfsu.edu

General University Information

President: Leslie E. Wong
Dean of Graduate School: Ann Hallum
University website: http://www.sfsu.edu/
Control: Public
Setting: Urban
Total Faculty: 1,180
Total Graduate Faculty: 820
Total number of Students: 25,000
Total number of Graduate Students: 5,500

Department Information

Department Chairman: Susan M. Lea, Chair
Department Contact: Susan Lea, Professor and Department Chair
Total full-time faculty: 12
Total number of full-time equivalent positions: 12
Full-Time Graduate Students: 55
First-Year Graduate Students: 22
Female First-Year Students: 3
Total Post Doctorates: 2

Department Address

1600 Holloway Ave
San Francisco, CA 94132
Phone: (415) 338-1659 (C)
Fax: (415) 338-2178
E-mail: physics@sfsu.edu
Website: http://physics.sfsu.edu

ADMISSIONS

Admission Contact Information

Address admission inquiries to: Registrar's Office and/or Department of Physics and Astronomy
Phone: (415) 338-1659
E-mail: physics@sfsu.edu

Admissions website: http://www.sfsu.edu/~gradstdy/main-domestic.htm

Application deadlines

Fall admission:
U.S. students: June 1   *Int'l. students*: June 1
Spring admission:
U.S. students: November 1   *Int'l. students*: November 1

Application fee

U.S. students: $55   *Int'l. students*: $55

Admissions information

For Fall of 2012:
Number of applicants: 42
Number admitted: 31
Number enrolled: 22

Admission requirements

Bachelor's degree requirements: Bachelors' degree in physics, engineering, or mathematics is required.
Minimum undergraduate GPA: 3.0

GRE requirements

The GRE is required.
We have no formal minimum score. We look at the whole application package.

Advanced GRE requirements

The Advanced GRE is not required.
Physics GRE is useful if score is good and GPA is not.

TOEFL requirements

The TOEFL exam is required for students from non-English-speaking countries.
PBT score: 550
iBT score: 80

Other admissions information

Additional requirements: Bachelor's degrees in other disciplines may be acceptable. Consult department.
Undergraduate preparation assumed: Gregory, Classical Mechanics; Griffiths, Electrodynamics; Zemansky and Dittman,

Heat and Thermodynamics; Eisberg, *Quantum Physics for Atoms, Molecules, Solids, Nuclei, and Particles*; Boyce and DiPrima, *Ordinary Differential Equations*.

TUITION

Tuition year 2012–13:
Tuition for in-state residents
  Full-time students: $3,858 per semester
  Part-time students: $2,442 per semester
Tuition for out-of-state residents
  Full-time students: $372 per credit
  Part-time students: $372 per credit
Out-of-state per credit fee is in addition to the in-state fees.
Credit hours per semester to be considered full-time: 24
Deferred tuition plan: No
Health insurance: Not available.
Academic term: Semester
Number of first-year students who received partial tuition waivers: 8

Teaching Assistants, Research Assistants, and Fellowships

Number of first-year
  Teaching Assistants: 8
  Fellowship students: 2
Average stipend per academic year
  Teaching Assistant: $4,600
  Fellowship student: $5,000
Some students teach two sections, which doubles their compensation.

FINANCIAL AID

Application deadlines
Fall admission:
U.S. students: March 3

Loans
Loans are available for U.S. students.
Loans are not available for international students.
GAPSFAS application required: No
FAFSA application required: Yes

For further information
Address financial aid inquiries to: Financial Aid Office.
Phone: (415) 338-7000
E-mail: finaid@sfsu.edu
Financial aid website: http://www.sfsu.edu/~finaid/

HOUSING

Availability of on-campus housing
  Single students: Yes
  Married students: Yes

For further information
Address housing inquiries to: Housing Office.
Phone: (415) 338-1067
E-mail: housing@sfsu.edu
Housing aid website: http://www.sfsu.edu/~housing/

Table A—Faculty, Enrollments, and Degrees Granted

Research Specialty	2013-14 Faculty	Enrollment Fall 2012 Master's	Enrollment Fall 2012 Doctorate	Number of Degrees Granted 2012–13 (2008–13) Master's	Number of Degrees Granted 2012–13 (2008–13) Terminal Master's	Number of Degrees Granted 2012–13 (2008–13) Doctorate
Astronomy	4	15	–	–	4(16)	–
Astrophysics	2	4	–	–	1(3)	–
Biophysics	–	–	–	–	–(1)	–
Condensed Matter Physics	2	6	–	–	–(5)	–
Fluids, Rheology	1	2	–	–	–(1)	–
Nuclear Physics	–	–	–	–	–	–
Optics	2	8	–	–	–(6)	–
Particles and Fields	3	5	–	–	3(5)	–
Physics and other Science Education	2	4	–	–	–(2)	–
Relativity & Gravitation	1	1	–	–	–	–
Non-specialized	–	–	–	–	–(13)	–
Total	11	45	–	–	8(52)	–
Full-time Grad. Stud.	–	45	–	–	–	–
First-year Grad. Stud.	–	22	–	–	–	–

GRADUATE DEGREE REQUIREMENTS

Master's: A total of 30 semester units with a "B" average is required. Fifteen semester units in dynamics, electromagnetic theory, mathematical physics, statistical mechanics, and quantum mechanics and six semester units of other graduate physics courses must be included. Nine units of upper division and graduate level courses in mathematics, science, engineering, or other appropriate fields, selected with the approval of the graduate advisor, complete the course requirements. 24 semester units must be completed in residence. Either a Master's Comprehensive Oral Examination or a Master's Thesis with oral defense is required.
Thesis: Thesis may be written in absentia.

SPECIAL EQUIPMENT, FACILITIES, OR PROGRAMS

Instrumentation and computational facilities for elementary particle physics; nuclear sources and counters, Mössbauer spectrometer; optics laboratory with spectrometers, laser optics, and holography; vacuum deposition equipment with cleanroom and photolithography facilities; solid state laboratory with NMR and ESR apparatus, ultrahigh vacuum capability, and advanced instrumentation; low-temperature laboratory with temperatures to 20 mK, SQUID systems, high-field superconducting magnets; astronomical observatory, high-quality planetarium.

Table B—Separately Budgeted Research Expenditures by Source of Support

Source of Support	Departmental Research	Physics-related Research Outside Department
Federal government	$2,732,000	
State/local government		
Non-profit organizations		
Business and industry		
Other		
Total	$2,732,000	

Table C—Separately Budgeted Research Expenditures by Research Specialty

Research Specialty	No. of Grants	Expenditures ($)
Astrophysics	6	$1,324,000
Condensed Matter Physics	1	$50,000
Optics	4	$1,113,000
Particles and Fields	2	$245,000
Total	13	$2,732,000

FACULTY

Professor

Chen, Zhigang, Ph.D., Bryn Mawr College, 1995. *Atomic, Molecular, & Optical Physics, Optics.* Nonlinear photonics.

Cool, Andrienne, Ph.D., Harvard University, 1994. *Astronomy, Astrophysics, Physics and other Science Education.* Observational astronomy; stellar astrophysics.

Golterman, Maarten, Ph.D., University of Amsterdam, 1986. Graduate coordinator. *High Energy Physics, Particles and Fields, Theoretical Physics.* Theoretical physics; particle physics.

Greensite, Jeffrey P., Ph.D., University of California, Santa Cruz, 1981. Graduate admissions chair. *Computational Physics, Particles and Fields, Theoretical Physics.* Theoretical elementary particle physics. Computational physics.

Lea, Susan M., Ph.D., University of California, Berkeley, 1974. Department Chair. *Astrophysics, Fluids, Rheology, Physics and other Science Education, Plasma and Fusion, Theoretical Physics.* Theoretical astrophysics: accretion dynamics, x-ray astronomy, plasma astrophysics.

Marzke, Ronald, Ph.D., Harvard University, 1994. Associate Department Chair. *Astronomy, Astrophysics.* Observational cosmology; galaxy formation and evolution; universe structure.

Neuhauser, Barbara J., Ph.D., Stanford University, 1985. *Condensed Matter Physics, Low Temperature Physics, Solid State Physics.* Low-temperature physics; ultralow temperatures; superfluid 3He; neutrino detectors.

Associate Professor

Barranco, Joseph A., Ph.D., University of California, Berkeley, 2004. undergraduate advisor. *Astronomy, Astrophysics, Computational Physics, Fluids, Rheology, Plasma and Fusion.* Theoretical and Computational Astrophysics; astrophysical and geophysical fluid dynamics; accretion disks; Star and Planet formation.

Assistant Professor

Kane, Stephen, Ph.D., Univ. of Tasmania, 2000. *Astronomy, Astrophysics.* Exoplanets.

Mahdavi, Andisheh, Ph.D., Harvard University, 2001. *Astronomy, Astrophysics, Cosmology & String Theory, Relativity & Gravitation.* Observational and computational astrophysics; groups and clusters of galaxies; dark matter; plasma astrophysics; x-ray astronomy; galactic dynamics; gravitational lensing.

Man, Weining, Ph.D., Princeton University, 2005. *Atomic, Molecular, & Optical Physics, Condensed Matter Physics, Materials Science, Metallurgy, Optics.* Experimental and numerical soft condensed matter physics; colloid physics; colloidal suspension thin films; photonic band gap materials; quasicrystals; granular materials.

Unsal, Mithat, Ph.D., University of Washington, 2004. On leave spring 2014. *Particles and Fields, Quantum Foundations, Theoretical Physics.* Theoretical physics.

Emeritus

Bland, Roger W., Ph.D., University of California, Berkeley, 1968. Lecturer. *Acoustics, Marine Science/Oceanography, Particles and Fields.* Experimental particle physics; underwater acoustics.

Lockhart, James, Ph.D., Stanford University, 1976. *Condensed Matter Physics, Low Temperature Physics, Physics and other Science Education.* Experimental solid state physics; low-temperature physics; SQUID detectors; instrumentation.

Adjunct Professor

Adler, Ronald J., Ph.D., Stanford University, 1965. *Cosmology & String Theory, Low Temperature Physics, Particles and Fields, Relativity & Gravitation, Theoretical Physics.* Relativity theory; superconductivity.

Barsony, Mary, Ph.D., California Institute of Technology, 1989. *Astronomy, Astrophysics.* Observational astronomy; infrared and x-ray astronomy, star formation.

Fischer, Debra, Ph.D., University of California, Santa Cruz, 1998. *Astronomy, Astrophysics.* Observational astronomy; extrasolar planet searches, orbital dynamics, spectroscopy, spectral synthesis, interferometry.

Lipschultz, Fred, Ph.D., Cornell University, 1966. *Low Temperature Physics.* Low temperature physics; physics education.

Marcy, Geoffrey W., Ph.D., University of California, Santa Cruz, 1982. *Astronomy, Astrophysics, Planetary Science.* Stellar astrophysics; observational astronomy.

McCarthy, Chris, Ph.D., University of California, Los Angeles, 2001. Lecturer. *Astronomy, Astrophysics.* Brown dwarf stars; extra-solar planets; stellar spectroscopy.

DEPARTMENTAL RESEARCH SPECIALTIES AND STAFF

Theoretical

Astronomy. Barranco, Barsony, Cool, Fischer, Kane, Lea, Mahdavi, Marcy, Marzke, McCarthy.

Astrophysics. Barranco, Cool, Kane, Lea, Mahdavi, Marcy, Marzke, McCarthy.

Particles and Fields. Adler, Bland, Golterman, Greensite, Unsal.

Physics and other Science Education. Cool, Lea, Lockhart.

Experimental

Acoustics. Bland.

Astronomy. Barsony, Cool, Fischer, Kane, Mahdavi, Marcy, Marzke, McCarthy.

Condensed Matter Physics. Chen, Lipschultz, Man, Neuhauser.

Low Temperature Physics. Lipschultz, Lockhart, Neuhauser.

Optics. Chen, Man.

SAN JOSE STATE UNIVERSITY

DEPARTMENT OF PHYSICS AND ASTRONOMY

San Jose, California 95192-0106
http://www.physics.sjsu.edu/

General University Information
President: Dr. Mo Qayoumi
Dean of Graduate School: Dr. Pamela Stacks
University website: http://www.sjsu.edu
Control: Public
Setting: Urban
Total Faculty: 1,740
Total number of Students: 28,007
Total number of Graduate Students: 5,766

Department Information
Department Chairman: Michael Kaufman, Chair
Department Contact: Bertha Aguayo, Administrative
  Analyst/Specialist
  Total full-time faculty: 16
  Total number of full-time equivalent positions: 22
  Full-Time Graduate Students: 40
  First-Year Graduate Students: 11
  Female First-Year Students: 2

Department Address
One Washington Square
San Jose, CA 95192-0106
Phone: (408) 924-5210
Fax: (408) 924-2917
E-mail: Bertha.Aguayo@sjsu.edu
Website: http://www.physics.sjsu.edu/

ADMISSIONS

Admission Contact Information
Address admission inquiries to: Graduate Admissions Office
Phone: (408) 924-2480
E-mail: graduate@sjsu.edu
Admissions website: http://www.sjsu.edu/gape

Application deadlines
Fall admission:
U.S. students: April 1   *Int'l. students*: April 1

Application fee
U.S. students: $55

Admissions information
For Fall of 2011:
  Number of applicants: 28
  Number enrolled: 16

Admission requirements
Bachelor's degree requirements: A Bachelor's degree is required.
Minimum undergraduate GPA: 2.5

GRE requirements
The GRE is not required.

Advanced GRE requirements
The Advanced GRE is not required.

TOEFL requirements
The TOEFL exam is required for students from non-English-speaking countries.
PBT score: 550

Other admissions information
Additional requirements: The GRE Physics is not required for admission to the program, but it must be taken with a score of 550 no later than two semesters prior to graduation.
Undergraduate preparation assumed: Young and Freedman, University Physics; Marion, Classical Dynamics of Particles & Systems; Beiser, Concepts of Modern Physics; Griffiths, Introduction to Electrodynamics; Griffiths, Introduction to Quantum Mechanics.

TUITION

Tuition year 2013-14:
Tuition for in-state residents
  Full-time students: $4,486.5 per semester
  Part-time students: $2,941.5 per semester
Tuition for out-of-state residents
  Full-time students: $7,382.5 per semester
Academic term: Semester

Teaching Assistants, Research Assistants, and Fellowships
  Number of first-year
    Teaching Assistants: 5

FINANCIAL AID

Loans
Loans are available for U.S. students.
Loans are not available for international students.
GAPSFAS application required: No
FAFSA application required: No

HOUSING

Availability of on-campus housing
  Single students: Yes
  Married students: Yes

For further information
Address housing inquiries to: Attn.: Housing Office.
Housing aid website: http://housing.sjsu.edu

Table A—Faculty, Enrollments, and Degrees Granted

Research Specialty	2007–08 Faculty	Enrollment Fall 2007		Number of Degrees Granted 2007–08 (2000–07)		
		Master's	Doctorate	Master's	Terminal Master's	Doctorate
Acoustics	1	–	–	–	–	–
Applied Physics	1	–	–	–	–	–
Astronomy	4	–	–	–	–	–
Astrophysics	4	–	–	–	–	–
Biophysics	1	–	–	–	–	–
Condensed Matter Physics	3	–	–	–	–	–
Optics	3	–	–	–	–	–
Plasma and Fusion	1	–	–	–	–	–
Non-specialized	–	28	–	–	7(36)	–
Total	–	30	–	–	7(36)	–
Full-time Grad. Stud.	–	–	–	–	–	–
First-year Grad. Stud.	–	11	–	–	–	–

GRADUATE DEGREE REQUIREMENTS

Master's: A total of 30 semester units with a B average is required. Fifteen semester units in mathematical physics, advanced dynamics, electromagnetic theory, statistical physics, and quantum mechanics and at least six semester units of other graduate physics courses must be included. Twelve units of graduate level and/or upper division courses in mathematics, science, engineering, or other appropriate fields, selected with the approval of the graduate advisor, complete the course requirements. Twenty-four semester units must be completed in residence. All graduate students must attend weekly department seminars in at least one semester, must achieve a satisfactory score on the Physics portion of the GRE no later than two semesters prior to graduation, and must satisfy the English writing requirement prior to the semester of graduation. A comprehensive oral examination, literature review, or a thesis presentation is required.
Thesis: Optional.

SPECIAL EQUIPMENT, FACILITIES, OR PROGRAMS

The Department has academic options in optics and condensed matter and offers a concentration in computational physics. The Institute for Modern Optics coordinates optics research and collaboration with local industries. A 4,000-sf instruction and research facility contains equipment for laser spectroscopy, nonlinear optics, Fourier optics, holography, and optical metrology. A world-class center for novel laser materials research and modeling of laser systems is located here. The solid state laboratory includes equipment for measuring magnetic susceptibility over a 4–1,000 K temperature range. Computing equipment consists of on-campus workstations and internet access to remote facilities and databases. A 10,000-sf Nuclear Science Facility is used by all science departments. Equipment includes neutron and gamma irradiators; x-ray fluorescence, magnetic resonance, and Mössbauer spectrometers; and a range of radiation analyzing equipment. Several faculty carry out research with colleagues at NASA Ames Research Center. Located in "Silicon Valley" we have research and instructional collaborations with many local instrumentation, materials, semiconductor, and optics companies.

Table B—Separately Budgeted Research Expenditures by Source of Support

Source of Support	Departmental Research	Physics-related Research Outside Department
Federal government	$500,000	
State/local government		
Non-profit organizations		
Business and industry	$840,000	$820,000
Other	$4,843,000	$720,000
Total	$6,183,000	$1,540,000

Table C—Separately Budgeted Research Expenditures by Research Specialty

Research Specialty	No. of Grants	Expenditures ($)
Astronomy	2	$10,000
Astronomy	2	$100,000
Atmosphere, Space Physics, Cosmic Rays	4	$200,000
Materials Science, Metallurgy	2	$490,000
Physics and other Science Education	1	$75,000
Solid State Physics	1	$188,000
Other	6	$4,570,000
Total	18	$5,633,000

FACULTY

Professor

Bahuguna, Ramendra D., Ph.D., Indian Institute of Technology, Delhi, 1979. Holographic interferometry; laser speckle metrology; display holography; Fourier optics; fingerprint verification.
Boekema, Carolus, Ph.D., University of Groningen, Netherlands, 1977. Magnetism and superconductivity in solids; computational condensed matter physics; muon spin research: Mössbauer spectroscopy.
Garcia, Alejandro, Ph.D., University of Texas, Austin, 1984. Computational fluid mechanics; statistical mechanics.
Holmes, Brian W., Ph.D., Boston University, 1980. Musical acoustics; sports physics; physics education.
Kaufman, Michael J., Ph.D., Johns Hopkins University, 1995. Astrophysics: interstellar medium, interactions of young stars with molecular clouds, dynamics and chemistry of molecular shocks, infrared/submillimeter observations.
Lam, Lui, Ph.D., Columbia University, 1973. Histophysics; nonlinear physics; liquid crystals; pattern formation; complex systems.
Parvin, Kiumars, Ph.D., University of California, Riverside, 1978. Experimental solid state physics; magnetic materials.
Wharton, Kenneth B., Ph.D., University of California, Los Angeles, 1998. Plasma physics, laser-plasma interactions; subpicosecond x-ray sources, foundations of quantum mechanics, relativistic quantum mechanics.

Associate Professor

Batalha, Natalie M., Ph.D., University of California, Santa Cruz, 1997. Stellar astrophysics: variable stars, magnetic activity, T Tauri stars; Extra-solar planet detection.
Beyersdorf, Peter T., Ph.D., Stanford University, 2001. Gravitational wave detection, precision measurements and optical interferometry.
Kress, Monika, Ph.D., Rensselaer Polytechnic Institute, 1997. Astrophysics and planetary science. Computer modeling of protoplanetary disks, comet impacts, and planetary environments; meteorites; astrobiology.

Assistant Professor

Heindl, Ranko, Ph.D., University of South Florida, 2006. Magnetism, spintronics, and nonvolatile memories.

Paul, Cassandra, Ph.D., University of California, Davis, 2012. Physics Education Research, Astronomy Education Research.

Romanowsky, Aaron, Ph.D., Harvard University, 1999. Astrophysics (galaxies, dark matter, computation).

Emeritus

Anderson, Merlin F., Ph.D., Oregon State University, 1966. Gas dynamics; computer applications.

Becker, Joseph F., Ph.D., New York University, 1976. Spectroscopy; biophysics; optics, optoelectronic devices.

Bloomer, Iris L., Ph.D., University of London, 1976. General relativity; optical properties of materials.

Finkelstein, Jerome, Ph.D., University of California, Berkeley, 1967. Theoretical physics; elementary particles.

Gruber, John B., Ph.D., University of California, Berkeley, 1961. Engineering physics; solid state spectroscopy; rare-earth/transition-metal ion solid state lasers; chemical physics; quantum electronics.

Hamill, Patrick, Ph.D., University of Arizona, 1971. Atmospheric physics; aerosol physics; celestial mechanics.

Morris, Marvin L., Ph.D., University of Utah, 1966. High-energy cosmic rays; computer-aided instruction; physics education.

Strandburg, Donald L., Ph.D., Iowa State University, 1961. Low-temperature physics; magnetism.

Tomley, Leslie J., Ph.D., University of Washington, 1968. *Astrophysics*. Astrophysics.

Tucker, Allen B., Ph.D., Stanford University, 1965. Nuclear physics; cosmic rays; accelerator mass spectrometry; health physics.

Williams, Gareth T., Ph.D., University of Wales, 1960. Optics; holography.

Adjunct Professor

Bolton, Paul R., Ph.D., Yale University, 1982. Laser design and laser spectroscopy; laser ablation; strong optical field atomic and plasma physics; ultrafast laser-driven phenomena and diagnostics; laser applications to accelerator development; xuv/x-ray spectroscopy; electron photoinjectors.

Castellano, Timothy, Ph.D., University of California, Santa Cruz, 2001. Detection of extrasolar planets by transit method; stellar main sequence variability.

Freund, Friedemann, Ph.D., University of Marburg, 1959. Defects in crystals; proton conductivity.

Lecturer

Berman, Irina, Ph.D., Moscow State University, 1989. Superconductivity of conventional, high-temperature and disordered superconductors; localization effects and superconductivity.

Hubickyj, Olenka, Ph.D., City University of New York, 1983. RR Lyrae stars' instability strip; primordial atmosphere of Earth; convection turbulence in the Solar Nebula; formation and evolution of gas giant planets.

Kwok, Ray, Ph.D., University of California, Los Angeles, 1990. Condensed matter physics; solid state physics; phase transitions; electrical and thermal transport; Muon spin rotation; applied superconductivity RF and microwave communications; applied physics.

Mosqueira, Ignacio, Ph.D., Cornell University, 1995. Planetary formation.

Sauke, Todd B., Ph.D., University of Illinois at Urbana-Champaign, 1989. Tunable diode laser technology; development of planetary exploration applications for the measurement of isotopic ratios in planetary surface and atmospheric samples.

Sherman, Douglas, Ph.D., University of California, Berkeley, 1987. Hydrogen solubility and diffusivity in refractory ceramics.

DEPARTMENTAL RESEARCH SPECIALTIES AND STAFF

Theoretical

Astrophysics. Batalha, Hubickyj, Kaufman, Kress.
Condensed Matter Physics. Kwok, Parvin.

Experimental

Acoustics.
Applied Physics.
Astrophysics.
Biophysics.
Optics.
Plasma and Fusion.

View additional information about this department at
www.gradschoolshopper.com

STANFORD UNIVERSITY

DEPARTMENT OF APPLIED PHYSICS

Stanford, California 94305–4090
http://appliedphysics.stanford.edu

General University Information

President: John L. Hennessy
Dean of Graduate School: Richard P. Saller
University website: http://www.stanford.edu
Control: Private
Setting: Suburban

Total Faculty: 1,995
Total Graduate Faculty: N/A
Total number of Students: 15,870
Total number of Graduate Students: 8,871

Department Information
Department Chairman: Hideo Mabuchi, Chair
Department Contact: Paula P. Perron, Department Manager
  Total full-time faculty: 22
  Full-Time Graduate Students: 134
  First-Year Graduate Students: 19
  Female First-Year Students: 4
  Total Post Doctorates: 28

Department Address
348 Via Pueblo Mall
Stanford, CA 94305–4090
Phone: (650) 723-4027
E-mail: pperron@stanford.edu
Website: http://appliedphysics.stanford.edu

ADMISSIONS

Admission Contact Information
Address admission inquiries to: Graduate Admissions Office, 482
  Galvez Mall, Suite 120, Stanford, CA 94305-6032
Phone: (866) 432-7472
E-mail: gradadmissions@stanford.edu
Admissions website: http://gradadmissions.stanford.edu

Application deadlines
Fall admission:
U.S. students: December 17   *Int'l. students*: December 17

Application fee
U.S. students: $125   *Int'l. students*: $125
Application fee to be paid by credit/debit card only (Visa or MasterCard). Only one application to one department is allowed.

Admissions information
For Fall of 2013:
  Number of applicants: 209
  Number admitted: 40
  Number enrolled: 23

Admission requirements
Bachelor's degree requirements: Bachelor's degree in physics, mathematics, chemistry, or electrical engineering is required.

GRE requirements
The GRE is required.
We do not set minimum accepted GRE scores.

Advanced GRE requirements
The Advanced GRE is required.
We do not set minimum accepted GRE Physics scores.

TOEFL requirements
The TOEFL exam is required for students from non-English-speaking countries.
  PBT score: 600
  iBT score: 100
Refer to http://studentaffairs.stanford.edu/gradadmissions/applying/exams for Stanford's policy on TOEFL scores.

Other admissions information
Undergraduate preparation assumed: Intermediate undergraduate (junior and senior) courses in mechanics, electricity and magnetism, modern physics, statistical mechanics, and thermodynamics.

TUITION

Tuition year 2013-14:
  Full-time students: $9,250 per quarter

$9,250/quarter is the standard rate for graduate students with the exception that a few outside fellowships provide the University's full tuition rate of $14,230/quarter.
Deferred tuition plan: Yes
Health insurance: Available.
Other academic fees: Fees vary. Student health insurance is required.
Academic term: Quarter

Teaching Assistants, Research Assistants, and Fellowships
Number of first-year
  Research Assistants: 17
  Fellowship students: 6
Number of first-year fellowships refers to both internal and outside fellowships the first-year students have received. Stipends vary.

FINANCIAL AID

Application deadlines
Fall admission:
U.S. students: December 17   *Int'l. students*: December 17

Loans
Loans are available for U.S. students.
Loans are available for international students.
GAPSFAS application required: No
FAFSA application required: No

For further information
Address financial aid inquiries to: Paula P. Perron, Department Manager, Department of Applied Physics.
Phone: (650) 723-4027
E-mail: pperron@stanford.edu
Financial aid website: http://financialaid.stanford.edu

HOUSING

Availability of on-campus housing
  Single students: Yes
  Married students: Yes

For further information
Address housing inquiries to: Student Housing Office, 565 Cowell Lane, Stanford University, Stanford, CA 94305-8581.
Phone: (650) 725-1600
E-mail: studenthousing@stanford.edu
Housing aid website: http://studenthousing.stanford.edu

Table A—Faculty, Enrollments, and Degrees Granted

Research Specialty	2012-13 Faculty	Enrollment Fall 2012		Number of Degrees Granted 2012-13 (2008-13)		
		Master's	Doctorate	Master's	Terminal Master's	Doctorate
Applied Physics	48	1	133	5(30)	–(3)	27(117)
Total	48	1	133	5(30)	–(3)	27(117)
Full-time Grad. Stud.	–	1	133	–	–	–
First-year Grad. Stud.	–	1	18	–	–	–

GRADUATE DEGREE REQUIREMENTS

Master's: Subject matter: advanced mechanics, electrodynamics, and quantum mechanics. A "B" average is required. Total number of course units required is 45. There is no foreign language requirement, no comprehensive and/or qualification examination, and no thesis option. A terminal M.S. degree program is offered.

Doctorate: Subject matter: advanced mechanics, electrodynamics, quantum mechanics, statistical physics, and one advanced laboratory, with the remaining required courses to be distributed between major and minor fields. Total number of course units required is 135. A "B" average is required. Departmental qualification examination, fourth-year research progress report, dissertation, and oral defense of dissertation are required. There is no foreign language requirement.

SPECIAL EQUIPMENT, FACILITIES, OR PROGRAMS

The Applied Physics Department participates in the Honors Cooperative Program, which offers the opportunity to qualified engineers and scientists employed by companies in the general vicinity of the University to pursue graduate work on a part-time basis leading to the M.S. degree.

NOTE: No data for the separately budgeted research expenditures by source of support or by research specialty will be provided. That information is not available at the reporting unit's level.

FACULTY

Professor

Block, Steven M., Ph.D., California Institute of Technology, 1983. Biophysics.

Bucksbaum, Philip H., Ph.D., University of California, Berkeley, 1980. Atomic, molecular, and optical physics; ultrafast science.

Byer, Robert L., Ph.D., Stanford University, 1969. Nonlinear optics.

Doniach, Sebastian, Ph.D., University of Liverpool, 1958. Theory of cooperative phenomena; biophysics.

Fejer, Martin M., Ph.D., Stanford University, 1986. Quantum electronics; guided-wave optics; and optical materials.

Fisher, Daniel S., Ph.D., Harvard University, 1979. Theoretical condensed-matter physics; biophysics; evolutionary dynamics.

Fisher, Ian R., Ph.D., University of Cambridge, 1996. Condensed-matter physics; materials physics.

Hwang, Harold Y., Ph.D., Princeton University, 1997. Materials physics; emergent phenomena in oxide heterostructures; devices.

Kapitulnik, Aharon, Ph.D., Tel Aviv University, 1984. Theoretical and experimental low-temperature and condensed-matter physics; superconductivity.

Kasevich, Mark A., Ph.D., Stanford University, 1992. Atomic, molecular, and optical physics.

Mabuchi, Hideo, Ph.D., California Institute of Technology, 1998. Department Chair. Quantum optics; quantum information and control.

Moler, Kathryn A., Ph.D., Stanford University, 1995. Condensed-matter physics; materials physics; physics of small structures and of novel materials.

Petrosian, Vahé, Ph.D., Cornell University, 1967. Theoretical astrophysics and cosmology.

Quake, Stephen R., Ph.D., University of Oxford, 1994. Biophysics.

Shen, Zhi-Xun, Ph.D., Stanford University, 1989. Condensed-matter physics; electronic structure; photoelectron spectroscopy; synchrotron radiation.

Suzuki, Yuri, Ph.D., Stanford University, 1995. Condensed-matter physics; materials and device physics.

Yamamoto, Yoshihisa, Ph.D., University of Tokyo, 1978. Quantum optics; mesoscopic physics.

Associate Professor

Reis, David A., Ph.D., University of Rochester, 1999. Condensed-matter physics; ultrafast science.

Schnitzer, Mark J., Ph.D., Princeton University, 1999. Biophysics.

Assistant Professor

Ganguli, Surya, Ph.D., University of California, Berkeley, 2004. Theoretical neuroscience; biophysics.

Lev, Benjamin L., Ph.D., California Institute of Technology, 2005. Atomic, molecular, and optical physics; novel microscopy and imaging; condensed-matter physics; materials research.

Emeritus

Beasley, Malcolm R., Ph.D., Cornell University, 1968. Low-temperature and condensed-matter physics; superconductivity.

Bienenstock, Arthur I., Ph.D., Harvard University, 1962. Synchrotron radiation.

Fetter, Alexander L., Ph.D., Harvard University, 1963. Theoretical condensed-matter physics.

Geballe, Theodore H., Ph.D., University of California, Berkeley, 1949. Low-temperature and condensed-matter physics; superconductivity.

Harris, Stephen E., Ph.D., Stanford University, 1963. Quantum electronics and XUV lasers.

Harrison, Walter A., Ph.D., University of Illinois, 1956. Theoretical condensed-matter physics: electronic structure.

Quate, Calvin F., Ph.D., Stanford University, 1950. Scanning tunneling microscopy; imaging.

Sturrock, Peter A., Ph.D., University of Cambridge, 1951. Theoretical solar physics and astrophysics.

Wiedemann, Helmut, Ph.D., University of Hamburg, 1971. Accelerator physics; electron storage rings.

Winick, Herman, Ph.D., Columbia University, 1957. Synchrotron radiation.

Research Professor

Digonnet, Michel, Ph.D., Stanford University, 1984. Fiber optics; sensors; slow and fast light.

Courtesy Professor

Clemens, Bruce M., Ph.D., California Institute of Technology, 1983. Metal-metal multilayers; interfaces and interface reactions; magnetic thin films; x-ray diffraction.

Harris, James S., Ph.D., Stanford University, 1969. Optoelectronic device structures; quantum electronics.

Hesselink, Lambertus, Ph.D., California Institute of Technology, 1977. Nonlinear optics.

Miller, David A. B., Ph.D., Heriot-Watt University, 1979. Electro-optic wave devices; engineering physics.

Moerner, W. E., Ph.D., Cornell University, 1982. Quantum optics.

Osheroff, Douglas D., Ph.D., Cornell University, 1973. Low-temperature and condensed-matter physics.

Zhang, Shoucheng, Ph.D., Stony Brook University, 1987. Theoretical condensed-matter physics.

DEPARTMENTAL RESEARCH SPECIALTIES AND STAFF

Theoretical

Astrophysics.
Biophysics.
Condensed Matter Physics.
Neuroscience/Neuro Physics.

Experimental
Atomic, Molecular, & Optical Physics.
Biophysics.
Condensed Matter Physics. Materials research.
Materials Science, Metallurgy.

Nano Science and Technology.
Novel Microscopy and Imaging.
Optics.
Relativity & Gravitation.
Synchrotron Radiation.

View additional information about this department at
www.gradschoolshopper.com

STANFORD   UNIVERSITY

DEPARTMENT OF PHYSICS

Stanford, California 94305
http://physics.stanford.edu

General University Information
President: John Hennessy
Dean of Graduate School: Ann Arvin
University website: http://www.stanford.edu
Control: Private
Setting: Suburban
Total Faculty: 1,508
Total Graduate Faculty: 340
Total number of Students: 18,217
Total number of Graduate Students: 11,154

Department Information
Department Chairman: Peter Michelson, Chair
Department Contact: Maria Frank, Student Services Officer
   Total full-time faculty: 40
   Total number of full-time equivalent positions: 39
   Full-Time Graduate Students: 172
   First-Year Graduate Students: 40
   Female First-Year Students: 8
   Total Post Doctorates: 84

Department Address
382 Via Pueblo Mall
Stanford, CA 94305
Phone: (650) 723-4344
E-mail: phys-admissions@lists.stanford.edu
Website: http://physics.stanford.edu

ADMISSIONS

Admission Contact Information
Address admission inquiries to: Before calling or emailing Graduate Admissions, please look for answers to your questions on the Applying to Graduate Admissions page http://studentaffairs.stanford.edu/gradadmissions/applying, and the Frequently Asked Questions page http://studentaffairs.stanford.edu/gradadmissions/faq.
Phone: (866) 432-7472
E-mail: gradadmissions@stanford.edu
Admissions website: http://gradadmissions.stanford.edu

Application deadlines
Fall admission:
U.S. students: December 17   *Int'l. students*: December 17

Application fee
U.S. students: $125   *Int'l. students*: $125

Admissions information
For Fall of 2013:
   Number of applicants: 617
   Number admitted: 72
   Number enrolled: 32

Admission requirements
Bachelor's degree requirements: Bachelor's degree in physics (or a related field) is required.

GRE requirements
The GRE is required.
No minimum scores specified. We strongly advise you to take the general and physics GRE tests no later than September, so that your scores will be received by the application deadline or shortly thereafter.

Advanced GRE requirements
The Advanced GRE is required.
No minimum scores specified. We strongly advise you to take the general and physics GRE tests no later than September, so that your scores will be received by the application deadline or shortly thereafter.

TOEFL requirements
The TOEFL exam is required for students from non-English-speaking countries.
   PBT score: 600
   iBT score: 100
Adequate command of spoken and written English is required for admission. Applicants whose first language is not English must submit an official test score from the Test of English as a Foreign Language (TOEFL). Scores must be submitted from a test taken within the last 18 months. Exceptions are granted for applicants who have earned a U.S. bachelor's or master's degree from a college or university accredited by a regional accrediting association in the United States, or equivalent of either degree from a university of recognized standing in countries where all instruction is provided in English. For example, applicants with degrees from: Australia,

Canada (except Quebec), New Zealand, Singapore, United Kingdom (England, Scotland, Ireland, Wales) are exempt from taking the TOEFL. Applicants with degrees from countries where English is spoken but not all courses provided in English are not exempt from taking the TOEFL. U.S. citizenship does not automatically exempt an applicant from taking the TOEFL; if the applicant's first language is not English.

Other admissions information

Additional requirements: No minimum scores specified.
The average GRE scores for admitted students in 2013–14 were: Verbal–163, Quantitative–168; Analytical–4.24; Physics Subject–934.

TUITION

Tuition year 2013-14:
  Full-time students: $37,000 annual
Credit hours per semester to be considered full-time: 10
Deferred tuition plan: No
Health insurance: Available at the cost of $3,936 per year.
Other academic fees: Varies (student health insurance is required).
Academic term: Quarter
Number of first-year students who received full tuition waivers: 32

Teaching Assistants, Research Assistants, and Fellowships
Number of first-year

FINANCIAL AID

Loans
Loans are available for U.S. students.
Loans are not available for international students.
GAPSFAS application required: No
FAFSA application required: Yes

For further information
Address financial aid inquiries to: Financial Aid Office, Stanford University, Montag Hall, 355 Galvez Street, Stanford, CA 94305-6106.
Phone: (888) 326-3773
E-mail: financialaid@stanford.edu
Financial aid website: http://www.stanford.edu/dept/finaid/grad/

HOUSING

Availability of on-campus housing
Single students: Yes
Married students: Yes

For further information
Address housing inquiries to: R&DE Student Housing Assignments, 482 Galvez Mall, Suite 110, Stanford, CA 94305.
Phone: (650) 725-2810
Housing aid website: http://www.stanford.edu/dept/rde/cgi-bin/drupal/housing/

Table A—Faculty, Enrollments, and Degrees Granted

Research Specialty	2011–12 Faculty	Enrollment Fall 2012 Master's	Enrollment Fall 2012 Doctorate	Number of Degrees Granted 2012-13 Master's	Number of Degrees Granted 2012-13 Terminal Master's	Number of Degrees Granted 2012-13 Doctorate
Astrophysics	14	–	34	2(-)	–	5(-)
Atomic, Molecular, & Optical Physics	4	–	19	–	–	6(-)
Condensed Matter Physics	11	–	37	1(-)	–	5(-)
Particles and Fields	10	–	43	1(-)	–	13(-)
Non-specialized	–	–	36	2(-)	–	–
Other	1	–	3	1(-)	–	8(-)
Total	40	–	172	7(-)	–	–
Full-time Grad. Stud.	–	–	172	–	–	–
First-year Grad. Stud.	–	–	40	–	–	–

GRADUATE DEGREE REQUIREMENTS

Master's: The Physics Department does not offer a separate program for the master of science degree, but this degree may be awarded for a portion of the doctoral degree; one-year residency is required.
Doctorate: See website http://physics.stanford.edu/.
Thesis: Thesis may be written in absentia.

SPECIAL EQUIPMENT, FACILITIES, OR PROGRAMS

Access to SLAC National Accelerator Laboratory; Teaching Center-Science and Engineering Quad; Hansen Experimental Physics Laboratory; Edward L. Ginzton Laboratory; Kavli Institute for Particle Astrophysics and Cosmology; Center for Space Science and Astrophysics; Laboratory for Advanced Materials.

FACULTY

Professor
Blandford, Roger, Ph.D., Magdalene College, 1974. Theoretical Astrophysics and Cosmology.
Bucksbaum, Philip, Ph.D., University of California, Berkeley, 1980. Director, Ultrafast Science Center, SLAC National Accelerator Laboratory. Optics. Atomic, Molecular, and Optical physics.
Burchat, Patricia, Ph.D., Stanford University, 1986. Experimental and Observational Astrophysics and Cosmology; Experimental Particle Physics.
Cabrera, Blas, Ph.D., Stanford University, 1975. Experimental and Observational Astrophysics and Cosmology; Experimental Particle Physics.
Chu, Steven, Ph.D., University of California, Berkeley, 1976. Professor of Molecular and Cellular Physiology. Atomic, Molecular, and Optical Physics.
Church, Sarah E., Ph.D., University of Cambridge, 1991. Experimental and Observational Astrophysics and Cosmology.
Dimopoulos, Savas, Ph.D., University of Chicago, 1978. Theoretical Particle physics.
Doniach, Sebastian, Ph.D., University of Liverpool, 1958. Biophysics; Theoretical Condensed Matter Physics.
Drell, Persis, Ph.D., University of California, Berkeley, 1983. Professor of Physics and Particle Physics and Astrophysics, SLAC National Accelerator Laboratory. Experimental and Observational Astrophysics and Cosmology; Experimental Particle Physics.

Goldhaber-Gordon, David, Ph.D., Massachusetts Institute of Technology, 1999. Experimental Condensed Matter Physics.

Gratta, Giorgio, Ph.D., Laurea, University of Rome, 1986. Experimental Particle Physics.

Hayden, Patrick, Ph.D., Oxford University, 2001. Theoretical Physics. Quantum Information.

Irwin, Kent, Ph.D., Stanford University, 1995. Experimental and Observational Astrophysics and Cosmology; Experimental Particle Physics.

Kachru, Shamit, Ph.D., Princeton University, 1994. Theoretical Particle Physics.

Kahn, Steven M., Ph.D., University of California, Berkeley, 1980. Experimental and Observational Astrophysics and Cosmology.

Kallosh, Renata, Ph.D., Lebedev Physical Institute, Moscow, 1968. Theoretical Particle Physics.

Kapitulnik, Aharon, Ph.D., Tel Aviv University, 1984. Experimental Condensed Matter Physics.

Kasevich, Mark, Ph.D., Stanford University, 1992. Associate Chair. Atomic, Molecular and Optical Physics.

Kivelson, Steven A., Ph.D., Harvard University, 1979. Theoretical Condensed Matter Physics.

Laughlin, Robert, Ph.D., Massachusetts Institute of Technology, 1979. Theoretical Condensed Matter Physics.

Levin, Craig, Ph.D., Yale University, 1993. Prof. of Radiology and by courtesy, Physics, Electrical Engineering and Bioengineering.

Linde, Andrei, Ph.D., Lebedev Physical Institute, Moscow, 1974. Theoretical Particle Physics.

Macintosh, Bruce, Ph.D., UCLA, 1994. Experimental Astrophysics and Observational Cosmology.

Michelson, Peter, Ph.D., Stanford University, 1979. Current Department Chair. Experimental and Observational Astrophysics and Cosmology.

Moler, Kathryn A., Ph.D., Stanford University, 1995. Experimental Condensed Matter physics.

Petrosian, Vahe, Ph.D., Cornell University, 1967. Theoretical Astrophysics and Cosmology.

Quake, Stephen, Ph.D., University of Oxford, 1999. Professor of Bioengineering and by courtesy, Physics. Biophysics.

Romani, Roger, Ph.D., California Institute of Technology, 1987. Theoretical Astrophysics and Cosmology.

Shen, Zhi-Xun, Ph.D., Stanford University, 1989. Experimental Condensed Matter Physics.

Shenker, Stephen H., Ph.D., Cornell University, 1980. Theoretical Particle physics.

Silverstein, Eva, Ph.D., Princeton University, 1996. Theoretical Particle Physics.

Susskind, Leonard, Ph.D., Cornell University, 1965. Theoretical Particle Physics.

Wieman, Carl, Ph.D., Stanford University, 1977. Physics Education and Atomic and Molecular Physics.

Zare, Richard N., Ph.D., Harvard University, 1964. Professor of Chemistry and by courtesy, Physics.

Zhang, Shoucheng, Ph.D., Stony Brook University, 1987. Theoretical Condensed Matter Physics.

Associate Professor

Abel, Thomas, Ph.D., L. Maxemillian Univ. Munich, 2000. Acting Director of Kavli Institute of Particle Astrophysics and Cosmology. Theoretical Astrophysics and Cosmology.

Allen, Steven, Ph.D., University of Cambridge, 1994. Experimental and Observational Astrophysics and Cosmology.

Manoharan, Hari, Ph.D., Princeton University, 1997. Experimental Condensed Matter Physics.

Wechsler, Risa, Ph.D., University of California, Santa Cruz, 2001. Theoretical Astrophysics and Cosmology.

Assistant Professor

Das, Rhiju, Ph.D., Stanford University, 2005. Assistant Professor of Biochemistry and, by courtesy, Physics.

Graham, Peter, Ph.D., Stanford University, 2007. Theoretical Particle Physics.

Hartnoll, Sean, Ph.D., University of Cambridge, 2005. Theoretical Particle Physics.

Kuo, Chao-Lin, Ph.D., University of California, Berkeley, 2003. Experimental and Observational Astrophysics and Cosmology.

Lev, Benjamin, Ph.D., California Institute of Technology, 2005. Assistant Professor of Applied Physics and, by courtesy, Physics. Atomic, Molecular, and Optical Physics.

Qi, Xiaoliang, Ph.D., Tsinghua University, 2007. Theoretical Condensed Matter Physics.

Raghu, Srinivas, Ph.D., Princeton University, 2006. Theoretical Condensed Matter Physics.

Schleier-Smith, Monika, Ph.D., Massachusetts Institute of Technology, 2011. Atomic, Molecular and Optical Physics.

Senatore, Leonardo, Ph.D., Massachusetts Institute of Technology, 2006. Theoretical Particle Physics.

Professor Emeritus

Fetter, Alexander L., Ph.D., Harvard University, 1963. *Condensed Matter Physics.*

Lipa, John A., Ph.D., Univ. of Western Australia, 1969. Professor (Research) Emeritus. *Low Temperature Physics.*

Little, William A., Ph.D., Rhodes, 1955. *Low Temperature Physics.*

Osheroff, Douglas D., Ph.D., Cornell University, 1973. *Condensed Matter Physics, Low Temperature Physics.*

Ritson, David M., Ph.D., University of Oxford, 1948. *High Energy Physics.*

Schwettman, H. Alan, Ph.D., Rice University, 1962. *Low Temperature Physics.*

Smith, Todd I., Ph.D., Rice University, 1965. Professor (Research) Emeritus. Free Electron Laser Physics.

Sturrock, Peter, Ph.D., University of Cambridge, 1951. Professor (Emeritus) by Courtesy. *Astrophysics, Solar Physics.*

Taylor, Richard, Ph.D., Stanford University, 1962. *Particles and Fields.*

Turneaure, John, Ph.D., Stanford University, 1967. Professor (Research) Emeritus. *Relativity & Gravitation.*

Wagoner, Robert V., Ph.D., Stanford University, 1965. *Astrophysics, Relativity & Gravitation.*

Walecka, John D., Ph.D., Massachusetts Institute of Technology, 1958. *Theoretical Physics.*

Wojcicki, Stanley G., Ph.D., University of California, Berkeley, 1961. *High Energy Physics, Particles and Fields.*

Yearian, Mason, Ph.D., Stanford University, 1959. *High Energy Physics.*

Research Professor

Hollberg, Leo, Ph.D., University of Colorado Boulder, 1984. Atomic, Molecular, and Optical Physics.

Scherrer, Philip H., Ph.D., University of California, Berkeley, 1973. Observational astrophysics; solar physics.

DEPARTMENTAL RESEARCH SPECIALTIES AND STAFF

Theoretical

Theoretical Astrophysics and Cosmology. Abel, Blandford, Petrosian, Romani, Wechsler.

Theoretical Condensed Matter. Doniach, Kivelson, Laughlin, Qi, Raghu, Zhang.

Theoretical Particle Physics. Dimopoulos, Graham, Hartnoll, Hayden, Kachru, Kalkosh, Linde, Senatore, Shenker, Silverstein, Susskind.

Experimental

Atomic, Molecular, & Optical Physics. Bucksbaum, Chu, Hollberg, Kasevich, Lev, Schleier-Smith, Wieman.

Experimental Condensed Matter Physics. Goldhaber-Gordon, Kapitulnik, Manoharan, Moler, Shen.

Experimental Particle Physics. Burchat, Cabrera, Drell, Gratta.

Experimental and Observational Astrophysics and Cosmology. Allen, Burchat, Cabrera, Church, Drell, Irwin, Kahn, Kuo, Macintosh, Michelson, Scherrer.

View additional information about this department at www.gradschoolshopper.com

---

UNIVERSITY OF CALIFORNIA, BERKELEY

DEPARTMENT OF PHYSICS

Berkeley, California 94720-7300
http://www.physics.berkeley.edu

General University Information

President: Mark G. Yudof
Dean of Graduate School: Andrew J. Szeri
University website: http://www.grad.berkeley.edu/
Control: Public
Setting: Urban
Total Faculty: 2,082
Total number of Students: 36,142
Total number of Graduate Students: 10,257

Department Information

Department Chairman: Steven Boggs, Chair
Department Contact: Liza Chudnovsky, Academic HR Analyst
  Total full-time faculty: 60
  Total number of full-time equivalent positions: 47
  Full-Time Graduate Students: 246
  First-Year Graduate Students: 42
  Female First-Year Students: 11
  Total Post Doctorates: 44

Department Address

366 Le Conte Hall
MC 7300
Berkeley, CA 94720-7300
Phone: (510) 642-3316
Fax: (510) 643-8497
E-mail: liza.chudnovsky@berkeley.edu
Website: http://www.physics.berkeley.edu

ADMISSIONS

Admission Contact Information

Address admission inquiries to: Donna K. Sakima, Graduate Student Affairs, Physics Student Services, 370 LeConte Hall #7300, University of California, Berkeley, CA 94720-7300
Phone: (510) 642-0596
E-mail: sakima@berkeley.edu
Admissions website: http://www.grad.berkeley.edu/admissions/index.shtml

Application deadlines

Fall admission:
U.S. students: December 16   *Int'l. students*: December 16

Application fee

U.S. students: $80   *Int'l. students*: $100

Admissions information

For Fall of 2013:
  Number of applicants: 775
  Number admitted: 121
  Number enrolled: 47

Admission requirements

Bachelor's degree requirements: Bachelor's degree in physics and/or related field is required.
Minimum undergraduate GPA: 3.0

GRE requirements

The GRE is required.
The average percentiles of the General GRE scores for admission are: verbal–45% or better; quantitative–90% or better; and analytical writing–45% or better.

Advanced GRE requirements

The Advanced GRE is required.
  Minimum accepted Advanced GRE score: 730
  Mean Advanced GRE score range (25th–75th percentile): 890-990

TOEFL requirements

The TOEFL exam is required for students from non-English-speaking countries.
  PBT score: 570
  iBT score: 68
Students who do not speak English as a native language and do not hold a Bachelor's degree from a U.S. institution must demonstrate oral English proficiency to be appointed as a Graduate Student Instructor. Oral English proficiency can be demonstrated by a passing iBT speaking section score (26 or better).

Other admissions information

Additional requirements: Supervised undergraduate research is strongly encouraged.

Undergraduate preparation assumed: 3 semesters-General Physics, Giancoli; 1 semester-Mechanics, Taylor; 2 semesters-Electromagnetism and Optics, Griffiths, Pedrotti; 1 semester-Thermal/Statistical, Kittel & Kroemer; 2 semesters-Atomic Physics and Quantum Mechanics, Griffiths; 2 semesters-Advanced Undergraduate Laboratory, Horowitz & Hill, Sedra & Smith, Melissinos & Napolitano. Plus mathematics courses in vector calculus, linear algebra, ordinary and partial differential equations, complex variable. (Berkeley undergraduates have in addition 1 semester of physics electives; for example, solid state physics, plasma physics, nuclear and particle physics, relativity).

TUITION

Tuition year 2012–13:
Tuition for in-state residents
  Full-time students: $15,339 annual
Tuition for out-of-state residents
  Full-time students: $30,281 annual
Fees subject to change. 2013-2014 fees were not available at the time of survey.
Credit hours per semester to be considered full-time: 12
Deferred tuition plan: No
Health insurance: Yes, Included in above tuition.
Academic term: Semester
Number of first-year students who received full tuition waivers: 32

Teaching Assistants, Research Assistants, and Fellowships

Number of first-year
  Teaching Assistants: 32
  Research Assistants: 3
  Fellowship students: 20
Average stipend per academic year
  Teaching Assistant: $17,655
  Research Assistant: $28,499
  Fellowship student: $30,000

FINANCIAL AID

Application deadlines
Fall admission:
U.S. students: December 16   *Int'l. students*: December 16

Loans
Loans are available for U.S. students.
Loans are not available for international students.
GAPSFAS application required: No
FAFSA application required: Yes

For further information
Address financial aid inquiries to: Financial Aid and Scholarships Office, UC Berkeley, 201 Sproul Hall #1960, Berkeley, CA 94720-1960.
Phone: (510) 664-9181
E-mail: fao_grad@berkeley.edu
Financial aid website: http://students.berkeley.edu/finaid/

HOUSING

Availability of on-campus housing
Single students: Yes
Married students: Yes

For further information
Address housing inquiries to: Cal Rentals, 2610 Channing Way, 2nd Floor, University of California, Berkeley CA 94720-2272.
Phone: (510) 642-3644
E-mail: homeinfo@berkeley.edu

Housing aid website: http://www.housing.berkeley.edu/livingatcal/graduatestudents.html

Table A—Faculty, Enrollments, and Degrees Granted

Research Specialty	2012-13 Faculty	Enrollment Fall 2012 Master's	Enrollment Fall 2012 Doctorate	Number of Degrees Granted 2012-13 (2008–13) Master's	Number of Degrees Granted 2012-13 (2008–13) Terminal Master's	Number of Degrees Granted 2012-13 (2008–13) Doctorate
Astrophysics	12	–	31	–	–	6(39)
Atomic, Molecular, & Optical Physics	7	–	25	–	–	4(14)
Biophysics	6	–	15	–	–	2(11)
Condensed Matter Physics	16	–	73	–	–	14(57)
Nuclear Physics	2	–	1	–	–	–(3)
Particles and Fields	13	–	45	–	–	8(43)
Plasma and Fusion	4	–	10	–	–	1(7)
Non-specialized	–	–	52	–	–	–
Total	60	–	252	–	–	35(174)
Full-time Grad. Stud.	–	–	252	–	–	–
First-year Grad. Stud.	–	–	42	–	–	–

GRADUATE DEGREE REQUIREMENTS

Master's: Thirty-five semester units in approved program with satisfactory performance; comprehensive exam required; thesis not required; two semester residence requirement; no language requirement.

Doctorate: Thirty-eight graduate units in approved program with satisfactory performance, preliminary examination, candidacy qualifying examination, and dissertation required; four semester residency requirement, no language requirement. Interdepartmental research: Some graduate students are engaged in research problems involving interdepartmental collaboration of which the following are examples: (1) nuclear physics, in programs with the Chemistry Department or the Lawrence Berkeley National Laboratory; (2) astrophysics, with the Department of Astronomy, the Berkeley Center for Cosmological Physics, or the Space Sciences Laboratory; (3) solid state physics, with the Departments of Electrical Engineering and Computer Sciences, and Materials Science and Engineering; (4) plasma physics, with the Departments of Electrical Engineering and Computer Sciences and Nuclear Engineering or the Lawrence Berkeley National Laboratory; (5) Biophysics and medical physics. Interdisciplinary groups: There are a number of graduate Interdisciplinary Groups with Ph.D. programs separate from the Ph.D. in Physics, particle physics, with the Berkeley Center for Theoretical Physics.

SPECIAL EQUIPMENT, FACILITIES, OR PROGRAMS

In addition to our own research programs and facilities, our dynamic collaborations bring access to local research facilities such as: Lick Observatory (University of California, Mount Hamilton, San Jose CA); Space Sciences Laboratory (University of California, Berkeley, CA); Lawrence Berkeley National Laboratory/LBNL (DOE, Berkeley, CA); Advanced Light Source (LBNL); National Center for Electron Microscopy (LBNL); The Molecular Foundry, a nanoscience user research facility (LBNL); and Lawrence Livermore National Laboratory (Livermore, CA).

The following local, national, and international laboratories also make facilities available for graduate student research: Brookhaven National Laboratory (New York); Fermi National Accelerator Laboratory (Illinois); NASA Ames Center (California); Kitt Peak National Observatory (Arizona); SLAC National Accelerator Laboratory (California); University of California, Berkeley Micro/ Nanofabrication Facility (California); the University of California, Berkeley Radio Astronomy Laboratory (California); Argonne National Laboratory (Illinois); W. M. Keck Observatory (Hawaii); CERN (Headquarters Switzerland); Gran Sasso Underground Laboratory (Italy); Assergi (Italy); Kamioka Observatory (Japan); Institute for the Physics and Mathematics of the Universe/IPMU (Japan); Paul Scherrer Institute (Switzerland); and TRIUMF (Canada).

Table B—Separately Budgeted Research Expenditures by Source of Support

Source of Support	Departmental Research	Physics-related Research Outside Department
Federal government	$16,267,394	$21,222,783
State/local government	$153,906	
Non-profit organizations	$3,259,687	
Business and industry	$741,577	
Other	$1,638,115	
Total	$22,060,679	$21,222,783

Table C—Separately Budgeted Research Expenditures by Research Specialty

Research Specialty	No. of Grants	Expenditures ($)
Astrophysics	33	$4,808,722
Atomic, Molecular, & Optical Physics	44	$4,592,377
Biophysics	7	$183,870
Condensed Matter Physics	87	$8,568,550
Nuclear Physics	12	$682,764
Particles and Fields	24	$1,994,492
Plasma and Fusion	11	$992,500
Cosmology & String Theory	3	$237,404
Other	0	
Total	221	$22,060,679

FACULTY

Professor

Bale, Stuart, Ph.D., University of Minnesota, 1994. Experimental astrophysics.

Birgeneau, Robert J., Ph.D., Yale University, 1966. *Condensed Matter Physics*. Experimental condensed matter physics.

Boggs, Steven E., Ph.D., University of California, Berkeley, 1998. Chair, Physics Department. *Astrophysics*. Gamma ray spectroscopy. Experimental astrophysics.

Bousso, Raphael, Ph.D., University of Cambridge, 1997. *Particles and Fields*. Theoretical particle physics. Particle theory.

Budker, Dmitry, Ph.D., University of California, Berkeley, 1993. *Atomic, Molecular, & Optical Physics*. Atomic physics.

Bustamante, Carlos J., Ph.D., University of California, Berkeley, 1981. *Biophysics*. Biophysics.

Crommie, Michael, Ph.D., University of California, Berkeley, 1991. *Condensed Matter Physics*. Experimental condensed matter physics.

Davis, Marc, Ph.D., Princeton University, 1973. Professor, Astronomy Department. Theoretical astrophysics; extragalactic astronomy; cosmology.

Dynes, Robert C., Ph.D., McMaster University, 1968. Experimental condensed matter physics.

Fajans, Joel, Ph.D., Massachusetts Institute of Technology, 1985. Experimental plasma physics.

Falcone, Roger W., Ph.D., Stanford University, 1979. Quantum electronics and experimental atomic physics.

Genzel, Reinhard L., Ph.D., University of Bonn, 1978. Experimental astrophysics; infrared and microwave astronomy.

Hŏrava, Petr, Ph.D., Czech Academy of Sciences, 1981. Theoretical particle physics.

Hall, Lawrence J., Ph.D., Harvard University, 1981. Theory of elementary particles.

Haxton, Wick, Ph.D., University of California, Santa Cruz, 1971. Theoretical particle physics and astrophysics.

Heinemann, Beate, Ph.D., University of Hamburg, 1999. Experimental particle physics.

Hellman, Frances D., Ph.D., Stanford University, 1985. *Condensed Matter Physics*. Experimental condensed matter physics.

Holzapfel, William L., Ph.D., University of California, Berkeley, 1996. Experimental astrophysics.

Jacobsen, Robert G., Ph.D., Stanford University, 1991. Experimental elementary particle physics.

Knobloch, Edgar, Ph.D., Harvard University, 1978. Theoretical astrophysics; fluid dynamics; nonlinear dynamics.

Kolomensky, Yury, Ph.D., University of Massachusetts, 1997. BaBar, B hadron decays, with particular emphasis on understanding the origin of CD noninvariance.

Lanzara, Alessandra, Ph.D., University of Rome, 1999. *Condensed Matter Physics*. Experimental condensed matter physics.

Lee, Adrian, Ph.D., Stanford University, 1993. Cryogenic far-infrared and mm-wave detector development.

Lee, Dung-Hai, Ph.D., Massachusetts Institute of Technology, 1982. Theory of condensed matter; quantum phase transitions; strongly correlating electronic systems.

Leone, Stephen R., Ph.D., University of California, Berkeley, 1974. Gas phase laser spectroscopy.

Littlejohn, Robert G., Ph.D., University of California, Berkeley, 1980. Theoretical plasma physics and nonlinear dynamics.

Louie, Steven G., Ph.D., University of California, Berkeley, 1976. Theory of condensed matter.

Luk, Kam-Biu, Ph.D., Rutgers University, 1983. Experimental elementary particle physics.

Moore, Joel E., Ph.D., Massachusetts Institute of Technology, 2000. *Condensed Matter Physics*. Condensed matter and statistical physics-theory/biophysics theory.

Murayama, Hitoshi, Ph.D., University of Tokyo, 1991. Theory of elementary particles.

Nomura, Yasunori, Ph.D., University of Tokyo, 2000. *Particles and Fields*. Theoretical particle physics.

Orenstein, Joseph W., Ph.D., Massachusetts Institute of Technology, 1980. Experimental condensed matter physics.

Perlmutter, Saul, Ph.D., Harvard University, 1986. *Astrophysics*. Experimental astrophysics.

Qiu, Zi Qiang, Ph.D., Johns Hopkins University, 1990. Experimental condensed matter physics.

Ramesh, Ramamoorthy, Ph.D., University of California, Berkeley, 1987. *Condensed Matter Physics*. Experimental condensed matter physics.

Rokhsar, Daniel S., Ph.D., Cornell University, 1987. Statistical and many-body physics; biophysics.

Sadoulet, Bernard, Ph.D., d'Etude es Sciences, Orsay, 1971. Experimental cosmology.

Seljak, Uroš, Ph.D., Massachusetts Institute of Technology, 1995. *Astrophysics*. Cosmology and theoretical astrophysics.

Shapiro, Marjorie, Ph.D., University of California, Berkeley, 1984. Experimental elementary particle physics.

Siegrist, James L., Ph.D., Stanford University, 1979. Director, Physics Division, Lawrence Berkeley National Lab. Experimental elementary particle physics.

Stamper-Kurn, Dan, Ph.D., Massachusetts Institute of Technology, 1999. Bose-Einstein Condensation (BEC) of atoms.

White, Martin, Ph.D., Yale University, 1992. Theoretical astrophysics.

Wurtele, Jonathan S., Ph.D., University of California, Berkeley, 1979. *Plasma and Fusion*. Theoretical plasma physics.

Zettl, Alex, Ph.D., University of California, Los Angeles, 1983. Experimental condensed matter physics.

Associate Professor

Aganagic, Mina, Ph.D., California Institute of Technology, 1999. Theoretical particle physics.

Ganor, Ori, Ph.D., Tel Aviv University, 1996. *Particles and Fields.* Theoretical particle physics.

Liphardt, Jan T., Ph.D., University of Cambridge, 1999. Experimental biophysics.

Quataert, Eliot, Ph.D., Harvard University, 1999. *Astronomy, Astrophysics.* Astronomy/astrophysics.

Siddiqi, Irfan, Ph.D., Yale University, 2002. *Condensed Matter Physics.* Experimental condensed matter physics.

Vishwanath, Ashvin, Ph.D., Princeton University, 2001. Theoretical condensed matter physics.

Wang, Feng, Columbia University, 2004. *Condensed Matter Physics.* Experimental condensed matter physics.

Assistant Professor

Analytis, James, Ph.D., University of Oxford, 2006. *Condensed Matter Physics.* Experimental condensed matter physics.

DeWeese, Michael, Ph.D., Princeton University, 1995. Experimental biophysics.

Ginsberg, Naomi, Ph.D., Harvard University, 2007. Professor in the Department of Chemistry. *Atomic, Molecular, & Optical Physics, Chemical Physics.*

Häffner, Hartmut, Ph.D., University of Mainz, 2000. Experimental atomic molecular and optical physics.

Hallatschek, Oskar, Ph.D., Free University of Berlin, 2004. *Biophysics.* Biophysics.

Kasen, Daniel, Ph.D., University of California, Berkeley, 2004. Assistant Professor, Astronomy. Astrophysics.

Müller, Holger, Ph.D., Humboldt University, Berlin, 2004. Experimental atomic, molecular, and optical physics.

Orebi-Gann, Gabriel, Ph.D., University of Oxford, 2008. *Particles and Fields.* Partical/nuclear experimental physics.

Yildiz, Ahmet, Ph.D., University of Illinois, 2004. *Biophysics.* Experimental biophysics.

Emeritus

Arons, Jonathan, Ph.D., Harvard University, 1970. Professor of the Graduate School. *Plasma and Fusion.* Theoretical high-energy astrophysics; x-ray sources; plasma physics.

Bardakci, Korkut, Ph.D., University of Rochester, 1962. Theory of elementary particles.

Chew, Geoffrey F., Ph.D., University of Chicago, 1948. Theory of elementary particles.

Chinowsky, William, Ph.D., Columbia University, 1955.

Clarke, John, Ph.D., University of Cambridge, 1968. Professor of the Graduate School. Experimental superconductivity and low-temperature physics.

Cohen, Marvin L., Ph.D., University of Chicago, 1964. Professor of the Graduate School. *Condensed Matter Physics.* Theory of condensed matter.

Commins, Eugene D., Ph.D., Columbia University, 1958. Experimental atomic and nuclear physics.

Ely, Robert P., Ph.D., Massachusetts Institute of Technology, 1959.

Frazer, William R., Ph.D., University of California, Berkeley, 1959. *Particles and Fields.* Elementary particle theory; cosmology.

Gaillard, Mary K., Ph.D., University of Paris, 1968. Professor of the Graduate School. Theory of elementary particles.

Hahn, Erwin L., Ph.D., University of Illinois, 1949. Experimental condensed matter physics and modern optics.

Halpern, Martin B., Ph.D., Harvard University, 1964. Theory of elementary particles.

Jackson, J. D., Ph.D., Massachusetts Institute of Technology, 1949. Theory of elementary particles.

Kaufman, Allan N., Ph.D., University of Chicago, 1953. Plasma theory and nonlinear dynamics.

Kerth, Leroy T., Ph.D., University of California, Berkeley, 1957. Experimental elementary particle physics.

Kittel, Charles, Ph.D., University of Wisconsin-Madison, 1941.

Kunkel, Wulf B., Ph.D., University of California, Berkeley, 1951. Experimental plasma physics.

Mandelstam, Stanley, Ph.D., University of Birmingham, England, 1956. Theory of elementary particles.

Marrus, Richard, Ph.D., University of California, Berkeley, 1959. Experimental atomic physics; beam foil spectroscopy.

McKee, Christopher F., Ph.D., University of California, Berkeley, 1970. Professor of the Graduate School. Theoretical astrophysics; interstellar medium; high-energy astrophysics.

Mozer, Forrest S., Ph.D., California Institute of Technology, 1956.

Muller, R. A., Ph.D., University of California, Berkeley, 1969. Professor of the Graduate School. Astrophysics experiment; experimental physics.

Packard, Richard E., Ph.D., University of Michigan, 1969. Professor of the Graduate School. Experimental condensed matter physics; experimental low-temperature physics.

Price, P. Buford, Ph.D., University of Virginia, 1958. Professor of the Graduate School. Astrophysics experiment; cosmic radiation and relativistic nuclear physics; high-energy neutrino astrophysics; microbes in polar ice.

Reif, Frederick, Ph.D., Harvard University, 1953.

Richards, Paul L., Ph.D., University of California, Berkeley, 1960. Professor of the Graduate School. Experimental condensed matter physics; infrared spectroscopy; infrared astrophysics.

Rosenfeld, Arthur H., Ph.D., University of Chicago, 1954. Physics related to environmental problems; energy and conservation.

Sachs, Rainer K., Ph.D., Syracuse University, 1958.

Schwartz, Charles L., Ph.D., Massachusetts Institute of Technology, 1954.

Shank, Charles V., Ph.D., University of California, Berkeley, 1969. Professor of the Graduate School. *Condensed Matter Physics.* Experimental condensed matter physics.

Shen, Yuen-Ron, Ph.D., Harvard University, 1963. Professor of the Graduate School. Experimental condensed matter physics; quantum and nonlinear optics.

Shugart, Howard A., Ph.D., University of California, Berkeley, 1957. Experimental atomic physics and atomic beam studies; nuclear spins and moments.

Smoot, George F., Ph.D., Massachusetts Institute of Technology, 1970. Professor of the Graduate School. Experimental astrophysics.

Steiner, Herbert M., Ph.D., University of California, Berkeley, 1956. Experimental elementary particle physics.

Stevenson, M. Lynn, Ph.D., University of California, Berkeley, 1953. Experimental elementary particle physics.

Strovink, Mark W., Ph.D., Princeton University, 1970. Experimental elementary particle physics.

Suzuki, Mahiko, Ph.D., University of Tokyo, 1965. Theory of elementary particles.

Townes, Charles H., Ph.D., California Institute of Technology, 1939. Professor of the Graduate School. Experimental astrophysics; infrared and microwave astronomy.

Trilling, George H., Ph.D., California Institute of Technology, 1955. Experimental elementary particle physics.

Tripp, Robert D., Ph.D., University of California, Berkeley, 1955. Experimental elementary particle physics.

Wichmann, Eyvind H., Ph.D., Columbia University, 1956. Theory of elementary particles; mathematical physics.

Yu, Peter, Ph.D., Brown University, 1972. *Condensed Matter Physics.* Experimental condensed matter physics; semiconductor physics.

Zumino, Bruno, Ph.D., University of Rome, 1945. Professor of the Graduate School. Theory of elementary particles.

DEPARTMENTAL RESEARCH SPECIALTIES AND STAFF

Theoretical

Astrophysics. Interstellar medium; star formation; binary stars; stellar convection; pulsars; x-ray sources; active galactic nuclei and quasars; galaxy formation; cosmology. Arons, Haxton, Kasen, McKee, Seljak, White.

Condensed Matter and Materials Physics. Uncover new states of matter and understand their physical properties. Theoretical and computational studies of the behaviors of novel materials and nanostructures, including electronic, vibrational, optical, thermal, transport, magnetic and superconducting properties; emergent phenomena, quantum phase transitions, and strongly correlated electron systems; many-body effects in bulk, reduced-dimensional and nanostructured systems; surface, interface, phase transition, and alloy properties. Cohen, Dung-Hai Lee, Louie, Moore, Vishwanath.

Nonlinear Dynamics, Plasma and Beam Physics, and Complex Systems. Dynamics of neutral and nonneutral plasmas with applications to antihydrogen trapping, laser-plasma interaction and particle acceleration; Chaos and approach to chaos;bifurcation theory; pattern formation; fluid dynamics; semiclassical mechanics; climate change. Littlejohn.

Nonlinear Dynamics, Plasma and Beam Physics, and Complex Systems. Dynamics of ionized gases far from thermal equilibrium and dominated by long-range electromagnetic fields with applications to controlled fusion, astrophysics, and space science. Fajans, Knobloch, Littlejohn, Wurtele.

Particles and Fields. Gauge theory of weak and electromagnetic interactions and perturbative QCD; Theories of physics beyond the standard model, including grand unification, supersymmetry, supergravity and extra dimensions; quantum field theory; string theory; theories of gravity. Aganagic, Bousso, Gaillard, Ganor, Hŏrava, Hall, Haxton, Murayama, Nomura, Zumino.

Relativity & Gravitation. general relativity and quantum gravity. Bousso, Littlejohn.

Experimental

Astrophysics. Dark matter searches, astroparticle physics, observations of the cosmic microwave background, cosmology, studies of galaxy clusters, studies of star forming galaxies, study of dark energy using supernovae, galaxy counts, and Baryon acoustic oscillations, Studies of astrophysical neutrinos, solar physics, and high-energy neutrino astrophysics. Bale, Boggs, Davis, Genzel, Holzapfel, Adrian Lee, Muller, Perlmutter, Price, Richards, Sadoulet, Smoot, Townes.

Atmosphere, Space Physics, Cosmic Rays. Magnetospheric physics: space plasmas and fields; auroras; isotopic and elemental composition of cosmic rays; search for new particles and antimatter in cosmic rays; spectrum and anisotropy of the universal microwave radiation; infrared astronomical spectroscopy and spatial interferometry; millimeter and submillimeter spectra; the galactic center; star formation; new astronomical detectors; automated supernova search; x-ray spectroscopy and laboratory astrophysics; high-energy gamma-ray astrophysics; experimental cosmology including particle astrophysics.

Atomic, Molecular, & Optical Physics. Precision measurements, ultra-cold quantum gases, quantum information, attosecond physics, x-ray lasers, tests of fundamental symmetries, variation of fundamental constants, quantum phase transitions, ion trapping, hybrid quantum systems, anti-hydrogen trapping and spectroscopy, opto-mechanical systems, magnetometry, low-field NMR, frequency combs, precision spectroscopy of atomic systems near and in bulk materials, NV-centers, novel approaches to microscopy, energy transfer in complex molecules, non-linear interaction of light with matter, hot and dense plasmas. Budker, Falcone, Ginsberg, Häffner, Leone, Müller, Stamper-Kurn.

Biophysics. Molecular biophysics and structural biology. Application of atomic force spectroscopy and optical tweezers to biological problems. Bustamante, DeWeese, Hallatschek, Liphardt, Rokhsar, Yildiz.

Condensed Matter Physics. Synthesis and experimental investigation of novel condensed phase materials, including high-Tc superconductors, low-dimensional materials, strongly correlated materials, thin-film oxides, and topological insulators. Understand the fundamental electrical, magnetic, optical, superconducting, and superfluidic behavior using state of the art experimental techniques, including electrical and magnetic transport, scanning tunneling microscopy, in-situ transmission electron microscopy, ultrafast laser spectroscopy, angle-resolved photoemission and neutron scattering. Applications of superconducting devices for quantum measurements and quantum information; quantum bits; superconducting parametric amplifiers. Analytis, Birgeneau, Clarke, Crommie, Dynes, Hellman, Lanzara, Orenstein, Packard, Qiu, Ramesh, Shank, Shen, Siddiqi, Wang, Zettl.

Energy Sources & Environment. The Center for Building Science, in the Applied Science Division at Lawrence Berkeley National Laboratory.

Nonlinear Dynamics, Plasma and Beam Physics, and Complex Systems. Nonneutral plasmas and antihydrogen synthesis; laser-driven particle acceleration in plasmas; high brightness electron and ion beams. Fajans, Knobloch, Littlejohn, Rokhsar, Yildiz.

Nuclear Physics. Neutrino physics, including studies of neutrinos produced in solar, atmospheric and reactor interactions; Nuclear astrophysics; studies of symmetry breaking in nuclear systems; searches for neutrinoless double beta decay; weak interactions in nuclei; heavy ion collisions. Heinemann, Jacobsen, Kolomensky, Luk, Orebi-Gann, Shapiro, Siegrist.

Particles and Fields. Experiments utilizing particle accelerators such as electron-positron and hadron colliders, as well as fixed target machines to test and extend the standard model and to search for physics beyond the standard model; studies of neutrino physics using neutrinos produced in accelerators and reactors; searches for Dark Matter and other experimental studies in the field of particle astrophysics; development and fabrication of detectors appropriately matched to these goals. Heinemann, Jacobsen, Kolomensky, Luk, Orebi-Gann, Shapiro, Siegrist.

Plasma and Fusion. Plasma production and heating; magnetic confinement of high-temperature plasma; development and application of plasma diagnostic methods; atomic physics problems related to controlled fusion, accelerator research for heavy-ion driven pellet fusion; single species plasma. Fajans, Knobloch, Littlejohn, Wurtele.

View additional information about this department at **www.gradschoolshopper.com**

UNIVERSITY OF CALIFORNIA, DAVIS

DEPARTMENT OF PHYSICS

Davis, California 95616
http://www.physics.ucdavis.edu/

General University Information
Chancellor: Linda Katehi
Dean of Graduate School: Jeffery C. Gibeling
University website: http://www.ucdavis.edu/index.html
Control: Public
Setting: Suburban
Total Faculty: 1,878
Total Graduate Faculty: 1,878
Total number of Students: 33,300
Total number of Graduate Students: 5,203

Department Information
Department Chairman: Andreas Albrecht, Chair
Department Contact: Angela Sharma, Graduate Program
 Coordinator
 Total full-time faculty: 40
 Total number of full-time equivalent positions: 40
 Full-Time Graduate Students: 136
 First-Year Graduate Students: 19
 Female First-Year Students: 5
 Total Post Doctorates: 25

Department Address
One Shields Avenue
Physics Department
Davis, CA 95616
Phone: (530) 752-1501
Fax: (530) 752-4717
E-mail: grad-info@physics.ucdavis.edu
Website: http://www.physics.ucdavis.edu/

ADMISSIONS

Admission Contact Information
Address admission inquiries to: Graduate Program, Department
 of Physics, One Shields Avenue, Davis, CA 95616
Phone: (530) 752-1501
E-mail: grad-info@physics.ucdavis.edu
Admissions website: http://www.physics.ucdavis.edu/resources-
 _for_graduates/index.html

Application deadlines
Fall admission:
U.S. students: January 15   *Int'l. students*: January 15

Application fee
U.S. students: $80   *Int'l. students*: $100
January 15 is a priority deadline for applications. If room remains
 in the program, further applications are accepted until May
 31.

Admissions information
For Fall of 2013:
 Number of applicants: 364
 Number admitted: 104
 Number enrolled: 39

Admission requirements
Bachelor's degree requirements: Bachelor's degree from an accredited college or university with a grade-point average of
 3.0 or better (on a scale where A=4.0) in upper division
 coursework and in any graduate courses taken.
Minimum undergraduate GPA: 3.0

GRE requirements
The GRE is required.

Advanced GRE requirements
The Advanced GRE is recommended.

TOEFL requirements
The TOEFL exam is required for students from non-English-
 speaking countries.
 PBT score: 550
 iBT score: 80

Other admissions information
Additional requirements: No minimum score specified.
Undergraduate preparation assumed: Typical texts: Tipler,
 Physics for Scientists and Engineers; Serway, Moses, and
 Moyer, Modern Physics; Boas, Mathematical Methods in the
 Physical Sciences; Marion and Thornton, Classical Dynamics;
 Griffiths, Introduction to Electrodynamics; Reif, Fundamen-
 tals of Statistical and Thermal Physics; Griffiths, Quantum
 Mechanics.

TUITION
Tuition year 2012–13:
Tuition for in-state residents
 Full-time students: $15,387 annual
 Part-time students: $9,777 annual
Tuition for out-of-state residents
 Full-time students: $30,489 annual
 Part-time students: $17,328 annual
Credit hours per semester to be considered full-time: 6
Deferred tuition plan: No
Health insurance: Available
Other academic fees: None
Academic term: Quarter
Number of first-year students who received full tuition waivers: 18

Teaching Assistants, Research Assistants, and Fellowships
Number of first-year
 Teaching Assistants: 18
 Fellowship students: 8
Average stipend per academic year
 Teaching Assistant: $22,469
 Research Assistant: $23,823
 Fellowship student: $22,469
Most first-year fellowships are much smaller, since they sup-
plement rather than replace TA positions. In later years fel-
lowships are intended as full support. All numbers are for
12 months of support.

FINANCIAL AID

Application deadlines

Fall admission:

U.S. students: January 15   *Int'l. students*: January 15

Loans

Loans are available for U.S. students.
Loans are not available for international students.
GAPSFAS application required: No
FAFSA application required: Yes

For further information

Address financial aid inquiries to: Graduate Financial Aid Office, One Shields Avenue, Davis, CA 95616.
Phone: (530) 752-9246
E-mail: gradfinaid@ucdavis.edu
Financial aid website: http://financialaid.ucdavis.edu/graduate/index.html

HOUSING

Availability of on-campus housing

Single students: Yes
Married students: Yes

For further information

Address housing inquiries to: Student Housing Office, One Shields Avenue, Davis, CA 95616.
Phone: (530) 752-2033
E-mail: studenthousing@ucdavis.edu
Housing aid website: http://www.housing.ucdavis.edu/

Table A—Faculty, Enrollments, and Degrees Granted

Research Specialty	2012-13 Faculty	Enrollment Fall 2012 Master's	Enrollment Fall 2012 Doctorate	Number of Degrees Granted 2012-13 (2008-13) Master's	Number of Degrees Granted 2012-13 (2008-13) Terminal Master's	Number of Degrees Granted 2012-13 (2008-13) Doctorate
Astrophysics	10	–	21	–	–	4(19)
Biophysics	1	–	1	–	–	1(6)
Computational Physics	2	–	7	–	–	2(7)
Condensed Matter Physics	14	3	47	–	–(1)	9(36)
Nuclear Physics	3	1	11	–	–	2(5)
Particles and Fields	12	–	40	–	–	3(19)
Physics and other Science Education	1	–	3	–	–	1(3)
Relativity & Gravitation	1	–	6	–	–	–(3)
Non-specialized	–	–	–	14(62)	1(10)	–
Total	44	4	136	28(76)	1(11)	22(98)
Full-time Grad. Stud.	–	4	136	–	–	–
First-year Grad. Stud.	–	1	18	–	–	–

GRADUATE DEGREE REQUIREMENTS

Master's: Two master's programs are offered. Plan I requires 32 quarter hours of graduate and upper division coursework and a master's thesis. Plan II requires 36 quarter hours of graduate and upper division coursework, of which at least 18 hours must be at the graduate level, and passing the preliminary examination at the master's level. The preliminary examination covers senior undergraduate and first-year graduate level physics. Both plans require coursework in classical physics, quantum mechanics, and mathematical methods.

Doctorate: The Ph.D. degree requires a thorough understanding of the foundations of physics and mathematical methods as evidenced by performance on the preliminary examination and oral examination and submission of a dissertation, which

must include an original contribution to fundamental physics. Ph.D. students must also complete the graduate core courses in classical physics, statistical physics, quantum mechanics, and mathematical methods. The required curriculum can be tailored to fit the individual student's preparation and needs. Each graduate student selects a course of study in consultation with a graduate advisor. A student with weaknesses in preparation may be advised to audit or take for credit specific advanced undergraduate courses. A student entering with advanced preparation may skip the core courses by passing the written examination upon entrance to the program. Students are to take a cluster of advanced graduate courses determined by their field of specialization. For more information, please visit http://www.physics.ucdavis.edu.

Thesis: Thesis may be written in absentia.

Table B—Separately Budgeted Research Expenditures by Source of Support

Source of Support	Departmental Research	Physics-related Research Outside Department
Federal government	$6,981,923	
State/local government	$416,063	
Non-profit organizations	$1,905,967	
Business and industry		
Other	$636,628	
Total	$9,940,581	

Table C—Separately Budgeted Research Expenditures by Research Specialty

Research Specialty	No. of Grants	Expenditures ($)
Cosmology	37	$1,904,878
Biophysics	8	$1,611,691
Condensed Matter Physics	37	$2,087,404
Nuclear Physics	6	$401,393
High Energy	26	$3,244,491
Computational Physics	16	$690,724
Total	130	$9,940,581

FACULTY

Professor

Albrecht, Andreas, Ph.D., University of Pennsylvania, 1983. Department Chair. *Cosmology & String Theory*. Cosmology; high-energy theory.

Becker, Robert, Ph.D., University of Maryland, 1975. *Astronomy*. Creating new astronomical surveys; radioastronomy.

Calderón de la Barca Sánchez, Manuel, Ph.D., Yale University, 2001. *Nuclear Physics*. Relativistic heavy-ion physics.

Carlip, Steven, Ph.D., University of Texas, Austin, 1987. *Relativity & Gravitation*. Theoretical high-energy physics; quantum gravity.

Cebra, Daniel A., Ph.D., Michigan State University, 1990. *Nuclear Physics*. Relativistic heavy-ion physics.

Cheng, Hsin-Chia, Ph.D., University of California, Berkeley, 1996. *High Energy Physics*. Theoretical particle physics.

Chertok, Maxwell B., Ph.D., Boston University, 1997. Vice Chairperson of the Department. *High Energy Physics*. Experimental high-energy physics; hadron collider physics.

Chiang, Shirley, Ph.D., University of California, Berkeley, 1983. *Condensed Matter Physics, Surface Physics*. Experimental condensed-matter physics and surface physics.

Conway, John, Ph.D., University of Chicago, 1987. *High Energy Physics*. Experimental high-energy physics.

Cox, Daniel L., Ph.D., Cornell University, 1985. *Biophysics, Condensed Matter Physics*. Theoretical biological physics.

Crutchfield, James, Ph.D., University of California, Santa Cruz, 1983. *Computational Physics*. Nonlinear dynamics; condensed-matter physics; physics of computation; evolutionary dynamics; pattern discovery; dynamics of learning; distributed robotics.

Curro, Nicholas, Ph.D., University of Illinois at Urbana-Champaign, 1998. *Condensed Matter Physics*. Experimental condensed-matter physics with a focus on nuclear magnetic resonance of condensed matter electron systems, especially phenomena associated with magnetism, unconventional superconductivity, and heavy fermion physics.

Erbacher, Robin, Ph.D., Stanford University, 1998. *High Energy Physics*. Experimental high-energy particle physics: top quark physics; searches for new particles and phenomena; particle detector development.

Fadley, Charles S., Ph.D., University of California, Berkeley, 1970. *Condensed Matter Physics*. Advanced light source; surfaces and interfaces via angle-resolved photo-electron spectroscopy.

Fassnacht, Christopher D., Ph.D., California Institute of Technology, 1999. *Astronomy*. Observational cosmology; galaxy structure; cosmological parameter measurement through gravitational lensing.

Ferenc, Daniel, Ph.D., University of Zagreb, Croatia, 1992. *Astrophysics, Nuclear Physics*. Relativistic universe; high-energy astrophysics; gamma-ray astronomy; next-generation underground lab for proton decay and neutrino physics.

Fong, Ching-Yao, Ph.D., University of California, Berkeley, 1968. *Condensed Matter Physics*. Theoretical solid-state physics.

Galli, Giulia, Ph.D., International School for Advanced Studies, Trieste, Italy, 1986. *Computational Physics, Condensed Matter Physics*. Computational materials theory.

Gunion, John F., Ph.D., University of California, San Diego, 1970. *High Energy Physics*. Theoretical particle physics.

Kaloper, Nemanja, Ph.D., University of Minnesota, Minneapolis, 1992. *Cosmology & String Theory, High Energy Physics*. Cosmology; high-energy theory.

Knox, Lloyd D., Ph.D., University of Chicago, 1995. *Cosmology & String Theory*. Cosmology.

Liu, Kai, Ph.D., Johns Hopkins University, 1998. *Condensed Matter Physics*. Experimental condensed matter.

Lubin, Lori M., Ph.D., Princeton University, 1995. *Astronomy*. Observational cosmology.

Luty, Markus, Ph.D., University of Chicago, 1992. *High Energy Physics*. Theoretical particle physics; theoretical gravity.

Pickett, Warren E., Ph.D., Stony Brook University, 1975. *Condensed Matter Physics*. Theoretical condensed-matter physics; metals and superconductors.

Rundle, John B., Ph.D., University of California, Los Angeles, 1976. *Computational Physics, Geophysics*. Dynamics of complex nonlinear systems.

Savrasov, Sergey Y., Ph.D., Lebedev Physical Institute, Moscow, 1994. *Condensed Matter Physics*. Theoretical condensed-matter physics; electronic structure of solids; computational approaches to strongly correlated systems.

Scalettar, Richard T., Ph.D., University of California, Santa Barbara, 1986. *Condensed Matter Physics*. Theoretical condensed-matter physics; statistical mechanics; many-body theory; highly correlated systems.

Singh, Rajiv R. P., Ph.D., Stony Brook University, 1986. *Biophysics, Condensed Matter Physics*. Theoretical condensed-atter physics; spin glasses; critical phenomena; low-dimensional systems, biological physics.

Svoboda, Robert, Ph.D., University of Hawaii, 1985. *High Energy Physics*. Neutrino physics.

Terning, John, Ph.D., University of Toronto, 1990. *High Energy Physics*. Theoretical particle physics; electroweak symmetry breaking; supersymmetry; cosmology; extra dimensions; AdS/CFT correspondence.

Tripathi, S. Mani, Ph.D., University of Pittsburgh, 1986. *High Energy Physics*. Experimental high-energy physics; productions of direct photons in hadronic interactions.

Tyson, J. Anthony, Ph.D., University of Wisconsin-Madison, 1967. *Astrophysics, Cosmology & String Theory*. Observational cosmology; development of new astronomical surveys, detectors, astronomical instrumentation, and analysis algorithms.

Zhu, Xiangdong, Ph.D., University of California, Berkeley, 1989. *Biophysics, Condensed Matter Physics*. Experimental condensed-matter physics; nonlinear optics; laser studies of surfaces.

Zieve, Rena, Ph.D., University of California, Berkeley, 1992. Vice Chair, Graduate Affairs. *Condensed Matter Physics, Low Temperature Physics*. Experimental condensed-matter physics; low-temperature physics; unconventional superconductors.

Zimanyi, Gergely, Ph.D., Central Research Institute for Physics, Budapest, 1985. *Condensed Matter Physics*. Theoretical condensed-matter physics; localization; electron gas; low-dimensional system; superconductivity.

Associate Professor

Bradac, Marusa, Ph.D., University of Bonn, 2004. *Astronomy*. Cosmology gravitational lensing; first galaxies; dark matter.

Wittman, David M., Ph.D., University of Arizona, 1997. *Astronomy*. Observational cosmology; deep lens survey; weak lensing.

Assistant Professor

Mulhearn, Michael J., Ph.D., Massachusetts Institute of Technology, 2004. *High Energy Physics*. Experimental high-energy physics.

Yu, Dong, Ph.D., University of Chicago, 2005. *Condensed Matter Physics*. Experimental condensed-matter physics.

Emeritus

Potter, Wendell H., Ph.D., University of Illinois, 1970. *Physics and other Science Education*. Science education.

Professor Emeritus

Brady, F. Paul, Ph.D., Princeton University, 1960. *Nuclear Physics*. Experimental nuclear physics; neutron physics and relativistic nuclear collisions.

Cahill, Thomas A., Ph.D., University of California, Los Angeles, 1965. Experimental nuclear and atomic physics; analytical applications of accelerator beams; atmospheric physics.

Chau, Ling-Lie, Ph.D., University of California, Berkeley, 1966. Theoretical particle physics.

Coleman, Lawrence B., Ph.D., University of Pennsylvania, 1975. *Condensed Matter Physics*. Experimental condensed-matter physics; far infrared spectroscopy; phase transitions in solids; structure-property relationships.

Corruccini, Linton R., Ph.D., Cornell University, 1972. *Condensed Matter Physics, Low Temperature Physics*. Experimental low-temperature and condensed-matter physics; liquid helium.

Draper, James E., Ph.D., Cornell University, 1952. Experimental nuclear physics; particle-gamma and ion-beam atomic spectroscopy.

Erickson, Glen W., Ph.D., University of Minnesota, 1960. Theoretical physics; quantum field theory.

Garrod, Claude, Ph.D., New York University, 1963. *Statistical & Thermal Physics.* Theoretical physics; quantum-mechanical many-particle systems; statistical mechanics.

Jungerman, John A., Ph.D., University of California, Berkeley, 1949. *Nuclear Physics.* Experimental nuclear physics; nuclear arms control.

Kiskis, Joseph E., Ph.D., Stanford University, 1974. *High Energy Physics.* Theoretical particle physics; lattice gauge theory.

Klein, Barry M., Ph.D., New York University, 1969. *Condensed Matter Physics.* Theoretical condensed matter.

Knox, William J., Ph.D., University of California, Davis, 1951. Experimental nuclear physics.

Ko, Winston T., Ph.D., University of Pennsylvania, 1971. Dean, Division of Mathematical and Physical Sciences. *High Energy Physics.* Experimental particle physics.

Lander, Richard L., Ph.D., University of California, Berkeley, 1958. *High Energy Physics.* Experimental particle physics.

McColm, Douglas W., Ph.D., Yale University, 1961. Experimental atomic physics; molecular beam studies.

Peek, Neal F., Ph.D., University of California, Davis, 1966. Experimental nuclear physics.

Pellett, David E., Ph.D., University of Michigan, 1966. Experimental particle physics.

Pines, David, Ph.D., Princeton University, 1950. *Condensed Matter Physics.* Theoretical condensed-matter physics.

Yager, Philip M., Ph.D., University of California, San Diego, 1973. Experimental particle physics.

Adjunct Professor

de Roeck, Albert, Ph.D., University of Antwerp, Belgium. Experimental particle physics.

Radousky, Harry B., Ph.D., University of Illinois, 1982. *Condensed Matter Physics.* Experimental condensed-matter physics.

Vogt, Ramona, Ph.D., Stony Brook University, 1989. *Nuclear Physics.* Theoretical nuclear physics.

Lecturer with Rank of Professor

Boeshaar, Patricia, Ph.D., Ohio State University, 1976. *Astrophysics.* Physics and astronomy education.

Webb, David J., Ph.D., University of Maryland, 1983. *Physics and other Science Education.* Physics education.

Lecturer

Cole, Rodney W., Ph.D., University of Wyoming, 1978.

Harris, Randy R., Ph.D., University of California, Davis, 1985.

Professional Specialist

Klavins, Peter, M.S., Iowa State University, 1987. Experimental materials physics.

Associate Professional Specialist

Gee, Perry A., M.S., University of California, Davis, 2004. Cosmology.

Project Physicist

Cox, Peter Timothy, Ph.D., University of Michigan, 1980. High-energy experimental physics.

Jee, Myungkook (James), Ph.D., Johns Hopkins University, 2005. *Astrophysics.* Astrophysics; gravitational lensing; x-ray analysis of galaxy clusters.

Smith, John R., Ph.D., University of California, Davis, 1982. *High Energy Physics.* High-energy experimental physics.

Researcher

Breedon, Richard, Ph.D., Rockefeller University, 1988. *High Energy Physics.* High-energy experimental physics.

Gregg, Michael D., Ph.D., Yale University, 1985. *Astronomy, Astrophysics, Cosmology & String Theory.* Cosmology.

Richter, Matthew J., Ph.D., University of California, Berkeley, 1995. *Astronomy, Astrophysics.* Cosmology.

Stanford, Spencer Adam, Ph.D., University of Wisconsin-Madison, 1990. *Astronomy, Astrophysics.* Cosmology.

Associate Researcher

Marchesini, Stefano, Ph.D., University J. Fourier, Grenoble, 2000. Physics; X-ray imaging.

Assistant Researcher

Scranton, Ryan E., Ph.D., University of Chicago, 2002. *Astronomy, Astrophysics, Cosmology & String Theory.* Cosmology.

DEPARTMENTAL RESEARCH SPECIALTIES AND STAFF

Theoretical

Computational Physics. Computational simulations of complex physical and biological systems; data mining of massive and multidimensional data sets; understanding emergent patterns and coherent structures in nonlinear systems; prediction and forecasting; scaling; network and cluster formation and dynamics. Crutchfield, Galli, Rundle.

Condensed Matter Physics. Studies of macroscopic phases of matter, including metals, insulators, superconductors, superfluids, magnets, spin-glasses, supersolids, etc.; thermal and quantum-phase transitions and critical phenomena; microscopic theory and phenomenology of high-temperature superconductivity and of superconductivity coexisting with magnetism; microscopic studies of strongly correlated electron systems; Hubbard, t-J, and Heisenberg models; mesoscopic physics and nanostructures; electronic structure calculations. Quantum Monte Carlo and other advanced numerical methods are used to study alloy phase stability, semiconductor nanostructures, spintronic materials, and surface physics properties from first principles. Daniel Cox, Fong, Galli, Pickett, Pines, Savrasov, Scalettar, Singh, Zimanyi.

Cosmology & String Theory. Physics of the early universe and the formation of cosmic structure; cosmic inflation; dark-energy models and probes; cosmic microwave background; precision cosmology. Albrecht, Gregg, Kaloper, Lloyd Knox, Scranton.

Nuclear Physics. Relativistic heavy-ion collisions; quark-gluon plasma. Vogt.

Particles and Fields. Gauge theories of the electromagnetic, weak, and strong interactions; perturbative and nonperturbative analysis; lattice gauge theory; unification, supersymmetry; extra dimension; symmetry breaking; phenomenology; heavy quark systems; soluble models; quantum gravity. Cheng, Gunion, Kiskis, Luty, Terning.

Relativity & Gravitation. Conceptual issues in quantum gravity; low-dimensional models for quantum gravity; black hole thermodynamics; physical and mathematical foundations of general relativity. Carlip.

Experimental

Astrophysics. Physics of the early universe; evolution of active galactic nuclei (AGN); quasar absorption line systems; clusters of galaxies; gravitational lenses; large-scale surveys. Becker, Boeshaar, Bradac, Fassnacht, Ferenc, Gregg, Lubin, Stanford, Tyson, Wittman.

Condensed Matter Physics. Surfaces and interfaces; magnetism; low-temperature physics; lattice dynamics; quantum fluctuations; transport properties; high-temperature superconductivity; granular materials; light scattering on biological materials and nanostructured materials. Systems under study include atoms, molecules, and nanostructures on surfaces; nanoparticles and nanowires; epitaxial multilayer thin films; patterned nano-structures; organic and biological nanostructures; 4 He vortices; dipolar magnets and frustrated antiferromagnets; exotic superconductors; complex magnetic and ferroelectric materials; and disordered media. Sample synthesis and processing are by magnetron sputtering; thermal and e-beam evaporation; electrodeposition; high-temperature sintering and single-crystal growth; and photo- and e-beam lithography. Characterizations are by photoelectron spectroscopy, diffraction, and holography using laboratory sources and high-brightness synchrotron radiation at the nearby Lawrence Berkeley National Laboratory; x-ray diffraction; scanning electron microscopy; ultra-high-vacuum scanning tunneling and atomic force microscopy; low-energy electron microscopy and diffraction; surface magneto-optical Kerr effect; Raman spectroscopy; far infrared spectroscopy; SQUID and local Hall probe magnetometry; metallographic and thermal analyses; semi-adiabatic calorimetry from dilution refrigerator range to high temperatures; transport measurements; and vibrating wire studies. Chiang, Coleman, Corruccini, Curro, Fadley, Klavins, Liu, Yu, Zhu, Zieve.

Nuclear Physics. Study of relativistic heavy-ion collisions; creation of quark-antiquark pairs; discovery and study of quark matter (or quark-gluon plasma, as it is often called). Brady, Calderón de la Barca Sánchez, Cebra, Draper, Ferenc.

Particles and Fields. Experiments using particle accelerators and colliders around the world to probe the structure of matter at the deepest possible level; design and construction of neutrino and WIMP dark-matter detectors to search for dark matter and make precision measurements of neutrino properties; properties of and the forces between quarks and leptons and searches for new particles; extensive computer utilization for data analysis and design modeling; development of new detectors such as micron-size silicon pixel devices; design and fabricate read-out electronics using new techniques such as ASIC (application specific integrated circuit) and flip chip bump bonding. Breedon, Chertok, Conway, de Roeck, Erbacher, Ko, Lander, Mulhearn, Pellett, Smith, Svoboda, Tripathi.

Physics and other Science Education. The process of how students come to an understanding of physics/science concepts; instructional models; use of visual models in understanding atomic phenomena. Potter, Webb.

View additional information about this department at
www.gradschoolshopper.com

UNIVERSITY OF CALIFORNIA, IRVINE

DEPARTMENT OF PHYSICS AND ASTRONOMY

Irvine, California 92697-4575
http://www.physics.uci.edu/

General University Information
President: Chancellor Michael Drake
Dean of Graduate School: Frances Leslie
University website: http://www.uci.edu/
Control: Public
Setting: Urban
Total Faculty: 2,883
Total number of Students: 28,184
Total number of Graduate Students: 5,875

Department Information
Department Chairman: Peter Taborek, Chair
Department Contact: Kirsten Lodgard, Manager, Student Affairs
Total full-time faculty: 47
Full-Time Graduate Students: 133
First-Year Graduate Students: 22
Female First-Year Students: 3
Total Post Doctorates: 32

Department Address
4129 Frederick Reines Hall
Irvine, CA 92697-4575
Phone: (949) 824-6911
Fax: (949) 824-2174
E-mail: physics@uci.edu
Website: http://www.physics.uci.edu/

ADMISSIONS

Admission Contact Information
Address admission inquiries to: University of California, Irvine, Graduate Admissions Committee, Department of Physics and Astronomy, 4129 Frederick Reines Hall, Irvine, CA 92697-4575
Phone: (949) 824-3496
E-mail: physgrad@uci.edu
Admissions website: http://www.physics.uci.edu/graduate

Application deadlines
Fall admission:
U.S. students: January 1   *Int'l. students*: January 1

Application fee
U.S. students: $80   *Int'l. students*: $100

Admissions information
For Fall of 2013:
Number of applicants: 298
Number admitted: 75
Number enrolled: 26

Admission requirements
Bachelor's degree requirements: A Bachelor's degree is required.
Minimum undergraduate GPA: 3.0

GRE requirements
The GRE is required.

Advanced GRE requirements
The Advanced GRE is required.

TOEFL requirements
The TOEFL exam is required for students from non-English-speaking countries.
PBT score: 550
iBT score: 80

Other admissions information
Additional requirements: No minimum scores on the General GRE and Physics GRE are specified.
Undergraduate preparation assumed: The following texts (or their equivalents): Halliday & Resnick, General Physics; Marion, Mechanics; Griffiths, E and M; Liboff, Quantum Mechanics; Reif, Thermal and Statistical Physics; Arfken, Mathematics for Physics; and advanced undergraduate laboratory.

TUITION

Tuition year 2012–13:
Tuition for in-state residents
  Full-time students: $15,049.5 annual
  Part-time students: $9,439.5 annual
Tuition for out-of-state residents
  Full-time students: $30,151.5 annual
  Part-time students: $16,990.5 annual
Credit hours per semester to be considered full-time: 12
Deferred tuition plan: No
Health insurance: Yes, 2088.00.

Teaching Assistants, Research Assistants, and Fellowships
Number of first-year
  Teaching Assistants: 22
  Fellowship students: 20
Average stipend per academic year
  Teaching Assistant: $17,655
  Research Assistant: $17,244

FINANCIAL AID

Application deadlines
Fall admission:
U.S. students: March 2

Loans
Loans are available for U.S. students.
Loans are not available for international students.
GAPSFAS application required: No
FAFSA application required: Yes

For further information
Address financial aid inquiries to: Financial Aid and Scholarships office, 102 Aldrich Hall, Irvine, CA 92697-2825.
Phone: (949) 824-8262
E-mail: finaid@uci.edu
Financial aid website: http://www.ofas.uci.edu/content/default.aspx

HOUSING

Availability of on-campus housing
  Single students: Yes
  Married students: Yes

For further information
Address housing inquiries to: Housing Outreach Services, G465 Student Center, Irvine, CA 92697-5225.
Phone: (949) 824-6375

E-mail: housing@uci.edu
Housing aid website: http://www.housing.uci.edu

Table A—Faculty, Enrollments, and Degrees Granted

Research Specialty	2012-13 Faculty	Enrollment Fall 2012 Master's	Enrollment Fall 2012 Doctorate	Number of Degrees Granted 2012-13 (2008-13) Master's	Number of Degrees Granted 2012-13 (2008-13) Terminal Master's	Number of Degrees Granted 2012-13 (2008-13) Doctorate
Astrophysics	12	–	24	1(10)	2(4)	1(12)
Atmosphere, Space Physics, Cosmic Rays	–	–	–	–	–	–(3)
Biophysics	6	–	9	2(9)	–(2)	2(6)
Chemical Physics	–	–	39	3(16)	–	4(20)
Condensed Matter Physics	19	–	23	1(16)	–	7(17)
Medical, Health Physics	–	–	3	–(4)	–(2)	1(2)
Particles and Fields	16	–	26	3(20)	2(2)	4(24)
Plasma and Fusion	5	–	9	–(9)	–	–(10)
Total	58	–	133	10(84)	4(10)	19(94)
Full-time Grad. Stud.	–	–	133	–	–	–
First-year Grad. Stud.	–	–	22	–	–	–

GRADUATE DEGREE REQUIREMENTS

Master's: The requirements for the M.S. degree are: (1) at least three quarters of residence; (2) mastery of graduate course material, which must be demonstrated by passing, with a grade of "B" or better, a minimum of eight quarter courses including 211, 213A-B, 214A, 215A, 223, and two other courses approved by the graduate advisor, which can include undergraduate upper-division courses in related areas; and (3) either option A, a research project and written thesis, or option B, a comprehensive written examination. Students pursuing option A typically complete three quarters of research, enrolling in Physics 295 or 296. Students following option B should take Physics 215B. There is no foreign language requirement. For option A, the thesis need be of no specified length or format, but must report significant results in readable, meaningful form while at the same time revealing the student's general grasp of the field and awareness of related work. For option B, the comprehensive examination for the M.S. degree is identical to that for the Ph.D. degree. The level of performance required for the M.S. degree by examination is identical to the required level of performance for the Ph.D. degree.

Doctorate: The principal requirements for the Ph.D. degree are a minimum of six quarters of residence, passage of a written and an oral examination, and successful completion and defense of a dissertation reporting results of original research. In addition, the Ph.D. candidate must complete certain graduate course requirements. Experience in teaching is an integral part of the graduate program, and all Ph.D. students are required to participate in the teaching program for at least three quarters during their graduate careers. Foreign students must pass a campus-approved spoken English proficiency examination (required in order to be a teaching assistant) by the time they advance to candidacy. There is no foreign language requirement for the Ph.D. degree.

Thesis: The thesis need be of no specified length or format, but must report significant results in readable, meaningful form while at the same time revealing the student's general grasp of the field and awareness of related work.

Table B—Separately Budgeted Research Expenditures by Source of Support

Source of Support	Departmental Research	Physics-related Research Outside Department
Federal government		$11,216,733.25
State/local government		$97,766
Non-profit organizations		
Business and industry		$1,541,502.11
Other		
Total		$12,856,001.36

Table C—Separately Budgeted Research Expenditures by Research Specialty

Research Specialty	No. of Grants	Expenditures ($)
Astrophysics	99	$2,055,016
Condensed Matter Physics	90	$4,102,524
Medical, Health Physics	24	$742,832
Particles and Fields	73	$5,119,411
Plasma and Fusion	39	$1,729,722
Total	325	$13,749,505

FACULTY

Professor

Barth, Aaron, Ph.D., University of California, Berkeley, 1998. Observational astrophysics.

Barwick, Steven, Ph.D., University of California, Berkeley, 1986. Particle astrophysics.

Bullock, James, Ph.D., University of California, Santa Cruz, 1999. Theoretical cosmology; astrophysics; astronomy.

Buote, David, Ph.D., Massachusetts Institute of Technology, 1995. Observational astrophysics; cosmology.

Burke, Kieron, Ph.D., University of California, Santa Barbara, 1989. Theoretical condensed-matter physics.

Chanan, Gary A., Ph.D., University of California, Berkeley, 1978. Observational astrophysics.

Chernyshev, Alexander L., Ph.D., Novosibirsk, 1995. Theoretical condensed-matter physics.

Cooray, Asantha, Ph.D., University of Chicago, 2001. Theoretical cosmology; astrophysics; planetary science.

Dennin, Michael, Ph.D., University of California, Santa Barbara, 1995. Experimental condensed-matter physics; biological physics.

Feng, Jonathan L., Ph.D., Stanford University, 1995. Elementary particle theory; cosmology.

Fisk, Zachary, Ph.D., University of California, San Diego, 1969. Experimental condensed-matter physics.

Gratton, Enrico, Ph.D., University of Rome, 1969. Biomedical physics.

Gross, Steven, Ph.D., University of Texas, Austin, 1995. Experimental biological physics.

Hamber, Herbert, Ph.D., University of California, Santa Barbara, 1980. Elementary particle theory; general relativity.

Heidbrink, William, Ph.D., Princeton University, 1984. Experimental plasma physics.

Ho, Wilson, Ph.D., University of Pennsylvania, 1979. Experimental condensed-matter physics; chemistry.

Kaplinghat, Manoj, Ph.D., Ohio State University, 1999. Theoretical cosmology; astrophysics.

Kirkby, David P., Ph.D., California Institute of Technology, 1995. Experimental particle physics.

Lankford, Andrew, Ph.D., Yale University, 1978. Experimental particle physics.

Lin, Zhihong, Ph.D., Princeton University, 1996. Theoretical plasma physics.

McWilliams, Roger D., Ph.D., Princeton University, 1980. Experimental plasma physics.

Molzon, William, Ph.D., University of Chicago, 1979. Experimental particle physics.

Rajaraman, Arvind, Ph.D., Stanford University, 1998. Elementary particle theory.

Rutledge, James, Ph.D., University of Illinois, 1978. Experimental condensed-matter physics.

Shirman, Yuri, Ph.D., University of California, Santa Cruz, 1997. Theoretical particle physics.

Siwy, Zuzana, Ph.D., Silesian Univ. of Tech., 1997. Experimental biological physics; condensed-matter physics.

Sobel, Henry, Ph.D., Case Institute of Technology, 1968. Experimental particle physics.

Su, Lydia, Ph.D., University of California, Irvine, 1993. Radiological sciences; medical physics.

Taborek, Peter, Ph.D., California Institute of Technology, 1980. Experimental condensed-matter physics.

Tait, Timothy, Ph.D., Michigan State University, 1999. Theoretical particle physics.

Trimble, Virginia L., Ph.D., California Institute of Technology, 1968. Theoretical astronomy.

White, Steven, Ph.D., Cornell University, 1988. Theoretical condensed-matter physics.

Wu, Ruqian, Ph.D., Peking University, 1989. Theoretical condensed-matter physics.

Yu, Clare, Ph.D., Princeton University, 1984. Theoretical condensed-matter physics; biological physics.

Associate Professor

Abazajian, Kevork, Ph.D., University of California, San Diego, 2001. Theoretical cosmology.

Casper, David W., Ph.D., University of Michigan, 1990. Experimental particle physics.

Chen, Mu-Chun, Ph.D., University of Colorado Boulder, 2002. Theoretical particle physics.

Collins, Philip G., Ph.D., University of California, Berkeley, 1998. Experimental condensed-matter physics.

Gulsen, Gultekin, Ph.D., Bogazici University, 1999. Experimental biological physics.

Krivorotov, Ilya, Ph.D., University of Minnesota, 2002. Experimental condensed-matter physics.

Ritz, Thorsten, Ph.D., University of Ulm, 2001. Theoretical biological physics; theoretical condensed-matter physics.

Smecker-Hane, Tammy, Ph.D., Johns Hopkins University, 1993. Observational astrophysics; astronomy.

Taffard, Anyes, Ph.D., University of Liverpool, 2002. Experimental particle physics.

Whiteson, Daniel, Ph.D., University of California, Berkeley, 2003. Experimental particle physics.

Assistant Professor

Cooper, Michael, Ph.D., University of California, Berkeley, 2007. Observational astrophysics and cosmology.

Murgia, Simona, Ph.D., Michigan State University, 2002. *Astrophysics, Particles and Fields.* Experimental particle physics; astrophysics.

Xia, Jing, Ph.D., Stanford University, 2008. Experimental condensed-matter physics.

Research Professor

Chen, Liu, Ph.D., University of California, Berkeley, 1972. Theoretical plasma physics.

Maradudin, Alexei A., Ph.D., University of Bristol, 1957. Theoretical condensed-matter physics.

Newman, Riley, Ph.D., University of California, Berkeley, 1966. Experimental particle physics; gravitational physics.

Parker, William H., Ph.D., University of Pennsylvania, 1967. Experimental condensed-matter physics.

Rostoker, Norman, Carnegie Institute of Technology, 1950. Plasma physics.

Yodh, Gaurang, Ph.D., University of Chicago, 1955. Experimental particle astrophysics.

Adjunct Faculty

Tajima, Toshiki, Ph.D., University of Califorina, Irvine, 1975. Plasma physics.

Vagins, Mark, Ph.D., Yale University, 1994. Experimental particle physics.

Researcher

Boehmer, Heinrich, Ph.D., Muenster, 1961. Experimental plasma physics.

Bystritskii, Vitaly, Ph.D., Institute of High Current Electronics, 1977. Experimental plasma physics.

Garate, Eusebio, Ph.D., Dartmouth College, 1984. Experimental plasma physics.

Kropp, William, Ph.D., Case Institute of Technology, 1964. Experimental particle physics.

Assistant Researcher

Fang, Taotao, Ph.D., Massachusetts Institute of Technology, 2001. *Astrophysics*. Astrophysics.

Smy, Michael, Ph.D., Colorado State University, 1997. Experimental particle physics.

Other

Mine, Shunichi, Ph.D., University of Tokyo, 1996. Associate Project Scientist,. Experimental particle physics.

DEPARTMENTAL RESEARCH SPECIALTIES AND STAFF

Theoretical

Astrophysics. Galaxy formation and evolution; large-scale structure; cosmology; high energy astrophysics; dark matter; the early universe.

Biophysics.

Condensed Matter Physics. Surfaces, superlattices, and ultrathin films; electromagnetic interactions with solids; lattice dynamics; semiconductors.

Cosmology & String Theory.

High Energy Physics.

History & Philosophy of Physics/Science.

Medical, Health Physics.

Plasma and Fusion. Wave and particle dynamics in plasmas; non-linear theories and large-scale numerical simulations.

Experimental

Astrophysics. Optical, infrared, and x-ray astronomy; high-energy astrophysics; galaxy evolution; quasars and active galactic nuclei; cosmic background radiation; black holes and dark matter, large telescope optics.

Biophysics. Photosynthesis, magnetonavigation, nanopores, cancer, systems biology.

Condensed Matter Physics. Electronic, mechanical, and optical properties of solids and liquids, especially at interfaces or at the nanoscale; surface physics; materials science; superconductivity, magnetism, and topological phases of matter; spintronics; nanotechnology.

High Energy Physics.

Medical, Health Physics.

Plasma and Fusion. Relativistic electron beams; fast ion dynamics; plasma waves and turbulence; collective accelerators.

View additional information about this department at
www.gradschoolshopper.com

UNIVERSITY OF CALIFORNIA, LOS ANGELES

DEPARTMENT OF PHYSICS AND ASTRONOMY

Los Angeles, California 90095-1547
http://www.physics.ucla.edu

General University Information

President: Gene D. Block
Dean of Graduate School: Robin L. Garrell
University website: http://www.ucla.edu
Control: Public
Setting: Urban
Total Faculty: 2,564
Total number of Students: 41,341
Total number of Graduate Students: 13,400

Department Information

Department Chairman: James Rosenzweig, Chair
Department Contact: Cecile Chang, Academic Personnel Assistant
Total full-time faculty: 65

Total number of full-time equivalent positions: 62
Full-Time Graduate Students: 160
First-Year Graduate Students: 30
Female First-Year Students: 8
Total Post Doctorates: 47

Department Address

430 Portola Plaza
M/C 154705
Los Angeles, CA 90095-1547
Phone: (310) 825-3403
Fax: (310) 206-0864
E-mail: cfchang@physics.ucla.edu
Website: http://www.physics.ucla.edu

ADMISSIONS

Admission Contact Information
Address admission inquiries to: Physics and Astronomy Department, Graduate Affairs Office, 1-707B PAB, Los Angeles, CA 90095-1547
Phone: (310) 825-2307
E-mail: apply@Physics.ucla.edu
Admissions website: http://www.physics.ucla.edu

Application deadlines
Fall admission:
U.S. students: December 15   *Int'l. students*: December 15

Application fee
U.S. students: $80   *Int'l. students*: $100

Admissions information
For Fall of 2013:
  Number of applicants: 434
  Number admitted: 143
  Number enrolled: 44

Admission requirements
Bachelor's degree requirements: A Bachelor's degree in physics is required.
Minimum undergraduate GPA: 3.0

GRE requirements
The GRE is required.

Advanced GRE requirements
The Advanced GRE is required.

TOEFL requirements
The TOEFL exam is required for students from non-English-speaking countries.
  PBT score: 570
  iBT score: 88

Other admissions information
Additional requirements: There is no minimum established for GRE scores. Students from non-English speaking countries are required to demonstrate proficiency in UCLA entrance exams in English. Depending on score, student may be required to take one or more ESL (English as a second language) courses. TSE (Test of Spoken English) is strongly recommended.
Undergraduate preparation assumed: Marion & Thornton, Classical Dynamics of Particles and Systems; Griffiths, Introduction to Quantum Mechanics and/or Gasiorowicz, Quantum Physics; Arfken, Mathematical Methods for Physicists; Wong, Introduction to Mathematical Physics; Griffiths, Introduction to Electrodynamics; Wangness, Electromagnetic Fields; Kittel, Thermal Physics (texts in current use).

TUITION

Tuition year 2012–13:
Tuition for in-state residents
  Full-time students: $4,873 per quarter
Tuition for out-of-state residents
  Full-time students: $9,907 per quarter
Credit hours per semester to be considered full-time: 12
Deferred tuition plan: No
Health insurance: Available at the cost of 2243 per year.

Academic term: Quarter
Number of first-year students who received full tuition waivers: 22

Teaching Assistants, Research Assistants, and Fellowships
Number of first-year
  Teaching Assistants: 22
  Fellowship students: 22
Average stipend per academic year
  Teaching Assistant: $17,655
  Research Assistant: $17,994
  Fellowship student: $15,000

FINANCIAL AID

Loans
Loans are available for U.S. students.
Loans are not available for international students.
GAPSFAS application required: No
FAFSA application required: Yes

For further information
Address financial aid inquiries to:
Financial aid website: http://www.fao.ucla.edu

HOUSING

Availability of on-campus housing
  Single students: Yes
  Married students: Yes

For further information
Address housing inquiries to: www.housing.ucla.edu.
Phone: (310) 825-3401
Housing aid website: http://ask.housing.ucla.edu

Table A—Faculty, Enrollments, and Degrees Granted

Research Specialty	2012-13 Faculty	Enrollment Fall 2012 Master's	Enrollment Fall 2012 Doctorate	Number of Degrees Granted 2012-13 (2008-13) Master's	Number of Degrees Granted 2012-13 (2008-13) Terminal Master's	Number of Degrees Granted 2012-13 (2008-13) Doctorate
Accelerator	2	–	13	–	–	4(7)
Astronomy	–	–	23	4(15)	–(2)	4(22)
Astrophysics	21	–	6	–	–	2(12)
Atmosphere, Space Physics, Cosmic Rays	1	–	2	–	–	–
Atomic, Molecular, & Optical Physics	1	–	4	–	–	–(6)
Biophysics	6	–	8	–	–	–(6)
Condensed Matter Physics	13	–	31	–	–	5(32)
High Energy Physics	8	–	13	–	–	3(11)
Low Temperature Physics	1	–	2	–	–	2(2)
Nuclear Physics	1	–	3	–	–	2(7)
Particles and Fields	10	–	13	–	–	–(15)
Plasma and Fusion	8	–	17	–	–	3(20)
Non-specialized	–	–	23	16(67)	1(1)	–
Total	71	–	158	20(82)	1(3)	25(134)
Full-time Grad. Stud.	–	–	158	–	–	–
First-year Grad. Stud.	–	–	23	–	–	–

GRADUATE DEGREE REQUIREMENTS

Master's: The Department does not offer a terminal Master's program. The M.S. degree is awarded to students in the Ph.D. program after satisfying a minimum course requirement of 9–11 courses, 6 of which are graduate level physics or astronomy courses, and passing the comprehensive examination at the Master's level of achievement. The residence requirement for the M.S. degree is as follows: minimum period of residence for one academic year, of which at least two quarters must be spent on the Los Angeles campus. There is no foreign language requirement. A "B" average in physics or astronomy and an overall "B" average in all courses taken in graduate status are required.

Doctorate: Physics Doctorate A written comprehensive examination is required and must be passed at the Ph.D. level of performance. It is based on the contents of the six core courses*, 210A, 210B, 215A, 220, 221A, 221B. Students must also take two elective courses that my be upper division or at the graduate level and pass with "B" or better. A qualifying dissertation oral and final oral are required. The minimum residence requirement for the Ph.D. is two academic years (six quarters) at UCLA, one of which, ordinarily the second, must be spent in continuous residence. A "B" average in physics and an overall "B" average in all courses taken in graduate status are required. No language requirement. Astronomy Doctorate Students are expected to fulfill the usual university requirements for a dissertation, and to pass oral, preliminary, and final exams. Students are required to take 7 core courses and two elective courses, and one 2-quarter research project during the second year. The comprehensive exam occurs during spring of the second year. One year working as a teaching assistant is required. *Title of Physics Graduate Courses: 210A,B Electromagnetic Theory 215A Statistical Mechanics. 220 Classical Mechanics. 221A,B Quantum Mechanics.

Thesis: Thesis may not be written in absentia.

SPECIAL EQUIPMENT, FACILITIES, OR PROGRAMS

The Astronomy Division supports a major program to develop astronomical instrumentation using infrared techniques and encourages graduate student participation. The main off-campus facilities are the Lick Observatory, the Keck Observatory, and solar facilities at Mt. Wilson. Graduate students have also used the national observatories(Kitt Peak, CTIO, NRAO), other major observatories (OVRO, CSO), and satellites (IRAS, HST, ISO, COBE, and ROAST) to obtain data.

Table B—Separately Budgeted Research Expenditures by Source of Support

Source of Support	Departmental Research	Physics-related Research Outside Department
Federal government	$25,540,916	
State/local government		
Non-profit organizations	$5,137,648	
Business and industry	$54,264	
Other	$839,241	
Total	$31,572,069	

Table C—Separately Budgeted Research Expenditures by Research Specialty

Research Specialty	No. of Grants	Expenditures ($)
Astronomy	86	$4,272,028
Condensed Matter Physics	49	$3,375,929
Energy Sources & Environment	23	$14,130,826
Particles and Fields	24	$3,759,533
Plasma and Fusion	51	$6,033,752
Total	233	$31,572,068

FACULTY

Professor

Arisaka, Katsushi, Ph.D., University of Tokyo, 1985. *High Energy Physics, Particles and Fields*. High-energy experiment.

Ashour-Abdalla, Maha, Ph.D., Imperial College, 1971. *Astrophysics, Atmosphere, Space Physics, Cosmic Rays, Plasma and Fusion*. Plasma theory.

Bern, Zvi, Ph.D., University of California, Berkeley, 1986. *Particles and Fields*. Elementary particle theory.

Brown, Stuart, Ph.D., University of California, Los Angeles, 1988. *Condensed Matter Physics*. Condensed matter experiment.

Bruinsma, Robijn, Ph.D., University of Southern California, 1979. *Biophysics, Condensed Matter Physics*. Condensed matter theory.

Carter, Troy, Ph.D., Princeton University, 2001. *Plasma and Fusion*. Experimental plasma physics.

Chakravarty, Sudip, Ph.D., Northwestern University, 1976. *Condensed Matter Physics*. Condensed matter theory.

Cline, David, Ph.D., University of Wisconsin-Madison, 1965. *Astronomy, Astrophysics, High Energy Physics, Particles and Fields*. Elementary particle experiment.

Coroniti, Ferdinand, Ph.D., University of California, Berkeley, 1969. Associate Dean, Div. of Physical Sciences. *Astronomy, Astrophysics, Atmosphere, Space Physics, Cosmic Rays, Plasma and Fusion*.

Cousins, Robert, Ph.D., Stanford University, 1981. *High Energy Physics, Particles and Fields*. High-energy experiment.

D'Hoker, Eric, Ph.D., Princeton University, 1981. Vice Chair, Academic Affairs. *Particles and Fields*. Elementary particle theory.

Ferrara, Sergio, Ph.D., University of Rome, 1968. *Particles and Fields*. Elementary particle theory.

Fronsdal, Christian, Ph.D., University of California, Los Angeles, 1957. *Particles and Fields*. Field theory.

Furlanetto, Steven, Ph.D., Harvard University, 2003. *Astronomy, Astrophysics*. Theoretical Astrophysics.

Gekelman, Walter, Ph.D., Stevens Institute of Technology, 1972. *Plasma and Fusion*. Plasma experiment.

Gelmini, Graciela, Ph.D., Universidad Nacional de La Plata, 1981. *Particles and Fields*. Elementary particle theory.

Ghez, Andrea, Ph.D., California Institute of Technology, 1992. *Astronomy, Astrophysics*. Astrophysics.

Grüner, George, Ph.D., Hungarian Academy of Science, 1977. *Biophysics, Condensed Matter Physics*. Condensed matter experiment.

Gutperle, Michael, Ph.D., University of Cambridge, 1997. *Particles and Fields*. Elementary particle theory.

Hansen, Bradley, Ph.D., California Institute of Technology, 1996. *Astronomy, Astrophysics*. Theoretical astrophysics.

Hauser, Jay, Ph.D., California Institute of Technology, 1985. *High Energy Physics, Particles and Fields*. High energy experiment.

Holczer, Karoly, Ph.D., Eotvos Lorand University, 1977. *Condensed Matter Physics*. Condensed matter experiment.

Huang, Huan Z., Ph.D., Massachusetts Institute of Technology, 1990. *Nuclear Physics*. Nuclear experiment.

Jewitt, David, Ph.D., California Institute of Technology, 1983. *Astronomy, Astrophysics*. Astronomy, Planetary Sciences.

Jiang, Hong-Wen, Ph.D., Case Western Reserve University, 1989. *Condensed Matter Physics*. Condensed matter experiment.

Jura, Michael A., Ph.D., Harvard University, 1971. *Astronomy, Astrophysics*. Physics of interstellar medium; mass loss.

Kraus, Per, Ph.D., Princeton University, 1995. *Particles and Fields*. Elementary particle theory.

Kusenko, Alexander, Ph.D., State University of New York, 1994. *Particles and Fields*. Elementary particle theory.

Larkin, James, Ph.D., California Institute of Technology, 1995. *Astronomy, Astrophysics*.

Levine, Alexander, Ph.D., University of California, Los Angeles, 1996. *Biophysics*.

Malkan, Matthew A., Ph.D., California Institute of Technology, 1983. *Astronomy, Astrophysics*. Quasars and active galaxies: infrared, optical, ultraviolet, and x-ray observations of their line and continuum emission: models of accretion disks around massive black holes.

Margot, Jean-Luc, Ph.D., Cornell University, 1999. *Astronomy, Geophysics*.

Mason, Thomas, Ph.D., Princeton University, 1995. *Condensed Matter Physics*. Soft condensed matter.

McLean, Ian, Ph.D., University of Glasgow, 1974. Vice Chair, Astronomy & Astrophysics. *Astronomy, Astrophysics*.

Mehta, Mayank, Ph.D., Indian Institute of Science, Bangalore, 1993. *Biophysics*. Neurophysics.

Miao, Jianwei, Ph.D., Stony Brook University, 1999. *Condensed Matter Physics*. Soft condensed matter.

Morales, George, Ph.D., University of California, San Diego, 1973. *Atmosphere, Space Physics, Cosmic Rays, Plasma and Fusion*. Plasma theory.

Mori, Warren, Ph.D., University of California, Los Angeles, 1987. *Atmosphere, Space Physics, Cosmic Rays, Plasma and Fusion*. Plasma theory.

Morris, Mark R., Ph.D., University of Chicago, 1975. *Astronomy, Astrophysics*. Mass loss envelopes around red giants; The Galactic Center.

Newman, William, Ph.D., Cornell University, 1979. *Astronomy, Astrophysics*. Astrophysics.

Ong, Rene, Ph.D., Stanford University, 1987. *Astronomy, Astrophysics, High Energy Physics, Particles and Fields*. Astroparticle physics.

Patel, C. Kumar N., Ph.D., Stanford University, 1961. *Condensed Matter Physics*. Condensed matter experiment.

Peccei, Roberto, Ph.D., Massachusetts Institute of Technology, 1969. *Particles and Fields*. Elementary particle theory.

Putterman, Seth, Ph.D., Rockefeller University, 1970. *Condensed Matter Physics, Low Temperature Physics*. Condensed matter experiment.

Rosenzweig, James, Ph.D., University of Wisconsin-Madison, 1988. Chair of the Department. *Plasma and Fusion*.

Rudnick, Joseph, Ph.D., University of California, San Diego, 1970. Dean, Division of Physical Sciences. *Biophysics, Condensed Matter Physics*.

Saltzberg, David, Ph.D., University of Chicago, 1994. Vice Chair, Resources. *High Energy Physics, Particles and Fields*. Elementary particle experiment.

Shapley, Alice, Ph.D., California Institute of Technology, 2003. *Astronomy, Astrophysics*. Experimental Astrophysics.

Tomboulis, E. Terry, Ph.D., Massachusetts Institute of Technology, 1976. *Particles and Fields*. Elementary particle theory.

Tserkovnyak, Yaroslav, Ph.D., Harvard University, 2004. *Condensed Matter Physics*. Condensed matter theory.

Turner, Jean L., Ph.D., University of California, Berkeley, 1984. *Astronomy, Astrophysics*. Star formation; radio emission from starburst galaxies; molecular line studies of young stellar objects.

Vassiliev, Vladimir, Ph.D., University of Minnesota, 1997. *Astrophysics*. Astroparticle physics.

Williams, Gary, Ph.D., University of California, Berkeley, 1974. *Condensed Matter Physics, Low Temperature Physics*. Low-temperature experiment.

Wright, Edward L., Ph.D., Harvard University, 1976. *Astronomy, Astrophysics*. Cosmic background radiation; infrared astronomy; relativity.

Zocchi, Giovanni, Ph.D., University of Chicago, 1990. *Biophysics, Condensed Matter Physics*. Biophysics.

Associate Professor

Bozovic, Dolores, Ph.D., Harvard University, 2001. *Biophysics, Condensed Matter Physics*. Soft condensed matter.

Musumeci, Pietro, Ph.D., University of California, Los Angeles, 2004. *Particles and Fields, Plasma and Fusion*. Elementary Particle experiment/accelerator.

Niemann, Christoph, Ph.D., University of Technology, Darmstadt, 2002. *Plasma and Fusion*. Plasma physics.

Assistant Professor

Campbell, Wesley, Ph.D., Harvard University, 2008. *Atomic, Molecular, & Optical Physics*.

Fitzgerald, Michael, Ph.D., University of California, Berkeley, 2007. *Astronomy, Astrophysics*. Astrophysics.

Hudson, Eric, Ph.D., University of Colorado, 2006. *Atomic, Molecular, & Optical Physics*. Atomic, Molecular and Optical Physics.

Ni, Ni, Ph.D., Iowa State University, 2009. *Condensed Matter Physics*.

Regan, B. Chris, Ph.D., University of California, Berkeley, 2001. *Condensed Matter Physics*. Condensed matter.

Roy, Rahul, Ph.D., University of Illinois at Urbana-Champaign, 2007. *Condensed Matter Physics*.

Winslow, Lindley, Ph.D., University of California, Berkeley, 2008. *High Energy Physics, Nuclear Physics, Particles and Fields*.

DEPARTMENTAL RESEARCH SPECIALTIES AND STAFF

Theoretical

Astrophysics. Stellar structure; stellar atmospheres; plasma astrophysics; binary star evolution; high-energy astrophysics; interstellar and circumstellar processes. Ashour-Abdalla, Cline, Coroniti, Furlanetto, Ghez, Hansen, Jura, Malkan, Newman, Wright.

Biophysics. Bruinsma, Rudnick.

Condensed Matter Physics. Solid State and Statistical Mechanics. Bruinsma, Chakravarty, Roy, Rudnick, Tserkovnyak.

Low Temperature Physics. Putterman.

Experimental

Astrophysics. External galaxies and quasars; galactic nuclei; binary stars; star forming regions; nebulae; the Sun; x-rays; ultraviolet; optical; infrared; radio astronomical instrumentation; polarization. Fitzgerald, Ghez, Larkin, Malkan, McLean, Morris, Ong, Shapley, Turner, Vassiliev, Wright.

Atomic, Molecular, & Optical Physics. Campbell, Hudson.

Biophysics. Grüner, Mehta, Zocchi.

Condensed Matter Physics. Solid State Physics and Statistical Mechanics. Brown, Grüner, Holczer, Jiang, Mason, Miao, Ni, Patel, Putterman, Regan.

High Energy Physics. High Energy & Particle Physics. Arisaka, Cline, Cousins, Hauser, Ong, Saltzberg, Winslow.

Nuclear Physics. Huang, Winslow.

Other. Accelerator. Musumeci, Rosenzweig.

Particles and Fields. Bern.

Plasma Physics, Space Physics, Astrophysics. Ashour-Abdalla, Coroniti, Morales, Mori.

View additional information about this department at
www.gradschoolshopper.com

UNIVERSITY OF CALIFORNIA, RIVERSIDE

DEPARTMENT OF PHYSICS AND ASTRONOMY

Riverside, California 92521
http://www.physics.ucr.edu

General University Information
Chancellor: Jane Close Conoley
Dean of Graduate School: Joseph Childers
University website: http://www.ucr.edu
Control: Public
Setting: Suburban
Total Faculty: 658
Total Graduate Faculty: 658
Total number of Students: 21,005
Total number of Graduate Students: 2,466

Department Information
Department Chairman: Umar Mohideen, Chair
Department Contact: Derek Deving, Student Affairs Officer
  Total full-time faculty: 31
  Total number of full-time equivalent positions: 31
  Full-Time Graduate Students: 122
  First-Year Graduate Students: 28
  Female First-Year Students: 5
  Total Post Doctorates: 29

Department Address
900 University Avenue, PHYS 3047
Riverside, CA 92521
Phone: (951) 827-5332
E-mail: gophysics@ucr.edu
Website: http://www.physics.ucr.edu

ADMISSIONS

Admission Contact Information
Address admission inquiries to: Graduate Advisor, Department of Physics and Astronomy, University of California, Riverside, CA 92521
Phone: (951) 827-5332
E-mail: gophysics@ucr.edu
Admissions website: http://www.physics.ucr.edu/graduate

Application deadlines
Fall admission:
U.S. students: January 5   *Int'l. students*: January 5

Application fee
U.S. students: $80   *Int'l. students*: $100

Admissions information
For Fall of 2013:
  Number of applicants: 259
  Number admitted: 65
  Number enrolled: 29

Admission requirements
Bachelor's degree requirements: Bachelor's degree in physics or equivalent is required.
Minimum undergraduate GPA: 3.25

GRE requirements
The GRE is required.
QGRE minimum 300; QGRE 25%-75%: 308-321.

Advanced GRE requirements
The Advanced GRE is required.
  Minimum accepted Advanced GRE score: 600
  Mean Advanced GRE score range (25th–75th percentile): 633-870
Average PGRE is 750.

TOEFL requirements
The TOEFL exam is required for students from non-English-speaking countries.
  PBT score: 550
  IBT score: 80

Other admissions information
Undergraduate preparation assumed: Kittel, Thermal Physics; Fowles, Analytical Mechanics; Lorrain and Corson, Electromagnetic Fields and Waves; Liboff, Introductory Quantum Mechanics.

TUITION

Tuition year 2013-14:
Tuition for in-state residents
  Full-time students: $14,685 annual
Tuition for out-of-state residents
  Full-time students: $29,787 annual
Credit hours per semester to be considered full-time: 12
Deferred tuition plan: Yes
Health insurance: Available
Academic term: Quarter
Number of first-year students who received full tuition waivers: 26

Teaching Assistants, Research Assistants, and Fellowships

Number of first-year
Teaching Assistants: 26
Fellowship students: 26
Average stipend per academic year
Teaching Assistant: $11,770
Fellowship student: $11,211

FINANCIAL AID

Application deadlines
Fall admission:
U.S. students: January 5   *Int'l. students*: January 5

Loans
Loans are available for U.S. students.
Loans are not available for international students.
GAPSFAS application required: No
FAFSA application required: Yes

For further information
Address financial aid inquiries to: Graduate Advisor, Department of Physics and Astronomy.
Phone: (951) 827-5332
E-mail: gophysics@ucr.edu

HOUSING

Availability of on-campus housing
Single students: Yes
Married students: Yes

For further information
Housing aid website: http://www.housing.ucr.edu

Table A—Faculty, Enrollments, and Degrees Granted

Research Specialty	2013-14 Faculty	Enrollment Fall 2013 Master's	Enrollment Fall 2013 Doctorate	Number of Degrees Granted 2012-13 (2008-13) Master's	Number of Degrees Granted 2012-13 (2008-13) Terminal Master's	Number of Degrees Granted 2012-13 (2008-13) Doctorate
Astrophysics	5	–	15	–	–(9)	1(5)
Biophysics	3	–	2	–	–(2)	1(5)
Condensed Matter Physics	15	–	56	–	–(5)	13(32)
High Energy Physics	10	–	20	–	1(2)	7(17)
Non-specialized	–	–	29	–	–(2)	1(9)
Total	**30**	**–**	**122**	**–**	**–(20)**	**23(68)**
Full-time Grad. Stud.	–	–	122	–	–	–
First-year Grad. Stud.	–	–	29	–	–	–

GRADUATE DEGREE REQUIREMENTS

Master's: (1) Satisfactory completion of a minimum of 36 quarter units of approved physics courses taken for a letter grade after admission to graduate study; of these, at least 24 quarter units must be in the 200 series. (2) Either: (a) satisfactory completion of a thesis in a field of physics to be chosen in consultation with a faculty supervisor or (b) satisfactory performance on the comprehensive examination. See the general catalog for more details: http://www.catalog.ucr.edu.

Doctorate: (1) Coursework: Each course must be passed with a grade of "B-" or better. Each student must maintain an average of "B" or better for all courses. (2) Written comprehensive examination to be taken at the end of the student's first year.

In the event of a failure, a make-up examination is offered in the winter quarter of the second year. The examination covers mechanics, statistical and thermal physics, quantum mechanics, and electromagnetism. The comprehensive examination for students pursuing the astronomy specialization consists of an examination that covers mechanics, statistical and thermal physics, electromagnetism, and fundamental astrophysics. (3) Oral qualifying examination in the general area of the student's proposed research. This examination is conducted by a doctoral committee charged with general supervision of the student's research. (4) Dissertation containing a review of existing knowledge relevant to the area of the candidate's research and the results of the candidate's original research. This research must be of sufficiently high quality to constitute a contribution to knowledge in the subject area. (5) Final oral examination may be required. See the general catalog for more details: http://www.catalog.ucr.edu.

Table B—Separately Budgeted Research Expenditures by Source of Support

Source of Support	Departmental Research	Physics-related Research Outside Department
Federal government	$7,292,317	
State/local government	$526,978	
Non-profit organizations	$54,851	
Business and industry		
Other	$416,327	
Total	**$8,290,473**	

Table C—Separately Budgeted Research Expenditures by Research Specialty

Research Specialty	No. of Grants	Expenditures ($)
Astronomy	18	$571,019
Astrophysics	1	$39,177
Condensed Matter Physics	43	$4,863,134
Nuclear Physics	2	$884,948
Particles and Fields	2	$1,932,195
Total	**66**	**$8,290,473**

FACULTY

Distinguished University Professor

Hanson, Gail G., Ph.D., Massachusetts Institute of Technology, 1973. *Particles and Fields*. Experimental high-energy physics.

Varma, Chandra M., Ph.D., University of Minnesota, 1968. *Condensed Matter Physics*. Theoretical condensed-matter physics.

Professor

Barish, Kenneth N., Ph.D., Yale University, 1996. *Nuclear Physics*. Relativistic heavy ion and nuclear spin.

Bockrath, Marc, Ph.D., University of California, Berkeley, 1999. *Condensed Matter Physics, Nano Science and Technology*. Experimental condensed-matter physics.

Clare, Robert B., Ph.D., Massachusetts Institute of Technology, 1982. *Particles and Fields*. Experimental high-energy physics.

Ellison, John A., Ph.D., Imperial College of London, 1987. *Particles and Fields*. Experimental high-energy physics.

Gary, John W., Ph.D., University of California, Berkeley, 1985. *Particles and Fields*. Experimental high-energy physics.

Lau, Chun Ning (Jeanie), Ph.D., Harvard University, 2001. *Condensed Matter Physics, Nano Science and Technology.* Experimental condensed-matter physics.

Long, Owen, Ph.D., University of Pennsylvania, 1997. *Particles and Fields.* Experimental high-energy physics.

Ma, Ernest, Ph.D., University of California, Irvine, 1970. *Particles and Fields.* Theoretical particle physics.

Mills, Allen P., Ph.D., Brandeis University, 1967. *Atomic, Molecular, & Optical Physics, Condensed Matter Physics.* Experimental condensed-matter physics.

Mobasher, Braham, Ph.D., University of Durham, UK, 1988. *Astronomy, Astrophysics.* Observational astrophysics.

Mohideen, Umar, Ph.D., Columbia University, 1992. *Biophysics, Condensed Matter Physics, Optics.* Fundamental precision measurements; biophysics; laser spectroscopy and nonlinear optics.

Seto, Richard, Ph.D., Columbia University, 1983. *Nuclear Physics.* Experimental heavy ion physics.

Shi, Jing, Ph.D., University of Illinois, 1994. *Condensed Matter Physics, Nano Science and Technology.* Experimental condensed-matter physics.

Shtengel, Kirill, Ph.D., University of California, Los Angeles, 1999. *Condensed Matter Physics.* Theoretical condensed-matter physics.

Tom, Harry W. K., Ph.D., University of California, Berkeley, 1984. *Condensed Matter Physics, Nano Science and Technology, Optics.* Experimental laser spectroscopy and nonlinear optics; surface science.

Wilson, Gillian, Ph.D., University of Durham, 1996. *Astronomy, Astrophysics.* Observational astrophysics.

Wimpenny, Stephen J., Ph.D., Sheffield University (England), 1980. *Particles and Fields.* Experimental high-energy physics.

Wudka, Jose, Ph.D., Massachusetts Institute of Technology, 1986. *Particles and Fields.* Theoretical high-energy physics.

Yarmoff, Jory A., Ph.D., University of California, Los Angeles, 1985. *Condensed Matter Physics.* Experimental surface sciences.

Associate Professor

Beyermann, Ward, Ph.D., University of California, Los Angeles, 1988. *Condensed Matter Physics.* Experimental condensed-matter physics.

Canalizo, Gabriela, Ph.D., University of Hawaii, 2000. *Astronomy, Astrophysics.* Observational astrophysics.

Pryadko, Leonid P., Ph.D., Stanford University, 1996. *Condensed Matter Physics.* Theoretical condensed-matter physics.

Tsai, Shan-Wen, Ph.D., Brown University, 2000. *Condensed Matter Physics.* Theoretical condensed-matter physics.

Zandi, Roya, Ph.D., University of California, Los Angeles, 2001. *Biophysics, Condensed Matter Physics.* Theoretical condensed-matter physics.

Assistant Professor

Aji, Vivek, Ph.D., University of Illinois at Urbana-Champaign, 2002. *Condensed Matter Physics.* Theoretical condensed-matter physics.

Gabor, Nathaniel, Ph.D., Cornell University, 2010. *Condensed Matter Physics.*

Reddy, Naveen, Ph.D., California Institute of Technology, 2006. *Astronomy, Astrophysics.* Observational astrophysics.

Siana, Brian, Ph.D., University of California, San Diego, 2005. *Astronomy, Astrophysics.* Observational astrophysics.

Yu, Hai-Bo, Ph.D., University of Maryland, 2007. *Astrophysics, Theoretical Physics.* Cosmology.

Professor Emeritus

Desai, Bipin R., Ph.D., University of California, Berkeley, 1961. *Particles and Fields.* Theoretical high-energy physics.

Research Professor

MacLaughlin, Douglas E., Ph.D., University of California, Berkeley, 1966. Professor Emeritus. *Condensed Matter Physics.* Experimental condensed-matter physics: high-temperature superconductivity; heavy-fermion compounds; magnetic resonance techniques.

Zych, Allen D., Ph.D., Case Western Reserve University, 1968. Professor Emeritus. Experimental space physics; gamma-ray astronomy; atmospheric and solar gamma-rays and neutrons.

Research Physicist

Kawakami, Roland, Ph.D., University of California, Berkeley, 1999. *Condensed Matter Physics, Nano Science and Technology.* Experimental condensed-matter physics.

Nagamine, Kanetada, Ph.D., University of Tokyo, 1969. *Condensed Matter Physics.* Experimental muon physics.

Assistant Researcher

Zhu, Lijun, Ph.D., Rice University, 2005. *Condensed Matter Physics.* Theoretical condensed-matter physics.

DEPARTMENTAL RESEARCH SPECIALTIES AND STAFF

Theoretical

Biophysics. Virus structure and structure formation kinetics; elastic and Casimir-type interaction between particles embedded in bilayer lipid membranes. There is one postdoctoral physicist. Pryadko, Zandi.

Condensed Matter Physics. Research in quantum and statistical mechanics of many-body systems, including studies of quantum critical phenomena, novel phases, and phase transitions in strongly correlated matter; superconductivity; singular Fermi liquids; low-dimensional systems including graphene, quantum computation, Casimir interactions, and cold atom physics. There are seven postdoctoral physicists. Aji, Gabor, Pryadko, Shtengel, Tsai, Zandi.

Particles and Fields. Gauge theories; extensions of the standard model of strong, weak, and electromagnetic interactions; effective theories of electroweak interactions and their potential impact on future LHC and ILC data; neutrinos and related physics beyond the standard model; the problem of mass, specifically the possible connection between quark and lepton masses and their respective mixing matrices; higher symmetries such as SU(5) and SO(10), as well as models of extra spacetime dimensions. There is one postdoctoral physicist. Desai, Ma, Wudka, Yu.

Experimental

Astronomy. Astronomy/astrophysics/cosmology. The main research area in the astronomy group at UCR is observational extragalactic astronomy. Using data from space and ground-based facilities (i.e., the Hubble and Spitzer Space Telescopes, Keck and Gemini telescopes, etc.), we study the following topics: formation and evolution of galaxies at different look-back times, high redshift universe, large-scale structure and clustering of galaxies, gravitational lensing, active galactic nuclei, and multi-waveband galaxy surveys. The UCR astronomy group is an active member of a number of large collaborations aimed at studying the evolution of galaxies. There are 20 graduate students and five postdoctoral physicists. Canalizo, Mobasher, Reddy, Siana, Wilson.

Atomic, Molecular, & Optical Physics. Positronium atom physics; Bose-Einstein condensation of a dense collection of positronium atoms; measuring the energy structure of the positronium atom; making a Bose-Einstein condensed positronium annihilation gamma ray laser. There is one postdoctoral physicist. Mills.

Biophysics. Virus structure and structure formation kinetics; elastic and Casimir-type interactions between particles embedded in bilayer lipid membranes. There is one postdoctoral physicist. Mohideen.

Condensed Matter Physics. Strongly correlated electron systems such as heavy-fermion materials, biological materials, high-temperature superconductors, non-Fermi liquids, Bose-Einstein condensates, quantum magnets, magnetic superconductors, spin glasses, fullerenes, and impurity systems. Experimental techniques include nuclear magnetic resonance (NMR) and muon spin rotation (muSR), terahertz spectroscopy, and both elastic and inelastic neutron scattering. NMR experiments are carried out at UCR, and the group travels to "meson factory" facilities such as TRIUMF in Vancouver, Canada, to perform muSR experiments using accelerator-produced beams of muons. Transport and thermodynamic properties are also investigated as functions of temperature and magnetic fields. In some cases, experiments are conducted down to mK temperatures or in pulsed magnetic fields up to 60T. Some of this research is performed offsite at user facilities such as Paul Scherrer Institute, LANSCE, and NHMFL. Research is also being conducted on devising an analog neural network computer based on the interactions of DNA molecules. There are three postdoctoral physicists. Beyermann, MacLaughlin, Mills.

Environmental Physics. Environmental physics: surface physics of the air/water/solid interface and complex materials such as zeolites with application to atmospheric and soil sciences, chemical and bioremediation, and chemical and biological sensors. Tom, Yarmoff.

Fundamental Precision Measurements. Precision measurements and theory of the Casimir force and other effects of the zero point fields for different shapes of the interacting bodies made using microcantilever-based techniques such as in atomic force microscopy. There are six postdoctoral physicists. Mohideen, Zandi.

Joint Programs. A joint Ph.D. in physics and M.S. in soil physics, a joint Ph.D. in soil physics and M.S. in physics, and a Ph.D. in physics with emphasis on environmental physics are offered in conjunction with the Department of Soil and Environmental Sciences. Interdisciplinary training is offered in the area of chemical physics through collaboration with faculty members in the Department of Chemistry. Tom.

Nano Science and Technology. Properties of graphene and carbon nanotubes, including quantum transport, strain engineering, spin transport, thermoelectric, and nanomechanical properties; synthesis of oxide heterostructures, semiconductors, topological insulators, ultrathin magnetic films, large area graphene, and carbon nanotubes; optical probes of interface dynamics, magnetism, and spin coherence; spintronics in semiconductor, metal, and carbon-based materials; fundamental properties and fabrication of metal nanoclusters; thermoelectric and thermo-spintronic phenomena; nanoscale superconductivity; physics of information storage devices. Techniques include nanofabrication by electron beam lithography and focused ion beam milling, electron transport in He3 and dilution refrigerators, pulsed laser and nonlinear optical spectroscopy, scanning probe microscopy, molecular beam epitaxy (MBE), laser-MBE, muon spin relaxation, and SQUID magnetometry. There are four postdoctoral physicists. Bockrath, Lau, Nagamine, Shi, Tom, Yarmoff.

Nucleon Structure. Proton spin research: what are the gluon and anti-quark contributions to the proton spin? Spin is a property of particles as fundamental as charge and mass. The spin of the proton was first determined in 1927, yet we still do not understand the spin structure of the proton. We are investigating the longitudinal and transverse spin structure of the proton at Brookhaven National Laboratory. There is one postdoctoral physicist and two graduate students. Barish, Seto.

Optics. Laser spectroscopy and nonlinear optics: femtosecond and picosecond laser pulses are used to time-resolve chemical processes at surfaces, the rotational dynamics of liquids, electron-phonon coupling in solids, and the orientation of molecules at environmentally significant interfaces. Novel laser sources in the far infrared are being developed. Picosecond time-resolved nanoscale microscopy is used to study the electronic properties of quantum dot structures and the mobility of ferroelectric and ferromagnetic domain walls. There is one postdoctoral physicist. Mohideen, Tom.

Particles and Fields. Neutrino factory and muon collider research and development: muon beam R&D is being carried out as part of the Muon Accelerator Program (MAP) hosted by Fermilab and the Muon Ionization Cooling Experiment (MICE) at the Rutherford Lab in the U.K. with the goals of developing designs for a future neutrino factory and muon collider. In a neutrino factory, an intense, well-controlled beam of high-energy neutrinos is produced from the decays of circulating muons, which can be used to study neutrino oscillations and possible CP violation in neutrinos. In a muon collider, muons and antimuons collide, providing high-energy collisions of fundamental particles in a relatively small accelerator complex with the potential to advance our fundamental understanding of matter and energy in unique ways. The MICE experiment will provide an experimental demonstration of ionization cooling, the technique needed for muons. There are two postdoctoral physicists. Clare, Ellison, Gary, Hanson, Long, Wimpenny.

Particles and Fields. High-energy physics: research efforts are ongoing in the study of proton-antiproton collisions at Fermilab, proton-proton collisions at CERN, and electron-positron collisions at SLAC. At Fermilab, high-energy proton-antiproton collisions in the DZero detector at the Tevatron Collider are used to study the top and bottom quark production and decay, electroweak physics, and the search for the Higgs boson and other new phenomena. The CMS experiment at the CERN Large Hadron Collider is the highest-energy collider in the world. Our detector efforts focus on the end-cap muon chambers, the silicon strip tracker, and the hadron calorimeter, as well as upgrades for future high-luminosity running. Physics interests include tracking software, top quark physics, the search for the Higgs boson, and new physics beyond the Standard Model, such as supersymmetry. At SLAC, the data-taking phase of the BaBar experiment is now complete. Our analysis efforts include measurements of charmless B decays relevant for CP violation studies, searches for new physics in flavor-changing neutral current processes, and bottomonium spectroscopy. There are six postdoctoral physicists. Hanson, Wimpenny.

Relativistic heavy-ion physics. The experimental group studies relativistic nucleus-nucleus collisions at high energies to explore the behavior of extended nuclear matter under extreme conditions of density and temperature, i.e., the Quark Gluon Plasma (QGP). These studies use the PHENIX detector at the Brookhaven Relativistic Heavy Ion Collider (RHIC), where energies of 100 GeV per nucleon are available. The initial goal for creating the QGP has been accomplished: a strongly interacting plasma behaving like a liquid has been found. The main goal now is to characterize this new state of matter. There are two postdoctoral physicists, six graduate students, and two undergraduate students. Barish, Seto.

Surface Physics. The geometric, electronic, chemical, and magnetic properties of solid surfaces are investigated with a wide variety of modern surface spectroscopies. We have particular expertise in studying ion-surface and laser-surface interactions. In addition, the surface properties of novel materials are investigated. There are two postdoctoral physicists. Tom, Yarmoff.

UNIVERSITY OF CALIFORNIA, SAN DIEGO

DEPARTMENT OF PHYSICS

La Jolla, California 92093-0319
http://physics.ucsd.edu/

General University Information
Chancellor: Pradeep K. Khosla
Dean of Graduate School: Kim E. Barrett
University website: http://www.ucsd.edu/
Control: Public
Setting: Suburban
Total Faculty: 1,177
Total Graduate Faculty: 1,177
Total number of Students: 29,052
Total number of Graduate Students: 4,588

Department Information
Department Chairman: Dmitri Basov, Chair
Department Contact: Hilari Ford, Graduate Coordinator
  Total full-time faculty: 50
  Total number of full-time equivalent positions: 50
  Full-Time Graduate Students: 168
  First-Year Graduate Students: 43
  Female First-Year Students: 3
  Total Post Doctorates: 45

Department Address
9500 Gilman Drive
La Jolla, CA 92093-0319
Phone: (858)534-3293
E-mail: apply@physics.ucsd.edu
Website: http://physics.ucsd.edu/

ADMISSIONS

Admission Contact Information
Address admission inquiries to: Graduate Admissions Office, Department of Physics (0319), 9500 Gilman Dr., La Jolla, CA 92093-0319
Phone: (858)822-1074
E-mail: apply@physics.ucsd.edu
Admissions website: https://gradapply.ucsd.edu/

Application deadlines
Fall admission:
U.S. students: December 15   *Int'l. students*: December 15

Application fee
U.S. students: $80   *Int'l. students*: $100

Admissions information
For Fall of 2013:
  Number of applicants: 525
  Number admitted: 125
  Number enrolled: 32

Admission requirements
Bachelor's degree requirements: Entering graduate students are required to have a sound knowledge of undergraduate mechanics, electricity, and magnetism; to have had senior courses or their equivalent in atomic and quantum physics, nuclear physics, and thermodynamics; and to have taken upper division laboratory work.
Minimum undergraduate GPA: 3.5

GRE requirements
The GRE is required.
The GRE examinations must be taken no later than November in order for the scores to be considered by the Admissions Committee. Incomplete files and applications received after December 15th will be considered only if space is available.

Advanced GRE requirements
The Advanced GRE is required.

TOEFL requirements
The TOEFL exam is required for students from non-English-speaking countries.
The minimum TOEFL score required is 550 for paper-based (PBT), 213 for computer-based (CBT) and 80 for internet based (iBT). The minimum International English Language Testing System (IELTS) score required is Band Score 7. The Test for Spoken English (TSE) is highly recommended. International Students whose native language is not English will be required to demonstrate English language proficiency before they may serve as teaching assistants.

Other admissions information
Additional requirements: The average GRE scores for admitted students for 2012-2013 were verbal–158; quantitative–165; advanced–821.
Undergraduate preparation assumed: Griffiths, Introduction to Electrodynamics; Dubin, Numerical and Analytical Methods for Scientists and Engineers; Thornton and Marion, Classical Dynamics; Barnaal, Analog Electronics for Scientific Application; Horowitz and Hill, The Art of Electronics; Griffiths, Introduction to Quantum Mechanics; Gasiorowicz, Quantum Physics; Sakurai, Advanced Quantum Mechanics; Carter, Classical and Statistical Thermodynamics; Kittel and Zetti, Introduction to Solid State Physics; Luth and Ibach, Solid-State Physics: An Introduction to Principles of Materials Science.

TUITION

Tuition year 2013-14:
Tuition for in-state residents
  Full-time students: $15,604.5 annual $30,706.5
Based on a proposed graduate student fees/tuition. Amounts subject to change.
Credit hours per semester to be considered full-time: 12
Deferred tuition plan: No
Health insurance: Available at the cost of 957.00 per year.
Other academic fees: Health insurance is included in student fees.
Academic term: Quarter

Teaching Assistants, Research Assistants, and Fellowships
Number of first-year
  Teaching Assistants: 18
  Research Assistants: 7
Average stipend per academic year
  Teaching Assistant: $17,658
  Research Assistant: $18,041
  Fellowship student: $10,000
Only select students receive a fellowship stipend.

FINANCIAL AID

Loans
Loans are available for U.S. students.
Loans are not available for international students.
GAPSFAS application required: No
FAFSA application required: Yes

For further information
Address financial aid inquiries to: Student Financial Services, 0013 Graduate Division, 9500 Gilman Dr., La Jolla, CA 92093-0013.
Phone: (858) 534-4480
E-mail: finaid@ucsd.edu
Financial aid website: http://ogs.ucsd.edu/financial-support/index.html

HOUSING

Availability of on-campus housing
Single students: Yes
Married students: Yes

For further information
Address housing inquiries to: ARCH (Associated Residential Community Housing), 9500 Gilman Drive, Dept 0907, La Jolla, CA 92093-0907.
Phone: (858) 822-6274
Housing aid website: http://hdh.ucsd.edu/arch/gradhousing.asp

GRADUATE DEGREE REQUIREMENTS

Master's: A "B" average in 36 units of graduate work and a comprehensive written examination are required. A thesis is not required. There is no language requirement. Three quarters of residency are required. There is no terminal master's program.

Doctorate: A "B" average must be maintained in all coursework. A comprehensive departmental examination at the beginning of the second year, completion of five advanced courses, completion of teaching requirement, followed by an oral qualifying examination for advancement to candidacy are required. A dissertation and successful oral defense of dissertation are required. There is no language requirement. Six quarters residency are required.

Other Degrees: A Ph.D. in physics (biophysics) is also available. This option has the same requirements as the regular Ph.D., except that the departmental examination can be taken at beginning of the third year and five courses related to the life sciences are required. A Ph.D. with a specialization in computational science (CSME)is designed to allow students to obtain training in their chosen field of science, mathematics, or engineering, with additional training in computational science integrated into their graduate studies. Prospective students must apply and be admitted into the Ph.D. program in physics, and then be admitted into the CSME program.

Thesis: Thesis may not be written in absentia.

SPECIAL EQUIPMENT, FACILITIES, OR PROGRAMS

Departmental facilities include excellent electronics and machine shops, a liquid He facility, and extensive computing facilities. Additional computing support is available from the campus-based San Diego Supercomputing Center (SDSC).

Table A—Faculty, Enrollments, and Degrees Granted

Research Specialty	2012-13 Faculty	Enrollment Fall 2012 Master's	Enrollment Fall 2012 Doctorate	Number of Degrees Granted 2012-13 (2008-13) Master's	Number of Degrees Granted 2012-13 (2008-13) Terminal Master's	Number of Degrees Granted 2012-13 (2008-13) Doctorate
Acoustics	1	–	–	–	–	–(1)
Applied Mathematics	2	–	3	–	–	–(1)
Applied Physics	7	–	11	–	–	–(1)
Astrophysics/ Astronomy	15	–	21	–	–	3(19)
Atomic, Molecular, & Optical Physics	3	–	7	–	–	–(1)
Biophysics	12	–	22	–	–	6(19)
Computational Science	–	–	4	–	–	1(3)
Condensed Matter Physics	20	–	38	–	–	6(34)
Elementary Particles	–	–	3	–	–	1(-)
Fluids, Rheology	8	–	3	–	–	–(1)
High Energy Physics	–	–	22	–	–(1)	4(9)
Materials Science, Metallurgy	3	–	4	–	–(4)	1(3)
Nonlinear Dynamics	5	–	3	–	–	1(5)
Nuclear Physics	5	–	–	–	–	–
Physics Education	1	–	1	–	–	–
Physics of Beams	4	–	–	–	–	–(2)
Plasma and Fusion	5	–	16	–	–	2(8)
Statistical & Thermal Physics	7	–	–	–	–	–
Non-specialized	–	–	14	–	3(12)	1(6)
Total	–	–	172	–	3(17)	26(113)
Full-time Grad. Stud.	–	–	172	–	–	–
First-year Grad. Stud.	–	–	41	–	–	–

Table B—Separately Budgeted Research Expenditures by Source of Support

Source of Support	Departmental Research	Physics-related Research Outside Department
Federal government	$12,228,762	
State/local government	$1,928,475	
Non-profit organizations		
Business and industry		
Other	$541,116	
Total	**$14,698,353**	

Table C—Separately Budgeted Research Expenditures by Research Specialty

Research Specialty	No. of Grants	Expenditures ($)
Atomic, Molecular, & Optical Physics	1	$362,437
Biophysics	23	$7,500,854
Condensed Matter Physics	18	$3,342,618
Plasma and Fusion	2	$531,438
Total	**55**	**$14,698,353**

FACULTY

Distinguished University Professor

Abarbanel, Henry D. I., Ph.D., Princeton University, 1966. *Biophysics*. Nonlinear dynamics of fluids; optical systems and neural assemblies; geophysical fluid dynamics and physical oceanography.

Diamond, Patrick H., Ph.D., Massachusetts Institute of Technology, 1979. Theoretical plasma physics and astrophysics; nonlinear dynamics.

Grinstein, Benjamin, Ph.D., Harvard University, 1984. Elementary particle theory. Interested in mathematical models of the interactions of elementary particles: creating models, developing methods to analyze them and confronting them with experiment. Interested as well in formal aspects of Quantum Field Theory.

Kuti, Julius, Ph.D., Hungary, 1967. Elementary particles and fields.

Maple, M. Brian, Ph.D., University of California, San Diego, 1969. Superconductivity; magnetism, strongly correlated electron phenomena; high-pressure physics; surface science.

Norman, Michael L., Ph.D., University of California, Davis, 1980. *Astrophysics*. Computational astrophysics and cosmology.

Schuller, Ivan K., Ph.D., Northwestern University, 1976. *Condensed Matter Physics*. Experimental condensed-matter physics; materials science (thin films, heterostructures, magnetism, nanostructures, and superconductivity).

Surko, Clifford M., Ph.D., University of California, Berkeley, 1968. *Atomic, Molecular, & Optical Physics*. Experimental studies of nonlinear non-equilibrium phenomena, plasma physics using positrons and positron–matter interactions.

Professor

Arovas, Daniel P., Ph.D., University of California, Santa Barbara, 1986. *Condensed Matter Physics*. Condensed matter theory; statistical mechanics.

Basov, Dmitri N., Ph.D., Lebedev Institute, USSR, 1991. Chair, Department of Physics. *Condensed Matter Physics*. Experimental condensed matter. Novel electronic and magnetic materials; meta-materials; advanced methods for optical spectroscopy and nano-imaging.

Branson, James G., Ph.D., Princeton University, 1977. *High Energy Physics*. Experimental elementary particle physics. My group has recently found a new "Higgs-like" boson at the Large Hadron Collider in the decay mode into two photons. We have played a leading role in the search for the Higgs in this mode, where the strongest signal has been observed. Now we are working on measuring the properties of this new particle to determine whether it is simply a Standard Model Higgs or something much more interesting.

Butov, Leonid V., Ph.D., Institute of Solid State Physics, 1991. *Condensed Matter Physics, Optics*. Experimental condensed matter physics. Studies of basic properties of electron-hole systems in semiconductors and development of new methods for optoelectronic signal processing.

Di Ventra, Massimiliano, Ph.D., Ecole Polytechnique Federale de Lausanne, 1997. *Condensed Matter Physics*. Condensed matter physics. Interests are in the theory of non-equilibrium phenomena in nanoscale and biological systems, with particular focus on applicative contexts. Employs both analytical and numerical approaches to understand and predict the behavior of many-body systems out of equilibrium.

Dubin, Daniel H. E., Ph.D., Princeton University, 1984. *Computational Physics*. Theoretical plasma physics. I am a plasma theorist working mainly on nonneutral plasmas, such as pure electron plasmas or pure ion plasmas. In addition to plasma physics this work involves statistical physics, nonlinear dynamics and advanced computer simulation methods.

Fogler, Michael M., Ph.D., University of Minnesota, 1997. *Condensed Matter Physics*. Condensed matter theory. Theoretical study of low dimensional and nanoscale electron systems, in particular, graphene, semiconductor nanowires and quantum wells. Investigation of correlations, disorder, and quantum effects in transport and optical properties of such materials.

Fuller, George M., Ph.D., California Institute of Technology, 1981. *Astrophysics*. Astrophysics/Astronomy. My research centers on nuclear and particle astrophysics and a few issues in gravitation. Recent work by my group and me includes calculation of neutrino flavor transformation in supernovae and the early universe, neutrino mass in cosmology, sterile neutrino dark matter, and general relativistic instability of super-massive stars.

Griest, Kim, Ph.D., University of California, Santa Cruz, 1987. *Astrophysics*. Theoretical and observational astrophysic; theoretical elementary particle physics; dark matter.

Hirsch, Jorge E., Ph.D., University of Chicago, 1980. *Condensed Matter Physics*. Theoretical condensed-matter physics.

Holst, Michael, Ph.D., University of Illinois at Urbana-Champaign, 1993. Director: Mathematical and Computational Physics Research Group Co-Director: Center for Computational Mathematics. *Biophysics*. Biophysics, mathematical physics, general relativity.

Hwa, Terence T.-L., Ph.D., Massachusetts Institute of Technology, 1990. *Biophysics, High Energy Physics*. Statistical mechanics; biological physics; systems biology; molecular evolution; genomics; condensed-matter physics; dynamics of complex systems; polymer physics.

Intriligator, Kenneth A., Ph.D., Harvard University, 1992. *High Energy Physics*. Elementary particle physics. High-energy theory topics: quantum field theory, supersymmetry, string theory, and dualities.

Jenkins, Elizabeth, Ph.D., Harvard University, 1989. *Nuclear Physics*. Elementary particle physics.

Kleinfeld, David, Ph.D., University of California, San Diego, 1984. *Biophysics, Neuroscience/Neuro Physics*. Experimental and computational neuroscience and neurovascular studies at the cellular through systems level; Advancements in electrical instrumentation, molecular probes, and optical microscopy for imaging and manipulation.

Manohar, Aneesh V., Ph.D., Harvard University, 1983. *High Energy Physics*. Elementary particle physics. Theoretical High Energy Physics and Cosmology.

Paar, Hans P., Ph.D., Columbia University, 1974. *Astrophysics*. Observational Cosmology. I do Observational Cosmology as a member of the Polarbear Collaboration. We operate a microwave telescope in the Andes Mountains in Chile with which we study the properties of the Cosmic Microwave Background.

Sharma, Vivek A., Ph.D., Syracuse University, 1990. *High Energy Physics*. Prof. Sharma's group is engaged in Higgs boson physics with the CMS detector at the Large Hadron collider at CERN.

Sinha, Sunil K., Ph.D., University of Cambridge, 1964. *Condensed Matter Physics*. Neutron and x-ray scattering studies of condensed matter.

Tytler, David, Ph.D., University of London, 1982. *Astrophysics*. Cosmology and galaxy formation; quasars; ultraviolet, optical and infrared observations; telescopes and astronomical instrumentation; other planetary systems.

Wuerthwein, Frank, Ph.D., Cornell University, 1995. Elementary particle physics. Interested in search for new physics at the LHC. CMS SUSY convener for 2013/14.

Yagil, Avraham, Ph.D., Weizmann Institute of Science, 1989. *High Energy Physics*. Experimental high energy physics at the large hadron collider.

Associate Professor

Burgasser, Adam, Ph.D., California Institute of Technology, 2001. *Astrophysics*. Astrophysics/Astronomy. Observational astrophysics; low mass stars, brown dwarfs and extrasolar planets; near-infrared instrumentation.

Coil, Alison, Ph.D., University of California, Berkeley, 2004. *Astrophysics*. Astrophysics/Astronomy. Observer, studying distant, high-redshift galaxies and accreting supermassive black holes, as well as outflowing galactic winds.

Dudko, Olga K., Ph.D., Institute for Low Temperature Physics and Engineering, Kharkov, Ukraine, 2001. *Biophysics*. Theoretical biophysics.

Groisman, Alexander, Ph.D., Weizmann Institute of Science, 2001. *Biophysics*. Experimental biophysics, fluid mechanics, microfluidics, mechanobiology.

Keating, Brian, Ph.D., Brown University, 2000. *Astrophysics*. Astrophysics/Astronomy. Cosmic Microwave Background, Experimental cosmology, low noise electronics, and detector physics.

Murphy, Thomas M., Ph.D., California Institute of Technology, 2000. Vice Chair of Education, Physics. *Astrophysics*. Astrophysics/Astronomy. Tests fundamental gravitation by measuring the Earth-Moon separation to millimeter precision using an apparatus he and his group constructed for the Apache Point Observatory in New Mexico.

Shpyrko, Oleg, Ph.D., Harvard University, 2004. *Condensed Matter Physics*. Condensed matter physics. We use x-ray, light and neutron scattering, as well as scanning probes to understand and manipulate the behavior of materials at the nanoscale, including problems that range from soft to hard condensed matter.

Smith, Douglas E., Ph.D., Stanford University, 1999. *Biophysics*. Biophysics. We conduct research in experimental biophysics, primarily using optical tweezers to manipulate single DNA molecules. Topics include viral DNA packaging, molecular motors, and protein-DNA interactions.

Wu, Congjun, Ph.D., Stanford University, 2005. *Condensed Matter Physics*. Condensed matter physics. Wu's research interest covers the theoretical study of new states of matter in condensed matter systems, including unconventional magnetism and superconductivity, orbital physics, spin-orbit coupling and spintronics, excitons, quantum phase transitions and criticality, strongly correlated bosonic and fermionic systems with cold atoms, and numerical algorithms for two dimensional quantum systems.

Assistant Professor

John, McGreevy, Ph.D., Stanford University, 2002. *High Energy Physics*. High-energy theory.

Jun, Suckjoon, Ph.D., Simon Fraser University, 2004. *Biophysics*. Experimental biophysics.

Keres, Dusan, Ph.D., University of Massachusetts, Amherst, 2007. *Astrophysics*. Astrophysics/Astronomy. Galaxy formation and evolution; cosmological simulations.

Schoetz, Eva-Maria, Ph.D., Technical University of Dresden, 2007. *Biophysics*. Experimental biophysics.

Emeritus

McIlwain, Carl E., Ph.D., University of Iowa, 1960. Research Professor. Space physics; experimental and theoretical studies of planetary magnetospheres; observational and instrumental astrophysics.

Peterson, Laurence E., Ph.D., University of Minnesota, 1960. Research Professor. *Astrophysics*. X- and gamma-ray astronomy; cosmic rays; space physics; balloon and satellite instrumentation.

Vernon, Wayne, Ph.D., Princeton University, 1965. Properties of elementary particles and their interactions; neutrino physics and astrophysics; particle detectors and acceleration techniques; free-electron lasers; Compton backscattered x-ray production.

Professor Emeritus

Berkowitz, Ami E., Ph.D., University of Pennsylvania, 1953. Magnetic materials investigations; correlation of microstructures with magnetic behavior; surface effects; relaxation phenomena.

Brueckner, Keith A., Ph.D., University of California, Berkeley, 1950. Theoretical nuclear physics; statistical mechanics; plasma physics; interaction of lasers with matter; magnetohydrodynamics; theory of metals.

Burbidge, E. Margaret, Ph.D., London Observatory, 1943. Extragalactic studies; spectrophotometric and imaging; observational work on normal galaxies; galaxies with active nuclei, especially radio galaxies; quasars using Lick Observatory 3-M telescope and Keck Observatory 10-M telescope.

Chen, Joseph C.Y., Ph.D., University of Notre Dame, 1961. Theory of atomic and molecular structure and processes; history and philosophy of science.

Driscoll, C. Fred, Ph.D., University of California, San Diego, 1976. Experimental plasma physics; waves and transport in pure electron and pure ion plasmas; 2D fluid dynamics and turbulence.

Dynes, Robert C., Ph.D., McMaster University, 1968. Experimental condensed-matter physics; solid-state physics.

Feher, George, Ph.D., University of California, Berkeley, 1954. *Biophysics*. Biophysics; photosynthesis; magnetic resonance; mechanisms of crystallization of macromolecules.

Fredkin, Donald R., Ph.D., Princeton University, 1961. Solid-state theory; applied magnetics; biophysics.

Goldberger, Marvin L., Ph.D., University of Chicago, 1948. Elementary particle physics; quantum field theory; collision theory.

Goodkind, John M., Ph.D., Duke University, 1960. *Condensed Matter Physics*. Low-temperature experimental research; 2D electrons; solid He; geophysical and fundamental gravity; quantum computing.

Jones, Barbara, Ph.D., University of London, 1976. *Astrophysics*. Infrared astrophysics; galactic and extragalactic astronomy; astronomical instrumentation; research in physics education.

Liebermann, Leonard N., Ph.D., University of Chicago, 1940. Magnetism; propagation of underwater sound; molecular and chemical physics; extremely low-frequency electromagnetic waves.

Lovberg, Ralph H., Ph.D., University of Minnesota, 1955. Experimental plasma physics; geophysics.

Nguyen-Huu, Xuong, Ph.D., University of California, Berkeley, 1962. *Biophysics*. Biophysics; protein crystallography; electron microscopy, detectors for x-rays and electrons.

O'Neil, Thomas M., Ph.D., University of California, San Diego, 1965. Theoretical plasma physics.

Okamura, Melvin Y., Ph.D., Northwestern University, 1970. *Biophysics*.

Onuchic, José, Ph.D., California Institute of Technology, 1987. *Biophysics*. Theoretical biophysics and chemical physics; theoretical studies in electron-transfer reactions in chemical and biological systems and in the protein-folding problem; bioinformatics.

Schultz, Sheldon, Ph.D., Columbia University, 1960. *Condensed Matter Physics*. Negative index of refraction; meta-materials; photonic band gap structures; plasmon resonant particles; advanced instrumentation in biotechnology.

Sham, Lu Jeu, Ph.D., University of Cambridge, 1963. *Condensed Matter Physics*. Theoretical condensed-matter physics.

Suhl, Harry, Ph.D., University of Oxford, 1948. Theoretical solid-state physics, particularly superconductivity, magnetism, and surface kinetics; nonlinear dynamics.

Swanson, Robert A., Ph.D., University of Chicago, 1958. Experiments involving properties and interactions of elementary particles; interference and decay of neutral K-mesons; deep inelastic muon scattering; nucleon structure and fragmentation; rare kaon decays and CP violation.

Ticho, Harold, Ph.D., University of Chicago, 1949. Experimental elementary particle physics.

Wolfe, Arthur M., Ph.D., University of Texas, 1967. Endowed Chair, Chancellor's Associates Chair IV. *Astrophysics*. Observational cosmology; galaxy formation; star formation.

Wong, David Y., Ph.D., University of Maryland, 1958. *High Energy Physics*. Theoretical high-energy physics.

Adjunct Professor

Pathria, Raj K., Ph.D., University of Delhi, 1957. Statistical physics; quantum fluids; low-temperature physics.

Waltz, Ronald, Ph.D., University of Chicago, 1970. Theoretical plasma physics; numerical simulation of turbulence in plasma.

Adjunct Assistant Professor

Sharpee, Tatyana O., Ph.D., Michigan State University, 2001. *Biophysics*. Our group works on theoretical principles of how the brain processes information. We are interested in how sensory processing in the brain is shaped by the animal's need to create parsimonious representations of events in the outside world. Our approaches are often derived from methods in statistical physics, mathematics, and information theory.

Lecturer with Rank of Professor

Anderson, Michael G., Ph.D., University of California, Davis, 2006. *Physics and other Science Education*. Physics education.

DEPARTMENTAL RESEARCH SPECIALTIES AND STAFF

Theoretical

Acoustics. Abarbanel.

Astrophysics. Shapiro, Shu. Diamond, Fuller, Griest, Keres, Norman, Wolfe.

Biophysics. Abarbanel, Dudko, Hwa.

Condensed Matter Physics. Arovas, Di Ventra, Fogler, Hirsch, Hwa, Sham, Suhl, Wu.

High Energy Physics. McGreevy. Grinstein, Intriligator, Jenkins, Kuti, Manohar.

History & Philosophy of Physics/Science.

Nonlinear Dynamics and Complex Systems. Abarbanel, Diamond, Hwa, Suhl.

Plasma and Fusion. Diamond, Dubin.

Statistical & Thermal Physics. Arovas, Brueckner, Diamond, Dubin, Hwa, Suhl.

Experimental

Astronomy. Burbidge, Burgasser, Coil, Griest, Jones, Keating, McIlwain, Murphy, Paar, Peterson, Tytler, Vernon, Wolfe.

Biophysics. Feher, Groisman, Jun, Kleinfeld, Nguyen-Huu, Okamura, Schoetz, Sharpee, Smith.

Condensed Matter Physics. Basov, Berkowitz, Butov, Dynes, Goodkind, Liebermann, Maple, Schuller, Schultz, Sinha, Smith.

High Energy Physics. Branson, Sharma, Swanson, Vernon, Wuerthwein, Yagil.

Nuclear Physics. Vernon.

Plasma and Fusion. Driscoll, Lovberg, Surko.

Polymer Physics/Science. Hwa.

View additional information about this department at
www.gradschoolshopper.com

UNIVERSITY OF CALIFORNIA, SANTA CRUZ

DEPARTMENT OF PHYSICS

Santa Cruz, California 95064
http://physics.ucsc.edu

General University Information

President: Mark Yudof
Dean of Graduate School: Tyrus Miller
University website: http://www.ucsc.edu
Control: Public
Setting: Suburban
Total Faculty: 860
Total Graduate Faculty: 860
Total number of Students: 17,404
Total number of Graduate Students: 1,426

Department Information

Department Chairman: Michael Dine, Chair
Department Contact: Davina Walker, Graduate Adviser
  Total full-time faculty: 45
  Full-Time Graduate Students: 52
  First-Year Graduate Students: 8

Female First-Year Students: 3
Total Post Doctorates: 12

Department Address

1156 High Street
Santa Cruz, CA 95064
Phone: (831) 459-4122
Fax: (831) 459-5265
E-mail: gradadviser@physics.ucsc.edu
Website: http://physics.ucsc.edu

ADMISSIONS

Admission Contact Information

Address admission inquiries to: University of California, Santa Cruz, Graduate Application Processing, 1156 High Street, Santa Cruz, CA 95064

Phone: (831) 459-5905
E-mail: gradadm@ucsc.edu
Admissions website: http://graddiv.ucsc.edu

Application deadlines
Fall admission:
U.S. students: January 15   *Int'l. students*: January 15

Application fee
U.S. students: $80   *Int'l. students*: $100

Admissions information
For Fall of 2013:
  Number of applicants: 205
  Number admitted: 61
  Number enrolled: 18

Admission requirements
Bachelor's degree requirements: A bachelor's degree in physics is required.
Minimum undergraduate GPA: 3.0

GRE requirements
The GRE is required.

Advanced GRE requirements
The Advanced GRE is required.

TOEFL requirements
The TOEFL exam is required for students from non-English-speaking countries.
  PBT score: 550
  iBT score: 83

Other admissions information
Additional requirements: none.

TUITION

Tuition year 2012–13:
Tuition for in-state residents
  Full-time students: $5,391 per quarter
Tuition for out-of-state residents
  Full-time students: $10,425 per quarter
Credit hours per semester to be considered full-time: 12
Deferred tuition plan: No
Health insurance: Available at the cost of $2667 per year.
Academic term: Quarter
Number of first-year students who received full tuition waivers: 16

Teaching Assistants, Research Assistants, and Fellowships
Number of first-year
  Teaching Assistants: 18
Average stipend per academic year
  Teaching Assistant: $17,654
  Research Assistant: $22,992
  Fellowship student: $21,000

FINANCIAL AID

Application deadlines
Fall admission:
U.S. students: January 15   *Int'l. students*: January 15

Loans
Loans are available for U.S. students.
Loans are not available for international students.
GAPSFAS application required: No
FAFSA application required: Yes

For further information
Address financial aid inquiries to: Financial Aid and Scholarship Office, University of California, Santa Cruz, 1156 High Street, 205 Hahn Student Services Building, Santa Cruz, CA 95064.
Phone: (831) 459-2963
Financial aid website: http://www2.ucsc.edu/fin-aid

HOUSING

Availability of on-campus housing
Single students: Yes
Married students: Yes

For further information
Address housing inquiries to: Graduate Housing Office, University of California, Santa Cruz, 1156 High Street, Santa Cruz, CA 95064.
Phone: (831) 459-5712
E-mail: gradhsg@ucsc.edu
Housing aid website: http://housing.ucsc.edu

Table A—Faculty, Enrollments, and Degrees Granted

Research Specialty	2012-13 Faculty	Enrollment 2012-13 Master's	Enrollment 2012-13 Doctorate	Degrees Master's	Degrees Terminal Master's	Degrees Doctorate
Physics and other Science Education	45	2	57	12(47)	4(9)	16(71)
Total	45	2	57	12(47)	4(9)	16(71)
Full-time Grad. Stud.	–	2	57	–	–	–
First-year Grad. Stud.	–	2	16	–	–	–

GRADUATE DEGREE REQUIREMENTS

Master's: Coursework plus a master's thesis or qualifying examinations are required.
Doctorate: Coursework plus qualifying examinations and a dissertation are required.

SPECIAL EQUIPMENT, FACILITIES, OR PROGRAMS

Astronomy and Astrophysics graduate students have access to state-of-the-art instrument development and data reduction technology, the UCO/Lick Observatory computer network, and an unusually extensive astronomical library at the Lick Observatory headquarters on campus. Graduate students may conduct supervised research with selected telescopic facilities of the Lick Observatory on Mount Hamilton, 55 miles from Santa Cruz. The 10-meter Keck Telescope in Hawaii, the world's largest, is administered from the University of California, Santa Cruz campus and is used for frontier research by University of California astronomers. The Center for Adaptive Optics (CfAO) is also headquartered at UCSC. Education is central to the CfAOs mission, with a particular focus on graduate students. In addition to research facilities, the center provides access to an interdisciplinary nationwide network of scientists in astronomy and vision science.

FACULTY

Professor
Banks, Tom, Ph.D., Massachusetts Institute of Technology, 1973. String and particle theory; quantum gravity and cosmology.
Belanger, David P., Ph.D., University of California, Santa Barbara, 1981. Chair, Graduate Recruitment Committee (Phys-

ics). Experimental condensed matter physics; phase transitions.

Bernstein, Rebecca, Ph.D., California Institute of Technology, 1997. Vice-chair (Astronomy). *Astronomy, Astrophysics*. Formation and evolution of galaxies and stellar populations; astronomical instrumentation and optical design.

Bolte, Mike J., Ph.D., University of Washington, 1987. *Astronomy, Astrophysics*. Dynamics of star clusters; ages of star clusters; chemical enrichment history of the galaxy; observations of interacting galaxies.

Brodie, Jean P., Ph.D., University of Cambridge. *Astronomy, Astrophysics*. Galaxies; instrumentation.

Carter, Sue A., Ph.D., University of Chicago, 1993. Experimental condensed matter physics; polymer physics; molecular electronic; phase transitions; electronic and optical properties of materials.

Deutsch, Joshua, Ph.D., University of Cambridge, 1983. Condensed matter theory.

Dine, Michael, Ph.D., Yale University, 1978. Chair, Department of Physics. Particle theory.

Epps, Harland W., Ph.D., University of Wisconsin-Madison, 1964. *Astronomy, Astrophysics*. Astronomical optics and instrumentation.

Faber, Sandra M., Ph.D., Harvard University, 1972. Director, University of California Observatories. *Astronomy, Astrophysics*. Galaxies; stellar populations; cosmology; instrumentation.

GuhaThakurta, Puragra (Raja), Ph.D., Princeton University, 1989. *Astronomy, Astrophysics*. Faint blue galaxies, study of faint stars using multicolor CCD data; search for Kuiper belt comets; gravitational lensing by galaxy clusters; HST studies of dense globular cluster cores; near infrared Tully-Fisher diagram; galactic "cirrus" clouds; interacting galaxies and dwarf galaxies.

Haber, Howard, Ph.D., University of Michigan, 1978. Theory and phenomenology of fundamental particles and their interactions.

Illingworth, Garth D., Ph.D., Australian National University, 1973. *Astronomy, Astrophysics*. Astrophysics; stellar and galaxy dynamics; instrumentation.

Johnson, Robert P., Ph.D., Stanford University, 1986. Experimental high energy physics; astrophysics.

Koo, David C., Ph.D., University of California, Berkeley, 1981. *Astronomy, Astrophysics*. Cosmology; birth and evolution of galaxies and quasars.

Laughlin, Gregory, Ph.D., University of California, Santa Cruz, 1994. Chair (Astronomy). *Astronomy, Astrophysics*. Extrasolar planets; numerical astrophysics; astrophysical phenomena of the extremely distant future.

Lin, Douglas N.C., Ph.D., University of Cambridge. *Astronomy, Astrophysics*. Fluid dynamics; star formation; galactic structure; planetary systems; accretion disks.

Madau, Piero, Ph.D., International School for Advanced Studies (Trieste). Director, NEXSI. *Astronomy, Astrophysics*. Cosmology; high energy astrophysics.

Max, Claire, Ph.D., Princeton University. Director, Center for Adaptive Optics. *Astronomy, Astrophysics*. Adaptive optics; planetary science.

Narayan, Onuttom, Ph.D., Princeton University. Condensed matter theory.

Primack, Joel, Ph.D., Stanford University. Director, University of California Systemwide High Performance Astro-Computing Center. Cosmology; galaxy; formation and evolution; particle astrophysics; nature of dark matter; gamma-ray astronomy.

Prochaska, Jason, Ph.D., University of California, San Diego. Associate Director, Lick Observatory. *Astronomy, Astro-physics*. Damped Lya systems in quasars; Lyman limit systems; stellar abundances; thick disk imaging of the galaxy.

Ramirez-Ruiz, Enrico, Ph.D., University of Cambridge. Director, TASC. *Astronomy, Astrophysics*. Stellar explosions; gamma-ray bursts; accretion physics near compact stars.

Ritz, Steven, Ph.D., University of Wisconsin-Madison. Director, SCIPP. Particle physics and astrophysics.

Rockosi, Constance, Ph.D., University of Chicago. Associate Director, UCO, Technical Facilities. *Astronomy, Astrophysics*. Galactic structure; stellar populations; CCD detectors; astronomical instrumentation.

Schlesinger, Zack, Ph.D., Cornell University. Condensed matter experimental.

Schumm, Bruce, Ph.D., University of Chicago. Particle experimental.

Seiden, Abraham, Ph.D., University of California, Santa Cruz. Particle experimental physics.

Shastry, B. Sriram, Ph.D., Tata Institute of Fundamental Research. Condensed matter theory.

Smith, David M., Ph.D., University of California, Berkeley. Particle experiment; high energy astrophysics.

Smith, Graeme H., Ph.D., Australian National University. *Astronomy, Astrophysics*. Stellar populations; chromospheric activity among late-type stars.

Vogt, Steven S., Ph.D., University of Texas, Austin. *Astronomy, Astrophysics*. Stellar spectroscopy; instrumentation.

Woosley, Stanford E., Ph.D., Rice University. *Astronomy, Astrophysics*. Nuclear astrophysics; stellar structure.

Young, A. Peter, Ph.D., University of Oxford. Condensed matter theory.

Associate Professor

Aguirre, Anthony, Ph.D., Harvard University. Particle theory.

Fortney, Jonathan J., Ph.D., University of Arizona. *Astronomy, Astrophysics*. The physics of giant planet atmospheres and interiors, with a focus on exoplanets.

Krumholz, Mark R., Ph.D., University of California, Berkeley. *Astronomy, Astrophysics*. Studies star formation and the interstellar medium using both analytic and numerical techniques, with particular focus on how massive stars and star clusters form, the origin of the stellar initial mass function, the life cycles of molecular clouds, and regulation of the star formation rate on galactic scales.

Nielsen, Jason A., Ph.D., University of Wisconsin-Madison. Experimental high energy physics.

Profumo, Stefano, Ph.D., International School for Advanced Studies (Trieste). Chair, Graduate Committee (Physics). Theory of particle physics and particle astrophysics.

Assistant Professor

Conroy, Charlie, Ph.D., Princeton University. *Astronomy, Astrophysics*. Galaxy formation and evolution; stellar populations.

Gweon, Gey-Hong, Ph.D., University of Michigan. Condensed matter experimental.

Jeltema, Tesla, Ph.D., Massachusetts Institute of Technology. Large-scale structure; galaxy evolution; indirect detection of dark matter.

Sher, Alexander, Ph.D., University of Pittsburgh. *Biophysics*. Development of experimental techniques for recording and stimulation of activity at hundreds of neurons and use of these techniques to study neural function, structure, and development.

Emeritus

Nelson, Jerry E., Ph.D., University of California, Berkeley. *Astronomy*. Design and construction of large telescopes; project scientist for Keck telescope and Thirty Meter telescope.

Lecturer

Steinacker, Adriane, Ph.D., University of Bonn. *Astronomy, Astrophysics.* Magneto-hydrodynamical (MHD) simulations of protoplanetary accretion disks and the interaction between turbulent accretion disks and planetary cores.

DEPARTMENTAL RESEARCH SPECIALTIES AND STAFF

Theoretical

Astronomy. Research within astrophysical and planetary sciences ranges from the (relatively) pedestrian bodies of this solar system to the highly exotic and energetic events of the Big Bang. Planetary sciences covers the characterization, origin, and evolution of planetary bodies, in both this solar system and elsewhere (exoplanets). Moving slightly further afield, research into the Milky Way and environs covers the evolution of stars and stellar populations. More distant, more energetic and more ancient phenomena are covered by the High Energy & Particles and Extragalactic & Cosmology fields of research, including studies of quasars, black holes, supernovae, and gamma-ray bursts. These broad research categories are facilitated by instrumentation (telescopes, spacecraft) and computational research areas. Bernstein, Bolte, Brodie, Conroy, Epps, Faber, Fortney, GuhaThakurta, Illingworth, Koo, Krumholz, Laughlin, Lin, Madau, Max, Narayan, Prochaska, Ramirez-Ruiz, Rockosi, Graeme Smith, Vogt, Woosley.

View additional information about this department at
www.gradschoolshopper.com

UNIVERSITY OF SOUTHERN CALIFORNIA

DEPARTMENT OF PHYSICS AND ASTRONOMY

Los Angeles, California 90089-0484
http://dornsife.usc.edu/physics/

General University Information

President: C. L. Max Nikias
Dean of Graduate School: Sally (Sarah) Pratt, Vice Provost
University website: http://www.usc.edu/
Control: Private
Setting: Urban
Total Faculty: 5,286
Total Graduate Faculty: 5,286
Total number of Students: 38,000
Total number of Graduate Students: 20,500

Department Information

Department Chairman: Hans Bozler, Chair
Department Contact: Betty Byers, Graduate Coordinator
  Total full-time faculty: 40
  Total number of full-time equivalent positions: 40
  Full-Time Graduate Students: 74
  First-Year Graduate Students: 11
  Female First-Year Students: 2
  Total Post Doctorates: 5

Department Address

3620 Mc Clintock Avenue
SGM 408
Los Angeles, CA 90089-0484
Phone: (213) 740-0848 (C)
Fax: (213) 740-8094
E-mail: physics@college.usc.edu
Website: http://dornsife.usc.edu/physics/

ADMISSIONS

Admission Contact Information

Address admission inquiries to: Department of Physics and Astronomy
Phone: (213) 740-1117
E-mail: physicsgradadmis@college.usc.edu
Admissions website: http://www.usc.edu/admission/graduate/

Application deadlines

Fall admission:
U.S. students: December 1   *Int'l. students*: December 1

Application fee

U.S. students: $85   *Int'l. students*: $85

Admissions information

For Fall of 2013:
  Number of applicants: 147
  Number admitted: 28
  Number enrolled: 15

Admission requirements

Bachelor's degree requirements: Bachelor's degree in physics is required.
Minimum undergraduate GPA: 3.2

GRE requirements

The GRE is required.
  Quantitative score: 750
  Verbal score: 550
  Mean GRE score range (25th–75th percentile): V+Q=1300

Advanced GRE requirements

The Advanced GRE is required.
  Minimum accepted Advanced GRE score: 780

TOEFL requirements

The TOEFL exam is required for students from non-English-speaking countries.

PBT score: 600
iBT score: 80

Other admissions information

Additional requirements: The minimum acceptable score suggested for admission is verbal—550; quantitative—750; total—1,300.

The GRE Advanced is required, the minimum acceptable score suggested for admission is 780.

A minimum TOEFL score of 100 or higher with no less than 20 on each of the four individual sections of the Internet-based TOEFL (iBT) or (600 on paper-based/250 or higher on the computer-based TOEFL) is required for Teaching Assistants.

Undergraduate preparation assumed: Reitz and Milford, Foundations of Electromagnetic Theory; Eisberg, Quantum Physics for Atoms, Molecules, Solids, Nuclei, and Particles; Saxon, Elementary Quantum Mechanics; Reif, Foundation of Statistical and Thermal Physics; Boyce and DiPrima, Elementary Differential Equations and Boundary Value Problems.

TUITION

Tuition year 2013-14:
Full-time students: $18,432 per semester
Part-time students: $1,536 per credit
Credit hours per semester to be considered full-time: 6
Deferred tuition plan: Yes
Health insurance: Available at the cost of $1,519 per year.
Other academic fees: New Student Orientation Fee $35.00 (first semester) Student Programming Fee $37.50 per semester Student Services Fee $12.50 per semester Norman Topping Student Aid Fund $8.00 per semester.
Academic term: Semester

Teaching Assistants, Research Assistants, and Fellowships

Number of first-year
Teaching Assistants: 4
Fellowship students: 12
Average stipend per academic year
Teaching Assistant: $20,500
Research Assistant: $20,500
Fellowship student: $30,000
Additional summer support is available for Teaching and Research Assistants.

FINANCIAL AID

Application deadlines

Fall admission:
U.S. students: December 1   *Int'l. students*: December 1

Loans

Loans are available for U.S. students.
Loans are not available for international students.
GAPSFAS application required: No
FAFSA application required: No

For further information

Address financial aid inquiries to: Department of Physics and Astronomy.
Phone: (213) 740-1117
E-mail: physicsgradadmis@college.usc.edu
Financial aid website: http://dornsife.usc.edu/physics/graduate/graduate_financial_support.cfm

HOUSING

Availability of on-campus housing

Single students: Yes
Married students: Yes

For further information

Address housing inquiries to: Housing Services Office, USC, Los Angeles, CA 90089-1332.
Phone: (800) 872-4632
E-mail: housing@usc.edu
Housing aid website: http://housing.usc.edu/

Table A—Faculty, Enrollments, and Degrees Granted

Research Specialty	2012-13 Faculty	Enrollment Fall 2012		Number of Degrees Granted 2012-13 (2009-13)		
		Master's	Doctorate	Master's	Terminal Master's	Doctorate
Astronomy	3	–	–	–	–	–(1)
Astrophysics	2	–	3	–	–	1(3)
Atmosphere, Space Physics, Cosmic Rays	1	–	1	–	–	–
Atomic, Molecular, & Optical Physics	8	–	3	–	–	–(4)
Biophysics	3	–	8	–	–	–
Chemical Physics	3	–	1	–	–	–(3)
Computational Physics	1	–	5	–	–	1(2)
Condensed Matter Physics	13	–	15	–	–	2(10)
Cosmology & String Theory	6	–	9	–	–	–(4)
Electrical Engineering	2	–	2	–	–	–(1)
High Energy Physics	6	–	9	–	–	–(3)
Low Temperature Physics	3	1	1	–	–	–(2)
Materials Science, Metallurgy	1	–	1	–	–	–
Nano Science and Technology	2	–	4	–	–	–(2)
Plasma and Fusion	2	–	2	–	–	1(1)
Quantum Foundations	–	–	3	–	–	2(6)
Statistical & Thermal Physics	1	–	–	–	–	–
Non-specialized	–	–	13	3(12)	–(4)	–
Total	–	1	72	3(12)	–(4)	
Full-time Grad. Stud.	–	1	72	–	–	–
First-year Grad. Stud.	–	–	14	–	–	–

GRADUATE DEGREE REQUIREMENTS

Master's: The M.S. Physics degree requires satisfactory completion of seven courses of which no more than one course may be directed research. The M.A. Physics degree requires satisfactory completion of eight courses (exclusive of directed research) plus a comprehensive exam. For all master's degrees, a GPA of 3.0 and one-year residency is required; there is no language requirement.

Doctorate: A minimum of 11 courses exclusive of dissertation and directed research courses, taken at this university and elsewhere with a minimum GPA of 3.0; comprehensive exam, qualifying exam, dissertation, and dissertation exam required; one-year residency required; there is no language requirement.

Thesis: Thesis may not be written in absentia.

SPECIAL EQUIPMENT, FACILITIES, OR PROGRAMS

Molecular beam lab, low temperature labs, laser optics labs, biological physics labs, electron microscopy, nanofabrication labs, nanomaterials synthesis.

Large scale parallel computing facilities, quantum computing (D-Wave).

Table B—Separately Budgeted Research Expenditures by Source of Support

Source of Support	Departmental Research	Physics-related Research Outside Department
Federal government	$5,300,000	
State/local government		
Non-profit organizations		
Business and industry		
Other		
Total	$5,300,000	

Table C—Separately Budgeted Research Expenditures by Research Specialty

Research Specialty	No. of Grants	Expenditures ($)
Astrophysics	10	$1,280,000
Atomic, Molecular, & Optical Physics	1	$70,000
Biophysics	4	$400,000
Condensed Matter Physics	6	$495,000
Nano Science and Technology	2	$200,000
Particles and Fields	7	$355,000
Computational Physics	15	$2,500,000
Total	45	$5,300,000

FACULTY

Distinguished University Professor

Armstrong, Lloyd, Ph.D., University of California, Berkeley, 1966. Former Provost of USC.

Professor

Bars, Itzhak, Ph.D., Yale University, 1971. Theoretical elementary particle physics.

Bergmann, Gerd, Ph.D., University of Göttingen, 1963. Experimental condensed matter physics.

Bickers, Nelson E., Ph.D., Cornell University, 1986. Vice Provost for Undergraduate Programs. Theoretical condensed matter physics.

Bozler, Hans, Ph.D., Stony Brook University, 1972. Experimental low-temperature and condensed matter physics.

Dappen, Werner, Ph.D., ETH, Zurich, 1978. Theoretical atomic physics; astronomy.

Feinberg, Jack, Ph.D., University of California, Berkeley, 1977. Experimental laser physics; nonlinear optics.

Gould, Christopher, Ph.D., Cornell University, 1978. Experimental low-temperature and condensed matter physics.

Haas, Stephan, Ph.D., Florida State University, 1995. Theoretical condensed matter physics.

Johnson, Clifford V., Ph.D., University of Southampton. Theoretical quantum gravity; string theory.

Kalia, Rajiv, Ph.D., Northwestern University, 1976. Computational condensed matter physics.

Kresin, Vitaly, Ph.D., University of California, Berkeley, 1992. Physics of small clusters.

Nemeschansky, Dennis, Ph.D., Princeton University, 1984. Theoretical high-energy physics.

Pierpaoli, Elena, Ph.D., SISSA - ISAS, Trieste, Italy, 1998. Cosmology and theoretical astrophysics.

Pilch, Krzysztof, Ph.D., University of Wroclaw, 1979. Theoretical elementary particle physics.

Rhodes, Edward J., Ph.D., University of California, Los Angeles, 1977. *Astronomy.*

Saleur, Hubert, Ph.D., Universite Paris 6, 1987. Theoretical elementary particle physics.

Shakeshaft, Robin, Ph.D., University of Nebraska-Lincoln, 1972. Theoretical atomic physics.

Wagner, William G., Ph.D., California Institute of Technology, 1962. Theoretical quantum electronics; elementary particle physics.

Warner, Nicholas P., Ph.D., University of Cambridge, 1982. Theoretical elementary particle physics.

Zanardi, Paolo, Ph.D., Università di Roma, 1995. Theoretical physics and quantum information science.

Associate Professor

Lu, Jia Grace, Ph.D., Harvard University, 1997. Experimental condensed matter physics; nanoscale materials and electronics.

Thompson, Richard S., Ph.D., Harvard University, 1965. Theoretical superconductivity and low-temperature physics.

Assistant Professor

El-Naggar, Moh, Ph.D., California Institute of Technology, 2006. Biological nanostructure.

Haselwandter, Christoph A., Ph.D., Imperial College, 2007. *Biophysics.* Theoretical biological physics.

Professor Emeritus

Chang, Tu-nan, Ph.D., University of California, Riverside, 1972. Theoretical atomic physics.

Judge, Darrell L., Ph.D., University of Southern California, 1965. Experimental atomic and molecular physics; space physics.

Research Professor

Didkovsky, Leonid, Ph.D., Main Astronomical Observatory of Ukraine, 1990. *Astrophysics.*

Peters, Geraldine, Ph.D., University of California, Los Angeles, 1974. Stellar astrophysics.

Wu, Robert, Ph.D., University of Illinois, 1973. Emeritus. Chemical physics.

Research Assistant Professor

Campos Venuti, Lorenzo, Ph.D., University of Stuttgart, 2003. *Condensed Matter Physics, Quantum Foundations.*

Colombo, Loris, Ph.D., University of Milano, 2005. Cosmology.

Joint Appointment

Brun, Todd, Ph.D., California Institute of Technology, 1994. Quantum information theory.

Dapkus, Daniel, Ph.D., University of Illinois, 1970. Electrophysics and photonics.

Gundersen, Martin A., Ph.D., University of Southern California, 1972. Experimental laser physics.

Hellwarth, Robert W., Ph.D., University of Oxford, 1955. Theoretical and experimental laser physics; solid state and statistical physics.

Kunc, Joseph, Ph.D., Warsaw Technical University, 1974. Plasma physics.

Levi, Anthony F. J., Ph.D., University of Cambridge, 1983. Photonic devices.

Lidar, Daniel, Ph.D., Hebrew University of Jerusalem, 1997. Physical theoretical chemistry.

Madhukar, Anupam, Ph.D., California Institute of Technology, 1971. Experimental and theoretical quantum-well physics.

Nakano, Aiichiro, Ph.D., University of Tokyo, 1989. Computational physics.

Povinelli, Michelle, Ph.D., Massachusetts Institute of Technology, 2004. *Electrical Engineering.*

Takahashi, Susumu, Ph.D., University of Florida, 2005. *Biophysics, Chemical Physics, Medical, Health Physics.*

Tanguay, Armand, Ph.D., Yale University, 1977. Optics and photonics.

Vashishta, Priya, Ph.D., Indian Institute of Technology, Kampur, 1967. Computational physics.

Vilesov, Andrey, Ph.D., St. Petersburg State University, 1985. Physical chemistry.

DEPARTMENTAL RESEARCH SPECIALTIES AND STAFF

Theoretical

Astrophysics. Theoretical Solar Physics uses the Sun as a plasma physics laboratory. On the one hand, this work involves state-of-the-art solar modeling and an analysis of helioseismic data. On the other hand, helioseismology is the first accurate experiment that puts strong constraints on the thermodynamic quantities of the plasma of stellar interiors. Cosmology is interested in a variety of scientific issues including dark matter and dark energy models, the early Universe, cosmological parameter determination, data analysis and interpretation. The group offers the opportunity of collaborative research with scientists at the Jet Propulsion Laboratory on current and future space missions. Dappen, Pierpaoli.

Atomic, Molecular, & Optical Physics. Interactions of strong electromagnetic radiation (lasers) with matter; multiphoton processes; energy development related basic atomic transitions; many-body approach to atomic transitions; multiply excited atomic resonances; atomic optics; collective properties in atomic traps and confined Bose-Einstein condensates; atomic lithography; atoms and ions in dense stellar plasmas; interactions of high-power laser beams with matter leading to nonlinear optical phenomena; propagation of light in dense inhomogeneous plasmas; free-electron lasers; high-power unstable laser oscillators. Chang, Didkovsky, Feinberg, Gundersen, Hellwarth, Judge, Kresin, Kunc, Shakeshaft, Takahashi, Tanguay, Wu.

Biophysics. The area is characterized by the use of quantitative methods, modeling, and physics based experimental methods to make discoveries related to biological systems. This sub-discipline anticipates the movement of both basic biological science and medical applications to areas where physics has an ever greater impact. Haas, Kalia, Madhukar.

Computational Physics. Large-scale simulations of quantum spin liquids, atomic spectra, time-dependent atomic processes in intense fields, multi-scale hybrid simulations of materials, algorithm design, high performance programming environments for massively parallel machines, interactive three-dimensional scientific visualization. Brun, Campos Venuti, Haas, Kalia, Nakano.

Condensed Matter Physics. Superconductivity, superfluidity in 3He; nonlinear transport phenomena in reduced dimensionality; electronic structures, strongly correlated metals; nonlinear phenomena in metals; high-Tc superconductors; quantum magnetism. Bergmann, Bickers, Bozler, Brun, Campos Venuti, Dapkus, El-Naggar, Gould, Haas, Kalia, Kresin, Levi, Lidar, Lu, Madhukar, Nakano, Povinelli, Saleur, Thompson, Vashishta, Wagner, Zanardi.

High Energy Physics. Quantum field theory and unification of fundamental interactions; cosmology; superstring theory; M-theory; gauge theories; supersymmetry and supergravity; conformal field theory; statistical mechanics; mathematical physics; integrable models. Bars, Johnson, Nemeschansky, Pilch, Saleur, Warner.

Quantum Foundations. Use of quantum mechanical resources for computation, communication, and other information-processing tasks. Effect of noise and decoherence; quantum error correction and suppression, dynamical decoupling, and decoherence-free subspaces and subsystems; weak and continuous measurements and quantum trajectories; quantum random walks; quantification of entanglement; quantum information theory; algebraic description of quantum states and observables; quantum information and many-body physics; entanglement and quantum phase transitions; quantum algorithms for classical statistical physics; quantum process tomography; geometric phases; adiabatic, holonomic, and topological quantum computation; quantum computing implementations, using quantum dots, linear optics, and magnetic resonance force microscopy; quantum information. Brun, Campos Venuti, Haas, Lidar, Zanardi.

Experimental

Astrophysics. Study of the overall structure, composition, origins and evolution of the Universe; analysis of cosmic microwave background data and the study of dark matter and galaxy clusters. Study of the structure and dynamics of the solar atmosphere and interior using observations and theory of solar local and global oscillations; use of helioseismology to probe properties of dense plasmas. Didkovsky, Judge, Peters, Pierpaoli, Rhodes, Wu.

Atmosphere, Space Physics, Cosmic Rays. Laser spectroscopy, highly excited atomic states, study of planetary atmospheres from space flight experiments; photoabsorption and emission in planetary atmospheres; vacuum ultraviolet radiation interacting with gaseous plasmas. Didkovsky, Judge, Wu.

Biophysics. Research at the interface between biological and inorganic systems. Cell-surface interactions. Microbe-inorganic interactions. Bioenergy production in microbial fuel cells. Electronic and enzymatic activity of extracellular nanostructures. Intersection of information and biological sciences. Biochemical sensors and intelligent bio-mimetic coatings for neural prostheses. Optical imaging and spectroscopic studies of intra-cellular biochemical processes. El-Naggar.

Condensed Matter Physics. Electronic transport and quantum-interference in two-dimensional metals, superconducting films, magnetic surface impurities, anomalous Hall effect and spin-orbit scattering, quantum wires; quasi-one-dimensional conductors; localization; magnetoresistance. Superfluidity in 3He; transport properties of anisotropic conductors and thin alkali metal films; phase transitions in metals of low dimensionality; two-dimensional magnetism of 3He films; development of primary thermometry at ultralow temperatures. Electronic and thermal properties of size-quantized metal nanoclusters; photodissociation, transport, and particle growth in ultra-cold liquid helium clusters. Bergmann, Bozler, Feinberg, Hellwarth, Kresin, Levi, Lu, Povinelli, Wu.

Optics. Laser spectroscopy; nonlinear optical mixing; optical fibers and devices; phase-conjugation, photorefractive effect; Raman-induced Kerr effect spectroscopy; spectroscopy of glassy solids; laser plasma studies; photochemistry of simple molecular systems; interaction in wave guides. Feinberg, Hellwarth.

View additional information about this department at
www.gradschoolshopper.com

COLORADO SCHOOL OF MINES

DEPARTMENT OF PHYSICS

Golden, Colorado 80401
http://physics.mines.edu

General University Information
President: Myles W. Scoggins
Dean of Graduate School: Thomas M. Boyd
University website: http://www.mines.edu
Control: Public
Setting: Suburban
Total Faculty: 368
Total Graduate Faculty: 230
Total number of Students: 5,632
Total number of Graduate Students: 1,463

Department Information
Department Chairman: Thomas Furtak, Head
Department Contact: Barbara Johnson, Program Assistant
　Total full-time faculty: 37
　Total number of full-time equivalent positions: 25
　Full-Time Graduate Students: 72
　First-Year Graduate Students: 10
　Female First-Year Students: 2
　Total Post Doctorates: 17

Department Address
1523 Illinois Street
Meyer Hall, Room 325
Golden, CO 80401
Phone: (303) 273-3830
Fax: (303) 273-3919
E-mail: bpjohnso@mines.edu
Website: http://physics.mines.edu

ADMISSIONS

Admission Contact Information
Address admission inquiries to: David Wood
Phone: (303) 273-3853
E-mail: dmwood@mines.edu
Admissions website: http://www.mines.edu/graduate_admissions

Application deadlines
Fall admission:
U.S. students: January 15 　　*Int'l. students*: June 1
Spring admission:
U.S. students: October 15 　　*Int'l. students*: October 15

Application fee
U.S. students: $75 　　*Int'l. students*: $95
Fee is discounted for applications received prior to March 1: $50
　domestic, $75 international.

Admissions information
For Fall of 2013:
　Number of applicants: 89
　Number admitted: 15
　Number enrolled: 10

Admission requirements
Bachelor's degree requirements: Bachelor's degree in physics
　is required.
Minimum undergraduate GPA: 3.0

GRE requirements
The GRE is required.
The average score (V+Q) on the General Test of students recently, under the new scoring system, was 319.

Advanced GRE requirements
The Advanced GRE is recommended.
The average score on the Advanced (Physics Subject) Test of students admitted for F2013 was 720.

TOEFL requirements
The TOEFL exam is required for students from non-English-speaking countries.
　PBT score: 550
　iBT score: 79

Other admissions information
Additional requirements: The Advanced GRE Test is strongly urged and is required for financial aid.
Undergraduate preparation assumed: One semester of classical mechanics at the level of Marion, two semesters of electromagnetism at the level of Griffiths, one year of modern physics, and one semester each of thermodynamics, optics, mathematical physics, and electronics.

TUITION

Tuition year 2013-14:
Tuition for in-state residents
　Full-time students: $7,200 per semester
　Part-time students: $2,800 per semester
Tuition for out-of-state residents
　Full-time students: $15,170 per semester
　Part-time students: $5,900 per semester
Part-time tuition is for 1 to 4 credit hours.
Credit hours per semester to be considered full-time: 9
Deferred tuition plan: Yes
Health insurance: Available at the cost of $1,596 per year.
Other academic fees: $1,043 per academic term.
Academic term: Semester
Number of first-year students who received full tuition waivers: 9
Number of first-year students who received partial tuition waivers: 3

Teaching Assistants, Research Assistants, and Fellowships
Number of first-year
　Teaching Assistants: 6
　Fellowship students: 4
Average stipend per academic year
　Teaching Assistant: $22,838
　Research Assistant: $23,751
　Fellowship student: $22,838
All listed stipends are for initial appointments. More senior graduate students receive larger stipends.

FINANCIAL AID

Application deadlines
Fall admission:
U.S. students: January 15 　　*Int'l. students*: June 1
Spring admission:

U.S. students: October 15   *Int'l. students*: October 15

Loans

Loans are available for U.S. students.
Loans are not available for international students.
GAPSFAS application required: No
FAFSA application required: No

For further information

Address financial aid inquiries to: David Wood.
Phone: (303) 273-3853
E-mail: dmwood@mines.edu
Financial aid website: http://gradschool.mines.edu/
   GS-Assistantship-Policies

HOUSING

Availability of on-campus housing
Single students: Yes
Married students: Yes

For further information

Address housing inquiries to: Housing Office, Colorado School
   of Mines.
Phone: (800) 446-9488, x3350
Housing aid website: http://www.mines.edu/Housing_GRPR

Table A—Faculty, Enrollments, and Degrees Granted

Research Specialty	2013-14 Faculty	Enrollment Fall 2013		Number of Degrees Granted 2012-13 (2009-13)		
		Master's	Doctorate	Master's	Terminal Master's	Doctorate
Condensed Matter						
Physics	9	10	28	–	5(17)	7(17)
Nuclear Physics	4	1	13	–(2)	1(7)	1(8)
Optics	4	7	6	–	3(9)	2(4)
Other	7	2	5	–	–	–
Total	24	20	52	–(2)	9(33)	10(29)
Full-time Grad. Stud.	–	20	49	–	–	–
First-year Grad. Stud.	–	3	7	–	–	–

GRADUATE DEGREE REQUIREMENTS

Master's: Thirty-six semester hours in approved program with 3.0 GPA; thesis; no foreign language. 27 credit hours including thesis must be taken in residence.

Doctorate: Thirty-four semester hours coursework, 38 of research credit. Coursework includes 12 hours in a specialty area, which include programs in Optical Science and Engineering, Photovoltaics, Nanotechnology, Materials Physics, Electromagnetics, Nuclear Physics and Astroparticle Physics in addition to topic areas in the other degree programs at CSM. Ph.D. candidacy established by grades or oral examination. Two semesters in residence are required.

Other Degrees: Interdisciplinary research is organized under centers: Renewable Energy Materials Research Science and Engineering Center (REMRSEC), Center for Microintegrated Optics for Advanced Bioimaging and Control (MOABC), Nuclear Science and Engineering Center (NUSEC), Golden Energy Computing Organization (GECO). Special solar energy related research programs are available in conjunction with the nearby National Renewable Energy Laboratory (NREL). Interdisciplinary M.S. and Ph.D. degrees are granted in Materials Science and in Nuclear Engineering.

Thesis: Thesis may be written in absentia.

SPECIAL EQUIPMENT, FACILITIES, OR PROGRAMS

The Department specializes in applied physics. Available materials processing and lithography facilities provide extensive capabilities for making and characterizing nanocrystalline and amorphous semiconductors, patterned nanostructures, and self assembled nanostructures.

Capabilities include growth systems (e.g., PECVD, low pressure CVD, MOCVD, sputtering, and electrochemical deposition), transmission electron microscopy and field emission scanning electron microscopy, reactive ion etching, ion implantation, wet etching and cleaning stations, oxide and nitride growth and deposition, dopant diffusion, annealing furnaces, and a clean room with optical lithography.

Materials characterization capabilities include surface profilometers, visible and IR spectrometers, visible and FTIR spectroscopic ellipsometers, a complete x-ray analysis laboratory, an imaging x-ray photoelectron spectroscopy system, scanning Auger electron spectroscopy system, temperature dependent Seebeck and Hall effect systems, and characterization of photovoltaic devices and integrated circuits.

Also included are an electron microprobe, an atomic force microscope, near-field microscopy, confocal and conventional Raman spectroscopy, and various chemical techniques for determining elemental compositions. Laboratories are also equipped with electron paramagnetic resonance (EPR) and nuclear magnetic resonance (NMR) as well as instruments for temperature dependent electroluminescence and photoluminescence (PL), PL excitation spectroscopy, and microwave modulated PL.

The applied optics group maintains a state-of-the-art ultrafast (femtosecond) spectroscopy laboratory, including a 1 terawatt, 1 kHz, Ti:sapphire laser system, a 6 terawatt, 20 Hz, Ti:sapphire laser system used for development of novel x-ray sources, an extended cavity Ti:sapphire oscillator and a diode-pumped Nd:glass laser used for multiphoton imaging, and a fiber oscillator used for ultrafast spectroscopy and nonlinear imaging.

The department's microwave and millimeter wave laboratory includes a vector network analyzer with imaging capability and laser-based THz spectroscopy.

The subatomic physics laboratory includes a 180 keV ion accelerator and facilities for detector development as well as laser-based atmospheric monitoring instrumentation. Faculty participate in experiments at nuclear facilities in Oak Ridge (HRIBF), Argonne (ATLAS), and at TRIMF in Vancouver. Astroparticle physics of ultra-high energy cosmic rays is studied at the Pierre Auger Observatory in Argentina.

Computational physics is conducted on a massively 2144-core 268-node supercomputer with 16 GB of RAM, capable of a peak speed of 23 Tflop.

Table B—Separately Budgeted Research Expenditures by Source of Support

Source of Support	Departmental Research	Physics-related Research Outside Department
Federal government	$1,117,820	$4,350,283
State/local government	$101,841	$277,361
Non-profit organizations	$78,961	$43,334
Business and industry	$40,000	$1,459,100
Other		
Total	$1,338,622	$6,130,078

Table C—Separately Budgeted Research Expenditures by Research Specialty

Research Specialty	No. of Grants	Expenditures ($)
Condensed Matter Physics	40	$2,854,948
Nuclear Physics	20	$1,746,225
Optics	10	$1,190,134
Other	12	$1,677,393
Total	**82**	**$7,468,700**

FACULTY

Professor

Carr, Lincoln, Ph.D., University of Washington, 2001. *Atomic, Molecular, & Optical Physics, Computational Physics, Condensed Matter Physics, Nonlinear Dynamics and Complex Systems, Quantum Foundations, Theoretical Physics.* Theoretical condensed matter physics.

Collins, Reuben T., Ph.D., California Institute of Technology, 1984. REMRSEC Associate Director. *Applied Physics, Energy Sources & Environment, Materials Science, Metallurgy, Nano Science and Technology, Solid State Physics, Surface Physics.* Electronic materials and devices.

Furtak, Thomas E., Ph.D., Iowa State University, 1975. Department Head. *Applied Physics, Condensed Matter Physics, Energy Sources & Environment, Materials Science, Metallurgy, Nano Science and Technology, Optics, Surface Physics.* Raman spectroscopy; linear and nonlinear optical properties of materials.

Greife, Uwe, Ph.D., University of Bochum, 1994. NuSEC Director. *Applied Physics, Astrophysics, Nuclear Engineering, Nuclear Physics.* Experimental nuclear physics and astrophysics.

Kowalski, Frank V., Ph.D., Stanford University, 1978. *Applied Physics, Atomic, Molecular, & Optical Physics, Optics, Physics and other Science Education.* Experimental laser physics, technology-assisted learning.

Lusk, Mark T., Ph.D., California Institute of Technology, 1992. GECO Director. *Applied Physics, Computational Physics, Condensed Matter Physics, Energy Sources & Environment, Materials Science, Metallurgy, Nano Science and Technology, Solid State Physics, Surface Physics.* Theoretical and computational condensed matter physics.

Scales, John A., Ph.D., University of Colorado, 1984. *Applied Physics, Atomic, Molecular, & Optical Physics, Condensed Matter Physics, Electromagnetism, Theoretical Physics.* Mesoscopic materials, electromagnetics, wave phenomena.

Squier, Jeff, Ph.D., University of Rochester, 1992. MOABC Co-Director. *Applied Physics, Optics.* Ultrafast optics, nonlinear optics, biological imaging, microscopy, micromachining.

Taylor, P. Craig, Ph.D., Brown University, 1969. REMRSEC Director. *Applied Physics, Condensed Matter Physics, Energy Sources & Environment, Materials Science, Metallurgy, Nano Science and Technology, Solid State Physics.* Optical, electrical, and structural properties of crystalline and amorphous semiconductors.

Associate Professor

Durfee, Charles G., Ph.D., University of Maryland, 1994. *Applied Physics, Optics.* Generation, characterization of ultrashort laser pulses, interactions with atoms and plasmas.

Ohno, Timothy R., Ph.D., University of Maryland, 1989. *Applied Physics, Condensed Matter Physics, Energy Sources & Environment, Materials Science, Metallurgy, Solid State Physics, Surface Physics.* Experimental solid state physics, surface physics, photovoltaics.

Sarazin, Frederic, Ph.D., University of Caen, 1999. *Astrophysics, Atmosphere, Space Physics, Cosmic Rays, Nuclear Engineering, Nuclear Physics.* Experimental nuclear physics, astroparticle physics.

Wiencke, Lawrence, Ph.D., Columbia University, 1992. *Astrophysics, Atmosphere, Space Physics, Cosmic Rays.* Experimental particle astrophysics.

Wood, David M., Ph.D., Cornell University, 1981. *Computational Physics, Condensed Matter Physics, Materials Science, Metallurgy, Solid State Physics, Theoretical Physics.* Solid state theory.

Assistant Professor

Hager, Ulrike, Ph.D., University of Hamburg, 2007. *Astrophysics, Nuclear Engineering, Nuclear Physics.* Experimental nuclear astrophysics.

Sarkar, Susanta K., Ph.D., University of Oregon, 2006. *Biophysics, Nano Science and Technology, Optics.* Single molecule microscopy.

Toberer, Eric S., Ph.D., University of California, Santa Barbara, 2006. *Applied Physics, Condensed Matter Physics, Energy Sources & Environment, Materials Science, Metallurgy, Nano Science and Technology, Solid State Physics.* Experimental materials physics.

Wu, Zhigang, Ph.D., College of William and Mary, 2002. *Computational Physics, Condensed Matter Physics, Energy Sources & Environment, Materials Science, Metallurgy, Nano Science and Technology, Solid State Physics, Surface Physics, Theoretical Physics.* Theoretical and computational condensed matter physics, first principles electronic structure, atomistic model simulations.

Zimmerman, Jeramy D., Ph.D., University of California, Santa Barbara, 2008. *Applied Physics, Condensed Matter Physics, Energy Sources & Environment, Nano Science and Technology, Solid State Physics.* Experimental materials physics.

Professor Emeritus

Cecil, Edward, Ph.D., Princeton University, 1972. *Astrophysics, Nuclear Physics, Plasma and Fusion.* Experimental nuclear physics, high-temperature fusion plasma diagnostics.

McNeil, James A., Ph.D., University of Maryland, 1979. *Nuclear Physics, Theoretical Physics.* Archeological dating, biomechanics.

Williamson, Don L., Ph.D., University of Washington, 1971. *Applied Physics, Condensed Matter Physics, Energy Sources & Environment, Materials Science, Metallurgy, Solid State Physics.* Experimental solid state physics, Mössbauer effect, x-ray scattering.

Research Professor

Arnold, Gerald B., Ph.D., University of California, Los Angeles, 1977. *Condensed Matter Physics, Solid State Physics, Theoretical Physics.* Theoretical condensed matter physics.

Coffey, Mark W., Ph.D., Iowa State University, 1991. *Computational Physics, Condensed Matter Physics, Solid State Physics, Theoretical Physics.* Quantum computing algorithms, superconducting and magnetic systems.

Mace, Jonathan L., Ph.D., Washington State University, 1990. *Condensed Matter Physics.* Experimental and theoretical high energy density physics.

Shayer, Zeev, Ph.D., Tel-Aviv University, 1985. *Applied Physics, Nuclear Engineering, Nuclear Physics.* Nuclear physics and engineering.

Research Associate Professor

Beach, Joseph, Ph.D., Colorado School of Mines, 2002. *Applied Physics, Materials Science, Metallurgy.* Experimental materials physics, photovoltaics.

Bernard, James E., Ph.D., University of Delaware, 1984. *Computational Physics, Condensed Matter Physics, Solid State Physics.* Electronic materials theory.

Research Assistant Professor

Bradley, Scott, Ph.D., Massachusetts Institute of Technology, 2009. *Applied Physics, Atomic, Molecular, & Optical Physics, Materials Science, Metallurgy, Nano Science and Technology.* Experimental optoelectronics.

Flammer, P. David, M.S., Colorado School of Mines, 2001. *Applied Physics, Condensed Matter Physics, Electromagnetism, Nano Science and Technology, Theoretical Physics.* Experimental and theoretical condensed matter physics, plasmonics.

Kendrick, Chito E., Ph.D., University of Canterbury, 2008. *Applied Physics, Materials Science, Metallurgy.* Nanomaterials synthesis and characterization.

Tamboli, Adele C., Ph.D., University of California, Santa Barbara, 2009. *Applied Physics, Condensed Matter Physics, Engineering Physics/Science, Materials Science, Metallurgy, Nano Science and Technology, Solid State Physics.* Experimental materials physics.

Teaching Professor

Flournoy, Alex T., Ph.D., University of Colorado, 2003. *Cosmology & String Theory, Particles and Fields, Relativity & Gravitation, Theoretical Physics.* Theoretical particle physics, string theory, and cosmology.

Kohl, Patrick, Ph.D., University of Colorado, 2007. *Physics and other Science Education.* Physics education research.

Kuo, Hsia-Po Vincent, Ph.D., University of Minnesota, 2004. CEE Associate Director. *Physics and other Science Education.* Physics education research.

Ruskell, Todd G., Ph.D., University of Arizona, 1996. Assistant Department Head. *Applied Physics, Atomic, Molecular, & Optical Physics, Optics, Physics and other Science Education, Surface Physics.* Computer-assisted instruction.

Stone, Charles A., Ph.D., University of California, Los Angeles, 1990. *Applied Physics, Energy Sources & Environment, Nuclear Engineering, Physics and other Science Education.* Renewable energy science and technology.

Young, Matt, Ph.D., University of Rochester, 1967. *Applied Physics, Optics.* Optics, metrology.

Teaching Associate Professor

Callan, Kristine E., Ph.D., Duke University, 2013. *Applied Physics, Optics, Physics and other Science Education.*

DEPARTMENTAL RESEARCH SPECIALTIES AND STAFF

Theoretical

Condensed Matter Physics. Theoretical many body quantum and classical mechanics in application to ultracold quantum gases: quantum phase transitions; atomic and molecular superfluidity and superconductivity; atom lasers; nonlinear waves; fractals, solitons, and vortices; quantum information science; mathematical physics and inverse problems; novel semiconductor materials and structures; semiconductor alloys; phonon properties; surfaces and interfaces; nanostructures; plasmonic phenomena. Arnold, Bernard, Carr, Coffey, Flammer, Lusk, Wood, Wu.

Experimental

Condensed Matter Physics. Semiconductor science, electronic devices, optical properties of materials and interfaces, transport phenomena, soft condensed matter and liquid crystals, self-assembled monolayers, bio-inorganic composites, quantum nanostructures, surface physics and catalysis, transport phenomena, nonlinear optical properties of surfaces, amorphous materials. Beach, Bradley, Collins, Flammer, Furtak, Mace, Ohno, Scales, Tamboli, Taylor, Toberer, Williamson.

Energy Sources & Environment. Solar photovoltaics, third-generation photoconversion, nanostructures, thermoelectrics, artificial photosynthesis, organic photovoltaics, photoelectrochemistry, optical and electronic properties of crystalline and amorphous semiconductors, thin film photovoltaic materials, photoexcitation and relaxation in nanostructures. Beach, Collins, Furtak, Ohno, Tamboli, Toberer, Williamson.

Nuclear Physics. Nuclear astrophysics, low energy nuclear physics, astrophysics with radioactive beams, ultra-high energy cosmic ray physics, astroparticle physics, low-energy nuclear reactions, fusion diagnostics, nuclear engineering. Cecil, Greife, Hager, Sarazin, Shayer.

Optics. Laser physics and ultrafast optical phenomena, plasmonic electronic systems, nonlinear microscopy and micromachining, ultra-high intensity lasers, ultrafast x-ray diffraction, frequency shifted feedback lasers, precision measurement, radio-frequency wave propagation in random media, ultrasonics, mesoscopic phenomena, quantum chaos, biophotonics, single-molecule microscopy. Durfee, Sarkar, Scales, Squier, Young.

View additional information about this department at
www.gradschoolshopper.com

COLORADO STATE UNIVERSITY

DEPARTMENT OF PHYSICS

Fort Collins, Colorado 80523
http://www.physics.colostate.edu

General University Information

President: Tony Frank

Dean of Graduate School: Jodie Hanzlik, Interim Vice Provost for Graduate Affairs

University website: http://www.colostate.edu

Control: Public

Setting: Urban

Total Faculty: 1,450

Total Graduate Faculty: No separate graduate faculty
Total number of Students: 28,547
Total number of Graduate Students: 3,671

Department Information

Department Chairman: John L. Harton, Chair
Department Contact: Veronica Nicholson, Administrative
  Assistant III
  Total full-time faculty: 21
  Total number of full-time equivalent positions: 29
  Full-Time Graduate Students: 70
  First-Year Graduate Students: 9
  Total Post Doctorates: 5

Department Address

Campus Delivery 1875
Fort Collins, CO 80523
Phone: (970) 491-6206
Fax: (970) 491-7941
E-mail: physics_grad_admissions@mail.colostate.edu
Website: http://www.physics.colostate.edu

ADMISSIONS

Admission Contact Information

Address admission inquiries to: Graduate Admissions Commit-
  tee, Department of Physics, 1875 Campus Delivery, Colorado
  State University, Fort Collins, CO 80523
Phone: (970) 491-6207
E-mail: veronica.nicholson@colostate.edu
Admissions website: http://www.physics.colostate.edu/
  GraduateInfo/applying

Application deadlines

Fall admission:
U.S. students: February 1   *Int'l. students*: February 1

Application fee

U.S. students: $50   *Int'l. students*: $50
Late applications can be considered.

Admissions information

For Fall of 2013:
  Number of applicants: 33
  Number admitted: 9
  Number enrolled: 33

Admission requirements

Bachelor's degree requirements: Bachelor's degree in physics
  or a related field is required.
Minimum undergraduate GPA: 3.0

GRE requirements

The GRE is required.

Advanced GRE requirements

The Advanced GRE is required.

TOEFL requirements

The TOEFL exam is required for students from non-English-
  speaking countries.
  PBT score: 600
  iBT score: 100

Other admissions information

Additional requirements: Scores better than 700-quantitative and
  600-advanced are desirable.
Undergraduate preparation assumed: Mechanics: Marion, Classical Dynamics; Electromagnetism: Reitz and Milford, Foundations of Electromagnetic Theory; Thermal Physics: Kittel
  and Kroemer, Thermal Physics; Quantum Mechanics: Griffith,
  Introduction to Quantum Mechanics.

TUITION

Tuition year 2013-14:
Tuition for in-state residents
  Full-time students: $5,082.62 per semester
  Part-time students: $555.42 per credit
Tuition for out-of-state residents
  Full-time students: $11,172.72 per semester
  Part-time students: $1,232.12 per credit
These figures include the student fees.
Credit hours per semester to be considered full-time: 9
Deferred tuition plan: No
Health insurance: Yes, 1100.00.
Academic term: Semester
Number of first-year students who received full tuition waivers: 20

Teaching Assistants, Research Assistants, and Fellowships

Number of first-year
  Teaching Assistants: 20
  Fellowship students: 1
Average stipend per academic year
  Teaching Assistant: $15,975
  Research Assistant: $15,975
  Fellowship student: $15,000

FINANCIAL AID

Application deadlines

Fall admission:
U.S. students: February 15

Loans

Loans are available for U.S. students.
Loans are not available for international students.
GAPSFAS application required: Yes
FAFSA application required: Yes

For further information

Address financial aid inquiries to: Colorado State University,
  Student Financial Services, Campus Delivery 1065, Fort Col-
  lins, CO 80523.
Phone: (970) 491-6321
Financial aid website: http://sfs.colostate.edu

HOUSING

Availability of on-campus housing

Single students: Yes
Married students: Yes

For further information

Address housing inquiries to: Office of Housing and Residence
  Education, 8032 Campus Delivery, Colorado State Univer-
  sity, Fort Collins, CO 80523.
Phone: (970) 491-6511
Housing aid website: http://www.housing.colostate.edu

Table A—Faculty, Enrollments, and Degrees Granted

Research Specialty	2011–12 Faculty	Enrollment Fall 2011		Number of Degrees Granted 2011–12 (2007–11)		
		Master's	Doctorate	Master's	Terminal Master's	Doctorate
Acoustics	1	–	–	–	–	–
Astronomy	1	–	–	–	–(1)	–
Atomic, Molecular, & Optical Physics	5	2	18	2(8)	–(4)	–(3)
Condensed Matter Physics	6	4	8	2(6)	–(5)	–(9)
High Energy Physics	6	4	12	1(9)	1(2)	–(1)
Other	3	–	2	–(3)	–(2)	–(4)
Total	22	23	29	5(26)	1(14)	–(17)
Full-time Grad. Stud.	–	23	26	–	–	–
First-year Grad. Stud.	–	–	20	–	–	–

GRADUATE DEGREE REQUIREMENTS

Master's: Two options. Thesis Option: 30 credits of course work and research are required, 12 credits in core physics courses with another 6 in 500-level physics courses or above; a minimum of 24 credits must be earned at Colorado State University; no language requirement.

Doctorate: Seventy-two credits in course work and research in an approved program, 18 credits in core physics courses with another 6 in 500-level physics courses or above; a minimum of 32 credits must be earned at Colorado State University; one-year residency required; no language requirement. Oral examination to determine mastery of specialized field of proposed dissertation required. Dissertation and dissertation defense required.

Other Degrees: 32 credits of course work are required, 15 credits in core physics courses with another 6 in 500-level physics courses or above; a minimum of 24 credits must be earned at Colorado State University; no language requirement. Journal presentation required.

Thesis: Thesis may be written in absentia.

SPECIAL EQUIPMENT, FACILITIES, OR PROGRAMS

Major research facilities in the department include conventional and superconducting magnets, microwave spectrometer for ferromagnetic-resonance studies, ESR spectrometers, vibrating-sample magnetometer, ultrasonic spectrometers, x-ray-diffraction instrumentation, several pulsed and cw lasers, tunable dye lasers along with detection and signal-processing equipment for quantum-electronics research, sputter-induced resonant-ionization spectroscopy, facilities for semiconductor fabrication and analysis, very-high-pressure equipment, and high-speed work-stations. Many facilities are available within other departments, such as electron microscopes, microelectronic fabrication, molecular beam epitaxy, and nuclear magnetic resonance.

Table B—Separately Budgeted Research Expenditures by Source of Support

Source of Support	Departmental Research	Physics-related Research Outside Department
Federal government	$3,006,924	
State/local government		
Non-profit organizations	$22,514	
Business and industry	$8,019	
Other		
Total	$3,037,457	

Table C—Separately Budgeted Research Expenditures by Research Specialty

Research Specialty	No. of Grants	Expenditures ($)
Atomic, Molecular, & Optical Physics	13	$624,660
Condensed Matter Physics	24	$788,988
Particles and Fields	20	$1,623,809
Total	57	$3,037,457

FACULTY

Professor

Bradley, R. Mark, Ph.D., Stanford University, 1985. *Condensed Matter Physics, Nonlinear Dynamics and Complex Systems.* Condensed matter theory; pattern formation in nonequilibrium systems.

Culver, Roger B., Ph.D., Ohio State University, 1971. *Astronomy.* Astronomy; experimental astrophysics.

Fairbank, William M., Ph.D., Stanford University, 1974. Undergraduate Advisor. *Atomic, Molecular, & Optical Physics, Optics, Particles and Fields.* Tunable laser spectroscopy; ultrasensitive laser mass spectrometry.

Harton, John L., Ph.D., Massachusetts Institute of Technology, 1988. Department of Physics Chair. *Astrophysics, High Energy Physics, Particles and Fields.* Experimental particle physics.

Krueger, David A., Ph.D., University of Washington, 1967. Undergraduate Advisor. *Climate/Atmospheric Science, Optics.* Fluids; lidar studies of the atmosphere.

Lee, Siu Au, Ph.D., Stanford University, 1976. Graduate Admissions Committee. *Atomic, Molecular, & Optical Physics, Optics.* Laser spectroscopy; laser manipulation of atoms.

Leisure, Robert G., Ph.D., Washington University, 1967. Undergraduate Advisor. *Condensed Matter Physics.* Ultrasonics.

Lundeen, Stephen R., Ph.D., Harvard University, 1975. Graduate Advising Committee Chair. *Atomic, Molecular, & Optical Physics.* Atomic, molecular, and optical physics.

Patton, Carl E., Ph.D., California Institute of Technology, 1967. Graduate Admissions Committee Chair. *Condensed Matter Physics.* Magnetism and magnetic materials.

Rocca, Jorge J., Ph.D., Colorado State University, 1983. *Atomic, Molecular, & Optical Physics, Plasma and Fusion.* Lasers, plasmas, and quantum electronics.

Sites, James R., Ph.D., Cornell University, 1969. Associate Dean for Research, College of Natural Sciences. *Condensed Matter Physics.* Semiconductor physics; solar cells.

Toki, Walter H., Ph.D., Massachusetts Institute of Technology, 1976. Undergraduate Advisor. *High Energy Physics, Particles and Fields.* Experimental particle physics.

Wilson, Robert J., Ph.D., Purdue University, 1983. Graduate Advising Committee. *High Energy Physics, Particles and Fields.* Experimental particle physics.

Associate Professor

Eykholt, Richard E., Ph.D., University of California, Irvine, 1984. Society of Physics Students Faculty Sponsor. *Nonlinear Dynamics and Complex Systems.* Nonlinear dynamics; chaos; mathematical physics.

Field, Stuart B., Ph.D., University of Chicago, 1986. Graduate Advising Committee. *Condensed Matter Physics, Low Temperature Physics.* Vortices in superconductors; nonlinear dynamics.

Gelfand, Martin P., Ph.D., Cornell University, 1990. Undergraduate Key Advisor. *Computational Physics, Condensed Matter Physics, Statistical & Thermal Physics.* Condensed matter theory.

Mostafa, Miguel, Ph.D., Instituto Balseiro, Universidad Nacional de Cuyo, 2001. *Astrophysics, Particles and Fields.* Undergraduate Recruitment and Retention. Particle astrophysics.

Roberts, Jacob L., Ph.D., University of Colorado Boulder, 2001. Department of Physics Associate Chair. *Atomic, Molecular, & Optical Physics, Optics.* Laser spectroscopy; ultracold plasmas.

Robinson, Raymond S., Ph.D., Colorado State University, 1979. Department of Physics Associate Chair. *Plasma and Fusion.* Low-density plasmas; space electric propulsion; ion-beam applications, including surface microtexturing.

Assistant Professor

Berger, Bruce E., Ph.D., Cornell University, 2000. Graduate Admissions Committee. *High Energy Physics, Particles and Fields.* Experimental particle physics.

Buchanan, Kristen, Ph.D., University of Alberta, 2004. Graduate Admissions Committee. *Condensed Matter Physics, Nano Science and Technology.* Magnetics.

Buchanan, Norman, Ph.D., University of Alberta, 2003. *High Energy Physics, Particles and Fields.* Particle astrophysics.

Wu, Mingzhong, Ph.D., Huazhong University of Science and Technology, 1999. Undergraduate Advisor. *Condensed Matter Physics, Nano Science and Technology.* Magnetics.

Other

Camley, Robert, Ph.D., University of California, Irvine, 1979. Magnetism.

Craine, Eric, Ph.D., Ohio State University, 1973. *Astronomy.*

Kabos, Pavel, Ph.D., Slovak Technical University, 1979. Theoretical electrotechnique.

DEPARTMENTAL RESEARCH SPECIALTIES AND STAFF

Theoretical

Statistical & Thermal Physics. Nonlinear dynamical systems; chaos and fractals; mathematical physics; phase transitions; far-from-equilibrium behavior; fractal aggregates; highly disordered materials; percolation theory; polymers; lattice dynamics of molecular crystals and crystallites using phonon theories; equations-of-state of condensed phases; phase transitions of bulk solids and adsorbed molecular surfaces; computational physics. Bradley, Eykholt.

Experimental

Atomic, Molecular, & Optical Physics. Recent highlights include Presidential Early Career Award for cold atom research, and W. M. Keck Foundation grant for Si quantum computer research. Research topics include efficient creation of BEC with new non-evaporative cooling technique and tailored optical traps; study of quantum fluids as a function of confinement geometry; development of Si based quantum computer; laser cooled single atom on demand source for deterministic ion deposition; new techniques for quantum information; investigation of nanostructures fabricated by laser manipulation of cold atoms; Na fluorescence Lidar for both day and night measurements of temperature, zonal wind and meridional wind in the upper atmosphere; studies of atmospheric wave and global changes; fast beam high precision laser-RF spectroscopy of excited states of atoms and molecules; Rydberg atom studies; ultra-sensitive single-atom detection for tagging Ba+ ion in the neutrinoless double-beta decay (EXO collaboration); extreme ultraviolet laser development, nanolithography and imaging. Fairbank, Krueger, Lee, Lundeen, Roberts.

Condensed Matter Physics. Thin-film semiconductors; semiconductor surfaces; solar cells; magnetic thin films, multiferroics; nanomagnetism, magnetic vortex physics in nanodots; magnetic recording physics and materials, ferrite materials, magnetodynamics and magnetic relaxation; magnon Brillouin light scattering; magnetism in rare earth and actinide compounds; chaos and solitons in thin magnetic films; optical spectroscopy at high pressure; structure-physical property relations of novel low-dimensional materials; ultrasonic spectroscopy; metal-hydrogen systems; phase transitions; crystal growth dynamics; conducting polymers; superconducting vortex dynamics; scanning Hall probe microscopy. Kristen Buchanan, Field, Leisure, Patton, Sites, Wu.

High Energy Physics. Members of the High Energy Physics and Particle Astrophysics group work in the T2K, KamLAND, BaBar, and DRIFT experiments, and the Pierre Auger and HAWC Observatories. T2K and KamLAND both study neutrino oscillations, BaBar investigates the weak decays of bottom and charm mesons, DRIFT aims to detect dark matter particles, Auger detects ultra high energy cosmic rays, and HAWC will study TeV gamma rays. The KamLAND, BaBar, DRIFT, and Auger collaborations are analyzing data. R&D on electronics for an upgraded DRIFT detector is being done at CSU. Main components of the T2K P0D detector and HAWC water Cherenkov detectors are being constructed at CSU, and installed in Japan and Mexico, respectively. Berger, Kristen Buchanan, Harton, Mostafa, Toki, Wilson.

View additional information about this department at
www.gradschoolshopper.com

UNIVERSITY OF COLORADO, BOULDER

DEPARTMENT OF PHYSICS

Boulder, Colorado 80309
http://phys.colorado.edu/

General University Information
President: Bruce Benson
Dean of Graduate School: John A. Stevenson
University website: http://www.colorado.edu/
Control: Public
Setting: Urban
Total Faculty: 1,517
Total number of Students: 30,659
Total number of Graduate Students: 4,900

Department Information
Department Chairman: Paul D. Beale, Chair
Department Contact: Jeanne Nijhowne, Graduate Program
  Assistant
  Total full-time faculty: 67
  Total number of full-time equivalent positions: 51
  Full-Time Graduate Students: 235
  First-Year Graduate Students: 32
  Female First-Year Students: 6
  Total Post Doctorates: 48

Department Address
2000 Colorado Avenue
Boulder, CO 80309
Phone: (303) 735-0519
Fax: (303) 492-3352
E-mail: jeanne.nijhowne@colorado.edu
Website: http://phys.colorado.edu/

ADMISSIONS

Admission Contact Information
Address admission inquiries to: Jeanne Nijhowne, Graduate Pro-
  gram Assistant, 390 UCB, University of Colorado, Deptart-
  ment of Physics, Boulder, CO 80309
Phone: (303) 735-0519
E-mail: jeanne.nijhowne@colorado.edu
Admissions website: http://phys.colorado.edu/prospective-
  students/application-information

Application deadlines
Fall admission:
U.S. students: December 15   *Int'l. students*: December 1

Application fee
U.S. students: $50   *Int'l. students*: $70

Admissions information
For Fall of 2013:
  Number of applicants: 609
  Number admitted: 137
  Number enrolled: 32

Admission requirements
Bachelor's degree requirements: Bachelor's degree is required.
Minimum undergraduate GPA: 3.0

GRE requirements
The GRE is required.

Advanced GRE requirements
The Advanced GRE is required.
We do not set minimum score requirements.

TOEFL requirements
The TOEFL exam is required for students from non-English-
  speaking countries.
iBT score: 85

Other admissions information
Additional requirements: The average GRE scores for 2013–14
  for U.S. admissions were verbal-163; quantitative-165. The
  average GRE advanced physics score for 2013–14 U.S. admissions was 825. The average GRE scores for 2013-14 international admissions were verbal-160; quantitative-166. The
  average GRE advanced physics score for 2013-14 international admissions was 919.
Undergraduate preparation assumed: An undergraduate program
  for students entering graduate study in physics should typically include the following: ; Physics: 3 Semesters Introductory Physics; 1 Semester Advanced Classical Mechanics;
  1 Semester Quantum Mechanics; 1 Semester Statistical Mechanics; 2 Semesters Advanced Electricity and Magnetism;
  2 Semesters Advanced Laboratory Course/Project Work; 1
  Semester Advanced Course in modern Physics such as Condensed Matter, Geophysics,; Atomic, Nuclear, or Particle
  Physics; Math: 3 Semesters Calculus; 1 Semester Linear Algebra; 1 Semester Differential Equations; Computing: ; General knowledge.

TUITION

Tuition year 2013-14:
Tuition for in-state residents
  Full-time students: $9,918 annual
Tuition for out-of-state residents
  Full-time students: $26,712 annual
Credit hours per semester to be considered full-time: 9
Deferred tuition plan: Yes
Health insurance: Yes, $3,030.00.
Other academic fees: $908.84 Grad student fees for AY 2013-14
Academic term: Semester
Number of first-year students who received full tuition waivers: 32

Teaching Assistants, Research Assistants, and Fellowships
Number of first-year
  Teaching Assistants: 19
  Research Assistants: 7
  Fellowship students: 6
Average stipend per academic year
  Teaching Assistant: $16,881.56
  Research Assistant: $19,971
  Fellowship student: $19,971

FINANCIAL AID

Application deadlines
Fall admission:
U.S. students: December 15   *Int'l. students*: December 1

Loans
Loans are available for U.S. students.
Loans are not available for international students.
GAPSFAS application required: No
FAFSA application required: No

For further information
Address financial aid inquiries to: University of Colorado Boulder Office of Financial Aid, 556 UCB, Boulder, Colorado 80309-0556.
Phone: (303) 492-5091
E-mail: finaid@colorado.edu
Financial aid website: http://www.colorado.edu/finaid/grad.html

HOUSING

Availability of on-campus housing
Single students: Yes
Married students: Yes

For further information
Address housing inquiries to: University of Colorado Housing and Dining Services, 159 UCB, Boulder, CO 80309-0159.
Phone: (303) 492-6384
E-mail: familyhousing@colorado.edu
Housing aid website: http://housing.colorado.edu/residences/graduate-family

Table A—Faculty, Enrollments, and Degrees Granted

Research Specialty	2013-14 Faculty	Enrollment Fall 2013 Master's	Enrollment Fall 2013 Doctorate	Number of Degrees Granted 2012-13 (2008-13) Master's	Number of Degrees Granted 2012-13 (2008-13) Terminal Master's	Number of Degrees Granted 2012-13 (2008-13) Doctorate
Total	–	–	238	–	–	28(135)
Full-time Grad. Stud.	–	–	235	–	–	–
First-year Grad. Stud.	–	–	32	–	–	–

GRADUATE DEGREE REQUIREMENTS

Master's: We do not offer a stand-alone Master's program. All grad students are admitted directly into the Ph.D. program.
Doctorate: Students must complete five of the six required Comps I courses with a "B-" or better. Five additional courses are needed to complete the 30 hours of required coursework. Twenty-seven of the 30 hours must be physics courses. Students must have at least a 3.0 average. All students are required to take the Comps II examination. When students are ready, they take a Comps III examination and are admitted into candidacy. They then write a doctoral thesis that they must defend. Students have six years to complete their doctorate, although this limit may be extended.

SPECIAL EQUIPMENT, FACILITIES, OR PROGRAMS

Liquid Crystal Material Research Center, Extreme Ultraviolet Engineering Research Center, and the Colorado Center for Lunar Dust and Atmospheric Studies are used by our students. Programs at CERN, SLAC, and the Fermi National Accelerator Laboratory are also available. Experimental and theoretical research programs at JILA, the Cooperative Institute for Research in Environmental Sciences, the National Institute for Standards & Technology (NIST), the Laboratory for Atmospheric and Space Physics (LASP), the High Altitude Observatory (HAO), the National Center for Atmospheric Research (NCAR), the National Renewable Energy Laboratory (NREL), the National Oceanic and Atmospheric Administration (NOAA), the U.S. Geological Survey (USGS), and the Center for Opto-Electronic Computing Systems (COCS), Center for Integrated Plasma Studies (CIPS) in astrophysics; atomic, molecular and optical physics and geophysics are available.

Table B—Separately Budgeted Research Expenditures by Source of Support

Source of Support	Departmental Research	Physics-related Research Outside Department
Federal government	$49,458,213	$108,307,472
State/local government	$457,376	
Non-profit organizations	$35,109	
Business and industry		
Other		
Total	**$49,950,698**	**$108,307,472**

Table C—Separately Budgeted Research Expenditures by Research Specialty

Research Specialty	No. of Grants	Expenditures ($)
Astrophysics	18	$1,127,896
Atomic, Molecular, & Optical Physics	244	$32,841,669
Biophysics	3	$114,263
Condensed Matter Physics	72	$4,963,044
Geophysics	28	$916,556
Nuclear Physics	5	$544,302
Particles and Fields	16	$2,279,449
Physics and other Science Education	17	$1,410,482
Plasma and Fusion	46	$5,753,037
Total	**449**	**$49,950,698**

FACULTY

Professor
Anderson, Dana Z., Ph.D., University of Arizona, 1981. *Optics*. Experimental quantum optics.
Baker, Daniel N., Ph.D., University of Iowa, 1974. Joint appointment with LASP and APS. *Plasma and Fusion, Solar Physics*.
Beale, Paul D., Ph.D., Cornell University, 1982. *Condensed Matter Physics*. Theoretical condensed-matter physics.
Cary, John, Ph.D., University of California, Berkeley, 1979. *Physics of Beams, Plasma and Fusion*. Theoretical plasma physics.
Clark, Noel A., Ph.D., Massachusetts Institute of Technology, 1970. *Condensed Matter Physics*. Experimental condensed-matter physics.
Cumalat, John P., Ph.D., University of California, Santa Barbara, 1977. *High Energy Physics*. Experimental particle physics.
de Alwis, Senarath P., Ph.D., University of Cambridge, 1969. *High Energy Physics, Particles and Fields*. Theoretical elementary particle physics.
DeGrand, Thomas A., Ph.D., Massachusetts Institute of Technology, 1976. *High Energy Physics, Particles and Fields*. Theoretical particle physics.
Dessau, Daniel, Ph.D., Stanford University, 1992. *Condensed Matter Physics*. Experimental condensed-matter physics.
Finkelstein, Noah, Ph.D., Princeton University, 1998. *Physics and other Science Education*. Physics education research.
Ford, William T., Ph.D., Princeton University, 1967. *High Energy Physics*. Experimental elementary particle physics.

Franklin, Allan D., Ph.D., Cornell University, 1965. *History & Philosophy of Physics/Science.* History and philosophy of science.

Goldman, Martin, Ph.D., Harvard University, 1965. *Plasma and Fusion, Solar Physics.* Plasma physics.

Greene, Chris H., Ph.D., University of Chicago, 1980. *Atomic, Molecular, & Optical Physics.* Theoretical atomic physics.

Hamilton, Andrew J. S., Ph.D., University of Virginia, 1983. Joint appointment with APS. *Astrophysics.*

Hasenfratz, Anna, Ph.D., Lorand Eotvos University, 1982. *High Energy Physics, Particles and Fields.* Theoretical high-energy physics.

Holland, Murray, Ph.D., University of Oxford, 1994. *Atomic, Molecular, & Optical Physics.* Theoretical quantum optics; collision studies; Bose-Einstein condensation.

Horanyi, Mihaly, Ph.D., Lorand Eotvos University, Budapest, 1982. *Atmosphere, Space Physics, Cosmic Rays, Plasma and Fusion.* Dusty plasmas.

Kapteyn, Henry, Ph.D., University of California, Berkeley, 1989. *Atomic, Molecular, & Optical Physics, Optics.* Ultrafast nonlinear optics to generate coherent x-rays; XUV and x-ray lasers and the use of ultrashort x-ray pulses to probe dynamic processes in chemical and material systems.

Kinney, Edward R., Ph.D., Massachusetts Institute of Technology, 1988. *Nuclear Physics.* Experimental nuclear physics.

Mahanthappa, Kalyana T., Ph.D., Harvard University, 1961. *High Energy Physics, Particles and Fields.* Theoretical elementary particle physics.

Murnane, Margaret, Ph.D., University of California, Berkeley, 1989. *Atomic, Molecular, & Optical Physics, Optics.* Development and application of novel ultrafast coherent light sources in the visible and x-ray regions of the spectrum.

Nagle, Jamie, Ph.D., Yale University, 1996. *Nuclear Physics.* Experimental nuclear physics.

Parker, Scott, Ph.D., University of California, Berkeley, 1990. *Plasma and Fusion.* Computational plasma physics.

Peterson, Roy J., Ph.D., University of Washington, 1966. *Nuclear Physics.* Experimental nuclear physics; intermediate-energy physics.

Pollock, Steven, Ph.D., Stanford University, 1987. *Physics and other Science Education.* Physics education research.

Price, John C., Ph.D., Stanford University, 1986. *Condensed Matter Physics.* Experimental low-temperature physics.

Radzihovsky, Leo, Ph.D., Harvard University, 1993. *Condensed Matter Physics.* Theoretical condensed-matter physics.

Rankin, Patricia, Ph.D., University of London, 1982. *High Energy Physics.* Experimental elementary particle physics.

Ritzwoller, Michael H., Ph.D., University of California, San Diego, 1987. *Geophysics.* Theoretical geophysics.

Rogers, Charles T., Ph.D., Cornell University, 1987. *Condensed Matter Physics.* Experimental condensed-matter physics.

Wahr, John, Ph.D., University of Colorado, 1979. *Geophysics.* Theoretical geophysics.

Wieman, Carl E., Ph.D., Stanford University, 1977. *Atomic, Molecular, & Optical Physics.* Bose-Einstein condensation; AMO physics; physics education research.

Zhong, Shijie, Ph.D., University of Michigan, 1994. *Geophysics.* Theoretical geophysics.

Zimmerman, Eric, Ph.D., University of Chicago, 1998. *High Energy Physics.* Experimental particle physics.

Associate Professor

Becker, Andreas, Ph.D., Beilefeld University, 1997. *Atomic, Molecular, & Optical Physics, Optics.* Theoretical atomic physics.

Betterton, Meredith D., Ph.D., Harvard University, 2000. *Biophysics.* Biophysics; systems biology; bioinformatics; pattern formation.

DeWolfe, Oliver, Ph.D., Massachusetts Institute of Technology, 2000. *High Energy Physics, Particles and Fields.* Theoretical elementary particles and fields.

Glenn, Jason, Ph.D., University of Arizona, 1997. Joint appointment with APS. *Astrophysics.*

Gurarie, Victor, Ph.D., Princeton University, 1996. *Condensed Matter Physics.* Theoretical condensed-matter physics.

Halverson, Nils, Ph.D., University of California, Berkeley, 2002. Joint appointment with APS. *Astrophysics.* Observational CMB cosmology.

Lewandowski, Heather, Ph.D., University of Colorado, 2002. *Atomic, Molecular, & Optical Physics.* Experimental atomic and molecular physics.

Munsat, Tobin, Ph.D., Princeton University, 2001. *Atmosphere, Space Physics, Cosmic Rays, Plasma and Fusion.* Experimental plasma physics.

Piestun, Rafael, Ph.D., Israel Institute of Technology, 1998. Joint appointment with Electrical and Computer Engineering. *Computer Science.*

Raschke, Markus B., Ph.D., Technical University of Munich, 1999. *Atomic, Molecular, & Optical Physics, Chemical Physics, Nano Science and Technology, Optics.* Science of nano-optics.

Reznik, Dmitry, Ph.D., University of Illinois at Urbana-Champaign, 1993. *Condensed Matter Physics.* Experimental condensed-matter and materials physics.

Stenson, Kevin, Ph.D., University of Wisconsin-Madison, 1998. *High Energy Physics.* Experimental elementary particles and fields.

Assistant Professor

Gopinath, Juliet, Ph.D., Massachusetts Institute of Technology, 2005. Joint appointment with ECEE. *Optics.*

Hermele, Michael, Ph.D., University of California, Santa Barbara, 2005. *Condensed Matter Physics.* Theoretical condensed-matter physics.

Hough, Loren E., Ph.D., University of Colorado Boulder, 2007. *Biophysics.* Biophysics; NMR of biological processes in living cells.

Kempf, Sascha, Ph.D., Friedrich Schiller University, Jena, 1999. *Atmosphere, Space Physics, Cosmic Rays.* Dusty plasmas.

Lee, Minhyea, Ph.D., University of Chicago, 2004. *Condensed Matter Physics.* Experimental condensed-matter and materials physics.

Marino, Alysia, Ph.D., University of California, Berkeley, 2004. *High Energy Physics.* Experimental particle physics (neutrinos).

McElroy, Kyle P., Ph.D., University of California, Berkeley, 2005. *Condensed Matter Physics.* Experimental condensed-matter physics.

Popovic, Milos, Ph.D., Massachusetts Institute of Technology, 2007. Joint appointment with ECEE. *Optics.*

Regal, Cindy, Ph.D., University of Colorado, 2006. *Atomic, Molecular, & Optical Physics.* Atomic, molecular, and optical physics.

Romatschke, Paul, Ph.D., Technical University of Vienna, 2003. *Nuclear Physics.* Theoretical nuclear physics.

Schibli, Thomas, Ph.D., University of Karlsruhe, 2001. *Optics.* Experimental optics.

Smalyukh, Ivan, Ph.D., Kent State University, 2003. *Condensed Matter Physics.* Experimental condensed-matter physics.

Uzdensky, Dmitri A., Ph.D., Princeton University, 1998. *Plasma and Fusion, Solar Physics.* Theoretical plasma physics and plasma astrophysics.

Research Professor

Bohn, John, Ph.D., University of Chicago, 1995. *Atomic, Molecular, & Optical Physics*. Cold collisions; semiconductor devices; few-body physics.

Research Associate Professor

Rey, Ana Maria, Ph.D., University of Maryland, 2004. *Atomic, Molecular, & Optical Physics*. Optical lattices; quantum degenerate Fermi gases; and ultracold Boson-Fermion mixtures.

Adjunct Professor

Nesbitt, David, Ph.D., University of Colorado, 1981. *Chemical Physics*. Chemical physics, including high resolution laser spectroscopy, chemical deaction Dynamics, quantum nanostructures and single-molecule biophysics.

Perkins, Thomas T., Ph.D., Stanford University, 1997. *Biophysics*. Single molecule measurements of biological systems using applying high precision measurements based on optical traps and atomic force microscopes.

Robertson, Scott, Ph.D., Cornell University, 1972. *Atmosphere, Space Physics, Cosmic Rays, Plasma and Fusion*. Experimental plasma physics; fusion.

Professor Adjoint

Cornell, Eric, Ph.D., Massachusetts Institute of Technology, 1990. *Atomic, Molecular, & Optical Physics*. Bose-Einstein condensation; laser and magnetic trapping of neutral cesium using diode lasers and a vapor cell.

Cundiff, Steven, Ph.D., University of Michigan, 1992. *Optics*. Telecommunications; fiber optics; ultrafast optical studies of semiconductors.

Diddams, Scott, Ph.D., University of New Mexico, 1996. *Optics*. Experimental laser physics; femtosecond lasers and ultrafast phenomena; nonlinear optics; precision spectroscopy; optical frequency combs; metrology.

Hall, John, Ph.D., Carnegie Institute of Technology, 1961. *Atomic, Molecular, & Optical Physics, Optics*. Development of laser stabilization and measurement techniques that lead toward the creation of phase-stable optical frequency sources and their application to precision tests of fundamental principles.

Jin, Deborah, Ph.D., University of Chicago, 1995. *Atomic, Molecular, & Optical Physics*. Bose-Einstein condensation in a dilute atomic gas: collective excitations, thermodynamics at the phase transition, interaction effects, vortices, and other analogs of superfluidity.

Levine, Judah, Ph.D., New York University, 1966. *Geophysics*. Application of precision measurement techniques to problem of geophysical interest.

Wineland, David, Ph.D., Harvard University, 1970. *Atomic, Molecular, & Optical Physics*. Laser-cooled trapped ions in the areas of high-resolution spectroscopy, basic plasma physics, and quantum information.

Ye, Jun, Ph.D., University of Colorado, 1997. *Atomic, Molecular, & Optical Physics, Optics*. Atomic and optical physics.

Associate Professor Adjoint

Lehnert, Konrad, Ph.D., University of California, Santa Barbara, 1999. *Atomic, Molecular, & Optical Physics, Nano Science and Technology*. Physics.

Assistant Professor Adjoint

Thompson, James K., Ph.D., Massachusetts Institute of Technology, 2003. *Atomic, Molecular, & Optical Physics*. Precision measurement.

Professor Attendant Rank

Glaser, Matthew A., Ph.D., University of Colorado, 1991. *Computational Physics, Condensed Matter Physics*. Computer simulation techniques for problems in condensed-matter physics and statistical physics.

Maclennan, Joseph, Ph.D., University of Colorado, 1988. *Condensed Matter Physics*. Ferroelectric liquid crystals; freely suspended liquid crystal films; instrumentation.

Smith, James G., Ph.D., University of California, San Diego, 1975. *High Energy Physics*. Experimental high-energy physics.

Associate Professor Attendant Rank

Perkins, Katherine, Ph.D., Harvard University, 2000. *Physics and other Science Education*. Physics education research with a focus on the use of interactive simulations for teaching and learning physics, students' beliefs about physics (and chemistry), and sustainable course reform.

Wagner, Stephen, Ph.D., Johns Hopkins University, 1983. *High Energy Physics*. High-energy physics.

Instructor

Dubson, Michael, Ph.D., Cornell University, 1984. *Physics and other Science Education*. Physics education research.

Lecturer

Knill, Emanuel, Ph.D., University of Colorado, 1991. *Applied Mathematics*. Quantum information science.

Ullom, Joel, Ph.D., Harvard University, 1997. Experimental condensed-matter physics: superconductivity, low-temperature physics, and radiation detectors and their applications.

DEPARTMENTAL RESEARCH SPECIALTIES AND STAFF

Theoretical

Astrophysics. Astrophysics is concentrated in the Department of Astrophysical and Planetary Science, but many of our Ph.D. students work with faculty at APS, JILA, CASA, and LASP. Glenn, Halverson, Hamilton.

Atmosphere, Space Physics, Cosmic Rays. Solar, space, and atmospheric plasma studies and lunar surface environment. Horanyi, Kempf, Munsat.

Atomic, Molecular, & Optical Physics. Our AMO physics is concentrated in JILA, one of the leading research institutes in the world in this area. Research at JILA includes high-precision spectroscopy and precision measurement, ultracold cold atoms and molecules, ultrafast and ultra-high-power lasers, and micron- and nanometer-scale optics. Anderson, Becker, Bohn, Cornell, Greene, Hall, Holland, Jin, Kapteyn, Lehnert, Lewandowski, Murnane, Raschke, Regal, Rey, Thompson, Wieman, Wineland, Ye.

Biophysics. Molecular motors and motors that can change their track; DNA-protein interactions; self-assembly of cytoskeletal materials; liquid crystalline aggregates. Betterton, Glaser, Hough, Thomas Perkins.

Chemical Physics. Theory of ultracold chemical reactions; chemical reaction dynamics. Nesbitt, Raschke.

Condensed Matter Physics. Theoretical condensed-matter research on soft materials, strongly interacting quantum systems, fractional quantum Hall effect, exotic quantum states, and statistical mechanics of spin systems. Beale, Clark, Dessau, Glaser, Gurarie, Hermele, Lee, Maclennan, McElroy, Price, Radzihovsky, Reznik, Rogers, Smalyukh.

Geophysics. Physics of earth dynamics, including both gravitational and magnetic fields and tectonics and volcanism. Levine, Ritzwoller, Wahr, Zhong.

High Energy Physics. Non-perturbative QCD lattice gauge theories; grand unified theories; supersymmetry; string theory.

Cumalat, de Alwis, DeGrand, DeWolfe, Ford, Hasenfratz, Mahanthappa, Marino, Rankin, Smith, Stenson, Wagner, Zimmerman.

History & Philosophy of Physics/Science. History and philosophy of 20th-century physics, especially high-energy physics. Anderson, Franklin.

Nuclear Physics. Theory of highly relativistic ion collisions, relativistic fluids, and the quark-gluon plasma. Kinney, Nagle, Peterson, Romatschke.

Optics. Ultrafast and ultra-high-power lasers and interactions. Becker, Cundiff, Diddams, Gopinath, Hall, Kapteyn, Murnane, Popovic, Raschke, Schibli, Ye.

Other. Gravitational physics, including general relativistic effects.

Particles and Fields. Non-perturbative QCD lattice gauge theories; grand unified theories; supersymmetry; string theory. de Alwis, DeGrand, DeWolfe, Hasenfratz, Mahanthappa.

Physics and other Science Education. Uses of technology in physics education; assessments (conceptual, epistemological, and belief oriented); curricular and classroom materials at the middle- and upper-division levels; theoretical models of students learning physics; social and contextual foundations of student learning; examination of successful educational reforms and replication studies of such reforms; student problem-solving in physics. Dubson, Finkelstein, Katherine Perkins, Pollock.

Plasma and Fusion. Theory of space and laboratory plasmas; plasma turbulence; magnetic reconnection. The Center for Integrated Plasma Studies is used. Baker, Cary, Goldman, Horanyi, Munsat, Parker, Robertson, Uzdensky.

Statistical & Thermal Physics. Statistical mechanics of soft condensed-matter systems; phase transitions; spin systems.

Experimental

Applied Physics. Research in materials science; nanomechanical systems; nano-optics; ultrafast and ultra-high-power lasers.

Astrophysics. Astrophysics is concentrated in the Department of Astrophysical and Planetary Science, but many of our Ph.D. students work with faculty at APS, JILA, CASA, and LASP. Glenn, Halverson, Hamilton.

Atmosphere, Space Physics, Cosmic Rays. Solar, space, and atmospheric plasma studies; lunar surface environment. Horanyi, Kempf, Munsat.

Atomic, Molecular, & Optical Physics. Our AMO physics is concentrated in JILA, one of the leading research institutes in the world in this area. Research at JILA includes high-precision spectroscopy and precision measurement, ultracold cold atoms and molecules, ultrafast and ultra-high-power lasers, and micron- and nanometer-scale optics. Anderson, Becker, Bohn, Cornell, Greene, Hall, Holland, Jin, Kapteyn, Lehnert, Lewandowski, Murnane, Raschke, Regal, Rey, Thompson, Wieman, Wineland, Ye.

Biophysics. Atomic force spectroscopy of biological molecules; nanoassembly. Betterton, Hough.

Chemical Physics. Laser spectroscopy of molecules; ultracold molecules; state-resolved chemical reactions; nanometer/femtosecond measurements of electrons in molecules.

Condensed Matter Physics. Soft condensed-matter and liquid crystal physics; femtosecond optical, electron, and neutron spectroscopy on materials; nanoscale electronic structure studies of surfaces; electrical and mechanical properties of nanofabricated materials; low-temperature properties of exotic materials. Beale, Clark, Dessau, Glaser, Gurarie, Hermele, Lee, Maclennan, McElroy, Price, Radzihovsky, Reznik, Rogers, Smalyukh.

High Energy Physics. Our high-energy physics experimentalists are members of the CMS collaboration at the LHC, the T2K neutrino collaboration, and other high-energy experiments at FermiLab and SLAC. Cumalat, de Alwis, DeGrand, DeWolfe, Ford, Hasenfratz, Mahanthappa, Marino, Rankin, Smith, Stenson, Wagner, Zimmerman.

Low Temperature Physics. Low-temperature properties of exotic materials.

Materials Science, Metallurgy. Electronic and mechanical properties of nanostructures.

Nuclear Physics. Relativistic heavy ion collisions and medium-energy nuclear structure studies. Kinney, Nagle, Peterson, Romatschke.

Optics. Ultrafast and ultra-high-power lasers; femtosecond/nanoscale optics; nonlinear optics; integrated nano-optics. Becker, Cundiff, Diddams, Gopinath, Hall, Kapteyn, Murnane, Popovic, Raschke, Schibli, Ye.

Other. Gravitational physics, including measurements of general relativistic effects.

Physics and other Science Education. Uses of technology in physics education; assessments (conceptual, epistemological, and belief oriented); curricular and classroom materials at the middle- and upper-division levels; theoretical models of students learning physics; social and contextual foundations of student learning; examination of successful educational reforms and replication studies of such reforms; student problem-solving in physics. Dubson, Finkelstein, Katherine Perkins, Pollock.

Plasma and Fusion. Laboratory and space plasmas; measurement and assessment of turbulence and cross-field transport in magnetically confined plasmas; solar plasma observations; dusty plasmas; lunar surface environment. The Center for Integrated Plasma Studies, Colorado Center for Lunar Dust and Atmospheric Studies, is used. Baker, Cary, Goldman, Horanyi, Munsat, Parker, Robertson, Uzdensky.

Statistical & Thermal Physics. Soft condensed-matter and liquid crystal physics; femtosecond optical, electron, and neutron spectroscopy on materials; nanoscale electronic structure studies of surfaces; electrical and mechanical properties of nanofabricated materials; low-temperature properties of exotic materials.

View additional information about this department at
www.gradschoolshopper.com

UNIVERSITY OF DENVER

DEPARTMENT OF PHYSICS AND ASTRONOMY

Denver, Colorado 80208-2238
http://www.physics.du.edu

General University Information
Chancellor: Robert D. Coombe
Dean of Graduate School: Andrei Kutateladze, Dean of Natural Science and Mathematics (NSM)
University website: http://www.du.edu
Control: Private
Setting: Suburban
Total Faculty: 640
Total Graduate Faculty: N/A
Total number of Students: 11,362
Total number of Graduate Students: 6,301

Department Information
Department Chairman: Davor Balzar, Chair
Department Contact: Barbara Stephen, Assistant to Chair
  Total full-time faculty: 13
  Total number of full-time equivalent positions: 11
  Full-Time Graduate Students: 23
  First-Year Graduate Students: 4
  Female First-Year Students: 2
  Total Post Doctorates: 5

Department Address
2112 E. Wesley Ave
Denver, CO 80208-2238
Phone: (303) 871-2238
Fax: (303) 871-4405
E-mail: bstephen@du.edu
Website: http://www.physics.du.edu

ADMISSIONS

Admission Contact Information
Address admission inquiries to: Davor Balzar, Chair, Department of Physics and Astronomy
Phone: (303) 871-2238
E-mail: balzar@du.edu
Admissions website: http://www.physics.du.edu

Application deadlines
Fall admission:
U.S. students: February 1   *Int'l. students*: February 1

Application fee
U.S. students: $65   *Int'l. students*: $65

Admissions information
For Fall of 2013:
  Number of applicants: 57
  Number admitted: 13
  Number enrolled: 6

Admission requirements
Bachelor's degree requirements: A bachelor's degree in physics (or a physics and mathematics background equivalent to that required for a bachelor of science degree in physics) is required.
Minimum undergraduate GPA: 3.0

GRE requirements
The GRE is required.
  Quantitative score: 550
Quantitative: 146 is the minimum accepted score for the revised GRE.

Advanced GRE requirements
The Advanced GRE is recommended.
Physics subject test is strongly recommended.

TOEFL requirements
The TOEFL exam is required for students from non-English-speaking countries.
  PBT score: 550
  iBT score: 80
Graduate Teaching Assistants (GTAs) must demonstrate fluency in spoken English by scoring a 26 on the TOEFL (iBT) speaking section or 8.0 on the IELTS speaking section.

Other admissions information
Additional requirements: Applicants with a GPA of less than 3.0 may be considered as individual cases. Admission preference will be given to those submitting strong GRE Advanced Physics scores. All those who wish to become teaching assistants must demonstrate oral English proficiencies via the ibT (minimum acceptable score of 26 on the speaking section).
Undergraduate preparation assumed: Taylor, Classical Mechanics; Griffiths, Electromagnetism; Liboff, Introductory Quantum Mechanics; advanced undergraduate laboratory; mathematics including differential equations, vector calculus, and linear algebra.

TUITION

Tuition year 2013-14:
  Full-time students: $1,104 per credit
  Part-time students: $1,104 per credit
Full-time equivalent faculty teaching and research assistants receive full tuition waivers for normal academic loads; the health fee is paid by the university for full-time graduate assistants.
Credit hours per semester to be considered full-time: 8
Deferred tuition plan: Yes
Health insurance: Available at the cost of $2,420 per year.
Other academic fees: Health insurance provided at no cost for students with GTA or GRA support. Student technology fee: $4/credit hour; graduate activity fee: $50/quarter.
Academic term: Quarter
Number of first-year students who received full tuition waivers: 8

Teaching Assistants, Research Assistants, and Fellowships
Number of first-year
  Teaching Assistants: 3
  Fellowship students: 1
Average stipend per academic year
  Teaching Assistant: $20,000
  Research Assistant: $20,000
  Fellowship student: $20,000

FINANCIAL AID

Application deadlines

Fall admission:
U.S. students: February 1   *Int'l. students*: February 1

Loans

Loans are available for U.S. students.
Loans are not available for international students.
GAPSFAS application required: No
FAFSA application required: No

For further information

Address financial aid inquiries to: Davor Balzar, Chairman, Dept.
   of Physics and Astronomy.
Phone: (303) 871-2238
E-mail: phys-gradinfo@du.edu
Financial aid website: http://www.physics.du.edu

HOUSING

Availability of on-campus housing

Single students: Yes
Married students: Yes

For further information

Address housing inquiries to: Department of Housing, University
   of Denver.
Phone: (303) 871-2246
E-mail: housing@du.edu
Housing aid website: http://www.du.edu/housing/

Table A—Faculty, Enrollments, and Degrees Granted

Research Specialty	2012-13 Faculty	Enrollment Fall 2012 Master's	Enrollment Fall 2012 Doctorate	Number of Degrees Granted 2012-13 (2008-13) Master's	Number of Degrees Granted 2012-13 (2008-13) Terminal Master's	Number of Degrees Granted 2012-13 (2008-13) Doctorate
Astrophysics	3	–	8	–(1)	–	2(3)
Atmosphere, Space Physics, Cosmic Rays	1	–	1	–	–	–(1)
Biophysics	3	1	2	–	1(1)	1(1)
Condensed Matter Physics	5	1	12	1(1)	–(2)	4(5)
Physics and other Science Education	1	–	–	–	–	–
Other	1	–	–	–	–	–
Total	14	2	23	1(2)	1(3)	7(10)
Full-time Grad. Stud.	–	2	23	–	–	–
First-year Grad. Stud.	–	–	4	–	–	–

GRADUATE DEGREE REQUIREMENTS

Master's: Option I (research thesis): 45 quarter hours in an approved course of study, up to 10 hours of which may be in thesis research; an acceptable thesis; oral final examination (primarily a thesis defense). Option II (no thesis): 45 quarter hours in an approved course of study; oral final examination covering coursework. Common requirements are a comprehensive examination. Residence requirement is enrollment as a graduate student in the University for at least three quarters. There is no departmental foreign language requirement.

Doctorate: Minimum of three years of full-time study beyond the baccalaureate degree, with at least 90 quarter hours of approved graduate credit; acceptable dissertation; comprehensive examination. There is no departmental foreign language

requirement. Residence requirement is enrollment as a graduate student in the University for at least six quarters including at least two consecutive quarters of full-time attendance.
Other Degrees: M.S. (applied physics) for baccalaureate engineers. For those currently employed, research projects will be matched whenever possible to the employers' programs. Requirements similar to conventional M.S.
Thesis: Thesis may be written in absentia.

SPECIAL EQUIPMENT, FACILITIES, OR PROGRAMS

The Department has major research programs in: circumstellar material surrounding particularly massive stars; optical and infra-red astronomy using the department's observatory at the summit of Mt. Evans (4313 m) and other facilities and data sources; and condensed matter and materials physics focusing on using nano-fabrication techniques to control and measure thermal, magnetic, and electronic properties of thin films and nanostructures down to 300 mK; studies of high resolution microcalorimeter x-ray and gamma-ray detectors; research on transport and recombination effects of carriers in semiconductors, photoconductors, and low-dimensional structures such as nanoparticles and nanotubes; studies of organic semiconductors, in particular the development of organic photo-voltaic devices or "plastic solar cells" for low-cost solar energy harvesting; nanoscale research on ferroelectrics, nano- and biomaterials research using x-ray and neutron diffraction and AFM techniques; studies of mechanical properties of environmentally interesting materials (some graduate research assistantships may be available in cooperation with several nearby Federal laboratories, in particular the National Institute of Standards and Technology, NIST, and the National Renewable Energy Laboratory, NREL).

Table B—Separately Budgeted Research Expenditures by Source of Support

Source of Support	Departmental Research	Physics-related Research Outside Department
Federal government	$1,498,506	
State/local government		
Non-profit organizations		
Business and industry		
Other		
Total	$1,498,506	

Table C—Separately Budgeted Research Expenditures by Research Specialty

Research Specialty	No. of Grants	Expenditures ($)
Astronomy	7	$259,262
Atmosphere, Space Physics, Cosmic Rays	4	$257,484
Biophysics	3	$96,082
Condensed Matter Physics	12	$885,678
Total	26	$1,498,506

FACULTY

Professor

Stencel, Robert E., Ph.D., University of Michigan, 1977. Womble Chair in Astrophysics. *Astronomy, Astrophysics.*

Associate Professor

Balzar, Davor, Ph.D., University of Zagreb, Croatia, 1993. Department Chair. *Applied Physics, Condensed Matter Physics, Materials Science, Metallurgy, Nano Science and Technology.*

Calbi, M. Mercedes, Ph.D., University of Buenos Aires, Argentina, 2000. *Condensed Matter Physics, Nano Science and Technology.*

Shaheen, Sean, Ph.D., University of Arizona, 1999. Chair, Graduate Committee. *Biophysics, Condensed Matter Physics, Nano Science and Technology.*

Ueta, Toshiya, Ph.D., University of Illinois at Urbana-Champaign, 2002. *Astronomy, Astrophysics.*

Zink, Barry, Ph.D., University of California, San Diego, 2002. *Condensed Matter Physics, Materials Science, Metallurgy, Nano Science and Technology.*

Assistant Professor

Ghosh, Kingshuk, Ph.D., University of Massachusetts, Amherst, 2003. *Biophysics.*

Hoffman, Jennifer, Ph.D., University of Wisconsin-Madison, 2002. *Astrophysics.*

Loerke, Dinah, Ph.D., University of Göttingen, Germany, 2004. Chair, Undergraduate Committee. *Biophysics.*

Siemens, Mark, Ph.D., University of Colorado, 2009. *Condensed Matter Physics, Nano Science and Technology.*

Emeritus

Blatherwick, Ronald D., Ph.D., University of Denver, 1976. *Atmosphere, Space Physics, Cosmic Rays.* Atmospheric physics.

Goldman, Aaron, Technion-Israel Institute of Technology, 1965. John Evans Professor. *Atmosphere, Space Physics, Cosmic Rays, Atomic, Molecular, & Optical Physics.* Atmospheric physics; atomic and molecular spectroscopy.

Neumann, Herschel, Ph.D., University of Nebraska-Lincoln, 1965. *Physics and other Science Education, Other.* Physics education.

Olson, John R., Ph.D., Iowa State University, 1963. *Acoustics, Atmosphere, Space Physics, Cosmic Rays, Other.* Atmospheric physics; electronics.

van der Merwe, Alwyn J., Ph.D., University of Bern, Switzerland, 1971. *History & Philosophy of Physics/Science, Other.* Intermolecular forces; history and foundations of physics.

Van Zyl, Bert, Ph.D., University of Washington, 1963. *Atomic, Molecular, & Optical Physics.* Atomic and molecular physics.

Williams, Walter John, M.S., University of Denver, 1963. *Atmosphere, Space Physics, Cosmic Rays.* Upper atmospheric physics.

Research Professor

Amme, Robert C., Ph.D., Iowa State University, 1958. *Atomic, Molecular, & Optical Physics, Condensed Matter Physics, Materials Science, Metallurgy, Mechanics.* Atomic and molecular physics; environmental materials; pure and applied mechanics.

Ormes, Jonathan F., Ph.D., University of Minnesota, 1967. *Astrophysics, High Energy Physics, Other.*

Adjunct Professor

Mickle, Ronald, M.S., Swinburne University, 2008. *Astronomy.*

Lecturer

Iona, Steven, Ph.D., University of Denver, 1994. *Physics and other Science Education.*

DEPARTMENTAL RESEARCH SPECIALTIES AND STAFF

Theoretical

Biophysics. Complex systems and nonlinear behavior of dynamical networks. Shaheen.

Biophysics. Computational studies of protein-protein interactions. Ghosh.

Condensed Matter Physics. Gas adsorption on nanostructures. Calbi.

Experimental

Astronomy. Mickle, Stencel, Ueta.

Astrophysics. Hoffman, Ormes, Stencel, Ueta.

Biophysics. Molecular dynamics of actin. Loerke.

Condensed Matter Physics. Materials physics and science; applied physics; low-temperature physics; thin films and nanostructures; measurements of thermal, magnetic, electronic, and mechanical properties; organic semiconductors; spintronics; strains and defects; x-ray and neutron diffraction; materials structure; compaction of granular materials; ultra-fast laser optics. Amme, Balzar, Calbi, Shaheen, Siemens, Zink.

Other. Cosmic rays. Ormes.

View additional information about this department at
www.gradschoolshopper.com

SOUTHERN CONNECTICUT STATE UNIVERSITY

DEPARTMENT OF PHYSICS

New Haven, Connecticut 06515

http://www.southernct.edu/academics/schools/arts/departments/physics/

General University Information

President: Dr. Mary A. Papazian

Dean of Graduate School: Dr. Gregory Paveza

University website: http://www.southernct.edu

Control: Public

Setting: Urban

Total Faculty: 752

Total Graduate Faculty: 320

Total number of Students: 10,800

Total number of Graduate Students: 2,615

Department Information

Department Chairman: Christine C. Broadbridge, Chair

Department Contact: Elliott P. Horch, Graduate Program Coordinator
Total full-time faculty: 7
Total number of full-time equivalent positions: 8
Full-Time Graduate Students: 3
First-Year Graduate Students: 6
Female First-Year Students: 2

Department Address

501 Crescent Street
New Haven, CT 06515
Phone: (203) 392-6450
Fax: (203) 392-6466
E-mail: physics@southernct.edu
Website: http://www.southernct.edu/academics/schools/arts/ departments/physics/

ADMISSIONS

Admission Contact Information

Address admission inquiries to: Prof. Elliott P. Horch, Graduate Program Coordinator, Department of Physics, 501 Crescent Street, New Haven, CT 06515
Phone: (203) 392-6393
E-mail: horche2@southernct.edu
Admissions website: http://www.southernct.edu/physics/graduate programs/msdegree/admission

Application deadlines

Fall admission:
U.S. students: May 1   *Int'l. students*: March 1
Spring admission:
U.S. students: November 1   *Int'l. students*: October 1

Application fee

U.S. students: $50   *Int'l. students*: $50

Admissions information

For Fall of 2013:
Number of applicants: 12
Number admitted: 5
Number enrolled: 3

Admission requirements

Bachelor's degree requirements: A Bachelor's degree in science, engineering, or a related field is required.
Minimum undergraduate GPA: 3.0

GRE requirements

The GRE is not required.

Advanced GRE requirements

The Advanced GRE is not required.

TOEFL requirements

The TOEFL exam is required for students from non-English-speaking countries.
PBT score: 550
iBT score: 79

Other admissions information

Additional requirements: Undergraduate transcripts, two letters of recommendation, and a one-page personal statement are required to complete the admission package.
Undergraduate preparation assumed: An undergraduate preparation that includes at least six physics courses is assumed (the same as for a minor in physics at the undergraduate level at SCSU).

TUITION

Tuition year 2013-14:
Tuition for in-state residents
    Full-time students: $4,985 per semester
    Part-time students: $622 per credit
Tuition for out-of-state residents
    Full-time students: $10,712 per semester
    Part-time students: $644 per credit
The figures above include university fees for full-time students.
Credit hours per semester to be considered full-time: 9
Deferred tuition plan: No
Health insurance: Available at the cost of 1088 per year.
Academic term: Semester
Number of first-year students who received partial tuition waivers: 2

Teaching Assistants, Research Assistants, and Fellowships

Number of first-year
    Fellowship students: 2
Average stipend per academic year
    Fellowship student: $9,600

FINANCIAL AID

Application deadlines

Fall admission:
U.S. students: March 2   *Int'l. students*: March 2
Spring admission:
U.S. students: November 2   *Int'l. students*: November 2

Loans

Loans are available for U.S. students.
Loans are available for international students.
GAPSFAS application required: No
FAFSA application required: Yes

For further information

Address financial aid inquiries to: Financial Aid, Southern Connecticut State University, Wintergreen Building, 501 Crescent St., New Haven, CT 06515.
Phone: (203) 392-5222
Financial aid website: http://www.southernct.edu/admissions/ undergraduate/financial-aid/index.html

HOUSING

Availability of on-campus housing

Single students: Yes
Married students: No

For further information

Address housing inquiries to: Office of Residence Life, Southern Connecticut State University, 501 Crescent St. - Schwartz Hall RM 100, New Haven, CT 06515.
Phone: (203) 392-5870
E-mail: reslife@southernct.edu
Housing aid website: http://www.southernct.edu/student-life/ campus-life/residencelife/index.html/

Table A—Faculty, Enrollments, and Degrees Granted

Research Specialty	2012-13 Faculty	Enrollment Spring 2013		Number of Degrees Granted 2012-13 (2012-13)		
		Master's	Doctorate	Master's	Terminal Master's	Doctorate
Applied Physics	7	9	–	–	–	–
Total	7	9	–	–	–	–
Full-time Grad. Stud.	–	2	–	–	–	–
First-year Grad. Stud.	–	9	–	–	–	–

GRADUATE DEGREE REQUIREMENTS

Master's: The Master of Science degree in applied physics requires completion of a total of 36 credits with a "B" or better average. All students in the program must complete an interdisciplinary core consisting of the six courses with a "B" or better average. After the core, the student selects one of two tracks, nanotechology/materials science or optics/optical detectors. The student then completes two electives and a research capstone experience. An internship (not for class credit) with a local technology company is also required for graduation.

Thesis: A Master's thesis based on research acceptable to the department is one way to satisfy the research capstone requirement. For the thesis option, students must complete a thesis proposal and initial research, and then develop and write a thesis. A student must apply to the department for the thesis defense and provide a final draft of the completed thesis at least two weeks prior to the defense date.

SPECIAL EQUIPMENT, FACILITIES, OR PROGRAMS

The Department has significant nanotechnology facilities, including sample preparation stations, scanning electron microscopy, transmission electron microscopy, atomic force microscopy, and optical microscopy. Photonics and optical instrument development facilities exist in the Department that are engaged in fiber optics, spectroscopy, and high-resolution camera development projects. The Department also has high-speed computing facilities.

Table B—Separately Budgeted Research Expenditures by Source of Support

Source of Support	Departmental Research	Physics-related Research Outside Department
Federal government	$1,946,447	$2,152,018
State/local government	$4,464	
Non-profit organizations		
Business and industry		
Other		
Total	$1,950,911	$2,152,018

Table C—Separately Budgeted Research Expenditures by Research Specialty

Research Specialty	No. of Grants	Expenditures ($)
Astronomy	2	$187,911
Nano Science and Technology	1	$1,763,000
Total	3	$1,950,911

FACULTY

Professor

Bidarian, Akbar, Ph.D., University of Kentucky, 1990. *Applied Physics, Electrical Engineering, Engineering Physics/Science, Low Temperature Physics, Materials Science, Metallurgy, Polymer Physics/Science, Solid State Physics*. Properties of polymers; superconductors with high critical temperatures; robotics; automation and electronic control systems.

Broadbridge, Christine C., Ph.D., Brown University, 1993. Department Chairperson; Education Director, Center for Research on Interface Structures and Phenomena, Yale University/SCSU; Director, Connecticut State University Nanotechnology Center.*Applied Physics, Condensed Matter Physics, Materials Science, Metallurgy, Nano Science and Technology, Physics and other Science Education, Solid State Physics, Surface Physics, Systems Science/Engineering*. Advanced materials and nanostructures for microelectronics and optoelectronics; scanning probe and electron microscopy; x-ray diffraction and spectroscopy; best practices in professional development for STEM educators.

Cummings, Karen, Ph.D., State University of New York at Albany, 1996. *Physics and other Science Education*. Learning and teaching physics, including the effective use of technology, the development of problem-solving ability in introductory students, and the development of curricular materials for use in modern learning environments.

Dolan, James F., Ph.D., University of Connecticut, 1983. *Applied Physics, Optics*. Spectroscopy of solid-state lasers and effects of ultraviolet light on optical fibers; polarization effects in interferometric measurements. Fluorescence of biological specimens.

Associate Professor

Enjalran, Matthew, Ph.D., University of California, Davis, 2000. Associate Director, Connecticut State University Nanotechnology Center. *Applied Physics, Computational Physics, Condensed Matter Physics, Materials Science, Metallurgy, Nano Science and Technology, Solid State Physics, Surface Physics, Theoretical Physics*. Condensed-matter physics of strongly correlated many-body systems with a focus on ground state phases and phase transitions in magnetic and correlated electron systems; numerical techniques for many-body systems.

Horch, Elliott P., Ph.D., Stanford University, 1994. Graduate Program Coordinator; Adjunct Astronomer, Lowell Observatory. *Applied Physics, Astronomy, Astrophysics, Optics*. High-resolution imaging techniques for astronomy; design and construction of instrumentation for large telescopes; binary stars as probes of stellar and galactic evolution.

Assistant Professor

Schwendemann, Todd C., Ph.D., University of Virginia, 2006. Visiting Assistant Professor, Yale University Department of Mechanical Engineering. *Applied Physics, Chemical Physics, Condensed Matter Physics, Engineering Physics/Science, Materials Science, Metallurgy, Nano Science and Technology, Solid State Physics, Surface Physics, Systems Science/Engineering*. Nanoscale friction/adhesion (nanotriboloy) through the manipulation of nanoparticles; thin-film and nanoparticle deposition by matrix-assisted processing (pulsed laser deposition); high-resolution scanning probe microscopy.

DEPARTMENTAL RESEARCH SPECIALTIES AND STAFF

Theoretical

Condensed Matter Physics. Magnetism; correlated electrons; phase transitions; metal-insulator transition; numerical and analytic methods for many-body systems. Enjalran.

Experimental

Astronomy. High-resolution imaging techniques for astronomy; binary stars as probes of stellar and galactic evolution. Horch.

Condensed Matter Physics. Properties of polymers; superconductors with high critical temperatures; thin-film and nanoparticle deposition by matrix assisted processing (pulsed laser deposition). Bidarian, Broadbridge, Schwendemann.

Nano Science and Technology. Advanced materials and nano-structures for microelectronics and optoelectronics; scanning probe and electron microscopy; x-ray diffraction and spectroscopy; nanoscale friction/adhesion (nanotriboloy) through the manipulation of nanoparticles; high-resolution scanning probe microscopy. Broadbridge, Schwendemann.

Optics. Spectroscopy of solid-state lasers and effects of ultra-violet light on optical fibers; polarization effects in interferometric measurements. Fluorescence of biological specimens. Design and construction of optical instrumentation for large telescopes. Dolan, Horch.

Physics and other Science Education. Learning and teaching physics, including the effective use of technology, the development of problem-solving ability in introductory students, the development of curricular materials for use in the modern learning environment, and professional development for STEM educators. Broadbridge, Cummings.

View additional information about this department at
www.gradschoolshopper.com

UNIVERSITY OF CONNECTICUT

DEPARTMENT OF PHYSICS

Storrs, Connecticut 06269-3046
http://www.physics.uconn.edu

General University Information

President: Susan Herbst
Dean of Graduate School: Kent Holsinger
University website: http://www.uconn.edu
Control: Public
Setting: Rural
Total Faculty: 1,377
Total Graduate Faculty: 1,200
Total number of Students: 22,301
Total number of Graduate Students: 7,955

Department Information

Department Chairman: Douglas Hamilton, Head
Department Contact: Kim Giard, Administrative Assistant
  Total full-time faculty: 27
  Total number of full-time equivalent positions: 27
  Full-Time Graduate Students: 78
  First-Year Graduate Students: 10
  Total Post Doctorates: 3

Department Address

2152 Hillside Road
Storrs, CT 06269-3046
Phone: (860) 486-4924
Fax: (860) 486-3346
E-mail: kim.giard@uconn.edu
Website: http://www.physics.uconn.edu

ADMISSIONS

Admission Contact Information

Address admission inquiries to: Graduate School, Unit 1006
Phone: (860) 486-3617
E-mail: gradphysics@uconn.edu
Admissions website: http://www.grad.uconn.edu

Application deadlines

Fall admission:
U.S. students: January 15   *Int'l. students*: January 15

Application fee

U.S. students: $75   *Int'l. students*: $75

Admissions information

For Fall of 2013:
  Number of applicants: 115
  Number admitted: 35
  Number enrolled: 13

Admission requirements

Bachelor's degree requirements: A bachelor's degree in physics or a closely allied field with a sufficient concentration in physics is required.
Minimum undergraduate GPA: 3.0

GRE requirements

The GRE is required.
There are no minimum accepted scores, but GRE scores are considered as part of application.

Advanced GRE requirements

The Advanced GRE is recommended.
  Minimum accepted Advanced GRE score: 670
  Mean Advanced GRE score range (25th–75th percentile): 49th

TOEFL requirements

The TOEFL exam is required for students from non-English-speaking countries.
iBT score: 92

Other admissions information

Additional requirements: Non-degree students may register under the Extended Education Program. Transfer to the Graduate Program may be possible for those with adequate academic performance.

TUITION

Tuition year 2013–14:
Tuition for in-state residents
   Full-time students: $5,728 per semester
   Part-time students: $636 per credit
Tuition for out-of-state residents
   Full-time students: $114,870 per semester
   Part-time students: $1,652 per credit
Tuition is waived for graduate assistants, veterans, and other certain groups of individuals (see graduate catalog for details).
Credit hours per semester to be considered full-time: 9
Deferred tuition plan: No
Health insurance: Available
Other academic fees: General university fee, $1368; Activity fee, $26; Matriculation fee, $84; Maintenance fee, $468; Transit fee for on-campus residents to support campus shuttle, $110; Technology Fee, $150.
Academic term: Semester
Number of first-year students who received full tuition waivers: 13

Teaching Assistants, Research Assistants, and Fellowships

Number of first-year
   Teaching Assistants: 9
Average stipend per academic year
   Teaching Assistant: $19,400
   Research Assistant: $19,400
   Fellowship student: $9,700

FINANCIAL AID

Application deadlines

Fall admission:
U.S. students: January 15   *Int'l. students*: January 15

Loans

Loans are available for U.S. students.
Loans are not available for international students.
GAPSFAS application required: No
FAFSA application required: Yes

For further information

Address financial aid inquiries to: Douglas Hamilton, Head, U-3046.
Phone: (860) 486-4924
E-mail: gradphysics@uconn.edu
Financial aid website: http://financialaid.uconn.edu/index.php/Gradprocess

HOUSING

Availability of on-campus housing

Single students: Yes
Married students: Yes

For further information

Address housing inquiries to: Office of the Graduate School, U-1006.
Housing aid website: http://www.reslife.uconn.edu

Table A—Faculty, Enrollments, and Degrees Granted

Research Specialty	2012-13 Faculty	Enrollment Fall 2012 Master's	Enrollment Fall 2012 Doctorate	Number of Degrees Granted 2012–13 (2008–13) Master's	Number of Degrees Granted 2012–13 (2008–13) Terminal Master's	Number of Degrees Granted 2012–13 (2008–13) Doctorate
Astrophysics	1	–	3	1(1)	–	–
Atomic, Molecular, & Optical Physics	7	–	27	1(12)	–(4)	5(23)
Condensed Matter Physics	8	1	20	5(10)	1(-)	–(13)
Geophysics	1	–	3	–(1)	–(1)	–(1)
Nuclear Physics	4	–	17	1(9)	–(2)	3(4)
Particles and Fields	4	–	6	–(3)	–(1)	–(8)
Polymer Physics/Science	1	–	1	–	–	–
Relativity & Gravitation	1	–	1	–(1)	–(1)	1(1)
Total	27	1	78	8(37)	1(9)	9(50)
Full-time Grad. Stud.	–	–	78	–	–	–
First-year Grad. Stud.	–	–	10	–	–	–

GRADUATE DEGREE REQUIREMENTS

Master's: For Plan A, 15 credits of graduate courses and thesis are required. For Plan B, 24 credits of graduate courses are required, but a thesis is not. Transfer credits are not ordinarily accepted. Either of these degrees may (but need not) be part of a Ph.D. program. The courses submitted must be approved in advance by the student's Advisory Committee. An average of B or better must be maintained. A final examination is required; it may be written, oral, or both. There is no foreign language or residency requirement.

Doctorate: The student must complete a plan of study of extent and quality satisfactory to the student's Advisory Committee and the Dean of the Graduate School. Ordinarily, the program will include at least 20 to 24 credits beyond the master's degree. At least one year must be in residence. The General Examination is usually taken by the end of the fifth semester.

Thesis: Thesis may be written in absentia.

SPECIAL EQUIPMENT, FACILITIES, OR PROGRAMS

The Physics Department occupies an 81,350-square-foot physics building and 25,000 square feet in the basement of an adjacent Biological Sciences/Physics building that provides research facilities for theoretical and experimental atomic, molecular, and optical physics, condensed-matter physics, nuclear physics, theoretical particle physics, astrophysics, and general relativity. In addition, the Institute of Materials Science, adjacent to the Physics Department, provides research facilities for theoretical and experimental condensed matter physics.

The Physics Department's own computer research network includes three high-performance parallel clusters. The Department's computer laboratory provides students and faculty with access to Linux, Windows, and Mac workstations and a variety of software for numerical analysis, scientific visualization, symbolic processing, and program development.

Faculty from the Physics Department have developed joint research programs with Brookhaven National Laboratory, the Thomas Jefferson National Accelerator Facility, the NASA/Caltech Jet Propulsion Laboratory, the Lawrence Livermore National Laboratory, Oak Ridge National Laboratory, the National Institute for Science and Technology, and many other American and foreign universities and research institutions.

Prospective students should visit http://www.phys.uconn.edu/research.

Table B—Separately Budgeted Research Expenditures by Source of Support

Source of Support	Departmental Research	Physics-related Research Outside Department
Federal government	$4,423,564	
State/local government	$60,786	
Non-profit organizations		
Business and industry		
Other		
Total	**$4,484,350**	

Table C—Separately Budgeted Research Expenditures by Research Specialty

Research Specialty	No. of Grants	Expenditures ($)
Astrophysics	1	
Atomic, Molecular, & Optical Physics	13	$2,776,957
Condensed Matter Physics	4	$72,358
Geophysics	2	$176,107
Nuclear Physics	11	$1,129,150
Particles and Fields	2	
Polymer Physics/Science	2	$135,208
Course Enhancement	1	$5,000
Total	**36**	**$4,294,780**

FACULTY

Professor

Côté, Robin, Ph.D., Massachusetts Institute of Technology, 1995. *Atomic, Molecular, & Optical Physics.* Theoretical atomic and molecular physics; ultracold collisions; Bose-Einstein condensation.

Cormier, Vernon F., Ph.D., Columbia University, 1976. *Geology/Geochemistry, Geophysics.* Wave propagation in deep earth structures.

Dobrynin, Andrey V., Ph.D., Moscow Institute of Physics and Technology, 1991. *Condensed Matter Physics, Polymer Physics/Science.* Theoretical polymer physics; modeling of self-assembling polymers and polymer/nanoparticle mixtures.

Dunne, Gerald V., Ph.D., Imperial College, London, 1988. Associate Department Head for Graduate Research and Education. *Particles and Fields.* Theoretical particle physics; quantum field theory; gauge theory.

Dutta, Niloy K., Ph.D., Cornell University, 1978. *Condensed Matter Physics, Engineering Physics/Science, Optics.* Experimental condensed matter and optical physics; semiconductor laser technology; quantum wires; fiber-optic transmission systems.

Eyler, Edward E., Ph.D., Harvard University, 1982. *Atomic, Molecular, & Optical Physics.* Experimental atomic, molecular, and optical physics; precision laser spectroscopy.

Fernando, Gayanath W., Ph.D., Cornell University, 1985. *Condensed Matter Physics.* Theoretical condensed matter physics; properties of transition metals.

Gibson, George N., Ph.D., University of Illinois at Chicago, 1990. Associate Department Head for Administration. *Atomic, Molecular, & Optical Physics.* High-intensity, short-pulse laser physics; laser spectroscopy.

Gould, Phillip L., Ph.D., Massachusetts Institute of Technology, 1986. *Atomic, Molecular, & Optical Physics.* Experimental quantum optics; laser cooling and trapping of atoms.

Hamilton, Douglas S., Ph.D., University of Wisconsin-Madison, 1976. Department Head. *Condensed Matter Physics, Optics.* Experimental condensed matter physics; nonlinear optics; light scattering; solid state laser design; dynamics of ions in solids.

Javanainen, Juha, Ph.D., University of Helsinki, 1981. *Atomic, Molecular, & Optical Physics.* Theoretical quantum optics; interaction of light with atoms.

Joo, Kyungseon, Ph.D., Massachusetts Institute of Technology, 1997. *Nuclear Physics.* High energy nuclear physics.

Kharchenko, Vasili A., Ph.D., Ioffe Institute, St. Petersburg, 1977. *Astrophysics, Atomic, Molecular, & Optical Physics, Planetary Science.* Theoretical physics and x-ray astrophysics; hot atoms in planetary atmospheres; kinetics and theory of atomic collisions.

Kovner, Alex, Ph.D., Tel Aviv University, 1985. *Particles and Fields.* Theoretical particle physics; strongly coupled gauge theories.

Mallett, Ronald L., Ph.D., Pennsylvania State University, 1973. *Astrophysics, Relativity & Gravitation.* Theoretical physics; relativity and gravitation; relativistic quantum theory.

Mannheim, Philip D., Ph.D., Weizmann Institute of Science, 1970. *Astrophysics, Particles and Fields, Relativity & Gravitation.* Theoretical physics; elementary particle theory; general relativity; astrophysics.

Peterson, Cynthia W., Ph.D., Cornell University, 1964. Manages the planetarium and observatory. *Astronomy, Biophysics, Condensed Matter Physics.* Experimental condensed matter physics; vacuum UV reflection spectroscopy; UV photoemission spectroscopy; biophysics.

Stwalley, William C., Ph.D., Harvard University, 1969. *Atomic, Molecular, & Optical Physics.* Experimental atomic and molecular interactions; laser spectroscopy and dynamics of atoms and molecules; ultracold atoms and molecules.

Wells, Barrett O., Ph.D., Stanford University, 1992. Associate Department Head for Undergraduate Education. *Condensed Matter Physics.* Experimental condensed matter physics; neutron scattering; superconductivity; photoemission.

Yelin, Susanne F., Ph.D., Ludwig-Maximilians University (Munich), 1998. *Atomic, Molecular, & Optical Physics, Condensed Matter Physics, Optics.* Theoretical quantum optics and condensed matter physics; spin physics in semiconductors; quantum coherence and quantum information; nonlinear optics in atoms and semiconductors.

Associate Professor

Blum, Thomas, Ph.D., University of Arizona, 1995. *High Energy Physics, Particles and Fields.* Theoretical high energy physics; lattice gauge theory; quantum chromodynamics (QCD); electroweak physics.

Dormidontova, Elena, Ph.D., Moscow State University, 1994. *Biophysics, Condensed Matter Physics, Polymer Physics/Science, Theoretical Physics.* Theoretical soft condensed matter physics, biophysics; theoretical and computational soft matter physics, (bio)macromolecules, networks, associating systems, micelles, biomedical applications, polymer-modified nanoparticles, surfaces, phase separation.

Jones, Richard T., Ph.D., Virginia Polytechnic Institute and State University (Virginia Tech), 1988. *Nuclear Physics.* Experimental nuclear physics.

Sinkovic, Boris, Ph.D., University of Hawaii, 1986. *Condensed Matter Physics.* Experimental condensed-matter physics; magnetic properties of films, surfaces, and nanostructures.

Assistant Professor

Bezrukov, Fedor, Ph.D., Russian Academy of Sciences, 2003. *Particles and Fields*. Standard model; dark matter; neutrino oscillations.

Hancock, Jason, Ph.D., University of California, Santa Cruz, 2005. *Condensed Matter Physics*. Topological insulators; superconductivity; correlated materials.

Jain, Menka, Ph.D., University of Puerto Rico, 2004. *Condensed Matter Physics*. Experimental condensed matter physics.

Schweitzer, Peter, Ph.D., University of Bochum, 2001. *Nuclear Physics, Particles and Fields*. Theoretical nuclear physics.

Emeritus

Azaroff, Leonid V., Ph.D., Massachusetts Institute of Technology, 1954. *Condensed Matter Physics*. Experimental condensed matter physics; metal physics; x-ray crystallography.

Bartram, Ralph H., Ph.D., New York University, 1960. *Condensed Matter Physics*. Theoretical condensed matter physics; optical and magnetic properties of point imperfections in solids.

Best, Philip E., Ph.D., University of Western Australia, 1962. *Condensed Matter Physics*. Experimental surface physics; electron scattering.

Budnick, Joseph I., Ph.D., Rutgers University, 1955. *Condensed Matter Physics*. Experimental condensed matter physics; nuclear magnetic resonance; critical phenomena.

Gilliam, O. R., Ph.D., Duke University, 1950. *Condensed Matter Physics*. Experimental condensed matter physics; electron spin resonance.

Hahn, Yukap, Ph.D., Yale University, 1962. *Atomic, Molecular, & Optical Physics*. Theoretical atomic, molecular, and optical physics.

Hayden, Howard, Ph.D., University of Denver, 1967. *Condensed Matter Physics*. Experimental condensed matter physics.

Hines, William A., Ph.D., University of California, Berkeley, 1967. *Condensed Matter Physics*. Experimental condensed matter physics; nuclear magnetic resonance and magnetization studies of metals and alloys.

Islam, M. M., Ph.D., Imperial College, London, 1961. *High Energy Physics*. Theoretical physics; high energy scattering; nucleon structure.

Kappers, Lawrence A., Ph.D., University of Missouri, Columbia, 1970. *Condensed Matter Physics*. Experimental condensed matter physics; color centers; optical properties; radiation damage.

Kessel, Quentin, Ph.D., University of Connecticut, 1966. *Astrophysics, Atomic, Molecular, & Optical Physics*. Experimental atomic and molecular physics; ionization; x-rays; Auger electrons; laboratory astrophysics.

Markowitz, David, Ph.D., University of Illinois, 1963. *Condensed Matter Physics*. Theoretical condensed matter physics.

Otter, Fred A., Ph.D., University of Illinois, 1959. *Condensed Matter Physics*. Experimental characterization and modification of metal and semiconductor surfaces and thin films; structural, mechanical, electrical, and optical properties.

Pease, Douglas M., Ph.D., University of Connecticut, 1972. *Condensed Matter Physics*. Experimental condensed matter physics; x-ray studies of alloys.

Rawitscher, George, Ph.D., Stanford University, 1956. *Nuclear Physics*. Theoretical physics; nuclear reaction and electron-nucleus scattering; mathematical physics.

Russek, Arnold, Ph.D., New York University, 1953. *Atomic, Molecular, & Optical Physics*. Theoretical atomic, molecular, and optical physics.

Smith, Winthrop W., Ph.D., Massachusetts Institute of Technology, 1963. *Atomic, Molecular, & Optical Physics*. Experimental atomic physics; ion-atom collisions; beam-foil spectroscopy; laser spectroscopy; laboratory astrophysics.

Adjunct Professor

Bates, Stephen C., Ph.D., Massachusetts Institute of Technology, 1977. *Condensed Matter Physics*. Experimental condensed-matter physics.

Deveney, Edward F., Ph.D., University of Connecticut, 1993. *Atomic, Molecular, & Optical Physics*. Experimental atomic and molecular physics.

Kussow, Adil-Gerai, Ph.D., A.F. Ioffe Institute of Physics and Technology (St. Petersburg), 1977. *Condensed Matter Physics, Solid State Physics*. Solid state theory: electron excitations; magnetic properties of thin films; theory of defects and fracture; first-principle electronic structure calculations.

Affiliate Professor

Birge, Robert R., Ph.D., Wesleyan College, 1972. Molecular biophysics; biomolecular electronics; biomolecular spectroscopy.

Edson, James, Ph.D., Pennsylvania State University, 1989. *Climate/Atmospheric Science, Marine Science/Oceanography*. Boundary layer meteorology with a focus on surface layer turbulence and air-sea interaction.

Huber, Greg, Ph.D., Boston University, 1993. *Biophysics, Nonlinear Dynamics and Complex Systems*. Biological physics and mechanics; biocomplexity; soft matter physics; nonequilibrium and nonlinear dynamics.

Liu, Lambo, Ph.D., Stanford University, 1993. *Geophysics*. Applied and computational geophysics; continental tectonophysics.

Michels, H. Harvey, Ph.D., University of Delaware, 1960. *Atomic, Molecular, & Optical Physics, Chemical Physics*. Theoretical atomic and molecular physics.

Montgomery, John A., Ph.D., Columbia University, 1978. *Atomic, Molecular, & Optical Physics, Computational Physics*. Computational molecular physics.

O'Donnell, James, Ph.D., University of Delaware, 1986. *Fluids, Rheology, Marine Science/Oceanography*. Physics of the coastal ocean; environmental fluid dynamics; mathematical models of environmental processes.

Papadimitrakopoulos, Fotios, Ph.D., University of Massachusetts, 1993. *Biophysics, Condensed Matter Physics*. Self-assembly of organic, inorganic, biological, and hybrid nanostructures; devices and sensors; organic semiconductors; II-VI and Si nanocrystals; carbon nanotubes.

Roychoudhuri, Chandrasekhar, Ph.D., University of Rochester, 1973. *Optics*. Experimental optical physics; semiconductor laser technology.

Schweitzer, Jeffrey S., Ph.D., Purdue University, 1972. *Astrophysics, Nuclear Physics, Solar Physics*. Experimental nuclear physics; nuclear astrophysics; solar physics.

DEPARTMENTAL RESEARCH SPECIALTIES AND STAFF

Theoretical

Astrophysics. Dark matter, dark energy, and the cosmological constant problem; interplay of cosmology and particle physics; neutrino oscillations; applications of standard and alternate gravity theories to astrophysics and cosmology; astrophysical plasmas and neutral gas; physics of the heliosphere; new astrophysical sources of x-ray emission; applications of quantum collision theory to the physics of planetary atmospheres and interstellar gas. Bezrukov, Kharchenko, Mallett, Mannheim.

Atomic, Molecular, & Optical Physics. Atomic and molecular structure; interaction of light with atoms; Bose-Einstein condensation; ultracold collisions; quantum coherence and quantum information. Côté, Javanainen, Kharchenko, Yelin.

Condensed Matter Physics. Electronic properties of point imperfections in ionic solids; thermal and electronic properties of transition metals; thermal conduction by lattice waves and electrons; spin physics in semiconductors; light-matter interactions. Dormidontova, Fernando, Yelin.

High Energy Physics. Particle physics; lattice gauge theory, quantum chromodynamics, and quantum electrodynamics; neutrino oscillations. Bezrukov, Blum, Dunne, Kovner, Mannheim, Peter Schweitzer.

Nuclear Physics. Scattering and rearrangement reactions of projectiles on nuclei; saturation physics; theory and phenomenology of strong interactions. Jones, Kovner, Peter Schweitzer.

Polymer Physics/Science. Self-assembling polymers; polymer/nanoparticle mixtures. Dobrynin, Dormidontova.

Experimental

Atomic, Molecular, & Optical Physics. Cold ion-neutral collisions; quasimolecular processes in heavy ion collisions; laser spectroscopy and atomic and molecular interactions; laser cooling and trapping of atoms; high-intensity, short-pulse laser physics; precision laser spectroscopy; ultracold molecules and plasmas. Eyler, Gibson, Gould, Stwalley.

Condensed Matter Physics. Electronic structure of alloys; magnetically ordered systems; high-temperature superconductivity; surface modification; point defects in nonmetallic crystals via ESR; NMR and magnetization studies of metals and alloys; point defects in ionic crystals via optical methods; laser spectroscopy of solids; angular-dependent electron spectroscopy from surfaces; semiconductor laser technology and fiber-optic transmission systems; neutron scattering; photoemission; x-ray absorption spectroscopy; very high peak power ps diodes for material processing; broadly tunable diodes and novel mux/demux devices for DWDM. Dobrynin, Dutta, Hamilton, Hancock, Jain, Peterson, Sinkovic, Wells.

Geophysics. Deep earth and planetary structure; elastic wave propagation; earthquake source properties. Cormier.

Nuclear Physics. Nuclear astrophysics; nuclear structure; the structure of the nucleon; low-energy QCD; meson spectroscopy. Jones, Joo.

View additional information about this department at
www.gradschoolshopper.com

UNIVERSITY OF DELAWARE

DEPARTMENT OF PHYSICS AND ASTRONOMY AND BARTOL RESEARCH INSTITUTE

Newark, Delaware 19716
http://web.physics.udel.edu

General University Information

President: Patrick T. Harker
Dean of Graduate School: James Richards
University website: http://www.udel.edu
Control: Public
Setting: Suburban
Total Faculty: 1,128
Total number of Students: 21,856
Total number of Graduate Students: 3,654

Department Information

Department Chairman: Edmund Nowak, Chair
Department Contact: Maura Perkins, Academic Support Coordinator
  Total full-time faculty: 35
  Total number of full-time equivalent positions: 37
  Full-Time Graduate Students: 80
  First-Year Graduate Students: 15
  Total Post Doctorates: 12

Department Address

104 The Green/ 217 Sharp Lab
Newark, DE 19716
Phone: (302) 831-1995
Fax: (302) 831-1637
E-mail: physics@physics.udel.edu
Website: http://web.physics.udel.edu

ADMISSIONS

Admission Contact Information

Address admission inquiries to: Chair of Graduate Admissions Committee.
Phone: (302) 831-1995
E-mail: physics@physics.udel.edu
Admissions website: http://www.physics.udel.edu/graduate/apply

Application deadlines

Fall admission:
U.S. students: April 15   *Int'l. students*: April 15

Application fee

U.S. students: $75   *Int'l. students*: $75

Admissions information

For Fall of 2013:
  Number of applicants: 123
  Number admitted: 28
  Number enrolled: 10

Admission requirements

Bachelor's degree requirements: Admission to either the MS or Ph.D. program requires a Bachelor's degree in physics or a closely related field.
Minimum undergraduate GPA: 3.2

GRE requirements

The GRE is required.
  Quantitative score: 770

Advanced GRE requirements

The Advanced GRE is recommended.
Minimum accepted Advanced GRE score: 680

TOEFL requirements

The TOEFL exam is required for students from non-English-speaking countries.
PBT score: 600
iBT score: 100
IELTA score of 7.5 is acceptable in place of TOEFL.

Other admissions information

Additional requirements: Advanced GRE score expected for financial aid consideration.
Undergraduate preparation assumed: Electricity and Magnetism, Classical Mechanics, Quantum Mechanics, Thermodynamics.

TUITION

Tuition year 2013-14:
Tuition for in-state residents
Full-time students: $14,966 per semester
Part-time students: $1,578 per credit
Tuition for out-of-state residents
Full-time students: $14,966 per semester
Part-time students: $1,578 per credit
Teaching and Research Assistant tuition is waived.
Credit hours per semester to be considered full-time: 6
Deferred tuition plan: Yes
Health insurance: Available at the cost of $238 per year.
Other academic fees: $504 (Health Service) per year.
Academic term: Semester
Number of first-year students who received full tuition waivers: 14

Teaching Assistants, Research Assistants, and Fellowships

Number of first-year
Teaching Assistants: 14
Average stipend per academic year
Teaching Assistant: $26,000
Research Assistant: $26,000

FINANCIAL AID

Application deadlines

Fall admission:
U.S. students: February 15   *Int'l. students*: February 15

Loans

Loans are not available for U.S. students.
Loans are not available for international students.
GAPSFAS application required: No
FAFSA application required: No

For further information

Address financial aid inquiries to: Chair Graduate Admissions Committee.
E-mail: physics@physics.udel.edu

HOUSING

Availability of on-campus housing

Single students: Yes
Married students: Yes

For further information

Address housing inquiries to: Office of Housing and Residence Life, www.udrentals.com.
Housing aid website: http://www.udel.edu/has/grad/main.html

Table A—Faculty, Enrollments, and Degrees Granted

Research Specialty	2012-13 Faculty	Enrollment Fall 2012 Master's	Enrollment Fall 2012 Doctorate	Number of Degrees Granted 2012-13 (2007–13) Master's	Number of Degrees Granted 2012-13 (2007–13) Terminal Master's	Number of Degrees Granted 2012-13 (2007–13) Doctorate
Astrophysics	6	–	6	–	–(1)	2(6)
Atmosphere, Space Physics, Cosmic Rays	5	–	5	1(1)	–	1(5)
Atomic, Molecular, & Optical Physics	6	–	16	1(1)	–(1)	1(8)
Biophysics	1	–	2	–	–	–
Condensed Matter Physics	10	–	23	–(2)	–(2)	2(30)
Nuclear Physics	1	–	–	–	–(1)	–(1)
Particles and Fields	7	1	15	–	–(1)	3(9)
Non-specialized	–	–	13	1(1)	–(8)	–
Total	36	1	80	2(4)	7(14)	10(51)
Full-time Grad. Stud.	–	1	80	–	–	–
First-year Grad. Stud.	–	–	15	–	–	–

GRADUATE DEGREE REQUIREMENTS

Master's: Twenty-four credit hours of classroom courses plus six credits of M.S. thesis. Thirty credit hours for MS without thesis.

Doctorate: Thirty credit hours of classroom courses, passing the Ph.D. written and oral candidacy exam, Ph.D. thesis. Students entering the program with a Master's degree may follow the Ph.D. fast track which has a reduced course requirement of 12 credit hours.

Thesis: Thesis may be written in absentia.

SPECIAL EQUIPMENT, FACILITIES, OR PROGRAMS

The Department of Physics and Astronomy is housed in Sharp Laboratory, which has its own library, machine and electronics shops, as well as research and teaching laboratories, classrooms, and office space. The condensed matter and material science programs have in house scanning and transmission microscopes, a variety of magnetometers, x-ray diffractometers, differential scanning calorimeters, thin-film deposition systems and cryogenic facilities, and make use of accelerator based facilities for x-ray and neutron scattering. The atomic and molecular physics laboratories include femtosecond and high-power pulsed lasers for nonlinear optical studies and high resolution multiphoton spectroscopy. The astro-particle physics programs include high-altitude balloon flights and high-energy cosmic ray and neutrino experiments in Antarctica (ICECUBE and Anita). Space physics programs maintain a world-wide network of neutron monitors and are involved with MMS, the Magnetosphere MultiScale mission, and multi spacecraft missions such as Cluster-2, to study the magnetosphere and the solar wind. Opportunities are available for participation in several NASA missions: ACE, The Spitzer infrared telescope, the Chandra X-ray satellite and the Hubble Space Telescope. UD is the lead institution for the Whole Earth Telescope (WET). Further programs on campus are the Institute for Energy Conversion and the Center for Composite Materials.

Table B—Separately Budgeted Research Expenditures by Source of Support

Source of Support	Departmental Research	Physics-related Research Outside Department
Federal government	$7,397,602.21	
State/local government	$433,008.78	
Non-profit organizations	$22,516.3	
Business and industry	$142,492.49	
Other	$1,185,592.65	
Total	$9,181,212.43	

Table C—Separately Budgeted Research Expenditures by Research Specialty

Research Specialty	No. of Grants	Expenditures ($)
Astronomy	4	$142,996.37
Astrophysics	9	$1,147,083.61
Atmosphere, Space Physics, Cosmic Rays	24	$1,967,274.02
Atomic, Molecular, & Optical Physics	22	$872,594.12
Condensed Matter Physics	22	$3,246,816.7
Nuclear Physics	1	$94,019.12
Particles and Fields	8	$1,226,684.23
High Energy Physics	3	$483,744.26
Total	93	$9,181,212.43

FACULTY

Professor

Barr, Stephen, Ph.D., Princeton University, 1978. *Particles and Fields.* Elementary particle theory.

Bieber, John W., Ph.D., University of Maryland, 1977. *Atmosphere, Space Physics, Cosmic Rays.* Space plasma physics; cosmic rays physics.

Chui, Siu-Tat, Ph.D., Princeton University, 1972. *Condensed Matter Physics.* Condensed matter theory; low-dimensional and amorphous materials.

Evenson, Paul A., Ph.D., University of Chicago, 1972. *Atmosphere, Space Physics, Cosmic Rays.* Space physics; solar and cosmic-ray studies.

Gaisser, Thomas K., Ph.D., Brown University, 1967. *Particles and Fields.* Elementary particle theory; high-energy astrophysics.

Glyde, Henry, Ph.D., University of Oxford, 1964. *Condensed Matter Physics.* Condensed matter theory; neutron studies of condensed matter; liquid helium.

Hadjipanayis, George C., Ph.D., University of Manitoba, 1979. *Condensed Matter Physics.* Experimental condensed matter physics; magnetism; nanocrystalline materials.

Leung, Chung Ngoc, Ph.D., University of Minnesota, 1983. *Particles and Fields.* Elementary particle theory.

MacDonald, James, Ph.D., University of Cambridge, 1979. *Astrophysics.* Astronomy and astrophysics; white dwarfs; cataclysmic variables.

Matthaeus, William H., Ph.D., College of William and Mary, 1979. *Atmosphere, Space Physics, Cosmic Rays.* Space physics; plasma physics; turbulence theory; computational physics.

Mulders, Norbert, Ph.D., Delaware, 1991. *Condensed Matter Physics.* Quantum fluids and solids.

Mullan, Dermott J., Ph.D., University of Maryland, 1969. *Astrophysics.* Astrophysics; solar and stellar physics.

Nikolic, Branislav, Ph.D., Stony Brook University, 2000. *Condensed Matter Physics.* Theoretical and computational condensed matter physics; transport phenomena.

Nowak, Edmund R., Ph.D., University of Minnesota, 1994. *Condensed Matter Physics.* Experimental condensed physics, magnetism, superconductivity, granular materials.

Owocki, Stanley P., Ph.D., University of Colorado, 1982. *Astrophysics.* Astrophysics; stellar winds.

Pittel, Stuart, Ph.D., University of Minnesota, 1968. *Nuclear Physics.* Theoretical nuclear physics and nuclear astrophysics.

Safronova, Marianna, Ph.D., University of Notre Dame, 2001. *Atomic, Molecular, & Optical Physics.* Quantum computing with neutral atoms, Rydberg atoms.

Seckel, David, Ph.D., University of Washington, 1983. *Particles and Fields.* Particle astrophysics; cosmology.

Shafi, Qaisar, Ph.D., Imperial College, London, 1971. *Particles and Fields.* Elementary particle theory; cosmology.

Shah, Ismat, Ph.D., University of Illinois, 1986. *Condensed Matter Physics.* Material Science and Engineering. Thin film, surface, interface, and nanostructures.

Shipman, Harry L., Ph.D., California Institute of Technology, 1971. *Astronomy.* Astronomy and astrophysics; white dwarfs.

Stanev, Todor, Ph.D., Sofia, Bulgaria, 1977. *Particles and Fields.* Cosmic-ray physics; particle astrophysics.

Szalewicz, Krzysztof, Ph.D., University of Warsaw, 1977. *Atomic, Molecular, & Optical Physics.* Theoretical atomic and molecular physics.

Unruh, Karl M., Ph.D., Johns Hopkins University, 1983. *Condensed Matter Physics.* Condensed matter; finite size effects; thin films.

Walker, Barry, Ph.D., Stony Brook University, 1996. *Atomic, Molecular, & Optical Physics.* Light-matter interactions; condensed matter; experimental; optical physics.

Watson, George, Ph.D., University of Delaware, 1984. *Condensed Matter Physics.* Laser spectroscopy of condensed matter; photonic band structure.

Xiao, John Q., Ph.D., Johns Hopkins University, 1993. *Condensed Matter Physics.* Specialty–metallic thin films and multilayers, superconducting materials; granular materials.

Associate Professor

Clem, John, Ph.D., Vanderbilt, 1990. *Atmosphere, Space Physics, Cosmic Rays.* IceCube and IceTop Neutrino Observatory. ANITA detector for high energy neutrinos created by collisions between cosmic rays and the cosmic microwave photons.

DeCamp, Matthew F., Ph.D., University of Michigan, 2002. *Atomic, Molecular, & Optical Physics.* Atomic and molecular physics.

Gizis, John, Ph.D., California Institute of Technology, 1998. *Astrophysics.* Astronomy, subdwarfs, brown dwarfs.

Holder, Jamie, Ph.D., University of Durham, UK, 1997. *Particles and Fields.* Cosmic ray physics.

Ji, Yi, Ph.D., Johns Hopkins University, 2003. *Condensed Matter Physics.* Condensed matter and materials physics.

Morgan, John D., Ph.D., University of California, Berkeley, 1978. *Atomic, Molecular, & Optical Physics.* Theoretical atomic and molecular physics.

Shay, Michael, Ph.D., University of Maryland, College Park, 1998. *Atmosphere, Space Physics, Cosmic Rays.* Plasma physics, space physics, and astrophysics.

Assistant Professor

Gundlach, Lars, Ph.D., Free University of Berlin, 2005. *Atomic, Molecular, & Optical Physics.*

Lorenz, Virginia, Ph.D., University of Colorado Boulder, 2007. *Atomic, Molecular, & Optical Physics.* Atomic and molecular physics.

Lyman, Edward, Ph.D., Virginia Polytechnic Institute and State University (Virginia Tech), 2004. *Biophysics.* Computational Biophysics, soft condensed matter.

Provencal, Judith L., Ph.D., Newton Wellesley Hospital, 1993. Director of the Delaware Asteroseismic Research Center (DARC). *Astronomy.* Observational Astronomy and Asteroseismic Research.

DEPARTMENTAL RESEARCH SPECIALTIES AND STAFF

Theoretical
Astrophysics. MacDonald, Mullan, Owocki, Shay.

Atmosphere, Space Physics, Cosmic Rays. Bieber, Matthaeus, Shay, Stanev.

Atomic, Molecular, & Optical Physics. Lyman, Morgan, Safronova, Szalewicz.

Biophysics. Lyman.

Condensed Matter Physics. Chui, Glyde, Nikolic.

Nuclear Physics. Pittel.

Particles and Fields. Barr, Gaisser, Leung, Seckel, Shafi, Stanev.

Experimental
Astrophysics. Gizis, Holder, Shipman.

Atmosphere, Space Physics, Cosmic Rays. Bieber, Evenson, Holder.

Atomic, Molecular, & Optical Physics. DeCamp, Gundlach, Lorenz, Morgan, Szalewicz, Walker.

Condensed Matter Physics. Glyde, Hadjipanayis, Ji, Mulders, Nowak, Shah, Unruh, Watson, Xiao.

Optics. DeCamp, Walker, Watson.

View additional information about this department at www.gradschoolshopper.com

CATHOLIC UNIVERSITY OF AMERICA

DEPARTMENT OF PHYSICS

Washington, D.C. 20064
http://physics.cua.edu

General University Information
President: John H. Garvey

Dean of Graduate School: James J. Greene

University website: http://www.cua.edu

Control: Private

Setting: Urban

Total Faculty: 798

Total Graduate Faculty: 342

Total number of Students: 6,838

Total number of Graduate Students: 3,144

Department Information
Department Chairman: Steven B. Kraemer, Chair

Department Contact: Steven B. Kraemer, Chair

Total full-time faculty: 11

Total number of full-time equivalent positions: 11

Full-Time Graduate Students: 26

First-Year Graduate Students: 5

Female First-Year Students: 3

Total Post Doctorates: 34

Department Address
Hannan Hall - Room 200

620 Michigan Avenue, NE

Washington, DC 20064

Phone: (202) 319-4335/-5315

Fax: (202) 319-4448

E-mail: kraemer@cua.edu

Website: http://physics.cua.edu

ADMISSIONS

Admission Contact Information
Address admission inquiries to: Graduate Admissions, Physics Department, 200 Hannan Hall, CUA, 620 Michigan Avenue, NE, Washington, DC 20064

Phone: (202) 319-5315

E-mail: cua-physgrad@cua.edu

Admissions website: http://admissions.cua.edu/graduate

Application deadlines
Fall admission:

U.S. students: August 1   *Int'l. students*: July 1

Spring admission:

U.S. students: December 1   *Int'l. students*: November 15

Application fee
U.S. students: $55   *Int'l. students*: $55

Applications for financial assistance are due by February 1.

Admissions information
For Fall of 2013:

Number of applicants: 28

Number admitted: 16

Number enrolled: 5

Admission requirements
Bachelor's degree requirements: Bachelor's degree with a background in physics and mathematics equivalent to a physics major beginning the senior year.

Minimum undergraduate GPA: 3.0

GRE requirements
The GRE is required.

Quantitative score: 600

Verbal score: 400

Mean GRE score range (25th–75th percentile): Q 640-790, V 440-680

Advanced GRE requirements
The Advanced GRE is not required.

TOEFL requirements
The TOEFL exam is required for students from non-English-speaking countries.
PBT score: 580
iBT score: 92

Other admissions information
Additional requirements: Most graduate fellowships require a combined verbal and quantitative score of 1300.
Undergraduate preparation assumed: Equivalent to: Taylor, Classical Mechanics; Griffiths, Introduction to Electrodynamics; Griffiths, Introduction to Quantum Mechanics.

TUITION

Tuition year 2013-14:
  Full-time students: $39,000 annual
  Part-time students: $1,525 per credit
Students with teaching assistantships and research assistantships receive full tuition waivers.
Credit hours per semester to be considered full-time: 8
Deferred tuition plan: No
Health insurance: Available at the cost of 2,250 per year.
Other academic fees: New student fee of $425 (one time only in the first semester); activities fee of $50 per semester; technology fee of $150 per semester.
Academic term: Semester
Number of first-year students who received full tuition waivers: 4

Teaching Assistants, Research Assistants, and Fellowships
Number of first-year
  Teaching Assistants: 2
  Research Assistants: 2
  Fellowship students: 1
Average stipend per academic year
  Teaching Assistant: $18,000
  Research Assistant: $18,400
  Fellowship student: $18,000

FINANCIAL AID

Application deadlines
Fall admission:
U.S. students: February 1   *Int'l. students*: February 1

Loans
Loans are available for U.S. students.
Loans are available for international students.
GAPSFAS application required: No
FAFSA application required: Yes

For further information
Address financial aid inquiries to: Graduate Admissions, Physics Department, CUA, 200 Hannan Hall, 620 Michigan Ave. NE, Washington, DC 20064.
Phone: (202) 319-5315
E-mail: cua-physgrad@cua.edu
Financial aid website: http://physics.cua.edu

HOUSING

Availability of on-campus housing
  Single students: Yes
  Married students: No

For further information
Address housing inquiries to: Housing Services, CUA, 620 Michigan Ave. NE, 160 O'Connell Hall, Washington, DC 20064.
Phone: (202) 319-5615
E-mail: cua-housing@cua.edu
Housing aid website: http://housing.cua.edu

Table A—Faculty, Enrollments, and Degrees Granted

Research Specialty	2012-13 Faculty	Enrollment 2012-13 Master's	Enrollment 2012-13 Doctorate	Number of Degrees Granted 2012-13 (2007-13) Master's	Number of Degrees Granted 2012-13 (2007-13) Terminal Master's	Number of Degrees Granted 2012-13 (2007-13) Doctorate
Astrophysics	12	–	12	3(-)	1(-)	1(8)
Atmosphere, Space Physics, Cosmic Rays	16	–	3	–	–	–(1)
Biophysics	2	–	3	1(-)	–	–(1)
Condensed Matter Physics	7	–	4	–	–	–(3)
Energy Sources & Environment	4	–	–	–	–	–
Materials Science, Metallurgy	8	–	–	3(-)	1(-)	1(1)
Nano Science and Technology	4	–	5	–	–	–(1)
Nuclear Physics	3	–	3	1(-)	–	–(2)
Planetary Science	9	–	–	–	–	–(1)
Solar Physics	9	–	1	–	–	–
Non-specialized	–	–	3	3(12)	–(3)	–
Total	74	–	34	8(20)	2(5)	2(18)
Full-time Grad. Stud.	–	–	34	–	–	–
First-year Grad. Stud.	–	–	5	–	–	–

GRADUATE DEGREE REQUIREMENTS

Master's: Thirty graduate credits required with a GPA of 2.5, of which six may be dissertation guidance if the thesis option is selected. A master's comprehensive examination and one year of residency are required. There is no language requirement.

Doctorate: Fifty-three graduate credits are required, of which 35 must be in physics. Written and oral comprehensive examinations are given after completion of 35 credits and must be passed before beginning thesis work. Students must maintain a GPA of 3.0 in order to qualify to take the comprehensive examinations. Three years of residency are required. There is no language requirement.

Other Degrees: M.S. in nuclear environmental protection (see http://nep.cua.edu/); M.S. in materials science and engineering (see http://materialsscience.cua.edu/).

Thesis: Thesis may not be written in absentia.

SPECIAL EQUIPMENT, FACILITIES, OR PROGRAMS

The Vitreous State Laboratory has a large array of equipment for research in materials science, nanoscience, nuclear environmental protection, and condensed-matter physics. Students in astrophysics have access to the facilities of the NASA Goddard Space Flight Center. The Department has close ties with many Federal facilities in the Washington area, such as the Naval Research Laboratory, the Goddard Space Flight Center, and the National Institute for Standards and Technology, as well as with Jefferson Laboratory in Newport News, VA. CUA is a member of a 14-university consortium in the Washington, D.C., area. Courses may be taken at other universities in the consortium.

Table B—Separately Budgeted Research Expenditures by Source of Support

Source of Support	Departmental Research	Physics-related Research Outside Department
Federal government	$15,457,000	
State/local government		
Non-profit organizations	$237,000	
Business and industry	$886,000	
Other		
Total	**$16,580,000**	

Table C—Separately Budgeted Research Expenditures by Research Specialty

Research Specialty	No. of Grants	Expenditures ($)
Astrophysics	101	$6,167
Energy Sources & Environment	52	$8,031,000
Materials Science, Metallurgy	9	$1,125,000
Nano Science and Technology	3	$794,000
Nuclear Physics	3	$463,000
Total	**168**	**$10,419,167**

FACULTY

Professor

Kraemer, Steven B., Ph.D., University of Maryland, 1985. Chair, Physics Department. *Astrophysics.* Theoretical and observational astrophysics.

Pegg, Ian L., Ph.D., University of Sheffield, England, 1982. Director, Vitreous State Laboratory. *Chemical Physics, Condensed Matter Physics, Energy Sources & Environment, Materials Science, Metallurgy.* Experimental materials science.

Resca, Lorenzo, Ph.D., Scuola Normale, Superiore di Pisa, 1970. *Condensed Matter Physics, Materials Science, Metallurgy, Nano Science and Technology, Statistical & Thermal Physics, Other.* Theoretical physics; nanostructures and superlattices; ferro-, piezo-, and thermoelectricity; psychophysical measurements.

Sober, Daniel I., Ph.D., Cornell University, 1969. *Nuclear Physics.* Experimental hadronic and intermediate-energy nuclear physics; medium-energy physics.

Associate Professor

DeMello, Duilia, Ph.D., University of São Paulo, 1995. *Astrophysics.* Theoretical and observational astrophysics.

Dutta, Biprodas, Ph.D., Vanderbilt University, 1987. *Materials Science, Metallurgy, Nano Science and Technology.* Materials science; nanotechnology.

Klein, Franz J., Ph.D., University of Bonn, 1996. *Nuclear Physics, Particles and Fields.* Experimental hadronic and intermediate-energy nuclear physics; measurement of spin observables in photoproduction using polarized beams and targets.

Philip, John, Ph.D., Indian Institute of Science, Bangalore, 2001. *Materials Science, Metallurgy, Nano Science and Technology.* Materials science; nanotechnology.

Uritsky, Vadim M., Ph.D., St. Petersburg State University, 1998. *Astrophysics, Physics and other Science Education, Solar Physics, Other.*

Assistant Professor

Horn, Tanja, Ph.D., University of Maryland, 2006. *Computational Physics, Nuclear Physics, Particles and Fields.* Experimental hadronic physics in the transition region between QCD confinement and asymptotic freedom; quark and gluon de-

grees of freedom in exclusive meson production; timelike Compton scattering; development of the electron-ion collider.

Sarkar, Abhijit, Ph.D., University of Illinois at Chicago, 2002. *Biophysics, Chemical Physics, Computational Physics, Condensed Matter Physics, Statistical & Thermal Physics.* Experimental: single-molecule DNA-protein micromanipulation to study protein-DNA interactions. Theoretical: apply statistical mechanical techniques to biological systems, especially using ideas from polymer physics.

Emeritus

Überall, Herbert M., Ph.D., University of Vienna 1953 and Cornell University 1956. *Acoustics.* Theoretical physics; nuclear physics; acoustics. ear and acoustics.

Crannell, Hall L., Ph.D., Stanford University, 1964. Experimental nuclear and medium-energy physics.

Leibowitz, Jack R., Ph.D., Brown University, 1962. *Low Temperature Physics.* Low-temperature and condensed-matter physics.

Meijer, Paul H. E., Ph.D., Leiden University, 1950. *Statistical & Thermal Physics.* Theoretical physics; statistical mechanics; solid-state physics.

Werntz, Carl, Ph.D., University of Minnesota, 1960. Theoretical physics; nuclear astrophysics.

Research Professor

Aikin, Arthur, Ph.D., Pennsylvania State University, 1960. *Astrophysics, Atmosphere, Space Physics, Cosmic Rays.* Astrophysics.

Bruhweiler, Frederick, Ph.D., University of Texas, 1977. Director, Institute for Astrophysics and Computational Science (IACS). *Astrophysics.* Theoretical and observational astrophysics.

Krasnopolsky, Vladimir, Ph.D., University of Moscow, 1972. *Astrophysics, Atmosphere, Space Physics, Cosmic Rays.* Planetary and cometary atmospheres.

Ofman, Leon, Ph.D., University of Texas, Austin, 1992. *Astrophysics, Atmosphere, Space Physics, Cosmic Rays.* Solar physics; astrophysics.

Research Associate Professor

Brosius, Jeffrey, Ph.D., University of Delaware, 1985. *Solar Physics.* Solar physics.

Clark, Pamela, Ph.D., University of Maryland, 1979. *Astrophysics.* Geochemistry; remote sensing.

Kutepov, Alexander A., Ph.D., University of Leningrad, 1975. *Astrophysics, Atmosphere, Space Physics, Cosmic Rays.* Transport of radiation through atmospheres.

Smith, Myron A., Ph.D., University of Arizona, 1971. *Astrophysics.* Stellar astrophysics.

Starr, Richard D., Ph.D., University of Illinois, 1978. *Astrophysics.* Gamma-ray astronomy.

Wahlgren, Glenn, Ph.D., Ohio State University, 1986. *Astrophysics, Atomic, Molecular, & Optical Physics.* Astrophysics.

Research Assistant Professor

Chen, Peter, Ph.D., Case Western Reserve University, 1979. *Astrophysics, Optics.* Astronomy and advanced optics.

Moran, Thomas, Ph.D., Massachusetts Institute of Technology, 1988. *Astrophysics, Atmosphere, Space Physics, Cosmic Rays.* Solar physics.

Nielsen, Krister, Ph.D., Lund University, 2002. *Astrophysics, Atomic, Molecular, & Optical Physics.* Atomic spectroscopy.

Reginald, Nelson, Ph.D., University of Delaware, 2000. *Solar Physics.* Solar physics and instrumentation.

Adjunct Professor

Bell, Michael I., Ph.D., Brown University, 1972. *Condensed Matter Physics, Materials Science, Metallurgy.* Solid-state physics; spectroscopy.

Gopalswamy, Natchimuthukonar, Ph.D., Indian Institute of Science, 1982. *Astrophysics, Atmosphere, Space Physics, Cosmic Rays.* Solar physics.

Kondo, Yoji, Ph.D., University of Pennsylvania, 1965. *Astrophysics.* Astrophysics.

Muller, Isabelle, Ph.D., Pierre et Marie Curie University, 1986. *Chemical Physics, Materials Science, Metallurgy.* Physical chemistry.

Resta, Raffaele, Ph.D., University of Pisa, 1969. *Condensed Matter Physics.* Theoretical solid-state physics.

Adjunct Associate Professor

Disanti, Michael, Ph.D., University of Arizona, 1989. *Astrophysics.* Astrophysics and planetary science.

Pulkkinen, Antti A., Ph.D., University of Helsinki, 2003. *Astrophysics, Atmosphere, Space Physics, Cosmic Rays, Physics and other Science Education, Other.*

Senior Research Scientist

Reiner, Michael, Ph.D., University of Maryland, 1974. *Atmosphere, Space Physics, Cosmic Rays.* Heliospheric physics.

Associate Research Scientist

Airapetian, Vladimir, Ph.D., Byurakan Astrophysical Observatory, 1989. *Astrophysics.* Stellar astrophysics.

Research Associate

Bonev, Boncho, Ph.D., University Toledo, 2005. *Planetary Science.* Comets.

Villaneuva, Geronimo, Ph.D., Albert Ludwigs University, 2004. *Planetary Science.* Planetary atmospheres; Mars research.

Associate Researcher

Yashiro, Seiji, Ph.D., University of Tokyo, 2000. *Atmosphere, Space Physics, Cosmic Rays, Solar Physics.* Heliospheric physics.

DEPARTMENTAL RESEARCH SPECIALTIES AND STAFF

Theoretical

Astrophysics. MHD simulations of sun and solar environment; photoionization modeling; dynamic modeling of solar system; radiative transfer in upper atmospheres. Airapetian, Kraemer, Ofman, Uritsky.

Biophysics. Statistical mechanical modeling of DNA, proteins, and single-molecule experiments; statistical approaches to systems biology and biological processes. Sarkar.

Condensed Matter Physics. Semiconductors; insulators; superlattices. Bell, Resca, Resta.

Experimental

Astrophysics. Solar physics; observational cosmology and AGNs; stellar physics; upper atmosphere; exoplanets and young planetary systems; planetary science. Aikin, Bonev, Brosius, Bruhweiler, Chen, Clark, DeMello, Disanti, Gopalswamy, Kondo, Kraemer, Krasnopolsky, Kutepov, Moran, Nielsen, Reginald, Reiner, Smith, Starr, Villaneuva, Wahlgren, Yashiro.

Biophysics. Single-molecule DNA-protein interaction studies using magnetic tweezers; interaction of electromagnetic radiation with cells. Sarkar.

Energy Sources & Environment. Nuclear waste treatment technologies; energy recovery; thermoelectric ceramic heterostructures; solar cells; low-carbon impact materials. Dutta, Muller, Pegg.

Materials Science, Metallurgy. Structure and properties of novel glass-forming systems; thermoelectric ceramic oxides; semiconductor thin films; solar cells; magnetic thin films; thin film heterostructures; magnetic tunnel junctions; spin valves; fly ash activity; geopolymers. Dutta, Muller, Pegg.

Nano Science and Technology. Nanoelectronics; nanospintronics; growth of semiconducting and metallic nanostructures; nanoscale device physics; nanoscale sensors; MEMS (microelectromechanical systems); bioMEMS. Dutta, Philip.

Nuclear Physics. Nuclear and hadronic structure studies with intermediate-energy electron and photon beams. Horn, Klein, Sober.

View additional information about this department at
www.gradschoolshopper.com

GEORGE WASHINGTON UNIVERSITY

DEPARTMENT OF PHYSICS

Washington, D.C. 20052
http://phys.gwu.edu/

General University Information

President: Steven Knapp
Dean of Graduate School: Ben Vinson
University website: http://www.gwu.edu/
Control: Private
Setting: Urban
Total Faculty: 4,673
Total Graduate Faculty: N/A

Total number of Students: 20,327
Total number of Graduate Students: 9,935

Department Information

Department Chairman: Allena K. Opper, Chair
Department Contact: Allena K. Opper, Chair
  Total full-time faculty: 22
  Total number of full-time equivalent positions: 19

Full-Time Graduate Students: 35
First-Year Graduate Students: 10
Female First-Year Students: 3
Total Post Doctorates: 10

Department Address
Corcoran Hall, Suite 105
725 21st Street, NW
Washington, D.C., DC 20052
Phone: (202) 994-6275
E-mail: phys@gwu.edu
Website: http://phys.gwu.edu/

ADMISSIONS

Admission Contact Information
Address admission inquiries to: Prof. Weiqun Peng, Director of
  Graduate Programs, Department of Physics, 725 21st Street,
  NW, Washington, D.C. 20052
Phone: (202) 994-6275
E-mail: gradappl@gwu.edu
Admissionswebsite:http://www.gwu.edu/~ccas/grad/application.
  html

Application deadlines
Fall admission:
U.S. students: January 15   *Int'l. students*: January 15
Spring admission:
U.S. students: September 1   *Int'l. students*: September 1

Application fee
U.S. students: $75   *Int'l. students*: $75

Admissions information
For Fall of 2013:
  Number of applicants: 65
  Number admitted: 18
  Number enrolled: 6

Admission requirements
Bachelor's degree requirements: Bachelor's degree in physics
  or equivalent is preferred.
Minimum undergraduate GPA: 3.0

GRE requirements
The GRE is required.
We require GRE general with a good Q score and reasonable
  A and V scores.

Advanced GRE requirements
The Advanced GRE is recommended.
GRE subject is optional but strongly recommended to enhance
  your chances.

TOEFL requirements
The TOEFL exam is required for students from non-English-
  speaking countries.
  PBT score: 600
  iBT score: 100
Or a score of over 250 in the computer based test.

Other admissions information
Undergraduate preparation assumed: Subjects at the level of Symon, Mechanics; Lorrain and Corson, Electricity and Magnetism; Reif, Statistical and Thermal Physics; Eisberg and Park,
  Modern and Quantum Physics.

TUITION
Tuition year 2013-14:
  Full-time students: $1,340 per credit
  Part-time students: $1,340 per credit

Tuition rate effective since Fall 2012.
Credit hours per semester to be considered full-time: 9
Deferred tuition plan: Yes
Health insurance: Available at the cost of 1977 per year.
Other academic fees: Yes
Academic term: Semester
Number of first-year students who received full tuition waivers: 10

Teaching Assistants, Research Assistants, and Fellowships
Number of first-year
  Teaching Assistants: 8
  Research Assistants: 2
Average stipend per academic year
  Teaching Assistant: $23,000
  Research Assistant: $23,000
  Fellowship student: $25,000

FINANCIAL AID

Loans
Loans are available for U.S. students.
Loans are not available for international students.
GAPSFAS application required: No
FAFSA application required: Yes

For further information
Address financial aid inquiries to: Department of Physics.

HOUSING

Availability of on-campus housing
  Single students: Yes
  Married students: No

For further information
Address housing inquiries to: Director of Housing.

Table A—Faculty, Enrollments, and Degrees Granted

Research Specialty	2012-13 Faculty	Enrollment Fall 2012		Number of Degrees Granted 2012–13 (2008–13)		
		Master's	Doctorate	Master's	Terminal Master's	Doctorate
Astrophysics	8	–	3	–	–	–(1)
Biophysics	6	–	7	–	–(1)	1(5)
Medical, Health Physics	–	–	1	–	–	1(2)
Nuclear Physics	11	–	8	–(1)	–(2)	1(8)
Physics and other Science Education	4	–	1	–	–	–(1)
Non-specialized	–	1	13	–(1)	–(2)	–
Total	26	1	34	–(2)	–(4)	6(9)
Full-time Grad. Stud.	–	1	34	–	–	–
First-year Grad. Stud.	–	1	10	–	–	–

GRADUATE DEGREE REQUIREMENTS

Master's: M.A. degree with thesis or no thesis options: 30 semester-
  hours of course work in physics plus thesis, or 36 semester hours
  of course work in physics and mathematics, including a tool requirement in computer programming. A 3.0 GPA is required.
Doctorate: A minimum of 72 semester hours of approved
  courses for students with only a Baccalaureate. For students
  with a Master's degree, a minimum of 48 semester hours is
  required. Tool requirement: completion of a numerical methods course. A 3.0 GPA is required.
Thesis: Thesis may be written in absentia.

SPECIAL EQUIPMENT, FACILITIES, OR PROGRAMS

High-end central computing facility; several departmental computing facilities, including five high-end clusters, two CMP/biophysics research labs; machine shop; Virginia campus facilities contains labs for design, construction, and testing of particle and radiation detectors for use at major accelerator laboratories worldwide.

A new Science and Engineering Hall (SEH) is scheduled to open January 2015.

Table B—Separately Budgeted Research Expenditures by Source of Support

Source of Support	Departmental Research	Physics-related Research Outside Department
Federal government	$2,250,000	
State/local government		
Non-profit organizations		
Business and industry		
Other	$100,000	
Total	$2,350,000	

FACULTY

Professor

Briscoe, William, Ph.D., Catholic University of America, 1978. *Nuclear Physics*. Experimental nuclear physics and particle physics.

Eskandarian, Ali, Ph.D., George Washington University, 1967. *Astrophysics, Computational Physics, Nuclear Physics*. Theoretical nuclear physics; astrophysics.

Feldman, Gerald, Ph.D., University of Washington, 1987. Experimental nuclear physics; physics education research.

Lee, Frank X., Ph.D., Ohio University, 1993. *Computational Physics, Nuclear Physics*. Theoretical nuclear and particle physics.

Opper, Allena K., Ph.D., Indiana University, 1991. Experimental nuclear and particle physics; computational physics.

Reeves, Mark E., Ph.D., University of Illinois, 1989. Experimental condensed matter physics; biophysics; medical physics.

Zeng, Chen, Ph.D., Cornell University, 1994. Theoretical condensed matter physics; biophysics.

Associate Professor

Afanasev, Andrei V., Ph.D., Kharkov National University, Ukraine, 1990. Theoretical nuclear physics.

Dhuga, Kalvir S., Ph.D., University of Birmingham, 1980. Experimental nuclear physics; astrophysics.

Griesshammer, Harald, Ph.D., University of Erlangen-Nürnberg, 1996. Theoretical nuclear and particle physics.

Haberzettl, Helmut, Ph.D., University of Bonn, 1979. Theoretical nuclear and particle physics.

Peng, Weiqun, Ph.D., University of Illinois, 2001. Theoretical biophysics.

Assistant Professor

Alexandru, Andrei, Ph.D., Louisiana State University, 2001. Theoretical nuclear physics.

Cobb Kung, Bethany, Ph.D., Yale University, 2008. Also teaches in the University Honors Program. *Astrophysics*. Astrophysics (gamma-ray bursts, time-domain, astronomy).

Corsi, Alessandra, Ph.D., Rome University Sapienza, 2007. *Astrophysics*. Observational astrophysics.

Downie, Evangeline J., Ph.D., University of Gaslow, 2007. Experimental nuclear physics.

Kargaltsev, Oleg, Ph.D., Pennsylvania State University, 2004. Observational astrophysics.

Lan, Ganhui, Ph.D., Johns Hopkins University, 2008. Computational biophysics.

Qui, Xiangyun, Ph.D., Michigan State University, 2004. Experimental condensed-matter physics; biophysics.

Teodorescu, Raluca, Ph.D., George Washington University, 2009. *Physics and other Science Education*. Physics Education Research.

Emeritus

Lehman, Donald R., Ph.D., George Washington University, 1970. Theoretical nuclear physics.

Parke, William C., Ph.D., George Washington University, 1967. Theoretical nuclear physics; biophysics; astrophysics.

Research Professor

Maximon, Leonard C., Ph.D., Cornell University, 1952. Theoretical nuclear physics; mathematical physics; astrophysics.

Strakovsky, Igor, Ph.D., University of St. Petersburg, 1984. Experimental nuclear physics; phenomenology.

Research Associate Professor

Workman, Ron, Ph.D., University of British Columbia, 1987. Theoretical nuclear physics; phenomenology.

Research Assistant Professor

Wang, Guanyu, Ph.D., Goettingen University, 1998. Theoretical biophysics.

DEPARTMENTAL RESEARCH SPECIALTIES AND STAFF

Theoretical

Astrophysics. For further information, click on Astrophysics Group web pages. Eskandarian, Maximon, Parke.

Biophysics. The theoretical biophysics group currently consists of our four faculty members, and associated postdoctoral associates, graduate students, and undergraduate students. Current research interests of the faculty members are as follows: Ganhui Lan - Theoretical analysis of biochemical networks for cells to maintain their precise temporal and spatial regulation; computational modeling of intracellular assembling processes Weiqun Peng - Computation study of functional genomics, epigenomics, and gene regulation; bioinformatics; mathematical modeling of evolutionary dynamics Guanyu Wang - Physical Oncology, disease modeling, and bionetwork analysis Chen Zeng - Computational modeling and design of protein structures and numerical studies of bionetworks' robustness and evolvability To learn more, click on the Biophysics Group's website. Lan, Peng, Wang, Zeng.

Nuclear Physics. The theoretical nuclear physics research group aims to understand the structure and interactions of photons, hadrons, and nuclei at low and intermediate energy scales. It employs a variety of theoretical tools, such as lattice QCD and QCD sum rules, coupled-channels analysis, relativistic reaction theories, and effective field theories. For more information, click on the Theoretical Nulcear Physics website. Afanasev, Alexandru, Eskandarian, Griesshammer, Haberzettl, Lee, Lehman, Workman.

Experimental

Astrophysics. Members of the GW astrophysics group have done research in nuclear and particle physics for many years, going back to George Gamow, the developer of the hot Big Bang Theory of the Universe. With the rapid expansion of information in high-energy astrophysics due to new satellite data of unprecedented quality, and with the convergence of central themes in nuclear physics and astronomy in recent years, our group is now investigating new fundamental questions in astrophysics. Our current interests center on understanding the underlying physical processes occurring near extremely compact and dense objects such as neutron stars and black holes. We study these processes through the analysis of x-ray and gamma-ray data that have been collected by a number of space-borne telescopes. Gamma-ray bursts (GRBs) are associated with some of the most energetic and explosive events in the Universe. We explore the mechanisms that lead to the observed spectra of GRBs and the released energy. Neutron stars born with very strong magnetic fields, called magnetars, are also of special interest. Other areas of study include the evolution of very massive stars and their interactions with the interstellar medium. The GW group benefits extensively through collaborations with colleagues at NASA/Goddard Space Flight Center in Greenbelt, Maryland and around the world. For further information, click on the Astrophysics Group web pages. Cobb Kung, Corsi, Dhuga, Kargaltsev.

Biophysics. The experimental biophysics and condensed-matter group currently consists of faculty members Mark Reeves and Xiangyun Qiu, and their graduate and undergraduate students. The group features expertise in scanning probe-based near-field microscopy; detection of biomolecules by localized surface plasmon sensing; analysis of biomolecular structure, interaction, and functional relationships; X-ray and neutron scattering; and osmotic stress methods for mdofying celluar components. These techniques are being applied to the study of electronic materials, biomaterials, and to problems in cellular biological physics. Our expertise allows our students to study structural linkages in proteins and crystalline systems, and to study biological and electronic functionality through sub-wavelength length-scale probes of the electromagnetic response of materials. Collaborations with federal laboratories (NRL, ORNL, NIH, NIST) and with faculty in chemistry, biology, and in the medical school allow us to address a wide array of research questions. New approaches to investigating protein functionality are being developed, based on the electronic and optical response of self assembled nanoparticle systems. To learn more, click on the Biophysics Group's website. Qui, Reeves.

Nuclear Physics. The focus of the experimental nuclear physics group remains the understanding of the strong interaction in the nuclear medium. Our intention is to measure the elementary amplitudes for meson photoproduction and baryon excitation on the nucleon and see how they are modified in the nuclear medium, particularly in the light nuclei, where the nuclear density changes dramatically with very little change in nuclear size. To learn more, click on the GW Experimental Nuclear Physics Research Group's website. Briscoe, Downie, Feldman, Opper, Strakovsky.

Physics and other Science Education. Peer Instruction: developing and testing large collection of ConcepTests, organizing coherent sequences (ConcepModules), linking conceptual questions with numerical problems Thinking Skills Curriculum: taxonomy of Physics problems (based on Marzano); develop cognitive skills necessary for problem solving; problem-solving protocol (GW–ACCESS) Collaborative SCALEUP classroom: students work in cooperatively groups; full integration of lecture; recitation and lab; entirely focused on students (instructor as coach). Feldman, Reeves, Teodorescu.

View additional information about this department at www.gradschoolshopper.com

---

GEORGETOWN UNIVERSITY

DEPARTMENT OF PHYSICS

Washington, D.C. 20057
http://physics.georgetown.edu/

General University Information
President: John J. DeGioia
Dean of Graduate School: Gerald M. Mara
University website: http://www.georgetown.edu/
Control: Private
Total Faculty: 2,173
Total number of Students: 13,385
Total number of Graduate Students: 5,832

Department Information
Department Chairman: Amy Liu, Chair
Department Contact: Jim Freericks, Director of Graduate Studies
Total full-time faculty: 17
Full-Time Graduate Students: 24
First-Year Graduate Students: 6
Female First-Year Students: 1
Total Post Doctorates: 10

Department Address
37th & O Streets NW
Washington, DC 20057
Phone: (202) 687-5592
E-mail: graduatehelp@physics.georgetown.edu
Website: http://physics.georgetown.edu/

ADMISSIONS

Admission Contact Information
Address admission inquiries to: GSAS, Office of Graduate Admissions, Box 571004, Washington, DC 20057-1004
Phone: (202) 687-5568
E-mail: gradmail@georgetown.edu
Admissions website: http://grad.georgetown.edu/admissions/

Application deadlines
Fall admission:
U.S. students: December 15   *Int'l. students*: December 15

Application fee

U.S. students: $80   *Int'l. students*: $80

Admission requirements

Bachelor's degree requirements: Bachelor's degree in physics or related field is required, plus a personal statement, three letters of recommendation, and a resume/CV.

Minimum undergraduate GPA: 3.0

GRE requirements

The GRE is required.

Advanced GRE requirements

The Advanced GRE is required.

TOEFL requirements

The TOEFL exam is required for students from non-English-speaking countries.
PBT score: 600
iBT score: 100

Other admissions information

Additional requirements: GRE physics subject test is required.

Undergraduate preparation assumed: Intermediate-level courses in classical mechanics, quantum mechanics, electricity and magnetism, and statistical and thermal physics, as well as a working knowledge of an advanced computer language.

TUITION

Tuition year 2013-14:
Full-time students: $1,739 per credit
Credit hours per semester to be considered full-time: 9
Deferred tuition plan: No
Health insurance: Yes, Provided to full-time students.
Other academic fees: Mandatory Yates recreation fee: $348 per year.
Academic term: Semester
Number of first-year students who received full tuition waivers: 6

Teaching Assistants, Research Assistants, and Fellowships

Number of first-year
Teaching Assistants: 6
Average stipend per academic year
Teaching Assistant: $24,000
Research Assistant: $24,000
Fellowship student: $24,000

FINANCIAL AID

Application deadlines

Fall admission:
U.S. students: December 15   *Int'l. students*: December 15

Loans

Loans are available for U.S. students.
Loans are available for international students.
GAPSFAS application required: Yes
FAFSA application required: Yes

For further information

Address financial aid inquiries to: Director of the Graduate Studies, Department of Physics, 506 Reiss Science Building, Georgetown University, Washington, DC 20057.

HOUSING

Availability of on-campus housing

Single students: No
Married students: No

Table A—Faculty, Enrollments, and Degrees Granted

Research Specialty	2013-14 Faculty	Enrollment 2013-14 Master's	Enrollment 2013-14 Doctorate	Number of Degrees Granted 2012-13 (2004–13) Master's	Number of Degrees Granted 2012-13 (2004–13) Terminal Master's	Number of Degrees Granted 2012-13 (2004–13) Doctorate
Biophysics	3	–	2	–	–	1(1)
Condensed Matter Physics	4	–	2	1(1)	–	1(3)
Nano Science and Technology	2	–	3	–	–(2)	–(3)
Optics	2	–	4	–	–	2(9)
Physics and other Science Education	3	–	–	–	–	–
Statistical & Thermal Physics	3	–	4	–	–	1(1)
Non-specialized	–	–	9	–	1(9)	–
Total	17	1	24	1(1)	1(11)	5(17)
Full-time Grad. Stud.	–	1	24	–	–	–
First-year Grad. Stud.	–	–	6	–	–	–

GRADUATE DEGREE REQUIREMENTS

Master's: The thesis option requires 27 credits of satisfactory graduate coursework plus a thesis; the non-thesis option requires 30 credits of satisfactory graduate coursework.

Doctorate: The Ph.D. requires 34 credits of satisfactory graduate coursework and a dissertation. Graduate examinations include comprehensive and qualifying examinations and a dissertation defense.

Other Degrees: The Ph.D. program offers a traditional physics track and an Industrial Leadership in Physics (ILP) track. The latter is intended for students interested in scientific careers in industry. The curriculum of the ILP track includes business electives and a year-long internship in industry. Both tracks emphasize communication skills and teamwork.

Thesis: Thesis may be written in absentia.

SPECIAL EQUIPMENT, FACILITIES, OR PROGRAMS

Georgetown GNμLab

GNμLab is a micro fabrication and materials research facility that is wholly managed by the Department of Physics. GNμLab comprises a 2,000-square-foot facility including 1,200 square feet of clean room space and the following capabilities: optical and electron beam lithography, deposition, evaporation, and sputtering systems; etching: RIE/DRIE equipment and wet TMAH etching; high-temperature furnaces; measurement: stress, film thickness, and FESEM; and wire bonding. The Carbon Nanotube Synthesis facility (CVD) is located in GNμLab.

Soft Matter Lab

The soft matter lab has equipment for the synthesis, characterization, and manipulation of colloidal particles through the use of our wet chemistry facility. It also contains functional facilities for the preparation of biological materials such as cells and biopolymers. Materials are studied with a high-speed laser scanning Leica SP5 confocal microscope equipped with seven-laser lines ranging from UV to near IR. Mechanical measurements are performed with a customized Anton Paar MCR-301 rheometer that works in conjunction with the confocal microscope.

Georgetown Laser Laboratory

The Laser and Optical Characterization lab comprises laser systems for the characterization of nanoparticle formation, nonlinear optical effects, and spectrally resolved imaging. High-powered laser systems include a nanosecond pulsed laser (Quanta-Ray GCR03) pumping an optical parametric oscillator (GWU), an argon/dye continuous wave laser (Coherent Medical Lambda Plus), and an Ar-pumped femtosecond Ti:sapphire laser system. Nanoparticle formation is studied with an apparatus for photon and fluorescence correlation spectroscopy based on a hardware autocorrelator (ALV5000) and a photon-counting avalanche photodiode. Spectrally resolved fluorescence imaging capabilities are provided by a FALCON chemical imaging microscope (ChemImage). This equipment is augmented by advanced detection and signal processing instrumentation.

Dynamics Imaging Laboratory

The Dynamics Imaging Laboratory contains high-speed and high-resolution digital imaging systems; software for image acquisition, processing, and analysis; a high-speed confocal microscope; and a high-powered optical tweezer.

Computing Resources

The computational groups have one Beowulf cluster machine. The University has additional parallel computation resources.

Superconductivity and Nanoelectronics Laboratory

The Superconductivity and Nanoelectronics Laboratory has a room-temperature testing station (with four micromanipulators); a low-noise, low-temperature (to 1.5 K) transport measurement setup; a variable-temperature high-vacuum scanning microscope; and a JEOL JSPM-4210 scanning probe microscope capable of performing AFM and STM measurements under ambient conditions and in vacuum, controlled gas environments, and temperatures from 100 to 800 K.

Table B—Separately Budgeted Research Expenditures by Source of Support

Source of Support	Departmental Research	Physics-related Research Outside Department
Federal government	$2,869,950	
State/local government		
Non-profit organizations		
Business and industry	$168,395	
Other	$204,823	
Total	**$3,243,168**	

Table C—Separately Budgeted Research Expenditures by Research Specialty

Research Specialty	No. of Grants	Expenditures ($)
Biophysics	5	$823,402
Condensed Matter Physics	21	$2,419,766
Total	**26**	**$3,243,168**

FACULTY

Professor

Currie, John F., Ph.D., Cornell University, 1977. *Condensed Matter Physics.* Device microfabrication; biomedical devices.
Freericks, James K., Ph.D., University of California, Berkeley, 1991. *Computational Physics, Condensed Matter Physics.*
Liu, Amy Y., Ph.D., University of California, Berkeley, 1991. *Condensed Matter Physics.* Theoretical condensed matter physics.
Urbach, Jeffrey S., Ph.D., Stanford University, 1993. *Biophysics, Condensed Matter Physics.*

Associate Professor

Barbara, Paola, Ph.D., Technical University of Denmark, 1995. *Condensed Matter Physics.*
Chiao-Yap, Lydia, Ph.D., University of California, Berkeley, 1961. *Condensed Matter Physics.*
Egolf, David A., Ph.D., Duke University, 1994. *Computational Physics, Condensed Matter Physics.* Theoretical soft condensed matter physics; QCD.
Mathews, Wesley N., Ph.D., University of Illinois at Urbana-Champaign, 1966. *Condensed Matter Physics.*
Paranjape, Makarand, Ph.D., University of Alberta, 1993. Microelectromechanical systems; biomedical microdevices.
Rigol, Marcos, Ph.D., University of Stuttgart, 2004. *Computational Physics, Condensed Matter Physics.*
Van Keuren, Edward, Ph.D., Carnegie Mellon University, 1990. *Condensed Matter Physics, Optics.*

Assistant Professor

Blair, Daniel L., Ph.D., Clark University, 2003. *Biophysics, Condensed Matter Physics.*
Dzakpasu, Rhonda, Ph.D., University of Michigan, 2003. *Biophysics, Computational Physics.*

Emeritus

McClure, Joseph A., Ph.D., University of North Carolina, Chapel Hill, 1963. Physics education.
Serene, Joseph W., Ph.D., Cornell University, 1974. Theoretical condensed matter physics.

Research Professor

de Vincenz, Andre M., Ph.D., Georgetown University, 1987. *Materials Science, Metallurgy.*
Esrick, Mark A., Ph.D., Georgetown University, 1981. *Biophysics.*

Adjunct Professor

Clinton, Thomas W., Ph.D., University of Maryland, 1992. *Condensed Matter Physics.*
Lavine, James, Ph.D., University of California, Irvine, 1971. *Condensed Matter Physics.* Semiconductor physics.
Slakey, Francis, Ph.D., University of Illinois at Urbana-Champaign, 1992. Science policy.
Zlatic, Veljko, Ph.D., Imperial College of London, 1974. *Condensed Matter Physics.* Theoretical condensed-matter physics.

DEPARTMENTAL RESEARCH SPECIALTIES AND STAFF

Theoretical

Biophysics. Axonal chemotaxis. Dzakpasu, Urbach.
Hard condensed-matter physics. Superconductivity, magnetism, strongly correlated materials; structural, electronic, and transport properties; interaction of light with matter. Freericks, Liu, Mathews, Rigol.
Other. Ultracold gases. Freericks, Rigol.
Particles and Fields. Effective theories. Egolf.
Statistical & Thermal Physics. Nonequilibrium dynamical systems, both classical and quantum. Egolf, Freericks, Rigol.

Experimental

Biophysics. Cellular biophysics; pattern formation in neural systems; biomaterials. Blair, Dzakpasu, Urbach.

Hard Condensed Matter Physics. Superconductivity; superconducting devices; transport in nanostructures. Barbara.

Nano Science and Technology. Semiconductor technology; sensors and actuarors; nanotube devices; organic photovoltaic; applications to environmental monitoring, bioengineering, medical imaging. Barbara, Currie, Paranjape, Van Keuren.

Optics. Nanoparticle synthesis and characterization; imaging of soft materials; biomedical optics. Blair, Urbach, Van Keuren.

Soft condensed-matter physics. Soft glasses; colloidal and polymer physics; biomaterials; granular material; fluids; nonlinear dynamics. Blair, Urbach, Van Keuren.

View additional information about this department at
www.gradschoolshopper.com

HOWARD UNIVERSITY

DEPARTMENT OF PHYSICS AND ASTRONOMY

Washington, D.C. 20059
http://www.physics1.howard.edu

General University Information
President: Sidney A. Ribeau
Dean of Graduate School: Gary Harris
University website: http://www.howard.edu
Control: Private
Setting: Urban
Total Faculty: 1,082
Total Graduate Faculty: 503
Total number of Students: 10,594
Total number of Graduate Students: 3,411

Department Information
Department Chairman: Prabhakar Misra, Chair
Department Contact: Dr. Prabhakar Misra, Chairman
  Total full-time faculty: 20
  Total number of full-time equivalent positions: 20
  Full-Time Graduate Students: 17
  First-Year Graduate Students: 2
  Female First-Year Students: 1

Department Address
2355 6th Street, NW
Room 105
Washington, DC 20059
Phone: (202) 806-6245
Fax: (202) 806-5830
E-mail: pmisra@howard.edu
Website: http://www.physics1.howard.edu

ADMISSIONS

Admission Contact Information
Address admission inquiries to: Admissions: the Graduate School of Arts and Sciences.
Phone: (202) 806-6800
E-mail: gharris@howard.edu
Admissions website: http://www.gs.howard.edu/

Application deadlines
Fall admission:
U.S. students: April 1   *Int'l. students*: April 1
Spring admission:
U.S. students: September 1   *Int'l. students*: September 1

Application fee
U.S. students: $45   *Int'l. students*: $45

Admissions information
For Fall of 2012:
  Number of applicants: 2
  Number admitted: 2
  Number enrolled: 2

Admission requirements
Bachelor's degree requirements: Bachelor's degree in physics or a closely related field is required.
Minimum undergraduate GPA: 3.0

GRE requirements
The GRE is required.
  Quantitative score: 550
  Verbal score: 450
  Mean GRE score range (25th–75th percentile): 1000

Advanced GRE requirements
The Advanced GRE is recommended.

TOEFL requirements
The TOEFL exam is required for students from non-English-speaking countries.
  PBT score: 550
  iBT score: 79
Minimum accepted computer-based exam (CBT) score: 213.

Other admissions information
Additional requirements: The minimum acceptable score suggested for admission is verbal–450; quantitative–550.
Undergraduate preparation assumed: Marion, Mechanics; Reitz, Milford and Christy, Electricity and Magnetism; Jenkins and White, Optics; Zemansky, Thermodynamics; Tipler, Atomic Physics.

TUITION

Tuition year 2012–13:
Tuition for in-state residents
  Full-time students: per semester
  Part-time students: per credit
Tuition for out-of-state residents
  Full-time students: per semester
  Part-time students: per credit
  Full-time students: $29,090 annual
  Part-time students: $1,500 per credit
Tuition costs are for graduate. Undergraduate costs are different.
Credit hours per semester to be considered full-time: 9
Deferred tuition plan: Yes
Health insurance: Available
Academic term: Semester
Number of first-year students who received full tuition waivers: 2

Teaching Assistants, Research Assistants, and Fellowships

Number of first-year
  Teaching Assistants: 2
Average stipend per academic year
  Teaching Assistant: $16,000

FINANCIAL AID

Application deadlines
Fall admission:
U.S. students: April 1   *Int'l. students*: April 1
Spring admission:
U.S. students: September 1   *Int'l. students*: September 1

Loans
Loans are not available for U.S. students.
Loans are not available for international students.
GAPSFAS application required: Yes
FAFSA application required: No

For further information
Address financial aid inquiries to: Dr. Belay Demoz, Department of Physics and Astronomy, 2355 6th Street, NW, Washington, D.C. 20059.
Phone: (202) 806-6267
E-mail: bbdemoz@howard.edu
Financial aid website: http://www.howard.edu/financialaid/contacts/staff-finaid.htm

HOUSING

Availability of on-campus housing
Single students: Yes
Married students: No

For further information
Address housing inquiries to: Student Housing, Administration Building, 2400 6th Street, NW, Washington, D.C. 20059.
Phone: (202) 806-6131
E-mail: mdlee2@howard.edu
Housing aid website: http://www.howard.edu/residencelife/default.htm

Table A—Faculty, Enrollments, and Degrees Granted

Research Specialty	2012–13 Faculty	Enrollment Fall 2012		Number of Degrees Granted 2013 (2008–13)		
		Master's	Doctorate	Master's	Terminal Master's	Doctorate
Atmosphere, Space Physics, Cosmic Rays	4	–	–	2(3)	–	1(4)
Atomic, Molecular, & Optical Physics	4	1	3	–(3)	–	–(5)
Biophysics	1	–	1	–	–	–
Condensed Matter Physics	7	2	3	1(1)	–	–
Other	4	6	1	2(2)	–	1(2)
Total	20	9	8	5(9)	–	2(11)
Full-time Grad. Stud.	–	8	9	–	–	–
First-year Grad. Stud.	–	2	–	–	–	–

GRADUATE DEGREE REQUIREMENTS

Master's: Modern physics, mathematical methods in physics, and advanced laboratory are recommended; classical mechanics, electromagnetic theory, quantum mechanics, and statistical mechanics may be required; both thesis and nonthesis/comprehensive examination options are available.

Doctorate: A total of 72 hours, including the M.S. requirements and dissertation, are required; the student is required to pass a qualifying examination; other advanced courses, such as solid-state physics or astrophysics, are strongly recommended.

Thesis: Thesis may be written in absentia.

SPECIAL EQUIPMENT, FACILITIES, OR PROGRAMS

The Department conducts extensive atmospheric studies through the NASAURC and the NOAA Center for Atmospheric Sciences. The Department houses a modern laser spectroscopy laboratory, a string theory group, experimental condensed-matter facilities, and a computational physics laboratory.

Table B—Separately Budgeted Research Expenditures by Source of Support

Source of Support	Departmental Research	Physics-related Research Outside Department
Federal government	$2,610,985	
State/local government		
Non-profit organizations		
Business and industry		
Other		
Total	$2,610,985	

Table C—Separately Budgeted Research Expenditures by Research Specialty

Research Specialty	No. of Grants	Expenditures ($)
Atmosphere, Space Physics, Cosmic Rays	5	$2,188,418
Atomic, Molecular, & Optical Physics	2	$222,609
Condensed Matter Physics	3	$199,958
Particles and Fields	–	
Computational Physics	–	
Total	10	$2,610,985

FACULTY

Professor

Batra, Anand P., Ph.D., Rensselaer Polytechnic Institute, 1966. Assistant Chairman. *Condensed Matter Physics, Physics and other Science Education, Solid State Physics.* Solid-state physics.

Catchings, Robert M., Ph.D., Wayne State University, 1970. Associate Dean for the Natural Sciences, College of Arts and Sciences. *Condensed Matter Physics.* Solid-state physics.

Demoz, Belay, Ph.D., University of Nevada, Reno, 1985. Director of Graduate Studies. *Climate/Atmospheric Science.* Atmospheric physics.

Hubsch, Tristan, Ph.D., University of Maryland, College Park, 1987. *Particles and Fields, Theoretical Physics.* Elementary particle physics; field theory; strings.

Jenkins, Gregory, Ph.D., University of Michigan, 1991. *Atmosphere, Space Physics, Cosmic Rays, Climate/Atmospheric Science.* Atmospheric science.

Joseph, Everette, Ph.D., State University of New York (at Albany), 1997. Director, Howard University Program in Atmospheric Sciences (HUPAS). *Climate/Atmospheric Science.* Atmospheric physics.

Kushawaha, Vikram S., Ph.D., Benaras Hindu University (India), 1973. *Atomic, Molecular, & Optical Physics.* Atomic, molecular, and optical physics.

Lindesay, James, Ph.D., Stanford University, 1981. *Astrophysics, Computational Physics, Particles and Fields.* Field theory; computational physics.

Lowe, Walter P., Ph.D., Stanford University, 1983. *Condensed Matter Physics.* Condensed-matter physics; synchrotron radiation studies.

Misra, Prabhakar, Ph.D., Ohio State University, 1986. Chairman. *Atomic, Molecular, & Optical Physics, Chemical Physics, Nano Science and Technology, Optics.* Molecular spectroscopy; chemical physics; laser spectroscopy; optics.

Salu, Yehuda, Ph.D., University of Tel Aviv (Israel), 1973. *Biophysics.* Biophysics; medical physics.

Thorpe, Arthur N., Ph.D., Howard University, 1964. *Condensed Matter Physics.* Magnetism; solid-state physics.

Venable, Demetrius D., Ph.D., American University, 1974. *Atomic, Molecular, & Optical Physics, Climate/Atmospheric Science, Optics.* Optical physics; atmospheric physics.

Associate Professor

Alfred, Marcus, Ph.D., Howard University, 2000. *Computational Physics, Cosmology & String Theory, Particles and Fields.* Computational physics; particles and fields.

Gatica, Silvina, Ph.D., University of Buenos Aires, 1995. *Condensed Matter Physics, Nano Science and Technology.* Condensed matter and nanophysics; simulation and modeling of nanoscale phenomena.

Assistant Professor

Stancil, Kimani, Ph.D., Massachusetts Institute of Technology, 2002. *Condensed Matter Physics, Nano Science and Technology.* Condensed matter; nanophysics.

Adjunct Professor

Pass, Barry, Ph.D., Rutgers University, 1968. *Biophysics, Medical, Health Physics.* Radiation physics.

Romanyukha, Alex, Ph.D., Russian Academy of Sciences, 1985. *Atomic, Molecular, & Optical Physics, Biophysics, Medical, Health Physics, Optics.* Radiation physics.

Ting, Antonio, Ph.D., University of Maryland, 1984. *Atomic, Molecular, & Optical Physics, Condensed Matter Physics, Plasma and Fusion.* Plasma physics.

Lecturer

Connell, Rasheen, Ph.D., Howard University, 2010. *Climate/Atmospheric Science.* Atmospheric science.

Finch, Tehani, Ph.D., Howard University, 2008. *Cosmology & String Theory, Particles and Fields.* String theory.

Instiful, Peter, Ph.D., Howard University, 2007. *Climate/Atmospheric Science, Condensed Matter Physics.* Condensed-matter physics.

Weldegaber, Mengsteab, Ph.D., University of Maryland, Baltimore County, 2009. *Atmosphere, Space Physics, Cosmic Rays, Climate/Atmospheric Science.* Atmospheric physics.

DEPARTMENTAL RESEARCH SPECIALTIES AND STAFF

Theoretical

Climate Modeling. Weldegaber.
Computational Physics. Alfred, Gatica, Lindesay, Misra.
Condensed Matter Physics. Gatica, Lowe, Stancil, Thorpe.
Cosmology & String Theory. Finch, Hubsch, Lindesay.
LIDAR.
Modeling.
Neural Networks.
Spectroscopy.

Experimental

Atmospheric Measurements. Connell, Demoz, Instiful, Jenkins, Joseph, Venable, Weldegaber.
Laser Optics & Spectroscopy. Kushawaha, Misra, Pass, Romanyukha, Ting, Venable.
Solid State & Nanophysics. Batra, Catchings, Lowe, Stancil, Thorpe.

View additional information about this department at
www.gradschoolshopper.com

FLORIDA ATLANTIC UNIVERSITY

DEPARTMENT OF PHYSICS

Boca Raton, Florida 33431
http://www.physics.fau.edu/

General University Information
President: Mary Jane Saunders
Dean of Graduate School: Barry T. Rosson, Ph.D.
University website: http://www.fau.edu
Control: Public
Setting: Suburban
Total Faculty: 1,023
Total number of Students: 28,232
Total number of Graduate Students: 4,299

Department Information
Department Chairman: Warner A. Miller, Chair
Department Contact: Zia Smith, Secretary
  Total full-time faculty: 15
  Full-Time Graduate Students: 33
  First-Year Graduate Students: 1

Department Address
777 Glades Road
Boca Raton, FL 33431
Phone: (561) 297-3380
Fax: (561) 297-2662
E-mail: zasmith@fau.edu
Website: http://www.physics.fau.edu/

ADMISSIONS

Admission Contact Information
Address admission inquiries to: Andy Lau, Recruitment Chair
Phone: (561) 297-3380
E-mail: alau@fau.edu
Admissions website: http://www.physics.fau.edu/

Application deadlines
Fall admission:
U.S. students: July 1   *Int'l. students*: January 15
Spring admission:
U.S. students: November 1   *Int'l. students*: July 15

Application fee
U.S. students: $30

Admissions information
For Fall of 2012:
  Number of applicants: 72
  Number admitted: 10
  Number enrolled: 9

Admission requirements
Bachelor's degree requirements: Bachelor's degree in physics is required.
Minimum undergraduate GPA: 3.0

GRE requirements
The GRE is required.
There is no minimum GRE score required.

Advanced GRE requirements
The Advanced GRE is recommended.

TOEFL requirements
The TOEFL exam is required for students from non-English-speaking countries.
  PBT score: 550
  iBT score: 79

Other admissions information
Undergraduate preparation assumed: Reitz and Milford, Foundations of Electromagnetic Theory; Symon, Mechanics; Saxon, Quantum Mechanics;* Boyce and Deprima, Elementary Differential Equations and Boundary Value Problems; Reif, Statistical and Thermal Physics;* Boas, Mathematical Methods in the Physical Sciences.*; *May be taken during first year of graduate study.

TUITION

Tuition year 2013-14:
Tuition for in-state residents
  Full-time students: $369.82 per credit
  Part-time students: $369.82 per credit
Tuition for out-of-state residents
  Full-time students: $1,024.81 per credit
  Part-time students: $1,024.81 per credit
Credit hours per semester to be considered full-time: 9
Deferred tuition plan: No
Health insurance: Available at the cost of 1,450 per year.
Academic term: Semester

Teaching Assistants, Research Assistants, and Fellowships
Number of first-year
  Fellowship students: 2
Average stipend per academic year
  Teaching Assistant: $20,050
All first-year Ph.D. students are offered a TA stipend per academic year.

FINANCIAL AID

Application deadlines
Fall admission:
U.S. students: May 1
Spring admission:
U.S. students: November 28   *Int'l. students*: November 28

Loans
Loans are not available for U.S. students.
Loans are not available for international students.
GAPSFAS application required: No
FAFSA application required: Yes

For further information
Address financial aid inquiries to: Office of Student Financial Aid.
Phone: (561) 297-3530
Financial aid website: http://www.fau.edu/finaid/gradstudent.php

HOUSING

Availability of on-campus housing
Single students: Yes
Married students: No

For further information
Address housing inquiries to: Director of Student Housing.
Phone: (561) 297-2880
E-mail: housing@fau.edu
Housing aid website: http://www.fau.edu/housing/index.php

Table A—Faculty, Enrollments, and Degrees Granted

Research Specialty	2012–13 Faculty	Enrollment Spring 2013 Master's	Enrollment Spring 2013 Doctorate	Number of Degrees Granted 2012 Master's	Number of Degrees Granted 2012 Terminal Master's	Number of Degrees Granted 2012 Doctorate
Biophysics	2	–	9	–	–	–
Condensed Matter Physics	5	–	6	–	–	–
Medical, Health Physics	3	17	–	–	–	–
Optics	1	–	1	–	–	–
Quantum Foundations	–	–	5	–	–	–
Relativity & Gravitation	4	–	6	–	–	–
Total	–	17	27	–	–	–
Full-time Grad. Stud.	–	17	27	–	–	–
First-year Grad. Stud.	–	–	1	–	–	–

GRADUATE DEGREE REQUIREMENTS

Master's: 30 credits in approved program with a 3.0 sustained GPA, including 7 credits of thesis research. Students must be in residence for two semesters. Final thesis.

Doctorate: 50 credits in approved program with a 3.0 sustained GPA beyond the M.S., including 30 credits of dissertation research; comprehensive written examination covering mechanics, electromagnetism, quantum mechanics, and statistical mechanics. Dissertation and oral examination required.

Other Degrees: The MST in physics requires 30 credits with a 3.0 GPA which may include 6 thesis credits. In addition, a 6 credit internship requirement must be satisfied for students without teaching experience. The Ed.D. degree in Curriculum and Instruction is offered for junior college teachers with physics as a first or second teaching field. Master's in Medical Physics. 41 credits and Thesis in MSMP program. Clinical Training. The MSMP is recognized as a professional degree(PSM) by the CGS.

Thesis: Thesis may be written in absentia.

SPECIAL EQUIPMENT, FACILITIES, OR PROGRAMS

The Department is rapidly growing and is poised to meet the new challenges for the field of physics in today's environment at a research university.

We are augmenting and integrating our core expertise in condensed matter physics with the growing field of the Physics of Biological Systems. Additionally, we are in the process of forming the Center for Spacetime Physics. This center will pull together our current expertise in supernova astrophysics, cosmology as well as classical and quantum gravity. Our growing effort in the Physics of Biological Systems is tightly integrated with the expanding efforts here at FAU within the Charles E. Schmidt College of Science in the areas of Biomedical Science, Mathematics, Biology and Biochemistry. We will also be filling faculty lines in areas aligned with, and supporting, the charter of the Scripps Institute that will be establishing a new campus in Palm Beach County.

Our Spacetime Physics effort here already provides numerical and mathematical support for the DOE's TSI initiative as well the NSF's LIGO observatories. In addition to these three general-relativistic astrophysics areas, a major thrust of the center will be the search for the origin of the concept of space and time and will involve fundamental research in the foundations of gravity, and quantum theory.

Table B—Separately Budgeted Research Expenditures by Source of Support

Source of Support	Departmental Research	Physics-related Research Outside Department
Federal government	$900,000	
State/local government		
Non-profit organizations		
Business and industry		
Other	$250,000	
Total	$1,150,000	

FACULTY

Professor

Leventouri, Theodora, Ph.D., University of Athens, 1972. *Medical, Health Physics*. Experimental condensed matter physics, biophysics. X-ray diffraction of biocompatible materials.

Miller, Warner A., Ph.D., University of Texas, Austin, 1986. Classical and quantum gravity; general relativistic astrophysics, numerical relativity, foundations of quantum mechanics.

Qiu, Shen-Li, Ph.D., City University of New York, 1985. Experimental condensed matter; photoemission; electronic structure and magnetic behavior of metals and alloys.

Wille, Luc T., Ph.D., Ghent University, 1983. Theoretical condensed matter; alloys, high-Tc superconductivity.

Associate Professor

Beetle, Christopher, Ph.D., Pennsylvania State University, University Park, 2000. Classical and quantum gravity; numerical relativity.

Fuchs, Armin, Ph.D., University of Stuttgart, 1990. Nonlinear dynamical systems; complex systems and brain sciences.

Lau, Andy W. C., Ph.D., University of California, Santa Barbara, 2000. Theoretical soft condensed matter physics; biophysics and statistical mechanics.

Marronetti, Pedro, Ph.D., University of Notre Dame, 1999. Numerical relativity, relativistic astrophysics, neutron star binary systems, gravitational wave physics.

Tichy, Wolfgang, Ph.D., Cornell University, 2001. Numerical relativity; binary black hole systems; gravitational wave physics.

Assistant Professor

Engle, Jonathan, Ph.D., Pennsylvania State University, 2006. Loop quantum gravity.

Sorge, Korey D., Ph.D., University of Tennessee, Knoxville, 2002. Condensed matter physics.

Emeritus

Bruenn, Stephen W., Ph.D., Columbia University, 1968. Theoretical astrophysics; supernovae models; radiation transport.

Dean, Nathan W., Ph.D., University of Cambridge, 1968. Theoretical elementary particle physics; mathematical finance.

Faulkner, John S., Ph.D., Ohio State University, 1959. Theoretical physics; theory of alloys.

Jordan, Robin G., Ph.D., University of Sheffield, 1967. Experimental condensed matter; UV photoemission; alloys.

Lamborn, Bjorn N. A., Ph.D., University of Florida, 1962. Theoretical physics; nonlinear coupling of plasma waves; relativistic plasmas.

McGuire, James B., Ph.D., University of California, Los Angeles, 1963. Mathematical physics; three-body problem; statistical physics; quantum field theory.

Medina, Fernando D., Ph.D., Princeton University, 1975. Experimental condensed matter physics; spectroscopic studies of solids.

Research Associate Professor

Kreymerman, Grigoriy, Ph.D., Academy of Sciences, Soviet Union, 1989. *Optics*. Optics.

Instructor

Chen, De Huai, Ph.D., City University of New York, 1987.
Gross, Robert, Ph.D., Florida Atlantic University, 2002.

Martinez, Leonardo, M.S., State University of New York, Albany, 1971.

DEPARTMENTAL RESEARCH SPECIALTIES AND STAFF

Theoretical

Biophysics. Soft condensed matter physics, neuroscience.

Condensed Matter Physics. Theory of alloys; electronic structures; properties and photoemission; high-temperature superconductivity.

Statistical & Thermal Physics. Phase transitions; nonlinear phenomena, complex systems.

Experimental

Biophysics. Biomaterials physics, x-ray and neutron powder diffraction, magnetic nanomaterials.

Condensed Matter Physics. Electrical, magnetic, structural, and optical properties of solids; thin films; surface effects; alloys; high-temperature superconductors; bioceramics.

View additional information about this department at
www.gradschoolshopper.com

FLORIDA INSTITUTE OF TECHNOLOGY

DEPARTMENT OF PHYSICS AND SPACE SCIENCES

Melbourne, Florida 32901-6975
http://cos.fit.edu/pss/

General University Information

President: Anthony Catanese
Dean of Graduate School: Dr. Monica Baloga
University website: http://www.fit.edu/
Control: Private
Setting: Suburban
Total Faculty: 467
Total Graduate Faculty: 315
Total number of Students: 5,384
Total number of Graduate Students: 2,406

Department Information

Department Chairman: Terry Oswalt, Head
Department Contact: Terry Oswalt, Department Head
  Total full-time faculty: 14
  Total number of full-time equivalent positions: 14
  Full-Time Graduate Students: 33
  First-Year Graduate Students: 10
  Female First-Year Students: 4

Department Address

150 West University Boulevard
Physics and Space Sciences Department
Melbourne, FL 32901-6975
Phone: (321) 674-8098
Fax: (321) 674-7482
E-mail: toswalt@fit.edu
Website: http://cos.fit.edu/pss/

ADMISSIONS

Admission Contact Information

Address admission inquiries to: Cheryl A. Brown, Graduate Admissions
Phone: (321) 674-7581
E-mail: grad-admissions@fit.edu
Admissions website: http://www.fit.edu/grad

Application deadlines

Fall admission:
U.S. students: September 1   *Int'l. students*: September 1
Spring admission:
U.S. students: December 11   *Int'l. students*: December 11

Application fee

U.S. students: $50   *Int'l. students*: $60
$50.00 for master's and $60.00 for doctorate; alumni have application fees waived.

Admissions information

For Fall of 2012:
  Number of applicants: 90
  Number admitted: 22
  Number enrolled: 10

Admission requirements

Bachelor's degree requirements: A Bachelor's degree in physics, astronomy, or related subject is required.
Minimum undergraduate GPA: 3.0

GRE requirements
The GRE is recommended.
Recommended, but not required.

Advanced GRE requirements
The Advanced GRE is recommended.
Recommended, but not required.

TOEFL requirements
The TOEFL exam is required for students from non-English-speaking countries.
PBT score: 550
iBT score: 90

Other admissions information
Additional requirements: A score of 600 (250 computer based, 100 iBT) is necessary to qualify for a teaching assistantship.
Undergraduate preparation assumed: Halliday, Resnick, Walker, Fundamentals of Physics; Griffiths, Electrodynamics; Thornton, Classical Dynamics; Zemansky and Dittman, Heat and Thermodynamics; Griffith, Introduction to Quantum Mechanics.

TUITION

Tuition year 2013-14:
Full-time students: $1,123 per credit
Part-time students: $1,123 per credit
Credit hours per semester to be considered full-time: 9
Deferred tuition plan: No
Health insurance: Yes, 1107.00.
Academic term: Semester
Number of first-year students who received partial tuition waivers: 3

Teaching Assistants, Research Assistants, and Fellowships
Number of first-year
  Teaching Assistants: 3
Average stipend per academic year
  Teaching Assistant: $12,502
  Research Assistant: $12,502

FINANCIAL AID

Loans
Loans are not available for U.S. students.
Loans are not available for international students.
GAPSFAS application required: No
FAFSA application required: Yes

For further information
Address financial aid inquiries to: Terry Oswalt, Physics and Space Sciences.
Phone: (321) 674-8795
E-mail: toswalt@fit.edu
Financial aid website: http://www.fit.edu/financialaid/

HOUSING

Availability of on-campus housing
Single students: Yes
Married students: No

For further information
Address housing inquiries to: Campus Services Office.
Phone: (321) 674-8033
E-mail: askevakis@fit.edu
Housing aid website: http://www.fit.edu/housing/

Table A—Faculty, Enrollments, and Degrees Granted

Research Specialty	Fall 2012-13 Faculty	Enrollment Fall 2012 Master's	Enrollment Fall 2012 Doctorate	Number of Degrees Granted 2012-13 (2008-13) Master's	Number of Degrees Granted 2012-13 (2008-13) Terminal Master's	Number of Degrees Granted 2012-13 (2008-13) Doctorate
Astronomy	6	5	13	3(12)	–	–(1)
Atmosphere, Space Physics, Cosmic Rays	4	6	9	2(10)	–	1(3)
Condensed Matter Physics	2	–	–	–(1)	–	–(1)
Geophysics	4	1	2	–(3)	–	–(1)
High Energy Physics	3	2	4	2(5)	–	–(4)
Other	2	–	–	–(5)	–	–
Total	14	10	30	7(36)	–	1(10)
Full-time Grad. Stud.	–	7	26	–	–	–
First-year Grad. Stud.	–	4	6	–	–	–

GRADUATE DEGREE REQUIREMENTS

Master's: Thirty semester credit hours with a 3.0 GPA minimum are required; 2 semesters in residence; 2 comprehensive examinations, 1 written in first semester of residence, 1 oral or written at completion of master's program; thesis optional.

Doctorate: Forty-five semester credit hours beyond the master's or 75 beyond the bachelor's is required; 3.2 GPA minimum; at least 2 calendar years, including 1 year in doctoral research; comprehensive written examination before formal admission to candidacy; oral dissertation defense; dissertation must be submitted to a major physics journal.

Other Degrees: Master's and doctorate in space sciences: same general requirements as for physics.

Thesis: Thesis may be written in absentia.

SPECIAL EQUIPMENT, FACILITIES, OR PROGRAMS

Olin Physical Sciences Building: The Department is located in a 70,000-sq. ft. building, which includes specialized laboratories, an observatory, a 3500-sq. ft. high-bay laboratory, and a NASA-qualified clean room.

Astronomy and Astrophysics Laboratory: Astronomy faculty and students work on a wide variety of topics, including the evolution of white dwarf stars, simulations of cataclysmic variable systems, astrophysical fluid dynamics, accretion phenomena, the physics and evolution of active galactic nuclei and their jets, cosmology, exoplanets, and solar and stellar atmospheres. Observations are conducted from the radio to the gamma-rays, including observations with the Hubble Space Telescope, Chandra X-ray Observatory, and Spitzer Space Telescope.

Members of the group are involved in the development of instrumentation for the SuperNova Acceleration Probe (SNAP) and in the Canari-Cam Science Team, a guaranteed-time effort on the 10.4-m Gran Telescopio Canarias.

Resources in the laboratory include Linux computers and astronomical data reduction packages such as IRAF, AIPS, CIAO, Ftools, and HEADAS. The laboratory is also the control center for remote access to the SARA telescopes in Arizona and Chile and the 0.8-m Ortega telescopes on campus.

Geospace Physics Laboratory (GPL): Hosts the space physics research activities of Florida Tech's Space Sciences program. GPL operates a 10-site meridianal array of agnetometers up the east coast of the United States (MEASURE array). Measurements from various satellites are used for studying the magnetic wave energy propagation within the geospace environment and the dynamics of the earth's plasmasphere and the storm-time radiation belt.

Research includes the study of energetic particle acceleration and propagation within the heliosphere and in interstellar space involving energetic particle measurements from the Ulysses, ACE, Wind, and CRRES spacecraft and numerical modeling of particle transport processes.

High Bay Physics and Space Sciences Laboratory: This 2-story, 3500-sq. ft. hall is available for projects that require a large footprint and/or heavy utility access. Currently housed in this facility is a ground-link station for a planned experiment for the International Space Station, a 400-sq. ft. NASA-qualified clean room, a magnetic levitation track, the first node of the Florida Tech Grid Computing Project, and the starting payload for the first UNESCO satellite.

High-Energy Physics Laboratory (HEP): Presently, HEP research is focused on the commissioning of the CMS experiment at CERN and data analysis. Since 2001, the Florida Tech group has had responsibilities for calibration of the hadron calorimeters and precision alignment of the muon end-cap detectors. The physics analyses are initially focused on measurements of the properties of the top and bottom quarks and search for new gauge bosons. With anticipated higher luminosities, our physics program will switch to the search for the Higgs boson and more exotic phenomena at a multi-TeV energy scale.

Another main research area is the development and construction of a muon tomography system for detecting high-Z materials hidden in cargo based on advanced micropattern gas detectors such as gas electron multipliers.

The HEP laboratory houses a state-of-the-art Linux-based computing cluster with about 100 CPU cores used for muon tomography detector simulation work and serves as a Tier-3 site on the Open Science Grid for CMS data analysis. The group also conducts research and development on advanced particle detector technology for the Super-LHC upgrade programs.

In addition, Florida Tech is a member of the PHENIX experiment at BNL's Relativistic Heavy Ion Collider, which is searching for a new state of matter dubbed the "quarkgluon plasma" and the L3 collaboration at the LEP accelerator.

Maglev Facility: Houses a unique 43-foot magnetic levitation and propulsion track, the only such device at an academic institution. It supports research in controls,

aerodynamics, mechanical stability, super-conducting

technology, and electromagnetic acceleration levitation to

study the feasibility of maglev launch assist for future spacecraft.

SARA Observatories: Florida Tech is the administrative institution for the Southeastern Association for Research in Astronomy (SARA). SARA operates an automated 0.9-m telescope at Kitt Peak National Observatory near Tucson, AZ, and a 0.6-m telescope at Cerro Tololo Interamerican Observatory in Chile. Both are accessible via the Internet and are used primarily for CCD imaging and photometry. Ten percent of all observing time on these facilities are allocated to Florida Tech.

Ortega Observatory: A 0.8-m Ritchey-Chretien telescope was installed on the FIT campus in November 2007. The telescope is equipped with a wide-field CCD and medium-resolution spectrograph.

Scanning Probe Microscopy Laboratory: This facility provides researchers with the ability to image the surface structure of a solid and to probe the electronic surface properties of a material down to the atomic scale using a scanning tunneling microscope (STM). This laboratory is also used to investigate novel applications of the STM (e.g., in the field of electrochemistry) and in the development of other types of scanning probe microscopes.

Table B—Separately Budgeted Research Expenditures by Source of Support

Source of Support	Departmental Research	Physics-related Research Outside Department
Federal government	$1,927,618	
State/local government		
Non-profit organizations	$1,700	
Business and industry		
Other		
Total	**$1,929,318**	

Table C—Separately Budgeted Research Expenditures by Research Specialty

Research Specialty	No. of Grants	Expenditures ($)
Astrophysics	17	$881,779
Atmosphere, Space Physics, Cosmic Rays	11	$722,396
Condensed Matter Physics	2	$68,307
High Energy Physics	10	$397,421
Other	0	
Total	**40**	**$2,069,903**

FACULTY

Professor

Baarmand, Marc M., Ph.D., University of Wisconsin-Madison, 1987. *High Energy Physics*. Experimental high-energy particle physics at CERN (CMS Experiment); hadroproduction of top and bottom quarks in pQCD; Higgs physics; particle detector technology; grid computing.

Durrance, Samuel, Ph.D., University of Colorado, 1980. *Astronomy, Astrophysics, Biophysics, Geophysics, Medical, Health Physics*. Space science missions and human space exploration.

Dwyer, Joseph, Ph.D., University of Chicago, 1994. *Atmosphere, Space Physics, Cosmic Rays, Electromagnetism*. Space physics; solar/heliospheric energy particle observations.

McCay, T. Dwayne, Ph.D., Auburn University, 1974. Executive Vice President, and Chief Operating Officer. *Engineering Physics/Science, Materials Science, Metallurgy, Nano Science and Technology, Plasma and Fusion*. Materials processing in space; engineering physics.

Oswalt, Terry D., Ph.D., Ohio State University, 1981. *Astronomy, Astrophysics*. Observational astrophysics; stellar evolution, stellar luminosity functions, binary stars, stellar chromospheric activity, and minor planets.

Rassoul, Hamid K., Ph.D., University of Texas, Dallas, 1987. Dean, College of Science. *Atmosphere, Space Physics, Cosmic Rays, Electromagnetism, Geophysics*. Observation and modeling of auroras; photochemistry of the earth's upper atmosphere; solar wind-magnetosphere interactions.

Zhang, Ming, Ph.D., Massachusetts Institute of Technology, 1991. *Atmosphere, Space Physics, Cosmic Rays, Electromagnetism, Optics*. Cosmic radiation and interactions with the plasma and magnetic fields in the interstellar medium, the heliosphere, and magnetospheres.

Associate Professor

Hohlmann, Marcus, Ph.D., University of Chicago, 1997. *Computational Physics, High Energy Physics*. Experimental high-energy physics at CERN (L3, CMS); supersymmetry and Higgs searches; detector development.

Liu, Ningyu, Ph.D., Pennsylvania State University, 2006. *Atmosphere, Space Physics, Cosmic Rays, Electromagnetism*. Atmospheric electricity; space physics; computational electrodynamics; plasma physics; gas discharge phenomena; high-performance computing.

Perlman, Eric S., Ph.D., University of Colorado, 1994. *Astronomy, Astrophysics*. Active galactic nuclei; jets; observational cosmology; high-energy and multiwavelength astrophysics.

Turner, Niescja E., Ph.D., University of Colorado Boulder, 2000. *Atmosphere, Space Physics, Cosmic Rays, Geophysics*. Space physics; inner magnetosphere; ring current; energetics of magnetic storms; physics and astronomy education research.

Assistant Professor

Batcheldor, Daniel, Ph.D., University of Hertfordshire, 2005. *Astronomy, Astrophysics*. Active galactic nuclei; galactic dynamics, supermassive black holes.

Neish, Catherin, Ph.D., University of Arizona, 2008. *Planetary Science*.

Oluseyi, Hakeem M., Ph.D., Stanford University, 1999. *Astronomy, Astrophysics, Condensed Matter Physics, Materials Science, Metallurgy, Physics and other Science Education*. Solar/stellar atmospheres; cosmology; history of astronomy; physics education; instrumentation development.

Ragozzine, Darin, Ph.D., California Institute of Technology, 2009.

Sawyer, Benjamin M., M.S., Florida Institute of Technology, 1975. *Physics and other Science Education*. Physics education.

Yumiceva, Francisco, Ph.D., University of South Carolina, 2004. *Particles and Fields*.

Emeritus

Blatt, Joel H., Ph.D., University of Alabama, 1970. *Optics*. Applied optics and machine/human vision.

Patterson, James D., Ph.D., University of Kansas, 1962. *Condensed Matter Physics, Materials Science, Metallurgy*. Theoretical solid-state physics; magnetism; narrow gap semiconductors.

Research Professor

Foing, Bernard, Ph.D., LPSP/ENSET, Paris, 1983. Chief scientist of the European Space Agency Science Program. *Astronomy, Astrophysics, Geophysics*. Space missions (SMART 1, Mars Express, Huygens Probe).

Gamayunov, Konstantin, Ph.D., Institute of Terrestrial Magnetism, Ionosphere, and Radio Wave Propagation, Moscow, Russia, 1994. *Atmosphere, Space Physics, Cosmic Rays*. Space plasma physics; simulation of waves and energetic particles in the Earth's magnetosphere; energetic particle acceleration propagation in the heliosphere and interstellar medium.

Hannikainen, Diana, Ph.D., University of Helsinki, 1999. *Astronomy, Astrophysics*. High-energy astrophysics.

Mantovani, James G., Ph.D., Clemson University, 1985. NASA Kennedy Space Center. *Condensed Matter Physics, Materials Science, Metallurgy, Nano Science and Technology*. Condensed-matter physics, theoretical and experimental, particularly surface physics and electron microscopy; Mars and Moon environment.

Principe, Edward L., Ph.D., Pennsylvania State University, 1996. *Optics*. Materials processing; semiconductor physics; metrology; laboratory management.

Instructor

Gering, James A., M.S., Indiana University, 1984. Director of undergraduate laboratories. *Physics and other Science Education*. Physics education.

Researcher

Vodopiyanov, Igor, Ph.D., St. Petersburg Nuclear Physics Institute, 1995. *High Energy Physics*. High-energy physics at LEP and LHC.

DEPARTMENTAL RESEARCH SPECIALTIES AND STAFF

Theoretical

Astrophysics.

Atmosphere, Space Physics, Cosmic Rays. Multiscale modeling of atmospheric electricity, sprites, jets, and related phenomena. Dwyer, Liu, Rassoul.

Atomic, Molecular, & Optical Physics. Thermospheric-ionospheric coupling and space debris. Rassoul.

High Energy Physics. Modeling of heavy-quark pair production in hadron collisions and tests of perturbative QCD. Baarmand.

Experimental

Astronomy. Space-based observatories; high-contrast ratio astronomy. Batcheldor.

Astronomy. Development of next-generation ground and space-based observatories (e.g., the Large Synoptic Survey Telescope); multi-country research in ethnoastronomy in Africa. Oluseyi.

Astrophysics. Supermassive black holes. Batcheldor.

Astrophysics. Stellar evolution and rotation. Oswalt.

Astrophysics. Active galactic nuclei and quasars. Oluseyi, Perlman.

Astrophysics. Spectroscopy and photometry of binary stars, minor planets, and exoplanets. Durrance, Oluseyi, Oswalt.

Astrophysics. Age and evolution of the galaxy. Oswalt.

Atmosphere, Space Physics, Cosmic Rays. Study of the generation of the sun's corona and the sources and variability of solar EUV radiation using data from the SoHO, Hinode, TRACE, and STEREO satellites. Oluseyi.

Atmosphere, Space Physics, Cosmic Rays. Observational and experimental research in space physics, including planetary atmospheres, ionospheric and upper atmospheric physics (airglow and aurora, solar/heliospheric energetic particles, solar modulation of galactic and anomalous cosmic rays, and cosmic ray interstellar propagation. Dwyer, Hannikainen, Liu, Rassoul, Turner, Zhang.

Condensed Matter Physics. Carrier transport and generation in high-purity silicon; tests of optical, electrical, and scattering properties of two-dimensional nanostructures for next-generation detectors. Oluseyi.

Condensed Matter Physics. Scanning tunneling microscopy and optical spectroscopy of semiconductors; dielectric properties of insulators and granular materials; Mars/lunar research with NASA KSC. Durrance, Mantovani.

High Energy Physics. Search for Higgs and new phenomena at the eV energy scale; proton-proton collisions at LHC (CMS Experiment) at CERN; heavy-ion collisions (PHENIX Experiment) at BNL; gaseous detector development for muon radiography. Baarmand, Hohlmann, Yumiceva.

UNIVERSITY OF CENTRAL FLORIDA

DEPARTMENT OF PHYSICS

Orlando, Florida 32816-2385
http://physics.cos.ucf.edu/

General University Information
President: John C. Hitt
Dean of Graduate School: Dr. C. Ross Hinkle
University website: http://www.ucf.edu
Control: Public
Setting: Urban
Total Faculty: 1,600
Total Graduate Faculty: 1,000
Total number of Students: 58,900
Total number of Graduate Students: 9,321

Department Information
Department Chairman: Talat S. Rahman, Chair
Department Contact: Elizabeth Rivera, Graduate Program
  Admissions Specialist
  Total full-time faculty: 40
  Total number of full-time equivalent positions: 40
  Full-Time Graduate Students: 104
  First-Year Graduate Students: 12
  Female First-Year Students: 2
  Total Post Doctorates: 8

Department Address
4000 Central Florida Blvd.
Physical Sciences, Building 121, Room 430
Orlando, FL 32816-2385
Phone: (407) 823-2325
Fax: (407) 823-5112
E-mail: physics@ucf.edu
Website: http://physics.cos.ucf.edu/

ADMISSIONS

Admission Contact Information
Address admission inquiries to: Prof. H. Heinrich, Graduate Co-
  ordinator, Dept. of Physics, Univ. of Central Florida, Orlando,
  FL 32816-2385
Phone: 407 823 1884
E-mail: helge.heinrich@ucf.edu
Admissions website: http://www.graduate.ucf.edu

Application deadlines
Fall admission:
U.S. students: July 15   *Int'l. students*: January 15
Spring admission:
U.S. students: December 1   *Int'l. students*: July 1

Application fee
U.S. students: $30   *Int'l. students*: $30

Admissions information
For Fall of 2013:
  Number of applicants: 99
  Number admitted: 21
  Number enrolled: 12

Admission requirements
Bachelor's degree requirements: Bachelors degree in Physics is
  required.
Minimum undergraduate GPA: 3.0

GRE requirements
The GRE is required.
There is no minimum requirement.

Advanced GRE requirements
The Advanced GRE is not required.
A General GRE score of at least 1,000 on the combined verbal-
  quantitative sections of the Aptitude Test is recommended for
  admission to the Ph.D./M.S. program. Under the new GRE scoring system, a combined score of at least 310 is recommended.

TOEFL requirements
The TOEFL exam is required for students from non-English-
  speaking countries.
  PBT score: 550
  iBT score: 80

Other admissions information
Additional requirements: The SPEAK test is administered once
  the student arrives at the university. Minimum acceptable for
  the SPEAK test is 50.

TUITION
Tuition year 2011–12:
Tuition for in-state residents
  Full-time students: $367.35 per credit
Tuition for out-of-state residents
  Full-time students: $1,191.75 per credit
Students with assistantships receive tuition and general fee
  waiver and free health insurance. The department offers 10-12
  assistantships per year, pending funding. (Tuition rates may
  change after August 1).
Credit hours per semester to be considered full-time: 9
Deferred tuition plan: No
Health insurance: Available
Other academic fees: International Students will have to pay a
  $50 per semester fee for services provided by the International
  Student Center (ISC). In addition, all students must pay a
  yearly $10 fee for ID services. Additional fees may apply.
Academic term: Semester
Number of first-year students who received full tuition waivers: 15

Teaching Assistants, Research Assistants, and Fellowships
Number of first-year
  Teaching Assistants: 13
  Research Assistants: 2
  Fellowship students: 3
Average stipend per academic year
  Teaching Assistant: $20,600
  Research Assistant: $22,000

FINANCIAL AID

Loans
Loans are available for U.S. students.
Loans are not available for international students.
GAPSFAS application required: No
FAFSA application required: Yes

For further information

Address financial aid inquiries to: Office of Student Financial Assistance, 4000 Central Florida Blvd., Millican Hall, Room 120, Orlando, FL 32816-0113.
Phone: (407) 823-2827
E-mail: finaid@ucf.edu
Financial aid website: http://finaid.ucf.edu/index.htm

HOUSING

Availability of on-campus housing

Single students: Yes
Married students: No

For further information

Address housing inquiries to: Department of Housing & Residence Life, P.O. Box 163222, University of Central Florida, Orlando, FL 32816-3222.
Phone: (407) 823-4663
Housing aid website: http://www.housing.ucf.edu/

Table A—Faculty, Enrollments, and Degrees Granted

Research Specialty	2011–12 Faculty	Enrollment Fall 2012 Master's	Enrollment Fall 2012 Doctorate	Number of Degrees Granted Spring 2012 (2006-11) Master's	Number of Degrees Granted Spring 2012 (2006-11) Terminal Master's	Number of Degrees Granted Spring 2012 (2006-11) Doctorate
Atomic, Molecular, & Optical Physics	5	–	–	–	–	–
Computational Physics	6	–	–	–	–	–
Condensed Matter Physics	22	–	–	–	–	–
Planetary Science	9	1	–	–	–	–
Quantum Foundations	2	–	–	–	–	–
Total	–	–	–	–(150)	–	4(91)
Full-time Grad. Stud.	–	–	103	–	–	–
First-year Grad. Stud.	–	–	12	–	–	–

GRADUATE DEGREE REQUIREMENTS

Master's: A total of 30 semester credit hours is required. The student has the option of choosing courses specialized in General Physics, Condensed Matter Physics, and Optical Physics, with either a thesis or a non-thesis option. All students must take a set of core courses. The thesis option requires additional semester hours of electives plus 6 semester hours of thesis. The non-thesis option requires electives plus a comprehensive exit exam. The Physics M.S. program also offers a Planetary Science track.

Doctorate: Students have the option of choosing from three specializations: General Physics, Condensed Matter Physics, and Optical Physics. A total of 72 semester credit hours, of which 15 are required dissertation hours, are needed for the doctoral degree. The remaining 57 hours are divided into 18 hours of core courses, and a combination of specialization specific electives and research. Upon completion of the core, the student must take the written part of the Ph.D. candidacy examination. The Physics Ph.D. program also offers a Planetary Science track with specific core course requirements geared towards this track. For more information, please visit our website, http://physics.cos.ucf.edu.

Thesis: Thesis may be written in absentia.

SPECIAL EQUIPMENT, FACILITIES, OR PROGRAMS

The Department has recently established attosecond physics and NMR facility. Its new research building (inaugurated April 2011) is also equipped with Helium Liquefier, for research at very low temperatures. Two shared-facilities labs for microscopy and for UHV deposition techniques and spectrometry as well as a cleanroom for nanofabrication are available in the new building.

The Department operates the Surface Physics Laboratory, which includes a heavy ion backscattering spectrometer, 400-KEV ion implanter, imaging x-ray photoemission spectrometer, imaging secondary ion mass spectrometer, ultra-high vacuum atomic force microscope and scanning tunneling microscope, digital low energy electron diffraction, digital reflection high-energy electron diffraction, angle resolved ultraviolet photoemission spectrometer, and equipment for ultra-thin epitaxial film growth.

Other Department laboratories provide equipment for Mössbauer spectroscopy, magnetic susceptibility measurements, SQUID magnetometry, x-ray diffraction, Fourier spectroscopy, Raman and FTIR spectroscopy, inductively coupled plasma etching, e-beam thermal thin film evaporation, sub-micron optical lithography, high-sensitivity on-chip magnetometry, ultra-low temperature high magnetic field high frequency EPR and FMR spectroscopy, and high temperature incubation of bacterial cultures.

Equipment for performing biophysical research include an autoclave, a probe sonicator, a PCR thermocycler, a BioSafety hood, multiple centrifuges, and a soft-wall clean room.

The Department also has two ultrahigh vacuum systems with an e-beam evaporator, a quartz microbalance, a hybrid atom/ion plasma source, an argon sputter gun, a mass spectrometer, XPS and Auger spectrometers, UPS, TPD, and a variable temperature STM.

Computation facilities include several Linux Beowulf clusters with several hundred nodes each. UCF also has a state of the art high performance computer facility consisting of more than 1000 nodes is also available.

Department faculty participate in several campus research centers which provide access to additional specialized equipment. Research in advance materials and microelectronics is supported by the Materials Characterization Facility (MCF), located in the campus Research Park. Department faculty and students using MCF have access to Rutherford backscattering spectroscopy, imaging secondary ion mass spectroscopy, transmission and scanning electron microscopes, a focused ion beam system, a scanning Auger microprobe, x-ray photoelectron spectroscopy, and x-ray diffraction.

The Department also is responsible for operating the Robinson Observatory, an on-campus research and education facility that houses a 0.51-m f/8.2 Ritchey-Chretien astronomical telescope. The primary instrumentation on the telescope is a camera with a research-grade 3072-by-2048 pixel CCD that allows for multi-wavelength imaging.

Table B—Separately Budgeted Research Expenditures by Source of Support

Source of Support	Departmental Research	Physics-related Research Outside Department
Federal government		$5,252,953
State/local government		$25,720
Non-profit organizations		
Business and industry		$449,978
Other		
Total		$5,728,651

FACULTY

Distinguished University Professor

Chang, Zhenghu, Ph.D., Chinese Academy of Sciences, 1998. *Atomic, Molecular, & Optical Physics, Optics*. Atomic, Molecular, & Optical Physics, Optics and precision mechanics.

Rahman, Talat S., Ph.D., University of Rochester, 1977. *Computational Physics, Condensed Matter Physics, Surface Physics.* Chemical Physics, Computational Physics, Condensed Matter Physics, Nano Science and Technology, Surface Physics.

Professor

Britt, Daniel T., Ph.D., Brown University, 1991. Geophysics, Planetary Science.

Campins, Humberto, Ph.D., University of Arizona, 1982. *Planetary Science.* Planetary science; comets and asteroids.

Chernyak, Leonid, Ph.D., Weizmann Institute of Science, 1996. *Applied Physics, Condensed Matter Physics, Materials Science, Metallurgy, Nano Science and Technology.* Condensed Matter Physics, Materials Science, Nanostructure device physics.

Chow, Lee, Ph.D., Clark University, 1981. *Condensed Matter Physics, Nano Science and Technology.* Nano Science and Technology, Experimental condensed matter; carbon nanotubes and diamond-like carbon films.

Colwell, Joshua, Ph.D., University of Colorado Boulder, 1989. *Planetary Science.* Planetary science; planetary rings.

Johnson, Michael D., Ph.D., University of Virginia, 1986. *Condensed Matter Physics, Nano Science and Technology, Theoretical Physics.* Theoretical condensed matter; nanostructures.

Luo, Weili, Ph.D., University of California, Los Angeles, 1989. *Condensed Matter Physics, Fluids, Rheology.* Fluids, Experimental condensed matter; complex systems.

Mucciolo, Eduardo, Ph.D., Massachusetts Institute of Technology, 1994. *Condensed Matter Physics, Nano Science and Technology, Theoretical Physics.* Theoretical condensed matter; nanoelectronics.

Peale, Robert E., Ph.D., Cornell University, 1989. *Applied Physics, Condensed Matter Physics, Optics.* Experimental condensed matter; far-infrared semiconductor lasers.

Saha, Haripada, Ph.D., Calcutta, 1978. *Atomic, Molecular, & Optical Physics, Optics, Theoretical Physics.* Theoretical atomic, molecular, and optical physics.

Schulte, Alfons F., University of Munich, 1985. *Biophysics.* Biophysics; dynamics of proteins and disordered systems.

Associate Professor

Bhattacharya, Aniket, Ph.D., University of Maryland, 1992. *Condensed Matter Physics, Polymer Physics/Science, Theoretical Physics.* Polymer Physics/Science, Theoretical soft condensed matter; nonlinear dynamics.

del Barco, Enrique, Ph.D., Barcelona, 2001. *Condensed Matter Physics, Low Temperature Physics, Nano Science and Technology.* Nano Science and Technology; Experimental condensed matter; single-molecule magnets.

Efthimiou, Costas, Ph.D., Cornell University, 1996. *Physics and other Science Education, Theoretical Physics.* Mathematical physics; physics education research.

Fernández, Yanga R., Ph.D., University of Maryland, 1999. *Planetary Science.* Planetary science, comets, and asteroids.

Harrington, Joseph, Ph.D., Massachusetts Institute of Technology, 1994. *Atmosphere, Space Physics, Cosmic Rays, Planetary Science.* Planetary science; spectroscopy.

Heinrich, Helge H., Ph.D., ETH Zurich, 1994. *Condensed Matter Physics, Nano Science and Technology, Solid State Physics.* Condensed Matter Physics, Nano Science and Technology; Electron microscopy; nanomaterials.

Kara, Abdelkader, Ph.D., Lille, Saclay, 1985. *Chemical Physics, Computational Physics, Surface Physics.* Chemical Physics, Computational Physics, Condensed Matter Physics, Nano Science and Technology, Surface Physics.

Khondaker, Saiful I., Ph.D., University of Cambridge, 1999. *Condensed Matter Physics, Nano Science and Technology.* Experimental condensed matter; nanoelectronics.

Klemm, Richard, Ph.D., Harvard University, 1974. *Condensed Matter Physics, Nano Science and Technology, Theoretical Physics.* Theoretical condensed matter; superconductivity; nanomagnetism.

Kokoouline, Viatcheslav, Ph.D., St. Petersburg, 1999. *Atomic, Molecular, & Optical Physics, Theoretical Physics.* Theoretical atomic and molecular physics.

Leuenberger, Michael N., Ph.D., Basel, 2002. *Condensed Matter Physics, Quantum Foundations, Theoretical Physics.* Theoretical condensed matter; quantum computing.

Schelling, Patrick K., Ph.D., University of Minnesota, 1999. *Condensed Matter Physics, Materials Science, Metallurgy, Theoretical Physics.* Theoretical condensed matter and materials physics.

Stolbov, Sergei, Ph.D., Rostov, 1982. *Chemical Physics, Condensed Matter Physics, Surface Physics, Theoretical Physics.* Theoretical condensed physics, surface physics, chemical physics, material design.

Tatulain, Suren, Ph.D., St. Petersburg, 1979. *Biophysics.* Experimental bio-physics; catalysis.

Assistant Professor

Chanda, Debashis, Ph.D., University of Toronto, 2008. *Nano Science and Technology, Optics.* Research in his lab will focus on confining coherent/partially-coherent or incoherent light at nanoscale to enhance light-matter interactions for novel device applications as well as energy harvesting purposes. The emphasis will be design and development of high throughput, large scale and low cost fabrication of optical nanostructures for enhanced light-matter interactions in artificially structured metal/dielectric structures (metamaterials, plasmonic nanostructures), transformation optics for display/camouflage, strong coupling between photonic and plasmonic resonances and trapping light in thin film solar cells.

Chen, Bo, Ph.D., Northwestern University, 2007. *Biophysics, Nuclear Physics.* Biological Physics, Nuclear Magnetic Resonance.

Ishigami, Masahiro, Ph.D., California, 2004. *Condensed Matter Physics, Nano Science and Technology.* Condensed matter and materials physics.

Tetard, Laurene, Ph.D., University of Tennessee, 2010. staff scientist at the Oak Ridge National Laboratory. She received a B.S. in physics in 2004 and a M.S. in physics with concentration in nanotechnologies and nanoscience in 2006, from the University of Burgundy, France. She was awarded the Chancellor's Honors for Extraordinary professional promise in 2010, and was also one of the recipients of the R&D100 2010 award for her graduate work on subsurface imaging at the nanoscale. She held a Eugene P. Wigner Fellowship at the Oak Ridge National Laboratory from May 2011 to May 2013. *Biophysics, Nano Science and Technology.* development of high resolution microscopy and spectroscopy tools to advance the level of understanding of the behavior of materials and complex systems based on morphological, subsurface, physical and chemical properties at the nanoscale. The major applications of this work consist of lignocellulosic biomass, cancer cells, photovoltaic systems, proton conductors and polymers.

Emeritus

Bose, Subir K., Ph.D., Allahabad, 1967. *Optics, Statistical & Thermal Physics.* Statistical mechanics; quantum optics.

Brennan, J., Ph.D., Georgia Institute of Technology, 1968.

Research Professor

Liboff, Richard, New York University, 1961.

Research Assistant Professor

Hong, Sampyo, Ph.D., Kansas State University, 2005. *Computational Physics, Surface Physics.*

Turkowski, Volodymyr, Ph.D., Kiev University, Ukraine, 1998. *Computational Physics, Nano Science and Technology.* Development and application of the theoretical and numerical approaches to study electron-electron and electron-hole correlation effects in different bulk- and nano-systems; Ultrafast processes; Development of the dynamical mean-field theory (DMFT) approach for nanostructures, including the nonequilibrium theory; Extension of the time-dependent density functional theory approach (TDDFT) on the case of multiple excitations, especially electron-hole bound states (excitons, biexcitons etc.), plasmons and polarons and on the case of strongly correlated systems; Nanooptics; Nanomagnetism; Superconductivity in cuprates and graphene-based systems.

Zhao, Kun (Harry), Ph.D., Peking University, 2006. *Optics.* Attosecond optical pulse generation and characterization, ultrafast electron dynamics, ultrafast molecular dynamics, short and intense laser pulse generation and amplification, interaction of ultra-intense and ultra-short light pulse with plasmas, high-energy electron beam generation with laser wakefield acceleration.

Affiliate Professor

Bass, Michael, The College of Optics and Photonics. *Optics.*

Delfyett, Peter J., The College of Optics and Photonics. *Optics.*

Dogariu, Aristide, The College of Optics and Photonics. *Optics.*

Glebov, Leonid B., The College of Optics and Photonics. *Optics.*

Hagan, David J., The College of Optics and Photonics. *Optics.*

Hickman, James J., Ph.D., Burnett College of Biomolecular Sciences, and Electrical Engineering. Nanoscience Technology Center. *Medical, Health Physics, Nano Science and Technology.*

Kar, Aravinda, The College of Optics and Photonics. *Optics.*

Kik, Pieter, The College of Optics and Photonics. *Nano Science and Technology, Optics.*

Kuebler, Stephen, The College of Optics and Photonics. *Chemical Physics, Optics.*

Masunov, Artëm E., Ph.D., Nanoscience Technology Center. *Computational Physics, Nano Science and Technology.*

Richardson, Martin C., The College of Optics and Photonics. *Optics, Physics of Beams.*

Schoenfeld, Winston Vaughan, The College of Optics and Photonics. *Optics.*

Silfvast, William, The College of Optics and Photonics. *Optics.*

Soileau, M. J., The College of Optics and Photonics. *Optics.*

Stegeman, George I., The College of Optics and Photonics. *Optics.*

Su, Ming, Ph.D., Nanoscience Technology Center. *Medical, Health Physics, Nano Science and Technology.*

Van Stryland, Eric W., The College of Optics and Photonics. *Optics.*

Zeldovich, Boris Y., The College of Optics and Photonics. *Optics.*

Zhai, Lei, Ph.D., Nanoscience Technology Center. *Chemical Physics, Nano Science and Technology.*

Visiting Assistant Professor

Chini, Jacquelyn, Ph.D., Kansas State University, 2010. *Physics and other Science Education.* "physics education research". "implementing and assessing course transformation; physics teacher preparation".

Lecturer

Al Rawi, Ahlam, Ph.D., Kansas State University, 2001. *Biophysics, Computational Physics, Condensed Matter Physics.* Bio-

physics, computational surface physics and condensed matter physics.

Bindell, Jeffrey B., Ph.D., Brooklyn Poly, 1969. *Physics and other Science Education.* Physics education research.

Brueckner, Thomas J., Ph.D., Montana State University, 1997. *Astronomy, Physics and other Science Education.* Physics education research; astronomy.

Cooney, James H., Ph.D., University of Florida, 2004. *Cosmology & String Theory, Physics and other Science Education, Planetary Science.* Planetary Science; Physics education research; cosmology.

Dubey, Archana H., Ph.D., Bhavnagar, 1998. *Computational Physics, Fluids, Rheology.* Ferromagnetic fluids, computational physics.

Flitsiyan, Elena S., Ph.D., University of Moscow, 1975. *Condensed Matter Physics, Nuclear Physics.* Nuclear physics, experimental condensed matter physics.

Montgomery, Michele M., Ph.D., FIT, 2004. *Astrophysics, Planetary Science.* Planetary science; astrophysics.

Velissarius, Christos, Ph.D., University of Rochester, 1995. *High Energy Physics.* Experimental high-energy physics.

Zhmudsky, Oleksandr O., Ph.D., Kiev State, 1976. *Nonlinear Dynamics and Complex Systems, Theoretical Physics.* Theoretical physics; nonlinear dynamics.

Associate Research Scientist

Palotai, Csaba, Ph.D., University of Louisville. *Computational Physics.*

DEPARTMENTAL RESEARCH SPECIALTIES AND STAFF

Theoretical

Applied Mathematics. The Department of Physics at UCF has one regular faculty in Mathematical Physics and two distinguished affiliated faculty. As our university and our department grow, we are hoping to create a strong and prominent group. The research interests of the group are Conformal Field Theory, Integrable Models in Quantum Mechanics and Quantum Field Theory, Supersymmetry, String Theory, and Calabi-Yau manifolds. Several outstanding students have worked on the area and are currently pursuing doctorates in highly selective schools.

Atomic, Molecular, & Optical Physics. The theoretical Atomic, Molecular and Optical (AMO) physics group is engaged in research directed toward the elucidation of the fundamental details of the structure and dynamics of electrons and positrons interacting within atoms and molecules in the gas phase at thermal and ultra-cold energies. The primary aim of this group is to perform state-of-the-art calculation of (1) photoionization and photo-detachment of electrons from atoms and molecules, (2) electron-atom and electron molecule collisional processes, (3) the collisional processes in molecular plasmas with or without light emitted/absorbed, (4) elementary processes in ultra-cold atomic and molecular gases near the quantum degeneracy using completely ab-initio sophisticated theoretical approach.

Biophysics. The theoretical and computational Soft Condensed Matter and Biological Physics group engages in both classical and quantum calculations at various length and time scales using high-performance computing systems. There is a strong collaboration with the Biophysics group at the Physics department with several faculty at the College of Medicine.

Chemical Physics. Researchers in chemical physics study chemical processes from the point of view of physics. The focus is on understanding the microscopic factors that control the chemical characteristics of materials at the nanoscale. Size, shape, material, and support dependent properties of nanoparticles, for example, are examined using a range of state-of-art experimental, computational and theoretical techniques. Understanding the effect of the local atomic scale environment is the key towards designing materials for applications such as in solar cells, catalytic convertors, etc. Rahman, Stolbov.

Computational Physics. A variety of computational approaches are developed and used to address a wide array of physical problems. Focus extends from atomic-scale simulation of phenomena at surfaces and in nanomaterials, to heat, mass, and phonon transport phenomena at the nanoscale, to mesoscopic phenomena (DNA translocation through nanopores, self-assembly of amphiphilic peptides), to multiscale modeling, which extend the length and time scales accessible to simulation predictions to make more direct contact with experiment.

Condensed Matter Physics. The theoretical and computational condensed matter physics group focuses on fundamental physical phenomena, with a strong emphasis in the fields of nanomagnetism, nanotransport, nanocayalysis, energy related materials, surface science, vibrational dynamics at the nanoscale, superconductivity and semiconductors physics. These investigations will eventually lead to exciting novel applications in current technologies, e.g., catalysis, renewable energies, optoelectronics, medicine, homeland security, as well as potential applications in emerging technologies, e.g., quantum information and computation, coherent terahertz radiation, nanoelectronics or spintronics and high-sensitivity UV detection.

Physics and other Science Education. The Physics Education Research group (PER@UCF) is one of the newest research programs in PER in the nation. Formed by faculty originally trained in traditional areas of physics, our group initially started programs that promote science literacy, quantitative fluency, and information fluency. However, our group has expanded its projects to embrace successful educational reforms and replicate them smoothly at UCF. A partial list of interests for the members of our group is: Peer Instruction, Physics in Films, Video Analysis, Interactive Nanoscience Blogging, and Studio Physics.

Planetary Science. The UCF Planetary Sciences Group uses spacecraft data, images from the world's most powerful telescopes, meteorites and moon rocks, and supercomputer calculations to investigate fundamental questions like these: How did

our solar system form? What do the surfaces of other worlds tell us about their history? What happens when a comet hits a planet? What's going on in the atmospheres of extrasolar planets? Are there any other planetary systems like ours? Are we alone?.

Quantum Foundations. Research in this area engages in the study of quantum bits made from optical semiconductor quantum dots and explores the physical limits of quantum computation and communications and ways to mitigate the noise problem.

Experimental

Atomic, Molecular, & Optical Physics. The current AMO experimental group is focused on far/Terahertz spectroscopy and technology development, in particular, intracavity laser absorption spectroscopy for ultra-trace vapor detection. A large experimental effort is in Attosecond laser pulse generation, characterization and applications. The extremely short XUV pulses are used to probe and control electron dynamics in matter. Overlap between theory and experiment will form a critical component of our research program. Applications of this Attosecond laser pulse source to problems in condensed matter physics are underway.

Biophysics. The experimental Soft Condensed Matter and Biological Physics group involves various spectroscopic studies of proteins under high pressure and other extreme conditions, protein membrane interactions and light scattering studies. One recent technique is solid state NMR.

Condensed Matter Physics. The experimental condensed matter physics group focuses on the examination of many fundamental physical phenomena, with a strong emphasis in the fields of nanomagnetism, nanotransport, nanocayalysis, energy related materials, surface science, vibrational dynamics at the nanoscale, and semiconductors physics.

Planetary Science. The UCF Planetary Sciences Group uses spacecraft data, images from the world's most powerful telescopes, meteorites and moon rocks, and supercomputer calculations to investigate fundamental questions like these: How did our solar system form? What do the surfaces of other worlds tell us about their history? What happens when a comet hits a planet? What's going on in the atmospheres of extrasolar planets? Are there any other planetary systems like ours? Are we alone?.

View additional information about this department at www.gradschoolshopper.com

UNIVERSITY OF FLORIDA

DEPARTMENT OF PHYSICS

Gainesville, Florida 32611-8440
http://www.phys.ufl.edu/

General University Information
President: J. Bernard Machen
Dean of Graduate School: Henry T. Frierson
University website: http://www.ufl.edu
Control: Public
Setting: Urban
Total Faculty: 4,000
Total Graduate Faculty: 2,982

Total number of Students: 50,000
Total number of Graduate Students: 16,536

Department Information
Department Chairman: Kevin Ingersent, Chair
Department Contact: Pam Marlin, Program Assistant
Total full-time faculty: 48

Full-Time Graduate Students: 130
First-Year Graduate Students: 22
Total Post Doctorates: 22

Department Address

P.O. Box 118440
Gainesville, FL 32611-8440
Phone: (352) 392-0521 (C)
E-mail: studentservices@phys.ufl.edu
Website: http://www.phys.ufl.edu/

ADMISSIONS

Admission Contact Information

Address admission inquiries to: Graduate Affairs Office, P.O. Box 118440
Phone: (352) 392-1365
E-mail: webrequests@admissions.ufl.edu
Admissions website: http://www.admissions.ufl.edu/

Application deadlines

Fall admission:
U.S. students: February 1   *Int'l. students*: January 15

Application fee

U.S. students: $30

Admissions information

For Fall of 2013:
     Number of applicants: 265
     Number admitted: 22
     Number enrolled: 22

Admission requirements

Bachelor's degree requirements: Bachelor's degree in physics is required.
Minimum undergraduate GPA: 3.3

GRE requirements

The GRE is required.

Advanced GRE requirements

The Advanced GRE is required.
     Minimum accepted Advanced GRE score: 560

TOEFL requirements

The TOEFL exam is required for students from non-English-speaking countries.
PBT score: 550
iBT score: 80

Other admissions information

Additional requirements: The minimum acceptable GRE score required for admission is total–1200. The minimum acceptable GRE advanced score suggested for admission is 560. The average GRE scores for Fall 2010 admissions were verbal–470; quantitative–750; total–1220. The average GRE advanced score for Fall 2009 admissions was 780.
Undergraduate preparation assumed: Applicants are expected to have a solid knowledge at the level of the following textbooks or their equivalents: Griffiths, Introduction to Quantum Mechanics; Griffiths, Introduction to Electrodynamics; Thornton & Marion, Classical Dynamics; Zemansky and Kittel, Thermal Physics.

TUITION

Tuition year 2012–13:
Tuition for in-state residents
     Full-time students: $12,590 annual
Tuition for out-of-state residents
     Full-time students: $29,984 annual

Tuition fees for students employed as teaching assistants are paid the Department. Tuition fees for students employed as research assistants are paid by principal investigators on the grants that pay the students' stipends.
Credit hours per semester to be considered full-time: 9
Deferred tuition plan: No
Health insurance: Yes, Annual: $1,188.
Other academic fees: Graduate students pay their share of tuition fees, which cover University services, athletics, and public transportation.
Academic term: Semester
Number of first-year students who received full tuition waivers: 22

Teaching Assistants, Research Assistants, and Fellowships

Number of first-year
     Teaching Assistants: 22
     Fellowship students: 13
Average stipend per academic year
     Teaching Assistant: $22,000
     Research Assistant: $22,000
     Fellowship student: $25,000

FINANCIAL AID

Loans

Loans are available for U.S. students.
Loans are available for international students.
GAPSFAS application required: No
FAFSA application required: No

For further information

Address financial aid inquiries to: Graduate Affairs Office, P.O. Box 118440, Department of Physics, Gainesville, FL 32611-8440.
Phone: (352) 392-0181
Financial aid website: http://www.fa.ufl.edu/ufs/

HOUSING

Availability of on-campus housing

Single students: Yes
Married students: Yes

For further information

Address housing inquiries to: Director of Housing.
Housing aid website: http://www.housing.ufl.edu/

Table A—Faculty, Enrollments, and Degrees Granted

Research Specialty	2012-13 Faculty	Enrollment Summer 2012-Spring 2013		Number of Degrees Granted 2012-13 (2010-13)		
		Master's	Doctorate	Master's	Terminal Master's	Doctorate
Astrophysics	8	–	14	2(3)	–	2(3)
Biophysics	3	–	8	2(-)	–	–(2)
Chemical Physics	3	–	13	–(1)	1(-)	1(3)
Condensed Matter Physics	16	–	54	10(7)	1(-)	9(25)
Low Temperature Physics	4	–	12	2(1)	–	3(-)
Particles and Fields	12	–	29	4(5)	–(1)	5(8)
Non-specialized	–	–	–	–	–(6)	–
Total	46	–	130	20(18)	2(6)	20(40)
Full-time Grad. Stud.	–	–	130	–	–	–
First-year Grad. Stud.	–	–	22	–	–	–

GRADUATE DEGREE REQUIREMENTS

Master's: Thirty graduate credits in an approved program with satisfactory performance are required. Master's examination is required. A non-thesis option is available. One year of residence is required. There is no foreign language requirement.

Doctorate: Minimum of 90 hours and satisfactory performance in an approved course program are required. Full-time residency for two consecutive semesters is required. There is no foreign language requirement. Preliminary examination, qualifying examination, dissertation, and dissertation examination are required. Teaching experience, residence, a period of concentrated study, and a publishable thesis are required.

Other Degrees: Master of science in teaching degree requires 36 credits (half in graduate courses) with satisfactory performance in an approved program. Six credits of internship are required. No thesis is required. All candidates for advanced degrees must be registered during the semester in which the degree is awarded.

Thesis: Thesis may be written in absentia.

SPECIAL EQUIPMENT, FACILITIES, OR PROGRAMS

Helium liquefier and Microkelvin Laboratory located on campus; library and superb computer facilities include approximately 2000 processors running Linux 20 TB of disk. Graphics research is done on a 4-cpu SGI 4D/340 with video, Evans & Sutherland Freedom 3200. Mathematical Maple, Macsyma, PV-wave, Wave front, Mat lab, S-Plus, and Explorer software packages are available. Teaching laboratories include 100 Pentium PCs with data acquisition hardware and software. An eight-station electronics laboratory is used for training in analog and digital circuits, instrumentation design, and computer interfacing and programming. Highlights of the Advanced Physics Laboratory suite include FTIR spectrometer, pulsed NMR apparatus, x-ray diffracts meter, gamma ray spectrometer, helium cryostat, and holography room. Many workstations are distributed in faculty and staff offices. An undergraduate computer laboratory with 12 Linux PCs is provided for all student access. A full-service machine shop specializing in all forms of CNC and conventional machining, welding, and soldering is available. An electronics shop is also located in the building, specializing in design, prototype instrumentation, consultation, and repair. Cooperative research programs are conducted at Los Alamos National Laboratory, Oak Ridge National Laboratory, Florida Atlantic University, Florida International University, AT&T Bell Laboratories, Brookhaven National Laboratory, Fermilab, Cornell University, CERN, and the National High-Magnetic Field Laboratory (NHMFL). The University of Florida is a member of the consortium (UF, FSU, and LANL) responsible for the operation of the National High-Magnetic Field Laboratory.

Table B—Separately Budgeted Research Expenditures by Source of Support

Source of Support	Departmental Research	Physics-related Research Outside Department
Federal government	$8,300,000	
State/local government		
Non-profit organizations		
Business and industry	$405,500	
Other		
Total	**$8,705,500**	

Table C—Separately Budgeted Research Expenditures by Research Specialty

Research Specialty	No. of Grants	Expenditures ($)
Astrophysics	23	$2,126,293
Chemical Physics	2	$924,580
Condensed Matter Physics	16	$2,529,279
Low Temperature Physics	6	$2,116,360
Particles and Fields	18	$4,283,255
Total	**65**	**$11,979,767**

FACULTY

Distinguished University Professor

Hebard, Arthur F., Ph.D., Stanford University, 1970. *Applied Physics, Condensed Matter Physics*. Experimental condensed-matter physics, solid-state devices.

Mitselmakher, Guenakh, Ph.D., University of Moscow, 1974. *Accelerator, High Energy Physics, Particles and Fields*. Experimental high-energy physics.

Ramond, Pierre, Ph.D., Syracuse University, 1969. Elementary particle theory.

Sikivie, Pierre, Ph.D., Yale University, 1975. *Astrophysics*. Elementary particle theory; cosmology.

Tanner, David B., Ph.D., Cornell University, 1972. *Astrophysics, Condensed Matter Physics, Cosmology & String Theory*. Experimental condensed-matter physics; experimental astrophysics, LIGO.

Will, Clifford, Ph.D., California Institute of Technology, 1971. *Astrophysics, Cosmology & String Theory*. Experimental validation of Einstein's laws.

Professor

Acosta, Darin, Ph.D., University of California, San Diego, 1993. *High Energy Physics, Particles and Fields*. Experimental high-energy physics.

Avery, Paul R., Ph.D., University of Illinois, 1980. *Computational Physics, High Energy Physics, Particles and Fields*. Experimental high-energy physics, data management.

Cheng, Hai Ping, Ph.D., Northwestern University, 1988. *Chemical Physics, Computational Physics, Condensed Matter Physics*. Computational physics; molecules; clusters; nanostructures.

Detweiler, Steven L., Ph.D., University of Chicago, 1975. *Astrophysics, Cosmology & String Theory*. Relativistic astrophysics.

Field, Richard D., Ph.D., University of California, Berkeley, 1971. *Particles and Fields*. Elementary particle theory.

Fry, James N., Ph.D., Princeton University, 1979. *Astrophysics, Cosmology & String Theory*. Theoretical astrophysics; cosmology.

Hershfield, Selman P., Ph.D., Cornell University, 1989. *Condensed Matter Physics*. Theoretical condensed-matter physics.

Hirschfeld, Peter J., Ph.D., Princeton University, 1985. *Condensed Matter Physics*. Theoretical condensed-matter physics.

Ingersent, J. Kevin, Ph.D., University of Pennsylvania, 1990. Chair-elect. *Condensed Matter Physics*. Theoretical condensed-matter physics.

Korytov, Andrey, Ph.D., Dubna University, 1991. *Accelerator, High Energy Physics, Particles and Fields*. Experimental high-energy physics.

Kumar, Pradeep, Ph.D., University of California, San Diego, 1973. *Condensed Matter Physics, Low Temperature Physics*. Theoretical condensed-matter and low-temperature physics.

Lee, Yoonseok, Ph.D., Northwestern University, 1997. *Condensed Matter Physics, Low Temperature Physics*. Experimental condensed-matter and low-temperature physics.

Müller, Guido, Ph.D., Hannover, 1997. *Astrophysics, Cosmology & String Theory*. Experimental astrophysics.

Maslov, Dmitri, Ph.D., Landau Institute (Moscow), 1989. Graduate Coordinator. *Condensed Matter Physics*. Theoretical condensed-matter physics.

Matchev, Konstantin, Ph.D., Johns Hopkins University, 1997. *Cosmology & String Theory, High Energy Physics, Particles and Fields*. High-energy theory.

Meisel, Mark W., Ph.D., Northwestern University, 1983. *Biophysics, Condensed Matter Physics, Low Temperature Physics*. Low-temperature physics; experimental condensed-matter physics; biomagnetism.

Muttalib, Khandker A., Ph.D., Princeton University, 1982. *Condensed Matter Physics*. Theoretical condensed-matter physics.

Obukhov, Sergei, Ph.D., Landau Institute (Moscow), 1979. *Condensed Matter Physics, Polymer Physics/Science*. Theoretical condensed-matter and polymer physics.

Reitze, David, Ph.D., University of Texas, 1990. Presently on leave of absence as Executive Director of the LIGO Laboratory, California Institute of Technology, Pasadena, CA. *Astrophysics, Condensed Matter Physics, Optics*. ultrafast optical probes of solids; experimental astrophysics.

Rinzler, Andrew, Ph.D., University of Connecticut, 1991. *Applied Physics, Condensed Matter Physics*. Experimental condensed-matter physics; solid-state devices.

Stanton, Chris J., Ph.D., Cornell University, 1986. *Condensed Matter Physics*. Theoretical condensed-matter physics.

Stewart, Gregory R., Ph.D., Stanford University, 1975. *Condensed Matter Physics*. Experimental condensed-matter physics.

Sullivan, Neil S., Ph.D., Harvard University, 1972. *Condensed Matter Physics, Low Temperature Physics*. Experimental condensed-matter and low-temperature physics.

Takano, Yasumasa, Ph.D., University of Helsinki, 1978. *Condensed Matter Physics, Low Temperature Physics*. Experimental condensed-matter and low-temperature physics.

Thorn, Charles B., Ph.D., University of California, Berkeley, 1971. *High Energy Physics, Particles and Fields*. High-energy theory.

Whiting, Bernard F., Ph.D., University of Melbourne, 1979. *Astrophysics*. Theoretical astrophysics.

Woodard, Richard P., Ph.D., Harvard University, 1984. *Cosmology & String Theory, High Energy Physics, Particles and Fields*. Quantum gravity and quantum field theory.

Yelton, John M., Ph.D., University of Oxford, 1981. Chair. *High Energy Physics, Particles and Fields*. Experimental high-energy physics.

Associate Professor

Biswas, Amlan, Ph.D., University of Bangalore, India, 1999. Undergraduate Coordinator. *Condensed Matter Physics*. Experimental condensed-matter physics.

Furic, Ivan, Ph.D., Massachusetts Institute of Technology, 2004. *High Energy Physics, Particles and Fields*.

Hagen, Stephen, Ph.D., Princeton University, 1989. Associate Chair. *Biophysics, Condensed Matter Physics*. Experimental condensed-matter physics; molecular biophysics; optical spectroscopy.

Qiu, Zongan, Ph.D., University of Chicago, 1986. *Particles and Fields*. High-energy theory.

Saab, Tarek Khaled, Ph.D., Stanford University, 2002. *Astrophysics*. Experimental astrophysics.

Assistant Professor

Hamlin, James, Ph.D., Washington University, 2007. *Condensed Matter Physics*. Experimental condensed-matter physics.

Matcheva, Katia, Ph.D., Johns Hopkins University, 2000. *Atmosphere, Space Physics, Cosmic Rays, Planetary Science*. Planetary atmospheres.

Ray, Heather, Ph.D., University of Michigan, 2004. *Astrophysics, Particles and Fields*. Experimental neutrino physics.

Emeritus

Adams, E. Dwight, Ph.D., Duke University, 1960. *Low Temperature Physics*. Low-temperature physics.

Dufty, James W., Ph.D., Lehigh University, 1967. *Nonlinear Dynamics and Complex Systems, Polymer Physics/Science*. Nonequilibrium statistical mechanics.

Dunnam, F. Eugene, Ph.D., Louisiana State University, 1958. *Nuclear Physics*. Nuclear physics.

Hanson, Harold P., Ph.D., University of Wisconsin-Madison, 1948. Atomic physics; condensed-matter physics.

Ihas, Gary G., Ph.D., University of Michigan, 1971. *Low Temperature Physics*. Experimental low-temperature physics.

Ipser, James R., Ph.D., California Institute of Technology, 1969. *Astrophysics*. Relativistic astrophysics.

Klauder, John R., Ph.D., Princeton University, 1959. *Quantum Foundations*. Mathematical physics.

Monkhorst, Hendrik J., Ph.D., University of Groningen, 1968. *Chemical Physics, Plasma and Fusion*. Theoretical chemical physics; neutron-free fusion reactor development.

Sabin, John R., Ph.D., University of New Hampshire, 1966. *Accelerator, Atomic, Molecular, & Optical Physics*. Ion-beam interaction with solids.

Tobey, Frank L., Ph.D., University of Michigan, 1962.

Trickey, Samuel B., Ph.D., Texas A&M University, 1968. *Chemical Physics, Computational Physics, Condensed Matter Physics*. Condensed-matter theory and computation.

Affiliate Professor

Bartlett, Rodney J., Ph.D., University of Florida, 1971. *Atomic, Molecular, & Optical Physics, Chemical Physics, Computational Physics*. Many-electron theory of atoms, molecules, and solids.

Ford, Eric, Ph.D., Princeton, 2003. *Astronomy, Astrophysics*.

Mareci, Thomas, Ph.D., University of Oxford, 1982. *Biophysics, Medical, Health Physics*. Magnetic resonance imaging (MRI).

Micha, David A., Ph.D., Uppsala University, 1966. *Chemical Physics, Computational Physics*. Chemical physics; theoretical, molecular, and materials sciences.

Ohrn, N. Yngve, Ph.D., Uppsala University, 1963. *Chemical Physics, Computational Physics*. Quantum theory of matter.

Roitberg, Adrian, Ph.D., University of Chicago, 1992. *Biophysics, Chemical Physics, Computational Physics*. Chemical physics, theoretical biophysics.

Shabanov, Sergai, Ph.D., St. Petersburg State University, 1988. *Applied Mathematics, Quantum Foundations*. Mathematical physics.

Tan, Jonathan, Ph.D., University of California, Berkeley, 2001. *Astronomy, Astrophysics*.

Lab Personnel

Deserio, Robert, Ph.D., University of Chicago, 1981. Director of Undergraduate Laboratories. *Physics and other Science Education*. Physics education.

Scientist

Klimenko, Sergey, Ph.D., Novosibirsk State University, 1993. *Astrophysics, High Energy Physics*. Experimental astrophysics.

Konigsberg, Jacobo, Ph.D., University of California, Los Angeles, 1989. *Accelerator, High Energy Physics, Particles and Fields.* Experimental high-energy physics.

Associate Scientist

Andraka, Bohdan, Ph.D., Temple University, 1986. *Condensed Matter Physics.* Experimental condensed-matter physics.

Xia, Jian-Sheng, Ph.D., University of Science and Technology of China, 1989. *Low Temperature Physics.* Experimental low-temperature physics.

DEPARTMENTAL RESEARCH SPECIALTIES AND STAFF

Theoretical

Astrophysics. Stellar structure and evolution; variable stars; planetary atmospheres nonlinear dynamics and chaos; high-energy astrophysics; solar neutrinos; general relativity; black holes; gravitational waves; neutron stars; cosmology: early universe. Detweiler, Fry, Matcheva, Tan, Whiting.

Chemical Physics. Many-body scattering theory; intermolecular forces; radiation-molecule interactions; propagator methods; density functional theory and local density models; semiempirical molecular orbital methods: electronic structure computation; theory of quantum crystals, molecular crystals, biological molecules, and thin films; theory and application of new computational methods; neutron-free fusion reactor development. Bartlett, Cheng, Ingersent, Micha, Saab, Stanton.

Computational Physics. Quantum theory. Muttalib, Shabanov.

Condensed Matter Physics. Theory of highly correlated systems; liquid and solid 3He; high-Tc superconductivity and heavy fermions; glasses; disorder and localization; optical interactions in semiconductors; nonlinear phenomena in condensed matter; solitons; nuclear spin ordering in metals; pattern formation (Dorsey). Hershfield, Hirschfeld, Ingersent, Kumar, Muttalib, Obukhov, Stanton.

Nonlinear Dynamics and Complex Systems. Time series analysis; reconstruction of attractors, glasses, and complex fluids. Obukhov.

Particles and Fields. Standard Model of strong and electromagnetic interactions; grand unified theories; string unification; superstring theory; particle astrophysics. Field, Matchev, Qiu, Ramond, Sikivie, Thorn, Woodard.

Planetary Science. Modeling of atmospheric phenomena in planets. Matcheva.

Experimental

Biophysics. Protein conformational dynamic and folding; microfluids; in vivo and high-resolution MRI and NMR spectroscopy; bio-optically active processes (Petkova). Hagen, Mareci, Meisel.

Condensed Matter Physics. NMR of solids, liquids, and polymers; NMR imaging; optical properties; quantum magnetic excitations; conducting polymers; organic superconductors; composite materials; thermal and magnetic properties of superconductors; high-temperature superconductors; surface physics; synchrotron radiation; electronic properties of novel materials; ultrafast spectroscopy of novel materials; nonlinear nonequilibrium phenomena; ultrafast spectroscopy of novel materials; high-intensity laser-solid interactions.

Low Temperature Physics. Properties of macroscopic quantum systems, in particular liquid and solid 3He, solid H2, and electronic and magnetic systems down to 10 μK; thermodynamic, hydrodynamic, magnetic, and transport properties of materials are studied using NMR, ultrasound (P,V,T), and other probes. Ihas, Lee, Meisel, Sullivan, Takano, Xia.

View additional information about this department at
www.gradschoolshopper.com

UNIVERSITY OF MIAMI

DEPARTMENT OF PHYSICS

Coral Gables, Florida 33124
http://www.physics.miami.edu

General University Information

President: Donna Shalala
Dean of Graduate School: M. Brian Blake
University website: http://www.miami.edu
Control: Private
Setting: Suburban
Total Faculty: 3,027
Total Graduate Faculty: 2,557
Total number of Students: 15,657
Total number of Graduate Students: 5,289

Department Information

Department Chairman: Joshua Cohn, Chair
Department Contact: Joshua Cohn, Chair
  Total full-time faculty: 18
  Total number of full-time equivalent positions: 18
  Full-Time Graduate Students: 21
  First-Year Graduate Students: 9
  Total Post Doctorates: 2

Department Address

P.O. Box 248046
Coral Gables, FL 33124
Phone: (305) 284-7123

Fax: (305) 284-4222
E-mail: cohn@physics.miami.edu
Website: http://www.physics.miami.edu

ADMISSIONS

Admission Contact Information

Address admission inquiries to: Department of Physics, Graduate
    Program
Phone: (305) 284 7120
E-mail: grad_info@physics.miami.edu
Admissions website: http://web.physics.miami.edu/Graduate/

Application deadlines

Fall admission:
U.S. students: February 1   *Int'l. students*: February 1

Application fee

U.S. students: $50   *Int'l. students*: $50

Admissions information

For Fall of 2013:
   Number of applicants: 92
   Number admitted: 15
   Number enrolled: 7

Admission requirements

Bachelor's degree requirements: A bachelor's degree in physics
   or related field is required.
Minimum undergraduate GPA: 3.0

GRE requirements

The GRE is required.

Advanced GRE requirements

The Advanced GRE is recommended.

TOEFL requirements

The TOEFL exam is required for students from non-English-
   speaking countries.
PBT score: 550
iBT score: 80

Other admissions information

Additional requirements: Exceptions can be made at the dis-
   cretion of the committee on graduate student advising. The
   GRE is required for domestic applicants.
No minimum acceptable score for admission is specified.
Undergraduate preparation assumed: Griffiths, Introduction to
   Electrodynamics: Schroeder, Thermal Physics; Symon, Me-
   chanics; Griffiths, Quantum Mechanics.

TUITION

Tuition year 2013-14:
   Full-time students: $31,140 annual
   Part-time students: $1,730 per credit
Credit hours per semester to be considered full-time: 9
Deferred tuition plan: Yes
Other academic fees: $648 (includes student activity and athletic
   fees)
Academic term: Semester
Number of first-year students who received full tuition waivers: 7

Teaching Assistants, Research Assistants, and Fellowships

Number of first-year
   Teaching Assistants: 7
Average stipend per academic year
   Teaching Assistant: $22,000
   Research Assistant: $22,000

FINANCIAL AID

Application deadlines

Fall admission:
U.S. students: February 1   *Int'l. students*: February 1

Loans

Loans are available for U.S. students.
Loans are not available for international students.
GAPSFAS application required: No
FAFSA application required: No

For further information

Address financial aid inquiries to: Assistantships: Department of
   Physics, Graduate Program;, graduate school loans: Financial
   Aid Services.
Phone: (305) 284-7120
E-mail: grad_info@physics.miami.edu

HOUSING

Availability of on-campus housing

   Single students: No
   Married students: No

For further information

Address housing inquiries to: Department of Residence Halls,
   P.O. Box 248044, Miami, FL 33124.

Table A—Faculty, Enrollments, and Degrees Granted

Research Specialty	2012-13 Faculty	Enrollment Fall 2012		Number of Degrees Granted 2012-13 (2008-13)		
		Master's	Doctorate	Master's	Terminal Master's	Doctorate
Astrophysics	3	–	3	–	–	1(4)
Biophysics	1	–	–	–	–	–
Condensed Matter Physics	4	–	4	–	–	1(3)
Nonlinear Dynamics and Complex Systems	2	–	4	–	–	1(2)
Optics	3	–	4	–	–	1(4)
Particles and Fields	4	–	–	–	–	1(5)
Plasma and Fusion	1	–	–	–	–	–(1)
Non-specialized	–	–	8	1(12)	–(1)	–
Total	17	–	23	1(12)	–(1)	5(19)
Full-time Grad. Stud.	–	–	23	–	–	–
First-year Grad. Stud.	–	–	8	–	–	–

GRADUATE DEGREE REQUIREMENTS

Master's: Thirty graduate credits in approved program with a
B average, a score of S or better in the comprehensive departmental
exam, and two semesters of full-time study or equivalent in residence are required. There is no language requirement.

Doctorate: Sixty credits beyond the baccalaureate degree and 24
credits beyond the master's degree in an approved program with
a B average, a score of P in the comprehensive departmental
exam by the student's second yearly attempt at the latest, dissertation and dissertation exam, and one year of residency are
required. There is no foreign language requirement.

Thesis: Thesis may not be written in absentia.

SPECIAL EQUIPMENT, FACILITIES, OR PROGRAMS

Cooperative research programs are conducted with the Rosenstiel School of Marine and Atmospheric Sciences of the University of Miami and with the Atlantic Oceanographic and Meteorological Laboratory of NOAA and various domestic and foreign universities.

Table B—Separately Budgeted Research Expenditures by Source of Support

Source of Support	Departmental Research	Physics-related Research Outside Department
Federal government	$7,920,000	
State/local government		
Non-profit organizations		
Business and industry		
Other	$10,000	
Total	**$7,930,000**	

Table C—Separately Budgeted Research Expenditures by Research Specialty

Research Specialty	No. of Grants	Expenditures ($)
Astrophysics	11	$2,331,000
Condensed Matter Physics	1	$395,000
Optics	12	$3,900,000
Particles and Fields	4	$370,000
Plasma and Fusion	1	$615,000
Nonlinear Dynamics and Complex Systems	2	$898,000
Total	**31**	**$8,509,000**

FACULTY

Professor

Alexandrakis, George C., Ph.D., Princeton University, 1968. *Condensed Matter Physics*. Solid state experiment; transmission resonance; magneto-acoustic propagation in ferromagnetic metals.

Alvarez, Orlando, Ph.D., Harvard University, 1979. *Particles and Fields*. Theory of elementary particles.

Barnes, Stewart, Ph.D., University of California, Los Angeles, 1972. *Condensed Matter Physics*. Solid state theory; many-body theory; superconductivity and magnetism.

Cohn, Joshua L., Ph.D., University of Michigan, 1989. *Condensed Matter Physics*. Condensed matter experiment, materials physics, electronic and lattice transport.

Curtright, Thomas, Ph.D., California Institute of Technology, 1977. *Particles and Fields*. Theory of elementary particles.

Galeazzi, Massimiliano, Ph.D., University of Genoa, Italy, 1999. *Astrophysics*. X-ray astrophysics; studies of interstellar/intergalactic medium and x-ray sources; development of x-ray detectors.

Gordon, Howard R., Ph.D., Pennsylvania State University, 1965. *Optics*. Optical oceanography; light scattering; radiative transfer; remote sensing.

Gundersen, Joshua O., Ph.D., University of California, Santa Barbara, 1995. *Astrophysics*. Observational cosmology; experimental cosmology and astrophysics; cosmic microwave and infrared backgrounds.

Huerta, Manuel A., Ph.D., University of Miami, 1970. *Plasma and Fusion*. Statistical mechanics; plasma physics; numerical simulations in MHD.

Johnson, Neil F., Ph.D., Harvard University, 1989. *Nonlinear Dynamics and Complex Systems*. Theoretical condensed matter physics; biological complexity; social complexity; physical complexity.

Mezincescu, Luca, Ph.D., University of Bucharest, 1978. *Particles and Fields*. Theory of elementary particles.

Nepomechie, Rafael I., Ph.D., University of Chicago, 1982. Theory of elementary particles.

Voss, Kenneth J., Ph.D., Texas A&M University, 1984. Department Chair. *Marine Science/Oceanography, Optics*. Ocean optics; light scattering; atmospheric optics.

Zuo, Fulin, Ph.D., Ohio State University, 1988. Condensed matter experiment.

Associate Professor

Korotkova, Olga, Ph.D., University of Central Florida, 2003. *Optics*. Theoretical optics; wave propagation and scattering in random media.

Assistant Professor

Song, Chaoming, Ph.D., City University of New York, 2008. *Nonlinear Dynamics and Complex Systems*.

Syed, Sheyum, Ph.D., Columbia University, 2004. *Biophysics, Condensed Matter Physics, Nonlinear Dynamics and Complex Systems*. Biological physics with emphasis on understanding the circadian clock and related behaviors at the molecular and the organismal levels.

Emeritus

Faber, Shepard, Ph.D., University of Florida, 1959. Teaching techniques.

Hirschberg, Joseph G., Ph.D., University of Wisconsin-Madison, 1952. *Optics*. Physical optics; Fabry-Perot interferometry; plasma spectroscopy.

Nearing, James C., Ph.D., Columbia University, 1965. *Physics and other Science Education*. Theoretical physics; bifurcation theory in fully nonlinear plasma systems.

Perlmutter, Arnold, Ph.D., New York University, 1955. *Nuclear Physics, Particles and Fields*. Nuclear and particle physics.

Adjunct Professor

Ghandour, Ghassan, Ph.D., University of California, Berkeley, 1974. *Particles and Fields*. Elementary particles; mathematical physics.

Moore, William Franklin, Ph.D., University of Miami, 1980.

Van Vliet, Carolyne M., Ph.D., Free University of Amsterdam, 1956. *Statistical & Thermal Physics*. Equilibrium and nonequilibrium statistical mechanics; stochastic processes; quantum transport in solids.

DEPARTMENTAL RESEARCH SPECIALTIES AND STAFF

Theoretical

Astrophysics. Origin and evolution of the universe; cosmic microwave background; clustering and large-scale structure; dark matter and dark energy; intergalactic medium.

Condensed Matter Physics. Electronic structure of solids; many-body physics; high-temperature superconductivity; magnetism; linear and nonlinear quantum transport; reduced-dimensionality systems. Barnes.

Nonlinear Dynamics and Complex Systems. Emergent cooperative phenomena in biological and social systems; dynamical evolution; extreme behavior; prediction and soft control. Johnson, Song, Syed.

Optics. Transmission, propagation and scattering of electromagnetic waves in deterministic and random media. Korotkova.

Optics. Radiative transfer; remote determination of ocean chlorophyll concentrations; wave propagation and scattering in random media. Gordon.

Particles and Fields. Quantum field theory (especially integrable models); supersymmetry; supergravity; superstrings. Alvarez, Curtright, Mezincescu, Nepomechie.

Plasma and Fusion. Numerical simulations in plasmas and other systems. Huerta.

Experimental

Astrophysics. Studies of the cosmic microwave and infrared background; studies of the interstellar/intergalactic medium and x-ray sources; instrumentation for low-noise RF and mm-wave detectors and telescopes; development of high-resolution cryogenic microcalorimeters and bolometers cross-correlation studies; statistical analysis. Galeazzi, Gundersen.

Biophysics. Kinetics of and interactions between fruitfly clock proteins; behavioral and molecular studies exploring the roles of circadian clock and homeostasis in sleep; statistical models of fly behavior and underlying neurons. Syed.

Condensed Matter Physics. Ferromagnetic transmission resonance in metals; spin relaxation; exchange energy; phonon excitation and propagation; nonlinear phenomena; transport and magnetic properties of materials at low temperatures; transition metal oxides, high-temperature and organic superconductors, and reduced dimensional systems (e.g., layered systems and thin films); electrical and thermal conduction, thermoelectric effects; vortex dynamics; critical current; quantum tunneling; laser deposition of ferroelectric thin films. Alexandrakis, Cohn, Zuo.

Optics. Light scattering and absorption by marine particulates; instrumentation for measurement of optical properties of the ocean and atmosphere. Voss.

View additional information about this department at
www.gradschoolshopper.com

UNIVERSITY OF SOUTH FLORIDA

DEPARTMENT OF PHYSICS

Tampa, Florida 33620
http://physics.usf.edu/

General University Information
President: Judy L. Genshaft
Dean of Graduate School: Karen D. Liller
University website: http://usf.edu/
Control: Public
Setting: Urban
Total Faculty: 2,114
Total number of Students: 40,211
Total number of Graduate Students: 8,957

Department Information
Department Chairman: Pritish Mukherjee, Chair
Department Contact: Lilia M. Woods, Graduate Director
  Total full-time faculty: 32
  Full-Time Graduate Students: 87
  First-Year Graduate Students: 17
  Female First-Year Students: 6
  Total Post Doctorates: 10

Department Address
4202 E. Fowler Avenue, ISA 2019
Tampa, FL 33620
Phone: (813) 974-2871
Fax: (813) 974-5813
E-mail: lmwoods@usf.edu
Website: http://physics.usf.edu/

ADMISSIONS

Admission Contact Information
Address admission inquiries to: Director of Graduate Studies, Lilia M. Woods or Academic Program Specialist Candice Pietri (cpietri1@usf.edu)
Phone: (813) 974-2871
E-mail: lmwoods@usf.edu
Admissions website: http://physics.usf.edu/graduate/

Application deadlines
Fall admission:
U.S. students: February 1   *Int'l. students*: January 2
Spring admission:
U.S. students: September 1   *Int'l. students*: July 1

Application fee
U.S. students: $30   *Int'l. students*: $30
Submit the online Graduate Admission Application for the University of South Florida via http://www.grad.usf.edu/graduate-admissions.asp. During this process, you will be asked to pay a one-time application fee of $30.00 by credit card (Master Card, Visa, or Discover) or e-check.

Admissions information
For Fall of 2012:
  Number of applicants: 87
  Number admitted: 25
  Number enrolled: 17

Admission requirements
Bachelor's degree requirements: Bachelor's degree in physics or related fields is required.
Minimum undergraduate GPA: 3.0

GRE requirements

The GRE is required.

Advanced GRE requirements

The Advanced GRE is recommended.

TOEFL requirements

The TOEFL exam is required for students from non-English-speaking countries.

PBT score: 550
iBT score: 79

Other admissions information

Additional requirements: Test of English as a Foreign Language (TOEFL, http://www.toefl.org): Applicants whose native language is not English or who have not earned a degree in the United States must also submit TOEFL scores earned within two (2) years of the desired term of entry. A minimum total score of 79 on the internet-based test, 213 on the computer-based test, or 550 on the paper-based test are required. Applications submitted with TOEFL scores that do not meet the minimum requirements will be denied. The TOEFL requirement may be waived if the applicant meets one of the following conditions: (a) has scored 500 or higher on the GRE Verbal Test, (b) has earned a college degree at a U.S. institution of higher learning, (c) has earned a college degree from an institution whose language of instruction is English (must be noted on the transcript), or (d) Has scored 6.5 on the International English Language Testing System exam (IELTS, http://www.ielts.org).

TUITION

Tuition year 2012–13:
Tuition for in-state residents
  Full-time students: $430.85 per credit
  Part-time students: $430.85 per credit
Tuition for out-of-state residents
  Full-time students: $752.22 per credit
  Part-time students: $752.22 per credit
Credit hours per semester to be considered full-time: 9
Deferred tuition plan: Yes
Health insurance: Available at the cost of 1,806 per year.
Academic term: Semester
Number of first-year students who received full tuition waivers: 17

Teaching Assistants, Research Assistants, and Fellowships

Number of first-year
  Teaching Assistants: 17
Average stipend per academic year
  Teaching Assistant: $19,800
  Research Assistant: $19,800
  Fellowship student: $25,000

FINANCIAL AID

Application deadlines

Fall admission:
U.S. students: February 1   *Int'l. students*: January 1
Spring admission:
U.S. students: September 1   *Int'l. students*: July 1

Loans

Loans are available for U.S. students.
Loans are not available for international students.
GAPSFAS application required: Yes
FAFSA application required: Yes

For further information

Address financial aid inquiries to: Director of Financial Aid, SVC 1102.
Phone: (813) 974-3700
Financial aid website: http://usfweb2.usf.edu/finaid/

HOUSING

Availability of on-campus housing

Single students: Yes
Married students: Yes

For further information

Address housing inquiries to: Director of Housing and Food Service: Thomas Kane, RAR 229.
Phone: (813) 974-0001
Housing aid website: http://www.housing.usf.edu/

Table A—Faculty, Enrollments, and Degrees Granted

Research Specialty	2012–13 Faculty	Enrollment Fall 2012		Number of Degrees Granted 2012–13 (2008–13)		
		Master's	Doctorate	Master's	Terminal Master's	Doctorate
Atomic, Molecular, & Optical Physics	10	–	15	–(6)	–(1)	1(5)
Biophysics	4	–	25	–(5)	1(1)	–(3)
Condensed Matter Physics	9	–	47	3(7)	3(8)	4(19)
Total	23	–	87	3(18)	4(10)	5(27)
Full-time Grad. Stud.	–	–	70	–	–	–
First-year Grad. Stud.	–	–	17	–	–	–

GRADUATE DEGREE REQUIREMENTS

Master's: The Department offers Master's degrees in Physics and Applied Physics. There are 2 options for the M.S. in Physics: Thesis Option: minimum 30 semester hours in an approved program, six of which may be for thesis; final oral exam on the thesis required; no language requirement; one academic year of residence required. Non-Thesis Option: minimum 30 semester hours of course work in an approved program; no language requirement; one academic year of residence required. Dual Master's: The 51-semester-hour program culminates in the student receiving an M.S. degree in Physics and an M.S. Degree in Engineering Science; single thesis included; approximately three years to complete; 15 hours of course work each in Physics and Electrical Engineering (microelectronics option); 9 hours overlap courses; 6 thesis hours each in Physics and Electrical Engineering. Master of Science degree program in Materials Science and Engineering (MSE) at USF: Students require a minimum of 30 total credit hours to earn the MS degree in MSE. The degree may be completed within 12 months by taking 12 credit hours in each of the Fall and Spring semesters followed by 6 credit hours during the summer. Students must take 15 credit hours of core courses (including a maximum of 3 credit hours for an interdisciplinary Graduate Materials Seminar), and 9 hours of elective courses for the thesis option which requires an additional 6 hours of thesis research. In addition, 6 hours of elective courses would be required in lieu of thesis hours for the non-thesis option.

Doctorate: The Department offers a Ph.D. program in Applied Physics. This program encompasses the areas of biophysics; biomedical physics; atomic, molecular, and optical physics; solid state and materials physics; computational physics; and physics education. An integrated curriculum incorporating interdisciplinary training in applied physics overlapping these areas of emphasis, and research programs to prepare students for the increasingly technological workforce of the future are key components of the program. These aspects of the program and an industrial practicum provide a bridge between theoretical, fundamental concepts and their practical, engineering applications. These programs offer maximum flexibility and are tailored to suit the interests of students and his/her career objectives.

SPECIAL EQUIPMENT, FACILITIES, OR PROGRAMS

Experimental and theoretical research in the Department of Physics is conducted in the following laboratories/programs:

Cellular and Molecular Biophysics Research Laboratory

The Center is highly interdisciplinary, involving physics, physiology, cell biology, and molecular biology. The structural and functional relationship of membrane proteins, such as ion channels, membrane transporters, and electrogenic pump molecules, as well as their interfaces with external electromagnetic field, are currently studied. One of the projects is to study direct energy transform from inorganic energy to the living system by electrically activating the membrane electrogenic pump molecules, Na/K ATPases. Other projects include the study of percutaneous and targeted drug delivery, understanding the mechanisms underlying electrical injury, and the development of novel techniques for wound healing. The research involves both basic science and its practical application to biology and medicine. The research program is supported by the National Institutes of Health (NIH) since 1994, as well as by many other funding agencies. Our research is conducted and focused on cellular and molecular levels in nanoscales by using broad, state-of-arts techniques, including whole cell/patch clamps, various microscopic imagining systems such as multiple-laser confocal microscope and near field microscope, full line of cellular, and molecular biology technique. (wchen@usf.edu)

Nanomedicine and Nanobiotechnology Laboratory

This interdisciplinary and entrepreneurial research group applies principles of physics other subjects and utilizes analytical methods to study the structure and function of specific proteins and to design, fabricate and characterize nanostructures and polymer-based biomaterials, especially polypeptide-based materials. Proteins of interest play a key role in cell adhesion and migration. Materials of interest include coatings for in vitro/ex vivo cell and tissue culture, coatings for medical implant devices and coatings for cancer therapy. The research group has expertise in spectroscopic, acoustic, and microscopic methods for physical and chemical characterization, and bacterial and mammalian cell culture for biological characterization. Instrumentation in the laboratory includes a Fourier transform infrared spectrometer, an ultrasensitive dual-beam UV-visible wavelength spectrometer with Peltier temperature control, a fluorescence microscope equipped with a heated stage, gas control and low light level imaging device, a tissue culture suite, an electrospinning apparatus, a fume hood and various small items of equipment. We collaborate with investigators in several USF departments, the Moffitt Cancer Center, other universities and biotechnology companies. (dhaynie@usf.edu)

Soft Materials Physics Laboratory

Many mechanical, dynamical and structural properties of materials remain poorly understood for reasons independent of system-specific chemistry. Great advances in understanding these properties can be achieved through coarse-grained and multiscale simulations that are computationally efficient enough to access experimentally accessible spatiotemporal scales yet "chemically" realistic enough to capture the essential physics underlying the properties under study. I have and will continue to concentrate on explaining poorly-understood behaviors of polymeric, col-

loidal, and nanocomposite systems through coarse-grained modeling and concomitant development of analytic theories. The general theme is to do basic research on topics that are of high practical interest. Current work includes studies of polymer crystallization and polymer-nanocomposite mechanics. Facilities include a 32-core workstation and access to USF's 500-node CIRCE cluster. (rshoy@usf.edu)

Condensed matter/materials physics optical and laser spectroscopy:

This laboratory is currently under development and is designed to apply state-of-the-art optical and laser spectroscopy techniques to study the fundamental properties of advanced materials. Some of the most advanced optical and laser spectroscopy techniques will be available, including multidimensional ultrafast laser spectroscopy possibly spanning a wide spectral range, and single nanostructure photoluminescence and lifetime spectroscopy. The aforementioned optical techniques will be combined to explore the electronic, vibrational, and light-matter interactions in a variety of advanced nanomaterials, correlated-electron materials, and organic/inorganic hybrid materials, used in renewable energy applications for harvesting and storage. Understanding their intrinsic properties will contribute in improving the device performance of solar cells and batteries, important to a renewable-energy future. (karaiskaj@usf.edu)

Digital Holography & Microscopy Laboratory (DHML)

The main theme of our research activities is in the development of novel imaging technologies, with emphasis in holographic and interferographic microscopy. In digital holography (DH), the hologram is recorded by a CCD camera, instead of photographic plates, and the holographic images are calculated numerically using the electromagnetic diffraction theory. This gives direct access to the phase profile of the optical field and leads to a number of powerful imaging techniques that are difficult or impossible in real space holography. Transparent objects, such as many biological cells, thin film structures, and MEMS devices, can be imaged that reveal minute thickness variations with nanometer precision. Optical tomography by digital interference holography (DIH) yields cross-sectional images of biological tissues without actually cutting into them. Cellular motility can be studied by imaging the adhesion layers between a crawling cell and the substrate through the DH of total reflection, important in the study of embryogenesis, neuronal growth, and cancer cell metastasis. Furthermore, we are not only able to image cells and their components, but also manipulate them in full three dimensions, using patterns of light produced by holographic optical tweezers (HOT). Cells and organelles can be captured and tracked, coaxed into artificially patterned growth and motion, and operated on with micromanipulation and microsurgery. Students can expect to work on cutting-edge research topics and be trained extensively in advanced optical design and construction, digital image acquisition, computer programming, electronic instrumentation, and cellular and biomedical laboratory procedures. Digital holography is an emerging technology that has been experiencing exponential growth in the last decade, and has potential applications in wide-ranging areas including cellular microscopy, metrology, manufacturing processes and testing, medical imaging and diagnostics, biometry, environmental research, and food science, just to name a few. (mkkim@usf.edu)

Laboratory for Advanced Materials Science and Technology (LAMSAT)

Explores innovations in pulsed laser ablation and plasma processes for the growth of thin films of technologically significant materials, including super hard materials, magnetic materials, superconductors, and compound semiconductors for solar cells. Past NSF- and DOE-sponsored research projects have focused on the application of a dual-laser ablation process discovered in this laboratory to grow large-area, particulate-free films of

Cu(InGa)Se2 and ZnO for solar cell applications, and to fabricate diamond and diamond-like carbon structures for MEMS applications. One of the recently funded NSF projects focuses on an hybrid process where chemical self-assembly and physical vapor deposition techniques are combined to grow vertically aligned nano-grained films of superhard materials. Novel optical techniques for high resolution, in-situ plasma imaging, and development of new laser-assisted plasma growth processes are being researched. The research encompasses thin film growth, nanostructures, dynamic optical process diagnostics, thin film analysis, characterization and process modeling leading to the fabrication of single-layer and hetero-structure devices. (pritish@usf.edu, switanac@usf.edu)

Functional Materials Laboratory

This laboratory is equipped with experimental facilities for studying the electrical and magnetic properties of novel materials. Investigation of the material properties are done over a wide range in temperature ($2 K < T < 350 K$) and applied magnetic fields up to 7 Tesla. In addition, the frequency-dependent electromagnetic response is probed from DC to 6 GHz. A novel resonant radio-frequency (RF) method has been developed to accurately determine the magnetic anisotropy and switching in materials. Current research projects focus on studies of dynamic magnetic response and high-frequency impedance in nanoparticles, composites, thin films, magnetic semiconductors and multiferroic systems. These technologically important materials are promising candidates as building blocks for the next generation multifunctional device. Other interests include magnetocaloric effect in nanostructured materials, spin polarization studies, and physics of strongly correlated systems. Ongoing research support by NSF, DoD, and DOE. (sharihar@usf.edu)

The Bio-Nano Research Group

The work of this laboratory is the investigation of the structure/function relationship in biological systems ranging from the single molecule to the multicellular level. Molecular level structures determine the materials properties of the system, which in turn determines the macroscopic biological function. Using the expertise in atomic force microscopy, fluorescence microscopy, rheology, and other techniques found in the laboratory, the physical properties of single molecules and macromolecules are measured, and bulk models are developed and experimentally tested. These models are used to help explain the biology or pathology of systems. Example projects within the lab include investigations of cell surface and extracellular matrix glycoproteins and glycosaminoglycans through single molecule imaging and force spectroscopy. The data from these experiments is used to develop models for the viscoelastic properties of solutions of these biopolymers, which are then tested experimentally. These rheological properties are important for the function of tissues ranging from joint interstitial fluids to lung epithelium and will be used to understand the behaviors observed in these systems. The outcomes of the lab are geared to make significant contributions to biomedicine, and as such require a close collaboration with the Departments of Biology and Chemistry and with the School of Medicine. The work is inherently multidisciplinary, and students develop a broad range of skills from physics, biology, and chemistry. (garrettm@usf.edu)

Spintronics Lab

Facilities include multi-target sputtering system for thin film heterostructure growth, electron transport measurements in high magnetic fields (7T) and cryogenic temperatures ($\sim 1 K$). Current interests include spin-dependent tunneling, spin-injection, and spin-polarization measurements, and combine modeling to complement experiments when appropriate. (millercw@usf.edu)

Novel Materials Laboratory

The laboratory is designed for the synthesis and characterization (including structural, optical, electrical, thermal, and magnetic) of novel materials for technologically significant applications. The emphasis is understanding the structure-property relationships of material systems; that is, how crystal structure variations affect the electrical, thermal, optical, magnetic, and mechanical properties of materials. The laboratory applies this understanding towards the crystal growth and synthesis of new and novel materials for varying technologically significant applications. The Lab's research is based on new materials for energy-related technologies. Current materials research includes new semiconductors for electronics and optoelectronics applications, transport properties of "open structured" semiconductors, nanocrystal synthesis and self assembly approaches, and new magnetic materials. The research is supported by NSF, DOE, ONR, ARO, NASA, and industry. Close collaboration with industry is typical in this interdisciplinary Materials Physics research program that encompasses all aspects of physics and materials science. Students typically acquire a large variety of skill sets and apply these foundations to their applied physics research. (gnolas@usf.edu)

Materials Simulations Laboratory (MSL)

Director: Prof. I. I. Oleynik Associate Director: V. V. Zhakhovsky

The research program at MSL focuses on modeling the atomic, electronic, and chemical properties of systems of fundamental and technological importance using the powerful arsenal of first-principles density functional theory, tight-binding, and classical interatomic molecular dynamics techniques. The research thrusts are (1) energetic materials, shock physics, and materials at extreme conditions; (2) materials for information technology including molecular and spin electronics; (3) ultrafast laser-matter interactions in metals and semiconductors; (4) graphene nanomechanics and surface chemistry; (5) development of novel algorithms and computational methods for large-scale materials simulations using Tera- and Peta-Flop computers. The ultimate goal is to understand fundamental mechanisms and establish structure-property relationships that are difficult or sometimes impossible to obtain from experiment. Other areas of research at MSL include the development of analytic bond order potentials for large-scale atomistic simulations, surface and interface science, including surface chemistry of metal-oxide and metal-polymer interfaces in magnetic tunneling junctions. The research program at MSL is supported by NSF, DARPA, ONR, and ARO. MSL is one of the major users of the NSF funded Teragrid network of supercomputers. The computational facilities at MSL include a 400-CPU Beowulf cluster consisting of the latest 2.7 GHz AMD Shanghai quad-core nodes interconnected by an Infiniband switching fabric, a 12 Tb file server, and 20 linux workstations. MSL personnel include five graduate students, two undergraduate students, two postdoctoral associates, and a research associate professor. MSL is running a vibrant REU program supported by NSF that is specifically focused on attracting minority and female students. Further info can be obtained at http://msl.cas.usf.edu.

Condensed-Matter-Theory Research Group

This group works in condensed-matter theory, with current projects in crystallography, biological data analysis, and magnetic systems. In collaboration with Dr. Benji Fisher, we have reformulated Fourier-space crystallography into the language of cohomology of groups and applied the results to a wider class of structures than previously considered. Continuing work focuses on homological invariants of a new kind and their possible physical implications. In biology, we have been collaborating with Dr. Chun-Min Lo on analysis of electric cell-substrate impedance-sensing experiments. Looking only at statistical signatures of electrical noise, we can distinguish cancerous from non-cancerous cell cultures of the same type of cell and can detect physiological effects of the toxin cytochalasin-B at lower concentrations than possible with other techniques. Recent work in magnetic systems includes a statistical-mechanical model of helimagnetism in rare-earth heterostructures and a study of the ballistic-to-diffusive crossover in quantum wires. The latter may have applications in quantum computing. (davidra@ewald.cas.usf.edu)

Laboratory of Optical Biophysics

The overall research focus in our laboratory is on the basic physical principles that govern the phase separation and aggregation of proteins in solution. Depending on the specifics of the protein interactions and solution conditions, proteins can either stay soluble or undergo a variety of phase transitions. These phase transitions include crystallization, liquid-liquid phase separation, precipitation or formation of amyloid fibrils. We are using a variety of optical and spectroscopic techniques to study the thermodynamics and kinetics of these phase separation phenomena of proteins. Our current focus is on the mechanisms governing the self-assembly of a variety proteins into so-called amyloid fibrils. This aggregation process is the molecular hallmark for a large class of protein aggregation diseases, including Alzheimer's disease, Parkinson's disease and even type-II diabetes. Aside from these biomedical applications, amyloid fibril formation has also been recognized as a unique model for general mechanisms of self-assembly of biomolecules into highly ordered nano- and biomaterials with intriguing mechanical and optical properties. We are pursuing both the biomedical and nano-materials aspects of this assembly process in our current research. (mmuschol@usf.edu)

Computational Soft Matter Laboratory

Interest of our group lie in the areas of computational bio-physics and mathematical modeling of biological and social systems. In this context, we focus on molecular dynamics simulations, monte-carlo methods, multi-scale modeling based on mean field theory, dynamical systems with an emphasis on Hamiltonian systems, associated numerical methods, and underlying parallel processing issues. Our current focus is on the study of lipid bilayer systems, which form integral components of cellular membranes. These systems are studied using various computational modeling techniques and validated through close interactions with experimentalists. One aspect of our current work involves the study of heterogeneous model membrane systems. We study the interactions between various membrane components such as phospho- and sphingolipids, and cholesterol that give rise to stable structures such as "rafts" and caveolae. We intend to develop a multi-scale mean filed theory based model into a complete simulation methodology for membrane simulations. The models and methods developed will be used to study the structure and stability of membrane structures such as "raft" and caveolae, which are known to play critical roles in the activation of T-cells in immune response. (pandit@usf.edu)

Computational Condensed Matter Physics and Materials Science Program

Research interest include the area of theoretical condensed matter physics with a focus on computational nanoscience. The materials of interest include semiconductors, ferroelectrics, ferromagnets, and multiferroics in both bulk and low-dimensional forms. Examples are nanotubes, nanowires, nanodots, and thin films. An exciting feature of such nanoforms is the appearance of new properties and phenomena that do not exist in bulk. The purpose of my research is to identify these novel features, study their fundamental aspects, and explore their new functionalities for future applications in nanoscale devices. An example is utilizing a novel vortex structure that is a unique feature of ferroelectric nanodots in ultrahigh density memory that may increase the current memory capacitance by orders of magnitude. Another research focus is the development of computational techniques that will expand their capabilities beyond existing levels. Examples include the devel-

opment of first-principle-based techniques for new material forms (nanoscale ferroelectrics and multiferroics) and properties (dielectric loss and tunability). The ultimate research objective is the efficient design of new materials conducted in close collaboration with experimental groups. (iponomar@usf.edu)

Advanced Materials and Devices Theory Group

Our group is engaged in various problems related to theoretical modeling and description of structural, functional, and nanoscale materials and devices. We pursue two complementary routes - analytical and computational. Analytical techniques based on quantum mechanics, quantum electrodynamics, and many-body theory, are being developed. First principle density functional theory and tight binding models on high-performance supercomputers are being utilized. Currently, we are pursuing problems related to the Casimir effect in nanostructured materials, thermoelectric properties of materials with enhanced cooling and power generation performances, simulations of nanostructured materials properties, and related devices. The projects are funded by various national funding agencies. Our group maintains strong collaborations with experimental teams as well as other theoretical groups from the University of South Florida, other universities and national research laboratories. We are devoted to conducting leading edge research to advance our understanding of complex materials and devices using analytical and computational methods. (lmwoods@usf.edu)

The Nanophysics and Surface Science Laboratory

In this laboratory, we investigate condensed matter at the atomic scale. The surface of a material is where the action is; at a surface the material interacts with its environment and thus many chemical and physical processes occur at the interface between a solid and a different medium. Our goal is to understand the structural and electronic properties of surfaces and to tune these properties in order for the surface to perform new or improved functions. Currently investigated surface-functional materials are metal oxides for their use as solid state gas sensors and for solar energy conversion. Modification of surfaces with nanoclusters to improve their functionality is one approach to improve and create new functionalities.

Nanoclusters are aggregates of atoms in the realm between molecules and bulk materials. In this size range condensed matter exhibits new properties, which can be conveniently tuned by controlling their size. In our laboratory we assemble clusters atom by atom in the gas phase and subsequently place them on a support material. This allows investigating the cluster-support interaction and the cluster-size properties relationship. Most of the sample preparation and characterization is done under ultra high vacuum conditions to ensure the integrity of the samples under investigation. In addition to the in-house measurements, some supplementing photoemission and x-ray absorption studies are performed at synchrotron facilities. (mbatzill@usf.edu)

Soft Semiconducting Materials and Devices Lab

This lab studies soft semiconductors with low dimensional electronic structure and are solution processable. Some examples are organic semiconductor, colloidal quantum dots, carbon nanotubes, and graphene. One research focus is on systematic investigation of correlation between excitonic properties of thin films and electronic responses of pertinent devices. Technical approaches include optical study by linear (absorption, steady state and time-resolved photoluminescence) and various modulation (continuous wave photoinduced absorption, doping induced absorption &Electroabsorption) spectroscopies, and transport measurements using vertical (TOF, CELIV, SCLC, CV) and lateral (FET) device structures. Another major research line is to advance renewable energy technology by development of processing methodologies to integrate multifunctional materials onto soft substrates. One example is the fabrication of transparent and flexible organic solar module by solution-based techniques such as spray and printing. (xjiang@usf.edu)

Solid-State Quantum Optics Lab

This laboratory is equipped with experimental facilities for studying quantum optical phenomena in solid-state nanostructures such as quantum dots, nanocrystals, and impurity centers. The long-term goal of this research is to realize controlled light-matter interactions for use in quantum communication and quantum information science. Optical techniques employed include high-resolution spectroscopy, interferometry, and multi-photon correlation measurements in both ambient and low-temperature (liquid-helium) environments. Novel optical microcavities are being developed to enhance the interactions of light with single quantum emitters and mechanical resonators for harnessing cavity-electrodynamics and cavity-optomechanics phenomena. (mullera@usf.edu)

Quantum Nonlinear Photonics Lab

This laboratory is aimed to experimentally study quantum nonlinear photonics, a broadly defined field that investigates the physics and applications of the nonlinear and quantum aspects of photons and light-matter interaction. In specific, the lab is equipped with facilities to study photonic crystal and other nanophotonics devices on silicon-on-insulator chips and plasmonic nanostructures. Another part of the lab is aimed to develop new concepts of optical technologies that utilize nonclassical nature of light, especially to encode and retrieve information from an optical field in terms of color, amplitude, phase and polarization. (zhiminshi@usf.edu)

Microwave and Ultrafast Terahertz Research Lab

This group focuses on research in theoretical modeling and experimental characterization of photonic structures including metamaterials, surface plasmons and photonic crystals. Numerical modeling and simulation techniques based on the finite element method and the finite-difference time-domain method are used to design functional photonic structures. State-of-the-art microwave and ultrafast terahertz equipments provide the capability of characterizing photonic materials over a wide frequency range. Researchers in this group are devoted to conducting cutting-edge research in both theoretical and experimental aspects of various photonic phenomena. Current research activities include studying nonlinear properties of photonic metamaterials, RF and solar energy harvesting metamaterials and surface plasmons. This group maintains strong collaboration and interaction with several leading research groups from other universities and national laboratories. (jiangfengz@usf.edu)

General Support Facilities

Include a machine shop to build custom mechanical and vacuum parts and an electronics shop capable of custom design, repair, and fabrication of electronics and computer components.

Table B—Separately Budgeted Research Expenditures by Source of Support

Source of Support	Departmental Research	Physics-related Research Outside Department
Federal government	$3,679,808	
State/local government	$56,300	
Non-profit organizations		
Business and industry		
Other	$424,266	
Total	$4,160,374	

Table C—Separately Budgeted Research Expenditures by Research Specialty

Research Specialty	No. of Grants	Expenditures ($)
Atomic, Molecular, & Optical Physics	4	$482,909
Biophysics	4	
Condensed Matter Physics	9	$3,196,899
Other	4	$480,566
Total	21	$4,160,374

FACULTY

Professor

Chen, Wei, Ph.D., Temple University, 1988. Cellular and molecular biophysics, structure and function of membrane proteins, bioenergetics, new technique in synchronization modulation of the electrogenic pump molecules to electrically activate the pump functions, and its biomedical applications.

Kim, Myung K., Ph.D., University of California, Berkeley, 1986. *Applied Physics, Atomic, Molecular, & Optical Physics, Biophysics, Optics.* Digital holography; phase contrast microscopy; optical tomography; biomedical imaging; quantum optics; laser spectroscopy.

Mukherjee, Pritish, Ph.D., State University of New York at Buffalo, 1986. Chair of Physics Department. *Applied Physics, Atomic, Molecular, & Optical Physics, Chemical Physics, Condensed Matter Physics, Energy Sources & Environment, Materials Science, Metallurgy, Nano Science and Technology, Optics.* Picosecond lasers and applications; laser-assisted materials growth; nanostructures, thin films and heterostructures of semi-conductors and oxides.

Nolas, George S., Ph.D., Stevens Institute of Technology, 1994. *Applied Physics, Chemical Physics, Condensed Matter Physics, Energy Sources & Environment, Engineering Physics/Science, Materials Science, Metallurgy, Nano Science and Technology.* Experimental solid-state, materials, and condensed matter physics.

Oleynik, Ivan I., Ph.D., Russian Academy of Sciences, 1992. *Applied Physics, Chemical Physics, Computational Physics, Condensed Matter Physics, Nano Science and Technology.* Theoretical condensed matter and chemical physics; computational materials science.

Srikanth, Hariharan, Ph.D., Indian Institute of Science, 1993. *Applied Physics, Condensed Matter Physics, Materials Science, Metallurgy, Nano Science and Technology.* Experimental condensed matter; materials sciences.

Witanachchi, Sarath, Ph.D., State University of New York at Buffalo, 1989. Associate Chair of Physics Department. *Applied Physics, Atomic, Molecular, & Optical Physics, Chemical Physics, Condensed Matter Physics, Energy Sources & Environment, Materials Science, Metallurgy, Nano Science and Technology, Optics.* Laser ablation, plasma processing and chemical synthesis of films and nanostructures.

Associate Professor

Batzill, Matthias, Ph.D., University of Newcastle upon Tyne, UK, 1999. *Applied Physics, Chemical Physics, Condensed Matter Physics, Materials Science, Metallurgy, Nano Science and Technology.* Surface science; gas-surface interactions; structure and electronic properties of metal oxide surfaces; nanoclusters and quantum dots; solid state gas sensors; photocatalysis and photovoltaic for sustainable and renewable energy.

Haynie, Donald T., Ph.D., Johns Hopkins University, 1994. *Applied Physics, Biophysics, Materials Science, Metallurgy, Nano Science and Technology.* Nanomedicine and bionanotechnology.

Matthews, Garrett, Ph.D., University of North Carolina, 2001. *Applied Physics, Biophysics, Medical, Health Physics, Nano Science and Technology*. Biological macromolecules and macromolecular biopolymers.

Miller, Casey W., Ph.D., University of Texas, 2003. *Applied Physics, Condensed Matter Physics, Materials Science, Metallurgy, Nano Science and Technology*. Experimental condensed matter: Spin-dependent transport properties of novel materials and devices.

Muschol, Martin, Ph.D., City University of New York, 1992. *Applied Physics, Atomic, Molecular, & Optical Physics, Biophysics, Medical, Health Physics, Optics*. Neuronal plasticity; advanced optical techniques to probe cellular mechanisms; protein crystallization.

Rabson, David, Ph.D., Cornell University, 1991. *Applied Physics, Biophysics, Computational Physics, Condensed Matter Physics, Statistical & Thermal Physics*. Condensed matter theory.

Woods, Lilia, Ph.D., University of Tennessee, 1999. *Applied Physics, Condensed Matter Physics, Materials Science, Metallurgy, Nano Science and Technology*. Theoretical Condensed Matter Physics: theory and computation of nanostructures, dispersive interactions, thermoelectric transport.

Assistant Professor

Hoy, Robert S., Ph.D., Johns Hopkins University, 2008. *Applied Physics, Computational Physics, Theoretical Physics*. Soft matter physics, Theoretical and Computational modeling of polymeric, colloidal and nanocomposite systems, computational method development.

Jiang, Xiaomei, Ph.D., University of Utah, 2004. *Applied Physics, Atomic, Molecular, & Optical Physics, Chemical Physics, Condensed Matter Physics, Energy Sources & Environment, Materials Science, Metallurgy, Nano Science and Technology, Optics*. Organic electronic materials; fabrication and characterization of light emitting diodes and photovoltaic devices for solar cell applications.

Karaiskaj, Denis, Ph.D., Simon Fraser University, 2002. *Applied Physics, Atomic, Molecular, & Optical Physics, Materials Science, Metallurgy, Nano Science and Technology, Optics*. Two-dimensional spectroscopy on nanostructures, and proteins; optical spectroscopic studies of carbon nanotubes; ultrahigh resolution spectroscopy of semiconductors.

Muller, Andreas, Ph.D., University of Texas, Austin, 2007. *Applied Physics, Atomic, Molecular, & Optical Physics, Optics*. Experimental quantum optics of nanostructures; quantum dots, nanocrystals, impurity centers; cavity quantum electrodynamics and optomechanics for quantum communication and quantum information science.

Pandit, Sagar A., Ph.D., University of Pune, India, 1999. *Applied Physics, Biophysics, Computational Physics, Condensed Matter Physics, Statistical & Thermal Physics*. Computational Biophysics and mathematical modeling of biological and social systems.

Ponomareva, Inna, Ph.D., Russian Academy of Sciences, 2004. *Applied Physics, Computational Physics, Condensed Matter Physics, Materials Science, Metallurgy, Nano Science and Technology*. Condensed matter physics, numerical quantum chemistry, computational physics, nanoscience, developing and implementation of computational techniques.

Shi, Zhimin, Ph.D., University of Rochester, 2011. *Applied Physics, Atomic, Molecular, & Optical Physics, Optics, Quantum Foundations*. Quantum nonlinear photonics,Nanophotonics, silicon photonics, photonic crystal, Plasmonics, metamaterial. Optical methods utilizing non-classical nature of light.

Zhou, Jiangfeng, Ph.D., Iowa State University, 2008. *Applied Physics, Atomic, Molecular, & Optical Physics, Condensed Matter Physics, Electrical Engineering, Electromagnetism,*

Optics. Metamaterials, photonic crystals, plasmonics, numerical electromagnetics, and THz photonics.

Emeritus

Chang, Robert S. F., Ph.D., Cornell University, 1976. *Applied Physics, Atomic, Molecular, & Optical Physics, Condensed Matter Physics*. Solid state laser spectroscopy; energy transfer studies; crystal growth; fiber optics.

Djeu, Nicholas, Ph.D., Cornell University, 1970. *Applied Physics, Atomic, Molecular, & Optical Physics, Optics*. Laser physics and non-linear optics; fiberoptics.

Johnson, Dale E., Ph.D., University of Chicago, 1971. Graduate Director. *Applied Physics*. Electron microscopy.

Killinger, Dennis K., Ph.D., University of Michigan, 1978. Distinguished University Professor Emeritus. *Applied Physics, Atomic, Molecular, & Optical Physics, Optics*. Laser optics laser physics; laser remote sensing/LIDAR; Optical Transmission of the Atmosphere. physics; quantum electronics; laser spectroscopy.

Research Associate Professor

Zhakhovsky, Vasily V., Ph.D., Russian Academy of Science, 1997. *Applied Physics, Atomic, Molecular, & Optical Physics, Chemical Physics, Computational Physics, Condensed Matter Physics, Optics*. Atomic simulations of laser-matter interactions, shock wave physics, and materials at extreme conditions.

Research Assistant Professor

Datta, Anuja, Ph.D., Indian Association for the Cultivation of Science, 2008. *Nano Science and Technology*. Synthesis of inorganic nanostructured materials and characterization, optical properties (Static and time resolved spectroscopy) of nanomaterials, nanoscale thermoelectrics (Materials synthesis and transport properties), and structure-property relationships of new materials.

Lisenkov, Sergei, Ph.D., Russian Academy of Sciences, 2005. *Applied Physics, Computational Physics, Condensed Matter Physics, Materials Science, Metallurgy*. Finite-temperature properties of multiferroic materials, Perovskite superlattices and nanostructures, electronic and stability properties of nanotubes and fullerenes.

Phan, Manh-Huong, Ph.D., University of Bristol, 2006. *Applied Physics, Condensed Matter Physics, Materials Science, Metallurgy, Nano Science and Technology*. Nanomagnetism and magnetic materials, giant magnetoimpedance (GMI) materials, giant magnetocaloric (GMC) materials, colossal magnetoresistive (CMR) materials, nanoparticles and nanocomposites, multiferroic materials.

Instructor

Chabot, Michelle, Ph.D., University of Texas, Austin, 2001. *Applied Physics, Condensed Matter Physics, Physics and other Science Education*. Experimental condensed matter physics; undergraduate lab development; education.

Criss, Robert, Ph.D., University of Texas, Dallas, 1993. *Applied Physics, Atomic, Molecular, & Optical Physics, Physics and other Science Education*. Applied VUV-VIS spectroscopy; physics education.

Gobeille, Doug, Ph.D., Brandeis University, 2010. *Applied Physics, Astronomy, Astrophysics, Electromagnetism*. High-energy astrophysics, specifically regarding Active Galactic Nuclei (AGN), Quasar Morphology, AGN jet physics, and the role of AGN in galactic evolution. Primary research on these topics is conducted in the radio regime with panchromatic observations for complete spectral coverage. Special attention is given to jet physics using the Chandra X-ray space telescope.

Mackay, Kevin, Ph.D., Queen's University, Belfast, N. Ireland, 2000. *Applied Physics, Astronomy, Astrophysics.* Extra-solar planets, astronomy education, and thin-film magnetic materials.

Woods, Gerald, Ph.D., University of Tennessee, 2001. General Physics Lab Director. *Applied Physics, Condensed Matter Physics, Physics and other Science Education.* Experimental condensed matter.

DEPARTMENTAL RESEARCH SPECIALTIES AND STAFF

Theoretical
Applied Physics.
Atomic, Molecular, & Optical Physics.

Biophysics.
Medical, Health Physics.

Experimental
Applied Physics.
Atomic, Molecular, & Optical Physics.
Biophysics.
Condensed Matter Physics.
Medical, Health Physics.

> *View additional information about this department at*
> *www.gradschoolshopper.com*

EMORY UNIVERSITY

DEPARTMENT OF PHYSICS

Atlanta, Georgia 30322
http://www.physics.emory.edu

General University Information
President: James W. Wagner
Dean of Graduate School: Lisa Tedesco
University website: http://www.emory.edu
Control: Private
Total Faculty: 3,600
Total Graduate Faculty: 873
Total number of Students: 12,755
Total number of Graduate Students: 5,865

Department Information
Department Chairman: Eric Weeks, Chair
Department Contact: Calvin E. Jackson, Program Coordinator
  Total full-time faculty: 16
  Total number of full-time equivalent positions: 16
  Full-Time Graduate Students: 34
  First-Year Graduate Students: 10
  Female First-Year Students: 2
  Total Post Doctorates: 10

Department Address
400 Dowman Drive
Math and Science Center
Atlanta, GA 30322
Phone: (404) 727-6584
Fax: (404) 727-0873
E-mail: asc@physics.emory.edu
Website: http://www.physics.emory.edu

ADMISSIONS

Admission Contact Information
Address admission inquiries to: Director of Graduate Studies, Department of Physics
Phone: (404) 727-6584
E-mail: asc@physics.emory.edu
Admissions website: http://www.physics.emory.edu/graduate/

Application deadlines
Fall admission:
U.S. students: January 3   *Int'l. students*: January 3

Application fee
U.S. students: $75   *Int'l. students*: $75
Fee is waived for applications submitted by October 31. All applications have to be submitted online at http://www.gs.emory.edu/admissions/application.html.

Admissions information
For Fall of 2013:
  Number of applicants: 43
  Number admitted: 20
  Number enrolled: 10

Admission requirements
Bachelor's degree requirements: Bachelor's or higher degree in physics or related field is required. For all requirements, see http://www.physics.emory.edu/graduate/gradstudguide.html.
Minimum undergraduate GPA: 3.0

GRE requirements
The GRE is required.
  Quantitative score: 155
  Verbal score: 153

Analytical score: 3
Mean GRE score range (25th–75th percentile): 315-330
GRE scores should be supplied online at http://www.gs.emory.edu/admissions/application.html at least by February 15.

Advanced GRE requirements
The Advanced GRE is recommended.

TOEFL requirements
The TOEFL exam is required for students from non-English-speaking countries.
PBT score: 600

Other admissions information
Additional requirements: The minimum acceptable scores suggested for admission are verbal 153 (or 500) and quantitative 155 (or 700).

TUITION

Tuition year 2013–14:
Tuition for out-of-state residents
Full-time students: $33,800 annual
Part-time students: $33,800 annual
All admitted students receive a full tuition waiver and a stipend. More details on stipends and benefits provided can be found at http://www.physics.emory.edu/graduate/.
Credit hours per semester to be considered full-time: 9
Deferred tuition plan: No
Health insurance: Available
Other academic fees: Activity and athletic fee (optional). Information on available facilities can be found at https://wpec.emory.edu/wpec/facilities.cfm.
Academic term: Semester
Number of first-year students who received full tuition waivers: 10

Teaching Assistants, Research Assistants, and Fellowships
Number of first-year
Teaching Assistants: 10
Fellowship students: 2
Average stipend per academic year
Teaching Assistant: $23,000
Research Assistant: $25,000
Fellowship student: $25,500
Admitted students are automatically considered for various graduate school fellowships that range between $2,500 to $5,000 of additional support per year. Details are found at http://www.gs.emory.edu/financial_support/fellowships.html.

FINANCIAL AID

Application deadlines
Fall admission:
U.S. students: January 3   *Int'l. students*: January 3

Loans
Loans are available for U.S. students.
Loans are available for international students.
GAPSFAS application required: No
FAFSA application required: No

For further information
Address financial aid inquiries to: Graduate Coordinator, Department of Physics.
E-mail: asc@physics.emory.edu
Financial aid website: http://www.gs.emory.edu/admissions/assistance.html

HOUSING

Availability of on-campus housing
Single students: Yes
Married students: Yes

For further information
Address housing inquiries to: University Housing Office.
Phone: (404) 727-7631
E-mail: housinghelp@learnlink.emory.edu
Housing aid website: http://www.emory.edu/HOUSING/GRAD/gradhouse.html

Table A—Faculty, Enrollments, and Degrees Granted

Research Specialty	2013-14 Faculty	Enrollment Fall 2013 Mas-ter's	Enrollment Fall 2013 Doc-torate	Number of Degrees Granted 2012–13 (2008-13) Mas-ter's	Number of Degrees Granted 2012–13 (2008-13) Terminal Master's	Number of Degrees Granted 2012–13 (2008-13) Doc-torate
Biophysics	5	–	13	–	–(1)	2(13)
Condensed Matter Physics	4	–	15	–	–(4)	–(6)
Statistical & Thermal Physics	3	–	5	–	1(3)	1(2)
Total	12	–	33	–	1(8)	3(21)
Full-time Grad. Stud.	–	–	33	–	–	–
First-year Grad. Stud.	–	–	10	–	–	–

GRADUATE DEGREE REQUIREMENTS

Master's: Twenty-four semester hours, including 20 hours of coursework or seminar work with a GPA of "B" or better. Two semesters of residence are required. A thesis and oral defense of research work are required. For complete requirements, see http://www.physics.emory.edu/graduate/gradstudguide.html.
Doctorate: Forty-eight semester hours beyond the M.S. level; maximum of 36 hours in research or guided study with an average grade of "B−". A minimum of four semesters in residence are required beyond the M.S. level. Ph.D. candidacy is determined by a GPA of "B" in physics coursework, at least two research rotations, and a written and orally presented qualifier proposal (research proposal). A dissertation and oral defense are required. For complete requirements, see http://www.physics.emory.edu/graduate/gradstudguide.html. We offer a joint degree option with a Ph.D. in Physics and an M.S. in Computer Science, see http://www.physics.emory.edu/graduate/gradstudguide.html#joint_degree.
Thesis: A thesis is required for Ph.D. and M.S. degrees.

SPECIAL EQUIPMENT, FACILITIES, OR PROGRAMS

For the latest information on our facilities, see http://www.physics.emory.edu/facilities/.

Research: The Physics Department at Emory University hosts a number of research facilities for material and device preparation and deposition, nanofabrication, characterization, and measurements. Facilities listed as shared departmental equipment are available free of charge to all researchers affiliated with the Physics Department. Facilities located in individual research laboratories are often available to other researchers on a collaborative basis. Other research facilities available in close proximity include the Emory X-ray Crystallography Center, the Nanotechnology Research Center at Georgia Tech, and the Oak Ridge National Laboratory.

Computation: The Technical Team at the Department of Physics provides comprehensive technical support and services in a variety of areas, including software and desktop hardware support. Our department maintains or participates in several clusters for scientific computation, including the Computational Statistical Physics Cluster, The Cherry L. Emerson Center for Scientific Computation, and Emory's High Performance Computing Center.

Astronomy: The Physics Department houses three major astronomical facilities: A custom-designed planetarium featuring a Zeiss Skymater projector, a 24″ Ritchey-Crietien DFM Casegrain optical telescope, and a 25-foot diameter dish radio telescope. These facilities provide a comprehensive hands-on educational experience encompassing major areas of modern astronomy.

Education: Our classrooms and educational technologies help to enhance our teaching efforts. The introductory laboratories are kept small to emphasize a personalized approach. Up-to-date instrumentation and computer-based data acquisition prepare students for the challenges of modern science and technology. The advanced laboratories use custom-built experiments that have produced six education-related papers.

Machine Shop: In addition to the basic machining services, the Physics Machine Shop modifies commercial and custom-built research equipment to specification. We assist in the engineering, design, and construction of complex custom apparatuses. No matter what the need, the Machine Shop strives to be the department's one-stop destination for advanced research equipment needs.

Table B—Separately Budgeted Research Expenditures by Source of Support

Source of Support	Departmental Research	Physics-related Research Outside Department
Federal government	$1,600,000	
State/local government		
Non-profit organizations		
Business and industry		
Other	$500,000	
Total	**$2,100,000**	

Table C—Separately Budgeted Research Expenditures by Research Specialty

Research Specialty	No. of Grants	Expenditures ($)
Biophysics	6	$800,000
incl. Soft Matter and Polymer Physics	5	$400,000
incl. Computational and Nonlinear PhysicsPhysics	2	$400,000
Total	**13**	**$1,600,000**

FACULTY

Professor

Berland, Keith M., Ph.D., University of Illinois, 1995. Director of Undergraduate Studies. *Biophysics, Optics*. Experimental biophysics. Applications of fluorescence correlation spectroscopy (FCS) in biophysics, of high sensitivity fluorescence measurements in cellular biophysics. Investigation of the role of dynamic processes in biological function, and development and application of novel methods to quantify specific protein-protein interactions in living cells.

Family, Fereydoon, Ph.D., Clark University, 1974. *Biophysics, Computational Physics, Condensed Matter Physics, Nonlinear Dynamics and Complex Systems, Statistical & Thermal Physics*. Theoretical and Computational Studies of Nonequilibrium Phenomena. Development of mathematical models and computational techniques for studying nonequilibrium

phenomena, particularly in the areas of condensed matter physics and biology.

Finzi, Laura, Ph.D., University of New Mexico, 1990. *Biophysics*. Experimental biophysics. Single-molecule biophysics of transcriptional regulation. Molecular mechanisms of transcriptional regulation using single-molecule techniques, such as the tethered particle motion technique (TPM), magnetic tweezers (MT) and atomic force microscopy (AFM).

Hentschel, H. George E., Ph.D., University of Cambridge, 1978. *Biophysics, Condensed Matter Physics, Fluids, Rheology, Nonlinear Dynamics and Complex Systems, Statistical & Thermal Physics*. Nonequilibrium statistical mechanics, glasses, fracture, Laplacian growth, DLA, disordered systems, computional biophysics, morphogenesis, physics of cancer.

Warncke, Kurt, Ph.D., University of Pennsylvania, 1990. *Biophysics*. Experimental biophysics. Pulsed-EPR and optical studies of metallocenter- and radical-mediated enzyme and chemical catalyses, with a focus on time-resolve approaches.

Weeks, Eric, Ph.D., University of Texas, Austin, 1997. Department Chair. *Condensed Matter Physics, Fluids, Rheology, Nonlinear Dynamics and Complex Systems*. Experimental soft condensed matter. Microscopy of colloidal glasses, nonlinear dynamics, complex fluids, and granular media.

Associate Professor

Boettcher, Stefan, Ph.D., Washington University, 1993. Director of Graduate Studies. *Applied Mathematics, Computational Physics, Nonlinear Dynamics and Complex Systems, Statistical & Thermal Physics*. Statistical physics, critical phenomena, computational and theoretical problems in strongly disordered systems, with applications ranging from evolutionary dynamics and amorphous materials to combinatorial optimization.

Nemenman, Ilya, Ph.D., Princeton University, 2000. *Biophysics, Computational Physics, Neuroscience/Neuro Physics, Nonlinear Dynamics and Complex Systems, Statistical & Thermal Physics*. Theoretical and statistical physics. Theoretical biophysics, coarse-grained modeling in systems biology and neuroscience, information transduction in biological systems, learning and adaptation in molecular, neural, and evolutionary systems.

Roth, Connie, Ph.D., University of Guelph, 2004. *Condensed Matter Physics, Nano Science and Technology, Polymer Physics/Science*. Experimental soft condensed matter physics. Polymer materials, glass transition, physical aging, photophysics, miscibility and phase separation; Effect of nanoconfinement, surfaces and interfaces, external stresses, electric fields, and nanoparticles.

Urazhdin, Sergei, Ph.D., Michigan State University, 2002. *Condensed Matter Physics, Materials Science, Metallurgy, Nano Science and Technology, Solid State Physics*. Experimental condensed matter physics; spintronics, electronic and magnetic properties of surfaces and nanostructures, nonlinear dynamics in nanomagnetic systems, and strongly correlated materials.

Assistant Professor

Burton, Justin C., Ph.D., University of California, Irvine, 2006. *Condensed Matter Physics, Fluids, Rheology, Geophysics, Nonlinear Dynamics and Complex Systems*. Experimental condensed matter physics, soft matter physics, non-equilibrium physics, fluid mechanics, jamming and granular physics, geophysics.

Kim, Minsu, Ph.D., University of Illinois at Urbana-Champaign, 2008. *Biophysics, Nonlinear Dynamics and Complex Systems*. Experimental Biophysics. Advanced biophysical and conventional microbial techniques to characterize biological processes at the molecular and cellular levels. Interdisciplinary research at the intersection of physics, microbiology, synthetic biology, and theoretical biology.

Adjunct Professor

Malko, John A., Ph.D., Ohio University, 1970. Magnetic imaging.

Senior Lecturer

Williamon, Richard M., Ph.D., University of Florida, 1972. *Astronomy*. Astronomy.

Lecturer

Bing, Thomas, Ph.D., University of Maryland, 2007. *Physics and other Science Education*. Physics education.

Brody, Jed, Ph.D., Georgia Institute of Technology, 2003. Photovoltaics.

DEPARTMENTAL RESEARCH SPECIALTIES AND STAFF

Theoretical

Computational Physics. Computational biophysics; Simulations of disordered materials and growth processes; Multiscale simulations; Optimization. Boettcher, Family, Nemenman.

Nonlinear Dynamics and Complex Systems. Nonlinear dynamics; Complex Materials; Criticality, self-organization and emergence. Boettcher, Burton, Family, Hentschel, Weeks.

Statistical & Thermal Physics. Glassy and disordered systems; Nonlinear dynamics; Networks and information processing; Evolutionary processes; Nonequilibrium phenomena; Self-organization and emergence. Boettcher, Family, Hentschel, Nemenman.

Experimental

Biophysics. Molecular biophysics; Cellular biophysics; Theoretical and experimental systems and populations biophysics. We investigate how structure and dynamics at the molecular and cellular level contributes to the observed function of biological systems (proteins, nucleic acids, biological cells) and bio-inspired artificial systems. We probe spatial scales from bond lengths (Ångstroms, 10-10 m) to cells (micrometers, 10-6 m), and study dynamics on time scales corresponding to fast collisions of reactants (nanoseconds, 10-9 s) to intracellular fluxes (seconds or greater). Berland, Finzi, Kim, Nemenman, Warncke.

Condensed Matter Physics. Soft matter and fluids; Photonics, magnonics, and light-matter interactions; Spintronics, electronic and magnetic properties of surfaces and nanostructures, nonlinear dynamics in nanomagnetic systems, and strongly correlated materials. Burton, Roth, Urazhdin, Weeks.

Fluids, Rheology. Dynamics of fluid and granular flow; Friction; Self-organizion and pattern formation. Experimental and theoretical studies of far-from-equilibrium systems with complex interactions and emergent behavior over many different length and time scales. Burton, Hentschel, Weeks.

Materials for Energy. Bio-Inspired Solar Energy Conversion; Polymer materials; photophysics. Roth, Warncke.

Nano Science and Technology. Nanomaterials; Surfaces and interfaces; Polymers. Investigates new physical phenomena that emerge in nanoscale systems, at surfaces and interfaces of materials with different physical properties. The overarching goal is to develop fundamental understanding of the effects of confinement, interfaces, the resulting emerging interactions, and strongly nonequilibrium physical states that become possible to achieve only at nanoscale. Roth, Urazhdin.

View additional information about this department at
www.gradschoolshopper.com

GEORGIA INSTITUTE OF TECHNOLOGY

SCHOOL OF PHYSICS

Atlanta, Georgia 30332-0430
http://www.physics.gatech.edu

General University Information

President: G. P. Peterson
Dean of Graduate School: Rafael Bras
University website: http://www.gatech.edu
Control: Public
Setting: Urban
Total Faculty: 1,067
Total Graduate Faculty: 937
Total number of Students: 21,557
Total number of Graduate Students: 7,030

Department Information

Department Chairman: Paul Goldbart, Chair
Department Contact: James Sowell, Graduate Recruiter
  Total full-time faculty: 38
  Total number of full-time equivalent positions: 38

Full-Time Graduate Students: 142
First-Year Graduate Students: 22
Female First-Year Students: 1
Total Post Doctorates: 30

Department Address

837 State Street
Atlanta, GA 30332-0430
Phone: (404) 894-5200
Fax: (404) 894-9958
E-mail: jim.sowell@physics.gatech.edu
Website: http://www.physics.gatech.edu

ADMISSIONS

Admission Contact Information
Address admission inquiries to: Graduate Recruiter, School of
  Physics
Phone: (404) 385-1294
E-mail: jim.sowell@physics.gatech.edu
Admissions website: http://www.physics.gatech.edu

Application deadlines
Fall admission:
U.S. students: January 31   *Int'l. students*: January 31

Application fee
Int'l. students: $50

Admissions information
For Fall of 2013:
  Number of applicants: 197
  Number admitted: 79
  Number enrolled: 22

Admission requirements
Bachelor's degree requirements: Bachelor's degree in physics
  is preferred, with a minimum undergraduate GPA of 3.0 preferred for the M.S. program and 3.5 for the Ph.D. program.
Minimum undergraduate GPA: 3.5

GRE requirements
The GRE is required.
  Quantitative score: 155
  Verbal score: 153
  Analytical score: 3

Advanced GRE requirements
The Advanced GRE is required.
  Minimum accepted Advanced GRE score: 640

TOEFL requirements
The TOEFL exam is required for students from non-English-
  speaking countries.
  iBT score: 106

Other admissions information
Undergraduate preparation assumed: Classical Mechanics,
  Thornton & Marion; Electrodynamics, Griffiths; Quantum
  Mechanics, Griffiths; Thermal Physics, Schroeder.

TUITION

Tuition year 2013-14:
Tuition for in-state residents
  Full-time students: $13,000 annual
Tuition for out-of-state residents
  Full-time students: $29,000 annual
Credit hours per semester to be considered full-time: 12
Deferred tuition plan: No
Health insurance: Available
Other academic fees: Students employed as Graduate Research
  or Teaching Assistants pay $25 per semester tuition plus
  $1100 fee per semester.
Academic term: Semester
Number of first-year students who received full tuition waivers: 22

Teaching Assistants, Research Assistants, and Fellowships
Number of first-year
  Teaching Assistants: 21
  Research Assistants: 1
Average stipend per academic year
  Teaching Assistant: $24,060
  Research Assistant: $24,060

FINANCIAL AID

Loans
Loans are available for U.S. students.
Loans are available for international students.
GAPSFAS application required: No
FAFSA application required: Yes

HOUSING

Availability of on-campus housing
  Single students: Yes
  Married students: Yes

For further information
Address housing inquiries to: Housing Office.
Housing aid website: http://www.housing.gatech.edu

Table A—Faculty, Enrollments, and Degrees Granted

Research Specialty	2013-14 Faculty	Enrollment Fall 2013		Number of Degrees Granted 2012-13 (2008–13)		
		Master's	Doctorate	Master's	Terminal Master's	Doctorate
Applied Physics	2	–	–	–	–	–
Astrophysics	7	–	5	–	–	1(1)
Atomic, Molecular, & Optical Physics	8	–	4	–	–	2(13)
Biophysics	6	–	15	–	–	1(5)
Computational Physics	3	–	–	–	–	–(3)
Computer Science	2	–	–	–	–	–
Condensed Matter Physics	16	–	24	–	–	1(12)
Electrical Engineering	1	–	1	–	–	–(2)
Electromagnetism	1	–	–	–	–	–
Energy Sources & Environment	–	–	–	–	–	–
Fluids, Rheology	4	–	–	–	–	–
Low Temperature Physics	2	–	2	–	–	–(5)
Materials Science, Metallurgy	1	–	9	–	–	–(4)
Mechanics	1	–	–	–	–	–
Nano Science and Technology	16	–	12	–	–	1(9)
Nonlinear Dynamics and Complex Systems	6	–	8	–	–	1(6)
Nuclear Physics	–	–	–	–	–	–(2)
Optics	5	–	3	–	–	–(5)
Particles and Fields	3	–	–	–	–	–
Physics and other Science Education	1	–	–	–	–	–(2)
Quantum Optics	3	–	4	–	–	2(3)
Relativity & Gravitation	3	–	3	–	–	–
Statistical & Thermal Physics	9	–	–	–	–	–
Non-specialized	–	1	54	–(7)	2(6)	–
Other	1	–	–	–	–	–(2)
Total	101	1	144	–(7)	1(6)	8(74)
Full-time Grad. Stud.	–	1	144	–	–	–
First-year Grad. Stud.	–	–	22	–	–	–

GRADUATE DEGREE REQUIREMENTS

Master's: Thirty semester hrs. are required. Thesis is optional.
  2.7 GPA is required. One-year residency required. No language requirement.

Doctorate: The number of credit hours is not stipulated except 9 hrs. in minor with 2.9 GPA required. One year residency required. No comprehensive examination. Thesis and thesis examination are required.

Thesis: Thesis may be written in absentia.

SPECIAL EQUIPMENT, FACILITIES, OR PROGRAMS

Research programs are described at: http://www.physics.gatech.edu.

Table B—Separately Budgeted Research Expenditures by Source of Support

Source of Support	Departmental Research	Physics-related Research Outside Department
Federal government	$18,446,801	$632,328
State/local government		
Non-profit organizations		
Business and industry		
Other	$640,238	
Total	**$19,087,039**	**$632,328**

Table C—Separately Budgeted Research Expenditures by Research Specialty

Research Specialty	No. of Grants	Expenditures ($)
Applied Physics	–	
Astrophysics	11	$1,668,704
Atomic, Molecular, & Optical Physics	14	$1,857,078
Biophysics	11	$1,699,049
Chemical Physics	–	
Computer Science	–	
Condensed Matter Physics	23	$9,857,192
Electromagnetism	–	
Electrical Engineering	–	
Energy Sources & Environment	–	
Fluids, Rheology	–	
Low Temperature Physics	–	
Materials Science, Metallurgy	–	
Mechanics	–	
Nano Science and Technology	–	
Nuclear Physics	–	
Optics	5	$1,303,956
Particles and Fields	–	
Physics and other Science Education	1	$567,722
Relativity & Gravitation	–	
Statistical & Thermal Physics	–	
Nonlinear Dynamics and Complex Systems	11	$1,449,100
Other	5	$200,000
Total	**81**	**$18,602,801**

FACULTY

Chair Professor

Cvitanović, Predrag, Ph.D., Cornell University, 1973. Glen P. Robinson Chair in Nonlinear Sciences. *Particles and Fields, Statistical & Thermal Physics.* Nonlinear dynamics.

Landman, Uzi, Ph.D., Haifa, 1969. F.E. Callaway Chair in Computational Materials Science; Regents' and Institute Professor; Director, Center for Computational Materials Science. *Condensed Matter Physics, Nano Science and Technology.* Theoretical condensed matter physics; computational physics.

Professor

Bellissard, Jean, Ph.D., University of Provence, Marseille, 1974. *Condensed Matter Physics, Nano Science and Technology.* Mathematical physics.

Chapman, Michael S., Ph.D., Massachusetts Institute of Technology, 1995. *Atomic, Molecular, & Optical Physics, Optics.* Experimental quantum optics; atomic physics.

Chou, Mei-Yin, Ph.D., University of California, Berkeley, 1986. *Condensed Matter Physics, Nano Science and Technology.* Theoretical condensed matter physics; electronic structure of materials; computational materials physics.

Conrad, Edward H., Ph.D., University of Wisconsin-Madison, 1983. *Condensed Matter Physics, Nano Science and Technology.* Experimental surface physics.

de Heer, Walter A., Ph.D., University of California, 1984. Regents' Professor. *Condensed Matter Physics, Nano Science and Technology.* Experimental condensed matter physics; magnetic and electronic properties of clusters; carbon nanostructures.

First, Phillip, Ph.D., University of Illinois at Urbana-Champaign, 1988. *Condensed Matter Physics, Nano Science and Technology.* Experimental condensed matter physics.

Goldbart, Paul M., Ph.D., Imperial College, 1985. Dean - College of Sciences. *Condensed Matter Physics, Nano Science and Technology, Statistical & Thermal Physics.* Theoretical statistical and condensed matter physics.

Kennedy, T. A. Brian, Ph.D., Queen's Belfast University, 1986. Associate Chair for Undergraduate Program. *Atomic, Molecular, & Optical Physics, Optics.* Theoretical quantum optics.

Kuzmich, Alex, Ph.D., University of Rochester, 2000. *Atomic, Molecular, & Optical Physics, Optics.* Experimental atomic, molecular, and optical physics.

Laguna, Pablo, Ph.D., University of Texas, Austin, 1987. *Astrophysics, Computer Science, Relativity & Gravitation.* Numerical relativity.

Sá de Melo, Carlos, Ph.D., Stanford University, 1991. *Condensed Matter Physics, Nano Science and Technology.* Theoretical condensed matter physics.

Schatz, Michael F., Ph.D., University of Texas, 1991. *Fluids, Rheology, Statistical & Thermal Physics, Other.* Experimental nonlinear dynamics; fluid dynamics.

Trebino, Rick, Ph.D., Stanford University, 1983. *Applied Physics, Atomic, Molecular, & Optical Physics, Optics.* Ultrafast optics.

Uzer, Turgay, Ph.D., Harvard University, 1979. Regents' Professor. *Atomic, Molecular, & Optical Physics, Statistical & Thermal Physics.* Theoretical, molecular, and chemical physics; nonlinear dynamics.

Wiesenfeld, Kurt, Ph.D., University of California, Berkeley, 1985. *Biophysics, Statistical & Thermal Physics.* Theoretical nonlinear dynamics; biophysics.

Zangwill, Andrew, Ph.D., University of Pennsylvania, 1981. Associate Chair for Graduate Programs. *Condensed Matter Physics, Electromagnetism, Nano Science and Technology.* Theoretical condensed matter physics.

Associate Professor

Curtis, Jennifer E., Ph.D., University of Chicago, 2002. *Biophysics, Fluids, Rheology.* Experimental biophysics.

Davidovic, Dragomir, Ph.D., Johns Hopkins University, 1996. *Low Temperature Physics.* Mesoscopics and low-temperature physics.

Fenton, Flavio, Ph.D., Northeastern University, 1999. *Biophysics.* Non-linear dynamics, physics of the heart; computational biology, complex systems.

Goldman, Daniel I., Ph.D., University of Texas, Austin, 2002. *Biophysics, Fluids, Rheology, Mechanics, Statistical & Thermal Physics.* Experimental biophysics; nonlinear dynamics.

Grigoriev, Roman, Ph.D., California Institute of Technology, 1998. *Statistical & Thermal Physics.* Theoretical non-linear dynamics.

Kindermann, Markus, Ph.D., Universiteit Leiden, 2003. *Condensed Matter Physics, Nano Science and Technology*. Theoretical condensed matter physics.

Pustilnik, Michael, Ph.D., Bar-Ilan University, 1997. *Condensed Matter Physics, Nano Science and Technology, Statistical & Thermal Physics*. Theoretical condensed matter.

Raman, Chandra, Ph.D., University of Michigan, 1997. *Atomic, Molecular, & Optical Physics, Optics*. Experimental atomic physics.

Riedo, Elisa, Ph.D., University of Milan, 2000. *Biophysics, Condensed Matter Physics, Statistical & Thermal Physics*. Experimental condensed matter; biophysics.

Shoemaker, Deirdre, Ph.D., University of Texas, Austin, 1999. *Astrophysics, Computer Science, Relativity & Gravitation*. Numerical relativity.

Assistant Professor

Ballantyne, David R., Ph.D., University of Cambridge, 2002. *Astrophysics*. Theoretical astrophysics.

Bogdanovic, Tamara, Ph.D., Pennsylvania State University, 2006. *Astrophysics, Relativity & Gravitation*. Black hole astrophysics.

Fernandez de las Nieves, Alberto, Ph.D., University of Granada, 2000. Dunn Family Assistant Professor. *Condensed Matter Physics, Fluids, Rheology, Nano Science and Technology, Statistical & Thermal Physics*. Experimental condensed matter physics.

Gumbart, James, Ph.D., University of Illinois, 2009. *Biophysics*. Molecular dynamics, proteins, bacteria.

Jiang, Zhigang, Ph.D., Northwestern University, 2005. *Condensed Matter Physics, Low Temperature Physics, Nano Science and Technology*. Experimental condensed matter physics.

Kim, Harold, Ph.D., Stanford University, 2004. *Biophysics*. Experimental biophysics.

Otte, Nepomuk, Ph.D., Max Planck Institute, 2007. *Astrophysics, Particles and Fields*. Gamma-ray astrophysics.

Taboada, Ignacio, Ph.D., University of Pennsylvania, 2002. *Astrophysics, Particles and Fields*. Astrophysics.

Tan, Shina, Ph.D., University of Chicago, 2006. *Atomic, Molecular, & Optical Physics, Condensed Matter Physics, Nano Science and Technology*. Theoretical atomic and condensed matter physics.

Wise, John H., Ph.D., Stanford University, 2007. *Astrophysics, Computational Physics*. Early universe.

Professor Emeritus

Gole, James L., Ph.D., Rice University, 1971. *Applied Physics, Condensed Matter Physics, Materials Science, Metallurgy, Nano Science and Technology*. Experimental chemical and condensed matter physics; optics; material science.

Adjunct Faculty

Amini, Jason, Ph.D., University of California, Berkeley, 2006. Quantum Information Systems.

Brown, Kenton, Ph.D., University of Maryland, 2005. Quantum Information Systems.

Adjunct Professor

Bréchignac, Catherine, Ph.D., University of Paris-Sud, Orsay, 1977. *Atomic, Molecular, & Optical Physics, Condensed Matter Physics, Nano Science and Technology*. Molecular and cluster physics.

Brown, Kenneth, Ph.D., University of California, Berkeley, 2003. *Atomic, Molecular, & Optical Physics*. Atomic and molecular physics.

Harvey, Stephen C., Ph.D., Dartmouth College, 1971. *Biophysics*. Biophysics.

Kokkotas, Kostas, Ph.D., Thessalonike, 1988. *Astrophysics, Relativity & Gravitation*. Gravity wave astrophysics.

Orlando, Thomas, Ph.D., Stony Brook University, 1988. *Chemical Physics, Materials Science, Metallurgy*. Experimental physical, analytical and materials chemistry.

Wartell, Roger, Ph.D., University of Rochester, 1971. *Biophysics*. Experimental biophysics.

Weitz, Joshua, Ph.D., Massachusetts Institute of Technology, 2003. *Biophysics*. Theoretical biophysics.

Zhu, Cheng, Ph.D., Columbia University, 1988. *Biophysics*. Biophysics.

Adjunct Assistant Professor

Hu, David, Ph.D., Massachusetts Institute of Technology, 2005. Fluid mechanics.

Senior Research Scientist

Barnett, Robert N., Ph.D., University of Kansas, 1980. *Condensed Matter Physics, Nano Science and Technology*. Theoretical condensed matter physics.

Gao, Jianping, Ph.D., Brown University, 1989. *Condensed Matter Physics, Nano Science and Technology*. Theoretical condensed matter physics.

Luedtke, William D., Ph.D., Georgia Institute of Technology, 1984. *Condensed Matter Physics, Nano Science and Technology*. Theoretical condensed matter physics.

Yannouleas, C., Ph.D., University of Maryland, 1982. *Condensed Matter Physics, Nano Science and Technology, Nuclear Physics*. Theoretical condensed matter physics; theoretical nuclear physics.

Research Scientist

Berger, Claire, Ph.D., University Joseph Fourier, Grenoble, 1987. *Condensed Matter Physics, Nano Science and Technology*. Experimental condensed matter physics.

Ruan, Wen-Ying, Ph.D., Zhongshan University, 1992. *Condensed Matter Physics, Nano Science and Technology*. Theoretical condensed matter.

Yoon, Bokwon, Ph.D., University of Paris-Sud, Orsay, 1997. *Condensed Matter Physics, Nano Science and Technology*. Theoretical condensed matter physics.

Senior Academic Professional

Scherbakov, Andrew, Ph.D., Georgia Institute of Technology, 1997. *Condensed Matter Physics, Other*. Mesoscopic physics.

Sowell, James, Ph.D., University of Michigan, 1986. *Astronomy, Other*. Astronomy.

Academic Professional

Greco, Edwin, Ph.D., Georgia Institute of Technology, 2008. *Nonlinear Dynamics and Complex Systems, Physics and other Science Education, Other*. Physics Education.

Jarrio, Marty, Ph.D., Georgia Institute of Technology, 1996. *Nuclear Physics, Other*. Nuclear physics.

Murray, Eric, Ph.D., Cornell University, 1992. *Materials Science, Metallurgy, Other*. Materials science.

DEPARTMENTAL RESEARCH SPECIALTIES AND STAFF

Theoretical

Astrophysics. General relativity; gravitational wave patterns; gravitational interactions of compact binaries; theoretical and phenomenological astrophysics; galaxy and black hole evolution; high-energy particle astrophysics; accretion disks; numerical relativity; cosmology; gravitating systems; black holes; galaxy and black hole evolution; high-energy particle astrophysics; accretion disks; gravitational physics. Ballantyne, Bogdanovic, Kokkotas, Laguna, Shoemaker, Wise.

Atomic, Molecular, & Optical Physics. Three-body recombination; anti-hydrogen formation; cold collisions; collisional Stark mixing; Rydberg plasmas; classical-quantal correspondences; atomic Fermi gas transport; optical lattices; spin squeezing of atomic ensembles; Bose-Einstein condensate mixtures; quantum fluctuations; spatial solitary waves; nonlinear optical parametric processes; Rydberg atoms; light/matter interactions. Bréchignac, Kenneth Brown, Chapman, Kennedy, Kuzmich, Raman, Tan, Trebino, Uzer.

Biophysics. Energy transduction; chemosmosis; noise; protein biosynthesis; energy metabolism; ion channel fluctuations; molecular motors; Hodgkin-Huxley equations; chemo-mechanical energy conversion; energy driven rectification of Brownian motion; quantum mutations in DNA. Curtis, Goldman, Gumbart, Harvey, Kim, Riedo, Wartell, Weitz, Wiesenfeld, Zhu.

Computational Physics. Spatially extended non-equilibrium systems; chaotic mixing in fluids; thin liquid films; dynamics of solid surfaces; epitaxial growth processes; solid-liquid interfaces; melting; glasses; surface diffusion; atomic-scale friction and lubrication; confined complex fluids; electron localization; dynamics of small clusters; kinetic Monte Carlo and molecular dynamics simulations; density functional theory; quantum Monte-Carlo techniques; first-principles electronic structure; Landau-Lifshitz-Gilbert simulations; numerical relativity. Barnett, Fenton, Gumbart, Laguna, Landman, Shoemaker, Yannouleas.

Condensed Matter Physics. Nanoscience; phase transitions; mesoscopic physics; quantum interference effects; superconductors in high magnetic fields; Bose-Einstein superconductivity; macroscopic quantum phenomena; ferroelectrics; Sutherland-Calogero models; ferromagnets; spintronics; semiconductor quantum dots. Barnett, Bellissard, Berger, Bréchignac, Chou, Conrad, de Heer, Fernandez de las Nieves, First, Gao, Goldbart, Gole, Jiang, Kindermann, Landman, Luedtke, Pustilnik, Riedo, Ruan, Sá de Melo, Scherbakov, Tan, Yannouleas, Yoon, Zangwill.

Nonlinear Dynamics and Complex Systems. Molecular fluctuations; chaotic dynamics; quantum chaos; Husimi-Wigner wave packets; Lyapunov exponent; Rydberg states; trajectory analysis; massively coupled oscillators; chemical reaction dynamics; Hamiltonian flows. Bellissard, Cvitanović, Grigoriev, Schatz, Uzer, Wiesenfeld.

Optics. Classical and Quantum. Quantum optics; atomic Fermi gas transport in optical lattices; nonlinear optics and lasers; radiative interactions; squeezed states, quantum computing, cavity QED. Chapman, Kennedy, Kuzmich, Raman, Trebino.

Physics and other Science Education. Matter and Interactions curriculum. Schatz.

Experimental

Astrophysics. Neutrino and gamma-ray astrophysics. Otte, Taboada.

Atomic, Molecular, & Optical Physics. Fundamental properties of ultra-cold condensed gases; atom trapping; multi-atom entanglement; cavity QED; laser Raman and Brillouin scattering; chemical biosensors; photovoltaic devices; quantum memory; ultrafast optics. Kenneth Brown, Chapman, Kuzmich, Raman, Trebino.

Biophysics. Morphogenesis, noise; "g-jitter" thin organic films; nanotribology; gene expression; biomechanics. Curtis, Goldman, Gumbart, Kim, Riedo, Zhu.

Condensed Matter Physics. Nanoscience; soft matter; scanning tunneling microscopy; high-resolution x-ray scattering; magnetic heterostructures; graphene; Josephson tunneling; molecular clusters; thin-film magnetism; semiconductor nanostructures; atomic force microscopes; friction; nanowires; spintronics; liquid crystals; colloids. Conrad, de Heer, Fernandez de las Nieves, First, Gole, Jiang, Riedo.

Nonlinear Dynamics and Complex Systems. Spatiotemporal chaos; control/exploitation of chaos; pattern formation in fluids; weather-in-a-box; spontaneous and manipulated patterns; fluid instabilities; coupled mechanical oscillators; granular matter. Fenton, Goldman, Schatz.

Optics. Bose and Fermi condensed gases; atom and ion trapping; atom optics; multi-atom entanglement; quantum information; cavity QED; ultrafast optics; frequency-resolved optical gating (FROG); ultrashort laser pulses. Kenneth Brown, Chapman, Kuzmich, Raman, Trebino.

View additional information about this department at www.gradschoolshopper.com

---

GEORGIA STATE UNIVERSITY

DEPARTMENT OF PHYSICS AND ASTRONOMY

Atlanta, Georgia 30303
http://www.phy-astr.gsu.edu

General University Information

President: Mark P. Becker
Dean of Graduate School: MaryAnn Romski, Associate Dean for Research and Graduate Studies
University website: http://www.gsu.edu
Control: Public
Setting: Urban
Total Faculty: 1,886

Total number of Students: 32,009
Total number of Graduate Students: 7,919

Department Information

Department Chairman: D. Michael Crenshaw, Chair
Department Contact: Dr. D. Michael Crenshaw, Chair
Total full-time faculty: 21
Total number of full-time equivalent positions: 21

Full-Time Graduate Students: 57
First-Year Graduate Students: 8
Female First-Year Students: 3
Total Post Doctorates: 7

Department Address
25 Park Place
Suite 605
Atlanta, GA 30303
Phone: (404) 413-6033
Fax: (404) 413-5481
E-mail: dcrenshaw@gsu.edu
Website: http://www.phy-astr.gsu.edu

ADMISSIONS

Admission Contact Information
Address admission inquiries to: Dr. Xioachun He, Director of
Physics Graduate Program, xhe@gsu.edu, Dr. Todd Henry,
Director of Astronomy Graduate Program, thenry@chara.
gsu.edu
Phone: 404-413-6033
E-mail: xhe@gsu.edu
Admissions website: http://www.cas.gsu.edu/grad_services.html

Application deadlines
Fall admission:
U.S. students: July 1   *Int'l. students*: February 15

Application fee
U.S. students: $50   *Int'l. students*: $50
Deadlines depend on program (astronomy or physics). See http://
www.cas.gsu.edu/grad_services.html .

Admissions information
For Fall of 2013:
Number of applicants: 80
Number admitted: 16
Number enrolled: 8

Admission requirements
Bachelor's degree requirements: B.S. in Physics or related field.

GRE requirements
The GRE is required.
Quantitative score: 144
Verbal score: 144

Advanced GRE requirements
The Advanced GRE is required.

TOEFL requirements
The TOEFL exam is required for students from non-English-
speaking countries.
iBT score: 80

Other admissions information
Additional requirements: The GRE general test is required for
admission, and the GRE subject test is required for assis-
tantships.
Admission is based on the applicant's undergraduate record,
GRE scores and recommendations.
Georgia State University, a unit of the University System of
Georgia, is an equal opportunity/affirmative action educa-
tional institution.
Undergraduate preparation assumed: Eisberg and Resnick,
Quantum Physics: Griffiths, Introduction to Electrodynamics;
Mand1, Statistical Physics; Kreyszig, Advanced Engineering
Mathematics; Fowles, Analytic Mechanics.

TUITION
Tuition year 2013-14:
Tuition for in-state residents
Full-time students: $4,032 per semester
Part-time students: $336 per credit
Tuition for out-of-state residents
Full-time students: $14,400 per semester
Part-time students: $1,200 per credit
Tuition waived for students with assistantships or fellowships.
Credit hours per semester to be considered full-time: 12
Deferred tuition plan: No
Health insurance: Available at the cost of $366 per year.
Other academic fees: $1,064.00 per semester (not covered by
assistantship)
Academic term: Semester
Number of first-year students who received full tuition waivers: 8

Teaching Assistants, Research Assistants, and Fellowships
Number of first-year
Teaching Assistants: 7
Fellowship students: 1
Average stipend per academic year
Teaching Assistant: $20,000
Fellowship student: $22,000

FINANCIAL AID

Application deadlines
Fall admission:
U.S. students: July 1

Loans
Loans are available for U.S. students.
Loans are not available for international students.
GAPSFAS application required: No
FAFSA application required: Yes

For further information
Address financial aid inquiries to: Dr. X. He, Director of Physics
Graduate Program, or Dr. T. Henry, Director of Astronomy
Graduate Program, Department of Physics and Astronomy.
E-mail: xhe@gsu.edu
Financial aid website: http://www.gsu.edu/financialaid/

HOUSING

Availability of on-campus housing
Single students: Yes
Married students: Yes

For further information
Housing aid website: http://www.gsu.edu/housing/

Table A—Faculty, Enrollments, and Degrees Granted

Research Specialty	2013 Faculty	Enrollment Fall 2012		Number of Degrees Granted 2012-13		
		Master's	Doctorate	Master's	Terminal Master's	Doctorate
Astronomy	8	–	23	–	2(-)	4(-)
Atomic, Molecular, & Optical Physics	1	1	2	–	–	–
Biophysics	1	–	2	–	–	–
Condensed Matter Physics	6	1	18	–	–	2(-)
Neuroscience/Neuro Physics	1	–	5	–	–	–
Nuclear Physics	2	–	5	–	–	2(-)
Physics and other Science Education	2	1	2	–	–	–
Total	21	3	57	–	2(-)	8(-)
Full-time Grad. Stud.	–	3	57	–	–	–
First-year Grad. Stud.	–	–	8	–	–	–

GRADUATE DEGREE REQUIREMENTS

Master's: M.S. students must complete a minimum of 24 semester hours of course work, pass a comprehensive exam, and pass a foreign language exam. An alternate research skill, such as computer programming, may be substituted for the foreign language exam. M.S. students must either complete an acceptable thesis or complete 6 additional hours of course work.

Doctorate: The Ph.D. degrees each require a minimum of 71 semester hours (beyond the B.S.). Students must complete and defend an acceptable dissertation in physics or astronomy. Written qualifying and oral preliminary exams and reading exams in a foreign language are also required. Alternate research skills such as computer programming may be substituted for the foreign language exams.

Thesis: Thesis may not be written in absentia.

SPECIAL EQUIPMENT, FACILITIES, OR PROGRAMS

Research in the department currently is supported by the National Science Foundation, the Department of Energy, the National Institutes of Health, the National Aeronautics and Space Administration, the U.S. Army, and the U.S. Airforce.

Research apparatus within the department includes the following: X band and K band EPR/ENDOR spectrometers, x-ray diffraction apparatus, UV, X, and gamma irradiation facilities, a wide assortment of CAMAC and NIM modules, positron annihilation lifetime, closed-cycle refrigerator, Ge detector, CO2 Nd:YAB lasers, monochromators, a 32-channel logic analyzer with a dual channel digital scope, al-mil wire bonder, three FTIR spectrometers one for TR spec down to 10 ns, a steady-state and time-resolved photoluminescence spectrometer, a Raman spectrometers, a high-resolution absorption spectrometer, a high-power frequency agile laser system (200 nm–3000 nm) utilized in a variety of gas phase and materials characterization techniques, ultra high vacuum surface science apparatus including high-resolution electron energy loss spectrometer, Auger electron spectrometer, and low-energy electron diffraction apparatus.

Nuclear researchers utilize the particle accelerators at FermiLab and Brookhaven National Laboratories.

The Center for High Angular Resolution Astronomy (CHARA) research uses high-spatial resolution imaging techniques to attain image detail beyond that normally obtained with large telescopes. CHARA's major project is the CHARA Array, an optical and infrared interferometer at Mount Wilson Observatory in California. The CHARA Array consists of six telescopes of 1 m aperture that form a Y-shaped figure with baselines from 30 to 330 m. It is currently the premier instrument of its kind in the world for high angular resolution of stars and their environments.

GSU astronomers have guaranteed observing time with the Small and Medium Aperture Research Telescope System at the Cerro Tololo Interamerican Observatory in Chile, and the 1.8-m and smaller telescopes on Lowell Observatory's Anderson Mesa near Flagstaff, AZ.

GSU's Hard Labor Creek Observatory (HLCO) is located in a state park 80 kilometers east of Atlanta. There is an electronics shop and living quarters for observers. The principal telescopes at HLCO are new, research-quality 20-inch and 24-inch telescopes.

Research facilities on campus include a network of unix-based workstations used extensively for data analysis and image processing.

Table B—Separately Budgeted Research Expenditures by Source of Support

Source of Support	Departmental Research	Physics-related Research Outside Department
Federal government	$2,768,000	
State/local government		
Non-profit organizations		
Business and industry		
Other	$52,000	
Total	$2,820,000	

Table C—Separately Budgeted Research Expenditures by Research Specialty

Research Specialty	No. of Grants	Expenditures ($)
Astronomy	9	$1,308,000
Atomic, Molecular, & Optical Physics	3	$210,000
Biophysics	1	$300,000
Condensed Matter Physics	4	$850,000
Nuclear Physics	1	$152,000
Total	18	$2,820,000

FACULTY

Professor

Crenshaw, D. Michael, Ph.D., Ohio State University, 1985. Supermassive black holes, active galactic nuclei.

Dietz, Nikolaus, Ph.D., Technical University, Berlin, 1991.

Gies, Douglas R., Ph.D., University of Toronto, 1985.

He, Xiaochun, Ph.D., University of Tennessee, 1991.

Henry, Todd J., Ph.D., University of Arizona, 1991.

Manson, Steven T., Ph.D., Columbia University, 1966.

Perera, A. G. Unil, Ph.D., University of Pittsburgh, 1987.

Stockman, Mark I., Ph.D., Novosibirsk, 1989. Theoretical physics; electronic and optical properties of disordered systems; kinetic and non-linear optical effects in semiconductors.

Associate Professor

Apalkov, Vadym M., Ph.D., University of Utah, 1995. Condensed Matter Theory; disordered electronic systems; photonic crystals; quantum cascade lasers.

Hastings, Gary, Ph.D., Imperial College, London (U.K.), 1992. Experimental biophysics; static and time resolved infrared spectroscopies applied to biological systems; experimental physics; energy and electron transfer in photosynthetic organisms.

Lepine, Sebastien, Ph.D., University of Montreal, 1998. All-sky surveys, low-mass stars and brown dwarfs, galactic structure, extra-solar planets.

Mani, Ramesh, Ph.D., University of Maryland, 1990. Experimental condensed matter.

Thoms, Brian D., Ph.D., Cornell University, 1992. Experimental physics; surface science; wide band gap semiconductors; film growth mechanisms.

Assistant Professor

Baron, Fabien, Ph.D., University of Paris, 2005. *Astronomy, Astrophysics.* Stellar astrophysics, interferometry.

Bentz, Misty C., Ph.D., Ohio State University, 2007. Active galaxies and quasars; Black hole mass.

Dhamala, Mukesh, Ph.D., University of Kansas, 2000. Experimental biophysics; neuroimaging.

Kozhanov, Alexander, Ph.D., Moscow State University, 2006.

Kuzio de Naray, Rachel, Ph.D., University of Maryland, 2007. *Astronomy, Astrophysics.* Dark matter, low surface brightness galaxies.

Sarsour, Murad, Ph.D., University of Houston, 2002. Experimental nuclear physics.

Von Korff, Josh, Ph.D., University of California, Berkeley, 2010.

White, Russel, Ph.D., University of California, Los Angeles, 1999. Stellar astronomy.

Emeritus

McAlister, Harold A., Ph.D., University of Virginia, 1975. Observational astronomy; binary stars; speckle and long-baseline interferometry.

Miller, H. Richard, Ph.D., University of Florida, 1970. High-energy astrophysics, variability of AGN.

Nave, Carl R., Ph.D., Georgia Institute of Technology, 1966. Experimental physics; molecular structure; magnetic resonance; acoustics.

Wingert, David W., Ph.D., Princeton University, 1974. Observational astronomy; theoretical astrophysics.

Adjunct Professor

Francombe, Maurice H., Ph.D., University of London, 1958. Experimental physics; semiconductor thin films; magnetic alloys; ferroelectric oxides.

Liu, Hui C., Ph.D., University of Pittsburgh, 1987. Experimental physics; quantum transport in reduced dimensional semiconductor structures; high-frequency device applications.

Ridgway, Steven, Ph.D., Stony Brook University, 1972. Experimental infrared astronomy; high-resolution Fourier transform spectroscopy; imaging.

Tennakone, Kirti, Ph.D., University of Hawaii, 1972. Experimental physics; dye sensitized semiconductor nanostructure; solar cells.

Lecturer

Doluweera, Sumith, Ph.D., University of Cincinnati, 2008.

Evans, J., Ph.D., Georgia State University, 1998.

McGimsey, Ben, Ph.D., University of Florida, 1980.

Wang, Ruli, Ph.D., Georgia State University, 2005.

Research Scientist

Sturmann, Judit, Ph.D., Vanderbilt University, 1999. Astronomical instrumentation, optical long baseline interferometry.

Sturmann, Laszlo, Ph.D., Vanderbilt University, 1997. Astronomical instrumentation, optical long baseline interferometry.

ten Brummelaar, Theo A., Ph.D., University of Sydney, 1993. Associate Director of CHARA. Optical propagation in a turbulent atmosphere; long baseline optical stellar interferometry; observational astronomy.

Turner, Nils H., Ph.D., Georgia State University, 1998. Astronomical instrumentation, optical long baseline interferometry.

Academic Professional

Wilson, John, Ph.D., Georgia State University, 2004.

DEPARTMENTAL RESEARCH SPECIALTIES AND STAFF

Theoretical

Atomic, Molecular, & Optical Physics. Atomic and molecular structure and collisions. Manson.

Condensed Matter Physics. Electronic and optical properties of disordered systems; kinetic, electrical and transport phenomena in semiconductor devices, and optical effects in semiconductors, nano-plasmonics, nano-optics. Apalkov, Stockman.

Experimental

Astronomy. Active galaxies; Quasi-stellar objects; supermassive black holes; stellar populations, stellar masses, exoplanets, astrobiology, astroinformatics, hot and cool stars, variable stars; star formation; binary stars; star formation; young stars, Be stars, long baseline interferometry. Baron, Bentz, Crenshaw, Gies, Henry, Kuzio de Naray, McAlister, McGimsey, Miller, Ridgway, Judit Sturmann, Laszlo Sturmann, ten Brummelaar, Turner, White, Wilson.

Biophysics. Solar energy conversion in natural and artificial systems. Photosynthetic protein complexes. Hastings.

Condensed Matter Physics. Defects in solids, acoustical, linear and nonlinear optical, electrical and thermal properties of semiconductors; growth of III-N (high-pressure CVD) and waveguided, birefringent heterostructures (OMCVD; real-time growth diagnostics; optoelectronic semiconductor device applications. Dietz, Kozhanov, Mani, Perera, Thoms.

Neuroscience/Neuro Physics. Neuroimaging of the human brain. Dhamala.

Nuclear Physics. Studies of quark-gluon plasma in heavy-ion collisions at Brookhaven National Labs, proton spin structure. He, Sarsour.

Physics and other Science Education. Doluweera, Evans, Thoms, Von Korff, Wang, Wilson.

View additional information about this department at
www.gradschoolshopper.com

BOISE STATE UNIVERSITY

MATERIALS SCIENCE AND ENGINEERING

Boise, Idaho 83725-2090
http://coen.boisestate.edu/mse/

General University Information
President: Bob Kustra
Dean of Graduate School: Jack Pelton
University website: http://boisestate.edu
Control: Public
Setting: Urban
Total Faculty: 643
Total Graduate Faculty: 360
Total number of Students: 19,657
Total number of Graduate Students: 3,021

Department Information
Department Chairman: Peter Mullner, Chair
Department Contact: Dena Ross, Department Manager
  Total full-time faculty: 21
  Total number of full-time equivalent positions: 14
  Full-Time Graduate Students: 15
  First-Year Graduate Students: 13
  Female First-Year Students: 3
  Total Post Doctorates: 1

Department Address
1910 University Drive
Boise, ID 83725-2090
Phone: (208) 426-5600
Fax: (208) 426-4466
E-mail: materials@boisestate.edu
Website: http://coen.boisestate.edu/mse/

ADMISSIONS

Admission Contact Information
Address admission inquiries to: Chad Watson, Director for Academic and Technical Advancement.
Phone: (208) 426-4897
E-mail: chadwatson1@boisestate.edu
Admissions website: http://coen.boisestate.edu/mse/degreeprograms/

Application deadlines
Fall admission:
U.S. students: February 28   *Int'l. students*: February 28

Application fee
U.S. students: $55   *Int'l. students*: $85
visit us at coen/boisestate.edu/mse for more information.

Admissions information
For Fall of 2013:
  Number of applicants: 36
  Number admitted: 18
  Number enrolled: 15

Admission requirements
Bachelor's degree requirements: Bachelor of Science in Materials Science and Engineering or related field.
Minimum undergraduate GPA: 3.0

GRE requirements
The GRE is required.

The Boise State Materials Science and Engineering Department does not have a minimum GRE score for admission.

Advanced GRE requirements
The Advanced GRE is not required.

TOEFL requirements
The TOEFL exam is required for students from non-English-speaking countries.
  PBT score: 587
  iBT score: 95
Or IELTS score of 6.5

Other admissions information
Additional requirements: The graduate program application process includes an application form, an application fee, official transcripts, GRE scores, TOEFL scores (if necessary) a resume, a statement of purpose, and three letters of recommendation.
Undergraduate preparation assumed: Materials Science and Engineering is an interdisciplinary program and we accept applicants from a variety of engineering and science backgrounds. If you have a B.S. degree in a materials science and engineering-related field (materials, metallurgy, ceramics, or polymer engineering), other engineering field (mechanical, electrical, chemical, industrial, civil, aerospace, biological, environmental, computer, or nuclear), the physical sciences (physics, earth science, or chemistry), the life sciences, or mathematics, you should apply.

TUITION

Tuition year 2012–13:
Tuition for in-state residents
  Full-time students: $6,972 annual
  Part-time students: $312 per credit
Tuition for out-of-state residents
  Full-time students: $18,412 annual
Students accepted into the Graduate Program receive a tuition and fee waiver, research assistantship, and a generous salary.
Credit hours per semester to be considered full-time: 9
Deferred tuition plan: Yes
Health insurance: Available at the cost of $2,120 per year.
Other academic fees: The above health insurance fees for the graduate student are fully supported with the tuition and fee waiver.
Academic term: Semester
Number of first-year students who received full tuition waivers: 19

Teaching Assistants, Research Assistants, and Fellowships
Number of first-year
  Teaching Assistants: 3
  Research Assistants: 10
Average stipend per academic year
  Teaching Assistant: $26,000
  Research Assistant: $26,000

FINANCIAL AID

Application deadlines
Fall admission:
U.S. students: March 15

Loans
Loans are available for U.S. students.
Loans are available for international students.
GAPSFAS application required: No
FAFSA application required: Yes

For further information
Address financial aid inquiries to: Financial Aid Office.
Phone: (208) 426-1664
E-mail: faquest@boisestate.edu
Financial aid website: http://financialaid.boisestate.edu/

HOUSING

Availability of on-campus housing
Single students: Yes
Married students: Yes

For further information
Address housing inquiries to: University Housing Office.
Phone: (208) 447-1001
E-mail: bsuhousing@boisesetate.edu
Housing aid website: http://housing.boisestate.edu/

Table A—Faculty, Enrollments, and Degrees Granted

Research Specialty	2013-14 Faculty	Enrollment Spring 2013		Number of Degrees Granted Fall 2012		
		Master's	Doctorate	Master's	Terminal Master's	Doctorate
Materials Science and Engineering	21	21	18	5(-)	–	–
Total	21	21	18	5(-)	–	–
Full-time Grad. Stud.	–	21	18	–	–	–
First-year Grad. Stud.	–	5	10	–	–	–

GRADUATE DEGREE REQUIREMENTS

Master's: Please visit http://coen.boisestate.edu/mse/degree programs/master-of-science-program/m-s-degree-requirements/.

Doctorate: Please visit http://coen.boisestate.edu/mse/degree programs/phd-program/ph-d-in-materials-science-and-engineering-degree-requirements/. Master of Engineering, Materials Science and Engineering. Please visit http://coen. boisestate.edu/mse/degreeprograms/master-of-engineering-program/m-engr-degree-requirements/.

Thesis: Thesis is required for the Master of Science Degree. Dissertation is required for the Doctoral Degree. Comprehensive exam is required for the Master of Engineering Degree.

SPECIAL EQUIPMENT, FACILITIES, OR PROGRAMS

Electrochemical Energy Laboratory

Functional Ceramics Laboratory

Device Characterization Laboratory

NanoEngineering Laboratory

Integrated Nanomaterials Laboratory

Surface Science Laboratory

Nanophotonics Laboratory

Idaho Microfabrication Laboratory

Boise State Center for Materials Characterization

Advanced Materials Laboratory

Magnetic Materials Laboratory

Radiation Materials Science Laboratory

Materials Theory and Modeling Laboratory

We collaborate very closely with the Idaho National Laboratory and the Center for Advanced Energy Studies.

FACULTY

Professor

Butt, Darryl, Ph.D., Pennsylvania State University, 1991. Department Chair, Materials Science and Engineering Associate Director for the Center of Advanced Energy Studies. *Energy Sources & Environment, Materials Science, Metallurgy, Nuclear Engineering, Polymer Physics/Science, Other*. Ceramics, graphite, and high-temperature materials; Synthesis of powders and novel structures; Modeling and measurement of thermodynamics and kinetics; Materials processing and structure-property relations; Nuclear fuels and materials; Corrosion and materials durability; Failure analysis, brittle fracture and failure; Novel methods of joining; Environmentally friendly coatings; Carbon dioxide sequestration; Ion transport membranes.

Callahan, Janet, Ph.D., University of Connecticut at Storrs, 1990. Associate Dean, College of Engineering. *Materials Science, Metallurgy, Other*. Surface modification of materials, oxidation, engineering education.

Knowlton, Bill, Ph.D., University of California, Berkeley, 1998. Graduate Program Co-Coordinator. *Materials Science, Metallurgy, Nano Science and Technology, Other*. DNA nanotechnology; Nanophotonics; Ni-Mn-Ga magnetic shape memory alloys.

Moll, Amy, Ph.D., University of California, Berkeley, 1994. Interim Dean, College of Engineering. *Materials Science, Metallurgy, Other*. Materials Science. Microelectronic materials and packaging; Ceramic microfluidic and micro analytical systems, engineering education.

Mullner, Peter, Ph.D., ETH Zurich, 1994. *Materials Science, Metallurgy, Other*. Materials Science, Physical Metallurgy. Formation and characterization of microstructures, including phase transformations, defect characterization and modeling defect interaction, emphasis on magnetic shape-memory alloys.

Associate Professor

Frary, Megan, Ph.D., Massachusetts Institute of Technology, 2005. Undergraduate Academic Advisor. *Materials Science, Metallurgy, Other*. Research uses both experiments and computational modeling to study the relationship between microstructure and properties. Non-uniform properties associated with surfaces of different orientations or interfaces with different atomic structures. Current work is focused on the following projects with the goal of better understanding the processing-structure-properties relationship in metals and ceramics: (1) The role of grain boundary character on dynamic recrystallization. (2) Microstructural evolution during spark plasma sintering of metals, ceramic-metal composites.

Hughes, Will, Ph.D., Georgia Institute of Technology, 2006. *Materials Science, Metallurgy, Nano Science and Technology, Other*. Biomedicine; DNA nanotechnology; Synthetic biology; Engineering education.

Ubic, Rick, Ph.D., University of Sheffield, 1998. *Materials Science, Metallurgy, Other*. Structure-property relationships in functional ceramics, including microwave dielectrics, ferroelectrics, and ionic conductors; point defects; electron microscopy; and nuclear graphite.

Assistant Professor

Estrada, David, Ph.D., University of Illinois at Urbana-Champaign, 2013. *Nano Science and Technology, Other.* Nanoscale materials synthesis and characterization. Fundamental studies of thermal and electrical transport in nanoscale devices and materials. Applications of nanotechnology in energy harvesting/conversion and bioengineering. Environmental impact of engineered nanomaterials.

Graugnard, Elton, Ph.D., Purdue University, 2000. *Materials Science, Metallurgy, Nano Science and Technology, Other.* Nanoscale Science and Technology. Nucleic acid based nanoscale devices and molecular circuitry, Self-assembly, and Atomic Layer Deposition.

Li, Lan, Ph.D., University of Cambridge, 2006. *Materials Science, Metallurgy, Other.* Theoretical and Computational Materials Science. Develop and apply computer-based theoretical methodologies to capture structure-property-performance relationships and design materials with optimum properties and desired behavior. Current work focuses on (1) Development of materials-by-design for energy and gas separation applications, such as thermoelectrics, solar cells and carbon capture; (2) Fundamental studies of magnetism, transport and quantum phenomena in nanoscaled materials; (3) Development of multi-scale modeling techniques.

Wharry, Janelle, Ph.D., University of Michigan, 2012. *Materials Science, Metallurgy, Nuclear Engineering, Other.* Materials for current and next-generation nuclear power applications with focus on a mechanistic understanding the effects of irradiation on materials.

Xiong, Hui (Claire), Ph.D., Univeresity of Pittsburgh, 2007. *Materials Science, Metallurgy, Other.* Materials Science, Electrochemistry. The research is focused on the interdisciplinary areas of electrochemistry, materials science, and engineering, and surface chemistry. Current work is focused on the synthesis and characterization of new nano-architectured electrode materials for energy storage and conversion to a better understanding of the structure-property relationships for the development of advanced functional nanomaterials: (1) Electrochemical synthesis of nanostructured electrode materials with controlled size and shape; (2) Development of electrode materials for alternative battery.

Research Associate Professor

Allahar, Kerry, Ph.D., University of Florida, 2003. *Materials Science, Metallurgy.* Chemical engineering.

Wu, Yaqiao, Ph.D., Chinese Academy of Sciences, 2000. *Materials Science, Metallurgy, Other.* Materials design, synthesis, property analysis, nanoscale structure and chemistry characterization, establishing connections between structure, chemistry and property/behavior through combinational transmission electron microscopy (TEM) and Atom probe tomography(APT) techniques.

Research Assistant Professor

Hurley, Mike, Ph.D., University of Virginia, Charlottesville, 2007. *Materials Science, Metallurgy, Other.* Environmental degradation of materials through applied electrochemistry; Understanding and predicting corrosion in various complex applications and environments; Corrosion testing, materials characterization and design, predictive modeling, and sensor development to improve performance or enable prognostic health monitoring of structures and/or materials.

Lindquist, Paul, Ph.D., University of Illinois, 1988. *Materials Science, Metallurgy, Other.* Ni-Mn-Ga magnetic shape memory alloys.

Youngsman, John, Ph.D., Washington State University, 2009. *Materials Science, Metallurgy, Other.* Materials and structures for renewable energy and sensing.

Assistant Professor Adjoint

Lee, Jeunghoon, Ph.D., University of Connecticut, 2005. *Nano Science and Technology, Other.* Bottom-up synthesis and top-down fabrication of inorganic nanoparticle building blocks; assembly and manipulation of the nanoscale building blocks into functional structures using functional organic molecules and polymers; characterization and analysis of the nanostructures.

Affiliate Professor

Hanna, Charles, Ph.D., Stanford University. Department Chair, Physics. *Physics and other Science Education.* Interactions in quantum and in biophysical systems; Bose-Einstein Condensates; quantum effects in low-dimensional and nanoscale electron and boson systems; effects of interactions between electrons in semiconductor structures Quantum Hall effect, especially in multicomponent systems; many-body physics and statistical physics.

Oxford, Julia, Ph.D., Washington State. Molecular Mechanisms of Cellular Mechanoreception in Bone; Regulation of cell signaling by Col11a1 during craniofacial development in the zebrafish; Extracellular Matrix is a Key Factor in Cancer Progression; The Effects of Simulated Microgravity and Radiation on Articular Cartilage; Induction of Early Stages of Osteoarthritis after Exposure to Microgravity.

Punnoose, Alex, Ph.D., Aligarh Muslim University. *Nano Science and Technology, Physics and other Science Education.* Nanoscience, Nanotechnology, and Condensed Matter Physics. Physics of nanoscale materials; semiconductor materials and spintronics; nanomedicine; nanoparticles-based cancer and antibacterial therapies; nanotoxicology; nanosensors; catalysis; photonics; nanobiotechnology.

Russell, Dale, Ph.D., University of Arizona. Chemistry. Electroanalytical chemistry: electrochemical sensor development for strategic and environmentally important species including uranium, plutonium, mercury, and VOCs. Sensors for field applications, hand-held, autonomous and clandestine deployment, long-term data logging. Electrokinetic separation in non-polar media. Electrical Field Flow Fractionation of polymers, particles, and lipophilic (membrane) proteins.

Affiliate Associate Professor

Campbell, Kris, Ph.D., University of California, Davis. *Electrical Engineering.* Novel materials and electrical properties of devices including chalcogenide-based materials used as memristors, as nonvolatile memory, and reconfigurable electronics applications.

Kim, Byung, Ph.D., Seoul National University. *Physics and other Science Education.* Single molecular chiral recognition on metal surfaces; molecular scale investigation of bioactive surfaces; interfacial water structure and its application; Nanotribology: chemical modification of interfacial mechanical properties Magnetic-force microscopy of semiconducting magnetic thin films; rapid detection and analysis of biomolecular interactions.

Kuang, Wan, Ph.D., University of Southern California. *Electrical Engineering.* Nano-scale photonic devices; photonic bandgap material; electrodynamic numerical simulations.

Mitkova, Maria, Ph.D., Technological University Sofia. *Electrical Engineering.* Research on glass formation of semiconducting glasses; formation of microstructures on thin films; real-time optical recording on thin chalcogenide films; Conductive Bridge Memristive Devices (CBRAM); resonant frequency alteration; formation of elements for bio-lab on a chip; alteration of reflectance; Chalcogenide Glass Radiation Sensor and radiation induced effects in chalcogenide glasses.

Tenne, Dmitiri, Ph.D., Russian Academy of Sciences. *Physics and other Science Education*. Multifunctional oxide and semiconductor materials and nanostructures for electronic and optoelectronic applications; optical spectroscopy.

Affiliate Assistant Professor

Plumlee, Don, Ph.D., University of Idaho, 2007. *Other*. Microfluidic applications in Low Temperature Co-Fired Ceramics (LTCC).

Raghani, Pushpa, Ph.D., Jawaharlal Nehru Center for Advanced Scientific Research, 2006. *Nano Science and Technology, Physics and other Science Education*. Applications of ab initio Density Functional Theory, Density Functional Perturbation Theory and Nudged Elastic Band theory, in the following areas: Magnetic properties of transition metal atoms adsorbed on metal surfaces; Chemical reactivity of clusters and surfaces; Structural and dynamical properties of adsorbates on metal surfaces; Structural, thermal and dynamical properties of small atomic clusters; Reconstruction of homoepitaxial and heteroexpitaxial metal surfaces.

Lecturer

Donovan, Sean, Ph.D., University of Florida. *Materials Science, Metallurgy*. Electronic materials.

Research Fellow

Yurke, Bernard, Ph.D., Cornell University, 1982. *Materials Science, Metallurgy*. Materials science; Biomaterials; Soft condensed matter; Low temperature physics; Quantum optics; Liquid crystals, Biophysics; Microelectromechanical systems (MEMS); DNA nanotechnology. Participated in the first squeezed state experiments at optical and microwave frequencies and created the first DNA-based nanodevices powered by DNA. Current research is focused on DNA nanotechnology.

DEPARTMENTAL RESEARCH SPECIALTIES AND STAFF

Theoretical

Ceramic Micro-Electromechanical Systems. Boise State's research on ceramic microelectronic devices focuses on microsystems fabricated in low temperature co-fired ceramics. Current projects include micropropulsion and energy scavenging devices. Moll, Plumlee.

DNA Nanotechnology. Boise State's DNA nanotechnology is focused on both DNA origami and DNA catalytic networks. Graugnard, Hughes, Knowlton, Kuang, Lee, Yurke.

Electrochemistry. The research is focused on the interdisciplinary areas of electrochemistry, materials science and engineering, and surface chemistry. Current work is focused on the synthesis and characterization of new nano-architectured electrode materials for energy storage and conversion to a better understanding of the structure-property relationships for the development of advanced functional nanomaterials: (1) Electrochemical synthesis of nanostructured electrode materials with controlled size and shape; (2) Development of electrode materials for alternative battery systems (e.g. Na, Mg); (3) In-situ characterization of electrochemical processes at electrode materials. Xiong.

Electronic Materials. Collaborations include faculty members from Penn State, Oregon State, and University of Maryland. Examples of electronic materials research are high dielectric constant multilayer metal oxide semiconductor devices and nonvolatile memory materials. Campbell, Knowlton, Mitkova, Punnoose, Tenne, Ubic.

Integrated Nanomaterials. Nanoscale materials synthesis and characterization. Fundamental studies of thermal and electrical transport in nanoscale devices and materials. Applications of nanotechnology in energy harvesting/conversion and bioengineering. Environmental impact of engineered nanomaterials. Estrada.

Materials Theory and Modeling. Develop and apply computer-based theoretical methodologies to explore the structure-property-performance relationships and understand the fundamental science underlying observed properties and processes, as well as to manipulate and predict outcome for future experiments. Current projects includes the development of materials-by-design for energy and gas separation applications in collaboration with National Institute of Standards and Technology; fundamental studies of magnetism, transport and quantum phenomena in nanoscaled materials with The University of Texas at Dallas; and the development of multiscale modeling techniques. Li.

Materials for Extreme Environments. Boise State's research on materials for extreme environments involves collaborations through the Center for Advanced Energy Studies. The work focuses on processing and characterization of metals, ceramics, and composites for applications in extreme environments where high temperatures, corrosive gases and/or radiation damage are expected. Butt, Frary, Ubic.

Shape Memory Alloys. Boise State's research on ferromagnetic shape memory alloys involves collaborations with Northwestern University and other international partners. The work focuses on understanding the microstructure of these materials and their mechanical response. Lindquist, Mullner.

Experimental

Ceramic Micro-Electromechanical Systems. Boise State's research on ceramic microelectronic devices focuses on microsystems fabricated in low temperature co-fired ceramics. Current projects include micropropulsion and energy scavenging devices. Moll, Plumlee.

DNA Nanotechnology. Boise State's DNA nanotechnology is focused on both DNA origami and DNA catalytic networks. Graugnard, Hughes, Knowlton, Kuang, Lee, Yurke.

Electrochemistry. The research is focused on the interdisciplinary areas of electrochemistry, materials science and engineering, and surface chemistry. Current work is focused on the synthesis and characterization of new nano-architectured electrode materials for energy storage and conversion to a better understanding of the structure-property relationships for the development of advanced functional nanomaterials: (1) Electrochemical synthesis of nanostructured electrode materials with controlled size and shape; (2) Development of electrode materials for alternative battery systems (e.g. Na, Mg); (3) In-situ characterization of electrochemical processes at electrode materials. Xiong.

Electronic Materials. Collaborations include faculty members from Penn State, Oregon State, and University of Maryland. Examples of electronic materials research are high dielectric constant multilayer metal oxide semiconductor devices and nonvolatile memory materials. Campbell, Knowlton, Mitkova, Punnoose, Tenne, Ubic.

Integrated Nanomaterials. Nanoscale materials synthesis and characterization. Fundamental studies of thermal and electrical transport in nanoscale devices and materials. Applications of nanotechnology in energy harvesting/conversion and bioengineering. Environmental impact of engineered nanomaterials. Estrada.

Materials Theory and Modeling. Develop and apply computer-based theoretical methodologies to explore the structure-property-performance relationships and understand the fundamental science underlying observed properties and processes, as well as to manipulate and predict outcome for future experiments. Current projects includes the development of materials-by-design for energy and gas separation applications in collaboration with National Institute of Standards and Technology; fundamental studies of magnetism, transport and quantum phenomena in nanoscaled materials with The University of Texas at Dallas; and the development of multiscale modeling techniques. Li.

Materials for Extreme Environments. Boise State's research on materials for extreme environments involves collaborations through the Center for Advanced Energy Studies. The work focuses on processing and characterization of metals, ceramics, and composites for applications in extreme environments where high temperatures, corrosive gases and/or radiation damage are expected. Butt, Frary, Ubic.

Shape Memory Alloys. Boise State's research on ferromagnetic shape memory alloys involves collaborations with Northwestern University and other international partners. The work focuses on understanding the microstructure of these materials and their mechanical response. Lindquist, Mullner.

View additional information about this department at www.gradschoolshopper.com

---

DEPAUL UNIVERSITY

DEPARTMENT OF PHYSICS

Chicago, Illinois 60614

http://csh.depaul.edu/departments/physics/Pages/default.aspx

General University Information

President: Reverend Dennis H. Holtschneider C. M.
Dean of Graduate School: Margaret Silliker
University website: http://www.depaul.edu
Control: Private
Setting: Urban
Total Faculty: 1,842
Total number of Students: 25,072
Total number of Graduate Students: 7,795

Department Information

Department Chairman: Jesús Pando, Chair
Department Contact: Anuj P. Sarma, Associate Professor & Graduate Director
Total full-time faculty: 7
Full-Time Graduate Students: 10
First-Year Graduate Students: 2

Department Address

2219 N. Kenmore Avenue
Byrne Hall 211
Chicago, IL 60614
Phone: (773) 325-7330
E-mail: asarma@depaul.edu
Website: http://csh.depaul.edu/departments/physics/Pages/default.aspx

ADMISSIONS

Admission Contact Information

Address admission inquiries to: Dr. Anuj P. Sarma, Dept. of Physics, Byrne 211, De Paul University, 2219 N. Kenmore Avenue, Chicago, IL 60614-3504
Phone: (773) 325-1373
E-mail: asarma@depaul.edu
Admissions website: http://csh.depaul.edu/departments/physics/Pages/default.aspx

Application deadlines

Fall admission:
U.S. students: June 1   *Int'l. students*: June 1

Application fee

U.S. students: $40

Admissions information

For Fall of 2012:
 Number of applicants: 15
 Number admitted: 2
 Number enrolled: 2

Admission requirements

Bachelor's degree requirements: Bachelor's degree in physics, mathematics, chemistry, or engineering is required.
Minimum undergraduate GPA: 2.5

GRE requirements

The GRE is required.

Advanced GRE requirements

The Advanced GRE is not required.

TOEFL requirements

The TOEFL exam is required for students from non-English-speaking countries.

Other admissions information

Undergraduate preparation assumed: Tipler or Serway, General Physics; Tipler, Modern Physics; Fowles, Mechanics; Griffiths, Electricity and Magnetism; Schroeder, Thermal Physics; Boas, Mathematical Methods in The Physical Sciences; Griffiths, Quantum Mechanics.

TUITION

Tuition year 2012–13:
 Full-time students: $630 per credit
Credit hours per semester to be considered full-time: 8
Deferred tuition plan: Yes
Academic term: Quarter
Number of first-year students who received full tuition waivers: 3

Teaching Assistants, Research Assistants, and Fellowships

Number of first-year
Teaching Assistants: 2
Average stipend per academic year
Teaching Assistant: $11,000

FINANCIAL AID

Loans

Loans are available for U.S. students.
Loans are not available for international students.
GAPSFAS application required: No
FAFSA application required: Yes

For further information

Address financial aid inquiries to: Financial Aid Office, De Paul Center, Room 9100, 1 E. Jackson Blvd., Chicago, IL 60604-2287.
Phone: (312) 362-8610
E-mail: finaid1@depaul.edu
Financial aid website: http://www.depaul.edu/admission-and-aid/financial-aid/Pages/default.aspx

HOUSING

Availability of on-campus housing

Single students: Yes

For further information

Address housing inquiries to: Department of Housing Services, Centennial Hall, Suite 301, 2345 N. Shefield Ave., Chicago, IL 60614.
Phone: (773) 325-7196
Housing aid website: http://offices.depaul.edu/housing/Pages/default.aspx

Table A—Faculty, Enrollments, and Degrees Granted

Research Specialty	2012–13 Faculty	Enrollment Fall 2012		Number of Degrees Granted 2012-13 (2008-13)		
		Master's	Doctorate	Master's	Terminal Master's	Doctorate
Astrophysics	2	2	–	–	–(7)	–
Biophysics	–	3	–	–	1(4)	–
Condensed Matter Physics	2	2	–	–	1(3)	–
Fluids, Rheology	1	–	–	–	–(2)	–
Nonlinear Dynamics and Complex Systems	1	1	–	–	–(1)	–
Nuclear Physics	1	2	–	–	1(4)	–
Non-specialized	–	–	–	–	–	–
Total	**7**	**10**	**–**	**–**	**3(21)**	**–**
Full-time Grad. Stud.	–	8	–	–	–	–
First-year Grad. Stud.	–	2	–	–	–	–

GRADUATE DEGREE REQUIREMENTS

Master's: Eleven (4 quarter-hours each) courses, including thesis; minimum 2.75 grade point average on a scale of 4.0; no time residence requirement; no language requirement; oral thesis examination is required. Faculty expertise in astrophysics, biophysics, nonlinear dynamics, complex systems and computational physics, nuclear physics, fluids, laser physics, and condensed-matter physics.
Thesis: Thesis may be written in absentia.

SPECIAL EQUIPMENT, FACILITIES, OR PROGRAMS

Ultrafast optics laboratory including a Titanium:Sapphire 20 fs laser oscillator and a Q-switched Nd:YAG laser; high-temperature furnaces (maximum $T = 1500$ C) with different gas environments; four-point electrical conductivity probe; thermopower setup; hydraulic press (maximum applied load = 11 metric tons).

Table B—Separately Budgeted Research Expenditures by Source of Support

Source of Support	Departmental Research	Physics-related Research Outside Department
Federal government	$20,000	
State/local government		
Non-profit organizations		
Business and industry		
Other		$40,000
Total	**$20,000**	**$40,000**

Table C—Separately Budgeted Research Expenditures by Research Specialty

Research Specialty	No. of Grants	Expenditures ($)
Astrophysics	1	$20,000
Condensed Matter Physics	1	$40,000
Total	**2**	**$60,000**

FACULTY

Professor

Goedde, Christopher G., Ph.D., University of California, Berkeley, 1990. *Computational Physics, Nonlinear Dynamics and Complex Systems*. Nonlinear optics and dynamical systems; computational physics.

Associate Professor

Fischer, Susan M., Ph.D., University of Notre Dame, 1994. *Nuclear Physics*. Nuclear physics; gamma-ray spectroscopy.
Pando, Jesús, Ph.D., University of Arizona, 1997. *Astrophysics*. Astrophysics: cosmology and large-scale structure.
Sarma, Anuj, Ph.D., University of Kentucky, 2000. *Astronomy, Astrophysics*. Astrophysics: star formation and radio astronomy.

Assistant Professor

González Avilés, Gabriela, Ph.D., Northwestern University, 2003. *Materials Science, Metallurgy*. Materials science.
Kustusch, Mary Bridget, Ph.D., North Carolina State University, 2011. *Physics and other Science Education*. Physics Education Research.
Landahl, Eric, Ph.D., University of California, Davis, 2001. *Accelerator, Biophysics*. Ultra-fast physics.

Emeritus

Behof, Anthony F., Ph.D., University of Notre Dame, 1965. *Optics*. Optics.
El Saffar, Zuhair M., Ph.D., University of Wales, 1960. Solid-state physics.
Milton, John, M.S., Saint Louis University, 1960. Physics education.
Schillinger, Edwin J., Ph.D., University of Notre Dame, 1950. Physics education.
Stinchcomb, Thomas G., Ph.D., University of Chicago, 1951. Medical physics; radiation physics.
Van Ostenburg, Donald O., Ph.D., Michigan State University, 1956. Solid-state physics.

Lecturer

Corso, George, Ph.D., Northwestern University, 1975. *Astronomy*. Astronomy.

Laboratory Coordinator

Mihalcea, Gabi, M.S., Kansas State University.

DEPARTMENTAL RESEARCH SPECIALTIES AND STAFF

Theoretical

Astrophysics. Pando.
Computational Physics. Goedde.
Nonlinear Dynamics and Complex Systems. Goedde.

Experimental

Astronomy. Sarma.
Biophysics. Landahl.
Materials Science, Metallurgy. González Avilés, Landahl.
Nuclear Physics. Fischer.
Physics and other Science Education. Kustusch.

View additional information about this department at
www.gradschoolshopper.com

NORTHERN ILLINOIS UNIVERSITY

DEPARTMENT OF PHYSICS

DeKalb, Illinois 60115
http://www.physics.niu.edu/physics/

General University Information

President: Douglas D. Baker
Dean of Graduate School: Bradley Bond
University website: http://www.niu.edu
Control: Public
Setting: Suburban
Total Faculty: 1,185
Total Graduate Faculty: 768
Total number of Students: 22,671
Total number of Graduate Students: 5,365

Department Information

Department Chairman: Laurence Lurio, Chair
Department Contact: David Hedin, Professor, Director of Graduate Studies
  Total full-time faculty: 21
  Total number of full-time equivalent positions: 19
  Full-Time Graduate Students: 45
  First-Year Graduate Students: 13
  Female First-Year Students: 1
  Total Post Doctorates: 11

Department Address

Physics Department, 202 LaTourette Hall, Normal Road
Northern Illinois University
DeKalb, IL 60115
Phone: (815) 753-1772
Fax: (815) 753-8565
E-mail: askphysics@niu.edu
Website: http://www.physics.niu.edu/physics/

ADMISSIONS

Admission Contact Information

Address admission inquiries to: Graduate School
Phone: (815) 753-0395
E-mail: gradsch@niu.edu
Admissions website: http://www.niu.edu/grad/apply/

Application deadlines

Fall admission:
U.S. students: June 15   *Int'l. students*: April 15

Application fee

U.S. students: $40   *Int'l. students*: $40
International application fee can be waived if payment of the fee puts a significant financial strain on the applicant.

Admissions information

For Fall of 2013:
  Number of applicants: 59
  Number admitted: 19
  Number enrolled: 12

Admission requirements

Bachelor's degree requirements: Bachelor's degree in physics or a related discipline is required.
Minimum undergraduate GPA: 3.0

GRE requirements

The GRE is required.

Advanced GRE requirements

The Advanced GRE is recommended.
A good score helps.

TOEFL requirements

The TOEFL exam is required for students from non-English-speaking countries.
  iBT score: 80

TOEFL may be substituted by IELTS, for which the minimum acceptable score is 6.5.

Other admissions information
Additional requirements: GRE Physics is not required, but strongly recommended, especially for international students.
Undergraduate preparation assumed: Corson and Lorrain, Electricity and Magnetism; Fowles, Mechanics; Weidener and Sells, Modern Physics; Fowles, Optics.

TUITION

Tuition year 2013–14:
Tuition for in-state residents
Full-time students: $3,081 per semester
Tuition for out-of-state residents
Full-time students: $6,162 per semester
Graduate assistantships (TA/RA) include tuition waiver.
Credit hours per semester to be considered full-time: 9
Deferred tuition plan: No
Health insurance: Available at the cost of $1060 per year.
Other academic fees: Full-time—$1,591/sem. Graduate assistants—$1,591/sem.
Academic term: Semester
Number of first-year students who received full tuition waivers: 11

Teaching Assistants, Research Assistants, and Fellowships
Number of first-year
Teaching Assistants: 9
Research Assistants: 2
Average stipend per academic year
Teaching Assistant: $14,961
Research Assistant: $16,372
Fellowship student: $16,651
All of the above amounts are for nine months (fall+spring). Most students are able to find full support for the three summer months as RA's and a few as TA's at the same monthly rates.

FINANCIAL AID

Application deadlines
Fall admission:
U.S. students: February 15   *Int'l. students*: February 15
Spring admission:
U.S. students: October 15   *Int'l. students*: October 15

Loans
Loans are available for U.S. students.
Loans are not available for international students.
GAPSFAS application required: No
FAFSA application required: Yes

For further information
Address financial aid inquiries to: Student Financial Aid Office, Swen Parson Hall 245, Northern Illinois University, DeKalb, IL 60115.
Phone: 800-892-3050
E-mail: finaid@niu.edu
Financial aid website: http://www.niu.edu/fa/

HOUSING

Availability of on-campus housing
Single students: Yes
Married students: Yes

For further information
Address housing inquiries to: Housing & Dining, 101 East Neptune Hall, Northern Illinois University, DeKalb, IL 60115.
Phone: (815) 753-1525
E-mail: housingdining@niu.edu
Housing aid website: http://www.niu.edu/housing/

Table A—Faculty, Enrollments, and Degrees Granted

Research Specialty	2013–14 Faculty	Enrollment Fall 2013 Master's	Enrollment Fall 2013 Doctorate	Number of Degrees Granted 2011–12 (2007–12) Master's	Number of Degrees Granted 2011–12 (2007–12) Terminal Master's	Number of Degrees Granted 2011–12 (2007–12) Doctorate
Applied Physics	–	1	–	–	–	–(2)
Astrophysics	1	–	1	1(1)	–(2)	–
Condensed Matter Physics	14	4	17	4(18)	3(13)	2(7)
High Energy Physics	7	–	10	1(6)	3(8)	1(8)
Medical, Health Physics	1	1	2	–	–	–
Physics and other Science Education	1	2	–	–	–	–
Physics of Beams	4	–	9	–	–(2)	3(4)
Total	26	7	39	6(25)	6(25)	8(21)
Full-time Grad. Stud.	–	7	37	–	–	–
First-year Grad. Stud.	–	1	10	–	–	–

GRADUATE DEGREE REQUIREMENTS

Master's: 30 hours of course work with 24 in physics; thesis required for pure and applied physics specializations.

Doctorate: Students are required to complete 90 semester hours of graduate course work. This includes 15 hours in five out of six core courses covering classical and quantum mechanics, statistical physics, and electromagnetic theory, and twelve hours in two different areas of physics. A minimum of 24 hours dedicated to dissertation research is required. The remaining hours may include additional dissertation work or other graduate course work in physics and related fields. Students entering the program without a master's degree in physics are required to pass a qualifying examination, which is usually taken at the end of the first year. Successful completion of a candidacy examination based on the core courses and other graduate courses is required of all students in the Ph.D. program. Transfer credits for students entering with a master's degree or with graduate coursework from another institution are allowed, pending approval by the Graduate Studies Committee.

Thesis: Thesis may be written in absentia.

SPECIAL EQUIPMENT, FACILITIES, OR PROGRAMS

Students may specialize in five principal areas: condensed matter and materials physics, elementary particles and fields, accelerator physics, medical physics, and physics education.

The Department makes special efforts to accommodate the needs of students, such as employees of nearby industrial government laboratories and teachers employed in the region who wish to gain advanced degrees in physics on either a part-time or full-time basis.

On the departmental faculty are eight condensed matter experimenters and three theorists with whom graduate students may work on their thesis research. In addition there are joint and adjunct professors from Argonne National Laboratory and Fermi National Accelerator Laboratory.

The Physics Department is a member of NIU's Institute for Nano Science, Engineering, and Technology (InSET) and as such faculty and students have access to a class 100 clean room containing a wide array of fabrication and characterization instruments and deposition systems. Some faculty members of the department also base their research programs at Argonne National Laboratory, about a one hour-drive by car from NIU, where they utilize national user facilities such as the Advanced Photon Source, the Electron Microscopy Center, and the Center for Nanomaterials.

Among the faculty working on High Energy (Elementary Particle) Physics are seven experimenters and one theorist, several research scientists and a number of graduate students doing thesis research. At present, the experimenters participate in the the ATLAS proton-proton-collision experiment at CERN (Geneva, Switzerland) as well as the g-2 and the proposed Mu2E and experiment at the Fermi National Acceleration Laboratory (40 minutes by car). The group is also active on research and development of particle detection technologies and associated algorithms, with emphasis on use of scintillator detectors at a future linear collider.

Accelerator physics R&D are coordinated through the Northern Illinois Center for Accelerator and Detector Development (NICADD). Current areas include studies of intense electron sources at the Fermilab-NICADD Photoinjector Laboratory and beam diagnostics using resources at NIU, Argonne and Fermilab; Argonne Tandem Linear Accelerator System; muon-based accelerators.

Both the particles and accelerator groups collaborate closely with nearby Fermilab and Argonne National Lab where they have access to laboratory and computing facilities. They also have their own laboratories and high-end computing facilities at the university.

One faculty member works with a group of graduate students on medical physics, focusing on both diagnostic (proton-computed tomography) and therapeutic (proton therapy) aspects in close collaboration with Northwestern University and a nearby hospital.

A faculty member works closely with graduate students on methods of physics teaching and serves as a supervisor of their student teaching at selected nearby high schools.

Table B—Separately Budgeted Research Expenditures by Source of Support

Source of Support	Departmental Research	Physics-related Research Outside Department
Federal government	$1,700,488	
State/local government		
Non-profit organizations	$22,125	
Business and industry		
Other		
Total	**$1,722,613**	

FACULTY

Professor

Blazey, Gerald, Ph.D., University of Minnesota, 1984. Distinguished Research Professor. Director of the Northern Illinois Center for Accelerator and Detector Development. Currently on leave of absence to serve as the Assistant Director for Physical Sciences in the U.S. President's Office of Science and Technology Policy. *High Energy Physics.* Elementary Particles, Experiment.

Chakraborty, Dhiman, Ph.D., Stony Brook University, 1994. Presidential Research Professor. *High Energy Physics.* Elementary Particles, Experiment.

Dabrowski, Bogdan M., Ph.D., Northwestern University, 1987. Distinguished Research Professor. *Condensed Matter Physics, Materials Science, Metallurgy.* Experiment.

Hedin, David, Ph.D., University of Wisconsin-Madison, 1980. Board of Trustees Professor. Distinguished Research Professor. Director of Graduate Studies. Deputy Director of Northern Illinois Center for Accelerator and Detector Development. *High Energy Physics.* Elementary Particles, Experiment.

Lurio, Laurence, Ph.D., Harvard University, 1993. Chair. *Condensed Matter Physics.* Experiment.

Martin, Stephen, Ph.D., University of California, Santa Barbara, 1988. Distinguished Research Professor. Presidential Teaching Professor. *Particles and Fields.* Elementary Particles, Theory.

Mini, Susan, Ph.D., Southern Illinois University, 1991. Associate Provost. *Condensed Matter Physics.*

Piot, Philippe, Ph.D., University of Grenoble, 1999. Acting Director, Northern Illinois Center for Accelerator and Detector Development. Scientist, Fermi National Accelerator Laboratory. *Physics of Beams.* Accelerator physics, Experiment.

Thompson, Carol, Ph.D., University of Houston, 1987. *Condensed Matter Physics.* Experiment.

Van Veenendaal, Michel, Ph.D., Laboratory of Solid State Physics, RUG, Netherlands, 1994. Presidential Research Professor. Managing Director, Institute of Nano Science, Engineering, and Technology. *Condensed Matter Physics, Nano Science and Technology.* Theory.

Xiao, Zhili, Ph.D., University of Konstanz, 1996. Board of Trustees Professor. Presidential Research Professor. Physicist, Materials Science Division, Argonne National Laboratory. *Condensed Matter Physics, Materials Science, Metallurgy, Nano Science and Technology.* Experiment.

Associate Professor

Brown, Dennis, Ph.D., Stanford University, 1993. *Condensed Matter Physics.* Experiment.

Chmaissem, Omar W., Ph.D., University of Grenoble, France, 1992. *Condensed Matter Physics.* Experiment.

Coutrakon, George, Ph.D., Stony Brook University, 1983. *Medical, Health Physics.* Medical Physics, Experiment.

Erdelyi, Bela, Ph.D., Michigan State University, 2001. *Nonlinear Dynamics and Complex Systems, Physics of Beams.* Experiment and theory.

Fortner, Michael, Ph.D., Brandeis University, 1989. *High Energy Physics.* Experiment.

Glatz, Andreas, Ph.D., Cologne University, 2004. *Materials Science, Metallurgy, Nonlinear Dynamics and Complex Systems.* Theory.

Ito, Yasuo, Ph.D., University of Cambridge, 1996. Assistant Chair. *Condensed Matter Physics.* Experiment.

Winkler, Roland, Ph.D., University of Regensburg, 1994. *Condensed Matter Physics, Quantum Foundations.* Theory.

Assistant Professor

Eads, Michael, Ph.D., Northern Illinois, 2007. Physics Teaching, Elementary Particles, Experiment.

Shin, Young-Min, Ph.D., Seoul National University, 2006. *Physics of Beams.* Accelerator Physics, Experiment.

Emeritus

Albright, Carl H., Ph.D., Princeton University, 1960. *Particles and Fields.* Elementary Particles, Theory.

Kimball, Clyde W., Ph.D., St. Louis University, 1959. *Condensed Matter Physics, Nano Science and Technology.* Experiment.

Willis, Suzanne, Ph.D., Yale University, 1979. *Physics and other Science Education.* Physics education.

Adjunct Professor

Alp, Ercan, Ph.D., Southern Illinois University, 1984. *Condensed Matter Physics.* Condensed Matter, Experiment.

Bhat, Pushpa, Ph.D., Bangalore University, 1982. *High Energy Physics*. Elementay Particles, Experiment.

Cummings, MaryAnne, Ph.D., University of Michigan, 1990. *High Energy Physics*. Elementary Particles and accelerator physics, Experiment.

Welp, Ulrich, Ph.D., University of Konstanz, 1988. Experiment.

Zaluzec, Nestor, Ph.D., University of Illinois at Urbana-Champaign, 1973. *Condensed Matter Physics*. Experiment.

DEPARTMENTAL RESEARCH SPECIALTIES AND STAFF

Theoretical

Condensed Matter Physics. Liquid metals; magnetism and cooperative phenomena; many-body theory; optical properties of solids; electronic structure; multi-particle systems; quantum macrophysics; non-linear dynamics; transport phenomena; non-equilibrium systems; dynamics of disordered elastic systems. Glatz, Van Veenendaal, Winkler.

High Energy Physics. Weak interactions; gauge theory; phenomenology; super-symmetric theories. Albright, Martin.

Physics of Beams. Nonlinear dynamics, Applications of symplectic geometry in, and numerical methods for Hamiltonian dynamics. Erdelyi.

Experimental

Condensed Matter Physics. Mössbauer effect; superconductivity; lattice defects; optical and transport properties of amorphous and crystalline solids; synchrotron radiation; surface physics; magnetic properties of solids; low-temperature physics; x-ray crystallography; materials preparation, polymer physics, biophysics. Alp, Brown, Chmaissem, Dabrowski, Ito, Kimball, Lurio, Mini, Thompson, Xiao, Zaluzec.

High Energy Physics. LHC/ATLAS; Tevatron/D0: collider physics at the energy frontier: studies of top quark production and decay, studies of higgs boson(s), searches for new massive states, detector design and operations; Fermilab (Mu2E, g-2): searches for rare processes at the intensity frontier; R&D of detector technologies and algorithms. Bhat, Blazey, Chakraborty, Cummings, Eads, Hedin.

Physics of Beams. Simulation and operation of high brightness photoinjectors. Electron beam diagnostics. Muon and heavy nuclei accelerators. Erdelyi, Piot, Shin.

Radiation therapy and imaging. Proton-computed tomography, proton therapy. Blazey, Coutrakon, Hedin.

View additional information about this department at
www.gradschoolshopper.com

NORTHWESTERN UNIVERSITY

DEPARTMENT OF PHYSICS AND ASTRONOMY

Evanston, Illinois 60208
http://www.physics.northwestern.edu/index.html

General University Information

President: Morton Schapiro
Dean of Graduate School: Dwight A. McBride
University website: http://www.northwestern.edu/
Control: Private
Setting: Urban
Total Faculty: 2,500
Total Graduate Faculty: 1,000
Total number of Students: 19,129
Total number of Graduate Students: 9,663

Department Information

Department Chairman: Heidi Schellman, Chair
Department Contact: Grant Darktower, Graduate Assistant
  Total full-time faculty: 29
  Total number of full-time equivalent positions: 30
  Full-Time Graduate Students: 90
  First-Year Graduate Students: 14
  Female First-Year Students: 3
  Total Post Doctorates: 30

Department Address

2145 Sheridan Road
Evanston, IL 60208
Phone: (847) 491-3685

Fax: (847) 491-9982
E-mail: a-darktower@northwestern.edu
Website: http://www.physics.northwestern.edu/index.html

ADMISSIONS

Admission Contact Information

Address admission inquiries to: Graduate Admissions Committee Department of Physics & Astronomy, Northwestern University, Evanston, IL 60208
Phone: (847) 491-3685
E-mail: a-darktower@northwestern.edu
Admissions website: http://www.physics.northwestern.edu/graduate/

Application deadlines

Fall admission:
U.S. students: December 31   *Int'l. students*: December 31

Application fee

U.S. students: $75   *Int'l. students*: $75
Students can be admitted at times other than the fall under special conditions.

Admissions information

For Fall of 2012:
Number of applicants: 269
Number admitted: 36
Number enrolled: 14

Admission requirements

Bachelor's degree requirements: Students must have a bachelor's degree, preferably in physics, astronomy, or mathematics.
Minimum undergraduate GPA: 3.0

GRE requirements

The GRE is required.

Advanced GRE requirements

The Advanced GRE is required.
Minimum accepted Advanced GRE score: 700
Mean Advanced GRE score range (25th–75th percentile): 75%

TOEFL requirements

The TOEFL exam is required for students from non-English-speaking countries.
PBT score: 600
iBT score: 90

Other admissions information

Additional requirements: Students must submit results of the physics GRE examination and three letters of recommendation. For admission, equal weight is given to GPA, GRE scores, and letters.
Undergraduate preparation assumed: Symon, Mechanics; Reitz and Milford, E & M; Reif, Statistical Mechanics; Zemansky, Thermodynamics; and McGervy, Modern Physics.
  Math preparation should include: Ordinary Differential Equations; Partial Differential Equations; Boundary Value Problems; Complex Variable Theory; Linear Algebra.

TUITION

Tuition year 2013-14:
Tuition for out-of-state residents
  Full-time students: $44,708 annual
The full cost of tuition is included if we provide you with financial support.
Credit hours per semester to be considered full-time: 3
Deferred tuition plan: No
Health insurance: Available
Other academic fees: About $40 per quarter for using off-campus transportation and the athletic center.
Academic term: Quarter
Number of first-year students who received full tuition waivers: 14

Teaching Assistants, Research Assistants, and Fellowships

Number of first-year
  Fellowship students: 15
Average stipend per academic year
  Research Assistant: $23,000

FINANCIAL AID

Application deadlines

Fall admission:
U.S. students: December 31   *Int'l. students*: December 31

Loans

Loans are not available for U.S. students.
Loans are not available for international students.
GAPSFAS application required: No
FAFSA application required: Yes

For further information

Address financial aid inquiries to: Graduate Admissions Committee, Department of Physics & Astronomy, Northwestern University, Evanston, IL 60208.
Phone: (847) 491-3685
E-mail: a-darktower@northwestern.edu
Financial aid website: http://www.physics.northwestern.edu/graduate/finaid.html

HOUSING

Availability of on-campus housing

Single students: Yes
Married students: Yes

For further information

Address housing inquiries to: Graduate Student Housing Office, 1915 Maple Ave, Northwestern University, Evanston, IL 60208.
Phone: (847) 491-5127
E-mail: grad-housing@northwestern.edu
Housing aid website: http://www.northwestern.edu/gradhousing/

Table A—Faculty, Enrollments, and Degrees Granted

Research Specialty	2013–14 Faculty	Enrollment Fall 2013 Master's	Enrollment Fall 2013 Doctorate	Number of Degrees Granted 2012–13 Master's	Number of Degrees Granted 2012–13 Terminal Master's	Number of Degrees Granted 2012–13 Doctorate
Astrophysics	8	–	18	–	-(1)	2(-)
Atomic, Molecular, & Optical Physics	2	–	15	–	–	1(-)
Biophysics	2	–	6	–	–	2(-)
Condensed Matter Physics	9	–	22	–	–	2(-)
Particles and Fields	8	–	16	2(-)	–	1(-)
Physics and other Science Education	–	–	–	–	–	–
Non-specialized	-	–	12	–	–	–
Total	29	–	89	2(-)	1(-)	8(-)
Full-time Grad. Stud.	–	–	89	–	–	–
First-year Grad. Stud.	–	–	14	–	–	–

GRADUATE DEGREE REQUIREMENTS

Master's: Seven required quarter-courses in physics with a "B" average required. Minimum of three quarters of full-time study is required. Master's examination is required. No thesis is required. There is no foreign language requirement.

Doctorate: Minimum of two years of residency is required; 13 quarter-courses in physics and/or astronomy are required; "B" average is required. Departmental preliminary examination is required. There is no foreign language requirement.

Other Degrees: We do not offer a terminal master's degree. Only students admitted to the Ph.D. program can obtain a master's degree.

Thesis: Thesis may be written in absentia.

SPECIAL EQUIPMENT, FACILITIES, OR PROGRAMS

We have ties to many research centers at Northwestern, including the Center for Quantum Optics, the Materials Research Center, and the Center for Interdisciplinary Astrophysical Research (CIERA). Our faculty also has extensive access to major government research facilities, including Fermilab, CERN, Argonne National Laboratory, the National High-Magnetic Field Laboratory, and many ground-based and space-based astrophysical observatories. We have joint faculty in the Departments of Electrical Engineering, Chemistry, Materials Science, Molecular Biology, and Applied Physics.

Table B—Separately Budgeted Research Expenditures by Source of Support

Source of Support	Departmental Research	Physics-related Research Outside Department
Federal government	$5,062,665	$408,353
State/local government		
Non-profit organizations	$305,642	
Business and industry		
Other		
Total	**$5,368,307**	**$408,353**

Table C—Separately Budgeted Research Expenditures by Research Specialty

Research Specialty	No. of Grants	Expenditures ($)
Astrophysics	16	$1,023,711
Atomic, Molecular, & Optical Physics	3	$610,356
Biophysics	5	$731,431
Condensed Matter Physics	10	$1,851,019
Particles and Fields	10	$1,151,790
Total	**44**	**$5,368,307**

FACULTY

Professor

Chandrasekhar, Venkat, Ph.D., Yale University, 1989. Co-Director of Applied Physics. *Applied Physics, Condensed Matter Physics, Nano Science and Technology.* Mesoscopic systems; transport and magnetic properties of small particles.

Dutta, Pulak, Ph.D., University of Chicago, 1980. Co-Director of Applied Physics. *Applied Physics, Biophysics, Condensed Matter Physics, Fluids, Rheology.* Nanoscale order in soft materials; X-ray scattering studies of self-assembled molecular layers.

Ellis, Donald, Ph.D., Massachusetts Institute of Technology, 1966. *Applied Physics, Condensed Matter Physics, Materials Science, Metallurgy.* Numerical calculation of material properties, including composite materials, electroceramics, and bioceramics.

Freeman, Arthur, Ph.D., Massachusetts Institute of Technology, 1956. *Applied Physics, Condensed Matter Physics, Materials Science, Metallurgy.* First-principles simulations of complex materials, including structural, electronic, magnetic, optical, and mechanical properties.

Garg, Anupam, Ph.D., Cornell University, 1983. *Applied Physics, Atomic, Molecular, & Optical Physics, Condensed Matter Physics.* Spin tunneling in magnetic molecules; macroscopic quantum phenomena; quantum computing.

Halperin, William, Ph.D., Cornell University, 1974. *Applied Physics, Condensed Matter Physics, Low Temperature Physics.* Low-temperature physics; high-magnetic-field NMR studies of high-Tc superconductors; fluid transport in porous media.

Kalogera, Vassiliki, Ph.D., University of Illinois, 1997. Co-Director of CIERA. *Astrophysics, Relativity & Gravitation.* Gravitational waves; X-ray emission from compact binary objects; coalescence of neutron-star binaries.

Ketterson, John, Ph.D., University of Chicago, 1962. *Applied Physics, Atomic, Molecular, & Optical Physics, Biophysics, Condensed Matter Physics.* Superlattices; conventional and high-Tc super- conductors; organic films; scanning-electron and atomic-force microscopy; nonlinear optics.

Marko, John, Ph.D., Massachusetts Institute of Technology, 1989. Joint appointment with Molecular Biology. *Biophysics, Polymer Physics/Science, Statistical & Thermal Physics.* Applications of statistical mechanics and polymer physics to biophysical problems.

Meyer, David, Ph.D., University of California, Los Angeles, 1984. Co-Director of CIERA. *Astrophysics.* High signal-to-noise spectroscopy of interstellar and extragalactic absorption lines; small-scale structures in diffuse galactic clouds.

Motter, Adilson, Ph.D., UNICAMP (Brazil), 2002. *Biophysics, Statistical & Thermal Physics.* Theory of complex systems; nonlinear phenomena; statistical physics.

Novak, Giles, Ph.D., University of Chicago, 1988. *Astrophysics.* Large-scale galactic magnetic fields; magnetic fields in the vicinity of low-mass protostars; gas turbulence in molecular clouds.

Rasio, Frederic, Ph.D., Cornell University, 1991. *Astrophysics, Relativity & Gravitation.* Evolution of dense star clusters; massive black hole formation; coalescing compact binaries; gravity waves; extrasolar planets.

Sauls, James, Ph.D., Stony Brook University, 1980. Director of Graduate Studies. *Applied Physics, Condensed Matter Physics, Low Temperature Physics.* Theory of quantum fluids; strongly correlated metals; systems with disorder; complex symmetry breaking in superfluids and superconductors.

Schellman, Heidi, Ph.D., University of California, Berkeley, 1984. Chair of the Department of Physics & Astronomy. *Particles and Fields.* Interactions of quarks and leptons; decay of weak vector bosons; high-intensity neutrino experiments; cosmological high-energy physics.

Seth, Kamal, Ph.D., University of Pittsburgh, 1957. *Nuclear Physics, Particles and Fields.* Experimental study of exotic combinations of valence quarks and gluons, glueballs, and charmonium.

Taam, Ronald, Ph.D., Columbia University, 1973. *Astrophysics.* Hydrodynamics of gas flows; close binary-star systems; propagation of nuclear burning fronts in stars; X-ray bursts from neutron stars.

Ulmer, Melville, Ph.D., University of Wisconsin-Madison, 1970. *Astrophysics.* Development of X-ray mirrors and ultraviolet detectors; matter distribution in galactic clusters; transient hard X-ray sources.

Velasco, M ayda, Ph.D., Northwestern University, 1995. *Particles and Fields.* Matter-antimatter asymmetryl physics beyond the Standard Model; beam instrumentation; multi-TeV electron/positron and gamma/gamma colliders.

Yusef-Zadeh, Farhad, Ph.D., Columbia University, 1986. *Astrophysics.* Study of the black hole Sgr A* at the center of the galaxy; supernova remnant masers; stellar formation within evolved HII regions.

Associate Professor

de Gouvêa, André, Ph.D., University of California, Berkeley, 1999. *Particles and Fields.* Theory of neutrino oscillations; lepton number non-conservation in the Big Bang; dark matter.

Petriello, Frank, Ph.D., Stanford University, 2003. *Particles and Fields.* Application of perturbative QCD to collider physics; new phenomena at the TeV scale; new calculational techniques for quantum field theory.

Schmitt, Michael, Ph.D., Harvard University, 1991. *Particles and Fields.* Electroweak behavior in the Standard Model; searches for new elementary particles and new particle interactions; detector technology.

Assistant Professor

Dahl, Eric, Ph.D., Princeton University, 2009. *Particles and Fields*. Identifying and detecting dark matter in the galaxy.

Hahn, Kristian, Ph.D., University of Pennsylvania, 2006. *Particles and Fields, Physics of Beams*. Electroweak symmetry breaking; Higgs physics and searches for physics beyond the standard model.

Koch, Jens, Ph.D., Freie Universität, Berlin, 2006. *Applied Physics, Condensed Matter Physics*. Strongly correlated systems; quantum information processing with solid-state devices; theory of quantum transport.

Lithwick, Yoram, Ph.D., California Institute of Technology, 2002. *Astrophysics*. Planet formation; dynamics of planetary systems; accretion disks; MHD turbulence; cosmological halo formation; gamma-ray bursts.

Low, Ian, Ph.D., Carnegie Mellon University, 2000. Joint appointment with Argonne National Laboratory. *Particles and Fields*. Theoretical particle physics: electroweak symmetry breaking; properties of the Higgs boson; aspects of dark matter.

Odom, Brian, Ph.D., Harvard University, 2004. *Applied Physics, Atomic, Molecular, & Optical Physics*. Experiments on mK-trapped molecular ions to investigate changing fundamental constants, parity violation in chiral matter, and quantum effects in chemical reactions below 1 K.

Stern, Nathaniel, Ph.D., University of California, Santa Barbara, 2008. *Condensed Matter Physics*. Experimental optical condensed-matter physics; nanoscale photonics and magnetism; cavity QED with single atoms and photons; spin dynamics in the solid state.

Research Professor

Anastassov, Anton, Ph.D., Ohio State University, 2000. *Particles and Fields*. Experimental high-energy particle physics, primarily at the CMS experiment at the LHC.

Sesha, Ramakrishna, Ph.D., Indian Institute of Science, 1995. *Atomic, Molecular, & Optical Physics*. Theoretical and computational research on the ultrafast electronic, vibrational, and rotational dynamics of molecules.

Shafranjuk, Serhii, Ph.D., Kiev University, 1985. *Condensed Matter Physics, Nano Science and Technology*. Electromagnetic and far-infrared properties of graphene and carbon nanotube junctions.

Tomaradze, Amiran, Ph.D., Institute of High-Energy Physics, 1988. *Nuclear Physics, Particles and Fields*. Experimental studies of charmed quarks and antiquarks (charmonium).

Adjunct Professor

Bader, Sam, Ph.D., University of California, Berkeley, 1974. Primary appointment is at Argonne National Laboratory. *Condensed Matter Physics*. Experimental studies of magnetic surfaces and films; ferromagnetic-superconducting multilayers; bio-inspired self-assembly of magnetic nanostructures.

Boughezal, Radja, Ph.D., Institute Theoretische Physik, Universität Zurich, 2005. Primary appointment is at Argonne National Laboratory. *Particles and Fields*. Precision QCD predictions for collider observables; Higgs phenomenology; electroweak precision observables.

Patashinski, Alexander, Ph.D., Kharkov-Moscow University, 1968. Primary appointment is in Chemistry. *Condensed Matter Physics*. Condensed-matter physics; bulk and surface phenomena; phase and glass transitions; turbulence.

Roberts, Douglas, Ph.D., University of Oklahoma, 1986. *Astrophysics*. Radio astronomy, especially as used to study the center of the Milky Way galaxy.

Affiliate Professor

Bedzyk, Michael, Ph.D., SUNY, Albany, 1982. Primary appointment is in Materials Science. *Applied Physics, Condensed Matter Physics, Materials Science, Metallurgy*. Materials science; x-ray studies of semiconductor multilayers; characterization of thin films and water/crystalline interfaces.

Jacobsen, Chris, Ph.D., Stony Brook University, 1988. Primary appointment is at Argonne National Laboratory. *Applied Physics, Biophysics, Materials Science, Metallurgy*. X-ray microscopy applied to problems in biology, materials science, and environmental science.

Kumar, Prem, Ph.D., State University of New York at Buffalo, 1980. Primary appointment is in Electrical Engineering. *Applied Physics, Atomic, Molecular, & Optical Physics, Optics*. Quantum communications and computing; nonlinear and quantum optics; fiber-optic communications; all-optical networks.

Seideman, Tamar, Ph.D., Weizmann Institute of Science, 1990. Primary appointment is in Chemistry. *Atomic, Molecular, & Optical Physics*. Molecular electronics; current-driven nanochemistry; interaction of matter with intense fields; photomanipulation of molecular modes.

Shahriar, Selim, Ph.D., Massachusetts Institute of Technology, 1992. Primary appointment is in Electrical Engineering. *Applied Physics, Atomic, Molecular, & Optical Physics, Relativity & Gravitation*. Gravity wave detection; cooling and trapping of neutral atoms; quantum computing; optical communication; optical coherence tomography.

Solla, Sara, Ph.D., University of Washington, 1982. Primary appointment is in Neurology. *Biophysics, Statistical & Thermal Physics*. Neural networks; generalization abilities in adaptive systems; pattern recognition; motor control; advanced statistical mechanics.

Yuen, Horace P., Ph.D., Massachusetts Institute of Technology, 1970. Primary appointment is in Electrical Engineering. *Atomic, Molecular, & Optical Physics*. Theory of quantum optics and communication; the foundations of quantum physics; new quantum devices; quantum cryptography.

Lecturer

Brown, Deborah, Ph.D., Northwestern University, 1983. *Physics and other Science Education*. Physics education.

Rivers, Andrew, Ph.D., New Mexico Institute of Technology, 2000. Joint appointment with College of Arts and Sciences Advising Center. *Physics and other Science Education*. Physics education.

Schmidt, Arthur, Ph.D., University of Notre Dame, 1974. Director of Undergraduate Laboratories. *Physics and other Science Education*. Physics education.

Smutko, Michael, Ph.D., University of Chicago, 1998. Joint appointment with Adler Planetarium (Chicago). *Physics and other Science Education*. Physics education.

Taylor, David, Ph.D., University of Maryland, 1983. *Physics and other Science Education*. Physics education.

DEPARTMENTAL RESEARCH SPECIALTIES AND STAFF

Theoretical

Astrophysics. Neutron stars; formation and dynamics of multiple star systems; stellar atmospheres; Type-I supernovae; dynamics of dense stellar systems; hydrodynamic stellar interactions; X-ray binaries; gravitational waves; planetary formation and dynamics; accretion disks; MHD turbulence; cosmological halo formation; gamma-ray bursts. Kalogera, Lithwick, Meyer, Novak, Rasio, Smutko, Ulmer, Yusef-Zadeh.

Atomic, Molecular, & Optical Physics. Molecular electronics; current-driven nanochemistry; interaction of matter with intense fields; photomanipulation of molecular modes.

Biophysics. Nonlinear dynamics; chaos; neural networks; statistical physics of biological systems. Motter, Solla.

Condensed Matter Physics. Electronic structure of molecules and crystals; electronic, optical, and magnetic structure studies using accurate first principles methods; electronic properties of semiconductors and composite structures; electron-hole liquid at metallic densities and the electron-hole liquid in semiconductors; many-body physics including monomolecular films, chemisorption, and liquid and solid helium. Ellis, Freeman, Garg, Koch, Sauls.

Particles and Fields. Fundamental interactions of elementary particles; quantum chromodynamics; electroweak symmetry breaking; phenomenology of weak interactions; neutrino oscillations; astrophysical particle theory. Anastassov, Bader, Boughezal, de Gouvêa, Low, Petriello, Seth.

Experimental

Astrophysics. Optical/UV observations of interstellar gas/dust and quasar absorption line systems; sub-mm polarimetry of interstellar magnetic fields; radio/IR/X-ray observations of su-

pernova remnants; star formation regions and the Galactic Center; sub-mm/UV/X-ray astronomical instrumentation; gamma-ray bursts.

Atomic, Molecular, & Optical Physics. Precision measurements; ultracold chemistry; quantum control of trapped molecular ions; atom interferometry; Bose-Einstein condensates; plasmonic surface traps. Kumar, Odom, Shahriar, Stern.

Condensed Matter Physics. Ultra-low-temperature physics; properties of superfluid helium; magnetic, structural, and superconducting properties of composition modulated alloys; nuclear magnetic resonance; x-ray studies of monolayer and multilayer films; catalysis and properties of small metallic particles; semiconductor superlattices; light scattering. Chandrasekhar, Dutta, Halperin, Ketterson.

Nuclear Physics. Strong interactions in many-body systems; electron-, pion-, nucleon-, and heavy-ion-induced reactions and scattering on nuclei.

Particles and Fields. Studies of electroweak interactions; spectroscopy of heavy quark states; supersymmetry searches at the 2-TeV pp collider (TeV II) at Fermilab; searches for charmomium states at Fermilab; rare K decay studies (CERN); CMS preparation at CERN. Anastassov, Dahl, Hahn, Schmitt, Velasco.

View additional information about this department at www.gradschoolshopper.com

THE UNIVERSITY OF CHICAGO

DEPARTMENT OF ASTRONOMY AND ASTROPHYSICS

Chicago, Illinois 60637
http://astro.uchicago.edu

General University Information
President: Robert J. Zimmer
Dean of Graduate School: Robert Fefferman
University website: http://www.uchicago.edu
Control: Public
Setting: Urban
Total Faculty: 2,750
Total number of Students: 15,219
Total number of Graduate Students: 9,850

Department Information
Department Chairman: Angela Olinto, Chair
Department Contact: Laticia Rebeles, Graduate Student Affairs Administrator
Total full-time faculty: 24
Total number of full-time equivalent positions: 24
Full-Time Graduate Students: 28
First-Year Graduate Students: 3
Female First-Year Students: 1
Total Post Doctorates: 19

Department Address
5640 S. Ellis Avenue
Chicago, IL 60637
Phone: (773) 702-9808

Fax: (773) 702-8212
E-mail: lrebeles@oddjob.uchicago.edu
Website: http://astro.uchicago.edu

ADMISSIONS

Admission Contact Information
Address admission inquiries to: Director of Admissions, Department of Astronomy and Astrophysics, 5640 S. Ellis Avenue, Chicago, IL 60637
Phone: (773) 702-9808
E-mail: lrebeles@oddjob.uchicago.edu
Admissions website: http://astro.uchicago.edu

Application deadlines
Fall admission:
U.S. students: January 3   *Int'l. students*: January 3

Application fee
U.S. students: $55   *Int'l. students*: $55

Admissions information
For Fall of 2013:
Number of applicants: 141
Number admitted: 15
Number enrolled: 3

Admission requirements

Bachelor's degree requirements: Bachelor's degree, preferably in physics or astronomy, is required, but others will be considered.
Minimum undergraduate GPA: 3.0

GRE requirements

The GRE is required.

Advanced GRE requirements

The Advanced GRE is required.

TOEFL requirements

The TOEFL exam is required for students from non-English-speaking countries.
iBT score: 90

TUITION

Tuition year 2013-14:
Tuition for in-state residents
Full-time students: per quarter
Full-time students: $60,660 annual
Credit hours per semester to be considered full-time: 300
Deferred tuition plan: No
Health insurance: Available at the cost of $4,028 per year.
Other academic fees: University Student Medical Basic Plan, Student life fee 2013-2014, $1,338/qtr.
Academic term: Quarter
Number of first-year students who received full tuition waivers: 3

Teaching Assistants, Research Assistants, and Fellowships

Average stipend per academic year
  Teaching Assistant: $29,200
  Research Assistant: $29,200
  Fellowship student: $32,000
All of our students are always supported by either an RA or TA fellowship, or by a fellowship outside of the department.

FINANCIAL AID

Loans

Loans are available for U.S. students.
Loans are available for international students.
GAPSFAS application required: No
FAFSA application required: No

For further information

Address financial aid inquiries to:
Phone: (773) 702-8666
E-mail: college-aid@uchicago.edu
Financial aid website: http://collegeaid.uchicago.edu

HOUSING

Availability of on-campus housing

Single students: Yes

For further information

Address housing inquiries to: Neighborhood Student Apartments, 5316 South Dorchester Ave., Chicago, IL 60615-5360.
Phone: (773) 753-2218
Housing aid website: http://apartments.uchicago.edu/realtors/gradopts.html

Table A—Faculty, Enrollments, and Degrees Granted

Research Specialty	2012-13 Faculty	Enrollment Fall 2012 Master's	Enrollment Fall 2012 Doctorate	Number of Degrees Granted 2012-13 Master's	Number of Degrees Granted 2012-13 Terminal Master's	Number of Degrees Granted 2012-13 Doctorate
Astronomy	36	–	29	–	–	2(29)
Total	36	–	29	–	–	2(29)
Full-time Grad. Stud.	–	–	29	–	–	–
First-year Grad. Stud.	–	–	6	–	–	–

GRADUATE DEGREE REQUIREMENTS

Master's: Full-time registration. The candidate must complete a required sequence of courses with a 3.0 average and pass the Ph.D. candidacy examination.
Doctorate: Full-time registration. The candidate must complete a required sequence of courses with a 3.0 average, pass the Ph.D. candidacy examination, submit and defend-successfully a thesis, and have the thesis submitted to a recognized journal.
Thesis: Thesis may be written in absentia.

SPECIAL EQUIPMENT, FACILITIES, OR PROGRAMS

Research in astronomy and astrophysics at the University of Chicago covers a broad range of topics, including the Sun and solar-like stars, cosmic rays, the chemical origin of meteorites and comets, interstellar matter, the birth of stars, the death of stars and nucleosynthesis, high energy and relativistic astrophysics, the origins and dynamics of galaxies, and cosmology. The activities involve theoretical, experimental, and observational programs among a community of faculty members from the Departments of Astronomy and Astrophysics, Chemistry, Geophysical Sciences, Mathematics, and Physics, with connections to Argonne National Laboratory and Fermi National Accelerator Laboratory.

The students and faculty of the University of Chicago enjoy access to a wide range of observational facilities. The Departmental observational facility is the 3.5-meter aperture telescope at Apache Point Observatory in New Mexico. This telescope has been designed to permit routine remote observing and rapid changeover between instruments, and is instrumented to work from 0.35 mm to 2 microns. A very high-resolution Echelle Spectrograph built by a team led by faculty member Roger Hildebrand allows researchers to determine the composition of stars nearby and to probe the Universe at a time before stars and galaxies existed. Adaptive optics are being developed for the telescope, which will enable faint objects to be studied with a resolution of 0.1 arcsecond; this program is part of a larger NSF-funded effort at Chicago to bring a variety of adaptive-optics techniques to bear on improving the image quality of large-aperture reflecting telescopes. The instrumentation and the adaptive optics are being developed both at the Chicago campus and at the Yerkes Observatory in Williams Bay, Wisconsin. Yerkes serves as a laboratory for development of the instruments and techniques to be used on major telescopes, including the 3.5-meter telescope, the Stratospheric Observatory for Infrared Astronomy (SOFIA), and the Infrared Telescope Facility, and also provides a continuing observational program, with its famous 40-inch (1-meter) refractor and is 41-inch (1-meter) and 0.6-meter reflectors.

In addition, Chicago astronomers regularly use telescopes at the national observatories (Kitt Peak National Observatory, Gemini telescopes and the Cerro Tololo Inter-American Observatory), as well as at other observatories such as the CSO, JCMT, Keck I and II and UKIRT facilities on Mauna Kea, telescopes of the McDonald Observatory in Texas, the 200-inch Hale telescope at Palomar, and the Very Large Array (VLA), BIMA and OVRO radio arrays. Various active NASA satellites (and archives) are also used, including IRAS, Einstein, EXOSAT, IUE, HST, the Chandra X-ray Observatory, HETE-2, ROSAT, COBE, Compton GRO, Rossi XTE, and EUVE, as well as high-altitude balloons. Chicago astronomers will soon participate in observatories that are coming into operation, including the SubMillimeter Array (SMA), SOFIA, and the Gemini 8-meter telescopes.

In collaboration with Fermilab, Princeton University, the Institute for Advanced Study, Johns Hopkins University, a consortium of Japanese institutions, the US Naval Observatory, the Max-Planck-Institute for Astronomy and the University of Washington, we have built a 2.5-meter dedicated telescope, a half-billion-pixel CCD camera, and a 600-object spectrograph to study the large-scale structure of the Universe. The main scientific goals of the Sloan Digital Sky Survey are to construct a three-dimensional map of the Universe by obtaining redshifts for a million galaxies and 100,000 QSOs and accurate digital five-color photometry for 200 million objects. The 30-terabyte SDSS database will soon become the largest and most important astronomical database in existence.

Computing facilities for theoretical and numerical work at Chicago have recently been expanded in the area of visualization, as part of a major NASA-supported initiative in high-performance computing and communications. Much of the campus pioneering of computer networks, graphics, workstations, and telecommunications have been done in this department, which enjoys a close working relationship with the Argonne National Laboratory (ANL), managed by the University of Chicago for the Department of Energy, which has a particularly strong program in high-performance parallel computing.

This department is also the principal host of the Center for Astrophysical Thermonuclear Flashes ("Flash Center"), headed by Donald Lamb. This Center is one of five university-based-centers of excellence funded by the DOE Accelerated Strategic Computing Initiative (ASCI) program, and represents a large collaboration between some 35 University scientists, representing almost all of the Physical Sciences Division's departments and institutes, and scientists at Argonne National Laboratory, at Rensselaer Polytechnic Institute, and at the three DOE defense programs laboratories. The primary focus of the Center is to develop a new generation of computational tools for attacking the problem of nuclear burning on the surfaces of neutron stars and white dwarfs, and in the interior of white dwarfs. This development involves creation of new tools for computing on massively parallel computers; new algorithms for following the complex fluid behavior of astrophysical nuclear flames; and new methods for storing and displaying the resulting data.

A number of faculty in this department are active participants in the Kavali Institute for Cosmological Physics (KICP), a Physics Frontier Center funded by the National Science Foundation. Over the past two decades some of the most important discoveries both in physics and astronomy have come at the cosmology/particle physics boundary. These discoveries, as well as theoretical advances, raise a new set of questions that involve astronomy and particle physics in an indivisible way. This Center is devoted to exploiting the connections between physics at the smallest scale—interactions of the quarks and leptons—and at the largest scale—the constitution and birth of the cosmos itself. The key is an integrated approach—astronomers and physicists, theorists and experimentalists working together, using telescopes and accelerators.

The study of astronomical objects by researchers at Chicago begins nearby, with the solar system. Our proximity to the Sun allows detailed studies of this star. Studies of active regions provide clues to the nature and origin of its magnetic field, and numerical simulations of turbulent compressible-convection help us to understand the nature of energy, angular momentum, and magnetic field transport in its outer layers; tools of helioseismology, together with theory, are being used to probe the interior of the Sun. Observations of other, solar-like, stars are then used by Chicago scientists as a means by which ideas developed in the solar context can be tested: such stars thus become our laboratory.

We can now trace the history of the Universe back to within a fraction of a second of the beginning as well as tracing stars from birth to death. We are asking deep questions about how the Universe began, how stars explode, the origin of the chemical elements and the interworkings of black holes. The confluence of advances in our understanding of the Universe and leaps in technological capability have astrophysics poised for many exciting decades as the next millennium dawns. We are well prepared to participate in this most important and exciting endeavor.

FACULTY

Chair Professor

Olinto, Angela V., Ph.D., Massachusetts Institute of Technology, 1987. *Astrophysics, Cosmology & String Theory, Particles and Fields*. Particle and nuclear astrophysics; cosmology.

Professor

Carlstrom, John E., Ph.D., University of California, Berkeley, 1988. *Astronomy, Cosmology & String Theory*. Star formation and cosmology; observation and new instrumentation.

Cronin, James W., Ph.D., University of Chicago, 1955. *Astrophysics*. Ultra high-energy gamma-ray astrophysics.

Cudworth, Kyle M., Ph.D., University of California, Santa Cruz, 1974. *Astronomy*. Star clusters.

Frieman, Joshua A., Ph.D., University of Chicago, 1985. *Astrophysics, Cosmology & String Theory*. Cosmology; particle astrophysics.

Harper, Doyal A., Ph.D., Rice University, 1971. *Astronomy*. Infrared astronomy and the structure of active objects.

Hildebrand, Roger H., Ph.D., University of California, Berkeley, 1951. Far infrared and submillimeter astronomy.

Hobbs, Lewis M., Ph.D., University of Wisconsin-Madison, 1966. *Astronomy*. Interstellar matter and galactic structure.

Hogan, Craig, Ph.D., University of Cambridge, 1980. *Astrophysics*. Dark Energy; particle astrophysics.

Königl, Arieh, Ph.D., California Institute of Technology, 1980. *Astrophysics*. Theoretical high-energy astrophysics.

Khokhlov, Alexei, Ph.D., Moscow State University, 1984. *Astrophysics*. Fluid dynamics; supernovae jets; nucleosynthesis; experimental astrophysics.

Kibblewhite, E. J., Ph.D., University of Cambridge, 1971. *Optics*. Adaptive optics; high-resolution imaging.

Kolb, Edward W., Ph.D., University of Texas, 1978. *Astrophysics, Cosmology & String Theory, Particles and Fields*. Particle physics; cosmology; theoretical astrophysics.

Kravtsov, Andrey, Ph.D., New Mexico State University, 1999. *Cosmology & String Theory*. Cosmology; structure formation in the universe; numerical simulations.

Kron, Richard G., Ph.D., University of California, Berkeley, 1978. *Astronomy*. Director of Yerkes Observatory. Observational studies of active galaxies.

Lamb, Donald Q., Ph.D., University of Rochester, 1974. *Astrophysics*. Compact objects; high-energy astrophysics.

Meyer, Stephan S., Ph.D., Princeton University, 1979. *Astrophysics, Cosmology & String Theory*. Infrared astrophysics and observational cosmology.

Oka, Takeshi, Ph.D., University of Tokyo, 1960. Laser spectroscopy and interstellar molecules.

Palmer, Patrick E., Ph.D., Harvard University, 1968. *Astronomy.* Radio astronomy and interstellar molecules.

Privitera, Paolo, Ph.D., Laurea University of Applied Sciences, 1993. *Astronomy, Astrophysics, Atmosphere, Space Physics, Cosmic Rays.* Ultra-high energy cosmic rays; extragalactic astronomy; high energy astrophysics.

Rosner, Robert, Ph.D., Harvard University, 1975. *Astrophysics, Plasma and Fusion, Solar Physics.* Fluid and plasma dynamics; solar physics; high-energy astrophysics.

Truran, James W., Ph.D., Yale University, 1965. *Astrophysics.* Nuclear astrophysics; evolution of stars and galaxies; high-energy astrophysics.

Turner, Michael S., Ph.D., Stanford University, 1978. *Astrophysics, Cosmology & String Theory, Particles and Fields, Relativity & Gravitation.* Cosmology and particle physics; relativistic astrophysics.

Vandervoort, Peter O., Ph.D., University of Chicago, 1960. *Astronomy.* Analytical dynamics of galaxies.

York, Donald G., Ph.D., University of Chicago, 1970. *Cosmology & String Theory.* Interstellar and intergalactic matter; observational cosmology.

Associate Professor

Cattaneo, Fausto, Ph.D., University of Cambridge, 1984. *Astronomy, Astrophysics, Solar Physics.* Solar system astronomy, high energy and computational astrophysics.

Chen, Hsiao-Wen, Ph.D., Stony Brook University, 1999. Observational extragalactic astronomy.

Dodelson, Scott, Ph.D., Columbia University, 1988. *Cosmology & String Theory, Relativity & Gravitation.* Cosmology; gravitational lensing.

Gladders, Michael, Ph.D., University of Toronto, 2002. *Cosmology & String Theory.* Observational cosmology and instrumentation.

Gnedin, Nickolay Y., Ph.D., Princeton University, 1996. *Astronomy, Cosmology & String Theory.* Cosmology; galaxy formation; supercomputer simulations.

Hooper, Dan, Ph.D., University of Wisconsin-Madison, 2003. *Astrophysics.* Theoretical astrophysics.

Hu, Wayne, Ph.D., University of California, Berkeley, 1995. *Astronomy, Cosmology & String Theory.* Precision cosmology; CMB; galaxy surveys; weak lensing.

Kent, Stephen M., Ph.D., California Institute of Technology, 1980. *Astronomy.* Observational studies of galaxies.

Miller, Richard H., Ph.D., University of Chicago, 1957. *Astronomy.* Numerical experiments and the dynamical evolution of galaxies.

Assistant Professor

Bean, Jacob, Ph.D., University of Texas, 2007. *Astronomy, Astrophysics.* Extra solar planets; properties and physics of low-mass stars; high-precision spectroscopic, photometric, and astrometric methods.

Fabrycky, Daniel, Ph.D., Princeton University, 2007. *Astronomy.* Formation and evolution of planets.

DEPARTMENTAL RESEARCH SPECIALTIES AND STAFF

Theoretical
History & Philosophy of Physics/Science.

View additional information about this department at
www.gradschoolshopper.com

UNIVERSITY OF ILLINOIS AT CHICAGO

DEPARTMENT OF PHYSICS

Chicago, Illinois 60607
http://physicsweb.phy.uic.edu/

General University Information
President: Robert A. Easter
Dean of Graduate School: Karen J. Colley
University website: http://www.uic.edu
Control: Public
Setting: Urban
Total Faculty: 2,036
Total Graduate Faculty: 1,589
Total number of Students: 27,512
Total number of Graduate Students: 8,119

Department Information
Department Chairman: David Hofman, Head
Department Contact: James Nell, Graduate Advisor
  Total full-time faculty: 22
  Full-Time Graduate Students: 86
  First-Year Graduate Students: 20

Female First-Year Students: 3
Total Post Doctorates: 11

Department Address
845 West Taylor Street, Room 2236
SES, MC 273
Chicago, IL 60607
Phone: (312) 996-3400
Fax: (312) 996-9016
E-mail: physics@uic.edu
Website: http://physicsweb.phy.uic.edu/

ADMISSIONS

Admission Contact Information
Address admission inquiries to: Graduate Admissions, Department of Physics, M/C 273, 845 W. Taylor #2236 SES, Chicago, IL 60607-7059, or FAX: (312) 996-9016

Phone: (312) 996-3400
E-mail: physics@uic.edu
Admissions website: http://grad.uic.edu

Application deadlines
Fall admission:
U.S. students: May 15   *Int'l. students*: February 15

Application fee
U.S. students: $60   *Int'l. students*: $60
January 1 for consideration for the University Fellowship; February 15 priority deadline for domestic applicants.

Admissions information
For Fall of 2012:
Number of applicants: 145
Number admitted: 47
Number enrolled: 20

Admission requirements
Bachelor's degree requirements: Bachelor's degree is required. Prior academic work must include at least 20 semester hours of physics, including upper-level undergraduate electrodynamics, quantum mechanics, and classical mechanics.
Minimum undergraduate GPA: 2.75

GRE requirements
The GRE is required.

Advanced GRE requirements
The Advanced GRE is not required.

TOEFL requirements
The TOEFL exam is required for students from non-English-speaking countries.
PBT score: 550
iBT score: 80

Other admissions information
Additional requirements: A complete application will include an online application and payment of the application fee (http://www.uic.edu/depts/oar/grad/apply_grad_degree.html); electronic transcripts for previous post-secondary coursework and proof of any degrees earned (if in a language other than English, an electronic certified translated copy is also required); GRE and TOEFL or IELTS scores (minimum acceptable scores: TOEFL paper based 550; computer based statement 213; iBT-80; IELTS 6.5 total); 3 electronic letters of recommendation; an academic statement of purpose; an application for graduate appointment if applying for financial assistance.
Undergraduate preparation assumed: Thornton and Marion - Classical Dynamics; Griffiths - Introduction to Electrodynamics; Gould and Tobochnik - Statistical and Thermal Physics; Griffiths - Introduction to Quantum Mechanics.

TUITION
Tuition year 2012–13:
Tuition for in-state residents
Full-time students: $6,316 per semester
Part-time students: $4,210 per semester
Tuition for out-of-state residents
Full-time students: $12,310 per semester
Part-time students: $8,210 per semester
Tuition is waived for all students with assistantships. Currently, all physics graduate students have assistantships.
Credit hours per semester to be considered full-time: 12
Deferred tuition plan: Yes
Health insurance: Available at the cost of $802 per year.

Other academic fees: Between $1574-$1853/semester-dependent on number of credits (figures include the $401/sem. health insurance fee). All fees are waived for students with assistantships (currently, all physics graduate students have assistantships) except $425 general fee and a portion of $401 health insurance fee.
Academic term: Semester
Number of first-year students who received full tuition waivers: 20

Teaching Assistants, Research Assistants, and Fellowships
Number of first-year
Teaching Assistants: 20
Average stipend per academic year
Teaching Assistant: $15,885
Research Assistant: $16,203
Currently, all physics graduate students are supported on either TA or RA appointments. Outstanding physics graduate students have also received graduate college fellowships (http://grad.uic.edu/cms/?pid=1000893).

FINANCIAL AID

Application deadlines
Fall admission:
U.S. students: March 1

Loans
Loans are available for U.S. students.
Loans are not available for international students.
GAPSFAS application required: No
FAFSA application required: Yes

For further information
Address financial aid inquiries to: Graduate Admissions, Department of Physics, M/C 273, 845 W. Taylor #2236 SES, Chicago, IL 60607-7059.
Phone: (312) 996-3400
E-mail: physics@uic.edu
Financial aid website: http://www.uic.edu/depts/financialaid/

HOUSING

Availability of on-campus housing
Single students: Yes
Married students: No

For further information
Address housing inquiries to: University of Illinois at Chicago (M/C 579), 818 S. Wolcott St., Chicago, IL 60612.
Phone: (312) 355-6300
E-mail: housing@uic.edu
Housing aid website: http://www.housing.uic.edu

Table A—Faculty, Enrollments, and Degrees Granted

Research Specialty	2012–13 Faculty	Enrollment Fall 2011		Number of Degrees Granted 2012-13 (2007–13)		
		Master's	Doctorate	Master's	Terminal Master's	Doctorate
Atomic, Molecular, & Optical Physics	–	–	2	–(1)	–	1(2)
Biophysics	3	1	5	–(1)	–	1(8)
Condensed Matter Physics	7	–	23	–(15)	–	2(17)
Energy Sources & Environment	1	–	2	–	–	–
High Energy Heavy Ion Physics	5	–	6	–(3)	–	–(1)
High Energy Particle Physics, Particles and Fields	6	–	10	–(1)	–	4(8)
Non-specialized	–	8	29	7(19)	1(24)	–
Total	22	9	77	7(40)	1(24)	8(36)
Full-time Grad. Stud.	–	9	77	–	–	–
First-year Grad. Stud.	–	3	17	–	–	–

GRADUATE DEGREE REQUIREMENTS

Master's: The general requirement for the Master of Science is satisfactory completion of 32 semester hours of work in courses approved by the department. At least 20 of these hours must be at the 500 level; they must include Physics 501 and 502 (Electrodynamics) and Physics 511 and 512 (Quantum Mechanics), and may not include more than 4 hours of Physics 596 (Individual Study) or more than 8 hours of Physics 598 (Master's Thesis Research).

Doctorate: The minimum requirements for the Ph.D. are: (1) The satisfactory completion of 96 semester hours of course work approved by the department, including at least 36 hours of 500-level courses, exclusive of Physics 596 (Individual Study) and 599 (Thesis Research). These 36 hours must include the sequence Physics 501 and 502 (Electrodynamics), Physics 511 and 512 (Quantum Mechanics), Physics 561 (Statistical Mechanics), at least one complete sequence chosen from among the following: Physics 521 and 522 (Molecular and Laser Physics), Physics 531 and 532 (Solid State Physics), Physics 551 and 552 (Elementary Particle Physics), Physics 513 and 514 (Quantum Field Theory), and five semesters of the Graduate Seminar, Physics 595. (2) Satisfactory performance in a comprehensive qualifying examination consisting of 400-level problems on classical mechanics, electricity and magnetism, quantum mechanics, and thermodynamics and statistical mechanics. This examination may be repeated once but must be passed no later than January of the student's second year in residence. Details on this examination are available from the department office. (3) Satisfactory performance on an oral examination in the general area of the student's doctoral thesis research, which is to be taken within two years after passing the qualifying examination. The examination will normally start with a brief oral report by the student on his or her proposed research. If the performance is only marginally satisfactory, the student may be asked to retake the examination. (4) Satisfactory completion and defense of a doctoral dissertation. (5) Each student is required to serve as a teaching assistant for at least two semesters.

Other Degrees: For physics graduate students interested in science education careers, UIC offers an exciting Masters of Education Degree (MEd) in Instructional Leadership: Science Education (http://www.uic.edu/gcat/EDINLE.shtml). Many physics graduate students who have earned a master's in physics have also earned their MEd degree while at UIC.

Thesis: Thesis may not be written in absentia.

SPECIAL EQUIPMENT, FACILITIES, OR PROGRAMS

Significant onsite research laboratories include the University of Illinois at Chicago (UIC) Research Resources Center, the Laboratory of Atomic/Molecular Laser Physics, the Microphysics Laboratory, and the Silicon Detector Laboratory for Experimental Nuclear Physics. These research laboratories contain many notable resources including a new aberration corrected scanning transmission microscope (the UIC JEOL JEM-ARM200CF) located at the Research Resources Center that is the first such instrument in the United States with such a high level of capability. The Laboratory of Atomic/Molecular Laser Physics has the world's highest spectral brightness tunable ultraviolet laser, as well as dye laser systems, ti-sapphire, various excimer lasers, high-resolution spectrometers, and other computer-assisted optical equipment. The Microphysics Laboratory has molecular beam epitaxy (MBE), ultrahigh-vacuum growth chambers and surface analytical facilities - as well as helium-3 and dilution refrigerators, an ultrasonic spectrometer, Foner and SQUID magnetometers, a susceptibility balance, a Bruker NMR spectrometer, an atomic force microscopy apparatus, an x-ray diffraction system, an automated adiabatic calorimeter and on-line computer data handling systems with dedicated mini-and microcomputers. The Silicon Detector Laboratory for Experimental Nuclear Physics has testing equipment and facilities used in the assembly of silicon detectors for large particle and nuclear physics experiments.

Faculty and students are utilizing two major local research laboratories for a wide variety of research; Argonne National Laboratory and Fermilab (located only 22 miles and 35 west of UIC, respectively). Faculty and students are also engaged in materials, laser and energy research in collaboration with the Pacific Northwest National Laboratory and with the National Renewable Energy Laboratory; use the extensive experimental high-energy particle and nuclear physics facilities of Fermilab, Brookhaven National Laboratory in New York, and CERN in Switzerland; the solid state and the photon source facilities at Argonne National Laboratory and Brookhaven National Laboratory; and the neutron and biophysics related facilities of the NIST Center for Neutron Research, the ISIS facility at the Rutherford Appleton Laboratory in the UK and the ILL (Institut Laue-Langevin) facility in France.

Research is supported by several significant dedicated computer clusters hosted and administered within the department as well as dedicated facilities at the UIC Computer Center. UIC is also a major hub of the forefront STARLIGHT advanced research and education high speed network. This high-performance research network is connected to Europe, Asia, and the rest of the world. The UIC Computer Center also hosts campus-wide networks and has wide bandwidth connections to the supercomputer centers at the University of Illinois at Urbana-Champaign and nationwide.

The Department of Physics research program is supported by a large machine shop with ample equipment and instrument making expertise. The department maintains a sample preparation facility with an arc melter, zone refiner, a spark cutter, x-ray equipment, furnaces, dry boxes, polishing equipment, and other metallurgical instruments.

Table B—Separately Budgeted Research Expenditures by Source of Support

Source of Support	Departmental Research	Physics-related Research Outside Department
Federal government	$5,102,780	
State/local government		
Non-profit organizations		
Business and industry	$83,005	
Other	$2,613	
Total	**$5,188,398**	

Table C—Separately Budgeted Research Expenditures by Research Specialty

Research Specialty	No. of Grants	Expenditures ($)
Atomic, Molecular, & Optical Physics	1	$306,496
Biophysics	6	$582,673
Condensed Matter Physics	25	$2,143,936
Energy Sources & Environment	3	$255,144
High Energy Heavy Ion Physics	4	$756,266
High Energy Particle Physics, Particles and Fields	15	$1,143,883
Total	**54**	**$5,188,398**

FACULTY

Professor

Ansari, Anjum, Ph.D., University of Illinois at Urbana-Champaign, 1988. *Biophysics*. Biological physics.

Aratyn, Henrik, Ph.D., Copenhagen, N. Bohr Institute, 1984. Associate Dean, College of Liberal Arts and Sciences. *High Energy Physics*. Mathematical physics.

Campuzano, Juan C., Ph.D., University of Wisconsin-Milwaukee, 1978. *Condensed Matter Physics*. Experimental solid state physics.

Crabtree, George, Ph.D., University of Illinois at Chicago, 1974. *Condensed Matter Physics, Energy Sources & Environment*.

Gerber, Cecilia, Ph.D., Universidad de Buenos Aires, 1995. Director of Undergraduate Studies and Visiting Associate Head. *High Energy Physics*. Experimental high-energy particle physics.

Grein, Christoph, Ph.D., Princeton University, 1989. Director of Graduate Studies. *Condensed Matter Physics*. Theoretical condensed matter physics.

Hofman, David, Ph.D., Stony Brook University, 1994. Acting Head. *Nuclear Physics*. Experimental high-energy nuclear physics.

Keung, Wai-Yee, Ph.D., University of Wisconsin-Madison, 1980. *High Energy Physics*. High-energy physics theory and phenomenology.

Morr, Dirk, Ph.D., University of Wisconsin-Madison, 1997. *Condensed Matter Physics*. Theoretical condensed matter.

Ogut, Serdar, Ph.D., Yale University, 1995. Co-Director of Graduate Studies. *Computational Physics, Condensed Matter Physics*. Theoretical condensed matter physics.

Schlossman, Mark, Ph.D., Cornell University, 1987. *Biophysics, Condensed Matter Physics*. Experimental soft condensed matter physics, biological physics.

Schroeder, W. Andreas, Ph.D., University of London, Imperial College, 1987. *Condensed Matter Physics*. Ultrafast laser spectroscopy.

Sivananthan, Sivalingham, Ph.D., University of Illinois at Chicago, 1988. Director, Microphysics Laboratory. *Condensed Matter Physics, Materials Science, Metallurgy*. Experimental solid state physics, material science.

Stephanov, Mikhail, Ph.D., University of Oxford, 1994. Co-Director of Undergraduate Studies. *Nuclear Physics, Particles and Fields, Theoretical Physics*. Theoretical nuclear and high-energy physics.

Varelas, Nikos, Ph.D., University of Rochester, 1994. *High Energy Physics*. Experimental high-energy particle physics.

Associate Professor

Evdokimov, Olga, Ph.D., Joint Institute for Nuclear Research, Dubna, Russia, 1999. *Nuclear Physics*. Experimental high-energy nuclear physics.

Imbo, Tom, Ph.D., University of Texas, 1988. *High Energy Physics, Theoretical Physics, Other*. High-energy theory, mathematical physics, foundations of quantum physics.

Klie, Robert F., Ph.D., University of Illinois at Chicago, 2002. *Condensed Matter Physics, Surface Physics*. Atomic-resolution transmission electron microscopy of nanoscale materials systems.

Assistant Professor

Cavanaugh, Richard, Ph.D., Florida State University, 1999. *High Energy Physics*. Experimental high-energy particle physics.

Perez-Salas, Ursula, Ph.D., University of Maryland, 2000. *Biophysics*. Biological physics.

Ye, Zhenyu, Ph.D., University of Hamburg, 2006. *Nuclear Physics, Other*. Experimental high-energy nuclear physics, the development of silicon detector technology.

Yee, Ho-Ung, Ph.D., Yale University, 2003. *Nuclear Physics, Theoretical Physics*.

Emeritus

Adams, Mark R., Ph.D., Stony Brook University, 1981. *High Energy Physics*. Experimental high-energy particle physics.

Betts, R. Russell, Ph.D., University of Pennsylvania, 1972. *Nuclear Physics*. Nuclear experiment.

Boccara, Nino, Ph.D., Paris, 1961. Mathematical physics and modeling.

Bodmer, Arnold, Ph.D., University of Manchester, 1953. Theoretical nuclear physics.

Carhart, Richard, Ph.D., University of Wisconsin-Madison, 1965. Theoretical high-energy physics and environmental physics.

Claus, Helmut, Ph.D., Karlsruhe University, 1965. Experimental solid state physics.

Faurie, Jean-Pierre, Ph.D., University of Clermont-Ferrand, 1970. Experimental solid state physics.

Garland, James, Ph.D., University of Chicago, 1966. Theoretical solid state physics.

Goldberg, Howard, Ph.D., University of California, Berkeley, 1964. Experimental high-energy physics.

Halliwell, Clive, Ph.D., University of Manchester, 1971. *High Energy Physics*. High-energy particle and nuclear physics.

Licht, Arthur L., Ph.D., University of Maryland, 1963. Theoretical high-energy physics; astrophysics; many-body theory.

McLeod, Donald W., Ph.D., Cornell University, 1962. Experimental high-energy particle and nuclear physics.

McNeil, Edward, Ph.D., University of Illinois at Urbana-Champaign, 1951.

Montano, Pedro, Ph.D., Technion, 1972. Synchrotron radiation science.

Pagnamenta, Antonio, Ph.D., University of Maryland, 1965. Theoretical high-energy physics; microdosimetry; radiation physics.

Rhodes, Charles, Ph.D., Massachusetts Institute of Technology, 1969. *Atomic, Molecular, & Optical Physics*. Laser physics; atomic and molecular physics.

Sharma, Ram R., Ph.D., University of California, Riverside, 1965. Theoretical solid state physics; biophysics.

Solomon, Julius, Ph.D., University of California, Berkeley, 1963. Experimental high-energy physics.

Sukhatme, Uday, Ph.D., Massachusetts Institute of Technology, 1971. Theoretical high-energy particle physics.

Research Associate Professor

Borisov, Alexey, Ph.D., Moscow State University, 1985.

Chang, Yong, Ph.D., Shanghai Institute of Technical Physics, 1996.

Hahn, Suk-Ryong, Ph.D., Oregon Graduate Institute of Science and Technology, 1993.

Research Assistant Professor

Apanasevich, Leonard, Ph.D., Michigan State University, 2005.

Kouznetsov, Serguei, Ph.D., Semyonov Institute of Chemical Physics, 1994.

Phillips, Patrick, Ph.D., Ohio State University, 2012.

Adjunct Professor

Cho, Michael, Ph.D., Drexel University, 1991.

Dutta, Mitra, Ph.D., University of Cincinnati, 1981.

Kang, TaeWon, Ph.D., Dongguk University, 1982.

Kunde, Gerd, Ph.D., University of Frankfurt, 1994.

Lu, Hui, Ph.D., Beckman Institute, 1999.

Mueller, Mark, Ph.D., Stanford University, 1984.

Norris, James, Ph.D., Washington University, 1968.

Stroscio, Michael, Ph.D., Yale University, 1974.

Visiting Research Associate Professor

Gao, Wei, Ph.D., University of Abertay, 1996.

Visiting Research Assistant Professor

Bazterra, Victor, Ph.D., University of Buenos Aires, 2006.

Boguta, John, Ph.D., University of Illinois at Urbana-Champaign, 1968.

Espinoza, Randall, Ph.D., University of Illinois at Chicago, 2005.

Zhao, Jun, Ph.D., Shanghai Institute of Technical Physics, 1996.

Lecturer

Barkan, Adrian, Ph.D., University of Iowa, 1997.

Goeckner, Hans, Ph.D., University of Illinois at Chicago, 1995.

Tillotson, Andrew, Ph.D., University of Maryland, 2006.

Posdoctoral Research Associate

Adriano, Cris, Ph.D., Universidade Estadual de Campinas, Brazil, 2009.

Bu, Wei, Ph.D., Iowa State University, 2009.

Ciani, Anthony, Ph.D., University of Illinois at Chicago, 2008.

Kurt, Pelin, Ph.D., Cukurova University, Turkey, 2009.

Lacroix, Florent, Ph.D., Blaise Pascal University, 2008.

Moon, Dongho, Ph.D., Korea University, 2011.

Pandit, Yadav, Ph.D., Kent State University, 2012.

Schwaller, Pedro, Ph.D., University of Zurich, 2010.

Springer, Todd, Ph.D., University of Minnesota, 2009.

Strom, Derek, Ph.D., Northwestern University, 2009.

Wang, Yaping, Ph.D., Central China Normal University, 2008.

DEPARTMENTAL RESEARCH SPECIALTIES AND STAFF

Theoretical

Biophysics. Dynamics of nucleic acids, RNA folding, protein-DNA interactions, membrane protein-lipid interactions, structural and dynamics studies in model membranes. Ansari, Schlossman.

Computational Physics. Multifaceted activities in computational physics which incorporates: high power density plasmas; particle physics; heavy-ion nuclear physics; protein structure and dynamics; materials for energy; ab initio pseudopotential total energy calculations; molecular dynamics simulations. Aratyn, Evdokimov, Grein, Ogut.

Condensed Matter Physics. Metals, semiconductors, and insulators; Density functional electronic structure calculations; cooperative and critical phenomena and phase transitions; Density functional electronic structure calculations; magnetism in disordered systems; structural instabilities; high-Tc superconductivity; surfaces and thin films; ion implantation; thermodynamic and transport properties; optical properties from Raman scattering, ellipsometry, electrore-flectance, and photocapacitance; growth by molecular beam epitaxy of II-VI semiconducting epilayers and microstructures such as superlattices and tunneling structures; electronic properties of two-dimensional systems (Shubnikov-Dehaas; Quantum Hall Effect); processing and physics of electronic devices. Grein, Morr, Ogut, Schlossman.

High Energy Physics. Standard model phenomenology, strong and electroweak interactions, Higgs boson and new particle searches, top-quark physics, strings and integrable models, strong and electroweak gauge interactions, CP violation, algebraic and topological aspects of quantum field theory, the foundations of quantum mechanics, quantum information theory, exotic statistics. Aratyn, Imbo, Keung.

Nuclear Physics. Theory of strong interactions (Quantum Chromodynamics) and its applications, such as the physics of neutron stars, heavy-ion collisions, and the quark-gluon plasma. Theory of the tri-critical point in the nuclear phase diagram. Stephanov, Yee.

Experimental

Atomic, Molecular, & Optical Physics. X-ray microimaging and advanced forms of x-ray generation, ultrafast laser spectroscopy. Rhodes, Schroeder.

Biophysics. The primary techniques of x-ray and Neutron Surface Scattering, Laser Temperature-Jump, Single-Molecule FRET, and Fluctuation Correlation Spectroscopy are used to study the following: the dynamics of Nucleic Acids; RNA Folding Protein-DNA Interactions; Membrane Protein-Lipid Interactions; Structure and Dynamics of Lipid-Lipid and Lipid-Cholesterol Interactions in Membranes; Structure and electrostatic interactions at liquid surfaces and interfaces. Ansari, Perez-Salas, Schlossman.

Condensed Matter Physics. Metals, semiconductors, and insulators; Density functional electronic structure calculations; cooperative and critical phenomena and phase transitions; Density functional electronic structure calculations; magnetism in disordered systems; structural instabilities; high-Tc superconductivity; surfaces and thin films; ion implantation; thermodynamic and transport properties; optical properties from Raman scattering, ellipsometry, electrore-flectance, and photocapacitance; growth by molecular beam epitaxy of II-VI semiconducting epilayers and microstructures such as superlattices and tunneling structures; electronic properties of two-dimensional systems (Shubnikov-Dehaas; Quantum Hall Effect); processing and physics of electronic devices; Ultrafast laser spectroscopy; Materials for energy; Structure and electrostatic interactions at liquid surfaces and interfaces. Campuzano, Crabtree, Grein, Klie, Schroeder, Sivananthan.

High Energy Physics. Collider physics at the CMS experiment at the Large Hadron Collider in CERN and analysis of data from the D0 Experiment at Fermilab, precision measurements of strong and electroweak interactions, Higgs boson and new particle searches, top-quark physics, searches for new fundamental symmetries in nature and extra dimensions of space, trigger systems development, silicon microstrip tracking detectors; trigger systems and silicon tracker development for CMS detector at CERN LHC. Adams, Cavanaugh, Gerber, Varelas.

Nuclear Physics. Relativistic heavy ion collision physics, dense nuclear matter, and studies of the quark gluon plasma as mea- sured by the STAR experiment at the Relativistic Heavy Ion Collider (at Brookhaven National Laboratory) and the CMS experiment at the Large Hadron Collider (at CERN, in Geneva Switzerland). Evdokimov, Hofman, Ye.

View additional information about this department at
www.gradschoolshopper.com

UNIVERSITY OF ILLINOIS AT URBANA-CHAMPAIGN

DEPARTMENT OF PHYSICS

Urbana, Illinois 61801-3080
http://physics.illinois.edu

General University Information
President: Robert Easter
Dean of Graduate School: Debasish Dutta
University website: http://illinois.edu
Control: Public
Setting: Suburban
Total Faculty: 2,975
Total Graduate Faculty: 1,871
Total number of Students: 42,605
Total number of Graduate Students: 10,673

Department Information
Department Chairman: Dale J. Van Harlingen, Head
Department Contact: S. Lance Cooper, Associate Head for Graduate Programs
Total full-time faculty: 60
Total number of full-time equivalent positions: 60
Full-Time Graduate Students: 257
First-Year Graduate Students: 44
Female First-Year Students: 12
Total Post Doctorates: 64

Department Address
Loomis Laboratory of Physics
1110 West Green Street
Urbana, IL 61801-3080
Phone: (217) 333-3645
Fax: (217) 244-5073
E-mail: grad@physics.illinois.edu
Website: http://physics.illinois.edu

ADMISSIONS

Admission Contact Information
Address admission inquiries to: Graduate Records Secretary, Department of Physics, 1110 W. Green St., Urbana, IL 61801-3080
Phone: (217) 333-3645
E-mail: grad@physics.illinois.edu
Admissions website: http://physics.illinois.edu

Application deadlines
Fall admission:
U.S. students: January 15   *Int'l. students*: January 15

Application fee
U.S. students: $70   *Int'l. students*: $90
The application fee must be paid by a credit card at the time an application is submitted online. Payment is valid for only one semester and must be submitted before any action is taken on an application.

Admissions information
For Fall of 2013:
Number of applicants: 522
Number admitted: 45
Number enrolled: 44

Admission requirements
Bachelor's degree requirements: A bachelor's degree in physics or a related field is required. On the last 60 hours of work, 20 semester hours (30 quarter hours) of intermediate and advanced undergraduate physics is also required.
Minimum undergraduate GPA: 3.0

GRE requirements
The GRE is required.

Advanced GRE requirements
The Advanced GRE is required.
Minimum accepted Advanced GRE score: 550
Mean Advanced GRE score range (25th–75th percentile): 790-950
No definite minimum score is set for the GRE, but applicants must demonstrate mastery of upper-level undergraduate physics concepts. The average GRE Physics subject score for 2013 admissions was 870.

TOEFL requirements
The TOEFL exam is required for students from non-English-speaking countries.
PBT score: 610
iBT score: 102
The vast majority of students to whom we offer admission are also offered a teaching assistantship for financial support. To receive an appointment as a teaching assistant, an international graduate student is required to demonstrate proficiency

in spoken English. This proficiency can be demonstrated in one of four ways: by having a score of 24 or above on the speaking sub-section of the Internet Based TOEFL; by having a score of 8 or above on the speaking sub-section of the IELTS academic exam; by having a score of 50 or above on the TSE; or by having a score of 50 or above on the locally administered University of Illinois Speak Test.

Other admissions information

Additional requirements: Admission to our program is competitive. We consider your grade-point average, GRE scores, research experiences, and potential fit into our research programs. Admissions decisions are made by a committee of our senior faculty; please do not contact individual professors requesting admission to our program. No informal assessment of your chances for admission can be made.

Undergraduate preparation assumed: Although preparation will vary, we generally expect one year of upper-division mechanics, one year of electricity and magnetism, one semester of optics, one semester of statistical and thermal physics, and one year of quantum mechanics. One or two semesters of advanced laboratory courses are also expected.

TUITION

Tuition year 2013-14:
Tuition for in-state residents
  Full-time students: $16,754 annual
Tuition for out-of-state residents
  Full-time students: $30,246 annual
Appointment as a research or teaching assistant includes a full tuition waiver.
Credit hours per semester to be considered full-time: 8
Deferred tuition plan: Yes
Health insurance: Available
Other academic fees: A description of fees is posted at http://www.registrar.illinois.edu/financial/tuition_1314/Fall/index.html. Research and teaching assistantships include tuition and partial fee waivers.
Academic term: Semester
Number of first-year students who received full tuition waivers: 44

Teaching Assistants, Research Assistants, and Fellowships

Number of first-year
  Teaching Assistants: 39
  Research Assistants: 2
  Fellowship students: 5
Average stipend per academic year
  Teaching Assistant: $21,120
  Research Assistant: $21,120
  Fellowship student: $25,000 The rates quoted above are for the nine-month academic year; students typically are also supported for the two-month summer term. The Department of Physics makes every effort to ensure that eligible prospective students are not deterred from attending because of financial constraints, and we are proud of our tradition of providing continuing and adequate support for our students. In case of financial emergencies, short-term loans are available from the University's Office of Student Financial Aid.

FINANCIAL AID

Loans

Loans are not available for U.S. students.
Loans are not available for international students.
GAPSFAS application required: No
FAFSA application required: No

For further information

Address financial aid inquiries to: Graduate Records Secretary, Department of Physics, 1110 W. Green St., Urbana, IL 61801-3080.
Phone: (217) 333-3645
E-mail: grad@physics.illinois.edu
Financial aid website: http://physics.illinois.edu/grad/financial-aid.asp

HOUSING

Availability of on-campus housing

  Single students: Yes
  Married students: Yes

For further information

Address housing inquiries to: Housing Division, 200 Clark Hall, 1203 S. Fourth, Champaign, IL 61820.
Phone: (217) 333-7111
E-mail: housing@illinois.edu
Housing aid website: http://housing.illinois.edu

GRADUATE DEGREE REQUIREMENTS

Master's: See Academic information on website. Thirty-two hours of satisfactory (GPA 2.75/4.0) graduate course work required. All hours must be at the 400-level or higher. Sixteen of the 32 hours must be in Physics, with at least 8 hours of them at the 500-level. At most, 8 hours of individual study may be counted toward the master's degree. At least 16 hours must be in courses meeting on the Urbana-Champaign campus; credit for graduate work taken elsewhere is by petition only. There is no foreign language requirement.

Doctorate: Ninety-six hours of (2.75/4.0 GPA) satisfactory graduate work. Part of these hours must be thesis work. There is no specific residence requirement, but 64 hours must be taken on the Urbana-Champaign campus. The qualifying examination (the "qual") tests the candidate's broad understanding of basic physics and his or her preparation to proceed to thesis research. A student must take and pass the qual by the beginning of the third semester of enrollment in our graduate program. The preliminary examination (the "prelim") reviews the feasibility and appropriateness of a candidate's thesis research proposal. The prelim must be taken within the first two years of joining a research group. The thesis is a comprehensive publication describing the independent research project and its results. The final defense is an oral examination conducted by the candidate's thesis committee and based on the thesis, at which the candidate presents the results of his research. There are no foreign language requirements.

Other Degrees: The Medical Scholars Program, which allows students to earn joint MD/Ph.D. degrees, combines cutting edge research in physics with individualized clinical training in medicine. All graduate and medical training is done at the Urbana-Champaign campus. Only U.S. citizens and permanent residents are eligible for admission.

Thesis: Theses may be written in absentia.

SPECIAL EQUIPMENT, FACILITIES, OR PROGRAMS

The Department of Physics offers world-class research facilities in many research areas. For a complete description of physics facilities, please consult our website, http://physics.illinois.edu.

Table B—Separately Budgeted Research Expenditures by Source of Support

Source of Support	Departmental Research	Physics-related Research Outside Department
Federal government	$17,825,000	$1,425,000
State/local government	$56,000	
Non-profit organizations	$990,000	$83,000
Business and industry	$519,000	
Other	$1,100,000	
Total	**$20,490,000**	**$1,508,000**

Table C—Separately Budgeted Research Expenditures by Research Specialty

Research Specialty	No. of Grants	Expenditures ($)
Astrophysics	3	$513,000
Atomic, Molecular, & Optical Physics	13	$1,728,000
Biological Physics	25	$4,902,000
Condensed Matter Physics	44	$5,678,000
Low Temperature Physics	19	$3,431,000
Nuclear Physics	7	$3,314,000
Physics Education Research	3	$183,000
High Energy Physics	8	$2,249,000
Total	**122**	**$21,998,000**

FACULTY

Professor

Abbamonte, Peter, Ph.D., University of Illinois, 1999. *Condensed Matter Physics*. Experimental condensed matter physics; resonant soft x-ray scattering; electron self-organization; oxide devices; quantum phase transitions; collective excitations.

Beck, Douglas H., Ph.D., Massachusetts Institute of Technology, 1986. Principal investigator, Nuclear Physics Laboratory. *Nuclear Physics*. Experimental nuclear and particle physics; nucleon structure; fundamental symmetries; electric dipole moments.

Bezryadin, Alexey, Ph.D., J. Fourier Université, 1995. *Condensed Matter Physics, Nano Science and Technology*. Experimental condensed matter physics; nanometer-scale mescopic physics and molecular electronics; quantum phase transitions.

Ceperley, David M., Ph.D., Cornell University, 1976. Founder Professor of Engineering; Center for Advanced Study Professor of Physics. *Computational Physics, Condensed Matter Physics*. Theoretical condensed matter physics; electronic structue; superfluidity; Monte Carlo methods; physics at high pressure.

Cooper, S. Lance, Ph.D., University of Illinois, 1988. Associate Head for Graduate Programs. *Condensed Matter Physics*. Experimental condensed matter physics; optical spectroscopy; strongly correlated systems; superconductivity.

Dahmen, Karin A., Ph.D., Cornell University, 1995. *Biophysics, Condensed Matter Physics, Geophysics, Nonlinear Dynamics and Complex Systems*. Theoretical condensed matter physics; nonequilibrium dynamical systems; hysteresis; avalanches; earthquakes; population biology; disorder-induced critical behavior.

Eckstein, James N., Ph.D., Stanford University, 1978. *Condensed Matter Physics*. Experimental condensed matter physics; atomic layer-by-layer molecular beam epitaxy; colossal magnetoresistance.

El-Khadra, Aida X., Ph.D., University of California, Los Angeles, 1989. *High Energy Physics*. Theoretical high-energy physics; lattice field theory; phenomenology; quark flavor physics.

Errede, Steven M., Ph.D., Ohio State University, 1981. *High Energy Physics*. Experimental high-energy physics; interactions of the electroweak gauge bosons; physics of music.

Fradkin, Eduardo H., Ph.D., Stanford University, 1979. Director, Institute for Condensed Matter Theory. *Condensed Matter Physics*. Theoretical condensed matter physics; quantum Hall effects; strongly correlated systems; superconductors; critical phenomena; disordered systems; field theory.

Gammie, Charles F., Ph.D., Princeton University, 1992. Chair, Department of Astronomy. *Astrophysics, Computational Physics*. Theoretical and computational astrophysics; star formation; planet formation; relativistic accretion flows.

Giannetta, Russell W., Ph.D., Cornell University, 1980. *Condensed Matter Physics*. Experimental condensed matter physics; superconductivity; magnetic resonance; organic superconductors.

Goldenfeld, Nigel D., Ph.D., University of Cambridge, 1982. Swanlund Chair; Center for Advanced Study Professor of Physics; Director, Institute for Universal Biology; Theme Leader, Institute for Genomic Biology. *Biophysics, Condensed Matter Physics, Statistical & Thermal Physics*. Theoretical physics and complexity; pattern formation; statistical physics; microbiology and evolutionary biology; fluid mechanics; materials theory; quantitative finance.

Gollin, George D., Ph.D., Princeton University, 1980. *High Energy Physics*. Experimental high-energy physics; CP violation; lepton number violation; axion production; higher education policy.

Greene, Laura H., Ph.D., Cornell University, 1984. Swanlund Chair; Center for Advanced Study Professor of Physics. *Condensed Matter Physics*. Experimental condensed matter physics; thin-film growth and tunneling in novel superconducting materials.

Grosse Perdekamp, Matthias, Ph.D., University of California, Los Angeles, 1995. *Nuclear Physics*. Experimental high-energy nuclear physics; nucleon structure, including spin structure and nuclear effects; spin-dependent hadron fragmentation.

Ha, Taekjip, Ph.D., University of California, Berkeley, 1996. Howard Hughes Medical Investigator; Edward William and Jane Marr Gutgsell Endowed Professor; Theme Leader, Institute for Genomic Biology; Co-Director, Center for the Physics of Living Cells. *Biophysics*. Experimental biological physics; single molecule fluorescence microscopy and spectroscopy; DNA protein interactions; molecular biology.

Kwiat, Paul G., Ph.D., University of California, Berkeley, 1993. Bardeen Chair of Physics and of Electrical and Computer Engineering. *Atomic, Molecular, & Optical Physics, Quantum Foundations*. Experimental quantum optics; optical approaches to quantum information; foundations of quantum mechanics.

Leggett, Anthony J., University of Oxford, 1964. John D. and Catherine T. MacArthur Chair; Center for Advanced Study Professor of Physics; Nobel Laureate in Physics (2003). *Atomic, Molecular, & Optical Physics, Condensed Matter Physics, Low Temperature Physics, Quantum Foundations*. Foundations of quantum mechanics; superfluidity; high-temperature superconductivity; Bose–Einstein condensation; low-temperature properties of glasses.

Leigh, Robert G., Ph.D., University of Texas, Austin, 1991. *Condensed Matter Physics, Cosmology & String Theory, High Energy Physics*. Theoretical high-energy physics; quantum field theory, supersymmetric gauge theory; superstring theory.

Liss, Tony M., Ph.D., University of California, Berkeley, 1984. *High Energy Physics*. Experimental high-energy physics; production and decay of the top quark; Higgs studies.

Makins, Naomi C.R., Ph.D., Massachusetts Institute of Technology, 1994. *Nuclear Physics*. Experimental nuclear physics; proton and neutron spin.

Mestre, Jose, Ph.D., University of Massachusetts, 1979. Professor and Chair, Department of Educational Psychology. *Physics and other Science Education*. Physics education research; cognitive processes in learning; role and interaction of language in problem solving; educational technologies.

Mouschovias, Telemachos Ch, Ph.D., University of California, Berkeley, 1975. Professor of Astronomy. *Astrophysics*. Theoretical astrophysics; astrophysical magnetohydrodynamics; astrophysical fluid dynamics; cosmic magnetic fields; star formation; numerical astrophysics.

Nayfeh, Munir H., Ph.D., Stanford University, 1974. *Atomic, Molecular, & Optical Physics, Nano Science and Technology*. Experimental atomic, molecular, and optical physics; laser atomic spectroscopy; silicon nanotechnology.

Oono, Yoshitsugu, Ph.D., Kyushu University, 1976. *Statistical & Thermal Physics*. Nonequilibrium statistical physics/dynamical systems; system reduction/asymptotic analysis, including reduction of large data sets.

Peng, Jen-Chieh, Ph.D., University of Pittsburgh, 1975. *Nuclear Physics*. Experimental medium- and high-energy nuclear physics; parton structures of the nucleons and nuclei; neutrino physics.

Phillips, Philip W., Ph.D., University of Washington, 1982. *Condensed Matter Physics*. Theoretical condensed matter physics; strongly correlated electronic low-dimensional systems; quantum Hall effect; quantum critical phenomena; quantum magnetism.

Pitts, Kevin T., Ph.D., University of Oregon, 1994. Associate Head for Undergraduate Programs. *High Energy Physics*. Experimental high-energy physics; CP violation in bottom quark decays.

Schiffer, Peter E., Ph.D., Stanford University, 1993. Vice Chancellor for Research. *Condensed Matter Physics*. Experimental condensed matter physics; measurements of magnetic oxides, geometrically frustrated magnets, and magnetic nanostructures.

Schulten, Klaus J., Ph.D., Harvard University, 1974. Swanlund Chair; Center for Advanced Study Professor of Physics; Co-Director, Center for the Physics of Living Cells. *Biophysics, Computational Physics*. Theoretical and computational biological physics; physics of the living cell.

Selen, Mats A., Ph.D., Princeton University, 1988. *Physics and other Science Education*. Physics education research.

Selvin, Paul R., Ph.D., University of California, Berkeley, 1990. *Biophysics*. Experimental biological physics; structure and dynamics of biological macromolecules; fluorescence microscopy.

Shapiro, Stuart L., Ph.D., Princeton University, 1973. Professor of Astronomy; Senior Research Scientist, NCSA. *Astrophysics, Computational Physics, Relativity & Gravitation*. Theoretical astrophysics and general relativity; physics of black holes and neutron stars; gravitational collapse; generation of gravitational waves; stellar dynamics; magnetohydrodynamics; numerical relativity.

Song, Jun, Ph.D., Massachusetts Institute of Technology, 2001. Professor of Bioengineering. *Biophysics*. Computational biological physics; systems biology; biostatistics.

Stack, John D., Ph.D., University of California, Berkeley, 1965. *High Energy Physics*. Theoretical physics.

Stone, Michael, Ph.D., University of Cambridge, 1976. *Condensed Matter Physics*. Theoretical condensed matter physics; quantum Hall effect; superconductivity and superfluidity; mathematical physics.

Thaler, Jon J., Ph.D., Columbia University, 1972. *Astrophysics*. Observational cosmology, focusing on the properties of dark matter and dark energy, as well as neutrino masses and diverse phenomena.

Van Harlingen, Dale J., Ph.D., The Ohio State University, 1977. Department head; Willett Professor of Engineering; Center for Advanced Study Professor of Physics. *Condensed Matter Physics, Low Temperature Physics, Quantum Foundations*. Experimental condensed matter physics; superconductivity; superconductor device physics; foundations of quantum mechanics; quantum information.

Weaver, Richard L., Ph.D., Cornell University, 1977. *Acoustics, Nonlinear Dynamics and Complex Systems*. Condensed matter physics; stochastic waves, disordered and complex structures, quantum chaos, random matrix theory, ultrasonics, structural acoustics.

Willenbrock, Scott S., Ph.D., University of Texas, Austin, 1986. *High Energy Physics*. Theoretical high-energy physics; phenomenology; electroweak symmetry breaking; top quark physics; Higgs phenomena.

Wiss, James E., Ph.D., University of California, Berkeley, 1977. *High Energy Physics*. Experimental high-energy physics; photoproduction of charmed mesons; precision study of charmed mesons.

Associate Professor

Aksimentiev, Aleksei, Ph.D., Institute of Physical Chemistry, Warsaw, 1999. *Biophysics, Computational Physics*. Theoretical and computational biological physics; biomolecular modeling, molecular motors, mechanical proteins, silicon biotechnology, membrane transport.

Budakian, Raffi O., Ph.D., University of California, Los Angeles, 2000. *Condensed Matter Physics, Nano Science and Technology*. Experimental condensed matter physics; magnetic resonance force microscopy; micro- and nanomechanical devices.

Chemla, Yann R., Ph.D., University of California, Berkeley, 2001. *Biophysics*. Experimental biological physics; molecular motors; nucleic acid and protein translocases.

DeMarco, Brian, Ph.D., University of Colorado Boulder, 2001. *Atomic, Molecular, & Optical Physics, Condensed Matter Physics, Quantum Foundations*. Experimental atomic, molecular, and optical physics; quantum information science; atomic Bose–Einstein condensates and Fermi gases.

Hubler, Alfred W., Ph.D., Technical University of Munich, 1987. *Nonlinear Dynamics and Complex Systems*. Theoretical and experimental nonlinear dynamics and complex systems.

Mason, Nadya, Ph.D., Stanford University, 2001. *Condensed Matter Physics, Nano Science and Technology*. Experimental condensed matter physics; quantum properties of nanostructures; superconductivity; quantum phase transitions.

Neubauer, Mark, Ph.D., University of Pennsylvania, 2001. *High Energy Physics*. Experimental particle physics; particle astrophysics; neutrino physics; heavy flavor physics; Higgs boson; electroweak diboson physics.

Stelzer, Timothy J., Ph.D., University of Wisconsin-Madison, 1993. *High Energy Physics, Physics and other Science Education*. Theoretical high energy physics; standard model physics at hadron colliders; computational physics; physics education research.

Vishveshwara, Smitha, Ph.D., University of California, Santa Barbara, 2002. *Condensed Matter Physics*. Theoretical condensed matter physics; strongly correlated systems; phase transitions and critical phenomena; disorder and localization physics; superconductivity; quantum Hall systems; Luttinger liquids and edge states; nanophysics; topological systems; cold atom physics.

Assistant Professor

Adshead, Peter, Ph.D., Yale University, 2010. *Astrophysics, Cosmology & String Theory*. Theoretical physics; inflation and early universe cosmology; theoretical cosmology.

Clark, Bryan, Ph.D., University of Illinois at Urbana-Champaign, 2009. *Computational Physics, Condensed Matter Physics*. Computational condensed matter physics; many-body and strongly correlated physics.

Faulkner, Thomas, Ph.D., Massachusetts Institute of Technology, 2009. *Condensed Matter Physics, High Energy Physics.* Theoretical condensed matter physics; high-energy physics and string theory.

Hughes, Taylor L., Ph.D., Stanford University, 2009. *Condensed Matter Physics.* Theoretical condensed matter physics; topological insulators/superconductors; use of quantum information/entanglement techniques to characterize quantum condensed matter systems.

Kuhlman, Thomas E., Ph.D., University of California, San Diego, 2007. *Biophysics.* Experimental biological physics; regulation of gene expression in prokaryotes; role of chromosomal organization in transcriptional regulation.

MacDougall, Gregory, Ph.D., McMaster University, 2008. *Condensed Matter Physics.* Experimental condensed matter physics; neutron scattering and muon spin rotation measurements of unconventional superconductors, geometrically frustrated magnets, and multiferroics; single crystal growth of new materials.

Ryu, Shinsei, Ph.D., University of Tokyo, 2005. *Condensed Matter Physics.* Theoretical condensed matter physics; nanoscale physics; strongly correlated systems.

Shelton, Julia, Ph.D., Massachusetts Institute of Technology, 2006. *High Energy Physics.* Theoretical high-energy physics.

Yang, Liang, Ph.D., Harvard University, 2006. *Nuclear Physics.* Experimental low-energy nuclear physics; neutrino physics; low-background detectors; neutrinoless double beta decay. Fundamental properties of neutrinos and testing fundamental symmetries.

Professor Emeritus

Baym, Gordon, Ph.D., Harvard University, 1960. *Astrophysics, Atomic, Molecular, & Optical Physics, Condensed Matter Physics, History & Philosophy of Physics/Science, Nuclear Physics.* Theoretical physics; Bose–Einstein condensation in trapped atomic systems and excitons; superfluid helium; matter under extreme conditions; neutron stars.

Debevec, Paul T., Ph.D., Princeton University, 1972. *Energy Sources & Environment, Nuclear Physics.* Experimental nuclear physics; photonuclear interactions; precision muon physics; energy and the environment.

Gladding, Gary E., Ph.D., Harvard University, 1971. *Physics and other Science Education.* Physics education research.

Lamb, Frederick K., University of Oxford, 1970. *Astrophysics.* Theoretical astrophysics; plasma, magnetohydrodynamic, and high-energy processes.

Lamb, Susan A., University of Oxford, 1973. *Astrophysics, Computational Physics.* Theoretical astrophysics, computational astrophysics; galaxy collisions and star formation.

Nathan, Alan M., Ph.D., Princeton University, 1975. *Nuclear Physics, Other.* Experimental nuclear physics; physics of baseball.

Slichter, Charles P., Ph.D., Harvard University, 1949. *Condensed Matter Physics.* Experimental condensed matter physics; nuclear magnetic resonance.

Weissman, Michael B., Ph.D., University of California, San Diego, 1976. *Condensed Matter Physics.* Experimental condensed matter physics; 1/f noise, spin glasses, amorphous materials.

Research Professor

Chiang, Tai-Chang, Ph.D., University of California, Berkeley, 1978. *Condensed Matter Physics.* Experimental condensed matter physics; atomically uniform films; electronic properties of impurities, surfaces, and quantum structures.

Research Associate Professor

Errede, Deborah M., Ph.D., University of Michigan, 1987. *High Energy Physics.* High energy particle physics; precision measurements of the W mass.

Research Assistant Professor

Kuehn, Seppe, Ph.D., Cornell University, 2007. *Biophysics.* Experimental biological physics; microbial population dynamics in closed ecosystems; phenotypic variation of microbial behavior.

Riedl, Caroline K., Ph.D., Friedrich-Alexander-Universitaet Erlangen, 2005. *Nuclear Physics.* Experimental nuclear physics; nucleon structure; exclusive processes.

Yodh, Jaya G., Ph.D., Johns Hopkins University, 1993. Director of Education and Outreach, Center for the Physics of Living Cells. *Biophysics.* Experimental biological physics; single-molecule investigations of helicase and chromatin systems.

Research Scientist

Wagner, Lucas K., Ph.D., North Carolina State University, 2006. *Computational Physics, Condensed Matter Physics.* Theoretical and computational condensed matter physics; high-performance computing; quantum Monte Carlo.

DEPARTMENTAL RESEARCH SPECIALTIES AND STAFF

Theoretical

Astrophysics. Astrophysics at Illinois encompasses problems in star formation, planet formation, stellar dynamics, astrophysical fluid dynamics, the physics of compact objects, and theoretical and observational cosmology. Physics faculty in the astrophysics group work closely with colleagues in the high-energy physics group, the Department of Astronomy, the Department of Chemistry, the National Center for Supercomputing Applications, and the program in Computational Science and Engineering, and many hold joint appointments. Adshead, Baym, Gammie, Frederick Lamb, Susan Lamb, Mouschovias, Shapiro.

Atomic, Molecular, & Optical Physics. Theoretical research in ultracold atomic systems focuses on quantum many-body physics and intersections with electronic solids and high-density nuclear matter. Research topics include numerical simulations of lattice gases, the BEC-BCS crossover in Fermi gases, artificial gauge fields and rotating superfluids, and analogs with QCD and nuclear matter. Baym, Ceperley, Fradkin, Leggett, Vishveshwara.

Biological Physics. Theoretical and computational biological physics research at Illinois includes such topics as biomolecular modeling of molecular motors, multiscale modeling of pattern formation, cellular mechanics, multiscale modeling of cells, biocomplexity, and bionanotechnology. Aksimentiev, Dahmen, Goldenfeld, Oono, Schulten, Song.

Condensed Matter Physics. Theoretical research in condensed matter physics focuses on the collective properties of matter in its solid and liquid forms, the emergence of novel and unusual states, and the behavior of complex systems. Illinois has long been a leader in research on superconductivity, superfluidity, and strongly correlated systems, and it is known for its close and fruitful collaborations of theorists and experimentalists. Every area of modern-day condensed matter physics is represented at Illinois, together with numerous interdisciplinary projects in atomic, molecular and optical physics, quantum information, string theory, materials science, theoretical and applied mechanics, chemistry, biology, and computer science and engineering. Current topics include high-temperature superconductivity, nonequilibrium dynamical systems, pattern formation, Bose–Einstein condensation, quantum phase transitions and quantum critical phenomena, strongly correlated and low-dimensional systems, quantum entanglement, topological insulators and superconductors, and nanoscale physics. Baym, Ceperley, Clark, Dahmen, Faulkner, Fradkin, Goldenfeld, Hughes, Leggett, Oono, Phillips, Ryu, Stone, Vishveshwara, Weaver.

High Energy Physics. Theoretical research in high-energy physics at Illinois covers a very diverse set of topics, including lattice field theory and quark flavor physics, collider phenomenology and simulations, top quark and Higgs physics, as well as quantum field theory, duality, and string theory. There is close collaboration with the high-energy experimental group, as well as the astrophysics and condensed matter theory groups. There are also overlapping interests with the math department in string theory research as well as with the National Center for Supercomputing Applications in computational physics. El-Khadra, Faulkner, Leigh, Shelton, Stack, Stelzer, Willenbrock.

Nuclear Physics. Theoretical research in nuclear physics focuses on phase structure of ultrahot and dense hadronic matter; ultrarelativistic heavy ion collisions; hot nuclear matter, pairing in nuclear matter, and equation of state of nuclear matter, with applications to neutron stars; and transport properties of quantum fluids with application to experimental searches for a neutron electric dipole moment. Baym.

Relativity & Gravitation. The Illinois Relativity group focuses on the application of Einstein's theory of general relativity to forefront problems in relativistic astrophysics. The development and application of numerical relativity to tackle problems by computational means are major activities. The merger of binary compact objects (including binary black holes) leading to the generation of gravitational waves and, in some cases, electromagnetic radiation, are areas of great interest. Shapiro.

Experimental

Astrophysics. Experimental astrophysics research at Illinois aims to measure the properties of the universe and its constituents using methods of observational astronomy by designing and building instruments for the Dark Energy Survey and the Large Synoptic Survey Telescope collaborations. Our scientific goals are measurements of the properties of dark matter and dark energy (which comprise 96% of the universe), of the neutrino mass, and of the inflationary epoch (approximately the first $10-32$ s of the big bang). We make these measurements by studying supernovae, the time development of large-scale structure in the universe, and the properties of galaxy clusters. Theoretical astrophysics focuses on many of the same issues. Our group studies several topics in cosmology: 1) Observable consequences of, and constraints on, the inflationary epoch; 2) Extensions of cosmic microwave background (CMB) measurements, which are inherently two-dimensional, to the third dimension using 21-cm radiation and galaxy clustering; and 3) Probes of fundamental physics, physics beyond the standard model, and modified gravity. Adshead, Thaler.

Atomic, Molecular, & Optical Physics. Experimental AMO physics at Illinois focuses on two general areas: quantum information science using entangled photons, and quantum simulation using ultracold atoms trapped in optical lattices. Current research topics include experimental studies of quantum nonlocality and the development of advanced resources for quantum computation, quantum cryptography, and quantum metrology. We also study cooling, dynamics, and phase transitions in strongly correlated and disordered quantum gases, and we work closely with condensed matter colleagues at Illi-

nois to address outstanding problems in many-body physics and the foundations of quantum mechanics. DeMarco, Kwiat, Nayfeh.

Biological Physics. Experimental biological physics groups at Illinois employ a variety of single-molecule techniques, including single-molecule fluorescence microscopy and spectroscopy, optical trapping, and microfluidics, to investigate molecular motors, DNA-protein interactions, gene regulation, intracellular transport, and the structure and dynamics of biological macromolecules. Chemla, Ha, Kuehn, Kuhlman, Selvin, Yodh.

Condensed Matter Physics. Condensed matter experiment at Illinois ranges from the design and growth of new materials, to the development of novel methods to elucidate and control quantum phenomena, to the design and construction of ground-breaking new instruments for fundamental physics research. Experimentalists work closely with theorists and across disciplines to address outstanding problems in condensed matter physics. Examples of current projects include imaging electron dynamics in the attosecond regime, detecting nuclear spins with attonewton force sensitivity, engineering solid-state qubits, measuring and controlling the magnetic and superconducting properties of nanodevices and nanostructure arrays, growing epitaxial heterostructures and bulk single crystals of strongly correlated materials, and elucidating the novel phases of magnetic and superconducting materials using neutron, light, and electron spectroscopies. Illinois condensed matter researchers carry out experiments in state-of-the-art facilities at the Frederick Seitz Materials Research Laboratory, the Micro and Nanotechnology Laboratory, the Beckman Institute, and Argonne National Laboratory, as well as in their own well-equipped laboratories. Abbamonte, Bezryadin, Budakian, Chiang, Cooper, Eckstein, Giannetta, Greene, Hubler, MacDougall, Mason, Nayfeh, Schiffer, Slichter, Van Harlingen, Weissman.

High Energy Physics. High energy experiment at Illinois encompasses accelerator-based experiments at the Energy Frontier and the Intensity Frontier. At the former, the group works at the CDF experiment at Fermilab and the ATLAS experiment at the Large Hadron Collider, studying the properties of top and bottom quarks and the Higgs boson, measuring the CKM matrix elements, and searching for rare phenomena and physics beyond the standard model. At the Intensity Frontier, the group is involved in three planned experiments at Fermilab: g–2, which makes precision measurements of the muon g-factor; Mu2e, which will search for the forbidden lepton-number-violating decay of a muon into an electron; and ORKA, which will make a precision measurement of a rare kaon decay. Opportunities exist in all these projects for detector development and operation as well as data analysis. Deborah Errede, Steven Errede, Gollin, Liss, Neubauer, Pitts, Wiss.

Nuclear Physics. The Nuclear Physics Laboratory (NPL) at Illinois focuses on discovery in fundamental nuclear physics using advanced instrumentation and modern data analysis techniques that are developed and built at NPL. The group develops instruments for novel experimental approaches in four main areas of nuclear physics: the precision measurement of the electric dipole moment of the neutron, a broad program studying structure and formation of hadrons, the precise determination of $\sin \theta\_13$ through a ν_e disappearance experiment, and the search for neutrino-less double beta decay. Recent and current examples of instrumentation developed at Illinois include the large-volume superconducting spectrometer magnet for the G0 experiment at Jefferson Laboratory, the muLan and muCap spectrometers for precision muon lifetime and capture experiments at the Paul Scherrer Institute in Switzerland, the cryogenic He-4 target for the neutron EDM experiment at Oakridge National Laboratory, large-area RPC detector stations for the PHENIX W-trigger at Brookhaven National Laboratory, and the Drell-Yan muon trigger scintillator hodoscopes for the SeaQuest experiment at Fermi National Accelerator Laboratory. Beck, Grosse Perdekamp, Makins, Peng, Riedl, Yang.

Physics and other Science Education. Physics education research (PER) investigates the learning, understanding, and teaching of physics and the application of physics knowledge. The Illinois PER group has pioneered the application of technology to physics teaching, including development of the i-clicker® student-response system, web-based multi-media learning modules, and a personal, hand-held device that can measure acceleration, spatial orientation, magnetic fields, electrical signals, frequency spectra, and time constants and perform other introductory physics laboratory tasks. Research interests include the role of mathematics and reflection in physics learning, the organization and deployment of physics knowledge by experts and novices, transfer studies, the design and implementation of web-based instruction, curriculum reform, and the evaluation of educational assessments. Experimental techniques and analyses used include eye-tracking, video analysis, student interviews, web-based log data analysis, and analysis of exam data. Gladding, Mestre, Selen, Stelzer.

**View additional information about this department at
www.gradschoolshopper.com**

THE UNIVERSITY OF CHICAGO

DEPARTMENT OF PHYSICS

Chicago, Illinois 60637
http://physics.uchicago.edu/

General University Information
President: Robert J. Zimmer
Dean of Graduate School: N/A
University website: http://uchicago.edu
Control: Private
Setting: Urban
Total Faculty: 2,750
Total Graduate Faculty: N/A
Total number of Students: 15,219
Total number of Graduate Students: 9,850

Department Information
Department Chairman: Edward C. Blucher, Chair
Department Contact: Shadla Cycholl, Assistant to Chairman
   Total full-time faculty: 49
   Total number of full-time equivalent positions: 40
   Full-Time Graduate Students: 146
   First-Year Graduate Students: 35
   Female First-Year Students: 2
   Total Post Doctorates: 51

Department Address
5720 S. Ellis Avenue
Chicago, IL 60637
Phone: (773) 702-7006
Fax: (773) 702-2045
E-mail: physics@uchicago.edu
Website: http://physics.uchicago.edu/

ADMISSIONS

Admission Contact Information
Address admission inquiries to: Office of Graduate Admissions, Department of Physics, University of Chicago, 5720 S. Ellis Avenue, Chicago, IL 60637
Phone: (773) 702-7007

E-mail: physics@uchicago.edu
Admissions website: http://gradadmissions.uchicago.edu

Application deadlines
Fall admission:
U.S. students: December 28   *Int'l. students*: December 28

Application fee
U.S. students: $55   *Int'l. students*: $55
No mid-year admissions.

Admissions information
For Fall of 2013:
   Number of applicants: 650
   Number admitted: 95
   Number enrolled: 20

Admission requirements
Bachelor's degree requirements: Bachelor's degree in any physical science or engineering is required.

GRE requirements
The GRE is required.
   Mean GRE score range (25th–75th percentile): 0
Taking the GRE well in advance of submitting the application is strongly recommended so that an applicant can self-report the scores before submitting the application. Score verification is done electronically.

Advanced GRE requirements
The Advanced GRE is required.
   Minimum accepted Advanced GRE score: 0
   Mean Advanced GRE score range (25th–75th percentile): 0
Taking the October 19, 2013 Physics Subject Test is required. (Registration deadline: September 13, 2013.)

TOEFL requirements
The TOEFL exam is required for students from non-English-speaking countries.
   iBT score: 102

Minimum total score: 102. Minimum scores for the subsections: 26, 26, 24, and 26 (24 is the minimum required for the Speaking Section).

Other admissions information

Additional requirements: The average GRE Advanced score for 2013-2014 admissions was 879/990. Students from non-English-speaking countries are required to demonstrate proficiency in English via the TOEFL or the IELTS.

Undergraduate preparation assumed: Equivalent of Marion and Thornton, Classical Dynamics of Particles and Systems; Reif, Statistical and Thermal Physics; Wangsness, Electromagnetic Fields: Shankar, Principles of Quantum Mechanics; Eisberg and Resnick, Quantum Physics of Atoms, Molecules, Solids, Nuclei and Particles; Kittel, Introduction to Solid State Physics, 8th ed.

TUITION

Tuition year 2013-14:

Full-time students: $45,495 annual

Annual tuition charge per academic year is $45,495 (or $15,165 per quarter for three quarters). In the Summer Quarter regular Research Assistants register for a research course, and therefore the 4th quarter tuition becomes due. However, all RAs receive full tuition coverage.

Credit hours per semester to be considered full-time: 300

Deferred tuition plan: No

Health insurance: Available at the cost of $3,021 per year.

Other academic fees: $313 per quarter student life fee; $45 one-time transcript fee.

Academic term: Quarter

Number of first-year students who received full tuition waivers: 35

Teaching Assistants, Research Assistants, and Fellowships

Number of first-year
  Teaching Assistants: 24
  Research Assistants: 8
  Fellowship students: 3
Average stipend per academic year
  Teaching Assistant: $21,150
  Research Assistant: $21,150
  Fellowship student: $22,500

FINANCIAL AID

Application deadlines

Fall admission:
U.S. students: December 28   *Int'l. students*: December 28

Loans

Loans are available for U.S. students.
Loans are available for international students.
GAPSFAS application required: No
FAFSA application required: No

For further information

Address financial aid inquiries to: Graduate Admissions, Dept. of Physics, University of Chicago, 5720 S. Ellis Ave., Chicago, IL 60637.
Phone: (773) 702-7007
E-mail: physics@uchicago.edu
Financial aid website: http://physics.uchicago.edu

HOUSING

Availability of on-campus housing

Single students: Yes
Married students: Yes

For further information

Address housing inquiries to: Graduate Student Housing Office, 5555 S. Ellis Ave., Chicago, IL 60637.
Phone: (773) 753-2218
E-mail: rshousing@uchicago.edu
Housing aid website: http://rs.uchicago.edu

Table A—Faculty, Enrollments, and Degrees Granted

Research Specialty	2012-13 Faculty	Enrollment Fall 2012 Master's	Enrollment Fall 2012 Doctorate	Number of Degrees Granted 2012-13 (2008–13) Master's	Number of Degrees Granted 2012-13 (2008–13) Terminal Master's	Number of Degrees Granted 2012-13 (2008–13) Doctorate
Astrophysics	7	–	13	–	–	3(26)
Atomic, Molecular, & Optical Physics	3	–	6	–	–	–(2)
Biophysics	2	–	4	–	–	–(6)
Chemical Physics	–	–	–	–	–	–(6)
Condensed Matter Physics	14	–	49	–	–	3(35)
Nuclear Physics	1	–	2	–	–	1(1)
Particles and Fields	18	–	41	–	–	3(32)
Physics of Beams	1	–	–	–	–	1(1)
Relativity & Gravitation	1	–	4	–	–	1(3)
Non-specialized	–	–	27	10(110)	1(5)	–
Total	47	–	146	10(110)	1(5)	12(115)
Full-time Grad. Stud.	–	–	146	–	–	–
First-year Grad. Stud.	–	–	35	–	–	–

GRADUATE DEGREE REQUIREMENTS

Master's: Although students are not admitted to study for a master's, they may receive a master's degree while studying for the Ph.D. For the master's degree, there is a minimum residence requirement of three-quarters of full-time registration or the equivalent, nine quarter-length courses. In addition, a student must either pass the Candidacy Examination at the Master's level or pass nine approved graduate courses (six of which are the "core" graduate physics courses) and complete the experimental physics requirement, with a GPA of 2.5 or better overall. There is no thesis or foreign language requirement.

Doctorate: There is a minimum residence requirement of nine quarters of full-time registration. The candidate must pass the advanced physics laboratory course or participate in a first-year experimental research experience, and also pass six advanced physics courses. Four of these advanced courses must be selected from course offerings in either three or four general categories associated with active areas of contemporary physics research; the remaining two must be advanced, seminar-type elective courses. Other requirements include passing the Candidacy Examination at the Ph.D. level, convening a first/introductory meeting of the Ph.D. Committee, a pre-oral meeting to discuss the substantive issues of the dissertation, defending the dissertation before the candidate's Ph.D. Committee, and submitting a paper based on the dissertation to a recognized journal. Normally at the pre-oral meeting, preceding submission of the thesis, the Ph.D. Committee and the student decide whether the paper to be submitted will be a single-authored or multiple-authored paper.

Thesis: Electronic submission of the final thesis to the Dissertation Office is required of all students.

SPECIAL EQUIPMENT, FACILITIES, OR PROGRAMS

The Department of Physics at the University of Chicago offers Ph.D. programs in many areas of physics. Students' formal classwork takes place in the modern lecture halls, classrooms, and instructional laboratories of the Kersten Physics Teaching Center. This building also houses special equipment and support facilities for student experimental projects, departmental administrative offices, and meeting rooms. The Center is situated on the science quadrangle near the John Crerar Science Library, which holds over 1,000,000 volumes and provides modern literature search and data retrieval systems.

Student participation is crucial to virtually all research projects, and both graduate and undergraduate research and training are given high priority. Most of the experimental and theoretical research of Physics faculty and graduate students is carried out within the Enrico Fermi Institute, the James Franck Institute, and the Institute for Biophysical Dynamics. These research institutes provide close interdisciplinary contact, crossing the traditional boundaries between departments.

In the Enrico Fermi Institute, members of the Department of Physics carry out theoretical research in particle theory, string theory, field theory, general relativity, and theoretical astrophysics and cosmology. There are active experimental groups in high-energy physics, nuclear physics, astrophysics and space physics, infrared and optical astronomy, electron and ion microscopy, and atomic physics. Some of this research is conducted at the Fermi National Accelerator Laboratory, at Argonne National Laboratory, and at the European Organization for Nuclear Research (CERN) in Geneva, Switzerland.

Physics faculty in the James Franck Institute study chemical, solid state, condensed matter, and statistical physics. Fields of interest include chaos, chemical kinetics, critical phenomena, high Tc superconductivity, non-linear dynamics, low temperature, disordered and amorphous systems, the dynamics of glasses, fluid dynamics, surface and interface phenomena, non-linear and nanoscale optics, unstable and metastable systems, laser cooling and trapping, and polymer physics. Much of the research utilizes specialized facilities operated by the Institute, including a low temperature laboratory, a materials preparation laboratory, x-ray diffraction and analytical chemistry laboratories, laser equipment, a scanning-tunneling microscope, and extensive shop facilities. Some members of the faculty are involved in research at Argonne National Laboratory.

The Institute for Biophysical Dynamics includes members of both the Physical Sciences and Biological Sciences Divisions, and focuses on the physical basis for molecular and cellular processes. This interface between the physical and biological sciences is an exciting and rapidly developing area, with a bi-directional impact. Research topics include the creation of physical materials by biological self-assembly, the molecular basis of macromolecular interactions and cellular signaling, the derivation of sequence-structure-function relationships by computational means, and structure-function relationships in membranes.

In the areas of chemical, atomic, and biophysics, research toward the doctorate may be done in either the Physics or the Chemistry Department. Facilities are available for research in crystal chemistry, degenerate quantum gases, molecular physics, molecular spectra from infrared to far ultraviolet and Raman spectra, both experimental and theoretical, surface physics, statistical mechanics, radio chemistry, and quantum electronics.

Interdisciplinary research leading to a Ph.D. degree in physics may be carried out under the guidance of faculty committees including members of other departments in the Physical Sciences Division, such as Astronomy and Astrophysics, Chemistry, Computer Science, Geophysical Sciences, Mathematics, or related departments in the Biological Sciences Division.

Table B—Separately Budgeted Research Expenditures by Source of Support

Source of Support	Departmental Research	Physics-related Research Outside Department
Federal government	$18,966,529	$3,059,941
State/local government		
Non-profit organizations	$576,130	
Business and industry	$168,904	
Other	$1,862,120	
Total	$21,573,683	$3,059,941

Table C—Separately Budgeted Research Expenditures by Research Specialty

Research Specialty	No. of Grants	Expenditures ($)
Astrophysics	38	$9,873,604
Atmosphere, Space Physics, Cosmic Rays	9	$349,680
Atomic, Molecular, & Optical Physics	14	$1,000,551
Biophysics	10	$1,128,270
Condensed Matter Physics	43	$1,777,289
Particles and Fields	39	$645,374
Physics of Beams	10	$675,911
Relativity & Gravitation	9	$4,881,198
Statistical & Thermal Physics	2	$46,268
Solid State Physics	1	$1,192,586
Other	1	$2,950
Total	176	$21,573,681

FACULTY

Professor

Blucher, Edward C., Ph.D., Cornell University, 1988. Chairman of the Department of Physics. *High Energy Physics, Particles and Fields.* Experimental physics; particle physics.

Carena, Marcela, Ph.D., University of Hamburg, 1989. Scientist, Fermilab. *Particles and Fields.* Theoretical physics; elementary particles.

Carlstrom, John E., Ph.D., University of California, Berkeley, 1988. Professor of Astronomy & Astrophysics. *Astronomy, Astrophysics, Cosmology & String Theory.* Experimental physics and astrophysics; star formation and cosmology; observation and new instrumentation.

Chin, Cheng, Ph.D., Stanford University, 2001. *Atomic, Molecular, & Optical Physics.* Laser cooling; trapping, degenerate quantum gases.

Frisch, Henry, Ph.D., University of California, Berkeley, 1971. *Particles and Fields.* Experimental physics; particle physics.

Guyot-Sionnest, Philippe, Ph.D., University of California, Berkeley, 1987. Professor of Chemistry. Experimental physics; surface physics; nonlinear optical spectroscopy.

Harvey, Jeffrey A., Ph.D., California Institute of Technology, 1981. *Cosmology & String Theory, Particles and Fields.* Theoretical physics; particle physics; quantum field theory; superstring theory.

Isaacs, Eric, Ph.D., Massachusetts Institute of Technology, 1988. Director, Argonne National Laboratory. *Condensed Matter Physics.* Experimental physics; condensed matter physics.

Jaeger, Heinrich M., Ph.D., University of Minnesota, 1987. *Condensed Matter Physics.* Experimental condensed matter physics; mesoscopic physics; high-temperature superconductivity.

Kang, Woowon, Ph.D., Princeton University, 1992. *Condensed Matter Physics.* Experimental condensed matter physics; fractional quantum Hall effect; semiconductor physics.

Kim, Kwang-Je, Ph.D., University of Maryland, 1970. Senior Scientist, Argonne National Laboratory. *Physics of Beams.* Theoretical physics; beam physics.

Kim, Young-Kee, Ph.D., University of Rochester, 1990. Deputy Director, Fermilab. *High Energy Physics, Particles and Fields.* Experimental elementary particle physics.

Kutasov, David, Ph.D., Weizmann Institute of Science, 1989. *Cosmology & String Theory, Particles and Fields.* Theoretical physics; quantum field theory; string theory.

Levin, Kathryn, Ph.D., Harvard University, 1970. *Atomic, Molecular, & Optical Physics, Condensed Matter Physics.* Theoretical physics; solid-state physics.

Littlewood, Peter B., Ph.D., University of Cambridge, 1980. Associate Director, Argonne National Laboratory. *Condensed Matter Physics.* Condensed matter theory.

Lu, Zheng-Tian, Ph.D., University of California, Berkeley, 1994. Senior Physicist, Argonne National Laboratory. *Atomic, Molecular, & Optical Physics, Nuclear Physics.* Experimental physics; atomic physics.

Martinec, Emil J., Ph.D., Cornell University, 1984. Director, Enrico Fermi Institute. *Cosmology & String Theory, Particles and Fields.* Theoretical physics; string theory; quantum field theory; elementary particles.

Mazenko, Gene F., Ph.D., Massachusetts Institute of Technology, 1971. *Condensed Matter Physics, Systems Science/Engineering.* Theoretical physics; statistical physics.

Merritt, Frank S., Ph.D., California Institute of Technology, 1976. *Particles and Fields.* Experimental physics; particle physics.

Meyer, Stephan S., Ph.D., Princeton University, 1979. Professor of Astronomy & Astrophysics. *Astronomy, Astrophysics, Cosmology & String Theory.* Experimental astrophysics; infrared astrophysics; observational cosmology.

Nagel, Sidney R., Ph.D., Princeton University, 1974. *Condensed Matter Physics, Fluids, Rheology, Nonlinear Dynamics and Complex Systems.* Experimental physics; condensed matter physics; non-linear dynamics.

Oddone, Pier, Ph.D., Princeton University, 1970. Director, Fermilab. *High Energy Physics, Particles and Fields.* Experimental physics.

Oreglia, Mark J., Ph.D., Stanford University, 1980. *High Energy Physics, Particles and Fields.* Experimental physics; particle physics.

Privitera, Paolo, Ph.D., Karlsruhe University, 1993. Professor of Astronomy and Astrophysics. *Astronomy, Astrophysics.* Experimental physics; ultra-high energy cosmic rays.

Rosenbaum, Thomas F., Ph.D., Princeton University, 1982. Provost, University of Chicago. *Condensed Matter Physics, Low Temperature Physics.* Experimental physics; solid-state physics; low-temperature physics.

Rosner, Robert, Ph.D., Harvard University, 1976. Professor of Astronomy & Astrophysics. *Astrophysics, Fluids, Rheology, Plasma and Fusion.* Theoretical physics; fluid and plasma dynamics; solar physics; high-energy astrophysics.

Savard, Guy, Ph.D., McGill University, 1988. Senior Scientist, Argonne National Laboratory. *Nuclear Physics.* Experimental physics; nuclear physics.

Shochet, Melvyn J., Ph.D., Princeton University, 1972. *High Energy Physics, Particles and Fields.* Experimental particle physics.

Son, Dam Thanh, Ph.D., Institute for Nuclear Research, Moscow, 1995. Theoretical nuclear physics. *Theoretical Physics.*

Turner, Michael S., Ph.D., Stanford University, 1978. Professor of Astronomy & Astrophysics. Director, Kavli Institute for Cosmological Physics. *Astrophysics, Cosmology & String Theory.* Theoretical astrophysics; particle physics; cosmology.

Wagner, Carlos E. M., Ph.D., Hamburg, 1989. Physicist, Argonne National Laboratory. *High Energy Physics, Particles and Fields.* Theoretical physics; elementary particles; supersymmetric theories.

Wah, Yau W., Ph.D., Yale University, 1983. *High Energy Physics, Particles and Fields.* Experimental physics; particle physics.

Wald, Robert M., Ph.D., Princeton University, 1972. *Relativity & Gravitation.* Theoretical physics; general relativity.

Wiegmann, Paul B., Ph.D., Landau Inst., Moscow, 1978. Director, James Franck Institute. *Condensed Matter Physics.* Theoretical physics; condensed matter physics.

Associate Professor

Collar, Juan I., Ph.D., University of South Carolina, 1992. *Astrophysics, Nuclear Physics.* Experimental physics; neutrino and astroparticle physics.

Gardel, Margaret L., Ph.D., Harvard University, 2004. *Biophysics, Condensed Matter Physics.* Experimental biophysics.

Sethi, Savdeep S., Ph.D., Harvard University, 1996. *Cosmology & String Theory, High Energy Physics, Particles and Fields.* Theoretical physics; quantum field theory; string theory; particle physics.

Wakely, Scott P., Ph.D., University of Minnesota, 1999. *Astrophysics.* Experimental astroparticle physics, high-energy astrophysics.

Wang, LianTao, Ph.D., University of Michigan, 2002. *High Energy Physics, Particles and Fields.* Theoretical physics; elementary particles.

Zhang, Wendy W., Ph.D., Harvard University, 2001. *Condensed Matter Physics, Fluids, Rheology, Nonlinear Dynamics and Complex Systems.* Condensed matter theory.

Assistant Professor

Biron, David, Ph.D., Weizmann Institute of Science, 2004. *Biophysics.* Experimental biophysics.

Grandi, Luca, Ph.D., Universita degli Studi di Pavia, 2005. *Astrophysics.* Experimental physics; dark matter, and astroparticle physics.

Hill, Richard, Ph.D., Cornell University, 2002. *High Energy Physics, Particles and Fields.* Theoretical physics; elementary particles.

Holz, Daniel E., Ph.D., University of Chicago, 1998. *Astrophysics, Cosmology & String Theory, Relativity & Gravitation.* general relativity, astrophysics, cosmology.

Irvine, William T., Ph.D., University of California, Santa Barbara, 2006. *Condensed Matter Physics, Electromagnetism.* Experimental soft condensed matter; knotted fields.

Rust, Michael J., Ph.D., Harvard University, 2010. Assistant Professor of Molecular Genetics and Cell Biology. *Biophysics.*

Schmitz, David, Ph.D., Columbia University, 2008. Experimental particle physics; experimental neutrino physics.

Schuster, David I., Ph.D., Yale University, 2007. *Condensed Matter Physics.* Experimental condensed matter; quantum computing; superconducting circuits.

Simon, Jonathan, Ph.D., Harvard University, 2010. Experimental atomic, molecular, and optical physics.

Emeritus

Abella, Isaac D., Ph.D., Columbia University, 1963. *Atomic, Molecular, & Optical Physics, Optics, Physics and other Science Education.* Experimental physics; quantum optics; atomic physics; laser spectroscopy.

Cronin, James W., Ph.D., University of Chicago, 1955. *Astrophysics, Particles and Fields.* Experimental physics; particle physics; ultra-high energy γ-ray astronomy.

Eastman, Dean E., Ph.D., Massachusetts Institute of Technology, 1965. *Condensed Matter Physics.* Experimental physics; condensed matter physics.

Freund, Peter G. O., Ph.D., University of Vienna, 1960. *Particles and Fields*. Theoretical physics; particle physics; field theory.

Geroch, Robert P., Ph.D., Princeton University, 1967. *Relativity & Gravitation*. Theoretical physics; general relativity.

Hildebrand, Roger H., Ph.D., University of California, Berkeley, 1951. *Astronomy*. Experimental physics; infrared astronomy.

Kadanoff, Leo P., Ph.D., Harvard University, 1960. *Computational Physics, Condensed Matter Physics, Nonlinear Dynamics and Complex Systems, Statistical & Thermal Physics*. Theoretical physics; hydrodynamics; statistical physics.

Levi-Setti, Riccardo, Ph.D., University of Pavia, Italy, 1949. Experimental physics; ion microscopy; secondary ion mass spectrometry; ion-solid interaction.

Müller, Dietrich, Ph.D., University of Bonn, 1964. *Astrophysics*. Experimental physics; cosmic rays; high-energy astrophysics.

Nambu, Yoichiro, Tokyo, Japan, 1952. *Particles and Fields*. Theoretical physics; particle physics; field theory.

Parker, Eugene N., Ph.D., California Institute of Technology, 1951. *Astrophysics, Plasma and Fusion*. Theoretical physics; astrophysics; plasma physics; space physics.

Pilcher, James E., Ph.D., Princeton University, 1968. *High Energy Physics, Particles and Fields*. Experimental physics; particle physics.

Rosner, Jonathan L., Ph.D., Princeton University, 1965. *High Energy Physics, Particles and Fields*. Theoretical physics; particle physics; field theory.

Schiffer, John P., Ph.D., Yale University, 1954. Senior Physicist, Argonne National Laboratory. *Nuclear Physics*. Experimental physics; nuclear physics.

Witten, Thomas A., Ph.D., University of California, San Diego, 1971. *Condensed Matter Physics, Polymer Physics/Science*. Theoretical physics; weakly connected matter.

Lecturer

Gazes, Stuart B., Ph.D., Massachusetts Institute of Technology, 1983. Undergraduate Program Chair. *Nuclear Physics*. Experimental physics; nuclear physics.

Reid, David D., Ph.D., Wayne State University, 1995. Executive Officer. *Atomic, Molecular, & Optical Physics, Physics and other Science Education, Relativity & Gravitation*. Theoretical physics, discrete space-time, electron- and positron-gas scattering; physics pedagogy.

DEPARTMENTAL RESEARCH SPECIALTIES AND STAFF

Theoretical

Astrophysics & Cosmology. Cosmology and early universe particle physics. Big-bang nucleosynthesis. Tests of the Big Bang model. Ultra-high energy cosmic-ray processes. Baryogenesis and cosmological phase transitions. Topological defects. Inflationary cosmology. Cosmic microwave background radiation. Dark matter. Formation of structure in the universe. The cosmological constant and dark energy. Aspects of string cosmology. Solar and stellar astrophysics. Astrophysical fluid dynamics. Holz, Parker, Robert Rosner, Turner, Wagner, Wang.

Atomic, Molecular, & Optical Physics. Trapped Fermi and Bose gases. Ionization dynamics. Inner-shell physics of atoms, molecules, and clusters, strong-field and electron-correlation effects. Free-electron lasers. Ultrafast laser-induced phenomena. Electronic many-body theory. Non-Hermiticity in quantum mechanics. Levin.

Condensed Matter Physics. Macroscopic dynamics of materials, interfacial singularities, and non-linear processes. Turbulent, chaotic, and stochastic behavior in hydrodynamic and other dynamical systems. Spatial self-organization in polymers, surfactant monolayers, colloids and cell assemblies. Physics of magnetic and superconducting materials (systems) driven by a strong interaction. Physics in low dimensions. Fermi liquid and non-Fermi liquid states in many body systems. High temperature superconductivity. Quantum phase transitions. Phase ordering kinetics and defect dynamics. Non-perturbative phenomena in electronic systems; strongly correlated electronic systems, magnetism. Transition between jammed and fluid states in granular matter, glass-forming liquids, and magnetic flux lattices. Integrable models of statistical mechanics and quantum field theory. Stochastic processes. Irvine, Kadanoff, Levin, Littlewood, Mazenko, Wiegmann, Witten, Zhang.

Particle Physics & String Theory. String theory and unification, duality in gauge theory and string theory, solitons and topological structures, precision electroweak measurements, dark matter candidates, effective field theory, electroweak baryogenesis, low-energy supersymmetry, CP violation, heavy quark physics, confinement in QCD, quantum theory of black holes, large extra dimensions, fermion mass hierarchy, integrable systems. Carena, Freund, Harvey, Hill, Kutasov, Martinec, Nambu, Jonathan Rosner, Sethi, Wagner, Wang.

Relativity & Gravitation. Black holes. Asymptotic structure. Gravitational radiation. Mathematical aspects of general relativity. Quantum field theory in curved space-times. Quantum gravitation. Alternative theories. Geroch, Holz, Wald.

Experimental

Astrophysics. Studies of the cosmic microwave background radiation spectrum and anisotropy with ground and space-based detectors. Search for polarization in the cosmic background radiation. Measurements of the Sunyaev-Zel'dovich effect for clusters of galaxies. Measurements of intergalactic radiation fields. High-energy gamma-ray astrophysics with atmospheric Cherenkov telescopes. Development of giant air shower array (Auger Project) for investigation of the highest energy cosmic rays. Development of large detectors for high-energy cosmic rays on space and balloon payloads. Experimental investigations of cosmic ray electrons and of the elemental and isotopic abundances of cosmic-ray nuclei over a wide energy range. Investigations of solar, magnetospheric, and heliospheric phenomena with satellite and deep space missions. Cosmic dust studies. Development of instruments to detect polarization in the far-infrared emission from interstellar clouds. Investigation of the magnetic field structure of dense cloud cores. Airborne and mountain-top polarimetry. Direct searches for non-baryonic dark matter. Accelerator-based nuclear astrophysics experiments. Carlstrom, Collar, Cronin, Grandi, Hildebrand, Müller, Meyer, Privitera, Wakely.

Atomic, Molecular, & Optical Physics. Bose-Einstein condensation of molecules and fermionic superfluids. Laser cooling and trapping of atoms. Scalable quantum manipulation and quantum computation. Testing time-reversal symmetry in atoms and nuclei. Abella, Chin, Lu, Simon.

Biophysics. Cell migration and division, physical aspects of biological organization, mechanical behavior of cells, regulation of cell physiology, non-linear dynamics, computational biology, time-resolved fluorescence, confocal microscopy, protein-engineering, signal transduction, gene expression, mathematical modeling, large-scale simulations, stochastic and self-assembly processes, elasticity of polymer networks, optical and holographic traps, single-molecule biophysics, non-linear optics methods, noise and information in intraneuronal pathways and interneuronal communication, homeostatic regulation of single neuronal function and of the function of small neural circuits, design principles of biological networks, biophysics in vivo — quantifying behavior and physiological activity of neurons, high power computation (grid and parallel computing), biophysics of sleep. Biron, Gardel, Rust.

Condensed Matter Physics. Optical and electronic transport in normal and superconducting nanocrystals and arrays. Collective effects at ultra-low temperatures including (fractional) quantum Hall effect, vortex tunneling, metal-insulator transitions, and magnetic quantum critical points. Symmetry-breaking and fluctuations in heavy fermion, organic, and high-Tc superconductors. Nonlinear dynamics and flow properties of granular materials. Scaling behavior of liquid flow and droplet breakup. Mathematical analysis and computer simulation of singularity formation. Universal scaling behavior of relaxation phenomena in supercooled liquids and glasses. Microscopic kinetics and dynamics of phase transitions in colloidal suspensions. Manipulation by dynamic optical holographic traps. Molecular regulation within living cells. Self-assembly and morphology of ultrathin polymer films. Biological properties of the cytoskeleton of eukaryotic cells. The mechanical behavior of cells. Chin, Guyot-Sionnest, Irvine, Isaacs, Jaeger, Kang, Nagel, Rosenbaum, Schuster, Simon.

Nuclear Physics. Studies of the nuclear many-body system: Nuclear structure and interactions, nuclear reactions in astrophysics, nuclear matter under extreme conditions, precision measurements of critical information to nucleosynthesis along the r- and rp-process paths. Low-energy experiments in fundamental interactions and symmetries, exotic nuclear structure, double beta decay, coherent nuclear scattering. Production, cooling and trapping of rare isotopes, R&D for the Rare Isotope Accelerator (RIA) project. Non-nucleonic degrees of freedom in nuclei and phenomena requiring a quark description. Collar, Lu, Savard, Schiffer.

Particles and Fields. Measurements of properties of the top quark. Searches for supersymmetric particles, the Higgs boson, and other new physics. Precision tests of the standard model in W and Z decays. Studies of pp interactions at center-of-mass energies of 1800 GeV. High-precision measurement of CP violation parameters in K decays; high-sensitivity search for rare K decays and for CPT violation. High-precision measurements of hyperon rare decays. High-precision measurements of electroweak interactions at LEP, both near the Z0 and at center-of-mass energies up to 200 GeV. Searches for new physics including the Higgs boson and supersymmetry; precision measurement of Mw. Preparation for the ATLAS experiment at the LHC (high-energy pp interactions at 14 TeV). Research and development on muon colliders and neutrino factories. Use of facilities at Fermi National Accelerator Laboratory and at CERN. Blucher, Frisch, Young-Kee Kim, Merritt, Oddone, Oreglia, Pilcher, Jonathan Rosner, Schmitz, Shochet, Wah.

Physics of Beams. Investigation of particle and photon beams and their mutual interactions with the goal of developing novel accelerators or radiation devices. Some current topics are production and acceleration of high-brightness electron beams for linear colliders and free electron lasers; beam dynamics in ionization cooling for muon colliders and neutrino factories; self-amplified spontaneous emission for intense, coherent x-rays; miniature IR radiation source via Smith-Purcell process using electron microscope beams. Theoretical and experimental programs at the Enrico Fermi Institute on campus, at the Argonne National Laboratory Advanced Photon Source, and the A0 facility in Fermilab. Kwang-Je Kim.

View additional information about this department at
www.gradschoolshopper.com

BALL STATE UNIVERSITY

DEPARTMENT OF PHYSICS AND ASTRONOMY

Muncie, Indiana 47306
http://www.bsu.edu/physics

General University Information
President: Jo Ann Gora
Dean of Graduate School: Robert Morris
University website: http://www.bsu.edu
Control: Public
Setting: Urban
Total Faculty: 949
Total Graduate Faculty: 1,108
Total number of Students: 21,053
Total number of Graduate Students: 4,401

Department Information
Department Chairman: Thomas Jordan, Chair
Department Contact: Thomas Jordan, Chairperson
  Total full-time faculty: 16

Total number of full-time equivalent positions: 16
Full-Time Graduate Students: 18
First-Year Graduate Students: 10
Female First-Year Students: 1

Department Address
Department of Physics and Astronomy
Ball State University
Muncie, IN 47306
Phone: (765) 285-8860
Fax: (765) 285-5674
E-mail: physics@bsu.edu
Website: http://www.bsu.edu/physics

ADMISSIONS

Admission Contact Information
Address admission inquiries to: Chair, Graduate Committee, Department of Physics and Astronomy
E-mail: physics@bsu.edu
Admissions website: http://www.bsu.edu/physics

Application deadlines
Fall admission:
U.S. students: May 1   *Int'l. students*: March 1

Application fee
U.S. students: $55   *Int'l. students*: $55
Students who wish to be considered for assistantships should have applications complete by March 1, 2013 for assistantships for the 2013-2014 academic year. International students are encouraged to apply well before that deadline to permit time for communications and processing of international applications.

Admissions information
For Fall of 2013:
Number of applicants: 16
Number admitted: 14
Number enrolled: 8

Admission requirements
Bachelor's degree requirements: Bachelor's degree with a physics major or minor or equivalent from an accredited college or university is required.
Minimum undergraduate GPA: 2.75

GRE requirements
The GRE is recommended.
While we do not require the GRE for admission, it is strongly recommended and is considered in the award of assistantships. We can consider applications for probationary admission if the GPA is slightly below 2.75, but such admission does require GRE scores.

Advanced GRE requirements
The Advanced GRE is not required.

TOEFL requirements
The TOEFL exam is required for students from non-English-speaking countries.
PBT score: 550
iBT score: 79
Students may petition to be conditionally accepted with scores lower than these, but must complete extensive English language courses before being officially accepted into the Graduate School and physics program.

Other admissions information
Additional requirements: Additional requirements for Ed.D. in Science Education applicants: cumulative undergraduate GPA or cumulative GPA for last two years of undergraduate work of 3.0/4.0.
Undergraduate preparation assumed: Fowles, Mechanics; Krane, Modern Physics; Reitz and Milford, Electricity and Magnetism; Pedrotti and Pedrotti, Optics; Sears, Thermodynamics.

TUITION

Tuition year 2012–13:
Tuition for in-state residents
Full-time students: $299 per credit
Part-time students: $299 per credit
Tuition for out-of-state residents
Full-time students: $935 per credit
Part-time students: $935 per credit

Credit hours per semester to be considered full-time: 6
Deferred tuition plan: No
Other academic fees: Approximately $647 general fees, $550 course fee for typical 10 hours, $76 health fee, $168 technology fee, and $87 recreation fee/semester.
Academic term: Semester
Number of first-year students who received full tuition waivers: 5

Teaching Assistants, Research Assistants, and Fellowships
Number of first-year
Teaching Assistants: 4
Research Assistants: 1
Average stipend per academic year
Teaching Assistant: $13,084
Research Assistant: $13,084

FINANCIAL AID

Application deadlines
Fall admission:
U.S. students: March 1   *Int'l. students*: March 1

Loans
Loans are available for U.S. students.
Loans are not available for international students.
GAPSFAS application required: No
FAFSA application required: Yes

For further information
Address financial aid inquiries to: Chair, Graduate Committee, Department of Physics and Astronomy.
Financial aid website: http://www.bsu.edu/finaid

HOUSING

Availability of on-campus housing
Single students: Yes
Married students: Yes

For further information
Address housing inquiries to: Housing Office.
Phone: (765) 285-8000
E-mail: athome@bsu.edu
Housing aid website: http://www.bsu.edu/housing

Table A—Faculty, Enrollments, and Degrees Granted

Research Specialty	2013-2014 Faculty	Enrollment Fall 2012		Number of Degrees Granted 2012-13 (2008-13)		
		Master's	Doctorate	Master's	Terminal Master's	Doctorate
Astronomy	4	3	–	1(8)	–	–
High Energy Physics	2	–	–	1(2)	–	–
Medical, Health Physics	2	4	–	2(8)	–	–
Nano Science and Technology	5	6	–	2(10)	–	–
Nuclear Physics	1	–	–	–	–	–
Physics and other Science Education	1	1	–	2(6)	–	–
Non-specialized	2	–	–	–	–	–
Total	16	'13	–	8(34)	–	–
Full-time Grad. Stud.	–	15	–	–	–	–
First-year Grad. Stud.	–	7	–	–	–	–

GRADUATE DEGREE REQUIREMENTS

Master's: M.S. requirements: 33 semester hours; 3.0 GPA; formal thesis; oral defense of thesis; 22 semester hours in residency; no language requirement. M.A.: same as above but with 3-semester-hour research paper substituting for 6-semester-hour thesis.

Doctorate: Ed.D. in Science Education; 90 semester hours of graduate study, which may include one's master's degree work and other graduate work; a minimum of 48 hours must be completed at Ball State.

Other Degrees: M.A. in physics education (30 hours); M.A. in science education (30 hours).

Thesis: Thesis may be written in absentia.

SPECIAL EQUIPMENT, FACILITIES, OR PROGRAMS

The Nano Materials and Devices Research Laboratory has facilities for growth of carbon nanotubes and related materials, and for electron emission characterizations. The photonics laboratory including a CCD camera and a spectrometer to collect light spectra of thin film materials, two laser sources, an NdYAG laser with emission wavelength of 532 nm and a maximum power of 30 mW, and a 10 mW crystal laser with emission wavelength of 783 nm. Imaging Lab with Transmission Electron Microscope (TEM). Cooperative research with the STAR experiment at Brookhaven National Laboratory.

Beowulf parallel-processing computing cluster, 32 nodes with 16 processor Intel Xeon nodes; Gaussian, IRAF and Matlab software.

Center for Computational Nanoscience, collaborative research group for computational simulation related to nanotechnology.

Planetarium; planned construction of a new planetarium with a 52-foot diameter dome and GOTO projection technology which will be the largest in the state of Indiana (projected completion, 2014).

Astronomical Observatory, with five permanently mounted telescopes, two with CCD cameras attached. CCD astronomical image processing; member of SARA (Southeastern Association for Research in Astronomy) with telescopes on Kitt Peak in Arizona and Cerro Tololo in Chile.

State-of-the-art machine shop with professional staff machinist.

Table B—Separately Budgeted Research Expenditures by Source of Support

Source of Support	Departmental Research	Physics-related Research Outside Department
Federal government	$9,000	
State/local government		
Non-profit organizations		
Business and industry		
Other	$2,600	
Total	**$11,600**	

FACULTY

Professor

Grosnick, David, Ph.D., University of Chicago, 1986. *Particles and Fields, Physics and other Science Education.* Elementary particle physics.

Joe, Yong Suk, Ph.D., Ohio University, 1993. *Condensed Matter Physics.* Theoretical condensed-matter physics.

Kaitchuck, Ron, Ph.D., Indiana University, 1981. *Astronomy, Astrophysics.* Spectroscopy and photometry of interacting binary stars.

Khatun, Mahfuza, Ph.D., Ohio University, 1985. *Condensed Matter Physics, Nano Science and Technology, Statistical & Thermal Physics.* Theoretical condensed-matter physics.

Associate Professor

Bryan, Joel A., Ph.D., Texas A&M University, 2003. *Physics and other Science Education.* Curriculum and instruction; science education.

Islam, Saiful, Ph.D., Ohio University, 1986. *Nuclear Physics.* Experimental nuclear physics.

Jin, Feng, Ph.D., Wayne State University, 1998. *Nano Science and Technology.* Electrical engineering,; experimental nanoscience.

Jordan, Thomas, Ph.D., Oklahoma State University, 1979. *Astronomy, Physics and other Science Education.* Stellar atmospheres; astrophotography.

Maqbool, Muhammad, Ph.D., Ohio University, 2005. *Biophysics, Condensed Matter Physics, Medical, Health Physics.* Medical physics, health physics, photonics and experimental condensed matter physics.

Wijesinghe, Ranjith, Ph.D., Vanderbilt University, 1988. *Biophysics, Medical, Health Physics.* Medical physics.

Assistant Professor

Berrington, Robert, Ph.D., Indiana University, 2000. *Astronomy, Astrophysics, Computational Physics.* Numerical and observational extragalactic astronomy; structure and evolution of galaxies; cosmology.

Cancio, Antonio, Ph.D., University of Illinois, 1994. *Computational Physics, Condensed Matter Physics.* Theoretical condensed-matter physics.

Gonzalez, Guillermo, Ph.D., University of Washington, 1993. *Astronomy, Astrophysics.*

Hedin, Eric, Ph.D., University of Washington, 1986. *Plasma and Fusion.* Computational nano-electronics and Theoretical condensed-matter physics.

Maxin, James A., Ph.D., Texas A & M University, 2010. *Cosmology & String Theory, Particles and Fields.*

Nelson, Christopher B., Ph.D., University of Utah, 2003. *Condensed Matter Physics.* Condensed matter theory.

Emeritus

Cosby, Ronald M., Ph.D., University of Kentucky, 1971. *Condensed Matter Physics, Energy Sources & Environment.* Semiconductor physics; photovoltaics; solar energy.

Errington, Paul R., Ph.D., West Virginia University, 1966. *Electrical Engineering.* Electronics; microprocessor applications.

Howes, Ruth, Ph.D., Columbia University, 1971. *Nuclear Physics.* Women in physics; SPIN-UP.

Koltenbah, David E., Ph.D., Kent State University, 1968. *Chemical Physics.*

Ober, David R., Ph.D., Purdue University, 1968. *Nuclear Physics.* Nuclear spectroscopy; medium-energy nuclear reactions; radon study.

Place, Ralph L., Ph.D., University of Kentucky, 1968. *Computer Science, Nuclear Physics.* Microprocessor applications, expert systems, and computer vision.

Robertson, Thomas H., Ph.D., Case Western Reserve University, 1978. *Astronomy.* Galactic structure and kinematics; observational stellar astronomy.

Thomas, Gerald P., Ph.D., State University of New York at Buffalo, 1968. *Particles and Fields.* Elementary particle physics.

DEPARTMENTAL RESEARCH SPECIALTIES AND STAFF

Theoretical

Condensed Matter Physics. Optical properties of semiconductors and nanostructures; density functional theory; nanoscale systems; nanoscience/nanotechnology education and training; optical properties and electron transport in coupled quantum dots; electron transport in nanostructures; molecular nanoelectronics; quantum cellular automata. Cancio, Cosby, Hedin, Joe, Khatun.

Statistical & Thermal Physics. Low-dimensional lattice models. Khatun.

Experimental

Astrophysics. Observational stellar astronomy; galactic structure; extragalactic numerical and observational CCD imaging; astronomy education; interacting binary stars. Berrington, Jordan, Kaitchuck, Robertson.

Medical, Health Physics. Biomedical physics; EEG and MEG imaging; magnetic signals in nerves; radiation treatment planning. Wijesinghe.

Nano Science and Technology. Electronic and photonic materials and devices; gas discharge physics and devices. Jin.

Nuclear Physics. Nuclear structure and nuclear reactions for stellar processes. Errington, Howes, Islam, Ober, Place.

Particles and Fields. Elementary particles (Brookhaven, Fermi, and Argonne National Laboratories). Grosnick.

Physics and other Science Education. Woodrow Wilson Fellowship PhysTEC Noyce. Bryan.

View additional information about this department at www.gradschoolshopper.com

INDIANA UNIVERSITY, BLOOMINGTON

DEPARTMENT OF ASTRONOMY

Bloomington, Indiana 47405-7105
http://www.astro.indiana.edu/

General University Information

President: Michael A. McRobbie
Dean of Graduate School: James Wimbush
University website: http://www.indiana.edu/~grdschl/index.php
Control: Public
Setting: Suburban
Total Faculty: 1,941
Total Graduate Faculty: 1,344
Total number of Students: 42,133
Total number of Graduate Students: 9,762

Department Information

Department Chairman: John Salzer, Chair
Department Contact: Ross Wood, Senior Office Services Assistant
Total full-time faculty: 10
Total number of full-time equivalent positions: 12
Full-Time Graduate Students: 16
First-Year Graduate Students: 3
Female First-Year Students: 2
Total Post Doctorates: 2

Department Address

727 East 3rd St
Bloomington, IN 47405-7105
Phone: (812) 855-6911
Fax: (812) 855-8725
E-mail: astdept@indiana.edu
Website: http://www.astro.indiana.edu/

ADMISSIONS

Admission Contact Information

Address admission inquiries to: Dr. Liese van Zee, Graduate Advisor, Astronomy Department, Swain Hall West 319, 727 East Third St, Bloomington, IN 47405-7105
Phone: (812) 855-6911
E-mail: astdept@indiana.edu
Admissions website: http://www.astro.indiana.edu/admissions.shtml

Application deadlines
Fall admission:
U.S. students: January 15   *Int'l. students*: December 1

Application fee
U.S. students: $55   *Int'l. students*: $65

Admissions information
For Fall of 2013:
Number of applicants: 40
Number admitted: 10
Number enrolled: 2

Admission requirements
Bachelor's degree requirements: Bachelor's degree in physics, astronomy, astrophysics, or a related discipline is required.
Minimum undergraduate GPA: 3.0

GRE requirements
The GRE is required.

Advanced GRE requirements

The Advanced GRE is required.
Minimum accepted Advanced GRE score: 0
Mean Advanced GRE score range (25th–75th percentile): 540-730
There is no minimum GRE requirement.

TOEFL requirements

The TOEFL exam is required for students from non-English-speaking countries.
PBT score: 550

Other admissions information

Undergraduate preparation assumed: Physics and math background sufficient to handle the astronomy in the following text is assumed: Introduction to Modern Stellar Astrophysics by Carroll and Ostlie.

TUITION

Tuition year 2013-14:
Tuition for in-state residents
Full-time students: $331.56 per credit
Tuition for out-of-state residents
Full-time students: $994.28 per credit
Bloomington campus
Credit hours per semester to be considered full-time: 8
Deferred tuition plan: No
Health insurance: Available
Other academic fees: Approximately $1000 per semester for a full-time student.
Academic term: Semester
Number of first-year students who received full tuition waivers: 2

Teaching Assistants, Research Assistants, and Fellowships

Number of first-year
Teaching Assistants: 2
Average stipend per academic year
Teaching Assistant: $14,858
Research Assistant: $19,933
Fellowship student: $30,000

FINANCIAL AID

Application deadlines

Fall admission:
U.S. students: January 15   *Int'l. students*: December 1

Loans

Loans are available for U.S. students.
Loans are not available for international students.
GAPSFAS application required: No
FAFSA application required: Yes

For further information

Address financial aid inquiries to: Dr. Liese van Zee, Director of Graduate Studies, Astronomy Department, Swain Hall West 319 or to Office of Scholarships and Financial Aids.
E-mail: astdept@indiana.edu
Financial aid website: http://admit.indiana.edu/cost/index.shtml

HOUSING

Availability of on-campus housing

Single students: Yes
Married students: Yes

For further information

Address housing inquiries to: Halls of Residence, 801 N. Jordan, Bloomington, IN 47406.
Phone: (800) 817-6371
E-mail: housing@indiana.edu
Housing aid website: http://www.rps.indiana.edu/index.cfml

Table A—Faculty, Enrollments, and Degrees Granted

Research Specialty	2011 Faculty	Enrollment Fall 2011 Master's	Enrollment Fall 2011 Doctorate	Number of Degrees Granted 2011–12 (2007–12) Master's	Number of Degrees Granted 2011–12 (2007–12) Terminal Master's	Number of Degrees Granted 2011–12 (2007–12) Doctorate
Total	9	1	22	2(14)	-(3)	5(14)
Full-time Grad. Stud.	–	1	20	–	–	–
First-year Grad. Stud.	–	1	3	–	–	–

GRADUATE DEGREE REQUIREMENTS

Master's: The M.A. degree requires 30 hours with a minimum GPA of 3.0. There is no specific residence requirement. A thesis may be required at the discretion of the faculty. A final oral examination covering work for the degree is also required.

Doctorate: The Ph.D. in astronomy requires 90 hours with a minimum GPA of 3.0. There is no specific residency requirement but the student must be continuously enrolled after admission to candidacy. The qualifying examination consists of a written examination (normally after the fourth semester). The requirements for the Ph.D. in astrophysics include at least four physics courses not required by the Ph.D. in astronomy. Candidacy is attained by passage of a combination of tests administered by the Physics Department and the Astronomy Department.

Thesis: Thesis may be written in absentia.

SPECIAL EQUIPMENT, FACILITIES, OR PROGRAMS

The Department is a partner in the 3.5-m and the 0.9-m WIYN telescopes at Kitt Peak National Observatory near Tuscon, Arizona. Most data for thesis research are obtained at WIYN or with national facilities that provide telescope access for X-ray, UV, optical, infrared, radio, and space applications. Indiana University has superb centralized supercomputing facilities available for student and faculty research and the Department has its own computational systems for data processing and scientific computing. Small local telescopes are available for student training.

Table B—Separately Budgeted Research Expenditures by Source of Support

Source of Support	Departmental Research	Physics-related Research Outside Department
Federal government	$3,939,889	$672,565
State/local government		
Non-profit organizations		
Business and industry		
Other		
Total	$3,939,889	$672,565

Table C—Separately Budgeted Research Expenditures by Research Specialty

Research Specialty	No. of Grants	Expenditures ($)
Astronomy	29	$3,939,889
Total	29	$3,939,889

FACULTY

Professor

Cohn, Haldan N., Ph.D., Princeton University, 1979. *Astronomy, Astrophysics*. Dynamical evolution of dense stellar systems; high-performance N-body simulations; globular clusters structure and stellar content; x-ray binaries.

Friel, Eileen, Ph.D., University of California, Santa Cruz, 1986. *Astronomy, Astrophysics*. Formation and evolution of the Milky Way; galactic chemical evolution; star clusters; stellar evolution and nucleosynthesis; stellar populations.

Lugger, Phyllis M., Ph.D., Harvard University, 1982. *Astronomy, Astrophysics*. Dynamical evolution of globular clusters and other dense stellar systems; x-ray studies of compact binary stars.

Mufson, Stuart L., Ph.D., University of Chicago, 1974. *Astronomy, Astrophysics*. High-energy astrophysics; underground cosmic ray physics; neutrino physics.

Pilachowski, Catherine A., Ph.D., University of Hawaii, 1975. *Astronomy, Astrophysics*. Origin of the elements in the Milky Way; star clusters; stellar evolution; the compositions of stars; stellar populations; stellar seismology.

Salzer, John J., Ph.D., University of Michigan, 1987. *Astronomy, Astrophysics*. Galaxy evolution; active galactic nuclei; starburst galaxies; chemical evolution in galaxies multiwavelength studies of dwarf galaxies; emission-line surveys.

Associate Professor

Deliyannis, Constantine P., Ph.D., Yale University, 1990. *Astronomy, Astrophysics*. Stellar evolution; galactic evolution; primordial lithium; Big Bang nucleosynthesis.

Rhode, Katherine, Ph.D., Yale University, 2003. *Astronomy, Astrophysics*. Extragalactic globular clusters systems; galaxy formation; rotation and evolution of solar-type pre-main-sequence stars.

van Zee, Liese, Ph.D., Cornell University, 1996. *Astronomy, Astrophysics*. Galaxy evolution; element enrichment; star formation; extragalactic neutral hydrogen.

Assistant Professor

Vesperini, Enrico, Ph.D., Scuola Normale Superiore, Pisa, 1994. *Astronomy, Astrophysics*. Theoretical and computational stellar dynamics; dynamical evolution of globular clusters.

Emeritus

Burkhead, Martin S., Ph.D., University of Wisconsin-Madison, 1964. *Astronomy, Astrophysics*. Photoelectric photometry; star clusters; galaxies.

Durisen, Richard H., Ph.D., Princeton University, 1972. *Astronomy, Astrophysics*. Star formation; astrophysical fluid dynamics; stellar rotation; planetary rings; complex plasmas.

Honeycutt, R. Kent, Ph.D., Case Western Reserve University, 1968. *Astronomy, Astrophysics*. Stellar astronomy; instrumentation; accretion disks in cataclysmic variables and in other interacting binary stars.

Johnson, Hollis R., Ph.D., University of Colorado, 1960. *Astronomy, Astrophysics*. Model stellar atmospheres: theoretical stellar spectra; ultraviolet spectra of red-giant stars.

Research Professor

Salim, Samir, Ph.D., Ohio State University, 2002. *Astronomy, Astrophysics*. Galaxy evolution; star formation indicators; galaxy bimodality; SED fitting; galaxy surveys; data mining; UV astronomy.

Steiman-Cameron, Thomas Y., Ph.D., Indiana University, 1984. *Astronomy, Astrophysics*. Dynamics of nonplanar astrophysics disks; galaxy formation and evolution; structure of galactic halos; spiral structure of the Milky Way; accretion-driven compact x-ray binary stars.

Thornburg, Jonathan, Ph.D., University of British Columbia, 1993. *Astronomy, Astrophysics*. Numerical simulations of gravitational radiation from extreme-mass-ratio binary black hole inspirals/mergers; numerical simulations of binary black hole mergers; gravitational-wave astrophysics.

DEPARTMENTAL RESEARCH SPECIALTIES AND STAFF

Theoretical

Theoretical Astrophysics. Dynamical evolution of dense stellar systems; globular clusters; stellar rotation; planetary rings; complex plasma; x-ray studies; high-performance N-body simulations (Cohn, Durisen, Lugger, Steiman-Cameron, Vesperini). Cohn, Durisen, Lugger, Steiman-Cameron, Vesperini.

Experimental

Observational Astrophysics. Ground-based and space-based multi-wavelength astronomy; imaging and spectroscopy of stars, star clusters, and external galaxies; studies of steller abundances and evolution, galaxy evolution, and chemical evolution; accretion disks interacting binaries; X-ray binaries; dark energy. Cohn, Deliyannis, Friel, Honeycutt, Lugger, Mufson, Pilachowski, Rhode, Salzer, van Zee.

Observational Astrophysics. Instrumentation; CCD systems; spectrography design; telescope automation. Honeycutt, Pilachowski.

Observational Astrophysics. High-energy particle astrophysics; neutrino and muon astronomy. Mufson.

View additional information about this department at
www.gradschoolshopper.com

INDIANA UNIVERSITY, BLOOMINGTON

DEPARTMENT OF PHYSICS

Bloomington, Indiana 47405-7105
http://www.iub.edu/~iubphys

General University Information
President: Michael A. McRobbie
Dean of Graduate School: James Wimbush
University website: http://www.indiana.edu/
Control: Public
Setting: Suburban
Total Faculty: 1,344
Total Graduate Faculty: 1,581
Total number of Students: 42,133
Total number of Graduate Students: 9,762

Department Information
Department Chairman: Rob de Ruyter, Chair
Department Contact: Jordan Tillett, Fiscal Officer
  Total full-time faculty: 35
  Total number of full-time equivalent positions: 35
  Full-Time Graduate Students: 85
  First-Year Graduate Students: 23
  Female First-Year Students: 5
  Total Post Doctorates: 27

Department Address
727 E. Third St.
Bloomington, IN 47405-7105
Phone: (812) 855-1247
Fax: (812) 855-5533
E-mail: gradphys@indiana.edu
Website: http://www.iub.edu/~iubphys

ADMISSIONS

Admission Contact Information
Address admission inquiries to: Chairman, Graduate Admissions, Dept. of Physics, Swain Hall West Room 117, 727 E. Third St., Bloomington, IN 47405
Phone: (812) 855-3973
E-mail: gradphys@indiana.edu
Admissions website: http://www.indiana.edu/~iubphys/graduate/admissions shtml

Application deadlines
Fall admission:
U.S. students: January 15   *Int'l. students*: December 1
Spring admission:
U.S. students: October 1   *Int'l. students*: October 1

Application fee
U.S. students: $55   *Int'l. students*: $65

Admissions information
For Fall of 2013:
  Number of applicants: 188
  Number admitted: 74
  Number enrolled: 20

Admission requirements
Bachelor's degree requirements: Bachelor's degree in physics or a strong background in physics included in another degree is required.
Minimum undergraduate GPA: 3.0

GRE requirements
The GRE is required.
  Quantitative score: 160
  Verbal score: 154
  Analytical score: 3.5
Quantitative = 60%; Verbal = 30% and Analytical writing = 20%.

Advanced GRE requirements
The Advanced GRE is recommended.
  Minimum accepted Advanced GRE score: 610
  Mean Advanced GRE score range (25th–75th percentile): 65%
Minimum accepted advanced GRE score is 35 percentile.

TOEFL requirements
The TOEFL exam is required for students from non-English-speaking countries.
  PBT score: 550
  iBT score: 80
Minimum 213 computer based.

TUITION

Tuition year 2013-14:
Tuition for in-state residents
  Full-time students: $331.56 per credit
  Part-time students: $331.56 per credit
Tuition for out-of-state residents
  Full-time students: $994.28 per credit
  Part-time students: $994.28 per credit
Bloomington Campus
Credit hours per semester to be considered full-time: 12
Deferred tuition plan: No
Health insurance: Available at the cost of 2,776 per year.
Other academic fees: $645.03 mandatory fees
Academic term: Semester

Teaching Assistants, Research Assistants, and Fellowships
  Number of first-year
  Teaching Assistants: 19
  Research Assistants: 1
  Average stipend per academic year
  Teaching Assistant: $16,659
  Research Assistant: $22,114
  Fellowship student: $24,340
  Info from Fall 2012. Fall 2013 as yet unknown.

FINANCIAL AID

Application deadlines
Fall admission:
U.S. students: January 15   *Int'l. students*: December 1

Loans
Loans are available for U.S. students.
Loans are available for international students.
GAPSFAS application required: No
FAFSA application required: Yes

For further information
Address financial aid inquiries to: Chairman, Graduate Admissions, Physics Department.
Phone: (812) 855-3973
E-mail: gradphys@indiana.edu
Financial aid website: http://www.indiana.edu/~sfa/

HOUSING

Availability of on-campus housing
Single students: Yes
Married students: Yes

For further information
Address housing inquiries to: Halls of Residence, 801 N. Jordan, Indiana University, Bloomington, IN 47405.
Phone: (812) 855-1764
E-mail: housing@indiana.edu
Housing aid website: http://www.rps.indiana.edu

Table A—Faculty, Enrollments, and Degrees Granted

Research Specialty	2012-13 Faculty	Enrollment Fall 2012		Number of Degrees Granted 2012-13 (2008-13)		
		Master's	Doctorate	Master's	Terminal Master's	Doctorate
Astrophysics	3	–	2	–	–	–(3)
Biophysics	4	–	12	–	–	3(10)
Chemical Physics	–	–	–	–	–	–(1)
Condensed Matter Physics	10	–	13	–	–	3(5)
Nuclear Physics	9	–	18	–	–	1(16)
Particles and Fields	8	–	12	–	–	2(9)
Physics of Beams	1	–	10	–	–	–(10)
Other	–	4	22	3(33)	5(12)	–
Total	35	4	89	3(33)	5(12)	9(54)
Full-time Grad. Stud.	–	–	93	–	–	–
First-year Grad. Stud.	–	–	20	–	–	–

GRADUATE DEGREE REQUIREMENTS

Master's: Master of Science: 30 semester hours, at least 20 in physics, 14 of which must be in courses numbered P501 and higher passed with an average grade of "B" or higher. Physics courses numbered below P501 and passed with a grade of "B−" or lower do not count toward this degree. (Seminar, research, and reading courses may not be counted toward the 14 hour requirement.) Master of Science in Beam Physics and Technology (a national program in collaboration with the U.S. Particle Accelerator School, USPAS): 30 credit hours, including the following: P441 (or equivalent at another institution), P506 (or equivalent), P570, one course at the 500 level or above in laboratory techniques or computational methods, and a master's thesis course (P802). Four advanced courses in beam physics should be chosen from among the special topics courses P571, P671, and P672, with topics to be listed in a syllabus prepared jointly by the I.U. Physics Department and the USPAS. A grade point average of 3.0 or better must be maintained in the courses satisfying the 30 credit-hour requirement. In particular, both P441 and P506 (or equivalents) must be passed with a grade B (3.0) or above. Master's examination is required. Thesis is required. Either an oral defense of the thesis or a written final examination is required and should take place at Indiana University. The written examination may be substituted for the oral defense only with the permission of the thesis committee. Master of Science in Medical Physics: A total of 40 credit hours of which at least 18 credit hours must be in physics courses numbered 501 or above. Seminars, research, and reading courses may not be counted toward this 18 credit hour requirement. Required courses include: P576 Introduction to Medical Diagnostic Imaging, P572 Radiation Oncology Physics, P526 Principles of Health Physics and Dosimetry, P578 Radiation Biophysics, P683 Practicum in Medical Physics, and one course in scientific ethics. A580 Human Anatomy for Medical Imaging Evaluation, S520 Statistical Methods, and P551 Experiments in Modern Physics are also required if an equivalent course has not been completed previously. Either a research thesis or a written final examination is required.

Doctorate: 90 semester hours in course, reading, and research credits; a minimum of 9 credit hours per semester at the P501 level or above with an average grade of "B" or higher (first-year students are allowed a minimum of 7 credit hours at the P501 level or above); minor requirement can be met either outside of the Department of Physics or within Physics but outside of the student's area of thesis research; written qualifying examination; candidacy seminar; thesis; final oral examination; a minimum of two consecutive semesters in residence. All candidates are required to undertake supervised teaching as an associate instructor for at least one semester. All first-time teaching associate instructors must enroll in a one-hour graduate credit course "Practicum in Physics Laboratory." Associate instructors whose native language is not English are required to take an "Associate Instructor English Examination," which they must pass in order to be qualified to teach. This examination must be passed by the end of the second year of study.

Other Degrees: Master of Arts for Teachers: 36 credit hours with a minimum of 20 in physics. Ph.D. in Astrophysics: If in residence in the Physics Department, a student must pass specifically designated parts of the qualifying examinations of both departments; thesis; final oral examination. Ph.D. in Biophysics: Students must pass specifically designated parts of the regular departmental qualifying examination and a biophysics qualifying examination; thesis; final oral examination. Ph.D. in Chemical Physics: If in residence in the Physics Department, same qualifying examination as above; minor in chemistry with eight hours in designated courses; thesis; final oral examination. Ph.D. in Mathematical Physics: If in residence in the Physics Department, same qualifying examination as above, and a special qualifying examination in the Mathematics Department; thesis; final oral examination.

Thesis: Thesis may be written in absentia.

SPECIAL EQUIPMENT, FACILITIES, OR PROGRAMS

There is a large joint library for astronomy, computer science, math, and physics in the same building. An extensive machine shop now includes several programmable CNC milling machines and a CNC lathe, numerous other machines, and four full-time machinists. Expertise and high-bay area workshops provide support for assembly of large-scale instrumentation, such as high-energy physics detector assemblies. Capabilities for design and testing of electronics are provided in an Electronics Design Facility, staffed by an experienced electronics engineer and a technician. A 192-node parallel PC cluster is available for research computing. The University provides extensive supercomputing support, including Big Red II, a petaflop scale high-performance parallel computing system, and the Data Capacitor, a high-speed, 5 PB capacity storage facility for large data sets.

The newly established Center for the Exploration of Energy and Matter (CEEM) is a multipurpose research center that supports basic research in nuclear, particle, accelerator, and condensed-matter physics. This center shares the Integrated Science and Accelerator Technology (ISAT) Hall with IU Health's Proton Therapy Center and Radiation Effects Research program, which use proton beams to treat cancer and evaluate radiation effects in electronics, respectively.

Major facilities at CEEM include the Low Energy Neutron Source (LENS) and the Advanced eLectron and Photon Accelerator (ALPHA). LENS is the first pulsed cold neutron source located at a university and provides neutron beams for studying large-scale structure in materials and the development of advanced neutron instrumentation. ALPHA is a state-of-the-art facility for measuring radiation effects and is expected to provide brilliant and tunable hard-x-ray beams for materials research. CEEM includes specialized shops and experienced technical support for design and construction, cryogenic testing, an assembly area, dilution refrigerator, low-vibration room, polarized 3He laser laboratory, and testing of large or complex detector and electronics systems.

Condensed-matter and low-temperature equipment include two x-ray diffraction systems including SAXS; a multisource high-vacuum sputtering system; a pulsed-laser deposition system; and superconducting solenoids; and other low-temperature measurement equipment for measuring electron transport as a function of field (up to 14 T), temperature (down to 20 mK), and frequency (up to 20 GHz). The new Nanoscale Characterization Facility in Simon Hall also has a variety of instruments for patterning and characterizing materials on nanometer-length scales, including a JEOL JEM3200FS cryo TEM, SEMs with e-beam lithography capability, a focused ion beam instrument, an XPS instrument, various scanning probe microscopes, confocal microscopes, and various smaller instruments for optical lithography, thin film deposition, thermal gravimetric analysis, reactive ion etching, IR absorption, and other techniques.

Facilities for biophysics research include cell culture and incubation laboratories, a cell sorter, one-photon and two-photon scanning confocal microscopes, instrumentation for multi-electrical array recording and general neurophysics instrumentation, and access to shared core facilities at the Indiana Molecular Biology Institute.

The Center for Spacetime Symmetries (IUCSS) aims to promote and catalyze scientific progress in theoretical and experimental studies of spacetime and its symmetries, which are central ideas in gravity and particle theories. Studies of spacetime symmetries and their violations offer opportunities to explore foundational aspects of nature. For example, Einstein's theories of special and general relativity are based on spacetime symmetries. IUCSS researchers have proposed that the fundamental theory unifying gravity and quantum physics could lead to tiny violations of the laws of relativity, and many sensitive searches for these effects are underway.

Table B—Separately Budgeted Research Expenditures by Source of Support

Source of Support	Departmental Research	Physics-related Research Outside Department
Federal government	$11,328,824	$2,757,469
State/local government		
Non-profit organizations		
Business and industry	$4,553	
Other		
Total	**$11,333,377**	**$2,757,469**

Table C—Separately Budgeted Research Expenditures by Research Specialty

Research Specialty	No. of Grants	Expenditures ($)
Astrophysics	–	$1,041,147
Biophysics	–	$2,239,267
Condensed Matter Physics	–	$2,885,056
Nuclear Physics	–	$2,290,456
Particles and Fields	–	$2,186,604
Physics of Beams	–	$624,397
Other	0	$66,450
Total	**–**	**$11,333,377**

FACULTY

Distinguished University Professor

Gottlieb, Steven A., Ph.D., Princeton University, 1978. Director of PhD Minor in Scientific Computing. *Computational Physics, High Energy Physics, Particles and Fields, Theoretical Physics.* Lattice field theory.

Kostelecky, V. Alan, Ph.D., Yale University, 1982. *Astronomy, High Energy Physics, Particles and Fields, Theoretical Physics.* Spacetime symmetries; Lorentz and CPT violation.

Professor

Baxter, David V., Ph.D., California Institute of Technology, 1984. *Condensed Matter Physics, Low Temperature Physics, Nano Science and Technology, Solid State Physics, Surface Physics.* Condensed matter (experimental).

Berger, Michael S., Ph.D., University of California, Berkeley, 1991. *Particles and Fields.* Theoretical physics; elementary particles.

de Ruyter van Steveninck, Robert, Ph.D., University of Groningen, Netherlands, 1986. *Biophysics.* Biophysics (experimental).

Evans, Harold G., Ph.D., University of California, Los Angeles, 1991. *Particles and Fields.* Elementary particle physics (experimental).

Fertig, Herbert A., Ph.D., Harvard University, 1988. *Condensed Matter Physics.* Condensed-matter theory.

Glazier, James, Ph.D., University of Chicago, 1989. *Biophysics.* Biophysics (experimental).

Horowitz, Charles J., Ph.D., Stanford University, 1981. *Nuclear Physics.* Nuclear theory.

Jose, Jorge, Ph.D., National University of Mexico. Vice President for Research. *Biophysics, Condensed Matter Physics.* Condensed matter theory and theoretical biophysics.

Lee, Shyh-Yuan, Ph.D., Stony Brook University, 1972. *Physics of Beams.* Accelerator physics.

Londergan, J. Timothy, Ph.D., University of Oxford, 1969. *Nuclear Physics.* Theoretical physics; nuclear theory.

Messier, Mark, Ph.D., Boston University, 1999. *Astrophysics.* Astrophysics (experimental).

Musser, James A., Ph.D., University of California, Berkeley, 1984. *Astrophysics.* Astrophysics (experimental).

Ortiz, Gerardo, Ph.D., Swiss Federal Institute of Technology Theoretical Physics, 1992. *Computational Physics, Condensed Matter Physics, Low Temperature Physics, Quantum Foundations, Statistical & Thermal Physics, Theoretical Physics.* Quantum Information and Computation.

Pynn, Roger, Ph.D., University of Cambridge, 1969. *Condensed Matter Physics.* Condensed matter physics (experimental).

Snow, W. Michael, Ph.D., Harvard University, 1990. *Astrophysics, Condensed Matter Physics, Nuclear Physics, Particles and Fields.* Neutron physics; strong and electroweak interactions; nuclear precision measurements; neutron lifetime measurement for Big Bang Nucleosynthesis, searches for weakly coupled long-range interactions between nucleons; search for dark energy scalar fields; few body physics; polarized 3He neutron scattering; neutron source and instrumentation development.

Sokol, Paul E., Ph.D., Ohio State University, 1981. *Biophysics, Chemical Physics, Condensed Matter Physics, Energy Sources & Environment, Low Temperature Physics, Materials Science, Metallurgy, Nano Science and Technology, Physics of Beams, Polymer Physics/Science, Solid State Physics, Statistical & Thermal Physics.* Microscope structure and dynamics of condensed matter using x-ray and neutron scattering techniques; collective excitations in confined quantum liquids; momentum distribution of hydrogen on surfaces; microscopic structure of confined solids; wetting on nonstructural surfaces; dynamics of hydrogen in reduced dimensionality; hydrogen storage materials.

Szczepaniak, Adam P., Ph.D., University of Washington, 1990. Chair, Department of Physics. *Nuclear Physics.* Theoretical physics.

Urheim, Jon, Ph.D., University of Pennsylvania, 1990. *Astrophysics.* Astrophysics (experimental).

Van Kooten, Richard J., Ph.D., Stanford University, 1990. *Particles and Fields.* Heavy flavor physics; searches for new particles; Higgs boson properties; detector research and development.

Wissink, Scott W., Ph.D., Stanford University, 1986. Director of the IU Center for Exploration of Energy and Matter; Deputy Spokesperson for the STAR Experiment. *High Energy Physics, Nuclear Physics.* Experimental studies of the structure of the nucleon in terms of quarks and gluons with emphasis on spin and flavor dependencies.

Associate Professor

Beggs, John, Ph.D., Yale University, 1998. *Biophysics, Computational Physics, Nonlinear Dynamics and Complex Systems.* Experiments on living neural networks; computational models; statistical physics; network science; information theory; criticality; avalanches.

Carini, John P., Ph.D., University of Chicago, 1988. Associate Chair, Department of Physics. *Condensed Matter Physics.* Condensed matter physics (experimental).

Liu, Chen-Yu, Ph.D., Princeton University, 2002. *Nuclear Physics.* Nuclear physics (experimental).

Setayeshgar, Sima, Ph.D., California Institute of Technology, 1997. *Biophysics.* Biophysics (theoretical).

Shepherd, Matthew R., Ph.D., Cornell University, 2004. *Particles and Fields.* Elementary particle physics (experimental).

Tayloe, Rex, Ph.D., University of Illinois, 1995. *Nuclear Physics.* Nuclear physics (experimental).

Assistant Professor

Dermisek, Radovan, Ph.D., Ohio State University, 2002. *High Energy Physics, Particles and Fields.* Theoretical physics; elementary particles.

Kaufman, Lisa, Ph.D., University of Massachusetts, 2007. *Nuclear Physics.* Neutrinoless double beta decay; weak interactions; fundamental symmetries.

Lammers, Sabine, Ph.D., University of Wisconsin-Madison, 2004. *Particles and Fields.* Elementary particle physics (experimental).

Liao, Jinfeng, Ph.D., Stony Brook University, 2008. *Nuclear Physics.* Theoretical physics.

Long, Josh, Ph.D., Johns Hopkins University, 1997. *Nuclear Physics.* Nuclear physics (experimental).

Seradjeh, Babak H., Ph.D., Simon Fraser University, 2006. *Condensed Matter Physics.* Condensed matter physics (theoretical).

Strigari, Louis, Ph.D., The Ohio State University, 2005. *Astrophysics, Cosmology & String Theory, Theoretical Physics.*

Zhang, Shixiong, Ph.D., University of Maryland, 2007. *Condensed Matter Physics.* Condensed matter physics (experimental).

Emeritus

Alyea, Ethan D., Ph.D., California Institute of Technology, 1962. Astrophysics (experimental).

Bacher, Andrew D., Ph.D., California Institute of Technology, 1967. Intermediate-energy nuclear physics (experimental).

Brabson, Bennet, Ph.D., Massachusetts Institute of Technology, 1966. The physics of climate change.

Cameron, John M., Ph.D., University of California, Los Angeles, 1967. Nuclear physics (experimental).

Challifour, John L., Ph.D., University of Cambridge, 1963. Theoretical physics; mathematical physics; quantum field theory.

Crittenden, Ray R., Ph.D., University of Wisconsin-Madison, 1960. Elementary particle physics (experimental).

Dzierba, Alex R., Ph.D., University of Notre Dame, 1969. Elementary particle physics (experimental).

Goodman, Charles, Ph.D., University of Rochester, 1955. Nuclear physics (experimental).

Hake, Richard R., Ph.D., University of Illinois, 1955. Physics education.

Heinz, Richard M., Ph.D., University of Michigan, 1964. Astrophysics (experimental).

Hendry, Archibald W., Ph.D., University of Glasgow, 1962. Theoretical physics; elementary particles.

Kesmodel, Larry L., Ph.D., University of Texas, 1974. Condensed matter physics (experimental); surfaces.

Lichtenberg, Don B., Ph.D., University of Illinois, 1955. Elementary particle physics (theoretical).

Meyer, Hans Otto, Ph.D., Basel, Switzerland, 1970. Nuclear physics (experimental).

Miller, Daniel W., Ph.D., University of Wisconsin-Madison, 1951. Nuclear physics (experimental); nuclear reactions.

Nann, Herman, Ph.D., Goethe University, 1967. Intermediate energy nuclear physics (experimental).

Newton, Roger G., Ph.D., Harvard University, 1953. Theoretical and mathematical physics; scattering theory.

Ogren, Harold O., Ph.D., Cornell University, 1970. Elementary particle physics (experimental).

Olmer, Catherine, Ph.D., Yale University, 1976. Physics education.

Pollock, Robert E., Ph.D., Princeton University, 1963. Nuclear physics; nuclear reactions; cyclotron design.

Schaich, William L., Ph.D., Cornell University, 1970. Condensed matter physics (theoretical).

Schwandt, Peter, Ph.D., University of Wisconsin-Madison, 1967. Nuclear physics (experimental).

Swihart, James C., Ph.D., Purdue University, 1955. Condensed matter physics (theoretical).

Walker, George E., Ph.D., Case Western Reserve University, 1966. Nuclear physics (theoretical); intermediate energy physics.

Wills, John G., Ph.D., University of Washington, 1963. Theoretical physics.

Research Professor

Gagnon, Pauline, Ph.D., University of California, Santa Cruz, 1993. *Particles and Fields.* Elementary particle physics (experimental).

Klein, Susan, Ph.D., University of California, Berkeley, 1986. Director, Medical Physics M.S. Program. *Applied Physics, Biophysics, Medical, Health Physics.* Radiation physics.

Luehring, Frederick, Ph.D., Northwestern University, 1986. *Particles and Fields.* Elementary particle physics (experimental).

Lunghi, Enrico, Ph.D., International School of Advanced Studies, Italy, 2000. *Particles and Fields*. Theoretical physics; elementary particles.

Mitchell, Ryan, Ph.D., University of Tennessee, 2003. *Particles and Fields*. Elementary particle physics (experimental).

Nagao, Michihiro, Ph.D., University of Tokyo, 2001. *Condensed Matter Physics*. Condensed matter physics (experimental).

Sluka, James, Ph.D., California Institute of Technology, 1988. *Biophysics*. Biophysics (experimental).

Swat, Maciej, Ph.D., Indiana University, 2003. *Biophysics*. Biophysics (experimental).

Vossen, Anselm, Ph.D., University of Freiburg, 2008. *Nuclear Physics*. Nuclear physics (experimental).

Warren, Garfield, Ph.D., Tuskegee University, 1988. *Condensed Matter Physics*. Condensed matter physics (experimental).

Zieminska, Daria, Ph.D., Warsaw University, 1974. *Particles and Fields*. Elementary particle physics (experimental).

DEPARTMENTAL RESEARCH SPECIALTIES AND STAFF

Theoretical

Biophysics. Intracellular signaling networks; waves in excitable media; non-equilibrium systems; biocomplexity; theoretical neuroscience; information theory. Jose, Setayeshgar.

Condensed Matter Physics. Topological condensed matter systems; fractionalization and exotic quasiparticles; graphene; quantum Hall effect; superconductivity, spin transport, and magnetoresistance; nanoscale systems; soft matter; colloidal and biological materials; electron–phonon interaction in metals; optical and electrical properties of solids; collective excitations; many-body theory; surface electrodynamics; random alloys; quantum computation; correlated electronic materials; many-body physics, strongly correlated systems: high-Tc superconductivity, heavy fermions, fermions in high magnetic fields; exotic superconductors; magnetism and spin systems; quantum fluids and solids; ultracold Fermi and Bose gases; topologically quantum ordered systems; quantum statistical mechanics and field theory methods in condensed matter; quantum information and computation; quantum measurement theory. Fertig, Jose, Ortiz, Schaich, Seradjeh.

High Energy Physics. Phenomenology of elementary particle properties and interactions; quantum chromodynamics and electroweak interactions; lattice gauge field theory; solar neutrinos; grand-unified theories; supersymmetry; gravity and supergravity; superstring theory; CPT and Lorenz symmetry. Berger, Dermisek, Gottlieb, Lunghi.

Nuclear Physics. Nuclear structure; medium- and high-energy nuclear reactions; quantum chromodynamics; hadron spectra and structure; gluon dynamics; relativistic quantum hadrodynamics; neutron stars and nuclear astrophysics; stellar evolution; neutrino transport. Horowitz, Liao, Londergan, Szczepaniak.

Physics of Beams. Nonlinear beam dynamics; beam-beam interactions; transition energy problems; transverse and longitudinal coherent instabilities and Landau damping; bunched beam cooling; electron storage ring physics; spin motion in synchrotrons. Lee.

Experimental

Astrophysics. Magnetic monopoles; antimatter; supernovae; dark matter searches; bigbang cosmology; neutrino oscillations; dark energy; solar neutrino. Facilities include an assortment of computers, particle detectors, electronics development equipment, data acquisition systems, and spectrophotometers. Experiments are being performed at Fermi National Laboratory, Superkamiokande, and at a number of balloon launch facilities. Messier, Musser, Urheim.

Biophysics. Experimental and computational neuroscience; multielectrode records in vitro; intracellular and extracellular neural recording in vivo; experimental biocomplexity. Beggs, de Ruyter van Steveninck, Glazier, Sluka, Swat.

Chemical Physics. Self-assembly and nano-structured materials; optical properties of solids; low-temperature properties of solids; chemisorption and catalysis; confined fluids; nuclear chemistry; x-ray and neutron diffraction. Baxter, Carini, Pynn, Sokol.

Condensed Matter Physics. Confined complex and quantum fluids; x-ray and neutron scattering surfactant systems; atomic and electronic transport nanostructures; topological insulators; metal oxides; compositionally modulated thin films; thin film and exotic magnetism and magnetoresistance; metastable systems; surface studies: STM, AFM, EELS; topological insulators; dynamics of electrons in disordered metals and correlated electron systems. Experiments are conducted with a wide variety of in-house equipment (see facilities description) and at national user facilities. Baxter, Carini, Pynn, Sokol, Warren, Zhang.

High Energy Physics. Searches for new particles (Higgs bosons, supersymmetric particles, exotics, hybrid systems, glueballs); heavy quark physics (top, bottom, charm); light quarks; neutrino oscillation; and testing of fundamental symmetries. Detectors used include drift chambers, drift tubes, scintillating fibers, transition radiation detectors, Cerenkov counters, and calorimeters and trigger systems. IU facilities include data acquisition and large scale data analysis computer clusters, detector construction areas including a high-bay area and large class-10000 clean room, and an electronics design facility. Students work on DØ and MINOS and NOvA at Fermilab, ATLAS at CERN, experiments at Jefferson Laboratory, and in preparations for LBNE for the Linear Collider. Evans, Gagnon, Lammers, Luehring, Messier, Mitchell, Musser, Shepherd, Urheim, Van Kooten, Zieminska.

Nuclear Physics. Nucleon structure studies: gluon spin distributions, anti-quark and sea quark contributions to nucleon properties using the STAR detector at RHIC and the Belle detector at KEK; weak interaction studies with slow neutrons: precision measurements of neutron decay and neutron weak interactions at NIST, LANSCE, and SNS; methods for production of ultra-cold neutrons; studies of neutrino properties and interactions at Fermilab and WIPP; fundamental symmetry tests: searches for time-reversal violation via electric dipole moments of the electron and neutron; searches for exotic nucleon spin-dependent couplings, formation and decay of hot nuclei, damped collisions between heavy nuclei, and nuclear fission at MSU, ATLAS, and other laboratories. Kaufman, Liu, Long, Snow, Szczepaniak, Tayloe, Vossen, Wissink.

Physics of Beams. Nonlinear beam dynamics; electron cooling; properties of cooled beams; damping of transverse and longitudinal instabilities; spin motion of synchrotrons with spin rotators (snakes); overlapping spin resonances and snake resonance. Lee.

View additional information about this department at
www.gradschoolshopper.com

INDIANA UNIVERSITY — PURDUE UNIVERSITY INDIANAPOLIS

DEPARTMENT OF PHYSICS

Indianapolis, Indiana 46202-3273
http://physics.iupui.edu/

General University Information
President: Michael A. McRobbie
Dean of Graduate School: Sherry F. Queener
University website: http://www.iupui.edu/
Control: Public
Setting: Urban
Total Faculty: 3,041
Total number of Students: 30,300
Total number of Graduate Students: 8,200

Department Information
Department Chairman: Andrew Gavrin, Chair
Department Contact: Kimberly Wright, Administrative Assistant / Graduate Coordinator
Total full-time faculty: 16
Total number of full-time equivalent positions: 16
Full-Time Graduate Students: 20
First-Year Graduate Students: 2
Total Post Doctorates: 2

Department Address
402 N. Blackford St.
LD154
Indianapolis, IN 46202-3273
Phone: (317) 274-6900
Fax: (317) 274-2393
E-mail: physics@iupui.edu
Website: http://physics.iupui.edu/

ADMISSIONS

Admission Contact Information
Address admission inquiries to: Prof. Ricardo Decca, Director of Graduate Program
E-mail: rdecca@iupui.edu
Admissions website: http://physics.iupui.edu/graduate/graduate-admissions

Application deadlines
Fall admission:
U.S. students: March 15   *Int'l. students*: March 15
Spring admission:
U.S. students: November 15   *Int'l. students*: September 15

Application fee
U.S. students: $60   *Int'l. students*: $65
Dates are not hard deadlines; they are recommended dates to have applications submitted.

Admissions information
For Fall of 2012:
Number of applicants: 20
Number admitted: 2
Number enrolled: 20

Admission requirements
Bachelor's degree requirements: Bachelor's degree in physics or related areas is required.

GRE requirements
The GRE is recommended.

Advanced GRE requirements
The Advanced GRE is recommended.

TOEFL requirements
The TOEFL exam is required for students from non-English-speaking countries.
PBT score: 550
iBT score: 79

Other admissions information
Undergraduate preparation assumed: Symon, Mechanics; Corson and Lorrain, Electricity and Magnetism; Eisberg and Resnick, Quantum Physics; Rief, Thermodynamics.

TUITION

Tuition year 2012–13:
Tuition for in-state residents
   Full-time students: $345.3 per credit
Tuition for out-of-state residents
   Full-time students: $1,007 per credit
Credit hours per semester to be considered full-time: 8
Deferred tuition plan: No
Health insurance: Yes, 2,520.00.
Other academic fees: $600 - $1,000 for partial tuition and fees.
Academic term: Semester

Teaching Assistants, Research Assistants, and Fellowships
Number of first-year
   Teaching Assistants: 1
Average stipend per academic year
   Teaching Assistant: $19,000
   Research Assistant: $19,000
   Fellowship student: $19,000

FINANCIAL AID

Loans
Loans are available for U.S. students.
Loans are not available for international students.
GAPSFAS application required: No
FAFSA application required: No

For further information
Address financial aid inquiries to: Department of Physics Graduate Program.
Financial aid website: http://www.iupui.edu/~finaid

HOUSING

Availability of on-campus housing
Single students: Yes
Married students: Yes

For further information

Address housing inquiries to: Department of Physics, Graduate Program.

E-mail: physics@iupui.edu

Housing aid website: http://life.iupui.edu/housing

Table A—Faculty, Enrollments, and Degrees Granted

Research Specialty	2012-13 Faculty	Enrollment Fall 2011		Number of Degrees Granted 2009–10 (2005–10)		
		Master's	Doctorate	Master's	Terminal Master's	Doctorate
Atomic, Molecular, & Optical Physics	4	2	3	–(2)	–	–(1)
Biophysics	5	5	3	1(4)	1(6)	1(1)
Condensed Matter Physics	3	4	3	–(3)	1(1)	1(2)
Physics and other Science Education	1	2	–	–(1)	–	–(1)
Total	13	13	9	1(10)	2(9)	2(5)
Full-time Grad. Stud.	–	22	22	–	–	–
First-year Grad. Stud.	–	–	–	–	–	–

GRADUATE DEGREE REQUIREMENTS

Master's: Both thesis and non-thesis master's programs are available. For each program, the student must complete 30 credit hours and maintain a grade point average of 2.7. Twenty-four credit hours must be in physics/biophysics and 6 hours in mathematics. For the thesis master's program, six of the 24 hours are satisfied by completing the thesis. All students must pass a qualifying examination early in their program and an oral examination at the completion of their program. The minimum residence requirement is two semesters of full-time work or the equivalent in credits.

Doctorate: Qualified students may pursue the Ph.D. degree at IUPUI in areas in which a program has been arranged with Purdue University-West Lafayette. Students are usually expected to complete an M.S. degree before pursuing the Ph.D. degree. Currently, a Ph.D. program is available in the areas of biological physics, optics, and materials science.

Thesis: Thesis may be written in absentia.

SPECIAL EQUIPMENT, FACILITIES, OR PROGRAMS

The NMR facilities of the department consist of three multinuclear high-resolution FT spectrometers and one solid-state spectrometer. Included among the high-resolution instruments are narrow-bore 500 and 200 MHz spectrometers and a wide-bore 300 MHz spectrometer. The solid state instrument has broadline and MAS capabilities, operates at 4.2 T, and is home built. An x-band EPR spectrometer is used for biophysics research. Full facilities for the preparation and characterization of biological samples are available. Two optics research laboratories are equipped with argon laser-pumped, CW frequency-doubled Ti-sapphire lasers, diode lasers, and He–Ne lasers, among others. State-of-the-art data-acquisition equipment includes digital oscilloscopes, spectrum analyzers, and computer interfaces. High-finesse optical cavities and high-vacuum systems are used to study atomic behavior in laser fields. A thin-film sputter deposition system with 4 high-rate magnetron guns, rf and dc excitation, substrate heating to 1000 K, computer-controlled substrate and shutter motions is available. The chamber is cryopumped with background pressure of 10.

Table C—Separately Budgeted Research Expenditures by Research Specialty

Research Specialty	No. of Grants	Expenditures ($)
Biophysics	2	$215,000
Condensed Matter Physics	2	$190,000
Physics and other Science Education	2	$325,968.2
Total	6	$730,968.2

FACULTY

Professor

Decca, Ricardo S., Ph.D., Instituto Balseiro, 1994. Director of Graduate Program. *Condensed Matter Physics, Low Temperature Physics, Materials Science, Metallurgy, Nano Science and Technology.* Condensed matter; scanning probe microscopy; spectroscopy of low dimensional systems; corrections to Newtonian gravity; lipid bilayers.

Kemple, Marvin D., Ph.D., University of Illinois, 1971. *Biophysics, Medical, Health Physics.* Biological physics; magnetic resonance and fluorescence.

Ou, Zhe-Yu Jeff, Ph.D., University of Rochester, 1990. *Atomic, Molecular, & Optical Physics, Optics.* Quantum optics; nonlinear optics.

Rao, B. D. Nageswara, Ph.D., Aligarh Muslim University, 1961. *Biophysics, Medical, Health Physics.* Biophysics; magnetic resonance.

Vemuri, Gautam, Ph.D., Georgia Institute of Technology, 1990. *Atomic, Molecular, & Optical Physics, Nonlinear Dynamics and Complex Systems, Optics.* Nonlinear optics; laser physics.

Wassall, Stephen R., Ph.D., University of Nottingham, 1981. *Biophysics.* Solid-state NMR; biophysics; lipid membranes.

Associate Professor

Gavrin, Andrew D., Ph.D., Johns Hopkins University, 1992. Department Chair. *Materials Science, Metallurgy, Physics and other Science Education.* Just-in-Time Teaching; use of technology in education.

Joglekar, Yogesh, Ph.D., Indiana University, 2001. *Condensed Matter Physics, Nano Science and Technology.* Condensed matter; noise spectroscopy; graphene; non-Hermitian quantum mechanics.

Petrache, Horia, Ph.D., Carnegie Mellon University, 1998. *Biophysics.* X-ray scattering; membrane biophysics.

Assistant Professor

Cheng, Ruihua, Ph.D., University of Nebraska-Lincoln, 2002. *Condensed Matter Physics, Materials Science, Metallurgy, Nano Science and Technology.* Condensed matter; magnetic nanostructures.

Luo, Le, Ph.D., Duke University, 2008. *Atomic, Molecular, & Optical Physics, Optics.* BEC; optical trap; quantum computation; atomic physics (cold atom); quantum information; optics.

Presse, Steve, Ph.D., Massachusetts Institute of Technology, 2008. *Biophysics, Chemical Physics, Condensed Matter Physics.* Bio-organic Chemistry.

Zhu, Fangqiang, Ph.D., University of Illinois at Urbana-Champaign, 2004. *Biophysics, Computational Physics, Theoretical Physics.* Biomolecular simulation; membrane protein; protein conformation and assembly.

Senior Lecturer

Woodahl, Brian A., Ph.D., Purdue University, 1999. *Cosmology & String Theory, Particles and Fields.* Physics education; theoretical particle physics.

Lecturer

Rhoads, Edward, Ph.D., University of Minnesota, 2005. *Astronomy*. Astronomy.

Ross, John B., Ph.D., Boston University, 1993. *Physics and other Science Education, Other*. Physics education.

Associate Scientist

Ray, Bruce D., Ph.D., Indiana University, 1983. *Biophysics*. Biochemistry; isotope labeling; NMR.

DEPARTMENTAL RESEARCH SPECIALTIES AND STAFF

Theoretical

Biophysics. Theoretical aspects of magnetic resonance methods used in macromolecular structure studies; computer simulations of nuclear magnetic resonance spectra and electron paramagnetic resonance spectra in biological and nonbiological systems; nonlinear chemical exchange. Kemple, Petrache, Rao, Wassall.

Biophysics. Theoretical modeling of proteins; protein folding dynamics and thermodynamics; de novo protein design; protein structure prediction by computer simulations; inferring biological functions from simulations of large-scale motions in proteins and supramolecular assemblies. Kemple, Petrache, Rao, Wassall.

Biophysics. Theoretical modeling of membrane transport of water and non-ionic solutes and osmotic behavior; lipid-protein interactions; membrane dynamics and electrostatistics. Kemple, Petrache, Rao, Wassall.

Condensed Matter Physics. Strongly correlated systems including excitonic condensates in semiconductors, noise spectroscopy, graphene, and quantum Hall systems. Cheng, Decca, Presse.

Optics. Quantum optics in integrated structures; nonlinear dynamics of diode and solid-state lasers; atomic coherence effects; response to fluctuating fields; quantum noise and measurement; quantum fluctuations in nonlinear optical processes; quantum multiphoton interference. Luo, Ou, Vemuri.

Experimental

Biophysics. Small-angle scattering of membrane systems in solution to determine structure and molecular interactions relevant to biological functions. Zhu.

Biophysics. Macromolecular structure-function relationships in enzyme-substrate complexes of ATP-utilizing enzymes and alcohol dehydrogenase; internal motions of peptides, proteins, and their complexes; NMR and computer simulations of protein dynamics; broadline deuterium NMR of molecular order and dynamics in membranes; MAS NMR determination of peptide conformation; EPR membrane and cytoplasmic studies of reactive oxygen species and molecular order in model membranes. Zhu.

Condensed Matter Physics. Spatial and time-resolved spectroscopy in quantum systems; metal-insulator transition in superconductors; correlated electronic systems; Casimir interaction; spin-dependent transport. Decca, Joglekar, Presse.

Condensed Matter Physics. Near-field scanning optical microscopy; atomic force microscopy; probe-sample interaction effect; image analysis and deconvolution; single molecule detection and tracking. Decca, Joglekar.

Materials Science, Metallurgy. Artificially structured materials; ferromagnetic domains and giant magnetoresistive effects in granular metals; domain wall pinning in amorphous alloys; scanning electron microscopy with polarization analysis; fabrication of magnetic nanowires and nanodots. Decca, Joglekar.

Optics. Semiconductor lasers; non-linear optics; diode laser and amplifier statistical properties; laser instabilities; chaos and communication; cavity QED; nonlinear optical frequency conversion; photon statistics of nonclassical states; test of EPR nonlocality; multiphoton interference. Luo, Ou, Vemuri.

Physics and other Science Education. Just-in-Time-Teaching (JiTT) modulus course design. Gavrin.

View additional information about this department at
www.gradschoolshopper.com

PURDUE UNIVERSITY

DEPARTMENT OF PHYSICS

West Lafayette, Indiana 47907
http://www.physics.purdue.edu

General University Information

President: Mitchell E. Daniels, Jr.
Dean of Graduate School: Mark J. T. Smith
University website: http://www.purdue.edu
Control: Public
Setting: Suburban
Total Faculty: 1,807
Total number of Students: 39,256
Total number of Graduate Students: 8,163

Department Information

Department Chairman: Andrew S. Hirsch, Head
Department Contact: Sandy Formica, Graduate Secretary
  Total full-time faculty: 56
  Full-Time Graduate Students: 143
  First-Year Graduate Students: 21
  Female First-Year Students: 3
  Total Post Doctorates: 15

Department Address

525 Northwestern Avenue
West Lafayette, IN 47907
Phone: (765) 494-3099

Fax: (765) 494-0706
E-mail: physcontacts@purdue.edu
Website: http://www.physics.purdue.edu

ADMISSIONS

Admission Contact Information
Address admission inquiries to: Sandy Formica, Graduate Secretary
Phone: (765) 494-3099
E-mail: physcontacts@purdue.edu
Admissions website: http://www.physics.purdue.edu/academic_programs/graduate/admissions.shtml

Application deadlines
Fall admission:
U.S. students: January 15   *Int'l. students*: January 15

Application fee
U.S. students: $60   *Int'l. students*: $75

Admissions information
For Fall of 2013:
Number of applicants: 285
Number admitted: 71
Number enrolled: 23

Admission requirements
Bachelor's degree requirements: Bachelor's degree in physics is required.

GRE requirements
The GRE is required.

Advanced GRE requirements
The Advanced GRE is required.

TOEFL requirements
The TOEFL exam is required for students from non-English-speaking countries.
PBT score: 550
iBT score: 77

Other admissions information
Additional requirements: TOEFL (iBT) minimum required individual scores are 18 writing, 18 speaking, 14 listening, and 19 reading.
Undergraduate preparation assumed: A good preparation for entering students includes a sound knowledge of general physics, intermediate level mechanics, electricity and magnetism, optics, statistical and thermal physics, introductory atomic and nuclear physics including some principles of quantum mechanics. A corresponding mathematical background would include vector analysis, advanced calculus, ordinary differential equations, boundary value problems, and some knowledge of introductory complex analysis. Graduate credit courses are offered at two levels in mechanics, electricity and magnetism, thermal physics and modern physics; first-year students can be placed in courses that will supplement the undergraduate program and correct deficiencies. Strong undergraduate preparation would be provided by adequate study of textbooks at the level of: Marion, Classical Dynamics; Griffiths, Introduction to Electrodynamics; Reif, Statistical and Thermal Physics; Jenkins and White, Fundamentals of Optics; and Gasiorowicz, Quantum Physics.

TUITION

Tuition year 2013-14:
Tuition for in-state residents
Full-time students: $9,900 annual
Tuition for out-of-state residents
Full-time students: $28,702 annual

Fees for TAs, and RAs is $252 per semester.
Credit hours per semester to be considered full-time: 9
Deferred tuition plan: Yes
Health insurance: Available at the cost of $432 per year.
Academic term: Semester

Teaching Assistants, Research Assistants, and Fellowships
Number of first-year
Teaching Assistants: 22
Fellowship students: 4
Average stipend per academic year
Teaching Assistant: $21,700
Research Assistant: $21,700
Fellowship student: $25,500
Includes summer appointments.

FINANCIAL AID

Loans
Loans are available for U.S. students.
Loans are available for international students.
GAPSFAS application required: No
FAFSA application required: No

For further information
Address financial aid inquiries to: Sandy Formica, Graduate Secretary, Department of Physics.
Phone: (765) 494-3099
E-mail: physcontacts@purdue.edu
Financial aid website: http://www.purdue.edu/dfa

HOUSING

Availability of on-campus housing
Single students: Yes
Married students: Yes

For further information
Address housing inquiries to: Graduate Housing: ghapp@purdue.edu, Married and Family Housing: pvapp@purdue.edu.
E-mail: ghapp@purdue.edu
Housing aid website: http://www.purdue.edu/purdue/admissions/housing.html

Table A—Faculty, Enrollments, and Degrees Granted

Research Specialty	2012-13 Faculty	Enrollment Fall 2012		Number of Degrees Granted 2012-13 (2007-12)		
		Master's	Doctorate	Master's	Terminal Master's	Doctorate
Applied Physics	2	–	–	–	–	1(5)
Astrophysics	6	–	9	–	–(1)	3(13)
Atomic, Molecular, & Optical Physics	1	–	–	–	–	–
Biophysics	5	–	15	–	1(-)	1(21)
Condensed Matter Physics	17	–	52	–	–(2)	6(37)
Geophysics	2	–	7	–	–	1(3)
Nuclear Physics	8	–	14	–	–	1(8)
Particles and Fields	14	–	20	–	–	4(29)
Physics and other Science Education	2	–	1	–	–	–
Non-specialized	–	2	32	9(27)	2(23)	–
Total	57	2	151	9(27)	3(26)	17(116)
Full-time Grad. Stud.	–	2	151	–	–	–
First-year Grad. Stud.	–	–	22	–	–	–

GRADUATE DEGREE REQUIREMENTS

Master's: Non-thesis option: completion of a minimum of 30 credit hours with at least 24 hours of approved 500–600 level courses in physics, including one laboratory course, and 6 credit hours in 500–600 level mathematics courses, which may be replaced in whole or in part by Methods of Theoretical Physics I, II: grade in a 500-level physics course must be A or B, and grade in a 600-level physics or a mathematics course A, B, or C; minimum graduate grade average of 2.8/4.0; qualifying examination must be taken; written and oral final examinations are given or waived at discretion of student's advisory committee. More than half of the Purdue credits must be earned through the Purdue campus where the degree is conferred. Thesis option: thesis replaces 9 credit hours of physics requirement: final oral examination over thesis is required.

Doctorate: At least 90 hours of credit hours are required for the Ph.D. plan of study. Core requirements include statistical physics (one semester), advanced electricity and magnetism (one semester), quantum mechanics (two semesters), and three graduate-level specialty courses. A core course need not be taken at Purdue if its equivalent has been taken previously. A student entering with a B.S. degree and holding a teaching assistantship needs about 2 years to complete all courses. A master's degree or professional doctoral degree from any accredited institution may be considered to contribute up to 30 credit hours toward satisfying the 90 credits required for a Ph.D. degree. An average GPA of 3.0 is required in core courses. At the start of first semester, students are required to take a qualifying examination to demonstrate undergraduate knowledge of mechanics at the level of Marion, Classical Dynamics; of electricity and magnetism at the level of Griffiths, Introduction to Electrodynamics; and of modern physics at the level of Gasiorowicz, Quantum Physics. Students are formally admitted to candidacy for the Ph.D. degree only after they have passed the Ph.D. preliminary examination. The student is eligible to attempt this examination when he or she has completed the core courses with at least a B average. The Preliminary Examination Committee of a given student decides on the nature and coverage of that student's examination. The examination may have written and oral portions. There is no department-wide preliminary examination. After passing the preliminary examination, students can devote practically all of their time to the original research that will serve as the basis for their theses. The research must be of fully professional character and publishable quality. Completion of the Ph.D. requirements include the completion of the thesis, passing an oral examination in defense of the thesis, and preparation of the thesis material for publication.

Other Degrees: Computational Science and Engineering The Computational Science and Engineering (CS&E) Program at Purdue provides students with the opportunity to study a specific science or engineering discipline along with computing in a multidisciplinary environment. The aim of the program is to produce a student who has learned how to integrate computing with another scientific or engineering discipline and is able to make original contributions in both disciplines. The Physics Department is one of the original departments since the inception of this program. The participating departments now number 18 spread over 5 colleges. Physics CSE students must satisfy both the standard Physics departmental degree requirements and those of the CSE Program. Usually some of the math and specialty course requirements of the Physics Department can be met by courses which simultaneously contribute toward the satisfaction of the CSE requirements; however, generally, both the number of courses and grade requirement are higher for the physics students who elect to specialize in the CSE Program. M.S. graduates should be well prepared to join and make significant contributions to interdisciplinary research teams. Ph.D. graduates are expected to become leaders in research and development at the forefront of their fields, applying advanced computational techniques and theory to solve key problems.

Thesis: Thesis may be written in absentia if necessary.

SPECIAL EQUIPMENT, FACILITIES, OR PROGRAMS

Among the major facilities is PRIME Lab, a national center for accelerator mass spectrometry (AMS), which is based on an 8-MeV Tandem Van de Graaff accelerator. AMS is an ultra-sensitive analytical technique for measuring low levels of long-lived radio nuclides and rare trace elements, and has wide applications to the earth and space sciences, biological sciences, and materials sciences.

The Department has a class 10,000 clean room used for testing and assembling detectors for use in high energy physics experiments. Members of the Purdue High Energy Physics group have recently built a silicon vertex detector for use in experiments at the Cornell University electron-positron collider and are developing silicon sensors for the CMS Collaboration at the Large Hadron Collider at CERN.

The Physics Department Library has a seating capacity for 100 users and occupies approximately 11,000 square feet of space in the Physics Building. Its collection has in excess of 47,000 volumes and subscriptions of over 250 journals. Most of the journals are also available electronically.

The Physics Department has an Instrument Shop, a Faculty Machine Shop, and an Electronics Shop for building scientific apparatus. The Instrument Shop is staffed by professional machinists and features a CNC (Computer Numerical Control) milling machine and CNC lathe. Many undergraduate and graduate students receive training and practical experience in machining and electronics in the Faculty Machine Shop and Electronics Shop. Machining techniques and safety are taught to physics staff by a professional machinist in the Faculty Machine Shop, which is used by physics faculty, staff, and students. Electronics for both research and instruction are designed, built, and repaired in the Electronics Shop, which is staffed by an electrical engineer.

Purdue University has created Discovery Park for interdisciplinary research in both bio- and nano-science. The Birck Nanotechnology Center and the Bindley Bioscience Center are state-of-the-art facilities where Physics faculty, postdoctoral researchers, and graduate students join colleagues from other disciplines in performing ground-breaking research in nano-physics, biophysics, and Sensory Science & Technology. Condensed matter experimentalists make use of synchrotron radiation sources at Argonne National Laboratory and Brookhaven National Laboratory. Physicists in High Energy Particle and High Energy Nuclear physics are engaged in experiments at Brookhaven National Laboratory, Fermi National Acceleratory Laboratory, the Cornell Electron Storage Ring, and the CERN Laboratory. Astronomy and astrophysics researchers use facilities at the Whipple telescope at Kitt Peak National Observatory in Arizona, the Hubble Space Telescope, and a variety of other space-based instruments.

Table B—Separately Budgeted Research Expenditures by Source of Support

Source of Support	Departmental Research	Physics-related Research Outside Department
Federal government	$8,917,542	
State/local government	$2,922	
Non-profit organizations		
Business and industry	$354,849	
Other	$252,903	
Total	$9,528,216	

Table C—Separately Budgeted Research Expenditures by Research Specialty

Research Specialty	No. of Grants	Expenditures ($)
Applied Physics	3	$103,710
Astrophysics	17	$801,892
Biophysics	12	$888,010
Condensed Matter Physics	40	$2,585,841
Geophysics	11	$389,734
Nuclear Physics	5	$891,337
Particles and Fields	22	$3,231,706
Physics and other Science Education	18	$140,477
Other	10	$495,508
Total	138	$9,528,215

FACULTY

Distinguished University Professor

Bortoletto, Daniela, Ph.D., Syracuse University, 1989. Edward M. Purcell Distinguished Professor. *Particles and Fields.* Experimental particle physics at CDF with focus on searches for the standard model Higgs and particles expected in beyond the standard model theories. Silicon pixel detectors for CMS, the SLHC, and the linear collider. Study of WW production and Drell-Yan at the LHC with the CMS detector.

Greene, Chris, Ph.D., University of Chicago, 1980. *Atomic, Molecular, & Optical Physics.* Theoretical atomic, molecular, and optical physics. Ultracold few-body and many-body quantum systems. Electron-molecule collisions and dissociative recombination. Ultrafast laser interactions with atoms and molecules. Photofragmentation of atoms and molecules. Novel Rydberg molecules and multichannel Rydberg atoms.

Melosh, H. J., Ph.D., California Institute of Technology, 1972. University Distinguished Professor, Professor of Earth, Atmospheric, and Planetary Sciences. *Geophysics.* Ramifications of impact cratering, planetary tectonics, and the physics of earthquakes and landslides.

Ramdas, Anant K., Ph.D., Raman Research Institute, 1956. Lark-Horovitz Distinguished Professor. *Biophysics, Condensed Matter Physics, Optics.* Experimental condensed matter physics and materials sciences; laser optics; chemical physics; biophysics.

Shipsey, I. P. J., Ph.D., Edinburgh University, 1986. Julian Schwinger Distinguished Professor. *Particles and Fields.* Experimental elementary particle physics; heavy quark flavor physics and tests of lattice gauge theory with the CLEO-c experiment; the high energy frontier (LHC/LC); nuclear instrumentation, especially silicon and gas based detectors; dark energy and dark matter; science education and outreach; the natural physiology or hearing and cochlear implantation.

Professor

Barnes, Virgil E., Ph.D., University of Cambridge, 1962. *Particles and Fields.* CDF, pp collisions at the Collider Detector at the Fermilab, including search for supersymmetric (SUSY) particles. Design, construction and calibration of the CDF Run II scintillating tile-fiber Endplug Calorimeter, including early discovery of wavelength shifting fiber patterns with excellent response uniformity over the entire scintillating tile, and demonstration of precision calorimeter calibration using moving radioactive sources. Design, construction and calibration of the CMS Hadron Calorimeters, to probe beyond the Standard Model, including Higgs and SUSY particle searches, with 14 TeV pp collisions at the CERN Large Hadron Collider. Design R&D for gas Cherenkov calorimeters for use at Next Linear Collider and for possible ultra-high-luminosity "Super LHC" at CERN. Characterization and development of SiPM

(Silicon Photomultiplier) devices for use in High Energy Physics detectors.

Bryan, Lynn, Ph.D., Purdue University, 1997. Professor of Curriculum and Instruction. *Physics and other Science Education.* Science teacher education, physics education; sociocultural influences on teaching and learning, particularly in international and/or rural contexts; evidence-based inquiry and reflection in teacher education; teacher knowledge and beliefs; qualitative research methods.

Caffee, Marc, Ph.D., Washington University, St. Louis, 1986. Associate Department Head, Director of PRIME Lab. *Applied Physics, Geophysics.* Accelerator mass spectrometry; application of stable- and radio-nuclides to problems in the geosciences including quaternary landform evolution, cosmochronology, hydrology and atmospheric processes; the development of techniques to enable the measurement of new cosmogenic nuclides.

Carlson, Erica W., Ph.D., University of California, Los Angeles, 2000. *Condensed Matter Physics.* Condensed matter theory of strongly correlated electronic systems; liquid crystalline vortex matter in type II superconductors; theory of high temperature superconductivity; stripe phases in doped antiferromagnets; granular superconductors; analytic work and Monte Carlo simulations of the XY model; field theoretic calculation of spectral functions in quasi-one-dimensional superconductors; dimensional crossover; anisotropic bipolarons.

Clark, Thomas E., Ph.D., New York University, 1974. *Particles and Fields.* Quantum field theory and its application to elementary particle physics; the general structure of quantum field theoretic actions with particular attention paid to the realization of internal global and gauge symmetries and especially supersymmetry including higher dimensional space-time brane world models; Renormalization theory including renormalized perturbation theory as well as exact Wilson renormalization group equations and techniques.

Córdova, France, Ph.D., California Institute of Technology, 1979. President Emeritus, Purdue University. *Astrophysics.* High energy astrophysics.

Cui, Wei, Ph.D., University of Wisconsin-Madison, 1994. *Astrophysics.* Microquasars, active galactic nuclei, gamma ray bursts, x-ray binaries, and instrumentation for astronomical applications.

Durbin, Stephen M., Ph.D., University of Illinois, 1983. *Biophysics, Condensed Matter Physics.* Experimental condensed matter physics; biophysics. X-ray studies of vibrational modes in biomolecules. X-ray fluorescence imaging. X-ray holographic imaging. Sector Four at the Advanced Photon Source.

Elliott, Daniel S., Ph.D., University of Michigan, 1981. Professor of Electrical and Computer Engineering. *Atomic, Molecular, & Optical Physics.* Experimental atomic, molecular and optical physics; coherent and quantum optics.

Finley, John P., Ph.D., University of Wisconsin-Madison, 1990. *Astrophysics.* Optical, x-ray, and gamma ray studies of compact objects. The evolution of galactic supernova remnants and their impact on the interstellar medium. The origin of the soft x-ray background. The origin of the cosmic rays.

Fischbach, Ephraim, Ph.D., University of Pennsylvania, 1967. *Particles and Fields.* Theoretical physics; elementary particle theory.

Gutay, Laszlo J., Ph.D., Florida State University, 1964. *Particles and Fields.* Experimental high-energy physics; electroweak physics; W, Z pair production, Higgs, SS particles.

Hirsch, Andrew S., Ph.D., Massachusetts Institute of Technology, 1977. *Nuclear Physics, Physics and other Science Education.* Experimental exploration of the equation of state of state of nuclear matter. Physics Education Research projects: (1) Making Sense of Global Warming and Climate Change:

Model of Student Learning via Collaborative Research; (2) high energy nuclear physics project (STAR).

Khlebnikov, Sergei, Ph.D., Institute for Nuclear Research of the Academy of Sciences, Moscow, 1988. *Particles and Fields.* Elementary particle theory; cosmology; and quantum field theory.

Kim, Yeong E., Ph.D., University of California, Berkeley, 1963. *Applied Physics, Condensed Matter Physics, Nuclear Physics, Statistical & Thermal Physics.* Theoretical physics; nuclear theory; applied nuclear physics; nuclear astrophysics; condensed matter theory; quantum statistical mechanics.

Koltick, David S., Ph.D., University of Michigan, 1978. *Applied Physics, Nuclear Physics.* Experimental high-energy physics; applied nuclear physics.

Lister, Matthew L., Ph.D., Boston University, 1999. *Astrophysics.* High luminosity active galactic nuclei, astrophysical jets and shocks, quasars and BL Lacertae objects, very long baseline interferometry, special relativity.

Love, Sherwin T., Ph.D., Stanford University, 1978. *Particles and Fields.* Quantum field theory and its application to elementary particle physics with focus on aspects of dynamical symmetry breaking, supersymmetric field theories, and the renormalization group.

Manfra, Michael J., Ph.D., Boston University, 1999. William F. and Patty J. Miller Professor. *Condensed Matter Physics, Nano Science and Technology.* MBE growth of semiconductor nanostructures; transport properties of low dimensional correlated electron systems.

Miller, David H., Ph.D., Imperial College, London University, 1963. *Particles and Fields.* Discovery physics at the high energy frontier; physics of quarks.

Moffett, Thomas J., Ph.D., University of Texas, Austin, 1973. *Astronomy.* Distances to galaxies in the Local Group.

Muzikar, Paul, Ph.D., Cornell University, 1980. *Condensed Matter Physics, Other.* Various aspects of geochronology form the focus of research. Specific topics include: (1) the use of cosmogenic nuclides such as Be-10 and Al-26 to determine exposure ages, burial ages, and erosion rates; (2) radiocarbon dating in archaeology and the earth sciences; (3) the application of Bayesian statistics to issues in geochronology.

Nakanishi, Hisao, Ph.D., Harvard University, 1980. *Condensed Matter Physics.* Theoretical physics; condensed matter theory; statistical mechanics.

Nolte, David D., Ph.D., University of California, Berkeley, 1988. *Biophysics, Condensed Matter Physics, Optics.* The Pico-Science and BioPhotonics group, directed by Professor David D. Nolte, applies the ultimate sensitivity of laser interferometry to a broad range of topics that include solid state physics, plasmonics in gold films, graphene, semiconductor physics, biointerferometry in biological physics, protein surface chemistry and holographic imaging of living biological tissues. In all these areas, the picometer sensitivity of laser interferometry provides unprecedented sensitivity to study the optical properties of materials. Examples include the Bio-CD (Biological Compact Disks) that rely on diffraction of lasers from antibodies on spinning discs, to real-time video flythroughs of rat bone-cancer tumors using digital holography.

Pyrak-Nolte, Laura, Ph.D., University of California, Berkeley, 1988. *Geophysics.* Wave propagation in fractured and cracked media.

Reifenberger, Ronald G., Ph.D., University of Chicago, 1976. *Condensed Matter Physics, Nano Science and Technology.* Scanning probe techniques.

Ritchie, Kenneth P., Ph.D., University of British Columbia, 1998. *Biophysics.* Experimental biophysics; dynamics of the formation of signaling complexes in the plasma membrane of cells. Diffusion in lipid bilayer membranes with embedded mobile and immobile obstacles. Development of ultra-fast imaging techniques for observing individual molecules in living cells. Single molecule biophysics.

Robicheaux, Francis J., Ph.D., University of Chicago, 1991. *Atomic, Molecular, & Optical Physics.* Time dependent atomic phenomena, highly excited (Rydberg) atoms, electron scattering, strong fields, and ultracold plasmas.

Rokhinson, Leonid, Ph.D., Stony Brook University, 1996. *Condensed Matter Physics.* Experimental condensed matter physics; electron transport in mesoscopic systems; spintronics and spin interactions; quantum information processing; molecular electronics; nanofabrication; novel materials and devices.

Scharenberg, Rolf P., Ph.D., University of Michigan, 1955. *Nuclear Physics.* Phase diagram of hadronic matter; liquid-gas phase transitions; quark-gluon plasma.

Wang, Fuqiang, Ph.D., Columbia University, 1996. *Nuclear Physics.* High energy nuclear physics, STAR at Brookhaven National Laboratory RHIC.

Associate Professor

Chen, Yong, Ph.D., Princeton University, 2005. *Atomic, Molecular, & Optical Physics, Condensed Matter Physics, Nano Science and Technology.* Experimental condensed matter physics, experimental atomic, molecular and optical physics, nanoscience and nanotechnology.

Csathy, Gabor, Ph.D., Pennsylvania State University, 2001. *Condensed Matter Physics.* Experimental condensed matter physics; new physics in 2D electrons; BCS-like pairing of composite fermions; non-Abelian statistics and possible applications for quantum computing; solid phases in electronic systems; spin physics in low dimensional semiconductors; spectrally enhanced chemical detection with nanotube transistors.

Haugan, Mark P., Ph.D., Stanford University, 1978. Associate Department Head. *Astrophysics, Physics and other Science Education, Relativity & Gravitation.* Conceptual and empirical foundations of relativity and gravitation physics and physics education research and development.

Hu, Jiangping, Ph.D., Stanford University, 2002. *Condensed Matter Physics.* Condensed matter physics from active application like spintronics and nano-physics to experimental phenomenology of materials such as high-Tc superconductivity to more abstract topics like application of ideas borne in condensed matter theory to other fields of research in physics.

Jones, Matthew, Ph.D., Carleton University, 1997. *Particles and Fields.* Experimental high energy physics.

Kaufmann, Birgit, Ph.D., University of Bonn, 1999. Assistant Professor of Mathematics. *Condensed Matter Physics.* Condensed matter theory, correlated systems, non-equilibrium phenomena in quantum phase transitions; integrable models and scattering theories; Bethe Ansatz for quantum spin chains; reaction-diffusion systems.

Kruczenski, Martin, Ph.D., University of Buenos Aires, 1998. *Particles and Fields.* Theoretical high energy physics; string theory and its connections to gauge theory; string theory and blackhole physics.

Lyanda-Geller, Yuli, Ph.D., Ioffe Physico-Technical Institute, 1987. *Atomic, Molecular, & Optical Physics, Condensed Matter Physics.* Mesoscopic physics and interference phenomena; transport and optical phenomena in nanostructures; physics of quantum information.

Lyutikov, Maxim, Ph.D., California Institute of Technology, 1998. *Astrophysics.* Theoretical astrophysics; high energy astrophysics compact objects; extragalactic astrophysics; cosmic rays; plasma astrophysics.

Molnar, Dénes, Ph.D., Columbia University, 2002. *Nuclear Physics.* Properties of nuclear matter at extreme energy densities, physics of relativistic heavy-ion collisions and the quark-gluon plasma, transport theory.

Neumeister, Norbert, Ph.D., Vienna University of Technology, 1996. *Particles and Fields*. High energy particle physics; phenomenon of electro-weak symmetry breaking; the origin of the matter anti-matter asymmetry in the universe; the search for new physics beyond the established standard model of particle physics.

Peterson, John, Ph.D., Columbia University, 2003. *Astrophysics*. Observational cosmology: studies of dark energy and dark matter; high resolution x-ray spectroscopy; x-ray emission from clusters of galaxies; cooling flows in clusters of galaxies; surveys of clusters of galaxies; the Chandra and XMM-Newton X-ray Observatories; optical and x-ray astrophysics instrumentation; weak gravitational lensing of clusters and large scale structure; optical astrophysics simulation; advanced multivariate Monte Carlo data analysis techniques; the Large Synoptic Survey Telescope (LSST).

Rodriguez, Jorge, Ph.D., University of Illinois, 1995. *Biophysics*. Theoretical biophysics; computational electronic structure of active sites in metalloproteins; density functional theory of (bio)molecules; electronic structure and mesoscopic properties of (bio)molecular nanostructures; simulation of biological Mössbauer, EPR and x-ray spectra; (anti)ferromagnetism in molecular magnets and finite fermion systems.

Savikhin, Sergei, Ph.D., Tartu State University, 1991. *Biophysics*. Experimental biophysics; femtosecond optical studies of artificial and natural biological systems; membrane proteins: structure and function, structure-based computer modeling; Exciton kinetics in semiconductors, molecular crystals and biological structures; biomimetic devices; ultrafast experimental techniques.

Xie, Wei, Ph.D., Chinese Academy of Sciences, 1997. *Nuclear Physics*. Experimental high energy nuclear physics; quark-gluon plasma.

Yang, Chen, Ph.D., Harvard University, 2006. Assistant Professor of Chemistry. *Condensed Matter Physics, Nano Science and Technology*. Semiconductor nanowires.

Assistant Professor

Giannios, Dimitrios, Ph.D., University of Crete, 2005. *Astrophysics*. Theoretical astrophysics.

Lang, Rafael F., Ph.D., Max Planck Institut fur Physik, 2008. *Astrophysics, Particles and Fields*. Search for dark matter.

Lifton, Nathaniel A., Ph.D., University of Arizona, 1997. Assistant Professor of Earth, Atmospheric, and Planetary Sciences. *Geophysics*. Methods for using in situ cosmogenic nuclides to derive surface exposure ages and/or erosion rates for process oriented geomorphic studies.

Malis, Oana, Ph.D., Boston University, 1999. *Atomic, Molecular, & Optical Physics, Condensed Matter Physics, Nano Science and Technology*. Structural and optical properties of nanostructured materials.

Pushkar, Yulia, Ph.D., Freie Universität Berlin, 2003. *Biophysics, Energy Sources & Environment*. Experimental biophysics; energy research.

Todd, Brian, Ph.D., Case Western Reserve University, 2003. *Biophysics*. Single-molecule biophysics; intermolecular forces and biological recognition; DNA gymnastics and gene regulation.

Research Assistant Professor

Srivistava, Brijesh, Ph.D., Indian Institute of Technology, Kampur, 1975. Experimental high energy nuclear physics.

Courtesy Professor

Kais, Sabre, Ph.D., Hebrew University of Jerusalem, 1989. Professor of Chemistry. *Condensed Matter Physics*. Theoretical condensed matter physics.

Roychoudry, Anita, Ph.D., Indiana University, 1990. Associate Professor of Curriculum and Instruction. *Physics and other Science Education*. Teaching and learning of physics.

Shalaev, Vladimir M., Ph.D., Krasnoyarsk University, 1983. Robert and Anne Burnett Distinguished Professor of Electrical and Computer Engineering. *Electrical Engineering, Nano Science and Technology, Optics*. Fields and optics; biomedical imaging and sensing; communications, networking, signal and image processing; microelectronics and nanotechnology.

Wasserman, Adam, Ph.D., Rutgers University, 2005. Assistant Professor of Chemistry. *Condensed Matter Physics*. Theoretical condensed matter physics.

DEPARTMENTAL RESEARCH SPECIALTIES AND STAFF

Theoretical

Astrophysics. Cosmology; cosmic microwave background; extra dimensions; experimental tests of general relativity; gravitation; plasma and high energy astrophysics; Pulsars and Supernova remnants, active galactic nuclei; gamma ray bursts. Giannios, Lyutikov.

Atomic, Molecular, & Optical Physics. Ultracold atomic gases; electron-molecule collisions; laser-molecule interactions; time dependent atomic phenomena; highly excited (Rydberg) atoms; strong fields. Greene, Robicheaux.

Biophysics. Exploring the dynamics of large biomolecules (e.g., metalloproteins, DNA) using theoretical methods for normal mode analysis and simulations of Nuclear Resonant Vibrational Spectroscopy data. Theoretical and computational studies of the electronic structure and magnetic properties of metalloproteins by means of density functional theory (DFT), correlated ab inito methods, and phenomenological spin Hamiltonians. Development of genetic algorithms in conjunction with high-performance supercomputing for the simulation of biological resonant (Moesbauer, EPR) and XANES spectroscopies. Computational neuroscience and related studies of signal propagation in the brain. Rodriguez.

Condensed Matter Physics. Low-dimensional systems: Ground state properties of two-dimensional electron systems, quantum dots and quantum wires, phase diagrams of electron liquids, Quantum Hall effect, Wigner crystal, spin-orbit interactions and polarization of electronic states; transport and optical properties of quantum wells, wires and dots, Luttinger liquid, spin and charge density waves, electron-phonon interactions. Mesoscopics: quantum coherent phenomena in electronic transport, Aharonov-Bohm effect and Berry's phases, weak localization, universal conductance fluctuations, localization and metal-insulator transitions, Kondo effect, spin-dependent phenomena, spin relaxation and non-equilibrium spin polarization. Nano and Quantum physics: Quantum computation in quantum dots; Bose-Einstein condensation, optical lattices, decoherence and dissipation in nanostructures, Coulomb and spin blockade, electron spin resonance and nuclear spin resonance in quantum dots, coherent spin control. Superconductivity and Magnetism: transport in superconductors, Josephson effect, Vortices, BCS and unconventional superconductors. Strongly correlated electronic systems, electronic liquid crystals, high temperature superconductivity. Ferromagnetism and Antiferromagnetism, dilute magnetic semiconductors. Transport in ferromagnets, effects of domain walls, spin injection. Spintronics. Carlson, Hu, Kais, Kaufmann, Wasserman.

Geophysics. Ramifications of impact cratering; planetary tectonics; physics of earthquakes and landslides. Melosh.

Nuclear Physics. From keV to TeV energies; theory of three-particle bound and scattering states in nuclear and elementary particle physics; strong, weak, and electromagnetic interactions in nuclei; nuclear many-body problem including the structure of finite nuclei; rotational states of deformed nuclei; theory of nuclear fusion reactions; solar neutrino problem; nuclear astrophysics; study of nuclear matter at extreme conditions; dynamics in relativistic nucleus-nucleus collisions; properties of the quark-gluon plasma. Kim, Molnar.

Particles and Fields. Theory and phenomenology of the standard model of elementary particle interactions and its possible extensions; aspects of supersymmetry; neutrino oscillations and their application to astro-physical phenomena; dynamical symmetry breaking; renormalization group studies; cosmological phase transitions; inflationary models of the early universe; brane world models; string theory; string/gauge theory duality. Clark, Fischbach, Khlebnikov, Kruczenski, Love.

Statistical & Thermal Physics. Phase transitions and critical phenomena; phenomenology of first order phase transitions; Ising systems: percolation and other clustering phenomena; quantum percolation and other quantum transport; statistical properties of surfaces and interfaces; scaling in linear and branched polymer chains; kinetics of disorderly growth processes; Brownian motion. Kim.

Experimental

Applied Physics. Applications of nuclear physics to the detection of hazardous materials in commerce and public areas. Work in conjunction with the Center for Sensing Science and Technology and NSWC, Crane, Indiana to reduce terrorist threat from chemical agents, radiation threats, and explosives. Kim, Koltick.

Astrophysics. Studies of black holes, clusters of black holes, clusters of galaxies, dark matter, dark energy, neutron stars, Galactic structures, very metal-poor stars, horizontal-branch stars, stellar spectroscopy and abundance analysis, active galactic nuclei, relativistic jets, diffuse x-ray and infrared background, and interstellar medium; simultaneous photometry and photoelectric radial velocity observation of ultra-short-period pulsating variable stars. Satellite-based x-ray and gamma ray astronomy; ground-based very high energy gamma-ray experiments (Whipple and VERITAS); instrumentation for high energy astrophysics; radio astronomy and interferometry; optical survey telescopes (LSST). Córdova, Cui, Finley, Lang, Lister, Peterson.

Atomic, Molecular, & Optical Physics. Studies of two-pathway coherent control processes, including control of photoionization branching ratios, precision measurements of weak optical interactions, and control of photoelectron angular distributions and molecular processes. Coherent optical interactions in trapped ultra-cold atoms. Phase conjugate, four-wave mixing in two-level atomic systems. Bose-Einstein condensates. Quantum manipulation of atoms and molecules with lasers. Quantum information and quantum simulation of condensed matter problems with ultracold atoms and molecules. Coherence and decay in Bose-Einstein Condensates. Quantum interference effects in spin or Bose-Einstein Condensates. Chen, Elliott, Malis.

Biophysics. Modeling of real nervous systems containing small numbers of neutrons, and of learning and memory in simple neural systems. Measuring vibrational properties of metalloproteins and other biomolecules using many techniques, to explore how dynamics controls biological function. Nuclear Resonant Vibrational Spectroscopy conducted at the Argonnne x-ray synchrotron focuses on heme proteins (myoglobin, hemoglobin, cytochromes). Resonant Raman scattering and FTIR are applied to cytochromes and various model heme compounds. Terahertz time-delay spectroscopy is a new technique exploring vibrations of macromolecules beyond the far infrared. Single molecule spectroscopy is being developed to study photosynthetic complexes (PS I) using advanced laser spectroscopy methods. Live cell, single molecule imaging of membrane molecule dynamics and interactions. Durbin, Nolte, Pushkar, Ramdas, Ritchie, Savikhin, Todd.

Condensed Matter Physics. Optical absorption, Raman and Brillouin scattering, and photoluminescence of semiconductors and their sub-micron heterostructures; effects of uniaxial stress on vibrational and electronic levels; phonons and magnetic excitations in magnetic diluted semiconductors; solid state plasmas; electron spin resonance and electron-nucleus double resonance; graphene and carbon nanotubes; Mössbauer spectroscopy; nanoscience, nanomaterials and nanodevices; non-linear optics of semiconductors and their quantum well structures; metallic surfaces; resistivity of metals; x-ray studies of quasicrystals; x-ray synchrotron physics; phase problem; x-ray standing waves; superconductivity; magneto-optics; magnetic materials; studies of mesoscopic systems; quantum transport in GaAs/AlxGa1−xAs microstructures; transport studies of "quantum chaos" in both open and closed systems; transport in the fractional quantum Hall effect; scanning Hall probe microscopy; studies of Si/SiO2 interface roughness; electrical transport in one dimensional nanoscale structures and hetero-structures. Chen, Csathy, Durbin, Malis, Manfra, Nolte, Ramdas, Reifenberger, Rokhinson, Shalaev.

Geophysics. Rock mechanics and physics of rocks; physical acoustics of heterogeneous materials and discontinuities; volumetric nondestructive imaging of opaque materials; hydrology and percolation physics. Nolte, Pyrak-Nolte.

Nuclear Physics. Experimental studies of nuclear matter at high energy densities and temperatures, search for evidence of the formation of the quark-gluon plasma in relativistic nucleus-nucleus collisions by examining statistical and dynamical properties of the reaction products created at the Relativistic Heavy Ion Collider and the Large Hadron Collider. Hirsch, Koltick, Scharenberg, Wang, Xie.

Other. Methods of nuclear physics used to operate a tandem Van de Graaff accelerator and develop new techniques for measuring long-lived radionuclides and other rare particles. Applications are in physics (neutron transport, trace impurities, cross sections), earth science (dating and tracing processes and events, global change, environment), and biological science (drug metabolism, toxicity). Caffee, Lifton, Muzikar.

Particles and Fields. Current experiments in particle physics include the Collider Detector at the Fermilab which is searching for the Higgs particle and SUSY particles, and studying properties of the top quark, in proton anti-proton events at the Tevatron colliding beam accelerator; and CLEO, which is performing precision studies of the weak and strong interactions using charm quarks produced in e+e− annihilations at the Cornell Electron Storage Ring. Over the past fifteen years the research program wit the Compact Muon Solenoid experiment at the Large Hadron Collider at CERN in Switzerland has been considerably expanded. Extensive apparatus has been designed, constructed, and installed in CMS. The LHC will begin taking data in late 2009, allowing searches for the Higgs particle, dark matter particles, Supersymmetry, and extra dimensions of space and time. Particle astrophysics work includes contributions to the design, simulation and fabrication of the optical survey telescopy, LSST, and the study of dark energy. Local facilities include state of the art laboratories for the design, fabrication and evaluation of silicon and gas microstrip detectors for particle physics and particle astrophysics, and a major computing facility dedicated to the analysis of data from the LHC. (For more information http://www-.physics.purdue.edu/particle/). Barnes, Bortoletto, Gutay, Jones, Lang, Miller, Neumeister, Shipsey.

Physics and other Science Education. Focus on curricular components and pedagogical methods as it affects student under-

standing of physics concepts and problem solving strategies. Research on the effective means of dissemination of curriculum innovations. Bryan, Haugan, Hirsch, Roychoudry.

View additional information about this department at
www.gradschoolshopper.com

UNIVERSITY OF NOTRE DAME

DEPARTMENT OF PHYSICS

Notre Dame, Indiana 46556-5670
http://physics.nd.edu

General University Information
President: Rev. John I. Jenkins, C.S.C.
Dean of Graduate School: Christine Maziar
University website: http://www.nd.edu/
Control: Private
Setting: Suburban
Total Faculty: 1,364
Total number of Students: 12,004
Total number of Graduate Students: 3,552

Department Information
Department Chairman: Christopher Kolda, Chair
Department Contact: Shari Herman, Sr. Admin. Assistant, Graduate Student Programs
Total full-time faculty: 40
Total number of full-time equivalent positions: 41
Full-Time Graduate Students: 103
First-Year Graduate Students: 12
Female First-Year Students: 4
Total Post Doctorates: 18

Department Address
225 Nieuwland Science Hall
Notre Dame, IN 46556-5670
Phone: (574) 631-6386
Fax: (574) 631-5952
E-mail: physics@nd.edu
Website: http://physics.nd.edu

ADMISSIONS

Admission Contact Information
Address admission inquiries to: Chair, Graduate Admissions Committee, Dept. of Physics
Phone: (574) 631-6386
E-mail: physics@nd.edu
Admissions website: http://graduateschool.nd.edu/admissions/

Application deadlines
Fall admission:
U.S. students: January 15   *Int'l. students*: January 15

Application fee
U.S. students: $75   *Int'l. students*: $75

Admissions information
For Fall of 2013:
Number of applicants: 180
Number admitted: 36
Number enrolled: 12

Admission requirements
Bachelor's degree requirements: Bachelor's degree in physics is required.
Minimum undergraduate GPA: 3.2

GRE requirements
The GRE is required.

Advanced GRE requirements
The Advanced GRE is required.
Minimum accepted Advanced GRE score: 600

TOEFL requirements
The TOEFL exam is required for students from non-English-speaking countries.
PBT score: 600
iBT score: 100
Speaking score should be 23 or over.

Other admissions information
Additional requirements: The GRE physics test is required with a score of at least 600.
Undergraduate preparation assumed: Resnick, Halliday and Krane, Physics, Volumes 1 and 2; Taylor, Classical Mechanics; Griffiths, Quantum Mechanics; Griffiths, Introduction to Electrodynamics; Schroeder, Introduction to Thermal Physics.

TUITION

Tuition year 2013-14:
Tuition for out-of-state residents
    Full-time students: $44,380 annual
    Part-time students: $2,443 per credit
Credit hours per semester to be considered full-time: 9
Deferred tuition plan: Yes
Health insurance: Yes, $1,175.00.
Other academic fees: 75% of Health Insurance is subsidized by the Graduate School; $65 Graduate Student Union fee per year; $100 parking fee per year.
Academic term: Semester
Number of first-year students who received full tuition waivers: 12

Teaching Assistants, Research Assistants, and Fellowships

Number of first-year
Teaching Assistants: 10
Research Assistants: 2
Fellowship students: 3
Average stipend per academic year
Teaching Assistant: $25,055
Research Assistant: $25,055
Fellowship student: $30,000

FINANCIAL AID

Loans
Loans are available for U.S. students.
Loans are available for international students.
GAPSFAS application required: No
FAFSA application required: Yes

For further information
Address financial aid inquiries to: Director, Financial Aid.
Phone: (574) 631-6436
E-mail: finaid@nd.edu
Financial aid website: http://financialaid.nd.edu/grad/index.shtml

HOUSING

Availability of on-campus housing
Single students: Yes
Married students: Yes

For further information
Address housing inquiries to: Housing Office, Dean of Students.
Phone: (574) 631-5878
E-mail: orlh@nd.edu
Housing aid website: http://housing.nd.edu/graduate/

Table A—Faculty, Enrollments, and Degrees Granted

Research Specialty	2012-13 Faculty	Enrollment Fall 2012 Master's	Enrollment Fall 2012 Doctorate	Number of Degrees Granted 2011–12 (2008–12) Master's	Number of Degrees Granted 2011–12 (2008–12) Terminal Master's	Number of Degrees Granted 2011–12 (2008–12) Doctorate
Astrophysics	9	–	15	–	–(1)	1(5)
Atomic, Molecular, & Optical Physics	4	–	2	–	–(2)	–(1)
Biophysics	2	–	10	–	–(4)	1(5)
Condensed Matter Physics	11	–	21	–	–(7)	6(15)
High Energy Physics	13	–	17	–	–(5)	3(7)
Nuclear Physics	16	–	27	–	–(5)	3(17)
Total	55	–	92	–	2(23)	14(40)
Full-time Grad. Stud.	–	–	102	–	–	–
First-year Grad. Stud.	–	–	20	–	–	–

GRADUATE DEGREE REQUIREMENTS

Master's: The graduate program in the Department of Physics is research oriented. For that reason, the Department does not normally accept students who plan to terminate their studies with a master's degree. Under certain conditions, a non-research master's program is available. Students must complete 30 credit hours and maintain a grade point average of 3.0. The student must pass a comprehensive oral examination

in the major field. Applicants are cautioned that financial aid is normally restricted to students pursuing Ph.D. programs of study. The minimum residence is two successive semesters.

Doctorate: Students must complete 33 credit hours and maintain a grade point average of 3.0. The minimum residency requirement for the Ph.D. degree is full-time status for four consecutive semesters. The student will normally take a sequence of basic courses in a two-year core curriculum, followed by advanced courses and seminars in specialized areas of study. Included in the core curriculum are one semester of mathematical methods in physics, one semester of experimental methods in physics, one semester each of classical mechanics and statistical mechanics, two semesters of quantum mechanics, two semesters of electrodynamics, and one semester of a research-area course (astrophysics, atomic physics, biophysics, condensed-matter physics, elementary particle physics, or nuclear physics), and completion of two semesters of a breadth requirement in physics. Incoming students who have already successfully completed courses equivalent to any of those in the core curriculum will not be expected to take the corresponding core curriculum courses. However, all incoming students are required to take and pass a qualifying examination on undergraduate physics. The student is encouraged to become an active participant in research in the second semester of the first year of his/her graduate work. Prior to admission to candidacy for the Ph.D. degree, the student must pass comprehensive written and oral examinations. There is no foreign language requirement. Approval of the thesis by the research director and three readers and an oral defense of the thesis complete the requirements.

Thesis: Thesis may be written in absentia.

Table B—Separately Budgeted Research Expenditures by Source of Support

Source of Support	Departmental Research	Physics-related Research Outside Department
Federal government	$8,433,735	
State/local government	$136,517	
Non-profit organizations		
Business and industry		
Other		
Total	$8,570,252	

Table C—Separately Budgeted Research Expenditures by Research Specialty

Research Specialty	No. of Grants	Expenditures ($)
Astronomy	16	$810,988
Atomic, Molecular, & Optical Physics	3	$250,702
Condensed Matter Physics	9	$877,934
Nuclear Physics	7	$4,612,105
High Energy Physics	10	$1,882,006
Other	2	$136,517
Total	47	$8,570,252

FACULTY

Professor

Alber, Mark S., Ph.D., University of Pennsylvania, 1990. Concurrent Professor of Physics; Vincent J. Duncan Family Professor of Applied Mathematics and Concurrent Professor of Physics; Director, Center for the Study of Biocomplexity. *Applied Mathematics*. Dynamical systems treatment of nonlinear partial differential equations with applications to biology and nonlinear optics.

Aprahamian, Ani, Ph.D., Clark University, 1986. Frank M. Friemann Professor of Physics. *Nuclear Physics.* Experimental nuclear physics; gamma-ray spectroscopy; nuclear masses; lifetimes; astrophysics.

Berry, H. Gordon, Ph.D., University of Wisconsin-Madison, 1967. *Atomic, Molecular, & Optical Physics.* Experimental atomic physics.

Bigi, Ikaros I., Ph.D., Universität München, 1976. Grace-Rupley II Professor. *High Energy Physics.* Refining the standard model phenomenology for the decays of hadrons carrying the quantum numbers strangeness, charm and beauty and on electric dipole moments to use them as 'indirect' searches for new physics.

Blackstead, Howard A., Ph.D., Rice University, 1967. *Condensed Matter Physics.* Experimental physics; solid-state physics; magnetism and acoustics.

Bunker, Bruce A., Ph.D., University of Washington, 1980. Director, Materials Research Collaborative Access Team (MR-CAT), a multi-institutional consortium developing and using x-ray beamlines at the Advanced Photon Source, Argonne National Laboratory. *Condensed Matter Physics.* Experimental physics; X-ray, UV, and electron spectroscopy of condensed-matter and biological/environmental systems.

Crawford, Gregory P., Ph.D., Kent State University, 1991. William K. Warren II Foundation Dean College of Science. *Condensed Matter Physics.* Liquid crystal and polymer physics; optics; solid-state nuclear magnetic resonance.

Dobrowolska-Furdyna, Malgorzata, Ph.D., Polish Academy of Sciences, 1979. The Rev. John Cardinal O'Hara, C.S.C. Professor of Physics. *Condensed Matter Physics.* Experimental solid-state physics.

Frauendorf, Stefan G., Ph.D., Technical University Dresden, 1971. *Nuclear Physics.* Theoretical nuclear physics; atomic physics; mesoscopic systems.

Furdyna, Jacek K., Ph.D., Northwestern University, 1960. The Aurora and Thomas Marquez Chair in Information Theory and Computer Technology; Fellow, Nanovic Institute for European Studies; Director, Center for Material Fabrication & Nanotechnology. *Condensed Matter Physics, Materials Science, Metallurgy.* Experimental solid-state physics; man-made materials.

Garg, Umesh, Ph.D., Stony Brook University, 1978. Director of the Department of Physics Research Experience for Undergraduates (REU) program. *Nuclear Physics.* Experimental nuclear physics; nuclear structure; giant resonances; gamma-ray spectroscopy; high spin states.

Garnavich, Peter M., Ph.D., University of Washington, 1991. *Astrophysics.* Astrophysics/observational cosmology.

Hildreth, Michael D., Ph.D., Stanford University, 1995. *High Energy Physics.* Experimental high-energy elementary particle physics.

Hyder, Anthony K., Ph.D., Air Force Institute of Technology, 1976. *Nuclear Physics.* Experimental physics; space physics; nuclear physics.

Jankó, Boldizsár, Ph.D., Cornell University, 1996. Director, Institute for Theoretical Sciences. *Condensed Matter Physics.* Theoretical condensed-matter physics.

Jessop, Colin P., Ph.D., Harvard University, 1994. *Particles and Fields.* Experimental high-energy physics.

Kolata, James J., Ph.D., Michigan State University, 1969. *Nuclear Physics.* Experimental physics; nuclear structure; heavy-ion reactions; radioactive beam physics.

Kolda, Christopher F., Ph.D., University of Michigan, 1995. Chair, Department of Physics. *High Energy Physics.* Theoretical high-energy physics; supersymmetry.

Livingston, Eugene A., Ph.D., University of Alberta, 1974. *Atomic, Molecular, & Optical Physics.* Experimental physics; atomic physics; spectroscopy of highly ionized atoms.

LoSecco, John M., Ph.D., Harvard University, 1976. *Nuclear Physics, Particles and Fields.* Experimental and theoretical physics; high-energy elementary particle physics.

Mathews, Grant J., Ph.D., University of Maryland, 1977. Director, Center for Astrophysics at Notre Dame University (CANDU). *Astrophysics.* Theoretical astrophysics/cosmology; general relativity.

Newman, Kathie E., Ph.D., University of Washington, 1981. *Condensed Matter Physics, Mechanics, Statistical & Thermal Physics.* Theoretical physics; statistical mechanics; semiconductors.

Rettig, Terrence W., Ph.D., Indiana University, 1976. *Astronomy.* Observational astronomy: comets, solar system formation, and T Tauri stars.

Ruchti, Randal C., Ph.D., Michigan State University, 1973. *High Energy Physics.* Experimental physics; high-energy elementary particle physics.

Ruggiero, Steven T., Ph.D., Stanford University, 1981. *Condensed Matter Physics.* Experimental physics; condensed matter and low-temperature physics; superconductivity.

Sapirstein, Jonathan R., Ph.D., Stanford University, 1979. *Atomic, Molecular, & Optical Physics.* Theoretical physics; quantum electrodynamics.

Tanner, Carol E., Ph.D., University of California, Berkeley, 1985. *Atomic, Molecular, & Optical Physics.* A variety of precision measurements in atomic cesium that are motivated by the study of PNC, fundamental symmetries, and measurements of fundamental constants.

Toroczkai, Zoltan, Ph.D., Virginia Polytechnic Institute and State University (Virginia Tech), 1997. Concurrent Professor, Computer Science and Engineering. *Biophysics, Condensed Matter Physics.* Theoretical condensed-matter physics; biophysics; complex network theory.

Wayne, Mitchell R., Ph.D., University of California, Los Angeles, 1985. *High Energy Physics.* Experimental high-energy elementary particle physics.

Wiescher, Michael C.F., Ph.D., Universität Münster, 1980. Frank M. Freimann Professor of Physics; Director for the Joint Institute for Nuclear Astrophysics (JINA). *Nuclear Physics.* Experimental nuclear physics; nuclear astrophysics.

Associate Professor

Balsara, Dinshaw S., Ph.D., University of Illinois, 1990. Concurrent Associate Professor, Department of Applied and Computational Mathematics and Statistics. *Astrophysics, Computational Physics.* Theoretical and computational astrophysics.

Caprio, Mark A., Ph.D., Yale University, 2003. *Nuclear Physics.* Nuclear structure theory; many-body physics.

Collon, Philippe A., Ph.D., Universität Wien, 1999. Associate Chair and Director of Undergraduate Studies. Outreach Coordinator for the Underground Accelerator Collaboration DIANA at the new National Deep Underground Science and Engineering Laboratory (DUSEL). *Nuclear Physics.* Experimental nuclear physics; new techniques, AMS.

Delgado, Antonio, Ph.D., Universidad Autonoma de Madrid, 2001. *Particles and Fields.* The last building block left to be discovered within the Standard Model of Particle Physics; the Higgs boson.

Eskildsen, Morten R., Ph.D., University of Copenhagen, 1998. *Condensed Matter Physics.* Studies of superconductivity, especially in the vortices induced in type-II superconductors by an applied magnetic field.

Howk, Christopher J., Ph.D., University of Wisconsin-Madison, 1999. Associate Chair and Director of Graduate Studies. *Astro-*

physics. Observational astrophysics; interstellar and intergalactic media.

Peng, Jeffrey W., Ph.D., University of Michigan, 1993. Concurrent Associate Professor of Physics. *Biophysics.* Molecular biophysics using NMR spectroscopy.

Assistant Professor

Brodeur, Maxime, Ph.D., University of British Columbia, 2010. *Nuclear Physics.*

Couder, Manoel, Ph.D., Université Catholique de Louvain, 2004. *Nuclear Physics.* Design, simulate, develop and optimize new solutions and apply those solutions to real-world measurements of low-energy nuclear reactions of astrophysical interest.

Crepp, Justin R., Ph.D., University of Florida, 2008. Frank M. Freimann Assistant Professor of Physics. *Astronomy.* Developing new technologies and observational techniques to detect faint substellar companions that orbit nearby stars. Design and build instruments that operate at visible and near-infrared wavelengths to directly image and study brown dwarfs and extrasolar planets. Use the Doppler method to measure the radial velocity "wobble" of stars as they gravitationally interact with their planets.

Gomes, Kenjiro K., Ph.D., University of Illinois at Urbana-Champaign, 2008. Frank M. Freimann Assistant Professor of Physics. *Condensed Matter Physics.* Experimental condensed matter physics, scanning tunneling microscope.

Lannon, Kevin P., Ph.D., University of Illinois, 2003. *High Energy Physics.* Experimental high-energy particle physics.

Martin, Adam, Ph.D., Boston University, 2007. *High Energy Physics.*

Ptasinska, Sylwia, Ph.D., Leopold-Franzens-University, 2004. *Condensed Matter Physics.* Experimental studies on electron interaction with molecules and radiation damage to DNA and its component biomolecules.

Research Professor

Bennett, David P., Ph.D., Stanford University, 1986. *Astrophysics, Cosmology & String Theory.* Gravitational microlensing.

Berg, Georg P., Ph.D., University of Cologne, 1974. *Nuclear Physics.* Nuclear structure and its reaction mechanism mostly using spectrometers exploiting their high resolution and 0-degree capabilities. –ST. GEORGE recoil separator, a project of JINA funded by the National Science Foundation.

Gorres, Joachim, Ph.D., Universität Münster, 1983. *Nuclear Physics.*

Research Associate Professor

Lehner, Nicolas, Ph.D., The Queen's University of Belfast, 2000. *Astrophysics.* Understanding the physical processes that drive and regulate the growth of galaxies.

Liu, Xinyu, Ph.D., University of Notre Dame, 2003. *Condensed Matter Physics.*

Marinelli, Nancy, Ph.D., University of Bari, 1993. *High Energy Physics.* Actively engaged in the experimental program of the LHC (the Large Hadron Collider, also at CERN).

Research Assistant Professor

Karmgard, Daniel J., Ph.D., Florida State University, 1999. *High Energy Physics.* Research with the Compact Muon Solenoid (CMS) experiment at the European Organization for Nuclear Research (CERN).

Phillips, Lara Arielle, Ph.D., Princeton University, 2003. *Astrophysics.* Astrophysics and cosmology theory; the missing baryon problem.

Tan, Wanpeng, Ph.D., Michigan State University, 2002. *Nuclear Physics.*

Professional Specialist

Saddawi, Shafa D.J., Ph.D., Warsaw University of Technology, 1989. Integrated and fiber optics; electro-optical light control in lithium niobate for planar and stripe waveguides; measurement of thin film parameters with a prism coupler; magneto-optical sensor devices based on diluted magnetic semiconductors.

Associate Professional Specialist

Stech, Edward J., Ph.D., University of Notre Dame, 2004. *Nuclear Physics.* Study nuclear reactions important to the understanding of energy production and the origin of elements in stars and explosive stellar environments.

Assistant Professional Specialist

Zech, William F., Ph.D., University of Notre Dame, 2010. Director of Advanced Physics Laboratories. *Astrophysics.*

Concurrent Assistant Professional Specialist

Davis, Keith W., Ph.D., Clemson University, 2007. Director, Digital Visualization Theater. *Astronomy, Astrophysics.*

DEPARTMENTAL RESEARCH SPECIALTIES AND STAFF

Theoretical

Astrophysics. Inflationary cosmology; primordial nucleosynthesis; cosmic microwave background; galaxy formation and evolution; large-scale structure; stellar evolution and nucleosynthesis; black holes in a magnetic field; charged black holes; neutron stars; neutron star binaries; gravity waves; gamma-ray bursts; supernovae; numerical realtivity. Balsara, Mathews.

Atomic, Molecular, & Optical Physics. Quantum electrodynamics; weak interactions; atomic many-body theory; photoionization and photoexcitation. Sapirstein.

Biophysics. Biological networks; cellular and population dynamics; organogensis and tissue development; epidemics; endosymbiotic evolution. Alber, Toroczkai.

Condensed Matter Physics. Many-body problem; high-temperature superconductivity; superconductivity and magnetism on the nanoscale; tunneling phenomena; metal-metal interfaces; inhomogeneous and layered superconductors; hopping transport; studies of ordering in semiconductors, magnetic semiconductors. Jankó, Newman, Toroczkai.

Nuclear Physics. Many-body problem; nuclear reactions, few-body problem; boson expansions, structure of nuclei with momentum high angular momentum and exotic proton and neutron numbers. Caprio, Frauendorf.

Particles and Fields. Formal properties of quantum field theories; supersymmetry, grand unification, spontaneous symmetry breaking, Higgs physics, phenomenology of strong and weak processes, rare decays, and CP violation; lepton dynamics; supergravity; extra dimensions; new particles. Bigi, Delgado, Kolda, Martin.

Statistical & Thermal Physics. Complex networks, phase transitions; critical phenomena in fluids; networks; computer simulations. Newman, Toroczkai.

Experimental

Astrophysics. Spectra and images of comets; stellar nuclear reaction rates; high redshift supernovae; exoplanets; cosmological parameters. Bennett, Crepp, Davis, Garnavich, Howk, Lehner, Zech.

Atomic, Molecular, & Optical Physics. Atomic structure; parity violation; tests of fundamental symmetries; excitation mechanisms; radiative decays in neutral and ionized atoms; precision lifetimes. Berry, Livingston, Tanner.

Condensed Matter Physics. Low-temperature physics; superconducting microwave absorption; metal and semiconductor superlattices; magnetism; magnetic resonance; magnetoelastic effects; high-temperature superconductivity; heavy fermion superconductivity; unconventional superconductivity; optical and far-infrared spectroscopy of semiconductors; crystal growth and MBE of semiconductors; magnetostatic effects; layered superconductors; single-electron tunneling; scanning tunneling microscopy and spectroscopy; optical and infrared photoresponse; x-ray absorption spectroscopy and x-ray scattering; neutron scattering. Blackstead, Bunker, Crawford, Dobrowolska-Furdyna, Eskildsen, Furdyna, Gomes, Jankó, Liu, Newman, Ruggiero, Toroczkai.

High Energy Physics. CMS at CERN (Higgs boson properties, supersymmetry, top quark properties, electroweak physics, physics beyond the standard model); Double Chooz (neutrino oscillations); ILC (Beam monitoring, detector R&D); Fermilab D0 experiment (study of the top quark, bottom quark, W boson, and physics beyond the standard model); BaBar experiment at SLAC (CP violation in the b-quark system). Hildreth, Jessop, Karmgard, Lannon, LoSecco, Marinelli, Ruchti, Wayne.

Nuclear Physics. Nuclear structure; reaction energies; electromagnetic transitions; gamma-ray spectroscopy; high spin states; polarized particles; giant resonances; heavy-ion reactions; radioactive beam studies; nuclear astrophysics. Aprahamian, Berg, Brodeur, Collon, Couder, Garg, Gorres, Kolata, Stech, Tan, Wiescher.

View additional information about this department at www.gradschoolshopper.com

---

IOWA STATE UNIVERSITY

DEPARTMENT OF PHYSICS AND ASTRONOMY

Ames, Iowa 50011

http://www.physastro.iastate.edu

General University Information

President: Steven Leath
Dean of Graduate School: Williams Graves
University website: http://www.iastate.edu
Control: Public
Setting: Urban
Total Faculty: 1,845
Total Graduate Faculty: 1,396
Total number of Students: 31,040
Total number of Graduate Students: 5,195

Department Information

Department Chairman: Frank Krennrich, Chair
Department Contact: Diane Smith, Administrative Specialist
  Total full-time faculty: 42
  Total number of full-time equivalent positions: 53
  Full-Time Graduate Students: 88
  First-Year Graduate Students: 19
  Female First-Year Students: 3
  Total Post Doctorates: 15

Department Address

12 Physics Hall
Osborn Drive
Ames, IA 50011
Phone: (515) 294-0749
Fax: (515) 294-6027
E-mail: smithd@iastate.edu
Website: http://www.physastro.iastate.edu

ADMISSIONS

Admission Contact Information

Address admission inquiries to: Graduate Admission, Department of Physics and Astronomy
Phone: (515) 294-5870

E-mail: physastro@iastate.edu
Admissions website: http://www.admissions.iastate.edu/graduate/

Application deadlines

Fall admission:
U.S. students: September 1   *Int'l. students*: September 1
Spring admission:
U.S. students: October 15   *Int'l. students*: October 15

Application fee

U.S. students: $40   *Int'l. students*: $90

Admissions information

For Fall of 2012:
  Number of applicants: 179
  Number admitted: 55
  Number enrolled: 19

Admission requirements

Bachelor's degree requirements: A Bachelor's degree is required.
Minimum undergraduate GPA: 3.0

GRE requirements

The GRE is required.
  Analytical score: 1
  Mean GRE score range (25th–75th percentile): 54

Advanced GRE requirements

The Advanced GRE is required.
  Minimum accepted Advanced GRE score: 200
  Mean Advanced GRE score range (25th–75th percentile): 65%

TOEFL requirements

The TOEFL exam is required for students from non-English-speaking countries.
  PBT score: 550
  iBT score: 79

Other admissions information

Additional requirements: The average GRE percentage for 2012 admissions were verbal-64%; quantitative-87%; analytical-32%. The average GRE Physics score for admissions was 63%.

Undergraduate preparation assumed: Saxon, Elementary Quantum Mechanics; Marion, Classical Dynamics; Kittel, Thermal Physics; Lorrain and Corson, Electromagnetic Fields and Waves.

TUITION

Tuition year 2013-14:
Tuition for in-state residents
   Full-time students: $4,764 per semester
Tuition for out-of-state residents
   Full-time students: $10,920 per semester
Graduate students on 1/4 time or greater assistantships are assessed at the resident rate. In addition, a scholarship of 1/4 tuition is given to students with 1/4-time assistantships, and 1/2 tuition for students with 1/2-time assistantships.
Credit hours per semester to be considered full-time: 12
Deferred tuition plan: Yes
Health insurance: Not available.
Other academic fees: $849 per academic year.
Academic term: Semester
Number of first-year students who received full tuition waivers: 19

Teaching Assistants, Research Assistants, and Fellowships

Number of first-year
   Teaching Assistants: 19
Average stipend per academic year
   Teaching Assistant: $1,700
   Research Assistant: $1,800

FINANCIAL AID

Application deadlines

Fall admission:
U.S. students: March 1
Spring admission:
U.S. students: September 1   *Int'l. students*: September 1

Loans

Loans are available for U.S. students.
Loans are not available for international students.
GAPSFAS application required: No
FAFSA application required: Yes

For further information

Address financial aid inquiries to: Director of Student Financial Aids, 210 Beardshear.
Phone: (515) 294-2223
E-mail: financialaid@iastate.edu
Financial aid website: http://www.financialaid.iastate.edu

HOUSING

Availability of on-campus housing

Single students: Yes
Married students: Yes

For further information

Address housing inquiries to: Director of Residence, 2419 Friley.
Phone: (515) 294-2900
E-mail: housing@iastate.edu
Housing aid website: http://www.housing.iastate.edu

Table A—Faculty, Enrollments, and Degrees Granted

Research Specialty	2012-13 Faculty	Enrollment Fall 2012		Number of Degrees Granted 2012-13 (2007–13)		
		Master's	Doctorate	Master's	Terminal Master's	Doctorate
Astronomy	4	–	–	–	–	–
Astrophysics	3	–	6	–(7)	–(1)	1(7)
Biophysics	2	–	–	–	–	–
Condensed Matter Physics	15	2	34	–(2)	–(3)	7(40)
High Energy Physics	9	1	9	1(-)	–(1)	2(7)
Nuclear Physics	7	2	11	1(7)	2(2)	1(7)
Physics and other Science Education	2	–	23	–	–(1)	–(1)
Total	42	5	83	2(16)	2(8)	11(62)
Full-time Grad. Stud.	–	5	83	–	–	–
First-year Grad. Stud.	–	–	19	–	–	–

GRADUATE DEGREE REQUIREMENTS

Master's: The Master of Science degree is offered with and without thesis in various areas of physics (applied physics, atmospheric, high-energy, nuclear, condensed matter as examples) and astronomy. The minimum residential requirement is 30 credits, at least 21 of which must be in physics department graduate courses and 6 of which must be outside the major area. A "B" average (3.0 GPA) must be maintained. There is no foreign language requirement.

Doctorate: The Ph.D. degree in the same areas has a basic requirement of 72 credits, at least one-half of which must be earned at Iowa State University, and 12 of which must be outside the major area. A "B" average (3.0 GPA) must be maintained. There is no foreign language requirement. A qualifying examination given at the beginning of the student's second year, a preliminary oral examination, and a final thesis defense are the other major requirements.

Other Degrees: Close relationships exist with the Chemistry, Geological and Atmospheric Sciences, Electrical Engineering and Computer Engineering, Materials Science and Engineering, Computer Science, and Mathematics Departments, and joint programs are possible.

Thesis: Thesis may be written in absentia.

SPECIAL EQUIPMENT, FACILITIES, OR PROGRAMS

Five automated ultrasensitive SQUID magnetometers; helium dilution refrigerator millikelvin facility; two rotating anodes and several conventional x-ray diffraction facilities in combination with high and low temperature units and a new liquid surface reflectometer; ultra-high vacuum systems for surface physics studies using LEED, RHEED, and STMs; low temperature, high field magneto-optic spectrometer; microelectronics center for thin-film deposition (MBE, e-beam, etc.) and characterization (EELS, x-ray, Auger, ...); magnetic resonance spectrometers with superconducting solenoids for both high- and ultra low-temperature studies; precision spectrometers for neutron scattering and photoemission spectrometry (both carried out at national facilities); high-resolution Ge gamma-ray detectors; a campus-wide network of about 800 workstations linked with fiberoptic (FDDI) and Ethernet connections and running modified Athena software, over 60 of these fast computers are located in the Department, there are an additional 50 networked Unix, and SUN workstations, IBM RISC 6000 Models 550–590 and many MacIntoshes, PC's, and windowing terminals, the network is connected to all national labs and supercomputing centers through 45 Mbits/s(T3) lines to the NFSnet/Internet.

In addition, there are four multi processor Silicon Graphics Computers, and many PC-clusters ranging up to 128 processors. The clusters run with Linux and communicate via fast switches and either fast or gigabit-ethernet. Electronics and machine shop support; scanning Auger (600 Å resolution) and atomic resolution electron microscopes are available.

Research facilities are also utilized at the following laboratories: Fermilab (Batavia, IL), CERN (Geneva, Switzerland), BNL (Upton, NY), SRC (Stoughton, WI), ORNL (Oak Ridge, TN), Advanced Photon Source and Intense Pulsed Neutron Source at Argonne National Laboratory (Argonne, IL), and Stanford Linear Accelerator (Palo Alto, CA).

High-energy physics programs include collaboration on: the BABAR experiment at the PEPII B-factory detector, the DELPHI experiment at the CERN LEP collider, and the D0 experiment at the Fermilab Tevatron Collider, CMS, and ATLAS at the CERN LHC Collider. A facility for testing high-speed electronics components has been used to evaluate electronics for use in high-energy physics experiments and for applications outside high-energy physics, e.g., the human genome project.

The Nuclear Physics group constructed the Level-1 trigger for the PHENIX detector used at the RHIC accelerator (Brookhaven) to search for the quark-gluon plasma. It is also using PHENIX to study nuclear matter under extreme conditions of temperature and density and to probe the properties of the QCD vacuum. It constructed the Late-Energy Trigger for the E864 experiment at the AGS accelerator (Brookhaven) which was used in the search for strange quark matter. Studies on electromagnetic dissociation in relativistic heavy-ion reactions is being carried out.

Observational astronomy is pursued at all wavelengths at ISU. The gamma-ray astronomy group is collaborating with the Harvard-Smithsonian Center for Astrophysics, the University of Michigan, and several European groups in a program of very high-energy gamma-ray astronomy centered at the Smithsonian's Whipple Observatory. Optical data are obtained at ISU's Erwin W. Fick Observatory, which houses a 24-in. Cassegrain telescope. Observations are made with the coude spectrograph and direct imaging with the 800×800 pixel CCD camera. Optical and infrared data are also obtained with the Hubble Space Telescope, and telescopes at KPNO, CTIO, UKIRT and the Anglo Australian Observatory. Far-IR studies utilize the "Infrared Astronomical Satellite" (IRAS) database and new image processing tools. Ultraviolet and x-ray observations are obtained with the IUE satellite. Spectral line and radio continuum studies are performed with the Very Large Array (New Mexico) and the Australia Telescope National Facility (ATNF). Related research projects in planetary science and meteoritics are carried out in the Geological and Atmospheric Sciences Dept.

Table B—Separately Budgeted Research Expenditures by Source of Support

Source of Support	Departmental Research	Physics-related Research Outside Department
Federal government	$4,371,269	$7,143,252
State/local government	$4,991,468	
Non-profit organizations		
Business and industry		
Other	$153,333	
Total	$9,516,070	$7,143,252

Table C—Separately Budgeted Research Expenditures by Research Specialty

Research Specialty	No. of Grants	Expenditures ($)
Astronomy	9	$219,789
Astrophysics	7	$427,969
Biophysics	3	$416,559
Condensed Matter Physics	8	$7,932,426
Nuclear Physics	6	$1,413,177
Particles and Fields	11	$1,257,934
Physics and other Science Education	–	
Total	44	$11,667,854

FACULTY

Distinguished University Professor

Canfield, Paul C., Ph.D., University of California, Los Angeles, 1990. Distinguished Professor of Liberal Arts & Sciences. *Condensed Matter Physics, Physics and other Science Education.* Experimental physics; design, growth and characterization of new correlated electron materials.

Goldman, Alan I., Ph.D., Stony Brook University, 1984. *Condensed Matter Physics.* Experimental physics; x-ray and neutron scattering.

Harmon, Bruce N., Ph.D., Northwestern University, 1973. Deputy Director Ames Laboratory. *Condensed Matter Physics.* Computational materials science.

Ho, Kai-Ming, Ph.D., University of California, Berkeley, 1978. *Condensed Matter Physics, Theoretical Physics.* Properties of solids and surfaces.

Johnston, David C., Ph.D., University of California, San Diego, 1975. Divisional Associate Editor, Physical Review Letters. *Condensed Matter Physics.* Experimental solid state physics; high temperature superconductors, low-dimensional antiferromagnets, heavy fermion compounds, novel materials synthesis and measurement.

Soukoulis, Costas M., Ph.D., University of Chicago, 1978. *Condensed Matter Physics, Engineering Physics/Science, Nano Science and Technology.* Development of a theoretical understanding of the property of disordered systems, photonic crystals, metamaterials, random lasers, random magnetic systems, nonlinear systems, and amorphous semiconductors.

Willson, Lee Anne, Ph.D., University of Michigan, 1973. Vice President, American Astronomical Society. *Astronomy.* Mass loss, stellar winds, stellar evolution with mass loss.

Chair Professor

Krennrich, Frank, Ph.D., Ludwig Maximilian University, Munich, 1996. *Astrophysics, Particles and Fields.* Experimental particle astrophysics; gamma-ray astronomy.

Professor

Anderson, E. Walter, Ph.D., Columbia University, 1965. *High Energy Physics.*

Cochran, James H., Ph.D., Stony Brook University, 1993. *High Energy Physics.* Experimental physics.

Evans, James W., Ph.D., University of Adelaide, 1979. Scientist USDOE National Laboratory. *Condensed Matter Physics, Statistical & Thermal Physics, Theoretical Physics.* Surface science, epitaxial thin film evolution, far-from-equilibrium phenomena.

Hauptman, John M., Ph.D., University of California, Berkeley, 1974. *High Energy Physics.* Calorimetry, collider detectors.

Hill, John C., Ph.D., Purdue University, 1967. *Nuclear Physics.* Relativistic heavy ion physics and spin physics.

Kaminski, Adam, Ph.D., University of Illinois at Chicago, 2001. *Condensed Matter Physics.* Experimental condensed matter physics, superconductivity, angle-resolved photoelectron spectroscopy.

Kawaler, Steven D., Ph.D., University of Texas, Austin, 1986. Scientific Editor of 'The Astrophysical Journal.'. *Astronomy, Astrophysics.* Stellar astrophysics; asteroseismology.

Lajoie, John, Ph.D., Yale University, 1996. *Nuclear Physics.* Experimental nuclear physics; relativistic heavy-ion physics and nucleon spin.

Luban, Marshall, Ph.D., University of Chicago, 1962. *Condensed Matter Physics.* Molecular magnetism, statistical physics.

McQueeney, Rob, Ph.D., University of Pennsylvania, 1996. Scientist at Ames Laboratory. *Condensed Matter Physics.* Neutron scattering.

Ogilvie, Craig A., Ph.D., University of Birmingham, 1987. Assistant Dean of the ISU Graduate College. *Nuclear Physics.* Experimental nuclear physics; relativistic heavy-ion physics.

Prell, Soeren, Ph.D., Hamburg University, 1996. *High Energy Physics.* CP violation, B physics, standard model and beyond, BaBar, ATLAS.

Prozorov, Ruslan, Ph.D., Bar-Ilan University, 1998. Associate Scientist (US DOE Ames Laboratory). *Condensed Matter Physics, Low Temperature Physics.* Superconductivity, magnetism, nanoscience, bio-inspired materials.

Rosati, Marzia, Ph.D., McGill University, 1992. *Nuclear Engineering.* Experimental nuclear physics; relativistic heavy-ion physics.

Rosenberg, Eli I., Ph.D., University of Illinois, 1971. *High Energy Physics.*

Schmittmann, Beate, Ph.D., University of Edinburgh, 1984. Dean of the College of Liberal Arts & Sciences at Iowa State University. *Biophysics, Condensed Matter Physics.* Statistical mechanics, biological physics, complex networks.

Shinar, Joseph, Ph.D., Hebrew University, 1980. *Condensed Matter Physics.* Experimental Condensed Matter Physics. Optically detected magnetic resonance; optical properties of semiconductors, conducting polymers, and OLEDS.

Struck, Curtis, Ph.D., Yale University, 1981. Assistant Department Chair. *Astronomy, Astrophysics.* Theoretical astrophysics; galaxy formation and evolution; galaxy collisions.

Travesset, Alex, Ph.D., Universitat de Barcelona, 1997. *Biophysics, Condensed Matter Physics.* Self-assembly, protein-protein and protein-lipid interactions, new computation methods.

Tringides, Michael, Ph.D., University of Chicago, 1984. *Condensed Matter Physics.* Experimental physics: surface science; LEED; RHEED scanning tunneling microscopy.

Valencia, German, Ph.D., University of Massachusetts, 1988. *Particles and Fields, Theoretical Physics.* Particle phenomenology.

Vary, James P., Ph.D., Yale University, 1970. *Nuclear Physics, Particles and Fields.* Quantum many-particle and many-field systems.

Whisnant, Kerry L., Ph.D., University of Wisconsin-Madison, 1982. *Particles and Fields, Theoretical Physics.* Particle phenomenology.

Yu, Edward, Ph.D., University of Michigan, 1997. *Biophysics.* Protein x-ray crystallography.

Associate Professor

Furukawa, Yuji, Ph.D., Kobe University, 1995. *Condensed Matter Physics.* Nuclear magnetic resonance, nanoscale molecular magnet, low dimensional quantum spin systems, superconductors.

Kerton, Charles, Ph.D., University of Toronto, 2000. *Astrophysics.* Star formation, interstellar medium.

Marengo, Massimo, Ph.D., SISSA/ISAS, 2000. *Astrophysics.* Stellar astrophysics.

Sánchez, Mayly, Ph.D., Tufts University, 2003. *High Energy Physics.* Experimental high-energy physics. Neutrino physics.

Tuchin, Kurill, Ph.D., Tel Aviv University, 2001. *Nuclear Physics.* Nuclear and high-energy physics.

Wang, Jigang, Ph.D., Rice University, 2006. *Condensed Matter Physics.* Quantum processes and collective phenomena in condensed matter, light-matter interaction in nanoscale and complex materials, ultrafast spectroscopy and nonlinear optics, spin dynamics, coherence, magneto-optics and high magnetic field phenomena.

Assistant Professor

Chen, Chunhui, Ph.D., University of Pennsylvania, 2003. Experimental High Energy Physics.

Sivasankar, Sanjeevi, Ph.D., University of Illinois at Urbana-Champaign, 2001. *Biophysics, Nano Science and Technology.* Single molecule biophysics, physics at the bio-nano interface, molecular biotechnology.

Weinstein, Amanda, Ph.D., Stanford University, 2005. *Astrophysics, High Energy Physics, Particles and Fields.* Experimental particle astrophysics.

Emeritus

Clayton, Swenson, Ph.D., Oxford, 1949. *Condensed Matter Physics.* Experimental physics, thermodynamic properties at low temperatures.

Clem, John, Ph.D., Illinois, 1965. Distinguished Emeritus Professor. *Condensed Matter Physics.* Theoretical physics, superconductivity.

Finnemore, Douglas, Ph.D., Illinois, 1962. *Condensed Matter Physics.* Experimental physics, superconductivity and very low temperature phenomena.

Lynch, David, Ph.D., Illinois, 1958. *Condensed Matter Physics.* Experimental physics, optical properties and photoelectric spectroscopy of solids.

Research Associate Professor

Pieter, Maris, Ph.D., Groningen, 1993. *Nuclear Physics.* Theoretical nuclear and particle physics, strongly interaction quantum many-particle and many-field systems.

Adjunct Professor

Biswas, Rana, Ph.D., Cornell University, 1984. *Condensed Matter Physics.* Solar energy materials, photonics, atomistic simulations.

Vaknin, David, Ph.D., Hebrew University, 1987. *Condensed Matter Physics.* Low dimensional magnetism, superconductivity-magnetism, multiferroic materials, magnetic molecules, liquid services, and membrane physics.

Adjunct Associate Professor

Bud'ko, Sergey L., Ph.D., University of Moscow, 1986. Thermodynamic and transport properties of materials in multi-extreme conditions.

Tanatar, Makariy, Ph.D., Ukrainian Academy of Science, 1980. *Condensed Matter Physics.* Experimental condensed matter, transport phenomena.

Adjunct Assistant Professor

Kreyssig, Andreas, Ph.D., Dresden, 2001. *Condensed Matter Physics.* Experimental physics, condensed matter.

Affiliate Professor

Johnson, Duane, Ph.D., Cincinnati, 1985. *Condensed Matter Physics.* Condensed matter theory, materials physics, electronic-structure methods, computational materials science, thermodynamics and multiscale modeling.

Zia, Royce, Ph.D., MIT, 1968. *Condensed Matter Physics*. Theoretical physics, statistical and condensed matter, complex systems far from equilibrium, biological physics.

Senior Lecturer

Atwood, David, Ph.D., McGill University, 1989. *High Energy Physics, Theoretical Physics*.

Herrera-Siklody, Paula, Ph.D., Universitat de Barcelona, 1999. Laboratory Supervisor. *Physics and other Science Education*. Physics education.

Lecturer

Fretwell, Helen, Ph.D., University of Bristol, 1993. Lecturer in Physics. *Physics and other Science Education*. Physics education.

Courtesy Professor

Jernigan, Robert, Ph.D., Stanford, 1968. *Biophysics*. Biophysics, computational biology, bioinformatics, structural biology, systems biology, datamining, simulation science.

Shin, Yeon-Kyun, Ph.D., Seoul, 1990. *Biophysics*. Biochemistry, biophysics, electron paramagnetic resonance.

DEPARTMENTAL RESEARCH SPECIALTIES AND STAFF

Theoretical

Astronomy. Stellar evolution, stellar winds, and mass loss; pulsating and variable stars. Galaxy formation and evolution, galaxy collisions, and star formation and population evolution in galaxies. Kawaler, Kerton, Krennrich, Marengo, Struck, Willson.

Condensed Matter Physics. Superconducting vortex pinning; superconducting-normal proximity effects; Josephson junction arrays; electron-phonon interactions; magnetism. Harmon, Ho, Luban, Soukoulis.

Condensed Matter Physics. Photoemission; surface properties; optical properties; magnetic properties, electronic structure, lattice dynamics, critical phenomena; localization in disordered and quasiperiodic solids; spin glasses; quantum nanostructures; quantum computing photonic band gaps, many body theory, and simulations. Biswas, Bud'ko, Harmon, Ho, Luban.

Condensed Matter Physics. Phenomenology of the standard model and extensions; neutrino mass and oscillations; quantum chromodynamics; dynamical symmetry breaking and chiral perturbation theory. Biswas, Ho, Luban, Soukoulis, Travesset.

Nuclear Physics. Quark and gluon interactions in nuclei; relativistic heavy-ion and intermediate energy interactions; high-energy reactions of leptons and hadrons with nuclei; nuclear structure. Lajoie, Ogilvie, Tuchin, Vary.

Experimental

Astronomy. Radio, infrared, and optical studies of galaxies; image processing techniques; studies of galaxy clusters at high redshift; photoelectric stellar radial velocity measurements; multi-wavelength spectroscopy and photometry of variable stars; stellar seismology; identification and study of TeV gamma-ray sources. Kerton, Krennrich, Marengo, Weinstein.

Astrophysics. $Z°$ studies and b-production in e+ers- Aannihilations, CP violation in the b-quark; pp interactions at 1.8 TeV, Higgs particles, heavy leptons, exotic quarks, and top quark studies; quark and gluon jets; glueballs; high-energy neutrino astronomy. Experiments at CERN, SLAC, and Fermilab. Krennrich, Weinstein.

Condensed Matter Physics. Superconducting materials. Magneto-optic devices. Amorphous material devices. Semiconductors. Biswas, Bud'ko, Canfield.

Condensed Matter Physics. Optical properties; photoemission; neutron scattering; x-ray diffraction; magnetism; thermodynamic measurements; electrical properties; nuclear magnetic resonance; surface studies; thin films; transport properties; new materials design and growth; anisotropic superconductors; x-ray scattering studies of surfaces. Bud'ko, Canfield, Furukawa, Goldman, Johnston, Kaminski, McQueeney, Prozorov, Shinar, Tringides, Vaknin, Wang.

Energy Sources & Environment. Energy conservation; solar energy; radon in homes. Shinar.

Nuclear Physics. Search for quark-gluon plasma and strange quark matter with RHIC collider and AGS accelerator. Studies of nuclear matter under extreme conditions of density and temperature using the PHENIX detector. Studies of electromagnetic dissociation at the CERN-SPS. Hill, Lajoie, Ogilvie, Rosati.

View additional information about this department at
www.gradschoolshopper.com

THE UNIVERSITY OF IOWA

DEPARTMENT OF PHYSICS AND ASTRONOMY

Iowa City, Iowa 52242
http://www.physics.uiowa.edu

General University Information

President: Sally Mason
Dean of Graduate School: John C. Keller
University website: http://www.uiowa.edu
Control: Public
Setting: Urban
Total Faculty: 2,296
Total Graduate Faculty: 1,658

Total number of Students: 31,498
Total number of Graduate Students: 5,617

Department Information

Department Chairman: Mary Hall Reno, Chair
Department Contact: Mary Hall Reno, Professor & Chair
Total full-time faculty: 30
Full-Time Graduate Students: 73

First-Year Graduate Students: 14
Female First-Year Students: 2
Total Post Doctorates: 9

Department Address
203 Van Allen Hall
30 N. Dubuque St.
Iowa City, IA 52242
Phone: (319) 335-1686
Fax: (319) 335-1753
E-mail: admissions@newton.physics.uiowa.edu
Website: http://www.physics.uiowa.edu

ADMISSIONS

Admission Contact Information
Address admission inquiries to: Dean of Admissions, 107 Calvin
    Hall, The University of Iowa, Iowa City, IA 52242-1396 USA
Phone: (319) 335-3847
E-mail: admissions@uiowa.edu
Admissions website: http://www.uiowa.edu/admissions/index.
    html

Application deadlines
Fall admission:
U.S. students: January 15   *Int'l. students*: January 15
Spring admission:
U.S. students: January 15   *Int'l. students*: January 15

Application fee
U.S. students: $60   *Int'l. students*: $100
If you apply online, the $60 application fee ($100 for interna-
    tional students) is payable by Discover, MasterCard, or Visa.
    If you cannot pay by credit card, you may download and print
    an application and pay the fee by check or money order in
    U.S. currency made payable to "The University of Iowa."

Admissions information
For Fall of 2013:
Number of applicants: 160
Number admitted: 28
Number enrolled: 6

Admission requirements
Bachelor's degree requirements: Bachelor's degree in physics
    and/or astronomy is required with a minimum undergraduate
    GPA of "B".
Minimum undergraduate GPA: 3.0

GRE requirements
The GRE is required.

Advanced GRE requirements
The Advanced GRE is recommended.
75% of our offers of admission had GRE subject scores greater
    than 800.

TOEFL requirements
The TOEFL exam is required for students from non-English-
    speaking countries.
PBT score: 550
iBT score: 81
normally scores under 100 don't pass the on-campus English pro-
    ficiency exam required for graduate teaching assistantships

Other admissions information
Additional requirements: The minimum acceptable score is not
    specifically stated.
Undergraduate preparation assumed: Griffiths, Quantum Me-
    chanics; Griffiths, Introduction to Electrodynamics; Fowles
    and Cassiday, Analytical Mechanics; Reif, Statistical and
    Thermal Physics.

TUITION
Tuition year 2013-14:
Tuition for in-state residents
    Full-time students: $4,053 per semester
Tuition for out-of-state residents
    Full-time students: $12,345 per semester
Part-time is identified as 8 hours or less so tuition rates differ
    depending on the amount of credit hours you are taking (1-8
    credit hours).
Credit hours per semester to be considered full-time: 9
Deferred tuition plan: Yes
Health insurance: Yes, for a small fee.
Other academic fees: Technology Fee (FT) $231.50 Health Fee
    (FT) $118.50 Student Activity Fee (FT) $35.00 Student Ser-
    vices Fee (FT) $37.00 Student Union Fee (FT) $60.00 Build-
    ing Fee (FT) $61.50 Arts & Cultural Events Fee (FT) $12.00
    Recreation Fee (FT) $123.00 Professional Enhancement (FT)
    $30.00
Academic term: Semester

Teaching Assistants, Research Assistants, and Fellowships
Number of first-year
    Teaching Assistants: 13
    Fellowship students: 1
Average stipend per academic year
    Teaching Assistant: $19,650
    Research Assistant: $19,650
Average Fellowship Stipend (by fiscal year) $22,000.

FINANCIAL AID

Application deadlines
Fall admission:
U.S. students: January 15   *Int'l. students*: January 15
Spring admission:
U.S. students: January 15   *Int'l. students*: January 15

Loans
Loans are available for U.S. students.
Loans are not available for international students.
GAPSFAS application required: No
FAFSA application required: Yes

For further information
Address financial aid inquiries to: The University of Iowa, Office
    of Student Financial Aid, 208 Calvin Hall, Iowa City, IA
    52242-1315.
Phone: (319) 335-1450
E-mail: financial-aid@uiowa.edu
Financial aid website: http://www.uiowa.edu/financial-aid/index.
    shtml

HOUSING

Availability of on-campus housing
Single students: Yes
Married students: Yes

For further information
Address housing inquiries to: University Housing, 4141 Burge
    Hall, Iowa City, IA 52242.
Phone: (319) 335-3009
E-mail: housing@uiowa.edu
Housing aid website: http://housing.uiowa.edu/contactus.htm

Table A—Faculty, Enrollments, and Degrees Granted

Research Specialty	2012-13 Faculty	Enrollment Fall 2012		Number of Degrees Granted 2012-13 (2008-13)		
		Mas-ter's	Doc-torate	Mas-ter's	Terminal Master's	Doc-torate
Astronomy	6	4	–	1(3)	–(1)	2(2)
Atmosphere, Space Physics, Cosmic Rays	4	1	16	–(3)	–(2)	1(1)
Condensed Matter Physics	8	2	12	–(7)	–(2)	4(5)
Medical, Health Physics	3	–	1	–	–	–
Nuclear Physics	2	–	2	–	–	–
Particles and Fields	6	6	17	1(7)	–(1)	1(6)
Plasma and Fusion	5	–	12	–(2)	–(1)	2(1)
Non-specialized	–	–	–	–(1)	–(1)	–
Other	6	1	3	–	–	–
Total	40	13	60	2(23)	–(5)	10(14)
Full-time Grad. Stud.	–	13	60	–	–	–
First-year Grad. Stud.	–	1	13	–	–	–

GRADUATE DEGREE REQUIREMENTS

Master's: Thirty hours of coursework and research with a grade point average of at least 3.00 and an oral final examination are required. Thesis or critical essay options are available. No foreign language requirement is specified. The residence requirement may be fulfilled by completing a minimum of 24 semester hours under the auspices of The University of Iowa.

Doctorate: A minimum of 72 hours of coursework and research with a grade point average of at least 3.00, a qualifying examination, comprehensive examination, participation in advanced seminars, and original research are required for the Ph.D. A candidate for the degree will not be recommended until he/she has written the dissertation in proper form for formal publication and has submitted it, with the approval of the research advisor, for publication to a standard scientific journal of wide distribution. There is no specific foreign language requirement. Beyond the first 24 semester hours of graduate work, the residence requirement may be fulfilled by either: (1) enrollment as a full-time student (nine semester hours minimum) in each of two semesters or (2) enrollment for a minimum of six semester hours in each of three semesters during which time the student holds at least a one-third-time assistantship certified by the Department as contributing to the student's doctoral program.

Thesis: Thesis may be written in absentia.

SPECIAL EQUIPMENT, FACILITIES, OR PROGRAMS

Comprehensive facilities for design, construction, and testing of instruments for space flight and for the decoding, analysis, and display of flight data; automated 37-cm optical telescope at remote dark site, 3.0 and 4.5 m radio telescopes for instrumentation development; steady-state magnetized plasma devices (gas discharges and Q-machines) with 1-10 kG magnetic fields; a wide array of gas discharges and diagnostics for the study of plasmas containing charged dust grains; compact medical cyclotron for production of radionuclides; molecular-beam epitaxy machines for growth of state-of-the-art III-V semiconductor quantum wells, superlattices, quantum dots, and optoelectronic devices; a large computer cluster for analysis of high-energy nuclear data with direct connections to Fermilab and CERN; a state-of-the-art computer-controlled photomultiplier test station for high-energy and nuclear physics detectors; X-ray astronomical instrumentation facility for design of x-ray polarimeters and high-resolution timing instruments to be flown on x-ray space missions; high-performance computer cluster with 64 nodes for large-scale kinetic simulation of plasmas; and a large number of continuous-wave and pulsed (including ultrafast) lasers for spectroscopy.

Table B—Separately Budgeted Research Expenditures by Source of Support

Source of Support	Departmental Research	Physics-related Research Outside Department
Federal government	$16,680,838	
State/local government	$1,066,630	
Non-profit organizations		
Business and industry		
Other		
Total	$17,747,468	

Table C—Separately Budgeted Research Expenditures by Research Specialty

Research Specialty	No. of Grants	Expenditures ($)
Astronomy	26	$1,583,771
Atmosphere, Space Physics, Cosmic Rays	36	$8,460,591
Condensed Matter Physics	31	$3,722,498
Nuclear Physics	2	$106,114
Particles and Fields	27	$1,673,335
Plasma and Fusion	14	$1,290,514
Other	9	$910,645
Total	145	$17,747,468

FACULTY

Professor

Andersen, David R., Ph.D., Purdue University, 1986. Primary appointment in Electrical and Computer Engineering. *Condensed Matter Physics, Electrical Engineering, Materials Science, Metallurgy, Optics.* Nonlinear optics; quantum electronics; solid state; embedded systems.

Boggess, Thomas F., Ph.D., North Texas State University, 1982. Additional appointment in Electrical and Computer Engineering. *Applied Physics, Condensed Matter Physics, Electrical Engineering, Nano Science and Technology.* Nonlinear optics; ultrafast spectroscopy of semiconductor heterostructures.

Flatté, Michael E., Ph.D., University of California, Santa Barbara, 1992. Director of Optical Science Technology Center. *Applied Physics, Condensed Matter Physics, Materials Science, Metallurgy, Nano Science and Technology.* Condensed-matter physics; materials theory.

Goree, John A., Ph.D., Princeton University, 1985. *Condensed Matter Physics, Plasma and Fusion.* Experimental plasma physics; biomedical applications of plasmas; soft condensed-matter physics.

Gurnett, Donald A., Ph.D., University of Iowa, 1965. *Atmosphere, Space Physics, Cosmic Rays.* Experimental space plasma physics.

Hichwa, Richard, Ph.D., University of Wisconsin-Madison, 1981. Primary appointment in Department of Radiology. *Medical, Health Physics.* Medical physics.

Kaaret, Philip E., Ph.D., Princeton University, 1989. *Astronomy, Astrophysics.* X-ray and gamma-ray astronomy and instrumentation; black hole binaries; jet ejection from black holes.

Kleiber, Paul D., Ph.D., University of Colorado, 1981. Departmental Associate Chair. *Atmosphere, Space Physics, Cosmic Rays, Atomic, Molecular, & Optical Physics.* Atmospheric physics; chemical physics.

Kletzing, Craig A., Ph.D., University of California, San Diego, 1989. *Atmosphere, Space Physics, Cosmic Rays*. Experimental space plasma physics; laboratory plasma physics.

Madsen, Mark T., Ph.D., University of Wisconsin-Madison, 1979. Primary appointment in Department of Radiology. *Medical, Health Physics*. Medical physics.

Mallik, Usha, Ph.D., City University of New York, 1978. *Particles and Fields*. Experimental elementary particle physics.

Merlino, Robert L., Ph.D., University of Maryland, 1980. *Plasma and Fusion*. Experimental plasma physics.

Meurice, Yannick, Ph.D., UCL Louvain-la-Neuve, 1985. *Particles and Fields*. Theoretical elementary particle physics; lattice gauge theory; optical lattices.

Mutel, Robert L., Ph.D., University of Colorado, 1975. *Astronomy, Astrophysics, Atmosphere, Space Physics, Cosmic Rays*. Radio astronomy; space physics; plasma astrophysics.

Onel, Yasar, Ph.D., University of London, 1975. *Nuclear Physics, Particles and Fields*. Experimental elementary particle and nuclear physics.

Polyzou, Wayne N., Ph.D., University of Maryland, 1979. *Nuclear Physics*. Theoretical nuclear physics; mathematical physics.

Reno, Mary Hall, Ph.D., Stanford University, 1985. Departmental Chair. *Particles and Fields*. Theoretical elementary particle physics; astroparticle physics.

Rodgers, Vincent G. J., Ph.D., Syracuse University, 1985. *Particles and Fields*. Theoretical particle physics; string theory.

Scudder, Jack D., Ph.D., University of Maryland, 1975. *Atmosphere, Space Physics, Cosmic Rays, Plasma and Fusion*. Space plasma physics.

Skiff, Frederick N., Ph.D., Princeton University, 1985. *Plasma and Fusion*. Laser spectroscopy; plasma physics.

Smirl, Arthur L., Ph.D., University of Arizona, 1975. *Electrical Engineering, Optics*. Optical properties of semiconductors; ultrafast photonics; nonlinear optics; laser physics.

Spangler, Steven R., Ph.D., University of Iowa, 1975. *Astronomy, Astrophysics, Atmosphere, Space Physics, Cosmic Rays, Plasma and Fusion*. Radio astronomy; plasma astrophysics; space plasma physics.

Associate Professor

Gayley, Kenneth G., Ph.D., University of California, San Diego, 1990. *Astronomy, Astrophysics*. Radiative transfer; radiation hydrodynamics; spectral line diagnostics.

Howes, Gregory G., Ph.D., University of California, Los Angeles, 2004. *Computational Physics, Plasma and Fusion*. Theoretical and computational plasma physics.

Lang, Cornelia C., Ph.D., University of California, Los Angeles, 2000. *Astronomy, Astrophysics*. Radio astronomy; x-ray astronomy; observational study of interstellar medium and galactic center.

Nachtman, Jane M., Ph.D., University of Wisconsin-Madison, 1997. *Particles and Fields*. Experimental elementary particle physics.

Prineas, John P., Ph.D., University of Arizona, 2000. *Applied Physics, Condensed Matter Physics, Electrical Engineering, Materials Science, Metallurgy, Nano Science and Technology, Optics*. Experiemental semiconductor physics; growth and fabrication; spectroscopy; microscopy; semiconductor nanostructures; optoelectronics and photonics; III–V MBE growth; nonlinear optics.

Pryor, Craig, Ph.D., University of California, Santa Barbara, 1990. *Computational Physics, Condensed Matter Physics, Nano Science and Technology*. Theoretical condensed-matter and semiconductor nanostructures.

Sunderland, John J., Ph.D., University of Wisconsin-Madison, 1990. Primary appointment in Department of Radiology; Director of PET Imaging. *Medical, Health Physics*. Medical physics.

Wohlgenannt, Markus, Ph.D., University of Utah, 2000. *Condensed Matter Physics*. Experimental polymer physics.

Assistant Professor

Baalrud, Scott D., Ph.D., University of Wisconsin-Madison, 2010. *Plasma and Fusion, Theoretical Physics*. Basic and applied theoretical plasma physics.

Fu, Hai, Ph.D., University of Hawaii, 2008. *Astronomy, Astrophysics*. Extragalactic Astronomy.

Khodas, Maxim, Ph.D., Weizmann Institute of Science, 2006. *Condensed Matter Physics, Materials Science, Metallurgy*. Condensed-matter and materials physics.

McEntaffer, Randall L., Ph.D., University of Colorado Boulder, 2007. *Astronomy, Astrophysics*. X-ray astronomy and instrumentation; diffuse hot interstellar medium.

Siochi, R. Alfredo C., Ph.D., Virginia Polytechnic Institute and State University (Virginia Tech), 1990. Primary appointment in Radiation Oncology. *Medical, Health Physics*. Medical physics.

Emeritus

Klink, William H., Ph.D., Johns Hopkins University, 1964. Theoretical nuclear physics; mathematical physics.

Knorr, Georg, Ph.D., Ludwig Maximilian University, Munich, 1963. Theoretical plasma physics.

Lonngren, Karl E., Ph.D., University of Wisconsin-Madison, 1964. Experimental plasma physics.

McCliment, Edward R., Ph.D., University of Illinois, 1962. Elementary particle physics.

Neff, John S., Ph.D., University of Wisconsin-M adison, 1961. Observational optical astronomy.

Newsom, Charles R., Ph.D., University of Texas, 1980. *Particles and Fields*. Experimental elementary particle physics.

Norbeck, Edwin, Ph.D., University of Chicago, 1956. *Nuclear Physics, Particles and Fields*. Experimental nuclear and elementary particle physics.

Payne, Gerald L., Ph.D., University of California, San Diego, 1967. *Nuclear Physics*. Theoretical nuclear physics.

Schweitzer, John W., Ph.D., University of Cincinnati, 1966. Theoretical and experimental solid-state physics.

DEPARTMENTAL RESEARCH SPECIALTIES AND STAFF

Theoretical

Astrophysics. Radiation-driven winds from hot stars; plasma waves and turbulence in the interplanetary and interstellar media; physics of nonthermal radio sources. Gayley, Howes, Mutel, Spangler.

Atmosphere, Space Physics, Cosmic Rays. Physics of space magneto-plasmas and their kinetic properties; analytical and numerical solution of MHD, Vlasov, and Fokker-Planck equations; magnetic reconnection and plasma turbulence in the magnetotail; the solar corona and the solar wind. Howes, Scudder.

Atomic, Molecular, & Optical Physics. Gauge interactions on optical lattices. Meurice.

Condensed Matter Physics. Strong correlation problems and magnetic properties of materials; electrical, magnetic, and optical properties of nanostructures; spintronics; quantum computation. Flatté, Khodas, Pryor, Schweitzer.

Mathematical Physics. Coherent states in semiconductors; carrier dynamics in semiconductor lasers and detectors; nonlinear propagation phenomena. Andersen, Flatté, Khodas, Pryor.

Nuclear Physics. Numerical and theoretical studies of reactions, structure, and electroweak properties of few hadron and few quark systems using relativistic and non-relativistic quantum mechanics and quantum field theory. Klink, Payne, Polyzou.

Other. Mathematical methods with emphasis on group theory, operator algebras, infinite dimensional Lie algebras, and non-linear dynamics; Yang Mill' and Chern Simons theories. Klink, Meurice, Polyzou, Rodgers.

Particles and Fields. Particle phenomenology of colliders and neutrino detectors; astroparticle physics and dark matter; renormalization group methods; lattice gauge theory; heavy flavor physics; gauge/gravity duals; superstrings. Meurice, Reno, Rodgers.

Plasma and Fusion. Basic plasma physics; theoretical and computational studies of space; astrophysics; laboratory plasmas; turbulence in kinetic plasmas; development and implementation of high-performance gyrokinetic codes for first-principles simulation of kinetic plasmas; strongly coupled plasmas; high-energy-density plasmas; magnetic reconnection; sheaths; double layers; plasma-based electron sources. Baalrud, Howes.

Relativity & Gravitation. Superstring theory; string theory; gauge/gravity duality; supergravity; conformal field theory; cosmology; representations of diffeomorphisms. Klink, Rodgers.

Experimental

Astronomy. Radio, x-ray, and gamma-ray astronomy; very-long-baseline radio astronomy; interstellar radio scintillation; studies of the galactic center; radio continuum observations of galactic and extragalactic sources; radio-imaging spectroscopy; x-ray and gamma-ray observations of black holes, neutron stars, and the interstellar medium; jet ejection from black holes; plasma astrophysics, including radiation hydrodynamics, and astrophysical turbulence; X-ray and gamma-ray astronomical instrumentation; x-ray sounding rocket payloads; high-resolution x-ray spectroscopy. Gayley, Kaaret, Lang, McEntaffer, Mutel, Spangler.

Atmosphere, Space Physics, Cosmic Rays. Space plasmas and atmospheric physics; collisionless magnetic reconnection; energetic particles and waves in the radiation belts; electric and magnetic fields; plasma instabilities; wave phenomena and radio emissions at earth, in planetary magnetospheres, and in the interplanetary medium; auroral phenomena and magnetosphere-ionsphere coupling, including global imaging; chemistry and physics of atmospheric dust. Gurnett, Kleiber, Kletzing, Scudder.

Condensed Matter Physics. MBE growth and fabrication of III–V semiconductor nanomaterials and devices; synthesis and applications of organic semiconductors; electrical transport and magnetic properties of layered ternary transition metal sulfides; ultrafast optical and electronic properties of semiconductor quantum wells and superlattices; optical measurements of high-speed carrier dynamics, including transport, recombination, energy relaxation, and scattering in semiconductor structures and devices; soft condensed matter, including structure and waves in Coulomb crystals and 2D and 3D experiments under laboratory and microgravity conditions. Boggess, Goree, Prineas, Schweitzer, Smirl, Wohlgenannt.

Medical, Health Physics. Design of radiation detector systems for medical applications and nuclear medicine imaging; development of processing methodologies for analysis of medical images; development of high-speed electronics, associated hardware and application software for PET imaging devices; fabrication of nuclear targets and automated radiochemistry synthesis systems to produce PET radiopharmaceuticals. Hichwa, Madsen, Siochi, Sunderland.

Nuclear Physics. Study of Pb+Pb at 1000 TeV with CMS/LHC at CERN. Breakup of excited nuclei into large fragments using large national accelerators in the United States and Europe. Norbeck, Onel.

Other. Remotely controlled imaging telescope linked to regent's institutions. Mutel.

Other. Development of ultrafast optical sources and their use in spectroscopy of bulk and microstructure semiconductors; nonlinear optical properties of solid-state materials; coherent processes in semiconductors; semiconductor lasers, detectors, and other photonic devices. Andersen, Boggess, Prineas, Smirl, Wohlgenannt.

Particle Physics. Higgs, SUSY, and beyond the Standard Model searches at Fermilab, CMS/LHC, and ATLAS/LHC; Liquid Argon Calorimeter Upgrade; study of charm baryons at Fermilab; quartz fiber Cherenkov calorimetry development for collider physics at Iowa-HEP laboratory; silicon pixel and microstrip detector development at Fermilab and silicon pixel study in ATLAS/LHC; digital calorimetry, Compton polarimeter, and Particle Flow Algorithm development for a high-energy e+e– linear collider. Mallik, Nachtman, Norbeck, Onel.

Plasma and Fusion. Plasma waves and instabilities; nonlinear particle dynamics; negative ion plasmas; dusty plasma; interdisciplinary study of strongly coupled Coulomb crystal structure and dynamics; laboratory simulation of space and astrophysical plasmas; microgravity experiments; technological topics, including biomedical applications of plasmas, laser scattering and laser-induced fluorescence diagnostics of plasmas, plasma source for plasma processing, and particulate contamination in plasma processing. Goree, Kletzing, Lonngren, Merlino, Skiff.

View additional information about this department at
www.gradschoolshopper.com

KANSAS STATE UNIVERSITY

DEPARTMENT OF PHYSICS

Manhattan, Kansas 66506-2601
http://www.phys.ksu.edu

General University Information

President: Kirk Schulz
Dean of Graduate School: Carol Shanklin
University website: http://www.k-state.edu/
Control: Public

Setting: Rural
Total Faculty: 1,251
Total Graduate Faculty: 900
Total number of Students: 23,500
Total number of Graduate Students: 3,921

Department Information
Department Chairman: Amit Chakrabarti, Head
Department Contact: Michael O'Shea, Professor
  Total full-time faculty: 27
  Total number of full-time equivalent positions: 30
  Full-Time Graduate Students: 57
  First-Year Graduate Students: 11
  Female First-Year Students: 2
  Total Post Doctorates: 22

Department Address
Cardwell Hall
Manhattan, KS 66506-2601
Phone: (785) 532-6786
Fax: (785) 532-6806
E-mail: mjoshea@phys.ksu.edu
Website: http://www.phys.ksu.edu

ADMISSIONS

Admission Contact Information
Address admission inquiries to: Graduate Secretary, Department of Physics, Kansas State University, Manhattan, KS 66506-2601
Phone: (785) 532-6786
E-mail: graduate@phys.ksu.edu
Admissions website: http://www.phys.ksu.edu

Application deadlines
Fall admission:
U.S. students: December 16   *Int'l. students*: December 16

Application fee
U.S. students: $50   *Int'l. students*: $75

Admissions information
For Fall of 2012:
  Number of applicants: 120
  Number admitted: 34
  Number enrolled: 11

Admission requirements
Bachelor's degree requirements: Bachelor's degree in physics is required.
Minimum undergraduate GPA: 3.0

GRE requirements
The GRE is not required.

Advanced GRE requirements
The Advanced GRE is required.

TOEFL requirements
The TOEFL exam is required for students from non-English-speaking countries.
  PBT score: 550
  iBT score: 79

Other admissions information
Additional requirements: Candidates with engineering or mathematics degrees will also be considered. No minimum acceptable score for admission is specified.
Undergraduate preparation assumed: Mechanics (3 to 6 hours); Physics Lab (6 hours); Electricity and Magnetism (3 to 6 hours); Modern Physics (3 hours); Quantum Mechanics (3 hours); Mathematics through Differential Equations and Vector Analysis.

TUITION

Tuition year 2012–13:
Tuition for in-state residents
  Full-time students: $3,076 per semester
Tuition for out-of-state residents
  Full-time students: $6,584 per semester
Graduate assistants pay resident fees; tuition is waived for graduate teaching assistants.
Credit hours per semester to be considered full-time: 6
Deferred tuition plan: No
Health insurance: Available
Other academic fees: Full-time tuition rates include fees and required additional assessments. Part-time students are required to pay $65.00/semester special fees and have the option of using the student health center for a fee of $25/semester.
Academic term: Semester
Number of first-year students who received full tuition waivers: 11

Teaching Assistants, Research Assistants, and Fellowships
Number of first-year
  Teaching Assistants: 11
Average stipend per academic year
  Teaching Assistant: $14,200

FINANCIAL AID

Application deadlines
Fall admission:
U.S. students: April 1

Loans
Loans are not available for U.S. students.
Loans are not available for international students.
GAPSFAS application required: No
FAFSA application required: No

HOUSING

Availability of on-campus housing
  Single students: Yes
  Married students: Yes

For further information
Address housing inquiries to: Department of Housing, Pitmann Building, Kansas State University.
Phone: (785) 532-6453
E-mail: housing@ksu.edu
Housing aid website: http://www.housing.ksu.edu/

Table A—Faculty, Enrollments, and Degrees Granted

Research Specialty	2009–10 Faculty	Enrollment Fall 2010		Number of Degrees Granted 2010–11 (2006–10)		
		Master's	Doctorate	Master's	Terminal Master's	Doctorate
Atomic, Molecular, & Optical Physics	11	–	30	–(2)	–(2)	2(18)
Condensed Matter Physics	7	–	5	–(1)	–(4)	1(18)
Cosmology & String Theory	1	–	–	–	–	–(1)
High Energy Physics	4	–	9	–	–(1)	–(5)
Physics and other Science Education	3	–	8	–(5)	–(4)	2(7)
Non-specialized	–	–	5	–	–	–
Other	1	–	–	–	–	–
Total	27	–	57	–(8)	–(11)	5(49)
Full-time Grad. Stud.	–	–	57	–	–	–
First-year Grad. Stud.	–	–	7	–	–	–

GRADUATE DEGREE REQUIREMENTS

Master's: Thirty graduate credits in approved program with satisfactory performance and "B" average in coursework are required. Thesis is required for which up to six credits may be earned. Examination over thesis and one academic year of residence are also required.

Doctorate: Ninety graduate credits in approved program of study with satisfactory performance with a "B" average in coursework is required. Preliminary examination in area of specialization and related fields is required. Dissertation for which 30 credits may be earned, oral examination, and one full year of residency are also required.

Thesis: Thesis may be written in absentia.

SPECIAL EQUIPMENT, FACILITIES, OR PROGRAMS

Ultrafast Ti:Sapphire laser systems for studies of the interaction of high-intensity ultrafast electromagnetic pulses with matter, Cr:Forsterite, and fiber lasers for studies of optical frequency standards, a 7-MV tandem Van de Graff accelerator, a 150 keV accelerator with a CRYEDIS ion source, and a 30 keV ECR ion source for ion research in atomic, molecular, optical, and solid-state physics. Linux computer clusters are available for large-scale computational studies. State-of-the-art laser laboratories equipped with argon ion, Nd:YAG, and dye lasers and Raman, Fabry-Perot, and correlation spectrometers. Other facilities include laboratories for study of soft matter.

Table B—Separately Budgeted Research Expenditures by Source of Support

Source of Support	Departmental Research	Physics-related Research Outside Department
Federal government	$9,211,517	
State/local government		
Non-profit organizations		
Business and industry		
Other		
Total	$9,211,517	

Table C—Separately Budgeted Research Expenditures by Research Specialty

Research Specialty	No. of Grants	Expenditures ($)
Atomic, Molecular, & Optical Physics	15	$6,141,538
Condensed Matter Physics	11	$1,200,000
Particles and Fields	10	$1,268,252
Physics and other Science Education	6	$461,290
Other	1	$223,000
Total	43	$9,294,080

FACULTY

Professor

Ben-Itzhak, Itzik, Ph.D., Technion, Haifai, 1986. *Atomic, Molecular, & Optical Physics*. Experimental AMO physics; imaging the dissociation and ionization of molecular ions by ultrashort intense laser pulses; ionization and break-up of molecules by both fast and slow highly charged ions.

Bolton, Timothy, Ph.D., Massachusetts Institute of Technology, 1988. *High Energy Physics*. Experimental high-energy physics.

Chakrabarti, Amitabha, Ph.D., University of Minnesota, 1987. Department Head. *Condensed Matter Physics*. Theoretical and computational studies in soft-condensed matter and biological physics including self-assembly of nanoparticles, colloids, proteins, and aerosols.

De Paola, Brett David, Ph.D., University of Texas, Dallas, 1984. *Atomic, Molecular, & Optical Physics*. Experimental atomic physics; crossed beams; laser-beam interactions.

Esry, Brett D., Ph.D., University of Colorado Boulder, 1997. *Atomic, Molecular, & Optical Physics*. Theoretical atomic physics; ultracold three-body collisions; intense laser-molecule and laser atom interactions; few-body physics.

Law, Bruce M., Ph.D., University of Victoria, 1985. *Condensed Matter Physics*. Condensed matter interfaces; nonequilibrium liquids.

Lin, Chii-Dong, Ph.D., University of Chicago, 1974. *Atomic, Molecular, & Optical Physics*. Laser-molecule interactions; attosecond physics; few-body physics; atomic collisions; dynamic chemical imaging.

O'Shea, Michael, Ph.D., University of Sussex, 1981. *Condensed Matter Physics*. Experimental condensed matter physics.

Ratra, Bharat, Ph.D., Stanford University, 1986. *Cosmology & String Theory*. Cosmology; particle theory.

Rebello, N. Sanjay, Ph.D., Brown University, 1995. *Physics and other Science Education*. Research-based pedagogy; models of learning and transfer; problem solving; technology in teaching and learning.

Sorensen, Christopher M., Ph.D., University of Colorado, 1976. *Condensed Matter Physics*. Experimental condensed matter physics.

Thumm, Uwe, Ph.D., University of Freiburg, 1989. *Atomic, Molecular, & Optical Physics*. Theoretical atomic physics; electronic excitation, charge transfer, and fragmentation processes in interactions of photons, electrons, and ions with surfaces, nanostructures, molecules, clusters, and atoms.

Weaver, O. Lawrence, Ph.D., Duke University, 1970. *Theoretical Physics*. General theoretical physics.

Wysin, Gary M., Ph.D., Cornell University, 1985. *Condensed Matter Physics*. Condensed matter theory; Monte Carlo transport calculations; quantum Monte Carlo calculations.

Zollman, Dean A., Ph.D., University of Maryland, 1970. *Physics and other Science Education*. Inquiry methods; nontraditional learning experiences; audiovisual aids.

Associate Professor

Corwin, Kristan, Ph.D., University of Colorado Boulder, 1999. *Atomic, Molecular, & Optical Physics*. Experimental atomic physics; optical frequency metrology; ultrafast optics; laser development.

Flanders, Bret, Ph.D., University of Chicago, 1999. *Biophysics, Condensed Matter Physics*. Soft matter nanotechnology and biological physics.

Horton-Smith, Glenn, Ph.D., Stanford University, 1998. *High Energy Physics*. Experimental neutrino physics.

Maravin, Yurii, Ph.D., Southern Methodist University, 2002. *High Energy Physics*. Experimental high-energy physics.

Washburn, Brian R., Ph.D., Georgia Institute of Technology, 2002. *Atomic, Molecular, & Optical Physics*. Experimental atomic physics; nonlinear fiber optics; ultrafast optics; laser development.

Assistant Professor

Ivanov, Andrew G., Ph.D., University of Rochester, 2004. *High Energy Physics*. Experimental high-energy research aimed at pursuing searches for new phenomena at the Large Hadron Collider (LHC).

Kling, Matthias, Ph.D., University of Goettingen, 2002. *Atomic, Molecular, & Optical Physics*. Experimental atomic, molecular, and optical physics; 3D imaging of attosecond electron dynamics in molecules and nanostructures.

Kumarappan, Vinod, Ph.D., Tata Institute of Fundamental Research, 2002. *Atomic, Molecular, & Optical Physics*. Strong field alignment and orientation of molecules; experimental atomic, molecular, and optical physics.

Rudenko, Artem A., Ph.D., Moscow Institute of Physics & Technology, 2002. *Atomic, Molecular, & Optical Physics*. Experimental atomic, molecular, and optical physics; ultrafast laser and free electron laser interactions with atoms and molecules; 3D imaging of fragmentation processes.

Sayre, Eleanor, Ph.D., University of Maine, 2007. *Physics and other Science Education*. Improving and extending the theoretical bases for physics education research; investigating the interplay between physics and mathematics understanding.

Schmit, Jeremy, Ph.D., University of California, Santa Barbara, 2005. *Biophysics, Condensed Matter Physics*. Theoretical soft matter physics and biological physics; protein phase behavior, crystallization, and fibril formation.

Szoszkiewicz, Robert, Ph.D., Swiss Federal Institute of Technology, Lausanne, Switzerland, 2003. *Biophysics, Condensed Matter Physics*. Soft matter and biological physics.

Trallero, Carlos, Ph.D., Stony Brook University, 2007. *Atomic, Molecular, & Optical Physics*. Experimental atomic, molecular, and optical physics; strong-field molecular spectroscopy; higher-order harmonic generation; attosecond physics; coherent quantum control.

Emeritus

Bhalla, Chander P., Ph.D., University of Tennessee, 1960. *Atomic, Molecular, & Optical Physics*. Theoretical physics; atomic collisions; radiative and non-radiative transitions.

Cocke, C. Lewis, Ph.D., California Institute of Technology, 1967. *Atomic, Molecular, & Optical Physics*. Experimental atomic, molecular and optical physics; inner-shell excitations by beam foil processes.

Folland, Nathan O., Ph.D., Iowa State University, 1965. *Solid State Physics*. Theoretical solid-state physics.

Gray, Thomas J., Ph.D., Florida State University, 1967. *Atomic, Molecular, & Optical Physics*. Experimental atomic physics; inner-shell ionization in ion-atom collisions; fluorescence yields and lifetimes of highly ionized atomic and molecular species.

Hagman, Siegbert, Ph.D., University of Cologne, 1977. *Atomic, Molecular, & Optical Physics*. Experimental atomic physics.

Legg, James C., Ph.D., Princeton University, 1962. *Atomic, Molecular, & Optical Physics*. Experimental atomic physics; lifetimes of ionized molecules; breakup of molecular ions.

Rahman, Talat, Ph.D., University of Rochester, 1977. *Condensed Matter Physics, Solid State Physics*. Theoretical solid-state physics; surfaces and nanostructures.

Reay, Neville W., Ph.D., University of Minnesota, 1962. *High Energy Physics*. Experimental high-energy physics.

Richard, Patrick, Ph.D., Florida State University, 1964. *Atomic, Molecular, & Optical Physics*. Experimental atomic physics.

Stanton, Noel R., Ph.D., Cornell University, 1965. *High Energy Physics*. Experimental high-energy physics.

Research Associate Professor

Carnes, Kevin D., Ph.D., Purdue University, 1984. *Atomic, Molecular, & Optical Physics, Nuclear Physics*. Accelerator physics; computer controls.

Research Assistant Professor

Fehrenbach, Charles, Ph.D., University of Michigan, 1993. *Atomic, Molecular, & Optical Physics*. Experimental atomic physics.

Le, Auh-Thu, Ph.D., Belarussian State University, 1994. *Atomic, Molecular, & Optical Physics*. Theoretical atomic physics.

DEPARTMENTAL RESEARCH SPECIALTIES AND STAFF

Theoretical

Atomic, Molecular, & Optical Physics. Laser interactions with atoms, molecules, clusters, and surfaces; cold-atom collisions; atomic collisions; few-body problems; basis set expansion, grid propagation, and hyperspherical methods; attosecond physics; high-order harmonic generation; ultrafast chemical imaging with lasers. Esry, Le, Lin, Thumm.

Condensed Matter Physics. Magnetic models; statistical mechanics; spin dynamics; solitons; vortices; magnetic resonances; impurities (computer simulations including quantum Monte Carlo are used to study these systems). Wysin.

Cosmology & String Theory. Dark energy; dark matter; classical cosmological tests; cosmic microwave background anisotropies; inflation; cosmological magnetic fields; cosmological gravitational waves; large-scale structure. Ratra.

Soft Condensed Matter and Biological Physics. Theoretical and computational studies of various aspects of self-assembly and phase separation in soft-condensed matter and biological physics including: superlattice formation in various nanoparticle systems; kinetics of protein self-assembly from the dispersed phase; systematic approach to the theoretical study of certain diseases developing as a consequence of an undesired self-assembled protein phase; gelation in aerosols, colloids, and protein solutions (large-scale Monte Carlo and Brownian Dynamics simulations are used to study these systems). Chakrabarti, Schmit.

Experimental

Atomic, Molecular, & Optical Physics. Harmonic generation, AMO ultrafast, and attosecond science; collisions with MOT targets; collisions between highly charged ions and atomic and molecular targets; laser interactions with nano-structures and fast ion beams; laser metrology; fiber laser development; nonlinear fiber optics. Ben-Itzhak, Carnes, Corwin, De Paola, Fehrenbach, Kling, Kumarappan, Rudenko, Trallero, Washburn.

Biophysics. Soft matter nanotechnology and biophysics; nanoelectronic devices fabricated to measure electromechanical properties at selected sites on living cells. Flanders, Szoszkiewicz.

Biophysics. Soft matter and biological physics; mechanics and mechanochemical reactions of single molecules. Flanders, Szoszkiewicz.

High Energy Physics. Collider and neutrino physics; tests of electroweak theory and searches for new phenomena using the D0 detector at the Fermilab Tevatron and the CMS detector at the CERN Large Hadron Collider; measurements of neutrino oscillations using the KAMLAND detector and the double Chooz detector; development of next-generation pixel detectors, new algorithms for calorimetry, and advanced methods in detector and physics simulation; research in new directions for electron-positron physics, neutrino physics, and antihydrogen physics. Bolton, Horton-Smith, Ivanov, Maravin.

Physics and other Science Education. Pedagogy; models of learning and transfer; problem solving; development of traditional and nontraditional learning experiences; technology in teaching and learning. Rebello, Sayre, Zollman.

View additional information about this department at **www.gradschoolshopper.com**

PITTSBURG STATE UNIVERSITY

DEPARTMENT OF PHYSICS

Pittsburg, Kansas 66762
http://www.pittstate.edu/physics

General University Information
President: Steve Scott
Dean of Graduate School: Pahan Kahol
University website: http://www.pittstate.edu
Control: Public
Setting: Rural
Total Faculty: 411
Total Graduate Faculty: 301
Total number of Students: 7,289
Total number of Graduate Students: 1,123

Department Information
Department Chairman: David M. Kuehn, Chair
Department Contact: Desirae Tyler, Administrative Specialist
   Total full-time faculty: 6
   Total number of full-time equivalent positions: 6
   Full-Time Graduate Students: 6
   First-Year Graduate Students: 3
   Female First-Year Students: 2

Department Address
1701 S. Broadway Street
Pittsburg, KS 66762
Phone: (620) 235-4391
Fax: (620) 235-4050
E-mail: dtyler@pittstate.edu
Website: http://www.pittstate.edu/physics

ADMISSIONS

Admission Contact Information
Address admission inquiries to: Pittsburg State University, Office of Admissions
Phone: (800) 854-7488
E-mail: psuadmit@pittstate.edu
Admissions website: http://www.pittstate.edu/admission

Application deadlines
Fall admission:
U.S. students: July 1   *Int'l. students*: June 1

Spring admission:
U.S. students: December 1   *Int'l. students*: October 15

Application fee
U.S. students: $35   *Int'l. students*: $60
There is not a set application deadline. It is recommended that the application be submitted at least a month before the semester begins.

Admissions information
For Fall of 2012:
   Number of applicants: 10
   Number admitted: 5
   Number enrolled: 4

Admission requirements
Bachelor's degree requirements: A bachelor's degree in physics is required.
Minimum undergraduate GPA: 2.5

GRE requirements
The GRE is not required.

Advanced GRE requirements
The Advanced GRE is not required.

TOEFL requirements
The TOEFL exam is required for students from non-English-speaking countries.
PBT score: 520
iBT score: 68

Other admissions information
Undergraduate preparation assumed: B.S. in physics or equivalent is required.

TUITION

Tuition year 2012–13:
Tuition for in-state residents
   Full-time students: $3,082 per semester
   Part-time students: $261 per credit
Tuition for out-of-state residents
   Full-time students: $7,259 per semester
   Part-time students: $609 per credit

Gorilla Advantage:Students from bordering counties of Missouri, Oklahoma, and Northwest Arkansas may qualify for Gorilla Advantage tuition rate, which is tuition equal to in-state tuition and fees. Beginning Fall 2013 Gorilla Edge, for students with a permanent residency in Arkansas, Missouri, Oklahoma, or Texas who are not a part of our Gorilla Advantage program, we are offering a reduced tuition plan.

Credit hours per semester to be considered full-time: 9
Deferred tuition plan: No
Health insurance: Available at the cost of 338.00 per year.
Other academic fees: Campus privilege fees.
Academic term: Semester
Number of first-year students who received full tuition waivers: 3
Number of first-year students who received partial tuition waivers: 1

Teaching Assistants, Research Assistants, and Fellowships

Number of first-year
  Teaching Assistants: 3
  Research Assistants: 1
Average stipend per academic year
  Teaching Assistant: $10,000
  Research Assistant: $8,500

FINANCIAL AID

Application deadlines
Fall admission:
U.S. students: April 1   *Int'l. students*: April 1
Spring admission:
U.S. students: April 1   *Int'l. students*: April 1

Loans
Loans are available for U.S. students.
Loans are available for international students.
GAPSFAS application required: No
FAFSA application required: Yes

For further information
Address financial aid inquiries to: Pittsburg State University, Department of Student Financial Aid, 1701 S. Broadway, Pittsburg, KS 66762.
Phone: (620) 235-4240
E-mail: finaid@pittstate.edu
Financial aid website: http://www.pittstate.edu/office/financial_aid/

HOUSING

Availability of on-campus housing
Single students: Yes
Married students: No

For further information
Address housing inquiries to: Pittsburg State University, Department of University Housing, 1701 S. Broadway, Pittsburg, KS 66762.
Phone: (620) 235-4245
E-mail: house@pittstate.edu
Housing aid website: http://www.pittstate.edu/office/housing/

Table A—Faculty, Enrollments, and Degrees Granted

Research Specialty	2012-13 Faculty	Enrollment Fall 2012		Number of Degrees Granted 2012-13 (2008-13)		
		Master's	Doctorate	Master's	Terminal Master's	Doctorate
Astronomy	2	–	–	–(4)	–	–
Atomic, Molecular, & Optical Physics	1	–	–	–(1)	–	–
Condensed Matter Physics	1	1	–	1(4)	–	–
Nuclear Physics	1	1	–	1(4)	–	–
Physics and other Science Education	–	–	–	–	–	–
Solid State Physics	1	–	–	–	–	–
Other	–	4	–	1(6)	–	–
Total	6	6	–	3(19)	–	–
Full-time Grad. Stud.	–	6	–	–	–	–
First-year Grad. Stud.	–	3	–	–	–	–

GRADUATE DEGREE REQUIREMENTS

Master's: A minimum of 32 credit hours or 30 credit hours with thesis are required. Students must have maintained a 3.0 GPA at the time of graduation. There is no residency requirement. Qualifying examination is required.
Thesis: Thesis may be written in absentia.

SPECIAL EQUIPMENT, FACILITIES, OR PROGRAMS

Modern laboratory facilities provide invaluable hands-on experience in front-line research projects: observational astronomy at astrophysical observatory, ultracapacitors, gamma spectroscopy, surface analysis for nanoscale systems, and materials science, including microscopy (scanning electron, atomic force, and scanning tunneling microspectroscopy), and x-ray scatterometry.

Table B—Separately Budgeted Research Expenditures by Source of Support

Source of Support	Departmental Research	Physics-related Research Outside Department
Federal government	$28,331	
State/local government		
Non-profit organizations		
Business and industry		
Other		
Total	$28,331	

Table C—Separately Budgeted Research Expenditures by Research Specialty

Research Specialty	No. of Grants	Expenditures ($)
Physics and other Science Education	1	$28,331
Total	1	$28,331

FACULTY

Professor
Blatchley, Charles C., Ph.D., Louisiana State University, 1984. *Physics and other Science Education, Plasma and Fusion.* Particle accelerator applications.
Kuehn, David M., Ph.D., New Mexico State University, 1990. *Astronomy, Computational Physics, Energy Sources & En-*

vironment, Planetary Science. Planetary astronomy; computational physics; renewable energy.

Associate Professor

Butler, Rebecca, Ph.D., Ohio State University, 2002. *Atomic, Molecular, & Optical Physics, Computational Physics.* Molecular spectroscopy.

Uran, Serif, Ph.D., Illinois Institute of Technology, 2000. *Condensed Matter Physics, Materials Science, Metallurgy, Nano Science and Technology.* Condensed matter physics.

Assistant Professor

Tayo, Benjamin O., Ph.D., Lehigh University, 2012. *Computational Physics, Nano Science and Technology, Solid State Physics.* Many-body theory and computational modeling of the electronic and optical properties of carbon nanotubes; impurity scattering of excitons in semiconducting nanotubes; quantum physics of nanotube DNA hybrids.

Lecturer

Scarborough, Kyla, M.S., Pittsburg State University, 2005. *Astronomy, Planetary Science.* Planetary physics.

DEPARTMENTAL RESEARCH SPECIALTIES AND STAFF

Experimental

Condensed Matter. Nanotechnology; optical and electronic properties of solids; 3D holography; magnetic thin films; oxidation of metals. Uran.

Gamma Ray. Gamma-ray spectrometry; cosmogenic, environmental, and accelerator nuclides; tribology; wear and corrosion measurements. Blatchley.

Molecular Spectroscopy. Butler.

Nano Science and Technology. Many-body effects in the electronic properties of quasi one-dimensional systems, such as single-walled carbon nanotubes; modeling of optical spectra of carbon nanotubes; DNA-nanotube interaction. Tayo.

Planetary Science. Vertical structure modeling in multiple-scattering atmospheres, especially of the outer planets; positional and photometric observations of asteroids and comets using the PSU Greenbush Observatory; astronomical instrument development and testing, especially imagers and spectrometers that use acousto-optic tunable filters at near-infrared wavelengths, including custom control software development. Kuehn, Scarborough.

> *View additional information about this department at*
> *www.gradschoolshopper.com*

UNIVERSITY OF KANSAS

DEPARTMENT OF PHYSICS AND ASTRONOMY

Lawrence, Kansas 66045
http://www.physics.ku.edu

General University Information

Chancellor: Bernadette Gray-Little
Dean of Graduate School: Thomas Heilke
University website: http://www.ku.edu
Control: Public
Setting: Suburban
Total Faculty: 2,663
Total Graduate Faculty: unknown
Total number of Students: 27,939
Total number of Graduate Students: 7,988

Department Information

Department Chairman: Hume A. Feldman, Chair
Department Contact: Teri Leahy, Graduate Coordinator
  Total full-time faculty: 26
  Total number of full-time equivalent positions: 26
  Full-Time Graduate Students: 49
  First-Year Graduate Students: 15
  Female First-Year Students: 1
  Total Post Doctorates: 10

Department Address

Malott Hall
1251 Wescoe Hall Dr, Room 1082
Lawrence, KS 66045

Phone: (785) 864-1225
Fax: (785) 864-5262
E-mail: physics@ku.edu
Website: http://www.physics.ku.edu

ADMISSIONS

Admission Contact Information

Address admission inquiries to: Graduate Admissions, University of Kansas, Department of Physics and Astronomy, Malott Hall, 1251 Wescoe Hall Drive, Room 1082, Lawrence, Kansas 66045
Phone: (785) 864-1225
E-mail: physics@ku.edu
Admissions website: http://www.graduate.ku.edu/apply

Application deadlines

Fall admission:

U.S. students: December 1	*Int'l. students*: December 1

Spring admission:

U.S. students: October 1	*Int'l. students*: October 1

Application fee

U.S. students: $55	*Int'l. students*: $65

*International applicants not in possession of a U.S. Visa should apply earlier.

Admissions information

For Fall of 2013:
Number of applicants: 81
Number admitted: 27
Number enrolled: 13

Admission requirements

Bachelor's degree requirements: A Bachelor's degree in physics, astronomy, or a related field is desired.
Minimum undergraduate GPA: 3.0

GRE requirements

The GRE is recommended.
GRE General Scores: Not required, but highly recommended. A scanned version of score report is acceptable for review. However, an official score report is required before an admission can be offered.

Advanced GRE requirements

The Advanced GRE is recommended.
The Advanced GRE is recommended for all applicants. The exam is not required for admission; however, a Ph.D. student who sends in an original copy of a Physics GRE score of 650 or higher before enrollment will be excused from the Department's undergraduate certification process. A scanned version of score report is acceptable for review. However, an official score report is required before an admission can be offered.

TOEFL requirements

The TOEFL exam is required for students from non-English-speaking countries.
iBT TOEFL Test Score Requirements: -Regular Admission: A minimum score of 20 in each section (Reading, Listening, Writing, Speaking) is required. -Regular Admission with GTA (Graduate Teaching Assistant): A minimum score of 20 in Reading, Listening, Writing AND a score of 24 in the Speaking section. IELTS Test Score Requirements: -Regular Admission: A minimum overall score of 6.0, with no individual section having a score below 5.5. -Regular Admission with GTA: A minimum score of 6.0 in Reading, Listening, and Writing AND a score of 8.0 in the Speaking section.

Other admissions information

Additional requirements: International student applicants must take a TOEFL test with a speaking component/speaking score in order to be considered for a Teaching Assistantship (TA) position. The KU administered TSE exam is recommended for those applying for a TA and who have not taken the TOEFL iBT exam.
Undergraduate preparation assumed: Mechanics (at the level of the textbook by Marion and Thornton), Electrodynamics (level of D. J. Griffiths), Quantum Mechanics (level of Liboff), Laboratory (level of Melissinos or Brophy); at least two courses in mathematics beyond elementary calculus.

TUITION

Tuition year 2012–13:
Tuition for in-state residents
Full-time students: $329.8 per credit
Part-time students: $329.8 per credit
Tuition for out-of-state residents
Full-time students: $771.55 per credit
Part-time students: $771.55 per credit

Fees include a $10/credit hour technology fee.
Credit hours per semester to be considered full-time: 9
Deferred tuition plan: No
Health insurance: Yes, 1,248.00.
Other academic fees: There are also some additional miscellaneous campus fees of approximately $412.20/semester.
Academic term: Semester
Number of first-year students who received full tuition waivers: 6

Teaching Assistants, Research Assistants, and Fellowships

Number of first-year
Teaching Assistants: 6
Research Assistants: 4
Fellowship students: 2
Average stipend per academic year
Teaching Assistant: $18,336
Research Assistant: $18,336
Fellowship student: $32,000

FINANCIAL AID

Application deadlines

Fall admission:
U.S. students: December 31   *Int'l. students*: December 31
Spring admission:
U.S. students: November 15   *Int'l. students*: November 15

Loans

Loans are not available for U.S. students.
Loans are not available for international students.
GAPSFAS application required: No
FAFSA application required: No

For further information

Address financial aid inquiries to: Graduate Admissions, University of Kansas, Department of Physics and Astronomy, Malott Hall, 1251 Wescoe Hall Drive, Room 1082, Lawrence, KS 66045.
Phone: (785) 864-1225
E-mail: physics@ku.edu
Financial aid website: http://www.physics.ku.edu/

HOUSING

Availability of on-campus housing

Single students: Yes
Married students: Yes

For further information

Address housing inquiries to: KU Dept. of Student Housing, 422 West 11th St., Ste. DSH, Lawrence, KS 66045-3312.
Phone: (785) 864-4560
E-mail: housing@ku.edu
Housing aid website: http://www.housing.ku.edu/

Table A—Faculty, Enrollments, and Degrees Granted

Research Specialty	2012-13 Faculty	Enrollment Fall 2012		Number of Degrees Granted 2012-13 (2007-13)		
		Mas-ter's	Doc-torate	Mas-ter's	Terminal Master's	Doc-torate
Astronomy	4	–	3	–	–(1)	–
Astrophysics	5	–	2	2(5)	–(1)	2(7)
Atmosphere, Space Physics, Cosmic Rays	2	2	4	–	–(4)	3(5)
Biophysics	2	–	4	–	–(1)	–
Condensed Matter Physics	5	–	10	–	–(4)	2(13)
High Energy Physics	4	5	10	1(3)	–(3)	2(11)
Nuclear Physics	2	–	7	–	–(1)	2(2)
Physics of Beams	1	2	1	1(1)	1(2)	–(1)
Theoretical Physics	3	2	5	1(1)	–	1(1)
Non-specialized	–	5	2	–	–	–
Other	3	–	–	–(3)	–(2)	–(2)
Total	**35**	**4**	**33**	**5(15)**	**1(19)**	**12(42)**
Full-time Grad. Stud.	–	13	48	–	–	–
First-year Grad. Stud.	–	6	11	–	–	–

GRADUATE DEGREE REQUIREMENTS

Master's: Thirty hours of advanced courses and at least two hours of Master's research with satisfactory progress; no foreign language requirement; better than a B average required and a general examination in physics required. 30 hours of resident study is required, but up to six of these may be transferred from another accredited university.

Doctorate: Thirty-three hours of advanced lecture courses; course work should average better than a B; students with a cumulative average grade less than B will be placed on probation; three full academic years of residency are required, two semesters normally consecutive and excluding summer session subsequent to the first year of graduate study must be spent at the University of Kansas; no foreign language requirement; demonstrated skill in computer programming related to the student's field of study is required; undergraduate certification by the graduate committee and a comprehensive exam are required; a dissertation showing the results of original research is required.

Other Degrees: Computational Physics and Astronomy (M.S.).
Thesis: Thesis may be written in absentia.

SPECIAL EQUIPMENT, FACILITIES, OR PROGRAMS

The nearby campus Computer Center has a small supercomputer system (a 64-processor SGI Origin 2000) and several high-end Compaq Alpha servers (UNIX and Open VMS). The campus computer network has high-speed redundant connections to the Internet, and the University is a member of the Internet II Development Consortium (consisting of more than 120 universities), which provides the latest network connection technology.

Condensed matter physics facilities include an advanced materials research lab, a quantum electronics lab, and a semiconductor laser optics lab. These labs are well equipped with thin film deposition systems, a new scanning electron microscope, a unique UHV multi-probe scanning microscopy system, an X-ray diffractometer, SQUID magnetometers, a 6-mK dilution refrigerator, microwave synthesizers and a vector network analyzer, a Nd:YAG laser, and an optical parametric oscillator. A clean room with photo- and electron beam lithography, as well as wafer processing tools is also available for micro- and nano-fabrication of solid state devices and circuits. Professionally staffed machine shop and "student" shop.

High-energy and nuclear physics groups utilize experimental facilities at various universities and national laboratories as part of collaborative experiments.

The Astrobiophysics Working Group sponsors collaborations among the departments of Physics and Astronomy, Ecology and Evolutionary Biology, Geology, and the Biodiversity Research Center. This group has a large allocation of supercomputer time at the NSF Teragrid.

The Department shares with San Diego State University access to a new 1.25-m reflecting telescope located at Mt. Laguna Observatory in southern California. This research-quality instrument is located at an excellent site and is equipped with a CCD imager and a variety of filters. Observing is done remotely from on-campus.

Table B—Separately Budgeted Research Expenditures by Source of Support

Source of Support	Departmental Research	Physics-related Research Outside Department
Federal government	$4,409,928.12	
State/local government		
Non-profit organizations		
Business and industry		
Other		
Total	**$4,409,928.12**	

Table C—Separately Budgeted Research Expenditures by Research Specialty

Research Specialty	No. of Grants	Expenditures ($)
Astronomy	4	$82,549.73
Astrophysics	2	$150,390.37
Atmosphere, Space Physics, Cosmic Rays	10	$428,731.16
Biophysics	5	$150,389.67
Condensed Matter Physics	14	$478,199.66
Energy Sources & Environment	4	$855,075.02
Nuclear Physics	3	$272,726.56
Particles and Fields	4	$246,653.41
High Energy Physics	18	$1,745,212.54
Total	**64**	**$4,409,928.12**

FACULTY

Professor

Anthony-Twarog, Barbara J., Ph.D., Yale University, 1981. *Astronomy, Astrophysics*. Stellar evolution in open star clusters; CCD and photoelectric photometry; globular clusters; high resolution stellar spectroscopy.

Baringer, Philip S., Ph.D., Indiana University, 1985. Associate Chair. Director of Undergraduate Studies. *High Energy Physics*. Experimental physics; elementary particle physics.

Bean, Alice L., Ph.D., Carnegie Mellon University, 1987. *High Energy Physics*. Experimental physics; elementary particle physics.

Besson, David Z., Ph.D., Rutgers University, 1986. *High Energy Physics*. Experimental physics; elementary particle physics.

Cravens, Thomas E., Ph.D., Harvard University, 1975. *Atmosphere, Space Physics, Cosmic Rays*. Experimental, theoretical physics; astrophysics; space physics; plasma physics.

Feldman, Hume A., Ph.D., Stony Brook University, 1989. Department Chair. *Astrophysics, Cosmology & String Theory*. Astrophysics and cosmology; computational physics; particle Astrophysics.

Han, Siyuan, Ph.D., Iowa State University, 1986. *Condensed Matter Physics.* Experimental condensed matter physics; physics and application of Josephson junctions and SQUIDs; mesoscopic physics; quantum computing.

Hawley, Steven A., Ph.D., University of California, Santa Cruz, 1977. Director of Engineering Physics. *Astronomy, Astrophysics.* Observational astronomy; spectrophotometry of H II regions and planetary nebulae; astrobiology; human spaceflight.

Marfatia, Danny, Ph.D., University of Wisconsin-Madison, 2001. *High Energy Physics, Particles and Fields.* Theoretical particle physics; particle astrophysics.

Medvedev, Mikhail V., Ph.D., University of California, San Diego, 1996. *Atmosphere, Space Physics, Cosmic Rays.* Theoretical astrophysics; space physics; plasma physics; nonlinear dynamics; astrobiology.

Melott, Adrian L., Ph.D., University of Texas, 1981. *Astrophysics, Geophysics, Other.* Astrobiophysics.

Murray, Michael J., Ph.D., University of Pittsburgh, 1989. *Nuclear Physics.* Experimental nuclear physics, Astrobiophysics.

Ralston, John P., Ph.D., University of Oregon, 1980. *Astrophysics, High Energy Physics, Particles and Fields.* Theoretical physics; elementary particle physics; particle astrophysics.

Sanders, Stephen J., Ph.D., Yale University, 1977. Graduate Student Advisor. *Nuclear Physics.* Experimental nuclear physics.

Shandarin, Sergei F., Ph.D., Moscow Physical Technical Institute, 1975. *Cosmology & String Theory.* Astrophysics and cosmology; nonlinear dynamics; computational physics.

Shi, Jicong, Ph.D., University of Houston, 1991. *Nonlinear Dynamics and Complex Systems, Physics of Beams.* Theoretical physics; nonlinear dynamics; beam dynamics; accelerator physics; computational physics.

Twarog, Bruce A., Ph.D., Yale University, 1980. *Astronomy, Astrophysics.* Stellar nucleosynthesis; chemical evolution of galaxies; stellar photometry; high resolution stellar spectroscopy.

Wu, Judy Z., Ph.D., University of Houston, 1993. *Condensed Matter Physics, Energy Sources & Environment.* Experimental condensed matter physics; fabrication, characterization, and application of thin films and nanowires.

Associate Professor

Fischer, Christopher J., Ph.D., University of Michigan, 2000. *Biophysics.* Biophysics, Astrophysics.

Rudnick, Gregory H., Ph.D., University of Arizona, 2001. Director of Graduate Studies. *Astronomy, Astrophysics.* Astronomy; galaxy evolution; galaxy formation.

Wilson, Graham W., Ph.D., University of Lancaster, 1989. *High Energy Physics.* Experimental physics; elementary particle physics.

Assistant Professor

Antonik, Matthew, Ph.D., University of Maine, 1994. *Biophysics.* Biophysics.

Chan, Wai-Lun, Ph.D., Brown University, 2007. *Condensed Matter Physics, Materials Science, Metallurgy.* Renewable energy.

Chiu, Hsin-Ying, Ph.D., California Institute of Technology, 2009. *Condensed Matter Physics.* Carbon electronics; carbon optoelectronics; graphene electro-mechanics.

Kong, Kyoungchul K.C., Ph.D., University of Florida, Gainesville, 2006. *High Energy Physics, Particles and Fields.* Theoretical physics; elementary particle physics.

Zhao, Hui, Ph.D., Northern Jiaotong University, 2000. *Condensed Matter Physics.* Experimental condensed matter physics.

Emeritus

Armstrong, Thomas P., Ph.D., University of Iowa, 1966. *Atmosphere, Space Physics, Cosmic Rays.* Experimental, theoretical physics; astrophysics; space physics; plasma physics.

Bearse, Robert C., Ph.D., Rice University, 1964. *Nuclear Physics.* Nuclear physics; materials control and accounting; nuclear safeguards; computer database applications.

Davis, Robin E. P., Ph.D., University of Oxford, 1962. *High Energy Physics.* Experimental physics; elementary particle physics.

Eagleman, Joe R., Ph.D., University of Missouri, 1963. *Other.* Atmospheric science.

Friauf, Robert J., Ph.D., University of Chicago, 1953. Experimental condensed-matter physics, diffusion and color centers, molecular dynamics and Monte Carlo simulations.

Kwak, Nowhan, Ph.D., Tufts University, 1962. *High Energy Physics.* Experimental physics; elementary particle physics.

McKay, Douglas W., Ph.D., Northwestern University, 1968. Theoretical physics; elementary particle physics; particle astrophysics.

Munczek, Herman J., Ph.D., University of Buenos Aires, 1958. *Particles and Fields.* Theoretical physics; elementary particle physics.

Sapp, Richard C., Ph.D., Ohio State University, 1955. *Condensed Matter Physics.* Experimental physics; solid state and low-temperature; low-temperature magnetism.

Shawl, Stephen J., Ph.D., University of Texas, Austin, 1972. *Astronomy.* Observational astronomy; stellar astronomy; polarization; globular clusters; astronomy education.

Wiseman, Gordon G., Ph.D., University of Kansas, 1950. *Condensed Matter Physics.* Experimental physics; solid state physics; dielectrics; ferroelectricity.

Wong, Kai-Wai, Ph.D., Northwestern University, 1962. *Condensed Matter Physics.* Theoretical physics; many-body theory; superconductivity; liquid helium.

DEPARTMENTAL RESEARCH SPECIALTIES AND STAFF

Theoretical

Astrobiophysics. Melott.

Astrophysics. Dark matter; γ-ray bursts; particle astrophysics. Feldman, Medvedev, Melott, Ralston.

Atmosphere, Space Physics, Cosmic Rays. Space probes; trapped particles; solar wind; radiation belts. Cravens, Medvedev.

Cosmology & String Theory. Large-scale structure. Feldman, Shandarin.

Particles and Fields. Symmetry properties and dynamics of elementary particles. Kong, Marfatia, Ralston.

Experimental

Astrobiophysics. Fischer, Murray.

Astrophysics. Stellar astronomy; nebular astrophysics; galaxy evolution; polarization. Anthony-Twarog, Besson, Feldman, Hawley, Melott, Ralston, Rudnick, Twarog.

Biophysics. Kinetics and thermodynamics of protein-protein and protein-nucleic acid interactions. Antonik, Fischer.

Condensed Matter Physics. Quantum tunneling and coherence, superconducting and single electron devices, high-temperature superconductivity, electronic structure, semiconductors. Chan, Chiu, Han, Wu, Zhao.

High Energy Physics. Study of proton-antiproton collisions with the D0 experiment at the Fermilab Tevatron; study of electron-positron annihilation with CLEO at CESR; astrophysics research with RICE in Antarctica; study of proton-proton collisions with the CMS experiment at the CERN LHC; research and development work on a future linear electron-positron collider. Baringer, Bean, Besson, Kong, Marfatia, Ralston, Wilson.

Nuclear Physics. Heavy-ion reactions at RHIC, and nuclear structure. Murray, Sanders.

UNIVERSITY OF KENTUCKY

DEPARTMENT OF PHYSICS AND ASTRONOMY

Lexington, Kentucky 40506-0055
http://www.pa.uky.edu/

General University Information
President: Eli Capilouto
Dean of Graduate School: Jeannine Blackwell
University website: http://www.uky.edu/
Control: Public
Setting: Urban
Total Faculty: 2,291
Total Graduate Faculty: 1,221
Total number of Students: 28,928
Total number of Graduate Students: 7,207

Department Information
Department Chairman: Sumit Das, Chair
Department Contact: Sumit Das, Chair
  Total full-time faculty: 31
  Total number of full-time equivalent positions: 31
  Full-Time Graduate Students: 68
  First-Year Graduate Students: 9
  Female First-Year Students: 2
  Total Post Doctorates: 12

Department Address
177 Chemistry-Physics Building
600 Rose Street
Lexington, KY 40506-0055
Phone: (859) 257-6722
Fax: (859) 323-2846
E-mail: das@pa.uky.edu
Website: http://www.pa.uky.edu/

ADMISSIONS

Admission Contact Information
Address admission inquiries to: Graduate Admissions Committee, Department of Physics and Astronomy, University of Kentucky, 177 Chemistry-Physics Building, 600 Rose Street, Lexington, KY 40506-0055
Phone: (859) 257-6722
E-mail: gradapp@pa.uky.edu
Admissions website: http://www.pa.uky.edu/grad

Application deadlines
Fall admission:
U.S. students: February 1   *Int'l. students*: February 1

Application fee
U.S. students: $65   *Int'l. students*: $75

Admissions information
For Fall of 2012:
  Number of applicants: 122
  Number admitted: 19
  Number enrolled: 9

Admission requirements
Bachelor's degree requirements: A Bachelor's degree in physics, astronomy, or a related field is required.

GRE requirements
The GRE is required.

The GRE General Test is required for application. There is no official minimum score requirement.

Advanced GRE requirements
The Advanced GRE is not required.
Although the Physics Subject GRE is not required for application, we recommend submitting scores if available.

TOEFL requirements
The TOEFL exam is required for students from non-English-speaking countries.
  PBT score: 550
  iBT score: 79
The University of Kentucky requires a minimum TOEFL score of 213 computer-based (550 paper-based, iBT 79) or a minimum IELTS score of 6.5 for all international students whose first language is not English. Submitted scores must be no more than two years old. International students who receive college degrees from US universities and universities in other designated English-speaking countries may be exempted from taking the TOEFL test. For example, if you obtain an M.S. degree from a US University you do not need the TOEFL.

Other admissions information
Undergraduate preparation assumed: Substantial variations in preparation can be accommodated. Generally, the Department recommends one semester of advanced undergraduate mechanics, two semesters of advanced undergraduate electricity and magnetism, two semesters of advanced undergraduate quantum physics, and at least one semester of laboratory at the junior and senior level.

TUITION

Tuition year 2012–13:
Tuition for in-state residents
  Full-time students: $4,980 per semester
Tuition for out-of-state residents
  Full-time students: $10,524 per semester
Nearly all admitted graduate students are offered a Teaching Assistantship or Fellowship, both of which include a stipend in addition to a full-tuition scholarship.
Credit hours per semester to be considered full-time: 9
Deferred tuition plan: No
Health insurance: Available at the cost of $1,646 per year.
Academic term: Semester
Number of first-year students who received full tuition waivers: 8

Teaching Assistants, Research Assistants, and Fellowships
Number of first-year
  Teaching Assistants: 7
  Fellowship students: 1
Average stipend per academic year
  Teaching Assistant: $17,500
  Research Assistant: $17,500
  Fellowship student: $21,000
In addition, summer support is provided.

FINANCIAL AID

Loans

Loans are available for U.S. students.
Loans are not available for international students.
GAPSFAS application required: No
FAFSA application required: No

For further information

Address financial aid inquiries to: Office of Student Financial Aid, 128 Funkhouser Building, University of Kentucky, Lexington, KY 40506-0054.
Phone: (859) 257-3172
Financial aid website: http://www.uky.edu/financialaid/

HOUSING

Availability of on-campus housing

Single students: Yes
Married students: Yes

For further information

Phone: (859) 257-3721
E-mail: ukapthousing@email.uky.edu
Housing aid website: http://www.uky.edu/Housing/graduate/

Table A—Faculty, Enrollments, and Degrees Granted

Research Specialty	2012-13 Faculty	Enrollment Fall 2012 Master's	Enrollment Fall 2012 Doctorate	Number of Degrees Granted 2012-13 (2008-13) Master's	Number of Degrees Granted 2012-13 (2008-13) Terminal Master's	Number of Degrees Granted 2012-13 (2008-13) Doctorate
Astronomy	7	–	5	1(8)	1(3)	4(6)
Atomic, Molecular, & Optical Physics	2	–	5	–	–	–(1)
Condensed Matter Physics	9	–	10	5(10)	–	4(10)
Nuclear Physics	7	–	11	2(11)	2(3)	–(8)
Particles and Fields	6	–	5	3(3)	–	1(5)
Non-specialized	–	–	32	–(1)	3(12)	1(1)
Total	31	–	68	11(33)	6(18)	10(31)
Full-time Grad. Stud.	–	–	68	–	–	–
First-year Grad. Stud.	–	–	9	–	–	–

GRADUATE DEGREE REQUIREMENTS

Master's: PLAN A: 24 semester hours of graduate credit with satisfactory performance and a thesis. PLAN B: Same as Plan A except six hours of course work at the advanced level is substituted for the thesis. Minimum GPA of 3.0 for both plans.

Doctorate: Minimum GPA of 3.0 in approved course program, as determined by student's advisory committee. Students must pass the GRE Physics Subject Test with a score at the 50th percentile or above and also pass 6 graduate core physics courses with GPA of 3.0 or higher. 36 credit hours and two-year full-time residency before and one-year full-time residency after qualification exam.

Thesis: Written thesis required for Ph.D. degree.

SPECIAL EQUIPMENT, FACILITIES, OR PROGRAMS

On Campus: 6.5 MV van de Graaff accelerator, Center for Advanced Materials, Liquid Helium Liquefier, MacAdam Student Observatory, Center for Computational Sciences. Users at ORNL, LANL, NIST, Fermilab, PSI, TRIUMF, Lund, HIGS, RHIC, NRAO, JLab/CEBAF.

Table B—Separately Budgeted Research Expenditures by Source of Support

Source of Support	Departmental Research	Physics-related Research Outside Department
Federal government	$3,295,721	$4,782,069
State/local government		
Non-profit organizations		
Business and industry		
Other		
Total	$3,295,721	$4,782,069

FACULTY

Professor

Brill, Joseph W., Ph.D., Stanford University, 1978. *Condensed Matter Physics*. Experimental condensed matter physics.

Cao, Gang, Ph.D., Temple University, 1992. Director, Center for Advanced Materials. *Condensed Matter Physics*. Experimental condensed matter physics.

Cavagnero, Michael, Ph.D., University of Chicago, 1983. Chair, Department of Physics and Astronomy. *Atomic, Molecular, & Optical Physics*. Theoretical atomic physics.

Das, Sumit R., Ph.D., University of Chicago, 1983. *Particles and Fields*. Theoretical particle physics; string theory.

DeLong, Lance E., Ph.D., University of California, San Diego, 1977. *Condensed Matter Physics*. Experimental condensed matter physics.

Draper, Terrence, Ph.D., University of California, Los Angeles, 1984. *Particles and Fields*. Theoretical particle physics.

Eides, Michael I., Ph.D., Leningrad State University, 1977. *Particles and Fields*. Theoretical particle physics.

Elitzur, Moshe, Ph.D., Weizmann Institute of Science, 1971. *Astrophysics*. Theoretical astrophysics.

Ferland, Gary J., Ph.D., University of Texas, 1978. *Astrophysics*. Theoretical astrophysics.

Gardner, Susan V., Ph.D., Massachusetts Institute of Technology, 1988. *Nuclear Physics, Particles and Fields*. Theoretical nuclear and particle physics.

Gorringe, Tim P., Ph.D., University of Birmingham, 1984. Director of Graduate Studies. *Nuclear Physics*. Experimental nuclear physics.

Korsch, Wolfgang, Ph.D., University of Marburg, 1990. *Nuclear Physics*. Experimental nuclear physics.

Kovash, Michael A., Ph.D., Ohio State University, 1978. *Nuclear Physics*. Experimental nuclear physics.

Li, Bing An, Ph.D., Chinese Academy of Sciences, 1968. *Particles and Fields*. Theoretical particle physics.

Liu, Keh-Fei, Ph.D., Stony Brook University, 1975. *Particles and Fields*. Theoretical nuclear and particle physics.

Martin, Nicholas L. S., Ph.D., University of Oxford, 1977. *Atomic, Molecular, & Optical Physics*. Experimental atomic and molecular physics.

Murthy, Ganpathy, Ph.D., Yale University, 1987. *Condensed Matter Physics*. Theoretical condensed matter physics.

Ng, Kwok-Wai, Ph.D., Iowa State University, 1986. Director of Undergraduate Studies. *Condensed Matter Physics*. Experimental condensed matter physics.

Shapere, Alfred D., Ph.D., University of California, Santa Barbara, 1988. *Particles and Fields*. Theoretical particle physics; string theory.

Shlosman, Isaac, Ph.D., University of Tel-Aviv, 1985. *Astrophysics*. Theoretical astrophysics.

Straley, Joseph P., Ph.D., Cornell University, 1970. *Condensed Matter Physics, Physics and other Science Education*. Theoretical condensed matter physics; physics education.

Troland, Thomas H., Ph.D., University of California, Berkeley, 1980. *Astronomy*. Observational astronomy.

Associate Professor

Crawford, Christopher, Ph.D., Massachusetts Institute of Technology, 2005. *Nuclear Physics*. Experimental nuclear physics.
Fatemi, Renee, Ph.D., University of Virginia, 2002. *Nuclear Physics*. Experimental nuclear physics.
Plaster, Brad, Ph.D., Massachusetts Institute of Technology, 2004. *Nuclear Physics*. Experimental nuclear physics.
Wilhelm, Ronald, Ph.D., Michigan State University, 1995. *Astronomy*. Observational astronomy.

Assistant Professor

Kaul, Ribhu, Ph.D., Duke University, 2006. *Condensed Matter Physics*. Theoretical condensed matter physics.
Kocevski, Dale, Ph.D., University of Hawaii, 2006. *Astronomy*. Observational astronomy.
Seo, Sung, Ph.D., Seoul National University, 2007. *Condensed Matter Physics*. Experimental condensed matter physics.
Strachan, Doug, Ph.D., University of Maryland, 2002. *Condensed Matter Physics*. Experimental condensed matter physics.
Yan, Renbin, Ph.D., University of California, Berkeley, 2007. *Astronomy*. Observational astronomy.

Emeritus

Christopher, John E., Ph.D., University of Virginia, 1967. *Physics and other Science Education*. Physics education.
MacAdam, Keith B., Ph.D., Harvard University, 1971. *Atomic, Molecular, & Optical Physics*. Experimental atomic and molecular physics.
MacKellar, Alan D., Ph.D., Texas A&M University, 1966. *Nuclear Physics*. Experimental nuclear physics.
McEllistrem, Marcus T., Ph.D., University of Wisconsin-Madison, 1956. *Nuclear Physics*. Experimental nuclear physics.

Research Assistant Professor

Rudnev, Vladimir, Ph.D., St-Petersburg State University, 2000. *Atomic, Molecular, & Optical Physics*. Theoretical atomic physics.
Tishchenko, Vladimir, Ph.D., Joint Institute for Nuclear Research, Dubna, 2003. *Nuclear Physics*. Experimental nuclear physics.

DEPARTMENTAL RESEARCH SPECIALTIES AND STAFF

Theoretical

Theoretical Astrophysics. Interstellar masers; radiation transport; cosmology; stellar dynamics; high-z radio-galaxies; active galactic nuclei; cataclysmic variable stars; accretion disks; red giant atmospheres. Elitzur, Ferland, Shlosman.

Theoretical Atomic Physics. Few-body states of atoms and molecules. Collective effects in atoms, molecules and ultracold gases. Cold and ultra-cold dipolar gases. Elementary collision processes in atomic and molecular physics. Cavagnero, Rudnev.

Theoretical Condensed Matter Physics. Structure and electronic properties of clusters, defects, and surfaces; catalysis; statistical mechanisms of phase transitions; Josephson networks; quantum dots; quantum Hall effect; tunneling; superconductivity; mesoscopic systems, quantum Monte-Carlo, exact diagonalization, quantum criticality, nano-physics. Kaul, Murthy, Straley.

Theoretical Nuclear Physics. Fundamental symmetries; P-, T-, and CP-violation, dark matter; dark energy. Gardner.

Theoretical Particle Physics. Lattice QCD calculations; quark models; string theory; black holes; cosmology. Das, Draper, Eides, Li, Liu, Shapere.

Experimental

Experimental Atomic Physics. Electron impact excitation and ionization of atoms. (e,2e) spectroscopy of autoionizing levels. Non-dipole photoelectron studies on autoionizing levels. Laser assisted electron scattering. MacAdam, Martin.

Experimental Condensed Matter Physics. Physical properties of new and novel materials; organic metals, magnetic oxides, heavy fermion systems, charge density wave systems, superconducting materials; scanning tunneling electron microscopy, patterned thin films. Brill, Cao, DeLong, Ng, Seo, Strachan.

Experimental Nuclear Physics. Tests of fundamental symmetries, neutron beta-decay, muon capture, muon anomalous magnetic moment g-2, neutron electric dipole moment, radium electric dipole moment, nucleon electromagnetic structure, nucleon spin structure. Crawford, Fatemi, Gorringe, Korsch, Kovash, McEllistrem, Plaster, Tishchenko.

Observational Astronomy. Observational radio astronomy; interstellar material and star formation; interstellar magnetic fields; kinematics of ancient stars; active galactic nuclei; galaxy formation. Kocevski, Troland, Wilhelm, Yan.

View additional information about this department at
www.gradschoolshopper.com

LOUISIANA STATE UNIVERSITY

DEPARTMENT OF PHYSICS AND ASTRONOMY

Baton Rouge, Louisiana 70803
http://www.phys.lsu.edu

General University Information
President: F. King Alexander
Dean of Graduate School: Gary Byerly
University website: http://www.lsu.edu
Control: Public

Setting: Urban
Total Faculty: 1,035
Total Graduate Faculty: 1,250
Total number of Students: 24,631
Total number of Graduate Students: 4,525

Department Information

Department Chairman: Michael L. Cherry, Chair
Department Contact: Ilya Vekhter, Professor
  Total full-time faculty: 52
  Full-Time Graduate Students: 108
  First-Year Graduate Students: 21
  Female First-Year Students: 7
  Total Post Doctorates: 33

Department Address

202 Nicholson Hall
Baton Rouge, LA 70803
Phone: (225) 578-2261
E-mail: vekhter@phys.lsu.edu
Website: http://www.phys.lsu.edu

ADMISSIONS

Admission Contact Information

Address admission inquiries to: Assistantship Committee Chair,
  Department of Physics and Astronomy
Phone: (225) 892-2261
E-mail: vekhter@phys.lsu.edu
Admissions website: http://www.phys.lsu.edu

Application deadlines

Fall admission:
U.S. students: January 15   *Int'l. students*: January 15
Spring admission:
U.S. students: October 15   *Int'l. students*: October 15

Application fee

U.S. students: $75   *Int'l. students*: $75
Fee waived if application sent directly to department.

Admissions information

For Fall of 2012:
  Number of applicants: 220
  Number admitted: 56
  Number enrolled: 22

Admission requirements

Bachelor's degree requirements: Bachelor's degree in physics
  or a related field is required.
Minimum undergraduate GPA: 3.0

GRE requirements

The GRE is required.
  Mean GRE score range (25th–75th percentile): 1320
Minimum quantitative + verbal GRE score for admission 1150.

Advanced GRE requirements

The Advanced GRE is recommended.

TOEFL requirements

The TOEFL exam is required for students from non-English-
  speaking countries.
  PBT score: 600
  iBT score: 100

Other admissions information

Additional requirements: The Physics GRE is preferred but not
  required.
Undergraduate preparation assumed: Gasiorowicz, Quantum
  Physics; Griffiths, Introduction to Electrodynamics; Zemansky, Heat and Thermodynamics; Marion, Classical Dynamics
  of Particles and Systems.

TUITION

Tuition year 2013-14:
Tuition for in-state residents
  Full-time students: $9,499 per semester
Tuition for out-of-state residents
  Full-time students: $23,528 per semester
Students on a full-time graduate assistantship will receive a full
  exemption in tuition.
Credit hours per semester to be considered full-time: 9
Deferred tuition plan: No
Health insurance: Available
Other academic fees: None.
Academic term: Semester
Number of first-year students who received full tuition waivers: 21

Teaching Assistants, Research Assistants, and Fellowships

Number of first-year
  Teaching Assistants: 17
  Research Assistants: 2
  Fellowship students: 2
Average stipend per academic year
  Teaching Assistant: $22,000
  Research Assistant: $22,000
  Fellowship student: $28,000
Essentially all offers of admission are accompanied by an as-
  sistantship or fellowship offer.

FINANCIAL AID

Application deadlines

Fall admission:
U.S. students: January 15   *Int'l. students*: January 15
Spring admission:
U.S. students: October 15   *Int'l. students*: October 15

Loans

Loans are available for U.S. students.
Loans are not available for international students.
GAPSFAS application required: No
FAFSA application required: No

For further information

Address financial aid inquiries to: Assistantship Committee
  Chairman.
Phone: (225) 578-2261
E-mail: vekhter@phys.lsu.edu
Financial aid website: http://www.lsu.edu/financialaid/

HOUSING

Availability of on-campus housing

Single students: Yes
Married students: Yes

For further information

Address housing inquiries to: Department of Residential Life.
Phone: (225) 578-8663
E-mail: reslife@lsu.edu
Housing aid website: http://appl003.lsu.edu/slas/reslifeweb.nsf/
  index

Table A—Faculty, Enrollments, and Degrees Granted

Research Specialty	2012-13 Faculty	Enrollment Fall 2012 Master's	Enrollment Fall 2012 Doctorate	Number of Degrees Granted 2011–12 Master's	Number of Degrees Granted 2011–12 Terminal Master's	Number of Degrees Granted 2011–12 Doctorate
Astronomy	5	–	5	1(-)	–	2(-)
Astrophysics	5	–	8	2(-)	–	–
Atomic, Molecular, & Optical Physics	6	–	9	–	–	2(-)
Condensed Matter Physics	15	–	27	2(-)	–	1(-)
Medical, Health Physics	3	23	2	–	13(-)	–
Nuclear Physics	3	–	4	–	–	–
Particles and Fields	2	–	1	–	1(-)	–
Relativity & Gravitation	8	–	8	–	–	2(-)
Total	47	23	77	5(-)	14(-)	7(-)
Full-time Grad. Stud.	–	23	77	–	–	–
First-year Grad. Stud.	–	6	17	–	–	–

GRADUATE DEGREE REQUIREMENTS

Master's: The minimum course requirement is 36 semester hours without a thesis or 24 hours with a thesis. Minimum GPA is a "B" average. There is no minimum residence time or foreign language requirement. Thesis candidates have an oral thesis defense and non-thesis candidates must pass a written comprehensive examination.

Doctorate: Twelve hours of advanced courses beyond the core courses are required. The minimum GPA is a "B" average and the minimum time for residence is one year. Examinations are: (a) the qualifying exam, (b) the general examination, and (c) the dissertation defense. A substantial portion of the dissertation work must be published in an appropriate refereed professional journal.

Other Degrees: Master of Natural Science degree: requirements are similar to the M.S. degree except a broader scope of scientific courses (including science education) is emphasized. Thesis may be written in absentia. Master of Science in Medical Physics and Health Physics: The concentration in medical physics requires 29 hours of coursework, six hours (two semesters) of clinical training at Mary Bird Perkins Cancer center, and a minimum of six hours of thesis research. The concentration in health physics requires 33 hours of coursework and a minimum of six hours of thesis research. The degree requires a thesis of publishable quality with an oral defense. Minimum GPA is a "B" average. Program is accredited by the Commission on Accreditation of Medical Physics Educational Programs, Inc. (CAMPEP).

SPECIAL EQUIPMENT, FACILITIES, OR PROGRAMS

LSU holds the distinction of being one of 25 universities in the nation to hold both land-grant and sea-grant status; is a member of LaSPACE, the Louisiana consortium in NASA's Space Grant program; and is designated as a doctoral/research extensive institution by the Carnegie Foundation. The Hearne Institute for Theoretical Physics carries out interdisciplinary research in relativity and quantum theory.

Extensive in-house facilities include the Center for Advanced Microstructures and Devices (CAMD), a 1.5 GeV electron synchrotron light source currently providing x-rays up to 50 keV for microfabrication and for solid-state, surface science, condensed-matter, and materials science research. Other condensed matter instrumentation includes a low-temperature (5 mK) dilution refrigerator-high magnetic field (17.5 Tesla) facility used for studying high-temperature semiconductors and for materials research and extensive crystal growing and characterization facilities.

The NSF's Laser Interferometer Gravitational-wave Observatory (LIGO) is located 24 miles from campus. Astrophysical observation at LIGO is under way, as is advanced detector technology development.

Nuclear physics experiments are conducted at Oak Ridge, TRIUMF, and the National Superconducting Cyclotron Laboratory at Michigan State. The high-energy physics group participates in particle physics experiments at FermiLab (BooNE), the Sudbury Neutrino Observatory (SNO), KAMIOKANDE, and at JPARC/KEK in Japan and is involved in the deep underground experiments at DUSEL.

Members of the Astronomy, Astrophysics, and Space Science groups are presently conducting observations at Kitt Peak, Cerro Tololo, Lowell Observatory, the Hubble Space Telescope, SPITZER, and the Swift and Fermi satellite instruments. These groups are designing x-ray and cosmic ray experiments for long-duration balloon and space missions and are involved in the CALET high-energy electron experiment to be launched on the Space Station in 2014. Ultra-high-energy cosmic rays are being measured in Argentina (AUGER).

The Highland Road Park Observatory (HRPO) is located about 8 miles from campus and includes two fully computer-controlled reflecting optical telescopes with 20" and 16" diameter primary mirrors, plus a small 10' diameter radio telescope. The optical telescopes are equipped with CCD camera, filter wheel, and spectrograph, and the system is capable of imaging magnitude 19 stars with an exposure of a couple of minutes. The HRPO is used for teaching our undergraduate observational astronomy course, graduate observational techniques course, and for student research projects and public outreach.

Computational facilities include a large cluster of Unix workstations and Power PCs through which students and faculty gain access to a variety of high-performance computing facilities. These facilities include a 5,440-core, 50 TFlops Supercomputer and several smaller clusters operated by the Louisiana Optical Network Initiative. In addition, LSU's Center for Computation and Technology (CCT) and High Performance Computing Center (HPC) operate a 1440-core 15 TFlop machine. All of these facilities are used extensively for numerical calculations of general relativity, analysis of high-energy neutrino and gamma ray astronomy data, experimental calculations and simulations of star collisions, gravitational waves, calculations of strongly correlated materials, and simulations of biological materials.

Medical physics facilities at Mary Bird Perkins Cancer Center include Varian Clinac electron and x-ray beams, TomoTherapy HI-ART, BrainLab Novalis, GE PET/CT, HDR brachytherapy, comprehensive dosimetry laboratories, and Elekta Synergy and multi-vendor treatment planning laboratories.

Table B—Separately Budgeted Research Expenditures by Source of Support

Source of Support	Departmental Research	Physics-related Research Outside Department
Federal government	$6,403,899	$64,857
State/local government	$973,276	$394,782
Non-profit organizations		
Business and industry		
Other	$1,342,851	$55,922
Total	$8,720,026	$515,561

Table C—Separately Budgeted Research Expenditures by Research Specialty

Research Specialty	No. of Grants	Expenditures ($)
Astronomy	26	$1,028,844
Astrophysics	28	$1,934,570
Atomic, Molecular, & Optical Physics	6	$352,142
Condensed Matter Physics	19	$2,132,685
Medical, Health Physics	10	$614,072
Nuclear Physics	6	$651,126
Particles and Fields	6	$708,807
Relativity & Gravitation	21	$1,297,780
Total	**122**	**$8,720,026**

FACULTY

Professor

Adams, Philip W., Ph.D., Rutgers University, 1986. *Condensed Matter Physics, Low Temperature Physics, Materials Science, Metallurgy.* Transport properties in two-dimensional systems; transport properties of superconducting films.

Blackmon, Jeffrey, Ph.D., University of North Carolina, Chapel Hill, 1994. *Nuclear Physics.* Nuclear astrophysics; neutrino physics.

Browne, Dana, Ph.D., Stanford University, 1981. Associate Chair. *Condensed Matter Physics.* Phase transitions and self organization in non-equilibrium systems; electronic structure of materials.

Cherry, Michael L., Ph.D., University of Chicago, 1978. Department Chair. *Astrophysics.* Gamma ray astronomy; cosmic rays; high-energy astrophysics.

Clayton, Geoffrey C., Ph.D., University of Toronto, 1983. *Astronomy.* Astronomy and astrophysics; interstellar and extragalactic dust; circumstellar dust; R Coronae Borealis stars.

DiTusa, John, Ph.D., Cornell University, 1992. *Condensed Matter Physics.* Experimental condensed-matter physics.

Dowling, Jonathan P., Ph.D., University of Colorado, 1988. Hearne Research Chair; Co-Director, Hearne Institute for Theoretical Physics. *Atomic, Molecular, & Optical Physics, Quantum Foundations.* Quantum optics; quantum science and technologies; photonic crystal theory.

Draayer, Jerry P., Ph.D., University of Iowa, 1968. *Nuclear Physics.* Theoretical nuclear structure.

Frank, Juhan, Ph.D., University of Cambridge, 1978. *Astronomy.* Accretion in close binaries and active galactic nuclei.

Gaarde, Mette, Ph.D., University of Copenhagen, Denmark, 1997. *Atomic, Molecular, & Optical Physics.* Theory of atomic and optical physics; generation of high-order harmonics and attosecond pulses.

Giaime, Joseph, Ph.D., Massachusetts Institute of Technology, 1995. *Astrophysics, Relativity & Gravitation.* Gravitational-wave physics at the Laser Interferometer Gravitational-Wave Observatory (LIGO); very-low-noise instrumentation.

González, Gabriela, Ph.D., Syracuse University, 1995. *Astrophysics, Relativity & Gravitation.* Gravitational wave physics at LIGO (Laser Interferometer Gravitational-wave Observatory: instrumental commissioning, detector characterization, and data analysis.

Guzik, T. Gregory, Ph.D., University of Chicago, 1980. *Astrophysics.* Solar flares; particle interactions; accelerator experiments; cosmic rays.

Jarrell, Mark, Ph.D., University of California, Santa Barbara, 1987. *Computational Physics, Condensed Matter Physics.* Massively parallel simulations of strongly correlated electronic systems.

Jin, Rongying, Ph.D., Swiss Federal Institute of Technology, Zurich, 1997. *Condensed Matter Physics.* The development of novel complex materials with intriguing physical properties, such as new phases that exist on the edge of instabilities (un-

conventional superconductivity, quantum critical phenomena, heavy Fermion behavior, thermoelectricity, etc.).

Johnson, Warren W., Ph.D., Rutgers University, 1974. *Relativity & Gravitation.* Gravitational radiation detectors; Josephson devices, parametric transducers, and quantum nondemolition.

Kurtz, Richard L., Ph.D., Yale University, 1983. Associate Dean; Director, CAMD Synchrotron. *Condensed Matter Physics.* Surface science; experimental condensed-matter physics.

Matthews, James M., Ph.D., University of Wisconsin-Madison, 1984. *Astrophysics, Particles and Fields.* Experimental cosmic-ray research at extreme energy; Pierre Auger Observatory.

Newhauser, Wayne, Ph.D., University of Wisconsin-Madison, 1995. Director, Medical Physics Program. *Applied Physics, Medical, Health Physics.* Medical physics.

O'Connell, Robert F., Ph.D., University of Notre Dame, 1962. *Atomic, Molecular, & Optical Physics, Quantum Foundations, Relativity & Gravitation.* Dissipative and fluctuation phenomena in quantum physics, decoherence, and entanglement; general relativity.

Plummer, W. Ward, Ph.D., Cornell University, 1968. *Condensed Matter Physics.* Materials science; condensed-matter physics with emphasis on broken symmetry and reduced dimensionality.

Pullin, Jorge, Ph.D., Instituto Balseiro, Bariloche, Argentina, 1989. Hearne Research Chair in Theoretical Physics; Co-Director, Hearne Institute of Theoretical Physics. *Relativity & Gravitation.* Quantum gravity; classical, quantum mechanical and its astrophysical implications.

Rau, A. Ravi P., Ph.D., University of Chicago, 1970. *Atomic, Molecular, & Optical Physics, Quantum Foundations.* Theoretical atomic physics.

Schaefer, Bradley, Ph.D., Massachusetts Institute of Technology, 1983. *Astrophysics.* Gamma-ray bursts; novae; supernovae.

Schafer, Ken, Ph.D., University of Arizona, 1989. Graduate Advisor. *Atomic, Molecular, & Optical Physics.* Theory of high-intensity ultra-fast laser-matter interactions.

Seidel, Edward, Ph.D., Yale University, 1988. Director, Mathematical and Physical Sciences, National Science Foundation. *Computational Physics, Relativity & Gravitation.* General relativity, numerical relativity, black holes, and gravitational waves; relativistic astrophysics; computational physics, high-performance computing, and scientific and grid computing.

Sprunger, Phillip, Ph.D., University of Pennsylvania, 1993. *Condensed Matter Physics.* Condensed-matter physics.

Stadler, Shane, Ph.D., Tulane University, 1998. *Condensed Matter Physics.* Magnetocaloric systems; half-metallic spintronic systems; x-ray absorption spectroscopy.

Tohline, Joel E., Ph.D., University of California, Santa Cruz, 1978. Director, Center for Computation and Technology. *Astronomy, Computational Physics.* Astrophysics; star formation; galaxy dynamics.

Young, David P., Ph.D., Florida State University, 1998. Undergraduate Advisor. *Condensed Matter Physics.* Materials science.

Zhang, Jiandi, Ph.D., Syracuse University, 1994. *Condensed Matter Physics.* Condensed-matter physics; exploring novel properties of complex materials, such as transition-metal oxides manifested by broken symmetry and reduced dimensionality (e.g., at the surface, interfaces, and artificially structured multilayers).

Associate Professor

Hynes, Robert I., Ph.D., Open University, England, 1999. *Astronomy.* Multiwavelength observational astronomy of accreting objects.

Jia, Guang, Ph.D., The Ohio State University, 2006. *Medical, Health Physics.* Medical physics, radiology.

Kutter, Thomas, Ph.D., University of Heidelberg, Germany (University of Chicago), 1999. *Particles and Fields.* Experimental neutrino physics.

Lee, Hwang, Ph.D., Texas A&M University, 1998. *Atomic, Molecular, & Optical Physics, Quantum Foundations.* Quantum optics; quantum information science.

Matthews, Kenneth L., Ph.D., University of Chicago, 1997. *Applied Physics, Medical, Health Physics.* Radiological imaging; nuclear medical imaging.

Moreno, Juana, Ph.D., Rutgers University, 1997. *Computational Physics, Condensed Matter Physics.* Computational approaches to strongly correlated electron systems.

Sheehy, Daniel, Ph.D., University of Illinois at Urbana-Champaign, 2001. *Atomic, Molecular, & Optical Physics, Condensed Matter Physics.* High-temperature superconductivity; degenerate bosonic and fermionic atomic gases; vortices in condensed-matter and cold-atom contexts.

Stacy, J. Gregory, Ph.D., University of Maryland, 1980. *Astrophysics.* Gamma-ray detector development; galactic structure and star formation regions.

Vekhter, Ilya, Ph.D., Brown University, 1998. Chair, Graduate Admissions Committee. *Condensed Matter Physics.* Theoretical condensed-matter physics; unconventional superconductivity; strongly correlated electron systems.

Assistant Professor

Ajullo, ivan, Ph.D., University of Valencia, 2009. *Cosmology & String Theory, Relativity & Gravitation.* Theoretical cosmology, black hole physics, loop quantum gravity and quantum field theory.

Corbitt, Thomas, Ph.D., Massachusetts Institute of Technology, 2008. *Relativity & Gravitation.* Gravitational physics; LIGO.

Deibel, Catherine, Ph.D., Yale University, 2008. *Nuclear Physics.* Experimental nuclear physics.

Shikhaliev, Polad, Ph.D., Ioffe Physico-Technical Institute, 1998. *Applied Physics, Medical, Health Physics.* X-ray and CT imaging; photon counting/energy resolving x-ray detectors; applications to cancer and heart disease.

Singh, Parampreet, Pune, 2004. *Relativity & Gravitation.* Theoretical relativity and quantum gravity.

Tzanov, Martin M., University of Pittsburgh, 2005. *Particles and Fields.* Experimental neutrino physics.

Wilde, Mark, Ph.D., University of Southern California, 2008. *Computer Science, Quantum Foundations.* Quantum information theory, quantum computing.

Emeritus

Chan, Lai-Him, Ph.D., Harvard University, 1966. *Particles and Fields.* Theory of elementary particles and fields.

Hamilton, William O., Ph.D., Stanford University, 1963. *Relativity & Gravitation.* Gravitational radiation; cryogenics; infrared detectors.

Hogstrom, Kenneth R., Ph.D., Rice University, 1977. *Applied Physics, Medical, Health Physics.* Electron beam therapy; image-guided radiotherapy; x-ray capture therapy.

Landolt, Arlo U., Ph.D., Indiana University, 1962. *Astronomy.* Photometry; variable stars; eclipsing binaries.

Metcalf, William J., Ph.D., California Institute of Technology, 1974. *Particles and Fields.* Experimental high-energy physics.

Wefel, John, Ph.D., Washington University, St. Louis, 1971. Director, Louisiana Space Grant and NASA EPSCoR. *Astrophysics.* High-energy astrophysics; space science; LaSPACE.

Zganjar, Edward F., Ph.D., Vanderbilt University, 1966. *Nuclear Physics.* Heavy-ion reactions; nuclei far from stability; nuclear deformation.

Instructor

Grocholski, Aaron, Ph.D., University of Florida, 2006. *Astronomy.*

McElgin, William, Ph.D., University of Chicago, 2007. *Particles and Fields, Theoretical Physics.* String theory, conformal field theory, black hole physics.

Rupnik, Dubravka, Ph.D., Louisiana State University, 1994. *Nuclear Physics.*

Other

Diener, Peter, Ph.D., University of Copenhagen, 1997. *Computational Physics, Relativity & Gravitation.* Numerical relativity, astrophysical applications.

DEPARTMENTAL RESEARCH SPECIALTIES AND STAFF

Theoretical

Astrophysics. Star formation; galaxy dynamics; cataclysmic variables; white dwarfs, neutron stars, and black holes. Ajullo, Diener, Frank, Pullin, Schaefer, Singh, Tohline.

Atomic, Molecular, & Optical Physics. Electron, atom, and ion molecule scattering; intense magnetic fields; variational principles; ultra-short-pulse laser-atom interactions; dissipation and fluctuation in quantum physics; quantum optics and quantum computing. Dowling, Gaarde, Lee, O'Connell, Rau, Schafer, Wilde.

Condensed Matter Physics. Collective phenomena: magnetism, superconductivity, Bose condensation, and quantum critical phenomena; studies of novel materials, band structure, and model calculations. Browne, Jarrell, Moreno, Sheehy, Vekhter.

High Energy Physics. Experimental neutrino physics; high-energy cosmic rays. Chan, Kutter, James Matthews, McElgin, Tzanov.

Medical, Health Physics. Radiation transport; dose calculations; microdosimetry; radiation biology; medical imaging; cancer diagnosis and therapy. Hogstrom, Jia, Kenneth Matthews, Newhauser, Shikhaliev.

Nuclear Physics. Nuclear structure. Draayer.

Relativity & Gravitation. General relativity; numerical relativity; black hole formation; gravitational wave detection; gravitational wave theory; numerical analysis simulation and visualization; scientific computing for relativity; quantum gravity. Diener, O'Connell, Pullin, Seidel, Singh.

Experimental

Astronomy. Gamma-ray bursts; supernovae; x-ray binaries; dust; novae; photometry; stellar abundances. Clayton, Grocholski, Hynes, Landolt, Schaefer.

Astrophysics. Cosmic rays; neutrinos; gamma rays; high-energy nucleus–nucleus interactions; satellite, balloon, accelerator, air shower, and underground experiments. Cherry, Guzik, James Matthews, Stacy, Wefel.

Condensed Matter Physics. Electrons in metals and phase transitions; optical properties of solids; thermal conductivity; surface science; electron spectroscopies; tunneling microscopes; growing and characterizing novel materials. Adams, DiTusa, Jin, Kurtz, Plummer, Sprunger, Stadler, Young, Zhang.

Low Temperature Physics. Gravitational radiation; cryogenics; superconductivity; mesoscopic physics. Adams, Corbitt, DiTusa, Giaime, González, Hamilton, Jin, Johnson, Plummer, Sprunger, Young, Zhang.

Medical, Health Physics. Helical tomotherapy; image-guided radiotherapy; cardiovascular imaging; x-ray photon counting imaging; radioisotope imaging; synchrotron x-ray radiotherapy. Hogstrom, Kenneth Matthews, Newhauser, Shikhaliev.

Nuclear Physics. Heavy-ion reactions; nuclei far from stability; radioactive beams; nuclear astrophysics. Blackmon, Deibel, Zganjar.

Particles and Fields. Neutrino astronomy; neutrino oscillations; high-energy cosmic rays; experiments at Fermilab, the Sudbury Neutrino Observatory, Pierre Auger Observatory, JPARC, and KEK. Kutter, James Matthews, Tzanov.

Relativity & Gravitation. Search for gravitational waves with LIGO and development of advanced detectors. Corbitt, Giaime, González, Hamilton, Johnson.

View additional information about this department at
www.gradschoolshopper.com

LOUISIANA   TECH   UNIVERSITY

DEPARTMENT OF PHYSICS

Ruston, Louisiana 71272
http://www.phys.latech.edu

General University Information
President: Dr. Leslie Guice
Dean of Graduate School: Dr. Sheryl Shoemaker
University website: http://www.latech.edu
Control: Public
Setting: Suburban
Total Faculty: 369
Total Graduate Faculty: 321
Total number of Students: 11,360
Total number of Graduate Students: 2,386

Department Information
Department Chairman: Lee Sawyer, Head
Department Contact: Dr. Lee Sawyer, Director of Chemistry & Physics
  Total full-time faculty: 11
  Total number of full-time equivalent positions: 12
  Full-Time Graduate Students: 12
  First-Year Graduate Students: 1
  Female First-Year Students: 1
  Total Post Doctorates: 5

Department Address
P.O. Box 10348
600 W. Arizona Ave, Carson Taylor Hall 316
Ruston, LA 71272
Phone: (318) 257-4911
Fax: (318) 257-3823
E-mail: sawyer@phys.latech.edu
Website: http://www.phys.latech.edu

ADMISSIONS

Admission Contact Information
Address admission inquiries to: Graduate School, Louisiana Tech University, 1642 Wyly Tower, Railroad Avenue, P. O. Box 7923, Ruston, LA 71272
Phone: (318) 257-2924
E-mail: gschool@latech.edu
Admissions website: https://app.applyyourself.com/?id=latech-g

Application deadlines
Fall admission:
U.S. students: August 1   *Int'l. students*: June 1

Spring admission:
U.S. students: February 1   *Int'l. students*: December 1

Application fee
U.S. students: $40   *Int'l. students*: $40
Application deadline for Fall quarter admission is August 1 for domestic students and June 1 for international students.

Admissions information
For Fall of 2012:
  Number of applicants: 6
  Number admitted: 6
  Number enrolled: 3

Admission requirements
Bachelor's degree requirements: Bachelor's degree in Physics, Engineering Physics, or closely related engineering, science, or mathematics degree.
Minimum undergraduate GPA: 2.5

GRE requirements
The GRE is required.
No minimum GRE score is specified.

Advanced GRE requirements
The Advanced GRE is not required.

TOEFL requirements
The TOEFL exam is required for students from non-English-speaking countries.
  PBT score: 550
  iBT score: 80
International students have the option to attend the ELS Language Center located on the Louisiana Tech campus in Ruston, Louisiana. By enrolling in and successfully completing an ELS Level 112 intensive course, international students can meet the English language proficiency requirement.

Other admissions information
Undergraduate preparation assumed: Mechanics, Electromagnetism, Thermodynamics, Optics, Modern Physics and Quantum Mechanics, and Calculus through Ordinary Differential Equations. Students may be admitted with deficiencies in one or more areas, which must be made up before starting graduate courses.

TUITION

Tuition year 2013-14:

Tuition for in-state residents
Full-time students: $1,863 per quarter
Part-time students: $841 per quarter

Tuition for out-of-state residents
Full-time students: $3,895 per quarter
Part-time students: $841 per quarter

Credit hours per semester to be considered full-time: 6
Deferred tuition plan: Yes
Health insurance: Available.
Academic term: Quarter

Teaching Assistants, Research Assistants, and Fellowships

Number of first-year
Teaching Assistants: 1
Research Assistants: 1
Fellowship students: 2

Average stipend per academic year
Teaching Assistant: $14,400
Research Assistant: $20,000
Fellowship student: $30,000

All incoming PhD students in Fall, 2012, were supported.

FINANCIAL AID

Loans

Loans are available for U.S. students.
Loans are not available for international students.
GAPSFAS application required: Yes
FAFSA application required: Yes

For further information

Address financial aid inquiries to: Louisiana Tech University, College of Engineering and Science, Research and Graduate Studies, P.O. Box 10348, Ruston, LA 71272.
Phone: (318) 257-4314
E-mail: gradengr@coes.latech.edu
Financial aid website: http://coes.latech.edu/grad-programs/financial-aid.php

HOUSING

Availability of on-campus housing

Single students: Yes
Married students: Yes

For further information

Address housing inquiries to: Department of Residential Life, Louisiana Tech University, P.O. Box 3174 TS, Ruston, LA 71272.
Phone: (318) 257-4917
E-mail: housing@latech.edu
Housing aid website: http://www.latech.edu/students/residential-life/index.shtml

Table A—Faculty, Enrollments, and Degrees Granted

Research Specialty	12 Faculty	Enrollment 2013 Master's	Enrollment 2013 Doctorate	Number of Degrees Granted 2012-13 (2010-13) Master's	Number of Degrees Granted 2012-13 (2010-13) Terminal Master's	Number of Degrees Granted 2012-13 (2010-13) Doctorate
Masters in Applied Physics	12	3	–	4(11)	2(4)	–
PhD in Computational Analysis & Modeling	11	–	3	–	–	1(1)
PhD in Engineering Physics	11	–	15	–	–	2(3)
Total	12	3	18	4(11)	2(4)	3(4)
Full-time Grad. Stud.	–	3	18	–	–	–
First-year Grad. Stud.	–	1	4	–	–	–

GRADUATE DEGREE REQUIREMENTS

Master's: M.S. Applied Physics: Twelve semester hours of core courses including Mathematical Methods and 3 of the following: Electromagnetic Theory, Solid State Physics, Applied Nuclear & Particle Physics, Theoretical Mechanics, Quantum Mechanics, and Statistical Mechanics. Non-thesis Option: Additional 21 semester hours of directed elective courses plus 3 semester hours Research & Reporting course. Thesis Option: Additional 12 semester hours of directed elective courses plus 6 semester hours of Research & Thesis course culminating in a Masters thesis.

Doctorate: Ph.D. Engineering: Engineering Physics Track: Core courses include Mathematical Methods for Scientists and Engineers, Statistics for Engineering and Science, Numerical Solutions to PDEs, Electromagnetic Theory, Solid State Physics, Theoretical Mechanics, Quantum Mechanics, and Statistical Mechanics (24 semester hours total). Students pass written and qualifying exams. Additional 24 semester hours of doctoral seminar and directed study courses. Eighteen semester hours of research and dissertation culminating a doctoral dissertation. Ph.D. in Computational Analysis & Modeling, with research emphasis in Physics. See website at http://coes.latech.edu/cam/index.php for details.

Thesis: Dissertation required for Ph.D. Thesis and non-thesis options available for Master's.

SPECIAL EQUIPMENT, FACILITIES, OR PROGRAMS

Physics faculty and students work closely with several research centers on campus, including the Institute for Micromanufacturing (IfM), the Center for Applied Physics Studies (CAPS), the Center for Biomedical Engineering and Rehablitation Science (CBERS), the

Center for Secure Cyberspace (CSC), and the Louisiana Optical Network Initiative (LONI). IfM resources include nano-assembly, biotechnology, sensor, and characterization laboratories. See http://www.latech.edu/ifm/equipment.shtml for details. CAPS resources include detector and electronics development laboratories for high energy and nuclear physics experiments, electronics testing facilities, and nanopulse characterization equipment. LONI provides 12 high-performance computing clusters that are interconnected via the LONI fiber backbone and dedicated to furthering the efforts of researchers throughout Louisiana.

Table B—Separately Budgeted Research Expenditures by Source of Support

Source of Support	Departmental Research	Physics-related Research Outside Department
Federal government	$8,143,639	$1,400,000
State/local government	$1,908,346	
Non-profit organizations		
Business and industry	$101,719	
Other		
Total	**$10,153,704**	**$1,400,000**

FACULTY

Chair Professor

Lvov, Yuri M., Ph.D., Moscow State University, 1979. T. Pipes Eminent Endowed Chair on Micro and Nano Systems. *Applied Physics, Biophysics, Chemical Physics, Engineering Physics/Science, Nano Science and Technology*. Nanofabrication of ordered multilayers by alternate adsorption of polyions, nanoparticles, and proteins.

Professor

Greenwood, Zeno D., Ph.D., University of South Carolina, 1978. Nuclear Safety Officer. *High Energy Physics, Nuclear Physics, Particles and Fields*. Collider-based Experimental High Energy Physics, Grid Computing for Particle Physics Experiments, Top Quark Production at High Energy Colliders, Neutrino Physics.

Johnston, Kathleen, Ph.D., University of Houston, 1991. Physics Program Chair. *Nuclear Physics, Physics and other Science Education*. Experimental Nuclear Physics, Experimental Particle Physics, Physics Education.

Ramachandran, Bala, Ph.D., Kansas State University, 1987. Associate Dean for Research, College of Engineering & Science. *Chemical Physics, Computational Physics*. Computational study of metal oxide electrode coatings for lithium ion batteries, structure and magnetic properties of metal and metal oxide clusters, and chemical catalysis on metal and metal oxide surfaces.

Sawyer, Lee, Ph.D., Florida State University, 1991. Director, Chemistry & Physics. *High Energy Physics, History & Philosophy of Physics/Science, Nonlinear Dynamics and Complex Systems, Particles and Fields, Physics and other Science Education*. Collider-based Experimental High Energy Physics, Experimental Tests of Quantum Chromodynamics, Jet Production at High-Energy colliders, Nonlinear Dynamical Systems, Development of New Physics instructional models.

Simicevic, Neven, Ph.D., University of Zagreb, 1990. Director, Center for Applied Physics Studies. *Applied Physics, Biophysics, Computational Physics, Engineering Physics/Science, Nuclear Physics*. Nucleon structure, Experimental nuclear physics, Computations of Interactions of Radiation with Biological and Geological Materials.

Associate Professor

Bishop, Thomas C., Ph.D., University of Illinois, 1996. *Biophysics, Chemical Physics, Computational Physics*. Theoretical and Computational Molecular Biology, Molecular Modeling and Molecular Dynamics Simulations of Proteins and DNA. Multiscale Model of DNA, Nucleosomes and Chromatin.

Derosa, Pedro A., Ph.D., National University of Cordoba, 1997. *Chemical Physics, Computational Physics, Nano Science and Technology*. Charge and Molecular Transport in Nanostructures, Computational Nanosystems.

Wells, Steve P., Ph.D., Indiana University, 1993. *Nuclear Engineering, Particles and Fields, Theoretical Physics*. Nucleon Structure, Weak Interaction, Parity Violation, Two Photon Exchange Physics.

Wobisch, Markus, Ph.D., RWTH Aachen University, 2000. *High Energy Physics, Particles and Fields, Theoretical Physics*. Collider-based experimental high energy physics, experimental tests of Quantum Chromodynamics, jet production at high energy colliders, calculations of jet observables at hadron colliders.

Assistant Professor

Genov, Dentcho A., Ph.D., Purdue University, 2005. *Applied Physics, Computational Physics, Electrical Engineering, Electromagnetism, Theoretical Physics*. Electromagnetic Properties of Nano-structured Complex Media Including: Metamaterials and Negative Index Media, Electric and Magnetic Plasmons, Electromagnetic Cloaking.

Research Assistant Professor

Murray, Erica, Ph.D., Northwestern University, 1999. *Applied Physics, Energy Sources & Environment, Engineering Physics/Science, Materials Science, Metallurgy*. Electroceramic Materials, Gas Sensors and Solid Oxide Fuel Cells, Fostering university and industry research partnerships.

Lecturer

Shaw, John A., Ph.D., College of William and Mary, 1993. *Astronomy, Astrophysics, Atomic, Molecular, & Optical Physics, Nonlinear Dynamics and Complex Systems*. Closed-orbit theory calculations of photoabsorption spectra of atoms in external electric and magnetic fields, Atomic and Radiative physics, Variable Star Astronomy.

DEPARTMENTAL RESEARCH SPECIALTIES AND STAFF

Theoretical

Computational Chemistry. Computational studies of structure-energy relationships, thermodynamic properties and reactivity of molecules, liquids, solids, and surfaces using wave function methods, density functional theory, molecular dynamics, and Monte-Carlo methods. Bishop, Derosa, Ramachandran.

Computational Physics. Multiphysics multiscale computational simulations of interactions of electromagnetic fields with matter spanning length scales from subnanometer to several meters and time scales from femtoseconds to several hours. Bishop, Derosa, Genov, Simicevic.

Nano Science and Technology. Computational and experimental studies of properties and interactions of materials and structures at subnanometer to micron scales. Derosa, Genov, Lvov, Murray, Ramachandran.

Experimental

Astronomy. Variable star observations. Shaw.

High Energy Physics. Experimental collaborations include the D0 experiment at the Fermilab Tevatron collider and the ATLAS experiment at the CERN Large Hadron Collider. Detector development for future experiments. Top quark and QCD studies at hadron colliders. Development of distributed computing for high-energy physics experiments. Greenwood, Sawyer, Wobisch.

Medical, Health Physics. Collaborations with the LA Tech Center for Biomedical Research & Rehabilitation Science and area medical centers on medical imaging, radiation effects on biological samples, dosimetry, modeling of radiation in tissue. Greenwood, Sawyer, Simicevic.

Nano Science and Technology. Computational and experimental studies of properties and interactions of materials and structures at subnanometer to micron scales. Derosa, Genov, Lvov, Murray, Ramachandran.

Nuclear Physics. Experimental studies of nucleon structure, parity violation, and weak interactions at the Jefferson National Accelerator Facility in Virginia. Development of high speed electronics for nuclear experiments. Detector development for future experiments. Johnston, Simicevic, Wells.

View additional information about this department at
www.gradschoolshopper.com

UNIVERSITY OF MAINE

DEPARTMENT OF PHYSICS AND ASTRONOMY

Orono, Maine 04469-5709
http://www.physics.umaine.edu/

General University Information
President: Paul W. Ferguson
Dean of Graduate School: Dan H. Sandweiss
University website: http://www.umaine.edu
Control: Public
Setting: Suburban
Total Faculty: 825
Total Graduate Faculty: 736
Total number of Students: 10,901
Total number of Graduate Students: 1,656

Department Information
Department Chairman: David J. Batuski, Chair
Department Contact: David E. Clark, Lecturer
  Total full-time faculty: 14
  Total number of full-time equivalent positions: 12
  Full-Time Graduate Students: 35
  First-Year Graduate Students: 5
  Female First-Year Students: 2
  Total Post Doctorates: 5

Department Address
5709 Bennett Hall
Orono, ME 04469-5709
Phone: (207) 581-1040
Fax: (207) 581-3410
E-mail: physics@maine.edu
Website: http://www.physics.umaine.edu/

ADMISSIONS

Admission Contact Information
Address admission inquiries to: Dr. James McClymer, Department of Physics and Astronomy, Bennett Hall
Phone: (207) 581-1034

E-mail: JamesMcClymer@umit.maine.edu
Admissions website: http://www.umaine.edu/graduate/admissions/

Application deadlines
Fall admission:
U.S. students: September 1   *Int'l. students*: September 1

Application fee
U.S. students: $50
Application review begins in mid-February and continues until programs are filled. Early application is recommended if financial assistance is required.

Admissions information
For Fall of 2013:
  Number of applicants: 29
  Number admitted: 7
  Number enrolled: 7

Admission requirements
Bachelor's degree requirements: Bachelor's degree in physics is normally required.

GRE requirements
The GRE is required.

Advanced GRE requirements
The Advanced GRE is required.

TOEFL requirements
The TOEFL exam is required for students from non-English-speaking countries.
  PBT score: 600
  iBT score: 100

Other admissions information
Additional requirements: No minimum score for admission is required.
Undergraduate preparation assumed: Modem Physics: Beiser, Concepts of Modern Physics; Mechanics: Fowles and Cassiday, Analytical Mechanics; Electricity and Magnetism: Griffiths, Introduction to Electrodynamics; Mathematics: Paul, Differential Equations for Mathematics, Science, and Engineering.

TUITION

Tuition year 2012–13:

Tuition for in-state residents
  Full-time students: $418 per credit
  Part-time students: $418 per credit
Tuition for out-of-state residents
  Full-time students: $1,284 per credit
  Part-time students: $1,284 per credit
Credit hours per semester to be considered full-time: 6
Deferred tuition plan: Yes
Health insurance: Available at the cost of $2,500 per year.
Other academic fees: $1,200–3,400/year (activity fee, communication fee, unified fee, recreation center fee)
Academic term: Semester
Number of first-year students who received full tuition waivers: 5

Teaching Assistants, Research Assistants, and Fellowships

Number of first-year
  Teaching Assistants: 5
Average stipend per academic year
  Teaching Assistant: $15,300
  Research Assistant: $23,000

FINANCIAL AID

Loans

Loans are available for U.S. students.
Loans are available for international students.
GAPSFAS application required: No
FAFSA application required: No

For further information

Address financial aid inquiries to: Dr. James McClymer, Department of Physics and Astronomy, Bennett Hall.
Phone: (207)581-1034
E-mail: James.McClymer@umit.maine.edu
Financial aid website: http://www.umaine.edu/stuaid/applying-for-aid/

HOUSING

Availability of on-campus housing

Single students: Yes
Married students: Yes

For further information

Address housing inquiries to: Housing Services, Suite 101, 5734 Hilltop Commons, Orono, Maine 04469-5734.
Phone: (207) 581-4580
Housing aid website: http://umaine.edu/housing

Table A—Faculty, Enrollments, and Degrees Granted

Research Specialty	2012–13 Faculty	Enrollment Fall 2012		Number of Degrees Granted 2012–13 (2008-13)		
		Master's	Doctorate	Master's	Terminal Master's	Doctorate
Astronomy	2	–	–	–	–	–
Astrophysics	2	1	8	-(1)	–	1(2)
Biophysics	3	2	4	1(-)	-(1)	-(2)
Chemical Physics	4	–	1	–	–	–
Condensed Matter Physics	4	–	3	–	–	1(-)
Energy Sources & Environment	1	–	1	-(2)	1(-)	-(1)
Engineering Physics/Science	3	3	–	1(-)	2(3)	–
Fluids, Rheology	1	–	–	–	–	–
Geophysics	1	–	–	–	–	1(2)
Marine Science/ Oceanography	1	–	–	–	–	-(1)
Materials Science, Metallurgy	4	–	–	-(1)	-(2)	–
Nano Science and Technology	4	–	2	-(1)	-(2)	-(1)
Nuclear Physics	1	–	–	–	–	–
Physics and other Science Education	4	–	8	-(1)	-(1)	-(4)
Polymer Physics/Science	1	–	–	–	–	–
Relativity & Gravitation	1	–	–	–	–	–
Statistical & Thermal Physics	2	–	–	–	–	-(1)
Non-specialized	–	–	3	–	–	–
Total	–	6	30	2(8)	3(12)	2(17)
Full-time Grad. Stud.	–	4	26	–	–	–
First-year Grad. Stud.	–	1	4	–	–	–

GRADUATE DEGREE REQUIREMENTS

Master's: Master's: 30 graduate (semester) credits, with 24 devoted to courses in physics and allied fields; courses must be passed with a minimum grade of "C," and no more than six hours of a "C" grade may be applied toward the degree; at least 12 hours of coursework must be taken while a full-time student in residence. There are no foreign language or computer language requirements. No comprehensive or qualifying examination is required. A thesis is required.

Doctorate: Doctorate: 30 graduate (semester) credits in courses required; courses must be passed with a minimum grade of "C," and no more than six hours of a "C" grade may be applied toward the degree. Residence requirement satisfied by registering for a full program of study for two consecutive years following the baccalaureate, or for one year following the award of a Master's degree; no foreign language required, no computer language required; comprehensive examination required; thesis required.

Other Degrees: Other programs: through cooperative efforts of faculty in the Departments of Physics, Chemistry, Earth Sciences, Electrical and Computer Engineering, and Biochemistry, students may work for a M.S. or Ph.D. degree with concentration in applied physics, materials science, quaternary studies, and biophysics. M.S. in engineering (engineering physics) requirements are the same as for the M.S. in physics except that nine hours of the 30 must be selected from engineering courses and a non-thesis option is available and 36 course hours are required.

Thesis: Thesis may be written in absentia with permission.

SPECIAL EQUIPMENT, FACILITIES, OR PROGRAMS

The Laboratory for Surface Science and Technology (LASST) unites researchers from the Departments of Chemistry, Physics, Electrical and Computer Engineering, and Chemical and Biological Engineering in many projects spanning aspects of surface and interface science, thin films, sensors, microsystems, and nanotechnology. Current facilities include thin film synthesis, electron and optical spectroscopies, scanning probe microscopies, x-ray and electron diffraction, focused ion beam-scanning electron microscopy, fluorescence microscopy, device fabrication (Class-1000 clean room with photolithography, metallization, wet and dry etch, PECVD, sputtering, mask generation, and packaging), and sensor testing (gas delivery systems, electrical and microwave test equipment, and data acquisition/integrated electronic test suites).

The Physics Education Research Laboratory has several facilities and equipment for conducting research on the learning and teaching of physics: a classroom dedicated to curricular activities based on physics education research (PER) containing computer-equipped work areas for 24 people, including education software, video equipment, and items for hands-on inquiry-based activities; dedicated clinical interview space to ensure the anonymity and privacy of students participating in our research work (as required by our institutional review board for testing with human subjects); digital video cameras to record individual interviews and classroom interactions; and individual computers to help with video analysis, documentation, data management, communication, and other relevant activities.

Astrophysics: A Linux/PC workstation network is dedicated to research in galactic dynamics and radio and optical observational astronomy.

The Radon Measurement Laboratory contains a liquid scintillation counter, 3 HP Ge detectors, x-ray fluorescence, Wrenn detectors, and a portable NaI spectrometer interfaced to a Digidart portable multichannel analyzer.

Biophysics and Optics: Three laboratories include a super-resolution localization microscopy facility and four F-PALM microscopes, image processing computer cluster, tunable femtosecond pulsed Ti:Sapphire laser and optical parametric oscillator (OPO), cell culture facilities, polymerase chain reaction (PCR) thermal cycler and other equipment for molecular biology, confocal and two-photon laser-scanning microscopes, fluorescence correlation and cross-correlation microscope, fluorimeter, spectrophotometer, Krypton Argon and Argon ion lasers, numerous diode lasers spanning visible wavelengths from 400-700 nm, and optical tweezer.

Table B—Separately Budgeted Research Expenditures by Source of Support

Source of Support	Departmental Research	Physics-related Research Outside Department
Federal government	$1,825,720	$2,538,417
State/local government	$60,000	$70,000
Non-profit organizations	$12,000	$80,746
Business and industry Other	$2,111	$62,815
Total	$1,899,831	$2,751,978

Table C—Separately Budgeted Research Expenditures by Research Specialty

Research Specialty	No. of Grants	Expenditures ($)
Astrophysics	2	$15,000
Condensed Matter Physics	1	$12,000
Energy Sources & Environment	1	$5,000
Physics and other Science Education	7	$2,952,414
Surface Physics	8	$1,667,395
Total	19	$4,651,809

FACULTY

Professor

Astumian, R. D., Ph.D., University of Texas, Arlington, 1983. *Biophysics, Condensed Matter Physics, Nano Science and Technology.* Theoretical and experimental condensed-matter physics; biophysics of molecular motors and pumps.

Batuski, David J., Ph.D., New Mexico State University, 1986. *Astronomy, Astrophysics.* Observational cosmology; large-scale structure in the universe; radio sources in galaxy clusters.

Comins, Neil F., Ph.D., University College, Cardiff, 1978. *Astronomy, Astrophysics, Cosmology & String Theory, Physics and other Science Education, Relativity & Gravitation.* Observational and theoretical astrophysics; galactic evolution and stability; stellar systems; general relativity; astronomy education.

Hess, C. Thomas, Ph.D., Ohio University, 1967. *Energy Sources & Environment, Medical, Health Physics, Nuclear Physics.* Environmental nuclear physics; health physics; radioactivity studies.

Lad, Robert J., Ph.D., Cornell University, 1986. *Condensed Matter Physics, Nano Science and Technology.* Surface physics; thin films; sensor technology; materials science; ceramics; electronic materials; photovoltaics; material characterization.

McKay, Susan R., Ph.D., Massachusetts Institute of Technology, 1987. *Nonlinear Dynamics and Complex Systems.* Theoretical condensed-matter physics; nonlinear systems and transitions to chaos; phase transitions and critical phenomena; spin glasses; amorphous magnetism; quenched disorder; pattern formation; systems far from equilibrium; applications of network theory.

Unertl, William N., Ph.D., University of Wisconsin-Madison, 1973. *Condensed Matter Physics, Materials Science, Metallurgy, Nano Science and Technology.* Surface physics; surface analysis techniques; materials science; tribology; paper surface science.

Associate Professor

Hess, Samuel T., Ph.D., Cornell University, 2002. *Biophysics.* Experimental and theoretical biophysics; super-resolution fluorescence microscopy and spectroscopy; function and lateral organization of biomembranes; influenza virus infection; single-molecule fluorescence photophysics.

McClymer, James P., Ph.D., University of Delaware, 1986. *Optics, Other.* Digital imaging and light scattering from equilibrium and nonequilibrium transitions in liquid crystals and complex fluids.

Mountcastle, Donald B., Ph.D., University of Virginia, 1971. *Biophysics, Physics and other Science Education, Statistical & Thermal Physics.* Physics education; physics-mathematics interface; curriculum development.

Thompson, John R., Ph.D., Brown University, 1998. *Physics and other Science Education.* Physics education: research on teaching and learning, curriculum development and assessment, use of mathematics in physics, and teacher knowledge of student thinking.

Wittmann, Michael C., Ph.D., University of Maryland, 1998. *Physics and other Science Education.* Physics education: learning theory development, curriculum development and evaluation, use of mathematics in physics, and teacher knowledge of student thinking.

Assistant Professor

Meulenberg, Robert W., Ph.D., University of California, Santa Barbara, 2002. *Condensed Matter Physics, Materials Science, Metallurgy, Nano Science and Technology.* Experimental condensed-matter physics: electronic structure of nanoscale materials; surface and interfacial physics of nanostructures; magnetic materials; applications of synchrotron radiation to materials science.

Stetzer, MacKenzie R., Ph.D., University of Pennsylvania, 2000. *Physics and other Science Education.* Physics education: research on teaching and learning, curriculum development and assessment, analog electronics, and laboratory courses.

Emeritus

Kleban, Peter H., Ph.D., Brandeis University, 1970. *Condensed Matter Physics, Statistical & Thermal Physics.* Statistical mechanics; phase transitions and kinetics on surfaces; conformal field theory; instrumentation; electron and ion spectrometry.

DEPARTMENTAL RESEARCH SPECIALTIES AND STAFF

Theoretical

Biophysics. Membrane phase transitions; numerical simulations of optical systems; cooperative interactions. Astumian, Samuel Hess, Mountcastle.

Nano Science and Technology. Energy exchange processes; phase transitions; stepped surfaces; adhesion and friction. Astumian, Lad, Meulenberg, Unertl.

Physics and other Science Education. Learning theories; cognitive development in physics; metacognition and epistemology. Wittmann.

Relativity & Gravitation. Gravitational radiation detection data analysis; black hole theory; computer models of galactic dynamics. Comins.

Statistical & Thermal Physics. Theory of phase transitions; conformal field theory; population dynamics; spin glasses; amorphous magnetism; nonlinear systems; systems far from equilibrium; pattern formation. Kleban, Mountcastle.

Experimental

Astronomy. Identifying and remediating student misconceptions about the cosmos: graduate students observe at the Very Large Array in New Mexico, the Steward Observatory in Arizona, and the European Southern Observatory and Cerro Tololo Inter-American Observatory in Chile. Batuski, Comins.

Astrophysics. Simulations of spiral galaxy dynamics; analysis of large-scale structure in the universe. Batuski, Comins.

Biophysics. Ultra-high-resolution fluorescence microscopy; cell membrane biophysics; single-molecule photophysics; molecular motors and pumps. Astumian, Samuel Hess.

Engineering Physics/Science. Use of microprocessors and microcomputers for data acquisition; development of new analytical techniques for surface analysis; thin film development; sensor technology; heat transfer; nanotribology. Lad, McClymer, Meulenberg, Unertl.

Materials Science, Metallurgy. Electronic structure and optical properties of nanoparticles, magnetic materials, and doped and alloy materials. Lad, Meulenberg, Unertl.

Nano Science and Technology. Crystal structure and reactivity; adsorption at surfaces; surface modification; surface structure of nanoscale materials surface spectroscopy; electronic properties; phase transitions; liquid metal and polymer surfaces; ion-beam modification; model membranes; thin-film synthesis; friction and adhesion; mechanical properties of surfaces; sensors; tribology. Astumian, Lad, Meulenberg, Unertl.

Nuclear Physics. Studies of radon in homes, schools, and ground water; nuclear fallout; site characterization; radioactive contamination field studies; Lead-210 dating of sediment cores. C. Hess.

Physics and other Science Education. Empirical studies of conceptual understanding; development and assessment of instructional materials; identification of specific student difficulties and productive resources while learning physics concepts; teacher knowledge of student thinking about physics and physical science. Physics topics include thermal physics (in physics and in engineering courses), mechanics, energy, electronics, and wave physics. Data come from free-response written items, surveys, individual and group interviews, and classroom video. Additional emphasis is on the use and understanding of mathematics in upper-division physics and teaching and learning at the mathematics-physics interface. Populations under study include introductory, advanced, and general education students, and pre-service and in-service K-12 teachers. Comins, Mountcastle, Stetzer, Thompson, Wittmann.

Polymer Physics/Science. Metal polymer adhesion; surface structure; mechanical properties; wetting and spreading phenomena. Unertl.

View additional information about this department at
www.gradschoolshopper.com

JOHNS HOPKINS UNIVERSITY

THE HENRY A. ROWLAND DEPARTMENT OF PHYSICS AND ASTRONOMY

Baltimore, Maryland 21218
http://physics-astronomy.jhu.edu/

General University Information

President: Ronald J. Daniels
Dean of Graduate School: Katherine S. Newman
University website: http://www.jhu.edu/

Control: Private
Setting: Urban
Total Faculty: 540
Total Graduate Faculty: Not separated.

Total number of Students: 7,047
Total number of Graduate Students: 1,981

Department Information
Department Chairman: Daniel H. Reich, Chair
Department Contact: Carm King, Academic Affairs
  Administrator
Total full-time faculty: 29
Total number of full-time equivalent positions: 29
Full-Time Graduate Students: 116
First-Year Graduate Students: 17
Female First-Year Students: 3
Total Post Doctorates: 26

Department Address
3400 N. Charles Street
366 Bloomberg Center
Baltimore, MD 21218
Phone: (410) 516-7344
Fax: (410) 516-7239
E-mail: admissions@pha.jhu.edu
Website: http://physics-astronomy.jhu.edu/

ADMISSIONS

Admission Contact Information
Address admission inquiries to: The Henry A. Rowland Department of Physics and Astronomy, 366 Bloomberg Center, 3400 N. Charles Street, Baltimore, MD 21218
Phone: (410) 516-7344
E-mail: admissions@pha.jhu.edu
Admissions website: http://grad.jhu.edu/admissions/

Application deadlines
Fall admission:
U.S. students: December 15   *Int'l. students*: December 15

Application fee
U.S. students: $75   *Int'l. students*: $75

Admissions information
For Fall of 2013:
  Number of applicants: 238
  Number admitted: 49
  Number enrolled: 15

Admission requirements
Bachelor's degree requirements: Bachelor's degree in physics, astrophysics, or astronomy is required.

GRE requirements
The GRE is required.
No minimum score is set.

Advanced GRE requirements
The Advanced GRE is required.
No minimum score is set.

TOEFL requirements
The TOEFL exam is required for students from non-English-speaking countries.
  PBT score: 600
  iBT score: 100

Other admissions information
Additional requirements: Letters of recommendation, undergraduate transcripts, statement of purpose, and GRE scores are considered equally in granting admission. Students from non-English-speaking countries are required to demonstrate proficiency in English via the TOEFL examination; minimum score required is 600.

TUITION
Tuition year 2013-14:
  Full-time students: $45,470 annual
Deferred tuition plan: No
Health insurance: Yes, 1,848.00.
Other academic fees: $500 one-time matriculation fee for first-year graduate students.
Academic term: Semester
Number of first-year students who received full tuition waivers: 15

Teaching Assistants, Research Assistants, and Fellowships
Number of first-year
  Teaching Assistants: 11
  Fellowship students: 4
Average stipend per academic year
  Teaching Assistant: $23,000
  Research Assistant: $23,000
  Fellowship student: $23,000
Stipend is for 9 months.

FINANCIAL AID

Application deadlines
Fall admission:
U.S. students: April 1

Loans
Loans are available for U.S. students.
Loans are not available for international students.
GAPSFAS application required: No
FAFSA application required: Yes

For further information
Address financial aid inquiries to: Office of Student Financial Services, 146 Garland Hall, 3400 N. Charles Street, Baltimore, MD 21218.
Phone: (410) 516-8028
E-mail: fin_aid@jhu.edu
Financial aid website: http://www.jhu.edu/finaid/

HOUSING

Availability of on-campus housing
  Single students: No
  Married students: No

For further information
Address housing inquiries to: Housing Office, Wolman Hall.
Phone: (410) 516-8282
E-mail: incoming@hd.jhu.edu
Housing aid website: http://www.jhu.edu/hds/

Table A—Faculty, Enrollments, and Degrees Granted

Research Specialty	2012-13 Faculty	Enrollment Fall 2012 Master's	Enrollment Fall 2012 Doctorate	Number of Degrees Granted 2012-13 (2008–13) Master's	Number of Degrees Granted 2012-13 (2008–13) Terminal Master's	Number of Degrees Granted 2012-13 (2008–13) Doctorate
Astrophysics	12	–	45	5(22)	–(1)	10(26)
Condensed Matter Physics	8	–	42	5(27)	1(3)	5(23)
Particles and Fields	10	–	29	7(19)	–(1)	3(13)
Plasma and Fusion	–	–	–	–	–(1)	–
Total	30	–	116	17(68)	1(6)	17(62)
Full-time Grad. Stud.	–	–	116	–	–	–
First-year Grad. Stud.	–	–	17	–	–	–

GRADUATE DEGREE REQUIREMENTS

Doctorate: The principal goal of graduate study in physics and astronomy is to train the student to conduct original research. The first year of our graduate program emphasizes research and facilitates the transition between the classwork that characterizes the undergraduate education to the research that ultimately constitutes the bulk of the graduate-school experience. All entering graduate students are assigned an adviser who works closely with them during their first year. This first-year adviser meets regularly with the student to determine courses of study, familiarize them with the department, and help them find research opportunities. The first-year adviser works with the student until a thesis adviser has been appointed. Students are encouraged to find a thesis adviser near the beginning of the second year and no later than the beginning of the third year. A thesis committee, comprised of the thesis adviser and two other faculty members, is appointed to track the student's progress toward the degree. Details about this process are given at Graduate Orientation. Before the beginning of the first semester, incoming students take a one-day diagnostic written exam based on undergraduate physics curriculum. Based on the results of the exam, the first-year adviser may suggest a course of action (e.g., enrollment in a relevant undergraduate class or independent reading) to remedy any deficiencies. Every first-year student is required to get involved in research during the first and second semesters and during the summer after the first year. First-year students must find a member of the professorial faculty who agrees to advise them in some sort of research project. The first-semester project continues through Intersession in January. The spring-semester research project continues until the end of the spring semester. The summer project lasts from June through August. Students may continue with one adviser through the entire first year, or they may choose to cycle through several different research advisers. This research requirement continues until the end of the second year, or until the student finds a thesis adviser, whichever comes first. The nature of these first- and second-year research projects will vary from student to student and from one sub-field of physics to another. In some cases they may lead to published scientific papers. In other cases, they may be first steps in a longer-term research project. Alternatively, they may comprise reading or independent-study projects to develop background for subsequent research. The student must provide at the end of each semester a brief written summary of the research experience. A series of short research presentations are given by faculty during Orientation to introduce incoming graduate students to the various research programs and prospective faculty research advisers in the department (Fall Research Jamboree). At the beginning of the second year, each student takes a one-hour oral research exam that consists of a 30-minute presentation to a committee of three faculty members about the research they have carried out in their first year and questions from the committee about the research and related scientific background. There are four semesters of required classes for the physics PhD and for the astronomy and astrophysics PhD that must be passed with a grade of B- or better. Students are strongly encouraged to complete these requirements in their first year, although they may be deferred until the second year under certain circumstances. Some of the required classes may be waived, at the discretion of the first-year adviser, if the student has successfully completed a comparable class elsewhere. The required classes are as follows: PhD in Physics • 171.605-606 Quantum Mechanics • 171.603 Electromagnetism • 171.703 Advanced Statistical Mechanics PhD in Astronomy and Astrophysics • 171.611 Stellar structure and evolution • 171.612 Interstellar medium and astrophysical fluid dynamics • 171.613 Radiative astrophysics • 171.627 Astrophysical dynamics In addition to the courses above, physics students must take one semester of graduate seminar (172.631). In the seminars, students practice giving presentations and discuss current literature. Astrophysics students take 172.633 Language of Astrophysics. The department offers a complete range of advanced classes in physics and astrophysics, and students are strongly encouraged to take more classes beyond these required four semesters. The specific course of study should be tailored, under consultation with the first-year adviser and/or with prospective research advisers, to the needs of each student and to their research goals.

Thesis: Thesis may be written in absentia.

SPECIAL EQUIPMENT, FACILITIES, OR PROGRAMS

The high-energy physics group has facilities for constructing the electronics and detectors needed in modern experiments and also has independent computing capabilities that allow full analyses of massive amounts of data. Nuclear physics equipment includes facilities for relativistic heavy-ion collision studies and studies of nuclear interactions utilizing muons and pions with cryogenic targets. Among the diverse techniques used for studying condensed matter physics are SQUID (Superconducting Quantum Interference Device) magnetometry/susceptometry, vibrating sample magnetometry, atomic force and magnetic force microscopy, X-ray and electron diffraction, Auger spectroscopy, X-ray fluorescence spectroscopy, electron microscopy, and neutron scattering at the nearby NIST Center for Neutron Research and at other leading international facilities. A variety of cryostats, He3 refrigerators, and He3-He4 dilution refrigerators, and high-temperature ovens, electromagnets, and superconducting magnets allow measurements to be made from 0.05-1100 K and in magnetic fields up to 12 Tesla. Equipment for the preparation of samples include single-crystal growth vacuum furnaces, arc furnaces, several high-vacuum and ultra-high-vacuum chambers for thin film fabrication using evaporation, MBE, pulsed laser deposition, and sputtering. The department maintains a Class-1000 cleanroom for microfabrication and nanofabrication and supports an instrument design group with five full-time engineers and a machine shop with three full-time machinists.

Table B—Separately Budgeted Research Expenditures by Source of Support

Source of Support	Departmental Research	Physics-related Research Outside Department
Federal government	$17,283,084	
State/local government		
Non-profit organizations	$2,354,717	
Business and industry	$1,543,538	
Other	$61,033	
Total	$21,242,372	

Table C—Separately Budgeted Research Expenditures by Research Specialty

Research Specialty	No. of Grants	Expenditures ($)
Astrophysics	108	$11,303,604
Condensed Matter Physics	28	$3,664,543
Particles and Fields	7	$1,338,581
Plasma and Fusion	1	$305,417
High Energy Physics	8	$1,094,490
Other	2	$1,704,572
Intensive Data	10	$1,831,165
Total	164	$21,242,372

FACULTY

Professor

Bagger, Jonathan A., Ph.D., Princeton University, 1983. Vice Provost for Graduate and Post-doctoral Programs and Special Projects and Kreiger Eisenhower Professor of Physics and Astronomy. *Particles and Fields, Theoretical Physics*. Theoretical high-energy physics.

Barnett, Bruce A., Ph.D., University of Maryland, 1970. *High Energy Physics*. Experimental particle physics.

Bennett, Charles L., Ph.D., Massachusetts Institute of Technology, 1984. Alumni Centennial Professor of Physics and Astronomy. *Astronomy, Astrophysics*. Experimental cosmology.

Blumenfeld, Barry J., Ph.D., Columbia University, 1974. Experimental particle physics.

Broholm, Collin, Ph.D., University of Copenhagen, 1988. Gerhard H. Dieke Professor. *Nano Science and Technology*. Experimental condensed-matter physics.

Chen, Shiyi, Ph.D., Peking University, 1987. Primary appointment in Mechanical Engineering. Statistical theory and computation of fluid turbulence.

Chien, Chia-Ling, Ph.D., Carnegie Mellon University, 1972. Jacob L. Hain Professor; Director, Materials Research Science and Engineering Center. *Materials Science, Metallurgy*. Experimental condensed-matter physics; artificially structured solids.

Chien, Chih-Yung, Ph.D., Yale University, 1966. Experimental high-energy physics.

Eyink, Gregory, Ph.D., Ohio State University, 1987. Primary appointment in Applied Mathematics and Statistics. Mathematical physics; fluid mechanics; turbulence.

Giacconi, Riccardo, Ph.D., University of Milan, 1954. Nobel Laureate. Astrophysics.

Heckman, Timothy, Ph.D., University of Washington, 1978. Director, Center for Astrophysical Sciences; A. Hermann Pfund Professor. *Astronomy, Astrophysics*. Astrophysics; active galaxies and quasars.

Kamionkowski, Marc, Ph.D., University of Chicago, 1991. *Cosmology & String Theory*. Theoretical physics specializing in cosmology and particle physics.

Kaplan, David, Ph.D., University of Washington, 1999. Theoretical particle physics.

Krolik, Julian H., Ph.D., University of California, Berkeley, 1977. *Astrophysics*. Theoretical astrophysics.

Maksimovic, Petar, Ph.D., Massachusetts Institute of Technology, 1997. *High Energy Physics*. Experimental high-energy physics.

Morava, Jack, Ph.D., Rice University, 1968. Primary appointment in Department of Mathematics. Algebraic topology; mathematical physics.

Mountain, Charles Mattias, Ph.D., University of London, 1983. Director, Space Telescope Science Institute. Star formation in galaxies; capabilities of "second-generation telescope".

Neufeld, David A., Ph.D., Harvard University, 1987. Molecular astrophysics; submillimeter and infrared astronomy; interstellar medium.

Norman, Colin A., Ph.D., University of Oxford, 1973. *Astrophysics*. Theoretical astrophysics.

Reich, Daniel H., Ph.D., University of Chicago, 1988. Chair. *Condensed Matter Physics, Materials Science, Metallurgy*. Experimental condensed-matter physics.

Riess, Adam, Ph.D., Harvard University, 1996. Krieger Eisenhower Professor of Physics and Astronomy; Nobel Laureate. *Astronomy, Astrophysics*. Astrophysics.

Robbins, Mark O., Ph.D., University of California, Berkeley, 1983. Theoretical condensed-matter physics.

Searson, Peter, Ph.D., University of Manchester, Institute of Sci. and Tech., 1982. Primary appointment in Materials Science and Engineering. Electronic; nanophase; semiconductor materials.

Strobel, Darrell F., Ph.D., Harvard University, 1969. Primary appointment in Earth and Planetary Sciences. Planetary atmospheres and astrophysics.

Swartz, Morris L., Ph.D., University of Chicago, 1983. *High Energy Physics*. Experimental high-energy physics.

Szalay, Alexander S., Ph.D., University of Budapest, 1975. Alumni Centennial Professor and Director, IDIES. *Astrophysics, Computational Physics, Computer Science*. Theoretical astrophysics; galaxy formation.

Wyse, Rosemary F. G., Ph.D., University of Cambridge, 1982. *Astrophysics*. Theoretical astrophysics; galaxy formation and evolution.

Associate Professor

Falk, Michael, Ph.D., University of California, Santa Barbara, 1998. Theoretical and computational research.

Gritsan, Andrei, Ph.D., University of Colorado Boulder, 2000. *High Energy Physics*. Experimental high-energy physics.

Leheny, Robert L., Ph.D., University of Chicago, 1997. *Condensed Matter Physics*. Experimental soft condensed-matter physics.

Markovic, Nina, Ph.D., University of Minnesota, 1998. *Condensed Matter Physics*. Experimental condensed-matter physics.

Melnikov, Kirill, Ph.D., Mainz University, 1996. *Theoretical Physics*. Theoretical particle physics.

Tchernyshyov, Oleg, Ph.D., Columbia University, 1998. *Condensed Matter Physics, Theoretical Physics*. Theoretical condensed-matter physics; high-temperature superconductivity.

Assistant Professor

Armitage, N. Peter, Ph.D., Stanford University, 2001. *Condensed Matter Physics*. Experimental condensed-matter physics.

Kaplan, Jared D., Ph.D., Harvard, 2009. *Particles and Fields*. Effective field theory, particle physics, and cosmology, as well as formal aspects of scattering amplitudes, holography, and conformal field theory.

Marriage, Tobias, Ph.D., Princeton University, 2006. *Astronomy, Astrophysics*. Cosmology; astrophysics.

McQueen, Tyrel, Ph.D., Princeton University, 2009. *Condensed Matter Physics, Solid State Physics*. Solid-state chemistry; condensed-matter physics.

Menard, Brice, Ph.D., University of Paris, 2003. *Astronomy, Astrophysics*. Extragalactic astrophysics; cosmology; large surveys.

Zakamska, Nadia, Ph.D., Princeton University, 2005. *Astronomy, Astrophysics*. Astronomy; astrophysics.

Emeritus

Domokos, Gabor, Ph.D., Dubna University, 1963. Algebraic approaches to elementary particle physics; symmetries with application to field theories of particles; strong interactions at high energies.

Judd, Brian R., Ph.D., University of Oxford, 1955. Theoretical atomic and molecular physics; group theory; solid-state theory.

Kövesi-Domokos, Susan, Ph.D., University of Budapest, 1963. Theoretical high-energy physics; astroparticle physics.

Kim, Chung W., Ph.D., Indiana University, 1963. Nuclear theory; elementary particle theory; cosmology.

Lee, Yung-Keun, Ph.D., Columbia University, 1961. Nuclear physics.

Pevsner, Aihud, Ph.D., Columbia University, 1954. Joseph L. Hain Professor Emeritus. Experimental high-energy physics.

Walker, Calvin J., Ph.D., Princeton University, 1961. Experimental condensed-matter physics; thin films and surfaces; nuclear physics.

Research Professor

Bianchi, Luciana, Ph.D., University of Padua, 1978. Experimental astrophysics.

Blair, William, Ph.D., University of Michigan, 1981. Astrophysics; shockwaves; spectroscopy of plasmas.

Feldman, Paul D., Ph.D., Columbia University, 1964. Astrophysics; space physics; planetary and cometary atmospheres; spectroscopy.

Finkenthal, Michael, Ph.D., Hebrew University (Jerusalem). Plasma physics.

Ford, Holland, Ph.D., University of Wisconsin-Madison, 1970. Stellar dynamics; evolution of galaxies; active galactic nuclei; astronomical instrumentation.

Henry, Richard C., Ph.D., Princeton University, 1967. Director, Maryland Space Grant Consortium. Astronomy; astrophysics.

McCandliss, Stephan, Ph.D., University of Colorado Boulder, 1988. Experimental astrophysics; sounding rocket space astronomy in the far UV.

Moos, Warren H., Ph.D., University of Michigan, 1961. Astrophysics; plasma physics.

Murray, Stephan, Ph.D., California Institute of Technology, 1971. X-ray astronomy.

Silk, Joseph, Ph.D., Harvard University, 1968. Cosmology.

Weaver, Jr., Harold A., Ph.D., Johns Hopkins University, 1982. Ultraviolet, optical, infrared, Xray, and radio spectroscopy and imaging of comets, planets, and satellites.

Research Associate Professor

Budavari, Tamas, Ph.D., Eotvos Lorand University, 2001. Observational cosmology; large-scale structure; galaxy clustering; galaxy properties and evolution; statistical inference and embedding; data-intensive parallel computing.

Adjunct Professor

Allen, Ronald J., Ph.D., Massachusetts Institute of Technology, 1967. Spiral structure of galaxies; interstellar medium; radio; optical imaging.

Ferguson, Henry, Ph.D., Johns Hopkins University, 1990. Observational cosmology; space astronomy.

Hauser, Michael, Ph.D., California Institute of Technology, 1967. Cosmology, especially infrared background radiation.

Hornschemeier, Ann, Ph.D., Pennsylvania State University, 2002. Astronomy; astrophysics.

Petrovic, Cedomir, Ph.D., Florida State University, 2000. Condensed-matter physics.

Stiles, Mark, Ph.D., Cornell University, 1986. Condensed-matter theory.

van der Marel, Roeland, Ph.D., Leiden University, 1994. Astronomy.

Weaver, Kimberly, Ph.D., University of Maryland, 1993. Observational astrophysics.

Williams, Robert, Ph.D., Harvard University, 1986. Observational astronomy.

Adjunct Assistant Professor

Predrag, Nikolic, Ph.D., Massachusetts Institute of Technology, 2004. Adjunct Assistant Professor, George Mason University. Theoretical condensed-matter physics.

Joint Appointment

McCullough, Peter R., Ph.D., UC Berkeley, 1993. *Astronomy.* extrasolar transiting planets.

DEPARTMENTAL RESEARCH SPECIALTIES AND STAFF

Theoretical

Astrophysics. Cosmology and large-scale structure; active galaxies; stellar populations; the interstellar medium; astrophysical magnetohydrodynamics; dark matter; mass-transfer binaries; physics of accretion disks. Budavari, Krolik, Menard, Neufeld, Norman, Silk, Wyse, Zakamska.

Atomic, Molecular, & Optical Physics. Electronic transitions involving lanthanide ions in crystals and solutions; orthogonal operators for the energy levels of free atoms; group theory in atomic structure and icosahedral systems.

Condensed Matter Physics. Study of high-Tc superconductivity; quantum critical phenomena; superfluidity; magnetotransport and electrons in high magnetic fields; disordered systems; nonequilibrium processes such as growth and friction. Chen, Falk, Robbins, Stiles, Tchernyshyov.

High Energy Physics. Elementary fields and their interactions; gauge theories and superstring theory; experimental tests of the Standard Model and its extensions such as supersymmetry, technicolor, and theories with additional microscopic or mesoscopic spatial dimensions; dynamics of heavy quark systems; implications of particle physics for astrophysics and cosmology. Bagger, Domokos, Kövesi-Domokos, Kamionkowski, David Kaplan, Melnikov.

Experimental

Astrophysics. Ultraviolet astronomy; interstellar medium; infrared and radio astronomy; active galactic nuclei; galactic structure and dynamics; cosmology; planetary atmospheres and magnetospheres; stellar atmospheres and chromospheres; physics of comets; laboratory astrophysics. Allen, Bennett, Bianchi, Blair, Feldman, Ferguson, Giacconi, Hauser, Heckman, Henry, Hornschemeier, Marriage, McCandliss, Moos, Murray, Pevsner, Riess, Strobel, Weaver, Weaver, Jr., Williams.

Condensed Matter Physics. Magnetic thin films; nanostructured materials; superconductivity; strongly correlated electron systems; quantum magnetism; glass transitions; soft condensed matter; surface and interfaces. Armitage, Broholm, Chia-Ling Chien, Lee, Leheny, Markovic, Reich.

Cosmology & String Theory. Kamionkowski.

High Energy Physics. Strong, electromagnetic, and weak interaction using counter techniques. Research in progress involves: (1) the CDF experiment at the proton-antiproton Tevatron Collider at Fermilab, (2) a future experiment, CMS, at the Large Hadron Collider at CERN, and (3) a satellite experiment, AMS, looking for antimatter in cosmic rays. Barnett, Blumenfeld, Chih-Yung Chien, Gritsan, Jared Kaplan, Maksimovic, Swartz.

Nuclear Physics. Properties of relativistic heavy ion collisions.

Plasma and Fusion. Spectra of high-temperature plasmas used in fusion; highly ionized atoms. Finkenthal.

UNIVERSITY OF MARYLAND, BALTIMORE COUNTY

DEPARTMENT OF PHYSICS, APPLIED PHYSICS GRADUATE PROGRAM

Baltimore, Maryland 21250
http://www.physics.umbc.edu/

General University Information
President: Freeman A. Hrabowski
Dean of Graduate School: Janet C. Rutledge
University website: http://www.umbc.edu/
Control: Public
Setting: Suburban
Total Faculty: 1,017
Total Graduate Faculty: 651
Total number of Students: 13,637
Total number of Graduate Students: 2,684

Department Information
Department Chairman: L. Michael Hayden, Chair
Department Contact: Jennifer Salmi, Programs Specialist
Total full-time faculty: 19
Total number of full-time equivalent positions: 19
Full-Time Graduate Students: 46
First-Year Graduate Students: 7
Female First-Year Students: 2
Total Post Doctorates: 5

Department Address
1000 Hilltop Circle
Baltimore, MD 21250
Phone: (410) 455-2513
Fax: (410) 455-1072
E-mail: jen.salmi@umbc.edu
Website: http://www.physics.umbc.edu/

ADMISSIONS

Admission Contact Information
Address admission inquiries to: Dr. Laszlo Takacs, Graduate Admissions Coordinator, Dept. of Physics
Phone: (410) 455-2513
E-mail: takacs@umbc.edu
Admissions website: http://physics.umbc.edu/

Application deadlines
Fall admission:
U.S. students: January 1   *Int'l. students*: January 1

Application fee
U.S. students: $50   *Int'l. students*: $50

Admissions information
For Fall of 2013:
Number of applicants: 40
Number admitted: 20
Number enrolled: 8

Admission requirements
Bachelor's degree requirements: Bachelor's degree in physics, chemistry, math, or engineering is required.
Minimum undergraduate GPA: 3.0

GRE requirements
The GRE is required.

Advanced GRE requirements
The Advanced GRE is recommended.

TOEFL requirements
The TOEFL exam is required for students from non-English-speaking countries.
PBT score: 550
iBT score: 80

Other admissions information
Additional requirements: Letters of recommendation, undergraduate transcripts, personal statement.
Undergraduate preparation assumed: Young and Freedman, University Physics; Reif, Thermal Physics; Marion, Newtonian Dynamics; Tippler and Llewellyn, Introduction to Modern Physics; Hecht, Fundamentals of Optics; Griffiths, Quantum Mechanics; Weber and Arfken, Mathematical Methods for Physicists; Griffiths, Introduction to Electrodynamics.

TUITION

Tuition year 2012–13:
Tuition for in-state residents
   Full-time students: $530 per credit
Tuition for out-of-state residents
   Full-time students: $878 per credit
Credit hours per semester to be considered full-time: 9
Deferred tuition plan: No
Health insurance: Available
Other academic fees: $119/credit.
Academic term: Semester
Number of first-year students who received full tuition waivers: 7

Teaching Assistants, Research Assistants, and Fellowships
Number of first-year
   Teaching Assistants: 7
Average stipend per academic year
   Teaching Assistant: $23,000
   Research Assistant: $25,000
   Fellowship student: $28,000

FINANCIAL AID

Application deadlines
Fall admission:
U.S. students: April 15   *Int'l. students*: April 15

Loans
Loans are available for U.S. students.
Loans are available for international students.
GAPSFAS application required: No
FAFSA application required: No

For further information
Address financial aid inquiries to: Dr. Laszlo Takacs, Graduate Admissions Coordinator, Department of Physics.
E-mail: takacs@umbc.edu
Financial aid website: http://www.umbc.edu/financialaid/

HOUSING

Availability of on-campus housing
Single students: Yes
Married students: No

For further information
Address housing inquiries to: Office of Residential Life.
Housing aid website: http://www.umbc.edu/reslife/

Table A—Faculty, Enrollments, and Degrees Granted

Research Specialty	2012–13 Faculty	Enrollment Fall 2012		Number of Degrees Granted 2012–13 (2008–13)		
		Master's	Doctorate	Master's	Terminal Master's	Doctorate
Astrophysics	4	–	8	–(7)	–	–(1)
Atmospheric Physics	4	–	14	4(6)	–(6)	1(6)
Condensed Matter Physics	6	–	10	1(5)	1(3)	–(4)
Nonlinear/Ultrafast Optics	2	–	4	1(3)	1(3)	–(6)
Quantum Optics and Quantum Information	3	–	10	1(7)	–(1)	2(6)
Total	19	–	46	7(28)	2(13)	3(23)
Full-time Grad. Stud.	–	–	44	–	–	–
First-year Grad. Stud.	–	–	7	–	–	–

GRADUATE DEGREE REQUIREMENTS

Master's: Completion of 30 credit hours of coursework, including required core courses in quantum mechanics and mathematical physics. Overall competence must be demonstrated by an oral thesis defense (thesis option) or a written comprehensive examination (non-thesis option). For thesis option, six hours of the required 30 credit hours are for thesis research.

Doctorate: Minimum requirement is a total of 46 credit hours, with 28 credit hours of lecture courses at the 600 level or higher and 12 credit hours of doctoral research. All students will be at least required to complete a core curriculum consisting of quantum mechanics, statistical mechanics, mathematical physics, classical mechanics, and electromagnetic theory. In addition, students are required to take computational physics, quantum mechanics II, mathematical physics II, advanced electromagnetic theory, physics seminar, and at least one elective course. Students are required to pass a written examination in order to qualify for candidacy for the Ph.D. degree. Upon completion of the doctoral research, students are required to write and to defend a dissertation before a committee constituted in accordance with the graduate school regulations.

Thesis: Thesis may be written in absentia.

SPECIAL EQUIPMENT, FACILITIES, OR PROGRAMS

There are close working relationships between UMBC and several other research institutions in the Baltimore-Washington area. Formal cooperative agreements are in place between UMBC and the NASA Goddard Space Flight Center (GSFC)in the form of the Joint Center for Earth Systems Technology (JCET) [http://jcet.umbc.edu/] and the Center for Research and Exploration in Space Science and Technology (CRESST) [http://cresst.umd.edu/]. UMBC also hosts the Center for Advanced Studies in Photonics Research (CASPR) [http://www.umbc.edu/caspr/].

Table B—Separately Budgeted Research Expenditures by Source of Support

Source of Support	Departmental Research	Physics-related Research Outside Department
Federal government	$4,453,938	$256,851
State/local government		
Non-profit organizations		
Business and industry		
Other		
Total	$4,453,938	$256,851

Table C—Separately Budgeted Research Expenditures by Research Specialty

Research Specialty	No. of Grants	Expenditures ($)
Astrophysics	5	$366,472
Condensed Matter Physics	1	$105,968
Nonlinear/Ultrafast Optics	1	$389,227
Atmospheric Physics	19	$2,732,069
Quantum Optics and Quantum Information	6	$860,202
Total	32	$4,453,938

FACULTY

Professor
Franson, James D., Ph.D., California Institute of Technology, 1977. Quantum optics and quantum computing.
Hayden, L. Michael, Ph.D., University of California, Davis, 1987. Nonlinear optical properties of polymers; electro-optic techniques; photonic devices.
Hoff, Raymond M., Ph.D., Simon Fraser University, 1975. Atmospheric physics; LIDAR.
Johnson, Anthony M., Ph.D., City College of New York, 1981. Nonlinear optics; ultrafast optics; optoelectronics; ultrashort pulse propagation.
Shih, Yanhua, Ph.D., University of Maryland, 1987. Quantum optics; laser physics; nonlinear optics.
Turner, T. Jane, Ph.D., University of Leicester, 1988. Extragalactic astrophysics; x-ray astronomy.

Associate Professor
Georganopoulos, Markos, Ph.D., Boston University, 1999. Broad-band synchrotron emission from relativistic flows in active galaxies, galactic microquasars, and gamma-ray bursts.
George, Ian M., Ph.D., University of Leicester, 1988. Astrophysics; x-ray astronomy.
Gougousi, Theodosia, Ph.D., University of Pittsburgh, 1996. Nanoscience; interfaces.
Henriksen, Mark J., Ph.D., University of Maryland, 1986. Astrophysics; x-ray astronomy.
Kramer, Ivan, Ph.D., University of California, Berkeley, 1967. Mathematical modeling.
Martins, Vanderlei, Ph.D., University of São Paolo, 1999. Radiative effects of biomass burning and bio-aerosols.
Pittman, Todd B., Ph.D., University of Maryland, Baltimore County, 1996. Quantum optics and quantum computing.
Sparling, Lynn C., Ph.D., University of Texas, Austin, 1987. Atmospheric physics; modeling.
Takacs, Laszlo, Ph.D., Eotvos University, 1978. Amorphous and metastable crystalline alloys; energy-dispersive x-ray diffraction; magnetic susceptibility.
Worchesky, Terrance L., Ph.D., Georgetown University, 1984. Optical properties of semiconductors; photonics.

Assistant Professor

Kestner, Jason, Ph.D., University of Michigan, 2009. Condensed-matter theory; quantum information theory.

Pelton, Matthew, Ph.D., Stanford, 2002. Optical studies of nano-materials; nanophotonics.

Zhang, Zhibo, Ph.D., Texas A&M University, 2008. Remote sensing; aerosol-cloud-precipitation interactions; atmospheric physics.

Emeritus

Melfi, Harvey, Ph.D., College of William and Mary, 1970. Atmospheric LIDAR; remote sensing.

Rasera, Robert L., Ph.D., Purdue University, 1965. Perturbed gamma-ray angular correlation spectroscopy.

Reno, Robert C., Ph.D., Brandeis University, 1970. Hyperfine interactions in solids; electron microscopy; neutron diffraction measurement.

Rous, Philip J., Ph.D., Imperial College of Science and Technology, 1986. Theoretical physics: surfaces, interfaces, and nano-structures.

Rubin, Morton H., Ph.D., Princeton University, 1964. Theoretical physics; quantum optics.

Summers, Geoffrey P., Ph.D., University of Oxford, 1970. Radiation effects in semiconductors; defects in solids.

Wu, En-Shinn, Ph.D., Cornell, 1972. Optical studies of macromolecules.

Research Professor

Strow, L. Larrabee, Ph.D., University of Maryland, 1981. High-resolution infrared molecular spectroscopy; atmospheric radiative transfer.

Research Associate Professor

Peter, Kuchunov, Ph.D., University of Texas Health Science Center at San Antonio, 2001. MRI, quantitative imaging, imaging genetics.

Adjunct Professor

Fitelson, Michael, Ph.D., Pennsylvania State University, 1966. Advanced technologies.

Jacobs, Bryan, Ph.D., University of Maryland, Baltimore County, 2003. Quantum optics; quantum information.

Krotkov, Nickolay, Ph.D., Shirshov Institute, Russian Academy of Sciences, 1990. Atmospheric physics.

Kuchner, Marc, Ph.D., California Institute of Technology, 2000. Astrophysics.

Affiliate Professor

Remer, Lorraine, Ph.D., University of California, Davis, 1991. Climate change; remote sensing.

Affiliate Associate Professor

Davis, David, Ph.D., University of Maryland, College Park, 1994. Galaxy clusters; x-ray astronomy.

Hoban, Susan, Ph.D., University of Maryland, 1989. Planetary science, comets, dust in the solar system, STEM education.

Kundu, Prasun, Ph.D., University of Rochester, 1981. Satellite- and ground-based remote sensing.

McCann, Kevin J., Ph.D., Georgia Institute of Technology, 1974. LIDAR and atmospheric aerosols.

Olson, William, Ph.D., University of Wisconsin-Madison, 1987. Remote sensing of precipitation.

Pottschmidt, Katja, Ph.D., Universitaet Tuebingen, 2002. High energy astrophysics, accreting x-ray binary stars.

Varnai, Tamas, Ph.D., McGill University, 1996. Cloud physics and radiation transfer.

Affiliate Assistant Professor

De Souza-Machado, Sergio, Ph.D., University of Maryland, College Park, 1996. Infrared remote sensing and radiation transfer.

Engel, Don, Ph.D., University of Pennsylvania, 2005. Computational physics, molecular biophysics, statistical artificial intelligence.

Lecturer

Anderson, Eric, Ph.D., Arizona State University, 1993. Physics education.

Cui, Lili, Ph.D., Kansas State University, 2006. Physics education.

DEPARTMENTAL RESEARCH SPECIALTIES AND STAFF

Theoretical

Astrophysics. High energy astrophysics; active galactic nuclei; relativistic jets; quasars, x-ray astronomy. Georganopoulos, George, Henriksen, Turner.

Atmospheric Physics. Cloud physics; aerosol-cloud-precipitation; atmospheric dynamics. Sparling, Zhang.

Condensed Matter Physics. semi-conductor quantum dots; cold atoms; theoretical medical physics. Kestner, Kramer.

Quantum Optics and Quantum Information. Quantum foundations; entanglement; non-classical states; quantum imaging; photonic qubits; spin qubits; quantum information theory. Franson, Kestner, Shih.

Experimental

Astrophysics. X-ray astronomy; active galaxies; extragalactic astrophysics. Georganopoulos, George, Henriksen, Turner.

Atmospheric Physics. LIDAR; aerosols; clouds; instrument development; remote sensing. Hoff, Martins, Strow, Zhang.

Condensed Matter Physics. Thin films; surfaces and interfaces; atomic layer deposition; polymer physics; mechanical alloying; semi-conductors; optical studies of nanomaterials. Gougousi, Hayden, Pelton, Takacs, Worchesky.

Nonlinear/Ultrafast Optics. frequency conversion, THz generation and detection, ultrafast carrier dynamics, pulse propagation, low-power nonlinearities, nanophotonics. Franson, Hayden, Johnson, Pelton, Pittman, Shih, Worchesky.

Quantum Optics and Quantum Information. photonic quantum information; quantum imaging; entanglement; single photon physics; quantum foundations. Franson, Pelton, Pittman, Shih.

View additional information about this department at
www.gradschoolshopper.com

UNIVERSITY OF MARYLAND, BALTIMORE COUNTY

DEPARTMENT OF PHYSICS, ATMOSPHERIC PHYSICS GRADUATE PROGRAM

Baltimore, Maryland 21250
http://physics.umbc.edu

General University Information
President: Freeman A. Hrabowski
Dean of Graduate School: Janet C. Rutledge
University website: http://www.umbc.edu
Control: Public
Setting: Suburban
Total Faculty: 1,017
Total Graduate Faculty: 651
Total number of Students: 12,888
Total number of Graduate Students: 2,678

Department Information
Department Chairman: L. Michael Hayden, Chair
Department Contact: Jennifer Salmi, Business Services Specialist
  Total full-time faculty: 19
  Total number of full-time equivalent positions: 19
  Full-Time Graduate Students: 46
  First-Year Graduate Students: 7
  Female First-Year Students: 2
  Total Post Doctorates: 5

Department Address
1000 Hilltop Circle
Physics Building, room 220
Baltimore, MD 21250
Phone: (410) 455-2513
Fax: (410) 455-1072
E-mail: jen.salmi@umbc.edu
Website: http://physics.umbc.edu

ADMISSIONS

Admission Contact Information
Address admission inquiries to: Dr. Laszlo Takacs, Graduate Admissions Coordinator, Dept. of Physics
Phone: (410)455-2513
E-mail: takacs@umbc.edu
Admissions website: http://physics.umbc.edu

Application deadlines
Fall admission:
U.S. students: January 1   *Int'l. students*: January 1
Spring admission:
U.S. students: November 1   *Int'l. students*: May 1

Application fee
U.S. students: $50   *Int'l. students*: $50

Admissions information
For Fall of 2013:
  Number of applicants: 40
  Number admitted: 20
  Number enrolled: 8

Admission requirements
Bachelor's degree requirements: Bachelor's degree in physics, atmospheric sciences, chemistry, math, engineering, or related field is required.
Minimum undergraduate GPA: 3.0

GRE requirements
The GRE is required.

Advanced GRE requirements
The Advanced GRE is recommended.

TOEFL requirements
The TOEFL exam is required for students from non-English-speaking countries.
  PBT score: 550
  iBT score: 90

Other admissions information
Additional requirements: Letters of recommendation, undergraduate transcripts, personal statement.
Undergraduate preparation assumed: Young and Freedman. University Physics; Reif, Thermal Physics; Marion, Newtonian Dynamics; Tippler and Llewellyn, Introduction to Modern Physics; Griffiths, Introduction to Electrodynamics.

TUITION

Tuition year 2012–13:
Tuition for in-state residents
  Full-time students: $530 per credit
Tuition for out-of-state residents
  Full-time students: $878 per credit
Credit hours per semester to be considered full-time: 9
Deferred tuition plan: No
Health insurance: Available
Other academic fees: $119/credit.
Academic term: Semester
Number of first-year students who received full tuition waivers: 7

Teaching Assistants, Research Assistants, and Fellowships
Number of first-year
  Teaching Assistants: 7
Average stipend per academic year
  Teaching Assistant: $23,000
  Research Assistant: $26,400
  Fellowship student: $28,000

FINANCIAL AID

Application deadlines
Fall admission:
U.S. students: April 15   *Int'l. students*: April 15
Spring admission:
U.S. students: November 1   *Int'l. students*: November 1

Loans
Loans are available for U.S. students.
Loans are available for international students.
GAPSFAS application required: No
FAFSA application required: No

For further information
Address financial aid inquiries to: Dr. Laszlo Takacs, Graduate Admissions Coordinator, Department of Physics.
Phone: (410) 4552513

E-mail: takacs@umbc.edu
Financial aid website: http://www.umbc.edu/financialaid/

HOUSING

Availability of on-campus housing
Single students: Yes
Married students: No

For further information
Address housing inquiries to: Office of Residential Life.
Housing aid website: http://www.umbc.edu/reslife/

Table A—Faculty, Enrollments, and Degrees Granted

Research Specialty	2012–13 Faculty	Enrollment Fall 2012		Number of Degrees Granted 2012–13 (2008–13)		
		Mas-ter's	Doc-torate	Mas-ter's	Terminal Master's	Doc-torate
Astrophysics	4	–	8	–(7)	–	–(1)
Atmospheric Physics	4	–	14	4(6)	3(6)	2(6)
Condensed Matter Physics	6	–	10	1(5)	1(3)	–(4)
Nonlinear/Ultrafast Optics	2	–	4	1(3)	1(3)	–(6)
Quantum Optics and Quantum information	3	–	10	1(7)	–(1)	2(6)
Total	19	–	46	7(28)	2(13)	3(23)
Full-time Grad. Stud.	–	–	44	–	–	–
First-year Grad. Stud.	–	–	7	–	–	–

GRADUATE DEGREE REQUIREMENTS

Master's: Completion of 30 hours of course work, including required core courses in Atmospheric Physics I and II, quantum mechanics, and mathematical physics. Minimum acceptable GPA 3.0 ("A" = 4.0). Overall competence must be demonstrated by oral thesis defense (thesis option) or written comprehensive examination (non-thesis option). For thesis option, six hours of the required 30 are for thesis research.

Doctorate: Minimum requirement is a total of 46 credit hours, with 28 credit hours of lecture courses at the 600 level or higher and 12 credit hours of Doctoral Research. Overall GPA must be at least 3.0 ("A" = 4). All students will be at least required to complete a core curriculum consisting of Atmospheric Physics I and II, Mathematical Physics, Electromagnetic Theory, Quantum Mechanics, and Statistical Mechanics,. In addition, they are required to take Computational Physics, at least two specialized courses in Atmospheric Physics (e.g., Radiative Transfer, Remote Sensing, Dynamics, etc.), Professional Techniques, and Physics Seminar. After completion of the Ph.D. core curriculum, prospective Ph.D. students will be required to pass a written examination in order to qualify for candidacy for the Ph.D. degree. Upon completion of the doctoral research, the student will be required to write and to defend a dissertation before a committee constituted in accordance with the graduate school regulations.

Thesis: Thesis may be written in absentia.

SPECIAL EQUIPMENT, FACILITIES, OR PROGRAMS

There are close relationships between UMBC and several other research institutions in the Baltimore-Washington area. Formal cooperative agreements are in place between UMBC and NASA Goddard Space Flight Center (GSFC) in the form of the Joint Center for Earth Systems and Technology (JCET) http://jcet. umbc.edu, Center for Research and Exploration in Space Science and Technology (CRESST) http://cresst.umd.edu/, and the Goddard Planetary Heliophysics Institute (GPHI) http://gphi.umbc. edu. UMBC also hosts the Center for Advanced Studies in Photonics Research (CASPR) http://www.umbc.edu/caspr/.

Table B—Separately Budgeted Research Expenditures by Source of Support

Source of Support	Departmental Research	Physics-related Research Outside Department
Federal government	$4,453,938	$256,851
State/local government		
Non-profit organizations		
Business and industry		
Other		
Total	$4,453,938	$256,851

Table C—Separately Budgeted Research Expenditures by Research Specialty

Research Specialty	No. of Grants	Expenditures ($)
Astrophysics	5	$366,472
Condensed Matter Physics	1	$105,968
Nonlinear/Ultrafast Optics	1	$389,227
Atmospheric Physics	19	$2,732,069
Quantum Optics and Quantum Information	–	$860,202
Total	26	$4,453,938

FACULTY

Professor

Franson, James D., Ph.D., California Institute of Technology, 1977. Quantum optics and quantum computing.

Hayden, L. Michael, Ph.D., University of California, Davis, 1987. Nonlinear optical properties of polymers; electro-optic techniques; photonic devices.

Hoff, Raymond M., Ph.D., Simon Fraser University, 1975. Atmospheric physics; lidar, air quality, satellite remote sensing.

Johnson, Anthony, Ph.D., City College of New York, 1981. *Optics*.

Rous, Philip, Ph.D., Imperial College of Science and Technology, University of London, 1986. *Condensed Matter Physics*.

Shih, Yanhua, Ph.D., University of Maryland, 1987. Quantum optics; laser physics; nonlinear optics.

Associate Professor

George, Ian M., Ph.D., University of Leicester, 1988. Astrophysics, x-ray astronomy.

Henriksen, Mark J., Ph.D., University of Maryland, 1986. Astrophysics; x-ray astronomy.

Hoban, Susan, Ph.D., University of Maryland, College Park, 1989. scientific information systems, digital library technologies and information technologies for science, technology, engineering and mathematics (STEM) education.

Kramer, Ivan, Ph.D., University of California, Berkeley, 1967. Mathematical modeling.

Martins, Vanderlei J., Ph.D., University of Sao Paulo, 1999. Graduate Program Director, Atmospheric Physics. Aerosol and Cloud Physics, Radiative Transfer, Optics, satellite remote sensing, instrumentation development for laboratory, field, aircraft, and satellite measurements.

Pittman, Thomas, Ph.D., University of Maryland, Baltimore County, 1996. Quantum optics and quantum computing.

Reno, Robert C., Ph.D., Brandeis University, 1970. Hyperfine interactions in solids; electron microscopy; neutron diffraction measurement.

Sparling, Lynn C., University of Texas, Austin, 1987. Atmospheric physics; modeling.

Takacs, Laszlo, Ph.D., Eotvos University, 1978. Amorphous and metastable crystalline alloys; energy-dispersive x-ray diffraction; magnetic susceptibility.

Turner, T. Jane, Ph.D., University of Leicester, 1988. Extragalactic astrophysics; x-ray astronomy.

Worchesky, Terrance L., Ph.D., Georgetown University, 1984. Optical properties of semiconductors; photonics.

Assistant Professor

Georganopoulos, Markos, Ph.D., University of Thessaloniki, 1989. Broad-band synchrotron emission from relativistic flows in active galaxies, galactic microquasars and gamma-ray bursts.

Gougousi, Theodosia, Ph.D., University of Pittsburg, 1996. Nanoscience, interfaces.

Kestner, Jason, Ph.D., University of Michigan, 2009. Condensed Matter Theory, Quantum Information Theory.

Zhang, Zhibo, Ph.D., Texas A&M University, 2008. Remote sensing; aerosol-cloud-precipitation interactions; atmospheric physics.

Emeritus

Melfi, Harvey, Ph.D., College of William and Mary, 1970. Atmospheric lidar, remote sensing.

Rasera, Robert L., Ph.D., Purdue University, 1965. Perturbed gamma-ray angular correlation spectroscopy.

Rubin, Morton H., Ph.D., Princeton University, 1964. Theoretical physics, quantum optics.

Research Professor

Remer, Lorraine A., Ph.D., University of California, Davis, 1991. *Climate/Atmospheric Science*. Aerosol and cloud remote sensing. Cloud-Aerosol-Precipitation-Climate interactions.

Strow, L. Larrabee, Ph.D., University of Maryland, 1981. High-resolution infrared molecular spectroscopy; atmospheric radiative transfer.

Research Associate Professor

Davis, David, Ph.D., University of Maryland, College Park, 1994. Galaxy clusters, x-ray astronomy.

Kundu, Prasun, Ph.D., University of Rochester, 1981. Satellite and ground-based remote sensing.

McCann, Kevin J., Ph.D., Georgia Institute of Technology, 1974. Lidar and atmospheric aerosols.

Olson, William, Ph.D., University of Wisconsin-Madison, 1987. Remote sensing of precipitation.

Varnai, Tamas, Ph.D., McGill University, 1996. Cloud physics and radiation transfer.

Research Assistant Professor

De Souza-Machado, Sergio, Ph.D., University of Maryland, College Park, 1996. Infrared remote sensing, radiation transfer, spectroscopy, plasma physics.

Johnson, Benjamin, Ph.D., University of Wisconsin-Madison, 2007. *Climate/Atmospheric Science*. Precipitation cloud modeling, radiative transfer, and remote sensing.

Yuan, Tianle, Ph.D., University of Maryland College Park, 2008. *Climate/Atmospheric Science*. Aerosol-Cloud-Precipitation Interactions; remote sensing.

Lecturer

Anderson, Eric, Ph.D., Arizona State University, 1993. Physics education.

Cui, Lili, Ph.D., Kansas State University, 2006. Physics education.

DEPARTMENTAL RESEARCH SPECIALTIES AND STAFF

Theoretical

Astrophysics. High energy astrophysics; active galactic nuclei; relativistic jets; quasars; x-ray astronomy. Georganopoulos, George, Henriksen, Turner.

Atmospheric Physics. Atmospheric dynamics; hurricane; radiative transfer; aerosol-cloud-precipitation interactions; cloud physics. Benjamin Johnson, Martins, Sparling, Yuan, Zhang.

Condensed Matter Physics. Semi-conductor quantum dots; cold atoms; theoretical medical physics. Kestner, Kramer.

Optics. Radiative transfer in inhomogeneous media; light polarization; IR spectroscopy. De Souza-Machado, Benjamin Johnson, Martins, Strow, Yuan, Zhang.

Quantum Optics and Quantum Information. Quantum optics, quantum information processing, quantum theory. Franson, Kestner, Shih.

Experimental

Astrophysics. x-ray astronomy, active galaxies, extra galactic astrophysics. Georganopoulos, George, Henriksen, Turner.

Atmospheric Physics. Aerosol and cloud properties; atmospheric dynamics; remote sensing measurements; LIDAR; aerosol-cloud-precipitation interactions; air pollution; atmospheric radiative transfer; optics instrumentation. De Souza-Machado, Hoff, Benjamin Johnson, Kundu, Martins, Remer, Sparling, Strow, Varnai, Yuan, Zhang.

Condensed Matter Physics. Thin films; surface and interfaces; atomic layer deposition; polymer physics; mechanical alloying; semi-conductors; optical studies of nanomaterials. Gougousi, Hayden, Takacs, Worchesky.

Nonlinear/Ultrafast Optics. Frequency conversion; THz generation and detection; pulse propagation; low-power nonlinearities; nanophotonics. Franson, Hayden, Anthony Johnson, Pittman, Shih, Worchesky.

Optics. Light scattering and absorption by suspended particles; light polarization; instrument development. Hoff, Martins, Zhang.

Quantum optics and quantum information. Photonic quantum information; quantum imaging; entanglement; single photon physics; quantum foundations. Franson, Pittman, Shih.

View additional information about this department at
www.gradschoolshopper.com

UNIVERSITY OF MARYLAND

DEPARTMENT OF ASTRONOMY

College Park, Maryland 20742-2421
http://www.astro.umd.edu

General University Information
President: Wallace D. Loh
Dean of Graduate School: Dr. Charles Caramello
University website: http://www.umd.edu
Control: Public
Setting: Suburban
Total Faculty: 4,387
Total Graduate Faculty: 1,472
Total number of Students: 37,248
Total number of Graduate Students: 10,710

Department Information
Department Chairman: Stuart Vogel, Chair
Department Contact: Ms. MaryAnn Phillips, Coordinator
   Total full-time faculty: 74
   Full-Time Graduate Students: 40
   First-Year Graduate Students: 5
   Female First-Year Students: 2
   Total Post Doctorates: 29

Department Address
CSS 1204
College Park, MD 20742-2421
Phone: (301) 405-1505
Fax: (301) 314-9067
E-mail: maryann@astro.umd.edu
Website: http://www.astro.umd.edu

ADMISSIONS

Admission Contact Information
Address admission inquiries to: Graduate Entrance Committee,
   c/o MaryAnn Phillips
Phone: (301) 405-1505
E-mail: astr-grad@deans.umd.edu
Admissions website: http://www.astro.umd.edu/graduate/
   admissions.html

Application deadlines
Fall admission:
U.S. students: December 1   *Int'l. students*: December 1

Application fee
U.S. students: $75   *Int'l. students*: $75
Note that the Department of Astronomy accepts applications for
   the Ph.D. program only. (Admitted students typically receive
   an M.S. degree after their second year in the program.)

Admissions information
For Fall of 2013:
   Number of applicants: 110
   Number admitted: 22
   Number enrolled: 5

Admission requirements
Bachelor's degree requirements: An undergraduate degree in a
   related field (normally Astronomy or Physics) is required.
Minimum undergraduate GPA: 3.0

GRE requirements
The GRE is required.
There are no set minimums; however, a strong performance is
   helpful.

Advanced GRE requirements
The Advanced GRE is required.
There are no set minimums; however, a strong performance is
   helpful.

TOEFL requirements
The TOEFL exam is required for students from non-English-
   speaking countries.
   PBT score: 575
   iBT score: 84
For details and the most current information, see http://www.
   international.umd.edu/ies/97.

Other admissions information
Additional requirements: The Department of Astronomy relies on
   a combination of course grades and letters of recommendation. Students from non-English speaking countries are required to demonstrate proficiency in English via the TOEFL or IELTS exams.
Undergraduate preparation assumed: Students who enter the
   graduate program are normally expected to have strong backgrounds in astronomy, physics, and mathematics. A student with
   deficiencies in one of these areas may be admitted but will be
   expected to remedy such deficiencies as soon as possible.

TUITION

Tuition year 2013-14:
Tuition for in-state residents
   Full-time students: $573 per credit
Tuition for out-of-state residents
   Full-time students: $1,236 per credit
The Department guarantees funding (with tuition waivers) for
   a minimum of 6 years, assuming adequate progress toward
   the degree. Students in teaching or research assistantship posi-
   tions or on full fellowships receive 10 credits of tuition remis-
   sion per semester.
Credit hours per semester to be considered full-time: 8
Deferred tuition plan: No
Health insurance: Available
Other academic fees: Semester fees are listed at http://bursar.
   umd.edu/Tuitionfees.php
Academic term: Semester
Number of first-year students who received full tuition waivers: 5

Teaching Assistants, Research Assistants, and Fellowships
Number of first-year
   Teaching Assistants: 5
Average stipend per academic year
   Teaching Assistant: $25,250
   Research Assistant: $27,000
   Fellowship student: $25,000

FINANCIAL AID

Loans
Loans are not available for U.S. students.
Loans are not available for international students.
GAPSFAS application required: No
FAFSA application required: No

HOUSING

For further information
Housing aid website: http://www.och.umd.edu

Table A—Faculty, Enrollments, and Degrees Granted

Research Specialty	2012–13 Faculty	Enrollment Fall 2013 Master's	Enrollment Fall 2013 Doctorate	Number of Degrees Granted 2012–13 (2007–12) Master's	Number of Degrees Granted 2012–13 (2007–12) Terminal Master's	Number of Degrees Granted 2012–13 (2007–12) Doctorate
Astronomy	21	–	40	7(22)	2(5)	5(24)
Total	21	–	40	7(22)	2(5)	5(24)
Full-time Grad. Stud.	–	–	40	–	–	–
First-year Grad. Stud.	–	–	5	–	–	–

GRADUATE DEGREE REQUIREMENTS

Master's: Thirty credits (including six credits of Master's research) are required; 12 must be in the major area and 12 at an advanced graduate level. Both thesis and non-thesis options are available. Residence of one year of full-time study is required. A minimum GPA of 3.0 is required for graduation. No foreign language requirement. A research project and written comprehensive exam are required.

Doctorate: Six graduate Astronomy courses plus two courses in supporting areas plus a minimum of 12 credits of doctoral research must be completed. Three years of residency required. A research project and comprehensive exam must be completed prior to admission to candidacy, which must occur within four years of admission to the doctoral program. Dissertation and dissertation defense required (no less than one nor more than four years from admission to candidacy). Minimum GPA of 3.0; no foreign language required.

Thesis: Theses may be written in absentia.

SPECIAL EQUIPMENT, FACILITIES, OR PROGRAMS

In collaboration with four other excellent astronomy departments, the University of Maryland operates CARMA (Combined Array for Research in Millimeter-wave Astronomy), the most powerful millimeter-wave telescope in the northern hemisphere. CARMA is ideally suited for the study of planetary and star formation, the birth and evolution of galaxies, and the feeding of supermassive black holes that power active galactic nuclei. The Department has guaranteed access to the 4.3-meter Discovery Channel Telescope through a partnership with Lowell Observatory. Graduate students also observe with some of the largest telescopes in the United States and around the world, as well as a wide range of space telescopes covering the electromagnetic spectrum from gamma-rays to the submillimeter. Our planetary science team is heavily involved with space missions visiting solar system bodies, such as NASA's Deep Impact and EPOXI missions to study comets. Complementing its observational program, the department has a strong theory group, and there is also an important emphasis on the design and building of powerful new instruments.

A number of our students conduct research with distinguished scientists at the nearby NASA Goddard Space Flight Center. The university's scientific partnership with Goddard has recently been further strengthened via the creation of the Joint Space Science Institute (JSI). The first component of JSI is a black hole center, a close collaboration between the Departments of Astronomy and Physics and Goddard scientists that is unique in the world in involving all observational and theoretical aspects of black hole research.

An extensive department network provides seamless access to software and hardware on a variety of UNIX and LINUX platforms. The computational astrophysics group maintains and upgrades a Beowulf cluster for computation-intensive science projects and has additional access to a larger cluster maintained by the university.

The Department has recently established a partnership with Pontificia Universidad Catolica de Chile (PUC), one of the top two institutions for astronomy in Chile. PUC signed an agreement with UMD in 2010 that enables astronomy graduate students at both institutions to participate in a joint Ph.D. program starting in their third year. These students split their time between both locations and conduct their thesis research under the supervision of UMD and PUC co-advisors. UMD students gain improved access to Chilean observatories, which include many of the best telescopes in the world.

Table B—Separately Budgeted Research Expenditures by Source of Support

Source of Support	Departmental Research	Physics-related Research Outside Department
Federal government	$21,510,915	
State/local government		
Non-profit organizations		
Business and industry		
Other		
Total	$21,510,915	

Table C—Separately Budgeted Research Expenditures by Research Specialty

Research Specialty	No. of Grants	Expenditures ($)
Astronomy	178	$21,510,915
Total	178	$21,510,915

FACULTY

Professor

Deming, L. Drake, Ph.D., University of Illinois, 1976. *Astronomy, Planetary Science.* Extrasolar planets; supernovae.

Hamilton, Douglas P., Ph.D., Cornell University, 1994. *Astrophysics, Planetary Science.* Solar system dynamics; solar system origins.

Harris, Andrew I., Ph.D., University of California, Berkeley, 1986. *Astronomy.* Extragalactic astrophysics; radio astronomy; instrumentation.

Miller, M. Coleman, Ph.D., California Institute of Technology, 1990. Graduate Director. *Astrophysics, Relativity & Gravitation.* Theoretical modeling of neutron stars and black holes; gravitational lensing.

Mundy, Lee G., Ph.D., University of Texas, 1984. Director, Laboratory for Millimeter-wave Astronomy; Director, Center for Research and Exploration in Space Science and Technology. *Astronomy.* Millimeter-wave and IR astronomy; star and planet formation; interstellar matter; astrobiology.

Mushotzky, Richard, Ph.D., University of California, San Diego, 1976. *Astronomy, High Energy Physics*. High-energy astrophysics; x-ray astronomy; extragalactic astronomy.

Papadopoulos, Konstantinos, Ph.D., University of Maryland, 1968. *Atmosphere, Space Physics, Cosmic Rays*. Space plasma physics; Earth's radiation belts.

Reynolds, Christopher, Ph.D., University of Cambridge, 1996. *Astrophysics, High Energy Physics, Relativity & Gravitation*. High-energy astrophysics; black holes; AGN.

Richardson, Derek, Ph.D., University of Cambridge, 1994. *Astrophysics, Computational Physics, Planetary Science*. Asteroid evolution; granular dynamics; computational astrophysics.

Sunshine, Jessica, Ph.D., University of California, San Diego, 1993. *Astronomy, Planetary Science*. Comets; asteroids; space missions.

Veilleux, Sylvain, Ph.D., University of California, Santa Cruz, 1989. Director, Discovery Channel Telescope Partnership. *Astronomy*. Extragalactic astronomy; AGNs; formation and evolution of galaxies.

Vogel, Stuart, Ph.D., University of California, Berkeley, 1983. Department Chair. *Astronomy*. Millimeter-wave astronomy; interstellar medium; extragalactic astronomy.

Associate Professor

Bolatto, Alberto D., Ph.D., Boston University, 2000. *Astronomy*. Extragalactic astronomy; IR and radio astronomy.

Ricotti, Massimo, Ph.D., University of Colorado, 2001. *Astrophysics, Computational Physics, Cosmology & String Theory*. Theoretical cosmology; galaxy formation; computational astrophysics.

Assistant Professor

Boylan-Kolchin, Michael, Ph.D., University of California, Berkeley, 2006. *Astrophysics, Computational Physics, Cosmology & String Theory*. Cosmology and galaxy formation; computational astrophysics.

Gezari, Suvi, Ph.D., Columbia University, 2005. *Astronomy*. Time domain astrophysics; supermassive black holes and AGNs; supernovae.

Emeritus

A'Hearn, Michael F., Ph.D., University of Wisconsin-Madison, 1966. *Astronomy, Planetary Science*. Comets; asteroids; space missions.

Earl, James A., Ph.D., Massachusetts Institute of Technology, 1957. *Atmosphere, Space Physics, Cosmic Rays*. Cosmic rays.

Erickson, William C., Ph.D., University of Minnesota, 1956. *Astronomy*. Radio astronomy; extragalactic astronomy.

Harrington, J. Patrick, Ph.D., Ohio State University, 1967. *Astrophysics*. Planetary nebulae; interstellar matter; stellar atmospheres.

Leventhal, Marvin, Ph.D., Brown University, 1964. *Astronomy*. Gamma-ray astronomy.

DEPARTMENTAL RESEARCH SPECIALTIES AND STAFF

Theoretical

Atmospheric & Space Physics. Space plasma physics, terrestrial radiation belts. Papadopoulos.

Computational Astrophysics. Asteroid evolution. Richardson.

Computational Astrophysics. Simulations of cosmological halo evolution. Boylan-Kolchin, Ricotti.

Cosmology. Cosmology and galaxy formation. Boylan-Kolchin, Ricotti.

High Energy Astrophysics. Active galactic nuclei, black holes, neutron stars, gravitational radiation. Miller, Reynolds.

Planetary Science. Solar system dynamics: rings, asteroids, collisions; solar system origins. Hamilton, Richardson.

Relativity & Gravitation. Black hole astrophysics. Miller.

Experimental

CARMA Millimeter-Wave Astronomy. Star formation; interstellar medium; galactic structure, dynamics and evolution; protostellar disks; active galactic nuclei; instrumentation. Bolatto, Harris, Mundy, Vogel.

Extragalactic Astronomy. Optical, infrared, radio, and x-ray observations. Active galactic nuclei; jets; starbursts; star formation; galactic winds; intergalactic medium; galaxy clusters; dark matter; cosmology. Bolatto, Gezari, Mushotzky, Reynolds, Veilleux.

Planetary Science. Comets; asteroids; Deep Impact and EPOXI missions (A'Hearn, Sunshine). Extrasolar planets (Deming). A'Hearn, Deming, Sunshine.

View additional information about this department at
www.gradschoolshopper.com

UNIVERSITY OF MARYLAND

DEPARTMENT OF CHEMICAL PHYSICS

College Park, Maryland 20742
http://www.chemicalphysics.umd.edu/

General University Information

President: Wallace D. Loh
Dean of Graduate School: Dr. Charles Caramello
University website: http://umd.edu
Control: Public
Setting: Suburban

Total Faculty: 3,578
Total number of Students: 37,595
Total number of Graduate Students: 10,805

Department Information
Department Chairman: Michael Coplan, Director
Department Contact: Debbie Jenkins, Program Coordinator
Full-Time Graduate Students: 38
First-Year Graduate Students: 6

Department Address
Computer and Space Science Building
Room 4203
College Park, MD 20742
Phone: (301) 405-4780
Fax: (301) 314-9363
E-mail: dajenkin@umd.edu
Website: http://www.chemicalphysics.umd.edu/

ADMISSIONS

Admission Contact Information
Address admission inquiries to: Professor Michael A. Coplan,
Chemical Physics Program, Institute for Physical Science and
Technology
Phone: (301) 405-4858
E-mail: coplan@umd.edu
Admissions website: http://www.gradschool.umd.edu

Application deadlines
Fall admission:
U.S. students: February 1   *Int'l. students*: February 1
Spring admission:
U.S. students: June 1   *Int'l. students*: June 1

Application fee
U.S. students: $75   *Int'l. students*: $75
Application fee is nonrefundable.

Admissions information
For Fall of 2013:
Number of applicants: 22
Number admitted: 12
Number enrolled: 5

Admission requirements
Bachelor's degree requirements: Bachelor's degree in physics,
chemistry, engineering, or mathematics is required.
Minimum undergraduate GPA: 3.5

GRE requirements
The GRE is required.
Quantitative score: 776
Verbal score: 541
Mean GRE score range (25th–75th percentile): 70

Advanced GRE requirements
The Advanced GRE is not required.

TOEFL requirements
The TOEFL exam is required for students from non-English-
speaking countries.
iBT score: 100
IBT TOFEL: Score of 84-99 provisional range must take and
English proficiency placement exam upon their arrival on
campus.

Other admissions information
Additional requirements: IELTS minimum acceptable score for
admission is 6.5.
Undergraduate preparation assumed: Bachelor degree in physics, chemistry, engineering, or mathematics is required with
a minimum undergraduate GPA of 3.5.

TUITION
Tuition year 2010–11:
Tuition for in-state residents
Full-time students: $525 per credit
Part-time students: $525 per credit
Tuition for out-of-state residents
Full-time students: $1,077 per credit
Part-time students: $1,077 per credit
Credit hours per semester to be considered full-time: 10
Deferred tuition plan: Yes
Health insurance: Available
Other academic fees: $691.49 per semester, Mandatory Fee
Academic term: Semester

Teaching Assistants, Research Assistants, and Fellowships
Number of first-year
Teaching Assistants: 6
Fellowship students: 6
Average stipend per academic year
Teaching Assistant: $17,666
Research Assistant: $22,848
Fellowship student: $5,000

FINANCIAL AID

Application deadlines
Fall admission:
U.S. students: February 1   *Int'l. students*: February 1
Spring admission:
U.S. students: June 1   *Int'l. students*: June 1

Loans
Loans are available for U.S. students.
Loans are not available for international students.
GAPSFAS application required: Yes
FAFSA application required: Yes

For further information
Address financial aid inquiries to: Professor Michael A. Coplan,
Director, Chemical Physics Program, Institute for Physical
Science and Technology.
Phone: (301) 405-4780
E-mail: coplan@umd.edu
Financial aid website: http://www.chemicalphysics.umd.edu/
08_financial.htm

HOUSING

Availability of on-campus housing
Single students: No
Married students: No

For further information
Address housing inquiries to: University of Maryland Off-
Campus Housing, 1110 Stamp Student Union, College Park,
MD 20741.
Phone: (301) 314-3645
E-mail: och@umd.edu
Housing aid website: http://www.och.umd.edu/

Table A—Faculty, Enrollments, and Degrees Granted

Research Specialty	2012–13 Faculty	Enrollment Fall 2012		Number of Degrees Granted 2012–13		
		Master's	Doctorate	Master's	Terminal Master's	Doctorate
Chemical Physics	63	–	33	2(-)	–	6(6)
Total	63	–	33	2(-)	–	6(-)
Full-time Grad. Stud.	–	–	33	–	–	–
First-year Grad. Stud.	–	–	6	–	–	–

GRADUATE DEGREE REQUIREMENTS

Master's: MASTER'S WITH THESIS ● Written Masters Thesis ● B average ● Scholarly paper ● 30 graduate credits including: 1. Six credits of CHPH799 - (M.S. thesis research) 2. 24 course credits 3. Two credits of seminar, can be included in the non-course credits 4. Advanced laboratory course 5. Advanced course at the 600 level or above The Examining Committee consists of at least two faculty members, who will read the scholarly paper and attend the oral presentation. The paper should provide an informative review of the research topic selected by the candidate in consultation with his/her academic and research advisors. The bibliography is a particularly important part of the paper and should include the most significant references to the topic. The length of the paper is expected to be approximately 20 double space pages (12-point font) with 1-inch margins. The presentation is to last approximately one hour and can be part of regularly scheduled seminar series such as the Informal Statistical Mechanics Seminar or the Nonlinear Dynamics Seminar. Two faculty must be present and there should be sufficient time for questions and discussion. The Thesis Examining Committee is to consist of at least three faculty members including the research advisor. The Examination Committee will review the M.S. thesis, attend the oral presentation and participate in the defense of the thesis. The thesis is to consist of an introduction to the field of research with which the student is engaged, a clear statement of the problem under study, the objectives of the research, the approach taken, original results, interpretation, discussion, and conclusions. A concise review of the literature, and a bibliography of the most important literature should also be included. The M.S. thesis has no set length, but is typically 30 to 40 pages. The format of the thesis (font, margins, etc.) must follow the University of Maryland Thesis and Dissertation Style Guide. MASTER'S NON-THESIS ● Written Qualifying Examination passed at the M.S. level ● B average ● Scholarly paper ● 30 graduate credits of which 24 must be course credits including: 1. Advanced laboratory course 2. Two credits of seminar, can be included in the non-course credits 3. Advanced course at the 600 level or above The Examining Committee consists of at least two faculty members, who will read the scholarly paper and attend the oral presentation. The paper should provide an informative review of the research topic selected by the candidate in consultation with his/her academic and research advisors. The bibliography is a particularly important part of the paper and should include the most significant references to the topic. The length of the paper is expected to be approximately 20 double space pages (12-point font) with 1-inch margins. The presentation is to last approximately one hour and can be part of regularly scheduled seminar series such as the Informal Statistical Mechanics Seminar or the Nonlinear Dynamics Seminar. Two faculty must be present and there should be sufficient time for questions and discussion.

Doctorate: ● Written Qualifying Examination passed at the Ph.D. level and normally taken at the beginning of the second year ● B average ● Scholarly paper research presentation ● 24 graduate course credits including: 1. Two credits of seminar 2. Advanced laboratory course 3. Advanced course outside of the student's main field of study at the 600 level or above In order to advance to Ph.D. candidacy, the student must submit a scholarly paper and make an oral presentation. The paper and presentation are evaluated by a candidacy committee consisting of at least two faculty members, generally including the advisor and a member of the advisory committee. Students with a well-developed thesis topic and research results are expected to include these results together with further research plans in their paper and presentation. Students less far along with research will present background material and summaries of the research areas in which they will be working. A concise review of the literature is expected, along with a bibliography of the most important literature. The length of the paper is expected to be between approximately 20 double space pages (12-point font) with 1-inch margins. The paper is to be submitted to the candidacy committee at least two weeks before the date of the oral presentation. The presentation is to last approximately 50 minutes and can be part of regularly scheduled seminar series such as the Informal Statistical Mechanics Seminar or the Nonlinear Dynamics Seminar. Two members of the candidacy committee must be present and there should be sufficient time for questions and discussion. Within 12 to 18 months after beginning Ph.D. research, the student is to select a Ph.D. Thesis Examination Committee. 12 credits of CHPH899 (Ph.D. dissertation research, only available after advancement to Ph.D. candidacy) Written Ph.D. dissertation. The format of the dissertation (font, margins, etc.) must follow the University of Maryland Thesis and Dissertation Style Guide.

Other Degrees: QUALIFYING EXAMINATION: Students are usually advised to take the Qualifying Examination after their first summer, but some students require more than one year to prepare for the Qualifying Examination. The written examination has two parts: Chemical Physics and Quantum Mechanics. The examination questions are based on the course material in Physics 622 (Quantum Mechanics I), Physics 623 (Quantum Mechanics II), Chemistry 684 (Chemical Thermodynamics), Chemistry 687, (Statistical Mechanics) and Chemistry 691(Quantum Chemistry), Physics 601 (Classical Mechanics) with optional questions based on the material in Physics 606 (Electrodynamics), and Chemistry 601 (Inorganic Chemistry). The examination is two half-days in length held at the same time and in the same place as the Physics Qualifying Examination. The first day of the Qualifying Examination students choose either three or four questions from six questions that are based on the material in Physics 601, Physics 606, Chemistry 601, Chemistry 684, Chemistry 687, and Chemistry 691. On the second day the questions are identical to those on the Quantum part of the Physics Qualifying Examination. The results of the Qualifying Examination are: 1. Pass 2. Conditional Pass with an oral examination 3. Conditional Pass with conditions set by the Qualifying Examination Committee 4. Fail The purpose of the oral examination is to determine whether a student with satisfactory grades on most of the written examination but a deficiency in a limited area is qualified to pursue Ph.D. research without further course work or supplemental study. When the oral examination confirms the results of the written examination, the Qualifying Examination Committee, can propose remedies including additional course work, individual study, retaking portions of the written examination, or a second oral examination.

FACULTY

Distinguished University Professor

Alexander, Millard H., Ph.D., University of Paris, 1967. Molecular collisions and energy transfer.

Fisher, Michael E., Ph.D., King's College, London, 1957. Statistical mechanics; condensed matter theory; physical chemistry and associated mathematics.

Gupta, Ashwani K., Ph.D., University of Sheffield, 1973. Combustion, laser probes, and diagnostics.

Lorimer, George H., Ph.D., Michigan State University, 1972. Studies of the mechanism of chaperonin-assisted protein folding.

Ott, Edward, Ph.D., Polytechnic Institute of New York, 1967. Chaos in dynamical systems including fundamental aspects of chaotic dynamics and applications.

Phillips, William D., Ph.D., Massachusetts Institute of Technology, 1976. Interaction of atoms and photons.

Sengers, Jan, Ph.D., University of Amsterdam, 1962. Fluctuation phenomena in soft condensed matter.

Thirumalai, Devarajan, Ph.D., University of Minnesota, 1982. Problems in equilibrium and non-equilibrium statistical mechanics, aspects of the transition from liquid to amorphous state, theoretical study of polymer-colloid interactions, dynamics of protein folding.

Weeks, John D., Ph.D., University of Chicago, 1969. State and dynamic properties of interfaces; pattern formation during crystal growth; theories for the structure and dynamics of liquids.

Williams, Ellen D., Ph.D., California Institute of Technology, 1982. Experimental research on the properties of solid surfaces.

Professor

Anisimov, Mikhail, Ph.D., Moscow State University, 1968. Phase transitions and critical phenomena in fluids, fluid mixtures, liquid crystals, surfactant solutions and other "soft" condensed matter systems.

Briber, Robert M., Ph.D., University of Massachusetts, 1984. Thermodynamics and structure of polymers with an emphasis on the study of complex mixtures of polymers by neutron scattering.

Carton, James, Ph.D., Princeton University, 1983. Physical oceanography, ocean circulation, dynamics of the tropical Atlantic as it responds to meteorological forcing on seasonal and longer timescales.

Chellappa, Ramailingam, Ph.D., Purdue University, 1981. Image analysis and computer vision techniques to medical imaging.

Coplan, Michael A., Ph.D., Yale University, 1963. Electron and ion impact spectroscopy, space physics.

Dagenais, Mario, Ph.D., Rochester University, 1978. Nonlinear optical interactions in condensed matter, optical switching.

Davis, Christopher C., Ph.D., University of Manchester, 1970. Quantum electronics; molecular energy transfer, atmospheric trace monitoring.

Dickerson, Russell R., Ph.D., University of Michigan, 1980. Analytical techniques in atmospheric chemistry.

Ehrman, Sheryl, Ph.D., University of California, Los Angeles, 1997. Gas-phase synthesis routes to nanostructured materials, particle-surface interactions in chemical polishing, chemical transformation and transport of particulates, air pollution.

Eichhorn, Bryan, Ph.D., Indiana University, 1987. Materials chemistry, materials synthesis, heterogeneous catalysis, fuel cell research, transition metal main group clusters (Zintl ions), NMR spectroscopy, crystallography.

Einstein, Theodore L., Ph.D., University of Pennsylvania, 1973. Surface science, phase transitions of absorbates, surface spectroscopy.

Falvey, Daniel L., Ph.D., University of Illinois, 1989. Photochemical and photophysical behavior of biological molecules.

Fourkas, John, Ph.D., Stanford University, 1991. Ultra fast nonlinear optical spectroscopy of liquids, dynamics of nanoconfined liquids, nonlinear optical microscopy, nontraditional approaches to micro and nanofabrication, dynamics of single molecules and single nanoparticles.

Fushman, David, Ph.D., University of Kazan, 1985. Theoretical and experimental studies of structure, dynamics, and interactions of biomacromolecules.

Hill, Wendell T., Ph.D., Stanford University, 1980. Atomic and molecular structure, laser spectroscopy.

Jarzynski, Christopher, Ph.D., University of California, Berkeley, 1994. Statistical mechanics at the molecular level, foundations of nonequilibrium thermodynamics, application of statistical mechanics to biophysics, development of efficient numerical schemes for estimating thermodynamic properties of complex systems.

Kirkpatrick, Theodore R., Ph.D., Rockefeller University, 1981. Nonequilibrium statistical mechanics, quantum fluids, glasses, disordered electronic systems.

Kofinas, Peter, Ph.D., Massachusetts Institute of Technology, 1994. Synthesis and characterization of block polymers.

Lathrop, Daniel P., Ph.D., University of Texas, Austin, 1991. Nonlinear dynamics and chaos: turbulence, fluid dynamics. Singularities in liquid surface waves, the origin and dynamics of the magnetic field of the Earth.

Mignerey, Alice C., Ph.D., University of Rochester, 1975. Heavy-ion induced nuclear reactions.

Milchberg, Howard M., Ph.D., Princeton University, 1985. High-intensity laser-matter interactions.

Mullin, Amy, Ph.D., University of Colorado Boulder, 1991. Dynamics of collisions and reactions of high energy molecules, high resolution time-resolved optical probing of state-resolved molecular pathways, vibrationally enhanced chemical reactions, controlling molecular rotation using strong optical fields, reaction mechanisms of radicals and other transient species.

Reutt-Robey, Janice E., Ph.D., University of California, Berkeley, 1986. Transient surface chemical processes—energy transfer, diffusion.

Rolston, Steven, Ph.D., Stony Brook University, 1986. Ultracold neutral plasmas, Bose-Einstein condensation, quantum computation, non-linear atom optics, production of antihydrogen, atomic frequency standards, laser cooling.

Roy, Rajarshi, Ph.D., University of Rochester, 1981. Quantum electronics/optics, noise and nonlinear dynamics in optical systems, laser physics, semiconductor and solid state lasers, fiber and integrated optics, optical bistability, control of spatio-temporal systems, experimental statistical physics.

Salamanca-Riba, Lourdes, Ph.D., Massachusetts Institute of Technology, 1985. Multiferroic nanocomposites, DNA-based biosensors and radiation sensors on GaAs, semiconductor nanowires, and self-assembly for the formation of nanocomposite materials and structures.

Seo, Eun-Suk, Ph.D., Louisiana State University, 1991. Space-based experiments to cosmic-ray H, He, and heavier nuclei energy spectra at energies approaching 10 15 eV.

Sita, Lawrence, Ph.D., Massachusetts Institute of Technology, 1985. Transition and main group metal inorganic and organometallic chemistry, new synthetic methodology, catalyst development, polymers, chemically tailored surfaces and interfaces, molecular and mesoscopic self-assembly, new nanofabrication processes.

Vedernikov, Andrei, Ph.D., Kazan State University, 1986. Aerobic hydrocarbon functionalization, design ligands that allows tuning reactivity of transition metals for activation of kinetically inert bonds and molecules.

Zachariah, Michael, Ph.D., University of California, Los Angeles, 1986. Nanoparticles (aerosols) synthesis, characterization, application, and modeling.

Associate Professor

Cumings, John, Ph.D., University of California, Berkeley, 2002. Experimental condensed matter.

Dimitrakopoulos, Panagiotis, Ph.D., University of Illinois at Urbana-Champaign, 1998. Fluid mechanics of drops and bubbles in constrained geometries.

Lee, Sang Bok, Ph.D., Seoul National University, 1997. Electrochemical synthesis of nanotube-structured materials for electrochromics, supercapacitors, solar cells, MR imaging, drug delivery, and biosensors.

Losert, Wolfgang, Ph.D., City College of City University of New York, 1998. Equilibrium dynamics of granular flows, and the nonlinear dynamics of microstructure formation in alloys and biomaterials.

Martínez-Miranda, Luz J., Ph.D., Massachusetts Institute of Technology, 1985. Analysis and characterization of thin films and buried interfaces, using x-ray scattering and glancing incidence scattering techniques.

Ouyang, Min, Ph.D., Harvard University, 2001. Spin degree of freedom of electrons and nuclei within ordered nano-engineered architectures with an emphasis on synthetic methodologies for spin-based hybrid organic-inorganic nanostructures. Spin-charge interactions and spin transport within nanostructured systems using femtosecond optical spectroscopy, magnetotransport and low temperature scanning probe microscopy. New functional nanospintronic devices and technologies.

Papoian, Garegin, Ph.D., Cornell University, 1999. Complex biological phenomena based on fundamental physical and chemical principles. Current research interests include DNA packing in cells of higher organisms and cellular cytoskeleton and motility.

Raghavan, Srinivasa, Ph.D., North Carolina State University, 1998. Complex fluids and soft matter structured at the micro- or nano- scale; rheology, microscopy and scattering techniques.

Yu, Yihua Bruce, Ph.D., Johns Hopkins University, 1996. Biomaterials engineering focusing on structure function relationships.

Assistant Professor

La Porta, Arthur, Ph.D., University of California, San Diego, 1998. Development of optical techniques to be applied to problems in molecular biology. Single molecular techniques to investigate force, torque generation in molecular motors and to characterize proteins that bind and modify the structure of DNA.

Nie, Zhihong, Ph.D., University of Toronto, 2008. Microfluidics and microreactors; molecular and colloidal self-assembly; soft nanotechnology; nanoscience and nanochemistry; plasmonics and metamaterials; biomedicine and medical diagnostics; biomineralization and bioinspired materials.

Paglione, Johnpierre, Ph.D., University of Toronto, 2004. Strongly correlated electron systems, unconventional superconductivity, quantum criticality, crystal growth.

Rabin, Oded, Ph.D., Massachusetts Institute of Technology, 2004. Synthesis and characterization of nanoparticles and nanowires.

Upadhyaya, Arpita, Ph.D., University of Notre Dame, 2000. Mathematical modeling, quantitative imaging and genetic manipulation to uncover signaling networks and the physical properties of the cell and its surroundings.

Wang, Yu-Huang, Ph.D., Rice University, 2004. Nanomaterials assembly of nanostructures into ordered solids and functional networks.

White, Ian, Ph.D., Stanford University, 2002. Optical biosensors with ring resonator structures, microfluidic technologies and optical biosensors, integrated biosensing devices.

Research Associate Professor

Gudipati, Murthy, Ph.D., Indian Institute of Science, 1987. Astrophysics linked by atomic and molecular spectroscopy from far infrared into deep vacuum ultraviolet.

Adjunct Professor

Clark, Charles W., Ph.D., University of Chicago, 1979. NIST. Measurement for needed by emerging electronic and optical technologies, ultra cold atoms and molecules, Bose-Einstein condensation.

Nossal, Ralph J., Ph.D., University of Michigan, 1963. NIH. Laser inelastic light scattering, statistical mechanics of condensed media; phase changes and rheological properties of polymer networks and biological gels, membrane biophysics, cell motility and chemotaxis.

DEPARTMENTAL RESEARCH SPECIALTIES AND STAFF

Theoretical

Atomic, Molecular, & Optical Physics. Molecular collisions and energy transfer, molecular spectra, molecular quantum mechanics, photoelectron spectroscopy, many-body theory of quantum mechanical scattering, atomic and molecular processes at surfaces. Alexander, Clark.

Quantum Electronics and Optical Physics. Multiphoton processes, strong-field laser-atom interaction, Bose-Einstein condensation. Clark, Phillips, Rolston.

Statistical Physics. Nonequilibrium statistical mechanics, interfaces and surfaces, molecular hydrodynamics, phase transitions and critical phenomena, polymer science, disordered solids, thermophysical properties. Anisimov, Fisher, Jarzynski, Kirkpatrick, Ott, Weeks.

Experimental

Atmospheric Physics and Chemistry. Atmospheric spectroscopy, earth's magnetosphere, atmospheric adiation, aerosol physics, atmospheric photochemistry, laser sensing of atmospheric constituents. Dagenais, Davis, Dickerson.

Biophysics. Fushman, La Porta, Lorimer, Losert, Thirumalai, Upadhyaya, Weeks.

Condensed Matter Physics. Raman, Rayleigh, Brillouin scattering, neutron scattering, nuclear magnetic resonance, critical phenomena and phase transitions, surface science. Martínez-Miranda, Reutt-Robey, Williams.

Molecular Physics. Atomic and molecular spectroscopy, electron and ion impact spectroscopy, picosecond spectroscopy, infrared and Raman spectroscopy. Clark, Coplan, Davis, Fourkas, Hill, Mullin, Phillips, Rolston, Sita.

Photochemistry. Laser induced atomic and molecular collisions, ultraviolet and infrared laser photochemistry, photolysis rates in the atmosphere. Dickerson, Falvey.

Quantum Electronics and Optical Physics. Gas laser; light scattering, spectroscopy, nonlinear optics, ultrafast phenomena, laser plasmas, dye lasers and intense field effects. Davis, Hill, Mignerey, Ouyang, Phillips, Roy.

Space Physics. Solar and planetary atmospheres, solar wind, cosmic ray. Coplan, Seo.

Thermophysic. Energy conversion, thermodynamics properties, transport properties. Anisimov, Cumings, Dimitrakopoulos, Gupta, Raghavan.

UNIVERSITY OF MARYLAND

DEPARTMENT OF PHYSICS

College Park, Maryland 20742
http://www.umdphysics.umd.edu

General University Information
President: Wallace D. Loh
Dean of Graduate School: Charles Caramello
University website: http://www.umd.edu/
Control: Public
Setting: Suburban
Total Faculty: 3,996
Total number of Students: 37,631
Total number of Graduate Students: 10,805

Department Information
Department Chairman: Andrew Baden, Chair
Department Contact: Sarah Eno, Associate Chair for Graduate
  Education
  Total full-time faculty: 96
  Total number of full-time equivalent positions: 76
  Full-Time Graduate Students: 245
  First-Year Graduate Students: 39
  Female First-Year Students: 9
  Total Post Doctorates: 80

Department Address
1120 John S. Toll Physics
College Park, MD 20742
Phone: (301) 405-5982
E-mail: grad@umd.edu
Website: http://www.umdphysics.umd.edu

ADMISSIONS

Admission Contact Information
Address admission inquiries to: Ms. Linda O'Hara, Secretary,
  Graduate Entrance Committee, Department of Physics, Uni-
  versity of Maryland, College Park, MD 20742-4111
Phone: (301) 405-5982
E-mail: grad@physics.umd.edu
Admissions website: http://www.umdphysics.umd.edu/
  academics/graduate/how-to-apply.html

Application deadlines
Fall admission:
U.S. students: December 15   *Int'l. students*: December 15

Application fee
U.S. students: $75   *Int'l. students*: $75

Admissions information
For Fall of 2012:
  Number of applicants: 742
  Number admitted: 135
  Number enrolled: 39

Admission requirements
Bachelor's degree requirements: Bachelor's degree is required.
Minimum undergraduate GPA: 3.0

GRE requirements
The GRE is required.

Advanced GRE requirements
The Advanced GRE is required.

TOEFL requirements
The TOEFL exam is required for students from non-English-
  speaking countries.
  PBT score: 575
  iBT score: 100

Other admissions information
Additional requirements: Strong undergraduate academic record
  is required. Undergraduate research experience is highly val-
  ued. Strength of letters of recommendation is strongly considered, as is performance on the Advanced Physics GRE. All
  foreign students admitted with teaching responsibilities, or
  those who may assume teaching responsibilities at a later date,
  must participate in the Maryland English Institute (MEI) evaluation before beginning their first semester.

TUITION

Tuition year 2013-14:
Tuition for in-state residents
  Full-time students: $5,730 per semester
  Part-time students: $4,584 per semester
Tuition for out-of-state residents
  Full-time students: $12,360 per semester
  Part-time students: $9,888 per semester
Virtually all students receive an assistantship that includes a tu-
  ition waiver. For a schedule of costs, please see http://
  www.umd.edu/bursar/t_grd1112.html.
Credit hours per semester to be considered full-time: 9
Deferred tuition plan: No
Health insurance: Yes, $503-$855.
Other academic fees: Mandatory Fees: $723/full-time per semes-
  ter; $403.15 part-time per semester.
Academic term: Semester
Number of first-year students who received full tuition waivers: 38

Teaching Assistants, Research Assistants, and Fellowships
Number of first-year
  Teaching Assistants: 29
  Research Assistants: 11
  Fellowship students: 33
Average stipend per academic year
  Teaching Assistant: $23,980
  Research Assistant: $23,980
  Fellowship student: $5,000

FINANCIAL AID

Application deadlines
Fall admission:
U.S. students: December 15   *Int'l. students*: December 15

Loans
Loans are available for U.S. students.
Loans are not available for international students.
GAPSFAS application required: No
FAFSA application required: No

For further information

Address financial aid inquiries to: Graduate Entrance Committee, Department of Physics, University of Maryland, College Park, MD 20742.

Phone: (301) 405-3401

E-mail: grad@physics.umd.edu

HOUSING

Availability of on-campus housing

Single students: No

Married students: No

For further information

Address housing inquiries to: Office of off-campus housing, 1110 Stamp Student Union, University of Maryland, College Park, MD 20742.

Phone: (301) 314-3645

Housing aid website: http://www.umd.och101.com

Table A—Faculty, Enrollments, and Degrees Granted

Research Specialty	2012-13 Faculty	Enrollment Fall 2012 Master's	Enrollment Fall 2012 Doctorate	Number of Degrees Granted Fall 2012-Spring 2013 (2008-13) Master's	Number of Degrees Granted Fall 2012-Spring 2013 (2008-13) Terminal Master's	Number of Degrees Granted Fall 2012-Spring 2013 (2008-13) Doctorate
Astrophysics	–	–	–	–	–	–(7)
Atmosphere, Space Physics, Cosmic Rays	–	–	–	–	–	–(6)
Atomic, Molecular, & Optical Physics	–	–	–	–	–	2(15)
Biophysics	–	–	–	–	–	–(3)
Condensed Matter Physics	–	–	–	–	–	8(35)
Electrical Engineering	–	–	–	–	–	–(1)
Nano Science and Technology	–	–	–	–	–	–(3)
Nonlinear Dynamics and Complex Systems	–	–	–	–	–	2(23)
Nuclear Physics	–	–	–	–	–	2(10)
Particles and Fields	–	–	–	–	–	2(14)
Physics and other Science Education	–	–	–	–	–	1(4)
Physics of Beams	–	–	–	–	–	1(2)
Plasma and Fusion	–	–	–	–	–	3(13)
Relativity & Gravitation	–	–	–	–	–	–(11)
Statistical & Thermal Physics	–	–	–	–	–	–(2)
Total	–	–	–	–	–	21(149)
Full-time Grad. Stud.	–	–	213	–	–	–
First-year Grad. Stud.	–	–	34	–	–	–

GRADUATE DEGREE REQUIREMENTS

Master's: Master's Degree non-thesis: Minimum of 30 credits with 18 at graduate level. Must include at least four courses in general physics, also graduate lab. Student must pass at Master's level one part of the Ph.D. Qualifying Exam and prepare a scholarly paper. Student must maintain an overall "B" average in all courses taken. Residence for at least two semester is required. A five-year calendar limit is imposed for both full-time and part-time students. Master's Degree with thesis: Same minimum credits as above, which will include six credits of thesis and research. Passing one part of Ph.D. Qualifying Exam at Master's level and completing scholarly paper are not required. An oral examination in defense of thesis and covering all course material must be passed.

Doctorate: No minimum number of credits is required. However, a student must have at least 12 credits of thesis research, at least two credits of seminar, and six credits outside specialty. An overall B average must be maintained. At least two semesters of full-time residence are required. Ph.D. Qualifying Exam must be passed by end of second year. No foreign language is required. Each student must present a paper as evidence of scholarly writing ability as a condition for admission for candidacy. This paper must be written independently of and in excess of course requirements.

Thesis: Thesis may be written in absentia.

SPECIAL EQUIPMENT, FACILITIES, OR PROGRAMS

Owing to its excellent location near the NASA Goddard Space Flight Center, the National Institute of Standards and Technology, and other government research entities, the University of Maryland offers superb placements for its students. In addition, a state-of-the-art Physical Sciences Complex on campus will provide extraordinary laboratory space when it opens in 2013. http://www.cmns.umd.edu/psc.htm.

Table B—Separately Budgeted Research Expenditures by Source of Support

Source of Support	Departmental Research	Physics-related Research Outside Department
Federal government	$32,738,733.89	$12,726,897.26
State/local government		
Non-profit organizations		
Business and industry		
Other	$2,357,825.17	$1,369,984.46
Total	$35,096,559.06	$14,096,881.72

Table C—Separately Budgeted Research Expenditures by Research Specialty

Research Specialty	No. of Grants	Expenditures ($)
Atmosphere, Space Physics, Cosmic Rays	24	$3,655,724.83
Atomic, Molecular, & Optical Physics	22	$5,856,119.72
Biophysics	7	$514,792.04
Chemical Physics	6	$234,127.65
Condensed Matter Physics	41	$6,124,085
Electromagnetism	20	$2,202,210.11
Materials Science, Metallurgy	1	$1,451,034.63
Mechanics	5	$5,887,123.46
Nuclear Physics	14	$670,240.98
Physics and other Science Education	5	$250,981.04
Plasma and Fusion	47	$6,503,409.12
Relativity & Gravitation	13	$1,411,720.04
High Energy Physics	7	$2,024,118.34
Other	27	$12,306,753.82
Total	239	$49,193,440.78

FACULTY

Professor

Anderson, J. Robert, Ph.D., Iowa State University, 1963. *Condensed Matter Physics.* Experimental condensed matter physics; diluted magnetic semiconductors; electronic structures and Fermi surfaces of metals and semimetals.

Anlage, Steven, Ph.D., California Institute of Technology, 1988. *Condensed Matter Physics, Nano Science and Technology.*

Superconductivity-electromagnetic properties, proximity effect; near-field microwave microscopy; experimental chaos.

Antonsen, Thomas, Ph.D., Cornell University, 1977. *Plasma and Fusion*. Plasma physics; coherent sources of radiation.

Baden, Andrew R., Ph.D., University of California, Berkeley, 1986. Department chair. *High Energy Physics*. Experimental high-energy physics with accelerators; data acquisition; high-performance computing; data analysis.

Banavar, Jayanth, Ph.D., University of Pittsburgh, 1978. *Nonlinear Dynamics and Complex Systems*.

Beise, Elizabeth J., Ph.D., Massachusetts Institute of Technology, 1988. *Nuclear Physics*. Experimental nuclear physics-intermediate energy; electron scattering; polarization; few-nucleon and subnucleon systems.

Buonanno, Alessandra, Ph.D., University of Pisa, 1996. *Cosmology & String Theory, Relativity & Gravitation*. Gravity theory; gravitational waves.

Chen, Hsing-Hen, Ph.D., Columbia University, 1973. *Nonlinear Dynamics and Complex Systems*. Astrophysics; plasma physics; nonlinear dynamical systems.

Cohen, Thomas D., Ph.D., University of Pennsylvania, 1984. Associate chair for graduate education. *Nuclear Physics*. Nuclear theoretical physics; soliton models of baryons; chiral symmetry; effective low-energy models for QCD.

Das Sarma, Sankar, Ph.D., Brown University, 1979. Distinguished University Professor; Director, Condensed Matter Theory Center. *Condensed Matter Physics, Nano Science and Technology, Statistical & Thermal Physics*. Theoretical condensed matter; many body theory; semiconductor nanostructures; nonequilibrium statistical mechanics.

Dorland, William, Ph.D., Princeton University, 1993. *Plasma and Fusion*. Turbulence in magnetized plasma; computational physics.

Drake, James F., Ph.D., University of California, Los Angeles, 1975. *Plasma and Fusion, Solar Physics*. Plasma physics; magnetic reconnection; Tokamak transport.

Einstein, Theodore L., Ph.D., University of Pennsylvania, 1973. *Chemical Physics, Condensed Matter Physics, Materials Science, Metallurgy*. Theoretical condensed matter physics; surface physics; statistical and thermal physics.

Ellis, Richard F., Ph.D., Princeton University, 1970. *Plasma and Fusion*. Experimental plasma physics; plasma waves and instabilities; microwave and far infrared diagnostics for fusion plasmas; plasma probes and analyzers.

Eno, Sarah C., Ph.D., University of Rochester, 1990. *High Energy Physics*. Experimental high-energy physics with accelerators.

Fisher, Michael E., Ph.D., King's College, 1957. Statistical physics; condensed matter theory; theoretical chemistry; phase transitions and critical phenomena; associated mathematics.

Fuhrer, Michael S., Ph.D., University of California, Berkeley, 1998. *Condensed Matter Physics, Materials Science, Metallurgy, Nano Science and Technology*. Experimental condensed matter.

Galitski, Victor M., Ph.D., University of Minnesota, 2002. *Condensed Matter Physics*. Condensed matter theory.

Gates, S. James, Ph.D., Massachusetts Institute of Technology, 1977. *Cosmology & String Theory, Particles and Fields*. Elementary particles-supersymmetry, supergravity, and superstrings.

Goodman, Jordan A., Ph.D., University of Maryland, 1978. *Astrophysics, High Energy Physics*. Particle astrophysics.

Greenberg, Oscar W., Ph.D., Princeton University, 1956. *Particles and Fields*. Elementary particles and field theory.

Greene, Richard L., Ph.D., Stanford University, 1967. *Condensed Matter Physics*. Experimental condensed matter physics.

Hadley, Nicholas J., Ph.D., University of California, Berkeley, 1983. *High Energy Physics*. High-energy physics.

Hamilton, Douglas C., Ph.D., University of Chicago, 1977. *Atmosphere, Space Physics, Cosmic Rays, Solar Physics*. Experimental space physics; magnetospheric physics; solar wind, solar energetic particles; particle acceleration and transport.

Hassam, Adil B., Ph.D., Princeton University, 1978. *Plasma and Fusion, Solar Physics*. Plasma physics of the sun; thermonuclear fusion.

Hu, Bei-Lok, Ph.D., Princeton University, 1972. *Cosmology & String Theory, Relativity & Gravitation, Statistical & Thermal Physics*. General relativity; gravitation and cosmology; quantum field theory; statistical field theory.

Jacobson, Theodore A., Ph.D., University of Texas, Austin, 1983. *Quantum Foundations, Relativity & Gravitation*. Gravitation theory; quantum gravity; black hole thermodynamics.

Jawahery, Abolhassan, Ph.D., Tufts University, 1981. *High Energy Physics*. High-energy physics with accelerators.

Ji, Xiangdong, Ph.D., Drexel University, 1987. *Nuclear Physics, Quantum Foundations*. Theoretical nuclear physics; quantum chromodynamics; quark and gluon structure of hadrons.

Kirkpatrick, Theodore, Ph.D., Rockefeller University, 1981. *Condensed Matter Physics, Statistical & Thermal Physics*. Theoretical statistical mechanics; condensed matter theory.

Lathrop, Daniel P., Ph.D., University of Texas, 1991. *Fluids, Rheology, Nonlinear Dynamics and Complex Systems*. Nonlinear dynamics and chaos; turbulence; fluid dynamics.

Liu, Chuan Sheng, Ph.D., University of California, Berkeley, 1968. *Plasma and Fusion*. Plasma physics; fusion and space science.

Lobb, Christopher J., Ph.D., Harvard University, 1980. *Condensed Matter Physics*. Experimental superconductivity; superconducting devices; physics and applications of mesoscopic systems; condensed matter physics.

Mather, John C., Ph.D., University of California, Berkeley, 1974. *Astrophysics, Cosmology & String Theory*. Cosmology; far IR astronomy and instrumentation; Fourier transform spectroscopy.

Milchberg, Howard M., Ph.D., Princeton University, 1985. *Atomic, Molecular, & Optical Physics*. Atomic, molecular and optical physics; nonlinear optics, laser and optical physics.

Mohapatra, Rabindra N., Ph.D., University of Rochester, 1969. *Cosmology & String Theory*. Elementary particles; quantum field theory; cosmology.

Monroe, Christopher, Ph.D., University of Colorado Boulder, 1992. *Atomic, Molecular, & Optical Physics, Condensed Matter Physics, Quantum Foundations*. Cold atomic physics; quantum information science; ultrafast control of cold atoms; interface between atomic and condensed matter physics; foundations of quantum mechanics.

Orozco, Luis, Ph.D., University of Texas, Austin, 1987. *Optics*. Quantum optics; precision measurement; fundamental interactions.

Ott, Edward, Ph.D., Polytechical University., Brooklyn, 1967. *Plasma and Fusion*. Chaotic dynamics; plasmas.

Papadopoulos, Dennis, Ph.D., University of Maryland, 1968. *Plasma and Fusion*. Space plasma physics; lightning; photoconducting plasmas.

Phillips, William D., Ph.D., Massachusetts Institute of Technology, 1976. *Atomic, Molecular, & Optical Physics*. Laser cooling; atom trapping; atomic clocks; atomic and optical physics; cold collisions, photoassociative spectroscopy.

Redish, Edward F., Ph.D., Massachusetts Institute of Technology, 1968. *Physics and other Science Education, Other*. Physics education research and development.

Rolston, Steven L., Ph.D., Stony Brook University, 1986. *Plasma and Fusion, Quantum Foundations*. Laser cooling of neutral atoms; ultra cold plasmas; Bole-Einstein condensations quantum info.

Roy, Rajarshi, Ph.D., University of Rochester, 1981. *Nonlinear Dynamics and Complex Systems*. Nonlinear dynamics in optical systems; laser physics; wave propagation in optical fibers; coherence and stochastic processes.

Sagdeev, Roald Z., Siberian Branch, USSR Acad. of Sciences, 1962. *Astrophysics, Plasma and Fusion, Other*. Plasma physics; controlled fusion; space physics; planetary research and astrophysics; arms control; science policy; global security and environment.

Seo, Eun-Suk, Ph.D., Louisiana State University, 1991. *Atmosphere, Space Physics, Cosmic Rays*. Cosmic Ray physics.

Skuja, Andris, Ph.D., University of California, Berkeley, 1972. *High Energy Physics, Particles and Fields*. Experimental high-energy physics with accelerators; experimental particle physics.

Sprangle, Phillip, Ph.D., Cornell. *Applied Physics*.

Sreenivasan, Katepalli, Ph.D., Indian Instutute of Science.

Sullivan, Gregory W., Ph.D., University of Illinois, 1990. *Astrophysics*. Electroweak physics; Standard Model; top quark search.

Sundrum, Raman, Ph.D., Yale University, 1990. *Particles and Fields*. Particle theory.

Wellstood, Frederick, Ph.D., University of California, Berkeley, 1988. *Condensed Matter Physics*. Superconductivity-High Tc (YBCO) superconducting quantum interference devices; magnetic microscopy; Coulomb blockade electrometers.

Williams, Ellen D., Ph.D., California Institute of Technology, 1982. *Condensed Matter Physics, Statistical & Thermal Physics*. Condensed matter physics; surface science; scanning tunneling microscopy; statistical mechanics of surfaces.

Yakovenko, Victor M., Ph.D., Landau Institute for Theoretical Physics, Moscow, 1987. *Condensed Matter Physics*. Condensed matter theory; organic and high-Tc superconductors; the quantum Hall effect; effects of high magnetic fields.

Yorke, James A., Ph.D., University of Maryland, 1966. *Nonlinear Dynamics and Complex Systems, Statistical & Thermal Physics*. Chaos and non-linear dynamics; statistical physics.

Associate Professor

Agashe, Kaustubh, Ph.D., University of California, Berkeley, 1998. *Particles and Fields*. Particle phenomenology; collider signals.

Appelbaum, Ian, Ph.D., Massachusetts Institute of Technology, 2003. *Condensed Matter Physics*. Condensed matter.

Bedaque, Paulo, Ph.D., University of Rochester, 1994. *Nuclear Physics*. Nuclear theory.

Chacko, Zacharia, Ph.D., University of Maryland, 1999. *Particles and Fields*. Elementary particles.

Girvan, Michelle, Ph.D., Cornell University, 2003. *Nonlinear Dynamics and Complex Systems*. Nonlinear dynamics.

Hall, Carter, Ph.D., Harvard University, 2002. *Nuclear Physics*. Experimental nuclear.

Hoffman, Kara, Ph.D., Purdue University, 1998. *Particles and Fields*. Particle astrophysics.

Losert, Wolfgang, Ph.D., City College of the City University of New York, 1998. *Biophysics, Nonlinear Dynamics and Complex Systems*. Biophysics.

Ouyang, Min, Ph.D., Harvard University, 2001. *Condensed Matter Physics, Nano Science and Technology*. Condensed matter experiment; nanoelectronics.

Paglione, Johnpierre, Ph.D., University of Toronto, 2004. *Condensed Matter Physics*. Condensed Matter Experiment.

Roberts, Douglas A., Ph.D., University of California, 1994. *High Energy Physics*. High-energy physics with accelerators.

Assistant Professor

Belloni, Alberto, Ph.D., Massachusetts Institute of Technology. *High Energy Physics*.

Kim, Kiyong, Ph.D., University of Maryland, 2003. *Optics*. Laser and optical physics.

LaPorta, Arthur, Ph.D., University of California, San Diego, 1996. *Biophysics*. Biophysics, single molecule biophysics.

Levin, Michael, Ph.D., Massachusetts Institute of Technology, 2006. *Condensed Matter Physics*. Condensed Matter Theory.

Manucharyan, Vladamir, Ph.D., Yale.

McKinney, Jonathan C., Ph.D., University of Illinois, 2004. Accretion disk theory near central objects; numerical modeling; GR perturbation theory. *Relativity & Gravitation*.

Sau, Jay, Ph.D., University of California, Berkeley.

Shawhan, Peter, Ph.D., University of Chicago, 1999. *Relativity & Gravitation*. Gravitation experiment.

Tiglio, Manuel, Ph.D., University of Cordoba (Argentina), 2000. *Relativity & Gravitation*. Gravitation Theory.

Upadhyaya, Arpita, Ph.D., University of Notre Dame, 2000. *Biophysics*. Biophysics, force generation by actin polymerization, cell polarization, surface and elastic forces in developing tissues, mechanical properties of membrane nanotubes, biological springs.

Williams, Jimmy, Ph.D., Harvard University.

Emeritus

Roos, Philip, Ph.D., Massachusetts Institute of Technology. *Nuclear Physics*.

Professor Emeritus

Boyd, Derek, Ph.D., Stevens Institute of Technology.

Brill, Dieter, Ph.D., Princeton University.

Chang, Chia-Cheh, Ph.D., University of Southern California. *Nuclear Physics*.

Currie, Douglas, Ph.D., University of Rochester.

Dorfman, Jay Robert, Ph.D., Johns Hopkins.

Dragt, Alex J., Ph.D., University of California, Berkeley, 1963. *Mechanics, Nonlinear Dynamics and Complex Systems, Particles and Fields*. Elementary particles and field theory; mechanics; dynamical systems and accelerator theory; charged particle and light optics.

Dragt, Alex, University of California, Berkeley.

Drew, H. Dennis, Ph.D., Cornell University, 1967. *Condensed Matter Physics, Statistical & Thermal Physics*. Experimental condensed matter physics; statistical and thermal physics; semiconductor heterostructures; infrared properties of superconductors; near-field optical scanning microscopy.

Gloeckler, George, Ph.D., University of Chicago.

Griem, Hans, Ph.D., Universitat Kiel.

Griffin, James, Ph.D., Princeton University. *Nuclear Physics*.

Holmgren, Harry, University of Minnesota.

Kim, Young-Suh, Ph.D., Princeton University.

Misner, Charles, Ph.D., Princeton University.

Paik, Ho Jung, Ph.D., Stanford University, 1974. *Relativity & Gravitation*. Experimental general relativity; gravitational waves; precision tests of laws of gravity.

Park, Robert L., Ph.D., Brown University, 1964. *Condensed Matter Physics*. Experimental condensed matter physics; surface physics; science policy.

Pati, Jogesh, Ph.D., University of Maryland.

Richard, Jean-Paul, Ph.D., Universite de Paris.

Satindar, Bhagat, Ph.D., University of Delhi. *Condensed Matter Physics*.

Sucher, Joseph, Ph.D., Columbia University.

Research Professor

Wallace, Stephen J., Ph.D., University of Washington, 1971. *Relativity & Gravitation*. Theoretical physics—scattering theory;

nucleon-nucleon interactions, relativistic bound states; electron scattering.

Adjunct Professor

Bryant, Garnett, Ph.D., Indiana University, 1978.

Campbell, Gretchen, Ph.D., Massachusetts Institute of Technology, 2006. *Quantum Foundations.*

Clark, Charles W., Ph.D., University of Chicago, 1979. *Quantum Foundations.*

Julienne, Paul S., Ph.D., University of North Carolina, 1969. *Quantum Foundations.*

Lett, Paul D., Ph.D., University of Rochester, 1986. *Quantum Foundations.*

Lynn, Jeffrey W., Ph.D., Georgia Institute of Technology, 1974. *Condensed Matter Physics.* Condensed matter physics; neutron scattering; superconductivity; phase transitions and critical phenomena; magnetic materials.

McEnery, Julie, Ph.D., University College Dublin, 1997. *Astrophysics, Particles and Fields.* Particle astrophysics.

Migdall, Alan, Ph.D., Massachusetts Institute of Technology, 1984. *Quantum Foundations.*

Porto, James (Trey), Ph.D., Cornell University, 1996. *Quantum Foundations.*

Solomon, Glen, Ph.D., Stanford University, 1997. *Quantum Foundations.*

Spielman, Ian, Ph.D., California Institute of Technology, 2004. *Quantum Foundations.*

Taylor, Jacob, Ph.D., Harvard University, 2006.

Tiesinga, Eite, Ph.D., Eindhoven University of Technology, 1993. *Quantum Foundations.*

Tycko, Robert, Ph.D., University of California, Berkeley, 1984. *Nuclear Physics.* Solid state nuclear magnetic resonance (NMR) spectroscopy.

Williams, Carl J., Ph.D., University of Chicago, 1987. *Quantum Foundations.*

Affiliate Professor

Cumings, John, Ph.D., University of California, Berkeley, 2002. *Condensed Matter Physics, Materials Science, Metallurgy.* Condensed matter experiment; materials science.

Hill, Wendell T., Ph.D., Stanford University, 1980. *Atomic, Molecular, & Optical Physics.* Atomic, Molecular and Optical Physics.

O'Shea, Patrick, Ph.D., University of Maryland, 1986. *Particles and Fields, Physics of Beams.* Particles beams.

Oehrlein, Gottlei, Ph.D., State University of New York, Albany, 1981. *Condensed Matter Physics, Materials Science, Metallurgy.* Condensed Matter Experiment; Nanostructure fabrication, materials science, thin films.

Phaneuf, Raymond, Ph.D., University of Wisconsin-Madison, 1985. *Condensed Matter Physics.* Condensed Matter Experiment.

Takeuchi, Ichiro, Ph.D., University of Maryland, 1996. *Materials Science, Metallurgy, Nano Science and Technology.* Center for Nanophysics and Advanced Materials.

Weeks, John D., Ph.D., University of Chicago, 1969. *Condensed Matter Physics.* Condensed Matter Experiment; nanoelectronics, thin films.

Senior Research Scientist

Ipavich, Fred M., Ph.D., University of Maryland, 1972. *Astrophysics, Solar Physics.* Space physics; interplanetary physics; astrophysics; solar physics; magnetospheric physics.

Kane, Bruce, Ph.D., Princeton University, 1988. *Condensed Matter Physics.* Experimental quantum devices; silicon.

Kellogg, Richard G., Ph.D., Yale University, 1975. *High Energy Physics, Particles and Fields.* Experimental high energy and particle physics.

Moody, Martin Vol, Ph.D., University of Virginia, 1980. *Relativity & Gravitation.* Experimental general relativity; gravitational physics.

DEPARTMENTAL RESEARCH SPECIALTIES AND STAFF

Theoretical

Atomic, Molecular, & Optical Physics. Laser cooling and trapping of neutral atoms; quantum optics; cavity quantum electrodynamics; correlated photons; quantum feedback and control; spectroscopy and tests of discrete symmetries in atoms; ultracold atoms in optical lattices; linear and nonlinear atom optics (deBroglie-wave optics); atom interferometry; ultra-cold plasmas; collisions of atoms at ultra-low-energy; photoassociation of ultracold atoms; quantum-degenerate gases and Bose-Einstein condensation; quantum gases in low-dimensions; quantum transport of atoms in periodic potentials; quantum chaos; quantum information and quantum computing. http://umdphysics. umd.edu/research/theoretical/83-atomicmolecularoptical. html Monroe.

Condensed Matter Physics. Many-body theory relating to superconductivity, ferromagnetism, effects of critical fluctuations near phase transitions; properties of interfaces, surfaces and surface adsorbates and steps on surfaces; solitons; the quantum Hall effect; two-dimensional systems; theory of superlattices; transport and photoconductivity in microstructures; theory of crystal growth; quantum chaos; first principles band structure calculations; properties of clusters strongly correlated systems—quasi-one-dimensional and organic conductors, high-Tc superconductors, physical effects of high magnetic fields, quantum computation, econophysics. http://umdphysics.umd. edu/research/theoretical/84-condensedmatter.html Das Sarma, Einstein, Galitski, Levin, Yakovenko.

Elementary Particles. Unified gauge theories of elementary particle interactions; composite models of quarks and leptons; cosmology; supersymmetry; supergravity; superstrings; spontaneous symmetry breaking and mass generation; neutrino masses and mixings; bound-state models of hadrons; confinement of quarks and color in quantum chromodynamics; problems in quantum electrodynamics; quantum groups; small violations of particle statistics; hidden variables and EPR problems; field theory with finite boundary conditions; coherent and squeezed states; and representations of the Lorentz groups. http://umdphysics.umd.edu/research/theoretical/86-elementaryparticles.html Agashe, Chacko, Gates, Greenberg, Mohapatra, Sundrum.

Gravitational Theory. Gravity waves. Numerical relativity. Gravitation and cosmology; quantum field theory in curved spacetime; black hole thermodynamics; the early universe and quantum cosmology; nonequilibrium quantum fields; global and causal structures of spacetimes; topology change in quantum gravity; discrete models for spacetime. http://umdphysics. umd.edu/research/theoretical/87-gravitationaltheory.html Buonanno, Hu, Jacobson, McKinney, Tiglio.

Nonlinear Dynamics and Complex Systems. Chaos, complex systems, statistical physics, nonlinear dynamics,computer science and networks. http://umdphysics.umd.edu/research/ theoretical/88-nonlineardynamicschaos.html Banavar, Girvan, Ott, Roy, Yorke.

Plasma and Fusion. Research on a wide variety of topics in laboratory, magnetospheric, and solar plasma physics, including anomalous transport, plasma flows, the generation and dissipation of magnetic fields, nonlinear waves, nonlinear dynamics and chaos, and plasma interaction with intense radiation and high-energy beams. http://umdphysics.umd.edu/research/theoretical/89-plasmatheory.html. Dorland, Drake, Hassam, Liu, Sagdeev.

Quantum Coherence & Information Theory. http://umdphysics.umd.edu/research/theoretical/90-quantumcoherenceinformationtheory.html Hu.

Quarks, Hadron, and Nuclei. QCD and its connection to hadrons and nuclei, e.g., perturbative QCD, lattice QCD, large N QCD, effective field theory, structure of nucleon and hadrons and nucleon-nucleon interactions. http://umdphysics.umd.edu/research/theoretical/106-quarkshadronnuclei.html. Bedaque, Cohen, Ji.

Spintronics & Spin Quantum Computing. Maryland physicists are exploring the burgeoning field of spintronics, where active control of both spin and charge dynamics of electrons in solids leads to new phenomena and functionalities. The study of topics such as spin relaxation and spin transport is both answering fundamental physics questions and providing applications to the micro-electronics industry. http://umdphysics.umd.edu/research/theoretical/91-spintronicsspinquantumcomputing.html Das Sarma, Drew, Yakovenko.

Experimental

Astrophysics. Experimental study of ultrahigh-energy cosmic rays and nonaccelerator high-energy physics. Search for point sources of ultrahigh- and very high-energy gamma rays using the existing MILAGRO detector. Using the cubic kilometer IceCube detector located beneath the South Pole we will open the field of neutrino astronomy. We explore an energy region where the Universe is opaque to high energy gamma rays originating from beyond the edge of our own galaxy, and where cosmic rays do not carry directional information because of their deflection by magnetic fields. HAWC in Mexico. http://umdphysics.umd.edu/research/experimental/76-particleastrophysics.html. Goodman, Hoffman, McEnery, Sullivan.

Atmosphere, Space Physics, Cosmic Rays. Experimental investigations of high energy particles from outer space. Search for exotic matter such as antimatter and dark matter; precision measurements of galactic cosmic rays to understand their origin, acceleration, propagation and to explore the limit of supernova shock wave acceleration. Particle detector development. Balloon-borne and space-based experiments. Monte Carlo simulations.Experiments on satellites such as SOHO and ACE and deep space probes such as Voyager, Ulysses, and Cassini to study the compositions and distribution functions of space plasmas including the solar wind and solar, heliospheric, and magnetospheric energetic ions. http://umdphysics.umd.edu/research/experimental/71-cosmicray.html. Hamilton, Seo.

Atomic, Molecular, & Optical Physics. Laser cooling and trapping of neutral atoms; quantum optics; cavity quantum electrodynamics; correlated photons; quantum feedback and control; spectroscopy and tests of discrete symmetries in atoms; ultracold atoms in optical lattices; linear and non-linear atom optics (deBroglie-wave optics); atom interferometry; ultracold plasmas; collisions of atoms at ultra-low-energy; photoassociation of ultracold atoms; quantum-degenerate gases and Bose-Einstein condensation; quantum gases in low-dimensions; quantum transport of atoms in periodic potentials; quantum chaos; quantum information and quantum computing. http://umdphysics.umd.edu/research/experimental/68-atomicmolecularoptical.html. Bryant, Campbell, Clark, Julienne, Lett, Migdall, Monroe, Orozco, Phillips, Porto, Rolston, Solomon, Spielman, Taylor, Tiesinga, Carl Williams.

Biophysics. With outstanding scientists from the fields of biology, physics, chemistry and engineering, Maryland's burgeoning biophysics initiative is addressing complex questions in biology, biomedicine and bioengineering. Using novel theoretical and computational methods and next-generation experimental equipment (such as a holographic laser tweezer array and a two-photon confocal microscope), this interdisciplinary team is exploring biophysics at both the cellular and molecular levels. http://umdphysics.umd.edu/research/experimental/69-biophysics.html. LaPorta, Losert, Upadhyaya.

Condensed Matter Physics. Electrons in metals and semiconductors; surface properties of solids; spin dynamics of randomized systems; magnetization at low temperatures; ferromagnetism; spin glasses; re-entrant magnetism; diluted magnetic semiconductors; deHass-Van Alphen effect; properties of amorphous systems; electronic band structure of metals and alloys; microwave, far infrared and visible optical studies of superconductors and semiconductors in high magnetic fields and low temperatures; electron diffraction studies of 2D phase transitions; surface geometry and electronic structure; properties of epitaxial films; modulated optical studies of materials prepared by molecular beam epitaxy; studies of equilibrium and driven surface structures; scanning tunneling microscopy; low-energy electron microscopy; surface modification by laser annealing; preparation and study of artificially structured semiconductor and metal systems; low temperature physics of mesoscopic structures, universal conductance fluctuations, persistent currents in normal metal rings, single electron devices. Low temperature physics of mesoscopic structures, universal conductance fluctuations, persistent currents in normal metal rings, single electron devices. http://umdphysics.umd.edu/research/experimental/70-condensedmatterexperiment.html. Anderson, Anlage, Appelbaum, Drew, Fuhrer, Lobb, Lynn, Ouyang, Paglione, Wellstood, Ellen Williams.

Fluids, Rheology. Quantum turbulence, superfluidity, rotating turbulence, Coriolis-restored internal wave motions, inertial waves and related Rossby waves. Lathrop.

Geophysics. Self-generating magnetic fields (dynamos), turbulent flows of magnetized fluids. Lathrop.

High Energy Physics. Search for the fundamental constituents of matter and the properties of the interactions between these constituents using the techniques of high-energy physics; studies of electron-positron collisions with the OPAL detector at LEP in CERN, and pp interactions with the DO Detector at the Tevatron in Fermilab dominate the program. The Maryland HEP group is also a member of the Solenoidal Detector Collaboration at the SSCL. http://umdphysics.umd.edu/research/experimental/73-highenergyphysics.html. Baden, Belloni, Eno, Hadley, Jawahery, Roberts.

Nonlinear Dynamics and Chaos. Since the mid-1970s, the Chaos Group at Maryland has done extensive research in various areas of chaotic dynamics ranging from the theory of dimensions, fractal basin boundaries, chaotic scattering, controlling chaos, etc. http://umdphysics.umd.edu/research/experimental/74-nonlineardynamicsexperiment.html. Antonsen, Girvan, Lathrop, Losert, Ott, Roy, Yorke.

Nuclear Physics. The experimental Nuclear Physics Group at the University of Maryland is engaged in a program of experiments related to study of the QCD structure of nucleons and fundamental symmetries of nature. Much of our recent work has taken place at the Jefferson National Laboratory in Newport News, VA, where there is a polarized electron beam of energies up to 6 GeV. Our new projects include a search for neutrinoless double beta decay in liquid Xenon (EXO), as well as Dark Matter (LUX), symmetry experiments with cold neutrons at the NIST fundamental neutron beam line in Gaithersburg, MD, and new experiments to study the QCD properties of nucleons (E906). Our group is supported primarily by funds from the National Science Foundation. http://umdphysics.umd.edu/research/experimental/75-nuclearphysics.html Beise, Hall.

Physics Eduation. The University of Maryland Physics Education Research Group is a combined effort of the Physics Department and School of Education. We study the learning and teaching of physics at the high school and university level. Graduate degrees may be earned in physics or in education. http://umdphysics.umd.edu/research/experimental/77-physics educationresearch.html. Redish.

Plasma Physics. Experimental studies in plasma physics are primarily concerned with measurements that broaden understanding of basic properties of plasmas and of ions in the plasma environment: transport properties, fluctuations, and influences of plasma fields on the radiative properties of atoms and ions (plasma spectroscopy). Out of these studies come, in addition to basic data for comparison with plasma theory, new diagnostic techniques that may be applied in future fusion power and industrial techniques. http://umdphysics.umd.edu/research/experimental/78-plasmaphysics.html. Ellis, Hassam, Kiyong Kim, Milchberg.

Relativity & Gravitation. LIGO data interpretation. Antennas sensitive to gravitational radiation are being developed. These include multimode wide band antennas using quantum optics instrumentation. Superconducting transducers for millekelvin temperature antennas are also being developed. A sensitive superconducting gravity gradiometer is being developed for precision test of the inverse square law of gravitation; a highly stable laser clock is being developed. Precision tests of general relativity in space are being studied. http://umdphysics.umd.edu/research/experimental/72-generalrelativityexperiment.html. Shawhan.

Space Physics. The plasma and energetic particle observations carried out by the Space Physics Group require novel instrumentation carried on Earth-orbiting satellites and deep-space probes. Instruments are designed and constructed on campus by the group's technical staff, with participation by graduate as well as undergraduate students. The basic instrumentation technique is time-of-flight mass spectrometry, customized for rugged, lightweight systems as required by space missions. The group actively collaborates with other research teams in the United States and Europe. http://umdphysics.umd.edu/research/experimental/80-spacephysics.html Hamilton, Ipavich.

Superconducting Quantum Computing. The Maryland Superconducting Quantum Computing group has a large experimental and theoretical effort to develop a quantum computer based on superconducting devices. A quantum computer uses quantum mechanics to do calculations. While ordinary computers use a system based on either zeros or ones, quantum computers would use a logical system that is based on zeros, ones or a simultaneous combination of both. This would theoretically allow a quantum computer to do certain calculations, like finding prime factors of a large number, exponentially faster than conventional computers. Anderson, Alex Dragt, Lobb, Wellstood.

View additional information about this department at
www.gradschoolshopper.com

BOSTON COLLEGE

DEPARTMENT OF PHYSICS

Chestnut Hill, Massachusetts 02467-3811
http://www.bc.edu/physics

General University Information
President: Fr. William P. Leahy, S.J.
Dean of Graduate School: David Quigley
University website: http://www.bc.edu
Control: Private
Total Faculty: 708
Total number of Students: 14,796
Total number of Graduate Students: 4,960

Department Information
Department Chairman: Michael J. Naughton, Chair
Department Contact: Stephanie Zuehlke, Program Administrator
Total full-time faculty: 18
Total number of full-time equivalent positions: 18
Full-Time Graduate Students: 46
First-Year Graduate Students: 8
Female First-Year Students: 1
Total Post Doctorates: 3

Department Address
140 Commonwealth Avenue
Higgins 335
Chestnut Hill, MA 02467-3811

Phone: (617) 552-2195
E-mail: physics@bc.edu
Website: http://www.bc.edu/physics

ADMISSIONS

Admission Contact Information
Address admission inquiries to: Admissions, Graduate School of Arts and Sciences, Boston College, 140 Commonwealth Avenue, Chestnut Hill, MA 02467
Phone: 617-552-3268
E-mail: gsasinfo@bc.edu
Admissions website: http://www.bc.edu/gsas

Application deadlines
Fall admission:
U.S. students: January 2   *Int'l. students*: January 2

Application fee
U.S. students: $75   *Int'l. students*: $75

Admissions information
For Fall of 2013:
Number of applicants: 150

Number admitted: 12
Number enrolled: 11

Admission requirements

Bachelor's degree requirements: A bachelor's degree in physics is required.

GRE requirements

The GRE is required.

Advanced GRE requirements

The Advanced GRE is recommended.

TOEFL requirements

The TOEFL exam is required for students from non-English-speaking countries.
  PBT score: 600
  iBT score: 100

Other admissions information

Additional requirements: A statement of purpose, three letters of recommendation, and official transcripts are required.

Undergraduate preparation assumed: Reif, Fundamentals of Statistical and Thermal Physics; Lorrain and Corson, Electromagnetic Fields and Waves (2nd ed.); Symon, Mechanics; Rosenberg, The Solid State; Eisberg, Modern Physics.

TUITION

Tuition year 2013-14:
Tuition for out-of-state residents
  Full-time students: per credit
Deferred tuition plan: No
Health insurance: Available
Other academic fees: $422 (health services)
Academic term: Semester

Teaching Assistants, Research Assistants, and Fellowships

Number of first-year
  Teaching Assistants: 7
  Research Assistants: 1
Average stipend per academic year
  Teaching Assistant: $23,500
  Research Assistant: $24,000

FINANCIAL AID

Loans

Loans are not available for U.S. students.
Loans are not available for international students.
GAPSFAS application required: No
FAFSA application required: No

For further information

Address financial aid inquiries to: Admissions, Graduate School of Arts and Sciences.

HOUSING

Availability of on-campus housing
  Single students: Yes
  Married students: Yes

For further information

Address housing inquiries to: Housing Office.
Phone: 617.552.3060
E-mail: reslife@bc.edu
Housing aid website: http://www.bc.edu/offices/reslife/

Table A—Faculty, Enrollments, and Degrees Granted

Research Specialty	2012-13 Faculty	Enrollment Fall 2013		Number of Degrees Granted 2012-13		
		Master's	Doctorate	Master's	Terminal Master's	Doctorate
Condensed Matter						
Physics	17	3	42	–	3(-)	5(-)
Total	18	–	50	3(-)	-(1)	7(-)
Full-time Grad. Stud.	–	–	45	–	–	–
First-year Grad. Stud.	–	–	8	–	–	–

GRADUATE DEGREE REQUIREMENTS

Master's: Thesis and non-thesis options are available. They are typically two-year programs. Also offered is a Master of Science in Teaching.

Doctorate: For a doctorate degree, the following are required: (1) comprehensive exam, including general and special field, both written and oral parts; (2) thesis with oral defense; (3) course requirements, including advanced quantum mechanics, electromagnetic theory, and statistical physics; and (4) a distributional requirement of electives in four distinct areas of the graduate curriculum. One year of residency is required. There is no language requirement. All requirements must be completed within eight consecutive years.

SPECIAL EQUIPMENT, FACILITIES, OR PROGRAMS

The Physics Department has significant research facilities available to its researchers, including a state-of-the-art materials characterization facilities, low-temperature/high magnetic field, metamaterial, nanophotonics, photovoltaics, AFM, STM, and thermoelectric and optical spectroscopy research laboratories.

In addition, the Physics Department has established ties to many outside facilities, including Los Alamos and Argonne National Laboratories, NASA, and the National High Magnetic Field Laboratory at Florida State University.

Table B—Separately Budgeted Research Expenditures by Source of Support

Source of Support	Departmental Research	Physics-related Research Outside Department
Federal government	$2,500,000	
State/local government		
Non-profit organizations	$1,000,000	
Business and industry	$300,000	
Other		
Total	$3,800,000	

Table C—Separately Budgeted Research Expenditures by Research Specialty

Research Specialty	No. of Grants	Expenditures ($)
Condensed Matter Physics	30	$2,800,000
Plasma and Fusion	2	$220,000
Total	32	$3,020,000

FACULTY

Professor

Bedell, Kevin S., Ph.D., Stony Brook University, 1979. *Condensed Matter Physics, Solid State Physics*. Theoretical condensed matter physics.

Broido, David A., Ph.D., University of California, San Diego, 1985. Theoretical condensed matter physics.

Di Bartolo, Baldassare, Ph.D., Massachusetts Institute of Technology, 1964. Solid state spectroscopy; flash photolysis of gases and liquids.

Graf, Michael J., Ph.D., Brown University, 1987. Experimental condensed matter physics at low temperatures.

Kempa, Krzysztof, Ph.D., University of Wroclaw, 1980. Theoretical condensed matter physics.

Naughton, Michael J., Ph.D., Boston University, 1986. *Applied Physics, Computational Physics, Condensed Matter Physics, Electromagnetism, Energy Sources & Environment, Engineering Physics/Science, Low Temperature Physics, Medical, Health Physics, Nano Science and Technology, Neuroscience/Neuro Physics, Optics, Solar Physics, Solid State Physics*. Experimental condensed matter; nanophotonics; organic superconductors; energy physics.

Wang, Ziqiang, Ph.D., Columbia University, 1989. Theoretical condensed matter physics.

Associate Professor

Engelbrecht, Jan R., Ph.D., University of Illinois at Urbana-Champaign, 1993. Theoretical condensed matter physics; biological physics.

Madhavan, Vidya, Ph.D., Boston University, 2000. Experimental condensed matter physics.

Opeil, Cyril P., Boston College, 2004. Experimental condensed matter physics.

Padilla, Willie J., Ph.D., University of California, San Diego, 2004. Experimental condensed matter physics.

Uritam, Rein A., Ph.D., Princeton University, 1968. Particle theory; history and philosophy of science.

Assistant Professor

Ran, Ying, Ph.D., Massachusetts Institute of Technology, 2007. Theoretical condensed matter physics; correlated electrons.

Wilson, Stephen D., Ph.D., University of Tennessee, 2007. Experimental condensed matter physics; neutron diffraction.

Research Professor

Bakshi, Pradip, Ph.D., Harvard University, 1962. Mathematical physics; theoretical plasma physics; quantum field theory.

Kalman, Gabor, Israel Inst. of Tech., 1961. Theoretical plasma physics; many-body physics; astrophysics.

Research Associate Professor

Herczynski, Andrzej, Ph.D., Lehigh University, 1987. Fluid dynamics.

Scientist

Shepard, Stephen, Framingham State College. Nanofabrication Clean Room Facility Manager.

Research Physicist

McMahon, Gregory, Ph.D., University of Saarland, 1994. Focused Ion Beam Facility Manager.

DEPARTMENTAL RESEARCH SPECIALTIES AND STAFF

Theoretical

Condensed Matter Physics. Electronic, optical, magnetic and transport properties of nanoscale systems; thermoelectrics; strongly correlated electron systems; superconductivity; heavy fermion systems; Fermi liquid theory; electromagnetic response of metals; surfaces; topological insulators. Bakshi, Bedell, Broido, Engelbrecht, Herczynski, Kalman, Kempa, Ran, Wang.

History & Philosophy of Physics/Science. Selected topics in the foundations of physics and in the history and philosophy of science. Uritam.

Plasma and Fusion. Strongly coupled plasmas; plasma response function; dense plasmas; plasma phase transitions; dusty plasmas; plasma instabilities in inhomogeneous systems. Bakshi, Kalman.

Experimental

Condensed Matter Physics. Electronic, optical, magnetic, and transport properties of nanoscale systems; photovoltaics; nanophotonics; thermoelectrics; strongly correlated electron systems; superconductivity; heavy fermion systems; Fermi liquid theory; topological insulators; low-temperature physics; heavy-fermion systems; molecular organic conductors; physics in strong magnetic fields; neutron scattering; high-resolution STM and STS; fluorescence spectroscopy; nanoscale manipulation of light; nanomaterials for energy; metamaterials. Di Bartolo, Graf, Herczynski, Madhavan, McMahon, Naughton, Opeil, Padilla.

View additional information about this department at
www.gradschoolshopper.com

BOSTON UNIVERSITY

DEPARTMENT OF ASTRONOMY

Boston, Massachusetts 02215

http://www.bu.edu/astronomy

General University Information

President: Robert A. Brown
Dean of Graduate School: W. Jeffrey Hughes
University website: http://www.bu.edu/grs

Control: Private
Setting: Urban
Total Faculty: 3,622
Total Graduate Faculty: 567

Total number of Students: 31,766
Total number of Graduate Students: 13,232

Department Information

Department Chairman: Tereasa Brainerd, Chair
Department Contact: Paul Withers, Assistant Professor,
  Director of Graduate Admissions
Total full-time faculty: 15
Total number of full-time equivalent positions: 15
Full-Time Graduate Students: 34
First-Year Graduate Students: 5
Total Post Doctorates: 6

Department Address

725 Commonwealth Avenue
Boston, MA 02215
Phone: (617) 353-1531
E-mail: withers@bu.edu
Website: http://www.bu.edu/astronomy

ADMISSIONS

Admission Contact Information

Address admission inquiries to: Graduate Admissions Committee, Astronomy Department, Boston University, 725 Commonwealth Avenue, Boston, MA 02215
Phone: (617) 353-2696
E-mail: withers@bu.edu
Admissions website: http://www.bu.edu/astronomy/admissions/
  graduate-program-admission/

Application deadlines

Fall admission:
U.S. students: December 15   *Int'l. students*: December 15

Application fee

U.S. students: $80   *Int'l. students*: $80

Admissions information

For Fall of 2012:
  Number of applicants: 100
  Number admitted: 15
  Number enrolled: 5

Admission requirements

Bachelor's degree requirements: Bachelor's degree in physics or astronomy is required with a minimum cumulative undergraduate GPA equivalent to "B" or higher.
Minimum undergraduate GPA: 3.0

GRE requirements

The GRE is required.
The minimum acceptable score suggested for admission is not specified. The minimum acceptable score for admission is dependent upon the applicant's overall record.

Advanced GRE requirements

The Advanced GRE is required.
The minimum acceptable score suggested for admission is not specified. The minimum acceptable score for admission is dependent upon the applicant's overall record.

TOEFL requirements

The TOEFL exam is required for students from non-English-speaking countries.
  PBT score: 550
  iBT score: 85

Other admissions information

Undergraduate preparation assumed: Students are expected to have a strong background in physics and mathematics. Ideally, students should have two semesters each of E&M and quantum mechanics, as well as calculus through differential equations.

TUITION

Tuition year 2010–11:
  Full-time students: $43,970 annual
  Part-time students: $1,374 per credit
Credit hours per semester to be considered full-time: 12
Deferred tuition plan: Yes
Health insurance: Available
Other academic fees: Full time Student Service Fee $145.00 per semester; Health & Wellness fee $160.00 per semester.
Academic term: Semester
Number of first-year students who received full tuition waivers: 5

Teaching Assistants, Research Assistants, and Fellowships

Number of first-year
  Teaching Assistants: 4
  Research Assistants: 1
Average stipend per academic year
  Teaching Assistant: $20,000
  Research Assistant: $20,000
  Fellowship student: $19,300

FINANCIAL AID

Application deadlines

Fall admission:
U.S. students: December 15   *Int'l. students*: December 15

Loans

Loans are available for U.S. students.
Loans are available for international students.
GAPSFAS application required: Yes
FAFSA application required: Yes

For further information

Address financial aid inquiries to: Boston University, GRS Admissions, 725 Commonwealth Avenue, Room 112, Boston, MA 02215.
Phone: 617-353-2696

HOUSING

Availability of on-campus housing

Single students: Yes

For further information

Address housing inquiries to: Housing Director, Boston University.

Table A—Faculty, Enrollments, and Degrees Granted

Research Specialty	2012-13 Faculty	Enrollment Fall 2012		Number of Degrees Granted 2011–12 (2005-12)		
		Master's	Doctorate	Master's	Terminal Master's	Doctorate
Astronomy	–	–	8	–	3(-)	6(-)
Astrophysics	7	1	20	4(26)	1(7)	3(11)
Atmosphere, Space Physics, Cosmic Rays	8	–	17	3(26)	–(4)	5(18)
Total	16	–	37	7(52)	1(11)	8(29)
Full-time Grad. Stud.	–	–	37	–	–	–
First-year Grad. Stud.	–	–	5	–	–	–

GRADUATE DEGREE REQUIREMENTS

Master's: Eight semester courses passed with grades of "B−" or better. A Master's thesis or a passing grade on the Departmental comprehensive exam is required for a degree in Astronomy. The residence requirement is a minimum of two consecutive semesters of full-time graduate study at Boston University.

Doctorate: 64 credits (typically 13 classes plus research and journal club) passed with grades of "B−" or better; a passing grade on the Departmental comprehensive exam; an oral exam; a dissertation and a dissertation defense. Each student must satisfy a residence requirement of two consecutive semesters of full-time graduate study at Boston University.

Thesis: Thesis may be written in absentia.

SPECIAL EQUIPMENT, FACILITIES, OR PROGRAMS

The Department of Astronomy has a long-standing scientific partnership with Lowell Observatory (located in Flagstaff, AZ). Our partnership includes guaranteed access to all Lowell facilities, including the new 4.3 m Discovery Channel telescope. Departmental facilities on campus include a rooftop teaching observatory with 14″, 10″, and 8″ telescopes, a library, and a network of over 200 computers. Individual research groups are actively involved in various instrumentation projects to design and build equipment that will be used in both ground- and space-based observations.

Table B—Separately Budgeted Research Expenditures by Source of Support

Source of Support	Departmental Research	Physics-related Research Outside Department
Federal government	$27,186,283	
State/local government		
Non-profit organizations		
Business and industry		
Other		
Total	$27,186,283	

Table C—Separately Budgeted Research Expenditures by Research Specialty

Research Specialty	No. of Grants	Expenditures ($)
Astrophysics	27	$5,613.79
Atmosphere, Space Physics, Cosmic Rays	67	$25,000
Total	94	$30,613.79

FACULTY

Professor

Bania, Thomas M., Ph.D., University of Virginia, 1977. Radiospectroscopy; galactic structure; interstellar medium.

Brecher, Kenneth, Ph.D., Massachusetts Institute of Technology, 1969. Theoretical high-energy astrophysics; relativity and cosmology, neutron stars.

Clarke, John, Ph.D., Johns Hopkins University, 1980. Associate Chair. Director, Center for Space Physics. Planetary atmospheres; UV astrophysics; FUV instruments for remote observations.

Clemens, Dan P., Ph.D., University of Massachusetts, 1985. Director of Undergraduate Studies. Infrared, and optical astronomy; interstellar medium; galactic structure; star formation.

Fritz, Theodore A., Ph.D., University of Iowa, 1967. Space plasma physics; magnetospheric physics; solar wind; rockets and satellites.

Hughes, W. Jeffrey, Ph.D., University of London, 1974. Associate Dean of the Graduate School Director, Center for Integrated Space Weather Modeling. Magnetospheric physics; space physics.

Jackson, James, Ph.D., Massachusetts Institute of Technology, 1986. Associate Dean for Research and Outreach. Radio, infrared, and gamma-ray astronomy; interstellar medium; starburst galaxies; star formation; the Milky Way; Antarctic astronomy.

Marscher, Alan P., Ph.D., University of Virginia, 1977. Director, Institute for Astrophysical Research. Extragalactic astrophysics; quasars and active galaxies; radio infrared, x-ray, and gamma-ray astronomy.

Mendillo, Michael, Ph.D., Boston University, 1971. Space physics; solar system astronomy; planetary atmospheres.

Oppenheim, Meers, Ph.D., Cornell University, 1995. Director of Graduate Studies. Computational and theoretical space plasma physics; particle-wave interactions; meteor trails.

Associate Professor

Brainerd, Tereasa G., Ph.D., Ohio State University, 1992. Chair. Theoretical astrophysics; cosmology; computational astrophysics; galaxy formation and evolution; gravitational lensing.

Assistant Professor

Blanton, Elizabeth, Ph.D., Columbia University, 2000. Clusters of galaxies; the intercluster medium; Galactic and radio astronomy.

Opher, Merav, Ph.D., University of Sao Paulo, 1998. Computational and theoretical plasma physics in space physics and astrophysics.

West, Andrew, Ph.D., University of Washington, 2005. Kinematics and magnetic activity of M and L dwarfs; star formation in nearby galaxies.

Withers, Paul, Ph.D., University of Arizona, 2003. Director of Graduate Admissions. Behavior of the Martian upper atmosphere and ionosphere; Analysis of accelerometer data.

Professor Emeritus

Janes, Kenneth A., Ph.D., Yale University, 1972. Galactic structure; photometry and spectroscopy; star clusters.

DEPARTMENTAL RESEARCH SPECIALTIES AND STAFF

Theoretical

Astrophysics. High-energy astrophysics; neutron stars; quasars and active galaxies; cosmology; gravitational lensing; computational astrophysics and space physics. Brainerd, Brecher, Marscher, Opher, Oppenheim.

Atmosphere, Space Physics, Cosmic Rays. Magnetospheric physics; space plasma physics; hydromagnetic waves. Opher, Oppenheim.

Experimental

Atmosphere, Space Physics, Cosmic Rays. Solar-terrestrial relations; geomagnetic storms; plasma waves; ionospheric physics. Fritz, Hughes, Mendillo, Opher, Oppenheim.

Atmosphere, Space Physics, Cosmic Rays. Terrestrial, planetary, and cometary atmospheres; aeronomy; Fabry-Perot interferometry. Clarke, Mendillo, Withers.

Extragalactic Astronomy. Quasars and active galaxies; jets (radio, x-ray, gamma ray); clusters of galaxies; dark matter; kinematics of satellite galaxies. Blanton, Brainerd, Marscher.

Galactic Astronomy. Chemical history of the galaxy; composition of stars; star clusters; interstellar medium; galactic structure; star formation (optical, radio, infrared). Bania, Clemens, Jackson, Janes, West.

View additional information about this department at
www.gradschoolshopper.com

BOSTON UNIVERSITY

DEPARTMENT OF PHYSICS

Boston, Massachusetts 02215
http://physics.bu.edu

General University Information
President: Robert A. Brown
Dean of Graduate School: Virginia Sapiro
University website: http://www.bu.edu
Control: Private
Setting: Urban
Total Faculty: 4,047
Total Graduate Faculty: Unavailable
Total number of Students: 33,683
Total number of Graduate Students: 14,175

Department Information
Department Chairman: Professor Sid Redner, Chair
Department Contact: Mirtha Cabello, Administrative Coordinator
Total full-time faculty: 41
Total number of full-time equivalent positions: 41
Full-Time Graduate Students: 112
First Year Graduate Students: 19
Female First-Year Students: 4
Total Post Doctorates: 20

Department Address
590 Commonwealth Avenue
Room 255
Boston, MA 02215
Phone: (617) 353-2623
Fax: (617) 353-9393
E-mail: cabello@bu.edu
Website: http://physics.bu.edu

ADMISSIONS

Admission Contact Information
Address admission inquiries to: Graduate Admissions, Physics Department
Phone: (617) 353-2623

E-mail: cabello@bu.edu
Admissions website: http://physics.bu.edu

Application deadlines
Fall admission:
U.S. students: December 15   *Int'l. students*: December 15

Application fee
U.S. students: $80   *Int'l. students*: $80

Admissions information
For Fall of 2013:
   Number of applicants: 284
   Number admitted: 54
   Number enrolled: 16

Admission requirements
Bachelor's degree requirements: Bachelor's degree in physics or astronomy is required.
Minimum undergraduate GPA: 3.3

GRE requirements
The GRE is required.
The average scores of the entering class are: *Quantitative score*: 164 *Verbal score*: 159 *subject in physics score*: 844.

Advanced GRE requirements
The Advanced GRE is required.
   Minimum accepted Advanced GRE score: 730

TOEFL requirements
The TOEFL exam is required for students from non-English-speaking countries.
PBT score: 600
Minimum scores by section (iBT) Reading section: 21 Listening section: 18 Speaking section: 23 Writing section: 22.

Other admissions information
Additional requirements: Exceptional candidates from other fields will be considered.
Undergraduate preparation assumed: Students are expected to have taken junior/senior-level courses in classical mechanics

(at the level of Marion & Thornton or equivalent), electromagnetism (Griffiths), quantum or modern physics (Liboff or Griffiths), and statistical/thermal physics (Kittel or Reif).

TUITION

Tuition year 2012–13:
   Full-time students: $42,610 annual
A tuition and fees scholarship and medical insurance coverage is awarded to teaching fellows and research assistants.
Credit hours per semester to be considered full-time: 12
Deferred tuition plan: No
Health insurance: Available at the cost of 2141 per year.
Other academic fees: Student Services Fee $290/year ($145/sem) Health fees $320/year ($160/sem) These fees will be covered by financial aid for full-time students with fellowships.
Academic term: Semester
Number of first-year students who received full tuition waivers: 19

Teaching Assistants, Research Assistants, and Fellowships

Number of first-year
   Teaching Assistants: 14
   Research Assistants: 2
   Fellowship students: 3
Average stipend per academic year
   Teaching Assistant: $29,700
   Research Assistant: $29,700
   Fellowship student: $30,200

FINANCIAL AID

Loans

Loans are available for U.S. students.
Loans are available for international students.
GAPSFAS application required: Yes
FAFSA application required: Yes

For further information

Address financial aid inquiries to: Graduate School, Financial Aid Office, 705 Commonwealth Avenue, Boston, MA 02215.
Phone: (617) 353-2696
E-mail: grs@bu.edu
Financial aid website: http://www.bu.edu/cas/admissions/graduate/aid/

HOUSING

Availability of on-campus housing

Single students: Yes
Married students: Yes

For further information

Address housing inquiries to: Rental Property Mgmt., Boston Univ.
Phone: (617) 353-3511
E-mail: housing@bu.edu
Housing aid website: http://www.bu.edu/housing/living/graduate/

Table A—Faculty, Enrollments, and Degrees Granted

Research Specialty	2012-13 Faculty	Enrollment Fall 2012		Number of Degrees Granted 2012-13 (2009-12)		
		Master's	Doctorate	Master's	Terminal Master's	Doctorate
Biophysics	4	–	10	1(2)	–	2(4)
Computational Physics	1	–	2	–	–	–(1)
Condensed Matter Physics	8	–	34	4(8)	1(2)	4(14)
High Energy Physics	10	–	19	3(3)	–(2)	1(1)
Particles and Fields	6	–	4	1(4)	–(5)	–(7)
Statistical & Thermal Physics	8	–	43	4(16)	–	7(30)
Other	5	–	–	–	–	–
Total	42	–	112	13(33)	1(4)	14(57)
Full-time Grad. Stud.	–	–	112	–	–	–
First-year Grad. Stud.	–	–	19	–	–	–

GRADUATE DEGREE REQUIREMENTS

Master's: Eight semester courses (32 credits) required, including Advanced Lab, Mathematical Physics, Statistical Physics and Thermodynamics I, Electrodynamics I, Mathematical Methods, Quantum Mechanics I and II, and one elective course. A passing grade on the departmental comprehensive exam or a Master's thesis. All students must complete a "Scholarly Methods in Physics" course. Each student must satisfy a residency requirement of a minimum of two consecutive regular semesters of full-time graduate study at Boston University.

Doctorate: Eight semester courses (32 credits) beyond those used to fulfill the Master's degree requirements. These must include Advanced Lab, if not already taken to fulfill Master's requirements, and at least five lecture courses numbered between 500 and 850. Up to three non-lecture courses may count toward the total of eight courses, but no more than one directed study course and one seminar course. The five lecture courses must include at least two distribution courses from the category outside the student's area of specialization. (Category I includes elementary particle and mathematical physics and Category II includes biological physics and condensed matter physics.) Passing with an Honors grade on the departmental comprehensive exam; preliminary oral exam; a departmental seminar; a dissertation and a dissertation defense. Each student must satisfy a residency requirement of a minimum of two consecutive regular semesters of full-time graduate study at Boston University.

Other Degrees: Interdisciplinary Ph.D. is also available with many other departments, including Astronomy, Mathematics, Biology, Chemistry, and with departments in the College of Engineering.

Thesis: Thesis may be written in absentia.

SPECIAL EQUIPMENT, FACILITIES, OR PROGRAMS

The Scientific Instrument Facility employs three experimental machinists, and an experimental machinist/director who runs the facility. The facility houses a variety of CNC machine tools interfaced to a state-of-the-art CAD/CAM system, as well as manual lathes, milling machines, and grinders, high-vacuum welding equipment, leak detection and precision measurement equipment. It is capable of assisting with practically any scientific hardware need. The facility also has an assembly expert for fabrication and assembly of complex parts.

The Electronics Design Facility provides complete electronics design and assembly capabilities for research instrumentation. The facility is equipped with extensive CAD tools for analog and digital circuit design and simulation and PCB layout. Facility engineers have extensive experience in low-noise and high-speed circuit design, programmable logic, fiber-optic links and controlled-impedance PCB design.

In house research laboratories include central facilities, as well as individual group laboratories. Central facilities available in Physics, Photonics and Materials Science, and Engineering include STEM, SEM and AFM microscopy, focused ion beam (FIB), vapor-phase deposition, spectroscopy, and submicron UV photolithography and electron-beam lithographic nano-fabrication facilities. Individual research laboratories include the UHV elastic and inelastic He surface scattering laboratory; the electronic-structure laboratory using high-resolution electron and photon spectroscopies; the materials diffraction x-ray scattering laboratory; the low-temperature scanning-optical-microscopy laboratory; the MBE laboratory for growing wide-band-gap semiconductors; and the nano-optics laboratory with near-field solid immersion and subcellular florescent microscopies. They also include the nanoscale transport laboratory with He3 and dilution refrigerator; the low-temperature, high-frequency and high-magnetic field facilities; molecular biophysics and polymer labs, electron, atomic force, confocal, and near-field scanning probe microscopies, and Fourier transform infrared, Raman, UV-visible and ultrafast laser spectroscopy laboratories.

Computer Facilities: An extensive network of computational facilities supports the research activities of the Department. There are Linux cluster and Windows PC's available to departmental faculty, staff, and students. Additional Linux servers and workstations, as well as many PC's, are available to research groups. For computationally intensive applications, students have access to supercomputing resources supported through the Center for Computational Science and the Office of Information Services & Technology. At the high end, these currently consist of an Intel based Linux cluster with 1024 cores, a combined 9.26TB of memory, and combined perfomance of 10.6 Teraflops. These resources are integrated in a well-endowed distributed computing and visualization environment, which includes a high resolution, stereographic display wall; the Computer Graphics Laboratory; and InfiniBand, 10GigE, and Gigabit Ethernet networks. A vast and diverse array of optical fiber connections to the NoX, Metro Ring and commercial ISPs provide multiple Gb/s of bandwidth and connectivity to the Internet, Internet2, and international research networks. The Departmental Computer Facility supports a wide range of software applications for physics data collection, analysis, simulation, and visualization.

Table B—Separately Budgeted Research Expenditures by Source of Support

Source of Support	Departmental Research	Physics-related Research Outside Department
Federal government	$7,375,444.89	
State/local government		
Non-profit organizations		
Business and industry		
Other	$630,565.59	
Total	**$8,006,010.48**	

Table C—Separately Budgeted Research Expenditures by Research Specialty

Research Specialty	No. of Grants	Expenditures ($)
Biophysics	3	$575,291.26
Condensed Matter Physics	28	$2,274,528.41
Particles and Fields	48	$4,538,826.57
Physics and other Science Education	8	$527,810.28
Statistical & Thermal Physics	1	$89,553.96
Total	**88**	**$8,006,010.48**

FACULTY

Professor

Ahlen, Steven P., Ph.D., University of California, Berkeley, 1976. *Astrophysics, Particles and Fields.* Experimental Particle Physics and Astrophysics.

Bansil, Rama, Ph.D., University of Rochester, 1974. *Biophysics, Condensed Matter Physics.* Experimental Condensed Matter Physics and Biological Physics.

Bigio, Irving, Ph.D., University of Michigan, 1974. Joint Appointment with the College of Engineering. *Biophysics.* Biomedical and Biological Physics.

Bishop, David, Ph.D., Cornell University, 1978. Professor of Physics and ECE. Head of the BU Division of Materials Science and Engineering. *Condensed Matter Physics, Nano Science and Technology.* Condensed Matter Experimental including Superconductivity, Magnetic Vortices, MEMS devices for Optical Networks, Single atom MBE.

Brecher, Kenneth, Ph.D., Massachusetts Institute of Technology, 1969. Joint Appointment with the Astronomy Department. *Astronomy.* Theoretical Astrophysics.

Brower, Richard, Ph.D., University of California, Berkeley, 1969. Joint Appointment with the College of Engineering. *Theoretical Physics.* Theoretical Particle Physics.

Butler, John M., Ph.D., Stanford University, 1986. *Astrophysics, High Energy Physics.* Experimental Particle Physics and Astrophysics.

Campbell, David, Ph.D., University of Cambridge, 1970. *Condensed Matter Physics, Quantum Foundations.* Quantum Condensed Matter Theory.

Carey, Robert, Ph.D., Harvard University, 1989. Director of Undergraduate Studies. Experimental Particle Physics and High-Precision Measurements.

Castro Neto, Antonio H., Ph.D., University of Illinois at Urbana-Champaign, 1994. Quantum Condensed Matter Theory.

Chamon, Claudio, Ph.D., Massachusetts Institute of Technology, 1996. *Condensed Matter Physics, Quantum Foundations, Theoretical Physics.* Quantum Condensed Matter Theory.

Cohen, Andrew G., Ph.D., Harvard University, 1986. *Particles and Fields, Theoretical Physics.* Theoretical Particle Physics.

De Rújula, Alvaro, Ph.D., University of Madrid, 1968. Joint Appointment with CERN. Theoretical Particle Physics.

DeLisi, Charles, Ph.D., New York University, 1969. Joint Appointment with the College of Engineering. Biological Physics.

El-Batanouny, Maged, Ph.D., University of California, Davis, 1978. *Condensed Matter Physics.* Experimental Condensed Matter Physics.

Erramilli, Shyamsunder, Ph.D., University of Illinois, 1986. *Biophysics.* Biological Physics.

Evans, Evan, Ph.D., University of California, San Diego, 1970. Joint Appointment with the College of Engineering. Biological Physics.

Giles, Roscoe, Ph.D., Stanford University, 1975. Joint Appointment with the College of Engineering. Advanced Computer Architectures; Distributed and Parallel Computing; Computational Science.

Glashow, Sheldon, Ph.D., Harvard University, 1958. *Theoretical Physics.* Theoretical Particle Physics.

Goldberg, Bennett B., Ph.D., Brown University, 1987. *Nano Science and Technology.* Experimental Condensed Matter Physics.

Kearns, Edwards, Ph.D., Harvard University, 1990. *Astrophysics, Particles and Fields.* Experimental Particle Physics and Astrophysics.

Klein, William, Ph.D., Temple University, 1972. *Statistical & Thermal Physics.* Statistical Physics.

Kreimer, Dirk, Ph.D., Johannes Gutenberg Universitat, Mainz, Germany, 1992. Joint Appointment with the Mathematics Department. Quantum Field Theory; Mathematical Physics.

Lane, Kenneth D., Ph.D., Johns Hopkins University, 1970. Theoretical Particle Physics.

Ludwig, Karl F., Ph.D., Stanford University, 1986. Director of Academics. Experimental Condensed Matter Physics.

Miller, James P., Ph.D., Carnegie Mellon University, 1974. *Particles and Fields*. Experimental Particle Physics and High-Precision Measurements.

Mohanty, Pritiraj, Ph.D., University of Maryland, College Park, 1998. Experimental Condensed Matter Physics.

Moustakas, Theodore, Ph.D., Columbia University, 1974. Joint Appointment with the College of Engineering. Synthetic novel materials.

Pi, So-Young, Ph.D., Stony Brook University, 1974. *Particles and Fields*. Theoretical Particle Physics.

Rebbi, Claudio, Ph.D., International University College of Turin, 1967. Center for Computational Science Department Chair. Theoretical Particle Physics.

Redner, Sidney, Ph.D., Massachusetts Institute of Technology, 1977. Department Chair. *Statistical & Thermal Physics*. Statistical Physics.

Roberts, B. Lee, Ph.D., College of William and Mary, 1974. Experimental Particle Physics and High-Precision Measurements.

Rohlf, James, Ph.D., California Institute of Technology, 1980. *High Energy Physics*. Experimental Particle Physics and Astrophysics.

Rothschild, Kenneth, Ph.D., Massachusetts Institute of Technology, 1973. Experimental Biological Physics.

Sandvik, Anders, Ph.D., University of California, Santa Barbara, 1993. *Condensed Matter Physics*. Computational Quantum Condensed Matter Physics.

Schmaltz, Martin, Ph.D., University of California, San Diego, 1995. Theoretical Particle Physics.

Sergienko, Alexander, Ph.D., Moscow State University, 1987. Joint Appointment with the College of Engineering. Correlation Spectroscopy.

Skocpol, William J., Ph.D., Harvard University, 1974. Experimental Condensed Matter Physics.

Smith, Kevin, Ph.D., Yale University, 1988. *Condensed Matter Physics*. Experimental Condensed Matter Physics.

Stanley, H. Eugene, Ph.D., Harvard University, 1967. Statistical Physics; Director, Center for Polymer Studies.

Stone, James L., Ph.D., University of Michigan, 1977. Experimental Particle Physics and Astrophysics.

Sulak, Lawrence, Ph.D., Princeton University, 1970. Experimental Particle Physics and Astrophysics.

Tsui, Ophelia, Ph.D., Princeton University, 1996. *Condensed Matter Physics*. Biological Physics; Experimental Condensed Matter Physics.

Whitaker, J. Scott, Ph.D., University of California, Berkeley, 1976. Experimental Particle Physics and Astrophysics.

Associate Professor

Averitt, Richard, Ph.D., Rice University, 1998. Director of Graduate Studies. *Condensed Matter Physics*. Experimental Condensed Matter Physics.

Katz, Emanuel, Ph.D., Massachusetts Institute of Technology, 2001. Theoretical Particle Physics.

Meller, Amit, Ph.D., Weizmann Institute of Science, 1998. Biological Physics.

Mertz, Jerome, Ph.D., University of California, Santa Barbara, 1991. Joint Appointment with the College of Engineering. Biological Physics.

Polkovnikov, Anatoli, Ph.D., Yale University, 2003. Quantum Condensed Matter Theory.

Swan, Anna, Ph.D., Boston University, 1993. Joint Appointment with the College of Engineering. Experimental Condensed Matter Physics.

Assistant Professor

Black, Kevin, Ph.D., Boston University, 2005. *High Energy Physics*. Experimental Particle Physics and Astrophysics.

Bose, Tulika, Ph.D., Columbia University, 2006. *High Energy Physics*. Experimental Particle Physics and Astrophysics.

Mehta, Pankaj, Ph.D., Rutgers University, 2006. *Statistical & Thermal Physics*. Biological Physics; Statistical Physics.

Emeritus

Booth, Edward C., Ph.D., Johns Hopkins University, 1955. Biological Physics.

Chasan, Bernard, Ph.D., Cornell University, 1961. Biological Physics.

Cohen, Robert S., Ph.D., Yale University, 1948. Philosophy of Physics.

Corinaldesi, Ernesto, Ph.D., The University of Manchester, 1951. Quantum Mechanics.

Edmonds, Jr., Dean S., Ph.D., Massachusetts Institute of Technology, 1958. Electronics and Instrumentation.

Hellman, William S., Ph.D., Syracuse University, 1961. Elementary Particle Theory.

Shimony, Abner, Ph.D., Yale University, 1953. Foundations of Physics.

Willis, Charles R., Syracuse University, 1957. Biophysics; Nonlinear Physics; Statistical Physics.

Zimmerman, George O., Yale University, 1963. Low-Temperature Physics; Magnetism.

Professor Emeritus

Teich, Malvin C., Ph.D., Cornell University, 1966. Joint Appointment with the College of Engineering. Quantum Optics and Imaging.

Research Faculty

Hong, Mi Kyung, Ph.D., University of Illinois, 1988. *Biophysics*. Experimental Biophysics.

Ivanov, Plamen, Ph.D., Boston University, 1998. *Polymer Physics/Science*.

Krapivsky, Paul, Ph.D., Moscow Physical Technical Institute, 1991. *Condensed Matter Physics*. Theoretical Condensed Matter Physics.

Shank, James, Ph.D., University of California, Berkeley, 1988. *High Energy Physics*.

Youssef, Saul, Ph.D., Carnegie Mellon University, 1992. *High Energy Physics*.

Senior Lecturer

Duffy, Andrew, Ph.D., Queen's University at Kingston, 1995. Physics Education Research.

Lecturer

Jariwala, Manher, Ph.D., University of Maryland, 2004. Physics Education Research.

Joint Appointment

Ünlü, M. Selim, Ph.D., University of Illinois, 1997. Joint Appointment with the College of Engineering. Near-field Optical Microscopy and Spectroscopy.

DEPARTMENTAL RESEARCH SPECIALTIES AND STAFF

Theoretical

Condensed Matter Physics. Strongly interacting electron systems. Low-dimensional quantum magnetism and quantum antiferromagnets. High-temperature and organic superconductors and heavy electron systems. Fractional Quantum Hall effect. Surface physics; solitons on surfaces. Structural and vibrational properties of adsorbed atomic layers. Equilibrium and non-equilibrium properties of interacting many particle atomic systems. Quantum Monte Carlo algorithms. Graphene. Bose-Einstein condensates in optical lattices. Quantum adiabatic algorithm. Quantum quenches. 4 postdoctoral fellows. Campbell, Castro Neto, Chamon, Ivanov, Klein, Krapivsky, Mehta, Polkovnikov, Redner, Sandvik, Stanley, Teich.

Particles and Fields. Physical origin of electroweak and flavor symmetry breaking, including theoretical and phenomenological studies of technicolor, little Higgs, extra dimensions and supersymmetry. Quantum chromodynamics. Collider phenomenology. Numerical simulations of lattice gauge theories. Fundamental studies of quantum field theory. Theoretical astrophysics and cosmology; dark matter, inflation, baryogenesis, and the formation of large scale structure. 7 postdoctoral fellows. Brower, Andrew Cohen, Glashow, Katz, Lane, Pi, Schmaltz, Shank, Youssef.

Statistical & Thermal Physics. Kinetics of phase transitions and coarsening processes. Chemical reactions, stochastic processes, and the role of spatial fluctuations. Structure of heterogeneous networks; the dynamics, resilience, and failure mechanism of networks. Population biology models. Theoretical biology; collective biological behavior; biological computation and environmental response. Fluctuation dynamics, nonlinearity, and complexity in physiological systems. Neural control and coupling of organ systems. Network physiology. Dynamics of social systems. Econophysics. Mechanisms of nucleation and spinodal decomposition. Physics of disordered media; percolation models of disordered materials. Fractals and multifractals. Hydrogen-bonded network formation in liquid water. Dynamics of earthquake faults. Dynamics of materials damage mechanisms. Acceleration algorithms for Monte Carlo simulations. First-passage processes and their applications. Stochastic transport processes. Theoretical studies of polymers. 2 postdoctoral fellows. Campbell, Chamon, Ivanov, Klein, Krapivsky, Mehta, Polkovnikov, Redner, Stanley.

Experimental

Astrophysics. Atmospheric and solar neutrino studies and neutrino astrophysics with the Super-K experiment. Dark matter searches. 2 postdoctoral fellows. Kearns, Stone, Sulak, Whitaker.

Biophysics. Energy transduction, ion transport, and signal receptor studies of microbial and vertebrate rhodopsins and their complexes by FTIR, resonance Raman spectroscopy and their bioengineering for optogenetics. Biomembrane technology and molecular electronics. Structure and electrical properties of membranes. Ultrafast vibrational spectroscopy, STM, AFM imaging of macromolecular assemblies, membrane surfaces, and protein-lipid interactions. Gelation of mucin and mucus. Novel Fluorescent Imaging for subcellular microscopy. Dy-

namics of DNA. Signaling and Information Processing in Biochemicals networks. 1 postdoctoral fellow. Bansil, Erramilli, Goldberg, Rothschild, Tsui.

Condensed Matter Physics. Mesoscopic phenomena; quantum transport and quantum coherence phenomena in nanostructures. Advanced electronic materials. Nano-optics and spectroscopy of quantum dots, photonic bandgap systems, and carbon nanotubes. Single molecule spectroscopy and subcellular imaging. Studies of structural phase transitions in thin-film semiconductors. Synchrotron x-ray scattering studies of kinetics of nucleation, spinodal decomposition, and phase transitions. Growth of artificially structured materials using molecular beam epitaxy, sputtering, and chemical vapor deposition. Properties of high-Tc superconductor-normal interfaces. X-ray emission and photoemission studies of wide band gap semiconductors, organic superconductors, and low dimensional transition metal oxides. Terahertz spectroscopy and time-integrated and time-resolved optical spectroscopy of correlated electron materials. Investigation of the structural, dynamical, and magnetic properties of solid surfaces using neutral helium and metastable He beam scattering. Current interest is focused on the surfaces of topological insulators and multi-ferroics. High resolution photoemission and x-ray emission studies of metals, semiconductors, and oxides. Superconductivity, magnetic vortices, MEMS devices for optical networks, single-atom MBE. 4 postdoctoral fellows. Averitt, Bansil, Bishop, El-Batanouny, Erramilli, Goldberg, Hong, Ludwig, Mohanty, Rothschild, Smith, Tsui, Zimmerman.

Particles and Fields. Studies of the Higgs boson and electroweak symmetry breaking, W and Z bosons, top quarks, rare b quark decays and searches for new physics beyond the standard model at the CERN LHC. Study of neutrino properties using long baseline neutrino oscillation. Study of grand unified theories using proton decay. Precision measurements of the anomalous part of the muon magnetic dipole moment. Precision measurement of the muon lifetime and the Fermi constant. New limit on muon to electron conversion. New limit on the electric dipole moment of the muon and neutron. Precision measurement of muon capture on hydrogen and determination of the pseudoscalar coupling constant. Precision measurement of muon capture on deuterium. 8 postdoctoral fellows. Ahlen, Black, Booth, Bose, Butler, Carey, Kearns, Miller, Roberts, Rohlf, Sulak.

Polymer Physics/Science. X-ray, light-scattering, rheology and microscopy studies of the structure, dynamics, and phase separations of gels, block copolymers, polymer nanocomposites, polymer thin films, and interfaces. 2 postdoctoral fellows. Bansil, Stanley, Tsui.

Surface Physics. Investigation of the structural, dynamical, and magnetic properties of solid surfaces using neutral helium and metastable helium beam scattering. Current interest is focused on the surfaces of topological insulators and multiferroics. High resolution photoemission studies of metals, semiconductors, and oxides. 2 postdoctoral fellows. El-Batanouny, Smith.

View additional information about this department at
www.gradschoolshopper.com

BRANDEIS UNIVERSITY

DEPARTMENT OF PHYSICS

Waltham, Massachusetts 02454-9110
http://brandeis.edu/departments/physics/

General University Information
President: Frederick M. Lawrence
Dean of Graduate School: Malcolm W. Watson
University website: http://www.brandeis.edu
Control: Private
Setting: Suburban
Total Faculty: 350
Total number of Students: 3,588
Total number of Graduate Students: 923

Department Information
Department Chairman: Jané Kondev, Chair
Department Contact: Joan Thorne, Senior Department
  Coordinator
  Total full-time faculty: 17
  Total number of full-time equivalent positions: 17
  Full-Time Graduate Students: 46
  First-Year Graduate Students: 10
  Female First-Year Students: 2
  Total Post Doctorates: 9

Department Address
415 South Street
Waltham, MA 02454-9110
Phone: (781) 736-2800
Fax: (781) 736-2915
E-mail: physics@brandeis.edu
Website: http://brandeis.edu/departments/physics/

ADMISSIONS

Admission Contact Information
Address admission inquiries to: Chairman, Graduate Admissions
  Committee, Physics Department, Mailstop 057, Waltham,
  MA 02454-9110
Phone: (781) 736-2800
E-mail: physics@brandeis.edu
Admissions website: http://brandeis.edu/gsas/apply

Application deadlines
Fall admission:
U.S. students: January 15   *Int'l. students*: January 15

Application fee
U.S. students: $75   *Int'l. students*: $75
Graduate application fee required: $75 online/$100 paper.

Admissions information
For Fall of 2013:
  Number of applicants: 122
  Number admitted: 18
  Number enrolled: 9

Admission requirements
Bachelor's degree requirements: Bachelor's degree is required.

GRE requirements
The GRE is required.
Graduate Record Exam (GRE) general test is required.

Advanced GRE requirements
The Advanced GRE is recommended.
The GRE subject test is recommended, but not required. Our ETS
  institution code is 3092.

TOEFL requirements
The TOEFL exam is required for students from non-English-
  speaking countries.
  PBT score: 600
  iBT score: 100

Other admissions information
Additional requirements: No minimum acceptable score for admission is specified.
Undergraduate preparation assumed: Kleppner and Kolenkow,
  ; An introduction to Mechanics edition 1; Young and Freedman, Univ. Physics with Modern Physics 13th edition; M.
  Boas, Mathematical Methods in the Physical Sciences, 3rd
  edition; C. Meyer, Basic Electronics: An Introduction to Electronics for Science Students; D. Griffiths, Introduction to Electrodynamics (3rd ed); D. Griffiths, Introduction to Quantum
  Mechanics, 2nd edition; J. Jose & E. Saletan, Classical Dynamics: A Contemporary Approach.

TUITION

Tuition year 2013-14:
  Full-time students: $44,100 annual
  Part-time students: $22,050 annual
Credit hours per semester to be considered full-time: 12
Deferred tuition plan: No
Health insurance: Available at the cost of $1903 per year.
Other academic fees: QSHIP student + spouse $7,974, student
  + child $4,925, family $10,996. For more information on the
  plan go to www.bluecrossma.com then search Blue Care Elect
  Preferred or www.brandeis.edu/health under insurance in-
  formation. Optional Individual Health Center Fee $726.
Academic term: Semester

Teaching Assistants, Research Assistants, and Fellowships
Number of first-year
  Teaching Assistants: 7
  Research Assistants: 3
Average stipend per academic year
  Teaching Assistant: $24,480
  Research Assistant: $27,950

FINANCIAL AID

Application deadlines
Fall admission:
U.S. students: January 15

Loans
Loans are available for U.S. students.
Loans are not available for international students.
GAPSFAS application required: No
FAFSA application required: No

For further information

Address financial aid inquiries to: Graduate School of Arts and Sciences, 415 South Street, Waltham, MA 02453-2783.

Phone: (781) 736-3410

E-mail: gradschool@brandeis.edu

Financial aid website: http://www.brandeis.edu/gsas/financing/index.html

HOUSING

Availability of on-campus housing

Single students: No

Married students: No

For further information

Address housing inquiries to: Brandeis has extremely limited space for housing graduate students on campus, and space is limited to one year. We do have many resources available for helping students find off-campus housing.

Housing aid website: http://www.brandeis.edu/gsas/accepted/housing.html

Table A—Faculty, Enrollments, and Degrees Granted

Research Specialty	2012-13 Faculty	Enrollment Fall 2012 Master's	Enrollment Fall 2012 Doctorate	Number of Degrees Granted 2012–13 (2008–13) Master's	Number of Degrees Granted 2012–13 (2008–13) Terminal Master's	Number of Degrees Granted 2012–13 (2008–13) Doctorate
Astrophysics	2	–	1	–(2)	–(1)	–(2)
Biological Physics	4	–	19	–(1)	–	–(5)
Condensed Matter Physics	4	–	13	–(4)	1(4)	2(9)
Neuroscience/Neuro Physics	–	–	1	–	–	–
Particles and Fields	8	–	11	–	–(2)	–(4)
Non-specialized	–	–	1	–	–	–
Total	18	1	46	–(7)	1(7)	–(20)
Full-time Grad. Stud.	–	–	46	–	–	–
First-year Grad. Stud.	–	–	10	–	–	–

GRADUATE DEGREE REQUIREMENTS

Master's: One year's residence as a full-time student. Eight semester courses in physics numbered above 160 with a grade of at least "B-". A thesis on an approved topic may be accepted in place of a semester course. Satisfactory performance on the qualifying examination.

Doctorate: Three year's residence as a full-time student. Nine semester courses of advanced work in physics with a grade of at least "B." Outstanding performance on the qualifying examination. Passing of an Advanced Examination in topics related to the student's thesis subject. This examination will normally be taken after preparatory studies in the prospective field of research. Doctoral thesis and final oral examination.

Other Degrees: Doctoral program is available in biophysics.

Thesis: Thesis may be written in absentia.

Table B—Separately Budgeted Research Expenditures by Source of Support

Source of Support	Departmental Research	Physics-related Research Outside Department
Federal government	$4,637,131	
State/local government		
Non-profit organizations		
Business and industry	$353,123	
Other		
Total	$4,990,254	

Table C—Separately Budgeted Research Expenditures by Research Specialty

Research Specialty	No. of Grants	Expenditures ($)
Astrophysics	1	$71,938
Biophysics	4	$282,958
Condensed Matter Physics	4	$301,884
Materials Science, Metallurgy	12	$2,295,657
Particles and Fields	8	$1,171,927
Relativity & Gravitation	6	$865,890
Total	35	$4,990,254

FACULTY

Professor

Bensinger, James R., Ph.D., University of Wisconsin-Madison, 1970. Experimental high-energy physics.

Blocker, Craig A., Ph.D., University of California, Berkeley, 1980. Experimental high-energy physics.

Chakraborty, Bulbul, Ph.D., Stony Brook University, 1979. Theoretical condensed matter physics.

Fraden, Seth, Ph.D., Brandeis University, 1987. Physics of liquid crystals; colloids; macromolecules; microfluidics.

Kirsch, Lawrence E., Ph.D., Rutgers University, 1964. Experimental high-energy physics.

Kondev, Jané, Ph.D., Cornell University, 1995. Theoretical condensed matter physics; biological physics.

Roberts, David H., Ph.D., Stanford University, 1973. *Astrophysics*. Theoretical astrophysics; radio astronomy.

Schnitzer, Howard J., Ph.D., University of Rochester, 1960. Quantum theory of fields; string theory.

Wardle, John F. C., Ph.D., University of Manchester, England, 1969. *Astrophysics*. Radio astronomy; cosmology.

Associate Professor

Dogic, Zvonimir, Ph.D., Brandeis University, 2001. Soft condensed matter physics; biological physics.

Hagen, Michael F., Ph.D., University of California, Berkeley, 2003. Computation and theory in biological physics.

Lawrence, Albion, Ph.D., University of Chicago, 1996. String theory and its applications to particle physics and cosmology.

Sciolla, Gabriella, Ph.D., Torino University, 1996. High-energy experiment: physics beyond the Standard Model; Dark Matter detection with DMTPC.

Wellenstein, Hermann F., Ph.D., University of Texas, Austin, 1971. Experimental high-energy physics.

Assistant Professor

Baskaran, Aparna, Ph.D., University of Florida, 2006. Non-equilibrium statistical mechanics; biophysics.

Headrick, Matthew, Ph.D., Harvard University, 2003. String theory, quantum field theory, and geometry.

Samadani, Azadeh, Ph.D., Clark University, 2002. Experimental biological physics; soft condensed matter physics.

Professor Emeritus

Meyer, Robert B., Ph.D., Harvard University, 1970. Physics of liquid crystals and colloids.

DEPARTMENTAL RESEARCH SPECIALTIES AND STAFF

Theoretical

Astrophysics. http://www.brandeis.edu/departments/physics/research/astro.html. Roberts, Wardle.

Biological Physics. http://www.brandeis.edu/departments/physics/research/condt.html. Baskaran, Chakraborty, Hagen, Kondev.

Condensed Matter Physics. www.brandeis.edu/departments/physics/research/condt.html. Baskaran, Chakraborty.

Particles and Fields. www.brandeis.edu/departments/physics/research/hegt.html. Headrick, Lawrence, Schnitzer.

Experimental

Astrophysics. Radio astronomy of galaxies and quasars. www.brandeis.edu/departments/physics/research/astro.html. Roberts, Wardle.

Biological Physics. www.brandeis.edu/departments/physics/research/conde.html. Dogic, Samadani.

Condensed Matter Physics. www.brandeis.edu/departments/physics/research/conde.html. Fraden, Meyer.

Particles and Fields. Collider experiments. www.brandeis.edu/departments/physics/research/hep.html. Bensinger, Blocker, Kirsch, Sciolla, Wellenstein.

View additional information about this department at
www.gradschoolshopper.com

CLARK UNIVERSITY

DEPARTMENT OF PHYSICS

Worcester, Massachusetts 01610
http://physics.clarku.edu

General University Information

President: David Angel
Dean of Graduate School: William F. Fisher
University website: http://clarku.edu
Control: Private
Setting: Urban
Total Faculty: 197
Total Graduate Faculty: 197
Total number of Students: 3,235
Total number of Graduate Students: 960

Department Information

Department Chairman: Charles C. Agosta, Chair
Department Contact: Sujata Davis, Office Manager
   Total full-time faculty: 6
   Total number of full-time equivalent positions: 6
   Full-Time Graduate Students: 7
   First-Year Graduate Students: 1

Department Address

950 Main Street
Worcester, MA 01610
Phone: (508) 793-7169
Fax: (508) 793-8861
E-mail: physics@clarku.edu
Website: http://physics.clarku.edu

ADMISSIONS

Admission Contact Information

Address admission inquiries to: Graduate Admissions Office, Clark University, 950 Main Street, Worcester, MA 01610
Phone: (508) 793-7373

E-mail: gradadmissions@clarku.edu
Admissions website: http://www.clarku.edu/graduate

Application deadlines

Fall admission:
U.S. students: January 15   *Int'l. students*: January 15

Application fee

U.S. students: $75   *Int'l. students*: $75

Admissions information

For Fall of 2013:
   Number of applicants: 33
   Number admitted: 3
   Number enrolled: 1

Admission requirements

Bachelor's degree requirements: A Bachelor's degree in physics, chemistry, mathematics, or engineering with a minimum undergraduate GPA of "B-" specified is required.
Minimum undergraduate GPA: 2.7

GRE requirements

The GRE is recommended.

Advanced GRE requirements

The Advanced GRE is recommended.

TOEFL requirements

The TOEFL exam is required for students from non-English-speaking countries.

Other admissions information

Additional requirements: The GRE and GRE Advanced are recommended. There are no set minimum scores for GRE or GRE Advanced; each case is judged individually. Students from non-English-speaking countries are required to demonstrate proficiency in English via the TOEFL exam.

TUITION

Tuition year 2013-14:
Full-time students: $39,200 annual
Part-time students: $4,900 per credit
Credit hours per semester to be considered full-time: 9
Deferred tuition plan: No
Health insurance: Available at the cost of $1,711 per year.
Other academic fees: Health and accident insurance is required but is waived on evidence of other insurance. Graduate students are charged a $100 Graduate Enrollment Fee, in the semester they begin their graduate program. This is a one-time only fee.
Academic term: Semester
Number of first-year students who received full tuition waivers: 1

Teaching Assistants, Research Assistants, and Fellowships

Number of first-year
Teaching Assistants: 1
Average stipend per academic year
Teaching Assistant: $13,625
Research Assistant: $13,625

FINANCIAL AID

Loans

Loans are not available for U.S. students.
Loans are not available for international students.
GAPSFAS application required: No
FAFSA application required: No

For further information

Address financial aid inquiries to: Department of Physics, Clark University, 950 Main Street, Worcester, MA 01610.
Phone: (508) 793-7169
E-mail: physics@clarku.edu
Financial aid website: http://catalog.clarku.edu/content.php?catoid=4&navoid=144

HOUSING

Availability of on-campus housing

Single students: Yes
Married students: No

For further information

Address housing inquiries to: Department of Physics, Clark University, 950 Main Street, Worcester, MA 01610.
Phone: (508) 793-7169
E-mail: physics@clarku.edu
Housing aid website: http://www.clarku.edu/offices/housing/

Table A—Faculty, Enrollments, and Degrees Granted

Research Specialty	2013-14 Faculty	Enrollment Fall 2012 Master's	Enrollment Fall 2012 Doctorate	Number of Degrees Granted 2012-13 (2008–13) Master's	Number of Degrees Granted 2012-13 (2008–13) Terminal Master's	Number of Degrees Granted 2012-13 (2008–13) Doctorate
Condensed Matter Physics	5	–	6	2(8)	–(1)	1(7)
Physics and other Science Education	1	–	–	–	–	–
Statistical & Thermal Physics	2	–	–	–	–	–
Total	8	–	6	2(8)	–(1)	1(7)
Full-time Grad. Stud.	–	–	6	–	–	–
First-year Grad. Stud.	–	–	1	–	–	–

GRADUATE DEGREE REQUIREMENTS

Master's: Eight total courses are required, two of which may be transferred and one may be for thesis. A thesis (or Ph.D. candidacy) is required. There is no language requirement. Teaching experience is required.
Doctorate: One year in residence beyond the master's degree (eight courses) is required; four area qualification examinations and thesis proposal examination, teaching experience, and dissertation are also required.
Thesis: Thesis may be written in absentia.

SPECIAL EQUIPMENT, FACILITIES, OR PROGRAMS

The Department stresses research experience at the earliest possible time and encourages all students to enroll in a research apprenticeship during their first semester. Students work directly with several faculty members on specific research projects and can begin dissertation work substantially earlier than is usually the case. Clark's size affords a uniquely close association between faculty and students.

Table B—Separately Budgeted Research Expenditures by Source of Support

Source of Support	Departmental Research	Physics-related Research Outside Department
Federal government	$152,000	
State/local government		
Non-profit organizations		
Business and industry		
Other	$39,000	
Total	$191,000	

Table C—Separately Budgeted Research Expenditures by Research Specialty

Research Specialty	No. of Grants	Expenditures ($)
Biophysics	1	$33,345
Condensed Matter Physics	6	$157,655
Total	7	$191,000

FACULTY

Professor

Agosta, Charles C., Ph.D., Duke University, 1986. *Applied Physics, Astronomy, Computational Physics, Condensed Matter Physics, Energy Sources & Environment, Low Temperature Physics, Nano Science and Technology, Statistical & Thermal Physics*. Properties of organic superconductors and other materials in high, pulsed magnetic fields.
Kudrolli, Arshad, Ph.D., Northeastern University, 1995. *Applied Physics, Condensed Matter Physics, Engineering Physics/Science, Fluids, Rheology, Geophysics, Mechanics, Nonlinear Dynamics and Complex Systems, Statistical & Thermal Physics*. Experimental nonlinear physics; granular matter and soft condensed matter; self-assembly.
Landee, Christopher P., Ph.D., University of Michigan, 1975. *Condensed Matter Physics*. Experimental condensed-matter physics; quantum magnetism; magnetochemistry.

Associate Professor

Mukhopadhyay, Ranjan, Ph.D., California Institute of Technology, 1998. *Biophysics, Condensed Matter Physics, Statistical & Thermal Physics*. Theoretical condensed-matter physics; complex fluids; biophysics.

Assistant Professor

Boyer, Michael, Ph.D., Massachusetts Institute of Technology, 2008. *Condensed Matter Physics, Low Temperature Physics.* Experimental condensed-matter physics; correlated electron systems; scanning tunneling microscopy studies.

Emeritus

Andersen, Roy S., Ph.D., Duke University, 1952. *Condensed Matter Physics.* Experimental condensed-matter physics; molecular spectroscopy; history and philosophy of science.

Blatt, S. Leslie, Ph.D., Stanford University, 1965. *Astronomy, Nuclear Physics, Optics, Physics and other Science Education.* Experimental nuclear physics; college/pre-college curriculum development and teacher education.

Gould, Harvey A., Ph.D., University of California, Berkeley, 1966. *Computational Physics, Condensed Matter Physics, Statistical & Thermal Physics.* Statistical physics; computer simulation; phase transitions.

Kohin, Roger P., Ph.D., University of Maryland, 1961. *Condensed Matter Physics.* Experimental condensed-matter physics; electron spin resonance studies.

Lecturer

Colonna-Romano, Louis, M.S., Stephens Institute of Technology, 1971. *Computational Physics, Condensed Matter Physics, Statistical & Thermal Physics.* Computational physics, particularly, computer simulations of dynamical systems and the glass transition in supercooled liquids, and the evaluation and optimization of computational algorithms.

DEPARTMENTAL RESEARCH SPECIALTIES AND STAFF

Theoretical

Biophysics. Physical mechanisms underlying molecular and cellular processes; membranes; biopolymers; mechanical, biomechanical, and genetic networks. Mukhopadhyay.

Condensed Matter Physics. Soft matter, including polymers, liquid crystals, and gels; complex and disordered systems. Colonna-Romano, Gould, Mukhopadhyay.

Statistical & Thermal Physics. Computer simulation; phase transitions. Colonna-Romano, Gould, Mukhopadhyay.

Experimental

Condensed Matter Physics. Magnetism; low-dimensional conductors; organic superconductors; pulsed magnetic fields; scanning tunneling microscopy studies; active soft matter. Agosta, Boyer, Kudrolli, Landee.

Energy Sources & Environment. Risk assessment; energy policy. Agosta.

Nonlinear Dynamics and Complex Systems. Granular matter; pattern formation; soft matter. Kudrolli.

Physics and other Science Education. Teaching methods in science. Blatt.

View additional information about this department at www.gradschoolshopper.com

---

HARVARD UNIVERSITY

DEPARTMENT OF PHYSICS

Cambridge, Massachusetts 02138
http://www.physics.harvard.edu

General University Information

President: Drew Gilpin Faust
Dean of Graduate School: Xiao-Li Meng
University website: http://www.harvard.edu
Control: Private
Setting: Urban
Total Faculty: 2,436
Total Graduate Faculty: 2,436
Total number of Students: 21,049
Total number of Graduate Students: 13,804

Department Information

Department Chairman: Melissa Franklin, Chair
Department Contact: Lisa Cacciabaudo, Graduate Program Administrator
Total full-time faculty: 35
Total number of full-time equivalent positions: 35
Full-Time Graduate Students: 197
First-Year Graduate Students: 36
Female First-Year Students: 10
Total Post Doctorates: 100

Department Address

17 Oxford St.
Jefferson Physical Lab, Room 370
Cambridge, MA 02138
Phone: (617) 495-4327
Fax: (617) 495-0416
E-mail: gradinfo@physics.harvard.edu
Website: http://www.physics.harvard.edu

ADMISSIONS

Admission Contact Information

Address admission inquiries to: Graduate School of Arts and Sciences, Harvard University, Office of Admissions, Holyoke Center, 3rd Floor, 1350 Massachusetts Avenue, Cambridge, MA 02138
Phone: (617) 496-6100
E-mail: admiss@fas.harvard.edu
Admissions website: http://www.gsas.harvard.edu/prospective_students/admissions_overview.php

Application deadlines

Fall admission:

U.S. students: December 15   *Int'l. students*: December 15

Application fee

U.S. students: $105   *Int'l. students*: $105

All applicants to the Department of Physics are required to submit, in addition to the Abstract of Courses, a list of their four most advanced courses in physics and their two most advanced courses in mathematics, indicating textbooks (and authors) used in each course.

Admissions information

For Fall of 2013:

Number of applicants: 530

Number admitted: 63

Number enrolled: 33

Admission requirements

Bachelor's degree requirements: A Bachelor's degree (or equivalent) is required.

GRE requirements

The GRE is required.

Advanced GRE requirements

The Advanced GRE is required.

TOEFL requirements

The TOEFL exam is required for students from non-English-speaking countries.

iBT score: 80

Other admissions information

Undergraduate preparation assumed: The only specific requirements for admission are those stipulated by the Graduate School of Arts and Sciences. An undergraduate degree in physics is preferred but not assumed. Applicants with an undergraduate degree in engineering, mathematics, or chemistry have ideally taken introductory and intermediate physics courses.

TUITION

Tuition year 2013–14:

Tuition for out-of-state residents

Full-time students: $42,036 annual

Note: All physics graduate students receive fellowships that cover tuition, fees, health insurance, and stipend annually.

Health insurance: Yes. Included in tuition.

Academic term: Semester

Teaching Assistants, Research Assistants, and Fellowships

Average stipend per academic year

Teaching Assistant: $32,232

Research Assistant: $32,232

Fellowship student: $32,232

FINANCIAL AID

Loans

Loans are not available for U.S. students.

Loans are not available for international students.

GAPSFAS application required: No

FAFSA application required: No

For further information

Address financial aid inquiries to: N/A.

HOUSING

Availability of on-campus housing

Single students: Yes

Married students: Yes

For further information

Address housing inquiries to: Ashley Skipwith, Housing Coordinator, Patty Collyer, Housing Assistant, Dudley House B2.

Phone: (617) 495-5060

E-mail: gsashous@fas.harvard.edu

Housing aid website: http://www.gsas.harvard.edu/current_students/housing.php

Table A—Faculty, Enrollments, and Degrees Granted

Research Specialty	2010–2011 Faculty	Enrollment Fall 2010		Number of Degrees Granted 2010–2011		
		Master's	Doctorate	Master's	Terminal Master's	Doctorate
Applied Physics	–	–	1	–	–	–
Astronomy	–	–	–	–	–	–
Astrophysics	5	–	12	–	–	1(-)
Atomic, Molecular, & Optical Physics	10	–	47	–	–	3(-)
Biophysics	14	–	16	–	–	–
Chemical Physics	–	–	–	–	–	–
Condensed Matter Physics	19	–	59	–	–	6(-)
Cosmology & String Theory	5	–	6	–	–	1(-)
Electromagnetism	–	–	–	–	–	–
Energy Sources & Environment	2	–	–	–	–	–
High Energy Physics	8	–	20	–	–	6(-)
History & Philosophy of Physics/Science	1	–	–	–	–	–
Low Temperature Physics	–	–	–	–	–	–
Low Temperature Physics	–	–	18	–	–	–
Low Temperature Physics	8	–	18	–	–	–
Materials Science, Metallurgy	–	–	–	–	–	–
Mechanics	–	–	–	–	–	–
Nuclear Physics	1	–	–	–	–	–
Optics	–	–	–	–	–	1(-)
Particles and Fields	–	–	–	–	–	–
Relativity & Gravitation	–	–	–	–	–	–
Statistical & Thermal Physics	–	–	–	–	–	–
Total	–	–	196	–	–	–
Full-time Grad. Stud.	–	–	191	–	–	–
First-year Grad. Stud.	–	–	33	–	–	–

GRADUATE DEGREE REQUIREMENTS

Master's: The Department of Physics does not admit students whose sole purpose is to study for the master of arts degree. However, the A.M. degree is frequently taken by students who continue on for the Ph.D. degree. For those who do not attain the doctorate, the A.M. degree attests to the completion of a full year's study beyond the bachelor's degree, including 8 graduate courses.

Doctorate: Each student is required to demonstrate proficiency in a broad range of fields of physics by obtaining honor grades ("B-" or better) in at least eight half-courses. (A minimum of four core courses and an additional four elective courses.) Qualifying oral examination for Ph.D. candidacy. Thesis, and final oral examination thereon. No language requirement; laboratory experience is required for students not submitting experimental theses.

Thesis: The final examination, or thesis defense, includes a talk given by the thesis student defending the thesis. If the coursework does not indicate a wide proficiency in the field of the dissertation, the examination may be extended to test this proficiency as well.

SPECIAL EQUIPMENT, FACILITIES, OR PROGRAMS

Electronic Instrument Design Lab

The Lab provides work space, tools, parts, and technical assistance for Physics/SEAS students and faculty who need to design and build custom electronic instruments. The Lab staff can assist with schematic and circuit board design, help order parts, provide soldering/assembly instruction, and help debug circuits. Alternatively, we can start from your specifications and develop fully custom instrumentation to meet your needs.

Physics/SEAS Instructional Instrument Lab

The Machine Shop is set up to be primarily a teaching Shop. We have a state-of-the-art facility, complete with computerized machine tools and full arc welding capabilities.

Biophysics (PhD-Track Program)

The primary objective of the program is to educate and train individuals with background in physical or quantitative science, especially chemistry, physics, computer science, or mathematics, to apply the concepts and methods of the physical sciences to the solution of biological problems.

Center for Nanoscale Systems (CNS)

The Center's scientific focus is on how nanoscale components can be integrated into large and complex interacting systems. It brings together the disciplines of chemistry, physics, engineering, materials science, geology, biology, and medicine.

The Center for the Fundamental Laws of Nature

This interdisciplinary theoretical research center aims to advance our basic knowledge of the universe through the interactive collaboration of physicists, mathematicians, and cosmologists.

The Center for Ultracold Atoms (CUA)

A joint venture with MIT, the CUA encompasses experimental and theoretical research in the following areas:

- Bose-Einstein condensates: development of new methods for manipulating and probing condensed atomic gases, ultracold interactions, and collision dynamics.

- Atom optics: atom interferometry, atom waveguides, surface physics and quantum reflection, many body physics in lower dimensions.

- Cryogenic Sources for BEC: creation of large condensates of alkalis and other atoms, sympathetic cooling, novel condensates, creation of intense hydrogen sources, and optical techniques for ultracold hydrogen.

Engineering and Physical Biology Program (EPB) (PhD-Track Program)

A joint venture between Physics, Engineering, Chemistry, and Biology that focuses on determining how basic physical principles govern and explain biological processes.

Harvard-Smithsonian Center for Astrophysics (CfA)

The Center for Astrophysics combines the resources and research facilities of the Harvard College Observatory and the Smithsonian Astrophysical Observatory to study the basic physical processes that determine the nature and evolution of the universe. Some of its pioneering achievements include:

- Development of instrumentation for orbiting observatories in space.

- Ground-based gamma-ray astronomy.

- The application of computers to problems of theoretical astrophysics, particularly stellar atmospheres.

The Institute for Quantum Science and Engineering (IQSE)

The mission of the IQSE is to foster cross-disciplinary research and education in new areas at the intersection of nanoscience, atomic physics, device engineering and computer science, that in various ways seeks to apply principles of quantum mechanics to advanced technologies.

Institute for Theoretical Atomic and Molecular Physics (ITAMP)

The Institute for Theoretical Atomic, Molecular and Optical Physics was established in November 1988 at the Harvard-Smithsonian Center for Astrophysics in order to address the critical shortage of theorists in atomic and molecular physics at major universities throughout the nation.

Laboratory for Particle Physics and Cosmology (LPPC)

The Laboratory for Particle Physics and Cosmology carries out forefront programs in high energy physics research and provides first-rate educational opportunities for students. LPPC's experimental programs are carried out at the major accelerator centers throughout the world and address important questions both within and beyond the Standard Model.

Materials Research Science and Engineering Center (MRSEC)

The Materials Research and Engineering Center is the focus of Harvard's long tradition of interdisciplinary materials research.

Nanoscale Science Engineering Center (NSEC)

The Nanoscale Science and Engineering Center is a collaborative effort that combines "top down" and "bottom up" approaches to construct novel electronic and magnetic devices with nanoscale sizes and understand their behavior, including quantum phenomena.

Several collaborations and projects are also being carried out by Physics Department faculty and graduate students at centers outside of Cambridge: the Fermi National Accelerator Laboratory; the CERN in Geneva; the Stanford Linear Accelerator Center; the Soudan Mines in Northern Minnesota; and the National Institute of Standards and Technology.

Table B—Separately Budgeted Research Expenditures by Source of Support

Source of Support	Departmental Research	Physics-related Research Outside Department
Federal government	$24,043,756	
State/local government		
Non-profit organizations		
Business and industry		
Other	$2,929,465	
Total	$26,973,221	

Table C—Separately Budgeted Research Expenditures by Research Specialty

Research Specialty	No. of Grants	Expenditures ($)
Astrophysics	9	$5,041,707
Atomic, Molecular, & Optical Physics	8	$7,174,629
Biophysics	10	$1,369,545
Condensed Matter Physics	8	$9,490,817
Particles and Fields	9	$967,058
Total	44	$24,043,756

FACULTY

Professor

Berg, Howard C., Ph.D., Harvard University, 1964. *Biophysics.* Biophysics, motile behavior of bacteria.

Cohen, Adam, Ph.D., University of Cambridge, 2003. *Chemical Physics.* Experimental physical chemistry.

Demler, Eugene, Ph.D., Stanford University, 1998. *Atomic, Molecular, & Optical Physics, Condensed Matter Physics.* Theoretical condensed matter physics.

Doyle, John M., Ph.D., Massachusetts Institute of Technology, 1991. *Other.* Quantum Science and Particle Physics. Cold molecules, search for electron EDM.

Feldman, Gary, Ph.D., Harvard University, 1971. *High Energy Physics.* Experimental high-energy physics.

Franklin, Melissa, Ph.D., Stanford University, 1982. Chair, Department of Physics. *High Energy Physics.* Experimental high-energy physics.

Gabrielse, Gerald, Ph.D., University of Chicago, 1980. *Atomic, Molecular, & Optical Physics, Electromagnetism, High Energy Physics, Mechanics, Nuclear Physics, Optics, Particles and Fields.* Experimental atomic, optical, plasma, and elementary particle physics.

Galison, Peter, Ph.D., Harvard University, 1983. History of 20th century physics; particle physics.

Georgi, Howard M., Ph.D., Yale University, 1971. *Particles and Fields.* Field theory and particle physics.

Golovchenko, Jene A., Ph.D., Rensselaer Polytechnic Institute, 1972. *Applied Physics, Atomic, Molecular, & Optical Physics, Biophysics, Condensed Matter Physics, Electromagnetism, Materials Science, Metallurgy.*

Greiner, Markus, Ph.D., Ludwig Maximilians, 2003. *Atomic, Molecular, & Optical Physics, Condensed Matter Physics.* Bose-Einstein condensation; optical lattices; quantum simulation; quantum magnetism; quantum gas microscopy.

Halperin, Bertrand I., Ph.D., University of California, Berkeley, 1965. *Condensed Matter Physics, Low Temperature Physics, Nano Science and Technology, Statistical & Thermal Physics.* Condensed matter theory, especially properties of electrons and spins in nanoscale systems, electron systems in strong magnetic fields, effects of disorder and electron-electron interactions.

Hau, Lene Vestergaard, Ph.D., Aarhus, 1991. Experimental atomic physics; Bose-Einstein condensation; non-linear optics.

Heller, Eric J., Ph.D., Harvard University, 1973. *Acoustics, Applied Physics, Atomic, Molecular, & Optical Physics, Chemical Physics, Computational Physics, Condensed Matter Physics, Nonlinear Dynamics and Complex Systems.* Theory of electrons in two dimensions and imaging by scanning probe microscopy. Semiclassical methods for scattering theory, decoherence, quantum chaos. Wave propagation in random media. Molecular spectroscopy and photochemistry. Ultracold collisions of atoms and molecules. Random matrix and semiclassical approaches to statistical physics.

Huth, John, Ph.D., University of California, Berkeley, 1985. Experimental high-energy physics.

Jaffe, Arthur M., Ph.D., Princeton University, 1966. *Other.* Foundations of Theoretical Physics: Fundamental investigation of quyantum theory, relativity, and statistical physics, as well as other possible physical views of nature.

Kaxiras, Efthimios, Ph.D., Massachusetts Institute of Technology, 1987. *Computational Physics.* Condensed matter theory.

Lukin, Mikhail, Ph.D., Texas A&M University, 1998. *Atomic, Molecular, & Optical Physics, Condensed Matter Physics.* Quantum optics, quantum information science, many-body quantum dynamics.

Mahadevan, L., Ph.D., Stanford University, 1995. *Applied Mathematics.* Organismic and evolutionary biology.

Manoharan, Vinothan, Ph.D., University of California, Santa Barbara, 2004. *Applied Physics, Biophysics, Condensed Matter Physics, Fluids, Rheology, Optics, Statistical & Thermal Physics.* Light scattering, optical microscopy, spectroscopy, synthesis and other experimental techniques to understand the physics of self-organization.

Mazur, Eric, Ph.D., Leiden University, 1982. *Applied Physics, Materials Science, Metallurgy, Optics, Physics and other Science Education.* Interaction of short laser pulses with matter.

Morii, Masahiro, Ph.D., University of Tokyo, 1994. *High Energy Physics.* Experimental high-energy physics.

Murray, Cherry, Ph.D., Massachusetts Institute of Technology, 1978. Dean, School of Engineering and Applied Sciences. *Applied Physics.*

Narayanamurti, Venkatesh, Ph.D., Cornell University, 1965. *Applied Physics.* Science and Technology Policy. Experimental condensed matter physics.

Nelson, David R., Ph.D., Cornell University, 1975. *Applied Mathematics, Applied Physics, Biophysics, Chemical Physics, Condensed Matter Physics, Polymer Physics/Science, Statistical & Thermal Physics.* Statistical physics and condensed matter theory.

Park, Hongkun, Ph.D., Stanford University, 1996. *Atomic, Molecular, & Optical Physics, Biophysics, Condensed Matter Physics.* Experimental Optical Physics, Condensed Matter Physics and Biophysics. Quantum optoplasmonics, quantum information science, nano-bio interfaces for cell-circuit studies, brain-machine interfacing.

Pershan, Peter S., Ph.D., Harvard University, 1960. *Chemical Physics, Condensed Matter Physics.* Experimental Chemical Physics, Experimental Condensed Matter Physics.

Prentiss, Mara, Ph.D., Massachusetts Institute of Technology, 1986. *Biophysics.* Biologically inspired strategies for self-assembly; roles of mechanical force in biological systems; single molecule force experiments.

Randall, Lisa, Ph.D., Harvard University, 1987. *Cosmology & String Theory, High Energy Physics, Particles and Fields, Relativity & Gravitation.* Field theory; the Standard Model and beyond.

Sachdev, Subir, Ph.D., Harvard University, 1985. *Condensed Matter Physics.*

Samuel, Aravinthan, Ph.D., Harvard University, 1999. *Biophysics.* Biophysics, neurobiology, and animal behavior.

Shapiro, Irwin I., Ph.D., Harvard University, 1955. *Astrophysics.* Astrophysics; observational tests of general relativity.

Silvera, Isaac F., Ph.D., University of California, Berkeley, 1965. *Condensed Matter Physics.* High pressure and low-temperature physics: electrons in liquid helium, in particular multi-electron bubbles; high pressure: properties of hydrogen under pressure to multi- megabar pressures, aimed at metallization of hydrogen;nEDM-measurement of the electric dipole moment of the neutron.

Strominger, Andrew, Ph.D., Massachusetts Institute of Technology, 1982. *High Energy Physics.* String theory, black holes and topological strings, 2D quantum field theory.

Stubbs, Christopher, Ph.D., University of Washington, 1988. *Astronomy, Astrophysics, Relativity & Gravitation.* Experimental physics at the interface between particle physics, cosmology, and gravitation.

Vafa, Cumrun, Ph.D., Princeton University, 1985. *High Energy Physics.* String theory, interplay between geometry and physics, including applications to QFTs and black holes.

Weitz, David, Ph.D., Harvard University, 1978. *Condensed Matter Physics.* Soft condensed matter Physics, Biophysics, and Biotechnology.

Westervelt, Robert M., Ph.D., University of California, Berkeley, 1977. *Condensed Matter Physics, Low Temperature Physics, Medical, Health Physics, Nano Science and Technology, Nonlinear Dynamics and Complex Systems.* Experimental condensed matter physics: Cooled Scanning Probe Microscope imaging of electron motion in nanostructures; Biophysics: programmable Integrated Circuit Microfluidic chips for biomedicine.

Wu, Tai Tsun, Ph.D., Harvard University, 1956. *Electromagnetism, High Energy Physics, Statistical & Thermal Physics.* Electromagnetism; high-energy physics; statistical and thermal physics; quantum information processing.

Yacoby, Amir, Ph.D., Weizmann Institute of Science, 1994. *Condensed Matter Physics.* Experimental condensed matter physics; quantum computing; strongly correlated electrons; graphene; scan probe techniques; NV centers in diamond.

Zhuang, Xiaowei, Ph.D., University of California, Berkeley, 1996. *Biophysics.* Biophysics and Bioimaging. Single-molecule spectroscopy, super-resolution imaging, single-molecule biophysics, cellular biophysics.

Associate Professor

Finkbeiner, Douglas, Ph.D., University of California, Berkeley, 1999. *Astrophysics.* Astroparticle physics, interstellar medium, large-scale astronomical surveys.

Guimaraes da Costa, Joao, Ph.D., University of Michigan, 2000. *High Energy Physics, Particles and Fields.*

Hoffman, Jennifer, Ph.D., University of California, Berkeley, 2003. *Condensed Matter Physics, Low Temperature Physics.* Scanning probe microscopies of exotic electron materials. Scanning tunneling microscopy, force microscopy studies of correlated electron systems, superconductors, and topological materials.

Schwartz, Matthew, Ph.D., Princeton University, 2003. *Particles and Fields.* Cosmology, particles, and fields. Theoretical particle physics and cosmology with connections to experimental data.

Yin, Xi, Ph.D., Harvard University, 2006. *Atomic, Molecular, & Optical Physics, High Energy Physics, Particles and Fields, Relativity & Gravitation.* String theory.

Assistant Professor

Desai, Michael, Ph.D., Harvard University, 2006. *Biophysics, Nonlinear Dynamics and Complex Systems, Statistical & Thermal Physics.* Studies of genetic variation to develop methods to infer the evolutionary history of populations from the variation observed in sequence data. Primary focus on natural selection in asexual populations such as microbes and viruses.

Jafferis, Daniel L., Ph.D., Harvard University, 2007.

Kovac, John, Ph.D., University of Chicago, 2004. *Astronomy, Astrophysics, Cosmology & String Theory.* Observations of cosmic microwave background temperature and polarization anisotropies and their use to search for evidence of Inflation, constrain cosmological parameters, and study the evolution of structure. Cosmology using radio observations.

Levine, Erel, Ph.D., Weizmann Institute of Science, 2005. *Biophysics, Computer Science, Nonlinear Dynamics and Complex Systems, Statistical & Thermal Physics.* Collective dynamics in biological systems; multiscale approaches to gene regulation; statistical and information physics of embryonic and post-embryonic development.

Reece, Matthew, Ph.D., Cornell University, 2008. *High Energy Physics.* Physics beyond the Standard Model.

Emeritus

Glauber, Roy J., Ph.D., Harvard University, 1949. *Atomic, Molecular, & Optical Physics, Electromagnetism, High Energy Physics, Nuclear Physics, Optics, Particles and Fields, Statistical & Thermal Physics.* Theoretical physics.

Holton, Gerald, Ph.D., Harvard University, 1948. Experimental physics; history of 19th and 20th century physics.

Horowitz, Paul, Ph.D., Harvard University, 1970. Experimental astrophysics.

Wilson, Richard, University of Oxford, 1949. Experimental nuclear and elementary particle physics; energy related, environmental, and medical physics.

Senior Lecturer / Research Associate

Walsworth, Ronald, Ph.D., Harvard University, 1991. *Applied Physics, Astrophysics, Atomic, Molecular, & Optical Physics, Biophysics, Condensed Matter Physics, Low Temperature Physics, Medical, Health Physics, Nano Science and Technology, Optics.*

Senior Research Fellow

Yelin, Susanne, Ph.D., Ludwig-Maximilians University, 1998. Senior Research Fellow. *Atomic, Molecular, & Optical Physics.* Quantum optics, quantum information science, cold molecules.

DEPARTMENTAL RESEARCH SPECIALTIES AND STAFF

Theoretical

Atomic, Molecular, & Optical Physics.

Atomic, Molecular, & Optical Physics. Quantum Gases,Quantum Information Science, Quantum Optics. Demler, Doyle, Gabrielse, Glauber, Golovchenko, Greiner, Heller, Lukin.

Biophysics. Theoretical Problems in Soft Matter and Quantitative Biology: One focus is on natural selection in asexual populations such as microbes and viruses.

Computational Physics. The development of multiscale methods to couple disparate spatial and temporal scales for the study of complex physical phenomena, and the application of such methods to problems related to energy conversion and energy storage systems.

Condensed Matter Physics. Quantum Phase Transitions of Correlated Electrons and Atoms, Quantum Simulations of Condensed Matter Systems using Ultra-cold Atomic Gases.

Cosmology & String Theory. Studying the Nature of Time in deSitter Space and Exploring Models of Quantum Gravity.

Energy Sources & Environment.

Particles and Fields. Theoretical Particle physics and Cosmology with connections to experimental data from the Large Hadron Collider. Understanding Jets at The Large Hadron Collider and studying effective field theories and the study of unparticles.

Experimental

Astrophysics. The experimental astrophysics effort includes microwave background polarization measurements to constrain the physics of inflation, observations designed to better determine the nature of dark energy, dark matter searches using indirect detection techniques and precision tests of gravity in astrophysical systems. Additional projects include efforts to improve the precision of astronomical measurements and detector development for LSST. Finkbeiner, Kovac, Shapiro, Stubbs, Walsworth.

Atomic, Molecular, & Optical Physics. Ultracold molecules, physics in the quantum regime, advanced cold molecule electron electric dipole moment search, cold atom interferometry for inertial sensing, the production and atudy of antiprotons and cold antihydrogen, monlinear optics and ultrashort laser pulses, quantum optical techniques for solid-state quantum information processing. Quantum sensing, quantum optics, strongly correlated quantum gas with single site address-ability. Doyle, Gabrielse, Glauber, Golovchenko, Greiner, Heller, Lukin, Mazur, Park, Walsworth.

Biophysics. Biophysical approaches to complex navigational behaviors in larval Drosophila melanogaster, Subcellular surgery,Coherent Molecular Profiling Using Nano-Structured Environments Illegitimate Recombination by Drug Resistance Elements. Berg, Desai, Golovchenko, Levine, Manoharan, Mazur, Prentiss, Samuel, Walsworth, Westervelt, Zhuang.

Condensed Matter Physics. Mesoscopic quantum phenomena at the interface between micro and macro scales, Bulk Magnetometry, and Quantum Information Processing (QIP) focusing on Diamond Characterization and Quantum Physics. Interaction of ultrashort laser pulses with matter, Novel semiconductor materials, Light scattering, optical microscopy, spectroscopy, synthesis and other experimental techniques to understand the physics of self-organization. Multi-

Qubit Systems Based on Electron Spins in Coupled Quantum Dots, Development of Novel Quantum Devices Using Nitrogen Vacancy Centers in Diamond, Development of Solid-State Topological Quantum Computing, Spin Resolved Imaging of Correlated Electron Systems Including Cuprates and Pnictides, Metallization of Hydrogen. Doyle, Golovchenko, Greiner, Halperin, Hoffman, Manoharan, Mazur, Narayana-murti, Nelson, Pershan, Sachdev, Silvera, Walsworth, Weitz, Westervelt, Yacoby.

High Energy Physics. Collider physics with the ATLAS experiment at the Large Hadron Collider. Neutrino physics with the MINOS and NOVA experiments. Feldman, Franklin, Gabrielse, Guimaraes da Costa, Morii.

Low Temperature Physics. Study of multi-electron bubbles in superfluid helium, hydrogen at low temperature, measurement of the electric dipole mome. Halperin, Heller, Hoffman, Silvera, Walsworth, Westervelt.

View additional information about this department at
www.gradschoolshopper.com

HARVARD UNIVERSITY SCHOOL OF ENGINEERING AND APPLIED SCIENCES

APPLIED PHYSICS AREA

Cambridge, Massachusetts 02138

http://www.seas.harvard.edu/teaching-learning/areas/applied_physics

General University Information
President: Drew Faust
Dean of Graduate School: Xiao-Li Meng
University website: http://www.harvard.edu
Control: Private
Setting: Urban
Total Faculty: 1,732
Total number of Students: 21,041
Total number of Graduate Students: 13,796

Department Information
Department Chairman: Eric Mazur, Chair
Department Contact: Julie Holbrook, Director of Student Affairs
  Total full-time faculty: 43
  Total number of full-time equivalent positions: 29
  Full-Time Graduate Students: 143
  First-Year Graduate Students: 23
  Female First-Year Students: 6
  Total Post Doctorates: 32

Department Address
Pierce Hall
29 Oxford Street
Cambridge, MA 02138
Phone: (617) 495-2747
E-mail: holbrook@seas.harvard.edu
Website: http://www.seas.harvard.edu/teaching-learning/areas/applied_physics

ADMISSIONS

Admission Contact Information
Address admission inquiries to: Harvard University, Graduate School of Arts and Sciences, Admissions Office, Holyoke Center, 1350 Massachusetts Avenue, 3rd Floor, Cambridge, MA 02138
Phone: (617) 495-5315
E-mail: admiss@fas.harvard.edu
Admissions website: http://www.gsas.harvard.edu

Application deadlines
Fall admission:
U.S. students: December 14   *Int'l. students*: December 14

Application fee
U.S. students: $105

Admissions information
For Fall of 2012:
  Number of applicants: 202
  Number admitted: 35
  Number enrolled: 22

Admission requirements
Bachelor's degree requirements: Students with Bachelor's degrees in the natural sciences, mathematics, or engineering are invited to apply for admission.

GRE requirements
The GRE is required.

Advanced GRE requirements
The Advanced GRE is not required.

Massachusetts

Harvard U. School of Eng. and Appl. Sci., Appl. Phys.

TOEFL requirements

The TOEFL exam is required for students from non-English-speaking countries.

PBT score: 550
iBT score: 80

TUITION

Tuition year 2011–12:

Full-time students: $37,576 annual
Part-time students: $1,174 per credit

$37,576/years 1 and 2 $9,770/years 3 and 4 $2,402/after year 4 Students admitted for a Ph.D. program in Applied Physics can expect to receive some level of assistance (and for many cases, full assistance) to cover their tuition and other costs.

Deferred tuition plan: Yes
Health insurance: Available at the cost of $2,168 per year.
Other academic fees: $2,168 for health insurance; $930 annual for Student Health Fee. Required student fees = $3,098 per year.
Academic term: Semester
Number of first-year students who received full tuition waivers: 22

Teaching Assistants, Research Assistants, and Fellowships

Number of first-year
  Fellowship students: 22
Average stipend per academic year
  Teaching Assistant: $10,837.5
  Research Assistant: $30,228
  Fellowship student: $30,228

FINANCIAL AID

Application deadlines

Fall admission:
U.S. students: December 14

Loans

Loans are available for U.S. students.
Loans are not available for international students.
GAPSFAS application required: Yes
FAFSA application required: No

For further information

Address financial aid inquiries to: Graduate School of Arts and Sciences, Admissions Office, Holyoke Center, 1350 Massachusetts Avenue, 3rd Floor, Cambridge, MA 02138.
Phone: (617) 495-5315
E-mail: admiss@fas.harvard.edu
Financial aid website: http://www.gsas.harvard.edu

HOUSING

Availability of on-campus housing

Single students: Yes
Married students: No

For further information

Address housing inquiries to: The GSAS Office of Housing Services, Dudley House, Room B-2, Harvard Yard, Cambridge, MA 02138.
Phone: (617) 495-5060
E-mail: gsashous@fas.harvard.edu
Housing aid website: http://www.gsas.harvard.edu/prospective_students/gsas_housing_services.php

Table A—Faculty, Enrollments, and Degrees Granted

Research Specialty	2012–13 Faculty	Enrollment Fall 2012		Number of Degrees Granted		
		Master's	Doctorate	Master's	Terminal Master's	Doctorate
Applied Physics	43	–	143	20(-)	–	20(-)
Total	43	–	143	20(-)	–	20(-)
Full-time Grad. Stud.	–	–	143	–	–	–
First-year Grad. Stud.	–	–	22	–	–	–

GRADUATE DEGREE REQUIREMENTS

Master's: There is a Master of Science degree offered in Applied Physics that is open to enrolled Ph.D. students only which can be earned en-route to the Ph.D. degree. At least eight graduate courses are required for the Master of Science in the School of Engineering and Applied Sciences.

Doctorate: The Ph.D. requires a minimum academic residency of two years beyond the bachelor's degree. Programs are individually developed in consultation with a field advisor and a Committee on Higher Degrees (CHD) advisor and must be approved by the CHD, which also reviews any requests for exceptions to the requirements. Normally, students spend one-and-one-half to two years on coursework. The goal of our curriculum is to foster the education of PhD students so that they develop both the in-depth knowledge of their fields and the broader appreciation and skills that they will need after graduation. Students should work in close consultation with their advisors to develop an appropriate program of study which will contain a minimum of ten courses. Courses provide the background knowledge that is often needed to successfully complete research, and allow one to learn more broadly about a field or related fields in a structured fashion. Courses are not meant as and should not be seen as an impediment to research, but as a means of enhancing one's research ability and as part of the process of becoming a mature, well-rounded member of one's field. We emphasize that the 10 course requirement is considered a minimum, and not a goal; students are encouraged to take additional courses whenever appropriate. Of the ten required courses for the PhD degree: At least 8 courses will normally be disciplinary courses, i.e. courses that provide the scientific, mathematical, and technical depth that students need for our graduate programs in engineering and applied science. Up to two courses can normally be "298r" or "299r" courses, "Innovation" style courses that broaden a student's perspective, or relevant courses at a suitable level in other departments (e.g. economics) or schools (HKS, Business School, Medical School). Each course must be passed with a grade of B- or better, and a B average must be maintained. Academic, but not financial, credit may be granted for graduate work done elsewhere, but only if those courses are approved by the Committee on Higher Degrees as part of the degree program and justification for inclusion has been provided. Ordinarily, five such courses is the maximum number approved, and only four will count towards the eight disciplinary courses. In most cases fewer than five will be accepted. The first year is ordinarily spent principally on coursework, although some students may begin research. The second year is usually divided between coursework and research, with coursework completed during the third year if necessary.

Thesis: Upon successful completion of the qualifying examination, a committee usually consisting of three or four Harvard faculty members, is selected and chaired by the research supervisor and constituted to oversee the dissertation research. The committee must include at least two SEAS faculty members, one of whom must be a senior faculty member. In the student's sixth term, a progress report, which includes remarks by the student and comments by the committee members, must be submitted by the end of the reading period. Beginning with the eighth term, progress reports are due by the end of the reading period each term, and the committee is strongly urged to meet as a group with the student at least once each year to complete the progress report. Original research culminating in the dissertation is usually completed in the fourth or fifth year. The dissertation must, in the judgment of the research committee, meet the standards of significant and original research. No prospectus is required by the division. The dissertation should be a coherent document addressed to a broad audience in the subject area. A collection of manuscripts intended for publication as technical papers is not considered by SEAS to constitute an acceptable dissertation. When the dissertation is completed to the satisfaction of the research committee, generally in the fourth or fifth year and rarely later than the end of the student's sixth year, a final oral examination is scheduled at a time to which the committee has agreed. This public examination devoted to the field of the dissertation is conducted by the student's research committee. It consists of a presentation and defense of the dissertation itself and may also include more general questions relating to the field of the research. At the end of the examination, the committee may accept the dissertation, possibly subject to revisions, or specify further requirements. Three copies of the final dissertation, one bound and two unbound copies, must be delivered to the SEAS academic office prior to the degree meeting of the Committee on Higher Degrees, where recommendations on degrees are voted. An electronic copy of the dissertation must also be provided to the academic office. After a positive recommendation to grant the PhD is voted in SEAS, the dissertation is delivered to the Registrar's office.

SPECIAL EQUIPMENT, FACILITIES, OR PROGRAMS

We are closely linked with a variety of multidisciplinary and innovative education and research institutes, centers, and initiatives.

The University is also part of an integrated partnership, called the National Nanotechnology Infrastructure Network (NNIN), comprised of thirteen user facilities, led by Cornell and Stanford, that provide opportunities for nanoscience and nanotechnology research.

*BASF Advanced Research Initiative at Harvard University

*Center for Research in Computation and Society

*Institute for Applied Computational Science

*Kavli Institute for Bionano Science and Technology at Harvard University

*Materials Research Science and Engineering Center

*Center for Brain Science

*Center for Nanoscale Systems

*Harvard University Center for the Environment

*Institute for Quantum Science and Engineering

*The Microbial Sciences Initiative

*The Rowland Institute at Harvard

*Wyss Institute for Biologically Inspired Engineering at Harvard University

Based at SEAS BASF Advanced Research Initiative at Harvard, is set up as an integrated partnership among Harvard and BASF researchers, the BASF Advanced Research Initiative at Harvard benefits from having strong ties with departments and schools throughout the University. The decisive difference of this collaboration between academia and industry from most research initiatives is its more integrative nature: BASF researchers from Ger-

many are working closely with Harvard academic research teams, easing scientific exchange on the projects, as well as fostering broader interaction between the two institutions. This arrangement also gives the students the opportunity to benefit from a close interaction and early exposure to industry. Present projects focus on approaches to prevent biofilm formation and the use of colloidal techniques to develop formulations of pharmaceutical actives. For more information contact Marc Schroeder, North American Center for Research on Advanced Materials.

The Center for Research in Computation and Society (CRCS) was founded to develop a new generation of ideas and technologies designed to address some of society's most vexing problems. The Center brings computer scientists together with economists, psychologists, legal scholars, ethicists, neuroscientists, and other academic colleagues across the University and throughout the world, to address fundamental computational problems that cross disciplines, and to create new technologies informed by societal constraints to address those problems.

The Institute for Applied Computational Science was established in September 2010. It is charged with launching a unique interdisciplinary education and research program in computational science and engineering (CSE). The new Institute will create an intellectual home for faculty and students applying computational methods to major challenges in science and enhance existing courses in applied mathematics and computation and develop new computational science courses, activities and research opportunities for Harvard students from across the sciences. By establishing the Institute, SEAS has committed to fostering graduate training and research in applied computational science, infusing the curriculum with new courses and student research opportunities that will focus on the use of computation to power discovery and innovation.

The Kavli Institute for Bionano Science and Technology (KIBST) seeks to develop a deeper understanding of the functioning of life and biology at the nanoscale level. The Institute: brings together a wide range of scientists, including physicists, engineers, chemists, biologists as well as HMS clinicians to address fundamental questions about the behavior and functioning of biological systems; allows biologists, engineers, and clinicians to potentially use such knowledge to foster applications and new technologies; and provides a way for the tool-developers (physicists, engineers, computer scientists) to work with the tool-users (biologists, chemists, clinicians) in the early stages of scientific inquiry and encourage scientific collaboration at the innovation stage of tool development.

The Materials Research Science and Engineering Center (MRSEC) at Harvard is one of eleven such centers sponsored by the National Science Foundation. The Center is the focus of interdisciplinary research at the University. The participants of the MRSEC are drawn from five areas, including the SEAS; Chemistry and Chemical Biology (Chemistry), Physics; Earth and Planetary Sciences (EPS); and the Medical School (HMS). The center is organized into three Interdisciplinary Research Groups (IRGS): IRG 1: Multiscale Mechanics of Films and Interfaces; IRG 2: Engineering Materials and Techniques for Biological Studies at Cellular Scales; and IRG 3: Interface-Mediated Assembly of Soft Materials.

The Nanoscale Science and Engineering Center (NSEC) is a collaboration among Harvard University, the Massachusetts Institute of Technology, the University of California at Santa Barbara and the Museum of Science in Boston with participation by Delft University of Technology (Netherlands), the University of Tokyo (Japan), and Brookhaven National Laboratory, Oak Ridge National Laboratory and Sandia National Laboratory. This Center combines "top down" and "bottom up" approaches to construct novel electronic and magnetic devices with nanoscale sizes and understand their behavior, including quantum phenomena.

Based at FAS and/or University-wide Center for Brain Science (CBS), researchers are discovering the structure and function of neural circuits. They do this to understand:

-how these circuits govern behavior and vary between individuals.

-how they change during development and aging.

-how they underlie neurological and psychiatric disorders.

To accomplish this mission, CBS brings neuroscientists together with physical scientists and engineers to develop new tools for neuroscience. Members are drawn from the Faculty of Arts and Sciences, the Department of Neurobiology at the Harvard Medical School, the School of Engineering and Applied Sciences, and the Harvard-affiliated hospitals.

Center for Nanoscale Systems (CNS):

The inclusion of CNS in the National Nanotechnology Infrastructure Network (NNIN) in 2004 has expanded that function to include any and all other members of the larger research community both local and national, academic and non-academic who conduct research in any aspect of the large and growing field of nanoscale science.

Theoretical research is supported by computers within the School of Engineering and Applied Sciences(SEAS). Within SEAS, the computers consist of a generic 300 node High Performance Computing Cluster, an IBM BlueGene/L supercomputer with 4096 cores, and a 16 node NVidia GPU Cluster with 7000 cores. Students also have access to a large 12000 core commodity cluster located on campus in the Faculty of Arts and Sciences computing facility. Access to supercomputing centers at Federal National Labs are routine.

Course Work:

The basic core courses consist of the standard courses in classical mechanics, electrodynamics, quantum mechanics, statistical mechanics, and mathematical physics. In addition, students are required to take at least nine hours of electives in solid state physics, atomic and nuclear physics, physics of fluids, physics of surfaces, applied quantum mechanics, applied mathematics, and computer science. Students must take at least 10 courses for the Ph.D. These courses can be from the departments of Applied Physics, Applied Mathematics, Computer Science, Engineering Sciences, Physics, Mathematics, Statistics, Chemistry, or the life sciences.

Research:

Research training of students emphasizes development of analytical skills in mathematics and theoretical physics. This is attained by each student working on research projects of faculty members. Six hours of graduate credit may be awarded for the completion of a thesis.

Table B—Separately Budgeted Research Expenditures by Source of Support

Source of Support	Departmental Research	Physics-related Research Outside Department
Federal government	$13,049,732	$23,686,622
State/local government		
Non-profit organizations	$52,978	$541,879
Business and industry	$1,097,795	$2,035,657
Other	$221,430	$629,154
Total	$14,421,935	$26,893,312

FACULTY

Professor

Aizenberg, Joanna, Ph.D., Weizmann Institute of Science, 1996. *Applied Physics, Engineering Physics/Science.* Biophysics and Self-Assembly; Materials Science; Soft Condensed mat-

ter; Surface and Interface Science; Biomechanics and Motor Control.

Anderson, James G., Ph.D., University of Colorado, 1970. *Applied Mathematics, Engineering Physics/Science.* Atmospheric and Climate Modeling; Oceans and Geophysics; Observation and Field Testing for Environmental Systems.

Aziz, Michael J., Ph.D., Harvard University, 1983. *Applied Physics, Engineering Physics/Science.* Materials Science; Surface and Interface Science; Energy and Environmental Technologies.

Brenner, Michael P., Ph.D., University of Chicago, 1994. *Applied Mathematics, Applied Physics.* Biomechanics and Motor Control; Biophysics and Self-Assembly; Fluid Mechanics; Materials Science; Soft Condensed Matter.

Capasso, Federico, Ph.D., University of Rome, 1973. *Applied Physics.* Electromagnetic and Nano-Electronics; Nanophotonics; Quantum Devices.

Clarke, David, Ph.D., University of Cambridge, 1974. *Applied Physics, Engineering Physics/Science.* Electromagnetics and Nanoelectronics; Materials Science; Nanophotonics.

Cluzel, Phillippe, Ph.D., University of Paris VI, 1996. *Applied Physics.* Materials Science; Soft Condensed Matter; Surface and Interface Science; Biomechanics and Motor Control; Cell and Tissue Engineering and Biomaterials; Biophysics and Self-Assembly.

Farrell, Brian F., Ph.D., Harvard University, 1981. *Applied Physics, Engineering Physics/Science.* Modeling Physical/Biological Phenomena and Systems; Oceans and Geophysics.

Friend, Cynthia M., Ph.D., University of California, Berkeley, 1981. *Applied Physics.* Energy and Environmental technologies; Materials Science; Surface and Interface Science.

Golovchenko, Jene A., Ph.D., Rensselaer Polytechnic Institute, 1972. *Applied Physics.* Surface and Interface Science; Biophysics and Self-Assembly; Electromagnetics and Nanoelectronics; Nanophotonics; Soft Condensed Matter.

Ham, Donhee, Ph.D., California Institute of Technology, 2002. *Applied Physics, Engineering Physics/Science.* Control Theory and Stochastic Systems; Circuits and VLSI; Electromagnetics and Nanoelectronics; Nanophotonics; Quantum Devices.

Hau, Lene V., Ph.D., Aahrus, 1991. *Applied Physics.* Nanophotonics; Quantum Devices.

Horowitz, Paul, Ph.D., Harvard University, 1970. *Applied Physics, Electrical Engineering, Engineering Physics/Science.* Circuits and VLSI.

Hu, Eveyln, Ph.D., Columbia University, 1975. Area Dean for Electrical Engineering. *Applied Physics, Electrical Engineering, Engineering Physics/Science.* Electromagnetics and Nanoelectronics; Materials Science; Nanophotonics; Energy and Environmental Technologies.

Jacob, Daniel J., Ph.D., California Institute of Technology, 1985. *Applied Physics, Climate/Atmospheric Science, Engineering Physics/Science.* Atmospheric and Climate Modeling; Ocean and Geophysics.

Kaxiras, Efthimios, Ph.D., Massachusetts Institute of Technology, 1987. Director, Institute for Applied Computational Science; Area Dean for Applied Mathematics. *Applied Mathematics, Applied Physics.* Electromagnetics and Nanoelectronics; Materials Science; Surface and Interface Science; High Performance Computing; Modeling Physical/Biological Phenomena and Systems.

Keith, David, Ph.D., MIT, 1991. David Keith has worked near the interface between climate science, energy technology and public policy for twenty years. He took first prize in Canada's national physics prize exam, won MIT's prize for excellence in experimental physics, and was listed as one of TIME magazine's Heroes of the Environment 2009. *Applied Physics, Climate/Atmospheric Science, Energy Sources & Environ-*

ment. Engineering, Economic Development, and Resource Management.

Kuang, Zhiming, Ph.D., California Institute of Technology, 2003. *Applied Mathematics, Applied Physics, Engineering Physics/Science*. High Performance Computing; Modeling Physical/Biological Phenomena and Systems; Oceans and Geophysics.

Lewis, Jennifer A., Other, MIT, 1991. Prof. Lewis received her Sc.D. in Ceramics Science from MIT, 1991. *Energy Sources & Environment, Materials Science, Metallurgy, Mechanics*. Materials Science & Mechanical Engineering; Bioengineering

Lieber, Charles, Ph.D., Stanford University, 1985. *Applied Physics*. Electromagnetics and Nanoelectronics; Materials Science.

Loncar, Marko, Ph.D., California Institute of Technology, 2003. *Applied Physics, Engineering Physics/Science, Nano Science and Technology*. Electromagnetics and Nanoelectronics; Nanophotonics; Quantum Devices.

Mahadevan, L., Ph.D., Stanford University, 1995. *Applied Mathematics, Applied Physics*. Soft Condensed Matter; Solid Mechanics; Surface and Interface Science; Control Theory and Stochastic Systems; Modeling Physics/Biological Phenomena and Systems.

Manoharan, Vinothan, Ph.D., University of California, Santa Barbara, 2004. *Applied Physics*. Biophysics and Self-Assembly; Materials Science; Nanophotonics; Soft Condensed Matter; surface and interface science.

Martin, Scot T., Ph.D., California Institute of Technology, 1995. *Applied Physics, Engineering Physics/Science*. Atmospheric and Climate Modeling; Energy and Environmental Technologies; Environmental Chemistry and Microbiology; Oceans and Geophysics; Pollution Monitoring.

Mazur, Eric, Ph.D., Leiden University, 1981. Area Dean for Applied Physics. *Applied Physics*. Nanophotonics; Science and Engineering Education.

McElroy, Michael B., Ph.D., Queen's University, 1962. *Applied Physics, Engineering Physics/Science*. Energy, Environment and Sustainability; Atmospheric and Climate Modeling; Oceans and Geophysics; Engineering, Economic Development and Resource Management.

Murray, Cherry A., Ph.D., Massachusetts Institute of Technology, 1978. Dean of Harvard School of Engineering and Applied Science. *Applied Physics*. Nanophotonics; soft condensed matter; surface and interface science.

Narayanamurti, Venkatesh, Ph.D., Cornell University, 1965. *Applied Physics*. Electromagnetics and Nanoelectronics; Materials Science; Surface and Interface Science; Energy, Environment and Sustainability; Governance of Emerging Technologies/Innovation Policy; Science and Engineering Education; Social enterprise.

Nelson, David R., Ph.D., Cornell University, 1975. *Applied Physics*. Modeling Physical/Biological Phenomena and Systems; Biophysics and Self-Assembly; Soft Condensed Matter.

Parker, kevin K., Ph.D., Vanderbilt University, 1998. *Applied Physics, Engineering Physics/Science*. Biophysics and Self-assembly; Biomechanics and Motor Control; Cell and Tissue Engineering and Biomaterials.

Pershan, Peter S., Ph.D., Harvard University, 1960. *Applied Physics*. Soft Condensed Matter; Surface and Interface Science.

Ramanathan, Sharad, Ph.D., Harvard University, 1997. *Applied Physics*. Cell and Tissue Engineering and Biomaterials; Biophysics and Self-Assembly; Electromagnetics and Nanoelectronics; Materials Science; Surface and Interface Science.

Rice, James R., Ph.D., Lehigh University, 1964. *Applied Physics, Engineering Physics/Science*. Solid Mechanics; Fluid Mechanics; Oceans and Geophysics; Materials Science; Mod-

eling Physics/Biological Phenomena and Systems; Energy and Environmental Technologies.

Spaepen, Frans, Ph.D., Harvard University, 1975. *Applied Physics, Engineering Physics/Science*. Materials Science; Soft Condensed Matter; Surface and Interface Science.

Tziperman, Eli, Ph.D., Massachusetts Institute of Technology, 1987. *Applied Physics, Engineering Physics/Science*. Atmospheric and Climate Modeling; Oceans and Geophysics; Fluid Mechanics; Solid Mechanics; Modeling Physical/Biological Phenomena and Systems.

Weitz, David A., Ph.D., Harvard University, 1978. *Applied Physics*. Biomechanics and Motor Control; Biophysics and self-Assembly; Materials Science; Soft Condensed Matter; Surface and Interface Science; Start-ups and Technology Transfer.

Westervelt, Robert M., Ph.D., University of California, Berkeley, 1977. *Applied Physics*. Electromagnetics and Nanoelectronics; Circuits and VLSI; Biophysics and Self-Assembly.

Wofsy, Steven C., Ph.D., Harvard University, 1971. *Applied Physics, Engineering Physics/Science*. Atmospheric and Climate Modeling; Oceans and Geophysics; Observation and Field Testing.

Wu, Tai, Ph.D., Harvard University, 1956. *Applied Physics*. Modeling Physical/Biological Phenomena and Systems.

Yacoby, Amir, Ph.D., Weizmann Institute of Science, 1994. *Condensed Matter Physics, Electromagnetism, Nano Science and Technology, Quantum Foundations*. Nanophotonics and Quantum Devices.

Associate Professor

Crozier, Kenneth B., Ph.D., Stanford University, 2003. *Applied Physics, Engineering Physics/Science*. Nanophotonics.

Needleman, Daniel, Ph.D., University of California, Santa Barbara. *Applied Physics*. Biomechanics and Motor Control; Cell and Tissue Engineering and Biomaterials; Biophysics and Self-Assembly; Materials Science; Soft Condensed Matter; Surface and Interface Science.

Ramanathan, Shriram, Ph.D., Stanford University, 2002. *Applied Physics, Engineering Physics/Science, Materials Science, Metallurgy*. Energy and Environmental Technologies; Materials Science, Surface and Interface Science; Electromagnetics and Nanoelectronics.

DEPARTMENTAL RESEARCH SPECIALTIES AND STAFF

Theoretical

Other. Ph.D. students in SEAS/GSAS can elect to complete a secondary field in Computational Science and Engineering (CSE). SEAS welcomes applications for the Secondary Field in Computational Science and Engineering (CSE). This Secondary Field is available to any student enrolled in a Ph.D. program in the Graduate School of Arts and Sciences, upon approval of a plan of study by the CSE Program Committee and the student's home department Director of Graduate Studies. CSE is an exciting and rapidly evolving field that exploits the power of computation as an approach to major challenges on the frontiers of natural and social science and all engineering fields. In keeping with Harvard's emphasis on foundational knowledge, our program will focus on cross-cutting mathematical and computational principles important across disciplines. Completion of the Secondary Field will equip students with rigorous computational methods for approaching scientific questions. These approaches include mathematical techniques for modeling and simulation of complex systems; parallel programming and collaborative software development; and methods for organizing, exploring, visualizing, processing and analyzing very large data sets.

MASSACHUSETTS INSTITUTE OF TECHNOLOGY

DEPARTMENT OF EARTH, ATMOSPHERIC, AND PLANETARY SCIENCES

Cambridge, Massachusetts 02139
http://eapsweb.mit.edu

General University Information
President: L. Rafael Reif
Dean of Graduate School: Christine Ortiz
University website: http://web.mit.edu
Control: Private
Setting: Urban
Total Faculty: 1,685
Total Graduate Faculty: 1,685
Total number of Students: 10,500
Total number of Graduate Students: 6,300

Department Information
Department Chairman: Robert van der Hilst, Chair
Department Contact: Vicki McKenna, Academic Program
  Administrator
  Total full-time faculty: 36
  Total number of full-time equivalent positions: 39
  Full-Time Graduate Students: 169
  First-Year Graduate Students: 29
  Female First-Year Students: 15
  Total Post Doctorates: 64

Department Address
54-912
77 Massachusetts Avenue
Cambridge, MA 02139
Phone: (617) 253-3381
Fax: (617) 253-8298
E-mail: eapsinfo@mit.edu
Website: http://eapsweb.mit.edu

ADMISSIONS

Admission Contact Information
Address admission inquiries to: Education Office, Department of
  Earth, Atmospheric, and Planetary Sciences, Room 54-912,
  MIT, 77 Massachusetts Avenue, Cambridge, MA 02139
Phone: (617) 253-3381
E-mail: eapsinfo@mit.edu
Admissions website: http://eapsweb.mit.edu/academics/
  admissions/graduate

Application deadlines
Fall admission:
U.S. students: January 5   *Int'l. students*: January 5
Spring admission:
U.S. students: November 5   *Int'l. students*: November 5

Application fee
U.S. students: $75   *Int'l. students*: $75

Admissions information
For Fall of 2013:
  Number of applicants: 201
  Number admitted: 31
  Number enrolled: 15

Admission requirements
Bachelor's degree requirements: A strong undergraduate emphasis in math and science is necessary.

GRE requirements
The GRE is required.

Advanced GRE requirements
The Advanced GRE is not required.

TOEFL requirements
The TOEFL exam is required for students from non-English-
  speaking countries.
  PBT score: 577
  iBT score: 100

Other admissions information
Additional requirements: IELTS is the preferred examination,
  rather than Tthe OEFL, and 7.0 is the minimum acceptable
  score. The GRE subject test in chemistry or physics is required for the planetary science program.
Undergraduate preparation assumed: An undergraduate degree
  should have strong emphasis on math and science. Specific
  preparation will depend on the area of study chosen.

TUITION

Tuition year 2013-14:
Tuition for in-state residents
  Full-time students: per semester
  Full-time students: $21,605 per semester
  Part-time students: $670 per credit
Deferred tuition plan: Yes
Health insurance: Yes, $1,980 for AY13.
Other academic fees: $288 student life fee.
Academic term: Semester
Number of first-year students who received full tuition waivers: 26

Teaching Assistants, Research Assistants, and Fellowships
Number of first-year
  Teaching Assistants: 2
Average stipend per academic year
  Teaching Assistant: $24,561
  Research Assistant: $23,976
  Fellowship student: $24,561
All students accepted into our doctoral programs are provided
  with support that includes tuition, a stipend, and health in-
  surance.

FINANCIAL AID

Loans
Loans are available for U.S. students.
Loans are not available for international students.
GAPSFAS application required: No
FAFSA application required: No

For further information
Address financial aid inquiries to: Student Financial Services,
  MIT Room 11-320, 77 Massachusetts Avenue, Cambridge,
  MA 02139.
Phone: (617) 253-4971
E-mail: sfs@mit.edu

Financial aid website: http://web.mit.edu/sfs/financial_aid/index.html

HOUSING

Availability of on-campus housing
Single students: Yes
Married students: Yes

For further information
Address housing inquiries to: Graduate Housing Office, MIT Room W59-200, 77 Massachusetts Avenue, Cambridge, MA 02139.
Phone: (617) 253-5148
E-mail: graduatehousing@mit.edu
Housing aid website: http://housing.mit.edu/graduatefamily/graduate_family_housing

Table A—Faculty, Enrollments, and Degrees Granted

Research Specialty	36 Faculty	Enrollment 169		Number of Degrees Granted 40		
		Master's	Doctorate	Master's	Terminal Master's	Doctorate
Climate/Atmospheric Science	17	1	29	–	1(-)	7(-)
Geology/Geochemistry	13	–	20	–	2(-)	2(-)
Geophysics	10	2	30	–	3(-)	1(-)
Marine Science/ Oceanography	4	–	74	–	5(-)	15(-)
Planetary Science	9	1	12	1(-)	1(-)	2(-)
Total	36	4	165	1(-)	12(-)	27(-)
Full-time Grad. Stud.	–	4	165	–	–	–
First-year Grad. Stud.	–	3	26	–	–	–

GRADUATE DEGREE REQUIREMENTS

Master's: Sixty-six units of total credit, which include 42 units within a subject area and 44 units of higher-level graduate credit. A thesis is required.
Doctorate: Completion of the departmental program and a thesis are required.
Thesis: Required for all students.

SPECIAL EQUIPMENT, FACILITIES, OR PROGRAMS

Students have access to the Magellan telescopes at Las Campanas Observatory. Students in the MIT/WHOI Joint Program have access to the extensive oceanographic research facilities of the Woods Hole Oceanographic Institution. The Department gives students access to excellent computer and laboratory facilities. Students may also participate in a variety of field camps: geological, geophysical, astronomical, and oceanographic.

Table B—Separately Budgeted Research Expenditures by Source of Support

Source of Support	Departmental Research	Physics-related Research Outside Department
Federal government	$11,404,663	
State/local government	$15,372	
Non-profit organizations	$880,840	
Business and industry	$912,328	
Other	$15,723	
Total	**$13,228,926**	

Table C—Separately Budgeted Research Expenditures by Research Specialty

Research Specialty	No. of Grants	Expenditures ($)
Astronomy	5	$408,440
Geophysics	38	$3,133,095
Marine Science/Oceanography	5	$597,891
Climate/Atmospheric Science	60	$6,205,739
Geology/Geochemistry	61	$3,600,805
Planetary Science	31	$4,121,992
Total	200	$18,067,962

FACULTY

Professor

Binzel, Richard P., Ph.D., University of Texas, 1986. *Astronomy*. Planetary astronomy; collisional evolution of asteroids; physical parameters and surface features of the Pluto-Charon system.

Bowring, Samuel A., Ph.D., University of Kansas, 1985. *Geology/Geochemistry*. Origin and evolution of continental lithosphere using radiogenic isotopes including earliest history of the Earth; U-Pb geochronology and thermochronology of orogenic belts; high-precision U-Pb geochronology applied to problems in earth history.

Boyle, Edward A., Ph.D., Massachusetts Institute of Technology, 1976. *Geology/Geochemistry, Marine Science/Oceanography*. Paleoceanography and paleoclimatology; variability of the chemical composition of seawater; trace element chemistry of seawater, rivers, and estuaries.

Burchfiel, B. Clark, Ph.D., Yale University, 1961. *Geology/Geochemistry*. Origin, development, and structural evolution of the continental crust; active tectonics; studies of the older, more deeply eroded parts of the continental crust. Current studies focus on China and Eastern Europe.

Emanuel, Kerry A., Ph.D., Massachusetts Institute of Technology, 1981. *Climate/Atmospheric Science*. Relationship between cumulus convection and large-scale circulations; parametric representation of convection in large-scale weather forecast and climate models; the Hadley circulation; mesoscale dynamics of fronts and cyclones; tropical cyclone dynamics.

Evans, J. Brian, Ph.D., Massachusetts Institute of Technology, 1978. *Geophysics*. Strength of rocks; the effect of fluids and impurities on strength; recrystallization and grain growth; microstructures of naturally deformed rocks; applications of rock mechanics to tectonic problems; interrelationships of porosity, permeability, and plastic flow.

Ferrari, Raffaele, Ph.D., Scripps Institution of Oceanography, 2001. *Climate/Atmospheric Science, Marine Science/Oceanography*. Turbulence in the ocean and atmosphere using a combination of theory, models, and observations; role of the ocean on climate and on biological productivity.

Flierl, Glenn R., Ph.D., Harvard University, 1975. *Climate/Atmospheric Science*. Impacts of oceanic eddies upon the distribution of tracers and on the biology of the sea, including both transport and alterations in the reaction terms.

Grove, Timothy L., Ph.D., Harvard University, 1976. *Geology/Geochemistry, Planetary Science*. Igneous petrology; magma generation processes in island arc-continental settings and mid-ocean ridges; crystal growth and nucleation; phase transitions in minerals; diffusion in crystalline solids and silicate melts; thermal histories of geologic materials.

Hager, Bradford H., Ph.D., Harvard University, 1978. *Geophysics*. Physics of geologic processes; numerical modeling of mantle convection in terrestrial planets; numerical modeling of crustal deformation; GPS geodesy.

Herring, Thomas A., Ph.D., Massachusetts Institute of Technology, 1983. *Geophysics*. Techniques of space geodesy, including very Long Baseline interferometry and the use of the Global Positioning System; surface deformations related to plate tectonics and plate boundary zones; effects of whole-Earth dynamics on the nutation series.

Marshall, John C., Ph.D., Imperial College, 1980. *Climate/Atmospheric Science, Marine Science/Oceanography*. Dynamics and causes of the general circulation of the atmosphere and ocean; thermocline theory; geostrophic eddies; global-scale ocean modeling.

Morgan, Dale, Ph.D., Massachusetts Institute of Technology, 1981. *Geophysics*. Rock physics; geoelectromagnetism; inverse methods; applied seismology; environmental geophysics.

Plumb, R. Alan, Ph.D., University of Manchester, 1972. *Climate/Atmospheric Science*. Eddy transport processes in the atmosphere and ocean; dynamics of the stratosphere and mesosphere and their interaction with the lower atmosphere; large-scale tropospheric dynamics; transport of chemical constituents.

Prinn, Ronald G., Ph.D., Massachusetts Institute of Technology, 1971. *Climate/Atmospheric Science, Planetary Science*. Chemical-dynamical models of the atmosphere; measurement and modeling of the long-lived gases involved in the greenhouse effect and ozone depletion; atmospheric chemistry of carbon and sulfur compounds; integrated global system modelling that couples atmospheric, oceanic, and terrestrial physics, chemistry, and biology.

Rizzoli, Paola M., Ph.D., Scripps Institution of Oceanography, 1978. *Climate/Atmospheric Science, Marine Science/Oceanography*. Numerical modeling of the ocean general circulation with specific emphasis on the tropical Atlantic ocean, tropical/subtropical interactions, tropical instability waves, and coupled ocean-atmosphere modes; assimilation of oceanographic data into ocean numerical models through ensemble approaches and optimal design of fixed and adaptive observational arrays; physical-biochemical modeling of the Black Sea ecosystem.

Rothman, Daniel H., Ph.D., Stanford University, 1986. *Climate/Atmospheric Science, Geology/Geochemistry, Geophysics*. Dynamical organization of the past and present environment, including co-evolution of life and the physical environment; dynamics of the carbon cycle; geological fluid mechanics; geomorphology.

Royden, Leigh H., Ph.D., Massachusetts Institute of Technology, 1982. *Geology/Geochemistry, Geophysics*. Regional geology and geophysics; plate tectonics; thermal effects of continental deformation; mechanics of large-scale continental deformation; lithospheric flexure; continental extensions and sedimentary basin formation; uplift and erosion in mountain belts.

Seager, Sara, Ph.D., Harvard University, 1999. *Astronomy, Atmosphere, Space Physics, Cosmic Rays, Planetary Science*. Finding and characterizing earth-like exoplanets; theoretical models of atmospheres, interiors, and biosignatures of all kinds of exoplanets. Astrobiology.

Solomon, Susan, Ph.D., University of California, Berkeley, 1981. *Climate/Atmospheric Science*. Atmospheric chemistry and transport in the stratosphere and troposphere; climate change and its coupling to chemistry; comparative studies of environment and society.

Summons, Roger, Ph.D., University of New South Wales, 1972. *Climate/Atmospheric Science, Geology/Geochemistry*. Lipid chemistry of microbes; early biotic and environmental evolution; extinction and radiation events in earth history; biogeochemical fossils; petroleum; astrobiology.

van der Hilst, Robert D., Ph.D., Utrecht University, 1990. *Geophysics*. Seismic tomography; studies of the earth's structure with emphasis on mantle beneath convergent plate boundaries; tectonic evolution of subduction systems; mantle dynamics; structure and evolution of continental lithosphere; field studies with portable seismometers.

Wisdom, Jack, Ph.D., California Institute of Technology, 1981. *Planetary Science*. Solar system dynamics; long-term evolution of the orbits and spins of the planets and natural satellites; qualitative behavior of dynamical systems; chaotic behavior.

Zuber, Maria T., Ph.D., Brown University, 1986. *Geophysics, Planetary Science*. Theoretical modeling of geophysical processes; analysis of altimetry, gravity, and tectonics to determine the structure and dynamics of the earth and solid planets; development and implementation of spacecraft laser and radio-tracking experiments.

Associate Professor

Bosak, Tanja, Ph.D., California Institute of Technology, 2004. *Geology/Geochemistry*. Microbial sediments throughout geologic time as indicators of biological processes and environmental conditions; morphological and chemical biosignatures; early earth; astrobiology.

Cziczo, Daniel J., Ph.D., University of Chicago, 1999. *Climate/Atmospheric Science*. Chemical composition of atmospheric aerosols with an emphasis on their effect on cloud formation mechanisms, the arth's radiative budget, and meteoritic debris and launch vehicle emissions in the atmosphere.

Follows, MIchael, Ph.D., University of East Anglia, 1990. *Climate/Atmospheric Science, Marine Science/Oceanography*. Biogeochemical cycles of carbon and nutrients in the ocean. Use of numerical models to understand the combination of physical transport, chemical and biological processes that determine the distributions and fluxes of these elements in the ocean.

Heald, Colette L., Ph.D., Harvard University, 2005. *Climate/Atmospheric Science*. Atmospheric chemistry and composition; biosphere-atmosphere interactions; global modeling; satellite observations.

O'Gorman, Paul, Ph.D., California Institute of Technology, 2004. *Climate/Atmospheric Science*. Large-scale circulation of the atmosphere; interactions of moisture and baroclinic eddies; effect of climate change on the hydrological cycle; turbulence closure theories.

Weiss, Benjamin P., Ph.D., California Institute of Technology, 2003. *Geology/Geochemistry, Geophysics, Planetary Science*. Paleomagnetic studies of rocks from Mars, the moon, asteroids, and the earth; dynamo evolution, planetary histories and interiors; use and development of new magnetometry techniques.

Assistant Professor

Cahoy, Kerri, Ph.D., Stanford University, 2008. *Climate/Atmospheric Science, Planetary Science*. Planetary atmospheres; exoplanet atmospheres with optical direct imaging instruments (coronagraphs) onboard spacecraft; solar system planets with spacecraft radio occultation; earth with GNSS radio occultation.

Jagoutz, Oliver, Ph.D., ETH Zurich, 2004. *Geology/Geochemistry*. Field-related studies of igneous processes; crust mantle interaction; formation and evolution of the oceanic and continental lithosphere.

Malcolm, Alison, Ph.D., Colorado School of Mines, 2005. *Geophysics*. Wave propagation in complicated media; seismic imaging in the shallow earth; locating buried resources (primarily oil and gas); applications of microlocal analysis in imaging; nonlinear wave propagation; exploiting information in multiply scattered waves to infer earth properties.

McGee, David, Ph.D., Columbia University, 2009. *Climate/Atmospheric Science, Geology/Geochemistry.* Reconstruction of past climates using cave, lake, and marine deposits; U-Th dating of cave and lacustrine carbonates; U-series investigations of marine sediments; constant flux proxies; records of past atmospheric circulation and hydrology.

Ono, Shuhei, Ph.D., Pennsylvania State University, 2001. *Geology/Geochemistry.* Isotope biogeochemistry of sulfur and oxygen, water-rock-microbe interactions, seafloor hydrothermal deposits, deep biosphere, and global sulfur cycles.

Perron, J. Taylor, Ph.D., University of California, Berkeley, 2006. *Geology/Geochemistry, Planetary Science.* Measurement and modeling of physical processes that shape the surfaces of planets; river networks; biotic effects on landscape evolution; volatile cycling on Mars and Titan.

Prieto, Germån, Ph.D., University of California, San Diego, 2007. *Geophysics.* Use seismic records to understand the earthquake source, the interior of the Earth and how both affect the ground motions that we feel on the Earth's surface. Seismological observations are affected by the internal structure of the Earth, for example amplification of seismic waves in sedimentary basins. The nature of the earthquake source has also a significant impact on ground motions, and I am interested in a better understanding of earthquake ruptures.

Schlichting, Hilke, Ph.D., California Institute of Technology, 2009. *Astronomy.* My research interests span all aspects of planet formation theory, extrasolar planets and solar system dynamics. I am interested in our solar system, since it is the only place where we can examine the outcome of planet formation in detail and it provides a wealth of information about planet formation processes that will remain unattainable for extrasolar systems.

Selin, Noelle Eckley, Ph.D., Harvard University, 2007. *Climate/Atmospheric Science.* Atmospheric chemistry modeling; biogeochemical cycling of mercury (Hg); air pollution/climate interactions; air pollution health impacts; science-policy interactions.

DEPARTMENTAL RESEARCH SPECIALTIES AND STAFF

Theoretical

Climate/Atmospheric Science. Dynamics of the atmosphere and ocean; climate dynamics and modeling; theory of monsoons; coupled ocean-atmosphere models; chemical dynamical models of the atmosphere; inverse methods applied to global trace gas cycles; climatic effects of changes in carbon dioxide concentrations and solar constant; dynamics of planetary atmospheres; data assimilation and adaptive sampling; predictability and ensemble forecasting; mesoscale dynamics of fronts and cyclones; dynamics of tropical intraseasonal oscillations and tropical cyclones; modeling planetary atmospheres. Boyle, Cziczo, Emanuel, Ferrari, Flierl, Heald, Marshall, McGee, O'Gorman, Plumb, Prinn, Rizzoli, Rothman, Seager, Selin, Solomon, Summons.

Geophysics. Numerical models of nonlinear dynamical systems; fluid dynamics; theoretical models of rock physics; numerical methods for seismology; mantle dynamics; geodesy. Evans, Hager, Herring, Malcolm, Morgan, Rothman, Royden, van der Hilst, Weiss, Zuber.

Marine Science/Oceanography. Dynamics of thermohaline circulation of the ocean, numerical modeling of ocean-climate interactions, and analysis of oceanic data; modeling of the physics, chemistry, and biology of strongly nonlinear eddies and meandering jets; interactions between waves and vortices. Boyle, Ferrari, Flierl, Marshall, Rizzoli.

Planetary Science. Numerical experiments and theoretical studies of geophysical fluid dynamics; origins and evolution of planetary jet-stream wind profiles; solar system dynamics; long-term evolution of orbits and spins of planets and satellites; chaotic behavior; dynamics of planetary rings; planetary history, planetary gravity, and magnetic fields; geochemical and geophysical studies of meteorites, atmospheres, and interiors of exoplanets. Binzel, Cahoy, Grove, Perron, Prinn, Seager, Weiss, Wisdom, Zuber.

Experimental

Climate/Atmospheric Science. Ocean general circulation, paleo-oceanography, and paleo-climatology; abrupt climate change; development and application of trace element, organic geochemical, and stable isotopic techniques in oceanography and paleoclimatology; decadal-to-millennial scale climate change; origin of organic-rich sediment sequences in the marine environment; marine nitrogen cycle; acoustic tomography; hydrometeorology and hydroclimatology; global measurements of radiatively and chemically important trace gases; climate diagnostic studies; El Niño Southern Oscillation phenomenon; diagnostic studies of the general circulation; satellite observations of planetary atmospheres. Boyle, Cziczo, Emanuel, Ferrari, Flierl, Heald, Marshall, McGee, O'Gorman, Plumb, Prinn, Rizzoli, Rothman, Seager, Selin, Solomon, Summons.

Geology/Geochemistry. Rift magmatism, origin, and evolution of continental lithosphere using radiogenic isotopes, earth history, active tectonics, structural geology, metamorphic and igneous petrology, geochronology, and numerical simulations of depositional systems; magma generation processes in arcs, ocean ridges, ocean islands, and large igneous processes; mineralogy of the mantle and mantle processes controlling mantle geochemistry; mechanics and thermal effects of continental deformation, sediment transport by currents and waves, interpretation of ancient sedimentary environments, process geomorphology, debris-flow rheology, tectonic geomorphology, environmental monitoring of natural terrestrial and marine ecosystems, and the role of climate in the evolution of orogenic systems; petroleum systems; lipids of cultured microbes and microbial consortia; molecular signatures of hydrothermal ecosystems; signals of biochemical change through time. Bosak, Bowring, Boyle, Burchfiel, Grove, Jagoutz, McGee, Ono, Perron, Rothman, Royden, Summons, Weiss.

Geophysics. Application of rock mechanics to tectonic problems; mantle dynamics; numerical modeling of solid-state convection; space geodesy; plate tectonics; seismology; geodetic observation of surface deformation; thermal structure of oceanic lithosphere in the vicinity of hot-spot volcanoes; rock physics; environmental geophysics; seismic tomography for characterization of the earth's crust and petroleum reservoirs; tectonic evolution of subduction systems; structure and evolution of continental lithospheres. Evans, Hager, Herring, Malcolm, Morgan, Rothman, Royden, van der Hilst, Weiss, Zuber.

Marine Science/Oceanography. Numerical modeling of ocean general circulation; paleo-oceanography and paleo-climatology; trace element geochemistry; acoustic tomography. Boyle, Ferrari, Flierl, Marshall, Rizzoli.

Planetary Science. Collisional of asteroids; planetary atmospheric dynamics; Kuiper belt; stellar occultations at optical and infrared wavelengths; geodesy; radar and radio studies of physical properties of planets; the Pluto-Charon system; origins and evolution of eddy features (e.g., Jupiter's Great Red Spot); models of planetary lithosphere deformation and the physics of volcanism; development and implementation of space-based laser ranging systems; planetary paleomagnetism and geomagnetism; astrobiology and planetary history; extrasolar planets. Binzel, Cahoy, Grove, Perron, Prinn, Seager, Weiss, Wisdom, Zuber.

MASSACHUSETTS INSTITUTE OF TECHNOLOGY

DEPARTMENT OF PHYSICS

Cambridge, Massachusetts 02139
http://web.mit.edu/physics/index.html

General University Information
President: L. Rafael Reif
Dean of Graduate School: Christine Ortiz
University website: http://web.mit.edu
Control: Private
Setting: Urban
Total Faculty: 1,022
Total Graduate Faculty: 1,022
Total number of Students: 11,189
Total number of Graduate Students: 6,686

Department Information
Department Chairman: Edmund Bertschinger, Head
Department Contact: Matthew Cubstead, Administrative
  Officer
  Total full-time faculty: 75
  Total number of full-time equivalent positions: 75
  Full-Time Graduate Students: 228
  First-Year Graduate Students: 35
  Female First-Year Students: 9
  Total Post Doctorates: 99

Department Address
77 Massachusetts Avenue
MIT 4-315
Cambridge, MA 02139
Phone: (617) 253-4803
E-mail: cubstead@mit.edu
Website: http://web.mit.edu/physics/index.html

ADMISSIONS

Admission Contact Information
Address admission inquiries to: Graduate Admissions Officer,
  Dept. of Physics, Room 4-315
Phone: (617) 253-9703
E-mail: physics-grad@mit.edu
Admissions website: http://web.mit.edu/physics/prospective/
  graduate/index.html

Application deadlines
Fall admission:
U.S. students: December 15   *Int'l. students*: December 15

Application fee
U.S. students: $75   *Int'l. students*: $75

Admissions information
For Fall of 2013:
  Number of applicants: 784
  Number admitted: 96
  Number enrolled: 49

Admission requirements
Bachelor's degree requirements: Bachelor's degree in physics
  is required.

GRE requirements
The GRE is required.

Advanced GRE requirements
The Advanced GRE is required.

TOEFL requirements
Students from non-English speaking countries are required to
  demonstrate proficiency in English via the IELTS exam or
  the TOEFL exam.

Other admissions information
Undergraduate preparation assumed: Reif, Fundamentals of Statistical and Thermal Physics: Marion, Classical Dynamics,
  and Classical Electromagnetic Radiation; Griffiths, Quantum
  Mechanics.

TUITION

Tuition year 2013-14:
  Full-time students: $43,210 annual
  Part-time students: $670 per credit
(12 units/course) Definition of 'tuition waivers': all incoming
  students have their tuition fully paid through a fellowship,
  research assistantship, or teaching assistantship.
Credit hours per semester to be considered full-time: 36
Deferred tuition plan: No
Health insurance: Available at the cost of $2,088 per year.
Other academic fees: Student Life Fee $280.
Academic term: Semester
Number of first-year students who received full tuition waivers: 49

Teaching Assistants, Research Assistants, and Fellowships
Number of first-year
  Research Assistants: 7
  Fellowship students: 43
Average stipend per academic year
  Teaching Assistant: $32,752
  Research Assistant: $32,752
  Fellowship student: $32,752

FINANCIAL AID

Application deadlines
Fall admission:
U.S. students: December 15   *Int'l. students*: December 15

Loans
Loans are not available for U.S. students.
Loans are not available for international students.
GAPSFAS application required: No
FAFSA application required: No

For further information
Address financial aid inquiries to: Graduate Admissions Officer,
  Dept. of Physics, Room 4-315.
Phone: (617) 253-9703
E-mail: physics-grad@mit.edu
Financial aid website: http://web.mit.edu/physics

HOUSING

Availability of on-campus housing
Single students: Yes
Married students: Yes

For further information
Address housing inquiries to: Campus Housing Information W59-200.
Phone: (617) 253-5148
E-mail: graduatehousing@mit.edu
Housing aid website: http://housing.mit.edu/graduatefamily/ graduate_family_housing

Table A—Faculty, Enrollments, and Degrees Granted

Research Specialty	2013-14 Faculty	Enrollment Fall 2012 Master's	Enrollment Fall 2012 Doctorate	Number of Degrees Granted 2012-13 (2008-13) Master's	Number of Degrees Granted 2012-13 (2008-13) Terminal Master's	Number of Degrees Granted 2012-13 (2008-13) Doctorate
Astrophysics	17	–	28	–	–(1)	5(27)
Atomic, Molecular, & Optical Physics	4	–	38	–	1(3)	6(23)
Biophysics	6	–	14	–	–	2(14)
Condensed Matter Physics	15	–	56	–(1)	1(3)	6(40)
Nuclear Physics	10	1	18	1(1)	1(2)	9(15)
Particles and Fields	20	–	54	–	1(3)	5(50)
Plasma and Fusion	1	–	20	–	–	4(16)
Quantum Information and Quantum Computing	3	–	–	–	–	–
Total	76	1	228	1(2)	4(12)	37(185)
Full-time Grad. Stud.	–	2	228	–	–	–
First-year Grad. Stud.	–	–	35	–	–	–

GRADUATE DEGREE REQUIREMENTS

Master's: Approximately six graduate level courses in physics are required. A "B−" average must be maintained. Thesis required. Residence: one semester. There are no foreign language or comprehensive exam requirements.

Doctorate: Two academic years of full-time graduate work (including thesis) are required for the Ph.D. degree. Two courses are required inside, and two outside, the candidate's doctoral specialty. A "B−" average must be maintained. A general Doctoral examination consisting of two written parts and one oral part must be passed in the second or third year of graduate work. Original research, demonstrated through a thesis is required. The thesis and oral defense of the thesis complete the requirements for the doctorate.

Thesis: Thesis may be written in absentia (with special permission only).

Table B—Separately Budgeted Research Expenditures by Source of Support

Source of Support	Departmental Research	Physics-related Research Outside Department
Federal government	$98,228,505	
State/local government	$679,675	
Non-profit organizations	$485,237	
Business and industry	$1,616,137	
Other	$1,761,867	
Total	$102,771,421	

Table C—Separately Budgeted Research Expenditures by Research Specialty

Research Specialty	No. of Grants	Expenditures ($)
Astrophysics	–	$25,436,190
Atomic, Molecular, & Optical Physics	–	$7,988,991
Biophysics	–	$2,033,003
Condensed Matter Physics	–	$13,751,645
Nuclear Physics	–	$23,291,550
Plasma and Fusion	–	$29,688,655
Other	0	$581,387
Total	–	$102,771,421

FACULTY

Professor

Ashoori, Raymond, Ph.D., Cornell University, 1990. *Condensed Matter Physics*. Experimental and condensed matter physics.

Belcher, John W., Ph.D., California Institute of Technology, 1970. Associate Head for Education. *Astrophysics, Plasma and Fusion*. Theoretical physics; solar plasma.

Bertschinger, Edmund W., Ph.D., Princeton University, 1984. *Astrophysics*. Theoretical physics; astrophysics.

Canizares, Claude, Ph.D., Harvard University, 1972. Vice President for Research and Associate Provost. *Astronomy, Astrophysics*. Experimental physics; X-ray astronomy.

Chakrabarty, Deepto, Ph.D., California Institute of Technology, 1996. Division Head, Astrophysics. *Astronomy, Astrophysics*. Experimental physics; compact objects.

Chakraborty, Arup, Ph.D., University of Delaware, 1988.

Chen, Min, Ph.D., University of California, Berkeley, 1969. *Particles and Fields*. Experimental physics; high energy.

Chuang, Isaac, Ph.D., Stanford University, 1997. *Atomic, Molecular, & Optical Physics, Computational Physics*. Experimental physics; quantum computation.

Conrad, Janet, Ph.D., Harvard University, 1993. *Particles and Fields*. Experimental physics; high energy.

Farhi, Edward H., Ph.D., Harvard University, 1978. Division Head, Theoretical Nuclear and Particle Physics. *Computational Physics, Particles and Fields*. Theoretical physics; elementary particle, and quantum computation.

Fisher, Peter H., Ph.D., California Institute of Technology, 1988. Division Head, Experimental Nuclear and Particle Physics. *Particles and Fields*. Experimental physics; high energy.

Freedman, Daniel, Ph.D., University of Wisconsin-Madison, 1964. *Particles and Fields, Theoretical Physics*. Theoretical physics.

Greytak, Thomas, Ph.D., Massachusetts Institute of Technology, 1967. Interim Department Head. *Condensed Matter Physics*. Condensed matter physics.

Guth, Alan J., Ph.D., Massachusetts Institute of Technology, 1972. *Astronomy, Particles and Fields*. Theoretical physics; elementary particle physics; cosmology.

Hewitt, Jacqueline, Ph.D., Massachusetts Institute of Technology, 1986. *Astrophysics*. Experimental physics; gravitational lenses.

Jaffe, Robert, Ph.D., Stanford University, 1972. *Particles and Fields*. Theoretical physics; elementary particles.

Joannopoulos, John, Ph.D., University of California, Berkeley, 1974. *Condensed Matter Physics, Nano Science and Technology*. Theoretical physics; solid state, nanotechnology.

Joss, Paul C., Ph.D., Cornell University, 1971. *Astrophysics*. Theoretical physics; astrophysics.

Kardar, Mehran, Ph.D., Massachusetts Institute of Technology, 1986. *Biophysics, Condensed Matter Physics*. Condensed matter theory, biophysics.

Kastner, Marc, Ph.D., University of Chicago, 1972. Dean, MIT School of Science. *Condensed Matter Physics.* Experimental physics; semi-conductors.

Ketterle, Wolfgang, Ph.D., Ludwig-Maximilian University of Munich, 1986. *Atomic, Molecular, & Optical Physics.* Experimental physics; atomic resonance and scattering.

Kowalski, Stanley B., Ph.D., Massachusetts Institute of Technology, 1963. *Particles and Fields.* Experimental nuclear and particle physics.

Lee, Patrick, Ph.D., Massachusetts Institute of Technology, 1970. *Condensed Matter Physics.* Condensed matter theory.

Lee, Young, Ph.D., Massachusetts Institute of Technology, 2000. *Particles and Fields.* Experimental physics; X-ray scattering and neutron scattering.

Levitov, Leonid, Ph.D., Moscow, Physical Technical Institute, 1989. *Condensed Matter Physics.* Condensed matter theory.

Mavalvala, Nergis, Ph.D., Massachusetts Institute of Technology, 1997. *Astrophysics.* Astrophysics, gravity.

Milner, Richard, Ph.D., California Institute of Technology, 1984. *Nuclear Physics.* Experimental nuclear physics.

Negele, John, Ph.D., Cornell University, 1969. *Nuclear Physics, Particles and Fields.* Theoretical physics; nuclear structure.

Paus, Christoph M.E., Ph.D., III Phys. Institut RWTH Aachen, 1996. *Particles and Fields.* Experimental physics, high energy.

Porkolab, Miklos, Ph.D., Stanford University, 1967. *Plasma and Fusion.* Experimental physics; plasma physics.

Pritchard, David E., Ph.D., Harvard University, 1968. *Atomic, Molecular, & Optical Physics.* Experimental physics; atomic resonance and scattering.

Rajagopal, Krishna, Ph.D., Princeton University, 1993. *Condensed Matter Physics, Nuclear Physics, Particles and Fields.* Theoretical nuclear and particle physics.

Redwine, Robert, Ph.D., Northwestern University, 1973. *Nuclear Physics.* Experimental physics; nuclear structure.

Roland, Gunther, Ph.D., University of Frankfurt, 1993. *Nuclear Physics.* Experimental physics, relativistic heavy ion.

Schechter, Paul, Ph.D., California Institute of Technology, 1974. *Astronomy, Astrophysics.* Experimental physics; extragalactic astronomy.

Seager, Sara, Ph.D., Harvard University, 1999. *Astrophysics, Atmosphere, Space Physics, Cosmic Rays.* Theoretical physics; astrophysics.

Seung, Sebastian, Ph.D., Harvard University, 1990. *Biophysics, Computational Physics.* Computational neuroscience.

Soljacic, Marin, Ph.D., Princeton University, 2000. *Condensed Matter Physics, Optics.* Theoretical physics; nonlinear optics.

Stewart, Iain, Ph.D., California Institute of Technology, 1999. *Nuclear Physics, Particles and Fields.* Theoretical nuclear physics.

Taylor, Washington, Ph.D., University of California, Berkeley, 1993. *Particles and Fields.* Theoretical physics; string theory.

Tegmark, Max, Ph.D., University of California, Berkeley, 1994. *Astronomy, Astrophysics.* Theoretical physics; cosmology.

Ting, Samuel C. C., Ph.D., University of Michigan, 1962. *High Energy Physics, Particles and Fields.* Experimental physics; high energy.

Todadri, Senthil, Ph.D., Yale University, 1997. *Condensed Matter Physics.* Condensed matter theory.

van Oudenaarden, Alexander, Ph.D., Delft University of Technology, 1997. *Biophysics.* Biological physics.

Vuletic, Vladan, Ph.D., University of Munich, 1997. Division Head, Atomic, Biological, Condensed Matter, and Plasma Physics. *Atomic, Molecular, & Optical Physics, Low Temperature Physics.* Experimental physics; atomic.

Wen, Xiao-Gang, Ph.D., Princeton University, 1987. *Condensed Matter Physics.* Condensed matter theory.

Wilczek, Frank, Ph.D., Princeton University, 1974. *Particles and Fields.* Theoretical particle physics.

Wyslouch, Boleslaw, Ph.D., Massachusetts Institute of Technology, 1987. *Nuclear Physics, Particles and Fields.* Experimental nuclear physics; high energy.

Zwiebach, Barton, Ph.D., California Institute of Technology, 1983. *Particles and Fields.* Theoretical; elementary particle theory.

Zwierlein, Martin, Ph.D., Massachusetts Institute of Technology, 2007. *Atomic, Molecular, & Optical Physics.* Experimental physics; atomic.

Associate Professor

Adams, Allan, Ph.D., Stanford University, 2000. *Particles and Fields, Theoretical Physics.* Theoretical Physics; string theory.

Figueroa-Feliciano, Enectali, Ph.D., Stanford University, 2001. *Astronomy, Particles and Fields.* Experimental physics; X-ray, astronomy; cosmology.

Formaggio, Joseph, Ph.D., Columbia University, 2001. *Particles and Fields.* Experimental physics; high energy.

Gedik, Nuh, Ph.D., University of California, Berkeley, 2004. *Condensed Matter Physics, Optics.* Experimental condensed matter physics.

Hughes, Scott, Ph.D., California Institute of Technology, 1998. *Astrophysics.* Theoretical physics; gravitational physics.

Jarillo-Herrero, Pablo D., Ph.D., Delft University of Technology, 2005. *Condensed Matter Physics.* Experimental condensed matter physics.

Liu, Hong, Ph.D., Case Western Reserve University, 1997. *Particles and Fields.* Theoretical particle physics; string theory.

Mirny, Leonid, Ph.D., Harvard University, 1998. *Biophysics, Computational Physics, Medical, Health Physics.* Health sciences.

Simcoe, Robert, Ph.D., California Institute of Technology, 2003. *Astronomy, Astrophysics, Optics.* Experimental physics; optical.

Winn, Joshua, Ph.D., Massachusetts Institute of Technology, 2001. *Astronomy, Astrophysics, Optics.* Astrophysics; optical astronomy.

Assistant Professor

Detmold, William, Ph.D., University of Adelaide, 2002. *Nuclear Physics, Particles and Fields.* Theoretical Nuclear Physics: lattice QCD.

England, Jeremy L, Ph.D., Stanford University, 2009. *Biophysics.* Theoretical biophysics and statistical physics.

Evans, Matthew, Ph.D., California Institute of Technology, 2002. *Astrophysics.* Astrophysics, gravity.

Frebel, Anna L., Ph.D., Australian National University, 2006. *Astronomy.* Astrophysics.

Fu, Liang, Ph.D., University of Pennsylvania, 2009. *Condensed Matter Physics.* Theoretical condensed matter physics.

Gore, Jeff, Ph.D., University of California, Berkeley, 2005. *Biophysics.* Biological physics.

Harrow, Aram, Ph.D., Massachusetts Institute of Technology, 2005. *Computational Physics.* Quantum information and computing.

Klute, Markus, Ph.D., University of Bonn, 2004. *Particles and Fields.* Experimental physics; high energy.

Lee, Yen-Jie, Ph.D., Massachusetts Institute of Technology, 2011. *Nuclear Physics.* Heavy ion physics; QCD.

Monroe, Jocelyn, Ph.D., Columbia University, 2006. *Particles and Fields.* Experimental physics; high energy.

Slatyer, Tracy, Ph.D., Harvard University, 2010. *Particles and Fields.* Particle and fields; cosmology; astrophysics.

Thaler, Jesse, Ph.D., Harvard University, 2004. *Particles and Fields.* Theoretical physics; elementary particles.

Weinberg, Nevin N., Ph.D., California Institute of Technology, 2005. *Astrophysics.* Theoretical astrophysics.

Williams, Michael, Ph.D., Carnegie Mellon University, 2007. *Nuclear Physics, Particles and Fields.* Experimental nuclear physics.

Zuccon, Paolo, Ph.D., Perugia University, 2003. *Atmosphere, Space Physics, Cosmic Rays, Particles and Fields.* Experimental high energy physics.

Adjunct Professor

Moncton, David, Ph.D., Massachusetts Institute of Technology, 1975. *Condensed Matter Physics, Nuclear Physics.* Experimental condensed matter physics.

DEPARTMENTAL RESEARCH SPECIALTIES AND STAFF

Theoretical

Astrophysics. Belcher, Bertschinger, Hughes, Joss, Seager, Tegmark, Weinberg.

Atomic, Molecular, & Optical Physics.

Biophysics. England, Kardar.

Condensed Matter Physics. Fu, Joannopoulos, Kardar, Patrick Lee, Levitov, Soljacic, Todadri, Wen.

Nuclear Physics. Detmold, Negele, Rajagopal, Stewart, Todadri, Wen.

Particles and Fields. Adams, Farhi, Freedman, Guth, Jaffe, Liu, Negele, Rajagopal, Stewart, Taylor, Thaler, Wilczek, Zwiebach.

Plasma and Fusion. Belcher.

Quantum Information and Quantum Computing. Chuang, Farhi, Harrow.

Experimental

Astrophysics. Canizares, Chakrabarty, Frebel, Hewitt, Mavalvala, Schechter, Simcoe, Winn.

Atomic, Molecular, & Optical Physics. Chuang, Ketterle, Pritchard, Vuletic, Zwierlein.

Biophysics. Gore, Mirny, Seung, van Oudenaarden.

Condensed Matter Physics. Ashoori, Gedik, Jarillo-Herrero, Kastner.

Nuclear Physics. Kowalski, Milner, Moncton, Redwine, Roland, Williams, Wyslouch.

Particles and Fields. Chen, Conrad, Figueroa-Feliciano, Fisher, Formaggio, Klute, Kowalski, Young Lee, Monroe, Paus, Ting, Wyslouch, Zuccon.

Plasma and Fusion. Porkolab.

View additional information about this department at www.gradschoolshopper.com

---

NORTHEASTERN UNIVERSITY

DEPARTMENT OF PHYSICS

Boston, Massachusetts 02115
http://www.northeastern.edu/physics/

General University Information

President: Joseph E. Aoun
Dean of Graduate School: J. Murray Gibson
University website: http://www.northeastern.edu/
Control: Private
Setting: Urban
Total Faculty: 1,033
Total number of Students: 20,530
Total number of Graduate Students: 3,985

Department Information

Department Chairman: Paul Champion, Chair
Department Contact: Mark Williams, Professor and Graduate Coordinator
Total full-time faculty: 30
Total number of full-time equivalent positions: 30
Full-Time Graduate Students: 75
First-Year Graduate Students: 22
Female First-Year Students: 1
Total Post Doctorates: 21

Department Address

111 Dana Research Center
360 Huntington Avenue
Boston, MA 02115

Phone: (617) 373-4240
Fax: (617) 373-2943
E-mail: gradphysics@neu.edu
Website: http://www.northeastern.edu/physics/

ADMISSIONS

Admission Contact Information

Address admission inquiries to: Graduate Coordinator, Physics Department
Phone: (617) 373-4240
E-mail: gradphysics@neu.edu
Admissions website: http://www.physics.neu.edu

Application deadlines

Fall admission:
U.S. students: February 1   *Int'l. students*: February 1

Application fee

U.S. students: $75   *Int'l. students*: $75

The deadline for Ph.D. programs is February 1. The Priority date for master's programs for those interested in financial aid is February 1.

Admissions information

For Fall of 2013:
Number of applicants: 220
Number admitted: 45
Number enrolled: 17

Admission requirements

Bachelor's degree requirements: A bachelor's degree in physics or a related field is required.

GRE requirements

The GRE is required.

Advanced GRE requirements

The Physics subject test is required.

TOEFL requirements

The TOEFL exam is required for students from non-English-speaking countries.
PBT score: 600
iBT score: 100

Other admissions information

Additional requirements: The GRE General and Physics Subject Test are required for admission to the Ph.D. program. The minimum acceptable score suggested for admission is not specified.

Undergraduate preparation assumed: Although preparation will vary, a strong background in differential and integral calculus and differential equations is expected. Courses using Classical Mechanics (Marion), Electromagnetic Theory (Hayt and Buck), and Modern Physics (Serway) are assumed. It is also desirable, but not required, to have studied complex variables and linear algebra and to have an undergraduate background in most of the following areas: statistical physics and thermodynamics (Sears), optics (Hecht), solid state physics (Kittel), and quantum mechanics (Griffiths).

TUITION

Tuition year 2013–14:
Full-time students: $1,220 per credit
Credit hours per semester to be considered full-time: 10
Deferred tuition plan: Yes
Health insurance: Available.
Other academic fees: Full health insurance coverage provided for all graduate assistants.
Academic term: Semester
Number of first-year students who received full tuition waivers: 17

Teaching Assistants, Research Assistants, and Fellowships

Number of first-year
Teaching Assistants: 17
Average stipend per academic year
Teaching Assistant: $20,600
Research Assistant: $20,600
Average stipend is per "8 month" academic year, and the stipend includes full tuition coverage.

FINANCIAL AID

Application deadlines

Fall admission:
U.S. students: February 1

Loans

Loans are available for U.S. students.
Loans are not available for international students.
GAPSFAS application required: No
FAFSA application required: Yes

For further information

Address financial aid inquiries to: Student Financial Services.
Phone: (617) 373-3190
E-mail: sfs@neu.edu
Financial aid website: http://www.northeastern.edu/financialaid/index.html

HOUSING

Availability of on-campus housing

Single students: Yes
Married students: No

For further information

Address housing inquiries to: Housing and Residential Life.
Phone: (617) 373-2814
E-mail: housing@neu.edu
Housing aid website: http://www.northeastern.edu/reslife/

Table A—Faculty, Enrollments, and Degrees Granted

Research Specialty	2012-13 Faculty	Enrollment Fall 2012		Number of Degrees Granted 2012–13 (2008–13)		
		Master's	Doctorate	Master's	Terminal Master's	Doctorate
Biophysics	13	–	19	–	–	1(10)
Condensed Matter Physics	10	–	10	–	–	3(8)
Medical, Health Physics	1	–	–	–	–	–(2)
Nano Science and Technology	8	–	7	–	–	1(8)
Network Science	2	–	10	–	–	1(-)
Particles and Fields	10	–	14	–	–	1(5)
Non-specialized	–	1	21	1(41)	1(-)	–
Total	30	1	81	1(41)	1(-)	5(33)
Full-time Grad. Stud.	–	–	81	–	–	–
First-year Grad. Stud.	–	1	22	–	–	–

GRADUATE DEGREE REQUIREMENTS

Master's: Thirty-two semester hours, of which 24 are in specific courses, and a grade average of B are required. Time in residence is not stipulated. Foreign languages and comprehensive and/or qualifying examination are not required. Some options include a standard M.S. with/without an M.S. thesis or an M.S. with a concentration in applied physics, engineering physics, chemical physics, biophysics, materials physics, mathematical physics, or computational physics.

Doctorate: Forty-two semester hours and a grade average of B are required. Time in residence is one year after the qualifying examination. Foreign languages are not required. A qualifying examination is required after completion of one year of graduate courses (with a full undergraduate preparation, two years are needed with less undergraduate preparation). A thesis is required. M.S. degree may be earned while qualifying for Ph.D. degree.

SPECIAL EQUIPMENT, FACILITIES, OR PROGRAMS

Northeastern University is located in the Back Bay section of Boston, close to the Museum of Fine Arts, the Conservatory of Music, Symphony Hall, and historic Copley Square. It is an exciting, vibrant place to pursue graduate studies, because Greater Boston is home to more universities and research facilities than any other area in the world.

Thesis research can be undertaken in any one of the department's research specialties or in interdisciplinary areas, such as materials physics, mathematical physics, chemical physics, molecular biophysics, or applied engineering physics. An additional option allows cooperative research to be done at high-technology industrial, government, national or international laboratories, and at medical research institutions in the Boston area.

The department is housed in the Dana Research Center, with some optics, biological physics, and condensed matter physics laboratories also located in the Egan Research Center. There are ample modern research laboratories, department and student machine shops, an electronics shop, conference and seminar rooms, and faculty and graduate student offices. The Egan Center provides a direct interface with materials researchers in chemistry and engineering and includes extensive meeting space in the Technology Transfer Center. Numerous computational facilities are available on campus, including the Physics Department Computer Center in the Dana Research Center, and the newly developed Massachusetts Green High Performance Computational Center near Holyoke, Massachusetts (http://www.mghpcc.org/).

In addition to the research they do at campus facilities, faculty members and graduate students also work at research centers located in the United States and Europe. High energy physics experiments are under way at Fermilab (Batavia, Illinois) and CERN (Geneva, Switzerland). Astroparticle physics research is performed at the Pierre Auger Observatory in Argentina. Some groups use the synchrotron facilities at Brookhaven National Laboratory (Long Island, New York) and Argonne National Laboratory (Argonne, Illinois), and many faculty members have flourishing collaborations with scientists in Europe, Asia, and South America.

Table B—Separately Budgeted Research Expenditures by Source of Support

Source of Support	Departmental Research	Physics-related Research Outside Department
Federal government	$7,976,688	
State/local government		
Non-profit organizations		
Business and industry		
Other		
Total	**$7,976,688**	

Table C—Separately Budgeted Research Expenditures by Research Specialty

Research Specialty	No. of Grants	Expenditures ($)
Biophysics	19	$2,124,864
Condensed Matter Physics	10	$979,170
Medical, Health Physics	14	$706,752
Nano Science and Technology	12	$1,457,204
Particles and Fields	7	$466,303
Network Science	22	$1,492,314
Total	**84**	**$7,226,607**

FACULTY

Professor

Bansil, Arun, Ph.D., Harvard University, 1974. Theoretical Condensed Matter Physics.

Barabási, Albert-László, Ph.D., Boston University, 1994. Theoretical condensed matter physics and biological physics.

Champion, Paul M., Ph.D., University of Illinois, 1975. Experimental biological physics.

Gibson, J. Murray, Ph.D., University of Cambridge, 1978. Dean, College of Science. Nanophysics.

Goldberg, Haim, Ph.D., Massachusetts Institute of Technology, 1963. Theoretical particle physics.

Heiman, Donald, Ph.D., University of California, Irvine, 1975. Nanophysics.

Karma, Alain S., Ph.D., University of California, Santa Barbara, 1985. Theoretical condensed matter and biological physics.

Kravchenko, Sergey, Ph.D., Institute of Solid State Physics, Chernogolovka, 1988. Nanophysics.

Lowndes, Robert P., Ph.D., University of London, 1966. Condensed matter physics.

Markiewicz, Robert S., Ph.D., University of California, Berkeley, 1975. Condensed matter physics.

Nath, Pran, Ph.D., Stanford University, 1964. Theoretical particle physics.

Sokoloff, Jeffrey B., Ph.D., Massachusetts Institute of Technology, 1967. Theoretical condensed matter physics.

Sridhar, Srinivas, Ph.D., California Institute of Technology, 1983. Nanophysics.

Taylor, Tomasz, Ph.D., Warsaw University, 1981. Theoretical particle physics.

Vespignani, Alessandro, Ph.D., University of Rome, 1994. Theoretical condensed matter and biological physics.

Widom, Allan, Ph.D., Cornell University, 1967. Theoretical condensed matter physics.

Williams, Mark C., Ph.D., University of Minnesota, 1998. Experimental biological physics.

Wood, Darien, Ph.D., University of California, Berkeley, 1987. Experimental particle physics.

Associate Professor

Alverson, George O., Ph.D., University of Illinois, 1979. *Particles and Fields*. Experimental particle physics.

Barberis, Emanuela, Ph.D., University of California, Santa Cruz, 1996. Experimental particle physics.

Israeloff, Nathan, Ph.D., University of Illinois, 1991. Nanophysics.

Krioukov, Dimitri, Ph.D., Old Dominion University, 1998. Network science.

Menon, Latika, Ph.D., Tata Institute of Fundamental Research, 1998. Nanophysics.

Nelson, Brent, Ph.D., University of California, Berkeley, 2001. Theoretical particle physics.

Sage, J. Timothy, Ph.D., University of Illinois, 1986. Experimental biological physics.

Stepanyants, Armen, Ph.D., University of Rhode Island, 1999. Theoretical condensed matter and biological physics.

Swain, John D., Ph.D., University of Toronto, 1990. Experimental particle physics.

Assistant Professor

Feiguin, Adrian E., Ph.D., Facultad de Ciencias Exactas e Ingenieria. Universidad Nacional de Rosario, 2000. Theoretical condensed matter theory.

Kar, Swastik, Ph.D., Indian Institute of Science, 2004. Nanophysics.

Orimoto, Toyoko J., Ph.D., University of California, Berkeley, 2006. Experimental particle physics.

Wanunu, Meni, Ph.D., Weizmann Institute of Science, 2005. Experimental biological physics.

Whitford, Paul C., Ph.D., University of California, San Diego, 2009. Theoretical Condensed Matter and Biological Physics.

Emeritus

Aaron, Ronald, Ph.D., University of Pennsylvania, 1961. Medical physics.

Argyres, Petros N., Ph.D., University of California, Berkeley, 1954. Condensed matter theory.

Garelick, David A., Ph.D., Massachusetts Institute of Technology, 1963. Medical physics.

Glaubman, Michael J., Ph.D., University of Illinois, 1953. High energy experimental physics.

José, Jorge V., National University of Mexico, 1976.

Malenka, Bertram, Ph.D., Harvard University, 1951. Elementary particle theory.

Perry, Clive H., Ph.D., University of London, 1960. Condensed matter experimental physics.

Shiffman, Carl A., Ph.D., University of Oxford, 1956. Medical physics.

Srivastava, Yogendra, Ph.D., Indiana University, 1964. Condensed matter theory.

Vaughn, Michael T., Ph.D., Purdue University, 1960. Elementary particle theory.

von Goeler, Eberhard, Ph.D., University of Illinois, 1961. High energy experimental physics.

Wu, Fa-Yueh, Ph.D., Washington University, 1963. Condensed matter theory.

Adjunct Professor

Anchordoqui, Luis, Ph.D., Universidad Nacional de La Plata, Argentina, 1998. Elementary astroparticle physics.

Baublitz, Millard, Ph.D., Cornell University.

Chen, George Tze Yung, Ph.D., Brown University, 1972. Biomedical physics.

Dova, Maria-Teresa, Ph.D., Universidad Nacional de La Plata, 1989. High-Energy experimental physics.

Farmelo, Graham, Ph.D., University of Liverpool, 1977. High energy experimental physics.

Fenker, Howard, Ph.D., Vanderbilt University, 1978. High energy experimental physics.

Gongora-Trevino, Maria Araceli, Ph.D., University of Oxford, 1984. Condensed matter physics.

Kaprzyk, Stanislaw, Ph.D., Academy of Metallurgy (Krakow), 1981. Condensed matter theory.

Kern, Wolfhard, Ph.D., University of Bonn, 1958. High energy experimental physics and education.

Lindroos, Matti, Ph.D., Tampere University of Technology, Finland, 1979. Condensed matter theory.

Lu, Wentao, Ph.D., Northeastern University, 2001.

Mijnarends, Peter, Ph.D., Delft University of Technology, 1969. Condensed matter theory.

Morgan, Robert C., Ph.D., Massachusetts Institute of Technology, 1969. Condensed matter theory.

DEPARTMENTAL RESEARCH SPECIALTIES AND STAFF

Theoretical

Condensed Matter Theory. The group performs research on diverse topics that span forefront areas of hard/soft condensed matter physics and emerging areas at the intersection of physics and other disciplines. Specific research areas include the electronic structure and spectroscopy of high-temperature superconductors and other complex materials, nanotribology atomic-scale friction in crystalline and polymeric materials, network science with applications to technological, biological, and social networks, theoretical/computational materials science, cardiac nonlinear dynamics, and theoretical/computational neuroscience. Bansil, Barabási, Feiguin, José, Karma, Sokoloff, Stepanyants, Vespignani, Whitford, Widom, Wu.

High Energy Physics. The faculty and students in the theoretical particle physics group are actively exploring questions concerning supersymmetry SUSY, and more specifically its local extension to supergravity SUGRA, with a view to understanding the connection between the universe at very large and very small scales. This leads to the study of supersymmetry, and supergravity, possible extra dimensions beyond the usual four, and related exotic phenomena, such as mini-black holes, which may be produced at accelerators or by ultra high energy cosmic rays. Our formal investigations in superstring theory and M-theory are also conducted with the purpose of making connections between fundamental theory and experiment. The elementary particle theory group at NU initiated the PASCOS and SUSY series of conferences, which have become major conferences in high energy physics. Goldberg, Malenka, Nath, Nelson, Srivastava, Taylor.

Network Science. Complex network research is not a single discipline; it is highly interdisciplinary, seeking the answers to some fundamental questions about living, adaptable, and changeable systems. Several of the main disciplines are "network theory" involving the research areas of computer science, network science, and graph theory. Another is "network science (NS)" attempting to research engineered networks, information networks, biological networks, semantic networks, and social networks, whereas "dynamic network analysis (DNA)" will use traditional social network analysis, link analysis and multi-agent systems involving large amounts of electronic data. We should also add "complex adaptive systems," which is grounded in modern chemistry, biological views on adaption, expatriation, and evolution. In all of these and more network-related areas, the study of emergence and self-organization are fundamental. Although academic disciplines are hugely diverse in complex network research, here in the Department of Physics, disciplines in statistical analysis involving physics, mathematics, and computational analysis (data mining) are its primary focus. Barabási, Vespignani.

Experimental

Biological and Medical Physics. The group performs research on multiple levels from molecules DNA and proteins to cells regulatory and metabolic protein networks to tissue cell-to-cell signaling in heart muscle and brain. Eight faculty members have externally funded research programs in specific research areas, including single molecule DNA-protein interactions, vibrational dynamics of biomolecules, femtosecond protein dynamics, biological networks signaling, metabolic, and transcription-regulatory networks, cardiac nonlinear dynamics, and theoretical/computational neuroscience. Aaron, Barabási, Champion, Israeloff, José, Karma, Sage, Shiffman, Stepanyants, Vespignani, Wanunu, Williams.

Experimental Particle Physics. Experimental Particle Physics The Experimental Particle Physics group concentrates its efforts on the following three activities: CMS, D0, and the Pierre Auger Observatory. Compact Muon Solenoid at LHC The CMS detector recently began operations at the Large Hadron Collider (LHC), located near Geneva, Switzerland. The LHC is currently colliding protons at 8 TeV, the highest energy available in the world, and is scheduled to run at 13 TeV starting in 2014. At Northeastern we are supporting the end cap muon detector and the electromagnetic detector, are studying the newly-found Higgs boson, and are searching for leptoquarks (exotic particles with properties of both leptons and quarks), massive new gauge bosons (Stueckelberg Z-primes), and looking for other new physics. Post-doctoral fellows include Dr Andreas Massironi and Daniele Trocino. D0 The D0 (DZero) detector in the Tevatron collider at Fermi National Accelerator Laboratory in Illinois measured the products of 2 TeV head-on collisions of protons and antiprotons. The large data set at D0 is used for the study of such massive particles as the top quark and for searches for new heavy states. At Northeastern we continue to mine this trove of data for new physics. Pierre Auger Observatory The Pierre Auger Observatory makes use of the one accelerator bigger than either the Tevatron or the LHC-the one that gives us cosmic rays from intergalactic space. Currently taking data with a fully instrumented detector covering 3000 square kilometers in Argentina, the PAO observes the showers of particles coming from ultra-high-energy cosmic rays. Northeastern is providing the overall software infrastructure for data analysis and is participating in several research projects, including searches for exotic particles produced at extremely high energies. Research personnel: Dr Thomas Paul, Prof. John Swain. Alverson, Barberis, Orimoto, Swain, Wood.

Nanophysics. The faculty is actively pursuing research at the frontiers of nanoscience. The thrust areas in nanophysics include the following: left-handed metamaterials for photonic crystals, nanomedicine, spintronics, mesoscopic physics, low-dimensional electronic systems, nanomagnetism, and quantum chaos. Research is aimed at the synthesis of nanoscale materials and devices, as well as fundamental materials issues. Gibson, Heiman, Israeloff, Kar, Kravchenko, Menon, Perry, Sridhar.

View additional information about this department at www.gradschoolshopper.com

TUFTS UNIVERSITY

DEPARTMENT OF PHYSICS AND ASTRONOMY

Medford, Massachusetts 02155
http://ase.tufts.edu/physics

General University Information
President: Anthony P. Monaco
Dean of Graduate School: Lynne Pepall
University website: http://www.tufts.edu/
Control: Private
Setting: Suburban
Total Faculty: 1,315
Total number of Students: 9,857
Total number of Graduate Students: 3,141

Department Information
Department Chairman: Roger Tobin, Chair
Department Contact: Shannon Landis-Amerault, Department Manager
Total full-time faculty: 18
Full-Time Graduate Students: 31
First-Year Graduate Students: 8
Female First-Year Students: 2
Total Post Doctorates: 6

Department Address
Robinson Hall
212 College Ave.
Medford, MA 02155
Phone: (617) 627-5360
Fax: (617) 627-3878
E-mail: shannon.landis@tufts.edu
Website: http://ase.tufts.edu/physics

ADMISSIONS

Admission Contact Information
Address admission inquiries to: Graduate Student Committee, Physics Department
Phone: (617) 627-3395
E-mail: grasp@tufts.edu
Admissions website: http://gradstudy.tufts.edu/default.aspx

Application deadlines
Fall admission:
U.S. students: January 15   *Int'l. students*: December 15
Spring admission:
U.S. students: October 15   *Int'l. students*: September 15

Application fee
U.S. students: $75   *Int'l. students*: $75

Admissions information
For Fall of 2012:
Number of applicants: 87
Number admitted: 21
Number enrolled: 4

Admission requirements
Bachelor's degree requirements: A Bachelor's degree in physics is required with no minimum undergraduate GPA specified, although strong GPA and recommendations are crucial.

GRE requirements
The GRE is required.

Advanced GRE requirements
The Advanced GRE is recommended.

TOEFL requirements
The TOEFL exam is required for students from non-English-speaking countries.
PBT score: 550

TUITION

Tuition year 2012–2013:
Tuition for out-of-state residents
Full-time students: annual
Tuition is waived for students admitted with Teaching Assistant offers. Number of Teaching Assistant offer varies by year.
Deferred tuition plan: No
Health insurance: Available
Other academic fees: $48 activities fee
Academic term: Semester

Teaching Assistants, Research Assistants, and Fellowships

Number of first-year
Teaching Assistants: 8
Research Assistants: 1
Average stipend per academic year
Teaching Assistant: $23,000
Research Assistant: $23,000

FINANCIAL AID

Loans
Loans are not available for U.S. students.
Loans are not available for international students.
GAPSFAS application required: No
FAFSA application required: No

For further information
Address financial aid inquiries to: Graduate Student Committee, Physics Department.
E-mail: grasp@tufts.edu

HOUSING

Availability of on-campus housing
Single students: Yes
Married students: No

For further information
Address housing inquiries to: Office of Residential Life and Learning, South Hall, Tufts University, Medford, MA 02155.
Phone: (617) 627-3248
E-mail: reslife@tufts.edu
Housing aid website: http://ase.tufts.edu/reslife/

Table A—Faculty, Enrollments, and Degrees Granted

Research Specialty	2012-2013 Faculty	Enrollment 2012-2013 Master's	Enrollment 2012-2013 Doctorate	Number of Degrees Granted 2012-2013 (2007-13) Master's	Number of Degrees Granted 2012-2013 (2007-13) Terminal Master's	Number of Degrees Granted 2012-2013 (2007-13) Doctorate
Astrophysics	4	–	2	–	-(2)	–
Condensed Matter Physics	6	–	5	-(5)	–	1(8)
Particles and Fields	6	–	5	-(5)	-(1)	-(5)
Relativity & Gravitation	4	–	5	-(5)	-(1)	1(5)
Non-specialized	–	–	10	–	–	–
Other	–	–	4	-(1)	-(1)	2(2)
Total	21	–	–	-(16)	-(5)	4(20)
Full-time Grad. Stud.	–	–	31	–	–	–
First-year Grad. Stud.	–	–	8	–	–	–

GRADUATE DEGREE REQUIREMENTS

Master's: Eight graduate level courses are required in approved program with grades of B– or better; thesis optional; two semesters residence required; no language requirement; no examination requirements.

Doctorate: Eight graduate courses required. The student must demonstrate proficiency in classical physics and in quantum mechanics. Preliminary exam, dissertation, and dissertation exam are required; three academic years of study with at least one year in residence; no language requirement.

Thesis: Thesis may be written in absentia.

SPECIAL EQUIPMENT, FACILITIES, OR PROGRAMS

Cooperative research programs are carried out at the Arecibo Laboratory (National Astronomy and Ionospheric Center), Argonne National Laboratory, Brookhaven National Laboratory, European Center for Nuclear Research CERN, Fermi National Accelerator Laboratory, National Radio Astronomy Observatory (Socorro, NM), and the Soudan II Underground Laboratory. Biomedical research in cooperation with local hospitals is also possible.

Table B—Separately Budgeted Research Expenditures by Source of Support

Source of Support	Departmental Research	Physics-related Research Outside Department
Federal government	$1,576,594	
State/local government		
Non-profit organizations		
Business and industry		
Other	$460,739	
Total	$2,037,333	

FACULTY

Professor

Cebe, Peggy, Ph.D., Cornell University, 1984. *Chemical Physics, Condensed Matter Physics, Energy Sources & Environment, Materials Science, Metallurgy, Nano Science and Technology, Polymer Physics/Science.* Experimental condensed matter physics.

Ford, Lawrence H., Ph.D., Princeton University, 1974. *Relativity & Gravitation, Other.* General relativity and cosmology; quantum field theory.

Goldstein, Gary R., Ph.D., University of Chicago, 1968. *Particles and Fields.* Theoretical particle physics.

Lang, Kenneth R., Ph.D., Stanford University, 1969. *Astronomy, Astrophysics.* Astrophysics; radio astronomy.

Mann, W. Anthony, Ph.D., University of Massachusetts, 1970. *Particles and Fields.* Experimental particle physics.

Napier, Austin, Ph.D., Massachusetts Institute of Technology, 1978. *Particles and Fields.* Experimental particle physics.

Oliver, William P., Ph.D., University of California, Berkeley, 1969. *Particles and Fields.* Experimental particle physics.

Sliwa, Krzysztof, Ph.D., Jagiellonian, 1980. *Particles and Fields.* Experimental particle physics.

Tobin, Roger, Ph.D., University of California, Berkeley, 1985. Chair. *Chemical Physics, Condensed Matter Physics, Nano Science and Technology, Physics and other Science Education, Surface Physics.* Experimental condensed matter physics.

Vilenkin, Alexander, Ph.D., State University of New York at Buffalo, 1977. L. and J. Bernstein Chair in Evolutionary Science, Cosmology, general relativity, astrophysics. *Particles and Fields, Relativity & Gravitation, Other.* General relativity and cosmology.

Associate Professor

Blanco-Pillado, José, Ph.D., Tufts University, 2001. *Particles and Fields, Relativity & Gravitation.* Cosmology.

Gallagher, Hugh, Ph.D., University of Minnesota, 1996. *Particles and Fields, Physics and other Science Education.* High energy physics.

Assistant Professor

Atherton, Timothy, Ph.D., University of Exeter, 2007. *Condensed Matter Physics, Fluids, Rheology.* Theoretical soft condensed matter physics.

Beauchemin, Pierre-Hugues, Ph.D., McGill University, 2005. *High Energy Physics, Particles and Fields*. Experimental high energy physics.

Marchesini, Danilo, Ph.D., S.I.S.S.A.–I.S.A.S., 2004. *Astronomy, Astrophysics*. Astrophysics.

Sajina, Anna, Ph.D., University of British Columbia, 2006. *Astronomy, Astrophysics*. Astronomy, Astrophysics.

Staii, Cristian, Ph.D., University of Pennsylvania, 2005. *Biophysics, Condensed Matter Physics, Medical, Health Physics, Nano Science and Technology*.

Emeritus

Everett, Allen E., Ph.D., Harvard University, 1960. *Particles and Fields, Relativity & Gravitation*. Theoretical particle physics; cosmology.

Gunther, Leon, Ph.D., Massachusetts Institute of Technology, 1964. *Acoustics, Condensed Matter Physics, Statistical & Thermal Physics*. Theoretical condensed matter physics.

McCarthy, Kathryn A., Ph.D., Harvard University, 1957. *Solid State Physics*. Experimental solid state physics.

Milburn, Richard H., Ph.D., Harvard University, 1954. *Particles and Fields*. Experimental particle physics.

Mumford, George S., Ph.D., University of Virginia, 1954. *Astronomy, Astrophysics*.

Schneps, Jack, Ph.D., University of Wisconsin, 1956. *Particles and Fields*. Experimental particle physics.

Shapira, Yaacov, Ph.D., Massachusetts Institute of Technology, 1964. *Solid State Physics*. Experimental solid state physics.

Research Professor

Olum, Kenneth, Ph.D., Massachusetts Institute of Technology, 1997. *Cosmology & String Theory, Relativity & Gravitation*. General relativity and cosmology.

Adjunct Professor

Boghosian, Bruce, Ph.D., University of California, Davis, 1987. *Applied Mathematics, Fluids, Rheology, Nonlinear Dynamics and Complex Systems*. Mathematics, quantum computing, fluid dynamics.

Omenetto, Fiorenzo, Ph.D., Universita di Pavia, 1997. *Applied Physics, Biophysics, Electrical Engineering, Medical, Health Physics, Optics*. Ultrafast and nonlinear optics; biomedical engineering

Affiliate Professor

Hammer, David, Ph.D., University of California, Berkeley, 1991. *Physics and other Science Education, Other*. Science Education.

Thornton, Ronald, Ph.D., Brown University, 1976. Director, Center for Science and Mathematics Teaching. *Physics and other Science Education, Other*. Science and mathematics education.

Senior Lecturer

Willson, R., Ph.D., Tufts University, 1979. *Astronomy, Astrophysics, Other*. Radio astronomy.

DEPARTMENTAL RESEARCH SPECIALTIES AND STAFF

Theoretical

Condensed Matter Physics. Soft condensed matter, complex fluids. Atherton.

Particles and Fields. Quarks and quantum chromodynamics; electroweak theory; high-energy phenomenology. Goldstein.

Physics and other Science Education. Physics and science education. Hammer, Thornton, Tobin.

Relativity & Gravitation. Physical processes in the very early universe; cosmic strings; cosmological phase transitions; inflation, quantum gravity; quantum field theory in curved spacetime. Blanco-Pillado, Ford, Olum, Vilenkin.

Experimental

Astrophysics. Properties and evolution of galaxies in the early universe. Radio interferometry of the sun; x-ray and gamma ray studies of the sun; radio observations of active stars. Multiwavelength observations of stars, nebulae and galaxies. Lang, Marchesini, Sajina, Willson.

Condensed Matter Physics. Surface physics; polymers; biophysics. Cebe, Staii, Tobin.

Particles and Fields. High-energy neutrino physics, search for neutrino oscillations; top quark studies and search for Higgs particles; heavy quark spectroscopy. Beauchemin, Gallagher, Mann, Napier, Oliver, Sliwa.

View additional information about this department at
www.gradschoolshopper.com

UNIVERSITY OF MASSACHUSETTS, AMHERST

DEPARTMENT OF PHYSICS

Amherst, Massachusetts 01003
http://www.physics.umass.edu

General University Information

Chancellor: Kumble R. Subbaswamy
Dean of Graduate School: John McCarthy
University website: http://www.umass.edu/
Control: Public
Setting: Rural
Total Faculty: 1,472

Total Graduate Faculty: 1,356
Total number of Students: 28,236
Total number of Graduate Students: 6,308

Department Information

Department Chairman: Rory Miskimen, Head

Department Contact: Jane Knapp, Graduate Program
  Coordinator
  Total full-time faculty: 33
  Total number of full-time equivalent positions: 34
  Full-Time Graduate Students: 85
  First-Year Graduate Students: 10
  Female First-Year Students: 1
  Total Post Doctorates: 17

Department Address

710 N. Pleasant
Amherst, MA 01003
Phone: (413) 545-2548
Fax: (413) 545-0648
E-mail: jknapp@physics.umass.edu
Website: http://www.physics.umass.edu

ADMISSIONS

Admission Contact Information

Phone: (413) 545-2548
E-mail: gradmiss@physics.umass.edu
Admissions website: http://umass.edu/gradschool

Application deadlines

Fall admission:
U.S. students: January 15   *Int'l. students*: January 15

Application fee

U.S. students: $65   *Int'l. students*: $65
Please see website (www.umass.edu/gradschool/admissions).

Admissions information

For Fall of 2013:
  Number of applicants: 200
  Number admitted: 50
  Number enrolled: 14

Admission requirements

Bachelor's degree requirements: Bachelor's degree in physics
  or a related area is required.
Minimum undergraduate GPA: 3.0

GRE requirements

The GRE is required.

Advanced GRE requirements

The Advanced GRE is required.

TOEFL requirements

The TOEFL exam is required for students from non-English-
  speaking countries.
  iBT score: 90

Other admissions information

Undergraduate preparation assumed: Marion, Classical Mechanics; Marion, Electricity and Magnetism; Eisberg, Fundamentals of Modern Physics; Jenkins and White, Fundamentals of Optics; Reif, Statistical Mechanics.

TUITION

Tuition year 2013-2014:
Tuition for in-state residents
  Full-time students: $110 per credit
  Part-time students: $110 per credit

Tuition for out-of-state residents
  Full-time students: $414 per credit
  Part-time students: $414 per credit
Tuition and most fees are waived with TA/RA appointment. For
  updated tuition and fees, please see website (http://umass.edu/
  bursar).
Credit hours per semester to be considered full-time: 9
Deferred tuition plan: No
Health insurance: Yes, 171 for single plan.
Other academic fees: First year entering fee, $357 (one-time-only
  fee). Service fee, graduate senate fee. Health Insurance Family
  Plan, $851 per year.
Academic term: Semester
Number of first-year students who received full tuition waivers: 10

Teaching Assistants, Research Assistants, and Fellowships

Number of first-year
  Teaching Assistants: 7
  Research Assistants: 3
  Fellowship students: 1
Average stipend per academic year
  Teaching Assistant: $18,810
  Research Assistant: $18,810
  Fellowship student: $2,000

FINANCIAL AID

Application deadlines

Fall admission:
U.S. students: January 15   *Int'l. students*: January 15

Loans

Loans are available for U.S. students.
Loans are not available for international students.
GAPSFAS application required: No
FAFSA application required: No

For further information

Address financial aid inquiries to:
Phone: (413) 577-0555
E-mail: grads@finaid.umass.edu
Financial aid website: http://www.umass.edu/umfa/

HOUSING

Availability of on-campus housing

Single students: Yes
Married students: Yes

For further information

Phone: (413) 545-2100
Housing aid website: http://www.housing.umass.edu

Table A—Faculty, Enrollments, and Degrees Granted

Research Specialty	2012-13 Faculty	Enrollment Fall 2012 Master's	Enrollment Fall 2012 Doctorate	Number of Degrees Granted 2012-13 (2007-12) Master's	Number of Degrees Granted 2012-13 (2007-12) Terminal Master's	Number of Degrees Granted 2012-13 (2007-12) Doctorate
Biophysics	5	–	12	–	–	–
Condensed Matter Physics	10	–	31	–	–(1)	6(13)
Low Temperature Physics	1	–	–	–	–	–(2)
Medical, Health Physics	1	–	–	–	–	–(1)
Nuclear Physics	5	–	7	–	1(2)	1(3)
Particles and Fields	7	–	14	–(1)	–(3)	6(7)
Polymer Physics/Science	2	–	10	–	–	1(6)
Relativity & Gravitation	2	–	1	–	–	–(1)
Statistical & Thermal Physics	3	–	9	–	–(1)	–(5)
Non-specialized	2	–	9	5(26)	–(4)	–
Total	**37**	**–**	**85**	**5(27)**	**1(9)**	**14(38)**
Full-time Grad. Stud.	–	–	85	–	–	–
First-year Grad. Stud.	–	–	10	–	–	–

GRADUATE DEGREE REQUIREMENTS

Master's: Thirty credits in approved program with "B" average are required. Master's examination is required. Thesis is optional. There is no language or residency requirement. There is no terminal M.S. program.

Doctorate: Six core physics graduate courses are required. A qualifying examination, dissertation, and dissertation examination are required. There is no language requirement. Three research area courses are required, at least one of which must be outside of the dissertation research area.

Thesis: Thesis/dissertation may be written in absentia.

SPECIAL EQUIPMENT, FACILITIES, OR PROGRAMS

The Center for Hierarchical Manufacturing, an NSF Nanoscale Science and Engineering Center that provides expansive facilities for nanotechnology research and supports interdisciplinary graduate research; Nanotechnology Innovation: From Discovery to Product, an NSF IGERT program providing graduate fellowships in nanotechnology and educational experiences in innovation and entrepreneurship.

Table B—Separately Budgeted Research Expenditures by Source of Support

Source of Support	Departmental Research	Physics-related Research Outside Department
Federal government	$4,727,053	$244,052
State/local government	$148,959	
Non-profit organizations	$50,588	$17,556
Business and industry	$28,745	
Other	$7,456	$68,603
Total	**$4,962,801**	**$330,211**

Table C—Separately Budgeted Research Expenditures by Research Specialty

Research Specialty	No. of Grants	Expenditures ($)
Biophysics	5	$362,139
Condensed Matter Physics	26	$1,545,792
Low Temperature Physics	3	$179,848
Nuclear Physics	7	$920,753
Particles and Fields	6	$1,155,316
Relativity & Gravitation	5	$262,088
Statistical & Thermal Physics	3	$536,865
Total	**55**	**$4,962,801**

FACULTY

Professor

Candela, Donald, Ph.D., Harvard University, 1983. Experimental low-temperature physics; condensed-matter physics.

Donoghue, John F., Ph.D., University of Massachusetts, 1976. Theoretical high-energy physics.

Goldner, Lori, Ph.D., Cornell University, 1984. Biological physics.

Hallock, Robert B., Ph.D., Stanford University, 1969. Experimental low-temperature physics; condensed-matter physics.

Kumar, Krishna S., Ph.D., Syracuse University, 1990. Experimental nuclear physics.

Machta, Jonathan L., Ph.D., Massachusetts Institute of Technology, 1980. Theoretical statistical mechanics; condensed-matter physics.

Menon, Narayanan, Ph.D., University of Chicago, 1995. Experimental condensed-matter physics.

Miskimen, Rory A., Ph.D., Massachusetts Institute of Technology, 1983. Department Head. Experimental nuclear physics.

Parsegian, V. Adrian, Ph.D., Harvard, 1969. *Biophysics.*

Prokofiev, Nikolay, Ph.D., Kurchatov, 1987. Theoretical condensed-matter physics; computational physics.

Rabin, Monroe S. Z., Ph.D., Rutgers University, 1967. Associate Department Head. Medical physics.

Ramsey-Musolf, Michael J., Ph.D., Princeton, 1989. Theoretical High Energy Physics.

Svistunov, Boris V., Ph.D., Kurchatov, 1990. Theoretical condensed-matter physics; computational physics.

Traschen, Jennie, Ph.D., Harvard University, 1984. Theoretical high-energy physics; relativity; gravitation.

Tuominen, Mark, Ph.D., University of Minnesota, 1990. Experimental condensed-matter physics.

Willocq, Stéphane, Ph.D., Tufts University, 1992. Experimental high-energy physics.

Associate Professor

Blaylock, Guy, Ph.D., University of Illinois, 1986. Experimental high-energy physics.

Cadonati, Laura, Ph.D., Harvard University, 2002. Experimental gravitational physics.

Dallapiccola, Carlo J., Ph.D., University of Colorado, 1993. Graduate Program Director. Experimental high-energy physics.

Davidovitch, Benjamin, Ph.D., Weizmann Institute of Science, 2001. Theoretical condensed-matter physics.

Dinsmore, Anthony D., Ph.D., University of Pennsylvania, 1997. Experimental condensed-matter physics.

Kawall, David, Ph.D., Stanford University, 1996. Experimental nuclear physics.

Ross, Jennifer, Ph.D., University of California, Santa Barbara, 2004. Experimental biological physics.

Santangelo, Christian, Ph.D., University of California, Santa Barbara, 2004. Theoretical condensed-matter physics.

Sorbo, Lorenzo, Ph.D., (SISSA/ISAS) of Trieste, 2001. Theoretical high-energy physics.

Assistant Professor

Babaev, Egor, Ph.D., University of Uppsala, 2001. Theoretical condensed-matter physics.

Brau, Benjamin, Ph.D., Massachusetts Institute of Technology, 2002. Experimental high-energy physics.

Kilfoil, Maria, Ph.D., Memorial University, 2001. *Biophysics.* Biological physics.

Pocar, Andrea, Ph.D., Princeton University, 2003. Experimental neutrino physics.

Yan, Jun, Ph.D., Columbia, 2009. *Nano Science and Technology.* Experimental Condensed Matter Physics.

Adjunct Faculty

Aidala, Katherine, Ph.D., Harvard University, 2006. Experimental condensed-matter physics.

Barnes, Michael, Ph.D., Rice University, 1991. Single-molecule spectroscopy of nanoscale systems.

Decowski, Piotr, Ph.D., University of Warsaw, 1967. Experimental nuclear physics; hadron physics.

Friedman, Jonathan, Ph.D., City College of New York, 1996. Experimental condensed-matter physics.

Grason, Gregory, Ph.D., University of Pennsylvania, 2005. Statistical mechanics of macromolecular and soft-matter systems.

Lannert, Courtney, Ph.D., University of California, Santa Barbara, 2002. Theoretical condensed-matter physics.

Muthukumar, Murugappan, Ph.D., University of Chicago, 1978. Theoretical polymer physics.

Podgornik, Rudolf, Ph.D., University of Ljubljana, 1986. *Biophysics.*

Tewari, Shubha, Ph.D., University of California, Los Angeles, 1993. Theoretical condensed-matter physics.

Lecturer

Bourgeois, Paul, Ph.D., University of Massachusetts, 2005. Director of Physics Teaching Laboratories. Experimental nuclear physics.

Darnton, Nicholas, Ph.D., Princeton University, 2002. Biological physics.

Hatch, Heath, Other, University of Northern British Columbia, 1998.

Kastor, David, Ph.D., University of Chicago, 1988. Theoretical high-energy physics; relativity; gravitation.

DEPARTMENTAL RESEARCH SPECIALTIES AND STAFF

Theoretical

Condensed Matter Physics. Complex and disordered systems; phase transitions, dynamics, and transport; computational methods and computational complexity; classical and quantum Monte Carlo methods; polymers, liquid crystals, and poly-electrolytes; self-assembly and pattern recognition of nanostructures; packaging of chromosomal assemblies; transport of biological macromolecules through membranes; quantum fluids and solids: Bose-Einstein condensation and kinetic theory; quantum dissipation and decoherence: tunneling, qubits, nanomagnets, and low-dimensional conductors. Babaev, Davidovitch, Machta, Muthukumar, Prokofiev, Santangelo, Svistunov.

Particles and Fields. Gauge theories; CP violation; heavy-quark physics; structure of weak interactions; physics beyond the Standard Model; gravitation; string theory; cosmology. Donoghue, Kastor, Ramsey-Musolf, Sorbo, Traschen.

Experimental

Biophysics. Fundamental Biophysics Problems in: biomaterials, biomechanics, intramolecular forces, advanced single molecule imaging and instrumentation; Biopolymers such as: RNA, DNA, cytoskeletal filaments; Biological systems: membranes; pores; intracellular organization; mitosis; aggregation kinetics; cellulose machinery. Dinsmore, Goldner, Kilfoil, Muthukumar, Rabin, Ross.

Experimental Particles and Fields. At the LHC at CERN: search for physics beyond the Standard Model. At the Borexino experiment: solar neutrinos. At the DarkSide experiment: dark matter. At LIGO: gravitational wave physics. Brau, Cadonati, Dallapiccola, Willocq.

Low Temperature Physics. Quantum fluids and solids: superconductivity, spin-polarized systems, 3He-4He mixtures, weakbinding systems, helium films, and solid helium; phase transitions: wetting, 2D effects, restricted geometry, quenched disorder, localization, and nanostructures; third sound; NMR; microbalance; thermal techniques; high field/temperature ratio. Candela, Hallock, Tuominen.

Nano Science and Technology. Thin film and nanostructures: device fabrication, electron-beam lithography, time-resolved optical spectroscopy, single electron devices, superconductivity, mesoscopic quantum phenomena, plasmonics, liquid helium, superfluidity, physics in two dimensions, and liquid helium mixtures; magnetic and transport properties of nanostructure; electrical and optical properties of semiconductors, functional nanostructures, photonic crystal devices, and networks, electrical and optical properties of atomically thin two dimensional crystal. Candela, Dinsmore, Hallock, Tuominen, Yan.

Nuclear Physics. Electromagnetic interaction studies of the structure of hadrons using multi-GeV electron and proton beams. At Jlab: Determination of strange quark electromagnetic distributions of the proton by parity-violating electron scattering, precision measurement of the neutral pion lifetime and the chiral axial anomaly. At the Relativistic Heavy Ion Collider (RHIC): Studies of polarized proton-proton collisions to measure the contribution of gluons to the spin of the proton. Neutrino physics; neutrinoless double beta decay with EXO experiment; solar neutrinos with borexino experiment. At Fermilab: Precision measurement of the anomalous magnetic moment of the muon. Kawall, Kumar, Miskimen, Pocar.

Soft Condensed Matter Physics. Complex and disordered systems: fluids in porous media; diffusion and dispersion in random media; avalanche phenomena; flow and rheology of granular materials; glass transitions; fluid-solid interfacial phenomena; dendritic and fractal growth; polymers and macromolecules; complex fluids; chemical self-assembly; polymer nanostructures; molecular-scale devices; x-ray imaging; optical microscopy; light scattering; ellipsometry. Candela, Dinsmore, Menon, Tuominen.

View additional information about this department at
www.gradschoolshopper.com

UNIVERSITY OF MASSACHUSETTS, LOWELL

DEPARTMENT OF PHYSICS AND APPLIED PHYSICS

Lowell, Massachusetts 01854
http://www.uml.edu/physics

General University Information

Chancellor: Martin T. Meehan
Dean of Graduate School: Donald Pierson
University website: http://www.uml.edu
Control: Public
Setting: Urban
Total Faculty: 556
Total Graduate Faculty: N/A
Total number of Students: 16,294
Total number of Graduate Students: 4,007

Department Information

Department Chairman: Robert H. Giles, Chair
Department Contact: James J. Egan, Physics Graduate
  Coordinator
  Total full-time faculty: 24
  Total number of full-time equivalent positions: 24
  Full-Time Graduate Students: 70
  First-Year Graduate Students: 25
  Female First-Year Students: 5
  Total Post Doctorates: 6

Department Address

1 University Ave.
Lowell, MA 01854
Phone: (978) 934-3750
Fax: (978) 934-3068
E-mail: James_Egan@uml.edu
Website: http://www.uml.edu/physics

ADMISSIONS

Admission Contact Information

Address admission inquiries to: Prof. James J. Egan, Physics
  Graduate Coordinator, Dept. of Physics and Applied Physics
Phone: (978) 934-3774
E-mail: James_Egan@uml.edu
Admissions website: http://www.uml.edu/grad

Application deadlines

Fall admission:
U.S. students: March 15   *Int'l. students*: March 15
Spring admission:
U.S. students: November 15   *Int'l. students*: November 15

Application fee

U.S. students: $50   *Int'l. students*: $50

Admissions information

For Fall of 2013:
  Number of applicants: 76
  Number admitted: 47

Admission requirements

Bachelor's degree requirements: Bachelor's degree in physics
  or related area is required.
Minimum undergraduate GPA: 3.0

GRE requirements

The GRE is required.

Advanced GRE requirements

The Advanced GRE is required.
Physics GRE required for Ph.D. applicants but not for M.S. applicants.

TOEFL requirements

The TOEFL exam is required for students from non-English-speaking countries.
  PBT score: 547
  iBT score: 78

Other admissions information

Additional requirements: No minimum acceptable score is specified for General or Advanced GRE.
Undergraduate preparation assumed: Taylor, Mechanics; Wangsness or Griffiths, Electromagnetism; Liboff, Quantum Mechanics; Mandl, Statistical Mechanics.

TUITION

Tuition year 2012–13:
Tuition for in-state residents
  Full-time students: $11,229 annual
  Part-time students: $1,248 per credit
Tuition for out-of-state residents
  Full-time students: $20,774 annual
  Part-time students: $2,308 per credit
Credit hours per semester to be considered full-time: 9
Deferred tuition plan: Yes
Health insurance: Available at the cost of $1203 per year.
Academic term: Semester
Number of first-year students who received full tuition waivers: 17

Teaching Assistants, Research Assistants, and Fellowships

Number of first-year
  Teaching Assistants: 15
  Research Assistants: 2
Average stipend per academic year
  Teaching Assistant: $15,680
  Research Assistant: $15,680

FINANCIAL AID

Application deadlines

Fall admission:
U.S. students: March 15   *Int'l. students*: March 15
Spring admission:
U.S. students: November 15   *Int'l. students*: November 15

Loans

Loans are available for U.S. students.
Loans are not available for international students.
GAPSFAS application required: No
FAFSA application required: No

For further information

Address financial aid inquiries to: Prof. James J. Egan, Physics
  Graduate Coordinator, Dept. of Physics and Applied Physics.
Phone: (978) 934-3774

E-mail: James_Egan@uml.edu
Financial aid website: http://www.uml.edu/FinancialAid

HOUSING

Availability of on-campus housing
Single students: Yes
Married students: No

For further information
Address housing inquiries to: University Housing Officer.
Phone: (978) 934-5100
E-mail: StudentFinancialServ@uml.edu
Housing aid website: http://www.uml.edu/student-services/reslife/housing

Table A—Faculty, Enrollments, and Degrees Granted

Research Specialty	2012–13 Faculty	Enrollment Fall 2012–13		Number of Degrees Granted 2012–13 (2009–13)		
		Master's	Doctorate	Master's	Terminal Master's	Doctorate
Applied Physics	9	–	3	–	–	–
Astronomy	3	–	–	–(1)	–(1)	–
Astrophysics	3	–	2	–	–	–
Atmosphere, Space Physics, Cosmic Rays	1	–	6	–(1)	–(1)	–
Atomic, Molecular, & Optical Physics	7	–	7	–(8)	–(4)	–(14)
Biophysics	3	–	4	–(1)	–	–
Condensed Matter Physics	3	1	7	1(2)	–(1)	–
Engineering Physics/Science	2	–	–	–	–	–(1)
Materials Science, Metallurgy	7	–	–	–	–	–
Medical, Health Physics	4	20	15	5(45)	4(28)	3(5)
Nano Science and Technology	3	–	3	–(3)	–(2)	1(1)
Nuclear Physics	5	–	8	2(4)	1(2)	1(3)
Physics and other Science Education	2	–	–	–	–	–
Polymer Physics/Science	1	–	3	–(3)	–	2(3)
Non-specialized	–	7	3	–(3)	–	–
Total	–	27	61	8(71)	5(39)	7(27)
Full-time Grad. Stud.	–	9	61	–	–	–
First-year Grad. Stud.	–	3	22	–	–	–

GRADUATE DEGREE REQUIREMENTS

Master's: Thirty graduate credits with a "B" average are required. There is no foreign language requirement. There is no qualifying or comprehensive examination required. One year of residence is required. Thesis or project is required. A master's degree can be obtained en route to a Ph.D. without thesis or project by passing the Ph.D. comprehensive examination and earning 30 graduate credits.

Doctorate: Sixty graduate credits with a "B" average are required, 21 of which must be in specific courses. Comprehensive written and oral examinations are required; the doctoral research admission oral examination is taken after completion of a two-semester research project unless a master's thesis is completed. One-year residence is required. Dissertation is required.

SPECIAL EQUIPMENT, FACILITIES, OR PROGRAMS

Astronomy research involves the development of instruments used on suborbital sounding rockets to study the structure of galaxies and interstellar media and the investigation of atmospheres and environments of planets and exoplanets.

The Center for Advanced Materials is involved in the design, synthesis, characterization, and processing of materials for application in new technologies by bringing together state-of-the-art instrumentation, facilities, and expert personnel.

The Photonics Center forms a core of design and fabrication facilities to support various university initiatives requiring innovative semiconductor-based photonic and electronic device technologies, which primarily apply semiconductor, dielectric, and metallic nanomaterials for new robust photonic devices for defense, medical, and commercial applications. Equipment includes three molecular beam epitaxy machines and concomitant lithography and epilayer characterization facilities.

The Submillimeter-Wave Technology Laboratory is a leader in terahertz transmitter and receiver technologies, pioneering the design and fabrication of broadband solid-state multiplier sources, ultra-stable optically pumped lasers, and laser/microwave hybrid systems. A 20-member research team, with the aid of graduate and undergraduate students, builds and maintains a variety of high-performance solid-state and laser-based measurement systems to generate terahertz frequency radiation, resulting in the development of a wide range of materials characterization techniques and high-resolution imaging systems for industry, defense, and medical applications.

The Radiation Laboratory with a 1-MW research reactor, an intense Cobalt-60 gamma source, and 5.5-MV Van de Graaff accelerator, is a unique interdisciplinary facility for nuclear science and technology research. Applied nuclear research includes materials studies, fast neutron and fission spectroscopy, radiation damage, dosimetry, and aerosol transport. The heavy-ion spectroscopy (HI-SPIN) group carries out fundamental nuclear structure research, with experiments at national heavy-ion facilities with high-resolution detector arrays, as well as detector development with industry using analog and digital signal processing with multi-parameter data acquisition and analysis, for nuclear science, advanced nuclear energy research and development, medical imaging, and homeland security applications.

The Multiscale Electromagnetics Group combines theoretical and experimental physics, supporting investigations in solid nanomaterials, device design optimization, and interpretive material and device characterization studies.

The Advanced Biophotonics Laboratory provides fundamental expertise on the structural and functional characterization of pathology for exploratory efforts in medical and bioengineering applications. Integrating multiple optical imaging and spectroscopic approaches, researchers monitor biochemical and physiological processes in real time on a variety of spatially different scales.

The Laboratory for Nano-science and Laser Applications has developed a regenerative amplified femtosecond Ti:Sapphire laser facility to acquire femtosecond laser light-matter interaction data for investigation of material structures and chemical reactions at the molecular level. The laser technology is also used to facilitate the manufacture of micro- and nanostructure materials. Through the development of ultra-fast femtosecond optical spectroscopy measurement systems, nanometer scale spatial and temporal resolution material characterization studies are performed and nanostructures on solid surfaces are fabricated using intense femtosecond laser pulse irradiation.

Space physics research is carried out at the UMass Lowell Center for Atmospheric research, where research projects include solar wind-magnetosphere interaction modeling, magnetosphere-ionosphere-thermosphere coupling theory, plasmasphere sounding and modeling, plasmasphere depletion and refilling processes, ionosphere sounding and modeling, radiation belt wave-particle interactions, antenna-plasma interaction, antenna radiation theory and experiments, whistler mode wave propagation, and ionospheric coupling.

Table B—Separately Budgeted Research Expenditures by Source of Support

Source of Support	Departmental Research	Physics-related Research Outside Department
Federal government	$8,000,000	
State/local government	$102,000	
Non-profit organizations		
Business and industry	$2,000,000	
Other		
Total	$10,102,000	

Table C—Separately Budgeted Research Expenditures by Research Specialty

Research Specialty	No. of Grants	Expenditures ($)
Applied Physics	–	$5,000,000
Astronomy	–	$28,000
Astrophysics	–	$1,000,000
Atmosphere, Space Physics, Cosmic Rays	–	$850,000
Materials Science, Metallurgy	–	$450,000
Medical, Health Physics	–	$413,000
Nano Science and Technology	–	$130,000
Nuclear Physics	–	$2,000,000
Polymer Physics/Science	–	$231,000
Total	–	$10,102,000

FACULTY

Professor

Chakrabarti, Supriya, Ph.D., University of California, Berkeley, 1982. *Astronomy, Astrophysics.*

Chowdhury, Partha, Ph.D., Stony Brook University, 1979. *Materials Science, Metallurgy, Nuclear Physics.*

Egan, James J., Ph.D., University of Kentucky, 1969. *Nuclear Physics.* Experimental nuclear physics.

French, Clayton S., Ph.D., University of Massachusetts, Lowell, 1985. *Medical, Health Physics.* Radiological science.

Giles, Robert, Ph.D., University of Massachusetts, Lowell, 1986. *Applied Physics, Materials Science, Metallurgy, Optics.* Terahertz laser physics.

Kumar, Jayant, Ph.D., Rutgers University, 1983. *Biophysics, Condensed Matter Physics, Materials Science, Metallurgy, Optics, Polymer Physics/Science.*

Lister, Christopher J., Ph.D., University of Liverpool, 1977. *Nuclear Physics.*

Mittler, Arthur, Ph.D., University of Kentucky, 1970. *Physics and other Science Education.* Physics education.

Podolskiy, Viktor, Ph.D., New Mexico State University, 2002. *Computational Physics, Condensed Matter Physics, Optics.* Materials science; photonics; plasmonics.

Sajo, Erno, University of Massachusetts, Lowell, 1990. *Medical, Health Physics, Nuclear Engineering.* Aerosol science.

Schier, Walter, Ph.D., University of Notre Dame, 1964. *Nuclear Physics.* Experimental nuclear physics.

Sebastian, Kunnat J., Ph.D., University of Maryland, 1969. *Atomic, Molecular, & Optical Physics, Particles and Fields.* Theoretical elementary particle physics; theoretical atomic physics.

Song, Paul, Ph.D., University of California, Los Angeles, 1991. *Atmosphere, Space Physics, Cosmic Rays.*

Stimets, Richard W., Ph.D., Massachusetts Institute of Technology, 1969. *Astronomy, Condensed Matter Physics, Optics.* Image processing.

Associate Professor

Shen, Mengyan, Ph.D., University of Science and Technology of China, 1990. *Applied Physics, Nano Science and Technology, Optics.* Femtosecond laser physics.

Tries, Mark A., Ph.D., University of Massachusetts, Lowell, 1999. *Medical, Health Physics.* Radiological science.

Yaroslavsky, Anna, Ph.D., Saratov State University, 1999. *Biophysics, Medical, Health Physics, Optics.* Medical imaging.

Assistant Professor

Cook, Timothy A., Ph.D., University of Colorado, 1991. *Astronomy, Astrophysics.*

Guo, Wei, Ph.D., Brown University, 2008. *Applied Physics, Materials Science, Metallurgy, Optics, Solid State Physics.* Photonics and optoelectronics; molecular beam epitaxy.

Ngwa, Wilfred F., Ph.D., University of Leipzig, 2004. *Biophysics, Medical, Health Physics.*

Emeritus

Altman, Albert, Ph.D., University of Maryland, 1962. Physics education.

Goodhue, William D., Ph.D., University of Massachusetts, Lowell, 1982. *Applied Physics, Condensed Matter Physics, Materials Science, Metallurgy, Optics.* Photonics and optoelectronics; molecular beam epitaxy.

Hardy, F. Raymond, M.S., University of Massachusetts, Lowell, 1962. Physics education.

Karakashian, Aram S., Ph.D., University of Maryland, 1970. *Condensed Matter Physics, Optics.* Theoretical and experimental solid-state physics/optics.

Kegel, Gunter H. R., Ph.D., Massachusetts Institute of Technology, 1961. *Materials Science, Metallurgy, Nuclear Physics, Physics of Beams.*

Pullen, David J., University of Oxford, 1963. *Nuclear Physics.* Physics and science education.

Waldman, Jerry, Ph.D., Massachusetts Institute of Technology, 1970. *Applied Physics, Materials Science, Metallurgy, Optics.* Experimental laser physics.

Adjunct Faculty

Antal, John J., Ph.D., Saint Louis University, 1952. *Medical, Health Physics, Nuclear Engineering.* Neutron radiography.

Ascoli, Frank A., M.S., University of Cincinnati, 1976. *Medical, Health Physics.*

Baird, Christopher S., Ph.D., University of Massachusetts, Lowell, 2007. *Applied Physics, Optics.* Electromagnetic theory; terahertz imagery.

Bliss, David F., Ph.D., Stony Brook University, 2000. Materials science; substrate engineering; crystal growth.

Bobek, Leo, M.S., University of Massachusetts, Lowell, 1989. *Medical, Health Physics, Nuclear Engineering.* Radiological sciences.

Coulombe, Michael, University of Massachusetts, Lowell, 1989. *Systems Science/Engineering.* Microwave systems; terahertz physics.

DeMartinis, Guy B., Ph.D., University of Massachusetts, Lowell, 2008. *Applied Physics.* Terahertz technology.

Fox, Herbert L., Ph.D., University of Massachusetts, Lowell, 2001. *Physics and other Science Education*. Physics education.

Gatesman, Andrew, Ph.D., University of Massachusetts, Lowell, 1993. *Engineering Physics/Science, Optics, Systems Science/Engineering*. Radar signatures; IR, submillimeter, and millimeter wave optical systems.

Goyette, Thomas M., Ph.D., Duke University, 1990. *Applied Physics, Optics*. Laser systems; terahertz spectroscopy.

Guess, Carol J., Ph.D., Michigan State University, 2010. *Nuclear Physics*. Isomer decay of heavy nuclei; physics education.

Joseph, Cecil, Ph.D., University of Massachusetts, Lowell, 2010. *Applied Physics, Biophysics, Physics and other Science Education*. Biophysics; medical applications.

Lazewatsky, Joel L., Ph.D., Massachusetts Institute of Technology, 1979. *Biophysics, Medical, Health Physics*.

Li, Lian, Ph.D., University of Massachusetts, Lowell, 1993. *Optics, Polymer Physics/Science*. Nonlinear optics.

Maginnis, Thomas O., Ph.D., Adelphi University, 1973. *Applied Physics, Fluids, Rheology*.

Medich, David, Ph.D., University of Massachusetts, Lowell, 1997. *Medical, Health Physics*. Radiological health physics.

Menyhart, Gabor, M.S., University of Kentucky, 2002. *Medical, Health Physics*.

Montesalvo, Mary, M.S., University of Massachusetts, Lowell, 1985. *Medical, Health Physics*. Radiation dosimetry.

Mosurkal, Ravi, Ph.D., University of Hyderabad, 1998. *Polymer Physics/Science*. Polymer physics.

Mower, Herbert W., Other, Massachusetts Institute of Technology, 1972. *Medical, Health Physics*. Electrical engineering; radiation oncology.

Narayan, Chandrika, Ph.D., University of Massachusetts, Lowell, 1992. *Applied Physics, Materials Science, Metallurgy, Physics and other Science Education*. Materials physics; accelerator applications.

Pandian, Lakshmi S., Ph.D., Tata Institute of Fundamental Research, 2005. *Nuclear Physics*. Gamma-ray spectroscopy.

Pretorius, P. Hendrik, Ph.D., University of Orange Free State, 1994. *Biophysics, Medical, Health Physics*.

Regan, Thomas, M.S., University of Massachusetts, Lowell, 1994. *Nuclear Engineering, Nuclear Physics*. Nuclear engineering; neutron radiography; gamma-ray spectroscopy.

Rivard, Mark, Ph.D., Wayne State University, 1998. *Medical, Health Physics*. Medical physics.

Salesky, Edward T., Ph.D., University of Massachusetts, Lowell, 1978. *Astrophysics, Physics and other Science Education*. Theoretical applied physics; physics education.

Seco, Joao, Ph.D., University of London, 2001. *Medical, Health Physics*. Radiation oncology.

Sivjee, Abbas H., Ph.D., Johns Hopkins University, 1970. *Astronomy, Astrophysics*.

Snay, Steven, M.S., University of Massachusetts, Lowell, 2007. *Medical, Health Physics*. Radiological science.

Sullivan, Nancy L. B., Ph.D., University of Massachusetts, Lowell, 1993. *Medical, Health Physics*. Physics education.

Weintraub, Sheri M., M.S., University of Cincinnati, 1999. *Medical, Health Physics*. Radiation oncology.

Wong, Eric T., Other, Rutgers Medical School, 1989. M.D., Neuro-oncology.

Yang, Ke, Ph.D., University of Massachusetts, Lowell, 1999. *Materials Science, Metallurgy, Polymer Physics/Science*. Optics; materials science.

Lecturer

Danylov, Andriy, Ph.D., University of Massachusetts, Lowell, 2010. *Materials Science, Metallurgy, Optics, Physics and other Science Education*. Sub-millimeter wave technology.

Laycock, Silas, Ph.D., University of Southampton, 2002. *Astronomy, Astrophysics*. Physics Education.

Lepeshkin, Nikolay, Ph.D., New Mexico State University, 2001. *Optics*. Physics education.

DEPARTMENTAL RESEARCH SPECIALTIES AND STAFF

Theoretical

Atomic, Molecular, & Optical Physics. Radiative transitions in mesic atoms. Podolskiy, Sebastian.

Biophysics. Giles, Joseph, Yaroslavsky.

Optics. Quantum optics; dielectric waveguides; surface plasmons; ultraviolet and far-infrared spectra; electronic and vibrational cluster calculations. Baird, Danylov, Podolskiy.

Particles and Fields. Sebastian.

Experimental

Applied Physics. Radiation effects; Rutherford backscattering; PIXE; nuclear instrumentation; proton microbeam applications. Chowdhury, Egan, Kegel, Kumar, Lister.

Applied Physics. Development of coherent sources, receivers, and novel imaging systems for applications at terahertz frequencies. Baird, Coulombe, DeMartinis, Giles, Goyette, Joseph, Waldman.

Applied Physics. Materials tunable visible infrared and far-infrared lasers; opto-electronic materials and devices; image processing; surface plasmons; polymers and biological materials. Baird, Danylov, DeMartinis, Gatesman, Giles, Goyette, Guo, Joseph, Kumar, Li, Shen, Waldman.

Astronomy. Studies of x-ray binaries; galactic formation and structure: use of suborbital sounding rockets for astronomical observation; development of novel instruments and data analysis techniques; ultaviolet imaging and spectroscopy; exoplanet observations. Chakrabarti, Cook, Laycock.

Atmosphere, Space Physics, Cosmic Rays. Solar wind-magnetosphere-ionosphere interactions. Song.

Medical, Health Physics. Dosimetry; shielding; biological effects of radiation; radon monitoring studies; radiation safety and control; aerosol physics. Antal, French, Lazewatsky, Medich, Ngwa, Pretorius, Rivard, Sajo, Seco, Tries, Weintraub, Wong, Yaroslavsky.

Nano Science and Technology. Femtosecond laser surface interactions. Goodhue, Shen.

Nuclear Physics. Neutron cross-sections; fission reaction studies; inelastic neutron scattering; fission product studies; in-beam gamma-ray spectroscopy; high-spin nuclear structure; heavy-ion fusion reactions, detector development. Chowdhury, Egan, Guess, Kegel, Lister, Pandian, Schier.

Optics. Danylov, Gatesman, Giles, Guo, Kumar, Waldman, Yaroslavsky.

View additional information about this department at
www.gradschoolshopper.com

WORCESTER POLYTECHNIC INSTITUTE

DEPARTMENT OF PHYSICS

Worcester, Massachusetts 01609
http://www.wpi.edu/academics/physics.html

General University Information
President: Phillip Ryan
Dean of Graduate School: Richard Sisson
University website: http://www.wpi.edu
Control: Private
Setting: Urban
Total Faculty: 458
Total Graduate Faculty: 250
Total number of Students: 5,615
Total number of Graduate Students: 1,734

Department Information
Department Chairman: Germano S. Iannacchione, Head
Department Contact: Jackie Malone, Administrative Assistant
  Total full-time faculty: 18
  Total number of full-time equivalent positions: 18
  Full-Time Graduate Students: 18
  First-Year Graduate Students: 6
  Female First-Year Students: 3
  Total Post Doctorates: 2

Department Address
100 Institute Road
Worcester, MA 01609
Phone: (508) 831-5258
Fax: (508) 831-5886
E-mail: physics@wpi.edu
Website: http://www.wpi.edu/academics/physics.html

ADMISSIONS

Admission Contact Information
Address admission inquiries to: Graduate Admissions, Boynton Hall, 100 Institute Road, Worcester, MA 01609-2280
Phone: (508) 831-5301
E-mail: grad@wpi.edu
Admissions website: http://www.wpi.edu/admissions/graduate/

Application deadlines
Fall admission:
U.S. students: March 1   *Int'l. students*: March 1

Application fee
U.S. students: $70

Admissions information
For Fall of 2012:
  Number of applicants: 56
  Number admitted: 5
  Number enrolled: 3

Admission requirements
Bachelor's degree requirements: A bachelor's degree in physics or a related field is required.
Minimum undergraduate GPA: 3.0

GRE requirements
The GRE is recommended.
  Quantitative score: 150
  Verbal score: 150

  Analytical score: 4

Advanced GRE requirements
The Advanced GRE is recommended.
  Minimum accepted Advanced GRE score: 700
Specific subject matter is physics.

TOEFL requirements
The TOEFL exam is required for students from non-English-speaking countries.
  PBT score: 550

Other admissions information
Additional requirements: A statement of purpose highlighting a connection to department research is required.
Undergraduate preparation assumed: Taylor, Mechanics; Griffiths, Electromagnetism; Griffiths, Quantum Theory; Rief, Statistical Mechanics.

TUITION

Tuition year 2012–13:
  Full-time students: $1,300 per credit
  Part-time students: $1,300 per credit
There is a total of 247 full-time faculty plus 119 part-time faculty. There are 201 full-time graduate faculty plus 49 part-time graduate faculty.
Credit hours per semester to be considered full-time: 9
Deferred tuition plan: No
Health insurance: Available
Other academic fees: $85 service fee
Academic term: Semester

Teaching Assistants, Research Assistants, and Fellowships
Number of first-year
  Teaching Assistants: 11
  Research Assistants: 2
  Fellowship students: 2
Average stipend per academic year
  Teaching Assistant: $18,000
  Research Assistant: $21,000
  Fellowship student: $24,000

FINANCIAL AID

Loans
Loans are not available for U.S. students.
Loans are not available for international students.
GAPSFAS application required: No
FAFSA application required: No

For further information
Address financial aid inquiries to: Graduate Admissions, Boynton Hall, 100 Institute Road, Worcester, MA 01609-2280.
Phone: (508) 831-5301
E-mail: grad@wpi.edu
Financial aid website: http://www.wpi.edu/admissions/graduate.html

HOUSING

Availability of on-campus housing
Single students: No
Married students: No

For further information
Address housing inquiries to: Office of Residential Services, East
Hall, 100 Institute Road, Worcester, MA 01609-2280.
Phone: (508) 831-5645
E-mail: res-services@wpi.edu
Housing aid website: http://www.wpi.edu/offices/rso.html

Table A—Faculty, Enrollments, and Degrees Granted

Research Specialty	2012-13 Faculty	Enrollment Fall 2012 Master's	Enrollment Fall 2012 Doctorate	Number of Degrees Granted 2012-13 (2004–12) Master's	Number of Degrees Granted 2012-13 (2004–12) Terminal Master's	Number of Degrees Granted 2012-13 (2004–12) Doctorate
Atomic, Molecular, & Optical Physics	2	1	–	–	–	–
Biophysics	1	–	6	–	–	1(-)
Computational Physics	2	1	1	1(-)	1(-)	1(-)
Condensed Matter Physics	4	–	1	–	–	–
Electromagnetism	1	–	–	–	–	–
Low Temperature Physics	1	–	–	–	–	–
Medical, Health Physics	1	–	1	–	–	–
Nano Science and Technology	2	2	1	1(-)	1(-)	–
Optics	1	–	2	–	–	–
Statistical & Thermal Physics	3	1	2	1(-)	1(-)	–
Total	**17**	**–**	**–**	**3(-)**	**3(-)**	**2(-)**
Full-time Grad. Stud.	–	6	13	–	–	–
First-year Grad. Stud.	–	6	2	–	–	–

GRADUATE DEGREE REQUIREMENTS

Master's: Thirty semester-hour credits are required, 24 by approved courses and six by thesis or directed research. Thesis and non-thesis options are available. Fifteen credits in classical mechanics, quantum mechanics, electromagnetism, and statistical mechanics are required.

Doctorate: Ninety semester-hour credits beyond the baccalaureate degree are required, including a minimum of 42 in approved courses and directed study, 30 of dissertation research, and completion and defense of a Ph.D. dissertation. Passage of a qualifying examination is required. Completion of an M.S. degree is not required.

SPECIAL EQUIPMENT, FACILITIES, OR PROGRAMS

We offer a master's degree in Physics Education (MDPE) and Nuclear Science and Engineering.

Table B—Separately Budgeted Research Expenditures by Source of Support

Source of Support	Departmental Research	Physics-related Research Outside Department
Federal government	$154.81	
State/local government		
Non-profit organizations		
Business and industry		
Other	$251,801	$50,000
Total	**$251,955.81**	**$50,000**

Table C—Separately Budgeted Research Expenditures by Research Specialty

Research Specialty	No. of Grants	Expenditures ($)
Condensed Matter Physics	5	$406,614
Total	**5**	**$406,614**

FACULTY

Professor

Aravind, Padmanabhan K., Ph.D., Northwestern University, 1980. Associate Department Head. *Applied Mathematics, Atomic, Molecular, & Optical Physics, Computational Physics.* Quantum information theory; cryptography; and computing.

Phillies, George D. J., Ph.D., Massachusetts Institute of Technology, 1973. Light scattering spectroscopy and theory of complex fluids.

Ram-Mohan, L. Ramdas, Ph.D., Purdue University, 1971. Many-body theory and optical properties of solids.

Zozulya, Alex A., Ph.D., Lebedev Physics Institute of the Academy of Sciences of the U.S.S.R., 1984. *Atomic, Molecular, & Optical Physics, Computational Physics, Condensed Matter Physics, Optics.* Nonlinear optics; BEC; atomtronics.

Associate Professor

Burnham, Nancy A., Ph.D., University of Colorado Boulder, 1987. *Nano Science and Technology.* Mechanical properties of nanostructures; instrumentation and metrology for nanomechanics.

Iannacchione, Germano S., Ph.D., Kent State University, 1993. Department Head. *Biophysics, Chemical Physics, Condensed Matter Physics, Materials Science, Metallurgy, Nano Science and Technology, Polymer Physics/Science, Statistical & Thermal Physics.* Experimental soft matter; biophysics; liquid crystals.

Quimby, Richard S., Ph.D., University of Wisconsin-Madison, 1979. *Applied Physics, Atomic, Molecular, & Optical Physics, Materials Science, Metallurgy, Optics, Physics of Beams.* Optical properties of materials.

Assistant Professor

Medich, David C., Ph.D., UMass-Lowell, 1997. *Medical, Health Physics.* Active research programs in diagnostic and therapeutic medical physics, including functional neutron imaging, intensity modulated brachytherapy dosimetry, Monte Carlo radiation transport simulations, gamma spectroscopy, air kerma strength calibrations, and radiation shielding.

Stroe, Izabela, Ph.D., Clark University, 2005. *Biophysics, Chemical Physics, Medical, Health Physics, Nuclear Physics, Polymer Physics/Science, Statistical & Thermal Physics.* Experimental biophysics.

Tüzel, Erkan, Ph.D., University of Minnesota, 2006. *Biophysics, Chemical Physics, Computational Physics, Fluids, Rheology, Medical, Health Physics, Polymer Physics/Science.* Computational physics; statistical mechanics of biology and materials.

Wen, Qi, Ph.D., Brown University, 2005. *Biophysics, Chemical Physics, Mechanics, Optics, Polymer Physics/Science, Statistical & Thermal Physics.* Experimental biophysics.

Research Assistant Professor
Popovic, Marko B., Ph.D., Boston University, 2002. *Applied Physics, Biophysics, Mechanics, Medical, Health Physics.*

Teaching Assistant Professor
Dick, Frank A., Ph.D., Worcester Polytechnic Institute, 2007. *Astronomy, Astrophysics, Atmosphere, Space Physics, Cosmic Rays, Nuclear Engineering, Nuclear Physics, Particles and Fields, Physics of Beams, Plasma and Fusion.*

Kashuri, Hektor, Ph.D., Northeastern University, 2008. *Biophysics, Chemical Physics, Condensed Matter Physics, Materials Science, Metallurgy, Mechanics, Nano Science and Technology, Polymer Physics/Science.*

Sarkar, Sabyasachi, Ph.D., University of Nebraska-Lincoln, 2008. *Biophysics, Materials Science, Metallurgy, Nano Science and Technology, Optics, Polymer Physics/Science, Surface Physics.* Ellipsometry, polymers, and biomacromolecules; surface chemistry; separations; material synthesis and fabrication.

Tüzel, Vasfiye Hande, Ph.D., University of Minnesota, 2009. *Applied Mathematics, Applied Physics, Biophysics, Fluids, Rheology, Medical, Health Physics, Nano Science and Technology, Nonlinear Dynamics and Complex Systems.* Inverse problems; mathematical biology; continuum and fluid mechanics; mechanics of biopolymers; level-set method.

DEPARTMENTAL RESEARCH SPECIALTIES AND STAFF

Theoretical
Computational Physics. Cold atom Bose-Einstein condensation of bosons and fermions; atom wave guides and interferometers; quantum information: Bell's theorem quantum algorithms; wave function engineering nanostructures; finite-element modeling of quantum systems and wells; field theory; biophysics and cell mobility. Aravind, Burnham, Ram-Mohan, Erkan Tüzel, Vasfiye Hande Tüzel.

Medical/Health Physics. Functional neutron imaging; intensity modulated brachytherapy dosimetry; Monte Carlo radiation transport simulations; gamma spectroscopy; air kerma strength calibrations; radiation shielding. Medich.

Nano Science and Technology. Semiconductors: optical properties of superlattices, heterostructure laser design, spintronics in diluted magnetic semiconductors and devices; nanomechanics: adhesion of microsensor surfaces, compliance of tissue-growth substrates, analysis of force–curve data, atomic–force microscopy calibration. Aravind, Burnham, Ram-Mohan, Erkan Tüzel, Vasfiye Hande Tüzel.

Physics and other Science Education. Research in physics education focusing on aspects of teaching and learning physics, spanning a broad range of topics from psychology (studying student behaviors) to computer science (studying uses of new interactive technologies in learning). Dick, Iannacchione.

Experimental
Biophysics. Liquid crystals: thermotropic/lyotropic/colloidal systems, phase transitions and critical phenomena, cooperative behavior and self-assembly, quenched random disorder effects, calorimetry instrumentation; liquids: diffusion and transport properties, light scattering spectroscopy of liquids and polymer solutions, wetting phenomena, phase transitions and critical phenomena, superfluidity; polymers: molecular properties of small-sample volumes and single molecules, polymer and bio–macromolecular solutions, surfactants, and colloids; biophysics: proteins, dynamics and structure of self-assemblies of biomaterials, DNA, biomechanics, cellular functions. Iannacchione, Kashuri, Phillies, Sarkar, Stroe, Erkan Tüzel, Vasfiye Hande Tüzel, Wen.

Nano Science and Technology. Nanoscience, an interdisciplinary field that incorporates elements of physics, engineering, biotechnology, and chemistry, deals with structures that are very small in nature, generally those smaller than 100 nm, or approximately one ten-millionth of an inch. Nanoscience and nanotechnology involve the ability to see and control these tiny, individual atoms that make up everything on Earth. The food we eat, the clothes we wear, the houses we live in, and even the human body all consist of atoms. Aravind, Burnham, Popovic, Quimby, Ram-Mohan, Erkan Tüzel, Vasfiye Hande Tüzel, Zozulya.

Optics. Lasers: development of infrared fiber lasers and materials, mid-IR and FIR quantum cascade laser design, THz lasers; photonics: fiber amplifiers and optical communications; spectroscopy: laser spectroscopy of impurity ions in glasses and crystals. Iannacchione, Quimby, Sarkar, Wen, Zozulya.

View additional information about this department at www.gradschoolshopper.com

CENTRAL MICHIGAN UNIVERSITY

DEPARTMENT OF PHYSICS

Mt. Pleasant, Michigan 48859
http://www.cst.cmich.edu/physics

General University Information
President: George Ross
Dean of Graduate School: Roger Coles
University website: http://www.cmich.edu

Control: Public
Setting: Rural
Total Faculty: 1,017
Total Graduate Faculty: 650

Total number of Students: 28,389
Total number of Graduate Students: 6,777

Department Information
Department Chairman: Christopher Tycner, Chair
Department Contact: Christopher Tycner, Associate Professor and Chair
   Total full-time faculty: 15
   Total number of full-time equivalent positions: 18
   Full-Time Graduate Students: 17
   First-Year Graduate Students: 10
   Female First-Year Students: 3
   Total Post Doctorates: 3

Department Address
Dept of Physics/Dow 203
Ottawa Court
Mt. Pleasant, MI 48859
Phone: (989) 774-3321
Fax: (989) 774-2697
E-mail: tycne1c@cmich.edu
Website: http://www.cst.cmich.edu/physics

ADMISSIONS

Admission Contact Information
Address admission inquiries to: College of Graduate Studies, Central Michigan University, Foust Hall 100, Mount Pleasant, MI 48859
Phone: (989) 774-4723
E-mail: grad@cmich.edu
Admissions website: https://apply.cmich.edu/

Application deadlines
Fall admission:
U.S. students: May 1   *Int'l. students*: April 1

Application fee
There is no application fee required.
Applications from students seeking graduate assistantships should be received by February 1. Students are rarely admitted for a spring semester start.

Admissions information
For Fall of 2013:
   Number of applicants: 30
   Number admitted: 15
   Number enrolled: 10

Admission requirements
Bachelor's degree requirements: Bachelor's degree in physics or a closely related discipline from an accredited science or engineering program.
Minimum undergraduate GPA: 2.7

GRE requirements
The GRE is recommended.
GRE scores are recommended for all applicants, but especially for those seeking a graduate assistantship.

Advanced GRE requirements
The Advanced GRE is recommended.
A Physics GRE score is recommended for applicants seeking a graduate assistantship.

TOEFL requirements
The TOEFL exam is required for students from non-English-speaking countries.
   PBT score: 550
   iBT score: 79

Students with scores of 500 to 549 (PBT) or 61-78 (iBT) can be considered for conditional admission.

Other admissions information
Additional requirements: Three letters of recommendation are requested for applicants seeking an assistantship.
Undergraduate preparation assumed: Coursework that is similar to that offered in CMU's undergraduate physics major. See www.cst.cmich.edu/physics.

TUITION

Tuition year 2013-14:
Tuition for in-state residents
   Full-time students: $485 per credit
   Part-time students: $485 per credit
Tuition for out-of-state residents
   Full-time students: $766 per credit
   Part-time students: $766 per credit
All students receiving graduate assistantships receive tuition remission of up to 30 credit hours per calendar year. This is more than enough to cover all tuition needs.
Credit hours per semester to be considered full-time: 6
Deferred tuition plan: No
Health insurance: Available at the cost of 1,500 per year.
Other academic fees: A health plan is available for purchase.
Academic term: Semester
Number of first-year students who received full tuition waivers: 10

Teaching Assistants, Research Assistants, and Fellowships
Number of first-year
   Teaching Assistants: 10
   Research Assistants: 2
Average stipend per academic year
   Teaching Assistant: $12,500
   Research Assistant: $12,500
   Fellowship student: $12,500
Majority (more than 90%) of graduate students receive RA support during the summer that brings an additional $6,250 in support for a total of $18,750 per calendar year.

FINANCIAL AID

Loans
Loans are available for U.S. students.
Loans are not available for international students.
GAPSFAS application required: No
FAFSA application required: Yes

For further information
Address financial aid inquiries to: CMU Office of Scholarships and Financial Aid, Student Service Court, Mount Pleasant, MI 48859.
Phone: (989) 774-3674
E-mail: CMUOSFA@cmich.edu
Financial aid website: http://go.cmich.edu/financial_information/graduate/Pages/Financial_Aid.aspx

HOUSING

Availability of on-campus housing
   Single students: Yes
   Married students: Yes

For further information
Address housing inquiries to: Residence Life, Bovee University Center.
Phone: (989) 774-3111
Housing aid website: http://www.reslife.cmich.edu

Table A—Faculty, Enrollments, and Degrees Granted

Research Specialty	2012-13 Faculty	Enrollment Fall 2013		Number of Degrees Granted 2012-13		
		Master's	Doctorate	Master's	Terminal Master's	Doctorate
Astronomy	2	2	–	–	2(-)	–
Atomic, Molecular, & Optical Physics	1	1	–	–	1(-)	–
Condensed Matter Physics	5	5	2	–	2(-)	–
Materials Science, Metallurgy	2	2	2	–	1(-)	2(-)
Medical, Health Physics	–	2	–	–	1(-)	–
Nuclear Physics	5	5	–	–	–	–
Total	15	17	4	–	7(-)	2(-)
Full-time Grad. Stud.	–	17	4	–	–	–
First-year Grad. Stud.	–	10	2	–	–	–

GRADUATE DEGREE REQUIREMENTS

Master's: A total of 30 credit hours are required, plus a thesis.

Other Degrees: The Department of Physics participates in an interdisciplinary PhD program in the Science of Advanced Materials (SAM). See www.cst.cmich.edu/sam for details. Students may begin with the M.S. in Physics and enter the SAM Ph.D. program after earning the M.S.

Thesis: A written thesis is required, along with an oral thesis defense.

SPECIAL EQUIPMENT, FACILITIES, OR PROGRAMS

Laser spectroscopy laboratory, rheology laboratory, x-ray crystallography laboratory, astronomical observatory, polymer physics laboratory, three experimental nuclear physics labs. Faculty and students also access computational resources at the High Performance Computer Center at Michigan State University.

Table B—Separately Budgeted Research Expenditures by Source of Support

Source of Support	Departmental Research	Physics-related Research Outside Department
Federal government	$800,000	
State/local government		
Non-profit organizations	$40,000	
Business and industry	$10,000	
Other		
Total	$850,000	

Table C—Separately Budgeted Research Expenditures by Research Specialty

Research Specialty	No. of Grants	Expenditures ($)
Condensed Matter Physics	7	$595,000
Nuclear Physics	3	$255,000
Total	10	$850,000

FACULTY

Professor

Finck, Joseph E., Ph.D., Michigan State University, 1982. *Nuclear Physics*. Experimental nuclear physics; properties of neutron-rich nuclei near the neutron drip-line; MoNA and LISA neutron detectors.

Fornari, Marco, Ph.D., University of Trieste, 1998. *Condensed Matter Physics*. Electronic structure, thermo-electric materials, ferro- and piezo-electric materials; physics education.

Hirschi, Stanley, Ph.D., Massachusetts Institute of Technology, 1971. *Fluids, Rheology, Polymer Physics/Science*. Mechanical and electrical properties of materials, especially fluids and gels.

Horoi, Mihai, Ph.D., Institute of Atomic Physics, Bucharest, 1990. *Nuclear Physics*. Theoretical nuclear physics; nuclear shell structure; medical physics.

Jackson, Koblar A., Ph.D., University of Wisconsin-Madison, 1989. *Condensed Matter Physics*. Density functional theory-based methods for studying the properties of materials; theory of atomic clusters; chemical physics.

Petkov, Valeri G., Ph.D., University of Sofia, 1991. *Condensed Matter Physics*. X-ray diffraction of materials.

Sieradzan, Andrzej, Ph.D., Warsaw University, 1976. *Atomic, Molecular, & Optical Physics*. High-precision laser spectroscopy; optics; atomic traps.

Williams, Glen, Ph.D., University of Michigan, 1983. *Astrophysics*. Studies of hydrodynamics and radiation transfer in accretion disks of Cataclysmic Variable stars.

Associate Professor

Peralta, Juan E., Ph.D., University of Buenos Aires, 2002. *Condensed Matter Physics*. Magnetic phenomena in molecules and nanomaterials from first-principles; novel theoretical and computational methods for understanding the chemical and physical properties of new materials.

Tycner, Christopher, Ph.D., University of Toronto, 2004. *Astronomy, Astrophysics*. Study of circumstellar disks of hot stars using a variety of ground-based instruments, including long-baseline optical interferometry and spectroscopy.

Assistant Professor

Barone, Veronica, Ph.D., University of Buenos Aires, 2003. *Condensed Matter Physics*. Electronic structure calculations based on density functional theory with applications in energy storage, molecular magnetism, electronic devices, drug delivery, and characterization methods; nanoscience.

Mellinger, Axel P., Ph.D., Technical University Munich, 1995. *Condensed Matter Physics, Polymer Physics/Science*. Ferroelectret polymers and dielectric nanocomposites: new concepts for piezoelectric sensors and actuators; non-destructive 3D space-charge and polarization tomography; energy harvesting.

Perdikakis, Georgios, Ph.D., National Technical University of Athens, 2006. *Nuclear Physics*. Experimental nuclear physics and nuclear astrophysics, stellar nucleosynthesis, stellar energy production, nuclear structure and reactions, physics with rare isotopes.

Redshaw, Matthew, Ph.D., Florida State University, 2007. *Atomic, Molecular, & Optical Physics, Nuclear Physics*. Precision mass measurements using ions confined in a Penning trap; atomic mass measurements on stable and short-lived isotopes with applications in nuclear physics and nuclear astrophysics, neutrino physics, atomic physics, chemistry and metrology.

Wimmer, Kathrin, Ph.D., Technical University of Munich, 2010. *Nuclear Physics*. Nuclear structure and reaction studies using radioactive ion beams; the evolution of shell structure away from the valley of stability towards the most exotic nuclei.

Visiting Research Assistant Professor

Senkov, Roman, Ph.D., Budker Institute of Nuclear Physics. *Nuclear Physics*. Theoretical nuclear physics; nuclear shell structure.

DEPARTMENTAL RESEARCH SPECIALTIES AND STAFF

Theoretical

Computational materials physics. Electronic structure of materials using first-principles techniques. Barone, Fornari, Jackson, Peralta.

Nuclear structure physics. Spectroscopy and nuclear structure; shell model calculations. Finck, Horoi.

Physics of circumstellar disks. Observational studies of disks using optical interferometry; computational modeling of radiative transfer in disks. Tycner, Williams.

Rare isotope physics. High-precision mass determinations; transfer reactions. Perdikakis, Redshaw, Wimmer.

Experimental

Rare isotope physics. High-precision mass determinations; transfer reactions. Finck, Perdikakis, Redshaw, Wimmer.

materials characterization. X-ray characterization of materials; nanoparticles; disordered materials; rheology and polymer physics. Hirschi, Mellinger, Petkov.

View additional information about this department at www.gradschoolshopper.com

---

EASTERN MICHIGAN UNIVERSITY

DEPARTMENT OF PHYSICS AND ASTRONOMY

Ypsilanti, Michigan 48197
http://www.emich.edu/physics

General University Information

President: Susan W. Martin
Dean of Graduate School: Jeffrey D. Kentor
University website: http://www.emich.edu
Control: Public
Setting: Suburban
Total Faculty: 957
Total number of Students: 24,287
Total number of Graduate Students: 5,627

Department Information

Department Chairman: Alexandria Oakes, Head
Department Contact: J. Marshall Thomsen, Advisor
  Total full-time faculty: 12
  Total number of full-time equivalent positions: 14
  Full-Time Graduate Students: 10
  First-Year Graduate Students: 5
  Female First-Year Students: 1

Department Address

Physics and Astronomy
Ypsilanti, MI 48197
Phone: (734) 487-8794
Fax: (734) 487-0989
E-mail: aoakes@emich.edu
Website: http://www.emich.edu/physics

ADMISSIONS

Admission Contact Information

Address admission inquiries to: Office of Admissions, 401 Pierce Hall, Ypsilanti, MI 48197
Phone: (800) 468-6368
E-mail: graduate.admissions@emich.edu
Admissions website: http://www.emich.edu/graduate/admissions

Application deadlines

Fall admission:
U.S. students: March 15   *Int'l. students*: February 15

Application fee

U.S. students: $35
$25 online

Admission requirements

Bachelor's degree requirements: A Bachelor's degree in physics is required.
Minimum undergraduate GPA: 2.7

GRE requirements

The GRE is not required.

Advanced GRE requirements

The Advanced GRE is not required.

TOEFL requirements

The TOEFL exam is required for students from non-English-speaking countries.
PBT score: 550
iBT score: 85

Other admissions information

Additional requirements: Conditional admission may be granted to those without a standard undergraduate preparation in physics. A physics education degree, which includes a minor in physics and status as an in-service or prospective teacher, is available. A physical science degree, which includes a minor in a science with not less than 30 semester-hours in science and mathematics and status as an in-service or prospective teacher is also available. Students from non-English-speaking countries are required to demonstrate proficiency in English via the ELI exam; the minimum acceptable score for admission is 85 (admission with lower scores, but requiring special English courses, may be granted).

TUITION

Tuition year 2012–13:
Tuition for in-state residents
Full-time students: $449 per credit
Tuition for out-of-state residents
Full-time students: $885 per credit
Credit hours per semester to be considered full-time: 6
Deferred tuition plan: No
Health insurance: Yes, see website.
Other academic fees: $40/semester + $71/credit.

Teaching Assistants, Research Assistants, and Fellowships

Number of first-year
Teaching Assistants: 4
Average stipend per academic year
Teaching Assistant: $9,240

FINANCIAL AID

Application deadlines
Fall admission:
U.S. students: February 15   *Int'l. students*: February 15

Loans
Loans are available for U.S. students.
Loans are not available for international students.
GAPSFAS application required: No
FAFSA application required: No

For further information
Address financial aid inquiries to: Graduate Assistantships: Graduate School., Other: Director of Financial Aid.
Phone: (734) 487-0042
E-mail: graduate_school@emich.edu
Financial aid website: http://www.emich.edu/graduate/admissions/financialassistance/

HOUSING

Availability of on-campus housing
Single students: Yes
Married students: Yes

For further information
Address housing inquiries to: EMU Housing, Ypsilanti, MI 48197.
Phone: (734) 487-1300
E-mail: housing@emich.edu
Housing aid website: http://www.emich.edu/residencelife

Table A—Faculty, Enrollments, and Degrees Granted

Research Specialty	2012–13 Faculty	Enrollment Fall 2012 Master's	Enrollment Fall 2012 Doctorate	Number of Degrees Granted 2010–11 (2006–11) Master's	Number of Degrees Granted 2010–11 (2006–11) Terminal Master's	Number of Degrees Granted 2010–11 (2006–11) Doctorate
Physics and other Science Education	11	7	–	2(8)	–	–
Non-specialized	–	13	–	6(35)	–	–
Other	2	7	–	4(11)	–	–
Total	13	27	–	11(54)	–	–
Full-time Grad. Stud.	–	10	–	–	–	–
First-year Grad. Stud.	–	11	–	–	–	–

GRADUATE DEGREE REQUIREMENTS

Master's: Master's degree in physics: A "B" average in 30 semester hours of graduate credits on an advisor-approved program is required. There are no language requirements. 18 hours of program must be taken on campus. An oral examination and a written research report/thesis are required. Master's degree in physics education: The requirements are as specified for the Physics Master's program, except that the program emphasizes courses beneficial to secondary school teachers. Master's degree in physical science: A program for middle school science teachers. A "B" average in 30 semester hours of approved graduate credits. Eighteen hours of the program must be taken on campus. There are no language, oral examination, or thesis requirements.
Thesis: Thesis may be written in absentia.

SPECIAL EQUIPMENT, FACILITIES, OR PROGRAMS

The Department has among its facilities an observatory, workstations, a staffed machine shop, a plasma physics laboratory, a surface science and nano-tribology laboratory, and a modern optics laboratory.

Table B—Separately Budgeted Research Expenditures by Source of Support

Source of Support	Departmental Research	Physics-related Research Outside Department
Federal government	$1,402,000	
State/local government		
Non-profit organizations		
Business and industry	$60,000	
Other	$31,000	
Total	$1,493,000	

Table C—Separately Budgeted Research Expenditures by Research Specialty

Research Specialty	No. of Grants	Expenditures ($)
Atmosphere, Space Physics, Cosmic Rays	4	$431,000
Nano Science and Technology	3	$486,000
Physics and other Science Education	2	$555,000
Total	9	$1,472,000

FACULTY

Professor

Behringer, Ernest, Ph.D., Cornell University, 1994. *Condensed Matter Physics, Optics.*

Carroll, James, Ph.D., West Virginia University, 1997. Plasma physics.

Jacobs, Diane A., Ph.D., University of Texas, Austin, 1984. *Condensed Matter Physics, Plasma and Fusion.*

Oakes, Alexandria, Ph.D., Lehigh University, 1986. *Acoustics, Mechanics.*

Sharma, Natthi, Ph.D., Ohio University, 1982. *Atomic, Molecular, & Optical Physics, Condensed Matter Physics.*

Sheerin, James P., Ph.D., University of Michigan, 1980. *Atmosphere, Space Physics, Cosmic Rays, Nonlinear Dynamics and Complex Systems, Plasma and Fusion.*

Shen, Weidian, Ph.D., Wayne State University, 1988. *Condensed Matter Physics, Nano Science and Technology.*

Thomsen, Marshall, Ph.D., Michigan State University, 1984. *Condensed Matter Physics, Theoretical Physics.*

Wylo, Bonnie, Other, University of Michigan, 1993. Ed.D. *Physics and other Science Education.*

Associate Professor

Koehn, Patrick, Ph.D., University of Michigan, 2002. Space physics.

Assistant Professor

Kubitskey, Beth, Ph.D., University of Michigan, 2006. *Physics and other Science Education.* Teacher education.

Pawlowski, David J., Ph.D., University of Michigan, 2009. *Atmosphere, Space Physics, Cosmic Rays.*

DEPARTMENTAL RESEARCH SPECIALTIES AND STAFF

Theoretical

Acoustics. Oakes.
Atmosphere, Space Physics, Cosmic Rays. Koehn, Pawlowski, Sheerin.
Computational Physics. Pawlowski, Sheerin, Thomsen.
Condensed Matter Physics. Sharma, Thomsen.
Electromagnetism. Sharma.
Mechanics. Oakes.
Optics. Sharma.
Plasma and Fusion. Koehn, Pawlowski, Sheerin.

Experimental

Atmosphere, Space Physics, Cosmic Rays. Koehn, Sheerin.
Condensed Matter Physics. Behringer, Jacobs, Shen, Thomsen.
Fluids, Rheology. Jacobs.
Nonlinear Dynamics and Complex Systems. Jacobs, Sheerin.
Optics. Behringer, Sharma, Thomsen.
Physics and other Science Education. Carroll, Kubitskey, Wylo.
Plasma and Fusion. Carroll, Jacobs, Sheerin.
Surface Physics. Nano-tribology. Shen.

View additional information about this department at
www.gradschoolshopper.com

MICHIGAN STATE UNIVERSITY

DEPARTMENT OF PHYSICS AND ASTRONOMY

East Lansing, Michigan 48824-2320
http://www.pa.msu.edu

General University Information
President: LouAnna K. Simon
Dean of Graduate School: Karen Klomparens
University website: http://www.msu.edu
Control: Public
Setting: Suburban
Total Faculty: 4,950
Total Graduate Faculty: Not separated
Total number of Students: 48,906
Total number of Graduate Students: 10,247

Department Information
Department Chairman: Phillip Duxbury, Chair
Department Contact: Scott Pratt, Graduate Program Director
Total full-time faculty: 71
Total number of full-time equivalent positions: 51
Full-Time Graduate Students: 146
First-Year Graduate Students: 31
Female First-Year Students: 5
Total Post Doctorates: 48

Department Address
567 Wilson Road
1312 BPS Building
East Lansing, MI 48824-2320
Phone: (517) 884-5534
Fax: (517) 432-6191
E-mail: pratts@pa.msu.edu
Website: http://www.pa.msu.edu

ADMISSIONS

Admission Contact Information
Address admission inquiries to: Michigan State University, Department of Physics & Astronomy, Attn: Graduate Secretary, 567 Wilson Road, 1312 BPS Bldg., East Lansing, MI 48824
Phone: (517) 884-5532
E-mail: barratt@pa.msu.edu
Admissions website: http://www.pa.msu.edu

Application deadlines
Fall admission:
U.S. students: January 1   *Int'l. students*: January 1
Spring admission:
U.S. students: November 30   *Int'l. students*: November 30

Application fee
U.S. students: $50   *Int'l. students*: $50

Admissions information
For Fall of 2013:
Number of applicants: 315
Number admitted: 70
Number enrolled: 23

Admission requirements
Bachelor's degree requirements: Bachelor's degree in physics or astronomy is required.
Minimum undergraduate GPA: 3.0

GRE requirements
The GRE is required.

Advanced GRE requirements
The Advanced GRE is required.
Minimum accepted Advanced GRE score: 600

Mean Advanced GRE score range (25th–75th percentile): 1250-1450

TOEFL requirements

The TOEFL exam is required for students from non-English-speaking countries.
PBT score: 570
iBT score: 80
No subscore below 19 for reading, listening, and speaking; no subscore below 22 for writing.

Other admissions information

Additional requirements: Applicants with degrees in other fields will be considered and admitted on a provisional basis. The GRE general test and physics test are both required. The average GRE scores for 2011–12 admissions were verbal—590; quantitative—760; analytical—4-6. The average GRE Advanced score for 2011–12 admissions was 730.
Undergraduate preparation assumed: Bauer and Westfall, University Physics; Marion, Classical Dynamics; Pollack and Stump, Electromagnetism; Griffith, Introduction to Quantum Mechanics.

TUITION

Tuition year 2013-14:
Tuition for in-state residents
Full-time students: $621.25 per credit
Part-time students: $621.25 per credit
Tuition for out-of-state residents
Full-time students: $1,220.25 per credit
Part-time students: $1,220.25 per credit
Graduate assistants receive a full waiver of tuition, and all registration fees and are provided with health insurance.
Credit hours per semester to be considered full-time: 6
Health insurance: Available
Academic term: Semester
Number of first-year students who received full tuition waivers: 23

Teaching Assistants, Research Assistants, and Fellowships

Number of first-year
Teaching Assistants: 14
Research Assistants: 2
Fellowship students: 7
Average stipend per academic year
Teaching Assistant: $23,700
Research Assistant: $26,000
Fellowship student: $27,000

FINANCIAL AID

Application deadlines

Fall admission:
U.S. students: January 1
Spring admission:
U.S. students: November 30   *Int'l. students*: November 30

Loans

Loans are available for U.S. students.
Loans are available for international students.
GAPSFAS application required: Yes
FAFSA application required: Yes

For further information

Address financial aid inquiries to: Graduate Program Director, Michigan State University, Dept. of Physics & Astronomy, 567 Wilson Rd-1312 BPS Bldg., East Lansing, MI 48824-2320.
Phone: (517) 884-5532
E-mail: pratts@pa.msu.edu

HOUSING

Availability of on-campus housing

Single students: Yes
Married students: Yes

For further information

Address housing inquiries to: University Apartments, Michigan State University, East Lansing, MI 48824.
Phone: (517) 355-9550
Housing aid website: http://www.liveon.msu.edu

Table A—Faculty, Enrollments, and Degrees Granted

Research Specialty	2012-13 Faculty	Enrollment Fall 2012		Number of Degrees Granted 2012-13		
		Master's	Doctorate	Master's	Terminal Master's	Doctorate
Acoustics	1	–	1	–	–	–
Astronomy	8	–	13	1(-)	–	4(-)
Atomic, Molecular, & Optical Physics	–	–	–	–	–	–
Biophysics	2	–	8	1(-)	1(-)	–
Chemical Physics	–	–	5	–	–	1(-)
Condensed Matter Physics	15	–	34	1(-)	–	2(-)
Materials Science, Metallurgy	–	–	7	–	–	–
Nuclear Physics	23	–	50	1(-)	1(-)	9(-)
Particles and Fields	15	–	15	1(-)	–	–
Physics and other Science Education	2	–	2	1(-)	–	–
Physics of Beams	6	–	11	–	–	1(-)
Total	71	–	146	6(-)	2(-)	17(-)
Full-time Grad. Stud.	–	–	146	–	–	–
First-year Grad. Stud.	–	–	31	–	–	–

GRADUATE DEGREE REQUIREMENTS

Master's: Thirty semester credit hours with a 3.0 minimum GPA are required. Six credits earned on campus is the minimum University residence requirement. There is no foreign language requirement. There is a thesis option with between 4 and 10 credits for thesis research (PHY 899) counted toward the degree.

Doctorate: Physics doctorate: There are 8 specific course requirements and a 3.0 GPA must be maintained. Two consecutive semesters is the minimum residency requirement. Qualifying and candidacy examination must be passed at the Ph.D. level. A thesis and thesis oral examination are required. Astronomy and astrophysics doctorate: Must take 8 specific core courses with minimum 3.375 GPA, complete a 2-semester second-year research project, pass qualifying and candidacy examinations, and complete a Ph.D. thesis and thesis oral examination. A minimum of 24 credits of thesis research (PHY 999) is required.

Other Degrees: Degrees are also available in accelerator physics, atomic and molecular physics, chemical physics, biophysics, beam physics, and mathematical physics.

SPECIAL EQUIPMENT, FACILITIES, OR PROGRAMS

Extensive research facilities in solid-state and low-temperature physics, astronomy and astrophysics, nuclear physics, and high-energy and particle physics. These facilities include: two superconducting heavy-ion cyclotrons, K500 and K1200, with associated apparatus; a modern A1900 fragment separator allowing efficient production and inflight separation of rare isotopes; a high-resolution S800 superconducting magnetic spectrograph; high-resolution and high-efficiency gamma-ray, beta-ray, neutron, and charged particle arrays; 9-T ion trap; rf-fragment separators and laser spectroscopy system; x-ray diffraction apparatus employing x-ray cameras; a 12 kW rotating anode x-ray source; class 100 clean room; UV mask/aligner; scanning electron microscope; multi-pocket e-beam evaporator; photolithographic and electron-beam lithographic facilities for device fabrication with 50 nm resolution; cryogenic facilities, including three helium-3 refrigerators and two helium-3/helium-4 mixing refrigerators; two (55-kG and 105-kG) superconducting magnets; four electromagnets; AC and DC automated SQUID magnetometer; an ultra-high-vacuum four-gun sputtering system; a 32-cubic-meter anechoic room, 174-cubic-meter reverberation room, and 32-cubic-meter double-walled quiet enclosure; a liquid argon calorimeter test station for detector development in neutrino measurements, collider experiments and future high-energy accelerators; a high-energy physics laboratory, which is a state-of-the-art electronics design facility where detectors for experiments are developed, tested, and constructed a major grid-based computing facility based on the LHC Computing Grid, supporting research on the ATLAS experiment at the LHC; and numerous minicomputers and microcomputers in all research areas.

Michigan State University is one of the primary partners (along with Notre Dame and the University of Chicago) in the Joint Institute for Nuclear Astrophysics (JINA), which is a $10M program funded as "Physics Frontier Center" focusing on the intersection of nuclear physics and astrophysics by the National Science Foundation. It addresses open questions related to the origin of the elements, compact stellar objects, stellar evolution, and stellar explosions through nuclear physics experiments, astronomical observations, scientific computing, and theory. MSU is a partner in the new SOAR 4.1m telescope, which is located on a superb observing site in Chile and is operated remotely from East Lansing. SOAR is optimized for very high image quality and offers a wide range of optical and near-infrared imagers and spectrographs.

Table B—Separately Budgeted Research Expenditures by Source of Support

Source of Support	Departmental Research	Physics-related Research Outside Department
Federal government	$58,869,050	
State/local government		
Non-profit organizations	$45,126	
Business and industry	$188,418	
Other	$1,914,804	
Total	**$61,017,398**	

Table C—Separately Budgeted Research Expenditures by Research Specialty

Research Specialty	No. of Grants	Expenditures ($)
Acoustics	1	$207,935
Astronomy	34	$847,163
Astrophysics	3	$97,465
Condensed Matter Physics	46	$4,200,535
Nuclear Physics	38	$51,746,740
Particles and Fields	31	$3,798,277
Physics and other Science Education	3	$119,285
Total	**156**	**$61,017,400**

FACULTY

Professor

Bauer, Wolfgang, Ph.D., University of Giessen, 1987. University Distinguished Professor. *Computational Physics, Nuclear Physics*. Theoretical physics; intermediate heavy-ion reactions; nuclear transport theory; computational physics; nuclear astrophysics.

Berz, Martin, Ph.D., University of Giessen, 1986. *Computational Physics, Physics of Beams*. Beam physics; mathematical physics; computational physics.

Birge, Norman O., Ph.D., University of Chicago, 1986. Experimental condensed-matter physics.

Bollen, Georg, Ph.D., University of Kaiserslautern, 1989. *Atomic, Molecular, & Optical Physics, Nuclear Physics*. Nuclear physics; atomic physics; molecular physics.

Brock, Raymond L., Ph.D., Carnegie Mellon University, 1980. University Distinguished Professor. *Particles and Fields*. Experimental high-energy physics; gluon resummation phenomenology and applications.

Bromberg, Carl M., Ph.D., University of Rochester, 1974. *Particles and Fields*. Experimental high-energy physics; neutrino oscillations; particle detectors.

Brown, B. Alex, Ph.D., Stony Brook University, 1974. *Nuclear Physics*. Theoretical nuclear physics; shell model theory.

Chivukula, R. Sekhar, Ph.D., Harvard University, 1987. Associate Dean to the College of Natural Science. *Particles and Fields*. Theoretical high-energy physics.

Danielewicz, Pawel, Ph.D., Warsaw University, 1981. *Nuclear Physics*. Theoretical nuclear physics; relativistic heavy-ion physics.

Donahue, Megan, Ph.D., University of Colorado, 1990. Space and ground-based observational studies of galaxies and galaxy clusters.

Duxbury, Phillip M., Ph.D., University of New South Wales, 1983. Chairperson of the Department. *Condensed Matter Physics, Statistical & Thermal Physics*. Theoretical condensed-matter physics; statistical mechanics.

Dykman, Mark, Ph.D., University of Kiev, 1973. *Condensed Matter Physics*. Theoretical condensed-matter physics.

Gade, Alexandra, Ph.D., University of Cologne, 2001. *Nuclear Physics*. Experimental nuclear physics.

Gelbke, C. Konrad, Ph.D., Heidelberg University, 1973. University Distinguished Professor; Director of the National Superconducting Cyclotron Laboratory and the Facility for Rare Isotope Beams Laboratory. *Nuclear Physics*. Experimental nuclear physics; nuclear reactions of heavy ions.

Glasmacher, Thomas, Ph.D., Florida State University, 1992. University Distinguished Professor; Project Manager of the Facility for Rare Isotope Beams Laboratory. *Nuclear Physics*. Experimental nuclear physics.

Golding, Brage, Ph.D., Massachusetts Institute of Technology, 1966. *Condensed Matter Physics*. Experimental condensed-matter physics; mesoscopic physics; fluctuation phenomena.

Hartmann, William M., Ph.D., University of Oxford, 1965. *Acoustics, Condensed Matter Physics*. Theoretical and experimental condensed-matter physics.

Hjorth-Jensen, Morten, Ph.D., University of Oslo, 1993. Theoretical Nuclear Physics.

Huston, Joey W., Ph.D., University of Rochester, 1982. *Particles and Fields*. Experimental high-energy physics; direct photon production; experimental tests of QCD. Studies of the Higgs boson and electroweak symmetry breaking.

Leitner, Daniela, Ph.D., University of Vienna, 1995. NSCL Associate Director for Operations. *Nuclear Physics*. Accelerator physics.

Linnemann, James T., Ph.D., Cornell University, 1978. *Astrophysics, Particles and Fields*. Experimental high-energy physics; searches for physics beyond the Standard Model of particle physics; particle astrophysics; statistical techniques.

Loh, Edwin D., Ph.D., Princeton University, 1977. Astrophysics.

Lynch, William G., Ph.D., University of Washington, 1980. *Nuclear Physics*. Experimental nuclear physics.

Mittig, Wolfgang, Ph.D., Paris, 1971. Hannah Professor.

Nunes, Filomena, Ph.D., University of Surrey, 1995. NSCL Theoretical Nuclear Science Department Head. *Nuclear Physics*. Theoretical nuclear physics; few-body structures and reactions.

Piermarocchi, Carlo, Ph.D., Ecole Polytechnique Federale de Lausanne, Switzerland, 1998. *Condensed Matter Physics*. Theoretical condensed-matter physics.

Pope, Bernard G., Ph.D., Columbia University, 1971. *Particles and Fields*. Experimental high-energy physics; tests of quantum chromodynamics; Hadron collider physics.

Pratt, Scott, Ph.D., University of Minnesota, 1985. Associate Chairperson; Graduate Program Director. *Nuclear Physics*. Theoretical nuclear physics.

Pumplin, Jon, Ph.D., University of Michigan, 1968. Theoretical high-energy physics; strong interaction theory.

Repko, Wayne W., Ph.D., Wayne State University, 1967. *Particles and Fields*. Theoretical high-energy physics; quantum electrodynamics; strong interaction theory.

Saito, Kenji, Ph.D., Tohoku University, 1983. *Nuclear Physics*. Accelerator physics.

Schatz, Hendrik, Ph.D., Heidelberg University, 1997. *Astrophysics, Nuclear Physics*. Experimental and theoretical nuclear astrophysics.

Sherrill, Bradley, Ph.D., Michigan State University, 1985. University Distinguished Professor. NSCL Chief Scientist. *Nuclear Physics*. Experimental nuclear physics.

Simmons, Elizabeth, Ph.D., Harvard University, 1990. Dean of Lyman Briggs College. *Particles and Fields*. Theoretical high-energy physics.

Stump, Daniel R., Ph.D., Massachusetts Institute of Technology, 1976. *Particles and Fields*. Theoretical physics; strong interaction theory; quantum chromodynamics.

Syphers, Michael, Ph.D., University of Illinois at Chicago, 1987. *Nuclear Physics*. Accelerator physics.

Thoennessen, Michael, Ph.D., Stony Brook University, 1988. *Nuclear Physics*. Experimental nuclear physics.

Tomanek, David, Ph.D., Freie University, Berlin, 1983. *Computational Physics, Condensed Matter Physics*. Theoretical condensed-matter physics; computational physics.

Voit, G. Mark, Ph.D., University of Colorado, 1990. Associate Dean for Undergraduate Studies. *Astronomy, Astrophysics*. Theoretical astrophysics; evolution of galaxy clusters; AGN.

Wei, Jie, Ph.D., Stony Brook University, 1989. *Nuclear Physics*. Accelerator physics.

Westfall, Gary, Ph.D., University of Texas, Austin, 1975. University Distinguished Professor. *Nuclear Physics*. Experimental heavy-ion physics; reaction mechanisms.

Yamazaki, Yoshishige, Ph.D., University of Tokyo, 1974. *Nuclear Physics*. Accelerator physics.

Yuan, Chien-Peng, Ph.D., University of Michigan, 1988. Theoretical high-energy physics.

Zegers, Remco G. T., Ph.D., University of Groningen, The Netherlands, 1999. NSCL Associate Director for Experimental Nuclear Science. *Nuclear Physics*. Experimental nuclear physics.

Zelevinsky, Vladimir, Ph.D., Budker Institute of Nuclear Physics, 1974. *Nuclear Physics*. Theoretical nuclear physics.

Zepf, Stephen E., Ph.D., Johns Hopkins University, 1992. Associate Chair and Astronomy Program Director. *Astronomy, Astrophysics*. Observational and theoretical studies of evolution of galaxies.

Associate Professor

Bogner, Scott, Ph.D., Stony Brook University, 2002. *Nuclear Physics*. Theoretical nuclear physics.

Brown, Edward, Ph.D., University of California, Berkeley, 1999. *Astrophysics*. Theoretical astrophysics; compact objects; nuclear astrophysics.

Conway, Zachary, Ph.D., University of Illinois-Urbana Champaign, 2007. *Accelerator*.

Kortemeyer, Gerd, Ph.D., Michigan State University, 1997. *Physics and other Science Education*. Physics education.

Lapidus, Lisa, Ph.D., Harvard University, 1998. *Biophysics*. Experimental biophysics.

Naviliat-Cuncic, Oscar, Ph.D., Catholic University of Louvain, 1989. *Nuclear Physics*. Experimental nuclear physics.

Ruan, Chong-Yu, Ph.D., University of Texas, Austin, 2000. *Condensed Matter Physics, Optics*. Experimental condensed-matter physics; ultra-fast electron diffraction.

Schmidt, Carl., Ph.D., Harvard University, 1990. *Particles and Fields*. Theoretical high-energy physics.

Schwienhorst, Reinhardt, Ph.D., University of Minnesota, 2000. *Particles and Fields*. Experimental high-energy physics; single-top quark production; searches for new physics beyond the standard model of particle physics.

Tessmer, Stuart, Ph.D., University of Illinois at Urbana-Champaign, 1995. Associate Chairperson and Undergraduate Program Director. *Condensed Matter Physics*. Experimental condensed-matter physics.

Tollefson, Kirsten, Ph.D., University of Rochester, 1997. *Astrophysics, Particles and Fields*. Experimental high-energy physics; searches for new physics beyond the standard model at hadron colliders; particle astrophysics.

Assistant Professor

Caballero, Marcos D., Ph.D., Georgia Institute of Technology, 2011. *Physics and other Science Education*.

Chomiuk, Laura, Ph.D., University of Wisconsin-Madison, 2010. *Astronomy, Astrophysics*. Studies of Novae and supernovae.

Comstock, Matthew, Ph.D., University of California, Berkeley, 2008. Endowed Cowen Chair. *Biophysics, Condensed Matter Physics*. Single-molecule optical trapping and fluorescence.

Fisher, Wade C., Ph.D., Princeton University, 2004. *Particles and Fields*. Experimental high-energy physics; Studies of the Higgs boson and electroweak breaking; precision measurements of electroweak properties; statistical methods.

Iwasaki, Hironori, Ph.D., University of Tokyo, 2001. *Nuclear Physics*. Experimental nuclear physics.

Ke, Xianglin, Ph.D., University of Wisconsin-Madison, 2006. Cowen Endowed Chair. *Condensed Matter Physics*. Experimental condensed-matter physics.

Lai, Chih-Wei, Ph.D., University of California, Berkeley, 2004. Endowed Cowen Chair. *Condensed Matter Physics*. Experimental condensed-matter physics.

Levchenko, Oleksandr, Ph.D., University of Minnesota, 2009. *Condensed Matter Physics*.

McGuire, John A., Ph.D., University of California, Berkeley, 2004. *Condensed Matter Physics*. Experimental condensed-matter physics.

O'Shea, Brian, Ph.D., University of Illinois at Urbana-Champaign, 2005. *Astronomy, Astrophysics*. Computational astrophysics and astronomy.

Spyrou, Artemisia, Ph.D., University of Athens, 2007. *Astrophysics, Nuclear Physics*. Experimental nuclear physics; nuclear astrophysics.

Strader, Jay, Ph.D., University of California, Santa Cruz, 2007. *Astronomy*. Star clusters; stellar populations; galaxy formation; black holes.

Wrede, Christopher, Ph.D., Yale University, 2008. *Astrophysics, Nuclear Physics*. Nuclear physics; experimental nuclear astrophysics.

Zhang, Peng Peng, Ph.D., University of Wisconsin-Madison, 2006. *Condensed Matter Physics*. Experimental condensed-matter physics.

Professor Emeritus

Abolins, Maris A., Ph.D., University of California, San Diego, 1965. *Particles and Fields*.

Austin, Sam M., Ph.D., University of Wisconsin-Madison, 1960. University Distinguished Professor. Nuclear physics; nuclear astrophysics.

Baldwin, Jack A., Ph.D., University of California, Santa Cruz, 1974. *Astronomy, Atmosphere, Space Physics, Cosmic Rays*. Observational studies of quasars, planetary nebulae, and HII regions.

Bass, Jack, Ph.D., University of Illinois, 1964. *Condensed Matter Physics*.

Beers, Timothy C., Ph.D., Harvard University, 1983. University Distinguished Professor. *Astrophysics*. Observational studies of first- and second-generation stars; stellar cosmo-chronometry; galactic structure, kinematics, and dynamics; nuclear astrophysics.

Benenson, Walter, Ph.D., University of Wisconsin-Madison, 1962. University Distinguished Professor. *Nuclear Physics*.

Carlson, Edward H., Ph.D., Johns Hopkins University, 1959.

Galonsky, Aaron I., Ph.D., University of Wisconsin-Madison, 1954. *Nuclear Physics*.

Harrison, Michael, Ph.D., University of Chicago, 1960. *Condensed Matter Physics*.

Kaplan, Thomas A., Ph.D., University of Pennsylvania, 1954. *Condensed Matter Physics*.

Kashy, Edwin, Ph.D., Rice University, 1959. *Nuclear Physics*.

Kovacs, Julius S., Ph.D., Indiana University, 1955.

Mahanti, S. D., Ph.D., University of California, Riverside, 1968. *Condensed Matter Physics*. Theoretical condensed-matter physics.

McManus, Hugh, Ph.D., University of Birmingham, 1947.

Pollack, Gerald, Ph.D., California Institute of Technology, 1962. *Condensed Matter Physics*.

Pratt, William P., Ph.D., University of Minnesota, 1967. *Condensed Matter Physics*.

Schriber, Stan, Ph.D., McMaster University, 1966. *Nuclear Physics*.

Schroeder, Peter A., Ph.D., University of Bristol, 1955. *Condensed Matter Physics*.

Signell, Peter S., Ph.D., University of Rochester, 1958.

Smith, Horace A., Ph.D., Yale University, 1980. *Astronomy*. Chemical evolution of galaxies; pulsating variable stars.

Stein, Robert, Ph.D., Columbia University, 1966. *Astronomy*.

Thorpe, Michael F., University of Oxford, 1968. *Condensed Matter Physics*.

Weerts, Hendrick J., Ph.D., RWTH Aachen, 1981.

Research Professor

Ghosh, Ruby, Ph.D., Cornell University, 1991. *Condensed Matter Physics*. Solid-state and optical device physics.

Hauser, Reiner, Ph.D., Heidelberg University, 1994. *Particles and Fields*. Experimental high-energy physics.

Makino, Kyoko, Ph.D., Michigan State University, 1998. *Physics of Beams*. Beam physics; computational physics.

Adjunct Professor

Adami, Christoph, Ph.D., Stony Brook University, 1991. Theoretical physics.

Auerbach, Naftali, Ph.D., Weizmann Institute of Science, 1966.

Balasubramaniam, Shanker, Ph.D., Pennsylvania State University, 1993. Engineering science; mechanics.

Berger, Edmond, Ph.D., Princeton University, 1965.

Billinge, Simon, Ph.D., University of Pennsylvania, 1992. Materials science.

Chomiuk, Laura, Ph.D., University of Wisconsin-Madison, 2010. Jansky Fellow. *Astronomy, Astrophysics*. Astronomy; studies of novae and supernovae.

Dantus, Marcos, Ph.D., California Institute of Technology, 1991. Chemistry.

Doleans, Marc, Ph.D., Paris, 2003. *Nuclear Physics, Plasma and Fusion*. Accelerator physics.

Duguet, Thomas, Ph.D., Paris, 2002.

Fert, Albert, Ph.D., University of Paris-Sud at Orsay, 1970.

Fisher, Galen, Ph.D., Stanford University, 1966.

Gruebele, Martin, Ph.D., University of California, Berkeley, 1988. Chemistry.

Hartung, Walter, Ph.D., Cornell University, 1996. Nuclear physics.

Hussein, Mahir, Ph.D., Massachusetts Institute of Technology, 1971.

Izrailev, Felix, Ph.D., Budker Institute of Nuclear Physics, Russia, 1991.

Janssens, Robert, Ph.D., University of Catholique de Louvian-Belgium, 1978. *Nuclear Physics*. Nuclear physics.

Johnson, Ronald, Ph.D., University of Manchester, 1961. Theoretical physics.

Johnstone, Carol, Ph.D., University of Texas, Austin, 1984. *Nuclear Physics*. Experimental nuclear physics.

Kester, Oliver, Ph.D., University of Frankfurt, 1996. *Plasma and Fusion*. Accelerator physics.

Kuhn, Leslie, Ph.D., University of Pennsylvania, 1989. *Biophysics*. Biophysics.

Lambert, David, Ph.D., University of California, Berkeley, 1979. *Condensed Matter Physics*. Condensed-matter physics.

Langanke, Karlheinz, Ph.D., University of Muenster, Germany, 1985.

Mackay, Michael, Ph.D., University of Illinois at Urbana-Champaign, 1985. Chemical engineering.

Marti, Felix, Ph.D., Michigan State University, 1977. *Nuclear Physics*. Nuclear physics.

Morelli, Donald, Ph.D., University of Michigan, 1985. Chemical engineering; material science.

Ormand, William E., Ph.D., Michigan State University, 1985. *Nuclear Physics*. Nuclear physics.

Otsuka, Takaharu, Ph.D., University of Tokyo, 1979.

Platzman, P., Ph.D., California Institute of Technology, 1960.

Sinnis, Constantine, Ph.D., University of Hawaii, 1990.

Starosta, Krzysztof, Ph.D., University of Warsaw, 1996.

Tostevin, Jeffrey, Ph.D., University of Surrey, 1978.

Tsang, Manyee Betty, Ph.D., University of Washington, Seattle, 1980. Nuclear chemistry.

Wangler, Thomas, Ph.D., University of Wisconsin-Madison, 1964. Accelerator physics.

Wiescher, Michael, Ph.D., Universität Münster, 1980. *Nuclear Physics*. Nuclear physics.

DEPARTMENTAL RESEARCH SPECIALTIES AND STAFF

Theoretical

Astronomy. Theoretical and computational studies of structure formation, formation of the first stars, galactic chemical evolution, evolution of clusters of galaxies, and globular cluster formation and evolution; theoretical nuclear astrophysics: physics of compact objects and supervova explosions; petascale computing. There are three research associates and four graduate students. Edward Brown, O'Shea, Voit.

Atomic, Molecular, & Optical Physics. Nonlinear and quantum optics; coupled matter-light systems; ion-trap physics. There is one graduate student. Piermarocchi.

Biophysics. Control theory and optimization methods applied to systems biology; theoretical models of gene regulatory networks; information theory applied to biological systems; development of new statistical methods for the analysis of high-throughput experiments in cellular systems; combinatorial optimization in drug discovery. Duxbury, Piermarocchi.

Condensed Matter Physics. Electronic properties of surfaces, clusters, nanostructures, nanowires, nanotubes, and novel narrow band gap semiconductors, and exciton and polariton condensates; magnetism in strongly correlated systems; optical properties of quantum dots, quantum wires, and quantum wells; transport in correlated two-dimensional electron systems; electrons on helium; many-electron tunneling; activated processes in nonequilibrium systems; quantum computing using condensed-matter systems; atomic structure of complex materials, nanoparticles, and nanoparticle dispersion and their applications to energy and nanomedicine; photophysics; combinatorial optimization methods in statistical physics with applications to complex materials; multiscale modeling of ultrafast quantum and classical dynamics in correlated ilectron materials, and in materials under extreme conditions. There are three research associates and 15 graduate students. Duxbury, Dykman, Levchenko, Piermarocchi, Tomanek.

Nuclear Physics. Quantum many-body theory for nuclear structure, nuclear reactions, nuclear astrophysics, weak interactions physics, and properties of mesoscopic systems; nuclear structure including applications of ab initio, configuration interaction, collective model, and energy-density functional methods; nuclear reactions spanning energies from low- to ultra-relativistic and including quantum transport and coupled-channel methods. Applications are made to structure and reactions of rare isotopes studies at the NSCL and other facilities. For nuclear astrophysics, our work relates to the properties of neutron stars, stellar evolution, and nucleosynthesis. Applications for fundamental symmetries involve nuclear structure aspects of parity violations, time-reversal violation, and Standard Model tests from nuclear observables. At present there are 7 postdoctoral research associates, one FRIB Theory Fellow and 12 graduate students, along with several visiting faculty members. Bauer, Bogner, B. Brown, Danielewicz, Nunes, Zelevinsky.

Particles and Fields. Fundamental structure and interaction of elementary particles; global analysis of high-energy processes to determine the parton (quarks and gluons) structure of hadrons; stringent tests of quantum chromodynamics; origin of spontaneous symmetry breaking in the unified theory of electro-weak interactions; physics of the Higgs particle and the top quark; new phenomena at super high energies; quantum electrodynamics and quantum chromodynamics of bound systems; Monte Carlo methods in high-energy physics; astro-particle physics and constraints of dark energy. There are two postdoctoral research associates and three graduate students. Chivukula, Pumplin, Repko, Schmidt, Simmons, Stump, Yuan.

Physics of Beams. Nonlinear dynamics; beam physics; computational physics; self-validated numerical methods; non-Archimedian analysis; VU Beam program. There are three local graduate students and six remote graduate students in the VU Beam program. Berz, Makino.

Experimental

Accelerator. Design, construction and operation of superconducting linear accelerator and cyclotrons for heavy ions and isotopes; cutting-edge beam dynamics for accelerators at high beam-power frontier; ECR ion sources and radio-frequency quadrupole linac; high-gradient superconducting radio-frequency accelerating structures and technology; cryogenic system for large-scale superconducting RF cavities and magnets; mega-scale accelerator controls, beam diagnostics and instrumentations. One of the major world centers in low-beta superconducting radio-frequency physics and technology and in high-power hadron accelerators. There are 3 research associates and 5 graduate students. Leitner, Marti, Saito, Syphers, Wei, Yamazaki.

Acoustics. Signal processing by the human brain: physiologically based models and psychoacoustical experiments with an emphasis on binaural processing and sound localization in free field and complex room environments. There is one graduate student. Hartmann.

Astrophysics and Astronomy. Supernova remnants; the evolution of galaxies and galaxy clusters including globular clusters and globular cluster systems, quasars and active galaxies, and galaxies and hot gas in galaxy clusters; structure and evolution of the Milky Way Galaxy, including observational studies of first-and second-generation stars, and the dynamics of old stellar populations; observational cosmology; astronomical instrumentation. There are five research associates and six graduate students. Baldwin, Laura Chomiuk, Donahue, Loh, Strader, Zepf.

Atomic, Molecular, & Optical Physics. Traps and precision measurements and their applications to nuclear physics. There are 16 research associates, 30 graduate students, and many short-term visitors. Bollen.

Biophysics. Fast-folding studies of proteins and nucleic acids using optical spectroscopy and microfluidics; measurement and modeling of unfolded protein dynamics; single molecule optical trapping and fluorescence spectroscopy of DNA processing moleculer motors; scanning probe methods to study bioelectronic systems. There are seven graduate students. Comstock, Lapidus, Tessmer.

Condensed Matter Physics. Quantum phenomena in solids, including quantum transport, optical control of electron and nuclear spins, semiconductor quantum optics and electronic control of magnetization in ferromagnets; nanosciences and technologies, including many body effects in semiconductor nanostructures, ultrafast switching of photonic and electronic materials, scanning probe microscopy, single-electron capacitance spectroscopy of semiconductor nanostructures, and quantum Hall effects, studies of electronic and photovoltaic nanomaterials using scanning probe microscopy and manipulation of these materials via surface and interface engineering and characterization of devices made out of these materials, quantum dots: enhanced charge-charge and charge-spin interactions, transport in metals and alloys, including magnetic multilayers and the interplay between superconductivity and magnetism; ultrafast science and nonlinear optics, including surface nonlinear optics, ultrafast phenomena in solids, liquids and nanostructures, material transformations far from equilibrium, and femtosecond electron imaging and spectroscopy; device physics and fabrication, including diamond growth, dye-sensitized solar cells, charge transport and carrier enhancements, wide-band gap semiconductors at high temperatures, and chemical sensing techniques using fiber optics and solid-state devices. There are three research associates and 23 graduate students. Birge, Ghosh, Golding, Ke, Lai, McGuire, Ruan, Tessmer, Zhang.

Nuclear Physics. Test of nuclear models and studies of fundamental interactions; ion trapping for high precision mass measurements; study of rare isotopes with a large excess of neutrons and protons; search for new isotopes and exotic decay modes; exploration of structural changes of rare isotopes and resonance properties of nuclei; isospin dependence of nuclear reactions; nuclear astrophysics: stellar evolution, nova and supernova explosions, galactic chemical evolution, properties of neutron stars, and properties of the quark gluon plasma. There are 16 research associates, 30 graduate students, and many short-term visitors. Bollen, Gade, Gelbke, Glasmacher, Iwasaki, Lynch, Schatz, Sherrill, Spyrou, Thoennessen, Tsang, Westfall, Wrede, Zegers.

Particles and Fields. Hadron-induced high-energy and super-high-energy interactions; the inclusive production and study of jets and other hadronic final states; the inclusive production of leptons and photons; experimental tests of quantum chromodynamics; measurement of the quark and gluon structure of hadrons; studies of heavy flavor production (including studies of the top quark); experimental tests of electro-weak interactions including the production and study of top quarks; searches for supersymmetric particles and other new phenom-

ena beyond the Standard Model; neutrino properties, including interaction cross-sections and oscillation properties (mass hierarchy, mixing angles, search for CP violation); development of the liquid Argon time projection chambers for neutrino and proton-decay physics; particle astrophysics including cosmic ray physics; design principles for fast digitization, pipelined processing techniques, and computer software for high-energy event selection algorithms; design and research and development for experiments to be performed at the Fermi National Accelerator Laboratory (CDF, D0, and NOVA experiments) and the CERN Large Hadron Collider (ATLAS experiment). There are six research associates and 12 graduate students. Brock, Bromberg, Wade Fisher, Hauser, Huston, Linnemann, Pope, Schwienhorst, Tollefson.

View additional information about this department at www.gradschoolshopper.com

---

MICHIGAN TECHNOLOGICAL UNIVERSITY

DEPARTMENT OF PHYSICS

Houghton, Michigan 49931
http://www.phy.mtu.edu

General University Information
President: Glenn D. Mroz
Dean of Graduate School: Jacqueline E. Huntoon
University website: http://www.mtu.edu
Control: Public
Setting: Rural
Total Faculty: 485
Total Graduate Faculty: 609
Total number of Students: 6,945
Total number of Graduate Students: 1,322

Department Information
Department Chairman: Ravindra Pandey, Chair
Department Contact: Andrea S. Lappi, Departmental Coordinator
  Total full-time faculty: 19
  Total number of full-time equivalent positions: 19
  Full-Time Graduate Students: 39
  First-Year Graduate Students: 10
  Female First-Year Students: 4
  Total Post Doctorates: 4

Department Address
1400 Townsend Drive
Houghton, MI 49931
Phone: (906) 487-2086
Fax: (906) 487-2933
E-mail: physics@mtu.edu
Website: http://www.phy.mtu.edu

ADMISSIONS

Admission Contact Information
Address admission inquiries to: Dean of Graduate School.
Admissions website: http://www.mtu.edu/gradschool

Application deadlines
Fall admission:
U.S. students: February 1   *Int'l. students*: February 1

Application fee
There is no application fee required.

Admissions information
For Fall of 2012:
  Number of applicants: 111
  Number admitted: 16
  Number enrolled: 11

Admission requirements
Bachelor's degree requirements: A Bachelor's degree in physics is usually required; however, degree recipients in related areas often apply and are accepted.
Minimum undergraduate GPA: 3.0

GRE requirements
The GRE is required.
  Quantitative score: 720
  Analytical score: 3

Advanced GRE requirements
The Advanced GRE is not required.

TOEFL requirements
The TOEFL exam is required for students from non-English-speaking countries.
  PBT score: 570
  iBT score: 88

IELTS = Band of 6.5 or better.

Other admissions information

Undergraduate preparation assumed: Taylor, Classical Mechanics; Eisberg and Resnick, Quantum Physics; Griffiths, Introduction to Electrodynamics.

TUITION

Tuition year Fall, 2013, estimated:

Tuition for in-state residents

   Full-time students: $7,101 per semester

Tuition for out-of-state residents

   Full-time students: $7,101 per semester

$20 graduate fee for binding of thesis; any laboratory fees required in curriculum; student-voted fee of $113.10/year.

Credit hours per semester to be considered full-time: 9

Deferred tuition plan: No

Health insurance: Available at the cost of $368 per year.

Other academic fees: Variable

Academic term: Semester

Teaching Assistants, Research Assistants, and Fellowships

Number of first-year

   Teaching Assistants: 9

   Research Assistants: 1

Average stipend per academic year

   Teaching Assistant: $13,057

   Research Assistant: $13,057

FINANCIAL AID

Loans

Loans are available for U.S. students.

Loans are not available for international students.

GAPSFAS application required: No

FAFSA application required: Yes

For further information

Address financial aid inquiries to: Graduate Studies Chair, Physics Department/MTU, 1400 Townsend Dr., Houghton, MI 49931-1295.

E-mail: physics@mtu.edu

HOUSING

Availability of on-campus housing

Single students: Yes

Married students: Yes

For further information

Address housing inquiries to: Director of Housing.

Housing aid website: http://www.mtu.edu/housing

Table A—Faculty, Enrollments, and Degrees Granted

Research Specialty	2012-13 Faculty	Enrollment Fall 2012 Master's	Enrollment Fall 2012 Doctorate	Number of Degrees Granted 2012-13 (2008–13) Master's	Number of Degrees Granted 2012-13 (2008–13) Terminal Master's	Number of Degrees Granted 2012-13 (2008–13) Doctorate
Astronomy	1	2	1	–	–(2)	–
Astrophysics	3	–	4	–	1(1)	–
Atmosphere, Space Physics, Cosmic Rays	4	4	9	1(-)	1(2)	–(2)
Atomic, Molecular, & Optical Physics	3	–	2	–	–(3)	–
Biophysics	–	–	–	–	1(-)	–(1)
Condensed Matter Physics	5	–	6	–	–(1)	2(4)
Materials Science, Metallurgy	3	1	9	–	–(3)	2(9)
Total	19	7	31	1(-)	3(12)	4(16)
Full-time Grad. Stud.	–	7	31	–	–	–
First-year Grad. Stud.	–	4	7	–	–	–

GRADUATE DEGREE REQUIREMENTS

Master's: All physics graduate degree requirements include six core courses (15 credits) and two to three elective courses. M.S. degrees have a coursework option, project report option, and a thesis option.

Doctorate: Ph.D. in physics and engineering physics requires 30 course and/or research credits beyond the M.S. degree or 60 course/research credits beyond the Bachelor's degree and passing a written qualifying examination, dissertation proposal, research, and dissertation and final oral examination.

SPECIAL EQUIPMENT, FACILITIES, OR PROGRAMS

The department maintains a machine shop with a full-time machinist. Research laboratories include atmospheric physics, atomic and molecular laser spectroscopy, cloud physics, environmental optics, integrated photonics and materials integration, materials physics synthesis and characterization, nuclear magnetic resonance, quantum optics, and computational physics. For details see http://www.phy.mtu.edu/facilities.html. In addition, the university is well-equipped with numerous user facilities including microfabrication, x-ray diffraction, scanning and transmission electron microscopy, scanning probe microscopy, optical microscopy, focused ion beam system and more.

Table B—Separately Budgeted Research Expenditures by Source of Support

Source of Support	Departmental Research	Physics-related Research Outside Department
Federal government	$1,993,194	
State/local government	$4,907	
Non-profit organizations		
Business and industry		
Other	$158,192	
Total	$2,156,293	

Table C—Separately Budgeted Research Expenditures by Research Specialty

Research Specialty	No. of Grants	Expenditures ($)
Astronomy	1	$32,959
Astrophysics	6	$383,087
Atmosphere, Space Physics, Cosmic Rays	15	$994,202
Atomic, Molecular, & Optical Physics	5	$82,993
Biophysics	3	$82,879
Condensed Matter Physics	4	$143,242
Materials Science, Metallurgy	9	$273,832
Total	43	$1,993,194

FACULTY

Professor

Beck, Donald R., Ph.D., Lehigh University, 1968. Atomic & Molecular Physics. Computational atomic physics.

Borysow, Jacek I., Ph.D., University of Texas, Austin, 1986. Laser spectroscopy.

Fick, Brian E., Ph.D., Virginia Polytechnic Institute and State University (Virginia Tech), 1985. Experimental astro-particle physics.

Jaszczak, John A., Ph.D., Ohio State University, 1989. Computational solid-state physics; nanotechnology education.

Kostinski, Alexander, Ph.D., University of Illinois at Chicago, 1984. Radar meteorology; polarization optics; atmospheric physics.

Levy, Miguel, Ph.D., City University of New York, 1988. Surface physics.

Nemiroff, Robert J., Ph.D., University of Pennsylvania, 1987. Astronomy; astrophysics.

Nitz, David, Ph.D., University of Rochester, 1978. High-energy astrophysics.

Pandey, Ravindra, Ph.D., University of Manitoba, Canada, 1987. Materials theory.

Pati, Ranjit, Ph.D., University at Albany, State University of New York, 1998. *Atomic, Molecular, & Optical Physics, Condensed Matter Physics*. Theoretical condensed-matter physics; materials science.

Perger, Warren F., Ph.D., Colorado State University, 1987. Computational atomic and condensed-matter physics.

Seel, Maximillian J., Ph.D., University of Erlangen, West Germany, 1978. Provost. Theoretical chemistry.

Shaw, Raymond, Ph.D., Pennsylvania State University, 1998. *Atmosphere, Space Physics, Cosmic Rays*. Atmospheric sciences.

Suits, Bryan, Ph.D., University of Illinois, 1981. *Atomic, Molecular, & Optical Physics, Condensed Matter Physics*. Experimental condensed-matter physics; NMR.

Yap, Yoke Khin, Ph.D., Osaka University, 1999. *Condensed Matter Physics, Nano Science and Technology*. Materials science; materials and laser physics.

Associate Professor

Cantrell, Will, Ph.D., University of Alaska, 1999. *Atmosphere, Space Physics, Cosmic Rays*. Atmospheric sciences.

Mazzoleni, Claudio, Ph.D., University of Nevada, 2003. *Atmosphere, Space Physics, Cosmic Rays*. Atmospheric physics.

Moran, Peter D., Ph.D., University of Wisconsin-Madison. *Atomic, Molecular, & Optical Physics, Condensed Matter Physics, Materials Science, Metallurgy*. Materials physics; device fabrication and characterization.

Weidman, Robert S., Ph.D., University of Illinois, 1980. *Condensed Matter Physics*. Theoretical condensed-matter physics; electronic structure.

Assistant Professor

El-Ganainy, Ramy, Ph.D., University of Central Florida, 2009. *Optics*. Photonics; quantum optics.

Huentemeyer, Petra, Ph.D., University of Hamburg, Germany, 2001. *Astrophysics*. Particle astrophysics.

Lee, Kim Fook, Ph.D., Duke University, 2002. *Atomic, Molecular, & Optical Physics, Condensed Matter Physics*. Experimental quantum optics.

DEPARTMENTAL RESEARCH SPECIALTIES AND STAFF

Theoretical

Astronomy. All-sky monitoring and gamma ray burst detection. Nemiroff.

Astrophysics. Gravitational lensing; high-energy astrophysics; close binary stars; cosmology; cosmic rays. Fick, Huentemeyer, Nemiroff, Nitz.

Atmospheric Physics. Satellite meteorology; optics; digital image processing; cloud precipitation and nucleation. Kostinski.

Atomic, Molecular, and Condensed Matter Physics. Electronic structure of metals and oxides; biological molecules; point defects; surfaces; metal-insulator transitions; atomistic simulations of materials; relativistic and correlation effects in atomic structure. Beck, Jaszczak, Lee, Pandey, Pati, Perger, Seel.

Experimental

Astrophysics. High-energy cosmic rays; Pierre Auger Observatory; HAWC gamma rays. Fick, Huentemeyer, Nitz.

Atmospheric Physics. Optics; digital image processing; cloud precipitation and nucleation; climate; air quality. Cantrell, Mazzoleni, Shaw.

Atomic, Molecular, Optical, and Condensed Matter Physics. Laser spectroscopy; magnetic resonance; quantum optics. Borysow, El-Ganainy, Lee, Levy, Moran, Suits, Yap.

View additional information about this department at www.gradschoolshopper.com

OAKLAND UNIVERSITY

DEPARTMENT OF PHYSICS

Rochester, Michigan 48309-4487
http://www.oakland.edu/physics

General University Information
President: Gary Russi
Dean of Graduate School: Darlene Schott-Baer, Interim Vice Provost
University website: http://www.oakland.edu
Control: Public
Setting: Suburban
Total Faculty: 560
Total Graduate Faculty: 560
Total number of Students: 19,740
Total number of Graduate Students: 3,550

Department Information
Department Chairman: Andrei Slavin, Chair
Department Contact: Carol Searight, Administrative Secretary
Total full-time faculty: 12
Total number of full-time equivalent positions: 12
Full-Time Graduate Students: 23
First-Year Graduate Students: 3
Total Post Doctorates: 8

Department Address
190 Science and Engineering
Rochester, MI 48309-4487
Phone: (248) 370-3416
E-mail: searight@oakland.edu
Website: http://www.oakland.edu/physics

ADMISSIONS

Admission Contact Information
Address admission inquiries to: Graduate Study, 520 O'Dowd Hall, Rochester, MI 48309-4401
Phone: (248) 370-2700
E-mail: gradmail@oakland.edu
Admissions website: http://www.oakland.edu/grad

Application deadlines
Fall admission:
U.S. students: July 15   *Int'l. students*: May 1
Spring admission:
U.S. students: November 15   *Int'l. students*: September 1

Application fee
There is no application fee required.

Admissions information
For Fall of 2012:
Number of applicants: 61
Number admitted: 5
Number enrolled: 3

Admission requirements
Bachelor's degree requirements: A Bachelor's degree in physics is required with no minimum undergraduate GPA specified.

GRE requirements
The GRE is required.

Advanced GRE requirements
The Advanced GRE is required.

TOEFL requirements
The TOEFL exam is required for students from non-English-speaking countries.
PBT score: 550
iBT score: 79

Other admissions information
Additional requirements: The GRE and GRE Advanced are required for Ph.D. program only. The GRE Advanced is required for foreign students. Foreign students must meet minimum acceptable score for TOEFL or have a baccalaureate or more advanced degree from an institution in the United States.
Undergraduate preparation assumed: Fowles, Analytical Mechanics; Reitz and Milford, Electromagnetic Theory; Tipler, Modern Physics; Jenkins and White, Optics; Saxon, Elementary Quantum Mechanics.

TUITION

Tuition year 2012–13:
Tuition for in-state residents
   Full-time students: $595.25 per credit
   Part-time students: $595.25 per credit
Tuition for out-of-state residents
   Full-time students: $1,027 per credit
   Part-time students: $1,027 per credit
Credit hours per semester to be considered full-time: 8
Deferred tuition plan: No
Health insurance: Available.
Other academic fees: Health insurance cost varies per year. No other academic fees.
Academic term: Semester
Number of first-year students who received full tuition waivers: 1

Teaching Assistants, Research Assistants, and Fellowships
Average stipend per academic year
   Teaching Assistant: $10,250
   Research Assistant: $14,000
   Fellowship student: $14,000

FINANCIAL AID

Application deadlines
Fall admission:
U.S. students: January 1
Spring admission:
U.S. students: January 1   *Int'l. students*: January 1

Loans
Loans are available for U.S. students.
Loans are not available for international students.
GAPSFAS application required: No
FAFSA application required: Yes

For further information
Address financial aid inquiries to: Student Financial Services, 120 North Foundation Hall, 2200 N. Squirrel Road, Rochester, MI 48309.

Phone: (248) 370-2550
E-mail: finaid@oakland.edu
Financial aid website: http://www.oakland.edu/financialaid

HOUSING

Availability of on-campus housing
Single students: Yes
Married students: Yes

For further information
Address housing inquiries to: University Housing, 448 Hamlin Hall, Rochester, MI 48309.
Phone: (248) 370-3570
E-mail: housing@oakland.edu
Housing aid website: http://www.oakland.edu/housing/

Table A—Faculty, Enrollments, and Degrees Granted

Research Specialty	2012-13 Faculty	Enrollment Fall 2012		Number of Degrees Granted 2011–12 (2007–12)		
		Master's	Doctorate	Master's	Terminal Master's	Doctorate
Applied Physics	1	8	–	–	1(12)	–
Astrophysics	1	–	–	–	–	–
Condensed Matter Physics	7	–	–	–	–	–
Medical, Health Physics	3	–	15	–	–	2(12)
Statistical & Thermal Physics	–	–	–	–	–	–
Total	12	8	15	–	1(12)	2(12)
Full-time Grad. Stud.	–	8	15	–	–	–
First-year Grad. Stud.	–	3	–	–	–	–

GRADUATE DEGREE REQUIREMENTS

Master's: Thirty-six credits of graduate courses including four credits of PHY 673 (Quantum Mechanics), one credit of PHY 600 (Seminar), 23 credits of additional 400–500–600-level courses approved by the department, eight credits of research, including a thesis or a critical essay. No foreign language requirements.

Doctorate: Biomedical Sciences-Medical Physics. 90 semester hours of graduate credit including at least 30 hours of dissertation research, grade point average of 3.0 or higher, three full-time equivalent semesters (at least 8 credits/semester) in residence, qualifying examination, dissertation.

Thesis: Thesis may be written in absentia.

SPECIAL EQUIPMENT, FACILITIES, OR PROGRAMS

Research facilities in the high pressure optics laboratory include Raman spectrometers with single or multi-channel detectors, facilities for photoluminescence studies in the visible and infrared regions, argon ion and Ti: sapphire lasers, high-pressure cells capable of generating 10 GPa, and closed cycle helium refrigerators.

Research facilities in the condensed matter physics laboratories include a Faraday Magnetometer, an AC susceptometer, a ferromagnetic resonance spectrometer at x-band, a Scanning Probe Microscope, a Scanning Microwave Microscope, Vector Network Analyzers (1 kHz-110 GHz), a Philips x-ray diffractometer, and facilities for thin film deposition and fullerene preparation.

Research facilities in the NMR microscopy laboratory include a Bruker AVANCE II 300 NMR spectrometer with a 7-Tesla/89-mm bore superconducting magnet and micro-imaging accessories, PerkinElmer Spotlight 300 Fourier-transform infrared

microscope, Leica polarized light microscope, Skyscan 1174 micro-CT scanner, and EnduraTEC 3200 mechanical testing system. The department also has microwave device facilities in the frequency range from 100 kHz to 70 GHz. Supporting facilities include electronics and mechanical workshops staffed by experienced technical personnel. Computer facilities include a 200-node supercomputer cluster for high performance computing, plus a number of workstations, and numerous Macintosh and PC computers. Most research laboratories are located in the modern Science and Engineering Building on campus.

The Physics Department recently installed two modern computer clusters, the newest one, funded through the NSF Major Research Instrumentation program, is composed of 24 nodes containing two AMD Opteron 6136 "Magny-Cours", running at 2.4 GHz, with 8 cores and 64 gigabytes RAM each, for a total of 384 cores and more than 1.5 terabytes of RAM. In addition, the two head nodes combined provide 18 terabytes of hard disk space. Both clusters are located in modern computer rooms at the Oakland Information Technology Center.

Among research facilities in neighboring hospitals available to medical physics students are a 3.0-Tesla whole-body NMR system and two 7.0-Telsa/20-cm horizontal bore magnet NMR systems for imaging and in vivo spectroscopy, a 7-channel SQUID magnetometer, a 148-channel whole-head SQUID neuromagnetometer, a Zeiss LSM 510 two photon microscope, Leica LMD6000 laser microdissection system, Philips EM208 Transmission Electron Microscope, a nuclear medicine laboratory, radiology and CT scanning facilities, advanced modalities cancer therapy laboratory (including radiotherapy), diagnostic ultrasonic equipment, a laser surgery laboratory, and major hospital medical libraries.

Table B—Separately Budgeted Research Expenditures by Source of Support

Source of Support	Departmental Research	Physics-related Research Outside Department
Federal government	$1,473,602	
State/local government		
Non-profit organizations		
Business and industry		
Other	$83,226	
Total	$1,556,828	

Table C—Separately Budgeted Research Expenditures by Research Specialty

Research Specialty	No. of Grants	Expenditures ($)
Condensed Matter Physics	17	$911,537
Medical, Health Physics	5	$556,063
Relativity & Gravitation	2	$89,228
Total	24	$1,556,828

FACULTY

Distinguished University Professor

Chopp, Michael, Ph.D., New York University, 1975. *Solid State Physics*. My research efforts include basic and applied research into the development of neurorestorative therapies for the treatment of neurological diseases and injury, e.g. stroke, traumatic brain injury, multiple sclerosis, and peripheral neuropathy. These treatments are designed to remodel the central and peripheral nervous systems, to enhance neurological recovery resulting from a neurological disease or injury. In addition, my laboratory seeks to develop therapies for the treatment of brain tumors. Our studies range from research into fundamental molecular mechanisms of therapeutic action, to preclinical and clinical studies, using a vast array of biological and physical methods.

Slavin, Andrei, Ph.D., Leningrad Technical University, 1977. *Condensed Matter Physics.*

Srinivasan, Gopalan, Ph.D., Indian Institute of Technology, 1980. *Condensed Matter Physics, Materials Science, Metallurgy.* Current research projects are on the physics of composite multiferroics and applications for useful technologies. The primary focus is on the nature of magnetoelectric interactions in ferromagnetic piezoelectric composites. Studies involve measurements over 1 Hz–110 GHz in layered and nanocomposites. Applications-related efforts are on magnetic field sensors, microwave and millimeter devices and miniature antennas.

Professor

Elder, Ken, Ph.D., University of Toronto, 1989. *Condensed Matter Physics, Materials Science, Metallurgy, Statistical & Thermal Physics.* The main focus of my research is on understanding the formation of complex spatial morphologies that form in nature. This research has included studies of spinodal decomposition, eutectic solidification, order/disorder transitions, Rayleigh-Benard convection and the absorption of liquids by random media. Recently I developed, in collaboration with others, the "Phase Field Crystal" method of modeling elastic and plastic deformation in polycrystalline systems.

Garfinkle, David, Ph.D., University of Chicago, 1985. *Relativity & Gravitation.* I do research in general relativity, especially black holes, gravitational collapse, and singularities. My methods are mostly numerical. I also have interests in cosmology.

Roth, Bradley, Ph.D., Vanderbilt University, 1987. *Biophysics, Medical, Health Physics.* Theoretical and numerical modeling of bioelectric and biomagnetic phenomena.

Xia, Yang, Ph.D., Massey University, 1992. *Applied Physics, Biophysics, Fluids, Rheology, Mechanics, Medical, Health Physics, Polymer Physics/Science.* Concentrating research in the degradation of articular cartilage, which is a hallmark of clinical joint diseases such as osteoarthritis. - Uses multidisciplinary imaging techniques to study cartilage, including microscopic magnetic resonance imaging, polarized light microscopy, Fourier-transform infrared imaging, and microscopic x-ray tomography. - Aims to resolve subtle changes in the physical, biological, chemical, and morphological properties of cartilage at the early stage of the cartilage degradation non-destructively.

Associate Professor

Martins, George, Ph.D., Campinas State University, 1994. *Condensed Matter Physics.*

Rojo, Alberto, Ph.D., Instituto Balseiro, Bariloche, 1990. *Condensed Matter Physics.*

Assistant Professor

Khain, Evgeniy, Ph.D., Hebrew University of Jerusalem, 2005. *Condensed Matter Physics, Materials Science, Metallurgy, Other.* Other specialty: Biological physics. Modeling of collective cell behavior in biological systems (growth of malignant brain tumors, wound healing) - Population dynamics (mathematical biology) - Statistical physics far from equilibrium - Pattern formation and nonlinear dynamics - Driven granular gases, instabilities in granular flows.

Wang, Yuejian, Ph.D., Texas Christian University, 2006. *Applied Physics, Nano Science and Technology, Other.* My research focuses on the investigations of the new structures, properties, and applications of materials with various crystal

sizes under high pressures by using diamond anvil cells integrated with synchrotron X-ray, Raman, micro-photographic techniques. Our high-pressure research group also does synthesize novel materials with unique properties for practical application.

Emeritus

Liboff, Abraham R., Ph.D., New York University, 1964. *Biophysics, Medical, Health Physics.*

Tepley, Norman, Ph.D., Massachusetts Institute of Technology, 1963. *Low Temperature Physics, Medical, Health Physics.*

Research Associate Professor

Tyberkevych, Vasyl, Ph.D., University of Kyiv, 2001. *Applied Physics, Condensed Matter Physics, Materials Science, Metallurgy, Nano Science and Technology.* Static and dynamic properties of magnetic nano-dot arrays (ground states, multistability, dynamic excitations, parametric processes, etc.); dynamic phenomena induced by a spin-polarized current in magnetic nano-structures (magnetization switching, persistent microwave oscillations, etc.); spin waves in micro- and nano-sized magnetic particles and films (spin wave spectra, nonlinear spin wave interactions, etc.); practical applications of spin waves for microwave signal processing (isolators and circulators, passive and active delay lines, filters, oscillators, etc.).

Adjunct Professor

Bleil, Carl, Ph.D., University of Oklahoma, 1953.

Ewing, James R., Ph.D., Oakland University, 1992. *Medical, Health Physics.*

Gerhart, Grant R., Ph.D., Wayne State University, 1972. *Condensed Matter Physics.*

Marples, Brian, Ph.D., University College London and Gray Laboratory, 1991. *Medical, Health Physics.*

Sabbah, Hani, Ph.D., Oakland University, 1988. *Medical, Health Physics.*

Soltanian-Zadeh, Hamid, Ph.D., University of Michigan, 1992. *Medical, Health Physics.*

Venkatesan, Srinivasan, Ph.D., University of London, 1974.

Venkateswaran, Uma Devi, Ph.D., University of Missouri (Columbia), 1985.

Wilson, George D., Ph.D., University of Liverpool, 1980. *Medical, Health Physics.*

Yan, Di, Ph.D., Washington University, 1990. *Medical, Health Physics.*

Adjunct Associate Professor

Castoldi, Kapila Clara, Ph.D., University of Milan, 1976. *Particles and Fields.* Physics education.

Hammond, Robert L., Ph.D., Wayne State University, 2000. *Medical, Health Physics.*

Jiang, Quan, Ph.D., Oakland University, 1991. *Medical, Health Physics.*

Knight, Robert A., Ph.D., Oakland University, 1991. *Medical, Health Physics.*

Liang, Jian, Ph.D., Zhejiang University, 1994. *Medical, Health Physics.*

McDermott, Patrick N., Ph.D., University of Rochester, 1985. *Medical, Health Physics.*

Adjunct Assistant Professor

Bagher-Ebadian, Hassan, Ph.D., Oakland University, 2010. *Medical, Health Physics.*

Bowyer, Susan, Ph.D., Oakland University, 1998. *Medical, Health Physics.*

Ionascu, Dan, Ph.D., Northeastern University, 2005. *Medical, Health Physics.*

Jenrow, Kenneth, Ph.D., Oakland University, 1995. *Medical, Health Physics.*

Zhang, Tiezhi, Ph.D., University of Wisconsin-Madison, 1999. *Medical, Health Physics.*

Zhang, Zheng-Gang, Ph.D., Oakland University, 1995. *Medical, Health Physics.*

DEPARTMENTAL RESEARCH SPECIALTIES AND STAFF

Theoretical
Applied Physics. Tyberkevych.
Biological Physics. Khain.
Biophysics. Roth.
Condensed Matter Physics. Elder, Khain, Martins, Rojo, Slavin, Tyberkevych.
Materials Science, Metallurgy. Elder, Khain, Tyberkevych.
Medical, Health Physics. Roth.
Nano Science and Technology. Tyberkevych.

Relativity & Gravitation. Garfinkle.
Statistical & Thermal Physics. Elder.

Experimental
Applied Physics. Wang, Xia.
Atomic, Molecular, & Optical Physics. Xia.
Biophysics. Liboff, Xia.
Fluids, Rheology. Xia.
Geophysics. Xia.
High-pressure Physics. Wang.
Low Temperature Physics. Tepley.
Materials Science, Metallurgy. Srinivasan.
Mechanics. Xia.
Medical, Health Physics. Liboff, Tepley, Xia.
Nano Science and Technology. Wang.
Optics. Xia.
Polymer Physics/Science. Xia.
Solid State Physics. Chopp.

View additional information about this department at
www.gradschoolshopper.com

UNIVERSITY OF MICHIGAN

APPLIED PHYSICS PROGRAM

Ann Arbor, Michigan 48109-1040
http://www-applied.physics.lsa.umich.edu/

General University Information
President: Mary Sue Coleman
Dean of Graduate School: Janet A. Weiss
University website: http://www.umich.edu/
Control: Public
Setting: Urban
Total Faculty: 5,497
Total number of Students: 43,426
Total number of Graduate Students: 15,477

Department Information
Department Chairman: Cagliyan Kurdak, Director
Department Contact: Cynthia McNabb, Program Administrator
Total full-time faculty: 132
Full-Time Graduate Students: 72
First-Year Graduate Students: 19
Female First-Year Students: 5

Department Address
450 Church, 1425 Randall Laboratory
Ann Arbor, MI 48109-1040
Phone: (734) 936-0653 / (734) 764-4595
Fax: (734) 764-2193
E-mail: cyndia@umich.edu
Website: http://www-applied.physics.lsa.umich.edu/

ADMISSIONS

Admission Contact Information
Address admission inquiries to: Applied Physics Program, 450 Church, 1425 Randall Laboratory, University of Michigan, Ann Arbor, MI 48109-1040

Phone: (734) 764-4595
E-mail: csutton@umich.edu
Admissions website: http://www-applied.physics.lsa.umich.edu/

Application deadlines
Fall admission:
U.S. students: January 15   *Int'l. students*: January 15

Application fee
U.S. students: $65   *Int'l. students*: $75

Admissions information
For Fall of 2012:
Number of applicants: 192
Number admitted: 32
Number enrolled: 14

Admission requirements
Bachelor's degree requirements: Bachelor's degree is required in one of the following areas: physics, applied physics, engineering physics, or a related discipline with a strong physical science content.

GRE requirements
The GRE is required.

Advanced GRE requirements
The Advanced GRE is not required.

TOEFL requirements
The TOEFL exam is required for students from non-English-speaking countries.
PBT score: 560
iBT score: 84

Other admissions information

Additional requirements: At least 15 hours of introductory and intermediate-level physics are expected, including courses in electricity, magnetism, and quantum physics. Degree requirements equivalent to an American bachelor of arts or bachelor of science degree from an approved institution must have been completed by the time of first enrollment.

TUITION

Tuition year Fall 2011:
Tuition for in-state residents
    Full-time students: $9,717 per semester
Tuition for out-of-state residents
    Full-time students: $19,538 per semester
Tuition applies up to and including 9 or more credit hours. No additional fees for 10 hours and above up to 18 credit hours (rates shown are subject to change without notice).
Credit hours per semester to be considered full-time: 9
Deferred tuition plan: No
Health insurance: Available
Other academic fees: Registration fee: $97.19 each term.
Academic term: Semester
Number of first-year students who received full tuition waivers: 20

Teaching Assistants, Research Assistants, and Fellowships

    Number of first-year
        Fellowship students: 14
    Average stipend per academic year
        Fellowship student: $27,351

FINANCIAL AID

Loans

Loans are available for U.S. students.
Loans are not available for international students.
GAPSFAS application required: No
FAFSA application required: Yes

For further information

Address financial aid inquiries to: Mrs. Cynthia McNabb, 450 Church, 1425 Randall Laboratory, Applied Physics, Ann Arbor, MI 48109-1040.
Phone: (743) 936-0653
E-mail: cyndia@umich.edu
Financial aid website: http://www-applied.physics.lsa.umich.edu/doctoral.html

HOUSING

Availability of on-campus housing

    Single students: Yes
    Married students: Yes

For further information

Address housing inquiries to: University of Michigan Housing Office, 1500 Student Activities Building.
Phone: (734) 763-3164
E-mail: housing@umich.edu
Housing aid website: http://www.housing.umich.edu

Table A—Faculty, Enrollments, and Degrees Granted

Research Specialty	2012-2013 Faculty	Enrollment Fall 2012 Master's	Enrollment Fall 2012 Doctorate	Number of Degrees Granted 2012-13 (2008-13) Master's	Number of Degrees Granted 2012-13 (2008-13) Terminal Master's	Number of Degrees Granted 2012-13 (2008-13) Doctorate
Atomic, Molecular, & Optical Physics	7	–	10	–	–	–(2)
Biophysics	15	–	6	–	–	–(5)
Chemical Physics	10	–	9	–	–	–(3)
Computational Physics	6	–	2	–	–	–
Computer Science	6	–	1	–	–	–
Condensed Matter Physics	7	–	9	–	–	1(9)
Electrical Engineering	5	1	1	–	–	2(3)
Energy Sources & Environment	10	–	6	–	–	–(1)
Geophysics	5	–	–	–	–	–
Materials Science, Metallurgy	13	1	2	–	–	–(4)
Medical, Health Physics	3	1	–	–	–	–(3)
Nano Science and Technology	15	–	7	–	–	1(5)
Nuclear Physics	2	–	2	–	–	–(1)
Optics	9	–	9	–	–	–(14)
Plasma and Fusion	8	–	8	–	–	2(6)
Polymer Physics/Science	5	–	3	–	–	–
Statistical & Thermal Physics	4	–	2	–	–	–
Total	123	3	72	–	–	6(56)
Full-time Grad. Stud.	–	3	72	–	–	–
First-year Grad. Stud.	–	3	10	–	–	–

GRADUATE DEGREE REQUIREMENTS

Master's: Students must earn at least 30 credit hours of graduate-level courses of which at least 20 must be in the applied physics core curriculum at the 500 level or higher. A master's thesis is not required. Up to six credit hours may be transferred from other graduate programs or universities, subject to program approval. Students must maintain at least a "B" average overall. There is no language requirement.

Doctorate: In order to achieve candidacy and form a dissertation committee, the student must satisfy the following requirements: complete nine prescribed 500- and 600-level graduate courses comprising the applied physics core curriculum, plus two distribution courses; pass a qualifying examination on standard undergraduate material in basic physics; complete one four-credit hour course on non-thesis research (theoretical or experimental) under the supervision of a faculty member; complete any two 600-level advanced courses on special topics (may be done before or after candidacy); and pass a preliminary examination given by his/her dissertation committee on a prospective thesis topic. The student is expected to attain candidacy and be working on a thesis topic before the end of the fifth term. There is no foreign language proficiency requirement. Because applied physics is an interdisciplinary program offering a wide range of possibilities, the thesis research may be done under the supervision of any faculty member associated with the Applied Physics Program, or alternately under joint sponsorship with a faculty member outside of the program. Thesis may be written in absentia, but thesis research must be done on campus or at a designated location. At least two full terms of not less than nine credit hours each must be completed in residence on the campus at Ann Arbor.

SPECIAL EQUIPMENT, FACILITIES, OR PROGRAMS

The University of Michigan has extensive resources in laboratories and equipment.

FACULTY

Professor

Allen, James W., Ph.D., Stanford University, 1968. *Condensed Matter Physics.* Experimental condensed-matter physics.

Assanis, Dionissios N., Ph.D., Massachusetts Institute of Technology, 1986. Thermal and fluid sciences and their applications to automotive systems design; internal combustion engine processes and systems; development and validation of transient diesel engine simulation; in-cylinder CFD computations; experimental investigation of heat rejection; unburned hydrocarbon mechanisms and friction in spark-ignition engines.

Atzmon, Michael, Ph.D., California Institute of Technology, 1985. Experimental materials physics.

Banaszak-Holl, Mark M., Ph.D., Cornell University. Biophysical, surface, and organometallic chemistry.

Becchetti, Frederick D., Ph.D., University of Minnesota, 1969. Experimental nuclear physics; medical physics; experimental materials physics; short-lived nuclear beams; heavy ion collisions; nuclear instrumentation; radiation oncology and nuclear medicine.

Berman, Paul, Ph.D., Yale University, 1969. Theoretical atomic, molecular, and optical physics; spectroscopy of laser-cooled atoms; atom interference.

Bhattacharya, Pallab, Ph.D., University of Sheffield, 1978. Experimental solid-state electronics.

Carson, Paul L., Ph.D., University of Arizona, 1972. Diagnostic ultrasound; nuclear MRI; spectroscopy.

Chupp, Timothy E., Ph.D., University of Washington, 1983. Precision measurements in atomic, nuclear, and particle physics; biomedical NMR and MRI.

Clarke, Roy, Ph.D., Queen Mary College, London, 1973. Experimental condensed-matter physics.

Drake, R. Paul, Ph.D., Johns Hopkins University. Experimental astrophysics; nonlinear and radiation hydrodynamics; plasma waves in interplanetary and interstellar space.

Fessler, Jeffrey A., Ph.D., Stanford University (Departments of Electrical Engineering and Computer Science, Biomedical Engineering, Nuclear Medicine/Radiology and Applied Physics Program), 1990. Statistical aspects of medical imaging.

Forrest, Stephen R., Ph.D., University of Michigan, 1979. William Gould Dow Collegiate Professor of Electrical Engineering. Optoelectronic integrated circuits; organic thin film semiconductor and III-V semiconductor growth by molecular beam epitaxy; optoelectronic interconnections and phased array antenna systems.

Fowlkes, J. Brian, Ph.D., University of Mississippi (Radiology, Biomedical Engineering and Applied Physics Program), 1988. Diagnostic and therapeutic applications of medical ultrasound, including acoustic cavitation and contrast agents, vascular and 3D imaging, and MRI of acoustic wave propagation.

Gafni, Ari, Ph.D., Weizmann Institute of Science. Research focuses on mechanistic aspects of protein folding, on the identification of molecular interactions that stabilize the folded state, and on the characterization of those alterations in the folded structure that occur during aging and in some age-associated diseases.

Gallimore, Alex, Ph.D., Princeton University, 1992. Experimental plasma physics; plasma diagnostics in the areas of electric propulsion for spacecraft, anode sheath phenomena, cathode physics, and space plasma simulation.

Gianchandani, Yogesh B., Ph.D., University of Michigan, 1994. Design, fabrication, and packaging of micromachined sensors and actuators and their interface circuits.

Gidley, David W., Ph.D., University of Michigan, 1979. Atomic, molecular, and optical research; positron and positronium physics; materials research.

Gilchrist, Brian E., Ph.D., Stanford University, 1991. Investigations of electrodynamic tethered systems in the ionosphere and high-density electron beam effects; microwave diagnostics of dense, high-speed plasmas for space electric propulsion.

Gilgenbach, Ronald M., Ph.D., Columbia University, 1978. Interaction of intense laser and particle beams with plasmas.

Gland, John L., Ph.D., University of California, Berkeley (Department of Chemistry and Applied Physics Program), 1973. Experimental physical chemistry; spectroscopy of surface chemical reactions.

Glotzer, Sharon C., Ph.D., Boston University (Departments of Chemical Engineering, Materials Science and Engineering, Macromolecular Science and Engineering, Physics and Applied Physics Program), 1993. Assembly of nanoscale systems; supercooled and metastable liquids and complex fluids, colloids, and complex fluids; biomimetic materials design; computer simulation.

Goldman, Rachel S., Ph.D., University of California (Departments of Materials Science and Engineering, Electrical Engineering and Computer Science and Applied Physics Program), 1995. Atomic-scale design of electronic materials with a focus on the mechanisms of fundamental processes, including strain relaxation, alloy formation, and diffusion; correlations between microstructure and electronic and optical properties of semiconductor films, nanostructures, and heterostructures.

Goodson, Theodore G., Ph.D., University of Nebraska-Lincoln, 1996. Physical chemistry; organic materials; nonlinear and time-resolved spectroscopy; quantum optical effects in novel materials.

Green, Peter F., Ph.D., Cornell University, 1985. Structure, phase behavior, and dynamics of bulk and thin film polymer and polymer-based nanocomposite systems.

Islam, Mohammed, Ph.D., M.I.T. *Optics.* His current research interests include mid-infrared laser sources and their applications in fiber-to-the-home, advanced semiconductor process control, combustion monitoring, infrared counter-measures, chemical sensing and bio-medical selective laser ablation. Another area of this current work relates to ultra-high resolution imaging of automobile parts, such as transmissions. He also has on-going work in modulators and new architectures for fiber to the home systems.

Jones, J. Wayne, Ph.D., Vanderbilt University (Department of Materials Science and Engineering and Applied Physics Program), 1977. Mechanical behavior of structural metal alloys, specifically on fatigue and creep.

Kanicki, Jerzy, Ph.D., Free University of Brussels, Belgium (Department of Electrical Engineering and Computer Science), 1982. Molecular and organic electronics; amorphous and polycrystalline semiconductor (inorganic and organic) thin film devices and circuits; flat panel displays and sensor technology.

Kaviany, Massoud, Ph.D., University of California, Berkeley (Mechanical Engineering and Applied Physics Program), 1979. Thermal science, heat transfer, transport, reaction, and phase change in porous media.

Kieffer, John, Ph.D., Clausthal Technical University (Materials Science and Engineering), 1985. Materials for photonics, dielectrics, and structural mechanical applications.

Kopelman, Raoul, Ph.D., Columbia University (Department of Chemistry, Department of Physics), 1960. Laser spectroscopy and super-microscopy of molecular materials, membranes, nanostructures, and DNA sequencing; fractal chemical kinetics and supercomputer simulations.

Krushelnick, Karl, Ph.D., Princeton University, 1994. Ultra-high-intensity laser plasma interactions.

Kushner, Mark J., Ph.D., California Institute of Technology (Electrical Engineering and Computer Science), 1979. Computational plasma and plasma etching: development of computer simulations of low-temperature plasmas and technologically important devices that use low-temperature plasmas.

Larson, Ron, Ph.D., University of Minnesota (Chemical Engineering and Applied Physics Program), 1980. George Granger Brown Professor of Chemical Engineering. Complex fluids; polymers; fluid mechanics; surfactants; biomolecules; transport theory; rheology; instabilities; constitutive theory.

Lau, Yue-Ying, Ph.D., Massachusetts Institute of Technology (Department of Nuclear Engineering and Radiological Sciences and Applied Physics Program), 1973. Plasma and beam physics.

Mazumder, Pinaki, Ph.D., University of Illinois at Urbana-Champaign, 1987. Nanocircuits, nanoarchitectures, modeling and simulation tools for nano and quantum electronic circuits; ultra-fast circuit design with compound semiconductor devices; very deep sub-micron CMOS VLSI: mixed signal system design, testing methodology, and chip layout automation.

Merlin, Roberto D., Ph.D., University of Stuttgart (Department of Physics and Applied Physics Program), 1978. Experimental condensed-matter physics: inelastic light scattering, ultrafast lasers, coherent optical phenomena, low-dimensional semiconductor structures, superconductors, and magnetic materials.

Moghaddam, Mahta, Ph.D., University of Illinois at Urbana-Champaign, 1991. Radar remote sensing.

Munson, David C., Ph.D., Princeton University, 1977. Signal and image processing focused on radar imaging, passive millimeter-wave imaging, LIDAR imaging, tomography, interferometry, and high-precision GPS.

Nilton, Renno, Ph.D., Massachusetts Institute of Technology (Atmospheric, Oceanic and Space Sciences), 1992. Atmospheric convection; climate; planetary sciences; instrument development.

Nori, Franco M., Ph.D., University of Illinois at Urbana-Champaign, 1987. Quantum computing; nanoscience; theoretical condensed-matter physics; complex systems.

Norris, Theodore B., Ph.D., University of Rochester (Department of Electrical Engineering and Applied Physics Program), 1989. Ultrafast lasers and measurement techniques; ultrafast dynamics of carriers and excitons in semiconductor nanostructure; novel imaging techniques.

Orr, Bradford G., Ph.D., University of Minnesota (Department of Physics and Applied Physics Program), 1985. Condensed-matter physics; applied physics: molecular beam epitaxy film growing, surface science studies on semiconductors and metals, and AFM investigation of membranes and cells.

Pan, Xiaoqing, Ph.D., Universitaet des Saarlandes, Germany (Department of Materials Science and Engineering and Applied Physics Program), 1991. Structure/property relationships in both functional and structural resolution transmission electron microscopy (HRTEM) in combination with AEM utilized to study bulk ceramics, thin films of chemical sensors, ferroelectrics, and silicon-nitride-based materials.

Penner, Joyce E., Ph.D., Harvard University, 1977. Atmospheric chemistry and climate modeling.

Pipe, Kevin Patrick, Ph.D., Massachusetts Institute of Technology (Departments of Mechanical Engineering, Electrical Engineering and Computer Science and Applied Physics Program), 2004. MEMS (micro electromechanical systems); optics; photonics/optoelectronics; solid-state electronics.

Raithel, Georg, Ph.D., University of Munich, Germany (Department of Physics and Applied Physics Program), 1990. Experimental atomic, molecular, and optical physics: laser cooling, cold Rydberg atoms, cold plasmas, and atom lasers.

Rand, Stephen C., Ph.D., University of Toronto (Departments of Electrical Engineering and Computer Science; Department of Physics, Applied Physics Program), 1978. Experimental quantum electronics; nonlinear laser spectroscopy.

Ross, Marc Hansen, Ph.D., University of Wisconsin (Department of Physics and Applied Physics Program), 1948. Environmental physics: energy use and its impacts and the reduction of those impacts through efficiency and conservation.

Sander, Leonard M., Ph.D., University of California, Berkeley (Department of Physics and Applied Physics Program), 1968. Theoretical physics; condensed-matter physics.

Sension, Roseanne J., Ph.D., University of California, Berkeley (Departments of Chemistry, Physics and Applied Physics Program), 1986. Experimental atomic molecular, and optical physics; experimental optics; chemical physics.

Singh, Jasprit, Ph.D., University of Chicago (Department of Electrical Engineering and Computer Science and Applied Physics Program), 1980. Physics of semiconductor devices; use of heterostructures for high-performance electronic and optoelectronic devices.

Steel, Duncan G., Ph.D., University of Michigan (Departments of Electrical Engineering and Computer Science; Institute of Gerontology, Biophysics and Applied Physics Program), 1976. Optical physics in condensed matter and biomolecules: quantum optics in semiconductors (quantum computing) and laser spectroscopy of proteins.

Terry, Fred, Ph.D., Massachusetts Instittue of Technlogy (Department of Electrical Engineering and Computer Science and Applied Physics Program), 1985. Semiconductor manufacturing; solid-state electronics: electronic properties of materials and their effects on devices; physics of solid-state devices.

Thompson, Levi T., Ph.D., University of Michigan, 1986. Hydrogen Energy Technology Laboratory; catalysis and surface science.

Uher, Ctirad, Ph.D., University of New South Wales (Department of Physics and Applied Physics Program), 1975. Experimental solid-state physics.

Waite, Jack Hunter, Ph.D., University of Michigan, 1981. Outer planet aeronomy; mass spectrometry; planetary astronomy.

Winful, Herbert G., Ph.D., University of Southern California (Department of Electrical Engineering and Computer Science and Applied Physics Program), 1981. Nonlinear optics; semiconductor laser physics; nonlinear dynamics of coupled lasers; nonlinear fiber optics; integrated optics.

Winick, Kim Allen, Ph.D., University of Michigan (Department of Electrical Engineering and Computer Science and the Applied Physics Program), 1981. Glass and crystal integrated optics; nanophotonics; communications; information theory.

Yalisove, Steven M., Ph.D., University of Pennsylvania (Department of Materials Science and Engineering and Applied Physics Program), 1983. Researching the relationships between atomic structure and materials properties.

Yoon, Euisik, Ph.D., University of Michingan, 1990. My research group realizes self-contained microsystems that combine and process natural signals (such as bio, chemical, optical and thermal signals), as well as electrical signals on a single chip platform by integrating new MEMS/nano structures with low-power, wireless VLSI circuits and systems.

Zurbuchen, Thomas, Ph.D., University of Bern (Space Science and Aerospace Engineering), 1996.

Associate Professor

Al-Hashimi, Hashim, Ph.D., Yale University (Department of Chemistry and Department of Biophysics).

Becker, Udo, Ph.D., Virginia Technological University. Mineral surface chemistry and nanoscience; computational mineralogy.

Castro, M. Clara, Ph.D., University of Paris VI and Paris School of Mines, France (Department of Geological Sciences), 1995. Hydrogeology; paleoclimatology; mantle geochemistry.

Chen, Zhan, Ph.D., University of California-Berkeley (Department of Chemistry, Macromolecular Science and Engineering, Applied Physics Program and Optical Physics Interdisciplinary Lab). Integrating analytical chemistry, physical chemistry, materials chemistry, polymer science, surface science, molecular spectroscopy, microscopy, bioengineering, life sciences, laser techniques, nonlinear optics, and nanotechnology.

Currie, William S., Ph.D., University of New Hampshire Institute for the Study of Earth, Oceans and Space (School of Natural Resources and Environmental and Applied Physics Program), 1995. Ecosystem science; dynamic modeling.

Foster, John E., Ph.D., University of Michigan, Applied Physics (Nuclear Engineering and Radiological Sciences), 1996. Low-temperature plasma science, including propulsion plasmas, environmental plasmas, space and atmospheric plasma phenomena, energy conversion plasmas, and processing plasmas.

Garikipati, Krishnakumar R., Ph.D., Stanford University (Mechanical Engineering and Applied Physics Program), 1996. Nonlinear mechanics; materials physics; applied mathematics and numerical methods; theoretical and computational aspects of nonlinear continuum mechanics and inelasticity with a special emphasis on multi-scale material behavior and applications to processing of semiconductor material.

Guo, Lingjie J., Ph.D., University of Minnesota (Departments of Electrical Engineering and Computer Science and Applied Physics), 1997. Photonics/optoelectronics, semiconductor manufacturing, and solid-state electronics; nanofabrication technology and applications; polymer photonic devices and microresonator biosensors; nanoelectronics; nano-biotechnology; surface plasmon-polariton-based photonic structures.

Hunt, Alan J., Ph.D., University of Washington (Biomedical Engineering, Institute of Gerontology and Applied Physics Program), 1993. Cellular and molecular biomechanics: physical and mechanical properties of molecular motors that underlie biologic motility.

Jablonowski, Christiane, Ph.D., University of Michingan, 2004. *Atmosphere, Space Physics, Cosmic Rays.* Adaptive Mesh Refinement (AMR) and variable-resolution grid techniques for atmospheric General Circulation Models (GCMs). Development of test cases for dynamical cores of GCMs, model intercomparisons. Modern numerical methods and computational grids for dynamical cores of GCMs. Idealized simulations of the Quasi-Biennial Oscillation (QBO), stratospheric dynamics. Dynamics of tropical cyclones. Impact of climate change on tropical cyclone statistics in high-resolution GCM experiments. Subgrid-scale mixing in climate models. Cyberinfrastructure tools for the Earth System Sc.

Kurdak, Cagliyan, Ph.D., Princeton University, 1995. Condensed-matter research: transport in semiconductors and Josephson junctions, tunneling, and quantum coherence.

Markov, Igor L., Ph.D., University of California, Los Angeles (Electrical Engineering and Computer Science and Applied Physics Program), 2002. Combinatorial optimization with applications to the design and verification of integrated circuits and quantum logic circuits.

Mirecki-Millunchick, Joanna, Ph.D., Northwestern University (Materials Science and Engineering and Applied Physics Program), 1995. Materials and surface science of semiconductor thin film nucleation and epitaxy.

Mycek, Mary-Ann, Ph.D., University of California, Berkeley (Biomechanical Engineering Department, Applied Physics Program and Comprehensive Cancer Center Biomedical Optics), 1995. Development and application of optical science to probe noninvasively the complex living systems found in biology and medicine.

Newman, Mark E., Ph.D., University of Oxford, 1991. Statistical physics theory: networks, spin systems and percolation, and Monte Carlo methods.

Noll, Douglas C., Ph.D., Stanford University (Departments of Biomedical Engineering, Radiology and and Applied Physics Program), 1991. Data acquisition and processing for imaging brain function using MRI.

Violi, Angela, Ph.D., University of Naples Federico II, Italy (Mechanical Engineering, Biomedical Engineering, and Chemical Engineering), 1999. Nanoparticle growth and self-assembly; nanoparticle interactions with biomolecular systems; molecular modeling of complex systems using atomistic models; applied chemical kinetics; aerosols.

Wakefield, Gregory H., Ph.D., University of Minnesota (Department of Electrical Engineering and Computer Science and Applied Physics Program, Otolaryngology, School of Music), 1985. Perceptual acoustics; music processing; statistical signal processing; time-frequency distributions; signal analysis and synthesis; sound quality engineering; computational audition; psychoacoustics; auditory prosthetics; speech processing.

Walter, Nils, Ph.D., Technical University of Darmstadt, Germany (Department of Chemistry and Applied Physics Program), 1995. Chemical biology; folding and function of RNA enzymes; single-molecule fluorescence spectroscopy; biophysical chemistry of nucleic acids.

Assistant Professor

Aidala, Christine, Ph.D., Columbia University, 2005. *Nuclear Physics.* I do research at the boundary of nuclear and particle physics, studying nucleon structure and QCD dynamics. I have been on the PHENIX experiment at the Relativistic Heavy Ion Collider at Brookhaven National Laboratory since 2001 and on the E906/SeaQuest experiment at Fermilab since 2010.

Carmon, Tal, Ph.D., Israel Institute of Technology (Electrical Engineering and Computer Science). Optics, photonics, nonlinear optics, and MEMS; nanophotonics; photonic micro-electro-mechanical systems, visible on-chip emitters, and harnessing radiation pressure for opto-mechanical applications.

Deng, Hui, Ph.D., Stanford University (Department of Physics), 2006. Experimental quantum optics; quantum information processing; many-body physics; semiconductor physics.

Dunietz, Barry Dov, Ph.D., Columbia University (Department of Chemistry and Applied Physics Program), 2000. Theoretical and computational chemistry.

Gavini, Vikram, Ph.D., California Institute of Technology (Department of Mechanical Engineering), 2007. Developing computational and mathematical tools for electronic structure calculations at macroscopic scales; multi-scale modeling; analysis of approximation theories; electronic structure studies on defects in solids.

Geva, Eitan, Ph.D., Hebrew University of Jerusalem (Department of Chemistry and Applied Physics Program). Physical chemistry: theoretical and computational chemistry; chemical reactivity and spectroscopy in condensed matter.

Grbic, Anthony, Ph.D., University of Toronto (Electrical Engineering and Computer Science), 2005. Periodic structures (metamaterials, frequency selective surfaces, photonic crystals, and electromagnetic band-gap structures); planar antennas; microwave circuits; subwavelength optics; analytical electromagnetic modeling.

Huang, Xianglei, Ph.D., California Institute of Technology (Atmospheric, Oceanic and Space Sciences), 2004. Infrared radiative transfer and remote sensing.

Jarrahi, Mona, Ph.D., Stanford University (Electrical Engineering and Computer Science), 2007. Applied electromagnetics; optics and photonics; solid-state devices and nanotechnology; integrated circuit design and VLSI.

Kioupakis, Emmanouil (Manos), Ph.D., Univ. of California Berkeley, 2008. *Materials Science, Metallurgy.* I develop first-principles computational algorithms and employ high-performance computing resources to study the structural, electronic, and optical properties of solids, molecules, and nanostructures. I have worked on the microscopic energy loss mechanisms in nitride LEDs and lasers. My goal is to use first principles computational methods to understand, model, predict, and design the behavior of novel electronic, optoelectronic, photovoltaic, and thermoelectric materials.

Kripfgans, Oliver Daniel, Ph.D., University of Michigan (Department of Radiology and Applied Physics Program), 2002. Transcutaneous ultrasonic generation of bursts of vascular and urinary microbubbles for selective contrast imaging; beam aberration correction; occlusion therapy; 3D ultrasound imaging and vascular quantification techniques in the breast and prostate; 3D ultrasound image fusion.

Kubarych, Kevin Joel, Ph.D., Massachusetts Institute of Technology (Department of Chemistry and Applied Physics Program), 2004. Physical and biophysical chemistry: ultrafast nonlinear, multidimensional spectroscopy of biomolecules; optical control of biomolecular dynamics.

Leanhardt, Aaron E., Ph.D., Massachusetts Institute of Technology (Department of Physics), 2003. Experimental atomic, molecular, and optical physics: ultracold atoms and molecules, quantum degenerate gases, and precision measurement.

Lu, Wei, Ph.D., Rice University (Electrical Engineering and Computer Science and Applied Physics Program), 2003. Nanoelectronics; growth of semiconductor nanowires and nanoscale heterostructures; new device physics and structures for nano-quantum scale electronic devices; solid-state-based spintronics; nanoelectromechanical systems.

Maldonado, Stephen, Ph.D., University of Texas, Austin (Department of Chemistry). Heterogeneous charge transfer processes relevant to the fields of electronics, chemical sensing, and energy conversion/storage technologies.

Meiners, Jens Christian, Ph.D., Universität Konstanz, Germany (Departments of Physics, Biophysics Research Division and Applied Physics Program), 1997. Experimental biophysics: dynamics of single DNA molecules.

Ogilvie, Jennifer P., Ph.D., University of Toronto, Canada (Departments of Physics, Biophysics Research Division and Applied Physics Program), 2003. Experimental biophysics: protein dynamics, ultrafast and single-molecule spectroscopies, and nonlinear microscopy.

Shtein, Max, Ph.D., Princeton University, 2004. Organic semiconductors, organic-inorganic hybrid materials, and nanocomposites geared toward efficient energy conversion.

Sih, Vanessa, Ph.D., University of California, Santa Barbara, 2006. Experimental condensed-matter physics: electron spins in semiconductors and nanophotonics.

Thomas, Alex, Ph.D., Imperial College, London, UK (Nuclear Engineering and Radiological Sciences), 2006. Plasma physics; ultra-high-intensity laser-plasma interactions; compact laser-based particle accelerators; particle-in-cell simulation; radiation generation and back reaction; laser propagation in plasma at high intensity; inertial confinement fusion; Vlasov-Fokker-Planck modeling; nonlocal transport; magnetized plasmas; electromagnetic and electrothermal instabilities.

Thornton, Katsuyo S., Ph.D., University of Chicago (Department of Materials Science and Engineering, College and Applied Physics Program), 1997. Computational and theoretical investigations of the evolution of microstructures and nanostructures during processing and operation.

Van Der Ven, Anton, Ph.D., Massachusetts Institute of Technology (Materials Science and Engineering), 2000. Understanding and predicting equilibrium and non-equilibrium materials properties from first principles: combining electronic structure methods (density functional theory) with techniques from statistical mechanics to calculate thermodynamic and kinetic properties of new materials, including oxides and structures of assembled nanoparticles for battery and fuel cell components, metallic alloys, alloy surfaces for catalysis, and organic electronic materials.

Zochowski, Michal R., Ph.D., University of Warsaw (Departments of Physics, Biophysics Research Division and Applied Physics Program), 1995. Biophysics; complex systems; neuroscience: formation of spatiotemporal patterns, nonlinear systems, and neural integration.

Adjunct Professor

Torres-Isea, Ramon, M.S., Eastern Michigan University (Department of Physics and Applied Physics Program), 1983.

Assistant Research Scientist

Kuranz, Carolyn C., Ph.D., University of Michigan, Applied Physics (Atmospheric, Oceanic and Space Sciences), 2009. High-energy-density physic;, laboratory astrophysics; hydrodynamic instabilities; plasma physics; radiation hydrodynamics.

Spooner, Gregory J., Ph.D., University of California, Davis (Department of Electrical Engineering and Computer Science and Applied Physics Program), 1992. Ultrafast laser systems and procedures for ophthalmic applications.

DEPARTMENTAL RESEARCH SPECIALTIES AND STAFF

Theoretical

Materials Science, Metallurgy. Growth and stability of thin films and aggregates; quantum devices and nanostructures; molecular dynamics.

Plasma and Fusion. Intense charged particle beams; accelerator physics; laser interactions with matter.

Experimental

Atomic, Molecular, & Optical Physics. Ultrafast spectroscopy; phonon dynamics; atom trapping; high-intensity physics; atomic structure; materials properties.

Energy Sources & Environment. Energy systems: hydrogen energy technology; organic light-emitting diodes; fuel technologies: emissions and vehicle mass; solar and heliospheric physics; energy policy; thermoelectric energy conversion; photovoltaic energy conversion; nanostructured energy conversion devices; hybrid organic/inorganic devices; energy conversion plasmas; propulsion; nonlinear control and systems engineering.

Materials Science, Metallurgy. Semiconductor heterostructures; magnetic nanostructures; physics of optoelectronic devices; thin film structures; synchrotron radiation studies; molecular electronics; display technology.

Optics. Ultrafast optical science; ultrashort laser pulses; solid-state lasers; nonlinear optics; holography; chemical reaction dynamics; integrated optics; material interaction; glass and crystal integrated optics.

Plasma and Fusion. Interaction of intense laser beams and plasmas; ion propulsion systems; particle acceleration; experimental space research.

WAYNE STATE UNIVERSITY

DEPARTMENT OF PHYSICS AND ASTRONOMY

Detroit, Michigan 48201
http://physics.clas.wayne.edu

General University Information
President: M. Roy Wilson, M.D., M.S.
Dean of Graduate School: Ambika Mather, Interim Dean
University website: http://wayne.edu
Control: Public
Setting: Urban
Total Faculty: 1,806
Total Graduate Faculty: 1,150
Total number of Students: 30,765
Total number of Graduate Students: 8,032

Department Information
Department Chairman: Ratna Naik, Ph.D, Chair
Department Contact: Ratna Naik, PhD., Chair
  Total full-time faculty: 31
  Total number of full-time equivalent positions: 31
  Full-Time Graduate Students: 63
  First-Year Graduate Students: 10
  Female First-Year Students: 1
  Total Post Doctorates: 8

Department Address
666 West Hancock
Suite # 135 Physics Research Building
Detroit, MI 48201
Phone: (313) 577-2721 (C)
Fax: (313) 577-3932
E-mail: rnaik@wayne.edu
Website: http://physics.clas.wayne.edu

ADMISSIONS

Admission Contact Information
Address admission inquiries to: Graduate Admissions Committee, Dept. of Physics, Wayne State University, Detroit, MI 48201
Phone: (313) 577-2775
E-mail: ashis@wayne.edu
Admissions website: http://physics.clas.wayne.edu

Application deadlines
Fall admission:
U.S. students: July 1   *Int'l. students*: May 1

Application fee
U.S. students: $50   *Int'l. students*: $50

Admissions information
For Fall of 2013:
  Number of applicants: 96
  Number admitted: 14
  Number enrolled: 14

Admission requirements
Bachelor's degree requirements: Bachelor's degree in physics or related fields is required.
Minimum undergraduate GPA: 3.0

GRE requirements
The GRE is required.
  Quantitative score: 700

Advanced GRE requirements
The Advanced GRE is recommended.

TOEFL requirements
The TOEFL exam is required for students from non-English-speaking countries.
  PBT score: 550
550 or higher on paper based test. 213 or higher on Computer based test.

Other admissions information
Additional requirements: A GPA below 3.0 would require a probationary admission.
Undergraduate preparation assumed: J.R. Taylor, Classical Mechanics; D.J. Griffith, Introduction to Quantum Mechanics; Reitz, Milford and Christy Foundations of Electromagnetic Theory; A.H. Carter, Classical and Statistical Thermodynamics.

TUITION

Tuition year 2013-14:
Tuition for in-state residents
  Full-time students: $554.15 per credit
  Part-time students: $554.15 per credit
Tuition for out-of-state residents
  Full-time students: $1,200.35 per credit
  Part-time students: $1,200.35 per credit
Credit hours per semester to be considered full-time: 8
Deferred tuition plan: No
Health insurance: Not available.
Other academic fees: Registration Fee: $243.30 Omnibus Fee: 42.15 per cr. Fitness Maintenance Fee: $25.00
Academic term: Semester

Teaching Assistants, Research Assistants, and Fellowships
Number of first-year
  Teaching Assistants: 10
  Fellowship students: 3
Average stipend per academic year
  Teaching Assistant: $18,071
  Research Assistant: $18,071
  Fellowship student: $18,000

FINANCIAL AID

Loans
Loans are available for U.S. students.
Loans are not available for international students.
GAPSFAS application required: Yes
FAFSA application required: No

For further information

Address financial aid inquiries to: Office of Student Financial Aid, Wayne State University.
Phone: (313) 577-3378
E-mail: financialaid@wayne.edu
Financial aid website: http://finaid.wayne.edu

HOUSING

Availability of on-campus housing

Single students: Yes
Married students: Yes

For further information

Address housing inquiries to: Director, Housing & Residential Life, 598 Student Center Bldg., Detroit, MI 48202.
Phone: (313) 577-2116
E-mail: housing@wayne.edu
Housing aid website: http://www.housing.wayne.edu

Table A—Faculty, Enrollments, and Degrees Granted

Research Specialty	2012-13 Faculty	Enrollment Fall 2012 Master's	Enrollment Fall 2012 Doctorate	Number of Degrees Granted 2012-13 (2008-13) Master's	Number of Degrees Granted 2012-13 (2008-13) Terminal Master's	Number of Degrees Granted 2012-13 (2008-13) Doctorate
Applied Physics	–	–	–	–	–	–
Astrophysics	2	3	1	1(2)	1(2)	1(1)
Atomic, Molecular, & Optical Physics	1	–	1	–(6)	–(4)	3(3)
Biophysics	3	3	3	–	–	–
Condensed Matter Physics	12	3	25	4(26)	–(8)	9(22)
Low Temperature Physics	–	–	–	–	–	–
Materials Science, Metallurgy	–	–	–	–	–	–
Nuclear Physics	6	1	7	1(7)	1(3)	1(4)
Optics	–	–	–	–	–	–
Particles and Fields	5	1	15	1(10)	–(4)	1(7)
Other	1	–	–	–	–	–
Total	30	11	52	7(51)	2(21)	15(37)
Full-time Grad. Stud.	–	4	59	–	–	–
First-year Grad. Stud.	–	6	5	–	–	–

GRADUATE DEGREE REQUIREMENTS

Master's: The Master degree is offered with (M.S.) and without (M.A.) thesis in various areas of physics. Requirements for the M.S. degree are 24 credits of course work at the 5000 level or above plus an eight-credit thesis, while the M.A. degree requires 29 credits of course work at the 5000 level or above plus a three-credit essay. Both degrees require at least nine credits at the 7000 level or above with at least half of the course work in physics. Students must maintain a 3.0 GPA and must complete their degree within six years. A final oral exam over the thesis or essay is required of all students.

Doctorate: The Ph.D. degree has a basic requirement of 90 credits which include 30 dissertation credits. Courses at the graduate level in mathematical physics, mechanics and dynamics, quantum mechanics, electromagnetic theory, and statistical mechanics are required for all students as well as certain other courses depending on the area of concentration. A written Ph.D. qualifying exam usually taken after the end of the student's first year, a preliminary oral exam, and a final dissertation defense are the other major requirements. A 3.0 GPA must also be maintained. There is a seven-year time limit for completion of the degree.

Thesis: Thesis may be written in absentia.

SPECIAL EQUIPMENT, FACILITIES, OR PROGRAMS

The Department has numerous well-equipped research laboratories with concentrated efforts in the areas of high-energy nuclear and particle physics, applied physics, biophysics, and condensed matter physics.

The relativistic heavy-ion group participates in two major international collaborations: the STAR experiment at the Relativistic Heavy-Ion Collider (RHIC) at Brookhaven National Laboratory NY, and the ALICE experiment at the Large Hadron Collider (LHC) at CERN in Switzerland. On campus, the group operates laboratory facilities for the design, construction and testing of precision electromagnetic and hadronic calorimeters. Facilities for testing and development of high-density silicon drift detector arrays for use in nuclear physics experiments are also available.

The nuclear theory group is exploring a range of topics related to these experimental results, including the hydrodynamics of the quark-gluon plasma. The experimental particle physics groups are part of the CLEO collaboration CESR at Cornell University, the CDF collaboration at Fermilab, the CMS collaboration at CERN, and the Belle II collaboration at KEK (Japan) and have set up facilities on campus for the design and development of electronic systems for the particle detectors and accelerators. They also have a leadership role in a future high energy electron-positron linear collider.

The particle theory group works on understanding the fundamental properties of elementary particles, including phenomenology of quantum chromodynamics in heavy quark systems, studies of CP-violation and Dark Matter.

The astrophysics group is closely involved with the Sloan Digital Sky Survey to study type 1A supernovae.

Research programs in condensed matter physics have extensive materials characterization and synthesis facilities available for investigating problems ranging from superconductivity to magnetism to semiconductors to pattern formation to nanoconfined fluids. There is a strong emphasis on nanotechnology, with research projects including studies on carbon nanotubes and graphene, two-dimensional electron gas, nanoparticles, and thin films. The materials characterization tools available include systems for electrical, dielectric, thermodynamic, and optical studies, all of which can be performed under a range of temperatures and magnetic fields. Some specialized measurements available for these studies include micro Raman spectroscopy, atomic force spectroscopy, x-ray photoemission spectroscopy, and fluorescence correlation spectroscopy. The condensed matter theory group uses state-of-the-art computer facilities along with analytical calculations to investigate the dynamics of systems far from equilibrium, pattern formation, and positron interactions with biologically relevant molecules.

The biophysics group is interested in applying techniques from Physics to solve problems in medicine and biology. Active projects in the biophysics group include studies of molecular motors, such as myosin, protein-binding interactions using atomic force microscopy, and cancer detection using Raman spectroscopy.

The thermal wave imaging group uses several visible and infrared lasers, two infrared video cameras, and three image processing systems to investigate thermal properties and defects in materials. Argon ion and helium-neon lasers, a photon correlation spectrometer, and a polarizing microscope with an attached spectrometer and diode array detector are used to investigate properties of liquid crystals.

Information about the graduate program and other research activities in the department is also available at http://www.clas.wayne.edu/physics.

Table B—Separately Budgeted Research Expenditures by Source of Support

Source of Support	Departmental Research	Physics-related Research Outside Department
Federal government	$5,403,310	
State/local government	$575,723	
Non-profit organizations		
Business and industry		
Other	$53,769	
Total	**$6,032,802**	

Table C—Separately Budgeted Research Expenditures by Research Specialty

Research Specialty	No. of Grants	Expenditures ($)
Atomic, Molecular, & Optical Physics	–	
Biophysics	1	$250,232
Condensed Matter Physics	8	$876,650
Nuclear Physics	2	$2,913,789
Particles and Fields	8	$1,102,330
Other	0	
Total	**19**	**$5,143,001**

FACULTY

Professor

Cinabro, David A., Ph.D., University of Wisconsin-Madison, 1991. *Astrophysics, High Energy Physics, Particles and Fields.* Experimental high-energy particle physics; Astrophysics, Charm physics in e+e− collisions at CLEO; luminosity spectrum at linear collider.

Cormier, Thomas M., Ph.D., Massachusetts Institute of Technology, 1974. *Nuclear Physics.* Experimental nuclear physics; relativistic heavy-ion collisions; properties of quark matter and the QCD phase transition.

Gavin, Sean, Ph.D., University of Illinois, 1987. *Nuclear Physics, Theoretical Physics.* Theoretical nuclear; relativistic heavy ion physics, quark gluon plasma theory, QCD phenomenology.

Harr, Robert F., Ph.D., University of California, Berkeley, 1990. *Particles and Fields.* Experimental high-energy particle physics; CP violation, rare decays and Higgs particles at CDF.

Hoffmann, Peter M., Ph.D., Johns Hopkins University, 1999. Associate Dean, College of Liberal Arts and Sciences. *Biophysics, Condensed Matter Physics, Nano Science and Technology.* Experimental soft condensed matter physics, biophysics, nanomechanics; atomic force microscopy studies of interatomic and intermolecular forces.

Karchin, Paul E., Ph.D., Cornell University, 1982. *Particles and Fields.* Experimental particle physics; CDF and CMS experiments.

Keyes, Paul H., Ph.D., University of Maryland, 1972. *Condensed Matter Physics.* Experimental and theoretical condensed matter physics; liquid crystals; phase transitions.

Morgan, Caroline G., Ph.D., Princeton University, 1980. *Condensed Matter Physics, Theoretical Physics.* Theoretical condensed matter physics; Solid State Theory: Dynamic Processes at Surfaces/Defects in Semiconductors.

Nadgorny, Boris E., Ph.D., Stony Brook University, 1996. *Condensed Matter Physics.* Experimental condensed matter physics; Spin Polarization Mapping, Magnetism and Magnetic Materials.

Naik, Ratna, Ph.D., West Virginia University, 1982. Chair of the Department. *Condensed Matter Physics.* Experimental condensed matter physics; Materials Science, Magnetism and Magnetic Materials. Magnetic Nanoparticles; Sensor Materials.

Petrov, Alexey A., Ph.D., University of Massachusetts, Amherst, 1997. *Astrophysics, Particles and Fields.* Theoretical particle physics; heavy quark physics, CP violation, QCD, LHC phenomenology, effective field theories. Theoretical particle astrophysics: Dark matter.

Pruneau, Claude A., Ph.D., Universite Laval, Quebec, 1987. Director of Planetarium and Outreach. *High Energy Physics, Nuclear Physics.* Experimental nuclear physics; RHIC (relativistic heavy-ion collisions),LHC (Cern); Quark Gluon Plasma.

Voloshin, Sergei A., Ph.D., Moscow Engineering & Physics Institute, 1980. *High Energy Physics, Nuclear Physics.* Experimental nuclear physics; RHIC (relativistic heavy ion collisions); Phenomenology of multiparticle production.

Wadehra, Jogindra M., Ph.D., New York University, 1977. Associate Department Chair and Department Graduate Advisor. *Astrophysics, Atomic, Molecular, & Optical Physics, Theoretical Physics.* Theoretical atomic and molecular physics, astrophysics; the scattering of positrons (antiparticles of electrons) and electrons from various atoms and molecules.

Associate Professor

Bonvicini, Giovanni, Ph.D., Universita di Bologna, 1981. *Astrophysics, High Energy Physics, Particles and Fields.* Experimental high-energy particle, CLEO, and astrophysics.

Bowen, David, Ph.D., University of Pennsylvania, 1966. *Computer Science, Other.* Computers; science, Internet; creativity.

Huang, Zhi-Feng, Ph.D., Tsinghua University, 1999. *Condensed Matter Physics, Nano Science and Technology, Theoretical Physics.* Theoretical condensed matter physics; Nonequilibrium, nonlinear phenomena in complex dynamical systems. Nanostructures and defect dynamics in strained thin films; Mesophase dynamics of block copolymer films; nonlinear pattern formation and defect chaos.

Lawes, Gavin, Ph.D., Cornell University, 2001. *Condensed Matter Physics, Nano Science and Technology.* Experimental condensed matter physics; Multiferroics and magnetodielectrics; Magnetic semiconducting oxides and magnetic nanoparticles.

Majumder, Abhijit, Ph.D., McGill University, Montreal, 2002. *Nuclear Physics, Theoretical Physics.* Theoretical nuclear physics; study of extended systems of QCD matter, perturbative QCD calculations, lattice QCD simulations.

Mukhopadhyay, Ashis, Ph.D., Kansas State University, 2000. *Condensed Matter Physics.* Experimental soft condensed matter physics; Materials Science.

Padmanabhan, Karur R., Ph.D., University of Poona, 1975. *Condensed Matter Physics.* Experimental condensed matter physics; Materials Science, materials modification; ion-solid interaction and ion channeling.

Putschke, Joern H., Ph.D., Technical University of Munich, 2004. *High Energy Physics, Nuclear Physics.* Experimental high-energy nuclear physics; RHIC (relativistic heavy ion collisions).

Zhou, Zhixian, Ph.D., Florida State University, 2004. *Condensed Matter Physics, Nano Science and Technology.* Experimental condensed matter physics: Individual nanoscale materials and single organic molecules: synthesis and characterization, nanoscale device fabrication, electrical transport measurements.

Assistant Professor

Cackett, Edward M., Ph.D., University of St. Andrews, 2006. *Astronomy, Astrophysics.* Experimental astrophysics; compact objects (neutron stars and black holes), accretion across the mass scale; from neutron starts and black holes in X-ray binaries to Active Galactic Nuclei (AGN).

Chu, Xiang-Qiang, Ph.D., Massachusetts Institute of Technology, 2010. *Biophysics, Nano Science and Technology.* Experimental biophysics; probing the structure and dynamics of biomolecules, nanomaterials; protein structures and dynamics using neutron and x-ray scattering.

Huang, Jian, Ph.D., Michigan State University, 2001. *Condensed Matter Physics, Low Temperature Physics, Nano Science and Technology.* Experimental condensed matter physics; Quantum transport and low temperature physics; strongly correlated many-body quantum systems; interaction-driven in high-purity two-demensional systems.

Kelly, Christopher V., Ph.D., University of Michigan-Ann Arbor, 2009. *Biophysics.* Experimental biophysics; Subdiffraction-limited optics and biological membranes; spectroscopy.

Paz, Gil, Ph.D., Cornell University, 2006. *High Energy Physics, Particles and Fields, Theoretical Physics.* Theoretical particle physics; QCD; effective field theories; supersymmetry.

Sakamoto, Takeshi, Ph.D., Kanazawa University, 2001. *Biophysics, Medical, Health Physics.* Experimental biophysics: Mechanisms of myosin-dependent motility, protein-protein interactions, actin-myosin interactions and visualization using single molecule imaging techniques in vitro and in vivo.

Research Professor

Nazri, Gholam-Abbas, Ph.D., Case Western Reserve University, 1981. *Condensed Matter Physics.* Experimental: Energy storage and generation materials and systems.

Research Assistant Professor

Bao, Jianjun, Ph.D., Gunma University, 2002. Experimental biophysics: Cell biology and molecular/cellular imaging.

DEPARTMENTAL RESEARCH SPECIALTIES AND STAFF

Theoretical

Atomic, Molecular, & Optical Physics. Studying the scattering of positrons (antiparticles of electrons) and electrons from various atoms and molecules. Wadehra.

Condensed Matter Physics. Solid State Theory; Dynamic Processes at Surface/defects in Semiconductors. Nanostructured Systems Outside of Equilibrium. Zhi-Feng Huang, Morgan.

Nuclear Physics. QCD phenomenology; relativistic heavy ion collisions; quark gluon plasma. Extended systems of QCD matter at all temperatures and densities that are experimentally accessible. Gavin, Majumder.

Particles and Fields. Heavy quark physics, CP violation, electroweak physics, and QCD phenomenology; Higgs particles and W boson processes. Paz, Petrov.

Plasma and Fusion. Production and diagnostics of negative ion beams. Wadehra.

Experimental

Applied Physics. Thermal-wave imaging studies, including photo-acoustics, laser-induced ultrasound, and laser interferometry.

Astrophysics. Accretion onto neutron stars and black holes in X-ray Binaries & Active Galactic Nuclei (AGN), Relativistic Fe Lines in neutron star los-mass x-ray binaries, observational probes of the neutron star equation of state (neutron star radii and masses). Cackett, Cinabro.

Biophysics. Molecular Motor. Single molecule imaging studies with Total Internal Reflection Fluorescent (TIRF), microscopy, Molecular and cellular imaging, protein dynamics using neutron and x-ray scattering. Biological Physics - Subdiffraction-limited optics and biological membranes by using nanoscale engineering and biophysical techniques. Bao, Chu, Kelly, Sakamoto.

Condensed Matter Physics. Atomic force and scanning tunneling microscopy of surfaces; magnetic materials and device applications; conventional and high-temperature superconductivity; Andreev reflection; electron and Josephson tunneling; spin transport and spin polarization; spintronics; ion channeling; thin-film and materials research; surface studies and modification; energy storage and generation materials; liquid crystals; calorimetric and ultrasonic properties; Raman spectroscopy. Hoffmann, Jian Huang, Keyes, Lawes, Mukhopadhyay, Nadgorny, Nazri, Sakamoto, Zhou.

Nuclear Physics. Nuclear structure; nuclear lifetime determinators; relativistic heavy-ion collisions; monoenergetic e+e− pair production in super-heavy nucleus-nucleus collisions; quark gluon plasma physics. Cormier, Pruneau, Putschke, Voloshin

Other. Computers, networks, science, society and creativity. Bowen.

Particles and Fields. Bs mixing, heavy quark physics, CP violation and Higgs particles at CDF (Fermilab); charm physics at Cornell (CLEO); design studies for linear collider. Bonvicini, Cinabro, Harr, Karchin.

View additional information about this department at
www.gradschoolshopper.com

WESTERN MICHIGAN UNIVERSITY

DEPARTMENT OF PHYSICS

Kalamazoo, Michigan 49008-5252
http://tesla.physics.wmich.edu/

General University Information

President: John M. Dunn
Dean of Graduate School: Susan Stapleton
University website: http://www.wmich.edu/
Control: Public
Setting: Suburban

Total Faculty: 861
Total Graduate Faculty: 762
Total number of Students: 24,598
Total number of Graduate Students: 5,120

Department Information
Department Chairman: Kirk Korista, Chair
Department Contact: Lori Krum, Office Coordinator
 Total full-time faculty: 18
 Total number of full-time equivalent positions: 18
 Full-Time Graduate Students: 37
 First-Year Graduate Students: 8
 Female First-Year Students: 5
 Total Post Doctorates: 5

Department Address
1903 W Michigan Avenue
Kalamazoo, MI 49008-5252
Phone: (269) 387-4940
Fax: (269) 387-4939
E-mail: physics-department@wmich.edu
Website: http://tesla.physics.wmich.edu/

ADMISSIONS

Admission Contact Information
Address admission inquiries to: Graduate Advisor, Department
 of Physics.
Phone: (269) 387-4940
E-mail: physics-department@wmich.edu
Admissions website: http://www.wmich.edu/admissions/

Application deadlines
Fall admission:
U.S. students: February 15   *Int'l. students*: February 15

Application fee
U.S. students: $40   *Int'l. students*: $100
For full consideration for an assistantship, a completed file is
 needed by February 15.

Admissions information
For Fall of 2012:
 Number of applicants: 24
 Number admitted: 15
 Number enrolled: 8

Admission requirements
Bachelor's degree requirements: Bachelor's degree in physics
 or related discipline is required.
Minimum undergraduate GPA: 3.0

GRE requirements
The GRE is required.
 Quantitative score: 155
 Verbal score: 138
 Analytical score: 3
 Mean GRE score range (25th–75th percentile): 60th
GRE is not required for the master's degree; however, preference
 in admission is given to applicants who have taken and earned
 a good score on the exam.

Advanced GRE requirements
The Advanced GRE is recommended.
Preference in admission is given to applicants who have taken
 and earned a good score on the GRE Physics exam.

TOEFL requirements
The TOEFL exam is required for students from non-English-
 speaking countries.
 PBT score: 550
 iBT score: 80

Other admissions information
Undergraduate preparation assumed: Halliday and Resnick,
 Fundamentals of Physics; Sprott, Introduction to Modern
 Electronics; Fowles, Analytical Mechanics; Christy, Reitz,

and Milford, Electricity and Magnetism; Eisberg and Resnick,
Quantum Physics; Meyer-Arendt, Introduction to Classical
and Modern Optics; Sears, Thermodynamics, Kinetic Theory
and Statistical Thermodynamics.

TUITION
Tuition year 2013-14:
Tuition for in-state residents
 Full-time students: $496.69 per credit
 Part-time students: $496.69 per credit
Tuition for out-of-state residents
 Full-time students: $1,052.01 per credit
 Part-time students: $1,052.01 per credit
Master graduate assistants can receive up to 22 credit hours of
 tuition waiver. Doctoral associates and doctoral graduate as-
 sistants can receive up to 24 credit hours of tuition waiver.
Credit hours per semester to be considered full-time: 6
Deferred tuition plan: Yes
Health insurance: Available at the cost of $1,350 per year.
Other academic fees: $411.50/semester enrollment fee (5 or more
 credit hours); $26/semester student assessment fee; $8/semes-
 ter sustainability fee; $25 international student fee/semester;
 $300 one-time records initiation fee.
Academic term: Semester
Number of first-year students who received full tuition waivers: 5

Teaching Assistants, Research Assistants, and Fellowships
Number of first-year
 Teaching Assistants: 4
 Research Assistants: 1
Average stipend per academic year
 Teaching Assistant: $13,762
 Research Assistant: $14,125
Summer TA and RA assistantships are available.

FINANCIAL AID

Application deadlines
Fall admission:
U.S. students: February 15   *Int'l. students*: February 15

Loans
Loans are available for U.S. students.
Loans are not available for international students.
GAPSFAS application required: No
FAFSA application required: Yes

For further information
Address financial aid inquiries to: WMU Student Financial Aid,
 1903 W Michigan Ave, Kalamazoo, MI 49008-5337.
Phone: (269) 387-6000
E-mail: finaid-info@wmich.edu
Financial aid website: http://www.wmich.edu/finaid/

HOUSING

Availability of on-campus housing
 Single students: Yes
 Married students: Yes

For further information
Address housing inquiries to: WMU Campus Apartments, 1903
 W Michigan Ave, Kalamazoo, MI 49008-5312.
Phone: (269) 387-2175
E-mail: ca-info@wmich.edu
Housing aid website: http://www.wmich.edu/housing/

Table A—Faculty, Enrollments, and Degrees Granted

Research Specialty	2012-13 Faculty	Enrollment Fall 2012		Number of Degrees Granted 2012-13 (2008-13)		
		Master's	Doctorate	Master's	Terminal Master's	Doctorate
Astronomy	2	1	1	–	–(2)	–
Atomic, Molecular, & Optical Physics	5	1	7	–(3)	1(4)	1(7)
Condensed Matter Physics	5	–	8	–(6)	–(4)	4(7)
Nuclear Physics	4	–	7	–(2)	–(2)	–(4)
Physics and other Science Education	2	3	1	–	4(4)	–
Non-specialized	–	3	5	1(1)	–	–
Total	18	8	29	1(12)	5(16)	5(18)
Full-time Grad. Stud.	–	8	29	–	–	–
First-year Grad. Stud.	–	2	6	–	–	–

GRADUATE DEGREE REQUIREMENTS

Master's: Thirty semester hours of graduate credit with a GPA of 3.0/4.0 or better are required. Students may transfer only six hours from another institution. There is no residency requirement. 18 hours of coursework required in Mathematical Physics, Classical Mechanics, Electricity and Magnetism, Quantum Mechanics, Statistical Mechanics, and Research Seminar. In addition, six hours in physics, mathematics, computer science, or other departments chosen with the consent of the graduate advisor are required. Successful completion of Mathematical Physics, Quantum Mechanics I, Classical Mechanics, and Electricity and Magnetism I with a GPA of 3.0 or better or satisfactory completion of a master's thesis (6 credit hours) is also required.

Doctorate: Sixty semester hours of graduate credit with a GPA of 3.0 or better are required. Successful completion of Mathematical Physics, Quantum Mechanics I, Classical Mechanics, and Electricity and Magnetism I with a GPA of 3.0 is required.

Thesis: Thesis may not be written in absentia.

SPECIAL EQUIPMENT, FACILITIES, OR PROGRAMS

A 12-MV tandem Van de Graaff accelerator with the associated equipment and electronics and support staff is used for atomic, nuclear, and applied research. A well-equipped instrument shop and electronic shops with technical support staff are available. A computer laboratory is reserved for physics graduate students. The department has ready access to the University alpha-cluster and Sun systems. A superconducting NMR spectrometer and scanning electron microscopes are also available at the University.

Table B—Separately Budgeted Research Expenditures by Source of Support

Source of Support	Departmental Research	Physics-related Research Outside Department
Federal government	$1,176,669	
State/local government		
Non-profit organizations		
Business and industry		
Other		
Total	$1,176,669	

Table C—Separately Budgeted Research Expenditures by Research Specialty

Research Specialty	No. of Grants	Expenditures ($)
Astronomy	1	$173,221
Atomic, Molecular, & Optical Physics	3	$319,800
Condensed Matter Physics	3	$131,511
Nuclear Physics	4	$324,760
Physics and other Science Education	4	$227,377
Total	15	$1,176,669

FACULTY

Professor

Berrah, Nora, Ph.D., University of Virginia, 1987. *Atomic, Molecular, & Optical Physics.*

Burns, Clement, Ph.D., University of California, San Diego, 1993. *Condensed Matter Physics.*

Chung, Sung G., Ph.D., University of Tokyo, 1981. *Condensed Matter Physics.*

Gorczyca, Thomas, Ph.D., University of Colorado, 1990. *Atomic, Molecular, & Optical Physics.*

Halderson, Dean W., Ph.D., University of Kansas, 1974. *Nuclear Physics.* Intermediate energy nuclear theory.

Kamber, Emanuel Y., Ph.D., University of London, 1983. *Atomic, Molecular, & Optical Physics.*

Korista, Kirk, Ph.D., Ohio State University, 1990. Department Chair. *Astronomy.* Numerical simulations of photoionized gaseous nebulae.

McGurn, Arthur R., Ph.D., University of California, Santa Barbara, 1975. *Condensed Matter Physics.*

Pancella, Paul V., Ph.D., Rice University, 1987. *Nuclear Physics.*

Paulius, Lisa, Ph.D., University of California, San Diego, 1993. *Condensed Matter Physics.*

Tanis, John A., Ph.D., New York University, 1976. *Atomic, Molecular, & Optical Physics.*

Wuosmaa, Alan H., Ph.D., University of Pennsylvania, 1988. *Nuclear Physics.*

Associate Professor

Bautista, Manuel, Ph.D., Ohio State University, 1997. *Astronomy.* Theoretical modeling of plasma dynamics and spectra.

Famiano, Michael, Ph.D., Ohio State University, 2001. *Astrophysics, Nuclear Physics.*

Henderson, Charles, Ph.D., University of Minnesota, 2002. *Physics and other Science Education.*

Rosenthal, Alvin S., Ph.D., University of Colorado, 1978. Graduate . *Optics.* Theoretical nonlinear optics.

Schuster, David, Ph.D., University of Witwatersrand, 1972. *Physics and other Science Education.*

Professional Specialist

Kayani, Asghar, Ph.D., Ohio University, 2003. *Condensed Matter Physics.*

DEPARTMENTAL RESEARCH SPECIALTIES AND STAFF

Theoretical

Astronomy. Atomic data and spectral models for astrophysics; numerical simulations of spectra of photoionized gaseous nebulae. Bautista, Korista.

Atomic, Molecular, & Optical Physics. Electron-ion collisions; ion-atom collisions; many-body theory; R-matrix theory; atomic photoionization; photodetachment; dielectronic recombinations; nonlinear optics. Gorczyca, Rosenthal.

Condensed Matter Physics. Photonic crystals; Anderson localization; scattering of light by the localized surface polaritons of disordered media; inelastic neutron scattering from mixed Ising systems; computer simulations of the dynamics of Heisenberg magnetic alloys; fractons; nano-electronics; spintronics; quantum computing; superconductor-insulator transition; metal-insulator transition; novel many-body techniques. Chung, McGurn.

Nuclear Physics. The nuclear many-body problem; hypernuclear structure; reaction theory; strangeness-exchange reactions; high-energy photo-nuclear reactions; relativistic nuclear structure. Halderson.

Experimental

Astronomy. Acquisition and interpretation of spectroscopic data from active galactic nuclei, high-redshift quasars, and galactic emission line sources. Bautista, Korista.

Atomic, Molecular, & Optical Physics. Studies of strongly correlated electron systems including photon-atom, photon-molecule, photon-cluster, and photon negative ion interactions; spectroscopy with lasers and third-generation light sources; mechanisms of electronic excitation and charge-changing are investigated for collisions of ions with atomic and molecular targets. Berrah, Kamber, Tanis.

Condensed Matter Physics. Studies of highly correlated electron systems, including metal ammonia compounds and high-temperature superconductor parent compounds; development of inelastic x-ray scattering using synchrotron radiation; research on the properties of organic semiconductors; electrical and magnetic properties of high-temperature superconductors; flux vortex dynamics; metal-insulator transitions in rare-earth nickel oxides; electrical transport under high pressure; ion beam analysis of materials; development of solid oxide fuel cells and hydrogen storage materials. Burns, Kayani, Paulius.

Nuclear Physics. Two-nucleon problem; nucleon-nucleus phenomenology; direct capture of protons by light nuclei; experiments relevant to determination of spectroscopic factors; in-beam gamma ray spectroscopy; pion production; structure of exotic light nuclei; clustering phenomena in light nuclei; few nucleon transfer reactions for nuclear structure and astrophysics; relativistic heavy-ion collisions. Pancella, Wuosmaa.

Physics and other Science Education. Curriculum design, evaluation, and assessment; cognitive aspects of the teaching and learning of science, such as conceptual understanding, problem solving, and epistemology; teacher beliefs and teacher professional development. Henderson, Schuster.

View additional information about this department at www.gradschoolshopper.com

---

MINNESOTA STATE UNIVERSITY, MANKATO

DEPARTMENT OF PHYSICS AND ASTRONOMY

Mankato, Minnesota 56001
http://cset.mnsu.edu/pa

General University Information
President: Richard Davenport
Dean of Graduate School: Barry Ries
University website: http://www.mnsu.edu
Control: Public
Setting: Rural
Total Faculty: 453
Total Graduate Faculty: 395
Total number of Students: 15,640
Total number of Graduate Students: 1,981

Department Information
Department Chairman: Youwen Xu, Chair
Department Contact: Youwen Xu, Professor
  Total full-time faculty: 9
  Total number of full-time equivalent positions: 9
  Full-Time Graduate Students: 8
  First-Year Graduate Students: 5
  Female First-Year Students: 3

Department Address
141 Trafton Science Center, North
Mankato, MN 56001
Phone: (507) 389-5743
Fax: (507) 389-1095

E-mail: youwen.xu@mnsu.edu
Website: http://cset.mnsu.edu/pa

ADMISSIONS

Admission Contact Information
Address admission inquiries to: College of Graduate Studies, 115 Alumni Foundation Center, Mankato, MN 56001
Phone: (507) 389-2321
E-mail: grad@mnsu.edu
Admissions website: http://grad.mnsu.edu/home.html

Application deadlines
Fall admission:
U.S. students: July 1   *Int'l. students*: May 1
Spring admission:
U.S. students: November 1   *Int'l. students*: October 1

Application fee
U.S. students: $40   *Int'l. students*: $40

Admissions information
For Fall of 2012:
  Number of applicants: 5
  Number admitted: 3
  Number enrolled: 2

Admission requirements

Bachelor's degree requirements: A Bachelor's degree in physics or related field is required.

GRE requirements

The GRE is recommended.

Advanced GRE requirements

The Advanced GRE is not required.

TOEFL requirements

The TOEFL exam is required for students from non-English-speaking countries.
PBT score: 530
iBT score: 72

Other admissions information

Additional requirements: At least an undergraduate physics minor is required for admission into the M.S. program. A minimum score of TOEFL 82 (IBT) is required for teaching assistantship.
Undergraduate preparation assumed: Completion of at least two junior-senior level physics courses.

TUITION

Tuition year 2011–2012:
Tuition for in-state residents
   Full-time students: $345 per credit
   Part-time students: $345 per credit
Tuition for out-of-state residents
   Full-time students: $345 per credit
   Part-time students: $345 per credit
Tuition and fees to be determined for 2012-13. Students with an assistantship have tuition waived for up to 18 credit hours per academic year.
Credit hours per semester to be considered full-time: 6
Deferred tuition plan: No
Health insurance: Available at the cost of $1,650 per year.
Other academic fees: $36/credit
Academic term: Semester
Number of first-year students who received full tuition waivers: 1

Teaching Assistants, Research Assistants, and Fellowships

Number of first-year
   Teaching Assistants: 1
Average stipend per academic year
   Teaching Assistant: $9,000
Summer employment is possible.

FINANCIAL AID

Application deadlines

Fall admission:
U.S. students: March 15
Spring admission:
U.S. students: March 15   *Int'l. students*: March 15

Loans

Loans are available for U.S. students.
Loans are not available for international students.
GAPSFAS application required: No
FAFSA application required: No

For further information

Address financial aid inquiries to: Student Financial Services, Minnesota State University, Mankato, 120 Wigley Administration Center, Mankato, MN 56001.
Phone: (800) 722-0544
E-mail: campushub@mnsu.edu

Financial aid website: http://www.mnsu.edu/campushub/programs/

HOUSING

Availability of on-campus housing

Single students: Yes
Married students: No

For further information

Address housing inquiries to: Department of Residential Life, 111 Carkoski Commons, Minnesota State University, Mankato, MN 56001.
Phone: (800) 722-0544
E-mail: reslife@mnsu.edu
Housing aid website: http://www.mnsu.edu/reslife/

Table A—Faculty, Enrollments, and Degrees Granted

Research Specialty	2012-13 Faculty	Enrollment Fall 2012		Number of Degrees Granted 2012-13 (2008-13)		
		Master's	Doctorate	Master's	Terminal Master's	Doctorate
Astronomy	1	–	–	–	–	–
Astrophysics	1	1	–	–	–(2)	–
Condensed Matter Physics	3	2	–	–	–(5)	–
Medical, Health Physics	–	–	–	–	–	–
Nuclear Physics	2	–	–	–	1(2)	–
Physics and other Science Education	1	1	–	–	1(5)	–
Space phyiscs	1	–	–	–	–	–
Non-specialized	–	2	–	–	–	–
Total	9	6	–	–	2(14)	–
Full-time Grad. Stud.	–	5	–	–	–	–
First-year Grad. Stud.	–	2	–	–	–	–

GRADUATE DEGREE REQUIREMENTS

Master's: The Physics M.S. is the professional degree in physics. The thesis plan requires 30 semester hours with a minimum graduate GPA of 3.0 on a 4.0 scale. The non-thesis plan requires 34 semester hours. 15 semester hours must be earned in residency. Comprehensive written and oral exams are required with the thesis plan. The non-thesis plan requires a comprehensive written exam. The Master of Science in Physics Education is designed for individuals interested in strengthening their background in secondary school teaching. Previous teacher licensure is usually required. Students interested in teaching at a community college may elect the M.S. Community College Teaching track. Teacher certification is not required for this track.

Other Degrees: Master of Arts in Teaching (MAT). This degree requires a minimum of 34 semester hours which must include the 19 hours required for Minnesota licensure, a thesis, and internship. No foreign language is required. Graduate GPA of 3.0 on a 4.0 scale required. 15 semester hours must be earned as a resident student. Comprehensive written and oral exams required.

Thesis: Thesis may be written in absentia.

SPECIAL EQUIPMENT, FACILITIES, OR PROGRAMS

Some of the research equipment available includes an Auger spectrometer, scanning electron microscope, x-ray diffractometer, and facilities for microcomputer and materials research. A 0.5-meter telescope with photometers and a CCD camera is available. Facilities for fabrication and characterization of high-critical-temperature superconductors are available. A 400-kV Van de Graaff particle accelerator is available. A high performance computer with 140 64-bit parallel processors is available.

Table B—Separately Budgeted Research Expenditures by Source of Support

Source of Support	Departmental Research	Physics-related Research Outside Department
Federal government		$121,000
State/local government	$2,000	
Non-profit organizations		
Business and industry		
Other		
Total	$2,000	$121,000

Table C—Separately Budgeted Research Expenditures by Research Specialty

Research Specialty	No. of Grants	Expenditures ($)
Applied Physics	1	$2,000
Atmosphere, Space Physics, Cosmic Rays	2	$121,000
Total	3	$123,000

FACULTY

Professor

Eskridge, Paul, Ph.D., University of Washington, 1987. *Astrophysics*. Extragalactic astronomy.

Kipp, Steven L., Ph.D., University of Pittsburgh, 1980. *Astronomy*. Astrometry; galactic structure.

Kogoutiouk, Igor, Ph.D., Chernovtsy State University, Ukraine, 1981. *Condensed Matter Physics*. Highly correlated electronic materials, theory.

Palma, Russell L., Ph.D., Rice University, 1981. *Atmosphere, Space Physics, Cosmic Rays*. Space physics, cosmic materials.

Pickar, Mark A., Ph.D., Indiana University, 1982. *Nuclear Physics*. Nuclear physics, experiment.

Wu, Hai-Sheng, Ph.D., Iowa State University, 1988. *Condensed Matter Physics*. Semiconductors. Experiment.

Xu, Youwen, Ph.D., Iowa State University, 1987. Chair. *Condensed Matter Physics, Materials Science, Metallurgy*. Condensed matter, superconductivity, magnetism, experiment.

Associate Professor

Roberts, Andrew D., Ph.D., University of Wisconsin-Madison, 1995. *Medical, Health Physics, Nuclear Engineering, Nuclear Physics*. Applied nuclear physics and medical physics.

Assistant Professor

Brown, Thomas, Ph.D., Montana State University, 2003. Advisor for physics teaching majors. Liaison with College of Education. Graduate Coordinator. *Physics and other Science Education*. Physics education.

Professor Emeritus

Pierce, James N., Ph.D., Iowa State University, 1980. *Astrophysics*. Stellar astrophysics.

Schwartzkopf, Louis, Ph.D., University of California, Berkeley, 1974. *Condensed Matter Physics, Energy Sources & Environment*. Energy and environment.

DEPARTMENTAL RESEARCH SPECIALTIES AND STAFF

Theoretical

Theoretical Condensed Matter Physics. Theory of highly correlated electronic materials. Kogoutiouk.

Experimental

Applied Nuclear Science. The Department of Physics and Astronomy operates the Minnesota State Applied Nuclear Science Lab conveniently located in Trafton Center. This facility offers a unique opportunity for students at all levels to perform in depth, hands-on research in the broad field of applied nuclear physics. The lab is equipped with a vault shielded 400 kV Van de Graaff positive ion particle accelerator, a radioactive materials fume hood for nuclear chemistry work, and an extensive array of dedicated nuclear detectors and analysis electronics. Students use the facility to gain experience in a wide range of nuclear techniques including particle accelerator development, radiation safety, and remote analysis and control systems, and perform professionally recognized experiments in applied fields such as radioisotope production for nuclear imaging methods. Roberts.

Astrometry. The stellar luminosity function. Kipp.

Experimental Nuclear Physics. Pion production and radiative capture at medium energies for dew nucleon systems. Pickar.

Experimental Semiconductor Physics. Sermiconducting thin films; Thin film solar cells; Basic research on fabrications and characterizations of Organic Light Emitting Devices (OLEDs). Wu.

Experimental Space Physics. Use ultra-high vacuum mass spectrometry to measure noble gases and nitrogen in meteorites, lunar materials, interplanetary dust particles, tektites and spacecraft flown materials. Under the umbrella of investigating the origin and early evolution of the sun and solar system, research areas of current interest include participation in NASA/JPL's Stardust Mission and the connections between interplanetary dust particles and cometary material brought back by that mission. Palma.

Extra-galactic Astronomy. Understanding the resolved properties of galaxies in the nearby Universe as a means of studying the evolution of galaxies. The stellar content, the distribution of the various phases of the interstellar medium, and the morphology of galaxies are all essential observational outcomes of the process of galaxy evolution. This requires data from across the electromagnetic spectrum, including both ground- and space-based facilities. Eskridge.

Physics Education. Physics Education Research specialist. Specific interests include: prior knowledge states of introductory college Physics students, conceptual change and cognitive development, and curricular design and implementation. Brown.

Superconductivity and Magnetic Materials. Superconducting and magnetic materials. Material synthesize and characterization. Xu.

UNIVERSITY OF MINNESOTA, DULUTH

DEPARTMENT OF PHYSICS

Duluth, Minnesota 55812
http://www.d.umn.edu/physics

General University Information
President: Eric Kaler
Dean of Graduate School: Henning Schroeder
University website: http://www.d.umn.edu/
Control: Public
Setting: Urban
Total Faculty: 595
Total Graduate Faculty: 428
Total number of Students: 11,729
Total number of Graduate Students: 721

Department Information
Department Chairman: Alec Habig, Chair
Department Contact: Jay Austin, Director of Graduate Studies
  Total full-time faculty: 7
  Total number of full-time equivalent positions: 7
  Full-Time Graduate Students: 11
  First-Year Graduate Students: 7
  Female First-Year Students: 2

Department Address
371 Marshall W. Alworth Hall
1023 University Dr
Duluth, MN 55812
Phone: (218) 726-7124 (C)
E-mail: umdphys@d.umn.edu
Website: http://www.d.umn.edu/physics

ADMISSIONS

Admission Contact Information
Address admission inquiries to: Director of Graduate Studies,
  Physics Department
Phone: (218) 726-8773
E-mail: jaustin@d.umn.edu
Admissions website: http://www.d.umn.edu/physics/grad/index.
  html

Application deadlines
Fall admission:
U.S. students: July 15   *Int'l. students*: June 15

Application fee
U.S. students: $75   *Int'l. students*: $95

Admissions information
For Fall of 2013:
  Number of applicants: 20
  Number admitted: 6
  Number enrolled: 6

Admission requirements
Bachelor's degree requirements: Bachelor's degree in physics
  or the equivalent is required.
Minimum undergraduate GPA: 2.5

GRE requirements
The GRE is not required.

Advanced GRE requirements
The Advanced GRE is not required.

TOEFL requirements
The TOEFL exam is required for students from non-English-
  speaking countries.
PBT score: 550
iBT score: 79

Other admissions information
Undergraduate preparation assumed: Taylor, Classical Mechan-
  ics; Griffiths, Introduction to Electrodynamics; Schroeder,
  Thermal Physics; Griffiths, Quantum Mechanics.

TUITION

Tuition year 2013-14:
Tuition for in-state residents
  Full-time students: $12,044 annual
  Part-time students: $1,003 per credit
Tuition for out-of-state residents
  Full-time students: $19,042 annual
  Part-time students: $1,595 per credit
Tuition is waived for half-time assistants (up to 14 credits).
Credit hours per semester to be considered full-time: 6
Deferred tuition plan: No
Health insurance: Available at the cost of 1860 per year.
Other academic fees: $294 student services fee if registered for
  more than 6 credits; $6/credit computer network access fee;
  $150 technology fee.
Academic term: Semester
Number of first-year students who received full tuition waivers: 6

Teaching Assistants, Research Assistants, and Fellowships
Number of first-year
  Teaching Assistants: 6
Average stipend per academic year
  Teaching Assistant: $13,400
  Research Assistant: $13,400

FINANCIAL AID

Application deadlines
Fall admission:
U.S. students: March 15

Loans
Loans are available for U.S. students.
Loans are not available for international students.
GAPSFAS application required: No
FAFSA application required: No

For further information
Address financial aid inquiries to: Director of Graduate Studies,
  Physics Department.
Phone: (218) 726-8773
E-mail: jaustin@d.umn.edu
Financial aid website: http://www.d.umn.edu/onestop/student-
  finances/financial-aid/

HOUSING

Availability of on-campus housing
Single students: Yes
Married students: No

For further information
Address housing inquiries to: Housing Office, 149 Lake Superior Hall.
Phone: (218) 726-7170
E-mail: mbowman@d.umn.edu
Housing aid website: http://www.d.umn.edu/kirby/housing

Table A—Faculty, Enrollments, and Degrees Granted

Research Specialty	2010–11 Faculty	Enrollment Fall 2010 Master's	Enrollment Fall 2010 Doctorate	Number of Degrees Granted 2010–11 (2006-11) Master's	Number of Degrees Granted 2010–11 (2006-11) Terminal Master's	Number of Degrees Granted 2010–11 (2006-11) Doctorate
Condensed Matter Physics	1	1	–	–(3)	–	–
Marine Science/ Oceanography	2	2	–	1(3)	–	–
Optics	1	1	–	1(2)	–	–
Particles and Fields	3	5	–	–	–	–
Relativity & Gravitation	1	–	–	1(4)	–	–
Other	–	2	–	2(2)	–	–
Total	**7**	**11**	**–**	**3(14)**	**–**	**–**
Full-time Grad. Stud.	–	11	–	–	–	–
First-year Grad. Stud.	–	5	–	–	–	–

GRADUATE DEGREE REQUIREMENTS

Master's: There are two programs of study that are planned with faculty advisors to suit the needs and interests of students. A grade point average of 2.8 must be maintained (on a scale of 4.0). All students complete a common 14 semester credit core in classical and quantum physics, six credits in related fields, and a final examination. Plan A requires a master's thesis. Plan B requires 10 additional credits in approved electives and preparation of one or more papers.
Thesis: Thesis may be written in absentia.

SPECIAL EQUIPMENT, FACILITIES, OR PROGRAMS

Lasers and vacuum UV facilities; scanning probe microscopes; low-temperature facility; computing facilities; facilities for vacuum deposition of materials; opportunities in physical limnology and oceanography through Large Lakes Observatory; well-funded multidisciplinary Natural Resources Research Institute; participation in MINOS Nova and Minerva neutrino experiments.

Table B—Separately Budgeted Research Expenditures by Source of Support

Source of Support	Departmental Research	Physics-related Research Outside Department
Federal government	$281,000	$256,000
State/local government	$10,000	$50,000
Non-profit organizations		
Business and industry		
Other		
Total	**$291,000**	**$306,000**

Table C—Separately Budgeted Research Expenditures by Research Specialty

Research Specialty	No. of Grants	Expenditures ($)
Atomic, Molecular, & Optical Physics	2	$10,000
Particles and Fields	7	$281,000
Total	**9**	**$291,000**

FACULTY

Professor
Habig, Alec, Ph.D., Indiana University, 1996. *Astrophysics.* High-energy neutrinos.
Hiller, John R., Ph.D., University of Maryland, 1980. *Computational Physics.* Theoretical particle physics.
Jordan, Thomas F., Ph.D., University of Rochester, 1962. *Theoretical Physics.* General relativity and cosmology.
Sydor, Michael, Ph.D., University of New Mexico, 1964. *Condensed Matter Physics.* Optical characterization of semiconductor properties; remote sensing of particulates in water.

Associate Professor
Austin, Jay, Ph.D., Massachusetts Institute of Technology and Woods Hole Oceanographic Institution, 1999. Physical oceanography.
Gran, Richard, Ph.D., University of Minnesota, 2002. *High Energy Physics.*
Katsev, Serguei, Ph.D., University of Ottawa, 2002. Physical oceanography.

Assistant Professor
Maps, Jonathan, Ph.D., University of Massachusetts, 1982. *Condensed Matter Physics.*
Vanchurin, Vitaly, Ph.D., Tufts University, 2005. *Cosmology & String Theory.* Cosmic strings; quantum cosmology; cosmic inflation; dark energy.

DEPARTMENTAL RESEARCH SPECIALTIES AND STAFF

Theoretical
Marine Science/Oceanography. Dynamics of large and mesoscale circulation; numerical modeling of coastal shelves, estuaries, and large lakes; coupling between sediment and water column; sediment early diagenesis. Austin, Katsev.
Particles and Fields. Elementary particles; quark model calculations; nonperturbative quantum field theory. Hiller.
Relativity & Gravitation. General relativity; quantum mechanics; cosmology. Jordan.

Experimental
Atomic, Molecular, & Optical Physics. Optical characterization of suspended particles. Sydor.
Condensed Matter Physics. Scanning probe microscopy; surface states and excitons in alkali halides; resonance Raman spectroscopy; opto-electronic materials; device physics. Maps, Sydor.
Marine Science/Oceanography. Observations of the circulation dynamics of large lakes, estuaries, and coastal shelves. Austin.
Other. High-energy neutrinos. Gran, Habig.

MISSISSIPPI STATE UNIVERSITY

DEPARTMENT OF PHYSICS AND ASTRONOMY

Mississippi State, Mississippi 39762
http://www.msstate.edu/dept/physics

General University Information
President: Mark E. Keenum
Dean of Graduate School: Lori Bruce
University website: http://www.msstate.edu
Control: Public
Setting: Rural
Total Faculty: 1,346
Total Graduate Faculty: 982
Total number of Students: 20,365
Total number of Graduate Students: 3,975

Department Information
Department Chairman: Mark A. Novotny, Head
Department Contact: David L. Monts, Professor of Physics
 Total full-time faculty: 20
 Total number of full-time equivalent positions: 20
 Full-Time Graduate Students: 46
 First-Year Graduate Students: 10
 Female First-Year Students: 1
 Total Post Doctorates: 2

Department Address
355 Lee Boulevard
Hilbun Hall, Room 125
Mississippi State, MS 39762
Phone: (662) 325-2806
Fax: (662) 325-8898
E-mail: dlm1@ra.msstate.edu
Website: http://www.msstate.edu/dept/physics

ADMISSIONS

Admission Contact Information
Address admission inquiries to: Mississippi State University, Office of Graduate Studies, P.O. Box G, Mississippi State, MS 39762-5507
Phone: (662) 325-7400
Admissions website: http://www.grad.msstate.edu

Application deadlines
Fall admission:
U.S. students: February 1   *Int'l. students*: February 1
Spring admission:
U.S. students: September 1   *Int'l. students*: September 1

Application fee
U.S. students: $60   *Int'l. students*: $60

Admissions information
For Fall of 2012:
 Number of applicants: 75
 Number admitted: 15
 Number enrolled: 10

Admission requirements
Bachelor's degree requirements: A bachelor's degree in physics is required.
Minimum undergraduate GPA: 2.75

GRE requirements
The GRE is recommended.
The GRE is strongly recommended.

Advanced GRE requirements
The Advanced GRE is recommended.
The GRE is strongly recommended.

TOEFL requirements
The TOEFL exam is required for students from non-English-speaking countries.
 PBT score: 525
 iBT score: 69

Other admissions information
Additional requirements: Students from non-English-speaking countries are required to demonstrate proficiency in English via the TOEFL exam or IELTS examination. A minimum acceptable TOEFL score for admission to the M.S. program is 525 (or for 69 iBT) and 550 (or 79 for iBT) for the Applied Physics Ph.D. program. Minimum acceptable IELTS score for admission to the M.S. program is 6.0 and 6.5 for the Applied Physics Ph.D. program.
Undergraduate preparation assumed: Undergraduate major in physics; deficiencies may be corrected by additional coursework.

TUITION

Tuition year 2013–14:
Tuition for in-state residents
 Full-time students: $6,672 per semester
 Part-time students: $370.75 per credit
Tuition for out-of-state residents
 Full-time students: $16,860 per semester
 Part-time students: $936.75 per credit
Out-of-state portion of tuition is waived for assistantship holders.
Credit hours per semester to be considered full-time: 9
Deferred tuition plan: Yes
Health insurance: Yes, 1253 per year.
Other academic fees: $300 international student charges/year
Academic term: Semester
Number of first-year students who received partial tuition waivers: 9

Teaching Assistants, Research Assistants, and Fellowships
 Number of first-year
 Teaching Assistants: 8
 Research Assistants: 1
 Average stipend per academic year
 Teaching Assistant: $13,950
 Research Assistant: $14,100

FINANCIAL AID

Application deadlines
Fall admission:
U.S. students: February 1   *Int'l. students*: February 1
Spring admission:
U.S. students: September 1   *Int'l. students*: September 1

Loans

Loans are available for U.S. students.
Loans are not available for international students.
GAPSFAS application required: No
FAFSA application required: No

For further information

Address financial aid inquiries to: Mississippi State University, Department of Physics and Astronomy, Box 5167, Mississippi State, MS 39762-5167.
Phone: (662) 325-2806
E-mail: financialaid@saffairs.msstate.edu
Financial aid website: http://www.sfa.msstate.edu

HOUSING

Availability of on-campus housing

Single students: Yes
Married students: Yes

For further information

Address housing inquiries to: Housing and Residence Life, Box 9502, Mississippi State, MS 39762.
Phone: (662) 325-3555
E-mail: housing@saffairs.msstate.edu
Housing aid website: http://www.housing.msstate.edu

Table A—Faculty, Enrollments, and Degrees Granted

Research Specialty	2012–13 Faculty	Enrollment Fall 2012		Number of Degrees Granted 2012–13 (2008–13)		
		Master's	Doctorate	Master's	Terminal Master's	Doctorate
Astrophysics	2	2	1	1(4)	–	–
Atomic, Molecular, Plasma, & Optical Physics	6	12	10	4(18)	–	-(5)
Computational Physics	6	3	9	-(7)	–	2(10)
Condensed Matter Physics	4	–	–	–	–	–
Nuclear Physics	6	1	8	-(6)	–	-(3)
Physics and other Science Education	3	–	–	–	–	–
Total	27	18	28	5(35)	–	2(18)
Full-time Grad. Stud.	–	18	26	–	–	–
First-year Grad. Stud.	–	8	2	–	–	–

GRADUATE DEGREE REQUIREMENTS

Master's: A total of 30 credit hours is required with a minimum average grade of B. The residence requirement is a minimum of 30 weeks. For the thesis option, six credit hours of thesis research are required, as is an oral examination on the thesis. For the non-thesis option, written qualifying examinations on the physics core courses and an oral examination are required.

Doctorate: At least three academic years beyond the bachelor's degree are necessary. The number of credit hours will vary. A preliminary examination is required for admission to candidacy after completion of academic coursework. A minimum of 20 credit hours of research for the dissertation must be scheduled. An oral defense of the dissertation is required.

Thesis: Thesis may be written in absentia.

SPECIAL EQUIPMENT, FACILITIES, OR PROGRAMS

Faculty and graduate students in the Department of Physics and Astronomy are involved in research at the Center for Advanced Vehicular Systems (CAVS), the Center for Computational Sciences (CCS), the High Performance Computing Collaboratory (HPC2), and the Institute for Clean Energy Technology (ICET), all located on the Mississippi State University campus.

Table B—Separately Budgeted Research Expenditures by Source of Support

Source of Support	Departmental Research	Physics-related Research Outside Department
Federal government	$1,279,893	$1,235,049
State/local government	$208,774	
Non-profit organizations		
Business and industry	$7,534	
Other		
Total	$1,496,201	$1,235,049

Table C—Separately Budgeted Research Expenditures by Research Specialty

Research Specialty	No. of Grants	Expenditures ($)
Astrophysics	2	$554,868
Atomic, Molecular, Plasma, & Optical Physics	7	$312,013
Nuclear Physics	6	$579,284
Physics and other Science Education	3	$862,235
Computational Physics	7	$422,850
Total	25	$2,731,250

FACULTY

Professor

Afanasjev, Anatoli, Ph.D., Latvian Academy of Sciences, 1993. *Computational Physics, Nuclear Physics, Theoretical Physics*. Covariant (relativistic) density functional theory and its development and application to nuclei at normal and extreme conditions; rotating nuclei; nuclear fission; the crust of neutron stars.

Arnoldus, Hendrik F., Ph.D., Eindhoven University of Technology, The Netherlands, 1985. *Atomic, Molecular, & Optical Physics, Electromagnetism, Nano Science and Technology*. Theoretical optics with emphasis on near-field and nanoscale optics at interfaces.

Dunne, James A., Ph.D., American University, 1995. *Nuclear Physics*. Experimental medium energy nuclear physics; nucleon structure.

Harpole, Sandra H., Ph.D., Mississippi State University, 1986. *Physics and other Science Education*. Physics education and high school outreach.

Kim, Seong-Gon, Ph.D., Michigan State University, 1994. Director for the Center for Computational Sciences (CCS). *Computational Physics, Nano Science and Technology*. Application of first principles computational techniques of condensed matter physics and materials science to the study of the electronic and structural properties of nanostructures, semiconductors, and metals.

Ma, Wenchao, Ph.D., Vanderbilt University, 1985. *Nuclear Physics*. Experimental nuclear physics; nuclear structure at high spin states; nuclei far from stability; radiation detection and measurement technology.

Monts, David L., Ph.D., Columbia University, 1978. *Atomic, Molecular, & Optical Physics*. Molecular spectroscopy; environmental applications imaging.

Novotny, Mark A., Ph.D., Stanford University, 1978. Head. *Chemical Physics, Computational Physics, Condensed Matter Physics, Nano Science and Technology, Statistical & Thermal Physics*. Computational physics approaches to understanding the time dependence of classical and quantum models for materials.

Wang, Chuji, Ph.D., University of Science and Technology of China, 1998. *Atomic, Molecular, & Optical Physics, Biophysics, Engineering Physics/Science, Medical, Health Physics.* Develop and apply measuring, monitoring, and sensing technologies to address real problems in energy, environment, and biomedical engineering.

Winger, Jeffry A., Ph.D., Iowa State University, 1987. *Nuclear Physics.* Decay spectroscopy of neutron-rich nuclei; evolution of single-particle energies; application of nuclear physics to nuclear energy.

Associate Professor

Clay, R. Torsten, Ph.D., University of Illinois, 1999. *Computational Physics, Condensed Matter Physics.* Theoretical condensed matter physics; electronic and magnetic properties of strongly correlated materials; computational methods for strongly correlated systems.

Dutta, Dipangkar, Ph.D., Northwestern University, 1999. *Nuclear Physics.* Medium energy nuclear physics; precision measurement of the fundamental properties of nucleons.

Pierce, Donna M., Ph.D., University of Maryland, 2006. *Astrophysics.* Planetary astronomy; chemical composition of comets.

Ye, Jinwu, Ph.D., Yale University, 1993. *Atomic, Molecular, & Optical Physics, Computational Physics, Condensed Matter Physics.* Quantum systems both optical and condensed matter.

Assistant Professor

Berg, Matthew J., Ph.D., Kansas State University, 2008. *Applied Physics, Computational Physics, Electromagnetism, Optics.* Electromagnetic scattering.

Rupak, Gautam Lan Tai Moong, Ph.D., University of Washington, 2000. *Computational Physics, Nuclear Physics.* Nuclear structure and reactions using QCD and effective field theory.

Tanner, Angelle M., Ph.D., University of California, Los Angeles, 2004. *Astrophysics.* Exoplanets.

Adjunct Professor

Gaskell, David, Ph.D., Oregon State University, 2001. *Nuclear Physics.* Medium energy nuclear physics.

Lindner, Jeffrey S., Ph.D., Mississippi State University, 1985. Applied Physics, Atomic, Molecular, & Optical Physics, Polymer Physics/ScienceMolecular spectroscopy; polymer physics.

McIntyre, Dustin L., Ph.D., West Virginia University, 2007. *Atomic, Molecular, & Optical Physics.* Applications of laser spectroscopy.

Park, Brent K., Ph.D., Ohio University, 1991. Associate Laboratory Director at Oak Ridge National Laboratory. *Nuclear Physics.* Physics of national security issues.

Rykaczewski, Krzysztof P., Ph.D., Warsaw University, 1983. Decay spectroscopy for low energy nuclear structure.

Singh, J. P., Ph.D., Banares Hindu University, 1980. *Applied Physics, Atomic, Molecular, & Optical Physics.* Applications of laser spectroscopy.

Su, Yi, Ph.D., Wayne State University, 1996. *Applied Physics, Atomic, Molecular, & Optical Physics.* Applied spectroscopy and environmental sensing.

DEPARTMENTAL RESEARCH SPECIALTIES AND STAFF

Theoretical

Atomic, Molecular, Plasma, & Optical Physics. Computational modeling of electromagnetic scattering from complex single-particle and multiparticle systems and analytical investigations in fundamental electromagnetic theory and interactions with matter: (1) study of energy flow patterns in electromagnetic radiation fields at nanoscale (sub-wavelength) levels; (2) study of the quantum nature of electromagnetic radiation, including coherent states, squeezed states, correlations, photon counting statistics, and interactions with quantum phases of matter. Arnoldus, Berg, Ye.

Computational Physics. Computational methods to study a diverse range of physics topics; algorithm development and large-scale computational facilities at MSU are used to study the specific research specialization; research also involves collaborative efforts with many disciplines, including engineering, chemistry, mathematics, and computer science. Afanasjev, Berg, Clay, Kim, Novotny, Park, Rupak, Ye.

Condensed Matter Physics. Theoretical and computational condensed matter physics and materials science: (1) electronic and magnetic properties of strongly correlated materials; (2) computational methods for strongly correlated systems; (3) electronic and structural properties of nanostructures, semiconductors, and metals; and (4) time dependence of classical and quantum models for materials. Clay, Kim, Novotny, Ye.

Nuclear Physics. Theoretical and computational studies: (1) understanding nuclear structure and reactions in a manner that is consistent with quantum chromodynamics using effective field theory to construct the low energy nuclear theory; (2) properties of light nuclei and the properties of dense nuclear matter in an astrophysical context, such as neutron stars; and (3) use of relativistic mean field theory to understand the structure of medium to heavy mass nuclei. Afanasjev, Rupak.

Experimental

Astrophysics. Planetary astronomy focused on the study of asteroids and comets, particularly the study of cometary atmospheres; searches to discover exoplanets. Pierce, Tanner.

Atomic, Molecular, Plasma, & Optical Physics. Multidisciplinary research involving physics, chemistry, and optical engineering to develop and apply measuring, monitoring, and sensing technologies to address problems in energy, environment, and biomedical engineering; laser diagnostics in plasma and combustion, especially plasma-assisted combustion; time-domain fiber-optic sensor and sensor network for multifunction (physical, chemical, and biological) monitoring and sensing; cavity ring-down instrumentation for trace elements, isotopes, and volatile organic compounds; breath biomarkers for non-invasive disease diagnostics and metabolic status monitoring; in situ classification of airborne small particles. Berg, Lindner, McIntyre, Monts, Singh, Su, Wang.

Nuclear Physics. Low-energy nuclear physics (LENP) group studies both high-spin states (including triaxial super deformation) produced in fusion evaporation reactions and lower spin structures fed by various decay processes (including half-lives, delayed-neutron probabilities, shell closures, etc.); medium energy physics (MEP) group conducts high precision measurements that probe the fundamental nature of quarks inside atomic nuclei and precision tests of fundamental symmetries and the standard model and is also involved in building novel detectors, systems, and targets for the experimental program at the Thomas Jefferson National Accelerator Facility (JLab); research is performed at Argonne National Laboratory, JLab, Oak Ridge National Laboratory, and the National Superconducting Cyclotron Laboratory. Dunne, Dutta, Gaskell, Ma, Rykaczewski, Winger.

View additional information about this department at
www.gradschoolshopper.com

MISSOURI UNIVERSITY OF SCIENCE AND TECHNOLOGY

DEPARTMENT OF PHYSICS

Rolla, Missouri 65409
http://physics.mst.edu

General University Information
Chancellor: Cheryl B. Schrader
Dean of Graduate School: N/A
University website: http://mst.edu
Control: Public
Setting: Rural
Total Faculty: 492
Total Graduate Faculty: 343
Total number of Students: 7,647
Total number of Graduate Students: 1,804

Department Information
Department Chairman: G. Dan Waddill, Chair
Department Contact: Dan Waddill, Chairman
  Total full-time faculty: 18
  Total number of full-time equivalent positions: 19
  Full-Time Graduate Students: 51
  First-Year Graduate Students: 19
  Female First-Year Students: 4
  Total Post Doctorates: 1

Department Address
1315 N. Pine Street
Rolla, MO 65409
Phone: (573) 341-4781
Fax: (573) 341-4715
E-mail: waddill@mst.edu
Website: http://physics.mst.edu

ADMISSIONS

Admission Contact Information
Address admission inquiries to: Director of Admissions, 106
  Parker Hall, 300 W. 13th Street
Phone: (573) 341-4165
E-mail: graduate-admissions@mst.edu
Admissions website: http://futurestudents.mst.edu

Application deadlines
Fall admission:
U.S. students: September 1   *Int'l. students*: September 1
Spring admission:
U.S. students: November 1   *Int'l. students*: October 15

Application fee
U.S. students: $50   *Int'l. students*: $50

Admissions information
For Fall of 2012:
  Number of applicants: 29
  Number admitted: 11
  Number enrolled: 11

Admission requirements
Bachelor's degree requirements: A bachelor's degree in physics
  is required.
Minimum undergraduate GPA: 3.0

GRE requirements
The GRE is required.
  Quantitative score: 700
  Analytical score: 3
  Mean GRE score range (25th–75th percentile): 70

Advanced GRE requirements
The Advanced GRE is recommended.
  Minimum accepted Advanced GRE score: 700
  Mean Advanced GRE score range (25th–75th percentile): 70th

TOEFL requirements
The TOEFL exam is required for students from non-English-
  speaking countries.
  PBT score: 570
  iBT score: 80

Other admissions information
Additional requirements: The minimum acceptable score sug-
  gested for admission is 1100 (verbal plus quantitative).
  No minimum acceptable score is used.
Undergraduate preparation assumed: General Physics, Halliday;
  Fundamentals of Physics; Modern Physics, Thornton; Modern
  Physics for Scientists and Engineers; Mechanics, Marion;
  Classical Dynamics of Particles and Systems; Thermodynam-
  ics, Reif; Fundamentals of Statistical and Thermal Physics;
  Electricity and Magnetism, Griffiths; Introduction to Elec-
  trodynamics; Quantum Mechanics, Liboff; Introduction to
  Quantum Mechanics.

TUITION

Tuition year 2011–12:
Tuition for in-state residents
  Full-time students: $4,254 per semester
  Part-time students: $354.5 per credit
Tuition for out-of-state residents
  Full-time students: $10,982.4 per semester
  Part-time students: $915.2 per credit
Out-of-state fees are waived for Graduate Assistants.
Credit hours per semester to be considered full-time: 9
Deferred tuition plan: Yes
Health insurance: Yes, $1,656.00.
Other academic fees: $1543.50 academic year
Academic term: Semester

Teaching Assistants, Research Assistants, and Fellowships
Number of first-year
  Teaching Assistants: 2
  Research Assistants: 1
Average stipend per academic year
  Teaching Assistant: $16,650
  Research Assistant: $16,650

FINANCIAL AID

Application deadlines
Fall admission:
U.S. students: July 1   *Int'l. students*: June 15

Spring admission:
U.S. *students*: November 1   Int'l. *students*: October 15

Loans
Loans are available for U.S. students.
Loans are available for international students.
GAPSFAS application required: Yes
FAFSA application required: Yes

For further information
Address financial aid inquiries to: Student Financial Assistance
G1 Parker Hall, 300 W. 13th Street, Rolla, MO 65409.
Phone: (573) 341-4282
E-mail: sfa@mst.edu
Financial aid website: http://sfa.mst.edu

HOUSING

Availability of on-campus housing
Single students: Yes
Married students: Yes

For further information
Address housing inquiries to: Office of Residential Life, 205 W.
12th Street, Rolla, MO 65409.
Phone: (573) 341-4218
E-mail: reslife@mst.edu
Housing aid website: http://reslife.mst.edu

years (six semesters; those with a master's degree from the University of Minnesota, Rochester or other institution is two years, four semesters). A Ph.D. qualifying exam, dissertation, and dissertation exam are required. The language requirement includes passing an examination or an equivalent of one-year collegiate-level course work with a grade of B or better, with an overall requirement of B grades or better.
Thesis: Thesis may be written in absentia.

SPECIAL EQUIPMENT, FACILITIES, OR PROGRAMS

The Cloud and Aerosol Sciences Laboratory provides a wide range of special instrumentation for the study of atmospheric physics, including an assortment of aerosol generators, direct Aitken nuclei counters, condensation nuclei counters, diffusion cloud chambers, and expansion cloud chambers suitable for low-temperature applications. A major mobile laboratory facility is available for the characterization of gas turbine and rocket engine exhaust emissions and is suitable for ground test, airborne, and altitude chamber venues.

The Graduate Center for Materials Research provides accessibility to electron spectroscopy for chemical analysis, auger spectrometers, an scanning electron microscope, an automatic x-ray spectrometer, mass spectrometers, etc.

The Physics Department itself provides access to an ion energy loss spectrometer, an ion implantation accelerator system, electron spin resonance, a full range of lasers, and general research equipment.

Table A—Faculty, Enrollments, and Degrees Granted

Research Specialty	2012-13 Faculty	Enrollment Fall 2012 Master's	Enrollment Fall 2012 Doctorate	Number of Degrees Granted 2012-13 (2006–12) Master's	Number of Degrees Granted 2012-13 (2006–12) Terminal Master's	Number of Degrees Granted 2012-13 (2006–12) Doctorate
Astrophysics	–	–	2	–	–	1(1)
Atomic, Molecular, & Optical Physics	7	–	11	–	-(1)	-(7)
Biophysics	–	–	2	–	–	1(3)
Condensed Matter Physics	7	–	18	2(2)	-(1)	1(6)
Optics	–	–	–	–	–	–
Physics and other Science Education	2	–	2	–	–	–
Polymer Physics/Science	2	–	1	–	–	–
Statistical & Thermal Physics	2	–	–	–	–	–
Non-specialized	–	7	5	-(1)	1(2)	–
Other	4	–	3	1(-)	1(2)	1(-)
Total	23	7	44	3(3)	2(6)	4(17)
Full-time Grad. Stud.	–	7	44	–	–	–
First-year Grad. Stud.	–	4	10	–	–	–

Table B—Separately Budgeted Research Expenditures by Source of Support

Source of Support	Departmental Research	Physics-related Research Outside Department
Federal government	$1,232,466.38	$322,453.44
State/local government	$67,762.4	
Non-profit organizations		
Business and industry		
Other	$116,202.21	
Total	$1,416,430.99	$322,453.44

Table C—Separately Budgeted Research Expenditures by Research Specialty

Research Specialty	No. of Grants	Expenditures ($)
Atomic, Molecular, & Optical Physics	–	
Atomic, Molecular, & Optical Physics	15	$873,394.65
Condensed Matter Physics	–	$359,071.73
Condensed Matter Physics	6	
Physics and other Science Education	–	
Physics and other Science Education	4	$67,762.4
Other	–	
Other	3	$116,202.21
Total	28	$1,416,430.99

GRADUATE DEGREE REQUIREMENTS

Master's: Thirty graduate credit hours for a master's degree with thesis and 30 graduate credit hours for a non-thesis master's degree in an approved program with satisfactory performance are required. A thesis exam and a B average are required. There are no residence or language requirements.

Doctorate: A minimum of 72 hours with satisfactory performance is required. There is a residency requirement of three

FACULTY

Professor

Bieniek, Ronald J., Ph.D., Harvard University, 1975. *Atomic, Molecular, & Optical Physics, Physics and other Science Education*. Atomic and molecular collision processes; theoretical physics.

DuBois, Robert D., Ph.D., University of Nebraska-Lincoln, 1975. *Atomic, Molecular, & Optical Physics*. Ion-atom collisions-electron spectra; experimental physics.

Hagen, Donald E., Ph.D., Purdue University, 1970. *Other*. Cloud simulation models and experiments.

Hale, Barbara N., Ph.D., Purdue University, 1967. *Other*. Nucleation of water and ice; theoretical physics.

Madison, Don H., Ph.D., Florida State University, 1972. *Atomic, Molecular, & Optical Physics*. Electron-atom and ion-atom scattering; theoretical physics.

Parris, Paul E., Ph.D., University of Rochester, 1984. *Condensed Matter Physics, Polymer Physics/Science, Statistical & Thermal Physics*. Electron transport in disordered solids; theoretical physics.

Peacher, Jerry L., Ph.D., Indiana University, 1965. *Atomic, Molecular, & Optical Physics*. Atomic collisions; scattering; theoretical physics.

Pringle, Allan, Ph.D., University of Missouri, Columbia, 1981. *Condensed Matter Physics, Physics and other Science Education*. Neutron scattering; magnetic materials; experimental physics.

Schulz, Michael, Ph.D., University of Heidelberg, 1984. *Atomic, Molecular, & Optical Physics*. Ion-atom collisions; electron correlation; experimental physics.

Vojta, S. Thomas, Ph.D., University of Chemnitz, 1994. *Condensed Matter Physics*. Correlated electrons; quantum phase transitions; theoretical physics.

Waddill, G. Daniel, Ph.D., Indiana University, 1987. Chairman of the Department. *Condensed Matter Physics*. Characterization of metallic surfaces and interfaces by x-ray and ultraviolet photoelectron spectroscopy; experimental physics.

Wilemski, Gerald, Ph.D., Yale University, 1972. *Statistical & Thermal Physics, Other*. Nucleation theory; theoretical physics.

Associate Professor

Medvedeva, Julia E., Ph.D., Institute of Metal Physics, Russian Academy of Science, 2002. *Condensed Matter Physics*. Density functional theory; theoretical physics.

Schmitt, John L., Ph.D., University of Michigan, 1968. *Other*. Nucleation phenomena; experimental physics.

Story, J. Greg, Ph.D., University of Southern California, 1989. *Atomic, Molecular, & Optical Physics*. Studies of Rydberg atom properties; experimental physics.

Assistant Professor

Hor, Yew San, Ph.D., Rutgers University, 2004. *Condensed Matter Physics*. Growth and characterization of novel bulk and nanostructured materials; experimental physics.

Jentschura, Ulrich, Ph.D., University of Technology, Dresden, 1999. *Atomic, Molecular, & Optical Physics*. Quantum electrodynamic bound-state calculations; relativistic quantum dynamic processes in laser fields; analysis of high-precision experiments; theoretical physics.

Yamilov, Alexey, Ph.D., City University of New York, 2001. *Condensed Matter Physics*. Mesoscopic phenomena in light propagation; photonics; theoretical physics.

DEPARTMENTAL RESEARCH SPECIALTIES AND STAFF

Theoretical

Atomic, Molecular, & Optical Physics. Atomic collisions; scattering; primary ionization. Bieniek, Jentschura, Madison, Medvedeva, Parris, Peacher, Vojta, Yamilov.

Condensed Matter Physics. Electron transport in disordered solids; surface phenomena; electronic polymers; density functional theory; mesoscopic phenomena in light propagation; photonics; correlated electrons and quantum phase transitions.

Other. Cloud and Aerosol Sciences Laboratory; homomolecular and heteromolecular nucleation studies; properties of water clusters; surface nucleation; simulation of initial stages of cloud formation; condensational drop growth/evaporation; aerosol dynamics; neutron scattering by aerosols; density functional theory of inhomogeneous fluids. Hagen, Hale, Schmitt, Wilemski.

Experimental

Atomic, Molecular, & Optical Physics. Ion-atom and electron collisions; heavy-ion impact excitation; energy-loss spectrometry; lifetimes; photoionization; electron impact; recoil ion momentum spectroscopy; electron spectroscopy; quantum electronics; Penning reactions; lasers; crossed-beam molecular scattering. DuBois, Schulz, Story.

Condensed Matter Physics. Ion implantation; electron spin resonance; magnetism; ferroelectricity; surfaces; superconductivity; electronic ceramics. Bieniek, Pringle, Waddill.

Other. Cloud and Aerosol Sciences Laboratory; fundamental studies of aerosol generation, measurement of physical and chemical properties, and evolution in the atmosphere; current emphasis on the exhaust emissions from jet and rocket engines, with measurement campaigns conducted in situ with airborne facilities, in ground-based facilities in engine test stands, combustor rigs, altitude chambers, and in laboratory combustion facilities; the laboratory has an interdisciplinary flavor with strong interactions with chemistry and engineering. Hagen, Schmitt.

Physics and other Science Education. Development of traditional and nontraditional learning experiences; production of audiovisual aids. Bieniek, Pringle.

View additional information about this department at
www.gradschoolshopper.com

UNIVERSITY OF MISSOURI

DEPARTMENT OF PHYSICS AND ASTRONOMY

Columbia, Missouri 65211
http://physics.missouri.edu/

General University Information
President: Timothy M. Wolfe
Dean of Graduate School: George Justice
University website: http://www.missouri.edu
Control: Public
Setting: Urban
Total Faculty: 2,121
Total Graduate Faculty: 1,826
Total number of Students: 34,748
Total number of Graduate Students: 7,752

Department Information
Department Chairman: Peter Pfeifer, Chair
Department Contact: Carsten A. Ullrich, Director of Graduate
  Studies
  Total full-time faculty: 32
  Total number of full-time equivalent positions: 29
  Full-Time Graduate Students: 55
  First-Year Graduate Students: 12
  Female First-Year Students: 3
  Total Post Doctorates: 16

Department Address
223 Physics Building
Columbia, MO 65211
Phone: (573) 882-3335
Fax: (573) 882-4195
E-mail: umcasphysics@missouri.edu
Website: http://physics.missouri.edu/

ADMISSIONS

Admission Contact Information
Address admission inquiries to: Graduate School, University of
  Missouri, 210 Jesse Hall, Columbia, MO 65211
Phone: (800) 877-6312
E-mail: gradadmin@missouri.edu
Admissions website: http://gradschool.missouri.edu/admission/

Application deadlines
Fall admission:
U.S. students: March 15   *Int'l. students*: March 15
Spring admission:
U.S. students: October 1   *Int'l. students*: October 1

Application fee
U.S. students: $55   *Int'l. students*: $75

Admissions information
For Fall of 2013:
  Number of applicants: 103
  Number admitted: 14
  Number enrolled: 9

Admission requirements
Bachelor's degree requirements: A bachelor's degree in physics
  is required.
Minimum undergraduate GPA: 3.0

GRE requirements
The GRE is required.
  Quantitative score: 155
  Verbal score: 146
  Analytical score: 3
  Mean GRE score range (25th–75th percentile): 300-320

Advanced GRE requirements
The Advanced GRE is recommended.
  Minimum accepted Advanced GRE score: 500
  Mean Advanced GRE score range (25th–75th percentile): 500-
  800

TOEFL requirements
The TOEFL exam is required for students from non-English-
  speaking countries.
  PBT score: 550
  iBT score: 80

Other admissions information
Additional requirements: The minimum acceptable scores suggested for admission are as follows: verbal, 146; quantitative,
  155. The average GRE scores for admission are as follows:
  verbal, 150; quantitative, 160. The IELTS test with a minimum score of 6.5 is an acceptable alternative to the TOEFL.
  Students are exempt from this requirement if they have completed a year of full-time academic study in the United States.

TUITION

Tuition year 2013-14:
Tuition for in-state residents
  Full-time students: $342.2 per credit
  Part-time students: $342.2 per credit
Tuition for out-of-state residents
  Full-time students: $883.6 per credit
  Part-time students: $883.6 per credit
Credit hours per semester to be considered full-time: 9
Deferred tuition plan: Yes
Health insurance: Available at the cost of $1284 per year.
Other academic fees: Recreational facility fee, $69.35 per semes-
  ter (enrolled 6+ hours); activity fee, $98.40 per semester (en-
  rolled 8+ hours); health fee, $80.92 per semester (enrolled
  6+ hours); information technology fee, $12.20 per credit
  hour.
Academic term: Semester
Number of first-year students who received full tuition waivers: 13

Teaching Assistants, Research Assistants, and Fellowships
Number of first-year
  Teaching Assistants: 8
  Research Assistants: 5
  Fellowship students: 8
Average stipend per academic year
  Teaching Assistant: $18,000
  Research Assistant: $20,000
  Fellowship student: $21,000
TA students are eligible for up to $3,000 summer fellowship
support.

FINANCIAL AID

Application deadlines

Fall admission:
U.S. students: March 15   *Int'l. students*: March 15
Spring admission:
U.S. students: October 1   *Int'l. students*: October 1

Loans

Loans are available for U.S. students.
Loans are not available for international students.
GAPSFAS application required: Yes
FAFSA application required: Yes

For further information

Address financial aid inquiries to: University of Missouri-Columbia Office of Student Financial Aid, 11 Jesse Hall, Columbia, MO 65211.
Phone: (573) 882-7506
E-mail: finaidinfo@missouri.edu
Financial aid website: http://financialaid.missouri.edu/

HOUSING

Availability of on-campus housing

Single students: Yes
Married students: Yes

For further information

Address housing inquiries to: Department of Residential Life, University of Missouri, 0780 Defoe-Graham Hall, 901 Hitt Street, Columbia, MO 65211.
Phone: (573) 882-7275
E-mail: reslife@missouri.edu
Housing aid website: http://reslife.missouri.edu/

Table A—Faculty, Enrollments, and Degrees Granted

Research Specialty	2012-13 Faculty	Enrollment Fall 2012 Master's	Enrollment Fall 2012 Doctorate	Number of Degrees Granted 2011–12 (2007–12) Master's	Number of Degrees Granted 2011–12 (2007–12) Terminal Master's	Number of Degrees Granted 2011–12 (2007–12) Doctorate
Astronomy	4	1	9	–(2)	1(3)	1(2)
Biophysics	7	–	11	2(5)	–	1(6)
Condensed Matter Physics	11	1	20	3(16)	–(4)	2(10)
Energy Sources & Environment	2	–	8	1(4)	–(2)	2(4)
Low Temperature Physics	1	–	–	–	–	–
Nano Science and Technology	3	–	1	–	1(-)	–
Optics	1	–	3	–	–	1(3)
Physics and other Science Education	2	–	–	–	–	–
Relativity & Gravitation	2	–	–	–	–	1(1)
Total	33	2	52	6(27)	1(9)	7(26)
Full-time Grad. Stud.	–	2	52	–	–	–
First-year Grad. Stud.	–	1	12	–	–	–

GRADUATE DEGREE REQUIREMENTS

Master's: A master's degree requires completion of 30 credit hours beyond the bachelor's degree (at least 15 hours of those must be 8000 level courses) with a grade of 3.0 (B) or better. Completion of the departmental qualifying examination at the M.S. pass level or a written thesis is required. There are no foreign language or computing requirements.

Doctorate: A doctorate degree requires completion of a minimum of 18 hours beyond the master's degree with a grade of 3.0 (B) or better and completion of the department qualifying examination at the Ph.D. pass level. The degree candidate must also meet the residency requirements. A student is required to have taken a minimum of three full years of graduate work beyond the bachelor's degree. To be an official candidate, the student must pass the comprehensive examination for the Ph.D., which is based on graduate coursework in the department. A dissertation is required.

Thesis: Thesis may be written in absentia.

SPECIAL EQUIPMENT, FACILITIES, OR PROGRAMS

The University of Missouri Research Reactor at Columbia provides a unique opportunity for neutron scattering research. The thermal neutron flux at the beam port is 2×10^{14} neutrons/cm^{2}s, the highest of any university in the United States. Current projects/programs include studies of neutron and x-ray scattering, critical phenomena, surface and interfaces, lattice and liquid dynamics, magnetic materials, optoelectronics, theoretical, computational, and experimental biological physics, biomedical imaging and optics, theory of gravity and relativistic astrophysics, spintronics, and condensed matter theory.

Table B—Separately Budgeted Research Expenditures by Source of Support

Source of Support	Departmental Research	Physics-related Research Outside Department
Federal government	$21,385,491	
State/local government	$118,281	
Non-profit organizations	$340,000	
Business and industry	$1,310,067	
Other		
Total	$23,153,839	

Table C—Separately Budgeted Research Expenditures by Research Specialty

Research Specialty	No. of Grants	Expenditures ($)
Astronomy	12	$1,524,344
Biophysics	12	$7,970,854
Condensed Matter Physics	27	$8,373,957
Physics and other Science Education	7	$5,284,684
Total	58	$23,153,839

FACULTY

Professor

Chandrasekhar, H. R., Ph.D., Purdue University, 1973. *Condensed Matter Physics*. Optical properties of solids under pressure.

Chandrasekhar, Meera, Ph.D., Brown University, 1976. Curators' Distinguished Teaching Professor. *Condensed Matter Physics*. Optical spectroscopy of semiconductors and superconductors with an emphasis on high-pressure studies.

Chen, Shi-Jie, Ph.D., University of California, San Diego, 1994. *Biophysics*. Theoretical and computational biophysics.

Duncan, Robert, Ph.D., University of California, Santa Barbara, 1988. Vice Chancellor of Research. *Low Temperature Physics*. Low-temperature physics.

Forgacs, Gabor, Ph.D., Eotvos Roland University, Budapest, 1978. George H. Vineyard Distinguished Professor of Theoretical Physics. *Biophysics*. Physical mechanisms in cell and developmental biology.

Gangopadhyay, Shubhra, Ph.D., Indian Institute of Technology, Kharagpur, 1982. Professor of Electrical Engineering (joint appointment). *Condensed Matter Physics, Electrical Engineering, Nano Science and Technology.* Nanotechnology.

Godwin, Linda, Ph.D., University of Missouri, 1980. *Astronomy.* Astronomy; space science.

Guha, Suchi, Ph.D., Arizona State University, 1996. *Condensed Matter Physics.* Experimental condensed matter physics; light scattering and organic optoelectronics.

Hawthorne, M. Frederick, Ph.D., University of California, Los Angeles, 1953. Professor of Radiology and Director of Nanomedicine (joint position); Member of the National Academy of Sciences. *Medical, Health Physics, Nano Science and Technology.* Nanomedicine.

Katti, Kattesh, Ph.D., Indian Institute of Science, Bangalore, 1984. Director, University of Missouri Cancer Nanotechnology Platform; Curator's Distinguished Professor of Radiology and Physics (joint position). *Medical, Health Physics, Nano Science and Technology.* Development of site-specific radiopharmaceuticals; chemotherapeutic agents for cancer therapy; chemical and biomedical aspects of optical materials.

Kopeikin, Sergei, Ph.D., Space Research Institute of the Russian Academy of Science, Moscow, 1986. *Astrophysics, Relativity & Gravitation.* General relativity and cosmology; theoretical and experimental gravity.

Kosztin, Ioan, Ph.D., University of Illinois at Urbana-Champaign, 1997. *Biophysics.* Theoretical and computational biological physics.

Li, Aigen, Ph.D., Leiden University, 1998. *Astronomy, Astrophysics.* Theoretical astrophysics.

Mashhoon, Bahram, Ph.D., Princeton University, 1972. *Astrophysics, Relativity & Gravitation.* Theory of gravitation and relativistic astrophysics.

Miceli, Paul, Ph.D., University of Illinois at Urbana-Champaign, 1987. *Condensed Matter Physics.* Surfaces and interfaces of condensed matter investigated by x-ray and neutron scattering.

Pfeifer, Peter, Ph.D., ETH Zurich Swiss Federal Institute of Technology, 1980. Department Chair. *Condensed Matter Physics, Energy Sources & Environment.* Surface physics; fractals and quantum dynamics; porous media; energy storage.

Satpathy, Sashi, Ph.D., University of Illinois at Urbana-Champaign, 1982. *Condensed Matter Physics.* Theoretical condensed matter physics; electronic structure and magnetism in solids.

Speck, Angela, Ph.D., University College London, 1998. Director of Astronomy. *Astronomy, Astrophysics.* Astronomy, including stellar evolution, astromineralogy, and dust around evolved stars, galactic chemical evolution, and meteoritics.

Taub, Haskell, Ph.D., Cornell University, 1971. Director of Neutron Scattering. *Condensed Matter Physics.* Structure, phase transitions, and dynamics of absorbed films.

Ullrich, Carsten, Ph.D., University of Wuerzburg, 1995. Director of Graduate Studies. *Condensed Matter Physics.* Condensed matter theory; density-functional theory.

Vignale, Giovanni, Ph.D., Northwestern University, 1984. *Condensed Matter Physics.* Condensed matter theory.

Wexler, Carlos, Ph.D., University of Washington, 1997. *Condensed Matter Physics, Energy Sources & Environment.* Condensed matter theory.

Associate Professor

Hanuscin, Deborah, Ph.D., Indiana University, 2004. *Physics and other Science Education.* Physics education research.

Montfrooij, Wouter, Ph.D., University of Delft, 1990. *Condensed Matter Physics.* Phase transitions in condensed matter; neutron scattering.

Yu, Ping, Ph.D., Hong Kong University of Science and Technology, 1998. *Biophysics, Condensed Matter Physics, Optics.* Optoelectronics and biomedical imaging.

Zou, Xiaoqin, Ph.D., University of California, San Diego, 1993. *Biophysics.* Theoretical and computational biophysics.

Assistant Professor

Bompadre, Silvia, Ph.D., University of Washington, 1998. *Biophysics.* Biological physics; transmembrane proteins.

King, Gavin, Ph.D., Harvard University, 2004. *Biophysics.* Single-molecule biophysics; atomic force microscopy.

Singh, Deepak, Ph.D., University of Massachusetts, Amherst, 2006. *Condensed Matter Physics.* Nanoengineered materials; geometrically frustrated magnetic materials.

Yan, Haojing, Ph.D., Arizona State University, 2003. *Astronomy, Astrophysics.* Astrophysics; early galaxy formation.

Teaching Professor

Kosztin, Dorina, Ph.D., University of Illinois at Urbana-Champaign, 1998. Director of Undergraduate Studies. *Biophysics.* Theoretical biophysics.

Teaching Associate Professor

Zhang, Yun, Ph.D., University of California, San Diego, 1999. *Condensed Matter Physics.* Experimental condensed matter physics.

Teaching Assistant Professor

King, Karen, Ph.D., Dartmouth College, 2003. *Biophysics, Physics and other Science Education.* Medical imaging and biomechanical modeling; physics and engineering education.

DEPARTMENTAL RESEARCH SPECIALTIES AND STAFF

Theoretical

Biophysics. Molecular modeling transport through membranes; RNA folding and assembly; computational drug design. Chen, Karen King, Ioan Kosztin, Zou.

Condensed Matter Physics. Electronic structure of materials, magnetic devices, and spintronics; quantum many-body theory; density-functional theory; transport and optical excitations in semiconductors; quantum and classical statistical mechanics; fractals and phase transitions. Pfeifer, Satpathy, Ullrich, Vignale, Wexler.

Cosmology & String Theory. Origin and fate of the universe; gravitational radiation; post-Newtonian gravity; black holes. Kopeikin, Mashhoon.

Experimental

Astronomy. Cosmic dust; planetary and star formation and evolution; galactic chemical evolution; origin of molecules; early galaxy formation; science of space exploration. Godwin, Li, Speck, Yan.

Biophysics. Cellular biomechanics; physical mechanisms of cell and developmental biology; organ printing and tissue engineering; single-molecule atomic force microscopy; transmembrane proteins. Bompadre, Forgacs, Gavin King, Yu.

Condensed Matter Physics. Organic displays and photovoltaics; Raman scattering; ZnO-based optoelectronics; dielectrics; high pressure optical spectroscopy; magnetic fractals; quantum phase transitions; organic thin films; neutron and x-ray

scattering; epitaxial growth; alternative fuel research; hydrogen storage; surface science; geometrically frustrated magnetic systems. H. Chandrasekhar, Meera Chandrasekhar, Duncan, Gangopadhyay, Guha, Miceli, Montfrooij, Pfeifer, Singh, Taub, Yu.
Medical, Health Physics. Nanomedicine; drug delivery; cancer research. Hawthorne, Katti.

Optics. Biomedical imaging; nonlinear optics. Yu.
Physics and other Science Education. Writing-to-learn strategies; formative assessment tools; inquiry-based teaching methodologies. Hanuscin, Karen King.

View additional information about this department at
www.gradschoolshopper.com

UNIVERSITY OF MISSOURI–ST. LOUIS

DEPARTMENT OF PHYSICS AND ASTRONOMY

St. Louis, Missouri 63121
http://www.umsl.edu/~physics

General University Information
Chancellor: Thomas F. George
Dean of Graduate School: Judith Walker de Felix
University website: http://www.umsl.edu
Control: Public
Setting: Urban
Total Faculty: 467
Total Graduate Faculty: 350
Total number of Students: 16,719
Total number of Graduate Students: 2,747

Department Information
Department Chairman: Bernard Feldman, Chair
Department Contact: Bruce Wilking, Professor of Astronomy
Total full-time faculty: 9
Total number of full-time equivalent positions: 8
Full-Time Graduate Students: 18
First-Year Graduate Students: 5
Female First-Year Students: 2

Department Address
1 University Boulevard
503 Benton Hall
St. Louis, MO 63121
Phone: (314) 516-5023
Fax: (314) 516-6152
E-mail: bwilking@umsl.edu
Website: http://www.umsl.edu/~physics

ADMISSIONS

Admission Contact Information
Address admission inquiries to: Bruce A. Wilking
Phone: (314) 516-5081
E-mail: bwilking@umsl.edu
Admissions website: http://www.umsl.edu/divisions/graduate/

Application deadlines
Fall admission:
U.S. students: July 1   *Int'l. students*: May 1
Spring admission:
U.S. students: December 1   *Int'l. students*: October 1

Application fee
U.S. students: $35   *Int'l. students*: $40
Students seeking a teaching or research assistantship for fall should apply by March at the latest.

Admissions information
For Fall of 2013:
Number of applicants: 20
Number admitted: 8
Number enrolled: 4

Admission requirements
Bachelor's degree requirements: A bachelor's degree in physics or a related field is required, with an undergraduate and major field grade point average of B or better.
Minimum undergraduate GPA: 3.0

GRE requirements
The GRE is required.
Quantitative score: 148
Analytical score: 3
Minimum scores are required for Ph.D. applicants.

Advanced GRE requirements
The Advanced GRE is not required.

TOEFL requirements
The TOEFL exam is required for students from non-English-speaking countries.
PBT score: 570
iBT score: 88

Other admissions information
Additional requirements: To be admitted to the Ph.D. program, a student must have a GRE Quantitative score of at least 148 and an Analytical score of at least 3.0.
Undergraduate preparation assumed: Marion, Classical Dynamics; Wangsness, Electromagnetic Fields; Gasiorowicz, The Structure of Matter: A Survey of Modern Physics.

TUITION
Tuition year 2013-14:
Tuition for in-state residents
Full-time students: $409.1 per credit
Part-time students: $409.1 per credit

Tuition for out-of-state residents
Full-time students: $1,008.5 per credit
Part-time students: $1,008.5 per credit
Per credit hour amounts include tuition and mandatory fees.
Graduate Teaching and Research Assistants may receive
waivers for tuition and nonresident fees.
Credit hours per semester to be considered full-time: 9
Deferred tuition plan: Yes
Health insurance: Yes, 1783.00/12 months.
Academic term: Semester
Number of first-year students who received full tuition waivers: 4

Teaching Assistants, Research Assistants, and Fellowships

Number of first-year
Teaching Assistants: 4
Average stipend per academic year
Teaching Assistant: $14,500
Research Assistant: $16,000
Fellowship student: $16,000

FINANCIAL AID

Application deadlines
Fall admission:
U.S. students: April 1   *Int'l. students*: April 1
Spring admission:
U.S. students: October 1   *Int'l. students*: October 1

Loans
Loans are available for U.S. students.
Loans are not available for international students.
GAPSFAS application required: No
FAFSA application required: Yes

For further information
Address financial aid inquiries to: Student Financial Aid Office
327 Millennium Student Center, One University Boulevard,
St. Louis, MO 63121.
Phone: (314) 516-5526
E-mail: financialaid@umsl.edu
Financial aid website: http://www.umsl.edu/services/finaid/

HOUSING

Availability of on-campus housing
Single students: Yes
Married students: Yes

For further information
Address housing inquiries to: Residential Life and Housing,
C103 Provincial House and University Meadows, 2901 Uni-
versity Meadows Drive, St. Louis, MO 63121.
Phone: (314) 516-6877
E-mail: umslreslife@umsl.edu
Housing aid website: http://www.umsl.edu/reslife

Table A—Faculty, Enrollments, and Degrees Granted

Research Specialty	2013–14 Faculty	Enrollment Fall 2013 Master's	Enrollment Fall 2013 Doctorate	Number of Degrees Granted 2012–13 (2008–13) Master's	Number of Degrees Granted 2012–13 (2008–13) Terminal Master's	Number of Degrees Granted 2012–13 (2008–13) Doctorate
Astronomy	2	2	3	-(4)	-(2)	-(1)
Astrophysics	1	1	–	–	–	-(1)
Atomic, Molecular, & Optical Physics	–	–	–	–	–	–
Biophysics	1	1	4	1(4)	–	-(4)
Condensed Matter Physics	2	1	6	4(14)	-(7)	-(4)
Optics	–	–	–	–	–	–
Physics and other Science Education	1	–	–	–	–	–
Statistical & Thermal Physics	–	–	–	–	–	–
Non-specialized	1	1	1	–	-(5)	–
Total	8	6	14	5(27)	-(14)	-(10)
Full-time Grad. Stud.	–	4	13	–	–	–
First-year Grad. Stud.	–	3	1	–	–	–

GRADUATE DEGREE REQUIREMENTS

Master's: A student must complete 30 credit hours in graduate physics courses, with at least 15 of these at the 5000 or 6000 level. The writing of a thesis is optional. A maximum of 6 (3) credit hours of Research, P6490, may be counted toward the minimum 15 hours with (without) the thesis option. A comprehensive oral examination must be passed, which includes the defense of the thesis if the student has chosen to write one. A grade point average of a B or better must be maintained during each academic year. The requirements must be fulfilled within six years from the time of admission. At least two-thirds of required graduate credit must be taken in residence. There is no language requirement.

Doctorate: A minimum of 78 hours with satisfactory performance (i.e., B grade point average or better) is required for a Ph.D. in physics. There is a residency requirement of three years/six semesters (for those with Master's degree, two years/four semesters) at University of Missouri-St. Louis and/or cooperating Missouri University of Science and Technology campus. Ph.D. qualifying exam, dissertation, dissertation exam administered in cooperation with Missouri University of Science and Technology are required.

Thesis: Thesis may be written in absentia.

SPECIAL EQUIPMENT, FACILITIES, OR PROGRAMS

The William L. Clay Center for Nanoscience, which opened in 1996, is a facility bringing together both physicists and chemists for research in materials science. A focus of the center is to foster collaborations between its members and colleagues in industry. The center houses the Microscope Image and Spectroscopy Technology Laboratory, where research at the forefront of nanotechnology is conducted with transmission electron, scanning probe, and scanning electron microscopes in a building uniquely designed for such work.

The Center for Neurodynamics, established in 1995, conducts research at the interface between physics and biology, with a focus on the roles of noise and stochastic synchronization in neural processing. The center has an on-site high-speed (CCD) imaging system for studying the spatial dynamics of neural activity in the mammalian brain. Collaborations are ongoing with members of Department of Philosophy, who are involved in problems of neural computation and the philosophy of consciousness. Astronomers make use of national facilities at Kitt Peak, Cerro Tololo, and Mauna Kea Observatories.

The university provides E-mail and internet services through numerous student laboratories equipped with computers, flat-bed document scanners, and color printers with standard software packages for word and image processing.

Table B—Separately Budgeted Research Expenditures by Source of Support

Source of Support	Departmental Research	Physics-related Research Outside Department
Federal government	$226,000	
State/local government		
Non-profit organizations		
Business and industry	$149,000	
Other		
Total	$375,000	

Table C—Separately Budgeted Research Expenditures by Research Specialty

Research Specialty	No. of Grants	Expenditures ($)
Astronomy	4	$91,000
Biophysics	2	$93,000
Condensed Matter Physics	2	$125,000
Energy Sources & Environment	2	$66,000
Total	10	$375,000

FACULTY

Professor

Bahar, Sonya, Ph.D., University of Rochester, 1997. Director, Center for Neurodynamics. *Biophysics, Nonlinear Dynamics and Complex Systems*. Synchronization in neural systems; computational modeling of evolutionary dynamics.

Feldman, Bernard J., Ph.D., Harvard University, 1972. Chairperson. *Physics and other Science Education*.

Flores, Ricardo A., Ph.D., University of California, Santa Cruz, 1984. *Astrophysics, Particles and Fields*. Particle astrophysics; cosmology.

George, Thomas F., Ph.D., Yale University, 1970. Chancellor. *Condensed Matter Physics, Nano Science and Technology*. Laser/materials physics.

Henson, Bob L., Ph.D., Washington University, 1964. *Physics and other Science Education, Statistical & Thermal Physics*.

Wilking, Bruce A., Ph.D., University of Arizona, 1981. Associate Chairperson Graduate Coordinator. *Astronomy*. Star formation.

Associate Professor

Fraundorf, Phil B., Ph.D., Washington University, 1980. *Materials Science, Metallurgy, Nano Science and Technology, Physics and other Science Education*. Bayesian inference; pre-solar dust; physics content modernization.

Gibb, Erika, Ph.D., Rensselaer Polytechnic Institute, 2001. *Astronomy*. Astrochemistry in comets and young stars.

Majzoub, Eric, Ph.D., Washington University, 2000. *Computational Physics, Energy Sources & Environment, Nano Science and Technology*. Energy storage and conversion.

DEPARTMENTAL RESEARCH SPECIALTIES AND STAFF

Theoretical

Astrophysics. Particle astrophysics; structure of dark matter halos. Flores.

Biophysics. Computational studies of stochastic phase synchronization; spatiotemporal dynamics and information processing in neural ensembles. Bahar.

Condensed Matter Physics. Modeling of molecular clusters and nanostructures. George.

Energy Sources & Environment. Modeling of bulk and nanocrystalline materials for energy storage. Majzoub.

Nonlinear Dynamics and Complex Systems. Computational modeling of evolutionary dynamics. Bahar.

Experimental

Astronomy. Observations of pre-main sequence stars and protoplanetary disks. Gibb, Wilking.

Astronomy. Composition of cometary volatiles. Gibb.

Biophysics. In vivo imaging of neural dynamics in epilepsy; electrophysiological studies of the role of stochastic synchronization. Bahar.

Energy Sources & Environment. Characterization of bulk and nanocrystalline materials for energy storage. Majzoub.

Materials Science, Metallurgy. Ultrahigh temperature materials for hypersonic aerospace; defects in and on VLSI silicon and mica. Fraundorf.

View additional information about this department at
www.gradschoolshopper.com

MONTANA STATE UNIVERSITY

DEPARTMENT OF PHYSICS

Bozeman, Montana 59717-3840
http://www.physics.montana.edu

General University Information
President: Waded Cruzado
Dean of Graduate School: Ron W. Larsen, Assoc. Provost
University website: http://www.montana.edu

Control: Public
Setting: Rural
Total Faculty: 953
Total Graduate Faculty: N/A

Total number of Students: 14,660
Total number of Graduate Students: 1,888

Department Information
Department Chairman: Richard J. Smith, Head
Department Contact: Margaret Jarrett, Graduate Program
  Coordinator
Total full-time faculty: 17
Total number of full-time equivalent positions: 47
Full-Time Graduate Students: 54
First-Year Graduate Students: 10
Female First-Year Students: 2
Total Post Doctorates: 8

Department Address
264 Engineering/Physical Sciences Building
Bozeman, MT 59717-3840
Phone: (406) 994-3614 (C)
Fax: (406) 994-4452
E-mail: jarrett@physics.montana.edu
Website: http://www.physics.montana.edu

ADMISSIONS

Admission Contact Information
Address admission inquiries to: Prof. Rufus Cone, Physics
Phone: (406) 994-6175
E-mail: cone@physics.montana.edu
Admissions website: http://www.physics.montana.edu

Application deadlines
Fall admission:
U.S. students: January 1   *Int'l. students*: January 1

Application fee
U.S. students: $60   *Int'l. students*: $60
New graduates are only accepted for Fall term enrollment.

Admissions information
For Fall of 2012:
  Number of applicants: 75
  Number admitted: 10
  Number enrolled: 10

Admission requirements
Bachelor's degree requirements: Four-year Bachelor's degree in
  physics or a related field is required.
Minimum undergraduate GPA: 3.0

GRE requirements
The GRE is required.
  Quantitative score: 157
  Verbal score: 155
  Analytical score: 3.5

Advanced GRE requirements
The Advanced GRE is required.
  Minimum accepted Advanced GRE score: 570

TOEFL requirements
The TOEFL exam is required for students from non-English-
  speaking countries.
  PBT score: 600
  iBT score: 80
Overall score of 80 on iBT-TOEFL is required of all international
  students who wish to attend MSU. A score of 26 on the speaking portion is required for teaching assistantships.

Other admissions information
Additional requirements: Successful students in our program usually have GRE scores that exceed: verbal–70%, quantitative–80%, analytical–4.0, and physics–40%.

Undergraduate preparation assumed: Marion or Symon, Classical Mechanics; Griffiths, Electricity and Magnetism; Libof
  or Gasiorowicz, Quantum Mechanics; Reif, Statistical and
  Thermal Physics.

TUITION

Tuition year 2012–13:
Tuition for in-state residents
  Full-time students: $6,200 annual
Tuition for out-of-state residents
  Full-time students: $16,250 annual
Credit hours per semester to be considered full-time: 9
Deferred tuition plan: Yes
Health insurance: Available at the cost of $2,100 per year.
Academic term: Semester
Number of first-year students who received full tuition waivers: 10

Teaching Assistants, Research Assistants, and Fellowships
Number of first-year
  Teaching Assistants: 9
  Research Assistants: 1
Average stipend per academic year
  Teaching Assistant: $14,900
  Research Assistant: $14,900
  Fellowship student: $15,000

FINANCIAL AID

Loans
Loans are available for U.S. students.
Loans are not available for international students.
GAPSFAS application required: No
FAFSA application required: No

For further information
Address financial aid inquiries to: Dr. Rufus Cone.
Phone: (406) 994-6175
E-mail: cone@physics.montana.edu
Financial aid website: http://www.montana.edu/wwwfa/

HOUSING

Availability of on-campus housing
  Single students: Yes
  Married students: Yes

For further information
Address housing inquiries to: Residence Life Office.
Phone: (406) 994-3730
Housing aid website: http://www.montana.edu/fgh

Table A—Faculty, Enrollments, and Degrees Granted

Research Specialty	2012-13 Faculty	Enrollment Fall 2012		Number of Degrees Granted 2012-13 (2008-13)		
		Master's	Doctorate	Master's	Terminal Master's	Doctorate
Astrophysics	2	–	7	1(3)	–(2)	–(2)
Atmosphere, Space Physics, Cosmic Rays	1	–	–	–	–	–
Condensed Matter Physics	7	1	11	–(5)	1(5)	2(8)
Optics	6	1	7	2(6)	2(5)	–(3)
Physics and other Science Education	2	1	1	–(1)	–(1)	–(1)
Relativity & Gravitation	3	–	9	–(2)	–	–(4)
Solar Physics	8	–	16	1(8)	–(1)	–(8)
Total	29	–	–	4(25)	3(14)	2(26)
Full-time Grad. Stud.	–	3	51	–	–	–
First-year Grad. Stud.	–	1	9	–	–	–

GRADUATE DEGREE REQUIREMENTS

Master's: Twenty credits plus a thesis or 30 credits without a thesis in an approved program with satisfactory performance are required. An M.S. examination and two semesters of residency are required. There are no language requirements.

Doctorate: A minimum of 40 credits of acceptable course work; dissertation; satisfactory performance on comprehensive and dissertation examinations, and four semesters of residency are required. There are no language requirements.

Thesis: Thesis may be written in absentia.

SPECIAL EQUIPMENT, FACILITIES, OR PROGRAMS

Optical physics and laser spectroscopy laboratories, including lasers ranging from the ultrastable (few Hz linewidths) to the ultrafast (femtosecond pulses), for research in spectral hole burning phenomena, materials science, and devices, ultrafast holography, smart pixel sensors, Raman lasers, diode laser frequency control and noise characterization, optical frequency standards, and ultrastable optical lasers and cavities; Spectral Information Technology Laboratory (Spectrum Lab) and Optical Technology Center (OpTeC), fostering collaborations with the Departments of Chemistry and Biochemistry and Electrical and Computer Engineering, with local optics industries, and with several national and international laboratories and companies; millimeter-wave magneto-spectroscopy facility; magnetics laboratory; dielectrics laboratory; active materials development and testing laboratory; Magnetic Nanostructure Growth and Characterization Facility for synthesis and characterization of magnetic films, particles, and interfaces; Image and Chemical Analysis Laboratory; Montana Space Grant Consortium, a statewide program for research, education, and outreach in space science; Space Science and Engineering Laboratory with facilities for the design, development and testing of small satellite hardware and solar and space physics spaceflight instrumentation; public outreach programs in astrophysics, solar physics, and Mars exploration.

Table B—Separately Budgeted Research Expenditures by Source of Support

Source of Support	Departmental Research	Physics-related Research Outside Department
Federal government	$7,014,944	
State/local government		
Non-profit organizations		
Business and industry		
Other		
Total	$7,014,944	

Table C—Separately Budgeted Research Expenditures by Research Specialty

Research Specialty	No. of Grants	Expenditures ($)
Astrophysics	2	$36,789
Atmosphere, Space Physics, Cosmic Rays	10	$1,182,265
Condensed Matter Physics	6	$599,667
Optics	19	$1,956,264
Relativity & Gravitation	5	$306,130
Solar Physics	27	$2,933,829
Total	69	$7,014,944

FACULTY

Professor

Babbitt, William R., Ph.D., Harvard University, 1987. *Optics.* Optical physics; applied optics.

Carlsten, John L., Ph.D., Harvard University, 1974. *Optics.* Nonlinear optics; laser spectroscopy; atomic physics.

Cone, Rufus, Ph.D., Yale University, 1971. *Condensed Matter Physics, Optics.* Quantum optics; optical materials; lasers.

Cornish, Neil, Ph.D., University of Toronto, 1996. *Cosmology & String Theory, Relativity & Gravitation, Theoretical Physics.* Relativity theory; cosmology.

Francis, Gregory E., Ph.D., Massachusetts Institute of Technology, 1987. *Physics and other Science Education.* Physics education.

Idzerda, Yves, Ph.D., University of Maryland, 1986. *Condensed Matter Physics, Nano Science and Technology, Surface Physics.* Magnetic materials.

Link, Bennett, Ph.D., University of Illinois, 1991. *Astrophysics, Theoretical Physics.* Astrophysics.

Longcope, Dana, Ph.D., Cornell University, 1993. *Plasma and Fusion, Solar Physics, Theoretical Physics.* Plasma physics.

Neumeier, John J., Ph.D., University of California, San Diego, 1990. *Condensed Matter Physics, Low Temperature Physics.* Thermal expansion; phase transitions.

Rebane, Aleksander, Ph.D., University of Estonia, 1985. *Optics.* Optics; lasers.

Smith, Richard J., Ph.D., Iowa State University, 1975. Department Head. *Condensed Matter Physics, Materials Science, Metallurgy, Surface Physics.* Ion beams; surface physics.

Tsuruta, Sachiko, Ph.D., Columbia University, 1964. *Astrophysics, Theoretical Physics.* Astrophysics; compact objects.

Associate Professor

Kankelborg, Charles, Ph.D., Stanford University, 1996. *Solar Physics.* Solar physics.

Malovichko, Galina I., Ph.D., University of Kiev, 1987. *Condensed Matter Physics.* Optical materials.

Qiu, Jiong, Ph.D., Nanjing University, 1998. *Solar Physics, Theoretical Physics.*

Assistant Professor

Vorontsov, Anton, Ph.D., Northwestern University, 2004. *Condensed Matter Physics, Low Temperature Physics, Theoretical Physics*. Theoretical condensed-matter physics.

Willoughby, Shannon, Ph.D., Tulane University, 2003. *Astronomy, Condensed Matter Physics, Physics and other Science Education, Theoretical Physics*. Theoretical condensed-matter physics; physics and astronomy education.

Yunes, Nico, Ph.D., Pennsylvania State University, 2008. *Cosmology & String Theory, Relativity & Gravitation, Theoretical Physics*. Gravitational waves and quantum gravity.

Emeritus

Hermanson, John C., Ph.D., University of Chicago, 1966. *Condensed Matter Physics, Surface Physics, Theoretical Physics*. Surface physics theory.

Kirkpatrick, Larry, Ph.D., Massachusetts Institute of Technology, 1968. *High Energy Physics, Physics and other Science Education*. Science education.

Lapeyre, Gerald J., Ph.D., University of Missouri, Columbia, 1962. *Condensed Matter Physics, Surface Physics*. Photoemission; semiconductor materials; electron energy loss.

Schmidt, V. Hugo, Ph.D., University of Washington, 1961. *Condensed Matter Physics, Energy Sources & Environment*. Alternate energy.

Swenson, Robert, Ph.D., Lehigh University, 1961. *Statistical & Thermal Physics*. Statistical physics.

Wheeler, Gerald, Ph.D., Stony Brook University, 1972. *Nuclear Physics, Physics and other Science Education*. Experimental nuclear physics; science education.

Research Professor

Acton, Loren W., Ph.D., University of Colorado, 1965. *Solar Physics*. Solar physics.

Avci, Recep, Ph.D., University of Illinois, 1978. *Biophysics, Condensed Matter Physics, Surface Physics*.

Canfield, Richard C., Ph.D., University of Colorado, 1968. *Solar Physics*. Solar physics.

Craig, Alan, Ph.D., University of Arizona, 1982. *Condensed Matter Physics, Nano Science and Technology, Optics*. Coherent optics applications.

Hellings, Ronald, Ph.D., Montana State University, 1972. *Relativity & Gravitation, Theoretical Physics*. Relativity theory.

Klumpar, David M., Ph.D., University of New Hampshire, 1972. *Atmosphere, Space Physics, Cosmic Rays*. Experimental space physics; space instrumentation.

Martens, Petrus C., Ph.D., Utrecht University, 1983. *Solar Physics, Theoretical Physics*. Solar physics.

Research Associate Professor

Drobijev, Mikhail, Ph.D., Moscow Institute of Physics and Technology, 1986. *Biophysics, Optics*. Molecular biophysics; optics.

McKenzie, David E., Ph.D., University of Delaware, 1997. *Solar Physics*. Solar physics.

Research Assistant Professor

Leamon, Robert J., Ph.D., University of Delaware, 1999. *Solar Physics*. Solar physics.

Teaching Professor

Riedel, Carla M., Ph.D., University of Minnesota, 1996. *Nuclear Physics*. Nuclear physics.

Teaching Assistant Professor

Rugheimer, Paul P., Ph.D., University of Wisconsin - Madison, 2004. *Condensed Matter Physics, Surface Physics*. Thin film physics.

DEPARTMENTAL RESEARCH SPECIALTIES AND STAFF

Theoretical

Astrophysics. Neutron stars; active galactic nuclei; gamma-ray bursters. Link, Tsuruta.

Condensed Matter Physics. Correlated many-body (collective) effects such as superconductivity and superfluidity; influence of magnetic fields, impurities, and fluctuations on superconducting properties. Hermanson, Vorontsov.

Physics and other Science Education. Developing and implementing innovative programs for primary and secondary teacher education; developing techniques in education of non-science majors. Francis, Willoughby.

Relativity & Gravitation. Gravitational waves; black holes; quantum theory of gravity; early universe; experimental relativity. Cornish, Hellings, Yunes.

Experimental

Atmosphere, Space Physics, Cosmic Rays. Space instrumentation, including ultraviolet optics, to investigate the high-speed dynamics of magnetic reconnection in solar flares; solar magnetic activity; auroral physics; magnetospheric physics; development of space technologies and small satellites; heliophysics. Kankelborg, Klumpar.

Atomic, Molecular, & Optical Physics. Linear and nonlinear optical laser spectroscopy; coherent optical transients; optical hole burning; Raman scattering; solid-state laser material. Babbitt, Carlsten, Cone, Craig, Drobijev, Rebane.

Condensed Matter Physics. Measurements of physical properties to temperatures as low as 0.3 K; measurements of thermal expansion using a novel dilatometer capable of detecting sub-angstrom length changes of specimens to study phase transitions and critical phenomena; characterization of magnetic nanoparticles, thin films, and buried interfaces for spin transport devices; ceramics for fuel cells fabricated and tested for their electrical properties; characterization of defects in advanced materials at the atomic level using EPR, ENDOR, and optical spectroscopy. Avci, Cone, Idzerda, Malovichko, Neumeier, Schmidt, Smith.

View additional information about this department at www.gradschoolshopper.com

MONTANA TECH

GEOSCIENCES & ENGINEERING

Butte, Montana 59701

http://www.mtech.edu/admissions/graduate/graduate-programs.htm

General University Information
Chancellor: Donald M. Blackketter
Dean of Graduate School: Beverly Hartline
University website: http://www.mtech.edu
Control: Public
Setting: Suburban
Total Faculty: 217
Total Graduate Faculty: 63
Total number of Students: 2,701
Total number of Graduate Students: 163

Department Information
Department Chairman: Beverly Hartline, Dean
Department Contact: Beverly Hartline, Dean of Graduate
  School
  Total full-time faculty: 188
  Total number of full-time equivalent positions: 188
  Full-Time Graduate Students: 65
  First-Year Graduate Students: 17
  Female First-Year Students: 7

Department Address
1300 W. Park Street
Butte, MT 59701
Phone: (406) 496-4304
Fax: (406) 496-4710
E-mail: BHartline@mtech.edu
Website: http://www.mtech.edu/admissions/graduate/graduate-
  programs.htm

ADMISSIONS

Admission Contact Information
Address admission inquiries to: 1300 West Park Street Box 20,
  Butte, MT 59701-8997 USA
Phone: (406) 496-4304
E-mail: fsullivan@mtech.edu
Admissions website: http://www.mtech.edu/admissions/graduate/
  apply.htm

Application deadlines
Fall admission:
U.S. students: April 1   *Int'l. students*: April 1
Spring admission:
U.S. students: October 1   *Int'l. students*: October 1

Application fee
U.S. students: $30   *Int'l. students*: $30
These are priority admissions deadlines. Applications will be accepted after these dates. Late applications will be considered
  for admission, on a space-available basis.

Admissions information
For Fall of 2012:
  Number of applicants: 120
  Number admitted: 76
  Number enrolled: 63

Admission requirements
Bachelor's degree requirements: Bachelor's or International
  equivalent required. For some programs students with an undergraduate GPA between 2.5 and 2.75 can be admitted provi-
  sionally.
Minimum undergraduate GPA: 2.75

GRE requirements
The GRE is required.
General Test Scores are required for the following programs: ■
  Environmental Engineering ■ Industrial Hygiene Campus
  Program ■ Metallurgical/Mineral Processing ■ Technical
  Communications ■ Geoscience ■ Geochemistry ■ Geology
  ■ Geological Engineering ■ Geophysical Engineering ■
  Health Care Informatics ■ Hydrogeology ■ Hydrogeological
  Engineering

Advanced GRE requirements
The Advanced GRE is not required.

TOEFL requirements
The TOEFL exam is required for students from non-English-
  speaking countries.
  PBT score: 550
  iBT score: 75
A minimum IELTS score of 6.5 is also acceptable.

Other admissions information
Additional requirements: Electrical Engineering administers an
  entrance examination.

TUITION

Tuition year 2013-14:
Tuition for in-state residents
  Full-time students: $2,774 per semester
  Part-time students: $347 per credit
Tuition for out-of-state residents
  Full-time students: $8,525 per semester
  Part-time students: $986 per credit
Tuition amounts have mandatory fees included. Semester tuition
  is capped at the amount for 12 credits. Distance in-state tuition
  and fees are $302 per credit, and out-of-state is $547. Ph.D.
  program tuition is slightly higher.
Credit hours per semester to be considered full-time: 9
Deferred tuition plan: Yes
Health insurance: Available at the cost of $1900 per year.
Other academic fees: Books, supplies, building fees, insurance,
  lab fees.
Academic term: Semester
Number of first-year students who received full tuition waivers: 18
Number of first-year students who received partial tuition waivers: 12

Teaching Assistants, Research Assistants, and Fellowships
Number of first-year
  Teaching Assistants: 32
  Research Assistants: 5
Average stipend per academic year
  Teaching Assistant: $8,000
  Research Assistant: $10,000

FINANCIAL AID

Loans

Loans are available for U.S. students.
Loans are not available for international students.
GAPSFAS application required: Yes
FAFSA application required: No

For further information

Address financial aid inquiries to: Tressa Johnson.
Phone: 406-496-4465
E-mail: tjohnson@mtech.edu
Financial aid website: http://www.mtech.edu/admissions/graduate/financial-awards.htm

HOUSING

Availability of on-campus housing

Single students: Yes
Married students: Yes

For further information

Address housing inquiries to: Annie Telling.
Phone: 406-496-4425
E-mail: atelling@mtech.edu
Housing aid website: http://www.mtech.edu/admissions/graduate/apply-for-housing.htm

Table A—Faculty, Enrollments, and Degrees Granted

Research Specialty	Faculty	Enrollment Fall 2012		Number of Degrees Granted 53 (2012-13)		
		Master's	Doctorate	Master's	Terminal Master's	Doctorate
Electrical Engineering	5	4	–	5(-)	–	–
Geology/Geochemistry	7	13	–	18(-)	–	–
Geophysics	5	6	–	6(-)	–	–
Mechanical/ Environmental/Nano Engineering/Nano Technology	6	15	–	4(-)	–	–
Metallurgy, Energy, Biomaterials	15	9	–	5(-)	–	–
Total	**38**	**47**	**–**	**38(-)**	**–**	**–**
Full-time Grad. Stud.	–	46	–	–	–	–
First-year Grad. Stud.	–	35	–	–	–	–

GRADUATE DEGREE REQUIREMENTS

Master's: 31-37 semester credits depending on specialty, including 1 seminar and 1 writing class.

Doctorate: Collaborative program with University of Montana and Montana State University. Sixty semester credits at least 18 for dissertation and 32 for course work. Twenty credit core first year curriculum. Qualifying exam, dissertation, and dissertation defense.

Other Degrees: Master's programs have thesis and non-thesis options. A thesis can be accomplished by a published peer reviewed paper.

Thesis: Thesis must conform to a standard format. All theses are submitted electronically to ProQuest.

SPECIAL EQUIPMENT, FACILITIES, OR PROGRAMS

The Montana Bureau of Mines and Geologyhas the Ground Water Information Center (GWIC), the Ground Water Assessment Program GWAP), and the Ground Water Investigation Program (GWIP), and other studies that produce and make available extensive information on groundwater quality and quantity throughout Montana. The Geographic Information Systems (GIS) lab is fully equipped to produce and analyze digital maps and other spatial data, particularly geologic and hydrogeologic maps. MBMG's Analytical Laboratory is licensed by the state of Montana to analyze public water supplies and it has a QA/QC program meeting the criteria established by the U.S. Environmental Protection Agency (EPA) and the U.S. Geological Survey (USGS). Available instrumentation includes the following major instruments.

- Thermo ICAP inductively coupled argon plasma emission spectrophotometer (ICAPES)

- Thermo X-series® Inductively Coupled Plasma/Mass Spectrometer (ICP/MS) for trace-metal determination

- Dionex ion chromatographs for anion determination

- Agilent gas chromatograph with mass spectrometer detector (GS/MS) for organic compounds,

- Agilent gas chromatograph with electron capture detector (ECD) to measure low level pesticide and chlorinated compounds in water and soil

- ELISA testing in magnetic particle and 96 well plate formats for determining endocrine disrupting compounds in surface and ground waters

- Beckman scintillation counter for determination of radon in water

- Picarro water isotope analyzer

Metallurgical and Materials Engineering (M&ME)– Located in the Engineering Laboratory and Classroom (ELC) Building, where extensive laboratory facilities and equipment related to mineral, materials, and metallurgical processing, characterization, and testing are available in about twenty different laboratory rooms. The co-located Center for Advanced Mineral and Metallurgical Processing (CAMP) shares the research facilities and conducts research supporting the global mineral industry, including processing of minerals & metals and developing processes that minimize waste generation. The ELC Building has a loading dock for receiving large samples, along with a high bay, where Montana Tech's faculty/student machine shop is being installed.

Within the high bay on the first floor of the ELC Building is the Thermal Processing Area, housing a pilot plant roaster donated by Newmont Mining Company, and the Roasting/Calcining Lab with induction, box, and kiln furnaces as well as investment and traditional casters. The Materials Manufacturing Lab houses a freeform fabricator, two types of porosimeters, and an autoclave, as well as sample-preparation and wet-chemistry equipment. A nearby lab houses a ThermoFisher ICAP6000 inductive coupled plasma (ICP) spectrometer, an ion chromatograph and an Agilent 300A Micro Gas Chromatograph. The Separations/Recycling Lab is used for particulate processing and includes various chemical, density, electrostatic, hydrophobic, magnetic, and particle-size separators. The Comminution Lab contains equipment, such as crushers, grinding mills, pulverizers, splitters, and sieves.

The Physical Metallurgy Lab houses a cold pressure roll and vacuum furnace. Metallographic Lab 1 houses samples, microscopes, and sample-preparation equipment, as well as a macrohardness tester. Metallographic Lab 2 includes manual polishing wheels, submerged cut-off and diamond-blade saws. Metallographic Lab 3 contains automatic polishing equipment, as well as sample preparation benches. The Microhardness Lab houses three microhardness testers and a Neophot21. The Imaging Lab houses a computer-aided optical microscope and a tabletop imaging SEM. The X-ray Lab has an x-ray diffraction (XRD) instrument and a Bruker A20A4 x-ray fluorescence (XRF) spectrometer. The Scanning Electron Microscopy (SEM) Lab houses two instruments: a 1430VP SEM with two energy-dispersive x-ray analyzers (SEM/EDX) for mineral liberation analysis (MLA), and a Leo 1430 Upgrade SEM.

The three labs on the second floor of the ELC Building are the Materials Thermochemistry Lab, the Environmental Hydrometallurgy Lab, and the Corrosion and Special Projects Lab. The Thermochcemistry lab contains a Carbon/Sulfur analyzer, a TA Instruments Q5000 ThermoGravimetric Analyzer (TGA), a Netsch DIL 402C dilatometer, a TA Instruments Q800 dynamic mechanical analyzer (DMA), a scanning calorimeter, and several furnaces—for characterizing various types of materials, particularly polymers, ceramics, slags, composites, sulfide minerals and coals. The Hydrometallurgy Lab is basically a wet-chemistry lab, equipped additionally with water purification systems, drying ovens, and weighing scales. The Corrosion Lab houses potentiostats, galvanostats, plating equipment, and electrochemical cells and has space for short-duration research projects.

Geological Engineering – Has a Geomechanics Lab equipped with a $1-million 330,000-lb load frame with the capacity to test specimens under confinement pressure up to 20,000 psi and at temperatures up to 100 degrees C. The Lab also has a recently upgraded servo-controlled direct-shear apparatus and superfine surface grinding and polishing equipment for rapidly preparing specimens polished thin sections and slabs of geological and other materials for examination by transmitted and reflected light microscopy and/or scanning electron microscopy. A new Underground Mine Education Center on campus is being developed. This facility will be used by all departments and will allow Mining Engineering students in particular to conduct research in drilling, blasting, equipment, rock mechanics, ventilation, instrumentation and automation.

Environmental Engineering – Houses a Metal-Contamination Research Lab with instruments for analyzing and detecting trace amounts of mercury and other toxic heavy metals. The suite of equipment includes a Thermo inductively coupled argon plasma emission spectrometer (iCAPES), four different types of mercury analyzers, and a Leica microscope with imaging software to take pictures of mercury-contaminated specimens. The fully equipped sample preparation lab includes a microwave digester to prepare samples for the iCAPES. The lab benefits from co-ownership of the patent on a metal nanoparticle filter capable of removing trace metals from air and water.

General Engineering – Has several labs with devices to test and characterize asphalt, concrete, soil, timber, metals, composites, and ceramics. Research on nanomaterials, fluids and energy conversion is also done. A new Near-Field Scanning Optical Microscope (NSOM) is arriving during summer 2013. Equipment for materials testing includes standard machines for tensile/compression, impact/toughness, three-point bending, hardness and fatigue testing. The testing devices include a full range of systems for analysis of materials of many different descriptions. Characterization equipment for nondestructive evaluation includes dye-penetrant, magnetic-particle, eddy-current, ultrasonic, and radiographic testing, as well as optical microscopes, a probe station, and soil analysis equipment. There is also a range of manufacturing equipment in both subtractive (milling, lathe, and cutting equipment) and additive (welding, arc, laser, friction, stir, and 3D printing) forms. The standard range of specialized software is available, including Auto-CAD, Solidworks, Civil3D, MATLAB, MathCAD, Octave, and COMSOL.

Geophysical Engineering – The department of Geophysical Engineering has facilities that include a tank for developing small-scale models of soil structures, three field vehicles (including trucks and an ATV), and a full complement of seismic, magnetic, electric, and gravitational geophysics equipment to make field measurements. Equipment includes EM-31 and EM-34 devices (two of each) to actively image variations in subterranean electrical conductivity, as well as a GDP-32 multipurpose meter. Two IRIS resistivity meters and proton precession total field magnetometers with 1 nT and 10 nT sensitivity, round out the complement of electrical and magnetic measurement equipment.

The department also owns one Worden and two Lacost-Romberg gravimeters with 0.1 mGal sensitivity, for measuring density variations. Two vehicle-mounted accelerated weight drop machines (500 pound and 100 pound) and a 120-channel 24-bit digital recording system are used with a 96-geophone array to measure seismic activity. In addition, the Geophysical Engineering department has a Mala ground-penetrating radar system. All of the above equipment is fully complemented with modern software to interpret, process, and image the measurements. These softward packages include:

● Seismic

o Landmark Graphics Corporation ProMax seismic processing/interpretation software

o Kingdom Suite seismic interpretation software

o OpendeTect seismic interpretation software

o Vista seismic processing software

o SPW seismic processing software

o Rayfract refraction tomography software

o IXRefraX layered/GRM refraction software

o IXSEG2SEGY seismic utility software

o Hampson-Russel AVO and modeling software

o jTIPS well log modeling software

o MESA Expert seismic survey design software

o Surfseis surface wave analysis software

o Petrel reservoir and seismic modeling software

Table B—Separately Budgeted Research Expenditures by Source of Support

Source of Support	Departmental Research	Physics-related Research Outside Department
Federal government	$3,293	
State/local government	$4,817	
Non-profit organizations		
Business and industry	$1,289	
Other	$2,366	
Total	$11,765	

Table C—Separately Budgeted Research Expenditures by Research Specialty

Research Specialty	No. of Grants	Expenditures ($)
Electrical Engineering	6	$1,000,000
Geophysics	4	$1,000,000
Materials Science, Metallurgy	6	$1,700,000
Geology/Geochemistry	30	$6,000,000
Mechanical/Environmental/NanoEngineering	4	$900,000
Total	50	$10,600,000

FACULTY

Professor

Appleman, Rick, Ph.D., University of California at Irvine. *Other.* Environmental engineering.

Cameron, Douglas, Ph.D., University of California at Los Angeles. *Geology/Geochemistry.* Geochemistry.

Drury, Bill, Ph.D., Montana State University. *Other.* Bill Drury's research projects at Montana Tech have investigated passive biological systems for treating mine drainage containing metals and acid. These systems can be enclosed in a reactor, or a sub-surface flow wetland can be used to generate the same reactions. Bacteria utilize solid organic matter like compost or plant litter as electron donors. Sulfate reducing bacteria use sulfate in the mine water as their terminal electron acceptor, producing hydrogen sulfide that is a good precipitant for most of the metals in the water. Bicarbonate alkalinity is also generated that neutralized the acid. Specifically, the research has measured the kinetics of sulfide production and metals removal, and the hydraulic characteristics of treatment wetlands.

Gammons, Christopher, Ph.D., Penn State University. *Geology/Geochemistry.* Aqueous geochemistry at high and low temperatures, economic geology, acid mine drainage, stable isotopes.

Ganesan, Kumar, Ph.D., Washington State University Pullman. *Other.* Inventory of Mercury from historic mining and its impact on soil and watersheds in Montana. Director of EPA-EPSCoR funded research activities in Montana. Evaluating Montana coal for removing toxic metals from acid rock drains. Funded by DOE/EPSCoR. Southridge, SC. Field Evaluation of Montana coal for removing cadmium, zinc and iron from active acid mine drainage. Funded by DOE/EPSCoR.

Girard, Jim, Ph.D., University of Montana. *Geophysics.* Geophysical engineering.

Huang, Hsin-Hsiung, Ph.D., Stanford University. *Materials Science, Metallurgy.*

James, Rod, Ph.D., Montana State University. *Other.* Environmental engineering.

Link, Curtis, Ph.D., University of Houston. Department Head. *Geophysics.* Geophysical engineering.

MacLaughlin, Mary, Ph.D., UC Berkeley. *Geology/Geochemistry.* Geological engineering.

Parker, Steve, Ph.D., University of Montana. *Geology/Geochemistry.* Geochemistry.

Peterson, Holly, Ph.D., Washington State University Pullman. *Other.* Environmental engineering.

Speece, Marvin, Ph.D., The University of Wyoming. *Geophysics.* Geophysical engineering.

Trudnowski, Dan, Ph.D., Montana State University. Department Head. *Electrical Engineering.* Electrical engineering.

Wolfgram, Diane, Ph.D., UC Berkeley. *Geology/Geochemistry.* Geological engineering/geology.

Young, Courtney, Ph.D., University of Utah. *Materials Science, Metallurgy.* Materials science, metallurgical engineering.

Associate Professor

Donnelly, Matthew, Ph.D., Montana State University. *Electrical Engineering.*

Downey, Jerry, Ph.D., Colorado School of Mines. *Materials Science, Metallurgy.* Materials science and metallurgical engineering.

Gleason, William, Ph.D., University of Montana/Montana Tech. *Materials Science, Metallurgy.* Materials science and metallurgical engineering.

Kasinath, Rajendra, Ph.D., Nanyang Technological University. *Other.* Environmental engineering.

Meier, Al, Ph.D., Colorado School of Mines. *Materials Science, Metallurgy.*

Moon, Thomas, Ph.D., University of Washington. *Electrical Engineering.* Electrical engineering and geophysical engineering.

Morrison, John, Ph.D., University of Idaho. *Electrical Engineering.* Electrical engineering.

Smtih, Larry, Ph.D., University of New Mexico. *Geology/Geochemistry.* Stratigraphy, sedimentology, Quaternary geology,

subsurface and surface geologic mapping, and petroleum geology.

Sudhakar, K V., Ph.D., Indian Institute of Science. *Materials Science, Metallurgy.* Materials science, metallurgical engineering.

Zhou, Xiaobing, Ph.D., Universtiy of Alaska Fairbanks. *Geophysics.* Geophysical engineering.

Assistant Professor

Bayat, Jahan, Ph.D., University of Florida. *Other.* Mechanical engineering.

Hailer, Katie, Ph.D., University of Montana. *Geology/Geochemistry.* Geochemistry.

Hill, Bryce, Ph.D., University of Utah. *Electrical Engineering.* Electrical engineering.

Hugo, Bertete Aguirre, Ph.D., University of Utah. Geophysical engineering.

Klem, Michael, Ph.D., Iowa State University. *Geology/Geochemistry.* Geochemistry.

Shaw, Glenn, Ph.D., UC Merced School of Engineering. *Geology/Geochemistry.* Hydrogeology; groundwater-surface water interactions; contaminant transport; geochemical tracers in groundwater and surface water.

Skinner, Jack, Ph.D., UC Davis. *Materials Science, Metallurgy, Other.* ■ Energy systems ■ Microscale and nanoscale theory and design ■ Microfabrication and nanofabrication ■ Advanced materials ■ Plasmonics ■ Device characterization ■ Integration of microscale and nanoscale technologies.

DEPARTMENTAL RESEARCH SPECIALTIES AND STAFF

Experimental

Electrical Engineering. Appleman, Drury, Ganesan, James, Kasinath, Peterson.

Electrical Engineering. Appleman, Drury, Ganesan, James, Kasinath, Peterson.

Geochemistry. Geochemistry is an interdisciplinary field of study in which the science of chemistry is used to solve earth science problems. Areas of study include the full spectrum of topics from the determination of the thermodynamic properties of minerals, to the determination of the migration of pollutant species within a soil or hydrologic environment and the geobotanical/biogeochemical search for ore deposits. The geochemistry option is centered in the Chemistry and Geochemistry Department. Students acquire relatively strong backgrounds in chemistry and can choose thesis topics ranging among a large variety of geochemical/environmental topics. Cameron, Hailer, Klem, Parker.

Geological Engineering. Gammons, MacLaughlin, Shaw, Smtih, Wolfgram.

Geology. Gammons, Parker, Shaw, Smtih.

Geophysical Engineering. Oil Reservoir Characterization; Seismic Processing/Interpretation; Electrical Resistivity Studies; Neural Network Analysis/Application; Remote Sensing Analysis; Ground Penetrating Radar Studies; Shallow Seismic Investigations; Application of Artificial Neural Networks; Antarctic History; Gravity and Magnetic Studies. Hugo, Link, Moon, Speece, Zhou.

Hydrogeological Engineering. Gammons, Parker, Shaw, Smtih.

Materials Science & Engineering. Energy materials, biomaterials, materials synthesis, processing, characterization, and fabrication. Includes nanoscience and nanotechnology. Downey, Ganesan, Gleason, Kasinath, Klem, MacLaughlin, Meier, Skinner, Sudhakar, Young.

Metallurgical/Mineral Processing. Downey, Gleason, Huang, Meier, Sudhakar, Young.

CREIGHTON UNIVERSITY

DEPARTMENT OF PHYSICS

Omaha, Nebraska 68178
http://physicsweb.creighton.edu

General University Information
President: Timothy R. Lannon, S.J.
Dean of Graduate School: Gail M. Jensen
University website: http://www.creighton.edu
Control: Private
Setting: Urban
Total Faculty: 764
Total Graduate Faculty: 346
Total number of Students: 7,736
Total number of Graduate Students: 3,704

Department Information
Department Chairman: Janet E. Seger, Chair
Department Contact: Michael G. Nichols, Graduate Program
  Director
  Total full-time faculty: 9
  Total number of full-time equivalent positions: 9
  Full-Time Graduate Students: 11
  First-Year Graduate Students: 8
  Female First-Year Students: 1
  Total Post Doctorates: 2

Department Address
2500 California Plaza
Omaha, NE 68178
Phone: (402) 280-2159
Fax: (402) 280-2140
E-mail: mnichols@creighton.edu
Website: http://physicsweb.creighton.edu

ADMISSIONS

Admission Contact Information
Address admission inquiries to: Director, Physics Graduate Program, Department of Physics
Phone: (402) 280-2870
E-mail: gradsch@creighton.edu
Admissions website: http://www.creighton.edu/gradschool/admissioninformation/index.php

Application deadlines
Fall admission:
U.S. students: July 1   *Int'l. students*: July 1
Spring admission:
U.S. students: December 1   *Int'l. students*: December 1

Application fee
U.S. students: $50   *Int'l. students*: $50

Admissions information
For Fall of 2012:
  Number of applicants: 14
  Number admitted: 9
  Number enrolled: 6

Admission requirements
Bachelor's degree requirements: A Bachelors degree in physics or engineering-related field is required.

GRE requirements
The GRE is required.

Advanced GRE requirements
The Advanced GRE is not required.

TOEFL requirements
The TOEFL exam is required for students from non-English-speaking countries.
  PBT score: 550
  iBT score: 80

Other admissions information
Undergraduate preparation assumed: A typical student will have completed undergraduate courses in Classical Mechanics (Taylor, Marion, etc.), Quantum Mechanics (Griffiths, Townsend, etc.), Electricity and Magnetism (Griffiths, Corson and Lorrain, etc.), and Statistical Mechanics/Thermodynamics (Kittel, Reif, etc.).

TUITION

Tuition year 2013-14:
  Full-time students: $756 per credit
  Part-time students: $756 per credit
Teaching and Research Assistants do not pay tuition or semester.
Credit hours per semester to be considered full-time: 8
Deferred tuition plan: No
Health insurance: Available at the cost of $2182 per year.
Other academic fees: University Fee: $532/semester (full-time); $56 (part-time) University Technology Fee: $227/semester (full-time); $93/semester (part-time).
Academic term: Semester
Number of first-year students who received full tuition waivers: 6

Teaching Assistants, Research Assistants, and Fellowships
Number of first-year
  Teaching Assistants: 4
  Research Assistants: 1
Average stipend per academic year
  Teaching Assistant: $11,295
  Research Assistant: $11,295

FINANCIAL AID

Application deadlines
Fall admission:
U.S. students: July 23
Spring admission:
U.S. students: December 12   *Int'l. students*: December 12

Loans
Loans are available for U.S. students.
Loans are not available for international students.
GAPSFAS application required: No
FAFSA application required: Yes

For further information
Address financial aid inquiries to: Creighton University, Office of Financial Aid.
Phone: (402) 280-2731
E-mail: finaid@creighton.edu
Financial aid website: http://www.creighton.edu/financialaid

HOUSING

Availability of on-campus housing
Single students: Yes
Married students: Yes

For further information
Address housing inquiries to: Department of Residence Life, 136 Swanson Hall.
Phone: (402) 280-3900
E-mail: ResidenceLife@Creighton.edu
Housing aid website: http://www.creighton.edu/studentservices/ departmentofresidencelife/departmenthome/index.php

Table A—Faculty, Enrollments, and Degrees Granted

Research Specialty	2012–13 Faculty	Enrollment Fall 2012 Master's	Enrollment Fall 2012 Doctorate	Number of Degrees Granted 2012–13 (2008–13) Master's	Number of Degrees Granted 2012–13 (2008–13) Terminal Master's	Number of Degrees Granted 2012–13 (2008–13) Doctorate
Astrophysics	2	2	–	–	2(6)	–
Atomic, Molecular, & Optical Physics	1	–	–	–	–(2)	–
Biophysics	2	2	–	–	1(5)	–
Condensed Matter Physics	1	1	–	–	3(7)	–
Nano Science and Technology	1	–	–	–	–	–
Nuclear Physics	3	1	–	–	1(8)	–
Non-specialized	–	2	–	–	–	–
Total	10	8	–	–	7(28)	–
Full-time Grad. Stud.	–	8	–	–	–	–
First-year Grad. Stud.	–	5	–	–	–	–

GRADUATE DEGREE REQUIREMENTS

Master's: Plan A (thesis option) requires 30 credit hours of graduate-level courses, including six credit hours of Thesis Research. Plan B (non-thesis option) requires 33 credit hours of graduate-level courses, including three credit hours of Directed Independent Research for which a report is required. A minimum of fifteen credit hours must be in physics in either Plan. In both Plans, all students are required to take the core graduate courses in classical mechanics, electromagnetics, quantum mechanics, and statistical mechanics. Full-time students must enroll in Graduate Seminar each semester. A minimum grade average of 3.0 (B) is required, with no more than two grades of C. A three-part comprehensive exam offered three times each year on (1) Mechanics and Heat, (2) Electricity, Magnetism and optics, (3) Modern Physics is required to be passed. Each part can be taken separately. At least one part must be passed in the first year of study. Ordinarily a student must devote two semesters and a summer session to resident graduate study.
Thesis: Theses may be written in absentia.

SPECIAL EQUIPMENT, FACILITIES, OR PROGRAMS

Atomic, Molecular, and Optical Physics: 767 nm frequency-stabilized diode laser, frequency-stabilized HeNe laser, acousto-optic modulators, spectrum analyzer, rf sources, ultra-high vacuum equipment, x-ray fluorescence analyzers, high-resolution Si(li) detectors and a silicon PIN detector.

Condensed Matter Physics: Turbomolecular pump, 1500 C furnace, IR detector, glove box, laminar flow bench, muffle furnaces, vacuum oven, capacitance bridge, lock-in amplifier, photon correlation spectrometer, fluorescence correlation spectrometer, 5W CW 532-nm laser, 25W CW 1064-nm laser, CO_2 laser, x-ray diffractometer.

Biophysics: Femtosecond Ti:S laser, confocal laser scanning microscope, holographic microscope, Frequency Domain fluorimeter, fiber-optic UV-VIS-NIR absorption and fluorescence spectrometers, Cary 60 UV VIS spectrometer, 10W Nd:YAG laser, Optical stretcher, sterile cell culture facilities, inverted tissue culture microscopes, Zeiss 510 NLO LSM with FLIM electronics.

Computational molecular biophysics: AMD opteron and Intel Xeon multi-core workstations with 64-bit Linux operating system.

Astrophysics/Astro-particle Physics: Multi-node networked Intel-based workstations for simulations. Robotic Telescope facility.

Nuclear Physics: Compton-scattering spectrometer, gamma-gamma and beta-gamma angular correlation spectrometers, HPGe gamma-ray detector, NIM electronics, Computing cluster.

Nanoscience and technology: Agilent Atomic Force Microscope, Kurt J. Lesker high vacuum, three-source thermal evaporator, Advanced Research Systems (ARS) 7 K closed-cycle cryostat with optical access, GMW 1.2 T electromagnet, Keithley SourceMeter Unit (3A, 40V), Perkin-Elmer Fourier Transform Infrared Spectroscopy, ElVax X-ray Fluorescence Spectroscopy, Perkin-Elmer Differential Scanning Calorimeter, Newport Apex Monochromator light source, Ocean Optics UV-Vis Spectrometer, Admet 10 kN Universal Testing Machine, Bareiss Shore A and D Durometers, Filmetrics Spectral Reflectance, Headway Photoresist Spinner, PlasmaEtch Reactive Ion Etcher. Solvent induced ordering of block copolymers facility. Thin film deposition and testing facility.

Table B—Separately Budgeted Research Expenditures by Source of Support

Source of Support	Departmental Research	Physics-related Research Outside Department
Federal government	$436,700	
State/local government	$20,000	
Non-profit organizations		
Business and industry		
Other	$4,800	
Total	$461,500	

Table C—Separately Budgeted Research Expenditures by Research Specialty

Research Specialty	No. of Grants	Expenditures ($)
Astrophysics	4	$80,000
Biophysics	5	$86,500
Condensed Matter Physics	1	$80,000
Nuclear Physics	1	$215,000
Total	11	$461,500

FACULTY

Professor
Cherney, Michael G., Ph.D., University of Wisconsin-Madison, 1987. *High Energy Physics, Nuclear Physics*. Experimental high-energy nuclear physics (relativistic heavy ion physics); control systems; nuclear science education; energy sources and environment.
Seger, Janet E., Ph.D., University of Wisconsin-Madison, 1991. *High Energy Physics, Nuclear Physics*. Theoretical and experimental high-energy nuclear physics.

Associate Professor
Duda, Gintaras, Ph.D., University of California, Los Angeles, 2003. *Astrophysics, Cosmology & String Theory, Particles and Fields,*

Physics and other Science Education. Astro-particle physics; theoretical elementary particle physics; physics education research.

Nichols, Michael G., Ph.D., University of Rochester, 1996. *Biophysics, Medical, Health Physics, Optics.* Experimental biological physics; tissue spectroscopy and microscopy; photodynamic therapy of cancer; cellular mechanics.

Sidebottom, David L., Ph.D., Kansas State University, 1989. *Condensed Matter Physics.* Glass science; dynamic light scattering; optical spectroscopy; dielectric spectroscopy.

Assistant Professor

Baruth, Andrew G., Ph.D., University of Nebraska-Lincoln, 2009. *Condensed Matter Physics, Energy Sources & Environment, Nano Science and Technology.* Magnetic and electronic behavior of 3D nanostructures, block copolymer self assembly, magnetic heterostructures, solar cells, superconductivity.

Gabel, Jack R., Ph.D., Catholic University of America, 2000. *Astrophysics.* Observational astrophysics; UV, optical, IR spectroscopy; photoionization modeling of astrophysical plasmas.

McShane, Thomas S., M.S., Saint Louis University, 1956. Control systems; computer simulation; electronics.

Soto, Patricia, Ph.D., University of Groningen, 2004. *Biophysics.* Computational molecular biophysics, biomolecular modeling, structural bioinformatics, protein conformational dynamics, pathological protein folding.

Wrubel, Jonathan P., Ph.D., Cornell University, 2006. *Atomic, Molecular, & Optical Physics.* Atomic, Molecular, and Optical Physics, Optics. laser-cooling and trapping; ultracold atoms; Bose-Einstein condensates; spinors; nonlinear matter-waves.

Emeritus

Cipolla, Sam J., Ph.D., Purdue University, 1969. *Atomic, Molecular, & Optical Physics.* Experimental atomic physics; inner-shell ionization; response modeling of radiation detectors.

Zepf, Thomas H., Ph.D., Saint Louis University, 1963. *Condensed Matter Physics.* Condensed matter physics.

Instructor

Bruckman, Katie E., M.S., Creighton University, 2011. Sustainability Engineer and Assistant Program Director of the Energy Program. *Energy Sources & Environment, Engineering Physics/Science.*

Kriegler, David J., M.S., University of Nebraska-Lincoln, 1976. *Astronomy.* Astronomy.

Stuva, David R., M.S., Creighton University, 1983. General physics.

Watters, Kyle P., Ph.D., Stanford University, 2010. *Astrophysics.* Gamma ray pulsars; Fermi Gamma-ray Space Telescope.

Posdoctoral Research Associate

DeSilva, L. Chanaka, Ph.D., University of Houston, 2012. *High Energy Physics, Nuclear Physics.* Relativistic heavy ion collisions.

Nilsen, Bjorn, Ph.D., University of Minnesota, 1994. *High Energy Physics, Nuclear Physics.* Relativistic heavy ion collisions.

DEPARTMENTAL RESEARCH SPECIALTIES AND STAFF

Theoretical

Astrophysics. Characterization and detection of dark matter; prompt atmospheric lepton flux in high energy cosmic rays; high-energy cosmic rays beyond the GZK cutoff. Duda.

Biophysics. Protein structure, dynamics, and self-assembly; protein folding and misfolding; protein-membrane interactions. Soto.

Experimental

Astrophysics. Observations and analysis of active galactic nuclei using UV, optical, IR spectra from space-based and large ground-based observations; Photoionization modeling of astrophysical plasmas, Studies of energetic mass outflows from quasars. Gabel.

Atomic, Molecular, & Optical Physics. Laser-cooling of potassium atoms to produce Bose-Einstein condensates; Study of the spin dynamics (spinor physics) of ultracold atoms; nonlinear matter-wave physics in BECs. Wrubel.

Atomic, Molecular, & Optical Physics. Atomic inner-shell ionization. X-ray fluorescence. Cipolla.

Biophysics. Development and application of novel optical techniques to biology and medicine; Multiphoton and confocal laser scanning fluorescence microscopy; Molecular photophysics. Vis/Near-IR tissue spectroscopy; Photodynamic therapy (PDT) of Cancer. Studies of cellular mechanics use an optical stretcher apparatus and holographic microscope. Nichols.

Condensed Matter Physics. Ionic motion in glasses; dynamic light scattering of the glass transition. Evaluation of cryopreserving agents. Sidebottom.

Nano Science and Technology. Nanoscale systems involving novel magnetic and electronic phenomena at heterostructure interfaces; Solar Cells; self-assembly of block copolymers. Baruth.

Nuclear Physics. High-energy nuclear physics (relativistic heavy ion physics) in STAR collaboration at RHIC (Brookhaven National Laboratory) and ALICE collaboration at the LHC (CERN). Cherney, DeSilva, McShane, Nilsen, Seger.

View additional information about this department at www.gradschoolshopper.com

UNIVERSITY OF NEBRASKA–LINCOLN

DEPARTMENT OF PHYSICS AND ASTRONOMY

Lincoln, Nebraska 68588-0299
http://physics.unl.edu

General University Information

President: James B. Milliken
Dean of Graduate School: Dr. Lance Perez
University website: http://www.unl.edu

Control: Public
Setting: Urban
Total Faculty: 1,615
Total Graduate Faculty: 1,477

Total number of Students: 24,207
Total number of Graduate Students: 4,559

Department Information
Department Chairman: Daniel R. Claes, Chair
Department Contact: Marjorie Wolfe, Graduate Secretary
   Total full-time faculty: 28
   Total number of full-time equivalent positions: 43
   Full-Time Graduate Students: 76
   First-Year Graduate Students: 14
   Total Post Doctorates: 19

Department Address
Jorgensen Hall
855 N 16th Street
Lincoln, NE 68588-0299
Phone: (402) 472-2770
Fax: (402) 472-6148
E-mail: PAGrad@unl.edu
Website: http://physics.unl.edu

ADMISSIONS

Admission Contact Information
Address admission inquiries to: Chairman, Graduate Admissions
   Committee, Department of Physics and Astronomy
Phone: (402) 472-2770
E-mail: PAgrad@unl.edu
Admissions website: http://physics.unl.edu/grad/grad.shtml

Application deadlines
Fall admission:
U.S. students: January 31   *Int'l. students*: January 31

Application fee
U.S. students: $50   *Int'l. students*: $50
Steps for Admission website: http://physics.unl.edu/grad/grad_
   howtoapply.shtml

Admissions information
For Fall of 2013:
   Number of applicants: 130
   Number admitted: 40
   Number enrolled: 13

Admission requirements
Bachelor's degree requirements: Bachelor's degree is required.
Minimum undergraduate GPA: 3.0

GRE requirements
The GRE is required.
The preferred Quantitative score: greater than 150.

Advanced GRE requirements
The Advanced GRE is recommended.
The average GRE Physics score for the 2013-2014 admits was
   773.

TOEFL requirements
The TOEFL exam is required for students from non-English-
   speaking countries.
   PBT score: 550
   iBT score: 80
The minimum score for computer-based TOEFL is 213. The minimum score for the ILETS is 6.5

Other admissions information
Additional requirements: Each student is judged on his/her own
   merits. The average new Revised GRE scores for 2013–2014
   admissions were verbal–153; quantitative–162. The average
   GRE physics score for 2013-2014 admission was 773.

Undergraduate preparation assumed: G. Fowles and G. Cassiday, Analytical Mechanics; Quantum Mechanics D. Griffiths, Introduction to Electrodynamics, 4th Ed; D. Griffiths,
   Quantum Mechanics.

TUITION
Tuition year 2013-14:
Tuition for in-state residents
   Full-time students: $285 per credit
   Part-time students: $285 per credit
Tuition for out-of-state residents
   Full-time students: $791.75 per credit
   Part-time students: $791.75 per credit
Teaching and Research Assistants do not pay tuition.
Credit hours per semester to be considered full-time: 9
Deferred tuition plan: No
Health insurance: Available at the cost of $368 per year.
Other academic fees: Program and facilities fees: $520/semester
   if enrolled 7+ hours; library and technology fees; $11.50 per
   credit hour; registration fee $20. One-time new international
   student fee of $150, then $35/semester thereafter.
Academic term: Semester
Number of first-year students who received full tuition waivers: 13

Teaching Assistants, Research Assistants, and Fellowships
Number of first-year
   Teaching Assistants: 13
   Fellowship students: 4
Average stipend per academic year
   Teaching Assistant: $18,300
   Research Assistant: $18,300
All first-year graduate students have a Graduate Assistant ap-
   pointment.

FINANCIAL AID

Application deadlines
Fall admission:
U.S. students: April 1   *Int'l. students*: April 1

Loans
Loans are available for U.S. students.
Loans are not available for international students.
GAPSFAS application required: No
FAFSA application required: Yes

For further information
Address financial aid inquiries to: Scholarships and Financial
   Aid, University of Nebraska-Lincoln, 17 Canfield Admin
   Bldg, Lincoln, NE 68588-0411.
Phone: (800) 742-8800 (ext 2030)
E-mail: financialaid@unl.edu
Financial aid website: http://www.unl.edu/scholfa/

HOUSING

Availability of on-campus housing
   Single students: Yes
   Married students: Yes

For further information
Address housing inquiries to: University Housing, 1115 N. 16th
   St., Lincoln, NE 68588-0622.
Phone: (800) 742-8800 (ext 3561)
E-mail: housing@unl.edu
Housing aid website: http://housing.unl.edu/

Table A—Faculty, Enrollments, and Degrees Granted

Research Specialty	2012-13 Faculty	Enrollment Fall 2012		Number of Degrees Granted 2012-13 (2008-13)		
		Master's	Doctorate	Master's	Terminal Master's	Doctorate
Atomic, Molecular, & Optical Physics	9	2	28	2(9)	–(2)	–(4)
Condensed Matter Physics	11	6	31	5(19)	–	3(21)
High Energy Physics	5	1	6	–(2)	–(2)	1(2)
Non-specialized	1	1	–	–	–	–
Total	27	11	65	7(30)	–(4)	4(27)
Full-time Grad. Stud.	–	11	65	–	–	–
First-year Grad. Stud.	–	–	14	–	–	–

GRADUATE DEGREE REQUIREMENTS

Master's: Option I (for both terminal degree students and those continuing in the Ph.D. program): 30 credit hours of coursework with a minimum of 15 credits in physics plus a thesis. Option II (for students wishing an interdisciplinary degree program): 36 credit hours of coursework including a minimum of 18 hours in physics and a minimum of nine hours in each of one or more minor subject areas and no thesis required. Option III (for students continuing in the Ph.D. program only): 36 credit hours of coursework with a minimum of 18 hours in physics and no thesis required. There is no foreign language required for any option. Required examinations include: (1) an elementary examination on elementary physics (grade of "B" required) and (b) a comprehensive examination (written and/or oral) covering the student's program of study.

Doctorate: 90 credit hours including research credits. A maximum of 45 credit hours can be transferred. A grade of "B" or better grade point average is required. There is no foreign language requirement. 27 credit hours and 18 consecutive months must be completed in residence. Required examinations include: (1) elementary examination with a passing grade, (2) written advanced examinations, (3) comprehensive examination based on one week of intensive research on a topic approved by the supervisory committee, and (4) oral examination in defense of thesis.

Thesis: Thesis may be written in absentia.

SPECIAL EQUIPMENT, FACILITIES, OR PROGRAMS

Astronomy: Behlen Observatory has a 30-inch reflector telescope equipped with a CCD camera; both the telescope and camera are computer controlled. A 16-inch rooftop observatory equipped with a CCD camera is operational for student use. The 6-inch Minnich Solar Telescope equipped with an Hα filter provides safe views of the sun, allowing student studies of sunspots and solar prominences.

Condensed Matter and Materials Physics: The CMMP group operates laboratories for investigations of a broad range of phenomena and materials. The experimental systems include: two SQUID, alternating-gradient, and vibrating-sample magnetometers; several magneto-optic Kerr-effect (MOKE) systems; pump-probe and high-frequency (8 GHz) systems for fast magnetization dynamics; high-field superconducting and pulsed-field solenoids; high-sensitivity microcantilever torque magnetometer; electron-transport measurements including resistivity, magnetoresistance, and Hall effect; scanning-probe microscopes for magnetic and ferro-electric nanostructure studies; angle-resolved, angle-integrated, and inverse photoemission; x-ray photoemission; low-energy electron diffraction; several thin-film growth systems including molecular-beam epitaxy (MBE), sputtering, e-beam evaporation, chemical-vapor deposition, electrodeposition, and others; neutron accelerator source; optical and high-frequency dielectric measurements; collaborative studies at large-scale facilities including synchrotron light and neutron sources. Computational facilities include several workstations. Many other materials growth and structural characterization measurements are available in the Central Facilities operated by the Nebraska Center for Materials and Nanoscience (for more information, see http://www.unl.edu/ncmn/facilities.shtml).

Atomic, Molecular, Optical, and Plasma Physics: Behlen Laboratory houses the Extreme Light Laboratory featuring its ultra-high-intensity petawatt laser, DIOCLES, built to study the interactions of light with matter at the highest attainable field strengths. With the recent installation of new ten-times-higher-energy pump lasers, DIOCLES has a peak power of 1-petawatt, and operates at the highest duty cycle of any laser in its class. At its focus, it is capable of directly increasing an electron's relativistic mass by a factor of 20. Jorgensen Hall houses two amplified femtosecond laser systems, with repetition rates of 1 and 10 kHz and average powers of 2 and 10 W, respectively. These lasers are used to study ultrafast molecular dynamics and ionization using an ion mass spectrometer and a femtosecond electron diffractometer. Jorgensen Hall also houses an electron matter optics laboratory. In this laboratory, a 50-W Nd:YAG laser, 0-10 keV electron beam, nano-fabricated gratings at 100 nanometer periodicity, 2-D particle imaging systems, a state-of-the-art femtosecond electron pulse source, and an electron interferometer that is unique in the world today are available. A source of polarized electrons to study chiral molecules and fluorescence is also available.

High Energy Physics: The HEP group performs its research at the Fermilab Tevatron Collider in Batavia, Illinois; at the Large Hadron Collider at CERN in Geneva, Switzerland; at the Pierre Auger Observatory in Mendoza Province, Argentina; and at the Askaryan Radio Array experiment at the South Pole. The group manages a clean room facility for research and development on pixel detectors and operates a Tier-2 computing center on campus for CMS with an international user base for data analysis.

All: Excellent instrument (five full-time machinists) and electronics (two full-time technicians) shops are on site.

Library: Love Library houses an extensive collection of physics and astronomy periodicals, books, and monographs. Most research journals are available online, as are powerful journal databases. In addition, faculty participation in the Digital Commons, the University Libraries open access repository, demonstrates their commitment to open science and scholarly preservation.

Table B—Separately Budgeted Research Expenditures by Source of Support

Source of Support	Departmental Research	Physics-related Research Outside Department
Federal government	$9,876,244	$16,282,200
State/local government	$881,000	$310,651
Non-profit organizations		$5,709,860
Business and industry	$30,000	$324,888
Other		
Total	$10,787,244	$22,627,599

Table C—Separately Budgeted Research Expenditures by Research Specialty

Research Specialty	No. of Grants	Expenditures ($)
Atomic, Molecular, & Optical Physics	15	$2,857,975
Condensed Matter Physics	30	$4,856,223
Physics and other Science Education	3	$11,467
High Energy Physics	13	$3,061,579
Total	61	$10,787,244

FACULTY

Professor

Batelaan, Herman, Ph.D., University of Utrecht, 1991. *Atomic, Molecular, & Optical Physics.* Experimental atomic physics; laser cooling and trapping; matter optics and interferometry; ultrafast electron/photon physics.

Claes, Daniel R., Ph.D., Northwestern University, 1991. Department Chair. *Astrophysics, Particles and Fields.* Experimental high-energy physics; D0 at Fermilab; CMS experiment at CERN's large hadron collider; the cosmic ray observatory project (CROP) in Nebraska.

Dowben, Peter A., Ph.D., University of Cambridge, 1981. Charles Bessey Professor of Physics. *Chemical Physics, Condensed Matter Physics, Statistical & Thermal Physics.* Electronic structure of solids and surfaces; nonmetal to metal transitions; surface magnetism; surface ferroelectricity; organic molecular interfaces.

Ducharme, Stephen, Ph.D., University of Southern California, 1986. Vice Chair of Physics and Astronomy. *Condensed Matter Physics.* Experimental condensed-matter and optical physics; physics and applications of two-dimensional ferroelectric polymers; nanoscale ferroelectricity.

Eckhardt, Craig J., Ph.D., Yale University, 1967. *Chemical Physics, Condensed Matter Physics.* Experimental chemical physics; electronic and vibrational spectra of molecules and molecular crystals; dielectric properties of molecular crystals; detonation mechanism of energetic materials; optical properties of solids; lattice dynamics of molecular crystals; organized molecular monolayers; nanotribology.

Fabrikant, Ilya I., Ph.D., Riga, 1971. *Atomic, Molecular, & Optical Physics, Chemical Physics.* Theoretical, atomic, and molecular physics; atomic processes involving negative ions; electron-molecule and electron-cluster collisions; strong-field ionization of atoms and molecules.

Gay, Timothy J., Ph.D., University of Chicago, 1980. *Atomic, Molecular, & Optical Physics, Nuclear Physics.* Experimental atomic and nuclear physics; polarized electrons.

Gruverman, Alexei, Ph.D., Ural State University, 1990. *Condensed Matter Physics, Nano Science and Technology.* Fundamental studies of nanoscale physical phenomena in multiferroic and polar materials by means of scanning probe microscopy techniques; static and dynamic properties of ferroic domains; scaling behavior of ferroelectric-based devices; electronic properties of polar surfaces; SPM-assisted methods for fabrication of nanostructures; SPM studies of electromechanical and mechanical properties of biocompatible materials and biological systems.

Liou, Sy-Hwang, Ph.D., Johns Hopkins University, 1985. *Condensed Matter Physics.* Experimental condensed-matter physics; magnetic properties of thin films; magnetic domain imaging; magnetic sensors.

Sellmyer, David J., Ph.D., Michigan State University, 1965. University Professor and George Holmes Distinguished Professor; Director of the Nebraska Center for Materials and Nanoscience. *Condensed Matter Physics, Materials Science, Metallurgy, Nano Science and Technology.* Experimental studies of nanoscale self- and cluster-assembled magnetic structures; exchange-coupled and high-energy magnetic materials; extremely high-density magnetic recording media, new spintronic materials.

Snow, Gregory R., Ph.D., Rockefeller University, 1983. *Astrophysics, Particles and Fields, Physics and other Science Education.* Experimental high-energy physics; D0 experiment at Fermilab; CMS experiment at CERN's Large Hadron Collider; the Cosmic Ray Observatory Project (CROP) and the Action at a Distance project in Nebraska; the Pierre Auger Cosmic Ray Observatory in Argentina.

Starace, Anthony F., Ph.D., University of Chicago, 1971. George Holmes University Professor Director, Atomic, Molecular, Optical, and Plasma Physics Program of Excellence. *Atomic, Molecular, & Optical Physics, Computational Physics, Theoretical Physics.* Theory of intense laser-atom interactions and on processes that elucidate few-body dynamics, including attosecond and other ultrafast atomic and molecular processes; strong-field atomic processes such as above-threshold ionization/detachment, high-order harmonic generation, laser-assisted electron-atom scattering, and intense laser acceleration of electrons; photoionization and photo-detachment processes; atoms in strong external static fields; coherent control of atomic processes; entanglement and decoherence of spin-based quantum information systems.

Tsymbal, Evgeny Y., Ph.D., Russian Academy of Sciences, Moscow, 1988. George Holmes University Distinguished Professor Director, Materials Research Science and Engineering Center. *Condensed Matter Physics.* Theory of electronic, magnetic, ferroelectric, and transport properties of of materials and structures.

Umstadter, Donald P., Ph.D., University of California, Los Angeles, 1987. Leland J. and Dorothy H. Olson Chair of Physics. *Atomic, Molecular, & Optical Physics, Optics, Plasma and Fusion.* Extreme nonlinear optics of high-intensity laser interactions; relativistic plasmas; compact laser-driven particle accelerators and ultra-short-duration x-ray sources; high-energy density physics and extreme states of matter.

Associate Professor

Adenwalla, Shireen, Ph.D., Northwestern University, 1989. *Condensed Matter Physics, Materials Science, Metallurgy.* Experimental condensed-matter physics; magnetization dynamics in coupled systems; magneto-electric interactions in heterostructures; neutron detectors.

Belashchenko, Kirill, Ph.D., Kurchatov Institute, 1999. *Computational Physics, Condensed Matter Physics, Materials Science, Metallurgy.* Electronic theory of magnetism and magnetic materials, spin-dependent transport, alloy theory.

Binek, Christian, Ph.D., University of Duisburg, 1995. Coordinator of interdisciplinary research group of the Materials Research Science and Engineering Center (MRSEC) at the University of Nebraska; Leader of theme 1 in the Center for Nanoferroic Devices (CNFD), and PI in theme 1 of The Center for Spintronic Materials, Interfaces, and Novel Architectures (C-spin). *Applied Physics, Condensed Matter Physics, Materials Science, Metallurgy, Nano Science and Technology, Statistical & Thermal Physics.* Experimental condensed-matter physics; magnetic heterostructures and model systems; voltage-controlled magnetism, spintronics.

Bloom, Kenneth, Ph.D., Cornell University, 1997. *Particles and Fields.* Experimental high-energy physics focusing on top-quark physics, Higgs boson characterization and searches for new phenomena at the Large Hadron Collider; developments in high-throughput computing that support particle physics research.

Dominguez, Aaron, Ph.D., University of California, San Diego, 1998. *Particles and Fields*. Experimental high-energy physics.

Enders, Axel, Ph.D., Martin Luther University, 1998. *Condensed Matter Physics, Nano Science and Technology*. Experimental research on structure, electronic structure, magnetism and chemistry of low-dimensional metallic, organic, and oxidic materials and interfaces.

Shadwick, Bradley A., Ph.D., University of Texas at Austin, 1995. *Computational Physics, Computer Science, Mechanics, Physics of Beams, Plasma and Fusion*. Intense laser-plasma interactions; advanced accelerator concepts; plasma theory; Hamiltonian systems and methods; advanced numerical methods; cluster computing.

Uiterwaal, Kees, Ph.D., Utrecht University, 1994. Graduate Committee Chair. *Atomic, Molecular, & Optical Physics, Chemical Physics, Optics*. Photoexcitation and photoionization in ultrashort and intense laser pulses; molecular photochemistry; spatially resolved time-of-flight ion mass spectrometry; nonlinear optical effects; holographic spatial pulse shaping; optical vortices; optical orbital angular momentum.

Assistant Professor

Centurion, Martin, Ph.D., California Institute of Technology, 2005. *Atomic, Molecular, & Optical Physics, Optics, Physics of Beams*. Experimental ultrafast atomic, molecular, optical and plasma physics. Molecular imaging, ultrafast molecular dynamics, femtosecond electron sources, femtosecond electron diffraction.

Fuchs, Matthias, Ph.D., Ludwig-Maximilians University, Munich, 2010. *Applied Physics, Atomic, Molecular, & Optical Physics, Optics, Particles and Fields*. Experimental ultrafast and high-field X-ray science. Development of laser-driven compact ultrafast x-ray lightsource. Laser-wakefield acceleration. Ultrafast x-ray interactions. High-field interactions. Nonlinear x-ray optics.

Hong, Xia, Ph.D., Yale University, 2006. *Applied Physics, Condensed Matter Physics, Low Temperature Physics, Materials Science, Metallurgy, Nano Science and Technology*. Epitaxial growth and nanofabrication of complex oxide heterostructures; transport studies of nanoscale and low-dimensional electron systems.

Kovalev, Alexey, Ph.D., Technical University of Delft, 2006. *Condensed Matter Physics, Theoretical Physics*. Theoretical condensed-matter physics.

Kravchenko, Ilya, Ph.D., University of Kansas, 1999. *Atmosphere, Space Physics, Cosmic Rays, High Energy Physics, Particles and Fields*. Experimental high-energy physics.

Xu, Xiaoshan, Ph.D., Georgia Institute of Technology, 2007. *Condensed Matter Physics*. Experimental condensed-matter physics.

Emeritus

Burrow, Paul D., Ph.D., University of California, Berkeley, 1966. *Atomic, Molecular, & Optical Physics*. Experimental molecular physics; scattering of electrons from atoms and molecules; temporary negative ions; dissociative attachment.

Gallup, Gordon A., Ph.D., University of Kansas, 1953. *Atomic, Molecular, & Optical Physics*. Theoretical atomic and molecular physics; theory of vibrational excitation in polyatomic molecules upon electron impact; electron scattering from chiral molecules; atomic and molecular structure.

Hardy, Robert J., Ph.D., Lehigh University, 1962. *Condensed Matter Physics, Mechanics*. Theoretical physics; statistical mechanics; condensed matter; computer simulations.

Jaecks, Duane H., Ph.D., University of Washington, 1964. *Atomic, Molecular, & Optical Physics, History & Philosophy of Physics/Science*. Experimental study of unusual states of atoms and molecules formed in various dynamical processes, including photoionization and molecular dissociation using polarized photon-scattered particle correlation measurements and three-particle coincidence measurements; history of scientific instruments.

Jaswal, S. S., Ph.D., Michigan State University, 1964. *Condensed Matter Physics, Electromagnetism, Nano Science and Technology*. Theoretical condensed-matter physics; electron and magnetic properties of solids, surfaces, interfaces, and nanostructures.

Jones, C. Edward, Ph.D., University of California, Berkeley. *Particles and Fields*. Particle theory.

Kirby, Roger D., Ph.D., Cornell University, 1969. *Condensed Matter Physics, Nano Science and Technology, Optics*. Experimental condensed-matter physics; thin film magnetism; fabrication of nanostructural materials; magneto-optic properties; magneto-optic recording; spin dynamics measured using femtosecond laser pump-probe experiments.

Rudd, M. Eugene, Ph.D., University of Nebraska-Lincoln, 1962. *Atomic, Molecular, & Optical Physics, History & Philosophy of Physics/Science*. Experimental atomic collision physics; history of science and scientific instruments.

Schmidt, Edward G., Ph.D., Australian National University, 1970. *Astronomy, Astrophysics*. Astronomy and astrophysics; variable stars; spectroscopy and photometry; stellar interiors and evolution.

Research Professor

Skomski, Ralph, Ph.D., TU Dresden, 1991. *Applied Physics, Condensed Matter Physics, Materials Science, Metallurgy, Nano Science and Technology, Statistical & Thermal Physics*. Theoretical condensed matter physics; Magnetic nanostructures; Micromagnetics, intrinsic properties of permanent magnets.

Research Associate Professor

Bettis, Clifford L., Ph.D., University of Oklahoma, 1976. *Nonspecialized*. Educational physics.

Lee, Kevin M., Ph.D., University of Nebraska-Lincoln, 1997. *Physics and other Science Education*. Astronomy education; instructional technology; photometric observations of variable stars.

Liu, Yi, Ph.D., Tohoku University, 1988. *Applied Physics, Electromagnetism, Materials Science, Metallurgy, Nano Science and Technology, Physics and other Science Education*. High-resolution and analytical electron microscopy.

Research Assistant Professor

Banerjee, Sudeep, Ph.D., University of Mumbai, 2000. *Atomic, Molecular, & Optical Physics, Nuclear Physics, Optics, Physics of Beams*. Plasma physics, x-rays and gamma-rays, nuclear interrogation, ultra-intense lasers.

Burton, J. D., Ph.D., University of Nebraska-Lincoln, 2008. *Computational Physics, Condensed Matter Physics, Materials Science, Metallurgy*. Properties of advanced magnetic and ferroelectric nanostructures.

Chen, Shouyuan, Ph.D., University of Michigan. *Atomic, Molecular, & Optical Physics, Optics, Physics of Beams*. Experimental atomic, molecular and optical physics. High intensity laser plasma interaction. Laser wakefield acceleration. Nonlinear Thomson scattering. Quantum electrodynamics.

Kalmykov, Serguei, Ph.D., Joint Institute for High Temperatures of the Russian Academy of Sciences, 2001. *Computational Physics, Electromagnetism, Nonlinear Dynamics and Complex Systems, Optics, Physics of Beams*. Intense laser-plasma interactions, advanced particle accelerator concepts, compact plasma-based radiation sources, relativistic nonlinear optics, plasma theiry, Hamiltonian systems and methods, reduced numberical models of laser- and beam-plasma interactions.

Komesu, Takashi, Ph.D., University of Nebraska-Lincoln, 2002. *Condensed Matter Physics, Electromagnetism, Low Temperature Physics.*

Malik, Sudhir, Ph.D., University of Delhi, 1997. *Particles and Fields.* High-energy physics.

Sokolov, Andrei, Ph.D., Moscow State University, 1996. *Condensed Matter Physics, Materials Science, Metallurgy.* Experimental condensed-matter physics; magnetic and electronic properties of materials at reduced dimensions. Nanofabrication and thin-film syntheses.

Adjunct Professor

Boag, Neil, Ph.D., University of Bristol, 1980. Experimental condensed-matter physics.

Fridkin, Vladimir, Ph.D., Russian Academy of Sciences, 1958. Experimental condensed-matter physics; ferroelectricity.

Hadjipanayis, George C., Ph.D., University of Manitoba, 1979. Experimental condensed-matter physics; magnetic materials and applications.

Losovyj, Yaroslav, Ph.D., University of L'viv, 1984. *Condensed Matter Physics.* Condensed-matter physics.

Manakov, Nikolai L., Ph.D., Voronezh State University, 1971. *Atomic, Molecular, & Optical Physics.* Theoretical atomic physics; mathematical physics; theory of intense laser-atom interactions.

Mei, Wai-Ning, Ph.D., State University of New York at Buffalo, 1979. *Condensed Matter Physics.* Theoretical condensed-matter physics; surface physics and molecular dynamics simulations of molecular crystals.

Woollam, John A., Ph.D., Michigan State University, 1967. Professor, Electrical Engineering. *Condensed Matter Physics.* Experimental condensed-matter physics; optical and electrical properties of solids.

Zeng, Xiao Cheng, Ph.D., Ohio State University, 1989. Professor, Chemistry. *Computational Physics.* Computational and theoretical studies of equilibrium and kinetic properties of liquids, solids, and nanomaterials.

DEPARTMENTAL RESEARCH SPECIALTIES AND STAFF

Theoretical

Atomic, Molecular, & Optical Physics. Intense laser-matter inactions; high-energy-density physics; plasma-based light sources. Shadwick.

Atomic, Molecular, & Optical Physics. Theory of vibrational excitation in polyatomic molecules upon electron impact; electron scattering from chiral molecules; atomic and molecular structure. Gallup.

Atomic, Molecular, & Optical Physics. Theory of electron-atom and electron-molecule collisions; electron attachment to molecules and clusters, including systems of biological interest; strong-field ionization of atoms and molecules. Fabrikant.

Atomic, Molecular, & Optical Physics. Theory of intense laser-atom interactions and of processes that elucidate few-body dynamics, including attosecond and other ultrafast atomic and molecular processes, strong-field atomic processes such as above-threshold ionization/detachment, high-order harmonic generation, laser-assisted electron-atom scattering, and intense laser acceleration of electrons; photoionization and photo-detachment processes; atoms in strong external static fields; coherent control of atomic processes; and entanglement and decoherence of spin-based quantum information systems. Starace.

Condensed Matter Physics. Investigation of gas-liquid nucleation; understanding friction and lubrication between two solid surfaces; nanoscale materials. Zeng.

Condensed Matter Physics. Electronic, magnetic, and ferroelectric properties of solids, surfaces, interfaces, and nanostructures. Jaswal.

Condensed Matter Physics. Mathematical modeling of magnetic materials; development of rare-earth permanent magnets. Skomski.

Condensed Matter Physics. Theory of electronic, magnetic, ferroelectric, and transport properties of nanostructures. Tsymbal.

Condensed Matter Physics. Electronic theory of magnetism and magnetic materials, spin-dependent transport, alloy theory. Belashchenko.

Plasma. Intense short-pulse laser-plasma interactions with applications to advanced accelerators and light sources; kinetic theory of plasmas; Hamiltonian structure and methods; advanced numerical methods; large-scale simulations; computational methods. Shadwick.

Experimental

Astronomy. Photometric and spectroscopic study of binary stars; intrinsic variable stars; history of astronomy; spectroscopy and photometry of variable stars; stellar interiors and evolution. Schmidt.

Atomic, Molecular, & Optical Physics. Low-energy electron scattering in gases; temporary negative ion formation and dissociative attachment in bio-molecules. Burrow.

Atomic, Molecular, & Optical Physics. Vacuum field interaction of atoms and electrons; decoherence in matter interferometry. Batelaan.

Atomic, Molecular, & Optical Physics. Experimental ultrafast and high-field x-ray science. Development of laser-driven compact ultrafast x-ray lightsource. Laser-wakefield acceleration. Ultrafast X-ray interactions. High-field interactions. Nonlinear x-ray optics. Fuchs.

Atomic, Molecular, & Optical Physics. Photoexcitation and photoionization in intense, ultrashort laser pulses; molecular photochemistry; spatially resolved time-of-flight ion mass spectrometry; nonlinear optical effects; holographic spatial pulse shaping; optical vortices; optical orbital angular momentum. Uiterwaal.

Atomic, Molecular, & Optical Physics. Experimental ultrafast physics, molecular dynamics; laser-induced plasmas; nonlinear optics. Centurion.

Atomic, Molecular, & Optical Physics. Polarized electron collisions with atoms and molecules; development of polarized electron sources and electron polarimeters; neutrino mass measurements. Gay.

Atomic, Molecular, & Optical Physics. Electronic structure and magnetism of solids; spin-dependent transport; microstructural evolution in alloys; microstructure/property relationships in magnetic materials. Banerjee.

Atomic, Molecular, & Optical Physics. Nonlinear optics of high-intensity laser light in relativistic plasmas, with applications to compact particle accelerators and ultra-short duration x-ray sources. High-energy density physics and extreme states of matter. Umstadter.

Condensed Matter Physics. Magnetism in cluster-assembled nanostructure; magneto-electronic devices; fundamental limits on magnetic recording density; exchange-coupled and hybrid permanent-magnetic materials. Sellmyer.

Condensed Matter Physics. Electronic and magnetic properties of nanostructural materials; magnetic domain imaging applications of SPM; magnetic sensors. Liou.

Condensed Matter Physics. Physics and applications of two-dimensional ferroelectric polymers. Ducharme.

Condensed Matter Physics. Multiferroic and polar materials using SPM techniques; SPM-assisted methods for fabrication of nano-structures; studies of electromechanical properties of biocompatible materials. Gruverman.

Condensed Matter Physics. Magnetic and electronic properties of materials. Sokolov.

Condensed Matter Physics. Epitaxial growth and nanofabrication of complex oxide heterostructures; transport studies of nanoscale and low dimensional electron systems. Hong.

Condensed Matter Physics. Magnetic and electronic properties of materials. Sokolov.

Condensed Matter Physics. Magneto-optical Kerr effect in structured magnetic materials; optical pump-probe measurements of spin dynamics in magnetic thin films. Kirby.

Condensed Matter Physics. Magnetic/ferromagnetic/ferroelectric thin film interactions piezoelectric strain effects, exchange bias, solid state neutron detector development. Adenwalla.

Condensed Matter Physics. Molecular beam epitaxial growth of magnetic heterolayer structures, extrinsic control of exchange bias, model systems in statistical physics. Binek.

Condensed Matter Physics. Self-assembled magnetic and molecular nanostructures, low- and variable temperature STM; spin-polarized STM. Enders.

Condensed Matter Physics. Electronic structure of solids and surfaces; nonmetal to metal transitions; surface magnetism; surface ferroelectricity; organic molecular interfaces. Dowben.

High Energy Physics. The Askaryan Radio Array experiment at the South Pole. Kravchenko.

High Energy Physics. The Cosmic Ray Observatory Project (CROP)and the Action at a Distance Project are education and outreach projects to study extensive cosmic-ray air showers with particle detectors located at high schools throughout Nebraska. Claes, Snow.

High Energy Physics. The Compact Muon Solenoid (CMS) Experiment at CERN, Geneva, Switzerland studies the newly-observed Higgs boson, searches for new particles that are expected from physics beyond the standard model, and makes precision measurements of standard-model phenomena. The group has hardware, software and physics analysis responsibilities for the experiment, including the hosting of a Tier-2 computing facility on campus. The group has significant responsibilities in the construction of a future silicon pixel detector for CMS. Bloom, Claes, Dominguez, Kravchenko, Snow.

High Energy Physics. The Pierre Auger Observatory, the world's largest cosmic ray experiment in Argentina. Snow.

View additional information about this department at
www.gradschoolshopper.com

UNIVERSITY OF NEBRASKA–LINCOLN

NEBRASKA CENTER FOR MATERIALS AND NANOSCIENCE

Lincoln, Nebraska 68588-0298
http://www.unl.edu/ncmn/

General University Information
Chancellor: Harvey S. Perlman
Dean of Graduate School: Lance C. Perez
University website: http://www.unl.edu/
Control: Public
Setting: Urban
Total Faculty: 1,615
Total Graduate Faculty: 1,477
Total number of Students: 24,207
Total number of Graduate Students: 4,559

Department Information
Department Chairman: David J. Sellmyer, Director
Department Contact: Shelli Krupicka, Administrative Coordinator
  Total full-time faculty: 88
  Total number of full-time equivalent positions: 7
  Full-Time Graduate Students: 158
  Total Post Doctorates: 42

Department Address
N201 Voelte-Keegan Nanoscience Research Center
855 N. 16th Street
Lincoln, NE 68588-0298
Phone: (402) 472-7886

Fax: (402) 472-6148
E-mail: ncmn@unl.edu
Website: http://www.unl.edu/ncmn/

ADMISSIONS

Admission Contact Information
Address admission inquiries to: Admissions, Office of Graduate Studies, 1100 Seaton Hall, University of Nebraska, Lincoln, NE 68588-0619
Phone: (402) 472-2875
E-mail: graduate@unl.edu
Admissions website: http://www.unl.edu/gradstudies/

Application deadlines

Application fee
U.S. students: $50
Application deadlines vary by program. Please check directly with program of interest for most accurate application information.

Admissions information
For Spring of 2012:

Admission requirements

Bachelor's degree requirements: A bachelor's degree in physics, chemistry, materials science, engineering, or mathematics is required.

Minimum undergraduate GPA: 3.0

GRE requirements

The GRE is required.

Advanced GRE requirements

The Advanced GRE is recommended.

TOEFL requirements

The TOEFL exam is required for students from non-English-speaking countries.

Other admissions information

Additional requirements: Application deadlines and admission requirements vary by program. Please check directly with program of interest for additional and most accurate information.

TUITION

Tuition year 2013-14:

Tuition for in-state residents

   Full-time students: $285 per credit

Tuition for out-of-state residents

   Full-time students: $791.75 per credit

Teaching and Research Assistants do not pay tuition.

Credit hours per semester to be considered full-time: 9

Deferred tuition plan: No

Health insurance: Yes, $315 student cost.

Other academic fees: Financial aid and other deadlines vary by program. Please check with program of interest for accurate information.

Academic term: Semester

Teaching Assistants, Research Assistants, and Fellowships

   Number of first-year

FINANCIAL AID

Application deadlines

Fall admission:

U.S. students: April 1   *Int'l. students*: April 1

Loans

Loans are available for U.S. students.

Loans are not available for international students.

GAPSFAS application required: No

FAFSA application required: Yes

For further information

Address financial aid inquiries to: Scholarships & Financial Aid, 17 Canfield Administration Building, P.O. Box 880411, Lincoln, NE 68588-0411.

Phone: (402) 472-2030

E-mail: finaid2@unl.edu

Financial aid website: http://www.unl.edu/scholfa/

HOUSING

Availability of on-campus housing

   Single students: Yes

   Married students: Yes

For further information

Address housing inquiries to: Office of University Housing, 1115 N. 16th Street, Lincoln, NE 68588-0622.

Phone: (402) 472-3561

E-mail: housing@unl.edu

Housing aid website: http://housing.unl.edu/

Table A—Faculty, Enrollments, and Degrees Granted

Research Specialty	2012-13 Faculty	Enrollment Fall 2012 Master's	Enrollment Fall 2012 Doctorate	Number of Degrees Granted 2012–13 (Data Unavailable) Master's	Number of Degrees Granted 2012–13 (Data Unavailable) Terminal Master's	Number of Degrees Granted 2012–13 (Data Unavailable) Doctorate
Chemical Physics	20	21	42	–	–	–
Condensed Matter Physics	23	15	31	–	–	–
Materials Science, Metallurgy	39	15	30	–	–	–
Other	10	1	3	–	–	–
Total	92	52	106	–	–	–
Full-time Grad. Stud.	–	52	106	–	–	–
First-year Grad. Stud.	–	–	–	–	–	–

GRADUATE DEGREE REQUIREMENTS

Master's: Several options exist, including thesis and non-thesis. Generally, 30 or 36 hours of coursework are required, depending on the option chosen. See Graduate Studies Bulletin for details.

Doctorate: Ninety credit hours are required, with research credits included. A grade of B or better is required. No foreign language is required. Twenty-seven hours and 18 consecutive months must be completed in residence.

Thesis: Thesis may be written in absentia.

SPECIAL EQUIPMENT, FACILITIES, OR PROGRAMS

The Nebraska Center for Materials and Nanoscience operates and coordinates the following central service facilities: X-ray Structural Characterization, Electron Microscopy, Crystallography, Metallurgical and Mechanical Characterization, Scanning Probe Microscopy, Materials Preparation, Nanofabrication, and Cryogenics. A brief description of each facility is as follows.

The X-Ray Structural Characterization facility hosts (1) Bruker-AXS D8 Discover X-Ray Diffractometer for high resolution and General Area Detector Diffraction System (GADDS) applications, (2) Rigaku D/Max-B and Multiflex Diffractometers for powder diffraction applications, and (3) Bruker Smart Apex Single Crystal Diffractometer for X-Ray Crystallography. A new high-resolution diffractometer will be added to enhance the nanostructural characterization capabilities. The facility provides a Meiji optical microscope with a polarizer attachment for single crystal screening. Crystallographic databases from International Center for Diffraction Data (ICDD) and Cambridge Structural Database (CSD) are subscribed by the facility to assist users for phase identification, quantification, and molecular visualization needs. It also provides several analysis software suites to assist users for in-depth analysis of their powder, thin films, and single diffraction data.

The Electron Microscopy facility offers researchers the best tools for atomic-scale and submicron materials analysis. The facility houses four electron microscopes, including the latest 200 kV analytical (scanning) transmission electron microscope (FEI Tecnai Osiris) and a new high-resolution scanning electron microscope (FEI Nova NanoSEM450); the other two are JEOL JEM2010 transmission electron microscope and JEOL JSM840A scanning electron microscope. The instruments for sample preparation are well equipped.

The Cryogenics facility maintains a continuous supply of the liquid helium and liquid nitrogen necessary for on-campus use.

The Materials Preparation facility has various equipment for materials and sample preparation. Instruments available include a sputtering system, an e-beam evaporation system, and a pulsed-laser-deposition system to fabricate a variety of thin films, especially nanostructured films, including overlayers, multilayers, granular solids, clusters, etc. The sputtering system has 2 RF and 2 DC sputtering guns, and the Pulsed Laser Deposition (PLD) system has six pockets available for the targets. In situ reflection high energy electron diffraction (RHEED) is available for monitoring the layer-by-layer growth of the PLD films. Additionally, an arc-melting and a rapid solidification facilities will be soon available for the synthesis of novel materials. Furthermore, two tube furnaces (Lindberg 54233 and Lindberg 55332) together with vacuum pump stations are available for sample (or target) annealing, doping, and sintering.

The Mechanical and Materials Characterization facility contains a large variety of equipment to characterize the mechanical and physical properties of a wide range of materials. Primary equipment include an Olympus BX51 Microscope, Mettler Toledo FP900 Microscope Thermosystem, and Sartorius Cubis MSU2.7S-000-DM Microbalance. Sample preparation equipment includes several different saw and polishing systems, such as BUEHLER ISOMet 1000 Precision Saw and BUEHLER MiniMet 1000 grinder-polisher. Other equipment available are hardness testing equipment, such as Rockwell, Knoop, and Vickers testers, as well as tension/compression testing machines (MTS, Instron, Satec, and Tinius-Olson) and torsion testing equipment.

The Scanning Probe Microscopy (SPM) facility contains a Digital Instruments Dimension 3100 Scanning Probe Microscopy and a DI EnviroScope Atomic Force Microscope (ESCOPE) with environmental controls and vacuum chamber to image sample surface in air, vacuum, or a purged gas, as well as a heating environment. The SPM facility provides nanometer-scale surface characterization of materials by using scanning tunneling microscopy, atomic force microscopy, lateral force microscopy, magnetic force microscopy, electric force microscopy, and piezoresponse force microscopy.

The Nanofabrication facility operates a 5000 sq.ft. clean room with a full suite of leading edge tools for fabricating and characterizing nanoscale devices and structures. Equipment includes a Carl Zeiss field-emission microscope, Raith e-beam lithography, Suss mask aligner, Heidelberg laser write lithography system, FEI focused ion-beam workstation, Trion reactive ion etcher, wet etching benches and fume hoods, AJA e-beam evaporation system, Stylus profilometer, film thickness measurement system, optic microscope, Lucas resistivity measurement stand, ovens, and others.

The Specialized Research facilities include many state-of-the-art research facilities, including 14 Tesla NMR spectrometer equipped for solid-state NMR, NIMA Langmuir-Blodgett Trough for monolayer and multilayer films, atomic force microscopes, high-field superconductive solenoids, a SQUID magnetometer, angle-integrated photoemission and electron spin analysis facilities, Raman and Brillouin laser light-scattering facilities, a comprehensive laboratory for the study of magnetic materials, as well as high-temperature superconducting materials, a number of photoemission and inverse photoemission spectrometers (including spin-polarized inverse photoemission), and dedicated

minicomputers for theoretical calculations. In addition, pulsed laser facilities, atomic force microscope, and a comprehensive ellipsometer laboratory are available.

Table B—Separately Budgeted Research Expenditures by Source of Support

Source of Support	Departmental Research	Physics-related Research Outside Department
Federal government	$16,282,200	$9,876,244.86
State/local government	$310,651	$881,000.04
Non-profit organizations	$5,709,860	
Business and industry	$324,888	$30,000
Other		
Total	$22,627,599	$10,787,244.9

Table C—Separately Budgeted Research Expenditures by Research Specialty

Research Specialty	No. of Grants	Expenditures ($)
Chemical Physics	21	$2,303,006
Condensed Matter Physics	48	$15,624,465
Materials Science, Metallurgy	72	$4,700,128
Total	141	$22,627,599

FACULTY

Professor

Alexander, Dennis R., Ph.D., Kansas State University, 1976. Ultra fast lasers; nanoparticle production; surface modifications using fs lasers.

Barnes, Caren M., M.S., University of Missouri, Kansas City, 1974. The effects of polishing on dental materials and dental tissues.

Berkowitz, David B., Ph.D., Harvard University, 1990. Organic catalyst development and new methods for catalyst screening.

Brand, Jennifer I., Ph.D., University of California, San Diego, 1992. Ceramic thin films from cluster beams; supercritical fluid technology.

Chandra, Namas, Ph.D., Texas A&M University, 1986. Mechanics of nano, bio, and structural materials and structures; finite deformation; multiscale modeling and simulation; molecular dynamics; superplasticity; composites.

Di Magno, Stephen G., Ph.D., University of California, Berkeley, 1991. Synthesis and properties of organic solids.

Dowben, Peter A., Ph.D., University of Cambridge, 1981. Surface science; magnetic and ferroelectric films.

Ducharme, Stephen, Ph.D., University of Southern California, 1986. Two-dimensional ferroelectric polymers.

Dussault, Patrick H., Ph.D., California Institute of Technology, 1986. Synthesis and properties of organic materials.

Dzenis, Yuris A., Ph.D., University of Texas, Arlington, 1994. Composite materials and mechanics.

Eckhardt, Craig J., Ph.D., Yale University, 1967. Statics and dynamics of molecular crystals; solid state optical spectroscopy; Langmuir-Blodgett films; crystal engineering of organic solids.

Feng, Ruqiang, Ph.D., Johns Hopkins University, 1992. Experimental mechanics of materials, including high strain rate, shockwave, impact experiments, and the study of inelastic deformation and failure mechanisms of ceramics.

Forbes, Valery E., Ph.D., SUNY-Stony Brook, 1988. Fate and effects of engineered nanoparticles in the environment; nanosafety.

Gay, Timothy J., Ph.D., University of Chicago, 1980. Experimental atomic physics; state selected ion-atom collisions; spin-polarized sources.

Gruverman, Alexei, Ph.D., Ural State University, 1990. Ferroelectrics; piezoelectrics; scanning probe microscopy.

Hallbeck, Susan M., Ph.D., Virginia Polytechnic Institute and State University (Virginia Tech), 1990. Ergonomics, biomedical, and systems engineering.

Harbison, Gerard S., Ph.D., Harvard University, 1984. Solid state NMR spectroscopy; quantum chemical calculations on endohedral fullerenes and ionic crystalline solids.

Ianno, Natale (Ned) J., Ph.D., University of Illinois, 1981. Electronic properties of semiconductors and superconductors.

Langell, Marjorie, Ph.D., Princeton University, 1979. Surface chemistry; Auger and photoemission studies of transition metal oxides.

Larsen, Gustavo, Ph.D., Yale University, 1992. Surface chemistry and catalysis.

Liou, Sy-Hwang, Ph.D., Johns Hopkins University, 1985. Ultrafine particle magnetic films; high-temperature superconducting films.

Lu, Yongfeng, Ph.D., Osaka University, 1991. Laser material processing and characterization at micrometer and nanometer scales.

Negahban, Mehrdad, Ph.D., University of Michigan, 1988. Mechanical effects of phase transition in polymers under large strains.

Parkhurst, Lawrence J., Ph.D., Yale University, 1965. Optical properties of DNA; solid state protein thermodynamics.

Rajca, Andrzej T., Ph.D., University of Kentucky, 1985. High-spin organic polyradicals; synthesis of two-dimensional magnets from polyradicals.

Rajurkar, K. P., Ph.D., Michigan Technological University, 1982. Nonconventional machining at macro, micro, and nano scales; stochastic modeling of manufacturing processes and systems.

Redepenning, Jody G., Ph.D., Colorado State University, 1985. Electrochemistry; biocompatible materials.

Saraf, Ravi, Ph.D., University of Massachusetts, Amherst, 1987. Nanoscale material fabrication and devices; biophysics.

Sellmyer, David J., Ph.D., Michigan State University, 1965. Nanomagnetics and magnetoelectronics, including self- and cluster-assembled structures, dilute magnetic semiconductors and oxides, exchange-coupled nanocomposites, and fundamental limits on magnetic storage density.

Shield, Jeffrey E., Ph.D., Iowa State University, 1992. Microstructural evolution of materials; rapid solidification.

Subramanian, Anuradha, Ph.D., Virginia Polytechnic Institute and State University, 1995. Biomaterial development and additional use of these biomaterials in bioseparations and biomedical applications.

Takacs, James M., Ph.D., California Institute of Technology, 1981. Synthesis of novel polymers.

Timm, Delmar C., Ph.D., Iowa State University, 1967. Engineering properties of polymers.

Tsymbal, Evgeny Y., Ph.D., Russian Academy of Sciences, Moscow, 1988. Theory of electronic structure and spin-dependent transport in nanoscale magnetic systems.

Turner, Joseph A., Ph.D., University of Illinois at Urbana-Champaign, 1994. Wave propagation and vibrations, including ultrasonics, NDE, materials characterization, AFM dynamics, and nanoindentation.

Viljoen, Hendrik J., Ph.D., University of Pretoria, 1985. Synthesis and processing of materials with CVD.

Watkins, David K., Ph.D., Florida State University, 1984. Electron microscopy studies of minerals.

Woollam, John A., Ph.D., Michigan State University, 1967. Ellipsometric studies of oxide surfaces; magneto-optic films; optical coatings and electrochromics.

Yang, Jiashi, Ph.D., Princeton University, 1994. Electromechanical materials and devices.

Yang, Yiqi, Ph.D., Purdue University, 1991. Biopolymeric materials; biotextiles; industrial applications of agricultural by-products.

Zeng, Xiao Cheng, Ph.D., Ohio State University, 1989. Statistical mechanics of liquids and solids; phase transition; computer simulations.

Associate Professor

Adenwalla, Shireen, Ph.D., Northwestern University, 1989. Experimental condensed matter physics.

Baesu, Eveline, Ph.D., University of California, Berkeley, 1998. Continuum mechanics; plasticity; electrodynamics of continuum media.

Belashchenko, Kirill, Ph.D., Kurchatov Institute, 1999. Electronic structure theory; magnetic, transport, and structural properties of materials and nanostructures.

Binek, Christian, Ph.D., University of Duisburg, 1995. Magnetic heterostructures in basic research and spintronic applications.

Bobaru, Florin, Ph.D., Cornell University, 2001. Computational mechanics; numerical optimization of advanced materials and systems.

Cheung, Chin Li (Barry), Ph.D., Harvard University, 2002. Synthesis and characterization of materials at the nanoscale.

Covey, David A., M.S., University of Iowa, 1979. Restorative dental materials testing.

Enders, Axel, Ph.D., Max-Planck, Halle, 1999. Scanning probe studies of magnetic nanostructures.

Gu, Linxia, Ph.D., University of Florida, 2004. Soft-tissue mechanics; multiscale and multiphysics modeling.

Lai, Rebecca Y., Ph.D., University of Texas, Austin, 2003. Ligand-induced folding in biopolymers; electrochemical biosensors; surface plasmon resonance biosensors.

Pannier, Angela K., Ph.D., Northwestern University, 2007. Biomaterials for nonviral gene delivery and tissue engineering applications.

Schubert, Eva Franke, Ph.D., University of Leipzig, 1998. Ion beam processing; nanostructured thin-film fabrication for optical, electromechanical, and magnetic device applications.

Schubert, Mathias, Ph.D., Universität Leipzig, 1997. Condensed matter spectroscopy; ferroic semiconductor thin films and nanostructures.

Tan, Li, Ph.D., University of Michigan, 2002. Unconventional nanolithography; nanoimprint lithography; polymer thin films and devices/sensors.

Zhang, Zhaoyan, Ph.D., Pennsylvania State University, 2000. Experimental and theoretical study of laser material interactions.

Assistant Professor

Hong, Xia, Ph.D., Yale University, 2006. Epitaxial complex oxide thin films and nanostructures; two-dimensional electron systems.

Huang, Jinsong, Ph.D., University of California, Los Angeles, 2007. Organic solar cell; organic spintronics; ferroelectric polymers.

Kidambi, Srivatsan, Ph.D., Michigan State University, 2007. Tissue engineering and biomaterials; drug delivery; stem cells.

Li, Yusong, Ph.D., Vanderbilt University, 2005. Fate and transport of contaminants; numerical modeling.

Ndao, Sidy, Ph.D., Rensselaer Polytechnic Institute, 2010. Micro/nano systems energy conversion, storage and power generation; two-phase heat transfer in micro and nano domains; microfluidics and functional nanofluids; microscale combustion.

Othman, Shadi F., Ph.D., University of Illinois at Chicago, 2004. Microscopic magnetic resonance elastography; MRI.

Sinitskii, Alexander, Ph.D., Moscow State University, 2008. Carbon nanomaterials; graphene; nanoelectronics.

Zhang, Jian, Ph.D., University of Pittsburgh, 2008. The ability to control the shape, composition, and dynamics of matter on the nanoscale constitutes a major challenge of next decade. The Zhang group combines synthetic chemistry, physical chemistry, and theoretical modeling to provide a fundamental understanding of nanoscale interactions so that noble metal nanoparticles, metal-organic frameworks, and carbon-based nanomaterials can be rationally designed and synthesized to have desired electrocatalytic, gas sorption, and photocatalytic properties for clean energy-related applications.

Emeritus

Burrow, Paul D., Ph.D., University of California, Berkeley, 1966. Experimental molecular physics; scattering of electrons from atoms and molecules; temporary negative ions.

Diestler, Dennis J., Ph.D., California Institute of Technology, 1967. Theoretical studies of the fluid solid interface.

Hardy, Robert J., Ph.D., Lehigh University, 1962. Computer simulations of shock waves; statistical mechanics; theory of thermodynamic and transport properties of anharmonic solids.

Jaswal, Sitaram S., Ph.D., Michigan State University, 1964. Theory of electronic structure of magnetic compounds and multilayers.

Kirby, Roger D., Ph.D., Cornell University, 1969. Magneto-optic properties of thin films; light scattering.

Pearlstein, Edgar A., Other, Carnegie Institute of Technology, 1950. Doctor of Science. Defects in solids; electrical noise.

Robertson, Brian W., Ph.D., University of Glasgow, Scotland, 1979. Electron microscopy and techniques; nanofabrication; sensor materials and devices.

Soukup, Rodney J., Ph.D., University of Minnesota, 1969. Semiconductors; solar energy studies.

Williams, P. Frazer, Ph.D., University of Southern California, 1973. Electrical breakdown of semiconductors.

Research Professor

Skomski, Ralph, Ph.D., Dresden University of Technology, 1991. Theoretical solid state magnetism; micromagnetics; permanent magnetism.

Research Associate Professor

Liu, Yi, Ph.D., Tohoku University, 1988. High-resolution and analytical electron microscopy.

Research Assistant Professor

Fernandez-Ballester, Lucia, Ph.D., California Institute of Technology, 2007. Polymer physics; polymer processing and crystallization.

Hofmann, Tino, Ph.D., University of Leipzig, 2004. THz ellipsometry; magneto-optical materials characterization; three-dimensional nanostructures for optical sensor applications.

Sokolov, Andrei, Ph.D., Lomonosov Moscow State University, 1996. Magnetoelectronics.

Zhou, Yunshen, Ph.D., University of Science and Technology of China, 2005. Self-assembled monolayers and functionalization of C60.

Adjunct Professor

Darveau, Scott A., Ph.D., University of Chicago, 1998. Thin-film photovoltaic materials; Raman microscopy/laser spectroscopy.

Exstrom, Christopher, Ph.D., University of Minnesota, 1995. Characterization and development of novel solar cell film materials; preparation and characterization of novel solvatochromic transition-metal-based materials.

Fridkin, Vladimir, Ph.D., Institute of Crystallography, Academy of Sciences of the USSR, 1965. Ferroelectricity; ferroelectric polymers; phase transitions.

Hadjipanayis, George C., Ph.D., Manitoba, 1979. Experimental condensed matter physics; magnetic materials and applications.

Liu, J. Ping, Ph.D., University of Amsterdam, 1994. Magnetic materials.

Mei, Wai Ning, Ph.D., State University of New York at Buffalo, 1979. Surface structure determination using multiple-scattering theory, molecular dynamics simulations of alkali halides, and molecular solids.

Namavar, Fereydoon, Ph.D., Katholieke Universiteit Leuven, 1978. Nanotechnology; alternative surfaces for medical implants; smart surfaces; smart drugs and sensors for medical application.

Palencia, Hector, Ph.D., UNL/UNAM, 2005. Catalysis/nanocatalysts; biofuels; organic synthesis.

Reece, Timothy J., Ph.D., University of Nebraska-Lincoln, 2007. Brewster angle microscopy; Langmuir films; semiconductor devices.

Sabirianov, Renat, Ph.D., UTU, 1993. Defects and impurities in magnetic materials.

Smith, Robert W., Ph.D., Oregon State University, 1989. Synthesis, crystal growth, and structural characterization of new materials through x-ray diffraction techniques.

Visiting Professor

Boag, Neil, Ph.D., University of Bristol, 1981. Materials science and synthesis.

DEPARTMENTAL RESEARCH SPECIALTIES AND STAFF

Theoretical

Chemical Physics. Density-functional calculations of clusters, carbon nanotubes, layered nanostructures, magnetic systems. Eckhardt, Zeng.

Condensed Matter Physics. Structural and vibrational properties of crystalline and amorphous solids using the methods of statistical mechanics; giant magnetoresistance in magnetic metallic multilayers, spin-dependent tunneling in magnetic tunnel junctions, spin injection into semiconductors; permanent-magnet materials, magnetic nanostructures, time-dependent magnetization processes, spin structure of half-metallic ferromagnets, and quantum entanglement in nanodots. Belashchenko, Hardy, Jaswal, Skomski, Tsymbal.

Mechanics. Large deformation thermomechanical and mechanical response characterization; continuum thermodynamic modeling; fluid-solid interfacial phenomena. Baesu, Bobaru, Chandra, Feng, Negahban, Rajurkar, Tan, Timm, Viljoen, Zhaoyan Zhang.

Experimental

Chemical Physics. Synthesis and study of unnatural analogs of biological molecules, catalyst development; design and synthesis of inorganic/bio-organic nanoscaled components; development of new robust macrocyclic ligands; synthesis of peroxide-containing natural products, new methods for organic oxidations based on ozone, singlet oxygen, and hydrogen peroxide; nanotribology, mechanochemistry, energetic materials supramolecules, and organic ferroelectrics; influence of oxidation state on chemical and physical properties of metal-containing materials; structure and properties of solids, NMR; surface properties of transition-metal oxides and other metal compounds; interactions between DNA, RNA, and proteins; design and synthesis of stable high-spin polyradicals; electrochemistry, electrochemical deposition; structure function studies in supramolecular systems; organic synthesis and chemistry, piezoelectric devices. Berkowitz, Cheung, Di Magno, Dussault, Eckhardt, Harbison, Lai, Langell, Parkhurst, Rajca, Redepenning, Takacs, Jian Zhang.

Condensed Matter Physics. Structural characterization of materials, magnetic systems, polymers, development of solid state neutron detectors; exchange bias in magnetic metal/insulator heterosystems, matrix insulated magnetic nanoparticles; interplay between magnetic and electric properties of heterogeneous metallic magnetic systems; electronic band structure and the influence of electronics structure on various phase transitions; ferroelectric nanostructures and polymers; polarized electron physics; magnetism dynamics in thin films, superlattices; magnetic interactions in patterned nanostructures, quantum conductance and magnetoresistance in nanocontacts, nanofabrications; high-anisotropy magnetic nanocluster-assembled films, nanotube magnetism, spin-logic nanostructures, exchange-coupled nanocomposites. Adenwalla, Binek, Burrow, Dowben, Ducharme, Enders, Gay, Gruverman, Hong, Kirby, Liou, Yi Liu, Pearlstein, Sellmyer, Sokolov.

Materials Science, Metallurgy. Developing new materials and more efficient materials production processes for deposition of thin films, microfibers, and microparticles; novel devices; thin-film deposition, high-density plasma processing, nanoscale processing; catalysis, adsorption, and materials design; microscale and nanoscale laser material processing and characterization. Magnetic and electronic thin films, nanoscale wires and devices, and electron probe-based characterization; deposition and characterization of thin films of wear-resistant materials, piezoelectric oxide ceramics; optical properties of semiconductors and nanoscale materials, ellipsometry; deposition and study of semiconductor films; synthesis and development of novel biofunctional materials; microstructural evolution of materials during processing; composite materials comprising a polymeric matrix; piezoelectric materials; electrical breakdown of gases and semiconductors, plasma processing of semiconductors; laser-materials interactions. Alexander, Bobaru, Brand, Chandra, Dzenis, Feng, Fernandez-Ballester, Gu, Hallbeck, Hofmann, Huang, Ianno, Larsen, Li, Lu, Ndao, Negahban, Othman, Pannier, Rajurkar, Robertson, Saraf, Eva Schubert, Mathias Schubert, Shield, Soukup, Subramanian, Tan, Timm, Turner, Viljoen, Williams, Woollam, Jiashi Yang, Zhaoyan Zhang, Zhou.

Mechanics. Electromechanical effects, fiber networks, and biomechanics; mesh-free methods, structural and multidisciplinary optimization of solids; functional nanomaterials and nanomanufacturing; experimental and computational mechanics of materials; active materials, composites, micromechanics, and microstructure mechanics; stochastic wave propagation, experimental ultrasonics, structural acoustics; frequency stability of piezoelectric crystal resonators. Baesu, Bobaru, Brand, Chandra, Dzenis, Feng, Fernandez-Ballester, Gu, Kidambi, Li, Ndao, Negahban, Othman, Pannier, Rajurkar, Subramanian, Tan, Turner, Jiashi Yang, Zhaoyan Zhang, Zhou.

View additional information about this department at
www.gradschoolshopper.com

UNIVERSITY OF NEVADA, LAS VEGAS

DEPARTMENT OF PHYSICS AND ASTRONOMY

Las Vegas, Nevada 89154-4002
http://www.physics.unlv.edu

General University Information
President: Neal J. Smatresk
Dean of Graduate School: Tom Piechota
University website: http://go.unlv.edu
Control: Public
Setting: Urban
Total Faculty: 865
Total Graduate Faculty: 690
Total number of Students: 27,389
Total number of Graduate Students: 4,970

Department Information
Department Chairman: Stephen Lepp, Chair
Department Contact: Gail Michel-Parsons, Administrative Assistant IV
Total full-time faculty: 16
Total number of full-time equivalent positions: 16
Full-Time Graduate Students: 17
First-Year Graduate Students: 3

Female First-Year Students: 1
Total Post Doctorates: 5

Department Address
4505 S Maryland Pkwy
Las Vegas, NV 89154-4002
Phone: (702) 895-0868
Fax: (702) 895-0804
E-mail: gail.michel-parsons@unlv.edu
Website: http://www.physics.unlv.edu

ADMISSIONS

Admission Contact Information
Address admission inquiries to: Graduate Program Admissions, Department of Physics and Astronomy
Phone: (702) 895-3320
E-mail: gradcollege@unlv.edu
Admissions website: http://graduatecollege.unlv.edu/admissions/

Application deadlines
Fall admission:

U.S. students: August 1   Int'l. students: August 1

Spring admission:

U.S. students: October 1   Int'l. students: October 1

Application fee
U.S. students: $60   Int'l. students: $95

Admissions information
For Fall of 2012:

  Number of applicants: 20

  Number admitted: 11

  Number enrolled: 5

Admission requirements
Bachelor's degree requirements: Bachelor's degree is required.
Minimum undergraduate GPA: 2.75

GRE requirements
The GRE is required.

Advanced GRE requirements
The Advanced GRE is required.

  Minimum accepted Advanced GRE score: 520

Physics Graduate Program: The Advanced GRE is not required
for M.S. applicants in Physics. However, the Physics GRE
is required for Ph.D. applicants without an M.S. degree in
Physics. Astronomy and Astrophysics Graduate Program: The
Advanced GRE is not required for either the M.S. or the Ph.D.
applicants.

TOEFL requirements
The TOEFL exam is required for students from non-English-
speaking countries.

  PBT score: 550

Other admissions information
Additional requirements: The applicant must have completed 18
semester credits of upper division physics.
Undergraduate preparation assumed: Mechanics level—Marion,
Classical Dynamics; Electricity and Magnetism level—Wang-
sness, Electromagnetic Fields; Quantum Mechanics level—Grif-
fiths, Quantum Mechanics; Mathematics level—through Advanced
Calculus.

TUITION

Tuition year 2014:
Tuition for in-state residents

  Full-time students: $264 per credit

  Part-time students: $264 per credit

Tuition for out-of-state residents

  Full-time students: $264 per credit

  Part-time students: $264 per credit

Credit hours per semester to be considered full-time: 6
Deferred tuition plan: Yes
Health insurance: Available
Other academic fees: There will be additional fees.
Academic term: Semester
Number of first-year students who received full tuition waivers: 4

Teaching Assistants, Research Assistants, and Fellowships
Number of first-year

  Teaching Assistants: 4

FINANCIAL AID

Application deadlines
Fall admission:

U.S. students: August 1   Int'l. students: May 1

Spring admission:

U.S. students: October 1   Int'l. students: October 1

Loans
Loans are available for U.S. students.
Loans are not available for international students.
GAPSFAS application required: No
FAFSA application required: No

For further information
Address financial aid inquiries to: Graduate Program Admis-
sions, Department of Physics and Astronomy.
Phone: 702-895-3424
Financial aid website: http://www.unlv.edu/finaid

HOUSING

Availability of on-campus housing
  Single students: Yes

  Married students: No

For further information
Address housing inquiries to: Housing and Residential Life.
Phone: (702) 895-3489
E-mail: housing@unlv.edu
Housing aid website: http://housing.unlv.edu/

Table A—Faculty, Enrollments, and Degrees Granted

Research Specialty	2013–14 Faculty	Enrollment Fall 2014		Number of Degrees Granted 2012–13 (2005–13)		
		Master's	Doctorate	Master's	Terminal Master's	Doctorate
Astronomy	1	–	2	–	–	–(1)
Astrophysics	5	1	6	–(4)	–(5)	1(9)
Atomic, Molecular, & Optical Physics	2	4	1	–(2)	1(3)	–(3)
Chemical Physics	1	–	–	–(1)	–	–
Condensed Matter Physics	5	3	5	–(6)	1(5)	–(2)
Plasma and Fusion	1	–	–	–(1)	–(2)	–(1)
Polymer Physics/Science	1	–	–	–	–	–(1)
Total	17	8	14	–(14)	2(15)	1(17)
Full-time Grad. Stud.	–	8	14	–	–	–
First-year Grad. Stud.	–	1	4	–	–	–

GRADUATE DEGREE REQUIREMENTS

Master's: A minimum of 30 graduate credits is required for the
Master of Science degree in Physics or Astronomy, including
a minimum of 15 credits (excluding thesis) in 700-level
courses and six semester hours of research for thesis credit.
A final oral exam is required on course work and thesis except
for the satisfactory performance on an Astronomy Qualifying
Exam for the Astronomy Non-Thesis Option. A grade point
average of at least 3.0 is required in all course work that is
part of the degree program.

Doctorate: A minimum of 60 semester credits past the Bachelor's degree, including at least 36 graduate-level semester
credits in classroom courses in physics, astronomy, or related
fields and specified core courses. Course work used to satisfy
the requirement for a Master's degree may be included. A
minimum grade of B (3.00) is required in each course that
is used in the degree program. A minimum of 18 semester
credits of dissertation. Qualifying exam and oral defense of
the dissertation.

Other Degrees: Atomic, Molecular, and Optical Physics The AMO group consists of six faculty members: four experimentalists and two theorists. The research projects include nonlinear optics, studies of macromolecules, photon correlation spectroscopy, spectroscopy of molecular ions, studies of laser-produced low energy plasmas and trapped ions, atomic and molecular collisions, and modeling of molecular clouds in the interstellar medium. All four research laboratories are equipped with lasers, ultrahigh vacuums, and spectroscopy facilities. Modeling and calculations are conducted in the Computational Physics Laboratory, which is partially funded by the W. M. Keck Foundation. Astronomy/Astrophysics Faculty research interests include star formation in galaxies, active galactic nuclei, clusters of galaxies, gamma ray bursts, quasars, large scale structure of the universe, and variable stars. Faculty members have guaranteed access to SWIFT and successfully compete for time on other national facilities. Department facilities include an automated telescope in southern Arizona and access to the Lowell Observatory's 31-inch telescope in Flagstaff, Arizona, through UNLV's participation in the National Undergraduate Research Observatory. Condensed Matter and High Pressure Physics The UNLV High Pressure Science and Engineering Center (HiPSEC), recently established with support from the U.S. Department of Energy, brings together physicists, chemists, geoscientists, and mechanical engineers to consider fundamental experimental, computational, and engineering problems of materials under high pressures. Faculty and research staff study the equilibrium thermochemical properties, mechanical properties, reaction kinetics, and reaction products at static pressures using in situ x-ray diffraction, absorption, emission, and light-scattering spectroscopy from infrared to x-ray wavelengths, and other chemical and physical methods. Experiments are conducted at three in-house laboratories and in national laboratories, including an x-ray beam line in the Advanced Photon Source at the Argonne National Laboratory. In addition, along with experimental studies of complex fluids, state-of-the-art computational techniques are used to study highly correlated electron systems, including d- and f-band metals, clusters, thin films, quantum dots, and novel materials. Department Facilities and Funding Sources The Department resides in the Robert L. Bigelow Physics Building completed in 1994. Inside this 70,000 square-foot building, there are seven laboratories, other teaching and research laboratories, and supporting facilities, including two modern machine shops, a glass shop, and an electronics shop. The Computational Physics Laboratory, partially funded by the W. M. Keck Foundation, has a parallel/distributed computing system with a peak performance of about 8 GFlops. Research in the department is supported by NSF, DOD, DOE, NASA, EPA, the W. M. Keck Foundation, the Bigelow Foundation, and the UNLV Foundation.

Thesis: Thesis may be written in absentia with special approval only.

SPECIAL EQUIPMENT, FACILITIES, OR PROGRAMS

HiPSEC - High Pressure Science and Engineering Center.

Table B—Separately Budgeted Research Expenditures by Source of Support

Source of Support	Departmental Research	Physics-related Research Outside Department
Federal government	$18,674,609.58	
State/local government	$401,523.43	
Non-profit organizations		
Business and industry		
Other		
Total	$19,076,133.01	

Table C—Separately Budgeted Research Expenditures by Research Specialty

Research Specialty	No. of Grants	Expenditures ($)
Astronomy	13	$1,678,059.37
Condensed Matter Physics	7	$16,947,253.83
Physics and other Science Education	6	$450,819.81
Total	26	$19,076,133.01

FACULTY

Professor

Chen, Changfeng, Ph.D., Peking University, 1987. *Condensed Matter Physics*. Condensed matter theory.

Cornelius, Andrew L., Ph.D., Washington University, St. Louis, 1996. *Condensed Matter Physics*. Condensed matter experimental.

Farley, John W., Ph.D., Columbia University, 1977. *Atomic, Molecular, & Optical Physics*. Laser spectroscopy of molecular ions.

Kwong, Victor H. S., Ph.D., University of Toronto, 1979. *Atomic, Molecular, & Optical Physics*. Laser induced plasmas; ion charge transfer processes.

Lepp, Stephen H., Ph.D., University of Colorado Boulder, 1984. Department Chair. *Astrophysics, Atomic, Molecular, & Optical Physics*. Atomic and molecular theory.

Pang, Tao, Ph.D., University of Minnesota, 1989. *Condensed Matter Physics*. Condensed matter theory.

Selser, James C., Ph.D., University of California, Davis, 1975. *Polymer Physics/Science*. Static and dynamic light scattering from macromolecular systems.

Shelton, David P., Ph.D., University of Manitoba, 1979. *Atomic, Molecular, & Optical Physics*. Nonlinear optical properties of atoms and molecules.

Zhao, Yusheng, Ph.D., Stony Brook University, 1992. Director of High Pressure Science and Engineering Center(HiPSEC). *Condensed Matter Physics, Geophysics*.

Zygelman, Bernard, Ph.D., City University of New York, 1983. *Atomic, Molecular, & Optical Physics*. Atomic and molecular theory.

Associate Professor

Nagamine, Kentaro, Ph.D., Princeton University, 2001. *Astrophysics*. Cosmology, galaxy formation.

Pravica, Michael G., Ph.D., Harvard University, 1998. *Condensed Matter Physics*.

Proga, Daniel, Ph.D., Nicolaus Copernicus Astronomical Center, Warsaw, Poland, 1996. *Astrophysics*. High energy astrophysics.

Rhee, George, Ph.D., Leiden University, 1989. *Astronomy*. Extragalactic astronomy.

Spight, Lon D., Ph.D., University of Nevada, Reno, 1969. *Cosmology & String Theory*. Cosmology; interacting galaxies.

Zhang, Bing, Ph.D., Peking University, 1997. *Astronomy, Astrophysics*.

Emeritus

Cloud, Stanley D., Ph.D., University of Oregon, 1968. Nuclear physics; teaching instrumentation.

Smith, Diane Pyper, Ph.D., University of California, Santa Cruz, 1968. *Astronomy*. Stellar photometry and spectroscopy.

Weistrop, Donna E., Ph.D., California Institute of Technology, 1971. Extragalactic astronomy.

Zane, Leonard I., Ph.D., Duke University, 1970. Special Relativity.

Adjunct Faculty

Maxham, Amanda, Ph.D., University of Nevada Las Vegas, 2011. *Astronomy.*

Adjunct Professor

Kernan, Warnick, Ph.D., University of Rochester, 1989.

Naduvalath, Balakrishnan, Ph.D., Indian Institute of Technology, 1993.

Perry, Dale L., Ph.D., University of Houston, 1974.

Shaffer, David, Ph.D., California Institute of Technology, 1974.

Weck, Philippe, Ph.D., Paul Verlaine University, 2001.

DEPARTMENTAL RESEARCH SPECIALTIES AND STAFF

Theoretical

Astronomy. Astronomical phenomena of atomic and molecular systems; atomic and molecular collisions. Lepp, Nagamine, Proga, Rhee, Zhang.

Astrophysics. Interacting galaxies; digital image processing, gamma ray bursts. Lepp, Nagamine, Proga, Rhee, Zhang.

Condensed Matter Physics. High-temperature superconductivity; quantum liquids; disordered systems; correlated electron systems; spin systems. Chen, Pang.

Experimental

Astronomy. Galactic clustering and evolution. Photoelectric photometry, spectrophotometry, and spectroscopy of peculiar, upper main-sequence stars; observation and photometry of interacting galaxies. Rhee, Smith.

Chemical Physics. Dynamic behavior of macromolecular systems; nonlinear optics of atoms and molecules. Farley, Kwong, Shelton.

Condensed Matter Physics. Material characterization in high pressure. Cornelius, Farley, Kwong, Pravica, Selser, Zhao.

View additional information about this department at www.gradschoolshopper.com

UNIVERSITY OF NEVADA, RENO

DEPARTMENT OF PHYSICS

Reno, Nevada 89557-0058
http://www.physics.unr.edu

General University Information
President: Marc Johnson
Dean of Graduate School: Marsha Read
University website: http://www.unr.edu/
Control: Public
Setting: Urban
Total Faculty: 1,528
Total Graduate Faculty: 912
Total number of Students: 18,227
Total number of Graduate Students: 2,900

Department Information
Department Chairman: David Bennum, Chair
Department Contact: Jonathan Weinstein, Director of Graduate Studies
  Total full-time faculty: 26
  Total number of full-time equivalent positions: 15
  Full-Time Graduate Students: 48
  First-Year Graduate Students: 12
  Female First-Year Students: 2
  Total Post Doctorates: 3

Department Address
University of Nevada
MS 0220
Reno, NV 89557-0058
Phone: (775) 784-6792
Fax: (775) 784-1398
E-mail: weinstein@physics.unr.edu

Website: http://www.physics.unr.edu

ADMISSIONS

Admission Contact Information
Address admission inquiries to: Director of Graduate Studies, Department of Physics
Phone: (775) 784-6821
E-mail: weinstein@physics.unr.edu
Admissions website: http://www.unr.edu/grad/

Application deadlines
Fall admission:
U.S. students: March 1   *Int'l. students*: March 1
Spring admission:
U.S. students: November 1   *Int'l. students*: October 1

Application fee
U.S. students: $60   *Int'l. students*: $95

Admissions information
For Fall of 2013:
  Number of applicants: 27
  Number admitted: 11
  Number enrolled: 6

Admission requirements
Bachelor's degree requirements: A Bachelor's degree or Master's degree in physics is required.
Minimum undergraduate GPA: 3.0

GRE requirements

The GRE is required.

Physics GRE is strongly recommended.

Advanced GRE requirements

The Advanced GRE is not required.

TOEFL requirements

The TOEFL exam is required for students from non-English-speaking countries.

PBT score: 550

iBT score: 79

A minimum score of 24 on the speak portion of the iBT is required to guarantee financial support as a teaching assistant.

Other admissions information

Additional requirements: Applicants with degrees in related majors are considered, but may be required to take undergraduate physics courses based on coursework completed and interview.

Undergraduate preparation assumed: Boas, Mathematical Methods in the Physical Sciences; Marion and Thornton, Classical Dynamics; Griffiths, Introduction to Electrodynamics; Griffiths, Introduction to Quantum Mechanics; Hecht, Optics; Kittel, Thermal Physics.

TUITION

Tuition year 2013-14:

Tuition for in-state residents

Full-time students: $248 per credit

Tuition for out-of-state residents

Full-time students: $248 per credit

Graduate assistantships include a tuition fee waiver.

Credit hours per semester to be considered full-time: 6

Deferred tuition plan: No

Health insurance: Yes, $0 (all fees paid by department).

Academic term: Semester

Teaching Assistants, Research Assistants, and Fellowships

Average stipend per academic year

Teaching Assistant: $18,600

Assistantships provided for all first-year students. The current Teaching Assistant stipend is $1550 per month.

FINANCIAL AID

Loans

Loans are available for U.S. students.

Loans are not available for international students.

GAPSFAS application required: No

FAFSA application required: No

For further information

Address financial aid inquiries to: Jonathan Weinstein.

Phone: (775) 784-6821

E-mail: weinstein@physics.unr.edu

HOUSING

Availability of on-campus housing

Single students: Yes

Married students: Yes

For further information

Address housing inquiries to: Residential Life, Housing and Food Service.

Phone: (775) 784-1113

E-mail: housing@unr.edu

Housing aid website: http://www.unr.edu/housing

Table A—Faculty, Enrollments, and Degrees Granted

Research Specialty	2012-13 Faculty	Enrollment Fall 2012		Number of Degrees Granted 2010–11 (2006-11)		
		Master's	Doctorate	Master's	Terminal Master's	Doctorate
Astrophysics	2	–	–	–	–	–
Atomic, Molecular, & Optical Physics	9	–	–	–	–	–
Climate/Atmospheric Science	1	–	–	–	–	–
Materials Science, Metallurgy	3	–	–	–	–	–
Plasma and Fusion	14	–	–	–	–	–
Total	26	10	40	–	–	–
Full-time Grad. Stud.	–	8	40	–	–	–
First-year Grad. Stud.	–	4	9	–	–	–

GRADUATE DEGREE REQUIREMENTS

Master's: Plan A requires 30 graduate credits including six thesis credits. Plan B requires 32 graduate course credits. Minimum "B" average is required. There is no language requirement. A final oral examination on thesis work is required under Plan A; a final written and oral examination on coursework is required under Plan B. 24 credits under Plan A and 26 credits under Plan B must be earned in residence.

Doctorate: 48 course credits and 24 dissertation credits are required. Minimum "B" average is required. Written and oral comprehensive examinations and a final oral defense of dissertation are required. A minimum of six semesters beyond the bachelor's degree in residence are required, including at least two in succession.

Other Degrees: Master's and Ph.D. programs in atmospheric sciences are offered in association with the Desert Research Institute with courses and research topics in atmospheric physics. A Ph.D. program in chemical physics is offered in association with the Chemistry Department.

Thesis: Thesis may be written in absentia.

SPECIAL EQUIPMENT, FACILITIES, OR PROGRAMS

ECR multicharged ion source and electron-ion crossed-beams apparatus; photon-ion merged-beams end station at Advanced Light Source; negative ion beam facility; laser cooling atomic physics laboratory; cryogenic atomic physics laboratory; 2-TW Z-pinch pulsed-power device for plasma/x-ray physics and two 100 TW class lasers at the Nevada Terawatt Facility; well-equipped atmospheric physics laboratories on campus and at the Desert Research Institute; multiple computing clusters optimized for atomic physics and plasma physics/HED calculations.

Table B—Separately Budgeted Research Expenditures by Source of Support

Source of Support	Departmental Research	Physics-related Research Outside Department
Federal government	$5,200,000	
State/local government		
Non-profit organizations		
Business and industry		
Other		
Total	$5,200,000	

FACULTY

Professor

Arnott, Patrick, Ph.D., Washington State University, 1988. *Climate/Atmospheric Science*. Acoustic and optical sensing of the earth's atmosphere.

Bauer, Bruno, Ph.D., University of California, Los Angeles, 1992. *Plasma and Fusion*. Experimental research involving high-power lasers, Z-pinches linear and nonlinear plasma waves, and instabilities.

Derevianko, Andrei, Ph.D., Auburn University, 1996. *Atomic, Molecular, & Optical Physics, Theoretical Physics*. Theoretical atomic and molecular physics; many-body methods; tests of fundamental symmetries; cold atoms.

Mancini, Roberto C., Ph.D., University of Buenos Aires, 1983. *Atomic, Molecular, & Optical Physics, Plasma and Fusion*. Atomic and radiation physics of high-energy-density plasmas; stark-broadened line shapes; radiation transport; x-ray spectroscopy of plasmas; multi-objective spectroscopic data analysis.

Neill, Paul A., Ph.D., Queen's University, Belfast, 1984. Alignment and orientation studies in electron atom collisions; ionization and charge transfer in ion-atom collisions.

Phaneuf, Ronald A., Ph.D., University of Windsor, 1973. *Atomic, Molecular, & Optical Physics*. Measurements of interactions of photons and electrons with atomic and molecular ions; research using synchrotron radiation; atomic processes in plasmas and astrophysics; structure and dynamics of nanomolecules.

Sentoku, Yasuhiko, Ph.D., Osaka University, 1999. *Plasma and Fusion, Theoretical Physics*. Modeling the physics of high-energy density matter created by short laser pulses.

Thompson, J. S., Ph.D., University of Tennessee, 1989. Dean, College of Science. *Atomic, Molecular, & Optical Physics*. Photodetachment of negative ions.

Winkler, Peter, Ph.D., Erlangen-Nornberg, 1969. *Atomic, Molecular, & Optical Physics*. Theoretical description of atomic and molecular processes; many-body physics; quantum chemistry.

Winterberg, Friedwardt, Ph.D., Max Planck Institute, Goettingen, 1955. *High Energy Physics, Plasma and Fusion, Quantum Foundations, Relativity & Gravitation, Theoretical Physics*. Theoretical physics; elementary particle physics, in particular Planck scale physics; controlled nuclear fusion, in particular inertial confinement fusion; nuclear rocket propulsion.

Associate Professor

McCall, Katherine R., Ph.D., University of Massachusetts, Amherst, 1992. *Condensed Matter Physics*. Theoretical research on properties of hysteretic nonlinear materials and transport in disordered systems; rock physics.

Weinstein, Jonathan, Ph.D., Harvard University, 2002. *Atomic, Molecular, & Optical Physics*. Experimental research involving cryogenically cooled atoms and molecules.

Assistant Professor

Geraci, Andrew, Ph.D., Stanford University, 2007. *Atomic, Molecular, & Optical Physics*. Precision force measurements; laser-cooling and trapping of atoms and dielectric objects; hybrid atomic-microelectromechanical systems; quantum information science; Casimir forces; experimental gravitation.

Sawada, Hiroshi, Ph.D., University of Rochester, 2008.

Research Professor

Darling, Timothy, Ph.D., University of Melbourne, 1989. *Acoustics, Condensed Matter Physics, Plasma and Fusion*. Experimental studies of elasticity and stability in solid-state systems.

Kantsyrev, Victor L., Ph.D., Institute of Analytical Instrumentation, Moscow, Russia, 1992. *Plasma and Fusion*. X-ray spectroscopy of laser-produced plasmas and Z-pinch plasmas.

Safronova, Alla, Ph.D., Lebedev Physical Institute, Moscow, 1986. *Atomic, Molecular, & Optical Physics, Plasma and Fusion, Theoretical Physics*. Spectroscopy and modeling of hot, dense plasmas.

Safronova, Ulyana, Ph.D., Vilnus University, 1975. *Atomic, Molecular, & Optical Physics, Theoretical Physics*. High-precision relativistic electronic structure calculations; many-body theory.

Research Associate Professor

Ivanov, Vladimir, Ph.D., Lebedev Physical Institute, 1987. *Plasma and Fusion*. Experimental studies of Z-pinch and laser produced plasmas.

Presura, Radu, Ph.D., University of Bucharest, 1999. *Plasma and Fusion*. Laboratory studies of astrophysical plasmas.

Research Assistant Professor

Covington, Aaron, Ph.D., University of Nevada, Reno, 1997. *Atomic, Molecular, & Optical Physics, Plasma and Fusion*. Experimental studies of photon interactions with atoms, molecules, and solids; Nevada Terawatt Facility.

Esaulov, Andrey, Ph.D., Moscow Institute of Physics and Technology, 1996. *Plasma and Fusion, Theoretical Physics*. Theoretical modeling of plasmas.

Fuelling, Stephan, Ph.D., University of Nevada, Reno, 1991. *Plasma and Fusion*. Experimental plasma physics.

Paraschiv, Ioana, Ph.D., University of Nevada, Reno, 2007.

Wiewior, Piotr, Ph.D., Warsaw University, 1999.

Lecturer

Bach, Bernhar, Ph.D., College of William and Mary, 1995. *Optics*. Design and fabrication of optical systems and physics education research.

Bennum, David, Ph.D., University of Nevada, Reno, 1973. *Astronomy, Physics and other Science Education*. Observational investigation of superplanets and applications of astronomy research to education.

Rodrígue, Melodi, Ph.D., University of Nevada, Reno, 1998. *Physics and other Science Education*. Physics and astronomy education research.

DEPARTMENTAL RESEARCH SPECIALTIES AND STAFF

Theoretical

Atomic, Molecular, & Optical Physics. Precision calculations of atomic structure for tests of the Standard Model; research to improve the accuracy of atomic clocks; calculations and models of atomic interactions and spectroscopy in dense plasma environments; quantum computing; physics of ultracold degenerate gases. Derevianko, Alla Safronova, Ulyana Safronova, Winkler.

Condensed Matter Physics. The Materials Physics Group focuses on improving our understanding of the behavior of condensed matter. Materials of interest range from single crystals and aggregates of crystals such as minerals, metals and rocks, to amorphous and disordered materials such as polymers and glasses. Knowledge of the fundamental properties of these materials can be applied in diverse arenas, including solid state physics, geophysics and engineered materials. McCall.

Plasma and Fusion. The fields of Plasma Physics and High Energy Density Science are some of the department's core strengths, with active research in experiment, theory, and modeling. Research groups are carrying out theoretical, computational, experimental and applied investigations into many physical systems under extremes of pressure, temperature and density. Research areas include studies of the formation and time evolution of plasmas, and investigations of conditions ranging from the physics of the upper atmosphere to extreme pressures and temperatures found in astrophysical events. Mancini, Sentoku, Winterberg.

Experimental

Atmospheric Physics. Gas particle conversion; laboratory studies of nucleation and growth of particulates; cloud condensation nuclei; aerosol removal by scavenging; trace elements in snow; atmospheric remote sensing. Arnott.

Atomic, Molecular, & Optical Physics. Cold atomic and molecular collisions; laser cooling and trapping of atoms; quantum information science; photoionization and electron-impact ionization and fragmentation of atomic and molecular ions; collisional and photodetachment of negative ions; optomechanics and quantum precision sensors; hybrid quantum systems and quantum computing. Covington, Fuelling, Geraci, Neill, Phaneuf, Thompson, Weinstein.

Condensed Matter Physics. The Materials Physics group uses primarily acoustic and optical techniques to extract information from samples, including Resonant Ultrasound Spectroscopy (RUS), Raman Spectroscopy, and Fourier Transform Infrared (FTIR) spectroscopy. Typically, these tools are used to interrogate samples while they undergo changes in state variables such as temperature or pressure. Changes in the state variables can induce concomitant changes in the arrangement of atoms within the materials, which in turn can lead to changes in microscopic and macroscopic properties of the material. Covington, Darling.

Plasma and Fusion. The Physics Department has a number of distinctive research facilities that enhance the research activities in plasma physics and high-energy density science, including a 2 TW z-pinch accelerator, a 100 TW, 350 femtosecond laser, and a ninety-six node cluster computer. In addition to these rich on-campus resources, additional research is done through collaborations at national research facilities such as the Rochester's Laboratory for Laser Energetics, Sandia National Laboratory, and Livermore National Laboratory. Bauer, Covington, Darling, Esaulov, Fuelling, Ivanov, Kantsyrev, Mancini, Paraschiv, Presura, Sawada, Wiewior.

View additional information about this department at
www.gradschoolshopper.com

DARTMOUTH COLLEGE

DEPARTMENT OF PHYSICS AND ASTRONOMY

Hanover, New Hampshire 03755-3528
http://www.dartmouth.edu/~physics/

General University Information
President: Phillip J. Hanlon
Dean of Graduate School: F. Jon Kull
University website: http://www.dartmouth.edu
Control: Private
Setting: Rural
Total Faculty: 1,045
Total Graduate Faculty: Included above
Total number of Students: 6,277
Total number of Graduate Students: 2,084

Department Information
Department Chairman: James W. LaBelle, Chair
Department Contact: Judy Lowell, Department Administrator
  Total full-time faculty: 23
  Total number of full-time equivalent positions: 18
  Full-Time Graduate Students: 53
  First-Year Graduate Students: 9
  Female First-Year Students: 1
  Total Post Doctorates: 8

Department Address
6127 Wilder Laboratory
Hanover, NH 03755-3528
Phone: (603) 646-2854
Fax: (603) 646-1446

E-mail: physics.department@dartmouth.edu
Website: http://www.dartmouth.edu/~physics/

ADMISSIONS

Admission Contact Information
Address admission inquiries to: Chair of Graduate Admissions, Dept. of Physics and Astronomy, 6127 Wilder Laboratory
Phone: (603) 646-2854
E-mail: physics.department@dartmouth.edu
Admissions website: http://www.dartmouth.edu/~physics/academics/graduate/apply.html

Application deadlines
Fall admission:
U.S. students: January 15   *Int'l. students*: January 15

Application fee
U.S. students: $45   *Int'l. students*: $45

Admissions information
For Fall of 2013:
  Number of applicants: 128
  Number enrolled: 10

Admission requirements
Bachelor's degree requirements: Bachelor's degree in physics, astrophysics, or astronomy is normally required.

GRE requirements

The GRE is required.
The minimum acceptable GRE score is 1,200, combined verbal plus quantitative sections.

Advanced GRE requirements

The Advanced GRE is required.

TOEFL requirements

The TOEFL exam is required for students from non-English-speaking countries.
　PBT score: 600
　iBT score: 100
Minimum acceptable ILELTS Band score of 7.0; TWE score of 4.5

Other admissions information

Additional requirements: Mathematics majors with some physics training are considered in some cases.

TUITION

Tuition year 2013-14:
　Full-time students: $58,376 annual
Stipends carry full tuition scholarship; single student health insurance provided at no cost to student.
Credit hours per semester to be considered full-time: 8
Deferred tuition plan: No
Health insurance: Available
Other academic fees: Activity Fee: $50 Document Fee: $50 Health Access Fee: $300 International Student Fee: $300
Academic term: Quarter
Number of first-year students who received full tuition waivers: 10

Teaching Assistants, Research Assistants, and Fellowships

Number of first-year
　Teaching Assistants: 9
Average stipend per academic year
　Teaching Assistant: $27,240
　Research Assistant: $27,240
　Stipends carry full tuition scholarship; single student health insurance provided at no cost to student.

FINANCIAL AID

Loans

Loans are available for U.S. students.
Loans are available for international students.
GAPSFAS application required: No
FAFSA application required: Yes

For further information

Address financial aid inquiries to: Office of Graduate Studies 6062 Wentworth Hall, Room 305, Dartmouth College, Hanover, NH 03755.
Phone: (603) 646-2107
Financial aid website: http://graduate.dartmouth.edu/finaid/

HOUSING

Availability of on-campus housing

　Single students: Yes
　Married students: No

For further information

Address housing inquiries to: Dartmouth College Real Estate Office, 4 Currier Place, Suite 304, Hanover, NH 03755.
Phone: (603) 646-2446
E-mail: real.estate.office@dartmouth.edu
Housing aid website: http://www.dartmouthre.com

Table A—Faculty, Enrollments, and Degrees Granted

Research Specialty	2012-13 Faculty	Enrollment Fall 2012 Master's	Enrollment Fall 2012 Doctorate	Number of Degrees Granted 2012-13 (2008–13) Master's	Number of Degrees Granted 2012-13 (2008–13) Terminal Master's	Number of Degrees Granted 2012-13 (2008–13) Doctorate
Applied Mathematics	1	–	–	–	–	–
Applied Physics	2	–	–	–	–	–(2)
Astronomy	2	–	6	1(2)	–(2)	1(5)
Astrophysics	7	–	5	–	–	1(1)
Atmosphere, Space Physics, Cosmic Rays	8	2	13	1(3)	1(4)	1(7)
Atomic, Molecular, & Optical Physics	2	–	1	1(1)	–	–
Biophysics	1	–	–	–	–	–(3)
Chemical Physics	1	–	1	–	–	–
Computational Physics	2	–	1	1(1)	–	–
Computer Science	1	–	–	–	–	–
Condensed Matter Physics	4	–	9	–	–	3(7)
Cosmology & String Theory	3	–	4	–	–	1(1)
Engineering Physics/Science	1	–	–	–	–	–
Geophysics	2	–	–	–	–	–
High Energy Physics	1	–	–	–	–	–(1)
History & Philosophy of Physics/Science	1	–	–	–	–	–
Medical, Health Physics	–	–	4	–	–(1)	–
Nonlinear Dynamics and Complex Systems	1	–	1	–	–	–
Particles and Fields	2	–	1	–	–	–
Physics and other Science Education	1	–	–	–	–	–
Plasma and Fusion	5	1	3	–	–	–(2)
Quantum Foundations	1	–	–	–	–	–(1)
Quantum Information	1	–	1	–	–	–
Relativity & Gravitation	2	–	–	–	–	–(1)
Solar Physics	1	–	–	–	–	–
Statistical & Thermal Physics	2	–	–	–	–	–(1)
Theoretical Physics	2	–	–	–	–	–
Total	–	3	50	4(7)	1(7)	7(32)
Full-time Grad. Stud.	–	3	50	–	–	–
First-year Grad. Stud.	–	1	13	–	–	–

GRADUATE DEGREE REQUIREMENTS

Master's: Minimum of three consecutive terms in residence. Degree credit for eight graduate courses. Two of the eight may be Graduate Research. At least six of the eight should be in Physics and Astronomy. A satisfactory thesis must be completed and defended; or significant co-author of a publication submitted to a referred journal or referred conference proceedings, defended publicly; or passing the Ph.D. qualifying exam. Some teaching experience must be acquired. No foreign language requirement.

Doctorate: Minimum of six terms in residence. Degree credit for twelve graduate courses, exclusive of Graduate Research and teaching courses. Degree credit for at least two terms of supervised undergraduate teaching. A written qualifying examination must be passed. A thesis proposal must be presented and defended. A dissertation of substantial significance and publishable quality must be completed and defended before the faculty. No formal language requirement.

Thesis: Thesis may not be written in absentia.

SPECIAL EQUIPMENT, FACILITIES, OR PROGRAMS

Observational facilities including 1.3-m and 2.4-m reflecting telescopes at the MDM Observatory, Kitt Peak, Arizona, and the 10-m SALT telescope at Sutherland, South Africa.

Table B—Separately Budgeted Research Expenditures by Source of Support

Source of Support	Departmental Research	Physics-related Research Outside Department
Federal government	$3,064,300	
State/local government		
Non-profit organizations		
Business and industry		
Other	$73,000	
Total	$3,137,300	

Table C—Separately Budgeted Research Expenditures by Research Specialty

Research Specialty	No. of Grants	Expenditures ($)
Astronomy	5	$133,000
Astrophysics	7	$215,300
Atmosphere, Space Physics, Cosmic Rays	22	$1,756,500
Chemical Physics	1	$2,500
Condensed Matter Physics	4	$410,000
Physics and other Science Education	1	$60,000
Plasma and Fusion	2	$100,000
Nonlinear Dynamics and Complex Systems	1	$15,000
Cosmology & String Theory	1	$25,000
Quantum Information	5	$420,000
Total	49	$3,137,300

FACULTY

Professor

Blencowe, Miles P., Ph.D., University of London, 1989. Department Chair. *Condensed Matter Physics, Low Temperature Physics, Nano Science and Technology, Nonlinear Dynamics and Complex Systems, Particles and Fields, Relativity & Gravitation, Solid State Physics, Statistical & Thermal Physics, Theoretical Physics.* Condensed matter theory; mesoscopic physics; open quantum systems; quantum-classical correspondence.

Caldwell, Robert, Ph.D., University of Wisconsin-Milwaukee, 1992. *Cosmology & String Theory, Particles and Fields, Relativity & Gravitation, Theoretical Physics.* Theoretical cosmology; gravitation, particles, fields; the early universe; large scale structure.

Chaboyer, Brian C., Ph.D., Yale University, 1993. *Astronomy, Astrophysics.* Theoretical astrophysics; stellar evolution, stellar populations.

Fesen, Robert A., Ph.D., University of Michigan, 1981. *Astronomy, Astrophysics.* Observational astrophysics; supernovae and supernova remnants.

Gleiser, Marcelo, Ph.D., King's College, London, 1986. *Biophysics, Cosmology & String Theory, History & Philosophy of Physics/Science, Nonlinear Dynamics and Complex Systems, Particles and Fields, Physics and other Science Education, Relativity & Gravitation.* Field theory and cosmology; nonequilibrium phenomena; astrobiology; history and philosophy of science.

Hudson, Mary K., Ph.D., University of California, Los Angeles, 1974. *Atmosphere, Space Physics, Cosmic Rays, Plasma and Fusion.* Theoretical space plasma physics.

LaBelle, James W., Ph.D., Cornell University, 1985. *Atmosphere, Space Physics, Cosmic Rays, Plasma and Fusion.* Experimental space plasma physics.

Lynch, Kristina A., Ph.D., University of New Hampshire, 1992. *Applied Physics, Atmosphere, Space Physics, Cosmic Rays, Geophysics, Plasma and Fusion.* Experimental space plasma physics; auroral and mesospheric sounding rockets; thermal plasma laboratory experiments; small spacecraft design.

Rimberg, Alexander, Ph.D., Harvard University, 1992. *Condensed Matter Physics.* Experimental condensed matter physics; nanoscale physics; quantum phenomena and measurements; microwave and radio-frequency techniques; low-noise electrical measurements.

Rogers, Barrett N., Ph.D., Massachusetts Institute of Technology, 1991. *Plasma and Fusion.* Laboratory and space plasma theory and simulation; magnetic reconnection; plasma turbulence; magnetic fusion.

Thorstensen, John R., Ph.D., University of California, Berkeley, 1980. *Astronomy, Astrophysics.* Observational astronomy; interacting binary stars; astrometry.

Viola, Lorenza, Ph.D., University of Padova, 1996. *Condensed Matter Physics, Quantum Foundations, Theoretical Physics.* Theoretical quantum information processing; open quantum systems; entanglement; quantum phase transitions, topological phases of matter.

Wegner, Gary A., Ph.D., University of Washington, 1971. *Astronomy, Astrophysics.* Observational astrophysics.

Wybourne, Martin, Ph.D., University of Nottingham, 1980. Interim Provost. *Condensed Matter Physics.* Experimental physics; condensed matter.

Associate Professor

Alexander, Stephon, Ph.D., Brown University, 2000. *Cosmology & String Theory, Particles and Fields, Relativity & Gravitation.* Theoretical cosmology; early universe cosmology; large scale structure; string theory; loop quantum gravity; mathematics of music.

Millan, Robyn, Ph.D., University of California, Berkeley, 2002. *Atmosphere, Space Physics, Cosmic Rays.* Experimental space physics; radiation belt losses, x-ray instrumentation for space physics and atmospheric sciences.

Assistant Professor

Hickox, Ryan C., Ph.D., Harvard University, 2007. *Astronomy, Astrophysics.* Active galactic nuclei; galaxy evolution; large-scale structure of the Universe; the cosmic x-ray background.

Ramanathan, Chandrasekhar, Ph.D., Massachusetts Institute of Technology, 1996. *Chemical Physics, Condensed Matter Physics, Nano Science and Technology, Solid State Physics.* Experimental quantum information processing; magnetic resonance; many-body physics.

Wright, Kevin C., Ph.D., University of Rochester, 2009. *Atomic, Molecular, & Optical Physics.* Ultra-cold quantum gases, many-body physics, matter-wave interferometry; atom-photon interactions, quantum optics, microcavities.

Research Professor

Crane, Philippe, Ph.D., Yale University, 1969. *Astrophysics.* Astrophysics; extra-solar planets.

Denton, Richard E., Ph.D., University of Maryland, 1986. *Geophysics, Plasma and Fusion.* Computational plasma physics, wave phenomena, reconnection, magnetospheric density.

Lyon, John G., Ph.D., University of Maryland, 1972. *Computational Physics, Geophysics, Plasma and Fusion.* Space plasma physics and magnetospheric physics; numerical simulation and computational physics.

Research Associate Professor

Kress, Brian, Ph.D., Dartmouth College, 2002. *Atmosphere, Space Physics, Cosmic Rays, Computational Physics.* Space plasma physics, radiation belt dynamics, fluid and magnetofluid dynamics.

Müller, Hans, Ph.D., Dartmouth College, 1997. *Astrophysics, Atmosphere, Space Physics, Cosmic Rays.* Space plasma physics and heliospheric physics; cool stellar winds; numerical simulations.

Adjunct Professor

Fulton, Theodore, Ph.D., Cornell University, 1966. *Condensed Matter Physics.* Condensed matter physics, particularly single-electron devices and Josephson effects.

Lotko, William, Ph.D., University of California, Los Angeles, 1981. *Geophysics, Plasma and Fusion.* Geospace environment; space plasma physics, modeling, simulation; electromagnetic fields and waves.

Naumann, Robert A., Ph.D., Princeton University, 1953. *Nuclear Physics.* Nuclear spectroscopy.

Pogue, Brian W., Ph.D., McMaster University, 1996. *Medical, Health Physics.* Medical physics; medical imaging systems software and hardware; laser spectroscopy of tissue; biomedical optics.

Weaver, John B., Ph.D., University of Virginia, 1983. *Medical, Health Physics.* Medical imaging; magnetic nanoparticle sensing and imaging; MRI; MR elastography; radiation dosimetry.

Adjunct Associate Professor

Barnett, Alexander H., Ph.D., Harvard University, 2000. *Applied Mathematics.* Numerical analysis; partial differential equations; scientific computing; quantum chaos.

Brizard, Alain, Ph.D., Princeton University, 1990. *Plasma and Fusion.* Low-frequency nonlinear gyrokinetic theory, relativistic quasilinear transport driven by arbitrary-frequency electromagnetic fluctuations, variational formulations of exact and reduced, kinetic and fluid plasma equations, applications of Lie-transform methods in plasma physics.

Levey, Christopher G., Ph.D., University of Wisconsin-Madison, 1984. Microfabricataion technology; micro-optical and electromechanical systems (MEMS).

Smith, Timothy, Ph.D., University of Massachusetts, Lowell, 1990. *Nuclear Physics.* Experimental nuclear physics studying the quark structure of neutrons and protons.

Ukhorskiy, Aleksandr, Ph.D., University of Maryland, 2003. *Atmosphere, Space Physics, Cosmic Rays.* Energetic particles in space; magnetospheric plasma waves; nonlinear processes in space plasmas.

Adjunct Assistant Professor

Liu, Jifeng, Ph.D., Massachusetts Institute of Technology, 2006. Optoelectronic materials and devices; nanophotonics for solar cells and concentrated solar power (CSP); electronic-photonic integration.

Ticozzi, Francesco, Ph.D., University of Padova, 2007. *Condensed Matter Physics.* Quantum control; quantum information protection; stability theory; quantum communication.

DEPARTMENTAL RESEARCH SPECIALTIES AND STAFF

Theoretical

Astrophysics. Stellar structure and evolution; globular cluster ages; formation of the Milky Way and its galaxies. Chaboyer.

Biophysics. Astrobiology. Gleiser.

Computational Physics. Caldwell, Chaboyer, Denton, Gleiser, Kress, Lyon, Müller, Rogers.

Condensed Matter Physics. Quantum many-body systems, electronic and mechanical properties of mesoscopic systems; quantum measurement theory and decoherence; quantum phase transitions; topological quantum matter. Blencowe, Viola.

Cosmology & String Theory. Alexander, Caldwell, Gleiser.

Fundamental Quantum Physics and Quantum Information Science. Measurement, control, information processing at the quantum scale, entanglement. Blencowe, Ticozzi, Viola.

Particles and Fields. Alexander, Caldwell, Gleiser.

Plasma and Fusion. MHD and kinetic theory; computer simulations; waves and instabilities; radiation; transport properties; particle acceleration and heating; applications to space and laboratory plasmas. Brizard, Denton, Hudson, Kress, Lotko, Lyon, Müller, Rogers.

Relativity & Gravitation. Alexander, Blencowe, Caldwell, Gleiser.

Experimental

Astrophysics. Cataclysmic binaries; supernova remnants; white dwarfs, close binary stars, and clusters of galaxies; globular clusters; active galactic nuclei; supermassive black hole and galaxy evolution. Chaboyer, Crane, Fesen, Hickox, Thorstensen, Wegner.

Atomic, Molecular, & Optical Physics. Utra-cold quantum gases; many-body physics; matter-wave interferometry; Atomphoton interactions; quantum optics; microcavities. Wright.

Condensed Matter Physics. Phonon and electron transport in low-dimensional systems; cooling techniques for mechanical resonators; nanostructures; mesoscopic physics; spins in semiconductors; central spin problem; quantum magnetism; measurement, control, information processing at the quantum scale; quantum devices and sensors; ultra-cold quantum gases; many-body physics; matter-wave interferometry; atomphoton interactions; quantum optics. Fulton, Ramanathan, Rimberg, Wright, Wybourne.

Nuclear Physics. Using electron accelerators to measure the distribution of quarks in neutrons and protons. Naumann, Smith.

Plasma and Fusion. Plasma kinetics; microwave plasma interactions; plasma diagnostics; ionospheric and magnetospheric physics; plasma measurement in space; remote sensing of ionospheric plasma processes; auroral particle measurements; mesospheric dust particles; ionospheric and mesospheric sounding rockets; stratospheric balloons, radiation belt losses. LaBelle, Lynch, Millan.

Quantum Foundations. Measurement, control, information processing at the quantum scale. Ramanathan, Rimberg.

View additional information about this department at
www.gradschoolshopper.com

UNIVERSITY OF NEW HAMPSHIRE

DEPARTMENT OF PHYSICS

Durham, New Hampshire 03824
http://www.physics.unh.edu

General University Information
President: Mark Huddleston
Dean of Graduate School: Harry J. Richards
University website: http://www.unh.edu
Control: Public
Setting: Rural
Total Faculty: 985
Total Graduate Faculty: 578
Total number of Students: 14,761
Total number of Graduate Students: 2,196

Department Information
Department Chairman: Mark McConnell, Chair
Department Contact: Katie Makem-Boucher, Administrative
  Manager
  Total full-time faculty: 38
  Total number of full-time equivalent positions: 18
  Full-Time Graduate Students: 68
  First-Year Graduate Students: 13
  Female First-Year Students: 4
  Total Post Doctorates: 14

Department Address
9 Library Way
Durham, NH 03824
Phone: (603) 862-2669
E-mail: katie.makem@unh.edu
Website: http://www.physics.unh.edu

ADMISSIONS

Admission Contact Information
Address admission inquiries to: Graduate Admission Coordi-
  nator, Physics Department
Phone: (603) 862-2669
E-mail: physics.grad.info@unh.edu
Admissions website: http://www.gradschool.unh.edu/apply.html

Application deadlines
Fall admission:
U.S. students: February 1   *Int'l. students*: February 1

Application fee
U.S. students: $65   *Int'l. students*: $65
Please ensure that your letters of recommendation arrive in time,
  ideally by mid-January. Spring admission is by approval only.

Admissions information
For Fall of 2013:
  Number of applicants: 88
  Number admitted: 19
  Number enrolled: 8

Admission requirements
Bachelor's degree requirements: A bachelor's degree in physics
  or astronomy is usually required, but exceptions have been
  made.

GRE requirements
The GRE is required.

GRE is required. Request official test scores to be sent directly
to the Graduate School by the testing service. Test scores
more than five years old may not be acceptable. Student copies and photocopies of scores are not considered official. Our
CEEB code is 3918.

Advanced GRE requirements
The Advanced GRE is recommended.
Physics GRE is not required but highly recommended. If available, please enter your percentile score into the online application (test score information).

TOEFL requirements
The TOEFL exam is required for students from non-English-
  speaking countries.
  PBT score: 550
  iBT score: 80

Other admissions information
Undergraduate preparation assumed: The level of preparation
  should be comparable with that provided at University of New
  Hampshire for the bachelor's of science degree in Physics.
  This includes two semesters of advanced quantum mechanics
  (Griffiths, Introduction to Quantum Mechanics), two semesters of advanced E&M (Pollack & Stump, Electromagnetism),
  and advanced classical mechanics (Thornton & Marion, Classical Dynamics of Particles and Systems).

TUITION
Tuition year 2013-14:
Tuition for in-state residents
  Full-time students: $13,500 annual
  Part-time students: $750 per credit
Tuition for out-of-state residents
  Full-time students: $26,200 annual
  Part-time students: $1,100 per credit
The university reserves the right to adjust tuition and/or related
  expenses. Any changes will be announced as far in advance
  as possible.
Credit hours per semester to be considered full-time: 9
Deferred tuition plan: No
Health insurance: Available at the cost of $2180 per year.
Other academic fees: Total mandatory fees for full-time students
  (which includes health insurance) is $2326.
Academic term: Semester
Number of first-year students who received full tuition waivers: 13

Teaching Assistants, Research Assistants, and Fellowships
Number of first-year
  Teaching Assistants: 11
  Research Assistants: 1
  Fellowship students: 1
Average stipend per academic year
  Teaching Assistant: $15,100
  Research Assistant: $15,100
  Fellowship student: $15,100

Most first-year students serve as a TA and then move to an RA beginning with the summer after their first year. Students serving as either a TA or RA are provided with both a stipend and full tuition.

FINANCIAL AID

Application deadlines
Fall admission:
U.S. students: March 1   *Int'l. students*: March 1

Loans
Loans are available for U.S. students.
Loans are not available for international students.
GAPSFAS application required: No
FAFSA application required: Yes

For further information
Address financial aid inquiries to: Financial Aid Office, 11 Garrison Avenue, Stoke Hall, Durham, NH 03824.
Phone: (603) 862-3600
Financial aid website: http://financialaid.unh.edu/

HOUSING

Availability of on-campus housing
Single students: Yes
Married students: Yes

For further information
Address housing inquiries to: Department of Housing, 10 Academic Way, Durham, NH 03824.
Phone: (603) 862-2120
E-mail: housing.office@unh.edu
Housing aid website: http://www.unh.edu/housing/gradhousing/index.html

Table A—Faculty, Enrollments, and Degrees Granted

Research Specialty	2012-13 Faculty	Enrollment Fall, 2012 Master's	Enrollment Fall, 2012 Doctorate	Number of Degrees Granted 2012-13 (2008-13) Master's	Number of Degrees Granted 2012-13 (2008-13) Terminal Master's	Number of Degrees Granted 2012-13 (2008-13) Doctorate
Condensed Matter Physics	3	1	5	–	–	1(2)
High Energy Astrophysics	4	–	3	–	–	1(3)
High Energy Theory	3	–	2	–	–(1)	1(3)
Medical Imaging	1	–	1	–	–	–(1)
Nuclear Physics	3	1	3	–	–	–(1)
Optics	1	–	1	–	–	–
Physics Education	1	–	2	–	–	–(1)
Space Science	21	2	40	–	–(3)	4(13)
Total	37	4	57	–(5)	1(12)	7(24)
Full-time Grad. Stud.	–	5	63	–	–	–
First-year Grad. Stud.	–	–	14	–	–	–

GRADUATE DEGREE REQUIREMENTS

Master's: Students are required to satisfactorily complete an approved course program of five required courses plus: (1) nine additional credits of graduate coursework plus a master's thesis and an oral thesis defense or (2) 12 additional credits of graduate coursework plus a research project and an oral examination in the form of a seminar. There are no residency or foreign language requirements.

Doctorate: Students are required to complete satisfactorily an approved course program of eight required courses plus four additional courses at the graduate level, pass a written comprehensive examination on their undergraduate physics topics by the middle of their second year in the program, pass an oral qualifying examination in which a thesis proposal is presented and discussed, demonstrate proficiency in teaching (by service as a teaching assistant), complete a dissertation, and pass an oral dissertation examination. Students must satisfy a one-year residence requirement. Students can earn a master's degree after completing 30 graduate credits and passing the written comprehensive examination and the oral qualifying examination. There is no foreign language requirement.

SPECIAL EQUIPMENT, FACILITIES, OR PROGRAMS

Many faculty hold joint appointments in the Space Science Center (SSC), part of the Institute for the Study of Earth, Oceans, and Space (EOS). The SSC fosters research and graduate education in all of the space sciences, with studies ranging from the ionosphere to the Earth's magnetosphere, the local solar system, and out to the farthest reaches of the universe. Researchers and students have access to a CRAY XE6m-200 supercomputer with 4100 compute cores and 180 TB of storage. The SSC also maintains facilities that are used to design and fabricate space flight hardware, including clean rooms and thermal/vacuum chambers. The SSC maintains a close relationship with Southwest Research Institute (SwRI), which recently established (in 2013) a new department on the University of New Hampshire campus.

Faculty in the Condensed Matter Group are also part of the Materials Science Program, a Ph.D. program that offers research opportunities in the areas of science and engineering that cross traditional departmental boundaries, with an emphasis on the synthesis and characterization of nanoscale materials.

Several faculty members participate in the Applied Mathematics Program, a Ph.D. program designed to facilitate interdisciplinary research among graduate students and participating faculty. This interdisciplinary program gives students the opportunity to explore the frontier where the sciences meet cutting-edge mathematical analysis and high-performance computing.

Table B—Separately Budgeted Research Expenditures by Source of Support

Source of Support	Departmental Research	Physics-related Research Outside Department
Federal government	$30,311,872	
State/local government		
Non-profit organizations		
Business and industry		
Other		
Total	$30,311,872	

Table C—Separately Budgeted Research Expenditures by Research Specialty

Research Specialty	No. of Grants	Expenditures ($)
High Energy Astrophysics	12	$869,470
Space Science	159	$28,473,533
Condensed Matter Physics	3	$276,756
Medical Imaging	3	$71,620
Nuclear Physics	6	$423,024
Optics	3	$6,466
Physics Education	1	$27,087
High Energy Theory	4	$163,918
Total	191	$30,311,874

FACULTY

Professor

Balling, Ludwig C., Ph.D., Harvard University, 1965. *Atomic, Molecular, & Optical Physics.*

Berglund, Per, Ph.D., University of Texas, Austin, 1993. *Cosmology & String Theory, High Energy Physics, Particles and Fields, Theoretical Physics.* String theory.

Calarco, John R., Ph.D., University of Illinois at Chicago, 1969. *Accelerator, Nuclear Physics.* Investigating the structure of nucleons and light nuclei using the scattering of electrons or the absorption of gamma rays as the primary probe.

Chandran, Benjamin D. G., Ph.D., Princeton University, 1997. Member of Space Science Center; member of the Integrated Applied Mathematics Program. *Applied Mathematics, Astrophysics, Atmosphere, Space Physics, Cosmic Rays, Plasma and Fusion, Solar Physics, Theoretical Physics.* Studies of plasma turbulence; the role of turbulence in the solar corona and other astrophysical settings; the evolution of baryonic matter in clusters of galaxies; cosmic-ray propagation; particle acceleration in solar flares; the origin of astrophysical magnetic fields; space mission involvement includes Solar Probe Plus.

Echt, Olof, Ph.D., University of Konstanz, 1979. Member of the Materials Science Program. *Chemical Physics, Condensed Matter Physics, Materials Science, Metallurgy.* Pulsed-laser deposition of thin films; photophysics of fullerenes; implantation of ions into fullerenes; ion-molecule reactions in atomic clusters and doped helium nanodroplets.

Hersman, F. William, Ph.D., Massachusetts Institute of Technology, 1982. *Medical, Health Physics, Nuclear Physics, Optics.* Development of high-power wavelength-locked lasers for optical pumping with applications in defense technology, nuclear physics, and diagnostic medicine; investigation of the properties and utility of hyperpolarized xenon, particularly as a contrast agent in magnetic resonance imaging.

Kistler, Lynn M., Ph.D., University of Maryland, 1987. Director of the Space Science Center. *Atmosphere, Space Physics, Cosmic Rays, Solar Physics.* Magnetospheric physics, with specific emphasis on the sources, transport, and acceleration of magnetospheric particle populations, including the design, fabrication, and testing of state-of-the-art instrumentation for spacecraft and the analysis of the data collected by these instruments; space mission involvement includes FAST, Equator-S, CLUSTER, ACE, STEREO, Solar Orbiter, and SCOPE.

Lee, Martin A., Ph.D., University of Chicago, 1971. Member of Space Science Center; member of the Integrated Applied Mathematics Program. *Applied Mathematics, Astrophysics, Atmosphere, Space Physics, Cosmic Rays, Computational Physics, Solar Physics, Theoretical Physics.* Theoretical space physics, astrophysics, and plasma physics, including heliospheric plasmas, solar cosmic rays, and shock acceleration processes; space mission involvement includes: SOHO.

Möbius, Eberhard, Ph.D., Ruhr University Bochum, 1977. Member of Space Science Center. *Astrophysics, Atmosphere, Space Physics, Cosmic Rays, Solar Physics.* Acceleration of ions in the earth's magnetosphere, in interplanetary space, and in solar flares; interaction of interstellar gas with the solar wind and the study of the local interstellar medium; space mission involvement includes AMPTE, FAST, ACE, CLUSTER, Equator-S, STEREO, and IBEX.

McConnell, Mark L., Ph.D., University of New Hampshire, 1987. Chair of the Physics Department; member of Space Science Center. *Astrophysics, Atmosphere, Space Physics, Cosmic Rays, Solar Physics.* Experimental x-ray and gamma-ray astronomy using balloon and satellite platforms; x-ray and gamma-ray polarimetry of gamma-ray bursts and solar flares; space mission involvement includes CGRO, RHESSI, INTEGRAL, FERMI, LRO, and various balloon missions.

Pohl, Karsten, Ph.D., University of Pennsylvania, 1997. Member of the Materials Science Program. *Condensed Matter Physics, Nano Science and Technology, Surface Physics.* Experimental condensed matter physics and materials science focused on the study of the interplay of electronic, vibrational, and structural surface properties at the atomic scale.

Raeder, Joachim, Ph.D., University of Köln, 1989. Member of Space Science Center; member of the Integrated Applied Mathematics Program. *Applied Mathematics, Atmosphere, Space Physics, Cosmic Rays, Computational Physics, Theoretical Physics.* Space physics; space weather solar-terrestrial relationships; plasmas and magnetic fields in space; solar wind-magnetosphere-ionosphere-thermosphere coupling; geomagnetic activity; geomagnetic storms and substorms; large-scale modeling of magnetospheres; data assimilation; cometary physics; computational fluid dynamics; numerical methods; high-performance computing; space mission involvement includes GIOTTO and THEMIS.

Ryan, James M., Ph.D., University of California, Riverside, 1978. Member of Space Science Center. *Astrophysics, Atmosphere, Space Physics, Cosmic Rays, Solar Physics.* Terrestrial, solar, and astrophysical cosmic rays, neutrons, and gamma rays; space mission involvement includes SMM, CGRO, RHESSI, and various balloon experiments.

Spence, Harlan, Ph.D., University of California, Los Angeles, 1989. Director, Institute for the Study of Earth, Oceans and Space (EOS); member of Space Science Center. *Astrophysics, Atmosphere, Space Physics, Cosmic Rays, Planetary Science, Solar Physics.* Theoretical and experimental space plasma physics; cosmic rays and radiation belt processes; heliospheric, planetary magnetospheric, lunar, and auroral physics; space mission involvement includes POLAR, LRO, IBEX, TWINS, RBSP, FIREBIRD, and MMS.

Torbert, Roy B., Ph.D., University of California, Berkeley, 1979. Director of SwRI-EOS. *Atmosphere, Space Physics, Cosmic Rays.* Space plasma physics; physics of magnetospheres, aurora, and early solar system formation; space mission involvement includes WIND, POLAR, CLUSTER, Equator-S, MMS, and RBSP.

Associate Professor

Beane, Silas R., Ph.D., University of Texas, Austin, 1994. *High Energy Physics, Nuclear Physics, Particles and Fields, Theoretical Physics.* Nuclear and particle physics from first principles with lattice QCD and effective field theory methods; theory of atomic gases.

Connell, James, Ph.D., Washington University, 1988. Member of Space Science Center. *Astrophysics, Atmosphere, Space Physics, Cosmic Rays, Solar Physics.* The measurement of energetic particle radiation in space, including galactic cosmic rays, solar energetic particles, and the anomalous cosmic rays; space mission involvement includes Ulysses and GOES-R.

Holtrop, Maurik, Ph.D., Massachusetts Institute of Technology, 1995. *Accelerator, Nuclear Physics.* Experimental nuclear and particle physics using electron and photon scattering experiments at the Thomas Jefferson National Accelerator Facility.

Lessard, Mark, Ph.D., Dartmouth College, 1997. Member of Space Science Center. *Atmosphere, Space Physics, Cosmic Rays.* Experimental space plasma physics (rocket-borne and ground-based instruments), including measurements and analysis of auroral phenomena; space mission involvement includes various sounding rocket missions.

Meredith, Dawn C., Ph.D., California Institute of Technology, 1987. *Physics and other Science Education.* Physics education.

Schwadron, Nathan, Ph.D., University of Michigan, 1996. Member of Space Science Center. Heliospheric phenomena related to the solar wind, the heliospheric magnetic field, pickup ions, cometary x-rays, energetic particles, and cosmic rays; space mission involvement includes Ulysses, New Horizons, IBEX, and LRO.

Assistant Professor

Germaschewski, Kai, Ph.D., Heinrich-Heine University, 2001. Member of Space Science Center; member of the Integrated Applied Mathematics Program. *Applied Mathematics, Atmosphere, Space Physics, Cosmic Rays, Computational Physics, Plasma and Fusion.* Space physics; plasma physics; numerical simulations.

Slifer, Karl, Ph.D., Temple University, 2004. *Accelerator, Nuclear Physics.* Experimental nuclear and particle physics using electron- and photon-scattering experiments at the Thomas Jefferson National Accelerator Facility.

Solvignon, Patricia H., Ph.D., Temple University, 2006. *Accelerator, Nuclear Physics.* Experimental nuclear and particle physics using electron and photon scattering at Thomas Jefferson National Accelerator Laboratory; primarily focused on the structure of nuclei and the properties of the nuclear force.

Tang, Jian-Ming, Ph.D., University of Washington, 2001. Member of the Materials Science Program. *Condensed Matter Physics, Materials Science, Metallurgy, Solid State Physics, Surface Physics, Theoretical Physics.* Development of theoretical techniques to investigate the electronic properties of nanostructures near interfaces.

Professor Emeritus

Chupp, Edward L., Ph.D., University of California, Berkeley, 1954. Member of Space Science Center. *Astrophysics, Atmosphere, Space Physics, Cosmic Rays, Solar Physics.* Astrophysics (gamma-ray astronomy); solar flare particle acceleration; gamma-ray detector development; space mission involvement includes OSO-7 and SMM.

Dawson, John, Stanford University, 1962. *Nuclear Physics, Theoretical Physics.* Theoretical nuclear models; electron scattering; heavy ion collisions.

Forbes, Terry, Ph.D., University of Colorado Boulder, 1978. Member of Space Science Center. *Astrophysics, Atmosphere, Space Physics, Cosmic Rays, Solar Physics, Theoretical Physics.* Theories of solar flares, including magnetic reconnection; space mission involvement includes YOHKOH, SOHO, TRACE, and HINODE.

Heisenberg, Jochen, Ph.D., University of Hamburg, 1966. *Nuclear Physics, Theoretical Physics.* Theoretical nuclear physics.

Hollweg, Joseph, Ph.D., Massachusetts Institute of Technology, 1968. *Atmosphere, Space Physics, Cosmic Rays, Solar Physics.* Dynamics of the solar atmosphere and solar wind; waves in plasmas.

Kaufmann, Richard, Ph.D., Yale University, 1960. *Atmosphere, Space Physics, Cosmic Rays, Theoretical Physics.*

McKibben, R. Bruce, Ph.D., University of Chicago, 1972. Member of Space Science Center, Astrophysics, Atmosphere, Space Physics, Cosmic Rays, Solar Physics. *Astrophysics, Atmosphere, Space Physics, Cosmic Rays, Solar Physics.* Observational studies of galactic cosmic rays, solar energetic particles, and particles accelerated in interplanetary space or in planetary magnetospheres, using measurements from heliospheric and near-Earth spacecraft; space mission involvement includes Pioneer 10, Pioneer 11, and Ulysses.

Research Professor

Galvin, Antoinette, Ph.D., University of Maryland, 1982. Member of Space Science Center; Director of the NASA EPSCoR program in New Hampshire. *Atmosphere, Space Physics, Cosmic Rays, Solar Physics.* Studies of heliophysics particle populations (solar wind, magnetosphere, pickup ions, accelerated suprathermals), including their origins, evolution, and interactions in the inner heliosphere, and solar wind interactions with planetary bodies (space weather); the design, fabrication, and operation of state-of-the-art particle composition experiments for space missions; space mission involvement includes ISEE-1, ISEE-3, Ulysses, Geotail, SOHO, Wind, ACE, STEREO, Solar Orbiter, and SCOPE.

Isenberg, Philip A., Ph.D., University of Chicago, 1977. Member of Space Science Center. *Atmosphere, Space Physics, Cosmic Rays, Solar Physics, Theoretical Physics.* Theoretical space plasma physics; solar wind acceleration; solar wind interactions with neutral particles from comets and the interstellar medium; solar flares.

Smith, Charles, Ph.D., College of William and Mary, 1981. Member of Space Science Center. *Atmosphere, Space Physics, Cosmic Rays, Solar Physics.* A wide range of plasma physics investigations centered on the solar wind, its origin and evolution, particle acceleration, and the propagation of cosmic rays through the heliosphere; space mission involvement includes ISEE, Pioneer-Venus, Voyager, WIND, Ulysses, ACE, and RBSP.

Vasquez, Bernard, Ph.D., University of Maryland, 1992. Member of Space Science Center. *Atmosphere, Space Physics, Cosmic Rays.* Studies of the solar wind; waves; discontinuities; ion kinetics; and numerical simulations.

Research Associate Professor

Chen, Li-Jen, Ph.D., University of Washington, 2002. Member of Space Science Center. *Atmosphere, Space Physics, Cosmic Rays, Theoretical Physics.* Magnetic reconnection (including particle acceleration and structures of the diffusion region that affect the energy conversion rate and energy partition); dynamics of electrostatic structures and how they influence plasma bulk properties at current layers; propagation of dispersive Alfven waves and field line resonances; space mission involvement includes MMS.

Farrugia, Charles, Ph.D., University of Bern, 1984. Member of Space Science Center. *Atmosphere, Space Physics, Cosmic Rays, Solar Physics.* Magnetohydrodynamic flow of the solar wind around the magnetosphere; processes at the magnetopause and its boundary layers; magnetosphere-ionosphere coupling; interplanetary planar structures and magnetic clouds; the interaction of the solar wind with the magnetosphere; prediction of strong geomagnetic disturbances from the inner heliosphere; space mission involvement includes STEREO.

Kucharek, Harald, Ph.D., Technical University of Munich, 1989. Member of Space Science Center. *Atmosphere, Space Physics, Cosmic Rays, Solar Physics.* Kinetic numerical simulation (MHD, hybrid, and full-particle simulations) of physical processes in collisionless plasmas, in the earth's magnetosphere, and the heliosphere; solar wind composition and particle acceleration at the earth's bow shock and at interplanetary structures such as coronal mass ejections (CMEs) and corotating interaction regions (CIRs); space mission involvement includes SOHO, ACE, CLUSTER, IBEX, Solar Probe Plus, and SCOPE.

Lopate, Clifford, Ph.D., University of Chicago, 1989. Member of Space Science Center. *Astrophysics, Atmosphere, Space Physics, Cosmic Rays, Solar Physics.* Studies of heliospheric energetic particles in the 1-10,000 MeV energy range (the sources of these particles are cosmic rays, anomalous components, solar particle events, and planetary magnetospheres); space mission involvement includes Pioneer-10, Pioneer-11, IMP-8, and GOES-R.

Research Assistant Professor

Bloser, Peter F., Ph.D., Harvard University, 2000. Member of Space Science Center. *Astrophysics, Atmosphere, Space Physics, Cosmic Rays, Solar Physics.* Experimental gamma-ray astronomy and solar physics; space mission involvement includes various balloon experiments.

Bravar, Ulisse, Ph.D., New Mexico State University, 2001. Member of Space Science Center. *Astrophysics, Atmosphere, Space Physics, Cosmic Rays, Solar Physics.* Experimental high-energy astrophysics; space mission involvement includes PAMELA and various balloon experiments.

Ebrahimi, Fatima, Ph.D., University of Wisconsin-Madison, 2003. Member of Space Science Center. *Applied Mathematics, Astrophysics, Atmosphere, Space Physics, Cosmic Rays, Computational Physics, Plasma and Fusion.* Momentum transport from current-driven reconnection and flow-driven magnetorotational instability with application to both fusion and astrophysical plasmas.

Lugaz, Noe, Ph.D., University of Michigan, 2007. Member of the Space Science Center. *Atmosphere, Space Physics, Cosmic Rays, Solar Physics.* Solar-terrestrial physics with focus on solar eruptions and coronal mass ejections from the sun to the earth: initiation, propagation, interaction, and effect on Earth's magnetosphere; combining large, parallel numerical simulations; the analysis of remote-sensing observations from coronagraphs and heliospheric imagers and space plasma measurements.

Mattingly, David, Ph.D., University of Maryland, 2003. *Relativity & Gravitation, Theoretical Physics.* Development of new methods to test fundamental symmetries in general relativity and hypotheses about quantum gravity.

Affiliate Assistant Professor

Ruset, Iulian, Ph.D., University of New Hampshire, 2005. *Medical, Health Physics.* Development of instrumentation and techniques for medical imaging with polarized gases.

DEPARTMENTAL RESEARCH SPECIALTIES AND STAFF

Theoretical

Condensed Matter Physics. Electronic structure calculations of nanostructures; computational modeling of scanning tunneling microscopy; spin-dependent phenomena in semiconductors; quantum size effects in thin metallic films. Tang.

High Energy Physics. Theoretical research in quantum chromodynamics (QCD), field theory, string theory, and quantum gravity. Beane, Berglund, Dawson, Mattingly.

Nuclear Physics. Studies of theoretical nuclear models, electron scattering, heavy ion collisions, and condensed ultra-cold atomic systems. Dawson, Heisenberg.

Space Plasma Theory. The research activities of Space Plasma Theory Group encompass a broad range of subfields, in-cluding theoretical plasma physics, solar and heliospheric physics, magnetospheric physics, and plasma astrophysics. Chandran, Chen, Ebrahimi, Forbes, Germaschewski, Hollweg, Isenberg, Kaufmann, Lee, Lugaz, Raeder, Vasquez.

Experimental

Condensed Matter Physics. Research involving scanning tunneling microscopy and electron spectroscopies, thin-film deposition, and high-resolution mass spectrometry to investigate electronic and structural properties of nanoscale materials. Echt, Pohl.

Experimental Space Plasma. Investigations of the earth's environment in the solar system looking at space as a laboratory for plasma physics; satellite and sounding rocket investigations of the solar-terrestrial radiation environment. Active satellite programs include Cluster, ACE, STEREO, IBEX, and the Van Allen Probes (RBSP); upcoming missions for which University of New Hampshire is building instruments include FIREBIRD (launch date, 2013), MMS (launch date, 2014), and GOES-R (launch date, 2014). Farrugia, Galvin, Kistler, Kucharek, Lessard, Möbius, Schwadron, Smith, Spence, Torbert.

High Energy Astrophysics. Research involving high-energy radiations addressing a wide variety of astrophysical problems, including high energy emissions from solar flares, gamma-ray bursts, x-ray binaries, and pulsars; studies are conducted using x-ray detectors and gamma-ray detectors placed on suborbital balloon platforms and on orbital satellites (such as CGRO); recent work has also included the application of radiation detector technology to homeland security issues. Bloser, Bravar, Chupp, McConnell, Ryan.

Medical Imaging. Investigation of spin exchange optical pumping (SEOP) to identify new technologies for producing nuclear polarized gases; originally, these efforts were motivated by applications in fundamental physics, but more recently, these efforts are motivated by opportunities to apply hyperpolarized gas as a diagnostic tracer of inhaled lung gas with magnetic resonance imaging. Hersman, Ruset.

Nuclear Physics. Using electron- and photon-scattering experiments (conducted at various accelerator facilities, including the Jefferson Laboratory) to improve our understanding of the atomic nucleus. Calarco, Holtrop, Slifer, Solvignon.

Optics. Development of high-power laser systems for ballistic missile defense. Hersman.

Physics Education Research. Research to determine how students learn and developing teaching methodologies to improve student learning. Meredith.

Space Radiation. Research in high energy space particle radiation, including ions and electrons; space weather studies; measurements of cosmic ray and solar energetic particles, including detailed element and isotopic composition; recent work has included the High Energy Telescope (HET) for the Ulysses mission; currently providing energetic heavy ion sensors for the GOES-R mission; research also includes instrument development work in preparation for future space flight opportunities. Connell, Lopate, McKibben.

View additional information about this department at
www.gradschoolshopper.com

NEW JERSEY INSTITUTE OF TECHNOLOGY

DEPARTMENT OF PHYSICS

Newark, New Jersey 07102
http://physics.njit.edu/

General University Information
President: Joel Bloom
Dean of Graduate School: Sotirios Ziavras
University website: http://www.njit.edu/graduatestudies/
Control: Public
Setting: Urban
Total Faculty: 489
Total Graduate Faculty: 170
Total number of Students: 9,944
Total number of Graduate Students: 2,833

Department Information
Department Chairman: Andrei Sirenko, Chair
Department Contact: Andrei Sirenko, Professor & Chair
  Total full-time faculty: 20
  Total number of full-time equivalent positions: 20
  Full-Time Graduate Students: 16
  First-Year Graduate Students: 6
  Female First-Year Students: 2
  Total Post Doctorates: 4

Department Address
Tiernan Building # 463
161 Warren Street
Newark, NJ 07102
Phone: (973) 596-3562
Fax: (973) 596-5794
E-mail: andrei.sirenko@njit.edu
Website: http://physics.njit.edu/

ADMISSIONS

Admission Contact Information
Address admission inquiries to: Admissions, New Jersey Institute
  of Technology, Newark, NJ 07102-1982
Phone: 1-800-925-NJIT or 973-596-3300
Admissions website: http://www.njit.edu/admissions/
  contactadmissions.php

Application deadlines
Fall admission:
U.S. students: May 1   *Int'l. students*: May 1
Spring admission:
U.S. students: November 15   *Int'l. students*: October 15

Application fee
U.S. students: $65   *Int'l. students*: $65
The Department of Physics also runs the Interdisciplinary Pro-
  gram in Materials Science and Engineering at NJIT.

Admissions information
For Fall of 2013:
  Number of applicants: 35
  Number admitted: 10
  Number enrolled: 6

Admission requirements
Bachelor's degree requirements: Bachelor's degree in physics
  is required.
Minimum undergraduate GPA: 3.5

GRE requirements
The GRE is required.
  Quantitative score: 700
  Verbal score: 400

Advanced GRE requirements
The Advanced GRE is not required.

TOEFL requirements
The TOEFL exam is required for students from non-English-
  speaking countries.
  PBT score: 550
  iBT score: 80

Other admissions information
Additional requirements: Publications help in the admissions process.
Undergraduate preparation assumed: Symon, Mechanics; Scott,
  Electromagnetism; Mandl, Quantum Mechanics; Reif, Ther-
  modynamics; Kittel, Solid State Physics.

TUITION

Tuition year 2013-14:
Tuition for in-state residents
  Full-time students: $17,384 annual
  Part-time students: $945 per credit
Tuition for out-of-state residents
  Full-time students: $25,404 annual
  Part-time students: $1,341 per credit
Additional fees apply; tuition based on 9 credits per semester;
  see http://www.njit.edu/bursar/tuition/grad-tuition.php.
Credit hours per semester to be considered full-time: 9
Deferred tuition plan: Yes
Health insurance: Available at the cost of 715 per year.
Other academic fees: Per credit; see http://www.njit.edu/
  healthservices/health-insurance.php.
Academic term: Semester
Number of first-year students who received full tuition waivers: 3

Teaching Assistants, Research Assistants, and Fellowships
Number of first-year
  Teaching Assistants: 4
  Fellowship students: 2
Average stipend per academic year
  Teaching Assistant: $18,315
  Research Assistant: $26,000
  Fellowship student: $24,000

FINANCIAL AID

Application deadlines
Fall admission:
U.S. students: January 15   *Int'l. students*: January 15
Spring admission:
U.S. students: October 15   *Int'l. students*: October 15

Loans

Loans are available for U.S. students.

Loans are not available for international students.

GAPSFAS application required: No

FAFSA application required: Yes

For further information

Address financial aid inquiries to: Admissions, New Jersey Institute of Technology, Newark, NJ 07102-1982.

Phone: (973) 596-3300

Financial aid website: http://www.njit.edu/financialaid/

HOUSING

Availability of on-campus housing

Single students: Yes

Married students: No

For further information

Address housing inquiries to: Admissions Office, New Jersey Institute of Technology, Newark, NJ 07102-1982.

Phone: (973) 596-3039

E-mail: reslife@njit.edu

Housing aid website: http://www.njit.edu/reslife/

Table A—Faculty, Enrollments, and Degrees Granted

Research Specialty	2013-14 Faculty	Enrollment Fall 2013 Master's	Enrollment Fall 2013 Doctorate	Number of Degrees Granted 2013-14 (2002–07) Master's	Number of Degrees Granted 2013-14 (2002–07) Terminal Master's	Number of Degrees Granted 2013-14 (2002–07) Doctorate
Applied Physics	4	–	–	5(-)	–	7(-)
Astrophysics	6	1	6	1(-)	–	2(7)
Atomic, Molecular, & Optical Physics	2	–	–	–	–	1(-)
Biophysics	3	–	2	1(-)	–	–
Condensed Matter Physics	2	4	13	1(-)	–	2(8)
Materials Science, Metallurgy	3	–	7	5(-)	–	6(14)
Total	20	5	28	–(5)	–	6(45)
Full-time Grad. Stud.	–	–	28	–	–	–
First-year Grad. Stud.	–	–	5	–	–	–

GRADUATE DEGREE REQUIREMENTS

Master's: The interdisciplinary NJIT-Rutgers (Newark) joint M.S. degree in applied physics requires 30 credits; 24 credits are coursework, of which 18 credits are physics courses (including mathematical physics or applied mathematics) and 6 credits are electives. Four graduate physics courses, Classical Mechanics, Classical Electrodynamics I, Quantum Mechanics I, and Statistical Mechanics, are mandatory. Six credits are thesis research; with the approval of the academic advisor, the student can choose a 3-credit project plus an additional 3-credit course to replace the 6-credit thesis.

Doctorate: For entering students with B.S. or B.A. degrees, the interdisciplinary NJIT-Rutgers (Newark) joint Ph.D. degree in applied physics requires 75 credits (above the 600 level), of which 39 credits are coursework and 36 credits are dissertation research. Of the coursework, 24 credits are physics courses (including mathematical physics or applied mathematics) and 15 credits are electives. Of the 24 credits of physics courses, Classical Mechanics, 621 Classical Electrodynamics I and II, 631 Quantum Mechanics I and II, and Statistical Mechanics are mandatory For entering students with M.S. or M.A. degrees, the Joint NJIT-Rutgers (Newark) Ph.D. degree in applied physics requires 54 credits (above the 600 level), of which 18 credits are coursework and 36 credits are dissertation research. Of the coursework, 9 credits are physics courses (including mathematical physics or applied mathematics) and 9 credits are electives.

Thesis: Thesis is a requirement for all doctoral students.

SPECIAL EQUIPMENT, FACILITIES, OR PROGRAMS

Center for Solar-Terrestrial Research; Big Bear Solar Observatory (BBSO); Apollo CdTe Solar Cell Research Center; Space Weather Research Laboratory; Microelectronics Research Center with class 10 clean room silicon IC process facility and device research laboratories; Owens Valley Solar Array (OVSA); THz spectroscopy laboratory; laser spectroscopy laboratory; surface science laboratory; bio-sensor laboratory. Interdisciplinary applied physics research is carried out in collaboration with electrical engineering, chemistry, biomedical engineering, and biological sciences faculty and with the University of Medicine and Dentistry of New Jersey (UMDNJ). There is also extensive cooperative research with the National Solar Observatory, Brookhaven National Laboratory, National Renewable Energy Laboratory, US Army Research Lab, and other industrial and federal research laboratories. The Department of Physics at NJIT also operates the Interdisciplinary Program in Materials Science and Engineering. For information on this program, contact Prof. Ravindra at ravindra@adm.njit.edu or nmravindra@gmail.com.

Table B—Separately Budgeted Research Expenditures by Source of Support

Source of Support	Departmental Research	Physics-related Research Outside Department
Federal government	$15,000,000	$500,000
State/local government		
Non-profit organizations		
Business and industry	$500,000	
Other	$227,529	
Total	$15,727,529	$500,000

Table C—Separately Budgeted Research Expenditures by Research Specialty

Research Specialty	No. of Grants	Expenditures ($)
Applied Physics	6	$876,652
Biophysics	1	$50,944
Condensed Matter Physics	5	$832,274
Optics	2	$259,982
Solar Physics	40	$15,207,677
Total	54	$17,227,529

FACULTY

Professor

Chin, Ken K., Ph.D., Stanford University, 1986. Director, Apollo Center for CdTe Solar Cell Research. *Applied Physics, Condensed Matter Physics, Energy Sources & Environment, Materials Science, Metallurgy, Solid State Physics.* Semiconductors; solar cells; solid-state physics; power transmission.

Federici, John, Ph.D., Princeton University, 1989. Distinguished Professor/Professor Rank II. *Applied Physics, Biophysics, Materials Science, Metallurgy, Optics.* Terahertz spectroscopy; sensors; biophotonics.

Gary, Dale E., Ph.D., University of Colorado, 1982. Distinguished Professor/Professor Rank II. *Applied Physics, Astron-*

omy, *Astrophysics, Planetary Science, Solar Physics*. Radio solar physics.

Gatley, Ian, Ph.D., California Institute of Technology, 1978. *Astronomy, Astrophysics, Atmosphere, Space Physics, Cosmic Rays*. Infrared astronomy.

Gerrard, Andrew, Ph.D., Pennsylvania State University, 2002. *Astronomy, Astrophysics, Atmosphere, Space Physics, Cosmic Rays*. Upper atmospheric research.

Goode, Philip R., Ph.D., Rutgers University, 1968. Distinguished Professor/Professor Rank II. *Astronomy, Astrophysics, Atmosphere, Space Physics, Cosmic Rays, Solar Physics*. Astrophysics; solar physics.

Levy, Roland, Ph.D., Columbia University, 1973. Distinguished Professor/Professor Rank II. *Applied Physics, Materials Science, Metallurgy, Solid State Physics*. CVD; PVD; materials synthesis.

Murnick, E. Daniel, Ph.D., Massachusetts Institute of Technology, 1966. Distinguished Professor/Professor Rank II. *Atomic, Molecular, & Optical Physics, Biophysics, Medical, Health Physics*. Laser spectroscopy and applied physics.

Ravindra, N. M., Ph.D., Indian Institute of Technology, Roorkee, 1982. Program Director-Interdisciplinary Program in Materials Science and Engineering. *Applied Physics, Condensed Matter Physics, Materials Science, Metallurgy, Nano Science and Technology, Physics and other Science Education, Solid State Physics*.

Sirenko, Andrei, Ph.D., A. E. Ioffe Physical Technical Institute, Russia, 1993. Chair of the Physics Department. *Optics, Solid State Physics*. Optics; materials and device physics.

Spruch, Grace, Ph.D., New York University, 1955. *Applied Physics, Condensed Matter Physics*. Experimental condensed-matter physics.

Thomas, Gordon, Ph.D., University of Rochester, 1972. *Biophysics, Optics*. Optics; biophysics.

Tyson, Trevor, Ph.D., Stanford University, 1993. Distinguished Professor/Professor Rank II. *Condensed Matter Physics, Solid State Physics*. Condensed-matter physics.

Wang, Haimin, Ph.D., California Institute of Technology, 1988. Distinguished Professor/Professor Rank II. *Astronomy, Astrophysics, Atmosphere, Space Physics, Cosmic Rays, Solar Physics*. Solar physics; space weather.

Wu, Zhen, Ph.D., Columbia University, 1984. *Atomic, Molecular, & Optical Physics*. Atomic and molecular physics; laser spectroscopy and surface science.

Associate Professor

Ahn, Keun, Ph.D., Johns Hopkins University, 2000. *Computational Physics, Condensed Matter Physics, Solid State Physics*. Condensed-matter physics.

Cao, Wenda, Ph.D., Goettingen (Göttingen), Germany, 2001. *Optics, Solar Physics*. Solar physics.

Prodan, Camelia, Ph.D., University of Houston, 2005. *Biophysics*. Biophysics.

Russo, L. O., Ph.D., New Jersey Institute of Technology, 1975. *Solid State Physics*. Solid-state physics.

Schaden, Martin, Ph.D., University of Vienna, Austria, 1982. *Quantum Foundations*. Quantum field theory.

Zhou, Tao, Ph.D., Max Plank Institute for Solid State Research, 2004. *Applied Physics, Optics*. Optical spectroscopy.

Assistant Professor

Dias, Cristiano, Ph.D., McGill University. Computational Biophysics.

Research Professor

Abramenko, Valentyna, Ph.D., Ioffe Institute for Physics and Technology, St. Petersburg, Russia, 1990. Solar magnetic fields.

Farrow, Reginald, Ph.D., Stevens Institute of Technology, 1984. *Applied Physics, Nano Science and Technology, Solid State Physics*. Solid-state semiconductors.

Fiory, Anthony, Ph.D., Rutgers University, 1968. *Applied Physics, Solid State Physics*. Solid-state physics; semiconductors.

Jing, Ju, Ph.D., New Jersey Institute of Technology, 2005. Solar magnetic fields; solar activity.

Lanzerotti, Louis J., Ph.D., Harvard University, 1965. *Astronomy, Astrophysics, Atmosphere, Space Physics, Cosmic Rays*. Geophysics and space plasma physics.

Liu, Chang, Ph.D., New Jersey Institute of Technology, 2007. Solar flares and coronal mass ejections; solar magnetic fields.

Rimmele, Thomas, Ph.D., University of Freiburg, Germany, 1993.

Varsik, John, Ph.D., University of Hawaii, 1987. Solor polar magnetic fields.

Yurchyshyn, Vasyl, Ph.D., Astronomical Observatory, Kiev, Ukraine, 1998. Solar flares; solar coronal mass ejecta.

Research Associate Professor

Fleishman, Gregory, Ph.D., Ioffe Institute for Physics and Technology, St. Petersburg, Russia, 1998. *Astrophysics*. Nonthermal electromagnetic emission in structural astrophysical medium.

Research Assistant Professor

Nita, Gelu, Ph.D., New Jersey Institute of Technology, 2004. *Solar Physics*. Solar microwave radiation.

Visiting Research Professor

Deng, Na, Ph.D., New Jersey Institute of Technology, 2007. Space weather; solar-terrestrial research.

Sufian, Abedrabbo M., Ph.D., New Jersey Institute of Technology, 1998. *Applied Physics, Solid State Physics*. Solid-state physics; semiconductors.

Lecturer

Georgiou, George E., Ph.D., Columbia University, 1980. Semiconductors.

Gokce, Oktay, Ph.D., Montana State University, 1981. Chemical engineering.

Janow, Richard, Ph.D., City University of New York, 1977. *Condensed Matter Physics*.

Jerez, Andrez, Ph.D., Rutgers University, 1996. *Condensed Matter Physics*.

Maljian, Libarid A., M.S., Rutgers University, 2000. *Astronomy, Astrophysics*.

Opyrchal, Halina, Ph.D., Institute of Low Temperature and Structure Research, Polish Academy of Sciences, 1975. *Solid State Physics*.

Piatek, Slawomir, Ph.D., Rutgers University, 1994. *Astronomy, Astrophysics*. Solar physics.

Shneidman, Vitaly, Ph.D., Academy of Sciences, Ukraine. *Solid State Physics, Statistical & Thermal Physics*. Computational physics.

Laboratory Director

Maeng, Sung, Ph.D., New Jersey Institute of Technology. Materials Science and Engineering., 2005. Laboratory personnel.

Laboratory Coordinator

Zhou, Xuechong, Ph.D., City University of New York, 2005. Laboratory personnel. *Optics*. Semiconductors and optics.

DEPARTMENTAL RESEARCH SPECIALTIES AND STAFF

Experimental

Atomic, Molecular, & Optical Physics. Atomic, molecular, and applied laser physics. Two postdoctoral fellows. Federici, Murnick, Sirenko, Wu, Tao Zhou.

Biophysics. Biophysics-diagnostics and sensors. Four postdoctoral fellows. Federici, Murnick, Prodan, Thomas.

Materials Science, Metallurgy. Solid-state physics; materials science. Six postdoctoral fellows. Chin, Federici, Levy, Ravindra, Russo, Tyson, Wu.

Other. Semiconductors; solar cells. Three postdoctoral fellows. Chin, Levy, Ravindra.

Solar Physics. Solar Terrestrial Physics. Abramenko, Cao, Deng, Fleishman, Gary, Gerrard, Goode, Jing, Lanzerotti, Liu, Nita, Rimmele, Varsik, Wang, Yurchyshyn.

View additional information about this department at
www.gradschoolshopper.com

PRINCETON UNIVERSITY

DEPARTMENT OF ASTROPHYSICAL SCIENCES PLASMA PHYSICS SECTION

Princeton, New Jersey 08543
www.princeton.edu/plasma/

General University Information

President: Christopher L. Eisgruber
Dean of Graduate School: William B. Russel
University website: http://www.princeton.edu
Control: Private
Setting: Suburban
Total Faculty: 1,184
Total Graduate Faculty: Not separated
Total number of Students: 8,081
Total number of Graduate Students: 2,672

Department Information

Department Chairman: Nathaniel J. Fisch, Chair
Department Contact: Barbara Sarfaty, Graduate Program Administrator
  Total full-time faculty: 17
  Full-Time Graduate Students: 40
  First-Year Graduate Students: 7

Department Address

Forrestal Campus
P.O. Box 451, MS-30
Princeton, NJ 08543
Phone: (609) 243-2489 (office)
Fax: (609) 243-2662
E-mail: bsarfaty@pppl.gov
Website: www.princeton.edu/plasma/

ADMISSIONS

Admission Contact Information

Address admission inquiries to: Office of Graduate Admission, Princeton University, P.O. Box 270, Princeton, NJ 08544-0270
Phone: (609) 258-3034
E-mail: gs@princeton.edu
Admissions website: http://www.princeton.edu/gradschool/admission/

Application deadlines

Fall admission:
U.S. students: December 15   *Int'l. students*: December 15

Application fee

U.S. students: $90   *Int'l. students*: $90

Admissions information

For Fall of 2013:
  Number of applicants: 52
  Number admitted: 9
  Number enrolled: 6

Admission requirements

Bachelor's degree requirements: A Bachelor's degree, preferably in physics, mathematics, or engineering, is required.

GRE requirements

The GRE is required.
No minimum score required.

Advanced GRE requirements

The Advanced GRE is required.
Advanced GRE subject test in Physics required.

TOEFL requirements

The TOEFL exam is required for students from non-English-speaking countries.
  PBT score: 600
  iBT score: 108

Other admissions information

Additional requirements: The average GRE scores (in % for Fall 2013 entering class) for admissions were Verbal–91%, Quantitative–96%, writing-68%. The average GRE Advanced score for admissions was 89% in physics.
No minimum score required.

Undergraduate preparation assumed: A sound undergraduate education in physics and mathematics is assumed, including courses in electricity and magnetism, classical mechanics, thermodynamics and statistical mechanics, quantum mechanics, differential equations, and complex variables, is assumed.

TUITION

Tuition year 2013–14:
  Full-time students: $42,070 annual
Includes student health plan.
Deferred tuition plan: No
Health insurance: Yes, 0 (included in tuition).
Other academic fees: None

Academic term: Semester
Number of first-year students who received full tuition waivers: 6

Teaching Assistants, Research Assistants, and Fellowships
Number of first-year
Fellowship students: 6
Average stipend per academic year
Research Assistant: $25,170

FINANCIAL AID

Application deadlines
Fall admission:
U.S. students: December 15   *Int'l. students*: December 15

Loans
Loans are available for U.S. students.
Loans are available for international students.
GAPSFAS application required: No
FAFSA application required: No

For further information
Address financial aid inquiries to: Office of Graduate Admission, Princeton University, P.O. Box 270, Princeton, NJ 08544-0270.
Phone: (609) 258-3034
E-mail: gs@princeton.edu
Financial aid website: http://www.princeton.edu/gradschool/financial/

HOUSING

Availability of on-campus housing
Single students: Yes
Married students: Yes

For further information
Address housing inquiries to: Office of Graduate Admission, Princeton University, P.O. Box 270, Princeton, NJ 08544-0270.
Phone: (609) 258-3034
E-mail: gradhsg@princeton.edu
Housing aid website: http://www.princeton.edu/facilities/housing/graduate_info/

Table A—Faculty, Enrollments, and Degrees Granted

Research Specialty	2012–13 Faculty	Enrollment Fall 2012		Number of Degrees Granted 2012–13 (2005–11)		
		Mas-ter's	Doc-torate	Mas-ter's	Terminal Master's	Doc-torate
Plasma and Fusion	17	–	40	–	–	5(48)
Total	17	–	40	–	–	5(48)
Full-time Grad. Stud.	–	–	40	–	–	–
First-year Grad. Stud.	–	–	7	–	–	–

GRADUATE DEGREE REQUIREMENTS
Master's: The Department does not sponsor a specific Master's Degree program. However, a Master of Arts degree can be awarded as an incidental degree upon passing the General Examination.
Doctorate: There are no formal curriculum-based requirements in the doctoral program, and no course grades are given. Facility in a foreign language is encouraged, but a reading examination is no longer required. Students take the "Physics Prelims" in the Department of Physics, at the end of their first year. However, a minimum of one year in residence is required before taking the plasma physics portion (written and oral) of the General Examination.
Thesis: Thesis may be written in absentia.

SPECIAL EQUIPMENT, FACILITIES, OR PROGRAMS
Students normally request admission to either the astronomy section or the plasma physics section. The present report gives information only on the plasma physics program. As the name implies, this program has a strong physics orientation.

First-year studies typically include graduate courses given by the Department of Physics–quantum mechanics, electricity and magnetism, and statistical mechanics, together with the full-year introductory plasma physics course. Course offerings in later years cover plasma waves and instabilities, equilibrium and nonequilibrium statistical mechanics, theories of transport, structure and stability of plasma equilibria, magnetohydrodynamics, kinetic theory, nonlinear processes, and computational methods in plasma physics.

Applications of theory are made to astrophysical and geophysical plasmas as well as to the plasmas in controlled fusion research and in other areas of physics and engineering science. Studies of basic plasma physics and of the possibilities of controlled fusion power were initiated in Princeton in 1951. The principal fusion research devices at the Princeton Plasma Physics Laboratory currently include the National Spherical Torus Experiment (NSTX), the Current Drive Experiment-Upgrade (CDX-U), and the Magnetic Reconnection Experiment (MRX).

Laboratory scientists and engineers are also involved in the development of innovative confinement concepts, such as compact stellarators. In addition, Princeton Plasma Physics Laboratory scientists are collaborating with researchers on fusion science and technology at other facilities, both domestic and foreign. Through its efforts to build and operate magnetic fusion devices, PPPL has gained extensive capabilities in a host of disciplines including advanced computational simulations, lasers, microwave technology, vacuum technology, mechanics, materials science, electronics, computer technology, and high-voltage power systems. In addition, PPPL scientists and engineers are applying knowledge gained in fusion research to other theoretical and experimental areas including plasma thrusters for space propulsion, propagation of intense beams of ions, materials science, solar physics and manufacturing.

There are also experimental research opportunities on plasma thrusters, x-ray lasers, nonneutral plasmas, and plasma processing, as well as for theoretical and computational research in diverse areas of plasma physics, including magnetic confinement, plasma heating, transport and stability, plasma astrophysics, laser-plasma interactions, magneto-spheric physics and basic studies in plasma physics. This involves the innovative development of new calculation capabilities as well as the application of state of the art theoretical and computational tools to the interpretation of experimental results.

Local computation is carried out on workstations and personal computers with access to Cray Supercomputers and massively parallel computers. Graduate students from their first week participate in both the experimental and theoretical research programs at the Laboratory.

FACULTY

Professor
Bhattacharjee, Amitava, Ph.D., Princeton University, 1981. Head, PPPL Theory Department. *Astrophysics, Plasma and*

Fusion. Magnetic reconnection, turbulence, the dynamo effect, waves in weakly collisional plasmas, high-energy-density plasmas, dusty plasmas, free-electron lasers.

Fisch, Nathaniel J., Ph.D., Massachusetts Institute of Technology, 1978. Director, Graduate Studies, Program in Plasma Physics. *Plasma and Fusion.* Plasma waves; current drive; laser/plasma interaction.

Goldston, Robert J., Ph.D., Princeton University, 1977. *Plasma and Fusion.* Experimental plasma physics, plasma heating, transport.

Ji, Hantao, Ph.D., University of Tokyo, 1990. Magnetic reconnection; magnetorotational instability (MRI).

Prager, Stewart C., Ph.D., Columbia University, 1975. Director, Princeton Plasma Physics Laboratory. *Plasma and Fusion.* Experimental plasma physics; stability transport.

Affiliate Professor

Choueiri, Edgar, Ph.D., Princeton University, 1991. *Astrophysics, Plasma and Fusion.* Plasma propulsion; space plasma physics; turbulence in collisional plasmas; plasma dynamics and astronautics.

Szymon, Suckewer, Ph.D., Warsaw University, 1966. *Plasma and Fusion.* X-ray lasers development and applications; x-ray microscopy; powerful sub-picosecond lasers and their applications; applications of lasers to biology and medicine; plasma diagnostics; development of compact tokamak as a source of VUV and soft x-ray radiation.

Lecturer with Rank of Professor

Cohen, Samuel A., Ph.D., Massachusetts Institute of Technology, 1973. *Plasma and Fusion.* Experimental plasma physics.

Hammett, Gregory W., Ph.D., Princeton University, 1986. *Computational Physics, Plasma and Fusion.* Computational and analytical studies of plasma turbulence and RF heating.

Krommes, John A., Ph.D., Princeton University, 1975. *Plasma and Fusion.* Plasma stochasticity and turbulence.

Majeski, Richard P., Ph.D., Dartmouth College, 1979. *Plasma and Fusion.* Experimental plasma physics.

Phillips, Cynthia K., Ph.D., University of Wisconsin-Madison, 1982. RF heating.

Qin, Hong, Ph.D., Princeton University, 1998. *Particles and Fields, Physics of Beams.* Beam physics; gyrokinetic theory; particle simulation.

Reiman, Allan H., Ph.D., Princeton University, 1977. *Plasma and Fusion.* Theoretical plasma physics; non-linear dynamics; MHD.

Tang, William M., Ph.D., University of California, Davis, 1972. Lecturer with Rank of Professor. Kinetic stability theory.

White, Roscoe B., Ph.D., Princeton University, 1963. Lecturer with Rank of Professor. Resistive magnetohydrodynamic instabilities; parametric instabilities; transport.

Lecturer

Dodin, Ilya, Ph.D., Princeton University, 2005. *Particles and Fields, Plasma and Fusion.* Plasma physics; nonlinear dynamics; wave-particle interactions.

Efthimion, Philip C., Ph.D., Columbia University, 1977. *Plasma and Fusion.* Plasma physics; E&M theory and applications; instrumentation.

View additional information about this department at www.gradschoolshopper.com

---

PRINCETON UNIVERSITY

DEPARTMENT OF PHYSICS

Princeton, New Jersey 08544
http://princeton.edu/physics

General University Information
President: Christopher Eisgruber
Dean of Graduate School: William B. Russel
University website: http://www.princeton.edu
Control: Private
Total Faculty: 933
Total number of Students: 5,249
Total number of Graduate Students: 2,610

Department Information
Department Chairman: Lyman A. Page, Jr., Chair
Department Contact: Herman L. Verlinde, Director of Graduate Studies
Total full-time faculty: 38
Full-Time Graduate Students: 109
First-Year Graduate Students: 26
Female First-Year Students: 1
Total Post Doctorates: 25

Department Address
Washington Road
Princeton, NJ 08544
Phone: (609) 258-5585
E-mail: phydgs@princeton.edu
Website: http://princeton.edu/physics

ADMISSIONS

Admission Contact Information
Address admission inquiries to: Graduate Admissions Office, Clio Hall, Princeton University, Princeton, NJ 08544
Phone: (609) 258-3030
E-mail: gs@princeton.edu
Admissions website: http://www.princeton.edu/gradschool/

Application deadlines
Fall admission:
U.S. students: December 15   Int'l. students: December 15

Application fee
U.S. students: $90   Int'l. students: $90

Admissions information
For Fall of 2013:
  Number of applicants: 506
  Number admitted: 70
  Number enrolled: 26

Admission requirements
Bachelor's degree requirements: A bachelor's degree in physics is required.

GRE requirements
The GRE is recommended.

Advanced GRE requirements
The Advanced GRE is required.

TOEFL requirements
The TOEFL exam is required for students from non-English-speaking countries.

TUITION

Tuition year 2013-14:
  Full-time students: $42,070 annual
Deferred tuition plan: No
Health insurance: Available.

Teaching Assistants, Research Assistants, and Fellowships
Number of first-year
  Fellowship students: 26
Average stipend per academic year
  Teaching Assistant: $27,950
  Research Assistant: $25,170
  Fellowship student: $24,650

FINANCIAL AID

Loans
Loans are not available for U.S. students.
Loans are not available for international students.
GAPSFAS application required: No
FAFSA application required: No

For further information
Address financial aid inquiries to: Asst. Dean of Financial Affairs, Graduate School, 204 Nassau Hall, Princeton, NJ 08544.
Phone: (609) 258-3030
E-mail: gs@princeton.edu
Financial aid website: http://www.princeton.edu/gradschool/financial/

HOUSING

Availability of on-campus housing
Single students: Yes
Married students: Yes

For further information
Address housing inquiries to: Graduate Housing Department MacMillan Building, Princeton University, Princeton, NJ 08544.
Phone: (609) 258-4360
E-mail: gradhsg@princeton.edu
Housing aid website: http://www.princeton.edu/facilities/housing/graduate_info/

Table A—Faculty, Enrollments, and Degrees Granted

Research Specialty	2012 Faculty	Enrollment Fall 2012		Number of Degrees Granted 2012-13		
		Master's	Doctorate	Master's	Terminal Master's	Doctorate
Atomic, Molecular, & Optical Physics	2	–	2	–	–	–
Biophysics	5	–	12	–	–	3(-)
Computational Physics	2	–	2	–	–	1(-)
Condensed Matter Physics	9	–	35	–	–	–
Cosmology & String Theory	6	–	19	–	–	3(-)
High Energy Physics	14	–	34	–	–	–
Particles and Fields	2	–	5	–	–	3(-)
Total	40	–	109	–	–	10(-)
Full-time Grad. Stud.	–	–	109	–	–	–
First-year Grad. Stud.	–	–	26	–	–	–

GRADUATE DEGREE REQUIREMENTS

Master's: The master's degree is conferred only after passing a general examination. (Students who want to work toward a master's degree only will not be admitted.).

Doctorate: The formal course requirements include three core courses to be taken between the beginning of the first year of study and the end of the second year. (This requirement is part of the general examination.) Students taking these courses must achieve a grade of B or higher. In addition, one year of residency is required; general examination and the dissertation are required.

Thesis: Thesis may be written in absentia.

SPECIAL EQUIPMENT, FACILITIES, OR PROGRAMS

Theoretical research spans most of the central topics of modern physics. The department has decades-old traditions of excellence and leadership in these core areas of fundamental physics, and it is also rapidly building strength in newer areas, such as theoretical biology. In the newer areas, interaction between physics and other departments is critical, and major university-supported interdisciplinary initiatives provide a strong framework for this cooperation. There is also productive interaction between theorists in the department and those at the nearby Institute for Advanced Studies, although there is no formal connection between these institutions.

The high energy theory group works on quantum field theory, particle phenomenology and cosmology, string theory and quantum gravity models in various dimensions, and dualities between gauge theories and strings. Some members of the group are also interested in applications of quantum field theory and string theory to problems in statistical mechanics, the theory of turbulence, heavy-ion collisions, and condensed matter physics.

The cosmology theory group uses astrophysical, particle physics, and superstring theory combined with observations to study gravitation and the origin, composition, and evolution of the universe.

The theoretical condensed matter group works on quantum many-body theory of systems involving strong correlations and/or disorder, statistical mechanics, biological systems and systems far from equilibrium.

The mathematical physics group is concerned with problems in statistical mechanics, atomic and molecular physics, quantum field theory, and, in general, with the mathematical foundations of theoretical physics.

The theoretical biophysicists work on problems in statistical mechanics and information theory that arise in studying nervous systems, gene expression networks, the organization of genomes, and the mechanisms of evolution.

Experiments in high energy particle physics are directed toward understanding the fundamental interactions and particle structures at extremely small distances. The apparatus is designed and constructed in the physics shops in Jadwin Hall or at the nearby Elementary Particles Laboratory, which contains special facilities for the fabrication of detectors. The experiments are performed at large national and international laboratories, which currently include CERN (Switzerland), Fermilab (Illinois), and SLAC (California). The data are then analyzed at Princeton University.

The nuclear and particle astrophysics group is active in experimental studies of solar neutrinos and dark matter. The goal of the solar neutrino program is to explore neutrino oscillations and solar processes through a measurement of the low energy 7Be neutrino. Neutrinos will be detected with the Borexino liquid scintillation detector located in the Gran Sasso underground laboratory in Italy.

The dark matter group is designed to detect WIMPs in the galaxy by their collisions with either xenon or argon nuclei in a scintillation-ionization detector made of the rare gas atoms. Experiments are under development to provide a definitive search for rare WIMP collisions by combining the unique scintillation properties of the rare gas atoms with the low background methods developed for the Borexino solar neutrino experiment.f

Research in the condensed-matter physics group seeks to understand electronic behavior in novel low-dimensional solids in which interaction and correlation effects are dominant. Problems investigated have included the fractional and integer quantum Hall effects, high-temperature superconductivity, Kondo effect in quantum dots, spin-density-wave states in organic conductors, highly frustrated quantum-spin systems, and novel excitations in low-dimensional magnetic systems.

The research involves close collaborations between experimentalists and theorists, as well as with faculty in the Chemistry and Electrical Engineering Departments.

Experimental groups are also engaged in researching novel patterning techniques using diblock copolymers (with faculty in Chemical Engineering) and techniques for single-molecule detection and separation of biological molecules (with Molecular Biology and the Genomics Center).

In the experimental cosmology group, students often design and build specialized instrumentation to make unique and precise measurements, or analyze cosmological data. In recent years, experimental work has emphasized measurements of the anisotropy and polarization of the cosmic microwave background. Among other projects, Princeton is actively involved in all aspects of the WMAP satellite, is the lead institution for the ACT project, and is a collaborator on the QUIET experiment.

Research in atomic physics is primarily focused on spin-polarized gases, liquids, and solids, on their properties, interactions, and a wide range of applications. Among applications currently being developed are searches for violation of CP symmetry beyond the Standard Model, tests of Lorentz invariance, development of miniature atomic clocks, ultra-sensitive atomic magnetometers, and new biomedical techniques, such as lung imaging and mapping of the magnetic fields generated by the brain.

Biological physics spans a huge range of subjects, from neurobiology to genomics to fundamentals of protein action. Princeton has strengths in nearly all areas of modern biological physics. Many faculty with a strong physics background who are involved in biological physics are not solely in the Physics Department but have joint appointments with other departments or are completely in other departments. There is a strong community spirit to biological physics among these departments despite the vast range of subjects being studied.

For more information, please visit www.princeton.edu/physics.

Table B—Separately Budgeted Research Expenditures by Source of Support

Source of Support	Departmental Research	Physics-related Research Outside Department
Federal government	$19,402,090	$5,183,217
State/local government		
Non-profit organizations	$538,914	$766,076
Business and industry		
Other		
Total	$19,941,004	$5,949,293

Table C—Separately Budgeted Research Expenditures by Research Specialty

Research Specialty	No. of Grants	Expenditures ($)
Astrophysics	27	$3,455,901
Atomic, Molecular, & Optical Physics	9	$805,463
Biophysics	8	$988,644
Condensed Matter Physics	55	$6,590,873
Nuclear Physics	19	$4,425,444
Particles and Fields	81	$9,172,212
Other	3	$451,759
Total	202	$25,890,296

FACULTY

Professor

Aizenman, Michael, Ph.D., Belfer Graduate School of Science, Yeshiva Univ., 1975. Mathematical physics.

Austin, Robert, Ph.D., University of Illinois, 1975. *Biophysics*. Biophysics.

Bialek, William, Ph.D., University of California, Berkeley, 1983. *Biophysics*. Biophysics.

Calaprice, Frank, Ph.D., University of California, Berkeley, 1967. Nuclear physics.

Callan, Curtis G., Ph.D., Princeton University, 1964. Theoretical physics.

Groth, Edward, Ph.D., Princeton University, 1971. Cosmology; gravitation and relativity.

Gubser, Steven, Ph.D., Princeton University, 1998. Particle theory.

Haldane, F. Duncan M., Ph.D., University of Cambridge, 1978. Condensed matter.

Happer, William, Ph.D., Princeton University, 1964. Atomic physics.

Hasan, M. Zahid, Ph.D., Stanford University, 2001. *Condensed Matter Physics*. Condensed matter physics.

Huse, David A., Ph.D., Cornell University, 1983. Condensed matter physics.

Klebanov, Igor, Ph.D., Princeton University, 1986. Theoretical physics.

Lieb, Elliott, Ph.D., Massachusetts Institute of Technology, 1956. Mathematical physics.

Marlow, Daniel R., Ph.D., Carnegie Mellon University, 1981. High energy physics.

McDonald, Kirk, Ph.D., California Institute of Technology, 1972. High energy physics.

Meyers, Peter, Ph.D., University of California, Berkeley, 1983. High energy physics.

Olsen, James, Ph.D., University of Wisconsin-Madison, 1998. High energy physics.

Ong, Nai-Phuan, Ph.D., University of California, Berkeley, 1976. Condensed matter physics.

Page, Lyman, Ph.D., Massachusetts Institute of Technology, 1989. Cosmology; gravitation; relativity.

Polyakov, Alexandre, Ph.D., Landau Institute, USSR, 1969. Theoretical physics.

Pretorius, Frans, Ph.D., University of British Columbia, 2002. Theoretical cosmology.

Romalis, Michael, Ph.D., Princeton University, 1997. Atomic physics.

Smith, A. J. Stewart, Ph.D., Princeton University, 1966. High energy physics.

Sondhi, Shivaji Lal, Ph.D., University of California, Los Angeles, 1992. Condensed matter physics.

Staggs, Suzanne, Ph.D., Princeton University, 1993. Cosmology; gravitation; relativity.

Steinhardt, Paul, Ph.D., Harvard University, 1978. Cosmology.

Tully, Christopher, Ph.D., Princeton University, 1998. High energy physics.

Verlinde, Herman, Ph.D., Utrecht University, 1988. Particle theory.

Yazdani, Ali, Ph.D., Stanford University, 1995. Condensed matter.

Associate Professor

Galbiati, Cristiano, Ph.D., University of Milan, 1999. Nuclear physics.

Petta, Jason, Ph.D., Cornell University, 2003. Condensed matter.

Assistant Professor

Bakr, Waseem, Ph.D., Harvard University, 2011.

Bernevig, Bogdan an Andrei, Ph.D., Stanford University. Condensed matter.

Giombi, Simone, Ph.D., State University of New York at Stony Brook, 2007. *Nuclear Physics.*

Gregor, Thomas, Princeton University, 2005. *Biophysics.* Biophysics.

Jones, William C., Ph.D., California Institute of Technology, 2005. Cosmology.

Lisanti, Mariangela, Ph.D., Stanford University. *Particles and Fields.*

Pufu, Silviu, Ph.D., Princeton University, 2011. *High Energy Physics.*

Shaevitz, Joshua, Ph.D., Stanford University, 2004. Molecular biology.

Professor Emeritus

Nappi, Chiara, Ph.D., University of Naples, 1976. Theoretical physics.

Affiliate Professor

Bhatt, Ravindra, Ph.D., University of Illinois, 1976. Electrical engineering and computer science.

Car, Roberto, Ph.D., Laurea, Milan, 1971. Chemistry.

Shayegan, Mansour, Ph.D., Massachusetts Institute of Technology, 1983. *Electrical Engineering.* Electrical engineering.

Tank, David, Ph.D., Cornell University, 1983. *Biophysics.* Biophysics.

Torquato, Salvatore, Ph.D., Stony Brook University, 1980. Chemistry.

Visiting Professor

Adler, Stephen L., Ph.D., Princeton University, 1964. Theoretical physics.

Arkani-Harned, Nima, Ph.D., University of California, Berkeley, 1997. Particle theory.

Maldacena, Juan, Ph.D., Princeton University, 1996. Particle theory.

Seiberg, Nathan, Ph.D., Tel Aviv University, 1982. Particle theory.

Spergel, David N., Ph.D., Harvard University, 1985. Astrophysical sciences.

Witten, Edward, Ph.D., Princeton University, 1976. Visiting Lecturer with rank of Professor. Theoretical physics.

Posdoctoral Research Associate

Xu, Jingke, Ph.D., Princeton University, 2013.

Associate Faculty

Tsui, Dan, Ph.D., University of Chicago, 1967. Electrical engineering and computer science.

Wingreen, Ned, Ph.D., Cornell University, 1989. Molecular biology.

DEPARTMENTAL RESEARCH SPECIALTIES AND STAFF

Theoretical

Biophysics. The theoretical biophysicists work on problems in statistical mechanics and information theory that arise in studying nervous systems, gene expression networks, the organization of genomes, and the mechanisms of evolution. For more information, please visit www.princeton.edu/physics. Austin, Bialek, Gregor, Tank.

Condensed Matter Physics. The theoretical condensed matter group works on quantum many-body theory of systems involving strong correlations and/or disorder, statistical mechanics, biological systems, and systems far from equilibrium. For more information, visit www.princton.edu/physics. Hasan.

Relativity & Gravitation.

Relativity & Gravitation.

Experimental

Atomic, Molecular, & Optical Physics. Research in atomic physics is primarily focused on spin-polarized gases, liquids, and solids, on their properties, interactions, and a wide range of applications. Among applications currently being developed are searches for violation of CP symmetry beyond the standard model, tests of Lorentz invariance, development of miniature atomic clocks, ultra-sensitive atomic magnetometers, and new biomedical techniques, such as lung imaging and mapping of the magnetic fields generated by the brain. For more information, please visit www.princeton.edu/physics.

Biophysics. Biological physics spans a huge range of subjects, from neurobiology to genomics to fundamentals of protein action. Princeton has strengths in nearly all areas of modern biological physics. Many faculty with a strong physics background who are involved in biological physics are not solely in the Physics Department but have joint appointments with other departments or are completely in other departments. There is a strong community spirit to biological physics amongst these departments despite the vast range of subjects being studied. For more information, please visit www.princeton.edu/physics.

Condensed Matter Physics. Research in the condensed-matter physics group seeks to understand electronic behavior in novel low-dimensional solids in which interaction and correlation effects are dominant. Problems investigated have included the fractional and integer quantum Hall effects, high-temperature superconductivity, Kondo effect in quantum dots, spin-density-wave states in organic conductors, highly frustrated quantum-spin systems, and novel excitations in low-dimensional magnetic systems. The research involves close collaborations between experimentalists and theorists, as well as with faculty in the Chemistry and Electrical Engineering Departments. Experimental groups are also engaged in researching novel patterning techniques using diblock copolymers (with faculty in Chemical Engineering) and techniques for single-molecule detection and separation of biological molecules (with Molecular Biology and the Genomics Center). For more information, please visit www.princeton.edu/physics.

Nuclear Physics. The nuclear and particle astrophysics group is active in experimental studies of solar neutrinos and dark matter. The goal of the solar neutrino program is to explore neutrino oscillations and solar processes through a measurement of the low energy 7Be neutrino. Neutrinos will be detected with the Borexino liquid scintillation detector located in the Gran Sasso underground laboratory in Italy. For more information, please visit www.princeton.edu/physics.

Nuclear Physics. The nuclear and particle astrophysics group is active in experimental studies of solar neutrinos and dark matter. The goal of the solar neutrino program is to explore neutrino oscillations and solar processes through a measurement of the low energy 7Be neutrino. Neutrinos will be detected with the Borexino liquid scintillation detector located in the Gran Sasso underground laboratory in Italy. For more information, please visit www.princeton.edu/physics.

Particles and Fields. Experiments in high energy particle physics are directed toward understanding the fundamental interactions and particle structures at extremely small distances. The apparatus is designed and constructed in the physics shops in Jadwin Hall or at the Elementary Particles Laboratory a block away, which contains special facilities for the fabrication of detectors. The experiments are performed at large national and international laboratories, which currently include CERN (Switzerland), Fermilab (Illinois), KEK (Japan), and SLAC (California). The data are then analyzed at Princeton. For more information, please visit www.princeton.edu/physics.

Particles and Fields. Experiments in high energy particle physics are directed toward understanding the fundamental interactions and particle structures at extremely small distances. The apparatus is designed and constructed in the physics shops in Jadwin Hall or at the Elementary Particles Laboratory a block away, which contains special facilities for the fabrication of detectors. The experiments are performed at large national and international laboratories, which currently include CERN (Switzerland), Fermilab (Illinois), KEK (Japan), and SLAC (California). The data are then analyzed at Princeton. For more information, please visit www.princeton.edu/physics.

View additional information about this department at
www.gradschoolshopper.com

RUTGERS — THE STATE UNIVERSITY OF NEW JERSEY

DEPARTMENT OF PHYSICS AND ASTRONOMY

Piscataway, New Jersey 08854
http://www.physics.rutgers.edu

General University Information
President: Robert L. Barchi
Dean of Graduate School: Jerome J. Kukor
University website: http://www.rutgers.edu/
Control: Public
Setting: Suburban
Total Faculty: 3,000
Total Graduate Faculty: 6,500
Total number of Students: 58,000
Total number of Graduate Students: 8,200

Department Information
Department Chairman: Robert Bartynski, Chair
Department Contact: Ronald Gilman, Graduate Director
Total full-time faculty: 60
Total number of full-time equivalent positions: 60
Full-Time Graduate Students: 104
First-Year Graduate Students: 17
Female First-Year Students: 2
Total Post Doctorates: 50

Department Address
136 Frelinghuysen Road
Piscataway, NJ 08854
Phone: (732) 445-5500, x5489
Fax: (732) 445-4343
E-mail: graduate@physics.rutgers.edu
Website: http://www.physics.rutgers.edu

ADMISSIONS

Admission Contact Information
Address admission inquiries to: Dr. Ronald Gilman, Graduate Program Director, Department of Physics and Astronomy, 136 Frelinghuysen Road, Piscataway, NJ 08854-8019
Phone: (732) 445-5500 x 2502
E-mail: graduate@physics.rutgers.edu
Admissions website: http://gradstudy.rutgers.edu

Application deadlines
Fall admission:
U.S. students: January 1   *Int'l. students*: January 1
Spring admission:

U.S. students: November 1   *Int'l. students*: November 1

Application fee
U.S. students: $65   *Int'l. students*: $65
Later applications will be considered until July 15, depending on availability of positions.

Admissions information
For Fall of 2013:
> *Number of applicants*: 315
> *Number admitted*: 72
> *Number enrolled*: 17

Admission requirements
Bachelor's degree requirements: Bachelor's degree in physics or related field is required.
Minimum undergraduate GPA: 3.0

GRE requirements
The GRE is required.
> *Quantitative score*: 500
> *Verbal score*: 300
> *Analytical score*: 3
> *Mean GRE score range (25th–75th percentile)*: 50th-75th

Advanced GRE requirements
The Advanced GRE is required.
> *Minimum accepted Advanced GRE score*: 600
> *Mean Advanced GRE score range (25th–75th percentile)*: 50th-75th

TOEFL requirements
The TOEFL exam is required for students from non-English-speaking countries.
> *PBT score*: 560
> *iBT score*: 83

Other admissions information
Additional requirements: No minimum scores specified. The Advanced Physics average was 75% for students admitted. Students from non-English speaking countries are required to demonstrate proficiency in English via the TOEFL or IELTS exam.

TUITION
Tuition year 2012–2013:
Tuition for in-state residents
> *Full-time students*: $15,038 annual
> *Part-time students*: $626 per credit

Tuition for out-of-state residents
> *Full-time students*: $24,600 annual
> *Part-time students*: $1,025 per credit

Credit hours per semester to be considered full-time: 9
Deferred tuition plan: Yes
Health insurance: Available at the cost of 950 per year.
Other academic fees: $1937 annual computer fees, School fees and Campus fees.
Academic term: Semester
Number of first-year students who received full tuition waivers: 17

Teaching Assistants, Research Assistants, and Fellowships
Number of first-year
> *Teaching Assistants*: 11
> *Research Assistants*: 2
> *Fellowship students*: 4

Average stipend per academic year
> *Teaching Assistant*: $25,460
> *Research Assistant*: $25,460
> *Fellowship student*: $29,000

Teaching and research assistants receive full remission of fess and health insurance. Most fellowships provide modest health insurance but no fee remission.

FINANCIAL AID

Application deadlines
Fall admission:
U.S. students: January 1   *Int'l. students*: January 1
Spring admission:
U.S. students: November 1   *Int'l. students*: November 1

Loans
Loans are available for U.S. students.
Loans are not available for international students.
GAPSFAS application required: No
FAFSA application required: Yes

For further information
Address financial aid inquiries to: Prof. R. Gilman, Graduate Program Director, Rutgers University, Department of Physics and Astronomy, 136 Frelinghuysen Rd., Piscataway, NJ 08854-8019.
Phone: (732) 445-5500 x 2502
E-mail: graduate@physics.rutgers.edu
Financial aid website: http://studentaid.rutgers.edu/

HOUSING

Availability of on-campus housing
> *Single students*: Yes
> *Married students*: Yes

For further information
Address housing inquiries to: Graduate Student Housing, 581 Taylor Rd., Piscataway, NJ 08854.
Phone: 732-445-0750
E-mail: oncampus@rci.rutgers.edu
Housing aid website: http://housing.rutgers.edu/

Table A—Faculty, Enrollments, and Degrees Granted

Research Specialty	2012-13 Faculty	Enrollment Fall 2012 Master's	Enrollment Fall 2012 Doctorate	Number of Degrees Granted 2012-13 Master's	Number of Degrees Granted 2012-13 Terminal Master's	Number of Degrees Granted 2012-13 Doctorate
Astronomy	11	–	16	–	–	1(-)
Biophysics	6	–	11	–	–	1(-)
Condensed Matter Physics	28	–	43	–	–	4(-)
Nuclear Physics	7	–	4	–	–	–
Particles and Fields	18	–	28	1(-)	–	2(-)
Statistical & Thermal Physics	1	–	2	–	–	2(-)
Other	4	4	4	–	2(-)	–
Total	77	4	108	1(-)	2(-)	10(-)
Full-time Grad. Stud.	–	3	107	–	–	–
First-year Grad. Stud.	–	1	17	–	–	–

GRADUATE DEGREE REQUIREMENTS

Master's: The M.S. degree program is designed for part-time as well as full-time students. A comprehensive oral examination is required of all M.S. candidates. The M.S. degree requires 30 credits of which up to 12 may be in upperclass undergraduate courses (300–400 series). The candidate may choose to write a thesis (in which case, six of the 30 required credits may be devoted to this thesis research) or to submit an essay (which is to be based on material from a course he or she has taken). The thesis must be defended in the oral examination. There is no formal GPA requirement, but no more than three courses with grades of "C" can be counted toward the degree. There is no foreign language requirement.

Doctorate: The candidacy exam consists of an oral exam on a current topic in research. Candidates must present a written report on a current area of research, followed by an oral presentation and exam. The exam tests the candidate's ability to grasp the relevance, goals, techniques, and underlying physics of a current area of research. The exam is normally taken at the start of the second year. In addition, candidates are required to complete a set of core courses with grades of B or better, or pass an exam if exemption is requested based on previous course work. A dissertation of original research is required. There is a residence requirement of one year, but no foreign language requirement.

Other Degrees: Master of Science for Teachers (MST) degree: The MST degree is primarily a subject-matter oriented degree for practicing teachers, although others may be accepted. The requirements for the MST degree in physics consist of 30 credits, a comprehensive examination, and an essay or thesis. The courses are chosen in consultation with the departmental advisor to fit the needs of each individual student. The first aim is to give each candidate the opportunity to further his or her knowledge of physics. Both undergraduate and graduate courses may be used, depending on each person's previous experience.

Thesis: Thesis may be written in absentia.

SPECIAL EQUIPMENT, FACILITIES, OR PROGRAMS

The department has 60 faculty members. An additional 17 faculty members from other departments, primarily chemistry and mathematics, are members of the graduate program.

The department is housed in a modern, fully equipped research laboratory with networks of workstations and PCs that provide easy computer access for all students and faculty members.

The astrophysics group is focused on galactic dynamics and cosmology and has developed Fabry-Perot interferometers for observatories in Chile and South Africa. Rutgers astronomers have a 10% share of observing time at the 11 meter Southern African Large Telescope.

Condensed-matter theory faculty members study strongly correlated electron systems and electronic properties of materials. The multidisciplinary Laboratory for Surface Modification includes 7 physics faculty members and members of the chemistry, materials science, and engineering departments. Research in condensed matter experiment spans low temperature physics, mesoscopic electronics organic conductors, optical scattering spectroscopies, magnetic and multiferroic materials, and two-dimensional systems (e.g., graphene). New research initiatives focus on the synthesis of novel materials and their characterization using optical, scanning-probe, X-ray/neutron diffraction, and transport techniques.

High-energy theory research includes phenomenological studies and abstract approaches such as string theory and conformal field theories. High-energy experimentalists do research at CERN hadron collider. They search for supersymmetry, the Higgs particle and dark matter with leptons, photons, and jets and also study the top quark and gauge bosons. They are also developing detectors and detection technologies for high radiation environments.

Nuclear physics research in both theory and experiment span a broad range of questions, including the structure of the nucleon, the interaction of neutrinos with nuclei, the limits of angular momentum and stability in nuclei, and nucleosynthesis. Experiments are carried out at Jefferson Lab in Virginia, Fermilab Texas A&M, Argonne National Lab and Michigan State University.

Table B—Separately Budgeted Research Expenditures by Source of Support

Source of Support	Departmental Research	Physics-related Research Outside Department
Federal government	$11,000,000	
State/local government		
Non-profit organizations		
Business and industry		
Other		
Total	**$11,000,000**	

FACULTY

Professor

Andrei, Eva Y., Ph.D., Rutgers University, 1980. Member, National Academy of Science. *Condensed Matter Physics, Low Temperature Physics*. Experimental condensed matter physics.

Andrei, Natan, Ph.D., Princeton University, 1979. *Condensed Matter Physics, High Energy Physics*. Theoretical elementary particle/condensed matter physics.

Banks, Thomas, Ph.D., Massachusetts Institute of Technology, 1973. *Cosmology & String Theory, Particles and Fields*. Theoretical elementary particle physics.

Bartynski, Robert, Ph.D., University of Pennsylvania, 1986. Chair of the Department. *Condensed Matter Physics, Surface Physics*. Experimental condensed matter physics.

Bhanot, Gyan, Ph.D., Cornell University, 1979. *Biophysics*. Systems biology, cancer and population genetics.

Blumberg, Girsh, Ph.D., Estonian Academy of Sciences, 1987. *Condensed Matter Physics, Solid State Physics*. Experimental condensed matter physics.

Case, David, Ph.D., Harvard University, 1977. *Biophysics, Chemical Physics*. Theoretical chemistry of biomolecules.

Chandra, Premala, Ph.D., University of California, Santa Barbara, 1988. *Condensed Matter Physics*. Condensed matter theory.

Cheong, Sang-Wook, Ph.D., University of California, Los Angeles, 1989. *Condensed Matter Physics, Materials Science, Metallurgy, Solid State Physics*. Experimental condensed matter physics; material science.

Cizewski, Jolie A., Ph.D., Stony Brook University, 1978. *Nuclear Physics*. Experimental nuclear physics.

Coleman, Piers, Ph.D., Princeton University, 1984. *Condensed Matter Physics*. Theoretical condensed matter physics.

Croft, Mark, Ph.D., University of Rochester, 1977. *Condensed Matter Physics, Solid State Physics*. Experimental condensed matter physics.

Etkina, Eugenia, Ph.D., Moscow State Pedagogical University, 1997. Professor of Education. *Physics and other Science Education*. Physics education.

Feldman, Leonard, Ph.D., Rutgers University, 1967. *Condensed Matter Physics, Materials Science, Metallurgy, Solid State Physics*. Experimental condensed matter physics.

Friedan, Daniel, Ph.D., University of California, Berkeley, 1980. *Cosmology & String Theory, Particles and Fields, Theoretical Physics*. Theoretical elementary particle physics.

Garfunkel, Eric, Ph.D., University of California, Berkeley, 1983. *Chemical Physics, Solid State Physics, Surface Physics*. Experimental surface science.

Gershenson, Michael E., Ph.D., Institute of Radio Engineering and Electronics (Moscow), 1982. *Condensed Matter Physics, Solid State Physics*. Experimental condensed matter physics.

Gilman, Ronald, Ph.D., University of Pennsylvania, 1985. Associate Chair and Graduate Program Director. *Nuclear Physics*. Experimental nuclear physics.

Goldin, Gerald A., Ph.D., Princeton University, 1969. *Quantum Foundations, Theoretical Physics*. Mathematical physics.

Goldstein, Sheldon, Ph.D., Yeshiva University, 1974. *Quantum Foundations, Statistical & Thermal Physics, Theoretical Physics*. Statistical mechanics; foundations of quantum mechanics.

Gustafsson, Torgny, Ph.D., Chalmers University of Technology, Sweden, 1973. *Condensed Matter Physics, Solid State Physics, Surface Physics*. Experimental condensed matter physics; experimental surface physics.

Hughes,John,Ph.D.,ColumbiaUniversity,1984.*Astronomy,Astrophysics*. Observational astronomy.

Ioffe, Lev, Ph.D., Landau Inst. for Theoretical Physics, Russia, 1985. *Condensed Matter Physics, Solid State Physics*. Theoretical condensed matter physics.

Kiryukhin, Valery, Ph.D., Princeton University, 1997. *Condensed Matter Physics, Solid State Physics*. Experimental condensed matter physics.

Kloet, Willem M., Ph.D., Utrecht, Netherlands, 1973. *Nuclear Physics*. Theoretical nuclear physics.

Kojima, Haruo, Ph.D., University of California, Los Angeles, 1972. *Condensed Matter Physics, Low Temperature Physics*. Experimental condensed matter physics.

Kotliar, B. Gabriel, Ph.D., Princeton University, 1983. *Condensed Matter Physics*. Theoretical condensed matter physics.

Lebowitz, Joel, Ph.D., Syracuse University, 1956. Member National Academy of Science. *Statistical & Thermal Physics*. Theoretical statistical mechanics; math physics.

Levy, Ronald, Ph.D., Harvard University, 1976. *Biophysics*. Biological physics theory and simulation.

Matilsky, Terry A., Ph.D., Princeton University, 1971. *Astronomy, Physics and other Science Education*. Experimental astrophysics.

Mekjian, Aram, Ph.D., University of Maryland, 1968. *Nuclear Physics, Statistical & Thermal Physics*. Theoretical nuclear physics.

Moore, Gregory, Ph.D., Harvard University, 1986. *Cosmology & String Theory, Particles and Fields, Theoretical Physics*. Theoretical particle physics.

Murnick, Daniel E., Ph.D., Massachusetts Institute of Technology, 1966. *Applied Physics, Atomic, Molecular, & Optical Physics, Nuclear Physics*. Experimental nuclear and atomic physics.

Neuberger, Herbert, Ph.D., Tel Aviv University, 1979. *Particles and Fields*. Theoretical elementary particle physics.

Olson, Wilma, Ph.D., Stanford University, 1971. *Biophysics*. Biological physics theory and simulation.

Pryor, Carlton, Ph.D., Harvard University, 1982. Associate Chair for Undergraduate Education. *Astronomy*. Experimental astrophysics.

Rabe, Karin, Ph.D., Massachusetts Institute of Technology, 1987. Member National Academy of Science. *Condensed Matter Physics, Solid State Physics*. Theoretical condensed matter physics; theoretical surface physics.

Ransome, Ronald, Ph.D., University of Texas, Austin, 1981. *Nuclear Physics*. Experimental nuclear physics.

Schnetzer, Stephen R., Ph.D., University of California, Berkeley, 1981. *High Energy Physics, Particles and Fields*. Experimental elementary particle physics.

Sellwood, Jeremy, Ph.D., Manchester Univ., England, 1977. *Astronomy*. Theoretical astrophysics.

Sengupta, Anirvan, Ph.D., Bombay University, 1994. *Biophysics*. Biological physics.

Shapiro, Joel, Ph.D., Cornell University, 1967. *Particles and Fields, Physics and other Science Education*. Theoretical elementary particle physics.

Shinbrot, Troy, Ph.D., University o Maryland, 1992. *Biophysics, Fluids, Rheology, Statistical & Thermal Physics*. Computational bioengineering; self-assembly; mixing; chaos theory.

Soffer, Avraham, Ph.D., Tel-Aviv University, 1984. *Statistical & Thermal Physics, Theoretical Physics*. Theory of partial differential evolution equations; Schrödinger operators and scattering theory; general mathematical physics.

Somalwar, Sunil, Ph.D., University of Chicago, 1988. *High Energy Physics, Particles and Fields*. Experimental elementary particle physics.

Somerville, Rachel, Ph.D., University of California, Santa Cruz, 1997. *Astronomy, Astrophysics*. Astrophysics.

Thomas, Scott, Ph.D., University of Texas, Austin, 1993. *High Energy Physics, Particles and Fields*. Theoretical elementary particle physics.

Vanderbilt, David, Ph.D., Massachusetts Institute of Technology, 1981. Member National Academy of Science. *Condensed Matter Physics, Solid State Physics*. Theoretical condensed matter physics; theoretical surface physics.

Williams, Theodore B., Ph.D., California Institute of Technology, 1975. Director, South African Astronomical Observatories. *Astronomy, Astrophysics*. Experimental astrophysics.

Zamick, Larry, Ph.D., Massachusetts Institute of Technology, 1961. *Nuclear Physics*. Theoretical nuclear physics.

Zamolodchikov, Alexander, Ph.D., Inst. of Theoretical and Exp. Physics, Moscow, 1978. *Particles and Fields, Theoretical Physics*. Theoretical particle physics.

Associate Professor

Baker, Andrew, Ph.D., California Institute of Technology, 2000. *Astronomy*. Observational physics.

Diaconescu, Duiliu-Emanual, Ph.D., Rutgers University, 1998. *Cosmology & String Theory, Particles and Fields*. Theoretical high-energy physics.

Gawiser, Eric, Ph.D., University of California, Berkeley, 1999. *Astronomy, Astrophysics*. Observational astrophysics, cosmology.

Gershtein, Yuri, Ph.D., Moscow Inst. For Physics and Tech., 1996. *High Energy Physics, Particles and Fields*. Experimental high energy physics.

Halkiadakis, Eva, Ph.D., Rutgers University, 2001. *High Energy Physics, Particles and Fields*. Experimental particle physics.

Haule, Kristjan, Ph.D., University of Ljubljana, 2002. *Condensed Matter Physics, Solid State Physics*. Theoretical condensed matter physics.

Hinch, B. Jane, Ph.D., University of Cambridge, 1987. *Chemical Physics, Solid State Physics, Surface Physics*. Surface studies using atomic and molecular scattering.

Jha, Saurabh, Ph.D., Harvard University, 2002. *Astronomy, Astrophysics*. Observational cosmology.

Keeton, Charles, Ph.D., Harvard University, 1998. *Astronomy, Astrophysics*. Astronomy.

Lath, Amitabh, Ph.D., Massachusetts Institute of Technology, 1994. *High Energy Physics, Particles and Fields*. Experimental elementary particle physics.

Lukyanov, Sergei, Ph.D., Landau Institute, 1989. *Particles and Fields, Theoretical Physics*. Theoretical high-energy physics.

Morozov, Alexandre, Ph.D., University of Washington, 2003. *Biophysics*. Biophysics.

Oh, Seaongshik, Ph.D., University of Illinois, 2003. *Condensed Matter Physics, Materials Science, Metallurgy, Solid State Physics*. Experimental condensed matter physics.

Podzorov, Vitaly, Ph.D., Rutgers University, 2002. *Condensed Matter Physics, Solid State Physics*. Experimental condensed matter physics.

Wu, Weida, Ph.D., Princeton University, 2004. *Condensed Matter Physics, Materials Science, Metallurgy, Solid State Physics*. Experimental condensed matter physics.

Yuzbashyan, Emil, Ph.D., Princeton University, 2004. *Condensed Matter Physics*. Theoretical condensed matter physics.

Zimmermann, Frank M., Ph.D., Cornell University, 1995. *Condensed Matter Physics, Solid State Physics, Surface Physics*. Experimental surface science physics.

Assistant Professor

Brooks, Alyson, Ph.D., University of Washington, 2008. *Astronomy, Astrophysics*. Theoretical astrophysics.

Chou, John Paul, Ph.D., Harvard University, 2008. *High Energy Physics, Particles and Fields*. Experimental particle physics.

Salur, Sevil, Ph.D., Yale University, 2006. *High Energy Physics, Nuclear Physics*. Experimental nuclear physics.

Shih, David, Ph.D., Princeton University, 2006. *High Energy Physics, Particles and Fields*. Theoretical high energy physics.

Professor Emeritus

Kalelkar, Mohan S., Ph.D., Columbia University, 1975. *High Energy Physics, Particles and Fields, Physics and other Science Education*. Experimental elementary particle physics.

Koller, Noemie B., Ph.D., Columbia University, 1958. *Nuclear Physics*. Experimental nuclear physics.

Leath, Paul L., Ph.D., University of Missouri, Columbia, 1966. *Condensed Matter Physics*. Theoretical condensed matter physics.

Research Professor

Batson, Philip E., Ph.D., Cornell University, 1976. *Condensed Matter Physics, Materials Science, Metallurgy, Solid State Physics*. Experimental condensed matter physics.

DEPARTMENTAL RESEARCH SPECIALTIES AND STAFF

Theoretical

Astrophysics. Evolution, structure, and dynamics of galaxies, dark matter, gravitational lensing, gravitational N-body simulations. Brooks, Keeton, Sellwood, Somerville.

Biophysics. Bhanot, Case, Levy, Morozov, Olson, Sengupta, Shinbrot.

Condensed Matter Physics. Strongly correlated electron systems, novel superconductors, quantum phase transitions, quantum computing, electronic and structural properties of solids, dielectric and ferroelectric materials, magnetism and multiferroics, equilibrium and non-equilibrium statistical mechanics. Natan Andrei, Chandra, Coleman, Haule, Ioffe, Kotliar, Leath, Lebowitz, Rabe, Vanderbilt, Yuzbashyan, Zimmermann.

Mathematical and statistical physics. Statistical mechanics, foundations of quantum mechanics, mathematical physics. Goldin, Goldstein, Lebowitz, Soffer.

Nuclear Physics. Nuclear structure; quark dynamics; few-nucleon problem; relativistic heavy-ion reactions; dibaryon resonances; electron scattering; intermediate-energy, hadron scattering. Kloet, Mekjian, Zamick.

Particles and Fields. String theory, cosmology, high energy phenomenology, lattice gauge theory, conformal field theory. Natan Andrei, Banks, Diaconescu, Friedan, Lukyanov, Moore, Neuberger, Shapiro, Shih, Thomas, Zamolodchikov.

Experimental

Astrophysics. Galaxies and clusters of galaxies, cosmology and dark energy, supernovae, galaxies at high redshift, x-ray sources, imaging spectrophotometry, star clusters and dwarf galaxies. Baker, Gawiser, Hughes, Jha, Matilsky, Pryor, Williams.

Condensed Matter Physics. Surface physics: geometric structure, electronic structure, molecular adsorption, thin fills; superconductivity; electrical and thermal transport; superfluidity in helium; 2D electron gas; spin resonance; synchrotron radiation. Eva Andrei, Bartynski, Batson, Blumberg, Cheong, Croft, Feldman, Garfunkel, Gershenson, Gustafsson, Hinch, Kiryukhin, Kojima, Murnick, Oh, Podzorov, Wu, Zimmermann.

High Energy Physics. LHC physics. Chou, Gershtein, Halkiadakis, Kalelkar, Lath, Salur, Schnetzer, Somalwar.

Nuclear Physics. Nuclear structure; magnetic moments, nuclei far from stability, nuclear astrophysics; intermediate energy electron and proton scattering; neutrino scattering; relativistic heavy ion scattering. Cizewski, Gilman, Koller, Ransome, Salur.

View additional information about this department at
www.gradschoolshopper.com

STEVENS INSTITUTE OF TECHNOLOGY

DEPARTMENT OF PHYSICS AND ENGINEERING PHYSICS

Hoboken, New Jersey 07030
http://www.stevens.edu/ses/physics/

General University Information

President: Nariman Farvadin

Dean of Graduate School: Charles Suffel, Dean of Grad. Academics

University website: http://www.stevens.edu

Control: Private

Setting: Urban

Total Faculty: 450

Total Graduate Faculty: Not separated

Total number of Students: 5,950
Total number of Graduate Students: 3,600

Department Information

Department Chairman: Rainer Martini, Director

Department Contact: Diane E. Gioia, Administrative Assistant

Total full-time faculty: 9

Total number of full-time equivalent positions: 9

Full-Time Graduate Students: 23

Department Address
524 Burchard Bldg.
6th & River Streets
Hoboken, NJ 07030
Phone: (201) 216-5665
Fax: (201) 216-5638
E-mail: dgioia@stevens.edu
Website: http://www.stevens.edu/ses/physics/

ADMISSIONS

Admission Contact Information
Address admission inquiries to: Charles Suffel, Dean of Graduate
Academics
Phone: (210) 216-8031
E-mail: charles.suffel@stevens.edu
Admissions website: http://stevens.edu/admissions

Application deadlines
Fall admission:
U.S. students: March 31   *Int'l. students*: March 31
Spring admission:
U.S. students: November 30   *Int'l. students*: November 30

Application fee
U.S. students: $60

Admissions information
For Fall of 2013:
  Number of applicants: 45
  Number admitted: 32
  Number enrolled: 15

Admission requirements
Bachelor's degree requirements: Bachelor's degree in science or
engineering is required.
Minimum undergraduate GPA: 3.0

GRE requirements
The GRE is required.

Advanced GRE requirements
The Advanced GRE is required.

TOEFL requirements
The TOEFL exam is required for students from non-English-
speaking countries.
  PBT score: 530

Other admissions information
Additional requirements: No minimum acceptable score for admission is specified. The average GRE scores for admissions
were not calculated. The average GRE Advanced score for
admissions is not available.
Undergraduate preparation assumed: "Physics for Scientists and
Engineers: A Strategic Approach with Modern Physics
w/Mastering Physics" 5 volume boxed set. Author Randall
D. Knight, Publisher Pearson Addison Wesley.

TUITION
Tuition year 2013-14:
Tuition for out-of-state residents
  Full-time students: per credit
  Part-time students: per credit
  Full-time students: $1,220 per credit
  Part-time students: $1,220 per credit
Cost will probably go up 2013-2014.
Credit hours per semester to be considered full-time: 9
Deferred tuition plan: No
Health insurance: Available at the cost of $1,051 per year.

Other academic fees: $60 (application fee) World Wide Online
Enrollment 3,500.
Academic term: Semester

Teaching Assistants, Research Assistants, and Fellowships
Number of first-year
  Fellowship students: 1
Average stipend per academic year
  Teaching Assistant: $21,240
  Research Assistant: $21,240
  Fellowship student: $22,303

FINANCIAL AID

Loans
Loans are available for U.S. students.
Loans are available for international students.
GAPSFAS application required: Yes
FAFSA application required: Yes

For further information
Address financial aid inquiries to: Charles Suffel.
Phone: (201)-216-8031
E-mail: charles.suffel@stevens.edu
Financialaidwebsite:http://www.stevens.edu/sit/graduate/tuition/
index.cfm

HOUSING

Availability of on-campus housing
  Single students: Yes
  Married students: No

For further information
Address housing inquiries to: Student Housing, Trina Ballantyne,
Dean of Residence Life, Dining Services & Center Operation.
Phone: (201) 216-5128
E-mail: tballant@stevens.edu
Housing aid website: http://www.stevens.edu/housing

Table A—Faculty, Enrollments, and Degrees Granted

Research Specialty	2012-13 Faculty	Enrollment Fall 2012 Master's	Enrollment Fall 2012 Doctorate	Number of Degrees Granted 2012-13 (2008-13) Master's	Number of Degrees Granted 2012-13 (2008-13) Terminal Master's	Number of Degrees Granted 2012-13 (2008-13) Doctorate
Applied Physics	1	–	2	1(2)	–	1(2)
Atmosphere, Space Physics, Cosmic Rays	1	–	6	1(-)	–	1(10)
Atomic, Molecular, & Optical Physics	3	–	6	–(1)	–	1(8)
Condensed Matter Physics	3	–	4	1(1)	–	1(9)
Electromagnetism	–	–	–	–	–	–(1)
Optics	2	–	5	1(-)	–(1)	1(9)
Plasma and Fusion	–	–	–	–(1)	–	–(4)
Statistical & Thermal Physics	–	–	–	–	–	–(1)
Non-specialized	–	–	–	6(15)	–(18)	–
Total	10	–	23	10(20)	–(19)	3(44)
Full-time Grad. Stud.	–	–	23	–	–	–
First-year Grad. Stud.	–	–	–	–	–	–

GRADUATE DEGREE REQUIREMENTS

Master's: Thirty semester hour credits; 3.0 GPA in physics courses and overall; no residence requirement; no language or other comprehensive exams; thesis optional.

Doctorate: Ninety semester hour credits (including Master's credits) of which 50 minimum to be in courses, and 30 minimum to be dissertation research; 3.0 GPA in physics courses and overall; one year residence; no language requirement; comprehensive/qualifying exam (one combined exam); dissertation.

SPECIAL EQUIPMENT, FACILITIES, OR PROGRAMS

The Department has particular strength in areas of applied physics, such as optics, atomic and plasma physics, nanotechnology, and condensed matter physics.

Table B—Separately Budgeted Research Expenditures by Source of Support

Source of Support	Departmental Research	Physics-related Research Outside Department
Federal government	$2,276,337	
State/local government		
Non-profit organizations		
Business and industry		
Other		
Total	**$2,276,337**	

Table C—Separately Budgeted Research Expenditures by Research Specialty

Research Specialty	No. of Grants	Expenditures ($)
Applied Physics	–	
Atmosphere, Space Physics, Cosmic Rays	2	$792,270
Atomic, Molecular, & Optical Physics	1	$197,214
Condensed Matter Physics	1	$428,587
Electromagnetism	–	
Optics	1	$848,266
Total	**5**	**$2,266,337**

FACULTY

Professor

Carr, Wayne E., Ph.D., University of Illinois, 1967. *Computational Physics, Plasma and Fusion*. Plasma physics; electron+ion beams; computational physics.

Horing, Norman J., Ph.D., Harvard University, 1964. *Computational Physics, Condensed Matter Physics*. Quantum many-particle theory; solid state physics; surface physics; high-magnetic field effects.

Stamnes, Knut, Ph.D., University of Colorado, 1978. Director of the Physics/Engineering Physics Department. *Atmosphere, Space Physics, Cosmic Rays*. Electron transport and thermalization; kinetic theory; radiation transport; satellite remote

sensing; biophotonics for non-invasive diagnostic of biological tissue.

Whittaker, Edward A., Ph.D., Columbia University, 1982. *Optics*. Laser techniques; optical diagnosis of gas phase materials; processing reactors; Brillouin scattering; quantum optics.

Associate Professor

Malinovskaya, Svetlana, Ph.D., Novosibirsk State University, 1993. *Other*. Laser-matter interaction; coherent control; quantum optics.

Martini, Rainer, Rheinisch-Westfaelische Technischen Hochschule Aachen, Germany, 1999. Serves on numerous Institute Committees. *Optics, Other*. High-sensitivity laser spectroscopy.

Search, Christopher, Ph.D., University of Michigan, 2002. *Atomic, Molecular, & Optical Physics*. Bose-Einstein condensation; quantum optics; nonlinear optics.

Strauf, Stefan, Universität Bremen, Germany, 2001. *Nano Science and Technology, Optics, Other*. Nanophotonics; quantum optics.

Yu, Ting, Ph.D., Imperial College, University of London, UK, 1998. *Atomic, Molecular, & Optical Physics*. Atomic, Molecular and Optical Physics (AMO); Quantum Information and Quantum Optics.

Adjunct Associate Professor

Hutt, Marvin, Ph.D., New York University, 1987. *Optics*. Optical engineering.

Lenzing, Harry, Stevens Institute of Technology, 1962. *Optics*. Satellite-tracking systems; passive intermodulation (PIM).

Webb, Robert, M.S., New York University, 1966. *Optics*. Optics.

Visiting Professor

Supplee, James, Ph.D., University of Texas, Dallas, 1979. *Optics*. Spectroscopy and semiclassical optics.

DEPARTMENTAL RESEARCH SPECIALTIES AND STAFF

Theoretical

Electron Transport. Horing.

Ion Surface Interactions. Horing.

Laser Matter Interactions. Malinovskaya.

Quantum Information and Quantum Optics. Yu.

Quantum Optics. Malinovskaya, Search.

Semiconductor Solid State Theory. Horing.

Experimental

Atmosphere, Space Physics, Cosmic Rays. Radiation transport in planetary media including the coupled atmosphere-snow/ice-ocean system. Stamnes.

High Sensitivity Laser Spectroscopy. Martini, Strauf, Supplee, Whittaker.

Nanophotonics. Strauf.

Optical Control of Quantum Systems. Martini, Strauf.

Optical Diagnosis of Gas Phase Materials. Whittaker.

Other.

View additional information about this department at
www.gradschoolshopper.com

NEW MEXICO INSTITUTE OF MINING AND TECHNOLOGY

DEPARTMENT OF PHYSICS

Socorro, New Mexico 87801
http://physics.nmt.edu/

General University Information
President: Daniel Lopez
Dean of Graduate School: Lorie Liebrock
University website: http://www.nmt.edu
Control: Public
Setting: Rural
Total Faculty: 112
Total Graduate Faculty: 112
Total number of Students: 1,928
Total number of Graduate Students: 539

Department Information
Department Chairman: Michelle Creech-Eakman, Chair
Department Contact: Altagracia Lujan, Department Specialist
  Total full-time faculty: 12
  Total number of full-time equivalent positions: 12
  Full-Time Graduate Students: 32
  First-Year Graduate Students: 4
  Female First-Year Students: 2
  Total Post Doctorates: 2

Department Address
801 Leroy Place
Workman Center
Socorro, NM 87801
Phone: (575) 835-5328
Fax: (575) 835-5707
E-mail: physics@kestrel.nmt.edu
Website: http://physics.nmt.edu/

ADMISSIONS

Admission Contact Information
Address admission inquiries to: Dean of Graduate Studies, New
  Mexico Tech.
Phone: (575) 835-5513
E-mail: graduate@nmt.edu
Admissions website: http://www.nmt.edu/information

Application deadlines
Fall admission:
U.S. students: January 15   *Int'l. students*: January 15
Spring admission:
U.S. students: September 15   *Int'l. students*: September 15

Application fee
U.S. students: $45   *Int'l. students*: $45

Admissions information
For Fall of 2013:
  Number of applicants: 20
  Number admitted: 3
  Number enrolled: 2

Admission requirements
Bachelor's degree requirements: A Bachelor's degree in physics
  or closely related field is required.
Minimum undergraduate GPA: 3.0

GRE requirements
The GRE is required.

Advanced GRE requirements
The Advanced GRE is required.

TOEFL requirements
The TOEFL exam is required for students from non-English-
  speaking countries.

Other admissions information
Undergraduate preparation assumed: Traditional undergraduate
  physics program is assumed.

TUITION

Tuition year 2012–13:
Tuition for in-state residents
  Full-time students: $5,043 per semester
  Part-time students: $280.19 per credit
Tuition for out-of-state residents
  Full-time students: $16,682 other
  Part-time students: $926.8 per credit
Tuition is waived for students with assistantships (in most cases).
Credit hours per semester to be considered full-time: 12
Deferred tuition plan: Yes
Health insurance: Available at the cost of varies per year.
Academic term: Semester
Number of first-year students who received full tuition waivers: 2

Teaching Assistants, Research Assistants, and Fellowships
  Number of first-year
    Teaching Assistants: 1
    Research Assistants: 1
  Average stipend per academic year
    Teaching Assistant: $19,000
    Research Assistant: $21,000

FINANCIAL AID

Application deadlines
Fall admission:
U.S. students: September 15   *Int'l. students*: September 15
Spring admission:
U.S. students: January 15   *Int'l. students*: January 15

Loans
Loans are available for U.S. students.
Loans are not available for international students.
GAPSFAS application required: No
FAFSA application required: Yes

For further information
Address financial aid inquiries to: Graduate Office, New Mexico
  Tech (for TA/RA information: Dean of Graduate Studies,
  New Mexico Tech).
Phone: (575) 835-5513
E-mail: graduate@nmt.edu
Financial aid website: http://www.nmt.edu/financial-aid

HOUSING

Availability of on-campus housing
  Single students: Yes
  Married students: Yes

For further information

Address housing inquiries to: Auxiliary Services, New Mexico Tech.
Phone: (575) 835-5050
E-mail: Residential_Life@admin.nmt.edu
Housing aid website: http://www.nmt.edu/welcome-to-res-life

Table A—Faculty, Enrollments, and Degrees Granted

Research Specialty	2012-13 Faculty	Enrollment Fall 2012		Number of Degrees Granted 2012-13 (2007-13)		
		Master's	Doctorate	Master's	Terminal Master's	Doctorate
Astrophysics	6	5	20	1(15)	1(9)	1(4)
Climate/Atmospheric Science	6	2	3	1(4)	–	1(4)
Other	1	1	–	–	–	–(1)
Total	39	8	23	2(19)	1(9)	2(9)
Full-time Grad. Stud.	–	8	23	–	–	–
First-year Grad. Stud.	–	–	4	–	–	–

GRADUATE DEGREE REQUIREMENTS

Master's: With thesis: 24 hours of coursework plus six hours of thesis work; without thesis: 27 hours of coursework plus three hours of independent study, including a paper. Minimum GPA: 3.0, no grade less than C. Students should be in residence a minimum of 18 hours. There is no language requirement. Physics preliminary examination is required. Thesis topic is chosen after consultation with student's advisory committee.

Doctorate: Fifty hours of coursework minimum beyond the B.A. is required. Minimum GPA is 3.0/4.0. Students should be in residence a minimum of three semester hours; degree in absentia may be granted by petition. There is no foreign language requirement. Physics preliminary examination is required. Dissertation, oral defense, and paper submitted for publication to a recognized journal are also required.

Thesis: Thesis may be written in absentia, in part, by petition.

SPECIAL EQUIPMENT, FACILITIES, OR PROGRAMS

The Langmuir Laboratory for Atmospheric Research is located in the Magdalena Mountains at an elevation of 10,800 feet. The laboratory offers an unparalleled facility for atmospheric research, particularly in the areas of lightning physics, atmospheric electricity, and atmospheric chemistry. New Mexico Tech is also a member of the University Corporation for Atmospheric Research (UCAR), the institution that oversees the operation of the National Center for Atmospheric Research (NCAR).

New Mexico Tech's Magdalena Ridge Observatory (MRO) is under construction in the Magdalena Mountains. The optical interferometer will be a unique facility once completed, allowing for high-resolution imaging of stellar systems and complex astrophysical phenomena. In addition to the MRO, our astrophysics program takes advantage of the Very Large Array (VLA) and Very Long Baseline Array (VLBA) of the National Radio Astronomy Observatory (NRAO); its Domenici Science Operations Center is located on the Tech campus.

Table B—Separately Budgeted Research Expenditures by Source of Support

Source of Support	Departmental Research	Physics-related Research Outside Department
Federal government	$8,400,000	$3,000,000
State/local government	$20,000	
Non-profit organizations		
Business and industry		
Other	$69,000	
Total	$8,489,000	$3,000,000

Table C—Separately Budgeted Research Expenditures by Research Specialty

Research Specialty	No. of Grants	Expenditures ($)
Astrophysics	15	$4,000,000
Climate/Atmospheric Science	16	$4,489,000
Total	31	$8,489,000

FACULTY

Professor

Hofner, Peter, Ph.D., University of Wisconsin-Madison, 1995. *Astronomy, Astrophysics*. Star formation; interstellar medium; x-ray astronomy; extragalactic interstellar.

Minschwaner, Kenneth, Ph.D., Harvard University, 1992. *Climate/Atmospheric Science*. Radiative transfer and climate; physics of the upper atmosphere.

Raymond, David J., Ph.D., Stanford University, 1970. *Climate/Atmospheric Science*. Cloud physics; geophysical fluid dynamics; clouds and climate.

Romero, Van, Ph.D., SUNY, Albany, 1991. Energetic materials; shock phenomena; high-energy physics.

Westpfahl, David J., Ph.D., Montana State University, 1985. *Astrophysics*. Dynamics of spiral and dwarf galaxies.

Associate Professor

Creech-Eakman, Michelle, Ph.D., University of Denver, 1997. *Astrophysics*. Stellar astrophysics and mass-loss; optical/IR interferometry; IR instrumentation.

Eack, Kenneth, Ph.D., University of Oklahoma, 1997. *Climate/Atmospheric Science*. Thunderstorm electrification; atmospheric physics.

Morales Juberias, Raúl, Ph.D., University of the Basque Country, Vizcaya (Spain), 2002. *Atmosphere, Space Physics, Cosmic Rays*. Outer planets observations and atmospheric dynamics.

Sessions, Sharon L., Ph.D., University of Oregon, 2002. Atmospheric dynamics; statistical mechanics; many-body theory.

Sonnenfeld, Richard, Ph.D., University of California, Santa Barbara, 1987. Experimental physics.

Young, Lisa, Ph.D., University of New Mexico, 1987. *Astrophysics*. Astrophysics; elliptical galaxies.

Assistant Professor

Arendt, Paul, Ph.D., New Mexico Institute of Mining and Technology, 2003. *Astrophysics*. Elementary processes in extreme electromagnetic fields of astrophysics.

Meier, David, Ph.D., University of California, Los Angeles, 2002. *Astrophysics*. Molecular gas chemistry in star formation galaxies; earliest phases of starburst evolution in nearby galaxies.

Emeritus

Eilek, Jean, Ph.D., University of British Columbia, 1975. *Astrophysics*. Plasma astrophysics; quasars; radio galaxies; pulsars.

Hankins, Timothy H., Ph.D., University of California, San Diego, 1971. *Astrophysics*. Radio astronomy of pulsars; instrumentation; signal processing.

Krehbiel, Paul R., Ph.D., University of Manchester, 1982. *Climate/Atmospheric Science*. Lightning studies; radar meteorology; thunderstorm electrification.

LeFebre, Vernon G., Ph.D., University of Utah, 1963. Statistical physics; thermodynamics.

Schery, Stephen C., Ph.D., University of Colorado, 1973. Environmental radioactivity.

Winn, William P., Ph.D., University of California, Berkeley, 1966. *Climate/Atmospheric Science*. Atmospheric physics; electrical discharges in gases; instrumentation.

Adjunct Professor

Avramidi, Ivan, Ph.D., Moscow State University, Russia, 1987. Mathematical physics; analysis on manifolds; quantum field theory.

Buscher, David, Ph.D., University of Cambridge. Optical/IR interferometry; atmospheric seeing measurement; adaptive optics; early and late stages of stellar evolution.

Colgate, Stirling A., Ph.D., Cornell University, 1948. *Astrophysics, Plasma and Fusion*. Astrophysics; plasma physics; atmospheric physics.

Elvis, Martin, Ph.D., University of Leicester, 1978. Quasars and active galactic nuclei; x-ray astronomy.

Fuchs, Zeljka, Ph.D., New Mexico Institute of Mining and Technology. Atmospheric dynamics.

Goss, W. Miller, Ph.D., University of California, Berkeley, 1967. Radio astronomy; interstellar medium.

Haniff, Chris, Ph.D., University of Cambridge, 1988. Spatial interferometry at optical and near-infrared wavelengths; atmospheric turbulence; imaging theory; evolved stars.

Jurgenson, Colby, Ph.D., University of Denver, 2005. State-of-the-art infrared astronomical instrumentation.

Klinglesmith, Dan, Ph.D., Indiana University, 1967. Asteroids; robotic telescope operations.

Lopez-Carrillo, Carlos, Ph.D., New Mexico Institute of Mining and Technology, 2001. Doppler radar and data analysis; tropical dynamics.

Manney, Gloria, Ph.D., Iowa State University, 1988. Atmospheric science; stratospheric dynamics/transport; stratospheric polar processes and ozone loss.

Meason, John, Ph.D., University of Arkansas, 1965. Nuclear physics; nuclear and space radiation effects; electromagnetic radiation effects and directed energy.

Mihalas, Dimitri, Ph.D., California Institute of Technology, 1963. Theoretical astrophysics; stellar atmospheres.

Myers, Steven, Ph.D., California Institute of Technology, 1990. Cosmology; extragalactic radio astronomy; interferometric imaging algorithms.

Owen, Frazier, Ph.D., Radio galaxies.

Rison, William, Ph.D., University of Utah, 1980. Atmospheric electricity; radar meteorology; instrumentation.

Rupen, Michael, Ph.D., Princeton University, 1989. Gas and dust in galaxies; radio transients.

Taylor, Gregory, Ph.D., University of California, Los Angeles, 1991. Very long baseline radio astronomy; active galactic nuclei.

Teare, Scott, Ph.D., University of Guelph, 1991. Adaptive optics; instrumentation; astrophysics.

Thomas, Ronald J., Ph.D., Utah State University, 1970. Atmospheric physics; instrumentation.

Wrobel, Joan, Ph.D., University of Toronto, 1983. VLBI.

DEPARTMENTAL RESEARCH SPECIALTIES AND STAFF

Theoretical

Astrophysics. Arendt, Eilek, Owen.

Climate/Atmospheric Science. Fuchs, Raymond, Sessions, Winn.

Theoretical Physics. Arendt, Avramidi, Sessions.

Experimental

Astrophysics. Radio and x-ray astronomy; binary and variable stars; pulsars; cometary physics; young stars; evolved stars; astronomical instrumentation. Buscher, Colgate, Creech-Eakman, Elvis, Goss, Haniff, Hankins, Hofner, Jurgenson, Klinglesmith, Meier, Morales Juberias, Owen, Rupen, Taylor, Westpfahl, Wrobel, Young.

Climate/Atmospheric Science. Clouds and lightning; electrical phenomena; pressure waves and pulses; weather radar; precipitation mechanisms; tornadoes; atmospheric charges in gas; radiative transfer and climate; physics of the upper atmosphere; ice tribocharging; embedded instrumentation. Eack, Krehbiel, Lopez-Carrillo, Manney, Minschwaner, Morales Juberias, Raymond, Rison, Sessions, Sonnenfeld, Thomas, Winn.

Energy Sources & Environment. Geothermal energy and energetic materials. Meason, Romero.

View additional information about this department at
www.gradschoolshopper.com

NEW MEXICO STATE UNIVERSITY

DEPARTMENT OF ASTRONOMY

Las Cruces, New Mexico 88003
http://astronomy.nmsu.edu/

General University Information

President: Garrey Carruthers
Dean of Graduate School: Loui Reyes (interim)
University website: http://nmsu.edu
Control: Public

Setting: Suburban
Total Faculty: 1,154
Total Graduate Faculty: 665
Total number of Students: 18,024
Total number of Graduate Students: 3,529

Department Information
Department Chairman: Jon Holtzman, Head
Department Contact: Ofelia Ruiz, Inter. Admin. Assistant
  Total full-time faculty: 10
  Total number of full-time equivalent positions: 10
  Full-Time Graduate Students: 31
  First-Year Graduate Students: 4
  Female First-Year Students: 2
  Total Post Doctorates: 5

Department Address
PO Box 30001, MSC 4500
Las Cruces, NM 88003
Phone: (575) 646-4438
Fax: (575) 646-1602
E-mail: oruiz@nmsu.edu
Website: http://astronomy.nmsu.edu/

ADMISSIONS

Admission Contact Information
Address admission inquiries to: Dr. Nancy Chanover, Dept. of Astronomy, Box 30001/MSC 4500, Las Cruces, NM 88003-0001
Phone: (575) 646-2567
E-mail: gradapps@astronomy.nmsu.edu
Admissions website: http://astronomy.nmsu.edu/dept/html/academics.admissions.shtml

Application deadlines
Fall admission:
U.S. students: February 1   *Int'l. students*: February 1

Application fee
U.S. students: $40   *Int'l. students*: $50
NMSU uses an on-line application system, the Hobsons ApplyYourself system. Please email gradapps@astronomy.nmsu.edu if you experience any problems with this system.

Admissions information
For Fall of 2013:
  Number of applicants: 60
  Number admitted: 15
  Number enrolled: 4

Admission requirements
Bachelor's degree requirements: Bachelors degree in astronomy, physics, other science, or engineering is required.
Minimum undergraduate GPA: 3.0

GRE requirements
The GRE is required.

Advanced GRE requirements
The Advanced GRE is required.
The Physics subject GRE test is required for admission to the NMSU Astronomy graduate program.

TOEFL requirements
The TOEFL exam is required for students from non-English-speaking countries.
  PBT score: 530

Other admissions information
Additional requirements: No minimum score for admission is specified. Typical average scores for GRE verbal–570; quantitative–700.
Undergraduate preparation assumed: Math: differential equations; Physics: mechanics, modern physics, some of optics, electricity and magnetism, statistical mechanics, thermodynamics, etc.

TUITION
Tuition year 2013-14:
Tuition for in-state residents
  Full-time students: $3,256.8 per semester
  Part-time students: $271.4 per credit
Tuition for out-of-state residents
  Full-time students: $9,770.4 per semester
  Part-time students: $814.2 per credit
  Full-time students: per semester
  Part-time students: per credit
Credit hours per semester to be considered full-time: 9
Deferred tuition plan: Yes
Health insurance: Yes, $15.62/pay period.
Academic term: Semester
Number of first-year students who received partial tuition waivers: 5

Teaching Assistants, Research Assistants, and Fellowships
Number of first-year
  Teaching Assistants: 4
Average stipend per academic year
  Teaching Assistant: $16,500
  Research Assistant: $16,500
Most students usually get RA support in the summer, which would bring the total year stipend to around $22,000.

FINANCIAL AID

Loans
Loans are available for U.S. students.
Loans are not available for international students.
GAPSFAS application required: No
FAFSA application required: Yes

For further information
Address financial aid inquiries to: Department of Astronomy, Box 30001/MSC 4500, Las Cruces, NM 88003-8001.
Phone: (575) 646-4438
E-mail: gradapps@astronomy.nmsu.edu

HOUSING

Availability of on-campus housing
  Single students: Yes
  Married students: Yes

For further information
Address housing inquiries to: Housing Department, Box 30001/MSC 3BB, Las Cruces, NM 88003-0001.
Phone: (575) 646-3202
E-mail: housing@nmsu.edu
Housing aid website: http://www.nmsu.edu/~housing

Table A—Faculty, Enrollments, and Degrees Granted

Research Specialty	2012–13 Faculty	Enrollment Fall 2013		Number of Degrees Granted 2012–13 (2008–12)		
		Master's	Doctorate	Master's	Terminal Master's	Doctorate
Astronomy	13	12	17	5(-)	–	4(-)
Total	–	12	17	–(5)	–	4(4)
Full-time Grad. Stud.	–	12	17	–	–	–
First-year Grad. Stud.	–	4	–	–	–	–

GRADUATE DEGREE REQUIREMENTS

Master's: The M.S. program is closely geared to the Ph.D. program, and prospective students are requested to contact the department.

Doctorate: A minimum of 64 credits of graduate work in astronomy and related fields of which 33 are in formal courses. Qualification is ascertained during the student's third semester and a comprehensive oral given after formal course work is completed. Written exam evaluations are based on monthly cumulative examinations. A dissertation and a final oral examination on the dissertation is also required. The residence requirement is two consecutive semesters of full-time graduate work after the first 30 credits.

Thesis: Thesis may be written in absentia.

SPECIAL EQUIPMENT, FACILITIES, OR PROGRAMS

The Department is a member of the Astrophysical Research Consortium (ARC), which operates a state-of-the-art 3.5 m telescope at Apache Point, NM, and the Sloan Digital Sky Survey with a dedicated 2.5-m telescope. The Department also operates a 1m telescope at Apache Point. The department is home to NASA's Planetary Data System's Atmospheres Node archive of planetary atmosphere-related mission data.

Table B—Separately Budgeted Research Expenditures by Source of Support

Source of Support	Departmental Research	Physics-related Research Outside Department
Federal government	$1,920,115	
State/local government	$488,016	
Non-profit organizations		
Business and industry		
Other	$2,905,918	
Total	**$5,314,049**	

Table C—Separately Budgeted Research Expenditures by Research Specialty

Research Specialty	No. of Grants	Expenditures ($)
Astronomy	83	$5,314,049
Total	**83**	**$5,314,049**

FACULTY

Professor

Beebe, Reta F., Ph.D., Indiana University, 1969. *Astronomy, Planetary Science*. Planetary atmospheres; planetary physics; radiative transfer; cool star atmospheres; equation of state.

Holtzman, Jon, Ph.D., University of California, Santa Cruz, 1989. *Astronomy*. Stellar populations and chemical abundances. Spiral galaxies. Instrument development.

Klypin, Anatoly, Ph.D., University of Moscow, 1980. *Astronomy, Cosmology & String Theory*. Extragalactic astronomy; cosmology.

McNamara, Bernard J., Ph.D., University of California, Santa Cruz, 1975. *Astronomy*. Observational astronomy; photoelectric photometry; star formation; cluster photometry and membership.

Walterbos, Reinirus, Ph.D., Leiden University, 1986. *Astronomy*. Interstellar medium; stellar populations; extragalactic.

Webber, William R., Ph.D., University of Iowa, 1957. *Astronomy, Atmosphere, Space Physics, Cosmic Rays, Planetary Science*. Interplanetary physics; cosmic rays; delta-ray astronomy.

Associate Professor

Chanover, Nancy, Ph.D., New Mexico State University, 1997. *Astronomy, Planetary Science*. Planetary atmospheres. Astrobiology. Instrument Development.

Churchill, Chris, Ph.D., University of California, Santa Cruz, 1997. *Astronomy*. Quasar absorption lines galaxies and intergalactic medium.

Murphy, James, Ph.D., University of Washington, 1991. *Astronomy, Planetary Science*. Atmospheric sciences; planetary atmospheres, Mars exploration missions.

Vogt, Nicole, Ph.D., Cornell University, 1995. *Astronomy*. Galaxies; galaxy evolution.

Assistant Professor

Jackiewicz, Jason, Ph.D., Boston University, 2002. *Astronomy, Solar Physics*. Helioseismology.

McAteer, James, Ph.D., Queens University, Belfast, 2003. Solar physics. *Astronomy, Solar Physics*.

Emeritus

Beebe, Herbert, Ph.D., Indiana University, 1969. *Atmosphere, Space Physics, Cosmic Rays*. Atmospheres; spectral line formation.

Research Faculty

Harrison, Tom, Ph.D., University of Minnesota, 1989. *Astronomy*. Infrared astronomy.

Research Fellow

Johnson, Joni, Ph.D., University of Minnesota, 1990. *Astronomy*. Cataclysmic variables.

DEPARTMENTAL RESEARCH SPECIALTIES AND STAFF

Theoretical

Astrophysics. Klypin.

Cosmology & String Theory. Models of extragalactic objects; star formation. Klypin.

Solar Physics. Jackiewicz, McAteer.

Experimental

Astronomy. Herbert Beebe, Reta Beebe, Chanover, Churchill, Harrison, Holtzman, Jackiewicz, Johnson, Klypin, McAteer, McNamara, Murphy, Vogt, Walterbos, Webber.

Planetary Science. Reta Beebe, Chanover, Murphy.

Solar Physics. Jackiewicz, McAteer.

View additional information about this department at
www.gradschoolshopper.com

NEW MEXICO STATE UNIVERSITY

DEPARTMENT OF PHYSICS

Las Cruces, New Mexico 88003-8001
http://physics.nmsu.edu

General University Information
President: Garrey Carruthers
Dean of Graduate School: Loui Reyes
University website: http://www.nmsu.edu
Control: Public
Setting: Rural
Total Faculty: 1,153
Total Graduate Faculty: 809
Total number of Students: 17,651
Total number of Graduate Students: 3,375

Department Information
Department Chairman: Stefan Zollner, Head
Department Contact: Vassilios Papavassiliou, Graduate
 Program Head
 Total full-time faculty: 15
 Total number of full-time equivalent positions: 15
 Full-Time Graduate Students: 36
 First-Year Graduate Students: 7
 Female First-Year Students: 3
 Total Post Doctorates: 2

Department Address
1255 N. Horseshoe Drive
MSC 3D
Las Cruces, NM 88003-8001
Phone: (575) 646-3831 (C)
Fax: (575) 646-1934
E-mail: loretcha@nmsu.edu
Website: http://physics.nmsu.edu

ADMISSIONS

Admission Contact Information
Address admission inquiries to: Graduate Student Services, MSC
 3G, New Mexico State University, P.O. Box 30001, Las Cru-
 ces, NM 88003-8001
Phone: (575) 646-2736
E-mail: gradinfo@nmsu.edu
Admissions website: http://prospective.nmsu.edu/graduate/index.
 html

Application deadlines
Fall admission:
U.S. students: February 15   *Int'l. students*: February 15
Spring admission:
U.S. students: November 15   *Int'l. students*: September 1

Application fee
U.S. students: $40   *Int'l. students*: $50

Admissions information
For Fall of 2013:
 Number of applicants: 36
 Number admitted: 18
 Number enrolled: 10

Admission requirements
Bachelor's degree requirements: Bachelor's degree in physics
 or a related field is required.
Minimum undergraduate GPA: 3.0

GRE requirements
The GRE is required.
No minimum GRE scores have been established.

Advanced GRE requirements
The Advanced GRE is required.
Physics GRE is required for graduate assistantships. No min-
 imum score has been established.

TOEFL requirements
The TOEFL exam is required for students from non-English-
 speaking countries.
 PBT score: 550
 iBT score: 79
IELTS scores are also accepted as an alternative to TOEFL. Minimum IELTS score is 6.5 for regular admission. Applicants
 may be admitted to NMSU with scores lower than those above,
 but admission in such cases will be either 'tentative' or 'provisional;' see http://prospective.nmsu.edu/international/language/
 index.html for details.

Other admissions information
Additional requirements: A minimum GPA of 3.0 or equivalent
 is required for admission as a Ph.D. student or for regular
 (not provisional) admission as a Master's student.
Undergraduate preparation assumed: Physics: Marion, Classical
 Dynamics; Griffiths, Electromagnetism; Kittel and Kroemer,
 Thermal Physics; Townsend, Quantum Mechanics. Mathematics: Thomas, Calculus and Analytic Geometry; Boyce and
 di Prima, Boundary Value Problems.

TUITION

Tuition year 2011–12:
Tuition for in-state residents
 Full-time students: $279.5 per credit
 Part-time students: $279.5 per credit
Tuition for out-of-state residents
 Full-time students: $838.5 per credit
 Part-time students: $838.5 per credit
Out-of-state students awarded a graduate assistantship are also
 offered a waiver of out-of-state tuition; such students pay the
 in-state tuition rates.
Credit hours per semester to be considered full-time: 9
Deferred tuition plan: Yes
Health insurance: Available at the cost of 1166. per year.
Other academic fees: Tuition and fees are described in detail in
 http://www.nmsu.edu/~uar/Info%20Docs/TUITION%20
 AND%20FEES%20201240.pdf.
Academic term: Semester
Number of first-year students who received partial tuition waivers: 7

Teaching Assistants, Research Assistants, and Fellowships
 Number of first-year
 Teaching Assistants: 5
 Research Assistants: 2

Average stipend per academic year
Teaching Assistant: $18,300
Research Assistant: $18,500
Fellowship student: $18,100

FINANCIAL AID

Application deadlines
Fall admission:
U.S. students: February 15   *Int'l. students*: February 15
Spring admission:
U.S. students: November 15   *Int'l. students*: September 1

Loans
Loans are available for U.S. students.
Loans are not available for international students.
GAPSFAS application required: No
FAFSA application required: Yes

For further information
Address financial aid inquiries to: NMSU Financial Aid and scholarship Services, Educational Services Building, MSC 5100, P.O. Box 30001, Las Cruces, NM 88003-8001.
Phone: (575) 646-4105
E-mail: financialaid@nmsu.edu
Financial aid website: http://fa.nmsu.edu/index.html

HOUSING

Availability of on-campus housing
Single students: Yes
Married students: Yes

For further information
Address housing inquiries to: NMSU Department of Housing and Residential Life, Educational Services Center, Suite H, Housing and Residential Life, MSC 3BB, P.O. Box 30001, Las Cruces, NM 88003-8001.
Phone: (575) 646-3202
E-mail: housing@nmsu.edu
Housing aid website: http://www.nmsu.edu/~housing/

Table A—Faculty, Enrollments, and Degrees Granted

Research Specialty	Spring 2012 Faculty	Enrollment Spring 2012 Mas-ter's	Enrollment Spring 2012 Doc-torate	Number of Degrees Granted 2012-13 (2008-13) Mas-ter's	Number of Degrees Granted 2012-13 (2008-13) Terminal Master's	Number of Degrees Granted 2012-13 (2008-13) Doc-torate
Atomic, Molecular, & Optical Physics	1	–	3	–	2(2)	1(6)
Geophysics	2	–	2	–	2(2)	-(3)
Materials Science, Metallurgy	5	–	13	–	1(1)	1(9)
Nuclear Physics	6	1	8	–	–	1(12)
Physics and other Science Education	1	2	12	–	–	–
Total	15	3	38	–	5(5)	3(30)
Full-time Grad. Stud.	–	–	38	–	–	–
First-year Grad. Stud.	–	–	7	–	–	–

GRADUATE DEGREE REQUIREMENTS

Master's: Satisfactory completion of a minimum of 30 semester credits, of which at least 21 credits are in formal courses, including one laboratory course; a minimum residence requirement of two consecutive semesters of full-time graduate work; no foreign language required; overall GPA of at least

3.0 at time of application for admission to candidacy. Several program options available: (1) thesis, qualifying examination, and final oral exam; (2) successful completion of doctoral comprehensive exam; (3) course work option (at least 27 credits in formal courses) and successful completion of qualifying exam and final oral exam.

Doctorate: At least 36 credits in formal courses, and a minimum of 72 total credits of graduate work are required. These include 24 credits of graduate core courses and one laboratory course; foreign language not required; qualifying examination, comprehensive exam, dissertation, and final oral on dissertation required; a minimum residence requirement of two consecutive semesters of full-time graduate work.

Other Degrees: Space physics concentration for M.S. physics degree.

Thesis: Thesis may be written in absentia.

SPECIAL EQUIPMENT, FACILITIES, OR PROGRAMS

Cooperative research programs are conducted that involve using the particle accelerators at Fermilab and Brookhaven National Laboratory; with the Neutron Science Center (LANSCE) and the Earth and Environmental Sciences Division of Los Alamos National Laboratory; with Sandia National Laboratories; with the Physical Science Laboratory at NMSU; and with local industries. Specialized research equipment on campus includes 10-meter small angle x-ray diffractometer, liquids energy dispersive diffractometer, 2 two-crystal vacuum x-ray spectrometers, a high-resolution electron spectrometer, an energy dispersive x-ray spectrometer, 1 electron microscope, specimen preparation equipment, ultra high vacuum chambers, mass spectrometers, an x-ray photoemission spectrometer, low-energy electron diffraction, spherical and planar Fabry-Perot interferometers, a near-field optical microscope, an atomic force microscope, a femtosecond laser, a nanosecond optical parametric oscillator, a tunable diode laser, standalone Nd: YAG lasers, 1- and 2D detector arrays, a streak camera, a gravity meter, a magnetometer, and a broadband seismometer. Departmental computing facilities include Beowulf clusters and numerous Windows, MAC, and Linux workstations.

Table B—Separately Budgeted Research Expenditures by Source of Support

Source of Support	Departmental Research	Physics-related Research Outside Department
Federal government	$1,800,000	
State/local government		
Non-profit organizations		
Business and industry		
Other		
Total	$1,800,000	

FACULTY

Professor

Burkardt, Matthias, Ph.D., University of Erlangen-Nürnberg, 1989. Undergraduate physics program head. *High Energy Physics, Nuclear Physics, Particles and Fields, Theoretical Physics*. Quantum chromodynamics, nucleon structure.

Gibbs, William R., Ph.D., Rice University, 1961. Comprehensive exam chair. *Nuclear Physics, Particles and Fields, Theoretical Physics*. Hadronic interactions, quantum chromodynamics.

Nakotte, Heinrich, Ph.D., University of Amsterdam, 1994. Undergraduate advisor for engineering physics students. *Con-*

densed *Matter Physics, Solid State Physics*. Magnetic properties of materials, neutron scattering.

Ni, James F., Ph.D., Cornell University, 1984. *Geophysics*. Observational seismology, mantle dynamics, continental rifting.

Pate, Stephen F., Ph.D., University of Pennsylvania, 1987. Undergraduate advisor. *High Energy Physics, Nuclear Physics, Particles and Fields*. Nucleon structure, spin physics, relativistic heavy ion physics, electron and neutrino scattering.

Zollner, Stefan, Ph.D., Universitat Stuttgart, 1991. Department Head. *Applied Physics, Materials Science, Metallurgy, Nano Science and Technology, Optics, Solid State Physics*. Optical properties of materials, spectroscopic ellipsometry, semiconductor process integration.

Associate Professor

Engelhardt, Michael, Ph.D., University Erlangen, 1994. Undergraduate advisor. *Computational Physics, High Energy Physics, Nuclear Physics, Particles and Fields, Theoretical Physics*. Nucleon structure, quantum chromodynamics.

Hearn, Thomas M., Ph.D., California Institute of Technology, 1985. Undergraduate advisor. *Geophysics*. Seismic tomography, seismology.

Kanim, Stephen, Ph.D., University of Washington, 1999. Undergraduate advisor. *Other*. Physics education.

Kiefer, Boris, Ph.D., University of Michigan, 2003. Undergraduate advisor, SPS advisor. *Applied Physics, Computational Physics, Condensed Matter Physics, Energy Sources & Environment, Nano Science and Technology, Solid State Physics*. Computational material science, energy conversion technologies, earth and planetary materials.

Papavassiliou, Vassilios, Ph.D., Yale University, 1988. Graduate Program Head. *High Energy Physics, Nuclear Physics, Particles and Fields*. Nucleon structure, spin physics, relativistic heavy ion physics, electron and neutrino scattering.

Urquidi, Jacob, Ph.D., Texas Tech University, 2001. *Atomic, Molecular, & Optical Physics, Condensed Matter Physics, Fluids, Rheology*. Studies of materials using x-ray and neutron scattering.

Vasiliev, Igor V., Ph.D., University of Minnesota, 2000. *Computational Physics, Condensed Matter Physics, Nano Science and Technology, Solid State Physics, Theoretical Physics*. Theoretical condensed matter physics.

Assistant Professor

Fohtung, Edwin, Ph.D., Albert-Ludwig's University of Freiburg, 2010. *Applied Physics, Condensed Matter Physics, Nano Science and Technology, Solid State Physics*.

Wang, Xiaorong, Ph.D., Central China Normal University, 2002. *High Energy Physics, Nuclear Physics, Particles and Fields*. Nucleon structure, spin physics, relativistic heavy ion physics.

Emeritus

Armstrong, Robert, Ph.D., Johns Hopkins University, 1970. Optics and laser physics; atmospheric physics.

Burleson, George R., Ph.D., Stanford University, 1960. Nuclear and elementary particle physics.

Burr, Alex F., Ph.D., Johns Hopkins University, 1966. Physics education.

Chylek, Petr, Ph.D., Charles University. Light scattering; atmospheric physics.

Daw, Harold A., Ph.D., University of Utah, 1956. Physics education.

Goedecke, George H., Ph.D., Rensselaer Polytechnic Institute, 1961. Scattering theory; acoustics; optics; stochastic electrodynamics.

Goldman, Terrance, Ph.D., Harvard University, 1973. Theoretical nuclear physics.

Ingraham, Richard L., Ph.D., Harvard University, 1952. Foundations of quantum mechanics, quantum optics; nonlinear dynamics and chaos; general relativity and relativistic cosmology.

Kunz, Kaiser, Ph.D., University of Cincinnati, 1939. Electrodynamics; theoretical physics.

Kyle, Gary S., Ph.D., University of Minnesota, 1979. Nuclear and particle physics.

Liefeld, Robert J., Ph.D., Ohio State University, 1959. X-ray physics; electron, atomic, and solid state physics; physics education.

Ostashev, Vladimir E., Ph.D., University of Moscow, 1979. Acoustical and optical wave propagation in random media.

Stromberg, Thorsten F., Ph.D., Iowa State University, 1965. Physics education.

Strottman, Daniel, Ph.D., State University of New York, 1969. Theoretical nuclear physics.

Zund, Joseph, Ph.D., University of Texas, Austin, 1964. General Relativity.

Research Professor

Bruce, Charles, Ph.D., New Mexico State University, 1970. *Atmosphere, Space Physics, Cosmic Rays, Atomic, Molecular, & Optical Physics, Climate/Atmospheric Science, Condensed Matter Physics, Optics*. Aerosol physics, optical properties of nanoparticles.

Research Assistant Professor

Wei, Feng, Ph.D., Iowa State University, 2010. *Nuclear Physics*. Strong interaction, nucleon structure.

Teaching Associate Professor

Burkardt, Michaela, Ph.D., University of Erlangen-Nürnberg, 1992. *Physics and other Science Education*. Physics education.

DeAntonio, Michael, Ph.D., New Mexico State University, 1993. *Applied Physics, Climate/Atmospheric Science, Optics, Physics and other Science Education*. Polarimetry; ion mobility spectroscopy; remote sensing; atmospheric physics; nonlinear optics; physics education.

Adjunct Faculty

Pinnick, Ron, Ph.D., University of Wyoming, 1972. *Atmosphere, Space Physics, Cosmic Rays*. Atmospheric particulates.

Adjunct Professor

Higbie, Paul, Ph.D., Massachusetts Institute of Technology, 1968. *Atmosphere, Space Physics, Cosmic Rays*. Space physics.

Tompkins, Harland G., Ph.D., University of Wisconsin - Milwaukee, 1971. *Applied Physics, Optics, Surface Physics*. Spectroscopic ellipsometry, vacuum science, surface physics.

Laboratory Coordinator

Pennise, Christine, M.S., Johns Hopkins University, 1992.

Posdoctoral Research Associate

Hoellwieser, Roman, Ph.D., TU Vienna, 2009. *High Energy Physics, Nuclear Physics, Theoretical Physics*. Strong interaction, nucleon structure.

Miceli, Tia, Ph.D., University of California, Davis, 2013. *High Energy Physics*. Neutrino physics.

DEPARTMENTAL RESEARCH SPECIALTIES AND STAFF

Theoretical

Computational Physics. Engelhardt, Kiefer, Vasiliev, Wang.

Condensed Matter Physics. Fohtung, Kiefer, Nakotte, Vasiliev, Zollner.

Geophysics. Global seismic tomography; inversion methods; gravity; tectonophysics, mantle convection, fluids; planetary sciences. Hearn, Kiefer, Ni.

Nano Science and Technology. Fohtung, Kiefer, Nakotte.

Nuclear Physics. Hadron-hadron interactions; lepton scattering; nucleon structure; quantum chromodynamics. Matthias Burkardt, Engelhardt, Gibbs.

Optics. Scattering by particles of arbitrary shape and structure; single and multiple scattering; aerosol effects; inverse scattering theory; nonlinear effects; quantum optics. Bruce, DeAntonio, Goedecke, Pinnick.

Experimental

Applied Physics. Energy research; remote sensing. Bruce, Daw, DeAntonio, Kanim, Liefeld, Nakotte, Urquidi, Zollner.

Atomic, Molecular, & Optical Physics. Raman scattering; Rayleigh-Brillouin scattering; Doppler limited IR absorption; soft x-ray emission and absorption spectroscopy; x-ray line and continuum isochromats; photo and Auger electron spectroscopy; plasma and laser spectroscopy; spectroscopic ellipsometry. Armstrong, Bruce, DeAntonio, Urquidi, Zollner.

Geophysics. Seismic studies of the crust and mantle of Asia and North America. Hearn, Ni.

Materials Science, Metallurgy. Fabrication and properties of opto-electronic materials; optical and magneto-optical properties of nanostructured materials; electron, optical, and scanning probe microscopy. Fohtung, Nakotte, Urquidi, Zollner.

Neutrino physics. Neutrino interactions with nucleons and nuclei. Miceli, Papavassiliou, Pate.

Nuclear Physics. Quark-gluon structure of matter; spin physics; hadronic interactions; electron and neutrino scattering. Papavassiliou, Pate, Wang.

Optics. Laser Raman spectroscopy; laser absorption and emission spectroscopy; quantum optics, near-field optics; spectroscopic ellipsometry. Armstrong, Bruce, DeAntonio, Urquidi, Zollner.

Physics education. Michaela Burkardt, DeAntonio, Kanim.

Solid State Physics. Nakotte, Urquidi, Zollner.

View additional information about this department at
www.gradschoolshopper.com

UNIVERSITY OF NEW MEXICO

DEPARTMENT OF PHYSICS AND ASTRONOMY

Albuquerque, New Mexico 87131-0001
http://panda.unm.edu

General University Information

President: Robert Frank
Dean of Graduate School: Julie Coonrod
University website: http://www.unm.edu/
Control: Public
Setting: Urban
Total Faculty: 3,711
Total Graduate Faculty: Not separated
Total number of Students: 29,100
Total number of Graduate Students: 6,262

Department Information

Department Chairman: Wolfgang Rudolph, Chair
Department Contact: Alisa Gibson, Academic Programs Coordinator
  Total full-time faculty: 30
  Total number of full-time equivalent positions: 28
  Full-Time Graduate Students: 117
  First-Year Graduate Students: 11
  Total Post Doctorates: 16

Department Address

MSC07 4220
1 University of New Mexico
Albuquerque, NM 87131-0001
Phone: (505) 277-1514
Fax: (505) 277-1520
E-mail: agibson@unm.edu
Website: http://panda.unm.edu

ADMISSIONS

Admission Contact Information

Address admission inquiries to: Coordinator of Program Advisement, Department of Physics and Astronomy
Phone: (505) 277-1514
E-mail: pandainfo@phys.unm.edu
Admissions website: http://panda.unm.edu/pandaweb/graduate/app_adm_process.phtml

Application deadlines

Fall admission:
U.S. students: June 1   *Int'l. students*: January 15
Spring admission:
U.S. students: October 1   *Int'l. students*: August 1

Application fee

U.S. students: $50   *Int'l. students*: $50
A domestic student seeking an assistantship has the same application deadline as international students: January 15 for the Fall semester and August 1 for the Spring semester.

Admissions information

For Fall of 2013:
  Number of applicants: 146
  Number admitted: 40
  Number enrolled: 11

Admission requirements

Bachelor's degree requirements: Students wishing to enter the M.S. or the Ph.D. program in Physics must have an undergraduate degree in physics or its equivalent.

GRE requirements

The GRE is required.
Analytical score: 1

Advanced GRE requirements

The Advanced GRE is required.
Minimum accepted Advanced GRE score: 200

TOEFL requirements

The TOEFL exam is required for students from non-English-speaking countries.
PBT score: 550
iBT score: 79
We also accept the IELTS with a minimum score of 6.5

Other admissions information

Additional requirements: Admission to the Optical Science & Engineering graduate programs requires an undergraduate background that includes physics, optics, optical engineering, and/or optoelectronics. Additional information, specific criteria, application forms and instructions are available online at http://panda.unm.edu.
Undergraduate preparation assumed: 1 semester of Thermal Physics; 1 year of Mechanics and of E & M. 1 year of Quantum Physics strongly recommended.

TUITION

Tuition year 2013-14:
Tuition for in-state residents
Full-time students: $2,176.65 per semester
Part-time students: $301.85 per credit
Tuition for out-of-state residents
Full-time students: $7,995.78 per semester
Part-time students: $888.42 per credit
Credit hours per semester to be considered full-time: 9
Deferred tuition plan: No
Health insurance: Available at the cost of 993 per year.
Other academic fees: $50/year. Graduate and Professional Student Association $120 fee for international students assessed only for the first semester they start at UNM.
Academic term: Semester
Number of first-year students who received full tuition waivers: 15

Teaching Assistants, Research Assistants, and Fellowships

Number of first-year
Teaching Assistants: 14
Research Assistants: 3
Fellowship students: 1
Average stipend per academic year
Teaching Assistant: $15,196
Research Assistant: $16,436
Fellowship student: $30,000

FINANCIAL AID

Application deadlines

Fall admission:
U.S. students: January 15   *Int'l. students*: January 15
Spring admission:
U.S. students: August 1   *Int'l. students*: August 1

Loans

Loans are available for U.S. students.
Loans are not available for international students.
GAPSFAS application required: No
FAFSA application required: Yes

For further information

Address financial aid inquiries to: Student Financial Aid Office.
Phone: 1-800-CALLUNM
Financial aid website: http://finaid.unm.edu/

HOUSING

Availability of on-campus housing

Single students: Yes
Married students: Yes

For further information

Address housing inquiries to: Housing Office.
Phone: (505) 277-2606
E-mail: reshalls@unm.edu
Housing aid website: http://housing.unm.edu/

Table A—Faculty, Enrollments, and Degrees Granted

Research Specialty	2012-13 Faculty	Enrollment Fall 2013 Master's	Enrollment Fall 2013 Doctorate	Number of Degrees Granted 2012-13 (2008-13) Master's	Number of Degrees Granted 2012-13 (2008-13) Terminal Master's	Number of Degrees Granted 2012-13 (2008-13) Doctorate
Astronomy	7	–	–	–	–	1(4)
Astrophysics	3	–	–	–	–	–(2)
Atomic, Molecular, & Optical Physics	3	–	–	–	–	–(3)
Biophysics	5	–	–	–	–	3(8)
Computational Physics	1	–	–	–	–	–(1)
Condensed Matter Physics	6	–	–	–	–(1)	2(6)
Geophysics	1	–	–	–	–	–
High Energy Physics	5	–	–	–	–	–
Nonlinear Dynamics and Complex Systems	3	–	–	–	–	–
Nuclear Physics	3	–	–	–	–	–
Optics	10	–	–	2(-)	2(-)	3(16)
Particles and Fields	3	–	–	–	–	–(2)
Quantum Foundations	5	–	–	–	–(1)	1(11)
Relativity & Gravitation	–	–	–	–	–	–(1)
Statistical & Thermal Physics	2	–	–	–	–(1)	–
Total	49	–	–	12(35)	2(18)	10(56)
Full-time Grad. Stud.	–	7	128	–	–	–
First-year Grad. Stud.	–	1	18	–	–	–

GRADUATE DEGREE REQUIREMENTS

Master's: Master of Science in Physics and Master of Science in Optical Science and Engineering. Two plans are available for the M.S. Physics: I-A minimum of 24 hours of coursework with 16 hours of 500-level courses, six hours of thesis (599) credit; II-A minimum of 32 hours of coursework with 16 hours of 500-level courses. In either case a minimum GPA of "B" is expected, and only 6 hours may be transferred. No foreign language required. Three options are available for the M.S.-Optical Science and Engineering: thesis, non-thesis, or coursework plus, an industrial internship.

Doctorate: Ph.D. in Physics and in Optical Science and Engineering. The Ph.D. in Physics requires at least 48 semester hours (52 hours for Optical Sciences) exclusive of thesis and dissertation with 24 hours of physics in courses above 500, exclusive of problems and research courses. Formal requirements are: coursework; two semesters in residence; departmental qualifying procedure; application for and admission to candidacy; the doctoral comprehensive examination; the dissertation; "B" average.

Thesis: Thesis may be written in absentia, when approved.

SPECIAL EQUIPMENT, FACILITIES, OR PROGRAMS

Miscellaneous research opportunities available locally at Sandia National Laboratories, the Air Force Research Laboratory, the VLA (Very Large Array Radio Telescope), Los Alamos National Laboratory. Other facilities with cooperative agreements: Fermi National Accelerator Laboratory, Brookhaven National Laboratory. Some graduate courses are taught at Los Alamos at UNM's Graduate Center by UNM campus faculty and by Los Alamos staff members.

Table B—Separately Budgeted Research Expenditures by Source of Support

Source of Support	Departmental Research	Physics-related Research Outside Department
Federal government	$7,000,000	
State/local government	$350,000	
Non-profit organizations		
Business and industry		
Other		
Total	$7,350,000	

FACULTY

Professor

Ahluwalia, Harjit S., Ph.D., Gujarat University, 1960. *Atmosphere, Space Physics, Cosmic Rays, Plasma and Fusion, Other.* Cosmic-ray and high-energy astrophysics, solar and space physics, plasma physics; nuclear electronics.

Bassalleck, Bernd, Ph.D., Karlsruhe University, 1977. Department Chairman. *Nuclear Physics, Other.* Experimental subatomic physics, particularly strangeness nuclear physics, spin physics, and rare Kaon Decays.

Cahill, Kevin, Ph.D., Harvard University, 1967. *Biophysics, Medical, Health Physics, Particles and Fields, Other.* Particle theory, lattice-gauge theory. Medical physics.

Caves, Carlton M., Ph.D., California Institute of Technology, 1979. *Physics and other Science Education, Quantum Foundations, Other.* Quantum information theory and quantum computation, Theory of open quantum systems and decoherence; nonlinear dynamics and quantum chaos; theoretical quantum optics.

Deutsch, Ivan H., Ph.D., University of California, Berkeley, 1992. Associate Chair of Graduate Affairs. *Optics, Quantum Foundations, Other.* Quantum optics/atomic physics/quantum information theory: laser cooling and trapping, optical lattices, quantum computation, coherent control, quantum dissipative systems.

Diels, Jean-Claude, Ph.D., Brussels, 1973. *Nonlinear Dynamics and Complex Systems, Optics, Other.* Laser physics and nonlinear optics, ultrafast phenomena. High-resolution spectroscopy and imaging, adaptive optics and interferometry. Laser-induced discharges and plasmas. Laser gyros and related sensing of displacements, index of refraction, magnetic and electric fields. displacements; interaction of intense laser fields with matter; laser induced discharges.

Dunlap, David H., Ph.D., University of Rochester, 1987. Chair of Preliminary Exam Committee. *Condensed Matter Physics, Statistical & Thermal Physics, Other.* Transport and tunneling phenomena; hopping transport; disordered materials; molecular solids.

Gell-Mann, Murray, Ph.D., Massachusetts Institute of Technology, 1951. Complex systems; measures of complexity; decoherent histories in quantum mechanics.

Gold, Michael S., Ph.D., University of California, Berkeley, 1986. High-energy collider physics, fundamental interactions and supersymmetry, new particle searches, rare decays. Particle-physics instrumentation.

Henning, Patricia A., Ph.D., University of Maryland, 1990. Associate Chair of Undergraduate Affairs. Extragalactic astronomy; radio astronomy; galaxy clusters and superclusters; material content of cosmic voids.

Kenkre, V. M., Ph.D., Stony Brook University, 1971. Statistical physics; nonlinear science; mathematical biology and theory of the spread of epidemics; quantum transport and tunneling phenomena; Bose-Einstein Condensation; materials theory.

Malloy, Kevin J., Ph.D., Stanford University, 1984. Semiconductor physics; device physics.

Matthews, John A. J., Ph.D., University of Toronto, 1971. Cosmic-ray and high-energy astrophysics; high-energy collider physics; particle-physics instrumentation.

McGraw, John T., Ph.D., University of Texas, Austin, 1977. Adaptive optics and interferometry; galactic astronomy.

Prasad, Sudhakar, Ph.D., Harvard University, 1983. Adaptive optics and interferometric imaging; theoretical quantum optics; propagation in optical fibers.

Rand, Richard J., Ph.D., California Institute of Technology, 1991. Gaseous halos of galaxies, diffuse ionized gas, the disk-halo connection, superbubbles and chimneys; star formation processes in spirals, molecular gas, atomic gas, HII region populations; ring galaxies.

Rudolph, Wolfgang, Ph.D., University of Jena, 1985. High-resolution spectroscopy and imaging, laser physics and nonlinear optics, ultrashort light pulses; biophysics.

Seidel, Sally C., Ph.D., University of Michigan, 1987. High-energy collider physics, QCD, rare decays; particle-physics instrumentation.

Sheik-Bahae, Mansoor, Ph.D., State University of New York at Buffalo, 1987. Laser physics and nonlinear optics, ultrafast phenomena, solid-state physics.

Associate Professor

Allahverdi, Rouzbeh, Ph.D., University of Alberta, Edmonton, 2000. Particle physics and cosmology: physics beyond the standard model, supersymmetry, inflation, dark matter, lepto/baryogenesis.

Fields, Douglas, Ph.D., Indiana University, 1991. High energy-density nuclear physics, high-energy spin physics; collider instrumentation; physics education.

Lidke, Keith, Ph.D., University of Minnesota, 2002. Fluorescence imaging techniques: single particle tracking, single molecule imaging and superresolution techniques for measuring protein-protein interactions at the sub-cellular level.

Loomba, Dinesh, Ph.D., Boston University, 1998. Extragalactic astronomy, particle astrophysics, and dark matter.

Pihlström, Ylva M., Ph.D., Chalmers University of Technology, 2001. Radio astronomy: Active Galactic Nuclei, starburst galaxies, extragalactic and Galactic masers; radio interferometry techniques including Very Long Baseline Interferometry (VLBI).

Roy, Mousumi, Ph.D., Massachusetts Institute of Technology, 1997.

Taylor, Gregory B., Ph.D., University of California, Los Angeles, 1991. Clusters of galaxies; active galactic nuclei; jets; gamma-ray bursts; radio interferometry techniques.

Thomas, James L., Ph.D., Cornell University, 1991. *Biophysics.* Phospholipids are the amphiphilic molecules that are the structural foundation of essentially all cell membranes. In addition, they have found practical application in synthetic membranes for drug delivery vehicles, and as stabilizing monolayers for ultrasound contrast agents. Our laboratory is engaged in optical and fluorescence studies of phospholipid layers and cell membranes, in order to elucidate the dynamical properties of these important biophysical/biomedical materials and their constituent molecules.

Assistant Professor

Becerra, Francisco E., Ph.D., University of Maryland. *Atomic, Molecular, & Optical Physics, Optics, Quantum Foundations.* Experimental physics is quantum optics, quantum information science, atomic/molecular/optical. Studies of quantum communication, quantum memory, and computation.

Duan, Huaiyu, Ph.D., University of Minnesota, 2004. Neutrinos and nucleosynthesis in hot and dense matter.

Koch, Steven J., Ph.D., Cornell University, 2003. Biophysics. Experimental single-molecule biophysics: optical and magnetic tweezers, micro and nanofabricated devices for mechanical manipulation of biomolecules.

Miyake, Akimasa, Ph.D., University of Tokyo, 2004. *Quantum Foundations, Statistical & Thermal Physics.* Quantum Information Theory; many-body physics.

Research Professor

Gorelov, Igor V., Ph.D., ITEP, Moscow, 1991. Optical science and engineering.

Schwoebel, Paul, Ph.D., Cornell University, 1987. Surface physics, medical imaging, nuclear sensors. Crystal growth and nucleation phenomena; vacuum, gaseous and solid electrical breakdown; high electric field processes; ion sources, neutron generators, x-ray sources, radiation damage, plasma physics.

Thomas, Timothy L., Ph.D., University of Minnesota, 1995. Deputy Director, University of New Mexico Center for High Performance Computing.

Research Associate Professor

Atlas, Susan R., Ph.D., Harvard University, 1988. Chemical and molecular physics, materials theory; computational physics.

Boyd, Stephen T., Ph.D., University of California, Los Angeles, 1991. Low-temperature physics and critical phenomena. Transport phenomena.

Research Assistant Professor

Hasselbeck, Michael, Ph.D., University of Central Florida, 1995.

Hope, Douglas, Ph.D., University of New Mexico, 2004. Research involving the use of compressive sensing for space-object identification.

Nampoothiri, Vasudevan, Ph.D., Indian Institute of Technology, 1999.

Younus, Imran, Ph.D., Syracuse University, 2003.

Zimmer, Peter, Ph.D., University of New Mexico, 2004.

Adjunct Associate Professor

Landahl, Andrew, Ph.D., California Institute of Technology, 2002. Adjunct National Laboratory Associate Professor. Quantum information science. Specific interests include quantum computation, information theory, error correction, and control systems.

Affiliate Professor

Brueck, Steven R. J., Ph.D., Massachusetts Institute of Technology, 1971. Director, Center for High Technology Materials Primary appointment as Professor of Electrical Engineering and Computing Engineering. Solid-state devices; optical properties of solids.

Heintz, Philip H., Ph.D., University of Washington, 1971. Primary appointment as Professor of Radiology. Biomedical physics (Radiology).

Jain, Ravinder K., Ph.D., University of California, Berkeley, 1974. Primary appointment as Professor of Electrical Engineering and Computer Engineering. Solid-state devices, quantum electronics, optoelectronics. Electrical and computer engineering.

Moore, Christopher, Ph.D., Cornell University, 1991. Primary appointment as Assistant Professor of Computer Science. Phase transitions in NP-complete problems; quantum computation; computational complexity in statistical physics; analog computation; dynamical systems, cellular automata, recurrent neural networks; algebraic circuits; non-associative algebras (Quasigroups and Loops); glassy systems and slow relaxation; spin systems; Potts Models; random tilings; random networks; "Small Worlds"; Monte Carlo Algorithms; combinatorial games.

Osiński, Marek, Ph.D., Polish Academy of Sciences, 1979. Primary appointment as Professor of Electrical Engineering and Computer Engineering. Optoelectronic devices and materials, theory and experiment. Semiconductor lasers, vertical-cavity surface-emitting lasers, two-dimensional arrays; wide-bandgap materials and devices, group-III nitrides, light-emitting diodes, lasers from green to UV. Reliability and degradation physics; comprehensive computer simulation.

Posse, Stephan, Ph.D., University of Cologne, 1986. Primary appointment as Professor of Neurology. Creating and applying innovative MR imaging technologies toward a more comprehensive understanding of the human mind and body. Our current focus is the development of high-speed spectroscopic MR imaging and real-time functional MR imaging for applications in Cancer Research, Neurology and Psychiatry.

Lecturer

Odom, Boye M., M.S., University of Texas, El Paso, 1981.

Saul, Jeff, Ph.D., University of Maryland, 1998. Physics Education: Implementing, adapting, and assessing the effectiveness of activity-based instruction in high school and undergraduate physics classes.

DEPARTMENTAL RESEARCH SPECIALTIES AND STAFF

Theoretical

Astrophysics. Cosmology, dark matter. Duan.

Atomic, Molecular, & Optical Physics. AMO physics applied to quantum information science. Ultracold atoms. Caves, Deutsch.

Biophysics. Mathematical modeling of epidemics and related ecosystems, medical physics. Atlas, Cahill.

Computational Physics. Atlas.

Condensed Matter Physics. Transport and tunneling phenomena, disordered materials, organic materials, Materials theory, computational physics. Atlas, Dunlap, Kenkre.

Nonlinear Dynamics and Complex Systems. Interdisciplinary science connecting physics to biology, chemistry, and engineering. Deutsch, Gell-Mann, Kenkre.

Nuclear Physics. Neutrinos and nucleosynthesis in hot and dense matter. Duan.

Optics. Spectroscopy and imaging, laser physics and nonlinear optics, quantum optics. Caves, Deutsch, Prasad.

Particles and Fields. QCD, Lattice-gauge theory, dark matter. Allahverdi, Cahill, Duan.

Quantum Foundations. Quantum information science. Caves, Deutsch, Landahl, Miyake, Moore.

Statistical & Thermal Physics. Dunlap, Kenkre.

Experimental

Astronomy. Adaptive optics and interferometry, radio astronomy, galactic & extragalactic astronomy, solar and space physics. Ahluwalia, Henning, McGraw, Pihlström, Rand, Taylor, Zimmer.

Astrophysics. Cosmology, cosmic radiation, dark matter searches. Loomba, Matthews.

Atomic, Molecular, & Optical Physics. Metrology: Magnetometers, atom interferometers. Diels.

Biophysics. Biomedical imaging, cell and membrane imaging, single-molecule biophysics. Lidke, James Thomas.

Condensed Matter Physics. Low-temp. physics/critical phenomena, surface physics, nanotechnology. Boyd, Malloy, Schwoebel.

Geophysics. Roy.

High Energy Physics. Collider physics, instrumentation. Fields, Gold, Gorelov, Matthews, Seidel.

Nuclear Physics. RHIC Spin Physics, Rare Kaon decays, Strangeness Nuclear Physics. Bassalleck, Fields.

Optics. Ultrafast phenomena, laser physics and nonlinear optics, laser cooling of solids, nonlinear optics. Brueck, Diels, Jain, Malloy, Osiński, Rudolph, Sheik-Bahae.

View additional information about this department at
www.gradschoolshopper.com

BINGHAMTON UNIVERSITY (STATE UNIVERSITY OF NEW YORK)

DEPARTMENT OF PHYSICS, APPLIED PHYSICS, AND ASTRONOMY

Binghamton, New York 13902-6000
http://www2.binghamton.edu/physics

General University Information

President: Harvey Stenger
Dean of Graduate School: Susan Strehle
University website: http://www2.binghamton.edu
Control: Public
Setting: Suburban
Total Faculty: 866
Total Graduate Faculty: 496
Total number of Students: 15,308
Total number of Graduate Students: 2,952

Department Information

Department Chairman: Eric Cotts, Chair
Department Contact: Renee Farris, Administrator
　Total full-time faculty: 12
　Total number of full-time equivalent positions: 12
　Full-Time Graduate Students: 28
　First-Year Graduate Students: 9
　Female First-Year Students: 1
　Total Post Doctorates: 1

Department Address

4400 Vestal Parkway East
Science II, Room 256
Binghamton, NY 13902-6000
Phone: (607) 777-4609
Fax: (607) 777-2546
E-mail: physics@binghamton.edu
Website: http://www2.binghamton.edu/physics

ADMISSIONS

Admission Contact Information

Address admission inquiries to: C. Nelson, Graduate Director, Department of Physics, Applied Physics, and Astronomy
E-mail: cnelson@binghamton.edu
Admissions website: http://www2.binghamton.edu/grad-school/

Application deadlines

Fall admission:
U.S. students: February 15　　*Int'l. students*: February 15

Application fee

U.S. students: $75
Applications must be submitted online in addition to submitting documents as a single PDF to the department.

Admissions information

For Fall of 2013:
　Number of applicants: 20
　Number admitted: 15
　Number enrolled: 6

Admission requirements

Bachelor's degree requirements: Bachelor's degree is required.
Minimum undergraduate GPA: 3.0

GRE requirements

The GRE is required.
　Quantitative score: 156
　Verbal score: 153
　Analytical score: 3
Physics Subject GRE also required

Advanced GRE requirements

The Advanced GRE is required.
　Minimum accepted Advanced GRE score:

TOEFL requirements

The TOEFL exam is required for students from non-English-speaking countries.
　PBT score: 550
　iBT score: 100

Other admissions information

Additional requirements: Specialization in physics on the undergraduate level is desirable but not essential for admission. Two letters of reference are required.
No minimum acceptable score for admission is specified.

Undergraduate preparation assumed: One year of general physics; one year of Electromagnetic Theory (Marion and Heald, Classical Electromagnetic Radiation); one semester of Dynamics (Barger and Olsson, Classical Mechanics: A Modern Perspective); at least a semester of quantum mechanics (Cohen-Tannoudji, Quantum Mechanics); and mathematics through differential equations. Appropriate laboratory experience at the upper undergraduate levels is desirable.

TUITION

Tuition year 2012–2013:

Tuition for in-state residents
    Full-time students: $5,570 annual
    Part-time students: $464 per credit

Tuition for out-of-state residents
    Full-time students: $14,718 annual
    Part-time students: $674 per credit

Credit hours per semester to be considered full-time: 12

Deferred tuition plan: No

Other academic fees: $2,043 (maximum except international students)

Academic term: Semester

Number of first-year students who received full tuition waivers: 6

Teaching Assistants, Research Assistants, and Fellowships

Number of first-year
    Teaching Assistants: 6

Average stipend per academic year
    Teaching Assistant: $18,000
    Research Assistant: $18,000

FINANCIAL AID

Loans

Loans are available for U.S. students.
Loans are available for international students.

GAPSFAS application required: No

FAFSA application required: No

HOUSING

Availability of on-campus housing

Single students: Yes
Married students: Yes

For further information

Address housing inquiries to: Director of Graduate Housing.

Table A—Faculty, Enrollments, and Degrees Granted

Research Specialty	2012-13 Faculty	Enrollment Fall 2012		Number of Degrees Granted 2011–12 (2004–10)		
		Master's	Doctorate	Master's	Terminal Master's	Doctorate
Applied Physics	2	–	2	-(8)	–	–
Atomic, Molecular, & Optical Physics	–	–	–	–	–	–
Biophysics	1	–	1	–	–	–
Chemical Physics	–	–	–	–	–	–
Condensed Matter Physics	1	–	2	-(3)	–	–
Energy Sources & Environment	3	–	9	–	–	–
Engineering Physics/Science	–	–	–	–	–	–
Low Temperature Physics	–	–	–	–	–	–
Materials Science, Metallurgy	1	–	3	-(2)	–	–
Optics	2	–	6	-(1)	–	–
Statistical & Thermal Physics	1	–	–	-(2)	–	–
Non-specialized	1	1	–	–	–	–
Total	12	1	27	-(16)	–	–
Full-time Grad. Stud.	–	1	27	–	–	–
First-year Grad. Stud.	–	1	8	–	–	–

GRADUATE DEGREE REQUIREMENTS

Master's: 30 graduate credit hours with at least a "B" average. There is a two-semester residence requirement. Students have a choice of thesis or comprehensive examination.

Doctorate: At least 24 credit hours of course study (in residence) and 24 additional credit hours of dissertation work. Passing a written qualifying examination, in three parts, covering the core areas of physics and successful defense of dissertation are required.

Other Degrees: Master of Science in physics with a specialization in applied physics is designed for students seeking careers in applied physics. Emphasis is to provide a comprehensive education in fundamental physical principles and their applications to enhance the ability to evolve with changing technology and to avoid technical obsolescence. Student may study part-time and complete a degree in three years or complete a full-time graduate assistantship and complete the degree in two years or less. Thesis topics may be drawn from employment with the consent of the department and employer. M.A.T. and M.S.T. programs are designed for students who wish to teach physics at the secondary level. The M.A.T. program is designed for students with a physics background who need education courses; the M.S.T. program is designed for teachers who want to improve their physics background. The credits in professional education courses required for certification are offered, as well as additional work in physics and allied fields. Certified teachers may enroll in the M.S.T. program, in which almost all of the training involves substantive physics coursework.

Thesis: Thesis may be written in absentia.

SPECIAL EQUIPMENT, FACILITIES, OR PROGRAMS

AC and DC magnetic susceptibility bridges; x-ray diffractometers; 100,000 kilogauss superconducting magnet; 15″ iron core magnet; differential scanning calorimeters; sputtering equipment; vacuum deposition stations; dilution refrigerator; raman spectrometer; clean room; splat quencher; resistivity bridges; hydrator; cryo-cooler; dielectric analyzer; thermogravimetric analyzer; dynamic mechanical analyzer; thermomechanical analyzer; scanning electron microscope; high-pressure intensifier; squid magnetometer.

FACULTY

Professor

Cotts, Eric J., Ph.D., University of Illinois, 1983. Department Chair. *Solid State Physics.* Experimental solid-state physics.

Nelson, Charles A., Ph.D., University of Maryland, 1968. Graduate Director. *Applied Physics, High Energy Physics.* Theoretical high-energy physics.

Suzuki, Masatsugu, Ph.D., University of Tokyo, 1977. Undergraduate Director. *Solid State Physics.* Experimental solid-state physics.

Associate Professor

White, Bruce E., Ph.D., Cornell University, 1995. *Solid State Physics.*

Assistant Professor

DeSilva, Theja, Ph.D., University of Cincinnati, 2004. *Condensed Matter Physics.* Theoretical condensed-matter physics; spin-orbit physics.

Jang, Joon, Ph.D., University of Illinois at Urbana-Champaign, 2005. *Condensed Matter Physics, Optics.*

Kolmogorov, Aleksey, Ph.D., Penn State University, 2004. *Computational Physics.* Development and modeling of new materials with ab initio methods.

Lawler, Michael, Ph.D., University of Illinois at Urbana-Champaign, 2006.

Mativetsky, Jeffrey, Ph.D., McGill University, 2006. *Nano Science and Technology.* Relationships between nanoscale structure and electrical function in organic materials for solar cells and electronics.

Piper, Louis, Ph.D., University of Warwick, 2003. *Condensed Matter Physics.*

Shim, Bonggu, Ph.D., University of Texas @ Austin, 2006. *Nonlinear Dynamics and Complex Systems.* Nonlinear interactions with matter using high-power, ultrashort laser pulses.

Research Assistant Professor

Margine, Roxana, Ph.D., Penn State University, 2007. *Computational Physics.* Develop and apply ab initio computational methods for modeling of emerging materials with applications in energy transport and electronics.

Adjunct Professor

Poliks, Barbara, Ph.D., Jagiellonian University, 1982. Computer simulations of polymeric systems, including proteins and materials.

Pompi, Robert L., Ph.D., Cornell University, 1968. *Applied Physics.*

Sileo, Richard, Ph.D., Applied and Engineering Physics, Cornell University, 1974. *Physics and other Science Education.* Physics education.

Telesca, Andrew, M.S., State University of NY at Oneonta, 1976. *Astronomy, Physics and other Science Education.*

Wu, Tsu-Ming, Ph.D., University of Pennsylvania, 1966. *Applied Physics, Biophysics.*

DEPARTMENTAL RESEARCH SPECIALTIES AND STAFF

Theoretical
Applied Physics.

Experimental
Applied Physics.

Condensed Matter Physics. Low-temperature condensed-matter physics; localized magnetic moments in metallic crystals; induced valence changes in impurity doped metals; Raman spectroscopy; properties of disordered materials, amorphous metals, and layered materials.

View additional information about this department at www.gradschoolshopper.com

---

CLARKSON UNIVERSITY

DEPARTMENT OF PHYSICS

Potsdam, New York 13699-5820

http://www.clarkson.edu/physics

General University Information

President: Anthony G. Collins
Dean of Graduate School: Peter Turner, School of Arts & Sciences
University website: http://www.clarkson.edu
Control: Private
Setting: Rural
Total Faculty: 270
Total number of Students: 3,604
Total number of Graduate Students: 532

Department Information

Department Chairman: Dipankar Roy, Chair
Department Contact: Dipankar Roy, Chair
  Total full-time faculty: 15
  Total number of full-time equivalent positions: 10
  Full-Time Graduate Students: 14
  First-Year Graduate Students: 5
  Total Post Doctorates: 3

Department Address
Clarkson University
Science Center 273, Box 5820
Potsdam, NY 13699-5820
Phone: (315) 268-2396
Fax: (315) 268-7754
E-mail: bpichett@clarkson.edu
Website: http://www.clarkson.edu/physics

ADMISSIONS

Admission Contact Information
Address admission inquiries to: Chair, Physics Department,
Clarkson University, Potsdam, NY 13699-5820
Phone: (315) 268-2396
E-mail: samoy@clarkson.edu
Admissions website: http://www.clarkson.edu/admissions/
graduate_admission/apply.html

Application deadlines
Fall admission:
U.S. students: April 15   *Int'l. students*: April 15
Spring admission:
U.S. students: October 1   *Int'l. students*: October 1

Application fee
U.S. students: $25   *Int'l. students*: $35
Applications must be submitted online.

Admissions information
For Fall of 2013:
Number of applicants: 34
Number admitted: 15
Number enrolled: 4

Admission requirements
Bachelor's degree requirements: A Bachelor's degree in physics
is required.

GRE requirements
The GRE is required.
Minimum score required varies in different years.

Advanced GRE requirements
The Advanced GRE is recommended.

TOEFL requirements
The TOEFL exam is required for students from non-English-
speaking countries.
PBT score: 550
iBT score: 80

Other admissions information
Additional requirements: All applicants are required to take the
general Graduate Record Exam (GRE; there are NO minimum
scores specified for the GRE report; Clarkson's institution
code is 2084; international students should submit a TOEFL
or IELTS score; minimum scores are: TOEFL, paper-550,
Internet-80, CBT-213, IELTS-6.5; score is effective for 2
years).
Undergraduate preparation assumed: Mechanics, Symon or
Becker; Optics, Bennett or Hecht; Quantum Mechanics, Anderson or Merzbacher; Modern Physics, Krane; Thermal
Physics, Baierlein; Electrodynamics, Griffiths.

TUITION

Tuition year 2013-14:
Full-time students: $1,324 per credit
Clarkson University Payment Plan http://www.clarkson.edu/sas/
financial
Credit hours per semester to be considered full-time: 12
Deferred tuition plan: No

Other academic fees: $840
Academic term: Semester

Teaching Assistants, Research Assistants, and Fellowships
Number of first-year
Teaching Assistants: 12
Research Assistants: 3
Fellowship students: 1
Average stipend per academic year
Teaching Assistant: $23,329

FINANCIAL AID

Application deadlines
Fall admission:
U.S. students: April 15   *Int'l. students*: April 15
Spring admission:
U.S. students: October 1   *Int'l. students*: October 1

Loans
Loans are available for U.S. students.
Loans are not available for international students.
GAPSFAS application required: No
FAFSA application required: Yes

For further information
Address financial aid inquiries to: Graduate Studies Office;,
www.clarkson.edu/artsandsci/grad.
Phone: (315) 268-3802
E-mail: jreed@clarkson.edu
Financial aid website: http://www.clarkson.edu/sas/financial

HOUSING

Availability of on-campus housing
Single students: Yes
Married students: Yes

For further information
Address housing inquiries to: Dean of Student Life; Housing information may be obtained from the ISE Office at 315-268-
3856 or by requesting an electronic copy of the listing by
writing to isegrad@clarkson.edu.
Phone: (315) 268-3856
Housing aid website: http://www.clarkson.edu/ese/housing.html

Table A—Faculty, Enrollments, and Degrees Granted

Research Specialty	2012-13 Faculty	Enrollment Fall 2012		Number of Degrees Granted 2012-13 (2007–12)		
		Mas-ter's	Doc-torate	Mas-ter's	Terminal Master's	Doc-torate
Biophysics	3	1	4	2(2)	3(6)	1(3)
Chemical Physics	4	1	1	–	–	–
Condensed Matter						
Physics	7	1	4	2(2)	–(2)	–(4)
Optics	1	1	–	1(1)	–	1(1)
Physics and other						
Science Education	2	–	1	–(1)	–	–(2)
Statistical & Thermal						
Physics	3	1	–	–(1)	–(1)	–(2)
Total	15	5	10	5(7)	3(9)	2(12)
Full-time Grad. Stud.	–	3	10	–	–	–
First-year Grad. Stud.	–	1	2	–	–	–

GRADUATE DEGREE REQUIREMENTS

Master's: Thirty semester hours with a B average and 1 year of residence is required; no language or comprehensive examination; non-thesis option available.

Doctorate: Ninety semester hours with B average and 2 years of residence is required; master's degree may be accepted in lieu of a maximum of 30 hours; 90 hours must include 6 hours of seminar and a minimum of 39 hours of course work, 24 in major field, 9 in minor field, 6 out-of-department; there is no language requirement; examinations consist of a written comprehensive exam on undergraduate physics, an oral qualifying exam in the area of proposed research, and a defense of the dissertation.

Thesis: Thesis may be written in absentia.

Table B—Separately Budgeted Research Expenditures by Source of Support

Source of Support	Departmental Research	Physics-related Research Outside Department
Federal government	$230,686	
State/local government	$99,809	
Non-profit organizations		
Business and industry		
Other	$2,823	
Total	**$333,318**	

Table C—Separately Budgeted Research Expenditures by Research Specialty

Research Specialty	No. of Grants	Expenditures ($)
Biophysics	1	
Energy Sources & Environment	2	
Medical, Health Physics	1	
Nano Science and Technology	2	
Physics and other Science Education	4	
Other	2	
Total	**12**	

FACULTY

Distinguished University Professor

Forgacs, Gabor, Ph.D., Eötvös Roländ University, 1978. Executive and Scientific Director of the Shipley Center for Innovation Czanderna and Shirkey Professor of Physics. *Biophysics, Medical, Health Physics, Statistical & Thermal Physics.*

Professor

ben-Avraham, Daniel, Ph.D., Bar-Ilan University, 1985. *Biophysics, Chemical Physics, Computational Physics, Condensed Matter Physics, Statistical & Thermal Physics.*

Privman, Vladimir, Ph.D., Technion, 1982. Robert A. Plane Professo; Director, Center for Innovative Device Technologies. *Chemical Physics, Computational Physics, Condensed Matter Physics, Mechanics, Nano Science and Technology, Quantum Foundations, Statistical & Thermal Physics.* Self-healing materials; quantum information, foundations of quantum mechanics.

Roy, Dipankar, Ph.D., Rensselaer Polytechnic Institute, 1986. Departmental Chair. *Chemical Physics, Condensed Matter Physics, Energy Sources & Environment, Nano Science and Technology, Optics, Surface Physics.* Energy storage technologies.

Schulman, Lawrence, Ph.D., Princeton University, 1967. *Astrophysics, Condensed Matter Physics, Cosmology & String Theory, Quantum Foundations, Relativity & Gravitation, Statistical & Thermal Physics.*

Assistant Professor

Gracheva, Maria, Ph.D., Moscow State Engineering Physics Institute, 1998. *Biophysics, Condensed Matter Physics, Nano Science and Technology, Solid State Physics.*

Ramsdell, Michael, Ph.D., Clarkson University, 2004. Director First Year Physics, Team Design Program. *Engineering Physics/Science, Physics and other Science Education.* Science education, STEM.

Scrimgeour, Jan, D.Phil., University of Oxford, 2005. *Biophysics, Condensed Matter Physics, Nano Science and Technology, Optics.* Soft matter and polymer physics, cell physics, advanced optical imaging, biointerfaces.

Thomas, Joshua D., Ph.D., University of Toledo, 2012. Team Design Program, Clarkson University Reynold's Observatory. *Astronomy, Astrophysics, Computational Physics, Physics and other Science Education.* Observational Astronomy, Laboratory Astrophysics, and Spectroscopy.

Emeritus

Glasser, M. Lawrence, Ph.D., Carnegie Mellon University, 1962. *Applied Mathematics, Condensed Matter Physics, Nonlinear Dynamics and Complex Systems, Statistical & Thermal Physics.*

Affiliate Professor

Gorshkov, Vyacheslav, Ph.D., East Ukrainian University, 1992. *Condensed Matter Physics, Nano Science and Technology, Statistical & Thermal Physics.*

Sokolov, Igor, Ph.D., D.I. Mendeleev Metrology Institute, 1991. *Biophysics, Chemical Physics, Condensed Matter Physics, Nano Science and Technology, Statistical & Thermal Physics, Surface Physics.*

Affiliate Associate Professor

Wick, David, Ph.D., Clarkson University, 1996. *Fluids, Rheology, Nonlinear Dynamics and Complex Systems, Physics and other Science Education.*

Visiting Assistant Professor

Melnikov, Dmitriy, Ph.D., Lehigh University, 2001. *Biophysics, Condensed Matter Physics, Nano Science and Technology, Solid State Physics.*

Courtesy Professor of Physics

Bollt, Erik, Ph.D., University of Colorado, 1995. *Nonlinear Dynamics and Complex Systems.* Image processing, model reduction.

Cheng, Ming-Cheng, Ph.D., Polytechnic University, 1990. *Computational Physics, Condensed Matter Physics, Nano Science and Technology.*

DEPARTMENTAL RESEARCH SPECIALTIES AND STAFF

Theoretical

Astrophysics. Schulman, Thomas.

Biophysics. Modeling of protein dynamics; DNA translocation; cell biomechanics. ben-Avraham, Forgacs, Gracheva, Melnikov, Scrimgeour, Sokolov.

Chemical Physics. Reaction kinetics; energy storage materials; surface absorption. ben-Avraham, Privman, Roy.

Condensed Matter Physics. Quantum wells; semiconductors; ion mobility. ben-Avraham, Cheng, Glasser, Gorshkov, Gracheva, Melnikov, Privman, Roy, Schulman, Scrimgeour.

Nano Science and Technology. Quantum computing; device physics. Cheng, Gorshkov, Gracheva, Melnikov, Privman, Roy, Scrimgeour.

Optics. Optical sensors; nonlinear optics, microscopy. Roy, Scrimgeour.

Physics and other Science Education. Ramsdell, Thomas, Wick.

Statistical & Thermal Physics. Phase transitions; network statistics; scaling; finite size effects; percolation: self-avoiding walks. ben-Avraham, Forgacs, Glasser, Gorshkov, Privman, Schulman.

Experimental

Astronomy. Light scattering photoluminescence; stellar spectroscopy, and polarimetry. Thomas.

Biophysics. Quantitative localization miroscopy; Physical properties of tissues, cytoplasm, and cytoskeleton. Forgacs, Scrimgeour, Sokolov.

Physics and other Science Education. Physics Education Research, Assessment and Curriculum Development. Ramsdell.

Surface Physics. Scanning probe microscopy; nanomaterials; soft condensed matter. Scrimgeour, Sokolov.

Surface Physics. Interfacial electrochemistry of energy storage/conversion materials; kinetics of surface reactions; corrosion; catalysis. Roy.

View additional information about this department at
www.gradschoolshopper.com

COLUMBIA UNIVERSITY, FU FOUNDATION SCHOOL OF ENGINEERING AND APPLIED SCIENCE

DEPARTMENT OF APPLIED PHYSICS AND APPLIED MATHEMATICS

New York, New York 10027
http://www.apam.columbia.edu/

General University Information

President: Lee C. Bollinger
Dean of Graduate School: Mary C. Boyce
University website: http://www.columbia.edu
Control: Private
Setting: Urban
Total Faculty: 3,662
Total Graduate Faculty: 3,662
Total number of Students: 28,824
Total number of Graduate Students: 18,220

Department Information

Department Chairman: Ismail C. Noyan, Chair
Department Contact: Dina Amin, Department Administrator
  Total full-time faculty: 34
  Total number of full-time equivalent positions: 23
  Full-Time Graduate Students: 111
  First-Year Graduate Students: 53
  Female First-Year Students: 13
  Total Post Doctorates: 14

Department Address

500 West 120th Street
200 Seeley Mudd Bldg, MC 4701
New York, NY 10027
Phone: (212) 854-4457
Fax: (212) 854-8257
E-mail: seasinfo.apam@columbia.edu
Website: http://www.apam.columbia.edu/

ADMISSIONS

Admission Contact Information

Address admission inquiries to: Office of Graduate Student Services, The Fu Foundation of Engineering and Applied Science, 500 West 120th Street, Room 254, Engineering Terrace, Mail Code 4708, Columbia University, New York, NY 10027
Phone: (212) 854-6438
E-mail: seasgradmit@columbia.edu
Admissions website: http://www.gradengineering.columbia.edu/gradmissions

Application deadlines

Fall admission:
U.S. students: December 15   *Int'l. students*: December 15
Spring admission:
U.S. students: October 1   *Int'l. students*: October 1

Application fee

U.S. students: $85   *Int'l. students*: $85

December 15 is the deadline for doctoral applications; February 15 is the deadline for M.S., part-time, and non-degree applications. There are NO Spring doctoral admissions, and the Only Spring admissions for Masters is for Applied Mathematics.

Admissions information

For Fall of 2012:
  Number of applicants: 461
  Number admitted: 121
  Number enrolled: 53

Admission requirements

Bachelor's degree requirements: Bachelor's degree in sciences or engineering from an accredited institution is required.

GRE requirements
The GRE is required.
  Quantitative score: 700
No required minimums except for quantitative.

Advanced GRE requirements
The Advanced GRE is required.
  Minimum accepted Advanced GRE score: 680

TOEFL requirements
The TOEFL exam is required for students from non-English-speaking countries.
  PBT score: 619
  iBT score: 100
There are no required minimums; above is suggested minimum.

Other admissions information
Additional requirements: The minimum acceptable score suggested for admission is quantitative-700. The GRE Advanced is required for applicants to the Applied Physics Doctoral program. The GRE Advanced is strongly urged for doctoral applicants in Applied Mathematics and Materials Science and Engineering. The minimum acceptable score suggested for admission is 680.
Undergraduate preparation assumed: B.S. or B.A. degree in science or engineering from an accredited university.

TUITION

Tuition year 2012–13:
  Full-time students: $38,688 annual
  Part-time students: $1,578 per credit
Credit hours per semester to be considered full-time: 12
Deferred tuition plan: No
Health insurance: Available at the cost of $2,981 per year.
Other academic fees: approx. $4,000 (inclusive of health fees). See http://www.engineering.columbia.edu/tuition-fees-and-payments.
Academic term: Semester
Number of first-year students who received full tuition waivers: 16

Teaching Assistants, Research Assistants, and Fellowships
Number of first-year
  Teaching Assistants: 15
Average stipend per academic year
  Teaching Assistant: $23,750
  Research Assistant: $23,750
  Fellowship student: $24,000
Fellowships vary, depending on the source of funding.

FINANCIAL AID

Application deadlines
Fall admission:
U.S. students: December 15   *Int'l. students*: December 15

Loans
Loans are available for U.S. students.
Loans are not available for international students.
GAPSFAS application required: No
FAFSA application required: Yes

For further information
Address financial aid inquiries to: Office of Graduate Student Services, The Fu Foundation of Engineering and Applied Science, 500 West 120th Street, Room 254, Engineering Terrace, Mail Code 4708, Columbia University, New York, NY 10027.
Phone: (212) 854-3711
E-mail: engradfinaid@columbia.edu

Financial aid website: http://www.gradengineering.columbia.edu/financial-aid-4

HOUSING

Availability of on-campus housing
  Single students: Yes
  Married students: Yes

For further information
Address housing inquiries to: University Apartment Housing, 400 W. 119th Street, New York, NY 10027.
Phone: (212) 854-9300
E-mail: uah@columbia.edu
Housing aid website: http://facilities.columbia.edu/housing/overview

Table A—Faculty, Enrollments, and Degrees Granted

Research Specialty	2012-13 Faculty	Enrollment Fall 2012 Master's	Enrollment Fall 2012 Doctorate	Degrees Granted 2012-13 (2008-13) Master's	Degrees Granted 2012-13 (2008-13) Terminal Master's	Degrees Granted 2012-13 (2008-13) Doctorate
AM/climate research	–	2	4	2(10)	–(1)	2(8)
Applied Mathematics	12	16	16	14(53)	9(27)	1(13)
Applied Physics	14	5	–	4(18)	4(18)	–
Biophysics	–	–	–	–(1)	–	1(2)
Condensed Matter Physics	–	6	13	4(20)	–	–(8)
Materials Science, Metallurgy	7	19	12	10(63)	4(37)	3(14)
Medical, Health Physics	1	26	–	8(53)	8(51)	–(2)
Optics	–	3	4	3(11)	–	–(4)
Plasma and Fusion	–	2	10	2(20)	–	4(14)
Total	34	61	68	47(249)	25(134)	11(65)
Full-time Grad. Stud.	–	32	79	–	–	–
First-year Grad. Stud.	–	53	19	–	–	–

GRADUATE DEGREE REQUIREMENTS

Master's: The Master of Science (M.S.) degree is given for completion of 1 or more years of study beyond the Bachelor's degree. A minimum of 30 points (or semester hours) of residence credit of approved graduate course work completed at Columbia is required for the degree. A research report is required for the program in Materials Science and Engineering. A minimum grade point average of 2.5 must be maintained. The following are not required for the M.S. degree: knowledge of a foreign language, a comprehensive and/or qualifying examination, and a thesis. All degree requirements must be completed within 5 years of the beginning of graduate study. The Master of Philosophy (M.Phil.): all requirements for the Ph.D., except for the dissertation. The Program in Medical Physics, which leads to a 36-point M.S. degree and requires a comprehensive examination, is offered in collaboration with faculty from the College of Physicians and Surgeons and the Mailman School of Public Health. It prepares students for careers in medical physics and provides preparation for the ABR certification examination.

Doctorate: For both the Ph.D. and the Eng. Sc.D., candidates must successfully complete 30 points (or semester hours) or more of approved graduate course work beyond the Master's degree. A minimum grade point average of 3.0 must be maintained. Candidates must successfully pass both written and oral qualifying examinations. They must submit and successfully defend an approved dissertation. A knowledge of a foreign language is not required. A time limit of seven years is allowed for completion of the requirements for doctoral degrees. The residence requirement for the Ph.D. degree is satisfied by six Residence Units in the Graduate School of Arts and Sciences, two of which are granted for the Master's degree. The residence requirement for the Eng. Sc.D. degree is satisfied by 12 additional points of credit in APAM E9800 (Doctoral Research Instruction) in the School of Engineering and Applied Science.

Thesis: Thesis may be written in absentia.

SPECIAL EQUIPMENT, FACILITIES, OR PROGRAMS

Columbia's Plasma Physics Laboratory contains experiments to better understand and to control high-temperature magnetized plasma. Experiments include a toroidal high-beta tokamak, HBT-EP, a linear steady-state plasma experiment, a large laboratory terrella used to investigate space plasma physics, the CNT stellarator used to study non-neutral plasmas, and a variety of smaller devices used to produce pulsed and continuous plasmas for scientific investigations.

Graduate research opportunities also exist with one of several collaborative research programs between Columbia University and other national research projects. These include collaborations with the Princeton Plasma Physics Laboratory, MIT's Plasma Science and Fusion Center, and the Fusion Research Group at General Atomics. Free-electron laser (FEL) research opportunities also exist using the 40 MeV rf linac at the Brookhaven National Laboratory.

Research equipment in the Solid-State Physics and Optical Physics laboratories and the associated Columbia Center for Integrated Science and Engineering include extensive laser and spectroscopy facilities, a clean room that includes photolithography and thin film fabrication systems, ultra high-vacuum surface preparation and analysis chambers, direct laser writing stations, a molecular beam epitaxy machine, picosecond and femtosecond lasers, and diamond anvil cells.

Facilities and research opportunities also exist within the interdisciplinary DOE Energy Frontier Research Center which focuses on conversion of sunlight into electricity in nanometer-sized thin films, and the NSF Nanoscale Science and Engineering Center, which focuses on electron transport in molecular nanostructures.

Materials Science and Engineering facilities include transmission and scanning electron microscopes, x-ray diffraction, ellipsometry, x-ray photoelectron microscopy equipment, a clean room with photolithography, deposition, and etching capabilities, magnetic and electrical measurement equipment, laser spectroscopy, laser processing, and mechanical testing. Faculty members have research collaborations with local research and manufacturing centers including Lucent, Exxon, IBM, as well as major international research centers.

The National Synchrotron Light Source at Brookhaven National Laboratory is used for high-resolution x-ray diffraction and absorption measurements. There are also research opportunities in medical physics at the Columbia Presbyterian Medical Center, as well as at other medical institutes, employing state-of-the-art medical diagnostic imaging and treatment equipment.

The Applied Mathematics division is closely linked with the Lamont Doherty Earth Observatory (LDEO), with six faculty members sharing appointments in the Department of Earth and Environmental Sciences.

There are also close ties with the NASA Goddard Institute for Space Studies (GISS), with Columbia's Center for Computational Biology and Bioinformatics, and with Columbia's Center for Computational Learning Systems.

Ongoing joint research, instruction and supervision of graduate students with LDEO and NASA/GISS span the areas of atmosphere, ocean and climate modeling, geophysical/geological fluid dynamics, earthquake physics, remote sensing, and large-scale scientific computing. The research of the Plasma Lab is supported by a dedicated data acquisition/data analysis system.

Computational researchers have local access to the Department's SiCortex supercomputer with 1458 cores as well as to Columbia's 256-processor Linux cluster, and to supercomputer systems at the National Center for Atmospheric Research and the Lawrence Berkeley and Oak Ridge National Laboratories.

Table B—Separately Budgeted Research Expenditures by Source of Support

Source of Support	Departmental Research	Physics-related Research Outside Department
Federal government	$10,746,863	
State/local government		
Non-profit organizations		
Business and industry	$861,035	
Other	$689,696	
Total	**$12,297,594**	

Table C—Separately Budgeted Research Expenditures by Research Specialty

Research Specialty	No. of Grants	Expenditures ($)
Atmosphere, Space Physics, Cosmic Rays	55	$3,486,758
Biophysics	7	$198,342
Condensed Matter Physics	26	$1,980,982
Materials Science, Metallurgy	65	$2,552,373
Physics of Beams	2	$284
Plasma and Fusion	33	$2,749,260
Applied Mathematics	43	$1,329,595
Total	**231**	**$12,297,594**

FACULTY

Professor

Bal, Guillaume, Ph.D., University of Paris, 1997. *Applied Mathematics, Other.* Applied mathematics, partial differential equations with random coefficients, high frequency waves in random media, and application to time reversal, theory of inverse transport and applications in medical imaging and geophysical imaging.

Barmak, Katayun, Ph.D., Massachusetts Institute of Technology, 1989. *Applied Physics, Materials Science, Metallurgy.* Processing and structure (crystal structure and microstructure) relationships to electrical and magnetic properties of metal films; developing transmission electron microscopy automated orientation imaging techniques that can be applied to the study of nanostructured materials; use of differential scanning calorimetry for the study solid state reactions and phase transformations in thin films.

Bienstock, Daniel, Ph.D., Massachusetts Institute of Technology, 1985. (Joint with Industrial Engineering and Operations Research.). *Applied Mathematics.* Applied mathematics, methodology and high-performance implementation of optimization algorithms, applications of optimization: preventing national-scale blackouts, emergency management, approximate solution of massively large optimization problems, higher-dimensional reformulation techniques for integer programming, robust optimization.

Billinge, Simon J. L., Ph.D., University of Pennsylvania, 1992. *Applied Mathematics, Applied Physics, Materials Science, Metallurgy*. Nanoscale structure-property relationships in functional nanomaterials studied using novel x-ray and neutron scattering techniques coupled with advanced computing; solving the nanostructure problem.

Boozer, Allen H., Ph.D., Cornell University, 1970. *Applied Physics*. Plasma theory; theory of magnetic confinement for fusion energy; nonlinear dynamics.

Cane, Mark A., Ph.D., Massachusetts Institute of Technology, 1975. (Joint with Earth and Environmental Sciences.). *Applied Mathematics*. Climate dynamics; physical oceanography; geophysical fluid dynamics; computational fluid dynamics; impacts of climate on society, El Niño forecasting.

Chan, Siu-Wai, Ph.D., Massachusetts Institute of Technology, 1985. *Materials Science, Metallurgy*. Nanoparticles, electronic ceramics, grain boundaries and interfaces, oxide thin films.

Herman, Irving P., Ph.D., Massachusetts Institute of Technology, 1977. Chairman of Department. *Applied Physics*. Nanocrystals; laser diagnostics of thin-film processing; optical spectroscopy of nanostructured materials; mechanical properties of nanomaterials.

Im, James, Ph.D., Massachusetts Institute of Technology, 1989. *Materials Science, Metallurgy*. Laser-induced crystallization of thin films, phase transformations and nucleation in condensed systems.

Kim, Philip, Ph.D., Harvard University, 1999. (Joint with Physics.). *Applied Physics*. Experimental condensed matter physics, physical properties and applications of nanoscale low-dimensional materials, quantum thermal transport phenomena in 1-dimensional nanoscaled materials, mesoscopic thermoelectricity and thermoelectric applications of nanoscale materials, quantum transport in novel 2-dimensional materials, mesoscopic electron transport and thermodynamic processes for sensors and electric devices.

Mauel, Michael E., Ph.D., Massachusetts Institute of Technology, 1983. *Applied Physics*. Plasma physics, waves and instabilities, fusion and equilibrium control; space physics; plasma processing; international energy policy.

Navratil, Gerald A., Ph.D., University of Wisconsin-Madison, 1976. *Applied Physics*. Plasma physics; plasma diagnostics; fusion energy science.

Noyan, I. Cevdet, Ph.D., Northwestern University, 1984. *Materials Science, Metallurgy*. Characterization and modeling of mechanical and micromechanical deformation; residual stress analysis and nondestructive testing; x-ray and neutron diffraction, microdiffraction analysis.

Osgood, Richard M., Ph.D., Massachusetts Institute of Technology, 1973. (joint with Electrical Engineering). *Applied Physics*. Nanoscale optical and electronic phenomena, femtosecond lasers and laser probing, low-dimensional physics, integrated optics, nanofabrication and materials growth.

Pinczuk, Aron, Ph.D., University of Pennsylvania, 1969. *Applied Physics*. Spectroscopy of semiconductors and insulators; quantum structures, systems of reduced dimensions, atomic layers of graphene, electron quantum fluids.

Polvani, Lorenzo, Ph.D., Massachusetts Institute of Technology, 1988. *Applied Mathematics*. Atmospheric and climate dynamics, geophysical fluid dynamics, numerical methods for weather and climate modeling, planetary atmospheres.

Ruderman, Malvin A., Ph.D., California Institute of Technology, 1951. (Joint with Physics.). *Applied Physics*. Theoretical astrophysics; neutron stars; pulsars; early universe; cosmic gamma rays.

Scholz, Christopher H., Ph.D., Massachusetts Institute of Technology, 1967. (Joint with Earth and Environmental Sciences.). *Applied Mathematics*. Experimental and theoretical rock mechanics, especially friction, fracture and hydraulic transport properties, nonlinear systems, mechanics of earthquakes and faulting.

Sen, Amiya K., Ph.D., Columbia University, 1963. (Joint with Electrical Engineering.). *Applied Physics*. Plasma physics; fluctuation and anomalous transport in plasmas; control of plasma instabilities, plasma transport.

Sobel, Adam H., Ph.D., Massachusetts Institute of Technology, 1998. *Applied Mathematics*. Atmospheric science, geophysical fluid dynamics, tropical meteorology, climate dynamics.

Spiegelman, Marc W., Ph.D., University of Cambridge, 1989. *Applied Mathematics*. Coupled fluid/solid mechanics, reactive fluid flow, solid earth and magma dynamics, scientific computation/modeling.

Wang, Wen I., Ph.D., Cornell University, 1981. (Joint with Electrical Engineering.). *Applied Physics*. Heterostructure devices and physics; materials properties; molecular beam epitaxy.

Weinstein, Michael I., Ph.D., Courant Institute, New York University, 1982. *Applied Mathematics*. Applied mathematics, partial differential equations and analysis, waves in nonlinear, inhomogeneous and random media; dynamical systems; multiscale phenomena, applications to nonlinear optics, mathematical physics; fluid dynamics; geosciences.

Wuu, Cheng Shie, Ph.D., University of Kansas, 1985. (Joint with Radiation Oncology.). *Medical, Health Physics*. Microdosimetry, biophysical modeling, dosimetry of brachytherapy, gel dosimetry second cancers induced by radiotherapy, medical physics.

Associate Professor

Bailey, William E., Ph.D., Stanford University, 1999. *Materials Science, Metallurgy*. Nanoscale magnetic films and heterostructures, materials issues in spin-polarized transport, materials engineering of magnetic dynamics.

Marianetti, Chris A., Ph.D., Massachusetts Institute of Technology, 2004. *Applied Mathematics, Materials Science, Metallurgy*. Predicting materials properties from first-principles computations; density-functional theory; dynamical mean-field theory; transition-metal oxides; actinides.

Venkataraman, Latha, Ph.D., Harvard University, 1999. *Applied Physics*. Single molecule transport, single molecule force spectroscopy, electron transport in nanowires, scanning tunneling microscopy and spectroscopy.

Wiggins, Chris H., Ph.D., Princeton University, 1998. *Applied Mathematics*. Applied mathematics, mathematical biology, biopolymer dynamics, soft condensed matter, genetic networks and network inference, machine learning.

Assistant Professor

Cole, Andrew, Ph.D., University of Texas, Austin, 2006. *Applied Physics, Nuclear Physics, Plasma and Fusion, Theoretical Physics*. Plasma physics and nuclear fusion; particular focus on symmetry-breaking magnetic perturbations and their effect on plasma rotation in magnetically-confined fusion plasmas.

Letourneau, Pierre-David, Ph.D., Stanford University, 2013. *Applied Mathematics, Nonlinear Dynamics and Complex Systems*. Imaging in strongly-scattering and random media, Super-resolution, fast algorithms, numerical linear algebra numerical analysis.

Shaw, Tiffany A., Ph.D., University of Toronto, 2009. (Joint with Earth and Environmental Sciences.). *Applied Mathematics*. Atmospheric science, advection-diffusion of a passive scalar; Hamiltonian geophysical fluid dynamics; multiple scale asymptotics; wave-mean flow interaction.

Volpe, Francesco, Ph.D., Ernst Moritz Arndt University and International Max Planck Research School on Bounded Plasmas, 2003. *Applied Physics, Plasma and Fusion*. plasma physics and magnetic confinement fusion (tokamaks and stellarators) both experimentally and via numerical modeling, with emphasis on: (1) microwave heating, current drive and diagnostics and (2) magnetohydrodynamic instabilities and their control.

Yu, Nanfang, Ph.D., Harvard University, 2009. *Applied Physics, Atomic, Molecular, & Optical Physics, Nano Science and Technology, Optics.* Mid-infrared and far-infrared optics and optoelectronic devices, active plasmonics and metamaterials with gain media, reconfigurable metainterfaces based on phased optical antenna arrays, mid-infrared and terahertz quantum cascade lasers, semiconductor quantum wells and superlattices with largely tunable optical properties, infrared imaging and spectroscopy, nanophotonics, graphene optoelectronic devices.

Emeritus

Beshers, Daniel N., Ph.D., University of Illinois, 1956. Metallurgy and materials science.

Chu, C. K., Ph.D., Courant Institute, New York University, 1959. Applied mathematics.

Friedman, Morton B., Ph.D., New York University, 1953. (Joint with Civil Engineering.). Applied mathematics and mechanics; numerical analysis; parallel computing.

Gross, Robert A.; Ph.D., Harvard University, 1952. Plasma physics; energy; fluid dynamics.

Lidofsky, Leon, Ph.D., Columbia University, 1952. Radiation applications in medicine and technology, radiation shielding.

Marshall, Thomas C., Ph.D., University of Illinois, 1960. Accelerator concepts, free-electron lasers; relativistic beam dynamics and radiation.

Stormer, Horst, Ph.D., University of Stuttgart, 1977. Semiconductors; electronic transport; lower-dimensional physics; transport in nanostructures.

Other

Tippett, Michael, Ph.D., New York University, 1992. Lecturer in Discipline of Applied Mathematics. *Applied Mathematics, Computational Physics.* The predictability and variability of the climate system, with emphasis on the application of statistical methods to data from observations and numerical models.

DEPARTMENTAL RESEARCH SPECIALTIES AND STAFF

Theoretical

Applied Mathematics. Analytical and numerical analysis of PDEs, scientific computation, geophysical fluid dynamics, biomathematics, dynamical systems and chaos, as well as applications to various fields of physics including solid-state, plasma physics, nonlinear optics, medical imaging and the earth sciences and various fields of biology. In collaboration with the NASA Goddard Institute for Space Studies and the Lamont-Doherty Earth Observatory, atmospheric and oceanic modeling. Bal, Bienstock, Boozer, Cane, Friedman, Letourneau, Polvani, Scholz, Shaw, Sobel, Spiegelman, Tippett, Weinstein, Wiggins.

Materials Science, Metallurgy. Diffraction physics, dynamic and kinematic scattering of x-rays and neutrons, forward calculation of diffraction profiles and inverse calculation of material structures, theoretical analysis of heterogeneous and homogeneous nucleation kinetics. Barmak, Billinge, Chan, Im, Marianetti, Noyan.

Plasma and Fusion. Basic plasma theory; analytical and computational magnetohydrodynamics; toroidal high-beta equilibrium and stability; space/solar plasmas; linear and nonlinear interactions among waves and particles; plasma turbulence; plasma transport properties; free-electron laser theory. Boozer, Cole, Sen, Volpe.

Experimental

Condensed Matter Physics. Nanoscience and nanoparticles, the optical spectroscopy of semiconductor structures that are subjected to high pressure, electronic transport and inelastic light scattering in low-dimensional correlated electron systems, fractional quantum Hall effect, heterostructure physics and applications, molecular beam epitaxy, grain boundaries and interfaces, nucleation in thin films, and surface physics. Research opportunities also exist within the interdisciplinary NSF Nanoscale Science and Engineering Center, which focuses on electron transport in molecular nanostructures. Barmak, Billinge, Chan, Herman, Kim, Osgood, Pinczuk, Venkataraman, Wang, Yu.

Materials Science, Metallurgy. Polycrystalline semiconductor and metallic films, laser irradiation of materials, electronic ceramics, wide-band-gap semiconductors, plasma processing of materials and optical diagnostics of thin-film processing, ceramic nanocomposites, and magnetic films and spintronics, nondestructive testing of material systems, residual stress analysis. Bailey, Barmak, Chan, Im, Noyan.

Optics. Inelastic light scattering, the free-electron laser, accelerators, optical diagnostics of film processing, nonlinear optics, ultrafast optoelectronics, photonic switching, optical physics of surfaces and nanocrystals, laser-induced crystallization, and photon integrated circuits. Herman, Im, Osgood, Pinczuk, Wang, Yu.

Plasma and Fusion. High-beta tokamak equilibrium, stability, active feedback mode control and transport; HBT-EP Tokamak (R=0.92m, a=0.15m, B=0.5T); design of improved tokamak configurations; plasma heating at microwave and radio frequencies. Joint programs on advanced and innovative fusion confinement at the NSTX device at the Princeton Plasma Physics Laboratory, the DIII-D device at General Atomics investigating high-beta equilibrium and stability, and the LDX device at M.I.T. Plasma diagnostics; CO_2 laser scattering, multi-point Thomson scattering, and tomographic density reconstruction on high-beta plasmas; ultraviolet, optical, infrared, and microwave radiation and scattering. Waves in plasmas: trapped particle instabilities; space plasma physics, nonlinear and chaotic transport in planetary magnetospheres; feedback control of plasma instabilities; anomalous transport due to plasma instabilities. Intense relativistic electron beams: high-power millimeter wave sources-free electron laser; beam diagnostics; relativistic electron beam microwave production; inverse FEL accelerators; plasma processing of semiconductors using microwave sources, non-neutral plasmas. Mauel, Navratil, Sen, Volpe.

View additional information about this department at
www.gradschoolshopper.com

COLUMBIA UNIVERSITY

DEPARTMENT OF PHYSICS

New York, New York 10027
http://www.columbia.edu/cu/physics/

General University Information
President: Lee Bollinger
Dean of Graduate School: Carlos J. Alonso
University website: http://columbia.edu
Control: Private
Total Faculty: 3,566
Total Graduate Faculty: 700
Total number of Students: 25,459
Total number of Graduate Students: 15,067

Department Information
Department Chairman: William A. Zajc, Chair
Department Contact: Randy Torres, Senior Administrative
 Manager
 Total full-time faculty: 39
 Full-Time Graduate Students: 111
 First-Year Graduate Students: 25

Department Address
538 West 120th Street
704 Pupin Hall MC 5255
New York, NY 10027
Phone: (212) 854-3366
Fax: (212) 854-3379
E-mail: rtorres@phys.columbia.edu
Website: http://www.columbia.edu/cu/physics/

ADMISSIONS

Admission Contact Information
Address admission inquiries to: Office of Graduate Admissions,
 108 Low Library
Phone: (212) 854-8903
E-mail: gsas-admit@columbia.edu
Admissions website: http://gsas.columbia.edu/admissions

Application deadlines
Fall admission:
U.S. students: January 3   *Int'l. students*: January 3

Application fee
U.S. students: $105

Admissions information
For Fall of 2013:
 Number of applicants: 454
 Number admitted: 34
 Number enrolled: 12

Admission requirements
Bachelor's degree requirements: Bachelor's degree is required.

GRE requirements
The GRE is required.

Advanced GRE requirements
The Advanced GRE is required.

TOEFL requirements
The TOEFL exam is required for students from non-English-
 speaking countries.
 PBT score: 600

iBT score: 100

Other admissions information
Additional requirements: No minimum acceptable score is specified.

TUITION

Tuition year 2013–14:
 Full-time students: $39,852 annual
All regular graduate students in physics are given full financial
 aid, which includes exemption from all tuition and medical
 fees.
Deferred tuition plan: No
Other academic fees: $4,000/yr.
Academic term: Semester

Teaching Assistants, Research Assistants, and Fellowships
Number of first-year
 Teaching Assistants: 24
 Fellowship students: 24
Average stipend per academic year
 Teaching Assistant: $15,834
 Research Assistant: $23,750

FINANCIAL AID

Loans
Loans are not available for U.S. students.
Loans are not available for international students.
GAPSFAS application required: No
FAFSA application required: No

For further information
Address financial aid inquiries to: Thomas Tarduogno, Director
 of Financial Aid.
Phone: (212) 854-3809
E-mail: tt22@columbia.edu
Financial aid website: http://gsas.columbia.edu/financial-aid

HOUSING

Availability of on-campus housing
 Single students: Yes
 Married students: Yes

For further information
Address housing inquiries to: Office of Institutional Real Estate,
 University Apartment Housing, 400 W 119th Street, New
 York, NY 10027.
Phone: (212) 854-9423
E-mail: housing@columbia.edu
Housing aid website: http://housingservices.columbia.edu/

Table A—Faculty, Enrollments, and Degrees Granted

Research Specialty	2012-13 Faculty	Enrollment Fall 2012		Number of Degrees Granted 2012-13 (2008–13)		
		Master's	Doctorate	Master's	Terminal Master's	Doctorate
Applied Physics	2	–	3	–	–	–(8)
Astronomy	–	–	–	–	–	–
Astrophysics	8	–	12	–	–	6(24)
Atomic, Molecular, & Optical Physics	3	–	2	–	–	–
Biophysics	1	–	1	–	–	–
Chemical Physics	–	–	1	–	–	–
Condensed Matter Physics	9	–	14	–	–	3(-)
Engineering Physics/Science	–	–	–	–	–	–
High Energy Physics	5	–	8	–	–	2(13)
Nuclear Physics	2	–	4	–	–	2(8)
Theoretical Physics	5	–	12	–	–	5(14)
Non-specialized	1	–	49	46(84)	1(5)	–
Total	36	–	106	46(84)	1(5)	18(77)
Full-time Grad. Stud.	–	–	106	–	–	–
First-year Grad. Stud.	–	–	25	–	–	–

GRADUATE DEGREE REQUIREMENTS

Master's: M.A. 30 points of courses including at least 24 within the department at an appropriate level. Two semesters of residence. No requirements for grade average, foreign languages, comprehensive or qualifying examination, or thesis.

Doctorate: Master's degree plus four semesters of residence, qualifying examination, thesis. No requirement for foreign language. No course requirement beyond the Master's degree, except that courses must show competence in electromagnetic theory, quantum mechanics, and statistical mechanics. Grades must be satisfactory.

Other Degrees: M.Phil. All requirements for the Ph.D. except thesis and final departmental examination.

Thesis: Thesis may be written in absentia.

SPECIAL EQUIPMENT, FACILITIES, OR PROGRAMS

In addition to experimental facilities for Astrophysics, Atomic, Condensed Matter and Molecular in Pupin Laboratory on campus, extensive facilities for High-Energy, Nuclear Physics and Astrophysics, are located at the Nevis Laboratories on an estate at Irvington-on-Hudson, New York.

Table B—Separately Budgeted Research Expenditures by Source of Support

Source of Support	Departmental Research	Physics-related Research Outside Department
Federal government	$20,384,715	
State/local government		
Non-profit organizations		
Business and industry		
Other	$312,762	
Total	**$20,697,477**	

Table C—Separately Budgeted Research Expenditures by Research Specialty

Research Specialty	No. of Grants	Expenditures ($)
Astrophysics	25	$2,930,367
Condensed Matter Physics	43	$2,160,880
Nuclear Physics	7	$1,534,201
Particles and Fields	14	$12,505,472
Total	103	$21,527,641

FACULTY

Professor

Aleiner, Igor, Ph.D., University of Minnesota, 1996. Theoretical condensed matter physics.

Altshuler, Boris, Ph.D., Leningrad Institute for Nuclear Physics, 1979. Theoretical condensed matter physics.

Aprile, Elena, Ph.D., University of Geneva, 1982. Experimental high-energy astrophysics; gamma-ray astronomy.

Beloborodov, Andrei, Ph.D., Lebedev Physical Institute, Moscow, 1995. Theoretical astrophysics.

Christ, Norman, Ph.D., Columbia University, 1966. Theoretical particle physics; lattice field theory.

Cole, Brian, Ph.D., Massachusetts Institute of Technology, 1992. Experimental relativistic heavy-ion nuclear physics.

Denef, Frederik, Ph.D., Leuven University, 1999. Theoretical Physics.

Greene, Brian, Ph.D., University of Oxford, 1987. Theoretical high-energy physics; string theory.

Gyulassy, Miklos, Ph.D., University of California, Berkeley, 1972. Theoretical nuclear physics.

Hailey, Charles, Ph.D., Columbia University, 1983. Experimental high-energy astrophysics.

Halpin-Healy, Timothy, Ph.D., Harvard University, 1987. Theoretical statistical physics.

Heinz, Tony, Ph.D., University of California, Berkeley, 1982. Surface physics and nonlinear optics.

Hughes, Emlyn, Ph.D., Columbia University, 1987. Experimental particle physics.

Hui, Lam, Ph.D., Massachusetts Institute of Technology, 1996. Theoretical astrophysics.

Kim, Philip, Ph.D., Harvard University, 1999. Experimental condensed matter physics.

Mawhinney, Robert D., Ph.D., Harvard University, 1987. Theoretical particle physics; lattice field theory.

Miller, Amber, Ph.D., Princeton University, 2000. Experimental astrophysics.

Millis, Andrew, Ph.D., Massachusetts Institute of Technology, 1986. Theoretical condensed matter physics.

Mueller, Alfred H., Ph.D., Massachusetts Institute of Technology, 1965. Theoretical particle physics.

Mukherjee, Reshmi, Ph.D., Columbia University, 1993. Experimental astrophysics.

Parsons, John, Ph.D., University of Toronto, 1990. Experimental particle physics.

Pinczuk, Aron, Ph.D., University of Pennsylvania, 1969. Condensed matter physics.

Ruderman, Malvin A., Ph.D., California Institute of Technology, 1951. Theoretical astrophysics.

Shaevitz, Michael, Ph.D., Ohio State University, 1975. Experimental particle physics.

Tuts, M ichael, Ph.D., Stony Brook University, 1979. Experimental particle physics.

Uemura, Yasutomo J., Ph.D., University of Tokyo, 1982. Condensed matter physics.

Weinberg, Erick J., Ph.D., Harvard University, 1973. Theoretical high-energy physics.

Zajc, William, Ph.D., University of California, Berkeley, 1982. Experimental relativistic heavy-ion nuclear physics.

Associate Professor

Brooijmans, Gustaaf, Ph.D., Université Catholique de Louvian, 1998. Experimental high-energy particle physics.

Levin, Janna, Massachusetts Institute of Technology, 1993. Theoretical astrophysics.

Marka, Szabolcs, Ph.D., Vanderbilt University, 1999. Experimental astrophysics.

Nicolis, Alberto, Ph.D., Scuola Normale Superiore, Pisa, 2003. Theoretical cosmology.

Sahin, Ozgur, Ph.D., Stanford University, 2005. Biological physics.

Assistant Professor

Humensky, Thomas B., Ph.D., Princeton University, 2003. Gamma-ray astrophysics.

Johnson, Bradley R., Ph.D., University of Minnesota, 2004. Cosmic microwave background studies: SKIP (new project), EBEX, PIPER, APEX-SZ.

Metzger, Brian D., Ph.D., University of California, Berkeley, 2009. *Astrophysics, Theoretical Physics*. Theoretical high-energy astrophysics.

Pasupathy, Abhay, Ph.D., Cornell University. Experimental condensed matter physics.

Rosen, Rachel A., Ph.D., New York University, 2009. Theoretical Particle Physics and Cosmology.

Zelevinsky, Tanya, Ph.D., Harvard University, 2004. Atomic, molecular, and optical physics.

Adjunct Professor

Budick, Burton, Ph.D., University of California, Berkeley, 1962. Experimental particle physics.

May, Morgan, Ph.D., Columbia University, 1975. Experimental particle physics and nuclear physics.

Senior Lecturer

Dodd, Jeremy, Ph.D., University College London, 1990. Experimental particle physics.

DEPARTMENTAL RESEARCH SPECIALTIES AND STAFF

Theoretical

Astrophysics. Aprile, Beloborodov, Greene, Hailey, Hui, Humensky, Johnson, Levin, Marka, Metzger, Miller, Rosen, Ruderman.

Condensed Matter Physics. Aleiner, Altshuler, Kim, Millis, Pasupathy, Uemura.

Nuclear Physics. Nuclear physics and relativistic heavy-ion physics. Gyulassy.

Statistical & Thermal Physics. Statistical mechanics and low-temperature physics. Aleiner, Altshuler, Halpin-Healy, Millis.

Experimental

Astrophysics. UV, x-ray, and gamma-ray astronomy; dark matter searches, experiments using ground-based detectors, rockets, balloons, and satellites. Aleiner, Altshuler, Aprile, Hailey, Humensky, Johnson, Levin, Marka, Metzger, Miller, Mukherjee.

Atomic, Molecular, & Optical Physics. Heinz, Hughes, Zelevinsky.

Condensed Matter Physics. Heinz, Kim, Pasupathy, Uemura.

Nuclear Physics. Relativistic heavy-ion nuclear physics. Cole, Zajc.

View additional information about this department at
www.gradschoolshopper.com

CORNELL UNIVERSITY

DEPARTMENT OF PHYSICS

Ithaca, New York 14853
http://www.physics.cornell.edu

General University Information

President: David J. Skorton
Dean of Graduate School: Barbara Knuth
University website: http://www.cornell.edu
Control: Private
Setting: Rural
Total Faculty: 1,564
Total Graduate Faculty: 1,790
Total number of Students: 21,131
Total number of Graduate Students: 5,174

Department Information

Department Chairman: Jeevak Parpia, Chair
Department Contact: Kacey Acquilano, Graduate Field Assistant

Total full-time faculty: 42
Total number of full-time equivalent positions: 42
Full-Time Graduate Students: 186
First-Year Graduate Students: 32
Female First-Year Students: 5
Total Post Doctorates: 36

Department Address

117 Clark Hall
Ithaca, NY 14853
Phone: (607) 255-7561

E-mail: physics@cornell.edu
Website: http://www.physics.cornell.edu

ADMISSIONS

Admission Contact Information
Address admission inquiries to: Kacey Acquilano, Graduate Field Assistant
Phone: (607) 255-7561
E-mail: physics@cornell.edu
Admissions website: http://www.physics.cornell.edu/academics/graduate-program/how-to-apply/

Application deadlines
Fall admission:
U.S. students: December 15   *Int'l. students*: December 15

Application fee
U.S. students: $95   *Int'l. students*: $95

Admissions information
For Fall of 2013:
   Number of applicants: 456
   Number admitted: 100
   Number enrolled: 30

Admission requirements
Bachelor's degree requirements: Bachelor's degree from an accredited institution is required; major in physics is recommended, but not required.
Minimum undergraduate GPA: 3.45

GRE requirements
The GRE is required.
   Quantitative score: 154
   Verbal score: 146
   Analytical score: 3

Advanced GRE requirements
The Advanced GRE is required.
   Minimum accepted Advanced GRE score: 600
   Mean Advanced GRE score range (25th–75th percentile): 780-930

TOEFL requirements
The TOEFL exam is required for students from non-English-speaking countries.
   PBT score: 620
   iBT score: 105
Internet-based TOEFL minimums; writing score of 20, listening score of 15, reading score of 20, and speaking score of 22.

TUITION

Tuition year 2013-14:
   Full-time students: $29,500 annual
All regular graduate students in physics are normally appointed to positions that provide a stipend, health insurance, and full support for tuition.
Deferred tuition plan: No
Health insurance: Available
Other academic fees: $81 student activity per year.
Academic term: Semester
Number of first-year students who received full tuition waivers: 32

Teaching Assistants, Research Assistants, and Fellowships
   Number of first-year
   Teaching Assistants: 28
   Fellowship students: 4

Average stipend per academic year
   Teaching Assistant: $25,594
   Research Assistant: $23,587
   Fellowship student: $27,060

FINANCIAL AID

Application deadlines
Fall admission:
U.S. students: December 15   *Int'l. students*: December 15

Loans
Loans are available for U.S. students.
Loans are available for international students.
GAPSFAS application required: No
FAFSA application required: No

For further information
Address financial aid inquiries to: Kacey Acquilano, Graduate Field Assistant.
Phone: (607) 255-7561
E-mail: physics@cornell.edu
Financial aid website: http://www.physics.cornell.edu/academics/graduate-prgram/financial-aid/

HOUSING

Availability of on-campus housing
   Single students: Yes
   Married students: Yes

For further information
Address housing inquiries to: Housing & Dining Office.
Phone: (607) 255-5368
E-mail: housing@cornell.edu
Housing aid website: http://www.housing.cornell.edu

Table A—Faculty, Enrollments, and Degrees Granted

Research Specialty	2012-13 Faculty	Enrollment Fall 2012 Master's	Enrollment Fall 2012 Doctorate	Number of Degrees Granted 2012–13 (2008–13) Master's	Number of Degrees Granted 2012–13 (2008–13) Terminal Master's	Number of Degrees Granted 2012–13 (2008–13) Doctorate
Accelerator	6	–	–	1(8)	–	2(6)
Atomic, Molecular, & Optical Physics	2	–	10	2(6)	–	2(6)
Biophysics/Soft Matter	6	–	17	6(20)	–	5(27)
Condensed Matter Experiment	6	–	41	4(24)	–	5(26)
Condensed Matter Theory	5	–	1	6(27)	–	4(18)
Cosmology & String Theory	5	–	3	5(17)	–	4(16)
High Energy Physics	6	–	11	1(10)	–	5(16)
Particles and Fields	6	–	34	1(15)	–	4(13)
Non-specialized	–	–	8	–	1(4)	–
Total	42	–	186	26(127)	1(4)	31(128)
Full-time Grad. Stud.	–	–	186	–	–	–
First-year Grad. Stud.	–	–	32	–	–	–

GRADUATE DEGREE REQUIREMENTS

Master's: Four semesters of residence are required. Qualifying examination after one year is required. Thesis and/or master's examination is required. Maximum time allowed for degree is four years (minimum two years). There is no foreign language requirement. Master's degree is earned as part of admission to candidacy examination for Ph.D.

Doctorate: Six semesters of residence are required, at least four of which must be full-time at Cornell. Qualifying examination after one year is required. Admission to candidacy examination is required in third year. There are no fixed course requirements except one semester of Advanced Physics Lab. There is no foreign language requirement. Maximum time allowed for degree is seven years. Thesis and oral thesis examination are required. Three-person special committees direct the course program and supervise research individually for each student. Research in adjacent fields is permissible.

Thesis: Thesis may be written in absentia.

SPECIAL EQUIPMENT, FACILITIES, OR PROGRAMS

1.9+1.9 GeV e+e− CESR test accelerator; CESR facility also provides high-intensity, high-energy x-rays for NSF-funded synchrotron radiation facility Cornell High-Energy Synchrotron Source (CHESS) and NIH Research Resource MacCHESS (protein crystallographic studies); Center for Advanced Computing; Cornell Center for Materials Research (CCMR) and National Submicron Facility with specialized workshops and equipment; Electron and Optical Microscopy Facility; Nanobiotechnology Center; National Nanofabrication Facility; Kavli Institute at Cornell for Nanoscale Science; Atkinson Center for a Sustainable Future; Energy Materials Center at Cornell.

Table C—Separately Budgeted Research Expenditures by Research Specialty

Research Specialty	No. of Grants	Expenditures ($)
Laboratory of Atomic and Solid State Physics	–	$7,966,189
Laboratory of Elementary Particle Physics	–	$16,808,194
Total	–	$24,774,383

FACULTY

Professor

Alexander, James P., Ph.D., University of Chicago, 1985. *High Energy Physics.*

Arias, Tomas A., Ph.D., Massachusetts Institute of Technology, 1992. *Computational Physics, Condensed Matter Physics, Theoretical Physics.*

Csaki, Csaba, Ph.D., Massachusetts Institute of Technology, 1997. *High Energy Physics, Particles and Fields, Theoretical Physics.*

Davis, J.C. Seamus, Ph.D., University of California, Berkeley, 1989. *Condensed Matter Physics.*

Elser, Veit, Ph.D., University of California, Berkeley, 1984. *Computational Physics, Condensed Matter Physics, Theoretical Physics.*

Flanagan, Eanna, Ph.D., California Institute of Technology, 1993. *Astrophysics, Cosmology & String Theory.*

Ginsparg, Paul, Ph.D., Cornell University, 1981. *Computational Physics, Computer Science, Particles and Fields, Statistical & Thermal Physics, Theoretical Physics.* Digital knowledge networks.

Gruner, Sol M., Ph.D., Princeton University, 1977. *Biophysics, Condensed Matter Physics, Nano Science and Technology.*

Hartill, Donald L., Ph.D., California Institute of Technology, 1967. *Accelerator.*

Henley, Christopher L., Ph.D., Harvard University, 1983. *Biophysics, Condensed Matter Physics, Statistical & Thermal Physics, Theoretical Physics.*

Hoffstaetter, Georg, Ph.D., Michigan State University, 1994. *Accelerator.*

LeClair, Andre, Ph.D., Harvard University, 1987. *Particles and Fields, Theoretical Physics.*

Lepage, G. Peter, Ph.D., Stanford University, 1978. *Computational Physics, Particles and Fields, Theoretical Physics.*

McEuen, Paul, Ph.D., Yale University, 1991. *Biophysics, Condensed Matter Physics, Nano Science and Technology, Optics.*

Orlov, Yuri, Ph.D., Yerevan Physics Institute, 1958. *Accelerator.*

Parpia, Jeevak M., Ph.D., Cornell University, 1979. Department Chair. *Condensed Matter Physics, Low Temperature Physics, Nano Science and Technology, Optics.*

Patterson, J. Ritchie, Ph.D., University of Chicago, 1990. *High Energy Physics.*

Ralph, Daniel C., Ph.D., Cornell University, 1993. *Condensed Matter Physics, Low Temperature Physics, Nano Science and Technology.*

Rubin, David L., Ph.D., University of Michigan, 1983. *Accelerator.*

Sethna, James P., Ph.D., Princeton University, 1981. *Biophysics, Condensed Matter Physics, Statistical & Thermal Physics, Theoretical Physics.*

Sievers, Albert J., Ph.D., University of California, Berkeley, 1962. *Condensed Matter Physics.*

Teukolsky, Saul, Ph.D., California Institute of Technology, 1973. *Astrophysics, Computational Physics, Relativity & Gravitation, Theoretical Physics.*

Thorne, Robert E., Ph.D., University of Illinois, 1987. *Biophysics, Condensed Matter Physics, Fluids, Rheology, Statistical & Thermal Physics.*

Wang, Jane, Ph.D., University of Chicago, 1996. *Biophysics, Computational Physics, Statistical & Thermal Physics.* Biophysics/soft matter.

Wang, Michelle D., Ph.D., University of Michigan, 1993. *Biophysics, Optics, Polymer Physics/Science.*

Wasserman, Ira M., Ph.D., Harvard University, 1978. *Astrophysics, Relativity & Gravitation.*

Associate Professor

Bazarov, Ivan, Ph.D., Far Eastern State University, 2000. *Accelerator.*

Cohen, Itai, Ph.D., University of Chicago, 2001. *Biophysics, Condensed Matter Physics, Engineering Physics/Science, Fluids, Rheology, Polymer Physics/Science.*

Franck, Carl P., Ph.D., Princeton University, 1978. *Biophysics, Fluids, Rheology, Statistical & Thermal Physics.*

Gibbons, Lawrence K., Ph.D., University of Chicago, 1993. Director of Graduate Studies. *High Energy Physics.*

Grossman, Yuval, Ph.D., Weizmann Institute of Science, 1996. *Particles and Fields, Theoretical Physics.*

Liepe, Matthias, Ph.D., University of Hamburg, 2001. *Accelerator, Particles and Fields.*

McAllister, Liam, Ph.D., Stanford University, 2005. *Cosmology & String Theory, Theoretical Physics.*

Mueller, Erich, Ph.D., University of Illinois at Urbana-Champaign, 2001. *Atomic, Molecular, & Optical Physics, Condensed Matter Physics, Theoretical Physics.*

Perelstein, Maxim, Ph.D., Stanford University, 2000. *High Energy Physics, Particles and Fields, Theoretical Physics.*

Ryd, Anders, Ph.D., University of California, Santa Barbara, 1996. *High Energy Physics.*

Shen, Kyle, Ph.D., Stanford University, 2005. *Condensed Matter Physics, Surface Physics.*

Thom-Levy, Julia, Ph.D., University of Hamburg, 2001. *High Energy Physics.*

Wittich, Peter, Ph.D., University of Pennsylvania, 2000. *High Energy Physics.*

Assistant Professor

Kim, Eun-ah, Ph.D., University of Illinois at Urbana-Champaign, 2005. *Condensed Matter Physics, Theoretical Physics.*

Niemack, Michael, Ph.D., Princeton, 2008. *Cosmology & String Theory.*

Vengalattore, Mukund, Ph.D., Massachusetts Institute of Technology, 2005. *Atomic, Molecular, & Optical Physics.*

DEPARTMENTAL RESEARCH SPECIALTIES AND STAFF

Theoretical

Astrophysics. Relativistic astrophysics; numerical relativity; cosmology; neutron stars; black holes; exoplanets; dark energy; gravitational waves; accretion disks and jets. Flanagan, Teukolsky, Wasserman.

Atomic, Molecular, & Optical Physics. Bose Einstein condensation, degenerate Fermi gases, quantum simulation and emulation. Mueller.

Biophysics. Insect flight, symmetry breaking, viral capsids, mechanics of root growth, protein folding, biological networks, plant pathogens, mechanical properties of DNA, statistical mechanices of DNA, bioinformatics, variability. Elser, Henley, Sethna, Jane Wang.

Computational Physics. Ab initio quantum mechanical description of materials, multi-scale modeling, numerical relativity, numerical quantum field theories, lattice QCD, joint density functional theories, solvation and solid-liquid interfaces, quantum Monte Carlo methods, optimization, fluid mechanics, image reconstructions, phase retrieval algorithms. Arias, Elser, Lepage, Teukolsky, Jane Wang.

Condensed Matter Physics. Electronic structure of solids and liquids; graphene; high-Tc superconductors; quasicrystals; ultracold atoms. Arias, Elser, Henley, Kim, Mueller, Sethna.

Cosmology & String Theory. String theory, inflation, AdS/CFT, signatures of inflation in CMB spectrum, string compactifications, particle cosmology. Grossman, McAllister, Perelstein.

High Energy Physics. Phenomonology, electroweak symmetry breaking, supersymmetry, astroparticle physics, flavor physics, neutrinos, leptogenesis. Csaki, Grossman, Perelstein.

Particles and Fields. Quantum electrodynamics, gauge theories of strong interactions; spectroscopy of new heavy mesons, hadronic structure of the photon, numerical QCD, effective field theories, supersymmetric models. Csaki, Ginsparg, Grossman, Lepage, Perelstein.

Statistical & Thermal Physics. Turbulence, statistical mechanics of nonlinear fits, avalanches, shock waves, computational linguistics, quasicrystals, phase transitions. Ginsparg, Henley, Sethna, Jane Wang.

Experimental

Accelerator. Design, construction, and operation of particle accelerators; Cornell Electron Storage Ring (CESR): CESR Test-Accelerator; Energy Recovery Linear accelerator (ERL); muon g-2; International Linear Collider (ILC); linear and non-linear beam dynamics; microwave superconductivity; superconducting RF cavities; damping rings; photoinjectors; photocathode research and development; lasers; X-ray production; beam instrumentation. Bazarov, Hartill, Hoffstaetter, Liepe, Rubin.

Atomic, Molecular, & Optical Physics. Bose-enstein condensates, degenerate fermi gases, spinor gases, magnetic microscopy, correlated quantum matter. Vengalattore.

Biophysics. Single-molecule biophysics; mechanical manipulation; precision instrumentation; physics of cryopreservation. Franck, Gruner, McEuen, Thorne, Michelle Wang.

Condensed Matter Physics. Ultraviolet, visible, infrared, and Raman studies of solids; nuclear and electron spin resonance; electronic and thermal transport properties; phase transitions in one and two dimensions. Cohen, Gruner, McEuen, Parpia, Ralph, Shen, Sievers, Thorne.

Fluids, Rheology. Fluid dynamics in insect flight; wetting and conttact line pinning of fluids on surfaces; rheology of colloidal suspensions; drop breakup and dynamics. Cohen, Franck, Thorne.

High Energy Physics. Accelerator-based: collaboration in and detector development for experiments at proton-proton colliders (CMS experiment at CERN's Large Hadron Collider [LHC]), electron-positron colliders (b factories, International Linear Collider [ILC]) and in precision muon physics (muon g-2 at Fermilab). The work focuses on understanding nature's most fundamental particle content and particle interactions with an emphasis on the origin of mass and electroweak symmetry breaking and on the dark matter content of the universe, including probes of supersymmetry, extra dimensions, and composite states. Non-accelerator-based: studies of the Cosmic Microwave Background (CMB) radiation, currently utilizing the Atacama Cosmology Telescope (ACT) to explore the nature of the early universe. Alexander, Gibbons, Patterson, Ryd, Thom-Levy, Wittich.

Low Temperature Physics. Superfluidity in 3He and 4He; phase transitions in liquid 3He, 4He, and mixtures and in solid 3He. Hoffstaetter, Parpia, Ralph.

Nano Science and Technology. Dynamics of charge, spin, thermal, and mechanical degrees of freedom in nanoscale and single-molecule devices; new nanofabrication techniques; advanced microscopy/spectroscopy; single-molecule biophysics; graphene and other atomic-membrane materials; self-assembling materials. Gruner, McEuen, Parpia, Ralph.

Optics. Optical trapping; single-molecule fluorescence; nanophotonics. Hoffstaetter, McEuen, Parpia, Michelle Wang

Statistical & Thermal Physics. Phase transitions in water/ice, aqueous mixtures and colloids: phase behavior at high pressures; self-assembly of nanocomposites. Cohen, Franck, Gruner, Thorne.

View additional information about this department at
www.gradschoolshopper.com

CORNELL UNIVERSITY

SCHOOL OF APPLIED AND ENGINEERING PHYSICS

Ithaca, New York 14853-2501
http://www.aep.cornell.edu

General University Information
President: David J. Skorton
Dean of Graduate School: Barbara A. Knuth
University website: http://www.cornell.edu
Control: Private
Total Faculty: 1,564
Total Graduate Faculty: 1,564
Total number of Students: 22,254
Total number of Graduate Students: 7,004

Department Information
Department Chairman: Alexander L. Gaeta, Chair
Department Contact: Cynthia R. Reynolds, Academic
   Programs Coordinator
   Total full-time faculty: 15
   Full-Time Graduate Students: 107
   First-Year Graduate Students: 33
   Female First-Year Students: 6
   Total Post Doctorates: 30

Department Address
142 Science Drive
271 Clark Hall
Ithaca, NY 14853-2501
Phone: (607) 255-0638
Fax: (607) 255-7658
E-mail: aep_info@cornell.edu
Website: http://www.aep.cornell.edu

ADMISSIONS

Admission Contact Information
Address admission inquiries to: Academic Programs Coordina-
   tor, School of Applied & Engineering Physics, 261 Clark Hall,
   Cornell University, Ithaca, NY 14853
Phone: (607) 255-0638
E-mail: aep_info@cornell.edu
Admissions website: http://www.gradschool.cornell.edu/

Application deadlines
Fall admission:
U.S. students: January 6   *Int'l. students*: January 6

Application fee
U.S. students: $95   *Int'l. students*: $95
MS & M.ENG - May 1

Admissions information
For Fall of 2013:
   Number of applicants: 180
   Number admitted: 59
   Number enrolled: 33

Admission requirements
Bachelor's degree requirements: Bachelor's degree in science is
   required.

GRE requirements
The GRE is required.

Advanced GRE requirements
The Advanced GRE is recommended.

TOEFL requirements
The TOEFL exam is required for students from non-English-
   speaking countries.
   PBT score: 600
Minimums: Writing=20 Listening=15 Reading=20 Speak-
   ing=22

Other admissions information
Additional requirements: Last year, students with undergraduate
   majors in applied physics, electrical engineering, engineering
   physics, mathematics, materials science, and engineering, mechanical engineering, and physics were accepted.

TUITION

Tuition year 2013-14: $29,500
Credit hours per semester to be considered full-time: 12
Deferred tuition plan: No
Health insurance: Available at the cost of 2248 per year.
Other academic fees: $81 student fee.
Academic term: Semester

Teaching Assistants, Research Assistants, and Fellowships
Number of first-year
   Teaching Assistants: 6
   Research Assistants: 16
   Fellowship students: 6
Average stipend per academic year
   Teaching Assistant: $23,470
   Research Assistant: $23,470
   Fellowship student: $27,060

FINANCIAL AID

Loans
Loans are available for U.S. students.
Loans are not available for international students.
GAPSFAS application required: No
FAFSA application required: No

For further information
Address financial aid inquiries to: Graduate School, Caldwell
   Hall.
Phone: (607) 255-5820
Financial aid website: http://www.gradschool.cornell.edu/costs-
   and-funding

HOUSING

Availability of on-campus housing
Single students: Yes
Married students: Yes

For further information

Address housing inquiries to: Graduate Student Housing, Caldwell Hall.
Phone: (607) 255-5820
Housing aid website: http://www.campuslife.cornell.edu/campuslife/housing/gradhousing.cfm

Table A—Faculty, Enrollments, and Degrees Granted

Research Specialty	2013-14 Faculty	Enrollment Fall 2013		Number of Degrees Granted 2012-13		
		Master's	Doctorate	Master's	Terminal Master's	Doctorate
Applied Physics	15	9	22	16(-)	9(-)	15(-)
Total	15	23	91	16(-)	8(-)	15(-)
Full-time Grad. Stud.	–	25	69	–	–	–
First-year Grad. Stud.	–	11	22	–	–	–

GRADUATE DEGREE REQUIREMENTS

Master's: Master of Science in Applied Physics: Minimum residence time is two units. Full-time study for one semester with satisfactory accomplishment constitutes one residence unit. Requirements must be completed in four years from date of first registration. A written thesis and an oral presentation (e.g., group seminar) are required at the end of the project. Students must maintain a "C" average (2.0) or better to remain in good standing. There is no foreign language requirement. Master of Engineering in Engineering Physics: The Master of Engineering program is a terminal professional master's degree. The general requirement for the degree is a total of 30 credits for graduate-level courses or their equivalent, distributed as follows: (1) a design project in applied science or engineering (not less than six nor more than 12 credits), (2) an integrated program of graduate-level courses (14-20 credits), and (3) a required special topics seminar course (four credits).

Doctorate: Six units of residence are required. Full-time study for one semester with satisfactory accomplishment constitutes one residence unit. All requirements must be completed in seven years from the date of first registration. No more than two units may be earned in the summer. At least four of the six units must be earned as a full-time student. Two examinations (by the special committee) are required for doctoral degree: a uniform qualifying examination (oral and written) is taken no later than after three units of residence credit are accumulated and a final examination is taken after completion of the dissertation covering the subject matter of the dissertation. Publication of thesis by abstract and microfilm is required. No specific GPA is required. There is no foreign language requirement.

Other Degrees: Master of Engineering Physics.

Thesis: Thesis may be written in absentia.

SPECIAL EQUIPMENT, FACILITIES, OR PROGRAMS

Students in the Graduate Field of Applied Physics are able to access a diverse range of experimental facilities on the Cornell University campus, including the Physical Science Building, atomic and molecular physics laboratories, biophysics laboratories, electron microscopy laboratories, synchrotron x-ray laboratories, plasma physics laboratories, optical physics and photonics laboratories, geophysics laboratories, and condensed-matter physics laboratories (including materials physics, surface physics, device physics, and nanophysics).

Due to the interdisciplinary and multi-departmental nature of the graduate field of applied physics at Cornell, the research facilities available are much more extensive than those that could be provided by a single department.

The Department of Applied Physics is also closely associated with several major research facilities on the Cornell University campus: the Cornell Center for Materials Research (CCMR), the Center for Nanoscale Systems (CNS), the Nanobiotechnology Center (NBTC), the Cornell High Energy Synchrotron Source (CHESS), and the Cornell NanoScale Science & Technology Facility (CNF).

Other facilities available for applied physics research include the radar-radio observatory in Arecibo, Puerto Rico, the unique pulsed-power-driven high-energy plasma facilities operated by the Laboratory of Plasma Studies (LPS), and the National Institutes of Health Developmental Resource in Biophysical Imaging and Optoelectronics (DrBio).

Finally, close relationships with adjacent fields (i.e., astronomy, biochemistry, biophysics, molecular and cell biology, chemistry, electrical engineering, materials science, neurobiology, and physics) enable students to make use of facilities not formally represented in the School of Applied and Engineering Physics.

Table B—Separately Budgeted Research Expenditures by Source of Support

Source of Support	Departmental Research	Physics-related Research Outside Department
Federal government	$11,249,018	
State/local government	$29,717	
Non-profit organizations		
Business and industry	$228,614	
Other	$1,110,228	
Total	$12,617,577	

FACULTY

Professor

Brock, Joel D., Ph.D., Massachusetts Institute of Technology, 1987. High-resolution x-ray scattering studies of condensed matter.

Buhrman, Robert A., Ph.D., Cornell University, 1973. High-Tc superconductivity; nanostructure physics; quantum transport and tunneling studies; scanning tunneling microscopy.

Craighead, Harold G., Ph.D., Cornell University, 1980. Nanofabrication; ultrasmall solid-state devices and structures.

DiSalvo, F. J., Ph.D., Stanford University, 1971. Physical and chemical properties of solid-state compounds.

Gaeta, Alexander, Ph.D., University of Rochester, 1990. Quantum and nonlinear optics.

Giannelis, Emmanuel P., Ph.D., Michigan State University, 1985. Hybrid nanomaterials for energy, biomedical, transportation, and packaging.

Gruner, Sol., Ph.D., Princeton University, 1977. Biophysics; condensed-matter physics; advanced x-ray area detectors.

Hammer, David A., Ph.D., Cornell University, 1969. Plasma physics; nuclear fusion; high-power electron- and ion-beam physics.

Hines, Melissa, Ph.D., Stanford University, 1992.

Houck, James R., Ph.D., Cornell University, 1967. Astrophysics.

Kelley, Michael C., Ph.D., University of California, Berkeley, 1970. Space plasma physics; rocket and satellite instrumentation.

Kusse, Bruce R., Ph.D., Massachusetts Institute of Technology, 1969. Plasma physics; intense ion beams; inertial fusion.

Lindau, Manfred, Ph.D., Technical University of Berlin, 1983. Biophysics; optical and electrical studies of fusion membrane transport and exocytosis.

Lovelace, Richard V., Ph.D., Cornell University, 1970. Plasma physics theory; astrophysics.

McEuen, Paul L., Ph.D., Yale University, 1990. Nanostructures.

Muller, David, Ph.D., Cornell University, 1996. Condensed-matter physics; electronic microscopy.

Pollock, Clifford R., Ph.D., Rice University, 1981. Lasers; molecular spectroscopy; quantum electronics.

Ralph, Daniel, Ph.D., Cornell University, 1993. Nanoelectronics and magnetics.

Schlom, Darrell, Ph.D., Stanford University, 1990. Heteroepitaxial growth and characterization of oxide thin films; preparation of oxide superlattices and metastable phases by molecular-beam epitaxy (MBE).

Seyler, Charles E., Ph.D., University of Iowa, 1975. Plasma physics; thermonuclear fusion; high-power beams; space plasmas.

Shalloway, David I., Ph.D., Massachusetts Institute of Technology, 1975. Theoretical/computational analysis of protein structure and dynamics.

Tiwari, Sandip, Ph.D., Cornell University, 1980. Optical and electronic properties of semiconductor devices.

van Dover, Robert Bruce, Ph.D., Stanford University, 1980. Magnetic, dielectric, superconducting, and optical thin films.

Wang, Zheng Jane, Ph.D., University of Chicago, 1996. Biophysics; fluid dynamics; statistical physics; applied mathematics; physics of insect flight; modeling of biological organisms; computer-human interactions.

Webb, Watt W., Ph.D., Massachusetts Institute of Technology, 1955. Biological physics; chemical physics; cooperative phenomena; hydrodynamics; physical optics; fluctuation correlation spectroscopy.

Wise, Frank W., Ph.D., Cornell University, 1988. Ultrafast optics; semiconductor nanostructures.

Associate Professor

Erickson, David, Ph.D., University of Toronto, 2004. Fluid and thermal dynamics and optics; nanophotonics; nanofabrication; chemistry; biology.

Lal, Amit, Ph.D., University of California, Berkeley, 1996. Navigation; low-energy computation; bio-robotics; atomic microsystems.

Lipson, Michal, Ph.D., Israel Institute of Technology, 1998. Photonics.

Pollack, Lois, Ph.D., Massachusetts Institute of Technology, 1989. Low-temperature physics and biological physics.

Thompson, Michael O., Ph.D., Cornell University, 1984. Rapid phase transformations; nonequilibrium thermodynamics of semiconductor materials.

Xu, Chris, Ph.D., Cornell University, 1996. Optics.

Zipfel, Warren, Ph.D., Cornell University, 1993. Biophysics; biomedical engineering.

Assistant Professor

Fennie, Craig, Ph.D., Rutgers University, 2006. Condensed-matter physics; computational-matter physics.

Fuchs, Gregory, Ph.D., Cornell University, 2003. Nanomagnetics and nanophotonics.

Kourkoutis, Lena F., Ph.D., Cornell University, 2009. Biophysics, Condensed Matter Physics, Nano Science and *Technology*. Nanostructured Material, Biophysics,Condensed matter and Materials Physics,Electron Microscopy and Spectroscopy, Renewable energy materials.

Park, Jiwoong, Ph.D., University of California, Berkeley, 2003. Synthesis; assembly and characterization of nanoscale materials and devices.

Robinson, Richard D., Ph.D., Columbia University, 2004. Chemistry and physics of nanoparticles and their use in energy applications.

Emeritus

Fleischmann, Hans H., Technische Hochschule, Munich, 1962. Plasma physics; thermonuclear fusion.

Kostroun, Vaclav O., Ph.D., University of Oregon, 1968. Low-energy nuclear and atomic physics.

Silcox, John, Ph.D., University of Cambridge, 1961. Inelastic electron scattering; atomic resolution electron microscopy.

Adjunct Professor

Bilderback, Donald H., Ph.D., Purdue University, 1975. Synchrotron radiation instrumentation.

Hao, Quan, Ph.D., Chinese Academy of Sciences, 1988. Structural biology of proteins.

Heinekamp, Scott, Ph.D., Brown University. X-ray diffraction.

DEPARTMENTAL RESEARCH SPECIALTIES AND STAFF

Theoretical

Astrophysics. Lovelace.

Biophysics. Lindau.

Condensed Matter Physics.

Nano Science and Technology.

Optics. Gaeta.

Other. Bioimaging and multiphone microscopy high-resolution electron microscopy and spectroscopy; microfluids and nanofluidics. Nanophotonics. Nanostructured materials for energy conversion and energy storage. Quantum and nonlinear optics. X-ray diffraction.

Physics and other Science Education.

Plasma and Fusion. Kusse.

Quantum Foundations.

View additional information about this department at
www.gradschoolshopper.com

RENSSELAER POLYTECHNIC INSTITUTE

DEPARTMENT OF PHYSICS, APPLIED PHYSICS, AND ASTRONOMY

Troy, New York 12180-3590
http://www.rpi.edu/dept/phys/physics.html

General University Information
President: Shirley A. Jackson
Dean of Graduate School: Stanley M. Dunn
University website: http://www.rpi.edu/index.html
Control: Private
Total Faculty: 500
Total number of Students: 6,658
Total number of Graduate Students: 1,249

Department Information
Department Chairman: Angel E. Garcia, Head
Department Contact: Joan Perras, Assistant to the Department
  Head
  Total full-time faculty: 20
  Total number of full-time equivalent positions: 20
  Full-Time Graduate Students: 53
  First-Year Graduate Students: 5
  Female First-Year Students: 1
  Total Post Doctorates: 16

Department Address
110 8th Street
Troy, NY 12180-3590
Phone: (518) 276-6310
Fax: (518) 276-6310
E-mail: perraj@rpi.edu
Website: http://www.rpi.edu/dept/phys/physics.html

ADMISSIONS

Admission Contact Information
Address admission inquiries to: Paul Marthens, Vice President,
  Enrollment
Phone: (518) 276-6413
E-mail: marthp@rpi.edu
Admissions website: http://www.rpi.edu/dept/admissions/index.
  html

Application deadlines
Fall admission:
U.S. students: January 1   *Int'l. students*: January 1
Spring admission:
U.S. students: August 15   *Int'l. students*: August 15

Application fee
U.S. students: $75   *Int'l. students*: $75

Admissions information
For Fall of 2013:
  Number of applicants: 139
  Number admitted: 69
  Number enrolled: 22

Admission requirements
Bachelor's degree requirements: Bachelor's degree is required
  with courses and grades demonstrating ability and preparation
  adequate for graduate study in physics. Remedial courses
  available as needed.
Minimum undergraduate GPA: 3.2

GRE requirements
The GRE is required.
  Quantitative score: 146
  Verbal score: 156
  Analytical score: 4
The above are the suggested minimum.

Advanced GRE requirements
The Advanced GRE is required.
  Minimum accepted Advanced GRE score: 600
The above is suggested minimum.

TOEFL requirements
The TOEFL exam is required for students from non-English-
  speaking countries.
  PBT score: 600
  iBT score: 100
The above are the suggested minimum.

Other admissions information
Undergraduate preparation assumed: Students are normally expected to have taken intermediate-level courses in mechanics,
  electricity and magnetism, quantum physics, thermodynamics
  experimental physics. Typical texts are Marion and Thornton,
  Griffiths, Brehm and Mullin, Stowe, and Liboff. However, students may take a limited number of remedial courses after
  enrollment where inadequate preparation has been available,
  but where other courses and grade records indicate adequate
  ability.

TUITION

Tuition year 2013-2014:
Tuition for in-state residents
  Full-time students: annual
  Part-time students: per credit
  Full-time students: $45,100 annual
  Part-time students: $1,879 per credit
Credit hours per semester to be considered full-time: 12
Deferred tuition plan: No
Health insurance: Available at the cost of 828.00 per year.
Other academic fees: Other Fees per academic year: Health Cen-
  ter: $550.00 Dental Insurance: $188.00 Activity Fee: $334.00.
Academic term: Semester
Number of first-year students who received full tuition waivers: 20

Teaching Assistants, Research Assistants, and Fellowships
Number of first-year
  Teaching Assistants: 19
  Research Assistants: 1
Average stipend per academic year
  Teaching Assistant: $18,500
  Research Assistant: $18,500
  Fellowship student: $28,000
The fellowship includes summer stipend. The research as-
sistantship summer stipend minimum is $6,200.

FINANCIAL AID

Application deadlines
Fall admission:
U.S. students: February 1　　　　　*Int'l. students*: February 1
Spring admission:
U.S. students: November 1　　　　*Int'l. students*: November 1

Loans
Loans are available for U.S. students.
Loans are not available for international students.
GAPSFAS application required: No
FAFSA application required: Yes

For further information
Address financial aid inquiries to: Office of Financial Aid.
Phone: (518) 276-6813
E-mail: finaid@rpi.edu
Financial aid website: http://admissions.rpi.edu/aid/index.html

HOUSING

Availability of on-campus housing
Single students: Yes
Married students: Yes

For further information
Address housing inquiries to: Office of Residence Life.
Phone: (518) 276-6284
E-mail: res_life@rpi.edu
Housing aid website: http://reslife.rpi.edu/setup.do

Table A—Faculty, Enrollments, and Degrees Granted

Research Specialty	2012-13 Faculty	Enrollment Fall 2012 Master's	Enrollment Fall 2012 Doctorate	Number of Degrees Granted 2012-13 (2007–12) Master's	Number of Degrees Granted 2012-13 (2007–12) Terminal Master's	Number of Degrees Granted 2012-13 (2007–12) Doctorate
Astronomy and Astrophysics	4	–	6	2(3)	1(5)	–(1)
Condensed Matter Physics	8	4	31	6(9)	–(14)	7(32)
Optics	5	–	4	–(2)	–(6)	–(11)
Particles and Fields	3	1	7	2(2)	2(4)	2(6)
Total	20	5	48	10(16)	3(29)	9(50)
Full-time Grad. Stud.	–	5	48	–	–	–
First-year Grad. Stud.	–	3	2	–	–	–

GRADUATE DEGREE REQUIREMENTS

Master's: Thirty semester credit hours "B" average; one academic year (2 semesters) residence minimum. No foreign language, no exams. Thesis or research projects are required, but may be waived for students who pass Ph.D. candidacy exam. Maximum transfer credit 6 credit hours. Usually 6 to 9 credit hours for thesis or 3 credit hours for research project.

Doctorate: Seventy-two semester credit hours "B" average; three academic year (6 semesters) residence minimum; no foreign language requirement; qualifying examination (8 hours written) by end of first year covering advanced undergraduate level material (these may be waived for high physics GRES score or high performance in classes at Rensselaer); candidacy exam (oral, on physics related to proposed thesis research area); thesis dissertation required. Usually about 30 to 45 credit hours of research.

SPECIAL EQUIPMENT, FACILITIES, OR PROGRAMS

Astronomy and Astrophysics

Students' thesis research in astronomy and astrophysics enjoys access to world-class ground-based telescopes located at observing sites in the southern hemisphere and China. Our faculty cooperates with the international Large Sky Area Multi-Object Fiber Spectroscopic Telescope (LAMOST), the Sloan Digital Sky Survey and the immensely popular Milky Way @ Home project.

Students also access, analyze and interpret data from NASA's major satellite missions such as the Hubble Space Telescope, the Kepler Space Telescope, the Spitzer Space Telescope and the airborne SOFIA observatory.

A special program offered through the physics department is research on the origins of life on earth and beyond. This program is facilitated through the NASA sponsored New York Center for Astrobiology.

For students' education in observational astronomy and for public outreach the department maintains the Hirsch Observatory. It houses a Boller and Chivens 16″ Cassegrain Telescope, a Santa Barbara Instrument Group (SBIG) imaging camera, a SBIG spectrograph, and color filter wheel.

Biological Physics, Condensed Matter and Optics

State-of-the art equipment for graduate students' experimental research in optics and condensed matter physics is provided in the physics department. The equipment comprehends optical, electronic and cryogenic instruments, surface science techniques and materials growth equipment. Examples are Atomic Force Microscopy, Auger Electron Spectroscopy, Ellipsometry, High-Resolution Low Energy Electron Diffraction, Reflection High-Energy Electron Diffraction and X-Ray Crystallography. Also available for research are terahertz radiation sources and ultrafast laser systems. Also students engage in absorption, light scattering and photoluminescence spectroscopy using systems operating from the from the terahertz frequency band to the ultraviolet part of the electromagnetic spectrum.

Students interested in nanofabrication will find excellent facilities in Rensselaer's Nano- and Micro-Fabrication Clean Room (MNCR). This is a state-of-the-art, 10,000-square-foot, Class 100 multi-user facility. The MNCR offers infrastructure for end-to-end device fabrication, characterization, metrology and testing by the graduate student researcher. The facility processes 2″ - 8" diameter wafers for high-speed electronics, power devices, integrated circuits, and microsystems." In addition, the facility has a dedicated staff of five personnel to provide process solutions, training and teaching.

The Center for Biotechnology and Interdisciplinary Research on our campus offers extensive and high-quality facilities for students' experimental research in biological physics at the molecular, cellular and tissue level.

Students conducting research in theoretical biological and condensed matter physics use Rensselaer's own supercomputer. The Blue Gene/Q is one of the world's most powerful university-based supercomputers. Theoretical methods implemented by our students on this machine and other computers are density functional theory calculations, Monte-Carlo Simulations as well as classical and quantum mechanical molecular dynamics simulation.

In addition, our research activities in condensed matter physics and optics are affiliated with special programs of other research centers on campus: The Center for Integrated Electronics, the Center for Broadband Data Transport, Science and Technology and the MARCO Interconnect Focus Center.

Particle Physics

Students active in experimental particle physics research access, analyze and interpret data from the Daya Bay Reactor Neutrino Experiment and the Thomas Jefferson Linear Accelerator Facility (Jefferson Lab). Students pursuing thesis research in theoretical particle physics apply lattice field theories and implement the calculations on the Blue Gene/Q supercomputer.

Table B—Separately Budgeted Research Expenditures by Source of Support

Source of Support	Departmental Research	Physics-related Research Outside Department
Federal government	$5,661,929	
State/local government	$1,766,400	
Non-profit organizations	$40,066	
Business and industry	$101,536	
Other		
Total	$7,569,931	

Table C—Separately Budgeted Research Expenditures by Research Specialty

Research Specialty	No. of Grants	Expenditures ($)
Astronomy and Astrophysics	7	$732,125
Condensed Matter Physics	57	$5,354,313
Optics	19	$758,886
Particles and Fields	14	$724,607
Total	97	$7,569,931

FACULTY

Professor

Garcia, Angel E., Ph.D., Cornell University, 1987. Department Head; Professor; Senior Constellation Chaired Professor in Biocomputation and Bioinformatics. *Biophysics.* Theoretical and computational aspects of the structure, dynamics and stability of biological molecules.

Jackson, Shirley A., Ph.D., Massachusetts Institute of Technology, 1973. President, Rensselaer Polytechnic Institute. *Particles and Fields.* Theoretical elementary particle physics.

Korniss, Gyorgy, Ph.D., Virginia Tech, 1997. *Computational Physics, Computer Science, Condensed Matter Physics, Nonlinear Dynamics and Complex Systems, Statistical & Thermal Physics.* Statistical mechanics, dynamics in complex networks.

Lin, Shawn-Yu, Ph.D., Princeton University, 1992. Constellation Professor, Future Chips. *Optics.* Photonic crystals, plasmonics, nanophotonics, silicon photonics, solid state lighting, solar energy applications.

Lu, Toh-Ming, Ph.D., University of Wisconsin, Madison, 1976. Associate Director, Center for Integrated Electronics; Ray Palmer Baker Distinguished Professor. *Materials Science, Metallurgy, Nano Science and Technology, Solid State Physics, Surface Physics.* Materials physics, thin film morphology and texture, nanostructures for energy and electronics applications.

Morse, Jon A., Ph.D., University of North Carolina, Chapel Hill, 1992. *Astronomy, Astrophysics.* Studies of star formation, high-mass galaxies; shock waves and hydrodynamic flows; stellar and planetary systems and exoplanets; abundances; space missions and instrumentation; science policy.

Napolitano, James J., Ph.D., Stanford University, 1982. *Astrophysics, Computational Physics, High Energy Physics, Nuclear Engineering, Nuclear Physics, Particles and Fields,*

Physics and other Science Education, Quantum Foundations. Nuclear and particle physics.

Nayak, Saroj, Ph.D., Jawaharlal Nehru University, 1995. *Applied Mathematics, Biophysics, Chemical Physics, Computational Physics, Condensed Matter Physics, Energy Sources & Environment, Materials Science, Metallurgy, Nano Science and Technology, Physics and other Science Education, Solid State Physics, Surface Physics, Theoretical Physics.* Theoretical physics and first principles calculations.

Newberg, Heidi, Ph.D., University of California, Berkeley, 1992. Director, Hirsch Observatory. *Astronomy, Astrophysics, Computational Physics.* Structure and evolution of the Milky Way galaxy using stars as tracers; n-body simulationsusing Volunteer Computing with MilkyWay@home; large surveys including SDSS and LAMOST; the density distribution of dark matter in the Milky Way.

Persans, Peter D., Ph.D., University of Chicago, 1982. Department Associate Head. *Applied Physics, Condensed Matter Physics, Materials Science, Metallurgy, Optics, Physics and other Science Education, Solid State Physics.* Optical and optoelectronic materials. Introductory Physics education.

Roberge, Wayne G., Ph.D., Harvard University, 1981. *Astrophysics.* Astrophysics and Astrobiology. Evolution of ices in the interstellar medium and solar nebula; computer simulations of multifluid, magnetohydrodynamic (MHD) shock waves; analytic and numerical methods for multifluid MHD; physics of interstellar dust.

Schroeder, John, Ph.D., Catholic University of America, 1974. *Biophysics, Condensed Matter Physics, Solid State Physics.* Physics of Non-equilibrium Solids, High Pressure Physics. High pressure physics and biological physics.

Shur, Michael, Ph.D., A.F. Ioffe Institute, 1967. Patricia W. and C. Sheldon Roberts '48 Chaired Professor in Solid State Electronics; Professor of Electrical, Computer, and Systems Engineering; Professor of Physics, Applied Physics and Astronomy; Director, Center for Broadband Data Transport Science and Technology (Primary appointment with ECSE). *Applied Mathematics, Atomic, Molecular, & Optical Physics, Condensed Matter Physics, Energy Sources & Environment, Engineering Physics/Science, Low Temperature Physics, Materials Science, Metallurgy, Nano Science and Technology, Optics, Physics and other Science Education, Solid State Physics, Surface Physics, Theoretical Physics.* THz electronics. Physics of semiconductor materials and devices. Physics of color rendition. Deep Ultraviolet Light Emitting Diodes.

Stoler, Paul, Ph.D., Rutgers University, 1966. *Materials Science, Metallurgy, Nuclear Physics, Particles and Fields.* Experimental particle/nuclear physics, structure of hadrons.

Wang, Gwo-Ching, Ph.D., University of Wisconsin, Madison, 1978. Travelstead Institute Chair. *Applied Physics, Condensed Matter Physics, Nano Science and Technology, Solid State Physics, Surface Physics.* Growth mechanism of oblique angle deposited nanostructures and thin films for solar cell and hydrogen storage. Development of reflection high energy electron diffraction (RHEED) surface pole figure technique to study texture evolution and structures of nanostructures and thin films.

Wetzel, Christian M., Ph.D., Technical University, Munich, 1993. Wellfleet Career Development Constellation Professor, Future Chips. *Applied Physics, Condensed Matter Physics, Electrical Engineering, Energy Sources & Environment, Engineering Physics/Science, Low Temperature Physics, Materials Science, Metallurgy, Nano Science and Technology, Optics, Solid State Physics.* Electronic band and defect structure of wide band gap semiconductor materials and devices for energy efficiency by means of epitaxy and optical spectroscopy.

Whittet, Douglas C.B., Ph.D., University of St. Andrews, 1975. Director, New York Center for Astrobiology. *Astronomy, Astrophysics*. Origin and evolution of dust, ices and organic molecules in the interstellar medium; organic inventories of protoplanetary disks; infrared astronomy; astrobiology and the origins of habitable planets.

Zhang, Shengbai, Ph.D., University of California, Berkeley, 1989. Gail and Jeffrey L. Kodosky '70 Senior Constellation Professor of Physics, Information Technology, and Entrepreneurship. *Condensed Matter Physics*. First-principles structural and electronic properties of a broad range of solid-state materials from crystalline, amorphous semiconductors, metals, to various nanostructures.

Associate Professor

Giedt, Joel, Ph.D., University of California, Berkeley, 2002. *Computational Physics, Cosmology & String Theory, High Energy Physics, Particles and Fields, Theoretical Physics*. Particle phenomenology, lattice field theory, string compactifications, high energy mathematical and computational physics.

Lewis, Kim M., Ph.D., University of Michigan, 2004. *Applied Physics, Condensed Matter Physics, Low Temperature Physics, Nano Science and Technology, Solid State Physics*. Molecular Electronics. Molecular electronics, electron transport in molecules and nanostructures; scanning probe microscopy techniques; materials and device characterization using low-temperature techniques with magnetic fields.

Meunier, Vincent, Ph.D., University of Namur, 1999. Gail and Jeffrey L. Kodosky '70 Constellation Professor of Physics, Information Technology, and Entrepreneurship. *Chemical Physics, Computational Physics, Condensed Matter Physics, Materials Science, Metallurgy, Nano Science and Technology, Solid State Physics*. Theory, modeling and computer simulation in nanoscience, including energy storage, electronic transport properties, and materials by design development. Current activities focus on low dimensional systems, such as graphene-based materials and self-assembled structures on surfaces.

Wilke, Ingrid, Ph.D., ETH Zuerich, 1993. *Applied Physics, Optics, Solid State Physics*. Research on sources and detectors of terahertz (THz) radiation. Investigation of dielectric properties of materials at THz-frequencies and femtosecond laser applications.

Yamaguchi, Masashi, Ph.D., Hokkaido University, 1991. *Acoustics, Applied Physics, Chemical Physics, Condensed Matter Physics, Electrical Engineering, Electromagnetism, Low Temperature Physics, Materials Science, Metallurgy, Mechanics, Nano Science and Technology, Optics, Polymer Physics/Science, Solid State Physics, Statistical & Thermal Physics*. Structural and electronic dynamics in condensed matter; acoustic, thermal, and electric transport in nanoscale materials; THz spectroscopy of advanced materials; and THz science and technology.

Assistant Professor

Stark, Peter RH, Ph.D., Harvard University, 2010. Primary appointment with School of Architecture. *Condensed Matter Physics*. Building and energy physics, materials science and emerging technologies.

Research Assistant Professor

Herce, Henry D., Ph.D., North Carolina University, 2004. *Biophysics, Computational Physics, Statistical & Thermal Physics, Theoretical Physics*. Computational and experimental molecular biology.

Sun, Yiyang, Ph.D., National University of Singapore, 2004. *Computational Physics, Condensed Matter Physics, Energy Sources & Environment, Materials Science, Metallurgy, Nano Science and Technology, Solid State Physics, Surface Physics*. Computational materials science.

Other

Washington, Morris, Ph.D., New York University, 1976. Associate Director, Center for Integrated Electronics. *Condensed Matter Physics*. Photonic and electronic devices.

DEPARTMENTAL RESEARCH SPECIALTIES AND STAFF

Theoretical

Astrophysics. Theoretical projects include studies of shock waves in star forming regions, multifluid magnetohydrodynamics (MHD), MHD instabilities, the physics of dusty plasmas, and electromagnetic mechanisms for heating asteroids during the early solar system. We do analytical calculations and are also significant users of Rensselaer's supercomputing facility. We perform n-body simulations of the tidal disruption of dwarf galaxies in the Milky Way halo, using MilkyWay@home, a 0.5 PetaFLOPS volunteer computing platform built in-house. We compare the simulations to actual Milky Way data to determine the best parameters for the simulations, thus constraining the amount and distribution of dark matter in the halo. We are testing predictions of dark matter distribution for particular dark matter particles against the measured positions and motions of stars in the Milky Way. Newberg, Roberge.

Biophysics. Current research addresses theoretical and computational aspects of dynamics, and equilibrium and non-equilibrium statistical mechanics of biomolecular systems. The objectives are to understand the structure, dynamics, stability and function of biomolecules from physical principles. Protein folding, self-assembly, binding, and dynamics are important for understanding how proteins work and how they interact with other biomolecules. Knowledge gained from this research has applications in biotechnology, drug design, and biomaterials. Parallel computer simulation methods are being applied to study protein folding, and aggregation. Highly parallel computer simulations of the folding dynamics and thermodynamics of biomolecules in aqueous solutions are being performed. Other research interests include the hydrophobic effect, enzyme catalysis, nucleic acids, proteins, and membranes. Research in complex biological systems also addresses competition and invasion phenomena in large-scale ecological systems and population dynamics, and investigating epidemic spreading and contagion in social networks. On-campus collaborations and facilities include the Social Cognitive Network Academic Research Center. Garcia, Korniss.

Condensed Matter Physics. Theoretical and computational studies performed in condensed matter physics include the determination of the electronic structure of nanostructured material, the description of models for the structure and electronic properties of surfaces and interfaces and the binding and mobility of adsorbed atoms on metal surfaces. Significant effort is also devoted to investigating molecular electronics and spintronics, as well as developing understanding of far-from-equilibrium physics. Many other aspects of condensed matter physics, at the forefront of research are subjects of dedicated projects, such a studies devoted to light-material interactions for solar-energy harvesting, photo catalysis, energy conversion, sensing, and structural transformation in inorganic and organic semiconductors and in bio materials. Many-body interactions encountered in electron-phonon coupling for excited-state energy relaxation, superconductivity, heat management, and thermoelectricity are also parts of the research portfolio. The researchers in the condensed matter groups pay particular attention to emerging materials such as low-cost solar cell materials, topological insulators, two-dimensional layered structures, and van der Waals solids with exotic electronic structures and defect properties for applications in electronics, optoelectronics, spintronics, and beyond. Meunier, Zhang.

Particle Physics. Activities primarily focus on investigations on beyond the standard model applications of lattice field theory. This includes strongly coupled supersymmetric systems such as arise in hidden sector models of spontaneous supersymmetry breaking. We have also studied models of compositeness in the Higgs sector of the Standard Model, with electroweak symmetry broken by strong dynamics of a new gauge force. This has led us into developing software for the study of resonance properties from first principles, which is also useful for lattice quantum chromodynamics. Further investigations include dark matter cross sections based on calculations from lattice quantum chromodynamics, and the study of whether or not dark matter may have appreciable self-interactions. Much of our work has an eye toward string-inspired particle phenomenology, which we have worked on in the past. Giedt.

Stochastic Dynamics on Complex Networks. One of the major developments of the last two decades has been the ever-increasing interconnectivity of a broad class of information networks, including physical and data network types arising in telecommunication, social networks, and transportation and energy infrastructures. This interconnectivity has led to immense temporal and spatial complexity in modern networks and a critical need for basic mathematical theory and statistical modeling of complex interacting networks. Our current research in this direction includes structure and dynamics of social, information, and biological networks and applications to social dynamics, network vulnerability, epidemic models, and synchronization problems. On-campus collaborations and facilities are at the Social Cognitive Network Academic Research Center (SCNARC) and at the Network Science and Technology Center (NeST). Korniss.

Experimental

Astrophysics. Experimental research in the astrophysics group includes astrobiology, the chemistry of the interstellar medium, and Galactic structure. Research in astrobiology and interstellar chemistry describes how interstellar clouds evolve into new solar systems. Current interest focuses on spectroscopic detection of organic molecules in interstellar dust and gas and their contribution to the organic inventory of protoplanetary disks. Galactic structure research focuses on the outer structure of the Milky Way as revealed by millions of stars in the Sloan Digital Sky Survey. We are using spectroscopic data from the Chinese LAMOST project. The structure is used to constrain the processes by which the Milky Way galaxy formed and the distribution of the dark matter within it. The astrophysics group makes use of data from ground-based telescopes located at world class observing sites in the USA, Chile, and other major facilities around the world, and from large ground-based astronomy projects, including the Sloan Digital Sky Survey and the Two Micron All Sky Survey (2MASS). Rensselaer also has access to data from major space and airborne observatories, including the Hubble Space Telescope, Chandra, the Infrared Space Observatory, the Spitzer Space Telescope, and the Stratospheric Observatory for Infrared Astronomy. Morse, Newberg, Whittet.

Condensed Matter Physics. Experimental condensed matter research performed in the department distinguishes between the bulk of matter, its surface and interface, and proceeds in close partnership with theory and computational studies. Of interest are new concepts, materials, and techniques for nanotech-

nology and green technology such as renewable energy, energy conservation and conversion, storage, and delivery. In one dedicated effort, fundamental research is conducted on molecular electronics, which study the quantum transport of molecules that exhibit conductance switching and rectification behavior. These measurements are completed using scanning probe microscopy techniques and electromigrated nano junctions measured at temperatures of liquid He. Another project aims at improving our understanding of materials, their structure, and devices. The systems are prepared in thin film deposition and epitaxial growth. Their structural, electronic, spin, and optical properties are characterized and compared to theoretical investigations. Other studies include wide band gap semiconductors, photonic crystals, polymers, semiconductor nanoparticle composites, dielectrics, magnetic, and metallic thin films and nano structures. Works dedicated to structure growth focus on the relationships between noise and fractal, and the diffraction signature at fractals growth/etch front. The department makes use of state-of-the art characterization techniques such as electron, x-ray, ultraviolet, visible, infrared, terahertz, and scanning probe spectroscopies and microscopies. Local facilities include Micro and Nano Fabrication Clean Room, the Microelectronics Clean Room, and the Electron Microscope Laboratory. Lewis, Lu, Shur, Wang, Wetzel.

Optical Physics. Research in optical physics covers a wide range of activities related to photons and their interaction with various materials. Experimental and theoretical research is ongoing to provide innovative solutions to today's problems in both fundamental and application aspects of the research area. Particularly, the goals of these activities are directed towards the development of novel nanoelectronic and nanophotonic devices, creative solutions for homeland security, renewable energies, biological and biomedical investigations, solar harvesting, and smart lighting. Faculty research includes photonic crystals, plasmonics, photonic nanostructures, light emitting diodes, terahertz photonics, spectroscopy, imaging, chemical and biological sensing and identification, ultrafast and nonlinear phenomena, the development of novel ultrafast spectroscopic techniques, development of novel optical materials including wideband gap and narrow band gap semiconductors, metallic nanoparticles, nanowires and their arrays, semiconducting quantum dots and quantum wells, amorphous materials. Major facilities include various types of ultrafast lasers and ultrafast spectroscopy systems, terahertz imaging and spectroscopy systems, a micro and nanofabrication clean room for semiconductor processing, linear and nonlinear optical absorption, luminescence, Raman and Brillouin scattering, and various types of modulation spectroscopy systems. Lin, Persans, Schroeder, Wetzel, Wilke, Yamaguchi.

Particle Physics. The nature and structure of matter and energy remains one of mankind's leading research frontiers. The faculty members involved in this area are engaged in experimental and theoretical studies of the fundamental interactions of matter at sub-femtometer distances. This includes continued measurements of neutrino oscillations using a nuclear reactor complex in China, and R&D for a future neutrino experiments in the US and China. A longstanding program of experiments at the Thomas Jefferson National Accelerator Facility (JLab), examines the properties of the proton and its excited states, as well as novel investigations of nuclear matter properties with astrophysical significance. Measurements at Jefferson Laboratory, using parity violating electron scattering, are being made to determine the neutron radial distribution of key spherical nuclei. This work has direct implications for the equation of state of neutron stars and other dense astrophysical objects. Napolitano, Stoler.

447

ROCHESTER INSTITUTE OF TECHNOLOGY

RIT - CHESTER F. CARLSON CENTER FOR IMAGING SCIENCE

Rochester, New York 14623
http://www.cis.rit.edu

General University Information
President: Dr. William Destler
Dean of Graduate School: Dr. Hector Flores
University website: http://www.rit.edu
Control: Private
Setting: Suburban
Total Faculty: 1,536
Total number of Students: 17,950
Total number of Graduate Students: 2,865

Department Information
Department Chairman: Stefi Baum, Director
Department Contact: Dr. John Kerekes, Imaging Science
  Graduate Program Coordinator
  Total full-time faculty: 48
  Full-Time Graduate Students: 102
  First-Year Graduate Students: 22
  Total Post Doctorates: 6

Department Address
54 Lomb Memorial Drive
Rochester, NY 14623
Phone: (585) 475-6996
Fax: (585) 475-5988
E-mail: kerekes@cis.rit.edu
Website: http://www.cis.rit.edu

ADMISSIONS

Admission Contact Information
Address admission inquiries to: Rochester Institute of Technol-
  ogy, Office of Part-time & Graduate Enrollment, Bausch &
  Lomb Center - A130, 58 Lomb Memorial Drive, Rochester,
  NY 14623-5604
Phone: (585) 475-2229; (866) 260-3950
E-mail: gradinfo@rit.edu
Admissions website: http://www.rit.edu/emcs/ptgrad/grad/

Application deadlines
Fall admission:
U.S. students: January 15   *Int'l. students*: January 15

Application fee
U.S. students: $60   *Int'l. students*: $60
http://www.rit.edu/emcs/ptgrad/grad_checklist.html

Admissions information
For Fall of 2013:
  Number of applicants: 63
  Number admitted: 41
  Number enrolled: 26

Admission requirements
Bachelor's degree requirements: Graduate Application Instructions All applicants to RIT's graduate programs must
  hold—or currently be completing—a four-year baccalaureate
  degree, or the U.S. equivalent, granted by an accredited college or university. Decisions on graduate selection are made
  by the college offering the program.
Minimum undergraduate GPA: 3.0

GRE requirements
The GRE is required.

Advanced GRE requirements
The Advanced GRE is not required.

TOEFL requirements
The TOEFL exam is required for students from non-English-
  speaking countries.
  iBT score: 100

Other admissions information
Additional requirements: Application Options
  To Apply Online: https:/futurestudent.rit.edu/grad/create_ac-
  count.cfm. Submit only one application. It is NOT necessary
  to submit a complete new application. If you have previously
  applied, submit changes to your application in email:
  gradinfo@rit.edu or phone: (585) 475-2229 or fax: (585) 475-
  7164.
  To Check the Status of a Previous Submission: https:/future
  student.rit.edu/grad/login.cfm. You will need the login you
  created to gain access to your online application.
  To Check the Status of a Paper Application: https://
  futurestudent.rit.edu/grad/create_account.cfm
  If you choose to pay the application fee online, please pay
  the fee ONLY AFTER YOU HAVE SUBMITTED YOUR
  APPLICATION
  (online or in paper) so that your payment more easily can
  be matched to your application.

TUITION

Tuition year 2013-14:
  Full-time students: $1,552 per credit
  Part-time students: $1,552 per credit
International students requiring the services of our English Lan-
  guage Center may have additional tuition charges.
Credit hours per semester to be considered full-time: 9
Deferred tuition plan: Yes
Health insurance: Available.
Other academic fees: Student Activity Fee = $125 per semester.
Academic term: Semester
Number of first-year students who received full tuition waivers: 31
Number of first-year students who received partial tuition waivers: 9

Teaching Assistants, Research Assistants, and Fellowships
Number of first-year
  Teaching Assistants: 23
  Research Assistants: 3
Average stipend per academic year
  Teaching Assistant: $20,000
  Research Assistant: $20,000
  Fellowship student: $32,000

FINANCIAL AID

Application deadlines
Fall admission:
U.S. students: January 15   *Int'l. students*: January 15

Loans

Loans are available for U.S. students.

Loans are available for international students.

GAPSFAS application required: Yes

FAFSA application required: Yes

For further information

Address financial aid inquiries to: Rochester Institute of Technology, Office of Financial Aid and Scholarships, Bausch & Lomb Center, 56 Lomb Memorial Drive, Rochester, NY 14623-5604.

Phone: (585) 475-2186

E-mail: ritaid@rit.edu

Financial aid website: http://www.rit.edu/emcs/financialaid/graduate.html

HOUSING

Availability of on-campus housing

Single students: Yes

Married students: Yes

For further information

Address housing inquiries to: Our team is dedicated to assisting you with all your assignment and apartment maintenance needs and we welcome your comments and questions. Please feel free to stop in to our office or contact us at (585) 475-2572 or housing@rit.edu., Sincerely, Mary F. Niedermaier, Director, RIT Housing Operations, Main Office: Grace Watson Hall, Phone: (585) 475-2572, Fax: (585) 475-5050, Email: housing@rit.edu.

Phone: (585) 475-2572

E-mail: housing@rit.edu

Housing aid website: http://www.rit.edu/housing

Table A—Faculty, Enrollments, and Degrees Granted

Research Specialty	2012-13 Faculty	Enrollment 2012-13		Number of Degrees Granted 2012-13		
		Master's	Doctorate	Master's	Terminal Master's	Doctorate
Physics and other Science Education	48	30	79	10(-)	–	9(-)
Total	48	30	79	4(-)	–	8(-)
Full-time Grad. Stud.	–	30	79	–	–	–
First-year Grad. Stud.	–	8	14	–	–	–

GRADUATE DEGREE REQUIREMENTS

Master's: Master of Science Imaging Science The Master of Science curriculum emphasizes a systems approach to the study of imaging science, and with your background in science or engineering, this degree prepares you for positions in research, product development, and management in the imaging industry. The curriculum was developed in collaboration with industrial partners to emphasize skills needed by their scientists, engineers, and managers. You may concentrate on one of several "system tracks," or customize your own track. You may choose to complete either a research thesis or a project and a paper in the non-thesis option. You may also perform your thesis research at your place of employment. The program can be completed on a full or part-time basis. Click here to learn more about the curriculum and requirements of the on-campus MS program. Interested in pursuing your Master's degree online? Contact Dr. Roger Dube, Online Master's Graduate Program Coordinator, by email or at (585) 475-5836. To learn more about completing your Master's degree

online using distance learning, click here. Connections to the Colleges of Science, Engineering, and Computing - . The Chester F. Carlson Center has connections with faculty in many academic programs across RIT and beyond. You will study in an active research environment with students and scientists from many major industrial countries. Questions? More information can be found on the RIT Graduate Enrollment Services site. That department has several methods (online, chat, mail) for answering your general questions on registration and enrollment. For specific questions about course content and research, please contact Dr. Roger Dube, Online Master's Graduate Coordinator, (585) 475-5836, or Dr. John Kerekes, Imaging Science Graduate Program Coordinator, (585) 475-6996. Detailed information concerning Imaging Science Graduate Program policies and procedures is available in the Imaging Science Graduate Handbook.

Doctorate: Ph.D. Imaging Science The doctoral curriculum offers you a thorough course of study and research, structured and directed by experts in imaging science. You acquire the knowledge, skills, and experience to continue to expand the boundaries of the discipline and to meet current scholarly, industrial, and governmental demands. Fundamental understanding of the basis of imaging science. The course of study involves two years of course work beyond the baccalaureate and a research-based dissertation. The successful doctoral candidate completes course work, including the core curriculum, defined by a plan of study, passes the Ph.D. comprehensive and qualifying examinations, and completes a dissertation under the supervision of a research advisor and dissertation committee. Click here to learn more about the curriculum and requirements of the PhD program. Questions? Registration and enrollment questions should be directed to the RIT Offices of Part-time & Graduate Enrollment. For specific questions about course content and research, please contact Dr. John Kerekes, Imaging Science Graduate Program Coordinator, (585) 475-6996. Detailed information concerning Imaging Science Graduate Program policies and procedures is available in the Imaging Science Graduate Handbook.

Other Degrees: http://www.cis.rit.edu/grad.

SPECIAL EQUIPMENT, FACILITIES, OR PROGRAMS

Graduate Studies in Imaging Science http://www.cis.rit.edu/grad

CIS at RIT has offered the MS and Ph.D. in Imaging Science for over twenty years and is still the only PhD program in the United States in this field.

What is Imaging Science and why would you study it?

Imaging was named one of the top twenty engineering achievements of the 20th Century by the National Academies (greatachievements.org) and for good reason. In the last twenty years, imaging has made dramatic advances, transforming our ability as humans to see and understand a range of phenomenon, keeping us healthy, protecting our security, monitoring our earth, exploring our universe, uncovering and preserving our heritage, enhancing communication, and facilitating our every day lives.

Imaging Science encompasses the development, optimization, and application of imaging systems. Today, imaging systems use not only the visible light our eyes can see, but the full range of the electromagnetic spectrum from gamma rays through infrared and on down to the lowest radio frequencies. Modern ultrasound and electron microscopic imaging techniques transcend the realm of electromagnetic waves. Video and three-dimensional imaging further enhance our capabilities, and computers allow development of algorithms to extract information from databases of images, as well as visualizations techniques to display information so humans can utilize it.

Thus, the science and engineering of imaging encompasses a very wide range of subject areas, from the physics of energy sources to the psychophysics of high-level visual perception, from the engineering of optics and sensing systems to the design of algorithms. From how light is generated to how the world is perceived, imaging science addresses questions about every aspect of systems and techniques used to create, perceive, analyze, and optimize images. Application areas of imaging are equally diverse, including remote sensing, earth observation and monitoring, vision and perception, biomedical systems, astronomy, security, emergency response, video systems and printers, document reconstruction, data mining, to name some of the active areas of research within the Center. Thus, Imaging Science is both truly interdisciplinary in its content and multi-disciplinary in its applications.

Every year ~15 students graduate from CIS with PhD's in Imaging Science and 15 with MS degrees. These students will have carried out research theses and be ready to work in any number of applications areas. We offer stipends and tuition coverage for admitted PhD students and top MS students. Last year CIS was supported by 8 million dollars expended from external grants and contracts from the Federal and State agencies and Corporate America. We have over 40 faculty actively engaged with our students and programs.

You can explore active areas of research where you might pursue your thesis research here. Areas of active research within CIS where you might pursue your thesis research include:

* Remote Sensing - the use of satellite, airborne, or distributed sensor systems for purposes ranging from environmental science to national security.

* Biomedical Imaging – where imaging is used to non-invasively diagnose disease, to develop therapies and to track the success of treatment.

* Vision and visual perception – where imaging science study how humans use our own imaging system — our eyes and brain — to perceive the world around us.

* Imaging algorithms, data fusion and visualization – to allow full exploitation of imaging data.

* Video and 3-D Imaging

* Space Weather Alert Technologies and Ionospheric Monitoring

* Ancient Document Reconstruction - application of imaging techniques to the reading and reconstruction of ancient documents, such as the Archimedes Palimpsest.

* Imaging on the smallest scales – nano-imaging.

* Detector and sensor system development (hardware).

* Astronomical Imaging System Development.

* Emergency Response – where integrated imaging systems are developed to monitor and aide in critical response to natural disasters such as wildfires, floods, and hurricanes as well as man made disasters.

* Novel Optical System development

* Applications of Imaging Science to Green Energy, including use of lasers, optical techniques, and nano-materials research to develop improved fuel sources.

For more details, visit the specific pages for the Masters or Doctoral programs. Or Apply Right Now!

A complete list of past Ph.D. dissertation titles is available.

Table B—Separately Budgeted Research Expenditures by Source of Support

Source of Support	Departmental Research	Physics-related Research Outside Department
Federal government	$5,140,181	
State/local government	$193,882	
Non-profit organizations		
Business and industry		
Other	$834,870	
Total	**$6,168,933**	

Table C—Separately Budgeted Research Expenditures by Research Specialty

Research Specialty	No. of Grants	Expenditures ($)
Physics and other Science Education	143	$6,169,993
Total	**143**	**$6,169,993**

FACULTY

Professor

Baum, Stefi A., Ph.D., University of Maryland, 1987. Director, Chester F. Carlson Center for Imaging Science. *Astronomy*. Astronomical imaging systems and algorithms; extragalactic astrophysics: active galaxies, galaxy clusters, galaxy evolution, biomedical imaging.

Berns, Roy, Ph.D., Chemistry, Color Science Concentration, Rensselaer Polytechnic Institute, 1983. Richard S. Hunter Professor. Director, Munsell Color Science Laboratory. Spectral imaging, digital archiving, and reproduction; art conservation science; spectral-based color reproduction; multi-ink printing.

Dianat, Sohail, Ph.D., D.Sc., Electrical Engineering, George Washington University, 1981. Principal Scientist, Eastman Kodak Company (1996-2007); Principal Scientist, Carestream Health, Inc. (2007-2009). Digital signal and image processing; digital communication.

Easton, Roger, Ph.D., Optical Sciences, University of Arizona, 1986. Imaging of cultural heritage; Fourier imaging.

Farchild, Mark, Ph.D., Vision Science, University of Rochester, 1990. Color perception and imaging; color appearance modeling; image appearance; image preference; image perception; color measurement; image quality measurement; high-dynamic-range imaging; image and video rendering; science literacy through color.

Figer, Donald, Ph.D., University of California, Los Angeles, 1995. *Astronomy*. Imaging detectors; astronomical imaging.

Gaborski, Roger, Ph.D., Electrical Engineering, University of Maryland. Professor, Department of Computer Science. Director, Laboratory of Intelligent Systems. Visual and acoustic scene understanding; computer vision; video processing; artificial intelligence; blind source separation; machine learning.

Hornak, Joseph, Ph.D., Chemistry, University of Notre Dame, 1982. Magnetic resonance imaging (NMR and ESR); imaging science; physical chemistry; analytical chemistry; scientific publishing.

Kastner, Joel, Ph.D., Astronomy, University of California, Los Angeles, 1990. Details of research and interests are: star and planet formation, late stages of stellar evolution, x-ray, optical/IR, and radio spectroscopy and image processing algorithms and system.

Kerekes, John, Ph.D., Electrical Engineering, Purdue University, 1989. Imaging Science Graduate Program Coordinator. Research Details and Interests: remote sensing, system modeling and analysis, pattern recognition, medical imaging, image processing.

Kotlarchyk, Michael, Ph.D., Massachusetts Institute of Technology, 1984. Department Head - Physics. Professor of Physics. Radiation scattering techniques; laser light scattering; small-angle neutron and x-ray scattering; photon correlation spectroscopy; structure and interactions in complex fluids; optics and photonics.

Ninkov, Zoran, Ph.D., Astronomy, University of British Columbia, 1986. Novel 2-D CMOS detector arrays; fundamental limitations of visible and IR arrays; miniaturized multi-spectral systems.

Pelz, Jeff, Ph.D., Brain and Cognitive Science, University of Rochester, 1995. Visual Perception Lab web page. Eye tracking; visual perception.

Rao, Navalgund, Ph.D., Physics, University of Minnesota, 1979. Image and signal processing; medical imaging.

Rhody, Harvey, Ph.D., Electrical Engineering, Syracuse University, 1969. Imaging systems; remote sensing; imaging algorithms; image processing.

Saber, Eli, Ph.D., Electrical and Computer Engineering, University of Rochester, 1996. Associate Professor, Department of Electrical Engineering. Image segmentation; object tracking; text quality; defect classification; scene reconstruction.

Salvaggio, Carl, Ph.D., Environmental Resource Engineering, Syracuse University / State University of New York, 1994. Remote sensing; thermal infrared phenomenology; visible through longwave infrared optical properties measurements; scene modeling; small target restoration; bidirectional and polarized reflectance measurement and modeling; digital image processing; image compression.

Salvakis, Andreas, Ph.D., Electrical Engineering (Math Minor), North Carolina State University, 1991. Professor and Department Head, Department of Computer Engineering. Computer vision digital image processing, multimedia.

Smith, Bruce, Ph.D., Rochester Institute of Technology. Intel Professor of Microelectronic Engineering. Kate Gleason College Of Engineering Associate Dean. Immersion lithography; high NA and polarization; aberration metrology; UV/VUV thin films; high index fluids; optical extension & imaging theory.

Smith, Thomas, Ph.D., Chemistry, University of Michigan, 1973. Professor of Chemistry and Microsystems Engineering. Organic chemistry; polymer chemistry.

Associate Professor

Bajorski, Peter, Ph.D., Politechnika Wrocławska, 1990. Target detection and unmixing in hyperspectral images; multivariate analysis; regression analysis.

Cahill, Nathan, Ph.D., University of Oxford, 2009. Principal Scientist, Eastman Kodak Company (1996-2007). Principal Scientist, Carestream Health, Inc. (2007-2009). Image computing; image alignment and stitching; 3-D medical image registration; variational techniques and partial differential equations for image processing.

Esterman, Marcos, Ph.D., Stanford University, 2002. Associate Professor, Industrial & Systems Engineering. Sustainable print systems; systems engineering; product development; design robustness.

Ferwerda, James, Ph.D., Experimental Psychology, Cornell University, 1998. National Academies - Keck Futures Initiative Program (2010). High dynamic range imaging; perceptually-based rendering; material appearance; display systems; low vision and assistive technologies.

Geigel, Joe, Ph.D., D.Sc., Computer Science, George Washington University (2000) M.S. Computer Science, George Washington University, 2000. Associate Professor, Computer Science. *Computer Science*. Computer graphics; virtual photography; virtual theater.

Hailstone, Richard, Ph.D., Physical Chemistry, Indiana University, 1972. *Nano Science and Technology*. Research Details and Interests: nano-imaging, hyperspectral imaging, silver halide.

Helguera, Maria, Ph.D., Rochester Institute of Technology, 1999. *Medical, Health Physics*. Medical imaging system characterization; ultrasound tissue characterization; digital image processing; non-destructive evaluation techniques.

Hubbard, Seth, Ph.D., Electrical Engineering, University of Michigan, 2005. Assistant Professor, Department of Physics. *Electrical Engineering*. Next-generation photovoltaics devices; nanomaterials; epitaxial crystal growth; quantum confinement effects, novel and wide bandgap semiconductors, carbon nanotubes, gas sensor, semiconducting polymers and devices and high power electronic devices.

Qiao, Jie, Ph.D., University of Texas at Austin, 2001. *Optics*. Optical instrumentation, optical system design, optics and imaging quality testing and assessment, active optics and adaptive optics control.

Swartzlander, Grover, Ph.D., Johns Hopkins University, 1990. Associate Professor, with joint appointment in Imaging Science and the Department of Physics. Optical vortices: basic and applied optics; optical coronagraphs and high contrast imaging; pattern formation in linear and nonlinear optics; optical tweezers and optical lift; polarization holography. Optical coherence.

van Aardt, Jan, Ph.D., Forestry, Virginia Tech, 2004. Application of imaging spectroscopy and light detection and ranging (lidar) for spectral-structural characterization of natural systems (remote sensing of natural resources); forestry (vegetation) inventory and assessment; species diversity; foliar biochemistry; scaling of remote sensing estimates.

Vodacek, Anthony, Ph.D., Environmental Engineering, Cornell University, 1990. Remote sensing; integrating remote sensing and environmental models; data assimilation; Dynamic Data Driven Applications Systems (DDDAS; persistent simulation; imaging spectroscopy; passive and active remote sensing of water quality; remote detection of wildland fire. Feature extraction from hyperspectral data. Fourier analysis of landscape metrics.

Zanibbi, Richard, Ph.D., Computer Science, Queen's University. Theory and tools used to construct pattern recognition systems; document recognition and retrieval; optical character recognition (OCR; parsing diagrammatic notations (e.g. music, math); CAPTCHAs (particularly video-based CAPTCHAs).

Assistant Professor

Bhattacharya, Mishkat, Ph.D., University of Rochester, 2005. *Physics and other Science Education*. Quantum optics, mechanical systems in the quantum regime, superconducting quantum computing, quantum information, quantum measurement Laser cooling and trapping, ultracold atomic and molecular scattering, Feshbach resonances, quantum degenerate gases, level crossings.

Diaz, Gabriel, Ph.D., Rensselaer Polytechnic institute. *Optics*. Use of virtual reality, motion capture, and eye movements in complex visuo-motor tasks for predicting target location.

Gaborski, Thomas, Ph.D., Biomedical Engineering, University of Rochester, 2008. Nanomaterials and membrane fabrication, cellular biophysics, quantitative fluorescence imaging.

Hoffman, Matthew, Ph.D., Applied Mathematics and Scientific Computing, University of Maryland, 2009. Data Assimilation, applied mathematics, ocean and ecosystem modeling, Martian atmosphere and climate, breeding, ensemble Kalman filtering, scientific computation.

Tomaszewski, Brian, Ph.D., Geography, Pennsylvania State University, 2009. Geographic Information Science and Technology (GIS&T), geovisual analytics, geographic information retrieval, geocollaboration.

Research Professor

Dube, Roger, Ph.D., Physics, Princeton University, 1976. Native American Science Initiatives. Scholarships/Awards: RIT - Future Stewards Initiative Leadership and Mentoring Award (2010). Space weather; computer security; Ganondagan's Iroquois White Corn Project; cosmology; holographic data storage; preservation of Native American Intellectual Property; Native American entrepreneurship and job creation; quantum cryptography.

Kremens, Robert, Ph.D., New York University, 1981. Remote sensing.

Schott, John, Ph.D., Environmental Science, Remote Sensing, Syracuse University, 1980. Electro-optical systems; environmental systems and terrain classification; energy systems.

Research Associate Professor

Messinger, David, Ph.D., Physics, Rensselaer Polytechnic Institute, 1998. Associate Research Professor. Director, Digital Imaging and Remote Sensing Laboratory.Remote sensing image exploitation; Advanced mathematical approaches for spectral image processing; target detection in hyperspectral imagery.

Research Assistant Professor

Ientilucci, Emmett, Ph.D., Imaging Science, Rochester Institute of Technology, 2005. Remote sensing; hyperspectral image processing; multivariate statistics; target detection; radiometry.

Kushalnagar, Poorna, Ph.D., Developmental Psychology, Cognitive Neuroscience focus, University of Houston, 2007. Attention, cognition, and learning in population with sensory disabilities, eye tracking, and visual perception.

Noel-Storr, Jake, Ph.D., Astronomy, Columbia University, New York, 2004. Head of Insight Lab for Science Outreach and Learning Research. Supermassive Black Holes; active galactic nuclei; science education and learning; outreach.

Affiliate Professor

Frey, Franziska S., Ph.D., Natural Sciences, Federal Institute of Technology, Zurich, Switzerland, 1994. Associate Professor. McGhee Professor. Digitization of cultural heritage materials; digital fine art printing; digital asset management.

Garrett, Alfred, Ph.D., Civil Engineering, University of Texas at Austin, 1978, 1978. Remote sensing; computational fluid dynamics; thermodynamics; meteorology.

Johnson, Garrett, Ph.D., Imaging Science, Rochester Institute of Technology, 2003. High dynamic range imaging; color appearance modeling; image appearance modeling.

DEPARTMENTAL RESEARCH SPECIALTIES AND STAFF

Experimental

ASTRONOMY AND SPACE SCIENCE. With our collaborators in physics, mathematics, and computer science, we conduct research dedicated to understanding the nature and evolution of the universe in which we live, from the sun-earth environment to the earliest furthest reaches of the universe. Our research encompasses development of state-of-the-art instrumentation, observation and interpretation, theoretical physics and modelling using state-of-the art computation, and mining of large astronomical datasets. O'Dea (Physics), Robinson (Physics), Richmond (Physics) Related Laboratories, Centers and Programs RIDL Astrophysics Center for Computational Relativity and Gravitation Laboratory for Multiwavelength Astronomy Spaceweather Technologies and Research Lab (STAR). Baum, Dube, Figer, Kastner, Ninkov.

BIOMEDICAL IMAGING. We conduct research into applications of imaging to biomedical science, in conjunction with the Departments of Biology, Life Sciences and Engineering. Our research ranges from the development, enhancement and application of imaging techniques such as ultrasound, MRI, and hyperspectral imaging to medical and biology applications, to the development of algorithms for quantitative biomedical imaging and visualization of biomedical imagery. Vince Calhoun (Affiliate Professor, UNM) Related Laboratories Magnetic Resonance Imaging Biomedical and Materials Multimodal Imaging Lab. Baum, Cahill, Helguera, Hornak, Rao.

CULTURAL ARTIFACT AND DOCUMENT IMAGING. We conduct research aimed at deciphering ancient documents, preserving cultural heritage, and enhancing art reproduction and distribution. Related Laboratories, Centers and Programs Digital Image Capture of Cultural Heritage School of Print Media Historical Manuscript Imaging Munsell Color Science Lab Art Spectral Imaging. Berns, Easton, Frey.

DISASTER RESPONSE. We conduct research to develop and apply sensing information systems, in combination with geographic information systems and predictive modelling that inform preparation for and response to natural and manmade disasters. Associated Faculty and Staff: McKeown, Long, Krazeck, Related Laboratories, Centers and Programs Information Products for Emergency Response Agent Based Modeling and Crowd Response. Baum, Kremens, Tomaszewski, van Aardt, Vodacek.

NANOIMAGING & MATERIALS. We conduct research into the design, development, and application of imaging to material and biological systems on very small scales, in conjunction with the Departments of Chemistry, Physics, and Electrical Engineering.,Kotlarchyk (Physics), Thurston (Physics), Associated Laboratories Nanoimaging Laboratory Nanolithography Research Labs NanoPower Research Laboratory. Hailstone, Hubbard, Bruce Smith, Thomas Smith.

OPTICS. We conduct applied optics research in the use of optics from the smallest scales to the largest. Preble (Engineering), Associated Laboratories Nanophotonics Lab Optical Physics Optical Vortex Laboratory. Hubbard, Swartzlander.

REMOTE SENSING. We conduct remote sensing research focused on imaging the earth's environment in the visible, near infrared, and thermal infrared spectral regions. We use modeling tools, field measurements, and synthetic image generation to understand how remotely sensed data can be used to study environmental processes and provide security. Related Laboratories Digital Imaging & Remote Sensing Information Products Laboratory for Emergency Response. Garrett, Kerekes, Rhody, Salvaggio, Schott, van Aardt, Vodacek.

STEM PEDAGOGRY. We engage in research to develop and implement innovative K-12, family, undergraduate, and graduate education and outreach in science, technology, engineering, and mathematics. Associated Faculty and Staff: Pow, Noel Storr, Choate, Associated Laboratories Insight Lab Innovative Freshman Experience. Baum.

VISION. We conduct research to understand the human visual system and how it is used to help us perceive and interpret information from images and the environment. We develop algorithms, devices, and systems to aid humans in their use of vision, for impaired vision and for learning. Herbert (Psychology), Condry (Psychology), Schenkel (Psychology), Haake (Information Sciences and Technology, Related Laboratories Visual Perception laboratory Munsell Color Science Laboratory. Ferwerda, Pelz.

STATE UNIVERSITY OF NEW YORK AT ALBANY

DEPARTMENT OF PHYSICS

Albany, New York 12222
http://www.albany.edu/physics

General University Information

President: Robert Jones
Dean of Graduate School: Kevin Williams
University website: http://www.albany.edu
Control: Public
Setting: Suburban
Total Faculty: 1,000
Total number of Students: 17,040
Total number of Graduate Students: 4,936

Department Information

Department Chairman: Carolyn MacDonald, Chair
Department Contact: Carolyn MacDonald, Chair
  Total full-time faculty: 12
  Total number of full-time equivalent positions: 12
  Full-Time Graduate Students: 46
  First-Year Graduate Students: 10

Department Address

Physics 215
1400 Washington Avenue
Albany, NY 12222
Phone: (518) 442-4501
Fax: (518) 442-5260
E-mail: Physics@albany.edu
Website: http://www.albany.edu/physics

ADMISSIONS

Admission Contact Information

Address admission inquiries to: Paul Labate, Physics 215, Albany, NY 12222
Phone: (518) 442-4501
E-mail: physics@albany.edu

Application deadlines

Fall admission:
U.S. students: September 2   *Int'l. students*: September 2
Spring admission:
U.S. students: November 1   *Int'l. students*: November 1

Application fee

U.S. students: $75

Admissions information

For Fall of 2012:
  Number of applicants: 69
  Number admitted: 8
  Number enrolled: 8

Admission requirements

Bachelor's degree requirements: A bachelor's degree in physics or a related field is required.

GRE requirements

The GRE is not required.

Advanced GRE requirements

The Advanced GRE is not required.

TOEFL requirements

The TOEFL exam is required for students from non-English-speaking countries.
  PBT score: 600
  iBT score: 100
The minimum accepted computer-based exam (CBT) score is 250.

Other admissions information

Additional requirements: GPA, GRE, letters of recommendation, and research alignment with faculty specializations are all considered for admission.
Undergraduate preparation assumed: Symon, Mechanics; Griffiths, Introduction to Electrodynamics; Griffiths, Quantum Mechanics.

TUITION

Tuition year 2012–13:
Tuition for in-state residents
  Full-time students: $9,370 annual
  Part-time students: $390 per credit
Tuition for out-of-state residents
  Full-time students: $16,680 annual
  Part-time students: $695 per credit
Other academic fees: $75 application fee

Teaching Assistants, Research Assistants, and Fellowships

  Number of first-year

FINANCIAL AID

Loans

Loans are available for U.S. students.
Loans are not available for international students.
GAPSFAS application required: No
FAFSA application required: Yes

For further information

Address financial aid inquiries to: Student Financial Center, Campus Center G-26, 1400 Washington Avenue, Albany, NY 12222.
Phone: 518-442-3202
E-mail: sscweb@albany.edu
Financial aid website: http://www.albany.edu/studentservices

HOUSING

For further information

Address housing inquiries to: Residential Life, State University of New York at Albany, State Quad U-Lounge, 1400 Washington Avenue, Albany, NY 12222.
Housing aid website: http://www.albany.edu/housing/index.shtml

Table A—Faculty, Enrollments, and Degrees Granted

Research Specialty	2012–13 Faculty	Enrollment Fall 2012		Number of Degrees Granted 2012-13 (2005–11)		
		Master's	Doctorate	Master's	Terminal Master's	Doctorate
Biophysics	2	–	6	–	–	–
Computational Physics	2	–	5	–	–	1(8)
Condensed Matter Physics	2	–	6	–	–	–(5)
Optics	3	–	6	–	–	2(5)
Particles and Fields	3	–	13	–	–	1(-)
Total	12	11	42	–	–	4(31)
Full-time Grad. Stud.	–	9	38	–	–	–
First-year Grad. Stud.	–	5	7	–	–	–

GRADUATE DEGREE REQUIREMENTS

Master's: Thirty graduate course credits, at least 24 on campus, including core, elective, and research courses, are required. Master's thesis or passage of comprehensive examination is required.

Doctorate: Sixty credit hours beyond the bachelor's degree, including core, elective, and research courses with at least two full-time semesters, are required. Transfer credit of up to 30 hours is allowed. Students are required to pass a written comprehensive examination, followed by an oral qualifying examination. Dissertation and dissertation defense examinations are required. Dissertation research may be conducted off-campus in approved programs.

Thesis: Thesis may be written in absentia.

SPECIAL EQUIPMENT, FACILITIES, OR PROGRAMS

A transmission electron microscope laboratory studies defects and other materials properties. An EPR facility is used to study biological physics. X-ray research includes applications to materials and medicine. The experimental particle physics program is part of the ATLAS collaborations at CERN and the BABAR collaboration at the Stanford Linear Accelerator Center. A robotics laboratory is used to study intelligent behaviors. Cooperative programs have been established with nearby General Electric Research and Development Center (Watervliet Arsenal, NY), State Public Health, and IBM Watson Research Laboratories.

FACULTY

Professor

Alam, M. Sajjad, Ph.D., Indiana University, 1975. *Particles and Fields*. Experimental particle physics.

Caticha, Ariel, Ph.D., California Institute of Technology, 1985. Information physics; fundamental problems in quantum, statistical, and gravitational physics.

Das, Tara P., Ph.D., University of Calcutta, India, 1955. Theory of electronic structures and associated properties, including hyperfine interactions of atomic, molecular, biological, and condensed-matter systems.

Inomata, Akira, Ph.D., Rensselaer Polytechnic Institute, 1964. *Particles and Fields, Relativity & Gravitation*. Theoretical particle physics; general relativity.

Kimball, John C., Ph.D., University of Chicago, 1969. Theoretical condensed-matter physics; statistical mechanics.

Kuan, Tung-sheng, Ph.D., Cornell University, 1977. *Materials Science, Metallurgy*. Materials science; electron microscopy.

Lanford, William A., Ph.D., University of Rochester, 1972. *Applied Physics, Materials Science, Metallurgy*. Materials physics; applied physics; glasses; thin films; ion beam analysis.

MacDonald, Carolyn A., Ph.D., Harvard University, 1986. *Materials Science, Metallurgy, Medical, Health Physics, Optics*. X-ray optics; materials science; medical physics.

Associate Professor

Ernst, Jesse, Ph.D., University of Rochester, 1995. *Particles and Fields*. Experimental particle physics.

Knuth, Kevin, Ph.D., University of Minnesota, 1995. *Biophysics, Computational Physics*. Biophysics; space physics; computational physics.

Assistant Professor

Earle, Keith, Ph.D., Cornell University, 1994. *Biophysics*. Experimental biophysics.

Goyal, Philip, Ph.D., University of Cambridge, 2005. Information physics; fundamentals of quantum mechanics.

Jain, Vivek, Ph.D., University of Hawaii, 1988. *Particles and Fields*. Experimental particle physics.

Lunin, Oleg, Ph.D., Ohio State University, 2000. *Computational Physics, Particles and Fields*. String theory; computational particle physics.

Petruccelli, Jonathan, Ph.D., University of Rochester, 2010. *Optics*. Optics.

Emeritus

Benenson, Raymond E., Ph.D., University of Wisconsin-Madison, 1955. Particle–solid interactions; hydrogen in metals; teaching experiments.

Lanni, Robert P., M.A., SUNY, Albany, 1954. Physics education.

Marsh, Bruce B., Ph.D., University of Rochester, 1962. Experimental particle–solid interactions.

Roth, Laura M., Ph.D., Radcliffe Institute for Advanced Study, 1957. Theoretical condensed-matter physics; liquids and alloys; quantum mechanics.

Scholes, Charles P., Ph.D., Yale University, 1969. Experimental biophysics.

Scholz, Wilfried W., Ph.D., University of Freiburg, 1964. Atomic physics; inner shell phenomena; electron and ion collisions; nuclear physics.

Professor Emeritus

Garg, Jagadish B., Other, University of Paris, 1958. *Nuclear Physics*. Experimental nuclear physics; neutron resonances.

DEPARTMENTAL RESEARCH SPECIALTIES AND STAFF

Theoretical

Biophysics. Structure of hemoglobin and associated molecules; electronic structure and properties of chlorophyll and related systems. Earle.

Computational Physics. Caticha, Goyal, Knuth.

Mechanics. Caticha, Goyal, Knuth.

Optics. MacDonald.

Particles and Fields. Exotic particles; geometrical models for particles; string theory. Alam, Ernst, Jain.

String Theory. Lunin.

Experimental

Biophysics. ESR, EPR of bioinorganic molecules; x-ray imaging and therapy. Earle.

Medical, Health Physics. Multilayer thin-film, crystal, and capillary optics; materials and medical applications. MacDonald.

Optics. Surface/interface studies; x-ray crystallography; ion beam characterization of solids; electron microscopy. MacDonald.

Particles and Fields. Charm Baryon production from $e+e-$ collision. Alam, Ernst, Jain.

STONY BROOK UNIVERSITY, STATE UNIVERSITY OF NEW YORK

DEPARTMENT OF PHYSICS AND ASTRONOMY

Stony Brook, New York 11794-3800
http://www.physics.sunysb.edu

General University Information
President: Samuel Stanley
Dean of Graduate School: Charles Taber
University website: http://www.stonybrook.edu
Control: Public
Setting: Suburban
Total Faculty: 1,900
Total number of Students: 24,149
Total number of Graduate Students: 8,146

Department Information
Department Chairman: Laszlo Mihaly, Chair
Department Contact: Jacobus Verbaarschot, Graduate Program
  Director
  Total full-time faculty: 60
  Total number of full-time equivalent positions: 55
  Full-Time Graduate Students: 169
  First-Year Graduate Students: 38
  Female First-Year Students: 7
  Total Post Doctorates: 44

Department Address
100 Nicolls Road
Stony Brook, NY 11794-3800
Phone: (631) 632-8100
Fax: (631) 632-8176
E-mail: jacobus.verbaarschot@stonybrook.edu
Website: http://www.physics.sunysb.edu

ADMISSIONS

Admission Contact Information
Address admission inquiries to: Prof. Jacobus Verbaarschot, Director of Graduate Program, Department of Physics and Astronomy
Phone: (631) 732-8123
E-mail: jacobus.verbaarschot@stonybrook.edu
Admissions website: http://graduate.physics.sunysb.edu

Application deadlines
Fall admission:
U.S. students: January 15   Int'l. students: January 15
Spring admission:
U.S. students: October 15   Int'l. students: October 15

Application fee
U.S. students: $100   Int'l. students: $100
Spring admission, contact the Director of Graduate Program first.

Admissions information
For Fall of 2013:
  Number of applicants: 494
  Number admitted: 80
  Number enrolled: 24

Admission requirements
Bachelor's degree requirements: Bachelor's degree in physics or related area is required.
Minimum undergraduate GPA: 3.0

GRE requirements
The GRE is required.
  Quantitative score: 160
  Analytical score: 3
  Mean GRE score range (25th–75th percentile): 320-350
Used new GRE for Mean. It is the sum of verbal and quantitative.

Advanced GRE requirements
The Advanced GRE is recommended.
  Mean Advanced GRE score range (25th–75th percentile): 650-900
The Physics GRE is recommended but not required and is not very important in admissions decisions.

TOEFL requirements
The TOEFL exam is required for students from non-English-speaking countries.
  PBT score: 600
  iBT score: 90
The Speaking score should be at least 19.

Other admissions information
Numbers of students that applied, were admitted, and enrolled include PhD students only, and Master students are not included.
Undergraduate preparation assumed: Typical undergraduate preparation includes courses in classical mechanics, electrodynamics, quantum mechanics, and statistical mechanics/thermal physics.

TUITION

Tuition year 2013-14:
Tuition for in-state residents
  Full-time students: $9,370 annual
  Part-time students: $390 per credit
Tuition for out-of-state residents
  Full-time students: $16,680 annual
  Part-time students: $695 per credit
No differentiation of faculty, excluding health sciences.
Deferred tuition plan: Yes
Health insurance: Available at the cost of $1,884 per year.
Other academic fees: $607 per semester
Academic term: Semester
Number of first-year students who received full tuition waivers: 30

Teaching Assistants, Research Assistants, and Fellowships
Number of first-year
  Teaching Assistants: 24
  Fellowship students: 5
Average stipend per academic year
  Teaching Assistant: $20,000
  Research Assistant: $25,000
  Fellowship student: $24,000

FINANCIAL AID

Loans
Loans are available for U.S. students.
Loans are not available for international students.
GAPSFAS application required: No
FAFSA application required: Yes

HOUSING

Availability of on-campus housing
Single students: Yes
Married students: Yes

For further information
Address housing inquiries to: Regina Lagrasta, Assistant Director, Residence Life, Room 138, Administration Building.
Housing aid website: http://studentaffairs.stonybrook.edu/res/index.aspx

Table A—Faculty, Enrollments, and Degrees Granted

Research Specialty	2012–12 Faculty	Enrollment Fall 2013		Number of Degrees Granted 2012-13 (2007–12)		
		Master's	Doctorate	Master's	Terminal Master's	Doctorate
Astronomy	8	1	4	–	1(3)	2(7)
Atomic, Molecular, & Optical Physics	7	–	13	–	2(-)	3(20)
Biophysics	3	–	5	–	–	–(5)
Condensed Matter Physics	25	–	21	–	1(3)	5(25)
Cosmology & String Theory	7	2	12	–	2(3)	1(9)
High Energy Physics	10	–	9	–	–	1(13)
Nuclear Physics	11	1	24	–	1(3)	6(15)
Particles and Fields	7	–	7	–	–	1(6)
Physics of Beams	5	1	4	–	1(3)	–(4)
Statistical & Thermal Physics	4	–	8	–	–	1(5)
Theoretical Physics	41	–	58	–	–	9(-)
Non-specialized	–	9	50	–	3(5)	–
Total	87	14	157	–	11(20)	20(109)
Full-time Grad. Stud.	–	14	157	–	–	–
First-year Grad. Stud.	–	9	28	–	–	–

GRADUATE DEGREE REQUIREMENTS

Master's: Master of Science in Instrumentation (M.S.I.) requires a minimum of 2 years of study; one semester of teaching; minor project and Master's thesis. Master of Arts (M.A.) 30 graduate credits with average grade of B (GPA of 3.0) in approved program with satisfactory performance; either passing approved courses and exams or thesis is required; no residence or language requirements. Master of Arts in Teaching (M.A.T.) 41 graduate credits with 15 in physics, 20 in education and 6 in supervised teaching, as well as an approved teaching project. E-mail SPD@stonybrook.edu for more information.

Doctorate: Satisfactory performance in approved course program; successful completion of written comprehensive examination, dissertation, and dissertation examination. One year of teaching required. One year of residence required. No language requirements.

SPECIAL EQUIPMENT, FACILITIES, OR PROGRAMS

The Department of Physics and Astronomy participates with other departments in programs in physical biology, chemical physics, medical physics, and mathematical physics. A physics student in one of these programs may do thesis research under the supervision of a faculty member in the Physics and Astronomy Department or in one of the cooperating departments.

The C. N. Yang Institute for Theoretical Physics, affiliated with the Physics and Astronomy Department, carries on research in particle physics, nuclear physics, statistical mechanics, string and supergravity theory.

Cooperative programs are conducted at the Simons Center for Geometry and Physics, at the nearby Brookhaven National Lab, including research at the National Synchrotron Light Source, and in the Center for Functional Nanomaterials.

The Center for Accelerator Science and Education (CASE) coordinates efforts in accelerator physics. Experiments in particle physics are carried out at the Relativistic Heavy Ion Collider (BNL), at Super-Kamiokande (Japan), and at the Large Hadron Collider (CERN).

The astronomy group makes regular use of various large optical telescopes, large millimeter wave telescopes and arrays, and space observatories. It has a partnership with the Caltech Palomar Observatory, and it is a member of the SMARTs consortium operating several telescopes at Cerro Tololo, Chile. Faculty are affiliated with the New York Center for Computational Science NYCCS and use its 100 T flops supercomputer and other supercomputers for computational science.

Table B—Separately Budgeted Research Expenditures by Source of Support

Source of Support	Departmental Research	Physics-related Research Outside Department
Federal government	$12,247,917	
State/local government		
Non-profit organizations		
Business and industry		
Other	$2,749,569	
Total	**$14,997,486**	

Table C—Separately Budgeted Research Expenditures by Research Specialty

Research Specialty	No. of Grants	Expenditures ($)
Astronomy	9	$595,420
Astrophysics	3	$409,093
Atmosphere, Space Physics, Cosmic Rays	1	$112,117
Atomic, Molecular, & Optical Physics	8	$703,326
Biophysics	1	$40,000
Condensed Matter Physics	22	$2,517,757
Nuclear Physics	7	$1,732,962
Particles and Fields	2	$802,427
Physics and other Science Education	13	$484,277
Physics of Beams	3	$1,083,637
Statistical & Thermal Physics	3	$298,988
Cosmology & String Theory	1	$437,331
High Energy Physics	11	$3,030,582
Theoretical Physics	14	$2,749,569
Total	**98**	**$14,997,486**

FACULTY

Professor
Allen, Philip B., Ph.D., University of California, Berkeley, 1969. *Condensed Matter Physics*. Theoretical solid state physics.

Aronson, Meigan, Ph.D., University of Illinois, 1982. *Condensed Matter Physics*. Experimental solid state physics.

Averin, Dmitrii V., Ph.D., Moscow State University, 1987. *Solid State Physics*. Solid state theory.

Dill, Ken, Ph.D., University of California, San Diego, 1978. Director of the Laufer Center. *Biophysics*. Protein structure; computational biophysics.

Douglas, Michael, Ph.D., California Institute of Technology, 1988. *Cosmology & String Theory, Neuroscience/Neuro Physics*. String theory.

Drees, Axel, Ph.D., Heidelberg University, 1989. Associate Provost. *Nuclear Physics*. Relativistic heavy ions.

Engelmann, Roderich, Ph.D., Heidelberg University, 1966. *High Energy Physics*. Experimental high-energy physics.

Goldhaber, Alfred S., Ph.D., Princeton University, 1964. *Particles and Fields*. Theoretical physics; nuclear theory; particle physics.

Goldman, Vladimir J., Ph.D., University of Maryland, 1985. *Condensed Matter Physics*. Experimental solid state physics.

Gurvitch, Michael, Ph.D., Stony Brook University, 1978. *Condensed Matter Physics*. Experimental solid state physics.

Hemmick, Thomas, Ph.D., University of Rochester, 1989. Director of CASE. *Nuclear Physics*. Experimental relativistic heavy ion physics.

Hobbs, John, Ph.D., University of Chicago, 1991. *High Energy Physics*. Experimental high energy physics.

Jacak, Barbara, Ph.D., Michigan State University, 1984. Spokesperson of Phenix. *Nuclear Physics*. Relativistic Heavy Ion Collisions.

Jung, Chang Kee, Ph.D., Indiana University, 1986. *High Energy Physics*. Experimental High-Energy Physics, Neutrino Physics.

Kharzeev, Dmitri, Ph.D., Moscow State University, 1990. *Particles and Fields*. Heavy Ion Physics and Particle Theory.

Koch, Peter M., Ph.D., Yale University, 1974. *Atomic, Molecular, & Optical Physics*. Atomic Physics; Chaos.

Korepin, Vladimir, Ph.D., University of Leningrad, 1977. *Statistical & Thermal Physics*. Theoretical Physics, Quantum computing.

Lanzetta, Kenneth, Ph.D., University of Pittsburgh, 1988. *Astronomy*. Galactic Formation, Cosmology.

Lattimer, James M., Ph.D., University of Texas, 1976. *Astrophysics*. Nuclear Astrophysics.

Likharev, Konstantin K., Ph.D., Moscow State University, 1969. *Condensed Matter Physics, Neuroscience/Neuro Physics*. Mesoscopic Physics, Neuroscience.

McCarthy, Robert L., Ph.D., University of California, Berkeley, 1971. *High Energy Physics*. Experimental High-Energy Physics.

McCoy, Barry M., Ph.D., Harvard University, 1967. *Statistical & Thermal Physics, Theoretical Physics*. Theoretical Statistical Mechanics.

Mendez, Emilio, Ph.D., Massachusetts Institute of Technology, 1979. Director of the Center for Functional Nanomaterials. *Condensed Matter Physics*. Experimental Solid State Physics.

Metcalf, Harold J., Ph.D., Brown University, 1967. *Atomic, Molecular, & Optical Physics*. Experimental Atomic Physics; Laser Cooling.

Mihaly, Laszlo, Ph.D., Eotvos University, Budapest, 1977. Chair of the Department. *Condensed Matter Physics*. Experimental Solid State Physics.

Rijssenbeek, Michael, Ph.D., University of Amsterdam, 1974. *High Energy Physics*. Experimental High-Energy Physics.

Roček, Martin, Ph.D., Harvard University, 1979. *Cosmology & String Theory, Particles and Fields*. Theoretical Physics.

Shrock, Robert, Ph.D., Princeton University, 1975. *Particles and Fields*. Theoretical Physics, String Theory.

Shuryak, Edward, Ph.D., Novosibirsk, USSR, 1974. *Nuclear Physics, Particles and Fields*. Nuclear Theory, Quark gluon plasma, instantons.

Siegel, Warren, Ph.D., University of California, Berkeley, 1977. *Cosmology & String Theory, Particles and Fields*. Theoretical Physics.

Sprouse, Gene D., Ph.D., Stanford University, 1968. Editor-in-Chief of the American Physical Society. Experimental Nuclear Physics.

Stephens, Peter W., Ph.D., Massachusetts Institute of Technology, 1978. *Condensed Matter Physics*. Experimental Solid State Physics, Powder x-ray Scattering.

Sterman, George, Ph.D., University of Maryland, 1974. Director of the Yang Institute for Theoretical Physics. *Particles and Fields, Theoretical Physics*. Theoretical Physics.

van Nieuwenhuizen, Peter, Ph.D., Utrecht, 1971. *Cosmology & String Theory, Particles and Fields*. Theoretical Physics, supergravity.

Verbaarschot, Jacobus J., Ph.D., University of Utrecht, 1982. Graduate Program Director. *Nuclear Physics, Particles and Fields, Theoretical Physics*. Theoretical Physics, Strong Interactions, Random Matrix Theory.

Walter, Frederick M., Ph.D., University of California, Berkeley, 1981. *Astronomy*. Star Formation; Chromospheres and Coronae.

Zahed, Ismail, Ph.D., Massachusetts Institute of Technology, 1983. *Nuclear Physics, Particles and Fields, Theoretical Physics*. Theoretical Nuclear Physics.

Associate Professor

Abanov, Alexandre, Ph.D., University of Chicago, 1997. Deputy Director of the Simons Center. *Condensed Matter Physics, Statistical & Thermal Physics, Theoretical Physics*. Theoretical Condensed Matter.

Deshpande, Abhay, Ph.D., Yale University, 1995. Director of the Undergraduate Program. *Nuclear Physics*. Polarized photons, RHIC.

Gonzalez-Garcia, Concha, Ph.D., University of Valencia, 1991. *Particles and Fields*. Theoretical Physics.

Graf, Erlend H., Ph.D., Cornell University, 1968. *Physics and other Science Education*. Curriculum Development.

McGrew, Clark, Ph.D., University of California, Irvine, 1994. *High Energy Physics*. Experimental High Energy, Neutrino Physics.

Rastelli, Leonardo, Ph.D., Massachusetts Institute of Technology, 2000. *Cosmology & String Theory*. Theoretical Physics.

Schneble, Dominik, Ph.D., University of Konstanz, 2002. *Atomic, Molecular, & Optical Physics*. Experimental Atomic Physics.

Weinacht, Thomas, Ph.D., University of Michigan, 2000. *Atomic, Molecular, & Optical Physics*. Experimental Atomic Physics.

Zingale, Michael, Ph.D., University of Chicago, 2000. *Astrophysics*. Computational Astrophysics.

Assistant Professor

Allison, Thomas, Ph.D., Univ. of California at Berkeley, 2010. Atomic, Molecular and Optical Experiment.

Calder, Alan, Ph.D., Vanderbilt University, 1997. *Astrophysics*. Computational Astrophysics.

Dawber, Matthew, Ph.D., University of Cambridge, 2003. *Condensed Matter Physics*. Experimental Solid State Physics.

Du, Xu, Ph.D., University of Florida, 2004. *Condensed Matter Physics*. Experimental Condensed Matter, Graphene.

Essig, Rouven, Ph.D., Rutgers University, 2008. *Cosmology & String Theory, Particles and Fields*. Theoretical Particle Physics, Cosmology, Dark Matter.

Fernandez-Serra, Maria Victoria, Ph.D., University of Cambridge, 2005. *Condensed Matter Physics*. Computational Condensed Matter.

Fidkowski, Lukasz, Ph.D., Stanford University, 2007. Condensed Matter Theory.

Figueroa, Eden, Ph.D., Univ. of Calgary/Univ. of Konstanz, 2008. Atomic, Molecular and Optical Experiment.

Herzog, Chirstopher, Ph.D., Princeton University, 2002. *Cosmology & String Theory, Particles and Fields, Theoretical Physics*. Theoretical Physics, String Theory.

Kiryluk, Joanna, Ph.D., University of Warsaw, 2000. *Nuclear Physics, Particles and Fields*. Neutrino and Heavy Ion Physics.

Koda, Jin, Ph.D., University of Tokyo, 2002. *Astronomy*. Astronomy.

Meade, Patrick, Ph.D., Cornell University, 2006. *Particles and Fields*. Theoretical Physics, Particle Phenomenology.

Sehgal, Neelima, Ph.D., Rutgers University, 2008. *Astronomy, Cosmology & String Theory*. Dark Matter and Cosmology.

Teaney, Derek, Ph.D., Stony Brook University, 2001. *Nuclear Physics, Particles and Fields*. Theoretical Nuclear Physics, Relativistic Heavy Ion Physics.

Tsybychev, Dmitri, Ph.D., University of Florida, 2004. *High Energy Physics*. Experimental High Energy Physics.

Wei, Tzu-Chieh, Ph.D., University of Illinois at Urbana-Champaign, 2004. *Particles and Fields*. Quantum Computing.

Emeritus

de Zafra, Robert L., Ph.D., University of Maryland, 1958. *Atmosphere, Space Physics, Cosmic Rays*. Experimental Atmospheric Physics.

Kahn, Peter B., Ph.D., Northwestern University, 1960. *Nonlinear Dynamics and Complex Systems*. Theoretical Physics.

Kirz, Janos, Ph.D., University of California, Berkeley, 1963. *Optics*. X-ray Microscopy.

Kuo, T. T. S., Ph.D., University of Pittsburgh, 1964. *Nuclear Physics*. Nuclear Theory.

Lee, Linwood L., Ph.D., Yale University, 1955. *Nuclear Physics*. Nuclear physics.

Lukens, James, Ph.D., University of California, San Diego, 1968. *Condensed Matter Physics*. Condensed Matter Nanodevices.

McGrath, Robert L., Ph.D., University of Iowa, 1965. *Nuclear Physics*. Nuclear Physics.

Mould, Richard, Ph.D., Yale University, 1957. *Relativity & Gravitation*.

Paul, Peter, Ph.D., University of Freiburg, 1959. Associate Vice President for Brookhaven Affairs. *Nuclear Physics*. Experimental Nuclear Physics, Neutrino Physics.

Peterson, Deane M., Ph.D., Harvard University, 1968. *Astronomy*. Stellar Atmospheres.

Simon, Michal, Ph.D., Cornell University, 1967. *Astronomy*. Infrared Astronomy.

Smith, John, Ph.D., University of Edinburgh, 1963. *Particles and Fields*. High-Energy Theory.

Weisberger, William I., Ph.D., Massachusetts Institute of Technology, 1964. *Particles and Fields*. Theoretical Physics.

Yahil, Amos, Ph.D., California Institute of Technology, 1970. *Astronomy*. Cosmology.

Yang, Chen Ning, Ph.D., University of Chicago, 1984. *Particles and Fields*. Theoretical Physics.

Research Professor

Grannis, Paul D., Ph.D., University of California, Berkeley, 1965. *High Energy Physics*. Experimental High-Energy Physics.

Research Associate Professor

Swesty, Douglas, Ph.D., Stony Brook University, 1993. *Astrophysics*. Computational and Nuclear Astrophysics.

Yanagisawa, Chiaki, Ph.D., University of Tokyo, 1981. *High Energy Physics*. Experimental High-Energy Physics.

Distinguished Adjunct Professor

Ben-Zvi, Ilan, Ph.D., Weizmann Institute of Science, 1970. *Physics of Beams*. Accelerator Physics.

Litvinenko, Vladimir, Ph.D., Novosibirsk, 1989. *Physics of Beams*. Accelerator Physics.

Adjunct Professor

Bergeman, Thomas, Ph.D., Harvard University, 1971. *Atomic, Molecular, & Optical Physics*. Theoretical Atomic Physics.

Creutz, Michael, Ph.D., Stanford University, 1970. *Particles and Fields*. Theoretical Particle Physics, Lattice QCD.

Dawson, Sally, Ph.D., Harvard University, 1981. *Particles and Fields*. Theoretical Particle Physics.

Dierker, Steven, Ph.D., University of Illinois, 1983. Director of NSLS II. *Condensed Matter Physics*. Experimental Solid State Physics.

Forman, Miriam, Ph.D., Stony Brook University, 1972. *Astronomy*. Cosmic Rays.

Geller, Marvin, Ph.D., Massachusetts Institute of Technology, 1969. *Atmosphere, Space Physics, Cosmic Rays*. Atmospheric Physics.

Johnson, Peter, Ph.D., University of Warwick, 1978. *Condensed Matter Physics*. Expt. Condensed Matter.

Karsch, Frithjof, Ph.D., Bielefeld University, 1982. Professor at the University of Bielefeld. *Particles and Fields*. Lattice Gauge Theory.

Ku, Wei, Ph.D., University of Tennessee, 2000. *Condensed Matter Physics*. Theoretical Condensed Matter.

Maslov, Sergei, Ph.D., Stony Brook University, 1996. *Condensed Matter Physics*. Physical Biology, Soft Condensed Matter.

Ocko, Ben, Ph.D., Massachusetts Institute of Technology, 1984. *Condensed Matter Physics*. Expt. Condensed Matter.

Peggs, Stephen, Ph.D., Cornell University, 1981. *Physics of Beams*. Accelerator Physics.

Petrovic, Cedomir, Ph.D., Florida State University, 2000. *Condensed Matter Physics*. Experimental Condensed Matter.

Qiu, Jianwei, Ph.D., Columbia University, 1987. *Particles and Fields*. Theoretical Physics.

Semenov, Vasili, Ph.D., Moscow State University, 1975. *Condensed Matter Physics*. Experimental Condensed Matter.

Sivaramakrishnan, Anand, Ph.D., University of Texas, 1983. *Astronomy*. Astronomy.

Takai, Helio, Ph.D., University of Rio de Janeiro, 1986. *High Energy Physics*. Experimental particle physics.

Tolpygo, Sergey, Ph.D., Institute for Solid State Physics, Russian Acad. of Sciences, 1984. *Condensed Matter Physics*. Condensed Matter Nanodevices.

Tsvelik, Alexei, Ph.D., Kurchatov Institute of Atomic Energy Moscow, 1980. *Condensed Matter Physics*. Condensed Matter Theory.

Venugopalan, Raju, Ph.D., Stony Brook University, 1992. *Particles and Fields*. High-Energy Theory.

Zhu, Yimei, Ph.D., Nagoya University, 1987. *Optics*. Microscopy.

Affiliate Professor

Lacey, Roy A., Ph.D., Stony Brook University, 1987. *Nuclear Physics*. Experimental Heavy Ion Physics.

Oganov, Artem, Ph.D., University College, London, 2002. *Geophysics*. Computational Geosciences.

Wang, Jin, Ph.D., University of Illinois, 1991. Biological Physics.

Affiliate Associate Professor

Jia, Jiangyong, Ph.D., Stony Brook University, 2003. *Nuclear Physics*. Relativistic Heavy Ion Collisions.

DEPARTMENTAL RESEARCH SPECIALTIES AND STAFF

Theoretical

Astrophysics. Cosmology, galaxy formation, and evolution. Nuclear astrophysics, neutrino astrophysics. Neutron stars; equation of state of dense matter; stellar collapse. Dark matter, dark energy. Calder, Lattimer, Sehgal, Swesty, Zingale.

Atomic, Molecular, & Optical Physics. Theory of ultracold quantum gases, non-linear optics. Bose-Einstein condensation. Bergeman.

Condensed Matter Physics. Superconductivity; electron-phonon interactions; magnetic properties; optical properties; quantum computing, Coulomb blockade, quantum Hall effect, density-functional theory; properties of water; low-dimensional systems. Abanov, Allen, Averin, Fernandez-Serra, Korepin, Ku, Likharev, McCoy, Tsvelik, Wei.

Cosmology & String Theory. Quantum theory of gravitation; supergravity; supersymmetry; superstrings; cosmology, dark matter. Douglas, Essig, Goldman, Herzog, Rastelli, Roček, Siegel, van Nieuwenhuizen.

Nuclear Physics. Experimental Relativistic Heavy Ion Collisons; neutrino physics; nucleon-nucleon interaction; meson exchange currents and other mesonic effects in nuclei; effective interactions in nuclei and nuclear matter; heavy ion reactions; Fermi liquid theory; variational and extended semiclassical models of large systems; studies of dense nuclear matter; Random Matrix Theory. Kharzeev, Kuo, Shuryak, Teaney, Verbaarschot, Zahed.

Physical Biology. Protein structure; computational biophysics; bio-molecular networks; neuroscience. Dill, Likharev, Maslov, Wang.

Statistical & Thermal Physics. Mathematical studies of solvable models; relation between statistical mechanics and field theory; Ising models; lattice gauge fields; Random Matrix Theory. Abanov, Korepin, McCoy, Tsvelik, Verbaarschot, Wei.

Theoretical Particle Physics. Quantum field theory; unified gauge theory of weak, electromagnetic, and strong interactions; general gauge theory; nonperturbative effects in gauge theories; perturbative QCD; QCD at finite temperature and density. Creutz, Dawson, Douglas, Essig, Gonzalez-Garcia, Herzog, Karsch, Kharzeev, Meade, Rastelli, Roček, Shrock, Siegel, Sterman, van Nieuwenhuizen, Verbaarschot, Weisberger, Zahed.

Theoretical Physics. The largest theoretical physics group at Stony Brook is the Yang Institute for Theoretical Physics. About the same number of theoretical physicts work work in the physics department on condensed matter physics, astrophysics, nuclear physics, particle physics and statistical mechancis. Fernandez-Serra, Goldhaber, Herzog, Kahn, Karsch, Kharzeev, Korepin, Ku, Kuo, Lattimer, Likharev, Maslov, McCoy, Meade, Mould, Qiu, Rastelli, Roček, Sehgal, Shrock, Shuryak, Siegel, Sterman, Teaney, Tsvelik, van Nieuwenhuizen, Venugopalan, Verbaarschot, Wang, Wei, Yang, Zahed, Zingale.

Experimental

Astronomy. Cosmology; galactic structure and evolution; interstellar molecular clouds; quasar absorption lines; stellar astronomy; chromospheres; coronae; compact objects; star formation. Pre-main sequence objects; high mass star formation.; Exoplanets. Forman, Koda, Peterson, Simon, Sivaramakrishnan, Walter, Yahil.

Atomic, Molecular, & Optical Physics. Coherent control of molecules and atoms with tailored, ultrafast laser pulses. Cooling, trapping, and laser spectroscopy of atoms. Ultracold quantum gases. Quantum chaos studies with microwave systems and driven atoms. Metcalf, Schneble, Weinacht.

Condensed Matter Physics. Superconductivity; Josephson effect; x-ray scattering; single electronics; magnetic flux quantum devices; quantum wells, fractional quantum Hall effect; optical spectroscopy, ferroelectrics; graphene; Powder X-ray scattering. Dawber, Dierker, Du, Gurvitch, Johnson, Likharev, Lukens, Mendez, Mihaly, Oganov, Petrovic, Stephens, Tolpygo, Zhu.

High Energy Physics. Particle interactions in high energy collisions; properties of the top and bottom quarks and electroweak bosons; studies of the strong interaction; search for new particles. Rare kaon decays. Neutrino oscillations; Study of the Higgs boson. Engelmann, Grannis, Hobbs, Jung, McCarthy, McGrew, Paul, Rijssenbeek, Takai, Tsybychev, Yanagisawa.

Nuclear Physics. Relativistic heavy ion collisions; properties of quark-gluon plasma; polarized protons. Deshpande, Drees, Hemmick, Jacak, Kiryluk, Lee, Paul.

View additional information about this department at
www.gradschoolshopper.com

SYRACUSE UNIVERSITY

DEPARTMENT OF PHYSICS

Syracuse, New York 13244-1130
http://physics.syr.edu

General University Information

Chancellor: Nancy Cantor
Dean of Graduate School: Ben Ware
University website: http://www.syr.edu
Control: Private

Setting: Urban
Total Faculty: 1,563
Total Graduate Faculty: 1,563
Total number of Students: 21,029
Total number of Graduate Students: 6,231

Department Information
Department Chairman: A. Alan Middleton, Chair
Department Contact: Patricia Whitmore, Academic Program
   Coordinator
   Total full-time faculty: 31
   Total number of full-time equivalent positions: 31
   Full-Time Graduate Students: 68
   First-Year Graduate Students: 10
   Female First-Year Students: 1
   Total Post Doctorates: 12

Department Address
201 Physics Building
Syracuse, NY 13244-1130
Phone: (315) 443-5958
Fax: (315) 443-9103
E-mail: graduate@physics.syr.edu
Website: http://physics.syr.edu

ADMISSIONS

Admission Contact Information
Address admission inquiries to: Graduate Coordinator, Physics
   Department, 201 Physics Bldg.
Phone: (315) 443-5958
E-mail: graduate@phy.syr.edu
Admissions website: http://web.physics.syr.edu/graduate/

Application deadlines
Fall admission:
U.S. students: January 1   *Int'l. students*: January 1

Application fee
U.S. students: $75   *Int'l. students*: $75
Late applications accepted until available slots are filled.

Admissions information
For Fall of 2013:
   Number of applicants: 116
   Number admitted: 34
   Number enrolled: 17

Admission requirements
Bachelor's degree requirements: Bachelor's degree in physics
   is recommended but not required.
Minimum undergraduate GPA: 3.0

GRE requirements
The GRE is required.
   Quantitative score: 600
   Verbal score: 440

Advanced GRE requirements
The Advanced GRE is required.
   Minimum accepted Advanced GRE score: 600

TOEFL requirements
The TOEFL exam is required for students from non-English-
   speaking countries.
   PBT score: 600
   iBT score: 100

Other admissions information
Additional requirements: Letters of recommendation and per-
   sonal statement.
Undergraduate preparation assumed: Symon, Classical Mechan-
   ics; Reitz and Milford, Foundations of Electromagnetic
   Theory; Eisberg, Fundamentals of Modern Physics.

TUITION
Tuition year 2012–13:
   Full-time students: $1,294 per credit
   Part-time students: $1,294 per credit
Almost all department graduate students have their tuition cov-
   ered by an assistantship, except in rare cases.
Credit hours per semester to be considered full-time: 9
Deferred tuition plan: Yes
Health insurance: Available at the cost of $2,600 per year.
Other academic fees: $760 student fees.
Academic term: Semester
Number of first-year students who received full tuition waivers: 11

Teaching Assistants, Research Assistants, and Fellowships
Number of first-year
   Teaching Assistants: 8
   Fellowship students: 3
Average stipend per academic year
   Teaching Assistant: $21,866
   Research Assistant: $21,866
   Fellowship student: $22,460

FINANCIAL AID

Application deadlines
Fall admission:
U.S. students: January 1   *Int'l. students*: January 1

Loans
Loans are available for U.S. students.
Loans are available for international students.
GAPSFAS application required: Yes
FAFSA application required: Yes

For further information
Address financial aid inquiries to: Financial Aid Office, 200
   Archbold Gymnasium.
Phone: (315) 443-1513
E-mail: finmail@syr.edu
Financial aid website: http://financialaid.syr.edu

HOUSING

Availability of on-campus housing
   Single students: No
   Married students: No

For further information
Address housing inquiries to: Off-Campus and Commuter Ser-
   vices, 754 Ostrom Avenue, Syracuse, NY 13244.
Phone: (315) 443-5489
E-mail: offcampus@syr.edu
Housing aid website: http://occs.syr.edu/

Table A—Faculty, Enrollments, and Degrees Granted

Research Specialty	2012-13 Faculty	Enrollment Fall 2012 Master's	Enrollment Fall 2012 Doctorate	Number of Degrees Granted 2011–12 (2007–12) Master's	Number of Degrees Granted 2011–12 (2007–12) Terminal Master's	Number of Degrees Granted 2011–12 (2007–12) Doctorate
Biophysics	4	–	7	–(2)	–(1)	1(7)
Condensed Matter Physics	9	–	28	4(5)	1(1)	5(14)
Cosmology & String Theory	3	–	6	–	–	2(4)
First Year/Undecided	–	–	7	3(6)	3(6)	–
High Energy Physics	12	–	15	1(2)	–(1)	4(16)
Relativity & Gravitation	3	–	5	–(2)	–(1)	2(4)
Total	**31**	**–**	**68**	**8(17)**	**4(10)**	**14(45)**
Full-time Grad. Stud.	–	–	68	–	–	–
First-year Grad. Stud.	–	–	11	–	–	–

GRADUATE DEGREE REQUIREMENTS

Master's: Minimum of one year in residence is required. A student admitted to graduate work in the Department must take the comprehensive examination. The degree can be achieved in one of three ways: (1) 24 hours of coursework including a thesis, (2) 30 hours of coursework including a minor problem and passing the qualifying examination, or (3) 36 hours of coursework and passing the qualifying examination. A "B" average in coursework must be maintained to be eligible for a degree. There is no foreign language requirement.

Doctorate: Satisfactory performance in a course program approved by the student's research committee, which may include courses taken for the M.S. degree. Students must pass a written qualifying examination, a preliminary oral research examination, and a thesis defense. There is no language requirement. At least a "B" average is required.

Thesis: Thesis may be written in absentia.

SPECIAL EQUIPMENT, FACILITIES, OR PROGRAMS

The Department is a strong participant in the Syracuse Biomaterials Institute of Syracuse University. Some students are fellows in the Soft Interfaces Integrative Graduate Education and Research Traineeship (IGERT) program, which focuses on research on soft and biological interfaces and on the interaction between soft and hard materials. The research laboratories include a number of clean and shielded rooms and advanced cryostats served by a helium liquefier. The Department of Physics also hosts the University's Surface Characterization Facility. The high energy physics group is the major US group involved in the LHCb project at CERN in Geneva. Students also travel to a number of national facilities, including the Laser Interferometer Gravitational Wave Observatory (LIGO), Jefferson National Laboratory, and to the Cornell NanoScale Science Facility.

Table B—Separately Budgeted Research Expenditures by Source of Support

Source of Support	Departmental Research	Physics-related Research Outside Department
Federal government	$4,408,183	
State/local government	$60,698	
Non-profit organizations		
Business and industry	$155,019	
Other		
Total	**$4,623,900**	

Table C—Separately Budgeted Research Expenditures by Research Specialty

Research Specialty	No. of Grants	Expenditures ($)
Biophysics	6	$722,598
Condensed Matter Physics	8	$650,658
Low Temperature Physics	7	$1,065,709
Relativity & Gravitation	7	$1,188,375
High Energy Physics	9	$2,270,848
Other	4	$271,668
Total	**41**	**$6,169,856**

FACULTY

Professor

Artuso, Marina, Ph.D., Northwestern University, 1986. *Electrical Engineering, Particles and Fields*. Experimental elementary particles and fields.

Bowick, Mark, Ph.D., California Institute of Technology, 1983. *Condensed Matter Physics*. Theoretical condensed-matter physics.

Catterall, Simon, Ph.D., University of Oxford, 1988. Director of Graduate Studies. *Computational Physics, Particles and Fields*. Theoretical elementary particles physics; computational physics.

Foster, Kenneth, Ph.D., California Institute of Technology, 1972. *Particles and Fields*. Biological physics.

Lipson, Edward D., Ph.D., California Institute of Technology, 1971. *Applied Physics, Biophysics, Engineering Physics/Science, Medical, Health Physics*. Biological physics.

Marchetti, M. Cristina, Ph.D., University of Florida, 1982. Associate Director of Syracuse Biomaterials Institute; William R. Kenan, Jr. Professor. *Biophysics, Condensed Matter Physics, Fluids, Rheology, Statistical & Thermal Physics*. Theoretical condensed-matter physics.

Middleton, Alan, Ph.D., Princeton University, 1990. Associate Department Chair. *Applied Mathematics, Computational Physics, Condensed Matter Physics, Statistical & Thermal Physics*. Theoretical condensed-matter physics; computational physics.

Rosenzweig, Carl, Ph.D., Harvard University, 1972. *Particles and Fields*. Theoretical elementary particles and fields physics.

Saulson, Peter, Ph.D., Princeton University, 1981. Department Chair; Martin A. Pomerantz '37 Professor of Physics. *Relativity & Gravitation*. Experimental relativity and astrophysics.

Schechter, Joseph, Ph.D., University of Rochester, 1965. *Particles and Fields*. Theoretical elementary particles and fields physics.

Schiff, Eric, Ph.D., Cornell University, 1979. *Condensed Matter Physics, Solar Physics, Solid State Physics*. Experimental condensed-matter physics; solar energy.

Skwarnicki, Tomasz, Ph.D., Inst. of Nuclear Physics, Krakow, 1986. Chair of Graduate Admissions. *High Energy Physics, Particles and Fields*. Experimental elementary particles physics.

Souder, Paul A., Ph.D., Princeton University, 1971. *Particles and Fields*. Experimental elementary particles and fields physics; medium energy.

Stone, Sheldon, Ph.D., University of Rochester, 1972. *High Energy Physics, Particles and Fields*. Experimental particle physics.

Vidali, Gianfranco, Ph.D., Pennsylvania State University, 1982. *Astrophysics, Condensed Matter Physics, Surface Physics*. Laboratory astrophysics and surface science.

Associate Professor

Armendariz-Picon, Cristian, Ph.D., Ludwig-Maximilians-Universität, Munich, 2001. *Cosmology & String Theory, Particles and Fields, Relativity & Gravitation*. Cosmology; relativity; theoretical elementary particles physics.

Blusk, Steven, Ph.D., University of Pittsburgh, 1995. *High Energy Physics, Particles and Fields*. Experimental elementary particles physics.

Brown, Duncan, Ph.D., University of Wisconsin-Milwaukee, 2004. *Astrophysics, Computational Physics, Relativity & Gravitation*. Theoretical astrophysics; relativity.

Movileanu, Liviu, Ph.D., University of Bucharest, 1997. Member of Syracuse Biomaterials Institute. *Biophysics, Condensed Matter Physics*. Experimental biophysics.

Plourde, Britton, Ph.D., University of Illinois at Urbana-Champaign, 2000. *Condensed Matter Physics, Low Temperature Physics, Nano Science and Technology, Quantum Foundations*. Experimental condensed-matter physics.

Assistant Professor

Ballmer, Stefan, Ph.D., Massachusetts Institute of Technology, 2006. *Relativity & Gravitation*. Theoretical physics; gravitational waves.

Fan, JiJi, Ph.D., Yale University, 2009. *High Energy Physics, Particles and Fields*. Particle phenomenology, supersymmetry.

Forstner, Martin, Ph.D., University of Texas, Austin, 2003. *Biophysics, Condensed Matter Physics*. Experimental biological physics (e.g, cell membranes).

Hubisz, Jay, Ph.D., Cornell University, 2006. *Cosmology & String Theory, High Energy Physics, Particles and Fields*. Theoretical particle physics and cosmology.

LaHaye, Matthew, Ph.D., University of Maryland, College Park, 2005. *Condensed Matter Physics, Nano Science and Technology*. Experimental condensed-matter physics; nanomechanics.

Laiho, John, Ph.D., Princeton University, 2004. *Computational Physics, Particles and Fields, Theoretical Physics*. Theoretical elementary particle physics; Lattice QCD.

Manning, M. Lisa, Ph.D., University of California, Santa Barbara, 2008. *Biophysics, Condensed Matter Physics, Fluids, Rheology, Statistical & Thermal Physics*. Soft condensed matter; biophysics; granular materials; glasses.

Schnee, Richard, Ph.D., University of California, Santa Cruz, 1996. *Cosmology & String Theory*. Experimental/observational cosmology.

Schwarz, Jennifer, Ph.D., Harvard University, 2002. *Biophysics, Condensed Matter Physics, Statistical & Thermal Physics*. Theoretical condensed-matter physics.

Soderberg, Mitchell, Ph.D., University of Michigan, 2000. *Particles and Fields*. Experimental elementary particles physics: neutrinos.

Watson, Scott, Ph.D., Brown University, 2005. *Cosmology & String Theory, Particles and Fields*. Particle physics; cosmology.

Research Professor

Holmes, Richard, Ph.D., University of Maryland, 1985. *Particles and Fields*. Experimental elementary particles physics.

Mountain, Raymond, Ph.D., University of Notre Dame, 1992. *Particles and Fields*. Experimental elementary particles physics.

Saranak, Jureepan, Ph.D., Mount Sinai Medical School, 1981. *Biophysics*. Biophysics of cells.

Wali, Kameshwar, *History & Philosophy of Physics/Science, Particles and Fields*. Particles and fields; history of physics; physics of music.

Wang, Jianchun, Ph.D., Massachusetts Institute of Technology, 1997. *Particles and Fields*. Experimental elementary particles physics.

DEPARTMENTAL RESEARCH SPECIALTIES AND STAFF

Theoretical

Biophysics. Collective behavior of biological molecules, especially actins and motor proteins; interaction of living cells to form structure; pattern formation of active material; rheology of biological tissue. Bowick, Manning, Marchetti, Schwarz.

Computational Physics. Gravitational-wave data analysis and source modeling; grid computing; connections between algorithms and physical principles; study of condensed-matter order and optimal distributions on curved interfaces; analysis of phase transitions and phase structure in disordered systems; simulations of lattice quantum field theories; numerical simulations on parallel computers; technicolor and supersymmetric theories; models beyond Standard Model Physics. Bowick, Brown, Catterall, Laiho, Middleton.

Condensed Matter Physics. Soft condensed-matter physics; statistical mechanics; nonequilibrium dynamics including: two-dimensional matter, collective behavior of biological molecules, interaction of living cells, jamming in granular materials, superconductors, hysteresis in magnets, colloidal particles, topological defects, glassy materials, networks, relationship between algorithms and physics. Bowick, Manning, Marchetti, Middleton, Schwarz.

Cosmology & String Theory. Theoretical models of dark and cosmic acceleration; inflation and alternatives; origin and evolution of cosmological structures. Armendariz-Picon, Catterall, Fan, Hubisz, Watson.

Particles and Fields. Quantum gravity; supersymmetry; renormalization theory; chiral symmetries; monopoles and dyons in curved space-time; noncommutative geometry; random surfaces, electroweak theory; quantum chromodynamics; general quantum field theory; constrained field theories; geometric quantization; phenomenological particle dynamics; simulations of lattice QCD; supersymmetric field theories on space-time lattices; quark gluon plasma. Particle cosmology. Theories with extra dimensions. Simulations of lattice quantum field theories; technicolor and supersymmetric theories; holographic models of strings; models beyond Standard Model physics. Armendariz-Picon, Catterall, Fan, Hubisz, Rosenzweig, Schechter, Watson.

Experimental

Astrophysics. Laboratory studies of physical and chemical processes occurring in the interstellar medium and in planetary atmospheres, including formation of molecular hydrogen and hydrogenation and oxidation reaction on interstellar and/or planetary dust grain analogs. Vidali.

Biophysics. Single-molecule biophysics; membrane biophysics; bionanotechnology and biosensors; protein design; development of new optical technologies; photosensory transduction in microorganisms; bioinformatics; self-organized beating of cilia; phylogenetics and molecular clocks. Forstner, Foster, Lipson, Movileanu, Saranak.

Cosmology & String Theory. Dark matter detection and early universe cosmology: development of improved ultra-low-radioactivity environments and detectors of weakly interacting massive particles; analysis of data from dark matter searches. Schnee.

High Energy Physics. Experimental studies of the fundamental electroweak and strong interactions as manifested by the decays of beauty and charm quarks and the search for exotic particles; b & c quark decays are studied at the LHCb experiment at the CERN LHC hadron collider Geneva, Switzerland, concentrating on rare and CP violating decays; searches for exotic particle production, including unusual decays of the Higgs boson, are also done using LHCb; study of nucleon structure, including spin and quark components carried out at JLab; R&D into advanced silicon micro-pattern detectors, such as pixel sensors, and their related readout electronics. Members of the group have discovered several new particles, including the B, Ds, and Y(1D); made the first measurements of several very important decay modes of these objects; and is also starting an effort in neutrino physics. Artuso, Blusk, Holmes, Mountain, Skwarnicki, Soderberg, Souder, Stone, Wang.

Low Temperature Physics. Quantum coherent superconducting circuits; measurement and coupling of circuits for quantum computing; vortex dynamics in nanofabricated thin-film devices; superconducting microwave resonant circuits; nano-electromechanical systems (NEMS); quantum dynamics of mechanical systems; sensitive environmental gas and bio-sensors; measurements at millikelvin temperatures. LaHaye, Plourde.

Medical, Health Physics. Medical tomographic image reconstruction; biomedical molecular imaging; PET, SPECT, CT, and MRI; medical image registration; ultrafast laser-based x-ray source for radiological applications. Forstner, Lipson.

Nano Science and Technology. Much of the activity in this area is described under low-temperature physics (for example nanoscopic mechanical systems) and under biophysics (nano-pore technology). LaHaye, Movileanu, Plourde.

Nuclear Physics. Medium-energy physics: use of spin degrees of freedom to study quantum chromodynamics and the Standard Model at low energies. Experiments are under way at the Thomas Jefferson National Accelerator Facility (JLab). Souder.

Relativity & Gravitation. Gravitational-wave detection and astrophysics: searches for gravitational waves using the Laser Interferometer Gravitational Wave Observatory (LIGO); commissioning and technology development for advanced gravitational wave detectors; gravitational wave source modeling and phenomenology; developing tests of general relativity using gravitational waves. Ballmer, Brown, Saulson.

Solar Physics. Electronic and optical properties of unconventional semiconductors (e.g., amorphous silicon, porous titania, and silicon); solar cell device physics; thin-film growth (plasma, hot-wire); hybrid organic-inorganic semiconductor devices; v surface physics (i.e., structure, kinetics, dynamics, and reactions). Schiff.

View additional information about this department at
www.gradschoolshopper.com

UNIVERSITY AT BUFFALO, THE STATE UNIVERSITY OF NEW YORK

DEPARTMENT OF PHYSICS

Buffalo, New York 14260-1500
http://www.physics.buffalo.edu

General University Information
President: Satish K. Tripathi
Dean of Graduate School: John T. Ho
University website: http://www.buffalo.edu
Control: Public
Setting: Suburban
Total Faculty: 2,298
Total Graduate Faculty: 1,506
Total number of Students: 28,952
Total number of Graduate Students: 9,446

Department Information
Department Chairman: Hong Luo, Chair
Department Contact: Nicole D. Mercer, Assistant to the Chair
  Total full-time faculty: 28
  Total number of full-time equivalent positions: 28
  Full-Time Graduate Students: 76
  First-Year Graduate Students: 11
  Total Post-Doctorates: 13

Department Address
Fronczak Hall 239
North Campus
Buffalo, NY 14260-1500
Phone: (716) 645-2007
Fax: (716) 645-2507
E-mail: ubphysics@buffalo.edu
Website: http://www.physics.buffalo.edu

ADMISSIONS

Admission Contact Information
Address admission inquiries to: Director of Graduate Studies, Physics Department, Fronczak Hall 239, Buffalo, NY 14260-1500
Phone: (716) 645-2007
E-mail: ubphysics@buffalo.edu
Admissions website: http://www.physics.buffalo.edu/graduate.html

Application deadlines
Fall admission:
U.S. students: February 1   *Int'l. students*: February 1

Application fee
U.S. students: $75   *Int'l. students*: $75
For consideration of fellowship awards, an application and supporting materials must be completed and officially submitted by Wednesday, January 1, 2014. For frequently asked questions in physics, see http://www.physics.buffalo.edu/graduate/faq.html.

Admissions information
For Fall 2012:
  Number of applicants: 135
  Number admitted: 35
  Number enrolled: 13

Admission requirements
Bachelor's degree requirements: A bachelor's degree in physics with an average of B or above is required, although exceptions can be made in some circumstances. An applicant must satisfy the department of his/her ability to perform graduate work in physics.
Minimum undergraduate GPA: 3.0

GRE requirements
The GRE is required.
  Quantitative score: 162
  Verbal score: 152
  Analytical score: 3.5
  Mean GRE score range (25th–75th percentile): 65
Our institution code is 2925 and the Physics Department code is 0808.

Advanced GRE requirements
The Advanced GRE is required.
  Minimum accepted Advanced GRE score: 685
  Mean Advanced GRE score range (25th–75th percentile): 65
GRE advanced subject exam in physics is required for financial support and is otherwise highly recommended for all applicants. Our institution code is 2925 and the Physics Department code is 0808.

TOEFL requirements
The TOEFL exam is required for students from non-English-speaking countries.
  PBT score: 550
  iBT score: 79
If applying for a Teaching Assistantship, the minimum acceptable total score is 600 (paper-based) or 90 (internet-based). In lieu of TOEFL, IELTS is acceptable with a total score minimum of 6.5 (no individual score less than 6.0). Our institution code is 2925 and the Physics Department code is 76.

Other admissions information
Additional requirements: Foreign students applying for Teaching Assistantships must pass the Test of Spoken English (SPEAK) Exam before being assigned teaching duties. The exam is offered on campus at the beginning of the semester.
Undergraduate preparation assumed: Fowles and Cassiday, Analytical Mechanics; Griffiths, Introduction to Electrodynamics; Reif, Statistical and Thermal Physics; Bransden and Joachain, Quantum Mechanics.

TUITION
Tuition year 2013-14:
Tuition for in-state residents
  Full-time students: $4,935 per semester
  Part-time students: $411 per credit
Tuition for out-of-state residents
  Full-time students: $9,175 per semester
  Part-time students: $765 per credit

Additional costs, at full-time status, consist of comprehensive fees ($791.75), activity fees ($64), and academic excellence fees ($75). Tuition & fees are subject to change. Enrollment in a monthly payment plan is available. Additional information available at: Student Response Center http://sarfs.buffalo.edu/src.php. International tuition rates are available at http://wings.buffalo.edu/intadmit/formgeneral.htm.
Credit hours per semester to be considered full-time: 12
Deferred tuition plan: Yes
Health insurance: Available at the cost of $2,084 per year.
Other academic fees: Assistantships need nine credit hours for full-time. The deferred tuition plan provides assistance to veterans and dependents eligible under Chapter 35 (Veterans Deferred Tuition Payment Plan). Call 716-645-2271 for information. International fees are available at http://wings.buffalo.edu/intadmit.
Academic term: Semester
Number of first-year students who received full tuition waivers: 10

Teaching Assistants, Research Assistants, and Fellowships
Number of first-year
  Teaching Assistants: 10
  Fellowship students: 1
Average stipend per academic year
  Teaching Assistant: $17,500
  Research Assistant: $18,000
  Fellowship student: $19,000

FINANCIAL AID

Application deadlines
Fall admission:
U.S. students: March 1
Spring admission:
U.S. students: March 1   *Int'l. students*: March 1

Loans
Loans are available for U.S. students.
Loans are not available for international students.
GAPSFAS application required: No
FAFSA application required: Yes

For further information
Address financial aid inquiries to: Student Response Center, Office of Financial Aid, 232 Capen Hall, Buffalo, NY 14260-1631., If you are a U.S. citizen or permanent resident, you may apply for need-based aid in the form of loans, grants, and work study by completing a Free Application for Federal Student Aid (FAFSA), which is available at the Office of Financial Aid.
Phone: (716) 645-8232
E-mail: UBFA@buffalo.edu
Financial aid website: http://financialaid.buffalo.edu/

HOUSING

Availability of on-campus housing
  Single students: Yes
  Married students: Yes

For further information
Address housing inquiries to: University Residence Halls and Apartments, 106 Spaulding Quad, University at Buffalo, Buffalo, NY 14261-0054.
Phone: (716) 645-2171 / Toll Free: 866-285-8806
Housing aid website: http://www.ub-housing.buffalo.edu/

Table A—Faculty, Enrollments, and Degrees Granted

Research Specialty	2012-13 Faculty	Enrollment Fall 2012		Number of Degrees Granted 2012-13 (2006-13)		
		Master's	Doctorate	Master's	Terminal Master's	Doctorate
Acoustics	1	–	2	–	–	–
Applied Physics	7	1	5	–	1(2)	–(1)
Astrophysics	2	–	3	–	–	–(1)
Atmosphere, Space Physics, Cosmic Rays	1	–	–	–	–(2)	–
Atomic, Molecular, & Optical Physics	1	–	–	–	–	–
Biophysics	3	–	4	–(1)	–(4)	2(6)
Computational Physics	5	–	5	–	–(1)	1(1)
Condensed Matter Physics	18	1	23	1(2)	–(13)	6(36)
Cosmology & String Theory	2	–	4	–	–	1(3)
Engineering Physics/Science	1	–	–	–	–	–
Geophysics	1	–	2	–	–	–
High Energy Physics	6	–	9	–(1)	–(5)	–(4)
Low Temperature Physics	3	1	5	–(1)	–(1)	2(3)
Materials Science, Metallurgy	9	1	10	–	–	–(1)
Mechanics	1	–	–	–	–(1)	–
Medical, Health Physics	–	–	–	–(2)	–(4)	–(9)
Nonlinear Dynamics and Complex Systems	–	–	–	–	–	–(3)
Nuclear Physics	–	–	–	–	–(1)	–
Optics	2	–	–	–	–	–
Particles and Fields	5	–	2	–(1)	–(1)	1(3)
Physics and other Science Education	4	–	5	–	–	–
Statistical & Thermal Physics	3	–	–	–	–	–
Total	**28**	**4**	**79**	**1(8)**	**1(35)**	**13(71)**
Full-time Grad. Stud.	–	4	79	–	–	–
First-year Grad. Stud.	–	2	11	–	–	–

GRADUATE DEGREE REQUIREMENTS

Master's: Three options are available: (1) a thesis program, (2) a qualifying exam program, and (3) a project program. In the thesis program, a minimum of 30 credit hours is required. At least 15 credit hours are to be devoted to formal graduate course work, with the remaining hours culminating in a thesis. The qualifying exam program requires 30 credit hours of graduate courses plus passing the departmental qualifying exam at the M.S. level. The project program requires 30 credit hours: 15 credit hours of specified courses, 9 credit hours of other graduate courses and 6 credit hours of research (PHY 600).

Doctorate: A minimum of 72 semester hours of credit must be earned, with at least 36 in graduate physics lecture courses. To ensure breadth in the student's Ph.D. program, the department will evaluate his/her graduate work and may require the student to take specific courses in related fields. Within 18 months of enrollment, the qualifying exam must be passed. After an additional 18 months, a short defense of the proposed Ph.D. project must be presented to the candidate's Ph.D. committee. A doctoral dissertation is required, as well as an oral exam that consists of a defense of the dissertation and other topics determined by the candidate's committee.

Thesis: Thesis may be written in absentia.

SPECIAL EQUIPMENT, FACILITIES, OR PROGRAMS

The department has active research experimental and theoretical programs in condensed matter, biophysics, high energy and elementary particles, and programs in computational physics, photonics/biophotonics, statistical physics, and astrophysics and cosmology.

The department occupies a modern building with major research facilities on UB's North Campus, such as a helium liquefier, SQUID magnetometer, MBE systems, atomic and magnetic force microscopes, pulsed terahertz spectrometer, Raman spectrometer, and a 15 Tesla superconducting magnet.

Department members are users of several major national facilities, including the National Synchrotron Light Source at Brookhaven National Laboratories, the Tevatron at the Fermi National Accelerator Laboratory, the LHC at CERN, the National High Magnetic Field Laboratory (Tallahassee, Florida), the Center for Free Electron Laser Studies at the University of California, Santa Barbara, the W. M. Keck Foundation Free Electron Laser Center at Vanderbilt University (Nashville, Tennessee), and the Cornell Nanofabrication Facility.

The Physics Department has a joint B.S./M.S. in Computational Physics in cooperation with the Computer Science Department (CSE). A graduate student in physics can also pursue an Advanced Certificate in Computational Science or in Professional Science Management in Biophysics as additional credential.

Table B—Separately Budgeted Research Expenditures by Source of Support

Source of Support	Departmental Research	Physics-related Research Outside Department
Federal government	$3,007,276.82	
State/local government		
Non-profit organizations		
Business and industry		
Other	$368,742.72	
Total	**$3,376,019.54**	

Table C—Separately Budgeted Research Expenditures by Research Specialty

Research Specialty	No. of Grants	Expenditures ($)
Biophysics	7	$975,220.9
Condensed Matter Physics	29	$1,718,400.71
Low Temperature Physics	1	$107,975.36
Medical, Health Physics	–	
Particles and Fields	9	$511,637.58
Cosmology & String Theory	1	$62,784.99
Total	**47**	**$3,376,019.54**

FACULTY

Professor

Gasparini, Francis M., Ph.D., University of Minnesota, 1970. University at Buffalo Distinguished Professor; Moti Lal Rustgi Professor of Physics. *Condensed Matter Physics, Low Temperature Physics, Statistical & Thermal Physics.* Experimental low-temperature physics; phase transitions; quantum fluids.

Gonsalves, Richard J., Ph.D., Columbia University, 1976. *Computational Physics, High Energy Physics, Particles and Fields.* Theoretical high energy physics; computational physics.

Ho, John T., Ph.D., Massachusetts Institute of Technology, 1969. Vice Provost for Graduate Education and Dean of the Graduate School; SUNY Distinguished Professor. *Condensed Matter Physics.* Experimental condensed matter physics.

Hu, Xuedong, Ph.D., University of Michigan, 1996. Director of Graduate Studies. *Condensed Matter Physics*. Theoretical condensed matter physics; theoretical study of nanostructure physics; solid state quantum information processing.

Krotscheck, Eckhard, Ph.D., Universität zu Köln, 1974. Quantum many-body theory; theoretical condensed matter physics.

Luo, Hong, Ph.D., Purdue University, 1988. Department Chair. *Applied Physics, Condensed Matter Physics, Materials Science, Metallurgy*. Experimental condensed matter physics; molecular beam epitaxy; microscopy; spintronics; semiconductor nanostructures.

Markelz, Andrea, Ph.D., University of California, Santa Barbara, 1995. *Applied Physics, Atomic, Molecular, & Optical Physics, Biophysics, Condensed Matter Physics, Materials Science, Metallurgy, Optics*. Experimental protein dynamics using terahertz time domain spectroscopy and UV/Vis/IR ultrafast spectroscopy; experimental condensed matter physics: nanosystems spectroscopy and device development.

McCombe, Bruce D., Ph.D., Brown University, 1965. SUNY Distinguished Professor. *Applied Physics, Condensed Matter Physics, Low Temperature Physics, Materials Science, Metallurgy*. Experimental condensed matter physics; semiconductor physics; semiconductor nanostructures and nanoparticles; optical and infrared spectroscopy and magnetospectroscopy; electrical magneto-transport; spin effects in semiconductors.

Petrou, Athos, Ph.D., Purdue University, 1983. *Condensed Matter Physics*. Experimental solid state physics.

Prasad, Paras, Ph.D., University of Pennsylvania, 1971. SUNY Distinguished Professor; Department of Chemistry (joint appointment). *Atomic, Molecular, & Optical Physics, Condensed Matter Physics*. Theoretical photonics; ultrafast optical processes; nonlinear optics.

Ram, Michael, Ph.D., Columbia University, 1965. *Atmosphere, Space Physics, Cosmic Rays, Geophysics, Physics and other Science Education*. Atmospheric physics; climate change; theoretical physics.

Sen, Surajit, Ph.D., University of Georgia, 1990. *Acoustics, Biophysics, Computational Physics, Condensed Matter Physics, Engineering Physics/Science, Materials Science, Metallurgy, Mechanics, Physics and other Science Education, Statistical & Thermal Physics*. Theoretical non-equilibrium many-particle physics; nonlinear dynamics; granular materials; metamaterials; dust flow studies; battle problems; disease modeling; mathematical physics; science and math education at the middle school level.

Wackeroth, Doreen, Ph.D., University of Karlsruhe, 1995. Associate Chair. *High Energy Physics, Particles and Fields*. Theoretical particle physics; phenomenology of particle physics at present and future colliders; electroweak physics; perturbative quantum chromodynamics; supersymmetry.

Weinstein, Bernard A., Ph.D., Brown University, 1974. Director of Undergraduate Studies. *Applied Physics, Condensed Matter Physics, Materials Science, Metallurgy*. Experimental condensed matter physics; high-pressure properties of hard and soft solids; semiconductor physics; semiconductor nanostructures; oxides; optical spectroscopy; Raman scattering.

Associate Professor

Cerne, John, Ph.D., University of California, Santa Barbara, 1996. *Condensed Matter Physics, Materials Science, Metallurgy*. Experimental condensed matter physics; strongly correlated electronic materials; magnetic semiconductors; magnetic oxides; graphene; high-temperature superconductors; magneto-polarimetry; experimental biophysics.

Durbin, Steven M., Ph.D., Purdue University, 1994. *Condensed Matter Physics*. Experimental electronic properties and device applications of oxide and nitride semiconductors.

Ganapathy, Sambandamurthy, Ph.D., Indian Institute of Science, 2000. *Applied Physics, Condensed Matter Physics, Low Temperature Physics, Materials Science, Metallurgy*. Experi-

mental condensed matter physics; quantum transport in nanostructures; nanoelectronics; quantum phase transitions.

Han, Jong, Ph.D., Ohio State University, 1997. *Computational Physics, Condensed Matter Physics*. Theoretical condensed matter physics; quantum transport theory; strongly correlated systems; quantum simulations; nanoscale magnetism.

Iashvili, Ia, Ph.D., Humboldt University, Berlin, 2000. *High Energy Physics, Particles and Fields, Physics and other Science Education*. Experimental elementary particle physics; research, development, and construction of particle detectors; searches for Higgs and supersymmetric particles; precision measurements of particle properties at current and future accelerators.

Kharchilava, Avto, Ph.D., Tbilisi State University, 1990. *High Energy Physics, Particles and Fields, Physics and other Science Education*. Experimental elementary particle physics; research, development, and construction of particle detectors; searches for Higgs and supersymmetric particles; precision measurements of particle properties at current and future accelerators.

Kinney, William H., Ph.D., University of Colorado, 1996. *Astrophysics, Cosmology & String Theory, Particles and Fields*. Theoretical cosmology; high energy physics; astrophysics.

Stojkovic, Dejan, Ph.D., Case Western Reserve University, 2001. *Astrophysics, Cosmology & String Theory, High Energy Physics*. Theoretical cosmology; high energy physics; gravity; astrophysics.

Zeng, Hao, Ph.D., University of Nebraska-Lincoln, 2001. *Applied Physics, Condensed Matter Physics, Materials Science, Metallurgy*. Experimental condensed matter physics; nanoscale magnetism; spintronics; nanomaterial synthesis and self-assembly.

Zhang, Peihong, Ph.D., Pennsylvania State University, 2001. *Computational Physics, Condensed Matter Physics, Materials Science, Metallurgy*. Theoretical condensed matter physics; electronic structure theory; nanostructured materials; dilute magnetic semiconductors; wide gap semiconductors; electron-phonon renormalization in metals; quasi-particle properties in strongly correlated materials; high-performance computing.

Zutic, Igor, Ph.D., University of Minnesota, 1998. *Condensed Matter Physics*. Theoretical condensed matter physics; spin-polarized transport and spintronics; high-temperature and unconventional superconductivity; ferromagnetic semiconductors; quantum dots; theoretical nanoscience; computational physics.

Assistant Professor

Pralle, Arnd, Ph.D., Ludwig-Maximilians-University, Munich and European Molecular Biology Lab (EMBL), Heidelberg, 1999. *Applied Physics, Biophysics, Statistical & Thermal Physics*. Experimental biophysics; soft condensed matter physics; molecular and cellular mechanics and forces; spatiotemporal patterning; single-molecule spectroscopy.

Rappoccio, Salvatore, Ph.D., Harvard University, 2005. *High Energy Physics*. Experimental high energy physics.

Zheng, Wenjun, Ph.D., Stanford University, 2003. *Biophysics, Computational Physics*. Theoretical computational modeling of protein structures and dynamics.

Professor Emeritus

Fuda, Michael G., Ph.D., Rensselaer Polytechnic Institute, 1967. Relativistic quantum mechanics; pion-nucleon scattering; photo- and electro-production of mesons.

Fujita, Shigeji, Ph.D., University of Maryland, 1960. Statistical mechanics; many-body problems; quantum transport; superconductivity; quantum Hall effect.

Jain, Piyare L., Ph.D., University of Michigan, 1954. Experimental elementary particle; relativistic heavy-ion physics.

Lee, Yung Chang, Ph.D., University of Maryland, 1963. Condensed matter physics; many-body theory statistical mechanics; superconductivity; physics in confined systems.

Lin, Duo-Liang, Ph.D., Ohio State University, 1961. Theoretical condensed matter and optical physics.

Reichert, Jonathan F., Ph.D., University of Washington, 1962. Experimental condensed matter physics.

Research Professor

DeMarco, Michael, Ph.D., University of Cincinnati, 1981. Experimental condensed matter physics and Mossbauer effect.

Research Assistant Professor

Jones, Matthew D., Ph.D., University of Illinois, 1996. Theoretical condensed matter/computational physics.

Adjunct Professor

Bird, Jonathan P., Ph.D., University of Sussex, 1990. Experimental condensed matter physics; nanoelectronics; nanomaterials characterization.

Cartwright, Alexander N., Ph.D., University of Iowa, 1995. Experimental condensed matter physics; time-resolved optical spectroscopy and ultrafast optical measurements.

Dimock, Jonathan, Ph.D., Harvard University, 1971. Mathematical physics.

Mitin, Vladimir, Ph.D., Ukrainian Academy of Science, 1987. Condensed matter theory, modeling, and simulations; nano-electronic, microelectric, and optoelectronic devices and materials; transport and noise in heterostructures; thin films; quantum wells and quantum wires; material characterization; heat dissipation in low-dimensional structures and devices; particle, molecular dynamics, and Monte Carlo methods of simulation of thyristors; three-terminal lasers; photodetectors; terahertz generators and devices based on quantum wells, quantum wires, and wide-bandgap semiconductors.

Adjunct Associate Professor

Mac Isaac, Daniel L., Ph.D., Purdue University, 1994. *Physics and other Science Education*. Physics education.

Adjunct Assistant Professor

Wang, John, Ph.D., University of California, Berkeley, 2001. *Cosmology & String Theory*. Cosmology.

DEPARTMENTAL RESEARCH SPECIALTIES AND STAFF

Theoretical

Acoustics. Sen.
Astrophysics. Kinney, Stojkovic.
Atmosphere, Space Physics, Cosmic Rays. Ram.
Biophysics. Sen, Zheng.
Computational Physics. Gonsalves, Han, Sen, Zhang, Zheng.
Condensed Matter Physics. Han, Hu, Krotscheck, Prasad, Sen, Zhang, Zutic.
Cosmology & String Theory. Kinney, Stojkovic.
Engineering Physics/Science. Sen.
Geophysics. Ram.
High Energy Physics. Gonsalves, Stojkovic, Wackeroth.
Materials Science, Metallurgy. Sen, Zhang.
Mechanics. Sen.
Particles and Fields. Gonsalves, Kinney, Wackeroth.
Physics and other Science Education. Ram, Sen.
Statistical & Thermal Physics. Sen.

Experimental

Applied Physics. Ganapathy, Luo, Markelz, McCombe, Pralle, Weinstein, Zeng.
Atomic, Molecular, & Optical Physics. Markelz, Prasad.
Biophysics. Markelz, Pralle.
Condensed Matter Physics. Cerne, Durbin, Ganapathy, Gasparini, Ho, Luo, Markelz, McCombe, Petrou, Weinstein, Zeng.
High Energy Physics. Iashvili, Kharchilava, Rappoccio.
Low Temperature Physics. Ganapathy, Gasparini, McCombe.
Materials Science, Metallurgy. Cerne, Ganapathy, Luo, Markelz, McCombe, Weinstein, Zeng.
Optics. Markelz.
Particles and Fields. Iashvili, Kharchilava.
Physics and other Science Education. Iashvili, Kharchilava.
Statistical & Thermal Physics. Gasparini, Pralle.

View additional information about this department at
www.gradschoolshopper.com

UNIVERSITY OF ROCHESTER

DEPARTMENT OF PHYSICS AND ASTRONOMY

Rochester, New York 14627
http://www.pas.rochester.edu

General University Information

President: Joel Seligman
Dean of Graduate School: Margaret Kearney
University website: http://www.rochester.edu
Control: Private
Setting: Suburban
Total Faculty: 1,324
Total number of Students: 10,541
Total number of Graduate Students: 4,756

Department Information

Department Chairman: Dan Watson, Chair
Department Contact: Shirley Brignall, Administrative Assistant
Total full-time faculty: 26
Full-Time Graduate Students: 98
First-Year Graduate Students: 6
Female First-Year Students: 1
Total Post Doctorates: 24

Department Address

Bausch & Lomb Hall, Box 270171
Wilson Blvd.
Rochester, NY 14627
Phone: (585) 275-4344
Fax: (585) 273-3237
E-mail: grad@pas.rochester.edu
Website: http://www.pas.rochester.edu

ADMISSIONS

Admission Contact Information

Address admission inquiries to: Chair, Admissions Committee, Dept. of Physics and Astronomy
Phone: (585) 275-4356
E-mail: grad@pas.rochester.edu
Admissions website: http://www.pas.rochester.edu/graduate/applying.html

Application deadlines

Fall admission:
U.S. students: January 15   *Int'l. students*: January 15

Application fee

There is no application fee required.

Admissions information

For Fall of 2013:
Number of applicants: 301
Number admitted: 28
Number enrolled: 6

Admission requirements

Bachelor's degree requirements: A Bachelor's degree is required.

GRE requirements

The GRE is required.
No minimum required for application.

Advanced GRE requirements

The Advanced GRE is required.
No minimum required for application.

TOEFL requirements

The TOEFL exam is required for students from non-English-speaking countries.
No minimum required for application.

Other admissions information

Additional requirements: Our admission application can be found at: apply.grad.rochester.edu. We do ask for a personal statement, CV, 3 letters of recommendation submitted through the on-line application, GRE, GRE Physics and TOEFL for non-native English speakers.

Undergraduate preparation assumed: Classical Mechanics in J.B. Marion, Classical Dynamics of Particles and Systems; Electricity and Magnetism, for example, in D. J. Griffiths, Introduction to Electrodynamics; Thermodynamics, Kinetic Theory, and Statistical Mechanics, for example, by C. Kittel and H. Kroemer, Thermal Physics; Quantum Mechanics for example in R. L. Liboff, Introductory Quantum Mechanics. Mathematics, good knowledge of advanced calculus, ordinary differential equations; functions of complex variable, boundary value problems, modern algebra.

TUITION

Tuition year 2013-14:
Full-time students: $16,728 per semester
Full tuition scholarships are available. Applicants are considered automatically for scholarships; no additional scholarship application is required at this time.

Deferred tuition plan: Yes
Health insurance: Available at the cost of $2,616 per year.
Other academic fees: All active students pay a $10 per semester Graduate Organizing Group Fee. International Students pay an additional fee of $25.
Academic term: Semester

Teaching Assistants, Research Assistants, and Fellowships

Average stipend per academic year
Teaching Assistant: $25,000
Research Assistant: $25,000
All first-year graduate students will hold teaching assistant positions.

FINANCIAL AID

Loans

Loans are available for U.S. students.
Loans are not available for international students.
GAPSFAS application required: No
FAFSA application required: Yes

For further information

Address financial aid inquiries to: Financial Aid Office, Wallis Hall.
Phone: (585) 275-3226
Financial aid website: http://enrollment.rochester.edu/financial/grads

HOUSING

Availability of on-campus housing

Single students: Yes
Married students: Yes

For further information

Address housing inquiries to: University of Rochester, University Apartments Office, 020 Gates Wing, Susan B. Anthony Halls, Rochester, NY 14627.
Phone: (585) 275-5824
E-mail: uapts@reslife.rochester.edu
Housing aid website: http://www.rochester.edu/reslife

Table A—Faculty, Enrollments, and Degrees Granted

Research Specialty	2012-13 Faculty	Enrollment Fall 2012		Number of Degrees Granted 2012-13 (2006–13)		
		Master's	Doctorate	Master's	Terminal Master's	Doctorate
Astronomy	5	–	–	–	–	–
Astrophysics	4	–	13	3(4)	–	2(11)
Atomic, Molecular, & Optical Physics	10	–	30	3(33)	–	7(29)
Biophysics	2	–	3	–(3)	–	–(1)
Chemical Physics	–	–	–	–(3)	–(1)	–(4)
Condensed Matter Physics	12	–	12	1(7)	–	2(11)
Engineering Physics/Science	1	–	–	–(6)	–	–(3)
High Energy Physics	14	–	22	–(18)	–(1)	3(20)
Medical, Health Physics	–	–	–	–(1)	–	1(7)
Nuclear Physics	2	–	2	–(2)	–	1(1)
Plasma and Fusion	4	–	12	2(9)	–	1(8)
Non-specialized	–	–	24	2(13)	–	–
Other	–	–	1	–(12)	–(1)	–(1)
Total	54	–	119	11(111)	–(5)	17(96)
Full-time Grad. Stud.	–	–	119	–	–	–
First-year Grad. Stud.	–	–	22	–	–	–

GRADUATE DEGREE REQUIREMENTS

Master's: Master Degree Enroute to Ph.D. Only: 30 graduate credits; no minimum grade average required; no language or residency requirements; Master's exam and thesis required. Degrees awarded in physics only.

Doctorate: Ninety credit hours beyond Bachelor's degree in approved program required; no minimum GPA; minimum one year residency and full-time enrollment required; no language or computer language required; written preliminary-oral qualifying exams required; thesis and oral thesis exam required. Degrees available in physics or physics and astronomy.

Thesis: Thesis may be written in absentia with approval from Advisor and Dean of Graduate Studies.

Table B—Separately Budgeted Research Expenditures by Source of Support

Source of Support	Departmental Research	Physics-related Research Outside Department
Federal government	$7,635,046	
State/local government		
Non-profit organizations		
Business and industry		
Other	$26,198	
Total	**$7,661,244**	

Table C—Separately Budgeted Research Expenditures by Research Specialty

Research Specialty	No. of Grants	Expenditures ($)
Astrophysics	21	$1,210,274
Experimental Condensed Matter	1	$123,585
Theoretical Condensed Matter	2	$87,366
Nuclear Physics	1	$196,034
Experimental Quantum Optics	7	$867,258
Theoretical Quantum Optics	9	$814,033
Theoretical High Energy Physics	1	$212,191
Experimental High Energy Physics	22	$3,787,388
Physics and other Science Education	4	$363,115
Total	**68**	**$7,661,244**

FACULTY

Professor

Agrawal, G. P., Ph.D., Indian Institute of Technology, New Delhi, 1974. *Applied Physics, Optics.* Nonlinear photonics, lasers, optical communications.

Betti, R., Ph.D., Massachusetts Institute of Technology, 1992. *Engineering Physics/Science, Plasma and Fusion.* Theoretical plasma physics, nuclear and mechanical engineering, computational and plasma physics.

Bigelow, N. P., Ph.D., Cornell University, 1989. *Atomic, Molecular, & Optical Physics.* Experimental and theoretical quantum optics and quantum physics, studies of BEC, laser-cooled and trapped atoms.

Blackman, E. G., Ph.D., Harvard University, 1995. *Astronomy, Astrophysics, Biophysics, Plasma and Fusion.* Theoretical astrophysics, astrophysical plasmas and magnetic fields, accretion and ejection phenomena; relativistic and high-energy astrophysics.

Bocko, M. F., Ph.D., University of Rochester, 1984. *Electrical Engineering.* Superconducting electronics, quantum computing, musical acoustics, digital audio technology, sensors.

Bodek, A., Ph.D., Massachusetts Institute of Technology, 1972. *Nuclear Physics, Particles and Fields.* Experimental elementary particle physics, proton-antiproton collisions, QCD and structure functions, neutrino physics, electron scattering, and tile-fiber calorimetric detectors, Physics with W,Z and Higgs Bosons at the large Hadron Collider (LHC).

Boyd, R. W., Ph.D., University of California, Berkeley, 1977. *Atomic, Molecular, & Optical Physics, Optics.* Nonlinear optics.

Castner, T. G., Ph.D., University of Illinois, 1958. Emeritus. *Condensed Matter Physics, Electromagnetism, Low Temperature Physics, Statistical & Thermal Physics.* Experimental condensed matter physics, metal insulator transition.

Cline, D., Ph.D., University of Manchester, 1963. *Nuclear Physics.* Extreme states of nuclei pairing and shape correlations in nuclei.

Conwell, E., Ph.D., University of Chicago, 1948. Theoretical chemical physics, condensed matter physics, biological physics.

Das, A., Ph.D., Stony Brook University, 1977. *Particles and Fields.* Theoretical particle physics, finite temperature field theory, integrable systems, phenomenology, non-commutative field theory, and string/M theory.

Demina, R., Ph.D., Northeastern University, 1994. *High Energy Physics, Particles and Fields.* Experimental particle physics, proton-antiproton collisions, top and electroweak physics.

Douglass, D. H., Ph.D., Massachusetts Institute of Technology, 1959. *Applied Physics, Astrophysics, Atmosphere, Space Physics, Cosmic Rays, Condensed Matter Physics, Energy Sources & Environment, Geophysics, Low Temperature Physics, Materials Science, Metallurgy, Relativity & Gravitation.* Experimental condensed matter physics; climate change and pollution.

Duke, Charles B., Ph.D., Princeton University, 1963. *Condensed Matter Physics.* Theoretical condensed matter physics; geophysics and climate.

Eberly, J. H., Ph.D., Stanford University, 1962. *Atomic, Molecular, & Optical Physics.* Theoretical quantum optics, quantum entanglement, cavity QED, atoms in strong laser fields, nonlinear optical pulse propagation.

Ferbel, T., Ph.D., Yale University, 1963. *Particles and Fields.* Experimental elementary particle physics, studies of the top quark in hadronic collisions.

Forrest, W. J., Ph.D., University of California, San Diego, 1974. *Astronomy, Astrophysics, Electrical Engineering, Low Temperature Physics.* Observational astrophysics, infrared astronomy, stellar and planetary formation, low-mass stars and brown dwarfs, development of infrared detector arrays and instrumentation.

Foster, T. H., Ph.D., University of Rochester, 1990. Biological and medical physics.

Frank, A., Ph.D., University of Washington, Seattle, 1992. *Astronomy, Astrophysics, Computer Science, Plasma and Fusion.* Theoretical astrophysics, astrophysical plasmas, numerical hydrodynamics and magnetohydrodynamics.

Gao, Y., Ph.D., Purdue University, 1986. *Condensed Matter Physics.* Experimental condensed matter physics; surface physics.

Hagen, C. R., Ph.D., Massachusetts Institute of Technology, 1962. *Particles and Fields.* Theoretical elementary particle physics, quantum field theory, particularly 2+1 dimensional theories.

Helfer, H. L., Ph.D., University of Chicago, 1953. Emeritus. Theoretical astrophysics and plasma physics, high-energy astrophysics, dark matter in galactic halos.

Howell, John, Ph.D., Pennsylvania State University, 2000. *Applied Physics, Atomic, Molecular, & Optical Physics, Optics.* Experimental quantum optics and quantum physics, quantum cryptography and quantum computation.

Huizenga, J. R., Ph.D., University of Illinois, 1949. Emeritus. *Nuclear Physics*. Nuclear chemistry.

Knox, R. S., Ph.D., University of Rochester, 1958. Emeritus. *Atomic, Molecular, & Optical Physics, Biophysics, Chemical Physics, Condensed Matter Physics, Energy Sources & Environment, Statistical & Thermal Physics*. Theoretical biological physics and condensed matter physics; energy-balance models of climate.

Knox, W. H., Ph.D., University of Rochester, 1984. *Condensed Matter Physics, Medical, Health Physics, Optics*. Ultrafast sciences and technology, telecommunications, ultrafast biomedical optics, and optics education.

Koltun, D. S., Ph.D., Princeton University, 1961. Emeritus. *Nuclear Physics*. Theoretical nuclear physics, meson interactions with nuclei, many-body theory, electron scattering.

Manly, S. L., Ph.D., Columbia University, 1989. *Nuclear Physics, Particles and Fields*. Experimental relativistic heavy ion physics, experimental elementary particle physics.

McCrory, Robert L., Ph.D., Massachusetts Institute of Technology, 1973. *Accelerator, Particles and Fields, Plasma and Fusion*. Nuclear and mechanical engineering, computational hydrodynamics, physics of inertial fusion, National Nuclear Security policy.

McFarland, K. S., Ph.D., University of Chicago, 1994. *Nuclear Physics, Particles and Fields*. Experimental elementary particle physics, neutrino physics, electroweak unification, top quark properties.

Melissinos, A. C., Ph.D., Massachusetts Institute of Technology, 1958. *Particles and Fields*. Experimental particle physics, high intensity laser particle interactions, searches for relic gravitational radiation (retired).

Meyerhofer, D. D., Ph.D., Princeton University, 1987. *Plasma and Fusion*. Experimental plasma and laser physics; high-energy density physics and inertial confinement fusion; high-intensity laser-matter interaction experiments, quantum optics.

Milonni, Peter W., Ph.D., University of Rochester, 1974. *Atomic, Molecular, & Optical Physics, Electromagnetism, Optics*. Effects of the quantum vacuum, the Casimir force, laser-atom interactions, non-linear optics, and QED of dielectric materials.

Okubo, S., Ph.D., University of Rochester, 1958. Emeritus. *Particles and Fields*. Theoretical particle physics and mathematical physics; Lie and non-associative algebras.

Orr, L., Ph.D., University of Chicago, 1991. *Particles and Fields*. Theoretical elementary particle physics, phenomenology, quantum chromodynamics and electroweak physics.

Pipher, J. L., Ph.D., Cornell University, 1971. Emeritus. *Astronomy, Astrophysics*. Observational astrophysics, infrared astronomy, Galactic and extragalactic star formation, low-mass stars and brown dwarfs, development of infrared detector arrays and instrumentation, detector physics.

Quillen, Alice, Ph.D., California Institute of Technology, 1993. *Astronomy, Astrophysics*. Observational astrophysics, galactic structure and dynamics, active galactic nuclei, dynamics of planetary and protoplanetary systems, celestial mechanics.

Rajeev, S. G., Ph.D., Syracuse University, 1984. Math Physics. *Particles and Fields, Relativity & Gravitation, Statistical & Thermal Physics*. Theoretical particle physics; nonperturbative quantum field theory applied to strong interactions, mathematical physics.

Rothberg, L., Ph.D., Harvard University, 1983. *Chemical Physics*. Experimental chemical physics, organic electronics and biomolecular sensing.

Savedoff, M. P., Ph.D., Princeton University, 1957. Emeritus. Theoretical astrophysics, stellar interiors, interstellar matter, high-energy astrophysics.

Schröder, Wolf-Udo, Ph.D., Technical University of Darmstadt, 1971. *Nuclear Physics*. Experimental nuclear physics, dynamics of complex nuclear reactions, fundamental properties of nuclear matter, nuclear transmutation. Nuclear technology applications,nuclear plasma physics.

Shapir, Y., Ph.D., Tel Aviv University, 1981. *Applied Physics, Condensed Matter Physics, Nano Science and Technology, Polymer Physics/Science, Statistical & Thermal Physics*. Theoretical condensed matter physics, statistical mechanics; critical phenomena in ordered and disordered systems, fractal growth.

Simon, A., Ph.D., University of Rochester, 1950. Emeritus. *Energy Sources & Environment, Engineering Physics/Science, Nuclear Engineering, Plasma and Fusion*. Theoretical plasma physics, controlled thermonuclear fusion.

Slattery, P. F., Ph.D., Yale University, 1967. *Particles and Fields*. Experimental elementary particle physics, investigation of QCD via direct photon production, top quark studies and searches for new phenomena using high energy colliders.

Sobolewski, R., Ph.D., Polish Academy of Science, Warsaw, 1983. *Applied Physics, Condensed Matter Physics, Electrical Engineering, Electromagnetism, Engineering Physics/Science, Low Temperature Physics, Materials Science, Metallurgy, Nano Science and Technology, Optics*. Applied superconductivity, ultrafast electronics and optoelectronics.

Sproull, R. L., Ph.D., Cornell University, 1943. *Condensed Matter Physics*. Experimental condensed matter physics.

Stroud, Carlos R., Washington University, 1969. *Applied Physics, Atomic, Molecular, & Optical Physics, Optics*. Quantum optics, short-pulse excitation of atoms and molecules, quantum information, optical physics.

Tang, Ching W., Ph.D., Cornell University, 1975. *Applied Physics, Condensed Matter Physics, Energy Sources & Environment, Materials Science, Metallurgy, Nano Science and Technology*. Chemical and condensed matter physics, organic electronics.

Tarduno, J. A., Ph.D., Stanford University, 1987. *Astrophysics*. Geophysics, geomagnetism and geodynamics, plate tectonics and polar wander, geomagnetic reversals, fine particle magnetism, planetary astrophysics.

Teitel, S. L., Ph.D., Cornell University, 1981. *Condensed Matter Physics, Statistical & Thermal Physics*. Statistical and condensed matter physics.

Thomas, J. H., Ph.D., Purdue University, 1966. *Astrophysics, Fluids, Rheology*. Theoretical astrophysics, astrophysical plasmas, astrophysical fluid dynamics and magnetohydrodynamics, solar physics.

Thorndike, E. H., Ph.D., Harvard University, 1960. *Particles and Fields*. Experimental elementary particle physics, weak decays of bottom and charm quarks.

Van Horn, H. M., Ph.D., Cornell University, 1965. Emeritus. *Astrophysics*. Theoretical astrophysics, degenerate stars.

Watson, D. M., Ph.D., University of California, Berkeley, 1983. *Astronomy, Astrophysics*. Observational astrophysics, infrared astronomy, stellar and planetary formation, development of infrared detector arrays and instrumentation.

Wolf, E., Ph.D., University of Bristol, 1948. *Optics*. Theoretical optics. Statistical optics, theory of coherence and polarization, inverse scattering, diffraction tomography.

Wolfs, F. L. H., Ph.D., University of Chicago, 1987. *Astrophysics, Electrical Engineering, Nuclear Physics, Particles and Fields*. Experimental high-energy/nuclear physics, dark-matter searches.

Zhong, Jianhui, Ph.D., Brown University, 1988. *Biophysics, Medical, Health Physics*. Biological and medical physics. Advanced medical imaging, novel MRI techniques, physiological properties, biological tissues.

Associate Professor

Jordan, Andrew N., Ph.D., University of California, Santa Barbara, 2002. *Atomic, Molecular, & Optical Physics, Condensed Matter Physics, Statistical & Thermal Physics*. Theoretical quantum optics and condensed matter, quantum physics.

Ren, Chuang, Ph.D., University of Wisconsin-Madison, 1998. Associate Professor of Mechanical Engineering. *Plasma and Fusion*. Theoretical and computational plasma physics, controlled fusion.

Assistant Professor

Badolato, Antonio, Ph.D., University of California, Santa Barbara, 2005. *Applied Physics, Atomic, Molecular, & Optical Physics, Condensed Matter Physics*. Materials Science. Semiconductor quantum structures with emphasis on solid-state quantum optics and cavity quantum electrodynamics.

Dery, H., Ph.D., University of Technion, Haifa, Israel, 2004. *Applied Physics, Computer Science, Condensed Matter Physics, Electrical Engineering, Engineering Physics/Science, Nano Science and Technology*. Materials Science. Theory of semiconductor spin electronics.

Garcia-Bellido, Aran, Ph.D., Royal Holloway, University of London, 2002. *Particles and Fields*. Experimental particle physics, with interests in supersymmetry and physics of the top quark, and in particular electroweak production of single top quarks.

Mamajek, Eric E., Ph.D., University of Arizona, 2004. *Astronomy, Astrophysics*. Formation and evolution of stars, stellar groups, substellar objects, extrasolar planets, and circumstellar disks.

Visiting Professor

Visser, Taco D., Ph.D., University of Amsterdam, 1992. *Optics*. Theoretical optics, coherence theory, scattering and diffraction, surface plasmons.

DEPARTMENTAL RESEARCH SPECIALTIES AND STAFF

Theoretical

Astrophysics. Astrophysics. Astrophysical fluid dynamics and magnetohydrodynamics, astrophysical plasmas, computational astrophysics. Accretion disks and hypersonic outflows associated with young stars and degenerate objects. Evolution of protonplanetary disks. Celestial mechanics. Galactic dynamics. Stellar formation and death; planetary nebulae. High-energy astrophysics. Dark matter in galaxy haloes. Dynamo theory of magnetic-field generation in stars, galaxy disks, and planets. Origin and long-term behavior of Earth's magnetic field; field reversals and geodynamics. Physics of sunspots and solar magnetic flux tubes. Laboratory simulation of high-energy-density astrophysical plasmas. Blackman, Frank, Helfer, Quillen, Thomas. 2 postdoctoral research associates, collaborating faculty at the Laboratory of Laser Energetics. Blackman, Frank, Helfer, Quillen, Thomas.

Atomic, Molecular, & Optical Physics. Coherence phenomena in the interaction of light with matter. Subjects include coherent control, quantum entanglement, Bose-Einstein condensates, quantum dots, nanophotonics, atoms in intense laser fields, wave packet states of atoms and molecules, quantum imaging, single-cycle and half-cycle EM pulses, amplitude-coherent chemistry, correlation-induced spectral changes, solitons and inverse scattering theory, diffraction tomography. Six postdoctoral research associates. Agrawal, Badolato, Bigelow, Boyd, Eberly, Howell, Jordan, Milonni, Stroud, Visser, Wolf.

Condensed Matter Physics. Theory of thermodynamic and transport properties of disordered systems also near phase transitions. Theory of flux phases in type II superconductors and Josephson junctions. Scaling properties in clusters and polymers. Complex fluids, colloids, and biosystems. Theory of mesoscopic physics: electronic transport and noise properties. Quantum physics in the solid state – entanglement, measurement and information. Conwell, Dery, Duke, Jordan, Shapir, Teitel.

High Energy Physics. Nuclear Physics. String theory; matrix model, integrable models; Lagrangian field theory; thermofield theory; phenomenology, non-perturbative methods in field theory; structure functions of hadrons; supersymmetry, renormalization in quantum mechanics. One postdoctoral research associate. Das, Hagen, Koltun, Okubo, Orr, Rajeev.

Plasma and Fusion. Plasma Physics. Astrophysical plasmas in extreme environments. Fundamental processes common to laboratory and astrophysical plasmas. Space plasmas interaction of intense lasers with matter. High energy density physics with intense lasers. Hydrodynamic, magnetohydrodynamic and plasma instabilities; particle acceleration. Interaction of intense lasers with matter. Inertial confinement fusion and high-energy-density physics. Compression and heating of pellets to ignition relevant conditions. Blackman, Frank, Mamajek, McCrory, Meyerhofer, Simon.

Experimental

Astrophysics. Astrophysics. Observations with space-based and ground-based telescopes, and extensive archival work based upon our group's observations with the NASA Spitzer Space Telescope and ESA Herschel Space Observatory. Evolution of protostellar envelopes and protoplanetary disks, and the formation of stars and planets. Early stellar and planetary evolution. Brown dwarfs. Mineralogy of dust and chemistry of gases in protoplanetary and debris disks. Structure, dynamics, star formation histories, and chemical evolution of the Milky Way, other galaxies, and nearby young stellar associations. High resolution imaging of young stellar systems using adaptive optics. Evolution of stellar rotation, magnetic activity, solar wind, and interaction between Sun and magnetic field of early Earth. Origin and long-term behavior of Earth's magnetic field; field reversals and geodynamics. Development of infrared detector arrays and instruments for infrared astronomy. 2 postdoctoral research associates. Forrest, Mamajek, Pipher, Quillen, Tarduno, Watson.

Atomic, Molecular, & Optical Physics. Quantum interference effects and non-classical states of light, search for locality violation with photons, Bose-Einstein condensation, laser cooling and trapping of atoms and molecules, atom optics, generation of non-classical states of the atom, ultra-cold collisions, cold molecules, novel light sources. Nonlinear optics, quantum coherence, optical solitons. High intensity laser plasma and laser-atom interactions. Collaborating faculty at the Institute of Optics. One postdoctoral research associate. Bigelow, Boyd, Howell, W. Knox, Meyerhofer, Stroud.

Biological and Medical Physics. Biological and Medical Physics. Experimental and theoretical research in single molecule spectroscopy and manipulation, photodynamic therapy, diffusion tensor and functional MRI mechanisms and techniques, tissue optics, light scattering, biomolecular sensing, interactions of nanoparticles with biomolecules, and microscopy. Conwell, Foster, Rothberg, Zhong.

Condensed Matter Physics. Semiconductor heterojunctions; surface phenomena; synchrotron radiation photoemission; femtosecond time resolved photoemission; ultrafast dynamics in solids; interfaces in organic semiconductors; scanning tunneling microscopy; superconductivity and superconducting films; electron tunneling spectroscopy; metallic, magnetic, and superconducting nanowires; electron-beam lithography and mesoscopic structures. Badolato, Castner, Dery, Douglass, Duke, Gao, Jordan, R. Knox, W. Knox, Shapir, Sobolewski, Sproull, Tang, Teitel

High Energy Physics. Proton-antiproton colliding beam exper-
iment at the Fermi National Accelerator Laboratory—FNAL
(CDF and DZERO), proton-proton colliding beam experi-
ments at the CERN LHC Large Hadron Collider (LHC/CMS)
e+e−; colliding beams at Ithaca (LEPP/CLEO-C) and Beijing
(BES-III) Relativistic Heavy Ion Collisions at BNL (RHIC/
PHOBOS), Neutrino experiments (Fermilab/MINERvA) and
(Jparc/T2K. Electron scattering on nuclear targets (JLAB/JU-
PITER). R+D for future Linear Colliders. Dark matter search
at SUSEL (LUX). Thirteen senior scientists and research associ-
ates. Bodek, Demina, Ferbel, Garcia-Bellido, Manly, McFarland,
Melissinos, Slattery, Thorndike, Wolfs.

Nuclear Physics. Structure of exotic nuclei far from stability, relativistic heavy-ion physics, neutrino physics, electron scattering. Two postdoctoral research associate. Bodek, Cline, Huizenga, Koltun, Manly, Schröder, Wolfs.

Plasma and Fusion. Inertial confinement fusion and high-energy-
density physics with high energy and high intensity lasers.
Laser-plasma interactions and instabilities. Hydrodynamic instabilities. Equation of state at extreme conditions. Compression and heating of capsules to ignition relevant conditions. Plasma diagnostics. Meyerhofer.

View additional information about this department at
www.gradschoolshopper.com

UNIVERSITY OF ROCHESTER

THE INSTITUTE OF OPTICS

Rochester, New York 14627
http://www.optics.rochester.edu

General University Information
President: Joel Seligman
Dean of Graduate School: Margaret Kearney
University website: http://www.rochester.edu
Control: Private
Setting: Suburban
Total Faculty: 1,329
Total Graduate Faculty: Not separated.
Total number of Students: 10,111
Total number of Graduate Students: 2,723

Department Information
Department Chairman: Xi-Cheng Zhang, Chair
Department Contact: Lori Russell, Administrator
Total full-time faculty: 31
Total number of full-time equivalent positions: 20
Full-Time Graduate Students: 110
First-Year Graduate Students: 38
Female First-Year Students: 5
Total Post Doctorates: 8

Department Address
The Institute of Optics, University of Rochester,
275 Hutchison Road
Rochester, NY 14627
Phone: (585) 275-2322
Fax: (585) 244-4936
E-mail: gradadmissions@optics.rochester.edu
Website: http://www.optics.rochester.edu

ADMISSIONS

Admission Contact Information
Address admission inquiries to: Maria Schnitzler, Administrator,
Optics Graduate Admissions Committee
Phone: (585) 273-1155

E-mail: gradadmissions@optics.rochester.edu
Admissions website: http://www.optics.rochester.edu

Application deadlines
Fall admission:
U.S. students: February 1   *Int'l. students*: February 1

Application fee
U.S. students: $60   *Int'l. students*: $60

Admissions information
For Fall of 2013:
Number of applicants: 193
Number admitted: 101
Number enrolled: 38

Admission requirements
Bachelor's degree requirements: Bachelor's degree in physics
or engineering is required.

GRE requirements
The GRE is required.

Advanced GRE requirements
The Advanced GRE is not required.

TOEFL requirements
The TOEFL exam is required for students from non-English-
speaking countries.
PBT score: 620
iBT score: 105

Other admissions information
Additional requirements: The average GPA for admitted M.S.
students is 3.4/4.0. The average GPA for admitted Ph.D. students is 3.7/4.0. The average GRE scores for admitted students are 80th percentile in all categories.
The minimum accepted computer-based TOEFL score is 260.
IELTS is also accepted with a minimum score of 7.8.

TUITION

Tuition year 2013–14:
Tuition for out-of-state residents
  Full-time students: $45,372 annual
Tuition is waived for Ph.D. students.
Credit hours per semester to be considered full-time: 12
Deferred tuition plan: Yes
Health insurance: Available at the cost of $2124 per year.
Academic term: Semester
Number of first-year students who received full tuition waivers: 17
Number of first-year students who received partial tuition waivers: 21

Teaching Assistants, Research Assistants, and Fellowships

Number of first-year
  Teaching Assistants: 4
  Fellowship students: 4
Average stipend per academic year
  Teaching Assistant: $29,000
  Research Assistant: $29,000
  Fellowship student: $30,000

FINANCIAL AID

Application deadlines

Fall admission:
U.S. students: January 15

Loans

Loans are available for U.S. students.
Loans are not available for international students.
GAPSFAS application required: No
FAFSA application required: Yes

For further information

Address financial aid inquiries to: Administrator, Optics Graduate Admissions Committee.
Phone: (585) 275-3226
E-mail: elisabeth.carosa@rochester.edu
Financial aid website: http://www.enrollment.rochester.edu/financial

HOUSING

Availability of on-campus housing

Single students: Yes
Married students: Yes

For further information

Address housing inquiries to: University Apartments Office, 1351 Mt. Hope Ave., Rochester, NY 14620.
Phone: (585) 275-5824
E-mail: uapts@reslife.rocheter.edu
Housing aid website: http://www.rochester.edu/reslife

Table A—Faculty, Enrollments, and Degrees Granted

Research Specialty	2012-13 Faculty	Enrollment Fall 2012 Master's	Enrollment Fall 2012 Doctorate	Number of Degrees Granted 2012–13 (2006–13) Master's	Number of Degrees Granted 2012–13 (2006–13) Terminal Master's	Number of Degrees Granted 2012–13 (2006–13) Doctorate
Optics	32	18	94	12(63)	12(52)	13(56)
Total	32	18	94	26(114)	20(92)	25(99)
Full-time Grad. Stud.	–	11	93	–	–	–
First-year Grad. Stud.	–	12	13	–	–	–

GRADUATE DEGREE REQUIREMENTS

Master's: M.S. degrees require 30 hours of coursework, including 16 hours of required core courses. The M.S. degrees are normally completed in 9–12 months. A thesis-based MS degree is normally completed in 18-24 months. There are no residence or foreign language requirements. In a co-op program, students take the first semester of courses, work full-time for 12 months, and then return to campus for the final semester of classes.

Doctorate: General requirements: one year of full-time residence, 90 hours of graduate work (60 hours beyond the M.S.), two semesters of teaching assistantship, successful completion of a written preliminary examination and an oral qualifying examination, and completion and defense of a doctoral dissertation. There is no language requirement.

Thesis: Thesis may not be written in absentia.

SPECIAL EQUIPMENT, FACILITIES, OR PROGRAMS

Instruction is offered in optical instrumentation and design, quantum optics and electronics, laser engineering, optics of thin films, electro-optics, holography, interferometry, and most other areas of optical physics and engineering. Well-equipped laboratories allow student thesis research in such areas as ultra-high-resolution dye laser spectroscopy, semiconductor lasers, optical physics, nano-optics, optical communications, fiber optics, imaging, nonlinear optics, diffractive optics, gradient index optics, interferometry, image processing, optical materials, and high-power laser physics. In addition to extensive facilities within the Institute, thesis research may be carried out at the Laboratory for Laser Energetics, the School of Medicine and Dentistry, and the Center for Visual Science. Joint projects applying optical techniques in all of these areas are currently under way.

Table B—Separately Budgeted Research Expenditures by Source of Support

Source of Support	Departmental Research	Physics-related Research Outside Department
Federal government	$6,501,751.02	
State/local government	$91,612.62	
Non-profit organizations	$664,668.26	
Business and industry	$958,517.21	
Other	$113,886.81	
Total	$8,330,435.92	

FACULTY

Professor

Agrawal, Govind, Ph.D., Indian Institute of Technology, 1974. *Optics*. Semiconductor lasers and amplifiers; nonlinear optical phenomena; optical fiber communications.

Bigelow, Nicholas, Ph.D., Cornell University, 1989. Chair, Physics Department. *Atomic, Molecular, & Optical Physics, Optics*. Quantum optics and quantum physics.

Boyd, Robert W., Ph.D., University of California, Berkeley, 1977. *Atomic, Molecular, & Optical Physics, Optics*. Nonlinear optics; infrared detection and generation.

Brown, Thomas, Ph.D., University of Rochester, 1987. Director, Robert E. Hopkins Center for Optical Design and Engineering. *Nano Science and Technology, Optics*. Integrated optics; fiber optics: optical properties of solids; quantum electronics.

Eberly, Joseph H., Ph.D., Stanford University, 1962. *Atomic, Molecular, & Optical Physics, Optics*. Multiphoton processes; quantum electrodynamics; resonant interaction of light with atoms and molecules.

Fauchet, Philippe, Ph.D., Stanford University, 1984. *Electrical Engineering, Nano Science and Technology, Optics*. Optical processes in solids; femtosecond laser spectroscopy; materials science and device applications of light-emitting silicon.

Fienup, James R., Ph.D., Stanford University, 1975. *Optics*. Phase retrieval; unconventional imaging; image processing; wave-front sensing.

Foster, Thomas, Ph.D., University of Rochester. *Medical, Health Physics, Optics*. Medical optics; photodynamic therapy.

George, Nicholas, Ph.D., California Institute of Technology, 1959. *Electrical Engineering, Electromagnetism, Optics*. Optical systems; speckle; pattern recognition.

Jacobs, Stephen D., University of Rochester, 1976. *Optics*. Optical materials.

Knox, Wayne, Ph.D., University of Rochester, 1983. Associate Dean of Education and New Initiatives. Ultrafast science and technology; telecommunications; optoelectronics.

Moore, Duncan T., Ph.D., University of Rochester, 1974. Vice Provost for Entrepreneurship. *Optics*. Geometrical optics; optical instrumentation; gradient index glass; interferometry; medical optics.

Novotny, Lukas, Ph.D., Swiss Federal Institute of Technology. Chair, Graduate Admissions Committee. *Atomic, Molecular, & Optical Physics, Nano Science and Technology, Optics*. Nano-optics; light-matter interactions on the subwavelength scale.

Rolland, Jannick, Ph.D., University of Arizona, 1990. Director, Robert E. Hopkins Center for Optical Design and Engineering. *Optics*. Optical instrumentation and system engineering.

Stroud, Carlos R., Ph.D., Washington University, St. Louis, 1969. *Atomic, Molecular, & Optical Physics, Optics*. Quantum optics; short-pulse excitation of atoms and molecules.

Teegarden, Kenneth J., Ph.D., University of Illinois, 1954. *Optics*. Optical properties of materials.

Wicks, Gary, Ph.D., Cornell University, 1981. Director, The Institute of Optics. *Nano Science and Technology, Optics*. III–V semiconductors-epitaxial growth; optical properties; optical devices.

Williams, David, Ph.D., University of California, San Diego, 1979. Director, Rochester's Center for Visual Science. *Medical, Health Physics, Optics*. Sensitivity and resolution of the human visual system to patterns that are modulated in wavelength, space, and time.

Wolf, Emil, Ph.D., University of Bristol, 1948. *Optics*. Electromagnetic theory and physical optics; diffraction and theory of partial coherence.

Associate Professor

Alonso, Miguel, Ph.D., University of Rochester, 1996. Chair, Graduate Studies Committee. *Applied Mathematics, Optics*. Mathematical models for wave propagation; theory of partial coherence; connection between the ray and wave models.

Bentley, Julie, Ph.D., University of Rochester, 1995. *Optics*. Leng design.

Berger, Andrew, Ph.D., Massachusetts Institute of Technology. Chair, Undergraduate Studies Committee. *Medical, Health Physics, Optics*. Biomedical optics; Raman spectroscopy; optical analysis of blood and tissue.

Guo, Chunlei, Ph.D., University of Connecticut. *Atomic, Molecular, & Optical Physics, Optics*. High-intensity laser interactions with matter.

Krauss, Todd, Ph.D., Cornell University, 1998. *Chemical Physics, Nano Science and Technology, Optics*. Nanoscale materials and devices.

Marciante, John R., Ph.D., University of Rochester. *Optics*. Lasers, waveguide, and fiber optics.

Wolf, Seka, Ph.D., University of Texas. *Optics*. Lasers.

Yoon, Geunyoung, Ph.D., Osaka University. *Medical, Health Physics, Optics*. Biomedical and visual optics; adaptive optics.

Zavislan, James, Ph.D., University of Rochester. Director, Center for Institute Ventures. *Medical, Health Physics, Optics*. Optical engineering; medical optical instrumentation.

Assistant Professor

Ellis, Jonathan D., Ph.D., Delft University of Technology, 2010. *Optics*. High-precision optical metrology.

Lin, Qiang, Ph.D., University of Rochester, 2006. *Electrical Engineering, Nano Science and Technology, Optics*. Nonlinear optics; quantum optics; nanoscopic photonic structures.

Vamivakas, Nick, Ph.D., Boston University, 2007. *Atomic, Molecular, & Optical Physics, Nano Science and Technology, Optics*. Solid-state quantum optics and information science; nanoscale optics-based sensing.

DEPARTMENTAL RESEARCH SPECIALTIES AND STAFF

Theoretical

Optics. Agrawal, Alonso, Bentley, Berger, Bigelow, Boyd, Brown, Eberly, Ellis, Fauchet, Fienup, Foster, George, Guo, Jacobs, Krauss, Lin, Marciante, Moore, Novotny, Rolland, Stroud, Teegarden, Vamivakas, Wicks, Williams, Emil Wolf, Seka Wolf, Yoon, Zavislan.

Experimental

Atomic, Molecular, & Optical Physics. Berger, Bigelow, Boyd, Brown, Eberly, Fienup, Foster, Guo, Marciante, Moore, Novotny, Rolland, Stroud, Vamivakas, Wicks, Williams, Yoon, Zavislan.

View additional information about this department at
www.gradschoolshopper.com

APPALACHIAN STATE UNIVERSITY

DEPARTMENT OF PHYSICS AND ASTRONOMY

Boone, North Carolina 28608
http://physics.appstate.edu

General University Information
Chancellor: Kenneth E. Peacock
Dean of Graduate School: Edelma D. Huntley
University website: http://www.appstate.edu
Control: Public
Setting: Rural
Total Faculty: 898
Total Graduate Faculty: 629
Total number of Students: 16,168
Total number of Graduate Students: 992

Department Information
Department Chairman: Michael Briley, Chair
Department Contact: Dr. Sid Clements, Graduate Program
  Coordinator
  Total full-time faculty: 24
  Full-Time Graduate Students: 24
  First-Year Graduate Students: 16
  Female First-Year Students: 5
  Total Post Doctorates: 1

Department Address
525 Rivers Street
ASU Box 32106
Boone, NC 28608
Phone: (828) 262-2447
Fax: (828) 262-2049
E-mail: clementsjs@appstate.edu
Website: http://physics.appstate.edu

ADMISSIONS

Admission Contact Information
Address admission inquiries to: Office of Graduate Admissions,
  ASU Box 32068, Boone, NC 28608
Phone: (828) 262-2130
Admissions website: http://graduate.appstate.edu

Application deadlines
Fall admission:
U.S. students: July 1   *Int'l. students*: April 1
Spring admission:
U.S. students: November 1   *Int'l. students*: August 1

Application fee
U.S. students: $55

Admissions information
For Fall of 2013:
  Number of applicants: 18
  Number admitted: 9
  Number enrolled: 9

Admission requirements
Bachelor's degree requirements: Bachelor's degree in physics
  or a related discipline is required, with an undergraduate GPA
  of 3.0 or better preferred.

GRE requirements
The GRE is required.

Advanced GRE requirements
The Advanced GRE is not required.

TOEFL requirements
The TOEFL exam is required for students from non-English-
  speaking countries.
PBT score: 550

Other admissions information
Additional requirements: Recommended TOEFL score is 580 or
  better.

TUITION

Tuition year 2012–13:
Tuition for in-state residents
  Full-time students: $6,557 annual
  Part-time students: $879.95 other
Tuition for out-of-state residents
  Full-time students: $18,369 annual
  Part-time students: $2,356.45 other
In-state rate for 0-2 hours per semester: $879.95. Out-of-state
  rate for 0-2 hours per semester: $2,356.45.
Credit hours per semester to be considered full-time: 9
Deferred tuition plan: No
Health insurance: Not available.
Other academic fees: $ (included in above).
Academic term: Semester
Number of first-year students who received partial tuition waivers: 2

Teaching Assistants, Research Assistants, and Fellowships
Number of first-year
  Teaching Assistants: 11
  Research Assistants: 4
  Fellowship students: 1
Average stipend per academic year
  Teaching Assistant: $12,000
  Research Assistant: $15,000
  Fellowship student: $15,000

FINANCIAL AID

Application deadlines
Fall admission:
U.S. students: July 1   *Int'l. students*: April 1
Spring admission:
U.S. students: November 1   *Int'l. students*: August 1

Loans
Loans are available for U.S. students.
Loans are not available for international students.
GAPSFAS application required: No
FAFSA application required: Yes

For further information
Address financial aid inquiries to: Financial Aid Office, ASU Box
  32068, Boone, NC 28608.
Financial aid website: http://financialaid.appstate.edu

HOUSING

Availability of on-campus housing
Single students: Yes
Married students: Yes

For further information
Address housing inquiries to: Housing Office, P.O. Box 32111, Boone, NC 28608.
Housing aid website: http://housing.appstate.edu

Table A—Faculty, Enrollments, and Degrees Granted

Research Specialty	2012-13 Faculty	Enrollment Fall 2012		Number of Degrees Granted 2012-13 (2008-13)		
		Master's	Doctorate	Master's	Terminal Master's	Doctorate
Astronomy	5	1	–	1(2)	–	–
Astrophysics	2	1	–	–	–	–
Atmosphere, Space Physics, Cosmic Rays	2	3	–	2(4)	–	–
Atomic, Molecular, & Optical Physics	2	1	–	1(4)	–	–
Biophysics	2	–	–	–	–	–
Chemical Physics	1	–	–	1(1)	–	–
Computational Physics	3	1	–	1(1)	–	–
Computer Science	1	–	–	–	–	–
Condensed Matter Physics	3	1	–	-(2)	–	–
Electrical Engineering	2	1	–	1(1)	–	–
Energy Sources & Environment	2	1	–	1(1)	–	–
Engineering Physics/Science	5	4	–	2(8)	–	–
Fluids, Rheology	1	1	–	1(2)	–	–
Medical, Health Physics	1	–	–	-(1)	–	–
Nano Science and Technology	4	2	–	1(2)	–	–
Optics	3	3	–	1(2)	–	–
Physics and other Science Education	2	–	–	–	–	–
Total	40	20	–	13(31)	–	–
Full-time Grad. Stud.	–	23	–	–	–	–
First-year Grad. Stud.	–	9	–	–	–	–

GRADUATE DEGREE REQUIREMENTS

Master's: Master's in Engineering Physics - Minimum 36 credit hours, or 30 credit hours with thesis. Comprehensive exam is required. Professional Science Master's in Instrumentation & Automation or Nanoscience for Advanced Materials - 36 hours. Internship required. No comprehensive exam. Dual-degree: Professional Science Master's and Master's of Business Administration - Up to 18 hours of the 36-hour Professional Science Master's coursework is applied directly toward the 36-hour MBA program. 3 years, 2 internships. No comprehensive exam.

SPECIAL EQUIPMENT, FACILITIES, OR PROGRAMS

Students enjoy personal attention in an informal atmosphere, where the primary goal is quality teaching based on a collegial rapport between students and teachers. Modern laboratory facilities provide invaluable hands-on experience in cutting-edge research projects. Currently, these include observational astronomy and astrophysics at Dark Sky Observatory; ultrahigh vacuum technology and microscopy, including time-of-flight mass spectroscopy, time-of-flight secondary ion mass spectrometry, and an ion-storage facility; state-of-the-art optics laboratories, including AppalAir atmospheric optical studies, biophotonics, Raman spectroscopy, and optical tweezers; an applied electrostatics laboratory; surface analysis for nanoscale systems and materials science; a cryopumped thin film vacuum deposition system; and microscopy, including (SEM), (FIB), (AFM), (STM), and x-ray microanalysis.

Table B—Separately Budgeted Research Expenditures by Source of Support

Source of Support	Departmental Research	Physics-related Research Outside Department
Federal government	$154,915	$25,062
State/local government	$70,982	
Non-profit organizations		
Business and industry	$36,000	
Other		
Total	$261,897	$25,062

Table C—Separately Budgeted Research Expenditures by Research Specialty

Research Specialty	No. of Grants	Expenditures ($)
Astronomy	4	$64,748
Astrophysics	1	$49,253
Atmosphere, Space Physics, Cosmic Rays	1	$40,914
Biophysics	2	$67,258
Engineering Physics/Science	1	$36,000
Nano Science and Technology	3	$28,786
Total	12	$286,959

FACULTY

Professor

Allen, Patricia E., Ph.D., Iowa State University. Physics education pedagogy, surface physics.

Briley, Michael M., Ph.D., University of Maryland. Department Chair. Stellar spectroscopy/photometry, abundances and populations.

Calamai, Anthony G., Ph.D., North Carolina State University. Dean, College of Arts & Sciences. Experimental atomic, molecular, and optical physics; laboratory astrophysics.

Caton, Daniel B., Ph.D., University of Florida. Computer applications to astronomical instrumentation; photoelectric photometry of eclipsing binary stars.

Clements, J. Sid, Ph.D., Florida State University. Experimental applied electrostatics (aerospace and industrial); electrical discharges; electronic instrumentation.

Gray, Richard O., Ph.D., University of Toronto. Fellow, Royal Astronomical Society. Stellar spectroscopy/photometry.

Mamola, Karl C., Ph.D., Dartmouth College. Editor, *The Physics Teacher* magazine. Spectroscopy, thin film physics.

Pollock, Joseph T., Ph.D., University of Florida. Quasars; electronic imaging; asteroids.

Russell, Phillip E., Ph.D., University of Florida. Distinguished Professor of Science Education. Nanoscience, ion and electron microscopy.

Associate Professor

Coffey, Tonya S., Ph.D., North Carolina State University. Nanotribology; tribology; ultra-high vacuum technology; microscopy; microanalysis.

Sherman, James P., Ph.D., Colorado State University. Optics, laser physics, and applications to environmental physics.

Thaxton, Christopher S., Ph.D., North Carolina State University. Director of Environmental Studies Program. Geophysics, computational physics, sediment transport and erosion studies, electronics, computer interfacing.

Assistant Professor

Burris, Jennifer L., Ph.D., Colorado State University. Raman spectroscopy, luminescence, high-pressure spectroscopy.

Conrad, Brad R., Ph.D., University of Maryland. Organic electronics, photovoltaics, nanoscience.

Hester, Brooke C., Ph.D., University of Maryland. Optical trapping and extinction resonance.

Smith, Rachel L., Ph.D., University of California, Los Angeles. Director, Astronomy & Space Observation Research Laboratory. Nature Research Center, North Carolina Museum of Natural Sciences. Geochemistry, protostellar objects, cosmochemistry.

Emeritus

Rokoske, Thomas L., Ph.D., Auburn University. Electronics; microcomputer applications; robotics; electrical conduction mechanisms in thin films.

Visiting Professor

Allen, John E., Ph.D., University of Florida. Optical physics.

Germinario, Louis T., Ph.D., Catholic University of America. Physical chemistry, biophysics, microscopy and nanophysics.

Lecturer

Cecile, Danny J., Ph.D., Duke University. Quantum chromodynamics, lattice QCD modeling.

Genberg, Richard W., Ph.D., Case Institute of Technology. Superconductivity, low-temperature magnetism.

Sherman, Leah B., Ph.D., University of Michigan. Optoelectronics.

Postdoctoral Research Associate

Huston, Shawn M., Ph.D., North Carolina State University. Nanoscience, ion and electron microscopy, materials science.

View additional information about this department at
www.gradschoolshopper.com

DUKE UNIVERSITY

DEPARTMENT OF PHYSICS

Durham, North Carolina 27708-0305
http://www.phy.duke.edu

General University Information

President: Richard H. Brodhead
Dean of Graduate School: Paula D. McClain
University website: http://www.duke.edu
Control: Private
Setting: Suburban
Total Faculty: 2,303
Total Graduate Faculty: 1,301
Total number of Students: 14,621
Total number of Graduate Students: 7,977

Department Information

Department Chairman: Haiyan Gao, Chair
Department Contact: Randall S. Best, Administrative Manager
  Total full-time faculty: 45
  Total number of full-time equivalent positions: 57
  Full-Time Graduate Students: 76
  First-Year Graduate Students: 15
  Female First-Year Students: 2
  Total Post Doctorates: 22

Department Address

139 Physics Building, Box 90305
Science Drive, Duke University
Durham, NC 27708-0305
Phone: (919) 660-2500
Fax: (919) 660-2525
E-mail: rb37@phy.duke.edu
Website: http://www.phy.duke.edu

ADMISSIONS

Admission Contact Information

Address admission inquiries to: Shailesh Chandrasekharan, Director of Graduate Studies, Department of Physics, PO Box 90305, Duke University, Durham, NC 27708-0305
Phone: (919) 660-2593
E-mail: sch@phy.duke.edu
Admissions website: http://www.gradschool.duke.edu

Application deadlines

Fall admission:
U.S. students: December 31   *Int'l. students*: December 31

Application fee

U.S. students: $80   *Int'l. students*: $80

All application materials must be received by the deadline. Applications are normally reviewed by the Graduate Admissions Committee. Applications received earlier may be reviewed by other faculty and be recommended to the committee.

Admissions information

For Fall of 2013:
  Number of applicants: 244
  Number admitted: 52
  Number enrolled: 15

Admission requirements

Bachelor's degree requirements: A Bachelor's degree in physics or related subject is required.
Minimum undergraduate GPA: 3.0

GRE requirements

The GRE is required.

Mean GRE score range (25th–75th percentile): 0

While there are no official minimums for the GRE scores, students who obtain a scores below 160 in Quantitative, 140 in Verbal and 3.0 in Writing are rarely admitted and only under exceptional circumstances.

Advanced GRE requirements

The Advanced GRE is required.

Minimum accepted Advanced GRE score: 0

Mean Advanced GRE score range (25th–75th percentile): 0

While there is no official minimum for the Physics GRE score, students who receive a score of less than 600 are admitted only rarely and only under exceptional circumstances.

TOEFL requirements

The TOEFL exam is required for students from non-English-speaking countries.

PBT score: 560

iBT score: 83

Other admissions information

Additional requirements: The average GRE scores for 2013–14 admissions were verbal–158; quantitative–166; writing–4.0. The average GRE subject score for 2012–13 admissions was 859.

Undergraduate preparation assumed: Marion and Thornton, Classical Dynamics of Particles and Systems; Griffiths, Introduction to Electrodynamics; Kittel and Kroemer, Thermal Physics; Bernstein et al, Modern Physics; Shankar, Principles of Quantum Mechanics.

TUITION

Tuition year 2013–14:

Full-time students: $44,000 annual

Most students receive tuition waivers through the financial assistance package when admitted.

Credit hours per semester to be considered full-time: 9

Deferred tuition plan: Yes

Health insurance: Yes, No Charge.

Other academic fees: $40 (one time transcript charge). $309 per semester as health fee.

Academic term: Semester

Number of first-year students who received full tuition waivers: 15

Teaching Assistants, Research Assistants, and Fellowships

Number of first-year

Teaching Assistants: 15

Fellowship students: 4

Average stipend per academic year

Teaching Assistant: $28,773

Research Assistant: $28,773

Fellowship student: $31,773

FINANCIAL AID

Loans

Loans are available for U.S. students.

Loans are available for international students.

GAPSFAS application required: No

FAFSA application required: Yes

For further information

Address financial aid inquiries to: Ms. Lisa Alfman, Financial Aid Officer, Graduate School Office, 03 Allen Building.

Phone: (919) 681-3247

E-mail: lisa.alfman@duke.edu

Financial aid website: http://gradschool.duke.edu/financial_support/index.php

HOUSING

Availability of on-campus housing

Single students: Yes

Married students: Yes

For further information

Address housing inquiries to: Dept. of Housing Management, 218 Alexander, Apt. B, Durham, NC 27708-0451.

Phone: 919-684-4304

Table A—Faculty, Enrollments, and Degrees Granted

Research Specialty	2013–14 Faculty	Enrollment Fall 2013		Number of Degrees Granted 2012–13 (2008–13)		
		Master's	Doctorate	Master's	Terminal Master's	Doctorate
Atomic, Molecular, & Optical Physics	4	–	6	–	–	2(9)
Biophysics	5	–	3	–	–	–(1)
Cosmology & String Theory	4	–	1	–	–(1)	–
High Energy Physics	7	–	9	–	1(2)	1(3)
Medical, Health Physics	4	–	2	–	–	–
Nano Science and Technology	7	–	5	–	–(1)	–(5)
Nanophysics	5	–	2	–	–	1(3)
Nonlinear Dynamics and Complex Systems	10	–	5	–	–	4(7)
Nuclear Physics	5	–	10	–	–	–(15)
Particles and Fields	6	–	10	–	1(1)	–(4)
Photonics and Quantum Information	5	–	6	–	–	–
Physics of Beams	2	–	2	–	–	1(3)
Non-specialized	1	–	15	–	–	–
Total	65	–	76	–	2(5)	9(50)
Full-time Grad. Stud.	–	–	76	–	–	–
First-year Grad. Stud.	–	–	15	–	–	–

GRADUATE DEGREE REQUIREMENTS

Master's: Master of Arts (M.A.): At least 9 graduate courses (or sufficient placement exam performance), 3.0 GPA average. Oral final examination. Master of Science (M.S.): Same as M.A. plus written thesis. Final examination on thesis.

Doctorate: Doctor of Philosophy (Ph.D.): Same as M.A. plus preliminary oral exam before dissertation work. Written dissertation. Final examination on dissertation.

SPECIAL EQUIPMENT, FACILITIES, OR PROGRAMS

The world-renowned Duke Free Electron Laser laboratory is an integral part of the department. Duke nuclear physics plays a leading role as a part of the Triangle Universities Nuclear Laboratory (TUNL) located on the Duke Campus.

Table B—Separately Budgeted Research Expenditures by Source of Support

Source of Support	Departmental Research	Physics-related Research Outside Department
Federal government	$12,130,971	
State/local government		
Non-profit organizations		
Business and industry	$92,133	
Other		
Total	**$12,223,104**	

Table C—Separately Budgeted Research Expenditures by Research Specialty

Research Specialty	No. of Grants	Expenditures ($)
Atomic, Molecular, & Optical Physics	11	$776,760
Biophysics	3	$166,536
Nanophysics	4	$347,192
Nano Science and Technology	6	$64,558
Nuclear Physics	11	$4,728,974
Photonics and Quantum Information	11	$1,632,261
Particles and Fields	7	$869,662
Physics of Beams	11	$701,815
Nonlinear Dynamics and Complex Systems	16	$1,010,040
Cosmology & String Theory	1	$348
High Energy Physics	20	$1,924,958
Total	**101**	**$12,223,104**

FACULTY

Professor

Aspinwall, Paul, Ph.D., University of Oxford, 1988. Gravity and string theory (primary appointment: mathematics).

Baranger, Harold U., Ph.D., Cornell University, 1986. Theoretical condensed matter physics; nanophysics.

Bass, Steffen, Ph.D., J. W. Goethe University, 1997. *Computational Physics, Nuclear Physics, Particles and Fields, Statistical & Thermal Physics*. Theoretical nuclear and particle physics; relativistic heavy-ion collisions.

Behringer, Robert P., Ph.D., Duke University, 1975. *Condensed Matter Physics, Nonlinear Dynamics and Complex Systems*. Experimental condensed matter physics; statistical physics; granular materials; non-linear dynamics.

Beratan, David N., Ph.D., California Institute of Technology, 1985. *Biophysics, Chemical Physics*. Theoretical chemistry; theoretical molecular biophysics (primary appointment: chemistry).

Bray, Hubert L., Ph.D., Stanford University, 1997. *Astrophysics, Relativity & Gravitation*. General relativity and astrophysics (primary appointment: mathematics).

Chang, Albert Mien-Fu, Ph.D., Princeton University, 1983. *Condensed Matter Physics, Nano Science and Technology*. Experimental condensed matter physics; nanophysics.

Curtarolo, Stefano, Ph.D., Massachusetts Institute of Technology, 2003. *Condensed Matter Physics, Materials Science, Metallurgy, Statistical & Thermal Physics*. Solid state physics; thermodynamics of materials; computational materials science (primary appointment: mechanical engineering and materials science).

Edwards, Glenn S., Ph.D., University of Maryland, 1984. *Biophysics*. Biophysics.

Finkelstein, Gleb, Ph.D., Weizmann Institute of Science, 1998. *Condensed Matter Physics, Nano Science and Technology*. Experimental condensed matter physics; nanophysics.

Gao, Haiyan, Ph.D., California Institute of Technology, 1994. Chair of the Department. *Nuclear Physics*. Experimental medium energy nuclear physics.

Gauthier, Daniel J., Ph.D., University of Rochester, 1989. *Nonlinear Dynamics and Complex Systems, Optics*. Quantum information science; nonlinear and complex systems.

Goshaw, Alfred T., Ph.D., University of Wisconsin-Madison, 1966. *High Energy Physics*. Experimental elementary particle physics; instrumentation.

Greenside, Henry S., Ph.D., Princeton University, 1981. *Biophysics, Nonlinear Dynamics and Complex Systems*. Theoretical neuroscience.

Howell, Calvin, Ph.D., Duke University, 1984. *Nuclear Physics*. Nuclear physics; few-nucleon systems.

Johnson, Allan G., Ph.D., Duke University, 1974. *Medical, Health Physics*. Imaging physics; magnetic resonance imaging (primary appointment: radiology).

Katsouleas, Tom C., Ph.D., University of California, Los Angeles, 1984. *Electrical Engineering, Physics of Beams*. Plasma and accelerator physics (primary appointment: electrical and computer engineering).

Kotwal, Ashutosh V., Ph.D., Harvard University, 1995. *High Energy Physics*. Experimental elementary particle physics; instrumentation.

Liu, Jian-Guo, Ph.D., University of California, Los Angeles, 1990. *Computational Physics, Fluids, Rheology, Nonlinear Dynamics and Complex Systems, Other*. Computational physics, nonlinear and complex systems, fluid dynamics.

Mueller, Berndt, Ph.D., University of Frankfurt, 1973. *Nuclear Physics, Particles and Fields*. Theoretical nuclear and particle physics.

Oh, Seog, Ph.D., Massachusetts Institute of Technology, 1981. *Particles and Fields*. Experimental elementary particle physics.

Palmer, Richard G., Ph.D., University of Cambridge, 1973. *Condensed Matter Physics, Nonlinear Dynamics and Complex Systems*. Theoretical condensed matter physics; complex systems.

Petters, Arlie O., Ph.D., Massachusetts Institute of Technology, 1991. *Cosmology & String Theory, Relativity & Gravitation*. General relativity and cosmology; gravitational lensing (primary appointment: mathematics).

Samei, Ehsan, Ph.D., University of Michigan, 1997. *Medical, Health Physics*. Medical imaging (primary appointment: radiology).

Scholberg, Kate, Ph.D., California Institute of Technology, 1997. Director of Undergraduate Studies. *High Energy Physics*. Experimental elementary particle physics; neutrino physics.

Smith, David R., Ph.D., University of California, San Diego, 1994. *Electrical Engineering, Optics*. Quantum optics; photonic crystals; metamaterials (primary appointment: electrical engineering).

Socolar, Joshua E. S., Ph.D., University of Pennsylvania, 1987. *Condensed Matter Physics, Nonlinear Dynamics and Complex Systems*. Theoretical condensed matter physics; nonlinear systems; regulatory networks.

Springer, Roxanne P., Ph.D., California Institute of Technology, 1990. *Nuclear Physics, Particles and Fields*. Theoretical nuclear and particle physics; effective field theory.

Warren, Warren S., Ph.D., University of California, Berkeley, 1980. *Atomic, Molecular, & Optical Physics, Biophysics, Chemical Physics*. Optical physics, molecular, and biomolecular imaging (primary appointment: chemistry).

Wu, Ying, Ph.D., Duke University, 1995. *Physics of Beams*. Free electron laser physics, beam physics.

Yang, Weitao, Ph.D., University of North Carolina, Chapel Hill, 1986. *Chemical Physics, Computational Physics*. Density functional theory, electronic structures in nano and condensed systems (primary appointment: chemistry).

Associate Professor

Chandrasekharan, Shailesh, Ph.D., Columbia University, 1995. Director of Graduate Studies. *Computational Physics, Nuclear Physics, Particles and Fields*. Theoretical nuclear and particle physics; lattice field theory.

Dobbins, James T., Ph.D., University of Wisconsin at Madison, 1985. *Medical, Health Physics*. Medical physics; medical imaging; imaging physics (primary appointment: radiology).

Driehuys, Bastiaan, Ph.D., Princeton University, 1995. *Medical, Health Physics*. Medical and health physics; radiology (primary appointment: Department of Radiology and Biomedical Engineering).

Kim, Jungsang, Ph.D., Stanford University, 1999. photonics, quantum information (primary appointment: electrical and computer engineering).

Kruse, Mark, Ph.D., Purdue University, 1996. *High Energy Physics*. Experimental elementary particle physics.

Mehen, Thomas, Ph.D., Johns Hopkins University, 1998. *Nuclear Physics, Particles and Fields*. Theoretical nuclear and particle physics; effective field theory.

Plesser, M. Ronen, Ph.D., Harvard University, 1991. *Cosmology & String Theory, Particles and Fields, Relativity & Gravitation*. String theory; supersymmetry.

Teitsworth, Stephen W., Ph.D., Harvard University, 1986. *Condensed Matter Physics, Nano Science and Technology, Nonlinear Dynamics and Complex Systems*. Experimental condensed matter physics; stochastic nonlinear dynamics of electronic transport systems.

Walter, Christopher, Ph.D., California Institute of Technology, 1997. *High Energy Physics*. Experimental elementary particle physics; neutrino physics.

Assistant Professor

Arce, Ayana Tamu Holloway, Ph.D., Harvard University, 2006. *High Energy Physics*. Experimental high energy physics.

Barbeau, Philip, Ph.D., University of Chicago, 2009. *Astrophysics, Nuclear Physics*. Experimental particle and astroparticle physics, neutrinos, and dark matter.

Buchler, Nicolas Emile, Ph.D., University of Michigan, 2001. *Biophysics, Nonlinear Dynamics and Complex Systems*. Biophysics, non-linear dynamics and complex systems (joint appointment: biology).

Charbonneau, Patrick, Ph.D., Harvard University, 2006. *Chemical Physics, Condensed Matter Physics*. Theoretical condensed matter physics, chemical physics (primary appointment: chemistry).

Mikkelsen, Maiken H., Ph.D., University of California, Santa Barbara, 2009. *Condensed Matter Physics, Electrical Engineering, Nano Science and Technology*. Experimental condensed matter physics; nanophysics; spintronics; nano photonins and quantum information science (joint appointment: electrical and computer engineering).

Emeritus

Evans, Lawrence E., Ph.D., Johns Hopkins University, 1960. *Particles and Fields*. Theoretical elementary particle physics.

Han, Moo-Young, Ph.D., University of Rochester, 1963. *Particles and Fields*. Theoretical physics; elementary particle physics.

Meyer, Horst, Ph.D., University of Geneva, 1953. *Condensed Matter Physics, Low Temperature Physics*. Experimental low-temperature and solid state physics.

Roberson, Russell N., Ph.D., Johns Hopkins University, 1960. *Nuclear Physics*. Experimental nuclear physics.

Robinson, Hugh G., Ph.D., Duke University, 1954. *Atomic, Molecular, & Optical Physics*. Atomic and molecular physics.

Thomas, John E., Ph.D., Massachusetts Institute of Technology, 1979. *Atomic, Molecular, & Optical Physics, Condensed Matter Physics*. Experimental quantum optics; atomic and molecular collision physics.

Tornow, Werner, Ph.D., Universität zu Tübingen, 1974. *Nuclear Physics*. Experimental nuclear physics; neutrino physics.

Walter, Richard L., Ph.D., University of Notre Dame, 1959. *Nuclear Physics*. Experimental nuclear physics.

Weller, Henry R., Ph.D., Duke University, 1967. *Nuclear Physics*. Experimental nuclear physics; nuclear structure; gamma-ray studies.

Adjunct Professor

Ciftan, Mikael, Ph.D., Duke University, 1968. *Statistical & Thermal Physics*. Theoretical physics; solid state theory; statistical thermodynamics.

Everitt, Henry, Ph.D., Duke University, 1990. *Applied Physics, Condensed Matter Physics, Engineering Physics/Science, Optics*. Experimental condensed matter physics; molecular physics; quantum optoelectronics.

Guenther, Robert D., Ph.D., University of Missouri, 1968. *Applied Physics, Engineering Physics/Science, Optics*. Applied science; tera-hertz optics.

Lawson, Dewey T., Ph.D., Duke University, 1972. *Acoustics*. Acoustics.

Skatrud, David D., Ph.D., Duke University, 1984. *Electrical Engineering, Optics*. Quantum electronics; submillimeter/THz spectroscopy.

West, Bruce, Ph.D., University of Rochester, 1970. *Biophysics*. Biophysics.

Adjunct Associate Professor

Ahmed, Mohammad W., Ph.D., University of Houston, 1999. *Nuclear Physics*. Experimental nuclear physics.

Daniels, Karen E., Ph.D., Cornell University, 2002. *Nonlinear Dynamics and Complex Systems*. Experimental non-linear dynamics.

Tonchev, Anton, Ph.D., Joint Institute for Nuclear Research, Russia, 1995. *Nuclear Physics*. Experimental nuclear physics.

Lecturer

Brown, Robert G., Ph.D., Duke University, 1982. *Computational Physics, Statistical & Thermal Physics*. Theoretical physics; statistical mechanics and computational physics.

DEPARTMENTAL RESEARCH SPECIALTIES AND STAFF

Theoretical

Biophysics. Generation, storage, and learning of temporal sequences in songbirds in terms of experimentally known anatomy, physiology, and connectivity of neurons in a songbird's brain. Structure and function of complex dynamical networks, especially genetic regulatory networks. Information transfer between complex networks. Buchler, Greenside, Socolar.

Computational Physics. Computational Physics; Numerical techniques for solving nonlinear partial differential equations; Monte Carlo algorithms in field theory and statistical mechanics; molecular dynamics; networks; large scale computations on vector and parallel computers. Computational methods in fluid dynamics, material sciences, plasma physics, and geophysical flow; emergent behavior in flocking and swarming; numerical analysis and scientific computing. Baranger, Bass, Brown, Chandrasekharan, Charbonneau, Greenside, Liu, Palmer, Yang.

Cosmology & String Theory. String theory; geometry of space-time, supersymmetry and duality; mirror symmetry, general relativity. Aspinwall, Bray, Petters, Plesser.

Nanophysics. Coherence and correlations in nanoscale systems like quantum dots and carbon nanotubes; coulomb blockade; quantum impurity effects; quantum phase transitions; quantum computing quantum entanglement; quantum information; thermodynamics of materials; density functional theory. Baranger, Beratan, Chandrasekharan, Curtarolo, Yang.

Nonlinear Dynamics and Complex Systems. Nonlinear and Complex Systems; computational studies of nonlinear and biological systems including genetic networks, heart and brain dynamics; collective behavior in matter and dynamical systems; spin glasses and glasses; adaptive algorithms; static and dynamic critical behavior in optics and magnetism; granular materials network dynamics; fractal growth; granular matter; in- and out-of-equilibrium dynamical properties of materials self-assembly; microphase formation; protein aggregation; glass and gel formation;stochastic dynamics of far-from-equilibrium systems. Buchler, Charbonneau, Greenside, Liu, Palmer, Socolar, Teitsworth.

Particles and Fields. Quantum chromodynamics and weak interactions; heavy quark physics; quark-gluon plasma; heavy-ion collisions; effective field theories of particle and nuclear interactions; lattice field theories and Monte Carlo simulations; strongly coupled field theories, thermalization. Bass, Chandrasekharan, Mehen, Mueller, Springer.

Experimental

Atomic, Molecular, & Optical Physics. Slow, fast, and stored light; quantum optics; single photon switching, quantum information; optical noise, optoelectronics; new technologies for optical communication; electromagnetic properties of materials, photonic crystals and metamaterials; molecular and biomolecular imaging. Gauthier, Kim, Smith, Warren.

Biophysics. Emergent properties and tissue dynamics; fast thermodynamics in laser-tissue interactions; applications of free-electron lasers to biology and medicine; characterization and control of heart dynamics; stochastic processes in biological systems; optical analysis of molecular dynamics in single synapses; optical stimulation of single synapses; development of high resolution imaging techniques; evolution of bistable and oscillatory dynamics in gene networks. Buchler, Edwards, West.

High Energy Physics. Precision tests of the Standard Model using the top quark, W, Z and Higgs bosons; searches for new fundamental symmetries and extra dimensions; tests of the QCD hadron production models; studies of neutrino properties; searches for proton decay; neutrino astrophysics; research program based at Fermilab, CERN and in Japan; state-of-the-art wire chamber and silicon detector development and con-

struction; electronics design for high energy physics experiments. Arce, Goshaw, Kotwal, Kruse, Oh, Scholberg, Christopher Walter.

Medical, Health Physics. Biomedical imaging; magnetic resonance imaging; magnetic resonance microscopy, x-ray microscopy, tomography and microPET; X-ray imaging, breast tomosynthesis, dual-energy imaging, Monte Carlo simulation; radiation dose and image quality; imaging optimization. Dobbins, Driehuys, Johnson, Samei.

Nano Science and Technology. Electronic properties of carbon nanotubes, nano crystals, semiconductor quantum dots and self-assembled DNA structures; physics of Luttinger liquids; scanning tunneling; capacitance and atomic force microscopy; optoelectronic processes in semiconductor microstructures; sub-picosecond optical characterization of nanostructures; nanometer-scale photonic, plasmonic and phononic band engineering; solid state spintronics, quantum information science, nanophotonics;nonlinear electronic transport in semiconductor nanostructures. Chang, Everitt, Finkelstein, Gauthier, Mikkelsen, Smith, Teitsworth.

Nonlinear Dynamics and Complex Systems. Nonlinear and complex systems; granular materials; dynamics of granular flow; quantum chaos in classical wave systems; chaotic networks; pattern formation and spatio-temporal chaos in far-from-equilibrium fluids and electronic systems; Rayleigh-Bernard convection. Behringer, Buchler, Daniels, Gauthier, Teitsworth.

Nuclear Physics. QCD and weak interactions in nuclear physics; nucleon structure and nucleon-nucleon interactions; physics of few nucleon systems; electromagnetic nuclear physics; radiative capture reactions using polarized proton and deuteron beams; testing QCD with high intensity gamma-ray source; fundamental symmetry studies with ultra-cold neutrons and the search for neutron electric dipole moment; neutrino oscillations using detectors at KAMLAND; double beta-decay; nuclear astrophysics. Ahmed, Barbeau, Gao, Howell, Tonchev.

Photonics and Quantum Information. High data rate quantum key distribution; high brightness hyper-entangled sources; multimode quantum communication; multi-element photon counting detector development; Solid state spin qubits. Everitt, Gauthier, Kim, Mikkelsen, Skatrud, Smith.

Physics of Beams. Beam physics; FEL and novel light source development; high intensity gamma ray source; FEL applications; plasma accelerators. Katsouleas, Wu.

View additional information about this department at
www.gradschoolshopper.com

EAST CAROLINA UNIVERSITY

DEPARTMENT OF PHYSICS

Greenville, North Carolina 27858
http://www.ecu.edu/physics

General University Information

Chancellor: Dr. Steve M. Ballard
Dean of Graduate School: Dr. Paul Gemperline
University website: http://www.ecu.edu

Control: Public
Setting: Rural
Total Faculty: 1,700
Total Graduate Faculty: 827

Total number of Students: 25,008
Total number of Graduate Students: 5,324

Department Information
Department Chairman: John C. Sutherland, Chair
Department Contact: Dr. Michael Dingfelder, Assistant Chair
for Graduate Studies
Total full-time faculty: 19
Total number of full-time equivalent positions: 19
Full-Time Graduate Students: 40
First-Year Graduate Students: 10
Female First-Year Students: 3
Total Post Doctorates: 2

Department Address
Mailstop 563
Howell Science Complex
Greenville, NC 27858
Phone: (252) 328-6739
Fax: (252) 328-6314
E-mail: physics@ecu.edu
Website: http://www.ecu.edu/physics

ADMISSIONS

Admission Contact Information
Address admission inquiries to: Graduate School, 131 Ragsdale
Building, East Carolina University, Greenville, NC 27858-
4353
Phone: (252) 328-6012
E-mail: gradschool@ecu.edu
Admissions website: http://www.ecu.edu/gradschool

Application deadlines
Fall admission:
U.S. students: February 15   *Int'l. students*: February 15

Application fee
U.S. students: $70   *Int'l. students*: $70
The admission process is handled by Graduate School. Please
see website for details.

Admissions information
For Fall of 2012:
Number of applicants: 51
Number admitted: 12
Number enrolled: 10

Admission requirements
Bachelor's degree requirements: A Bachelor's degree in Physics
or related subject is required.
Minimum undergraduate GPA: 2.7

GRE requirements
The GRE is required.
Quantitative score: 147
Verbal score: 147
Analytical score: 3.5
Mean GRE score range (25th–75th percentile): 303-317
An average 30 percentile score of quantitative + verbal is the
absolute minimum.

Advanced GRE requirements
The Advanced GRE is not required.

TOEFL requirements
The TOEFL exam is required for students from non-English-
speaking countries.
PBT score: 550
iBT score: 80

An iBT TOEFL score of minimum 20 in each section is required,
with a total minimum score of 80.

Other admissions information
Additional requirements: Three letters of recommendation, a letter of intent, and a CV/Resume are required.
Minimum GPA for PhD program: 3.0.
Undergraduate preparation assumed: Mathematics: Calculus
through Differential Equations; Physics: Courses in Mechanics, Electricity and Magnetism, Thermodynamics, Modern
Physics (Atomic and Nuclear Physics), Intro to Quantum Mechanics, Advanced Laboratory.

TUITION
Tuition year 2013-14:
Tuition for in-state residents
Full-time students: $2,111.5 per semester
Tuition for out-of-state residents
Full-time students: $8,270 per semester
Reduced tuition rates for part-time available: 1-2 hours (25%),
3-5 hours (50%), 6-8 hours (75%). Tuition for out-of-state
students consists of in-state part and out-of-state part. Number
above is total.
Credit hours per semester to be considered full-time: 9
Deferred tuition plan: Yes
Health insurance: Yes, $1,290.00.
Other academic fees: Approximately $1,100 per semester for uni-
versity fees, educational/technology fees, and health services
fees.
Academic term: Semester
Number of first-year students who received full tuition waivers: 4

Teaching Assistants, Research Assistants, and Fellowships
Number of first-year
Teaching Assistants: 4
Average stipend per academic year
Teaching Assistant: $21,600
Research Assistant: $23,520
PhD Teaching Assistantship $21,600.00 plus tuition remission
plus health insurance. MS Teaching Assistantship $11,000.00
plus out-of-state tuition remission.

FINANCIAL AID

Application deadlines
Fall admission:
U.S. students: March 1

Loans
Loans are available for U.S. students.
Loans are not available for international students.
GAPSFAS application required: No
FAFSA application required: No

For further information
Address financial aid inquiries to: Financial Aid Office, Graduate
School, 131 Ragsdale, East Carolina University, Greenville,
NC 27858-4353.
Phone: (252) 328-6610
E-mail: faques@ecu.edu
Financial aid website: http://www.ecu.edu/financial

HOUSING

Availability of on-campus housing
Single students: Yes
Married students: No

For further information

Address housing inquiries to: University Housing Service, Jones Residence Hall, East Carolina University, Greenville, NC 27858-4353.

Phone: (252) 328-4663

E-mail: housing@ecu.edu

Housing aid website: http://www.ecu.edu/studentlife/campusliving

Table A—Faculty, Enrollments, and Degrees Granted

Research Specialty	2012-13 Faculty	Enrollment Spring 2013 Master's	Enrollment Spring 2013 Doctorate	Number of Degrees Granted 2012-13 (2008-13) Master's	Number of Degrees Granted 2012-13 (2008-13) Terminal Master's	Number of Degrees Granted 2012-13 (2008-13) Doctorate
Acoustics	1	–	–	–	–	–
Atomic, Molecular, & Optical Physics	4	1	5	–(2)	–	1(1)
Biophysics	2	–	3	–(1)	–	1(2)
Computational Physics	4	1	1	1(2)	–	–
Laser	4	–	5	–	–	1(4)
Medical, Health Physics	7	15	9	2(3)	5(38)	2(4)
Solar Physics	1	–	–	–	–	–
Total	**23**	**17**	**23**	**3(8)**	**5(38)**	**5(11)**
Full-time Grad. Stud.	–	17	23	–	–	–
First-year Grad. Stud.	–	6	4	–	–	–

GRADUATE DEGREE REQUIREMENTS

Master's: A minimum of 34 semester hours is required for the Applied Physics (AP) option, a minimum of 39 semester hours is required for the Health Physics (HP) option, and a minimum of 39 semester hours is required for the Medical Physics (MP) option. A major field test is administered upon entrance into the program; for the AP option, candidates must write and defend a thesis based on original research; a thesis is not required for MP and HP students.

Doctorate: A minimum of 48 semester hours beyond the master's degree is required; a master's degree in physics or related area is preferred; students entering with a BS in physics or a related area will follow the AP master's curriculum; doctoral written and oral exams covering biomedical physics curriculum; thesis required; dissertation required.

Other Degrees: Integrated PhD and MS-MP concentration.

Thesis: Thesis may be written in absentia; Ph.D. Program: 5 consecutive semester residence requirement.

SPECIAL EQUIPMENT, FACILITIES, OR PROGRAMS

The Department operates in addition to classroom, offices, and well-equipped teaching and research laboratories, as well as electronic and machine shops to provide service to faculty and students throughout the University. The Department maintains state-of-the-art computing facilities, including a recently upgraded student computing laboratory with Mac and PCs. Research laboratories include the Accelerator Laboratory, the Biophysics Spectroscopy Laboratory, the Biomedical Laser Laboratory, the Raman Spectroscopy Laboratory, the Radiation Instrumentation Laboratory, and high-performance scientific computing equipment. Equipment in these laboratories includes, but is not limited to, the following: Accelerator Laboratory: a 2 MV tandem light ion accelerator; Biophysics Spectroscopy Laboratory: circular and linear dichroism spectroscopy, photospectrometer, MALDI-TOF mass spectrometer; Biomedical Laser Laboratory: high-power Nd:YAG Q-switched laser system, nanosecond N2 and dye laser system, cw diode-pumped Nd:YAG laser, diode-pumped nd:

YVO4 Q-switched laser, harmonic generators with nonlinear crystals, multichannel fluorescence decay time measurement system, UV-VIS-IR spectrophotometer, high-resolution (0.03-nm) spectrometer; Raman Spectroscopy Laboratory: optical tweezers, Raman spectroscopy, confocal Raman imaging; computing: 60 processor parallel cluster with Linux operating system.

Table B—Separately Budgeted Research Expenditures by Source of Support

Source of Support	Departmental Research	Physics-related Research Outside Department
Federal government	$709,392	
State/local government	$65,850	
Non-profit organizations		
Business and industry		
Other	$25,000	
Total	**$800,242**	

Table C—Separately Budgeted Research Expenditures by Research Specialty

Research Specialty	No. of Grants	Expenditures ($)
Medical, Health Physics	5	$315,000
Total	**5**	**$315,000**

FACULTY

Professor

Hu, Xin-Hua, Ph.D., University of California, Irvine, 1991. Director, Biomedical Laser Laboratory. *Biophysics, Optics, Other*. Biomedical optics; Experimental.

Joyce, James M., Ph.D., University of Pennsylvania, 1967. *Biophysics*. Biomedical physics.

Kempf, Ruth, Ph.D., Rensselaer Polytechnic Institute. *Medical, Health Physics, Other*. Security Studies, Health Physics, Radiation Biology.

Lapicki, Gregory, Ph.D., New York University, 1975. *Atomic, Molecular, & Optical Physics, Computational Physics, Physics of Beams*. Theoretical atomic physics.

Li, Yong-qing, Ph.D., Academia Sinica, Shanghai, 1989. Director, Biomedical Optics Laboratory. *Biophysics, Optics, Other*. Experimental optical physics and biomedical physics.

Seykora, Edward, Ph.D., North Carolina State University, 1968. *Astrophysics, Solar Physics*. Solar physics.

Shinpaugh, Jefferson L., Ph.D., Kansas State University, 1990. Director, ECU Accelerator Laboratory. *Accelerator, Atomic, Molecular, & Optical Physics, Physics of Beams*. Experimental atomic and molecular physics.

Sutherland, John C., Ph.D., Georgia Institute of Technology, 1967. Chair, Physics Department. *Biophysics, Other*. Biophysics, Spectroscopy.

Associate Professor

Bier, Martin, Ph.D., Clarkson University, 1990. *Biophysics, Neuroscience/Neuro Physics, Nonlinear Dynamics and Complex Systems, Theoretical Physics*. Mathematics, modeling, computational physics.

Day, Orville W., Ph.D., Brigham Young University, 1973. *Computational Physics, Relativity & Gravitation, Theoretical Physics*. Quantum mechanics of atoms and molecules; General Relativity.

Dingfelder, Michael, Ph.D., Eberhard-Karls University, Tübingen, 1995. Assistant Chair for Graduate Studies. *Atomic, Mo-*

lecular, & Optical Physics, Theoretical Physics, Other. Theoretical physics, radiation physics and modeling.

Justiniano, Edson L.B., Ph.D., Kansas State University, 1982. *Accelerator, Atomic, Molecular, & Optical Physics, Computational Physics.* Experimental atomic physics.

Kenney, John M., Ph.D., Stony Brook University, 1985. Assistant Chair for Undergraduate Studies. *Biophysics, Chemical Physics, Optics, Other.* Fibril amyloid-like structures; CD-Spechroscopy.

Lin, Zi Wei, Ph.D., Columbia University, 1996. *Atmosphere, Space Physics, Cosmic Rays, Computational Physics, High Energy Physics, Theoretical Physics.* Theoretical physics and radiation modeling.

Lu, Jun Qing, Ph.D., University of California, Irvine, 1991. *Computational Physics, Condensed Matter Physics, Theoretical Physics.* Theoretical condensed matter and biomedical physics.

Sprague, Mark W., Ph.D., University of Mississippi, 1994. Chair of Faculty. *Acoustics, Atomic, Molecular, & Optical Physics, Other.* Bio-acoustics. Sound propagation in condensed media, especially water. Sound of fish.

Assistant Professor

DeWitt, Regina, Ph.D., University of Heidelberg, 2002. *Accelerator, Applied Physics, Atomic, Molecular, & Optical Physics, Medical, Health Physics, Other.* Radiation dosimetry, radiation physics.

Jung, Jae Won, Ph.D., Texas A&M University, 2007. *Medical, Health Physics, Other.* Health Physics.

Emeritus

Bissinger, George, Ph.D., University of Notre Dame. *Acoustics, Other.* Acoustics of music instruments.

Mumtaz, Dinno A., Ph.D., *Biophysics.*

Sayetta, Thomas C., Ph.D., University of South Carolina, 1964. *Non-specialized.*

Toburen, Larry H., Ph.D., Vanderbilt University. *Accelerator, Atomic, Molecular, & Optical Physics, Other.* Experimental Radiation Physics.

Adjunct Professor

Huang, Zhibin, Ph.D., Case Western Reserve University, 2008. *Medical, Health Physics.* Medical Physics, Clinical Radiation therapy.

McLawhorn, Robert A., Ph.D., East Carolina University, 2008. *Medical, Health Physics.* Clinical Medical Physics.

Rasmussen V, Karl H., Ph.D., University of Wisconsin - Madison, 2009. *Medical, Health Physics.* Clinical Medical Physics.

Rice, John R., Ph.D., University of Wisconsin - Madison, 1992. *Medical, Health Physics.* Clinical Medical Physics.

DEPARTMENTAL RESEARCH SPECIALTIES AND STAFF

Theoretical

Biophysics. Theoretical study of biological processes on a molecular level; motor proteins; modulated barriers; power line issue; electroporation. Bier.

Computational Physics. Quantum mechanics of atoms and molecules; density functional theory; light scattering from random media; characterization of inhomogeneity in tissue using non-contact laser speckle techniques; Monte Carlo modeling of charged particle track structure; nuclear radiation transport models and codes; biomedical applications of radiation transport models; space radiation physics; mathematical theory of plasma oscillations in black holes; application to general relativity. Day, Dingfelder, Lin, Lu.

Solar Physics. High-resolution optical imaging of the solar disc through the earth's atmosphere. Seykora.

Theoretical Radiation Physics. Atomic and molecular collisions; penetration of charged particles through matter; inner shell ionization and stopping power. Dingfelder, Lapicki.

Experimental

Acoustics. Atmospheric and underwater acoustics; computational studies on convective effects on sound propagation; acoustic characterization of Sciaenid fish calls. Sprague.

Biophysics. Electrical activity of the stomach; biological physics of membrane; development of medical devices. Joyce.

Experimental Radiation Physics. Experimental atomic physics; ion-atom collisions; charge transfer; recombination and excitation in electron-ion collisions; laser assisted electron-ion and ion-atom collisions; application of atomic physics to biological physics and trace-element analysis; measurement of cross-sections for collisions involving ions and neutral particles and application of these data in the study of charged particle track structure. DeWitt, Justiniano, Shinpaugh.

Medical, Health Physics. Radiation detection and measurement; dosimetric studies of brachytherapy seeds; dose reconstruction in radiation therapy; dose reconstruction; optically stimulated luminescence; thermoluminescence; MRI physics; accelerator physics; imaging analysis; low-dose-rate brachytherapy; image-guided radiation therapy; robotic brachytherapy; tumor tracking; clinical medical physics. DeWitt, Huang, Jung, Kempf, McLawhorn, Rasmussen V, Rice.

Other. Lasers, photonics, and spectroscopy: experimental and theoretical study of light interaction with biological tissues and cells; optical imaging of turbid medium; flow cytometry; circular and linear dichroism (CD) spectroscopy; spider silk; structure, function, and dynamics of DNA, proteins, and other biological materials; optical tweezers; Raman spectroscopy; confocal Raman imaging. Hu, Kenney, Li, Sutherland.

View additional information about this department at
www.gradschoolshopper.com

NORTH CAROLINA STATE UNIVERSITY, RALEIGH

DEPARTMENT OF PHYSICS

Raleigh, North Carolina 27695-8202
http://www.physics.ncsu.edu/index.html

General University Information
President: Randy Woodson
Dean of Graduate School: Rebecca Rufty
University website: http://www.ncsu.edu/
Control: Public
Setting: Urban
Total Faculty: 2,068
Total Graduate Faculty: 2,400
Total number of Students: 34,767
Total number of Graduate Students: 9,591

Department Information
Department Chairman: John M. Blondin, Head
Department Contact: John M. Blondin, Professor and Head
  Total full-time faculty: 53
  Total number of full-time equivalent positions: 43
  Full-Time Graduate Students: 114
  First-Year Graduate Students: 23
  Female First-Year Students: 4
  Total Post Doctorates: 17

Department Address
2401 Stinson Drive
Raleigh, NC 27695-8202
Phone: (919) 515-2521
Fax: (919) 515-6538
E-mail: john_blondin@ncsu.edu
Website: http://www.physics.ncsu.edu/index.html

ADMISSIONS

Admission Contact Information
Address admission inquiries to: Director of Graduate Programs, Physics Department, Box 8202, NCSU, Raleigh, NC 27695-8202
Phone: (919) 515-8706
E-mail: py-grad-program@ncsu.edu
Admissions website: http://www.physics.ncsu.edu

Application deadlines
Fall admission:
U.S. students: January 9   *Int'l. students*: January 9
Spring admission:
U.S. students: October 15   *Int'l. students*: August 15

Application fee
U.S. students: $65   *Int'l. students*: $75
Late applications will be processed for exceptional applicants if spots are still available.

Admissions information
For Fall of 2012:
  Number of applicants: 180
  Number admitted: 63
  Number enrolled: 22

Admission requirements
Bachelor's degree requirements: Bachelor's degree in physics (or related field) is required.
Minimum undergraduate GPA: 3.0

GRE requirements
The GRE is required.

Advanced GRE requirements
The Advanced GRE is required.
  Minimum accepted Advanced GRE score: 650
  Mean Advanced GRE score range (25th–75th percentile): 770-970
Minimum accepted Advanced GRE: 650.

TOEFL requirements
The TOEFL exam is required for students from non-English-speaking countries.
  PBT score: 560
  iBT score: 80
Minimum TOEFL scores for admission are as follows: Listening-20; Reading-20; Writing-20, Speaking-20. For a teaching assistantship, a minimum of 23 on the speaking portion is required. – The paper-based test requires a score of 560 or higher (with scores of 50 on at least two of the three sections and no section score below 45).

Other admissions information
Undergraduate preparation assumed: Griffiths, Electromagnetic Fields and Waves; Hand and Finch, Analytical Mechanics; Gasiorowicz, Quantum Physics. (or equivalent) Thermal physics.

TUITION

Tuition year 2012–13:
Tuition for in-state residents
  Full-time students: $6,888 annual
Tuition for out-of-state residents
  Full-time students: $18,930 annual
Part-time prorated.
Credit hours per semester to be considered full-time: 9
Deferred tuition plan: Yes
Health insurance: Available
Other academic fees: Required student fees are about $2,050/ year.
Academic term: Semester
Number of first-year students who received full tuition waivers: 22

Teaching Assistants, Research Assistants, and Fellowships
Number of first-year
  Teaching Assistants: 19
  Research Assistants: 1
  Fellowship students: 2
Average stipend per academic year
  Teaching Assistant: $22,000
  Research Assistant: $23,000
  Fellowship student: $23,000

FINANCIAL AID

Application deadlines
Fall admission:
U.S. students: March 1

Loans

Loans are available for U.S. students.
Loans are available for international students.
GAPSFAS application required: No
FAFSA application required: No

For further information

Address financial aid inquiries to: Financial Aid Office, 2016 Harris Hall, Box 7302.
Phone: (919) 515-2421
Financial aid website: http://www.ncsu.edu/finaid

HOUSING

Availability of on-campus housing

Single students: Yes
Married students: Yes

For further information

Address housing inquiries to: Housing Assignments Office, 1112 Pullen Hall, Box 7315.
Phone: (919) 515-2440
E-mail: housing@ncsu.edu
Housing aid website: http://www.ncsu.edu/housing

Table A—Faculty, Enrollments, and Degrees Granted

Research Specialty	2011–12 Faculty	Enrollment Fall 2011 Master's	Enrollment Fall 2011 Doctorate	Number of Degrees Granted 2009–10 (2005–10) Master's	Number of Degrees Granted 2009–10 (2005–10) Terminal Master's	Number of Degrees Granted 2009–10 (2005–10) Doctorate
Astrophysics	5	1	5	–(1)	–	1(4)
Atomic, Molecular, & Optical Physics	7	–	5	–(1)	–(1)	1(2)
Biophysics	4	–	9	2(3)	–(1)	1(3)
Computational Physics	5	–	15	–(1)	–	2(5)
Condensed Matter Physics	10	1	35	1(6)	2(6)	7(23)
Nuclear Physics	14	–	23	1(1)	–	3(7)
Physics and other Science Education	3	–	7	2(1)	–	2(4)
Non-specialized	–	4	13	4(20)	–	–
Other	5	–	8	–	–	–(1)
Total	**51**	**6**	**120**	**8(34)**	**2(8)**	**17(50)**
Full-time Grad. Stud.	–	6	118	–	–	–
First-year Grad. Stud.	–	4	16	–	–	–

GRADUATE DEGREE REQUIREMENTS

Master's: Thirty semester-hours; 3.0/4.0 overall GPA; two semesters residence; no foreign language; no computer language; Option A-comprehensive oral exam and thesis required, or Option B-comprehensive written exam required.
Doctorate: Eight semesters beyond the baccalaureate with a 3.0/4.0 overall GPA; no computer language; comprehensive written and oral exams; thesis required.
Thesis: Thesis may be written in absentia.

SPECIAL EQUIPMENT, FACILITIES, OR PROGRAMS

The majority of the department is located in 55,000 sq. ft. of the recently renovated Riddick Hall. It houses a number of modern laboratories and teaching facilities, including a clean room and shared experimental user facilities. The adjacent NC State Centennial Campus houses several departmental nano-science/materials laboratories in several buildings. Major facilities for nuclear physics research are provided by the Triangle Universities Nuclear Laboratory. The optical physics and nanoscience/materials laboratories are well equipped with an atomic force microscope, various laser systems, spectrometers, electron microscopes, materials preparation systems, and extensive data acquisition equipment. Both a student machine shop and a professionally staffed machine shop are available for custom equipment fabrication. In addition, collaboration exists with other departments and with various industrial and governmental laboratories. The computer facilities are state of the art. At the department and research group level, powerful graphics workstations serve as graphics and communications nodes. The University ranks highly in high performance computing (HPC) and has been awarded a very high bandwidth (Internet2) connectivity to National Supercomputing Centers and other nationally prominent Research Universities.

Table B—Separately Budgeted Research Expenditures by Source of Support

Source of Support	Departmental Research	Physics-related Research Outside Department
Federal government	$7,359,418	
State/local government	$1,791,983	
Non-profit organizations		
Business and industry		
Other	$194,215	
Total	**$9,345,616**	

Table C—Separately Budgeted Research Expenditures by Research Specialty

Research Specialty	No. of Grants	Expenditures ($)
Astrophysics	19	$616,426
Atomic, Molecular, & Optical Physics	2	$331,651
Biophysics	14	$704,024
Condensed Matter Physics	43	$2,628,590
Nuclear Physics	18	$1,758,630
Physics and other Science Education	3	$299,450
Computational Physics	17	$1,020,646
Other	5	$1,986,199
Total	**121**	**$9,345,616**

FACULTY

Professor

Ade, Harald, Ph.D., Stony Brook University, 1990. *Condensed Matter Physics, Nano Science and Technology, Polymer Physics/Science*. Experimental physics.
Aspnes, David E., Ph.D., University of Illinois, 1965. *Condensed Matter Physics*. Experimental physics; optical physics/materials.
Beichner, Robert J., Ph.D., State University of New York at Buffalo, 1989. Physics education research.
Bernholc, Jerzy, Ph.D., Lund University, 1977. Theoretical physics; nanoscience/materials.
Blondin, John M., Ph.D., University of Chicago, 1987. Theoretical physics; astrophysics.
Brown, J. David, Ph.D., University of Texas, 1985. Theoretical physics; astrophysics and relativity.
Cotanch, Stephen R., Ph.D., Florida State University, 1973. Theoretical physics; nuclear physics.
Ellison, Donald C., Ph.D., Catholic University of America, 1982. Theoretical physics; astrophysics.
Golub, Robert, Ph.D., Massachusetts Institute of Technology, 1968. Experimental physics; ultra-cold neutrons.
Gould, Christopher R., Ph.D., University of Pennsylvania, 1969. Experimental physics; nuclear physics.
Haase, David G., Ph.D., Duke University, 1975. Experimental physics; solid state physics; low-temperature physics.

Hallen, Hans, Ph.D., Cornell University, 1991. Experimental physics; optics.

Huffman, Paul, Ph.D., Duke University, 1995. Experimental physics; low temperature and nuclear physics.

Ji, Chueng, Ph.D., Korea Advanced Institute of Science & Technology (KAIST), 1982. Theoretical physics; nuclear physics.

Krim, Jacqueline, Ph.D., University of Washington, 1984. Experimental physics; nano-science/materials.

Lee, Dean, Ph.D., Harvard University, 1998. *Nuclear Physics, Theoretical Physics.*

Lucovsky, Gerald, Ph.D., Temple University, 1960. Experimental physics; nano-science/materials.

McLaughlin, Gail, Ph.D., University of California, San Diego, 1996. Theoretical physics; nuclear physics; astrophysics.

Mitas, Lubos, Ph.D., Slovak Academy of Sciences, 1989. Theoretical physics; nano-science/materials.

Paesler, Michael A., Ph.D., University of Chicago, 1975. Experimental physics; nano-science/materials.

Reynolds, Stephen P., Ph.D., University of California, Berkeley, 1980. Theoretical physics; astrophysics.

Roland, Christopher M., Ph.D., McGill University, 1989. Theoretical physics; nano-science/materials.

Sagui, Celeste, Ph.D., University of Toronto, 1995. Theoretical physics; biophysics.

Schaefer, Thomas, Ph.D., University of Regensburg, 1992. Theoretical nuclear/particle physics.

Thomas, John E., Ph.D., Massachusetts Institute of Technology, 1979. Atomic physics.

Young, Albert, Ph.D., Harvard University, 1995. Experimental physics; atomic and nuclear physics.

Associate Professor

Clarke, Laura, Ph.D., University of Oregon, 1998. Experimental physics; nano-science/materials; molecular physics.

Daniels, Karen E., Ph.D., Cornell University, 2002. Experimental physics; non-equilibrium granular and fluid systems.

Lazzati, Davide, Ph.D., Universita degli Studi, Milan, 2001. Theoretical physics; astrophysics.

Weninger, Keith, Ph.D., University of California, Los Angeles, 1997. Experimental physics; biophysics.

Assistant Professor

Dougherty, Daniel, Ph.D., University of Maryland, 2004. *Condensed Matter Physics.* Experimental physics; nanoscale science and technology.

Frohlich, Carla, Ph.D., University of Basel, 2007. Theoretical physics; nuclear physics; astrophysics.

Gundogdu, Kenan, Ph.D., University of Iowa, 2004. Experimental physics; nanoscale science and technology.

Kneller, James, Ph.D., Ohio State University, 2001. Theoretical physics; nuclear physics; astrophysics.

Lim, Shuang Fang, Ph.D., University of Cambridge, 2004. *Biophysics, Nano Science and Technology.*

Riehn, Robert, Ph.D., University of Cambridge, 2003. Experimental physics; biophysics.

Wang, Hong, Ph.D., University of North Carolina, Chapel Hill, 2003. *Biophysics.*

Research Professor

Borkowski, Kazimirez, Ph.D., University of Colorado, 1988. Experimental physics; astrophysics.

Mitchell, Gary E., Ph.D., Florida State University, 1962. Experimental nuclear physics.

Research Associate Professor

Kelley, John H., Ph.D., Michigan State University, 1995. Experimental physics; nuclear physics.

Lu, Wenchang, Ph.D., Fudan University, 1994. *Condensed Matter Physics.* Theoretical physics; nano-science/materials.

Research Assistant Professor

Bochinski, Jason, Ph.D., University of Oregon, 2000. Molecular physics; nano-science/materials.

Heyward, Keith, Ph.D., North Carolina State University, 2006.

Hodak, Miroslav, Ph.D., University of Pennsylvania, 2002.

Teaching Professor

Fortner, Brand, Ph.D., University of Illinois, 1993. Theoretical physics; astrophysics.

Senior Lecturer

Egler, Robert E., M.A., 2002

Warren, C. Keith, M.S., Appalachian State University, 1993.

DEPARTMENTAL RESEARCH SPECIALTIES AND STAFF

Theoretical

Astrophysics. Theoretical modeling and numerical simulations of supernovae and remnants, gamma-ray bursts, gravitational radiation from supernovae and colliding black holes, accretion onto compact objects, and planetary nebulae; shock waves, particle acceleration, and cosmic rays; neutrinos and nucleosynthesis in supernovae and gamma-ray bursts.

Condensed Matter Physics. Large scale simulations of real materials, bio-molecular processes, semiconductors, nanotubes and related nanoscale structures; quantum Monte Carlo simulations; O(N) and multiscale methods; quantum transport; nanostructured materials; phase separation; ferrofluids liquid-state theory; interfaces; diffusion; neural networks; pattern formation; electronic properties of transition-metal oxides and silicates.

Nuclear Physics. Electromagnetic structure studies of hadrons; relativistic quark models; light-cone quantization; B-physics; glueball and hybrid meson spectroscopy; application to astrophysics and cosmology; neutrino phenomenology; nonperturbative vacuum effects and mixing; CP violation; extra dimensions and physics beyond the standard model; QCD-based description of hadronic interactions; BCS methods and chiral symmetry breaking; Hartree-Fock techniques; lattice gauge theory; many body phenomena and computational algorithms, instantons, finite density QCD, superconductivity; nuclear lattice simulations; effective field theory.

Experimental

Astronomy. Radio, infrared, optical, and x-ray observations of supernova remnants, pulsar-wind nebulae, and planetary nebulae.

Atomic, Molecular, & Optical Physics. Laser polarization of atomic vapors, atom trapping, Femi-gas.

Biophysics. Dynamics of polymers on surfaces and in constrained thin films; nanoprobe tools for signal pathway investigations in cellular biology; single molecule techniques for dynamic and structural studies of proteins and cells; interfacial instabilities; thin film flows; statistical mechanics of granular materials.

Condensed Matter Physics. Nanotribology, micro/nano electromechanical systems, and liquid wetting phenomena; real time spectroscopy and microscopy of nanostructure growth and dynamics; subwavelength optical probesatomically-precise materials preparation and characterization including ultra thin films and related device structures; remote plasma enhanced chemical vapor deposition of semiconductors and insulators; x-ray spectromicroscopy of carbon-based electronics materials; molecular motion on surfaces; organic dielectrics; conduction and polarization of molecular and macromolecular assemblies; sound propagation in granular materials; electronic and magnetic properties of metal-organic interfaces; preparation, characterization, and control of graphene and its interfaces; nonlinear optical characterization of surfaces; ultra-fast laser spectroscopy of dynamical processes in solids.

Nuclear Physics. Tests of fundamental symmetries; neutron beta decay and electric dipole moments; neutrinos and neutrino oscillations; ultracold neutrons; studies of few nucleon systems; quantum chaos; statistical properties of nuclei; polarized nuclear targets.

Optics. Near-field optical microscopy and spectroscopy; nano-Raman spectroscopy; optical characterization of electronic materials; linear and non-linear optical spectroscopy of materials, films, surfaces and interfaces under static and dynamic conditions. Microscopy and spectroscopy of complex systems studied with synchrotron radiation.

Physics and other Science Education. Reexamination and redesign of modes of instruction and content for large enrollment courses; assessment of student understanding; role of computers including simulation, visualization, computer-based experiments, and student programming; distance learning.

View additional information about this department at
www.gradschoolshopper.com

UNIVERSITY OF NORTH CAROLINA, CHAPEL HILL

DEPARTMENT OF PHYSICS AND ASTRONOMY

Chapel Hill, North Carolina 27599-3255
http://www.physics.unc.edu/

General University Information
President: Thomas Ross
Dean of Graduate School: Steve Matson
University website: http://www.unc.edu/
Control: Public
Setting: Urban
Total Faculty: 3,608
Total Graduate Faculty: 2,039
Total number of Students: 29,278
Total number of Graduate Students: 8,262

Department Information
Department Chairman: Chris Clemens, Chair
Department Contact: Greg Smith, Executive Assistant
  Total full-time faculty: 31
  Total number of full-time equivalent positions: 46
  Full-Time Graduate Students: 116
  First-Year Graduate Students: 12
  Female First-Year Students: 4
  Total Post Doctorates: 7

Department Address
CB# 3255, Phillips Hall
Chapel Hill, NC 27599-3255
Phone: (919) 843-4815
Fax: (919) 962-0480
E-mail: g-admit@physics.unc.edu
Website: http://www.physics.unc.edu/

ADMISSIONS

Admission Contact Information
Address admission inquiries to: Graduate Admissions, Department of Physics and Astronomy, UNC-Chapel Hill, CB# 3255, Phillips Hall, Chapel Hill, NC 27599-3255
Phone: (919) 962-4703
E-mail: g-admit@physics.unc.edu

Admissions website: http://www.physics.unc.edu/

Application deadlines
Fall admission:
U.S. students: January 1   *Int'l. students*: January 1

Application fee
U.S. students: $100

Admissions information
For Fall of 2012:
  Number of applicants: 130
  Number admitted: 35
  Number enrolled: 12

Admission requirements
Bachelor's degree requirements: Bachelor's degree in physics is required.
Minimum undergraduate GPA: 3.0

GRE requirements
The GRE is required.
  Quantitative score: 780
  Verbal score: 600
  Analytical score: 4
  Mean GRE score range (25th–75th percentile): > 50

Advanced GRE requirements
The Advanced GRE is not required.

TOEFL requirements
The TOEFL exam is required for students from non-English-speaking countries.
  PBT score: 550
  iBT score: 95

Other admissions information
Additional requirements: A score of 550 or better on the TOEFL is required; a score of 600 or better is strongly preferred. A score of 95 or better on the internet based TOEFL is required; a score of 100 or better is strongly preferred.

Undergraduate preparation assumed: Symon, Mechanics; Martin, Elements of Thermodynamics; Hecht and Zajac, Optics; Griffiths, Introduction to Electrodynamics; Liboff, Introductory Quantum Mechanics, or comparable.

TUITION

Tuition year 2012–13:
Tuition for in-state residents
  Full-time students: $7,834 annual
Tuition for out-of-state residents
  Full-time students: $23,924 annual
In-state residents Full-time—$7,834.00 Part-time—0–2.9 hrs./ $1,958.50 3–5.9 hrs./$3,917.00 6–8.9 hrs./$5,875.50 Out-of-state residents Full-time—$23,924.00 Part-time—0–2.9 hrs./$5,981.00 3–5.9 hrs./$11,962.00 6–8.9 hrs./$17,943.00.
Credit hours per semester to be considered full-time: 9
Deferred tuition plan: No
Health insurance: Yes, $2693.64.
Other academic fees: In-state: $927.74. Out-of-state: $927.74.
Academic term: Semester

Teaching Assistants, Research Assistants, and Fellowships

Number of first-year
  Teaching Assistants: 12
  Research Assistants: 3
Average stipend per academic year
  Teaching Assistant: $22,800
  Research Assistant: $22,800

FINANCIAL AID

Application deadlines

Fall admission:
U.S. students: March 1   *Int'l. students*: March 1

Loans

Loans are available for U.S. students.
Loans are available for international students.
GAPSFAS application required: No
FAFSA application required: Yes

For further information

Address financial aid inquiries to: Graduate Admissions Committee, Department of Physics and Astronomy, CB #3255, Phillips Hall, UNC, Chapel Hill, NC 27599-3255.
Phone: (919) 962-4703
E-mail: g-admit@physics.unc.edu
Financial aid website: http://studentaid.unc.edu

HOUSING

Availability of on-campus housing

Single students: Yes
Married students: Yes

For further information

Address housing inquiries to: UNC Housing Student Academic Services Bldg, 450 Ridge Road, CB5500 Chapel Hill, NC 27599.
Phone: (919) 962-5401
E-mail: housing@unc.edu
Housing aid website: http://housing.unc.edu

Table A—Faculty, Enrollments, and Degrees Granted

Research Specialty	2012-13 Faculty	Enrollment Fall 2012 Master's	Enrollment Fall 2012 Doctorate	Number of Degrees Granted 2012-13 Master's	Number of Degrees Granted 2012-13 Terminal Master's	Number of Degrees Granted 2012-13 Doctorate
Applied Physics	1	–	2	–	–	–
Astronomy	11	–	6	2(-)	–	1(-)
Astrophysics	11	–	16	1(-)	–	2(-)
Biophysics	4	–	9	–	–	–
Condensed Matter Physics	18	–	25	–	–	1(-)
Cosmology & String Theory	1	–	3	–	–	–
Nano Science and Technology	1	–	4	1(-)	–	1(-)
Nuclear Physics	7	–	15	1(-)	–	1(-)
Particles and Fields	3	–	8	–	–	–
Relativity & Gravitation	2	–	4	–	–	–
Other	5	–	–	–	–	1(-)
Total	–	–	92	5(-)	–	7(-)
Full-time Grad. Stud.	–	–	92	–	–	–
First-year Grad. Stud.	–	–	16	–	–	–

GRADUATE DEGREE REQUIREMENTS

Master's: Thirty semester hours required of which three to six may be for thesis or Master's project (if non-thesis option elected). One comprehensive written examination, which also serves as qualifying examination for Ph.D. Oral examination on thesis, or on a Master's project (if non-thesis option is elected), is required. Residency is required.

Doctorate: No specific graduate course credit requirements, but satisfactory completion of approved sequence of courses; a doctoral written examination covering "core" curriculum; a preliminary oral examination for the dissertation, a dissertation, and oral examination on dissertation. Residency is required.

Thesis: Thesis may be written in absentia. Courses taken as a graduate student before enrolling at Chapel Hill may afford students the opportunity to opt out of one or more first-year core courses and may be eligible for transfer course credit. However, students are required to pass the doctoral written examination, which is offered at the end of the student's first year and covers core material in quantum mechanics, electromagnetic theory, dynamics, and statistical mechanics.

SPECIAL EQUIPMENT, FACILITIES, OR PROGRAMS

The University of North Carolina, together with NC State and Duke Universities, administers the Triangle Universities Nuclear Laboratory's three nuclear accelerators in Durham. Special astronomy facilities of UNC, Chapel Hill, include a 61-cm telescope and a 68-foot planetarium dome with full-dome digital video. Astronomical facilities: observational facilities include the SOAR 4.1-meter telescope in Chile 61 nights time share, 11-meter Southern African Large Telescope (SALT) in South Africa (10 nights), the University's PROMPT robotic telescope array six 0.4-meter telescopes in Chile, one 0.8-meter telescope in Chile under construction, and four 0.4-meter telescopes in Australia under construction. The department is also home to one-of-a-kind microscopes, x-ray technology, and computational facilities. Laboratory facilities include the Goodman Laboratory for Astronomical Instrumentation and the Chapel Hill Analytical and Nanofabrication Laboratory (CHANL).

Table B—Separately Budgeted Research Expenditures by Source of Support

Source of Support	Departmental Research	Physics-related Research Outside Department
Federal government	$7,502,698	$1,613,427
State/local government	$384,382	
Non-profit organizations		
Business and industry	$150,694	
Other	$2,544,731	
Total	**$10,582,505**	**$1,613,427**

Table C—Separately Budgeted Research Expenditures by Research Specialty

Research Specialty	No. of Grants	Expenditures ($)
Astronomy	32	$2,738,451
Biophysics	23	$3,834,072
Condensed Matter Physics	17	$1,053,416
Nuclear Physics	15	$4,536,493
Particles and Fields	1	$33,500
Total	**88**	**$12,195,932**

FACULTY

Professor

Carney, Bruce W., Ph.D., Harvard University, 1978. Executive Vice Chancellor & Provost. Observational astronomy; history of the Milky Way galaxy.

Cecil, Gerald N., Ph.D., University of Hawaii, 1987. Observational astronomy; active galactic nuclei.

Champagne, Arthur E., Ph.D., Yale University, 1982. Experimental nuclear physics.

Clegg, Thomas B., Ph.D., Rice University, 1965. Experimental nuclear physics; development of polarized beams.

Clemens, J. Christopher, Ph.D., University of Texas, Austin, 1994. Department Chair. Observational astronomy; astrophysics; astronomical instrumentation.

Dolan, Louise A., Ph.D., Massachusetts Institute of Technology, 1976. Particle theory.

Engel, Jonathan H., Ph.D., Yale University, 1986. Theoretical nuclear physics; violation of fundamental symmetries in nuclei; the role of nuclear effects in solar neutrino and dark matter experiments; r-process nucleosynthesis and quark effects in nuclei.

Evans, Charles R., Ph.D., University of Texas, Austin, 1984. General relativity; numerical hydrodynamics; astrophysics.

Frampton, Paul H., Ph.D., University of Oxford, 1968. Particle theory; grand unification; cosmology.

Iliadis, Christian, Ph.D., University of Notre Dame, 1993. Experimental nuclear astrophysics.

Karwowski, Hugon J., Ph.D., Indiana University, 1980. Experimental nuclear physics with polarized nuclei.

Khveshchenko, Dmitri, Ph.D., Landau Institute for Theoretical Physics, 1989. Theoretical condensed matter physics; quantum transport in strongly correlated systems.

Lu, Jianping, Ph.D., City University of New York, 1988. High-Tc superconductors; fullerenes and fullerides; strongly correlated electron systems; electronic structure and computational physics; disordered system and quantum transports; complex systems; x-ray imaging.

McNeil, Laurie E., Ph.D., University of Illinois, 1982. Optical studies of materials.

Ng, Yee Jack, Ph.D., Harvard University, 1974. Theoretical particle physics; gravitation.

Qin, Lu-Chang, Massachusetts Institute of Technology, 1994. Electron microscopy; materials science; nanotechnology.

Reichart, Daniel E., Ph.D., University of Chicago, 2000. Gamma ray bursts; early universe; interstellar extinction; galaxy clusters.

Superfine, Richard, Ph.D., University of California, Berkeley, 1991. Scanning probe microscopy of biological structures.

Tsui, Frank, Ph.D., University of Illinois at Urbana-Champaign, 1992. Molecular beam epitaxy of transition metals; scanning tunneling microscopy.

Washburn, Sean, Ph.D., Duke University, 1982. Effects of quantum-mechanical coherence in charge transport in small systems; ballistic transport in semiconductors; nano-scale electro-mechanical systems.

Wilkerson, John, Ph.D., University of North Carolina, Chapel Hill, 1982. Neutrino physics and neutrino astrophysics.

Wu, Yue, Ph.D., Catholic University of Leuven, 1987. Nuclear magnetic resonance; electron spin resonance in solids.

Zhou, Otto, Ph.D., University of Pennsylvania, 1992. Experimental materials science; structure/property relationships in solids.

Associate Professor

Henning, Reyco, Ph.D., Massachusetts Institute of Technology, 2003. Experimental neutrino physics.

Kannapan, Sheila, Ph.D., Harvard University, 2001. Observational astronomy; galaxy formation and evolution.

López, René, Ph.D., Vanderbilt University, 2002. Nanotechnology; nano-optics; ultra short laser/matter interaction; thin film science.

Mersini-Houghton, Laura, Ph.D., University of Wisconsin-Madison, 2000. Theoretical cosmology.

Assistant Professor

Branca, Rosa Tamara, Ph.D., La Sapienza, University of Rome, 2006. *Condensed Matter Physics.*

Drut, Joaquin E., Ph.D., University of Washington, 2008. Quantum many-body physics; nuclear, atomic and condensed matter physics; non-perturbative computational approaches.

Heitsch, Fabian, Ph.D., University of Heidelberg, 2001. Turbulence and Fragmentation in Molecular Clouds.

Oldenburg, Amy, Ph.D., University of Illinois, 2001. *Biophysics.* Biophysics.

Emeritus

Briscoe, Charles V., Ph.D., Rice University, 1958.

Choi, Sang-il, Ph.D., Brown University, 1961.

Christiansen, Wayne A., Ph.D., University of California, 1968.

Dy, Kian S., Ph.D., Cornell University, 1967.

Hernandez, John P., Ph.D., University of Rochester, 1967.

Hooke, William M., Ph.D., Princeton University, 1958.

Hubbard, Paul S., Ph.D., Harvard University, 1958.

Kessemeier, Horst, Ph.D., University of Washington (St. Louis), 1964.

Ludwig, Edward J., Ph.D., Indiana University, 1963.

Macdonald, J. Ross, University of Oxford, 1950.

Rose, James, Ph.D., Yale University, 1977.

Rowan, Lawrence G., Ph.D., University of California, Berkeley, 1963. Solid state experiments using EPR.

Schroeer, Dietrich, Ph.D., Ohio State University, 1965.

Shafroth, Stephen M., Ph.D., Johns Hopkins University, 1953.

Slifkin, Lawrence M., Ph.D., Princeton University, 1950.

Thompson, William J., Ph.D., Florida State University, 1967.

York, James W., Ph.D., North Carolina State University, 1966.

Research Professor

Sen, Pabitra, Ph.D., University of Chicago, 1972. Physics.

Taylor, Russell, Ph.D., University of North Carolina, Chapel Hill, 1994. Scientific Visualization, Distributed Virtual Worlds, Haptic Display, and Interactive 3D Computer Graphics.

Research Associate Professor

Parikh, Nalin R., Ph.D., McMaster University, 1985. Ion beam modification of materials; ion beam analysis, microwave plasma deposition/etching.

Research Assistant Professor

Falvo, Michael, Ph.D., University of North Carolina, Chapel Hill, 1997. Nanoscience.

Kleinhammes, Alfred, Ph.D., Clark University, 1990. Condensed matter physics; nuclear magnetic resonance.

O'Brien, E. Timothy, Ph.D., University of California, Santa Barbara, 1974. Cell biology; microscopy. arikh.

Adjunct Professor

Chaffee, Fred, Ph.D., 1968Astronomy. Astronomy.

Chang, Sha Xiao, Ph.D., Clark University, 1988. Solid state physics.

Lee, Yueh.

Rohm, Ryan M., Ph.D., Princeton University, 1985. Quantum field theory; theoretical elementary particle physics.

Rutland, Jonathan, Ph.D., University of North Carolina, 1988. Physics.

Silverstone, Murray, Ph.D., University of California, Los Angeles, 2000. *Astronomy*. Astronomy.

Tang, Jie, Ph.D., Osaka University, 1993. Materials science, nanotechnology.

Tonchev, Anton P., Ph.D., Flerov Laboratory of Nuclear Reactions, Joint Institute for Nuclear Research, 1995. Nuclear resonance; fluorescence; nuclear spectroscopy.

Zhang, Jian, Ph.D., University of North Carolina, Chapel Hill, 2005. Experimental physics.

Lecturer

Churukian, Alice, Ph.D., Kansas State University, 2002. Physics education.

Deardoff, Duane L., Ph.D., North Carolina State University, 2001. Undergraduate laboratories; physics education.

Smith, David, Ph.D., Dublin City University, 2005. Physics Education.

DEPARTMENTAL RESEARCH SPECIALTIES AND STAFF

Theoretical

Astrophysics. Astronomy and Astrophysics: Gravity, astrophysical gas dynamics. Evans, Heitsch.

Condensed Matter Physics. Many-body theory; fullerenes; high-temperature superconductors; correlated electron systems. Drut, Lu.

Nuclear Physics. Nucleosynthesis in stars and supernovae; violation of fundamental symmetries in nuclei; the role of nuclear effects in solar neutrino and dark matter experiments. Drut, Engel.

Particles and Fields. Field Theory, Particle Theory, Gravitation and Relativity. Quantum theory of fields and particles; superstring theory; cosmology; relativistic dynamics; general relativity; gravitational fields; quantum and supergravity. Dolan, Evans, Frampton, Mersini-Houghton, Ng.

Experimental

Astrophysics. Astronomy and Astrophysics: Radio, optical, and x-ray astronomy; stellar spectroscopy and photometry; nucleosynthesis. Carney, Cecil, Champagne, Clemens, Iliadis, Kannapan, Reichart.

Biophysics. Biological and medical physics. Nanoscale properties of biomolecules, DNA, and motors, biophysical properties of fluids, cells and tissues, biomedical imaging technology development in x-ray, optical, and magnetic resonance. Lu, Oldenburg, Superfine, Washburn, Zhou.

Condensed Matter Physics. Condensed Matter Physics, Materials Science, microelectronics. NMR studies of kinetic and thermodynamic properties of metallic supercooled liquids and glasses and interactions and functions of adsorbed molecules in nanomaterials and biological systems; Raman and Brillouin scattering from organic semiconductors; AFM and surface science in virtual reality interfaces to microscopy; low-temperature physics and quantum transport; growth and magnetic susceptibility of magnetic semiconductors; electrical and mechanical interaction of carbon nanotubes with surfaces and other tubes; chaos in biological systems, Optical properties of nanoscale materials, organic and hybrid solar cell materials. López, McNeil, Parikh, Qin, Superfine, Tsui, Washburn, Wu, Zhou.

Nuclear Physics. Nuclear astrophysics; radioactive beams; neutrino physics; spin polarization in nuclear reactions; few body physics. Champagne, Clegg, Henning, Iliadis, Karwowski, Wilkerson.

View additional information about this department at
www.gradschoolshopper.com

WAKE FOREST UNIVERSITY

DEPARTMENT OF PHYSICS

Winston-Salem, North Carolina 27109
http://www.physics.wfu.edu/

General University Information

President: Nathan O. Hatch
Dean of Graduate School: Bradley T. Jones, Dean
University website: http://www.wfu.edu/

Control: Private
Setting: Suburban
Total Faculty: 1,699
Total Graduate Faculty: 540

Total number of Students: 7,432
Total number of Graduate Students: 2,617

Department Information
Department Chairman: Keith D. Bonin, Chair
Department Contact: Heather Chapman, Staff Assistant
   Total full-time faculty: 15
   Total number of full-time equivalent positions: 32
   Full-Time Graduate Students: 43
   First-Year Graduate Students: 7
   Female First-Year Students: 1
   Total Post Doctorates: 10

Department Address
1834 Wake Forest Road
P.O. Box 7507 Reynolda Station
Winston-Salem, NC 27109
Phone: (336) 758-3223
Fax: (336) 758-6142
E-mail: chapmahn@wfu.edu
Website: http://www.physics.wfu.edu/

ADMISSIONS

Admission Contact Information
Address admission inquiries to: Dean of Graduate School,Wake
   Forest University, 1834 Wake Forest Rd., Room 124, Reyn-
   olda Hall,PO Box 7487 Reynolda Station, Winston-Salem,
   NC 27109-7487
Phone: (336) 758-5301
E-mail: gradschl@wfu.edu
Admissions website: http://graduate.wfu.edu/admissions/
   onlineapp.html

Application deadlines
Fall admission:
U.S. students: January 15   *Int'l. students*: January 15
Spring admission:
U.S. students: November 15   *Int'l. students*: November 15

Application fee
U.S. students: $75   *Int'l. students*: $75
Typically 15 students are admitted of which about 50% enroll.

Admissions information
For Fall of 2013:
   Number of applicants: 83
   Number admitted: 10
   Number enrolled: 7

Admission requirements
Bachelor's degree requirements: A Bachelor's degree in physics
   is required.

GRE requirements
The GRE is required.
   Quantitative score: 700
   Verbal score: 450
   Mean GRE score range (25th–75th percentile): 1200-1370

Advanced GRE requirements
The Advanced GRE is recommended.
   Minimum accepted Advanced GRE score: 570
   Mean Advanced GRE score range (25th–75th percentile): 630-
   870

TOEFL requirements
The TOEFL exam is required for students from non-English-
   speaking countries.
   PBT score: 575
   iBT score: 79

Other admissions information
Additional requirements: The verbal and analytical GRE scores
   quoted above are for domestic students. Occasionally, international students with lower verbal and analytical GRE scores
   may be admitted.
   A minimum quantitative GRE score of 700 is required for
   domestic and international students. Quantitative GRE scores
   are frequently well above 700.
Undergraduate preparation assumed: Mechanics—Symon, Mechanics; Electricity and Magnetism—Griffiths, Introduction to
   Electrodynamics; Quantum Mechanics—Gasiorowicz, Quantum Physics; Thermodynamics—Kittel and Kroemer, Thermal Physics.

TUITION
Tuition year 2012–13:
   Full-time students: $33,174 annual
   Part-time students: $1,181 per credit
Usually, all admitted students will receive a tuition waiver. All
   admitted students will also receive a fully loaded, state-of-the
   art laptop computer.
Credit hours per semester to be considered full-time: 9
Deferred tuition plan: Yes
Health insurance: Available
Other academic fees: Free health insurance is provided. There
   is a $150 health Center fee and a $30 graduate fee each sum-
   mer. Free satellite parking (plus shuttle) is available. $300 near-
   campus parking and $500 on-campus parking is available.
Academic term: Semester

Teaching Assistants, Research Assistants, and Fellowships
Number of first-year
   Teaching Assistants: 7
   Research Assistants: 2
   Fellowship students: 1
Average stipend per academic year
   Teaching Assistant: $20,000
   Research Assistant: $20,000
   Fellowship student: $22,000

FINANCIAL AID

Application deadlines
Fall admission:
U.S. students: January 15
Spring admission:
U.S. students: January 15   *Int'l. students*: January 15

Loans
Loans are available for U.S. students.
Loans are not available for international students.
GAPSFAS application required: No
FAFSA application required: No

For further information
Address financial aid inquiries to: Dean of the Graduate School,
   Wake Forest University, 1834 Wake Forest Rd., Room 124
   Reynolda Hall, P.O.Box 7487 Reynolda Station, Winston Sa-
   lem, NC 27109-7487.
Phone: (336) 758-5301
E-mail: gradschl@wfu.edu
Financial aid website: http://graduate.wfu.edu/

HOUSING

Availability of on-campus housing
Single students: No
Married students: No

For further information
Address housing inquiries to: Residence Housing, P.O. Box 7749, Wake Forest Univ., Winston-Salem, NC 27109.
Phone: (336) 758-7777
E-mail: housing@wfu.edu
Housing aid website: http://www.rlh.wfu.edu/

Table A—Faculty, Enrollments, and Degrees Granted

Research Specialty	2012-13 Faculty	Enrollment Fall 2012 Master's	Enrollment Fall 2012 Doctorate	Number of Degrees Granted 2012-13 (2006–11) Master's	Number of Degrees Granted 2012-13 (2006–11) Terminal Master's	Number of Degrees Granted 2012-13 (2006–11) Doctorate
Atomic, Molecular, & Optical Physics	–	–	4	–	–(1)	1(3)
Biophysics	–	–	12	–	–	–(10)
Condensed Matter Physics	5	–	7	–	–	1(5)
Medical, Health Physics	3	–	5	1(2)	–	2(2)
Nano Science and Technology	3	1	6	–	1(2)	3(6)
Relativity & Gravitation	3	–	4	–	–(1)	1(3)
Total	18	1	38	1(2)	1(4)	6(30)
Full-time Grad. Stud.	–	1	38	–	–	–
First-year Grad. Stud.	–	–	9	–	–	–

GRADUATE DEGREE REQUIREMENTS

Master's: Thirty semester hours of graduate credit; of those, at least 24 credits must be classes or seminars, and 6 credits can be research. 12 credits hours must be at the 700 level. Courses must include Phys 711, 712, 741 (Math Methods & Classical Mechanics, Electrodynamics, Quantum Mechanics-I). Participation at the departmental seminar is required. Minimum of 12 months full-time in residence. An oral defense of the thesis and a 3.0 average on courses are required.

Doctorate: Courses must include Physics 711, 712, 741, 742, and 770 (Math Methods & Classical Mechanics, Electrodynamics, Quantum Mechanics-I and -II, Statistical Physics) and three more elective courses at the 600 or 700 level (of which one must be in Physics). A written General Exam at the level of material normally covered in the first year of graduate study serves as the preliminary examination. Within 12 months of completing the preliminary examination, the students submits to her/his advisory committee and defends orally a dissertation research plan. An oral defense of the dissertation, and a 3.0 average on courses are required.

Thesis: Thesis may be written in absentia.

SPECIAL EQUIPMENT, FACILITIES, OR PROGRAMS

Wake Forest is among the top 5% of tier-1 national doctoral universities (US News and World Report), despite its small size. We take pride in being able to provide the personal attention of a liberal arts college and having significant resources usually attributed to a large research university.

The research in our department is focused on the following areas: experimental and computational biophysics; atomic, molecular and optical physics; experimental and computational condensed matter physics; computational and theoretical relativity and gravitation; medical and health physics; and nanophysics. All re-

search labs contain state-of-the-art instrumentation; computational physicists have access to the large deacon cluster (over 2000 nodes).

Table B—Separately Budgeted Research Expenditures by Source of Support

Source of Support	Departmental Research	Physics-related Research Outside Department
Federal government	$1,669,551.64	
State/local government		
Non-profit organizations		
Business and industry	$795,699.39	
Other		
Total	$2,465,251.03	

Table C—Separately Budgeted Research Expenditures by Research Specialty

Research Specialty	No. of Grants	Expenditures ($)
Biophysics	15	$1,090,441
Condensed Matter Physics	7	$381,410
Nano Science and Technology	6	$955,178
Relativity & Gravitation	1	$36,208
Total	29	$2,463,237

FACULTY

Professor

Anderson, Paul R., Ph.D., University of California, Santa Barbara, 1983. *Astronomy, Astrophysics, Relativity & Gravitation*. General relativity; quantum field theory in curved space.

Bonin, Keith D., Ph.D., University of Maryland, 1984. Department Chair. *Atomic, Molecular, & Optical Physics, Biophysics, Nano Science and Technology, Optics*. Atomic physics, nanophysics, biophysics, optics.

Carroll, David L., Ph.D., Wesleyan University, 1993. Director, Nanotechnology Center. *Nano Science and Technology*. Nanostructures and nanotechnology.

Fetrow, Jacquelyn, Ph.D., Pennsylvania State University, 1986. Dean of College. *Biophysics, Computational Physics, Computer Science*. Computational biophysics, computational drug discovery, cheminformatics, systems biology.

Holzwarth, N. A. W., Ph.D., University of Chicago, 1975. *Condensed Matter Physics*. Theoretical solid state physics; electronic structure of bulk solids, surfaces, and molecules.

Kim-Shapiro, Daniel, Ph.D., University of California, Berkeley, 1993. Director, Translational Science Center. *Biophysics*. Biophysics.

Matthews, Eric G., Ph.D., University of North Carolina, 1977. Assoc. Provost of Information Systems. *Condensed Matter Physics*. Thermally stimulated depolarization of defects in insulators; ab initio calculations of defect properties.

Williams, R. T., Ph.D., Princeton University, 1974. *Atomic, Molecular, & Optical Physics, Condensed Matter Physics, Optics*. Femtosecond laser studies of defects and electrons in solids.

Associate Professor

Carlson, Eric, Ph.D., Harvard University, 1988. *Astrophysics, Particles and Fields, Relativity & Gravitation*. Astrophysics and particle physics.

Cook, Gregory B., Ph.D., University of North Carolina, 1990. Undergraduate Advisor. *Astrophysics, Relativity & Gravitation*. General relativity and relativistic astrophysics.

Guthold, Martin, Ph.D., University of Oregon, 1997. Director of Physics graduate program. *Biophysics, Nano Science and Technology.* Biophysics.

Macosko, Jed, Ph.D., University of California, Berkeley, 1999. *Biophysics.* Biophysics of molecular motors and biopolymers.

Salsbury, Fred, Ph.D., University of California, Berkeley, 1999. Undergraduate Advisor. Director, Interdisciplinary Program in Structural and Computational Biophysics. *Biophysics, Computational Physics.* Computational biophysics.

Assistant Professor

Cho, Samuel, Ph.D., University of California, San Diego, 2007. *Biophysics, Computational Physics, Computer Science.* Computational biophysics, protein and RNA folding, biomolecular assembly, molecular machines, GPU-based programming.

Dostal, Jack, Ph.D., Montana State University, 2009. Lecturer. *Physics and other Science Education.* Physics education.

Jurchescu, Oana, Ph.D., University of Groningen, Netherlands, 2006. *Condensed Matter Physics, Nano Science and Technology.* Nanostructures and nanotechnology.

Thonhauser, Timo, Ph.D., Karl-Franzens University, Austria, 2001. *Computational Physics, Condensed Matter Physics.* Density functional theory.

Emeritus

Shields, Howard, Ph.D., Duke University, 1956. *Biophysics.*

Research Professor

Holzwarth, G. M., Ph.D., Harvard University, 1964. *Biophysics.*

Kerr, William C., Ph.D., Cornell University, 1967. *Statistical & Thermal Physics.* Theoretical solid state and statistical physics; structural phase transitions.

Research Associate Professor

Basu, Swati, Ph.D., University of Illinois at Urbana-Champaign, 1994. *Biophysics.*

Ucer, K. Burak, Ph.D., University of Rochester, 1997. *Atomic, Molecular, & Optical Physics, Optics.* Ultrafast lasers and spectroscopy.

Adjunct Professor

Bourland, J. D., Ph.D., University of North Carolina. *Medical, Health Physics.* Radiation oncology.

Miller, Timothy, Ph.D., Vanderbilt University, 2002. *High Energy Physics, Medical, Health Physics, Nuclear Physics.* High-energy nuclear and parallel computing.

Santago, Peter, Ph.D., North Carolina State University, 1986. Chair, Computer Science. *Medical, Health Physics.* Image enhancement.

DEPARTMENTAL RESEARCH SPECIALTIES AND STAFF

Theoretical

Biophysics. Computational and theoretical biophysics, computational systems biology, protein structure/function relationships, biological network modeling, signal transduction network modeling, molecular physics, drug discovery. Basu, Bonin, Cho, Fetrow, Guthold, G. Holzwarth, Kim-Shapiro, Macosko, Salsbury, Shields.

Condensed Matter Physics. Computational materials physics: simulation and prediction of energy storage materials, development of "first principles" simulation methods, condensed matter theory, semi-classical electron dynamics, Berry phase effects, non-linearity, computational and theoretical materials science, condensed matter physics, solid state physics, density functional theory, first-principles calculations, NMR, van der Waals forces, magnetization. N. Holzwarth, Jurchescu, Matthews, Thonhauser, Williams.

Gravitational. Gravitational physics, general relativity, numerical relativity, black holes, neutron stars, compact binaries, initial data, gravitational waves, quantum field theory in curved space, particle physics. Anderson.

Experimental

Atomic, Molecular, & Optical Physics. Electron spin resonance in irradiated organic solids, transport properties, semiconductor trapping centers, laser materials, ultrafast spectroscopy, excitons, scintillators, energy research, optics, optical trapping, mechanical effects of light, optogenetics, optical microscopy. Bonin, Ucer, Williams.

Biophysics. Optics, optical trapping, mechanical effects of light, optogenetics, optical microscopy, motor proteins, kinesin, optical and electron paramagnetic spectroscopy, hemoglobin, nitric oxide, nitrite, sickle cell disease, cardiovascular disease, nanobiotechnology, atomic force and optical microscopy, single molecule experiments, protein-DNA interactions, thrombosis and hemostasis, nanofibers, electrospinning, tissue engineering, drug discovery, Center for Translational Science. Basu, Bonin, Cho, Fetrow, Guthold, G. Holzwarth, Kim-Shapiro, Macosko, Salsbury, Shields.

Condensed Matter Physics. Organic and flexible electronics, transport properties, semiconductor trapping centers, laser materials, ultrafast spectroscopy, excitons, scintillators, energy research. N. Holzwarth, Jurchescu, Matthews, Thonhauser, Williams.

Energy Sources & Environment. The Center is engaged in a broad range of projects from the development of medical technologies, to green energy technologies, to the understanding of the environmental and ethical implications of such nano-based technologies, material design and synthesis, carbon nanotubes, metal nanoparticles, quantum dots, polymers, cage structures, solar cells, biofuels, batteries, high efficiency organic transistors, new lighting systems, antibiotic resistance, wound healing, tissue regeneration.

Nano Science and Technology. Nanostructures and nanotechnology, nanomotors, solar cells, meta materials and negative index materials, organic electronics, thin-film transistors, field-effect transistors, organic semiconductors, single crystals, microstructure. Bonin, Carroll, Guthold, Jurchescu.

View additional information about this department at
www.gradschoolshopper.com

UNIVERSITY OF NORTH DAKOTA

DEPARTMENT OF PHYSICS AND ASTROPHYSICS

Grand Forks, North Dakota 58202-7129
http://arts-sciences.und.edu/physics-astrophysics

General University Information
President: Robert Kelley
Dean of Graduate School: Wayne Swisher
University website: http://und.edu/
Control: Public
Setting: Urban
Total Faculty: 851
Total Graduate Faculty: 643
Total number of Students: 14,194
Total number of Graduate Students: 3,055

Department Information
Department Chairman: Ju H. Kim, Chair
Department Contact: Ms. Connie Cicha, Department Secretary
  Total full-time faculty: 10
  Total number of full-time equivalent positions: 10
  Full-Time Graduate Students: 12
  First-Year Graduate Students: 3
  Female First-Year Students: 1

Department Address
101 Cornell Street
Witmer Hall, Room 213
Grand Forks, ND 58202-7129
Phone: (701) 777-2911
Fax: (701) 777-3523
E-mail: physics@und.edu
Website: http://arts-sciences.und.edu/physics-astrophysics

ADMISSIONS

Admission Contact Information
Address admission inquiries to: Dr. Kanishka Marasinghe, Physics and Astrophysics Dept., 101 Cornell St., Stop 7129, University of North Dakota, Grand Forks, ND 58202
Phone: (701) 777-3560
E-mail: kanishka.marasinghe@und.edu
Admissions website: http://graduateschool.und.edu/my-gradspace.cfm

Application deadlines
Fall admission:
U.S. students: March 1   *Int'l. students*: March 1

Application fee
U.S. students: $35   *Int'l. students*: $35

Admissions information
For Fall of 2013:
  Number of applicants: 12
  Number admitted: 3
  Number enrolled: 3

Admission requirements
Bachelor's degree requirements: Bachelor's degree in physics is required.
Minimum undergraduate GPA: 3.0

GRE requirements
The GRE is recommended.

Advanced GRE requirements
The Advanced GRE is recommended.

TOEFL requirements
The TOEFL exam is required for students from non-English-speaking countries.
  PBT score: 550
  iBT score: 79
IELTS (Min: 6.5) or iBT (Min: Total 79, Speaking 26, Reading 19, Listening 19, Writing 15)REQUIRED to be considered for a GTA.

TUITION

Tuition year 2013–14:
Tuition for in-state residents
  Full-time students: $332 per credit
  Part-time students: $332 per credit
Tuition for out-of-state residents
  Full-time students: $793 per credit
  Part-time students: $793 per credit
Almost all GTAs receive tuition waivers.
Credit hours per semester to be considered full-time: 6
Deferred tuition plan: Yes
Health insurance: Available at the cost of $600 per year.
Other academic fees: A major fraction of health insurance costs are covered by the Department.
Academic term: Semester
Number of first-year students who received full tuition waivers: 3

Teaching Assistants, Research Assistants, and Fellowships
Number of first-year
  Teaching Assistants: 3
  Research Assistants: 2
Average stipend per academic year
  Teaching Assistant: $18,158
  Research Assistant: $18,158
Above stipends are for academic year 13/14. Stipend for M.S. candidates for AY 13/14 is $15,328.

FINANCIAL AID

Application deadlines
Fall admission:
U.S. students: April 30   *Int'l. students*: April 30

Loans
Loans are available for U.S. students.
Loans are not available for international students.
GAPSFAS application required: No
FAFSA application required: No

For further information
Address financial aid inquiries to: Dr. Kanishka Marasinghe, Department of Physics and Astrophysics.
Phone: (701) 777-3560
E-mail: kanishka.marasinghe@und.edu
Financial aid website: http://arts-sciences.und.edu/physics-astrophysics

HOUSING

Availability of on-campus housing
Single students: Yes
Married students: Yes

For further information
Address housing inquiries to: Housing Office, University of North Dakota.
Phone: (701) 777-4251
E-mail: housing@und.edu
Housing aid website: http://und.edu/student-life/housing/

Table A—Faculty, Enrollments, and Degrees Granted

Research Specialty	2013-14 Faculty	Enrollment Fall 2012–13		Number of Degrees Granted 2012–13 (2006–12)		
		Master's	Doctorate	Master's	Terminal Master's	Doctorate
Astrophysics	3	1	3	–(2)	–	–
Condensed Matter Physics	7	1	7	1(10)	–	–(3)
Total	9	2	10	1(12)	–	–(3)
Full-time Grad. Stud.	–	2	10	–	–	–
First-year Grad. Stud.	–	1	2	–	–	–

GRADUATE DEGREE REQUIREMENTS

Master's: Thirty graduate credits are required. Thesis is required. There is no language requirement. A minimum of two semesters of full-time study in residence is required.

Doctorate: Ninety graduate credits, including a minimum of 9 credits in a related field in an approved program, are required. Qualifying examination, comprehensive examination, dissertation, and final oral examination are required. Two years of residency are required. There is no language requirement.

SPECIAL EQUIPMENT, FACILITIES, OR PROGRAMS

Scanning tunneling microscope, low-energy electron diffraction and AUGER spectrometer, ultra vacuum system, PerkinElmer DTA/TGA/DSC/TMA/DMA suite, Phillips X'Pert PRO X-ray diffractometer, ThermoNIcolet NEXUS 460 FTIR/FTRaman spectrometers, Bruker Vertex 70v FTIR spectrometers equipped with options for time-resolved spectroscopy (TRS), Advanced Research Systems (ARS) cryogen-free 4K optical cryostat, OXFORD magnetic AC susceptometer, tetrahedral anvil high-pressure/temperature press (up to 6 GPa and 1400 C), materials processing facilities, 6 teraflop HPC Linux cluster, 144TB high-availability storage appliance, TOAST (Transient Object Automated Search Telescope) 12" robotic robodome telescope, 20" plane wave telescope on Mathis fork mount, mobile launch facility for 250-lb research rocket, planetarium dome.

Table B—Separately Budgeted Research Expenditures by Source of Support

Source of Support	Departmental Research	Physics-related Research Outside Department
Federal government	$357,000	
State/local government	$38,000	
Non-profit organizations		
Business and industry		
Other		
Total	$395,000	

Table C—Separately Budgeted Research Expenditures by Research Specialty

Research Specialty	No. of Grants	Expenditures ($)
Condensed Matter Physics	1	$357,000
Total	1	$357,000

FACULTY

Chair Professor
Kovacs, Geza, Ph.D., Etovos University, Budapest, 1981. *Astronomy, Astrophysics*. Variable stars; data analysis; exo-solar planets.

Professor
Dewar, Graeme A., Ph.D., Simon Fraser University, 1980. *Condensed Matter Physics, Electromagnetism, Optics, Solid State Physics*. Solid-state physics; magnetoelasticity; magnetic materials; negative index of refraction.

Kim, Ju H., Ph.D., University of Chicago, 1990. Chair. *Condensed Matter Physics*. Condensed-matter theory; strongly correlated systems; superconductivity; superconductor quantum devices.

Marasinghe, Kanishka, Ph.D., University of Missouri, Rolla, 1993. *Condensed Matter Physics, Materials Science, Metallurgy, Nano Science and Technology, Solid State Physics*. Atomic structure-property relationships of complex materials; vitreous optical media and novel magnetic materials; x-ray and neutron diffraction; X-ray absorption spectroscopy; IR and Raman spectroscopies.

Schwalm, William A., Ph.D., Montana State University, 1978. *Condensed Matter Physics, Nonlinear Dynamics and Complex Systems, Solid State Physics, Theoretical Physics*. Condensed-matter theory; mathematical physics.

Associate Professor
Barkhouse, Wayne, Ph.D., University of Toronto, 2003. *Astrophysics*. Astrophysics; galaxy clusters; dark energy.

Young, Timothy R., Ph.D., University of Oklahoma, 1994. *Astrophysics*. Supernovae and supernova remnants; hydrodynamics; numerical simulations of radiation and neutrino transport.

Assistant Professor
Loh, Yen Lee, Ph.D., University of Cambridge, 2005. *Atomic, Molecular, & Optical Physics, Computational Physics, Condensed Matter Physics*. Condensed-matter theory (strongly correlated systems, superconductivity, magnetism, frustration); ultracold atomic gases; topological insulators; computational physics (quantum Monte Carlo, etc.).

Oncel, Nuri, Ph.D., University of Twente, Netherlands, 2007. *Condensed Matter Physics*. Solid-state physics; nanowires; quantum dots; molecular electronics; scanning tunneling microscopy.

Tung, Li-Chun, Ph.D., University of California, Riverside, 2005. *Applied Physics, Atomic, Molecular, & Optical Physics, Condensed Matter Physics, Energy Sources & Environment, Low Temperature Physics, Optics, Solid State Physics*. Experimental condensed-matter physics; infrared optics.

Emeritus
Lykken, Glenn I., Ph.D., University of North Carolina, 1966. *Medical, Health Physics*. Health physics; trace element analysis.

Rao, Sesh B., Ph.D., Pennsylvania State University, 1963. *Atomic, Molecular, & Optical Physics, Optics*. Optics; spectroscopy; atomic and molecular physics.

Adjunct Professor

Crockett, Scott, M.S., University of North Dakota, 2001. *Computational Physics*. Material science; condensed-matter physics; equation of state; shock physics.

Delhommelle, Jerome, Ph.D., University of Paris XI-Orsay, 2000. *Nonlinear Dynamics and Complex Systems, Statistical & Thermal Physics*. Molecular simulation of nonequilibrium processes.

Jones, Michael, Ph.D., University of North Dakota, 1978. Inorganic materials under high temperatures.

Lefevre, Russell, Ph.D., University of California, Santa Barbara, 1976. *Electrical Engineering, Physics and other Science Education*. Engineering and physics education.

Moritz, Brian, Ph.D., University of North Dakota, 2000. *Condensed Matter Physics, Solid State Physics, Surface Physics*. Theoretical condensed-matter physics.

Schwalm, Mizuho, Ph.D., Montana State University, 1978. *Condensed Matter Physics, Physics and other Science Education*. Theoretical solid-state physics; physics education.

DEPARTMENTAL RESEARCH SPECIALTIES AND STAFF

Theoretical

Astrophysics. Dark energy; supernova remnants; cool stars; radiative transfer; hydrodynamics. Barkhouse, Young.

Condensed Matter Physics. High-temperature superconductivity; magnetic properties; strongly correlated systems; disordered and nonlinear systems; topological insulators. Kim, Loh, Moritz, William Schwalm.

Experimental

Astrophysics. Gamma ray bursters; extra-solar planets; galaxy clusters; supernova light curves; exoplanet transits; GRB afterglows; asteroid astrometry; light curves. Barkhouse, Young.

Chemical Physics. Surface physics; phase transitions; trace elements in lead. Oncel.

Condensed Matter Physics. Transport phenomena; thin films; superconductivity; magnetism; high-pressure physics; negative index of refraction; magnetoelasticity; low-dimensional physics; scanning tunneling microscopy/spectroscopy; nanowires; molecular electronics; vitreous optical media. Dewar, Marasinghe, Oncel, Tung.

View additional information about this department at www.gradschoolshopper.com

---

BOWLING GREEN STATE UNIVERSITY

DEPARTMENT OF PHYSICS AND ASTRONOMY

Bowling Green, Ohio 43403
http://physics.bgsu.edu

General University Information
President: Mary Ellen Mazey
Dean of Graduate School: Michael Ogawa
University website: http://bgsu.edu
Control: Public
Setting: Rural
Total Faculty: 844
Total Graduate Faculty: 514
Total number of Students: 20,222
Total number of Graduate Students: 2,909

Department Information
Department Chairman: John B. Laird, Chair
Department Contact: John B. Laird, Chair
  Total full-time faculty: 15
  Total number of full-time equivalent positions: 15
  Full-Time Graduate Students: 20
  First-Year Graduate Students: 7

Department Address
104 Overman Hall
Bowling Green, OH 43403
Phone: (419) 372-2421

Fax: (419) 372-9938
E-mail: laird@bgsu.edu
Website: http://physics.bgsu.edu

ADMISSIONS

Admission Contact Information
Address admission inquiries to: Graduate College Admissions Office, Bowling Green State University, Bowling Green, OH 43403
Phone: (419) 372-2791
E-mail: gradcol@bgsu.edu
Admissions website: http://www.bgsu.edu/colleges/gradcol/

Application deadlines
Fall admission:
U.S. students: February 28   *Int'l. students*: February 28

Application fee
U.S. students: $45

Admissions information
For Fall of 2013:
  Number of applicants: 61
  Number enrolled: 7

Admission requirements

Bachelor's degree requirements: B.S. with major in physics or minor in physics with major in cognate field required.

GRE requirements

The GRE is required.

Advanced GRE requirements

The Advanced GRE is not required.

TOEFL requirements

The TOEFL exam is required for students from non-English-speaking countries.
PBT score: 550
iBT score: 79

Other admissions information

Additional requirements: Transcripts and three letters of recommendation are required.

TUITION

Tuition year 2012–13:
Tuition for in-state residents
  Full-time students: $6,784 annual
  Part-time students: $424 per credit
Tuition for out-of-state residents
  Full-time students: $11,664 annual
  Part-time students: $729 per credit
Credit hours per semester to be considered full-time: 8
Deferred tuition plan: No
Health insurance: Available
Other academic fees: $496/semester (full-time).
Academic term: Semester
Number of first-year students who received full tuition waivers: 7

Teaching Assistants, Research Assistants, and Fellowships

Number of first-year
  Teaching Assistants: 6
  Research Assistants: 1
Average stipend per academic year
  Teaching Assistant: $11,000
  Research Assistant: $16,000

FINANCIAL AID

Application deadlines

Fall admission:
U.S. students: February 28   *Int'l. students*: February 28

Loans

Loans are not available for U.S. students.
Loans are not available for international students.
GAPSFAS application required: No
FAFSA application required: No

For further information

Address financial aid inquiries to: Graduate Coordinator, Department of Physics and Astronomy.
Phone: (419) 372-2421
E-mail: fulcher@bgsu.edu
Financial aid website: http://www.bgsu.edu/gradcoll/

HOUSING

Availability of on-campus housing

Single students: Yes
Married students: No

For further information

Address housing inquiries to: Off-Campus Student Services.
Phone: (419) 372-2843

Table A—Faculty, Enrollments, and Degrees Granted

Research Specialty	2013-14 Faculty	Enrollment Fall 2013 Master's	Enrollment Fall 2013 Doctorate	Number of Degrees Granted 2012-13 (2008–13) Master's	Number of Degrees Granted 2012-13 (2008–13) Terminal Master's	Number of Degrees Granted 2012-13 (2008–13) Doctorate
Acoustics	1	–	–	–	-(2)	–
Astronomy	7	1	–	–	-(4)	–
Computational Physics	1	1	–	–	-(3)	–
Nano Science and Technology	3	7	5	–	4(16)	-(2)
Physics and other Science Education	1	–	–	–	-(11)	–
Solid State Physics	2	–	1	–	–	–
Non-specialized	-	6	–	–	2(3)	–
Total	15	15	6	–	6(39)	-(2)
Full-time Grad. Stud.	–	15	6	–	–	–
First-year Grad. Stud.	–	7	–	–	–	–

GRADUATE DEGREE REQUIREMENTS

Master's: Plan I: 30 semester hours: 3.0 GPA; formal thesis; oral examination on thesis; 24 semester hours in residence; no language requirement. Plan II: 32 semester hours; 3.0 GPA; no thesis, but comprehensive examination and scholarly paper required; 24 semester hours in residence; no language requirement.

Other Degrees: MAT degree: 22–26 semester hours in academic major, 8–14 semester hours in education; 3.0 GPA, research paper required.

Thesis: Thesis may be written in absentia.

SPECIAL EQUIPMENT, FACILITIES, OR PROGRAMS

Ph.D.-preparatory or career-oriented master's degree available. Students can pursue a Ph.D. through the Center for Photochemical Sciences or through a cooperative program with the University of Toledo. Computer access includes many workstations and a high-speed cluster in the Department and remote access to the Ohio Supercomputer Center. Laboratory facilities are available for research in nanoscale and solid-state materials, including quantum dots, light-emitting devices, photovoltaics, photonics, spintronics, materials defects, and magnetic resonance. Additional facilities available for graduate student research include electron microscopes and a 0.5-m telescope with a CCD camera.

Table B—Separately Budgeted Research Expenditures by Source of Support

Source of Support	Departmental Research	Physics-related Research Outside Department
Federal government	$678,000	
State/local government	$196,000	
Non-profit organizations		
Business and industry		
Other		
Total	$874,000	

Table C—Separately Budgeted Research Expenditures by Research Specialty

Research Specialty	No. of Grants	Expenditures ($)
Nano Science and Technology	2	$678,000
Physics and other Science Education	2	$196,000
Total	4	$874,000

FACULTY

Professor

Fulcher, Lewis P., Ph.D., University of Virginia, 1969. *Acoustics, Fluids, Rheology, Mechanics, Nuclear Physics.* Human phonation; modeling the human larynx.

Laird, John B., Ph.D., Yale University, 1983. *Astronomy, Astrophysics, Physics and other Science Education.* Stellar astronomy; stellar populations; astronomy and physics education.

Layden, Andrew C., Ph.D., Yale University, 1993. *Astronomy, Astrophysics.* Stellar astronomy; variable stars; stellar populations.

Smith, Dale W., Ph.D., University of Washington, 1978. *Astronomy, Physics and other Science Education.* Public education.

Associate Professor

Xi, Haowen, Ph.D., Lehigh University, 1993. *Computational Physics.* Computational physics; convection; econophysics.

Zamkov, Mikhail A., Ph.D., Kansas State University, 2003. *Condensed Matter Physics, Materials Science, Metallurgy, Nano Science and Technology, Solid State Physics.* Nanosystems; thin films; photovoltaic cells.

Assistant Professor

Selim, Farida A., Ph.D., Alexandria University, 1999. *Condensed Matter Physics, Materials Science, Metallurgy, Nano Science and Technology, Solid State Physics.* solid state materials; structural, optical, electronic, and magnetic properties of oxides.

Sun, Liangfeng, Ph.D., University of Texas, 2006. *Condensed Matter Physics, Materials Science, Metallurgy, Nano Science and Technology, Solid State Physics.* Nanomaterials; surface nonlinear optics.

Zayak, Alexey T., Ph.D., University of Duisburg-Essen, 2004. *Computational Physics, Condensed Matter Physics, Materials Science, Metallurgy, Nano Science and Technology, Solid State Physics.* Condensed-matter theory; nanoscale systems.

Emeritus

Boughton, Robert I., Ph.D., Ohio State University, 1968. *Condensed Matter Physics, Low Temperature Physics, Materials Science, Metallurgy, Physics and other Science Education, Solid State Physics.* Low-temperature solid-state physics; physics education.

Crandall, A. Jared, Ph.D., Michigan State University, 1967. *Solar Physics.* Microcomputer interfacing; solar energy research.

Shirkey, Charles T., Ph.D., Ohio State University, 1969.

Instructor

Attygalle, Lilani, Ph.D., University of Toledo, 2008. *Materials Science, Metallurgy, Nano Science and Technology, Physics and other Science Education.* Materials physics.

Blanton, Miles C., Ph.D., University of North Carolina, 2008. *Astronomy, Astrophysics.* Radioastronomy.

Dellenbusch, Kate E., Ph.D., University of Wisconsin-Madison, 2008. *Astronomy, Astrophysics, Physics and other Science Education.* Extragalactic astronomy; astronomy education.

Mandell, Eric S., Ph.D., University of Missouri, 2007. *Astronomy, Astrophysics, Materials Science, Metallurgy, Nano Science and Technology, Physics and other Science Education, Solid State Physics.* Physics education; carbon nanostructures and star dust.

Rogel, Allen B., Ph.D., Indiana University, 2005. *Astronomy, Astrophysics.* Stellar astronomy; cataclysmic variables.

Lecturer

Tiede, Glenn P., Ph.D., Ohio State University, 1997. *Astronomy, Astrophysics.* Stellar populations.

DEPARTMENTAL RESEARCH SPECIALTIES AND STAFF

Theoretical

Acoustics. Human phonation; modeling the human larynx. Fulcher.

Computational Physics. Electronic structure; convection; econophysics. Xi, Zayak.

Nano Science and Technology. Nanostructures and condensed-matter physics; light-matter interactions. Zayak.

Experimental

Acoustics. Measurements of air flow and pressure in the human larynx. Fulcher.

Astronomy. Stellar spectroscopy and abundances; stellar populations; chemical evolution of galaxies; variable stars; cataclysmic variables; interstellar matter; extragalactic astronomy. Blanton, Dellenbusch, Laird, Layden, Mandell, Rogel, Tiede.

Nano Science and Technology. Nanosystems; nonlinear optics and devices; thin film devices; photovoltaic cells. Attygalle, Mandell, Sun, Zamkov.

Physics and other Science Education. Active learning strategies and materials; teacher preparation; public education. Attygalle, Blanton, Dellenbusch, Laird, Mandell, Smith, Tiede.

Solid State Physics. thin films; optical, electronic, magnetic properties oxides; solid-state lighting; photovoltaics. Attygalle, Mandell, Selim, Sun, Zamkov.

View additional information about this department at
www.gradschoolshopper.com

CASE WESTERN RESERVE UNIVERSITY

DEPARTMENT OF PHYSICS

Cleveland, Ohio 44106-7079
http://physics.cwru.edu

General University Information
President: Barbara Snyder
Dean of Graduate School: Charles E. Rozek
University website: http://www.cwru.edu
Control: Private
Setting: Urban
Total number of Students: 10,026
Total number of Graduate Students: 5,640

Department Information
Department Chair: Kathleen Kash, Chair
Department Contact: Corbin E. Covault, Director of the
   Graduate Program
   Total full-time faculty: 24
   Total number of full-time equivalent positions: 24
   Full-Time Graduate Students: 54
   First-Year Graduate Students: 10
   Female First-Year Students: 3
   Total Post Doctorates: 19

Department Address
Rockefeller Building
2076 Adelbert Road
Cleveland, OH 44106-7079
Phone: (216) 368-4000
Fax: (216) 368-4671
E-mail: admissions@phys.case.edu
Website: http://physics.cwru.edu

ADMISSIONS

Admission Contact Information
Address admission inquiries to: Admissions Director, Physics
Phone: (216) 368-8779
E-mail: admissions@phys.case.edu
Admissions website: http://www.phys.case.edu/grad/apply.php

Application deadlines
Fall admission:
U.S. students: January 15   *Int'l. students*: January 15

Application fee
U.S. students: $8   *Int'l. students*: $8
First-year graduate students are generally not considered for
   Spring semester admission. Advanced transfer students who
   intend to matriculate in January are treated on a case-by-case
   basis.

Admissions information
For Fall of 2013:
   Number of applicants: 179
   Number admitted: 54
   Number enrolled: 13

Admission requirements
Bachelor's degree requirements: Bachelor's degree in physics,
   mathematics, or related field is required.
Minimum undergraduate GPA: 3.0

GRE requirements
The GRE is recommended.
All applicants are strongly encouraged to complete the GRE General Test and to arrange to have scores submitted directly to CWRU. The inclusion of GRE scores will considerably strengthen any application.

Advanced GRE requirements
The Advanced GRE is recommended.
All applicants are strongly encouraged to complete the GRE Physics Subject Test and to arrange to have scores submitted directly to CWRU. The inclusion of GRE scores will considerably strengthen any application. GRE Physics subject test scores are expected from all domestic applicants who are applying with bachelor's degrees only. Advanced students may submit a record of graduate coursework and graduate level research experience in lieu of GRE Physics subject test results.

TOEFL requirements
The TOEFL exam is required for students from non-English-
   speaking countries.
   PBT score: 557
   iBT score: 90

Other admissions information
Additional requirements: No minimum acceptable GRE score is
   specified.
Undergraduate preparation assumed: Taylor, Classical Mechanics; Griffiths, Electrodynamics; Kittel, Thermal Physics; Griffiths, Quantum Mechanics; or equivalent textbooks; one or two years of advanced laboratory courses.

TUITION
Tuition year 2013-14:
   Full-time students: $27,828 annual
   Part-time students: $1,546 per credit
Credit hours per semester to be considered full-time: 9
Deferred tuition plan: Yes
Health insurance: Available at the cost of $1,452 per year.
Other academic fees: Effective Spring 2014, health insurance will
   be made available at no cost to incoming graduate students.
Academic term: Semester
Number of first-year students who received full tuition waivers: 10

Teaching Assistants, Research Assistants, and Fellowships
Number of first-year
   Teaching Assistants: 10
   Research Assistants: 3
Average stipend per academic year
   Teaching Assistant: $22,860
   Research Assistant: $22,860
   Fellowship student: $22,860
   TAs, RAs and Fellows in the PhD program are also eligible
   for tuition support.

FINANCIAL AID

Application deadlines
Fall admission:
U.S. students: January 15

Loans

Loans are available for U.S. students.
Loans are available for international students.
GAPSFAS application required: No
FAFSA application required: No

For further information

Address financial aid inquiries to: Director of Admissions.
E-mail: admissions@phys.case.edu

HOUSING

Availability of on-campus housing

Single students: No
Married students: No

For further information

Address housing inquiries to: Dean, Graduate Studies.
Housing aid website: http://gradstudies.case.edu/prospect/area/housing.html

Table A—Faculty, Enrollments, and Degrees Granted

Research Specialty	2013-2014 Faculty	Enrollment Spring 2013		Number of Degrees Granted 2013 (2005–13)		
		Master's	Doctorate	Master's	Terminal Master's	Doctorate
Applied Physics	2	3	–	–	1(8)	–(6)
Astrophysics	8	–	21	–	–(3)	–(18)
Atomic, Molecular, & Optical Physics	–	–	–	–	–	–
Biophysics	1	–	1	–	–	–(1)
Condensed Matter Physics	12	–	28	–(6)	1(5)	2(22)
Fluids, Rheology	1	–	4	–	–	–(1)
Low Temperature Physics	–	–	–	–(2)	–	–(2)
Medical, Health Physics	2	–	5	–(1)	–(10)	–
Nano Science and Technology	2	–	8	–	–	1(3)
Optics	5	–	15	–(2)	–	1(9)
Polymer Physics/Science	1	–	2	–	–	–(1)
Relativity & Gravitation	4	–	7	–	–	–(5)
Statistical & Thermal Physics	2	–	2	–	–	–
Surface Physics	1	–	–	–	–	–(2)
Non-specialized	–	1	–	–	–	–
Other	–	–	–	–	–	–
Total	23	4	54	–(8)	2(12)	2(56)
Full-time Grad. Stud.	–	4	54	–	–	–
First-year Grad. Stud.	–	2	10	–	–	–

GRADUATE DEGREE REQUIREMENTS

Master's: Twenty-seven graduate credit hours in approved program including six required hours; Master's exam required; thesis option; no residence or language requirement.

Doctorate: Up to 36 hours of coursework is required (may be reduced by graduate coursework done elsewhere); comprehensive and topical exams, dissertation, and dissertation exam required; one year residency; no language exam required. See http://www.phys.cwru.edu/grad/phd.php.

Thesis: Thesis may be written in absentia.

SPECIAL EQUIPMENT, FACILITIES, OR PROGRAMS

A wide variety of facilities and programs is available within the department, and in addition there are collaborative programs with other departments, including Macromolecular Science, Chemistry, Astronomy, Materials Science, and the Medical School.

In astrophysics research, experiments in collaboration with other universities are being performed to determine the nature of elementary particle dark matter in the universe by direct detection in underground detectors, while other experiments are performed to search for high-energy gamma rays and to explore the Cosmic Microwave Background. High-energy physics experiments are undertaken at various National Laboratories. Theoretical work on astrophysics and cosmology, as well as particle, condensed matter physics, and quantum computing, covers a large number of research topics.

Condensed matter studies include measurements of dielectric; optical and nonlinear optical properties; thin-film properties; nanoscopic physics; quantum computing; liquid crystal and complex fluid properties; semiconductor crystal growth; quantum wells, wires, and dots; other nanoscopic structures; spintronics; organic electronics; and photovoltaics.

A wide range of facilities is available in surface physics and in optics. Among the collaborative programs are experimental and theoretical studies of phase transitions in polymers and of liquid crystals, photovoltaic materials, surface physics, the physics of imaging, fluid physics, dark matter detection, and measurements of fundamental parameters in cosmology.

Departmental computing facilities are extensive and are used in both research and courses. Weekly specialized seminars in particle/astrophysics and condensed matter physics take place, in addition to a weekly departmental colloquium.

The Physics Department has been recognized six times by the U.S. Dept. of Education as meeting vital national needs. Special graduate fellowships are available.

In addition to a traditional physics program, the Department maintains a Physics Entrepreneurship Masters degree program. The program is designed to empower physicists as entrepreneurs and to enable students and graduates to build on their physics skills to start new high-tech businesses or to launch new product lines in existing companies.

Special Programs Center for Education and Research in Cosmology and Astrophysics: A new center created in collaboration with the Cleveland Museum of Natural History's Shafran Planetarium and CWRU's Astronomy Department to promote research and education in cosmology and astrophysics. http://cerca.case.edu

Institute for Advanced Materials: The Institute for Advanced Materials brings together internationally recognized faculty researchers to engage in multi-disciplinary efforts on a broad range of materials that not only are ubiquitous in everday life, but are cornerstones to many key technology areas. Specifically, IAM focuses on strategic research that impacts national needs in human health, energy, and the environment. The four focus areas are: Fundamental Materials Research, Materials for Human Health, Materials for Energy, and Materials for Sustainability. http://iam.case.edu

The Institute for the Science of Origins ISO is a collaborative team of faculty members and researchers from diverse scientific disciplines seeking to understand how complex systems emerge and evolve, from the universe to the mind, from microbes to humanity http://www.case.edu/origins

The Michelson Postdoctoral Lectureship is an annual prize sponsored by Case Western Reserve University. It is awarded to an outstanding recent Physics Ph.D. based on an international competition. The winner spends one week in residence in the Department, and delivers several seminars and a departmental colloquium on his/her research.

Physics Entrepreneurship Masters Degree: To empower physicists as entrepreneurs and enable graduate students to build on their physics skills to start new high-tech businesses or to launch new product lines in existing companies.

Workshops and Conferences: The Department regularly holds national and international meetings on a variety of topics. Recent conferences have included: The Future of Cosmology, Future Physics and Future Facilities, The Cosmic Microwave Background, Great Lakes Cosmology Workshop, the American Vacuum Society Conference, International Workshop on MRI, Einstein's Legacy, and Confronting Gravity.

Outreach: The Department works with high school teachers and students to improve science education locally and nationally. The Department also hosts a web site of a national program in astronomy education called Ask an Astronomer.

Recent Books by Faculty include Magnetic Resonance Imaging: Physical Properties and Sequence Design Robert Brown, A Quantum Approach to Condensed Matter Physics Philip L. Taylor.

International Programs: The department spearheaded three university-wide student and faculty exchange programs with the University of Calabria Italy, Nagaoka University of Science and Technology Japan and Universite´ Pierre et Marie Curie U. Paris 6, France.

Table B—Separately Budgeted Research Expenditures by Source of Support

Source of Support	Departmental Research	Physics-related Research Outside Department
Federal government	$19,350,081	
State/local government	$3,288,461	
Non-profit organizations	$5,556,840	
Business and industry	$2,031,767	
Other	$449,267	
Total	**$30,676,416**	

Table C—Separately Budgeted Research Expenditures by Research Specialty

Research Specialty	No. of Grants	Expenditures ($)
Astrophysics	34	$14,877,718
Condensed Matter Physics	23	$12,437,237
Medical, Health Physics	6	$1,296,976
Other	8	$2,064,484
Total	**71**	**$30,676,415**

FACULTY

Professor

Akerib, Daniel S., Ph.D., Princeton University, 1991. *Astronomy, Astrophysics, Cosmology & String Theory*. Experimental particle astrophysics, dark matter, low temperature detectors, particle physics.

Alexander, Iwan, Ph.D., Washington State University, 1981. Joint appointment with mechanical engineering. Fluid physics, microgravity.

Brown, Robert W., Ph.D., Massachusetts Institute of Technology, 1968. *Medical, Health Physics, Particles and Fields*. Theoretical physics; elementary particles; imaging physics.

Chottiner, Gary S., Ph.D., University of Maryland, 1980. *Condensed Matter Physics, Surface Physics*. Experimental condensed matter physics; surface physics.

Covault, Corbin, Ph.D., Harvard University, 1991. *Astrophysics, Atmosphere, Space Physics, Cosmic Rays*. Experimental high-energy astrophysics, particle interactions, cosmic rays.

Kash, Kathleen, Ph.D., Massachusetts Institute of Technology, 1982. Experimental condensed matter physics; optics; mesoscopic physics.

Kowalski, Kenneth L., Ph.D., Brown University, 1963. *Particles and Fields, Theoretical Physics*. Theoretical physics; nuclear and elementary particle physics.

Lambrecht, Walter R. L., Ph.D., Ghent University, 1980. *Condensed Matter Physics, Theoretical Physics*. Theoretical condensed matter physics; electronic structure of materials.

Luck, Earle, Ph.D., University of Texas, 1977. Joint appointment with astronomy. *Astronomy*. Stellar and galactic chemical evolution; stellar abundance analysis; spectrum synthesis techniques.

Mihos, Christopher, Ph.D., University of Michigan, 1992. Joint appointment with astronomy. *Astronomy, Astrophysics, Computational Physics, Cosmology & String Theory*. Observational and computational astrophysics; galactic dynamics; galaxy clusters; galaxy evolution.

Morrison, Heather, Ph.D., Australian National University, 1988. Joint appointment with astronomy. *Astronomy, Astrophysics*. Galaxy structure, formation, and evolution, especially Milky Way and Local Group.

Petschek, Rolfe G., Ph.D., Harvard University, 1981. *Condensed Matter Physics, Statistical & Thermal Physics, Theoretical Physics*. Theoretical physics; statistical physics; condensed matter physics.

Rosenblatt, Charles, Ph.D., Harvard University, 1978. *Condensed Matter Physics, Fluids, Rheology, Optics*. Experimental condensed matter physics; liquid crystals and complex fluids; optics; microgravity; fluid physics.

Ruhl, John, Ph.D., Princeton University, 1993. *Astrophysics, Cosmology & String Theory*. Experimental particle astrophysics, cosmic microwave background.

Shutt, Thomas, Ph.D., University of California, Berkeley, 1993. *Astrophysics, Cosmology & String Theory*. Experimental particle astrophysics, dark matter, neutrino physics.

Singer, Kenneth D., Ph.D., University of Pennsylvania, 1981. *Condensed Matter Physics, Nano Science and Technology, Optics*. Experimental physics; nonlinear optics; organic electronics; photovoltaics. Nanophysics.

Starkman, Glenn, Ph.D., Stanford University, 1988. *Astrophysics, Cosmology & String Theory, Particles and Fields, Theoretical Physics*. Theoretical physics, cosmology, particle physics, and astrophysics.

Strangi, Guiseppe, Ph.D., University of Calabria. *Condensed Matter Physics, Nano Science and Technology, Optics*. Experimental Condensed Matter Physics Optics and Photonics of Soft Condensed Matter DFB and Random Lasing Opto-Plasmonics in Nanostructured Metamaterials.

Taylor, Cyrus C., Ph.D., Massachusetts Institute of Technology, 1984. *Applied Physics, Particles and Fields, Theoretical Physics*. Theoretical physics, theoretical and experimental elementary particle physics; physics of entrepreneurship.

Taylor, Philip L., Ph.D., University of Cambridge, 1962. *Condensed Matter Physics, Theoretical Physics*. Theoretical condensed matter physics; physics of polymers and liquid crystals.

Associate Professor

Gao, Xuan, Ph.D., Columbia University, 2003. *Condensed Matter Physics, Nano Science and Technology*. Experimental condensed matter physics; Applied physics; electronic properties of low dimensional nanostructures; semiconductor nanowires; nanosensors.

Jankowsky, Eckhard, Ph.D., Dresden Institute of Technology, 1996. Joint appointment with biochemistry. Experimental biophysics, single molecule fluorescence, enzyme kinetics.

Martens, Michael A., Ph.D., Case Western Reserve University, 1991. *Medical, Health Physics*. Imaging physics.

Mathur, Harsh, Ph.D., Yale University, 1994. *Astrophysics, Condensed Matter Physics, Cosmology & String Theory, The-*

oretical Physics. Theoretical condensed matter physics; localization and mesoscopic physics. Cosmology and particles.

Shan, Jie, Ph.D., Columbia University, 2001. *Condensed Matter Physics, Optics*. Experimental condensed matter physics; ultrafast spectroscopy; terahertz time-domain spectroscopy.

Zehavi, Idit, Ph.D., Hebrew University of Jerusalem, 1999. Joint appointment with astronomy. *Cosmology & String Theory*. Theoretical astrophysics; cosmology; large-scale structure; galaxy and structure formation.

Assistant Professor

Berezovsky, Jesse, Ph.D., University of California, Santa Barbara, 2007. *Condensed Matter Physics, Quantum Foundations*. Experimental condensed matter. Transport, quantum coupling of spins and photons, quantum information.

de Rham, Claudia, Ph.D., University of Cambridge, 2005. *Astrophysics, Cosmology & String Theory, Relativity & Gravitation, Theoretical Physics*. Theoretical cosmology and particle physics.

Tolley, Andrew J., Ph.D., University of Cambridge, 2003. *Astrophysics, Cosmology & String Theory, Relativity & Gravitation, Theoretical Physics*. Theoretical cosmology and particle physics. Dark energy.

DEPARTMENTAL RESEARCH SPECIALTIES AND STAFF

Theoretical

Electronic Properties of Metals and Semiconductors. Electronic properties of metals and semiconductors; photovoltaics, crystal growth; transport properties in ordered and disordered materials; band structure; deformation potentials; localization, thermo-electricity; interface and surface physics; lattice vibrations. Lambrecht, Mathur, Petschek, Philip Taylor.

Imaging Physics. Algorithm development; bio-data acquisition and analysis; rf coil theory; inverse scattering theory; diagnostic imaging. Brown, Martens.

Liquid Crystals. Phase transitions, dynamics, symmetry and surface effects, nonlinear behavior. Petschek, Philip Taylor.

Particle Astrophysics. Cosmology and Gravitational Physics. Neutrino astrophysics; early universe cosmology; dark matter; dark energy; large scale structure; gravitational lensing; black hole evaporation; stellar evolution; cosmic strings; cosmic microwave background. de Rham, Mathur, Starkman, Cyrus Taylor, Tolley.

Particle Physics. Electroweak theory; standard model; cosmology; black hole physics; superstring theory SSC physics; supersymmetry, field theories at finite temperature; quark-gluon plasma; diffractive excitation mechanisms. de Rham, Kowalski, Mathur, Starkman, Cyrus Taylor, Tolley.

Polymer Physics/Science. Equations-of-state; phase transitions; dynamical behavior; piezoelectric effects; polymer liquid crystals. Petschek, Philip Taylor.

Statistical & Thermal Physics. Statics and dynamics of phase transitions; pattern formation and dendritic growth; liquid crystals, polymeric liquid crystals, complex fluids; oscillatory chemical reactions; membrane noise. Petschek, Philip Taylor.

Experimental

Electronic Strucure of Materials. Electronic structure of metals and alloys; surfaces; crystal growth; thin films; amorphous films; dielectric and cohesive properties; dielectric and mechanical relaxation; organic electronics; transport properties of nano-structures, quantum wells, mesoscopic systems, fuel cells; soft matter. Berezovsky, Gao, Kash, Strangi.

Experimental Particle Astrophysics. Dark matter; low-temperature detectors; neutrino experiments; cosmic microwave background; high-energy cosmic rays; gamma ray astrophysics. Akerib, Covault, Ruhl, Shutt.

Fluid Physics. Interface instabilities, magnetic levitation. Rosenblatt.

High Energy Particle Physics. Collider physics. Hadronic interactions. Akerib, Covault, Cyrus Taylor.

Liquid Crystals and Complex Fluids. Phase transitions; optical, magnetic, and electrical properties; microgravity; nanostructured LCs, symmetry effects. Chottiner, Rosenblatt.

Nanoscopic Physics. Quantum dots, wires, molecular electronics, nanoscopic surface modification, nanowires, and sensors. Berezovsky, Gao, Kash, Rosenblatt, Shan, Singer, Strangi.

Optical Properties of Materials. Linear and nonlinear optical properties of organic, polymeric materials, and mesoscopic systems, photovoltaics, ultrafast spectroscopy. Berezovsky, Kash, Rosenblatt, Shan, Singer, Strangi.

Polymer Physics/Science. Phase transformations; dielectric properties; magnetic and electric field effects; optical mechanical properties. Rosenblatt, Shan, Singer.

Surface Physics. Surface magnetization; secondary electron emission; surface analysis; physi- and chemisorption. Chottiner.

View additional information about this department at
www.gradschoolshopper.com

MIAMI UNIVERSITY

DEPARTMENT OF PHYSICS

Oxford, Ohio 45056
http://www.MiamiOH.edu/physics

General University Information

President: David Hodge
Dean of Graduate School: James Oris, Dean
University website: http://www.MiamiOH.edu
Control: Public

Setting: Rural
Total Faculty: 1,500
Total Graduate Faculty: 800
Total number of Students: 17,557
Total number of Graduate Students: 2,476

Department Information
Department Chairman: Herbert Jaeger, Chair
Department Contact: Dr. Samir Bali, Graduate Director
Total full-time faculty: 15
Total number of full-time equivalent positions: 15
Full-Time Graduate Students: 21
First-Year Graduate Students: 11
Female First-Year Students: 3

Department Address
620 E. Spring Street
133 Culler Hall
Oxford, OH 45056
Phone: (513) 529-5635
Fax: (513) 529-5629
E-mail: physics@MiamiOH.edu
Website: http://www.MiamiOH.edu/physics

ADMISSIONS

Admission Contact Information
Address admission inquiries to: Graduate School, 102 Roudebush
   Hall, Oxford, OH 45056
Phone: (513) 529-3734
E-mail: gradschool@MiamiOH.edu
Admissions website: http://www.miamioh.edu/graduate-studies/
   admission/index.html

Application deadlines
Fall admission:
U.S. students: March 15   *Int'l. students*: March 15

Application fee
U.S. students: $50   *Int'l. students*: $50
Late applications may be considered until all positions are filled.

Admissions information
For Fall of 2013:
   Number of applicants: 39
   Number admitted: 18
   Number enrolled: 11

Admission requirements
Bachelor's degree requirements: A bachelor's degree in physics
   or related areas is required. Consult our Graduate Student Adviser about the appropriateness of related area.
Minimum undergraduate GPA: 2.75

GRE requirements
The GRE is recommended.

Advanced GRE requirements
The Advanced GRE is recommended.

TOEFL requirements
The TOEFL exam is required for students from non-English-
   speaking countries.
   PBT score: 550
   iBT score: 80

Other admissions information
Additional requirements: A brief statement (one page or
   less) of research interest must be included with the
   application.
Undergraduate preparation assumed: Classical Mechanics, Symon; Electromagnetism, Griffiths; Quantum Mechanics, Griffiths; Statistical and Thermal Physics, Reif; courses in linear
   algebra and differential equations.

TUITION
Tuition year 2012–13:
Tuition for in-state residents
   Full-time students: $12,444 annual
Tuition for out-of-state residents
   Full-time students: $27,483 annual
Credit hours per semester to be considered full-time: 9
Deferred tuition plan: No
Health insurance: Available at the cost of $862 per year.
Other academic fees: Miami Metro (bus), Student Technology
   fee, and facilities fee total per academic year is $528. Some
   individual courses may charge special course fees.
Academic term: Semester
Number of first-year students who received full tuition waivers: 8

Teaching Assistants, Research Assistants, and Fellowships
Number of first-year
   Teaching Assistants: 8
Average stipend per academic year
   Teaching Assistant: $14,911
   Research Assistant: $14,911

FINANCIAL AID

Application deadlines
Fall admission:
U.S. students: March 1   *Int'l. students*: March 1

Loans
Loans are available for U.S. students.
Loans are available for international students.
GAPSFAS application required: No
FAFSA application required: No

For further information
Address financial aid inquiries to: Graduate School.
Phone: (513) 529-3734
Financial aid website: http://www.miamioh.edu/admission/finaid/
   graduate/index.html

HOUSING

For further information
Address housing inquiries to: Graduate School.
Phone: (513) 529-3734
Housing aid website: http://miamioh.edu/graduate-studies/student-
   life/housing.html

Table A—Faculty, Enrollments, and Degrees Granted

Research Specialty	2012-13 Faculty	Enrollment Fall 2012 Master's	Enrollment Fall 2012 Doctorate	Number of Degrees Granted 2012-13 (2009–13) Master's	Number of Degrees Granted 2012-13 (2009–13) Terminal Master's	Number of Degrees Granted 2012-13 (2009–13) Doctorate
Applied Physics	4	1	–	–	–(2)	–
Astrophysics	1	1	–	–	1(3)	–
Atomic, Molecular, & Optical Physics	2	7	–	–	4(12)	–
Biophysics	1	3	–	–	2(6)	–
Condensed Matter Physics	4	5	–	–	2(9)	–
Nano Science and Technology	3	1	–	–	1(6)	–
Physics and other Science Education	3	–	–	–	–(2)	–
Quantum Foundations	3	1	–	–	1(6)	–
Other	2	1	–	–	1(2)	–
Total	15	20	–	–	12(48)	–
Full-time Grad. Stud.	–	21	–	–		
First-year Grad. Stud.	–	11	–	–		

GRADUATE DEGREE REQUIREMENTS

Master's: For the thesis option, a minimum of 30 semester hours of graduate course work, research, and thesis credit is required. You must write a thesis proposal and defend it before your thesis committee. Subsequent completion and defense of the thesis are required. For the non-thesis option, a minimum of 36 semester hours of graduate credit is required. A comprehensive examination must also be passed. The thesis option is strongly recommended. For either the thesis or non-thesis option, you are expected to show proficiency in the areas of quantum physics, classical mechanics, electromagnetism, statistical physics, and mathematical methods used in physics. Evidence of proficiency means successful completion of courses at the graduate level. Graduate course work is selected in consultation with the thesis director (thesis option) and graduate program director.

SPECIAL EQUIPMENT, FACILITIES, OR PROGRAMS

Faculty qualified to direct graduate student research maintain or have access to the following: cold atom trap and optical lattice with tunable diode lasers; pulsed laser polarization spectrometer; nanosecond time-resolved fluorescence spectrometer and sectioning microscope; cell culture facility of laminar flow-hood, CO_2 incubators; near-field scanning optical microscope; electron beam lithography system; gamma-ray spectrometer; quantum design physical properties measurement system; ferromagnetic resonance spectrometer; grid cluster for computation; and Class 1000 clean room to fabricate nanodevices.

Table B—Separately Budgeted Research Expenditures by Source of Support

Source of Support	Departmental Research	Physics-related Research Outside Department
Federal government	$294,276	
State/local government		
Non-profit organizations		
Business and industry	$10,000	
Other		
Total	$304,276	

Table C—Separately Budgeted Research Expenditures by Research Specialty

Research Specialty	No. of Grants	Expenditures ($)
Atomic, Molecular, & Optical Physics	2	$95,000
Biophysics	1	
Condensed Matter Physics	7	$80,441
Physics and other Science Education	1	$136,148
Total	11	$311,589

FACULTY

Professor

Jaeger, Herbert, Ph.D., Oregon State University, 1987. Department Chair. *Applied Physics, Condensed Matter Physics.* Experimental solid state.

Rice, Perry, Ph.D., University of Arkansas, 1988. *Quantum Foundations.* Theoretical quantum optics and quantum information.

Taylor, Beverley A. P., Ph.D., Clemson University, 1978. *Physics and other Science Education.*

Yarrison-Rice, Jan M., Ph.D., University of Arkansas, 1990. Adjunct Professor of Physics, University of Cincinnati. *Condensed Matter Physics, Energy Sources & Environment, Nano Science and Technology.* Fabrication and characterization of nanodevices used for biosensing, photovoltaics, and tunable solar cells; exploration of the basic science of semiconducting nanowires, nanosheets, and gold nanoparticles grown in different configurations via optical spectroscopies, lithography, photocurrent analysis, and FDTD modeling.

Associate Professor

Alexander, Stephen, Ph.D., Pennsylvania State University, 1990. Chief Departmental Adviser. *Astronomy, Astrophysics, Computational Physics.* Planetary dynamics.

Bali, Samir, Ph.D., University of Rochester, 1994. Graduate Director. *Atomic, Molecular, & Optical Physics, Quantum Foundations.* Experimental quantum optics.

Bayram, S. Burcin, Ph.D., Old Dominion University, 1998. *Atomic, Molecular, & Optical Physics, Chemical Physics.* Experimental atomic spectroscopy.

Blue, Jennifer, Ph.D., University of Minnesota, 1997. *Physics and other Science Education.*

Clemens, James, Ph.D., University of Oregon, 2003. *Computational Physics, Quantum Foundations.* Theoretical quantum optics and quantum information.

Urayama, Paul, Ph.D., Princeton University, 2001. *Biophysics.* Metabolic sensing; piezophysiology; high-pressure biotechnology.

Assistant Professor

Eid, Khalid, Ph.D., Michigan State University, 2002. *Applied Physics, Condensed Matter Physics, Nano Science and Technology.* Nanotechnology; magnetism; semiconductors.

Khan, Mahmud, Ph.D., Southern Illinois University, 2007. *Condensed Matter Physics, Energy Sources & Environment.* Magnetism; materials for energy applications.

Lecturer

Beer, Christopher P., Ph.D., Ball State University, 2010. *Physics and other Science Education.*

DEPARTMENTAL RESEARCH SPECIALTIES AND STAFF

Theoretical

Astrophysics. Numerical simulations with gravitational N-body codes. Alexander.

Computational Physics. Scientific visualization, image processing and analysis, and the use of computer graphics and animation to enhance comprehension of physical phenomena. Clemens.

Quantum Optics. Theoretical and computational modeling of light and matter for generating nonclassical or entangled states of light and atoms. Clemens, Rice.

Experimental

Atomic, Molecular, & Optical Physics. Polarization spectroscopy and electron-imaging spectroscopy of alkali-rare gas collisions; line narrowing of high-power broad-area diode lasers; dynamics of laser-cooled atoms in magneto-optical traps and optical lattices; electromagnetically induced transparency and absorption in atoms; imaging and optical sensing in turbid media for biological and environmental applications. Bali, Bayram.

Biophysics. High-pressure studies; methods of fluorescence-based metabolic sensing and microscopy imaging applied to piezophysiological studies. Urayama.

Condensed Matter Physics. Fabrication and optical characterization of nanoscale materials and devices, including biosensors, using a variety of methods such as electron beam lithography and photolithography, magnetoresistance in nanodevices at cryogenic temperatures, nanoscale magnetodynamics in reduced dimensional systems, angular correlation spectroscopy of ceramic materials, positron annihilation lifetime spectroscopy, electronic and thermal properties of novel solid state materials, magnetic and structural transitions, giant magnetorisistance, giant magnetocaloric effect, exchange bial effect, shape memory effects, and permanent magnetic materials. Eid, Jaeger, Khan, Yarrison-Rice.

Physics and other Science Education. Strategies for teaching scientific reasoning and problem-solving skills in introductory physics classes; elementary school and K-12 science education. Beer, Blue, Taylor.

**View additional information about this department at
www.gradschoolshopper.com**

OHIO UNIVERSITY

DEPARTMENT OF PHYSICS AND ASTRONOMY

Athens, Ohio 45701
http://www.phy.ohiou.edu

General University Information
President: Roderick McDavis
Dean of Graduate School: Joseph Shields
University website: http://www.ohio.edu
Control: Public
Setting: Rural
Total Faculty: 914
Total number of Students: 21,724
Total number of Graduate Students: 4,204

Department Information
Department Chairman: David Ingram, Chair
Department Contact: Candy Dishong, Assistant Department Administrator
Total full-time faculty: 26
Total number of full-time equivalent positions: 26
Full-Time Graduate Students: 85
First-Year Graduate Students: 16
Female First-Year Students: 2
Total Post Doctorates: 7

Department Address
1 Ohio University
Clippinger Laboratories
Athens, OH 45701
Phone: (740) 593-1709
Fax: (740) 593-0433
E-mail: dishong@ohio.edu
Website: http://www.phy.ohiou.edu

ADMISSIONS

Admission Contact Information
Address admission inquiries to: Chair of Graduate Admissions Committee
E-mail: gradapp@helios.phy.ohiou.edu
Admissions website: http://www.phy.ohiou.edu

Application deadlines
Fall admission:
U.S. students: February 1   *Int'l. students*: February 1

Application fee
U.S. students: $55   *Int'l. students*: $55
Application fee is waived if submitted through the online application system.

Admissions information
For Fall of 2012:
Number of applicants: 102
Number admitted: 16
Number enrolled: 16

Admission requirements
Bachelor's degree requirements: Bachelors degree in science, mathematics, or engineering is required.
Minimum undergraduate GPA: 3.0

GRE requirements
The GRE is not required.
GRE scores are considered if submitted.

Advanced GRE requirements
The Advanced GRE is not required.

TOEFL requirements

The TOEFL exam is required for students from non-English-speaking countries.
PBT score: 600
iBT score: 100

Other admissions information

Additional requirements: The degree from an institution outside the United States must be equivalent to a four-year program in the United States.

For teaching assistantships, the Test of Spoken English is also required for all international students. An examination in English is given upon arrival, and students may be required to enroll in English language instruction.

Undergraduate preparation assumed: Students entering with a B.S. degree in physics are normally assumed to have completed studies in the following basic subjects to the levels indicated. Texts: Mechanics—J. B. Marion and S. T. Thornton, "Classical Dynamics of Particles and Systems", fourth edition; Electricity and Magnetism—D. J. Griffiths, Introduction to Electrodynamics; E. M. Purcell, Electricity and Magnetism; Modern Physics—Weidner and Sells, Elementary Modern Physics; Quantum Mechanics—D. J. Griffiths, "Introduction to Quantum Mechanics", second edition; Thermophysics—F. Reif, "Fundamentals of Statistical and Thermal Physics"; Mathematics—M. L. Boas, "Mathematical Methods in the Physical Sciences", calculus, including differential equations, Fourier series, complex variables, vector operators, basic algebra, including matrices and determinants.

TUITION

Tuition year 2012–13:
Tuition for in-state residents
Full-time students: $4,094 per semester
Tuition for out-of-state residents
Full-time students: $8,090 per semester
Tuition waiver for TAs and RAs covers all tuition, including out-of-state surcharge.
Credit hours per semester to be considered full-time: 15
Deferred tuition plan: No
Health insurance: Available at the cost of $1,370 per year.
Other academic fees: General fee $628/semester. Technology fee $97/semester.
Academic term: Semester
Number of first-year students who received full tuition waivers: 14

Teaching Assistants, Research Assistants, and Fellowships

Number of first-year
Teaching Assistants: 14
Average stipend per academic year
Teaching Assistant: $22,639
Research Assistant: $24,493

FINANCIAL AID

Application deadlines

Fall admission:
U.S. students: February 1   *Int'l. students*: February 1

Loans

Loans are available for U.S. students.
Loans are not available for international students.
GAPSFAS application required: No
FAFSA application required: Yes

For further information

Address financial aid inquiries to: Graduate Appointments Committee Chair.
E-mail: gradapp@helios.phy.ohiou.edu

HOUSING

Availability of on-campus housing

Single students: Yes
Married students: No

For further information

Address housing inquiries to: Office of Residential Housing.
Phone: (740) 593-4090
E-mail: housing@ohio.edu
Housing aid website: http://www.ohio.edu/housing

Table A—Faculty, Enrollments, and Degrees Granted

Research Specialty	2012-13 Faculty	Enrollment Fall 2012		Number of Degrees Granted 2012-13		
		Master's	Doctorate	Master's	Terminal Master's	Doctorate
Astrophysics	3	–	9	1(-)	–	1(-)
Biophysics	3	–	5	–	–	1(-)
Condensed Matter Physics	8	–	23	2(-)	–	2(-)
Nano Science and Technology	3	–	15	1(-)	–	1(-)
Nuclear Physics	8	–	18	–	–	2(-)
Non-specialized	–	–	15	–	–	–
Total	25	–	85	4(-)	–	7(-)
Full-time Grad. Stud.	–	–	85	–	–	–
First-year Grad. Stud.	–	–	16	–	–	–

GRADUATE DEGREE REQUIREMENTS

Master's: Thirty semester hours minimum; 3.0 ("B") average minimum; thesis optional; residence requirement not specified; no exams except in courses and for thesis; no foreign language required.

Doctorate: There is no specified number of course hours but a series of core courses are required; a minimum 3.0 ("B") average must be maintained. A comprehensive review, two semesters continuous residence minimum, and dissertation are required. No foreign language required.

Other Degrees: Interdepartmental studies (e.g., communications, education) and other special programs available by arrangement.

Thesis: Thesis may be written in absentia.

SPECIAL EQUIPMENT, FACILITIES, OR PROGRAMS

The Physics Department occupies two wings of Clippinger Research Laboratories, the renovated Edwards Accelerator Laboratory building, which contains Ohio University's 4.5 MV high-intensity tandem Van de Graaff accelerator; and the Surface-Science Research Laboratory. The Department has a well-equipped and staffed machine shop in addition to specialized research equipment. The computer facilities are excellent and include networked PCs, workstations, a Beowulf cluster and the CRAY T-90 and T3E at the Ohio Supercomputer Center. All are available to students over high speed networks (Internet II).

Table B—Separately Budgeted Research Expenditures by Source of Support

Source of Support	Departmental Research	Physics-related Research Outside Department
Federal government	$4,063,946	
State/local government		
Non-profit organizations		
Business and industry	$66,400	
Other		
Total	**$4,130,346**	

Table C—Separately Budgeted Research Expenditures by Research Specialty

Research Specialty	No. of Grants	Expenditures ($)
Astrophysics	8	$773,509
Biophysics	1	$187,498
Condensed Matter Physics	2	$135,000
Nano Science and Technology	10	$1,121,922
Nuclear Physics	16	$1,905,417
Other	1	$7,000
Total	**38**	**$4,130,346**

FACULTY

Professor

Brune, Carl R., Ph.D., California Institute of Technology, 1994. Experimental nuclear astrophysics.

Drabold, David A., Ph.D., Washington University, 1989. Theoretical condensed matter; computational methodology for electronic structure; theory of topologically disordered material.

Elster, Charlotte, University of Bonn, 1986. Nuclear and intermediate-energy theory.

Govorov, Alexander O., Ph.D., Novosibirsk (Russia), 1991. Theoretical condensed matter; semiconductor nanostructures; nanoscience.

Grimes, Steven M., Ph.D., University of Wisconsin-Madison, 1968. Nuclear physics.

Hicks, Kenneth H., Ph.D., University of Colorado, 1984. Nuclear and intermediate energy physics.

Hla, Saw-Wai, Ph.D., University of Ljubljana (Slovenia), 1997. Experimental nanoscience.

Ingram, David C., Ph.D., University of Salford (UK), 1980. Thin films; atomic collisions in solids; surface physics.

Jung, Peter, Ph.D., University of Ulm (Germany), 1985. Non-equilibrium statistical physics; non-linear stochastic processes; pattern formation; biophysics.

Kordesch, Martin E., Ph.D., Case Western Reserve University, 1984. Surface physics; wide gap materials.

Phillips, Daniel, Ph.D., Flinders University (Australia), 1995. Theoretical nuclear and particle physics.

Prakash, Madappa, Ph.D., Bombay University, 1979. Theoretical nuclear astrophysics.

Shields, Joseph C., Ph.D., University of California, Berkeley, 1991. Astrophysics; interstellar medium; active galactic nuclei.

Smith, Arthur R., Ph.D., University of Texas, 1995. Experimental semiconductor physics; thin films.

Statler, Thomas S., Ph.D., Princeton University, 1986. Astrophysics; galactic structure and dynamics.

Ulloa, Sergio E., Ph.D., State University of New York at Buffalo, 1984. Theoretical condensed matter.

Associate Professor

Castillo, Horacio E., Ph.D., University of Illinois, 1998. Theoretical condensed matter.

Chen, Gang, Ph.D., Lehigh University, 2004. Experimental condensed matter physics.

Clowe, Douglas, Ph.D., University of Hawaii, 1998. Observation astrophysics.

Lucas, Mark, Ph.D., University of Illinois, 1995. Experimental nuclear and particle physics.

Neiman, Alexander, Ph.D., Saratov (Russia), 1998. Biophysics and non-linear stochastic processes.

Roche, Julie, Ph.D., University B. Pascal (France), 1998. Nuclear and intermediate energy physics.

Sandler, Nancy, Ph.D., University of Illinois, 1998. Theoretical condensed matter.

Stinaff, Eric, Ph.D., Iowa State University, 2002. Experimental nanoscience.

Tees, David F. J., Ph.D., McGill University, 1995. *Biophysics.* Biophysics.

Assistant Professor

Frantz, Justin, Ph.D., Columbia, 2005. Experimental nuclear physics.

Adjunct Professor

Boettcher, Markus, Ph.D., University of Bonn, 1997. High-energy astrophysics.

DEPARTMENTAL RESEARCH SPECIALTIES AND STAFF

Theoretical

Astrophysics. Studies of galaxies and galaxy clusters, with emphasis on galaxy structure and dynamics, cluster central galaxies and intergalactic medium, quasars, and supermassive black holes in galaxy nuclei. High-energy astrophysics related to accretion onto compact objects, relativistic jets, and gamma-ray bursts. Nebular astrophysics applied to active galactic nuclei and starbursts. Investigations into these topics employ multiwavelength observations with national facilities (Hubble Space Telescope, Chandra X-Ray Observatory, MMT Observatory) as well as theoretical efforts including analytic calculations and large-scale numerical simulations.

Biophysics. Current projects include computational modeling of complex cellular signaling networks, especially intracellular and intercellular calcium signaling, modeling of neural and glial functions in healthy and epileptic tissue, stochastic modeling of electro-receptors in paddle fish, modeling of the neuronal circuitry of the cat's retina, stochastic and coherence resonance in excitable biologic systems, nano-scale ion channel and receptor clusters, modeling slow axonal transport.

Computational Physics. Quantum simulations, ab initio calculations, and visualization of many-body and few-body systems in condensed matter and nuclear physics. Numerical methods and algorithmic development for high performance vector and parallel computers. Analytical and algorithmic studies in differential and integral equations, probability theory, and series expansions.

Condensed Matter Physics. Statistical mechanics and nonequilibrium dynamics of disordered systems and glassy materials. Some areas of interest include: nanoscale-sized dynamical heterogeneities in glassy materials, slow activation-controlled motion of topological defects (e.g.: vortices, dislocations), and disordered electronic systems. Methodology of first principles simulation: development of local basis density functional methods, time-dependent density functional theory and efficient computation of Wannier functions, and the single particle density matrix. Theory of disordered insulators: Anderson transition, photostructural response, novel schemes for structural modeling of glasses and studies of pressure-induced polyamorphism. Optical and transport phenomena in nanoscale systems, including quantum dots, rings, and channels. Recent activity covers excitons in quantum rings, spin transport in nanocrystals, and quantum acoustoelectric interactions on nanoscale. Other nanoscience problems of interest include electronic transport in complex molecule systems, the role of controlled disorder on the metallic or insulating nature of one- and two-dimensional systems, and the role of collective effects on the optical and transport properties of quantum dot arrays. Studies of low-dimensional strongly correlated electron systems, disordered electronic systems and quantum Hall effect physics.

Nuclear Physics. Research in theoretical nuclear and particle physics at Ohio University has as its major component the modeling of processes involving atomic nuclei with mass numbers 1, 2, 3, and 4. We attempt to reveal aspects of the forces that are at work inside the nucleus by examining data obtained when targets made of hydrogen and helium isotopes are bombarded with photons, electrons, neutrons, and pions. In order to understand the dynamics of the nucleus we build theoretical descriptions of these reactions and compare our predictions to the experimental results. "Light nuclei" nuclei containing up to four nucleons are particularly useful in this regard because once the nuclear dynamics is specified the Schrodinger equation for these systems can be solved exactly. A recent focus of our group has been the application of effective field theory techniques to such reactions. Charge-symmetry breaking is of particular current interest, since here nuclear reactions such as the production of neutral pions in deuterium-deuterium collisions reveal aspects of Quantum Chromodynamics associated with the difference between up and down quarks. Lastly, we have also worked on providing reliable predictions for processes of relevance to astrophysics and cosmology.

Experimental

Astrophysics. Spectroscopic observations of stellar motions and stellar populations in elliptical galaxies, and evidence for dark matter. Ionized gas in galaxies. Gravitational lens in x-ray studies of galaxy clusters. Nuclear physics applied to astrophysics.

Biophysics. Stochastic resonance in psychophysics and animal behavior. Studies of stochastic non-linear dynamics in paddlefish electroreceptors. Experimental determination of the response of single cell adhesion molecules to applied forces using a microcantilever device. Lipid bilayer tether pulling on leukocytes and platelets using micropipette aspiration. Studies of cell adhesion in pressure gradients in micropipettes. Determination of cell membrane mechanical properties. Optical studies of biomolecules at the single molecule level using Total Internal Reflection Fluorescence microscopy and Fluorescence Resonance Energy Transfer. Biomineralization. Studies of ice-modifying antifreeze proteins. Studies of DNA-protein interactions using optical methods.

Condensed Matter Physics. Current projects encompass various areas in nanoscale science probe techniques, materials characterization by ion beams, synthesis and characterization of photonic and electronic materials, spin electronics and low-temperature mesoscopic physics. Relevant projects are illustrated by the following list: Thin film growth (by molecular beam epitaxy) and characterization (using scanning probe microscopy techniques, inc. spin-polarized) of the structural, electronic, and magnetic properties of transition metal nitride layers and magnetic-doped nitride semiconductors. Single atom/molecule manipulation using ultra high vacuum low temperature scanning tunneling microscopy, development of single molecule electronics and mechanical devices, molecular and metal thin films, surface science, and Microscopy techniques. Amorphous semiconductors and their photonic properties. MeV ion beam analysis of materials, and measurement of relevant cross-sections. Ion beam and plasma deposition of materials and their characterization. Synthesis of zeolite-related materials with catalytic and novel nanoelectronic properties, and their characterization by x-ray diffraction and electronic measurements at variable temperatures. Synthesis and characterization of nanowires and molecular wires from complex inorganic precursors.

Nuclear Physics. Contemporary research in experimental nuclear physics involves collaboration with scientists from many different institutions and heavy use of specialized accelerator facilities around the world. Ohio University nuclear physicists are recognized leaders in a variety of experimental programs spanning a broad energy domain. At higher energies our faculty are leading research programs at Jefferson Laboratory in Virginia. These include: the study of electromagnetic production of strange baryons and the search for new exotic baryons in Hall B; precision measurements of the weak charge of the proton in Hall C; the study of the nature of the gluonic flux tube in Hall D. An active program studying the photo-excitation of the nucleon is also ongoing at the SPring-8 experiment in Japan. At lower energies, our faculty are directing research programs in several distinct areas, including: fundamental symmetries in nuclear reactions via precision tests of charge symmetry breaking at TRIUMF in Canada; exotic nuclei far from the line of stability at GANIL and the Hahn-Meitner Institute in Europe; measurements of neutron cross sections at Los Alamos in New Mexico; studies of nuclear level densities at the Holifield Radioactive Ion Beam Facility in Tennessee; studies of pion photoproduction and QCD sum rules with the LEGS facility at Brookhaven National Laboratory in New York.

View additional information about this department at
www.gradschoolshopper.com

THE OHIO STATE UNIVERSITY

DEPARTMENT OF PHYSICS

Columbus, Ohio 43210
http://www.physics.osu.edu

General University Information
President: Joseph A. Alluto (interim)
Dean of Graduate School: Patrick Osmer
University website: http://www.osu.edu
Control: Public
Setting: Urban
Total Faculty: 6,509
Total Graduate Faculty: 2,903
Total number of Students: 63,058
Total number of Graduate Students: 10,297

Department Information
Department Chairman: James Beatty, Chair
Department Contact: Kris Dunlap, Program Coordinator
  Total full-time faculty: 64
  Full-Time Graduate Students: 195
  First-Year Graduate Students: 39
  Female First-Year Students: 5
  Total Post Doctorates: 37

Department Address
191 West Woodruff Avenue
Columbus, OH 43210
Phone: (614) 292-5713
E-mail: gradstudies@physics.osu.edu
Website: http://www.physics.osu.edu

ADMISSIONS

Admission Contact Information
Address admission inquiries to: Physics Graduate Studies, 191
  W. Woodruff Ave., Columbus, OH 43210
Phone: (614) 292-7675
E-mail: gradstudies@physics.osu.edu
Admissions website: https://physics.osu.edu/graduate-
  admissions-application

Application deadlines
Fall admission:
U.S. students: December 15   *Int'l. students*: November 30

Application fee
U.S. students: $5   *Int'l. students*: $5
To be considered for a fellowship, please submit all applications
  materials by December 15. Please note that the online application system for the Autumn 2014 academic year will reference
  a fee of $70 for domestic and $80 for international students,
  but the system will only process $5 for either when you go
  to submit the final payment information.

Admissions information
For Fall of 2013:
  Number of applicants: 374
  Number admitted: 154
  Number enrolled: 48

Admission requirements
Bachelor's degree requirements: Bachelor's degree in physics
  or related field is required.
Minimum undergraduate GPA: 3.0

GRE requirements
The GRE is required.

Advanced GRE requirements
The Advanced GRE is required.
Physics GRE is REQUIRED by our program.

TOEFL requirements
The TOEFL exam is required for students from non-English-
  speaking countries.
  PBT score: 650
  iBT score: 79
The average range of TOEFL scores for admission offers generally is between 90 and 105, which usually requires 1 year
  of English as a Second Language courses to receive teaching
  certification.

Other admissions information
Additional requirements: No minimum score set, but the minimum acceptable score suggested for admission is 600. The
  average advanced GRE score for 2010–11 admissions was
  736.
  Graduate Teaching Associates must demonstrate their fluency
  in spoken English before they will be allowed to assume classroom teaching duties.
Undergraduate preparation assumed: Symon, Mechanics; Lorrain and Corson, Electromagnetic Fields and Waves; Eisberg
  and Resnick, Quantum Physics: Reif, Statistical and Thermal
  Physics.

TUITION

Tuition year 2012–13:
Tuition for in-state residents
  Full-time students: $14,630 annual
Tuition for out-of-state residents
  Full-time students: $36,270 annual
The 2013-2014 9-month academic year is August 16 to May 15
  and 12-month year is August 16, 2013 to August 15, 2014.
  Tuition is based on a 12-month 3-semester system. First-year
  students with fellowship, teaching, or research appointments
  with minimum qualifications receive a nonresident tuition
  waiver in addition to in-state tuition paid by grant or state
  funding. Students are still responsible for the additional fees
  that are not eligible for coverage.
Credit hours per semester to be considered full-time: 8
Deferred tuition plan: No
Health insurance: Available at the cost of 345 per year.
Other academic fees: Recreation ($123), activity ($37.50), city
  bus ($13.50), student union ($27.50), legal fee ($40 paid AU
  term only).
Academic term: Semester
Number of first-year students who received full tuition waivers: 39

Teaching Assistants, Research Assistants, and Fellowships
Number of first-year
  Teaching Assistants: 31
  Research Assistants: 1
  Fellowship students: 7

Average stipend per academic year
Teaching Assistant: $24,540
Research Assistant: $25,488
Fellowship student: $26,000

FINANCIAL AID

Application deadlines
Fall admission:
U.S. students: December 15   *Int'l. students*: November 30

Loans
Loans are not available for U.S. students.
Loans are not available for international students.
GAPSFAS application required: No
FAFSA application required: No

For further information
Address financial aid inquiries to: Department of Physics, Graduate Studies Office, 191 W. Woodruff Ave., Columbus, OH 43210.
E-mail: gradstudies@physics.osu.edu
Financial aid website: http://www.sfa.osu.edu/

HOUSING

Availability of on-campus housing
Single students: Yes
Married students: Yes

For further information
Address housing inquiries to: Director, Graduate Student Housing, 350 Morrill Tower, 1910 Cannon Drive, Columbus, OH 43210.
Housing aid website: http://housing.osu.edu/

Table A—Faculty, Enrollments, and Degrees Granted

Research Specialty	2012-13 Faculty	Enrollment AU2012 Master's	Enrollment AU2012 Doctorate	Number of Degrees Granted 2012-13 (2008–12) Master's	Number of Degrees Granted 2012-13 (2008–12) Terminal Master's	Number of Degrees Granted 2012-13 (2008–12) Doctorate
Astrophysics	8	–	9	–	–	4(14)
Atomic, Molecular, & Optical Physics	6	–	28	–	–	3(10)
Biophysics	5	–	8	–	–	3(9)
Condensed Matter Physics	17	–	51	–	3(-)	15(61)
High Energy Physics	11	–	13	–	–	3(19)
Nuclear Physics	8	–	14	–	–	2(9)
Physics and other Science Education	3	–	5	–	1(-)	1(7)
Non-specialized	–	–	74	–	1(-)	–(1)
Total	58	–	–	–	–	30(130)
Full-time Grad. Stud.	–	–	202	–	–	–
First-year Grad. Stud.	–	–	39	–	–	–

GRADUATE DEGREE REQUIREMENTS

Master's: Plan A (with thesis): Students must maintain an overall GPA of 3.0, complete 30 credit hours of graduate-level coursework over a period of at least 2 semesters, including work on thesis and a final oral examination on the thesis. Plan B (no thesis): Students must complete 30 credit hours of graduate-level coursework and demonstrate competence in individual research work, experimental or theoretical. There is no language requirement for either plan. The master's degree can be earned as part of the general (candidacy) examination for the Ph.D., which the department requires a GPA of 3.3 in the Physics core curriculum.

Doctorate: Students must complete 80 credit hours of graduate-level coursework including research for the dissertation; of this amount, 30 hours may be transferred from a master's degree at OSU. Students must pass have previously passed the candidacy examination and have a GPA of 3.3 in the Physics core curriculum, satisfactorily complete an oral final examination on the dissertation and submit the dissertation. There is no language requirement.

Other Degrees: M.S. and Ph.D. in chemical physics are also available.

SPECIAL EQUIPMENT, FACILITIES, OR PROGRAMS

Laboratory facilities and equipment: At the OSU Department of Physics, each research group possesses a unique scientific capability including techniques of scanning tunneling microscopy, spatially resolved ferromagnetic resonance imaging, electron spin resonance characterization, etc. While these capabilities are available through contact with the individual research groups, there are a number of excellent departmental facilities that are shared by all. These facilities include a well-staffed machine shop, an electro-mechanical shop (specializing in electronics design and fabrication, low-temperature system construction, and optics support), and liquid helium and nitrogen facilities. Undergraduate and graduate students also have access to an efficient student machine shop upon successful completion of mandatory supervised training classes. In addition, the Department of Physics is a home to NanoSystems Laboratory (NSL), which is a campus-wide user facility providing the OSU material science community with cutting-edge characterization equipment. The instruments available at NSL include, but are not limited to: a state-of-the-art dual beam Focused Ion Beam/Scanning Electron Microscope (FIB/SEM) with capabilities for electron beam lithography, in situ nanomanipulation. and EDS X-ray microanalysis, a high-resolution triple axis X-ray Diffractometer, two Superconducting Quantum Interference Device (SQUID) Magnetometers, two Atomic Force/Magnetic Force Microscopes (AFM/MFM), a new Chemical Vapor Deposition System for diamond growth, three Physical Property Measurement Systems (PPMS) for high-magnetic-field electric and magnetic measurements, one of which has a PPMS-compatible cryogenic AFM/MFM, two Terahertz Spectrometers, an Electron Paramagnetic Resonance Spectrometer, and a rapid turnaround system for low-temperature magneto-transport measurements. In addition, NSL operates a Class 1000 clean room with instruments such as a Langmuir-Blodgett trough for molecular monolayer deposition, a combined magnetron sputtering/electron beam thin film deposition system, and a laser writer for mask-free photo lithography.

Interdisciplinary programs: The Ohio State University Center for Cosmology and AstroParticle Physics (CCAPP) is supported by a $5M award from the Provost's Targeted Investment in Excellence Program, by a $5M Exploration of Space endowment, and other private endowments. CCAPP's mission is to support collaborative research between the OSU Departments of Astronomy and Physics in areas where OSU can make fundamental impact: dark energy, dark matter, and multi-messenger particle astrophysics. Through the Center, OSU has joined some of the world's leading research efforts: DES, GLAST, AUGER, Ice-Cube, and SDDSS-III. Of particular importance is the Center's identity as a collection point for the world's best young researchers in the areas of cosmology and particle astrophysics. The Center houses approximately 20 faculty members, seven CCAPP postdocs, and other mission specific postdocs, graduate students, and staff. It typically hosts 100 visitors a year, along with 10 mini-workshops and several collaboration meetings. CCAPP researchers receive federal support from the Department of Energy, NASA, and the National Science Foundation. Department of Physics faculty members involved with CCAPP are: Beacom, Beatty, Honscheid, Hughes, Kass, Steigman, Walker, and Winer. Please visit http://www.ccapp.osu.edu for more information.

The Center for Exploration of Novel Complex Materials (ENCOMM) is supported by a $4.1M award from the Provost's Targeted Investment in Excellence Program. ENCOMM builds on broad strength at OSU in electronic, magnetic, and organic materials to address cutting-edge challenges in understanding and developing complex multicomponent materials. These problems are inherently multidisciplinary and require state-of-the-art facilities. ENCOMM's mission is to build and nurture the teams that can compete effectively for multidisciplinary block-funded centers, to create the environment in which these theoretical and experimental teams can form and interact, and to provide the infrastructure needed to perform the research that will define this field. ENCOMM membership extends across the Departments of Physics, Chemistry, Mechanical Engineering, Materials Science and Engineering, Electrical and Computing Engineering, and Biomedical Engineering. ENCOMM formed and nurtured the Interdisciplinary Research Groups that successfully competed for our new NSF-funded Materials Research Science and Engineering Center (MRSEC). ENCOMM includes 44 faculty members from six departments; Department of Physics faculty members are: Hammel, Johnston-Halperin, Brillson, Epstein, Gramila, Gupta, Jayaprakash, Lemberger, Pelz, Poirier, Putikka, Randeria, Sooryakumar, Stroud, Trevedi, Wilkins, and Yang. Please visit http://www.physics.ohio-state.edu/ENCOMM for more information.

Our Materials Research Science and Engineering Center (MRSEC), also called the Center for Emergent Materials, is supported by an NSF award of $10.8M over six years with an additional $6.2M in institutional support. The CEM is comprised of two interdisciplinary research groups with activities focusing on the general area of magnetoelectronics. The first, "Towards Spin Preserving, Heterogeneous Spin Networks," is focused on broadening the application of spintronics to new materials through collaborative development of innovative characterization and investigational tools and new materials growth and testing approaches. The second, "Double Perovskite Interfaces and Heterostructures," is creating new functionality in oxide materials by combining novel perovskite compounds into composite structures with controlled interfaces. By tuning magnetic and electronic properties of the constituent materials and exploiting controlled strain at interfaces, new properties emerge. The Center includes 27 graduate research fellows, five postdoctoral researchers, 20 undergraduate students, three staff members, and 21 faculty members; Department of Physics faculty members are: Johnston-Halperin, Gupta, Hammel, Heckler, Epstein, Pelz, Stroud, Brillson, Lemberger, Randeria, Trivedi, and Yang. Please visit http://www.cem.osu.edu for more information.

Table B—Separately Budgeted Research Expenditures by Source of Support

Source of Support	Departmental Research	Physics-related Research Outside Department
Federal government	$10,877,130	$3,787,476
State/local government	$281,498	$180,241
Non-profit organizations	$408,180	$4,059,220
Business and industry	$1,103,699	
Other	$1,180,758	$13,082
Total	**$13,851,265**	**$8,040,019**

Table C—Separately Budgeted Research Expenditures by Research Specialty

Research Specialty	No. of Grants	Expenditures ($)
Astrophysics	6	$1,199,813
Atomic, Molecular, & Optical Physics	27	$3,781,028
Biophysics	16	$792,819
Condensed Matter Physics	43	$3,254,705
Nuclear Physics	7	$1,508,582
Physics and other Science Education	10	$682,203
High Energy Physics	13	$3,132,115
Total	122	$14,351,265

FACULTY

Professor

Agostini, Pierre, Ph.D., University AIX Marseille, 1967. *Atomic, Molecular, & Optical Physics.* Experimental atomic, molecular, and optical physics.

Andereck, C. David, Ph.D., Rutgers University, 1980. *Condensed Matter Physics.* Atomic, molecular, and optical physics.

Bao, Lei, Ph.D., University of Maryland, 1999. *Physics and other Science Education.* Physics education.

Beacom, John, Ph.D., University of Wisconsin-Madison, 1997. Director of Center for Cosmology and Astro-Particle Physics (CCAPP). *Astrophysics, Theoretical Physics.* Theoretical astrophysics and cosmology.

Beatty, James, Ph.D., University of Chicago, 1986. *Astrophysics.* Experimental astrophysics.

Braaten, Eric, Ph.D., University of Wisconsin-Madison, 1981. *High Energy Physics.* Theoretical high-energy physics.

Brillson, Leonard, Ph.D., University of Pennsylvania, 1972. *Condensed Matter Physics, Electrical Engineering.* Experimental condensed-matter physics; electrical engineering.

Bundschuh, Ralf, Ph.D., University of Porsdam, 1996. *Biophysics.* Theoretical biophysics.

De Lucia, Frank C., Ph.D., Duke University, 1969. *Atomic, Molecular, & Optical Physics.* Experimental atomic, molecular, and optical physics.

DiMauro, Louis F., Ph.D., University of Connecticut, 1980. *Atomic, Molecular, & Optical Physics.* Experimental atomic, molecular, and optical physics.

Durkin, L. Stanley, Ph.D., Stanford University, 1981. *High Energy Physics.* Experimental high-energy physics.

Freeman, Richard R., Ph.D., Harvard University, 1973. *Atomic, Molecular, & Optical Physics.* Experimental atomic, molecular, and optical physics.

Furnstahl, Richard J., Ph.D., Stanford University, 1985. *Nuclear Physics.* Theoretical nuclear physics.

Gan, K. K., Ph.D., Purdue University, 1985. *High Energy Physics.* Experimental high-energy physics.

Gruzberg, Ilya A., Ph.D., Yale University, 1998. *Condensed Matter Physics.* Condensed Matter.

Hammel, P. Chris, Ph.D., Cornell University, 1984. Director of Center for Emergent Materials. *Condensed Matter Physics.* Experimental condensed-matter physics.

Harata, Chris, Ph.D., Princeton University, 2005. *Astrophysics.*

Heinz, Ulrich, Ph.D., Goethe Univ., 1980. *Nuclear Physics.* Nuclear theory.

Ho, Tin-Lun (Jason), Ph.D., Cornell University, 1977. *Condensed Matter Physics.* Theoretical condensed-matter physics.

Honscheid, Klaus, Ph.D., University of Bonn, 1988. *High Energy Physics.* Experimental high-energy physics.

Hughes, Richard E., Ph.D., University of Pennsylvania, 1992. *High Energy Physics.* Experimental high-energy physics.

Humanic, Thomas J., Ph.D., University of Pittsburgh, 1979. *Nuclear Physics*. Experimental nuclear physics.

Jayaprakash, Ciriyam, Ph.D., University of Illinois, 1978. *Condensed Matter Physics*. Theoretical condensed-matter physics.

Jeschonnek, Sabine, Ph.D., University of Bonn, 1996. *Nuclear Physics*. Nuclear theory.

Kagan, Harris P., Ph.D., University of Minnesota, 1979. *High Energy Physics*. Experimental high-energy physics.

Kass, Richard, Ph.D., University of California, Davis, 1978. *High Energy Physics*. Experimental high-energy physics.

Kovchegov, Yuri, Ph.D., Columbia University, 1998. *Nuclear Engineering, Nuclear Physics*. Theoretical nuclear physics.

Lemberger, Thomas R., Ph.D., University of Illinois, 1978. *Condensed Matter Physics*. Experimental condensed-matter physics.

Lisa, Michael A., Ph.D., Michigan State University, 1993. *Nuclear Physics*. Experimental nuclear physics.

Mathur, Samir, Ph.D., University of Bombay, 1987. *High Energy Physics*. High-energy theory.

Pelz, Jonathan P., Ph.D., University of California, Berkeley, 1987. Vice Chair of Graduate Studies and Research. *Condensed Matter Physics*. Experimental condensed-matter physics.

Perry, Robert J., Ph.D., University of Maryland, 1984. Vice Chair for Undergraduate Studies. *Nuclear Physics*. Theoretical nuclear physics.

Putikka, William O., Ph.D., University of Wisconsin-Madison, 1988. *Condensed Matter Physics*. Theoretical condensed-matter physics.

Raby, Stuart A., Ph.D., Tel Aviv University, 1976. *High Energy Physics*. Theoretical high-energy physics.

Randeria, Mohit, Ph.D., Cornell University, 1987. *Condensed Matter Physics*. Theoretical condensed-matter physics.

Shigemitsu, Junko, Ph.D., Cornell University, 1978. Theoretical high-energy physics.

Sooryakumar, R., Ph.D., University of Illinois at Urbana-Champaign, 1980. *Biophysics, Condensed Matter Physics*. Experimental condensed-matter physics.

Sugarbaker, Evan R., Ph.D., University of Michigan, 1976. Vice Chair of Administration. *Nuclear Physics*. Experimental nuclear physics.

Trivedi, Nandini, Ph.D., Cornell University, 1987. *Condensed Matter Physics*. Theoretical condensed-matter physics.

Van Woerkom, Linn D., Ph.D., University of Southern California, 1987. Associate Provost; Director, Honors & Scholars. *Atomic, Molecular, & Optical Physics*. Experimental atomic, molecular, and optical physics.

Walker, Terrence P., Ph.D., Indiana University, 1987. *Astrophysics*. Theoretical astrophysics; cosmology.

Wilkins, John W., Ph.D., University of Illinois, 1963. *Condensed Matter Physics*. Theoretical condensed-matter physics.

Winer, Brian L., Ph.D., University of California, Berkeley, 1991. *High Energy Physics*. Experimental high-energy physics.

Zhong, Dongping, Ph.D., California Institute of Technology, 1999. *Biophysics*. Experimental biophysics.

Associate Professor

Gramila, Thomas, Ph.D., Cornell University, 1989. *Condensed Matter Physics*. Experimental condensed-matter physics.

Gupta, Jay, Ph.D., University of California, Santa Barbara, 2002. *Condensed Matter Physics*. Experimental condensed-matter physics.

Hill, Christopher S., Ph.D., University of California, Davis, 2001. *High Energy Physics*. Experimental high-energy physics.

Johnston-Halperin, Ezekiel, Ph.D., University of California, Santa Barbara, 2003. Director of ENCOMM. *Condensed Matter Physics*. Experimental condensed-matter physics.

Lafyatis, Gregory P., Ph.D., Harvard University, 1982. *Atomic, Molecular, & Optical Physics*. Experimental atomic, molecular, and optical physics.

Poirier, Michael, Ph.D., University of Chicago, 2001. *Biophysics*. Experimental biophysics.

Schumacher, Douglass, Ph.D., University of Michigan, 1995. *Atomic, Molecular, & Optical Physics*. Experimental atomic, molecular, and optical physics.

Yang, Fengyuan, Ph.D., Johns Hopkins University, 2001. *Condensed Matter Physics*. Experimental condensed-matter physics.

Assistant Professor

Carpenter, Linda, Ph.D., Johns Hopkins University, 2006. *High Energy Physics*.

Connolly, Amy L., Ph.D., University of California, Berkeley, 2003. *High Energy Physics*. Experimental high-energy physics.

Heckler, Andrew, Ph.D., University of Washington, 1994. *Physics and other Science Education*. Physics education.

Kural, Comert, Ph.D., University of Illinois at Urbana-Champaign, 2007. *Biophysics*.

Peter, Annika, Ph.D., Princeton University, 2008. *Astrophysics*.

Emeritus

Aubrecht, Gordon J., Ph.D., Princeton University, 1976. *High Energy Physics, Physics and other Science Education*. Theoretical high-energy physics.

Epstein, Arthur J., Ph.D., University of Pennsylvania, 1971. *Condensed Matter Physics*. Experimental condensed-matter physics.

Ling, Ta-Yung, Ph.D., University of Wisconsin-Madison, 1971. *High Energy Physics*. Experimental high-energy physics.

Patton, Bruce R., Ph.D., Cornell University, 1971. *Condensed Matter Physics*. Theoretical condensed-matter physics.

Steigman, Gary, Ph.D., New York University, 1968. *Astrophysics*. Theoretical astrophysics and cosmology.

Stroud, David G., Ph.D., Harvard University, 1969. Theoretical condensed-matter physics.

Research Assistant Professor

Akli, Kramer, Ph.D., University of Cailfornia, 2006. *High Energy Physics*. High Energy.

Chowdhury, Enam, Ph.D., University of Delaware, 2004. *High Energy Physics*. High Energy.

Teaching Associate Professor

Kilcup, Gregory P., Ph.D., Harvard University, 1986. *High Energy Physics*. Theoretical high-energy physics.

Adjunct Professor

Weilhammer, Peter, Ph.D., University of Munich, CERN, 1969. *High Energy Physics*. Experimental high-energy physics.

Winnewisser, Brenda, Ph.D., Duke University, 1965. *Atomic, Molecular, & Optical Physics*. Experimental atomic, molecular, and optical physics.

Winnewisser, Manfred, Technical University of Karlsruhe, 1960. *Atomic, Molecular, & Optical Physics*. Experimental atomic, molecular, and optical physics.

DEPARTMENTAL RESEARCH SPECIALTIES AND STAFF

Theoretical

Biological Physics:. Genetic network analysis; immunological modeling; ecological modeling; biological sequence analysis; modeling of single-molecule experiments; RNA folding. Bundschuh, Jayaprakash.

Condensed Matter Theory. Electronic structure theory; quantum Monte Carlo and molecular dynamics; strongly correlated electrons; magnetism in correlated oxides; high-temperature superconductivity; quantum Hall effect and topological phases; metal-insulator and superconductor insulator transitions; transport in disordered systems; graphene; spintronics; ultracold atoms; strongly interacting Fermi gases; bosons and fermions in optical lattices; spinor condensates. Ho, Jayaprakash, Patton, Randeria, Stroud, Trivedi, Wilkins.

Cosmology & Astroparticle Physics Theory. Early universe theories and big-bang cosmology; large-scale structure and the cosmic microwave background; dark matter and dark energy; primordial and stellar nucleosynthesis; particle and nuclear astrophysics; star formation and supernovae; neutrinos, gamma rays, and cosmic rays; connections between theory and experiment. Beacom, Harata, Peter, Steigman, Walker.

High Energy Theory:. Phenomenology of the strong and electroweak interactions; nonperturbative methods in quantum field theory; effective field theories; lattice gauge theory; heavy quark physics; tests of the Standard Model; Beyond the Standard Model theories; supersymmetry; string theory; grand unified models derived from string theory; quantum gravity; black hole physics. Braaten, Carpenter, Kilcup, Mathur, Raby, Shigemitsu.

Nuclear Theory. Theory and phenomenology of ultra-relativistic heavy-ion collisions, relativistic dynamics, and transport properties of strongly coupled quantum field systems; quantum chromodynamics (QCD) in high-energy scattering and applications of AdS/CFT correspondence to QCD; electron scattering from light nuclei; effective field theory, renormalization group, and large-scale computational methods for low-energy nuclear physics. Furnstahl, Heinz, Jeschonnek, Kovchegov, Perry.

Experimental

Atomic, Molecular, & Optical Physics. Interaction of matter with extremely high-intensity lasers; laser-plasma physics; generation and use of atto-second pulses of UV light; femtosecond laser spectroscopy; X-ray free electron laser physics; submillimeter/terahertz spectroscopy and sensors; molecular astrophysics; chemical physics. Agostini, De Lucia, DiMauro, Freeman, Lafyatis, Schumacher, Van Woerkom.

Biological Physics. Protein/enzyme dynamics; femtosecond spectroscopy; biological macromolecular hydration; chromatin structure; DNA repair; DNA mechanics; single-molecule experiments. Kural, Poirier, Sooryakumar, Zhong.

Condensed Matter Experiment. Spin-injection, spin-transport and magnetism in novel materials including graphene, semiconducting nanostructures/nanowires, complex oxides, molecular and organic materials; time and energy resolved spectroscopy and magnetic/electronic/optical properties at low temp and high magnetic field; nano- and micro-scale device fabrication; MBE and CVD growth of complex oxide heterostructures, metallic and magnetic multilayers, diamond, nanowires, and graphene; characterization/ manipulation of atomic and molecular nanostructures using low-temperature scanning tunneling microscopy; surface/tip-enhanced near-field optical microscopy; scanned probe magnetic resonance imaging, ultrasensitive force detection, nm-scale electronic, optical, and magnetic properties of surfaces, thin-films, interfaces, and nanocontacts; X-ray and UV photoemission spectroscopies, atomic force/Kelvin probe/surface photovoltage and depth-resolved CL microscopies/spectroscopies of electronic materials; metallic/semiconducting polymers, molecular ferromagnets; tunneling and transport effects in superconductors; superconductor-to-insulator quantum transition fluctuations; mobile magnetic tweezers for self-assembly, biomolecule/cell manipulation and interrogation; micro- and nano-fluidic devices; structural and elastic properties of patterned nanostructures, solid-state membranes, and biological tissues via laser light scattering. Brillson, Epstein, Gramila, Gupta, Hammel, Johnston-Halperin, Lemberger, Pelz, Sooryakumar, Yang.

Experimental Cosmology and Astroparticle Physics. Highest-energy cosmic rays and neutrinos; cosmic ray spectrum, composition, and anisotropy; radio detection of cosmic rays and neutrinos; high-energy neutrino astronomy; connecting neutrino and cosmic ray measurements with astrophysics; dark energy; very-high-energy gamma-ray astronomy with FGST. Beatty, Connolly, Honscheid, Hughes, Winer.

Experimental High Energy. Proton-anti-proton scattering using the CDF detector at Fermilab; studies of the top quark and Higgs boson production; proton-proton collisions with the ATLAS and CMS detectors at CERN's LHC; measurement of the properties of the Higgs boson; search for extensions beyond the Standard Model; dDark energy studies using BOSS, DES, and LSST; instrumentation and detector development for high-energy physics experiments. Durkin, Gan, Hill, Honscheid, Hughes, Kagan, Kass, Winer.

Nuclear Physics. Relativistic heavy ion collisions; boson interferometry; LHC Alice Experiment; RHIC Star Experiment; silicon drift detector development. Humanic, Lisa, Sugarbaker.

Physics Education Research. The improvement of instruction in physics has always been an integral part of the work of the academic physics community. The establishment of research programs in this area has recently brought it formal recognition as one of the subfields of physics. Physics education research approaches its work in the context of the specific problems posed by instruction and learning in physics while drawing on relevant aspects of knowledge from cognitive psychology and pedagogy. Physicists, with their intimate knowledge of physics and the ability to use this knowledge to address complex problems, are in a unique position to make contributions to this field. In recent years, physics education research has ranged widely from the study of student behaviors and cognitive skills to the uses of new interactive technologies for learning physics. The research often leads to development of materials that improve student learning and that are then used to evaluate that learning. Members of the OSU Physics Education Research Group are especially interested in the following areas: curriculum development; evaluation and assessment; professional development; analysis of student problem-solving strategies; construction of conceptual models; the use of technology to enhance learning; and system change research. Aubrecht, Bao, Heckler.

View additional information about this department at
www.gradschoolshopper.com

THE UNIVERSITY OF AKRON

DEPARTMENT OF PHYSICS

Akron, Ohio 44325-4001
http://www.uakron.edu/physics

General University Information
President: Luis M. Proenza
Dean of Graduate School: George R. Newkome
University website: http://www.uakron.edu/
Control: Public
Setting: Urban
Total Faculty: 1,907
Total Graduate Faculty: 958
Total number of Students: 27,911
Total number of Graduate Students: 4,103

Department Information
Department Chairman: Michael Taschner, Chair
Department Contact: Dr. Ang Chen, Associate Professor,
  Graduate Committee Chair
  Total full-time faculty: 8
  Total number of full-time equivalent positions: 8
  Full-Time Graduate Students: 15
  First-Year Graduate Students: 7
  Female First-Year Students: 6

Department Address
250 Buchtel Commons
Akron, OH 44325-4001
Phone: (330) 972-7078
Fax: (330) 972-6918
E-mail: ang@uakron.edu
Website: http://www.uakron.edu/physics

ADMISSIONS

Admission Contact Information
Address admission inquiries to: Dr. Ang Chen, Department of
  Physics
Phone: (330) 972-7078
E-mail: ang@uakron.edu
Admissions website: http://www.uakron.edu/physics/academics/
  graduate/index.dot

Application deadlines
Fall admission:
U.S. students: March 15   *Int'l. students*: March 15

Application fee
U.S. students: $40   *Int'l. students*: $60

Admissions information
For Fall of 2013:
  Number of applicants: 30
  Number admitted: 15
  Number enrolled: 7

Admission requirements
Bachelor's degree requirements: A bachelor's degree in physics
  is preferred, but applicants with a bachelor's degrees in another field of science, engineering, mathematics, or science
  education will be considered on an individual basis.
Minimum undergraduate GPA: 2.75

GRE requirements
The GRE is not required.

Advanced GRE requirements
The Advanced GRE is not required.

TOEFL requirements
The TOEFL exam is required for students from non-English-
  speaking countries.
  PBT score: 550
  iBT score: 79

Other admissions information
Undergraduate preparation assumed: Ohanian, Modern Physics;
  Fowles and Cassiday, Mechanics; Griffiths, Introduction to
  Electrodynamics; Ohanian, Principles of Quantum Mechan-
  ics; Kittel and Roemer, Thermal Physics; Kittel, Introduction
  to Solid State Physics.

TUITION

Tuition year 2010–11:
Tuition for in-state residents
  Full-time students: $425.2 per credit
Tuition for out-of-state residents
  Full-time students: $703.7 per credit
Credit hours per semester to be considered full-time: 9
Deferred tuition plan: Yes
Health insurance: Yes, depends on plan.
Other academic fees: $200 (approximately).
Academic term: Semester

Teaching Assistants, Research Assistants, and Fellowships
Number of first-year
  Teaching Assistants: 6

FINANCIAL AID

Loans
Loans are available for U.S. students.
Loans are available for international students.
GAPSFAS application required: No
FAFSA application required: Yes

For further information
Address financial aid inquiries to: Dr. Ang Chen, Department
  of Physics.

HOUSING

Availability of on-campus housing
  Single students: No
  Married students: No

For further information
Address housing inquiries to: Graduate School, The University
  of Akron, Akron, OH 44325-2101.

Table A—Faculty, Enrollments, and Degrees Granted

Research Specialty	2010–11 Faculty	Enrollment Fall 2010		Number of Degrees Granted 2010–11 (2006–11)		
		Master's	Doctorate	Master's	Terminal Master's	Doctorate
Chemical Physics	5	–	1	–	–	–(1)
Condensed Matter Physics	6	7	–	–	4(20)	–
Low Temperature Physics	2	–	–	–	–	–
Physics and other Science Education	–	–	–	–	–(3)	–
Polymer Physics/Science	3	3	1	–	1(8)	–
Statistical & Thermal Physics	3	–	–	–	1(3)	–
Other	2	4	–	–	2(-)	–
Total	–	14	2	–	8(34)	–(1)
Full-time Grad. Stud.	–	14	2	–	–	–
First-year Grad. Stud.	–	7	–	–	–	–

GRADUATE DEGREE REQUIREMENTS

Master's: 30 semester credits are required; 3.0 GPA is required; thesis in a current area of active interest is required.

Other Degrees: The interdisciplinary Ph.D. program in polymer science includes an option in polymer physics. Three faculty members of the physics department have some affiliation with this internationally recognized program. One Physics Department faculty member has a joint appointment and directs Ph.D. research in polymer science and one in polymer engineering. Five members of the Physics Department have joint appointments in the Department of Chemistry. The Chemistry Department offers a chemical physics Ph.D. program in collaboration with the Physics Department. Students may enter the chemical physics program and choose one of these five physics faculty members as their Ph.D. advisor.

Thesis: Thesis may be written in absentia.

SPECIAL EQUIPMENT, FACILITIES, OR PROGRAMS

The Department of Physics is housed in Ayer Hall with space and facilities for instruction and research. An NMR laboratory provides facilities for research on molecular motions of large polymer molecules in the viscous liquid phase in solutions and in rubbery materials including composites. A class-100 clean room houses a laminar flow hood with wet/dry capabilities and an in-house-modified atomic force microscope for surface characterization and nanolithography. Surface science laboratories are dedicated to the study of surfaced-related phenomena and are equipped with several high-vacuum systems for the thermal and sputter deposition of thin films. The morphology and chemical reactivity of these thin films are characterized ex situ by atomic force and scanning tunneling spectroscopy, inelastic electra tunneling spectroscopy, and FTIR. Ultra-high vacuum systems are also available for the in situ study of surface phenomena. Measurement techniques include scanning tunneling microscopy, Auger electron spectroscopy, X-ray photoelectronic spectroscopy, optical spectroscopy, low-energy electron diffraction, temperature-programmed desorption, and mass spectroscopy. Facilities available for computational research include dual quad core processor workstations, an SGI workstation, a quad core workstation, and small clusters. These facilities are used to support instruction and theoretical research in condensed matter, polymer physics, statistical mechanics, critical phenomena, exact enumerations, Monte Carlo simulations, etc. Studies of the phys-

ical properties of polymeric materials also utilize the extensive facilities of the College of Polymer Science and Polymer Engineering.

Table B—Separately Budgeted Research Expenditures by Source of Support

Source of Support	Departmental Research	Physics-related Research Outside Department
Federal government	$62,780	
State/local government		
Non-profit organizations	$10,000	
Business and industry		
Other		
Total	$72,780	

Table C—Separately Budgeted Research Expenditures by Research Specialty

Research Specialty	No. of Grants	Expenditures ($)
Biophysics	2	$58,780
Optics	1	$14,000
Total	3	$72,780

FACULTY

Professor

Gujrati, P. D., Ph.D., Columbia University, 1979. *Computational Physics, Statistical & Thermal Physics*. Theoretical physics; phase transitions and critical phenomena; polymer physics; combinatorics and graph theory; renormalization group and field theory.

Hu, Ben Yu-Kuang, Ph.D., Cornell University, 1990. *Nano Science and Technology*. Graphene, Frictional drag in coupled electronic bilayers; transport in mesoscopic and nanoscale systems.

Lyuksyutov, Sergei, Ph.D., USSR Academy of Sciences, 1991. *Chemical Physics, Optics*. Experimental surface physics; nanolithography; photorefractive optics; small-angle neutron scattering, graphene.

Mallik, Robert R., Ph.D., Leicester Polytechnic, 1985. *Applied Physics, Chemical Physics, Low Temperature Physics, Nano Science and Technology*. Low-temperature physics; surface physics; electron tunneling; scanning probe microscopy.

Ramsier, Rex D., Ph.D., University of Pittsburgh, 1994. *Chemical Physics, Physics and other Science Education*. Surface science; surface functionalization; nanotechnology; nanofibers; physics education.

Associate Professor

Buldum, Alper, Ph.D., Bilkent University, 1998. *Atomic, Molecular, & Optical Physics, Computational Physics, Condensed Matter Physics*. Condensed-matter theory, nanoscience, and nanotechnology; carbon nanotubes; quantum transport; molecular electronics.

Chen, Ang, Ph.D., Zhejiang University, 1994. *Chemical Physics, Materials Science, Metallurgy*. Experimental condensed-matter physics; materials physics; ferroelectric physics; ferroelectric/piezoelectric physics; ferroelectric/piezoelectric oxides/polymers and devices; nanotechnology; characterization of dielectric/ferroelectric properties in a wide temperature and frequency range.

Dordevic, Sasa, Ph.D., University of California, San Diego, 2002. *Atomic, Molecular, & Optical Physics, Chemical Physics, Low Temperature Physics*. Experimental condensed-matter physics; high-temperature superconductivity; heavy Fermions; low-dimensional metals; vortex dynamics; strongly correlated electron systems; infrared, optical, and magneto-optical spectroscopy.

Luettmer-Strathmann, Jutta, Ph.D., University of Maryland, College Park, 1994. *Biophysics, Chemical Physics, Computational Physics, Statistical & Thermal Physics.* Statistical mechanics; polymer physics; static and dynamic properties of simple and complex fluids; phase transitions and critical phenomena; mode-coupling, integral-equation, and simulation techniques.

Emeritus

von Meerwall, Ernst D., Ph.D., Northwestern University, 1969. *Chemical Physics, Electromagnetism.* Condensed-matter physics, mainly polymers; NMR, diffusion, and structure-property relations; numerical methods and computation.

DEPARTMENTAL RESEARCH SPECIALTIES AND STAFF

Theoretical

Condensed Matter Physics. Transport in mesoscopic systems; two-dimensional electron gases; superlattices; electron tunneling. Buldum, Gujrati, Hu, Luettmer-Strathmann.

Nano Science and Technology. Monte Carlo simulations; nanotechnology; atomic-scale modeling. Buldum, Hu, Lyuksyutov, Ramsier.

Polymer Physics/Science. Glass transitions; branched polymers; collapsed phase; dynamical and topological properties; phase separation; transport properties. Gujrati, Luettmer-Strathmann, von Meerwall.

Statistical & Thermal Physics. Phase transitions and critical phenomena; static and dynamic properties of simple and complex fluids. Gujrati, Luettmer-Strathmann.

Experimental

Condensed Matter Physics. FTIR; AFM nanolithography; ferroelectric/piezoelectrics; heavy fermions; low-dimensional materials; vortex dynamics of strongly correlated electron systems; optical and magneto-optical spectroscopy. Chen, Dordevic, Lyuksyutov, Mallik, Ramsier.

Low Temperature Physics. Electron-tunneling spectroscopy; high-temperature superconductivity. Dordevic, Mallik.

Optics. Photorefractive optics; fiber Bragg gratings. Lyuksyutov.

Polymer Physics/Science. Deformation and fracture processes; crystallization; adhesion; NMR spectroscopy, pulsed NMR diffusion, and AFM lithography in polymer materials. von Meerwall.

View additional information about this department at
www.gradschoolshopper.com

UNIVERSITY OF CINCINNATI

DEPARTMENT OF PHYSICS

Cincinnati, Ohio 45221-0011
http://www.physics.uc.edu

General University Information

President: Santa J. Ono
Dean of Graduate School: Robert Zierolf
University website: http://www.uc.edu
Control: Public
Setting: Urban
Total Faculty: 2,094
Total Graduate Faculty: 1,284
Total number of Students: 32,283
Total number of Graduate Students: 9,834

Department Information

Department Chairman: Kay Kinoshita, Chair
Department Contact: Rohana Wijewardhana, Graduate Director
Total full-time faculty: 30
Total number of full-time equivalent positions: 30
Full-Time Graduate Students: 62
First-Year Graduate Students: 15
Female First-Year Students: 5
Total Post Doctorates: 13

Department Address

UC Physics Dept, 400 Geology/Physics
345 Clifton Court
Cincinnati, OH 45221-0011
Phone: (513) 556-0501
Fax: (513) 556-3425
E-mail: physics.grad@uc.edu
Website: http://www.physics.uc.edu

ADMISSIONS

Admission Contact Information

Address admission inquiries to: Graduate Program Director, Department of Physics, P.O. Box 210011
Phone: (513) 556-0512
E-mail: physics.grad@uc.edu
Admissions website: http://www.physics.uc.edu

Application deadlines

Fall admission:
U.S. students: April 15   *Int'l. students*: April 15

Application fee

U.S. students: $65   *Int'l. students*: $70
Review of applicants begins in early February.

Admissions information

For Fall of 2012:
Number of applicants: 115
Number admitted: 22
Number enrolled: 14

Admission requirements

Bachelor's degree requirements: Bachelor's degree in physics or a related science or engineering discipline is required.
Minimum undergraduate GPA: 3.0

GRE requirements

The GRE is required.

Advanced GRE requirements

The Advanced GRE is recommended.

TOEFL requirements

The TOEFL exam is required for students from non-English-speaking countries.
PBT score: 520
iBT score: 68
TOEFL can be replaced by other forms of proof of English proficiency. Go to www.isso.uc.edu for details.

Other admissions information

Undergraduate preparation assumed: Symon, Mechanics; Lorrain and Corson, Electromagnetic Fields and Waves; Kittel, Thermal Physics; Fermi, Thermodynamics; Anderson, Modern Physics and Quantum Mechanics.

TUITION

Tuition year 2012–13:
Tuition for in-state residents
Full-time students: $14,182 annual
Part-time students: $710 per credit
Tuition for out-of-state residents
Full-time students: $25,969 annual
Part-time students: $1,285 per credit
Includes General and Campus Life fees.
Credit hours per semester to be considered full-time: 10
Deferred tuition plan: Yes
Health insurance: Yes, $1587, but $1000 is paid by award.
Other academic fees: All full-time students are required to be covered by health insurance. Foreign students must purchase University insurance - grants are available to cover $1000 of the cost. ITIE fee is $348/year.
Academic term: Semester
Number of first-year students who received full tuition waivers: 12

Teaching Assistants, Research Assistants, and Fellowships

Number of first-year
Teaching Assistants: 12
Average stipend per academic year
Teaching Assistant: $20,000
Research Assistant: $21,000
Fellowship student: $22,000

FINANCIAL AID

Application deadlines

Fall admission:
U.S. students: April 15   *Int'l. students*: April 15

Loans

Loans are available for U.S. students.
Loans are not available for international students.
GAPSFAS application required: No
FAFSA application required: No

For further information

Address financial aid inquiries to: Graduate Program Director, Department of Physics, P.O. Box 210011, Cincinnati, OH.
Phone: (513) 556-0511
E-mail: physics.grad@uc.edu
Financial aid website: http://www.physics.uc.edu

HOUSING

Availability of on-campus housing

Single students: No
Married students: No

For further information

Address housing inquiries to: 2921 Scioto Street, PO Box 210045, Cincinnati, OH 45221-0045.
Phone: (513) 556-0682
E-mail: ucgradfa@uc.edu
Housing aid website: http://www.uc.edu/uchousing/graduate_housing.html

Table A—Faculty, Enrollments, and Degrees Granted

Research Specialty	2012-13 Faculty	Enrollment Spring 2013		Number of Degrees Granted 2012-13 (2006–13)		
		Master's	Doctorate	Master's	Terminal Master's	Doctorate
Astrophysics	2	–	1	–	–(2)	–(2)
Chemical Physics	1	–	4	–	–	–
Condensed Matter Physics	12	2	18	–	–	–(21)
Engineering Physics/Science	–	–	–	–	–(2)	–
Medical, Health Physics	–	–	5	–	–(1)	–(4)
Particles and Fields	12	2	13	–	–	–(15)
Physics and other Science Education	3	1	–	–	–(1)	–
Non-specialized	–	–	16	14(27)	–(2)	–
Total	30	5	57	11(27)	–(8)	8(45)
Full-time Grad. Stud.	–	5	57	–	–	–
First-year Grad. Stud.	–	2	14	–	–	–

GRADUATE DEGREE REQUIREMENTS

Master's: There are two options available. Both options require a one-year set of courses and attendance at colloquium and seminar. The thesis option requires graduate research resulting in a thesis. The non-thesis option requires an additional coursework and a passing grade on the department's qualifying exam at the master's level.

Doctorate: Satisfactory completion of 90 semester credits of graduate level course work, including seminars and research work on thesis topic. Pass written Qualifying Examination. Oral examination covering chosen field of research. Dissertation. Public presentation and defense of dissertation. Residency requirement of 30 semester credits. Teaching requirement. No language requirement.

Other Degrees: Numerous interdisciplinary research opportunities exist. Recent thesis research projects have been in such areas as Medical Physics, Chemistry, or Physics applied to various problems in the Engineering disciplines. The Department participates in the Interdisciplinary Graduate Degree Program of the Division of Graduate Education and Research. The Program allows custom tailoring of interdisciplinary studies to the individual student's interests.

Thesis: Thesis may be written in absentia.

SPECIAL EQUIPMENT, FACILITIES, OR PROGRAMS

A wide variety of facilities and programs are available to us, both within the department and elsewhere on campus. We have a modern research laboratory and office building that house our on-campus research activities. We have strong research ties with the Departments of Chemistry, Electrical Engineering, Chemical and Materials Engineering, and Radiology. In particle physics, work is undertaken at the national and international accelerator laboratories (SLAC, Fermilab, and KEK). Our condensed matter laboratories are well equipped with the normal variety of ultrasensitive measurement, analytical and sample preparation equipment. Special items include dilution refrigerators for the milli-Kelvin range, argon, and carbon dioxide lasers for Brillouin and Raman studies, and an excellent photolithographic and microelectronics laboratory.

Table B—Separately Budgeted Research Expenditures by Source of Support

Source of Support	Departmental Research	Physics-related Research Outside Department
Federal government	$2,400,000	
State/local government	$500,000	
Non-profit organizations		
Business and industry		
Other	$100,000	
Total	**$3,000,000**	

Table C—Separately Budgeted Research Expenditures by Research Specialty

Research Specialty	No. of Grants	Expenditures ($)
Astrophysics	2	$140,626
Condensed Matter Physics	7	$707,262
Particles and Fields	11	$1,630,936
Physics and other Science Education	2	$521,176
Total	**22**	**$3,000,000**

FACULTY

Professor

Argyres, Philip C., Ph.D., Princeton University, 1989. *Astrophysics, Cosmology & String Theory, Particles and Fields, Relativity & Gravitation*. Theoretical particle physics.

Beck, Thomas, Ph.D., University of Chicago, 1987. Joint Faculty with Chemistry Department. *Biophysics, Chemical Physics, Computational Physics*. Computational chemical and biophysics.

Endorf, Robert J., Ph.D., Carnegie Mellon University, 1971. *Physics and other Science Education*. Physics education.

Esposito, F. Paul, Ph.D., University of Chicago, 1971. *Particles and Fields, Relativity & Gravitation*. Theoretical physics; general relativity and astrophysics.

Hanson, Margaret M., Ph.D., University of Colorado, 1995. *Astronomy, Astrophysics*. Experimental astronomy and astrophysics.

Jackson, Howard E., Ph.D., Northwestern University, 1971. *Condensed Matter Physics, Nano Science and Technology, Optics*. Experimental condensed matter physics; laser light scattering studies.

Johnson, Randy A., Ph.D., University of California, Berkeley, 1975. *High Energy Physics*. Experimental high-energy physics.

Kim, Young H., Ph.D., University of Florida, 1986. *Condensed Matter Physics, Nano Science and Technology*. Experimental condensed matter physics.

Kinoshita, Kay, Ph.D., University of California, Berkeley, 1981. *High Energy Physics*. Experimental particle physics.

Ma, Michael, Ph.D., University of Illinois, 1983. *Condensed Matter Physics*. Theoretical condensed matter physics.

Meadows, Brian T., Ph.D., University of Oxford, 1966. *High Energy Physics*. Experimental high-energy physics.

Pinski, Frank J., Ph.D., University of Minnesota, 1977. *Computational Physics, Condensed Matter Physics*. Theoretical and computational condensed matter physics.

Scanio, Joseph J. G., Ph.D., University of California, Berkeley, 1967. *Particles and Fields*. Theoretical particle physics.

Schwartz, Alan, Ph.D., Harvard University, 1988. *High Energy Physics*. Experimental particle physics.

Sitko, Michael, Ph.D., University of Wisconsin-Madison, 1980. *Astronomy, Astrophysics*. Experimental astronomy; cosmic dust; protostellar disks; comets.

Smith, Leigh M., Ph.D., University of Illinois, 1988. *Condensed Matter Physics, Nano Science and Technology, Optics*. Experimental condensed matter physics.

Sokoloff, Michael D., Ph.D., University of California, Berkeley, 1983. *High Energy Physics*. High-energy experimental physics.

Wijewardhana, L. C. R., Ph.D., Massachusetts Institute of Technology, 1984. *Astrophysics, Cosmology & String Theory, Particles and Fields*. Theoretical particle physics.

Associate Professor

Kagan, Alexander L., Ph.D., University of Chicago, 1989. *Particles and Fields*. Theoretical particle physics.

Koenig, Kathleen, Ph.D., University of Cincinnati, 2004. Joint Faculty with Department of Curriculum and Instruction. *Physics and other Science Education*. Acquisition and transfer of scientific reasoning abilities, teaching and learning K-16, retention of STEM majors.

Kogan, Andrei B., Ph.D., Duke University, 2000. *Biophysics, Condensed Matter Physics, Low Temperature Physics, Nano Science and Technology*. Experimental condensed matter physics and biophysics.

Mast, David B., Ph.D., Northwestern University, 1982. *Condensed Matter Physics, Low Temperature Physics, Medical, Health Physics, Nano Science and Technology*. Experimental condensed matter physics.

Plano Clark, Mark, Ph.D., University of North Carolina-Chapel Hill, 1988. *Atomic, Molecular, & Optical Physics*.

Serota, Rostislav A., Ph.D., Massachusetts Institute of Technology, 1987. *Condensed Matter Physics, Nonlinear Dynamics and Complex Systems*. Theoretical condensed matter physics.

Wagner, Hans-Peter A., Ph.D., University of Regensburg, 1991. *Condensed Matter Physics, Nano Science and Technology, Optics*. Experimental condensed matter physics.

Assistant Professor

Bolech, Carlos J., Ph.D., Rutgers University, 2002. *Atomic, Molecular, & Optical Physics, Computational Physics, Condensed Matter Physics*. Theoretical condensed matter physics.

Shah, Nayana B., Ph.D., Rutgers University, 2003. *Condensed Matter Physics*. Theoretical condensed matter physics.

Sousa, Alexandre, Ph.D., Tufts University, 2005. *High Energy Physics*. Experimental high energy physics, neutrino physics.

Zupan, Jure, Ph.D., University of Ljubljana, 2002. *Astrophysics, Cosmology & String Theory, Particles and Fields*. Theoretical particle physics.

Emeritus

Chow, William S., Ph.D., Case Institute of Technology, 1964. Theoretical condensed matter physics; theory of semiconductors.

Fenichel, Henry, Ph.D., Rutgers University, 1964. Experimental condensed matter physics; fluids; optics; holography.

Goodman, Bernard, Ph.D., University of Pennsylvania, 1955.

Jha, Shacheenatha, Ph.D., University of Edinburgh, 1950. Experimental condensed matter physics.

Joiner, William C. H., Ph.D., Rutgers University, 1962. Experimental condensed matter physics; superconductivity.

Russell, James E., Ph.D., Yale University, 1958. Theoretical physics; exotic atoms; chemical physics.

Suranyi, Peter, Ph.D., Joint Institute for Nuclear Research, Dubna, 1964. Theoretical physics; high-energy theory; statistical mechanics.

Witten, Louis, Ph.D., Johns Hopkins University, 1951. Theoretical physics; general relativity.

Zhang, Fu-Chun, Ph.D., Virginia Polytechnic Institute and State University (Virginia Tech), 1983. Theoretical condensed matter physics.

DEPARTMENTAL RESEARCH SPECIALTIES AND STAFF

Theoretical

Astrophysics. Dark matter; baryogenesis; brane-world cosmology; black holes. Argyres, Esposito, Scanio, Suranyi, Wijewardhana, Witten, Zupan.

Computational Physics. Computational studies of equilibrium and non-equilibrium properties of quantum and classical condensed matter systems, chemical and biological systems. Beck, Bolech, Pinski.

Condensed Matter Physics. Stongly correlated systems; quantum disordered systems; nanophysics; mesoscopic physics and open quantum systems; superconductivity; non-equilibrium physics and transport; spintronics; quantum phase transitions; quantum chaos; critical phenomena; competing phases. Topological insulators and superconductors; dynamical response properties of metal electrons; quantum and classical fluids; quasi-one-dimensional conductors: phase transitions and critical phenomena; gas dynamics; high-Tc and unconventional superconductivity; metal-insulator transitions; random magnets; heavy Fermions; Oxides; quantum impurities and dots; nanoscale superconductors; cold atoms and optical lattices. Bolech, Goodman, Ma, Pinski, Serota, Shah.

Cosmology & String Theory. Dark matter; baryogenesis; brane-world cosmology; supersymmetric gauge theories; supergravity and string/M-theory. Argyres, Esposito, Scanio, Suranyi, Wijewardhana, Witten, Zupan.

Particles and Fields. Flavor physics; electroweak symmetry breaking and fermion mass generation; symmetry and constituent models of hadrons and leptons; grand unified theories; string interaction phenomenology and dynamics; nonperturbative methods in field theory; lattice field theories; phase transitions in quantum field theories; conformal field theories; supersymmetric gauge theories. Argyres, Esposito, Kagan, Scanio, Suranyi, Wijewardhana, Witten, Zupan.

Relativity & Gravitation. Black holes; supergravity; quantum gravity. Argyres, Esposito, Scanio, Suranyi, Wijewardhana, Witten, Zupan.

Experimental

Astronomy. Near-IR spectroscopy of massive young stars; star formation and galactic structure; structure and evolution of young planet-building protostellar disks, inner disk dynamics, and planet-disk interactions; small solar system bodies, solar system formation and evolution. Hanson, Sitko.

High Energy Physics. At Fermilab: Studying muon neutrino disappearance and muon to electron neutrino oscillations over short distances with the MiniBooNE and MicroBooNE detectors, and over long distances with the NOvA and MINOS+ experiments; developing the Liquid Argon Time Projection Chamber technology for future neutrino experiments. In China: Using the Daya Bay Nuclear Reactor complex to measure the final angle in the neutrino mass mixing matrix. At SLAC: heavy quark physics, especially CP violations in B-meson decays using the Babar detector, and charm meson decays. At KEK (Japan): studies of CP violation, mixing, and rare processes in B-meson and D-meson decays, using the Belle detector. At CERN: studies of heavy quark physics including CP violation, mixing, and rare processes in hadrons with b- and c-quarks, using the LHCb detector. Johnson, Kinoshita, Meadows, Schwartz, Sokoloff, Sousa.

Interdisciplinary Research. Econophysics, quantitative finance, cognitive psychology, biological systems. Serota.

Low Temperature Physics. Superconductors; electronic correlation phenomena in semiconductor quantum dots, non-equilibrium Kondo effect, photon-assisted transport, dynamic conductance spectroscopy, coherent manipulation of spin-based quantum systems; Luttinger liquids. Kogan, Mast.

Medical, Health Physics. Development of micro-miniature microwave and surface acoustic wave devices for biochemical sensors; investigation of membrane structures using near-field scanning microwave microscopy and ultrasonic acoustics. Mast.

Optics. Laser light scattering including Raman, Brillouin and photoluminescence; femtosecond and picosecond spectroscopy; electronic and optical properties of low dimensional semiconductors including quantum dots, nanowires, nanowire heterostructures and plasmonic nanostructures; electro-optical phenomena in nanostructured materials; study of linear and nonlinear optical properties in organic/semiconductor nanostructures and waveguides: two-photon absorption; nonlinear refractive index; phase coherent photorefractive effect in quantum wells; optical coherence imaging; near-field microwave scanning of HTS materials, composite ceramics, layered semiconductors and biological samples; phase coherent photorefractive effect in quantum wells; optical coherence imaging; near-field microwave scanning of HTS materials, composite ceramics, layered semiconductors and biological samples; near field spectroscopy of single wall carbon nanotubes; far-infrared (FIR) studies of metal-insulator transition in highly correlated layered systems, FIR charge dynamics in superconducting cuprates, and FIR absorption by small metal particles. Jackson, Kim, Smith, Wagner.

Physics and other Science Education. Improving introductory physics courses with the implementation of active learning and inquiry; improving the training of graduate teaching assistants; professional development programs for primary, middle school, and high school teachers; collaboration with other science departments, the College of Education, Criminal Justice, and Human Services and local school districts to improve K-16 science teaching. Endorf, Koenig, Plano Clark.

View additional information about this department at
www.gradschoolshopper.com

UNIVERSITY OF DAYTON

ELECTRO-OPTICS PROGRAM

Dayton, Ohio 45469-2951

http://www.udayton.edu/engineering/electrooptics_grad/index.php

General University Information
President: Daniel J. Curran
Dean of Graduate School: Paul Vanderburgh
University website: http://www.udayton.edu/
Control: Private
Setting: Urban
Total Faculty: 937
Total number of Students: 10,909
Total number of Graduate Students: 3,502

Department Information
Department Chairman: Partha P. Banerjee, PhD, Chair
Department Contact: Partha P. Banerjee, PhD, Professor and
    Director of the Electro-Optics Program
Total full-time faculty: 7
Total number of full-time equivalent positions: 13
Full-Time Graduate Students: 50
First-Year Graduate Students: 12
Female First-Year Students: 3
Total Post Doctorates: 2

Department Address
300 College Park
Dayton, OH 45469-2951
Phone: (937) 229-2797
Fax: (937) 229-2097
E-mail: pbanjeree1@udayton.edu
Website: http://www.udayton.edu/engineering/electroop-
    tics_grad/index.php

ADMISSIONS

Admission Contact Information
Address admission inquiries to: Electro-Optics Program, Univer-
    sity of Dayton, 300 College Park, CPC572, Dayton, OH
    45469-2951
Phone: (937) 229-2797
E-mail: nancy.wilson@notes.udayton.edu
Admissions website: http://www.udayton.edu/gradschool/

Application deadlines
Fall admission:
U.S. students: September 1   *Int'l. students*: September 1

Application fee
Int'l. students: $50
Paper applications are no longer accepted for domestic or interna-
    tional students.

Admissions information
For Fall of 2013:
    Number of applicants: 50
    Number admitted: 33
    Number enrolled: 12

Admission requirements
Bachelor's degree requirements: Bachelor's degree in physics,
    optics, or engineering is required.
Minimum undergraduate GPA: 3.25

GRE requirements
The GRE is not required.

Advanced GRE requirements
The Advanced GRE is not required.

TOEFL requirements
The TOEFL exam is required for students from non-English-
    speaking countries.
    PBT score: 550
    iBT score: 80

Other admissions information
Additional requirements: Students who have degrees in physics
    or electrical engineering are encouraged to apply. Students
    from other scientific or engineering fields may be required
    to take a limited amount of undergraduate work.
Undergraduate preparation assumed: Major in physics or in
    electrical engineering is assumed.

TUITION

Tuition year 2012–13:
    Full-time students: $788 per credit
    Part-time students: $788 per credit
Tuition year 2012-13 Engineering Master per credit hour $788
    Engineering PhD per credit hour $858
Credit hours per semester to be considered full-time: 6
Deferred tuition plan: No
Health insurance: Available
Academic term: Semester

Teaching Assistants, Research Assistants, and Fellowships
Number of first-year
    Teaching Assistants: 7
    Research Assistants: 17
Average stipend per academic year
    Teaching Assistant: $13,000
    Research Assistant: $21,000

FINANCIAL AID

Loans
Loans are available for U.S. students.
Loans are not available for international students.
GAPSFAS application required: Yes
FAFSA application required: Yes

For further information
Address financial aid inquiries to: Office of Scholarships and Fi-
    nancial Aid, University of Dayton, 300 College Park,
    CPC572, Dayton, OH 45469-2951.
Financial aid website: http://www.udayton.edu/live/financial_
    aid/

HOUSING

Availability of on-campus housing
Single students: Yes
Married students: No

For further information
Address housing inquiries to: Housing Services, Gosiger Hall 212, University of Dayton, Dayton, OH 45469-0950.
Housing aid website: http://housing.udayton.edu/

Table A—Faculty, Enrollments, and Degrees Granted

Research Specialty	2012–13 Faculty	Enrollment Fall 2012		Number of Degrees Granted 2012-13 (2007-2012)		
		Master's	Doctorate	Master's	Terminal Master's	Doctorate
Optics	7	32	20	8(30)	–	1(15)
Total	7	32	20	8(30)	–	1(15)
Full-time Grad. Stud.	–	24	19	–	–	–
First-year Grad. Stud.	–	11	2	–	–	–

GRADUATE DEGREE REQUIREMENTS

Master's: M.S. in electro-optics: 24 semester hours of courses (including 18 semester hours of core courses, three semester hours of elective, and three 1-semester-hour EO laboratory courses) and six semester hours of thesis in the case of thesis option. A non-thesis option is available in which the thesis is replaced by six semester hours of approved electives.

Doctorate: Ph.D. in electro-optics: A minimum of 90 semester hours beyond the bachelor's degree is required, including 12 semester hours of 600-level EO courses, 30 semester hours of dissertation, and six semester hours of mathematics. Comprehensive examination and residency are also required. All entering Ph.D. students are expected to have completed an M.S. in electro-optics or its equivalent.

Other Degrees: A Ph.D. in electrical engineering is available to students who would like to emphasize electro-optics in their dissertation research and meet all other electrical engineering requirements.

Thesis: Thesis may be written in absentia.

SPECIAL EQUIPMENT, FACILITIES, OR PROGRAMS

Electro-optics facilities include a total of 25 research laboratories dedicated to ellipsometry, optical processing, optical metrology, pattern recognition, nonlinear optics, spectroscopy, and nanophotonics fabrication and characterization. A wide range of optical and optical mounting equipment is available, including a variety of pulsed lasers and continuous wave lasers. The LADAR and Optical Communications Institute is a federally funded center that is part of the EO Program. The university is located within minutes of the Wright-Patterson Air Force Base, where opportunities are often available to EO students to work at one of the many government laboratories at this facility, including a graduate co-op program.

Table B—Separately Budgeted Research Expenditures by Source of Support

Source of Support	Departmental Research	Physics-related Research Outside Department
Federal government	$2,000,000	
State/local government		
Non-profit organizations		
Business and industry	$500,000	
Other		
Total	**$2,500,000**	

FACULTY

Professor

Banerjee, Partha, Ph.D., University of Iowa, 1983. *Nano Science and Technology, Optics*. Nonlinear optics; metamaterials; optoelectronic materials; digital holography.

Chatterjee, Monish, Ph.D., University of Iowa, 1985. *Optics*. Acousto-optics.

Duncan, Bradley D., Ph.D., Virginia Polytechnic Institute and State University (Virginia Tech), 1991. Fiber optics; LADAR imaging; EO sensors; Fourier optics.

Hardie, Russell C., Ph.D., University of Delaware, 1992. *Optics*. IR signal processing.

Haus, Joseph W., Ph.D., Catholic University of America, 1975. Program Director. *Nano Science and Technology, Optics*. Nonlinear optics; metamaterials; digital holography.

Loomis, John S., Ph.D., University of Arizona, 1980. *Optics*. Optical design; geometrical and physical optics; image processing.

Powers, Peter E., Cornell University, 1994. *Optics*. Nonlinear optics.

Sarangan, Andrew, Ph.D., University of Waterloo, 1996. *Nano Science and Technology, Optics*. Semiconductor lasers; fiber optics.

Vorontsov, Mikhail A., Ph.D., Lomonosov Moscow State University, 1989. *Atmosphere, Space Physics, Cosmic Rays, Nonlinear Dynamics and Complex Systems, Optics*. Adaptive optics; laser beam control; imaging through turbulence; wavefront sensing and control; optical communications.

Zhan, Qiwen, Ph.D., University of Minnesota, 2002. *Nano Science and Technology, Optics*. Ellipsometry; polarization engineering; plasmonics and nanoscale characterization and imaging.

Assistant Professor

Chong, Andy, Ph.D., Cornell University, 2008. *Optics*. Femtosecond lasers; pulse propagation; nonlinear optics.

Emeritus

Yaney, Perry, Ph.D., University of Cincinnati, 1963. *Nano Science and Technology, Optics*. Biomaterials and photonic devices.

Research Professor

McManamon, Paul, Ph.D., Ohio State University, 1977. *Optics*. Active sensing and digital holography.

Research Assistant Professor

Deng, Cong, Ph.D., University of Dayton, 2005. *Optics*. Optical design; nonlinear optics and imaging.

Polnau, Ernst, Ph.D., University of Bern, 1999. *Atmosphere, Space Physics, Cosmic Rays, Optics.* Atmospheric optics.

Weyrauch, Thomas, Ph.D., Technische Hochschule Darmstadt, 1997. *Atmosphere, Space Physics, Cosmic Rays, Optics.* Adaptive optics.

Adjunct Faculty

Dierking, Matt, Ph.D., University of Dayton, 2009. *Optics.* Laser radar and IR systems.

Grote, James G., Ph.D., University of Dayton, 1994. *Electrical Engineering, Nano Science and Technology, Optics.* Polymer and bio-related optoelectronic materials.

Schepler, Ken, Ph.D., University of Michigan, 1975. *Optics.* Laser physics.

Watson, Edward A., Ph.D., University of Rochester, 1991. *Optics.* Active imaging systems; nonmechanical beam steering.

DEPARTMENTAL RESEARCH SPECIALTIES AND STAFF

Theoretical

Optics. Image processing; beam propagation; nonlinear optical phenomena. Banerjee, Chatterjee, Chong, Grote, Hardie, Haus, Loomis, McManamon, Powers, Watson.

Experimental

Optics. Adaptive optics; ellipsometry; guides wave/fiber optics; spectroscopy; nonlinear optics; nanophotonics (materials, fabrication, and characterization); optical design; ultrafast lasers and pulse dynamics; optical metrology; optical processing/digital holography; optical systems and devices; optoelectronic materials. Banerjee, Chatterjee, Chong, Deng, Dierking, Duncan, Grote, Hardie, Haus, Loomis, McManamon, Polnau, Powers, Sarangan, Schepler, Vorontsov, Watson, Weyrauch, Yaney, Zhan.

View additional information about this department at www.gradschoolshopper.com

UNIVERSITY OF TOLEDO

DEPARTMENT OF PHYSICS AND ASTRONOMY

Toledo, Ohio 43606
http://www.physics.utoledo.edu

General University Information

President: Dr. Lloyd A. Jacobs
Dean of Graduate School: Dr. Patricia Komuniecki
University website: http://www.utoledo.edu
Control: Public
Setting: Urban
Total Faculty: 1,151
Total Graduate Faculty: 563
Total number of Students: 23,064
Total number of Graduate Students: 4,924

Department Information

Department Chairman: Lawrence S. Anderson-Huang, Chair
Department Contact: Dr. Lawrence S. Anderson-Huang, Department Chair
  Total full-time faculty: 23
  Total number of full-time equivalent positions: 23
  Full-Time Graduate Students: 68
  First-Year Graduate Students: 11
  Female First-Year Students: 4
  Total Post Doctorates: 9

Department Address

2801 W. Bancroft Street
McMaster Hall 2017, Mail Stop 111
Toledo, OH 43606
Phone: (419) 530-5165
Fax: (419) 530-2723
E-mail: lawrence.anderson@utoledo.edu
Website: http://www.physics.utoledo.edu

ADMISSIONS

Admission Contact Information

Address admission inquiries to: Graduate Admissions Committee Department of Physics and Astronomy
Phone: (419) 530-2648
E-mail: bo.gao@utoledo.edu
Admissions website: http://www.utoledo.edu/nsm/physast/programs/grad/admission.html

Application deadlines

Fall admission:
U.S. students: January 15   *Int'l. students*: January 15
Spring admission:
U.S. students: May 9   *Int'l. students*: May 9

Application fee

U.S. students: $45   *Int'l. students*: $75
For assistantship, application must be recieved by January 15 to be considered in the first round.

Admissions information

For Fall of 2013:
  Number of applicants: 61
  Number admitted: 12
  Number enrolled: 11

Admission requirements

Bachelor's degree requirements: Bachelor's degree in physics or closely related field is required.
Minimum undergraduate GPA: 2.7

GRE requirements

The GRE is required.

Advanced GRE requirements
The Advanced GRE is recommended.

TOEFL requirements
The TOEFL exam is required for students from non-English-speaking countries.
PBT score: 550
iBT score: 80
To qualify for a teaching assistantship, applicants must demonstrate TOEFL speaking at the level of 22 or higher.

Other admissions information
Undergraduate preparation assumed: Wangsness, Electricity and Magnetism; Marion and Thornton, Mechanics; Townsend, A Modern Approach to Quantum Mechanics; Halliday and Resnick, Fundamentals of Physics; Kittel, Introduction to Solid State Physics; Eisberg and Resnick, Quantum Physics of Atoms, Molecules, Solids, Nuclei, Particles.

TUITION

Tuition year 2013-14:
Tuition for in-state residents
Full-time students: $12,600 annual
Part-time students: $525 per credit
Tuition for out-of-state residents
Full-time students: $22,800 annual
Part-time students: $951 per credit
2014-15 may be slightly higher.
Credit hours per semester to be considered full-time: 12
Deferred tuition plan: Yes
Health insurance: Available at the cost of $2000 per year.
Other academic fees: General, full-time: $595/semester General, part-time; $50/credit hr Technology fee: $13/credit hr Lab fees: variable
Academic term: Semester
Number of first-year students who received full tuition waivers: 11

Teaching Assistants, Research Assistants, and Fellowships
Number of first-year
Teaching Assistants: 8
Average stipend per academic year
Teaching Assistant: $20,000
Research Assistant: $23,000

FINANCIAL AID

Application deadlines
Fall admission:
U.S. students: March 1   *Int'l. students*: March 1

Loans
Loans are available for U.S. students.
Loans are not available for international students.
GAPSFAS application required: Yes
FAFSA application required: Yes

For further information
Address financial aid inquiries to: Director of Financial Aid Office.
Phone: (419) 530-8700
E-mail: utfinaid@utnet.utoledo.edu
Financial aid website: http://www.utoledo.edu/financialaid/index.html

HOUSING

Availability of on-campus housing
Single students: Yes
Married students: No

For further information
Phone: (419) 530-2941
E-mail: reslife@utoledo.edu
Housing aid website: http://www.utoledo.edu/studentaffairs/reslife/index/

Table A—Faculty, Enrollments, and Degrees Granted

Research Specialty	2013-14 Faculty	Enrollment Fall 2013 Master's	Enrollment Fall 2013 Doctorate	Number of Degrees Granted 2012–13 (2008–13) Master's	Number of Degrees Granted 2012–13 (2008–13) Terminal Master's	Number of Degrees Granted 2012–13 (2008–13) Doctorate
Astronomy	6	–	12	–(2)	–(3)	2(10)
Atomic, Molecular, & Optical Physics	3	–	5	–	–(1)	1(3)
Biophysics	1	–	1	–(1)	–	–(1)
Computational	2	–	3	–(1)	–(1)	–(3)
Lecturers	2	–	2	–	–	–
Photovoltaics	6	7	18	–(1)	4(4)	3(15)
Phys and Astro Dept	1	–	5	–	–	2(7)
Radiatn Oncolgy Dept	3	–	–	–	–	–
Theory	3	–	3	–(2)	–(1)	1(5)
Undecided	–	–	12	–	–	–
Total	–	7	61	–(7)	4(10)	9(44)
Full-time Grad. Stud.	–	7	61	–	–	–
First-year Grad. Stud.	–	3	8	–	–	–

GRADUATE DEGREE REQUIREMENTS

Master's: Requires 30 hours, including at least 26 in physics, research thesis, and oral final examination.

Doctorate: Requires 90 hours, including 30 to 48 for research. Requires qualifying, comprehensive, and final oral examination. No language requirement.

Other Degrees: Professional Science Master in Photovoltaics (PSM-PV): designed as a two-year terminal Master's program, requiring 37 hours in a combination of business and photovoltaics-oriented courses, and including research and internship experience. Materials Science option, involving courses from physics, chemistry, and engineering. Astrophysics Option (Ph.D.) Master of Science and Education offered in cooperation with College of Education. Requires 32 hours, with 18 hours in physics, and a thesis or project in either physics or education. Medical Physics option (Ph.D.), offered jointly with the Radiation Oncology dept in the Medical College, CAMPEP accredited.

SPECIAL EQUIPMENT, FACILITIES, OR PROGRAMS

Thin-film materials laboratories include high- and ultrahigh-vacuum deposition systems using glow-discharge and hot-wire deposition, metal and metal-oxide sputtering, plasma-enhanced chemical vapor deposition (PECVD), MBE, nanomaterials solution-sourced layer-by-layer and spray deposition, thermal and electron-beam evaporation of metals; many deposition chambers incorporate in situ spectroscopic ellipsometry for real-time monitoring of film properties.

Other materials and device characterization include the magnetooptical Kerr effect, Raman spectroscopy, steady-state and time-resolved, photoluminescence, UV-to-mid-infrared ultrafast transient absorption spectroscopy, AFM/STM, SEM/EDS, quantum efficiency, current-voltage dependence under solar simulation.

Ritter Observatory houses a 1-meter reflecting telescope that is used for studies of variable stellar spectra.

Full partnership in the Discovery Channel Telescope located on the high desert plateau in Arizona, managed by Lowell Observatory.

Most UT astronomers' research programs are based on observations made at external ground- and space-based facilities.

Atomic physics research is done with 300-keV heavy-ion and 80-keV negative-ion accelerators.

Lasers are also used for thin-film scribing and thin-film index-of-refraction measurements.

Computing facilities include UNIX workstations and three cluster computing systems. Supercomputer access is provided through the Ohio Supercomputing Center via Internet 2.

Table B—Separately Budgeted Research Expenditures by Source of Support

Source of Support	Departmental Research	Physics-related Research Outside Department
Federal government	$5,500,000	
State/local government	$1,000,000	
Non-profit organizations		
Business and industry	$200,000	
Other	$130,000	
Total	$6,830,000	

Table C—Separately Budgeted Research Expenditures by Research Specialty

Research Specialty	No. of Grants	Expenditures ($)
Astrophysics	12	$1,000,000
Atomic, Molecular, & Optical Physics	2	$50,000
Materials Science, Metallurgy	19	$5,700,000
Other	1	$80,000
Total	34	$6,830,000

FACULTY

Distinguished University Professor

Bjorkman, Karen S., Ph.D., University of Colorado Boulder, 1989. Dean, College of Natural Sciences and Mathematics. *Astronomy, Astrophysics.* Hot stars; pre-main sequence stars; circumstellar disks.

Collins, Robert W., Ph.D., Harvard University, 1982. NEG Endowed Chair Faculty member of the Wright Center for Photovoltaics Innovation and Commercialization (PVIC). *Condensed Matter Physics, Materials Science, Metallurgy, Optics.* Materials science, thin films, optics of solids, ellipsometry and polarimetry.

Professor

Amar, Jacques G., Ph.D., Temple University, 1985. Associate Department Chair. *Computational Physics, Condensed Matter Physics, Materials Science, Metallurgy, Theoretical Physics.* Condensed matter physics and materials science; dynamics of thin-film and epitaxial growth; computational physics.

Anderson-Huang, Lawrence S., Ph.D., University of California, Berkeley, 1977. Department Chair. *Astronomy, Astrophysics.* Astronomy and astrophysics; stellar atmosphere theory.

Bjorkman, Jon E., Ph.D., University of Wisconsin-Madison, 1992. *Astronomy, Astrophysics.* Stellar winds; radiation transfer; young stellar objects.

Deng, Xunming, Ph.D., University of Chicago, 1990. *Condensed Matter Physics, Materials Science, Metallurgy.* Photovoltaics; thin-film semiconducting materials; applied physics.

Federman, Steven R., Ph.D., New York University, 1979. *Astronomy, Atomic, Molecular, & Optical Physics.* Astrophysics; interstellar chemistry; laboratory astrophysics.

Gao, Bo, Ph.D., University of Nebraska-Lincoln, 1989. Graduate Admission Chair for Physics and Astronomy, beginning Fall 2012. *Atomic, Molecular, & Optical Physics.* Theoretical atomic physics.

Heben, Michael, Ph.D., California Institute of Technology, 1990. PVIC Endowed Chair in Photovoltaics Faculty member of the Wright Center for Photovoltaics Innovation and Commercialization (PVIC). *Condensed Matter Physics, Materials Science, Metallurgy.* Synthesis and materials physics and chemistry of nanotube materials for energy and hydrogen storage.

Karpov, Victor G., Ph.D., Leningrad Polytechnic Institute, 1979. *Condensed Matter Physics, Materials Science, Metallurgy, Theoretical Physics.* Condensed matter; theoretical physics.

Kvale, Thomas J., Ph.D., University of Missouri, Rolla, 1984. *Atomic, Molecular, & Optical Physics, Medical, Health Physics.* Negative ion spectroscopy, ion-atom collisions; medical physics.

Lee, Scott A., Ph.D., University of Cincinnati, 1983. *Biophysics.* Ultra-high pressure physics; phase transitions of DNA; biological physics.

Palmer, James F., M.S., University of Florida, 1977. *Medical, Health Physics.* Health physics.

Yan, Yanfa, Ph.D., Wuhan University, 1993. Ohio Research Scholar Endowed Chair Faculty member of the Wright Center for Photovoltaics Innovation and Commercialization. *Condensed Matter Physics, Materials Science, Metallurgy.* Thin film solar cells; solar water splitting for hydrogen production; electron microscopy; materials physics.

Associate Professor

Chandar, Rupali, Ph.D., Johns Hopkins University, 2000. *Astronomy, Astrophysics.* Extragalactic astronomy; galaxy formation and evolution, star clusters, stellar populations.

Cheng, Song, Ph.D., Kansas State University, 1991. *Atomic, Molecular, & Optical Physics.* Experimental physics; ion-atom/molecule collisions; beam-foil spectroscopy.

Ellingson, Randall J., Ph.D., Cornell University, 1994. Faculty member of the Wright Center for Photovoltaics Innovation and Commercialization. *Condensed Matter Physics, Materials Science, Metallurgy, Optics.* Basic physics of nanostructured materials for solar energy conversion; transient laser spectroscopy.

Khare, Sanjay V., Ph.D., University of Maryland, 1996. *Computational Physics, Condensed Matter Physics, Materials Science, Metallurgy, Theoretical Physics.* Theoretical and computational; materials science, mechanical, structural and electronic properties.

Megeath, S. Thomas, Ph.D., Cornell University, 1993. *Astronomy, Astrophysics.* Star and planet formation; planet detection, infrared and astronomy millimeter-wave.

Smith, J. D., Ph.D., Cornell University, 2001. *Astronomy, Astrophysics.* Extragalactic astronomy; infrared astrophysics; astrophysical dust; active galactic nuclei; galaxy evolution.

Assistant Professor

Cushing, Michael C., Ph.D., University of Hawaii, 2004. Director, Ritter Planetarium. *Astronomy, Astrophysics.* Low-mass stars and brown dwarfs; infrared astronomy.

Podraza, Nikolas, Ph.D., University of Toledo, 2008. *Condensed Matter Physics, Materials Science, Metallurgy.* Thin films; photovoltaics; condensed matter physics.

Emeritus

Bagley, Brian G., Ph.D., Harvard University, 1968. *Optics.* Development and analysis of solid-state devices for optical computing and communications.

Bohn, Randy G., Ph.D., Ohio State University, 1969. *Condensed Matter Physics.* Thermal properties of solid hydrogen.

Compaan, Alvin D., Ph.D., University of Chicago, 1971. *Condensed Matter Physics, Materials Science, Metallurgy.* Photovoltaics, semiconductor physics; thin-film growth and characterization.

Curtis, Larry J., Ph.D., University of Michigan, 1963. *Atomic, Molecular, & Optical Physics.* Atomic structure and lifetimes; laboratory astrophysics.

Deck, Robert T., Ph.D., University of Notre Dame, 1961. *Atomic, Molecular, & Optical Physics, Optics.* Nonlinear optics and photonics, design of all-optical light signal processing elements.

Ellis, David G., Ph.D., Cornell University, 1964. *Atomic, Molecular, & Optical Physics.* Theory of atomic structure and spectra; correlation in multi-electron wavefunctions.

Witt, Adolf N., Ph.D., University of Chicago, 1967. *Astronomy, Astrophysics.* Interstellar dust, radiative transfer, photoluminescence by interstellar nanoparticles.

Lecturer

Shan, Kathy, Ph.D., University of Toledo, 2013. *Physics and other Science Education.*

DEPARTMENTAL RESEARCH SPECIALTIES AND STAFF

Theoretical

Astrophysics. Cosmochemistry; interstellar molecular gas; interstellar dust; spectroscopy, polarimetry, and theory of stellar winds, disks and envelopes; star formation; cluster evolution; radiation hydrodynamics; computational methods. Anderson-Huang, Jon Bjorkman, Karen Bjorkman, Chandar, Cushing, Federman, Megeath, Smith, Witt.

Atomic, Molecular, & Optical Physics. Theory of atomic structure and spectra, including high-performance computational techniques; quantum theories of two-atom, few-atom, and many-atom systems, including Bose-Einstein condensates; plasma discharges, atomic structure, Rydberg states, radioactive transitions, and photoionization; theory and design of optical integrated circuits, components, and devices. Bagley, Curtis, Deck, Ellis, Gao.

Condensed Matter Physics. Computational study of materials, surfaces and interfaces; disordered systems; non-equilibrium systems and thin-film growth; phase transition kinetics and thin-film photovoltaics; quantum many-atom systems. Amar, Karpov, Khare.

Experimental

Astronomy. Interstellar gas, molecular cloud chemistry, cosmochemistry, laboratory astrophysics; interstellar dust; light scattering, photoluminescence; stellar spectroscopy and polarimetry, Be stars, Herbig Ae/Be stars, circumstellar disks and envelopes; extragalactic star clusters, star formation; star formation, infrared and millimeter observations; extrasolar planet detection. Anderson-Huang, Jon Bjorkman, Karen Bjorkman, Chandar, Cushing, Federman, Megeath, Smith, Witt.

Atomic, Molecular, & Optical Physics. Beam-foil spectroscopy; atomic lifetime measurements; ion-atom collisions, excitation/ionization, charge transfer, secondary emission of electrons from surfaces, photodetachment; semi-empirical techniques for structure of highly excited molecules and atoms; integrated optics and non-linear optics; spectroscopic ellipsometry and polarimetry. Cheng, Collins, Curtis, Ellingson, Federman, Podraza.

Biophysics. DNA structure and bonding to cancer drugs, phase transitions in hyaluronic acid. Lee.

Condensed Matter Physics. Photovoltaic materials and devices: amorphous silicon, CdTe/CdS, CuInGaS2/CdS, CZTS, PbS and FeS2 nanocrystalline materials, low- and room-temperature deposition routes from solution sources; photoelectrochemical H2 generation; real time metrology; laser scribing of photovoltaic thin films. Bohn, Collins, Compaan, Deng, Ellingson, Heben, Podraza, Yan.

Medical, Health Physics. Applied accelerator-based physics; applications to radiation therapy. Radiation oncology (Health Science Campus): hyperbaric medicine; treatment of gastrointestinal cancer; radiation beam modeling with Monte Carlo simulation techniques, optimization in IMRT delivered external beam radiotherapy, stereotactic radiosurgery, intra-operative radiation therapy, and three-dimensional dosimetric analysis and quantitative bremsstrahlung SPECT imaging for gamma-emitting radiopharmaceuticals. Medical physics—diagnostic radiology: (Health Science Campus): Tomosynthesis imaging techniques in mammography, perfusion techniques for functional MRI and BOLD functional MRI, MR proton spectroscopy, diagnostic imaging system performance testing. Cheng, Kvale, Palmer.

View additional information about this department at
www.gradschoolshopper.com

WRIGHT STATE UNIVERSITY

DEPARTMENT OF PHYSICS

Dayton, Ohio 45435
http://www.wright.edu/cosm/departments/physics/

General University Information
President: David R. Hopkins, P.E.D
Dean of Graduate School: R. William Ayres
University website: http://wright.edu/
Control: Public
Total Faculty: 891

Total Graduate Faculty: 891
Total number of Students: 16,780
Total number of Graduate Students: 2,513

Department Information
Department Chairman: Doug Petkie, Chair
Department Contact: Doug Petkie, Chair
   Total full-time faculty: 13
   Total number of full-time equivalent positions: 16
   Full-Time Graduate Students: 12
   First-Year Graduate Students: 6
   Female First-Year Students: 3
   Total Post Doctorates: 3

Department Address
3640 Colonel Glenn Highway
Dayton, OH 45435
Phone: (937) 775-2955
Fax: (937) 775-2222
E-mail: physics@wright.edu
Website: http://www.wright.edu/cosm/departments/physics/

ADMISSIONS

Admission Contact Information
Address admission inquiries to: Wright State University, Graduate School, 3640 Colonel Glenn Highway, 344 Student Union, Dayton, OH 45435
Phone: 1-800-452-4723
E-mail: wsugrad@wright.edu
Admissions website: http://www.wright.edu/graduate-school/admissions/apply-now

Application deadlines
Fall admission:
U.S. students: June 30   *Int'l. students*: May 30
Spring admission:
U.S. students: November 3   *Int'l. students*: August 30

Application fee
U.S. students: $40   *Int'l. students*: $40

Admissions information
For Fall of 2012:
   Number of applicants: 20
   Number admitted: 5
   Number enrolled: 5

Admission requirements
Bachelor's degree requirements: Bachelor's degree in physics or a closely related field is required with a minimum undergraduate GPA of 2.7/4.0.
Minimum undergraduate GPA: 2.7

GRE requirements
The GRE is not required.
The GRE is not required but recommended.

Advanced GRE requirements
The Advanced GRE is recommended.
The GRE Physics is not required but recommended.

TOEFL requirements
The TOEFL exam is required for students from non-English-speaking countries.
   PBT score: 550
   iBT score: 79

Other admissions information
Additional requirements: A GPA of 2.5 will be acceptable but with a 3.0 or better in the major field or in the latter half of undergraduate work.
   Those applying for teaching assistantships are also required to take the TSE or OPT exam.
Undergraduate preparation assumed: Krane, Modern Physics; Griffiths, Introduction to Electromagnetics; Griffiths, Introduction to Quantum Mechanics; Reif, Fundamentals of Statistical and Thermal Physics; Fowles and Cassiday, Analytical Mechanics; Diefenderfer and Holton, Principles of Electronic Instrumentation Melissinos and Napolitano, Experiments in Modern Physics.

TUITION
Tuition year 2013-14:
Tuition for in-state residents
   Full-time students: $6,257 per semester
   Part-time students: $577 per credit $10,629 $983
Credit hours per semester to be considered full-time: 6
Deferred tuition plan: Yes
Health insurance: Yes, $1,340.00.
Academic term: Semester
Number of first-year students who received full tuition waivers: 7

Teaching Assistants, Research Assistants, and Fellowships
Number of first-year
   Teaching Assistants: 6
   Research Assistants: 1
Average stipend per academic year
   Teaching Assistant: $13,600
   Research Assistant: $13,600

FINANCIAL AID

Loans
Loans are available for U.S. students.
Loans are available for international students.
GAPSFAS application required: No
FAFSA application required: No

For further information
Address financial aid inquiries to: Wright State University, Raider Connect, 3640 Colonel Glenn Highway, 108 Student Union, Dayton, OH 45435.
Phone: (937) 775-4000
E-mail: RaiderConnect@wright.edu
Financial aid website: https://www.wright.edu/raider-connect

HOUSING

Availability of on-campus housing
Single students: Yes
Married students: Yes

For further information
Address housing inquiries to: Wright State University, Office of Residence Services, 3640 Colonel Glenn Highway, Community Building, Dayton, OH 45435.
Phone: (937) 775-4172
E-mail: housing@wright.edu
Housing aid website: http://webapp1.wright.edu/housing/index.php

Table A—Faculty, Enrollments, and Degrees Granted

Research Specialty	2012-13 Faculty	Enrollment Fall 2012		Number of Degrees Granted 2012-13 (2007-12)		
		Master's	Doctorate	Master's	Terminal Master's	Doctorate
Basista, Tosa, Rowley	2	–	–	3(-)	–	–
Brown, Farlow	2	1	–	3(1)	–	–
Deibel	1	–	–	3(-)	–	–
Fox	1	–	–	–	–	–
Foy	1	1	–	–(8)	–	–
Hunt, Skinner	2	1	–	–	–	–
Kozlowski	1	6	–	3(8)	–	–
Medvedev, Clark	2	2	–	1(2)	–	–
Petkie, Deibel	2	2	–	–(7)	–	–
Tebbens, Gerschenzon	2	–	–	–	–	–
Total	16	13	–	13(26)	–	–
Full-time Grad. Stud.	–	10	–	–	–	–
First-year Grad. Stud.	–	4	–	–	–	–

Table B—Separately Budgeted Research Expenditures by Source of Support

Source of Support	Departmental Research	Physics-related Research Outside Department
Federal government	$77,262	$340,784
State/local government	$379,563	
Non-profit organizations		
Business and industry		
Other		
Total	$456,825	$340,784

Table C—Separately Budgeted Research Expenditures by Research Specialty

Research Specialty	No. of Grants	Expenditures ($)
Planetary Science	1	$180,000
Geophysics	2	$155,000
Physics Education	2	$370,000
THz Sensors	0	
Total	5	$705,000

GRADUATE DEGREE REQUIREMENTS

Master's: Complete 30 semester-hours of graduate courses with 3.0/4.0 GPA minimum, including PHY 6800, 6810, 6730, 6830, 7100, 7110, 8000 (up to 4 CH) and a minimum of 6 CH of PHY 8990 (Research) with only 9 CH counting towards the degree. Residence requirement is two semesters. There are no foreign language requirements. Qualifying and comprehensive exams are not required. Students must present an approved thesis and pass a final exam.

Other Degrees: The Selected Graduate Studies Masters format may be used to develop an individual interdisciplinary course of study. It has been used, for example, to provide an electro-optics option through a combination of engineering and physics courses. Master of Science in Teaching Physics (M.S.T.) is for licensed teachers who have completed two years of college physics.

Thesis: A thesis is required.

SPECIAL EQUIPMENT, FACILITIES, OR PROGRAMS

- Ultrafast and fiber lasers
- Terahertz, millimeter-wave, and microwave laboratories and systems
- DLTS apparatus;
- TSC, PITS 2 MeV electron Van de Graaff
- 120 keV ion implanter
- Photoreflectence system
- 252-core, 2+ TB RAM Computational Cluster
- Nano-profilometric and diagnostic tools
- Evanescent microwave microscopy
- Atomic force microscopy
- Machine shop
- Vacuum systems
- Clean room (lithography, etching, . . .)

FACULTY

Professor

Brown, Elliott, Ph.D., California Institute of Technology. Endowed Chair of Terahertz Sensor Physics. *Applied Physics, Biophysics, Computational Physics, Condensed Matter Physics, Electromagnetism, Materials Science, Metallurgy, Solid State Physics, Systems Science/Engineering.* Terahertz Science and Technology; ultrafast lasers.

Hunt, Allen, Ph.D., University of California, 1983. *Geophysics, Nonlinear Dynamics and Complex Systems.* Theoretical Geophysics and complex systems.

Kozlowski, Gregory, Ph.D., Wroclaw, 1969. *Condensed Matter Physics, Materials Science, Metallurgy, Nano Science and Technology.* Materials science; magnetism; superconductivity; nanoscience and nanotechnology.

Skinner, Thomas E., Ph.D., Johns Hopkins University, 1984. *Biophysics, Computational Physics, Geophysics, Nonlinear Dynamics and Complex Systems.* Nuclear magnetic resonance spectroscopy; Optimal Control Theory.

Associate Professor

Basista, Beth, Ph.D., Cincinnati, 1994. *Physics and other Science Education.* Physics education research and K-12 teacher education.

Clark, Jerry D., Ph.D., University of Texas, 1982. *Atomic, Molecular, & Optical Physics, Materials Science, Metallurgy, Optics, Plasma and Fusion, Solid State Physics.* Atomic, molecular, and optical physics; plasma physics; quantum electronics.

Deibel, Jason, Ph.D., University of Michigan, 2004. *Applied Physics, Computational Physics, Materials Science, Metallurgy, Nano Science and Technology, Optics, Solid State Physics.* Ultrafast laser; THz spectroscopy and imaging.

Farlow, Gary C., Ph.D., North Carolina, Chapel Hill, 1982. Chairs the Physics Graduate Studies Committee. *Accelerator, Solid State Physics.* Ion-solid interactions; electron and proton irradiation-induced defects in solids.

Foy, Brent D., Ph.D., Massachusetts Institute of Technology, 1991. Director of the Medical Physics Program. *Biophysics, Computational Physics, Medical, Health Physics.* Computational modeling of biological systems.

Petkie, Doug, Ph.D., Ohio State University, 1996. Department Chair. *Applied Physics, Atomic, Molecular, & Optical Physics, Chemical Physics, Engineering Physics/Science.* Molecu lar spectroscopy; THz imaging and spectroscopy; high-frequency radar signatures and phenomenology.

Tebbens, Sarah., Ph.D., Columbia University, 1994. *Geophysics, Nonlinear Dynamics and Complex Systems*. Marine geology; geophysics.

Assistant Professor

Medvedev, Ivan, Ph.D., The Ohio State University, 2005. Faculty advisor to the Society of Physics Students. *Atomic, Molecular, & Optical Physics, Chemical Physics*. Molecular spectroscopy.

Tosa, Sachiko, Ph.D., University of Massachusetts Lowel, 2009. *Physics and other Science Education*. Physics education research and K-12 teacher education.

Research Faculty

Fox, Jane L., Ph.D., Harvard University, 1978. *Chemical Physics, Planetary Science*. Atmospheric physics; planetary science.

Gerschenzon, Naum I., Ph.D., University of Moscow, 1984. *Geophysics, Nonlinear Dynamics and Complex Systems*. Geophysics.

Lecturer

Rowley, Eric, Ph.D., University of Iowa, 2006. *Physics and other Science Education*.

Research Physicist

Look, Davod, Ph.D., University of Pittsburgh, 1966. Semiconductor Research Center. *Solid State Physics*.

DEPARTMENTAL RESEARCH SPECIALTIES AND STAFF

Theoretical

Biophysics. Nuclear magnetic resonance spectroscopy and simulations/modeling of biophysical systems. Foy, Skinner.

Complexity. Modeling and simulations of complex systems including topics in geophysics, biomedical physics, nuclear magnetic resonance spectroscopy, and planetary atmospheres. Fox, Foy, Gerschenzon, Hunt, Skinner, Tebbens.

Geophysics. seismo-electromagnetism, hydrology, percolation theory, transport phenomena. Gerschenzon, Hunt, Tebbens.

Physics and other Science Education. Physics education research (K-16). Teacher preparation. Basista, Rowley, Tosa.

Experimental

Atomic, Molecular, & Optical Physics. Atomic and molecular gas phase spectroscopy in the terahertz and optical regions. Clark, Medvedev, Petkie.

Solid State Physics. Electrical and optical properties of semiconductors; irradiation damage; thin films and ion implantation; nanostructures; nechanical properties of metals and alloys. Clark, Farlow, Kozlowski, Look.

Terahertz Science and Technology. Spectroscopy, imaging, and electromagnetic studies in the millimeter-wave and terahertz spectral regions. Basic and applied research of THz phenomenology and the development of new technologies and techniques. Brown, Deibel, Medvedev, Petkie.

View additional information about this department at www.gradschoolshopper.com

OKLAHOMA STATE UNIVERSITY

DEPARTMENT OF PHYSICS

Stillwater, Oklahoma 74078-3072
http://www.physics.okstate.edu

General University Information

President: V. Burns Hargis
Dean of Graduate School: Sheryl Tucker
University website: http://www.okstate.edu
Control: Public
Setting: Rural
Total Faculty: 1,400
Total Graduate Faculty: 1,300
Total number of Students: 20,000
Total number of Graduate Students: 5,000

Department Information

Department Chairman: John W. Mintmire, Head
Department Contact: Susan Cantrell, Administrative Assistant
Total full-time faculty: 23
Total number of full-time equivalent positions: 23

Full-Time Graduate Students: 46
First-Year Graduate Students: 12
Female First-Year Students: 1
Total Post Doctorates: 3

Department Address

145 PS-II
Stillwater, OK 74078-3072
Phone: (405) 744-5796 (C)
Fax: (405) 744-6811
E-mail: susan.cantrell@okstate.edu
Website: http://www.physics.okstate.edu

ADMISSIONS

Admission Contact Information
Address admission inquiries to: Graduate Coordinator, Department of Physics
Phone: (405) 744-5796
E-mail: physics.grad.coordinator@okstate.edu
Admissions website: http://www.physics.okstate.edu

Application deadlines
Fall admission:
U.S. students: March 1   *Int'l. students*: March 1

Application fee
U.S. students: $40   *Int'l. students*: $75

Admissions information
For Fall of 2013:
  Number of applicants: 120
  Number admitted: 24
  Number enrolled: 12

Admission requirements
Bachelor's degree requirements: Bachelor's degree in physics (or closely related field) required.
Minimum undergraduate GPA: 3.0

GRE requirements
The GRE is recommended.
  Mean GRE score range (25th–75th percentile): 300-320
  While not strictly required, the GRE (at least the General Test) is strongly recommended.

Advanced GRE requirements
The Advanced GRE is not required.
(See above)

TOEFL requirements
The TOEFL exam is required for students from non-English-speaking countries.
  PBT score: 610
  iBT score: 90
iBT strongly preferred over pBT (or IELTS). Also, iBT Speaking Section score must be at least 19.

Other admissions information
Additional requirements: 3 Letters of Reference.
  All college transcripts.
  1-2 page Personal Statement.

TUITION

Tuition year 2012–13:
Tuition for in-state residents
  Full-time students: $178 per credit
Tuition for out-of-state residents
  Full-time students: $709 per credit
GTAs and GRAs are eligible for full waiver of all eligible tuition, up to the nominal limits of degree program (e.g. 30 crdt-hr for the M.S.; up to 90 crdt-hr for the Ph.D.)
Credit hours per semester to be considered full-time: 6
Health insurance: Available
Other academic fees: Approx. $170/crdt-hr.
Academic term: Semester
Number of first-year students who received full tuition waivers: 12

Teaching Assistants, Research Assistants, and Fellowships
  Number of first-year
    Teaching Assistants: 11
    Research Assistants: 1
    Fellowship students: 1

Average stipend per academic year
  Teaching Assistant: $17,600
  Research Assistant: $22,000
  Fellowship student: $5,000
Typically, we recommend admission only if we intend financial support; thus, nearly all students are supported.

FINANCIAL AID

Application deadlines
Fall admission:
U.S. students: March 1   *Int'l. students*: March 1

Loans
Loans are available for U.S. students.
Loans are not available for international students.
GAPSFAS application required: Yes
FAFSA application required: No

For further information
Address financial aid inquiries to: Graduate College.
Phone: (405) 744-6368
E-mail: grad-i@okstate.edu
Financial aid website: http://gradcollege.okstate.edu

HOUSING

Availability of on-campus housing
  Single students: Yes
  Married students: Yes

For further information
Address housing inquiries to: Manager, University Apartments, or Residence Halls Housing, Iba Hall.
Phone: (405) 744-5592
E-mail: reslife@okstate.edu
Housing aid website: http://reslife.okstate.edu

Table A—Faculty, Enrollments, and Degrees Granted

Research Specialty	2010–11 Faculty	Enrollment Fall 2010		Number of Degrees Granted 2012-13 (2006-11)		
		Master's	Doctorate	Master's	Terminal Master's	Doctorate
Astrophysics	1	–	–	–	–	–
Atomic, Molecular, & Optical Physics	2	1	4	–	–	–
Biophysics	3	–	7	–	–	–
Chemical Physics	2	–	2	–	–	1(1)
Condensed Matter Physics	9	1	7	–	–	1(-)
Fluids, Rheology	2	–	–	–	–	–
Materials Science, Metallurgy	4	–	–	–	–	–
Optics	3	3	1	–	–	–(2)
Polymer Physics/Science	2	–	–	–	–	–
Statistical & Thermal Physics	1	–	–	–	–	–
Non-specialized	–	1	–	–	–	–
Other	1	–	1	–	–	–
Other	1	–	1	–	–	1(3)
Other	4	2	7	1(1)	–	1(1)
Other	6	–	–	–	–	–
Total	–	7	41	2(-)	–	8(29)
Full-time Grad. Stud.	–	7	41	–	–	–
First-year Grad. Stud.	–	–	6	–	–	–

GRADUATE DEGREE REQUIREMENTS

Master's: Twenty-four semester hours of approved physics courses plus 6 hours thesis. Options in optics and photonics medical physics. No language requirement. Last 8 semester hours and 21 total semester hours must be completed in residence. At least a "B" average is required. In addition, a "Professional" M.S. in Physics is offered as a 32 credit hour (Report) plan.

Doctorate: Ninety hours of approved courses (including thesis research) beyond Bachelor's degree. Departmental preliminary exam required. Minimum 30 semester hours and one of last two years in residence. Qualifying exam and dissertation defense. No language requirement. At least a "B" average is required.

SPECIAL EQUIPMENT, FACILITIES, OR PROGRAMS

Materials Growth and Characterization Laboratory; MBE lab; Rubidium and sodium BEC labs; 2-MeV Van de Graaff accelerator; ESR; Solid State NMR; femtosecond spectroscopy; optical absorption and fluorescence spectroscopy, Radiation Dosimetry Laboratory, Brillouin Scattering Laboratory, Powder XRD Core Facility. Mendenhall Observatory (0.6 m RC robotic telescope); Multiple faculty Beowolf PC clusters for their research programs. Physics, along with Electrical Engineering, participates in the multidisciplinary Ph.D. Photonics programs. The types of experimental techniques in progress include the following: photon correlation, Raman scattering, Brillouin scattering, whispering gallery modes, four-wave mixing, time-resolved site selection, holographic gratings, picosecond pulse-probe, and multiphoton excitation. The instrumentation includes a variety of solid state, liquid, and gas lasers, nonlinear optical crystals autocorrelators, streak cameras, FTIR, optical multichannel analyzers, boxcar integrators, and signal averagers, along with the standard monochromators, spectrum analyzers, detectors, and cryogenic equipment required for conventional spectroscopy.

Table B—Separately Budgeted Research Expenditures by Source of Support

Source of Support	Departmental Research	Physics-related Research Outside Department
Federal government	$680,795	
State/local government	$796,622	
Non-profit organizations		
Business and industry		
Other	$143,461	
Total	**$1,620,878**	

Table C—Separately Budgeted Research Expenditures by Research Specialty

Research Specialty	No. of Grants	Expenditures ($)
Biophysics	1	$205,000
Condensed Matter Physics	17	$714,731
Optics	1	$347,500
Particles and Fields	5	$242,307
Statistical & Thermal Physics	1	$111,334
Total	**25**	**$1,620,872**

FACULTY

Professor

Ackerson, Bruce J., Ph.D., University of Colorado, 1976. Dynamic light scattering, colloids; critical phenomena in fluids.

Agarwal, Girish, Ph.D., University of Rochester, 1969. Quantum Optics; Nonlinear Optics; Quantum Information Science and Foundations of Quantum Mechanics; Surface Optics-nano photonics.

Babu, K. S., Ph.D., University of Hawaii, 1986. Grand unification model building, fermion mass mixing; neutrinos.

Bandy, Donna K., Ph.D., Drexel, 1984. Theoretical laser physics; instabilities, nonlinear behavior, optical devices.

Harmon, H. James, Ph.D., Purdue University, 1974. Biophysics; high-resolution high-speed optical spectroscopy; spectroscopy determination of enzyme kinetic intermediates; design of solid state chemical sensors; photochemical reaction of porphyrins.

McKeever, Stephen W. S., Ph.D., University College of N. Wales (Bangor), 1975. Experimental solid state physics; thermoluminescence; thermally stimulated polarization currents; radiation dosimetry; semiconductors.

Mintmire, John W., Ph.D., University of Florida, 1980. Computational materials physics, electronic structure theory, nanostructured materials.

Nandi, Satyanarayan, Ph.D., University of Chicago, 1975. Theoretical High-Energy Physics, grand unification, supersymmetry, extra dimensions, physics at LHC.

Perk, Jacques H. H., Ph.D., Leiden, 1979. Theoretical physics; exactly solvable models in statistical mechanics.

Rosenberger, Albert T., Ph.D., University of Illinois, 1979. Experimental and theoretical optical physics; microresonator optics and plasmonics.

Wicksted, James P., Ph.D., City University of New York, 1983. Experimental solid-state physics; Raman and Brillouin scattering, nonlinear-optics, rare-earth doped glasses, nanomaterials.

Xie, Aihua, Ph.D., Carnegie Mellon University, 1987. Biophysics; structural dynamics of proteins; molecular mechanism of receptor activation in signal transduction; biomedical application of lasers.

Associate Professor

Benton, Eric, Ph.D., Dublin, 2004. Ionizing radiation dosimetry, radiation protection, effects of radiation on living organism.

Hauenstein, Robert J., Ph.D., California Institute of Technology, 1987. *Applied Physics, Condensed Matter Physics, Materials Science, Metallurgy, Solid State Physics*. Experimental semiconductor physics, molecular beam epitaxial growth; heterostructures.

Rizatdinova, Flera, Ph.D., Moscow State University, 1994. Experimental high-energy physics. Mental solid state, radiation dosimetry, thermoluminescence.

Shull, Peter O., Ph.D., Rice University, 1982. Supernova remnants, exoplanets, near-Earth asteroids.

Summy, Gilford, Ph.D., Griffith (Australia), 1995. BEC, quantum chaos, atom optics.

Yukihara, Eduardo, Ph.D., Sao Paulo (Brazil), 2001. Experimental solid state, radiation dosimetry, thermoluminescence.

Assistant Professor

Borunda, Mario, Ph.D., Texas A&M, 2008. *Atomic, Molecular, & Optical Physics, Computational Physics, Condensed Matter Physics*. Developing theoretical and computational techniques to model and analyze the transport properties of atomic, molecular, mesoscopic and macroscopic systems.

Haley, Joseph, Ph.D., Princeton University, 2009. To uncover the physics behind the pattern of fundamental particles and unification of the forces.

Khanov, Alexander, Ph.D., University of Rochester, 2004. Experimental high-energy physics.

Liu, Yingmei, Ph.D., University of Pittsburg, 2004. Atom, molecular, and optical physics.

Zhou, Donghua, Ph.D., College of William and Mary, 2003. *Biophysics*. Biophysics.

Emeritus

Dixon, G., Ph.D., University of Georgia, 1967. Physics.
Lange, J., Ph.D., Pennsylvania State University, 1964. Solid-state physics; acoustics.
Martin, J., Ph.D., Iowa State University, 1967. Physics.
Westhaus, P., Ph.D., Washington University, 1966. Theoretical physics.
Wilson, T., Ph.D., University of Florida, 1966. Theoretical physics; electronic structure of point defects.

Adjunct Professor

Akselrod, M., Ph.D., Urals State Technical University, 1983. Solid-state physics.
Chen, W., Ph.D., University of Oregon, 1988. Physics.
Lucas, A., Ph.D., Oklahoma State University, 2003. Physics.
Perk, H., Ph.D., Stony Brook University, 1973. Theoretical physics.

DEPARTMENTAL RESEARCH SPECIALTIES AND STAFF

Theoretical

Atomic, Molecular, & Optical Physics. Density functional theory of electronic structure. Borunda, Mintmire.
Condensed Matter Physics. Electronic structure of disordered systems; density functional theory, low-dimensional materials, dielectric response theory; optical properties of defects; vibronics; semiconductor molecular beam epitaxy; transport in semiconductors; quantum Hall effect. Borunda, Hauenstein, Mintmire.
Optics. Nonlinear behavior of laser systems; modeling of optical instabilities, quantum optics, nonlinear optics. Bandy, Rosenberger.
Particles and Fields. Grand unification, supersymmetry, extra dimension, physics at LHC, fermion masses and mixings, neutrinos. Babu, Nandi.
Physics and other Science Education. Various instructional strategies. Ackerson.

Quantum Foundations. Coherent control; ultraslow light, super-resolution; quantum imaging, quantum entanglement; decoherence; quantum optics of semiconductor dots; integrated structures and nano-mechanical quantum devices. Agarwal.
Statistical & Thermal Physics. Ising model, stochastic processes, exactly solvable models; low-dimensional systems, quasicrystals. H. Perk, Jacques Perk.

Experimental

Astrophysics. Supernova remnants; exoplanets; near-Earth asteroids; solar physics. Shull.
Atomic, Molecular, & Optical Physics. Liu, Summy.
Biophysics. Laser effects on biological materials; high resolution high-speed optical spectroscopy; enzyme kinetics; photochemical reactions of porphyrins; protein dynamics; protein structure; membrane proteins; amyloid proteins; solid-state NMR. Chen, Harmon, Xie, Zhou.
Chemical Physics. Photocatalysis, photoenergy conversion, monolayer surfaces, solid-state catalysts, photoreductive chemistry. Harmon.
Condensed Matter Physics. Optical, electrical, thermal, acoustical, structural and mechanical properties of solids; laser materials; ESR; energy transfer; epitaxial growth, nanoparticles and nanotubes. Harmon, Hauenstein, McKeever, Wicksted, Yukihara.
Energy Sources & Environment. Dosimetry materials and measurements, radiation effects. Akselrod, Benton, Lucas, McKeever, Yukihara.
Fluids, Rheology. Light scattering; phase transitions in colloids; dynamics of flow systems. Ackerson.
High Energy Physics. Haley, Khanov, Rizatdinova.
Optics. Response of materials to coherent excitation. Optical gain, frequency mixing, multiphoton effects, dephasing times, nonlinear susceptibility measurements involving free and trapped carriers, optical switching, and storage. Microresonators and plasmonics. Ackerson, Rosenberger, Wicksted.
Quantum Foundations. Spin squeezing. Liu.
Surface Physics. Electrode surfaces, corrosion systems, material surface properties, carbon fiber surfaces thin films, and plasmonics. Rosenberger.

View additional information about this department at
www.gradschoolshopper.com

UNIVERSITY OF OKLAHOMA

HOMER L. DODGE DEPARTMENT OF PHYSICS AND ASTRONOMY

Norman, Oklahoma 73019
http://www.nhn.ou.edu/

General University Information

President: David L. Boren
Dean of Graduate School: T. H. Lee Williams
University website: http://www.ou.edu/
Control: Public
Setting: Suburban
Total Faculty: 1,501
Total Graduate Faculty: 1,463

Total number of Students: 27,149
Total number of Graduate Students: 6,643

Department Information

Department Chairman: Gregory A. Parker, Chair
Department Contact: Debbie Barnhill, Program Coordinator
Total full-time faculty: 30
Total number of full-time equivalent positions: 30

Full-Time Graduate Students: 72
First-Year Graduate Students: 16
Female First-Year Students: 3
Total Post Doctorates: 12

Department Address

440 West Brooks St
Norman, OK 73019
Phone: (405) 325-3961
Fax: (405) 325-7557
E-mail: inquiry@physics.ou.edu
Website: http://www.nhn.ou.edu/

ADMISSIONS

Admission Contact Information

Address admission inquiries to: Graduate Recruiting and Se-
lection Committee, Homer L. Dodge Department of Physics
and Astronomy, University of Oklahoma, 440 West Brooks
St, Norman, OK 73019
Phone: (405) 325-3961
E-mail: inquiry@physics.ou.edu
Admissions website: http://www.nhn.ou.edu/grad/apply_OU_
PhyAst.shtml

Application deadlines

Fall admission:
U.S. students: February 15   *Int'l. students*: February 15
Spring admission:
U.S. students: September 1   *Int'l. students*: September 1

Application fee

There is no application fee required.
Send all application materials directly to the Department of Physics and Astronomy to avoid the application fee (see admissions website above). Applications submitted online through
the OU Admissions web site actually slow down consideration of your application and require the application fee.

Admissions information

For Fall of 2013:
Number of applicants: 169
Number admitted: 16
Number enrolled: 16

Admission requirements

Bachelor's degree requirements: Bachelor's degree in physics
and/or astronomy is required.
Minimum undergraduate GPA: 3.0

GRE requirements

The GRE is recommended.
Although not strictly required, we ask for the GRE because it
is an objective and standardized way for us to compare you
with other applicants; it forms part of an overall picture of
you. If you do not have a GRE score, this will put you at
a disadvantage compared with other applicants.

Advanced GRE requirements

The Advanced GRE is recommended.
Although not strictly required, we ask for the physics subject
GRE because it is an objective and standardized way for us
to compare you with other applicants; it forms part of an overall picture of you. If you do not have a physics GRE score,
this will put you at a disadvantage compared with other applicants.

TOEFL requirements

The TOEFL exam is required for students from non-English-
speaking countries.
PBT score: 600

iBT score: 100
Our department has a stated minimum TOEFL of 600/250/100
for paper/computer/IB or IELTS of 7.0. There is some flexibility for candidates who are exceptional in other areas. However, applicants who do not meet the Graduate College's minimum scores of 550/213/79 or 6.5 will not be considered.

Other admissions information

Undergraduate preparation assumed: Marion, Classical Dynam-
ics of Particles and Systems; French, Vibrations and Waves;
Griffiths, Introduction ot Electrodynamics; Saxon, Elementary
Quantum Mechanics; Kittel, Thermal Physics.

TUITION

Tuition year 2011–12:
Tuition for in-state residents
Full-time students: per semester
Our students are typically supported on teaching and research
assistantships that include a tuition waiver for up to 90 hours.
Credit hours per semester to be considered full-time: 9
Deferred tuition plan: No A student health plan is provided as
part of TA and RA support.
Other academic fees: $1548/semester.
Academic term: Semester
Number of first-year students who received full tuition waivers: 17

Teaching Assistants, Research Assistants, and Fellowships

Number of first-year
Teaching Assistants: 15
Research Assistants: 1
Average stipend per academic year
Teaching Assistant: $21,600
Research Assistant: $21,600
Fellowship student: $26,000

FINANCIAL AID

Loans

Loans are available for U.S. students.
Loans are not available for international students.
GAPSFAS application required: No
FAFSA application required: No

HOUSING

Availability of on-campus housing

Single students: Yes
Married students: Yes

For further information

Address housing inquiries to: Housing Office, 1406 Asp Ave.,
Norman, OK 73019.
Phone: (405) 325-2511
E-mail: housinginfo@ou.edu
Housing aid website: http://www.ou.edu/housingandfood.html

Table A—Faculty, Enrollments, and Degrees Granted

Research Specialty	2012–13 Faculty	Enrollment Fall 2012		Number of Degrees Granted 2011–12 (2009–12)		
		Mas-ter's	Doc-torate	Mas-ter's	Terminal Master's	Doc-torate
Astrophysics	7	1	17	5(7)	2(4)	1(3)
Atomic, Molecular, & Optical Physics	7	–	14	–(3)	–	3(9)
Condensed Matter Physics	8	2	15	1(4)	–(2)	5(9)
Engineering Physics/Science	–	–	–	1(4)	–(1)	–
High Energy Physics	8	2	19	2(5)	–(1)	3(8)
Physics and other Science Education	1	–	–	–	–	–
Total	–	5	65	9(23)	2(8)	12(29)
Full-time Grad. Stud.	–	5	65	–	–	–
First-year Grad. Stud.	–	–	17	–	–	–

GRADUATE DEGREE REQUIREMENTS

Master's: A student must complete 30 hours of coursework with a thesis or 32 hours of coursework without a thesis taken in accordance with the general rules of the Graduate College. The allowable minimum number of credits in physics and astronomy is 18 hours, six hours of which must be at the 5000 level or above.

Doctorate: The student must complete a minimum of 36 hours of coursework at the 5000 level or above, excluding the credit hours granted for preparation of the thesis or dissertation describing original research. These hours include 21 hours of specific required courses. Another 54 hours of graduate coursework is required as appropriate for the student's field of research specialization, including research hours. The qualifying examination is offered semiannually and is usually taken at the end of the first year of graduate study. The general examination for the Ph.D. degree consists of a written report and an oral examination, including a presentation of a topic related to the field of specialization and a probing of the student's knowledge of general principles, and is taken before the student begins dissertation research. The Ph.D. in physics may include an emphasis in astronomy or astrophysics.

Other Degrees: An advanced degree (M.S.) in engineering physics is also offered. Specialization areas include astrophysics; atomic, molecular, optical physics; condensed-matter physics; high-energy physics; and others.

Thesis: Thesis and dissertation may be written in absentia.

SPECIAL EQUIPMENT, FACILITIES, OR PROGRAMS

The Homer L. Dodge Department of Physics and Astronomy has access to many well-equipped facilities for experimental research in atomic, molecular, and optical physics, condensed-matter physics, and materials characterization. These include UHV chambers for laser cooling and trapping, laser spectrometers, molecular beam epitaxy systems, a clean room for nanofabrication and characterization, thin-film deposition and characterization facilities, optical and electron beam lithography, atomic force microscopes, scanning tunneling microscopes, infrared spectrometers, systems for transport and magneto-optic measurements at high magnetic field and low temperature, and an independent facility for scanning and transmission electron microscopy.

Our experimental research programs regularly make use of external facilities such as the National High Magnetic Field Laboratory, Los Alamos National Laboratory, the ATLAS detector for the Large Hadron Collider at CERN, the DØ detector at Fermilab, and Oak Ridge National Laboratories. OU is a Tier 2 Data Collection Center for ATLAS. Our astrophysics group routinely has access to Kitt Peak Observatory, the Very Large Array, the MDM Observatory, and the Cerro Tololo Inter-American observatories, as well as data from the Hubble Space Telescope, the Spitzer Space Telescope, and the Chandra x-ray telescope. An instrument and machine shop within the Department with three full-time machinists also supports the experimental research efforts.

Theoretical work is supported by departmental computing resources and the OU Supercomputing Center for Education and Research. For astronomical research, AIPS, IRAF, and IDL software are available.

Table B—Separately Budgeted Research Expenditures by Source of Support

Source of Support	Departmental Research	Physics-related Research Outside Department
Federal government	$4,397,778	
State/local government		
Non-profit organizations		
Business and industry		
Other		
Total	$4,397,778	

Table C—Separately Budgeted Research Expenditures by Research Specialty

Research Specialty	No. of Grants	Expenditures ($)
Astrophysics	15	$402,365
Atomic, Molecular, & Optical Physics	12	$678,904
Condensed Matter Physics	20	$1,681,910
Particles and Fields	15	$1,608,040
Total	63	$4,397,778

FACULTY

Professor

Abbott, Braden, Ph.D., Purdue University, 1994. *High Energy Physics, Particles and Fields*. Experimental high-energy physics; DØ experiment; CERN-ATLAS experiment.

Baer, Howard, Ph.D., University of Wisconsin, 1984. *Astrophysics, Cosmology & String Theory, Particles and Fields, Theoretical Physics*. High-energy theory; supersymmetry; dark matter; LHC physics.

Baron, Edward A., Ph.D., Stony Brook University, 1985. *Astrophysics, Computational Physics*. Radiative transfer; stellar evolution; supernovae; numerical astrophysics.

Furneaux, John E., Ph.D., University of California, Berkeley, 1979. *Atomic, Molecular, & Optical Physics, Optics*. Precision molecular spectroscopy.

Gutierrez, Phillip, Ph.D., University of California, Riverside, 1983. *High Energy Physics*. Experimental high-energy physics; Fermilab DØ experiment; CERN-ATLAS Experiment.

Henry, Richard C., Ph.D., University of Michigan, 1983. *Astrophysics*. Chemical evolution of galaxies; chemical abundances in nebulae; evolution of intermediate mass stars.

Johnson, Matthew B., Ph.D., California Institute of Technology, 1989. *Condensed Matter Physics, Engineering Physics/Science, Nano Science and Technology, Solid State Physics*. Experimental semiconductor and surface physics; scanning tunnelling microscopy.

Kantowski, Ronald, Ph.D., University of Texas, Austin, 1966. *Relativity & Gravitation*. Gravitational lens theory.

Kao, Chung, Ph.D., University of Texas, Austin, 1990. *Astrophysics, Cosmology & String Theory, Particles and Fields, Theoretical Physics.* Particle theory; electroweak symmetry breaking; supersymmetry and unification; CP violation; dark matter; extra dimensions.

Leighly, Karen, Ph.D., Montana State University, 1991. *Astronomy, Astrophysics.* Active Galactic Nuclei (AGN).

Milton, Kimball, Ph.D., Harvard University, 1971. *High Energy Physics, Particles and Fields, Theoretical Physics.* High-energy theory, particularly the development of nonperturbative methods to be applied to quantum chromodynamics and other field theories; physics of the quantum vacuum.

Mullen, Kieran, Ph.D., University of Michigan, 1989. *Condensed Matter Physics, Low Temperature Physics, Solid State Physics, Statistical & Thermal Physics, Theoretical Physics.* Theoretical solid-state physics.

Parker, Gregory, Ph.D., Brigham Young University, 1976. *Atomic, Molecular, & Optical Physics, Chemical Physics, Computational Physics, Theoretical Physics.* Theoretical molecular physics specializing in rearrangement collisions.

Santos, Michael, Ph.D., Princeton University, 1992. *Applied Physics, Condensed Matter Physics, Engineering Physics/Science, Materials Science, Metallurgy, Nano Science and Technology, Solid State Physics.* Experimental semi-conductor and surface physics; MBE growth of narrow gap systems.

Shaffer, James P., Ph.D., University of Rochester, 1999. *Atomic, Molecular, & Optical Physics, Computational Physics, Low Temperature Physics, Optics.* Atomic, molecular, and optical physics; hybrid quantum systems; quantum gases; laser cooling and trapping; atom-based sensing.

Skubic, Patrick, Ph.D., University of Michigan, 1977. *High Energy Physics, Particles and Fields.* Experimental high-energy physics; Fermilab-DØ experiment; CERN-ATLAS experiment.

Strauss, Michael, Ph.D., University of California, Los Angeles, 1988. *High Energy Physics, Particles and Fields.* Experimental high-energy physics; Fermilab DØ experiment; CERN-ATLAS Experiment.

Wang, Yun, Ph.D., Carnegie Mellon University, 1991. *Astrophysics, Cosmology & String Theory.* Cosmology; probing dark energy with galaxy clustering and supernovae; early universe physics.

Watson, Deborah K., Ph.D., Harvard University, 1977. *Atomic, Molecular, & Optical Physics, Theoretical Physics.* Theoretical atomic and molecular physics; many-body systems; group theory methods; Bose-Einstein condensates.

Associate Professor

Abraham, Eric, Ph.D., Rice University, 1996. *Atomic, Molecular, & Optical Physics, Chemical Physics, Low Temperature Physics, Optics.* Experimental atomic, molecular, and optical physics: ultracold atoms and molecules; ultracold collisions; atomic clocks; quantum degenerate gases.

Bumm, Lloyd A., Ph.D., Northwestern University, 1991. *Applied Physics, Condensed Matter Physics, Engineering Physics/Science, Materials Science, Metallurgy, Nano Science and Technology, Surface Physics.* Experimental condensed-matter physics; nanophysics; surface physics and chemistry; self-assembly; scanning tunneling microscopy; surface spectroscopy; molecular plasmonics; development of novel instrumentation.

Mason, Bruce A., Ph.D., University of Maryland, 1985. *Physics and other Science Education.* Educational digital libraries; technology for physics education; faculty and teacher development.

Murphy, Sheena, Ph.D., Cornell University, 1991. *Condensed Matter Physics, Nano Science and Technology, Solid State Physics.* Experimental semiconductor and superconductor physics; low-temperature physics.

Assistant Professor

Capogrosso-Sansone, Barbara, Ph.D., University of Massachusetts, Amherst, 2008. *Atomic, Molecular, & Optical Physics, Computational Physics, Theoretical Physics.* Many-body systems; phase transitions; ultracold atoms and polar molecules in optical lattices; superfluidity; quantum Monte Carlo simulations.

Dai, Xinyu, Ph.D., Pennsylvania State University, 2004. *Astronomy, Astrophysics.* Observational cosmology: gravitational lensing; galaxy clusters; galaxy evolution; high-energy astrophysics; x-ray astronomy; AGNs; gamma-ray bursts.

Kilic, Mukremin, Ph.D., University of Texas, Austin, 2006. *Astronomy, Astrophysics.* Observational astronomy; supernovae Ia progenitors; white dwarfs; merger systems; gravitational waves; planets; debris disks; galactic cosmochronology.

Marino, Alberto, Ph.D., University of Rochester, 2006. *Atomic, Molecular, & Optical Physics, Optics, Quantum Foundations.* Quantum optics; atomic physics; quantum information; quantum metrology.

Sellers, Ian, Ph.D., University of Sheffield, 2004. *Condensed Matter Physics, Energy Sources & Environment, Engineering Physics/Science, Nano Science and Technology, Solid State Physics.* Next-generation photovoltaics; optical and optoelectronic spectroscopy of semiconductor quantum dots; magneto-photoluminescence; solar cell physics.

Uchoa, Bruno, Ph.D., State University of Campinas, 2004. *Condensed Matter Physics.* Quantum critical systems; physics of graphene; low-dimensional systems; unconventional quasiparticles; strongly correlated systems.

Wisniewski, John P., Ph.D., University of Toledo, 2005. *Astronomy.* Circumstellar disks; extrasolar planets; astronomical polarimetry.

Professor Emeritus

Branch, David, Ph.D., University of Maryland, 1969. Spectroscopic astrophysics; supernovae.

Cowan, John J., Ph.D., University of Maryland, 1976. Stellar evolution and nucleosynthesis; supernovae; cosmology.

Doezema, Ryan E., Ph.D., University of Maryland, 1971. Experimental solid-state physics; 2D electron systems in semiconductors; superconductivity.

Herczeg, Tibor, Ph.D., Bonn University, 1959.

Morrison, Michael, Ph.D., Rice University, 1976. Theoretical atomic and molecular physics, particularly electron and positron collisions and near-threshold excitations.

Romanishin, William, Ph.D., University of Arizona, 1980. Extragalactic astronomy; clusters of galaxies; active galactic nuclei.

Ryan, Stewart, Ph.D., University of Michigan, 1971. Applied physics; materials characterization.

Adjunct Professor

Beasley, William, Ph.D., University of Texas, Dallas, 1974. Meteorology.

Crompton, Robert, Ph.D., University of Adelaide, 1954. Electron and ion diffusion.

Feldt, Andrew N., Ph.D., University of Oklahoma, 1980. Atomic and molecular theory.

MacGorman, Donald, Ph.D., Rice University, 1978. Atmospheric electricity.

Rust, David, Ph.D., New Mexico Institute of Mining and Technology, 1973. Atmospheric and plasma physics.

Snow, Joel, Ph.D., Yale University, 1983.

DEPARTMENTAL RESEARCH SPECIALTIES AND STAFF

Theoretical

Astronomy. Cosmology; extragalactic astronomy; nucleosynthesis; stellar atmospheres; stellar evolution; supernovae; gravitational lensing; active galactic nuclei. Baron, Henry, Wang.

Atomic, Molecular, & Optical Physics. Atomic and molecular collisions; ultracold physics; coherent control of bimolecular collisions; dimensional perturbation theory; electron molecule collision; large-scale Monte Carlo simulations; molecular bosonic gases; optical lattices; computational physics; conical intersections. Capogrosso-Sansone, Parker, Watson.

Condensed Matter Physics. Graphene; carbon nanotubes; topological insulators; low-dimensional quantum systems; strongly correlated materials. Mason, Mullen, Uchoa.

High Energy Physics. Quantum field theory; particle physics phenomenology; general relativity; particle physics; Casimir effect; cosmology; dark matter. Baer, Kantowski, Kao, Milton.

Experimental

Astronomy. Binary and variable stars; extragalactic astronomy; extragalactic H regions; supernovae; white dwarfs; gravitational lensing; active galactic nuclei. Dai, Kilic, Leighly, Wisniewski.

Atomic, Molecular, & Optical Physics. Atomic and molecular scattering; laser spectroscopy; multiphoton ionization; reactive scattering; precision measurement; cooling and trapping; quantum optics; Bose-Einstein condensation; nonlinear optics. Abraham, Marino, Shaffer.

Condensed Matter Physics. Molecular beam epitaxy; narrow-gap semiconductors; scanning probe microscopy (AFM & STM); electron microscopy (SEM & TEM); nanofabrication; surface physics; molecular plasmonics; spin transport; photovoltaics; quantum cascade lasers; magneto-optics; topological insulators. Bumm, Furneaux, Johnson, Murphy, Santos, Sellers.

High Energy Physics. New states of matter found at ATLAS and DØ. Abbott, Gutierrez, Skubic, Strauss.

View additional information about this department at
www.gradschoolshopper.com

OREGON STATE UNIVERSITY

DEPARTMENT OF PHYSICS

Corvallis, Oregon 97331-6507
http://www.physics.oregonstate.edu

General University Information

President: Edward J. Ray
Dean of Graduate School: Brenda McComb
University website: http://oregonstate.edu/
Control: Public
Setting: Rural
Total Faculty: 3,481
Total Graduate Faculty: 1,931
Total number of Students: 24,977
Total number of Graduate Students: 3,068

Department Information

Department Chairman: Henri J. F. Jansen, Chair
Department Contact: Talley Richardson, Graduate Coordinator
Total full-time faculty: 18
Total number of full-time equivalent positions: 18
Full-Time Graduate Students: 39
First-Year Graduate Students: 10
Female First-Year Students: 3
Total Post Doctorates: 4

Department Address

301 Weniger Hall
Oregon State University
Corvallis, OR 97331-6507
Phone: (541) 737-6708
Fax: (541) 737-1683
E-mail: talley.richardson@oregonstate.edu
Website: http://www.physics.oregonstate.edu

ADMISSIONS

Admission Contact Information

Address admission inquiries to: Department of Physics, Oregon State University, 301 Weniger Hall, Corvallis, OR 97331
Phone: (541) 737-6708
E-mail: gradinfo@physics.orst.edu
Admissions website: http://physics.oregonstate.edu/

Application deadlines

Fall admission:
U.S. students: January 15   *Int'l. students*: January 15

Application fee

U.S. students: $65   *Int'l. students*: $65

Admissions information

For Fall of 2012:
Number of applicants: 81
Number admitted: 29
Number enrolled: 10

Admission requirements

Bachelor's degree requirements: A bachelor's degree is required.
Minimum undergraduate GPA: 3.0

GRE requirements

The GRE is recommended.

Advanced GRE requirements

The Advanced GRE is not required.

TOEFL requirements

The TOEFL exam is required for students from non-English-speaking countries.
PBT score: 550
iBT score: 100

Other admissions information

Additional requirements: The average GRE scores for 1999–2000 admissions were as follows: verbal, 481; quantitative, 659; and analytical, 631.

Undergraduate preparation assumed: Halliday and Resnick, Fundamentals of Physics; Krane, Modern Physics; Boas, Mathematical Methods in the Physical Sciences; Marion, Classical Dynamics of Particles and Systems; Brophy, Basic Electronics for Scientists; Griffiths, Introduction to Electrodynamics; Liboff, Quantum Physics; Hecht, Optics; and Kittel and Kroemer, Thermal Physics.

TUITION

Tuition year 2011–12:
Tuition for in-state residents
   Full-time students: $12,844 annual
Tuition for out-of-state residents
   Full-time students: $19,756 annual
See the Graduate Catalog.
Credit hours per semester to be considered full-time: 12
Deferred tuition plan: Yes
Health insurance: Available
Other academic fees: $546/term; $1,638/annual
Academic term: Quarter
Number of first-year students who received full tuition waivers: 7

Teaching Assistants, Research Assistants, and Fellowships

Number of first-year
   Teaching Assistants: 7
Average stipend per academic year
   Teaching Assistant: $16,722
   Research Assistant: $16,722

FINANCIAL AID

Application deadlines

Fall admission:
U.S. students: January 15

Loans

Loans are available for U.S. students.
Loans are not available for international students.
GAPSFAS application required: No
FAFSA application required: No

For further information

Address financial aid inquiries to: Office of Financial Aid and Scholarships, 218 Kerr Administration Building, Corvallis, OR 97331-2120.
Phone: (541) 737-2241
E-mail: financial.aid@oregonstate.edu
Financial aid website: http://oregonstate.edu/financialaid/

HOUSING

Availability of on-campus housing

Single students: Yes
Married students: Yes

For further information

Address housing inquiries to: University Housing and Dining Services, Oregon State University, 102 Buxton Hall, Corvallis, OR 97331-1317.
Phone: (541) 737-4711
Housing aid website: http://oregonstate.edu/uhds

Table A—Faculty, Enrollments, and Degrees Granted

Research Specialty	2012-13 Faculty	Enrollment Fall 2012 Master's	Enrollment Fall 2012 Doctorate	Number of Degrees Granted 2012-13 (2006–13) Master's	Number of Degrees Granted 2012-13 (2006–13) Terminal Master's	Number of Degrees Granted 2012-13 (2006–13) Doctorate
Atomic, Molecular, & Optical Physics	4	2	9	–	1(2)	1(8)
Biophysics	2	1	6	–	–	–
Computational Physics	4	1	7	–	-(5)	1(2)
Condensed Matter Physics	5	1	11	–	2(5)	1(4)
Nuclear Physics	–	–	–	–	–	1(1)
Particles and Fields	1	–	–	–	–	-(1)
Physics and other Science Education	2	1	1	–	1(-)	–
Non-specialized	–	1	8	–	-(1)	–
Total	**18**	**6**	**42**	**–**	**2(13)**	**3(16)**
Full-time Grad. Stud.	–	6	36	–	–	–
First-year Grad. Stud.	–	–	11	–	–	–

GRADUATE DEGREE REQUIREMENTS

Master's: Forty-five term hours of credit, with a 3.0 grade point average (minimum), with approximately two-thirds of the credit in the major and the remaining one-third in a minor, if chosen. The optional minor is ordinarily completed within the physics department; however, students seeking a more flexible program may plan a minor in another discipline. Residence requirements include one academic year full-time load (15 hours per term) or fair equivalent. A maximum of 15 term hours of transferred credit may be applied toward the residence requirement. There is no foreign language or computer language requirement. Thesis is optional. Completion of a project is required if the non-thesis option is chosen. Satisfactory performance is also required in a two-hour oral examination on the major and minor subjects.

Doctorate: The doctor of philosophy degree is granted primarily for creative attainments, broadly viewed as the departmental requirements, advancement to candidacy, and completion of a thesis. There is no rigid credit requirement; however, the equivalent of at least three years of full-time graduate work beyond the bachelor's degree (at least 108 graduate credits) is required. All graduate student programs must consist of, at a minimum, 50% graduate stand-alone courses. After admission into the doctoral program, a minimum of one full-time academic year (at least 36 graduate credits) should be devoted to the preparation of the thesis. The equivalent of one full-time academic year of regular non-blanket course work (at least 36 graduate credits) must be included on a doctoral program. Advancement to Ph.D. candidacy requires satisfactory performance on written and oral comprehensive examinations. A thesis and oral defense of the thesis are required.

Thesis: Thesis may be written in absentia.

SPECIAL EQUIPMENT, FACILITIES, OR PROGRAMS

A laboratory for preparation and characterization of thin solid films includes facilities for deposition by evaporation, radio frequency sputtering, and pulsed laser techniques. Films are characterized by alternating current and direct current transport, optical transmission and reflection, and Hall effect measurements. Neutron diffraction studies of magnetic semiconductors are conducted at the facilities of the National Institute of Standards and Technology. Spectrometers and magnetic fields up to 8 T are used for solid state nuclear magnetic resonance and nuclear quadrupole resonance studies of semiconductors, conducting oxides and related materials. Laser cooling and atom trapping studies are being done using diode lasers. Cold atoms are used in studies of atom interferometry. Nonlinear optical studies of thin films and interfaces are conducted in a picosecond pulsed laser laboratory. A titanium-sapphire femtosecond oscillator and 532-nm diode-pumped solid state laser are used for terahertz studies of semiconductors.

Computational physics research is conducted in several areas, including nuclear and solid state physics. The Oregon State Physics Department is one of the few in the United States with a degree in computational physics. The department maintains a 20-node Solaris Beowulf cluster for research and advanced computational projects, as well as a 30-node cluster for computational physics education. Laboratory courses offer instruction in interfacing computers for laboratory experiments, as well as practical experience in computational physics. The department maintains an electronics shop and has access to on-campus machine shops.

Physics education research is conducted in the context of the "Paradigms in Physics" program, a unique curriculum developed for upper-division physics instruction, and in the implementation of interactive teaching techniques in large, lower-division courses. In the "Paradigms in Physics" curriculum, course content has been rearranged to better reflect the way professional physicists think about their field and to incorporate pedagogy that assigns to the students more responsibility for their own learning.

Cooperative arrangements permit students to pursue advanced physics degrees through research in other departments, including chemistry and electrical engineering.

Table B—Separately Budgeted Research Expenditures by Source of Support

Source of Support	Departmental Research	Physics-related Research Outside Department
Federal government	$1,095,941	$10,836,890
State/local government		
Non-profit organizations		
Business and industry		
Other	$86,263	$53,950
Total	$1,182,204	$10,890,840

Table C—Separately Budgeted Research Expenditures by Research Specialty

Research Specialty	No. of Grants	Expenditures ($)
Atomic, Molecular, & Optical Physics	4	$410,207
Condensed Matter Physics	19	$441,423
Physics and other Science Education	3	$145,265
Computational Physics	12	$140,244
Total	38	$1,137,139

FACULTY

Professor

Jansen, Henri J. F., Ph.D., Groningen, 1981. Chair of Physics Department. Theoretical solid state physics.

Krane, Kenneth S., Ph.D., Purdue University, 1970. Experimental nuclear and solid state physics.

Landau, Rubin H., Ph.D., University of Illinois, 1970. *Computational Physics, Nuclear Physics, Particles and Fields*. Theoretical nuclear and particle physics; computational physics.

Manogue, Corinne A., Ph.D., University of Texas, 1984. *Particles and Fields, Physics and other Science Education*. Theoretical particle physics; physics education research.

McIntyre, David H., Ph.D., Stanford University, 1987. *Atomic, Molecular, & Optical Physics*. Experimental atomic and optical physics.

Tate, Janet, Ph.D., Stanford University, 1988. Experimental solid state physics.

Associate Professor

Giebultowicz, Tomasz M., Ph.D., University of Warsaw, 1975. Experimental solid state physics.

Hetherington, William M., Ph.D., Stanford University, 1977. Experimental optical and chemical physics.

Lee, Yun-Shik, Ph.D., University of Texas, Austin, 1997. Experimental optical and solid state physics.

Ostroverkhova, Oksana G., Ph.D., Case Western Reserve University, 2001. Experimental optical and chemical physics.

Assistant Professor

Demaree, Dedra, Ph.D., Ohio State University, 2006. Physics education research.

Michael, Zwolak, Ph.D., California Institute of Technology, 2007. *Nano Science and Technology*. Nanoscale electronics.

Minot, Ethan D., Ph.D., Cornell University, 2004. Experimental condensed matter physics.

Roundy, David, Ph.D., University of California, Berkeley, 2001. Theoretical condensed matter physics.

Schneider, Guenter, Ph.D., Oregon State University, 1999. Theoretical condensed matter physics.

Sun, Bo, Ph.D., New York University, 2010.

Emeritus

Stetz, Albert W., Ph.D., University of California, Berkeley, 1968. Experimental intermediate energy nuclear physics.

Warren, William W., Ph.D., Washington University, 1965. Experimental solid state physics.

DEPARTMENTAL RESEARCH SPECIALTIES AND STAFF

Theoretical

Condensed Matter Physics. Electronic structure of solids; magnetism, metal clusters and reduced dimensionality; aqueous interfaces and solutions; and density functional theory.

Experimental

Atomic, Molecular, & Optical Physics. Nonlinear optical studies of surfaces and interfaces; ultrafast processes and nonlinear optics in organic materials; laser cooling and trapping of atoms; atom interferometry; laser spectroscopy; quantum optics; terahertz spectroscopy and ultrafast carrier dynamics in semiconductors using femtosecond lasers; photonic crystals.

Condensed Matter Physics. Electrical transport and optical studies of transparent; conducting thin films; neutron diffraction studies of magnetic semiconductor heterostructures for "spintronics"; nuclear magnetic resonance and nuclear quadrupole resonance studies of electron dynamics in conducting oxides; carbon nanotube biosensors.

PORTLAND STATE UNIVERSITY

DEPARTMENT OF PHYSICS

Portland, Oregon 97207
http://www.pdx.edu/physics/

General University Information
President: Wim Wiewel
Dean of Graduate School: Margaret Everett
University website: http://www.pdx.edu
Control: Public
Setting: Urban
Total Faculty: 1,601
Total number of Students: 29,703
Total number of Graduate Students: 6,481

Department Information
Department Chairman: John L. Freeouf, Chair
Department Contact: Alexandria Christy, Graduate Program
   Director
   Total full-time faculty: 17
   Total number of full-time equivalent positions: 18
   Full-Time Graduate Students: 47
   First-Year Graduate Students: 11
   Female First-Year Students: 1
   Total Post Doctorates: 5

Department Address
P.O. Box 751
Portland, OR 97207
Phone: (503) 725-3812
E-mail: physics@pdx.edu
Website: http://www.pdx.edu/physics/

ADMISSIONS

Admission Contact Information
Address admission inquiries to: Department of Physics
Phone: (503) 725-3812
E-mail: physics@pdx.edu
Admissions website: http://www.pdx.edu/physics/graduate-
   application-procedure

Application deadlines
Fall admission:
U.S. students: March 1   *Int'l. students*: March 1

Application fee
U.S. students: $50   *Int'l. students*: $50
can't be waived

Admissions information
For Fall of 2013:
   Number of applicants: 50
   Number admitted: 20
   Number enrolled: 5

Admission requirements
Bachelor's degree requirements: Bachelor's degree is required.
Minimum undergraduate GPA: 2.75

GRE requirements
The GRE is required.
   Mean GRE score range (25th–75th percentile): 50% - 70%
Actually based on percentage in year taken; should be above 60%
   of population.

Advanced GRE requirements
The Advanced GRE is not required.

TOEFL requirements
The TOEFL exam is required for students from non-English-
   speaking countries.
   PBT score: 550
   iBT score: 80
Minimum accepted Computer-based exam (CBT) score is 213.
   The internet-based TOEFL requirement (iBT) is a minimum
   score of 80 with minimum sub-scores of 18 each in reading
   and writing.

Other admissions information
Undergraduate preparation assumed: Resnick and Halliday,
   Fundamentals of Physics; Tipler, Modern Physics; Fowles,
   Analytical Mechanics; Hecht, Modern Optics; Reif, Statistical
   Physics; Griffiths, Intro to Electrodynamics.

TUITION

Tuition year 2014-15:
Tuition for in-state residents
   Full-time students: $10,000 annual
   Part-time students: $359 per credit
Tuition for out-of-state residents
   Full-time students: $14,000 annual
   Part-time students: $527 per credit
Includes fees per credit/term incidental, Recreation Center building fee.
Credit hours per semester to be considered full-time: 9
Deferred tuition plan: Yes
Health insurance: Available
Other academic fees: $180
Academic term: Quarter
Number of first-year students who received full tuition waivers: 2

Teaching Assistants, Research Assistants, and Fellowships
   Number of first-year
   Teaching Assistants: 2
   Research Assistants: 1
   Average stipend per academic year
   Teaching Assistant: $15,000
   Research Assistant: $15,000

FINANCIAL AID

Application deadlines
Fall admission:
U.S. students: March 1   *Int'l. students*: March 1

Loans
Loans are available for U.S. students.
Loans are available for international students.
GAPSFAS application required: No
FAFSA application required: Yes

For further information
Address financial aid inquiries to: Financial Aid Office.
Financial aid website: http://www.pdx.edu/finaid/

HOUSING

Availability of on-campus housing
Single students: Yes
Married students: Yes

For further information
Address housing inquiries to: University Housing.
Phone: (503) 725-4333
E-mail: housing@pdx.edu
Housing aid website: http://www.pdx.edu/housing

Table A—Faculty, Enrollments, and Degrees Granted

Research Specialty	2012-13 Faculty	Enrollment Fall 2012 Master's	Enrollment Fall 2012 Doctorate	Number of Degrees Granted 2012-13 (2008-13) Master's	Number of Degrees Granted 2012-13 (2008-13) Terminal Master's	Number of Degrees Granted 2012-13 (2008-13) Doctorate
Atmospheric Physics	3	–	6	1(3)	–	3(4)
Atomic, Molecular, & Optical Physics	4	1	5	1(3)	–	–(2)
Biophysics	2	1	7	1(4)	–	2(3)
Condensed Matter Physics	3	1	6	2(4)	–	–(2)
Materials Science	3	1	3	1(3)	1(1)	–(1)
Nano Science and Technology	3	1	15	3(9)	2(-)	1(3)
Total	**19**	**5**	**42**	**9(37)**	**1(1)**	**2(10)**
Full-time Grad. Stud.	–	3	35	–	–	–
First-year Grad. Stud.	–	1	7	–	–	–

GRADUATE DEGREE REQUIREMENTS

Master's: Program approval required. Forty-five quarter credit hours required. A GPA of 3.0/4.0 or better continuously is required. 30 quarter credit hours must be taken in residence. A final oral examination is required.

Doctorate: Program approval required. A minimum GPA of 3.0/4.0 continuously. A minimum of 3 consecutive quarters in residence is required. Qualifying examinations and oral defense of dissertation is required.

Thesis: Master's thesis may be written in absentia.

SPECIAL EQUIPMENT, FACILITIES, OR PROGRAMS

In addition to individual labs having research equipment, the Center for Electron Microscopy (CEMN) has multiple state-of-the-art instruments for performing research. The department also hosts an NSF REU (Research Experience for Undergraduates) program focused on the education and use of advanced microscopy techniques in research.

Table B—Separately Budgeted Research Expenditures by Source of Support

Source of Support	Departmental Research	Physics-related Research Outside Department
Federal government	$2,450,000	
State/local government		
Non-profit organizations		
Business and industry	$574,000	
Other	$173,000	
Total	**$3,197,000**	

Table C—Separately Budgeted Research Expenditures by Research Specialty

Research Specialty	No. of Grants	Expenditures ($)
Atmosphere, Space Physics, Cosmic Rays	1	$669,000
Atomic, Molecular, & Optical Physics	2	$245,000
Biophysics	2	$223,000
Materials Science, Metallurgy	2	$403,000
Nano Science and Technology	9	
Total	**16**	**$1,540,000**

FACULTY

Professor

Abramson, Jonathan J., Ph.D., University of Rochester, 1975. *Biophysics*. Biophysics, rational drug design.

Bodegom, Erik, Ph.D., Catholic University, 1982. Complex systems; charge-coupled devices.

Carruthers, John, Ph.D., University of Toronto, 1967. *Computational Physics*. Materials science and engineering.

Freeouf, John L., Ph.D., University of Chicago, 1973. *Nano Science and Technology*. Optical studies of semiconductors.

Jiao, Jun, Ph.D., University of Arizona, 1997. *Nano Science and Technology*. Electron microscopy; electron field emission of nanomaterials and carbon nanoclusters.

Könenkamp, Rolf, Ph.D., Tulane University, 1984. Solid state physics; electron optics.

Khalil, M. Aslam Khan, Ph.D., University of Texas, 1976. *Atmosphere, Space Physics, Cosmic Rays*. Atmospheric physics and chemistry.

Leung, Pui-Tak, Ph.D., State University of New York at Buffalo, 1982. *Atomic, Molecular, & Optical Physics*. Atomic, optical, and surface physics.

Mitchell, Drake C., Ph.D., University of Oregon, 1987. Chairman of the Department. *Biophysics*. Membrane biophysics, signal transduction.

Solanki, Raj, Ph.D., Colorado State University, 1982. *Nano Science and Technology*. Semiconductors, graphene.

Associate Professor

La Rosa, Andres H., Ph.D., North Carolina State University, 1996. Opto/ultrasonic near-field microscopy; optical MEMS.

Möck, Peter, Ph.D., Humboldt-University of Berlin, 1991. Electron microscopy; quantum dots; x-ray diffraction.

Rice, Andrew, Ph.D., University of California, Irvine, 2002. *Atmosphere, Space Physics, Cosmic Rays*. Atmospheric physics and chemistry.

Sánchez, Erik J., Ph.D., Portland State University, 1999. *Atomic, Molecular, & Optical Physics, Computational Physics, Electromagnetism*. Microscopy, lasers, nanotechnology.

Straton, Jack, Ph.D., University of Oregon, 1986. *Atomic, Molecular, & Optical Physics*. Quantum scattering theory.

Assistant Professor

Butenhoff, Chris, Ph.D., Portland State University, 2010. *Atmosphere, Space Physics, Cosmic Rays*. Atmospheric physics.

Widenhorn, Ralf, Ph.D., Portland State University, 2005. Thermally activated processes, charge coupled devices.

Emeritus

Dash, John, Ph.D., Pennsylvania State University, 1966. Metallurgy.

Howard, Donald G., Ph.D., University of California, Berkeley, 1964.

Semura, Jack, Ph.D., University of Wisconsin-Madison, 1972. Statistical mechanics.

Smejtek, Pavel K., Ph.D., Czechoslovak Academy, 1965. *Biophysics.*

DEPARTMENTAL RESEARCH SPECIALTIES AND STAFF

Theoretical

Atmosphere, Space Physics, Cosmic Rays. Climate modeling; atmospheric dispersion and chemistry; global emissions inventories of greenhouse gases. Butenhoff, Khalil, Rice.

Atomic, Molecular, & Optical Physics. Stopping power theory; sum rule calculations; near-field optics; surface effects. Leung, Sánchez, Straton.

Computational Physics. Computing nanoarchitectures. Carruthers, Sánchez.

Electromagnetism. Modeling of electromagnetic fields. Leung, Sánchez.

Nano Science and Technology. Molecular fluorescence; metallic nanoparticles and plasmonics. Freeouf, Jiao, Könenkamp, La Rosa, Sánchez, Solanki.

Experimental

Atmosphere, Space Physics, Cosmic Rays. Global change science, design of field experimentation, instrumental analysis of air and water samples. Butenhoff, Khalil, Rice.

Atomic, Molecular, & Optical Physics. AFM, STM, charge-coupled devices. Bodegom, La Rosa, Sánchez, Widenhorn.

Biophysics. Rational drug design, cardiac function, membrane structure/function, nano-calorimetry, biosensors. Abramson, Mitchell, Smejtek.

Condensed Matter Physics. AFM, STM, quantum dots, crystallographic identification of compounds. Freeouf, La Rosa, Möck, Sánchez, Solanki.

Materials Science, Metallurgy. Electronic device physics. Freeouf, Jiao, Könenkamp, Solanki.

Nano Science and Technology. Carbon nanotubes and nanoclusters, graphene, biosensors, nanometrology, nanoelectronics. Freeouf, Jiao, Möck, Sánchez, Solanki.

Optics. Electron optics, near-field optics, biological imaging, electron microscopy. Freeouf, Jiao, Könenkamp, La Rosa, Möck, Sánchez.

View additional information about this department at
www.gradschoolshopper.com

UNIVERSITY OF OREGON

DEPARTMENT OF PHYSICS

Eugene, Oregon 97403-1274
http://physics.uoregon.edu

General University Information

President: Michael Gottfredson
Dean of Graduate School: Kimberly Andrews Espy
University website: http://www.uoregon.edu
Control: Public
Setting: Urban
Total Faculty: 1,478
Total number of Students: 24,591
Total number of Graduate Students: 3,702

Department Information

Department Chairman: Raymond Frey, Head
Department Contact: Jodi Myers, Undergraduate and Graduate Programs Coordinator
Total full-time faculty: 31
Total number of full-time equivalent positions: 31
Full-Time Graduate Students: 81
First-Year Graduate Students: 16
Female First-Year Students: 3
Total Post Doctorates: 13

Department Address

1274 University of Oregon
Eugene, OR 97403-1274
Phone: (541) 346-4751
Fax: (541) 346-5861
E-mail: physgradinfo@uoregon.edu
Website: http://physics.uoregon.edu

ADMISSIONS

Admission Contact Information

Address admission inquiries to: Department of Physics, Graduate Selection Committee
Phone: (541) 346-4751
E-mail: physgradinfo@uoregon.edu
Admissions website: http://physics.uoregon.edu/~dsteck/application/

Application deadlines

Fall admission:
U.S. students: January 15   *Int'l. students*: January 15

Application fee

U.S. students: $50   *Int'l. students*: $50

Admissions information

For Fall of 2013:
Number of applicants: 160
Number admitted: 64
Number enrolled: 23

Admission requirements

Bachelor's degree requirements: A bachelor's degree in physics or a related subject is required.
Minimum undergraduate GPA: 3.0

GRE requirements

The GRE is required.

No minimum acceptable GRE score for admission is specified. The median GRE scores for fall 2013 admission with teaching assistantships were as follows: verbal, 163; quantitative, 164; and physics subject, 750.

Advanced GRE requirements
The Advanced GRE is required.

TOEFL requirements
The TOEFL exam is required for students from non-English-speaking countries.
PBT score: 600
iBT score: 100
Minimum acceptable TOEFL (paper-based test)score is 500 for admission and 600 for teaching assistants. Supplemental English training after arrival may be required for TOEFL scores below 575. Therefore, scores above 600 are given preference.

Other admissions information
Undergraduate preparation assumed: Familiarity with material at a level found in the following text books is assumed: Classical Mechanics, Chow; Analytical Mechanics: Electricity and Magnetism, Griffiths; Introduction to Electrodynamics; Statistical and Thermal Physics, Schroeder; Introduction to Thermal Physics; Modern Physics, Griffiths; Introduction to Quantum Mechanics, Griffiths.

TUITION

Tuition year 2012–13:
Tuition for in-state residents
Full-time students: $15,306 annual
Each GTF receives a tuition waiver for 9-16 credit hours per term. For current information, see http://gradschool.uoregon.edu/gtf/salary-benefits.
Credit hours per semester to be considered full-time: 9
Deferred tuition plan: Yes
Health insurance: Yes, 5% of premium.
Other academic fees: GTFs have access to GTF-specific insurance during each term of appointment. For current info, see http://gradschool.uoregon.edu/gtf/salary-benefits.
Academic term: Quarter
Number of first-year students who received full tuition waivers: 16

Teaching Assistants, Research Assistants, and Fellowships
Number of first-year
Teaching Assistants: 16
Average stipend per academic year
Teaching Assistant: $17,328
Research Assistant: $17,328

FINANCIAL AID

Application deadlines
Fall admission:
U.S. students: January 15   *Int'l. students*: January 15

Loans
Loans are available for U.S. students.
Loans are available for international students.
GAPSFAS application required: No
FAFSA application required: Yes

For further information
Address financial aid inquiries to: Office of Financial Aid & Scholarships.
Phone: (541) 346-3221
E-mail: financialaid@uoregon.edu
Financial aid website: http://financialaid.uoregon.edu/

HOUSING

Availability of on-campus housing
Single students: Yes
Married students: Yes

For further information
Address housing inquiries to: University Housing, Walton Hall, University of Oregon.
Phone: (541) 346-4277
E-mail: housing@uoregon.edu
Housing aid website: http://housing.uoregon.edu/

Table A—Faculty, Enrollments, and Degrees Granted

Research Specialty	2012-13 Faculty	Enrollment Fall 2012 Master's	Enrollment Fall 2012 Doctorate	Number of Degrees Granted 2012-13 Master's	Number of Degrees Granted 2012-13 Terminal Master's	Number of Degrees Granted 2012-13 Doctorate
Applied Physics	–	26	–	17(-)	–	–
Astrophysics	4	–	4	–	–	1(-)
Atomic, Molecular, & Optical Physics	9	–	24	1(-)	1(-)	8(-)
Biophysics	3	–	9	–	1(-)	–
Condensed Matter Physics	6	–	10	–	–	2(-)
Particles and Fields	8	–	8	–	–	3(-)
Non-specialized	5	1	25	9(-)	2(-)	–
Total	35	27	80	27(-)	4(-)	14(-)
Full-time Grad. Stud.	–	27	80	–	–	–
First-year Grad. Stud.	–	1	15	–	–	–

GRADUATE DEGREE REQUIREMENTS

Master's: Forty-five credit hours of graduate level courses, including 32 credits of physics and at least 24 credits of University of Oregon-graded courses, are required. These must include at least one three-term sequence in physics at the 600 level and an approved sequence in mathematics. A maximum of 15 hours of credits earned in another accredited graduate school with a grade of B or better may be counted. A minimum GPA of 3.0 must be maintained. Command of a foreign language is recommended but not required. A master's final examination, a thesis, or a certain course requirement has to be satisfactorily completed. The department offers an Applied Physics Master's Program that leads to a professional M.S. degree. This degree is an alternative to the research-based Ph.D. and is more oriented toward the needs of industrial physicists than the traditional master's degree. This program includes an internship component. The Applied Physics Master's Program is offered through the University of Oregon Materials Science Institute (see http://internship.uoregon.edu).

Doctorate: The student must pass a qualifying examination covering advanced undergraduate physics in mechanics, electricity and magnetism, quantum mechanics, statistical mechanics, and thermal physics. Students generally must complete core graduate courses in mechanics, statistical physics, electromagnetic theory, and quantum mechanics, although they can be excused based on previous study. In addition, students must take a total of six other one-quarter courses, chosen from the following areas: condensed matter physics; nuclear and particle physics; atomic physics and molecular physics; astronomy and early universe physics; experimental and theoretical techniques; and interdisciplinary. An oral comprehensive examination and a thesis are required. Proficiency in a foreign language is recommended but not required. Three years work beyond the bachelor's degree is required, of which three consecutive terms must be on the Eugene, Oregon campus.

Thesis: Thesis may be written in absentia.

SPECIAL EQUIPMENT, FACILITIES, OR PROGRAMS

The University of Oregon has several interdisciplinary institutes in which many physics faculty members participate.

The Materials Science Institute (MSI; http://www.uoregon.edu/~msiuo/) focuses much of its efforts on the creation and study of new materials and devices but also addresses more abstract questions in experiment and theory. The MSI has a wide range of fabrication and characterization capabilities located in both individual laboratories and common facilities. An important mission of the MSI is education, and in this connection it promotes integrated research between various departments and conducts Summer Industrial Internship programs in semiconductor processing, polymer technology, and, with the Oregon Center for Optics (OCO), optics and photonics. The MSI is a founding member and partner of the Oregon Nanoscience and Microtechnologies Institute (http://www.onami.us/).

The OCO (http://oco.uoregon.edu/) promotes and facilitates research and education in the sciences at the University of Oregon whenever optical science is involved in an essential manner in either its fundamental aspects or its technological applications. The OCO has a broad range of state-of-the-art lasers and spectroscopy equipment located in individual laboratories and also in common facilities.

The Institute of Theoretical Science (http://www.uoregon.edu/~its/) is a center for theoretical research in overlapping areas of physics, chemistry, and mathematics. It provides an environment in which theorists can share common themes and mathematical approaches.

The Institute of Molecular Biology (http://www.molbio.uoregon.edu/) comprises biologists, chemists, and physicists pursuing a molecular-level understanding of living systems. It runs a weekly seminar series and operates common facilities to assist with imaging, cell culture, and analytic characterization.

The Center for High Energy Physics (http://uoregon.edu/~chep/) supports experimental and theoretical high energy physics research activities at the University of Oregon and at various external laboratories, including CERN, Fermilab, the SLAC National Accelerator Laboratory, and LIGO.

The Pine Mountain Observatory (http://pmo-sun.uoregon.edu/) houses several telescopes and is equipped with charge-coupled device cameras for remote data collection.

The Center for Advanced Materials Characterization at Oregon (http://www.uoregon.edu/)houses capital-intensive equipment for microanalysis, surface analysis, electron microscopy, semiconductor device fabrication, as well as traditional chemical characterization for users from inside and outside the university. The staff members who run the facilities are experienced in sample preparation, data collection, and data analysis.

The Shared Laser Facility (SLF) is a multidisciplinary laboratory available to the university community and others by arrangement. Faculty members may either set up long-term experiments in the SLF or use shared equipment for short-term experiments. SLF personnel also provide expertise in setting up experiments in user laboratories.

The Technical Services Administration maintains professional and student machine shops and an electronics shop.

Table B—Separately Budgeted Research Expenditures by Source of Support

Source of Support	Departmental Research	Physics-related Research Outside Department
Federal government	$7,753,755	
State/local government	$77,506	
Non-profit organizations	$286,604	
Business and industry		
Other	$58,336	
Total	**$8,176,201**	

Table C—Separately Budgeted Research Expenditures by Research Specialty

Research Specialty	No. of Grants	Expenditures ($)
Astrophysics	1	$39,499
Biophysics	10	$611,324
Condensed Matter Physics	11	$1,151,400
Energy Sources & Environment	10	$210,553
Low Temperature Physics	–	
Optics	21	$1,803,552
Particles and Fields	11	$2,842,045
Physics and other Science Education	3	$1,517,826
Total	**67**	**$8,176,199**

FACULTY

Professor

Belitz, Dietrich, Ph.D., University of Munich, 1982. Director of Institute of Theoretical Physics (ITS). *Condensed Matter Physics*. Many-body theory; quantum phase transitions.

Bothun, Gregory D., Ph.D., University of Washington, 1981. *Astrophysics, Energy Sources & Environment*. Astronomy; properties of galaxies; observational cosmology; sustainable energy.

Brau, James E., Ph.D., Massachusetts Institute of Technology, 1978. Director of Center for High Energy Physics (UOCHEP). *High Energy Physics, Particles and Fields*. Experimental elementary particle physics; electroweak symmetry breaking (ATLAS and ILC); gravitational radiation (LIGO).

Csonka, Paul, Ph.D., Johns Hopkins University, 1963. *High Energy Physics, Theoretical Physics*. Elementary particle theory; accelerator theory; conservation laws in quantum mechanics.

Deshpande, Nilendra G., Ph.D., University of Pennsylvania, 1965. *High Energy Physics, Theoretical Physics*. Elementary particle theory; electroweak interactions; grand unification; neutrino physics.

Deutsch, Miriam, Ph.D., Hebrew University, 1996. Director of Oregon Center for Optics (OCO). *Optics*. Quantum optics; photonics.

Donnelly, Russell J., Ph.D., Yale University, 1956. *Fluids, Rheology*. Classical and superfluid hydrodynamics; low-temperature physics.

Frey, Raymond E., Ph.D., University of California, Riverside, 1984. Department Head. *Astrophysics, High Energy Physics, Particles and Fields, Relativity & Gravitation*. Experimental elementary particle physics; electroweak force; gravity waves.

Haydock, Roger, Ph.D., University of Cambridge, 1972. *Solid State Physics*. Solid state theory; electronic structure and processes at surfaces; computational physics; quantum chaos.

Imamura, James N., Ph.D., Indiana University, 1981. *Astrophysics*. Astrophysics; accretion disks; dense fluids; x-ray timing observations.

543

Kevan, Stephen D., Ph.D., University of California, Berkeley, 1980. *Solid State Physics*. Experimental solid state physics; thin-film and surface physics; magnetism.

Raymer, Michael G., Ph.D., University of Colorado, 1979. *Optics*. Quantum optics; quantum information; quantum control; semiconductor optical physics; nonlinear optics.

Schombert, James, Ph.D., Yale University, 1984. *Astrophysics*. Astronomy; galaxy surveys; evolution and properties of galaxies.

Soper, Davison E., Ph.D., Stanford University, 1971. *High Energy Physics, Theoretical Physics*. Elementary particle theory; quantum chromodynamics; calculational tools for particle physics.

Strom, David, Ph.D., University of Wisconsin-Madison, 1986. *High Energy Physics, Particles and Fields*. Experimental elementary particle physics; new physics with tau decays in proton-proton collisions; detectors and electronics for linear colliders; precision electroweak measurements; instrumentation for the detection of gravity waves; ATLAS; LIGO.

Taylor, Richard, Ph.D., University of Nottingham, 1988. Director of Materials Science Institute. *Biophysics, Condensed Matter Physics*. Experimental solid state physics; nanoelectronics; retinal implants; solar cells; visual science of fractals.

Toner, John, Ph.D., Harvard University, 1981. *Condensed Matter Physics, Theoretical Physics*. Condensed matter theory; flocking; liquid crystal and superconducting glasses; novel phases of Josephson junction arrays; quantum whistling; supersolids.

Torrence, Eric, Ph.D., Massachusetts Institute of Technology, 1997. *High Energy Physics, Particles and Fields*. Experimental high energy physics.

van Enk, Stephen, Ph.D., Leiden University. *Optics, Theoretical Physics*. Theoretical optical physics; quantum communication; entanglement; coherence and decoherence.

Wang, Hailin, Ph.D., University of Michigan, 1990. *Condensed Matter Physics*. Quantum optics; optical properties of semiconductor nanostruture.

Associate Professor

Gregory, Stephen, Ph.D., University of Waterloo, 1975. Director of Graduate Studies. *Solid State Physics*. Experimental condensed matter; tunneling microscopy and spectroscopy; molecular electronics.

Kribs, Graham, Ph.D., University of Michigan, 1998. *High Energy Physics, Particles and Fields, Theoretical Physics*. Theoretical high energy physics; effective field theory of models beyond the standard model; particle astrophysics; early universe cosmology.

Nöckel, Jens, Ph.D., Yale University, 1997. *Optics, Theoretical Physics*. Optical physics; quantum chaos and semiclassical physics in microactivity optics; optical and transport properties of mesoscopic systems.

Parthasarathy, Raghuveer, Ph.D., University of Chicago, 2002. *Biophysics, Condensed Matter Physics*. Experimental solid state and biophysics; material properties of biological membranes; mechanism of protein organization; advanced microscopy techniques.

Steck, Daniel, Ph.D., University of Texas, 2001. *Optics*. Experimental and theoretical optical physics; quantum and atom optics; quantum nonlinear dynamical systems; laser cooling and trapping atoms.

Assistant Professor

Chang, Spencer, Ph.D., Harvard University, 2004. *High Energy Physics, Theoretical Physics*. Theoretical high energy physics; beyond the standard model physics; electroweak symmetry breaking; dark matter; cosmology.

Corwin, Eric, Ph.D., University of Chicago, 2007. *Biophysics, Condensed Matter Physics*. Experimental soft condensed matter and biophysics; atomic force microscopy of single molecule proteins; structure and dynamics of jammed athermal packings.

Majewski, Stephanie, Ph.D., Stanford University, 2007. *High Energy Physics, Particles and Fields*. Experimental particle physics.

McMorran, Benjamin, Ph.D., University of Arizona, 2009. *Condensed Matter Physics, Optics*. Experimental condensed matter; optical physics; free electron physics and interferometry; matter wave optics; electron microscopy; magnetic materials.

Instructor

Livelybrooks, Dean, Ph.D., University of Oregon, 1990. Director of Undergraduate Studies, Tenured Senior Instructor. *Geophysics*. Science education and outreach; magnetotellurics; dynamic margin model constraints.

DEPARTMENTAL RESEARCH SPECIALTIES AND STAFF

Theoretical

Astrophysics. Astrophysical flows; accretion disks; dense fluids; x-ray timing observations. Imamura.

Condensed Matter Physics. Metal insulator transitions; localization; phases in complex fluids; quantum critical phenomena. Belitz, Haydock, Toner.

High Energy Physics. Electroweak symmetry breaking; dark matter; cosmology; early universe physics; particle astrophysics. Chang, Csonka, Deshpande, Kribs, Soper.

Optics. Quantum optics; quantum information and nanophotonics. Nöckel, Steck, van Enk.

Solid State Physics. Haydock.

Experimental

Astrophysics. Properties of galaxies; observational cosmology; evolution and properties of galaxies; gravitational waves. Bothun, Frey, Imamura, Schombert.

Biophysics. X-ray crystallography of proteins; membrane biophysics; biomolecule mechanics. Corwin, Parthasarathy, Taylor, Toner.

Condensed Matter Physics. Phonon and electron transport in low-dimensional systems; electronic properties of amorphous semiconductors; surface physics. Corwin, Deutsch, Donnelly, Gregory, Kevan, Parthasarathy, Taylor, Wang.

High Energy Physics. Experimental elementary particle physics; electroweak symmetry breaking (ATLAS and ILC); gravitational radiation (LIGO). Brau, Frey, Majewski, Strom, Torrence.

Optics. Quantum optics and quantum information; cold atoms; plasmonics. Deutsch, Gregory, McMorran, Raymer, Steck, Wang.

Solid State Physics. Photovoltaic materials; magnetism and magnetic materials; thin-film physics, plasmonics; optical properties of semiconductors. Deutsch, Gregory, Kevan, Taylor, Wang.

View additional information about this department at
www.gradschoolshopper.com

BRYN MAWR COLLEGE

DEPARTMENT OF PHYSICS

Bryn Mawr, Pennsylvania 19010-2899
http://www.brynmawr.edu/physics

General University Information
President: Jane Dammen McAuliffe
Dean of Graduate School: Mary Osirim
University website: http://www.brynmawr.edu/
Control: Private
Setting: Suburban
Total Faculty: 158
Total number of Students: 1,863
Total number of Graduate Students: 362

Department Information
Department Chairman: Elizabeth McCormack, Chair
Department Contact: Xumei Cheng, Assistant
  Professor/Graduate Advisor
  Full-Time Graduate Students: 6
  First-Year Graduate Students: 1
  Female First-Year Students: 1

Department Address
101 North Merion Ave
Bryn Mawr, PA 19010-2899
Phone: (610) 526-5358
Fax: (610) 526-7469
E-mail: xcheng@brynmawr.edu
Website: http://www.brynmawr.edu/physics

ADMISSIONS

Admission Contact Information
Address admission inquiries to: Dean of Graduate Studies
Phone: (610) 526-5072
Admissions website: http://www.brynmawr.edu/gsas/
  Admissions/

Application deadlines
Fall admission:
U.S. students: January 2   *Int'l. students*: January 2

Application fee
U.S. students: $50   *Int'l. students*: $50

Admissions information
For Fall of 2013:
  Number of applicants: 9

Admission requirements
Bachelor's degree requirements: Bachelor's degree in physics
  or a closely related field is required.

GRE requirements
The GRE is required.

Advanced GRE requirements
The Advanced GRE is required.

TOEFL requirements
The TOEFL exam is required for students from non-English-
  speaking countries.
  PBT score: 600
  iBT score: 100

Other admissions information
Additional requirements: Students from non-English-speaking
  countries are required to demonstrate proficiency in English via the TOEFL or IELTS examination; minimum
  IELTS score 7.

TUITION

Tuition year 2012–13:
Tuition for out-of-state residents
  Full-time students: $35,280 annual
  Part-time students: $5,880 other
1 academic unit: $5,880 and 1 unit of supervised work: $940.
  Full load is 3 units/semester and 6 units/year.
Deferred tuition plan: No
Health insurance: Yes, $3,380 domestic/$1294 international.
Other academic fees: (Financial support package includes a sub-
  sidy for health insurance.)
Academic term: Semester
Number of first-year students who received full tuition waivers: 1

Teaching Assistants, Research Assistants, and Fellowships
Number of first-year
  Teaching Assistants: 1
Average stipend per academic year
  Teaching Assistant: $23,400
  Research Assistant: $23,400
TA/RA stipends expected to increase to $25,200 in 2013-14.

FINANCIAL AID

Application deadlines
Fall admission:
U.S. students: January 2   *Int'l. students*: January 2

Loans
Loans are available for U.S. students.
Loans are not available for international students.
GAPSFAS application required: Yes
FAFSA application required: Yes

For further information
Address financial aid inquiries to: Financial Aid Office.
Phone: (610) 526-5245
E-mail: finaid@brynmawr.edu
Financial aid website: http://www.brynmawr.edu/financialaid/

HOUSING

Availability of on-campus housing
Single students: No
Married students: No

For further information
Address housing inquiries to: Angie Sheets, Director of Residential Life.
Phone: (610) 526-7334
E-mail: asheets@brynmawr.edu
Housing aid website: http://www.brynmawr.edu/residentiallife/

Table A—Faculty, Enrollments, and Degrees Granted

Research Specialty	2013-14 Faculty	Enrollment Fall 2013		Number of Degrees Granted 2012–13 (2004–13)		
		Master's	Doctorate	Master's	Terminal Master's	Doctorate
Atomic, Molecular, & Optical Physics	–	–	4	–	-(2)	-(2)
Biophysics	–	–	–	–	–	-(2)
Chemical Physics	1	–	–	-(1)	–	–
Condensed Matter Physics	1	–	1	–	–	–
Cosmology & String Theory	1	–	1	–	–	–
Other	1	–	–	–	–	–
Total	–	–	6	-(1)	-(2)	-(4)
Full-time Grad. Stud.	–	–	6	–	–	–
First-year Grad. Stud.	–	–	1	–	–	–

GRADUATE DEGREE REQUIREMENTS

Master's: At least six units of work with satisfactory performance, including at least one full year in residence; master's thesis and oral examination are required.

Doctorate: At least 12 units of work with satisfactory performance, including at least three full years in residence; written and oral preliminary examinations are required; dissertation and oral examination are required.

Thesis: Thesis may be written in absentia.

SPECIAL EQUIPMENT, FACILITIES, OR PROGRAMS

Cooperative agreements with the University of Pennsylvania and Drexel University allow Bryn Mawr graduate students to pursue work in special field areas not available at Bryn Mawr. State-of-the-art laboratory facilities include an atomic force microscope, electrochemical deposition system, AJA sputtering thin film deposition system, class 1000 soft-curtain clean room, vibrating sample magnetometer, x-ray diffractometer, solid-state NMR spectrometer, various tunable pulsed and CW laser systems, molecular beam apparatus, two ultrahigh vacuum systems for laser cooling and trapping, and a machine and instrument shop. Students have access to user facilities at national laboratories, including the Advanced Photon Source (APS) at Argonne National Laboratory and the Center for Functional Nanomaterials (CFN) at Brookhaven National Laboratory. The Collier Science Library offers extensive information technology and library resources. The college and the department offer computing facilities for data acquisition, modeling, and data analysis, as well as high-speed computer links to the national and international physics communities.

Table B—Separately Budgeted Research Expenditures by Source of Support

Source of Support	Departmental Research	Physics-related Research Outside Department
Federal government	$335,000	
State/local government		
Non-profit organizations		
Business and industry		
Other	$70,000	
Total	$405,000	

Table C—Separately Budgeted Research Expenditures by Research Specialty

Research Specialty	No. of Grants	Expenditures ($)
Atomic, Molecular, & Optical Physics	1	$125,000
Condensed Matter Physics	2	$230,000
Cosmology & String Theory	1	$50,000
Total	4	$405,000

FACULTY

Professor
Beckmann, Peter A., Ph.D., University of British Columbia, 1975. *Chemical Physics.* Solid-state dynamic nuclear magnetic resonance.

McCormack, Elizabeth F., Ph.D., Yale University, 1989. Department Chair. *Atomic, Molecular, & Optical Physics, Optics.* Molecular spectroscopy and dynamics.

Noel, Michael W., Ph.D., University of Rochester, 1996. *Atomic, Molecular, & Optical Physics.* Ultracold Rydberg atoms.

Associate Professor
Schulz, Michael B., Ph.D., Stanford University, 2002. *Cosmology & String Theory.* Theoretical physics with a focus on string theory.

Assistant Professor
Cheng, Xuemei May, Ph.D., Johns Hopkins University, 2006. *Condensed Matter Physics, Nano Science and Technology.* Nanomaterials, spintronics, and spin dynamics in nanomagnetic materials.

Visiting Assistant Professor
Kim Pechkis, Hyewon, Ph.D., University of Connecticut, 2010. NIST/JQI (2010-2013). *Atomic, Molecular, & Optical Physics.* Atomic, molecular and optical physics, Ultracold molecules and spinor Bose-Einstein Condensates.

Pechkis, Joseph A., Ph.D., University of Connecticut, 2010. NIST/JQI (2012-2013), Naval Research Lab (2010-2012). *Atomic, Molecular, & Optical Physics.* Atomic, molecular, and optical physics, Ultracold alkali and alkaline-earth gases and atom guiding in hollow fibers.

Lecturer
Matlin, Mark D., Ph.D., University of Maryland, 1991. Laboratory Coordinator. *Relativity & Gravitation.* General relativity.

DEPARTMENTAL RESEARCH SPECIALTIES AND STAFF

Theoretical
Cosmology & String Theory. String theory and its applications to quantum field theory, cosmology, and particle physics. Schulz.

Experimental

Atomic, Molecular, & Optical Physics. Resonant energy transfer in ultracold samples of highly excited atoms using laser cooling and trapping techniques to prepare and manipulate the atomic sample and study the extremely long-range many-body interactions that result when the atoms are excited to weakly bound states. Noel.

Atomic, Molecular, & Optical Physics. Laser-based studies of atomic and molecular excited-state structure and decay dynamics, including photoionization, autoionization, predissociation, and photodissociation; nonlinear optical techniques, including multiphoton excitation and detection, laser-induced grating spectroscopy, degenerate four-wave mixing, and vacuum ultraviolet light generation. McCormack.

Chemical Physics. Nuclear spin relaxation in solids (NMR) using H-1 and F-19 solid-state NMR relaxation studies in organic molecular solids and modeling the motion with knowledge of the equilibrium structure. Collaborators are at the University of California at San Diego (X-ray diffraction) and Chengdu, China (electronic structure calculations). Beckmann.

Condensed Matter Physics. Fabrication, characterization, and application of nanoscale materials, including: templated electrochemical deposition of nanoscaled materials for energy and medical applications, time-resolved photoemission electron microscopy imaging of spin dynamics in magnetic nanostructures, and x-ray magnetic circular dichroism study of multiferroic materials. Synchrotron x-ray based experiments are carried out at the Advanced Photon Source at Argonne National Laboratory. Cheng.

View additional information about this department at
www.gradschoolshopper.com

CARNEGIE MELLON UNIVERSITY

DEPARTMENT OF PHYSICS

Pittsburgh, Pennsylvania 15213
http://www.cmu.edu/physics

General University Information

President: Dr. Subra Suresh
Dean of Graduate School: Fred Gilman
University website: http://www.cmu.edu
Control: Private
Setting: Urban
Total Faculty: 1,436
Total Graduate Faculty: 1,436
Total number of Students: 12,569
Total number of Graduate Students: 6,290

Department Information

Department Chairman: Gregg Franklin, Chair
Department Contact: Heather Corcoran, Student Programs Coordinator
  Total full-time faculty: 33
  Total number of full-time equivalent positions: 37
  Full-Time Graduate Students: 75
  First-Year Graduate Students: 21
  Female First-Year Students: 4
  Total Post Doctorates: 40

Department Address

5000 Forbes Avenue
Pittsburgh, PA 15213
Phone: (412) 268-2849
Fax: (412) 681-0648
E-mail: physgrad@andrew.cmu.edu
Website: http://www.cmu.edu/physics

ADMISSIONS

Admission Contact Information

Address admission inquiries to: Graduate Studies, Department of Physics, Carnegie Mellon University, Pittsburgh, PA 15213
Phone: (412) 268-2849
E-mail: physgrad@andrew.cmu.edu
Admissions website: http://www.cmu.edu/physics/graduate-program/admission

Application deadlines

Fall admission:
U.S. students: January 1   *Int'l. students*: January 1

Application fee

There is no application fee required.

Admissions information

For Fall of 2012:
  Number of applicants: 338
  Number admitted: 97
  Number enrolled: 21

Admission requirements

Bachelor's degree requirements: A bachelor's degree in physics or related field is required.

GRE requirements

The GRE is required.

Advanced GRE requirements

The Advanced GRE is required.

TOEFL requirements

The TOEFL exam is required for students from non-English-speaking countries.

Other admissions information

Additional requirements: No minimum scores are specified.

Undergraduate preparation assumed: A typical student will have completed intermediate courses in mechanics (Marion), electricity and magnetism (Griffiths or Wangsness), modern physics (Eisberg and Resnick), wave mechanics (Townsend), thermodynamics and statistical mechanics (Reif or Swendsen), and modern physics laboratory (Melissinos).

TUITION

Tuition year 2013-14:
  Full-time students: $40,000 annual
  Part-time students: $556 per credit
Deferred tuition plan: No
Health insurance: Available at the cost of $1090 per year.
Other academic fees: $688/year
Academic term: Semester
Number of first-year students who received full tuition waivers: 21

Teaching Assistants, Research Assistants, and Fellowships

Number of first-year
  Teaching Assistants: 19
  Fellowship students: 2
Average stipend per academic year
  Teaching Assistant: $20,025
  Research Assistant: $20,025
  Fellowship student: $20,025

FINANCIAL AID

Loans

Loans are not available for U.S. students.
Loans are not available for international students.
GAPSFAS application required: No
FAFSA application required: No

HOUSING

Availability of on-campus housing

Single students: No
Married students: No

Table A—Faculty, Enrollments, and Degrees Granted

Research Specialty	2013-14 Faculty	Enrollment Fall 2012 Master's	Enrollment Fall 2012 Doctorate	Number of Degrees Granted 2012-13 (2008–13) Master's	Number of Degrees Granted 2012-13 (2008–13) Terminal Master's	Number of Degrees Granted 2012-13 (2008–13) Doctorate
Applied Physics	10	–	2	–	–	1(7)
Astrophysics	10	–	7	–	–	1(7)
Biophysics	9	–	11	–	–	2(7)
Computational Physics	13	–	–	–	–	–
Condensed Matter Physics	13	–	14	–	–	2(8)
Nuclear Physics	6	–	7	–	–	–(9)
Particles and Fields	10	–	12	–	–	3(7)
Quantum Foundations	1	–	1	–	–	–(3)
Statistical & Thermal Physics	4	–	1	–	–	1(2)
Non-specialized	–	–	20	12(73)	–	–
Total	–	–	75	12(73)	–	10(50)
Full-time Grad. Stud.	–	–	75	–	–	–
First-year Grad. Stud.	–	–	21	–	–	–

GRADUATE DEGREE REQUIREMENTS

Master's: Thirty-two semester hours (96 units) of course work with grade average of B or above are required. Four semester hours (12 units) of experimental work are required. There are no thesis or foreign language requirements. Written qualifying examination is required. One year of residence is required.

Doctorate: Satisfactory performance in an approved program. Additional course requirements will depend on level of preparation. Comprehensive oral and written qualifying examinations, annual research reviews, thesis, and final thesis defense are required. One year of residence as a full-time student required. There is a teaching requirement for the Ph.D. degree.

Thesis: Thesis may be written in absentia.

SPECIAL EQUIPMENT, FACILITIES, OR PROGRAMS

Astrophysics research is integrated within the Bruce and Astrid McWilliams Center for Cosmology, which brings together astrophysicists, particle physicists, computer scientists, and statisticians to advance our understanding of dark matter and dark energy which dominate the universe. Observational astrophysics is performed using a variety of space-based and ground-based telescopes. Computation for astrophysics research uses in-house clusters, including two clusters that together have over 1500 cores.

The department maintains facilities for condensed matter and biological physics research, including apparatus for x-ray diffraction and reflection, laser spectroscopies, calorimetry, magnetic and electrical transport measurements, optical characterization of interfaces, scanning tunneling and atomic force microscopies, low energy electron microscopy, and sample preparation. Scattering experiments are performed at an in-house x-ray facility, including fix tube and rotating anode sources as well as at national synchrotron and neutron facilities. Computation facilities for these groups include five multicore, multinode, high-performance clusters. Collaborations with other departments provide access to additional facilities, including clean-room facilities, electron microscopies, optical microscopies, magnetic measurements, and fluids and interface characterization.

High energy research is performed by faculty using facilities at the Fermi National Accelerator Laboratory (Chicago, Illinois), CERN (Geneva, Switzerland), and IHEP (Beijing, China). A data analysis laboratory is maintained on campus, as are laboratories for the development of detection systems.

The medium energy physics group builds and performs experiments at the Thomas Jefferson National Accelerator Facility (JLab) in Virginia. Present work uses the ongoing JLab energy upgrade and includes the GlueX exotic meson search and the Hall A spin-physics program. The group uses a 1000-core computer cluster for computational studies.

Departmental facilities include machine shops, numerous computer clusters, and a stock room. The University Computing Center operates an extensive system of networked scientific workstations and microcomputers with central file servers for research and educational applications. Access to a Cray XT3 MPP supercomputer as well as sets of SMP machines are available through the Pittsburgh Supercomputing Center. The Physics Department is located in Wean Hall, which also houses the science and engineering library.

Table B—Separately Budgeted Research Expenditures by Source of Support

Source of Support	Departmental Research	Physics-related Research Outside Department
Federal government	$6,393,826	$262,000
State/local government		
Non-profit organizations		
Business and industry	$163,566	
Other	$2,197,524	
Total	**$8,754,916**	**$262,000**

Table C—Separately Budgeted Research Expenditures by Research Specialty

Research Specialty	No. of Grants	Expenditures ($)
Astrophysics	20	$2,341,124
Biophysics	23	$1,826,970
Condensed Matter Physics	31	$1,157,331
Nuclear Physics	12	$1,410,520
Particles and Fields	13	$1,830,047
Statistical & Thermal Physics	3	$166,426
Quantum Foundations	4	$22,498
Total	**106**	**$8,754,916**

FACULTY

Professor

Briere, Roy A., Ph.D., University of Chicago, 1995. *Particles and Fields*. Experimental high energy physics; BES at Beijing.

Feenstra, Randall M., Ph.D., California Institute of Technology, 1982. *Condensed Matter Physics*. Experimental condensed matter physics; semiconductor surfaces.

Ferguson, Thomas A., Ph.D., University of California, Los Angeles, 1978. *Particles and Fields*. Experimental high energy physics; CMS at CERN.

Franklin, Gregg B., Ph.D., Massachusetts Institute of Technology, 1980. Department Head. *Nuclear Physics*. Experimental medium energy/nuclear physics; production and interactions of strange hadrons; strange sea quarks in the nucleon.

Garoff, Stephen, Ph.D., Harvard University, 1977. *Condensed Matter Physics*. Experimental condensed matter physics; surfaces and interfaces.

Gilman, Frederick, Ph.D., Princeton University, 1965. Dean of Mellon College of Science. *Particles and Fields*. Theoretical elementary particle physics; CP violation, heavy quarks, and leptons.

Griffiths, Robert B., Ph.D., Stanford University, 1962. *Quantum Foundations*. Theoretical physics; foundations of quantum mechanics.

Holman, Richard F., Ph.D., Johns Hopkins University, 1982. *Particles and Fields*. Theoretical particle physics and cosmology; inflation, dark energy.

Lösche, Mathias, Ph.D., Technical U. of Munich, 1986. *Biophysics*. Experimental biological physics; molecular and membrane biophysics.

Levine, Michael J., Ph.D., California Institute of Technology, 1963. Director of the Pittsburgh Supercomputer Center. *Particles and Fields*. Theoretical elementary particle physics.

Majetich, Sara A., Ph.D., University of Georgia, 1987. *Condensed Matter Physics*. Experimental condensed matter physics; semiconductor and magnetic nanoparticles.

Meyer, Curtis A., Ph.D., University of California, Berkeley, 1987. Associate Dean of Mellon College of Science. *Nuclear Physics*. Experimental medium-energy/nuclear physics; me-son spectroscopy; search for gluonic excitations with GlueX at JLab.

Morningstar, Colin J., Ph.D., University of Toronto, 1991. *Nuclear Physics*. Theoretical medium-energy physics; nonperturbative phenomena in quantum field theories.

Paulini, Manfred, Ph.D., University of Erlangen, 1993. *Particles and Fields*. Experimental high-energy physics; CDF at Fermilab; CMS at CERN.

Peterson, Jeffrey B., Ph.D., University of California, Berkeley, 1985. *Biophysics*. Experimental astrophysics; observational cosmology.

Quinn, Brian P., Ph.D., Massachusetts Institute of Technology, 1984. *Nuclear Physics*. Experimental medium energy/nuclear physics; production and interaction of strange hadrons; strange sea quarks in the nucleon.

Rothstein, Ira Z., Ph.D., University of Maryland, College Park, 1992. *Particles and Fields*. Theoretical particle physics and cosmology; LHC theory; gravity waves.

Russ, James S., Ph.D., Princeton University, 1966. *Particles and Fields*. Experimental high-energy physics; CDF at Fermilab; particle astrophysics.

Schumacher, Reinhard A., Ph.D., Massachusetts Institute of Technology, 1983. *Nuclear Physics*. Experimental medium energy/nuclear physics; production and interactions of strange hadrons; strange sea quarks in the nucleon.

Suter, Robert M., Ph.D., Clark University, 1978. *Condensed Matter Physics*. Experimental condensed matter physics; x-ray and neutron scattering studies.

Swendsen, Robert H., Ph.D., University of Pennsylvania, 1971. *Condensed Matter Physics*. Theoretical condensed matter physics; computer simulations; statistical mechanics of phase transitions and biological molecules.

Tristram-Nagle, Stephanie, Ph.D., University of California, Berkeley, 1981. *Biophysics*. Experimental biophysics; membrane biophysics.

Vogel, Helmut, Ph.D., University of Erlangen, 1979. *Particles and Fields*. Experimental high energy physics; CMS at CERN.

Widom, Michael, Ph.D., University of Chicago, 1983. *Condensed Matter Physics*. Theoretical condensed matter physics; metal alloys; crystallography; biophysics.

Associate Professor

Croft, Rupert, Ph.D., University of Oxford, 1995. *Astrophysics*. Theoretical astrophysics/cosmology; simulations of the evolution of the universe.

Deserno, Markus, Ph.D., University of Mainz, 2000. *Biophysics*. Theoretical condensed matter and biophysics; membrane structure and properties.

Di Matteo, Tiziana, Ph.D., University of Cambridge, 1998. *Astrophysics*. Theoretical astrophysics/cosmology; cosmological simulations.

Evilevitch, Alex, Ph.D., Lund University, 2001. *Biophysics*. Experimental biological physics; physics of viruses.

Kahniashvili, Tina, Ph.D., Space Research Institute, Moscow, 1988. *Astrophysics*. Theoretical cosmology/astrophysics, Theory of gravity; studying physical processes in the early universe.

Assistant Professor

Flauger, Raphael, Ph.D., University of Texas, 2009. *Astrophysics, Particles and Fields*. Cosmology and particle physics; phenomenology; QFT; early universe/inflation.

Heinrich, Frank, Ph.D., University of Leipzig, 2005. *Biophysics*. Experimental biological physics; neutron scattering.

Ho, Shirley, Ph.D., Princeton University, 2008. *Astrophysics*. Observational astrophysics/cosmology; Baryon acoustic oscillations; cosmic microwave background; study of large-scale structure.

Mandelbaum, Rachel, Ph.D., Princeton University, 2006. *Astrophysics*. Observational astrophysics/cosmology; lensing studies of galaxies and large-scale structure.

Trac, Hy, Ph.D., University of Toronto, 2004. *Astrophysics*. Theoretical astrophysics/cosmology; evolution of the dark matter, baryons, and stars.

Walker, Matthew, Ph.D., University of Michigan, 2007. *Astrophysics*. Dark matter; galactic dynamics; near-field cosmology.

Xiao, Di, Ph.D., University of Texas, Austin, 2007. *Condensed Matter Physics*. Theoretical condensed matter physics; quantum transport; Berry phase.

Emeritus

Berger, Luc, Ph.D., University of Lausanne, 1960. *Condensed Matter Physics*. Experimental and theoretical condensed matter physics; studies of metallic ferromagnets.

Edelstein, Richard M., Ph.D., Columbia University, 1960. *Particles and Fields*. Experimental high energy physics; dynamics of strong interactions.

Engler, Arnold, Ph.D., University of Berne, 1953. *Particles and Fields*. Experimental high energy physics; colliding beams techniques.

Fetkovich, John G., Ph.D., Carnegie Mellon University, 1959. Special Assistant to the President for Academic Affairs.

Kisslinger, Leonard S., Ph.D., Indiana University, 1956. *Nuclear Physics*. Theoretical nuclear and particle physics; nonperturbative QCD.

Kraemer, Robert W., Ph.D., Johns Hopkins University, 1962. *Particles and Fields*. Experimental high energy physics; colliding beams techniques.

Li, Ling-Fong, Ph.D., University of Pennsylvania, 1970. *Particles and Fields*. Theoretical elementary particle physics; unified theories of particle interactions.

Nagle, John F., Ph.D., Yale University, 1965. *Biophysics*. Experimental and theoretical biological physics; statistical mechanics of phase transitions; biomembranes.

Rayne, John A., Ph.D., University of Chicago, 1954. *Condensed Matter Physics*. Experimental condensed matter physics; electronic and magnetic properties of metals and alloys; ultrasonic absorption in solids.

Schumacher, Robert T., Ph.D., University of Illinois, 1955. Musical acoustics; magnetic resonance in solids.

Sekerka, Robert F., Ph.D., Harvard University, 1965. *Condensed Matter Physics*. Theoretical condensed matter physics; problems in materials science.

Vander Ven, Ned S., Ph.D., Princeton University, 1962. *Condensed Matter Physics*. Experimental condensed matter physics; electron and nuclear spin resonance in solids.

Wolfenstein, Lincoln, Ph.D., University of Chicago, 1949. *Particles and Fields*. Theoretical elementary particle physics; weak interactions and symmetry principles.

Young, Hugh D., Ph.D., Carnegie Mellon University, 1959. *Physics and other Science Education*. Physics education.

Faculty by Courtesy

Anna, Shelley, Ph.D., Harvard University, 2000. Dynamic of soft matter; fluid mechanics.

Greve, David, Ph.D., Lehigh University, 1979. Physics and development of novel sensors.

Islam, Mohammad, Ph.D., Lehigh University, 2000. Structure, dynamics, and self-assembly of soft matter; properties of nanoscale structures.

Maloney, Craig, Ph.D., University of California, Santa Barbara, 2005. Mechanical response of solid-like materials.

Mandal, Maumita, Ph.D., University of Hyderabad, 2004. RNA structure and conformational rearrangements.

McHenry, Michael, Ph.D., Massachusetts Institute of Technology, 1988. Magnetic properties of materials.

Rollett, Anthony, Ph.D., Drexel University, 1987. Microstructure of polycrystalline materials.

Zhu, Jian-Gang, Ph.D., University of California, San Diego, 1983. Magnetic data storage technologies.

DEPARTMENTAL RESEARCH SPECIALTIES AND STAFF

Theoretical

Astrophysics. The largest scale simulations of the structure formation of the universe yet performed; the evolution of galaxies, including galaxy mergers, and the associated supermassive black holes; the nature of dark matter and dark energy; the cosmology-particle physics interface; early universe/inflationary physics. Croft, Di Matteo, Flauger, Kahniashvili, Trac.

Biophysics. Theoretical analysis of biomembranes; Monte Carlo simulations of proteins; elastic continuum theory and differential geometry of fluid membranes, statistical physics, and coarse-grained molecular dynamics simulations of membranes and peptides; structure of viruses and nucleic acids. Deserno, Nagle, Widom.

Computational Physics. Computational physics at Carnegie Mellon is an umbrella that encompasses a rapidly growing and highly interdisciplinary set of activities that are taking place in all areas of the department. Croft, Deserno, Di Matteo, Levine, Meyer, Morningstar, Paulini, Rollett, Suter, Swendsen, Trac, Widom, Xiao.

Condensed Matter Physics. Topological insulators and Berry phases; Monte Carlo studies of complex fluids, biological molecules, disordered solids and phase transitions; modeling of quasicrystals, ferromagnets, incommensurate phases, and quantum transport. Maloney, Nagle, Rollett, Swendsen, Widom, Xiao, Zhu.

Nuclear Physics. Strong and weak nuclear force; formation of hadrons, confinement, exotic forms of matter; Markov-chain and Monte Carlo computation of QCD; lattice gauge theory; QCD sum rules. Kisslinger, Morningstar.

Particles and Fields. Quantum gauge field theories and their applications to experiments; weak interaction phenomenology; CP violation; heavy quark physics; inflationary universe dynamics; topological defects and their applications in cosmology; gravity wave physics; LHC phenomenology. Flauger, Gilman, Holman, Rothstein, Wolfenstein.

Quantum Foundations. Reformulation of quantum theory using consistent histories and decoherence and application of quantum mechanics in computing. Griffiths.

Experimental

Astrophysics. Astrophysics research is integrated within the Bruce and Astrid McWilliams Center for Cosmology, which brings together physicists, computer scientists, and statisticians to advance our understanding of dark matter and dark energy. Institutional member of the Sloan Digital Sky Survey and the Large Synoptic Survey Telescope collaborations. Individuals participate in a number of other ongoing observational cosmology experiments, including those in 21 cm cosmology, development of high-sensitivity receivers at millimeter wavelengths, studies of weak lensing and large-scale structure, early evolution and formation of galaxies, and dark matter via dynamics of dwarf galaxies (near-field cosmology). Ho, Mandelbaum, Peterson, Walker.

Biophysics. Structure and function of biomembranes; NMR studies of the structure of proteins and optical microscopic studies of cell structure; protein-membrane interactions; biofluid mechanics of lung airways; biopolymer dynamics; physics of viruses. Evilevitch, Garoff, Heinrich, Lösche, Nagle, Tristram-Nagle.

Condensed Matter Physics. Properties and applications of nanoparticles and nanostructures; structure of thin organic and metal solid films; structure and properties of liquid/solid interfaces; wetting of fluids on solids; structure of semiconductor and metal surfaces; structure and properties of graphene; influence of surface properties on semiconductor devices; magnetic films for data storage; x-ray scattering from thin films and surfaces; x-ray microscopy for characterization of grain structure and growth in metals; microfluidics; interfacial fluid mechanics; properties and application of nanotubes and nanorods; many of these activities are performed in active collaboration with other departments, institutes, and centers in the science and engineering colleges. Anna, Feenstra, Garoff, Greve, Islam, Majetich, McHenry, Suter, Zhu.

Nuclear Physics. Strong QCD; the spectrum of excited baryons; gluonic excitations of mesons and quark confinement using GlueX at JLab; fundamental form factors of the proton and neutron; strangeness content of the nucleon; electromagnetic interactions with hadronic systems in Hall A at JLab. Franklin, Meyer, Quinn, Reinhard Schumacher.

Particles and Fields. Operation and data analysis with the CMS detector at the LHC collider at CERN; ultra high energy cosmic ray neutrinos; study of heavy quark production and decay properties with CDF at Fermilab; CP violations and quarkonia spin alignment measurements; studies of the properties of charm quarks using the BES experiment in Beijing, China; search for super symmetry fourth-generation quarks and heavy quark production. Briere, Ferguson, Paulini, Russ, Vogel.

View additional information about this department at www.gradschoolshopper.com

---

DREXEL UNIVERSITY

DEPARTMENT OF PHYSICS

Philadelphia, Pennsylvania 19104
http://www.drexel.edu/physics

General University Information

President: John A. Fry
Dean of Graduate School: Teck-Kah Lim, Associate Vice Provost
University website: http://www.drexel.edu
Control: Private
Setting: Urban
Total Faculty: 1,500
Total number of Students: 24,860
Total number of Graduate Students: 9,813

Department Information

Department Chairman: Michel Vallières, Head
Department Contact: Michael Vogeley, Director of Graduate Studies
  Total full-time faculty: 27
  Total number of full-time equivalent positions: 21
  Full-Time Graduate Students: 42
  First-Year Graduate Students: 11
  Female First-Year Students: 3
  Total Post Doctorates: 2

Department Address

3141 Chestnut Street
Philadelphia, PA 19104
Phone: (215) 895-2708
Fax: (215) 895-5934
E-mail: physics@drexel.edu
Website: http://www.drexel.edu/physics

ADMISSIONS

Admission Contact Information

Address admission inquiries to: Office of Graduate Admissions, Main Building, Room 212, 3141 Chestnut Street, Philadelphia, PA 19104
Phone: 1-800-2-DREXEL
E-mail: enroll@drexel.edu
Admissions website: http://www.drexel.edu/grad/programs/coas/physics/

Application deadlines

Fall admission:
U.S. students: January 1   *Int'l. students*: January 1

Application fee

U.S. students: $75   *Int'l. students*: $75
Application Fee: waived for online applications.

Admissions information

For Fall of 2013:
  Number of applicants: 167
  Number admitted: 31
  Number enrolled: 10

Admission requirements

Bachelor's degree requirements: Bachelor's degree in an approved program is required.
Minimum undergraduate GPA: 3.0

GRE requirements

The GRE is required.
  Quantitative score: 150

Verbal score: 150 0 GRE Physics Subject Test is required for PhD applicants to be considered for assistantships (no minimum score).

TOEFL requirements

The TOEFL exam is required for students from non-English-speaking countries.
PBT score: 600
iBT score: 100
IELTS scores may be submitted in lieu of TOEFL scores - minimum band score 7.0

Other admissions information

Additional requirements: Teaching assistants educated in non-English-speaking countries must complete a special English program.
An essay, a resume, and two letters of recommendations are required for all applicants.
Undergraduate preparation assumed: Advanced undergraduate coursework in classical mechanics, electromagnetism, statistical physics, and quantum mechanics. Mathematics coursework in differential equations and linear algebra.

TUITION

Tuition year 2013-14:
Full-time students: $1,085 per credit
Part-time students: $1,085 per credit
Average cost per class: $3,255.
Credit hours per semester to be considered full-time: 9
Deferred tuition plan: No
Health insurance: Available.
Other academic fees: Health insurance is covered by Drexel for full-time Ph.D. program students with assistantships. General Fees: $280 full-time (per term) $140 part-time (per term) $35 immunization fee.
Academic term: Quarter
Number of first-year students who received full tuition waivers: 8

Teaching Assistants, Research Assistants, and Fellowships

Number of first-year
Teaching Assistants: 8
Fellowship students: 5
Average stipend per academic year
Teaching Assistant: $22,500
Research Assistant: $23,000
Fellowship student supplement: $5,000.

FINANCIAL AID

Application deadlines

Fall admission:
U.S. students: January 1   *Int'l. students*: January 1

Loans

Loans are available for U.S. students.
Loans are not available for international students.
GAPSFAS application required: No
FAFSA application required: Yes

For further information

Address financial aid inquiries to: Student Financial Services, 3141 Chestnut Street, Main Building, Room 222, Philadelphia, PA 19104.
Phone: (215) 895-1445
Financial aid website: http://www.drexel.edu/grad/financing/

HOUSING

Availability of on-campus housing

Single students: Yes
Married students: No

For further information

Address housing inquiries to: Office of University Housing, 101 N. 34th Street, Philadelphia, PA 19104.
Phone: (215) 895-6155
E-mail: gradhousing@drexel.edu
Housing aid website: http://www.drexel.edu/dbs/university Housing/graduateHousing/graduateHousingApplication/

Table A—Faculty, Enrollments, and Degrees Granted

Research Specialty	2012-13 Faculty	Enrollment Fall 2012-13 Master's	Enrollment Fall 2012-13 Doctorate	Number of Degrees Granted 2012-13 (2008–13) Master's	Number of Degrees Granted 2012-13 (2008–13) Terminal Master's	Number of Degrees Granted 2012-13 (2008–13) Doctorate
Astrophysics	4	–	13	2(11)	–(1)	1(4)
Biophysics	5	–	13	3(8)	–(4)	2(10)
Condensed Matter Physics	3	–	4	1(2)	–	–
Nonlinear Dynamics and Complex Systems	1	–	3	–(1)	–	1(5)
Particles and Fields	4	1	4	–(2)	1(2)	2(3)
Non-specialized	–	2	3	–	–(2)	–
Other	4	–	–	–	–	–
Total	21	3	40	6(24)	1(9)	6(22)
Full-time Grad. Stud.	–	3	38	–	–	–
First-year Grad. Stud.	–	2	9	–	–	–

GRADUATE DEGREE REQUIREMENTS

Master's: The requirement for the Master's degree is 45 quarter credits. The student is required to maintain at least a 3.0 GPA. There is no thesis or foreign language requirement for the M.S. degree. There is no specific residence requirement for the M.S. degree. There are no examinations required for the M.S. degree.

Doctorate: In addition to required graduate-level coursework in physics, the successful Ph.D. candidate must: (a) pass the Ph.D. candidacy examinations, both written and oral; and (b) perform original research, write a satisfactory thesis describing that research, and defend the thesis in an oral examination. There is no foreign language requirement.

Thesis: Thesis may be written in absentia.

SPECIAL EQUIPMENT, FACILITIES, OR PROGRAMS

Students in the Graduate program are able to access a diverse range of experimental facilities, including the following:

(1) Astrophysics Facilities:

Numerical Astrophysics Facility, primarily networked LINUX and Mac OS X workstations emphasizes theoretical and numerical studies of stars, star clusters, the early Universe, galaxy distributions, cosmology modeling, and gravitational lensing. The facility also employs special purpose high-performance computers, such as the Gravity Pipeline Engine (GRAPE), a new Beowulf cluster (128 processors, 128G RAM, 2TB RAID disk), and a system using Graphics Processing Units to achieve computational speeds of up to a trillion floating point operations per second. The Joseph R. Lynch Observatory houses a 16-inch Mead Schmidt-Cassegrain telescope equipped with SBIG CCD camera.

Drexel is a participant in the Sloan Digital Sky Survey, which operates a 2.5m telescope at Apache Point, NM, and the Large Synoptic Survey Telescope to be built in Chile (first light 2020).

(2) Biophysics Facilities:

(a) Modulated excitation kinetics laboratory uses frequency domain techniques to follow internal dynamics of biological molecules.

(b) Spatially resolved kinetics laboratory uses simultaneously resolved spatio-temporal data at microscopic resolution to follow biological self-assembly processes, such as polymerization of sickle hemoglobin.

(c) Atomic Force Microscope (AFM) facility to study the structure and interaction of macromolecule via imaging, and to investigate the mechanical and kinetic properties of individual protein molecules via nanomanipulation.

(d) Computational Biophysics facility including two Beowulf clusters (44-node dual-core Xeon, 43-node dual quad-core Xeon [344 cores]), 24TB RAID disk server, and ten Linux workstations connected through a gigabit network.

(e) Preparative laboratory provides facilities for biological sample purification and characterization.

(3) Condensed Matter Facilities:

(a) Ultra-low temperature laboratory has a dilution refrigerator, 3He and 4He cryostats and microwave sources to study quantum phenomena in nano- and microscale devices, superconducting qubits, nanostructures and quantum fluids and solids.

(b) Energy Materials Research Laboratory including Variable Temperature UHV Scanning Probe Microscope, installed in an STC-50 rated acoustic chamber.

(c) Magnetic material laboratory conducts research on amorphous magnetic thin films, fiber optical sensors.

(d) Surface science laboratory has scanning probe microscopy to study surface structure interfaces at the atomic level.

(4) Particle Physics Facilities:

Detector development laboratory provides experimental support for an international research program in nonaccelerator particle and nuclear physics performing tests of invariance principles and conservation laws, and neutrino oscillations.

(5) Laboratory for High-Performance Computational Physics: we have a computer lab built upon 15 powerful workstations-each with Intel Core i5 3570 running at 3.4 Ghz, 16 Gb RAM, and an nVidia GTX 650 graphics card. They are running Ubuntu 13.04 operating system. Each workstation has a 24 inch screen monitor. These world-class workstations are connected to our main file server via the highest quality gigabyte network connectors.

(6) General Support Facilities:

Include an electronics shop capable of custom design and fabrication of electronics and computer components, and a machine shop to assist in the design, construction, and repair of mechanical component.

Table B—Separately Budgeted Research Expenditures by Source of Support

Source of Support	Departmental Research	Physics-related Research Outside Department
Federal government	$1,022,979.25	
State/local government		
Non-profit organizations		
Business and industry		
Other		
Total	$1,022,979.25	

Table C—Separately Budgeted Research Expenditures by Research Specialty

Research Specialty	No. of Grants	Expenditures ($)
Astrophysics	14	$629,206.83
Biophysics	3	$174,978.82
Condensed Matter Physics	–	
Particles and Fields	5	$150,792.26
Nonlinear Dynamics and Complex Systems	1	$68,001.34
Other	0	
Total	23	$1,022,979.25

FACULTY

Professor

Bose, Shymalendu, Ph.D., University of Maryland, 1967. *Condensed Matter Physics*. High-temperature superconductors; nanoshells.

DiNardo, N. John, Ph.D., University of Pennsylvania, 1982. Vice Provost for Academic Affairs. Studies of surfaces and interfacial phenomena in solids.

Ferrone, Frank, Ph.D., Princeton University, 1974. Associate Vice Provost for Research. *Biophysics*. Protein assembly, exemplified by sickle hemoglobin; nucleation theory; molecular crowding; light scattering; kinetic methods.

Finegold, Leonard X., Ph.D., University of London, 1959. *Biophysics*. Biophysics; granular physics.

Gilmore, Robert, Ph.D., Massachusetts Institute of Technology, 1967. *Nonlinear Dynamics and Complex Systems*. Group theory and its applications to atomic, nuclear, and condensed matter physics; Catastrophe theory, nonlinear dynamics, chaos, quantum mechanics.

Goldberg, David M., Ph.D., Princeton University, 2000. Director of Undergraduate Studies. *Astrophysics*. Gravitational lensing; cosmic microwave background; cosmology; computational physics.

Lane, Charles C., Ph.D., California Institute of Technology, 1987. *Particles and Fields*. Nonaccelerator-based particle physics. Solar neutrinos and neutrino oscillations (Projects CHOOZ and KamLAND).

Lim, Tech-Kah, Ph.D., University of Adelaide, 1968. Associate Vice Provost for Graduate Studies. Physics education.

McMillan, Stephen L. W., Ph.D., Harvard University, 1983. *Astrophysics*. Stellar dynamics; large-scale computations of stellar systems.

Steinberg, Richard I., Ph.D., Yale University, 1969. Experimental tests of invariance principles and conservation laws; solar neutrinos and neutrino oscillations.

Tyagi, Somdev, Ph.D., Brigham Young University, 1976. *Condensed Matter Physics*. Physics of high-temperature superconductivity; magnetic properties of thin-sputtered films of amorphous metallic alloys; fiber optical sensors giant magnetoresistive (GMR) materials.

Vallières, Michel, Ph.D., University of Pennsylvania, 1972. Department Head. Large-scale shell-model calculations; computer architecture for nuclear physics problems.

Vogeley, Michael S., Ph.D., Harvard University, 1993. Director of Graduate Studies. *Astrophysics*. Cosmology, formation of structure in the universe, galaxies, active galactic nuclei, Sloan Digital Sky Survey.

Yuan, Jian-Min, Ph.D., University of Chicago, 1973. *Biophysics*. Theoretical and computational biophysics; biological pathways and networks; protein folding and stability; protein aggregation; systems biology; nonlinear dynamics.

Associate Professor

Cruz Cruz, Luis, Ph.D., Massachusetts Institute of Technology, 1994. *Biophysics*. Molecular dynamics of proteins; spatial correlations; cellular automata.

Karapetrov, Goran, Ph.D., Oregon State University, 1996. *Condensed Matter Physics, Nano Science and Technology*. Experimental solid state physics; scanning probe microscopy; nanoscale catalysis; mesoscopic superconductivity.

Richards, Gordon, Ph.D., University of Chicago, 2000. *Astrophysics*. Quasars; quasars absorption lines; gravitational lensing; galaxy evolution; Sloan Digital Sky Survey.

Urbanc, Brigita, Ph.D., University of Ljubljana, 1994. *Biophysics*. Computational biophysics; protein folding and assembly.

Assistant Professor

Dolinski, Michelle J., Ph.D., University of California, Berkeley, 2008. *Particles and Fields*. Application of traditional nuclear physics techniques to the search for physics beyond the Standard Model.

Ma, Hairong, Ph.D., University of Illinois at Urbana-Champaign, 2005. *Biophysics*. High-throughput characterization and optimization of genetically-encoded metal sensors with microfluidics technology.

Research Professor

Olson, Kevin, Ph.D., University of Massachusetts. *Astrophysics*. Development of parallel, numerical algorithms for Astrophysics applications.

Teaching Assistant Professor

Aprelev, Alexey, Ph.D., St. Petersburg University, 1995. *Biophysics*. Experimental biophysics.

Visiting Professor

Spicer, Daniel, Ph.D., University of Maryland, 1976. *Astrophysics*. Space and solar plasma physics; magnetohydrodynamics and numerical 3D.

Visiting Research Professor

Mordecai-Mark, Mac Low, Ph.D., University of Colorado, 1989. *Astrophysics*. Structure of the Interstellar Medium and Molecular Clouds, Blast Waves, Star Formation, Circumstellar Nebulae, Planetary Impacts, Computational Gas Dynamics and Magnetohydrodynamics.

Posdoctoral Research Associate

Jampani, Srinivasa, Ph.D., University of Hyderabad, 2010. *Biophysics*. Computational biophysics.

Williams, Thomas L., Ph.D., University of Bath, 2008. *Biophysics*.

DEPARTMENTAL RESEARCH SPECIALTIES AND STAFF

Theoretical

Astrophysics. Cosmology, gravitational lensing, numerical simulation of dense stellar systems, and high-performance computing. Goldberg, McMillan, Mordecai-Mark, Olson, Richards, Spicer, Vogeley.

Biophysics. Protein folding and self-assembly, neurodegenerative diseases, systems biology and bio-network. Aprelev, Cruz Cruz, Ferrone, Jampani, Urbanc, Williams, Yuan.

Condensed Matter Physics. Electronic and optical properties of nanoshells, graphene, carbon nanotubes and high-Tc superconductors. Bose, Karapetrov, Tyagi.

Nonlinear Dynamics and Complex Systems. Topological analysis of non-linear systems, driven molecular systems, chaotic scattering, and quantum- classical correspondence. Gilmore.

Experimental

Astrophysics. Large-scale structure and cosmology, galactic and extragalactic astronomy, galaxy surveys (Sloan Digital Sky Survey), active galactic nuclei/quasars, black holes, dynamics of star clusters and galactic nuclei. Richards, Vogeley.

Biophysics. Phase transitions in biology, force transduction in muscle, dynamics of biomolecules. Aprelev, Ferrone, Ma, Urbanc, Williams.

Condensed Matter Physics. Experimental solid state physics, scanning probe microscopy, Nanoscale Catalysis, Mesoscopic superconductivity. Ultra-low temperature studies and simulations of entanglement and coherence in superconducting qubits, enhanced Raman scattering and use of nanoparticles for biomedical applications. Bose, Karapetrov, Tyagi.

Particles and Fields. Experimental neutrino properties and oscillation, solar neutrinos, geoneutrinos and neutrino applications to nuclear non-proliferation. Dolinski, Lane.

View additional information about this department at
www.gradschoolshopper.com

LEHIGH UNIVERSITY

DEPARTMENT OF PHYSICS

Bethlehem, Pennsylvania 18015
http://www.physics.lehigh.edu

General University Information
President: Alice P. Gast
Dean of Graduate School: Donald E. Hall
University website: http://www.lehigh.edu
Control: Private
Setting: Suburban

Total Faculty: 480
Total Graduate Faculty: 480
Total number of Students: 7,080
Total number of Graduate Students: 2,197

Department Information
Department Chairman: Volkmar Dierolf, Chair
Department Contact: Lois Groff, Physics Graduate Coordinator
Total full-time faculty: 19
Total number of full-time equivalent positions: 19
Full-Time Graduate Students: 41
First-Year Graduate Students: 9
Female First-Year Students: 4
Total Post Doctorates: 8

Department Address
16 Memorial Drive, East
Bethlehem, PA 18015
Phone: (610) 758-3930
Fax: (610) 758-5730
E-mail: lg00@lehigh.edu
Website: http://www.physics.lehigh.edu

ADMISSIONS

Admission Contact Information
Address admission inquiries to: Graduate Admissions Officer, Lehigh University, Dept. of Physics, 16 Memorial Drive, East, Bethlehem, PA 18015
Phone: (610) 758-3931
E-mail: LG00@lehigh.edu
Admissions website: http://cas.cas2.lehigh.edu/content/graduate-students

Application deadlines
Fall admission:
U.S. students: March 15   *Int'l. students*: March 15

Application fee
U.S. students: $75   *Int'l. students*: $75
Review of graduate applications begins on January 15th of each year.

Admissions information
For Fall of 2012:
Number of applicants: 110
Number admitted: 20
Number enrolled: 9

Admission requirements
Bachelor's degree requirements: Bachelor's degree in physics or a related field is required.
Minimum undergraduate GPA: 3.0

GRE requirements
The GRE is required.

Advanced GRE requirements
The Advanced GRE is recommended.

TOEFL requirements
The TOEFL exam is required for students from non-English-speaking countries.
iBT score: 85

Other admissions information
Undergraduate preparation assumed: Intermediate mechanics, electricity and magnetism, atomic and quantum physics, thermodynamics, and laboratory experience. Mathematics through partial differential equations. Typical texts include Symon, Griffiths, Eisberg and Resnick, Merzbacher, Reif, and Van Ness. Able students with inadequate preparation may take a limited number of remedial courses after enrollment.

TUITION

Tuition year 2013–14:
   Full-time students: $1,300 per credit
   Part-time students: $1,300 per credit
Tuition waivers are part of each RA/TA/GA support package.
Credit hours per semester to be considered full-time: 10
Deferred tuition plan: Yes
Health insurance: Available at the cost of $990 per year.
Academic term: Semester
Number of first-year students who received full tuition waivers: 9

Teaching Assistants, Research Assistants, and Fellowships
Number of first-year
   Teaching Assistants: 6
   Fellowship students: 3
Average stipend per academic year
   Teaching Assistant: $26,200
   Research Assistant: $26,200
   Fellowship student: $26,300

FINANCIAL AID

Application deadlines
Fall admission:
U.S. students: March 15   *Int'l. students*: March 15

Loans
Loans are not available for U.S. students.
Loans are not available for international students.
GAPSFAS application required: No
FAFSA application required: No

For further information
Address financial aid inquiries to: Graduate Admissions Officer, Lehigh University, Dept. of Physics, 16 Memorial Drive, East, Bethlehem, PA 18015.
Phone: (610) 758-3931
E-mail: LG00@lehigh.edu
Financial aid website: http://cas.cas2.lehigh.edu/content/financial-aid

HOUSING

Availability of on-campus housing
   Single students: Yes
   Married students: Yes

For further information
Address housing inquiries to: Residence Operations Office, Lehigh University, Rathbone Hall, 63 University Drive, Bethlehem, PA 18015.
Phone: (610) 758-3500
E-mail: inrsd@lehigh.edu
Housing aid website: http://www4.lehigh.edu/housing/contactus/staff.aspx

Table A—Faculty, Enrollments, and Degrees Granted

Research Specialty	2012–13 Faculty	Enrollment Fall 2012		Number of Degrees Granted 2012–13 (2008–13)		
		Master's	Doctorate	Master's	Terminal Master's	Doctorate
Astrophysics	4	–	3	1(3)	–	–(2)
Atomic, Molecular, & Optical Physics	2	–	3	–(3)	–	1(5)
Biophysics	1	–	3	1(6)	–(1)	1(4)
Condensed Matter Physics	6	–	10	2(10)	1(1)	1(9)
Nano Science and Technology	1	–	4	–(2)	–	1(2)
Optics	1	–	6	1(6)	1(1)	2(7)
Plasma and Fusion	2	–	1	–(1)	–	–(2)
Polymer Physics/Science	1	–	–	–(2)	–	1(2)
Statistical & Thermal Physics	1	–	2	1(3)	–	1(1)
Non-specialized	–	–	9	–(2)	–(1)	–
Total	19	–	41	6(38)	2(4)	8(34)
Full-time Grad. Stud.	–	–	41	–	–	–
First-year Grad. Stud.	–	–	9	–	–	–

GRADUATE DEGREE REQUIREMENTS

Master's: Thirty credit hours required, including a research project. No minimum grade point average, but more than four grades below B cause a student to become ineligible for further graduate work. All work for M.S. must be done in residence at Lehigh. No foreign language requirement and no requirement for comprehensive/qualifying examination or thesis.

Doctorate: Nine credits of coursework beyond M.S. required, but Ph.D. programs usually include 20 or more credits beyond M.S. Minimum time requirement of two years with at least one year in residence. Qualifying examination, general examination, and thesis defense required. No departmental or university language requirements.

Other Degrees: The additional interdepartmental areas of research include materials science, surface science, photonics, and geophysics.

Thesis: Thesis may be written in absentia.

SPECIAL EQUIPMENT, FACILITIES, OR PROGRAMS

Research facilities are housed in the Sherman Fairchild Center for the Physical Sciences, containing the Lewis Laboratory, the Sherman Fairchild Laboratory for Solid State Studies, and a large connecting research wing. Well-equipped laboratory facilities are available for experimental investigations in research areas at the frontiers of physics. Instruments used for experimental studies include a wide variety of laser systems ranging from femtosecond and picosecond pulsed lasers to stabilized single-mode cw Ti-sapphire and dye lasers. There is also a Fourier-transform spectrometer, cryogenic equipment that achieves temperatures as low as 0.05 K, and magnetic fields up to 9 Tesla, a facility for luminescence microscopy, and a laser-tweezers system for studies of complex fluids. The Fairchild Laboratory also contains a processing laboratory where advanced Si devices can be fabricated and studied. All laboratories are well furnished with electronic instrumentation for data acquisition and analysis.

Table B—Separately Budgeted Research Expenditures by Source of Support

Source of Support	Departmental Research	Physics-related Research Outside Department
Federal government	$1,699,570	$2,067,010
State/local government	$21,430	$24,860
Non-profit organizations	$92,340	$10,280
Business and industry	$78,360	$70,270
Other	$186,170	
Total	$2,077,870	$2,172,420

Table C—Separately Budgeted Research Expenditures by Research Specialty

Research Specialty	No. of Grants	Expenditures ($)
Astronomy	3	$220,070
Atomic, Molecular, & Optical Physics	2	$167,460
Biophysics	5	$310,160
Condensed Matter Physics	17	$488,790
Nano Science and Technology	3	$294,930
Optics	3	$110,060
Plasma and Fusion	5	$441,220
Statistical & Thermal Physics	1	$45,180
Total	39	$2,077,870

FACULTY

Professor

Biaggio, Ivan, Ph.D., ETH-Zurich, 1993. *Condensed Matter Physics, Optics*. Experiment.

DeLeo, Gary G., Ph.D., University of Connecticut, 1979. *Astrophysics, Physics and other Science Education*. Theory.

Dierolf, Volkmar, Ph.D., University of Utah, 1992. *Condensed Matter Physics, Optics*. Experiment.

Gunton, James D., Ph.D., Stanford University, 1967. *Condensed Matter Physics, Statistical & Thermal Physics*. Theory.

Hickman, A. Peet, Ph.D., Rice University, 1973. *Atomic, Molecular, & Optical Physics*. Theory.

Huennekens, John P., Ph.D., University of Colorado, 1982. *Atomic, Molecular, & Optical Physics*. Experiment.

Kanofsky, Alvin S., Ph.D., University of Pennsylvania, 1966. *Particles and Fields*. Experiment.

Kim, Yong W., Ph.D., University of Michigan, 1968. *Atomic, Molecular, & Optical Physics, Fluids, Rheology, Plasma and Fusion, Statistical & Thermal Physics*. Experiment.

Kritz, Arnold H., Ph.D., Yale University, 1961. *Plasma and Fusion*. Theory.

McCluskey, George E., Ph.D., University of Pennsylvania, 1965. *Astrophysics*. Theory.

Ou-Yang, H. Daniel, Ph.D., University of California, Los Angeles, 1985. *Biophysics, Polymer Physics/Science*. Experiment.

Rickman, Jeffrey M., Ph.D., Carnegie Mellon University, 1989. *Condensed Matter Physics*. Theory.

Stavola, Michael J., Ph.D., University of Rochester, 1980. *Condensed Matter Physics*. Experiment.

Toulouse, Jean, Ph.D., Columbia University, 1981. *Condensed Matter Physics, Optics*. Experiment.

Associate Professor

Licini, Jerome C., Ph.D., Massachusetts Institute of Technology, 1987. *Condensed Matter Physics*. Experiment.

McSwain, M. Virginia, Ph.D., Georgia State University, 2004. *Astronomy, Astrophysics*. Experimental/Observational.

Rotkin, Slava, Ph.D., Ioffe Institute, St. Petersburg, 1997. *Condensed Matter Physics, Nano Science and Technology*. Theory and Experiment.

Vavylonis, Dimitrius, Ph.D., Columbia University, 2000. *Biophysics*. Theory.

Assistant Professor

Pepper, Joshua A., Ph.D., Ohio State Univ., 2007. *Astronomy, Astrophysics*. Theoretical and Observational.

Emeritus

Folk, Robert T., Ph.D., Lehigh University, 1958. *Mechanics, Nuclear Physics*. Theory.

Fowler, W. Beall, Ph.D., University of Rochester, 1963. *Condensed Matter Physics*. Theory.

Shaffer, Russell A., Ph.D., Johns Hopkins University, 1962. *Particles and Fields*. Theory.

Adjunct Professor

Cereghetti, Paola, Ph.D., Swiss Federal Institute of Technology, 2000. Condensed Matter.

Glueckstein, Jon, Ph.D., University of Wisconsin-Madison, 1997. Condensed Matter.

Loomis, John, M.S., University of Massachusetts, 1973. Astronomy.

Lucic, Dragan, M.S., University of Colorado, 1997.

Tupa, Peter R., Ph.D., Lehigh University, 2011. Astrophysics.

Visiting Professor

Cohen, Joel, Ph.D., Univ. of Illinois, 1968. Biophysics.

Fu, Jinxin, Ph.D., Institute of Physics, Chinese Academy of Sciences, 2011. Nanoscience.

Iolin, Eugene, Ph.D., Estonian Academy of Sciences, 1978. Condensed Matter.

Nemilentsau, Andrei, Ph.D., Institute of Physics, NAS Minsk, 2009. Optics.

Rafiq, Tariq, Ph.D., Chalmers University of Technology, 2004. Plasma.

Ryan, Gillian, Ph.D., Dalhousie University, 2010. Biophysics.

Tangri, Varun, Ph.D., Devi Ahyiliya Univ., 2006. Plasma.

Yusuf, Eddy, Ph.D., Florida State Univ., 2006. Biophysics.

DEPARTMENTAL RESEARCH SPECIALTIES AND STAFF

Theoretical

Astrophysics. Ultraviolet spectroscopy and gas dynamics of interacting binary systems; orbits of binary stars; N-body dynamics. DeLeo, McCluskey.

Atomic, Molecular, & Optical Physics. Charge exchange collisions; fine-structure changing collisions; optical processes in gases; molecular hyperfine spectroscopy. Hickman.

Biophysics. Physical and engineering principles involved in the assembly of actin proteins into filaments and larger scale structures; statistical mechanics and soft matter physics applied to actin protein assemblies and the emergent collective properties. Vavylonis.

Condensed Matter Physics. Electronic and vibrational properties of defects in semiconductors and insulators. Fowler, Gunton, Rickman.

Nano Science and Technology. Quantum mechanics of one-dimensional systems; many-body effects in carbon nanotubes; physics of nanotube/nanowire devices; electron transport in molecular systems; modeling of interactions between nano and biological systems. Rotkin.

Plasma and Fusion. Integrated modeling codes developed and used to predict temperature, momentum, and density profiles in magnetically confined controlled fusion plasma experi-

ments; theoretically derived physics models used in these codes and detailed comparisons between simulations and experimental data to understand the physics of transport and confinement in plasmas. There are active collaborations with theory and experimental groups, both nationally and internationally. Kritz.

Statistical & Thermal Physics. Pattern formation in nonlinear, non-equilibrium systems; kinetics of first-order phase transitions focusing on crystallization of globular proteins; cell-cell communication via calcium oscillations. Gunton.

Experimental

Astronomy. Observational studies to understand the formation and evolution of stars: young open clusters, binary stars, x-ray binaries and pulsars, the formation of disks in Be stars, and the origin of magnetic fields in massive stars. Lehigh has a significant amount of telescope access as a partner in the SMARTS Consortium (McSwain). Theoretical modeling for the discovery of transiting exoplanets. Operation of the KELT small-telescope transit survey. Variable- and eclipsing-binary star studies, and large astronomical surveys and developments in the field of astroinformatics (Pepper).

Atomic, Molecular, & Optical Physics. Collisional processes in atomic vapors including excitation transfer and "energy pooling" line-broadening, quenching, diffusion, resonance exchange, and velocity-changing collisions; molecular spectroscopy of bound singlet and triplet states of alkali diatomics, photodissociation, predissociation, and bound-free emission. Huennekens, Kim.

Biophysics. Application of optical imaging, trapping, and manipulation for cell mechanics studies. Ou-Yang.

Condensed Matter Physics. Charge transport in insulators and semiconductors; nonlinear optical spectroscopy (Biaggio); point defects in insulating materials with ferroelectric domain walls and other dopants; optical spectroscopy under application of hydrostatic pressure and magnetic fields; carrier localization in wide band gap semiconductors (Dierolf); quantum transport behavior of electrons; conduction in ultrasmall silicon MOSFETs and gallium-arsenide devices and carbon nanotubes at low temperature and high magnetic field (Licini); defects in semiconductors; defect complexes that contain light-element impurities such as H, C, O, and N; vibrational spectroscopy and uniaxial stress techniques to elucidate microscopic properties (Stavola); Raman and neutron scattering; dielectric and ultrasonic spectroscopies; collective vibrational dynamics of disordered ferroelectrics and glasses (Toulouse).

Fluids, Rheology. Nonlinear dynamics in fluid systems; dynamics of small particle suspensions; light scattering loss spectroscopy; instabilities of interfaces. Kim.

Nano Science and Technology. Quantum mechanical theory of carbon nanotubes; DNA nanotube hybrids; optics, optoelectronics, and electronics of nanotube devices; nanotube-organics complexes. Rotkin.

Optics. Multiple orders of light-matter interactions; time-resolved spectroscopy of second- and third-order nonlinear optical effects in organic and inorganic materials; optical frequency conversion and all-optical switching (Biaggio); fiber optics; nonlinear effects in optical fibers and waveguides (Dierolf, Toulouse).

Plasma and Fusion. Collisional and collisionless phenomena of very dense plasmas in or near a local thermodynamic equilibrium; anomalies in radiation transport properties; lowering of ionization potentials in dense plasmas; laser-produced plasmas. Kim.

Polymer Physics/Science. Soft condensed matter and complex fluids. Ou-Yang.

Statistical & Thermal Physics. Intrinsic fluctuations in fluids under external forcing such as Brownian motion; chaotic transitions; light scattering from fractals; 1/f-dynamics of granular avalanches. Kim.

PENNSYLVANIA STATE UNIVERSITY

DEPARTMENT OF PHYSICS

University Park, Pennsylvania 16802
http://www.phys.psu.edu

General University Information
President: Rodney A. Erickson
Dean of Graduate School: Henry Foley
University website: http://www.psu.edu
Control: Public
Setting: Suburban
Total Faculty: 3,173
Total Graduate Faculty: 2,800
Total number of Students: 45,351
Total number of Graduate Students: 6,159

Department Information
Department Chairman: Nitin Samarth, Head
Department Contact: Carol Deering, Graduate Program
  Coordinator
  Total full-time faculty: 49
  Total number of full-time equivalent positions: 42
  Full-Time Graduate Students: 126
  First-Year Graduate Students: 24
  Female First-Year Students: 3
  Total Post Doctorates: 23

Department Address
104 Davey Laboratory
University Park, PA 16802
Phone: (814) 865-7533
Fax: (814) 865-0978
E-mail: graduate-admissions@phys.psu.edu
Website: http://www.phys.psu.edu

ADMISSIONS

Admission Contact Information
Address admission inquiries to: The Pennsylvania State University, Department of Physics, 104 Davey Laboratory, University Park, PA 16802
Phone: (814) 865-7534
E-mail: graduate-admissions@phys.psu.edu
Admissions website: http://phys.psu.edu/graduate/apply/

Application deadlines
Fall admission:
U.S. students: January 15   *Int'l. students*: January 15
$65 fee upon matriculation.

Admissions information
For Fall of 2012:
  Number of applicants: 676
  Number admitted: 92
  Number enrolled: 24

Admission requirements
Bachelor's degree requirements: Bachelor's degree in physics or a related field is required.
Minimum undergraduate GPA: 3.0

GRE requirements
The GRE is required.
The minimum acceptable score required for admission is not fixed.

Advanced GRE requirements
The Advanced GRE is required.
The minimum acceptable score required for admission is not fixed.

TOEFL requirements
The TOEFL exam is required for students from non-English-speaking countries.
  PBT score: 550
  iBT score: 80

Other admissions information
Additional requirements: The best-qualified applicants will be accepted up to the number of spaces available. The average GRE scores for 2012–13 admissions were quantitative–89%; physics–71%.
Undergraduate preparation assumed: Marion & Thornton, Classical Dynamics of Particles and Systems; Griffiths, Introduction to Electrodynamics; Griffiths, Introduction to Quantum Mechanics; Reif, Fundamentals of Statistical and Thermal Physics.

TUITION

Tuition year 2012–13:
Tuition for in-state residents
  Full-time students: $18,672 annual
  Part-time students: $778 per credit
Tuition for out-of-state residents
  Full-time students: $31,512 annual
  Part-time students: $1,313 per credit
The Department of Physics admits graduate students with an assistantship or fellowship that fully covers tuition. There is no separate application for financial aid.
Credit hours per semester to be considered full-time: 9
Deferred tuition plan: No
Health insurance: Available at the cost of 427.40 per year.
Other academic fees: Thesis fees: M.S., $25; Ph.D., $95.
Academic term: Semester
Number of first-year students who received full tuition waivers: 24

Teaching Assistants, Research Assistants, and Fellowships
  Number of first-year
    Teaching Assistants: 19
    Research Assistants: 2
    Fellowship students: 3
  Average stipend per academic year
    Teaching Assistant: $21,685
    Research Assistant: $24,759

FINANCIAL AID

Application deadlines
Fall admission:
U.S. students: January 15   *Int'l. students*: January 15

Loans

Loans are not available for U.S. students.

Loans are not available for international students.

GAPSFAS application required: No

FAFSA application required: No

HOUSING

Availability of on-campus housing

Single students: Yes

Married students: Yes

For further information

Address housing inquiries to: The Pennsylvania State University, The Assignment Office for Campus Residences, 201 Johnston Commons, University Park PA 16802.

Phone: (814) 865-7501

E-mail: assignmentoffice@psu.edu

Housing aid website: http://www.hfs.psu.edu/housing/graduates/lease.shtml

Table A—Faculty, Enrollments, and Degrees Granted

Research Specialty	2012-13 Faculty	Enrollment Fall 2012 Master's	Enrollment Fall 2012 Doctorate	Number of Degrees Granted 2012-13 (2008–13) Master's	Number of Degrees Granted 2012-13 (2008–13) Terminal Master's	Number of Degrees Granted 2012-13 (2008–13) Doctorate
Astrophysics	–	–	–	–	–	–
Atomic, Molecular, & Optical Physics	5	–	21	–	–	3(4)
Biophysics	5	–	13	–	–	–(4)
Condensed Matter Physics	21	3	60	–	1(8)	8(50)
Particles and Fields	12	–	18	–	–	1(13)
Relativity & Gravitation	6	–	14	–	–(2)	1(14)
Total	**49**	**3**	**126**	**–**	**1(10)**	**13(85)**
Full-time Grad. Stud.	–	2	124	–	–	–
First-year Grad. Stud.	–	–	24	–	–	–

GRADUATE DEGREE REQUIREMENTS

Master's: Thirty credits (semester-equivalent), at least 18 of which are at the graduate (not dual) level, are required. A minimum 3.0 grade point average is required. There is no residence or foreign language requirement. There are no examinations. A thesis or research paper may be submitted.

Doctorate: There is no fixed number of total credits; 21 graduate course credits are specified, a minimum of 12 additional course credits are required. An overall 3.0 grade point average is required. A minimum of two semesters of residence after candidacy are required. There is no foreign language requirement. Candidacy examination is taken during the first year and a comprehensive examination usually during the second year. A thesis is required.

Other Degrees: M.Ed. is also available.

Thesis: Thesis may be written in absentia.

SPECIAL EQUIPMENT, FACILITIES, OR PROGRAMS

Several of our condensed-matter faculty members are associated with the Materials Research Institute, which coordinates, administers, and supports state-of-the-art research, fabrication, and characterization facilities at Penn State, including a site of the National Nanotechnology Infrastructure Network. Faculty members specializing in high-energy physics, particle astrophysics, and gravitational physics are also members of large external collaborations such as LIGO, Amanda, Ice Cube, Auger, CREAM,

HAWC, CREST, and STAR at RHIC (Brookhaven). Several faculty members also belong to the Institute for Gravitation and the Cosmos, a multidisciplinary institute dedicated to the study of the most fundamental structure and constituents of the universe.

Table B—Separately Budgeted Research Expenditures by Source of Support

Source of Support	Departmental Research	Physics-related Research Outside Department
Federal government	$10,569,700	
State/local government		
Non-profit organizations		
Business and industry	$14,982	
Other		
Total	**$10,584,682**	

Table C—Separately Budgeted Research Expenditures by Research Specialty

Research Specialty	No. of Grants	Expenditures ($)
Atomic, Molecular, & Optical Physics	11	$1,764,000
Biophysics	4	$320,000
Condensed Matter Physics	27	$5,550,000
Relativity & Gravitation	6	$847,700
High Energy Physics	15	$2,088,000
Total	**63**	**$10,569,700**

FACULTY

Professor

Albert, Reka, Ph.D., University of Notre Dame, 2001. Professor of Biology. *Biophysics, Nonlinear Dynamics and Complex Systems, Systems Science/Engineering*. Statistical mechanics; network theory; systems biology.

Anderson, James, Ph.D., Princeton University, 1963. Evan Pugh Professor of Chemistry. *Chemical Physics*. Quantum chemistry by Monte Carlo methods; predicting rare events in chemical dynamics; developing high efficiency in fluorescent lamps; kinetics of enzyme catalysis.

Ashtekar, Abhay V., Ph.D., University of Chicago, 1974. Eberly Professor of Physics. *Relativity & Gravitation*. Gravity; geometry; physics.

Bojowald, Martin, Ph.D., RWTH Aachen University, 2000. *Relativity & Gravitation*. Quantum gravity; classical and quantum cosmology; classical and quantum aspects of black holes; general aspects of quantization; Poisson geometry and noncommutative geometry.

Castleman, A. W., Ph.D., Polytechnic Institute of New York, 1969. Evan Pugh Professor. Professor of Chemistry. *Chemical Physics, Condensed Matter Physics, Nano Science and Technology*. Laser photo physics; clusters; quantum confinement effects.

Chan, Moses H. W., Ph.D., Cornell University, 1974. Evan Pugh Professor. *Condensed Matter Physics, Low Temperature Physics*. Low-temperature physics; phase transitions in two and three dimensions; superfluid and liquid-vapor transitions in random media; low-temperature transport studies of superconducting nanowires.

Cole, Milton W., Ph.D., University of Chicago, 1970. Distinguished Professor. *Condensed Matter Physics*. Theoretical surface physics; superfluids; statistical mechanics.

Collins, John, Ph.D., University of Cambridge, 1975. Distinguished Professor. *Biophysics, Particles and Fields*. Quantum field theory; perturbative methods and factorization in QCD; renormalization theory; neuroscience.

Coutu, Stephane, Ph.D., California Institute of Technology, 1993. Professor of Astronomy and Astrophysics. *Atmosphere, Space Physics, Cosmic Rays, High Energy Physics, Particles and Fields.* Cosmic rays; high-energy physics; particles and fields.

Cowen, Douglas, Ph.D., University of Wisconsin-Madison, 1990. Professor of Astronomy and Astrophysics. *Astrophysics, High Energy Physics, Particles and Fields.* Astrophysics; particles and fields.

Crespi, Vincent, Ph.D., University of California, Berkeley, 1994. Distinguished Professor; Professor of Materials Science and Engineering; Professor of Chemistry. *Computational Physics, Condensed Matter Physics, Materials Science, Metallurgy, Nano Science and Technology.* Theory of superconducting; transport; electronic and structural/mechanical properties of novel materials.

Diehl, Renee, Ph.D., University of Washington (Seattle), 1982. Associate Department Head. *Condensed Matter Physics.* Surface physics; surface structure, dynamics and phase transitions; quasicrystals; fullerenes; complex materials; low-energy electron diffraction.

Fichthorn, Kristen A., Ph.D., University of Michigan, 1989. Professor of Chemical Engineering. *Computational Physics, Condensed Matter Physics, Nano Science and Technology, Statistical & Thermal Physics.* Statistical mechanics and computer simulation; surface physics; colloidal nanostructures; structural evolution of materials; computer simulation.

Finn, Lee Samuel, Ph.D., California Institute of Technology, 1987. Professor of Astronomy and Astrophysics. *Astrophysics, Relativity & Gravitation.* Gravitational wave astronomy; sources of gravitational radiation; relativistic astrophysics; numerical relativity.

Günaydin, Murat, Ph.D., Yale University, 1973. *High Energy Physics, Particles and Fields.* Theoretical physics; superstrings; super-gravity; Kaluza-Klein theories.

Gibble, Kurt, Ph.D., University of Colorado JILA, 1990. *Atomic, Molecular, & Optical Physics.* Microwave and optical frequency atomic clocks; space-based atomic clocks; ultracold atom-atom scattering; ultra-stable lasers; laser cooling; precision measurements; experimental atomic physics.

Heppelmann, Steven F., Ph.D., University Minnesota, 1981. *High Energy Physics, Particles and Fields.* Experimental high-energy physics.

Jain, Jainendra K., Ph.D., Stony Brook University, 1985. Erwin W. Mueller Professor; Evan Pugh Professor. *Condensed Matter Physics.* Low dimensional systems; fractional quantum Hall effect; composite fermions; topological states; Luttinger liquids.

Larson, Daniel J., Ph.D., Harvard University, 1971. Verne M. Willaman Dean. *Atomic, Molecular, & Optical Physics.* Atomic physics; molecules; optical physics.

Li, Qi, Ph.D., Peking University, 1989. *Condensed Matter Physics.* Spintronics; multiferroic multilayer structures; superconductivity and strong correlated systems; magnetic nanostructures.

Liu, Ying, Ph.D., University of Minnesota, 1991. *Condensed Matter Physics.* Experimental condensed-matter physics; unconventional superconductivity, topological quantum states of matter, nanophysics.

Mahan, Gerald, Ph.D., University of California, Berkeley, 1964. Distinguished Professor. *Condensed Matter Physics, Nano Science and Technology, Statistical & Thermal Physics.* Condensed-matter theory; transport and optical properties; solid-state devices.

Mallouk, Thomas E., Ph.D., University of California, Berkeley, 1983. Evan Pugh Professor of Materials Chemistry and Physics. *Condensed Matter Physics, Nano Science and Technology.* Photocatalysis; molecular electronics; electrochemical energy conversion; environmental remediation; chemical sensing; motion on the nanoscale.

Meszaros, Peter, Ph.D., University of California, Berkeley, 1972. Eberly Chair of Astronomy & Astrophysics; Director, Center for Particle Astrophysics. *Astrophysics, Relativity & Gravitation.* Theoretical high-energy astrophysics; gamma-ray burst sources; ultra-high-energy neutrinos and photons; cosmology; neutron stars and black holes; active galactic nuclei.

Owen, Benjamin, Ph.D., California Institute of Technology, 1998. *Astrophysics, Relativity & Gravitation.* Gravitational waves; neutron stars.

Robinett, Richard W., Ph.D., University of Minnesota, 1981. Associate Department Head; Director of Undergraduate Studies; Director of Graduate Studies. *Quantum Foundations.* Mathematical physics; time-dependent quantum systems; pedagogical issues related to quantum mechanics.

Roiban, Radu, Ph.D., Stony Brook University, 2001. *High Energy Physics, Particles and Fields.* String theory; gauge theories; quantum field theory.

Samarth, Nitin, Ph.D., Purdue University, 1986. George A. and Margaret M. Downsbrough Department Head. *Applied Physics, Condensed Matter Physics, Materials Science, Metallurgy, Nano Science and Technology, Solid State Physics, Surface Physics.* Spintronics; topological insulators; quantum information processing; semiconductor, superconductor, and magnetic nanostructures; magneto-optical studies of artificial spin ice.

Schiff, Steven, Ph.D., Duke University Medical Center, 1985. M.D., Duke University Medical Center, 1985. Director, Penn State Center for Neural Engineering; Brush Chair Professor of Engineering; Professor of Neurosurgery; Professor of Engineering Science and Mechanics. *Biophysics, Neuroscience/Neuro Physics.* Neural engineering, control theory.

Sofo, Jorge, Ph.D., Inst. Balseriro, Bariloche, 1991. Professor of Materials Science and Engineering. *Applied Physics, Atomic, Molecular, & Optical Physics, Chemical Physics, Computational Physics, Condensed Matter Physics, Energy Sources & Environment, Geophysics, Low Temperature Physics, Materials Science, Metallurgy, Nano Science and Technology, Statistical & Thermal Physics.* Theoretical and computational methods to link properties and structures of materials.

Strikman, Mark, Ph.D., Leningrad Nuclear Physics Institute, 1978. Distinguished Professor. *High Energy Physics, Particles and Fields.* High energy quantum chromodynamics; hard processes with nuclei; microscopic nuclear structure.

Terrones, Mauricio, Ph.D., University of Sussex, 1998. Professor of Materials Science & Engineering; Professor of Chemistry; Distinguished Professor (Shinsu University, Japan). *Applied Physics, Chemical Physics, Condensed Matter Physics, Materials Science, Metallurgy, Nano Science and Technology.* Production of nanomaterials; electron microscopy techniques for analysis; molecular simulations.

Weiss, David, Ph.D., Stanford University, 1993. Associate Department Head. *Atomic, Molecular, & Optical Physics.* Optical lattices; Bose-Einstein condensation; 1D gases; quantum computing; precision measurements; laser cooling.

Willis, Roy F., Ph.D., University of Cambridge, 1967. *Condensed Matter Physics.* Solid-state physics; surface physics; electron spectroscopy of thin films and surfaces; synchrotron source radiation physics.

Associate Professor

DeYoung, Tyce, Ph.D., University of Wisconsin-Madison, 2001. *Astrophysics, High Energy Physics.* High-energy particle astrophysics, especially neutrino and gamma ray astronomy.

Hudson, Eric, Ph.D., University of California, Berkeley, 1999. *Condensed Matter Physics, Low Temperature Physics, Nano Science and Technology, Physics and other Science Education.* Scanning tunneling microscopy; high-temperature superconductors and other complex materials.

Jin, Dezhe, Ph.D., University of California, San Diego, 1999. *Biophysics, Neuroscience/Neuro Physics.* Theory of biological neural networks and computational models of neurobiological functions.

Mocioiu, Irina, Ph.D., Stony Brook University, 2002. *Astrophysics, High Energy Physics, Particles and Fields.* High-energy physics and its connections to astrophysics and cosmology; neutrinos.

Mostafa, Miguel, Ph.D., Universidad Nacional de Cuyo. Instituto Balseiro, 2001. *Astrophysics, High Energy Physics, Particles and Fields.* High-energy particle astrophysics, especially neutrino and gamma ray astronomy.

O'Hara, Kenneth, Ph.D., Duke University, 2000. *Atomic, Molecular, & Optical Physics.* Experimental atomic, molecular, and optical physics; condensed-matter physics.

Rigol, Marcos, Ph.D., University of Stuttgart, 2004. *Atomic, Molecular, & Optical Physics, Condensed Matter Physics.* Statistical mechanics; strong correlations; non-equilibrium quantum dynamics; ultracold gases; magnetism; disorder; superconductivity physics.

Zhu, Jun, Ph.D., Columbia University, 2002. *Applied Physics, Condensed Matter Physics, Low Temperature Physics, Materials Science, Metallurgy, Nano Science and Technology.* Experimental condensed-matter physics; low-temperature transport and scanned probe experiments; mesoscopic and nanoscale systems.

Assistant Professor

Bai, Lu, Ph.D., Cornell University, 2007. Assistant Professor of Biochemistry and Molecular Biology. *Biophysics, Optics, Statistical & Thermal Physics.* Single-molecule/single-cell fluorescence microscopy.

Gemelke, Nathan, Ph.D., Stanford University, 2007. Downsbrough Career Development Professorship. *Atomic, Molecular, & Optical Physics.* Bose-Einstein condensation; quantum gases; quantum information and computation.

Kozhevnikov, Alexey A., Ph.D., Yale University, 2001. *Biophysics.* Biological physics; neural computations; novel experimental techniques in neuroscience.

Liu, Chaoxing, Ph.D., Tsinghua University, 2009. *Condensed Matter Physics.* Theoretical condensed-matter physics; spintronics; topological insulators and superconductors; electronic and transport properties of low-dimensional systems.

Shandera, Sarah, Ph.D., Cornell University, 2006. *Cosmology & String Theory, Particles and Fields.* Theory of the very early universe.

Stasto, Anna, Ph.D., Institute of Nuclear Physics at Polish Academy of Science, 1999. *High Energy Physics, Particles and Fields.* Elementary particle physics; phenomenology of strong interactions; hadronic interactions at high energies.

Professor Emeritus

Maynard, Julian D., Ph.D., Princeton University, 1974. Distinguished Professor. *Acoustics, Low Temperature Physics.* Acoustics; liquid and solid helium; interface and two-dimensional phenomena.

Sommers, Paul, Ph.D., University of Texas, Austin, 1973. Professor of Astronomy and Astrophysics. *Astrophysics, High Energy Physics, Particles and Fields.* High-energy cosmic rays; astrophysics; general relativity.

DEPARTMENTAL RESEARCH SPECIALTIES AND STAFF

Theoretical

Atomic, Molecular, & Optical Physics. Nonequilibrium quantum dynamics; bosons and fermions in low dimensions; superfluidity and Bose-Einstein condensation; optical lattices; quantum criticality and precision measurements. Gibble, Rigol, Sofo.

Biophysics. Computational models for birdsong experiments; modeling of neural mechanisms by which high-level memories are coded; biophysically realistic models of rhythm generation, seizures, and Parkinson's disease; prototype feedback control systems; molecular-level interaction networks among genes and proteins; complex systems; statics and dynamics of ecological communities. Albert, Collins, Jin, Schiff.

Condensed Matter Physics. Fractional quantum Hall effect and composite fermions; Abelian and non-Abelian anyons; 1D Luttinger liquid; strongly correlated systems; many-body phenomena and electron/spin transport in low-dimensional solids (nanowires, nanotubes, quantum dots, 2DEGs, and graphene); topological insulators and topological superconductors; semiconductor spintronics; superconductivity in 1D, 2D, and 3D systems; first-principles density functional theory; dynamical mean field theory; empirical interatomic potentials; photonic band structures; effective continuum theories (Landau-Ginzburg); phase transitions in quantum solids and quantum fluids; wetting transitions of thin films; ultraweak forces binding atoms to surfaces; superfluidity in films; quasi-one-dimensional fluids within nanotubes; theory of classical frustrated magnetic systems; materials design; materials informatics; structural energetics of periodic biological systems; electromechanical response of nanostructures. Cole, Crespi, Fichthorn, Jain, Chaoxing Liu, Mahan, Rigol, Sofo.

Particles and Fields. Application of relativistic quantum theory to internal structure of subnuclear particles; quantum chromodynamics and quantum electrodynamics; quantum chromodynamics in nuclei and in ultra-high-energy hadronic collisions; collider and supercollider phenomenology; electroweak interactions; neutrino physics; supersymmetry; conformal field theory and applications for helicity amplitudes; M/Superstring theory; supergravity. Collins, Günaydin, Mocioiu, Roiban, Shandera, Stasto, Strikman.

Relativity & Gravitation. Classical and quantum theories of gravitation; nonperturbative approaches to quantum field theory and quantum gravity, including loop quantum gravity; gravitational waves; astrophysical applications of general relativity; numerical relativity; mathematical physics; cosmology. Ashtekar, Bojowald, Finn, Meszaros, Owen, Shandera.

Experimental

Atomic, Molecular, & Optical Physics. Bose-Einstein condensation; degenerate Fermi gases; laser-cooled atomic clocks; the structure and dynamics of atomic and molecular clusters; laser cooling and trapping of atoms; tests of fundamental symmetries; quantum computation; quantum scattering of cold atoms; optical lattices. Castleman, Gemelke, Gibble, O'Hara, Weiss.

Biophysics. Neural organization of complex motor sequences; neural activity in functioning neural circuits. Bai, Kozhevnikov, Schiff.

Condensed Matter Physics. Low-temperature studies of superfluidity and supersolid behavior; quantum and mesoscopic transport in low-dimensional solids such as graphene, carbon nanotubes, metal nanowires, and semiconductor nanowires; superconductivity in nanowires; optical spectroscopy of carbon-based and semiconductor nanostructures; physics of spintronic devices; exotic superconductors; quantum transport in topological insulator thin films, heterostructures, and nanowires; low-temperature scanning tunneling microscopy; high-resolution LEED measurements of surfaces. Chan, Diehl, Hudson, Jain, Li, Ying Liu, Mallouk, Samarth, Terrones, Willis, Zhu.

Materials Science, Metallurgy. Synthesis of carbon-based nanostructures; molecular beam epitaxy of semiconductor and topological insulator heterostructures; pulsed laser deposition of complex oxide thin films; electrochemical synthesis of metallic, ferromagnetic, superconducting, and semiconducting nanowires. Chan, Li, Ying Liu, Mallouk, Samarth, Terrones.

Particles and Fields. Particle astrophysics: studies of the properties of quarks and gluons in hadronic interactions; studies of nuclear color transparency with hard hadronic elastic collisions; spin physics in polarized proton-proton collisions at RHIC; direct measurements of high-energy cosmic rays and cosmic antimatter particles; studies of the highest-energy cosmic rays; neutrino physics; neutrino astrophysics; gamma-ray astrophysics. Coutu, Cowen, DeYoung, Heppelmann, Mostafa, Sommers.

View additional information about this department at
www.gradschoolshopper.com

UNIVERSITY OF PENNSYLVANIA

PHYSICS AND ASTRONOMY

Philadelphia, Pennsylvania 19104-6396
http://www.physics.upenn.edu

General University Information
President: Amy Gutmann
Dean of Graduate School: Ralph Rosen
University website: http://www.upenn.edu/
Setting: Urban
Total Faculty: 4,246
Total number of Students: 24,832
Total number of Graduate Students: 11,028

Department Information
Department Chairman: Larry Gladney, Chair
Department Contact: Millicent Minnick, Department
  Administrator
Total full-time faculty: 39
Full-Time Graduate Students: 102
First-Year Graduate Students: 13
Female First-Year Students: 2
Total Post Doctorates: 34

Department Address
209 South 33rd Street
Philadelphia, PA 19104-6396
Phone: (215) 898-3125
Fax: (215) 898-2010
E-mail: mminnick@physics.upenn.edu
Website: http://www.physics.upenn.edu

ADMISSIONS

Admission Contact Information
Address admission inquiries to: Physics and Astronomy, University of Pennsylvania, 209 S. 33rd Street, Philadelphia, PA 19104-6396
Phone: (215) 898-3125
E-mail: admiss@physics.upenn.edu
Admissions website: http://www.physics.upenn.edu/graduate

Application deadlines
Fall admission:
U.S. students: December 31   *Int'l. students*: December 31

Application fee
U.S. students: $80   *Int'l. students*: $80
In payment of the application fee, international applicants should send an international postal money order for the amount. Checks drawn on a foreign bank will be returned and cause a delay processing of the application will be delayed. Discover, MasterCard, and Visa are also accepted.

Admissions information
For Fall of 2013:
Number of applicants: 427
Number admitted: 55
Number enrolled: 23

Admission requirements
Bachelor's degree requirements: A Bachelor's degree in physics, astronomy, or a related science is required. If the Bachelor's degree is not in physics or astronomy, a strong physics minor is necessary. Prior research experience is strongly encouraged. The GRE general test is required. proficiency in English.

GRE requirements
The GRE is required.
The GRE subject test in physics is also required.

Advanced GRE requirements
The Advanced GRE is required.

TOEFL requirements
The TOEFL exam is required for students from non-English-speaking countries.
iBT score: 100
We will accept IELTS in hard copy in lieu of TOEFL scores. You must have them sent directly from IELTS.

TUITION
Tuition year 2013-14:
Tuition for in-state residents
   Full-time students: $31,260 annual $31,260
Ph.D. tuition is based on Fall/Spring and general fees.
Credit hours per semester to be considered full-time: 3
Deferred tuition plan: No
Health insurance: Yes, No cost to student.
Academic term: Semester

Teaching Assistants, Research Assistants, and Fellowships
Number of first-year
   Teaching Assistants: 23
Average stipend per academic year
   Teaching Assistant: $28,443
   Research Assistant: $28,443
All first-year students are teaching assistants.

FINANCIAL AID

Loans
Loans are not available for U.S. students.
Loans are not available for international students.
GAPSFAS application required: No
FAFSA application required: No

HOUSING

Availability of on-campus housing
Single students: Yes
Married students: Yes

For further information
Housing aid website: http://www.upenn.edu/life-at-penn/housing_dining.php

Table A—Faculty, Enrollments, and Degrees Granted

Research Specialty	Faculty	Enrollment Fall 2012		Number of Degrees Granted (2012-13) (2008-13)		
		Master's	Doctorate	Master's	Terminal Master's	Doctorate
Physics and other Science Education	–	–	–	4(-)	5(-)	12(-)
Total	–	–	–	4(-)	5(-)	12(-)
Full-time Grad. Stud.	–	3	99	–	–	–
First-year Grad. Stud.	–	3	13	–	–	–

GRADUATE DEGREE REQUIREMENTS
Master's: Grades of "B" or higher in each of Math Methods (Physics 500), Electrodynamics (Physics 516), and Quantum I and II (Physics 531/532). Satisfactory completion of four additional graduate-level courses with scores of B or better, at least one of which must be offered by the Physics and Astronomy Department. Upto three courses may be from other departments, provided they are in subjects related to Physics and Astronomy. There is no language requirement, but there is a one-year residency requirement.

Doctorate: 1. A grade of "B+" or higher in each of Math Methods (Physics 500), Electrodynamics (Physics 516) and Quantum I and II (Physics 531/532). A student who fails to attain a B+ in any course may re-take the course the following year in order to have the grade changed. 2. Successful completion of 20 graduate-level courses (including no more than 11 course credits for research and reading courses). These courses must include the four courses in requirement 1, Statistical Mechanics (611), and one course outside the student's field of specialization. ● Students may receive credit for graduate courses taken at other institutions, though no more than 8 credits may be transferred. If the Graduate Chair determines that an equivalent graduate course has been taken, then the student must go to the current instructor for that course for a standardized evaluation of the instructor's design. If the instructor determines that the student knows the course material, then that course will be waived and, if appropriate, credit will be given. 3. An Oral Candidacy Exam must be taken within 18 months of the successful completion of the four required courses in requirement 1. 4. Annual progress reports. From the time of successful completion of the Candidacy exam, until successful defense of the Ph.D. dissertation, the student and the three-member committee must provide annual progress reports to the Associate Chair for Graduate Affairs.

FACULTY

Professor
Balasubramanian, Vijay, Ph.D., Princeton University.
Beier, Eugene, Ph.D., University of Illinois, 1966.
Bernstein, Gary, Ph.D., University of California, Berkeley, 1989.
Cvetic, Mirjam, Ph.D., University of Maryland, College Park, 1984.
Devlin, Mark, Ph.D., University of California at Berkeley, 1993.
Drndic, Marija, Ph.D., Harvard University, 2000.
Durian, Douglas, Ph.D., Cornell University, 1989.
Fortune, H. Terry, Ph.D., Florida State Universtiy, 1967.
Gladney, Larry, Ph.D., Stanford University, 1985.
Goulian, Mark, Ph.D., Harvard University, 1990.
Heiney, Paul A., Ph.D., Massachusetts Institute of Technology, 1982.
Hollebeek, Robert, Ph.D., University of California, Berkeley, 1974.
Jain, Bhuvnesh, Ph.D., Massachusetts Institute of Technology, 1994.
Johnson, Charles A.T., Ph.D., Harvard University, 1990.
Kamien, Randall, Ph.D., Harvard University, 1992.
Kane, Charles, Ph.D., Massachusetts Institute of Technology, 1989.
Kikkawa, Jay, Ph.D., University of California, Santa Barbara, 1997.
Klein, Josh, Ph.D., Princeton University, 1994.
Kroll, Joseph I., Ph.D., Harvard University, 1989.
Lande, Kenneth, Ph.D., Columbia University, 1958.
Liu, Andrea, Ph.D., Cornell University, 1989.
Lubensky, Tom, Ph.D., Harvard University, 1969.

Mele, Eugene, Ph.D., Massachusetts Institute of Technology, 1978.
Nelson, Philip, Ph.D., Harvard University, 1984.
Ovurt, Burt A., Ph.D., University of Chicago, 1978.
Sheth, Ravi, Ph.D., University of Cambridge, 1994.
Trodden, Mark, Ph.D., Brown University, 1995.
Williams, Hugh H., Ph.D., Stanford University, 1972.
Yodh, Arjun, Ph.D., Harvard University, 1986.

Associate Professor

Bernardi, Mariangela, Ph.D., Ludwig-Maximilians-Universitaet, Munich, 1995.

Khoury, Justin, Ph.D., Princeton University, 2002.
Sako, Masao, Ph.D., Columbia University, 2001.
Thomson, Evelyn, Ph.D., University of Glasgow, 1998.

Assistant Professor

Aguirre, James, Ph.D., University of Chicago, 2003.
Blake, Cullen, Ph.D., Harvard University, 2009.
Lidz, Adam, Ph.D., Columbia University, 2004.
Lipeles, Elliot, Ph.D., California Institute of Technology, 2004.
Sweeney, Allison, Ph.D., Duke University, 2007.

Professor Emeritus

Balamuth, David, Ph.D., Columbia University, 1968.

View additional information about this department at
www.gradschoolshopper.com

UNIVERSITY OF PENNSYLVANIA

MASTER OF MEDICAL PHYSICS

Philadelphia, Pennsylvania 19104-6396
http://www.sas.upenn.edu/lps/graduate/mmp

General University Information
President: Amy Gutmann
Dean of Graduate School: Ralph M. Rosen
University website: http://www.upenn.edu/
Control: Private
Setting: Urban
Total Faculty: 4,049
Total number of Students: 24,107
Total number of Graduate Students: 9,853

Department Information
Department Chairman: Stephen Avery, Director
Department Contact: Glenn Fechner, Program Coordinator
  Total full-time faculty: 35
  Full-Time Graduate Students: 19
  First-Year Graduate Students: 10
  Female First-Year Students: 4

Department Address
209 S. 33rd Street
University of Pennsylvania
Philadelphia, PA 19104-6396
Phone: (215) 898-6105
Fax: (215) 573-2053
E-mail: mmp-info@sas.upenn.edu
Website: http://www.sas.upenn.edu/lps/graduate/mmp

ADMISSIONS

Admission Contact Information
Address admission inquiries to: Admissions Coordinator, College of Liberal and Professional Studies
Phone: 215-898-7326
E-mail: mmp-info@sas.upenn.edu

Admissions website: http://www.sas.upenn.edu/lps/graduate/mmp/application

Application deadlines
Fall admission:
U.S. students: January 15   *Int'l. students*: January 15

Application fee
U.S. students: $70

Admissions information
For Fall of 2013:
  Number of applicants: 64
  Number admitted: 23
  Number enrolled: 10

Admission requirements
Bachelor's degree requirements: Bachelor's degree in either physics or a related science is required.
Minimum undergraduate GPA: 3.5

GRE requirements
The GRE is required.

Advanced GRE requirements
The Advanced GRE is not required.

TOEFL requirements
The TOEFL exam is required for students from non-English-speaking countries.

Other admissions information
Additional requirements: If the degree is not in physics, a strong physics minor is necessary. No minimum undergraduate GPA is specified.
No minimum score is specified.
Undergraduate preparation assumed: A typical student will have completed intermediate and advanced courses in mechanics (Marion, Becker, etc.); electricity and magnetism (Reitz and

Milford, Corson and Lorrain); quantum mechanics (Saxon, etc.); undergraduate laboratory.

TUITION

Tuition year $35,182:
Full-time students: $5,026 per credit
Fifteen credit units are required for successful completion of masters degree. Total tuition for degree is approximately $75,390.
Credit hours per semester to be considered full-time: 3
Deferred tuition plan: No
Health insurance: Available at the cost of $2,642 per year.
Other academic fees: General Fee: $302.
Academic term: Semester

Teaching Assistants, Research Assistants, and Fellowships

Number of first-year
Research Assistants: 4
Average stipend per academic year
Teaching Assistant: $1,683
Research Assistant: $2,500

FINANCIAL AID

Loans

Loans are available for U.S. students.
Loans are not available for international students.
GAPSFAS application required: No
FAFSA application required: Yes

For further information

Address financial aid inquiries to: Student Financial Services.
Phone: 215-898-1988
E-mail: sfsmail@sfs.upenn.edu
Financial aid website: http://www.sfs.upenn.edu

HOUSING

Availability of on-campus housing

Single students: Yes
Married students: Yes

For further information

Address housing inquiries to: Graduate Housing Office, 3702 Spruce Str.
Phone: (215) 898-3547
Housing aid website: http://www.upenn.edu/provost/graduate_admissions/moving/housing/

Table A—Faculty, Enrollments, and Degrees Granted

Research Specialty	2012-13 Faculty	Enrollment Fall 2012		Number of Degrees Granted 2012-13		
		Master's	Doctorate	Master's	Terminal Master's	Doctorate
Medical, Health Physics	30	11	–	11(-)	11(-)	–
Total	–	11	–	11(-)	11(-)	–
Full-time Grad. Stud.	–	19	–	–	–	–
First-year Grad. Stud.	–	10	–	–	–	–

GRADUATE DEGREE REQUIREMENTS

Master's: Fifteen medical physics course units (CUs) at the graduate level will be required for the Master of Medical Physics (MMP) degree. With the exception of submatriculants ad-

mitted from Penn, students may not apply any graduate-level courses taken as undergraduates against that 15 CU requirement. The program will normally be completed full-time in four semesters, not including summer sessions. Exceptions for part-time study may be granted by the Program Director.
Other Degrees: The University of Pennsylvania's Certificate Program in Medical Physics offers a pathway for individuals who have earned a Ph.D. in physics or a related field (e.g., engineering, computer science, or physical chemistry with a strong physics minor) to enter a CAMPEP-accredited residency program in medical physics that will provide the clinical experience required for certification by the American Board of Radiology. Penn's CAMPEP-accredited Medical Physics Certificate Program will offer participants a curriculum based on the recommendations of AAPM report #197S. Coursework will include radiological physics, radiation protection, medical imaging, medical ethics/government regulation, anatomy and physiology, radiobiology, and the physics of radiation therapy. Students will complete a total of 6 CUs (18 semester hours) over the course of two semesters.

SPECIAL EQUIPMENT, FACILITIES, OR PROGRAMS

Students in the Master of Medical Physics program utilize the facilities of the Departments of Radiology and Radiation Oncology of the medical school and the Hospital of the University of Pennsylvania.

FACULTY

Chair Professor

Gladney, Larry, Ph.D., Stanford University, 1985. Physics Department Chair, Faculty Advisory Committee Chair. *Astrophysics*. Elementary particle experiments.

Professor

Ashmanskas, Bill, Ph.D., University of California, Berkeley, 1998. Course instructor. *Medical, Health Physics*. High-energy physics.
Gressman, Phillip, Ph.D., Princeton University, 2005. Course Instructor. *Applied Mathematics*. Harmonic analysis and geometry, including the study of geometric averaging operators (generalizing the Radon transform), oscillatory integral operators, sublevel set estimates, the Fourier restriction problem, and related objects and applications. Applications of harmonic analysis to PDEs, specifically the Boltzmann equation and the Gross-Pitaevskii Hierarchy.
Hollebeek, Robert, Ph.D., University of California, Berkeley, 1974. Admissions Committee Member. *High Energy Physics*. Elementary particle experiments.
Kamien, Randall D., Ph.D., Harvard University, 1992. *Theoretical Physics*. Condensed matter theory.
Karp, Joel, Ph.D., Massachusetts Institute of Technology, 1980. Faculty Advisory Committee. *Medical, Health Physics*. Radiology.
Kroll, Joseph I., Ph.D., Harvard University, 1989. Faculty Advisory Committee. *High Energy Physics*. Elementary particle experiment.
Maughan, Richard, Ph.D., University of Birmingham, 1974. Vice Chair and Director, Division of Medical Physics in Radiation Oncology; Faculty Advisory Committee Member; Admissions committee member. *Medical, Health Physics*.
Mele, Eugene J., Ph.D., Massachusetts Institute of Technology, 1978. Course instructor. *Condensed Matter Physics*. Condensed matter theory.
Nelson, Philip, Ph.D., Harvard University, 1984. Faculty Advisory Committee; Course instructor. *Condensed Matter Physics*. Condensed matter theory.

Ovrut, Burt, Ph.D., University of Chicago, 1978. *High Energy Physics*. High-energy particle physics.

Trodden, Mark, Ph.D., Brown University, 1995. Course instructor. *Particles and Fields*. Theoretical particle cosmology.

Williams, Hugh, Ph.D., Stanford University, 1972. Course instructor. *High Energy Physics*. Elementary particles.

Yodh, Arjun, Ph.D., Harvard University, 1986. *Condensed Matter Physics, Medical, Health Physics, Optics*. Condensed matter experiments.

Zhu, Timothy, Ph.D., Brown University, 1991. Course instructor. *Medical, Health Physics*. Photodynamic therapy.

Associate Professor

Kassaee, Alireza, Ph.D., State University of NY at Buffalo, 1989. Clinical Practicum Mentor. *Medical, Health Physics*. Clinical radiation.

Koumenis, Costas, Ph.D., University of Houston, 1994. Course instructor. *Medical, Health Physics*. Ionizing radiation.

Maidment, Andrew, Ph.D., University of Toronto, 1993. Course instructor. *Medical, Health Physics*.

McDonough, James, Ph.D., Temple University, 1987. Course instructor. *Medical, Health Physics*. Proton therapy.

Tsourkas, Andrew, Ph.D., Georgia Tech University, 2002. Course instructor. *Medical, Health Physics, Nano Science and Technology*. Nanosensors.

Assistant Professor

Ainsley, Christopher, Ph.D., University of Cambridge, 2003. Course instructor. *Medical, Health Physics*. Proton therapy equipment and techniques; Monte Carlo simulation; Optimization algorithms.

Avery, Stephen, Ph.D., Hampton University, 2002. Director, Master of Medical Physics Program; Course instructor; Admissions and Faculty Advisory Committee Member; Clinical Practicum Mentor. *Medical, Health Physics*. Small Animal Radiation Research Platform (SARRP); Quality assurance and safety in proton therapy treatment delivery.

Both, Stefan, Ph.D., Babes-Bolyai University, 2005. Clinical Practicum Mentor. *Medical, Health Physics*. Advanced treatment planning techniques development and mitigating toxicities in photon and proton radiotherapy. Motion and range uncertanties management in proton radiotherapy. Investigate dose volume histogram and outcome relationship.

Dorsey, Jay, Ph.D., University of South Florida, 2003. Course instructor. *Medical, Health Physics*. Oncology.

Finlay, Jarod, Ph.D., University of Rochester, 2004. Course instructor; clinical practicum . *Medical, Health Physics*. Fluorescence and reflectance spectroscopy of tissue. Tissue optics and modelling of light propagation; Photodynamic

therapy dosimetry and modelling; Monte Carlo simulation; Theoretical modelling of photochemical processes.

Lin, Liyong, Ph.D., University of Wisconsin, 2006. Course instructor. *Medical, Health Physics*. Independent dose calculation engine for Pencil Beam Scanning Treatment, in vivo delivery verification of Proton therapy treatment.

Munbodh, Reshma, Ph.D., Yale University, 2004. Course instructor. *Medical, Health Physics*. Medical image analysis, medical imaging physics, image registration, 2D-3D image registration, image segmentation, statistical estimation theory, and optimization techniques. Expertise in finite element analysis, feature extraction, signal processing, information theory, stochastic processes, decision theory and Bayesian inference, numerical methods and computer vision. Other interests include pattern classification and deformable registration with applications to image guided therapy and surgery, computer-aided diagnosis and visualization. Spatial and morphological modelling of complex radia.

Stripp, Diana, Ph.D., Robert Wood Johnson Medical School, 1998. Course instructor. *Medical, Health Physics*.

Teo, Kevin, Ph.D., University of Michigan, 2002. Course instructor. *Medical, Health Physics*.

Research Assistant Professor

Dahmane, Nadia, Ph.D., University of Paris, 1996. Course instructor. *Medical, Health Physics*. Neurosurgery.

Instructor

Posh, John, Other.Course instructor. *Medical, Health Physics*. Radiologic technology.

Spillane, Kate, Ph.D., Georgia Institute of Technology. Course instructor. *Medical, Health Physics*.

Staff Physicist

Bieda, Michael, M.S., Chief, Clinical Medical Physics; Clinical Practicum Mentor. *Medical, Health Physics*.

Goldberg, Craig, M.S., University of Pennsylvania, 2006. Course instructor. *Medical, Health Physics*.

Mooij, Rob, Ph.D., Course instructor. *Medical, Health Physics*.

Reddin, Janet, Ph.D., Course instructor. *Medical, Health Physics*.

Stambaugh, Michael, M.S., Course instructor. *Medical, Health Physics*.

DEPARTMENTAL RESEARCH SPECIALTIES AND STAFF

Experimental

Medical Physics Reserach. Research that emphasizes the connection between physics and medicine. Areas of research include therapeutic radiology, diagnostic imaging, medical health, and nuclear medicine.

View additional information about this department at
www.gradschoolshopper.com

UNIVERSITY OF PITTSBURGH

DEPARTMENT OF PHYSICS AND ASTRONOMY

Pittsburgh, Pennsylvania 15260
http://www.physicsandastronomy.pitt.edu

General University Information
Chancellor: Mark A. Nordenberg
Dean of Graduate School: N. John Cooper
University website: http://www.pitt.edu
Control: Public
Setting: Urban
Total Faculty: 4,981
Total Graduate Faculty: 1,368
Total number of Students: 28,769
Total number of Graduate Students: 10,340

Department Information
Department Chairman: David Turnshek, Chair
Department Contact: Andrew Zentner, Director of Graduate Studies
  Total full-time faculty: 37
  Full-Time Graduate Students: 97
  First-Year Graduate Students: 20
  Female First-Year Students: 1
  Total Post Doctorates: 17

Department Address
3941 O'Hara Street
Room 100 Allen Hall
Pittsburgh, PA 15260
Phone: (412) 624-9000
Fax: (412) 624-9163
E-mail: pagrad@pitt.edu
Website: http://www.physicsandastronomy.pitt.edu

ADMISSIONS

Admission Contact Information
Address admission inquiries to: Admissions Officer, Department of Physics and Astronomy, 3941 O'Hara Street, Rm. 100 Allen Hall, Univ. of Pittsburgh, Pittsburgh, PA 15260
Phone: (412) 624-9066
E-mail: pagrad@pitt.edu
Admissions website: http://www.physicsandastronomy.pitt.edu/prospective_students

Application deadlines
Fall admission:
U.S. students: January 31   *Int'l. students*: January 31

Application fee
Int'l. students: $50
Late applications and supporting materials are accepted on the basis of space availability.

Admissions information
For Fall of 2013:
  Number of applicants: 242
  Number admitted: 63
  Number enrolled: 20

Admission requirements
Bachelor's degree requirements: Bachelor's degree in one of the physical sciences, mathematics, or engineering with relevant physics courses or research experience)is required.

GRE requirements
The GRE is recommended.

Advanced GRE requirements
The Advanced GRE is recommended.
University of Pittsburgh GRE school code is 2927; Department code for Physics and Astronomy is 0808

TOEFL requirements
The TOEFL exam is required for students from non-English-speaking countries.
  iBT score: 90
Effective for 2013-14 applicants, required a score of 90 (with at least a score of 22 in all of the four sections of speaking, reading, listening, writing. The IELTS* minimum score is 7, with at least a 6.5 score in each of its four sections. Students admitted to the University with these scores are required to take an English Language Proficiency Test during orientation.

Other admissions information
Additional requirements: IELTS International Students See http://www.ois.pitt.edu/.

TUITION

Tuition year 2012–13:
Tuition for in-state residents
  Full-time students: $9,668 per semester
  Part-time students: $782 per credit
Tuition for out-of-state residents
  Full-time students: $15,829 per semester
  Part-time students: $1,295 per credit
Fees are per term and subject to change.
Credit hours per semester to be considered full-time: 9
Deferred tuition plan: Yes
Health insurance: Yes, $0 if eligible appointment.
Other academic fees: Mandatory fees per term Full-time $370 Part-time $200 Student Health Service—$85 (full-time) Activities—$20 (full-time)/$10 (part-time) Computing & Network Services—$175 (full-time)/$100 (part-time) Security, Safety & Transportation—$90 (full-time and part-time)
Academic term: Semester
Number of first-year students who received full tuition waivers: 20

Teaching Assistants, Research Assistants, and Fellowships
Number of first-year
  Teaching Assistants: 17
  Fellowship students: 3
Average stipend per academic year
  Teaching Assistant: $24,450
  Research Assistant: $24,450
  Fellowship student: $29,559
All newly admitted doctoral students receive funding in the form of TA, GSR, or Fellowship support. Our fellowship offers exceed the number show accepting. Rates for 2013-14 are not available until mid-summer 2013.

FINANCIAL AID

Application deadlines

Fall admission:

U.S. students: January 31   *Int'l. students*: January 31

Loans

Loans are available for U.S. students.

Loans are not available for international students.

GAPSFAS application required: No

FAFSA application required: No

For further information

Address financial aid inquiries to: In reference to loan information:, Financial Aid Office, 4227 Fifth Ave., Alumni Hall, University of Pittsburgh, Pittsburgh, PA 15260. Otherwise, please see the link to the Department's Graduate Program on the Department's website.

Phone: 412-624-7488

E-mail: oafa@pitt.edu

Financial aid website: http://www.oafa.pitt.edu/fahome.aspx

HOUSING

Availability of on-campus housing

Single students: No

Married students: No

For further information

Address housing inquiries to: (For Off Campus), Dept. Property Management, 127 N. Bellefield Ave., Pittsburgh, PA 15213, (For On Campus), Panther Central, Litchfield Towers Lobby, 412-648-1100, pc@bc.pitt.edu.

Phone: (412) 624-9900

Housing aid website: http://www.pitt.edu/~property

Table A—Faculty, Enrollments, and Degrees Granted

Research Specialty	2012-13 Faculty	Enrollment Fall 2011–12		Number of Degrees Granted 2011–12 (2007–11)		
		Master's	Doctorate	Master's	Terminal Master's	Doctorate
Astrophysics	13	1	20	–	1(2)	2(10)
Condensed Matter Physics	23	–	42	–	2(22)	4(21)
Particles and Fields	19	–	15	–	–	1(8)
Physics and other Science Education	4	2	3	–	2(2)	2(5)
Total	59	3	80	–	3(21)	9(44)
Full-time Grad. Stud.	–	3	80	–	–	–
First-year Grad. Stud.	–	–	16	–	–	–

GRADUATE DEGREE REQUIREMENTS

Master's: Candidates for the M.S. degree must satisfy the Preliminary Evaluation, which requires the successful completion of at least one course in each of the following core subjects: Dynamical Systems, Statistical Mechanics and Thermodynamics, Electricity and Magnetism, and Quantum Mechanics, with a final examination score of at least 50% for courses at the graduate level or 75% for courses at the advanced undergraduate level. M.S. candidates may elect one of three alternative options to earn the degree: (1) Submit a thesis and successfully complete at least six courses (at least four must be at the graduate level and the balance at the advanced undergraduate level); (2) Submit no thesis but successfully complete at least eight courses (at least four must be at the graduate level and the balance at the advanced undergraduate level);

(3) Submit no thesis but successfully complete at least six courses at the graduate level. M.S. students must maintain a grade point average of at least 3.00, which corresponds to a B average in all of their courses. The core subject courses require a minimum grade of B. There is no foreign language requirement. There is a residence requirement of two full terms with a total of 24 credits.

Doctorate: Ph.D. students must successfully complete the following six graduate-level core courses: Dynamical Systems (one term), Statistical Mechanics and Thermodynamics (one term), Classical Electricity and Magnetism (one term), Mathematical Methods (one term), and Non-relativistic Quantum Mechanics (two terms). Exemptions from any of these courses may be granted if a student has successfully completed an equivalent course elsewhere. Students must complete these core courses with a grade point average of at least 3.00, which corresponds to a B average; they must also maintain a GPA of at least 3.00 in all of their graduate courses. In order to satisfy the Ph.D. Comprehensive Examination requirement, students must achieve a score of at least 60% on the final examination in each of the six core courses. This requirement must be fulfilled within the first two years unless an extension is granted. After passing the Ph.D. Comprehensive Examination, the student must find a research advisor and begin the process that leads to Admission to Candidacy and ultimately to the preparation and defense of a satisfactory dissertation. All Ph.D. students are required to serve for two terms as a Teaching Assistant in introductory undergraduate laboratories or recitations. An exemption may be granted if a student has substantial prior teaching experience. There is no foreign language requirement. There is a residence requirement of six full terms, with a total of 72 credit hours. Under some circumstances prior graduate work may be transferred from another institution.

Other Degrees: Interdisciplinary research programs may be arranged on a case-by-case basis. There have been Physics Doctorates awarded for work done in collaboration with the faculty members in the Chemistry Department, the Mathematics Department, the Materials Science Department, the Electrical and Chemical Engineering Departments, the Department of Biological Sciences, the Department of Computational Biology and the Department of Radiology in the School of Medicine, among others.

SPECIAL EQUIPMENT, FACILITIES, OR PROGRAMS

The Department of Physics and Astronomy is located on the University of Pittsburgh main campus in a complex of five interconnecting buildings. The department is currently undergoing a large-scale renovation project that will both develop new high, quality laboratories and improve existing laboratories. In addition to new laboratories, new departmental infrastructure for faculty, staff, and students is being created and improved. This includes areas to house our newly created PITTsburgh Particle Physics, Astrophysics, and Cosmology Center (PITT PACC) and a new student resource room and Physics Exploration Center (PEC). The department facilities include three professionally staffed shops (a machine shop, an electronics shop, and a glass-blowing shop). The department has its own extensive computer resources, a professional computer consultant, and access to University computer resources, which include the Center for Simulation and Modeling (SAM, www.sam.pitt.edu). Departmental students have access to the facilities and expertise available at the Gertrude E. and John M. Peterson Institute of NanoScience and Engineering (PINSE, www.nano.pitt.edu) and its Nano Fabrication and Characterization Facility (NFCF, www.nano.pitt.edu/nfcf.html), as well as the Pittsburgh Supercomputing Center (PSC, www.psc.edu). Other local facilities include University of Pittsburgh's Allegheny Observatory (AO, www.pitt.edu/~aobsvtry).

Experiments in particle physics are carried out at national and international facilities such as Fermilab near Chicago, CERN in Switzerland and J–PARC in Japan. This includes, for example: the Large Hadron Collider ATLAS experiment at CERN and various neutrino experiments (MINOS, MINERvA, and T2K).

Similarly, observational programs in astrophysics and cosmology are conducted at national and international ground-based observatories located at, for example: Kitt Peak and Mount Hopkins, in Arizona, Cerro Tololo in Chile, Mauna Kea in Hawaii, and Apache Point in New Mexico for collection of Sloan Digital Sky Survey data. Faculty also makes use of space-based telescopes, for example: the Hubble Space Telescope; the Chandra X-Ray Telescope; and the GALEX UV Telescope. University of Pittsburgh faculty are also members of several current and/or future large-telescope consortia: the Sloan Digital Sky Survey (SDSS, www.sdss.org), the Atacama Cosmology Telescope (ACT, www.physics.princeton.edu/act/), the Panoramic Survey Telescope & Rapid Response System (Pan-STARRS, www.ps1sc.org), and the Large Synoptic Survey Telescope (LSST, www.lsst.org).

Table B—Separately Budgeted Research Expenditures by Source of Support

Source of Support	Departmental Research	Physics-related Research Outside Department
Federal government	$5,577,003	
State/local government		
Non-profit organizations	$500,446	
Business and industry		
Other		
Total	$6,077,449	

Table C—Separately Budgeted Research Expenditures by Research Specialty

Research Specialty	No. of Grants	Expenditures ($)
Astrophysics	23	$855,377
Condensed Matter Physics	34	$3,637,545
Particles and Fields	22	$1,387,838
Other	4	$196,689
Total	83	$6,077,449

FACULTY

Professor

Boudreau, Joseph, Ph.D., University of Wisconsin-Madison, 1991. *High Energy Physics, Particles and Fields.* Experimental particle physics.

Boyanovsky, Daniel, Ph.D., University of California, Santa Barbara, 1982. *Cosmology & String Theory, Particles and Fields, Relativity & Gravitation, Statistical & Thermal Physics.* Theoretical condensed matter physics, particle astrophysics, astrophysics, and cosmology.

Coalson, Rob, Ph.D., Harvard University, 1984. *Chemical Physics.*

Duncan, H. E. Anthony, Ph.D., Massachusetts Institute of Technology, 1975. *High Energy Physics, Particles and Fields.* Theoretical particle physics.

Dytman, Steven A., Ph.D., Carnegie Mellon University, 1978. *High Energy Physics, Particles and Fields.* Experimental particle physics, experimental neutrino physics.

Han, Tao, Ph.D., University of Wisconsin-Madison, 1990. Director of the PITTsburgh Particle physics, Astrophysics and Cosmology Center (PITT PACC). *Particles and Fields.* Theoretical particle physics.

Hillier, D. John, Ph.D., Australian National University, 1984. *Astrophysics, Computational Physics.* Theoretical and observational astrophysics.

Jasnow, David M., Ph.D., University of Illinois, 1969. *Biophysics, Condensed Matter Physics, Fluids, Rheology, Nonlinear Dynamics and Complex Systems, Polymer Physics/Science, Statistical & Thermal Physics.* Theory of phase transitions, statistical physics, biological physics.

Kosowsky, Arthur, Ph.D., University of Chicago, 1994. Associate Director of the PITTsburgh Particle physics, Astrophysics and Cosmology Center (PITT PACC). *Astrophysics, Cosmology & String Theory, Relativity & Gravitation.* Theoretical and experimental cosmology and astrophysics.

Levy, Jeremy, Ph.D., University of California, Santa Barbara, 1993. *Applied Physics, Condensed Matter Physics, Low Temperature Physics, Materials Science, Metallurgy, Nano Science and Technology, Optics.* Experimental condensed matter physics, nanoscience, quantum information.

Maher, James V., Ph.D., Yale University, 1969. *Condensed Matter Physics, Statistical & Thermal Physics.* Experimental solid state physics, critical phenomena, physics of fluids.

Petek, Hrvoje, Ph.D., University of California, Berkeley, 1985. Co-Director PINSE (Peterson Institute of Nanoscience and Engineering). *Atomic, Molecular, & Optical Physics, Chemical Physics, Condensed Matter Physics, Nano Science and Technology.* Experimental condensed matter/AMO physics, nanoscience, solid-state physics.

Roskies, Ralph Z., Ph.D., Princeton University, 1966. Co-Director of the Pittsburgh Supercomputing Center. *Computational Physics, High Energy Physics, Particles and Fields.* Theoretical particle physics, use of computers in theoretical physics.

Schulte-Ladbeck, Regina, Ph.D., Heidelberg University, 1985. *Astronomy, Astrophysics, Cosmology & String Theory.* Extragalactic astronomy, observational cosmology.

Singh, Chandralekha, Ph.D., University of California, Santa Barbara, 1993. *Physics and other Science Education.* Polymer physics, physics education research.

Snoke, David W., Ph.D., University of Illinois, 1990. *Applied Physics, Atomic, Molecular, & Optical Physics, Biophysics, Condensed Matter Physics, Low Temperature Physics, Nano Science and Technology, Optics, Statistical & Thermal Physics.* Experimental condensed matter physics, solid state physics, nanoscience.

Turnshek, David A., Ph.D., University of Arizona, 1981. Chair of the Department, Director of Allegheny Observatory. *Astronomy, Astrophysics.* Extragalactic astronomy, observational cosmology.

Wu, Xiao-Lun, Ph.D., Cornell University, 1987. *Biophysics.* Experimental condensed matter physics, experimental biological physics.

Yang, Judith, Ph.D., Cornell University, 1993. *Materials Science, Metallurgy.* Materials science and engineering.

Associate Professor

Devaty, Robert P., Ph.D., Cornell University, 1983. Chair of the Graduate Admissions Committee. *Condensed Matter Physics.* Solid state physics, semiconductor physics.

Leibovich, Adam, Ph.D., California Institute of Technology, 1997. Associate Chair of the Department. *High Energy Physics, Particles and Fields, Relativity & Gravitation.* Theoretical particle physics.

Liu, W. Vincent, Ph.D., University of Texas, Austin, 1999. *Condensed Matter Physics, Low Temperature Physics.* Theoretical condensed matter physics, cold atoms.

Mueller, James A., Ph.D., Cornell University, 1989. Director of the Undergraduate Program. *High Energy Physics, Particles and Fields.* Experimental particle physics.

Naples, Donna, Ph.D., University of Maryland, 1993. *Particles and Fields*. Experimental neutrino physics.

Newman, Jeffrey, Ph.D., University of California, Berkeley, 2000. *Astronomy, Astrophysics, Cosmology & String Theory*. Extragalactic astronomy, observational cosmology.

Paolone, Vittorio, Ph.D., University of California, Davis, 1990. *Particles and Fields*. Experimental particle physics, experimental neutrino physics.

Savinov, Vladimir, Ph.D., University of Minnesota, 1996. *High Energy Physics, Particles and Fields*. Experimental particle physics.

Swanson, Eric, Ph.D., University of Toronto, 1991. *Particles and Fields*. Theoretical particle physics.

Zentner, Andrew, Ph.D., Ohio State University, 2003. Director of the Graduate Program. *Astronomy, Astrophysics, Cosmology & String Theory, Particles and Fields, Relativity & Gravitation*. Theoretical cosmology.

Assistant Professor

Badenes, Carles, Ph.D., Universitat Politecnicade de Catalunya, 2004. *Astronomy, Astrophysics, Cosmology & String Theory*. Type 1a supernovae, supernova remnants, large astronomical data bases, extragalactic astronomy, observational cosmology.

D'Urso, Brian, Ph.D., Harvard University, 2003. *Condensed Matter Physics, Nano Science and Technology*. Experimental condensed matter physics, nanoscience.

Daley, Andrew, Ph.D., University of Innsbruck, 2005. *Atomic, Molecular, & Optical Physics, Computational Physics, Condensed Matter Physics*. Theoretical many-body physics, quantum optics, AMO.

Dutt, Gurudev, Ph.D., University of Michigan, 2004. *Condensed Matter Physics, Nano Science and Technology*. Quantum optics, quantum information.

Freitas, Ayres, Ph.D., University of Hamburg, 2002. *Cosmology & String Theory, High Energy Physics, Particles and Fields*. Theoretical particle physics.

Frolov, Sergey, Ph.D., University of Illinois, 2005. *Condensed Matter Physics, Nano Science and Technology*. Experimental condensed matter physics, quantum nanowires, Majorana fermions in nanowires, and nanowire quantum bits.

Pekker, David, Ph.D., University of Illinois @ Urbana-Champaign, 2007. *Atomic, Molecular, & Optical Physics, Computational Physics, Condensed Matter Physics*.

Salman, Hanna, Ph.D., Weizmann Institute of Science, 2002. *Applied Physics, Biophysics, Condensed Matter Physics, Nonlinear Dynamics and Complex Systems, Statistical & Thermal Physics*. Experimental biological physics.

Wood-Vasey, Michael, Ph.D., University of California, Berkeley, 2004. *Astronomy, Astrophysics, Cosmology & String Theory*. Extragalactic astronomy, observational cosmology.

Emeritus

Cleland, Wilfred E., Ph.D., Yale University, 1964. *High Energy Physics, Particles and Fields*. Experimental particle physics.

Engels, Eugene, Ph.D., Princeton University, 1962. *High Energy Physics*. Experimental particle physics.

Gerjuoy, Edward, Ph.D., University of California, Berkeley, 1942. *Atomic, Molecular, & Optical Physics*. Quantum computing/information.

Goldburg, Walter I., Ph.D., Duke University, 1955. *Condensed Matter Physics, Fluids, Rheology, Nonlinear Dynamics and Complex Systems, Statistical & Thermal Physics*.

Janis, Allen I., Ph.D., Syracuse University, 1957. *Relativity & Gravitation, Other*. History and philosophy of science.

Johnsen, Rainer, Ph.D., University of Kiel, 1966. *Atmosphere, Space Physics, Cosmic Rays, Atomic, Molecular, & Optical Physics, Chemical Physics, Fluids, Rheology, Plasma and Fusion*.

Koehler, Peter F. M., Ph.D., University of Rochester, 1967. *High Energy Physics, Particles and Fields, Physics and other Science Education*. Experimental particle physics, physics education research.

Newman, Ezra T., Ph.D., Syracuse University, 1956. *Relativity & Gravitation*.

Shepard, Paul, Ph.D., Princeton University, 1969. *High Energy Physics, Particles and Fields*. Experimental particle physics.

Vincent, C. Martin, Ph.D., University of the Witwatersrand, South Africa, 1966. *Nuclear Physics*.

Winicour, Jefferey, Ph.D., Syracuse University, 1964. *Astrophysics, Relativity & Gravitation*. General relativity, numerical relativity.

Research Professor

Choyke, W. James, Ph.D., Ohio State University, 1952. *Condensed Matter Physics*. Experimental solid state physics, defect states in semiconductors, large-bandgap spectroscopy.

Pratt, Richard H., Ph.D., University of Chicago, 1959. *Atomic, Molecular, & Optical Physics*. Theoretical atomic physics.

Rao, Sandhya, Ph.D., University of Pittsburgh, 1994. *Astronomy, Astrophysics, Cosmology & String Theory*. Extragalactic astronomy, observational cosmology.

Research Assistant Professor

Danko, Istvan, Ph.D., Vanderbilt University, 2001. *Particles and Fields*. Experimental neutrino physics.

Feng, Min, Ph.D., Chinese Academy of Sciences, 2005. *Condensed Matter Physics*. Experimental condensed matter physics.

Irvin, Patrick, Ph.D., University of Pittsburgh, 2009. *Condensed Matter Physics*. Experimental condensed matter physics.

Adjunct Associate Professor

Zhao, Jin, Ph.D., University of Science and Technology of China, 2003. *Condensed Matter Physics, Nano Science and Technology*.

Lecturer

Nero, David, Ph.D., University of Toledo, 2010. *Astronomy, Physics and other Science Education*.

Lecturer / Lab Supervisor

Broccio, Matteo, Ph.D., University of Messina, 2005. *Biophysics, Physics and other Science Education*. Physics education, experimental biophysics.

Clark, Russell, Ph.D., Louisiana State University, 1997. *Physics and other Science Education*. Neutrino physics.

DEPARTMENTAL RESEARCH SPECIALTIES AND STAFF

Theoretical

Astrophysics. Astrophysics and Cosmology. Early universe physics; dark matter and dark energy; theoretical and numerical cosmology; model stellar atmospheres; massive stars; supernovae; gravitational lensing; General Relativity and gravitation; numerical relativity; gravitational radiation; black hole physics; plasma physics. Badenes, Boyanovsky, Hillier, Kosowsky, Ezra Newman, Turnshek, Winicour, Zentner.

Condensed Matter Physics. Phase transitions; disordered systems; nonequilibrium behavior; polymer physics; biological physics; atomic cold gases; superconductivity; quantum kinetics, atomic, molecular, and optical physics. Boyanovsky, Coalson, Daley, Gerjuoy, Jasnow, Liu, Pratt, Zhao.

Particles and Fields. Gauge field theories; lattice calculations; nonperturbative effects; weak interaction models and phenomenology; heavy-quark physics; supersymmetry; QCD modeling. Boyanovsky, Duncan, Freitas, Han, Leibovich, Roskies, Swanson.

Experimental

Astrophysics. Astronomy, Astrophysics, and Cosmology. Local and distant galaxies; active galactic nuclei and quasars; studies of the interstellar medium, circumgalactic medium, and intergalactic medium using quasar absorption line systems; statistical analysis of the properties of galaxies; clustering and large-scale structure; dark matter and dark energy; cosmic microwave background; supernovae; massive stars; stellar atmospheres. Observations take place with ground-based telescopes around the world and with space telescopes. Badenes, Hillier, Kosowsky, Jeffrey Newman, Rao, Schulte-Ladbeck, Turnshek, Wood-Vasey, Zentner.

Condensed Matter Physics. Nanoscience; quantum information; quantum optics; quantum states of matter; semiconductor physics; soft condensed matter physics; superconductivity and superfluidity; ultrafast optics; atomic, molecular, and optical physics; biological physics; turbulence. Experimental work takes place on campus in the individual labs of faculty members, at the Peterson Institute for Nanoscience and Engineering (PINSE), and at the Nano Fabrication and Characterization Facility (NFCF). Choyke, D'Urso, Devaty, Dutt, Feng, Frolov, Goldburg, Irvin, Johnsen, Levy, Maher, Petek, Salman, Snoke, Wu, Yang.

Particles and Fields. Particle Physics. Origin of mass and flavor; search for new symmetries of nature; neutrino physics; CP violation; heavy quarks; leptoquarks; supersymmetry; extra dimensions; baryogenesis. Studies take place at the Tevatron proton-antiproton collider, located at the Fermi National Accelerator Laboratory, and at the Large Hadron Collider ATLAS detector, located at CERN. Studies at the LHC may uncover the elusive Higgs boson as well as a spectrum of new particles arising from "supersymmetry." Studies of fundamental properties of neutrinos, such as oscillations, mass differences, and neutrino-nucleus interactions take place at a variety of locations. Boudreau, Cleland, Danko, Dytman, Mueller, Naples, Paolone, Savinov, Shepard.

Physics and other Science Education. Physics Education Research. Identification of sources of student difficulties in learning concepts in both introductory and advanced-level physics courses; design, implementation, and outcome assessment of changes in curricular offerings; pedagogical methods that are designed to reduce learning difficulties. Broccio, Clark, Koehler, Singh.

View additional information about this department at
www.gradschoolshopper.com

BROWN UNIVERSITY

DEPARTMENT OF PHYSICS

Providence, Rhode Island 02912
http://www.brown.edu/academics/physics/

General University Information
President: Christina Hull Paxson
Dean of Graduate School: Peter Weber
University website: http://brown.edu
Control: Private
Setting: Urban
Total Faculty: 713
Total Graduate Faculty: 713
Total number of Students: 8,540
Total number of Graduate Students: 1,947

Department Information
Department Chairman: James M. Valles, Jr., Chair
Department Contact: Barbara Dailey, Student Affairs & Programs Manager
Total full-time faculty: 27
Total number of full-time equivalent positions: 27
Full-Time Graduate Students: 93
First-Year Graduate Students: 17
Female First-Year Students: 3
Total Post Doctorates: 13

Department Address
182 Hope Street
Providence, RI 02912
Phone: (401) 863-1434
Fax: (401) 863-2024
E-mail: Barbara_Dailey@Brown.edu
Website: http://www.brown.edu/academics/physics/

ADMISSIONS

Admission Contact Information
Address admission inquiries to: Graduate School: admission_Graduate@brown.edu, Physics:, physics_admission@brown.edu
Phone: 401-863-2600
E-mail: admission_Graduate@brown.edu
Admissions website: http://www.brown.edu/academics/gradschool/application-information

Application deadlines
Fall admission:
U.S. students: January 6   *Int'l. students*: January 6

Application fee
U.S. students: $75   *Int'l. students*: $75

The deadline listed is for the PhD program. The ScM program has rolling admissions (allowing for a possible Spring semester start) with an academic year application deadline of May 1.

Admissions information
For Fall of 2013:
Number of applicants: 326
Number admitted: 52
Number enrolled: 17

Admission requirements
Bachelor's degree requirements: Bachelor's degree in physics or related field is required. Applicants are expected to have a strong background in physics or closely related subjects.

GRE requirements
The GRE is required.
For PhD program, GRE General is required. For ScM program, GRE General is recommended.

Advanced GRE requirements
The Advanced GRE is recommended.
For both PhD and ScM program, GRE Subject is recommended.

TOEFL requirements
The TOEFL exam is required for students from non-English-speaking countries.
PBT score: 577
iBT score: 90
TOEFL is not required for one whose native language is not English but who has received a degree from a university in which English is the primary language of instruction. More information: http://www.brown.edu/academics/gradschool/application-information/international-applicants/language-proficiency-toefl-or-ielts

Other admissions information
Additional requirements: The Ph.D. program provides students with opportunities to perform independent research in some of the most current and dynamic areas of physics. Three letters of recommendation are required.
The ScM program is suitable as both a means for professional development and preparation for further graduate study. The program offers enough flexibility to allow for completion of the degree in two, three, or four semesters of full time enrollment, depending on a student's background. Two recommendation letters are required.
Undergraduate preparation assumed: Undergraduate requirements flexible to some extent; preference given for strong upper-class study in mechanics, E&M, wave theory, modern physics, and mathematics through partial differential equations. Purcell, Electricity and Magnetism; Schey, Div, Grad, Curl and All That; Feynman, The Feynman Lectures on Physics, Vol. II recommended; French and Taylor, An Introduction to Quantum Physics; Marion, Classical Electromagnetic Radiation; Gasiorowicz, Quantum Physics; Reif, Fundamentals of Statistical and Thermal Physics; French, Vibrations and Waves; Kibble, Classical Mechanics (2nd. ed. or later).

TUITION
Tuition year 2013-14:
Tuition for in-state residents
Full-time students: annual
Part-time students: per credit
Full-time students: $44,608 annual
Part-time students: $5,576 per credit
Average stipends listed below are for AY13-14 and do not include guaranteed summer funding. The Graduate School offers incoming doctoral students five years of guaranteed financial support, including stipend, tuition remission, health-services fee, and health-insurance subsidy. More information:

http://www.brown.edu/academics/gradschool/financing-support/phd-funding The majority of students enrolled in master's programs are self-supported. Students may be eligible for federal direct student loans and other loans administered through the Office of Financial Aid. Master's students are also eligible for conference travel funds. More information: http://www.brown.edu/academics/gradschool/financing-support/masters-funding More information on graduate student financing and support at Brown: http://www.brown.edu/academics/gradschool/financing-support
Credit hours per semester to be considered full-time: 4
Deferred tuition plan: Yes
Health insurance: Available
Other academic fees: $54 Activity Fee $64 Recreation Fee
Academic term: Semester
Number of first-year students who received full tuition waivers: 11

Teaching Assistants, Research Assistants, and Fellowships
Number of first-year
Teaching Assistants: 9
Research Assistants: 1
Fellowship students: 1
Average stipend per academic year
Teaching Assistant: $22,200
Research Assistant: $22,200
Fellowship student: $22,200
All Ph.D. students are supported as fellows, TA's or RA's regardless of the number of students enrolled in any particular year. Average stipends listed are for AY13-14 and do not include guaranteed summer funding. Summer stipends range from $2,500 to 1/3 of the academic year stipend, depending on a number of factors.

FINANCIAL AID

Loans
Loans are available for U.S. students.
Loans are available for international students.
GAPSFAS application required: No
FAFSA application required: Yes

For further information
Address financial aid inquiries to: Office of Financial Aid, Brown University, Box #1827, Providence, RI 02912.
Phone: (401) 863-2721
E-mail: GS_Financial_Aid@brown.edu
Financial aid website: http://www.brown.edu/about/administration/financial-aid/contact-information

HOUSING

Availability of on-campus housing
Single students: Yes
Married students: Yes

For further information
Address housing inquiries to: Off-Campus:, Office of Auxiliary Housing, Brown University, Box 1902, Providence, RI 02912, (401) 863-2541, On-campus:, Office of Residential Life, Box 1864, Brown University, Providence, RI 02912, Telephone: (401) 863-3500, Res_Life@brown.edu.
Phone: (401) 863-3500
E-mail: Res_Life@brown.edu
Housing aid website: http://www.brown.edu/academics/gradschool/graduate/housing

Table A—Faculty, Enrollments, and Degrees Granted

Research Specialty	Faculty	Enrollment Fall 2012		Number of Degrees Granted 2012-13 (2008-13)		
		Master's	Doctorate	Master's	Terminal Master's	Doctorate
Astrophysics	4	–	15	–	–	2(6)
Biophysics	3	–	11	–	–	–(13)
Condensed Matter						
Physics	10	2	34	–	–	3(37)
Particles and Fields	9	1	19	–	–	5(21)
Non-specialized	–	–	11	17(-)	2(-)	–
Other	–	–	–	–	–	–(1)
Total	26	3	90	17(64)	2(4)	10(78)
Full-time Grad. Stud.	–	3	90	–	–	–
First-year Grad. Stud.	–	–	9	–	–	–

GRADUATE DEGREE REQUIREMENTS

Master's: Approved sequence of eight semester courses. Of the eight required courses, four will be selected from the six core courses (PHYS2010, 2030, 2040, 2050, 2060, 2140). Because preparation of a Master's thesis is highly recommended, as it forms an important pillar of the professional training, one of the eight required courses may be Thesis Preparation. Three additional credits at the 2000 level are required.

Doctorate: Equivalent of three full-time years. Six one-semester core courses: quantum mechanics (2), classical physics (mechanics and electricity & magnetism) (2), experimental techniques (1), and statistical mechanics (1). Also, four advanced courses in area of research specialization. Required exams: qualifying (Sem. 3), preliminary (Sem. 6); also required: thesis defense (oral).

Thesis: Preparation of a Master's thesis is highly recommended. Ph.D. written dissertation and oral defense is required.

SPECIAL EQUIPMENT, FACILITIES, OR PROGRAMS

(a) Brown is a member of Universities Research Association, Inc. (URA), part of the Fermilab Research Alliance, which operates the Fermi National Accelerator Laboratory (FNAL) in Batavia, Illinois, and other facilities. Brown physicists are involved in the D0 experiment at the 2 TeV Tevatron collider located at FNAL, as well as in the CMS experiment at the 14 TeV Large Hadron Collider (LHC) located at CERN, in Geneva, Switzerland. Brown leads experimental collaborations which operate rare particle search experiments at international underground laboratories at Gran Sasso, Italy, and Sanford Lab, Homestake Mine, South Dakota. Brown in involved with ground-based, balloon-borne, and satellite-based cosmology and astrophysics experiments. Researchers at Brown are collaborating on telescope projects in Arizona and Chile, and balloon flights launched from Texas and the Antarctic. Using equipment designed and built at Brown, as well as the National Laboratories, data are recorded in experimental runs and analyzed at Brown with extensive use of computer systems.

(b) The Physics Department is an active participant in Brown's Institute for Molecular and Nanoscale Innovation (IMNI), an umbrella organization that supports centers and collaborative research teams in targeted areas of the molecular and nanosciences. IMNI is a "polydisciplinary" venture, with 60 faculty participants representing nine departments across campus. IMNI serves as a focal point for interaction with industry, government, and affiliated hospitals.

(c) Physics is also associated with the Institute for Brian and Neural Systems and the Center for Biomedical Engineering at Brown.

(d) Extensive computer facilities are available. These include a variety of Windows and Linux/UNIX workstations within the department, all of which are connected via Ethernet. In addition, the department has several powerful Zinux clusters for dealing with problems needing large scale computation. Several high-speed network links provide worldwide access to experimental facilities and enable extensive and efficient use of national supercomputing centers. A department web server provides access to personal home pages of faculty, staff, and students, as well as general departmental information.

Table B—Separately Budgeted Research Expenditures by Source of Support

Source of Support	Departmental Research	Physics-related Research Outside Department
Federal government	$6,893,007	
State/local government		
Non-profit organizations	$214,688	
Business and industry	$542,745	
Other	$1,953,253	
Total	$9,603,693	

Table C—Separately Budgeted Research Expenditures by Research Specialty

Research Specialty	No. of Grants	Expenditures ($)
Astrophysics	17	$1,651,317
Biophysics	10	$811,723
Condensed Matter Physics	19	$1,794,245
Particles and Fields	31	$5,346,408
Total	77	$9,603,693

FACULTY

Chair Professor

Valles, James M., Ph.D., University of Massachusetts, 1988. Experimental condensed matter physics, biological physics.

Professor

Cooper, Leon N., Ph.D., Columbia University, 1954. Neural studies; theoretical condensed matter physics.

Cutts, David, Ph.D., University of California, Berkeley, 1968. Experimental high-energy physics.

Gaitskell, Richard, Ph.D., University of Oxford, 1993. Experimental high-energy physics.

Guralnik, Gerald S., Ph.D., Harvard University, 1964. Theoretical high-energy physics.

Heintz, Ulrich, Ph.D., Stony Brook University, 1991. Experimental high-energy physics.

Jevicki, Antal, Ph.D., City University of New York, 1976. Theoretical high-energy physics.

Kosterlitz, J. Michael, Ph.D., University of Oxford, 1969. Theoretical condensed matter physics.

Landsberg, Greg L., Ph.D., Stony Brook University, 1994. Experimental high-energy physics.

Ling, Xinsheng, Ph.D., University of Connecticut, 1992. Experimental condensed matter physics, biological physics.

Maris, Humphrey J., Ph.D., Imperial College, 1963. Experimental and theoretical condensed matter physics.

Marston, J. Bradley, Ph.D., Princeton University, 1989. Theoretical condensed matter physics.

Narain, Meenakshi, Ph.D., Stony Brook University, 1991. Experimental high-energy physics.

Pelcovits, Robert A., Ph.D., Harvard University, 1978. Theoretical condensed matter physics.

Tan, Chung-I, Ph.D., University of California, Berkeley, 1968. Theoretical high-energy physics.

Tucker, Gregory S., Ph.D., Princeton University, 1991. Experimental astrophysics.

Xiao, Gang, Ph.D., Johns Hopkins University, 1988. Experimental condensed matter physics.

Ying, See-Chen, Ph.D., Brown University, 1968. Theoretical condensed matter physics, biological physics.

Associate Professor

Dell'Antonio, Ian P., Ph.D., Harvard University, 1995. Experimental astrophysics.

Feldman, Dmitri, Ph.D., Landau Institute for Theoretical Physics, 1998. Theoretical condensed matter.

Lowe, David A., Ph.D., Princeton University, 1993. Theoretical high-energy physics.

Mitrovic, Vesna, Ph.D., Northwestern University, 2001. Experimental condensed matter.

Spradlin, Marcus, Ph.D., Harvard University, 2001. Theoretical high-energy physics.

Stein, Derek, Ph.D., Harvard University, 2003. Experimental condensed matter physics, biological physics.

Tang, Jay X., Ph.D., Brandeis University, 1995. Experimental condensed matter, biological physics.

Volovich, Anastasia, Ph.D., Harvard University, 2002. Theoretical high-energy physics.

Assistant Professor

Koushiappas, Savvas, Ph.D., Ohio State University, 2004. Experimental astrophysics.

Emeritus

Elbaum, Charles, Ph.D., University of Toronto, 1954. Neural Studies, experimental condensed matter physics.

Fried, Herbert M., Ph.D., Stanford University, 1957. Theoretical high-energy physics.

Lanou, Robert E., Ph.D., Yale University, 1957. Experimental high-energy physics.

Seidel, George M., Ph.D., Purdue University, 1958. Experimental condensed matter physics.

Research Professor

Oldenbourg, Rudolf, Ph.D., University of Konstanz, 1981. Condensed matter physics.

Research Assistant Professor

Piperov, Stefan, Ph.D., Bulgarian Academy of Science, 2010.

Speer, Thomas, Ph.D., University of Geneva, 2000. Experimental high-energy physics.

Adjunct Professor

Lawandy, Nabil, Ph.D., Johns Hopkins University, 1980. Experimental quantum optics.

Nurmikko, Arto W., Ph.D., University of California, Berkeley, 1971. Experimental semiconductor physics.

Powers, Thomas R., Ph.D., University of Pennsylvania, 1995.

Stratt, Richard M., Ph.D., University of California, Berkeley, 1979. Theoretical condensed matter physics.

Xu, Jimmy, Ph.D., University of Minnesota, 1987. Experimental condensed matter physics.

Zaslavsky, Alexander, Ph.D., Princeton University, 1991. Experimental condensed matter physics.

Adjunct Associate Professor

Dickerson, II, James H., Ph.D., SUNY at Stony Brook, 2002.

Targan, David, Ph.D., University of Minnesota, 1988. *Astronomy.* Astronomy.

Adjunct Assistant Professor

Adetunji, Oludurotimi O., Ph.D., Ohio State University, 2008.

Zia, Rashid, Ph.D., Stanford University, 2005.

Senior Research Scientist

Fiorucci, Simon, Ph.D., Universite Paris-XI Orsay and CEA Saclay, 2005.

Korotkov, Andrei, Ph.D., Institute of Applied Physics, Russian Academy of Sciences, 1995.

DEPARTMENTAL RESEARCH SPECIALTIES AND STAFF

Theoretical

Astrophysics and Cosmology. Cosmological models for structure formation and particle-physics predictions of the nature of the dark matter are analyzed. Computational and analytic tools are used to predict the distribution of matter on sub- and super-galaxy scales and to aid in the design of the next generation of cosmological experiments. Koushiappas.

Neural Science. The major goal of the research is to elucidate the biological mechanisms that underlay learning and memory: to find principles of organization that can account both for experimental data on the cellular level. Among the detailed objectives are the following: to clarify the dependence of learning on synaptic modification; to elucidate the principles that govern synapse formation or modification; to use principles of organization that can account for observations on a cellular level to construct network models that can learn, associate and reproduce such higher level cognitive acts as abstraction, computation, and language acquisition. Cooper.

Physics of Condensed Matter. Research problems currently under investigation include interference and interaction in mesoscopic systems including the quantum Hall effect, quantum wires and quantum phase transitions; strongly correlated electrons in layered materials; development of a non-equilibrium statistical mechanics of planetary climates; non-equilibrium transport in nanostructures; modeling actinide complexes in aqueous solution; liquid crystal physics and its interface with biology; dynamics of biopolymers in nanochannels; strain relaxation and dynamics of heteroepitaxial nanostructures; microscopic theories of friction; out of equilibrium systems in the presence of weak stochastic noise; ultrafast dynamics in liquids. Feldman, Kosterlitz, Marston, Pelcovits, Stratt, Ying.

Physics of Elementary Particles. Current activities include studies in quantum field theory, quantum chromodynamics, gauge/gravity duality, nonperturbative methods in field theory, solitons, monopoles, spontaneous symmetry breaking, lattice field theories, renormalization group, field theoretic approaches to condensed matter, gauge theories of weak and electromagnetic interactions, grand unification theory and phenomenology, phenomenology of scattering and production processes, the quantum theory of gravitation, super symmetry, supergravity, superstrings, and cosmology. Guralnik, Jevicki, Lowe, Spradlin, Tan, Volovich.

Experimental

Astrophysics and Cosmology. The origins and evolution of the universe are being measured. Topics of research include the following: Studies of the Cosmic Microwave Background from satellite, balloon-borne and ground-based missions, measurements of the CMB to measure properties of the early Universe. A parallel effort in sub-millimeter cosmology is being carried out using the BLAST balloon-borne observatory to study the epoch of formation of the fist galaxies and the dynamics of star formation in our own galaxy. Wide-field optical and near-imaging surveys are being carried out with telescopes in Arizona and Chile to map out the gravitational lensing signal and measure the shear correlation function and the growth of clustering over cosmic time to measure the evolution of the dark energy equation of state. Studies of mass substructure from gravitational lensing maps of clusters of galaxies taken with HST and ground-based telescopes are being used to measure the clustering properties of Dark Matter. Investigations using the next generation of wide-field survey instruments to map the galaxy gropu and luster distribution out to high redshift are being planned. Studies of the galaxy interaction and star formation properties through optical photometry, spectroscopy, NIR photometry and radio spectral line observations are being carried out. Dell'Antonio, Gaitskell, Koushiappas, Tucker.

Biological Physics. Research problems currently under investigation include: development of single-molecule DNA sequencing technology using solid-state nanopores; electro-fluidics for single molecule biophysics; electronic DNA barcode sequencing; electrokinetic energy harvesting in the presence of hydrodynamic slip; probing the sequence and dynamics of single DNA molecules using solid-state nanopores, optical tweezers, and binding proteins; DNA sequencing using nanopore mass spectrometry; biophysical mechanism of bacterial swimming and adhesion; biomechanics of actin networks regulated by physical mechanisms; mechanics of intracellular pathogens and biomimetic systems propelled by actin comet tails; neutrophil mechano-sensing using traction microscopy; swimming and force sensing on microorganisms. Ling, Stein, Tang, Valles.

Physics of Condensed Matter. Studies of magnetoconductive, optical, and mechanical properties of amorphous metals and semiconductors, magnetic solids and high Tc superconductors; mesoscopic superconducting arrays and colloidal model systems; flux lattices in type-II superconductors; low-temperature scanning probe techniques and devices; giant and colossal magnetoresistance effects in magnetic superlattices, granular solids and oxides; electron-electron interactions in two dimensional electron systems at low temperatures, electronic and magnetic properties of artificial superlattices, quantum wires and dots; the quantum Hall effect; nonlinear optical phenomena and plasma dynamics studies in semiconductors using picosecond and femtosecond laser pulses; studies of ultrasonic and thermal properties of solids using pico-second laser pulses; properties of liquid and solid 4He and 3He, including elementary excitations and their interactions, cavitation, and levitation of superfluid 4He; NMR studies of the structure and bonding of crystalline and glassy solids; nuclear quadrupole resonance (NQR) studies of electronic distributions in molecules of biological importance and in inorganic solids. Ling, Maris, Mitrovic, Nurmikko, Stein, Tang, Valles, Xiao, Xu, Zaslavsky.

Physics of Elementary Particles. The properties of elementary particles and their interactions are being investigated, with current effort focused on the study of proton-proton collisions at the highest available energy with the CMS experiment at the Large Hadron Collider at CERN and proton-antiproton collisions at the previous energy frontier facility: the DO experiment at Fermilab Tevatron accelerator. The CMS program is focused on searches for new particles, forces, and properties of space-time, beyond the predictions of the Standard Model of particle physics. That includes searches for supersymmetry and other heavy partners of the known particles, extra spatial dimensions, and new forces. In addition to this avenue to discovery, the CMS and DO programs include precision measurements of the properties of electroweak and strong interactions, in particular measurement of the top-quark properties. An important component of the current DO and near-future CMS program is the search for the last missing piece of the Standard Model-the Higgs boson. High-performance LHC Computing Grid networking and videoconferencing facilities provide tight links between Brown, CERN, and Fermilab. Local computer cluster connected to the Grid allows for massive parallel computing support of the D0 and CMS physics program. Cutts, Heintz, Landsberg, Narain.

View additional information about this department at www.gradschoolshopper.com

---

CLEMSON UNIVERSITY

DEPARTMENT OF PHYSICS AND ASTRONOMY

Clemson, South Carolina 29634-0978
http://www.clemson.edu/ces/physics-astro/index.html

General University Information
President: James F. Barker
Dean of Graduate School: Karen Burg
University website: http://www.clemson.edu
Control: Public
Setting: Urban
Total Faculty: 1,245
Total number of Students: 20,768
Total number of Graduate Students: 4,206

Department Information
Department Chairman: Mark Leising, Chair
Department Contact: Risé Sheriff, Office Manager
Total full-time faculty: 26

Total number of full-time equivalent positions: 26
Full-Time Graduate Students: 63
First-Year Graduate Students: 14
Female First-Year Students: 2
Total Post Doctorates: 7

Department Address
118 Kinard Laboratory
Clemson, SC 29634-0978
Phone: (864) 656-3416
Fax: (864) 656-0805
E-mail: risem@clemson.edu
Website: http://www.clemson.edu/ces/physics-astro/index.html

ADMISSIONS

Admission Contact Information
Address admission inquiries to: Dr. Murray Daw, Graduate Student Recruiter, Dept. of Physics and Astronomy
Phone: (864) 656-3419
E-mail: daw@clemson.edu
Admissions website: http://www.grad.clemson.edu/prospective Students.php

Application deadlines
Fall admission:
U.S. students: January 15   *Int'l. students*: January 15

Application fee
U.S. students: $80   *Int'l. students*: $90
International students should apply directly to the Graduate School for a special self-managed application package.

Admissions information
For Fall of 2013:
  Number of applicants: 90
  Number admitted: 28
  Number enrolled: 14

Admission requirements
Bachelor's degree requirements: Bachelor's degree is required.

GRE requirements
The GRE is required.

Advanced GRE requirements
The Advanced GRE is not required.

TOEFL requirements
The TOEFL exam is required for students from non-English-speaking countries.
  PBT score: 570
  iBT score: 90

Other admissions information
Additional requirements: Usual preparation is undergraduate major in physics. Students from other fields will have an opportunity to make up deficiencies.
Undergraduate preparation assumed: Courses are based upon texts such as Hecht, Optics; Griffiths, Introduction to Electrodynamics; Marion and Thornton, Mechanics, Eisberg; Modern Physics, Stowe, Introduction to Statistical Mechanics and Thermodynamics; Griffiths and Townsend, Quantum Physics. Included with these standard areas of study should be an advanced undergraduate laboratory in experimental physics, mathematics including differential equations, complex variable, Fourier analysis, and operational mathematics. Some knowledge of computer programming including standard methods using Mathematica, Maple, MatLab, etc., will also be helpful.

TUITION
Tuition year 2013-14:
Tuition for in-state residents
  Full-time students: $4,184 per semester
  Part-time students: $529 per credit
Tuition for out-of-state residents
  Full-time students: $8,335 per semester
  Part-time students: $1,060 per credit
Per semester is full-time (12 credit hours or more).
Credit hours per semester to be considered full-time: 12
Deferred tuition plan: Yes
Health insurance: Yes, $863 annually.
Other academic fees: Graduate assistantship fees, $1,030 per semester (fall and spring).
Academic term: Semester
Number of first-year students who received full tuition waivers: 14

Teaching Assistants, Research Assistants, and Fellowships
Number of first-year
  Teaching Assistants: 13
  Research Assistants: 1
  Fellowship students: 2
Average stipend per academic year
  Teaching Assistant: $19,000
  Research Assistant: $23,000
  Fellowship student: $24,000

FINANCIAL AID

Application deadlines
Fall admission:
U.S. students: March 1   *Int'l. students*: March 1

Loans
Loans are available for U.S. students.
Loans are not available for international students.
GAPSFAS application required: No
FAFSA application required: Yes

For further information
Address financial aid inquiries to: Clemson University, Financial Aid Counselor, G-01 Sikes Hall, Box 345123, Clemson, SC 29634.
Phone: (864) 656-2280
E-mail: finaid@clemson.edu
Financial aid website: http://www.clemson.edu/financial-aid/index.html

HOUSING

Availability of on-campus housing
  Single students: Yes
  Married students: No

For further information
Address housing inquiries to: University Housing, 200 Mell Hall, Box 344075, Clemson, SC 29634-4075.
Phone: (864) 656-2295
E-mail: housinginfo-1@clemon.edu
Housing aid website: http://www.clemson.edu/campus-life/housing/

Table A—Faculty, Enrollments, and Degrees Granted

Research Specialty	2012-13 Faculty	Enrollment Fall 2012-13 Master's	Enrollment Fall 2012-13 Doctorate	Degrees Master's	Degrees Terminal Master's	Degrees Doctorate
Astrophysics	6	–	17	–	7(2)	2(5)
Atmosphere, Space Physics, Cosmic Rays	4	–	6	–	–	–
Atomic, Molecular, & Optical Physics	1	–	–	–	–	–
Biophysics	3	–	8	–	–	3(1)
Condensed Matter Physics	8	–	25	–	5(3)	3(4)
Quantum Foundations	1	–	2	–	–	–
Non-specialized	–	–	5	–	–	–
Total	23	–	63	–	12(5)	8(10)
Full-time Grad. Stud.	–	–	63	–	–	–
First-year Grad. Stud.	–	–	14	–	–	–

GRADUATE DEGREE REQUIREMENTS

Master's: Thirty semester credits of coursework are required, including six credits of thesis research with an oral defense. For non-thesis degrees, 36 credits are needed. A grade point average of 3.0 is required. There are no foreign language requirements.

Doctorate: A core of six graduate physics courses is required, except for students who have a M.S. degree in physics. All students complete four advanced graduate courses related to their areas of research. Students must pass a set of Ph.D. qualifying examinations and a dissertation topic defense. Original research culminates in a dissertation that is defended before a faculty advisory committee. A 3.0 grade point average is required. There is no foreign language requirement.

Thesis: Thesis may be written in absentia.

SPECIAL EQUIPMENT, FACILITIES, OR PROGRAMS

The department is housed in the four-story 64,000 sq. ft. physics and astronomy building. A fully equipped research/instrument lab and computing facilities is available, along with a state-of-the-art planetarium. Office space is provided for graduate students. Extensive research facilities include an electron beam ion trap facility for highly charged ion beam production;an atomic molecular and optical physics lab; a scanning tunneling microscope nanomaterial processing laboratory with electric arc discharge, pulsed laser vaporization, and CVD synthesis capabilities; bulk and thin film thermoelectric materials growth facilities; Raman scattering, infrared/visible spectroscopy, electron microscopy, atomic force microscopy, and electrical transport measurements are used extensively for characterizing carbon nanotubes, nanodiamond, semiconducting oxide nanobelts and nanowires. Access is also available to the SARA and Super Lotis telescopes.

Table B—Separately Budgeted Research Expenditures by Source of Support

Source of Support	Departmental Research	Physics-related Research Outside Department
Federal government	$2,341,098	
State/local government		
Non-profit organizations	$10,337	
Business and industry		
Other	$221,238	
Total	**$2,572,673**	

Table C—Separately Budgeted Research Expenditures by Research Specialty

Research Specialty	No. of Grants	Expenditures ($)
Astrophysics	18	$512,383
Atmosphere, Space Physics, Cosmic Rays	12	$790,800
Biophysics	5	$493,486
Condensed Matter Physics	9	$776,004
Total	**44**	**$2,572,673**

FACULTY

Professor

Alexov, Emil, Ph.D., Sofia University, 1990. *Biophysics, Computational Physics*. Developing methods for modeling electrostatics in biological systems (DelPhi package); predicting effects of nsSNP on human health.

Daw, Murray S., Ph.D., California Institute of Technology, 1981. *Condensed Matter Physics, Solid State Physics, Theoretical Physics*. Solid-state theory; structure and dynamics of defects in solids.

Hartmann, Dieter H., Ph.D., University of California, Santa Cruz, 1989. *Astronomy, Astrophysics*. Gamma-ray astronomy; nucleosynthesis; galactic structure.

King, J. R., Ph.D., University of Hawaii, 1993. *Astronomy, Astrophysics*. Stellar abundances; stellar atmospheres; galactic populations; high-resolution spectroscopy.

Larsen, M. F., Ph.D., Cornell University, 1979. Atmosphere, Space Physics.

Leising, Mark D., Ph.D., Rice University, 1987. Department Chair. *Astronomy, Astrophysics*. Gamma-ray astronomy; supernovae.

Marinescu, D. C., Ph.D., Purdue University, 1996. *Condensed Matter Physics, Theoretical Physics*. Condensed-matter theory.

Meriwether, J. W., Ph.D., University of Maryland, 1970. *Optics*. Atmosphere, space physics; optics.

Meyer, Bradley S., Ph.D., University of Chicago, 1989. *Astronomy, Astrophysics, Theoretical Physics*. Nuclear astrophysics; supernova theory; cosmology.

Rao, Apparao M., Ph.D., University of Kentucky, 1989. *Condensed Matter Physics, Nano Science and Technology, Solid State Physics*. Condensed-matter physics; nanomaterial synthesis; mechanical properties; chem-bio sensing; solid-state spectroscopy.

Tritt, Terry M., Ph.D., Clemson University, 1985. *Condensed Matter Physics*. Experimental condensed-matter and materials physics: electrical and thermal transport phenomena.

Valentini, Antony, Ph.D., International School for Advanced Studies (SISSA), Trieste, 1992. *Astrophysics, Quantum Foundations, Theoretical Physics*. Foundations of quantum mechanics and astrophysics; cosmology; black holes.

Associate Professor

Brittain, Sean, Ph.D., University of Notre Dame, 2004. *Astronomy, Astrophysics*. Astrophysics; planet formation; circumstellar disks; spectroscopy.

Flower, Phillip J., Ph.D., University of Washington, Seattle, 1976. *Astronomy, Astrophysics*. Stellar evolution; star clusters.

Lehmacher, Gerald, Ph.D., University of Bonn, 1993. Atmosphere, Space Physics; Atmospheric physics; turbulence in the mesosphere and lower thermosphere, suborbital rocket instrumentation.

Oberheide, Jens, Ph.D., Wuppertal University, 2000. *Climate/Atmospheric Science*. Atmosphere,Space Physics; Atmospheric and geospace physics; climate and weather of the sun-earth system.

Sosolik, Chad E., Ph.D., Cornell University, 2001. *Condensed Matter Physics, Physics of Beams, Solid State Physics*. Experimental surface and highly charged ion beam physics.

Assistant Professor

Ding, Feng, Ph.D., Boston University, 2004. *Biophysics, Computational Physics*. Multiscale modeling of biomolecules and molecular complexes; Understanding the interface between nanomaterials and biology; and designing functional biomolecules, including proteins and RNA.

He, Jian, Ph.D., University of Tennessee, 2004. *Condensed Matter Physics*. Condensed-matter physics; single crystal growth and characterizations; functional nanocomposite materials.

Kang, Hye Jung, Ph.D., University of Tennessee, 2005. *Condensed Matter Physics*. Condensed-matter physics; experimental neutron scattering.

Marler, Joan P., Ph.D., University of California, San Diego, 2005. *Atomic, Molecular, & Optical Physics*. Experimental low temperature atomic and molecular ion physics; laser trapping and cooling; cold chemistry.

Tewari, Sumanta, Ph.D., University of California, Los Angeles, 2003. *Condensed Matter Physics, Theoretical Physics.* Condensed-matter theory; topological quantum computation; high-temperature superconductivity; cold atomic physics in optical traps and lattices.

Research Professor

Skove, Malcolm J., Ph.D., University of Virginia, 1960. Alumni; Professor Emeritus of Physics. *Condensed Matter Physics.* Superconductivity; transport effects in whiskers.

Lecturer

Brown, Jason, Ph.D., Clemson University, 1999. *Physics and other Science Education.*

Pope, Amy, Ph.D., Clemson University, 2002. *Condensed Matter Physics, Physics and other Science Education.* Condensed-matter physics; physics education.

The, Lih-Sin, Ph.D., University of Arizona, 1989. *Astronomy, Astrophysics, Physics and other Science Education.* Gamma-ray astronomy; supernova remnants; stellar nucleosynthesis.

DEPARTMENTAL RESEARCH SPECIALTIES AND STAFF

Theoretical

Astrophysics. Nucleosynthesis; space astrophysics; stellar atmospheres; cosmic rays; gamma-ray bursts; supernova theory; origin of solar system; stellar evolution; cosmology. Brittain, Flower, Hartmann, King, Leising, Meyer, The, Valentini.

Atmosphere, Space Physics, Cosmic Rays. Atmospheric wave dynamics, propagation, and interaction with chemical and airglow processes; ionospheric electrodynamics: plasma; climate and weather of the sun-earth system. Larsen, Lehmacher, Meriwether, Oberheide.

Biophysics. DNA repair mechanisms; quantum biology. Alexov, Ding.

Computational Physics. Computational biophysics and bioinformatics; developing methods for modeling electrostatics in biological systems (DelPhi package); understanding the effects of missense mutations causing mental disorders and intellectual disability; predicting protein-protein interactions and 3D structures of the protein-protein complexes; computer simulations of protein-protein interactions modeling the role of conformation changes, pH, and salt concentration on biological function. Alexov, Ding.

Condensed Matter Physics. Surface phenomena, including scattering; anharmonic effects in crystal lattices; magnetic, optic, and transport properties of semiconductor mesoscopic structures (quantum wells and superlattices); broken symmetry states; charge and spin density waves; non-equilibrium superconductivity; topological computation; high-temperature superconductivity; cold atomic physics in optical traps and lattices. Daw, He, Kang, Marinescu, Rao, Skove, Sosolik, Tewari, Tritt.

Experimental

Astronomy. Gamma-ray astronomy; observational astronomy; stellar evolution; stellar atmospheres; circumstellar evolution; planet formation; stellar accretia; abundance determinations; galactic chemical evolution; star clusters; close binary star systems. Brittain, Flower, Hartmann, King, Leising, Meyer, The.

Atmosphere, Space Physics, Cosmic Rays. Rocket, radar, and spacecraft studies of ionospheric dynamics, electrodynamics, and plasma physics; studies of atmospheric dynamics and composition with LIDAR and Fabry-Perot systems; sounding rocket instrumentation, density, temperature, and turbulence in the mesosphere and lower thermosphere; satellite data analysis of vertical coupling processes. Larsen, Lehmacher, Meriwether, Oberheide.

Atomic, Molecular, & Optical Physics. Experimental low temperature atomic and molecular ion physics; laser trapping and cooling; cold chemistry; physics with highly charged ions. Marler, Sosolik.

Biophysics. Biomedical optical imaging; fluorescence tomography; optical spectroscopy; microwave imaging; ultrasound tomography; bioluminescence tomography; x-ray tomosynthesis; spectroscopy; DNA repair mechanisms; structure of biological molecules; biological effects of radiation damage; mechanisms of carcinogenesis; mechanisms for single-event effects in microelectronics; microdosimetry using microelectronic technology; structural biology of RNA; biomolecular structure function relationships; NMR spectroscopy; fluorescence spectroscopy; single-molecule biophysics; nanoscience.

Condensed Matter Physics. Atomic and molecular beam interactions at surfaces; formation and characterization of surface nanostructures with an energetic beam and scanning tunneling microscope. He, Kang, Marinescu, Rao, Skove, Sosolik, Tewari, Tritt.

Condensed Matter Physics. Thermoelectric materials and applied physics; thermophysical properties of novel materials, including investigations of low-temperature heat capacity and thermal transport; investigations of thermal, magnetic, and electronic transport properties of exotic systems, low-dimensional conductors, strongly electron correlated materials, and phase transition materials; high-temperature thermophysical properties of novel materials; synthesis of thermoelectric nanomaterials and composite thermoelectrics. Daw, He, Kang, Marinescu, Rao, Skove, Sosolik, Tewari, Tritt.

Nano Science and Technology. Synthesis of nanostructured materials using electric arc discharge, pulsed laser vaporization, neutron scattering, and CVD methods; optical characterization of novel materials by Raman scattering, infrared/visible, and fluorescence spectroscopy; mechanical properties of one-dimensional materials and chem-bio sensing using harmonic detection of resonance technique; superconducting nanotubes. Rao, Skove.

View additional information about this department at
www.gradschoolshopper.com

UNIVERSITY OF SOUTH CAROLINA

DEPARTMENT OF PHYSICS AND ASTRONOMY

Columbia, South Carolina 29208
http://www.physics.sc.edu

General University Information
President: Harris Pastides
Dean of Graduate School: Lacy Ford
University website: http://www.sc.edu
Control: Public
Setting: Urban
Total Faculty: 1,604
Total number of Students: 31,288
Total number of Graduate Students: 6,238

Department Information
Department Chairman: Milind Purohit, Chair
Department Contact: Beth Powell, Administrative Coordinator
  Total full-time faculty: 27
  Total number of full-time equivalent positions: 26
  Full-Time Graduate Students: 34
  First-Year Graduate Students: 9
  Female First-Year Students: 4
  Total Post Doctorates: 9

Department Address
712 Main Street
Columbia, SC 29208
Phone: (803) 777-8105
Fax: (803) 777-3065
E-mail: mepowell@sc.edu
Website: http://www.physics.sc.edu

ADMISSIONS

Admission Contact Information
Address admission inquiries to: Graduate Director, Department
  of Physics and Astronomy
Phone: (803) 777-4121
E-mail: davisaa@mailbox.sc.edu
Admissions website: http://www.physics.sc.edu

Application deadlines
Fall admission:
U.S. students: March 1   *Int'l. students*: March 1
Spring admission:
U.S. students: November 15   *Int'l. students*: November 15

Application fee
U.S. students: $50   *Int'l. students*: $50

Admissions information
For Fall of 2012:
  Number of applicants: 77
  Number admitted: 24
  Number enrolled: 9

Admission requirements
Bachelor's degree requirements: Bachelor's degree is required.

GRE requirements
The GRE is required.
  Verbal score: 350

Advanced GRE requirements
The Advanced GRE is not required.

TOEFL requirements
The TOEFL exam is required for students from non-English-
  speaking countries.
  PBT score: 570
  iBT score: 80

Other admissions information
Additional requirements: Usual preparation is an undergraduate
  major in physics. Students from other fields will have an opportunity to make up deficiencies.
Undergraduate preparation assumed: One semester advanced
  undergraduate courses in Mechanics, Electromagnetism,
  Modern Physics, Quantum Theory, and Experimental Physics,
  as well as mathematics through Differential Equations and
  Advanced Calculus, are assumed. Provision is made for students to make up deficiencies in these areas.

TUITION
Tuition year 2012–13:
Tuition for in-state residents
  Full-time students: $5,636 per semester
  Part-time students: $470 per credit
Tuition for out-of-state residents
  Full-time students: $12,098 per semester
  Part-time students: $1,008 per credit
Credit hours per semester to be considered full-time: 6
Deferred tuition plan: Yes
Health insurance: Available at the cost of 1349 per year.
Other academic fees: $500 foreign student enrollment fee.
Academic term: Semester
Number of first-year students who received full tuition waivers: 9

Teaching Assistants, Research Assistants, and Fellowships
Number of first-year
  Teaching Assistants: 7
Average stipend per academic year
  Teaching Assistant: $16,250
  Research Assistant: $16,704
  Fellowship student: $17,000

FINANCIAL AID

Loans
Loans are available for U.S. students.
Loans are not available for international students.
GAPSFAS application required: No
FAFSA application required: Yes

For further information
Address financial aid inquiries to: University of South Carolina,
  Office of Student Financial Aid and Scholarships, 1714 Col-
  lege Street, Columbia, SC 29208.
Phone: (803) 777-8134
E-mail: USCFAID@sc.edu
Financial aid website: http://www.sc.edu/financialaid

HOUSING

Availability of on-campus housing
Single students: Yes
Married students: Yes

For further information
Address housing inquiries to: University Housing, McBryde
   1309 Blossom Street, Columbia, SC 29208.
Phone: (803) 777-4283
E-mail: housing@sc.edu
Housing aid website: http://www.housing.sc.edu/

Table A—Faculty, Enrollments, and Degrees Granted

Research Specialty	2012-13 Faculty	Enrollment Fall 2012 Master's	Enrollment Fall 2012 Doctorate	Number of Degrees Granted 2012-13 (2008-13) Master's	Number of Degrees Granted 2012-13 (2008-13) Terminal Master's	Number of Degrees Granted 2012-13 (2008-13) Doctorate
Astronomy	1	–	2	–	–(1)	–(3)
Condensed Matter Physics	9	–	8	–(1)	–(10)	–(2)
High Energy Physics	6	1	8	–	–(9)	1(1)
Nuclear Physics	7	–	11	–	–(9)	–(7)
Theoretical Physics	4	–	2	–	–	1(2)
Non-specialized	–	1	8	–	–	–
Total	27	2	39	–(1)	3(22)	2(20)
Full-time Grad. Stud.	–	2	35	–	–	–
First-year Grad. Stud.	–	–	7	–	–	–

GRADUATE DEGREE REQUIREMENTS

Master's: The requirements for the Master of Science degree include 30 semester hours of course work, a thesis, and an oral comprehensive examination. The minimum residence is two semesters. There is no foreign language requirement.

Doctorate: The requirements for the degree of Doctor of Philosophy include 60 semester hours of advanced course work (or 30 semester hours beyond the Master's), written and oral examination for admission to candidacy, a reading knowledge of one foreign language, and a dissertation. Three years of residence are required, at least one of which is at the University of South Carolina.

Other Degrees: The MAT program includes 30 semester hours of graduate work with a distribution of graduate credit of 6 to 15 credits in professional education and 15 to 24 credits in the teaching area.

Thesis: Thesis may be written in absentia.

SPECIAL EQUIPMENT, FACILITIES, OR PROGRAMS

The Department is housed in the eight-story Physical Sciences Center, which also contains a machine shop and computer terminal facilities. Offices are provided for graduate students. Equipment for experimental research includes high- and low-temperature electron spin resonance spectrometers (cw and pulsed), two Mössbauer spectrometers, and a superconducting quantum interference device susceptometer. In addition to on-campus research facilities, University of South Carolina faculty and graduate students are utilizing the Thomas Jefferson National Accelerator Facility (JLab), Fermi National Accelerator Laboratory (Fermilab), Stanford Linear Accelerator Center (SLAC), and Instituto Nazionale di Fisica Nucleare (INFN) in Italy.

Table B—Separately Budgeted Research Expenditures by Source of Support

Source of Support	Departmental Research	Physics-related Research Outside Department
Federal government	$3,104,770	
State/local government	$209,744	
Non-profit organizations		
Business and industry		
Other	$197,356	
Total	$3,511,870	

Table C—Separately Budgeted Research Expenditures by Research Specialty

Research Specialty	No. of Grants	Expenditures ($)
Astronomy	6	$137,647
Condensed Matter Physics	17	$699,673
Nuclear Physics	9	$749,494
Particles and Fields	17	$1,773,187
Theoretical Physics	2	$151,869
Other	0	
Total	51	$3,511,870

FACULTY

Professor

Avignone, Frank T., Ph.D., Georgia Institute of Technology, 1965. *Astrophysics, Particles and Fields.*

Creswick, Richard J., Ph.D., University of California, Berkeley, 1981. *Computational Physics, Condensed Matter Physics, Statistical & Thermal Physics.*

Datta, Timir, Ph.D., Tulane University, 1979. *Condensed Matter Physics.*

Djalali, Chaden, Ph.D., University of Paris, 1984. Department Chair. *Nuclear Physics.*

Gothe, Ralf W., Ph.D., University of Mainz, 1990. *Nuclear Physics.*

Gudkov, Vladimir, Ph.D., Leningrad Nuclear Physics Institute, 1984. Graduate Director. *Computational Physics, Nuclear Physics.*

Kubodera, Kuniharu, Ph.D., University of Tokyo, 1970. *Astrophysics, Nuclear Physics.*

Kulkarni, Varsha P., Ph.D., University of Chicago, 1996. *Astronomy.*

Kunchur, Milind N., Ph.D., Rutgers University, 1988. *Condensed Matter Physics, Nano Science and Technology.*

Mazur, Pawel O., Ph.D., Jagellonian University, 1982. *Theoretical Physics.*

Mishra, Sanjib R., Ph.D., Columbia University, 1986. *High Energy Physics, Particles and Fields.*

Myhrer, Fred, Ph.D., University of Rochester, 1973. *Astrophysics, Nuclear Physics.*

Purohit, Milind V., Ph.D., California Institute of Technology, 1983. *High Energy Physics, Particles and Fields.*

Rosenfeld, Carl, Ph.D., California Institute of Technology, 1977. *High Energy Physics, Particles and Fields.*

Tedeschi, David J., Ph.D., Rensselaer Polytechnic Institute, 1993. *Medical, Health Physics, Nuclear Physics.*

Webb, Richard A., Ph.D., University of California, San Diego, 1973. *Condensed Matter Physics, Nano Science and Technology.*

Associate Professor

Altschul, Brett D., Ph.D., Massachusetts Institute of Technology, 2003. *Theoretical Physics.*

Bazaliy, Yaroslaw, Ph.D., Stanford University, 2000. *Condensed Matter Physics.*

Crawford, Thomas M., Ph.D., University of Colorado, 1992. *Condensed Matter Physics, Nano Science and Technology.*

Ilieva, Yordanka Y., Ph.D., Bulgaria Academy of Sciences, 2001. *Nuclear Physics.*

Petti, Roberto, Ph.D., Pavia University, 1998. *High Energy Physics, Particles and Fields.*

Strauch, Steffen, Ph.D., Darmstadt University, 1998. *Nuclear Physics.*

Wilson, Jeffrey R., Ph.D., Purdue University, 1985. Undergraduate Director. *High Energy Physics, Particles and Fields.*

Assistant Professor

Crittenden, Scott R., Ph.D., Purdue University, 2004. *Biophysics, Condensed Matter Physics, Nano Science and Technology.*

Pershyn, Yuriy, Ph.D., University of Konstanz, 2002. *Computational Physics, Nano Science and Technology.*

Schindler, Matthias R., Ph.D., University of Mainz, 2007. *Computational Physics, Nuclear Physics, Other.*

Wu, Yanwen, Ph.D., University of Michigan, 2007. *Condensed Matter Physics, Nano Science and Technology.*

DEPARTMENTAL RESEARCH SPECIALTIES AND STAFF

Theoretical

Computational Physics. Lie group applications. Creswick, Gudkov, Pershyn, Schindler.

Condensed Matter Physics. Monte Carlo calculations of the properties of spin glasses; magnetic resonance; theory of high-Tc superconductivity. Bazaliy, Creswick, Datta.

Nuclear Physics. Effective field theory, astrophysics, quark models. Gudkov, Kubodera, Myhrer.

Other. Generalized renormalization phenomena; effects of potentials and related topological concepts; generalized gauge invariance and pseudoperturbations. Altschul, Mazur.

Statistical & Thermal Physics. Symmetry properties and thermal properties of matter and radiation; nonlinear effects; critical phenomena and processes far from equilibrium. Creswick.

Experimental

Astronomy. Optical, infrared, and ultraviolet studies of quasars and distant galaxies to investigate chemical and morphological evolution of galaxies, interstellar/intergalactic matter, and intergalactic background radiation. Kulkarni.

Condensed Matter Physics. Theory of critical phenomena; numerical methods in statistical and quantum physics; Monte Carlo simulation; bulk and nano-structured thermoelectric materials; Aharonov-Bohm charge spectroscopy; colossal magnetic resistance; electron-spin resonance; biophysics; domain dynamics in ferroelectrics; superconductivity - lead inverse opals, high-dissipation phenomena, flux-vortex dynamics, metal-insulator transitions, magnetic penetration depth; magnetic nanoparticles and ultrasonic cavitation. Crawford, Crittenden, Datta, Kunchur, Webb, Wu.

High Energy Physics. High-energy proton-proton collisions are used to search for new particles indicative of physics beyond the standard model: the Higgs boson, supersymmetry, gravitons, black holes, extra dimensions, etc. Accelerator-based neutrino experiments study neutrino oscillations, masses and angles, electroweak parameters, as well as neutrino-induced reactions at new levels of precision. Mishra, Petti, Purohit, Rosenfeld, Wilson.

Nuclear Physics. The intermediate energy nuclear physics group has been playing a leadership role in research addressing the question of the structure and interaction of nucleons and nuclei in terms of quantum chromodynamic and has been making significant contributions. The research program uses multi-GeV photon and electron beams in Halls A and B from the Continuous Electron Beam Accelerator Facility (CEBAF) located at Thomas Jefferson National Laboratory (JLab) and are co-spokespersons of nine JLab experiments pursuing this quest in three interwoven areas: the study of medium modifications of hadrons, the study of the excited states of the nucleon, and the study of the transition from quark-gluon to pion-nucleon degrees of freedom in exclusive processes. Djalali, Gothe, Ilieva, Strauch, Tedeschi.

View additional information about this department at
www.gradschoolshopper.com

UNIVERSITY OF SOUTH DAKOTA

PHYSICS

Vermillion, South Dakota 57069
www.usd.edu/physics

General University Information

President: James W. Abbott
Dean of Graduate School: Laura Jenski
University website: http://www.usd.edu/
Control: Public
Setting: Rural

Total number of full-time equivalent positions: 6
Full-Time Graduate Students: 15
First-Year Graduate Students: 6
Female First-Year Students: 1
Total Post Doctorates: 4

Department Information

Department Chairman: Christina Keller, Director
Department Contact: Professor Christina Keller, Director
Total full-time faculty: 12

Department Address

414 E. Clark Street
Vermillion, SD 57069
Phone: (605) 677-6125

E-mail: Tina.Keller@usd.edu
Website: www.usd.edu/physics

ADMISSIONS

Admission Contact Information
Address admission inquiries to: Graduate School, McKusick Room 211, University of South Dakota, 414 E. Clark St., Vermillion SD 57069
Phone: 1-800-233-7937
E-mail: grad@usd.edu
Admissions website: www.usd.edu/graduate-school/future-students.cfm

Application deadlines
Fall admission:
U.S. students: February 1   *Int'l. students*: February 1

Application fee
U.S. students: $35   *Int'l. students*: $35

Admissions information
For Fall of 2013:
 Number of applicants: 8
 Number admitted: 6
 Number enrolled: 5

Admission requirements
Bachelor's degree requirements: B.S. or B.A. in Physics or related fields.
Minimum undergraduate GPA: 3.0

GRE requirements
The GRE is required.

Advanced GRE requirements
The Advanced GRE is required.
Physics GRE is required for Ph.D. applicants only.

TOEFL requirements
The TOEFL exam is required for students from non-English-speaking countries.
 PBT score: 550
 iBT score: 79
For score requirements of other tests, see Graduate School website.

TUITION

Tuition year 2013-14:
Tuition for in-state residents
 Full-time students: $210.4 per credit
Tuition for out-of-state residents
 Full-time students: $445.3 per credit
Other academic fees: Fees of $128.60/cr are charged in addition to tuition.
Academic term: Semester
Number of first-year students who received partial tuition waivers: 8

Teaching Assistants, Research Assistants, and Fellowships
Number of first-year
 Teaching Assistants: 5
 Research Assistants: 5
Average stipend per academic year
 Teaching Assistant: $24
 Research Assistant: $24
9-month stipends, summer support available.

FINANCIAL AID

Application deadlines
Fall admission:
U.S. students: March 15

Loans
Loans are available for U.S. students.
Loans are not available for international students.
GAPSFAS application required: No
FAFSA application required: Yes

For further information
Address financial aid inquiries to: Office of Financial Aid, Belbas Center, University of South Dakota, 414 E. Clark St., Vermillion SD 57069.
Phone: 605-677-5446
Financial aid website: www.usd.edu/finaid

HOUSING

Availability of on-campus housing
 Single students: Yes
 Married students: No

For further information
Address housing inquiries to: Office of Student Life/Housing, Muenster University Center 219, University of South Dakota, 414 E. Clark St., Vermillion SD 57069.
Phone: 605 677 5666
E-mail: housing@usd.edu
Housing aid website: www.usd.edu/campus-life/student-services/university-housing/

Table A—Faculty, Enrollments, and Degrees Granted

Research Specialty	Faculty	Enrollment		Number of Degrees Granted		
		Master's	Doctorate	Master's	Terminal Master's	Doctorate
Nuclear Physics	5	9	–	–	–	–
Total	–	–	–	4(-)	–	–
Full-time Grad. Stud.	–	9	–	–	–	–
First-year Grad. Stud.	–	4	–	–	–	–

SPECIAL EQUIPMENT, FACILITIES, OR PROGRAMS

The University of South Dakota has a close affiliation with the Sanford Underground Research Facility and experiments searching for dark matter and the nature of the neutrino.

FACULTY

Professor
Corey, Robert, Ph.D., Washington University, St. Louis. Teaching faculty at SDSMT that contributes to courses and serves on committees.
Keller, Christina, Ph.D., North Dakota State University, 2013. *Condensed Matter Physics*. Condensed Matter Physics.
Petukhov, Andre, Ph.D., St. Petersburg Technical University. Teaching faculty at SDSMT that contributes to courses and serves on committees.
Sobolev, Vladimir L., Ph.D., The Kharkov State University, Ukraine. Teaching faculty at SDSMT that contributes to courses and serves on committees.

Associate Professor
Mei, Dongming, Ph.D., University of Alabama, 2003. *Astrophysics, Nuclear Physics*. Nuclear physics, astrophysics.
Sun, Yongchen, Ph.D., Montana State University, 1993. *Condensed Matter Physics, Optics*. Laser spectroscopy, condensed matter physics.

Assistant Professor
Bai, Xinhua, Ph.D., Peking University. Teaching faculty at SDSMT that contributes to courses and serves on committees.
Martin, Ryan, Ph.D., Queen's University, 2009. *Astrophysics, Nuclear Physics*. Nuclear physics, astrophysics.

Sander, Joel, Ph.D., University of California, Santa Barbara, 2007. *Astrophysics, Particles and Fields*. Particle physics, astrophysics.

Research Assistant Professor
Zhang, Chao, Ph.D., Institute of High Energy Physics, 2007. *Atmosphere, Space Physics, Cosmic Rays, Nuclear Physics, Particles and Fields*. Dark matter and neutrino physics.

Visiting Assistant Professor
Khizar, Muhammad, Ph.D., UNC Charlotte, 2007. *Nano Science and Technology*. Photonics, nanoelectronics.

View additional information about this department at www.gradschoolshopper.com

UNIVERSITY OF TENNESSEE, KNOXVILLE

DEPARTMENT OF PHYSICS AND ASTRONOMY

Knoxville, Tennessee 37996-1200
http://www.phys.utk.edu

General University Information
President: Dr. Joe DiPietro
Dean of Graduate School: Carolyn R. Hodges
University website: http://www.utk.edu
Control: Public
Setting: Urban
Total Faculty: 1,406
Total number of Students: 27,107
Total number of Graduate Students: 6,101

Department Information
Department Chairman: Hanno Weitering, Head
Department Contact: Chrisanne Romeo, Administrative Assistant, Graduate Program
Total full-time faculty: 31
Total number of full-time equivalent positions: 26
Full-Time Graduate Students: 119
First-Year Graduate Students: 29
Female First-Year Students: 3

Department Address
1408 Circle Drive
401 Nielsen Physics Building
Knoxville, TN 37996-1200
Phone: (865) 974-3342
Fax: (865) 974-7843
E-mail: physics@utk.edu
Website: http://www.phys.utk.edu

ADMISSIONS

Admission Contact Information
Address admission inquiries to: Graduate School, University of Tennessee, 111 Student Services Building, Knoxville, TN 37996-0211
Phone: (865) 974-2475

E-mail: gradschool@utk.edu
Admissions website: http://gradschool.utk.edu/

Application deadlines
Fall admission:
U.S. students: July 31   *Int'l. students*: February 1
Spring admission:
U.S. students: December 20   *Int'l. students*: September 1

Application fee
U.S. students: $35   *Int'l. students*: $60

Admissions information
For Fall of 2013:
Number of applicants: 178
Number admitted: 77
Number enrolled: 12

Admission requirements
Bachelor's degree requirements: Bachelor's degree in physics, mathematics, or engineering is required.
Minimum undergraduate GPA: 2.7

GRE requirements
The GRE is recommended.
Subject GRE Required for financial support.

Advanced GRE requirements
The Advanced GRE is recommended.

TOEFL requirements
The TOEFL exam is required for students from non-English-speaking countries.
PBT score: 550
iBT score: 80
For financial consideration, speaking score must be greater than 20 OR an International English Language Testing System (IELTS) score of 6.5.

Other admissions information

Additional requirements: A minimum undergraduate GPA of 3.0 for international students.

Undergraduate preparation assumed: Griffiths, Introduction to Electrodynamics; Marion and Thornton, Classical Dynamics; Griffiths, Introduction to Quantum Mechanics or equivalent.

TUITION

Tuition year 2012–13:

Tuition for in-state residents
   Full-time students: $5,140 per semester

Tuition for out-of-state residents
   Full-time students: $14,384 per semester

Below listed Fees included in cost.

Credit hours per semester to be considered full-time: 9

Deferred tuition plan: Yes

Health insurance: Available at the cost of $1,272 per year.

Other academic fees: Program & Services, Library, Technology, Facilities, and Transportaton fees.

Academic term: Semester

Number of first-year students who received full tuition waivers: 21

Teaching Assistants, Research Assistants, and Fellowships

Number of first-year
   Teaching Assistants: 17
   Research Assistants: 5
Average stipend per academic year
   Teaching Assistant: $20,000
   Research Assistant: $22,000

FINANCIAL AID

Loans

Loans are available for U.S. students.
Loans are not available for international students.

GAPSFAS application required: No

FAFSA application required: Yes

For further information

Address financial aid inquiries to: Financial Aid Office, 115 Student Services Building, University of Tennessee, Knoxville, TN 37996-0210.

Phone: 865-974-3131

E-mail: finaid@utk.edu

Financial aid website: http://gradschool.utk.edu/gradfund.shtml

HOUSING

Availability of on-campus housing

Single students: Yes

Married students: No

For further information

Address housing inquiries to: University Housing, 405 Student Services Building and UT Rental Property, 472 South Stadium Hall, Knoxville, TN 37996.

Phone: 865-974-2426

E-mail: housing@utk.edu

Housing aid website: http://uthousing.utk.edu/tnliving/current/housing-options.shtml

Table A—Faculty, Enrollments, and Degrees Granted

Research Specialty	2012-13 Faculty	Enrollment 2012-13 Master's	Enrollment 2012-13 Doctorate	Number of Degrees Granted 2012-13 (2011–12) Master's	Number of Degrees Granted 2012-13 (2011–12) Terminal Master's	Number of Degrees Granted 2012-13 (2011–12) Doctorate
Astrophysics	2	3	9	–	–	1(1)
Atomic, Molecular, & Optical Physics	1	–	5	–	–(2)	1(-)
Biophysics	2	–	1	–	–(1)	–
Chemical Physics	1	1	1	–	–	1(-)
Condensed Matter Physics	8	–	32	–(2)	–(1)	8(2)
High Energy Physics	5	–	12	1(-)	–(1)	1(-)
Medical, Health Physics	–	–	1	–	–	1(-)
Nuclear Physics	8	1	14	–	–(1)	–(2)
Particles and Fields	–	1	6	–	–(1)	2(1)
Physics and other Science Education	3	1	31	–	2(-)	–
Total	29	7	112	1(2)	2(6)	15(7)
Full-time Grad. Stud.	–	7	112	–	–	–
First-year Grad. Stud.	–	4	24	–	–	–

GRADUATE DEGREE REQUIREMENTS

Master's: Thesis Option: The course requirements include 24 semester hours of physics courses, of which at least 12 semester hours are taken from Physics 506, 513–514, 521–522, 531, 541, 571, 573. Each candidate must present an acceptable thesis, 6 hours of 500, and pass an oral examination on course material and thesis. The department offers an M.S. thesis program with a concentration in geophysics. Program requirements are: 12 hours from Physics 506, 513–514, 521–522, 531, 541, 571, 573; a minimum of 12 additional hours in geology, geophysics, and/or physics, as approved by the student's committee; and the presentation of an acceptable thesis, 6 hours of Physics 500, an advanced seminar, and the passing of an oral examination on course material and thesis. Project Option: The course requirements include a minimum of 30 hours of graduate credit in courses composed of Physics 506, 513–514; 6 hours from Physics 593, 594 for a Project in Lieu of Thesis; 9 hours from general physics: 411–412, 421, 431–432, 461–462, 507, 508, 521–522, 531, 541, 555, 571, 573 (at least 3 hours above the 500-level); and 6 hours from a single minor field outside of the physics department, such as computer science, mathematics, engineering, chemistry, biology, education, business, or law. The candidate must pass an oral examination on course material and on the Project representing the culmination of an original research project completed by the student. A written report must be approved and accepted by the Physics Graduate Committee and the Department Head. An electronic version of the written report must also be submitted to the permanent electronic archive of the Physics Department available to the Internet. Non-Thesis Option: Students seeking the non-thesis option must apply to the department's graduate committee for permission to enroll under this program. The requirements are the satisfactory completion of 30 hours of course work composed of 18 semester hours from Physics 513–514, 521–22, 531–32, 541–42, and 571–72; 6 semester hours in a minor field; and 6 semester hours from other courses numbered above 400 (preferably of advanced laboratory nature). At least 20 hours must be taken at the 500-level or above. In addition, the candidate must pass a written examination administered by his/her committee.

Doctorate: All students are expected to take the graduate core curriculum in physics consisting of the following courses: Physics 521–22, 531, 541, 551, and 571. Students specializing in chemical physics may substitute Chemistry 572 for Physics 551, and should complete at least 6 semester hours from Chemistry 530, 570, 571, 573, 595, 630, 670, and 690. Students must take a minimum of 15 hours of 600-level courses, with 6 of these hours in their area of specialization. Physics 601–02 are normally required of students specializing in atomic physics; Physics 621–22 of students in nuclear physics; Physics 626–27 of students in elementary particle physics (and/or Physics 613–14 for students specializing in theoretical high-energy physics); and Physics 671–72 of students in condensed matter and surface physics. Students concentrating in nanomaterials must take a minimum 15 hours of 600-level courses, of which at least 6 hours are offered by the department and at least 6 hours are from a list of courses offered by several departments which are appropriate for a concentration in nanomaterials. To be admitted to Ph.D. candidacy students must: a) fulfill all general requirements by the Graduate School, b) pass the qualifying examination, c) have at least a 3.0 GPA on the graduate core curriculum in physics, d) form a doctoral committee, and e) pass the comprehensive examination. The qualifying examination is designed to test the student's general knowledge of the fundamentals of physics. The performance needed to pass this examination corresponds to a mature command of the material typically included in the undergraduate physics major curriculum. The qualifying examination should be passed after the student's first year of study. Based on the student's performance on a) the qualifying examinations, b) the course work, c) the GRE scores and d) optional research participation, the faculty will decide if the student will be allowed to continue in the Ph.D. program. Students are required to find a research advisor and form a doctoral committee before the end of the second year of study. This committee is responsible for advising the student and monitoring his/her progress toward the doctoral degree. The comprehensive examination is designed to test the students on a) specific knowledge and skills in the areas essential to the student's research program, b) capability to successfully complete the doctoral dissertation and c) general knowledge of the graduate core curriculum. The most essential component of this examination is the presentation and defense of an original research proposal. The comprehensive examination must be passed before the end of the third year of study. It contains both a written and an oral component and is conducted by the student's doctoral committee and an additional faculty member appointed by the department head. The dissertation topic will be chosen with reference to one of the fields in which research facilities can be made available either at The University of Tennessee laboratories in Knoxville; The University of Tennessee Space Institute at Tullahoma, Tennessee; the Oak Ridge National Laboratory, Oak Ridge, Tennessee; or at other research facilities used by the University faculty.

SPECIAL EQUIPMENT, FACILITIES, OR PROGRAMS

Special experimental facilities include semiconductor- and oxide molecular beam epitaxy growth facilities, various ultrahigh vacuum systems with provisions for photoelectron spectroscopy (UV and x-ray), time-of-flight mass spectrometry, ultrahigh resolution scanning tunneling microscopy, QPlus atomic force microscopy, low energy electron diffraction, reflection high energy electron diffraction, secondary ion mass spectrometry, angle-resolved photo electron spectroscopy, pump and probe photo emission. Other facilities include: optical floating zone furnaces; SQUID magnetometers; physical properties measurement systems; 15 crucible furnaces; a large computer-controlled, high-resolution molecular spectrometer; computer-controlled nuclear data analysis systems; scintillator detector test facility; and a complete biophysics laboratory.

The department has fully staffed instrument- and electronics shops. Research facilities are also available at Oak Ridge National Laboratory in astrophysics, nuclear, and condensed matter physics.

A Ph.D. program in Chemical Physics is conducted jointly with the Department of Chemistry. The department also conducts a resident Ph.D. program in physics at the University of Tennessee Space Institute in Tullahoma. Facilities there support research in atomic and molecular physics, laser physics, chemical physics, quantum optics, infrared spectroscopy, and laser scattering. A cooperative arrangement permits research appointments for selected students at the nearby Air Force Arnold Engineering Development Center research operation.

Table B—Separately Budgeted Research Expenditures by Source of Support

Source of Support	Departmental Research	Physics-related Research Outside Department
Federal government	$7,512,526	
State/local government	$1,667,550	
Non-profit organizations		
Business and industry		
Other	$137,701	
Total	**$9,317,777**	

Table C—Separately Budgeted Research Expenditures by Research Specialty

Research Specialty	No. of Grants	Expenditures ($)
Applied Physics	1	$21,351
Astrophysics	3	$28,954
Atomic, Molecular, & Optical Physics	5	$384,877
Biophysics	1	$45,587
Condensed Matter Physics	51	$3,643,122
Nuclear Physics	25	$4,121,240
Particles and Fields	9	$1,002,964
Physics and other Science Education	4	$69,682
Total	**99**	**$9,317,777**

FACULTY

Distinguished University Professor

Dagotto, Elbio, Ph.D., Bariloche, 1985. Distinguished Scientist. *Condensed Matter Physics.* Theoretical condensed matter physics.

Macek, Joseph H., Ph.D., Rensselaer Polytechnic Institute, 1964. *Atomic, Molecular, & Optical Physics.* Theoretical atomic physics.

Nazarewicz, Witold, Ph.D., University of Warsaw, 1981. *Nuclear Physics.* Theoretical nuclear physics.

Quinn, John J., Ph.D., University of Maryland, 1958. *Condensed Matter Physics.* Theoretical solid state physics.

Professor

Breinig, Marianne, Ph.D., University of Oregon, 1979. *Atomic, Molecular, & Optical Physics, Physics and other Science Education.* Atomic and molecular physics.

Compton, Robert N., Ph.D., University of Tennessee, 1964. *Atomic, Molecular, & Optical Physics, Chemical Physics.* Experimental chemical physics.

Dai, P., Ph.D., University of Missouri, 1993. *Condensed Matter Physics.* Experimental condensed-matter physics.

Efremenko, Yuri, Ph.D., ITEP, Moscow, Russia, 1989. *Particles and Fields.* High energy physics.

Eguiluz, Adolfo, Ph.D., Brown University, 1976. *Condensed Matter Physics.* Condensed matter.

Elston, Stuart, Ph.D., University of Massachusetts, 1975. *Atomic, Molecular, & Optical Physics.* Atomic and molecular physics.

Greene, Geoffrey L., Ph.D., Harvard University, 1977. *Nuclear Physics*. Experimental nuclear physics.

Grzywacz, Robert, Ph.D., University of Warsaw, 1997. *Nuclear Physics*. Experimental nuclear physics.

Guidry, Michael W., Ph.D., University of Tennessee, 1974. *Astrophysics*. Nuclear astrophysics.

Handler, Thomas, Ph.D., Rutgers University, 1974. *Particles and Fields*. Higher energy physics.

Kamychkov, Yuri A., Ph.D., University of Moscow, 1970. *Particles and Fields*. Experimental elementary particle physics.

Levin, Jon C., Ph.D., University of Oregon, 1986. *Atomic, Molecular, & Optical Physics*. Experimental atomic physics.

Moreo, Adriana, Ph.D., Bariloche, 1985. *Condensed Matter Physics*. Theoretical condensed matter physics.

Read, Kenneth F., Ph.D., Cornell University, 1987. *Nuclear Physics*. Experimental nuclear physics.

Riedinger, Leo L., Ph.D., Vanderbilt University, 1969. *Nuclear Physics*. Experimental nuclear physics.

Siopsis, George, Ph.D., California Institute of Technology, 1987. *Particles and Fields*. High energy physics.

Sorensen, Soren, Ph.D., Niels Bohr Institute, 1981. *Nuclear Physics*. Experimental nuclear physics.

Weitering, Harm H., Ph.D., University of Groningen, 1991. *Condensed Matter Physics, Nano Science and Technology, Solid State Physics, Surface Physics*. Experimental condensed matter physics.

Associate Professor

Hix, Raph W., Ph.D., Harvard University, 1995. *Astrophysics, Nuclear Physics*. Theoretical nuclear astrophysics.

Jones, Kate, Ph.D., University of Surrey, 2000. *Nuclear Physics*. Experimental nuclear physics.

Papenbrock, Thomas, Ph.D., Max-Planck Institute, 1996. *Nuclear Physics*. Theoretical nuclear physics.

Spanier, Stefan M., Ph.D., University Mainz, 1994. *Particles and Fields*. Experimental high energy physics.

Assistant Professor

Joo, Jaewook, Ph.D., Rutgers University, 2004. *Biophysics*. Theoretical biophysics.

Mannella, Norman, Ph.D., University of California, Davis, 2003. *Condensed Matter Physics*. Experimental condensed matter.

Mannik, Jaan, Ph.D., Stony Brook University, 2003. *Biophysics*. Experimental biophysics.

Nattrass, Christine, Ph.D., Yale University, 2009. *Nuclear Physics*. Experimental nuclear physics.

Zhou, Haidong, Ph.D., University of Texas at Austin, 2005. *Condensed Matter Physics*. Experimental condensed matter physics.

Professor Emeritus

Bingham, Carroll R., Ph.D., University of Tennessee, 1965. Experimental nuclear physics.

Blass, William E., Ph.D., Michigan State University, 1963. Experimental and theoretical molecular spectroscopy.

Bugg, William M., Ph.D., University of Tennessee, 1959. Experimental elementary particle physics.

Georghiou, Solon, Ph.D., University of Manchester, 1968. *Biophysics*. Biophysics.

Shih, Chia C., Ph.D., Cornell University, 1967. Theoretical elementary particle and atomic physics; experimental medical physics.

Thompson, James R., Ph.D., Duke University, 1969. Experimental solid state.

Research Professor

Bertrand, Fred E., Ph.D., Louisiana State University, 1968.

Blankenship, James L., Ph.D., Yale University, 1967.

Ferrell, Thomas L., Ph.D., Clemson University, 1969. Experimental surface physics.

Halbert, Melvyn L., Ph.D., University of Rochester, 1955.

Harvey, John, Ph.D..

Kerman, Arthur K., Ph.D., Massachusetts Institute of Technology, 1953.

Ovchinnikov, Serguei, Ph.D., Technical Institute of Lenigrad, 1985. Atomic physics.

Saltmarsh, Michael J., Ph.D..

Stone, Nicholas J., Ph.D., University of Oxford, 1963. Nuclear physics.

Wong, Cheuk-Yin, Ph.D., Princeton University, 1966.

Zucker, Alexander, Ph.D., Yale University, 1950.

Research Assistant Professor

Cheney, Christine P., Ph.D., Vanderbilt University, 2001. *Condensed Matter Physics*.

Tselev, Alexander, Ph.D., Dresden University of Technology, 2000.

Adjunct Professor

Aytug, Tolga, Ph.D., University of Kansas, 2000.

Bardayan, Daniel, Ph.D., Yale University, 1999.

Burgdoerfer, Joachim, Ph.D., Freie Univ., Berlin, 1982.

Calarco, John, Ph.D., University of Illinois at Urbana-Champaign, 1969.

Cooke, John, Ph.D., Georgia Institute of Technology, 1965.

Datskos, Panos, Ph.D., University of Tennessee, 1988. Molecular and chemical physics.

Dean, David, Ph.D., Vanderbilt University, 1991.

Egami, Takeshi, Ph.D., University of Pennsylvania, 1971.

Fernández-Baca, Jaime, Ph.D., University of Maryland, 1986.

Galindo-Uribarri, Alfredo, Ph.D., University of Toronto, 1991.

Garrett, William R., Ph.D., University of Alabama, 1963.

Grice, Warren, Ph.D., University of Rochester, 1997.

Henderson, Stuart, Yale University, 1991.

Holmes, Jeffrey, Ph.D., California Institute of Technology, 1976.

Katsaras, John, Ph.D., University of Guelph, 1991.

Keppens, Veerle, Ph.D., Katholieke Universiteit Leuven, Belgium, 1995.

Kristic, Predrag, Ph.D., City College of City University of New York, 1982.

Mandrus, David G., Ph.D., Stony Brook University, 1992.

Mason, Thom E., Ph.D., McMaster University.

Melnichenko, Yuri B., Ph.D., Kiev State University, 1985.

Mezzacappa, Anthony, Ph.D., University of Texas, Austin, 1988. Nuclear astrophysics.

Nagler, Stephen, Ph.D., University of Toronto, 1982.

Passian, Ali, Ph.D., University of Tennessee, 2000.

Ramsey, Chester, Ph.D., University of Tennessee, 2000.

Shen, Jian, Ph.D., Institut Mikrostrukturphysik, 1996.

Singh, David, Ph.D., University of Ottawa, 1985.

Smith, Michael, Ph.D., Yale University, 1990.

Sokolov, Alexei, Ph.D., Russian Academy of Sciences, 1986.

Stone, Jirina, Ph.D., Charles University, Prague, 1975.

Sun, Yang, Ph.D., Technische Universität, München, 1991.

Van Berkel, Gary, Ph.D., Washington State University, 1987.

Ziock, Klaus-Peter, Ph.D., Stanford University.

Adjunct Associate Professor

Thomas, Maier, Ph.D., University of Regensburg, Germany, 2001.

Adjunct Assistant Professor

Hagen, Gaute, Ph.D., University of Bergen, Norway.

Li, An-Ping, Ph.D., Peking University, 1997.

Lecturer with Rank of Professor

Parks, James E., Ph.D., University of Kentucky, 1970. Experimental atomic physics.

DEPARTMENTAL RESEARCH SPECIALTIES AND STAFF

Theoretical

Astrophysics. Core collapse supernovae, novae, computational modeling, nuclear astrophysics. Guidry, Hix, Mezzacappa.

Atomic, Molecular, & Optical Physics. Atomic scattering theory, interaction of electromagnetic radiation with atoms and electron correlation with atoms. Blass, Breinig, Burgdoerfer, Compton, Datskos, Elston, Ferrell, Garrett, Grice, Kristic, Levin, Macek, Ovchinnikov, Parks, Passian, Saltmarsh, Van Berkel.

Biophysics. Information flow in cellular systems. Studiess of cellular organization in E. coli through nanofabricated environments. Georghiou, Joo, Mannik, Sokolov.

Condensed Matter Physics. Strongly correlated electron systems, colossal magnetoresistance, high Tc superconductivity, transport properties, nanostructures, many-body excitations, thin film growth, surface physics, quantum Hall effect, quantum magnetism. Cooke, Dagotto, Dai, Egami, Eguiluz, Fernández-Baca, Li, Mandrus, Mannella, Mason, Moreo, Quinn, Shen, Singh, Sun, Thompson, Weitering, Zhou.

High Energy Physics. Neutrino physics. Neutron oscillations. CP-violation in the heavy quark sector. Strings and quantum gravity. Efremenko, Handler, Kamychkov, Siopsis, Spanier.

Nuclear Physics. Nuclear structure far from stability and at high spins. Hot and dense nuclear matter. Quark-Gluon Plasma. Neutron physics. Nuclear structure, many body problems, physics of open systems, quantum chaos, and random matrices. Bardayan, Bertrand, Bingham, Blankenship, Calarco, Dean, Galindo-Uribarri, Greene, Halbert, Jones, Kerman, Nazarewicz, Papenbrock, Wong, Zucker.

Physics and other Science Education. Breinig, Elston, Guidry, Levin.

Experimental

Accelerator. Accelerator physics studies utilizing the ORNL Spalation Neutron Source. Henderson, Holmes.

Biophysics. Nanotechnology-based probes to interrogate single molecules/molecular complexes within living cells. Mannik.

Chemical Physics. Laser spectroscopy, negative ions, chirality, synchrotron spectroscopy of atoms. Compton.

Condensed Matter Physics. Neutron scattering, high Tc superconductivity, magnetism and lattice effects in colossal magnetoresistance, physics of novel materials and complex electron systems, low dimensional materials, nanostructures, surface and interface physics, x-ray spectroscopy, thin-film materials. Dai, Mannella, Weitering, Zhou.

High Energy Physics. CP violation measurements, meson spectroscopy, electron neutrino detection, neutrino oscillation. Efremenko, Handler, Kamychkov, Spanier.

Nuclear Physics. Decay spectroscopy, experimental nuclear astrophysics, gamma ray spectroscopy, relativistic heavy ion physics, hot and dense nuclear matter, neutron physics. Greene, Grzywacz, Jones, Read, Sorensen.

View additional information about this department at
www.gradschoolshopper.com

VANDERBILT UNIVERSITY

DEPARTMENT OF PHYSICS AND ASTRONOMY

Nashville, Tennessee 37235-1807
http://www.vanderbilt.edu/physics/

General University Information

President: Nicholas S. Zeppos
Dean of Graduate School: Dennis Hall
University website: http://www.vanderbilt.edu/
Control: Private
Setting: Urban
Total Faculty: 3,309
Total Graduate Faculty: 972
Total number of Students: 12,745
Total number of Graduate Students: 5,949

Department Information

Department Chairman: Robert J. Scherrer, Chair
Department Contact: Robert J. Scherrer, Professor & Dept. Chair
  Total full-time faculty: 30
  Total number of full-time equivalent positions: 30
  Full-Time Graduate Students: 76
  First-Year Graduate Students: 16

Female First-Year Students: 1
Total Post Doctorates: 19

Department Address

6301 Stevenson Center
VU Station B #351807
Nashville, TN 37235-1807
Phone: (615) 322-2828
Fax: (615) 343-7263
E-mail: physics-astronomy@vanderbilt.edu
Website: http://www.vanderbilt.edu/physics/

ADMISSIONS

Admission Contact Information

Address admission inquiries to: Director of Graduate Studies, Physics & Astronomy Dept., 6301 Stevenson Center Vanderbilt University, Nashville, TN 37240
Phone: (615) 322-2828

E-mail: physics-astronomy@vanderbilt.edu

Admissions website: http://www.vanderbilt.edu/physics/grad/grad-home.php

Application deadlines

Fall admission:

U.S. students: January 15   *Int'l. students*: January 15

Application fee

There is no application fee required.

Online application is free for all students. No spring admits except under rare conditions.

Admissions information

For Fall of 2013:

Number of applicants: 222

Number admitted: 31

Number enrolled: 12

Admission requirements

Bachelor's degree requirements: Bachelor's degree in physics is required.

Minimum undergraduate GPA: 3.0

GRE requirements

The GRE is required.

Quantitative score: 550

Verbal score: 550

Advanced GRE requirements

The Advanced GRE is required.

Minimum accepted Advanced GRE score: 500

Mean Advanced GRE score range (25th–75th percentile): 500-700

For the Physics Subject test.

TOEFL requirements

The TOEFL exam is required for students from non-English-speaking countries.

PBT score: 550

iBT score: 88

Other admissions information

Additional requirements: The minimum acceptable score suggested for admission is verbal-550; quantitative-550; total-1100. Applicants in physics should have quantitative scores greater than 650. The Graduate School's minimum acceptable score suggested for admission is 650.

Undergraduate preparation assumed: Resnick and Halliday, Physics; Eisberg and Resnick, Quantum Physics of Atoms, Molecules, Solids, Nuclei, and Particles; Reitz, Milford, Christy, Foundations of Electromagnetic Theory; Zemansky,Heat and Thermodynamics; Reif, Fundamentals of Statistical and Thermal Physics; Symon, Mechanics; Saxon, Elementary Quantum Mechanics.

TUITION

Tuition year 2012–13:

Full-time students: $1,712 per credit

Part-time students: $1,712 per credit

Excludes Medical and Law Schools. Physics degree. Masters in Medical Physics administered through the School of Medicine.

Credit hours per semester to be considered full-time: 9

Deferred tuition plan: Yes

Health insurance: Available at the cost of 2382 per year.

Other academic fees: $404 for annual recreation center fee; $30 one-time transcript fee for all new students.

Academic term: Semester

Number of first-year students who received full tuition waivers: 12

Teaching Assistants, Research Assistants, and Fellowships

Number of first-year

Teaching Assistants: 12

Research Assistants: 1

Fellowship students: 2

Average stipend per academic year

Teaching Assistant: $20,200

Research Assistant: $24,500

Fellowship student: $8,000

Awarded on academic merit at acceptance to program and are typically multi-year.

FINANCIAL AID

Loans

Loans are available for U.S. students.

Loans are not available for international students.

GAPSFAS application required: No

FAFSA application required: Yes

For further information

Address financial aid inquiries to: Office of Student Financial Aid and Undergraduate Scholarships, 2309 West End Ave., Nashville, TN 37203-1725.

Phone: (615) 322-3591

E-mail: finaid@vanderbilt.edu

Financial aid website: http://www.vanderbilt.edu/financialaid/

HOUSING

Availability of on-campus housing

Single students: No

Married students: No

For further information

Address housing inquiries to: Office of Housing and Residential Education, Vanderbilt University, VU #351677, Station B, Nashville, TN 37235.

Phone: (615) 322-2591

E-mail: resed@vanderbilt.edu

Housing aid website: http://www.vanderbilt.edu/ResEd/main/

Table A—Faculty, Enrollments, and Degrees Granted

Research Specialty	2013-14 Faculty	Enrollment Fall 2013		Number of Degrees Granted 2012-13 (2008-13)		
		Master's	Doctorate	Master's	Terminal Master's	Doctorate
Astrophysics	4	–	17	3(8)	–	2(6)
Atomic, Molecular, & Optical Physics	2	–	6	–	1(2)	–(3)
Biophysics	3	–	9	2(4)	–(2)	3(8)
Condensed Matter Physics	7	–	10	1(9)	–(2)	1(10)
Medical, Health Physics	2	10	–	2(10)	3(4)	2(9)
Nuclear Physics	8	–	9	–(8)	–(2)	3(9)
Particles and Fields	4	–	3	1(6)	–	–(12)
Non-specialized	–	–	22	1(2)	–	–
Total	**30**	**10**	**76**	**10(47)**	**4(12)**	**11(57)**
Full-time Grad. Stud.	–	–	76	–	–	–
First-year Grad. Stud.	–	–	16	–	–	–

GRADUATE DEGREE REQUIREMENTS

Master's: Master's in Physics: 24 semester-hours of coursework plus research thesis. A "B" average is required. Degree can be completed in one year but often requires three semesters. There is no foreign language requirement and no comprehensive examination. Master's is also awarded without thesis on the basis of 42 hours of coursework, Ph.D.-qualifying examination, and some research experience. Master's in Astronomy: Non-thesis option is not available. This degree normally requires 4 semesters. Oral examination is required. Master's in Medical Physics: Thesis and non-thesis option available. Oral examination is required.

Doctorate: Ph.D. in Physics: 72 semester-hours of coursework is required, up to 36 of which can be research. A "B" average is required in formal coursework. Completion of core coursework and oral qualifying examination after first or second year establishes Ph.D. candidacy. Dissertation in physics or astronomy is required. One year of residency is required.

Other Degrees: Master of Science in Medical Physics, MAT, and interdisciplinary graduate program in materials science are also available. Interdisciplinary work is encouraged and is tailored to fit the needs of the individual student. Master of Science at Fisk University with a Ph.D. at Vanderbilt University is also available.

Thesis: Thesis may be written in absentia.

SPECIAL EQUIPMENT, FACILITIES, OR PROGRAMS

The Nanofabrication Laboratory, a collaborative activity with the School of Engineering, has a focused ion beam, state-of-the-art pulsed laser thin-film deposition system, atomic-force microscope, and a variety of diagnostics for studying nanometer-scale structures and materials. This laboratory is a centerpiece of the Vanderbilt Institute of Nanoscale Science and Engineering.

Vanderbilt University is a member of the University Radioactive Ion Beam (UNIRIB) Consortium at Oak Ridge National Laboratory that is in charge of a new generation recoil mass spectrometer. The new RMS was initiated by Vanderbilt for use with the new Holifield Radioactive Ion Beam Facility (HRIB). The Joint Institute for Heavy-Ion Research at Oak Ridge, Tennessee, is operated by Vanderbilt University, the University of Tennessee, and Oak Ridge National Laboratory to support users of the Holifield HRIB and sponsors a visitor's program that brings distinguished scientists for research and lecturing.

Members of the nuclear structure physics group have additional cooperative research programs at Argonne National Laboratory, Lawrence Berkeley Laboratory, Idaho National Engineering Laboratory, the Joint Institute for Nuclear Research in Russia, the Universities of Frankfurt, Tsinghua, and Bucharest, and access to a number of supercomputers around the country.

The Vanderbilt Relativistic Heavy Ion (RHI) nuclear group plays a senior management role in the PHENIX relativistic heavy-ion experiment at RHIC and the heavy-ion program at the CMS in CERN. The RHI group studies nuclear matter under extreme conditions of temperature and density and searches for exotic new states of matter.

The Atomic, Molecular, Optical and Surface (AMOS) groups study the dynamics of surface and interface processes under a wide variety of conditions and has collaborators in the Department of Chemistry and in the School of Engineering. The AMOS resources include femosecond Ti-Sapphire lasers, OPAs, linear and ring dye lasers, excimer and excimer-pumped lasers, frequency doublers, ultrafast tunable lasers, low-energy ion electron and atomic beam sources, 300 kV ion implanter, and visible and vacuum-ultraviolet spectrometers. Members of the group are involved in cooperative research programs at the Max-Planck-Institute in Garching, the Synchrotron Radiation Center in Madison, Wisconsin, and Universities in Vienna, Berlin, and Krakow.

Members of the biological physics group investigate cellular electric, magnetic, and mechanical phenomena. Included in the laboratories are a magnetic imaging facility with high-resolution SQUID magnetometers and microscopes, scanning stages and magnetic shields, numerous video fluorescence microscopes, and a confocal microscope with a coupled laser microsurgery system. Studies under way are examining the nonlinear electrodynamics of cardiac tissue, intracellular and paracrine signaling in cellular biosystems, coupling among genetics, morphologic change and the mechanics of soft condensed matter, and nanoparticle and nanocluster labeling and spectroscopy. The Project to Instrument and Control the Single Cell is developing tools and techniques for cellular biophysics and wide-bandwidth metabolic measurements. The group, in conjunction with the Vanderbilt Institute for Integrative Biosystems Research and Educations (VIBRE), operates the BioMEMS Fabrication Facility that includes three class-100 clean rooms for photolithography, soft lithography, biomicrofluidics, e-beam and ion-etch fabrication of metal microelectrode arrays, and extensive cellular biophysics instrumentation. The group uses infrared, visible, and ultraviolet spectroscopic techniques for investigating the dynamics of biopolymers and laser-tissue interactions.

The high-energy physics group at Vanderbilt University has research projects at the CMS detector at CERN.

The astronomy group actively uses national and international ground-based and space-platform observatories. Vanderbilt is a partner in the SMARTS (Small and Medium Aperture Research Telescope System) that operates telescopes at the Cerro Tololo Inter-American Observatory in Chile. Vanderbilt is a member of the Sloan Digital Sky Survey (SDSS) III. Vanderbilt University is a charter member of the Extreme Universe Space Observatory (EUSO) Consortium, formed of three U.S. universities, two national laboratories, and seven European nations. The energy, direction, and composition of nature's most energetic particles will be measured.

The Department has exceptionally good research computing facilities at the Advanced Computing Center for Research and Education (ACCRE). The facility is one of the top university supercomputing resources in the United States. There is also a wide variety of modern computational facilities within each of the research groups. Vanderbilt University is a participant in the Internet II network and is thus involved in developing and using the next-generation network.

The materials and nanoscience physics group focuses on the growth and analysis of thin films with enhanced electronic and optical properties. Resources include a 2.0 MeV Van de Graaff accelerator, 4 Kelvin optical cryostate, a 300 KeV accelerator for ion-scattering analysis and ion implantation, an Si-based molecular beam epitaxy (MBE) system, a combined focused ion beam-pulsed laser deposition system for nanostructure fabrication, and various apparatuses for growth, automated nanocrystalline thin film deposition system, annealing, and film measurement. Members of the group collaborate with Oak Ridge National Laboratory, the Engineering School, and other centers for materials science in an extensive interdisciplinary program.

The theoretical condensed-matter physics group and the nuclear theory group have joint research activities with Oak Ridge National Laboratory, where the computational facilities include several massively parallel computers.

Table B—Separately Budgeted Research Expenditures by Source of Support

Source of Support	Departmental Research	Physics-related Research Outside Department
Federal government	$10,525,512	
State/local government		
Non-profit organizations	$164,877	
Business and industry	$619,864	
Other		
Total	$11,310,253	

Table C—Separately Budgeted Research Expenditures by Research Specialty

Research Specialty	No. of Grants	Expenditures ($)
Astronomy	6	$258,371
Atomic, Molecular, & Optical Physics	3	$487,891
Biophysics	6	$6,144,123
Condensed Matter Physics	17	$2,191,523
Nuclear Engineering	3	$795,338
Particles and Fields	6	$1,433,007
Total	41	$11,310,253

FACULTY

Professor

Brau, Charles A., Ph.D., Harvard University, 1965. Theoretical and experimental physics; atomic and molecular physics; lasers; electron accelerators.

Ernst, David J., Ph.D., Massachusetts Institute of Technology, 1970. Nuclear theory; intermediate-energy nuclear reactions; neutrino oscillation phenomenology; hadronic structure.

Gore, John C., Ph.D., University of London, 1976. Development and application of imaging; magnetic resonance imaging and spectroscopic techniques.

Greene, Senta V., Ph.D., Yale University, 1992. Experimental physics; relativistic heavy ion collisions.

Haglund, Richard F., Ph.D., University of North Carolina, 1975. Experimental physics; nanoscale nonlinear optics and phase transitions; laser modification of surfaces and films; free-electron laser applications including polymer thin-film deposition and biomolecular mass spectroscopy.

Hamilton, Joseph H., Ph.D., Indiana University, 1958. Experimental nuclear physics; nuclear structure and reactions with heavy ions; fission processes.

Kephart, Thomas W., Ph.D., Northeastern University, 1981. Theoretical physics; elementary particles; field theory; cosmology.

Maguire, Charles F., Ph.D., Yale University, 1973. Experimental physics; high-energy and relativistic heavy-ion collisions; studies of the nuclear equation of state and the quark-gluon plasma.

Oberacker, Volker E., Ph.D., University of Frankfurt, 1977. Theoretical nuclear physics; computational physics; structure of exotic nuclei.

Pantelides, Sokrates T., Ph.D., University of Illinois, 1973. Theoretical physics; semiconductor physics; first principles atomic-scale dynamics; mesoscopic dynamics in complex solids; interactions of light with matter.

Ramayya, Akunuri V., Ph.D., Indiana University, 1964. Experimental nuclear physics; nuclear structure and reactions with heavy ions; fission processes.

Scherrer, Robert J., Ph.D., University of Chicago, 1986. Theoretical astrophysics: physics of the early universe and the large-scale structure of the universe, including studies of primordial nucleosynthesis, dark energy, the cosmic microwave background, and particle physics.

Sheldon, Paul D., Ph.D., University of California, Berkeley, 1986. Experimental physics; high-energy particles.

Stassun, Keivan, Ph.D., University of Wisconsin-Madison, 2000. Observations and modeling of star formation; science pedagogy; diversity issues.

Tolk, Norman H., Ph.D., Columbia University, 1966. Experimental physics; inelastic interactions with surfaces; particle-solid interactions; quantum-mechanical phase-interference effects; free-electron laser applications.

Umar, Sait A., Ph.D., Yale University, 1985. Theoretical computational physics; nuclear theory; heavy-ion nuclear and atomic physics; models of supernovae.

Velkovska, Julia, Ph.D., Stony Brook University, 1997. Experimental physics; relativistic heavy ion collisions.

Weiler, Thomas J., Ph.D., University of Wisconsin-Madison, 1976. Theoretical physics; elementary particles; high-energy astrophysics; cosmology.

Weintraub, David A., Ph.D., University of California, Los Angeles, 1989. Observational x-ray infrared and submillimeter astronomy; pre-main-sequence stars.

Wikswo, John P., Ph.D., Stanford University, 1975. Biological physics; biomedical engineering; cardiac and cellular electrophysiology; cellular instrumentation and control; complex matter; electromagnetism; nondestructive evaluation, non-linear dynamics, and non-equilibrium behavior; SQUID magnetometry.

Associate Professor

Csorna, Steven E., Ph.D., Columbia University, 1974. Experimental physics; high-energy particles; detector research and development.

Dickerson, James H., Ph.D., Stony Brook University, 2002. Experimental physics; optical and electro-optical spectroscopy of nanostructures; fundamental phenomena of semiconductors and insulators; nanostructured thin films.

Hutson, M. Shane, Ph.D., University of Virginia, 2000. Experimental biophysics.

Johns, Will, Ph.D., University of Colorado Boulder, 1995. Experimental physics; high-energy physics.

Varga, Kalman, Ph.D., University of Debrecen, 1996. Theoretical and computational research on multiscale modeling of materials.

Assistant Professor

Berlind, Andreas, Ph.D., Ohio State University, 2001. Large-scale structure and galaxy formation; ultra-high cosmic energy cosmic rays.

Bolotin, Kirill I., Ph.D., Cornell University, 2006. Experimental condensed matter physics; electron and spin transport in nanoscale systems (graphene).

Holley-Bockelmann, Kelly, Ph.D., University of Michigan, 1999. Galaxy dynamics; N-body simulations; supermassive black holes; gravitational waves.

Rericha, Erin, Ph.D., University of Texas, 2004. Experimental biological physics; effect of cellular environments on cell migration.

Xu, Yaqiong, Ph.D., Rice University, 2006. Interaction between single-walled carbon nanotubes and DNA.

Research Professor

O'Dell, Robert C., Ph.D., University of Wisconsin-Madison, 1962. Optical imaging and spectroscopy; protoplanetary disks; planetary nebula.

Webster, Medford S., Ph.D., University of Washington, 1959. Experimental physics; high-energy physics and astrophysics; photon JETS; high-energy neutrinos.

Research Associate Professor

Hmelo, Anthony B., Ph.D., Stony Brook University, 1987. Materials physics; surface and thin films science; low-dimensional materials.

Idrobo, Juan Carlos, Ph.D., University of California, Davis, 2004. Atomic-scale structure-property relationships of defects and interfaces in complex materials.

Mendenhall, Marcus H., Ph.D., California Institute of Technology, 1983. Experimental physics; free electron laser.

Oxley, Mark, Ph.D., University of Melbourne, 1999. Theoretical physics; first principles of atomic-scale dynamics; semiconductor physics.

Research Assistant Professor

Avanesyan, Sergey, Ph.D., Moscow State University, 1987. Experimental physics; laser modification of surfaces and films.

Bradshaw, Leonard, Ph.D., Vanderbilt University, 1995. Living state physics.

Gabella, William E., Ph.D., University of Colorado, 1991. Experimental; free electron laser.

Hebb, Leslie, Ph.D., The Johns Hopkins University, 2006. Eclipsing binary stars.

Huang, Shengli, Ph.D., University of Science and Technology of China, 2004. Experimental physics; relativistic heavy ion collisions.

Hwang, Jae-Kwang, Ph.D., Yonsei University, 1992. Nuclear structure, experimental low energy physics, nuclear fission dynamics.

Ivanov, Borislav, Ph.D., University of Chemical Technology & Metallurgy, Bulgaria, 1994. Experimental biophysics.

Jarvis, Jonathan, Ph.D., Vanderbilt University, 2009. High-brightness electron sources for free electron lasers.

Puzyrev, Yevgeniy, Ph.D., Florida Atlantic University, 2006. *Condensed Matter Physics.*

Sinha, Manodeep, Ph.D., Pennsylvania State University, 2008. Metal enrichment of early universe.

Stroud, Dina, Ph.D., Vanderbilt University, 2001. Executive Director, Fisk-Vanderbilt Masters-to-PhD Bridge Program. Cardiac repolarization; cardiac function and formation.

Tackett, Alan, Ph.D., Wake Forest University, 1998. Research computing systems administration.

Velkovsky, Momchil, Ph.D., Stony Brook University, 1997. Experimental biological physics.

Posdoctoral Research Associate

Bellovary, Jillian, Ph.D., University of Washington, 2010. Black holes, cosmological simulation of galaxy formations.

Bird, Jonathan, Ph.D., The Ohio State University, 2012. Galaxy formation theory, near field cosmology, numerical galaxy simulations.

Cargile, Phillip, Ph.D., Vanderbilt University, 2010. Stellar formation and evolution through studies of young stars.

DeLee, Nathan, Ph.D., Michigan State University, 2008. Formation and distribution of brown dwarfs in the local galaxy.

Dhital, Saurav, Ph.D., Vanderbilt University, 2012. Binary stars; low-mass stars.

Gregory, Justin, Ph.D., Vanderbilt University, 2013. *Atomic, Molecular, & Optical Physics.*

Ho, Chiu-Man, Ph.D., University of Pittsburg, 2007. *Particles and Fields.*

Krzyzanowska, Halina, Ph.D., Marie Curie-Sklodowska University, 2001. *Condensed Matter Physics, Optics.*

Misra, Rohan, Ph.D., The Ohio State University, 2012. Structure-property correlations in complex oxides, 2D layered materials and group III-V compounds using a combination of STEM and atomistic modeling.

Montez, Rodolfo, Ph.D., Rochester Institute of Technology, 2010. Astrophysics.

Newaz, AKM, Ph.D., SUNY-Stony Brook, 2006. *Condensed Matter Physics.*

Paegert, Martin, Ph.D., Ruhr Universitat Bochum, 2007. Astronomy.

Paxton, William, Ph.D., Vanderbilt University, 2013. Thermal energy conversion via thermionic emission utilizing diamond film.

Sharma, Monika, Ph.D., Panjab University, 2008. CMS.

Shen, Xiao, Ph.D., Stony Brook University, 2009. *Condensed Matter Physics.*

Sherrod, Stacy, Ph.D., Texas A&M University, 2008. Experimental and computational research in the automated inference of mathematical models of metabolic and signaling dynamics of small populations of cells.

Sherrod, Stacy, Ph.D., Texas A&M University, 2008. *Biophysics.*

Shipra, FNU, Ph.D., Jawaharlal Nehru Center for Advanced Scientific Research, 2012. Superconducting thin films.

Wang, Bin, Ph.D., Ecole Normale Superieure de Lyon, 2010. *Condensed Matter Physics.*

Ziegler, Jed, Ph.D., Vanderbilt University, 2011. *Atomic, Molecular, & Optical Physics, Optics.*

DEPARTMENTAL RESEARCH SPECIALTIES AND STAFF

Theoretical

Astronomy. Large-scale structure; galactic dynamics; black holes. Bellovary, Berlind, DeLee, Holley-Bockelmann.

Condensed Matter Physics. Equilibrium atomic configurations and atomic-scale dynamics of bulk defects, surfaces, and interfaces; growth process; grain boundary dynamics; interaction of radiation with materials; many-body effects in nanostructures; ultrafast dynamics of strongly correlated systems. Idrobo, Misra, Oxley, Pantelides, Shipra, Varga.

Cosmology & String Theory. Dark matter; dark energy; primordial nucleosynthesis; physics of the early universe. Kephart, Scherrer, Weiler.

High Energy Physics. Electroweak symmetry breaking; supersymmetry; unification of forces; super-strings; cosmology and particle astrophysics; highest-energy cosmic rays; particle and dark matter phenomenology.

Nuclear Physics. Nuclear structure; intermediate-energy and high-energy nuclear reactions; hadronic structure; supernovae modeling; neutron oscillation phenomenology. Ernst, Oberacker, Umar.

Experimental

Atomic, Molecular, & Optical Physics. Interactions of ions, atoms, electrons, and synchrotron and laser photons with surfaces, interfaces, and thin films, with emphasis on electronic processes; carrier and spin dynamics; damage processes; desorption induced by electronic transitions; quantum interference effects; modification of surface electronic structure. Avanesyan, Brau, Gabella, Gore, Haglund, Ivanov, Jarvis, Tolk.

Biophysics. Action potential propagation, shock response, and non-linear dynamics in excitable system such as cardiac and smooth muscle; magnetic imaging of bioelectric currents, magnetic markers, and remanent geomagnetism; laser-tissue interactions; cellular development; differentiation morphogenesis; fluorescence imaging; instrumenting and controlling single cells; membrane transporters and channels. Bradshaw, Gore, Hutson, Rericha, Stacy Sherrod, Stroud, Wikswo.

Condensed Matter Physics. Growth and characterization of nanostructures; electrophoretic deposition; magnetic and luminescent nanocrystals; carbon nanotubes; graphene. Bolotin, Dickerson, Haglund, Hmelo, Ivanov, Jarvis, Oxley, Pantelides, Tolk, Xu.

High Energy Physics. CMS experiment at CERN. Greene, Huang, Johns, Maguire, Sharma, Sheldon, Shipra, Velkovska.

Nuclear Physics. Nuclear structure; Coulomb excitation; fission processes; in-beam gamma-ray spectroscopy of heavy-ion reactions; isotope separator and recoil mass spectrometer work to study nuclei far from line of stability; behavior of nuclear matter at extreme temperatures; relativistic heavy ion physics at RHIC and CERN; the search for the quark gluon plasma and the colored glass condensate. Greene, Hamilton, Huang, Idrobo, Johns, Maguire, Oxley, Ramayya, Sharma, Shipra, Velkovska.

BAYLOR UNIVERSITY

DEPARTMENT OF PHYSICS

Waco, Texas 76798
http://www.baylor.edu/physics/

General University Information
President: Kenneth W. Starr
Dean of Graduate School: Larry Lyon
University website: http://www.baylor.edu/
Control: Private
Setting: Suburban
Total Faculty: 942
Total Graduate Faculty: 813
Total number of Students: 15,364
Total number of Graduate Students: 2,446

Department Information
Department Chairman: Gregory A. Benesh, Chair
Department Contact: Gregory A. Benesh, Chairman
  Total full-time faculty: 18
  Total number of full-time equivalent positions: 20
  Full-Time Graduate Students: 28
  First-Year Graduate Students: 7
  Female First-Year Students: 2
  Total Post Doctorates: 5

Department Address
One Bear Place, #97316
Waco, TX 76798
Phone: (254) 710-2511
Fax: (254) 710-3878
E-mail: chava_baker@baylor.edu
Website: http://www.baylor.edu/physics/

ADMISSIONS

Admission Contact Information
Address admission inquiries to: Graduate Admissions Office
Phone: (254) 710-3588
E-mail: gerald_cleaver@baylor.edu
Admissions website: http://www.baylor.edu/graduate/index.php?id=42273

Application deadlines
Fall admission:
U.S. students: February 1   *Int'l. students*: February 1

Application fee
U.S. students: $40   *Int'l. students*: $40

Admissions information
For Fall of 2013:
  Number of applicants: 76
  Number admitted: 15
  Number enrolled: 7

Admission requirements
Bachelor's degree requirements: Bachelor's degree in physics is required.
Minimum undergraduate GPA: 3.0

GRE requirements
The GRE is required.
A minimum acceptable score of 1,000 is required on the verbal plus quantitative portions of the General Test.

Advanced GRE requirements
The Advanced GRE is required.

TOEFL requirements
The TOEFL exam is required for students from non-English-speaking countries.
PBT score: 550

Other admissions information
Undergraduate preparation assumed: 32 semester hours of undergraduate physics, including 8 semester hours at senior level; 18 semester hours of undergraduate math, including differential equations.

TUITION

Tuition year 2013-2014:
  Full-time students: $24,426 annual
Deferred tuition plan: Yes
Health insurance: Available
Other academic fees: $142/hr.—11 hrs. or less Student service fees/semester.
Academic term: Semester
Number of first-year students who received full tuition waivers: 7

Teaching Assistants, Research Assistants, and Fellowships
  Number of first-year
    Teaching Assistants: 7
  Average stipend per academic year
    Teaching Assistant: $23,000
    Research Assistant: $23,000
    Fellowship student: $23,000

FINANCIAL AID

Application deadlines
Fall admission:
U.S. students: September 1   *Int'l. students*: September 1

Loans
Loans are not available for U.S. students.
Loans are not available for international students.
GAPSFAS application required: No
FAFSA application required: No

For further information
Address financial aid inquiries to: Dr. Gerald Cleaver, Dept. of Physics, One Bear Place #97316, Waco, TX 76798-7316.
Phone: (254) 710-2283
E-mail: gerald_cleaver@baylor.edu
Financial aid website: http://www.baylor.edu/physics/

HOUSING

Availability of on-campus housing
Single students: Yes
Married students: Yes

For further information

Address housing inquiries to: Campus Living and Learning, Baylor University, One Bear Place #97076, Waco, TX 76798-7076.

Phone: (254) 710-1766

E-mail: living_learning@baylor.edu

Housing aid website: http://www.baylor.edu/graduate/index.php?id=42521

Table A—Faculty, Enrollments, and Degrees Granted

		Enrollment Fall 2012		Number of Degrees Granted 2012-13 (2008–12)		
Research Specialty	2011–12 Faculty	Master's	Doctorate	Master's	Terminal Master's	Doctorate
Atmosphere, Space Physics, Cosmic Rays	2	–	6	–(5)	–	2(3)
Atomic, Molecular, & Optical Physics	1	–	1	–	–	–
Condensed Matter Physics	6	–	6	–(1)	–	1(1)
Cosmology & String Theory	2	–	8	–(2)	–	1(7)
Particles and Fields	4	–	7	1(1)	–	2(5)
Non-specialized	–	–	2	–(1)	–	–
Total	15	–	30	1(10)	–	6(16)
Full-time Grad. Stud.	–	–	30	–	–	–
First-year Grad. Stud.	–	–	7	–	–	–

GRADUATE DEGREE REQUIREMENTS

Master's: 30 semester hours with thesis; 36 semester hours without thesis; minimum GPA of 3.0; residence requirements: two full semesters or three full summers; comprehensive oral examination required for thesis and nonthesis degree; thesis requirements: under supervision of thesis director and three graduate faculty members. Completed draft of research thesis must meet approval of all committee members.

Doctorate: 78 semester hours with dissertation; minimum GPA of 3.0; residence requirements: at least two consecutive semesters (summer does not count as a full semester); preliminary (qualifying) examination: 12 hours written examination covering quantum mechanics, classical mechanics, electricity and magnetism, mathematical physics, statistical mechanics; and other topics (must be taken prior to admission to candidacy for Ph.D.); final oral exam: given after all course, research, and dissertation requirements have been fulfilled; dissertation requirements: covers program of original research, the results of which reveal scholarly competence and are publishable in AIP journals or equivalent.

Thesis: Thesis may be written in absentia.

SPECIAL EQUIPMENT, FACILITIES, OR PROGRAMS

Scanning tunneling microscope (STM) system. Metalorganic chemical vapor deposition (MOCVD) system. Equipment for x-ray diffraction studies. Surface analysis system (XSAM 800) allowing for the characterization of surface and atomic electronic structure(s) using ARXPS, ISS, AES, AM, and LEED. Hypervelocity accelerator lab including dust particle accelerator, light gas gun accelerators, laser gas cell accelerator system and laser hypervelocity impact simulation system, 2 GEC RF/DC Reference Cells used for complex (dusty) plasma and colloidal plasma physics, and an inductively coupled plasma generator (IPG). Low field MRI. Spectra Nd:YAG laser, optical parametric oscillator FTIR spectrometer.

Table B—Separately Budgeted Research Expenditures by Source of Support

Source of Support	Departmental Research	Physics-related Research Outside Department
Federal government	$591,675	
State/local government		
Non-profit organizations	$280,141	
Business and industry		
Other		
Total	$871,816	

Table C—Separately Budgeted Research Expenditures by Research Specialty

Research Specialty	No. of Grants	Expenditures ($)
Atmosphere, Space Physics, Cosmic Rays	2	$111,405
Condensed Matter Physics	6	$269,577
Particles and Fields	4	$270,140
Physics and other Science Education	1	$16,400
Cosmology & String Theory	4	$104,805
Total	17	$772,327

FACULTY

Professor

Benesh, Gregory A., Ph.D., Northwestern University, 1980. *Condensed Matter Physics, Surface Physics*. Theoretical condensed matter physics.

Cleaver, Gerald, Ph.D., California Institute of Technology, 1993. *Cosmology & String Theory, Particles and Fields*. Superstring/M theory.

Hyde, Truell, Ph.D., Baylor University, 1988. *Astrophysics, Plasma and Fusion*. Theoretical and experimental space physics.

Wang, Anzhong, Ph.D., University of Ioannina, 1991. *Cosmology & String Theory, Relativity & Gravitation*. Theoretical gravity and cosmology.

Ward, Bennie, Ph.D., Princeton University, 1973. *Cosmology & String Theory, Particles and Fields*. Theoretical particle physics and quantum general relativity.

Wilcox, Walter M., Ph.D., University of California, Los Angeles, 1981. *Particles and Fields*. Theoretical elementary particle physics.

Associate Professor

Ariyasinghe, Wickramasinghe, Ph.D., Baylor University, 1987. *Atomic, Molecular, & Optical Physics*. Experimental atomic, molecular, and solid state physics.

Dittmann, Jay R., Ph.D., Duke University, 1998. *Particles and Fields*. Experimental high energy particle physics.

Matthews, Lorin, Ph.D., Baylor University, 1998. *Atmosphere, Space Physics, Cosmic Rays, Plasma and Fusion*. Theoretical and experimental space physics.

Olafsen, Jeffrey, Ph.D., Duke University, 1994. *Nonlinear Dynamics and Complex Systems*. Experimental nonlinear dynamics.

Olafsen, Linda, Ph.D., Duke University, 1997. *Condensed Matter Physics*. Experimental condensed matter physics.

Park, Kenneth, Ph.D., University of Rochester, 1993. *Condensed Matter Physics, Surface Physics*. Experimental surface physics.

Russell, Dwight, Ph.D., Vanderbilt University, 1986. *Condensed Matter Physics, Surface Physics*. Experimental surface physics.

Assistant Professor

Hatakeyama, Kenichi, Ph.D., Rockefeller University, 2003. *Particles and Fields.* Experimental high energy particle physics.

Zhang, Zhenrong, Ph.D., Chinese Academy of Sciences, 2002. *Condensed Matter Physics, Surface Physics.* Experimental condensed matter physics.

Lecturer

Bolton, Daniel R., Ph.D., University of Washington, 2011. *Nuclear Physics.* Theoretical nuclear physics.

Kinslow, Linda, Ph.D., Baylor University, 1979. *Theoretical Physics.* Many body theory.

Vasut, John, Ph.D., Baylor University, 2001. *Atmosphere, Space Physics, Cosmic Rays, Plasma and Fusion.* Theoretical dusty plasmas.

DEPARTMENTAL RESEARCH SPECIALTIES AND STAFF

Theoretical

Atmosphere, Space Physics, Cosmic Rays. Charging processes and dust dynamics in space (planetary rings, cometary comas, interplanetary/interstellar dust, protostellar/protoplanetary clouds) and laboratory environments. Hyde, Matthews.

Condensed Matter Physics. Embedding problems; electronic structure of surfaces; surface energies, magnetism, and catalysis. Benesh.

Cosmology & String Theory. Cleaver, Wang. Particles and Fields. Ward, Wilcox.

Experimental

Atmosphere, Space Physics, Cosmic Rays. Hypervelocity impact phenomena as related to orbital debris and fusion devices, dusty plasmas, laser physics, shock physics, in situ instrumentation. Hyde, Matthews.

Atomic, Molecular, & Optical Physics. Heavy-ion-induced Auger electron studies, chemical binding effects on Auger electrons and energy-loss mechanisms, intermediate and high-energy electron scattering. Ariyasinghe.

Condensed Matter Physics. Surface atomic and electronic structure of transition metals and compounds, adsorbate-induced surface modifications, atomically resolved imaging of surface catalytic reactions on model catalytic systems, material characterization under ultrahigh vacuum. Park, Russell, Zhang.

Condensed Matter Physics. Optical and electronic properties of III-V semiconductors; infrared semiconductor lasers. Linda Olafsen.

Nonlinear Dynamics and Complex Systems. Chemical, granular, and soft condensed matter physics; insect biomechanics; dissipative and dynamical systems. Jeffrey Olafsen.

Particles and Fields. Studies of high-energy hadron collisions with the CDF experiment at the Fermi National Accelerator Laboratory and the CMS experiment at CERN. Dittmann, Hatakeyama.

View additional information about this department at
www.gradschoolshopper.com

RICE UNIVERSITY

PROFESSIONAL MASTER'S IN NANOSCIENCES AND SPACE STUDIES

Houston, Texas 77005
http://www.profms.rice.edu

General University Information

President: David Leebron
Dean of Graduate School: Dan Carson
University website: http://rice.edu
Control: Private
Setting: Urban
Total Faculty: 670
Total Graduate Faculty: 510
Total number of Students: 6,400
Total number of Graduate Students: 2,400

Department Information

Department Chairman: Barry Dunning, Director
Department Contact: Dagmar Beck, Program Director
  Total full-time faculty: 36
  Full-Time Graduate Students: 10
  First-Year Graduate Students: 6

Department Address

6100 Main Street
MS-103
Houston, TX 77005
Phone: (713) 348-3188
Fax: (713) 348-3121
E-mail: profms@rice.edu
Website: http://www.profms.rice.edu

ADMISSIONS

Admission Contact Information

Address admission inquiries to: Professional Master's Program, MS-103, Rice University, P.O. Box 1892, Houston, TX 77251-1892
Phone: (713) 348-3188
E-mail: profms@rice.edu
Admissions website: http://www.profms.rice.edu/content.aspx?id=60

Application deadlines
Fall admission:
U.S. students: April 30   *Int'l. students*: April 30
Spring admission:
U.S. students: July 25   *Int'l. students*: July 25

Application fee
There is no application fee required.

Admissions information
For Fall of 2013:
  Number of applicants: 18
  Number admitted: 10
  Number enrolled: 6

Admission requirements
Bachelor's degree requirements: Nanoscience: Bachelor's degree in physics, or related field; and Space Studies: Bachelor's degree in physics, life science, related engineering.
Minimum undergraduate GPA: 3.0

GRE requirements
The GRE is required.
  Quantitative score: 150
  Verbal score: 150
  Analytical score: 3.5

Advanced GRE requirements
The Advanced GRE is not required.
Optional.

TOEFL requirements
The TOEFL exam is required for students from non-English-speaking countries.
  PBT score: 500
  iBT score: 95

Other admissions information
Additional requirements: The average GRE scores for admissions were verbal 150; quantitative 153.
Undergraduate preparation assumed: Nanoscience: Undergraduate courses in quantum mechanics, differential equations, and calculus; and ; Space Studies: science/engineering, advanced math.

TUITION
Tuition year 2012–13:
  Full-time students: $28,000 annual
  Part-time students: $1,400 per credit
This program will cost three semesters of graduate tuition plus a nominal fee for the semester when the student is on internship. Internships are usually paid and income will help with finances.
Credit hours per semester to be considered full-time: 9
Deferred tuition plan: Yes
Health insurance: Yes, 1500.00.
Other academic fees: $560 miscellaneous student fees.
Academic term: Semester

Teaching Assistants, Research Assistants, and Fellowships
Number of first-year
  Teaching Assistants: 3
  Research Assistants: 1
Corporate scholarships are made available to students as they are being received by our corporate supporters.

FINANCIAL AID

Loans
Loans are available for U.S. students.
Loans are available for international students.
GAPSFAS application required: No
FAFSA application required: No

For further information
Address financial aid inquiries to: Office of Financial Aid-MS 12, Rice University, P.O. Box 1892, Houston, TX 77251-1892.
Phone: (713) 348-2139
E-mail: fina@rice.edu
Financial aid website: http://financialaid.rice.edu/

HOUSING

Availability of on-campus housing
  Single students: Yes
  Married students: Yes

For further information
Address housing inquiries to: Graduate House 1515 Bissonnet Houston, TX 77005-1813.
Phone: (713) 348-4723/–4050
E-mail: gradapts@rice.edu
Housing aid website: http://campushousing.rice.edu/rga/

Table A—Faculty, Enrollments, and Degrees Granted

Research Specialty	2011–12 Faculty	Enrollment Fall 2012		Number of Degrees Granted 2012-13		
		Master's	Doctorate	Master's	Terminal Master's	Doctorate
Atomic, Molecular, & Optical Physics	5	–	–	–	–	–
Biophysics	4	–	–	–	–	–
Condensed Matter Physics	1	–	–	–	–	–
Materials Science, Metallurgy	2	–	–	–	–	–
Nano Science and Technology	5	4	–	5(-)	–	–
Total	17	4	–	5(-)	–	–
Full-time Grad. Stud.	–	4	–	–	–	–
First-year Grad. Stud.	–	3	–	–	–	–

GRADUATE DEGREE REQUIREMENTS

Master's: The **Master of Science degree in Nanoscale Physics** will provide students the knowledge to successfully navigate the emerging field of nanoscale science and technology. **The Master of Science in Space Studies** is geared to help individuals increase their knowledge of space engineering, science, program management and policy. The 21-month program begins with two semesters of coursework at Rice followed by a three- to six-month industrial internship. After the internship, students will return to Rice for a final semester of coursework. For the **Nanoscience track** all students are required to complete Nanostructures and Nanotechnology I and II, Methods of Experimental Physics I and II, Characterization and Fabrication at the Nanoscale, and Computational Physics. In addition to these six core physics courses, students will enroll in one management course, one policy course, and a seminar jointly with students involved in the other professional master's tracks. Students will be able to choose four elective courses, for a minimum of 40 total credit hours for the program. At least 24 of these credit hours must be taken at Rice. No thesis is required, however students must present their internship project in both oral and written form. **For the Space Studies track** students will take advanced engineering, biological and physical science classes and introduces students to economics, public policy, and management disciplines, which impact space commercialization and national policy. This program focuses on training scientists and engineers interested in program management providing them with the tools to face the complex challenges inherent in US space policy, human and robotic space exploration, and the role of science in space exploration and technology development.

Other Degrees: The Professional Science Master's Program at Rice University also offers M.S. degrees in Bioscience and Health Policy, Environmental Analysis and Decision Making, and Subsurface Geoscience and SPACE STUDIES.

Thesis: No thesis is required; however, students have to participate in a 3- to 6-month internship.

FACULTY

Professor

Alexander, David, Ph.D., University of Glasgow, 1988. Professor in Physics and Astronomy; Track Director for Space Studies track; Director of Rice Space Institute. *Solar Physics*. Solar activity, sunspots, flares and coronal mass ejections.

Barron, Andrew R., Ph.D., Imperial College of Science and Technology, University of London, 1986. Co-teaching of Management course for Scientists and Engineers. Applications of inorganic chemistry to the materials science of aluminum, gallium, and indium.

Carson, Dan, Ph.D., Temple University, 1979. Dean of the Wiess School of Natural Science Head of PSM Program. Biochemistry and cell biology.

Colvin, Vicki L., Ph.D., University of California, Berkeley, 1994. Nanocrystals, confined liquids and glasses, porous solids, and photonic band gap materials.

Dunning, F. Barry, Ph.D., University College London, 1969. Track Director of Nanoscale Physics program. Experimental atomic and molecular physics; surface physics; spin dependent phenomena; surface magnetism; chemical physics; optics and instrumentation.

Hafner, Jason H., Ph.D., Rice University, 1998. Carbon nanotube synthesis: chemical kinetics and device fabrication. Lipid bilayer substrates for biological atomic force microscopy.

Killian, Thomas C., Ph.D., Massachusetts Institute of Technology, 1999. Atomic, molecular, and optical physics: cold collisions, Bose-Einstein condensation, fundamental measurements, high-resolution spectroscopy, atom-photon interactions, and low temperature plasmas.

Meade, Andrew, Ph.D., University of California. Chair, Department of Mechanical Engineering and Materials Science. *Nonlinear Dynamics and Complex Systems*. Experimental and numerical aerodynamics.

Natelson, Douglas, Ph.D., Stanford University, 1998. Track Adviser for Nanoscale Physics. Nanoscale physics, in particular the electrical and magnetic properties of systems with characteristic dimensions approaching the single-nm scale.

Toffoletto, Frank R., Ph.D., Rice University, 1987. Magnetospheric physics; numerical simulations; space weather.

DEPARTMENTAL RESEARCH SPECIALTIES AND STAFF

Experimental

Astrophysics. Alexander, Meade.

Nano Science and Technology. Dunning, Hafner, Killian, Natelson, Toffoletto.

View additional information about this department at
www.gradschoolshopper.com

TEXAS CHRISTIAN UNIVERSITY

DEPARTMENT OF PHYSICS AND ASTRONOMY

Fort Worth, Texas 76129
http://www.phys.tcu.edu

General University Information

Chancellor: Victor Boschini
Dean of Graduate School: Phil Hartman
University website: http://www.tcu.edu/
Control: Private
Setting: Urban
Total Faculty: 544
Total number of Students: 9,725
Total number of Graduate Students: 1,269

Department Information

Department Chairman: William R.M. Graham, Chair
Department Contact: Marilyn Yates, Administrative Assistant
  Total full-time faculty: 9
  Total number of full-time equivalent positions: 9
  Full-Time Graduate Students: 12

Department Address

TCU Box 298840
Fort Worth, TX 76129
Phone: (817) 257-7375
Fax: (817) 257-7742

E-mail: physics@tcu.edu
Website: http://www.phys.tcu.edu

ADMISSIONS

Admission Contact Information
Address admission inquiries to: Department of Physics & Astronomy Texas Christian University, TCU Box 298840, Fort Worth, TX 76129
Phone: (817) 257-7375
E-mail: physics@tcu.edu
Admissions website: http://www.phys.tcu.edu

Application deadlines
Fall admission:
U.S. students: February 1   *Int'l. students*: February 1
Spring admission:
U.S. students: October 1   *Int'l. students*: October 1

Application fee
U.S. students: $60

Admissions information
For Fall of 2013:
  Number of applicants: 30
  Number admitted: 4
  Number enrolled: 4

Admission requirements
Bachelor's degree requirements: A Bachelor's degree in physics is required.
Minimum undergraduate GPA: 3.0

GRE requirements
The GRE is required.

Advanced GRE requirements
The Advanced GRE is not required.

TOEFL requirements
The TOEFL exam is required for students from non-English-speaking countries.
  PBT score: 550

Other admissions information
Undergraduate preparation assumed: A B.A. or B.S. with a physics major or 24-semester-hour equivalent, including intermediate or advanced undergraduate courses in mechanics, electricity and magnetism, atomic and nuclear or modern physics, or their equivalents is needed. Twelve semester hours must be of junior or senior level. Mathematics through differential equations and a course in general chemistry are required.

TUITION

Tuition year 2013-14:
  Full-time students: $1,125 per credit
  Part-time students: per credit
Holders of assistantships have free tuition.
Credit hours per semester to be considered full-time: 8
Deferred tuition plan: Yes
Health insurance: Available at the cost of 1,774 per year.
Academic term: Semester

Teaching Assistants, Research Assistants, and Fellowships
Average stipend per academic year
  Teaching Assistant: $19,500
  Research Assistant: $19,500

FINANCIAL AID

Application deadlines
Fall admission:
U.S. students: February 1   *Int'l. students*: February 1
Spring admission:
U.S. students: October 1   *Int'l. students*: October 1

Loans
Loans are available for U.S. students.
Loans are not available for international students.
GAPSFAS application required: No
FAFSA application required: No

For further information
Address financial aid inquiries to: Department of Physics & Astronomy, Texas Christian University, TCU Box 298840, Fort Worth, TX 76129.
Phone: (817) 257-7375
E-mail: physics@tcu.edu
Financial aid website: http://www.phys.tcu.edu

HOUSING

Availability of on-campus housing
  Single students: No
  Married students: No

For further information
Address housing inquiries to: Housing and Residential Living Office.
Housing aid website: http://www.wholewideworld.tcu.edu/apartments.asp

Table A—Faculty, Enrollments, and Degrees Granted

Research Specialty	2012-13 Faculty	Enrollment Spring 2013 Master's	Enrollment Spring 2013 Doctorate	Number of Degrees Granted 2012-13 (2007-13) Master's	Number of Degrees Granted 2012-13 (2007-13) Terminal Master's	Number of Degrees Granted 2012-13 (2007-13) Doctorate
Astronomy	2	–	3	–	–	1(3)
Atomic, Molecular, & Optical Physics	4	–	4	–	–(3)	1(9)
Biophysics	2	–	2	–	–	–
Nano Science and Technology	1	–	1	–	–	1(4)
Statistical & Thermal Physics	1	–	2	–	–	–(1)
Total	10	–	12	–	–	3(17)
Full-time Grad. Stud.	–	–	12	–	–	–
First-year Grad. Stud.	–	–	–	–	–	–

GRADUATE DEGREE REQUIREMENTS

Master's: The M.S. degree requires 30 approved semester hours with a thesis or 36 semester hours without a thesis. Course requirements for the degree are: Quantum Mechanics I & II, three courses from Classical Mechanics, Electrodynamics I & II, Solid State Physics or Statistical Physics, plus a minimum of 6 additional semester hours in Physics. An oral examination over coursework and thesis, if any, is required.

Doctorate: The Ph.D. degree is available on a physics, astrophysics, or biophysics track. An M.B.A. degree may also be earned in combination with the Ph.D. Course work: The following core of courses are required and are normally completed during the first four semesters of graduate study. The core courses must be completed with an average grade of 2.75 (out of 4.0) or better. Physics Track: Quantum Mechanics I

& II; 12 semester hours selected from Classical Mechanics, Statistical Physics, Solid State Physics, or Electrodynamics I or II, and 9 semester hours of Research Problems in Physics. Additional coursework may be required to ensure adequate preparation for the specified courses. Astrophysics Track: Astrophysics, Quantum Mechanics I, Electrodynamics I; 12 semester hours selected from Quantum Mechanics II, Classical Mechanics, Statistical Physics or Electrodynamics II; and 9 semester hours of Research Problems in Astronomy. Students may also be required to take Advanced Topics in Astrophysics to ensure an adequate background for their dissertation research. Biophysics Track: Quantum Mechanics I and II; 12 semester hours selected from Nonlinear Dynamics with Applications, Optical Spectroscopy and Fluorescence, Electrodynamics I and II, Statistical Physics; and 9 semester hours of Research Problems in Biophysics. Students may be required to take one or more additional courses from Topics in Biophysics, Experimental Methods in Biochemistry and Biophysics or Advanced Topics in Biophysics to ensure an adequate background for their dissertation research. Each full-time student is required to participate in graduate seminars. The course requirements for any course other than Research Problems may also be met by satisfactory performance on a written examination or by transfer of credit in an equivalent course from another institution. Pre-dissertation examination: This examination is normally taken during the fourth semester of graduate study, and consists of three parts: first, a written report submitted to the advisory committee on a research project either completed or proposed for a dissertation; second, a colloquium based on the written research report; and third, an oral examination over the research report given by the advisory committee and faculty. Successful completion of the predissertation examination and the required core work constitute admission to candidacy for the Ph.D. degree. Unsuccessful completion of the predissertation examination may result in the student being advised to complete the requirements for a Master of Science degree. Dissertation: Completion of a dissertation consisting of an original research project directed by a faculty member at TCU. A final oral examination in defense of the dissertation is required and a paper based on the dissertation research must be submitted for publication in an appropriate scientific journal. Teaching Requirements: Each full-time graduate student pursuing a degree in physics is required to participate in the undergraduate teaching function of the department. The faculty are committed to effective teaching and believe that experience in the teaching of physics is an integral part of graduate education. This requirement is met by assisting in undergraduate laboratories, giving laboratory instructions, grading papers, conducting problems sessions, or offering tutorial help. No more than 8 hours per week. The Ph.D. in Physics is also available with an M.B.A. option. Students entering the Ph.D. program with a B.S. degree are normally expected to compete the Ph.D. requirements within five years. At the end of the fourth year of graduate studies, a candidate for the Ph.D. degree who has demonstrated sufficient progress in dissertation research may apply for the M.B.A. option. During the fifth year the student is expected to continue with the dissertation on a reduced scale, and, if on Departmental Teaching Assistantship, to perform designated departmental teaching duties. Students entering the Ph.D. program with advanced standing (M.S. degree or more) can request an accelerated program. In addition to the coursework, qualifying examinations, and dissertation requirements specified above for the Ph.D. degree in Physics or Astrophysics, students electing to take the M.B.A. option will take 18 hours of M.B.A. coursework during two consecutive semesters as outlined in the TCU Graduate Studies Bulletin. Students are required to attend the Team Building and Skills workshop conducted by the School of Business. Students are assessed a fee for the workshop. The results of the GRE will be accepted in lieu of the GMAT. Students who wish to continue their studies in the program after their first year of business courses and pursue the M.B.A. degree will be required to complete such additional coursework as required of other M.B.A. students and as outlined in the TCU Graduate Studies Bulletin. The maximum term of fellowship or assistantship support through the Department of Physics is 5 years for the Ph.D. Degree or the Ph.D. with M.B.A. Option. Support for M.B.A. courses from the TCU Physics Department fellowships or assistantship is limited to 18 hours. Financial support for the additional 24 hours required for completion of the M.B.A. degree is the student's responsibility. However, students are eligible to apply for financial aid from the School of Business.

Thesis: Thesis may be written in absentia.

SPECIAL EQUIPMENT, FACILITIES, OR PROGRAMS

The physics and astronomy research facilities are housed in spacious, well-equipped laboratories with specialized equipment, including TEM and SEM (transmission and scanning electron microscopes); FTIR (Fourier transform) spectrometers; a Nicolet FTIR spectrometer equipped with infrared microscope; x-ray powder diffractometer closed-cycle cryogenic refrigerators operating at 3 and 10 K for the preparation of condensed-matter samples; Nd-YAG, dye, and Ar lasers; Raman spectrometers with CCD detectors; a confocal microscope, gas adsorption apparatus, and time-resolved spectroscope; a quadrupole mass spectrometer; a CAMAC system for multiparameter coincidence experiments; and a positron lifetime and Doppler broadening apparatus.

Table B—Separately Budgeted Research Expenditures by Source of Support

Source of Support	Departmental Research	Physics-related Research Outside Department
Federal government	$130,000	
State/local government		
Non-profit organizations		
Business and industry		
Other	$35,362	$29,500
Total	$165,362	$29,500

FACULTY

Professor

Graham, W. R. M., Ph.D., York University, Toronto, 1971. *Astrophysics, Atomic, Molecular, & Optical Physics, Chemical Physics, Solid State Physics.* Molecular physics, laboratory astrophysics.

Gryczynski, Z. K., Ph.D., University of Gdansk, 1986. *Biophysics, Medical, Health Physics, Nano Science and Technology, Optics.* Fluorescence.

Kouris, Demitris, Ph.D., Northwestern University, 1987. Dean. *Engineering Physics/Science, Other.* Nanomechanics.

Miller, Bruce Neil, Ph.D., Rice University, 1969. *Nonlinear Dynamics and Complex Systems, Statistical & Thermal Physics.* Structure formation in the expanding universe; self-trapping of quantum particles; chaos in real-world billiards.

Rittby, Magnus, Ph.D., University of Stockholm, 1985. *Atomic, Molecular, & Optical Physics, Computational Physics.*

Zerda, Tadeusz W., Ph.D., Silesian University, 1978. *Atomic, Molecular, & Optical Physics, Materials Science, Metallurgy, Nano Science and Technology.* Nanoscience; molecular physics; high-pressure physics.

Associate Professor

Strzhemechny, Yuri, Ph.D., City University of New York, 2000. Graduate Program Director. *Atomic, Molecular, & Optical Physics, Condensed Matter Physics, Materials Science, Metallurgy, Nano Science and Technology, Optics.* Surface science; optical spectroscopy.

Assistant Professor

Dobrovolny, Hana M., Ph.D., Duke University, 2008. *Biophysics.* Modeling influenza and drug treatments.

Frinchaboy, Peter, Ph.D., University of Virginia, 2006. *Astronomy.*

Instructor

Ingram, Douglas, Ph.D., University of Washington, 1996. *Astronomy.*

DEPARTMENTAL RESEARCH SPECIALTIES AND STAFF

Theoretical

Nonlinear Dynamics and Complex Systems. Miller.
Statistical & Thermal Physics. Miller.

Experimental

Astronomy. Frinchaboy, Ingram.
Astrophysics. Molecular clusters in solids. Graham.
Atomic, Molecular, & Optical Physics. Graham, Rittby, Strzhemechny.
Biophysics. Dobrovolny, Gryczynski.
Nano Science and Technology. High-pressure materials science. Zerda.

View additional information about this department at
www.gradschoolshopper.com

TEXAS STATE UNIVERSITY–SAN MARCOS

DEPARTMENT OF PHYSICS

San Marcos, Texas 78666
http://www.txstate.edu/physics/

General University Information

President: Dr. Denise Trauth
Dean of Graduate School: Dr. J. Michael Willoughby
University website: http://www.txstate.edu/
Control: Public
Setting: Suburban
Total Faculty: 1,588
Total Graduate Faculty: 858
Total number of Students: 34,087
Total number of Graduate Students: 5,128

Department Information

Department Chairman: Thomas Myers, Chair
Department Contact: Dr. Ir. Wilhelmus J. Geerts, Graduate Advisor
Total full-time faculty: 12
Total number of full-time equivalent positions: 14
Full-Time Graduate Students: 10
First-Year Graduate Students: 6
Female First-Year Students: 1
Total Post Doctorates: 4

Department Address

601 University Drive
San Marcos, TX 78666
Phone: (512) 245-1821
Fax: (512) 245-8233
E-mail: wjgeerts@txstate.edu
Website: http://www.txstate.edu/physics/

ADMISSIONS

Admission Contact Information

Address admission inquiries to: Dr.Ir. Wilhelmus J. Geerts, Graduate Advisor, Department of Physics, 601 University Dr., San Marcos, TX 78666
Phone: (512) 245-1821
E-mail: wjgeerts@txstate.edu
Admissions website: http://www.gradcollege.txstate.edu/

Application deadlines

Fall admission:
U.S. students: June 15   *Int'l. students*: June 1
Spring admission:
U.S. students: October 15   *Int'l. students*: October 1

Application fee

U.S. students: $40   *Int'l. students*: $50

Admissions information

For Fall of 2012:
Number of applicants: 12
Number admitted: 8
Number enrolled: 6

Admission requirements

Bachelor's degree requirements: Bachelor's degree in physics is required.
Minimum undergraduate GPA: 3.0

GRE requirements

The GRE is recommended.
If applicants have a GPA below a 3.0 on junior and senior level physics courses and have taken the Graduate Record Exam (GRE) prior to admission with a preferred score of 302, they may be considered for admission.

Advanced GRE requirements
The Advanced GRE is not required.

TOEFL requirements
The TOEFL exam is required for students from non-English-speaking countries.
PBT score: 550
iBT score: 78

Other admissions information
Undergraduate preparation assumed: Halliday, Resnick and Walker, Fundamentals of Physics, Thornton and Marion, Classical Mechanics, Liboff, Introductory Quantum Mechanics, Griffiths, Introduction to Electrodynamics.

TUITION

Tuition year 2013-14:
Tuition for in-state residents
  Full-time students: $6,581.34 annual
  Part-time students: $365.63 per credit
Tuition for out-of-state residents
  Full-time students: $12,953.34 annual
  Part-time students: $719.63 per credit
Credit hours per semester to be considered full-time: 9
Deferred tuition plan: Yes
Health insurance: Available at the cost of $3,012 per year.
Academic term: Semester
Number of first-year students who received full tuition waivers: 3

Teaching Assistants, Research Assistants, and Fellowships
Number of first-year
  Teaching Assistants: 6
  Research Assistants: 2
Average stipend per academic year
  Teaching Assistant: $13,536
  Research Assistant: $15,000

FINANCIAL AID

Application deadlines
Fall admission:
U.S. students: March 15   *Int'l. students*: March 15
Spring admission:
U.S. students: March 15   *Int'l. students*: March 15

Loans
Loans are available for U.S. students.
Loans are available for international students.
GAPSFAS application required: No
FAFSA application required: Yes

For further information
Address financial aid inquiries to: Financial Aid and Scholarships, J. C. Kellam Building, Suite 240, Texas State University-San Marcos, 601 University Dr., San Marcos, Texas 78666–4602.
Phone: (512) 245-2315
E-mail: finaid@txstate.edu
Financial aid website: http://www.finaid.txstate.edu/

HOUSING

Availability of on-campus housing
Single students: Yes
Married students: Yes

For further information
Address housing inquiries to: Housing and Residential Life, J.C. Kellam, Suite 320/380, 601 University Drive, San Marcos, TX 78666.
Phone: (512) 245-2382
E-mail: reslife@txstate.edu
Housing aid website: http://www.reslife.txstate.edu/

Table A—Faculty, Enrollments, and Degrees Granted

Research Specialty	2012-13 Faculty	Enrollment Spring 2013 Master's	Enrollment Spring 2013 Doctorate	Number of Degrees Granted 2012-13 (2007-11) Master's	Number of Degrees Granted 2012-13 (2007-11) Terminal Master's	Number of Degrees Granted 2012-13 (2007-11) Doctorate
Astronomy	2	–	–	–	–	–
Astrophysics	1	–	–	–	–	–
Atomic, Molecular, & Optical Physics	1	1	–	–	-(4)	–
Computational Physics	2	2	–	–	1(1)	–
Condensed Matter Physics	4	4	–	–	3(15)	–
Nano Science and Technology	2	2	–	–	2(3)	–
Optics	1	1	–	–	1(-)	–
Physics Education Research	3	1	–	–	–	–
Relativity & Gravitation	1	–	–	–	–	–
Non-specialized	–	1	–	–	-(2)	–
Total	12	12	–	–	7(25)	–
Full-time Grad. Stud.	–	12	–	–	–	–
First-year Grad. Stud.	–	6	–	–	–	–

GRADUATE DEGREE REQUIREMENTS

Master's: Physics Master's A Masters of Science degree with thesis requires 15–18 semester hours of physics coursework, 6–9 hours of coursework in another science, and a minimum of 6 hours of thesis supported by a thesis oral defense. A Master of Science degree without a thesis is available which requires 6 additional hours of physics coursework in lieu of the thesis, and 6 hours of additional coursework. Materials Physics Master's A Master of Science degree requiring 35 semester hours of physics coursework. The curriculum emphasizes topics of interest to the microelectronics industry with a required thesis and industrial internship. In all degrees, an oral thesis defense is required. A minimum of 3.0/4.0 GPA is required for graduation, and a minimum of 24 hours of coursework must be completed on campus. There is no foreign language requirement.

Doctorate: Texas State also offers a Ph.D. in Materials Science, Engineering and Commercialization. More details on this program can be found at: http://www.gradcollege.txstate.edu/msec.html.

Thesis: Thesis may be written in absentia.

SPECIAL EQUIPMENT, FACILITIES, OR PROGRAMS

The Department has several research concentrations.

The materials physics and thin film solid state groups are focused toward preparing Master's graduates for professional employment, including the semiconductor industry and materials high tech industry, or further graduate study in a doctoral program. Thesis research may utilize thin film sputtering (magnetron and dual ion beam), molecular beam epitaxy, scanning electron microscopy, energy dispersive x-ray spectroscopy, infrared spectroscopy, high resolution x-ray diffraction/reflectivity, scanning probe microscopy (AFM and STM), magnetometry, resistivity, high temperature furnaces/ovens, ellipsometry, electric transport measurements, deep level transient spectroscopy, impedance spectroscopy, and photoluminescence. Competitive opportunities for industry internships are available.

The physics education research (PER) group focuses on embodied and participationist models of learning, including gesture, conceptual metaphor, conceptual blending, communities of practice, relational discourse, and identity development. These research areas are pursued through qualitative analysis of video records of interactions between students, environment, and teachers; quantitative analysis of standard conceptual and attitudinal surveys; and hybrid analysis of student written and graphical artifacts. Graduate study in PER prepares students for careers in K-14 physics education, and further graduate study in a doctoral program in either physics or education.

Other research groups focus on instrumentation development and theory. Theoretical focus is on the study of the physical properties of materials through computational simulations, either using first principles methods from density functional theory (DFT) or approximation methods within the effective mass theory, with an emphasis on semiconductors and oxides. Texas State also offers a Ph.D. in Materials Science, Engineering and Commercialization which has many collaborations with these areas.

A final research area consists of computational modeling of historical events in astronomy. The department maintains a small astronomical observatory (16″ reflector) for student use.

Table B—Separately Budgeted Research Expenditures by Source of Support

Source of Support	Departmental Research	Physics-related Research Outside Department
Federal government	$775,000	
State/local government		
Non-profit organizations		
Business and industry	$200,000	
Other		
Total	$975,000	

Table C—Separately Budgeted Research Expenditures by Research Specialty

Research Specialty	No. of Grants	Expenditures ($)
Condensed Matter Physics	10	$800,000
Physics Education Research	1	$175,000
Polymer Physics/Science	1	
Total	12	$975,000

FACULTY

Professor

Donnelly, David, Ph.D., University of California, Santa Barbara, 1990. Undergraduate Advisor. Society of Physics Students Advisor. *Condensed Matter Physics, Physics and other Science Education.*

Droopad, Ravi, Ph.D., Imperial College, London, 1989. *Condensed Matter Physics, Materials Science, Metallurgy, Nano Science and Technology.* Experimental condensed matter.

Holz, Mark, Ph.D., Virginia Polytechnic Institute, 1987. *Atomic, Molecular, & Optical Physics, Condensed Matter Physics, Nano Science and Technology, Solid State Physics.*

Myers, Tom, Ph.D., North Carolina State University. Interim department chair Physics. Director of the Materials Science Engineering and Commercialization. Associate Dean College of Science. *Condensed Matter Physics, Materials Science, Metallurgy, Optics.*

Olson, Donald W., Ph.D., University of California, Berkeley, 1975. *Astronomy, Astrophysics, Computational Physics, Relativity & Gravitation.* Astrophysics; general relativity; computational astronomy.

Piner, Edwin, Ph.D., North Carolina State University, 1998. *Condensed Matter Physics, Materials Science, Metallurgy, Nano Science and Technology.* GaN Devices.

Associate Professor

Geerts, Wilhelmus J., Ph.D., University of Twente, Enschede, The Netherlands, 1992. *Applied Physics, Condensed Matter Physics, Optics, Polymer Physics/Science, Systems Science/Engineering.* Nanostructured magnetic & semiconductor materials: electrical & optical characterization; Instrumentation development; solar technology.

Spencer, Gregory W., Ph.D., University of Florida, 1986. Undergraduate Advisor. *Condensed Matter Physics, Materials Science, Metallurgy, Nano Science and Technology.* Advanced electronic devices and materials.

Assistant Professor

Close, Hunter G., Ph.D., University of Washington, 2005. *History & Philosophy of Physics/Science, Physics and other Science Education.*

Lee, Byounghak, Ph.D., Indiana University, 2002. *Computational Physics, Condensed Matter Physics, Materials Science, Metallurgy, Nano Science and Technology.* First principles simulations of electronic structure of nanomaterials.

Theodoropoulou, Nikoleta, Ph.D., University of Florida, 2002. *Condensed Matter Physics, Materials Science, Metallurgy, Nano Science and Technology.* Properties of dilute magnetic nanostructures.

Senior Lecturer

Close, Eleanor, Other, Seattle Pacific University, 2009. Learning Assistant Program Director. *Physics and other Science Education.*

Lecturer

Doescher, Russell L., M.S., Southwest Texas State University, 1992. *Astronomy, Astrophysics.* Computational and observational astronomy; physics education.

Mount, Jennifer, Ph.D., University of Texas - Austin, 2012. *Physics and other Science Education.*

Scolfaro, Luisa, Ph.D., University of Sao Paulo, 1988. *Computational Physics, Condensed Matter Physics, Materials Science, Metallurgy.*

DEPARTMENTAL RESEARCH SPECIALTIES AND STAFF

Theoretical

Computational Physics. Theoretical focus is on the study of the physical properties of materials through computational simulations, either using first principles methods from density functional theory (DFT) or approximation methods within the effective mass theory, with an emphasis on semiconductors and oxides. Lee, Scolfaro.

Physics Education Research. The physics education research (PER) group focuses on embodied and participationist models of learning, including gesture, conceptual metaphor, conceptual blending, communities of practice, relational discourse, and identity development. These research areas are pursued through qualitative analysis of video records of interactions between students, environment, and teachers; quantitative analysis of standard conceptual and attitudinal surveys; and hybrid analysis of student written and graphical artifacts. Graduate study in PER prepares students for careers in K-14 physics education, and further graduate study in a doctoral program in either physics or education. Eleanor Close, Hunter Close, Donnelly, Mount.

Experimental

Adaptive Optics. The use of adaptive optics, including spatial light modulators, in Kerr microscopoy and optical lithography is investigated. Geerts.

Condensed Matter Physics. Dilute magnetic nanostructures, giant magnetoresistance. Theodoropoulou.

Forensic Astronomy. Computational modeling of historical events in astronomy. Doescher, Olson.

Materials Science, Metallurgy. We operate and maintain a large array of laboratory equipment that allows us to conduct research across the entire process of growing and characterizing, semiconductor and oxide materials. We are focused on the development of novel III-V and II-VI materials for next generation semiconductor device and oxide based multiferroic devices. Our principal focus is on detailed understanding of materials properties and their interaction with the electrical performance in devices. Our goal is to enable higher-k materials and other oxides for use in manufacturing process flows. We partner with leading companies in all areas of manufacturing these devices in order to accelerate the process of innovation. Droopad, Holz, Myers.

Materials Science, Metallurgy. MOCVD of GaN Devices grown on Silicon substrates for high power electronics and high electron mobility transistors. Piner.

Micro mechanics. Study of MEMs device including the use of various etching technologies to create them. Spencer.

Nano Science and Technology. 1. Optical properties of semiconductors, properties of aluminum-rich III-Nitride based devices, self-assembly in metallic thin-films, nanophotonic structures, growth and fundamental properties of semiconductor nanowires, nanoscale materials properties related to combustion. 2. Properties of silicon nanoparticles formed by annealing of thin films. Holz, Spencer.

Optical Materials Properties. The optical properties of bulk materials, thin films, and micro- and nanostructured materials are studied by Raman spectroscopy, ellipsometry, Magneto-Optical spectroscopy, FTIR, and photoluminescence experiments. Donnelly, Droopad, Geerts, Holz, Myers.

Physics Education Research. The physics education research (PER) group focuses on embodied and participationist models of learning, including gesture, conceptual metaphor, conceptual blending, communities of practice, relational discourse, and identity development. These research areas are pursued through qualitative analysis of video records of interactions between students, environment, and teachers; quantitative analysis of standard conceptual and attitudinal surveys; and hybrid analysis of student written and graphical artifacts. Graduate study in PER prepares students for careers in K-14 physics education, and further graduate study in a doctoral program in either physics or education. Eleanor Close, Hunter Close, Donnelly, Mount.

View additional information about this department at www.gradschoolshopper.com

---

UNIVERSITY OF HOUSTON

DEPARTMENT OF PHYSICS

Houston, Texas 77204-5005
http://www.phys.uh.edu/

General University Information

President: Renu Khator
Dean of Graduate School: Dmitri Litvinov
University website: http://www.uh.edu/
Control: Public
Setting: Urban

Department Information

Department Chairman: Gemunu Gunaratne, Chair
Department Contact: Dr. Gemunu Gunaratne, Chair
  Total full-time faculty: 30
  Total number of full-time equivalent positions: 55
  Full-Time Graduate Students: 104
  First-Year Graduate Students: 27
  Female First-Year Students: 5
  Total Post Doctorates: 21

Department Address

4800 Calhoun Road
617 Science & Research, Building 1
Houston, TX 77204-5005
Phone: (713) 743-3534
Fax: (713) 743-3589
E-mail: gemunu@uh.edu
Website: http://www.phys.uh.edu/

ADMISSIONS

Admission Contact Information

Address admission inquiries to: 617 Science & Research Building 1, 4800 Calhoun Road, Houston, TX 77204-5005
Phone: (713) 743-3550
Admissions website: http://www.uh.edu

Application deadlines

Fall admission:
U.S. students: February 1   *Int'l. students*: February 1

Application fee
Int'l. students: $75

Admissions information
For Fall of 2013:
Number of applicants: 107
Number admitted: 27
Number enrolled: 24

Admission requirements
Bachelor's degree requirements: Bachelor's degree is required.
Minimum undergraduate GPA: 3.0

GRE requirements
The GRE is required.

Advanced GRE requirements
The Advanced GRE is recommended.
No minimum requirement.

TOEFL requirements
The TOEFL exam is required for students from non-English-speaking countries.
iBT score: 79

Other admissions information
Additional requirements: For international students please follow the link: http://phys.uh.edu/graduate/application-intl/index.php
For domestic student requirements please follow the link: http://phys.uh.edu/graduate/application-domestic/index.php.

TUITION

Tuition year 2012–13:
Tuition for in-state residents
Full-time students: $320 per credit
Part-time students: $320 per credit
Tuition for out-of-state residents
Full-time students: $671 per credit
Part-time students: $671 per credit
Credit hours per semester to be considered full-time: 9
Deferred tuition plan: Yes
Health insurance: Available at the cost of 1108 per year.
Other academic fees: Please refer to website for additional fees: http://www.uh.edu/financial/graduate/tuition-fees/required-fees/index.php
Academic term: Semester
Number of first-year students who received full tuition waivers: 27

Teaching Assistants, Research Assistants, and Fellowships
Number of first-year
Teaching Assistants: 23
Research Assistants: 2
Fellowship students: 2
Average stipend per academic year
Teaching Assistant: $23,400
Research Assistant: $22,200
Fellowship student: $24,600

FINANCIAL AID

Application deadlines
Fall admission:
U.S. students: September 1   *Int'l. students*: September 1
Spring admission:
U.S. students: January 20   *Int'l. students*: January 20

Loans
Loans are available for U.S. students.
Loans are not available for international students.
GAPSFAS application required: Yes
FAFSA application required: Yes

For further information
Address financial aid inquiries to: Office of Scholarships and Financial Aid, University of Houston, 31 E Cullen Building, Houston,TX 77204-2010.
Phone: 713-743-1010
E-mail: sfa@uh.edu
Financial aid website: http://www.uh.edu/financial/

HOUSING

Availability of on-campus housing
Single students: Yes
Married students: Yes

For further information
Address housing inquiries to: On-Campus Housing Office, 4401 Wheeler Street Room 15J, Houston, TX 77204-3018.
Phone: 713-743-6000
Housing aid website: http://housing.uh.edu/

Table B—Separately Budgeted Research Expenditures by Source of Support

Source of Support	Departmental Research	Physics-related Research Outside Department
Federal government	$14,839,822.21	
State/local government	$594,016	
Non-profit organizations	$758,660	
Business and industry	$4,056,187	
Other	$2,648,143.3	
Total	$22,896,828.51	

FACULTY

Distinguished University Professor
Chu, Wei-Kan, Ph.D., Baylor University, 1969.

Chair Professor
Chu, Ching, Ph.D., University of California, San Diego, 1968.
Hungerford, Ed, Ph.D., Georgia Tech, 1967.
Ignatiev, Alex, Ph.D., Cornell University, 1972.
Kouri, Donald, Ph.D., University of Wisconsin, 1965.
Ren, Zhifeng, Ph.D., Chinese Academy of Sciences, 1990.
Weglein, Arthur, Ph.D., City College of the City - University of New York, 1975.

Professor
Bassler, Kevin E., Ph.D., Carnegie Mellon University, 1990.
Bellwied, Rene, Ph.D., Gutenberg University, Mainz, 1989.
Bering, Edgar, Ph.D., University of California, Berkeley, 1974.
Bittner, Eric, Ph.D., University of Chicago, 1994.
Gunaratne, Gemunu, Ph.D., Cornell University, 1986.
Hu, Bambi, Ph.D., Cornell University, 1974.
Lau, Kwong, Ph.D., University of Maryland, 1981.
Miller, John, Ph.D., University of Illinois at Urbana-Champaign, 1985.
Pan, Shuheng H., Ph.D., University of Texas at Austin, 1991.
Pinsky, Lawrence S., Ph.D., University of Rochester, 1973.
Reiter, George, Ph.D., Stanford University, 1967.
Selvamanickam, Venkat, Ph.D., University of Houston.
Sharma, Pradeep, Ph.D., University of Maryland, College Park, 2000.
Su, Wu-Pei, Ph.D., University of Pennsylvania, 1981.
Ting, Chin-Sen, Ph.D., University of California at San Diego, 1970.

Wood, Lowell, Ph.D., University of Texas at Austin, 1968.

Associate Professor

Cheung-Wyker, Margaret, Ph.D., University of California at San Diego, 2003.

Curran, Seamus A., Ph.D., Trinity College Dublin, 1995.

Hor, Pei-Herng, Ph.D., University of Houston, 1990.

Lubchenko, Vassiliy, Ph.D., University of Illinois at Urbana-Champaign, 2002.

Ordonez, Carlos, Ph.D., University of Texas, 1986.

Stokes, Donna, Ph.D., University of Houston, 1998.

Varghese, Oomman, Ph.D., Indian Institute of Technology, Delhi.

Assistant Professor

Chen, Shuo, Ph.D., Boston College, 2006. Energy conversion, Energy Storage, Transmission electron microscopy, Synthesis of nano material.

Das, Mini, Ph.D., Indian Institute of Technology, Delhi.

Li, Liming, Ph.D., California Institute of Technology, 2006.

Peng, Haibing, Ph.D., Harvard University, 2004.

Timmins, Anthony, Ph.D., University of Birmingham, 2008. Relativistic heavy ion physics.

Whitehead, Lisa, Ph.D., Stony Brook University, 2007.

DEPARTMENTAL RESEARCH SPECIALTIES AND STAFF

Theoretical

Biological & Medical Physics. Bassler, Cheung-Wyker, Das, Gunaratne, Su.

Econophysics. Bassler, Gunaratne, Reiter.

High Energy, Medium Energy & Heavy Ion Physics. Ordonez.

Medical Imaging Physics. Das.

Nanophysics. Cheung-Wyker.

Network Science. Bassler, Gunaratne.

Planetary Science. Li.

Scattering: X-ray, Neutron Light. Das, Reiter.

Seismic Physics. Weglein.

Signal & Image Processing. Das.

Solar (Renewable) Energy. Cheung-Wyker, Curran.

Space Radiation Physics. Pinsky.

Statistical & Non-linear Physics & Complexity. Bassler, Bittner, Cheung-Wyker, Gunaratne, Hu, Lubchenko.

Superconductivity. Bassler, Reiter, Su, Ting.

Experimental

Biological & Medical Physics. Curran, Das, Ignatiev, Miller.

High Energy, Medium Energy & Heavy Ion Physics. Lau, Pinsky, Whitehead.

Medical Imaging Physics. Das, Wood.

Nanophysics. Curran, Ignatiev, Peng, Varghese.

Planetary Science. Li.

Scattering: X-ray, Neutron Light. Das.

Signal & Image Processing. Das, Kouri.

Solar (Renewable) Energy. Curran, Selvamanickam, Stokes, Varghese.

Space Physics. Bering.

Statistical & Non-linear Physics & Complexity. Curran.

Superconductivity. Ching Chu, Wei-Kan Chu, Hor, Miller, Pan.

Surface Physics. Ignatiev, Stokes, Varghese.

View additional information about this department at
www.gradschoolshopper.com

UNIVERSITY OF NORTH TEXAS

DEPARTMENT OF PHYSICS

Denton, Texas 76203
http://www.physics.unt.edu

General University Information

President: Dr. V. Lane Rawlins
Dean of Graduate School: Dr. Mark Wardell
University website: http://www.unt.edu/
Control: Public
Setting: Suburban
Total Faculty: 1,047
Total Graduate Faculty: 887
Total number of Students: 35,694
Total number of Graduate Students: 7,412

Department Information

Department Chairman: David Schultz, Chair

Department Contact: Michelle Rumer, Administrative Coordinator
Total full-time faculty: 28
Total number of full-time equivalent positions: 43
Full-Time Graduate Students: 67
First-Year Graduate Students: 3
Total Post Doctorates: 4

Department Address

1155 Union Circle #311427
Denton, TX 76203
Phone: (940) 565-2626
Fax: (940) 565-2515

E-mail: physics@unt.edu
Website: http://www.physics.unt.edu

ADMISSIONS

Admission Contact Information
Address admission inquiries to: Admissions Office, University of North Texas, 1155 Union Circle, #311067, Denton, TX 76203
Phone: (940) 565-3255
E-mail: michelle.rumer@unt.edu
Admissions website: http://phys.unt.edu/

Application deadlines
Fall admission:
U.S. students: July 1   *Int'l. students*: March 15
Spring admission:
U.S. students: November 15   *Int'l. students*: September 15

Application fee
U.S. students: $60   *Int'l. students*: $95

Admissions information
For Fall of 2012:
   Number of applicants: 55
   Number admitted: 16
   Number enrolled: 7

Admission requirements
Bachelor's degree requirements: Bachelor's degree in physics is required. The transcripts must show you have earned a bachelor's degree from a regionally accredited institution and earned the required grade point average: GPA requirement for UNT consideration: 2.8 GPA or higher on your undergraduate degree OR 3.0 GPA on the last 60 hours of your undergraduate degree OR 3.4 GPA on a completed master's degree. Students not meeting GPA requirements may be eligible for admission under certain conditions or if they are seeking a second bachelor's degree. See the Graduate Catalog for details.
Minimum undergraduate GPA: 2.8

GRE requirements
The GRE is required.
   Quantitative score: 640
   Analytical score: 4

Advanced GRE requirements
The Advanced GRE is required.

TOEFL requirements
The TOEFL exam is required for students from non-English-speaking countries.
   PBT score: 550
   iBT score: 79

Other admissions information
Additional requirements: The average GRE scores for admissions are quantitative-710 and analytical writing-4.0.
Undergraduate preparation assumed: Cutnell & Johnson, Physics; Krane, Modern Physics; Fowles, Analytical Mechanics; Arfken, Essential Mathematical Methods for Physicists; Serway & Jewett, Physics for Scientists & Engineers; Griffith, Introduction to Quantum Mechanics; Bowley, Introductory Statistical Mechanics.

TUITION

Tuition year 2013–14:
Tuition for in-state residents
   Full-time students: $3,467.65 per semester
   Part-time students: $586.05 per credit
Tuition for out-of-state residents

   Full-time students: $6,653.65 per semester
   Part-time students: $940.05 per credit
Eighteen credit hours/calendar year out-of-state tuition waived for employees such as teaching assistants and research assistants (nine credit hours/long semester).
Credit hours per semester to be considered full-time: 9
Deferred tuition plan: Yes
Health insurance: Not available.
Other academic fees: Health insurance is available for TAs.
Academic term: Semester
Number of first-year students who received full tuition waivers: 1
Number of first-year students who received partial tuition waivers: 1

Teaching Assistants, Research Assistants, and Fellowships
   Number of first-year
      Teaching Assistants: 3
   Average stipend per academic year
      Teaching Assistant: $17,836
      Research Assistant: $17,836

FINANCIAL AID

Application deadlines
Fall admission:
U.S. students: March 31
Spring admission:
U.S. students: August 15   *Int'l. students*: August 15

Loans
Loans are available for U.S. students.
Loans are not available for international students.
GAPSFAS application required: No
FAFSA application required: No

For further information
Address financial aid inquiries to: Student Financial Aid, 1155 Union Circle, #311370, ESSC Room 228, Denton, TX 76203-5017.
Phone: (940) 565-2302
Financial aid website: http://financialaid.unt.edu/basics

HOUSING

Availability of on-campus housing
   Single students: Yes
   Married students: No

For further information
Address housing inquiries to: Housing Dept., University of North Texas, P.O. Box 13617, Denton, TX 76203-3617.
Phone: (940) 565-2610
E-mail: housinginfo@unt.edu
Housing aid website: http://reslife.unt.edu/

Table A—Faculty, Enrollments, and Degrees Granted

Research Specialty	2012-2013 Faculty	Enrollment Fall 2012		Number of Degrees Granted Fall 2012		
		Master's	Doctorate	Master's	Terminal Master's	Doctorate
Physics and other **Science Education**	28	5	67	2(-)	–	5(-)
Total	–	3	64	–	–	–
Full-time Grad. Stud.	–	5	67	–	–	–
First-year Grad. Stud.	–	3	3	–	–	–

GRADUATE DEGREE REQUIREMENTS

Master's: Option 1: M.A. requiring 30 semester credit hours, including six hours of thesis work. Option 2: M.S. requiring 33 semester credit hours, including six hours of problems in lieu of thesis, which are independent but not necessarily original studies that may be experimental, computational, tutorial, bibliographic, pedagogic, or a combination of these. As part of the requirements for each problems course, the student must present a formal written report of the work done in the course, which must be approved by an advisory committee. Option 3: M.S. requiring 36 semester credit hours with a coursework option.

Doctorate: A Ph.D. candidate must pass a qualifying examination over the core material of the graduate curriculum and receive approval of a research proposal from an advisory committee or complete six (6) core courses with a minimum of three "A's" and three "B's" and receive approval from a research proposal from an advisory committee to be accepted into candidacy. Coursework and research amounting to the equivalent of two academic years beyond the master's degree or three years beyond the bachelor's degree may be considered a minimum.

Thesis: Thesis may be written in absentia.

SPECIAL EQUIPMENT, FACILITIES, OR PROGRAMS

Atomic scattering physics laboratory.

Ion beam modification and analysis laboratory.

Nanoscale materials synthesis and characterization laboratory.

Polymer gels and hydrogels laboratory.

Precision atomic physics measurements laboratory.

Remote access and public astronomical observatories.

Scanning tunneling microscopy laboratory.

Semiconductor materials and devices characterization laboratory.

Sputter-initiated resonance ionization spectroscopy laboratory.

Two- and three-dimensional photonic materials synthesis and characterization laboratories.

Ultrafast spectroscopy and nanophotonics laboratory.

Table B—Separately Budgeted Research Expenditures by Source of Support

Source of Support	Departmental Research	Physics-related Research Outside Department
Federal government	$3,100,250	
State/local government	$121,000	
Non-profit organizations	$600,000	
Business and industry		
Other		
Total	**$3,821,250**	

Table C—Separately Budgeted Research Expenditures by Research Specialty

Research Specialty	No. of Grants	Expenditures ($)
Astronomy	4	$130,450
Atomic, Molecular, & Optical Physics	6	$550,000
Condensed Matter Physics	2	$350,000
Nonlinear Dynamics and Complex Systems	8	$2,000,000
Solid State Physics	2	$600,000
Total	**22**	**$3,630,450**

FACULTY

Professor

Buongiorno Nardelli, Marco, Ph.D., International School for Advanced Studies, 1993. *Computational Physics, Theoretical Physics.* Theoretical and computational materials physics; materials informatics; computational materials and high-performance simulations; materials and processes for energy and environment applications; nano-catalysis; molecular electronics at the nanoscale and quantum electronic and thermal transport in molecules and nanoscale materials; design of novel electronic devices; physics and chemistry at interfaces and surfaces; theoretical developments of ab initio DFT-based methods and multiscale computational techniques.

Glass, Gary A., Ph.D., University of Tennessee, 1984. *Applied Physics.* High-energy focused ion beam (HEFIB) microscopy and microfabrication; HEFIB technology development; ion beam analysis and modification.

Grigolini, Paolo, Ph.D., University of Pisa, 1969. *Biophysics, Nonlinear Dynamics and Complex Systems.* Science of complexity; joint action of order and randomness as a source of long-range correlation; self-organization in physics, biology, and material science; from dynamics to thermodynamics and from quantum to classical physics: the anomalous versus the ordinary statistical mechanical perspective.

Kobe, Donald H., Ph.D., University of Minnesota, 1961. *Low Temperature Physics, Solid State Physics.* Quantum mechanics; quantum field theory; electromagnetic theory; classical mechanics.

Krokhin, Arkadii, Ph.D., Kiev State University, Ukraine, 1983. *Nonlinear Dynamics and Complex Systems, Solid State Physics.* Solid-state physics; dynamical systems; quantum chaos.

Littler, Chris, Ph.D., North Texas State University, 1984. *Solid State Physics.* Solid state physics; semiconductor materials growth; scattering mechanisms and influence of impurities; transport coefficients; laser spectroscopy.

Matteson, Samuel E., Ph.D., Baylor University, 1976. *Applied Physics, Astronomy.* Ion-solid interactions; ion-beam analysis of materials; semiconductor materials; acoustics; ion optics.

McDaniel, Floyd, Ph.D., University of Georgia, 1971. Regents Professor. *Applied Physics.* Semiconductor materials; materials science; accelerator mass spectrometry; ion-atom interactions; nuclear physics.

Mueller, Dennis W., Ph.D., University of Nebraska-Lincoln, 1988. Atomic and molecular physics; nano technology; non-destructive inspection (NDI); instrumentation and measurement science; microelectronics.

Neogi, Arup, Ph.D., Yamagata University, Japan, 1992. *Nano Science and Technology, Optics.* Semiconductor nanostructures/semiconductor physics; lasers and nonlinear optics; ultrafast optics and optical spectroscopy; optical communication; plasmonic nanoscience.

Ordonez, Carlos A., Ph.D., University of Texas, Austin. *Plasma and Fusion.* Penning trap-related research including antimatter studies; cryogenic heat engine research; plasma space-charge research, including plasma sheath-related studies.

Perez, Jose M., Ph.D., University of California, Berkeley, 1983. *Solid State Physics.* Nanotechnology; carbon nanotubes; quantum dots; spintronics; scanning tunneling microscopy.

Quintanilla (Ward), Sandra, Ph.D., University of London, 1986. *Atomic, Molecular, & Optical Physics.* Theoretical atomic physics, in particular three-body (Coulomb) problems; low- and high-energy positron collisions; electron collisions; photodetachment of negative ions; variational principles; modified effective range theories; formulation of atomic theory.

Roberts, James A., Ph.D., University of Oklahoma, 1967. Microwave interaction with matter; molecular and atomic spectroscopy; plasma physics; electronic systems; astronomy.

Schultz, David R., Ph.D., University of Missouri, Rolla, 1989. *Astrophysics, Atomic, Molecular, & Optical Physics, Plasma and Fusion*. Interactions of photons, electrons, ions, and atoms with atoms and molecules; atomic fusion and plasma science; development of electrostatic, atomic, and molecular ion storage rings for basic physical, chemical, and biological science.

Shiner, David, Ph.D., University of Michigan, 1988. *Atomic, Molecular, & Optical Physics, Optics*. Atomic physics; laser spectroscopy; precision measurement.

Associate Professor

Drachev, Vladimir, Ph.D., Novosibirsk State University, Institute of Semiconductor Physics Russian Academy of Sciences, 1995. *Nano Science and Technology, Optics*. Optics and spectroscopy, nonlinear optics, nanotechnology, nanophotonics, ultrafast optics, nanomaterials.

Kowalski, Jacek M., Ph.D., Poland, 1973. *Nonlinear Dynamics and Complex Systems*. Statistical physics; quantum theory of magnetism; artificial and biological neuronal networks.

Lin, Yuankun, Ph.D., University of British Columbia, 2000. *Optics*. Nano-photonic; laser optics; optical micro-/nano-fabrication; photonic band-gap materials; plasmonic devices.

Rout, Bibhudutta, Ph.D., Institute of Physics, Bhubaneswar, India, 2001. *Solid State Physics*. Growth, characterization, and thermal behavior of epitaxial metallic layers on semiconductors and their self-assembled microstructures.

Weathers, Duncan, Ph.D., California Institute of Technology, 1989. *Applied Physics*. Ion-solid interactions, with an emphasis on sputtering, accelerator-based materials characterization techniques, ion optics, resonance ionization spectroscopy, and instrumentation.

Assistant Professor

Philipose, Usha, Ph.D., University of Toronto, 2006. *Nano Science and Technology, Solid State Physics*. Growth and characterization of semiconductor nanowires.

Reinert, Tilo, Ph.D., University of Leipzig, 2001. *Applied Physics*. Ion-beam physics.

Rostovtsev, Yuri, Russian Academy of Sciences, 1991. EIT; quantum coherence; FELS.

Shemmer, Ohad, Ph.D., Tel Aviv University, 2004. *Astronomy*. Black hole mass, accretion rate, and metal abundances in active galactic nuclei.

Emeritus

Duggan, Jerome L., Ph.D., Louisiana State University, 1961. Regents Professor. *Applied Physics*. Materials analysis with ion beams; study of low-energy nuclear reactions; innershell ionization studies; nuclear electronics and measurement; nuclear physics experiments for physics undergraduates.

DEPARTMENTAL RESEARCH SPECIALTIES AND STAFF

Theoretical

Atomic, Molecular, & Optical Physics. Electron and positron scattering; heavy particle scattering; quantum optics plasma and astrophysics applications. Quintanilla (Ward), Rostovtsev, Schultz, Shiner.

Low Temperature Physics. Many-body theory; quantum fluids. Kobe.

Nano Science and Technology. Computational material science. Chemical systems and materials simulation. Buongiorno Nardelli.

Nonlinear Dynamics and Complex Systems. Biophysics. Grigolini, Kowalski.

Plasma and Fusion. Penning trap; antimatter studies; cryogenic heat engine research. Ordonez.

Solid State Physics. Optical properties of solids; transport processes; theoretical and computational materials physics; materials informatics. Buongiorno Nardelli, Kobe, Kowalski, Krokhin, Philipose.

Solid State Physics. Optical properties of solids; transport processes; theoretical and computational materials physics; materials informatics. Buongiorno Nardelli, Kobe, Kowalski, Krokhin, Philipose.

Statistical & Thermal Physics. Complexity. Grigolini, Kowalski, Krokhin.

Statistical & Thermal Physics. Complexity. Grigolini, Kowalski, Krokhin.

Experimental

Acoustics. Matteson.

Applied Physics. Ion-implantation and trace analysis in semiconductors; trace analysis; accelerator; mass spectrometry. Duggan, Glass, Matteson, McDaniel, Reinert, Rout, Weathers.

Astronomy. Ultraviolet evolution of galaxy; atomic and molecular physics; ionization processes for inner shell electrons; ion penetration; sputtering. Matteson, Shemmer, Weathers.

Atomic, Molecular & Optical Physics. Experimental atomic and molecular physics; precision measurements; atomic and molecular spectroscopy. Mueller, Roberts, Shiner.

Nano Science and Technology. Metamaterials. Nanowire synthesis and characterization. Optical spectroscopy of nanobiophotonic systems. Drachev, Neogi, Philipose.

Optics. Laser technology and applications of lasers. Drachev, Lin, Neogi, Shiner.

Polymer Physics/Science. Hydrogels; biomaterials; light scattering.

Solid State Physics. Semiconductor materials growth; scattering mechanisms and influence of impurities; transport coefficients; laser spectroscopy. Littler, Neogi, Perez, Philipose, Rout.

View additional information about this department at
www.gradschoolshopper.com

UNIVERSITY OF TEXAS AT ARLINGTON

DEPARTMENT OF PHYSICS

Arlington, TX, Texas 76019
http://www.uta.edu/physics

General University Information
President: Vistasp M. Karbhari
Dean of Graduate School: Philip Cohen
University website: http://www.uta.edu/
Control: Public
Setting: Urban
Total Faculty: 1,302
Total Graduate Faculty: 775
Total number of Students: 33,806
Total number of Graduate Students: 7,000

Department Information
Department Chairman: Alex Weiss, Chair
Department Contact: Margaret Jackymack, Administrative
  Assistant
  Total full-time faculty: 22
  Total number of full-time equivalent positions: 22
  Full-Time Graduate Students: 45
  First-Year Graduate Students: 7
  Female First-Year Students: 4
  Total Post Doctorates: 14

Department Address
502 Yates Street
108 Science Hall
Arlington, TX, TX 76019
Phone: (817) 272-9041
Fax: (817) 272-3637
E-mail: jackymack@uta.edu
Website: http://www.uta.edu/physics

ADMISSIONS

Admission Contact Information
Address admission inquiries to: Physics Dept., Box 19059
Phone: (817) 272-9041
E-mail: jackymack@uta.edu
Admissions website: http://www.uta.edu/physics

Application deadlines
Fall admission:
U.S. students: April 15   *Int'l. students*: February 3
Spring admission:
U.S. students: October 15   *Int'l. students*: September 15

Application fee
U.S. students: $40   *Int'l. students*: $70
Applications are accepted throughout the year.

Admissions information
For Fall of 2013:
  Number of applicants: 50
  Number admitted: 7
  Number enrolled: 7

Admission requirements
Bachelor's degree requirements: Bachelor's degree in physics or in related fields with a minimum 3.0 GPA on a 4.0 scale is required.
Minimum undergraduate GPA: 3.0

GRE requirements
The GRE is required.

Advanced GRE requirements
The Advanced GRE is not required.

TOEFL requirements
The TOEFL exam is required for students from non-English-speaking countries.
PBT score: 550
iBT score: 79

Other admissions information
Additional requirements: The applicant should have a score of minimum 1,000 (verbal+quantitative) in GRE.
  The average total GRE score for 2009–10 admissions in Physics was 1150.
  The Doctor of Philosophy in Physics and Applied Physics program is designed to prepare broadly trained physicists who would be engaged primarily in research, teaching and development in academia or industry. To be admitted to this program, an applicant must satisfy the general requirements of the Graduate School, and have a Masters degree or 30 credit hours of graduate courses in physics or related fields.
Undergraduate preparation assumed: Junior and senior physics at the level suggested by the following textbooks: Carter, Statistical Physics; Wangsness, Electromagnetic Fields; Gasiorowicz, Quantum Mechanics; Marion, Classical Dynamics of Particle Systems; Riley, Hobson and Bence, Mathematical Methods for Physics and Engineering.

TUITION

Tuition year 2012–13:
Tuition for in-state residents
  Full-time students: $4,144 per semester
Tuition for out-of-state residents
  Full-time students: $8,500 per semester
Teaching or research assistants employed half-time are charged the same rate as in-state residents See Graduate Catalog for more detailed breakdown.
Credit hours per semester to be considered full-time: 9
Deferred tuition plan: Yes
Academic term: Semester
Number of first-year students who received partial tuition waivers: 7

Teaching Assistants, Research Assistants, and Fellowships
Number of first-year
  Teaching Assistants: 7
  Fellowship students: 7
Average stipend per academic year
  Teaching Assistant: $21,600
  Research Assistant: $21,600
  Fellowship student: $21,600

FINANCIAL AID

Application deadlines
Fall admission:

608

U.S. students: February 3   *Int'l. students*: February 3

Loans

Loans are available for U.S. students.
Loans are not available for international students.
GAPSFAS application required: No
FAFSA application required: No

For further information

Address financial aid inquiries to: Chairman, Graduate Admissions Committee, Department of Physics.

HOUSING

Availability of on-campus housing

Single students: Yes
Married students: Yes

For further information

Address housing inquiries to: Housing Office, 210 University Center.
Phone: 817-272-2791
E-mail: housing@uta.edu
Housing aid website: http://www.uta.edu/housing/index.php

Table A—Faculty, Enrollments, and Degrees Granted

Research Specialty	2012-13 Faculty	Enrollment Spring 2013		Number of Degrees Granted 2012-13 (2008–13)		
		Master's	Doctorate	Master's	Terminal Master's	Doctorate
Astrophysics	3	–	8	1(4)	–(1)	–(3)
Atmosphere, Space Physics, Cosmic Rays	2	–	3	1(4)	–(1)	2(5)
Condensed Matter Physics	9	–	20	1(8)	1(2)	1(8)
High Energy Physics	6	–	12	1(4)	–(1)	2(4)
Surface Physics	2	–	2	1(4)	–(1)	–(2)
Total	22	–	45	5(24)	1(6)	5(22)
Full-time Grad. Stud.	–	–	45	–	–	–
First-year Grad. Stud.	–	–	7	–	–	–

GRADUATE DEGREE REQUIREMENTS

Master's: The Master of Science in physics requires a minimum of 30 credit hours, of which 24 hours, including six hours of thesis, will be in physics, and six may be selected from physics, mathematics, chemistry, geology, biology or engineering as approved by the Graduate Advisor. The completion of this degree normally takes two years. Foreign language, comprehensive, and qualifying exams are not required. However, a grade point average of 3.0 (on a scale of 4.0 maximum) must be maintained for all work undertaken as a graduate student. The student must conduct research leading to a thesis which must be defended in an oral exam. Thesis may be written in absentia. Non-thesis option is also available.

Doctorate: For the completion of the degree, the student must 1) demonstrate competence in a minimum of 39 credit hours of core courses in physics, chosen under the guidance of the supervising committee, and approved in advance by the Graduate Studies Committee, 2) complete 9 credit hours of internship or six credit hours of research with a written report and three hours of Applied Physics course, 3) pass qualifying and comprehensive examinations, and 4) conduct research leading to a dissertation which must be defended in an oral exam. Dissertation may be written in absentia.

SPECIAL EQUIPMENT, FACILITIES, OR PROGRAMS

Angular correlation of annihilation radiation (ACAR) system, four-probe electrical conductivity measurement system with closed-cycle liquid helium cryostat, high-pressure diamond anvil cell, low-energy positron beams, magnetic resonance spectrometer, photoemission spectrometer, photoluminescence spectrometer, positron annihilation induced Auger electron spectroscopy, Raman spectrometer, scanning Auger microbe, thin-film deposition systems, variable temperature vibrating sample magnetometers, scanning electron microscopy with polarization analysis (SEMPA), alternating gradient magnetometer (AGM), magnetic properties measurement system (MPMS) SQUID magnetometer and arc melting furnace, thermal particle analyzer, rapid thermal processor (RTP) and high-energy physics detector construction facility.

Computational facilities at UTA include high performance computing environment that combines multiple independent systems connected via a private high-speed network. The servers at this facility feature Alpha, Intel IA-32, and Itanium architectures running Compaq's Tru64 UNIX and Redhat Linux. The high-energy physics group of the physics department has a farm of Linux PC's that participate in high level simulation and a distributed and parallel computing cluster facility run jointly with Computer Science and Engineering department. This facility has 160 CPUs and over 70 TB of disk space available. The HEP group is also establishing a 5000 processor grid from which will support analysis of data remote ATLAS experiment at CERN. University-wide parallel computing facilities are also available through Beowulf Cluster. Separately, the physics department has a computing laboratory with multiple PCs, printers and scanners. There is also currently a 32 node Beowulf-based computational physics cluster in the Department. In addition, research groups in the department have PCs and workstations.

Table C—Separately Budgeted Research Expenditures by Research Specialty

Research Specialty	No. of Grants	Expenditures ($)
Astrophysics	6	$300,000
Atmosphere, Space Physics, Cosmic Rays	4	$300,000
Condensed Matter Physics	6	$1,500,000
High Energy Physics	5	$2,600,000
Total	21	$4,700,000

FACULTY

Professor

Brandt, Andrew, Ph.D., University of California, Los Angeles, 1992. *High Energy Physics*. Experimental high-energy physics.

Chen, Wei, Ph.D., Peking University, 1992. *Biophysics, Condensed Matter Physics*. Nanobiophysics.

Cuntz, Manfred, Ph.D., Heidelberg University, 1988. *Astronomy, Astrophysics, Plasma and Fusion, Solar Physics*. Theoretical astrophysics; observational astronomy, astrobiology.

De, Kaushik, Ph.D., Brown University, 1988. *High Energy Physics*. Experimental high-energy physics.

Koymen, Ali R., Ph.D., University of Michigan, 1984. *Solid State Physics, Surface Physics*. Surface physics; surface magnetism; positron physics.

López, Ramón E., Ph.D., Rice University, 1986. *Climate/Atmospheric Science, Physics and other Science Education, Planetary Science, Plasma and Fusion*. Space physics; physics and science education.

Liu, Ping, Ph.D., University of Amsterdam, 1994. *Condensed Matter Physics.* Condensed matter physics; magnetic materials; nano-materials.

Musielak, Zdzislaw, Ph.D., University of Gdansk, Poland, 1980. *Cosmology & String Theory, Mechanics, Particles and Fields, Planetary Science, Relativity & Gravitation, Solar Physics.* Theoretical astrophysics; cosmology; chaos and nonlinear physics.

Ray, Asok, Ph.D., Texas Tech University, 1977. *Computational Physics, Condensed Matter Physics.* Condensed matter theory; clusters; electron transport theory.

Sharma, Suresh C., Ph.D., Brandeis University, 1976. *Condensed Matter Physics, Surface Physics.* Positron physics; high-pressure physics; surface science; nano-materials.

Weiss, Alex, Ph.D., Brandeis University, 1983. *Condensed Matter Physics, Physics of Beams, Solid State Physics, Surface Physics.* Positron physics; surface physics.

White, Andrew P., Ph.D., University of London, 1972. *High Energy Physics.* Experimental high-energy physics.

Yu, Jaehoon, Ph.D., Stony Brook University, 1993. *High Energy Physics.* Experimental high-energy physics.

Zhang, Qiming, Ph.D., SISSA, Trieste, Italy, 1989. *Computational Physics, Condensed Matter Physics.* Theoretical condensed matter.

Associate Professor

Farbin, Amir, Ph.D., University of Maryland, 2004. *High Energy Physics.* Experimental high-energy physics.

Fazleev, Nail, Ph.D., Kazan University, 1978. *Computational Physics, Condensed Matter Physics.* Theoretical condensed matter physics.

Assistant Professor

Deng, Yue, Ph.D., University of Michigan, 2006. *Atmosphere, Space Physics, Cosmic Rays.* Space physics.

Huda, Muhammad, Ph.D., University of Texas, Arlington. *Chemical Physics, Computational Physics, Condensed Matter Physics.* Condensed matter theory.

Jackson, Christopher, Ph.D., Florida State University, 2005. *Cosmology & String Theory, High Energy Physics.* Theoretical high-energy physics.

Mohanty, Samarendra, Ph.D., Indian Institute of Science, 2006. *Biophysics, Condensed Matter Physics, Optics.*

Ngai, Joseph, Ph.D., University of Alberta, 2001. Experimental condensed matter physics.

Park, Sangwook, Ph.D., Purdue University, 1998. *Astronomy, Astrophysics.* Astrophysics.

DEPARTMENTAL RESEARCH SPECIALTIES AND STAFF

Theoretical
Astrophysics. Cuntz, Musielak, Park.
Atmosphere, Space Physics, Cosmic Rays. Deng, López.
Biophysics. Chen, Mohanty.
Computational Physics. Fazleev, Huda, Ray, Zhang.
Condensed Matter Physics. Chen, Fazleev, Huda, Koymen, Liu, Mohanty, Ngai, Ray, Sharma, Weiss, Zhang.
Cosmology & String Theory. Cuntz, Jackson, Musielak.
High Energy Physics. Brandt, De, Farbin, Jackson, White, Yu.
Surface Physics. Koymen, Ngai, Weiss.

Experimental
Optics. Mohanty, Sharma.

View additional information about this department at
www.gradschoolshopper.com

THE UNIVERSITY OF TEXAS AT AUSTIN

DEPARTMENT OF PHYSICS

Austin, Texas 78712
http://www.ph.utexas.edu

General University Information
President: William Powers, Jr.
Dean of Graduate School: Judith H. Langlois
University website: http://https://www.utexas.edu
Control: Public
Setting: Urban
Total Faculty: 3,081
Total Graduate Faculty: 1,956
Total number of Students: 52,186
Total number of Graduate Students: 11,123

Department Information
Department Chairman: Richard Hazeltine, Chair
Department Contact: Matthew Ervin, Graduate Coordinator
   Total full-time faculty: 56

Full-Time Graduate Students: 228
First-Year Graduate Students: 39
Female First-Year Students: 7
Total Post Doctorates: 41

Department Address
2515 Speedway, C1600
Austin, TX 78712
Phone: (512) 471-1153
Fax: (512) 471-9637
E-mail: graduate@physics.utexas.edu
Website: http://www.ph.utexas.edu

ADMISSIONS

Admission Contact Information

Address admission inquiries to: Admissions Coordinator, Department of Physics, RLM 5.208, University of Texas at Austin, 2515 Speedway, C1600, Austin, TX 78712-1081
Phone: (512) 471-1153
E-mail: admissions@physics.utexas.edu
Admissions website: http://www.ph.utexas.edu/grad-admissions.php

Application deadlines

Fall admission:
U.S. students: December 1   *Int'l. students*: December 1
Spring admission:
U.S. students: October 1   *Int'l. students*: October 1

Application fee

U.S. students: $65   *Int'l. students*: $90

Admissions information

For Fall of 2013:
   Number of applicants: 410
   Number enrolled: 38

Admission requirements

Bachelor's degree requirements: Bachelor's degree is required.
Minimum undergraduate GPA: 3.0

GRE requirements

The GRE is required.
   Analytical score: 1
   Mean GRE score range (25th–75th percentile): 1200-1380

Advanced GRE requirements

The Advanced GRE is required.
   Minimum accepted Advanced GRE score: 400
   Mean Advanced GRE score range (25th–75th percentile): 700-960

TOEFL requirements

The TOEFL exam is required for students from non-English-speaking countries.
   PBT score: 550
   iBT score: 120

Other admissions information

Additional requirements: The GRE Physics Subject Test is required. The average GRE advanced score for 2011–12 admission was 819. The TOEFL is absolutely required for foreign applicants and cannot be waived, substituted, or delayed. Foreign students who accept teaching assistantships must pass an English language proficiency assessment before any appointment can be made.
Undergraduate preparation assumed: Mechanics at the level of Halliday, Resnick, and Krane, Physics, Vol. 1; electricity and magnetism at the level of Halliday, Resnick, and Krane, Physics, Vol. 2; thermodynamics at the level of Kittel and Kroemer, Thermal Physics; atomic physics at the level of Morrison, Estle, and Lane, Quantum States of Atoms, Molecules and Solids; quantum mechanics at the level of Morrison, Understanding More Quantum Physics.

TUITION

Tuition year 2012–13:
Tuition for in-state residents
   Full-time students: $8,350 annual

Tuition for out-of-state residents
   Full-time students: $16,454 annual
All required fees included in the above amounts. Other fees may vary.
Credit hours per semester to be considered full-time: 9
Deferred tuition plan: Yes
Health insurance: Available
Academic term: Semester
Number of first-year students who received full tuition waivers: 37

Teaching Assistants, Research Assistants, and Fellowships

Number of first-year
   Teaching Assistants: 30
   Research Assistants: 6
   Fellowship students: 4
Average stipend per academic year
   Teaching Assistant: $17,080
   Fellowship student: $18,670

FINANCIAL AID

Application deadlines

Fall admission:
U.S. students: December 1   *Int'l. students*: December 1

Loans

Loans are available for U.S. students.
Loans are available for international students.
GAPSFAS application required: No
FAFSA application required: No

For further information

Address financial aid inquiries to: Admissions Coordinator, Department of Physics, The University of Texas at Austin, RLM 5.208, Austin, TX 78712-1081.
E-mail: admissions@physics.utexas.edu

HOUSING

Availability of on-campus housing

   Single students: Yes
   Married students: No

For further information

Address housing inquiries to: Division of Housing and Food Service, P.O. Box 7666, The University of Texas at Austin, Austin, TX 78712-7666.
Phone: (512) 471-3136
Housing aid website: http://www.utexas.edu/student/housing/

Table A—Faculty, Enrollments, and Degrees Granted

Research Specialty	2012-13 Faculty	Enrollment Fall 2012-13		Number of Degrees Granted 2012–13 (2008–12)		
		Master's	Doctorate	Master's	Terminal Master's	Doctorate
Acoustics	–	–	–	–	–(1)	–
Atomic, Molecular, & Optical Physics	8	13	32	–(11)	1(2)	7(20)
Biophysics	3	9	6	–(3)	–(1)	3(8)
Condensed Matter Physics	13	22	43	–(8)	1(1)	13(41)
Cosmology & String Theory	6	8	18	1(2)	–	2(8)
High Energy Physics	6	5	5	–	–	–(11)
Nonlinear Dynamics and Complex Systems	2	4	6	–(1)	2(-)	–(5)
Nuclear Physics	3	2	4	–(1)	–	3(2)
Particles and Fields	5	–	4	–(2)	–	1(8)
Plasma and Fusion	7	14	26	2(2)	1(2)	2(13)
Relativity & Gravitation	1	–	7	–(1)	–(1)	–(6)
Statistical & Thermal Physics	2	1	1	–	–	–(5)
Non-specialized	–	1	–	–	–	–
Total	56	79	152	3(31)	5(8)	31(127)
Full-time Grad. Stud.	–	78	150	–	–	–
First-year Grad. Stud.	–	2	37	–	–	–

GRADUATE DEGREE REQUIREMENTS

Master's: Master of Arts: The time required for the degrees will average about one calendar year plus one semester for a student with a strong undergraduate background. Requirements include 30 semester hours with a "B" average. Eighteen to 24 semester hours, including the thesis, must be in the major program. The minor, which is obligatory, consists of a minimum of six hours in a supporting subject or subjects outside the major program. Each program must include at least 30 semester hours of graduate work, including the thesis. All completed work included in the degree program at the time of admission to candidacy must have been taken within the previous six years. The Master of Science in Applied Physics: This degree is designed to provide students with a broad background in physics and related fields, with an emphasis on those aspects of the science most used in an industrial setting. The required physics courses include PHY 380N, 387K, and 389K, a course in the physics of sensors, and a technical seminar. A thesis is also required. The supporting work must be in engineering, chemistry, or geological sciences.

Doctorate: A student must fulfill the following requirements to be admitted to candidacy for the Ph.D. degree in physics: (1) fulfill the core course requirements described below; (2) show evidence of exposure to modern methods of experimental physics–this exposure may be gained in a senior-level laboratory course taken by the student as an undergraduate and approved by the graduate adviser and the chairman of the Graduate Studies Committee by previous participation in an experimental program or in Physics 380N; and (3) fulfill the oral examination requirement described below. Core courses: During the first two years of graduate studies, the student must take four core courses: Classical Mechanics (385K), Statistical Mechanics (385L), Electromagnetism I (387K) or Electromagnetism II(387L), and Quantum Mechanics I (389K) or Quantum Mechanics II (389L). The student must earn an official grade of at least "B" in each course and must maintain a grade point average of at least 3.30 in the four courses. The

student may ask for the grade he or she earns in Physics 380N to be substituted for the grade in one of the core courses when the average is computed. A well-prepared student may seek to fulfill the core course requirement by earning satisfactory grades on the final examinations for some of these courses rather than by registering for them; in this case, the student does not receive graduate credit for these courses and the grade is not counted toward the required average. Oral qualifying examination: After satisfying the first two requirements above, and within 27 months of entering the program, the student must take an oral qualifying examination. The examination consists of a presentation before a committee of four physics faculty members, one of whom is a member of the Graduate Studies Subcommittee. The presentation is open to all interested parties. It is followed by a question-and-answer period restricted to the student and the committee. The questions during this session are directed to clarifying the presentation and determining whether the student has a solid grasp of the basic material needed for research in his or her specialization. The student passes the examination by obtaining a positive vote from at least three of the four faculty members on the oral qualifying committee. Each program of work for the doctoral degree must include at least four advanced courses in physics; a list of acceptable courses is maintained by the Graduate Studies Subcommittee. The program must also include three courses outside of the student's area of specialization; one of these must be an advanced physics course, another must be outside of the Department of Physics, and the third may be either an advanced physics course or a course outside of the Department of Physics. A dissertation is required of every candidate, followed by a final oral examination covering the dissertation and the general field of the dissertation.

SPECIAL EQUIPMENT, FACILITIES, OR PROGRAMS

Modern facilities for graduate study and research include a large-scale cryogenic laboratory; nuclear magnetic and electron paramagnetic resonance laboratories; extensive facilities for tunneling and force microscopy and nanostructure characterization, SQUID magnetometry, and electron spectroscopy; well-equipped laboratories in optical spectroscopy, quantum optics, femtosecond spectroscopy and diagnostics, and electron-atom and surface scattering; and facilities including a table-top 100-terawatt laser for strong-field physics studies for turbulent flow and nonlinear dynamics experiments and two petawatt lasers (one Ti-sapphire providing 30J in 30fs and another glass laser at 200J in 150fs).

Plasma physics experiments are conducted at the major national tokamaks in Boston and San Diego and on the local machine, the Helimak. Experiments in high-energy heavy ion nuclear and particle physics are conducted at large accelerator facilities such as Brookhaven National Laboratory (New York), Fermi National Laboratory (Illinois), and Germany's Deutsches Electron Synchrotron.

Theoretical work in plasma physics, condensed-matter physics, acoustics, nonlinear dynamics, relativity, astrophysics, statistical mechanics, and particle theory is conducted within the Department of Physics.

Students have access to excellent computer and library facilities, including Ranger, the 10th fastest computer at 504 Tflops.

The Department maintains and staffs a machine shop, student workshop, low-temperature and high-vacuum shop, and an electronics design and fabrication shop.

Table B—Separately Budgeted Research Expenditures by Source of Support

Source of Support	Departmental Research	Physics-related Research Outside Department
Federal government	$12,468,055	
State/local government	$3,825,793	
Non-profit organizations	$1,248,692	
Business and industry	$67,461	
Other		
Total	$17,610,001	

Table C—Separately Budgeted Research Expenditures by Research Specialty

Research Specialty	No. of Grants	Expenditures ($)
Atomic, Molecular, & Optical Physics	19	$1,202,651
Biophysics	7	$416,395
Condensed Matter Physics	32	$3,973,510
Nuclear Physics	5	$932,685
Particles and Fields	15	$1,407,548
Physics and other Science Education	8	$761,772
Plasma and Fusion	14	$3,543,280
Relativity & Gravitation	1	$45,278
Statistical & Thermal Physics	4	$73,534
Nonlinear Dynamics and Complex Systems	5	$511,684
Cosmology & String Theory	8	$1,094,281
High Energy Physics	26	$3,647,383
Total	144	$17,610,001

FACULTY

Professor

Böhm, A., Ph.D., Universität Marburg, 1966. *Particles and Fields*. Particle phenomena in terms of algebraic and group-theoretical methods.

Bengtson, R. D., Ph.D., University of Maryland, 1968. *Plasma and Fusion*. Experimental plasma physics; atomic reactions in plasmas.

Berk, H. L., Ph.D., Princeton University, 1964. Minority Student Liaison. *Plasma and Fusion*. Theoretical plasma physics; computer simulation of plasmas.

Chelikowsky, J., Ph.D., University of California, Berkeley, 1975. Director, Institute for Computational Engineering Sciences (ICES). *Condensed Matter Physics*. Solid-state physics; computational materials science.

Chiu, C. B., Ph.D., University of California, Berkeley, 1965. *Particles and Fields, Physics and other Science Education*. Theoretical particle physics, particularly in quantum chromodynamics; confinement problems; subquark and sublepton models; theories in hadron collisions.

Coker, W. R., Ph.D., University of Georgia, 1966. *Nuclear Physics*. Theoretical nuclear physics, with emphasis on scattering and reactions of hadrons and nuclei at medium energies.

de Lozanne, A. L., Ph.D., Stanford University, 1982. *Condensed Matter Physics*. Low-temperature vacuum-tunneling microscopy.

Demkov, A., Ph.D., University of Arizona, 1995. Chair, Graduate Recruitment Committee. *Condensed Matter Physics*. Condensed-matter theory; physics of electronic materials, surfaces, and interfaces; thin films and devices; novel materials; quantum transport.

Dicus, D. A., Ph.D., University of California, Los Angeles, 1968. *Particles and Fields*. Field theory of strong, weak, and electromagnetic interactions; astrophysical implications of the weak force.

Distler, J., Ph.D., Harvard University, 1987. *Cosmology & String Theory*. High-energy theory; mathematical physics; string theory.

Ditmire, T., Ph.D., University of California, Davis, 1995. Director, Texas Center for High-Intensity Laser Science. *Atomic, Molecular, & Optical Physics, Plasma and Fusion*. Intense ultrafast laser interactions.

Downer, M. C., Ph.D., Harvard University, 1983. *Atomic, Molecular, & Optical Physics, Condensed Matter Physics*. Atomic and molecular physics; atomic physics; femtosecond spectroscopy; condensed-matter surfaces; high-field atomic and plasma physics.

Erskine, J. L., Ph.D., University of Washington, 1973. *Condensed Matter Physics*. Experimental solid-state physics; surface physics; magnetism.

Fink, M., Ph.D., Technische Hochschule Karlsruhe, 1966. *Atomic, Molecular, & Optical Physics*. Electron diffraction.

Fischler, W., Ph.D., Université Libre de Bruxelles, 1976. *Cosmology & String Theory, Relativity & Gravitation*. Theoretical physics; particle theory; invisible axion and supersymmetry.

Fitzpatrick, R., Ph.D., University of Sussex, 1988. *Plasma and Fusion*. Magnetic reconnection and gross plasma instabilities in fusion, terrestrial, and astrophysical contexts.

Gentle, K. W., Ph.D., Massachusetts Institute of Technology, 1962. *Plasma and Fusion*. Experimental plasma physics.

Gleeson, A. M., Ph.D., University of Pennsylvania, 1965. Associate Chair for Development and Outreach. *Acoustics, Particles and Fields, Physics and other Science Education*. Field theory of strong interactions and the physics of superdense matter.

Hazeltine, R., Ph.D., University of Michigan, 1968. Department Chair. *Plasma and Fusion*. Theoretical plasma physics.

Heinzen, D., Ph.D., Massachusetts Institute of Technology, 1988. *Atomic, Molecular, & Optical Physics*. Atomic and molecular physics; laser cooling and atom trapping; Bose-Einstein condensation.

Hoffmann, G. W., Ph.D., University of California, Los Angeles, 1971. Associate Chair for Operations. *Nuclear Physics*. Experimental nuclear physics and chemistry using medium-energy projectiles.

Kaplunovsky, V., Ph.D., Tel Aviv University, 1983. *Cosmology & String Theory*. Particle theory; string phenomenology.

Keto, J. W., Ph.D., University of Wisconsin-Madison, 1972. Associate Chair for Graduate Affairs, Graduate Studies Committee Chair, and Graduate Adviser. *Atomic, Molecular, & Optical Physics*. Reactions and radiative processes of excited atoms and molecules; laser spectroscopy high-power lasers.

Lang, K., Ph.D., University of Rochester, 1985. *High Energy Physics*. Rare decay of the K-meson.

MacDonald, A. H., Ph.D., University of Toronto, 1978. *Condensed Matter Physics*. Condensed-matter theory with emphasis on electron-electron interactions.

Marder, M. P., Ph.D., University of California, Santa Barbara, 1986. Co-Director of UTeach. *Condensed Matter Physics, Nonlinear Dynamics and Complex Systems, Physics and other Science Education*. Nonlinear dynamics; statistical physics of solids.

Markert, J. T., Ph.D., Cornell University, 1987. *Condensed Matter Physics*. Experimental condensed-matter physics; crystal growth; high-Tc materials; magnetic materials; magnetic resonance; magnetic microscopies.

Matzner, R. A., Ph.D., University of Maryland, 1967. Director, Center for Relativity. *Relativity & Gravitation*. General relativity and cosmology; manifolds with little symmetry; kinetic theory; conservation laws in general relativity; black hole physics and gravitational radiation.

Morrison, P. J., Ph.D., University of California, San Diego, 1979. Chair, Graduate Welfare Committee. *Nonlinear Dynamics and Complex Systems, Plasma and Fusion.* Plasma physics.

Niu, Q., Ph.D., University of Washington, 1985. *Condensed Matter Physics.* Field theory of condensed matter; theory of superconductivity; mesoscopic physics; quantum transport and diffusion.

Orbach, Raymond L., Ph.D., University of California - Berkley, 1960. Cockrell Family Chair in Engineering #12, Director, UT-Austin's Energy Institute, Fellow of the American Physical Society and the American Association for the Advancement of Science. *Energy Sources & Environment, Physics and other Science Education, Theoretical Physics.* Energy related challenges.

Raizen, M., Ph.D., University of Texas, Austin, 1989. *Atomic, Molecular, & Optical Physics, Nonlinear Dynamics and Complex Systems.* Atomic, molecular, and optical physics; atom optics; quantum chaos.

Reichl, L. E., Ph.D., University of Denver, 1969. Director, Center for Complex Quantum Systems. Associate Chair for Undergraduate Affairs. *Statistical & Thermal Physics.* Nonequilibrium quantum statistical mechanics; Brownian motion; nonlinear dynamics.

Ritchie, J. L., Ph.D., University of Rochester, 1983. *High Energy Physics.* High-energy/nuclear physics.

Schwitters, R. F., Ph.D., Massachusetts Institute of Technology, 1971. *High Energy Physics.* Experimental high-energy physics detector development and B-physics studies.

Shih, C. K., Ph.D., Stanford University, 1988. *Condensed Matter Physics.* Condensed matter; study of surface properties of microelectronic materials.

Shvets, G., Ph.D., Massachusetts Institute of Technology, 1995. *Condensed Matter Physics, Plasma and Fusion.* Theory and simulations: laser-plasma interactions; plasma-based accelerators; photonics; nano-plasmonics. Experimental: phonon-assisted nanolithography; compact surface-wave accelerators.

Sitz, G. O., Ph.D., Stanford University, 1987. Undergraduate Advisor. *Atomic, Molecular, & Optical Physics.* Experimental atomic and molecular physics; oriented molecules; surface scattering.

Sudarshan, E. C. G., Ph.D., University of Rochester, 1958. Dirac Medalist. *Particles and Fields.* Elementary particle physics; quantum optics; quantum field theory; classical mechanics; foundations of physics.

Swinney, H. L., Ph.D., Johns Hopkins University, 1968. Director, Center for Nonlinear Dynamics. *Nonlinear Dynamics and Complex Systems.* Equilibrium and nonequilibrium phase transitions; dynamics of nonlinear systems.

Weinberg, S., Ph.D., Princeton University, 1957. Nobel Laureate; Director, The Weinberg Theory Group. *Cosmology & String Theory, Relativity & Gravitation.* Theoretical physics.

Associate Professor

Florin, E.-L., Ph.D., Technische Universität Munchen, 1990. *Biophysics, Nonlinear Dynamics and Complex Systems.* Experimentalist, nonlinear dynamics.

Hegelich, Bjorn, Ph.D., Ludwig-Maximilians-Universität München, 2002. *Atomic, Molecular, & Optical Physics, Plasma and Fusion.* Interaction of ultra-intense electromagnetic fields with matter; high-energy density physics; laser-particle acceleration.

Kopp, S., Ph.D., University of Chicago, 1994. Associate Dean, Curriculum and Programs, College of Natural Sciences; Director, Center for Inquiry in Mathematics and Sciences. *High Energy Physics, Physics and other Science Education.* CP violation; weak decays of heavy quarks; neutrino oscillations.

Li, X., Ph.D., University of Michigan, 2003. *Atomic, Molecular, & Optical Physics, Condensed Matter Physics.* Experimental condensed-matter physics; femtosecond spectroscopy; phase-sensitive nonlinear optical interactions in semiconductors.

Markert, C., Ph.D., Johann Wolfgang Goethe Universität, 2001. *Nuclear Physics.* Nuclear physics; relativistic heavy-ion physics; the quark-gluon plasma (QGP) phase.

Paban, S., Ph.D., Universidad de Barcelona, 1988. *Cosmology & String Theory.* Quantum mechanics; particle phenomenology; string theory.

Tsoi, M., Ph.D., Universität Konstanz, 1998. *Condensed Matter Physics.* Experimental condensed-matter physics; nanostructures; spintronics.

Turner, J. S., Ph.D., Indiana University, 1969. *Physics and other Science Education, Statistical & Thermal Physics.* Nonequilibrium statistical mechanics and thermodynamics; theoretical and experimental studies of chemical instabilities; dynamics of nonlinear systems.

Yao, Z., Ph.D., Harvard University, 1997. *Condensed Matter Physics.* Nanostructures and mesoscopic physics; condensed-matter physics; experimental physics.

Assistant Professor

Fiete, G. A., Ph.D., Harvard University, 2003. *Condensed Matter Physics.* Theory of quantum matter and correlated electrons at the nanoscale.

Gordon, Vernita, Ph.D., Harvard University, 2003. *Biophysics, Nonlinear Dynamics and Complex Systems.* Experimental biological physics; multicellular systems; the role of physics and spatial structure in developmental and evolutionary systems; biological physics and engineering of membranes.

Kilic, Can, Ph.D., Harvard University, 2006. *Particles and Fields.* Theoretical particle physics; extensions of the Standard Model; collider phenomenology; dark matter models and searches.

Lai, Keji, Ph.D., Princeton University, 2006. *Condensed Matter Physics.* Experimental condensed matter physics; nanoscale electromagnetic imaging; complex oxides; nano-materials; transport in low-dimensional systems.

Onyisi, Peter, Ph.D., Cornell University, 2008. *Computational Physics, High Energy Physics.* Experimental investigation of electroweak symmetry breaking and searches for new particles and interactions; computing with large datasets of structured data.

Shubeita, G. T., Ph.D., Université de Lausanne, 2002. *Biophysics, Nonlinear Dynamics and Complex Systems.* Biophysics-integrating biophysical approaches with molecular biology, genetics, and proteomics.

Emeritus

Antoniewicz, P. R., Ph.D., Purdue University, 1965. *Condensed Matter Physics.* Theoretical investigation of electromagnetic wave propagation and transport properties in solids and liquids; investigation of the properties of atoms and molecules on metal surfaces.

de Wette, F. W., Ph.D., Universiteit Utrecht, 1959. *Condensed Matter Physics.* Theoretical study of structural, thermodynamics, and scattering properties of crystal surfaces.

DeWitt-Morette, C., Ph.D., Doctorat d'Etat, Universite de Paris, 1947. Officer in the French Legion of Honor. *Relativity & Gravitation.* General relativity; mathematical physics; Feynman path integrals.

Drummond, W. E., Ph.D., Stanford University, 1958. *Plasma and Fusion.* Theoretical plasma physics.

Frommhold, L. W., Ph.D., Universität Hamburg, 1959. *Atomic, Molecular, & Optical Physics.* Atomic and molecular physics.

Gavenda, J. D., Ph.D., Brown University, 1959. *Condensed Matter Physics*. Study of properties of conduction electrons in metals using ultrasonic and electromagnetic waves.

Griffy, T. A., Ph.D., Rice University, 1961. *Acoustics*. Theoretical medium-energy physics; underwater acoustics.

Horton, C. W., Ph.D., University of California, San Diego, 1967. *Plasma and Fusion*. Theoretical plasma physics.

Kleinman, L., Ph.D., University of California, Berkeley, 1960. *Condensed Matter Physics*. Solid-state theory; electronic structure of solids, surfaces, and clusters; chemisorption.

McCormick, W. D., Ph.D., Duke University, 1959. *Condensed Matter Physics, Nonlinear Dynamics and Complex Systems*. Experimental low-temperature and solid-state physics; phase transitions in solids (critical phenomena); instabilities in nonequilibrium systems.

Moore, C. F., Ph.D., Florida State University, 1964. *Nuclear Physics*. Detection and measurement of the interactions and involvement of the nuclear continuum in scattering experiments; atomic interactions in highly ionized atoms.

Nolle, A. W., Ph.D., Massachusetts Institute of Technology, 1947. *Acoustics, Condensed Matter Physics*. Musical-instrument physics; underwater acoustics; visoelastic phenomena; fluid oscillations.

Oakes, M. E., Ph.D., University of Florida, 1964. *Plasma and Fusion*. Theoretical and experimental studies of wave propagation in plasmas.

Riley, P. J., Ph.D., University of Alberta, 1962. Associate Dean for Research and Facilities, College of Natural Sciences. *Nuclear Physics*. Experimental studies of the nucleon-nucleon interaction at medium energy; actions in decaying plasmas; environmental effects on spectra.

Schieve, W. C., Ph.D., Lehigh University, 1959. *Statistical & Thermal Physics*. Nonequilibrium statistical mechanics; quantum optics; stochastic processes.

Swift, J. B., Ph.D., University of Illinois, 1968. *Condensed Matter Physics, Nonlinear Dynamics and Complex Systems*. Studies of nonlinear dynamics; phase transitions.

Thompson, J. C., Ph.D., Rice University, 1956. *Condensed Matter Physics*. Studies of electronic states in disordered systems (metallic and semiconducting) by galvanomagnetic parameters; optical properties; photoemission; the metal-nonmetal transition.

Udagawa, T., Ph.D., Tokyo University of Education, 1962. *Nuclear Physics*. Theoretical nuclear physics.

DEPARTMENTAL RESEARCH SPECIALTIES AND STAFF

Theoretical

Condensed Matter Physics. Ab-initio electronic structure calculations of the physical, electronic, and magnetic (including noncolinear magnetic systems) properties of solids, surfaces, interfaces and liquids; molecular dynamics calculations of properties of solids, liquids, and crack propagation; density functional theory; Berry phases in polarization theory and spinwave theory; Block electrons in magnetic fields, quantum Hall effect; quantum theory of thin film growth and surface diffusion; theory of mesoscopic phenomena, phonon calculations and lattice dynamics for high Tc superconductors; theory of atom surface interactions, physisorption, chemisorption. Antoniewicz, Chelikowsky, de Wette, Demkov, Fiete, Gavenda, Kleinman, MacDonald, Marder, McCormick, Niu, Nolle, Shvets, Swift, Thompson.

Nonlinear Dynamics and Complex Systems. Dynamics of materials, especially fracture and dislocation dynamics; instabilities and turbulence in fluids, granular media, liquid crystals, and chemical reaction-diffusion systems; chaos in low-dimensional dynamical systems. Marder, McCormick, Morrison, Swift, Swinney.

Nuclear Physics. Scattering and reactions of hadrons and nuclei at medium energies; nuclear structure in the low-energy region using neutron-scattering techniques; nuclear structure and reaction mechanism. Coker, Moore, Riley, Udagawa.

Particles and Fields. Phenomenological studies of the properties of matter ranging from medium-energy physics; symmetries in elementary particle physics; field theory of strong interactions and the physics of superdense matter; quantum chromodynamics; confinement problems; subquark and sublepton models; supersymmetry; quantum optics, basic quantum field theory and quantum mechanics; classical mechanics; particle phenomena in terms of algebraic and group-theoretical methods; electromagnetic interactions. Böhm, Chiu, Dicus, Gleeson, Sudarshan.

Physics and other Science Education. Curriculum development and evaluation at the university level; science teacher preparation program; computer-based education. Chiu, Gleeson, Kopp, Marder, Orbach, Turner.

Plasma and Fusion. Kinetic theory and transport theory; turbulent heating; collisionless shock waves; plasma turbulence; computer simulation of plasmas; stability theory controlled fusion; plasma dynamics. Berk, Drummond, Fitzpatrick, Hazeltine, Horton, Morrison, Oakes, Shvets.

Relativity & Gravitation. Quantum theory of space time; techniques of quantization in curved space-time; string theory; path integration; stochastic processes; critical phenomena in gravitational collapse; computational relativity; cosmology; exact solutions in general relativity; conformal properties of space time; manifolds with little symmetry; kinetic theory; conservation laws in general relativity; black hole physics; black hole interactions; gravitational radiation; interaction of matter with gravitation. DeWitt-Morette, Fischler, Matzner, Schieve, Weinberg.

Statistical & Thermal Physics. Nonequilibrium statistical physics; thermodynamic processes; nonequilibrium quantum statistical mechanics; quantum chaos; mesoscopic physics; nonlinear dynamics; complex systems theory; Brownian motion. Reichl, Schieve, Turner.

The Weinberg Theory Group. Research spans the range from studies of physics at the most fundamental level to exploration of phenomenologically relevant current issues in elementary particle physics. On the more fundamental level, the work continues in gravity and quantum cosmology, conformal field theories, superstring theories, and M theory, with special attention to the links between these topics and to the implication of superstring and M theory for effective field theories at accessible energies. Such theories offer the hope of uniting all forces including gravitation in a theory of superstrings. So far, it seems that these theories allow for the first time a satisfactory elimination of the infinities that have plagued all earlier quantum theories of gravitation. Distler, Fischler, Kaplunovsky, Kilic, Paban, Weinberg.

Experimental

Atomic, Molecular, & Optical Physics. Atom optics; quantum transport in optical lattices; quantum chaos with ultracold atoms; ultracold collisions; Bose-Einstein condensation; search for atomic electric dipole moment; state-resolved molecular-surface scattering and gas-surface dynamics; Raman spectroscopy; electron diffraction; neutrino rest mass experiments; laser spectroscopy of nanoparticles; development of new materials; molecular collision and sonoluminescence; femtosecond spectroscopy; high-power lasers; wake-field accelerators; terawatt lasers; optical properties of nanostructured plasmas at high fields. Ditmire, Downer, Fink, Frommhold, Hegelich, Heinzen, Keto, Li, Raizen, Sitz.

Biophysics. Elastic properties of cells; motility of cells; bacterial competition; dynamics of swimming organisms; biofilms; spatial structures formed through intercellular interactions; adhesion phenomena; cell mechanics; cargo transport in cells; molecular motors (dynamics and regulation); membranes; assembly of biological complexes; diffusive and ballistic Brownian motion; biopolymers; characterization of single biomolecules; microtubule mechanics; yeast mechanics; membrane fusion; thermal noise imaging. Florin, Gordon, Shubeita.

Condensed Matter Physics. Surface and thin-film magnetism; dynamics of magnetization reversal; magnetic switching; Barkhansen noise; domain dynamics; magnetic and electronic effects in ultrathin films multilayers and nanostructures; normal and superconducting properties of high-temperature superconductors; nonlinear optical response of solids; femtosecond spectroscopy of solid-state systems; nanostructure fabrication and characterization based on scanning tunneling microscopy; intrinsic phenomena at surfaces and interfaces studied by electron diffraction, spectroscopy, atom surface scattering, linear and nonlinear optical spectroscopy; scanning probe techniques, including near-field optical microscopy; thin-film nucleation and growth; cluster physics, mesoscopic phenomena in solids; materials synthesis including novel magnetic and superconducting materials; transport and magnetic characterization; strongly correlated electron systems; mechanical properties of materials including fracture. de Lozanne, Demkov, Downer, Erskine, Lai, Li, Marder, J. Markert, Shih, Shvets, Tsoi, Yao.

High Energy Physics. Properties of elementary particles, particularly kaons, B-mesons, and neutrinos; rare decays of the kaons; tests of conservation laws and CP violation; B-meson decays; information on CP violation; neutrino oscillation measurements; information on neutrino mass; detector development; applications of particle detectors to medical imaging. Experiments are conducted at national and international accelerator laboratories. Kopp, Lang, Onyisi, Ritchie, Schwitters.

Nonlinear Dynamics and Complex Systems. Pattern formation and chaotic dynamics of diverse systems; planetary fluid dynamics (especially internal gravity waves in the oceans); viscous fingering; crack propagation in amorphous and crystalline solids; rupture in rubber; friction; control of atomic and molecular motion; trapping of different isotopes; trapping and cooling of macroscopic particles (microspheres); dynamics of Brownian motion; stretching and wrinkling of thin sheets and graphene; physics education research (people dynamics). See also biophysics. Florin, Gordon, Marder, McCormick, Morrison, Raizen, Shubeita, Swinney.

Nuclear Physics. The research focuses on two experiments: (1) E896 (using the AGS at the Brookhaven National Laboratory), a definitive search for the short-lived HO di-baryon, a strangeness$=-2$, 6-quark object predicted by bag models. E896 also searches for other short-lived objects composed of strange hadrons that may be produced in high-energy nucleus-nucleus collisions. (2) STAR [Solenoidal Tracker at RHIC (Relativistic Heavy Ion Collider)] at the Brookhaven National Laboratory to study primordial matter at conditions of extreme temperature and pressure. Such matter is produced through central collisions of circulating beams of Au ions of momenta 100 GeV/c per nucleon (total center-of-momentum energy$=40$ TeV). STAR searches for evidence of the formation of a quark-gluon plasma (a phase of nuclear matter in which quarks and gluons are not confined within nucleons or mesons) and for evidence of the restoration of the fundamental chiral symmetry of the strong interaction at high temperature. Both experiments explore the most fundamental physics and chemistry of nature as it may have existed during the early evolution of the Universe (about $10-7-10-6$ seconds after the Big Bang). Hoffmann, C. Markert, Moore, Riley.

Plasma and Fusion. Plasma turbulence and transport; plasma heating; plasma propulsion; plasma spectroscopy; plasma diagnostics; plasma processing; atomic reactions in plasmas. Bengtson, Ditmire, Downer, Gentle, Hegelich, Shvets.

View additional information about this department at
www.gradschoolshopper.com

THE UNIVERSITY OF TEXAS AT BROWNSVILLE

DEPARTMENT OF PHYSICS AND ASTRONOMY

Brownsville, Texas 78520
http://phys.utb.edu

General University Information

President: Dr. Juliet V. García
Dean of Graduate School: Dr. Charles Lackey
University website: http://www.utb.edu
Control: Public
Setting: Urban
Total Faculty: 321
Total Graduate Faculty: 212
Total number of Students: 12,228
Total number of Graduate Students: 1,040

Department Information

Department Chairman: Dr. Soma Mukherjee, Chair
Department Contact: Dr. Teviet Creighton, Assistant Professor, Chair Graduate Committee
Total full-time faculty: 18
Total number of full-time equivalent positions: 18
Full-Time Graduate Students: 33
First-Year Graduate Students: 4
Female First-Year Students: 1
Total Post Doctorates: 4

Department Address

One West University Boulevard
Cavalry 104
Brownsville, TX 78520
Phone: (956) 882-6651
Fax: (956) 882-6726
E-mail: teviet@phys.utb.edu
Website: http://phys.utb.edu

ADMISSIONS

Admission Contact Information

Address admission inquiries to: Graduate Program Coordinator, Dept. of Physics and Astronomy, The University of Texas at Brownsville, One West University Blvd, Brownsville, TX 78520
Phone: (956) 882-6651
E-mail: teviet@phys.utb.edu
Admissions website: http://www.utb.edu/vpaa/graduate/Pages/ApplicationMaterials.aspx

Application deadlines

Fall admission:
U.S. students: July 1   *Int'l. students*: July 1
Spring admission:
U.S. students: November 1   *Int'l. students*: November 1

Application fee

U.S. students: $30   *Int'l. students*: $45
International applicants are recommended to apply at least three months in advance of these deadlines.

Admissions information

For Fall of 2013:
Number of applicants: 12
Number admitted: 4
Number enrolled: 4

Admission requirements

Bachelor's degree requirements: Bachelor's degree in physics or a related field is required.
Minimum undergraduate GPA: 3.0

GRE requirements

The GRE is required.

Advanced GRE requirements

The Advanced GRE is not required.

TOEFL requirements

The TOEFL exam is required for students from non-English-speaking countries.
PBT score: 550
iBT score: 77

Other admissions information

Additional requirements: Two letters of recommendation from people familiar with the applicant's undergraduate scholastic record are required. Applicants who do not meet the above criteria can apply for conditional admission into the program. Until Spring 2013, students were admitted into a cooperative Ph.D. program between the UTB and UT San Antonio (UTSA). Students enrolled in the cooperative Ph.D. program between UTB and UTSA reside at UTB and perform their dissertation research under the supervision of a graduate faculty member of the UTB Physics and Astronomy Department. Admission and graduation requirements have been the same as those established for the UTSA Ph.D. program in Physics.

TUITION

Tuition year 2014-15:
Tuition for in-state residents
Full-time students: $2,180 per semester
Part-time students: $880 per semester
Tuition for out-of-state residents
Full-time students: $4,997 per semester
Part-time students: $1,819 per semester
Teaching or Research Assistants are charged the same rate as in-state residents. See Graduate Catalog for more detailed breakdown.
Credit hours per semester to be considered full-time: 9
Deferred tuition plan: No
Health insurance: Available at the cost of $1,240 per year.
Academic term: Semester
Number of first-year students who received full tuition waivers: 8

Teaching Assistants, Research Assistants, and Fellowships

Number of first-year
Teaching Assistants: 6
Average stipend per academic year
Teaching Assistant: $15,000

FINANCIAL AID

Application deadlines

Fall admission:
U.S. students: June 1
Spring admission:
U.S. students: November 1   *Int'l. students*: November 1

Loans

Loans are available for U.S. students.
Loans are not available for international students.
GAPSFAS application required: No
FAFSA application required: No

For further information

Address financial aid inquiries to: Graduate Program Coordinator, Dept. of Physics and Astronomy, The University of Texas at Brownsville, One West University Blvd., Brownsville, TX 78520.
Phone: (956) 882-6651
E-mail: teviet@phys.utb.edu
Financial aid website: http://www.utb.edu/em/fa/Pages/default.aspx

HOUSING

Availability of on-campus housing

Single students: Yes
Married students: No

For further information

Address housing inquiries to: Residential Life Office, 1915 University Blvd., Brownsville, TX 78520.
Phone: (956) 882-7191
E-mail: housing@utb.edu
Housing aid website: http://www.utb.edu/sa/residential/Pages/default.aspx

Table A—Faculty, Enrollments, and Degrees Granted

Research Specialty	2012-13 Faculty	Enrollment Spring 2013		Number of Degrees Granted 2012-13		
		Master's	Doctorate	Master's	Terminal Master's	Doctorate
Astrophysics	3	5	2	1(-)	–	–
Atomic, Molecular, & Optical Physics	2	5	1	1(-)	–	–
Biophysics	2	3	1	–	1(-)	–
Nano Science and Technology	2	2	1	2(-)	–	–
Physics and other Science Education	1	–	–	–	–	–
Relativity & Gravitation	5	6	6	1(-)	–	–
Non-specialized	–	–	–	–	–	–
Total	15	21	11	–	–	–
Full-time Grad. Stud.	–	21	11	–	–	–
First-year Grad. Stud.	–	8	–	–	–	–

GRADUATE DEGREE REQUIREMENTS

Master's: Students can choose a thesis or non-thesis option within this program. Both options require 30 semester hours of credit for successful completion. The thesis option requires 18 hours of course work, 6 hours of Graduate Research and 6 hours of Thesis. The non-thesis option requires 27 hours of course work and a 3-hour course, Research Problems in Physics. A M.S. with Emphasis in Applied Physics is available. Students taking this emphasis will specialize in advanced optics, electromagnetics and nano-photonics.

Doctorate: Students were admitted into a cooperative Ph.D. program with UT-San Antonio until Spring 2013. All requirements for the UTSA Ph.D. program in Physics apply. The doctoral students reported are the ones currently in the UTB-UTSA cooperative Ph.D. pipeline.

SPECIAL EQUIPMENT, FACILITIES, OR PROGRAMS

UTB/TSC is a member institution of the LIGO-Virgo Collaboration (LVC), which operates four large scale detectors of gravitational waves (three in the United States and one in Europe). As members of the LVC, faculty members in the Physics department conduct research in the areas of gravitational wave data analysis, detector characterization and instrumentation. Graduate and undergraduate students participating in LVC activities regularly visit the LIGO facilities located in Washington state and Louisiana. Members of the LVC have access to the computing resources of the LIGO Scientific Collaboration Grid for data analysis and detector characterization tasks.

ARCC—the Arecibo Remote Command Center—is a facility for remote observations using the Arecibo Radio Telescope at Puerto Rico. The research conducted with this facility includes searches and studies of radio pulsars. The ARCC also enables research based on observations made with the Green Bank and Parkes radio telescopes.

For biophysics research, a Nanoscope IV Atomic Force Microscope is available for imaging biological nanostructures and doing force spectroscopy. Optical tweezers with TIRF fluorescence for single molecule studies and a Cell Culture Facility, including a minus 80 freezer and laminar flow fume hoods, for cancer cell studies are available.

The department operates two computer clusters "Funes" and "Futuro", for astrophysical computations, data analysis, and modeling of photonic crystals with FDTD. Futuro is a 120 processor (2.4 GHz, 12 GB RAM/processor) IBM cluster and Funes is the older cluster with 70 processors.

The department is also involved in research related to LISA and other proposed space-based gravitational wave detectors, in collaboration with NASA and ESA. The research involves operation of FPGA-based phase meters and realization of the optical transmission and detection setup in the optics lab. The laboratory includes 2 soft-wall clean-room enclosures, a highly stabilized 10-W Nd:YAG MOPA laser (Lightwave) and other infrared and visible-range lasers with optics and optical-beam diagnostic equipment.

Table B—Separately Budgeted Research Expenditures by Source of Support

Source of Support	Departmental Research	Physics-related Research Outside Department
Federal government	$5,574,000	
State/local government		
Non-profit organizations		
Business and industry		
Other		
Total	$5,574,000	

Table C—Separately Budgeted Research Expenditures by Research Specialty

Research Specialty	No. of Grants	Expenditures ($)
Astrophysics	3	$550,000
Biophysics	–	$263,000
Nano Science and Technology	2	$637,970
Optics	1	$623,860
Relativity & Gravitation	3	$3,499,170
Total	9	$5,574,000

FACULTY

Professor

Benacquista, Matthew, Ph.D., Montana State University, 1988. *Astrophysics*. Relativistic astrophysics and gravitational wave physics. Disk population of close white dwarf binaries and globular cluster populations of relativistic binaries.

Díaz, Mario, Ph.D., University of Córdoba, 1987. *Relativity & Gravitation*. Gravitational wave astronomy; instrumentation and modeling.

Price, Richard H., Ph.D., California Institute of Technology, 1971. *Astronomy, Astrophysics, Relativity & Gravitation*. General relativity; astrophysics.

Romano, Joseph, Ph.D., Syracuse University, 1991. *Relativity & Gravitation*. Gravitational wave detection; data analysis.

Associate Professor

Dukes, Phillip R., Ph.D., Brigham Young University, 1996. *Physics and other Science Education*. Java applets for physics education, visualization of relativistically moving objects and stereoscopic visualization. Radio electronics. Teacher development.

Guevara, Natalia V., Ph.D., Moscow State University, 1989. *Biophysics*. Biophysics.

Hanke, Andreas, Ph.D., University of Wuppertal, 1998. *Biophysics, Nano Science and Technology*. Statistical mechanics, soft condensed matter physics, biophysics, nanoscience.

Jenet, Fredrick, Ph.D., California Institute of Technology, 2001. *Astrophysics*. Gravitational wave detection with Pulsar timing.

Martirosyan, Karen, Ph.D., Russian Academy of Sciences and SEUA - Chemical Engineering, 1991. *Condensed Matter Physics, Medical, Health Physics, Nano Science and Technology.* Structure and functions of nanoenergetic composites, rechargeable batteries, magnetic and ferroelectric devices, multifunctional contrast agent markers for biomedical imaging and therapy.

Mohanty, Soumya D., Ph.D., Inter-University Centre for Astronomy and Astrophysics, Pune University, 1998. *Relativity & Gravitation.* Gravitational wave astronomy; data analysis; computational methods.

Mukherjee, Soma, Ph.D., University of Calcutta, 1991. *Computational Physics, Relativity & Gravitation.* Gravitational wave astronomy; data analysis and detector characterization.

Assistant Professor

Creighton, Teviet, Ph.D., California Institute of Technology, 2000. *Astronomy, Astrophysics, Relativity & Gravitation.* Astrophysics.

Quetschke, Volker, Ph.D., University of Hannover, 2003. *Optics.* Laser physics; space-based experiments.

Rakhmanov, Malik, Ph.D., California Institute of Technology, 2000. *Optics, Relativity & Gravitation.* Experimental optics; nanophotonics.

Touhami, Ahmed, Ph.D., University of Pierre & Marie Curie, 1993. *Biophysics.* Biophysics; experimental optics.

DEPARTMENTAL RESEARCH SPECIALTIES AND STAFF

Theoretical

Astrophysics. Relativistic binaries; pulsar searches, pulsar timing. Benacquista, Creighton, Jenet, Price.

Atomic, Molecular, & Optical Physics. Lasers, Optics, Nanophotonics. Quetschke, Rakhmanov.

Biophysics. Modeling in molecular cell biology; conformational dynamics of DNA. Guevara, Hanke, Touhami.

Computational Physics. Gravitational wave data analysis, astroinformatics. Benacquista, Mohanty, Mukherjee.

Nano Science and Technology. Martirosyan, Rakhmanov, Touhami.

Physics and other Science Education. Java applets for physics education; visualization of relativistically moving objects. Dukes.

Relativity & Gravitation. Relativistic astrophysics, gravitational wave data analysis, computational methods and algorithm development. Creighton, Díaz, Jenet, Mohanty, Mukherjee, Price, Rakhmanov, Romano.

Experimental

Astrophysics. Radio observations with Arecibo, Green Bank, and Parkes radio telescopes, Gravitational wave interferometric instrumentation related Creep experiment. Díaz, Jenet.

Atomic, Molecular, & Optical Physics. Adaptive control of laser beam; adaptive optics, phasefront sensing; laser frequency stabilization; optical tweezers and wrenches; optical readout of AFM; photonic crystals. Quetschke, Rakhmanov.

Biophysics. Structure and functions of lipoproteins; ESR and fluorescence spectroscopy in studies of lipid-protein interactions; Atomic Force Microscopy. Guevara, Hanke, Touhami.

Nano Science and Technology. Martirosyan, Rakhmanov, Touhami.

View additional information about this department at
www.gradschoolshopper.com

UNIVERSITY OF TEXAS, DALLAS

PHYSICS DEPARTMENT

Richardson, Texas 75080-3021
http://www.utdallas.edu/physics/

General University Information

President: David E. Daniel
Dean of Graduate School: Austin J. Cunningham
University website: http://www.utdallas.edu/
Control: Public
Setting: Suburban
Total Faculty: 1,071
Total Graduate Faculty: 582
Total number of Students: 19,727
Total number of Graduate Students: 7,300

Department Information

Department Chairman: Robert Glosser, Head
Department Contact: Barbara Burbey, Graduate Support Assistant

Total full-time faculty: 20
Total number of full-time equivalent positions: 20
Full-Time Graduate Students: 67
First-Year Graduate Students: 12
Female First-Year Students: 4
Total Post Doctorates: 6

Department Address

800 West Campbell Road
EC 36
Richardson, TX 75080-3021
Phone: (972) 883-2835
Fax: (972) 883-2843
E-mail: bburbey@utdallas.edu
Website: http://www.utdallas.edu/physics/

ADMISSIONS

Admission Contact Information
Address admission inquiries to: Barbara Burbey, Graduate Support Assistant
Phone: (972) 883-2835
E-mail: bburbey@utdallas.edu
Admissions website: http://www.utdallas.edu/dept/physics

Application deadlines
Fall admission:
U.S. students: August 1   *Int'l. students*: May 1
Spring admission:
U.S. students: November 1   *Int'l. students*: September 1

Application fee
U.S. students: $50   *Int'l. students*: $100

Admissions information
For Fall of 2013:
   Number of applicants: 63
   Number admitted: 21

Admission requirements
Bachelor's degree requirements: Bachelor's degree in physics or a related field is required.
Minimum undergraduate GPA: 3.0

GRE requirements
The GRE is required.
   Quantitative score: 155
   Verbal score: 153
   Mean GRE score range (25th–75th percentile): 310-325

Advanced GRE requirements
The Advanced GRE is recommended.
Will be required for fall 2014 applicants.

TOEFL requirements
The TOEFL exam is required for students from non-English-speaking countries.
   iBT score: 80

Other admissions information
Additional requirements: Students must have a minimum of 155 on the quantitative and 153 on the verbal. Applicants with lower scores will be considered on an individual basis.
Undergraduate preparation assumed: It is assumed that the student has an undergraduate background that includes the following courses at the level indicated by texts referred to: mechanics at the level of Symon, Mechanics; electromagnetism at the level of Reitz and Milford, Foundations of Electromagnetic Theory; thermodynamics at the level of Kittel, Thermal Physics; quantum mechanics at the level of Griffiths, Introduction to Quantum Mechanics (chapters 1-4), some upper-division course(s) in modern physics, and atomic physics.

TUITION

Tuition year 2013-14:
Tuition for in-state residents
   Full-time students: $6,382 per semester
   Part-time students: $1,343 per credit
Tuition for out-of-state residents
   Full-time students: $11,549 per semester
   Part-time students: $1,918 per credit
Tuition and fees are waived for Teaching Assistants and Research Assistants. International students pay $100.00/semester as a fee.
Credit hours per semester to be considered full-time: 9
Deferred tuition plan: Yes
Health insurance: Available.

Other academic fees: International student orientation fee (one-time assessment)$50.00
Academic term: Semester
Number of first-year students who received full tuition waivers: 12

Teaching Assistants, Research Assistants, and Fellowships
Number of first-year
   Teaching Assistants: 11
   Research Assistants: 2
   Fellowship students: 1
Average stipend per academic year
   Teaching Assistant: $17,000
   Research Assistant: $20,400
   Fellowship student: $25,000

FINANCIAL AID

Loans
Loans are available for U.S. students.
Loans are not available for international students.
GAPSFAS application required: No
FAFSA application required: Yes

For further information
Address financial aid inquiries to: Barbara Burbey, Graduate Secretary, Mail Station EC36.
Phone: (972) 883-2835
E-mail: bburbey@utdallas.edu
Financial aid website: http://www.utdallas.edu/student/finaid/

HOUSING

Availability of on-campus housing
   Single students: Yes
   Married students: Yes

For further information
Address housing inquiries to: reslife@utdallas.edu.
Phone: (972)-883-5561
E-mail: reslife@utdallas.edu
Housing aid website: http://www.utdallas.edu/housing/

Table A—Faculty, Enrollments, and Degrees Granted

Research Specialty	2012-13 Faculty	Enrollment Fall 2012		Number of Degrees Granted 2012-13 (2008-13)		
		Master's	Doctorate	Master's	Terminal Master's	Doctorate
Applied Physics	1	–	2	2(5)	–	–
Astrophysics	2	–	5	–(3)	–	–(2)
Atmosphere, Space Physics, Cosmic Rays	4	–	6	–(3)	–	2(11)
Atomic, Molecular, & Optical Physics	1	–	4	–(4)	–	–(1)
Computational Physics	1	–	1	–	–	–
Condensed Matter Physics	4	–	1	–	–	–
High Energy Physics	2	–	2	–(5)	–	1(-)
Relativity & Gravitation	1	–	–	–	–(8)	–
Other	3	–	16	–(18)	2(-)	3(10)
Total	19	–	37	2(23)	–	6(24)
Full-time Grad. Stud.	52	6	61	–	–	–
First-year Grad. Stud.	17	3	12	–	–	–

GRADUATE DEGREE REQUIREMENTS

Master's: For the M.S., all students must complete at least 30 hours of graduate physics courses, including a 12-hour "core". The degree is completed either by six hours of research, including a thesis, or by six hrs of additional graduate courses.

Doctorate: The Ph.D. students must complete the 24-hour core, a minimum of 3 elective courses, 1 from within his/her area of specialization and 2 selected from different areas within the department plus whatever his/her committee requires. A Ph.D. candidate must pass a written qualifying exam that is presented twice each academic year. Once a dissertation topic has been selected and a faculty committee formed, the student presents a dissertation proposal to his/her committee for approval, presents a seminar, and is given an oral examination on the dissertation topic and related subjects. The student must then complete an acceptable dissertation and present a seminar. A successful defense of the dissertation concludes the requirements for the Ph.D. degree.

Thesis: Thesis may be written in absentia.

Table B—Separately Budgeted Research Expenditures by Source of Support

Source of Support	Departmental Research	Physics-related Research Outside Department
Federal government	$5,596,433	
State/local government	$50,000	
Non-profit organizations	$51,740	
Business and industry	$27,222	
Other		
Total	$5,725,395	

Table C—Separately Budgeted Research Expenditures by Research Specialty

Research Specialty	No. of Grants	Expenditures ($)
Atmosphere, Space Physics, Cosmic Rays	20	$3,668,466
Condensed Matter Physics	20	$1,594,188
Particles and Fields	3	$344,807
Relativity & Gravitation	2	$117,934
Total	45	$5,725,395

FACULTY

Professor

Anderson, Phillip C., Ph.D., University of Texas, Dallas, 1990. Graduate Advisor. *Atmosphere, Space Physics, Cosmic Rays.* Ionospheric and magnetospheric electrodynamics; space weather; space environment effects on human systems, properties of materials.

Cunningham, Augustine J., Ph.D., Queen's Belfast University, 1969. Graduate Dean. *Atomic, Molecular, & Optical Physics, Condensed Matter Physics, Solid State Physics.* Ion–electron recombination processes; ion–molecule reactions; high-temperature and pressure gas kinetics; ultraviolet spectroscopy; plasma etching, e-beam lithography.

Glosser, Robert, Ph.D., University of Chicago, 1967. Head, Department of Physics. *Condensed Matter Physics, Medical, Health Physics, Solid State Physics.* Optical properties of solids and biological materials; Raman, modulation, and fluorescence spectroscopies.

Heelis, Roderick A., Ph.D., University of Sheffield, 1973. Director of the William B. Hanson Center for Space Sciences. Endowed chair. *Atmosphere, Space Physics, Cosmic Rays.* Plasma processes and electrodynamics in planetary atmospheres and ionospheres; space flight instrumentation.

Hoffman, John H., Ph.D., University of Minnesota, 1958. *Atmosphere, Space Physics, Cosmic Rays, Planetary Science.* Ionospheric composition; planetary atmospheres; mass spectroscopy; stratospheric cluster ion composition.

Izen, Joseph M., Ph.D., Harvard University, 1982. *High Energy Physics, Particles and Fields.* Elementary particles, charm, bottom, and τ decay, e+e−; collider experiments, high-energy physics computing.

Lee, Mark, Ph.D., Stanford University, 1991. *Applied Physics, Condensed Matter Physics, Electromagnetism, Engineering Physics/Science, Low Temperature Physics, Nano Science and Technology, Solid State Physics.* Pure and applied condensed matter physics, science and engineering of novel electronic and optical materials and electronic and photonic device engineering.

Lou, Xinchou, Ph.D., State University of New York at Albany, 1989. *High Energy Physics, Particles and Fields.* Elementary particles physics, bottom and charm physics, e+e− colliders, offline software and distributed computing.

Rindler, Wolfgang, Ph.D., University of London, 1956. *Astrophysics, Cosmology & String Theory, Relativity & Gravitation.* Special and general relativity; cosmology; spinors.

Salamon, Myron B., Ph.D., University of California, Berkeley, 1966. *Condensed Matter Physics, Low Temperature Physics, Nano Science and Technology, Solid State Physics, Statistical & Thermal Physics.* Experimental studies of unconventional superconductors, manganites and layered magnetic materials. Low-temperature physics, neutron and x-ray scattering.

Zakhidov, Anvar, Ph.D., Institute of Spectroscopy, U.S.S.R. Academy of Sciences, 1981. Deputy Director of the Nano-Tech Institute. *Applied Physics, Condensed Matter Physics, Low Temperature Physics, Nano Science and Technology, Solid State Physics.* Nanotechnology; photonic crystals; carbon nanotubes; organic molecular crystals.

Associate Professor

Gartstein, Yuri, Ph.D., Institute for Spectroscopy, USSR Academy of Sciences, 1988. *Condensed Matter Physics, Nano Science and Technology, Solid State Physics, Theoretical Physics.* Condensed matter physics with emphasis on nanoscience; electronic, optical, and transport properties of organic materials.

Ishak-Boushaki, Mustapha, Ph.D., Queen's University, 2002. *Astrophysics, Computational Physics, Cosmology & String Theory, Relativity & Gravitation.* Classical and modern cosmology; relativity; gravitational lensing (cosmic shear); cosmological models; computer algebra systems applied to relativity.

King, Lindsay J., Ph.D., University of Manchester, 1995. *Astronomy, Astrophysics, Computational Physics, Cosmology & String Theory, Relativity & Gravitation.* Physical cosmology using tools such as gravitational lensing to understand dark matter and dark energy. Computational and theoretical work as well as observations with large telescopes.

Lary, David J., Ph.D., University of Cambridge, 1991. *Applied Physics, Atmosphere, Space Physics, Cosmic Rays, Computational Physics.* Computational and information systems to facilitate discovery and decision support in earth system science.

Zhang, Chuanwei, Ph.D., The University of Texas, Austin, 2005. *Atomic, Molecular, & Optical Physics, Computational Physics, Condensed Matter Physics, Low Temperature Physics, Materials Science, Metallurgy, Nano Science and Technology, Nonlinear Dynamics and Complex Systems, Solid State Physics, Theoretical Physics.* Topological superfluids, superconductors and insulators; ultra-cold atomic gases; quantum computation; graphene.

Assistant Professor

Chen, Xingang, Ph.D., Columbia University, 2003. *Astrophysics, Cosmology & String Theory, Relativity & Gravitation.* String Theory, The Early Universe.

Kesden, Michael H., Ph.D., California Institute of Technology, 2005. *Astrophysics, Cosmology & String Theory, Relativity & Gravitation.* Black holes, lensing.

Malko, Anton V., Ph.D., New Mexico State/Los Alamos National Labs, 2002. *Applied Physics, Condensed Matter Physics, Nano Science and Technology, Optics.* Femtosecond laser spectroscopy of Nanomaterials such as semiconductor quantum dots, wires and wells; photoluminescence spectroscopy and microscopy; quantum optics; photoluminescence spectroscopy of single nanoparticles; solid state physics; laser physics.

Rodrigues, Fabiano, Ph.D., Cornell University, 2008. *Atmosphere, Space Physics, Cosmic Rays, Solar Physics.* Atmosphere, Space Physics, Cosmic Rays, Radio remote sensing of the upper atmosphere/ionosphere, ionospheric electrodynamics and irregularities, space weather.

Slinker, Jason D., Ph.D., Cornell University, 2007. *Applied Physics, Biophysics, Condensed Matter Physics, Nano Science and Technology.* Organic optoelectronic devices and laboratory assays. Devices include light emitting electrochemical cells and electrochemical biosensors with DNA-modified electrodes.

Emeritus

Fenyves, Ervin J., Ph.D., University of Budapest, 1950. *Nuclear Physics, Particles and Fields.* Elementary particles; cosmic rays; gamma-ray astrophysics; gamma-ray and neutrino detectors.

Heikkila, Walter J., Ph.D., University of Toronto, 1954. *Atmosphere, Space Physics, Cosmic Rays, Plasma and Fusion, Solar Physics.* Magnetospheric physics, solar wind, auroral substorms.

Tinsley, Brian, Ph.D., University of Canterbury, 1963. *Atmosphere, Space Physics, Cosmic Rays.* Airglow; aurora; theoretical research in aeronomy; instrumentation for atmospheric spectroscopy.

Senior Lecturer

MacAlevey, Paul J., Ph.D., University of Texas, Dallas, 1996. *Physics and other Science Education, Relativity & Gravitation.*

Rasmussen, Beatrice, M.S., University of Texas, Dallas, 1996. *Atmosphere, Space Physics, Cosmic Rays, Biophysics, Com-* *putational Physics, Physics and other Science Education.* A study on Equatorial Spread F in the earth's ionosphere.

DEPARTMENTAL RESEARCH SPECIALTIES AND STAFF

Theoretical

Nano Science and Technology. Nanoscience, electronic, optical, and transport properties of organic materials. Calculation of electronic structure by LCAO. Gartstein, Lee, Salamon, Zakhidov, Zhang.

Relativity & Gravitation. Gravitational radiation; exact solutions of Einstein's field equations. Classical and modern cosmology; gravitational lensing (cosmic shear); cosmological models; computer algebra systems applied to relativity. Ishak-Boushaki, King, MacAlevey, Rindler.

Remote Sensing for Atmospheric Physics. Computational and information systems to facilitate discovery and decision support in earth system science. Heikkila, Lary, Rasmussen, Rodrigues.

Experimental

Atmosphere, Space Physics, Cosmic Rays. Aeronomy; thermospheric, ionospheric and magnetospheric physics; planetary atmospheres. Instrumentation and data analysis for various satellites and deep space probes; microphysics of clouds, climate. Atmospheric electricity. Thermal properties of airless planetary regoliths, distribution of volatiles in the Martian crust, misconceptions in physics and astronomy education, space science and physics educational outreach programs. Anderson, Heelis, Heikkila, Hoffman, Lary, Rasmussen, Rodrigues, Tinsley.

Condensed Matter Physics. Raman, photoluminescence, and modulation spectroscopy of solids. Unconventional superconductivity. Magnetism; disordered and nanoscale magnets. Femtosecond laser spectroscopy of materials, photoluminescence, absorption spectroscopy Novel electronic and optical materials and electronic and photonic device engineering. Organic optoelectronic devices and laboratory assays. Light emitting electrochemical cells and electrochemical biosensors with DNA-modified electrodes. Cunningham, Glosser, Lee, Malko, Salamon, Slinker, Zakhidov, Zhang.

Low Temperature Physics. Experimental studies of unconventional superconductors, manganites. and layered magnetic materials. Low-temperature physics, neutron and x-ray scattering. Salamon, Zakhidov.

Nano Science and Technology. Quantum semiconductor nanostructure, optical properties. Glosser, Lee, Malko, Salamon, Slinker, Zakhidov, Zhang.

Optics. Quantum and nonlinear optics, single and multiphoton emission processes. Ultrafast laser spectroscopy. Malko.

Particles and Fields. Charm, bottom, and τ decays at e+e− colliders; simulation of fixed target detectors for b physics. Fenyves, Izen, Lou.

View additional information about this department at
www.gradschoolshopper.com

UNIVERSITY OF TEXAS, EL PASO

DEPARTMENT OF PHYSICS

El Paso, Texas 79968
http://www.utep.edu/physics/

General University Information
President: Diana Natalicio
Dean of Graduate School: Benjamin Flores
University website: http://www.utep.edu
Control: Public
Setting: Urban
Total Faculty: 1,158
Total number of Students: 20,198
Total number of Graduate Students: 3,097

Department Information
Department Chairman: Vivian Incera, Chair
Department Contact: Juan Camacho, Office Supervisor
  Total full-time faculty: 17
  Total number of full-time equivalent positions: 17
  Full-Time Graduate Students: 16
  First-Year Graduate Students: 9
  Female First-Year Students: 3
  Total Post Doctorates: 4

Department Address
500 W. University Avenue
El Paso, TX 79968
Phone: (915) 747-5715
Fax: (915) 747-5447
E-mail: physics@utep.edu
Website: http://www.utep.edu/physics/

ADMISSIONS

Admission Contact Information
Address admission inquiries to: Dr. Vivian Incera, Physics Department, 500 W. University Avenue, El Paso, TX 79968-0515
Phone: (915) 747-5715
E-mail: vincera@utep.edu
Admissions website: http://utep.edu/admit/

Application deadlines
Fall admission:
U.S. students: August 1   *Int'l. students*: March 17
Spring admission:
U.S. students: November 1   *Int'l. students*: September 16

Application fee
U.S. students: $45   *Int'l. students*: $80

Admissions information
For Fall of 2013:
  Number of applicants: 26
  Number admitted: 18
  Number enrolled: 16

Admission requirements
Bachelor's degree requirements: Bachelor's degree in physics or equivalent is required.
Minimum undergraduate GPA: 3.0

GRE requirements
  The GRE is not required.

Advanced GRE requirements
  The Advanced GRE is not required.

TOEFL requirements
  The TOEFL exam is required for students from non-English-speaking countries.
  PBT score: 550
  iBT score: 79

Other admissions information
Undergraduate preparation assumed: Symon, Mechanics; Griffiths, Introduction to Electrodynamics; Park, Introduction to Quantum Theory.

TUITION

Tuition year 2013-14:
Tuition for in-state residents
  Full-time students: $411.3 per credit
Tuition for out-of-state residents
  Full-time students: $787.3 per credit
Teaching assistants pay in-state tuition. See Graduate Catalog for detailed breakdown.
Deferred tuition plan: Yes
Health insurance: Yes, $982.00.
Other academic fees: Go to www.sbs.utep.edu for additional fees.

Teaching Assistants, Research Assistants, and Fellowships
  Average stipend per academic year
    Teaching Assistant: $12,150

FINANCIAL AID

Loans
Loans are available for U.S. students.
Loans are available for international students.
GAPSFAS application required: No
FAFSA application required: Yes

For further information
Address financial aid inquiries to: Financial Aid, UTEP, Academic Services Building, Room 204, El Paso, TX 79902.
Phone: 915-747-5204
Financial aid website: http://academics.utep.edu

HOUSING

Availability of on-campus housing
  Single students: Yes
  Married students: Yes

For further information
Address housing inquiries to: Residence Life, UTEP, 2401 N. Oregon Street, El Paso, TX 79902.
Phone: (915) 747-5352
Housing aid website: http://sa.utep.edu/housing

Table A—Faculty, Enrollments, and Degrees Granted

Research Specialty	2012-13 Faculty	Enrollment Spring 2013		Number of Degrees Granted 2012-13 (2005–09)		
		Master's	Doctorate	Master's	Terminal Master's	Doctorate
Atmosphere, Space Physics, Cosmic Rays	1	–	2	–	–	1(-)
Atomic, Molecular, & Optical Physics	1	2	–	–	–	–
Computational Physics	3	1	3	1(-)	–	2(-)
Materials Science, Metallurgy	2	2	–	–	1(-)	–
Medical, Health Physics	1	3	–	–	4(-)	–
Nuclear Physics	3	3	–	–	2(-)	–
Physics and other Science Education	1	–	–	–	–	–
Non-specialized	–	2	–	–	–	–
Total	12	13	5	1(-)	7(-)	3(-)
Full-time Grad. Stud.	–	12	5	–	–	–
First-year Grad. Stud.	–	5		–	–	–

GRADUATE DEGREE REQUIREMENTS

Master's: The Department offers a program of courses and research leading to the degree of M.S. in Physics (http://science.utep.edu/physics/graduate/master-of-science-in-physics). Two routes may be taken. Plan 1 requires 30 semester hours of credit: 24 hours of course work plus a 6 hour thesis (Physics 5398 and 5399). Plan 2 requires 36 hours of course work, including a successful completion of a research problem (Physics 5391) being substituted for a thesis. A grade average of B is required. There are no qualifying exam requirements.

Doctorate: The Department of Physics participates in three interdisciplinary doctoral programs: materials science and engineering, computational science, and environmental science. Information about these programs may be obtained directly from the programs' links in the UTEP website. A Master of Science degree in geophysics (http://science.utep.edu/physics/graduate/master-of-science-in-geophysics), a Master of Science in Medical Physics (http://science.utep.edu/physics/graduate/master-of-science-in-medical-physics), and a Master of Science in Computational Science (http://science.utep.edu/computationalscience/), this last one required as part of the PhD in Computational Science program, are other options available.

Thesis: Thesis may be written in absentia, but enrollment on Thesis II is required during the semester of graduation.

SPECIAL EQUIPMENT, FACILITIES, OR PROGRAMS

Biophotonic Lab with two-photon fluorescence microscopy, femtosecond laser, and Raman spectroscopy for molecular and cellular imaging in medicine research.

Optical Spectroscopy & Microscopy

Lab with Raman scattering,infrared

absorpion, photoluminescence,atomic

force microscopy,and near-field

optical microscopy for organic-inorganic material characterizations.

Structural and Magnetic Characterization Lab with Quantum Design Physical Property Measurement system and Siemens D-5000 X-ray diffractometer to conduct materials science studies.

Surface Physics Lab with Auger spectroscopy, XPS spectroscopy, SIMS spectroscopy, SEM microscopy, and XRF spectroscopy to conduct studies of elemental composition of surfaces of metals, ceramics, etc.

FACULTY

Professor

Ferrer, Efrain J., Ph.D., P.N. Lebedev Physical Institute, Moscow, USSR, 1988. *High Energy Physics, Nuclear Physics, Theoretical Physics*. Theoretical nuclear and high energy physics.

Incera, Vivian, Ph.D., P.N. Lebedev Physical Institute, Moscow, USSR, 1988. *High Energy Physics, Nuclear Physics, Theoretical Physics*. Theoretical nuclear and high energy physics.

López, Jorge A., Ph.D., Texas A&M University, 1986. *Nuclear Physics*. Theoretical physics; nuclear physics.

Associate Professor

Botez, Cristian E., Ph.D., University of Missouri, Columbia, 2002. *Condensed Matter Physics, Materials Science, Metallurgy*. Experimental condensed matter.

Fitzgerald, Rosa M., Ph.D., U.C. Riverside, 1992. *Atmosphere, Space Physics, Cosmic Rays, Electromagnetism, Energy Sources & Environment*. Atmospheric physics.

Hagedorn, Eric A., Ph.D., University of Wisconsin-Milwaukee. *Other*. Testing and measurement in physics education.

Manciu, Felicia, Ph.D., State University of New York at Buffalo, 2004. *Materials Science, Metallurgy*. Experimental optics and material science.

Manciu, Marian, Ph.D., State University of New York at Buffalo, 2000. *Biophysics, Medical, Health Physics*. Biophysics and Medical Physics.

Ravelo, Ramon, Ph.D., Boston University, 1990. *Computational Physics, Condensed Matter Physics*. Computational physics; condensed matter; nonlinear phenomena.

Assistant Professor

Baruah, Tunna, Ph.D., University of Pune, 2000. *Chemical Physics, Computational Physics, Condensed Matter Physics, Materials Science, Metallurgy*. Computational chemical physics and material structures.

Li, Chunqiang, Ph.D., Princeton University, 2006. *Condensed Matter Physics, Optics*. Biophotonics.

Zope, Rajendra, Ph.D., University of Pune, 2000. *Chemical Physics, Computational Physics, Condensed Matter Physics*. Density functional theory.

DEPARTMENTAL RESEARCH SPECIALTIES AND STAFF

Theoretical
Biophysics. Marian Manciu.
Computational Physics. Baruah, Ravelo, Zope.
High Energy Physics. Ferrer, Incera.
Materials Science, Metallurgy. Baruah, Ravelo, Zope.
Nuclear Physics. Ferrer, Incera, López.
Physics and other Science Education. Hagedorn.

Experimental
Atmosphere, Space Physics, Cosmic Rays. Fitzgerald.
Atomic, Molecular, & Optical Physics. Li, Felicia Manciu.
Condensed Matter Physics. Botez, Li, Felicia Manciu.
Materials Science, Metallurgy. Botez, Li, Felicia Manciu.

THE UNIVERSITY OF TEXAS AT SAN ANTONIO

DEPARTMENT OF PHYSICS AND ASTRONOMY

San Antonio, Texas 78249
http://physics.utsa.edu

General University Information

President: Ricardo Romo
Dean of Graduate School: Dorothy Flannagan
University website: http://www.utsa.edu
Control: Public
Setting: Suburban
Total Faculty: 1,113
Total Graduate Faculty: 614
Total number of Students: 30,474
Total number of Graduate Students: 4,495

Department Information

Department Chairman: Miguel Jose Yacaman, Chair
Department Contact: Nakia Scott, Senior Administrative
   Associate
   Total full-time faculty: 22
   Total number of full-time equivalent positions: 22
   Full-Time Graduate Students: 80
   First-Year Graduate Students: 13
   Female First-Year Students: 3
   Total Post Doctorates: 1

Department Address

One UTSA Circle
San Antonio, TX 78249
Phone: (210) 458-5451
E-mail: nakia.scott@utsa.edu
Website: http://physics.utsa.edu

ADMISSIONS

Admission Contact Information

Address admission inquiries to: Dr. Lorenzo Brancaleon
Phone: (210) 458-5694
E-mail: lorenzo.brancaleon@utsa.edu
Admissions website: http://physics.utsa.edu/Graduate%20Studies/
   Admission_Support.html

Application deadlines

Fall admission:
U.S. students: January 1   *Int'l. students*: January 1
Spring admission:
U.S. students: October 1   *Int'l. students*: October 1

Application fee

U.S. students: $45   *Int'l. students*: $80
Late fee for late application (this fee is applicable to all) $10

Admissions information

For Fall of 2013:
   Number of applicants: 42
   Number admitted: 16
   Number enrolled: 15

Admission requirements

Bachelor's degree requirements: A B.S. in Physics or equivalent
   degree or a minimum of 12 credit hours of upper division
   Physics classes required.
Minimum undergraduate GPA: 3.0

GRE requirements

The GRE is required.
   Mean GRE score range (25th–75th percentile): 313
Mean Q-GRE 159

Advanced GRE requirements

The Advanced GRE is recommended.

TOEFL requirements

The TOEFL exam is required for students from non-English-
   speaking countries.
   PBT score: 550
   iBT score: 79
IELTS - 6.5

Other admissions information

Additional requirements: There are no minimum scores to be eligible to apply. For international students, the TOEFL or the
   IELTS score is required. The minimum required TOEFL
   score is 79 for Master's and PhD applicants. The minimum
   required IELTS score is 6.5 for Master's and Ph.D. applicants.
Undergraduate preparation assumed: A minimum of 12 hours
   of upper division Physics classes, which includes fundamentals of Classical Mechanics, Statistical Thermodynamics,
   Quantum Mechanics and Electrodynamics.

TUITION

Tuition year 2013-14:
Tuition for in-state residents
   Full-time students: $8,696 annual
   Part-time students: $242.75 per credit
Tuition for out-of-state residents
   Full-time students: $13,288 annual
   Part-time students: $868.75 per credit
Credit hours per semester to be considered full-time: 9
Deferred tuition plan: Yes
Health insurance: Available
Academic term: Semester

Teaching Assistants, Research Assistants, and Fellowships

Number of first-year
   Teaching Assistants: 4
   Research Assistants: 2
   Fellowship students: 2
Average stipend per academic year
   Teaching Assistant: $27,000
   Research Assistant: $25,000
   Fellowship student: $25,000

FINANCIAL AID

Loans

Loans are available for U.S. students.
Loans are available for international students.
GAPSFAS application required: Yes
FAFSA application required: Yes

For further information
Address financial aid inquiries to: One UTSA Circle, Attn: Financial Aid Office, San Antonio, TX 78249.
Phone: (210) 458-8000
Financial aid website: http://utsa.edu/financialaid/index.html

HOUSING

Availability of on-campus housing
Single students: Yes
Married students: No

For further information
Address housing inquiries to: One UTSA Circle, Attn: Housing division, San Antonio, TX 78249.
Phone: (210) 458-6200
Housing aid website: http://utsa.edu/housing/index2.html

Table A—Faculty, Enrollments, and Degrees Granted

Research Specialty	2012-13 Faculty	Enrollment Fall 2012 Master's	Enrollment Fall 2012 Doctorate	Number of Degrees Granted 2012-13 (2008-13) Master's	Number of Degrees Granted 2012-13 (2008-13) Terminal Master's	Number of Degrees Granted 2012-13 (2008-13) Doctorate
Astrophysics	9	2	17	–	–	–
Atmosphere, Space Physics, Cosmic Rays	10	4	13	–	2(4)	2(3)
Atomic, Molecular, & Optical Physics	4	–	5	–	–	–(4)
Biophysics	7	2	12	–	–(1)	–
Computational Physics	2	–	2	–	–	–
Condensed Matter Physics	4	–	4	–	–(2)	1(4)
Nano Science and Technology	4	3	14	–	–	1(1)
Non-specialized	–	2	–	–	–(4)	–
Total	40	13	67	–	2(11)	4(12)
Full-time Grad. Stud.	–	10	63	–	–	–
First-year Grad. Stud.	–	5	13	–	–	–

GRADUATE DEGREE REQUIREMENTS
Master's: The Master of Science program requires the successful completion of a minimum of 30 semester credit hours, including 12 hours of core classes, 6 hours of Directed Research, and 3 hours of research seminars.

Doctorate: The doctoral degree requires a minimum of 81 semester credit hours beyond the baccalaureate degree. The coursework in the Program of Study includes a Core Curriculum (12 semester credit hours) and Advanced Electives (27 semester credit hours), including graduate courses offered by other departments with the approval of the student's Graduate Advisor. Research hours, including Research Seminar (3 semester credit hours), Directed and Doctoral Research (27 semester credit hours) and Dissertation (12 semester credit hours), totaling at least 42 semester credit hours, complete the Program of Study.

Other Degrees: Master's students have a non-thesis option.
Thesis: All Ph.D. students must provide an original dissertation. M.S. students who opt for a thesis option must write a thesis that describes their research.

SPECIAL EQUIPMENT, FACILITIES, OR PROGRAMS
Our Department has access to an atomic layer deposition, sputterer, two clean rooms, surface profiler, XRD, thin film stress measurement, vector network analyzer, hot embosser, ellipsometer, tensile tester, porosimeter, three atomic force microscopes, FTIR spectrometers, inverted microscopes, micro-Raman spectrometer, two systems for pulsed laser deposition, barrel etcher, LAM etcher, three scanning electron microscopes, two transmission electron microscopes, two mass spectrometers, two facilities for femtosecond spectroscopy and imaging, circular dichroism spectrometer, fluorescence spectrometers, UV-Vis absorption spectrometers, system for picoseconds fluorescence lifetime detection, Raman microscope, Spectropolarimeter for circular dichroism spectroscopy, and a terahertz laser.

Table B—Separately Budgeted Research Expenditures by Source of Support

Source of Support	Departmental Research	Physics-related Research Outside Department
Federal government		
State/local government		
Non-profit organizations		
Business and industry		
Other	$1,669,102	
Total	**$1,669,102**	

FACULTY

Professor
Ayon, Arturo, Ph.D., Cornell University, 1996. *Materials Science, Metallurgy, Nano Science and Technology*. Sensor arrays; micro-chemical reactors; demonstration of miniaturized muon sensors; micropropulsion employing solid fuels; negative index of refraction materials for imaging and other photonic applications; utilization of MEMS actuators on phase array antennas; CMOS-compatible microwave varactors and other radio-frequency projects.

Chen, Chonglin, Ph.D., Pennsylvania State University, 1994. *Condensed Matter Physics, Materials Science, Metallurgy, Nano Science and Technology*. Electronic thin films; surface science and interface phenomena; nanostructures and nanophenomena; advanced materials; nano/micro scale characterizations; novel device fabrications.

Chen, Liao Y., Ph.D., Chinese Academy of Sciences, 1988. *Biophysics, Chemical Physics, Computational Physics*. Biological physics; chemical physics; condensed matter physics.

Jose-Yacaman, Miguel, Ph.D., National University of Mexico, 1973. *Biophysics, Materials Science, Metallurgy, Nano Science and Technology*. Structure and properties of nanoparticles including metals; semiconductors; and magnetic materials. Synthesis and characterization of new materials most of them nanoparticles, surfaces and interfaces, defects in solids, electron diffraction and imagining theory, quasicrystals, archaeological materials, and catalysis.

Sardar, Dhiraj K., Ph.D., Oklahoma State University, 1980. *Biophysics, Nano Science and Technology, Optics*. Optical properties of a variety of technologically important materials using high resolution laser spectroscopy techniques, optical properties of biological tissues as well as studying the laser-tissue interaction, an exciting area of biophysics.

Schlegel, Eric M., Ph.D., Indiana University, 1983. *Astronomy, Astrophysics.* X-ray line emission from cataclysmic variables, supernovae.

Whetten, Robert L., Ph.D., Cornell University. *Applied Physics, Chemical Physics, Nano Science and Technology.*

Associate Professor

Brancaleon, Lorenzo, Ph.D., University of Parma, Italy, 1996. *Biophysics.* Molecular biophysics; optical spectroscopy; biomaterials; photobiophysics.

Chabanov, Andrey A., Ph.D., City University of New York, 2002. *Atomic, Molecular, & Optical Physics, Materials Science, Metallurgy, Nonlinear Dynamics and Complex Systems.* Microwave properties of magnetic photonic crystals and their applications in antennas; propagation and localization of microwaves in random cavities and waveguides, fabrication and optical properties of photonic band gap materials for photonics applications; photon localization and lasing in disordered microstructures.

Koinov, Zlatko, Ph.D., St. Petersburg Electrotechnical University, Russia, 1999. *Computational Physics, Condensed Matter Physics.* Nanophysics: Optical properties of single and couple quantum wells. Bose-Einstein condensation: Formation of Bose-Einstein condensate of excitons; polariton condensation in microcavities: the effects of the symmetry-breaking disorder on the condensate; collective excitations in superconductors Magnetooptics: Quantum-well excitons in high magnetic fields; Bethe-Salpeter equations for magnetoexcitons; Confinement of magnetoexcitons in quantum wells due to inhomogeneous magnetic fields. Strongly correlated systems: Hubbard model; High-temperature superconductivity.

Assistant Professor

Lopez-Lozano, Xochitl, Ph.D., Universidad Autonoma de Puebla, Mexico, 2005. *Computational Physics, Nano Science and Technology.* Structural, electronic, optical, and catalytic properties of materials at the nanometer scale through semiempirical and ab initio density-functional theory calculations.

Marucho, Marcelo, Ph.D., National University of La Plata, Argentina, 2002. *Biophysics, Computational Physics.* Theoretical and computational research in chemical physics, biophysics, and polymer physics, mathematical physics, with particular emphasis on the development of new analytical and numerical tools.

Nash, Kelly, Ph.D., University of Texas, San Antonio, 2009. *Nano Science and Technology, Optics.* Applications of fluorescent polymer composites, diodes, lighting displays, optical biosensors, drug delivery.

Packham, Chris, Ph.D., University of Hertfordshire. *Astronomy, Astrophysics.*

Peralta, Xomalin G., Ph.D., University of California, Santa Barbara, 2002. *Atomic, Molecular, & Optical Physics, Biophysics, Materials Science, Metallurgy, Optics.* Terahertz spectroscopy; biophysics; nanoparticles.

Ponce Pedraza, Arturo, Ph.D., UNAM. *Materials Science, Metallurgy, Nano Science and Technology.* Nanomaterials.

Adjunct Professor

Allegrini, Frédéric, Ph.D., University of Bern, Switzerland. *Atmosphere, Space Physics, Cosmic Rays.*

Boice, Daniel, Ph.D., New Mexico State University. *Atmosphere, Space Physics, Cosmic Rays.*

Desai, Mihir I., Ph.D., University of Birmingham. *Atmosphere, Space Physics, Cosmic Rays.*

Goldstein, Jerry, Ph.D., Dartmouth College. *Atmosphere, Space Physics, Cosmic Rays.*

Jahn, Jorg-Micha, Ph.D., Dartmouth College. *Atmosphere, Space Physics, Cosmic Rays.*

Livi, Stefano, Ph.D., University of Florence (Italy). *Atmosphere, Space Physics, Cosmic Rays.*

McComas, David J., Ph.D., University of California, Los Angeles. *Atmosphere, Space Physics, Cosmic Rays.*

Pollock, Craig, Ph.D., University of New Hampshire. *Atmosphere, Space Physics, Cosmic Rays.*

Valek, Phillip, Ph.D., Auburn University. *Atmosphere, Space Physics, Cosmic Rays.*

Waite, Jack H., Ph.D., *Atmosphere, Space Physics, Cosmic Rays.*

Lecturer

Konno, Ishiro, Ph.D., Arizona State University. *Astronomy.*

Koynova, Aeta, Ph.D., Ioffe Physico-Technical Institute, St. Petersburg, Russia. *Condensed Matter Physics, Energy Sources & Environment.*

Lopez-Mobilia, Rafael, Ph.D., University of Texas, Austin. *Particles and Fields.*

DEPARTMENTAL RESEARCH SPECIALTIES AND STAFF

Theoretical

Astrophysics. Lopez-Mobilia, Packham, Schlegel.

Atmosphere, Space Physics, Cosmic Rays. Allegrini, Boice, Desai, Goldstein, Jahn, Livi, McComas, Pollock, Schlegel, Valek, Waite.

Atomic, Molecular, & Optical Physics. Chabanov, Sardar.

Biophysics. Brancaleon, Jose-Yacaman, Marucho, Nash, Peralta, Sardar.

Computational Physics. Liao Chen, Lopez-Lozano, Marucho.

Condensed Matter Physics. Chonglin Chen, Koinov.

Materials Science, Metallurgy. Ayon, Chabanov, Chonglin Chen, Jose-Yacaman, Lopez-Lozano, Peralta, Ponce Pedraza, Whetten.

Nano Science and Technology. Liao Chen, Jose-Yacaman, Lopez-Lozano, Nash, Peralta, Ponce Pedraza, Sardar, Whetten.

Optics. Sardar.

Other.

Experimental

Optics.

View additional information about this department at
www.gradschoolshopper.com

BRIGHAM YOUNG UNIVERSITY

DEPARTMENT OF PHYSICS AND ASTRONOMY

Provo, Utah 84602
http://www.physics.byu.edu/Graduate/default.aspx

General University Information
President: Cecil O. Samuelson
Dean of Graduate School: Wynn Stirling
University website: http://byu.edu
Control: Private
Setting: Suburban
Total Faculty: 1,283
Total Graduate Faculty: 1,084
Total number of Students: 32,955
Total number of Graduate Students: 3,355

Department Information
Department Chairman: Richard R. Vanfleet, Chair
Department Contact: Shelena Shamo, Graduate Secretary
  Total full-time faculty: 32
  Total number of full-time equivalent positions: 32
  Full-Time Graduate Students: 35
  First-Year Graduate Students: 9
  Female First-Year Students: 2
  Total Post Doctorates: 1

Department Address
Brigham Young University
N234 ESC
Provo, UT 84602
Phone: (801) 422-9299
Fax: (801) 422-0101
E-mail: graduatesecretary@physics.byu.edu
Website: http://www.physics.byu.edu/Graduate/default.aspx

ADMISSIONS

Admission Contact Information
Address admission inquiries to: Graduate Studies, Brigham
  Young University, 105 FPH, Provo, UT 84602
Phone: (801) 422-4091
E-mail: gradstudies@byu.edu
Admissions website: http://graduatestudies.byu.edu/

Application deadlines
Fall admission:
U.S. students: January 15   *Int'l. students*: January 15

Application fee
U.S. students: $50   *Int'l. students*: $50

Admissions information
For Fall of 2012:
  Number of applicants: 23
  Number admitted: 13
  Number enrolled: 9

Admission requirements
Bachelor's degree requirements: Bachelor's degree in physics
  or astronomy is required.
Minimum undergraduate GPA: 3.0

GRE requirements
The GRE is required.

Advanced GRE requirements
The Advanced GRE is required.

TOEFL requirements
The TOEFL exam is required for students from non-English-
  speaking countries.
  PBT score: 580
  iBT score: 85

Other admissions information
Undergraduate preparation assumed: Fowles, Analytical Me-
  chanics; Griffiths, Foundations of Electromagnetic Theory;
  Schroder, Thermal Physics; Griffiths, Quantum Mechanics;
  Hecht, Optics.

TUITION

Tuition year 2013-14:
  Full-time students: $6,130 annual
  Part-time students: $340 per credit
$12,260 if non-member of the Church of Jesus Christ of Latter-
  day Saints for full-time; $680 per credit if non-member of
  the Church of Jesus Christ of Latter-day Saints for part-time.
Credit hours per semester to be considered full-time: 2
Deferred tuition plan: No
Health insurance: Available.
Academic term: Semester
Number of first-year students who received full tuition waivers: 9

Teaching Assistants, Research Assistants, and Fellowships
Number of first-year
  Teaching Assistants: 6
  Research Assistants: 3
Average stipend per academic year
  Teaching Assistant: $19,680
  Research Assistant: $20,640

FINANCIAL AID

Application deadlines
Fall admission:
U.S. students: January 15

Loans
Loans are available for U.S. students.
Loans are available for international students.
GAPSFAS application required: No
FAFSA application required: No

For further information
Address financial aid inquiries to: Director of Financial Aid,
  A-41 ASB.
Phone: (801) 422-4104
Financial aid website: http://financialaid.byu.edu/

HOUSING

Availability of on-campus housing
Single students: Yes
Married students: Yes

For further information
Address housing inquiries to: Director of Housing, 100 SASB.
Phone: (801) 422-2611
Housing aid website: http://housing.byu.edu/

Table A—Faculty, Enrollments, and Degrees Granted

Research Specialty	2012-2013 Faculty	Enrollment Winter 2013		Number of Degrees Granted 2012-13 (2008-13)		
		Mas-ter's	Doc-torate	Mas-ter's	Terminal Master's	Doc-torate
Acoustics	4	7	2	2(9)	–	1(3)
Astrophysics	6	3	3	1(5)	–	–(3)
Atomic, Molecular, & Optical Physics	6	3	1	–(4)	–	–(2)
Condensed Matter Physics	8	3	2	3(8)	–	1(6)
Nuclear Physics	1	–	–	–	–	–
Plasma and Fusion	3	–	1	2(3)	–	–
Relativity & Gravitation	2	–	1	1(3)	–	–
Other	2	1	1	1(1)	–	–(1)
Total	32	17	11	10(33)	–	2(15)
Full-time Grad. Stud.	–	17	11	–	–	–
First-year Grad. Stud.	–	5	4	–	–	–

GRADUATE DEGREE REQUIREMENTS

Master's: Minimum of 24 hours of course credit (7.5 may be transferred) plus thesis (6 credit hour minimum); thesis required; oral examination in support of thesis required; residence of 20 semester hours taken on Provo campus required; minimum GPA is 3.0; there is no language requirement. Before admission to candidacy, a student must be accepted as a research student by a department faculty member and submit a proposed study list.

Doctorate: Semester hours requirement: 36 hours approved coursework (exclusive of graduate seminars) plus dissertation (18 hour minimum); minimum GPA is 3.0. Ordinarily, two years of full-time coursework must be taken on the Provo campus, of which two full-time semesters must be consecutive. Written qualifying examination in the first year plus oral examination in support of a dissertation are required. Before admission to candidacy, a student must be accepted as a research student by a departmental faculty member and submit a proposed study list.

Other Degrees: The Ph.D. in physics and astronomy has the same requirements as for the Ph.D. in physics except the course requirements are appropriate to astronomy and astrophysics.

Thesis: Thesis may be written in absentia.

SPECIAL EQUIPMENT, FACILITIES, OR PROGRAMS

Mountain observatory with 0.3, 0.5, and 0.9-meter telescopes nd robotic 0.4-meter telescope; ultra-short pulse laser system; laser cooling system; XUV reflectometers; on-campus supercomputers; atomic force, magnetic force, and scanning tunneling microscopes; anechoic chamber; reverberation chamber.

Table B—Separately Budgeted Research Expenditures by Source of Support

Source of Support	Departmental Research	Physics-related Research Outside Department
Federal government	$1,528,163	
State/local government		
Non-profit organizations		
Business and industry	$287,657	
Other		
Total	$1,815,820	

Table C—Separately Budgeted Research Expenditures by Research Specialty

Research Specialty	No. of Grants	Expenditures ($)
Acoustics	6	$203,630
Astrophysics	2	$1,512
Atomic, Molecular, & Optical Physics	4	$474,001
Condensed Matter Physics	6	$566,861
Plasma and Fusion	–	
Relativity & Gravitation	1	$120,000
Other	1	$449,816
Total	20	$1,815,820

FACULTY

Professor

Allred, David D., Ph.D., Princeton University, 1977. *Atomic, Molecular, & Optical Physics*. Lasers; x-rays; thin-film physics.

Bergeson, Scott D., Ph.D., University of Wisconsin-Madison, 1995. *Atomic, Molecular, & Optical Physics*. Laser cooling.

Berrondo, Manuel, Ph.D., University of Uppsala, 1973. *Theoretical Physics*. Theoretical solid-state physics.

Davis, Robert C., Ph.D., University of Utah, 1996. *Condensed Matter Physics*. Experimental condensed-matter physics; nanophysics.

Hart, Gus, Ph.D., University of California, Davis, 1999. *Condensed Matter Physics*. Computational and theoretical condensed-matter physics.

Migenes, Victor, Ph.D., University of Pennsylvania, 1989. *Astrophysics*. Astrophysics.

Moody, Joseph Ward, Ph.D., University of Michigan, 1986. *Astrophysics*. Experimental astrophysics.

Peatross, Justin, Ph.D., University of Rochester. *Atomic, Molecular, & Optical Physics*. High-intensity laser physics.

Rees, Lawrence B., Ph.D., University of Maryland, 1983. *Nuclear Physics*. Medium- and low-energy nuclear physics.

Sommerfeldt, Scott D., Ph.D., Pennsylvania State University, 1989. *Acoustics*. Acoustics and structural vibrations.

Spencer, Ross L., Ph.D., University of Wisconsin-Madison, 1979. *Plasma and Fusion*. Theoretical plasma physics and gas dynamics.

Turley, R. Steven, Ph.D., Massachusetts Institute of Technology, 1984. *Atomic, Molecular, & Optical Physics*. Computational electromagnetics and atomic physics.

Vanfleet, Richard, Ph.D., University of Illinois, 1999. *Condensed Matter Physics*. Experimental condensed-matter physics; electron microscopy.

Associate Professor

Campbell, Branton J., Ph.D., University of California, Santa Barbara, 1999. *Condensed Matter Physics*. Experimental and theoretical condensed-matter physics.

Christensen, Clark G., Ph.D., California Institute of Technology, 1972. *Astrophysics*. Astrophysics.

Colton, John S., Ph.D., University of California, Berkeley, 2000. *Condensed Matter Physics*. Experimental condensed-matter physics.

Durfee, Dallin S., Ph.D., Massachusetts Institute of Technology, 1999. *Atomic, Molecular, & Optical Physics*. Atomic physics.

Gee, Kent L., Ph.D., Pennsylvania State University, 2005. *Acoustics*. Acoustics.

Hart, Grant W., Ph.D., University of Maryland, 1983. *Plasma and Fusion*. Nonneutral plasma physics.

Hess, Bret C., Ph.D., Iowa State University, 1988. *Condensed Matter Physics*. Experimental condensed-matter physics.

Hintz, Eric G., Ph.D., Brigham Young University, 1995. *Astrophysics*. Observational astrophysics.

Hirschmann, Eric W., Ph.D., University of California, Santa Barbara, 1996. *Relativity & Gravitation*. Theoretical and computational physics.

Leishman, Timothy W., Ph.D., Pennsylvania State University, 2000. *Acoustics*. Acoustics.

Neilsen, David, Ph.D., University of Texas, Austin, 1999. *Relativity & Gravitation*. Theoretical and computational physics.

Van Huele, Jean-Francois, Ph.D., Brussels Free University, 1987. *Quantum Foundations*. Theoretical quantum electrodynamics.

Ware, Michael, Ph.D., Brigham Young University, 2001. *Atomic, Molecular, & Optical Physics*. Quantum optics.

Assistant Professor

Chesnel, Karine, Ph.D., Joseph Fourier University, 2002. *Condensed Matter Physics*. Experimental condensed-matter physics.

Stephens, Denise, Ph.D., New Mexico State University, 2002. *Astrophysics*. Astrophysics.

Transtrum, Mark K., Ph.D., Cornell University, 2011. *Computational Physics, Condensed Matter Physics, Theoretical Physics*. Condensed matter theory, extracting models from complex data.

Research Professor

Joner, Michael, Ph.D., Brigham Young University, 2011. *Astrophysics*. Stellar photometry (observational).

Research Associate Professor

Peterson, Bryan G., Ph.D., Brigham Young University, 1983. *Plasma and Fusion*. Nonneutral plasma physics.

DEPARTMENTAL RESEARCH SPECIALTIES AND STAFF

Theoretical

Acoustics. Active control of sound and vibration; aeroacoustics; architectural acoustics; audio acoustics; sound-structure interaction; acoustical treatments; nonlinear acoustics. Gee, Leishman, Sommerfeldt.

Atomic, Molecular, & Optical Physics. Computational study of the nonlinear interaction of intense waves with gases; computational studies of the reflection of electromagnetic waves from rough surfaces; quantum electrodynamics; ultra-cold plasmas; x-ray optics. Turley.

Condensed Matter Physics. Material simulations; first-principles calculations; alloy modeling; electronic structure and excitations in nanosystems; the group-theoretical description of phase transitions; incommensurately modulated structures. Campbell, Gus Hart, Hess, Transtrum.

Plasma and Fusion. Basic plasma studies focusing on non-neutral plasmas and particle simulations. Spencer.

Quantum Foundations. Quantum; relativity; and complexity; quantum foundations and information; spintronics; field theory; parallel computing; chaos; complex behavior; consensus/frustration models. Berrondo, Van Huele.

Relativity & Gravitation. Large-scale computation and numerical evolution of the dynamic behavior of very strong gravitational fields, including the description of gravitational collapse and the formation and evolution of black holes. Hirschmann, Neilsen.

Experimental

Acoustics. Active control of sound and vibration; aeroacoustics; architectural acoustics; audio acoustics; sound-structure interaction; acoustical treatments; electroacoustics; nonlinear acoustics; passive noise control. Gee, Leishman, Sommerfeldt.

Astronomy. Narrow-band photometry and spectrophotometry; broad-band photometry; galactic clusters; luminosity function of galaxies; metal abundances; variable stars; interstellar reddening; radio astronomy. Hintz, Joner, Migenes, Moody, Stephens.

Atomic, Molecular, & Optical Physics. Physics of laser-cooled gases and plasmas; laser photochemistry and plasma diagnostics; VUV optics and sources; ultrafast lasers; high harmonic generation; matter interferometry. Allred, Bergeson, Durfee, Peatross, Rees, Turley, Ware.

Condensed Matter Physics. Nanofabrication; carbon nanotube microstructures; biomolecule-templated electronics; electron microscopy; semiconductor spin dynamics; optical spectroscopy of nanostructures; nanomagnetism; nanostructure-property relations; magnetometry; magnetic x-ray spectroscopy; x-ray diffuse scattering; XUV optical thin films. Campbell, Chesnel, Colton, Davis, Vanfleet.

Nuclear Physics. Particle-induced x-ray emission (PIXE) spectroscopy and related ion beam analysis techniques. Rees.

Plasma and Fusion. Basic plasma studies focusing on non-neutral plasmas. Grant Hart, Peterson.

View additional information about this department at
www.gradschoolshopper.com

UNIVERSITY OF UTAH

DEPARTMENT OF PHYSICS AND ASTRONOMY

Salt Lake City, Utah 84112
http://www.physics.utah.edu

General University Information
President: David Pershing
Dean of Graduate School: David Kieda
University website: http://www.utah.edu/
Control: Public
Setting: Urban
Total Faculty: 3,582
Total Graduate Faculty: 1,558
Total number of Students: 32,388
Total number of Graduate Students: 7,548

Department Information
Department Chairman: Carleton DeTar, Chair
Department Contact: Wayne Springer, Director of Graduate
  Studies
  Total full-time faculty: 38
  Total number of full-time equivalent positions: 38
  Full-Time Graduate Students: 96
  First-Year Graduate Students: 26
  Female First-Year Students: 9
  Total Post Doctorates: 23

Department Address
115 South 1400 East #201
Salt Lake City, UT 84112
Phone: (801) 585-1390
Fax: (801) 581-4801
E-mail: springer@physics.utah.edu
Website: http://www.physics.utah.edu

ADMISSIONS

Admission Contact Information
Address admission inquiries to: Graduate Secretary, Department
  of Physics and Astronomy, 115 South 1400 East #201, Salt
  Lake City, UT 84112-0830
Phone: (801) 581-6861
E-mail: admissions@physics.utah.edu
Admissions website: http://www.physics.utah.edu/index.php/
  graduate-program

Application deadlines
Fall admission:
U.S. students: October 1   *Int'l. students*: September 14

Application fee
U.S. students: $55   *Int'l. students*: $65
http://www.physics.utah.edu/index.php?option=com_content&
  view=article&id=216

Admissions information
For Fall of 2013:
  Number of applicants: 143
  Number admitted: 40
  Number enrolled: 18

Admission requirements
Bachelor's degree requirements: A Bachelor's degree in physics
  is required or a Bachelor's degree closely related field (science, engineering) with a strong Physics and Math background.
Minimum undergraduate GPA: 3.0

GRE requirements
The GRE is recommended.
  Mean GRE score range (25th–75th percentile): 25th-98th
There is no minimum required score.

Advanced GRE requirements
The Advanced GRE is required.
  Minimum accepted Advanced GRE score: 0
  Mean Advanced GRE score range (25th–75th percentile):
  25th-97th
No minimum subject GRE score required.

TOEFL requirements
The TOEFL exam is required for students from non-English-
  speaking countries.
  PBT score: 500
  iBT score: 61
IELTS score of 6.0 or better is also accepted.

Other admissions information
Additional requirements: Bachelor's degree from an accredited
  school.
Undergraduate preparation assumed: Marion, Electricity and
  Magnetism; Saxon, Quantum Mechanics; Marion, Classical
  Dynamics; Tipler, Modern Physics.

TUITION
Tuition year 2013-14:
Tuition for in-state residents
  Full-time students: $3,050 per semester
  Part-time students: $1,209 per credit
Tuition for out-of-state residents
  Full-time students: $9,704 per semester
  Part-time students: $3,530 per credit
Tuition Waiver Benefits available to qualified students in good
  academic standing.
Credit hours per semester to be considered full-time: 9
Deferred tuition plan: No
Health insurance: Yes, $145/semester.
Other academic fees: Mandatory University fees. Some labs re-
  quire special fees.
Academic term: Semester
Number of first-year students who received full tuition waivers: 24

Teaching Assistants, Research Assistants, and Fellowships
  Number of first-year
    Teaching Assistants: 24
  Average stipend per academic year
    Teaching Assistant: $20,641
    Research Assistant: $23,500

FINANCIAL AID

Application deadlines
Fall admission:
U.S. students: February 1   *Int'l. students*: February 1
Spring admission:
U.S. students: February 1   *Int'l. students*: February 1

Loans
Loans are available for U.S. students.
Loans are not available for international students.
GAPSFAS application required: No
FAFSA application required: Yes

For further information
Address financial aid inquiries to: Graduate Secretary, Department of Physics and Astronomy, 115 South 1400 East #201, Salt Lake City, UT 84112-0830.
Phone: (801) 581-6861
E-mail: admissions@physics.utah.edu
Financial aid website: http://www.physics.utah.edu/graduate/application_financial_assistance.pdf

HOUSING

Availability of on-campus housing
Single students: Yes
Married students: Yes

For further information
Address housing inquiries to: Office of Residential Living, 5 Heritage Center, Salt Lake City, UT 84112-2036.
Phone: (801) 587-2002
Housing aid website: http://www.housing.utah.edu/graduate-housing.html

Table A—Faculty, Enrollments, and Degrees Granted

Research Specialty	2012-2013 Faculty	Enrollment Fall 2012 Master's	Enrollment Fall 2012 Doctorate	Number of Degrees Granted 2011-2013 (2003–10) Master's	Number of Degrees Granted 2011-2013 (2003–10) Terminal Master's	Number of Degrees Granted 2011-2013 (2003–10) Doctorate
Acoustics	1	–	2	–(3)	–	–(1)
Applied Physics	6	–	5	–(3)	1(-)	2(2)
Astronomy	5	–	10	–(18)	–	1(-)
Astrophysics	6	–	6	–	–	1(-)
Atmosphere, Space Physics, Cosmic Rays	4	1	3	–(1)	1(-)	3(4)
Biophysics	3	–	7	–	–	2(-)
Chemical Physics	1	–	1	–	–	–(1)
Condensed Matter Physics	11	–	28	2(-)	1(-)	7(-)
Medical, Health Physics	10	–	5	1(6)	1(-)	2(9)
Particles and Fields	3	–	6	–	–	–(1)
Physics and other Science Education	–	–	–	–(1)	–	–(5)
Polymer Physics/Science	–	–	2	–(2)	–	–
Relativity & Gravitation	1	–	2	1(17)	–	–(22)
Non-specialized	–	–	24	–	–	–
Total	38	1	102	2(69)	4(-)	11(46)
Full-time Grad. Stud.	–	1	102	–	–	–
First-year Grad. Stud.	–	1	24	–	–	–

GRADUATE DEGREE REQUIREMENTS

Master's: Thirty graduate semester hours required with a 3.0 grade average in an approved program with satisfactory performance on Departmental Common Exam. Either thesis or non-thesis M.S. available. Master's of Instrumentation: 30 graduate semester hours with a 3.0 grade average. Nine to fifteen hours will be related to the instrumentation project. No language required. For admission, a Bachelor's degree in engineering, biology, chemistry, or some related field may be substituted for a degree in physics.

Doctorate: Forty-five graduate semester hours required. Satisfactory performance on Departmental Common Exam or GRE Physics required for admission to PhD program (no set minimum). Satisfactory performance (3.0 average) in an approved course program is required. Qualifying exam, dissertation, and dissertation exam required. Teaching experience required, and one of last two years must be in residence. No language requirement.

Other Degrees: Interdisciplinary studies available in chemical physics, and a variety of other areas by special arrangement. The PhD program in Medical Physics is an interdisciplined program in which complex medical and biological systems are studied using physics-based techniques and models. M.A. requirements are the same as M.S., except proficiency in one foreign language is required. M.Phil. requirements the same as Ph.D., except no dissertation required.

Thesis: Thesis may be written in absentia.

SPECIAL EQUIPMENT, FACILITIES, OR PROGRAMS

The Department maintains extensive facilities for teaching and research. The INSCC Building (Intermountain Network & Scientific Computation Center) houses 7 state-of-the-art laser labs on the first floor run by several condensed matter and biophysics experimental groups.

The research of one group focuses on time-resolved and steady state (cw) investigations of photo-excitations in solids, particularly in semiconductors. A state of the art Ti:sapphire laser gives time resolution of 10 fs. Current efforts in this femtosecond lab include pulse amplification with a Nd:YAG laser and the generation of continuum pulses from the near-infrared to the UV.

In the picosecond lab, two tunable synchronously pumped dye lasers are used for further photoexcitation studies. A 2D streak camera is used to measure the photoluminescence spectrum evolution with picosecond resolution. The laser laboratory has also been used to study optical non-linear spectra in electronic polymers, solids and other semiconductors. This includes spectra of two-photon absorption, non-linear refractive index and third harmonic generation. Light absorption is measured from the UV to the far IR using self-contained commercial instruments (Cary 17 DX and Bruker IFS88, both recently upgraded with modern electronics), operated either by researchers or as a service provided by members of the technical staff.

The single molecule spectroscopy group runs a low temperature laser microscopy lab centering around a helium cryostat and a one-box femtosecond laser system with wide (680nm- 1080nm) automated tunability. An FEI NovaNano Field emission Scanning Electron Microscope with 1.0 nm resolution (1.6 nm at 1 keV or low vacuum) is widely used for imaging. EDS analysis and e-beam lithography, a Leo 440i SEM is used for images requiring extremely large depth of field as well as for teaching. The Scanning Probe Microscopy group has many scanning probe microscopes, including several Atomic Force Microscopes, a Scanning Tunneling Microscope, two Near-field Optical Microscopes, a Scanning Capacitance Microscope and an ultra-high vacuum AFM/STM system.

Two new biophysics labs are under construction. The first lab is located in the INSCC building. The focus of the group is on single molecule studies of molecular motor activity and other protein interactions. Equipment includes (or will shortly include) a high-resolution optical microscope with optical trapping and fluorescence capabilities, as well as auxiliary biological research equipment (e.g. low-temperature refrigeration facilities and a Beckman TL-100 ultracentrifuge). The second Biophysics lab is under construction in the James Fletcher Building. This lab focuses on understanding the mechanism of enveloped virus budding using single molecule, fluorescence spectroscopy and high resolution live cell imaging technologies. A new iMIC digital microscope would be installed that is capable of confocal, TIRF, live cell imaging and fluorescence correlation spectroscopy.

In addition, the Department operated a fully equipped Opto-Electronic Materials Laboratory for chemical synthesis (including organic semiconductors not commercially available), purification, growth of single crystals, vacuum/controlled atmosphere annealing, sample cutting and polishing, thin-film deposition via thermal evaporation of rf sputtering, as well as a wide variety of techniques for chemical and physical characterization. A low-temperature AFM/STM (5 Kelvin) from Omicron Nanotechnology should arrive by the end of 2010.

The Astronomy and Astrophysics Group consists of nine full-time faculty members who are leading research programs at world-class astronomical facilities. As full institutional members of the Third Sloan Digital Sky Survey, the astronomy research group pursues an active research program in the BOSS, APGEE, and SEGUE surveys. The astronomy group also pursues observational research using facilities in Chile, Hawaii, and the Soutwestern United States. The astronomy group is a key member of the proposed BigBOSS Observatory, a stage IV baryon acoustic oscillation survey designed to elucidate the formation of galaxies in the early universe, and properties of dark energy and dark matter. The Department operates the 32″ Willard L. Eccles Observatory on Frisco Peak, Utah, approximately 200 miles from Salt Lake City. This high altitude (9600 ft a.s.l.) observatory is being developed for IR spectroscopy and imaging surveys. The department operates a pair of 3-meter interferometric telescopes at StarBase Utah, approximately 35 miles west of Salt Lake City. The South Physics observatory on campus houses the department 14″ fully automated telescope and several others with CCD photometers, spectrometers, and other accessories. The gamma-ray research group pursues astrophysics research with the Very Energetic Radiation Imaging Telescope Array System (VERITAS) located near Tucson, Arizona. Its four 10m telescopes make stereoscopic measurements of TeV gamma rays from black holes, supernova remnants, pulsars, and active galactic nuclei. Faculty members also pursue cosmic ray and gamma-ray research at the High Altitude Water Cherenkov (HAWC) observatory, located on Sierra Negra, Mexico. The University of Utah is the host institution for the Telescope Array (TA) and Telescope Array Low Energy extension (TALE) Projects, located 125 miles from Salt Lake City in the west-central Utah desert. Its ground array has more than 500 scintillation detectors covering 750 square kilometers, accompanied by three air-fluorescence detectors. Both TA and TALE are designed to study the highest energy particles known, and both experiments make extensive use of the air-fluorescence technique first successfully employed at Utah by the Fly's Eye Experiment (1976-1991) here at University of Utah, Degree programs in astronomy are currently offered.

The department has a robust wired and wireless local network designed with growth and flexibility in mind. The local network is integrated into a cutting edge university network dedicated to providing premier Internet services. Core user and computational services are provided by a dozen Sun Fire and Sun Enterprise servers accompanied by several powerful Linux and Windows servers. Data storage and backup are provided by a growing storage area network (SAN) currently totaling roughly one terabyte of disk and ten terabytes of tape storage. We provide access to a large suite of programs for departmental use including Maple, Matlab, Mathematica, LabVIEW, Microsoft software, and educational software. In addition, the University provides deeply discounted prices on hundreds of software titles through their office of software licensing. There are numerous open access terminals, desktops, and printers in the department library, study areas, and the five open computer labs. Individual workstations are a mixture of Windows, Linux, Macintosh, and UNIX. The department also supports several research groups that have various computational computers and multiple terabyte size data arrays, usually based on UNIX type architectures. Research groups also have access to large computational clusters through the University's Center for High Performance Computing (CHPC), totaling well over 1500 processors in various clusters.

Research is also supported by a professional Research Machine Shop, as well as a Student Shop, the latter open to all faculty, staff, and students who have completed a training course. Both shops are equipped with state-of-the-art CNC lathes and mills, as well as cutting, drilling, and welding equipment. The well-equipped wood shop allows fabrication of non-magnetic supports and shipping containers. The ample stockroom saves time and effort in procuring both common and hard-to-find materials and supplies.

Table B—Separately Budgeted Research Expenditures by Source of Support

Source of Support	Departmental Research	Physics-related Research Outside Department
Federal government	$3,802,171	$1,252,815
State/local government		
Non-profit organizations		
Business and industry		
Other		
Total	**$3,802,171**	**$1,252,815**

Table C—Separately Budgeted Research Expenditures by Research Specialty

Research Specialty	No. of Grants	Expenditures ($)
Astrophysics	12	$1,356,222
Atmosphere, Space Physics, Cosmic Rays	2	$1,252,815
Condensed Matter Physics	13	$1,728,368
Medical, Health Physics	2	$494,632
High Energy Physics	3	$222,949
Total	32	$5,054,986

FACULTY

Professor

Ailion, David C., Ph.D., University of Illinois, 1964. *Atomic, Molecular, & Optical Physics, Biophysics, Condensed Matter Physics, Medical, Health Physics.* Experimental condensed matter.

Boehme, Christoph, Ph.D., University of Philipps, Marburg, 2002. Associate Chair. *Atomic, Molecular, & Optical Physics, Condensed Matter Physics, Materials Science, Metallurgy, Nano Science and Technology, Optics.* Experimental condensed matter.

Bromley, Benjamin C., Ph.D., Dartmouth College, 1994. *Astronomy, Astrophysics, Relativity & Gravitation.* Theoretical astrophysics.

Cassiday, George L., Ph.D., Cornell University, 1968. *Astronomy, Astrophysics, Atmosphere, Space Physics, Cosmic Rays.* Cosmic rays.

DeFord, John W., Ph.D., University of Illinois, 1962. *Physics and other Science Education.* Physics education.

DeTar, Carleton E., Ph.D., University of California, Berkeley, 1970. *Astrophysics, Particles and Fields.* Elementary particle theory.

Efros, Alexei L., Ph.D., Ioffe Physico Technical Institute, 1972. *Condensed Matter Physics.* Theoretical condensed matter.

Gondolo, Paolo, Ph.D., University of California, Los Angeles, 1991. *Astronomy, Astrophysics, Particles and Fields.* Cosmology; Dark matter.

Harris, Frank E., Ph.D., University of California, Berkeley, 1954. *Chemical Physics.*

Jui, Charles C., Ph.D., Stanford University, 1992. *Atmosphere, Space Physics, Cosmic Rays, History & Philosophy of Physics/Science, Plasma and Fusion.* Cosmic rays.

Kieda, David B., Ph.D., University of Pennsylvania, 1989. Department Chair. *Astronomy, Astrophysics, Atmosphere, Space Physics, Cosmic Rays.* Experimental high-energy astrophysics.

Mishchenko, Eugene, Ph.D., Landau Institute for Theoretical Physics. Moscow, 1998. *Condensed Matter Physics.* Theoretical condensed matter.

Raikh, Mikhail, Ph.D., Ioffe Physico Technical Institute, 1981. *Condensed Matter Physics.* Theoretical condensed matter.

Saam, Brian T., Ph.D., Princeton University, 1995. Associate Dean, College of Science. *Atomic, Molecular, & Optical Physics, Biophysics, Condensed Matter Physics, Medical, Health Physics.* Experimental atomic/molecular.

Sokolsky, Pierre V., Ph.D., University of Illinois, 1973. Dean, College of Science. *Astrophysics, Atmosphere, Space Physics, Cosmic Rays, Plasma and Fusion.* Cosmic rays; high-energy physics.

Starykh, Oleg, Ph.D., Russian Academy of Science, 1991. *Condensed Matter Physics.* Theoretical condensed matter.

Symko, Orest G., Ph.D., University of Oxford, 1967. *Acoustics, Low Temperature Physics.* Thermoacoustics.

Thomson, Gordon, Ph.D., Harvard University, 1972. *Astrophysics, Atmosphere, Space Physics, Cosmic Rays, Plasma and Fusion.* Cosmic rays; high-energy physics.

Vardeny, Zeev V., Ph.D., Technion, Israel, 1979. *Atomic, Molecular, & Optical Physics, Biophysics, Condensed Matter Physics, Energy Sources & Environment, Materials Science, Metallurgy, Nano Science and Technology, Optics.* Experimental condensed matter.

Williams, Clayton C., Ph.D., Stanford University, 1984. *Atomic, Molecular, & Optical Physics, Condensed Matter Physics, Electrical Engineering, Nano Science and Technology, Optics.* Experimental condensed matter.

Wu, Yong-Shi, Ph.D., Chinese Academy of Sciences, 1965. *Astrophysics, Condensed Matter Physics, Particles and Fields.* High-energy theory.

Associate Professor

Bergman, Douglas, Ph.D., Yale University, 1997. *Astrophysics, Atmosphere, Space Physics, Cosmic Rays, Plasma and Fusion.* Cosmic rays; high-energy physics.

Gerton, Jordan, Ph.D., Rice University, 2001. *Biophysics, Condensed Matter Physics, Medical, Health Physics, Nano Science and Technology, Optics.* Experimental condensed matter.

LeBohec, Stephan, Ph.D., Paris XI University, 1992. *Astronomy, Astrophysics, Particles and Fields.* Experimental high-energy astrophysics.

Rogachev, Andrey, Ph.D., Nagoya University, 2000. *Condensed Matter Physics, Nano Science and Technology, Optics.* Experimental condensed matter physics.

Springer, Wayne R., Ph.D., University of Maryland, 1991. *Astronomy, Astrophysics, Atmosphere, Space Physics, Cosmic Rays, Computer Science, Particles and Fields, Plasma and Fusion.* Experimental astrophysics.

Assistant Professor

Bolton, Adam, Ph.D., Massachusetts Institute of Technology, 2005. *Astronomy.* Astronomy.

Dawson, Kyle, Ph.D., University of California, Berkeley, 2004. *Astronomy.*

Deemyad, Shanti, Ph.D., Washington University, St. Louis, 2004. *Condensed Matter Physics, Low Temperature Physics.* Experimental condensed matter.

Deshpande, Vikram V., Ph.D., California Institute of Technology, 2008. *Applied Physics, Condensed Matter Physics.* Synthesis, nanofabrication and combined electrical, mechanical, and optical measurements of ultra-high quality graphene and carbon nanotube devices.

Ivans, Inese, Ph.D., University of Texas, Austin, 2002. *Astronomy.* Astronomy.

Li, Yan (Sarah), Ph.D., University of California, Riverside, 2010. *Condensed Matter Physics, Materials Science, Metallurgy.* Spin noise spectroscopy of electrons and holes in semiconductor quantum dots.

Pesin, Dmytro, Ph.D., University of Washington, 2009. *Condensed Matter Physics.* Theoretical condensed matter.

Saffarian, Saveez, Ph.D., Washington University, St. Louis, 2003. *Biophysics.* Biophysics.

Sandick, Pearl, Ph.D., University of Minnesota, 2008. *Astrophysics, Particles and Fields.*

Seth, Anil, Ph.D., University of Washington, 2006. *Astronomy.*

Vershinin, Michael, Ph.D., University of Illinois, 2004. *Biophysics.*

Zheng, Zheng, Ph.D., Ohio State University, 2004. *Astronomy.*

Emeritus

Ball, James S., Ph.D., University of California, Berkeley, 1960. *Astrophysics, Particles and Fields.* Elementary particle theory.

Bergeson, Haven E., Ph.D., University of Utah, 1962. *Astrophysics, Atmosphere, Space Physics, Cosmic Rays.* Cosmic rays.

Dick, B. Gale, Ph.D., Cornell University, 1958. *Condensed Matter Physics.* Theoretical condensed matter.

Fowles, Grant R., Ph.D., University of California, Berkeley, 1950. *Optics.*

Johnson, Owen W., Ph.D., University of Utah, 1962. *Condensed Matter Physics.* Experimental condensed matter.

Kuchar, Karel V., Ph.D., Charles University in Prague, 1966. *Relativity & Gravitation.* Relativity.

Luty, Fritz W., Ph.D., University of Stuttgart, 1955. *Condensed Matter Physics.* Experimental condensed matter.

Mattis, Daniel C., Ph.D., University of Illinois, 1957. *Condensed Matter Physics.* Theoretical condensed matter.

Ohlsen, William D., Ph.D., Cornell University, 1961. *Condensed Matter Physics.* Experimental condensed matter.

Price, Richard, Ph.D., California Institute of Technology, 1971. *Relativity & Gravitation.*

Rudolph, Sidney, Ph.D., University of Utah, 1986. Physics education.

Sutherland, T. Bill, Ph.D., Stony Brook University, 1968. *Condensed Matter Physics.* Theoretical condensed matter.

Taylor, P. Craig, Ph.D., University of Brown, 1969. *Condensed Matter Physics.* Experimental condensed matter.

Williams, George A., Ph.D., University of Illinois, 1956. *Condensed Matter Physics.* Experimental condensed matter.

Research Professor

Gellermann, Werner, Ph.D., Tech. Universitate, Hanover, 1978. *Biophysics, Condensed Matter Physics, Medical, Health Physics.* Experimental condensed matter.

Lupton, John M., Ph.D., University of Durham, London, 2001. *Atomic, Molecular, & Optical Physics, Condensed Matter Physics, Materials Science, Metallurgy, Nano Science and Technology.* Experimental condensed matter.

Matthews, John N., Ph.D., Rutgers University, 1995. *Astrophysics, Atmosphere, Space Physics, Cosmic Rays.* Cosmic rays.

Worlock, John M., Ph.D., Cornell University, 1962. *Condensed Matter Physics.* Condensed matter.

Research Associate Professor

Belz, John, Ph.D., Temple University. *Astrophysics, Atmosphere, Space Physics, Cosmic Rays.* Cosmic rays.

Laicher, Gernot, Ph.D., University of Utah, 1994. *Atomic, Molecular, & Optical Physics, Condensed Matter Physics, Medical, Health Physics.* Experimental condensed matter.

Research Assistant Professor

AbuZayyad, Tareq, Ph.D., University of Utah. *Astrophysics, Atmosphere, Space Physics, Cosmic Rays, Plasma and Fusion.* Cosmic rays.

Brownstein, Joel R., Ph.D., University of Waterloo, 2009. *Astronomy.* Galaxy evolution, BOSS, lensing.

Cady, Robert, Ph.D., University of Utah, 1983. *Astrophysics, Atmosphere, Space Physics, Cosmic Rays.* Cosmic rays.

Lupton, Elizabeth, Ph.D., University of Edinburgh, 2001. *Condensed Matter Physics, Materials Science, Metallurgy.*

Teaching Professor

Ingebretsen, Richard J., Ph.D., University of Utah, 1989. MD. *Medical, Health Physics, Physics and other Science Education.* Physics education.

Teaching Associate Professor

Pantziris, Anthony, Ph.D., Brown University, 1987. *Physics and other Science Education.* Physics education.

Stone, Christopher, Ph.D., University of Utah, 1992. *Physics and other Science Education.* Physics education.

Teaching Assistant Professor

Buehler, Tabitha C., Ph.D., Brigham Young University, 2011. *Astronomy.*

Nyawelo, Tino S., Ph.D., National Institute for Nuclear Physics and High Energy Physics, 2004. *Astrophysics, High Energy Physics, Particles and Fields, Theoretical Physics.*

Adjunct Faculty

Nahata, Ajay, Ph.D., Columbia University, 1997. *Electrical Engineering, Nano Science and Technology, Optics.*

Adjunct Professor

Bjorken, James D., Ph.D., Stanford University, 1959. *Astrophysics, Atmosphere, Space Physics, Cosmic Rays.* High-energy theory.

Chubukov, Andrey, Ph.D., Moscow State University, 1985. *Condensed Matter Physics.* Condensed matter.

Ehrenfreund, Eitan, Ph.D., Hebrew University, 1970. *Condensed Matter Physics.* Experimental condensed matter.

Facelli, Julio C., Ph.D., University of Buenos Aires, Argentina, 1981. *Atomic, Molecular, & Optical Physics, Computer Science.* Nuclear magnetic resonance.

Gaisser, Thomas K., Ph.D., Brown University, 1967. *Astrophysics, Atmosphere, Space Physics, Cosmic Rays, Particles and Fields.* Cosmic rays; elementary particle physics.

Johnson, Christopher R., Ph.D., University of Utah, 1989. *Applied Mathematics, Computer Science.* Physics.

Ormes, Jonathan F., Ph.D., University of Minnesota, 1967. *Atmosphere, Space Physics, Cosmic Rays.* Cosmic rays.

Parker, Dennis, Ph.D., University of Utah, 1978. *Biophysics, Medical, Health Physics.* Medical biophysics; computing.

Shahbazyan, Tigran, Ph.D., University of Utah, 1995. *Condensed Matter Physics.* Theoretical condensed matter.

Shapiro, Boris, Ph.D., USSR Academy of Science, 1970. *Condensed Matter Physics.* Theoretical condensed matter.

Stringfellow, Gerald, Ph.D., Stanford University, 1967. *Electrical Engineering.* Semiconductor physics.

Adjunct Associate Professor

Bartl, Michael, Ph.D., Karl-Franzens-University, Graz, Austria, 2000. *Chemical Physics, Nano Science and Technology.*

Blair, Steven, Ph.D., University of Colorado Boulder, 1998. *Electrical Engineering, Nano Science and Technology.* Nano-photonics.

Jeong, Eun-Kee, Ph.D., Washington University, St. Louis, 1991. *Medical, Health Physics.* MRI.

Liu, Feng, Ph.D., Virginia Commonwealth University, 1990. *Chemical Physics, Materials Science, Metallurgy.*

Martens, Kai, Ph.D., Heidelberg University, 1994. *Astrophysics, Atmosphere, Space Physics, Cosmic Rays.* Cosmic Rays.

Adjunct Assistant Professor

Huentemeyer, Petra, Ph.D., University of Hamburg, 2001. *Astrophysics.* High-energy astrophysics.

McCamey, Dane, Ph.D., University of New South Wales, Sydney, 2007. *Condensed Matter Physics.* Experimental condensed matter.

Instructor

Higgs, Lynn B., M.S., University of Utah, 1972. Undergraduate Advisor. Physics education.

DEPARTMENTAL RESEARCH SPECIALTIES AND STAFF

Theoretical

Astronomy. Bromley, Gondolo, Ivans, Pantziris, Zheng.

Astrophysics. Bromley, DeTar, Gondolo, Kuchar, Pantziris, Price, Sandick, Wu, Zheng.

Atmosphere, Space Physics, Cosmic Rays. AbuZayyad, Belz, Cady, Cassiday, Gaisser, Jui, Martens, Matthews, Ormes, Sokolsky, Thomson.

Chemical Physics. Harris.

Condensed Matter Physics. Chubukov, Efros, Mattis, Mishchenko, Raikh, Shahbazyan, Shapiro, Starykh, Sutherland, Taylor, Worlock, Wu.

Particles and Fields. DeTar, Gondolo.

Plasma and Fusion.

Relativity & Gravitation. Kuchar, Price.

Experimental

Acoustics. Symko.

Astronomy. Bolton, Dawson, Seth.

Astrophysics. AbuZayyad, Belz, Bergman, Cady, Cassiday, Gaisser, Jui, Kieda, LeBohec, Martens, Matthews, Ormes, Seth, Sokolsky, Springer, Thomson.

Biophysics. Ailion, Gellermann, Gerton, Ingebretsen, Jeong, Laicher, Parker, Saam, Saffarian, Vardeny, Vershinin.

Condensed Matter Physics. Ailion, Bergeson, Boehme, Deemyad, Ehrenfreund, Gellermann, Gerton, Jeong, Laicher, Liu, Elizabeth Lupton, John Lupton, Luty, McCamey, Nahata, Ohlsen, Rogachev, Saam, Taylor, Vardeny, Clayton Williams, George Williams.

Low Temperature Physics.
Materials Science, Metallurgy.
Medical, Health Physics.
Nano Science and Technology.
Optics.

View additional information about this department at
www.gradschoolshopper.com

UTAH STATE UNIVERSITY

DEPARTMENT OF PHYSICS

Logan, Utah 84322-4415
http://physics.usu.edu

General University Information
President: Stan Albrecht
Dean of Graduate School: Mark R. McLellan
University website: http://www.usu.edu
Control: Public
Setting: Rural
Total Faculty: 870
Total number of Students: 15,612
Total number of Graduate Students: 1,721

Department Information
Department Chairman: Jan J. Sojka, Chair
Department Contact: Karalee Ransom, Academic Advisor
  Total full-time faculty: 15
  Total number of full-time equivalent positions: 15
  Full-Time Graduate Students: 24
  First-Year Graduate Students: 5
  Female First-Year Students: 2
  Total Post Doctorates: 2

Department Address
Department of Physics
Utah State University
Logan, UT 84322-4415
Phone: (435) 797-2857
Fax: (435) 797-2492
E-mail: physics@usu.edu
Website: http://physics.usu.edu

ADMISSIONS

Admission Contact Information
Address admission inquiries to: Karalee Ransom, Department of Physics, 4415 Old Main Hill, Logan, UT 84322-4415
Phone: (435) 797-4021
E-mail: karalee.ransom@usu.edu
Admissionswebsite:http://physics.usu.edu/htm/students/graduate-program

Application deadlines
Fall admission:
U.S. students: January 15   *Int'l. students*: January 15

Application fee
U.S. students: $55   *Int'l. students*: $55
Official applications can be found on the School of Graduate Studies website: http://usu.edu/graduateschool.

Admissions information
For Fall of 2012:
  Number of applicants: 20
  Number admitted: 12
  Number enrolled: 5

Admission requirements
Bachelor's degree requirements: Bachelor's degree in physics or a related field.
Minimum undergraduate GPA: 3.0

GRE requirements
The GRE is required.

Advanced GRE requirements
The Advanced GRE is recommended.
  Minimum accepted Advanced GRE score: 200

TOEFL requirements
The TOEFL exam is required for students from non-English-speaking countries.
  PBT score: 550
  iBT score: 79

TUITION

Tuition year 2011–12:
Tuition for in-state residents
  Full-time students: $2,659.11 per semester
  Part-time students: $1,978.32 per semester
Tuition for out-of-state residents
  Full-time students: $8,297.68 per semester
  Part-time students: $6,013.87 per semester

Amounts shown are for tuition and fees. Full-time was determined by a 9-credit semester. Part-time was determined by a 6-credit semester.

Credit hours per semester to be considered full-time: 9

Deferred tuition plan: Yes

Health insurance: Available at the cost of $237 per year.

Other academic fees: International students pay a $100 per semester student fee. Health insurance is mandatory for all graduate students. The insurance is available through the University and is paid by both the student ($237 per academic year) and the Department ($941 per academic year).

Academic term: Semester

Number of first-year students who received full tuition waivers: 8

Teaching Assistants, Research Assistants, and Fellowships

Number of first-year
  Teaching Assistants: 5
  Research Assistants: 2
  Fellowship students: 1
Average stipend per academic year
  Teaching Assistant: $15,000
  Research Assistant: $17,500
  Fellowship student: $15,000

FINANCIAL AID

Application deadlines

Fall admission:
U.S. students: June 30
Spring admission:
U.S. students: June 30   *Int'l. students*: June 30

Loans

Loans are available for U.S. students.
Loans are not available for international students.
GAPSFAS application required: Yes
FAFSA application required: Yes

For further information

Address financial aid inquiries to: USU Financial Aid Office, 1800 Old Main Hill, Logan, UT 84322-1800.
Phone: (435) 797-0173
Financial aid website: http://usu.edu/finaid

HOUSING

Availability of on-campus housing

Single students: Yes
Married students: Yes

For further information

Address housing inquiries to: USU Housing and Residence Life, 8600 Old Main Hill, Logan, UT 84322-8600.
Phone: (800) 863-1085
E-mail: info@housing.usu.edu
Housing aid website: http://usu.edu/housing

Table A—Faculty, Enrollments, and Degrees Granted

Research Specialty	2013-14 Faculty	Enrollment Fall 2012		Number of Degrees Granted 2010–11 (2006–11)		
		Master's	Doctorate	Master's	Terminal Master's	Doctorate
Atmosphere, Space Physics, Cosmic Rays	6	3	10	–	–(3)	3(16)
Condensed Matter Physics	2	3	4	–	2(2)	–(3)
Nano Science and Technology	1	–	2	–	–	–(1)
Nonlinear Dynamics and Complex Systems	1	–	–	–	–	–
Plasma and Fusion	2	–	2	–	2(3)	–
Relativity & Gravitation	2	–	2	–	1(1)	–
Total	14	6	20	–	–	–
Full-time Grad. Stud.	–	6	25	–	–	–
First-year Grad. Stud.	–	–	5	–	–	–

Table B—Separately Budgeted Research Expenditures by Source of Support

Source of Support	Departmental Research	Physics-related Research Outside Department
Federal government	$700,834	$1,095,745
State/local government		
Non-profit organizations		
Business and industry		
Other		
Total	$700,834	$1,095,745

Table C—Separately Budgeted Research Expenditures by Research Specialty

Research Specialty	No. of Grants	Expenditures ($)
Atmosphere, Space Physics, Cosmic Rays	30	$1,095,745
Condensed Matter Physics	3	$300,834
Plasma and Fusion	3	$250,000
Nonlinear Dynamics and Complex Systems	2	$150,000
Total	38	$1,796,579

FACULTY

Professor

Dennison, J. R., Ph.D., Virginia Polytechnic Institute and State University (Virginia Tech), 1985. *Materials Science, Metallurgy, Solid State Physics, Surface Physics.* Solid-state and surface physics.

Edwards, W. Farrell, Ph.D., California Institute of Technology, 1960. *Atmosphere, Space Physics, Cosmic Rays, Electromagnetism, Geophysics, Plasma and Fusion, Theoretical Physics.* Electromagnetic theory.

Fejer, Bela, Ph.D., Cornell University, 1974. *Atmosphere, Space Physics, Cosmic Rays, Geophysics.* Space plasma physics.

Held, Eric D., Ph.D., University of Wisconsin-Madison, 1999. *Plasma and Fusion.* Plasma physics.

Peak, David, Ph.D., University at Albany, 1969. *Biophysics, Nonlinear Dynamics and Complex Systems.* Complex materials and dynamics.

Schunk, Robert W., Ph.D., Yale University, 1970. *Atmosphere, Space Physics, Cosmic Rays, Geophysics*. Space plasma physics.

Shen, T. C., Ph.D., University of Maryland, 1985. *Nano Science and Technology, Solid State Physics*. Surface science.

Sojka, Jan J., Ph.D., University College London, 1976. *Atmosphere, Space Physics, Cosmic Rays, Geophysics*. Space plasma physics.

Taylor, Michael, Ph.D., University of Southampton, 1986. *Atmosphere, Space Physics, Cosmic Rays, Geophysics*. Atmospheric physics.

Torre, Charles, Ph.D., University of North Carolina, 1985. Assistant Department Head. *Applied Mathematics, Particles and Fields, Relativity & Gravitation, Theoretical Physics*. Gravitational physics; field theory; mathematical physics.

Wickwar, Vincent, Ph.D., Rice University, 1971. *Atmosphere, Space Physics, Cosmic Rays, Optics*. Atmospheric physics.

Associate Professor

Riffe, D. Mark, Ph.D., Cornell University, 1989. *Condensed Matter Physics, Surface Physics*. Optical studies of surfaces.

Scherliess, Ludger, Ph.D., Utah State University, 1997. *Atmosphere, Space Physics, Cosmic Rays, Geophysics*. Space physics.

Wheeler, James T., Ph.D., University of Chicago, 1986. *Particles and Fields*. Relativity; particle physics.

Research Professor

Ji, Jeong-Young, Ph.D., *Plasma and Fusion*.

Wilkerson, Thomas D., Ph.D., University of Michigan, 1962. *Atmosphere, Space Physics, Cosmic Rays*. Atmospheric physics.

Research Associate Professor

Singh, Ajay Kumar, Ph.D., Institute for Plasma Research, 1993. *Plasma and Fusion*. Experimental Tokamak physics; plasma physics.

Zhu, Lie, Ph.D., University of Alaska, 1990. *Atmosphere, Space Physics, Cosmic Rays, Geophysics*. Space plasma physics.

Research Assistant Professor

Yuan, Titus, Ph.D., Colorado State University, 2004. *Atmosphere, Space Physics, Cosmic Rays, Optics*.

Lecturer

Triplett, Tonya, M.S., Utah State University, 2003. *Physics and other Science Education*. Physics education.

DEPARTMENTAL RESEARCH SPECIALTIES AND STAFF

Theoretical

Applied Mathematics. Torre.

Atmosphere, Space Physics, Cosmic Rays. Edwards, Fejer, Scherliess, Schunk, Sojka, Zhu.

Computational Physics. Held, Ji.

Condensed Matter Physics. Riffe.

Electromagnetism. Edwards.

High Energy Physics.

Nonlinear Dynamics and Complex Systems. Peak.

Particles and Fields. Torre, Wheeler.

Physics and other Science Education. Triplett.

Plasma and Fusion. Edwards, Held, Ji.

Relativity & Gravitation. Torre.

Experimental

Atmosphere, Space Physics, Cosmic Rays. Edwards, Fejer, Taylor, Wickwar, Wilkerson, Yuan.

Condensed Matter Physics. Dennison, Riffe, Shen.

Plasma and Fusion. Edwards, Singh.

View additional information about this department at
www.gradschoolshopper.com

UNIVERSITY OF VERMONT

DEPARTMENT OF PHYSICS

Burlington, Vermont 05405-0125

http://www.uvm.edu/physics

General University Information

President: E. Thomas Sullivan

Dean of Graduate School: Cynthia Forehand, Ph.D.

University website: http://www.uvm.edu

Control: Public

Setting: Urban

Total Faculty: 1,471

Total Graduate Faculty: 654

Total number of Students: 13,097

Total number of Graduate Students: 1,886

Department Information

Department Chairman: Dennis P. Clougherty, Chair

Department Contact: Denise M. Fontaine, Department Administrative Coordinator

Total full-time faculty: 12

Total number of full-time equivalent positions: 12

Full-Time Graduate Students: 13

First-Year Graduate Students: 5

Female First-Year Students: 1

Total Post Doctorates: 3

Department Address
82 University Place
Cook Science Building, Room A405
Burlington, VT 05405-0125
Phone: (802) 656-2664
Fax: (802) 656-0817
E-mail: physics@uvm.edu
Website: http://www.uvm.edu/physics

ADMISSIONS

Admission Contact Information
Address admission inquiries to: Department of Physics, University of Vermont, Cook Science Building, Room A405, 82 University Place, Burlington, VT 05405-0125
Phone: (802) 656-2644
E-mail: physics@uvm.edu
Admissions website: http://uvm.edu/~gradcoll/

Application deadlines
Fall admission:
U.S. students: February 15   *Int'l. students*: February 15
Spring admission:
U.S. students: November 15   *Int'l. students*: November 15

Application fee
U.S. students: $40   *Int'l. students*: $40
Online

Admissions information
For Fall of 2013:
Number of applicants: 28
Number admitted: 7
Number enrolled: 5

Admission requirements
Bachelor's degree requirements: Bachelor's degree in physics is required.

GRE requirements
The GRE is required.

Advanced GRE requirements
The Advanced GRE is not required.

TOEFL requirements
The TOEFL exam is required for students from non-English-speaking countries.
PBT score: 550
iBT score: 80
For student funded, an internet-based score of 100 is required, and a paper-based score of 600 is required.

Other admissions information
Additional requirements: The General GRE cannot be waived. The Advanced GRE can be waived in special cases. No minimum acceptable score is specified. The average GRE internet-based scores for 2013-2014 admissions were as follows: verbal, 153.0; quantitative, 156.7; analytical writing, 3.5.
Undergraduate preparation assumed: Taylor, Classical Mechanics; Griffiths, Introduction to Electrodynamics; Kittel and Kroemer, Thermal Physics; Griffiths, Introduction to Quantum Mechanics.

TUITION
Tuition year 2013-14:
Tuition for in-state residents
   Full-time students: $572 per credit
Tuition for out-of-state residents
   Full-time students: $1,444 per credit
Credit hours per semester to be considered full-time: 10
Deferred tuition plan: No
Health insurance: Available.
Other academic fees: $980 comp fees per semester
Academic term: Semester

Teaching Assistants, Research Assistants, and Fellowships
Number of first-year
   Teaching Assistants: 5
Teaching Assistant: $15,500 (M.S.) $17,250 (Ph.D, MATS)
Research Assistant: $15,500 (M.S.) $17,250 (Ph.D, MATS)

FINANCIAL AID

Application deadlines
Fall admission:
U.S. students: January 15   *Int'l. students*: January 15
Spring admission:
U.S. students: January 15   *Int'l. students*: January 15

Loans
Loans are available for U.S. students.
Loans are available for international students.
GAPSFAS application required: No
FAFSA application required: Yes

For further information
Address financial aid inquiries to: Student Financial Services, 223 Waterman Building, 85 South Prospect Street, Burlington, VT 05405.
Phone: (802) 656-5700
E-mail: sfs@uvm.edu
Financial aid website: http://www.uvm.edu/~stdfinsv/

HOUSING

Availability of on-campus housing
Single students: Yes
Married students: Yes

For further information
Address housing inquiries to: Office of Family Housing, Fort Ethan Allen, 36 Catamount Lane, Colchester, VT 05446.
Phone: (802) 654-1735
E-mail: familyhs@uvm.edu
Housing aid website: http://www.uvm.edu/~rlweb/graduate_students/

Table A—Faculty, Enrollments, and Degrees Granted

Research Specialty	2012-13 Faculty	Enrollment Fall 2012		Number of Degrees Granted 2012-13 (2003–09)		
		Master's	Doctorate	Master's	Terminal Master's	Doctorate
Acoustics	1	–	1	–	–(2)	–
Astronomy	3	2	–	–(1)	–	–
Astrophysics	–	–	–	–	–	–
Biophysics	2	–	–	–(2)	–(5)	–
Condensed Matter Physics	6	2	8	1(1)	–(6)	2(-)
History & Philosophy of Physics/Science	–	–	–	–	–(2)	–
Low Temperature Physics	–	–	–	–	–	–
Materials Science, Metallurgy	–	–	–	–	–	–
Optics	–	–	–	–	–	–
Polymer Physics/Science	–	–	–	–	–	–
Total	12	4	9	1(1)	–(15)	2(-)
Full-time Grad. Stud.	–	4	9	–	–	–
First-year Grad. Stud.	–	2	3	–	–	–

GRADUATE DEGREE REQUIREMENTS

Master's: A minimum of 30 semester hours of graduate credit is required. Of the 30 hours, 21 must be completed in residence. At least six must be in thesis research and nine in other courses numbered above 300 (graduate students only). No more than 15 hours of thesis research may be included in the degree program. The candidate must pass a written and oral comprehensive examination, as well as an oral examination on the thesis. The graduate student must maintain a B average. There are no foreign language requirements.

Other Degrees: For the Materials Science Program, Master's of Science and the Doctor of Philosophy degrees are offered in this interdisciplinary program. The faculty are drawn from the departments of Chemistry, Electrical Engineering, Mechanical Engineering, and Physics. The program is committed to educating the students in the application of basic sciences and engineering to promote understanding of the properties of materials, their development and applications, and to perform advanced and stimulating research in these areas. (The research program pursued in Materials Science at the University of Vermont has two areas of specialization: Electronic Materials and Bio/Polymeric Materials.) Each student must meet the general requirements or admissions as described at http://www.uvm.edu/~gradcoll. Students in the program are sponsored by the participating department that best reflects the student's background and interest. The degree of Doctor of Philosophy requires a minimum of 75 credit hours earned in courses and in dissertation research, of which a minimum of 51 hours must be earned in residence.

Thesis: Thesis may be written in absentia.

SPECIAL EQUIPMENT, FACILITIES, OR PROGRAMS

Research is concentrated in areas of astrophysics, biological physics, polymer physics, materials science, and condensed matter physics. Collaboration is feasible with other departments of science, engineering, and medicine on this geographically small campus. There is especially close cooperation with the School of Engineering, the College of Medicine, and the Department of Chemistry. The department shares a building with the Department of Chemistry.

Table B—Separately Budgeted Research Expenditures by Source of Support

Source of Support	Departmental Research	Physics-related Research Outside Department
Federal government	$1,525,758	
State/local government		
Non-profit organizations		
Business and industry		
Other		
Total	$1,525,758	

Table C—Separately Budgeted Research Expenditures by Research Specialty

Research Specialty	No. of Grants	Expenditures ($)
Astronomy	2	$160,987
Biophysics	3	$505,168
Condensed Matter Physics	6	$693,577
Physics and other Science Education	1	$166,026
Total	12	$1,525,758

FACULTY

Professor

Clougherty, Dennis P., Ph.D., Massachusetts Institute of Technology, 1989. Chair. *Atomic, Molecular, & Optical Physics, Condensed Matter Physics, Theoretical Physics*. Materials Science.

Headrick, Randall, Ph.D., University of Pennsylvania, 1988. *Condensed Matter Physics, Nano Science and Technology*. Materials Science. Molecular beam epitaxy; x-ray scattering surface processing.

Rankin, Joanna M., Ph.D., University of Iowa, 1970. *Astronomy, Astrophysics*. Radio astrophysics; history of science.

Wu, Jun-Ru, Ph.D., University of California, Los Angeles, 1985. *Acoustics, Biophysics, Condensed Matter Physics, Medical, Health Physics*. Experimental condensed matter physics and ultrasound.

Associate Professor

Chu, Kelvin, Ph.D., University of Illinois, 1995. *Biophysics, Low Temperature Physics*. Experimental biophysics; protein dynamics; low-temperature physics.

Furis, Madalina, Ph.D., State University of New York at Buffalo, 2004. *Condensed Matter Physics, Nano Science and Technology, Optics*. Materials science; ultrafast spectroscopy; time-resolved photoluminescence.

Spartalian, Kevork, Ph.D., Carnegie Mellon University, 1974. *Biophysics, Physics and other Science Education*. Mössbauer spectroscopy; biological physics; physics education.

Yang, Jie, Ph.D., Princeton University, 1987. *Biophysics*. Experimental biophysics; atomic force microscopy.

Assistant Professor

Del Maestro, Adrian, Ph.D., Harvard University, 2008. *Computational Physics, Condensed Matter Physics*. Theoretical condensed matter physics.

Kotov, Valeri, Ph.D., Clarkson University, 1996. *Condensed Matter Physics, Theoretical Physics*. Theoretical condensed matter physics.

Emeritus

Arns, Robert G., Ph.D., University of Michigan, 1960. *History & Philosophy of Physics/Science, Nuclear Physics*. History of science.

Brown, John S., Ph.D., Rutgers University, 1967. *Condensed Matter Physics*. Theoretical condensed matter physics.

Detenbeck, Robert W., Ph.D., Princeton University, 1962. Physics education.

Smith, David Y., Ph.D., University of Rochester, 1962. *Condensed Matter Physics, Optics*. Optical and x-ray properties of matter.

Adjunct Professor

Ohanian, Hans C., Ph.D., Princeton University, 1968. *Relativity & Gravitation*. Relativity.

Lecturer

Manley, Don, M.A., University of Oregon, 1954. *Astronomy*. Physics.

Pepe, Jason, M.S., University of Vermont, 2003. Physics education.

Perry, John, Ph.D., University of Rochester, 1992. *Astrophysics*. Astrophysics.

Sanders, Malcolm, Ph.D., Yale University, 1984. *Nonlinear Dynamics and Complex Systems*. Applied physics; nonlinear systems; chaos.

DEPARTMENTAL RESEARCH SPECIALTIES AND STAFF

Theoretical

Biophysics. Physical mechanisms for biological effects of ultrasound. Wu.

Condensed Matter Physics. Electronic and transport properties of metals, random alloys, and liquid metals; lattice dynamics; order-disorder phase transitions in alloys; superconductivity; superfluidity; strongly correlated electron systems; electronic properties of graphene; ultracold atom-surface scattering; Berry-phase effects in condensed matter systems. Clougherty, Del Maestro, Kotov.

Experimental

Acoustics. Physical mechanisms for biological effects of ultrasound. Wu.

Astrophysics. Pulsar radio-frequency emission; pulsars as probes of the interstellar medium. Rankin.

Biophysics. Spectroscopy of proteins and nucleic acids and the use of biomolecules in nanotechnological applications. Chu.

Condensed Matter Physics. Spin-polarized magneto-optical spectroscopy studies of nitride semiconductors; the time-resolved spectroscopy of nitride emitters and semiconductor nanocrystals; magneto-optical Kerr rotation spectroscopy of ferromagnetic nanostructures. Furis.

Materials Science, Metallurgy. Kinetics of thin-film growth and etching; real-time x-ray and electron diffraction studies of materials growth and surface evolution. Headrick.

View additional information about this department at
www.gradschoolshopper.com

COLLEGE OF WILLIAM AND MARY

DEPARTMENT OF PHYSICS

Williamsburg, Virginia 23187-8795
http://www.wm.edu/as/physics

General University Information

President: W. Taylor Reveley III
Dean of Graduate School: Virginia J. Torczon
University website: http://www.wm.edu/
Control: Public
Setting: Suburban
Total Faculty: 595
Total Graduate Faculty: n/a
Total number of Students: 8,200
Total number of Graduate Students: 2,087

Department Information

Department Chairman: David S. Armstrong, Chair
Department Contact: Paula C. Perry, Coordinator, Physics Graduate Program
Total full-time faculty: 32
Total number of full-time equivalent positions: 28
Full-Time Graduate Students: 71
First-Year Graduate Students: 15
Female First-Year Students: 3

Total Post Doctorates: 12

Department Address

P. O. Box 8795
Williamsburg, VA 23187-8795
Phone: (757) 221-3502
Fax: (757) 221-3540
E-mail: grad@physics.wm.edu
Website: http://www.wm.edu/as/physics

ADMISSIONS

Admission Contact Information

Address admission inquiries to: Director, Graduate Admissions, Department of Physics
Phone: (757) 221-3502
E-mail: grad@physics.wm.edu
Admissions website: http://www.wm.edu/as/physics

Application deadlines

Fall admission:

U.S. students: January 15   *Int'l. students*: January 15

Application fee
U.S. students: $45   *Int'l. students*: $45

Admissions information
For Fall of 2012:
Number of applicants: 108
Number admitted: 41
Number enrolled: 15

Admission requirements
Bachelor's degree requirements: A bachelor's degree in physics or a related field is required.
Minimum undergraduate GPA: 2.5

GRE requirements
The GRE is required.

Advanced GRE requirements
The Advanced GRE is required.
No minimum acceptable score for admissions is specified.

TOEFL requirements
The TOEFL exam is required for students from non-English-speaking countries.

Other admissions information
Undergraduate preparation assumed: Marion and Thornton, Mechanics; Griffiths, Quantum Physics; Griffiths, Electricity and Magnetism; Kittel and Kroemer, Thermal Physics.

TUITION

Tuition year 2013-14:
Tuition for in-state residents
   Full-time students: $11,884 annual
   Part-time students: $405 per credit
Tuition for out-of-state residents
   Full-time students: $26,960 annual
   Part-time students: $1,050 per credit
The tuition and fees of students on a graduate assistantship are paid by the department.
Credit hours per semester to be considered full-time: 9
Deferred tuition plan: No
Health insurance: Yes, paid by dept.
Other academic fees: None
Academic term: Semester
Number of first-year students who received full tuition waivers: 15

Teaching Assistants, Research Assistants, and Fellowships
Number of first-year
   Teaching Assistants: 14
   Research Assistants: 1
   Fellowship students: 3
Average stipend per academic year
   Teaching Assistant: $17,100
   Research Assistant: $17,100
   Fellowship student: $3,000
A graduate assistantship is a 12-month appointment of $22,800 plus paid health insurance premium.

FINANCIAL AID

Loans
Loans are available for U.S. students.
Loans are available for international students.
GAPSFAS application required: No
FAFSA application required: No

For further information
Address financial aid inquiries to: Director, Graduate Admissions, Department of Physics, P. O. Box 8795, College of William and Mary, Williamsburg, Virginia 23187-8795.
Phone: (757) 221-3502
E-mail: grad@physics.wm.edu

HOUSING

Availability of on-campus housing
Single students: Yes
Married students: No

For further information
Address housing inquiries to: Office of Residence Life.
Phone: (757) 221-4314
E-mail: living@wm.edu
Housing aid website: http://www.wm.edu/offices/residencelife

Table A—Faculty, Enrollments, and Degrees Granted

Research Specialty	2013–14 Faculty	Enrollment Fall 2012 Master's	Enrollment Fall 2012 Doctorate	Number of Degrees Granted 2012–13 (2008–13) Master's	Number of Degrees Granted 2012–13 (2008–13) Terminal Master's	Number of Degrees Granted 2012–13 (2008–13) Doctorate
Applied Physics	1	–	1	–(1)	–(1)	–(2)
Atmosphere, Space Physics, Cosmic Rays	1	–	–	–	–	–(1)
Atomic, Molecular, & Optical Physics	8	–	15	3(14)	–(2)	1(5)
Computational Physics	7	–	–	–	–	–(1)
Condensed Matter Physics	9	–	16	5(15)	–	1(6)
High Energy Physics	7	–	13	2(12)	–	2(7)
Nonlinear Dynamics and Complex Systems	4	–	4	1(3)	–	–(3)
Nuclear Physics	14	–	23	2(13)	–(1)	1(11)
Physics of Beams	2	–	1	–	–	–
Plasma and Fusion	3	–	1	–	–	–(1)
Total	41	–	74	13(58)	–(4)	5(37)
Full-time Grad. Stud.	–	–	73	–	–	–
First-year Grad. Stud.	–	–	15	–	–	–

GRADUATE DEGREE REQUIREMENTS

Master's: For the M.S. degree, the requirements are taking the Ph.D. qualifying exam and 32 satisfactory credits of graduate work with a B average. A student progressing toward the Ph.D. degree will usually satisfy the M.S. requirements en route. At least one semester must be spent in residence, and a minimum of one semester of teaching is required for all candidates. There are no foreign language or thesis requirements.

Doctorate: For the Ph.D., required courses include Classical Mechanics, Mathematical Physics, Quantum Mechanics I and II, Classical Electricity and Magnetism I and II, Field Theory and Relativistic Quantum Mechanics, and Statistical Physics and Thermodynamics. In addition, two semesters of Colloquium, Teaching Physics, at least one elective from inside and at least one outside the student's field of study may be required. The candidate must, in addition to passing the qualifying exam, demonstrate a mastery of the material in the first- and second-year courses, by either doing well in these courses or individual examinations. A student must maintain a B average for all course work. There is a one-year residence minimum for the degree. The research must be a significant original contribution. The dissertation must be approved by the candidate's faculty committee and must be successfully defended in a public oral examination. A Ph.D. candidate must teach a minimum of two semesters. There are no foreign language requirements.

Thesis: Thesis may be written in absentia.

SPECIAL EQUIPMENT, FACILITIES, OR PROGRAMS

The Department is housed in the William Small Physical Laboratory, which contains its own library, machine shop, and other support facilities in addition to research and teaching laboratories, classrooms, and offices. Small was entirely renovated in 2011, and the research space in the department was doubled by adding two wings to the building. The Physics Department has many high-performance computing clusters and access to supercomputers through national and international networks. Extensive computational resources are available through the Center for Piezoelectric Design and the nuclear/particle group. The high field solid state Nuclear Magnetic Resonance Laboratory houses a number of NMR machines, including a 17 T magnet. A 6-GeV continuous electron beam accelerator facility, Thomas Jefferson National Accelerator Facility, and Applied Research Center (ARC) is located in nearby Newport News. Faculty and graduate students are engaged in experiments at Fermilab (Batavia, Illinois), The Jefferson Laboratory, NASA (Langley, Virginia), Mainz Institute for Theoretical Physics (Mainz, Germany), TRIUMF (Vancouver, British Columbia, Canada), Soudan Underground Laboratory (Soudan, Minnesota), Ash River Neutrino Laboratory (Ash River, Minnesota), CERN (Geneva, Switzerland), and Joint Institute for Nuclear Research (Dubna, Russia).

Table B—Separately Budgeted Research Expenditures by Source of Support

Source of Support	Departmental Research	Physics-related Research Outside Department
Federal government	$4,278,512	
State/local government	$1,049,357	
Non-profit organizations	$40,880	
Business and industry	$556,586	
Other		
Total	$5,925,335	

Table C—Separately Budgeted Research Expenditures by Research Specialty

Research Specialty	No. of Grants	Expenditures ($)
Atomic, Molecular, & Optical Physics	9	$288,711
Chemical Physics	6	$428,591
Condensed Matter Physics	18	$887,470
Energy Sources & Environment	2	$574,698
Medical, Health Physics	3	$102,929
Nuclear Physics	25	$1,659,076
Plasma, Fusion, and Nonlinear Dynamics	3	$144,637
High Energy Physics	14	$937,503
Other	7	$901,720
Total	87	$5,925,335

FACULTY

Professor

Armstrong, David S., Ph.D., University of British Columbia, 1989. *Nuclear Physics, Particles and Fields.* Electroweak interactions; electron scattering; parity violation; muon capture; experiment.

Averett, Todd D., Ph.D., University of Virginia, 1995. *Atomic, Molecular, & Optical Physics, Nuclear Physics.* Nucleon structure; polarized nuclear targets; spin exchange optical pumping; experiment.

Carlson, Carl E., Ph.D., Columbia University, 1968. *Nuclear Physics, Particles and Fields.* Theory.

Carone, Christopher D., Ph.D., Harvard University, 1994. *High Energy Physics, Particles and Fields, Theoretical Physics.* Quantum field theoretic extensions of the standard model; theory.

Cooke, William E., Ph.D., Massachusetts Institute of Technology, 1976. *Atomic, Molecular, & Optical Physics, Energy Sources & Environment.* Experiement.

Delos, John B., Ph.D., Massachusetts Institute of Technology, 1970. *Atomic, Molecular, & Optical Physics, Chemical Physics, Medical, Health Physics, Nonlinear Dynamics and Complex Systems, Theoretical Physics.* Chaos; Theory.

Griffioen, Keith A., Ph.D., Stanford University, 1984. *Nuclear Physics, Particles and Fields.* Nucleon structure; dark matter searches; experiment.

Hoatson, Gina L., Ph.D., University of East Anglia, 1980. *Chemical Physics, Condensed Matter Physics.* Solid state NMR spectroscopy; Experiment.

Krakauer, Henry, Ph.D., Brandeis University, 1975. *Computational Physics, Condensed Matter Physics, Solid State Physics.* Theory.

Lukaszew, R. Alejandra, Ph.D., Wayne State University, 1996. *Condensed Matter Physics, Electromagnetism, Nano Science and Technology, Optics, Solid State Physics, Surface Physics.* Thin films and nanostructures; Experiment.

Manos, Dennis M., Ph.D., Ohio State University, 1976. Vice Provost for Research and Graduate/Professional Studies. *Applied Physics, Biophysics, Nano Science and Technology, Plasma and Fusion, Surface Physics.* Experiment.

McKeown, Robert D., Ph.D., Princeton University, 1979. Deputy Director for Science at the Thomas Jefferson National Accelerator Facility (JLAB). *Nuclear Physics, Particles and Fields.* Electron scattering and reactor neutrinos; experiment.

Pennington, Michael R., Ph.D., Westfield College, London, 1971. Associate Director for Theoretical and Computational Physics; Theory Center Director at the Thomas Jefferson National Accelerator Facility (JLAB). *Nuclear Physics, Particles and Fields.* Hadronic physics; theory.

Perdrisat, Charles F., Ph.D., Swiss Federal Institute of Technology, 1961. *High Energy Physics, Nuclear Physics, Particles and Fields.* Measurements of the proton structure from elastic electron scattering; experiment.

Sher, Marc T., Ph.D., University of Colorado, 1980. *Astrophysics, High Energy Physics, Particles and Fields, Theoretical Physics.* Theory.

Tracy, Eugene R., Ph.D., University of Maryland, 1984. *Nonlinear Dynamics and Complex Systems, Plasma and Fusion.* Phase space methods in plasma wave theory; nonlinear modeling and time series analysis; theory.

Vahala, George M., Ph.D., University of Iowa, 1972. *Computational Physics, Nonlinear Dynamics and Complex Systems, Plasma and Fusion.* Lattice Boltzmann; quantum lattice algorithms; theory.

Zhang, Shiwei, Ph.D., Cornell University, 1993. *Computational Physics, Condensed Matter Physics, Solid State Physics.* Quantum many-body simulations; strongly correlated systems; Theory.

Associate Professor

Aubin, Seth A. M., Ph.D., Stony Brook University, 2003. *Atomic, Molecular, & Optical Physics.* Ultracold quantum gases; precision measurements; laser cooling and trapping; experiment.

Erlich, Joshua, Ph.D., Massachusetts Institute of Technology, 1999. *High Energy Physics, Particles and Fields, Theoretical Physics.* Theory.

Kordosky, Michael A., Ph.D., University of Texas, Austin, 2004. *High Energy Physics, Nuclear Physics, Particles and Fields.* Experiment.

Nelson, Jeffrey K., Ph.D., University of Minnesota, 1994. *High Energy Physics, Particles and Fields.* Neutrino oscillations; neutrino scattering; experiment.

Novikova, Irina, Ph.D., Texas A&M University, 2003. *Atomic, Molecular, & Optical Physics, Optics.* Quantum optics; experiment.

Orginos, Konstantinos N., Ph.D., Brown University, 1998. *Applied Mathematics, Computational Physics, Nuclear Physics, Particles and Fields, Theoretical Physics.* Hadron structure; lattice QCD; nuclear physics; algorithms; theory.

Vahle, Patricia L., Ph.D., University of Texas, Austin, 2004. *High Energy Physics, Particles and Fields.* Neutrino oscillations; development of future neutrino facilities; experiment.

Assistant Professor

Deconinck, Wouter, Ph.D., University of Michigan, 2008. *Nuclear Physics, Particles and Fields.* Parity; parity violation; electron scattering; fundamental symmetries; Jefferson Laboratory; experiment.

Mikhailov, Eugeniy E., Ph.D., Texas A&M University, 2003. *Atomic, Molecular, & Optical Physics.* Quantum optics; experiment.

Qazilbash, M. Mumtaz, Ph.D., University of Maryland, College Park, 2004. *Condensed Matter Physics, Low Temperature Physics, Nano Science and Technology, Optics, Solid State Physics.* Infrared and optical spectroscopy; near-field infrared nanospectroscopy; metal-insulator transitions; structural instabilities; superconducting and density-wave transitions; experiment.

Rossi, Enrico, Ph.D., University of Texas, Austin, 2005. *Condensed Matter Physics, Nano Science and Technology, Solid State Physics, Theoretical Physics.* Electronic properties of materials; graphene; two-dimensional electron systems; strongly correlated electron systems; theory.

Walker-Loud, Andre P., Ph.D., University of Washington-Seattle, 2006. *Astrophysics, Computational Physics, High Energy Physics, Nuclear Physics, Particles and Fields, Theoretical Physics.* Nonperturbative QCD; implications of the standard model; theory.

Emeritus

Champion, Roy L., Ph.D., University of Florida, 1966. *Atomic, Molecular, & Optical Physics.* Experiment.

Eckhause, Morton, Ph.D., Carnegie Mellon University, 1962. *Nuclear Physics, Particles and Fields.* Experiment.

Gross, Franz L., Ph.D., Princeton University, 1963. *Nuclear Physics, Particles and Fields.* Theory.

Kane, John R., Ph.D., Carnegie Mellon University, 1964. *Nuclear Physics, Particles and Fields.* Experiment.

Kossler, William J., Ph.D., Princeton University, 1964. *Condensed Matter Physics, Nuclear Physics, Physics of Beams.* Experiment.

McKnight, John L., Ph.D., Yale University, 1957. *History & Philosophy of Physics/Science.*,Foundations of quantum theory.

Petzinger, Kenneth G., Ph.D., University of Pennsylvania, 1971. *Condensed Matter Physics.* Theory.

Remler, Edward A., Ph.D., University of North Carolina, 1963. *Nuclear Physics, Particles and Fields.* Theory.

Schone, Harlan E., Ph.D., University of California, Berkeley, 1960. *Condensed Matter Physics.* Experiment.

von Baeyer, Hans C., Ph.D., Vanderbilt University, 1964. *Particles and Fields, Theoretical Physics.* Theory; public understanding of science.

Walecka, J. Dirk, Ph.D., Massachusetts Institute of Technology, 1958. *High Energy Physics, Nuclear Physics, Particles and Fields.* Theory.

Welsh, Robert E., Ph.D., Pennsylvania State University, 1960. *Nuclear Physics, Particles and Fields.* Experiment.

Research Professor

Venkataraman, Malathy D., Ph.D., University of Kerala, 1968. *Chemical Physics.* Spectroscopy.

Research Associate Professor

Benner, D. Chris, Ph.D., University of Arizona, 1979. *Chemical Physics, Electromagnetism, Theoretical Physics.* Molecular spectroscopy.

Adjunct Professor

Bosted, Peter, Ph.D., Massachusetts Institute of Technology, 1980. *Nuclear Physics.* Experiment.

Carlini, Roger D., Ph.D., University of New Mexico, 1978. *Nuclear Physics, Particles and Fields.* Experiment.

Danehy, Paul M., Ph.D., Stanford University, 1995. *Atomic, Molecular, & Optical Physics.* Laser-based measurement techniques for NASA applications; experiment.

Detmold, William, Ph.D., University of Adelaide, 2002. *Computational Physics, Nuclear Physics.* Hadronic theory and lattice field theory.

Osborne, Alfred R., Ph.D., University of Houston, 1974. *Fluids, Rheology, Nonlinear Dynamics and Complex Systems.* Nonlinear phenomena in fluids; physical oceanography.

Reilly, Anne C., Ph.D., University of Michigan, 1996. *Atomic, Molecular, & Optical Physics, Condensed Matter Physics.* Experiment.

Richards, David, Ph.D., University of Cambridge, 1984. *Computational Physics, Nuclear Physics, Particles and Fields.* Theory.

Williams, Gwyn P., Ph.D., University of Sheffield, 1971. *Condensed Matter Physics, Physics of Beams.* Synchrotron radiation; free-electron lasers; experiment.

Wolf, Stuart A., Ph.D., Rutgers University, 1969. *Applied Physics, Condensed Matter Physics.* Magnetism; superconductivity and nanotechnology; experiment.

Academic Professional

Hancock, A. Dayle, Ph.D., University of Houston, 1981. Director of Teaching Labs. *History & Philosophy of Physics/Science, Physics and other Science Education.*

DEPARTMENTAL RESEARCH SPECIALTIES AND STAFF

Theoretical

Condensed Matter Physics. Electronic properties of materials; positive muons in solids; surface physics; high-temperature superconductivity ferroelectrics; ultra-cold atoms; graphene; two-dimensional systems; strongly correlated electron systems; electron-phonon dynamics; computational physics. Krakauer, Rossi, Zhang.

High Energy Physics. Electroweak phenomenology and symmetry breaking; extensions of the standard model; supersymmetry, grand unification and extra dimensions; string theory; cosmology. Carone, Erlich, Sher, Walker-Loud.

Nonlinear Dynamics and Complex Systems. Order and chaos in classical and quantum systems; atoms in strong fields; atomic and molecular collisions. Delos, Osborne, Tracy, Vahala.

Nuclear Physics. Perturbative and nonperturbative QCD; lattice gauge theory; effective field theories for hadrons. Carlson, Detmold, Orginos, Pennington, Richards, Walker-Loud.

Plasma and Fusion. Magnetohydrodynamics; kinetic theory; turbulence; numerical simulation of plasmas; applications to fusion; nonlinear dynamics and chaotic signal process; ocean waves; developing type II quantum computer algorithms for MHD; supercomputers are used at DoE-NERSC, DoD-NAVO, DoD-ERDC, and Earth Simulator (Japan). Tracy, Vahala.

Experimental

Atomic, Molecular, & Optical Physics. Ion-atom and ion-molecule collisions; collisional detachment; inelastic scattering; collisions of ions with surfaces; interactions of lasers with atoms; ultra-cold quantum gases; quantum optics; studies of nonlinear processes; electron-phonon dynamics. Aubin, Cooke, Danehy, Mikhailov, Novikova, Reilly, Williams.

Condensed Matter Physics. Infrared and optical spectroscopy; near-field infrared nanospectroscopy; phase transitions (metal-insulator, structural, magnetic, superconducting, and density wave); nuclear magnetic resonance; piezoelectrics and high dielectric microwave ceramics; thin films (metallic, magnetic, superconducting for SRF applications, correlated-electron-materials) and nanostructures; electron-phonon dynamics; plasmonics and magneto-plasmonics. Hoatson, Lukaszew, Qazilbash, Reilly, Williams, Wolf.

High Energy Physics. Experiments at Fermilab and the Soudan Underground Laboratory; neutrino masses and mixing; CP violation in neutrinos; neutrino interactions on nucleons and nuclei; structure of the weak current; reactor and long baseline neutrino oscillation experiments. Kordosky, McKeown, Nelson, Vahle.

Nuclear Physics. Intermediate energy experiments at Jefferson Laboratory and other facilities; measurements of the structure of nucleons and nuclei via electromagnetic and electroweak interactions; precision tests of the standard model and dark matter searches with electron scattering; hadron spectroscopy; hyperpolarized nuclear targets. Armstrong, Averett, Bosted, Carlini, Deconinck, Griffioen, Kordosky, McKeown, Perdrisat.

Physics of Beams. Particle accelerator physics; relativistic electron beams; synchrotron radiation; free-electron lasers; relativistic electrodynamics; superconducting RF cavity surface physics; design and optimization of future accelerator facilities. Lukaszew, Nelson, Williams.

Plasma and Fusion. Properties of glow discharges; glow discharge effects on surfaces. Manos.

View additional information about this department at
www.gradschoolshopper.com

UNIVERSITY OF VIRGINIA

DEPARTMENT OF PHYSICS

Charlottesville, Virginia 22904-4714
http://www.phys.virginia.edu/

General University Information
President: Teresa Sullivan
Dean of Graduate School: Philip Zelikow
University website: http://www.virginia.edu/
Control: Public
Total Faculty: 7,979
Total number of Students: 21,095
Total number of Graduate Students: 6,454

Department Information
Department Chairman: Joseph Poon, Chair
Department Contact: Helen McLaughlin, Physics and Educational Outreach Assistant
  Total full-time faculty: 35
  Full-Time Graduate Students: 95
  First-Year Graduate Students: 23
  Female First-Year Students: 3
  Total Post Doctorates: 16

Department Address
382 McCormick Road
Charlottesville, VA 22904-4714
Phone: (434) 924-3781

E-mail: grad-info-request@physics.virginia.edu
Website: http://www.phys.virginia.edu/

ADMISSIONS

Admission Contact Information
Address admission inquiries to: Graduate Admissions Advisor
Phone: (434) 924-3781
E-mail: Grad-Info-Request@physics.virginia.edu
Admissions website: http://gsas.virginia.edu/admission/requirements

Application deadlines
Fall admission:
U.S. students: January 15   *Int'l. students*: January 15

Application fee
U.S. students: $60

Admissions information
For Fall of 2013:
  Number of applicants: 209
  Number admitted: 63
  Number enrolled: 15

645

Admission requirements
Bachelor's degree requirements: Bachelor's degree is required.

GRE requirements
The GRE is required.
Quantitative score: 750
Verbal score: 500
Analytical score: 4

Advanced GRE requirements
The Advanced GRE is required.

TOEFL requirements
The TOEFL exam is required for students from non-English-speaking countries.
iBT score: 90

Other admissions information
Undergraduate preparation assumed: Marion and Thornton, Classical Dynamics of Particles and Systems (mechanics); Kittel, Thermal Physics (statistical physics); Fermi, Thermodynamics (statistical physics); Marion, Classical Electromagnetic Radiation (electromagnetism); Gasiorowicz, Quantum Physics (quantum mechanics).

TUITION

Tuition year 2013-14:
Tuition for in-state residents
Full-time students: $16,270 annual
Tuition for out-of-state residents
Full-time students: $26,276 annual
Please note that the tuition amount includes fees.
Credit hours per semester to be considered full-time: 12
Deferred tuition plan: Yes
Health insurance: Available at the cost of 2574 per year.
Other academic fees: $50 - Per semester international fee applied to all international students.

Teaching Assistants, Research Assistants, and Fellowships
Number of first-year
Teaching Assistants: 11
Fellowship students: 4
Average stipend per academic year
Teaching Assistant: $18,600
Research Assistant: $18,600
Fellowship student: $18,600

FINANCIAL AID

Application deadlines
Fall admission:
U.S. students: January 15   *Int'l. students*: January 15

Loans
Loans are available for U.S. students.
Loans are not available for international students.
GAPSFAS application required: No
FAFSA application required: No

For further information
Address financial aid inquiries to: Graduate Admissions Advisor.
Financial aid website: http://www.phys.virginia.edu/GraduateBrochure/FinancialAid.asp

HOUSING

Availability of on-campus housing
Single students: Yes
Married students: Yes

For further information
E-mail: housing@virginia.edu
Housing aid website: http://www.virginia.edu/housing/grad.php

Table A—Faculty, Enrollments, and Degrees Granted

Research Specialty	2012-13 Faculty	Enrollment Fall 2012 Master's	Enrollment Fall 2012 Doctorate	Number of Degrees Granted 2012–13 (2008–13) Master's	Number of Degrees Granted 2012–13 (2008–13) Terminal Master's	Number of Degrees Granted 2012–13 (2008–13) Doctorate
Atomic, Molecular, & Optical Physics	7	2	19	1(4)	–	–(15)
Condensed Matter Physics	11	–	13	–(2)	–	–(10)
Materials Science, Metallurgy	–	–	2	–	–	–(1)
Medical, Health Physics	2	–	4	–(1)	–	2(5)
Nuclear Physics	12	–	21	1(1)	–(2)	4(8)
Particles and Fields	9	–	19	–(4)	–(3)	3(20)
Non-specialized	–	–	17	–(1)	–	–
Other	2	10	–	–(1)	8(71)	–
Total	44	12	95	2(14)	8(76)	9(59)
Full-time Grad. Stud.	–	2	95	–	–	–
First-year Grad. Stud.	–	–	23	–	–	–

GRADUATE DEGREE REQUIREMENTS

Master's: 24 graduate credits in approved program with satisfactory performance, thesis, and thesis exam; no language requirement.

Doctorate: 54 graduate credits required; satisfactory performance in an approved course program is required; comprehensive exam, dissertation, and dissertation exam; two semesters residency required; no language required.

Other Degrees: The Department also offers M.S. and Ph.D. degrees in Engineering Physics in cooperation with the School of Engineering and Applied Science. A Ph.D. in Biophysics is available through an interdisciplinary program associated with the physics department and other science departments of the University. A Master of Arts in Physics Education (MAPE) degree is offered to High School teachers.

Thesis: Thesis may be written in absentia.

SPECIAL EQUIPMENT, FACILITIES, OR PROGRAMS

Department facilities include machine and electronics shops, a physics library, and extensive computing resources. Most on-site research is carried out in the J. W. Beams Laboratory building, with additional facilities on campus for work in laser science, materials science, nanoscale fabrication, and particle detector development. The department is active at many off-site facilities as well, including major particle accelerators, neutron sources, and synchrotron sources in the US and abroad.

Table B—Separately Budgeted Research Expenditures by Source of Support

Source of Support	Departmental Research	Physics-related Research Outside Department
Federal government	$6,140,269.55	
State/local government	$1,454,802.43	
Non-profit organizations	$1,416.01	
Business and industry		
Other	$491,112	
Total	$8,087,599.99	

Table C—Separately Budgeted Research Expenditures by Research Specialty

Research Specialty	No. of Grants	Expenditures ($)
Atomic, Molecular, & Optical Physics	26	$1,387,757.17
Condensed Matter Physics	30	$1,897,577.76
Nuclear Physics	40	$3,061,689.85
Particles and Fields	30	$1,544,263.85
Other	2	$192,311.36
Total	128	$8,083,599.99

FACULTY

Distinguished University Professor

Cardman, Lawrence S., Ph.D., Yale University, 1972. Experimental nuclear and particle physics.

Professor

Arnold, Peter B., Ph.D., Stanford University, 1986. Theoretical particle physics.

Bloomfield, Louis A., Ph.D., Stanford University, 1983. Experimental atomic and solid state physics.

Cates, Gordon D., Ph.D., Yale University, 1987. Experimental nuclear and atomic physics.

Cox, Bradley B., Ph.D., Duke University, 1967. Experimental high-energy particle physics.

Dukes, Edmond C., Ph.D., University of Michigan, 1984. Experimental elementary particle physics.

Fendley, Paul, Ph.D., Harvard University, 1990. Theoretical condensed matter and particle physics.

Fowler, Michael, Ph.D., St. John's College, Cambridge University, 1962. Theoretical physics; field theory and solid state theory; physics education.

Gallagher, Thomas F., Ph.D., Harvard University, 1971. Collisions and spectroscopy of atoms and molecules.

Hirosky, Robert J., Ph.D., University of Rochester, 1994. Experimental particle physics.

Hung, Pham Q., Ph.D., University of California, Los Angeles, 1978. Theoretical particle physics; cosmology.

Jones, Robert R., Ph.D., University of Virginia, 1990. Experimental atomic molecular and optical physics.

Lee, Seung-Hun, Ph.D., Johns Hopkins University, 1996. Experimental condensed matter physics.

Lehmann, Kevin, Ph.D., Harvard University, 1983. Experimental chemical physics; experimental atomic, molecular, and optical physics.

Louca, Despina A., Ph.D., University of Pennsylvania, 1997. Experimental condensed matter.

Norum, Blaine E., Ph.D., Massachusetts Institute of Technology, 1979. Experimental nuclear and particle physics.

Pfister, Olivier, Ph.D., University of Paris-North, 1993. Experimental atomic, molecular, and optical physics.

Počanić, Dinko, Ph.D., Zagreb, 1981. Experimental intermediate-energy nuclear and particle physics.

Poon, S. Joseph, Ph.D., California Institute of Technology, 1978. Experimental solid state physics; nanostructured materials; quasicrystals; thermoelectric compounds.

Thacker, Harry B., Ph.D., University of California, Los Angeles, 1973. Elementary particle physics and quantum field theory.

Thornton, Stephen T., Ph.D., University of Tennessee, 1967. Experimental nuclear physics; physics education.

Wolf, Stuart, Ph.D., Rutgers University, 1969. Experimental condensed matter physics.

Associate Professor

Kolomeisky, Eugene B., Ph.D., Academy of Sciences of the USSR, Moscow, 1988. Theoretical condensed matter.

Liyanage, Nilanga, Ph.D., Massachusetts Institute of Technology, 1999. Experimental nuclear and particle physics.

Paschke, Kent D., Ph.D., Carnegie Mellon University, 2001. Experimental nuclear and particle physics.

Sackett, Charles A., Ph.D., Rice University, 1998. Experimental atomic, molecular and optical physics.

Shivaram, Bellave S., Ph.D., Northwestern University, 1984. Experimental solid state physics.

Yoon, Jongsoo, Ph.D., Pennsylvania State University, 1997. Experimental condensed matter physics.

Zheng, Xiaochao, Ph.D., Massachusetts Institute of Technology, 2002. Experimental nuclear and particle physics.

Assistant Professor

Baeßler, Stefan, Ph.D., University of Heidelberg, 1996. Experimental nuclear and particle physics.

Chatterjee, Utpal, Ph.D., University of Illinois at Chicago, 2007. Experimental condensed matter physics.

Group, R. Craig, Ph.D., University of Florida, 2006. Experimental high energy physics.

Klich, Israel, Ph.D., Israel Institute of Technology, 2004. Theoretical condensed matter; theoretical mathematical physics.

Neu, Christopher, Ph.D., Ohio State University, 2003. Experimental high energy physics.

Vaman, Diana, Ph.D., Stony Brook University, 2001. Theoretical nuclear and particle physics.

Emeritus

Brill, Arthur S., Ph.D., University of Pennsylvania, 1956. Experimental biophysics; proteins and transition metal ions.

Celli, Vittorio, Ph.D., Pavia, 1958. Theoretical solid state physics; surface studies.

Conetti, Sergio, Ph.D., Trieste, 1967. Experimental high-energy particle physics.

Deaver, Bascom S., Ph.D., Stanford University, 1962. Experimental solid state physics: superconducting devices.

Fishbane, Paul M., Ph.D., Princeton University, 1967. Theoretical physics; elementary particles.

Hess, George B., Ph.D., Stanford University, 1967.

Minehart, Ralph C., Ph.D., Harvard University, 1962. Experimental nuclear and particle physics.

Ritter, Rogers C., Ph.D., University of Tennessee, 1961. Gravitation; precision measurements; medical physics.

Ruvalds, John, Ph.D., University of Oregon, 1967. Theoretical solid state physics.

Schnatterly, Stephen, Ph.D., University of Illinois, 1965. Experimental solid state physics; soft x-ray and inelastic electron scattering spectroscopy of solids, atoms, and molecules.

Sobottka, Stanley E., Ph.D., Stanford University, 1960. X-ray crystallography; x-ray detector development; experimental nuclear physics.

Weber, Hans J., Ph.D., University of Frankfurt, 1965. Theoretical nuclear and particle physics.

Research Professor

Crabb, Donald G., Ph.D., University of Southampton, 1967. Experimental nuclear and particle physics.

Day, Donal B., Ph.D., University of Virginia, 1979. Experimental nuclear and particle physics.

Gillies, George T., Ph.D., University of Virginia, 1980. Engineering Physics.

Lindgren, Richard A., Ph.D., Yale University, 1969. Experimental nuclear and particle physics, physics education.

Research Associate Professor

Liuti, Simonetta, Ph.D., Universitád: Roma, 1989. Theoretical nuclear and particle physics.

Williams, Mark B., Ph.D., University of Virginia, 1990. Medical Physics.

Affiliate Professor

Herbst, Eric, Ph.D., Harvard University, 1972. Theoretical chemical physics.

Visiting Professor

Imlay, Richard, Ph.D., Princeton University, 1967. Experimental high energy.

DEPARTMENTAL RESEARCH SPECIALTIES AND STAFF

Theoretical

Condensed Matter Physics. Field theoretic models for solid state systems; many-body physics in ultracold atomic gases; quantum Hall effect; Bethe Ansatz systems; topological quantum computation; phase transitions and renormalization group methods in statistical physics; Bose-Einstein condensation; theory of macroscopic quantum phenomena; pattern formation; nonperturbative statistical mechanics. Fendley, Klich, Kolomeisky.

High Energy Physics. Theoretical studies of high-energy physics including properties of quantum chromodynamics; lattice gauge theory; string/gauge duality; high-temperature field theory; electroweak interactions; grand unified theories; supersymmetry; neutrino physics including models of neutrino masses; dark matter; dark energy; cosmology. Arnold, Hung, Thacker, Vaman.

Nuclear Physics. Lattice gauge theory; inclusive and exclusive deep inelastic electron and neutrino scattering on nucleons and nuclei; the spin composition of quarks and gluons within hadrons; the role of QCD in hadronic structure. Liuti, Vaman.

Experimental

Atomic, Molecular, & Optical Physics. Laser manipulation and spectroscopy of atoms, ions, small molecules, and clusters, including Bose-Einstein condensation in dilute vapors; atom interferometry; quantum optics; quantum information; optical interferometry; dipole-dipole interactions between cold Rydberg atoms; ultracold plasmas; observation and control of electronic wavepackets in Rydberg atoms using microwave, THz, and optical fields; dynamics of atoms and molecules in intense femtosecond laser pulses; high-order harmonic generation in gases; spectroscopy of single and doubly excited Rydberg atoms; studies of magnetic properties of clusters; photo-detachment and photoionization; development of new techniques in laser spectroscopy; noble gas hyper polarization via spin exchange with optically pumped alkali atoms, optical control of chemical processes; cavity ring-down spectroscopy; spectroscopy using helium nanodroplet isolation; investigation of highly excited vibrational states; microwave-optical double resonance. Cates, Gallagher, Jones, Lehmann, Pfister, Sackett.

Condensed Matter Physics. The experimental condensed matter physics groups at UVa explore the structural, electronic, magnetic, and superconducting properties of different types of amorphous and crystalline solids including thin films. The groups are equipped with state-of-the-art equipment. Activities include the synthesis and characterization of amorphous alloys, quantum magnets, frustrated spin systems, multiferroics, high temperature superconductors, and strongly correlated systems. Several groups perform research at national and international neutron and synchrotron facilities. There are joint research programs with the Engineering School. Facilities accessible to the groups include photolithography lab and x-ray diffraction and electron microscopes, as well as national labs where high magnetic fields sources are available. Celli, Lee, Louca, Poon, Shivaram, Wolf, Yoon.

High Energy Physics. The experimental group participates in major research collaborations at the world's leading particle accelerators in the U.S. and in Europe where we are able to study the most fundamental interactions of matter to elicit the inner workings of the natural world. The group is housed in its own building a short walk from the main physics building. This superb laboratory has an electronics lab, mechanical shop, a large assembly area, and powerful computing capabilities. Cox, Dukes, Group, Hirosky, Neu.

Physics Education. The department has an outreach professional development program offering local and distance learning courses to K-12 teachers that include physical/Earth science teachers and high school physics teachers. Courses are offered at sites throughout Virginia, in residential summer workshops in Charlottesville, and distance learning, web-based courses on the Internet. Teachers take these courses for endorsement, recertification, or simply to increase their physics content. The Masters of Arts in Physics Education degree (MAPE) program is active with a rolling application deadline. Lindgren, Thornton.

View additional information about this department at **www.gradschoolshopper.com**

VIRGINIA COMMONWEALTH UNIVERSITY

DEPARTMENT OF PHYSICS

Richmond, Virginia 23284-2000
http://www.vcu.edu/hasweb/phy

General University Information

President: Michael Rao
Dean of Graduate School: F. Douglas Boudinot
University website: http://www.vcu.edu
Control: Public

Setting: Urban
Total Faculty: 2,048
Total Graduate Faculty: Not available
Total number of Students: 31,752
Total number of Graduate Students: 6,094

Department Information
Department Chairman: Robert H. Gowdy, Chair
Department Contact: Robert H. Gowdy, Chair
  Total full-time faculty: 17
  Total number of full-time equivalent positions: 18
  Full-Time Graduate Students: 12
  First-Year Graduate Students: 6
  Female First-Year Students: 1
  Total Post Doctorates: 11

Department Address
701 West Grace Street (Laurel Entrance)
P. O. Box 842000
Richmond, VA 23284-2000
Phone: (804) 828-1818 (C) 1821
Fax: (804) 828-7073
E-mail: rhgowdy@vcu.edu
Website: http://www.vcu.edu/hasweb/phy

ADMISSIONS

Admission Contact Information
Address admission inquiries to: School of Graduate Studies Virginia Commonwealth University, P.O. Box 843051, Richmond, VA 23284-3051
Phone: (804) 828-2233
E-mail: gradmail@vcu.edu
Admissions website: http://www.graduate.vcu.edu/

Application deadlines
Fall admission:
U.S. students: May 1   *Int'l. students*: May 1

Application fee
U.S. students: $50   *Int'l. students*: $50

Admissions information
For Fall of 2012:
  Number of applicants: 12
  Number admitted: 10
  Number enrolled: 7

Admission requirements
Bachelor's degree requirements: Bachelor's degree in physics or engineering is recommended.
Minimum undergraduate GPA: 2.7

GRE requirements
The GRE is required.
  Quantitative score: 145
  Verbal score: 145
  Analytical score: 3

Advanced GRE requirements
The Advanced GRE is not required.

TOEFL requirements
The TOEFL exam is required for students from non-English-speaking countries.
  PBT score: 550
  iBT score: 80

Other admissions information
Undergraduate preparation assumed: A typical student will have completed intermediate and/or advanced courses using Classical Mechanics, Marion and Thornton; Electricity and Magnetism, Reitz, Milford, and Christy; and Modern Physics, Eisberg and Resnick. Deficiencies in advanced courses may be made up while a graduate student.

TUITION
Tuition year 2013–14:
Tuition for in-state residents
  Full-time students: $9,911 annual
  Part-time students: $632.45 per credit
Tuition for out-of-state residents
  Full-time students: $20,378 annual
  Part-time students: $1,029.26 per credit
Credit hours per semester to be considered full-time: 9
Deferred tuition plan: Yes
Health insurance: Available
Other academic fees: Fees - In-state, Full-time $2091.46 Fees - Out-of-state Full-time $2,703.46
Academic term: Semester

Teaching Assistants, Research Assistants, and Fellowships
Number of first-year
  Teaching Assistants: 5
  Average stipend per academic year
  Teaching Assistant: $12,500
  Research Assistant: $12,500

FINANCIAL AID

Application deadlines
Fall admission:
U.S. students: March 1

Loans
Loans are available for U.S. students.
Loans are available for international students.
GAPSFAS application required: No
FAFSA application required: Yes

For further information
Address financial aid inquiries to: UES/Financial Aid, 1015 Floyd Avenue, Richmond, VA 23284-3026.
Phone: (804) 828-6669
Financial aid website: http://www.vcu.edu/enroll/finaid/

HOUSING

Availability of on-campus housing
  Single students: Yes
  Married students: No

For further information
Address housing inquiries to: Residential Life & Housing, 301 West Cary Street, Richmond, VA 23284.
Phone: (804) 828-7666
Housing aid website: http://www.housing.vcu.edu/

Table A—Faculty, Enrollments, and Degrees Granted

Research Specialty	2012-13 Faculty	Enrollment Fall 2012		Number of Degrees Granted 2012-13 (2008-12)		
		Master's	Doctorate	Master's	Terminal Master's	Doctorate
Chemical Physics	9	–	1	–	–	-(2)
Condensed Matter Physics	9	14	–	–	6(23)	–
Nano Science and Technology	9	–	5	–	–	2(3)
Relativity & Gravitation	1	–	–	–	-(2)	–
Total	10	14	6	–	6(23)	2(3)
Full-time Grad. Stud.	–	12	6	–	–	–
First-year Grad. Stud.	–	5	1	–	–	–

GRADUATE DEGREE REQUIREMENTS

Master's: Completion of 30 approved graduate credits with at least 15 credits of didactic or laboratory coursework and successful completion of a master's thesis. Each student will choose an advisor during the first semester and propose a plan of study to fulfill the student's individual career goals.

Thesis: Thesis may not be written in absentia.

SPECIAL EQUIPMENT, FACILITIES, OR PROGRAMS

The department has facilities for surface and material physics research, including an atomic force microscope and equipment for Raman and photoluminescence. Other analytical equipment (SEM, XPS, TEM) is available in a shared facility.

Table B—Separately Budgeted Research Expenditures by Source of Support

Source of Support	Departmental Research	Physics-related Research Outside Department
Federal government	$1,115,000	
State/local government		
Non-profit organizations		
Business and industry		
Other		
Total	$1,115,000	

Table C—Separately Budgeted Research Expenditures by Research Specialty

Research Specialty	No. of Grants	Expenditures ($)
Condensed Matter Physics	13	$1,115,000
Total	13	$1,115,000

FACULTY

Professor

Baski, Alison A., Ph.D., Stanford University, 1991. Executive Associate Dean, College of Humanities and Science. *Chemical Physics, Condensed Matter Physics, Nano Science and Technology*. Semiconductor surface studies.

Jena, Purusottam, Ph.D., University of California, Riverside, 1970. *Atomic, Molecular, & Optical Physics, Chemical Physics, Condensed Matter Physics, Energy Sources & Environment, Materials Science, Metallurgy, Nano Science and Technology, Solid State Physics, Surface Physics, Theoretical Physics*. Electronic structure theory of metals and alloys, semiconductors, intermetallics, and insulators; atomic clusters and cluster assembled materials.

Khanna, Shiv N., Ph.D., University of Delhi, 1976. *Chemical Physics, Condensed Matter Physics, Materials Science, Metallurgy, Nano Science and Technology, Solid State Physics*. Theoretical solid-state physics; electronic structure and magnetic properites of amorphous metals, small atomic clusters, and cluster assembled materials.

Associate Professor

Bertino, Massimo F., Ph.D., MPI-Germany, 1996. *Chemical Physics, Condensed Matter Physics, Nano Science and Technology*. Production of nanostructures by photo-lithographic methods.

Bishop, Marilyn F., Ph.D., University of California, Irvine, 1976. *Acoustics, Biophysics, Chemical Physics, Condensed Matter Physics, Medical, Health Physics, Nano Science and Technology, Optics, Solid State Physics, Statistical & Thermal Physics*. Charge density waves; superconductivity; semiconductors, biophysics.

Gowdy, Robert H., Ph.D., Yale University, 1968. Department Chair. *Relativity & Gravitation*. General relativity and cosmology.

Reshchikov, Michael A., Ph.D., Ioffe Physical-Technical Institute, 1989. *Condensed Matter Physics, Nano Science and Technology, Solid State Physics, Surface Physics*. Defects in semiconductors; photoluminescence.

Assistant Professor

Demchenko, Denis, Ph.D., South Dakota School of Mines & Technology, 2002. *Condensed Matter Physics, Nano Science and Technology, Solid State Physics*. Theoretical and computational nanoscience; electronic structure theory of semiconductor nanocrystals, defects in semiconductors.

Reed, Jason C., Ph.D., New York University, 2007. *Biophysics, Nano Science and Technology*. Applications of nanotechnology to biological systems: Cell-mass measurements by interference Microscopy; Gene expression profiling on Atomic Force Microscopy.

Reiner, Joseph, Ph.D., Stony Brook University, 2003. *Chemical Physics, Condensed Matter Physics, Nano Science and Technology*. Single-molecule biophysics nanopores; fluorescence; optical tweezers.

Ye, Dexian, Ph.D., Rensselaer Polytechnic Institute, 2006. *Chemical Physics, Condensed Matter Physics, Nano Science and Technology*. Fabrication and characterization of nanostructured surfaces.

Instructor

AbuEideh, Wael, M.S., Southern Illinois University, 2010. Physics Lab Coordinator. Photonics.

Ameen, David B., Ph.D., Virginia Commonwealth University, 2000. Undergraduate Academic Advisor. *Chemical Physics, Condensed Matter Physics, Nano Science and Technology*. Theoretical and computational nanoscience; electronic structure theory of semiconductor nanocrystals.

McMullen, J. Thomas, Ph.D., Queen's College, 1968. *Acoustics, Biophysics, Chemical Physics, Condensed Matter Physics, Medical, Health Physics, Nano Science and Technology, Optics, Solid State Physics, Statistical & Thermal Physics*. Charge density waves; superconductivity; semiconductors, biophysics.

Reveles, J. Ulises, Ph.D., Cinvestav, Mexico City, 2004. *Chemical Physics, Condensed Matter Physics, Nano Science and Technology*. Theoretical solid-state physics; electronic structure and magnetic properites of amorphous metals, small atomic clusters, and cluster assembled materials.

Skrobiszewski, John L., M.S., Virginia Commonwealth University, 2003. *Condensed Matter Physics, Nano Science and Technology*.

Woodworth, Patrick H., M.S., Virginia Commonwealth University, 2007. *Condensed Matter Physics, Nano Science and Technology*. Semiconductor surface studies.

DEPARTMENTAL RESEARCH SPECIALTIES AND STAFF

Theoretical

Acoustics. Bishop, McMullen.

Atomic, Molecular, & Optical Physics. Jena.

Biophysics. Bishop, McMullen, Reed.

Chemical Physics. Properties of metal clusters and cluster-assembled materials. Ameen, Baski, Bertino, Bishop, Jena, Khanna, McMullen, Reveles, Ye.

Energy Sources & Environment. Jena.

Materials Science, Metallurgy. Jena, Khanna.

Medical, Health Physics. Bishop, McMullen.

Nano Science and Technology. Electronic structure of defects and defect complexes; electronic and magnetic properties of multilayer thin films, dilute magnetic semiconductors, metal oxides, and hydrogen storage materials in bulk and nanostructured forms; transport theory for simple metals and charge density waves; electronic, structural, and elastic properties of semiconductors; nanostructures and nanoscale photovoltaics. Ameen, Baski, Bertino, Bishop, Demchenko, Jena, Khanna, McMullen, Reiner, Reshchikov, Reveles, Skrobiszewski, Woodworth, Ye.

Optics. Bishop, McMullen.

Relativity & Gravitation. Dynamical structure of gravitational theories; gravitational waves; exact solutions of Einstein's equations; quantum gravity. Gowdy.

Solid State Physics. Bishop, Demchenko, Jena, Khanna, McMullen, Reshchikov.

Statistical & Thermal Physics. Bishop, McMullen.

Surface Physics. Jena, Reshchikov.

Theoretical Physics. Jena.

Experimental

Condensed Matter Physics. Growth and characterization of semiconductor and metal systems using scanning probe microscopy techniques, modulation spectroscopy, Raman, and photoluminescence; photo-lithographic fabrication of nanocomposites for structural and energetic applications; directed self-assembly of nanostructures. Ameen, Baski, Bertino, Bishop, Demchenko, Jena, Khanna, McMullen, Reiner, Reshchikov, Reveles, Skrobiszewski, Woodworth, Ye.

View additional information about this department at www.gradschoolshopper.com

---

VIRGINIA TECH

DEPARTMENT OF PHYSICS

Blacksburg, Virginia 24061
http://www.phys.vt.edu

General University Information
President: Charles W. Steger
Dean of Graduate School: Karen DePauw
University website: http://www.vt.edu/
Control: Public
Setting: Rural
Total Faculty: 2,326
Total number of Students: 28,836
Total number of Graduate Students: 4,620

Department Information
Department Chairman: Leo E. Piilonen, Chair
Department Contact: Leo E. Piilonen, Department Chair
  Total full-time faculty: 33
  Full-Time Graduate Students: 81
  First-Year Graduate Students: 15
  Female First-Year Students: 4
  Total Post Doctorates: 13

Department Address
Robeson Hall (MC 0435)
850 West Campus Drive
Blacksburg, VA 24061
Phone: (540) 231-6544 (C) –7472
Fax: (540) 231-7511
E-mail: info@phys.vt.edu
Website: http://www.phys.vt.edu

ADMISSIONS

Admission Contact Information
Address admission inquiries to: Graduate Committee Chairman, Physics Department
Phone: (540) 231-8728

E-mail: gradphys@vt.edu
Admissions website: http://www.phys.vt.edu

Application deadlines
Fall admission:
U.S. students: January 15   *Int'l. students*: January 15

Application fee
U.S. students: $75   *Int'l. students*: $75

Admissions information
For Fall of 2013:
  Number of applicants: 146
  Number admitted: 68
  Number enrolled: 15

Admission requirements
Bachelor's degree requirements: Bachelor's degree in physics with a minimum undergraduate GPA of 3.0 in physics/math during the last two years of undergraduate study or, if the bachelor's degree is in a subject other than physics, 18 semester hours in intermediate mechanics, electromagnetism, and quantum mechanics, excluding general physics, are required. The GRE general and advanced physics examinations are required. Students from non-English-speaking countries are required to demonstrate proficiency in English via the TOEFL examination, with a minimum acceptable score of 550.
Minimum undergraduate GPA: 3.0

GRE requirements
The GRE is required.

Advanced GRE requirements
The Advanced GRE is required.

TOEFL requirements

The TOEFL exam is required for students from non-English-speaking countries.
PBT score: 550
iBT score: 80

Other admissions information

Undergraduate preparation assumed: *Undergraduate preparation assumed:* Thornton, Marion, *Classical Mechanics*; Reitz, Milford, and Christy, *Foundations of Electromagnetic Theory*; Griffiths, *Electrodynamics*; Hecht, *Optics*; Kittel, Kroemer, *Thermal Physics*; Griffiths, *Quantum Mechanics*; Liboff, *Quantum Mechanics*.

TUITION

Tuition year 2013-14:
Tuition for in-state residents
  Full-time students: $5,635.5 per semester
  Part-time students: $1,886 per semester
Tuition for out-of-state residents
  Full-time students: $11,116 per semester
  Part-time students: $3,712.25 per semester
Tuition for part-time is for three credit hours. Assistantships come with proportional tuition waivers.
Credit hours per semester to be considered full-time: 9
Deferred tuition plan: No
Health insurance: Yes, 90% subsidy of University negotiated plan.
Other academic fees: In-state residents, $876 ($276.75, part-time/three credits) per semester. Out-of-state residents: $1,178 ($427.75 part-time/three credits)per semester.
Academic term: Semester
Number of first-year students who received full tuition waivers: 15

Teaching Assistants, Research Assistants, and Fellowships

Number of first-year
  Teaching Assistants: 15
Average stipend per academic year
  Teaching Assistant: $15,000
  Research Assistant: $15,000
All assistantships come with tuition waivers. Average stipend per academic year is $15,000.

FINANCIAL AID

Application deadlines

Fall admission:
U.S. students: January 15   *Int'l. students*: January 15

Loans

Loans are not available for U.S. students.
Loans are not available for international students.
GAPSFAS application required: No
FAFSA application required: No

For further information

Address financial aid inquiries to: Graduate Program Coordinator, Physics Department.
E-mail: gradphys@vt.edu
Financial aid website: http://www.finaid.vt.edu/

HOUSING

Availability of on-campus housing

Single students: Yes
Married students: No

For further information

Address housing inquiries to: Housing Residence and Life, 144 New Hall West, Blacksburg, VA 24061-0428.
Phone: (540) 231-6205
Housing aid website: http://www.housing.vt.edu

Table A—Faculty, Enrollments, and Degrees Granted

Research Specialty	2013-14 Faculty	Enrollment Fall 2013 Master's	Enrollment Fall 2013 Doctorate	Number of Degrees Granted 2012–13 (2008-13) Master's	Number of Degrees Granted 2012–13 (2008-13) Terminal Master's	Number of Degrees Granted 2012–13 (2008-13) Doctorate
Astrophysics	3	–	6	–(1)	–(1)	1(3)
Biophysics	4	–	–	–	–	–
Condensed Matter						
Physics	13	–	37	4(17)	–(6)	7(27)
Medical, Health Physics	4	–	–	–	–	–
Nuclear Physics	5	–	11	3(6)	–(2)	2(6)
Particles and Fields	7	–	12	2(4)	–(2)	1(3)
Non-specialized	–	2	13	–	–	–
Total	–	2	79	9(28)	–(11)	11(39)
Full-time Grad. Stud.	–	2	79	–	–	–
First-year Grad. Stud.	–	2	13	–	–	–

GRADUATE DEGREE REQUIREMENTS

Master's: Both thesis and nonthesis options are available. For the thesis option, 24 hours of coursework and 6 hours of research are required. A written thesis must be submitted and defended at an oral final examination. For the nonthesis option, 30 hours of coursework are required. An oral final examination must be passed. For both options, a minimum 3.0 grade point average must be maintained. There is no foreign language requirement.

Doctorate: A Ph.D. candidate must pass an oral qualifying and preliminary examination covering classical mechanics, electromagnetism, and nonrelativistic quantum mechanics. The preliminary examination covers the proposed thesis research. 90 hours total (minimum) must be completed, including coursework (30 hours minimum) and research while maintaining a minimum 3.0 grade point average. A written dissertation is required and must be defended at an oral final examination.

SPECIAL EQUIPMENT, FACILITIES, OR PROGRAMS

The faculty in Virginia Tech's Physics Department conducts research in astronomical, mathematical, medical, nuclear, elementary particle, and condensed-matter physics. Medical and neuroscience research is conducted at sites in Arlington and Roanoke, Virginia. Much of the research activity in astronomy and experimental nuclear and particle physics utilizes off-campus facilities, while most of the instrumentation and data analysis are performed on-campus. These facilities include Brookhaven National Laboratory, Daya Bay, Fermilab, KEK, LANL, ORNL, TJNAF, NRAO, Gran Sasso, and the nearby Kimballton Underground Research Facility (KURF). Telescopes used by the astronomy group include the Hubble Space Telescope, the Very Large Telescope, the Chandra X-ray satellite, the Spitzer IR satellite, and the XMM-Newton X-ray satellite.

Experimental facilities in condensed-matter physics include low-temperature facilities and variable-temperature high-magnetic-field magneto-transport systems, low-temperature optical systems, pulsed near- and mid-infrared lasers, visible-ultraviolet lasers, spectrometers, confocal microscopy and related optical characterization facilities, nanofabrication systems, thin film materials deposition systems, materials synthesis, room-temperature and low-temperature scanning tunneling microscopy, and various other microscopy systems. More analytical and nanofabrication systems (e.g., x-ray, Auger, TEM, AFM, SIMS, SQUID, FIB) are housed in on-campus facilities. Research is also performed off-campus, for example, at the National High Magnetic Field Laboratory.

Housed in Robeson Hall is the University's Center for Neutrino Physics (CNP). Many theorists are members of the University Center for Statistical Mechanics, Mathematical Physics, and Theoretical Chemistry, comprised of faculty from the Departments of Chemistry, Physics, and Mathematics.

Virginia Tech University computing offers multiple high-performance computing systems. The Physics Department has two dedicated clusters and a distributed collection of about 200 limited-availability nodes, all running Linux. Access to supercomputers is available through national and international networks.

The Physics Department operates a professional machine shop, a computer shop, and a student shop.

Table B—Separately Budgeted Research Expenditures by Source of Support

Source of Support	Departmental Research	Physics-related Research Outside Department
Federal government	$2,645,694.49	
State/local government	$79,971.88	
Non-profit organizations		
Business and industry	$634,804.03	
Other	$8,311.33	
Total	**$3,368,781.73**	

FACULTY

Professor

Chang, Lay Nam, Ph.D., University of California, Berkeley, 1967. Dean, College of Science. Theoretical particle physics.

Heflin, James R., Ph.D., University of Pennsylvania, 1990. Experimental condensed-matter physics; biophysics.

Heremans, Jean J., Ph.D., Princeton University, 1994. Experimental condensed-matter physics.

Montague, Read P., Ph.D., University of Alabama, 1988. Neuroscience; medical physics; biophysics.

Mun, Seong K., Ph.D., SUNY, Albany, 1979. Neuroscience; medical physics.

Piilonen, Leo E., Ph.D., Princeton University, 1985. Chair of the Department of Physics; William E. Hassinger, Jr Senior Faculty Fellow in Physics. Experimental nuclear and particle physics.

Pitt, Mark, Ph.D., Princeton University, 1992. Experimental nuclear and particle physics.

Schmittmann, Beate, Ph.D., University of Edinburgh, 1984. Theoretical condensed-matter physics.

Simonetti, John H., Ph.D., Cornell University, 1985. Astrophysics.

Täuber, Uwe C., Ph.D., Technische Universität München, 1992. Theoretical condensed-matter physics.

Vogelaar, R. Bruce, Ph.D., California Institute of Technology, 1989. Experimental nuclear and particle physics.

Associate Professor

Arav, Nahum, Ph.D., University of Colorado, Boulder, 1994 Astrophysics.

Huber, Patrick, Ph.D., Technische Universität München, 2003. Theoretical particle physics.

Khodaparast, Giti A., Ph.D., University of Oklahoma, 2001. Experimental condensed-matter physics; biophysics.

Link, Jonathan M., Ph.D., University of California, Davis, 2001. Experimental nuclear and particle physics.

Minic, Djordje, Ph.D., University of Texas, Austin, 1993. Theoretical particles physics.

Park, Kyungwha, Ph.D., Princeton University, 2000. Theoretical condensed-matter physics.

Pleimling, Michel J.F., Ph.D., Universität des Saarlandes, 1996. Theoretical condensed-matter physics.

Robinson, Hans D., Ph.D., Boston University, 2000. Experimental condensed-matter physics.

Sharpe, Eric R., Ph.D., Princeton University, 1998. Theoretical particle physics.

Soghomonian, Victoria, Ph.D., Syracuse University, 1995. Experimental condensed-matter physics.

Takeuchi, Tatsu, Ph.D., Yale University, 1989. Theoretical particle physics.

Assistant Professor

Anderson, Lara B., Ph.D., Oxford University, 2008. Theoretical particle physics.

Cheng, Shengfeng, Ph.D., Johns Hopkins University, 2010. Experimental condensed-matter physics.

Farrah, Duncan, Ph.D., Imperial College London, 2002. Astrophysics.

Gray, James, Ph.D., University of Sussex, 2001. Theoretical particle physics.

Mariani, Camillo, Ph.D., University of Rome, 2008. Experimental nuclear and particle physics.

Mather, William, Ph.D., Georgia Institute of Technology, 2007. Biophysics.

Nguyen, Vinh, Ph.D., University of Amsterdam, Zeeman Intstitute, 2004. Experimental condensed-matter physics.

Scarola, Vito W., Ph.D., Pennsylvania State University, 2002. Theoretical condensed-matter physics.

Tao, Chenggang, Ph.D., University of Maryland, 2007. Experimental condensed-matter physics.

Research Faculty

Özcan, Alpay, Ph.D., Washington University, St. Louis, 2000. Neuroscience; medical physics.

Wong, Kenneth, Ph.D., University of California, Berkeley/San Francisco, 2002. Neuroscience; medical physics.

DEPARTMENTAL RESEARCH SPECIALTIES AND STAFF

Theoretical

Condensed Matter/Statistical Physics. Theoretical investigations of a wide range of systems, both in thermal equilibrium and driven far from equilibrium, are being carried out using both analytical techniques and computational approaches. Research interests include phase transitions, critical phenomena, electronic, transport and optical properties of a variety of physical systems. Examples include universal properties and scaling behavior in magnetic systems, topological matter, structural phase transitions, boson localization, driven diffusive systems, branching and annihilating random walks, vortex transport and flux pinning in superconductors, chemical reactions, population dynamics, and percolation problems. Research is also carried out on electronic, transport and optical properties of materials, interfaces, semiconductor heterostructures, molecular devices, biological systems, and ultracold quantum gases. Analytical approaches include classical Landau-Ginzburg theory, as well as modern techniques such as coherent-state path-integrals and field theoretic renormalization group analysis are used to study problems in quantum mechanics; molecular dynamics; dynamical systems; equilibrium and non-equilibrium statistical mechanics. Computational approaches include numerical solutions of Master and Langevin equations, Monte Carlo simulations of model systems and first-principle approaches for ground state and transport problems within density functional theory. Collaborations with numerous members in other departments on campus and at many American and foreign universities and industrial laboratories exist. Cheng, Park, Pleimling, Scarola, Schmittmann, Täuber.

Particles and Fields. Analysis of high-energy particle physics phenomenology and precision tests within and beyond the standard model framework. One special focus is neutrino phenomenolgy in close collaboration with the Center for Neutrino Physics and includes internationally well-known efforts like the development of the GLoBES software package. Neutrinos are also investigated in astrophysical settings. Another special focus is on string theory and M theory, especially string compactifications, supersymmetric field theories, and mathematical aspects of string theory. Research is also carried out on QCD and other gauge theories, supersymmetric and otherwise, in three and four dimensions. Anderson, Chang, Gray, Huber, Minic, Sharpe, Takeuchi.

Experimental

Astrophysics. The group at Virginia Tech is active in extragalactic astronomy and studies of radio transients. Current extragalactic research is concerned with measuring stellar and supermassive black hole mass assembly history in galaxies from multiwavelength surveys and the observation and interpretation of mass outflow from active galactic nuclei (AGNs). This work has impact on studies of the formation of galaxies and galaxy clusters and the way these structures trace the underlying dark matter distribution. Searches for radio transients are under way in collaboration with searches for gravity wave signals (e.g., by LIGO, the Laser Interferometer Gravitational Wave Observatory). This work has impact on the study of high-energy or explosive astrophysical events (e.g., supernovae, mergers of compact objects, and the explosion of primordial black holes) and implications for work at the frontier of fundamental physics (e.g., the existence of gravitational radiation and extra-spatial dimensions). Research facilities currently used include the Hubble Space Telescope, the Herschel Space Observatory, the Spitzer Space Telescope, the Chandra X-Ray Observatory, the Very Large Telescope, the Long Wavelength Array (LWA), and the Eight-meter-wavelength Transient Array (ETA). Arav, Farrah, Simonetti.

Biophysics. Topics include biosensors using ionic self-assembled multilayers on fiber gratings; targeted delivery of functionalized nanoparticles using laser techniques in nanomedicine; voltametric chemical detection methods for sub-second measurements of neurotransmitters in the human brain during active decision-making; and experimental, analytical, and computational investigations of synthetic molecular circuits in living biological cells, particularly in nonlinear, stochastic, and far-from-equilibrium systems. Examples of the latter include pattern formation in dense bacterial colonies, cross-talk due to shared processing pathways, oscillatory gene networks, and population dynamics. Experimental approaches include near-infrared laser techniques, self-assembly techniques, optical characterization, voltametric methods, temporally resolved fluorescence microscopy, custom microfluidic devices, and molecular biology techniques. Theoretical approaches include queuing theory, stability analysis of continuum models, automated image analysis for single-cell tracking, Monte Carlo simulation of molecular networks, discrete element simulations, and partially automated model inference. Heflin, Khodaparast, Mather, Montague.

Condensed Matter Physics. Research is being carried out on semiconductors, heterostructures, oxides, polymers, self-assembled nanostructures, lithographic nanostructures, metallic nanoparticles, biological systems, new quantum states of matter, and quantum mesoscopic systems, using nonlinear optics, terahertz science, ultrafast dynamics, transport, scanning probes, and low-temperature physics techniques. Topics addressed include nonlinear optical response in self-assembled organic materials; optoelectronic applications and photovoltaics of semiconducting polymers; plasmonic enhancement of nonlinear optical and photovoltaic effects; spintronics; mesoscopic physics, spin physics, and quantum physics of metals, semimetals and semiconductors; quantum transport, low-temperature physics, and magnetic properties; quantum and spin coherence effects in the solid-state; quantum information processing architectures; nanoscience and nanofabrication techniques; energy storage and conversion; gigahertz and terahertz spectroscopy of biological systems; ultrafast dynamics of quantum systems. Experimental facilities include Raman scattering, ultrafast spectroscopies, scanning probe microscopy, infrared and visible-ultraviolet spectroscopy, variable-temperature electronic transport and magnetotransport systems, nanofabrication systems, thin film materials deposition and materials synthesis systems, and laser systems for experiments in nonlinear optics. Collaborations exist with groups in the Departments of Chemistry. Cheng, Heflin, Heremans, Khodaparast, Nguyen, Robinson, Soghomonian, Tao.

Neuroscience and Medical Physics. Topics include computational models of cognitive functions to gain insight into healthy and injured brain cognition and the characterization of cognitive phenotypes, both supported by magnetic resonance imaging; the use of medical physics to study sleep; the transitions between wake and sleep states in the brainstem; the interplay between sleep and stress on brain networks; multisource-multimodal data analysis methods, including but not limited to medical imaging and bioinformatics, with initial focus on prostate cancer and multiple sclerosis; development of new diffusion magnetic resonance imaging methods for assessment of brain white matter integrity; development of mobile health systems for military medics development of open source electronic health record architectures. Experimental efforts use functional magnetic resonance imaging, positron emission tomography, and electroencephalography. A study of interacting subjects uses new models of social exchange and uses the new technique of hyperscanning. Özcan, Montague, Mun, Wong.

Nuclear and Particle Physics. Research in experimental nuclear and particle physics is primarily carried out as a major activity of the Department's Center for Neutrino Physics. Current efforts in neutrino physics include the measurement of neutrino mixing angles with the reactor neutrino experiments Daya Bay in China and Double Chooz in France and with the accelerator neutrino experiment MicroBooNE (liquid Argon based detector) at Fermilab. Solar neutrinos are studied with Borexino at Gran Sasso in Italy and LENS at the local Kimballton Underground Research Facility (KURF). Future experiments are being planned to study sterile neutrinos and fundamental neutrino parameters (LBNE). Heavy-flavor physics (b and c quarks and tau leptons) is studied to probe CP violation and other phenomena at the Belle and Belle II experiments at KEK in Japan. Fundamental properties of the neutron are measured with the UCNA experiment at Los Alamos. Tests of the Standard Model are carried out with the parity-violating electron scattering experiments Qweak and MOLLER at the Jefferson Laboratory. The potential role of accelerators in nuclear energy production is being explored. The department has laboratory space and machine/electronic shop support for significant equipment contributions to these experiments. The Kimballton Underground Research Facility (KURF) is a low-background laboratory at a 1700-foot depth in a nearby limestone mine; it is managed by the Department and supports several active VT and external experiments. Link, Mariani, Piilonen, Pitt, Vogelaar.

UNIVERSITY OF WASHINGTON

DEPARTMENT OF ASTRONOMY

Seattle, Washington 98195-1580
http://www.astro.washington.edu

General University Information
President: Dr. Michael Young
Dean of Graduate School: Dr. David Eaton
University website: http://www.washington.edu/
Control: Public
Setting: Urban
Total Faculty: 4,300
Total number of Students: 43,000
Total number of Graduate Students: 11,500

Department Information
Department Chairman: Scott Anderson, Chair
Department Contact: Main Office, Astronomy
  Total full-time faculty: 17
  Total number of full-time equivalent positions: 12
  Full-Time Graduate Students: 30
  First-Year Graduate Students: 4
  Female First-Year Students: 2
  Total Post Doctorates: 13

Department Address
Box 351580
Seattle, WA 98195-1580
Phone: (206) 543-2888
E-mail: office@astro.washington.edu
Website: http://www.astro.washington.edu

ADMISSIONS

Admission Contact Information
Address admission inquiries to: Graduate Program Advisor, Department of Astronomy, Box 351580, Seattle, WA 98195
Phone: (206) 543-2888
E-mail: grad@astro.washington.edu
Admissions website: http://www.astro.washington.edu/grad/

Application deadlines
Fall admission:
U.S. students: December 31   *Int'l. students*: December 31

Application fee
U.S. students: $75   *Int'l. students*: $75

Admissions information
For Fall of 2013:
  Number of applicants: 130
  Number admitted: 10
  Number enrolled: 5

Admission requirements
Bachelor's degree requirements: A Bachelor's degree in astronomy, physics, mathematics, or other field related to astronomy, is required.
Minimum undergraduate GPA: 3.0

GRE requirements
The GRE is required.
  Quantitative score: 146
  Verbal score: 140
  Analytical score: 2

Advanced GRE requirements
The Advanced GRE is required.
  Minimum accepted Advanced GRE score: 440

TOEFL requirements
The TOEFL exam is required for students from non-English-speaking countries.
  PBT score: 580
  iBT score: 92

Other admissions information
Undergraduate preparation assumed: Undergraduate preparation assumed allows for a range of backgrounds of incoming graduate students. However, the equivalent of an undergraduate physics program is typical.

TUITION

Tuition year 2012–13:
Tuition for in-state residents
  Full-time students: $14,337 annual
Tuition for out-of-state residents
  Full-time students: $26,768 annual
For students with appointments, tuition/fees are reduced to about $1,200 annually. Students with appointments that meet minimum qualifications receive a non-resident tuition waiver in addition to in-state tuition paid by grant/state funding. Students are still responsible for a portion of tuition/fees that cannot be covered.
Credit hours per semester to be considered full-time: 10
Deferred tuition plan: No
Health insurance: Available.
Academic term: Quarter
Number of first-year students who received full tuition waivers: 5

Teaching Assistants, Research Assistants, and Fellowships
Number of first-year
  Teaching Assistants: 5
Average stipend per academic year
  Teaching Assistant: $22,000
  Research Assistant: $23,700
  Fellowship student: $30,000

FINANCIAL AID

Loans
Loans are available for U.S. students.
Loans are not available for international students.
GAPSFAS application required: No
FAFSA application required: No

For further information
Address financial aid inquiries to: Office of Student Financial Aid.
Phone: (206) 543-6101
E-mail: osfa@u.washington.edu
Financial aid website: http://www.washington.edu/students/osfa

HOUSING

Availability of on-campus housing
Single students: Yes
Married students: Yes

For further information
Address housing inquiries to: Housing Office.
Phone: (206) 543-4059
Housing aid website: http://hfs.washington.edu

Table A—Faculty, Enrollments, and Degrees Granted

Research Specialty	2012-13 Faculty	Enrollment Fall 2012		Number of Degrees Granted 2012-13 (2005-13)		
		Master's	Doctorate	Master's	Terminal Master's	Doctorate
Astronomy	17	–	30	5(46)	–	4(37)
Total	17	–	30	5(46)	–	4(37)
Full-time Grad. Stud.	–	–	30	–	–	–
First-year Grad. Stud.	–	–	5	–	–	–

GRADUATE DEGREE REQUIREMENTS

Master's: With Thesis: 36 approved credits, of which 18 are in astronomy courses at the 500 level or above, and nine are thesis research. Without Thesis: 36 approved credits, of which 18 are in astronomy courses at the 500 level or above.

Doctorate: Admission Requirements: Entering students are expected to have a strong background in physics and mathematics. Graduation Requirements: Passage of the departmental qualifying examinations. Master's degree in astronomy or equivalent knowledge; at least three quarters of teaching experience in astronomy; dissertation and final examination. Students interested in work in theoretical astrophysics may take additional courses in physics and mathematics. Students working on other topics may take certain courses in related fields, such as astrobiology, astronautics, atmospheric sciences, geophysics, or computer science.

Thesis: Thesis may be written in absentia.

SPECIAL EQUIPMENT, FACILITIES, OR PROGRAMS

The Department owns, in consortium with several other universities, a 3.5-meter telescope at Apache Point, NM, and receives 25% of the observing time on this facility. It is operated largely remotely over the internet and used heavily for graduate student dissertation research. UW is also a participant in the Sloan Digital Sky Survey, a project making a digital photometric and spectroscopic map of 25% of the celestial sphere, using a special purpose 2.5-meter telescope also on Apache Point. The Department is a founding member of the future Large Synoptic Survey Telescope. The Department also operates a 0.8-meter telescope in the Cascade Mountains of Washington, for the use of its students. Additional facilities in Seattle include an electron microscopy laboratory for analysis of cosmic dust particles and laboratories for developing astronomical telescopes and instrumentation. Members of the faculty are on teams that supplied instrumentation for the Hubble Space Telescope. Faculty and students are also extensive users of other national ground- and space-based observatories at a variety of wavelengths, and of national supercomputing facilities. The Department operates a large network of Linux workstations in support of all of these efforts.

FACULTY

Professor

Anderson, Scott, Ph.D., University of Washington, 1985. Chair. *Astronomy*. Quasars; compact binaries; high-energy phenomena.

Balick, Bruce B., Ph.D., Cornell University, 1971. *Astronomy*. Planetary nebulae and late stages of stellar evolution; gas dynamics; active nuclei and their impact on galactic structure.

Brownlee, Donald E., Ph.D., University of Washington, 1971. *Astronomy*. Interplanetary dust; comet physics; meteoritics; origin of the solar system.

Connolly, Andrew, Ph.D., University of London, 1993. *Astronomy*. Formation and evolution of galaxies; cosmology; astronomical surveys.

Dalcanton, Julianne, Ph.D., Princeton University, 1995. *Astronomy*. Galaxy evolution and formation; cosmology; galactic dynamics.

Hawley, Suzanne, Ph.D., University of Texas, 1989. Associate Chair. *Astronomy*. Low mass stars; variable stars; star clusters; dwarf galaxies; galactic structure.

Ivezic, Zeljko, Ph.D., University of Kentucky, 1995. *Astronomy*. Deep sky surveys; quasars; stellar populations; asteroids; origin of interstellar dust.

Quinn, Tom, Ph.D., Princeton University, 1986. *Astronomy, Astrophysics*. Astrophysical dynamics on a wide range of scales, from asteroids to clusters of galaxies; solar system studies.

Sullivan, Woodruff T., Ph.D., University of Maryland, 1971. *Astronomy, History & Philosophy of Physics/Science*. Astrobiology; galaxies; clusters of galaxies; distance scale; history of radio astronomy.

Szkody, Paula, Ph.D., University of Washington, 1975. *Astronomy*. Cataclysmic variables; white dwarfs.

Associate Professor

Agol, Eric, Ph.D., University of California, Santa Barbara, 1997. *Astronomy, Astrophysics*. Relativistic astrophysics and gravity; black holes; active galaxies; accretion disks; extrasolar planets.

Meadows, Victoria, Ph.D., University of Sydney, 1994. *Astronomy*. Planetary atmospheres; astrobiology.

Professor Emeritus

Böhm, Karl Heinz, Ph.D., University of Kiel, 1954. *Astronomy, Astrophysics*. Stellar atmospheres and interiors; nebulae, circumstellar matter.

Böhm-Vitense, Erika, Ph.D., Christian Albrechts University, 1951. *Astronomy, Astrophysics*. Stellar atmospheres; convection; stellar chromospheres and coronae; magnetic stars; cepheids.

Hodge, Paul, Ph.D., Harvard University, 1960. *Astronomy*. Galaxies; the Magellanic clouds.

King, Ivan, Ph.D., Harvard University, 1952. *Astronomy, Astrophysics*. Stellar populations; star clusters; structure & dynamics of globular clusters.

Lutz, Julie, Ph.D., University of Illinois, 1972. *Astronomy*. Planetary nebulae and symbiotic stars; astronomy education.

Wallerstein, George, Ph.D., California Institute of Technology, 1958. *Astronomy*. Spectra of variable stars; chemical composition of stellar atmospheres; interstellar lines.

Research Professor

Governato, Fabio, Ph.D., University of Rome, 1995. *Astronomy, Astrophysics*. Galaxy and clusters; cosmic structure formation; planet formation.

Research Associate Professor

Becker, Andrew, Ph.D., University of Washington, 2000. *Astronomy*. Time domain science; techniques of massive survey astronomy; data mining.

Adjunct Professor

Burnett, Toby, Ph.D., University of California, San Diego, 1968. *Astrophysics, Particles and Fields*. High-energy astrophysics; gamma-ray sources.

Morales, M iguel, Ph.D., University of California, Santa Cruz, 2002. *Astronomy, Astrophysics, Electromagnetism*. Epoch of reionization; cosmology; radio astronomy surveys and instrumentation; radio transients.

Rosenberg, Leslie, Ph.D., Stanford University, 1985. *Astrophysics, Nuclear Physics, Particles and Fields*. Searches for axonic dark matter; surveys of dark matter and energy in the universe; novel particle and nuclear instrumentation; ultra low noise electromagnetic amplification.

Affiliate Professor

DeBattista, Victor, Ph.D., Rutgers University, 1998. *Astronomy, Astrophysics*. Cosmology; computational astrophysics.

Hughes Clark, Joanne, Ph.D., University of London, 1989. *Astronomy*. Observational astronomy; astrophysics of dwarf galaxies and globular clusters.

Linnell, Albert, Ph.D., Harvard University, 1950. *Astronomy, Astrophysics*. Modeling of accretion disks; cataclysmic variables.

Murphy, Thomas, Ph.D., California Institute of Technology, 2000. *Astrophysics, Relativity & Gravitation*. Solar-system tests of general relativity; energy and the environment.

Affiliate Assistant Professor

McQuinn, Matthew, Ph.D., Harvard, 2009. *Astrophysics*. Theoretical modeling of the intergalactic medium, galaxy formation and cosmology.

Senior Lecturer

Larson, Ana, Ph.D., University of British Columbia, 1996. *Astronomy*. Modeling stellar atmospheres; precise radial velocities of stars.

Smith, Toby, Ph.D., University of Washington, 1995. *Astronomy*. Terrestrial impact craters; meteoritics.

Lecturer

Fraser, Oliver, Ph.D., University of Washington, 2008. *Astronomy*. Variable stars; teaching methods.

Laws, Christopher, Ph.D., University of Washington, 2004. *Astronomy*. Extrasolar planetary systems; stellar evolution; chemical evolution.

Silvestri, Nicole, Ph.D., Florida Institue of Technology, 2002. *Astronomy*. Low mass stars, white dwarfs.

DEPARTMENTAL RESEARCH SPECIALTIES AND STAFF

Theoretical

Astrophysics. *Active Galaxies and Quasars: ejecta; lensing; accretion. *Astrobiology and Planetary Studies: formation and evolution of planetary systems; extrasolar planets; planetary atmospheres and surfaces. *Clusters of Stars: formation; structure; evolution. *Compact Objects: degenerate stars; black holes. *Computational Astrophysics: N-body simulations. *Cosmology: large-scale structure formation and evolution. *Galactic Nebulae: H II regions and planetary nebulae; Herbig-Haro objects. *Galaxies: structure and dynamics; formation; dark matter. *Stars: magnetic activity; magnetic activity, chromospheres, coronae, flares; convection; equation of state. Agol, Burnett, DeBattista, Governato, McQuinn, Morales, Quinn, Rosenberg.

Experimental

Astronomy. *Active Galaxies and Quasars: nuclear properties; luminosity functions and evolution; absorption lines; BL Lacs and other radio sources; high-energy phenomena. *Astrobiology and Planetary Studies: extrasolar planet detection and characterization; planetary atmospheres; asteroids and comets; meteorites. *Clusters of Stars: evolution; abundance determinations; statistical properties. *Compact Objects: degenerate stars; black holes; cataclysmic variables and other compact binaries. *Cosmology: intergalactic medium; large-scale structure formation and evolution; cosmological parameters; dark matter and dark energy. *Galactic Nebulae: supernovae; hot and cool components in the interstellar medium: H II regions and planetary nebulae. *Galaxies: structure and dynamics; formation and evolution; dark matter; gaseous and stellar content; internal motions; extragalactic distance scale; clusters of galaxies; properties of star-forming regions. *History of Astronomy. *Stars: chemical composition; magnetic activity and flares; circumstellar material; variable stars; low mass stars and brown dwarfs. *Survey Science: data mining; imaging, spectroscopic, and time-domain astronomical surveys. Anderson, Böhm, Böhm-Vitense, Balick, Becker, Brownlee, Connolly, Dalcanton, Fraser, Hawley, Hodge, Hughes Clark, Ivezic, King, Larson, Laws, Linnell, Lutz, Meadows, Silvestri, Smith, Sullivan, Szkody, Wallerstein.

View additional information about this department at
www.gradschoolshopper.com

UNIVERSITY OF WASHINGTON

DEPARTMENT OF PHYSICS

Seattle, Washington 98195
http://www.phys.washington.edu

General University Information
President: Michael Young
Dean of Graduate School: David Eaton
University website: http://www.washington.edu/
Control: Public
Setting: Urban
Total Faculty: 4,300
Total number of Students: 50,745
Total number of Graduate Students: 12,211

Department Information
Department Chairman: Blayne Heckel, Chair
Department Contact: Front desk, Receptionist
  Total full-time faculty: 57
  Total number of full-time equivalent positions: 43
  Full-Time Graduate Students: 140
  First-Year Graduate Students: 27
  Female First-Year Students: 4
  Total Post Doctorates: 30

Department Address
3910 15th Ave. NE
Seattle, WA 98195
Phone: (206) 543-2771
Fax: (206) 685-0635
E-mail: reception@phys.washington.edu
Website: http://www.phys.washington.edu

ADMISSIONS

Admission Contact Information
Address admission inquiries to: Graduate Program Assistant, Department of Physics, University of Washington Box 351560, Seattle, WA 98195-1560
Phone: (206) 543-2488
E-mail: grad@phys.washington.edu
Admissions website: http://www.phys.washington.edu/phd_admissions.htm

Application deadlines
Fall admission:
U.S. students: January 5   *Int'l. students*: January 5

Application fee
U.S. students: $75   *Int'l. students*: $75

Admissions information
For Fall of 2013:
  Number of applicants: 511
  Number admitted: 100
  Number enrolled: 27

Admission requirements
Bachelor's degree requirements: Bachelor's degree in physics is required.
Minimum undergraduate GPA: 3.0

GRE requirements
The GRE is required.

Advanced GRE requirements
The Advanced GRE is required.

TOEFL requirements
The TOEFL exam is required for students from non-English-speaking countries.
  PBT score: 580
  iBT score: 92
The following minimum English language proficiency test scores will exempt students from Academic English Program (AEP) requirements: 92 on the TOEFL iBT An applicant may be admitted with an English proficiency test score within the range of scores listed below. In such cases, the applicant will be required, when starting the graduate degree program, to take designated Academic English Program (AEP) courses through UW English Language Programs, unless an exempting test score is subsequently submitted. 61-91 TOEFL iBT.

Other admissions information
Additional requirements: There are no minimum GRE scores requirement but considerable weight is given to these scores. The average GRE advanced score of the first-year class is typically around 820.
Undergraduate preparation assumed: Undergraduate courses in quantum mechanics, electricity and magnetism, classical mechanics, statistical mechanics, and mathematical physics and one senior-level survey course.

TUITION

Tuition year 2012–13:
Tuition for in-state residents
  Full-time students: $14,337 annual
Tuition for out-of-state residents
  Full-time students: $26,748 annual
This tuition rate is for the academic year (9 months). Students holding assistantships of half-time or more receive a tuition waiver.
Credit hours per semester to be considered full-time: 10
Deferred tuition plan: No
Health insurance: Available
Other academic fees: $250/quarter student fees.
Academic term: Quarter
Number of first-year students who received full tuition waivers: 27

Teaching Assistants, Research Assistants, and Fellowships
Number of first-year
  Teaching Assistants: 22
  Research Assistants: 5
  Fellowship students: 4
Average stipend per academic year
  Teaching Assistant: $15,489
  Research Assistant: $17,271
  Fellowship student: $10,000

FINANCIAL AID

Application deadlines
Fall admission:
U.S. students: February 15

Loans
Loans are available for U.S. students.
Loans are available for international students.
GAPSFAS application required: No
FAFSA application required: No

For further information
Address financial aid inquiries to: Office of Student Financial Aid, University of Washington, Box 355880, Seattle, WA 98195-5880.
Phone: (206) 543-6101
E-mail: osfa@u.washington.edu
Financial aid website: http://www.washington.edu/students/osfa

HOUSING

Availability of on-campus housing
Single students: Yes
Married students: Yes

For further information
Address housing inquiries to: Student Services Office, Housing and Food Services, University of Washington, 301 Schmitz Hall, Box 355842, Seattle, WA 98195–5842.
Phone: (206) 543-4059
E-mail: hfsinfo@u.washington.edu
Housing aid website: http://hfs.washington.edu

Table A—Faculty, Enrollments, and Degrees Granted

Research Specialty	2012-2013 Faculty	Enrollment Fall 2013 Master's	Enrollment Fall 2013 Doctorate	Number of Degrees Granted 2012-13 Master's	Number of Degrees Granted 2012-13 Terminal Master's	Number of Degrees Granted 2012-13 Doctorate
Applied Physics	–	42	–	11(-)	11(-)	2(-)
Astrophysics	6	–	–	–	–	2(-)
Atomic, Molecular, & Optical Physics	6	–	–	–	–	1(-)
Biophysics	10	–	–	–	–	1(-)
Condensed Matter Physics	21	–	–	–	–	1(-)
High Energy Physics	8	–	–	–	–	2(-)
Nuclear Physics	22	–	–	–	–	5(-)
Particles and Fields	5	–	–	–	–	2(-)
Physics and other Science Education	3	–	–	–	–	–
Non-specialized	–	–	–	–	–	–
Total	78	49	138	11(-)	11(-)	16(-)
Full-time Grad. Stud.	–	–	138	–	–	–
First-year Grad. Stud.	–	19	27	–	–	–

GRADUATE DEGREE REQUIREMENTS

Master's: Minimum of 36 approved credits are required, 18 of which must in courses numbered 500 or above, including a minimum of three credits in Physics 600 research. At least 18 credits must be in graded courses. No thesis is required. No foreign language is required. Students must pass qualifying examination (Ph.D. program). Must submit project report and pass a final oral examination (evening master's degree program). A minimum of three full-time quarters of residency are required. Part-time quarters may be accumulated to meet this requirement. A grade point average of 3.0 is required.

Doctorate: Grade point average above 3.0 is required. A sequence of required courses must be taken. Qualifying examination, general examination, and a final examination, which is usually a defense of the dissertation, are required. 18 graded credits at the University of Washington are required. A minimum of three academic years of resident study are required. A minimum of 27 credits of dissertation over period of at least three quarters and some teaching experience are required. There is no language examination required. Students must be registered for the quarter that they receive their degree.

Thesis: Thesis may be written in absentia.

SPECIAL EQUIPMENT, FACILITIES, OR PROGRAMS

At our Center for Experimental Nuclear Physics and Astrophysics, an FN tandem Van de Graaff accelerator can be used with negative-ion injection to reach energies of 18 MeV for protons and higher energies for heavier ions. For nuclear astrophysics experiments with a terminal ion source, the proton (or helium ion) beam energy can be as low as 100 keV, with currents up to 30 microamps. The Center includes active research programs in neutrino physics and muon physics.

Our High-Energy Laboratory maintains facilities for the preparation and analysis of experiments performed at off-campus accelerators. We have a 3He dilution refrigerator for extremely low-temperature research, a laser facility for generating tunable optical radiation at precisely controlled frequencies, and an extended x-ray absorption fine structure (XAFS) facility for determining the atomic structure of condensed matter.

Facilities for research into nanostructure include two atomic force microscopes, one at room temperature and the other at low temperature in extra-high vacuum, and a 14-Tesla superconducting magnet.

The Department also houses the Institute for Nuclear Theory (INT), a national facility funded by the Department of Energy to host visitor programs for the exploration of current topics in nuclear theory. The INT is closely integrated with the Physics Department both physically and intellectually, with INT senior fellows supervising thesis research in physics and INT seminars frequently attended by members of the Physics Department.

Computing facilities include numerous modern workstations and server machines plus access to both University mainframes and national supercomputing centers. Since 1994, the Department has been located in a recently constructed building with state-of-the-art facilities for instruction and research.

Table B—Separately Budgeted Research Expenditures by Source of Support

Source of Support	Departmental Research	Physics-related Research Outside Department
Federal government	$15,514,576	
State/local government	$1,023,505	
Non-profit organizations		
Business and industry		
Other	$158,350	
Total	$16,696,431	

Table C—Separately Budgeted Research Expenditures by Research Specialty

Research Specialty	No. of Grants	Expenditures ($)
Atomic, Molecular, & Optical Physics	8	$1,594,480
Condensed Matter Physics	20	$2,190,758
Nuclear Physics	21	$6,406,752
Particles and Fields	11	$2,753,363
Physics and other Science Education	2	$724,724
Relativity & Gravitation	7	$2,002,849
Total	69	$15,672,926

FACULTY

Professor

Andreev, Anton, Ph.D., Massachusetts Institute of Technology, 1996. *Condensed Matter Physics*. Theoretical condensed-matter physics.

Bertsch, George F., Ph.D., Princeton University, 1965. *Nuclear Physics*. Theoretical physics; quantum many-body physics with application to nuclei.

Bulgac, Aurel, Ph.D., Leningrad Nuclear Physics Institute, 1977. *Nuclear Physics*. Theoretical nuclear physics.

Chaloupka, Vladimir, Ph.D., University of Geneva, 1975. *Particles and Fields*. Experimental elementary particle physics.

Cobden, David H., Ph.D., University of Cambridge, 1991. *Condensed Matter Physics*. Experimental condensed-matter physics; nanodevice physics.

den Nijs, Marcel, Ph.D., Radboud University Nijmegen, 1979. *Condensed Matter Physics, Statistical & Thermal Physics*. Theoretical condensed-matter physics; statistical physics; neural science.

Ellis, Stephen D., Ph.D., California Institute of Technology, 1971. *Particles and Fields*. Theoretical elementary particle physics.

Garcia, Alejandro, Ph.D., University of Washington, 1991. *Nuclear Physics*. Experimental nuclear physics.

Goussiou, Anna, Ph.D., University of Wisconsin-Madison, 1995. *Particles and Fields*. Experimental particle physics.

Gundlach, Jens, Ph.D., University of Washington, 1990. *Biophysics, Nuclear Physics*. Experimental nuclear physics and biophysics.

Heckel, Blayne, Ph.D., Harvard University, 1981. Chairman Physics Department. *Atomic, Molecular, & Optical Physics*. Experimental atomic physics.

Heron, Paula, Ph.D., University of Western Ontario, 1995. *Physics and other Science Education*. Physics education.

Hertzog, David W., Ph.D., William & Mary, 1983. *Astrophysics, Nuclear Physics*. Experimental nuclear physics; astrophysics.

Kaplan, David B., Ph.D., Harvard University, 1985. Director INT. *Nuclear Physics*. Theoretical nuclear physics.

Karch, Andreas, Ph.D., Humbold University, 1998. *Particles and Fields*. Theoretical particle physics.

Lubatti, Henry J., University of California, Berkeley, 1966. *Particles and Fields*. Experimental elementary particle physics.

McDermott, Lillian C., Ph.D., Columbia University, 1959. *Physics and other Science Education*. Physics education.

Miller, Gerald A., Ph.D., Massachusetts Institute of Technology, 1972. *Nuclear Physics*. Theoretical nuclear physics.

Nelson, Ann E., Ph.D., Harvard University, 1984. *Particles and Fields*. Theoretical particle physics.

Olmstead, Marjorie A., Ph.D., University of California, Berkeley, 1985. *Condensed Matter Physics, Nano Science and Technology*. Experimental condensed-matter physics; nanotechnology.

Reddy, Sanjay, Ph.D., Massachusetts Institute of Technology, 1998. Theoretical nuclear physics; astrophysics.

Rehr, John J., Ph.D., Cornell University, 1972. *Condensed Matter Physics*. Theoretical condensed-matter physics.

Robertson, R. G. Hamish, Ph.D., McMaster University, 1971. *Nuclear Physics*. Experimental nuclear physics; neutrino physics.

Rosenberg, Leslie, Ph.D., Stanford University, 1985. *Astrophysics*. Experimental astrophysics.

Savage, Martin J., Ph.D., California Institute of Technology, 1990. *Computational Physics, Nuclear Physics*. Theoretical nuclear physics.

Schick, Michael, Ph.D., Stanford University, 1967. *Biophysics, Condensed Matter Physics, Statistical & Thermal Physics*. Theoretical statistical physics and biophysics.

Seidler, Gerald T., Ph.D., University of Chicago, 1993. *Condensed Matter Physics*. Experimental condensed-matter physics.

Shaffer, Peter S., Ph.D., University of Washington, 1993. *Physics and other Science Education*. Physics education.

Sharpe, Stephen R., Ph.D., University of California, Berkeley, 1983. *Particles and Fields*. Theoretical elementary particle physics.

Sorensen, Larry B., Ph.D., University of Illinois, 1979. *Condensed Matter Physics*. Experimental condensed-matter physics; neural science.

Spivak, Boris, Ph.D., Leningrad Politecknical Institute, 1978. *Condensed Matter Physics*. Theoretical condensed-matter physics.

Watts, Gordon T., Ph.D., University of Rochester, 1994. *Particles and Fields*. Experimental elementary particle physics.

Wilkes, Richard J., Ph.D., University of Wisconsin-Madison, 1974. *High Energy Physics*. Experimental high-energy physics; space science.

Yaffe, Laurence G., Ph.D., Princeton University, 1980. *Particles and Fields*. Theoretical elementary particle physics.

Associate Professor

Blinov, Boris B., Ph.D., University of Michigan, 2000. *Atomic, Molecular, & Optical Physics*. Experimental atomic physics; quantum information.

Assistant Professor

Beane, Silas, Ph.D., University of Texas, Austin, 1994. *Computational Physics, Nuclear Physics*.

Detwiler, Jason, Ph.D., Stanford University, 2005. Neutrino physics.

Fu, Kai-Mei C., Ph.D., Stanford University, 2007. *Condensed Matter Physics*. Condensed matter experiment; single-impurity optoelectronics.

Gupta, Subhadeep, Ph.D., Massachusetts Institute of Technology, 2003. *Atomic, Molecular, & Optical Physics*. Experimental atomic physics.

Hsu, Shih-Chieh, Ph.D., University of California, San Diego, 2008. High-energy physics.

Laumann, Christopher, Ph.D., Princeton, 2009. *Condensed Matter Physics*. Theoretical Condensed Matter Physics.

Morales, Miguel F., Ph.D., University of California, Santa Cruz, 1992. *Astrophysics, Cosmology & String Theory*. Experimental astrophysics; cosmology.

Tolich, Nikolai, Ph.D., Stanford University, 2005. *Nuclear Physics*. Experimental nuclear physics.

Wiggins, Paul A., Ph.D., California Institute of Technology, 2005. *Biophysics*. Biophysics.

Xu, Xiaodong, Ph.D., University of Michigan, 2008. *Condensed Matter Physics*. Experimental condensed-matter physics; nanoscale optoelectronics.

Emeritus

Adelberger, Eric G., Ph.D., California Institute of Technology, 1967. *Nuclear Physics, Relativity & Gravitation*. Experimental nuclear physics; gravitation.

Baker, Marshall, Ph.D., Harvard University, 1958. *Particles and Fields*. Theoretical elementary particle physics.

Bardeen, James M., Ph.D., California Institute of Technology, 1965. *Astrophysics, Relativity & Gravitation*. Theoretical astrophysics and relativity.

Boulware, David G., Ph.D., Harvard University, 1962. *Astrophysics, Particles and Fields, Relativity & Gravitation*. Theoretical physics; astrophysics; relativity; elementary particles.

Boynton, Paul E., Ph.D., Princeton University, 1967. *Astronomy, Astrophysics, Relativity & Gravitation*. Astronomy; astrophysics; gravitation.

Brown, Frederick C., Ph.D., Harvard University, 1950. *Condensed Matter Physics*. Experimental condensed-matter physics.

Brown, Lowell S., Ph.D., Harvard University, 1961. *Particles and Fields*. Theoretical elementary particle physics.

Burnett, Thompson H., Ph.D., University of California, San Diego, 1968. *Particles and Fields*. Experimental elementary particle physics.

Cook, Victor, Ph.D., University of California, Berkeley, 1962. *Particles and Fields*. Experimental elementary particle physics.

Cramer, John G., Ph.D., Rice University, 1961. *Nuclear Physics*. Experimental nuclear physics.

Dehmelt, Hans G., Ph.D., Goettingen University, 1950. *Atomic, Molecular, & Optical Physics*. Experimental atomic physics.

Forston, E. Norval, Ph.D., Harvard University, 1963. *Atomic, Molecular, & Optical Physics*. Experimental atomic physics.

Halpern, Isaac, Ph.D., Massachusetts Institute of Technology, 1948.

Haxton, Wick C., Ph.D., Stanford University, 1976. *Nuclear Physics, Particles and Fields*. Theoretical nuclear physics.

Henley, Ernest M., Ph.D., University of California, Berkeley, 1952. *Nuclear Physics*. Theoretical nuclear physics.

Ingalls, Robert L., Ph.D., Carnegie Mellon University, 1962. *Condensed Matter Physics*. Experimental condensed-matter physics.

Mockett, Paul M., Ph.D., Massachusetts Institute of Technology, 1965. *High Energy Physics, Particles and Fields*. Experimental high-energy physics.

Puff, Robert D., Ph.D., Harvard University, 1960. *Nuclear Physics*. Theoretical nuclear and statistical physics.

Rothberg, Joseph E., Ph.D., Columbia University, 1963. *Particles and Fields*. Experimental elementary particle physics.

Snover, Kurt A., Ph.D., Stanford University, 1969. *Nuclear Physics*. Experimental nuclear physics.

Stern, Edward A., Ph.D., California Institute of Technology, 1955. *Condensed Matter Physics*. Experimental condensed-matter physics.

Storm, Derek W., Ph.D., University of Washington, 1970. *Nuclear Physics*. Experimental nuclear physics.

Thouless, David J., Ph.D., Cornell University, 1958. *Condensed Matter Physics*. Theoretical condensed-matter physics.

Van Dyck, Robert S., Ph.D., University of California, Berkeley, 1971. *Atomic, Molecular, & Optical Physics*. Experimental atomic physics.

Vilches, Oscar E., Ph.D., Universidad Nacional de Cuyo, Argentina, 1966. *Condensed Matter Physics, Low Temperature Physics*. Experimental low-temperature physics.

Weitkamp, William G., Ph.D., University of Wisconsin-Madison, 1965. *Nuclear Physics*. Experimental nuclear physics.

Williams, Robert W., Ph.D., Massachusetts Institute of Technology, 1948. *High Energy Physics*.

Research Professor

Doe, Peter J., Ph.D., Durham University, 1977. *Nuclear Physics*. Experimental nuclear physics.

Kammel, Peter, Ph.D., University of Vienna, 1982. *Nuclear Physics*. Nuclear physics.

Trainor, Thomas A., Ph.D., University of North Carolina, 1973. *Nuclear Physics*. Experimental nuclear physics.

Research Associate Professor

Zhao, Tianchi, Ph.D., Columbia University, 1986. *Particles and Fields*. Experimental elementary particle physics.

Research Assistant Professor

Enomoto, Sanshiro, Ph.D., Tohoku University, 2005. *Nuclear Physics*.

Forbes, Michael, Ph.D., Massachusetts Institute of Technology, 2005. Nuclear theory.

Gray, Rybka, Ph.D., Massachusetts Institute of Technology, 2007. Experimental astrophysics.

Lin, Huey-Wen, Ph.D., Columbia University, 2006. *Nuclear Physics*. Nuclear theory.

Schlamminger, Steven, Ph.D., University of Zurich, 2002. *Relativity & Gravitation*. Gravitational physics.

Steiner, Andrew, Ph.D., Stony Brook University, 2002. *Nuclear Physics*.

Adjunct Professor

Agol, Eric, Ph.D., University of California, Santa Barbara, 1997. *Astrophysics*. Theoretical astrophysics.

Baker, David, Ph.D., University of California, Berkeley, 1989. *Biophysics*. Computational biological physics; protein folding.

Buck, Warren W., Ph.D., College of William and Mary, 1976. *Nuclear Physics, Particles and Fields*. Theoretical nuclear and elementary particle physics.

Campbell, Charles, Ph.D., University of Texas, Austin, 1979. *Chemical Physics*. Chemical physics.

Dalcanton, Julianne, Ph.D., Princeton University, 1995. *Astrophysics*. Astrophysics.

Drobny, Gary, Ph.D., University of California, Berkeley, 1981. Biophysical chemistry.

Dunham, Scott T., Ph.D., Stanford University, 1985. Modeling and simulation of microfabrication processes and device behavior.

Fairhall, Adrienne, Ph.D., Weizmann Institute of Science, 1998. *Biophysics*. Theoretical neuroscience.

Fine, Arthur I., Ph.D., University of Chicago, 1963. *Other*. Foundations of physics.

Ginger, David S., Ph.D., University of Cambridge, 2001. *Chemical Physics, Condensed Matter Physics*. Experimental nanophysics.

Hawley, Suzanne L., Ph.D., University of Texas, Austin, 1989. *Astrophysics*. Theoretical astrophysics.

Hochberg, Michael, Ph.D., California Institute of Technology, 2006. *Condensed Matter Physics, Nano Science and Technology*. Condensed-matter physics; nanophotonics.

Holzworth, Robert, Ph.D., University of California, Berkeley, 1977. *Geophysics*. Geophysics.

Jarboe, Thomas R., Ph.D., University of California, Berkeley, 1974. *Plasma and Fusion*. Plasma physics.

Keller, Sarah L., Ph.D., Princeton University, 1995. *Biophysics*. Biophysics.

Krishnan, Kannon M., Ph.D., University of California, Berkeley, 1984. *Condensed Matter Physics, Materials Science, Metallurgy*. Material science.

Kutz, Jose N., Ph.D., Northwestern University, 1990. *Applied Mathematics*. Dynamical systems; nonlinear differential equations; bifurcation theory.

Lin, Lih, Ph.D., University of California, Los Angeles, 1996. *Condensed Matter Physics, Engineering Physics/Science, Nano Science and Technology*. Nanoscale photonic devices.

Ohuchi, Fumio, Ph.D., University of Florida, 1981. *Materials Science, Metallurgy, Nano Science and Technology*. Materials and surface science.

Quinn, Thomas R., Ph.D., Princeton University, 1986. *Astrophysics*. Astrophysics.

Reinhardt, William P., Ph.D., Harvard University, 1968. *Chemical Physics*. Chemistry; Bose-Einstein condensates.

Rieke, Frederick M., Ph.D., University of California, Berkeley, 1991. *Biophysics*. Neural science; vision sensory signal processing and computation.

Thompson, LuAnne, Ph.D., Massachusetts Institute of Technology, 1990. *Marine Science/Oceanography*. Oceanography.

Winglee, Robert, Ph.D., University of Sydney, 1985. *Geophysics*. Geophysics.

Affiliate Professor

Alberg, Mary A., Ph.D., University of Washington, 1974. *Nuclear Physics*. Theoretical nuclear physics.

Bacon, Dave, Ph.D., University of California, Berkeley, 2001. *Computer Science, Other*. Quantum computation.

Balantekin, A. Baha, Ph.D., Yale University, 1982. *Nuclear Physics*. Theoretical nuclear physics.

Barrett, Bruce R., Ph.D., Stanford University, 1967. *Nuclear Physics*. Nuclear many-body theory.

Bichsel, Hans, Ph.D., University of Basel, 1951. *Nuclear Physics*. Experimental nuclear physics.

Bowles, Thomas J., Ph.D., Princeton University, 1978. *Nuclear Physics*. Experimental nuclear physics.

Cahn, John, Ph.D., University of California, Berkeley, 1953. *Condensed Matter Physics*. Theoretical condensed-matter physics.

Cleveland, Bruce T., Ph.D., Johns Hopkins University, 1970. Experimental neutrino physics.

Elliot, Steven R., Ph.D., University of California, Irvine, 1987. *Nuclear Physics*. Experimental nuclear physics.

Friedman, William A., Ph.D., Massachusetts Institute of Technology, 1966. *Nuclear Physics*. Nuclear physics.

Gardner, Jeffrey P., Ph.D., University of Washington, 2000. *Computational Physics, Computer Science*.

Habig, Alec T., Ph.D., Indiana University, 1996. *Astrophysics, Nuclear Physics*.

Hoyle, Charles D., Ph.D., University of Washington, 2001. *Relativity & Gravitation*. Laboratory tests of gravity.

Magierski, Piotr A., Ph.D., Warsaw University of Technology, 1995. *Nuclear Physics*. Theoretical nuclear physics.

Mandula, Jeffrey E., Ph.D., Harvard University, 1966. *Particles and Fields*. Theoretical particle physics.

Nordtvedt, Kenneth, Ph.D., Stanford University, 1964. *Relativity & Gravitation*. General relativity.

Raab, Frederick J., Ph.D., Stony Brook University, 1980. *Relativity & Gravitation*. Experimental gravitational physics.

Raschke, Markus B., Ph.D., Max-Planck Institute for Quantum Optics and Technology University, Munich, 1999. *Nano Science and Technology*. Chemistry and nanoparticles.

Riedel, Eberhard K., Ph.D., München, 1966. Condensed Matter Physics, Statistical & Thermal Physics Theoretical condensed-matter and statistical physics.

Strassler, Matthew, Ph.D., Stanford University, 1993. *Particles and Fields*. Quantum field theory; string theory; particle physics.

Stubbs, Christopher W., Ph.D., University of Washington, 1988. *Cosmology & String Theory*. Experimental and observational cosmology.

Van Bibber, Karl, Ph.D., Massachusetts Institute of Technology, 1976. *High Energy Physics*. High-energy physics.

van Kolck, Ubirijara, Ph.D., University of Texas, Austin, 1993. *Nuclear Physics*. Theoretical nuclear physics.

Wettlaufer, John S., Ph.D., University of Washington, 1991. Ice physics.

Wilkerson, John F., Ph.D., University of North Carolina, 1982. *Nuclear Physics*. Experimental neutrino physics.

Affiliate Assistant Professor

Miller, Michael L., Ph.D., Yale University, 2004. *Nuclear Physics, Particles and Fields*. Experimental nuclear and particle physics.

DEPARTMENTAL RESEARCH SPECIALTIES AND STAFF

Theoretical

Astrophysics. Agol, Bardeen, Quinn.

Biophysics. David Baker, den Nijs, Fairhall, Schick.

Condensed Matter Physics. Andreev, Cahn, den Nijs, Dunham, Rehr, Reinhardt, Schick, Spivak, Thouless.

Nuclear Physics. Bertsch, Bulgac, Haxton, Henley, Kaplan, Huey-Wen Lin, Gerald Miller, Savage, Steiner.

Particles and Fields. Marshall Baker, Ellis, Nelson, Sharpe, Thompson, Yaffe.

Experimental

Astrophysics. Adelberger, Boynton, Gundlach, Heckel, Morales, Rosenberg, Wilkes.

Atomic, Molecular, & Optical Physics. Blinov, Dehmelt, Forston, Gupta, Heckel, Kutz, Van Dyck.

Biophysics. Gundlach, Keller, Huey-Wen Lin, Rieke, Sorensen, Wiggins.

Condensed Matter Physics. Campbell, Cobden, Drobny, Fu, Ginger, Hochberg, Ingalls, Keller, Krishnan, Ohuchi, Olmstead, Seidler, Sorensen, Stern, Vilches, Xu.

High Energy Physics. Burnett, Goussiou, Lubatti, Mockett, Rosenberg, Watts, Zhao.

Nuclear Physics. Adelberger, Cramer, Doe, Enomoto, Garcia, Gundlach, Heckel, Hertzog, Kammel, Gerald Miller, Robertson, Rosenberg, Schlamminger, Snover, Storm, Tolich, Trainor, Weitkamp.

Physics and other Science Education. Heron, McDermott, Shaffer.

Relativity & Gravitation. Adelberger, Gundlach, Heckel.

View additional information about this department at
www.gradschoolshopper.com

WASHINGTON STATE UNIVERSITY

DEPARTMENT OF PHYSICS AND ASTRONOMY

Pullman, Washington 99164-2814
http://www.physics.wsu.edu

General University Information
President: Elson S. Floyd
Dean of Graduate School: William Andrefsky, Jr.
University website: http://www.wsu.edu
Control: Public
Setting: Rural
Total Faculty: 2,128
Total Graduate Faculty: 2,128
Total number of Students: 27,679
Total number of Graduate Students: 3,714

Department Information
Department Chairman: Matt McCluskey, Chair
Department Contact: Sabreen Yamini Dodson, Principal
   Assistant to the Chair
   Total full-time faculty: 17
   Total number of full-time equivalent positions: 26
   Full-Time Graduate Students: 68
   First-Year Graduate Students: 17
   Female First-Year Students: 5
   Total Post Doctorates: 8

Department Address
1405 NE College Avenue
P.O. Box 642814
Pullman, WA 99164-2814
Phone: (509) 335-9532
Fax: (509) 335-7816
E-mail: physics@wsu.edu
Website: http://www.physics.wsu.edu

ADMISSIONS

Admission Contact Information
Address admission inquiries to: Chair, Graduate Studies, Department of Physics and Astronomy.
Phone: (509) 335-9532
E-mail: physics@wsu.edu
Admissions website: http://www.physics.wsu.edu

Application deadlines
Fall admission:
U.S. students: March 31   *Int'l. students*: March 31

Application fee
U.S. students: $75   *Int'l. students*: $75

Admissions information
For Fall of 2012:
   Number of applicants: 89
   Number admitted: 40
   Number enrolled: 17

Admission requirements
Bachelor's degree requirements: A Bachelor's degree is required.
Minimum undergraduate GPA: 3.0

GRE requirements
The GRE is recommended.
   Mean GRE score range (25th–75th percentile): 60-85%
   GRE is not required but strongly recommended.

Advanced GRE requirements
The Advanced GRE is recommended.
Subject GRE is not required but strongly recommended.

TOEFL requirements
The TOEFL exam is required for students from non-English-speaking countries.
   PBT score: 550
   iBT score: 80
TOEFL must be less than two years old at time of admission.

Other admissions information
Additional requirements: Subject GRE test in physics is highly recommended for all applicants. No minimum acceptable scores are specified.
Undergraduate preparation assumed: Symon, Mechanics; Reitz, Milford, and Christy, Foundations of Electromagnetic Theory; Zemansky, Heat and Thermodynamics; Liboff, Introductory Quantum Mechanics; Boas, Mathematical Methods in the Physical Sciences; Eisberg and Resnick, Quantum Physics of Atoms, Molecules, Solids, Nuclei, and Particles. Preparation in optics, solid-state physics, nuclear physics, and/or acoustics is encouraged.

TUITION

Tuition year 2013-14:
Tuition for in-state residents
   Full-time students: $10,890 annual
   Part-time students: $587 per credit
Tuition for out-of-state residents
   Full-time students: $23,794 annual
   Part-time students: $1,259 per credit
Out-of-state residents with half-time appointments may pay in-state tuition rates.
Credit hours per semester to be considered full-time: 14
Deferred tuition plan: Yes
Health insurance: Available at the cost of 1,792 per year.
Other academic fees: $864
Academic term: Semester
Number of first-year students who received full tuition waivers: 2
Number of first-year students who received partial tuition waivers: 13

Teaching Assistants, Research Assistants, and Fellowships
Number of first-year
   Teaching Assistants: 14
   Research Assistants: 2
   Fellowship students: 1
Average stipend per academic year
   Teaching Assistant: $15,075
   Research Assistant: $16,165
   Fellowship student: $18,744

FINANCIAL AID

Application deadlines
Fall admission:
U.S. students: February 15

Loans

Loans are available for U.S. students.
Loans are not available for international students.
GAPSFAS application required: No
FAFSA application required: Yes

For further information

Address financial aid inquiries to: Chair, Graduate Studies, Department of Physics and Astronomy.
Phone: (509) 335-9532
E-mail: physics@wsu.edu
Financial aid website: http://www.finaid.wsu.edu

HOUSING

Availability of on-campus housing

Single students: Yes
Married students: Yes

For further information

Address housing inquiries to: Housing and Dining Financial Services, Washington State University, Pullman, WA 99164-1722.
Phone: (509) 335-7732
E-mail: housing@wsu.edu
Housing aid website: http://housing.wsu.edu/

Table A—Faculty, Enrollments, and Degrees Granted

Research Specialty	2012-2013 Faculty	Enrollment Fall 2012		Number of Degrees Granted 2012-2013 (2008–13)		
		Master's	Doctorate	Master's	Terminal Master's	Doctorate
Acoustics	3	–	4	–(2)	–(2)	1(7)
Astronomy	2	–	–	–	–	–
Astrophysics	6	–	7	1(5)	2(3)	1(3)
Atomic, Molecular, & Optical Physics	2	–	8	–(3)	–	2(4)
Biophysics	1	–	2	1(1)	–	–(1)
Chemical Physics	6	–	–	–	–	–
Computational Physics	1	–	–	–	–	–
Condensed Matter Physics	12	–	2	–(2)	–(1)	–(1)
Cosmology & String Theory	1	–	–	–	–	–
Fluids, Rheology	1	–	–	–	–	–
High Energy Physics	1	–	–	1(1)	–	–
Low Temperature Physics	4	–	–	–	–	–
Materials Science, Metallurgy	7	2	7	1(4)	–(1)	4(12)
Nano Science and Technology	2	–	1	–	–	3(3)
Nonlinear Dynamics and Complex Systems	1	–	–	–	–	–
Nuclear Physics	2	–	–	–	–	–
Optics	7	1	7	–(5)	–(1)	2(3)
Polymer Physics/Science	–	–	–	–	–	–
Quantum Foundations	1	–	2	–	–	–
Relativity & Gravitation	2	–	1	–	–	–
Shock Physics	1	–	7	1(2)	–	–(1)
Solid State Physics	7	–	1	–	–	–(2)
Surface Physics	4	–	–	–	–	1(1)
Theoretical Physics	5	–	1	–	–	–
Non-specialized	–	2	13	–	1(4)	–
Total	79	5	63	4(25)	3(12)	14(38)
Full-time Grad. Stud.	–	5	63	–	–	–
First-year Grad. Stud.	–	1	16	–	–	–

GRADUATE DEGREE REQUIREMENTS

Master's: For the M.S. degree without a thesis, a minimum of 30 semester hour credits for graded courses is required; coursework must include a specified core curriculum. For the M.S. degree with a thesis, a minimum of 30 credits is required, of which 21 must be for graded courses. A minimum 3.0 GPA must be maintained. One academic year of residence is required. No foreign language is required. Must pass oral final examination. Students making normal progress toward a Ph.D. satisfy the requirements for a non-thesis M.S. degree at the end of the second year of study.

Doctorate: For the Ph.D. degree, a minimum of 72 semester hour credits is required, of which 36 credits must be for graduate-level graded coursework in physics, astrophysics, or related fields. Coursework must include a specified core curriculum. A minimum 3.0 GPA must be maintained. Minimum period of study is three years, of which two years must be in residence, including a minimum of two continuous semesters in an academic year. No foreign language is required. Qualifying examination, preliminary examinations, and final examination are required. Thesis is required.

Other Degrees: An interdepartmental doctoral program in materials science is available.

Thesis: Thesis may be written in absentia.

SPECIAL EQUIPMENT, FACILITIES, OR PROGRAMS

The Department occupies a modern, 96,000-sq. ft. building. Surface and solid-state physics laboratories are equipped for STM, LEED, Auger, molecular beam, optical and resonance spectroscopy, laser studies, and perturbed angular correlation and Mössbauer spectroscopies. The optical physics laboratories are equipped with several high-power ultrafast femto-, pico-, nano-sec, and continuous-wave lasers, as well as a wide assortment of detection systems. The physical acoustics research laboratory is equipped with a 6,000-gallon water tank for scattering and nonlinear acoustics experiments. Several computer systems are available for use in research and teaching. There is local access to a nuclear reactor, ESR and NMR spectrometers, x-ray spectrometers, and electron microscopes. The Institute for Shock Physics (ISP) is housed in a separate, new state-of-the-art building (~30,000 sq. ft.). Unique among academic institutions, ISP is equipped with an impact laboratory, laser shock laboratory, pulsed power facility, and static high-pressure laboratory with which to study materials under extreme conditions.

Table B—Separately Budgeted Research Expenditures by Source of Support

Source of Support	Departmental Research	Physics-related Research Outside Department
Federal government	$1,902,100	$7,681,297
State/local government		
Non-profit organizations		
Business and industry	$40,152	$620,618
Other		
Total	$1,942,252	$8,301,915

Table C—Separately Budgeted Research Expenditures by Research Specialty

Research Specialty	No. of Grants	Expenditures ($)
Acoustics	13	$319,146
Astrophysics	6	$245,015
Atomic, Molecular, & Optical Physics	5	$350,476
Condensed Matter Physics	1	$253
Materials Science, Metallurgy	6	$378,285
Nano Science and Technology	2	$152,526
Optics	4	$348,594
Theoretical Physics	4	$6,552
Other	2	$141,405
Total	43	$1,942,252

FACULTY

Professor

Blume, Doerte, Ph.D., Georg-August-Universität zu Göttingen, Germany, 1998. *Atomic, Molecular, & Optical Physics, Chemical Physics, Low Temperature Physics, Theoretical Physics.* Theoretical atomic and molecular physics.

Bose, Sukanta, Ph.D., University of Wisconsin-Milwaukee, 1996. *Astrophysics, Theoretical Physics.* Theoretical astrophysics.

Collins, Gary S., Ph.D., Rutgers University, 1976. *Condensed Matter Physics, Materials Science, Metallurgy.* Nuclear hyperfine interactions; PAC.

Dickinson, J. Thomas, Ph.D., University of Michigan, 1968. Regents Professor. *Chemical Physics, Nano Science and Technology, Solid State Physics, Surface Physics.* Solid-state physics; surface physics; chemical physics.

Gupta, Yogendra M., Ph.D., Washington State University, 1973. Director, Institute of Shock Physics; Regents Professor. *Condensed Matter Physics, Optics, Solid State Physics.* Shockwave physics; condensed-matter physics; optics; solid-state physics; materials science.

Kuzyk, Mark G., Ph.D., University of Pennsylvania, 1985. Regents Professor. *Chemical Physics, Optics.* Nonlinear optics.

Lynn, Kelvin G., Ph.D., University of Utah, 1974. Director, Center for Materials Research. *Materials Science, Metallurgy.* Materials science.

Marston, Philip L., Stanford University, 1976. *Acoustics, Optics.* Wave propagation and scattering; acoustics; optics; fluid mechanisms; microgravity.

McCluskey, Matthew D., Ph.D., University of California, Berkeley, 1997. Chair of the Department. *Materials Science, Metallurgy.* Materials science.

Miller, Michael D., Ph.D., Northwestern University, 1974. *Condensed Matter Physics, Low Temperature Physics, Theoretical Physics.* Theoretical physics.

Tomsovic, Steven L., Ph.D., University of Rochester, 1987. *Acoustics, Condensed Matter Physics, Nonlinear Dynamics and Complex Systems, Theoretical Physics.* Theoretical physics.

Associate Professor

Dexheimer, Susan, Ph.D., University of California, Berkeley, 1990. Ultrafast spectroscopy.

Engels, Peter, Ph.D., University of Hannover, Germany, 2000. *Atomic, Molecular, & Optical Physics, Low Temperature Physics.* Experimental atomic; molecular and optical physics.

Gu, Yi, Ph.D., Columbia University, 2004. *Condensed Matter Physics, Nano Science and Technology.* Experimental physics; nanomaterials.

Worthey, Guy, Ph.D., University of California, Santa Cruz, 1992. *Astronomy.* Astronomy.

Assistant Professor

Duez, Mathew, Ph.D., University of Illinois at Urbana-Champaign, 2005. *Astrophysics, Relativity & Gravitation.* Astrophysics; relativity and gravitation.

Emeritus

Donaldson, Edward E., Ph.D., Washington State University, 1953. Surface physics.

Dresser, Miles J., Ph.D., Iowa State University, 1964. Solid-state physics; surface physics.

Park, James L., Ph.D., Yale University, 1967. Theoretical physics.

Adjunct Faculty

Selim, Farida, Ph.D., Alexandria University, 1999. *Nano Science and Technology, Nuclear Physics.* Positron Annihilation Spectroscopy, Laser Matter Interactions, Material Characterization and Defect Studies, X-ray Fluorescence and Scattering, Particle Scattering and Channeling, Beam Optics of Charged Particles.

Adjunct Professor

Anderson, Roger H., Ph.D., University of Washington, 1961. Theoretical condensed-matter physics.

Blakeslee, John, Ph.D., Massachusetts Institute of Technology, 1997. Astronomy and cosmology.

Chaudhuri, Santanu, Ph.D., Stony Brook University, 2003. *Condensed Matter Physics, Surface Physics.* Condensed-matter physics; surface physics.

Clays, Koen J., Ph.D., University of Leuven, Belgium, 1989. Chemistry; nonlinear optics.

Eilers, Hergen, Ph.D., University of Hamburg, 1993. *Chemical Physics, Optics, Solid State Physics.* Optics; solid-state physics; chemical physics.

Kouzes, Richard T., Ph.D., Princeton University, 1974. Nuclear physics.

Lytel, Rick, Ph.D., Stanford University, 1980. Theoretical high-energy physics.

Mendell, Gregory, Ph.D., Montana State University, 1991. Astrophysics; gravitational waves.

Raab, Frederick J., Ph.D., Stony Brook University, 1980. Experimental physics.

Wang, Lai-Sheng, Ph.D., University of California, Berkeley, 1989. Metal clusters; photoelectron spectroscopy.

Zacate, Matthew O., Ph.D., Oregon State University, 1997. Computational materials science; experimental physics.

DEPARTMENTAL RESEARCH SPECIALTIES AND STAFF

Theoretical

Acoustics. Optics. Marston.

Astronomy. Astronomy and astrophysics; stellar populations; gravitational waves; numerical general relativity; black holes; neutron stars. Bose, Duez, Worthey.

Astrophysics. Astronomy and astrophysics; stellar populations; gravitational waves; numerical general relativity; black holes; neutron stars. Bose, Duez, Worthey.

Atomic, Molecular, & Optical Physics. Atomic physics; quantum clusters; Bose-Einstein condensates. Blume.

Condensed Matter Physics. Phase transitions in liquid mixtures; nonlinear dynamics; mesoscopic systems; low-temperature and many-body physics. Miller, Tomsovic.

Nonlinear Dynamics. Quantum chaos and semiclassical theory. Tomsovic.

Optics. Production and study of laser-induced plasmas and subpicosecond x-ray pulses; dynamics of excited states in solids and molecules; nonlinear optical properties of doped polymers; light scattering and Fourier optics; clusters; optome-chanical effects; all-optical devices; time-resolved optical spectroscopy; atomic spectroscopy. Dexheimer, Gupta, Kuzyk, Lytel, Marston.

Shock Wave and High Pressure Physics. Equations of state; finite amplitude wave propagation; material models of mechanical and thermal behavior. Gupta, Marston.

Experimental

Acoustics. Nonlinear acoustics; radiation pressure and scattering. Marston.

Atomic, Molecular, & Optical Physics. Atomic, molecular, and optical physics; Bose-Einstein condensates; degenerate Fermi gases. Engels.

Chemical Physics. Molecular interactions on surfaces; problems in catalysis. Dickinson.

Materials Physics. Low-energy positron beam studies of inter-faces and layered structures. Lynn, McCluskey.

Nanomaterials. Synthesis and device design. Gu.

Nuclear Solid State Physics. Local atomic environments in sol-ids; point defects in metals; perturbed angular correlation and Mössbauer spectroscopy. Collins.

Optics. Production and study of laser-induced plasmas; dynamics of excited states in solids and molecules; nonlinear optical

properties of doped polymers; light scattering and Fourier op-tics; clusters; optomechanical effects; all-optical devices; time-resolved optical spectroscopy; atomic spectroscopy. Dexheimer, Gupta, Kuzyk, Marston, Wang.

Shock Wave and High Pressure Physics. Structural and chem-ical changes in condensed materials; time-resolved optical spectroscopy; optical properties of semiconductors; nonlin-ear wave propagation; inelastic deformation. Gupta, Mc-Cluskey.

Solid State Physics. Solid-state physics; fracture of solids (Dick-inson); mechanical and optical properties at extreme condi-tions (Gupta); defects in semiconductor materials (Lynn); dy-namics of electronic excitations (Dexheimer); photoelectron spectroscopy in clusters (Wang); wide band gap semiconduc-tors (Lynn, McCluskey). Dexheimer, Gupta, Lynn, McClus-key.

Surface Physics. Molecular and atomic interactions and characterization of surfaces; reactive etching of surfaces; transmission electron microscopy on small metal clusters; photoelectric and thermal emission microscopy. Dickinson.

View additional information about this department at
www.gradschoolshopper.com

WEST VIRGINIA UNIVERSITY

DEPARTMENT OF PHYSICS

Morgantown, West Virginia 26506
http://physics.wvu.edu/

General University Information
President: Jim Clements
Dean of Graduate School: Robert Jones
University website: http://www.wvu.edu
Control: Public
Setting: Urban
Total Faculty: 1,709
Total Graduate Faculty: 1,100
Total number of Students: 29,707
Total number of Graduate Students: 6,880

Department Information
Department Chairman: Earl E. Scime, Chair
Department Contact: Earl Scime, Chair
  Total full-time faculty: 22
  Total number of full-time equivalent positions: 22
  Full-Time Graduate Students: 70
  First-Year Graduate Students: 11
  Female First-Year Students: 4
  Total Post Doctorates: 10

Department Address
135 Willey Street
P.O. Box 6315
Morgantown, WV 26506

Phone: (304) 293-3422
E-mail: escime@wvu.edu
Website: http://physics.wvu.edu/

ADMISSIONS

Admission Contact Information
Address admission inquiries to: Admissions Committee, De-partment of Physics, P.O. Box 6315, Morgantown, WV 26506
Phone: (304) 293-3422
Admissions website: http://www.physics.wvu.edu

Application deadlines
Fall admission:
U.S. students: February 15   *Int'l. students*: February 15

Application fee
U.S. students: $60
A personal statement and resume are required.

Admissions information
For Fall of 2013:
  Number of applicants: 100
  Number admitted: 20
  Number enrolled: 10

Admission requirements

Bachelor's degree requirements: A bachelor's degree in physics is required.

Minimum undergraduate GPA: 3.0

GRE requirements

The GRE is required.

Advanced GRE requirements

The Advanced GRE is required.

TOEFL requirements

The TOEFL exam is required for students from non-English-speaking countries.

PBT score: 550

iBT score: 79

The minimum accepted computer-based exam (CBT) score is 213. The minimum accepted IELTS score is 6.5.

Other admissions information

Additional requirements: No minimum score is specified.

Undergraduate preparation assumed: Intermediate mechanics, electricity and magnetism, atomic and quantum physics, thermodynamics, and mathematics through partial differential equations. Typical physics texts include Davis (mechanics), Wangsness (electricity and magnetism), Saxon (quantum mechanics), and Sears and Salinger (thermodynamics).

TUITION

Tuition year 2012–13:

Tuition for in-state residents

Full-time students: $3,405 per semester

Part-time students: $378 per credit

Tuition for out-of-state residents

Full-time students: $9,754 annual

Part-time students: $1,083 per credit

Credit hours per semester to be considered full-time: 12

Deferred tuition plan: Yes

Health insurance: Available

Other academic fees: Fees are included in tuition numbers cited above.

Academic term: Semester

Number of first-year students who received full tuition waivers: 70

Teaching Assistants, Research Assistants, and Fellowships

Number of first-year

Teaching Assistants: 26

Fellowship students: 5

Average stipend per academic year

Teaching Assistant: $22,500

Research Assistant: $22,500

Fellowship student: $28,000

FINANCIAL AID

Application deadlines

Fall admission:

U.S. students: February 15   *Int'l. students*: March 15

Spring admission:

U.S. students: November 1   *Int'l. students*: November 1

Loans

Loans are available for U.S. students.

Loans are not available for international students.

GAPSFAS application required: No

FAFSA application required: Yes

For further information

Address financial aid inquiries to: Graduate Program Committee, Department of Physics.

Phone: (304) 293-3422

E-mail: escime@wvu.edu

Financial aid website: http://grad.wvu.edu/financial_assistance

HOUSING

Availability of on-campus housing

Single students: Yes

Married students: Yes

For further information

Address housing inquiries to: Housing Office, Evansdale Campus.

Phone: (304) 293-4491

E-mail: wvuhousing@mail.wvu.edu

Housing aid website: http://housing.wvu.edu/graduate_student_faculty_and_staff_housing

Table A—Faculty, Enrollments, and Degrees Granted

Research Specialty	2013-2013 Faculty	Enrollment Fall 2012		Number of Degrees Granted 2012-2103 (2008-2013)		
		Master's	Doctorate	Master's	Terminal Master's	Doctorate
Astrophysics	5	–	16	2(7)	–	1(2)
Chemical Physics	1	–	–	–	–	1(1)
Condensed Matter Physics	9	1	33	9(19)	–(4)	3(24)
Fluids, Rheology	1	–	2	1(2)	–	1(1)
Medical, Health Physics	2	–	–	1(2)	–	–
Plasma and Fusion	6	–	17	1(12)	–(1)	–(8)
Statistical & Thermal Physics	2	–	1	–(1)	–	1(1)
Total	–	1	60	14(4)	–(5)	7(37)
Full-time Grad. Stud.	–	–	69	–	–	–
First-year Grad. Stud.	–	–	11	–	–	–

GRADUATE DEGREE REQUIREMENTS

Master's: Approved courses with a minimum GPA of 3.0 is required. There is no residence or language requirement. For a degree with thesis, 24 credits are required. For a degree without a thesis, 30 credits are required.

Doctorate: A minimum of 36 hours of course work in an approved program with a minimum GPA of 3.0 are required. A written comprehensive exam, oral research exam, dissertation, and oral dissertation defense are required.

Thesis: Thesis may be written in absentia.

SPECIAL EQUIPMENT, FACILITIES, OR PROGRAMS

The department and associated instrument and electronics shops are housed in White Hall, a six-story building located on the downtown campus. The building renovation was completed in 2011. The building houses a 60-seat planetarium, a roof-top observatory, a small radio telescope, and 23 state-of-the-art research laboratories.

The plasma facilities include a triple plasma source, a Q-machine for generating space-like plasmas and waves, two helicon plasma sources, a space simulation chamber, a plasma processing test facility, four laser facilities dedicated to plasma diagnosis, a toroidal experiment for turbulence studies, and a pulsed high-velocity plasma source.

The condensed matter physics facilities include four molecular beam epitaxy (MBE) growth facilities, magnetic resonance laboratory (EPR, ENDOR), microimaging MRI, SQUID magnetometer with magneto-resistance probe, rotating anode x-ray source, x-ray diffractometers, an e-beam writer, a scanning probe microscope, an atomic force microscope, Hall effect apparatus, an optical spectrophotometer, an FTIR spectrophotometer, a high-temperature graphite furnace, ultrasonic, thermogravimetry, and differential scanning calorimetry; characterization capabilities for thermoluminescence, optical absorption, photoreflectance, photoconductance and photoluminescence of materials, two-dimensional Fourier transform spectroscopy, second harmonic generation system for interface studies, and a sputtering system for thin-film deposition.

Laser facilities include four cw argon ion lasers, three dye lasers, three tunable diode lasers, three cw and Q-switched Nd:YAG lasers, and three femtosecond lasers.

Departmental computing facilities include two dedicated cluster facilities for development of new computational resources and two large computer clusters. Cooperative research programs with National Energy Technology Laboratory and Pittsburgh Supercomputing Center are possible. Also included are a joint facility with the Chemical Engineering Department for Auger and XPS studies and a joint facility with the Computer Science and Electrical Engineering Department for materials and device processing.

Table B—Separately Budgeted Research Expenditures by Source of Support

Source of Support	Departmental Research	Physics-related Research Outside Department
Federal government	$5,000,000	
State/local government	$500,000	
Non-profit organizations	$100,000	
Business and industry		
Other		
Total	**$5,600,000**	

Table C—Separately Budgeted Research Expenditures by Research Specialty

Research Specialty	No. of Grants	Expenditures ($)
Astrophysics	10	$1,500,000
Condensed Matter Physics	18	$2,600,000
Plasma and Fusion	10	$2,000,000
Total	**38**	**$6,100,000**

FACULTY

Professor

Abdul-Razzaq, Wathiq, Ph.D., University of Illinois at Chicago, 1986. *Condensed Matter Physics, Physics and other Science Education*. Experimental solid state; magnetism of nanoparticles; particulate matter in the environment.

Golubovic, Leonardo, Ph.D., University of Belgrade, 1987. *Condensed Matter Physics, Statistical & Thermal Physics*. Condensed matter theory and statistical physics.

Koepke, Mark E., Ph.D., University of Maryland, 1984. *Atmosphere, Space Physics, Cosmic Rays, Plasma and Fusion*. Experimental plasma physics; nonlinear dynamics.

Lederman, David, Ph.D., University of California, Santa Barbara, 1992. *Condensed Matter Physics, Materials Science, Metallurgy, Nano Science and Technology*. Experimental solid state physics; magnetic materials; superconductors.

Leslie-Pelecky, Diandra, Ph.D., Michigan State University, 1991. *Biophysics, Condensed Matter Physics, Materials Science, Metallurgy, Nano Science and Technology*. Experimental condensed matter physics.

Scime, Earl E., Ph.D., University of Wisconsin-Madison, 1992. *Applied Physics, Atmosphere, Space Physics, Cosmic Rays, Plasma and Fusion*. Experimental plasma physics.

Associate Professor

Cassak, Paul, Ph.D., University of Maryland, 2006. *Plasma and Fusion*. Theoretical plasma physics.

Lewis, James, Ph.D., Arizona State University, 1996. *Condensed Matter Physics*. Computational physics.

Lorimer, Duncan R., Ph.D., University of Manchester, 1994. *Astronomy, Astrophysics*. Radio astronomy; astrophysics.

McLaughlin, Maura A., Ph.D., Cornell University, 2001. *Astronomy, Astrophysics*. Radio astronomy; astrophysics.

Romero, Aldo H., Ph.D., University of California, San Diego, 1998. *Computational Physics, Condensed Matter Physics, Nano Science and Technology, Theoretical Physics*.

Assistant Professor

Anderson, Loren, Ph.D., Boston University, 2009. *Astronomy, Astrophysics*. Observational astrophysics.

Bristow, Alan, Ph.D., University of Sheffield, 2003. *Condensed Matter Physics, Optics*. Experimental condensed matter physics; optics.

Cen, Cheng, Ph.D., University of Pittsburgh, 2010. *Applied Physics, Condensed Matter Physics, Materials Science, Metallurgy, Nano Science and Technology*. Novel complex oxide-based material systems.

Flagg, Edward, Ph.D., University of Texas-Austin, 2008. *Nano Science and Technology, Optics, Quantum Foundations*. Quantum optics.

Holcomb, Mickel, Ph.D., University of California, Berkeley, 2009. *Condensed Matter Physics, Materials Science, Metallurgy, Optics*. Experimental condensed matter physics.

Ma, Renmin, Ph.D., Peking University, 2009. *Condensed Matter Physics*. Experimental condensed matter physics.

McWilliams, Sean, Ph.D., University of Maryland, 2008. General Relativity. *Astronomy, Astrophysics*. Gravitational wave theory.

Miller, Paul, Ph.D., West Virginia University, 2009. *Physics and other Science Education, Plasma and Fusion*. Physics education.

Pisano, Daniel J., Ph.D., University of Wisconsin-Madison, 2001. *Astronomy, Astrophysics*. Radio astronomy, astrophysics.

Schulze, Julian, Ph.D., Ruhr-University Bochum, 2009. *Plasma and Fusion*. Experimental plasma physics.

Stanescu, Tudor, Ph.D., University of Illinois at Urbana-Champaign, 2002. *Condensed Matter Physics*. Theoretical condensed matter physics.

Emeritus

Ferer, Martin V., Ph.D., University of Illinois, 1971. *Statistical & Thermal Physics*. Theory; statistical physics; applied physics; critical phenomena.

Halliburton, Larry E., Ph.D., University of Missouri, Columbia, 1971. *Applied Physics, Condensed Matter Physics, Materials Science, Metallurgy, Optics*. Optical and magnetic properties of point defects.

Pavlovic, Arthur S., Ph.D., Pennsylvania State University, 1966. *Condensed Matter Physics*. Solid state experiments.

Treat, Richard P., Ph.D., University of California, Riverside, 1967. Quantum field theory.

Weldon, H. Arthur, Ph.D., Massachusetts Institute of Technology, 1974. *Particles and Fields*. Particle theory.

Research Professor

Demidov, Vladimir, Ph.D., St. Petersburg State University, 1981. *Plasma and Fusion*. Experimental plasma physics.

Keesee, Amy M., Ph.D., West Virginia University, 2006. *Atmosphere, Space Physics, Cosmic Rays, Plasma and Fusion*. Space plasma physics; optical diagnostics of plasmas.

Seehra, Mohindar S., Ph.D., University of Rochester, 1969. *Applied Physics, Condensed Matter Physics*. Solid state experiment; x-ray scattering; applied physics; magnetism.

Vasiliadis, Dimitris, Ph.D., University of Maryland, 1992. *Atmosphere, Space Physics, Cosmic Rays*. Space plasma physics.

Adjunct Professor

Frayer, D., Ph.D., University of Virginia, 1996. Extragalactic astronomy.

Ganguli, Gurudas, Ph.D., Boston College, 1980. *Plasma and Fusion*. Plasma physics theory.

Lockman, Felix J., Ph.D., University of Massachusetts, 1979. Galactic and extragalactic radio astronomy.

O'Neil, K., Ph.D., University of Oregon, 1997. Extragalactic radio astronomy.

Raylman, Raymond R., Ph.D., University of Michigan, 1991. *Medical, Health Physics, Nuclear Physics*. Medical physics; radiology; imaging.

Rosen, R., Ph.D., University of North Carolina, 2007. White dwarf and Pulsar astronomy.

Smith, Duane, Ph.D., University of Chicago, 1970. *Statistical & Thermal Physics*. Statistical and applied physics; fluids.

DEPARTMENTAL RESEARCH SPECIALTIES AND STAFF

Theoretical

Applied Physics. Photocatalytic materials, new materials design, fracture formation. Ferer, Lewis, Romero.

Astrophysics. Interstellar medium; galactic structure; stellar evolution; compact objects; general relativity; pulsars. Anderson, Lorimer, McLaughlin, McWilliams, Pisano.

Condensed Matter Physics. Surface and interface phenomena; lattice stability and relaxation; molecular dynamics; properties of disordered materials; biomaterials; complex fluids and membranes; fracture; transport in random media; thin-film growth. Golubovic, Lewis, Romero, Stanescu.

Plasma and Fusion. Plasma instabilities; simulations applicable to space and laboratory plasmas; low-temperature plasmas; fusion diagnostics; space plasma instrumentation; space plasma modeling and data analysis. Cassak, Ganguli, Keesee, Koepke, Schulze, Scime, Vasiliadis.

Statistical & Thermal Physics. Fractals; percolation theory; chaos; phase transitions and critical phenomena; nonequilibrium growth and pattern formation. Ferer, Golubovic, Smith.

Experimental

Applied Physics. Preparation and characterization of nanoparticles; iron-based catalysts; properties of air-borne particulate matter; coal-based high purity carbons and carbon fibers; electrochemical detection of Hg and other trace metals using boron-doped diamond films; visible and UV light emitters and sensors; nonlinear optical and photorefractive materials. Bristow, Flagg, Halliburton, Holcomb, Lederman, Leslie-Pelecky, Ma, Seehra.

Astrophysics. Radio astronomy; x-ray astronomy; pulsars; tests of strong-field gravity; digital signal processing; computational astrophysics. Anderson, Frayer, Lockman, Lorimer, McLaughlin, McWilliams, O'Neil, Pisano, Rosen.

Condensed Matter Physics. Electronic structure and magnetic properties of artificially grown surfaces and superlattices and nanoscale particles; spin transport; properties of magnetic ions and clusters; elementary excitations in antiferromagnets; magnetic susceptibility; magnetostriction; electrical, structural, and electro-optic properties of semiconductors; optical and magnetic resonance characterization of point defects. Bristow, Cen, Flagg, Halliburton, Holcomb, Lederman, Leslie-Pelecky, Ma, Pavlovic, Seehra.

Materials Science, Metallurgy. X-ray scattering from disordered systems; Auger and x-ray photoelectron spectroscopy deposition physics; molecular beam epitaxy; properties of monolayer and multilayer thin films; optical properties of quantum confined systems and semiconductors. Holcomb, Lederman, Leslie-Pelecky, Ma, Seehra.

Nano Science and Technology. Nanostructured materials; nano toxicology; biological sensors; nanomagnetism. Cen, Holcomb, Lederman, Leslie-Pelecky, Ma, Seehra.

Physics and other Science Education. K-12 teacher training; development of GTA training programs; curriculum development. Abdul-Razzaq, Keesee, Miller.

Plasma and Fusion. Plasma waves and instabilities; nonlinear interactions; turbulence and chaos; space plasma instrument design; space plasma data analysis and instrument (sensor) development; magnetic reconnection; plasma processing. Cassak, Demidov, Ganguli, Keesee, Koepke, Miller, Schulze, Scime, Vasiliadis.

View additional information about this department at
www.gradschoolshopper.com

UNIVERSITY OF WISCONSIN, MADISON

DEPARTMENT OF ASTRONOMY

Madison, Wisconsin 53706-1582
http://www.astro.wisc.edu

General University Information
President: University of Wisconsin System: Kevin Reilly
Dean of Graduate School: Martin T. Cadwallader
University website: http://www.wisc.edu
Control: Public
Setting: Urban
Total Faculty: 2,173
Total Graduate Faculty: (not separated)
Total number of Students: 42,820
Total number of Graduate Students: 9,183

Department Information
Department Chairman: Ellen Zweibel, Chair
Department Contact: Sharon Pittman, Graduate Program Coordinator
  Total full-time faculty: 12
  Total number of full-time equivalent positions: 12
  Full-Time Graduate Students: 28
  First-Year Graduate Students: 7
  Female First-Year Students: 3
  Total Post Doctorates: 9

Department Address
475 North Charter Street
Madison, WI 53706-1582
Phone: (608) 262-3071 (C)
Fax: (608) 263-6386
E-mail: grading@astro.wisc.edu
Website: http://www.astro.wisc.edu

ADMISSIONS

Admission Contact Information
Address admission inquiries to: Graduate Coordinator, Department of Astronomy, 475 N. Charter Street, Madison, WI 53706
Phone: (608) 890-3775
E-mail: gradinq@astro.wisc.edu
Admissions website: http://www.astro.wisc.edu

Application deadlines
Fall admission:
U.S. students: December 30   *Int'l. students*: December 30

Application fee
U.S. students: $56   *Int'l. students*: $56
Application is submitted online through the UW Graduate School at https://www.gradsch.wisc.edu/eapp/eapp.pl.

Admissions information
For Fall of 2013:
  Number of applicants: 105
  Number admitted: 12
  Number enrolled: 7

Admission requirements
Bachelor's degree requirements: Applicants must have undergraduate preparation that includes at least three years of college physics and mathematics through differential equations.

GRE requirements
The GRE is required.

Advanced GRE requirements
The Advanced GRE is required.

TOEFL requirements
The TOEFL exam is required for students from non-English-speaking countries.
  PBT score: 600
  iBT score: 100

Other admissions information
Additional requirements: Applicants are judged on the basis of their previous academic record, letters of recommendation, personal statement, research plans, and Graduate Record Examination (GRE) scores.
Undergraduate preparation assumed: Applicants must have undergraduate preparation that includes at least three years of college physics and mathematics through differential equations.

TUITION

Tuition year 2013-14:
Tuition for in-state residents
  Full-time students: $5,919 per semester
Tuition for out-of-state residents
  Full-time students: $12,583 per semester
Assistantships and fellowships provide tuition remission.
Credit hours per semester to be considered full-time: 8
Deferred tuition plan: Yes
Health insurance: Available at the cost of $510 per year.
Other academic fees: None
Academic term: Semester
Number of first-year students who received partial tuition waivers: 7

Teaching Assistants, Research Assistants, and Fellowships
Number of first-year
  Teaching Assistants: 1
  Research Assistants: 6
  Fellowship students: 2
Average stipend per academic year
  Teaching Assistant: $14,088
  Research Assistant: $23,305
  Fellowship student: $23,305
Teaching Assistant: $14,088-16,264 Research Assistant: $23,305-28,451 TA's are supplemented to the RA rate.

FINANCIAL AID

Application deadlines
Fall admission:
U.S. students: December 30   *Int'l. students*: December 30

Loans
Loans are available for U.S. students.
Loans are not available for international students.
GAPSFAS application required: No
FAFSA application required: No

For further information

Address financial aid inquiries to: Graduate Coordinator, Department of Astronomy, 475 N. Charter Street, Madison, WI 53706.

Phone: (608) 890-3775

E-mail: grading@astro.wisc.edu

Financial aid website: http://uwoffr.wordpress.com/assistantships/

HOUSING

Availability of on-campus housing

Single students: Yes

Married students: Yes

For further information

Address housing inquiries to: University Apartments Office, University Apartments Community Center, 611 Eagle Heights, Madison, WI 53705.

Phone: (608) 262-3407

E-mail: universityapartments@housing.wisc.edu

Housing aid website: http://www.housing.wisc.edu/universityapartments

Table A—Faculty, Enrollments, and Degrees Granted

Research Specialty	Fall 2012-13 Faculty	Enrollment Fall 2012-13 Master's	Enrollment Fall 2012-13 Doctorate	Number of Degrees Granted 2012-13 Master's	Number of Degrees Granted (2012-13) Terminal Master's	Number of Degrees Granted Doctorate
Astronomy	12	–	25	7(-)	2(-)	2(-)
Total	12	–	25	7(-)	–	2(-)
Full-time Grad. Stud.	–	–	25	–	–	–
First-year Grad. Stud.	–	–	4	–	–	–

GRADUATE DEGREE REQUIREMENTS

Doctorate: Preliminary examination, complete minor requirements, 32 credits (equal to 4 semesters) including required coursework, all with grades of "B" or better; Ph.D. thesis.

Thesis: Theses may be written in absentia.

SPECIAL EQUIPMENT, FACILITIES, OR PROGRAMS

The Department has a 26% share in the WIYN 3.5-m Telescope, an advanced technology optical telescope in Arizona. Remote observing is done from the Department or students travel to WIYN to make their own observations. We also have a share in the nearby 0.9-m telescope.

UW-Madison is also a significant partner in the Southern African Large Telescope (SALT), an 11-m spectroscopic telescope. SALT's Robert Stobie Spectrograph (RSS) was designed and built in our department. A near-infrared addition to this spectrograph is currently under construction in our instrumentation labs.

On Cerro Tololo in Chile is the remotely operable Wisconsin H-Alpha Mapper (WHAM) observatory dedicated to studies of the diffuse interstellar medium.

Astronomy at Wisconsin combines strong traditions in observational, instrumental, and theoretical research. The analysis and interpretation of astronomical data require specialized tools and the Department operates a powerful network of image processing workstations. Theorists use both national high-performance computing facilities and several large departmental computer clusters.

We also house the extensive Woodman Astronomy Library.

Table B—Separately Budgeted Research Expenditures by Source of Support

Source of Support	Departmental Research	Physics-related Research Outside Department
Federal government	$158,490,415	$7,171,618
State/local government		
Non-profit organizations	$3,292,110	
Business and industry		
Other	$7,712,584	
Total	$169,495,109	$7,171,618

Table C—Separately Budgeted Research Expenditures by Research Specialty

Research Specialty	No. of Grants	Expenditures ($)
Astronomy	120	$33,525,327
Total	120	$33,525,327

FACULTY

Professor

Barger, Amy, Ph.D., University of Cambridge, 1997. *Astronomy*. Observational cosmology; distant galaxies and supermassive black holes; star formation and accretion histories of the universe.

Bershady, Matthew A., Ph.D., University of Chicago, 1994. *Astronomy*. Galaxy kinematics; stellar populations; galaxy and quasar evolution; optical and IR spectra and instrumentation.

Gallagher, John S., Ph.D., University of Wisconsin-Madison, 1972. *Astronomy*. Multi-wavelength observational investigations of evolutionary processes in galaxies; stellar populations; classical novae.

Lazarian, Alex, Ph.D., University of Cambridge, 1995. *Astronomy, Astrophysics*. Theoretical astrophysics, in particular plasma processes, properties of magnetic turbulence and techniques of its observational studies, magnetic reconnection, cosmic ray physics, star formation, physics of dusty plasmas, physics of microwave foregrounds, techniques for observational studies of astrophysical magnetic fields.

Mathieu, Robert D., Ph.D., University of California, Berkeley, 1983. Department Chair. *Astronomy*. Observational study of star formation, binary stars, and open star clusters; high-resolution optical and infrared spectroscopy; optical, infrared, and sub-mm imaging and photometry.

Wilcots, Eric M., Ph.D., University of Washington, 1992. *Astronomy*. Studies of the structure and evolution of galaxies through 21 cm HI, optical, and infrared observations; extended gas around galaxies; distribution and kinematics of the interstellar medium in nearby galaxies; structure and evolution of classical HII regions.

Zweibel, Ellen, Ph.D., Princeton University, 1977. *Astronomy, Astrophysics*. Theoretical plasma astrophysics; generation and evolution of astrophysical magnetic fields; interstellar astrophysics; star formation; stellar physics.

Associate Professor

Heinz, Sebastian, Ph.D., University of Colorado, 2000. *Astronomy*. Relativistic jets; black holes; AGN; X-ray binaries; galaxy clusters; gamma ray bursts; interstellar and intergalactic medium; numerical methods.

Stanimirovic, Snezana, Ph.D., University of Western Sydney Nepean, 1999. Undergraduate Advisor. *Astronomy*. Galactic disk/halos; dust properties in low-metallicity environments; physics of the ISM; radio techniques and applications.

Assistant Professor

D'Onghia, Elena, Ph.D., University of Milan, 2003. *Astronomy, Astrophysics*. Cosmology: nature of dark matter, large-scale structure formation, dynamics and galaxy formation.

Townsend, Richard, Ph.D., University College, London, 1997. Graduate Committee Chair. *Astronomy, Astrophysics*. Stellar astrophysics; magnetic fields; stellar winds, massive stars.

Tremonti, Christy, Ph.D., Johns Hopkins University, 2003. *Astronomy*. Galaxy and AGN co-evolution; galactic chemical evolution.

Emeritus

Cassinelli, Joseph P., Ph.D., Washington University, 1970. *Astronomy*. Structure of stellar winds; high-resolution X-ray observations; effects of rotation and magnetic fields on the circumstellar envelopes of hot stars.

Churchwell, Edward B., Ph.D., Indiana University, 1970. *Astronomy*. Star formation; hot molecular cores; UC HII regions; atomic abundances; radio and infrared astronomy.

Nordsieck, Kenneth H., Ph.D., University of California, Santa Cruz, 1972. *Astronomy*. Stellar and extragalactic optical/ultraviolet spectropolarimetry; ground-based instrument control; space astronomy.

Reynolds, Ronald J., Ph.D., University of Wisconsin-Madison, 1971. *Astronomy*. High-resolution spectroscopy of diffuse sources; development of high-throughput spectrometers; physics of the interstellar medium.

Savage, Blair D., Ph.D., Princeton University, 1968. *Astronomy*. Physical properties of the interstellar medium; gas in galactic halos and the intergalactic medium; high-resolution ultraviolet spectroscopy.

Scientist

Haffner, Matt, Ph.D., University of Wisconsin-Madison, 1999. *Astronomy*. Milky Way structure and dynamics; physics of the interstellar medium; extended galactic halos; diffuse emission-line spectroscopy; remote observing.

Orio, Marina, Ph.D., University of Technion, Haifa, Israel, 1987. *Astronomy, Astrophysics*. Stellar evolution and compact objects, particularly close binary stars, classical and recurrent novae, supersoft X-ray sources, cataclysmic variables, low-mass X-ray binaries, ionization nebulae.

Percival, Jeffrey, Ph.D., University of Wisconsin-Madison, 1979. *Astronomy*. Instrument control software; telescope control systems; guidance and navigation for suborbital rockets.

Wakker, Bastiaan, Ph.D., Gronigen University, 1990. *Astronomy*. High-velocity clouds and low-redshift intergalactic medium.

Whitney, Barbara, Ph.D., University of Wisconsin-Madison, 1989. *Astronomy*. Radiative transfer models of planets, forming stars, and galaxies; infrared surveys of our galaxy and the Magellanic Clouds.

DEPARTMENTAL RESEARCH SPECIALTIES AND STAFF

Theoretical

Extragalactic astronomy and cosmology. Barger, Bershady, D'Onghia, Gallagher, Tremonti, Wilcots.

Instrumentation. Bershady, Nordsieck, Percival, Reynolds.

Interstellar and intergalactic media. Haffner, Heinz, Reynolds, Savage, Stanimirovic, Wakker.

Plasma astrophysics and magnetic fields. Cassinelli, Lazarian, Nordsieck, Zweibel.

Stellar star formation, young stars. Cassinelli, Churchwell, Mathieu, Nordsieck, Orio, Townsend.

View additional information about this department at
www.gradschoolshopper.com

UNIVERSITY OF WISCONSIN, MADISON

DEPARTMENT OF PHYSICS

Madison, Wisconsin 53706
http://physics.wisc.edu

General University Information

President: Kevin Reilly
Dean of Graduate School: Martin Cadwallader
University website: http://www.wisc.edu
Control: Public
Setting: Urban
Total Faculty: 2,027
Total number of Students: 42,595
Total number of Graduate Students: 8,817

Department Information

Department Chairman: Robert Joynt, Chair

Department Contact: Renee Lefkow, Graduate Student Coordinator
Total full-time faculty: 44
Total number of full-time equivalent positions: 450
Full-Time Graduate Students: 177
First-Year Graduate Students: 37
Female First-Year Students: 5
Total Post Doctorates: 80

Department Address

1150 University Avenue
Chamberlin Hall
Madison, WI 53706

Phone: (608) 262-9678
Fax: (608) 263-0800
E-mail: physgrad@physics.wisc.edu
Website: http://physics.wisc.edu

ADMISSIONS

Admission Contact Information

Address admission inquiries to: Graduate Coordinator, Department of Physics, 1150 University Avenue, University of Wisconsin-Madison, Madison, WI 53706
Phone: (608) 262-9678
E-mail: physgrad@physics.wisc.edu
Admissions website: http://physics.wisc.edu

Application deadlines
Fall admission:
U.S. students: December 31   *Int'l. students*: December 31
Spring admission:
U.S. students: November 1   *Int'l. students*: November 1

Application fee
U.S. students: $56

Admissions information
For Fall of 2013:
  Number of applicants: 450
  Number admitted: 100
  Number enrolled: 37

Admission requirements
Bachelor's degree requirements: A bachelor's degree in physics or related field is required.
Minimum undergraduate GPA: 3.0

GRE requirements
The GRE is required.

Advanced GRE requirements
The Advanced GRE is required.

TOEFL requirements
The TOEFL exam is required for students from non-English-speaking countries.
  PBT score: 580
  iBT score: 92

Other admissions information
Additional requirements: The average GRE scores for admissions were as follows: verbal, 580; quantitative, 780; total, 1395. The average GRE subject score for admissions was 794. All international students who are admitted as teaching assistants will be required to take and pass the SPEAK test when they arrive on campus, as well as participate in the six-week Summer Orientation Program before the fall semester for which they have been admitted. An admitted applicant may be required to take the English Placement exam on arrival and register for the recommended English as a second language (ESL) course.
Undergraduate preparation assumed: Classical mechanics; electromagnetic fields; electric circuits and elementary electronics; waves and optics; thermal physics; quantum physics; and laboratory experience in classical and atomic physics.

TUITION

Tuition year 2012–13:
Tuition for in-state residents
  Full-time students: $5,919.4 per semester
  Part-time students: $742.55 per credit
Tuition for out-of-state residents
  Full-time students: $12,582.84 per semester
  Part-time students: $1,575.48 per credit

Segregated Fees are included in the tuition cost.
Credit hours per semester to be considered full-time: 8
Deferred tuition plan: No
Health insurance: Available
Academic term: Semester
Number of first-year students who received full tuition waivers: 37

Teaching Assistants, Research Assistants, and Fellowships
Number of first-year
  Teaching Assistants: 31
  Research Assistants: 6
Average stipend per academic year
  Teaching Assistant: $14,087
  Research Assistant: $16,668
  Fellowship student: $22,440

FINANCIAL AID

Loans
Loans are available for U.S. students.
Loans are not available for international students.
GAPSFAS application required: No
FAFSA application required: Yes

For further information
Address financial aid inquiries to: Office of Student Financial Aid, 333 E. Campus Mall, Madison, WI 53715.
Phone: (608) 262-3060
E-mail: finaid@finaid.wisc.edu
Financial aid website: http://finaid.wisc.edu

HOUSING

Availability of on-campus housing
  Single students: Yes
  Married students: Yes

For further information
Housing aid website: http://www.housing.wisc.edu

Table A—Faculty, Enrollments, and Degrees Granted

Research Specialty	2013-14 Faculty	Enrollment Fall 2012 Master's	Enrollment Fall 2012 Doctorate	Number of Degrees Granted 2011–12 (2007–12) Master's	Number of Degrees Granted 2011–12 (2007–12) Terminal Master's	Number of Degrees Granted 2011–12 (2007–12) Doctorate
Astrophysics	1	–	16	1(1)	-(4)	-(12)
Atmosphere, Space Physics, Cosmic Rays	2	–	7	-(6)	1(3)	-(6)
Atomic, Molecular, & Optical Physics	5	–	22	1(5)	1(11)	2(16)
Biophysics	1	–	1	–	–	1(1)
Condensed Matter Physics	12	–	36	2(5)	-(1)	1(24)
Engineering Physics/Science	–	–	–	-(3)	-(3)	1(1)
Materials Science, Metallurgy	1	–	1	–	-(1)	1(7)
Medical, Health Physics	–	–	–	-(1)	–	-(5)
Nuclear Physics	1	–	11	-(1)	-(1)	1(1)
Particles and Fields	15	–	58	-(8)	2(2)	4(35)
Plasma and Fusion	6	–	26	1(3)	-(1)	5(16)
Total	44	2	177	5(33)	4(27)	16(124)
Full-time Grad. Stud.	–	2	177	–	–	–
First-year Grad. Stud.	–	–	37	–	–	–

GRADUATE DEGREE REQUIREMENTS

Master's: For the Master of Science degree, the department requires that at least 12 of the 18 credits be in physics courses (other than research) numbered above 500. A degree is awarded to a student who has (1) satisfied the graduate-level credit and course requirements, (2) passed the qualifying exam, and (3) completed a master's project, including a thesis.

Doctorate: The following are required for a doctorate degree: a qualifying exam, a preliminary exam, complete minor requirements, 32 credits (equivalent to four semesters), including required course work, with grades of B or better, and a Ph.D. thesis.

Thesis: Thesis work may be done in absentia.

Table B—Separately Budgeted Research Expenditures by Source of Support

Source of Support	Departmental Research	Physics-related Research Outside Department
Federal government	$26,051,023	
State/local government		
Non-profit organizations		
Business and industry		
Other	$2,162,832	
Total	**$28,213,855**	

Table C—Separately Budgeted Research Expenditures by Research Specialty

Research Specialty	No. of Grants	Expenditures ($)
Atmosphere, Space Physics, Cosmic Rays	18	$2,688,462
Atomic, Molecular, & Optical Physics	19	$2,352,404
Condensed Matter Physics	27	$5,593,463
Nuclear Physics	8	$672,116
Particles and Fields	66	$9,514,138
Plasma and Fusion	15	$7,393,272
Total	**153**	**$28,213,855**

FACULTY

Professor

Balantekin, A. Baha, Ph.D., Yale University, 1982. *Nuclear Physics.* Theoretical physics at the interface of nuclear physics, particle physics, and astrophysics; mathematical physics; neutrino physics; astrophysics; fundamental symmetries; nuclear structure physics.

Barger, Vernon, Ph.D., Pennsylvania State University, 1963. Theory and phenomenology of elementary particle physics; neutrino physics; electroweak gauge models; heavy quarks; supersymmetry; cosmology.

Carlsmith, Duncan L., Ph.D., University of Chicago, 1984. *Particles and Fields, Plasma and Fusion.* High energy and fundamental particle physics at the Tevatron and LHC.

Chubukov, Andrey, Ph.D., Moscow State University, 1985. *Condensed Matter Physics, Electromagnetism.* Condensed matter theory; low-D magnetism; frustrated antiferromagnets; fermi liquid theory; high-temperature superconductivity.

Coppersmith, Susan N., Ph.D., Cornell University, 1983. *Biophysics, Condensed Matter Physics, Other.* Theoretical condensed matter physics; nonlinear dynamics; quantum computation and information; biomineralization.

Dasu, Sridhara R., Ph.D., University of Rochester, 1988. *Particles and Fields, Plasma and Fusion.* High energy physics; LHC; CMS; BaBar; SLAC National Accelerator Laboratory.

Eriksson, Mark, Ph.D., Harvard University, 1997. *Condensed Matter Physics, Nano Science and Technology, Other.* Condensed matter physics; nanoscience; semiconductor membranes; semiconductor nanostructures; quantum dots; quantum computing; thermoelectric materials.

Forest, Cary B., Ph.D., Princeton University, 1992. *Plasma and Fusion, Other.* Experimental plasma physics; liquid metal magnetohydrodynamics with applications to astrophysics and magnetic confinement of fusion plasmas.

Gilbert, Pupa, Ph.D., First University of Rome "La Sapienza", 1987. *Biophysics.* Biophysics, specializing in biomineralization, nanobiology, and synchrotron spectromicroscopy.

Halzen, Francis, Ph.D., University of Louvain, 1969. *Astrophysics, Particles and Fields.* Theory and phenomenology of particle physics; particle astrophysics; neutrino astronomy.

Herndon, M., Ph.D., University of Maryland, 1998. *Particles and Fields, Plasma and Fusion.* Fundamental particle physics involving high energy hadron collisions with the Collider Detector at Fermilab at the Tevatron and the CMS at LHC; research topics include rare decay of B hadrons, diboson physics, Higgs physics, and searches for fundamental new particles; detector and algorithm development involving muon triggers and tracking detectors.

Himpsel, Franz J., Ph.D., University of Munich, 1977. *Condensed Matter Physics, Nano Science and Technology.* Experimental condensed matter physics; synchrotron radiation techniques; nanoscience.

Joynt, R. J., Ph.D., University of Maryland, 1982. Department Chair. *Condensed Matter Physics.* Theory of superconductivity and heavy fermion systems; quantum Hall effect; magnetism; high-temperature superconductivity; quantum computing.

Karle, A., Ph.D., University of Munich, 1994. *Astronomy, Astrophysics, Atmosphere, Space Physics, Cosmic Rays, Particles and Fields.* Experimental particle astrophysics; high-energy neutrino astronomy; neutrino physics; cosmic rays.

Lagally, M. G., Ph.D., University of Wisconsin-Madison, 1968. *Materials Science, Metallurgy.* Surface physics; structure and disorder; electronic materials; thin-film growth.

Lawler, J. E., Ph.D., University of Wisconsin-Madison, 1978. *Astrophysics, Atomic, Molecular, & Optical Physics.* Experimental atomic physics; laser spectroscopy; gas discharges; laboratory astrophysics.

Lin, C. C., Ph.D., Harvard University, 1955. *Atomic, Molecular, & Optical Physics.* Atomic and molecular physics; atomic collisions.

McCammon, D., Ph.D., University of Wisconsin-Madison, 1971. *Astrophysics.* Astrophysics; x-ray astronomy; interstellar and intergalactic medium; x-ray detectors.

Onellion, Marshall, Ph.D., Rice University, 1984. *Condensed Matter Physics, Nano Science and Technology.* Experimental solid state, synchrotron radiation, and ultra-fast optical techniques; nanomaterials.

Rzchowski, Mark S., Ph.D., Stanford University, 1988. *Condensed Matter Physics, Materials Science, Metallurgy, Nano Science and Technology.* Experimental condensed matter physics; magnetic heterostructures and nanostructures; low-temperature scanning tunneling spectroscopy; superconductivity in novel materials; thin-film growth and fabrication.

Saffman, M., Ph.D., University of Colorado Boulder, 1994. *Atomic, Molecular, & Optical Physics, Optics.* Atomic physics; quantum computing with neutral atoms; quantum optics; entanglement; nonlinear optics; solitons; pattern formation.

Sarff, John S., Ph.D., University of Wisconsin-Madison, 1988. *Plasma and Fusion.* Plasma physics; magnetic confinement; instabilities and turbulence.

Schnack, Dalton, Ph.D., University of California, Davis, 1977. *Plasma and Fusion.* Computational plasma physics.

Shiu, Gary, Ph.D., Cornell University, 1998. *Particles and Fields*. String theory; theoretical physics; elementary particle physics; cosmology.

Smith, Wesley H., Ph.D., University of California, Berkeley, 1981. *Particles and Fields, Plasma and Fusion*. High energy and fundamental experimental particle physics and collisions at LHC, CERN, Geneva, Switzerland.

Terry, P. W., Ph.D., University of Texas, Austin, 1981. *Astrophysics, Plasma and Fusion*. Theory of turbulent plasmas and neutral fluids; plasma theory; anomalous transport and turbulence in fusion plasmas; plasma astrophysics.

Timbie, Peter T., Ph.D., Princeton University, 1985. *Astrophysics, Low Temperature Physics*. Observational astrophysics and cosmology; measurements of the 2.7 K cosmic microwave background radiation; 21-cm hydrogen tomography; microwave detectors; cryogenics.

Walker, T., Ph.D., Princeton University, 1988. *Atomic, Molecular, & Optical Physics, Low Temperature Physics*. Laser trapping of atoms; collisions between ultra-cold atoms; neutral atom quantum computing; spin-exchange optical pumping; biomagnetometry.

Westerhoff, Stefan, Ph.D., University of Wuppertal, 1996. *Astronomy, Astrophysics, Particles and Fields*. Experimental particle astrophysics; high energy neutrino astronomy; ultra high energy cosmic rays.

Winokur, Michael J., Ph.D., University of Michigan, 1985. *Condensed Matter Physics*. Condensed matter physics; structure of novel materials; phase transitions.

Wu, Sau Lan, Ph.D., Harvard University, 1970. *Electromagnetism, Particles and Fields, Plasma and Fusion*. High energy and elementary particle physics; weak, electromagnetic, and strong interactions; Higgs boson, CERN, Geneva, Switzerland.

Zweibel, Ellen, Ph.D., Princeton University, 1977. *Astronomy, Astrophysics*. Theoretical astrophysics; plasma astrophysics; origin and evolution of astrophysical magnetic fields.

Associate Professor

Boldyrev, S., Ph.D., Princeton University, 1999. *Plasma and Fusion*. Plasma theory.

Chung, Daniel J.H., Ph.D., University of Chicago, 1998. Theoretical cosmology; high energy physics; quantum field theory in curved space time.

Everett, Lisa, Ph.D., University of Pennsylvania, 1998. *Particles and Fields*. Theoretical elementary particle physics; superstring phenomenology; supersymmetry.

Hashimoto, Akikazu, Ph.D., Princeton University, 1997. *Particles and Fields*. String theory; black hole physics; quantum field theory; theoretical physics.

McDermott, Robert, Ph.D., University of California, Berkeley, 2002. *Condensed Matter Physics*. Experimental condensed matter physics; quantum computing.

Pan, Yibin, Ph.D., University of Wisconsin-Madison, 1991. *Particles and Fields, Plasma and Fusion*. High energy experimental particle physics.

Perkins, Natalia, Ph.D., Moscow State University, 1997. *Condensed Matter Physics*. Condensed matter theory; strongly correlated electron systems; orbital physics; frustrated magnetism; Kondo physics.

Vavilov, Maxim, Ph.D., Cornell University, 2001. *Condensed Matter Physics, Nano Science and Technology*. Condensed matter theory; nanoscale- and low-dimensional systems.

Yavuz, Deniz, Ph.D., Stanford University, 2003. *Atomic, Molecular, & Optical Physics*. Experimental atomic, molecular, and optical physics.

Assistant Professor

Bai, Yang, Ph.D., Yale University, 2007. *High Energy Physics, Particles and Fields*. Collider physics; dark matter; electroweak symmetry breaking; B physics; topological interactions.

Egedal, Jan, Ph.D., Oxford University, 1998. *Plasma and Fusion*. Magnetic reconnection.

Vandenbroucke, Justin, Ph.D., University of California, Berkeley, 2009. *Astrophysics, Atmosphere, Space Physics, Cosmic Rays*. Experimental particle astrophysics; cosmogenic neutrinos; cosmology; gamma-ray astronomy.

Emeritus

Anderson, L. W., Ph.D., Harvard University, 1960. Atomic and molecular physics; atomic collisions; lasers.

Bincer, A.M., Ph.D., Massachusetts Institute of Technology, 1956. Theoretical physics; quantum field theory; group theory.

Bruch, Ludwig W., Ph.D., University of California, San Diego, 1964. *Chemical Physics, Condensed Matter Physics, Statistical & Thermal Physics*. Theoretical condensed matter; statistical and chemical physics.

Callen, J. D., Ph.D., Massachusetts Institute of Technology, 1968. Plasma physics; theory of confinement and heating of magnetically confined plasmas, primarily for controlled thermonuclear fusion.

Camerini, Ugo, Ph.D., São Paulo, 1946. High energy and fundamental particle physics; cosmic rays; astrophysics.

Cox, Donald P., Ph.D., University of California, San Diego, 1970. Astrophysics and space physics; theoretical studies of interstellar matter; cosmic-ray acceleration.

Dexter, R. N., Ph.D., University of Wisconsin-Madison, 1955. Plasma physics; diagnostics of high-temperature plasma physics; fluctuation and turbulence studies.

Durand, Bernice, Ph.D., Iowa State University, 1971. Associate Vice Provost. Theoretical high energy physics; use of algebras in theoretical physics.

Durand, L., Ph.D., Yale University, 1957. Theoretical physics; elementary particle physics; electroweak interactions; scattering processes; mathematical physics; special functions and group theory.

Ebel, Marvin E., Ph.D., Iowa State University, 1953. Theoretical physics; high energy physics and interactions of elementary particles.

Friedman, William A., Ph.D., Massachusetts Institute of Technology, 1966. Nuclear physics; nuclear theory, including reaction theory and collective effects in nuclear models; heavy-ion reactions.

Goebel, C. J., Ph.D., University of Chicago, 1954. Theoretical physics; quantum field theory, including high energy interactions, elementary particles, and dispersion theory; general relativity.

Haeberli, Willy, Ph.D., Basel, 1952. Nuclear physics; polarized particle physics; polarized ion sources; polarized gas targets; tests of fundamental symmetries in hadronic and weak interactions.

Huber, David L., Ph.D., Harvard University, 1964. Condensed matter theory; magnetic and optical properties of solids.

Knutson, Lynn, Ph.D., University of Wisconsin-Madison, 1973. Nuclear physics with polarized particles; properties of few-nucleon systems; medium energy physics.

March, R. H., Ph.D., University of Chicago, 1960. High energy and fundamental particle physics; high energy astrophysics.

Morse, R., Ph.D., University of Wisconsin-Madison, 1969. High energy particle astrophysics; gamma-ray and neutrino astronomy.

Olsson, M. G., Ph.D., University of Maryland, 1964. Theory and phenomenology of fundamental particle physics; nonperturbative quantum chromodynamics; quark confinement; chiral dynamics; B-meson.

Pondrom, L. G., Ph.D., University of Chicago, 1958. High energy and fundamental particle physics; hadronic interactions.

Prager, Stewart, Ph.D., Columbia University, 1975. Plasma physics; magnetic confinement; instabilities and turbulence.

Prepost, R., Ph.D., Columbia University, 1961. High energy and fundamental particle physics; weak and electromagnetic interactions.

Quin, Paul A., Ph.D., University of Notre Dame, 1969. Nuclear reactions with polarized particles; weak and electromagnetic interactions; fundamental symmetries.

Reeder, D. D., Ph.D., University of Wisconsin-Madison, 1966. High energy and fundamental particle physics; weak and electromagnetic interactions; electron-positron colliders; cosmic rays.

Roesler, F. L., Ph.D., University of Wisconsin-Madison, 1962. Astrophysics; aeronomy; optical spectroscopy; interference spectroscopy.

Scherb, Frank, Ph.D., Massachusetts Institute of Technology, 1958. Astrophysics and space physics; space plasma physics; high-resolution astrophysical spectroscopy.

Sprott, J. C., Ph.D., University of Wisconsin-Madison, 1969. Plasma physics; computational nonlinear dynamics; chaos; complex systems.

Symon, K. R., Ph.D., Harvard University, 1948. Particle accelerators; plasma physics; nonlinear mechanics; particle orbit theory.

Webb, M. B., Ph.D., University of Wisconsin-Madison, 1956. Solid state physics; surface studies.

DEPARTMENTAL RESEARCH SPECIALTIES AND STAFF

Theoretical

Atomic, Molecular and Optical Theory. Scattering theory; electron-electron and electron-atom collisions; atomic collisions; molecular Rydberg states. Lin, Saffman, Walker.

Biophysics Theory. Modeling of a variety of complex biological systems. Coppersmith.

Condensed Matter Theory. Magnetism; optical properties; energy band structure; many-body problems; superconductivity; heavy fermion systems; quantum Hall effect; quantum algorithms; studies of decoherence; studies of novel experimental architectures. Chubukov, Coppersmith, Eriksson, Gilbert, Joynt, McDermott, Perkins, Vavilov.

High Energy Theory. Quantum field theory; particle astrophysics; string theory; mathematical physics; phenomenology of particle physics; collider physics; standard model and extensions; cosmology. Bai, Balantekin, Barger, Chung, Everett, Hashimoto, Shiu.

Materials Science, Metallurgy. Many-body problems; disordered systems; thin films. Lagally.

Neutrino and Astroparticle Theory. Early universe cosmology; dark matter and energy; baryogenesis; cosmic microwave background radiation; modified gravity; string cosmology; interstellar medium; supernova remnants; gas dynamics and radiation; cosmic-ray acceleration; neutrino and gamma-ray astronomy. Balantekin, Barger, Chung, Everett, Shiu, Zweibel.

Nuclear Theory. Reaction theory; scattering theory; nuclear structure; many-body theory; symmetry principles; heavy ions and intermediate energies; high energy nuclear physics; nuclear astrophysics. Balantekin.

Plasma Theory. Stability theory; plasma confinement; turbulence theory; anomalous transport; heating theory; computer simulation. Boldyrev, Terry, Zweibel.

Experimental

Atomic, Molecular, and Optical Experiment. Atomic collisions; lasers; atomic oscillator strengths; high-resolution spectroscopy; trapped atoms; weakly ionized plasmas; Rydberg atom flux qubit; semiconductoring architectures; very high-resolution studies of atomic, molecular, and astrophysical phenomena; spectral line-strength determinations. Lin, Saffman, Walker, Yavuz.

Biophysics Experiment. Photoelectron spectromicroscopy of biological systems; cancer therapy. Coppersmith, Gilbert.

Condensed Matter Experiment. Mesoscopic systems; scanning force microscopies; strongly correlated magnetic materials; high-temperature superconductivity; magnetic nanostructures and heterostructures; structural properties of polymers; synchrotron radiation studies of strongly correlated systems. Coppersmith, Eriksson, Gilbert, Himpsel, Lagally, McDermott, Onellion, Rzchowski, Winokur.

High Energy Experiment. CMS; CDF; ATLAS LBNE; Neutrino Physics at Daya Bay; weak, electromagnetic, and strong interactions; search for Higgs bosons and new physics phenomena; study of B-meson and neutrino oscillation physics; study of leptonic CV violation; dark matter searches; study of proton structure. Carlsmith, Dasu, Halzen, Herndon, Karle, Pan, Smith, Vandenbroucke, Wu.

Neutrino and Astroparticle Experiment. X-ray, gamma-ray, and neutrino astronomy; solar neutrinos; dark matter searches; observational cosmology; cosmic background radiation and spectroscopy. Balantekin, Halzen, Karle, McCammon, Timbie, Vandenbroucke, Westerhoff.

Nuclear Experiment. Polarization phenomena at low and intermediate energies; properties of few-nuclear systems; electromagnetic and weak interactions; tests of fundamental symmetries; neutrino physics; double beta decay and rare event searches; development of polarized-ion sources and polarized targets. Balantekin, Karle, Knutson, Vandenbroucke, Westerhoff.

Plasma Experiment. Toroidal confinement, instabilities, turbulence, and anomalous transport; reversed field pinch and tokamak. Boldyrev, Forest, Sarff, Schnack, Terry, Zweibel.

Quantum Computing. Coppersmith, Eriksson, McDermott, Saffman, Walker.

View additional information about this department at
www.gradschoolshopper.com

UNIVERSITY OF WISCONSIN, MILWAUKEE

DEPARTMENT OF PHYSICS

Milwaukee, Wisconsin 53201
http://www4.uwm.edu/letsci/physics/

General University Information
President: Kevin P. Reilly
Dean of Graduate School: David Yu (Interim)
University website: http://www4.uwm.edu/
Control: Public
Setting: Urban
Total Faculty: 826
Total Graduate Faculty: 826
Total number of Students: 29,145
Total number of Graduate Students: 4,946

Department Information
Department Chairman: Valerica Raicu, Chair
Department Contact: Kate Valerius, Graduate Program
  Assistant
  Total full-time faculty: 22
  Full-Time Graduate Students: 45
  First-Year Graduate Students: 8
  Female First-Year Students: 1
  Total Post Doctorates: 36

Department Address
P.O. Box 413
Milwaukee, WI 53201
Phone: (414) 229-4474
Fax: (414) 229-5589
E-mail: physgradadmissions@uwm.edu
Website: http://www4.uwm.edu/letsci/physics/

ADMISSIONS

Admission Contact Information
Address admission inquiries to: UW-Milwaukee, Department of
  Physics, Attn: Graduate Admissions, P.O. Box 413, Mil-
  waukee, WI 53201
Phone: (414) 229-4474
E-mail: physgradadmissions@uwm.edu
Admissions website: http://www4.uwm.edu/letsci/physics/
  graduate/admissions.cfm

Application deadlines
Fall admission:
U.S. students: February 15   *Int'l. students*: February 15

Application fee
U.S. students: $56   *Int'l. students*: $96
Applicants are encouraged to fully review all information on the
  admissions portion of our web page. The department has its
  own application process which is in addition to the required
  on-line application with the UWM Graduate School.

Admissions information
For Fall of 2012:
  Number of applicants: 58
  Number admitted: 19
  Number enrolled: 8

Admission requirements
Bachelor's degree requirements: For admission to the graduate
  program, a Bachelor's degree in physics or a related field is
  required, with a minimum undergraduate GPA of 2.75 specified. Applicants are required to take the GRE general test
  and are encouraged to take the GRE Physics subject test.
Minimum undergraduate GPA: 2.75

GRE requirements
The GRE is required.

Advanced GRE requirements
The Advanced GRE is recommended.

TOEFL requirements
The TOEFL exam is required for students from non-English-
  speaking countries.
  PBT score: 550
  iBT score: 79
These are minimum scores accepted. Minimum IELTS score accepted is 6.5.

Other admissions information
Additional requirements: For financial support, the minimum acceptable TOEFL score is 580 (237 for computer based test).
Undergraduate preparation assumed: Mechanics, Barger, Olsson, and Marion; Thermodynamics, Zemansky and Dittman;
  E.M., Corson, Lorrain, and Griffiths; Quantum Physics, Eisberg, Resnick, and Gasiorowitz.

TUITION

Tuition year 2012–13:
Tuition for in-state residents
  Full-time students: $7,780 annual
Tuition for out-of-state residents
  Full-time students: $17,140 annual
Additional amount/fee of approximately $1,000/AY, which are
  student segregated fees. These fees are not included in the
  amount quoted for full-time students (pro-rated for part-time
  students) and are the responsibility of the student.
Credit hours per semester to be considered full-time: 6
Deferred tuition plan: Yes
Health insurance: Available
Other academic fees: Special notes regarding stipends: 1) These
  are non-doctoral rates, 2) Non-resident portion of tuition is
  waived, 3) All amounts are for AY of 9 months, 4) Awards
  rarely exceed 50% of the full stipend amount indicated, and
  5) Additional awards are given competitively.
Academic term: Semester
Number of first-year students who received full tuition waivers: 8

Teaching Assistants, Research Assistants, and Fellowships
  Number of first-year
    Teaching Assistants: 7
    Research Assistants: 1
  Average stipend per academic year
    Teaching Assistant: $23,200
    Research Assistant: $34,000
    Fellowship student: $12,500

FINANCIAL AID

Application deadlines
Fall admission:

U.S. students: January 1   *Int'l. students*: January 1

Loans
Loans are available for U.S. students.

Loans are available for international students.

GAPSFAS application required: No

FAFSA application required: No

For further information
Address financial aid inquiries to: UW-Milwaukee, Dept. of Financial Aid, P. O. Box 469, Milwaukee, WI 53201.

Phone: (414) 229-4541

Financial aid website: http://www4.uwm.edu/financialaid/

HOUSING

Availability of on-campus housing
Single students: Yes

Married students: No

For further information
Address housing inquiries to: UW-Milwaukee, Attn. Housing/ Sandburg Hall, 3400 N. Maryland Ave., Milwaukee, WI 53211.

Phone: (414) 229-4065

E-mail: contract-group@uwm.edu

Housing aid website: http://www4.uwm.edu/housing/

Table A—Faculty, Enrollments, and Degrees Granted

Research Specialty	2012-13 Faculty	Enrollment Fall 2012 Master's	Enrollment Fall 2012 Doctorate	Number of Degrees Granted 2012-13 (2008-12) Master's	Number of Degrees Granted 2012-13 (2008-12) Terminal Master's	Number of Degrees Granted 2012-13 (2008-12) Doctorate
Astrophysics	3	–	4	–	1(-)	–(1)
Biophysics	4	–	7	–	–(3)	1(1)
Condensed Matter Physics	8	–	18	–	–(4)	2(7)
Medical, Health Physics	1	–	–	–(1)	–(2)	–
Optics	–	–	–	–	–	–(2)
Particles and Fields	1	–	1	–	–	–
Relativity & Gravitation	4	–	15	–(1)	–(1)	1(9)
Non-specialized	1	–	–	–	–	–
Total	22	–	45	–(2)	1(10)	4(20)
Full-time Grad. Stud.	–	–	45	–	–	–
First-year Grad. Stud.	–	–	8	–	–	–

GRADUATE DEGREE REQUIREMENTS

Master's: The Master's graduate must complete 24 graduate credits (12 credits must be earned at UWM), and pass a comprehensive examination. The student may choose to engage in research and present a thesis for 6 credits of the total needed. The Graduate School requires an average of at least 3.0 (4.0 basis) for all graduate work.

Doctorate: The PhD graduate is required to complete 54 graduate credits (27 credits must be earned in residence at UWM) beyond the Bachelor's degree and pass a written comprehensive examination. A dissertation reporting the results of original and independent research investigation must be written and defended. The Graduate School requires an average of at least 3.0 (4.0 basis) for all graduate work.

Thesis: Thesis may be written in absentia.

SPECIAL EQUIPMENT, FACILITIES, OR PROGRAMS

Certain faculty members of the department also participate in Laboratory for Surface Studies, an interdepartmental unit (with members from physics, chemistry, material science, etc.) dedicated to the study of surfaces on a microscopic scale.

Laboratory equipment in the Physics Building includes high and ultra high vacuum systems with facilities for the following analytical techniques: high- and low-energy-electron diffraction (RHEED & LEED), Auger spectroscopy, x-ray energy dispersive spectrometry (EDS), electron spectroscopy for chemical analysis (ESCA), infrared spectroscopy (IR), scanning tunneling microscopy (STM), atomic force microscopy (AFM), high resolution transmission electron microscopy (HRTEM) and molecular beam epitaxy (MBE).

Experiments using synchrotron radiation are carried out at the nearby Wisconsin Synchrotron Radiation center and at Brookhaven National Laboratory and Argonne National Laboratory.

Additional experimental condensed matter research facilities include a 3He-4He dilution refrigerator, several superconducting magnets (8–9.5T), pulsed-ultrasonic equipment (1 MHz–4 GHz), and a thin film deposition facility-sputtering, evaporation, microphotolithography and clean-room high temperature furnaces. Oxide synthesis in infrared image furnace and E-beam furnace for single crystal growth. Physical property measurement equipment.

Ultra high power femtosecond pulsed laser is a key component of the optics lab.

Biophysics laboratories are equipped with all necessary tools to grow bacteria, to purify and crystallize proteins, to collect x-ray diffraction data and to analyze them for structure determination; also equipment for Förster Resonance Energy Transfer (FRET) experiments.

Experimental collaborations exist with, among others, Argonne National Laboratory, Univ. of Illinois, Lucent Labs., Univ. of Hamburg, Illinois Institute of Technology, Univ. of Chicago, Science Univ. of Tokyo, Northwestern University and the National High Magnetic field Laboratory at Tallahassee and at Los Alamos.

Table B—Separately Budgeted Research Expenditures by Source of Support

Source of Support	Departmental Research	Physics-related Research Outside Department
Federal government	$4,359,101	
State/local government		
Non-profit organizations	$56,000	
Business and industry	$145,798	
Other		
Total	$4,560,899	

Table C—Separately Budgeted Research Expenditures by Research Specialty

Research Specialty	No. of Grants	Expenditures ($)
Biophysics	12	$1,522,857
Condensed Matter Physics	6	$699,973
Optics	–	
Relativity & Gravitation	19	$2,201,069
Other	2	$137,000
Total	39	$4,560,899

FACULTY

Professor

Agterberg, Daniel, Ph.D., University of Toronto, 1996. *Condensed Matter Physics*. Theoretical condensed matter physics (superconductivity and magnetism).

Brady, Patrick, Ph.D., University of Alberta, 1993. *Relativity & Gravitation*. Gravitational wave detection; detection of periodic sources, coalescing binary search. Gravitational wave astronomy. Numerical relativity; flux conservative formalisms; intermediate black hole coalescence. Dynamics of gravitational collapse, quantum effects in gravitational collapse.

Gajdardziska-Josifovska, Marija, Ph.D., Arizona State University, 1991. *Condensed Matter Physics, Solid State Physics*. Experimental solid state physics; transmission and reflection electron microscopy of solid surfaces and interfaces; electron holography; reflection high energy electron diffraction; nanodiffraction; x-ray and electron-energy-loss spectroscopy; growth and optical properties of thin films.

Hirschmugl, Carol J., Ph.D., Yale University, 1994. *Biophysics, Condensed Matter Physics, Optics, Surface Physics*. Experimental surface physics. Infrared synchrotron radiation studies of surfaces; photoelectron diffraction studies of hydrocarbons on metals; reflection-absorption spectroscopy; infrared microspectroscopy.

Li, Lian, Ph.D., Arizona State University, 1995. *Surface Physics*. Experimental surface physics. Molecular beam epitaxy, scanning tunneling microscopy.

Lyman, Paul, Ph.D., University of Pennsylvania, 1991. *Condensed Matter Physics, Surface Physics*. Experimental surface science. Semiconductor adsorbates and thin film growth using x-ray scattering, x-ray standing waves, photoemission, and reflection high energy electron diffraction.

Ourmazd, Abbas, Ph.D., University of Oxford, 1980. *Biophysics, Condensed Matter Physics*. Semiconductor physics, electron microscopy and holography, protein crystallography.

Saldin, Dilano K., Ph.D., University of Oxford, 1975. *Biophysics, Condensed Matter Physics, Surface Physics*. Theoretical surface science; crystallography of ordered and disordered surfaces; protein crystallography; theory of electron diffraction and microscopy; x-ray absorption; electron energy-loss spectroscopy.

Sarma, Bimal, Ph.D., Northwestern University, 1980. *Low Temperature Physics*. Low-temperature physics; heavy-fermion superconductivity; high-Tc superconductivity; ultrasonics; neutron diffraction, and physical phenomena at high magnetic fields.

Weinert, Michael, Ph.D., Northwestern University, 1982. *Condensed Matter Physics*. Theoretical condensed matter physics, magnetism and first-principles electronic structure.

Associate Professor

Anchordoqui, Luis, Ph.D., Universidad Nacional de La Plata, 1998. *Astrophysics, Cosmology & String Theory, Particles and Fields*. Cosmic ray astrophysics and neutrino astrophysics with applications to particle physics and cosmology.

Creighton, Jolien, Ph.D., University of Waterloo, 1996. *Relativity & Gravitation*. General relativistic theory of gravity. Physics of black holes and gravitational radiation.

Guptasarma, Prasenjit, Ph.D., Tata Institute of Fundamental Research, 1993. *Condensed Matter Physics*. Experimental condensed matter physics. Highly correlated electronic materials, unconventional superconductivity, floating-zone single crystal growth, physical properties of materials at low temperatures.

Patch, Sarah K., Ph.D., University of California, Berkeley, 1994. *Medical, Health Physics*. Image reconstruction, primarily for medical imaging. Current focus: thermo/opto-acoustic tomography, requiring inversion of the spherical Radon transform.

Raicu, Valerica, Ph.D., University of Bucharest, 1997. *Biophysics, Condensed Matter Physics, Optics*. Studies of biological systems by optical and electrical methods.

Schmidt, Marius, Ph.D., Technical University of Munich, 1996. *Biophysics*. Structural biophysics.

Siemens, Xavier, Ph.D., Tufts University, 2002. *Relativity & Gravitation*. Gravitational Physics.

Wiseman, Alan, Ph.D., Washington University, St. Louis, 1992. *Relativity & Gravitation*. Gravitational wave source modeling. Computer analysis of gravitational-wave data.

Assistant Professor

Chang, Philip, Ph.D., University of California, Santa Barbara, 2005. *Astrophysics*. Plasma astrophysics, theory of compact objects (neutron stars and white dwarfs), and astrophysical fluid dynamice.

Erb, Dawn, Ph.D., California Institute of Technology, 2005. *Astrophysics*. Galaxy formation and evolution, the kinematics, chemical evolution and stellar populations of galaxies at high redshift, feedback processes in starburst galaxies, and the evolution of the intergalactic medium at high redshift.

Kaplan, David, Ph.D., California Institute of Technology, 2004. *Astrophysics*. Observations of compact objects at all wavelengths, with an emphasis on connections to basic physics. Currently also working on the development of new facilities for observations of the transient radio sky.

Emeritus

Beck, Donald, Ph.D..
Chow, Yutze, Ph.D..
Dittman, Richard, Ph.D..
Friedman, John, Ph.D..
Greenler, Robert, Ph.D..
Levy, Moises, Ph.D..
Lubkin, Elihu, Ph.D..
McQuistan, Richmond, Ph.D..
Parker, Leonard, Ph.D..
Schmieg, Glenn, Ph.D..
Shurman, Michael, Ph.D..
Snider, Dale, Ph.D..
Sorbello, Richard, Ph.D..
Suchy, Raymond.
Walters, William, Ph.D..

Adjunct Associate Professor

Wood, Robert, Ph.D., University of Oxford, 1979. *Condensed Matter Physics*. Electron-nuclear double resonance (ENDOR) applied to studies of transferred hyperfine structure of dilute paramagnetic impurities in insulating crystals with the fluorite structure.

DEPARTMENTAL RESEARCH SPECIALTIES AND STAFF

Theoretical

Astrophysics. Galaxy formation and evolution, observations of compact objects at all wavelengths with an emphasis on connections to basic physics; also, the evolution of the intergalactic medium at high redshift, cosmic ray astrophysics (ultrahigh energy cosmic rays), radio astronomy, optical astronomy, and neutrino astronomy with applications to particle physics and cosmology. 3 postdoctoral fellows. Anchordoqui, Chang, Erb, Kaplan, Siemens.

Biophysics. Single molecule crystallography, direct methods. 2 senior scientists; 3 postdoctoral fellows. Ourmazd, Saldin.

Condensed Matter Physics. Transport phenomena; electron and atom transport in solids and low-dimensional systems, magnetism, phase stability, electronic structure, electromigration and superconductivity. 1 postdoctoral fellow. Agterberg, Weinert.

Medical, Health Physics. Image reconstruction and data corrections/pre-processing. Work relies heavily on Fourier analysis, and inversion of Radon transforms. Patch.

Particles and Fields. Elementary particle phenomenology related to experimental searches for new physics at the Large Hadron Collider; hadronic interaction models through analyses of extensive air showers recorded by the Pierre Auger Observatory. 1 postdoctoral fellow. Anchordoqui.

Relativity & Gravitation. Gravitational-wave astronomy; collaboration on the construction for LIGO (Laser-Interferometric Gravitational-wave Observatory) of templates to detect the gravity-wave background of the early universe and the waves emitted in the coalescence of binary systems of neutron stars (and of black holes). Early cosmology; use of renormalization-group methods to study the physics of inflationary universe models; gravitational waves from a network of cosmic strings. Quantum field theory on curved spacetime; blackhole evaporation, particle production in the early universe; field theory on spacetimes that are not globally hyperbolic. Aspects of quantum gravity associated with small-scale noneuclidean topology. Relativistic astrophysics; structure and stability of relativistic stars; limits on the mass and rotation of pulsars, and implied constraints on the equation of state of matter above nuclear density. 4 senior scientists; 10 postdoctoral fellows. Brady, Creighton, Friedman, Siemens, Wiseman.

Surface Physics. Crystallography of ordered and disordered surfaces, theory of electron diffraction and microscopy, x-ray absorption, electron-energy-loss spectroscopy, electron holography, dynamical processes at surfaces and interfaces, surface phonons, atom-solid bonding, photoemission, inelastic low-energy electron diffraction, and surface electronic structure of materials reconstruction. Saldin, Weinert.

Experimental

Biophysics. Protein crystallography. Development of infrared microspectroscopy, confocal microscopy, fluorescence, and Raman spectroscopy methods for characterization of living cells. Iron and toxic metal storage in organisms; atomic structure of botanical nanocrystalline iron biominerals; biomagnetism; biogenic self assembly; life in extreme environments. 1 senior scientist; 6 postdoctoral fellows. Gajdardziska-Josifovska, Hirschmugl, Ourmazd, Raicu, Schmidt.

Condensed Matter Physics. Studies of the electron-phonon interaction in superconductors, surface wave interaction with superconducting and magnetic films and spin-phonon interaction. The principal technique used in these studies is ultrasonic attenuation, and other physical property measurements at low temperatures. Areas of interest are high-Tc superconductivity, heavy fermion superconductivity, milliKelvin temperature physics and neutron diffraction. Single crystal growth by floating-zone and other techniques, oxide and intermetallic compounds, manganite and cuprate oxides, ferroelectric oxides. 3 postdoctoral fellows. Guptasarma, Sarma.

Optics. Infrared and visible microspectroscopy. Hirschmugl, Raicu.

Surface Physics. Physical methods for synthesis of advanced materials with reduced dimensionality, including thin films and nanocrystals of wide band gap semiconductors and high k-dielectrics. Development and applications of electron and photon based imaging, diffraction and spectroscopy methods for determination of atomic structures, with emphasis on solid surfaces, interfaces and nanocrystals. Effects of surface structure on controlled epitaxial growth of single crystal films; Stabilization mechanisms for polar oxide surfaces and interfaces; Low energy dynamics and structure of water-oxide interfaces, electronic friction at adsorbate-superconductor interfaces. 2 postdoctoral fellows. Gajdardziska-Josifovska, Hirschmugl, Li, Lyman.

View additional information about this department at
www.gradschoolshopper.com

UNIVERSITY OF WYOMING

DEPARTMENT OF PHYSICS AND ASTRONOMY

Laramie, Wyoming 82071-3905
http://www.uwyo.edu/physics

General University Information

President: Robert Sternberg
Dean of Graduate School: N/A
University website: http://www.uwyo.edu
Control: Public
Setting: Rural
Total Faculty: 749
Total number of Students: 13,922
Total number of Graduate Students: 2,923

Department Information

Department Chairman: Daniel Dale, Chair
Department Contact: Daniel Dale, Department Chairman
Total full-time faculty: 15
Total number of full-time equivalent positions: 15
Full-Time Graduate Students: 34
First-Year Graduate Students: 12
Female First-Year Students: 5
Total Post Doctorates: 6

Department Address

1000 E University Avenue
Dept 3905
Laramie, WY 82071-3905
Phone: (307) 766-6150
Fax: (307) 766-2562
E-mail: physics@uwyo.edu
Website: http://www.uwyo.edu/physics

ADMISSIONS

Admission Contact Information

Address admission inquiries to: Director of Graduate Studies, University of Wyoming, Department of Physics & Astronomy, 1000 E. University, Dept. 3905, Laramie, WY 82071
Phone: (307) 766-6150
E-mail: physics@uwyo.edu
Admissions website: http://www.uwyo.edu/admissions

Application deadlines

Fall admission:
U.S. students: January 22   *Int'l. students*: January 22
Spring admission:
U.S. students: January 22   *Int'l. students*: January 22

Application fee

There is no application fee required.

Admissions information

For Fall of 2013:
  Number of applicants: 85
  Number admitted: 11
  Number enrolled: 8

Admission requirements

Bachelor's degree requirements: Bachelor's degree in physics is required.
Minimum undergraduate GPA: 3.0

GRE requirements

The GRE is required.
  Analytical score: 1

Advanced GRE requirements

The Advanced GRE is required.
  Minimum accepted Advanced GRE score: 200

TOEFL requirements

The TOEFL exam is recommended for students from non-English-speaking countries.
  PBT score: 540
  iBT score: 76

Other admissions information

Undergraduate preparation assumed: Undergraduate preparation in physics and mathematics equivalent to that specified for a physics major.

TUITION

Tuition year 2013-14:
Tuition for in-state residents
  Full-time students: $3,780 annual
  Part-time students: $210 per credit
Tuition for out-of-state residents
  Full-time students: $11,322 annual
  Part-time students: $629 per credit $11,320 $5,660
Credit hours per semester to be considered full-time: 9
Deferred tuition plan: Yes
Health insurance: Available at the cost of 1,011 per year.
Academic term: Semester
Number of first-year students who received full tuition waivers: 7

Teaching Assistants, Research Assistants, and Fellowships

Number of first-year
  Teaching Assistants: 4
  Research Assistants: 2
  Fellowship students: 2
Average stipend per academic year
  Teaching Assistant: $15,795
  Research Assistant: $15,795
  Fellowship student: $15,795

FINANCIAL AID

Application deadlines

Fall admission:
U.S. students: August 30   *Int'l. students*: August 30

Loans

Loans are available for U.S. students.
Loans are not available for international students.
GAPSFAS application required: No
FAFSA application required: No

For further information

Address financial aid inquiries to: Student Financial Aid, Department 3335, 1000 E. University Avenue, Laramie, WY 82071.
Phone: (307) 766-2116
E-mail: finaid@uwyo.edu
Financial aid website: http://www.uwyo.edu/sfa/

HOUSING

Availability of on-campus housing

  Single students: Yes
  Married students: Yes

For further information

Address housing inquiries to: Residence Life & Dining, Dept. 3394, 1000 E. University Ave., Laramie, WY 82071.
Phone: (866) 653-0212
E-mail: reslife-dining@uwyo.edu
Housing aid website: http://www.uwyo.edu/reslife-dining/

Table A—Faculty, Enrollments, and Degrees Granted

Research Specialty	2012-13 Faculty	Enrollment Fall 2013 Master's	Enrollment Fall 2013 Doctorate	Number of Degrees Granted 2011–12 (2006-11) Master's	Number of Degrees Granted 2011–12 (2006-11) Terminal Master's	Number of Degrees Granted 2011–12 (2006-11) Doctorate
Astrophysics	6	–	10	–(5)	–(2)	2(7)
Biophysics	1	–	–	–	–	–
Condensed Matter						
Physics	5	–	12	3(3)	1(3)	–(1)
Plasma and Fusion	1	–	–	–	–	–
Other	1	7	1	–	1(2)	–
Total	14	7	23	3(8)	2(5)	2(8)
Full-time Grad. Stud.	–	–	30	–	–	–
First-year Grad. Stud.	–	1	8	–	–	–

GRADUATE DEGREE REQUIREMENTS

Master's: Thirty hours of graduate course work. Thesis planning, development, and production guided by the committee chair and graduate committee. 3.0 GPA (on 4.0 scale) required.
Doctorate: Forty-two hours of course work at the graduate level, 30 hours of research. Dissertation planning, development, and

production guided by the committee chair and graduate committee. 3.0 GPA (on 4.0 scale) required. Comprehensive exam required.

SPECIAL EQUIPMENT, FACILITIES, OR PROGRAMS

Wyoming Infrared Observatory-The University of Wyoming's 2.3-meter telescope is located at the Wyoming Infrared Observatory (WIRO) about 25 miles southwest of Laramie, WY, on the summit of Jelm Mt. and at an altitude of 9,656 ft. (2,943 m). The design of the WIRO 2.3-meter telescope is a classical Cassegrain. This includes a large, concave primary mirror with a parabolic surface and a smaller convex secondary mirror with a hyperbolic shape. The NCAR Supercomputer, located in Cheyenne, will contain some of the world's most powerful supercomputers dedicated to improving scientific understanding of climate change, severe weather, air quality, and other vital atmospheric science and geoscience topics. The center will also house a premier data storage and archival facility that holds irreplaceable historical climate records and other information.

Table B—Separately Budgeted Research Expenditures by Source of Support

Source of Support	Departmental Research	Physics-related Research Outside Department
Federal government	$4,813,000	
State/local government	$2,182,555	
Non-profit organizations		
Business and industry	$120,000	
Other	$2,234,000	
Total	$9,349,555	

FACULTY

Professor

Dahnovsky, Yuri, Ph.D., Russian Academy of Sciences, 1983. *Condensed Matter Physics.* Computational and theoretical physics: molecular electronic devices, solar cells, electronic properties of surfaces, nonequilibrium Green functions, photon-assisted tunneling and electron transfer reactions.

Dale, Daniel, Ph.D., Cornell University, 1998. *Astrophysics.* Ground- and space-based multi-wavelength studies of galaxies; clusters of galaxies; observational cosmology.

Johnson, Paul, Ph.D., University of Washington, 1979. *Biophysics.* Biophysics; detection of pathogenic micro-organisms.

Tang, Jinke, Ph.D., Iowa State University, 1989. *Condensed Matter Physics.* Experimental condensed matter physics and materials science: spintronics and optoelectronics; magnetic semiconductors; half-metals; magnetic, optical and thermoelectric properties of nanomaterials; tunneling magnetoresistance; thin films.

Thayer, David, Ph.D., Massachusetts Institute of Technology, 1983. *Plasma and Fusion.* Theoretical studies of plasmas, fusion, turbulence, nonlinear dynamics, global change, and quantum mechanics.

Associate Professor

Brotherton, Michael, Ph.D., University of Texas, 1996. *Astrophysics.* Multi-wavelength observations of quasars and active galaxies; quasar/galaxy mutual evolution.

Kobulnicky, Chip, Ph.D., University of Minnesota, 1997. *Astrophysics.* Chip is an observational astronomer, using ground-based and space-based telescopes at radio, infrared, optical, and x-ray wavelengths to study stars, star formation, the chemical composition of interstellar material, and the evolution of galaxies.

Michalak, Rudi, Ph.D., Physics, Ruhr-Universität Bochum, 1993. *Condensed Matter Physics.* Experimental condensed matter physics; nuclear magnetic resonance; science education.

Pierce, Michael, Ph.D., University of Hawaii, 1988. *Astrophysics.* Galaxies, clusters of galaxies, large-scale structure of the universe, observational cosmology, astronomical instrumentation.

Assistant Professor

Barrans, Rich, Ph.D., Caltech, 1993. Organic chemistry.

Chien, TeYu, Ph.D., University of Tennessee. *Condensed Matter Physics, Nano Science and Technology.* Solar cell devices; Complex oxides electronic devices.

Jang-Condell, Hannah, Ph.D., Harvard University, 2004. *Astrophysics.* Theoretical planetary formation.

Koncel, Ed, Other, University of Wyoming, 1992. Science education.

Myers, Adam, Ph.D., Durham University, 2004. *Astronomy.* Quasars; cosmology.

Wang, Wenyong, Ph.D., Yale University, 2004. *Condensed Matter Physics.* Experimental condensed matter physics; nanotechnology.

DEPARTMENTAL RESEARCH SPECIALTIES AND STAFF

Theoretical
Astrophysics. Jang-Condell.
Condensed Matter Physics. Dahnovsky.
Materials Science, Metallurgy. Dahnovsky.
Plasma and Fusion. Thayer.

Experimental
Astrophysics. Brotherton, Dale, Kobulnicky, Myers, Pierce.
Biophysics. Johnson.
Condensed Matter Physics. Chien, Michalak, Tang, Wang.
Materials Science, Metallurgy. Chien, Michalak, Tang, Wang.
Nano Science and Technology. Chien, Tang, Wang.

View additional information about this department at
www.gradschoolshopper.com

PART II

INTERNATIONAL

Geographic Listing of Graduate Programs

RYERSON UNIVERSITY

DEPARTMENT OF PHYSICS

Toronto, Canada
http://www.ryerson.ca/physics/

General University Information
President: Sheldon Levy
Dean of Graduate School: Jennifer Mactavish
University website: http://www.ryerson.ca
Total Faculty: 830
Total Graduate Faculty: 500
Total number of Students: 41,250
Total number of Graduate Students: 2,300

Department Information
Department Chair: Ana Pejović-Milić, Chair
Department Contact: Cynthia Dy, Graduate Program
  Administrator
  Total full-time faculty: 16
  Total number of full-time equivalent positions: 16
  Full-Time Graduate Students: 34
  First-Year Graduate Students: 20
  Female First-Year Students: 13
  Total Post Doctorates: 7

Department Address
350 Victoria Street
Toronto, ON M5B 2K3
CANADA
Phone: (416) 979-5000 ext. 4760
Fax: (416) 979-5343
E-mail: biomed@ryerson.ca
Website: http://www.ryerson.ca/physics/

ADMISSIONS

Admission Contact Information
Address admission inquiries to: Graduate Studies Admissions
  Office, 11th floor, 1 Dundas Street West, Toronto, ON, Can-
  ada
Phone: 416-979-5150
E-mail: gradadmit@ryerson.ca
Admissions website: http://www.ryerson.ca/gradstudies/
  admissions

Application deadlines
Fall admission:
Domestic students: February 1   *Int'l. students*: June 15

Application fee: $110
Applications for Fall admission are reviewed starting Feb 1 until
  the programs are full.

Admissions information
For Fall of 2012:
  Number of applicants: 70
  Number admitted: 21
  Number enrolled: 21

Admission requirements
Bachelor's degree requirements: Physics, engineering, or a related field.

GRE requirements
The GRE is not required.

Advanced GRE requirements
The Advanced GRE is not required.

TOEFL requirements
The TOEFL exam is required for students from non-English-
  speaking countries.
iBT score: 93
TOEFL is only required for students who have completed undergraduate education in a non-English speaking institution

TUITION
Tuition year 2012–13:
Tuition for in-state residents
  Full-time students: $3,436.06 per semester
Tuition for out-of-state residents
  Full-time students: $6,757.82 per semester
please see attached tuition fee schedule for 2013-2014.
Deferred tuition plan: No
Health insurance: Available at the cost of $295 per year.
Academic term: Semester

Teaching Assistants, Research Assistants, and Fellowships
Number of first-year
  Teaching Assistants: 12
  Fellowship students: 2
Average stipend per academic year
  Teaching Assistant: $9,000
  Research Assistant: $10,000
  Fellowship student: $6,300
salary rates for Teaching Assistants are driven by Ryerson
University's current CUPE 3 collective agreement.

FINANCIAL AID

For further information
Address financial aid inquiries to: Ryerson University Financial
  Aid Office, POD59, 350 Victoria Street, Toronto, ON M5B
  2K3. Canada.
Phone: 416-979-5113
E-mail: finaid@ryerson.ca
Financial aid website: http://www.ryerson.ca/currentstudents/
  financialaid/

HOUSING

Availability of on-campus housing
Single students: No
Married students: No

For further information
Address housing inquiries to: Ryerson University, through the
  Student Housing Services Office, provides an off-campus
  Housing Registry that may assist you in finding accommo-
  dation. There are no on campus housing facilities available
  to Graduate Students at this time. Please consult the Registry.
Phone: (416) 979-5284
E-mail: offcamp@ryerson.ca
Housing aid website: http://www.ryerson.ca/offcampushousing

Table A—Faculty, Enrollments, and Degrees Granted

Research Specialty	2012-13 Faculty	Enrollment Fall 2012		Number of Degrees Granted 2012-13 (2008-13)		
		Master's	Doctorate	Master's	Terminal Master's	Doctorate
Astronomy	1	–	–	–	–	–
Biomedical Physics	13	24	12	–(27)	–(18)	–
Nuclear Physics	1	–	–	–	–	–
Physics and other Science Education	1	–	–	–	–	–
Total	16	24	12	–(27)	–(18)	–
Full-time Grad. Stud.	–	24	12	–	–	–
First-year Grad. Stud.	–	14	7	–	–	–

GRADUATE DEGREE REQUIREMENTS

Master's: BP8102 Medical Diagnostic Techniques BP8103 Fundamentals of Radiation Physics BP82January 2 Master's seminar I/II Two (2) credits from elective list.

Doctorate: BP9101 Science Communication BP9201/2/3/4 Doctoral Seminar I/II/III/IV Two (2) credits from the elective list.

Other Degrees: Option in Medical Physics: BP8104 Radiation Therapy BP8113 Advanced Imaging BP8107 Radiation Protection and Dosimetry BP8112 Radiobiology BP8114 Anatomy and Physiology for Medical Physicists.

Thesis: Master's Thesis (Milestone) Doctoral Candidacy Examination Doctoral Dissertation.

SPECIAL EQUIPMENT, FACILITIES, OR PROGRAMS

MEDICAL IMAGING AND TREATMENT MODALITIES: Digitizers; Preamplifiers; Variable 400 volt square wave pulser/receivers with 35 MHz bandwidth; Two-dimensional raster scanner; Permanent magnet that can generate a magnetic field of 0.1-0.2 T within a volume of a 4 inch-cube; Different digital and analog function generators.

ADVANCED BIOMEDICAL ULTRASOUND IMAGING AND THERAPY: HIFU transducers and systems; Histotripsy transducers and systems; High-end ultrasound imaging systems (scanners); Micro-positioning systems for acoustic field measurements; Calibrated ultrasound hydrophones (standard and HIFU); RF power amplifiers; Digital oscilloscopes and impedance meters; Water conditioner system (degasser and deionizer); Thermocouple-based thermometry system; Nerve electrophysiology system; Ultrasound simulation software packages; High resolution thermal camera; 4-channel fluoroptic thermometry system; Acoustic camera; 4-channel laser and delivery system Arbitrary signal generator; Ultrasound transducers; Power amplifier; Lock-in amplifier; Digital oscilloscope.

ULTRASOUND BIOMICROSCOPY: Visualsonics VS40B high frequency ultrasound imager; 2.6GS arbitrary waveform generator (Textronix model AWG610); Power amplifier that can be used to achieve 56 dB gains up to 250 MHz (Empower RF systems model BBS0D3FRR); Various oscilloscopes (National Instruments, computer based) and high end oscilloscope (touch screen Textronix TDS5052); 3-D scanning apparatus with calibrated hydrophones (Precision Acoustics Ltd., England); Several high frequency ultrasound transducers (at 20, 30, 40, 55 MHz); Several heating transducers.

OPTOACOUSTIC IMAGING: Laser optoacoustic imaging system (LOIS) (Fairway Medical Technologies, Inc. Houston, Tx); Optical parametric oscillator (OPO - from Optotek Vibrant) delivering light pulses with duration 6ns, maximum energy 60 mJ/pulse and covering a spectral range from 680 to 950 nm; Double

integrating spheres for measuring different optical properties; Compact spectrometers capable of detection over the visible and NIR range of the spectrum; Interferometry setup around a solid state laser with cw output at 532 nm.

COMPUTATIONAL AND MATHEMATICAL PHYSICS: Linux based IBM high performance computational cluster containing six Opteron dual processor computational nodes, 40 gigabytes of RAM and a full license of the ANSYS Multiphysics numerical analysis software, and 30 licenses for COMSOL Multiphysics.

TRACE ELEMENT DETECTION: S2 Ranger Energy-Dispersive X-ray spectrometer (Bruker-AXS, Madison, WI,USA); S2 PicoFox Total Reflection X-ray spectrometer (Bruker-AXS, Madison, WI,USA); Tracer III-SD Handheld spectrometer (Bruker-AXS, Madison, WI,USA); Gamble Technologies SLP - 16220 - P: Silicon (Li) X-ray Detectors DSPEC-PLUS Digital Gamma-Ray Spectrometer with MAESTRO-32 Software; Canberra G:2020R HPGe Low Energy Photon Spectrometer; DSA-2000 Digital Gamma-Ray Spectrometer with Genie 2000 software; Ball mill; Pellett press.

FACULTY

Professor

Buckby, Margaret, Ph.D., University of Toronto. *Astronomy, Astrophysics, Nuclear Physics.* Nuclear reactions occurring during supernova explosions.

Carvalho, M. Juliana, Ph.D., University of Toronto. *Nuclear Physics, Particles and Fields, Physics and other Science Education.* Collective Motion in Nuclear Physics, Algebraic Models, Schur Function Formalism. MAPLE as a tool for teaching Physics.

Goldman, Pedro, Ph.D., University of Windsor. *Medical, Health Physics, Physics and other Science Education.* Radiation Therapy of Tumours, Fast Inverse Dose Optimizations for Intensity Modulated Radiation Therapy (IMRT) and Tomotherapy, Alternative Methods for Efficient CT Image Reconstruction, Physics Education.

Kolios, Michael C., Ph.D., University of Toronto. *Acoustics, Applied Physics, Biophysics, Computational Physics, Nano Science and Technology, Nonlinear Dynamics and Complex Systems, Optics.* Ultrasound imaging and therapeutics, Ultrasound imaging, Heat transfer in tissue, Thermal Therapies.

Pejović-Milić, Ana, Ph.D., McMaster University. Interim Chair. *Chemical Physics, Medical, Health Physics.* Medical physics, trace elements analysis in humans, bone strontium, aluminum, manganese, and magnesium, nuclear analytical methods for medical applications, X-ray fluorescence (XRF), in vivo neutron activation analysis (IVNAA).

Associate Professor

Antimirova, Tetyana, Ph.D., Institute for Problems of Materials Science, National Academy of Sciences of Ukraine Kyev, 1990. Undergraduate Program Director. *Condensed Matter Physics, Physics and other Science Education.* Physics Education. Physical Chemistry.

Beauchemin, Catherine, Ph.D., University of Alberta, 2005. Co-op Program Faculty Advisor. *Biophysics, Computational Physics, Computer Science, Medical, Health Physics.* Computational/mathematical modelling of self-organizing systems. Modelling dynamics of infectious diseases within a host and host/immune interactions.

Cordes, Dietmar, Ph.D., University of Nevada-Reno. *Computational Physics, Medical, Health Physics.* MR Physics, MR Imaging, Echoplanar Imaging and Functional MRI (fMRI),

fMRI Data Analysis and Mathematical Methods, Ultra-short Echo-time Imaging (UTE).

Douplik, Alexandre, Ph.D., Russian State Medical University, 1996. *Applied Physics, Biophysics, Chemical Physics, Medical, Health Physics, Nano Science and Technology, Optics.* Biospectroscopy for clinical diagnostics, Bioimaging for Cancer and Atherosclerotic Diagnostics, Laser Surgery Navigation and Monitoring, Photodynamic Therapy, Nanobiophotonics, Perception and reaction intelligent virtual training.

Karshafian, Raffi, Ph.D., University of Toronto, 2009. *Acoustics, Applied Physics, Biophysics, Medical, Health Physics, Nano Science and Technology, Nonlinear Dynamics and Complex Systems.* Ultrasound imaging and therapeutics, ultrasound contrast agents.

Kumaradas, J. Carl, Ph.D., University of Toronto, 2002. Graduate Program Director. *Acoustics, Applied Physics, Biophysics, Computational Physics, Electromagnetism, Medical, Health Physics, Nano Science and Technology, Optics, Physics and other Science Education.* Medical Physics, Thermal Therapy, Electromagnetism, Heat Transfer, Numerical Analysis.

Tavakkoli, Jahan, Ph.D., University of Lyon 1. *Acoustics, Medical, Health Physics.* Biomedical Ultrasound (Therapeutic and Diagnostic). Image-guided Ultrasound Surgery. Nonlinear Acoustic Modeling and Simulation. Ultrasound Signal and Image Processing. Medical Devices and Technologies.

Toronov, Vladislav, Ph.D., Saratov State University. *Medical, Health Physics, Optics.* Optical and Magnetic Resonance Biomedical Imaging Non-linear dynamics of the brain.

Xu, Yuan, Ph.D., Texas A&M University, 2003. *Acoustics, Computational Physics, Electromagnetism, Medical, Health Physics.* Ultrasound imaging. Novel imaging reconstruction algorithms. Multi-wave imaging methods. Medical imaging.

Assistant Professor

Chithrani, Basnagge Devika, Ph.D., University of Toronto. *Biophysics, Medical, Health Physics, Nano Science and Technology.* Medical physics, synthesis and characterization of nanoparticles, development of nanoparticle based systems for multimodal imaging and therapeutics, nanoparticle based radiosensitizers, drug delivery, intracellular fate of nanoparticles.

Heath, Emily, Ph.D., McGill University, 2007. *Computational Physics, Medical, Health Physics.* Radiation Therapy, Image Registration, Robust Treatment Plan Optimization, Monte Carlo simulation.

DEPARTMENTAL RESEARCH SPECIALTIES AND STAFF

Theoretical

Computational and Mathematical Physics. Physical Modelling in Biology, Immunology, and Ecology (phymbie). Computational Biomedical Physics Laboratory. Simulated treatment courses using Monte Carlo techniques. Beauchemin, Cordes, Heath, Kolios, Kumaradas, Tavakkoli, Toronov.

Physics Education. The investigation of the effectiveness and impact of using technology in both large and small undergraduate physics classrooms. The technologies under investigation include interactive peer response systems (clickers), microcomputer based laboratory equipment such as Logger Pro or interactive computer simulations like the ones created by the Physics Educational Technology Team at the University of Colorado at Boulder. We are also interested in studying and addressing the issue of motivation and attitudes through the analysis of undergraduate physics courses. Antimirova, Carvalho, Goldman, Kumaradas.

Experimental

Medical Imaging and Treatment Modalities. Ultrasound-mediated imaging. Advanced biomedical ultrasound imaging and therapy. Minimally invasive thermal therapy. Ultrasound biomicroscopy. Optoacoustic imaging. Magnetic Resonance Imaging and near infrared spectroscopy. Robust Treatment Planning. Clinical feedback for laser surgery. Nanoparticles for improved therapeutics and imaging in cancer therapy. Ultrasound and microbubble mediated therapeutic applications. Treatment optimization for radiation therapy and image reconstruction. Chithrani, Cordes, Douplik, Goldman, Heath, Karshafian, Kolios, Kumaradas, Tavakkoli, Toronov, Xu.

Trace Element Detection. Human trace element detection. X-ray fluorescence. Pejović-Milić.

View additional information about this department at www.gradschoolshopper.com

FUDAN UNIVERSITY

DEPARTMENT OF PHYSICS

Shanghai, China
http://www.physics.fudan.edu.cn/

General University Information

President: Yuliang Yang (杨玉良)
Dean of Graduate School: Fang Lu (陆昉)
University website: http://www.fudan.edu.cn/englishnew/
Control: Public
Setting: Urban
Total Faculty: 2,346
Total Graduate Faculty: 2,400

Total number of Students: 30,893
Total number of Graduate Students: 13,851

Department Information

Department Chairman: Jian Shen (沈健), Chair
Department Contact: E Xu (徐娥), Administrative Assistant
Total full-time faculty: 67
Total number of full-time equivalent positions: 105

Full-Time Graduate Students: 200
First-Year Graduate Students: 85
Female First-Year Students: 20
Total Post Doctorates: 20

Department Address
220 Handan Road
Shanghai, 200433
CHINA
Phone: 86-21-65642360
Fax: 86-21-65104949
E-mail: xue@fudan.edu.cn
Website: http://www.physics.fudan.edu.cn/

ADMISSIONS

Admission Contact Information
Address admission inquiries to: Graduate Admissions Working
   Group, Department of Physics, Fudan University, 220 Handan
   Road, Shanghai 200433, China
Phone: 86-021-65642364
E-mail: xhyan@fudan.edu.cn
Admissions website: http://www.physics.fudan.edu.cn/chinese/
   doku.php?id=graduate

Application deadlines
Fall admission: December 31

Application fee
Int'l. students: RMB 600

Admissions information
For Fall of 2013:
   Number admitted: 84
   Number enrolled: 84

Admission requirements
Bachelor's degree requirements: Bachelor's degree in physics
   or a related field.
Minimum undergraduate GPA: 2.8

TUITION
Tuition year 2012–13:
Tuition for in-state residents
   Full-time students: RMB 30,000 annual

Teaching Assistants, Research Assistants, and Fellowships
Average stipend per academic year
   Teaching Assistant: RMB 20,000
   Research Assistant: RMB 30,000
   Fellowship student: RMB 8,000

HOUSING

Availability of on-campus housing
   Single students: Yes
   Married students: Yes

Table A—Faculty, Enrollments, and Degrees Granted

Research Specialty	2012-13 Faculty	Enrollment Fall 2013 Master's	Enrollment Fall 2013 Doctorate	Number of Degrees Granted 2007-12 Master's	Number of Degrees Granted 2007-12 Terminal Master's	Number of Degrees Granted 2007-12 Doctorate
Condensed Matter						
Physics	36	–	–	–	–	–
Optics	7	–	–	–	–	–
Theoretical Physics	24	–	–	–	–	–
Total	67	–	–	–	–	–
Full-time Grad. Stud.	–	–	–	–	–	–
First-year Grad. Stud.	–	–	–	–	–	–

GRADUATE DEGREE REQUIREMENTS

Master's: **Time to degree:** - 3 years. **Coursework requirements:** - Master's students shall earn 31 degree credits with
a minimum of 17 coursework credits. **Other requirements:**
- Comprehensive exam in the third semester - Mid-term evaluation by the Graduate Steering Group in the fourth semester
- Publish at least one paper in a journal indexed by SCI.

Doctorate: **Time to degree** - 3 years with a master's degree
- 5 years without a master's degree **Coursework requirements:** - Doctoral students with a master's shall earn
12 degree credits for coursework - Doctoral students without a master's degree shall earn at least 26 credits for
coursework. **Other requirements** - Qualifying exam -
Three-year doctoral program students must publish 2 papers in SCI-indexed journals, with a combined impact factor
greater than 2. - Five-year doctoral program students must publish
3 papers in SCI-indexed journals, with a combined impact factor
greater than 3.

SPECIAL EQUIPMENT, FACILITIES, OR PROGRAMS

The Department runs 3 key labs: the State Key Laboratory
of Surface Physics, Key Laboratory of Matter Computational
Sciences, and Key Laboratory of Micro-/Nano-Photonic Structure.

Sponsored by the State Planning Commission, the State Key
Laboratory of Surface Physics (SKLSP), Fudan University
was established in 1990. SKLSP is based on the multi-
discipline intergradations of computational condensed matter
physics, semiconductor surface and interface physics, semiconductor optoelectronic physics and ultra-thin magnetic film
physics, majoring in the directions of semiconductor surface,
interface and optoelectronic physics to contribute to the development of new generation information science and technology
of China.

The laboratory is mainly engaged in research in the following
four subjects:

1. the novel properties of surface and interface;

2. optical physics and application of surface and micro-structure;

3. theory and computational s of surface and interface;

4. interface problems related to the soft matter and bio-physics.

Each research activity is cooperatively carried out by several research groups.

SKLSP is equipped with a large number of sophisticated equipment, of which 26 are large-scale apparatuses. The Lab stresses
and strengthens the cooperation among the research groups, and
most of the large-scale instruments are opened to and shared with
other research groups.

The Laboratory attaches great importance to the international and domestic academic collaboration and exchange. So far, many international and domestic cooperative programs have been carried out with universities and laboratories home and abroad by the programs of "Opening Project" and "Senior Visiting Scholar." Now SKLSP is making all efforts to reach its goal of contributing to the development of basic subjects in China.

FACULTY

Professor

Che, Jingguang, (车静光) Ph.D., Muenster University, 1992. *Computational Physics, Condensed Matter Physics, Solid State Physics, Surface Physics*. Computational models for predicting geometric and electronic structure evolution in surfaces and interfaces of metals and semiconductors.

Chen, Jiyao, (陈暨耀) Ph.D., Fudan University, 1991. *Biophysics, Medical, Health Physics, Optics*. Photobiology and laser applications in medicine.

Chen, Yan, (陈焱) Ph.D., Nanjing University, 1998. *Condensed Matter Physics, Quantum Foundations, Theoretical Physics*. Theoretical studies on many body physics in complex quantum systems. In particular, electronic states in high temperature superconductors, exotic superfluidity in ultrcold atoms, quantum phase transitions and quantum entanglement.

Chen, Zhanghai, (陈张海) Ph.D., Shanghai Institute of Technical Physics, Chinese Academy of Sciences, 1997. *Condensed Matter Physics, Nano Science and Technology, Optics, Quantum Foundations, Solid State Physics*. Optical spectroscopy of condensed matters, including (1) Light-matter coupling in semiconductor nano/micro-structures and (2) Quantum chaos of electrons in solid-state environment.

Feng, Donglai, (封东来) Ph.D., Stanford University, 2001. Fudan-Haoqing Chair Professor. *Condensed Matter Physics*. Experimental condensed matter physics. Study complex quantum materials, including correlated systems, cuprate and iron based superconductors, charge/spin/orbital ordered systems, and oxide interfaces, etc. with synchrotron & laser-based spectroscopy and scattering techniques.

Gong, Xingao, (龚新高) Ph.D., Institute of Solid State Physics, Chinese Academy of Sciences, 1993. *Computational Physics, Condensed Matter Physics, Nano Science and Technology*. Theoretical study of nano-particle and nano-structure; computational design of new energy materials; structure and dynamic properties of surfaces and intersurfaces; development of computational method on electron structure of complex systems.

Hao, Bailin, (郝柏林) Kharkov State University, 1963. Academician, Chinese Academy of Sciences. Academician, Third World Academy of Sciences. *Computational Physics, Theoretical Physics*. Prokaryote phylogeny and taxonomy based on their complete genomes, the Composition Vector (CVTree) approach; statistics, combinatorics, language theory and graph theory inspired by the study of K-tuples in symbolic sequences of biological origin.

Hou, Xiaoyuan, (侯晓远) Ph.D., Fudan University, 1987. *Condensed Matter Physics*. Organic functional device, organic semiconductor / metal interface interaction.

Huang, Jiping, (黄吉平) Ph.D., The Chinese University of Hong Kong, 2003. *Theoretical Physics*. Soft matter and econophysics.

Jiang, Zuimin, (蒋最敏) Ph.D., University of Science and Technology of China, 1988. *Condensed Matter Physics*. Si molecular epitaxy and low dimensional Si based materials.

Jin, Xiaofeng, (金晓峰) Ph.D., Fudan University, 1989. *Condensed Matter Physics*. Experimental condensed matter physics. Surface and ultrathin film magnetism, spin-dependent transport in low dimensional systems.

Li, Shiyan, (李世燕) Ph.D., University of Science and Technology of China, 2002. *Condensed Matter Physics*. Ultra-low temperature properties of superconductors, magnets, and quantum critical systems.

Lin, Zhifang, (林志方) Ph.D., Fudan University, 1990. *Optics, Statistical & Thermal Physics, Theoretical Physics*. Optical micromanipulation, electromagnetic metamaterials.

Liu, Weitao, (刘韡韬) Ph.D., University of California, Berkeley, 2008. *Condensed Matter Physics, Optics, Surface Physics*. Nonlinear optics.

Liu, Xiaohan, (刘晓晗) Ph.D., Fudan University, 1999. *Condensed Matter Physics*. Optical properties of condensed matter, photonic crystals.

Lu, Fang, (陆昉) Ph.D., Fudan University, 1995. *Condensed Matter Physics*. Deep levels in semiconductor, electric properties low dimensional quantum structures, semiconductor solar cell.

Ma, Shihong, (马世红) Ph.D., Fudan University, 1995. *Optics*. Functional ultrathin films physics and device, surface nonlinear optics, organic functional materials, teaching study of physics experiment.

Ma, Yongli, (马永利) Ph.D., Fudan University, 1993. Theoretical condensed matter physics; ensemble theory in quantum statistics; matter waves; quantum condensation phenomena and superfluidity.

Qiao, Shan, (乔山) Ph.D., Tokyo University, 1997. *Theoretical Physics*. Applications of synchrotron radiation; development of multiple channel electron spin polarimeter; electron spin and orbital-related physics.

Shen, Jian, (沈健) Ph.D., Max Planck Institute of Microstructure Physics, 1996. Department Chair. *Condensed Matter Physics*. Investigation of emerging phenomena at surface, in reduce dimensionality, and at nanometer scale. Specific interest includes magnetism and electronic transport of nanostructured materials, and their underlying physical mechanism.

Sheng, Weidong, (盛卫东) Ph.D., Chinese Academy of Sciences, 1997. *Condensed Matter Physics*. Physics of low-dimensional structures: electronic structure, optical properties, and many-body effects.

Shi, Yu, (施郁) Ph.D., Nanjing University, 1994. *Quantum Foundations, Theoretical Physics*. Quantum entanglement and its uses in condensed matter physics and particle physics.

Tan, Yanwen, (谭砚文) Ph.D., Columbia University, 2007. *Biophysics*. Experimental biological physics; single-molecule spectroscopy.

Tao, Ruibao, (陶瑞宝) Ph.D., Fudan University, 1964. *Theoretical Physics*. Theory of Condensed Matters: Current focus: (1) Quantum Spin systems and its dynamics, (2) Transport theory of change and spin in mesoscopic system. (3) Quantum evolution and manipul.

Tian, Chuanshan, (田传山) Ph.D., Fudan University, 2006. *Optics*. Characterization of material with (nonlinear) optical spectroscopy, exotic optical effects and processes, and chemical physics. Research topics include: (1) Molecular structure of water surface and interfaces: (2)Pre-melting of ice interfaces below freezing point; (3) Interfacial structure of photocatalytic metal oxides; (4) Microscopic process of catalytc cracking of hydrocarbon on metal surface.

Wang, Xun, (王迅) Fudan University, 1960. *Condensed Matter Physics, Surface Physics*. Semiconductor physics.

Wei, Guanghong, (韦广红) Ph.D., Fudan University, 1998. *Biophysics, Condensed Matter Physics*. Computation study of peptide folding, misfolding, aggregation; and ordered nanostructure formation; peptide-membrane iteractions; protein-nanoparticle interactions.

Wu, Changqin, (吴长勤) Ph.D., Fudan University, 1987. *Theoretical Physics*. Charge/spin transport and photoelectric conversion in organic materials and devices; Quantum charge/ spin and heat transport through nano-structures; Orders, phase transitions, and excitations in 1d correlated systems.

Wu, Hua, (吴骅) Ph.D., Institute of Solid State Physics, Chinese Academy of Sciences, 1999. *Computational Physics, Condensed Matter Physics, Surface Physics*. Computational condensed matter physics, correlated oxides, magnetic semiconductor, surface and interface.

Wu, Shiwei, (吴施伟) Ph.D., University of California, 2007. Scanning probe microscopy and spectroscopy; surface science; nano optics; ultrafast dynamics; ultrafast dynamics; scientific instrumentation.

Wu, Yizheng, (吴义政) Ph.D., Fudan University, 2001. *Condensed Matter Physics*. Experimental condense matter physics; magnetism; spintronics.

Wu, Yongshi, (吴咏时) M.S., Peking University, 1965. Distinguished Professor, Department of Physics and Astronomy, University of Utah. *Condensed Matter Physics, Particles and Fields, Quantum Foundations, Statistical & Thermal Physics, Theoretical Physics*. Prof. Wu is interested in topological, geometric, and algebraic structures underlying the fundamental laws in physics that unifies all matter and forces in Nature, as well as emergent phenomena in fundamental physics and strongly correlated systems. His fields over quantum field theory, particle physics, statistical physics, String/M theory, topological matter and topological quantum computation, an emergent interdisciplinary research frontier of physics, mathematics and computer science.

Xiang, Hongjun, (向红军) Ph.D., University of Science and Technology of China, 2006. *Computational Physics, Condensed Matter Physics, Nano Science and Technology*. Computational condensed matter physics; multiferroics, nanomaterials, photovoltaic materials, development of new methods.

Xiao, Yanhong, (肖艳红) Ph.D., Harvard University, 2004. *Optics*. Experimental Atomic Molecular and Optical (AMO) physics: coherent interaction between atoms and light, precision spectroscopy, quantum memory, magnetometry, atomic frequency standards and quantum optics.

Xiu, Faxian, (修发贤) Ph.D., University of California, Riverside, 2007. *Condensed Matter Physics*. Study of novel topological thin films and nanostructures by molecular beam epitaxy and tube furnaces. Integration of magnetic materials with topological insulators for novel device functionalities. Exploration of high Curie temperature dilute magnetic semiconductors and their spintronic application.

Yang, Zhongqin, (杨中芹) Ph.D., Fudan University, 2000. *Condensed Matter Physics*. Theoretical Condensed Matter Physics (1) Quantum charge and spin transport in nanojunctions. (2) Electronic states of novel materials. (3) Effects of spin-out coupling.

You, Jianqiang, (游建强) Ph.D., Institute of Solid State Physics, Chinese Academy of Sciences, 1997. *Theoretical Physics*. (1) Solid-state quantum computing (2) Quantum coherence and transport in quantum devices and confined systems (3) Quantum noise and quantum open systems.

Zhang, Xinyi, (张新夷) Ph.D., Universite Pierre et Marie Curie, 1981. *Condensed Matter Physics*. Luminescence dynamics and novel materials. Synchrotron radiation applications.

Zhang, Yuanbo, (张远波) Ph.D., Columbia University, 2006. *Condensed Matter Physics*. Experimental condensed matter physics, particularly the electronic properties of graphene and other low dimensional electron systems.

Zhao, Jun, (赵俊) Ph.D., University of Tennessee, 2010. Using various neutron and x-ray scattering techniques to study the strongly correlated electron systems. Specific interest includes high Tc superconductors, multiferroics and other transition metal oxides.

Zhao, Li, (赵利) Ph.D., Harbin Institute of Technology, 1994. *Optics*. Black silicon, microstructured metal films.

Zhong, Zhengyang, (钟振扬) Ph.D., Institute of Physics, Chinese Academy of Sciences, 2001. *Condensed Matter Physics*. Controlled formation of varieties of nanostructures on Si substrates; exploration of the unique properties and the applications of Si-based nanostructures.

Zhou, Lei, (周磊) Ph.D., Fudan University, 1997. *Optics, Theoretical Physics*. Metamaterials, nanophotonics, magnetism.

Zhou, Luwei, (周鲁卫) Ph.D., Temple University, 1986. *Condensed Matter Physics*. Soft matter physics.

Zi, Jian, (资剑) Ph.D., Fudan University, 1991. *Condensed Matter Physics, Optics*. Photonic crystals, plasmonics, metamaterials, natural photonic structures and structural colors, liquid surface waves propagating in periodic structures.

Associate Professor

An, Zhenghua, (安正华) Ph.D., Shanghai Institute of Microsystem and Information Technology, Chinese Academy of Sciences, 2004. *Nano Science and Technology, Optics*. Nanofabrication; low dimensional semi-conductors; optoelectronics; nano-photonics; subwavelength metallic plasmonics.

Bambi, Cosimo, Ph.D., Ferrara University, 2007. *Astronomy, Astrophysics, Theoretical Physics*. Theoretical physics in the areas of gravity, cosmology, and high-energy astrophysics. The research focuses on the possibility of testing the Kerr-nature of astrophysical black hole candidates by studying the electromagnetic radiation emitted by the gas in the accretion disk.

Cai, Qun, (蔡群) Ph.D., Fudan University, 1996. *Condensed Matter Physics, Surface Physics*. Surface and interface structured studied with scanning probe microscpy; formation and properties of metal nano-structures on semiconductor surfaces; kinetics of surface atomic processes; ultrathin dielectric films on semiconductor surfaces.

Chen, Junyi, (陈骏逸) Fudan University, 1985.

Chen, Wei, (陈唯) Ph.D., Institute of Physics, Chinese Academy of Sciences, 2001. *Biophysics, Condensed Matter Physics*. Soft matter: collodial physics; biophysics: cell motility.

Du, Side, (杜四德) Ph.D., Nanjing University, 1995. *Theoretical Physics*.

Ji, Min, (冀敏) Henan Normal University, 1981.

Lv, Jinglin, (吕景林) Jilin univeristy of Technology, 1983.

Modesto, Leonardo, Ph.D., Torino University, 2004. *Quantum Foundations, Relativity & Gravitation, Theoretical Physics*. Currently, I am working in modified theories of gravity with particular interest in nonlocal theories and/or a particular class of super-renormalizable theories of gravity in any dimension. I am especially interested in the connection between these theories and string theory and precisely with string field theory. In the past, I earned my Ph.D. in string theory introducing a new supersymmetric formalism for D-branes, but in the last years I worked in a non-perturbative approach to quantum gravity. I have applied non perturbative ideas coming from loop quantum gravity to the black hole physics, including phenomenology.

Shu, Lei, (殳蕾) Ph.D., University of California, Riverside, 2008. Muon spin relaxation/rotation, quantum criticality; non-Fermi-liquid phenomena, magnetism and superconductivity in correlated electron system, high temperature superconductivity.

Xiao, Jiang, (肖江) Ph.D., Georgia Institute of Technology, 2006. *Condensed Matter Physics*. Theoretical condensed matter physics, spintronics, magnetism.

Xu, Jianjun, (徐建军) Ph.D., Fudan University, 1988.

Xu, Xiaohua, (徐晓华) Ph.D., Nanjing University, 1996.

Yang, Xinju, (杨新菊) Ph.D., Fudan University, 1994. *Condensed Matter Physics*. Quantum structure fabrications and nanoscale electrical property studies.

Yin, Lifeng, (殷立峰) Ph.D., Fudan University, 2007. *Condensed Matter Physics*. Spintronics, Nanomagnetism, Complex Oxides under Spatial Confinement, Surface and Interface of Complex Oxides.

Zhang, Tong, (张童) Ph.D., Institute of Physics, Chinese Academy of Sciences, 2010. *Condensed Matter Physics*.

Assistant Professor

Yao, Yao, (姚尧) Ph.D., Fudan University, 2009. *Condensed Matter Physics*.

Zuo, Guanghong, (左光宏) Ph.D., Fudan University, 2010. *Biophysics, Theoretical Physics*. Biophysics,Theoretical Physics.

Adjunct Professor

Shen, Xuechu, (沈学础) Fudan University, 1958. *Nano Science and Technology*. Solid State Spectroscopy, methods and physics. Recently interested in electronic states and quantum interactions in nano-sized single quantum structures, such as wires and dot, also chaotic motion of electrons in solid state environment.

Shen, Yuanrang, (沈元壤) Ph.D., Harvard University, 1963. *Condensed Matter Physics, Optics, Surface Physics*. Molecular physics, nonlinear optics, laser spectroscopy, liquid crystals, and surface sciences. More recently developed novel nonlinear optical techniques to probe surfaces and interfaces, which can provide molecular-level information about interfacial water structures and orientations and conformation of adsorbates. Other interests include thee study of ice and oxide interfaces, development of a sensistive nonlinear optical spectroscopic technique to probe molecular chirality, and optical spectroscopy of nanostructures composite structures, and exotic materials.

Yang, Fujia, (杨福家) Fudan University, 1958. *Atomic, Molecular, & Optical Physics, Nuclear Physics*. Atomic physics, beam physics.

DEPARTMENTAL RESEARCH SPECIALTIES AND STAFF

Theoretical

Condensed Matter Physics. Quantum spin system with spin-dependent scattering, low-dimensional correlated electron systems, wave propagation of photonic crystals and electromagnetic exotic materials, first-principles study of low-dimensionaland nano structures, quantum information and computing, electronic transmission in nano-structure, ultracold atomic systems. Jingguang Che, Xingao Gong, Ruibao Tao, Xun Wang, Hua Wu, Yizheng Wu, Hongjun Xiang, Jiang Xiao, Zhongqin Yang, Xinyi Zhang.

Gravitation, Cosmology and Black Hole Physics. General relativity, black hole physics and cosmology, quantum physics and quantum information, specific media electromagnetic physics, soft matter theory. Cosimo Bambi, Leonardo Modesto, Yongshi Wu.

Partical Physics. Particle Physics and Field Theory. Yu Shi, Yongshi Wu.

Theoretical Physics. Yan Chen, Bailin Hao, Jiping Huang, Zhifang Lin, Yongli Ma, Yu Shi, Ruibao Tao, Changqin Wu, Jianqiang You, Lei Zhou, Guanghong Zuo.

Experimental

Condensed Matter and Material Physics. Semiconductor surface and interface, semiconductor quantum dots and nano-structure, low-dimensional magnetism and spintronics, strongly correlated electron systems, photonic crystals and electromagnetic metamaterials, synchrotron radiation applications, organic light-emitting and solar cells. Zhanghai Chen, Donglai Feng, Xiaoyuan Hou, Zuimin Jiang, Xiaofeng Jin, Shiyan Li, Weitao Liu, Xiaohan Liu, Fang Lu, Shan Qiao, Jian Shen, Weidong Sheng, Lei Shu, Xun Wang, Yizheng Wu, Faxian Xiu, Yao Yao, Lifeng Yin, Tong Zhang, Xinyi Zhang, Yuanbo Zhang, Jun Zhao, Zhengyang Zhong, Jian Zi.

Optics. Applications of quantum dots and nano-materials in biological systems, surface photophysics, femtosecond micromachining and surface modification, nano-materials, study of ultrafast photophysical and nonlinear optical properties in new materials like organic materials and biological systems. Zhanghai Chen, Zhifang Lin, Weitao Liu, Xiaohan Liu, Weidong Sheng, Yanwen Tan, Chuanshan Tian, Shiwei Wu, Yanhong Xiao, Xinju Yang, Li Zhao, Lei Zhou, Jian Zi.

Physical Biology. Physical biology, bioinformatics and computational biology, economical physics. Jiyao Chen, Wei Chen, Yanwen Tan, Guanghong Wei, Guanghong Zuo.

View additional information about this department at
www.gradschoolshopper.com

NANJING UNIVERSITY

SCHOOL OF PHYSICS

Nanjing, China
http://physics.nju.edu.cn/

General University Information
President: Jun CHEN (陈骏)
Dean of Graduate School: Jian Lv (吕建)
University website: http://www.nju.edu.cn/
Control: Public

Setting: Urban
Total number of Students: 50,000
Total number of Graduate Students: 13,581

Department Information
Department Chairman: Jianxin Li (李建新), Dean
Department Contact: Xiangyun Kong (孔祥云), Administrative
  Assistant
Total full-time faculty: 221
Total number of full-time equivalent positions: 221
Full-Time Graduate Students: 606
First-Year Graduate Students: 224
Total Post Doctorates: 65

Department Address
School of Physics
Nanjing University
Nanjing, 210093
CHINA
Phone: 86-25-83592870
Fax: 86-25-83326028
E-mail: kongxy@nju.edu.cn
Website: http://physics.nju.edu.cn/

ADMISSIONS

Admission Contact Information
Address admission inquiries to: Graduate School, Nanjing University, 22 Hankou Road, Nanjing, China 210093
Phone: 86-25-83594535
E-mail: issd@nju.edu.cn
Admissions website: domestic student:http://grawww.nju.edu.cn/content/zs.asp

Application deadlines
Fall admission:
Chinese students: December 31   *Int'l. students*: May 20
International student should submit application at http://www.studyinnju.com

Admissions information
For Fall of 2013:
  Number admitted: 134
  Number enrolled: 134

Admission requirements
Bachelor's degree requirements: Bachelor's degree in physics or a related field.

TUITION

Health insurance: Available.
Academic term: Semester

Teaching Assistants, Research Assistants, and Fellowships
Average stipend per academic year
  Teaching Assistant: RMB 4,800

HOUSING

Availability of on-campus housing
Single students: Yes
Married students: Yes

Table A—Faculty, Enrollments, and Degrees Granted

Research Specialty	2012-13 Faculty	Enrollment 2013		Number of Degrees Granted 2012-13		
		Master's	Doctorate	Master's	Terminal Master's	Doctorate
Acoustics	–	23	–	20	–	–
Biophysics	–	7	–	8	–	1
Condensed Matter Physics	–	46	–	41	–	19
Electrical Engineering	–	–	–	29	–	13
Low Temperature Physics	–	1	–	2	–	–
Optics	–	9	–	5	–	5
Particles and Fields	–	7	–	8	–	6
Theoretical Physics	–	19	–	16	–	5
Total	–	112	107	129	–	49
Full-time Grad. Stud.	–	–	–	–	–	–
First-year Grad. Stud.	–	–	–	–	–	–

GRADUATE DEGREE REQUIREMENTS

Master's: 3 years full-time study. Coursework requirements: Student with a Bachelor's degree in physics shall earn 32 degree credits; otherwise should earn 36 degree credits. Other requirements: Pass midterm evaluation; finish the degree thesis; publish at least one paper in a core journal as first author or single author.

Doctorate: 3 years or 5 years full-time study with a master degree. 5 years without a master degree. Coursework requirements: Complete the courses and credits required by the degree. Other requirements: Pass midterm evaluation; finish the degree thesis; publish two papers as first author in core journals. One of them should be in SCI or EI-indexed journal as first author.

SPECIAL EQUIPMENT, FACILITIES, OR PROGRAMS

The National Laboratory of Solid State Microstructures (NLSSM) was established in 1984, among the first group of State Key Laboratories established in China. The NLSSM is affiliated with Nanjing University. As a fundamental research laboratory, the missions of NLSSM are to design and fabricate new artificial microstructured materials via designing and tailoring the energy band structures in reciprocal space, find out new macroscopic and microscopic quantum effects and principles, develop new theories and methodology based on quantum physics, and endeavor to meet the scientific challenges and technological requirements encountered in the post-Moore era and the post-petroleum era.

As one of the major research centers in condensed matter physics in China, research projects currently being carried out in NLSSM cover many important branches in artificial microstructure physics, quantum physics and the associated electronics, nanostructure properties and devices, soft condensed matter physics and computational physics. Currently, 27 principal investigators are leading their groups working mainly in the Cyrus Tang Building with a total laboratory space of over 20,000 square meter. The major state-of-the-art equipments in the Laboratory include field-emission transmission electron microscope (Philips F20); spin-resolved ultrahigh vacuum scanning tunneling microscopy (Omicron); field-emission scanning electronic microscope (LEO 1530VP); solid state nuclear magnetic resonance system (Bruker Arance 300); and superconducting quantum interference device (SQUID) (MPMSXL-7). In addition, there are numerous thin film/bulk material fabrication facilities, as well as structural and spectrum characterization facilities. An advanced nanofabrication center, equipped with electron beam lithography system

(Raith e-Line) and focusing ion-beam fabrication facility (Philips FB201), has been established, which will serve the needs for fabrications of nano/micro sized structures.

In the period of 2005-2009, the Laboratory has published 1858 peer-reviewed research papers, among which there are 39 in Physical Review Letters, 209 in Physical Review series, and 178 in Applied Physics Letters. During this period the number of authorized patents reaches 90, in addition to 86 new applications.

NLSSM is also a public platform for international exchanges and cooperation. The Laboratory has accommodated many internship students, postdoctoral research associates and visitors, from domestic and abroad. The Laboratory has collaboration agreements with research institutes and universities in US, Japan, France, Germany, Italy, etc. Many joint research papers have been published in Science, Nature Photonics, Advanced Materials, and Physical Review Letters. The Laboratory sincerely welcomes international cooperation in areas of material design, quantum physics, nanooptics and plasmonics, clean energy and environment-related studies.

National Laboratory of Microstructures of Nanjing is under construction. (Budget is 300 millions)

FACULTY

Professor

An, Jin, (安晋) Ph.D., Nanjing University, 2001. *Condensed Matter Physics.* Strongly correlated electronic systems, unconventional superconductivity, topological insulators.

Cao, Yi, (曹毅) Ph.D., The University of British Columbia, 2009. *Biophysics.* Single molecule force spectroscopy; atomic force microscope; protein folding and dynamics; protein engineering; self-assembly.

Chen, Shenjian, (陈申见) Ph.D., Purdue University. *Theoretical Physics.*

Chen, Weizhong, (陈伟中) Ph.D., Nanjing University, Acoustics.

Cheng, Jianchun, (程建春) Ph.D., Nanjing University, 1992. NSFC Distinguished Young Scholar,. *Acoustics.*

Cheung, Yeuk-Kwan Edna, (张若筠) Ph.D., Princeton University, 2000. *Theoretical Physics.* String theory; early universe physics; beyond standard model physics.

Ding, Hai Feng, (丁海峰) Ph.D., Martin-Luther University and Max-Planck-Institute of microstructure physics, 2001. Member of the technical committee of IEEE Magnetic Society. *Condensed Matter Physics.* Scanning tunneling microscopy, low-dimensional magnetism, spintronics.

Ding, Jianping, (丁剑平) Ph.D., Institute of Applied Physics, Tsukuba University, 1995. Optical Information Processing, Optical Interferometry, Digital and Optical Holography, Manipulation of Light Beams, and Optical Tweezers.

Du, Jun, (杜军) Ph.D., Nanjing University, 1998. *Condensed Matter Physics.* Spintronics, Magnetism.

Du, You-Wei, (都有为) B.S., Nanjing University, 1957. Member of Chinese Academy of Science. *Condensed Matter Physics.* Magnetism and magnetic materials, spintronics.

Fang, Guiyin, (方贵银) Ph.D., Nanjing University, 1989.

Feng, Duan, (冯端) B.S., Nanjing University, 1946. Member of Chinese Academy of Science. *Condensed Matter Physics.*

Feng, Xiaoning, (冯小宁) B.S., Nanjing University, 1980.

Gong, Chang-de, (龚昌德) M.S., Fudan University, 1952. Member of Chinese Academy of Science. *Condensed Matter Physics.* Strong correlated system, superconductivity, topological insulators, and low dimensional condensed matter physics.

Gu, MIn, (顾民) Ph.D., Nanjing University, 1998. *Condensed Matter Physics.* Nuclear magnetic resonance spectroscopy, dielectric spectroscopy, relaxation and phase transition, struc-

ture and physical property of nanoparticle, interaction of nanoparticle with cell.

Huang, Runsheng, (黄润生) M.S., Nanjing University, 1986.

Huang, Yineng, (黄以能) Ph.D., Nanjing University, 1990.

Jiang, Zhengsheng, (蒋正生) Ph.D., Nanjing University, 1995.

Jin, Guojun, (金国钧) M.S., Nanjing University, 1982. *Theoretical Physics.*

Ju, Guoxing, (鞠国兴) Ph.D., Nanjing University, 1993.

Li, Jian-Xin, (李建新) Ph.D., Wuhan University, 1993. Cheung Kong Scholar, NSFC Distinguished Young Scholar,. *Theoretical Physics.* Strongly correlated electron system, Unconventional superconductivity(especially cuprates, iron-based superconductors), quantum magnetism.

Li, Jun, (李俊) Ph.D., Tsinghua University.

Li, Shao-Chun, (李绍春) Ph.D., Chinese Academy of Science, 2004. China Recruitment Program of Global Experts (C plan, 2012), Professor of Physics. *Condensed Matter Physics.* Scanning Tunneling Microscopy, Molecular Beam Epitaxy, Oxide, Surface Physics and Chemistry.

Liu, Feng, (刘锋) Ph.D., Nanjing University, Biophysics., 1998.

Liu, Hui, (刘辉) Ph.D., Nanjing University, 2003. Associate Director of the State Key Laboratory of Solid State Microstructures. *Nano Science and Technology.* Metamaterials, surface plasmons, photonic crystals.

Liu, Jun-Ming, (刘俊明) Ph.D., Northwestern Polytechnical University, 1989. Cheung Kong Scholar, NSFC Distinguished Young Scholar, Professor of Physics with Nanjing University. *Condensed Matter Physics.* Physics of multiferroics, strongly correlated electron systems, ferroelectricity, magnetism of oxides, and Monte Carlo simulation.

Liu, Xiao-Jun, (刘晓峻) Ph.D., Nagoya University, 2000. *Acoustics.* Acoustic artificial materials, photoacoustics.

Liu, Xiaozhou, (刘晓宙) Ph.D., Nanjing University, 1999. Director of Key Lab. of Modern Acoustics,council member of the Acoustical Society of China. *Acoustics.* Biomedical acoustics, nonlinear acoustics, ultrasonic nondestructive evaluation, acoustic signal processing.

Lv, Xiao-Mei, (吕笑梅) Ph.D., Nanjing University, 1995. NSFC Distinguished Young Scholar. *Condensed Matter Physics.* Physics of dielectrics and ferroelectrics; PFM investigation of ferroelectrics.

Ma, Yuqiang, (马余强) Ph.D., Nanjing University, 1993. *Condensed Matter Physics.* Soft condensed matter physics, biophysics, tatistical physics of active matters.

Miao, Feng, (缪峰) Ph.D., University of California, Riverside, 2009. *Condensed Matter Physics.* Mesoscopic physics, electron transport, nanoelectronics, graphene, memristor.

Min, Naiben, (闵乃本) Ph.D., Tohoku University, 1987. Member of Chinese Academy of Sciences.

Peng, Ruwen, (彭茹雯) Ph.D., Nanjing University, 1998. NSFC Distinguished Young Scholar,. *Condensed Matter Physics.*

Qiu, Xiaojun, (邱晓军) Ph.D., Nanjing University, 1995. *Acoustics.* Audio engineering, active noise control, electroacoustics, communication acoustics, noise control.

Ren, Zhongzhou, (任中洲) Ph.D., Nanjing Univrersity, 1988. Cheung Kong Scholar, NSFC Distinguished Young Scholar,. Properies of unstable nuclei, alpha-decay model, electron scattering with nuclei, beta-decay model, superheavy nuclei.

Sang, Hai, (桑海) Ph.D., Nanjing University, 1996. *Condensed Matter Physics.* Microstructural, magnetic, and electric properties in nanostructured materials; Exchange bias in FM/AF bilayer; perpendicular anisotropy and extraordinary Hall effect in magnetic multilayer; magnetoelectronics.

Shen, Rui, (沈瑞) Ph.D., Nanjing University, 1996.

Shen, Yong, (沈勇) Ph.D., Nanjing University, Acoustics.

Sheng, Li, (盛利) Ph.D., Nanjing University, 1996.

Shu, Da-Jun, (舒大军) Ph.D., Chinese Academy of Science, 2001. *Condensed Matter Physics.* Physics at Surfaces and Interfaces, Density Functional Theory Calculations, Molecular Dynamics Simulation.

Su, Wei-Ning, (苏为宁) Ph.D., Nanjing University, 1989. *Condensed Matter Physics*. growth of low-dimensional nanostructures and quantum physics.

Tang, Shao-Long, (唐少龙) Ph.D., Chinese Academy of Science, 1997. *Condensed Matter Physics*. Magneto-opical effect of magnteoplasmonic nanostructure; functional magnetic films; Permanent magnetic materials; growth of low-dimensional magnetic nanostructure.

Tang, Tao, (唐涛) Ph.D., Nanjing University, 2000. *Condensed Matter Physics*. Magnetism and magnetic materials.

Wan, Jian-Guo, (万建国) Ph.D., Nanjing University of Aeronautics and Astronautics, 1997. *Condensed Matter Physics*. Multiferroic materials and physics, cluster physics, low-dimensional nanostructures and physics.

Wan, Xiangang, (万贤纲) Ph.D., Nanjing University, 2000. *Condensed Matter Physics*. Condensed matter theory; topological insulator theory; strongly correlated electronic; density function theory.

Wang, Bogen, (王伯根) Ph.D., The University of HongKong, 2003. Cheung Kong Scholar, NSFC Distinguished Young Scholar,. *Theoretical Physics*.

Wang, Dunhui, (王敦辉) Ph.D., Nanjing University, 2004. New Century Excellent Talents of China. *Condensed Matter Physics*. Magnetism, magnetic phase transition material, multiferroic, spintronics.

Wang, Guanghou, (王广厚) B.S., Beijing Normal University, 1963. Academician, Chinese Academy of Sciences. *Condensed Matter Physics*. Atomic cluster physics, nanoscience, and nanotechnology.

Wang, Huitian, (王慧田) Ph.D., Chinese Academy of Science, 1994. Cheung Kong Scholar, NSFC Distinguished Young Scholar,. *Optics*.

Wang, Jun, (王骏) Ph.D., Nanjing University, 2001. *Biophysics*. Protein folding, allosteric motion, protein-protein interaction, simulation of biomolecules.

Wang, Mu, (王牧) Ph.D., Nanjing University, 1991. Cheung Kong Scholar, NSFC Distinguished Young Scholar,. *Condensed Matter Physics*.

Wang, Qiang-Hua, (王强华) Ph.D., Nanjing University, 1993. Cheung Kong Scholar, NSFC Distinguished Young Scholar,. *Theoretical Physics*. Strongly correlated electron systems, topological insulators/superconductors, quantum information and quantum computing.

Wang, Sihui, (王思慧) Ph.D., Nanjing University, 2001.

Wang, Wei, (王炜) Ph.D., Nanjing University, 1990. Cheung Kong Scholar, NSFC Distinguished Young Scholar,. *Biophysics, Condensed Matter Physics*. Protein folding, dynamics of biological networks, and self-assembling of biomolecules and nanoparticles.

Wang, Xiaoyong, (王晓勇) Ph.D., University of Arkansas at Fayetteville, 2004. *Optics*. Single molecule spectroscopy, ultrafast spectroscopy, nano optics.

Wang, Xinlong, (王新龙) Ph.D., Nanjing University, 1991. Cheung Kong Scholar, NSFC Distinguished Young Scholar,. *Acoustics*. Linear and Nonlinear Acoustics, Nonlinear Waves and Solitons, Nonlinear Signal Processing.

Wang, Zhen-Lin, (王振林) Ph.D., Nanjing University, 1996. Cheung Kong Scholar, NSFC Distinguished Young Scholar, Vice Chairman of the Optical Society of Jiangsu Province. *Optics*. Photonic crystals, plamonics, surface enhanced Raman scattering, light-matter coupling, self-assembled functional materials.

Wang, Zhihe, (王智河) Ph.D., Nanjing University, 1982.

Wen, Hai-Hu, (闻海虎) Ph.D., Chinese Academy of Science, 1993. Cheung Kong Scholar, NSFC Distinguished Young

Scholar,. *Condensed Matter Physics*. Condensed matter physics, superconductivity.

Wu, Di, (吴镝) Ph.D., Fudan University, 2001. *Condensed Matter Physics*. Spintronics, magnetic nanostructures.

Wu, Hao-Dong, (吴浩东) Ph.D., Nanjing University, 2001. Member of Chinese Academy of Acoustics. *Acoustics*. Ultrasonic transducer, SAW.

Wu, Xiao-Shan, (吴小山) Ph.D., Beijing Normal University, 1994. Member of Chinese Physics Society, Life Member of American Physics Society;Chair of Nanjing Chapter of IEEE Magnetic Society, Vice Dean of School of Physics, Nanjing University. *Condensed Matter Physics*. X-ray diffraction; low dimension transition metal oxides; photovoltage effect; synchrotron radiation.

Wu, Xing-Long, (吴兴龙) Ph.D., Nanjing University, 1995. Cheung Kong Scholar, NSFC Distinguished Young Scholar,. *Condensed Matter Physics*. Semoconductor nanostructured materials and physics, Raman scattering.

Xiao, Min, (肖敏) Ph.D., University of Texas at Austin, 1988. *Optics*. Quantum optics, atomic physics, nonlinear optics, ultrafast laser spectroscopy, and micro/nano photonics.

Xiao, Mingwen, (肖明文) Ph.D., Nanjing University, Acoustics, 1991.

Xing, Dingyu, (邢定钰) M.S., Nanjing University, 1987. Member of Chinese Academy of Science. *Theoretical Physics*.

Xu, Xiaonong, (徐小农) Ph.D., Nanjing University, 1997. *Condensed Matter Physics*.

Yang, Shao-Guang, (杨绍光) Ph.D., The University of Science and Technology of China, 1998. *Condensed Matter Physics*. One-dimensional nanostructures, magnetic materials.

Yang, Yue-tao, (杨跃涛) Ph.D., University of Science and Technology of China, 1999. *Acoustics*. Photoacoustic science, ultrasonic preparation of nanomaterials.

Yu, Tao, (于涛) Ph.D., Nanjing University, 1997. Materials Research Society (MRS) member, Chinese Physical Society (CPS) member. *Condensed Matter Physics*. Micro/nanostructured thin films and hetero junctions of semiconductive oxide material physics, photocatalysis, dye sensitized solar cells, ferroelectric and dielectric material physics.

Yu, Yang, (于扬) Ph.D., University of Kansas, 2002. Cheung Kong Scholar, NSFC Distinguished Young Scholar. *Condensed Matter Physics*. Superconducting quantum devices, quantum computing, macroscopic quantum phenomenon.

Zhang, Dong, (章东) Ph.D., Nanjing University, 1995. *Acoustics*. Medical ultrasound, nonlinear acoustics.

Zhang, Fengming, (张凤鸣) Ph.D., The University of Newcastle, 1997. Director of the Photovoltaic Engineering Center, Nanjing University. *Condensed Matter Physics*. Photovoltaics, spintronics.

Zhang, Jian, (张建) Ph.D., Nanjing Universit, 2002. *Biophysics*. Large-scale simulation of biological molecules, nanobiotechnologies.

Zhang, Shuyi, (张淑仪) M.S., Nanjing University, 1987. Member of Chinese Academy of Sciences.

Zhang, Weiyi, (章维益) Ph.D., Helsinki University of Technology, 1983. NSFC Distinguished Young Scholar,. *Condensed Matter Physics*.

Zhong, Wei, (钟伟) Ph.D., Nanjing University, 1989.

Zhou, Jin, (周进) M.S., Nanjing university, 1986. Professor of Physics and Director of fundamental physics experiment Center School of Physics Nanjing University. *Optics*. Optical information processing, binary optics.

Zhou, Yong, (周勇) Ph.D., University of Science and Technology of China, 2000. Editorial board memeber of current nanoscience. *Condensed Matter Physics*. Design of nanomaterials for clean renewable energy.

Zhu, Shining, (祝世宁) Ph.D., Nanjing University, 1996. Member of Chinese Academy of Science. *Optics*.

Zhu, Xin-hua, (朱信华) Ph.D., Xi'an Jiaotong University, 1996. Alexander von Humboldt (AvH) Research Scholar of Germany, Academic Consultant in School of Physical Sci. and Eng., King Abdullah University of Science & Technology (KAUST), Kingdom of Saudi Arabia, Editor Board Members of New Journal of Glass and Ceramics, Nanotechnology, and Journal of Chinese Electron Microscopy Society. *Condensed Matter Physics*. High-resolution (scanning) transmission electron microscopy, growth of low-dimensional multiferroic nanostructures and quantum physics, nano-scale physical property characterizations.

Zhu, Yongyuan, (朱永元) Ph.D., Nanjing University, 1991. *Condensed Matter Physics*. Dielectric superlattices, nonlinear optics, metamaterials, plasmonic crystals.

Zong, Hongshi, (宗红石) Ph.D., Nanjing University, 1994. *Nuclear Physics*.

Zou, Zhigang, (邹志刚) Ph.D., the University of Tokyo, 1996. Cheung Kong Scholar, NSFC Distinguished Young Scholar,. *Condensed Matter Physics*.

Associate Professor

Cao, Qingqi, (曹庆琪) Ph.D., Nanjing University, 1996.

Chen, Yan-bin, (陈延彬) Ph.D., University of Michigan, 2008. *Condensed Matter Physics*. Transmission electron microscopy; electron energy loss spectroscopy; correlated electron systems; Structure-property relationship.

Chen, Zhuo, (陈卓) Ph.D., University of Exeter, 2008. *Optics*. Near-field optics, surface plasmons, nonlinear optics.

Cheng, Li-Ping, (程利平) M.S., Nanjing University of Science and Technology, 1999. *Acoustics*. Surface acoustic wave actuators and sensors, acoustic streaming.

Fa, Wei, (法伟) Ph.D., Nanjing University, 1999.

Fan, Li, (范理) Ph.D., Nanjing University, 2007. *Acoustics*. Thermoacoustics, SAW sensors, acoustic metamaterials.

Gong, Xun, (公勋) Ph.D., Nanjing University, Acoustics.

Gu, Jun, (顾军) Ph.D., Nanjing University, 2002.

Guo, Xia-sheng, (郭霞生) Ph.D., nstitute of Acoustics, Nanjing University, 2008. *Acoustics*. Ultrasound in Medicine and Biology, Ultrasonic nondestructive testing.

Hu, Xiao-Peng, (胡小鹏) Ph.D., Nanjing University, 2007. *Condensed Matter Physics*. Nonlinear optics, quasi-phasematching materials, all-solid-state laser techniques, mode-locking lasers, guide wave optics.

Huang, Feng-Zhen, (黄凤珍) Ph.D., Nanjing University, 2008. *Condensed Matter Physics*. Dielectric, ferroelectric, and multiferroic materials; the transporting properties of charge defeats in dielectric materials.

Ju, Yan, (鞠艳) Ph.D., Nanjing University.

Li, Wenfei, (李文飞) Ph.D., Chinese Academy of Science, 2004. *Biophysics*. coupled protein structure and dynamics, including protein folding, aggregation.

Liang, Bin, (梁彬) Ph.D., Nanjing University, 2007. *Acoustics*. Acoustic metamaterials, acoustic diode, acoustic cloak, acoustic propagation in soft bubbly media.

Liu, Jie-Hui, (刘杰惠) Ph.D., Institute of Acoustics, Nanjing University, 1999. *Acoustics*. Nonlinear acoustics, ultrasonics.

Lu, Jing, (卢晶) Ph.D., Nanjing University, 2004. Member of Chinese Institute of Acoustics, Member of Chinese Institute of Electronics. *Acoustics*. Communication acoustics, acoustic signal processing, noise control.

Mao, Yiwei, (毛一葳) Ph.D., Nanjing University, Acoustics.

Qin, Meng, (秦猛) Ph.D., Nanjing Unversity. *Biophysics*.

Ren, Chun-Lai, (任春来) Ph.D., Nanjing University, 2006. Theoretically modeling on self-assembly of complex systems, drug delivery systems, biocompatible materials and intelligent materials. *Condensed Matter Physics*.

Shao, Lubing, (邵陆兵) Ph.D., Nanjing University, 2009.

Shi, Rong-Zhang, (石荣章) M.S., Nanjing Southeast University, 1993. *High Energy Physics*. High-energy microwave electronics, eiectron vacuum, eiectron-beam optics, physical electronics, and optical electronics.

Song, Fengqi, (宋凤麒) Ph.D., Nanjing University, 2005. *Condensed Matter Physics*. Atomic clusters; topological insulators; mesoscopic transport; scanning transmission electron microscopy; nanophysics.

Sun, Weimin, (孙为民) Ph.D., Nanjing University, 2002.

Tang, Nu-Jiang, (汤怒江) Ph.D., Nanjing University, 2005. *Condensed Matter Physics*. Magnetic properties of graphene-based materials, chemical doping of graphene and magnetism, magnetic field effect of photoluminescence of graphene-based materials, spintronics.

Tao, Chao, (陶超) Ph.D., Nanjing University, 2004. Associate Professor, MOE Key Laboratory of Modern Acoustics, Institute of Acoustics, School of Physics, Nanjing University. *Acoustics*. Photoacoustic imaging, voice production, nonlinear acoustics.

Tu, Juan, (屠娟) Ph.D., University of Washington, 2006. Member of Acoustical Society of China, Member of Acoustical Society of America. *Acoustics*. Medical ultrasound, nonlinear acoustics, physical acoustics.

Xu, Chang, (许昌) Ph.D., Nanjing University. *Theoretical Physics*.

Xu, Jianyi, (许坚毅) Ph.D., Nanjing University.

Xu, Ping, (徐平) Ph.D., Nanjing Univerisity, 2007. Assoc. Prof., New Century Excellent Talents in University. *Optics*. Photon entanglement and quantum information processing, quantum computing and quantum simulation, nonlinear optics, guided optics and integrated optics, ferroelectric materials.

Xu, Xi-Bin, (徐锡斌) Ph.D., nstitute of Physics, Nanjing University, 2006. Member of Chinese Academy of Refrigeration. *Condensed Matter Physics*. Superconducting vortex physics, soft matter, nanophysics, solid state, and complex systems.

Xu, Xiaodong, (徐晓东) Ph.D., Nanjing University, 2003. *Acoustics*. Nondestructive test, relationship of the microstructure and properties in materials, optical instruments, laser ultrasonics.

Yang, Huan, (杨欢) Ph.D., Chinese Academy of Science, 2009. *Condensed Matter Physics*. Superconductivity, scanning tunneling microscopy.

Yang, Jing, (杨京) Ph.D., Nanjing University, Acoustics.

Ying, Xuenong, (应学农) Ph.D., Nanjing University, 1999.

You, Biao, (游彪) Ph.D., Nanjing University, 1992.

Yu, Shun-Li, (于顺利) Ph.D., Nanjing University, 2009. *Theoretical Physics*. Correlated electron systems, Quantum magnetism and Superconductivity.

Zang, Wencheng, (臧文成) M.S., Nanjing University, 1985.

Zhan, Peng, (詹鹏) Ph.D., Nanjing University, 2008.

Zhang, Chunfeng, (张春峰) Ph.D., Fudan University, 2007. *Optics*.

Zhang, Hui, (张辉) Ph.D., Nanjing University, 2006. *Acoustics*. Acoustic transducers and resonators, Micro-ultrasonic actuators and sensors, Acoustic scanning microscopy.

Zhang, Jianhui, (章建辉) Ph.D., University of Science and Technology of China, 2000. *Condensed Matter Physics*. growth of novel nanostructures, fluorescence enhancement, surface plasmon resonance, solar cells, spintronics.

Zhang, Zhiyong, Ph.D., Nanjing University, 1996.

Zhang, Zhiyong, (张志勇) Ph.D., Nanjing University, 1996.

Zheng, Zhiming, (郑之明) Ph.D., Nanjing University, 1998.

Zhou, Yuan, (周苑) Ph.D., Nanjing University, 2004. *Condensed Matter Physics*. Strong correlated system, superconductivity, and topological insulators.

Zou, Xinye, (邹欣烨) Ph.D., Nanjing University, Acoustics.

Lecturer

Chen, Jianjun, (陈建军) Ph.D., Nanjing University, 2009.

Chen, Kai, (陈锴) Ph.D., Nanjing University, 2008. *Acoustics*. Acoustics signal processing, active noise control, wave field synthesis.

Cheng, Ying, (程营) Ph.D., Nanjing University, 2010. *Acoustics*. Acoustic metamaterials, ultrasonics.

Gao, Wenli, (高文莉) M.S., Nanjing University, 1993.

Lin, Zhibin, (林志斌) Ph.D., Nanjing University.

Sun, Liang, (孙亮) Ph.D., Nanjing University, 2006. *Condensed Matter Physics*. Magnetism.

Tao, Jiancheng, (陶建成) Ph.D., Nanjing University, 2008.

Xiong, Xiang, (熊翔) Nanjing University, 2011. *Condensed Matter Physics*. Plasmonics, metamaterials, nano fabrication.

DEPARTMENTAL RESEARCH SPECIALTIES AND STAFF

Theoretical

Biophysics. Yi Cao, Yuqiang Ma, Meng Qin, Wei Wang.

Condensed Matter Physics. Jian-Xin Li, Qiang-Hua Wang, Dingyu Xing, Shun-Li Yu.

Particles and Fields. Shenjian Chen, Yeuk-Kwan Cheung, Wenfei Li, Zhongzhou Ren, Hongshi Zong.

Theoretical Physics. Guojun Jin, Chun-Lai Ren, Jian Zhang.

Experimental

Acoutistical Science and Engineering. Li-Ping Cheng, Ying Cheng, Li Fan, Xia-sheng Guo, Bin Liang, Jie-Hui Liu, Xiao-Jun Liu, Xiaozhou Liu, Jing Lu, Xiaojun Qiu, Chao Tao, Juan Tu, Xinlong Wang, Hao-Dong Wu, Dong Zhang.

Condensed Matter Physics. Jin An, Hai Feng Ding, Jun Du, You-Wei Du, Duan Feng, MIn Gu, Feng-Zhen Huang, Shao-Chun Li, Jun-Ming Liu, Xiao-Mei Lv, Hai Sang, Da-Jun Shu, Fengqi Song, Liang Sun, Nu-Jiang Tang, Shao-Long Tang, Jian-Guo Wan, Xiangang Wan, Bogen Wang, Dunhui Wang, Guanghou Wang, Di Wu, Shao-Guang Yang.

Optics. Jianping Ding, Huitian Wang, Mu Wang, Xiaoyong Wang, Zhen-Lin Wang, Xiao-Shan Wu, Xing-Long Wu, Min Xiao, Xiang Xiong, Ping Xu, Yang Yu, Fengming Zhang, Shining Zhu, Xin-hua Zhu, Yongyuan Zhu, Zhigang Zou.

View additional information about this department at
www.gradschoolshopper.com

SHANGHAI JIAO TONG UNIVERSITY

DEPARTMENT OF PHYSICS AND ASTRONOMY

Shanghai, China
http://www.physics.sjtu.edu.cn/en/

General University Information

President: Jie Zhang (张杰)
Dean of Graduate School: Zheng Huang (黄震)
University website: http://en.sjtu.edu.cn/
Control: Public
Setting: Suburban
Total Faculty: 2,873
Total Graduate Faculty: 1,858
Total number of Students: 36,307
Total number of Graduate Students: 25,696

Department Information

Department Chairman: Xiangdong Ji (季向东), Chair
Department Contact: Xue Ying (薛颖), Graduate Educational Administrator
  Total full-time faculty: 69
  Total number of full-time equivalent positions: 127
  Full-Time Graduate Students: 244
  First-Year Graduate Students: 74
  Female First-Year Students: 23
  Total Post Doctorates: 16

Department Address

615A Physics Building
800 Dongchuan Road
Shanghai, 200240
CHINA
Phone: 86-21-5474-2662

Fax: 86-21-5474-1040
E-mail: xueying@sjtu.edu.cn
Website: http://www.physics.sjtu.edu.cn/en/

ADMISSIONS

Admission Contact Information

Address admission inquiries to: The Department of Physics and Astronomy, 615A Physics Building, 800Dongchuan Road, Shanghai, Peoples' Republic of China
Phone: 86-21-54742964
E-mail: catherinecherry@sjtu.edu.cn
Admissions website: http://www.sie.sjtu.edu.cn/ctrler.asp? action=list3_en&tp=00048

Application deadlines

Fall admission:
Domestic or Chinese students: March 31   *Int'l. students*: March 31

Application fee: RMB 800

Admissions information

For Fall of 2013:
  Number admitted: 110
  Number enrolled: 110

Admission requirements

Bachelor's degree requirements: Bachelor's degree in Physics or a related field. Those who apply for Ph.D. Program should have Master's Degrees.

Other admissions information

Additional requirements: Academic record for all universities an applicant has attended.

Recommendations from three faculty members who have supervised your coursework or research.

Statement of Purpose for your graduate study objectives.

Personal statement of any special circumstances you would like us to be aware of (optional).

TUITION

Tuition year 2014-15:

Full-time students: RMB 45,500 annual

Health insurance: Yes, ￥600.

Academic term: Semester

Teaching Assistants, Research Assistants, and Fellowships

Number of first-year

Teaching Assistants: 45

Research Assistants: 42

Fellowship students: 10

Average stipend per academic year

Fellowship student: RMB 8,000

FINANCIAL AID

For futher information

Address financial aid inquiries to: School of International Education, SJTU, No. 1954, Hua Shan RD. Shanghai, China 200030.

Phone: 86-21-62820638

E-mail: iso@sjtu.edu.cn

Financial aid website: http://www.sie.sjtu.edu.cn/ctrler.asp?action=detail_en&s=1109250004

HOUSING

Availability of on-campus housing

Single students: Yes

Married students: Yes

For further information

Address housing inquiries to: School of International Education, SJTU, No. 1954, Hua Shan RD. Shanghai, China 200030.

Phone: 86-21-62820638

E-mail: iso@sjtu.edu.cn

Housing aid website: http://www.sie.sjtu.edu.cn/ctrler.asp?action=detail_en&s=1207050001

Table A—Faculty, Enrollments, and Degrees Granted

Research Specialty	Faculty	Enrollment Master's	Enrollment Doctorate	Number of Degrees Granted Master's	Number of Degrees Granted Terminal Master's	Number of Degrees Granted Doctorate
Astronomy	7	–	–	–	–	–
Condensed Matter Physics	16	–	–	–	–	–
Optics	12	–	–	–	–	–
Particles and Fields	14	–	–	–	–	–
Plasma and Fusion	12	–	–	–	–	–
Theoretical Physics	8	–	–	–	–	–
Total	69	–	–	–	–	–
Full-time Grad. Stud.	–	86	158	–	–	–
First-year Grad. Stud.	–	43	31	–	–	–

GRADUATE DEGREE REQUIREMENTS

Master's: Time to degree: 2.5 years. Coursework requirements: Master's students shall earn 30 degree credits with a minimum of 19 coursework credits. Other requirements: Publish at least one paper on journals indexed by SCI as the first author.

Doctorate: Time to degree: 3 years with a master's degree.-5 years without a master's degree. Coursework requirements: Doctoral students shall earn 16 degree credits. Other requirements: Publish at least two papers on core journals indexed by SCI as the first author.

SPECIAL EQUIPMENT, FACILITIES, OR PROGRAMS

The Department of Physics and Astronomy, SJTU established in 1928, is one of the oldest physics departments in China. It currently has 67 full-time faculty members, 282 full-time undergraduates and 244 postgraduate students. Among those faculty members, there are 6 Academicians of Chinese Academy of Sciences and Engineering, 6 distinguished experts of "National 1000-Talent Search Program", 10 "Cheung Kong" chair professors and 16 NSFC Outstanding Junior Investigators.

The Department of Physics and Astronomy, SJTU has 6 research institutes, including the Institute of Theoretical and Interdisciplinary Physics, the Institute of Particle Physics and Cosmology, the Institute of Condensed Matter Physics, the Research Center of Optical Science and Engineering, Key Laboratory for Laser Plasma, and Center for Astronomy and Astrophysics. The department also has 4 ministry-level laboratories and more than 20 research teams.

Key Laboratories:

Key Laboratory on Artificial Structures and Quantum Management (Ministry of Education)

Key Laboratory for Laser Plasma (Ministry of Education)

State Key Laboratory on Fiber Optic Local Area Communication Networks and Advanced Optical Communication Systems

Shanghai Key Laboratory on Particle Physics and Cosmology

There is 2 National "973" Key Basic Research Project whose Chief Scientist is from the department, 10 research topics of "973" Project, 1 key project supported by "863" Program, and 2 National Science and Technology Major Projects. The department undertakes 1 Innovative Group of NSFC, with 10 projects supported by Major Program (Major International Joint Research Program) of NSFC and 3 projects supported by National Natural Science Funds for Distinguished Young Scholar. In the past five years, the department has seen a swift growth in research funds, reaching 79.18 million RMB in 2012. The Department of Physics and Astronomy also leads a Coordination and Innovation Center and 2 Innovative Groups, Ministry of Education.

Innovative Groups:

Innovative Group of National Natural Science Foundation on "Cutting-edge Studies of High-Energy Density Physics" (Led by Academician Zhang Jie)

Innovative Group of Ministry of Education on "Quantum Semiconductor Structures and Quantum Process Control" (Led by Professor Shen Wenzhong)

Innovative Group of Ministry of Education on "Physics with extremely high intensity lasers" (Led by Professor Sheng Zhengming)

The Department of Physics and Astronomy owns a state-level teaching base and national teaching fellowship. A wide range of disciplines, including particle and nuclear physics, astrophysics and cosmology, condensed matter physics, theoretical physics, optics, laser plasma physics are taught here. The department has established an all-English teaching system, cooperating with University of Maryland and Duke University to develop "2+2" Program and "3 +1" Program which gives students opportunities to gain a diploma abroad. The Department of Physics and Astronomy attaches great importance to cultivating students' innovative abilities, with a great number of students publishing papers on world-known journals. 2 PhD Dissertations have been included in the "National 100 Excellent Doctoral Dissertations" and 1 has been nominated.

Table B—Separately Budgeted Research Expenditures by Source of Support

Source of Support	Departmental Research	Physics-related Research Outside Department
Federal government	RMB60,000,000	
State/local government	RMB10,000,000	
Non-profit organizations		
Business and industry	RMB10,000,000	
Other		
Total	RMB80,000,000	

Table C—Separately Budgeted Research Expenditures by Research Specialty

Research Specialty	No. of Grants	Expenditures (RMB)
Condensed Matter Physics	75	RMB43,035,500
Optics and Optical Enginnering	32	RMB 7,488,300
Particle Physics, Cosmology and Astronomy	35	RMB 9,911,800
Laser Plasma	44	RMB15,253,500
Theoretical Physics	14	RMB 2,839,000
Other	5	RMB 660,000
Total	205	RMB79,188,100

FACULTY

Professor

Cai, Shen'ou, (蔡申瓯) Ph.D., Northwestern University, 1989. Director of the Institute of Natural Science National Chair Professor of "1000-Plan". *Biophysics, Theoretical Physics*. Theoretical and computational neuroscience, network dynamics, applied dynamical systems, applied stochastic processes, wave turbulence; development of theoretical and computational tools for their applications in physics, biology, and neuroscience.

Chen, Liewen, (陈列文) Ph.D., Institute of Modern Physics, Chinese Academy of Sciences, 2000. *Nuclear Physics*. (1) Theoretical Nuclear Physics (2) Heavy-Ion Collisions (3) Quark Gluon Plasma (4) Compact Stars.

Chen, Xianfeng, (陈险峰) Ph.D., Shanghai Jiao Tong University, 1999. Deputy Chair of the Department of Physics and Astronomy. NSFC Outstanding Junior Investigator. *Optics*. (1) Quasi-phase-matching nonlinear optics and its applications (2) Coherent control of ultra-fast laser pulse and interaction with matters (3) Micro-, Nano- photonics materials and applications.

Deng, Xiaoxu, (邓晓旭) Ph.D., Harbin Institute of Technology, 1999. *Applied Physics, Optics*. (1) Electro-optic waveguide devices (2) Micro/nano photonic devices (3) Experimental study on silicon nitride passivation film of solar cell (4) Study on laser peen forming of sheet metal.

Dong, Bing, (董 兵) Ph.D., Shanghai Institute of Metallurgy, 1997. *Condensed Matter Physics, Solid State Physics, Theoretical Physics*. (1) Quantum Transport and fluctuation in Nanosystems (2) Electron Correlation in Nanosystems (3)

Many-Particle Theory (4) Strongly-Correlated Electron Systems.

Fan, Dianyuan, (范滇元) Ph.D., Institute of Electronics, Chinese Academy of Sciences, 1996. Academician, Chinese Academy of Engineering. *Optics, Plasma and Fusion*. Intense Laser.

Giboni, Karl Ludwig, (Karl Ludwig Giboni) Ph.D., RWTH Aachen, 1980. *High Energy Physics, Particles and Fields*.

He, Xiaogang, (何小刚) Ph.D., University of Hawaii, 1987. National Chair Professor of "1000 Plan". *Cosmology & String Theory, Particles and Fields, Quantum Foundations*. Particle physics, quantum mechanics, and cosmology.

Ji, Xiangdong, (季向东) Ph.D., Drexel University, 1987. Chair of the Department of Physics and Astronomy National Chair Professor of "1000 Plan". *Particles and Fields*. (1) Dark Matter Detection (2) Quantum Chromodynamics (3) Hadron Physics (4) New Physics Beyond Standard Model.

Jia, Jinfeng, (贾金锋) Ph.D., Peking University, 1992. Yangtze Chair Professor. NSFC Outstanding Junior Investigator. *Condensed Matter Physics*. (1) Topological Insulators and new quantum materials (2) Low Temperature Scanning Tunneling Microscopy/Spectroscopy/ARPES (3) Thin film growth and its atomistic processes of various materials (semiconductor, magnetic multilayers and dielectric materials etc.) by molecular beam epitaxy (4) Novel properties of low-dimensional structures and interfaces.

Jing, Yipeng, (景益鹏) Ph.D., Scuola Internazionale Superiore di Studi Advanzati, 1992. Director of the Centre for Astronomy and Astrophysics NSFC Outstanding Junior Investigator. *Astrophysics, Cosmology & String Theory*.

Lei, Xiaolin, (雷啸霖) Ph.D., Peking University, 1963. Academician, Chinese Academy of Science. *Condensed Matter Physics, Theoretical Physics*. Electron transport and optics in semiconductors.

Li, Jiaming, (李家明) Ph.D., University of Chicago, 1974. Academician, Chinese Academy of Science. *Atomic, Molecular, & Optical Physics, Computational Physics, Theoretical Physics*.

Li, Yijie, (李贻杰) Ph.D., Shanghai Institute of Microsystem And Information Technology, Chinese Academy of Sciences, 1991. Chair Professor. . . *Applied Physics, Condensed Matter Physics*. High-Tc superconductivity research.

Liu, Huichun, (刘惠春) Ph.D., University of Pittsburgh, 1987. National Chair Professor of "1000 Plan". *Condensed Matter Physics*. Research of semiconductor, optoelectronics.

Liu, Ying, (刘荧) Ph.D., Univerisity of Minnesota, 1991. Director of the Institute of Condensed Matter Physics. National Chair Professor of "1000 Plan". *Condensed Matter Physics*.

Pan, Jianwei, (潘建伟) Ph.D., Universität Wien, 1999. Academician, Chinese Academy of Science. *Optics, Quantum Foundations*. (1) Quantum physics (2) Quantum optics.

Qian, Liejia, (钱列加) Ph.D., Shanghai Institute of Optics and Fine Mechanics, Chinese Academy of Sciences, 1989. Director of the Key Laboratory for Laser Plasma, Ministry of Education. NSFC Outstanding Junior Investigator. *Applied Physics, Optics, Plasma and Fusion*. Ultrafast nonlinear optics and high-intensity lasers.

Shen, Wenzhong, (沈文忠) Ph.D., Shanghai Institute of Technical Physics of the Chiese Academy of Sciences, 1995. Yangtz Chair Professor. NSFC Outstanding Junior Investigator. *Applied Physics, Condensed Matter Physics*. Solid spectrum and optoelectronic device applications.

Sheng, Zhengming, (盛政明) Ph.D., Shanghai Institute of Optics and Fine Mechanics, Chinese Academy of Sciences, 1993. Yangtze Chair Professor. NSFC Outstanding Junior Investigator. *High Energy Physics, Plasma and Fusion*. (1)High intensity laser-plasma interaction (2) advanced concepts of particle acceleration and radiation sources from relativistic laser-plasmas (3) beam-plasma interaction, physics of inertial confined fusion (4) large-scale numerical simulation (PIC, Vlasov-Fokker-Planck, etc.).

Sun, Hong, (孙 弘) Ph.D., Shanghai Jiao Tong University, 1987. Chair Professor. *Computational Physics, Condensed Matter Physics, Theoretical Physics*. First principles calculation studies of material physics with focuses on the studies and designs of super-hard materials in high pressures and high temperatures.

Sun, Yang, (孙 扬) Ph.D., University of Munich, 1991. Chair Professor. *Nuclear Physics*. (1) Theoretical Nuclear Physics (2) Nuclear Astrophysics (3) Computational Many-Body Physics (4) High-Temperature Superconductivity.

Wang, Bin, (王 斌) Ph.D., Fudan University, 1998. Yangtze Chair Professor. NSFC Outstanding Junior Investigator. *Cosmology & String Theory*. (1) General Relativity (2) Black hole physics (3) Cosmology.

Wang, Changshun, (王长顺) Ph.D., Jilin University, 1999. *Optics*. (1) Optical storage (2) Nonlinear optics of organic polymers (3) Synchrotron radiation stimulated etching.

Wang, Hui, (王 辉) Ph.D., Fudan University, 1999. *Optics*. New photoelectric effect and new photoelectric materials, spin electronics materials and devices.

Wang, Xijie, (王西杰) Ph.D., University of California, Los Angeles, 1992. National Chair Professor of "1000 Plan". *Accelerator, Plasma and Fusion*. (1) Laser acceleration (2) Ultra-fast physics and technology research.

Wu, Xiangping, (武向平) Ph.D., National Astronomical Observatories, CAS, 1989. Academician, Chinese Academy of Science. *Astrophysics, Cosmology & String Theory, Nuclear Physics*. (1) Cosmology (2) Low-frequency radio astronomy.

Xiaohu, Yang, (杨小虎) Ph.D., University of Science and Technology of China, 2002. NSFC Outstanding Junior Investigator. *Astronomy, Astrophysics, Cosmology & String Theory*.

Xing, Xiangjun, (邢向军) Ph.D., University of Colorado at Boulder, 2003. Chair Professor. *Theoretical Physics*. (1) Elasticity of complex elastomeric materials (2) Liquid Crystals and Membranes (3) Statistical Mechanics of charged systems (4) Generalized elasticity theory (5) Topological Defects.

Xu, Haiguang, (徐海光) Ph.D., Shanghai Jiao Tong University, 1998. Associate Chair of the Department of Physics and Astronomy NSFC Outstanding Junior Investigator. *Astronomy, Astrophysics*. X-ray imaging spectroscopic study of galaxies, galaxy groups, and galaxy clusters; Dark matter distributions; First generation objects in the universe and cosmic reionization; Low-frequency radio experiment and observation.

Xu, Jianqiu, (徐剑秋) Ph.D., Shanghai Institute of Optics and Fine Mechanics, Chinese Academy of Sciences, 1999. *Optics, Plasma and Fusion*. (1) High-power solid-state laser (2) MID-IR solid-state laser (3) New pattern solid-state laser.

Yang, Haijun, (杨海军) Ph.D., Institute of High Energy Physics, Chinese Academy of Sciences, 2000. *High Energy Physics, Particles and Fields*. (1) Collider Physics (2) Electroweak Physics (3) Neutrino Physics (4) Higgs, SUSY, Beyond Standard Model Physics (5) Dark Matter/Energy Search.

Yao, Xin, (姚 忻) Ph.D., University of Liverpool, 1993. Yangtze Chair Professor. *Condensed Matter Physics*. High-temperature superconductors crystal growth theory.

Zhan, Li, (詹 黎) Ph.D., City University of Hongkong, 2002. *Optics*. (1) Fiber optics and optical communication (2) Fiber laser and physics (3) Slow/fast light and superluminal signal processing (4) Laser spectrometry and its application (5) Single molecular technology and DNA sequencing.

Zhang, Jie, (张 杰) Ph.D., Institute of Theoretical Physics, Chinese Academy of Sciences, 1988. President of Shanghai Jiao Tong University. Academician, Chinese Academy of Science Foreign associate of American National Academy of Sciences (NAS). Member of Germany Academy of Sciences Leopoldina, Foreign member of Royal Academy of Engineering (FREng). Fellow of the Third World Academy of Sciences (TWAS). *Plasma and Fusion*. (1) High field physics (2) Laser plasma physics (3) ultrafast electron diffraction physics. (4) laboratory astrophysics.

Zhang, Pengjie, (张鹏杰) Ph.D., University of Toronto, 2003. NSFC Outstanding Junior Investigator. *Astronomy, Astrophysics, Atmosphere, Space Physics, Cosmic Rays*. Large-scale structure of the universe, the cosmological tests of general relativity, weak gravitational lensing, SZ effect.

Zhao, Yumin, (赵玉民) Ph.D., Nanjing University, 1995. NSFC Outstanding Junior Investigator. *Nuclear Physics*. (1) Theoretical Nuclear Physics (2) Many-body systems interacting by random interactions (3) Nucleon pair approximation of the shell model (4) Systematics of nuclear properties.

Zheng, Hang, (郑 杭) Ph.D., Shanghai Jiao Tong University, 1985. NSFC Outstanding Junior Investigator. *Condensed Matter Physics*. (1) Quantum coherence and manipulation (2) Physics of strongly correlated systems.

Zheng, Maojun, (郑茂俊) Ph.D., Institute of Solid State Physics, Chinese Academy of Sciences, 2001. *Condensed Matter Physics*. Preparation of nanostructured semiconductor materials, physical properties and their applications in solar and hydrogen; preparation and application of grapheme.

Zhu, Kadi, (朱卡的) Ph.D., Shanghai Jiao Tong University, 1991. Associate chair of the Department of Physics and Astronomy. *Condensed Matter Physics*. (1) Condensed Matter Optomechanics (2) Solid-State Quantum Computation and Quantum Information (3) Solid-State Quantum Optics et al.

Associate Professor

Chen, Jie, (陈洁) Ph.D., University of California, Irvine, 2008. *Plasma and Fusion*. Ultrafast optical spectroscopy, ultrafast x-ray spectroscopy, ultrafast electron diffraction.

Chen, Min, (陈民) Ph.D., Institute of Physics, Chinese Academy of Sciences, 2007. *Plasma and Fusion*. Laser plasma interaction, laser plasma wakefield acceleration, laser plasma ion acceleration, laser plasma radiation source, plasma theory and simulation.

Chen, Yuping, (陈玉萍) Ph.D., Shanghai Jiao Tong University, 2002. *Optics*. (1) QPM nonlinear optics and quantum nonlinear optics; (2) ultrafast optical interactions with matter; (3) photonic materials and devices and all-optical signal processing.

Dai, Dechang, (戴德昌) Ph.D., Case Western Reserve University, 2007. *Astrophysics, Cosmology & String Theory*.

Fu, Changbo, (符长波) Ph.D., Texas Tech University, 2007. *Nuclear Physics*. (1) PandaX project: Searching heavy dark matter particles with liquid Xe detector (2) Searching for Axion-like low mass dark matter by using polarized noble gas (3) Developing high efficient pumping and storage methods for Laser Pumped Polarized Noble Gases (4) Polarized noble gas for Medial Lung Imaging (5) Radioactive Ion beam physics, and Nuclear Astrophysics.

Gao, Chunlei, (高春雷) Ph.D., Max-Planck-Institute for Microstructure Physics, 2006. *Condensed Matter Physics*. (1) Topological Insulators and new quantum materials (2) Low Temperature Scanning Tunneling Microscopy/Spectroscopy/ARPES (3) Thin film growth and its atomistic processes of various materials (semiconductor, magnetic multilayers and dielectric materials etc.) by molecular beam epitaxy (4) Novel properties of low-dimensional structures and interfaces.

Hafz, Nasr A., (Nasr A. Mohamed Hafz) Ph.D., University of Tokyo, 2001. *Plasma and Fusion*. Laser-plasma acceleration (LPA) of electron beams.

He, Feng, (何 峰) Ph.D., Shanghai Institute of Optics and Fine Mechanics, Chinese Academy of Sciences, 2005. *Plasma and Fusion*. (1) Molecular image (2) Quantum coherent control of the chemical reaction (3) Ccreation of the new light source (4) Super-intense laser-matter interaction.

Jin, Xianmin, (金贤敏) Ph.D., University of Science and Technology of China, 2008. *Optics*. Ultrafast quantum optics and optical metrology.

Li, Liang, (李亮) Ph.D., University of Wisconsin, Madison, 2005. *Nuclear Physics, Particles and Fields*. Collider Physics; Intensity Frontier Physics; Standard Model and Beyond Standard Model Physics; Electroweak Physics; Top Quark Physics; QCD Physics; Higgs Physics; SUSY; Extra Dimension; Dark Matter/Energy.

Liu, Canhua (刘灿华) Ph.D., The University of Tokyo, 2006. *Condensed Matter Physics*. (1) Topological Insulators and new quantum materials (2) Low Temperature Scanning Tunneling Microscopy/Spectroscopy/ARPES (3) Thin film growth and its atomistic processes of various materials (semiconductor, magnetic multilayers and dielectric materials etc.) by molecular beam epitaxy (4) Novel properties of low-dimensional structures and interfaces.

Liu, Jianglai, (刘江来) Ph.D., University of Maryland, 2006. *Particles and Fields*. (1) Neutrino Experiment (2) Dark Matter Experiment.

Liu, Shiyong, (刘世勇) Ph.D., Shanghai Institute of Microsystem and Information Technology, Chinese Academy of Sciences, 2002. *Theoretical Physics*.

Liu, Xiang, (刘湘) Ph.D., Columbia University, 2003. *Particles and Fields*. (1) Neutrino physics (2) Dark matter (3) High-purity germanium detector.

Loach, James, (James Loach) Ph.D., Oxford University, 2008. *Nuclear Physics, Particles and Fields*. (1) Neutrinoless double beta decay (2) Neutrino oscillations (3) WIMP dark matter. (4) Low-radioactivity techniques.

Luo, Weidong, (罗卫东) Ph.D., University of California, Berkeley, 2004. *Condensed Matter Physics, Theoretical Physics*. (1) First-principles studies of complex oxides (2) Topological insulators.

Ni, Kaixuan, (倪凯旋) Ph.D., Columbia University, 2006. *Nuclear Physics, Particles and Fields*. Dark matter and neutrino physics, particle detector development.

Qian, Dong, (钱冬) Ph.D., Fudan University, 2003. *Condensed Matter Physics*. 1. Low-dimensional physics 2. Magnetism.

Shun, Wang, (王顺) Ph.D., University of Minnesota, 2010. *Condensed Matter Physics*.

Ulmschneider, Jakob, (Jakob Ulmschneider) Ph.D., Yale University, 2004. *Biophysics, Theoretical Physics*. (1) Computational Biophysics (2) Multi-scale algorithms for ion transport (3) Voltage-gated sodium channels (4) Monte Carlo algorithms (5) Parallel algorithms for large-scale simulations (6) Predicting the native fold of membrane proteins (7) Mesoscopic phenomena of peptide adsorption on membranes.

Wan, Wenjie, (万文杰) Ph.D., Princeton University, 2010. *Optics*. 1. Nonlinear optics 2. Nano-optics Quantum Optics 3. Plasmonics.

Wang, Yujie, (王宇杰) Ph.D., Massachusetts Institute of Technology, 2001. *Condensed Matter Physics*. Synchrotron radiation imaging.

Xie, Guoqiang, (谢国强) Ph.D., Fudan University, 2008. *Applied Physics, Plasma and Fusion*. (1) Ultrafast laser (2) Laser materials and devices (3) Mid-infrared laser.

Ye, Fangwei, (叶芳伟) Ph.D., University of Science and Technology of China, 2005. *Optics*. (1) Nonlinear Optics: Formation and evolution of optical localized states, Optical Spatial Solitons (2) Nanophotonics: Interaction between light and nanostructured materials (3) Interaction of light with complex physical systems, such as nonlinear materials, periodic/disordered systems, matematerials, etc.

Zhang, Hepeng, (张何朋) Ph.D., City University of New York, 2004. *Biophysics, Condensed Matter Physics*. (1) Bacterial Collective Motion (2) Competition between colonies (3) Swarming Dynamics (4) Internal Gravity Waves (5) Toughening Crystallites (6) Dynamics of Static Friction.

Zhang, Jie, (张洁) Ph.D., University of Pittsburgh, 2005. *Condensed Matter Physics*. (1) Statistical description of granular materials (2) Jamming transition of granular media under shear and compression (3) Mixing and segregation in granular materials (4) Geological implications of granular physics (5) Thermal convection and 2D turbulence (6) Interfacial physics: Instabilities in thin liquid films.

Zhang, Jun, (张骏) Ph.D., Columbia University, 2006. *Astrophysics, Cosmology & String Theory*. I am interested in the formation of the large scale structure in our Universe, and the underlying physical rules. Theoretically, my research focuses on understanding the properties of dark matter, dark energy, cosmic microwave background, galaxies, clusters, cosmic reionization, etc. Observationally, we use the gravitational lensing effect to directly probe the density distribution of our Universe, and to test the theory of gravity on the cosmic scales.

Zhong, Xiaoxia, (钟晓霞) Ph.D., Fudan University, 1998. *Nano Science and Technology, Optics, Plasma and Fusion*. Plasma nanotechnology and its applications.

Research Professor

Huang, Meizhen, (黄梅珍) Ph.D., Zhejiang University, 2001. *Optics*.

DEPARTMENTAL RESEARCH SPECIALTIES AND STAFF

Theoretical

Astronomy and Astrophysics. Center for Astronomy and Astrophysics (CAA) was founded in September 2012. The center has been attracting a team of excellent researchers who work in international advanced scientific research areas. CAA focuses on basic scientific studies of early universe, dark matter, dark energy, reionization, large-scale structure, galaxy formation and evolution, black hole and high energy astrophysics. Major Projects in CAA: SKA Dome-A CMB polarisation and optical telescopes AS3—After Sloan 3:e-BOSS, MANGA, APOGEE BigBOSS—The first stage IV Dark energy experiment Theoretical Cosmology and Black Hole Physics Observational Cosmology Galaxies and Clusters of Galaxies Yipeng Jing, Bin Wang, Xiangping Wu, Yang Xiaohu, Haiguang Xu, Jun Zhang, Pengjie Zhang.

Condensed Matter Physics. Research activities at the Institute of Condensed Matter Physics focuses on topological insulators and topological superconductors, semiconductors, high-Tc superconductors, theoretical and computational condensed matter physics, soft condensed matter physics, spectroscopies, and opto-electronic device physics. Hepeng Zhang, Jie Zhang, Hang Zheng, Kadi Zhu.

Nuclear Physics. The nuclear theory group focuses its research on the many-body origin of regular properties of the atomic nuclei, structure of heavy nuclei, and the hot and dense nuclear matter. Liewen Chen, Yang Sun, Yumin Zhao.

Particles and Fields. The particle theory group is working on models for massive neutrinos, Higgs production, and models for dark matter physics. There is also an active research in the area of quantum chromodynamics and properties of hadrons. Dechang Dai, Xiaogang He, Xiangdong Ji.

Plasma and Fusion. Min Chen, Feng He, Jiaming Li, Zhengming Sheng.

Theoretical Physics. Electronic transport in condensed matter and superconducting materials Non-linear physics Quantum optics and quantum information Soft condensed matter physics

Complex systems Theoretical biophysics Statistical physics Fluid dynamics Transformation optics and metamaterials Low-dimensional condensed matter physics Quantum field theory and phase transition. Shen'ou Cai, Bing Dong, Xiaolin Lei, Hong Sun, Jakob Ulmschneider, Xiangjun Xing.

Experimental

Astronomy and Astrophysics. Major Projects in CAA: SKA Dome-A CMB polarisation and optical telescopes AS3— After Sloan 3:e-BOSS, MANGA,APOGEE BigBOSS—The first stage IV Dark energy experiment. Xiangping Wu, Haiguang Xu.

Condensed Matter Physics. HTS Laboratory, STJTU owns the world-class cutting-edge research facilities. HTS Laboratory is the only institute in China able to fabricate 100 meters of second generation high-temperature superconducting tapes, with a rare earth oxide superconducting layer. Laboratory of low-dimensional physics and interface engineering boasts first-class equipments and a group of dynamic young researchers. Over 260 dissertations are involved in SCI, 3 theses are published on Science and Nature respectively. Papers are cited more than 3000 times. Condensed matter spectroscopy and opto-electronic physics is engaged in the study of controllable preparation and optical and electrical properties of nanocrystalline silicon films, and solar cell applications. Team members have made innovative achievements in large-scale solar cell applications. Chunlei Gao, Jinfeng Jia, Yijie Li, Canhua Liu, Huichun Liu, Ying Liu, Dong Qian, Wenzhong Shen, Xin Yao, Hang Zheng, Maojun Zheng, Kadi Zhu.

Optics and Optical Enginerring. The researches of the Center of Optical Science and Engineering include the basic scientific research in nonlinear optics, nanophotonics, and quantum optics, and applied basic research in fiber, optical waveguide, optical instruments and solar cell, etc. Xianfeng Chen, Yuping Chen, Xiaoxu Deng, Meizhen Huang, Xianmin Jin, Jianwei Pan, Wenjie Wan, Changshun Wang, Hui Wang, Fangwei Ye, Li Zhan, Xiaoxia Zhong.

Particles and Fields. Particle physics experiment team leads dark matter detection experiments — PandaX, and will carry out researches on dark matters at underground lab in Jinping, Sichuan Province. They also participate in other experimental projects include DayaBay neutrino experiment, LHC Collide Physics (ATLAS), Majorana double beta decay. Karl Giboni, Xiangdong Ji, Liang Li, Jianglai Liu, Xiang Liu, James Loach, Kaixuan Ni, Haijun Yang.

Plasma and Fusion. Laboratory of Laser Plasma fully dedicates to the frontier studies of high energy density physics and the relevant key technologies, as well as their applications in the advanced particle acceleration and radiation sources driven by intense lasers, new concepts of inertial confinement fusion, laboratory astrophysics, and ultrafast structural dynamics of matter. Jie Chen, Dianyuan Fan, Nasr Hafz, Liejia Qian, Zhengming Sheng, Xijie Wang, Guoqiang Xie, Jianqiu Xu, Jie Zhang.

View additional information about this department at
www.gradschoolshopper.com

CITY UNIVERSITY OF HONG KONG

DEPARTMENT OF PHYSICS AND MATERIALS SCIENCE

Kowloon, Hong Kong
http://www.ap.cityu.edu.hk

General University Information

President: Professor Way Kuo
Dean of Graduate School: Professor Horace Ip
University website: http://www.cityu.edu.hk/
Total Faculty: 787
Total number of Students: 19,250
Total number of Graduate Students: 6,215

Department Information

Department Chairman: Professor Xun-Li Wang, Head
Department Contact: Professor Ruiqin Zhang, Professor
Total full-time faculty: 28
Total number of full-time equivalent positions: 28
Full-Time Graduate Students: 64
First-Year Graduate Students: 16
Female First-Year Students: 4
Total Post Doctorates: 12

Department Address

G6702, 6/F, Green Zone, Academic 1,
City University of Hong Kong, 83 Tat Chee Avenue,
Kowloon, 852
HONG KONG
Phone: (852) 3442-7849
Fax: (852) 3442-0538
E-mail: aprqz@cityu.edu.hk
Website: http://www.ap.cityu.edu.hk

ADMISSIONS

Admission Contact Information

Address admission inquiries to: Block 1 To Yuen Building, Chow Yei Ching School of Graduate Studies, City University of Hong Kong, Kowloon, Hong Kong
Phone: (852) 3442-9076
E-mail: sg@cityu.edu.hk
Admissions website: http://www.sgs.cityu.edu.hk/

Application deadlines
Fall admission: December 10

Application fee: HK$200

Admissions information
For Fall of 2012:
Number of applicants: 98
Number admitted: 24
Number enrolled: 16

Admission requirements
Bachelor's degree requirements: Bachelor's degree with first class honours (or equivalent qualification) from a recognised university.
Equivalent qualifications mentioned above include relevant professional qualifications or other scholarly achievements recognised by the University.
Minimum undergraduate GPA: 3.5

GRE requirements
The GRE is not required.

Advanced GRE requirements
The Advanced GRE is not required.

TOEFL requirements
The TOEFL exam is required for students from non-English-speaking countries.
PBT score: 550
iBT score: 79
Applicants whose entrance qualification is obtained from an institution where the medium of instruction is NOT English should fulfill the following minimum English proficiency requirement: 550 (paper-based test) or 213 (computer-based test) or 79 (internet-based test).

Other admissions information
Additional requirements: English Proficienty Requirement:
Besides TOEFL, applicants from an institution where the language of teaching is not English may also take one of the following tests to satisfy the minimum English proficiency requirements:
-a minimum overall band score of 6.5 in IELTS; or
-a score of 490 in the Chinese mainland's College English Test Band 6; or
-other test scores that may be regarded as equivalent to TOEFL 550 (paper-based) or 213 (computer-based) or 79 (internet-based).
Other Entrance Requirements:
Apart from the first-class honours bachelor's degree holder, the University also accepts the following qualifications for minimum entrance requirements for admission to PhD:
- hold a higher degree by research (or equivalent qualification) from a recognised university; or
- hold a Master's degree (or equivalent qualification) from a recognised university
Equivalent qualifications mentioned above include relevant professional qualifications or
other scholarly achievements recognised by the University.

TUITION

Tuition year 2014-15:
For Government-funded Students:
Tuition fee: HK$3,508 per month (Full-time) or HK$1,754 per month (Part-time)

Continuation fee: HK$877 per month (Full-time) or HK$439 per month (Part-time)
For Self-financing Students:
Tuition fee: HK$7,016 per month (Full-time) or HK$3,508 per month (Part-time)
Continuation fee: HK$1,754 per month (Full-time) or HK$877 per month (Part-time)
Remarks: The above tuition fee is NON-REFUNDABLE.
Deferred tuition plan: No
Health insurance: Available.
Other academic fees:
Acceptance Fee (Non-refundable; payable upon acceptance of admission offer): HK$7,016
Membership Fee for CityU Postgraduate Association (One-off payment): HK$300 (Full-time) or HK$150 (Part-time)
(Re)Examination Fee (Non-refundable): HK$1,000
Graduation Fee (Refundable upon withdrawal/termination): HK$400
Academic term: Semester
Number of first-year students who received full tuition waivers: 4

Teaching Assistants, Research Assistants, and Fellowships
Number of first-year
Teaching Assistants: 15
Fellowship students: 3
Average stipend per academic year
Teaching Assistant: HK$172,800
Fellowship student: HK$240,000
Postgraduate Studentship (not applicable to self-financing research students) is granted on the basis of academic merit. Eligible new full-time students will be considered for the award of the Studentship as part of their application for admission to PhD. The Studentship is normally granted on a yearly basis, for a maximum period of 3 years. Postgraduate Studentship holder is required to perform some teaching duties as part of their postgraduate training.
Applicants who could demonstrate outstanding qualities of academic performance, research ability / potential, communication and interpersonal skills, and leadership abilities are encouraged to apply for admission through 'Hong Kong PhD Fellowship Scheme'(http://www.sgs.cityu.edu.hk/prospective/rpg/hkphd).

HOUSING

Availability of on-campus housing
Single students: Yes
Married students: Yes

For further information
Address housing inquiries to: Student Residence Office, City University of Hong Kong, 22 Cornwall Street, Kowloon Tong, Hong Kong.
Phone: (852) 3442-1200
E-mail: sropga@cityu.edu.hk
Housing aid website: http://www.cityu.edu.hk/sro/

Table A—Faculty, Enrollments, and Degrees Granted

Research Specialty	2012-13 Faculty	Enrollment Fall 2012		Number of Degrees Granted 2011–12		
		Master's	Doctorate	Master's	Terminal Master's	Doctorate
Applied Mathematics	1	–	–	–	–	–
Applied Physics	12	–	6	–	–	–
Astronomy	1	–	–	–	–	–
Atomic, Molecular, & Optical Physics	4	–	–	–	–	–
Biophysics	3	–	–	–	–	–
Chemical Physics	5	–	5	–	–	–
Computational Physics	6	–	1	–	–	–
Condensed Matter Physics	11	–	5	–	–	–
Electrical Engineering	3	–	–	–	–	–
Electromagnetism	1	–	1	–	–	–
Energy Sources & Environment	5	–	3	–	–	–
Engineering Physics/Science	4	–	–	–	–	–
Fluids, Rheology	1	–	–	–	–	–
Geology/Geochemistry	1	–	–	–	–	–
Geophysics	1	–	–	–	–	–
Low Temperature Physics	1	–	–	–	–	–
Materials Science, Metallurgy	18	–	7	–	–	–
Mechanics	3	–	–	–	–	–
Medical, Health Physics	2	–	–	–	–	–
Nano Science and Technology	19	–	10	–	–	–
Nonlinear Dynamics and Complex Systems	2	–	–	–	–	–
Nuclear Engineering	1	–	–	–	–	–
Nuclear Physics	1	–	–	–	–	–
Optics	4	–	1	–	–	–
Physics of Beams	2	–	–	–	–	–
Plasma and Fusion	1	–	–	–	–	–
Polymer Physics/Science	5	–	–	–	–	–
Solid State Physics	12	–	4	–	–	–
Statistical & Thermal Physics	3	–	–	–	–	–
Surface Physics	5	–	5	–	–	–
Theoretical Physics	3	–	1	–	–	–
Other	1	–	–	–	–	–
Total	142	–	–	–	–	–
Full-time Grad. Stud.	–	–	64	–	–	–
First-year Grad. Stud.	–	–	16	–	–	–

GRADUATE DEGREE REQUIREMENTS

Doctorate:

Coursework Requirement:

-14 credit units (including core course(s) of at least 4 credit units which shall include at least 2 credit units of research methodology or foundation course at postgraduate level) and

-1 credit unit compulsory course: Teaching Students: First Steps (SG8001)

To be recommended for the award of PhD degree, students must successfully complete the coursework requirements, and satisfy the examiners in respect of the thesis submitted, in an oral examination on the thesis and area of study concerned, and in any written or practical examinations as required.

Thesis: The thesis examination should include thesis assessment, an oral examination, and any other assessment arrangements that may be required by the Panel of Examiners. If the thesis is confirmed to be of the required academic standard by the Panel of Examiners, an oral examination will be arranged. The oral examination is compulsory. It should normally take place in Hong Kong and be conducted in English.

SPECIAL EQUIPMENT, FACILITIES, OR PROGRAMS

The Physics & Materials Science Laboratories are well equipped with sophisticated up-to-date equipment for teaching and research.

The Radiation Laboratories contain systems for Alpha, Beta and Gamma Spectrometry, Liquid Scintillation Counting, Tritium-in-air Detection, Radon Counting, X-Ray Fluorescence spectrometry and Thermo-Luminescence Dating.The Physics and Opto-electronics Laboratories contain several laser systems. (Argon ion, Carbon Dioxide, Excimer, YAG and dye lasers), and systems for Brillouin Spectrometry, Thermal Vediography, Laser Thickness Measurement, Image Analysis, Sputter Coating and Noise Analysis.

Also installed is state-of-the-art equipment for Materials Sciences including systems for Creep Testing, Thermo-Gravimetric Analysis, Differential Scanning Calorimetry, High Speed Photography, X-Ray Diffractometry as well as instrumentation for measuring true density, surface area, gas-sorption and particle size. In addition, systems exist for Atomic Absorption Spectrometry, and Fourier Transform IR Spectrometry. Scanning Electron Microscopes with windowless EDS and an Inter-institutional Transmission Electron Microscope are likewise available for microstructural examination of materials. Besides, the Department has a wide range of thin film deposition system including two Hot Filament and two Microwave CVD Diamond Deposition Systems. Ion Beam, Electron Beam and Magnetron Sputtering Deposition Systems, all with UHV design, are also available. Analysis of the deposition products can be done with the sophisticated surface science system which includes LEED, RHEED, Auger Electron spectroscopy and Photoemission.

The Plasma Laboratory has two plasma immersion ion implanters as well as ancillary sample preparation and materials characterization equipment.

In addition, a well-equipped mechanical workshop is providing support to both the teaching and research activities in the department.

FACULTY

Chair Professor

Chu, Paul Kim Ho, Ph.D., Cornell University. *Applied Physics, Biophysics, Condensed Matter Physics, Electrical Engineering, Energy Sources & Environment, Materials Science, Metallurgy, Mechanics, Nano Science and Technology, Physics of Beams, Plasma and Fusion, Polymer Physics/Science, Solid State Physics, Surface Physics.*

Lai, Ki Leuk Joseph, Ph.D., City University. *Materials Science, Metallurgy.*

Lee, Chun Sing, Ph.D., University of Hong Kong. Director of Centre of Super-Diamond and Advanced Films. *Applied Physics, Chemical Physics, Energy Sources & Environment, Engineering Physics/Science, Nano Science and Technology, Solid State Physics, Surface Physics.*

Rogatch, Andrei, Ph.D., Belarusian State University. Director of Centre for Functional Photonics. *Atomic, Molecular, & Optical Physics, Condensed Matter Physics, Materials Science, Metallurgy, Nano Science and Technology.*

Wang, Xun-Li, Ph.D., Iowa State University. Head of Department. *Applied Physics, Condensed Matter Physics, Materials*

Science, Metallurgy, Mechanics, Nano Science and Technology, Solid State Physics, Statistical & Thermal Physics.

Woo, Chung Ho, Ph.D., University of Waterloo. *Applied Physics, Computational Physics, Condensed Matter Physics, Engineering Physics/Science, Mechanics, Nano Science and Technology, Nonlinear Dynamics and Complex Systems, Nuclear Engineering, Solid State Physics, Statistical & Thermal Physics.*

Professor

Chan, Kwok Sum, Ph.D., University of Hong Kong. *Condensed Matter Physics.*

Li, Kwok Yiu Robert, Ph.D., Dublin University. Associate Dean of College of Science and Engineering. *Materials Science, Metallurgy, Nano Science and Technology, Polymer Physics/Science.*

Shek, Chan Hung, Ph.D., University of Hong Kong. Associate Head of Department. Assistant Dean of College of Science and Engineering. *Materials Science, Metallurgy.*

Tjong, Sie Chin, Ph.D., University of Manchester. *Materials Science, Metallurgy, Nano Science and Technology, Polymer Physics/Science.*

Wu, Chi Man Lawrence, Ph.D., University of Bristol. *Atomic, Molecular, & Optical Physics, Computational Physics, Geology/Geochemistry, Geophysics, Materials Science, Metallurgy, Polymer Physics/Science.*

Yu, Kwan Ngok Peter, Ph.D., University of Hong Kong. *Biophysics, Medical, Health Physics, Nuclear Physics, Physics of Beams.*

Zhang, Ruiqin, Ph.D., Shandong University. *Applied Physics, Atomic, Molecular, & Optical Physics, Chemical Physics, Computational Physics, Condensed Matter Physics, Energy Sources & Environment, Materials Science, Metallurgy, Nano Science and Technology, Solid State Physics, Surface Physics, Theoretical Physics.*

Zhang, Wenjun, Ph.D., Lanzhou University. *Materials Science, Metallurgy, Nano Science and Technology, Surface Physics.*

Associate Professor

Chu, Sai Tak, Ph.D., University of Waterloo. *Astronomy, Computational Physics, Electrical Engineering, Electromagnetism, Optics.*

Chung, Chi Yuen, Ph.D., University of Hong Kong. *Chemical Physics, Materials Science, Metallurgy, Medical, Health Physics, Nano Science and Technology.*

Vellaisamy, Arul Lenus Roy, Ph.D., Nagpur University. *Applied Physics, Solid State Physics.*

Xu, Zhengkui, Ph.D., University of Illinois at Urbana-Champaign. *Materials Science, Metallurgy.*

Zapien, Juan Antonio, Ph.D., The Pennsylvania State University. *Applied Physics, Atomic, Molecular, & Optical Physics, Nano Science and Technology, Optics, Solid State Physics, Surface Physics.*

Assistant Professor

Chen, Xianfeng, D.Phil., University of Oxford. *Nano Science and Technology.*

Fan, Jun, Ph.D., Princeton University. *Applied Mathematics, Biophysics, Computational Physics, Materials Science, Metallurgy, Nano Science and Technology.*

Ho, Johnny Chung Yin, Ph.D., University of California, Berkeley. *Applied Physics, Chemical Physics, Condensed Matter Physics, Electrical Engineering, Energy Sources & Environment, Engineering Physics/Science, Materials Science, Metallurgy, Nano Science and Technology, Solid State Physics.*

Li, Yangyang, Ph.D., University of California, San Diego. *Applied Physics, Energy Sources & Environment, Engineering*

Physics/Science, Nano Science and Technology, Optics, Polymer Physics/Science.

Mavila Chathoth, Suresh, Ph.D., Technical University of Munich. *Applied Physics, Chemical Physics, Condensed Matter Physics, Fluids, Rheology, Materials Science, Metallurgy, Other.*

Ruotolo, Antonio, Ph.D., University of Naples (IT) "Federico II". *Applied Physics, Condensed Matter Physics, Low Temperature Physics, Materials Science, Metallurgy, Nano Science and Technology, Nonlinear Dynamics and Complex Systems, Solid State Physics.*

Wang, Feng, Ph.D., Zhejiang University. *Condensed Matter Physics, Nano Science and Technology, Optics, Solid State Physics.*

Zhi, Chunyi, Ph.D., Institute of Physics, Chinese Academy of Sciences. *Materials Science, Metallurgy, Nano Science and Technology.*

Professor Emeritus

Rudowicz Czeslaw, Zygmunt, Ph.D., Adam Mickiewicz University. *Condensed Matter Physics, Materials Science, Metallurgy, Solid State Physics, Theoretical Physics.*

DEPARTMENTAL RESEARCH SPECIALTIES AND STAFF

Theoretical

Applied Mathematics. Jun Fan.

Applied Physics. Paul Kim Ho Chu, Chung Ho Woo, Ruiqin Zhang.

Atomic, Molecular, & Optical Physics. Chi Man Lawrence Wu, Ruiqin Zhang.

Biophysics. Jun Fan, Kwan Ngok Peter Yu.

Chemical Physics. Ruiqin Zhang.

Computational Physics. Sai Tak Chu, Jun Fan, Chung Ho Woo, Chi Man Lawrence Wu, Ruiqin Zhang.

Condensed Matter Physics. Kwok Sum Chan, Zygmunt Rudowicz Czeslaw, Chung Ho Woo, Ruiqin Zhang.

Electrical Engineering. Sai Tak Chu.

Electromagnetism. Sai Tak Chu.

Energy Sources & Environment. Ruiqin Zhang.

Engineering Physics/Science. Chung Ho Woo.

Materials Science, Metallurgy. Paul Kim Ho Chu, Jun Fan, Ki Leuk Joseph Lai, Kwok Yiu Robert Li, Zygmunt Rudowicz Czeslaw, Xun-Li Wang, Zhengkui Xu, Ruiqin Zhang.

Mechanics. Chung Ho Woo.

Medical, Health Physics. Kwan Ngok Peter Yu.

Nano Science and Technology. Paul Kim Ho Chu, Jun Fan, Kwok Yiu Robert Li, Yangyang Li, Chung Ho Woo, Ruiqin Zhang.

Nonlinear Dynamics and Complex Systems. Antonio Ruotolo, Chung Ho Woo.

Nuclear Engineering. Chung Ho Woo.

Nuclear Physics. Kwan Ngok Peter Yu.

Optics. Sai Tak Chu, Yangyang Li.

Physics of Beams. Kwan Ngok Peter Yu.

Plasma and Fusion. Paul Kim Ho Chu.

Polymer Physics/Science. Kwok Yiu Robert Li.

Solid State Physics. Zygmunt Rudowicz Czeslaw, Chung Ho Woo, Ruiqin Zhang.

Statistical & Thermal Physics. Chung Ho Woo.

Surface Physics. Paul Kim Ho Chu, Ruiqin Zhang.

Theoretical Physics. Zygmunt Rudowicz Czeslaw, Ruiqin Zhang.

Experimental

Applied Physics. Paul Kim Ho Chu, Johnny Chung Yin Ho, Chun Sing Lee, Yangyang Li, Suresh Mavila Chathoth, Antonio Ruotolo, Arul Lenus Roy Vellaisamy, Xun-Li Wang, Juan Antonio Zapien, Ruiqin Zhang.

Astronomy. Sai Tak Chu.

Atomic, Molecular, & Optical Physics. Andrei Rogatch, Chi Man Lawrence Wu, Juan Antonio Zapien.

Biophysics. Paul Kim Ho Chu, Kwan Ngok Peter Yu.

Chemical Physics. Chi Yuen Chung, Johnny Chung Yin Ho, Chun Sing Lee, Suresh Mavila Chathoth.

Condensed Matter Physics. Paul Kim Ho Chu, Johnny Chung Yin Ho, Suresh Mavila Chathoth, Andrei Rogatch, Antonio Ruotolo, Feng Wang, Xun-Li Wang.

Electrical Engineering. Paul Kim Ho Chu, Sai Tak Chu, Johnny Chung Yin Ho.

Electromagnetism. Sai Tak Chu.

Energy Sources & Environment. Paul Kim Ho Chu, Johnny Chung Yin Ho, Chun Sing Lee, Yangyang Li, Ruiqin Zhang.

Engineering Physics/Science. Johnny Chung Yin Ho, Chun Sing Lee, Yangyang Li.

Fluids, Rheology. Suresh Mavila Chathoth.

Geology/Geochemistry. Chi Man Lawrence Wu.

Geophysics. Chi Man Lawrence Wu.

Low Temperature Physics. Antonio Ruotolo.

Materials Science, Metallurgy. Paul Kim Ho Chu, Chi Yuen Chung, Johnny Chung Yin Ho, Ki Leuk Joseph Lai, Kwok Yiu Robert Li, Suresh Mavila Chathoth, Andrei Rogatch, Antonio Ruotolo, Chan Hung Shek, Sie Chin Tjong, Xun-Li Wang, Chi Man Lawrence Wu, Zhengkui Xu, Ruiqin Zhang, Wenjun Zhang, Chunyi Zhi.

Mechanics. Paul Kim Ho Chu, Xun-Li Wang.

Medical, Health Physics. Chi Yuen Chung, Kwan Ngok Peter Yu.

Nano Science and Technology. Xianfeng Chen, Paul Kim Ho Chu, Chi Yuen Chung, Johnny Chung Yin Ho, Chun Sing Lee, Kwok Yiu Robert Li, Yangyang Li, Andrei Rogatch, Antonio Ruotolo, Sie Chin Tjong, Feng Wang, Xun-Li Wang, Juan Antonio Zapien, Ruiqin Zhang, Wenjun Zhang, Chunyi Zhi.

Nonlinear Dynamics and Complex Systems. Antonio Ruotolo.

Nuclear Physics. Kwan Ngok Peter Yu.

Optics. Sai Tak Chu, Yangyang Li, Feng Wang, Juan Antonio Zapien.

Other. Suresh Mavila Chathoth.

Physics of Beams. Paul Kim Ho Chu, Kwan Ngok Peter Yu.

Plasma and Fusion. Paul Kim Ho Chu.

Polymer Physics/Science. Paul Kim Ho Chu, Kwok Yiu Robert Li, Yangyang Li, Sie Chin Tjong, Chi Man Lawrence Wu.

Solid State Physics. Paul Kim Ho Chu, Johnny Chung Yin Ho, Chun Sing Lee, Antonio Ruotolo, Arul Lenus Roy Vellaisamy, Feng Wang, Xun-Li Wang, Juan Antonio Zapien.

Statistical & Thermal Physics. Xun-Li Wang.

Surface Physics. Paul Kim Ho Chu, Chun Sing Lee, Juan Antonio Zapien, Ruiqin Zhang, Wenjun Zhang.

View additional information about this department at
www.gradschoolshopper.com

THE CHINESE UNIVERSITY OF HONG KONG

DEPARTMENT OF PHYSICS

Shatin, Hong Kong
http://www.phy.cuhk.edu.hk/

General University Information

President: Joseph J. Y. Sung
Dean of Graduate School: Wing Shing Wong
University website: http://www.cuhk.edu.hk

Department Information

Department Chairman: Ke-Qing Xia, Chair
Department Contact: Pui Yee Ho, Miss
Total full-time faculty: 28
Total number of full-time equivalent positions: 28
Full-Time Graduate Students: 128
First-Year Graduate Students: 49

Department Address

Room 108, 1st Floor, Science Centre North Block, Department of Physics, The Chinese University of Hong Kong
Shatin, N.T, 852, Hong Kong
Phone: (852) 39436339
Fax: (852) 26035204
E-mail: pyho@phy.cuhk.edu.hk
Website: http://www.phy.cuhk.edu.hk/

ADMISSIONS

Admission Contact Information

Address admission inquiries to: Miss P. Y. Ho
Phone: (852) 39436339
E-mail: pyho@phy.cuhk.edu.hk
Admissions website: http://www.phy.cuhk.edu.hk/

Application deadlines

Fall admission: January 31

Application fee: HK$300

Deadline for applicants who apply for PhD program through the Hong Kong PhD Fellowship Scheme (http://cerg1.ugc.edu.hk/hkpfs/index.html) is 1 December 2013. Deadline for applicants who apply for MSc program is 28 February 2014.

Admission requirements

Bachelor's degree requirements: Bachelor's degree in Physics (for Physics Program) or Materials Science and Engineering (for Materials Science and Engineering Program) or related disciplines, normally with Second Class Honours or overall average result of B or above.

GRE requirements

The GRE is not required.

Advanced GRE requirements
The Advanced GRE is recommended.

TOEFL requirements
The TOEFL exam is recommended for students from non-English-speaking countries.
PBT score: 550
iBT score: 79

Other admissions information
Additional requirements: Qualified applicants will be invited to take an entrance examination which consists of written examination and interview. However, applicants who have satisfactory score of the GRE Physics Subject Test taken within the past three years may be exempted from the written examination.

TUITION

Tuition year 2014-15:
Tuition for in-state residents
Full-time students: HK$42,100 annual
Tuition for out-of-state residents
Full-time students: HK$42,100 annual
Annual tuition fee for MSc program: Full-time: HK$76,800 Part-time: HK$38,400
Health insurance: Available.

Teaching Assistants, Research Assistants, and Fellowships
Average stipend per academic year
Teaching Assistant: HK$168,000
Postgraduate Studentship is a form of financial assistance provided to full-time postgraduate students registered in research degree programs. Separate application is not required.

FINANCIAL AID

For further information
Address financial aid inquiries to:
Financial aid website: http://www5.cuhk.edu.hk/oafa/index.php/scholarships

HOUSING

Availability of on-campus housing
Single students: Yes
Married students: Yes

For further information
Address housing inquiries to: Jockey Club Postgraduate Hall, General Office, The Chinese University of Hong Kong, Shatin, N.T., Hong Kong.
Phone: (852) 39433000
E-mail: enquiry@pgh.cuhk.edu.hk
Housing aid website: http://www.pgh.cuhk.edu.hk/

GRADUATE DEGREE REQUIREMENTS

Master's: MPhil in Physics/Materials Science and Engineering: Coursework: 12 units, Thesis research: 6 units each semester. MSc in Physics: Coursework: 24 units.
Doctorate: Ph.D.in Physics/Materials Science and Engineering: Coursework: 12 units, Guided study: 1 unit each semester, Thesis research: 6 units each semester during pre-candidacy period and 12 units each semester during post-candidacy period. Students must pass the candidacy examination within 24 months from first entry.

Thesis: In the final year of study, MPhil/Ph.D. student must submit a thesis and pass an oral examination defending his/her thesis.

SPECIAL EQUIPMENT, FACILITIES, OR PROGRAMS

Thin Film Deposition Facilities:

High vacuum coating system; low pressure chemical vapour deposition systems; sputtering thin film coating system; spin coater; pulsed-laser deposition system.

Materials Characterization:

High resolution transmission electron microscope (TEM) with an electron energy loss spectrometer; scanning electron microscopes (SEM) with energy dispersive X-ray (EDX) spectrometers, one with field emission gun, one equipped with a cathodoluminescence system, and one suitable for variable pressure imaging; X-ray diffractometers (XRD); physical property measuring system (PPMS) that provides an ultracold (0.35K) and highly magnetized (14T) environment, capable of measuring resistivity, Hall effect, magnetic moment, AC susceptibility, etc.; atomic force microscope (AFM) / scanning tunneling microscope (STM); surface profiler; differential scanning calorimeter; thermomechanical analysis system; thermogravimetric analysis system; dynamic mechanical analysis system; microhardness tester.

Equipment for Optical Measurement:

Stereo, inverted and ordinary optical microscopes; photoluminescence microscope; Raman and microRaman spectrometers; FTIR spectrometer; UV-VIS-IR double beam spectrometer; fluorescence lifetime imaging microscope system; equipped with a confocal microscope, an ultrafast femtosecond pulsed laser, and single-photon detectors, which allow the spatial and temporal imaging of optical materials down to diffraction limit and nanoseconds.

Computational Facilities:

The Department provides a powerful computational environment for various research fields through its three well-equipped computational physics laboratories. As of 2012, there are over 80 high speed UNIX workstations, servers and many PCs, including 63 high-speed Linux workstations and a 20-nodes Linux cluster for parallel calculations. Other computer accessories include notebook computers, networked B/W, colour laserjet printers, scanners, digital camera and photocopiers. All hardwares are networked and provide a full range of application softwares for scientific computing and graphics.

Others:

10 K closed cycle refrigerating system, thermoluminescence dating system, 20 GHz digitizing oscilloscope, Tektronix logic system, pulse-echo ultrasonic detection system, laser doppler velocimeter.

FACULTY

Distinguished University Professor
Yang, Chen Ning, (楊振寧) Ph.D., University of Chicago, 1948. Particle Physics, Statistical Mechanics, Conceptual History of Theoretical Physics.

Professor
Ching, Emily S. C., (程淑姿) Ph.D., University of Chicago, 1991. Non-equilibrium Systems, Turbulence, Complex Networks, Biophysics.
Chu, Ming Chung, (朱明中) Ph.D., California Institute of Technology, 1987. Astroparticle Physics, Cosmology, Neutrino Physics.

Hui, Pak Ming, (許伯銘) Ph.D., The Ohio State University, 1987. Complex Systems, Condensed Matter Physics.

Kui, Hin Wing, (瞿顯榮) Ph.D., Harvard University, 1986. Materials Science.

Leung, Pui Tang, (梁培燈) Ph.D., The Chinese University of Hong Kong, 1988. Theoretical Physics.

Li, Quan, (李 泉) Ph.D., Northwestern University, 2001. Materials Science.

Xia, Ke-Qing, (夏克青) Ph.D., University of Pittsburgh, 1987. Complex Fluids, Turbulence.

Xiao, Xudong, (肖旭東) Ph.D., University of California at Berkeley, 1992. Surface Science, Nano Science.

Young, Kenneth, (楊綱凱) Ph.D., California Institute of Technology, 1972. Theoretical Physics.

Yu, Kin Wah, (余建華) Ph.D., University of California, 1984. Condensed Matter Physics.

Associate Professor

Law, Chi Kwong, (羅志光) Ph.D., University of Rochester, 1994. Quantum Optics.

Liu, Renbao, (劉仁保) Ph.D., Chinese Academy of Sciences, 2000. Condensed Matter Physics, Quantum Physics, Optics.

Lo, Chi Fai, (羅志輝) Ph.D., Massachussetts Institute of Technology, 1989. Theoretical and Mathematical Physics, Quantitative Finance.

Ng, Dickon H. L., (吳恆亮) Ph.D., University of Manitoba, 1989. Materials Science.

Ong, Daniel H. C., (王福俊) Ph.D., Northwestern University, 1996. Materials Science.

Wang, Jianfang, (王建方) Ph.D., Harvard University, 2002. Nanomaterials and Nanophotonics.

Wong, King Young, (黃景揚) Ph.D., University of Pennsylvania, 1986. Nonlinear Optics.

Assistant Professor

Goh, Swee Kuan, (吳瑞權) Ph.D., University of Cambridge, 2009. Experimental Condensed Matter Physics, High Pressure Techniques.

Li, Hua-bai, (李華白) Ph.D., Northwestern University, 2006. Astrophysics.

Wang, Dajun, (王大軍) Ph.D., University of Connecticut, 2007. Atomic, Molecular, and Optical Physics.

Wang, Yi, (王一) Ph.D., University of Illinois at Urbana-Champaign, 2008. Computational Biophysics.

Wu, Yilin, (吳藝林) Ph.D., University of Notre Dame, 2009. Biophysics and Quantitative Biology.

Xu, Lei, (徐 磊) Ph.D., The University of Chicago, 2006. Soft Condensed Matter, Fluid Mechanics, Complex Fluids.

Zhou, Qi, (周 琦) Ph.D., The Ohio State University, 2009. Quantum Gases, Condensed Matter Physics.

Zhu, Junyi, (朱駿宜) Ph.D., University of Utah, 2009. Computational Material Physics.

Research Associate Professor

Gu, Shijian, (顧世建) Ph.D., Zhejiang University, 2002. Condensed Matter Physics, Quantum Information, and Quantum Physics.

Research Assistant Professor

Chen, Tao, (陳濤) Ph.D., Nanyang Technological University, 2010. Energy Materials, Organic/inorganic Hybrid Solar Cells.

Gong, Ming, (龔 明) Ph.D., University of Science and Technology of China, 2010. Condensed Matter Physics.

Lu, Xinhui, (路新慧) Ph.D., Yale University, 2010. Soft Matter, Photovoltaic Devices, X-ray Scattering.

DEPARTMENTAL RESEARCH SPECIALTIES AND STAFF

Theoretical

Biophysics and Quantitative Biology. Study of biological macromolecules using computational methods. Current research interests include molecular dynamics simulation of lipids and proteins, modeling of mycobacterial cell wall, and the prediction of drug-receptor binding affinity. Yi Wang.

Computational Physics. Active research has been devoted to computational studies of quantum and classical systems. For quantum systems, research projects include phase separation in highly correlated systems, superconductivity and magnetism in two-dimensional triangular lattice, issues in quantum Monte Carlo simulations, magnetic properties of transition metals and quantum entanglement and quantum phase transition. For classical systems, substantial efforts have been devoted to soft condensed matter physics. Projects include correlation effects in colloidal suspensions, electrorheological fluids and biological cell suspensions. Useful tools for investigation include molecular dynamics and Monte Carlo methods. Kin Wah Yu.

Condensed Matter Physics. Magnetic and electronic properties of disordered photonic crystals, density functional theory, geophysical systems, kinetic glass transition of amorphous materials, physics of nonlinear composites, plasmonics, dynamic ER effects in complex fluids, quantum spin systems, interacting electron-phonon systems, fermion systems, nonlinear optical properties of conjugated organic molecules and semiconductor nanostructures, bosons in confining potentials, spin dynamics in semiconductor nanostructures, topological effects in condensed matters. Shijian Gu, Pak Ming Hui, Renbao Liu, Chi Fai Lo, King Young Wong, Kin Wah Yu, Junyi Zhu.

Gravitational Waves. Quasinormal modes of gravitating systems, late-time behavior of gravitational waves. Frequency spectrum of gravitational waves, inversion of the internal structure of neutron stars. Pui Tang Leung, Kenneth Young.

Optics. Laser interaction with micrometer-sized droplets, open and dissipative systems, generalized Jaynes-Cummings models, interaction between photons and quantum dots, squeezed states, quantum optics, optical properties of novel quantum matters. Chi Kwong Law, Pui Tang Leung, Renbao Liu, Chi Fai Lo.

Other Theoretical Projects. Quasinormal modes, dissipative quantum systems, method of complex scaling, data compression in information science, S matrix calculation of the scattering of light by an atom near an absorption edge, interpretation of EM radiation by non-inertial observers, supersymmetric quantum mechanics, geometric phases, study of quantum anharmonic oscillators by state-dependent diagonalization, Lie algebraic approach to quantum systems, nonlinear stochastic systems, forces between biomolecules. Pui Tang Leung, Chi Fai Lo, Kenneth Young, Kin Wah Yu.

Quantitative Finance. Option pricing, risky bond pricing, financial risk modeling. Chi Fai Lo.

Quantum Information and Quantum Physics. Investigation of entanglement in many-body systems; quantification of entanglement, entanglement manipulation, quantum measurement decoherence. Shijian Gu, Chi Kwong Law, Renbao Liu.

Relativity and Astrophysics. Cosmology. Relativistic astrophysics, neutrino astrophysics. Ming Chung Chu, Pui Tang Leung.

Semiconductor Doping. To understand doping mechanisms, investigate electronic structures, and simulate thermal and kinetic processes of doping, density functional theory (DFT) calculations will be applied. Our goals are: investigation of defects and doping of semiconductors and their alloys (InGaN, AlGaP, CZTS, CIGS, SiC, diamond), which are very important solid state lighting, photo voltaic, and information materials. Defects and dopant formation energies and transition energies will be calculated. New strategies of tuning defects and dopants will be proposed. Junyi Zhu.

Sonoluminescence. Hydrodynamics, optical emission, plasma spectroscopy. Pui Tang Leung.

Theoretical Studies of Surface and Interface Physics. Tuning surface properties can be critical in thin film growth and device properties. We'll apply DFT calculations and classic molecular dynamics calculations to study surface phenomena. Our goals are: investigating the surface reconstructions, surface passivation, surface diffusion, surfactant effects, surface effects on doping in many different thin films and nano-materials, including CZTS, InGaN, CIGS, AlGaP, diamond, SiC, ScTiO3, various topological insulators and super conductors. Junyi Zhu.

Turbulence and Complex Systems. Scaling and intermittency, statistics and coherent structures. Drag reduction by polymers. Effect of polymers on heat transport. Glass transition. Agent-based models of complex systems, physics of complex networks, dynamical processes in complex networks. Emily Ching, Pak Ming Hui.

Experimental

Biophysics and Quantitative Biology. Study of how living things function, adapt and evolve, using bacteria as a proxy. Current research interests include bacterial motility, bacterial adaption to complex environments, and multicellular dynamics in microbial communities. Yilin Wu.

Fluid Turbulence. Study of fluid turbulence, in particular convective thermal turbulence. Main focus are heat transport, flow dynamics, energy and entropy cascades, and coherent structures. Ke-Qing Xia.

Fluid and Complex Fluids Experiment. Study the origin of liquid drop splashing, multiphase flow in porous medium, colloidal crystallization and colloidal glass. Lei Xu.

Functional Nanomaterials and Their Electronic Structures. Research focus on designing low-dimensional functional nanostructures, with emphasis on their microstructure/ electronic structures and the corresponding mechanical, magnetic, electrical, and optical properties. Quan Li.

Materials: Synthesis, Characterization, and Engineering. (a) Metal matrix composites: Aluminium metal matrix composites are synthesized by using appropriate internal and external oxidation techniques. The microstructures and properties of composites are studied. (b) Ceramic matrix composites: Chemical synthesis techniques are employed to produce various fine ceramic powders which are used to make new ceramic matrix composites. (c) Fabrication and characterization of 1-D and 2-D nanostructured materials. (d) Biomorphic materials: conversion of bio-organic materials to carbon-based composites and functional ceramics. Synthesis of catalysts by a bimorphic approach. Dickon Ng.

Neutrino Oscillation Experiment. The project aims at measuring the $\theta 13$ mixing angle of neutrino, using neutrinos from the Daya Bay Nuclear Power Plant. A satellite laboratory at the Aberdeen Tunnel has been setup to study the cosmic rays background. Ming Chung Chu.

Nonlinear Optics. Investigation of nonlinear optical properties of molecules using hyper-Rayleigh scattering. All-optical poling of glassy polymers. Nonlinear optical properties of organic materials and polymers. King Young Wong.

Optical Properties of Semiconductors. Spectroscopic methods are used to probe the electronic structures of semiconductors. Artificially structured materials are prepared and their novel optical properties studied. Daniel Ong.

Plasmonics. Study of metal nanoparticles and their interactions with metals and semiconductors. Study of surface plasmons by various far-field and near-field optical techniques. Daniel Ong, Jianfang Wang.

Solar Cells. Development of solar energy materials, device fabrication and physics of dye-sensitized solar cells and inorganic thin film solar cells. Tao Chen.

Surface Science and Materials Characterization. Surface analysis and structure analysis using XPS, UPS, SIMS, AES/SEM/ EBIC, TEM and Raman. Quan Li.

Surface Science and Nano Science. Metal and semiconductor surfaces, and nano structures fabricated on surfaces. Focus is on their atomic and electronic structures and their surface dynamics. The major techniques are scanning tunneling microscopy and spectroscopy at low temperature or variable temperatures. Xudong Xiao.

Ultracold Atoms and Molecules. Production and investigation of molecular quantum gases with strong dipolar interactions. Studies of a double-species Bose-Einstein condensate with tunable interactions. Dajun Wang.

Undercooled Liquids, Glass Formation and Nanostructures. Work is focused on novel materials, nucleation problem, glass and nanostructure formation, physical properties of liquid in its stable and metastable regime and microstructures of undercooled semiconductors and metals. Hin Wing Kui.

View additional information about this department at
www.gradschoolshopper.com

THE UNIVERSITY OF HONG KONG

DEPARTMENT OF PHYSICS

Pokfulam, Hong Kong
http://www.physics.hku.hk/

General University Information
President: Lap-Chee Tsui
Dean of Graduate School: Paul K.H. Tam
University website: http://hku.hk

Control: Public
Setting: Urban
Total Faculty: 1,041
Total Graduate Faculty: 1,041

Total number of Students: 23,033
Total number of Graduate Students: 11,543

Department Information
Department Chairman: K.S. Cheng, Head
Department Contact: K.S. Cheng, Head and Chair Professor
 Total full-time faculty: 20
 Total number of full-time equivalent positions: 20
 Full-Time Graduate Students: 57
 First-Year Graduate Students: 17
 Female First-Year Students: 6
 Total Post Doctorates: 26

Department Address
Room 518, Chong Yuet Ming Physics Building, The University of Hong Kong, Pokfulam Road, Hong Kong
CHINA
Phone: (852) 2859 2368
Fax: (852) 2559 9152
E-mail: hrspksc@hkucc.hku.hk
Website: http://www.physics.hku.hk/

ADMISSIONS

Admission Contact Information
Address admission inquiries to: Graduate School, Room P403, Graduate House, The University of Hong Kong, Pokfulam Road, Hong Kong
Phone: (852) 2857 3470
E-mail: gradsch@hku.hk
Admissions website: http://www.gradsch.hku.hk/gradsch/web/apply/index_ps.htm

Application deadlines
Fall admission: December 1

Application fee: HK$400
*Our Research Postgraduate (RPg) Programmes are open for application all year round. The application dates for admission in 2013-14 could be found at http://www.gradsch.hku.hk/gradsch/web/apply/. *For online payment by VISA or MASTERCARD, the application fees is HK$150 *List of documents to support an application refers to http://www.gradsch.hku.hk/gradsch/rola/app_doc.htm

Admission requirements
Bachelor's degree requirements: A good honors degree from the University of Hong Kong or an equivalent qualification from a comparable institution.

GRE requirements
The GRE is not required.

Advanced GRE requirements
The Advanced GRE is not required.

TOEFL requirements
The TOEFL exam is required for students from non-English-speaking countries.
 PBT score: 550
 iBT score: 80
TOEFL is required for applicants from institutions outside Hong Kong where the language of instruction and examination is not entirely in English. TOEFL can be replaced by other designated language examinations, including (IELTS) (taken

within a two-year period), Overseas General Certificate of Education (GCE), International General Certificate of Secondary Education (IGCSE) or Cambridge Test of Proficiency in English Language (CPE).

TUITION
Tuition year 2012–13:
Tuition for in-state residents
 Full-time students: HK$42,100 annual
 Part-time students: HK$53,100 annual
The tuition fee for local (in-state) and non-local (out-of-state) students are the same.
Deferred tuition plan: Yes
Health insurance: Available.
Academic term: Semester

Teaching Assistants, Research Assistants, and Fellowships
Number of first-year
 Fellowship students: 17
Average stipend per academic year
 Fellowship student: HK$13,600
All entitled full-time research postgraduate students are offered fellowships, namely Postgraduate Scholarship (PGS). The monthly PGS for the 2012-2013 academic year is HK$13,600 and subject to adjustment upon renewal. Currently PGS holders whose PhD probation has been confirmed will receive a higher PGS rate at $14,070.

FINANCIAL AID

Application deadlines
Fall admission: December 1

For further information
Address financial aid inquiries to: Graduate School.
Phone: (852) 28573470
E-mail: gradsch@hku.hk
Financial aid website: http://www.gradsch.hku.hk/gradsch/web/apply/index_ps4.htm

HOUSING

Availability of on-campus housing
 Single students: Yes
 Married students: Yes

For further information
Address housing inquiries to: Centre of Development and Resources for Students (CEDARS), 3/F, Meng Wah Complex, The University of Hong Kong, Pokfulam Road, Hong Kong.
Phone: (852) 2859 2305
E-mail: cedars@hku.hk
Housing aid website: http://apps.cedars.hku.hk/pg_housing/

Table A—Faculty, Enrollments, and Degrees Granted

Research Specialty	2012-13 Faculty	Enrollment 2012-13 Master's	Enrollment 2012-13 Doctorate	Number of Degrees Granted 2011–12 (2009-12) Master's	Number of Degrees Granted 2011–12 (2009-12) Terminal Master's	Number of Degrees Granted 2011–12 (2009-12) Doctorate
Astrophysics	4	5	2	7(19)	–	1(2)
Condensed Matter Experiments	3	5	7	3(14)	–	4(14)
Condensed Matter Theory	6	1	17	–(7)	–	5(14)
Environmental Radioactivity	–	–	–	–(2)	–	–(3)
Health Physics	–	–	–	–(1)	–	–
Material Science	5	7	9	7(21)	–	4(12)
Particle Physics	1	3	–	–(4)	–	2(2)
Quantum Information	1	1	–	1(4)	–	–(1)
Total	**20**	**22**	**35**	**19(72)**	**–**	**16(48)**
Full-time Grad. Stud.	–	22	35	–	–	–
First-year Grad. Stud.	–	9	8	–	–	–

GRADUATE DEGREE REQUIREMENTS

Master's: Requirements for conferment of the degree include course work, thesis, and oral examination. Details can be found at http://www.gradsch.hku.hk/gradsch/web/resources/handbooks/12/gshdbk1213.pdf.

Doctorate: Requirements for conferment of the degree include course work, thesis, oral examination, and public seminar. Details can be found at http://www.gradsch.hku.hk/gradsch/web/resources/handbooks/12/gshdbk1213.pdf.

SPECIAL EQUIPMENT, FACILITIES, OR PROGRAMS

The Department offers both M.Phil. and Ph.D. programs for full-time postgraduate students. Most of our researches are in condensed matter and material physics and in astrophysics and astronomy. In condensed matter and related fields, our interests include superconductivity, topological insulators, correlated electron systems, solid-state quantum computation, surface physics, material sciences, quantum transport in nanoscale and spintronics, semiconductor physics, optics, and positron physics. In the field of astrophysics and astronomy, our research covers neutron stars, gamma-ray bursts, pulsars, cosmological models, supernovae, planetary nebulae, interstellar chemistry, and neutrino physics.

The Department houses a number of state-of-art research facilities for multi-disciplinary researches in condensed matter physics and astrophysics.

● Surface Science Laboratory

Multi-chamber ultrahigh vacuum (UHV) systems are available for material synthesis and characterization by the techniques of molecular beam epitaxy (MBE), scanning tunneling microscopy (STM), low and high energy electron diffraction (LEED/RHEED), Ultraviolet photo-electron spectroscopy (UPS) and Auger electron spectroscopy (AES).

● Semiconductor Physics Laboratory

Facilities are available for investigating defect states in semiconductors. These include photovoltage, internal photoemission, photoconductivity, deep level optical spectroscopy (DLOS), and conventional deep-level transient spectroscopy (DLTS). Positron annihilation techniques are also employed in this laboratory.

● Positron-beam Laboratory

A variable mono-energetic positron beam is available for depth profiling of defect studies in thin films and at surface. Coincidence Doppler Broadening Spectroscopy (CDBS) is also available.

● Material Physics Laboratory

Deep level transient spectroscopy system; liquid nitrogen optical cryostat; Electrical characterization equipments: semiconductor parameter analyzer, multi-frequency LCR meter, pico-ammeter, electrometer, and etc.; 10K liquid He free optical cryostat; Photoluminescence system: 30mW HeCd laser, 500mm monochrometer, PMT and CCD detecting system; UV-visible spectrophotometer; Radio frequency magnetron sputtering system; Pulsed laser deposition system; Thermal evaporator; Tube furnace and box furnace.

● Thin-film Laboratory

Thin-films deposition techniques include two laser ablation systems, ion-beam and e-beam deposition chambers, two magnetron sputtering systems. The characterization tools include a cryofree superconducting magnet (10 Tesla), a x-ray diffractometer, a scanning probe microscope, and a clean room for photolithography and etching.

● Optoelectronics and Nanomaterials Laboratory

The laboratory is equipped with fume cupboards, tube furnaces, spin-coater, two thermal evaporators, and E-beam/sputtering deposition system. The characterization facilities include UV/Vis/NIR spectrometers for LED characterization and setups for power conversion efficiency and external quantum efficiency measurements for solar cells.

● Nanostructure Characterization Laboratory

We focus on optical and electrical properties of nanostructures and emerging semiconductors. The laboratory is equipped with home-made confocal spectroscopy, time-resolved spectroscopy and electric charactering system.

● Laser Spectroscopy Laboratory

The laboratory is equipped with variable-temperature (4.2K-300K) photoluminescence(PL), variable-temperature (1.5K-300K) magneto-PL (up to 7T) with super high spectral resolution, confocal micro-Raman micro/spectroscopy, broadband emission/ absorption spectroscopy, sub-ps time-resolved PL system, and near field scanning microscope. The laser sources include He-Cd laser, He-Ne laser, Ar-Kr mixed gas laser, high-energy YAG pulse laser, Ar laser pumped dye laser, and femtosecond broadband laser.

● Underground Laboratory

The underground laboratory in Aberdeen Tunnel is setup for study of cosmic ray and high energy particle physics. Currently, it has a Gd doped liquid scintillator neutron detector sandwiched between a muon tracker for measuring muon-induced neutrons. The laboratory also serves as a satellite laboratory for the Daya Bay Reactor Neutrino Experiment.

● For theoretical studies, besides the central computing facility of the university, staff and students of the department have at their disposal a 100-CPU Linux computer cluster solely dedicated to research.

● Our facilities in observational astrophysics include a 40 cm-diameter reflector telescope located in the main campus equipped with charged couple device (CCD) imager and spectrometer, and a 2.3m diameter Small Radio Telescope. We also have access to international telescope facilities on the ground (e.g. the Gemini Telescopes, Beijing Astronomical Observatories) and in space (e.g. Hubble Space Telescope, Chandra X-Ray Observatory) for our research projects.

Table B—Separately Budgeted Research Expenditures by Source of Support

Source of Support	Departmental Research	Physics-related Research Outside Department
Federal government		
State/local government	HK$10,494,887	
Non-profit organizations		
Business and industry		
Other	HK$777,000	
Total	**HK$11,271,887**	

Table C—Separately Budgeted Research Expenditures by Research Specialty

Research Specialty	No. of Grants	Expenditures (HK$)
Astrophysics	11	HK$1,073,779
Condensed Matter Experiment	4	HK$814,025
Condensed Matter Theory	15	HK$5,152,500
Material Science	17	HK$3,573,283
Quantum Information	1	HK$700,000
Total	**48**	**HK$11,313,587**

FACULTY

Professor

Chau, Hoi Fung, Ph.D., The University of Hong Kong, 1992. *Quantum Foundations*. Quantum cryptography and quantum information processing.

Cheng, Kwong Sang, Ph.D., Columbia University, 1984. Head & Chair Professor, FAPS. *Astrophysics*. Theoretical astrophysics, neutron star physics, pusars, gamma-ray bursts, strange stars.

Djurišić, Aleksandra B., Ph.D., University of Belgrade, 1997. *Materials Science, Metallurgy*. Nanomaterials, optoelectronics, solar energy, photocatalysis.

Fung, Stevenson, Ph.D., University of Oxford, 1980. *Condensed Matter Physics*. Semiconductor physics, Schottky barriers, positron annihilation spectroscopy, deep level states, wide band gap semiconductors.

Gao, Ju, Ph.D., Universiteit Twente, 1992. *Materials Science, Metallurgy*. Experimentalist in condensed matter physics, superconductivity, thin films, devices.

Kwok, Sun, Ph.D., University of Minnesota, 1974. Dean of Faculty of Science, Chair Professor. *Astronomy, Astrophysics*. Stellar evolution, interstellar chemistry, space astronomy.

Shen, Shunqing, Ph.D., Fudan University, 1992. *Condensed Matter Physics*. Topological insulators, and quantum transport.

Wang, Jian, Ph.D., University of Pennsylvania, 1988. Chair Professor. *Condensed Matter Physics*. General quantum transport theory, spintronics, nanoelectronics.

Wang, Zidan, Ph.D., Nanjing University, 1988. Chair Professor. *Condensed Matter Physics*. Theoretical condensed matter, quantum computation, and quantum information.

Xie, Maohai, Ph.D., University of London, 1994. *Materials Science, Metallurgy*. Materials science, surface science.

Zhang, Fuchun, Ph.D., Virginia Polytechnic Institute and State University, 1983. Zhou Guangzhao Professor in Natural Sciences, Chair Professor, FAPS. *Condensed Matter Physics*. Theoretical condensed matter physics, in particular Strongly Correlated Electron Systems, including high temperature superconductivity.

Associate Professor

Cui, Xiaodong, Ph.D., Arizona State University, 2001. *Condensed Matter Physics*. Nano-electronics.

Leung, John, Ph.D., The University of Hong Kong, 1983. *Energy Sources & Environment, Medical, Health Physics, Particles and Fields*. Radiation protection, radiation dosimetry, high-energy and particle physics.

Lim, Jeremy, Ph.D., Macquarie University, 1992. *Astrophysics*. Astrophysics - star formation, stellar activities, evolved stars, external galaxies, radio interferometry.

Ling, Francis, Ph.D., The University of Hong Kong, 1996. *Materials Science, Metallurgy*. Defects in semiconductors, electrical and optical properties of semiconductors, carrier transport in semiconductors, semiconductor junctions, positron annihilation spectroscopy, deep level transient spectroscopy.

Wu, Huasheng, Ph.D., Montana State University, 1994. *Materials Science, Metallurgy*. Experimental surface physics: Electrical and geometric properties of surfaces and interfaces, low-energy electron positron diffraction and holography.

Xu, Shijie, Ph.D., Xi'an Jiaotong University, 1993. *Condensed Matter Physics*. Optical properties including nonlinear optical properties and ultrafast phenomena of semiconductor nanostructures. Luminescence and imaging of individual quantum dots are also our interest.

Assistant Professor

Ng, Stephen, Ph.D., Stanford University, 2006. *Astrophysics*. Radio and x-ray observations of neutron stars, pulsar wind nebulae, and supernova remnants.

Yao, Wang, Ph.D., University of California, San Diego, 2006. *Condensed Matter Physics*. Theoretical condensed matter physics, quantum physics, and optical physics.

Zhang, Shizhong, Ph.D., University of Illinois at Urbana-Champaign, 2009. *Condensed Matter Physics*. Physics of very degenerate quantum gases and strongly correlated electronic system.

DEPARTMENTAL RESEARCH SPECIALTIES AND STAFF

Theoretical

Astrophysics. The major research areas are related to neutron stars and pulsars, which are rapidly spinning and magnetized neutron stars, including x-ray and gamma-ray emission mechanisms, stellar structure, stellar cooling and heating mechanisms and the internal activities, e.g. sudden unpinning of superfluid vortices. In addition to topics related to pulsars and neutron stars, we also study topics related to gamma-ray bursts, in particular the central engine problem, and high energy phenomena resulting from the stellar capture processes by supermassive black holes in the galactic center. Kwong Sang Cheng, Sun Kwok, Jeremy Lim, Stephen Ng.

Condensed Matter Theory. Our current research interest includes: 1) strongly correlated electron systems; 2) theories for high temperature superconducting cuprates and iron pnictides; 3) quantum computation; 4) quantum magnetism; 5) spintronics and quantum transport. Shunqing Shen, Jian Wang, Zidan Wang, Wang Yao, Fuchun Zhang, Shizhong Zhang.

Quantum Information. Our research works focus on quantum information processing. On the application side, we work on the proofs of unconditional security of various quantum cryptographic protocols and primitives. And on the theoretical side, we study how ideas used in quantum information processing and quantum information theory can be used to tackle fundamental problems in quantum mechanics. Hoi Fung Chau.

Experimental

Condensed Matter Experiment. 1) Characterization of defects in semiconductors (such as SiC, GaN, GaAs and InP) using Positron Annihilation Spectroscopy (PAS) and various conventional optical/electrical techniques. 2) Optical properties, including nonlinear optical properties, electronic structures, electron-phonon interactions, and ultrafast phenomena, in semiconductor nanostructures (e.g., quantum wells, dots and nanocrystals). 3)

Characterizations and applications of nanomaterials, such as carbon nanotubes, nanowires and molecular nanostructures, for fundamental physics and potential applications. 4) Spintronics research on electron spin related phenomenon, particularly spin current generation, detection and manipulation in nonmagnetic semiconductors. Xiaodong Cui, Stevenson Fung, Shijie Xu.

Materials Science. 1) Thin films and multilayer structures of advanced functional materials, such as high Tc superconductors, ABO3 compounds, and various perovskite transition metal oxides. 2) Ultrathin films and heterostructures of semiconductors and topological insulators grown by molecular-beam epitaxy. Studies of surface growth kinetics and surface structural and electronic properties by scanning probing microscopy, electron diffraction and holography, photoemission electron spectroscopy and Auger electron spectroscopy, etc. 3) Properties and effects of defects on transport and optical behavior of semiconductors, such as GaN and ZnO. 4) Fabrication and characterization of organic, organic/inorganic nanocomposite, inorganic optoelectronic devices (LEDs and solar cells), as well as fabrication and characterization of various nanostructures (mostly metal oxides and/or wide band gap semiconductors). Aleksandra Djurišić, Ju Gao, Francis Ling, Huasheng Wu, Maohai Xie.

Particle Physics. Participating in the Daya Bay Reactor Neutrino Experiment in determining precisely the value of $\theta 13$, the mixing angle of electron anti-neutrinos emitted from the Daya Bay nuclear power plants. The experiment is a joint force of 38 institutions from China, Czech Republic, Hong Kong, Russia, Taiwan and USA. A small underground laboratory inside the Aberdeen Tunnel in Hong Kong is now being used for studying the production of neutrons from cosmic-ray muons. The results of our work have provided useful radiation background information for the Daya Bay Experiment. John Leung.

Radiation Protection and Environmental Radioactivity. Development of a plume dispersion model for use in the events of nuclear accident that may occur in the Daya Bay nuclear power plants, production of Hong Kong radon potential map, measurement of radon in air; measurement of radon and radium isotopes in underground water; radioactive waste management in Siu A Chau Low-Level Radioactive Wastes Facility. John Leung.

**View additional information about this department at
www.gradschoolshopper.com**

BENEMÉRITA UNIVERSIDAD AUTÓNOMA DE PUEBLA

INSTITUTO DE FISICA LUIS RIVERA TERRAZAS

Puebla, Pue. 72570, Mexico
http://www.ifuap.buap.mx

General University Information
President: M. A. José Alfonso Esparza Ortíz
Dean of Graduate School: Ph.D. María del Rosario Huesca
University website: http://www.buap.mx
Control: Public
Setting: Urban
Total number of Students: 62
Total number of Graduate Students: 315

Department Information
Department Chairman: Ph.D., Juan Francisco Rivas Silva, Chair
Department Contact: Ph.D., Umapada Pal, Materials Science
 Coordinator, Ph.D.,Felipe Pérez Rodríguez, Academic
 Secretary
Total full-time faculty: 37
Total number of full-time equivalent positions: 37
Total Present Post Doctorates: 4

Department Address
Av San Claudio y Blvd. 18 Sur
C.U., Col. San Manuel
Puebla, PU 72570
MEXICO
Phone: (+52-222) 229-5610
Fax: (+52-222) 229-5611

E-mail: upal@ifuap.buap.mx, fperez@ifuap.buap.mx
Website: http://www.ifuap.buap.mx

ADMISSIONS

Admission Contact Information
Address admission inquiries to: Instituto de Física, UAP, Apdo.
 Postal J-48, 72570 Puebla, Puebla, México
Phone: (+52-222) 229-5610
E-mail: rivas@ifuap.buap.mx,sacad@ifuap.buap.mx
Admissions website: http://www.ifuap.buap.mx

Application deadlines
Fall admission deadlines for International students:
 (for both levels, M.S. ad Ph.D) January 6, 2014
Spring admission deadlines for all students:
 (for Ph.D level) December 1, 2014

Application fee
There is no application fee required.

Admissions information
For Fall of 2013:
 Number of applicants: 36
 Number admitted: 20
 Number enrolled: 18

Admission requirements

Bachelor's degree requirements: Bachelor's degree in Physics or related field is required. Physics–Spring or Summer Programs consist of (B.S.-level) intensive courses on Modern Physics, Classical Mechanics, Electromagnetism and Mathematical Methods of Physics are offered as a prerequisite to enter the M.S. program. Materials Science–Spring and Summer Programs consist of (B.S.-level) intensive courses on Mathematical Methods, General Physics, General Chemistry and Thermal Physics are offered as a prerequisite to enter the M.S. program.

Minimum undergraduate GPA: 8.0

GRE requirements

The GRE is not required.

Advanced GRE requirements

The Advanced GRE is not required.

TOEFL requirements

The TOEFL exam is not required for students from non-English-speaking countries.

Other admissions information

Additional requirements: Students from non-Spanish-speaking countries are required to demonstrate proficiency in Spanish. Passing an admission exam and/or attending and passing prerequisite courses are required for the Master's degree. Passing an admission exam is required for the Doctorate degree.

Undergraduate preparation assumed: Physics–Undergraduate preparation assumed: Reif, Statistical and Thermal Physics; Marion, Classical Dynamics of Particles and Systems; Reitz, Milford, and Christy, Foundations of Electromagnetic Theory; Arfken, Mathematical Methods for Physicists, 2nd ed. Merzbacher, Quantum Mechanics. Materials Science–Undergraduate preparation assumed; M.R. Spiege, Teoría y Problemas de Mateméticas Superiores para Ingenieros y Científicos (McGraw Hill, 1971); F. Reif, Fundamentos de la Física Estadística y Térmica (McGraw-Hill, 1968); Berkeley Physics Course, Mechanics Vol. 1 (Reverté), Berkeley Physics Course, Electromagnetism, Vol. 2 (Reverté), G.M. Bonder, Chemistry and Experimental Science, 2nd Edition (John Wiley, 1995).

TUITION

Tuition year 2013-14:
Tuition for in-state residents
 Full-time students: $100 per semester
Tuition for out-of-state residents
 Full-time students: $100 per semester
Mexican pesos. Credit hours per semester.
Credit hours per semester to be considered full-time: 6
Deferred tuition plan: No
Health insurance: Available
Other academic fees: None.
Academic term: Semester

Teaching Assistants, Research Assistants, and Fellowships

Number of first-year
Fellowships are provided by National Science Council (CONACyT-Federal Government), after requirements are fulfilled, for national and international students.

FINANCIAL AID

For further information

Address financial aid inquiries to: (Physics) Ph.D. Felipe Pérez Rodríguez, Instituto de Física, UAP, Apdo. Postal J-48, 72570 Puebla, Puebla, México, sacad@ifuap.buap.mx, (Material Sciences) Ph.D. Umapada Pal. UAP, Apdo. Postal J-48, 72570 Puebla, Puebla, México, upal@fuap.buap.mx.

Phone: (+52 222) 229-5610
E-mail: rivas@ifuap.buap.mx
Financial aid website: http://www.ifuap.buap.mx

HOUSING

Availability of on-campus housing
 Single students: No
 Married students: No

Table A—Faculty, Enrollments, and Degrees Granted

Research Specialty	Faculty	Enrollment		Number of Degrees Granted		
		Master's	Doctorate	Master's	Terminal Master's	Doctorate
Total	37	–	–	236	–	79
Full-time Grad. Stud.	–	10	27	–	–	–
First-year Grad. Stud.	–	11	14	–	–	–

GRADUATE DEGREE REQUIREMENTS

Master's: Master's in Physics: The Master's degree in Physics requires 9 courses (with a minimum combined grade average of 8, scale from 0 to 10). During the second year, the student must pass a qualifying examination. Alternatively, a student may write a thesis under the supervision of a faculty member. The program must be completed within 2 years. The M.Sc. is a terminal program. Master's in Materials Science: The Master's in Materials Science requires 10 courses (with a minimum combined grade average of 8, scale from 0 to 10). During the second year, the student must pass a qualifying examination. Alternatively, the student may write a thesis under the supervision of a faculty member. The program must be completed within 2 years. The M.Sc. is a terminal program.

Doctorate: Doctorate in Physics: 3 advanced courses are required. During the second year of the program, the student must pass a preliminary examination. The student must write a dissertation based on original research and must have a paper accepted for publication in a recognized journal prior to the final examination. Doctorate in Materials Science: 3 advanced courses are required. During the second year of the program, the student must pass a preliminary examination. The student must write a dissertation based on original research and must have a paper accepted for publication in a recognized journal prior to the final examination.

Thesis: Thesis may be written in absentia.

SPECIAL EQUIPMENT, FACILITIES, OR PROGRAMS

Library: Monographic material: 6,202 items; 5,758 books; 134 research magazines titles and, 319 thesis exemplars.

Services; data-base, books catalogues and electronic magazines, through pages web: "http://www.bibliocatalogo.buap.mx" and "http://conricyt.mx", copies, agreements with libraries within the country, like UNAM, UDLAP, CINVESTAV, INAOE, etc.

Laboratories, 17 specialized laboratories, 1 electronic shop and, 1 mechanic shop.

Computer Center, Workstations: 1WS SGI 02, CPU MIPS R12000 300 MHz, 1 WS Microway, CPU ALPHA 21264 600 MHz, 1 WS HP Visualize J5600, 2 CPU parisc 8500, 1 ws sgi Octane/SE, CPU MIPS R12000 400 MHz, 1 Server SGI 1400l, 4 CPU Xeon 500 MHz, 1 SGI 2200, 8 CPU MIPS R12000 400 MHz, 1 Cluster Beowulf, 16 alpha nodes 21164 600 MHz, 5PC Linux Fedora Core, Pentium 4 HT 3 GHz, etc.

FACULTY

Research Professor

Arriaga-Rodríguez, J. Jesús, Ph.D., SC CSIC. Madrid, 1992. *Applied Physics*. Condensed matter, solid state: New materials, electronic properties, physics of surfaces and interfaces. Photonic and phononic crystals.

Calixto Rodríguez, Ma. Estela, Ph.D., UNAM, 2001. *Applied Physics*. Physics of surfaces and interfaces; optical properties and acoustics of periodic artificial systems. Photonic and phononic crystals.

Carrillo-Estrada, José Luis, Ph.D., UNAM, México City, México, 1984. *Nano Science and Technology, Statistical & Thermal Physics*. Semiconductor physics; statistical physics.

Cartas Fuentevilla, Roberto, Ph.D., IFUAP, Puebla, México, 1999. Mathematical physics; experimental particle physic; astroparticle and astrophysics.

De la Peña Seaman, Omar, Ph.D., CINVESTAV-Merida, México, 2008. *Electromagnetism*. Advanced materials; superconductivity and magnetism; physical properties of advanced materials.

Dossetti Romero, Victor, Ph.D., IFUAP, BUAP, Puebla, México, 2005. Intelligent and complex materials; manostructured materials.

Escalante Hernández, Alberto, Ph.D., IFUAP, BUAP, Puebla, Mexico, 2005. *Particles and Fields*. Mathematical physics; experimental particle physics; astroparticle and astrophysics.

Flores-Riveros, Antonio, Ph.D., Uppsala University, 1986. *Condensed Matter Physics*. Quantum chemistry; optical properties.

García-Vázquez, Valentín, Ph.D., University of Arizona, 1992. Advanced materials; artificially structured materials; superconductivity and magnetism; physical properties of advanced materials.

González Melchor, Minerva, Ph.D., CINVESTAV-IPN, México City, México, 2002. Advanced materials; molecular simulations; structural properties; dynamics and thermodynamics. Polyelectrolytes; molecular dynamics; Brownian dynamics; dissipative particle dynamics; surface tension; liquid-gas coexistence.

González Ronquillo, Ana Lilia, Ph.D., IFUNAM-México, 2007. *Electromagnetism*. Optical properties of nanostructured systems; superconductividad and magnetism.

Gracia y Jiménez, Justo Miguel, Ph.D., UAP, Puebla, México, 1993. 0ptical properties of solids.

Hernández-Cocoletzi, Gregorio, Ph.D., UNAM, México City, México, 1991. *Condensed Matter Physics*. Optical properties of solids.

Hernández-Tejeda, Pedro Hugo, Ph.D., University of Michigan, 1987. Kinetics of phase transitions; x-ray diffraction.

Izrailev, Felix, Ph.D., Novosibirsk, URSS, 1969. *Nonlinear Dynamics and Complex Systems*. Quantum chaos.

López-Crúz, José Elías, Ph.D., CINVESTAV, México City, México, 1979. Electrical and optical properties of insulators and semiconductors, physics of surfaces and interfaces. Photonic and phononic crystals.

Luna-Acosta, Germán Aurelio, Ph.D., New Mexico University, USA, 1984. *Nonlinear Dynamics and Complex Systems*. Theoretical physics; classical quantum.

Márquez Beltrán, César, Ph.D., University De Paris-XI, Orsay, Francia, 2004. *Nano Science and Technology*. Nanoparticles and nanocomposites; complex and intelligent materials.

Méndez Bermúdez, José Antonio, Ph.D., IFUAP, Puebla, México, 2003. *Nonlinear Dynamics and Complex Systems*. Chaotic systems; quantum chaos; random matrices.

Méndez Blas, Antonio, Ph.D., UAM, Madrid, Spain, 2003. *Applied Physics*. Optical characterizations, UV-VIS and infrared, physics of surfaces and interfaces. Photonic and phononic crystals.

Martínez-Montes, Gerardo, Ph.D., University of Arizona, 1985. *Condensed Matter Physics*. Optical properties of semiconductors.

Mendoza-Álvarez, María Eugenia, Ph.D., Geneva, Switzerland, 1985. *Nano Science and Technology*. Crystal growth of ferroic materials; structural characterization.

Meza-Espinoza, Luis Octavio, Ph.D., CIO Guanajuato, México, 2011. Luminescent nanomaterials for bioimaging and displays, design of experiment.

Meza-Montes, Lilia, Ph.D., UAP, Puebla, México, 1993. *Nano Science and Technology*. Semiconductor physics; statistical mechanics.

Pérez-Rodríguez, Felipe, Ph.D., Kharkov, USSR, 1989. Advanced materials; metal physics, acoustic and optical properties of solids, superconductividad and magnetism.

Pal, Umapada, Ph.D., 1. T. Karagpur, India, 1991. *Nano Science and Technology*. Condensed matter physics; optical and electrical properties of solids.

Palma-Almendra, Alejandro, Ph.D., Uppsala University, 1976. Quantum chemistry.

Pando-Lambruschini, Carlos Leopoldo, Ph.D., Lomonosov Univ., Moscow, USSR, 1990. Quantum and nonlinear optics.

Quiroga-González, Enrique, Ph.D., University of Kiel, Germany, 2010.

Reyes Ayona, Edgar, Ph.D., IFUAP, México, 2006. *Applied Physics*. Solid state physics; acoustic properties.

Reyes-Coronado, Alejandro, Ph.D., UNAM, 2007. Advanced materials; nanoplasmonics & nanophotonics; metamaterials; effective medium theories.

Rivas-Silva, Juan Francisco, Ph.D., UAP, Puebla, México, 1991. *Computational Physics, Condensed Matter Physics*. Atomic and molecular physics.

Rosado-Sánchez, Alfonso, Ph.D., CINVESTAV, México City, México, 1984. *Particles and Fields*. Elementary particle physics.

Sánchez Mora, Enrique, Ph.D., UAM-Iztapalapa, México City, México, 2000. Thin-film growth, optical properties of thin films.

Sadurní-Hernández, Emerson Leao, Ph.D., IFUNAM, 2007. *Nonlinear Dynamics and Complex Systems*. Quantum dynamics; relativistic; quantum mechanics.

Saldaña-Saldaña, Xóchitl, Ph.D., IFUAP, México, 1995. Advanced material; optical properties of solids, superconductivity and magnetism.

Silva-González, Nicolás Rutilo, Ph.D., TU Dresden, GRD, 1988. Photocatalytic materials; electric and thermal transport in semiconductor thin films.

Soto-Manríquez, José, Ph.D., University of Arizona, 1983. Advanced materials; optical sciences, superconductivity, and magnetism.

DEPARTMENTAL RESEARCH SPECIALTIES AND STAFF

Theoretical

Adanced Materials.. This departmental of Research is experimental too. See the details bellow.

Applied Physics. J. Jesús Arriaga Rodríguez, S.N.I. II, (Leader) Antonio Méndez Blas, S.N.I. I, Luis Octavio Meza González S.N.I. I, Elías López Cruz, S.N.I. II, Ma. Estela Calixto Rodríguez, S.N.I. I, Edgar Reyes Ayona, S.N.I. I. Fields of Research: Surface and interface physics - Optical and acoustic properties of periodic artificial systems - Photonic and Phononic crystals.

Complex Systems. Felix M. Izrailev, S.N.I. III, José Antonio Méndez Bermúdez, S.N.I. II, (Leader) German Aurelio Luna Acosta, S.N.I. II, Emerson Leao Sadurní Hernández, S.N.I. I, Fields of Research: Complex Systems.

Complex, Intelligent and Nanostructured Materials. José Luis Carrillo Estrada, S.N.I. II, Ma. Eugenia Mendoza Álvarez, S.N.I. II,(Leader), Lilia Meza Montes, S.N.I. II, Pal Umapada,

S.N.I. III, César Márquez Beltran, S.N.I. I, Victor Dossetti Romero, S.N.I I. Fields of Research: Complex and intelligent materials - Nanoparticles and Nanocomposites.

Computational Physics of the Condensed Matter. Minerva González Melchor, S.N.I. II, Gregorio Hernaández Cocoletzi, S.N.I. III, (Leader) Pedro Hugo Hernández Tejeda, S.N.I. I, Gerardo Martínez Montes, Juan Francisco Rivas Silva, S.N.I. II. Antonio Flores Riveros, S.N.I. II. Alejandro Palma Almendra, S.N.I. III. Fields of Research: Ab initio calculation of the electronic structure of atoms, molecules and solids - Variational methods and their applications to confined systems - Optical properties - Molecular simulation of liquids.

Particles, Fields and General Relativity (FCFM-BUAP). Alfonso Rosado Sánchez, S.N.I. III, Alberto Escalante Hernández, S.N.I. I, Roberto Cartas Fuentevilla, S.N.I. II, Fields of Research: Experimental Particle physics, astrophysics and astropaticles-Theory and phenomenology of fundamental interactions.

Experimental

Advanced Materials. Valentín García Vazquez, S.N.I. I, Estela de Lourdes Juárez Ruíz, S.N.I. I, Felipe Pérez Rodríguez, S.N.I. II, (Leader) Xochitl Inés Saldaña Saldaña, S.N.I. I, Enrique Sánchez Mora, S.N.I. I. Omar De la Peña Seaman, S.N.I. I. Alejandro Reyes Coronado, S.N.I. I. Ana Lilia González Ronquillo, S.N.I. C. José Soto Manríquez. Fields of Research: Superconductivity and magnetism - Physics properties of advanced materials.

Applied Physics. This departmental of research is theoretical too, see the details above.

Complex, Intelligent and Nanostructured Materials. This departmental of research is theoretical too, see the details above.

Photocatalytic and Photoconductive Materials. Estela Gómez Barojas, S.N.I. I, Justo Miguel Gracia y Jiménez, S.N.I. II, Nicolás Rutilo Silva González, S.N.I. II, (Leader),Enrique Quiroga González, S.N.I. I. Fields of Research: Morphological and chemical properties of Materials - Photocatalytic, luminescent and photoelectric properties of materials.

View additional information about this department at
www.gradschoolshopper.com

CENTRO DE INVESTIGACIÓN Y DE ESTUDIOS AVANZADOS DEL INSTITUTO POLITÉCNICO NACIONAL

DEPARTMENT OF PHYSICS

Mexico City, Mexico
http://www.fis.cinvestav.mx

General University Information

President: Pablo Rene Asomoza y Palacios
Dean of Graduate School: Juan Manuel Mendez Nonell
University website: http://www.cinvestav.mx
Control: Public
Setting: Urban
Total Faculty: 623
Total Graduate Faculty: 623
Total number of Students: 2,000
Total number of Graduate Students: 2,000

Department Information

Department Chairman: Maximo Lopez-Lopez, Head
Department Contact: Diana Garcia Sotelo, Secretary
  Total full-time faculty: 49
  Total number of full-time equivalent positions: 49
  Full-Time Graduate Students: 109
  First-Year Graduate Students: 35
  Female First-Year Students: 9
  Total Post Doctorates: 10

Department Address

Av. IPN 2508
Col. San Pedro Zacatenco
Mexico DF, DF 07360

MEXICO
Phone: (525) 57473836
Fax: (525) 57473388
E-mail: admision@fis.cinvestav.mx
Website: http://www.fis.cinvestav.mx

ADMISSIONS

Admission Contact Information

Address admission inquiries to: Coordinacion de Admision., Departamento de Fisica., Av. IPN 2508, Col. Zacatenco Mexico DF 07360. Mexico.
Phone: (525) 57473836
E-mail: admision@fis.cinvestav.mx
Admissions website: http://www.fis.cinvestav.mx/ADMISION

Application deadlines

Fall admission:
Mexican students: May 5   *Int'l. students*: March 2
Spring admission:
Mexican students: February 3   *Int'l. students*: November 3

Application fee

There is no application fee required.
Non-Mexican applicants are encouraged to apply for a Mexican student VISA at least three months in advance to the application dead line.

Admissions information
For Fall of 2012:
Number of applicants: 97
Number admitted: 36
Number enrolled: 35

Admission requirements
Bachelor's degree requirements: MSc Program is open to students with a Bachelor's degree in Physics, Mathematics, Chemistry, or related areas of Engineering or with an equivalent preparation. The candidate must present and pass a level examination and/or take propaedeutic courses offered by our Department. To qualify for a fellowship, a minimum GPA of 8/10 is needed. Ph.D. PROGRAM: We offer two programs for a Ph.D. degree: Traditional and Direct Ph.D. The traditional program is open to students with a Master's degree in physics or with an equivalent preparation. The candidates are required to take and pass a general qualifying examination (predoctoral) on Classical Electrodynamics, Classical Mechanics, Equilibrium Statistical Mechanics and Quantum Mechanics. The candidates are required to present three letters of recommendation and, it is responsibility of the applicant to select and contact a faculty member who would be willing to serve as the student's advisor. Direct Ph.D program is open to students with Bachelor's Degree in Physics or related areas. In addition to the requirements for the MsC program, the candidates for Direct Ph.D. must have outstanding GPA and high score in the level examination.
Minimum undergraduate GPA: 8.0

GRE requirements
The GRE is not required.

Advanced GRE requirements
The Advanced GRE is recommended.
For the Ph.D. program, the foreign students who have presented and passed satisfactorily the GRE in physics, can miss out our general qualifying examination, depending on thier GRE score. 0 0 Students from non-Spanish-speaking, and/or non-English-speaking countries are required for at least 500-TOEFL record.

Other admissions information
Additional requirements: MSc: There are two ways to be accepted in the MSc program: a) The candidate should present and satisfactorily pass a level examination in classical mechanics, electromagnetism, thermodynamics and mathematical methods. b) The candidate must take and pass with a minimum GPA of 8.0/10 propaedeutic courses offered by the Department. According to our experience, the last method allows the student adapting to the academic life of our Department, and provides time to show academic skills in a better way.
PhD: The acceptance to the traditional PhD program is after the general qualifying examination. This examination may be repeated only once. During the year there are four dates for predoctoral: March, June, September and December. The foreign students who have presented the GRE in physics, depending on their score, may not be required for qualifying examination.
The students admitted to the Direct PhD program, have to take postgraduated courses in Classical and Quantum mechanics, Electrodynamics, Mathematical methods and Statistical mechanics. Afterwards they are required to present a general qualifying examination.
Undergraduate preparation assumed: MSc PROGRAM: Electromagnetism at level of Foundations of Electromagnetic Theory, Reitz, Milford and Christy, Addison-Wesley. Thermodynamics at the level of F. Reif, Fundamentals of Statistical and Thermal Physics, McGraw-Hill; H. B. Callen, Thrmody-

namics, Wiley & Sons. Mathematical Methods at the level of G. Arfken, Mathematical Methods for Physicist (Third edition), Academic Press, 1985. Classical Mechanics at the level of Ch. Kittel, Mecánica, Berkeley Physics Course, Vol. 1 and/or Classical Dynamics of particles and systems, Marion. PhD. PROGRAM: Classical Mechanics, Herbert Goldstein Addison-Wesley. Mathematical Methods for Physicists, George Arfken, Academic Press, New York, 1970. Classical Electrodynamics, John David Jackson, John Wiley & Sons, New York. Quantum Mechanics: Quantum Mechanics Vol.I C. Cohen-Tannoudji, B. Diu, F. Laloe, Wiley-Interscience, 2006. Quantum Mechanics, A. Messiah, Dover Publications 1999. Modern Quantum Mechanics, J. J. Sakurai, Addison Wesley, 1993.

TUITION
Tuition year 2014-15:
Our programs do not have tuition costs.
Credit hours per semester to be considered full-time: 150
Deferred tuition plan: No
Health insurance: Not available.
Other academic fees: There are no additional fees. Moreover, all our students are accepted after an exam and, if successful, they are granted the tuition and will receive a fellowship. In this sense, our local system is completely different from the standard U.S. graduate school system.
Academic term: Semester

Teaching Assistants, Research Assistants, and Fellowships
Number of first-year
Fellowship students: 40
Average stipend per academic year
Fellowship student: $7,450
All our domestic and international students qualify for a fellowship from "Consejo Nacional de Ciencia y Tecnologia (CONACYT)", a Federal Government Agency for Science and Technolgy.

FINANCIAL AID

Application deadlines
Fall admission: August 1
Spring admission: August 1

For further information
Address financial aid inquiries to: Coordinacion de Admision, Departamento de Fisica.
Phone: (525) 5747 3838
E-mail: admision@fis.cinvestav.mx
Financial aid website: http://www.fis.cinvestav.mx

HOUSING

Availability of on-campus housing
Single students: No
Married students: No

Table A—Faculty, Enrollments, and Degrees Granted

Research Specialty	2012 Faculty	Enrollment 2012		Number of Degrees Granted 2012-13 (2008-13)		
		Master's	Doctorate	Master's	Terminal Master's	Doctorate
High Energy Physics	13	6	5	6(17)	–	9(23)
Relativity & Gravitation	12	4	3	3(22)	–	–(5)
Solid State Physics	17	6	4	1(18)	–	3(9)
Statistical & Thermal Physics	7	4	2	1(13)	–	–(5)
Total	49	19	14	11(70)	–	12(42)
Full-time Grad. Stud.	–	53	65	–	–	–
First-year Grad. Stud.	–	35	15	–	–	–

GRADUATE DEGREE REQUIREMENTS

Master's: The MSc program consists of five semesters. From the 1st to the 3rd, the students are required to take nine advanced graduate courses in classical mechanics, mathematical methods, classical electrodynamics, quantum mechanics, statistical mechanics and advanced physics lab. During the 4th and 5th semester, the students are required to take an elective course and engage in their thesis research topic for a final defense of this thesis.

Doctorate: The Ph.D. program must be completed within four years. The Ph.D. students are required to be full-time students. They should take three elective advanced graduate courses and must participate in at least three teaching assistantships and engage in a research work, which involves the presentation of a plan of a research project and doctoral thesis that must be developed during four years. During the developing of the thesis research, the students are required to present a set of seminars related to the advancements in their investigations. Once the students have completed their thesis, they are required to publish at least one article in an international journal in the field, and they are asked to present and defend the thesis in front of a suitable committee.

Other Degrees: Direct Ph.D. Program. The Direct Ph.D. program consists of ten semesters. The Ph.D. students are required to be full-time students. From the 1st to the 3rd, semesters the students are required to take nine advanced graduate courses in classical mechanics, mathematical methods, classical electrodynamics, quantum mechanics, statistical mechanics, and advanced physics lab. During the 3rd semester, the candidates are required to take predoctorals exams; this examination may be repeated only once. In the remaining part of the program, the students must take three additional elective advanced graduate courses and must participate in at least three teaching assistantships. By the fourth semester, the students must be engaged in their research work, which involves the presentation of a plan for a research project and doctoral thesis. During the developing of the thesis research, the students are required to present a set of seminars related to the advance in their investigations. Once the students have completed their thesis, they are required to publish at least one article in an international journal in the field, they are asked to present and defend the thesis in front of a suitable committee.

Thesis: Obligatory. It must contain original research. The goal is to form independent researchers and encourage the resercher skills in areas of interest for the students.

SPECIAL EQUIPMENT, FACILITIES, OR PROGRAMS

Fifty specialized Labs for R&D in areas such as: Solid State Physics, High Energy Physics, Soft Condensed Matter. Some Examples: Molecular Beam Epitaxy, x-Ray Difractometers, Electron microscopies (HR-TEM & SEM), AFM, Videomicroscopy, Small Angle x-ray Scattering, Vibrating Sample Magnetometer and Transport Properties Measurements system (PPMS); Raman, UV-VIS and Photoluminscense spectroscopies; Particle Detectors, Photo-Lithography.

Library: 11,000 specialized books in physics an mathematics. 436 periodical journals in physics and mathematics. Additionally, ~1200 journals from other Departments of Cinvestav. Most of those journals can be consulted via on-line.

FACULTY

Chair Professor

LÓPEZ LÓPEZ, MÁXIMO, Ph.D., Toyohashi University of Technology, 1992. Chairman of Physics Department. *Condensed Matter Physics, Nano Science and Technology, Solid State Physics*. Growth and characterization of low dimensional systems. Molecular beam epitaxy.

Professor

AYÓN BEATO, ELOY, Ph.D., Cinvestav-IPN, 2000. *Cosmology & String Theory, Relativity & Gravitation*. Black Hole Physics, Higher Dimensional Gravity, Gravitational Aspects of String Theory.

BAQUERO PARRA, RAFAEL, Ph.D., CINVESTAV-IPN, 1972. *Solid State Physics*. superconductivity, surface science.

BRETÓN BÁEZ, NORA EVA, Ph.D., Cinvestav-IPN, 1986. *Relativity & Gravitation*. Relativity and Gravitation (T): General Relativity, Exact solutions in black holes, cosmlgical models, non linear electrodynamics.

CAPOVILLA CHIARIGLIONE, RICCARDO, Ph.D., Universidad de Maryland, 1991. *Relativity & Gravitation*. Relatividad y gravitación (T): Field theory, Geometrical methods for soft condensed matter.

CARBAJAL TINOCO, MAURICIO DEMETRIO, Ph.D., Universidad Autónoma de San Luis Potosí, 1997. *Condensed Matter Physics, Polymer Physics/Science, Statistical & Thermal Physics*. Statistical physics: polymers, coloidal systems.

CASTILLA VALDEZ, HERIBERTO, Ph.D., Cinvestav-IPN, 1991. *Accelerator, High Energy Physics, Particles and Fields*. Particles and Fields: experimental meassurements of the quark b. colisions p-pbar Dzero (Fermilab).

CASTRO HERNÁNDEZ, JORGE JAVIER, Ph.D., Oxford University, 1972. *Atmosphere, Space Physics, Cosmic Rays, Solid State Physics*. Nanoparticle, atmospheric physics.

CASTRO ROMAN, FRANCISCO, Ph.D., Université Montpellier II, 1999. *Biophysics, Statistical & Thermal Physics*. Biophyscis, soft condensed matter.

CONDE, AGUSTIN, Ph.D., CINVESTAV-IPN, 1995. ADMISSION COORDINATOR. *Solid State Physics*. Superconductivity, spintronics luminiscent materials.

CRUZ OREA, ALFREDO, Ph.D., Universidad Estadual de Campinas, Campinas SP, 1994. *Solid State Physics, Statistical & Thermal Physics*. Thermal properties of materials.

DE LA CRUZ BURELO, EDUARD, Ph.D., Cinvestav-IPN, 2005. *High Energy Physics, Particles and Fields*. Experimental High-Energy Physics. Heavy hadron physics (Physics of heavy quarks).

FALCONY GUAJARDO, CIRO, Ph.D., Universidad de Lehigh, 1980. *Solid State Physics*. MOS devices, semiconductor films, HTc supercconductors, photoluminiscent materials.

FERNANDEZ CABRERA, DAVID JOSE, Ph.D., Cinvestav-IPN, 1988. *Quantum Foundations, Theoretical Physics*. Quantum foundations, mathematical methods in quantum mechanics.

GARCÍA ROCHA, MIGUEL, Ph.D., Cinvestav-IPN, 1995. *Nano Science and Technology, Solid State Physics, Surface Physics.* Semiconductors physics, Growth and characterization of II-VI semiconductors, ultra-fast spectroscopy.

GARCIA COMPEAN, HECTOR HUGO, Ph.D., Cinvestav-IPN, 1994. *Cosmology & String Theory.* Mathematical aspects of string theory and field theory.

GODINA NAVA, JUAN JOSÉ, Ph.D., Cinvestav-IPN, 1994. *Medical, Health Physics.* High Energy Physics, Field Theory, Médical Physics.

GONZÁLEZ DE LA CRUZ, GERARDO, Ph.D., Universidad Estatal de Campinas, 1981. *Nano Science and Technology, Solid State Physics.* Electrical properties of low dimensional systems.

GONZÁLEZ MOZUELOS, PEDRO, Ph.D., Cinvestav-IPN, 1992. *Biophysics, Statistical & Thermal Physics.* Molecular liquids, electrostatic interactions in liquids.

GUREVICH GENRIJOVICH, YURI, Ph.D., Academy of Science of the USSR, 1980. *Nano Science and Technology, Solid State Physics.* Transport properties in semiconductors.

HERNÁNDEZ CALDERÓN, ISAAC, Ph.D., Universidad Estatal de Campinas, 1981. *Nano Science and Technology, Solid State Physics, Surface Physics.* Optical, electrical, and structural characterization of semiconductor heterostructures and low dimensional systems. Epitaxial growth (MBE, ALE, SPBE). Surface and interface physics.

HERNÁNDEZ CONTRERAS, MARTÍN, Ph.D., Universidad Autónoma de San Luis Potosí, 1995. *Fluids, Rheology, Polymer Physics/Science, Statistical & Thermal Physics.* Soft condensed matter: Colloids diffusion and their structural properties. Liquid crystal dynamics. Electrolytes. Use is made of non-equilibrium statistical thermodynamic approaches for their study and computer simulations.

HERRERA CORRAL, GERARDO, Ph.D., Universidad de Dortmund, 1991. *Accelerator, High Energy Physics, Medical, Health Physics, Particles and Fields.* Hadron production. E-791 experiment (Fermilab), ALICE detector for hevy ions (CERN).

KIELANOWSKI, PIOTR, Ph.D., Universidad de Varsovia, 1972. *High Energy Physics, Particles and Fields.* Phenomenological models for elmentary particles.

LÓPEZ CASTRO, GABRIEL, Ph.D., Université catholique de Louvain, 1988. *High Energy Physics, Particles and Fields.* Physics of heavy flavors, neutrino physics, precision tests of the electroweak theory, effective theories of strong interactions.

LÓPEZ FERNÁNDEZ, RICARDO, Ph.D., Université Joseph Fourier (Grenoble), 2001. *Accelerator, High Energy Physics, Particles and Fields.* Experimental High Energy Physics. Heavy Quark Production in Colliders. Accelerator Physics. Direct Detection of Dark Matter.

MÉNDEZ ALCARAZ, JOSÉ MIGUEL, Ph.D., Universidad de Constanza, 1993. *Condensed Matter Physics, Statistical & Thermal Physics.* Soft condensed matter.

MATOS CHASSIN, TONATIUH, Ph.D., Universidad F. Schiller-Jena, 1987. *Astrophysics, Particles and Fields, Relativity & Gravitation.* Dark matter, Dark energy, Gallaxy formations.

MELÉNDEZ LIRA, MIGUEL ÁNGEL, Ph.D., Cinvestav-IPN, 1993. *Nano Science and Technology, Solid State Physics, Systems Science/Engineering.* Raman spectroscopy, photoluminiscence, electroreflectance. Optical properties of semiconductors.

MENDOZA ÁLVAREZ, JULIO GREGORIO, Ph.D., Universidad Estadual de Campinas, 1979. *Nano Science and Technology, Solid State Physics.* Growth of semiconductors by liquid phase epitaxy and r.f. sputtering, sol-gel. Optical characterization by photoluminiscense and Raman spectroscopies. Optoelectrónics devices and III/V heterostructures like AlGaAs/GaAs, GaSb/InGaAsSb, GaSb/InGaAs, InP/InGaAs/InAs. QDs.

MIRANDA ROMAGNOLI, OMAR GUSTAVO, Ph.D., Cinvestav-IPN, 1997. *High Energy Physics, Particles and Fields.* Neutrins physics.

MONTAÑO ZETINA, LUIS MANUEL, Ph.D., Cinvestav-IPN, 1998. *High Energy Physics, Medical, Health Physics.* Elementary particles detection. Instrumentation for elemntary particle and x-ray detectors. Detectors for Medical physics.

MONTESINOS VELÁSQUEZ, MERCED, Ph.D., Cinvestav-IPN, 1997. *Quantum Foundations, Theoretical Physics.* Geometry and Gravitation (T): Quantum Gravity, Standard Theories, Canonical Quantization and Mathematical Physics.

OLGUÍN MELO, RITO DANIEL, Ph.D., Cinvestav-IPN, 1996. *Solid State Physics.* Theoretical calculations of optical and electronic properties of semiconductors, supercocnductors and metals.

PÉREZ ANGÓN, MIGUEL ÁNGEL, Ph.D., Cinvestav-IPN, 1972. *High Energy Physics, Particles and Fields.* Phenomenological gauge models.

PÉREZ LORENZANA, ABDEL, Ph.D., Cinvestav-IPN, 1998. *Cosmology & String Theory, High Energy Physics, Particles and Fields, Theoretical Physics.* Model building for Particle Physics and Beyond the Standard Theory. Phenomenology of models for neutrino masses and mixings, and new and extended symmetries. Models for Cosmology and the Universe dark components. Particle Physics in the early Universe. Astroparticle Physics.

ROJAS OCHOA, LUIS FERNANDO, Ph.D., University of Fribourg, Switzerland., 2004. Academic Coordinator. *Condensed Matter Physics, Nano Science and Technology, Optics, Polymer Physics/Science, Statistical & Thermal Physics.* Soft Condensed Matter, Statistical Optics, Photonic materials.

ROSAS ORTIZ, JOSÉ OSCAR, Ph.D., Cinvestav-IPN, 1997. *Quantum Foundations, Theoretical Physics.* Mathematical Physics, Quantum Control, Geometry of Quantum States and Supersymmetry.

SÁNCHEZ HERNÁNDEZ, ALBERTO, Ph.D., Cinvestav-IPN, 1997. Technical Coordinador. *High Energy Physics, Particles and Fields.* Quark-c properties. Experiemntal work in FOCUS experiment(Fermilab). Quark-b measurements at DZero (Fermilab) and CMS (CERN).

SANTOYO SALAZAR, JAIM E, Ph.D., IIM-UNAM, 2006. *Nano Science and Technology, Solid State Physics.* Physical properties of nanoparticles. Magnetic nanoparticles for cancer treatments.

SEMIONOVICH MANKO, VLADIMIR, Ph.D., Universidad de la Amistad de los Pueblos, 1986. *Relativity & Gravitation, Theoretical Physics.* Exact solutions in general relativity and dilatonic gravitation.

TOMÁS VELÁZQUEZ, SERGIO ARMANDO, Ph.D., Cinvestav-IPN, 1996. *Biophysics, Solid State Physics, Systems Science/Engineering.* Optical characterization by photothermal techniques. Low concentration gas detection by phothermal and infrared techniques.

TORRES VEGA, GABINO, Ph.D., Cinvestav-IPN, 1987. *Quantum Foundations, Theoretical Physics.* The time in quantum mechanics, nonlineal quantum mechanics, geometrical mechanics.

VÁZQUEZ LÓPEZ, CARLOS, Ph.D., Cinvestav-IPN, 1979. *Nano Science and Technology, Solid State Physics.* Optical and electrical properties of semiconductor materials. Atomic force microscopy. Detection of Nuclear traces by AFM.

ZELAYA ÁNGEL, ORLANDO, Ph.D., Cinvestav-IPN, 1985. *Solid State Physics.* Semiconductor materials.

Emeritus

GARCÍA DÍAZ, ALBERTO, Ph.D., Universidad Lomonosov, 1990. *Relativity & Gravitation, Theoretical Physics.* Exact solutions in general relativity and gravitation.

MIELNIK, BOGDAN, Ph.D., Cinvestav-IPN, 1964. *Quantum Foundations, Theoretical Physics*. Quantum foundations. Movility in nonlineal systems, theory of dynamic quantum states manipulation.

SÁNCHEZ SINENCIO, FELICIANO, Ph.D., Universidad de Sao Paulo, 1970. *Solid State Physics*. Semiconductor materials, gas detections.

ZEPEDA DOMINGUEZ, ARNULFO, Ph.D., Cinvestav-IPN, 1970. *Particles and Fields*. Astroparticles, cosmic rays.

DEPARTMENTAL RESEARCH SPECIALTIES AND STAFF

Theoretical

Cosmology & String Theory. Background Field (BF) Theories, Cosmology & The Dark Energy Problem, Extended objects, Lagrangian and Hamiltonian Methods, Numerical Relativity, Theoretical Astrophysics, Quantum Gravity. AYÓN BEATO, GARCIA COMPEAN, MATOS CHASSIN, MONTESINOS VELÁSQUEZ, PÉREZ LORENZANA.

High Energy Physics. Phenomenological Aspects of Particle Physics, Radiative Corrections, Precision Tests of the Standard Model, Effective Lagrangians for Strong and Electroweak Interactions, Neutrino Physics, Heavy Flavor physics, Cosmic Ray Physics, and on Phsics Beyond the Standard Model. KIELANOWSKI, LÓPEZ CASTRO, MIRANDA ROMAGNOLI, PÉREZ ANGÓN, PÉREZ LORENZANA.

Particles and Fields. Phenomenological Aspects of Particle Physics, Radiative Corrections, Precision Tests of the Standard Model, Effective Lagrangians for Strong and Electroweak Interactions, Neutrino Neutrino Physics, Heavy Flavor physics, Cosmic Ray Physics, and on Phsics Beyond the Standard Model. KIELANOWSKI, LÓPEZ CASTRO, LÓPEZ FERNÁNDEZ, MIRANDA ROMAGNOLI, PÉREZ ANGÓN, PÉREZ LORENZANA, ZEPEDA DOMINGUEZ.

Quantum Foundations. Quantum control and manipulation, Mobility of dynamic systems, Geometrical phases, coherent and compressed states, Quantum mechanics phase space representation, Supersymmetric quantum mechanics, Fundamentals of quantum mechanics, Quantum computation. FERNANDEZ CABRERA, MIELNIK, ROSAS ORTIZ, TORRES VEGA.

Relativity & Gravitation. Black Hole Physics, Exact Solutions, Gauge/Gravity Duality, Higher & Lower Dimensional Gravity. AYÓN BEATO, BRETÓN BÁEZ, CAPOVILLA CHIARIGLIONE, GARCÍA DÍAZ, GARCIA COMPEAN, MATOS CHASSIN, MONTESINOS VELÁSQUEZ, SEMIONOVICH MANKO.

Solid State Physics. Thermoelectricity, Transport properties in low dimensional systems, Semiconductor materials, Superconductivity, Band structure calculations, AB Initio Calculations. BAQUERO PARRA, CASTRO HERNÁNDEZ, GUREVICH GENRIJOVICH, OLGUÍN MELO.

Statistical & Thermal Physics. Liquid Theory Integral equations, Langevin dynamics, Smoluchowski and Fokker-Planck dynamics. Nonequilibrium statistical mechanics approaches. CARBAJAL TINOCO, CASTRO ROMAN, GONZÁLEZ MOZUELOS, HERNÁNDEZ CONTRERAS, MÉNDEZ ALCARAZ, ROJAS OCHOA.

Experimental

High Energy Physics. The D0 and E-831 Collaborations at Fermilab. The ALICE Collaboration at LHC (CERN). The Pierre Auger Project for detection of Ultrahigh Energy Cosmic Rays. They also develop instrumentation and detector techniques with potential applications in medical physics. CASTILLA VALDEZ, DE LA CRUZ BURELO, HERRERA CORRAL, LÓPEZ FERNÁNDEZ, MONTAÑO ZETINA, SÁNCHEZ HERNÁNDEZ, ZEPEDA DOMINGUEZ.

Nano Science and Technology. Quantum wells, nanowires, nanodots. Magnetic nanoparticles for medical applications. GARCÍA ROCHA, GONZÁLEZ DE LA CRUZ, HERNÁNDEZ CALDERÓN, LÓPEZ LÓPEZ, SANTOYO SALAZAR, ZELAYA ÁNGEL.

Solid State Physics. Semiconductors Materials, Low dimensional system: Quantum wells, nanowires, nanodots. Dielectrics Materials, Luminescent Materials, conventional and unconventional Superconductor Materials, Magnetic Materials, Photo-thermal Phenomena, Biophysics, Medical Physics, Nanostructures. CONDE, CRUZ OREA, FALCONY GUAJARDO, GARCÍA ROCHA, GONZÁLEZ DE LA CRUZ, HERNÁNDEZ CALDERÓN, LÓPEZ LÓPEZ, MELÉNDEZ LIRA, MENDOZA ÁLVAREZ, SÁNCHEZ SINENCIO, SANTOYO SALAZAR, TOMÁS VELÁZQUEZ, VÁZQUEZ LÓPEZ, ZELAYA ÁNGEL.

Statistical & Thermal Physics. Digital video microscopy, Small angle x-ray scattering, Light scattering. Colloidal suspensions, Polymeric Solutions, Liquid Crystals, Biomolecular materials. CARBAJAL TINOCO, CASTRO ROMAN, ROJAS OCHOA.

View additional information about this department at
www.gradschoolshopper.com

NATIONAL TSING HUA UNIVERSITY

DEPARTMENT OF PHYSICS AND INSTITUTE OF ASTRONOMY

Hsinchu, Taiwan
http://www.phys.nthu.edu.tw/en/

General University Information

President: Lih-Juann Chen
Dean of College of Science: Rai-Shung Liu
University website: http://www.nthu.edu.tw/english/index.php

Control: Public
Setting: Urban
Total Faculty: 640
Total Graduate Faculty: 640

Total number of Students: 12,059
Total number of Graduate Students: 5,844

Department Information
Department Chairman: Ci-Ling Pan, Chair
Department Contact: Daw-Wei Wang, Prof.
   Total full-time faculty: 38
   Total number of full-time equivalent positions: 38'
   Full-Time Graduate Students: 214
   First-Year Graduate Students: 63
   Female First-Year Students: 9
   Total Post Doctorates: 34

Department Address
101 Section 2 Kuang Fu Road
Vice Chair
Hsinchu, 30013
Taiwan
Phone: 886-3-574-2511
Fax: 886-3-574-3052
E-mail: clpan@phys.nthu.edu.tw
Website: http://www.phys.nthu.edu.tw/en/

ADMISSIONS

Admission Contact Information
Address admission inquiries to: International Students Division, Office of International Affairs (international student admission), National Tsing Hua University, 101, Kuang Fu Rd. Sec. 2, 30013 Hsinchu, Taiwan, R.O.C., Phone number: +886-3-5162461, E-mail address: hcchan@mx.nthu.edu.tw, Admissions website: http://oga.nthu.edu.tw/
Phone: +886-3-5162461
E-mail: hcchan@mx.nthu.edu.tw
Admissions website: http://oga.nthu.edu.tw/

Application deadlines
Fall admission: March 15
Spring admission: November 1

Application fee
There is no application fee required.

Admissions information
For Fall of 2013:
   Number of applicants: 129
   Number admitted: 64
   Number enrolled: 62

Admission requirements
Bachelor's degree requirements: Updated details (instructions and requirements) are given on http://oga.nthu.edu.tw/cont.php?id=91&m=m26&mm=mm41&tc=5&lang=en. BSc: High School Degree and evaluation by Department Language: Chinese Language Proficiency is not required but desirable for the beginning but it is expected from students to participate the offered language class.
Minimum undergraduate GPA: 1.0

GRE requirements
The GRE is not required.

Advanced GRE requirements
The Advanced GRE is not required.

TOEFL requirements
The TOEFL exam is not required but desirable for students from non-English-speaking countries.

Other admissions information
Additional requirements: Updated details (instructions and requirements) are given on http://oga.nthu.edu.tw/cont.php?id=91&m=m26&mm=mm41&tc=5&lang=en.

TUITION

Tuition year 2014-15:
Tuition for in-state residents
   Full-time students: NT$66,000 annual
Tuition for out-of-state residents
   Full-time students: NT$66,000 annual
Credit hours per semester to be considered full-time: 1
Health insurance: Yes, 8988 NTD.
Other academic fees: Dormitory expense is about TWD 10,700 to 16,000 per semester for 2-bed rooms. Off-campus housing costs about TWD 4,000~6,000 per month. Internet User Fee for dormitory is TWD 200 (per semester).
Academic term: Semester

Teaching Assistants, Research Assistants, and Fellowships
Number of first-year
   Teaching Assistants: 105
Average stipend per academic year
   Teaching Assistant: NT$60,000
Additional RA positions are available within individual research groups. Fellowships are offered from the College and University to qualified students. Special Fellowship are offered to qualified international students. All incoming Ph.D. students receive full tuition waivers for the first two years.

FINANCIAL AID

For further information
Address financial aid inquiries to: International Students Division, Office of International Affairs (international student admission), National Tsing Hua University, 101, Kuang Fu Rd. Sec. 2, 30013 Hsinchu, Taiwan, R.O.C.
Phone: +886-3-5162463
E-mail: cyuliu@mx.nthu.edu.tw
Financial aid website: http://oga.nthu.edu.tw/

HOUSING

Availability of on-campus housing
Single students: Yes
Married students: Yes

For further information
Address housing inquiries to: 101, Section 2, Kuang-Fu Road, Hsinchu, Taiwan 30013, R.O.C.
Phone: +886-3-5715416
E-mail: housing@my.nthu.edu.tw
Housing aid website: http://sthousing.web.nthu.edu.tw/bin/home.php

Table A—Faculty, Enrollments, and Degrees Granted

Research Specialty	2012-13 Faculty	Enrollment 2013		Number of Degrees Granted 2012-13		
		Master's	Doctorate	Master's	Terminal Master's	Doctorate
Astrophysics and Astronomy	6	1	5	–	–	2
Atomic, Molecular, & Optical Physics	7	36	15	11	–	2
Condensed Matter Experiment	11	31	31	13	–	2
Condensed Matter Theory	4	6	13	3	–	1
Field Theory	4	10	11	3	–	1
others	6	33	23	4	–	9
Total	38	116	98	35	–	17
Full-time Grad. Stud.	–	116	98	–	–	–
First-year Grad. Stud.	–	58	14	–	–	–

Table B—Separately Budgeted Research Expenditures by Source of Support

Source of Support	Departmental Research	Physics-related Research Outside Department
Government agencies	NT$100,853,200	
State/local government		
Non-profit organizations		
Business and industry	NT$14,972,509	
Other	NT$4,900,788	
Total	NT$120,726,497	

Table C—Separately Budgeted Research Expenditures by Research Specialty

Research Specialty	No. of Grants	Expenditures (TW$)
Non-specialized	19	NT$100,853,200
Total	19	NT$100,853,200

GRADUATE DEGREE REQUIREMENTS

Master's: 24 credit hours of course work and a thesis.
Doctorate: 18 credit hours of course work beyond those of the Master program (waiver possible for qualified students); qualifying exam and Ph.D. proposal exam, dissertation. . .

SPECIAL EQUIPMENT, FACILITIES, OR PROGRAMS

Facilities in the department:

- Liquid He Factory

- Machine Shop

- Ultrafast Spectroscopy Instrumentation

- Epitaxial Growth Facility

- Micro, Millimeter, and THz wave instrumentation

- Cluster of experimental research activities on cold atoms ranging from optical spectroscopy to Bose-Einstein Condensate

- Surface Science Cluster: Scanning Tunneling Microscopy and Spectroscopy from 5 K to above room temperature/Photoemission Spectroscopy

External Facilities:

- NTHU Nanotechnology Center

- National nanoDevices Laboratory

- National High-Speed Computing Center

- National Synchrotron Radiation Research Center, which the department has our own beam line and end stations (http://www.nsrrc.org.tw/)

- National Center for Theoretical Sciences on the Campus (http://www.cts.nthu.edu.tw/main.php) with our own computer cluster

- Industrial Technology Research Institute, Hsinchu (http://www.itri.org.tw/eng/)

- Industrial Collaborations with local industry (World-renowned Hsinchu Science-Based Industrial Park, featuring companies like TSMC - world second largest semiconductor company.)

FACULTY

Professor

Cazalilla, Miguel A., (米格爾) Ph.D., University of the Basque Country, 1999. *Condensed Matter Physics, Theoretical Physics*. Ultra-cold atomic gases, non-equilibrium physics.

Chang, Hsiang-Kuang, (張祥光) Ph.D., Bonn University, 1994. Associate Vice President for Academic Affairs. *Astronomy, Astrophysics*. Astronomical instrumentation, high-energy astrophysics, compact stars.

Chang, Shih-Lin, (張石麟) Ph.D., Polytechnic Institute of Brooklyn, New York, 1975. *Condensed Matter Physics*. Diffraction physics, x-ray crystallography (x-rays, synchrotron radiation, instrumentation) Condensed matter physics (thin films, interfaces, multilayers, polymer/liquid crystals).

Chang, Tsun-Hsu, (張存續) Ph.D., National Tsing Hua University, 1999. *Astronomy, Astrophysics*. Coherent Radiation, Electron Cyclotron Maser Instability, Nonlinear Plasma Dynamics, High-Frequency Electronics.

Cheung, Kingman, (張敬民) Ph.D., University of Wisconsin - Madison, 1992. *Particles and Fields*. Higgs boson, supersymmetry, beyond the standard model, dark matter.

Chou, Dean-Yi, (周定一) Ph.D., California Institute of Technology, 1986. *Astronomy, Astrophysics*. Astrophysics, Solar Physics, helioseismology, asteroseismology.

Chou, Ya-Chang, (周亞謙) Ph.D., Pittsburg, 1981. *Condensed Matter Physics*. Surface Physics.

Chu, Chong-Sun, (朱創新) Ph.D., UC Berkeley, 1996. *Theoretical Physics*. String theory, quantum gravity and quantum field theory.

Geng, Chao-Qiang, (耿朝強) Ph.D., Virgina Tech, 1987. *Cosmology & String Theory, High Energy Physics*.

Gwo, Shangjr, (果尚志) Ph.D., University of Texas at Austin, 1993. *Condensed Matter Physics*. Scanning Probe Microscopy/Spectroscopy, Nanostructure Physics, Surface Physics, Molecular beam epitaxy of nitride semiconductors.

He, Xiao-Gang, (何小剛) Ph.D., University of Hawaii, 1987. *Particles and Fields*. Particle phenomenology, cosmology, and quantum mechanics.

Jeng, Horng-Tay, (鄭弘泰) Ph.D., National Tsing-Hua University, 1996. *Computational Physics, Condensed Matter Physics*. transition-metal oxides, charge/orbital orderings, strong correlations, layer structures, nano-structures.

Kong, Albert, (江國興) Ph.D., University of Oxford, 2000. *Astronomy, Astrophysics*. Observational Astronomy; High-energy Astrophysics.

Kou, Chwung-Shan, (寇崇善) Ph.D., University of California LA, 1991. *Plasma and Fusion.* plasma physics and applications, microwave physics and applications, atmospheric plasma source, transformer coupled plasma source, microwave surface wave plasma source, material plasma processing.

Ku, Huan-Chiu, (古煥球) Ph.D., University of California, SD, 1980. *Condensed Matter Physics.* Experimental Condensed Matter Physics, Low Temperature Physics, Superconductivity, Magnetism, Strongly Correlated Electron Systems, Material Sciences, and Nanomagnetism.

Kwo, J. Raynien, (郭瑞年) Ph.D., Stanford University, 1981. *Condensed Matter Physics.* Nano-electronics, spintronics, superconductivity, topological insulators, advanced thin film growth.

Lin, Deng-Sung, (林登松) Ph.D., o University of Illinois at Urbana-Champaign, 1994. *Applied Physics, Chemical Physics, Condensed Matter Physics, Nano Science and Technology, Surface Physics.* Experimental surface physics/chemistry and nano-film science using both atomic resolved microscopy and synchrotron-radiation based spectroscopy technique.

Lin, Hsiu-Hau, (林秀豪) Ph.D., University of California, Santa Barbara, 1998. *Condensed Matter Physics, Statistical & Thermal Physics, Theoretical Physics.* Low dimensional physics, superconductivity, quantum biology.

Mou, Chung-Yu, (牟中瑜) Ph.D., Caltech, 1993. *Theoretical Physics.* Strongly correlated electronic systems, topological matters, superconductivity, biophysics.

Pan, Ci-Ling, (潘犀靈) Ph.D., Colorado State University, 1979. Chair of the department. *Applied Physics, Atomic, Molecular, & Optical Physics, Engineering Physics/Science, Nano Science and Technology, Optics.* Laser Science, Nonlinear Optics, Ultrafast and THz Photonics, Liquid Crystal and Fiber Photonics.

Shy, Jow-Tsong, (施宙聰) Ph.D., Optical Sciences Center, University of Arizona, 1982. *Atomic, Molecular, & Optical Physics.* Precision laser spectroscopy of simple atoms and molecules.

Soo, Yun-Liang, (蘇雲良) Ph.D., State University of New York at Buffalo, 1995. *Condensed Matter Physics.* Materials Physics, Synchrotron Radiation Applications.

Tai, Ming-Fong, (戴明鳳) Ph.D., National Tsing Hua University, 1989. *Other.* Nano- and micro-size magnetic particles for biomedical applications, New magnetic superconducting systems with strong electron correlation interaction, Magnetic materials applied in microwave devices or microwave absorption.

Wang, Daw-Wei, (王道維) Ph.D., University of Maryland, 2000. *Condensed Matter Physics, Theoretical Physics.* Strongly correlated physics, ultracold atoms.

Yu, Ite, (余怡德) Ph.D., MIT, 1993. *Atomic, Molecular, & Optical Physics.* Laser trapping and cooling, cold atoms and Bose-Einstein condensation, quantum optics and quantum information experiments.

Associate Professor

Chang, We-Fu, (張維甫) Ph.D., National Tsing Hua University, 2000. *High Energy Physics, Particles and Fields.*

Chen, Huei-Ru, (陳惠茹) Ph.D., University of California at Berkeley, 2004. *Astronomy, Astrophysics.* Radio Astronomy, Star Formation, Interstellar Medium.

Chen, Jeng-Chung, (陳正中) Ph.D., Purdue University, 2003. *Condensed Matter Physics.* Quantum transport in low-dimensional system.

Chen, Po-Chung, (陳柏中) Ph.D., University of California, San Diego, 2002. *Computational Physics, Condensed Matter Physics.* Numerical method for strongly correlated systems.

Hoffmann, Germar, (霍夫曼) Ph.D., University of Kiel, 2002. *Chemical Physics, Condensed Matter Physics, Low Temperature Physics, Nano Science and Technology, Solid State Physics, Surface Physics.* Molecular Magnetism, Molecular Spintronics, Molecular Electronics, Single-Molecule Spectroscopy, Scanning Tunneling Microscopy.

Jiang, Ing-Guey, (江瑛貴) Ph.D., University of Oxford, 1999. *Astronomy, Astrophysics.* Galaxies, Planetary Systems.

Lai, Shih-Ping, (賴詩萍) Ph.D., University of Illinois at Urbana-Champaign, 2001. *Astronomy.* Star formation, interstellar magnetic fields, radio astronomy, infrared astronomy.

Liu, Yi-Wei, (劉怡維) Ph.D., Oxford University, 1999. *Atomic, Molecular, & Optical Physics.* Cold collision, cold molecule, test fundamental physics with precision laser spectroscopy.

Tang, Shu-Jung, (唐述中) Ph.D., University of Tennessee, Knoxville, 2003. *Surface Physics.* Electronic and lattice structure of 2D system such as metal or semiconductor surfaces, metal thin films, and organic thin films.

Assistant Professor

Chuu, Chih-Sung, (褚志崧) Ph.D., University of Texas at Austin, 2006. *Applied Physics, Atomic, Molecular, & Optical Physics, Quantum Foundations.* Quantum Information Science, Quantum Optics.

Lin, Yi-Ping, (林怡萍) Ph.D., State University of New York at Stony Brook, 2002. *Condensed Matter Physics.* Spintronics, Low-dimensional Physics, Semiconductor Physics.

Lo, Rong-Li, (羅榮立) Ph.D., National Taiwan University, 1997. *Nano Science and Technology, Surface Physics.* Scanning Probe Microscopy, Atom-scale Surface Reactions.

Wang, Li-Bang, (王立邦) Ph.D., University of Illinois at Urbana-Champaign, 2005. *Atomic, Molecular, & Optical Physics.* Laser spectroscopy, optical metrology, and precision measurement.

Wu, Kuo-An, (吳國安) Ph.D., Northeastern University, Boston, 2006. *Computational Physics, Condensed Matter Physics, Nonlinear Dynamics and Complex Systems, Polymer Physics/Science, Theoretical Physics.* My research interests lie in theoretical understanding of morphology and pattern formation in non-equilibrium systems.

DEPARTMENTAL RESEARCH SPECIALTIES AND STAFF

Theoretical

Condensed Matter Physics. Miguel Cazalilla, Po-Chung Chen, Hsiu-Hau Lin, Chung-Yu Mou, Daw-Wei Wang.

Field Theory and High. We-Fu Chang, Kingman Cheung, Chong-Sun Chu, Chao-Qiang Geng.

Experimental

Astrophysics and Solar Physics. Hsiang-Kuang Chang, Tsun-Hsu Chang, Huei-Ru Chen, Dean-Yi Chou, Ing-Guey Jiang, Albert Kong, Shih-Ping Lai.

Atomic, Molecular, & Optical Physics. Bose-Einstein Condensate - Precision Laser Spectroscopy - Quantum Information, Ultrafast and THz Photonics. Chih-Sung Chuu, Yi-Wei Liu, Ci-Ling Pan, Jow-Tsong Shy, Li-Bang Wang, Ite Yu.

Condensed Matter Physics. Scanning Probe Microscopy, Low Temperature Physics, Quantum Transport, Novel Two-Dimensional Quantum Systems (topological insulators, graphene, . . .). Shih-Lin Chang, Jeng-Chung Chen, Ya-Chang Chou, Shangjr Gwo, Germar Hoffmann, Huan-Chiu Ku, J. Raynien Kwo, Deng-Sung Lin, Yi-Ping Lin, Rong-Li Lo, Yun-Liang Soo, Shu-Jung Tang.

Appendix I
Geographic Listing of Departments
UNITED STATES

Appendix II
Alphabetical Listing of Departments